中国工程院院士文集

李坚文集

（再续）

李坚◎著

中国林業出版社
CFPH
China Forestry Publishing House

图书在版编目(CIP)数据

李坚文集：再续／李坚著．—北京：中国林业出版社，2022.1
ISBN 978-7-5219-1501-3

Ⅰ.①李…　Ⅱ.①李…　Ⅲ.①木材学-文集　Ⅳ.①S781-53

中国版本图书馆CIP数据核字(2021)第281408号

责任编辑：于界芬　于晓文　徐梦欣　　电话：(010)83143542

出版发行　中国林业出版社(100009　北京市西城区德内大街刘海胡同7号)
　　　　　http://www.forestry.gov.cn/lycb.html
印　　刷　北京博海升彩色印刷有限公司
版　　次　2022年1月第1版
印　　次　2022年1月第1次印刷
开　　本　889mm×1194mm　1/16
印　　张　88　　彩插　12面
字　　数　3880千字
定　　价　688.00元

著名木材学家 教育家

李坚院士

李坚，男，1943年2月生，汉族，辽宁省阜新市人，东北林业大学教授，中国工程院院士，著名木材学家和教育家。1966年毕业于东北林学院木材利用系。1981年获工学硕士学位，1987年获工学博士学位。1989—1990年在英国北威尔士大学院合作研究。自1981年在东北林业大学从事木材科学与技术的教学、研究及教育管理工作，历任讲师、副教授、教授和教务处副处长、校长助理、副校长、校长。2011年当选为中国工程院院士，2013年当选为国际木材科学院院士。

李坚院士曾兼任教育部科技委委员、教育部学部委员、教育部奖励委员会委员、国务院学位委员会林业工程学科评议组召集人、教育部林业工程类教学指导委员会主任委员、黑龙江省政协委员、黑龙江省欧美同学会副会长等职务。现任东北林业大学材料科学与工程学院教授、博士生导师、林业工程一级学科（世界一流建设学科）带头人、生物质材料科学与技术教育部重点实验室学术委员会主任、木质新型材料教育部工程研究中心学术委员会主任、中国 3D 打印材料理事会荣誉主席、国家林业和草原局第六届科技委常务委员、国际林业研究组织联盟（IUFRO）会员、国际林产品学会联合会（IAWPS）执行理事和国际木材解剖学家协会（IAWA）会员。

李坚院士一生向大自然学习，立足木材科学与技术学科发展的理论和技术前沿，对接木材工业发展的战略需求，聚焦木材工业发展的技术瓶颈，在木材理学与人居环境、木材碳学与绿色制造、木材异质复合材料与增材制造、木材保护与界面修饰、木材仿生与智能响应、木材绿色胶接与涂饰等前沿领域开展了大量具有前瞻性和开创性的研究工作，引领一流学科发展，促进行业技术进步。他是木塑复合材料、纤维素气凝胶、纳米纤维素等生物质基新材料的开拓者，是无胶胶合、无卤阻燃、3D 打印、仿生智能化等木材加工新技术的先行者，是我国林业工程学科发展的领航人。

李坚院士主持完成 30 余项国家、省部级重大科研项目，获国家技术发明二等奖 1 项、国家科技进步二等奖 3 项、何梁何利基金科学与技术进步奖 1 项、黑龙江省重大科技效益奖 1 项、首届黑龙江省教材建设奖优秀教材特等奖 1 项，黑龙江省科学技术一等奖 8 项、省部级科学技术二等奖 9 项。发表学术论文 560 余篇，出版教材和专著 38 部，授权发明专利 30 余件。获“做出突出贡献的中国博士学位获得者”“国家级有突出贡献的中青年专家”“全国优秀林业科技工作者”“全国木材保护行业特殊贡献专家”“全国高校黄大年式教师团队”“全国工人先锋号”等荣誉称号，荣获国务院政府特殊津贴、2012 海峡两岸林业敬业奖励基金和“庆祝中华人民共和国成立七十周年”纪念章。

李坚院士一向倡导“立德树人、教科相长”的教育理念，以德立身，以德施教，在校内外设立了“勉学育英”和“救助贫困”励学金，并得以实施。长期坚持为本科新生主讲《专题报告》，为研究生开设院士金课《木材 · 人类 · 环境》和《林业工程前沿进展》，主编《木材科学》《木材波谱学》全国研究生教学用书和《木材保护学》《生物质复合材料学》等国家规划教材，不断地将最新科研成果引入教材和课堂。先后培养了国际木材科学院院士、全国百篇优秀博士学位论文获得者、长江学者特聘教授、国家优秀青年科学基金获得者、龙江学者特聘教授等一大批优秀人才，为我国林业教育事业呕心沥血，做出了杰出的贡献。

李坚院士在 1996—2006 年担任东北林业大学校长期间，发扬“团结拼搏、自我激励、发挥优势、争创一流”的东林精神，凝练“学参天地、德合自然”的东林校训，潜心改革，锐意创新，用现代教育理念办学治校，引领学校跨越式发展。2000 年 2 月学校由林业部划转教育部管理，2005 年 9 月学校正式成为国家“211 工程”重点建设高校。目前，东北林业大学是国家“双一流”建设高校。李坚院士为东北林业大学的建设和发展做出了历史性的贡献，谱写了浓墨重彩的新篇章。

李坚，木材科学家，中国工程院院士，东北林业大学教授。主要研究方向为纳米纤丝化纤维素与气凝胶、木材仿生与智能化、木材碳学与绿色加工、木材保护与功能修饰等。

建党百年，生辉地天；
科技盛世，绿色发展；
抢占先机，谱写新篇；
笃志立新，领跑前沿；
中华万岁，人民金安。

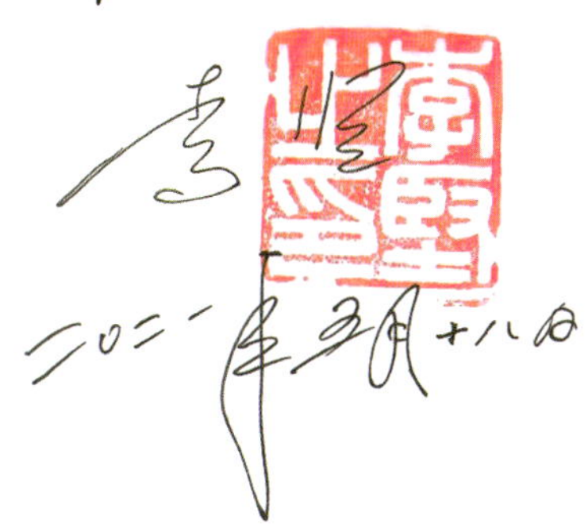

李坚院士于建党百年为《科技导报》期刊题词

李坚院士出席国家科学技术奖励大会并领奖

李坚院士为本科新生作专题报告

李坚院士录制研究生“院士金课”《木材·人类·环境》

李坚院士与学生交流讨论

李坚院士与研究生畅谈学习和人生

李坚院士与德华兔宝宝装饰新材股份有限公司技术人员交流

李坚院士出席新港集团院士工作站揭牌仪式

李坚院士与新港集团技术人员交流

李坚院士与久盛地板有限公司技术人员交流

李坚院士为东北林业大学材料科学与工程学院题词

李坚院士出席“盛和木业”研究生励学金颁奖典礼

李坚院士为本溪红叶家俬制造有限公司题词

李坚院士在2020年2月12日疫情期间
发给学生们的手机长信息

在读的和已毕业的李坚弟子们：

你们好！

1. 一丝不苟地执行习总书记和党中央的指示，做到：科学严密防疫情，安全有序抓复工！

2. 每位在家要做到：保护好家中所有的人不受病毒感染，身体无恙！根据每个人的发展和一流学科建设需要，静心，净心，凝心，用心设计写好各类人才项目的申报材料、国家自然科学基金申请书等。今年毕业答辩和博士后出站的各位撰写好博士学位论文和出站报告……

3. 严格遵守教育部、省教育厅和学校的各项规定，任何人不得违反！我们学科是正在建设中的世界一流学科，时刻不能放松学科建设！我们要顽强拼博，做到：一流学科，一流人才，一流队伍，一流成果，一流的教科相长，立德树人！

切切！

李坚

致我的学生们

《中国工程院院士文集》总序

二○一二年暮秋，中国工程院开始组织并陆续出版《中国工程院院士文集》系列丛书。《中国工程院院士文集》收录了院士的传略、学术论著、中外论文及其目录、讲话文稿与科普作品等。其中，既有早年初涉工程科技领域的学术论文，亦有成为学科领军人物后，学术观点日趋成熟的思想硕果。卷卷《文集》在手，众多院士数十载辛勤耕耘的学术人生跃然纸上，透过严谨的工程科技论文，院士笑谈宏论的生动形象历历在目。

中国工程院是中国工程科学技术界的最高荣誉性、咨询性学术机构，由院士组成，致力于促进工程科学技术事业的发展。作为工程科学技术方面的领军人物，院士们在各自的研究领域具有极高的学术造诣，为我国工程科技事业发展做出了重大的、创造性的成就和贡献。《中国工程院院士文集》既是院士们一生事业成果的凝练，也是他们高尚人格情操的写照。工程院出版史上能够留下这样丰富深刻的一笔，余有荣焉。

我向来以为，为中国工程院院士们组织出版《院士文集》之意义，贵在"真善美"三字。他们脚踏实地，放眼未来，自朴实的工程技术升华至引领学术前沿的至高境界，此谓其"真"；他们热爱祖国，提携后进，具有坚定的理想信念和高尚的人格魅力，此谓其"善"；他们治学严谨，著作等身，求真务实，科学创新，此谓其"美"。《院士文集》集真善美于一体，辩而不华，质而不俚，既有"居高声自远"之澹泊意蕴，又有"大济于苍生"之战略胸怀，斯人斯事，斯情斯志，令人阅后难忘。

读一本文集，犹如阅读一段院士的"攀登"高峰的人生。让我们翻开《中国工程院院士文集》，进入院士们的学术世界。愿后之览者，亦有感于斯文，体味院士们的学术历程。

徐匡迪

二○一二年

序 一

木材天地续辉煌

《李坚文集》和《李坚文集（续）》曾分别于2004年5月和2012年1月出版发行。文集收录了李院士在2011年以前发表的重要学术论文200余篇，内容涵盖木材保护学、木质环境学、生物木材学、木材超微构造学、木材加工化学、木材波谱学、木质复合材料、木质材料无胶胶合、木质功能材料、木材强化等诸多方面，集中反映了李院士在此期间的突出学术成就，记载了李院士对林业工程事业所做出的卓越贡献，已成为大家学习和参考的一份重要而珍贵的林业工程文献资料。文集的出版发行对于宣传李院士的创新学术思想、弘扬李院士的科学探索精神、梳理林业工程学科学术谱系、促进林业工程学科人才成长成才、推进林业工程学科文化传承与创新、引领林业工程学科争创一流发挥了重要作用。

2022年2月将迎来我们敬爱的李院士八十寿诞。为了集中反映李院士自新时代以来所取得的创新学术成就，配合东北林业大学林业工程一流学科学术谱系与文化传承研究，传李院士教科相长之学风，奏新时代立德树人之华章，特编辑出版《李坚文集（再续）》，作为全体学生献给恩师的一份生日礼物，以表达我们的崇敬、祝福和庆贺之情。

《李坚文集（再续）》是《李坚文集》和《李坚文集（续）》的再延续，主要精选了李院士2012年以来发表的重要学术论文100余篇，内容涵盖木基功能材料、木材仿生与智能响应、生物质基气凝胶、纳米纤维素等前沿研究方向。从《李坚文集（再续）》中，我们既可以饱览自新时代以来李院士和他的团队获得的创新学术成就，也可以学习李院士教科相长的严谨学风和立德树人的宗师风范。

这正是：

扶微索隐拓耕勤，育李培桃硕果殷。
道法天然多创意，学参宙宇渐登新。
博学笃志寻科秘，守正明德铸教魂。
老骥嘶风逾万里，辉煌再续又逢春。

我们诚挚地感谢中国工程院的大力资助以及中国林业出版社的鼎力支持和帮助！感谢东北林业大学材料科学与工程学院领导对本文集出版的关怀、支持和帮助！同时，也对李院士的合作者和同仁们的深情和贡献表示崇高的敬意！

本文集由我们的师兄弟钱学仁博士策划和编辑，甘文涛博士、王立娟博士校对，深致谢意！

李坚院士的学生们

2021 年 3 月

序　二

李坚：一生都向自然学

初次来到李坚家中的客人，常常惊讶于这位德高望重的院士竟然居住在一栋老楼里，面积不大、装饰朴素。李坚就在这栋楼里坦然地生活。他最爱吃蘸酱菜，闲暇时最喜欢到操场上散步，看看充满活力的年轻学生。他说："房子、票子都是身外物，不应迷恋。自然地生活最好。"虽然李坚对于物质生活没有高追求，但他却一直把为祖国富强而工作放在科学研究的首位，通过不断地向自然学习，开启了木材领域的新视界，为祖国的林业事业作出了贡献。

坚毅：特殊材料制成的院士

"种子哪怕落在石缝中，只要努力，都会生长。"

每天早上 7 点多，76 岁的李坚都会出现在办公室。只要他没有出差，这个时间基本没有变过，就连过年也不例外。李坚的开山弟子、和他已经共事 30 多年的刘一星教授用"特殊材料制成的院士"评价李坚，因为"不管工作多么繁重劳累，从来看不到他疲倦，总是精力旺盛的样子"。

制成李坚的"特殊材料"，就是他的执着和坚强。"你看自然界的种子，不管环境多么恶劣，哪怕落在石头缝儿里，只要自己努力，也会生根、发芽。人要向自然学习，只要有坚毅的精神，就没有什么克服不了的困难。"李坚说。

20 世纪 60 年代初是新中国成立后经济快速发展的一段时期，意气风发的李坚梦想着穿梭于高山峻岭、为美化祖国山河大展拳脚，所以在高考的时候，选择了当时的东北林学院木材利用系，从此与木材科学结下了不解之缘。

那个时代的大学生是真正的凤毛麟角。成绩优秀的李坚相信，一定有一个美好的未来在等待着自己。可是原本可以预知的美好，却在1966年毕业的时候发生了逆转。

“文革”的到来，使成绩最好的李坚被分配到最为艰苦的黑龙江省泰康县造纸厂。从天之骄子到基层工人，李坚拿到派遣证的时候，尽管也会因为难以施展抱负而有一丝失落，但他却没有说一个“不”字。“那个时候真的是完全服从组织分配。党让去哪里，就去哪里。”李坚收拾好行囊，到泰康县开始了三班倒的生活。

在泰康的日子里，李坚不仅认真工作，每年都把先进的奖状捧回家，他更利用休息时间偷偷地读书。在工厂埋头苦干的11年间，他开展了大量的技术革新工作，大大提高了工厂的生产效率和经济效益，60年代末，他就以芦苇为原料，首次试制成功了黏胶纤维。

如果说一腔热血可以让生命在逆境中燃烧，那么十几年的默默坚守，考验的则是一个人的毅力。1978年，恢复研究生考试第一年，李坚就以他坚持不懈的积累成为当时东北林学院木材科学专业的硕士研究生。

刚入学时，一直学习俄语的李坚根本听不懂英文的任何一个单词，这样的零基础却要直接挑战研究生的课程。有一位以前一直学习英语的同学不相信李坚能够跟上这样的节奏，不服气的李坚和他打赌：“毕业时，我的英语成绩一定比你好。”这场看似不可能成功的打赌，却以李坚的获胜告终。“除了刻苦，没有捷径。”那时的李坚，每天兜里都揣着单词卡片，晚上为了抵抗困意，他在脚下放了一盆凉水，“困了就把脚搁里面一会儿，精神了再接着学”。

绳锯木断，水滴石穿。在李坚的人生历程中，努力一直是抹不去的底色。正是这一抹底色，氤氲了人生，让这颗撒在石缝里的种子绽放精彩。

实践：学识当为社会服务
“我们是林业工作者，那里一定需要我们。”

1984年，李坚带学生到广西柳州木材防腐厂实习，工厂请李坚帮忙解决一个问题：如何使马尾松的性能更稳定。原来，广西生长着很多马尾松，可是由于这种木材易开裂、好变形，一直难以存放和应用，已经给企业造成了很多损失。李坚通过实验，提出了化学处理与干燥工艺相结合的《马尾松木材改性综合处理技术》，当年工厂就盈利300余万元。

1987年，大兴安岭发生新中国成立以来最严重的一次森林火灾，过火面积114万公顷、过火原木1500万立方米。6月初，火灾刚刚扑灭，虽然没有接到任何上级的指示，但李坚还是带着东北林业大学的老师、学生奔赴现场——“我们是林业工作者，那里一定需要我们。”

而现场的景象更让李坚终生难忘——被抢运下来的火烧原木堆积如山，虫害率高达98%以上。如果不及时处理虫害，不仅木材会受损失，居民的健康也会受到影响。云杉小黑天牛和落叶松八齿小蠹是主要害虫，它们其实不难对付，难的是传统的水浸、喷药等方法根本不能适用于这么多、这么巨大的原木。

经过现场实地试验，李坚筛选出适于大规模作业的高效灭虫药剂，还提出了可以处理千立方米以上楞垛的熏蒸法。这一方法不仅为大兴安岭火灾减少了上亿元的经济损失，还成为后来国际大规模保存火烧原木的通用做法。“这个方法看似简单，却体现了李坚的厚积薄发。”和李坚一起到大兴安岭工作的刘一星感慨道。

正是因为看到了大兴安岭的火灾，李坚开始更加关注木材的阻燃问题。2000年，李坚科研小组研发的新型木材阻燃剂FRW突破了国际上未能逾越的技术禁区，全面超越国际王牌产品，在多个国家和地区推广应用。

而这一项目的成功与李坚对团队工作的运筹密不可分。FRW负责人王清文在接手项目之初，有点发蒙：以目前的基础、数据根本无法支撑起这样的科研，能做吗？“可是做下来，你会发现，我们团队其他课题组的科研成果就会给FRW提供必要的支撑。这说明，在李老师的心里有一盘很大的棋，他会为了一个长远的目标调兵遣将、排兵布阵。”王清文说。

如果说一个人的冲锋需要勇气和胆识，那指挥千军万马则更需要智慧、韬略和魄力。一个学科的发展固然离不开单打独斗的科研工作者，但更离不开运筹帷幄的领头人。

“李坚把我国木材科学领域的科研带入了黄金期。”中国工程院院士、植物生理学教授尹伟伦这样评价。这个“黄金期”名副其实，这不仅缘于在重视生态文明的今天，人们的环境意识正在逐渐提高，更因为全国从事木材科研的机构已经在李坚的带领下，破除了壁垒、形成了合力。每年全国木材科学年会上，李坚都会在报告中畅谈他凝练出的科研方向，让同行们大呼“解渴”。近些年来，他提出的“追寻木材的碳足迹”“木材表面化学镀”“纳米纤维素与气凝胶”“木材仿生与智能响应”和“多元材料混合制造的3D打印”等方向的研究均得到全国同仁的首肯和响应。其研究成果和学术水平在世界范围内已成为并行者或领跑者。李坚也先后获得“做出突出贡献的中国博士学位获得者”“国家级中青年有突出贡献专家”等众多荣誉。

前沿：开启木材领域的新视界

“有时种子埋在地下好多年，是为了积蓄破土的力量。”

第八次全国森林资源清查结果显示，我国森林覆盖率只有21.63%。作为一个缺少森林的国家，在经济发展的过程中却不得不从森林中获取木材。所以，李坚一直有一个观点：保护和高效利用木材，就是不植树的造林，等于未增加森林面积而扩大了木材资源。

作为生物木材学的开拓者，自20世纪80年代开始，李坚就把传统木材科学研究从木材解剖、木材性质和木材缺陷扩大到生物木材学、木质环境学、木材基纳米复合材料、木材纤维素纳米纤维制备、木材基电磁屏蔽等多个领域，开辟了木材科学研究的新视界。

“他有着超前的眼光、一直走在时代的前面。”李坚的学生、全国优秀博士学位论文获得者王清文说。从20世纪80年代起，李坚就提倡学科的交叉融合，注意吸引化学、数学、自动化、艺术设计等专业的科研人员进入木材科学领域，正是这种远见，让李坚和他的团队创造出诸多个第一，给“超前”做了注解：

第一部应用现代波谱分析技术揭示木材内部及作用原理的专著——《木材波谱学》；

第一次运用有序聚类分析、计算机视觉技术以及优化统计理论解决成熟材和幼龄材界定难题的专著——《生物木材学》；

我国第一篇论述木材视觉、触觉、调湿等特性与人类和室内环境关系的论文——《木材、人类与环境》，成为唤醒人们生态意识的开篇；

第一次提出利用生物矿化原理制备无机纳米复合材料；

第一次采用超临界技术制备以木粉为原料的纤维素气凝胶；

第一次将采伐剩余物、木材加工剩余物、废旧木质材料这样的“三剩物”，以及秸秆、竹子等数量巨大却被弃之不用的材料纳入科学研究再利用的范畴；

……

李坚常常告诉身边的科研人员：陈旧的研究方法、陈旧的课题就不要再做了，搞科研必须要有新的思维、新的方法，否则就是误人子弟、误国发展。

这样的思路一以贯之。2011年，李坚当选院士后，仅4年多的时间，他已经带领全国木材科研人员对多项新课题开始了研究，其中很多都是向大自然学习所获得的启示：荷叶可以滴水不沾，我们能做出这样的木材吗？棉花轻柔飘逸，我们可以构筑相似的木质基仿生材料吗？这些木材仿生与智能响应、异质复合材料与智能制造、为3D打印提供支撑的生物质材料等很多课题，现在听起来还像天方夜谭，甚至让做这些科研的师生都有些怀疑——真能做出来吗？就算能做出来，得多少年才能出成果？可是李坚却很坚定：“科学研究一定是前沿性的，有些基础研究不可能在短期看到效益，做科研不能急功近利，就像一颗沉寂多年的种子，埋在地下，是为了积蓄破土的力量。”

育人：事业需要一代代人的努力
“我现在只希望能有更多的小树成为栋梁。”

或许一个人懂得越多，就越会发现知识的浩博。李坚常说的一句话是：“木材科学未被探寻的领域太多，需要一代代科研工作者的不懈努力。”

现在李坚依然保持着每天阅读文献、追踪前沿科技的习惯。有人问他累不累，他说：“能够给学生提供足够的营养，累也幸福。”让李坚觉得最为幸福的，并不是自己取得的任何成绩，而是培养了一大批在全国各地从事木材科学研究的栋梁。现在他的学生中，有的已经成长为长江学者、教授、学科带头人、首席专家，甚至很多“徒孙”辈的学生都已经成为博士生导师。所以，“李家军”到底有多少人，连李坚自己都说不清。

尽管身为院士，李坚的工作十分繁多，可是现在他仍然在招收硕士、博士研究生。目前在读的 11 名研究生他都亲自指导。“因为我给他们的研究方向，都是从来没人做过的课题，这种‘开第一枪’的研究，具有高度的冒险性。我直接指导，会让孩子们少走弯路。”李坚说。

可是要想成为李坚的学生，却并不是一件容易的事情，因为他至今依然保持着一个“让人看起来好笑，但却一直坚持”的选择学生标准——又红又专。“我觉得这个词一点都不过时。‘红’就是热爱党、热爱人民；‘专’就是雄厚的专业知识。没有热爱哪来的激情，没有知识哪来的力量。”李坚解释说。

做李坚的学生很累，因为他不仅每周都会去实验室看看科研进展，还要求学生们不能光做实验，更要养成总结归纳的习惯，每天至少要写 500 字的总结。而多年来，他本人每天写总结的字数则保持在 800 至 1000 字。尽管很累，可是很多李坚的学生都说，和李坚结识，自己的人生发生了改变。

现任生物质材料科学与技术教育部重点实验室副主任的许民是木塑复合材料方面的专家。许民考取博士的时候已经 38 岁。之前因为父亲去世，她心力交瘁，对人生也心灰意冷。当李坚鼓励她考博士时，她说自己已经评了副教授，不想再辛苦地读博士，因为那样“会累、会变老”。李坚说：“你不辛苦，就不会变老吗？人的一生怎样都会度过，你虚度光阴也好、刻苦努力也罢，一天的时间都是 24 个小时。可是不同的是你人生的意义。只有用积极的态度面对，才能收获一个无憾的人生。”这段话许民牢牢记在心里，正是这段话，使她的人生轨迹继续上扬。

高丽坤是一名研究生。高丽坤读本科的时候，母亲患上了骨癌，由于父亲很早就离开了她们，所以高丽坤只能和双胞胎姐姐一边读书、一边给母亲治疗，“那种压力让我看不到人生的希望，有时站到 7 楼宿舍，真想跳下去算了”。李坚知道后，一面帮她母亲联系医院，一面鼓励她“不管发生什么，都要努力、不能放弃”。李坚还和同事、学生为高丽坤捐款。李坚的鼓励和帮助，让高丽坤母亲的病情得到好转，也让高丽坤从自卑、自闭变得坚强、乐观。现在高丽坤不仅获得了国家奖学金、企业奖学金，并且已经发表了 11 篇 SCI 论文，成为保送的博士生。

如果说教给学生探索知识奥秘的钥匙是“授业”，那么教会学生面对人生的挑战、学会做人，则是“传道”。李坚常说“知识是力量、良知是方向”，他也一直用自己尊重他人、友善待人、低调做人的行动，影响着所有与他相识的人。

每次出差，尽管有学生、同事陪同，李坚也总是自己拿着行李，从不麻烦别人；凡是发给李坚的短信，就连过年时的拜年短信，他都会认真地一一回复……就连学校打字复印社的打字员都说，李坚老师特别随和。说起这些，李坚不好意思地笑着说：“‘我’字和‘你’字都是 7 画，人和人都是一样的，应该尊重别人。”

东北林业大学党委宣传部　孟姝轶

（发表于《东北林业大学报》第 573 期，2017 年 7 月 14 日）

自　序

从大自然中寻找灵感

在科学家座谈会上，习近平总书记发表了重要讲话，强调“我们必须走出适合国情的创新路子，特别是要把原始创新能力提升摆在更加突出的位置，努力实现更多‘从0到1’的突破。”总书记的讲话，对提升原始创新能力提出了更高要求。

如何提升原始创新能力？结合我的研究领域，有一些体会。自然界中的生物体各自有着独有的形貌、结构和行为，具备与外部环境相适应的独特性能或功能，观察它们，能为发明创造新型材料提供启发。

我和团队开展的很多科学研究，科研灵感均源于大自然。比如，“木材仿生智能科学研究”就是通过研究自然界生物的结构、性状、机理、行为及相互作用，为木材的各类加工技术提供新理念、新设计、新构成，赋予木材新的功能。

大家知道，荷叶是不沾水的，水珠要么完整地“躺”在荷叶表面，要么沿着荷叶滑落。根据这一现象，我和团队猜想：如果在木材上涂上一层纳米材料，阻止它吸收水分，是否可以延长家具等木制品的使用寿命？于是我们进行了木材防水防湿的研究，经过技术处理在木材表面形成一层无机纳米晶层，使得木材的拒水率至少在80%以上。

还有一次，我在海边散步，发现贻贝、牡蛎等贝类生物能够很牢固地黏附在坚硬的石壁上，并能在一定程度上经受住海浪的冲击。为什么它们之间具有如此强的黏附力呢？这种现象蕴含的原理，能不能用到木质人造板中呢？基于此，我们提出了木材仿生胶黏剂的概念，利用仿生海洋贝类代谢物的黏附性和层积效应，制备新型胶黏剂。

木材是重要的资源，与绿色环境和人体健康息息相关。要保证国家木材安全，让人民用上绿色清洁的木材制品，就要坚持自主创新，突破传统构架，拓展研究空间，进而推动林产工业转型升级。

作为一名教育工作者，我深感人才培养的重要性，对林业人才的培养要做到四个注重：注重培养学生创新思维和个性发展、注重培养学生的行动力、注重培养学生的坚定信念、注重引导学生服务于经济社会，为我国林业建设贡献智慧和力量。

李坚

（发表于《人民日报》2020 年 11 月 17 日　第 13 版）

目 录

第一部分 木基功能材料

第二部分　木材仿生与智能响应

第三部分　生物质基气凝胶

第四部分 纳米纤维素及其他

01

第一部分

木基功能材料

Electromagnetic Interference Shielding Material from Electroless Copper Plating on Birch Veneer*

Lili Sun, Jian Li, Lijuan Wang

Abstract: Copper coating on birch veneer substrate was conducted by electroless deposition to prepare electromagnetic interference shielding material. In the process, Pd^{2+} ions were chemically adsorbed on the wood surface modified with chitosan. Then they were reduced and dipped into a plating bath where copper film was successfully initiated. The coatings were characterized by SEM-EDS, XPS and XRD. The metal deposition, surface resistivity, and electromagnetic shielding effectiveness were measured. The morphology of the coating observed by SEM is uniform, compact and continuous. EDS, XPS and XRD results showed that the coating consists of Cu^0 with crystalline structure. Moreover, the copper films firmly adhere to the wood surface. Birch veneers plated with crystalline copper film exhibits high electro - conductivity with surface resistivity of 119.1 $m\Omega \cdot cm^{-2}$ and good electromagnetic shielding effectiveness of over 60 dB in frequencies ranging from 10 MHz to 1.5 GHz.
Keywords: Wood-particles; Chitosan; Pretreatment; APTHS

1 Introduction

Conductive wood, which is coated with nickel and copper, is a kind of wood-based composite for electromagnetic interference (EMI) shielding. It has also attracted many researchers' interests (Nagasawa and Kumagai 1989; Nagasawa et al., 1990, 1991a, b, 1992, 1999; Huang and Zhao 2004; Wang et al., 2006a, b, 2008; Wang and Li, 2007; Sun and Wang, 2008). Preparing this kind of composite often involves electroless plating technology with which uniformly and compactly continuous coatings on metallic and nonmetallic material can be obtained. Nonmetallic material should be catalyzed by using activators prior to the plating process. Palladium colloid is a conventional and commercial activator. However, palladium colloid contains many tin (Ⅱ) ions and easily precipitates. Furthermore, it can only be physically absorbed on the activated surface so that a small amount of palladium particles may drop off the substrate and cause self-decomposition of the plating solution used in the subsequent plating process. Pd (Ⅱ) is an extremely stable ion in aqueous solution and can be used as a tin-free activator. However, it is very difficult to directly absorb sufficient Pd (Ⅱ) for catalyzing without chemical bonding. Organosilanes have been used in many researches to promote adhesion between the coatings and a wide variety of substrates (Liu et al., 2005; Ikeda et al., 2009; Lu, 2009). Liu et al. (2010) and Liu and Wang (2010) have successfully used aminosilane to modify birch and Fraxinus *mandshurica* veneers and plated nickel coating on the modified veneers. However, the

---

* 本文摘自 Wood Science & Technology, 2012, 46: 1061-1071.

hydrolysis of aminosilane needs more time in the modification process. Aminosilane is also sensitive and easily reacts in storage. To overcome the problem, chitosan was employed to replace aminosilane.

Chitosan is a very abundant biopolymer from the alkaline deacetylation of chitin (a polymer made up of acetylglucosamine units) (Fig. 1) (Rani et al., 2010), which is present in the exoskeletons of crustaceans, the cuticles of insects, and the cell walls of most fungi. Chitosan is actually a heteropolymer containing both glucosamine units and acetylglucosamine units. The presence of amine groups offers its affinity for metal ions (Guibal, 2005; Renbutsu et al., 2008; Pillai et al., 2009). Moreover, it has the merits of non-toxicity, good film formers, and easy operation. Several researchers have utilized chitosan for pretreatment of fabric or ABS plastic in electroless nickel plating (Tang et al., 2008; Yu et al., 2011). Here, an activation process was developed to prepare electroless copper-plated wood veneers for EMI shielding using chitosan pretreatment. In this process (Fig. 2), chitosan formed a very thin membrane on the wood surface, which bonded wood and Pd (Ⅱ) via a hydrogen bond and N-Pd coordination bond. The metal deposition, surface resistance, and electromagnetic shielding effectiveness of the copper-coated wood veneer were measured, and the chitosan pretreatment in the activation process was also investigated. The coating on wood veneer was characterized by scanning electron microscopy (SEM-EDXA), X-ray photo electron spectroscopy (XPS), and X-ray diffraction (XRD).

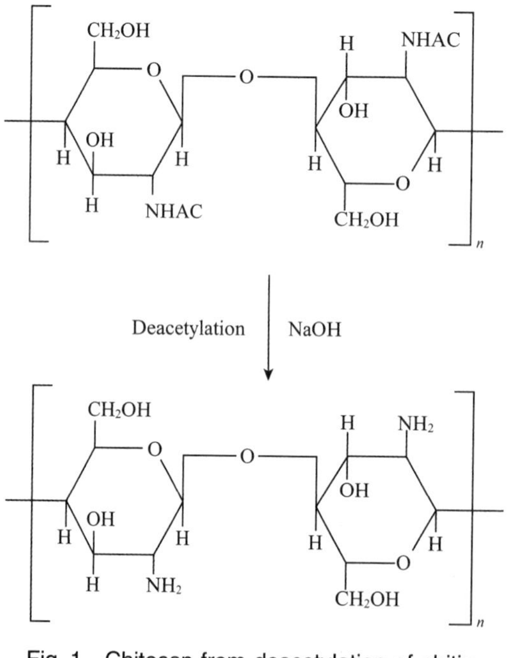

Fig. 1 Chitosan from deacetylation of chitin

2 Experimental

2.1 Materials

The substrates used were birch wood veneers as previously described (Wang et al., 2011). Chitosan was obtained from Jinan Haidebei Marline Bioengineering Company Limited. Palladium chloride ($PdCl_2$, >99.5%) was obtained from Shenyang Jinke Reagent Factory. All other chemicals were of analytical grade.

2.2 Pretreatment of wood veneer

Birch wood veneers were polished with 120-mesh emery paper to remove fine fibers and dust. In a general plating process, roughening is required in order to obtain higher bonding strength between the coating and the substrate. Because wood is porous, its surface, with its natural irregularities, promotes electroless plating without roughening procedures.

The samples were immersed in chitosan solution containing 1% acetic acid solution for 8 min at room temperature and removed from the solution; excess solution was removed and the samples were dried with hot air. The samples were coated with a chitosan membrane.

The wood-chitosan was immersed in a solution of palladium chloride ($PdCl_2 \cdot 2H_2O$: 0.2 $g \cdot L^{-1}$, HCl: 2 $g \cdot L^{-1}$) at 40 ℃ for 8 min, rinsed, and then reduced in sodium hypophosphite solution of 2 $g \cdot L^{-1}$ for 8 min. Thus, wood-chitosan-Pd was formed.

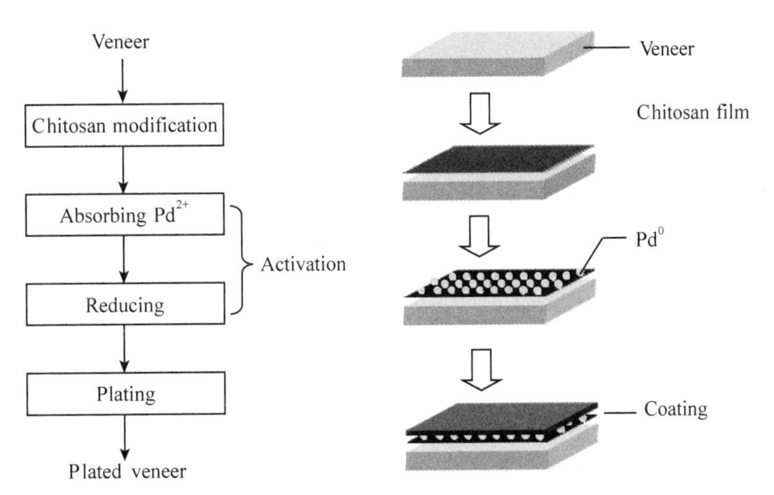

Fig. 2 Schematic illustration of electroless copper plating process on wood surface

2.3 Electroless copper plating of wood veneer

Wood-chitosan-Pd sample was put into the plating solution. The composition of electroless bath and the operating conditions are listed in Table 1. Distilled water was used to prepare the solutions. NaOH was used to adjust the pH of the plating solution. After plating, the samples were carefully rinsed with distilled water and dried in an oven at around 100 ℃ to constant weight.

Table 1 Composition and operation conditions of electroless copper plating

Chemical	Concentration ($g \cdot L^{-1}$)
$CuSO_4 \cdot 5H_2O$	20
$EDTANa_2 \cdot 2H_2O$	40
Glyoxylic	7.5
2, 2′-dipyridyl	0.01
Potassium ferrocyanide	0.015
pH	12
T	55 ℃

2.4 Measurement of metal deposition

The wood veneers were dried at 103±2 ℃ to constant weight (G_0). The copper-coated veneers were also dried to constant weight (G_1), and the metal deposition was calculated as:

$$\text{Metal deposition}\ (\%) = (G_1 - G_0)/G_0 \times 100\% \tag{1}$$

2.5 Measurement of surface resistance and shielding effectiveness

The measurement of surface resistance and shielding effectiveness was performed according to Li et al. (2010).

2.6 Characterization methods

The surface morphology and elemental compositions of the coatings were determined by scanning electron microscopy (SEM, Quanta 200) and X-ray energy dispersive spectrometer (EDS). Specimens were not coated with gold prior to analysis. XPS was used for chemical state analysis of the copper coating. XPS signals were recorded with a K-Alpha XPS Analyzer (Thermo Fisher Scientific Company) using an Al K_α source. In addition, the phase structure of the coating was investigated by X-ray diffraction (XRD, Rigaku D/max2200 diffractometer) using a Cu K_α radiation generator operated at 1200 W (40 kV×30 mA).

2.7 Bonding strength measurement

A tensile test was employed to measure the adhesion between the electroless copper coating and the wood by using a vertical pulling method. A schematic illustration of the test was used as in Li et al. (2010).

3 Results and discussion

3.1 The effect of chitosan concentration

Electroless copper plating requires a catalytic substrate for nucleation and growth. Therefore, activation is necessary for electroless plating on a noncatalytic substrate. In this study, chitosan was used for wood modification to absorb Pd (Ⅱ). This is the key process for the plating. The effects of chitosan concentration on metal deposition and surface resistivity are shown in Table 2. Metal deposition increased with chitosan concentration increasing from 0.2% to 0.8% and then changed slightly. The surface resistivity is negatively related to the metal deposition. Electrical conductivity is the best and the surface resistivity reached 119.1 $m\Omega \cdot cm^{-2}$ at a chitosan concentration of 0.8%. As is know, more chitosan will absorb more Pd (Ⅱ); therefore, more Pd from reduction can catalyze the copper plating reaction, making the copper to deposit quickly.

Table 2 The effects of chitosan concentration on plated veneers

Concentration($\% \ Wt \cdot Vol^{-1}$)	0.2	0.4	0.6	0.8	1.0
Metal deposition (%)	16.4	14.5	19.7	20.0	19.6
Surface resistivity ($m\Omega \cdot cm^{-2}$)	360.2	342.3	181.2	119.1	162.1

3.2 SEM-EDAX analysis

Fig. 3 shows SEM photographs of pristine, chitosan-modified and plated wood veneers. It is shown in Fig. 3a that many structures such as fibers, vessel, perforation plates, and pits exist on the wood surface showing the porous nature of wood. Fig. 3b shows the surface of chitosan-modified wood veneer. No obvious differences to Fig. 3a can be observed. In raw wood, the perforation plates are like ribs and separate each other. However, the chitosan membrane can be seen in the photograph at high magnification (Fig. 3c), especially on perforation plates indicating the uniform membrane obtained in the pretreatment. It can be found in Fig. 3d that the surface is entirely covered by the coating and has metallic sheen, and the coating deposited in the pores and the different structures (Fig. 3e). Because the coating is very thin, the porous structures are still preserved. Therefore, beautiful wood grain on the surface of plated wood can be observed with the unaided eye. It can be also observed in Fig. 3f that the coating is composed of small cells, whose diameters are less than 1 μm. These cells have been deposited together closely, so that the coating is very smooth and continuous. The color of the coated sample is copper-bright shown in Fig. 4, which is closely related to the compact structure of the coating.

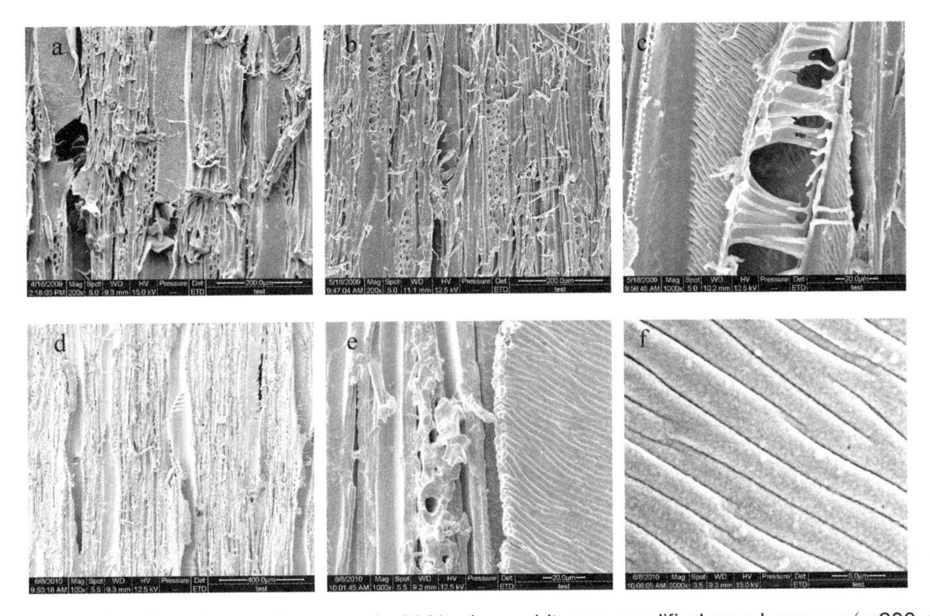

Fig. 3 SEM photographs of a raw wood veneer (×200); b, c chitosan-modified wood veneer (×200 and ×1,000); d-f plated veneers with 100, 1,000, and 10,000 times magnification, respectively

Fig. 4 Photo of birch veneer plated with copper film

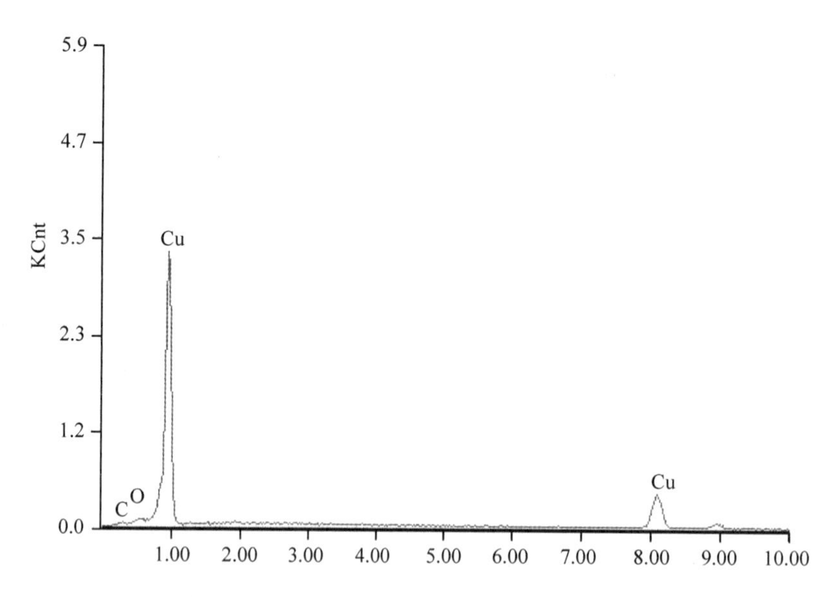

Fig. 5 EDS spectrum of plated wood veneer

Fig. 5 is the EDS analysis and results of plated wood veneer. It indicates that the coating is composed of Cu element, and little C and O is from the wood substrate or the contaminants and oxygen absorbed in wood pores. Other possible ingredients such as N and Pd are not detected, which indicates X-ray did not penetrate through the coating.

3.3 XPS analysis

XPS measurement was employed to provide further informationon the chemical state of the copper coating. A typical XPS wide spectrum is shown in Fig. 6, which indicates that only copper element with a small amount of O and C was detected. The peaks at 934, 123, and 77 eV are attributed to Cu 2p, Cu 3s, and Cu 3p, respectively. The results indicate that Cu element in the coating exists as metallic copper (Lu,

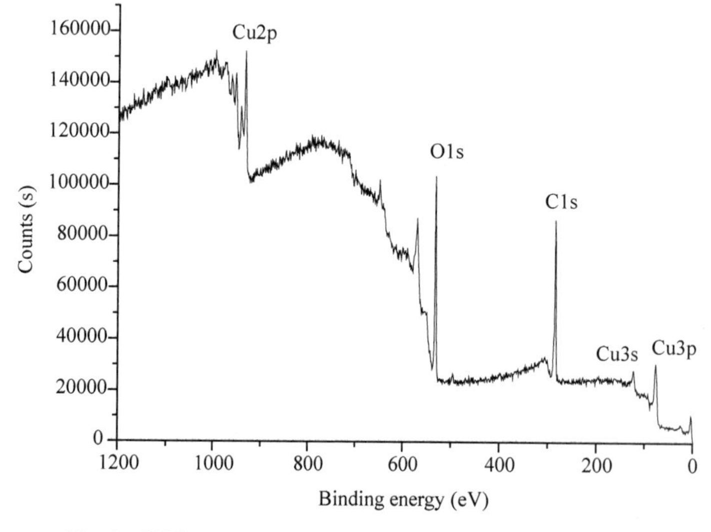

Fig. 6 XPS spectrum of copper coating on wood veneer

2010). Cu^{2+} has been reduced to Cu^0 in the plating process written as follow:

$$CuEDTA^{2-} + 2CHOCOOH + 4OH^- \rightarrow Cu\downarrow + 2C_2O_4^{2-} + 2H_2O + 2H_2\uparrow + EDTA^{4-} \quad (2)$$

As is known, XPS can only detect up to a depth of several nanometers less than the thickness of the coating. The fact that O and C were detected indicates that the contaminants and oxygen existed on or in the plated veneer (Wang et al., 2002).

3.4 XRD analysis

XRD spectra for birch veneers before and after plating are shown in Fig. 7. The strong diffraction peaks at $2\theta = 16.11°$ and $22.45°$ are characteristic peaks of cellulose in wood. The peaks at $2\theta = 43.43°$, $50.55°$, $74.21°$, and $90.11°$ are attributed to Cu (111), Cu (200), Cu (220), and Cu (311), respectively, which indicates the face-centered cubic phase of copper (JCPDS: 04-0836) and the crystalline nature of the copper coating. Compared with Fig. 7a, the characteristic peaks of wood in Fig. 7b obviously became weaker. This result indicates that the continuous and compact coating covered the wood surface entirely. In addition, no peaks of impurity and copper oxide were observed, which further proved the above obtained results from XPS analysis.

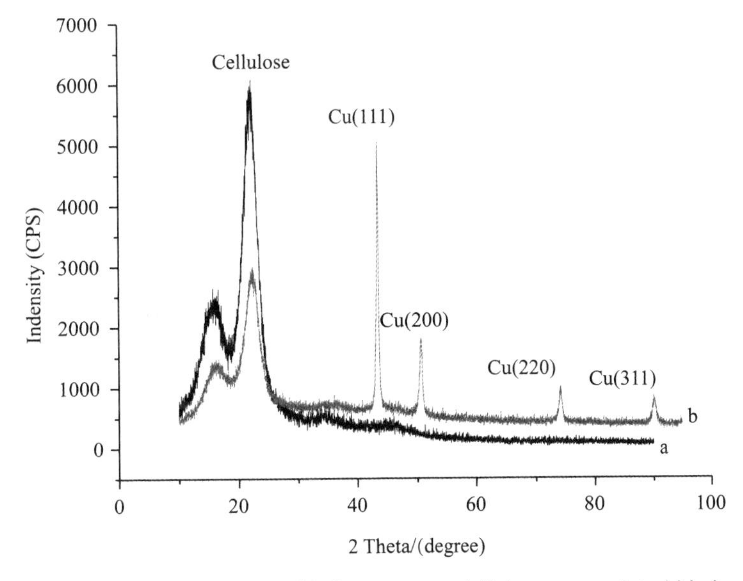

Fig. 7 XRD patterns of (a) raw birch veneer and (b) copper-plated birch veneer

3.5 Electromagnetic shielding property

The electromagnetic shielding results of the samples are shown in Fig. 8. The shielding effectiveness of raw birch veneer fluctuates around 0 dB; therefore, it has no shielding performance at all. The plated veneer has a shielding effectiveness which is higher than 60 dB in frequencies ranging from 10 MHz to 1.5 GHz, which indicates that the plated wood veneer has a better shielding effectiveness and can become a promising lightweight composite applied in the anti-EMI area.

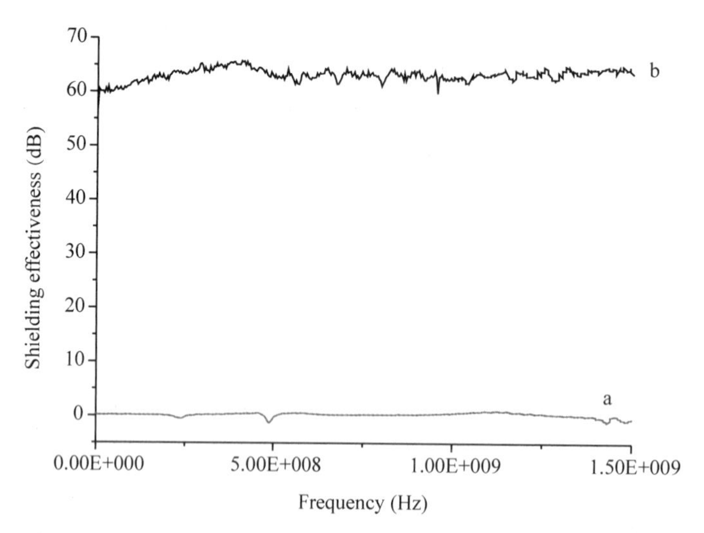

Fig. 8 Electromagnetic shielding effectiveness of wood veneers (a) before and (b) after copper plating

3.6 Adhesion of copper coatings to wood

The results of the tests of bonding strength between the coating and the wood surface are listed in Table 3. The samples measured were prepared under the same conditions. In the tests, some samples were damaged in the wood part or some in the adhesive layer or both. Therefore, the data listed in Table 3 cannot entirely represent the bonding strength between the coating and the wood surface; they can show only the strength of either the adhesive layer or the wood part. The experimental phenomena show that the coatings are firmly bonded to the surface of wood. Wood is naturally porous and has irregularities suitable for mechanical anchorage. A layer deposited on the surface of wood can bond firmly because of a twist lock effect. Furthermore, chitosan membrane acted as a coupling material firmly linking wood and copper film. In short, the adhesion strength can meet common application needs.

Table 3 Test results for adhesion between coating and wood surface

Sample no.	Bonding strength(MPa)	Rupture phenomenon observed
1	0.98	Wood rupture
2	2.45	Adhesive-layer rupture
3	1.55	Wood and adhesive-layer rupture
4	1.79	Wood rupture

4 Conclusions

(1) Electrolesscopper plating was successfully carried out on wood veneer modified with chitosan using Pd (Ⅱ) as activator.

(2) Chitosan concentration affects the performance of plated veneer. The surface resistivity of the plated birch veneer reached 119.1 $m\Omega \cdot cm^{-2}$ when the chitosan concentration was 0.8%.

(3) The color of the coated veneer is copper-bright. SEM analysis shows that the coatings plated on the birch veneers are very smooth, continuous, compact, and also crystalline.

(4) The EDS and XPS results indicate that Cu element in the coating exists as Cu^0.

(5) The electromagnetic shielding effectiveness of the plated birch veneers is higher than 60 dB in the frequency range from 10 MHz to 1.5 GHz.

(6) The copper coatings are strongly bonded to the birch veneer.

Acknowledgment

The authors gratefully acknowledge the Natural Science Foundation of Heilongjiang Province (C200906), and the Program for NCET (08-0752).

References

[1] Guibal E. Heterogeneous catalysis on chitosan-based materials: A review Prog[J]. Polym. Sci., 2005, 30(1): 71-109.

[2] Huang J T, Zhao G J. Electroless plating of wood[J]. Beijing For. Univ., 2004, 26(3): 88-92.

[3] Ikeda A, Sakamoto A, Hattori R, Kuroki Y. Electroless Ni-B plating on SiO_2 with 3-aminopropyl-triethoxysilane as a barrier layer against Cu diffusion for through-Si via interconnections in a 3-dimensional multi-chip package[J]. Thin Solid Films., 2009, 517(5): 1740-1745.

[4] Li J, Wang L J, Liu H B. A new process for preparing conducting wood veneers by electroless nickel plating[J]. Surf Coat Technol., 2010, 204(8): 1200-1205.

[5] Liu H B, Wang L J. Electroless nickel deposition on Fraxinus Mandshurica veneer modified with APTHS for EMI shielding[J]. BioResources., 2010, 5(4): 2040-2050.

[6] Liu Z C, Heb Q G, Hou P, Xiao P F, He N Y, Lu Z H. Electroless plating of copper through successive pretreatment with silane and colloidal silver[J]. Colloids Surf A Physicochem. Eng. Asp., 2005, 257-258(5): 283-286.

[7] Liu H B, Li J, Wang L J. Electroless nickel plating on APTHS modified wood veneer for EMI shielding[J]. Appl. Surf. Sci., 2010, 257(4): 1325-1330.

[8] Lu Y X. Electroless copper plating on 3-mercaptopropyltriethoxysilane modified PET fabric challenged by ultrasonic washing[J]. Appl. Surf. Sci., 2009, 255(20): 8430-8434.

[9] Lu Y X. Improvement of copper plating adhesion on silane modified PET film by ultrasonic-assisted electroless deposition[J]. Appl. Surf. Sci., 2010, 256(11): 3554-3558.

[10] Nagasawa C, Kumagai Y. Electromagnetic shielding particleboards with nickel-plated wood particle[J]. J. Wood Sci., 1989, 35: 1092-1099.

[11] Nagasawa C, Kumagai Y, Urabe K. Electromagnetic shielding effectiveness particles board containing nickel-metalized wood particles in the core layer[J]. J. Wood Sci., 1990, 36(7): 531-537.

[12] Nagasawa C, Kumagai Y, Urabo K. Electro-conductivity and electromagnetic shielding effectiveness of nickel-plated veneer[J]. J Wood Sci., 1991 a, 37(2): 158-163.

[13] Nagasawa C, Umehara H, Koshizaki N. Effects of wood species on electroconductivity and electromagnetic shielding properties of electrolessly plated sliced veneer with nickel[J]. J. Wood Sci., 1991 b, 40(10): 1092-1099.

[14] Nagasawa C, Kumagai Y, Koshizaki N, Kanbe T. Changes in electromagnetic shielding properties of particlesboards made of nickel-plated wood particles formed by various pre-treatment processes[J]. J. Wood Sci., 1992, 38(3): 256-263.

[15] Nagasawa C, Kumagai Y, Urabe K, Shinagawa S. Electromagnetic shielding particle board with nickel-plated wood particles[J]. J. Porous Mater., 1999, 6(3): 247-254.

[16] Pillai C K S, Paul W, Sharma C P. Chitin and chitosan polymers: Chemistry, solubility and fiber formation[J]. Prog. Polym. Sci., 2009, 34(7): 641-678.

[17] Rani M, Agarwal A, Negi Y S. Chitosan based hydrogel polymeric beads - as drug delivery system[J]. BioResources., 2010, 5(4): 2765-2807.

[18]Renbutsu E, Okabe S, Omura Y, Nakatsubo F, Minami S, Shigemasa Y, Saimoto H. Palladium adsorbing properties of UV-curable chitosan derivatives and surface analysis of chitosan-containing paint[J]. Int. J. of Bio Macromol., 2008, 43(1): 62-68.

[19]Sun B, Wang Z X. Study on the preparation technology of timber-metal composite materials[J]. J. of Zhongyuan Univ. of Technol., 2008, 19(6): 23-25.

[20]Tang X J, Cao M, Bi C L, Yan L J, Zhang B G. Research on a new surface activation process for electroless plating on ABS plastic[J]. Mater Lett., 2008, 62(6-7): 1089-1091.

[21] Wang L J, Li J. Electroless deposition of Ni-Cu-P alloy on Fraxinus mandshurica veneer[J]. Sci. Silv. Sinica., 2007, 43(11): 89-92.

[22] Wang W C, Kang E T, Neoh K G. Electroless plating of copper on polyimide films modified by plasma graft copolymerization with 4-vinylpyridine[J]. Appl Surf Sci., 2002, 199(1-4): 52-66.

[23] Wang L J, Li J, Liu Y X. Electroless nickel plating on poplar veneer[J]. Fine Chem., 2006 a, 23(3): 230-233.

[24] Wang L J, Li J, Liu Y X. Preparation of electromagnetic shielding wood-metal composite by electroless nickel plating[J]. J. For. Res., 2006 b, 17(001): 53-56.

[25] Wang L J, Li J, Liu Y X. Study on preparation of electromagnetic shielding composite by electroless copper plating on Fraxinus mandshurica veneer[J]. J. Mater Eng., 2008, 4: 56-60.

[26] Wang L J, Li J, Liu H B. A simple process for electroless plating nickel-phosphorus film on wood veneer[J]. Wood Sci. Technol., 2011, 45: 161-167.

[27]Yu D, Wang W, Wu J W. Preparation of conductive wool fabrics and adsorption behaviour of Pd (Ⅱ) ions on chitosan in the pre-treatment[J]. Synth. Met., 2011, 161(1-2): 124-131.

中文题目：桦木单板表面化学镀铜制备电磁屏蔽材料

作者：孙丽丽，李坚，王立娟

摘要：通过化学沉积法在桦木单板表面获得铜镀层制备了电磁屏蔽材料。制备过程中，利用壳聚糖改性木单板表现单吸附 Pb^{2+}，经还原后浸入镀液，成功引发化学镀铜反应。采用 SEM-EDS、XPS 和 XRD 对镀层进行表征，测定了化学镀木材的金属沉积率、表面电阻率和电磁屏蔽效能、SEM 观察发现镀层均匀、致密、连续，EDS、XPS 和 XRD 结果表明，镀层由晶态单质 Cu 组成，而且 Cu 镀层与木材表面结合牢固。镀铜桦木单板展现很高的导电性和电磁屏蔽性能，表面电阻率达到 119.1 $m\Omega \cdot cm^{-2}$，在 10 MHz~1.5 GHz 频段，电磁屏蔽效能超过 60 dB。

关键词：木单板；壳聚糖；预处理；APTHS

Hydrothermal Synthesis of Zirconium Dioxide Coating on the Surface of Wood with Improved UV Resistance*

Caichao Wan, Yun Lu, Qingfeng Sun, Jian Li

Abstract: Nano-ZrO_2 aggregations were successfully layer-by-layer doposited on the wood surface by a simple mild one-pot hydrothermal method. The resulting ZrO_2/wood nanocomposite was characterized by scanning electron microscopy (SEM), energy dispersive X-ray (EDX) spectrometer, X-ray diffraction (XRD), Fourier transform infrared spectroscopy (FTIR), X-ray photoelectron spectroscopy (XPS) and thermogravimetric analyzer (TGA). The results indicate that the amorphous ZrO_2 formed strong hydrogen bonds with the hydroxide radicals of wood surface, and the strong interaction contributes to the enhancement of the nanocomposite heat stability. Moreover, compared with the original wood, the ZrO_2/wood shows more superior UV-resistant ability through a 600-hour QUV accelerated aging test.

Keywords: Hydrothermal synthesis; Inorganic compound; Photodegradation; Nanoparticle; Layer-by-layer self-assembly

1 Introduction

Wood is the oldest material utilized by humans for construction after stone. Despite its complex chemical nature, wood has abundant excellent properties which lend themselves to human use. Indeed, wood is inexhaustible, reproducible and readily available. Besides, wood has not only easy machining and exceptionally strong relative to its weight, but also good heat and electrical insulator and aesthetic properties. Nowadays, wood has been intensively applied in various fields like building, shipbuilding, railway, papermaking, furniture, interior decoration, etc. As a complex natural biopolymer principally composed by cellulose, hemicellulose and lignin, wood has inferior resistance to degradation from environmental agencies including fire, water, light and microorganisms than many artificial materials. Especially for exposure in the sun, sharp decreases in surface color and mechanical strength would occur due to strong ultraviolet light absorption ability of lignin leading to radical induced depolymerisation of wood compositions. There is no doubt that these unfavorable declines would reduce utilization value, service life and aesthetic feeling of wood adverse to the long-term development and comprehensive utilization of wood.

To date, many effective approaches to preventing wood products from photodegradation or minimizing UV irradiation effect have been widely reported such as finishing, treatment with dilute aqueous solutions of inorganic salts and depositing hybrid inorganic-organic thin films on wood substrates. Currently, chemically bonding or grafting stabilising chemicals including UV absorbers or antioxidants is considered as one of the

---

* 本文摘自 Applied Surface Science, 2014, 321: 38-42.

most effective methods of stabilizing photo-labile polymers when subjected to exterior exposure or severe environments. Meanwhile, it is worth noting that this class of absorbers or antioxidants must be colorless or light-colored and avirulent or harmfulless, and do not significantly damage the wood surface. According to the previous reports, some chemically stable and nontoxic inorganic nanoparticles such as TiO_2, ZnO and ZrO_2 hold great prospects as photoprotective agents for wood owing to their UV irradiation absorption or scatter ability. The cases that TiO_2 and ZnO serve as ultraviolet resistant agents have been extensively studied, whereas the literatures aiming at ZrO_2 for UV resistance application is not abundant. Actually, zirconia is one of the important ceramic which is widely used as structural materials, thermal barrier coating, optical coating, solid oxide fuel cell electrolytes, semiconductor materials and catalysis or catalytic supports because of its superior biocompatibility, high mechanical strength and fracture toughness, high melting point, high refractivity, stable photochemical properties and corrosion resistance. Recently, Smirnov et al. reported a new type of ZrO_2/SiO_2 interference mirrors capable of efficiently protecting against UV radiation prepared by spin-coating-assisted layer-by-layer deposition of colloidal suspensions of nanoparticles of ZrO_2 and SiO_2. Wu et al. manufactured a kind of ultraviolet-resistant nano-ZrO_2 composite polyester functional fiber. Sun et al. fabricated cosmetics containing ZrO_2 with improved optical activity, which could better resist full-waveband ultraviolet radiation. Motivated by these studies, adopting a versatile hydrothermal method to deposit ZrO_2 on wood surface might receive an excellent UV resistance effect. Hydrothermal method is regarded as an efficient and mild way of synthesizing inorganic/wood hybrid materials, which has been demonstrated by our previous researches. Moreover, to the best of our knowledge, this is the first attempt to utilize ZrO_2 as wood photoprotective agent by hydrothermal method.

In this work, a simple mild one-pot hydrothermal method was introduced to prepare ZrO_2/wood nanocomposite. SEM observation suggests that the nano-ZrO_2 aggregations have been uniformly deposited on wood surface layer-by-layer, and the following characterizations including XPS, XRD, FTIR and EDX indicate the formation of ZrO_2 and the strong interaction between ZrO_2 and wood substrate. Besides, the thermal stability of ZrO_2/wood significantly improved compared with that of original wood. Under prolonged strong UV irradiation, the nanocomposite shows extraordinary UV resistance with only tiny discoloration. Meanwhile, we also proposed a possible schematic diagram of the deposition of ZrO_2 on wood surface according to the experimental results.

2 Materials and methods

2.1 Materials

Polar wood slices were cut with the sizes of 20 mm (longitudinal) × 20 mm (tangential) × 10 mm (radial), and the slices were subsequently ultrasonically washed in deionized water for 30 min and dried at 80 ℃ for 24 h in a vacuum. Zirconium oxychloride octahydrate ($ZrOCl_2 \cdot 8H_2O$) and ammonia solution (NH_4OH) used for this experiment were supplied by Tianjin Kermel chemical Co. Ltd., and used without further purification.

2.2 Preparation of ZrO_2/wood

The preparation of ZrO_2/wood was implemented as follows: firstly, $ZrOCl_2 \cdot 8H_2O$ (0.78 g) was dissolved in 100 mL of deionized water with magnetic stirring for 30 min, and then 0.6 mL NH_4OH was dropwise added into the mixed aqueous solution with continuous stirring to form an emulsion. Subsequently, the dried wood slice and the above mixed solution were transferred to a Teflon-lined stainless-steel autoclave, and the autoclave was sealed and heated to 90 ℃ for 4 h. After reaction, the wood specimen was took out and ultrasonically rinsed with deionized water for 30 min. Finally, the sample was dried at 60 ℃ for 24 h in vacuum, and the following ZrO_2/wood nanocomposite was obtained.

2.3 Characterizations

The as-prepared ZrO_2/wood was observed using FEI Quanta 200 SEM attached with EDX spectrometer unit. XRD patterns were taken with XRD technique (Rigaku, D/MAX 2200) using Ni-filtered Cu $K\alpha$ radiation ($\lambda = 1.5406$ Å) at 40 kV and 30 mA. Scattered radiation was detected in the range of $2\theta = 5° - 80°$ at a scan rate of 4°/min. FTIR spectra were recorded in the range of 400-4000 cm^{-1} on Thermo Electron Corp (Nicolet Magna 560) FTIR spectrometer. XPS spectra were recorded in the range of 0-1200 eV by using Thermo Escalab 250Xi XPS spectrometer (Germany). The thermal stabilities were investigated by thermogravimetric analyzer (TGA, TA Q600) from 28 to 700 ℃ with a heating rate of 10 ℃ · min^{-1} under nitrogen atmosphere.

2.4 Accelerated aging test

A QUV accelerated weathering tester (Atlas, Chicago, IL, USA) was applied to induce photo-discoloration of the nanocomposite, which could reproduce the damage caused by sunlight, rain and dew. The samples were placed under a 340 nm fluorescent UV lamps (UV-B region) and underwent continuous light irradiation at 60 ℃ for 2.5 h and following water spray for 0.5 h as well as succedent condensation at 45 ℃ for 24 h. The whole weathering schedule lasted for 600 h, and the surface color was measured about every 24 h for 0-200 h, 48 h for 200-400 h, and 96 h for 400-600 h.

The color changes induced by UV irradiation were determined using a portable spectrophotometer (NF-333, Nippon Denshoku Company, Japan) with CIELAB system in accordance with the ISO-2470 standard. The CIELAB system was characterized by three parameters L^*, a^*, and b^*. The L^* axis represents lightness, and L^* varies from 100 (white) to 0 (black); a^* and b^* are the chromaticity coordinates, and $+a^*$ is for red, $-a^*$ for green, $+b^*$ for yellow, $-b^*$ for blue. The change of L^*, a^*, and b^* are calculated using Eqs. (1)-(3).

$$\Delta L^* = L_2 - L_1 \quad (1)$$

$$\Delta a^* = a_2 - a_1 \quad (2)$$

$$\Delta b^* = b_2 - b_1 \quad (3)$$

Where ΔL^*, Δa^*, and Δb^* represent the differences of initial and final values of L^*, a^* and b^*; L_1, a_1 and b_1 are the initial color parameters; L_2, a_2 and b_2 are the ultimate color parameters after UV irradiation, respectively.

The overall color change (ΔE^*) was calculated using the Eq. 4, and lower ΔE^* value represents less significant color change.

$$\Delta E^* = \sqrt{(L_2^* - L_1^*)^2 + (a_2^* - a_1^*)^2 + (b_2^* - b_1^*)^2} \quad (4)$$

All parameters were measured at eight locations on each sample, and the average values were calculated as the final decision.

3 Results and discussion

Fig. 1a shows the survey broad scan XPS spectrum of ZrO_2/wood, in which elements of Zr, C and O were detected on the powder surface. Futhermore, XPS also provides an efficient way to study the surface oxidation states. As shown in Fig. 1b, two Zr 3d peaks were observed at the binding energies of 182. 5 eV and 184. 9 eV, which are attributed to ZrO_2. Fig. 1c shows the XRD patterns of ZrO_2/wood and original wood. Apart from the characteristic peaks of wood at around 16. 2° and 22. 3°, there are no other obvious diffaction peaks corresponding to ZrO_2 for the ZrO_2/wood indicating that the oxide coating is exclusively constituted of amorphous particles, which is similar to some previous reports that manufactured ZrO_2 materials at such similar low temperatures. For further exploring the surface chemical compositions differences before and after hydrothermal treatment, the FTIR measurements of ZrO_2/wood and original wood were performed, and the results are shown in Fig. 1b. It is not hard to find that a new strong adsorption peak attributed to Zr–O–Zr band at around 486 cm^{-1} occurs in the FTIR specturm of the nanocomposite, revealing that the ZrO_2 nanoparticles were successfully deposited on the wood surface by hydrothermal process. In addition, the broad adsorption band at around 3334 cm^{-1} assigned to the stretching vibration of hydroxyl group was shifted to lower wavenumber in the spectrum of ZrO_2/wood, which indicating the strong interaction between hydroxyl groups of wood surface and ZrO_2 nanoparticles.

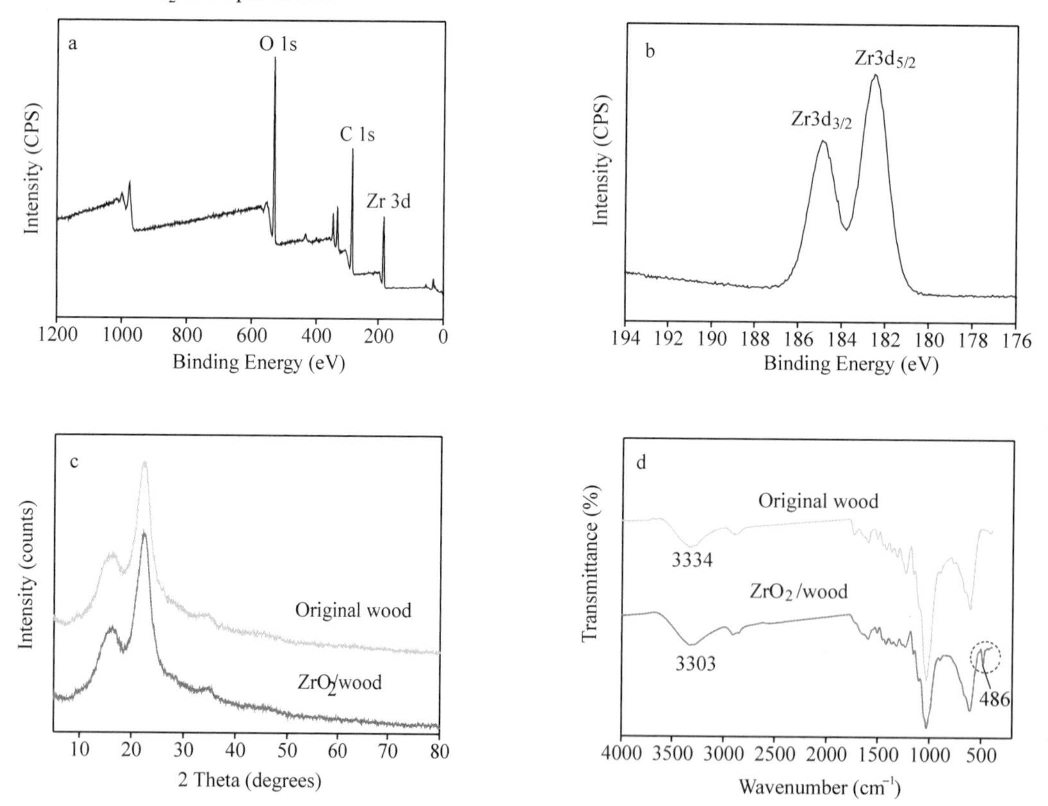

Fig. 1 Broad scan (a) and high resolution (b) XPS spectra of ZrO_2/wood. XRD patterns (c) and FTIR spectra (d) of ZrO_2/wood and original wood

The micrograph (Fig. 2a) obtained by SEM backscattering electron imaging, for the ZrO_2/wood nanohybrid, clearly show large-scale even-distributed aggregations of the oxide particles, appearing as bright points on the wood surface. These bright points were subsequently proved to be nano-ZrO_2 aggregations by EDX (Fig. 2c); the Cu peak originated from the coating layer used for electric conduction during the SEM observation, and all carbon and partial oxygen derived from the wood matrix. Fig. 2b shows the SEM image of ZrO_2/wood nanocomposite under higher magnification. Clearly, plentiful irregularly shaped ZrO_2 aggregations were layer-by-layer adhered to the wood substrate surface as indicated by the white arrow, and the driving force of layer-by-layer self-assembly behaviors of the aggregations is primarily due to the electrostatic adsorption. The increasing thickness of protective layer containing zirconium might be beneficial to resist ultraviolet radiation.

Fig. 2 SEM images at (a) low magnification and (b) high magnification of the ZrO_2/wood nanocomposite

Based on the above investigation of the experimental parameters for the ZrO_2/wood nanocomposite, its possible formation mechanism could be described as follows (Fig. 3). Firstly, the precipitation reaction, which took place as a result of mixing the $ZrOCl_2 \cdot 8H_2O$ aqueous solution and ammonia, might refer to the Eqs. (5). Then, the obtained $Zr(OH)_4$ nanoparticles were gradually reacted to generate nano-ZrO_2 with the help of hydrothermal energy [Eqs. (6)], and the resulting nano-ZrO_2 produced strong interaction with the hydroxyl groups of wood surface leading to formation of plentiful hydrogen bonds. Finally, owing to the electrostatic adsorption force, more ZrO_2 nanoparticles were attracted and deposited on the previous ZrO_2 or wood surfaces layer-by-layer, performing as a effective multilayered nano-uvioresistant agent.

$$ZrOCl_2 \cdot 8H_2O+NH_4OH \longrightarrow Zr(OH)_4+NH_4Cl+H_2O \qquad (5)$$

$$Zr(OH)_4 \xrightarrow{\text{Hydrothermal process}} ZrO_2 \qquad (6)$$

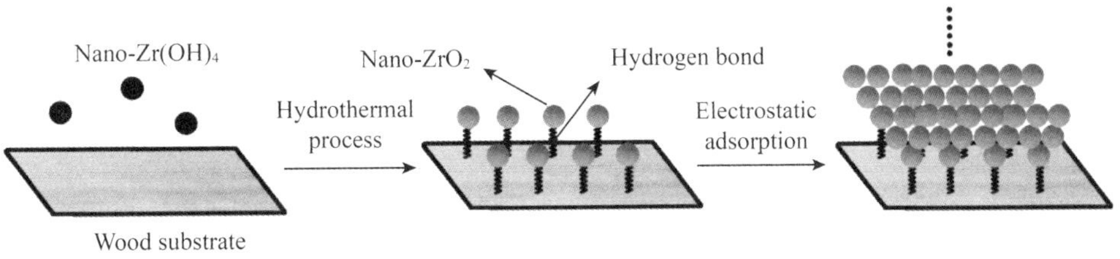

Nano-Zr(OH)₄ Nano-ZrO₂ Hydrogen bond
Hydrothermal process Electrostatic adsorption
Wood substrate

Fig. 3 Schematic illustrations for fabricating the ZrO_2/wood using the hydrothermal method

Fig. 4 represents the TG and DTG curves of ZrO_2/wood and original wood, respectively. The small initial drops occurring before 150 ℃ in both cases are due to the evaporation of retained moisture. The original wood started to degrade at around 192 ℃, whereas the ZrO_2/wood began to decompose at 223 ℃, respectively. At 50% weight loss, the decomposition temperature occurred at 339 ℃ for native wood and 381 ℃ for the nanocomposite. Moreover, the original wood reached maximum degradation rate at 357 ℃, which is 27 ℃ earlier than the ZrO_2/wood (ca. 384 ℃). These results imply that the thermal stability of ZrO_2/wood is higher than that of original wood. The reason of this enhancement is probably due to the strong interaction between wood substrate and ZrO_2. It is consistent with the shift of FTIR adsorption peak of ZrO_2/wood at 3334 cm^{-1} to lower wavenumber in Fig. 1b.

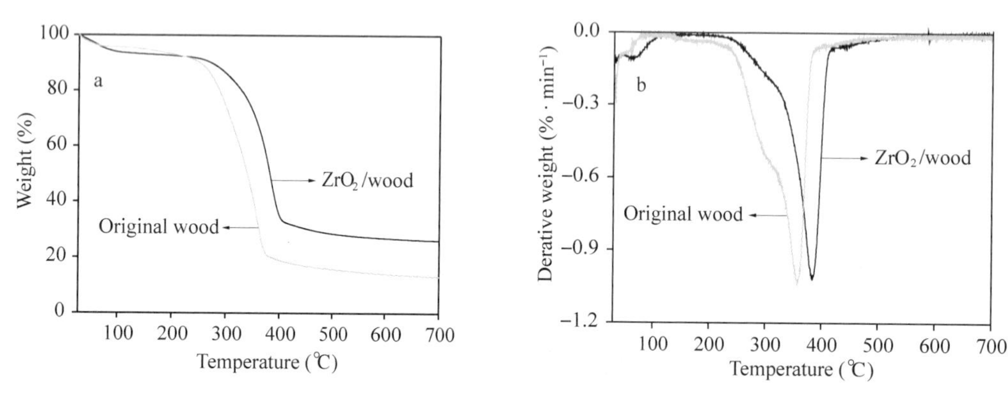

Fig. 4 TG (a) and DTG (b) curves of ZrO_2/wood and original wood, respectively

The color stability of the two material classes, the ZrO_2/wood and original wood, was examined by means of the change tendency of ΔL^*, Δa^*, Δb^* and ΔE^* (Fig. 5). It could be observed in Fig. 5 that all color feature parameters changes of original wood from 0 to 600 h are more significant than those of ZrO_2/wood notwithstanding the two materials both show similar change tendency of ΔL^*, Δa^*, Δb^* and ΔE^*, which suggests that the color stability of the ZrO_2/wood composite is more superior under UV light. Furthermore, the higher Δa^* and Δb^* and lower ΔL^* of original wood at the end of the 600-hour treatment (Fig. 5a-c) represent that the light-colored sample turned deeper shade of red, dark yellow and black, respectively. In addition, the total color change (ΔE^*) of the original wood is also obvious (Fig. 5d), nevertheless the change in the composite ΔE^* is much less (approximately 2/3). Therefore, the untreated wood with limited UV resistance suffered serious surface damage under UV light; hydrothermal treated wood exhibites more extraordinary UV protection, which also reveals that the ZrO_2 nanocoating is an effective UV protected agent.

4 Conclusions

The ZrO_2/wood nanocomposite was prepared by a simple mild one-pot hydrothermal method. The amorphous ZrO_2 aggregations, which were deposited on the surface of wood substrate with the help of hydrothermal energy and electrostatic adherence, generated strong hydrogen bonds with the hydroxide radicals of wood surface leading to a improvement of thermal stability for the nanocomposite. Through the 600-hour QUV accelerated aging test, the original wood suffered serious discoloration, whereas the nanocomposite shows

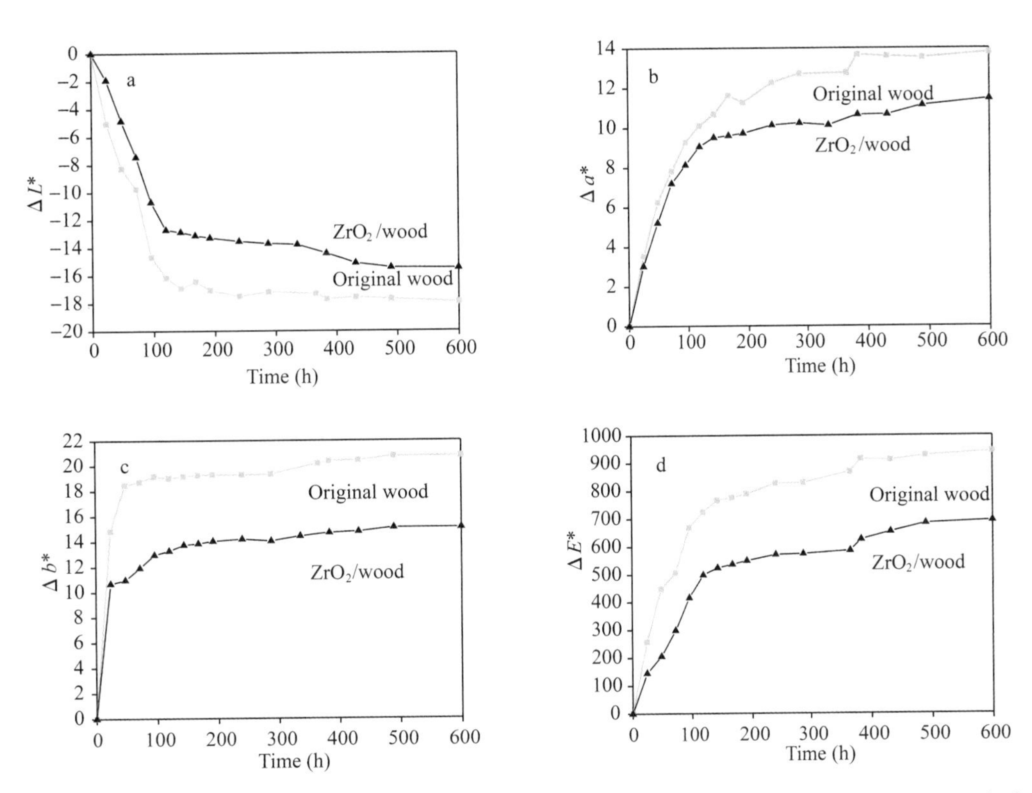

Fig. 5 Change tendency of ΔL^*, Δa^*, Δb^* and ΔE^* of ZrO_2/wood and original wood, respectively

superior anti-ultraviolet capability.

Acknowledgments

This work was financially supported by the National NaturalScience Foundation of China (grant no. 31270590).

References

[1] Ncube E, Meincken M. Surface characteristics of coated soft-and hardwoods due to UV-B ageing[J]. Applied Surface Science, 2010, 256(24): 7504-7509.

[2] Temiz A, Yildiz U, Aydin I, et al. Surface roughness and color characteristics of wood treated with preservatives after accelerated weathering test[J]. Applied Surface Science, 2005, 250(1-4): 35-42.

[3] Wang X, Ren H. Comparative study of the photo-discoloration of moso bamboo (Phyllostachys pubescens Mazel) and two wood species[J]. Applied Surface Science, 2008, 254(21): 7029-7034.

[4] Klaassen R K W M, van Overeem B S. Factors that influence the speed of bacterial wood degradation[J]. Journal of Cultural Heritage, 2012, 13(3): S129-S134.

[5] Raberg U, Terziev N, Daniel G. Degradation of Scots pine and beech wood exposed in four test fields used for testing of wood preservatives[J]. International Biodeterioration & Biodegradation, 2013, 79: 20-27.

[6] Rosu D, Teaca C A, Bodirlau R, et al. FTIR and color change of the modified wood as a result of artificial light irradiation[J]. Journal of Photochemistry and Photobiology B: Biology, 2010, 99(3): 144-149.

[7] Tolvaj L, Molnar Z, Nemeth R. Photodegradation of wood at elevated temperature: Infrared spectroscopic study[J]. Journal of Photochemistry and Photobiology B: Biology, 2013, 121: 32-36.

[8] Madey T, Faradzhev N, Yakshinskiy B, et al. Surface phenomena related to mirror degradation in extreme ultraviolet (EUV) lithography[J]. Applied Surface Science, 2006, 253(4): 1691-1708.

[9] Chang T C, Lin H Y, Wang S Y, et al. Study on inhibition mechanisms of light-induced wood radicals by Acacia confusa heartwood extracts[J]. Polymer Degradation and Stability, 2014, 105: 42-47.

[10] Pogge D J, Lonergan S M, Hansen S L. Influence of supplementing vitamin C to yearling steers fed a high sulfur diet during the finishing period on meat color, tenderness and protein degradation, and fatty acid profile of the longissimus muscle [J]. Meat Science, 2014, 97(4): 419-427.

[11] Nzokou P, Kamdem D P, Temiz A. Effect of accelerated weathering on discoloration and roughness of finished ash wood surfaces in comparison with red oak and hard maple[J]. Progress in Organic Coatings, 2011, 71(4): 350-354.

[12] Khelfa A, Bensakhria A, Weber J V. Investigations into the pyrolytic behaviour of birch wood and its main components: Primary degradation mechanisms, additivity and metallic salt effects [J]. Journal of Analytical and Applied Pyrolysis, 2013, 101: 111-121.

[13] Liu L, Sun J, Cai C, et al. Corn stover pretreatment by inorganic salts and its effects on hemicellulose and cellulose degradation[J]. Bioresource Technology, 2009, 100(23): 5865-5871.

[14] Van den Bulcke J, Van Acker J, Saveyn H, et al. Modelling film formation and degradation of semi-transparent exterior wood coatings[J]. Progress in Organic Coatings, 2007, 58(1): 1-12.

[15] Chou P L, Chang H T, Yeh T F, et al. Characterizing the conservation effect of clear coatings on photodegradation of wood[J]. Bioresource Technology, 2008, 99(5): 1073-1079.

[16] Forsthuber B, Müller U, Teischinger A, et al. Chemical and mechanical changes during photooxidation of an acrylic clear wood coat and its prevention using UV absorber and micronized TiO_2[J]. Polymer Degradation and Stability, 2013, 98(7): 1329-1338.

[17] Aloui F, Ahajji A, Irmouli Y, et al. Inorganic UV absorbers for the photostabilisation of wood-clearcoating systems: Comparison with organic UV absorbers[J]. Applied Surface Science, 2007, 253(8): 3737-3745.

[18] Cristea M V, Riedl B, Blanchet P. Enhancing the performance of exterior waterborne coatings for wood by inorganic nanosized UV absorbers[J]. Progress in Organic Coatings, 2010, 69(4): 432-441.

[19] Kiguchi M, Evans P D. Photostabilisation of wood surfaces using a grafted benzophenone UV absorber[J]. Polymer Degradation and Stability, 1998, 61(1): 33-45.

[20] Forsthuber B, Müller U, Teischinger A, et al. A note on evaluating the photocatalytical activity of anatase TiO_2 during photooxidation of acrylic clear wood coatings by FTIR and mechanical characterization[J]. Polymer Degradation and Stability, 2014, 105: 206-210.

[21] Kallio T, Alajoki S, Pore V, et al. Antifouling properties of TiO_2: Photocatalytic decomposition and adhesion of fatty and rosin acids, sterols and lipophilic wood extractives[J]. Colloids and Surfaces A: Physicochemical and Engineering Aspects, 2006, 291(1-3): 162-176.

[22] Salla J, Pandey K K, Srinivas K. Improvement of UV resistance of wood surfaces by using ZnO nanoparticles[J]. Polymer Degradation and Stability, 2012, 97(4): 592-596.

[23] Rodič P, Iskra J, Milošev I. Study of a sol-gel process in the preparation of hybrid coatings for corrosion protection using FTIR and 1H NMR methods[J]. Journal of Non-crystalline Solids, 2014, 396: 25-35.

[24] Nallis K, Katsumata K I, Isobe T, et al. Preparation and UV-shielding property of $Zr_{0.7}Ce_{0.3}O_2$-kaolinite nanocomposites[J]. Applied Clay Science, 2013, 80-81(aug.): 147-153.

[25] Pian X, Fan B, Chen H, et al. Preparation of m-ZrO_2 compacts by microwave sintering[J]. Ceramics International, 2014, 40(7): 10483-10488.

[26] Smirnov J R C, Calvo M E, Míguez H. Selective UV reflecting mirrors based on nanoparticle multilayers[J]. Advanced Functional Materials, 2013, 23(22): 2805-2811.

[27] Zhou J, Wu F X, Chen Y B, et al. Structural stability of layered n-$LaFeO_3$-$Bi_4Ti_3O_{12}$, $BiFeO_3$-$Bi_4Ti_3O_{12}$, and $SrTiO_3$-$Bi_4Ti_3O_{12}$ thin films[J]. Journal of Materials Research, 2012, 27(23): 2956-2964.

[28]X. Shi, C. Cui, C. Di, Y. Zou, M. Xu, Google Patents, 2011.

[29] Sun Q, Lu Y, Zhang H, et al. Hydrothermal fabrication of rutile TiO_2 submicrospheres on wood surface: An efficient method to prepare UV-protective wood[J]. Materials Chemistry and Physics, 2012, 133(1): 253-258.

[30] Li J, Yu H, Sun Q, et al. Growth of TiO_2 coating on wood surface using controlled hydrothermal method at low temperatures[J]. Applied Surface Science, 2010, 256(16): 5046-5050.

[31] Shukla S, Seal S, Vanfleet R. Sol-gel synthesis and phase evolution behavior of sterically stabilized nanocrystalline zirconia[J]. Journal of Sol-Gel Science and Technology, 2003, 27(2): 119-136.

[32] Botzakaki M A, Xanthopoulos N, Makarona E, et al. ALD deposited ZrO_2 ultrathin layers on Si and Ge substrates: A multiple technique characterization[J]. Microelectronic Engineering, 2013, 112: 208-212.

[33] Cai J, Kimura S, Wada M, et al. Nanoporous cellulose as metal nanoparticles support[J]. Biomacromolecules, 2009, 10(1): 87-94.

[34] Cai J, Zhang L. Rapid dissolution of cellulose in LiOH/urea and NaOH/urea aqueous solutions[J]. Macromolecular Bioscience, 2005, 5(6): 539-548.

[35] Zhao C, Richard O, Bender H, et al. E. Young, W. Tsai, G. Roebben and O. Van Der Biest, S. Haukka[J]. J. Non-Crys. Solids, 2002, 303: 144.

[36] Filho U P R, Gushikem Y, Fujiwara F Y, et al. Zirconium Dioxide Supported on. alpha. -Cellulose: Synthesis and Characterization[J]. Langmuir, 1994, 10(11): 4357-4360.

[37] Zink N, Emmerling F, Häger T, et al. Low temperature synthesis of monodisperse nanoscaled ZrO_2 with a large specific surface area[J]. Dalton Transactions, 2013, 42(2): 432-440.

[38] Lan W, Liu C F, Yue F X, et al. Ultrasound-assisted dissolution of cellulose in ionic liquid[J]. Carbohydrate Polymers, 2011, 86(2): 672-677.

[39] Olsson A M, Salmén L. The association of water to cellulose and hemicellulose in paper examined by FTIR spectroscopy[J]. Carbohydrate Research, 2004, 339(4): 813-818.

[40] Rana R, Langenfeld-Heyser R, Finkeldey R, et al. FTIR spectroscopy, chemical and histochemical characterisation of wood and lignin of five tropical timber wood species of the family of Dipterocarpaceae[J]. Wood Science and Technology, 2010, 44(2): 225-242.

[41] Kacurakova M, Capek P, Sasinkova V, et al. Carbohydr[J]. Polym. , 2000, 43, 195-203.

[42] Khan S B, Alamry K A, Marwani H M, et al. Synthesis and environmental applications of cellulose/ZrO_2 nanohybrid as a selective adsorbent for nickel ion[J]. Composites Part B Engineering, 2013, 50(jul.): 253-258.

[43] Lucovsky G, Rayner G B. Microscopic model for enhanced dielectric constants in low concentration SiO_2-rich noncrystalline Zr and Hf silicate alloys[J]. Applied Physics Letters, 2000, 77(18): 2912-2914.

[44] Dun H, Zhang W, Wei Y, et al. Layer-by-layer self-assembly of multilayer zirconia nanoparticles on silica spheres for HPLC packings[J]. Analytical Chemistry, 2004, 76(17): 5016-5023.

[45] Zhou H, Tian R, Ye M, et al. Highly specific enrichment of phosphopeptides by zirconium dioxide nanoparticles for phosphoproteome analysis[J]. Electrophoresis, 2010, 28(13): 2201-2215.

[46] Li J, Lu Y, Yang D, et al. Lignocellulose aerogel from wood-ionic liquid solution (1-allyl-3-methylimidazolium chloride) under freezing and thawing conditions[J]. Biomacromolecules, 2011, 12(5): 1860-1867.

[47] Shafizadeh F, Fu Y L . Pyrolysis of cellulose[J]. Carbohydrate Research, 1973, 29(1): 113-122.

[48] Yang H, Yan R, Chen H, et al. Characteristics of Hemicellulose, Cellulose and Lignin Pyrolysis[J]. Fuel, 2007, 86(12-13): 1781-1788.

[49] Valiokas R, Östblom M, Svedhem S, et al. Thermal stability of self-assembled monolayers: Influence of lateral

hydrogen bonding[J]. The Journal of Physical Chemistry B, 2002, 106(40): 10401-10409.

[50] Haurie L, Fernández A I, Velasco J I, et al. Thermal stability and flame retardancy of LDPE/EVA blends filled with synthetic hydromagnesite/aluminium hydroxide/montmorillonite and magnesium hydroxide/aluminium hydroxide/montmorillonite mixtures[J]. Polymer Degradation and Stability, 2007, 92(6): 1082-1087.

[51] Sun Q, Lu Y, Zhang H, et al. Improved UV resistance in wood through the hydrothermal growth of highly ordered ZnO nanorod arrays[J]. Journal of Materials Science, 2012, 47(10): 4457-4462.

中文题目：水热法在木材表面合成具有抗紫外线性能的二氧化锆涂层

作者：万才超，卢芸，孙庆丰，李坚

摘要：通过简单的温和的一锅水热法，纳米二氧化锆聚集体已成功地逐层沉积在木材表面上。通过扫描电子显微镜(SEM)，能量色散X射线光谱(EDX)，X射线衍射(XRD)，傅里叶变换红外光谱(FTIR)，X射线光电子能谱(XPS)对得到的二氧化锆/木材纳米复合材料进行表征 和热重分析(TGA)。结果表明，无定形二氧化锆与木材表面的氢氧根之间形成了很强的氢键，强相互作用促进了纳米复合材料的热稳定性。此外，与原始木材相比，二氧化锆/木材通过600 h QUV加速老化测试显示出更优异的抗紫外线能力。

关键词：水热合成；无机化合物；光降解；纳米粒子；层层自组装

Electromagnetic Shielding Wood-based Composite from Electroless Plating Corrosion Resistant Ni-Cu-P Coatings on *Fraxinus mandshurica* Veneer*

Bin Hui, Jian Li, Lijuan Wang

Abstract: A novel and simple electroless Ni-Cu-P plating process was used for preparing corrosion resistant and electromagnetic interference shielding wood-based composite. The effects of $CuSO_4 \cdot 5H_2O$ concentration, pH value in the plating solution and operation temperature on the metal deposition, surface resistivity, chemical composition, corrosion resistance and surface morphology of the composite were investigated. The surface morphologies were observed by using scanning electron microscopy (SEM) and analyzed by X-ray energy dispersive spectrometer (EDS). The electromagnetic shielding effectiveness (ESE) was measured by spectrum analyzer. The corrosion resistance was evaluated by potentiodynamic corrosion measurement. The results show that metal deposition increases with pH value and temperature increase, however, decreases with $CuSO_4 \cdot 5H_2O$ concentration increase. The corrosion resistance of the plated Ni-Cu-P coatings obviously depends on the total content of Cu and P in the coating. Higher total content of Cu and P leads to higher corrosion resistance. The optimum conditions are as follows: $CuSO_4 \cdot 5H_2O$ concentration of 1.0 $g \cdot L^{-1}$, pH value of 9.5, and operation temperature of 90 ℃. The obtained coating contains 77.41% Ni, 8.96% Cu, and 13.63% P. The wood-based composite exhibits higher corrosion resistance and *ESE* of around 60 dB in frequencies ranging from 9 kHz to 1.5 GHz. In this paper, a promising process for corrosion resistance and electromagnetic shielding wood-based composite was developed.

Keywords: Fraxinus mandshurica veneer; Electroless Ni-Cu-P plating; Corrosion resistance; Surface morphology; Electromagnetic shielding effectiveness

1 Introduction

In recent years, many kinds of electrical appliances have been commonly used. All electrical devices give off electromagnetic radiation which can cause serious and harmful electromagnetic interference (EMI). Electromagnetic shielding is an effective method for preventing EMI. So, electromagnetic shielding materials are widely used for reducing the electromagnetic field in a space by blocking the field. Those materials must be conductive or magnetic.

Wood is a natural and renewable biomass with many prominent properties. However, wood is not conductive, therefore, can not shield any EMI. To endow conductivity and electromagnetic shielding

---

* 本文选自 Wood Science & Technology, 2014, 48(5): 961-979.

performance to wood, conductive wood-based composite has attracted many attentions. Many researchers prepared conductive wood-based composite by using electroless nickel or copper plating on the surface to wood (Nagasawa et al., 1990, 1991a, b, 1992, 1999; Huang and Zhao, 2004; Wang et al., 2006a, b, 2008, 2011a, b, c; Wang and Li, 2007; Sun et al., 2012). As we all know, copper has high electric conductivity but poor corrosion resistance. It is easily oxidized to lose conductivity. Nickel has anticorrosion ability in some extent. But its conductivity is weaker. Therefore, the deposition of coatings with high conductivity and corrosion resistance has become a hot topic. Recently, it has been reported that the Ni-Cu-P coating has high antioxidant properties and excellent corrosion resistance (Yu et al., 2001). In addition, Abdel Hameed and Fekry (2010) reported that the Ni-Cu-P deposition results in higher corrosion resistance and lower corrosion current density (Icorr) than Ni-P deposition under various conditions. Ashassi Sorkhabi et al., (2002) thought that Ni-Cu alloys with Cu content around 30% are highly corrosion resistant under corrosion environments such as in halide solution like seawater.

Nonmetallic materials need to be activated prior to plating process due to no catalytic nuclei for initiating the autocatalytic stage. Traditionally, palladium colloid is used as catalyst in electroless plating because of its high activity. However, Pd is very expensive, and the colloid is not always stable. In recent years, $PdCl_2$ has been used as activator for electroless nickel or copper plating because of its controllability (Sun et al., 2012; Liu et al., 2010). But the operation process is complicated due to the fact that the coupling agents such as aminosilane or chitosan are used for combining Pd^{2+} with wood surface. With the objective of low-cost and simple activation process, many attempts have been made to employ Ni activation. Two methods are widely studied. One is that Ni^{2+} was loaded on the substrate and Ni^0 produced by thermal treatment at higher temperature (Li et al., 2006; Shao et al., 2007; Li and An 2008); the other is that Ni^{2+} was loaded on substrate and reduced to Ni^0 by $NaBH_4$ in two separate steps (Lai et al., 2006; Hu et al., 2006; Gao et al., 2007). In our lab, a simple activation method was used for nickel or copper plating (Li et al., 2010; Wang et al., 2011a) and Ni-Cu-P coatings plated on birch veneer (Hui et al., 2013, 2014). In order to investigate the universality of the activation process for different wood substrate, here, it was used for electroless plating Ni-Cu-P coatings on Fraxinus mandshurica veneer to prepare corrosion resistant and electromagnetic shielding wood-based composite. In the process, the wood veneer was immersed in $NaBH_4$ solution for loading $NaBH_4$, and then was directly placed in plating solution for activation and plating in one bath. Compared with previous electroless Ni-Cu-P plating, the process is greatly shortened, and the cost is decreased. Those provide a powerful foundation for the large-scale production and application of the functional wood-based composite.

2 Experimental

2.1 Materials

The substrates used were Fraxinus mandshurica veneers with thickness of 0.6 mm. All the chemicals were of analytical grade.

The veneers were polished by emery papers to remove fine fibers on the surface. The samples for corrosion studies were cut into squares in size of 30 mm×30 mm. The samples for electromagnetic shielding effectiveness and electric conductivity were cut according to Li et al. (2010).

2.2 Preparation of wood-based Ni-Cu-P composite

The preparing process of the wood-based composite is shown schematically in Fig. 1. First, the samples were dipped in $NaBH_4$ solution for a certain time; next, they were put in air for 1 min for $NaBH_4$ diffusing to the inner pores, and then they were placed in plating bath for activation and plating in one bath. After the samples were coated for a specified time, they were removed from bath, washed and air dried. The compositions of the plating solution and operating conditions used in this experiment were listed in Table 1. The solution pH was adjusted using $NH_3 \cdot H_2O$.

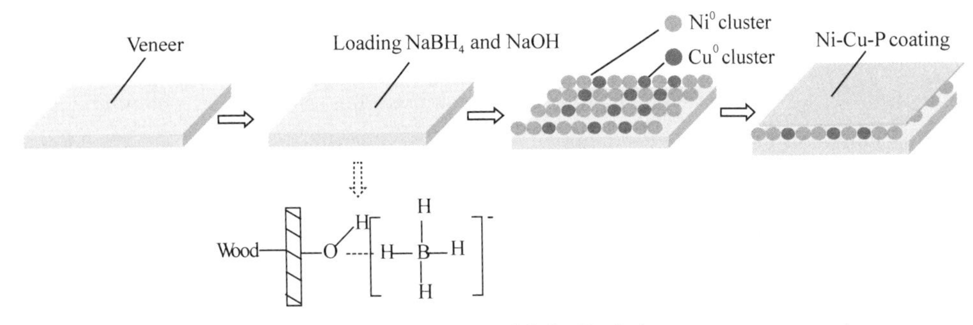

Fig. 1 Schematic illustration of electroless Ni-Cu-P plating process on wood veneer

Table 1 Compositions and operating conditions of electroless Ni-Cu-P plating

Chemicals	Content ($g \cdot L^{-1}$)
$NiSO_4 \cdot 6H_2O$	30
$CuSO_4 \cdot 5H_2O$	0.4-2.4
$NaH_2PO_2 \cdot H_2O$	40
Sodium citrate	50
Buffering agent	35
pH	7.5~10.5
Temperature (℃)	70~95

2.3 Measurement of metal deposition

The raw wood veneers were dried at 103±2 ℃ to constant weight (G_0). The Ni-Cu-P-coated veneers were also dried at 103±2 ℃ to constant weight (G_1). The metal deposition was calculated as:

$$\text{Metal deposition}(g/m^2) = (G_1 - G_0)/0.005 \tag{1}$$

where 0.005 is the total area of the sample (m^2).

2.4 Measurement of surface resistance and shielding effectiveness

The surface resistance of the metallized wood veneers was measured through the YD2511A-type smart low direct-current (DC) resistance tester, according to the Chinese National Military Standard GJB 2604-96. The detailed calculation was reported previously (Liu et al. 2010).

The electromagnetic shielding effectiveness (*ESE*) of the metallized wood veneers was measured with an Angilent E4402B spectrum analyzer and standard butt coaxial cable line with flange based on the Chinese Industrial Standard SJ20524-95. The *ESE* value was calculated as:

$$ESE\ (\mathrm{dB}) = -10\times\lg\left(\frac{P_{out}}{P_{in}}\right) \tag{2}$$

where P_{out} and P_{in} are the incidence and transmitted power, respectively.

2.5 Corrosion resistance

Potentiodynamic corrosion tests were employed to evaluate the anticorrosion properties of the Ni-Cu-P coatings. Tafel curves were obtained by using LK2005A electrochemical work station installed with software. A conventional three-electrode system was utilized in the measurements, where saturated calomel electrode (SCE) and platinum foil were used as reference and counter electrodes, respectively. The test sample was used as working electrode, and the exposed area to the corrosion medium was 30 mm×30 mm. Prior to the test, the sample was immersed in 3.5% NaCl solution in order to establish the open circuit potential (E_{ocp}) or steady state potential. The corrosion potential (E_{corr}) and corrosion current density (I_{corr}) were calculated from the intersection of the cathodic and anodic Tafel curves.

2.6 Characterization methods

In this study, the surface morphology and chemical composition of the coating were characterized by scanning electron microscopy (SEM, Quanta 200) and X-ray energy dispersive spectrometer (EDS). Samples were not sprayed with gold prior to analysis.

3 Results and discussion

3.1 Effect of $CuSO_4 \cdot 5H_2O$ concentration in the plating solution

$CuSO_4 \cdot 5H_2O$ concentration is a key factor for the deposition of Ni-Cu-P coatings. At pH value of 9.5 and operation temperature of 90 ℃, the effects of $CuSO_4 \cdot 5H_2O$ concentration on the metal deposition and surface resistivity are shown in Fig. 2. With increasing $CuSO_4 \cdot 5H_2O$ concentration from 0.6 to 2.2 $g \cdot L^{-1}$, the metal deposition decreased from 98.72 to 49.52 $g \cdot m^{-2}$ and the surface resistivity increased from 294.6 to 601.7 $m\Omega \cdot cm^{-2}$. In this process, nickel acts as activator for initiating the electroless plating. Once one film of nickel coating is formed, the reaction can proceed automatically. The addition of $CuSO_4 \cdot 5H_2O$ suppresses nickel deposition and blocks the activation process, therefore, leading to the decrease of metal deposition.

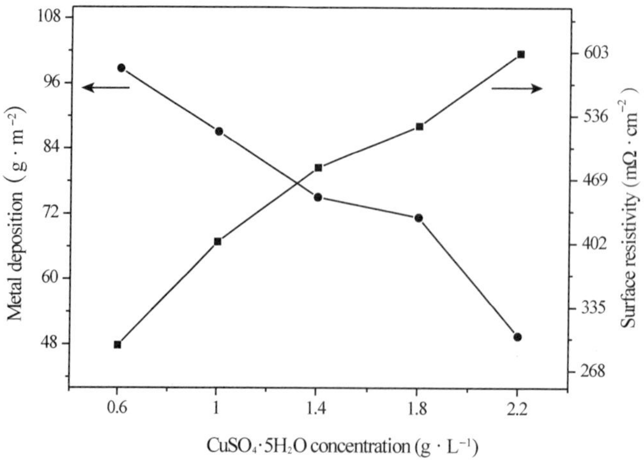

Fig. 2 Effect of $CuSO_4 \cdot 5H_2O$ concentration on metal deposition and surface resistivity (pH 9.5; temperature 90 ℃)

The effects of $CuSO_4 \cdot 5H_2O$ concentration on the chemical composition of coatings are shown in Fig. 3a. With increasing $CuSO_4 \cdot 5H_2O$ concentration from 0.6 to 1.4 g · L^{-1}, Cu content in the coating increased from 7.27 to 10.83%, while the Ni content slightly decreased from 78.58 to 77.52% and P content decreased from 14.15 to 11.65%. For the decrease of Ni and P content in the coating, it can be explained by the inhibition effect of nickel reduction by copper. This is due to the fact that the Cu is preferentially deposited because of its greater reduction potential than that of the Ni, as follows (Eqs. 3, 4) (Mallory and Hajdu1990; Bard and Faulkner 1980).

$$Ni^{2+}+2e^{-} \rightarrow Ni[E^{\ominus}(NHE)=-0.257\ V] \quad (3)$$

$$Cu^{2+}+2e^{-} \rightarrow Cu[E^{\ominus}(NHE)=0.340\ V] \quad (4)$$

As shown in Fig. 3b and Table 2, with increasing $CuSO_4 \cdot 5H_2O$ concentration from 0.6 to 1.4 g · L^{-1}, the E_{corr} of the Ni-Cu-P coating was positively shifted from −0.7746 to −0.6200 V and the I_{corr} was significantly reduced. At $CuSO_4 \cdot 5H_2O$ concentration of 1.4 g · L^{-1}, the I_{corr} reached the lowest value of 2.20×10^{-9} A · cm^{-2}, which is decreased by nearly 99% compared with the coating prepared at $CuSO_4 \cdot 5H_2O$ concentration of 0.6 g · L^{-1}. Therefore, the corrosion resistance is positively correlated with $CuSO_4 \cdot 5H_2O$ concentration in the plating solution. For the mechanism of corrosion resistance of the coating, it has been reported that this is ascribed to the formation of a P-rich surface film at the alloy/solution interface, blocking the dissolution of Ni and diffusion of Ni^{2+} toward bulk solution (Diegle et al. 1988; Krolikowski and Wiecko 2002; Liu et al. 2013). Meanwhile the incorporation of Cu deposition can reduce the free energy of Ni-Cu-P coatings due to the higher thermodynamic stability of Cu than that of Ni. In addition, the incorporated Cu deposition inhibits the cathodic reactions by increasing the over-potential of hydrogen evolution on Ni-Cu-P coatings (Georgieva and Armyanov 2007). By chemical composition analysis, we can find out that the total content of Cu and P in the coating increased with increasing $CuSO_4 \cdot 5H_2O$ concentration. The result proves the corrosion resistant sequence of coatings.

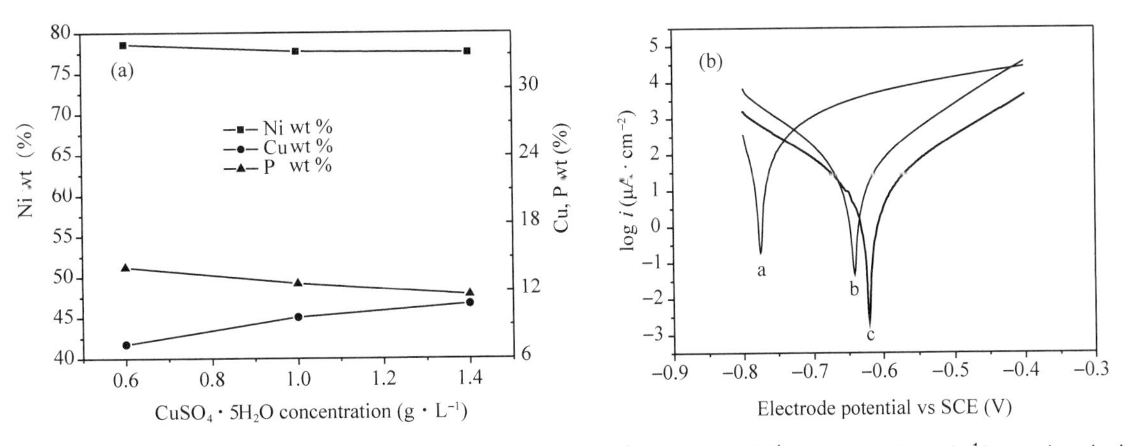

Fig. 3 Effect of $CuSO_4 \cdot 5H_2O$ concentration (a 0.6 g · L^{-1}, b 1.0 g · L^{-1}, and c 1.4 g · L^{-1}) on chemical composition and corrosion behavior (pH 9.5; temperature 90 ℃)

Table 2 Corrosion characteristics of electroless Ni-Cu-P plating

Sample	Scanning range (V)	E_{corr}(V vs. SCE)	I_{corr}($A \cdot cm^{-2}$)
a	−0. 80 ~ −0. 40	−0. 7746	2.00×10^{-7}
b	−0. 80 ~ −0. 40	−0. 6434	5.03×10^{-8}
c	−0. 80 ~ −0. 40	−0. 6200	2.20×10^{-9}

The following two samples were prepared at $CuSO_4 \cdot 5H_2O$ concentrations of 0. 6 and 1. 4 $g \cdot L^{-1}$, respectively. The surface morphology of Ni-Cu-P coatings with various Cu and P content was observed using SEM. The corresponding images of Ni−7. 27% Cu−14. 15% P and Ni−10. 83% Cu−11. 65% P plated on wood veneers are showed in Figs. 4, 5. The smoother structure can be observed for the higher Cu content sample (Figs. 5a, b). The influence of Cu on Ni-Cu-P morphology can be understood, where Cu^{2+} activates natural nucleation sites and retards the nodule growth, resulting in the smoother surface. After corrosion tests in 3. 5% NaCl solution, it is evident that Ni−10. 83% Cu−11. 65% P exhibits better corrosion resistance (Figs. 5c, d). This is probably ascribed to the more total content of Cu and P in the coating. In order to further dissect the relation between the corrosion resistance and surface morphology, we compared the surface resistivity of coatings before and after corrosion tests. It is shown in Table 3 that the added surface resistivity of the coating prepared at $CuSO_4 \cdot 5H_2O$ concentration of 1. 4 $g \cdot L^{-1}$ is lower than that at $CuSO_4 \cdot 5H_2O$ concentration of 0. 6 $g \cdot L^{-1}$. The results are in good agreement with that obtained previously by Abdel Aal and Shehata Aly (2009) on open cell stainless steel foam.

Table 3 Surface resistivity of coatings before and after corrosion tests

Surface resistivity ($m\Omega \cdot cm^{-2}$)	$CuSO_4 \cdot 5H_2O$ concentration ($g \cdot L^{-1}$)	
	0. 6	1. 4
Before corrosion	294. 6	481. 3
After corrosion	503. 5	642. 7
Added value	208. 9	161. 4

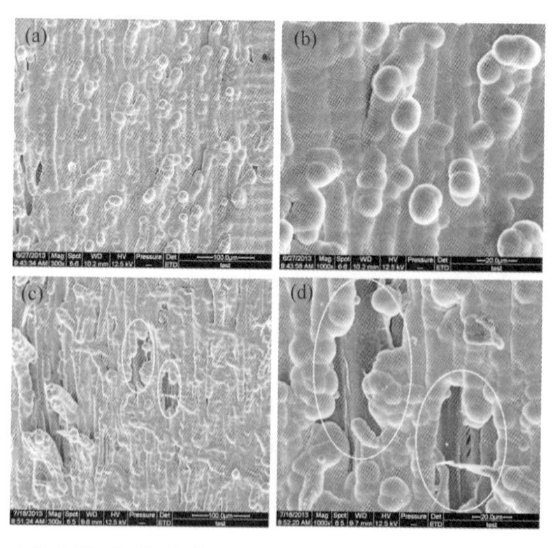

Fig. 4 Surface morphology of plated Ni−7. 27% Cu−14. 15% P on wood veneer (a and b) before and (c and d) after corrosion tests ($CuSO_4 \cdot 5H_2O$ 0. 6 $g \cdot L^{-1}$; pH 9. 5; temperature 90 ℃)

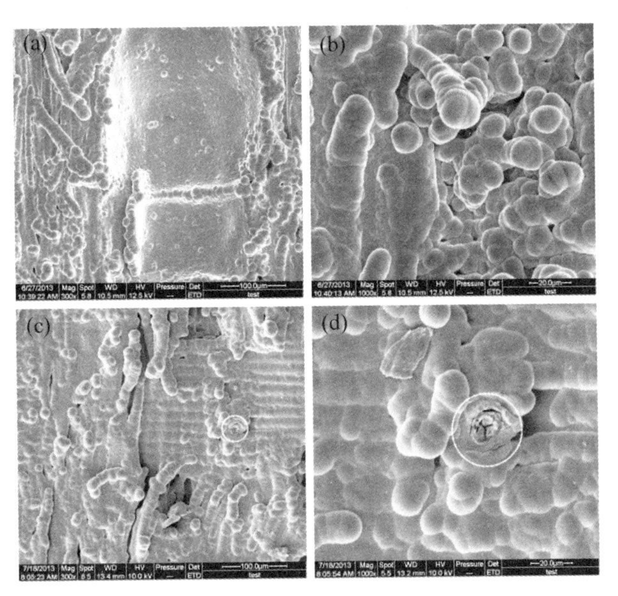

Fig. 5 Surface morphology of plated Ni–10. 83% Cu–11. 65% P on wood veneer (a and b) before and (c and d) after corrosion tests ($CuSO_4 \cdot 5H_2O$ 1. 4 g · L^{-1}; pH 9. 5; temperature 90 ℃)

3. 2 Effect of pH in the plating solution

pH plays an important role in the deposition of Ni-Cu-P coatings. At $CuSO_4 \cdot 5H_2O$ concentration of 1. 0 g · L^{-1} and operation temperature of 90 ℃, the effects of pH value on the metal deposition and surface resistivity are shown in Fig. 6. With the increase of pH from 8. 5 to 9. 5, the metal deposition increased from 71. 00 to 99. 98 g · m^{-2} and the surface resistivity decreased from 479. 9 to 242. 1 mΩ · cm^{-2}. An increase of solution pH can produce more Ni and Cu, whereas repress P deposition (Eqs. 5, 6, 7). As well known, the P deposition can significantly affect the conductive properties of coatings. The less the P content is, the higher electrical conductivity is. However, when further increasing pH to 10. 0, the metal deposition decreased to 76. 58 g · m^{-2} and the surface resistivity sharply increased to 679. 7 mΩ · cm^{-2} because the excess OH^- could make Ni^{2+} and Cu^{2+} transform to $Ni(OH)_2$ and $Cu(OH)_2$ which destroyed the continuity of the Ni-Cu-P coating. The autocatalytic stage is expressed as follows:

$$2H_2PO_2^- + 4OH^- + Ni^{2+} \rightarrow 2HPO_3^{2-} + 2H_2O + Ni \downarrow + H_2 \uparrow \quad (5)$$

$$2H_2PO_2^- + 4OH^- + Cu^{2+} \rightarrow 2HPO_3^{2-} + 2H_2O + Cu \downarrow + H_2 \uparrow \quad (6)$$

$$2H_2PO_2^- + 6H^+ + 4H^- \rightarrow 4H_2O + 2P \downarrow + 3H_2 \uparrow \quad (7)$$

Fig. 7a shows the chemical composition of Ni-Cu-P coatings in the pH range from 9. 0 to 10. 0. With increasing pH from 9. 0 to 10. 0, Ni content in the coating increased from 77. 37% to 82. 76% and Cu content slightly increased from 7. 80% to 8. 74%, while the P content decreased from 14. 83% to 8. 50%. Those results can be explained by Eqs. 5, 6, and 7. An increase of OH^- concentration can accelerate the reaction to produce more Ni and Cu and inhibit P deposition. As shown in Fig. 7b and Table 4, with increasing pH from 9. 0 to 10. 0, the E_{corr} of the coating deceased from −0. 6330 to −0. 6938 V. From the perspective of thermodynamics, the more positive the potential is, the stronger the corrosion resistance tends to be. The anticorrosion ability of coatings prepared at pH 9. 0 or 9. 5 was higher than that at pH 10. 0. The corrosion resistance is probably

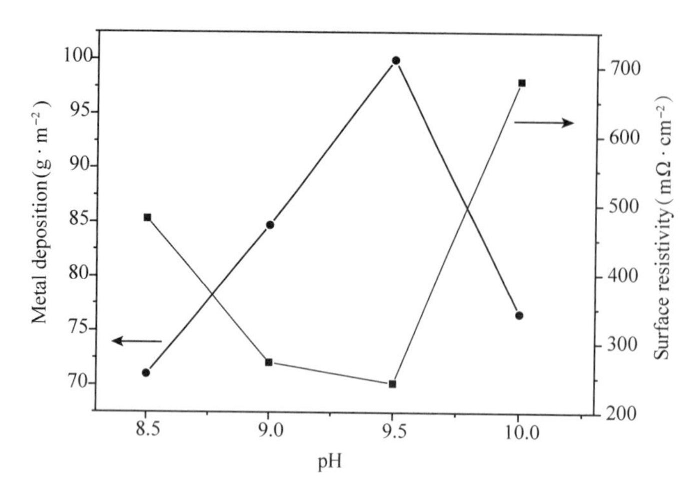

Fig. 6 Effect of pH on metal deposition and surface resistivity ($CuSO_4 \cdot 5H_2O$ 1.0 g · L^{-1}; temperature 90 ℃)

related to the total content of Cu and P in the coating. Chemical composition analysis shows that the total content of Cu and P in the coating prepared at pH 9.0 or 9.5 is more than that at pH 10.0.

Table 4 Corrosion characteristics of electroless Ni-Cu-P plating

Sample	Scanning range (V)	E_{corr}(V vs. SCE)	I_{corr}(A · cm^{-2})
a	−0.80 — −0.40	−0.6330	3.93×10^{-8}
b	−0.80 — −0.40	−0.6420	5.50×10^{-8}
c	−0.80 — −0.40	−0.6938	5.69×10^{-8}

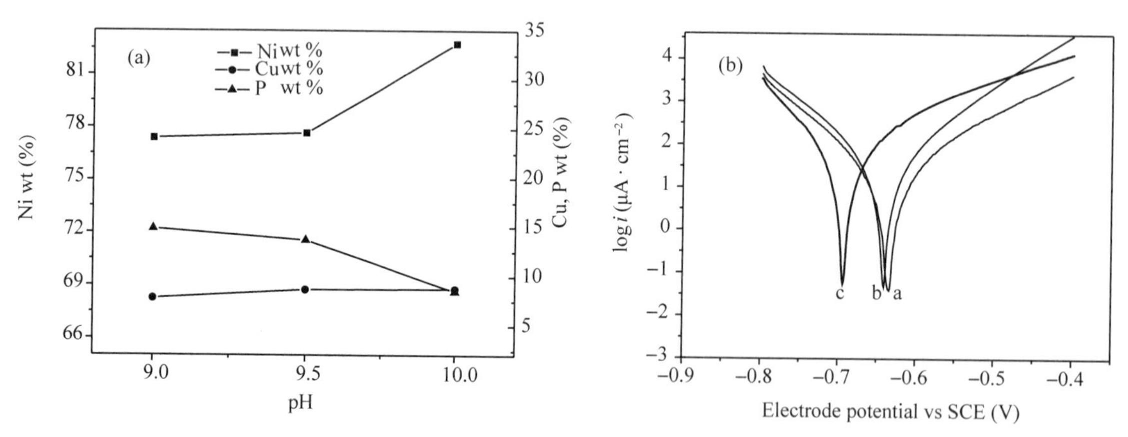

Fig. 7 Effect of pH (a 9.0, b 9.5, and c 10.0) on chemical composition and corrosion behavior ($CuSO_4 \cdot 5H_2O$ 1.0 g · L^{-1}; temperature 90 ℃)

The following two samples were prepared at solution pH values of 9.0 and 10.0, respectively. The corresponding surface morphologies are similar (Figs. 8a, b; Figs. 9a, b). This is because of the approximate Cu content in both coatings. After corrosion tests, delaminated areas of plated Ni−7.80% Cu−14.83% P on wood veneer (Figs. 8c, d) are smaller than those of plated Ni−8.74% Cu−8.50% P veneer (Figs. 9c, d). As well known that wood substrate is not conductive. The electrical conductivity of the composite comes from the alloy coating and relies on the continuity of the coating. After corrosion test, delaminated areas damaged the continuity of the coating, leading to the increase of the surface resistivity. Therefore, the change of surface resistivity can reflect the corrosion resistance in its entirety. It is shown in Table 5 that the surface resistivity of the coating prepared at pH 9.0 increased less than that at pH 10.0 after corrosion, indicating that the former coating has higher corrosion resistance. This is mainly attributed to the higher content of Cu and P combined in the coating.

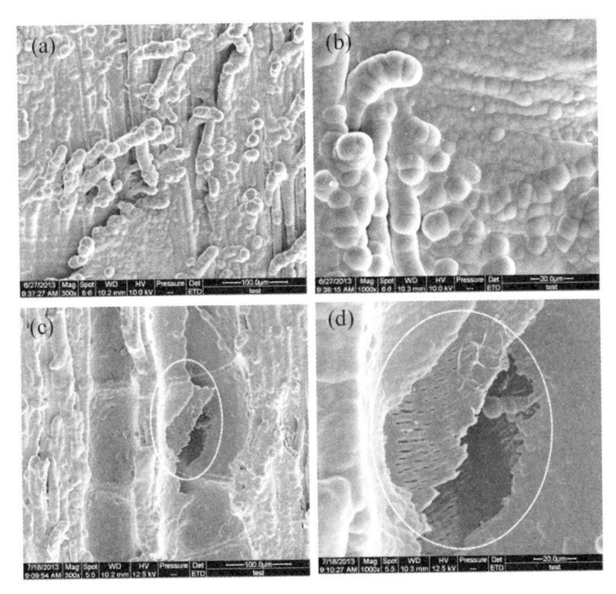

Fig. 8 Surface morphology of plated Ni-7.80% Cu-14.83% P on wood veneer (a and b) before and (c and d) after corrosion tests ($CuSO_4 \cdot 5H_2O$ 1.0 g · L^{-1}; pH 9.0; temperature 90 ℃)

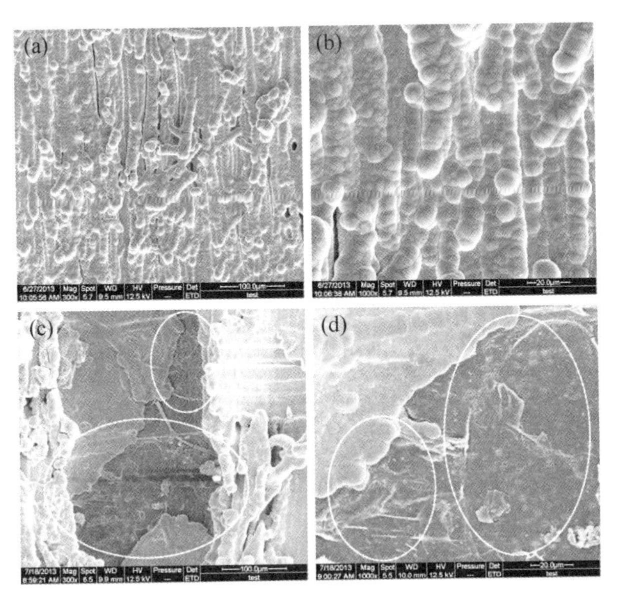

Fig. 9 Surface morphology of plated Ni-8.74% Cu-8.50% P on wood veneer (a and b) before and (c and d) after corrosion tests ($CuSO_4 \cdot 5H_2O$ 1.0 g · L^{-1}; pH 10.0; temperature 90 ℃)

Table 5 Surface resistivity of coatings before and after corrosion tests

Surface resistivity ($m\Omega \cdot cm^{-2}$)	pH	
	9.0	10.0
Before corrosion	271.40	679.70
After corrosion	382.83	843.52
Added value	111.43	163.82

3.3 Effect of the operation temperature

The operation temperature can significantly affect the metal deposition. At $CuSO_4 \cdot 5H_2O$ concentration of 1.0 $g \cdot L^{-1}$ and pH value of 9.5 in the plating solution, the effects of operation temperature on the metal deposition and surface resistivity are shown in Fig. 10. With increasing temperature from 75 to 85 ℃, the metal deposition obviously increased from 64.92 to 101.24 $g \cdot cm^{-2}$ and the surface resistivity decreased from 404.5 to 232.6 $m\Omega \cdot cm^{-2}$. Further increasing temperature did not markedly affect the metal deposition and the surface resistivity. As well known, the operation temperature can increase the migration velocity of ions, leading to the accelerated reactions. However, when temperature is much higher, the migration velocity of ions reaches to saturation. In addition, the plating solution is not stable for much higher temperature. Therefore, the appropriate operation temperature is of great importance for fabricating a superior coating.

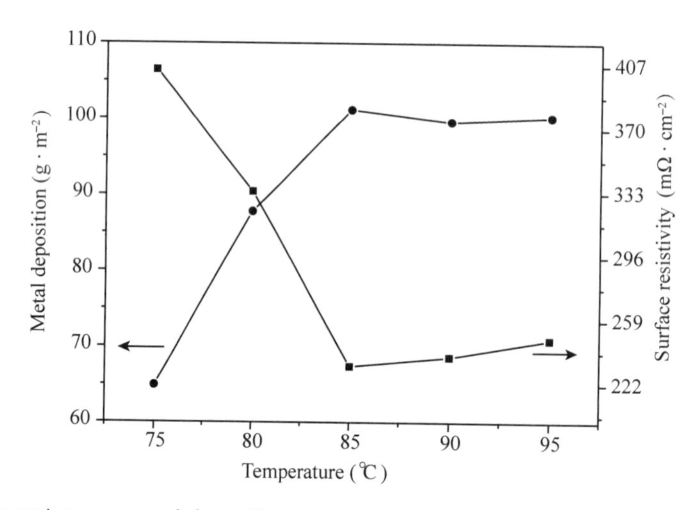

Fig. 10 Effect of temperature on metal deposition and surface resistivity ($CuSO_4 \cdot 5H_2O$ 1.0 $g \cdot L^{-1}$; pH 9.5)

Fig. 11a shows the variation of chemical composition at different operation temperatures. With temperature increase from 80 to 90 ℃, the Cu content in the coating slightly increased from 8.40% to 8.56% and P content increased from 9.39% to 14.54%, while the Ni content decreased from 82.21% to 76.90%. One of the reasons could be that the migration velocity of copper and phosphorus ions significantly increases with temperature increase, resulting in the increase of total content of Cu and P in the coating. As shown in Fig. 11b and Table 6, with increasing temperature from 80 to 90 ℃, the E_{corr} of the coating increased from −0.7116 to −0.6406 V, showing that the anticorrosion ability of the coating was improved. For the higher total content of Cu and P in the coating, this leads to the improvement of the corrosion resistance.

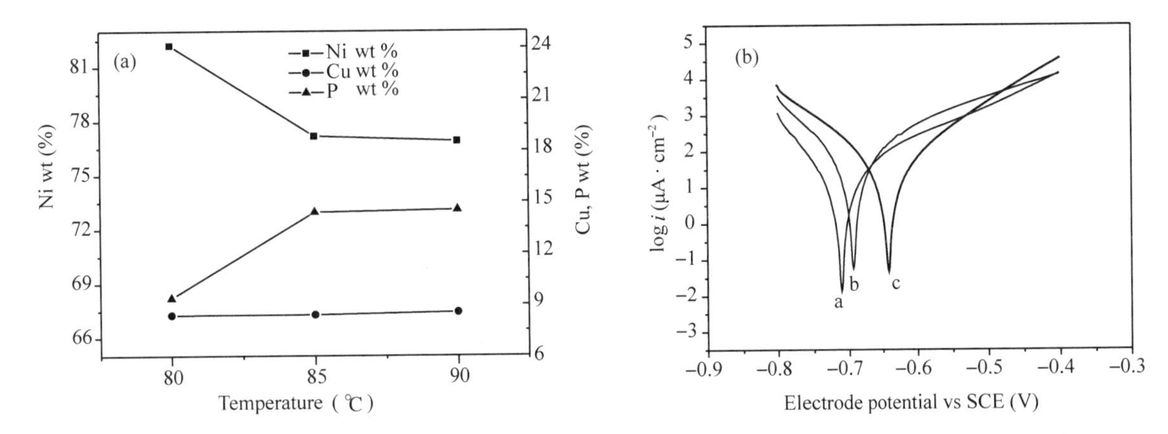

Fig. 11 Effect of temperature (a 80, b 85, and c 90 ℃) on chemical composition and corrosion behavior ($CuSO_4 \cdot 5H_2O$ 1.0 g · L^{-1}; pH 9.5)

Table 6 Corrosion characteristics of electroless Ni-Cu-P plating

Sample	Scanning range (V)	E_{corr}(V vs. SCE)	I_{corr}($A \cdot cm^{-2}$)
a	−0.80 ~ −0.40	−0.7116	1.47×10^{-8}
b	−0.80 ~ −0.40	−0.6927	5.90×10^{-8}
c	−0.80 ~ −0.40	−0.6406	5.03×10^{-8}

The following two samples were prepared at operation temperatures of 80 and 90 ℃, respectively. With the approximate Cu content in the coating, the nodular characteristics of deposited particles had no visible difference (Figs. 12a, b; Figs. 13a, b). However, after corrosion tests in 3.5% NaCl solution, the ratio between delaminated areas of the coatings at 80 ℃ and 90 ℃ was 1.48. In addition, it is shown in Table 7 that the added values of surface resistivity of the coatings prepared at 80 ℃ and 90 ℃ were 162.7 $m\Omega \cdot cm^{-2}$ and 148.3 $m\Omega \cdot cm^{-2}$, respectively. The results demonstrate that the corrosion resistance improves with temperature increase in some extent. This is mainly ascribed to the higher total content of Cu and P in the coating prepared at higher temperature.

Table 7 Surface resistivity of coatings before and after corrosion tests

Surface resistivity($m\Omega \cdot cm^{-2}$)	Temperature (℃)	
	80	90
Before corrosion	333.9	238.0
After corrosion	496.6	386.3
Added value	162.7	148.3

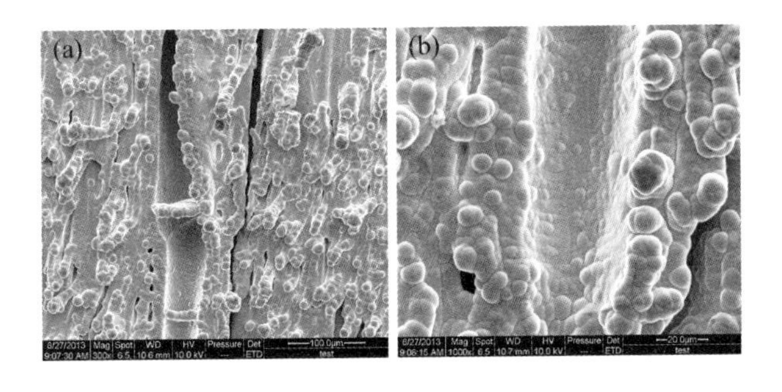

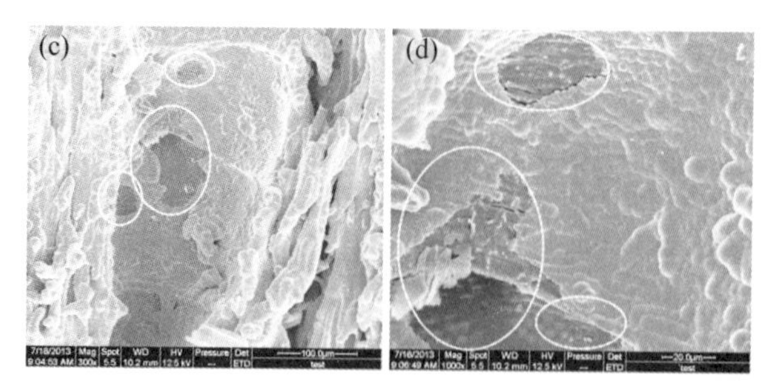

Fig. 12 Surface morphology of plated Ni-8.40% Cu-9.39% P on wood veneer (a and b) before and (c and d) after corrosion tests ($CuSO_4 \cdot 5H_2O$ 1.0 g $\cdot L^{-1}$; pH 9.5; temperature 80 ℃)

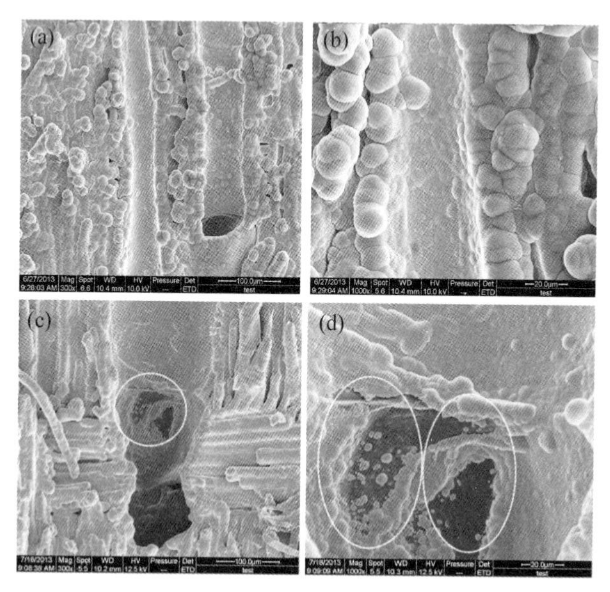

Fig. 13 Surface morphology of plated Ni-8.56% Cu-14.54% P on wood veneer (a and b) before and (c and d) after corrosion tests ($CuSO_4 \cdot 5H_2O$ 1.0 g $\cdot L^{-1}$; pH 9.5; temperature 90 ℃)

3.4 Shielding effectiveness of the wood-based composite

The sample of *ESE* was prepared at operating parameters, including $CuSO_4$ concentration of g $\cdot L^{-1}$, pH value of 9.5, and temperature of 90 ℃. Figures 14a, b shows the electromagnetic shielding results of the pristine and Ni-Cu-P-coated veneer, respectively. It is found that the pristine veneer fluctuated around 0 dB, indicating that it has no shielding effectiveness; however, the *ESE* of the Ni-Cu-P-coated veneer is around 60 dB in frequencies from 9 kHz to 1.5 GHz. Generally, it is regarded as an effective shield if the *ESE* value is over 35 dB. Therefore, the Ni-Cu-P-coated veneer is practically useful for some EMI shielding application and can meet special needs, such as advanced electronic products and national defense field.

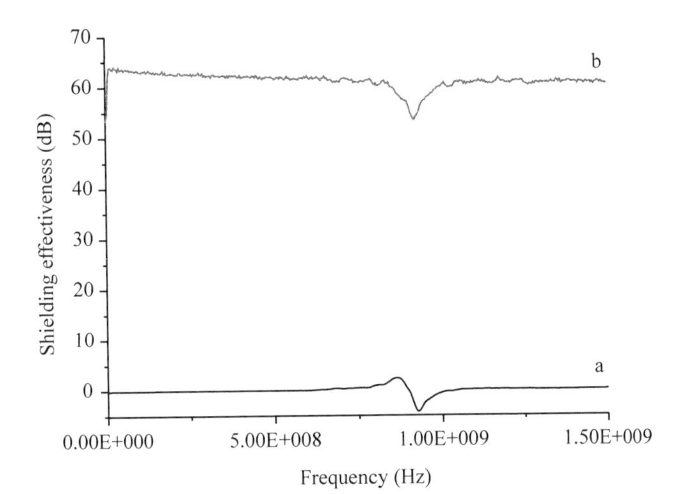

Fig. 14 Electromagnetic shielding effectiveness of wood veneers (a) before and (b) after plating ($CuSO_4 \cdot 5H_2O$ 1.0 g · L^{-1}; pH 9.5; temperature 90 ℃)

4 Conclusions

The operating parameters, including $CuSO_4 \cdot 5H_2O$ concentration, pH value in the plating solution and operation temperature, had significant effects on the metal deposition, surface resistivity, chemical composition, corrosion resistance, and surface morphology of the composite. The conclusions are summarized as follows:

(1) Metal deposition is negatively correlated with $CuSO_4 \cdot 5H_2O$ concentration, while positively correlated with pH value and temperature. Surface resistivity is negatively correlated with metal deposition.

(2) EDS analysis results show that, with the increase of $CuSO_4 \cdot 5H_2O$ concentration in the plating solution, Cu content in the coating increases significantly, whereas Ni and P content decrease. With increasing pH value, Ni and Cu content increase, while P content decreases.

(3) With the increase of operation temperature, Cu and P content in the coating increase, while Ni content decreases.

(4) The higher content of Cu and P combined in the coating favors the improvement of the corrosion resistance, and the images of the corresponding plated veneers are superior after corrosion in 3.5% NaCl solution.

(5) The optimum parameters to obtain corrosion resistant and EMI shielding wood-based composite is $CuSO_4 \cdot 5H_2O$ concentration of 1.0 g · L^{-1}, pH value of 9.5, and temperature of 90 ℃.

(6) The coating obtained under optimum conditions contains 77.41% nickel, 8.96% copper, and 13.63% phosphorus.

(7) The electromagnetic shielding effectiveness of the plated *Fraxinus mandshurica* veneer is around 60 dB in frequencies ranging from 9 kHz to 1.5 GHz.

Acknowledgements

The research is supported by " the Fundamental Research Funds for the Central Universities"

(2572014EB02-01).

References

[1] Abdel Aal A, Shehata Aly M. Electroless Ni-Cu-P plating onto open cell stainless steel foam[J]. Appl Surf. Sci., 2009, 255(13): 6652-6655.

[2] Abdel Hameed R M, Fekry A M. Electrochemical impedance studies of modified Ni-P and Ni-Cu-P deposits in alkaline medium[J]. Electrochim Acta, 2010, 55(20): 5922-5929.

[3] Ashassi Sorkhabi H, Dolati H, et al. Electroless deposition of Ni-Cu-P alloy and study of the influences of some parameters on the properties of deposits[J]. Appl Surf. Sci., 2002, 185(3): 155-160.

[4] Bard A J, Faulkner L R. Electrochemical Methods: Fundamentals and Applications[J]. Wiley, New York, 1980.

[5] Diegle R B, Sorensen N R, Clayton CR, Helfand MA, Yu YC. An XPS investigation into the passivity of an amorphous Ni-20P alloy[J]. J. Electrochem Soc., 1988, 135(5): 1085-109.

[6] Gao G Q, Huang J T. The effect factor analysis of the nickel activation techniques on the uniformity of the plating layer in wood electroless nickel plating[J]. J. Inner Mongolia Agric Univ., 2007, 28: 95-98.

[7] Georgieva J, Armyanov S. Electroless deposition and some properties of Ni-Cu-P and Ni-Sn-P coatings[J]. J. Solid State Electr., 2007, 11(7): 869-876.

[8] Hu G H, Wu H H, Yang F Z. Electroless nickel plating on carbon nanotube with non-palladium activation process [J]. Electrochemistry, 2006, 12(1): 25-28.

[9] Huang J T, Zhao G J. Electroless plating of wood[J]. J. Beijing For. Univ., 2004, 3: 88-92.

[10] Hui B, Li J, Wang L J. Preparation of EMI shielding and corrosion-resistant composite based on electroless Ni-Cu-P coated wood[J]. BioResour, 2013, 8(4): 6097-6110.

[11] Hui B, Li J, Zhao Q, Liang T Q, et al. Effect of $CuSO_4$ content in the plating bath on the properties of composites from electroless plating of Ni-Cu-P on birch veneer[J]. BioResour, 2014, 9(2): 2949-2959.

[12] Krolikowski A, Wiecko. A Impedance studies of hydrogen evolution on Ni-P alloys[J]. Electrochim Acta., 2002, 47 (13): 2065-2069.

[13] Lai D Z, Chen W X, Yao Y F, et al. Novel activation method using chemical plating on the fabric[J]. J. Textile Res, 2006, 27(1): 34-37.

[14] Li J, Wang L J, Liu H B. A new process for preparing conducting wood veneers by electroless nickel plating[J]. Surf. Coat Technol., 2010, 204(8): 1200-1205.

[15] Li L B, An M Z. Electroless nickel-phosphorus plating on SiCp/Al composite from acid bath with nickel activation [J]. J. Alloy and Compd., 2008, 461(1): 85-91.

[16] Li L B, An M Z, Wu G H. A new electroless nickel deposition technique to metallise SiCP/Al composites[J]. Surf. Coat Technol., 2006, 200(16): 5102-5112.

[17] Liu G C, Huang Z X, Wang L D, Sun W, Wang SL, Deng XL. Effects of Ce4+ on the structure and corrosion resistance of electroless deposited Ni-Cu-P coating[J]. Surf. Coat Technol., 2013, 222: 25-30.

[18] Liu H B, Li J, Wang L J. Electroless nickel plating on APTHS modified wood veneer for EMI shielding[J]. Appl Surf. Sci., 2010, 257(4): 1325-1330.

[19] Mallory G O, Hajdu J B. Electtoless Plating: Fundamentals and Applications[J]. American Electroplaters and Surface Finish Society, Orlando, Florida, 1990.

[20] Nagasawa C, Kumagai Y, Koshizaki N, et al. Changes in electromagnetic shielding properties of particlesboards made of nickel-plated wood particles formed by various pre-treatment processes[J]. J. Wood Sci., 1992, 38: 256-263.

[21] Nagasawa C, Kumagai Y, Urabe K. Electromagnetic shielding effectiveness particles board containing nickel-metalized wood particles in the core layer[J]. J. Wood Sci., 1990, 36(7): 531-537.

[22] Nagasawa C, Kumagai Y, Urabe K. Electro-conductivity and electromagnetic shielding effectiveness of nickel-plated

veneer[J]. J. Wood Sci., 1991a, 37(2): 158-163.

[23]Nagasawa C, Kumagai Y, Urabe K, et al. Electromagnetic shielding particle board with nickel-plated wood particles [J]. J. Porous Mater, 1999, 6: 247-254.

[24]Nagasawa C, Umehara H, Koshizaki N. Effects of wood species on electroconductivity and electromagnetic shielding properties of electrolessly plated sliced veneer with nickel[J]. J. Wood Sci., 1991b, 40(10): 1092-1099.

[25]Shao Q, Yang Y X, Ge S S, et al. Study on electroless nickel plating activated without palladium on the surface of cenospheres[J]. J. Funct. Mater, 2007, 38(12): 2001.

[26] Sun L L, Li J, Wang L J. Electromagnetic interference shielding material from electroless copper plating on birch veneer[J]. J. Wood Sci., 2012, 46(6): 1061-1071.

[27]Wang L J, Li J. Electroless deposition of Ni-Cu-P alloy on Fraxinus mandshurica veneer[J]. Sci. Silv. Sinica., 2007, 43(11): 89-92.

[28]Wang L J, Li J, Liu H B. A simple process for electroless plating nickel-phosphorus film on wood veneer[J]. Wood Sci. Technol., 2011a, 45(1): 161-167.

[29]Wang L J, Li J, Liu Y X. Electroless nickel plating on poplar veneer[J]. Fine Chem., 2006a, 3: 230-233.

[30]Wang L J, Li J, Liu Y X. Preparation of electromagnetic shielding wood-metal composite by electroless nickel plating [J]. J. For. Res., 2006b, 17: 53-56.

[31]Wang L J, Li J, Liu Y X. Study on preparation of electromagnetic shielding composite by electroless copper plating on Fraxinus mandshurica veneer[J]. J. Mater Eng., 2008, (4): 56-60.

[32]Wang L J, Liu H B. Electroless nickel plating on chitosan-modified wood veneer[J]. BioResour, 2011c, 6(2): 2045-2054.

[33]Wang L J, Sun L L, Li J. Electroless copper plating on Fraxinus mandshurica veneer using glyoxylic acid as reducing agent[J]. BioResour, 2011b, 6(3): 3493-3504.

[34]Yu H S, Luo S F, Wang Y R. A comparative study on the crystallization behavior of electroless Ni-P and Ni-Cu-P deposits[J]. Surf. Coat Technol., 2001, 148(2): 143-148.

中文题目：水曲柳单板化学镀 Ni-Cu-P 制备耐腐蚀性木质电磁屏蔽复合材料

作者：惠彬，李坚，王立娟

摘要：采用简单的新型化学镀 Ni-Cu-P 工艺制备了耐腐蚀的木质电磁屏蔽复合材料。研究了 $CuSO_4 \cdot 5H_2O$ 浓度、镀液 pH 值和温度对复合材料的金属沉积率、表面电阻率、化学组成、耐蚀性和表面形态的影响。采用扫描电子显微镜和 X 射线能量色散谱仪分别观察了表面形貌和分析了化学组成，通过频谱分析仪测量了电磁屏蔽效率(ESE)，利用动电位腐蚀测量评价了耐腐蚀性。结果表明，金属沉积率随着 pH 值和温度的升高而增加，但随 $CuSO_4 \cdot 5H_2O$ 浓度的增加而降低。Ni-Cu-P 镀层的耐蚀性明显与镀层中 Cu 和 P 总含量有关。Cu 和 P 的总含量越高，耐腐蚀性越强。研究获得的最佳施镀条件：$CuSO_4 \cdot 5H_2O$ 浓度 1.0g·L^{-1}、pH=9.5、温度 90 ℃，镀层含有 Ni 77.41%、Cu 8.96%和 P 13.63%。这种木质复合材料具有较高的耐腐蚀性，在 9 kHz 至 1.5 GHz 的频率范围内 ESE 约 60 dB。该研究开发了一种应用前途广阔的耐腐蚀性木质电磁屏蔽复合材料的制备方法。

关键词：水曲柳胶合板；化学镀镍-铜-磷；耐蚀性；表面形态；电磁屏蔽效能

Enhancement of Photo-catalytic Degradation of Formaldehyde through Loading Anatase TiO_2 and Silver Nanoparticle Films on Wood Substrates*

Likun Gao, Wentao Gan, Shaoliang Xiao, Xianxu Zhan, Jian Li

Abstract: Since formaldehyde is considered as a potential risk for human health, its emission must be eliminated in an effective way. In this study, anatase TiO_2 particles with silver nanoparticles were doped on wood substrates through a two-step method and used as a photo-catalyst for formaldehyde degradation. The effect of Ag dopant on the formaldehyde degradation of the TiO_2-treated wood was investigated. The results showed that the formaldehyde response of TiO_2 film was drastically improved by doping with Ag nanoparticles under the visible-light irradiation. In this heterostructured system, because the Fermi level of Ag was lower than the conduction band of TiO_2, Ag nanoparticles can act as electron scavenging centers for causing electron and hole pair separation, leading to an enhanced photo-catalytic activity of TiO_2.

Keywords: Silver nanoparticles; Anatase TiO_2; Formaldehyde degradation; Visible light; Wood

1 Introduction

As a significant volatile organic compounds (VOCs), Formaldehyde (HCHO) is considered as one of the most important pollutants in indoor environment.[1,2] wood-based products are under the category of natural materials due to their friendly properties of health, aesthetic and environmental aspects. However, because the urea-formaldehyde adhesives in wooden panels could produce HCHO gas during the application, the fear of the formaldehyde emission limits the use-value of wooden panels.[3] Although the elimination of formaldehyde is very difficult, the photo-decomposition was proved to be a promising method to remove gaseous pollutants.[4,5] Furthermore, due to the increasing concern on HCHO in the indoor environment, the abatement of HCHO is of significant practical interest at low temperature, especially at room temperature. Very recently, many studies on the reduction of HCHO by adsorbents or catalysts were reported.[6-8] However, the reclamation of the catalysts, such as powder and particles, plays a critical role in the practical applications, which would influence the development of environmental technology for decontamination, purification, and deodorization of the atmospheres. Based on above consideration, the wood-based products composited with a film of catalysts were proposed. The planned treatment would not only produce a catalyst for decomposition of the toxic organic pollutants, but also fundamentally modified the wood substrates.

Since Fujishima reported UV irradiation induced redox chemistry on TiO_2, photo-catalysis has attracted

* 本文摘自 RSC Advances, 2015, 5(65): 52985-52992.

the most interest as a green and sustainable solution for energy and environmental issues.[9, 10] As an *n*-type semiconductor oxide, TiO_2 is widely used as photo-catalyst in the applications such as self-cleaning, self-degradation, etc., because it is relatively inexpensive and can degrade toxic contaminants without requiring other reagents.[11, 12] Up to now, an increasing attention has been paid to improve its physical and chemical properties, which would further facilitate the profound understanding of the dependence of material properties. Driven by this requirement, great amount of innovative approaches have been conducted to develop novel semiconductor-based composites with desired properties and functions for special applications. In order to endow the TiO_2-based materials with extraordinary properties and superb photocatalytic performance in practical applications, the method of metal doped TiO_2, such as Ag/TiO_2 composites, has become one of the most attractive candidate solution.[7, 13, 14] Among numerous metals, such as Ru,[15] Au,[16, 17] Bi,[18, 19] Pt,[20, 21] etc., Ag is particularly suitable for industrial applications because of its low cost and nontoxic properties compared to other noble metals.[14,22] In addition, Ag has a good light absorption capability, extending the response of TiO_2 to visible light.[23, 24]

In this study, because formaldehyde was considered as a main indoor pollutant, it was chosen as the aim pollutant, and wood was selected as substrates coated by TiO_2 particles and silver nanoparticles. The photo-catalytic degradation characters of HCHO by the TiO_2-treated wood and the Ag-doped TiO_2/wood under visible-light irradiation was characterized by the scanning electron microscopy (SEM), transmission electron microscopy (TEM), X-ray diffraction (XRD), X-ray photoelectron spectroscopy (XPS), and UV-vis diffuse reflectance spectroscopy (UV-vis DRS). The analogue experiments of formaldehyde degradation were performed and the mechanism was explored simultaneously.

2 Materials and methods

2.1 Materials

All chemicals supplied by Shanghai Boyle Chemical Company Limited were of analytical reagent-grade quality and used without further purification. The Formaldehyde used in our experiment was an analytical reagent. Deionized water was used throughout the study. The wood blocks of 20 mm (*R*)×20 mm (*T*)×30 mm (*L*) were obtained from the sapwood sections of poplar (*Populus ussuriensis* Kom), which is one of the most common tree species in the northeast of China. After ultrasonically rinsing in deionized water for 30 min, the wood specimens were oven-dried (24 h, 103 ℃) to a constant weight, and the weights were determined.

2.2 Preparation of anatase TiO_2 film on the wood surface

The Ammonium fluorotitanate (0.4 M) and boric acid (1.2 M) were dissolved in the distilled water in a 500 mL glass container at a room temperature with vigorous magnetic stirring. A solution of 0.3 M hydrochloric acid was added until the pH value reached approximately 3. Then, the 75 mL adjusted solution was transferred into a 100 mL Teflon container. Wood specimens were subsequently placed into the solution, separately. The autoclave was sealed and maintained at 90 ℃ for 5 h, and then naturally cooled to room temperature. Finally, the treated samples were removed from the solution, ultrasonically rinsed with deionized water for 30 min, and dried at 45 ℃ for more than 24 h in a vacuum chamber. For the comparative purpose, un-treated wood samples were also examined. As a result, TiO_2 surfaces on the treated wood contained OH groups, which improved the surface hydrophilicity through the TiO_2 hydroxylation.

2.3 Loading of the TiO_2-treated wood surface with Ag nanoparticles

Aqua ammonia (28 wt. %) was added dropwise into a 0.5 M $AgNO_3$ aqueous solution with stirring until a transparent colourless $[Ag(NH_3)_2]^+$ solution was formed. The TiO_2-treated wood samples were soaked to the above solution for 1 h, then transferred to a 0.1 M glucose stock solution. After 5 min, the residual $[Ag(NH_3)_2]^+$ solution was also poured into the glucose solution. The reaction was continued for 15 min. Finally, the prepared samples were removed from the solution, ultrasonically rinsed with deionized water for 30 min, and dried at 45 ℃ for more than 24 h in a vacuum chamber. The detailed procedure for the synthesis of the Ag-doped TiO_2/wood heterostructures is illustrated in Fig. 1. In this work, the TiO_2 particles with abundant hydrophilic groups (OH) on the wood surface was dipped into the silver ammonia complex solution, the complexes of $[Ag(NH_3)_2]^+$ were easily and abundantly absorbed on the TiO_2 particles. And then the silver ammonia complex loaded TiO_2-treated wood was transferred into the solution of glucose and $[Ag(NH_3)_2]^+$ was in situ reduced into silver seeds on the TiO_2 particles. With the addition of $[Ag(NH_3)_2]^+$ and glucose, more and more silver ions were absorbed onto the TiO_2-treated wood surface and reduced into silver. The silver seeds grew larger into silver nanoparticles (Ag NPs) and attached to each other forming a compact film on the TiO_2-treated wood. [25]

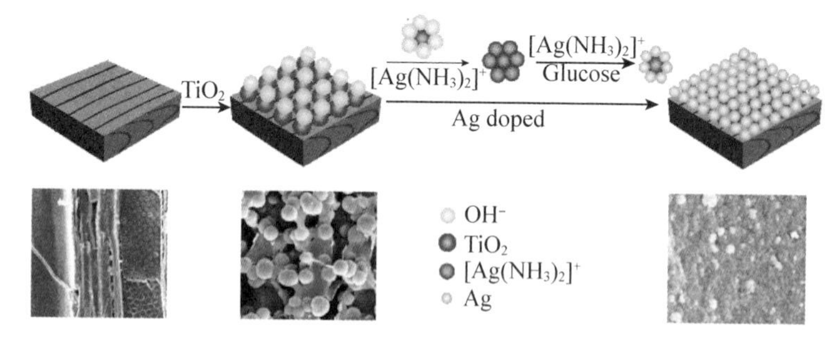

Fig. 1 Schematic representation of the formation of the Ag-doped TiO_2/wood

2.4 Characterizations

The morphology of the wood surfaces was observed through the field emission scanning electron microscopy (SEM, Quanta 200, FEI, Holland) operating at 12.5 kV in combination with EDS (Genesis, EDAX, Holland). The transmission electron microscopy (TEM) experiment was performed on a Tecnai G20 electron microscope (FEI, USA) with an acceleration voltage of 200 kV. Carbon-coated copper grids were used as the sample holders. The X-ray diffraction (XRD, Bruker D8 Advance, Germany) was employed to analyze the crystal structures of all samples applying graphite monochromatic with Cu $K\alpha$ radiation ($\lambda = 1.5418$ Å) in the 2θ range from 5° to 80° and a position-sensitive detector using a step size of 0.02° and a scan rate of 4° min^{-1}. Further evidence for the composition of the product was inferred from the X-ray photoelectron spectroscopy (XPS, Thermo ESCALAB 250XI, USA), using an ESCALab MKII X-ray photoelectron spectrometer with Mg-$K\alpha$ X-rays as the excitation source. Optical properties of the materials were characterized by the UV-vis diffuse reflectance spectroscopy (UV-vis DRS, Beijing Purkinje TU-190, China) equipped with an integrating sphere attachment, which $BaSO_4$ was the reference.

2.5 Photo-catalytic degradation of HCHO

A schematic diagram of the experimental system for photo-oxidation is shown in Fig. 2. The experiments were performed in an obturator with the size of 500 mm (L)×300 mm (W)×300 mm (H). A flat-type LED-light (wavelength is in the range of 400-760 nm, light intensity is approximately 3.6 mW · cm^{-2}) installed in the open central region was used to simulate visible light. Initially, there was no HCHO in the obturator, which was detected by a formaldemeter. Then, the desired amount of HCHO was injected into the obturator. The gaseous formaldehyde was allowed to reach adsorption and desorption equilibrium for 30 min. After equilibrium, the initial concentration of formaldehyde in the obturator was controlled in the region of approximately 1.5-2.5 mg · m^{-3}. Then, the LED light was turned on to irradiate the sample. Each set of degradation experiment under LED irradiation lasted for 10 h. HCHO concentration in the obturator was determined by the formaldemeter.

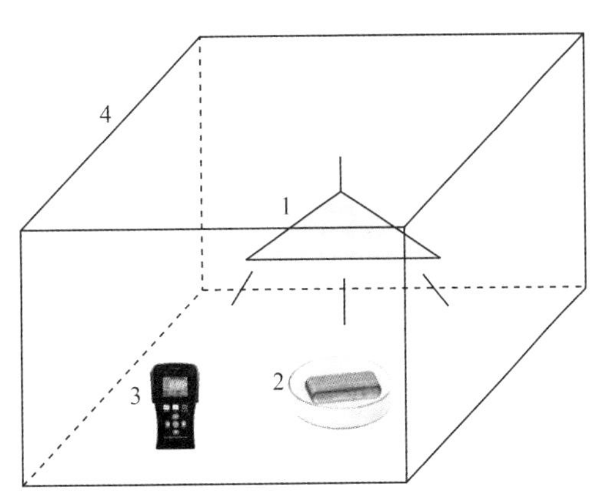

Fig. 2 Schematic diagram of experimental set-up [1. a flat-type LED-light; 2. the treated wood sample; 3. formaldemeter; 4. obturator (500 mm (L)×300 mm (W)×300 mm (H)]

3 Results and discussion

The SEM images of the untreated and treated wood samples are presented in Fig. 3. Fig. 3a shows the micrographs of untreated poplar wood surface, which clearly reveals that the poplar wood is a type of heterogeneous and porous material. Fig. 3b shows that the TiO_2 particles deposit on the wood cell walls and fill into the pits of the wood surface. The TiO_2 particles with the average diameter of approximately 1.5 μm are adhered onto wood surface through the chemical bonds between hydroxyl groups of wood surface and TiO_2 particles. As a comparison of the morphology of pristine TiO_2/wood with that of the Ag nanoparticles doped one, it was observed that a compact layer composed by Ag nanoparticles deposited on its surface (Fig. 3c). The magnified image (Fig. 3d) of the Ag-doped TiO_2/wood demonstrates that the film is approximately 1.8 μm thick and the sphere particles stacked over the wood surface is approximately 130 nm in diameter. The surface chemical elemental compositions of the treated wood are determined via the energy-dispersive X-ray spectroscopy (EDS), and the results are presented in Fig. 3e and Fig. 3f. The evidence of only fluorine, titanium, gold, and carbon elements could be detected in Fig. 3e, while the additional presence of silver element exists in Fig. 3f. The gold element is originated from the coating layer used for SEM observation, the fluorine element may be originated from the precursor ammonium fluorotitanate, and carbon element was from the wood substrate. Since no other elements were detected, the composition of the microspheres on the TiO_2-treated wood sample was confirmed to be TiO_2 as well as revealing the loading of metallic Ag on the Ag-doped TiO_2/wood sample.

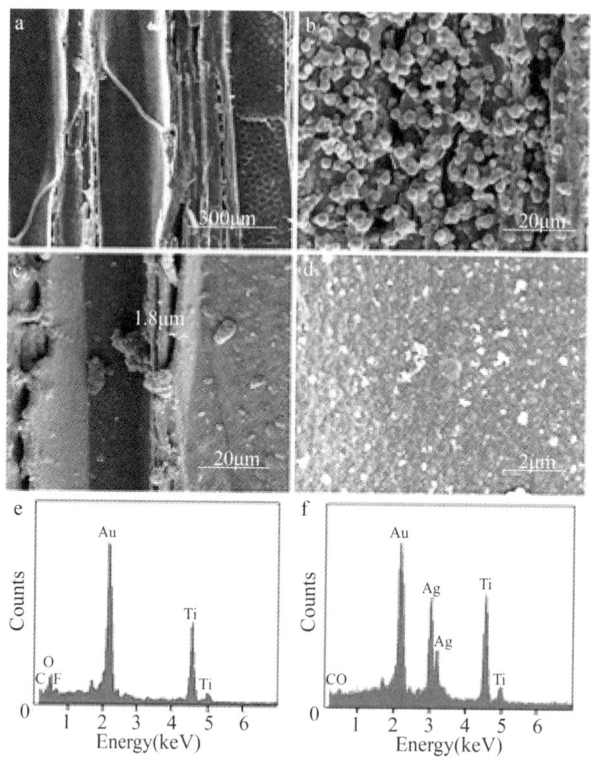

Fig. 3 SEM images of the surfaces of (a) the pristine wood, (b) the TiO_2-treated wood, (c, d) the Ag-doped TiO_2/wood at different magnifications, and (e, f) EDS spectrums of the TiO_2-treated wood and the Ag-doped TiO_2/wood, respectively

Fig. 4 shows the representative TEM image, the selected electron diffraction (SAED) pattern and high-resolution transmission electron microscopy (HRTEM) image of the Ag-doped TiO_2/wood. As shown in Fig. 4a, the TEM image shows that a large quantity of Ag nanoparticles are attached onto the surface of the TiO_2-treated wood matrix, and the size distribution of Ag nanoparticles is wide. The corresponding selected area electron diffraction (SAED) pattern of the Ag-doped TiO_2/wood is shown in Fig. 4b. The SAED pattern of TiO_2 microparticles in anatase phase and Ag nanoparticles in cubic phase shows that the diffraction patterns and rings are co-existent, and the different crystal planes are identified as (101), (200) and (211) diffraction planes of anatase TiO_2 as well as (111), (200), (220) and (311) diffraction planes of cubic Ag, respectively.[26] To further investigate the distribution of Ag and TiO_2 particles, the HRTEM image is displayed in Fig. 4c. The clear lattice fringe of $d=0.35$ nm matches that of the (101) plane of anatase phase of TiO_2, while the fringe of $d=0.24$ nm corresponds to the (111) plane of cubic phase of Ag.[7] No other impurities such as silver oxides are detected.

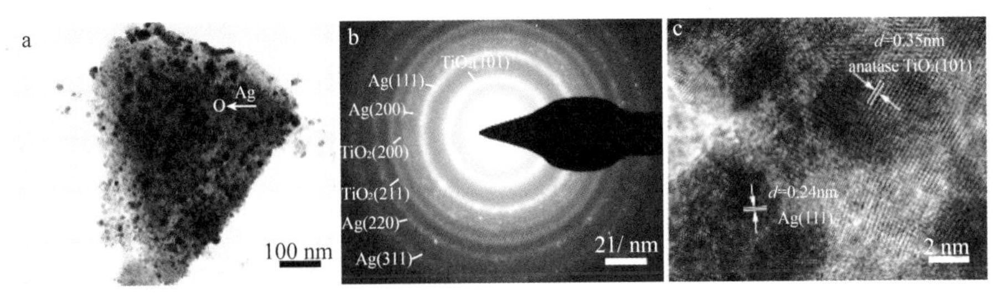

Fig. 4 (a) TEM image, (b) the selected electron diffraction (SAED) patterns, and (c) HRTEM image of the Ag-doped TiO_2/wood

Fig. 5 shows the XRD patterns of the pristine wood, the TiO_2-treated wood and the Ag-doped TiO_2/wood. Apparently, the diffraction peaks at 14. 8° and 22. 5° belonging to the (101) and (002) crystal planes of cellulose in the wood are observed in both the spectrum of the pristine wood and the treated wood samples.[27] It can be found that the diffraction peaks are well indexed to the standard diffraction pattern of anatase phase TiO_2(JCPDS file No. 21-1272) and the face-centered cubic (fcc) phase of Ag (JCPDS file No. 04-0783), indicating that the present synthesis strategy successfully achieves Ag/TiO_2 hierarchical heterostructures with high crystallinity on wood substrate. The curve in Fig. 5b shows that the diffraction peaks of pure anatase TiO_2 particles center at 2θ=25. 2°, 38. 0°, 47. 8°, 54. 2°, 62. 5°, 68. 8°, and 74. 9°, which agree with (101), (004), (200), (211), (204), (116) and (215) crystal planes of anatase TiO_2, respectively.[14] After doped with Ag nanoparticles (Fig. 5c), all of the new diffraction peaks at 38. 1°, 44. 2°, 64. 4°, and 77. 4°, except the diffraction peaks of TiO_2, can be perfectly indexed to (111), (200), (220) and (311) reflections of the face-centered cubic (fcc) phase of Ag,[7] which is in agreement with SAED pattern results in Fig. 4b.

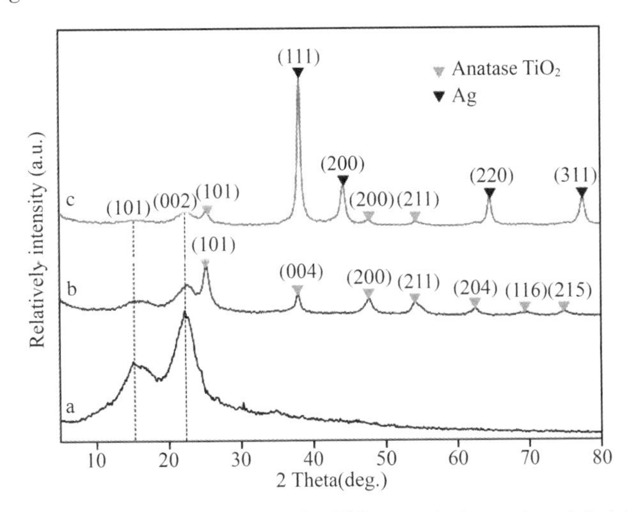

Fig. 5 XRD patterns of (a) the pristine wood, (b) the TiO_2-treated wood and (c) the Ag-doped TiO_2/wood

Fig. 6 ascertains more detailed information concerning the elemental and chemical state of the resulting samples. The fully scanned spectra (Fig. 6a) shows that the Ti, O, and C elements exist on the surface of the pure TiO_2-treated wood sample, while Ti, O, Ag, and C elements exist on the surface of the Ag-doped TiO_2/

wood sample. The C element can be ascribed to the wood substrate or the adventitious carbon-based contaminant.

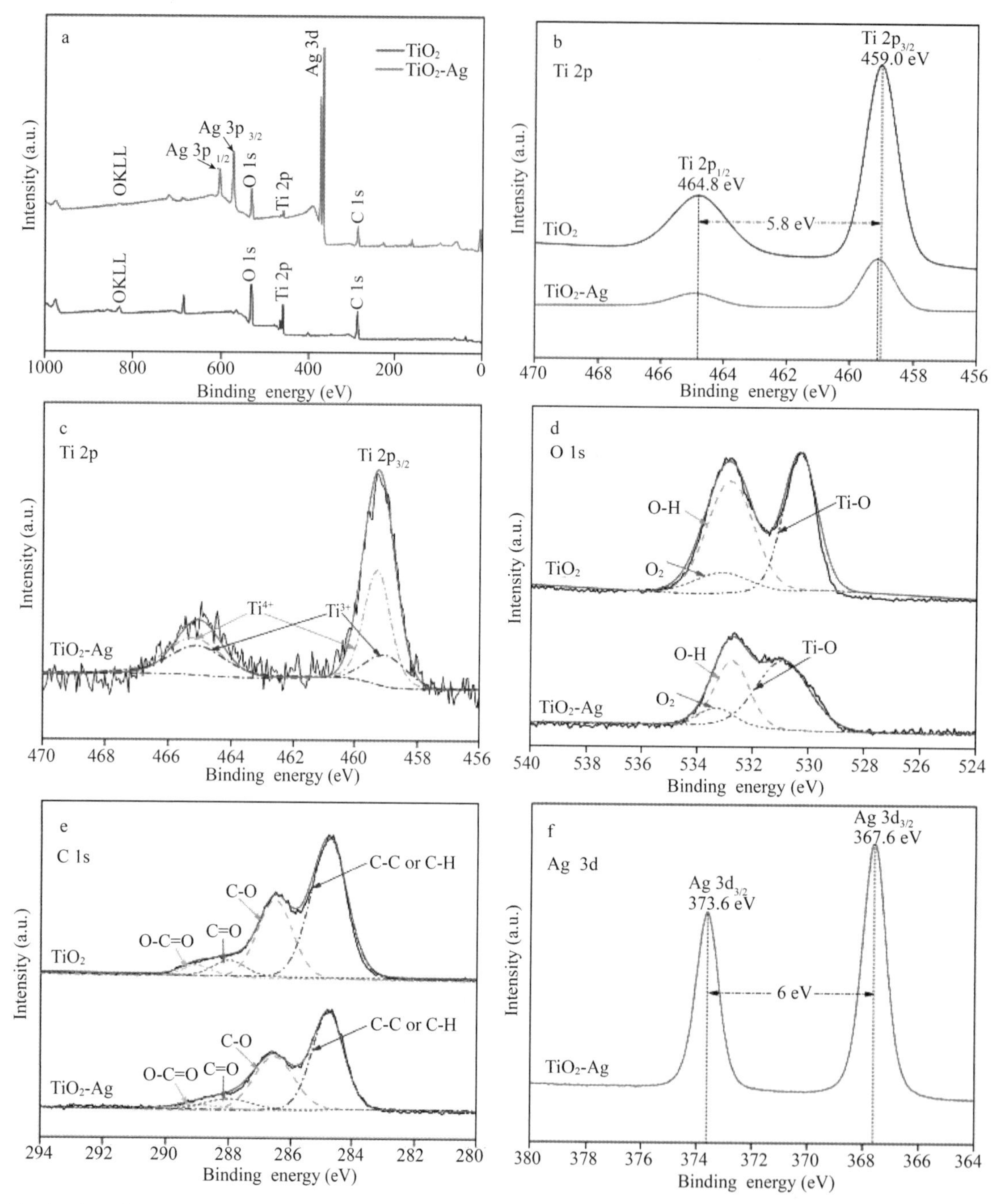

Fig. 6 (a) Survey scan and (b) Ti 2p XPS spectra of TiO_2-treated wood and the Ag-doped TiO_2/wood, (c) peaking-fitting results of Ti 2p XPS spectra of the Ag-doped TiO_2/wood, (d) O 1s and (e) C 1s XPS spectra of TiO_2-treated wood and the Ag-doped TiO_2/wood, (f) Ag 3d XPS spectra of the Ag-doped TiO_2/wood

From Fig. 6b, the Ti 2p XPS spectra of the TiO_2-treated wood with two peaks at binding energies of 459. 0 eV and 464. 8 eV, corresponding to the Ti $2p_{3/2}$ and Ti $2p_{1/2}$ peaks, respectively. The gap of 5. 8 eV between the two peaks indicates the existence of the Ti^{4+} oxidation state. [28] However, the peak position for Ti 2p in the Ag-doped TiO_2/wood sample shifts to a higher binding energy band than that in the pure TiO_2-treated wood sample. This confirms a lower electron density of the Ti atoms after doped with Ag nanoparticles, and there is a strong interaction between metallic Ag and TiO_2 in the Ag-doped TiO_2/wood. In addition, the Ti $2p_{3/2}$ XPS peak of the Ag-doped TiO_2/wood sample can be fitted into two components, one located at 459. 3 eV, attributed to a Ti^{4+} species, and the other located at 458. 9 eV, assigned to a Ti^{3+} species (in Fig. 6c), further indicating the strong interaction formed between Ag and TiO_2 species. Certainly, the presence of Ti^{3+} oxide with narrow band gap and energy level located between the valence and the conduction band of TiO_2 may be advantageous to the higher photocatalytic activity of the Ag-doped TiO_2/wood heterostructures driven by visible light. Moreover, the defect sites of TiO_2 surface induced by light, that is, Ti^{3+} species, are necessary for adsorption and photo-activation of oxygen, which are necessary for the photo-oxidation of pollutants.

The high resolution XPS spectra of O 1s in the TiO_2-treated wood and the Ag-doped TiO_2/wood are provided in Fig. 6d. The wide and asymmetric O 1s spectra suggests that there would be more than one component. Using the XPS Peak fitting program, version 4. 1, the spectra can be further fitted into three peaks including crystal lattice oxygen attributed to the peak of O_{Ti-O} at 530. 3 eV, surface hydroxyl groups assigned to the peak of O_{O-H} at 532. 8 eV, and adsorbed O_2 in molecular water or C-O bonds (at 533. 2 eV). [29]

Fig. 6 e represents the C 1s XPS spectra of the TiO_2-treated wood and the Ag-doped TiO_2/wood as well as their fitting results. The spectrum is de-convoluted into four separated Gaussian distributions. The peak around 294. 8 eV is assigned to C—C or C—H bonds, and the peak observed at 286. 5 eV corresponds to C—O bonds. The other two peaks around 288. 0 eV and 289. 0 eV are attributed to C=O and O—C=O bonds, respectively, which may be typical for adsorbed carbon on the samples. [30]

As shown in Fig. 6f, the Ag 3d XPS spectra with two peaks at binding energies of 367. 6 eV and 373. 6 eV, corresponding to the Ag $3d_{5/2}$ and Ag $3d_{3/2}$ peaks, respectively. The gap of 6. 0 eV between the two peaks is also indicative of metallic Ag, and there is no evidence for the presence of Ag^+, indicating that the Ag ions in the composites are fully reduced to metallic Ag. Additionally, it is obvious that the peaks of Ag 3d shift to the lower position compared with bulk Ag (368. 3 eV for Ag $3d_{5/2}$, and 374. 3 eV for Ag $3d_{3/2}$), suggesting that the electrons may migrate from the TiO_2 particles to metallic Ag, [31] which reveals that there is a strong interaction between Ag nanoparticles and TiO_2 particles in the interface of heterostructures.

In brief, it can be demonstrated that Ag nanoparticles are grown on the TiO_2-treated wood matrix, and there is a strong interaction between Ag and TiO_2 in the interface of the Ag-doped TiO_2/wood heterostructures. Furthermore, Ag nanoparticles can act as electron acceptors and be advantageous for separating the photo-excited electron-hole pairs, the heterostructures can inhibit the recombination of excited electrons and holes and then enhance the photo-catalytic activity of the products.

The probable scheme diagram for the photo-catalytic process of the Ag-doped TiO_2/wood sample and a simplified reaction scheme for the catalytic oxidation of HCHO on the surface of the Ag-doped TiO_2/wood sample are illustrated in Fig. 7. As shown in Figure. 7a, during visible-light irradiation, the Ag nanoparticles are photo-excited owing to the plasmon resonance, which the photo-excited electrons are subsequently

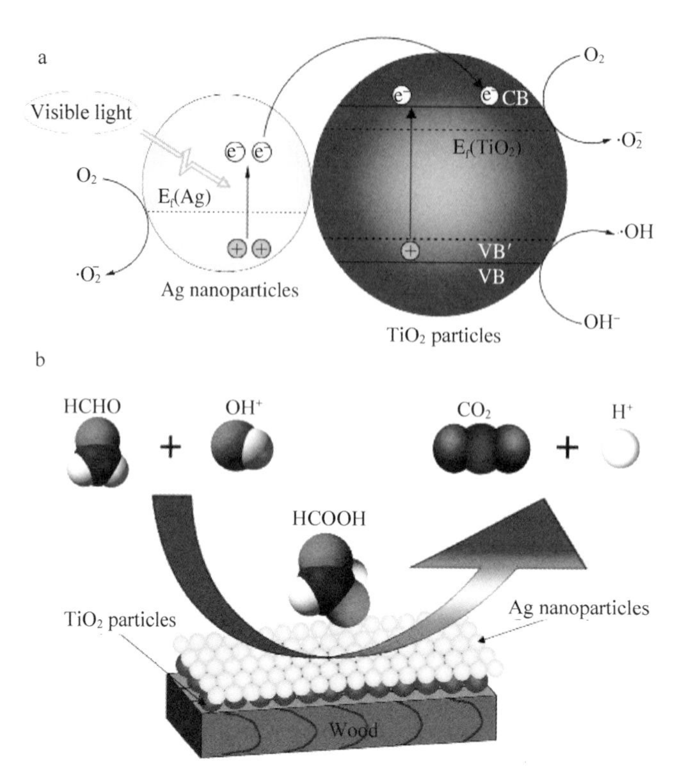

Fig. 7 (a) Proposed photo-induced charge separation and migration process in the Ag-doped TiO_2/wood sample under visible-light irradiation, and (b) possible scheme for HCHO adsorption and surface oxidation to CO_2 on the surface of Ag-doped TiO_2/wood sample

migrated from the surface of the Ag nanoparticles to the conduction band (CB) of TiO_2 particles. In addition, due to the high crystallinity of the Ag nanoparticles, the resistance of the electron migration is decreased leading to a reduction in the recombination of excited electrons and holes. The electrons on the conduction band (CB) of TiO_2 are scavenged by dissolved oxygen molecules (O_2) and the holes on the valence band (VB) of TiO_2 absorb water molecules (H_2O) or hydroxyl ions (OH^-), which yields abundant of highly oxidative species including superoxide radical anion ($\cdot O_2^-$) and hydroxyl radical ($\cdot OH$). This causes an improvement of the photo-catalytic performance and aggressive oxidation of the surface-adsorbed toxic organic pollutants (HCHO) to convert into CO_2 and water. Furthermore, as revealed by the XPS results, the production of Ti^{3+} species is ascribed to the strong interaction between Ag and TiO_2. The energy level of Ti^{3+} species is located between the valence band (VB) and conduction band (CB) of TiO_2, which is highly advantageous to promote the electrons in the new valence band (VB′) to be excited to the conduction band (CB) of TiO_2. Therefore, the Ag nanoparticles can be regarded as electrons trapping centers in the conduction band of TiO_2, leading to a decrease in the concentration of photo-induced charge carriers and an enhancement for the photo-catalytic activity of the samples. That is, the migration of charges is enhanced and the recombination of excited electrons and holes is suppressed as a result of the well-known Schottky barrier effect.[14] The proposed initial elementary reactions are listed in Eqs. (1)-(4) as the followings:

$$TiO_2/Ag + h\nu \rightarrow e^- + h^+ \quad (1)$$

$$TiO_2 + h\nu \rightarrow e^- + h^+ \quad (2)$$

$$e^- + O_2 \rightarrow \cdot O_2^- \quad (3)$$

$$h^+ + OH^- \rightarrow \cdot OH + H^+ \quad (4)$$

Moreover, published literatures have described that the mechanism of TiO_2 film on decomposition of HCHO. [1, 32, 33] Based on the discussion above, a scheme for HCHO adsorption and surface oxidation to CO_2 on the surface of Ag-doped TiO_2/wood sample is postulated in Scheme 3b. During the HCHO degradation in the photo-catalytic process, formic acid (HCOOH) is identified as intermediate. The related reactions of HCHO destruction are shown with Eqs. (5)-(9) as the followings:

$$HCHO + e^- \rightarrow H\cdot + HCO\cdot \quad (5)$$

$$HCHO + O\cdot \rightarrow \cdot OH + HCO\cdot \quad (6)$$

$$HCHO + \cdot OH \rightarrow HCO\cdot + H_2O \quad (7)$$

$$HCO\cdot + \cdot OH \rightarrow HCOOH \quad (8)$$

$$HCOOH + h^+ \rightarrow CO_2 + 2H\cdot \quad (9)$$

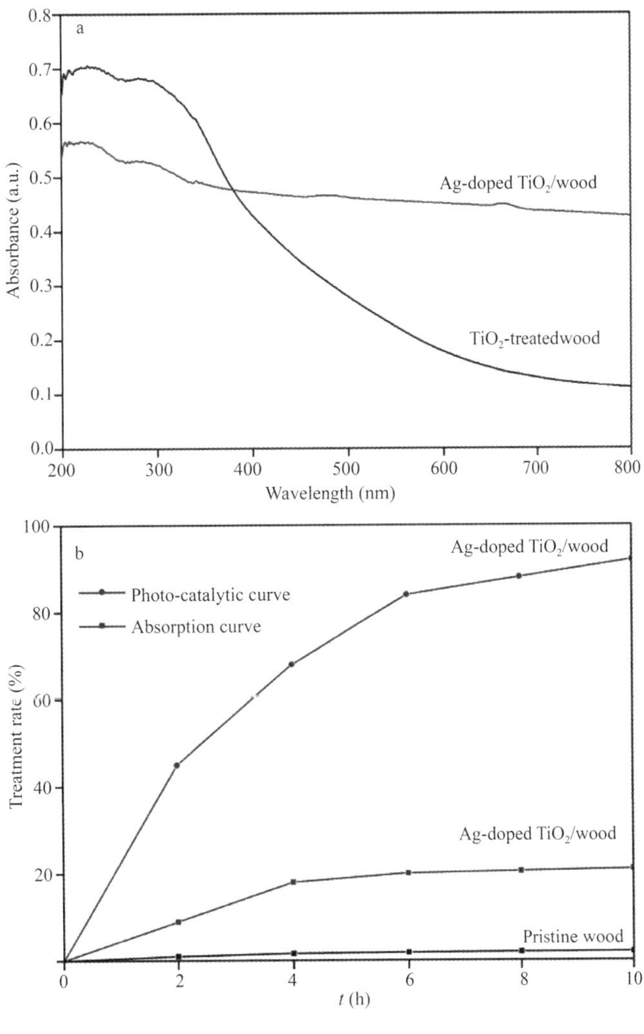

Fig. 8 (a) UV-vis absorption spectra of the TiO_2-treated wood and the Ag-doped TiO_2/wood, and (b) comparison of adsorption curve and photo-catalytic decomposition curve of HCHO over the pristine wood and the Ag-doped TiO_2/wood

In order to investigate the light absorbance of the samples, the UV-vis diffuse reflection spectra of the TiO_2-treated wood and the Ag-doped TiO_2/wood are depicted in Fig. 8a. As for the pure TiO_2-treated wood, it presents prominent adsorption below 350 nm wavelength. Whereas the Ag-doped TiO_2/wood exhibits a much higher absorption in the region 380-800 nm, indicating the absorption of Ag-doped TiO_2/wood is significantly red-shifted to the visible due to the loading of the Ag nanoparticles onto the surface of the TiO_2-treated wood. Theoretically, the wide visible absorption should be attributed to the characteristic absorption of surface plasmon resonance originating from the metallic Ag nanoparticles in the Ag-doped TiO_2/wood samples. Moreover, it is worth noting that the Ag-doped TiO_2/wood heterostructures with absorptions in the visible region may implicate higher activity in photo-catalysis.

Fig. 8b presents comparison of adsorption curve and photo-catalytic degradation curve of HCHO over the pristine wood and the Ag-doped TiO_2/wood for every 2 hours. For the pristine wood, both the adsorption and photo-catalytic degradation rate of HCHO is about 2% after 10 h, possibly because the effect of the porous structure of the cell wall of wood is similar to that of active carbon. For the Ag-doped TiO_2/wood, in the first two hours, the concentrations of HCHO decreased fast in both cases, and the treatment rate of HCHO by photo-catalytic degradation is still superior to that of HCHO by absorption. Finally, the treatment rate of HCHO by photo-catalytic degradation is 94.9%, while the treatment rate of HCHO by absorption is 21%. The obvious difference between two curves confirms that the photo-catalytic reaction plays an important role in the degradation of HCHO. After 10 h, the absorption curve variation of HCHO is less than 1% and tends to zero. The result indicates the absorption of HCHO over the Ag-doped TiO_2/wood reached the adsorption equilibrium. Furthermore, the initial concentrations of the HCHO are 2.14 $mg \cdot m^{-3}$ for the Ag-doped TiO_2/wood in the photo-catalytic degradation reaction. It is obvious that, for the Ag-doped TiO_2/wood, the degradation rate of HCHO is dramatically increased, after 6 hours reaction, the degradation rate is slightly decreased but still rising, indicating the photo-catalytic reaction of HCHO is continuing. As well-known, the national regulation illustrates that, for the indoor formaldehyde standard, the HCHO of public places should be below 0.12 $mg \cdot m^{-3}$, and the HCHO of residential indoors should be below 0.08 mg/m^3. Apparently, the results indicate that the removal efficiency of HCHO by the Ag-doped TiO_2/wood under visible-light irradiation could reach the national regulation for the indoor formaldehyde standard of public places, and approximate to the national regulation for indoor formaldehyde standard of residential indoor.

In addition, we further investigated the analogue experiment of formaldehyde degradation by the Ag-doped TiO_2/wood under the ultraviolet-germicidal-lamp illumination (365 nm wavelength and approximately 1.5 $mW \cdot cm^{-2}$ light intensity), as shown in Supplementary Material. Most interestingly, the experimental results present that, for the Ag-doped TiO_2/wood, the degradation rate of HCHO driven by UV light is dramatically improved, and the HCHO of 0.72 $mg \cdot m^{-3}$ in the obturator could be completely eliminated in approximately 1.02 min, elucidating an enhanced photo-catalytic activity of the Ag-doped TiO_2/wood under the UV irradiation. Such an important and useful property for the Ag-doped TiO_2/wood would greatly promote its application in a fast elimination of HCHO.

We have also evaluated the reusability of the Ag-doped TiO_2/wood for photo-catalytic degradation of HCHO, as shown in Fig. 9. After five cycled runs of the photo-catalytic degradation of HCHO, the photo-catalytic activity of the Ag-doped TiO_2/wood shows great decreases with the degradation rate decreased from

94.9% to 2.2%. It clearly demonstrates that the samples are unstable, because the stability of the wood substrate is not optimistic as the results of the damage of light, moisture, dirt, bacterium, etc.. Hence, further modification with the low-surface-energy materials is necessary to lead to a new durable photo-catalytic material with multifunction, such as superhydrophobicity, superoleophobicity and antibacterial performance. That is, the present study will be a basis of future photo-catalyst development.

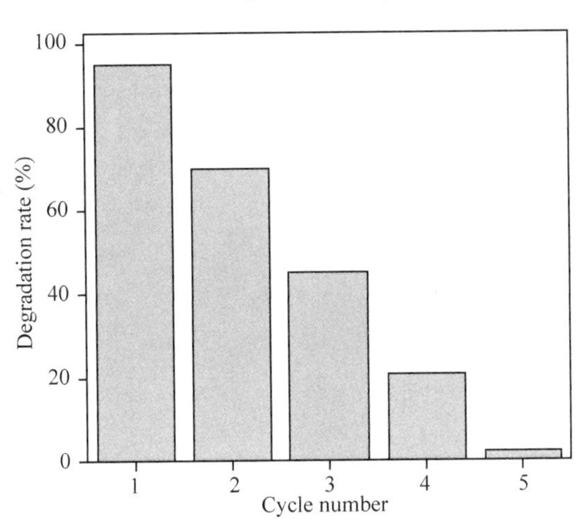

Fig. 9 Five cycles of photo-catalytic degradation of HCHO over the Ag-doped TiO_2/wood

4 Conclusions

The Heterostructured Ag-doped TiO_2/wood was prepared by a two-step method including the hydrothermal synthesis of TiO_2 particles on the wood substrate and silver mirror reaction of the Ag nanoparticles dopant on the surface of TiO_2/wood. The TiO_2-treated wood and the Ag-doped TiO_2/wood were used in an obturator for HCHO decomposition under visible light. The results indicate that the Ag-doped TiO_2/wood exhibits an enhanced photo-catalytic activity in the decomposition of HCHO driven by the visible light, which is mainly ascribed to the strong interaction between TiO_2 and Ag, and the surface plasmon resonances of Ag nanoparticles excited by visible-light irradiation. Meanwhile, the presence of Ti^{3+} in the Ag-doped TiO_2/wood is advantageous to higher photo-catalytic activity, which extends the photo-response of the products to the visible light. Since the Ag-doped TiO_2/wood enhanced the photo-catalytic degradation of formaldehyde, it would greatly promote the application of the modified wood in a fast elimination of HCHO. However, the stability of the photo-catalyst based on wood substrate requires more exploration in the future researches.

Acknowledgments

This work was financially supported by the National Natural Science Foundation of China (grant no. 31470584).

References

[1] Liang W, Li J, Jin Y. Photo-catalytic degradation of gaseous formaldehyde by TiO_2/UV, Ag/TiO_2/UV and Ce/TiO_2/UV[J]. Build Environ., 2012, 51: 345-350.

[2] Prado Ó J, Veiga M C, Kennes C. Biofiltration of waste gases containing a mixture of formaldehyde and methanol[J].

Appl Microbiol. Biot., 2004, 65: 235-242.

[3] Bulian F, Battaglia R, Ciroi S. Formaldehyde emission from wood based panels[J]. Holz Roh. Werkst, 2003, 61(3): 213-215.

[4] Wen Z, Tian-Mo L, De-Jun L. Formaldehyde gas sensing property and mechanism of TiO_2-Ag nanocomposite[J]. Physica B., 2010, 405(19): 4235-4239.

[5] Liao Y, Xie C, Liu Y, et al. Comparison on photocatalytic degradation of gaseous formaldehyde by TiO_2, ZnO and their composite[J]. Ceram. Int., 2012, 38(6): 4437-4444.

[6] Yuan Q, Wu Z, Jin Y, et al. Photocatalytic cross-coupling of methanol and formaldehyde on a rutile TiO_2(110) surface[J]. J. Am. Chem. Soc., 2013, 135, 135(13): 5212-5219.

[7] Wang S, Qian H, Hu Y, et al. Facile one-pot synthesis of uniform TiO_2-Ag hybrid hollow spheres with enhanced photocatalytic activity[J]. Dalton T., 2013, 42(4): 1122-1128.

[8] Shan Z, Wu J, Xu F, et al. Highly effective silver/semiconductor photocatalytic composites prepared by a silver mirror reaction[J]. J. Phys. Chem. C., 2008, 112(39): 15423-15428.

[9] Fujishima A, Rao T N, Tryk D A. Titanium dioxide photocatalysis[J]. J. Photoch. Photobio. C., 2000, 1(1): 1-21.

[10] Niu F, Zhang L S, Chen C Q, et al. Hydrophilic TiO_2 porous spheres anchored on hydrophobic polypropylene membrane for wettability induced high photodegrading activities[J]. Nanoscale, 2010, 2(8): 1480-1484.

[11] Gao L, Zhan X, Lu Y, et al. pH-dependent structure and wettability of TiO_2-based wood surface[J]. Mater Lett, 2015, 142: 217-220.

[12] Gao L, Lu Y, Zhan X, et al. A robust, anti-acid, and high-temperature-humidity-resistant superhydrophobic surface of wood based on a modified TiO_2 film by fluoroalkyl silane[J]. Surf. Coat. Tech., 2015, 262: 33-39.

[13] Xie Y, Meng Y. SERS performance of graphene oxide decorated silver nanoparticle/titania nanotube array[J]. RSC Adv., 2014, 4(79): 41734-41743.

[14] Su C, Liu L, Zhang M, et al. Fabrication of Ag/TiO_2 nanoheterostructures with visible light photocatalytic function via a solvothermal approach[J]. Cryst Eng Comm., 2012, 14(11): 3989-3999.

[15] Chu S Z, Wada K, Inoue S, et al. Fabrication and structural characteristics of ordered TiO_2-Ru (-RuO_2) nanorods in porous anodic alumina films on ITO/glass substrate[J]. J. Phys. Chem. B., 2003, 107(37): 10180-10184.

[16] Zhang P, Guo J, Zhao P, et al. Promoting effects of lanthanum on the catalytic activity of Au/TiO_2 nanotubes for CO oxidation[J]. RSC Adv., 2015, 5(16): 11989-11995.

[17] Sahu G, Gordon S W, Tarr M A. Synthesis and application of core-shell Au-TiO_2 nanowire photoanode materials for dye sensitized solar cells[J]. RSC Adv, 2012, 2(2): 573-582.

[18] Bessekhouad Y, Robert D, Weber J V. Bi_2S_3/TiO_2 and CdS/TiO_2 heterojunctions as an available configuration for photocatalytic degradation of organic pollutant[J]. J. Photoch. Photobio. A., 2004, 163(3): 569-580.

[19] Lin X, Xing J, Wang W, et al. Photocatalytic activities of heterojunction semiconductors Bi_2O_3/$BaTiO_3$: a strategy for the design of efficient combined photocatalysts[J]. J. Phys. Chem. C., 2007, 111(49): 18288-18293.

[20] Yu J, Qi L, Jaroniec M. Hydrogen production by photocatalytic water splitting over Pt/TiO_2 nanosheets with exposed (001) facets[J]. J. Phys. Chem. C., 2010, 114(30): 13118-13125.

[21] Chen S, Li Y, Wang C. Visible-light-driven photocatalytic H_2 evolution from aqueous suspensions of perylene diimide dye-sensitized Pt/TiO_2 catalysts[J]. RSC Adv., 2015, 5(21): 15880-15885.

[22] Zhang Q, Ye J, Tian P, et al. Ag/TiO_2 and Ag/SiO_2 composite spheres: synthesis, characterization and antibacterial properties[J]. RSC Adv., 2013, 3(25): 9739-9744.

[23] Cozzoli P D, Fanizza E, Comparelli R, et al. Role of metal nanoparticles in TiO_2/Ag nanocomposite-based microheterogeneous photocatalysis[J]. J. Phys. Chem. B., 2004, 108(28): 9623-9630.

[24] Seery M K, George R, Floris P, et al. Silver doped titanium dioxide nanomaterials for enhanced visible light photocatalysis[J]. J. Photoch. Photobio. A., 2007, 189(2-3): 258-263.

[25] Jin C, Li J, Han S, et al. Silver mirror reaction as an approach to construct a durable, robust superhydrophobic surface of bamboo timber with high conductivity[J]. J. Alloy. Compd., 2015, 635: 300-306.

[26] Yuan T, Zhao B, Cai R, et al. Electrospinning based fabrication and performance improvement of film electrodes for lithium-ion batteries composed of TiO_2 hollow fibers[J]. J. Mater Chem., 2011, 21(38): 15041-15048.

[27] Andersson S, Serimaa R, Paakkari T, et al. Crystallinity of wood and the size of cellulose crystallites in Norway spruce (Picea abies)[J]. J. Wood Sci., 2003, 49(6): 531-537.

[28] Sarma B K, Pal A R, Bailung H, et al. A hybrid heterojunction with reverse rectifying characteristics fabricated by magnetron sputtered TiO_x and plasma polymerized aniline structure[J]. J. Phys. D.: Appl Phys., 2012, 45(27): 275401.

[29] Zhang L, Chen L, Chen L, et al. A facile synthesis of flower-shaped TiO_2/Ag microspheres and their application in photocatalysts[J]. RSC Adv., 2014, 4(97): 54463-54468.

[30] Siuzdak K, Sawczak M, Klein M, et al. Preparation of platinum modified titanium dioxide nanoparticles with the use of laser ablation in water[J]. Phys. Chem. Chem. Phys., 2014, 16(29): 15199-15206.

[31] Wang C, Yifeng E, Fan L, et al. CdS-Ag nanocomposite arrays: Enhanced electro-chemiluminescence but quenched photoluminescence[J]. J. Mater Chem., 2009, 19(23): 3841-3846.

[32] Sekine, Y. Oxidative decomposition of formaldehyde by metal oxides at room temperature[J]. Atmos. Environ., 2002, 36(35): 5543-5547.

[33] Mao C F, Vannice M A. Formaldehyde oxidation over Ag catalysts[J]. J. Catal., 1995, 154(2): 230-244.

中文题目：以木材为基质负载锐钛矿型二氧化钛和银纳米薄膜用于提高光催化降解甲醛

作者：高丽坤，甘文涛，肖少良，詹先旭，李坚

摘要：甲醛对人体健康具有潜在的危害，必须有效地抑制其排放。本研究通过两步法将锐钛矿型二氧化钛和银纳米颗粒负载在木材基质上，并用作光催化剂降解甲醛。其中，我们探索了掺杂的银在二氧化钛/木材降解甲醛中的作用。结果表明，银掺杂极大提高了可见光照射下二氧化钛薄膜对甲醛的降解。在该异质结构体系中，由于银的费米能级比二氧化钛的导带低，银纳米颗粒可作为电子捕获中心，引发电子和空穴对的分离，从而提高二氧化钛的光催化效率。

关键词：木材；二氧化钛；银纳米颗粒；光催化

Durable Superamphiphobic Wood Surfaces from Cu_2O Film Modified with Fluorinated Alkyl Silane*

Likun Gao, Shaoliang Xiao, Wentao Gan, Xianxu Zhan, Jian Li

Abstract: A simple hydrothermal process with further hydrophobization was developed for fabricating durable superamphiphobic film of cuprous oxide (Cu_2O) microspheres on wood substrate. With the advantages of simple operation, low cost, short reaction time, and environmental friendliness, the present method can be well adopted to fabricate Cu_2O microstructures on the wood surfaces. Meanwhile, the wood coated with a hydrolysis product from long chain fluoroalkyl silane of (heptadecafluoro-1, 1, 2, 2-tetradecyl) trimethoxysilane has a durable superhydrophobic and superoleophobic surface and the coating shows excellent durability to acid, high temperature and humidity, and abrasion. The coatings effectively protect the substrate from damages, expanding the wood application fields. The functional coating may have a broad prospect of applications from the bridges and buildings to automobiles and other possible aspects.

Keywords: Durable; Superamphiphobic; Wood; Cuprous oxide (Cu_2O); Fluoroalkyl silane

1 Introduction

Recently bioinspired materials with extraordinary properties and functions provide a wide range of applications from photo-catalytic clothing to corrosion-resistant, pollutant-degrading surfaces.[1] The studies of lotus and other living things with unusual wetting characteristic of super-liquid-repellent surfaces have attracted much attention in both scientific and industrial areas.[2] Super-liquid-repellent surfaces by combining micro-and nanoscaled structures with low-surface-energy materials are often deemed superhydrophobic and superoleophobic up to the repelled liquid. Superhydrophobic surfaces with the water contact angle larger than 150° offer emerging applications including separation of oil from water, nonsticking, self-cleaning, anti-contamination, protection of devices, etc. Moreover, superoleophobic surfaces are defined as the structured surfaces that resist wetting of liquids with the surface tension below 35 $mN \cdot m^{-1}$, such as hexadecane (γ_{lv}-27.5 $mN \cdot m^{-1}$).[3-5] In addition, superoleophobic surfaces have great potential applications in preventing the substrates from fouling of hazard chemicals and biological contaminants. Being called as superamphiphobic surfaces, the combination of the superhydrophobicity and superoleophobicity on the same substrate, have an impact on a wide range of phenomena, including biofouling by marine organisms, anti-icing, anti-corrosion, self-cleaning and biomedical applications.[6]

However, no material would remain its properties for long time. The daily used materials will lose their efficacy in three ways: Aging, abrasion and rupture. Moreover, the durability of artificial super-liquid-

* 本文摘自 RSC Advances, 2015, 5: 98203-98208.

repellent coatings faced enormous threatens and challenges when exposed to acid rain, high-temperature-humidity conditions, or scratched away by sand in the wind or by animals. Redepositing the low-surface-energy materials is a common method for recovering the super-liquid-repellent properties of artificial coatings[7], which is not convenient. Durable function is defined as that plants maintain their super-liquid-repellent properties by regenerating the epicuticular wax layer after they are damaged. Recently, it is believed that the endowing artificial coatings with durable ability provide an efficient way to solve the problem of the poor durability. Inspired by the durable super-liquid-repellent properties of living plants, the study presents an artificial way through the two key surface parameters, roughness and surface energy, to fabricate durable superamphiphobic coatings. Wang et al. recently reported a durable superamphiphobic surface on anodized alumina that was prepared by filling the intrinsic pores with a low-surface energy liquid[8]. Wang et al. also reported that the durable superhydrophobic and superoleophobic surfaces could be obtained from the fluorinated-decyl polyhedral oligomeric silsesquioxane and hydrolyzed fluorinated alkyl silane[9].

As a natural organic material, wood is one of the widely used structural materials for various applications due to its satisfactory performance, such as low density, thermal expansion, renewability and aesthetically pleasing appearance.[10-12] Wood products with durable superamphiphobic properties would be greatly appreciated by a more discerning and demanding consumer market as high-value-added products.

Hence, our strategy is to design durable superamphiphobic coatings on wood surfaces as shown in Scheme 1. The first step (Fig. 1a) is to build microscaled cuprous oxide (Cu_2O) hierarchical structures on the wood substrate through a simple hydrothermal synthesis for only two hours. To date, most researchers have reported the superhydrophobic surfaces were successfully fabricated based on TiO_2-or ZnO-treated substrates. However, as a reddish p-type semiconductor, Cu_2O has attracted more and more attention by researchers due to its direct band gap of 2~2.2 eV as a result of the presence of Cu vacancies, which formed an acceptor level 0.4 eV above the valence band.[13] Cu_2O with unique chemical and physical properties has been widely used in electronics, catalysis, optical devices, biosensing, gas sensors, and antifouling. With its numerous attractive advantages, such as, low toxicity, good environmental acceptability, inexpensive, plentiful and readily available, Cu_2O may become a substitution for TiO_2 and ZnO.[14, 15] Therefore, we selected Cu_2O film as a rough coating on the wood surfaces.

The second step (Fig. 1b) is the fabrication of long-chain polymer coatings [(heptadecafluoro 1, 1, 2, 2-tetradecyl) trimethoxysilane, hereafter abbreviated as FAS-17] with low surface energy. After the deposition of a FAS-17 coating, the wood surface become superamphiphobic because of the formation of a covalently attached fluoroalkyl silane layer. As a most important property, the superamphiphobic coatings can preserve a large number of reacted fluoroalkyl silane moieties as healing agents. When the top fluoroalkyl silane layer of the superamphiphobic coating is decomposed or scratched, the preserved healing agents can migrate to the coating surface to heal the superamphiphobicity like a living plant. Moreover, due to the toxicity of fluoroalkyl silane, the study still needs deep investigation to develop an environmental-friendly and multifunctional material in the future. In the study, we designed harsh conditions, such as acid, high temperature and humidity, and abrasion, to explore the durable properties of the products. The results indicated that the coatings showed excellent durability to acid, high temperature and humidity, and abrasion, and the coatings could effectively protect the substrate from damages.

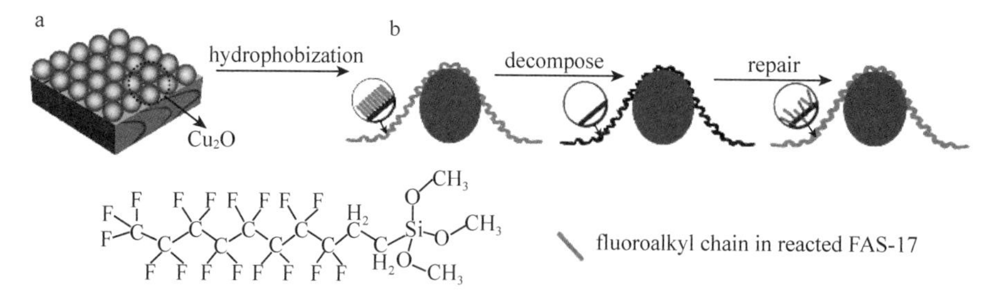

Fig. 1 Perform mechanism of the durable superamphiphobic wood surfaces: (a) the pristine wood coating with Cu_2O particles provided a rough microstructure; (b) the wood surfaces became superamphiphobic after hydrophobization with FAS-17, and the working principle of the durable process after decomposition

2 Materials and methods

2.1 Materials

All chemicals supplied by Shanghai Boyle Chemical Company, Limited were of analytical reagent-grade quality and used without further purification. Deionized water was used throughout the experiments. Wood blocks were obtained from the sapwood sections of poplar (*Populus ussuriensis* Kom), which is one of the most common tree species in the northeast of China. The wood specimens were oven-dried (24 h, 103±2 ℃) to a constant weight after ultrasonically rinsing in deionized water for 30 min, and oven-dried weight were determined.

2.2 Preparation of Cu_2O film on the wood surface

A typical synthesis of the Cu_2O film on wood substrate was performed as follows. $Cu(CH_3COO)_2 \cdot H_2O$ (0.5989 g) was dissolved in 30 mL deionized water, then D-glucose (0.5 g) and PVP (K-30, 0.3 g) were added under continuous stirring. After 1 h, the resulted solution was transferred into a 50 mL Teflon-lined autoclave and the cleaned wood blocks were immersed in the solution. The autoclave was kept in a vacuum oven at 180 ℃ for 2 h. After cooling to room temperature in air, the wood loaded with Cu_2O was removed, washed several times with deionized water and ethanol, and dried at 45 ℃ for more than 24 h in vacuum.

2.3 Hydrophobization of wood surfaces with FAS-17

A 20 mL methyl alcohol solution of 0.2 mL (heptadecafluoro-1, 1, 2, 2-tetradecyl) trimethoxysilane [$CF_3(CF_2)_7CH_2CH_2Si(OCH_3)_3$, hereafter denoted as FAS-17] was hydrolyzed by the addition of 60 mL water at room temperature. Then, 75 mL of the adjusted solution was transferred into a 100 mL Teflon container. The Cu_2O-treated wood samples were subsequently placed into the above reaction solution. The autoclave was sealed and maintained at 75 ℃ for 5 h, then allowed to naturally cool to room temperature. Subsequently, the samples were washed with ethyl alcohol to remove any residual chemicals and allowed to dry in air at room temperature. They were then dried at 45 ℃ for more than 24 h in vacuum. Thus, a superhydrophobic wood surface was obtained.

2.4 Characterization

The morphology of the wood surfaces was observed using the field emission scanning electron microscopy

(SEM, Quanta 200, FEI, Holland) operating at 12.5 kV in combination with EDS (Genesis, EDAX, Holland). The X-ray diffraction (XRD, Bruker D8 Advance, Germany) was employed to analyze the crystal structures of all samples applying graphite monochromatic with Cu $K\alpha$ radiation ($\lambda = 1.5418$ Å) in the 2θ range from 5° to 80° and a position-sensitive detector using a step size of 0.02° and a scan rate of 4° min^{-1}. The transmission electron microscopy (TEM) experiment was performed on a Tecnai G20 electron microscope (FEI, USA) with an acceleration voltage of 200 kV. Carbon-coated copper grids were used as the sample holders. FTIR spectra were obtained on KBr tablets and recorded using a Magna-IR 560 spectrometer (Nicolet) with a resolution of 4 cm^{-1} by scanning the region between 4000 and 400 cm^{-1}. Thermogravimetric and Differential Thermal Analysis (TG-DTA): SDT Q600 thermogravimetric analyzer (TA Instruments, USA), 10 mg sample size, N_2 as carrier gas (150 mL · min^{-1}), 10 ℃ min^{-1}, 25-800 ℃. Water contact angles (WCAs) and oil (hexadecane) contact angles (hereafter defined as OCAs) were measured on an OCA40 contact angle system (Dataphysics, Germany) at room temperature. In each measurement, a 5 μL drop of deionized water was injected onto the surfaces of the wood samples and the contact angles were measured at five different points of each sample. The final values of the contact angles were obtained by averaging the five measurement values. Furthermore, the roll-off angle (α) could be expressed as the difference between advancing and receding contact angle. The roll-off angle (α) is minimized on the surface expected to induce the drop roll off. The roll-off angle measurements were made at room temperature following tilted plate methodology.[16, 17] In this method, a water or hexadecane drop of volume ~10 μL was suspended with the needle and brought in contact with the superamphiphobic surfaces using a computer controlled device. The roll-off angle (α) was measured by inclining objective table when drops land on the sample surface. And roll-off angle (α) was calculated as the angle between the horizontal plane and the inclined plane when the drop started to sliding.

3 Results and discussion

The SEM images of the pristine wood and the Cu_2O-treated wood are presented in Fig. 2. The pristine poplar wood has a quite smooth surface and it is observed that the polar wood is a type of heterogeneous and porous material (Fig. 2a). After the wood is immersed in the hydrothermal system for 2 h, high density particles on the surface is formed (Fig. 2b). The corresponding high magnification SEM image reveals that the wood cell walls and the pits of the wood surface are coated with irregular particles with the size ranging from 3.9 μm to 2.7 μm (Fig. 2c). In the system, the microscaled particles and the heterogeneous wood substrate form a rough hierarchical morphology, which helps to endow the wood with improved wettability. The surface chemical elemental compositions of the Cu_2O-treated wood are determined using the energy-dispersive X-ray spectroscopy (EDS), and the results are presented in Fig. 2d. The evidence of only carbon, oxygen, copper, and gold elements could be detected. The gold element is originated from the coating layer used for SEM observation, and carbon element is from the wood substrate. Since no other elements were detected, the Cu_2O-treated wood is composed of Cu and O elements.

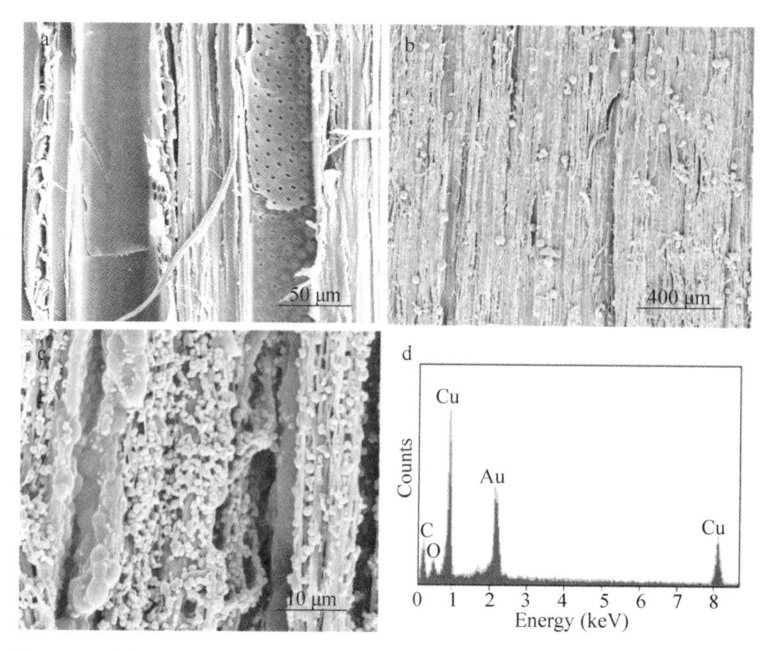

Fig. 2 SEM images of the surfaces of (a) the pristine wood, (b, c) the Cu_2O-treated wood at different magnifications, and (d) EDS spectrums of the Cu_2O-treated wood

Fig. 3 shows the XRD patterns of the pristine wood and the Cu_2O-treated wood, and the selected electron diffraction (SAED) pattern, the representative TEM image and high - resolution transmission electron microscopy (HRTEM) image of the Cu_2O-treated wood. In Fig. 3a, the diffraction peaks at 14.8° and 22.5° belonging to the (101) and (002) crystal planes of cellulose in the wood are observed in both the spectrum of the pristine wood and the Cu_2O-treated wood.[18, 19] For the Cu_2O-treated wood, all reflectance peaks are indexed to cuprite Cu_2O (JCPDS card no. 05-0667) and Cu species (JCPDS card no. 65-9026).[20, 21] The peaks at 2θ values of 29.5°, 36.4°, 42.3°, 61.6°, 73.8° and 77.6° correspond to (110), (111), (200), (220), (311) and (222) planes of pure cuprite Cu_2O crystal, respectively. In addition, the diffraction peaks located at 43.4° and 50.5° can be indexed to (111) and (200) planes of Cu, which may be from the intermediate material in the process of preparing the Cu_2O film. In view of the observation and analysis mentioned above, the Cu^{2+} ions are reduced first to Cu^+ by D-glucose. The possible chemical reaction for the formation of Cu_2O film is described with the following equations[22]:

$$Cu^+ \rightarrow Cu^0 \quad (1)$$

$$2Cu^+ \rightarrow Cu^0 + Cu^{2+} \quad (2)$$

$$2Cu^+ + 2OH^- \rightarrow 2CuOH \rightarrow Cu_2O + H_2O \quad (3)$$

When the D-glucose powders are added into the Cu^{2+} salt system, the D-glucose powder are firstly touched with organic additives (PVP) and reaches the reaction zone, then the Cu^{2+} ions are reduced to Cu^+ by D-glucose. Meanwhile, due to the existence of enough OH^- ions on the wood surfaces and in the aqueous solution, the main reaction tendency of Eq. (3) becomes dominant.

Moreover, the corresponding selected area electron diffraction (SAED) pattern of the Cu_2O-treated wood is shown in Fig. 3b. The different crystal planes are identified as (111), (200), (220), (311) and (222) diffraction planes of cuprite Cu_2O, which is in agreement with XRD pattern results in Fig. 3a. As shown in Fig. 3c, the TEM image shows that a large quantity of cuprite Cu_2O particles are attached onto the surface of the wood substrate. To further investigate the distribution of Cu_2O particles, the HRTEM image is displayed in Fig. 3d. The clear lattice fringe of $d-0.25$ nm matches that of the (110) plane of Cu_2O.[23] No other impurities such as CuO are detected.

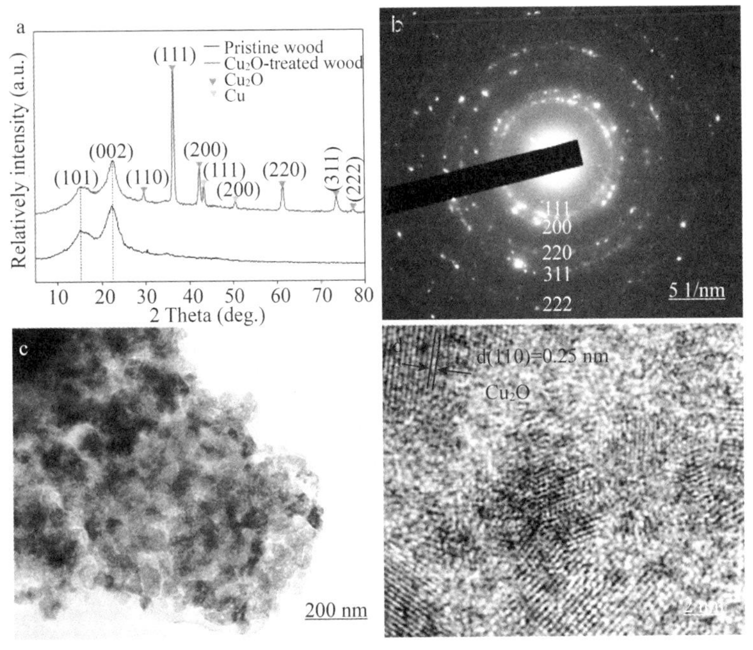

Fig. 3 (a) XRD patterns of the pristine wood and the Cu_2O-treated wood, and (b) the selected electron diffraction (SAED) patterns, (c) TEM image and (d) HRTEM image of the Cu_2O-treated wood, respectively

The TG profiles and DTG profiles of the pristine wood and the hydrophobized Cu_2O-treated wood are shown in Fig. 4. In Fig. 4b, the DTG curves of both the samples show an initial peak between 50 ℃ and 160 ℃, which correspond to a mass loss of absorbed moisture less than 5%. After this peak, the DTG curve of the pristine wood shows two decomposition steps[24]: ① the first decomposition shoulder peak at about 295 ℃ is attributed to thermal depolymerisation of hemicellulose or pectin (weight loss 15%); ② the major second decomposition peak at about 375 ℃ is attributed to cellulose decomposition (weight loss 62%). However, the DTG curve of the hydrophobized Cu_2O-treated wood shows only one major decomposition peak at about 380 ℃, which may be due to the protection effect of modifying agent (FAS-17). In addition, the hydrophobized Cu_2O-treated wood exhibits higher maximum decomposition temperature, because the composition of the sample that has the hydrophobic property may be responsible for the thermal insulation of cellulose degradation. As shown in Fig. 4a, it can be observed from the TG curves that carbon residues with the weight of 16.4% for the pristine wood remains, while the hydrophobized Cu_2O-treated wood is able to keep 22.8% of the weight left. The results indicate that the film combined with Cu_2O and FAS-17 provides wood with the protection.

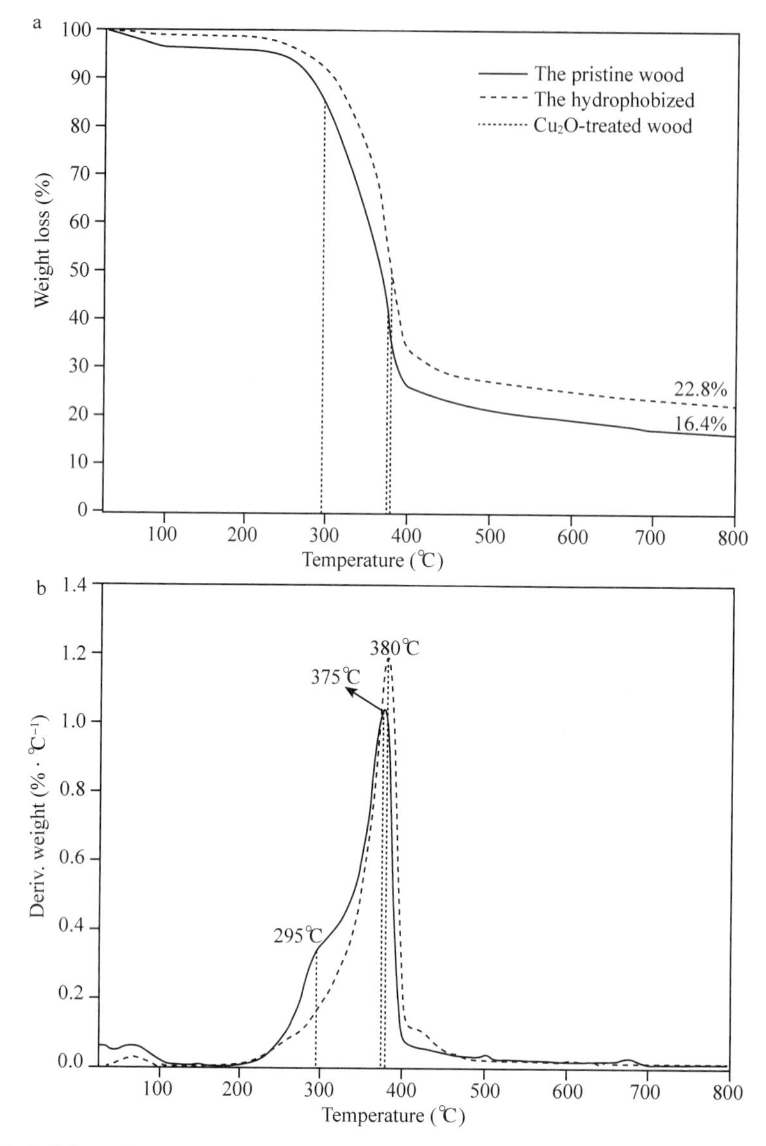

Fig. 4 (a) TG profiles and (b) DTG profiles of the pristine wood and the hydrophobized Cu_2O-treated wood, respectively

The wetting properties of the Cu_2O-treated wood before and after the hydrophobization are investigated by measuring the drop CAs for both water and hexadecane as shown in Fig. 5. The combined effect of CA and α determines the super-liquid-repellent properties of the surfaces. A surface owning CA more than 150° and α less than 10° shows super-liquid-repellent properties. The Cu_2O-treated wood presents hydrophobicity with the WCA of 130. 6° and the α of 13. 7°. (Fig. 5a), and superoleophilicity with the OCA of 0° (Fig. 5c). After being modified with FAS-17, the hydrophobicity of the Cu_2O-treated wood surface is raised to superhydrophobicity, and the WCA reaches 153. 8° and the α is 3. 6° (less than 10°). At the same time, the hydrophobized Cu_2O-treated wood possesses a superoleophobicity with OCA of 152. 1° and the α of 4. 5° (less than 10°). The results indicate that the superamphiphobic properties of wood surfaces are significantly increased.

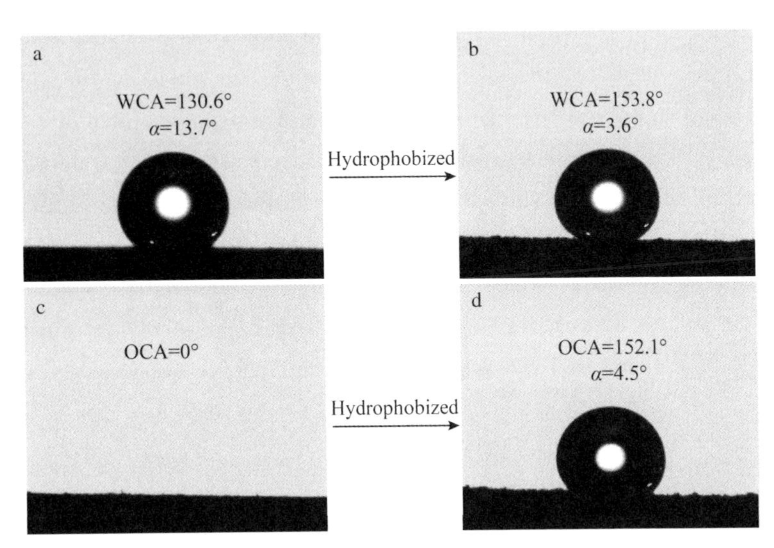

Fig. 5 (a, b) WCAs and roll-off angles (α) of the Cu_2O-treated wood before and after hydrophobization, (c, d) OCAs and roll-off angles (α) of the Cu_2O-treated wood before and after hydrophobization

In outdoor applications, it needs that superamphiphobic wood surfaces can survive harsh conditions. To investigate the mechanical resistance of the as-prepared coating, sand abrasion test is performed in Fig. 6a. Sand grains 100 μm to 300 μm in diameter are introduced to impact the surfaces from a height of 20 cm. Then, the contact angles (WCA and OCA) of the as-prepared superamphiphobic wood surfaces after sand impact tests are measured to estimate the physical and mechanical stability of the coatings as shown in Figs. 6b and c. It is obvious that the as-prepared superamphiphobic wood surfaces still remain superhydrophobicity with the WCA of 153.2°, and superoleophobicity with OCA of 151.8°.

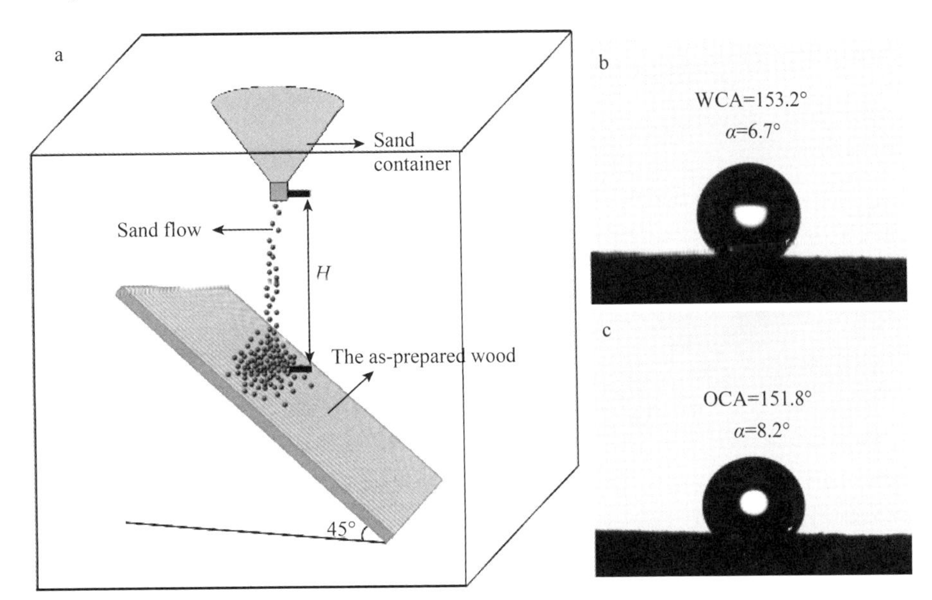

Fig. 6 Mechanical resistance quantified by sand abrasion. (a) Schematic drawing of a sand abrasion experiment. (b) Water drop and (c) hexadecane drop deposited on the as-prepared wood surface after 20 g of sand abrasion from 20 cm height

To assess the durable ability, the superamphiphobic wood surfaces is damaged artificially by sand impact tests for 100 times. Fig. 7a demonstrates water drop and hexadecane drop on the superamphiphobic wood surface after 100 cycles sand abrasion tests. It can be seen clearly that the water drop and hexadecane drop appear as spheres. Fig. 7d shows the change of the WCAs and OCAs with the abrasion cycles. The results indicate that the superamphiphobic wood surfaces can withstand at least 100 cycles of abrasion damages without changing its super-repellent feature. As shown in Fig. 7e, after immersed in an aqueous HCl solution (pH=1) for 24 h, the superamphiphobic wood surface results in a slight reduction in water and hexadecane contact angles, being reduced from 153.8° to 151.6° and 152.1° to 150.1°, respectively. Furthermore, Figs. 6b and c demonstrate snapshots of the captured videos of the superamphiphobic wood surface while being separately exposed to streams of water and hexadecane. It can be noted that the surface is able to effectively shed off the

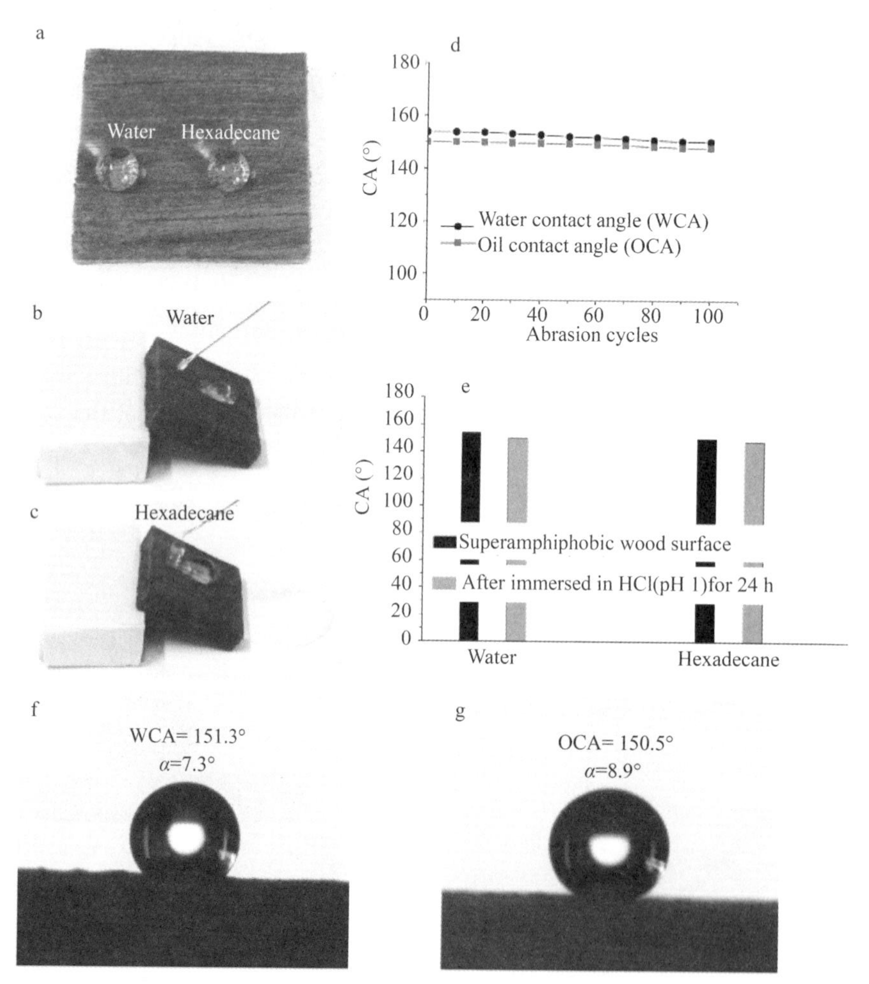

Fig. 7 (a) Digital photographs of water drop and hexadecane drop on the superamphiphobic wood surface after sand abrasion tests. (b, c) the surfaces exposed to the stream of water and hexadecane showing the chemical and physical roughness of surface in shedding the liquids, respectively. (d) CAs changes depending on the abrasion cycles. (e) CAs of the superamphiphobic wood surface before and after immersed in HCl (pH 1) for 24 h. (f) Water drop and (g) hexadecane drop deposited on the as-prepared wood surface after boiling at 100 ℃ for 30 h liquids without getting wet, which indicates that the superamphiphobic wood surface has the self-cleaning property that would scour off the dust during raining. In Fig. 6f and g, it could be observed that the global water and hexadecane drops on their surfaces with the WCA of 151.3° and the OCA of 150.5° after boiling at 100 ℃ for 30 h. The result powerfully proved that the thin film made of Cu_2O and FAS-17, as a barrier, effectively protected the wood substrate from acid and other degradations. Thus, the wood coated with the film shows robust anti-acid, mechanical resistance, and high temperature-humidity-resistant superamphiphobic property. All the above results indicate that the superamphiphobic wood surfaces possess excellent durable superamphiphobicity.

4 Conclusions

In summary, this study demonstrates that wood surfaces coated with hydrophobized Cu_2O films have a remarkable superamphiphobic surface with excellent durability against acid exposure, severe abrasion and high-temperature-humidity conditions. Such the functional coating may be useful for the development of innovative protective clothing for various applications, such as bridges, buildings, automobile, etc.

Acknowledgments

This work was financially supported by the National Natural Science Foundation of China (grant no. 31470584).

References

[1] Youngblood J P, Sottos N R. Bioinspired materials for self-cleaning and self-healing[J]. Mrs Bull., 2008, 33(8): 732-741.

[2] Li Y, Li L, Sun J. Bioinspired self-healing superhydrophobic coatings[J]. Angew. Chem. Int. Edit., 2010, 122 (35): 6265-6269.

[3] Tuteja A, Choi W, Ma M, et al. Designing superoleophobic surfaces[J]. Science, 2007, 318(5856): 1618-1622.

[4] Tuteja A, Choi W, McKinley G H, et al. Design parameters for superhydrophobicity and superoleophobicity[J]. Mrs Bull., 2008, 33(8): 752-758.

[5] Liu M, Wang S, Wei Z, et al. Bioinspired design of a superoleophobic and low adhesive water/solid interface[J]. Adv. Mater., 2009, 21(6): 665-669.

[6] Khedir K R, Saifaldeen Z S, Demirkan T M, et al. Robust superamphiphobic nanoscale copper sheet surfaces produced by a simple and environmentally friendly technique[J]. Adv. Eng. Mater., 2015, 17(7): 982-989.

[7] Risse G, Matys S, Böttcher H. Investigation into the photo-induced change in wettability of hydrophobized TiO_2 films [J]. Appl. Surf. Sci., 2008, 254(18): 5994-6001.

[8] Wang X, Liu X, Zhou F, et al. Self-healing superamphiphobicity[J]. Chem. Commun., 2011, 47(8): 2324-2326.

[9] Wang H, Xue Y, Ding J, et al. Durable, self-healing superhydrophobic and superoleophobic surfaces from fluorinated-decyl polyhedral oligomeric silsesquioxane and hydrolyzed fluorinated alkyl silane[J]. Angew. Chem. Int. Ed., 2011, 123 (48): 11635-11638.

[10] Gao L, Lu Y, Zhan X, et al. A robust, anti-acid, and high-temperature-humidity-resistant superhydrophobic surface of wood based on a modified TiO_2 film by fluoroalkyl silane[J]. Surf. Coat. Tech., 2015, 262: 33-39.

[11] Gao L, Gan W, Xiao S, et al. Enhancement of photo-catalytic degradation of formaldehyde through loading anatase TiO_2 and silver nanoparticle films on wood substrates[J]. RSC Adv., 2015, 5(65): 52985-52992.

[12] Gao L, Lu Y, Li J, et al. Superhydrophobic conductive wood with oil repellency obtained by coating with silver

nanoparticles modified by fluoroalkyl silane[J]. Holzforschung, 2016, 70(1): 63-68.

[13]Shoeib M A, Abdelsalam O E, Khafagi M G, et al. Synthesis of Cu_2O nanocrystallites and their adsorption and photocatalysis behavior[J]. Adv. Powder Technol., 2012, 23(3): 298-304.

[14]Kou T, Jin C, Zhang C, et al. Nanoporous core-shell Cu@ Cu_2O nanocomposites with superior photocatalytic properties towards the degradation of methyl orange[J]. RSC Adv., 2012, 2(33): 12636-12643.

[15]Chen C, Xu H, Xu L, et al. One-pot synthesis of homogeneous core-shell Cu_2O films with nanoparticle-composed multishells and their photocatalytic properties[J]. RSC Adv., 2013, 3(47): 25010-25018.

[16]Pierce E, Carmona F J, Amirfazli A. Understanding of sliding and contact angle results in tilted plate experiments [J]. Colloid. Surface. A, 2008, 323(1-3): 73-82.

[17]Extrand C W. Model for contact angles and hysteresis on rough and ultraphobic surfaces[J]. Langmuir, 2002, 18 (21): 7991-7999.

[18]Gao L, Zhan X, Lu Y, et al. pH-dependent structure and wettability of TiO_2-based wood surface[J]. Mater. Lett., 2015, 142: 217-220.

[19]Gao L, Lu Y, Cao J, et al. Reversible photocontrol of wood-surface wettability between superhydrophilicity and superhydrophobicity based on a TiO_2 film[J]. J. Wood Chem. Technol., 2015, 35(5): 365-373.

[20]Liu X. Cu_2O microcrystals: a versatile class of self-templates for the synthesis of porous Au nanocages with various morphologies[J]. RSC Adv., 2011, 1(6): 1119-1125.

[21]Lv D, Ou J, Hu W, et al. Superhydrophobic surface on copper via a one-step solvent-free process and its application in oil spill collection[J]. RSC Adv., 2015, 5(61): 49459-49465.

[22]Chen S J, Chen X T, Xue Z, et al. Solvothermal preparation of Cu_2O crystalline particles[J]. J. Cryst. Growth, 2002, 246(1-2): 169-175.

[23]Shang Y, Sun D, Shao Y, et al. A facile top-down etching to create a Cu_2O jagged polyhedron covered with numerous {110} edges and {111} corners with enhanced photocatalytic activity[J]. Chem. -Eur. J., 2012, 18(45): 14261-14266.

[24]Ouajai S, Shanks R A. Composition, structure and thermal degradation of hemp cellulose after chemical treatments [J]. Polym. Degrad. Stabil., 2005, 89(2): 327-335.

中文题目：基于氟硅烷改性的氧化亚铜薄膜制备耐久性超双疏木材表面

作者：高丽坤，肖少良，甘文涛，詹先旭，李坚

摘要：以木材为基质，通过简单的水热合成法和进一步疏水处理制备耐久性的超双疏氧化亚铜微球薄膜。该方法具有简单易操作、价廉、反应时间短及环境友好等优势，可广泛用于在木材表面制备氧化亚铜微米结构。同时，木材表面负载一层长链氟硅烷(十七氟葵基三甲氧基硅烷)的水解产物后，具有耐久的超疏水和超疏油表面特性，该表面对酸、高温高湿和磨损表现出优异的耐久性。该薄膜有效地保护了木材基质不受破坏，拓展了木材的应用领域。该功能性薄膜在桥梁和建筑乃至自动化和其它可能性领域中均具有广阔的应用前景。

关键词：木材；超双疏；氧化亚铜；耐久性

A Robust, Anti-acid, and High-temperature-humidity-resistant Superhydrophobic Surface of Wood Based on A Modified TiO_2 Film by Fluoroalkyl Silane*

Likun Gao, Yun Lu, Xianxu Zhan, Jian Li, Qingfeng Sun

Abstract: A two-steps method containing low-temperature hydrothermal synthesis with TiO_2 precursor solution and subsequent modification with fluoroalkyl silane on wooden substrates was investigated. Scanning electron microscopy (SEM) images showed that the TiO_2-treated wood substrate was covered with uniform TiO_2 particles, which generated a roughness on the wood surface favoring the formation of the superhydrophobic surface, and the decoration of (heptadecafluoro-1, 1, 2, 2-tetradecyl) trimethoxysilane on TiO_2-based wood surface acted as a crucial role in improving the repellency toward water. The superhydrophobic wood surface with the water contact angle (WCA) of 152.9°, maintained superhydrophobic property with the WCA larger than 150° after impregnating in 0.1 M hydrochloric acid solution for one week, irradiating under UV light for 24 h, or boiling at 150 ℃ for 10 h. The prepared wood surface showed multi-functions including super repellency toward water, anti-acid, and high-temperature-humidity-resistant. On the basis of the results, the functional coating on the wood surface provided an enlarged field of the wood works, such as historic structures protection.

Keywords: Anti-acid; Fluoroalkyl silane; TiO_2 film; Superhydrophobic; High-temperature-humidity-resistant

1 Introduction

Nowadays, wooden products are extensively used in a varied of daily applications such as indoor decoration, wooden bridge, pavement and so on[1, 2]. And efforts have been devoted to prolong the lifetime of wooden products especially for the place of historic structures that made of wood. However, the natural environmental factors cause discoloration and degradation to these wooden products by light[3, 4], moisture[3-5], microorganisms[6, 7], and acidic rain[8]. To solve this crucial and inevitable problem, researchers have carried out many studies on it, and found that reducing moisture adsorption was a key factor in improving the durability of the wooden products[2, 5]. Methods to prepare superhydrophobic surface by low-surface-energy materials involve one-step processes such as coaxial electrospinning[9], laser/plasma/chemical etching[10] and lithography[11], which are simple, but the excellent hydrophobicity of these superhydrophobic surfaces gradually degrades after a long-time outdoor exposure[12]. The superhydrophobic surface can also be fabricated

* 本文摘自 Surface & Coatings Technology, 2015, 262: 33-39.

by a two-steps method based on preparing a rough film on the substrate first, which is then modified by appending a low-surface-energy thin layer[12-14].

As an inorganic nano material, TiO_2 is frequently used to develop a roughness due to its low cost and nontoxicity[15-17], and a subsequent modification by silane or fluorine-containing polymers, thus leading to super water repellency on the surface[18-21], which separates the wettability from the native properties of the substrate[22]. Previous study has proved that the film of TiO_2 subsequent coating with fluoroalkyl silane maintained higher water contact angle than the film without TiO_2 after longtime outdoor exposure[22]. Recently, artificially constructed superhydrophobic surfaces possessing some interesting characters, such as low water and snow adhesion, low friction in surface dragging and easy removal of dirt, have become a hot issue in various fields[19, 20, 23, 24]. Kujawa et al. has reported that the hydrophobic titania ceramic membranes prepared by grafting of the long-chain polymer could be stable after 4 years of exposure to open air[25].

In this paper, we report a two-steps method for preparation of a superhydrophobic film on the wood surface which consists of fabricating a TiO_2 film through a low-temperature hydrothermal method and subsequent modification by (heptadecafluoro-1, 1, 2, 2-tetradecyl) trimethoxysilane. The durability of the superhydrophobic wood surface is demonstrated by impregnating in 0.1 M hydrochloric acid solution for one week, irradiating under UV light for 24 h, and boiling at 150 ℃ for 10 h. As a result, the prepared wood surface maintained hydrophobicity with the WCA larger than 150°, suggesting multi-functions including super repellency toward water, anti-acid, and high-temperature-humidity-resistant. Wood with such a superhydrophobic surface could possess these unique functions, increase the durability of wood, and enlarge the field of wood applications such as self-cleaning, and the historic structures protection.

2 Materials and methods

2.1 Materials

All chemicals were supplied by Shanghai Boyle Chemical Company Limited. and used as received. Each piece of poplar wood (*Populus ussuriensis* Kom) was carefully cut into an area of 20 mm×20 mm, and the thickness of the wooden chips was controlled within 5 mm. Then the wood samples were ultrasonically rinsed in deionized water for several times and oven dried at 105 ℃ until stabilization of their mass (approximately 48 h) before determination of their anhydrous weights.

2.2 Preparation of TiO_2 film on the wood surface via a hydrothermal method

Ammonium fluorotitanate (0.4 M) and boric acid (1.2 M) were dissolved in distilled water in a 500 mL glass container at room temperature under vigorous magnetic stirring. Then, a solution of 0.3 M hydrochloric acid (HCl) was added until the pH reached approximately 3. The adjusted solution was transferred into a Teflon container. Wood specimens were subsequently placed into the above reaction solution. The autoclave was sealed and maintained at 90 ℃ for 5 h, then cooled to room temperature naturally. Finally, the prepared samples were removed from the solution, ultrasonically rinsed with deionized water for 3 times, blown to dry by N_2 and dried in an oven at 60 ℃ for 24 h. For comparative studies, the blank wood samples were also selected.

2.3 Establishment of a robust, anti-acid, and high-temperature-humidity-resistant superhydrophobic wood surface

A methyl alcohol solution of (heptadecafluoro-1, 1, 2, 2-tetradecyl) trimethoxysilane ($CF_3(CF_2)_7CH_2CH_2Si(OCH_3)_3$, hereafter denoted as FAS-17) was hydrolyzed by the addition of a 3-fold molar excess of water at room temperature. The TiO_2-treated wood samples was placed into a Teflon container containing the prepared solution, then the Teflon container was put in an oven at 80 ℃ for 5 h to enable the silane group of FAS-17 vapor to react with the hydroxide group of titania particles completely. Subsequently, the samples were washed with ethyl alcohol (EtOH) to remove any residual chemicals, and they were then blown to dry by N_2 and dried in an oven at 60 ℃ for 24 h. Thus, a superhydrophobic wood surface was obtained. The durability test of superhydrophobic wood surfaces was carried out by impregnating the prepared wood samples in a solution of 0.1 M hydrochloric acid (HCl) for one week, irradiating under a 36 W UV light with a wavelength of 356 nm for 24 h, and boiling at 150 ℃ for 10 h, separately. The samples were dried at 45 ℃ for 2 h and then their repellency toward water droplets was measured every 1 h.

2.4 Characterization

The surface morphology of the samples was characterized using scanning electron microscope (SEM, Quanta 200, FEI) operating at 12.5 kV in combination with X-ray spectroscopy (EDS, Genesis, EDAX). Atomic Force Microscopy (AFM) images were obtained using a Multimode Nanoscope Ⅲa controllor (Veeco Inc, USA) with a silicon tip operated in tapping mode to characterize the surface morphology and roughness. X-ray diffraction studies (XRD, Bruker D8 Advance, Germany) were performed using Cu $K\alpha$ radiation (λ = 1.5418 Å) at a scan rate of 4°/min, 40 kV, 40 mA ranging from 5° to 70°. The transmission electron microscopy (TEM) experiment was performed on a Tecnai G20 electron microscope (FEI, USA) with an acceleration voltage of 200 kV. Carbon-coated copper grids were used as the sample holders. Thermogravimetric analysis was performed using a simultaneous thermal analyzer (DSC-TGA; SDT Q600, TA Instruments, USA) in a temperature range of 25-700 ℃ (10 ℃/min) under a dynamic N_2 atmosphere. The WCAs were measured using an OCA40 contact angle system (Dataphysics, Germany) at room temperature. For each measurement, a 5 μL droplet of deionized water was injected onto the radial section of each wood sample at a dosing rate of 0.5 $\mu L \cdot s^{-1}$, and the contact angles were measured at five different points on each sample. The final value for the WCA was obtained as an average of these five measurements.

3 Results and discussion

Fig. 1 shows the SEM images of the original wood, TiO_2-treated wood, and wood treated with TiO_2 and FAS-17. In Fig. 1a, the pristine surface of the poplar wood, which contained some pits, appeared to be free of any other materials. In Fig. 1b, the pristine wood surface was covered by a dense and even film of TiO_2 microparticles. The surface chemical elemental compositions of the TiO_2-treated wood are determined via EDS spectrum, and the results are presented as an insert in Fig. 1b. Evidence of only oxygen, titanium, gold, and carbon elements could be detected in the spectra. The gold originated from the coating layer used for SEM observation and carbon element was from the wood substrate. No other elements were detected, thus confirming the composition of the microspheres to be TiO_2. Fig. 1c shows the wood surface treated with TiO_2 and FAS-17, exhibiting a microrelief similar to that of the plant surface, which mainly combined with wax crystalloids and cuticular folds and provided a water-repellent surface[26, 27]. The wax crystalloids and cuticular folds are evident from the higher magnification images shown in Fig. 1d, which presents all over the surface providing the system with such a structure. The presence of such a rough binary structure on the wood surface is the main requirements for superhydrophobicity. While, The EDS analysis as an insert in Fig. 1c confirmed the existence of Si and F elements on the treated wood sample. It is reasonable that the appearance of Si and F elements originated from FAS-17, which verified the TiO_2-based wood surface was successfully modified by FAS-17.

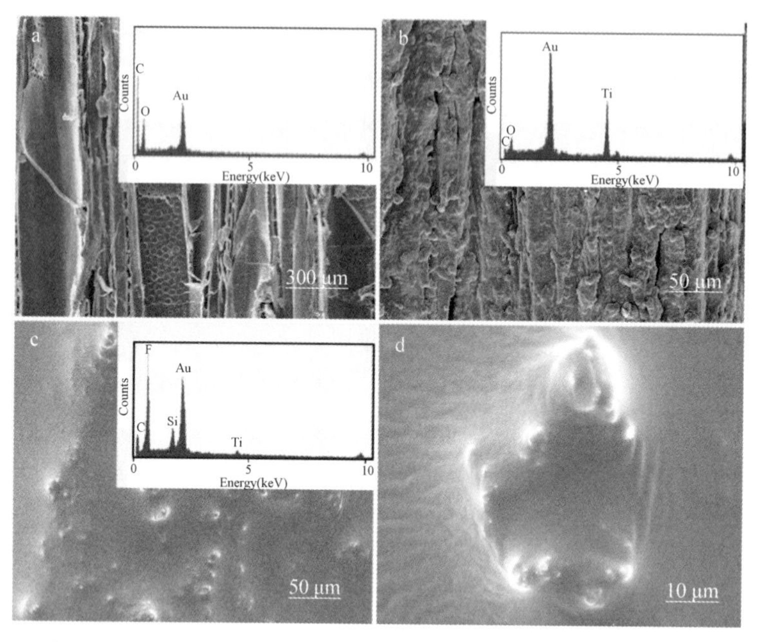

Fig. 1 SEM images of the surfaces of (a) the original wood, (b) the TiO_2-treated wood, and (c, d) the wood treated by TiO_2 and FAS-17 at different magnifications (the inserted EDS spectrums of the wood samples according to the Figures 1a, 1b and 1c, respectively)

In Fig. 2 top view and three-dimensional surface plots (300 nm×300 nm) AFM representations are presented for the original wood, the TiO_2-treated wood, and the wood treated by TiO_2 and FAS-17, respectively. Compared with the original wood surface (Fig. 2a), the TiO_2-treated wood surface (Fig. 2b) exhibits a fine microstructure and consisted of small and uniform grains forming a rather flat but more complex surface texture which is consistent with a rough topography. The surface of the wood treated by TiO_2 and FAS-17 (Fig. 2c) are composed of high "mountains" and deep "valleys". The difference of the surface roughness can be reflected in the rms roughness values (rms=root mean square=the standard deviation of the Z value, Z being the total height range analyzed) of the three surfaces, which were measured 23.5 nm (the original wood), 48.1 nm (the TiO_2-treated wood), and 94.8 nm (the wood treated by TiO_2 and FAS-17), respectively. Therefore, it can be concluded that the modified surface topology is responsible for the film generated hydrophobic characteristics, due to a change in surface roughness.

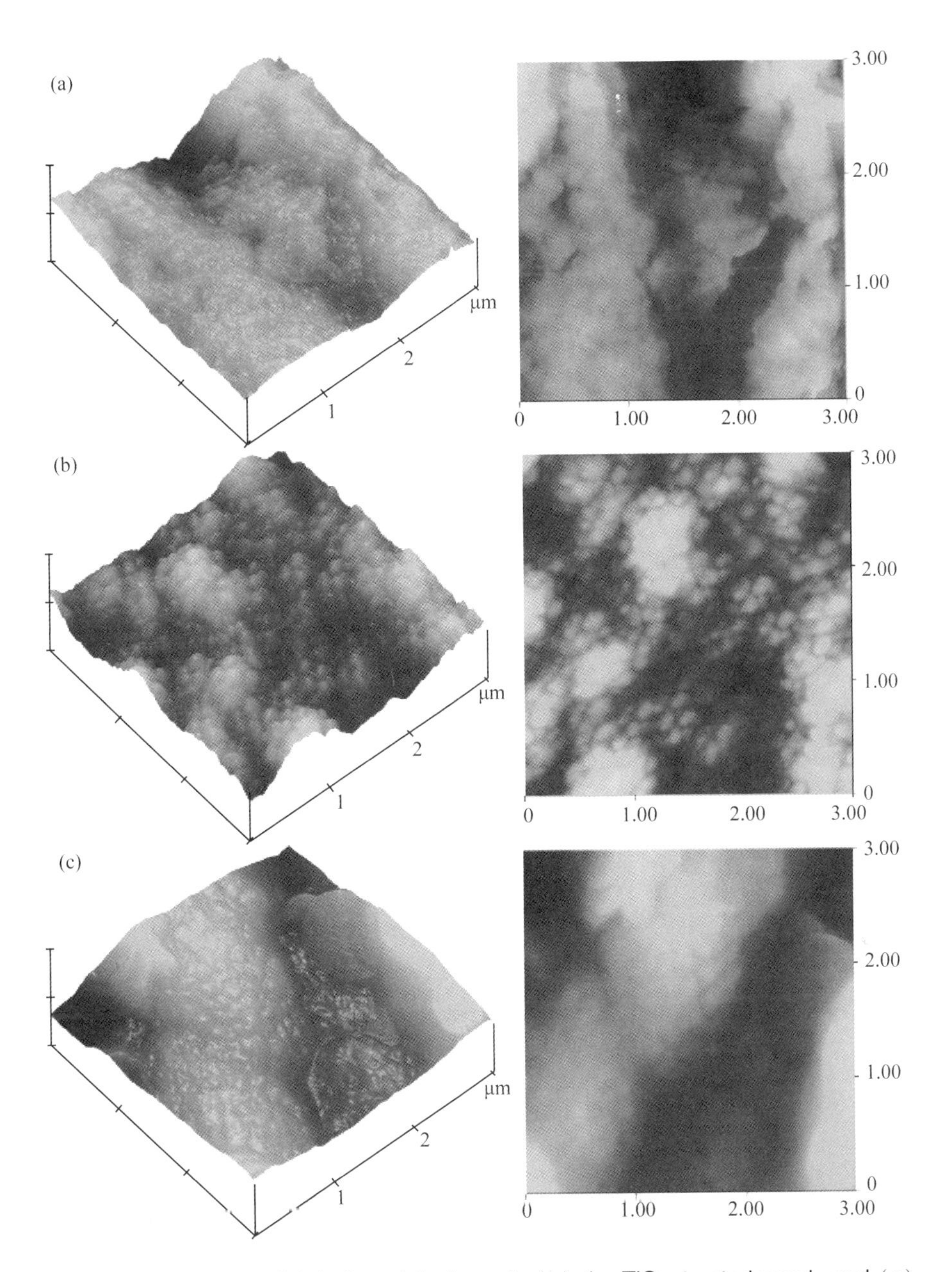

Fig. 2 AFM images of (a) the original wood, (b) the TiO_2-treated wood, and (c) the wood treated by TiO_2 and FAS-17

Fig. 3 presents the XRD patterns of the original and TiO_2-treated woods. Apparently, compared with the original wood, except for the diffraction peaks at 14.8° and 22.5° belonging to characteristic diffraction peaks of cellulose in the wood[28, 29], some new strong diffraction peaks are observed in the spectrum of the TiO_2-treated wood sample, indicating the formation of a new crystal structure on the wood surface. As shown in Fig. 3b, the diffraction peaks at 25.3°, 37.8°, 48.0°, 54.2°, and 62.5° are assigned to the (101), (004), (200), (105), and (204) diffraction planes of anatase TiO_2 (JCPDS card no. 21-1272), respectively. Moreover, the absence of other peaks indicates the purity of the anatase-TiO_2 coating, consistent with the EDS results.

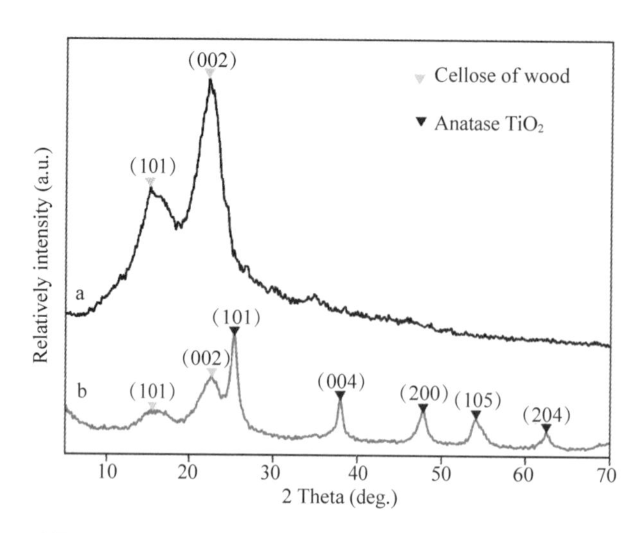

Fig. 3 XRD patterns of (a) the original, and (b) TiO_2-treated woods

The morphology and crystallinity of the TiO_2 - treated wood is further confirmed by the TEM observations. Fig. 4 shows TEM, HRTEM images and SAED pattern of TiO_2-treated wood. TEM dark field images of TiO_2 microparticles in anatase phases are shown in Fig. 4a. It could be observed that the film is composed of the accumulation of TiO_2 microparticles. The corresponding selected area electron diffraction (SAED) pattern of TiO_2 microparticles in anatase phases is shown in Fig. 4b. The SAED patterns of TiO_2 microparticles in anatase phase shows that the diffraction pattern and rings are co-existent. It is in agreement with XRD results in Fig. 3b that the different crystal planes was identified as (101), (004), (200), (105) and (204) diffraction planes of anatase TiO_2, respectively. From the HRTEM image in Fig. 4c, it is seen that the lattice fringes corresponding to (101) (d_{101}=0.35 nm) crystallographic plane of anatase is most frequently observed, which indicates the high crystallinity of the TiO_2 microparticles[30].

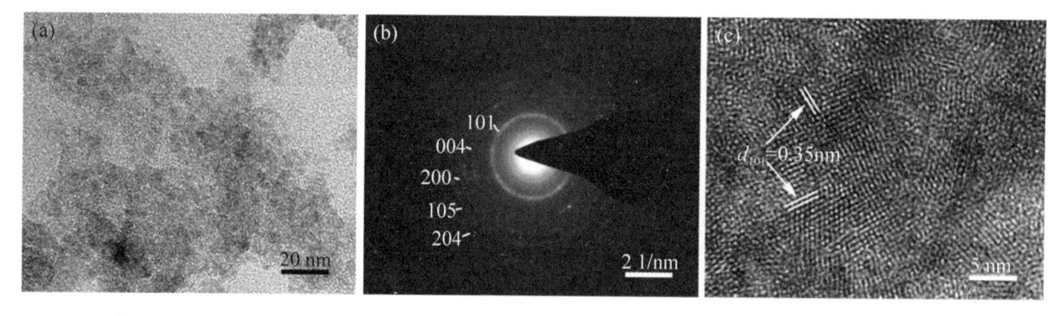

Fig. 4 (a) TEM images, (b) the selected electron diffraction (SAED) patterns, and (c) HRTEM images of the TiO_2-treated wood

The TG and DTG curves of the original wood, the TiO_2-treated wood, and the wood treated by TiO_2 and FAS-17 are shown in Fig. 5. Since the measurements performed on the original and treated woods show a similar course below 100 ℃, which is caused by the removal residual water before pyrolysis[31]. The three stages of the thermal degradation of the original wood were clearly visible. At stage one (190-250 ℃), the pyrolysis rate was low with the weight loss of approximately 10.1%, which mainly due to the partial degradation of hemicellulose[32]. Stage two (250 – 380 ℃) was mainly caused by cellulose degradation, accompanied with continuous degradation of lignin, whose maximum pyrolysis rate occurred at 375 ℃ and the weight loss reached 78.0%[33]. At stage three (380–700 ℃), all the components of wood degraded gradually leading to aromatization and carbonization. Lignin was the most difficult one to decompose. Its decomposition happened slowly and kept on along the whole calcining process[34, 35]. At last, carbon residues with the weight of 13.2% were left, as observed from the TG curve (Fig. 5a). In Fig. 5a, the DTG curve shows the first decomposition shoulder peak at about 295 ℃ is attributed to thermal depolymerisation of hemicellulose, and the major second decomposition peak at about 375 ℃ is attributed to cellulose decomposition[36]. And the DTA curve (Fig. 5b and Fig. 5c) presented strong and sharp endothermic peaks at a minimum of 349 ℃ for the TiO_2-treated wood and 340 ℃ for the wood treated by TiO_2 and FAS-17, respectively, corresponding to the weight losses shown in the TG curve. Due to the decomposition of cellulose and lignin, which the maximum degradation rates of the TiO_2-treated wood and the wood treated by TiO_2 and FAS-17 shifted to lower than the original wood. This might due to the catalysis of TiO_2, which generated an accelerated pyrolysis action on wood components[37]. Moreover, the maximum degradation rate of the wood treated by TiO_2 and FAS-17 was the lowest (Fig. 5c), which might due to the decomposition of FAS-17, as a barrier, effectively protected the wood sample. Computable weight loss along the whole process were about 84.8% for the original wood, 72.5% for the TiO_2-treated wood, and 65.6% for the wood treated by TiO_2 and FAS-17, respectively.

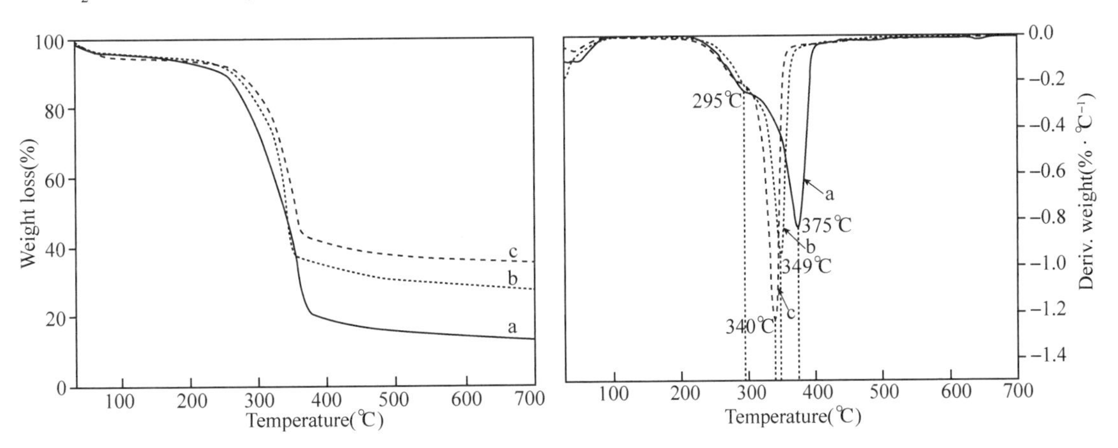

Fig. 5 Thermal analysis (TG/DTG) profiles of (a) the original wood, (b) the TiO_2-treated wood, and (c) the wood treated by TiO_2 and FAS-17

Fig. 6 demonstrates the variation of the water contact angles, and photographs of the TiO_2-treated wood (Fig. 6a), the wood treated by TiO_2 and FAS-17 (Fig. 6b), and the prepared superhydrophobic wood treated in different conditions (Fig. 6c–e). The TiO_2-treated wood surface presents superhydrophilicity with the WCA of 9.6°, and the water level could hardly be seen on its surface from the inserted photograph in Fig. 6a. As shown in Fig. 6b, the superhydrophilic wood surface converted into superhydrophobic one with the WCA of 152.9°, the global water drops stood on the surface observed from the inserted photograph.

Generally, for an absolutely smooth surface, the wettability can be described by the famous Young equation[38]:

$$\cos\theta = (\sigma_{sv} - \sigma_{sl}) / \sigma_{lv} \tag{1}$$

where σ_{sv}, σ_{sl}, and σ_{lv} are the interfacial tensions of the solid-vapor, solid-liquid, and liquid-vapor phases, respectively. To the actual surfaces, they are not absolutely smooth, thus Wenzel modified Eq. (1) and proposed a well-known equation for the contact angle θ_w of liquids on a rough homogeneous surface:

$$\cos\theta_w = r\ (\sigma_{sv} - \sigma_{sl}) / \sigma_{lv} = r\cos\theta_w \quad (2)$$

where r is the roughness factor defined as the ratio of the actual surface area of the rough surface to the geometric projected area and it is always larger than 1. In order to further understand the wettability of the superhydrophobic wood surface, the Cassie equation, which is generally applicable to heterogeneous roughness and low surface energy surfaces which exhibit superhydrophobic, was also employed[39-41]:

$$\cos\theta_\gamma = f_1\cos\theta - f_2 \quad (3)$$

where f_1 and f_2 were the fractional areas estimated for the solid and air on the surface, respectively, that is, $f_1 + f_2 = 1$, θ_γ and θ represent the water contact angle on rough and smooth surfaces, respectively. θ (100°) was the water contact angle on the smooth surface modified by FAS-17[42]; f_1 and f_2 were the fractional areas estimated for the solid and air on the surface, respectively, that is, $f_1 + f_2 = 1$. Here, the WCA value of the wood treated by TiO_2 and FAS-17 was 152.9° (θ_γ), the f_1 calculated using Cassie-Baxter equation was 0.13, which indicated that about 87% of the surface was occupied by air. The surface allowed air to be trapped more easily underneath the water droplets, so that the water droplets essentially rested on a layer of air. Therefore, WCAs of the wood treated by TiO_2 and FAS-17 increased significantly.

To evaluate the durable property of the superhydrophobic wood surface, the prepared wood samples were taken into the following environment, impregnating in 0.1 M HCl solution for one week (Fig. 6c), irradiating under a 36 W UV light with a wavelength of 356 nm for 24 h (Fig. 6d) and boiling at 150 ℃ for 10 h (Fig. 6e), separately. The result showed that the WCAs of the prepared wood surfaces were still more than 150° and each photograph was shown as an insertion in the corresponding figure. It could observed that the sample floated atop the solution after impregnating in 0.1 M HCl solution for one week, and the global water droplets on their surfaces both after irradiating under UV light for 24 h and boiling at 150 ℃ for 10 h. The result powerfully proved that the thin film made of TiO_2 and FAS-17, as a barrier, effectively protected the wood substrate from acid and other degradations, thus, the wood coated with such a film showed robust anti-acid, and high-temperature-humidity-resistant superhydrophobic property.

Fig. 7 explained the schematic illustration of the synthetic process of the superhydrophobic wood treated by TiO_2 and FAS-17. As shown in Fig. 7b, the TiO_2 particles settled onto the pristine wood surface (Fig. 7a) via a low-temperature method, the plentiful hydrogen bands between the surfaces of the TiO_2 microspheres, and the wood reacted with each other, and a TiO_2 film arose on the surface. The dense TiO_2 particles covered the wood surface and roughed the wood surface, which might also provide sites for fluorosilanization via covalent bonding between hydrolyzed silane coupling agents and hydroxyl groups (OH) available on the surface of TiO_2 coated layer[43]. Subsequently (Fig. 7c), in the modification progress, by hydrolysis FAS-17 reacts with water to form silicon hydroxide as a reactive group at the end of the molecule, which is responsible for the chemical bond with the substrate. The fluoroalcyl group is the hydrophobic part of the molecule[44]. Brownian motion helps the long-chain silanes to have an in-plane lateral mobility which leads to an in-plane reorganization and the formation of densely packed monolayer of vertical chains. Then, the hydroxyl groups of Si-OH groups can form hydrogen bond to the surface hydroxyl groups of TiO_2 and eventually the covalent bonds of Si-O-Ti can

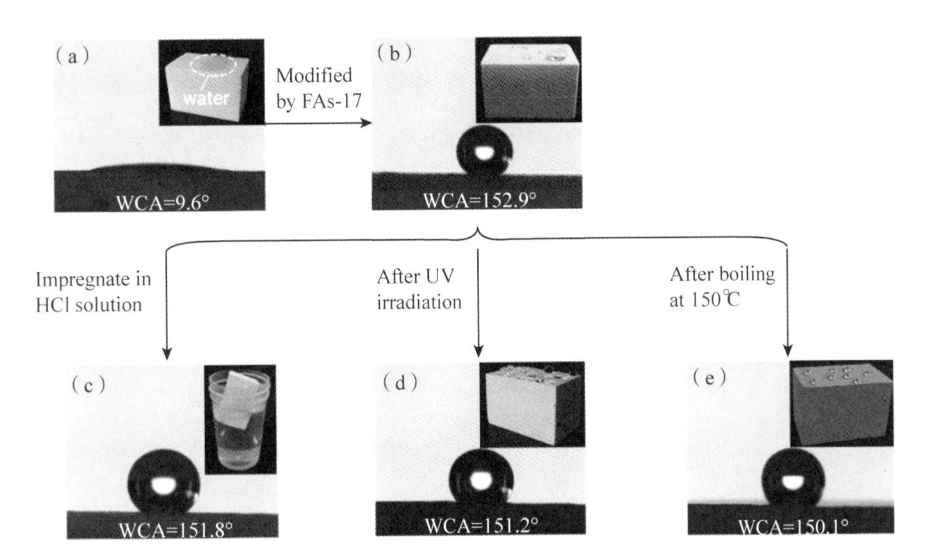

Fig. 6　WCAs of (a) the TiO_2-treated wood sample, (b) the superhydrophobic wood treated by TiO_2 and FAS-17, (c) the prepared superhydrophobic sample after impregnating in 0. 1 M HCl solution for one week, (d) the prepared superhydrophobic sample after irradiating under a 36 W UV light with a wavelength of 356 nm for 24 h, and (e) the prepared superhydrophobic sample after boiling at 150 ℃ for 10 h

take place[43]. Finally, under the combined effects of TiO_2 particles and FAS-17, it may lead to a dense, robust and water repellent layer on wood surface.

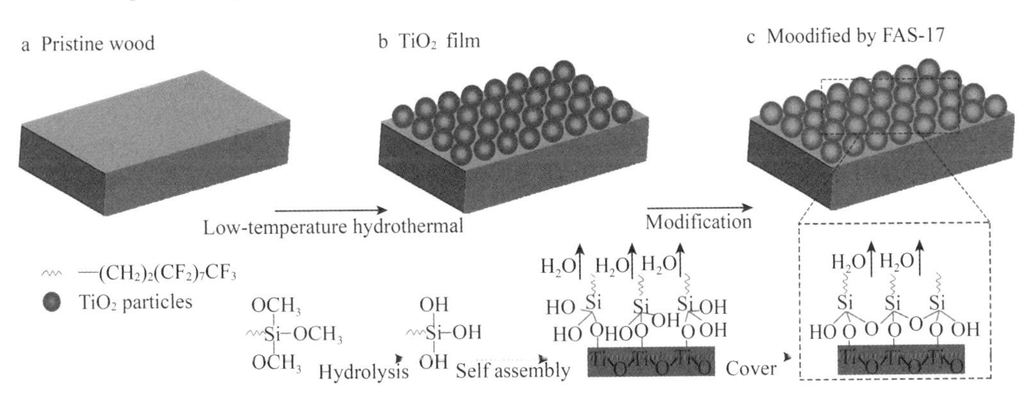

Fig. 7　Schematic illustration of the synthetic process of the superhydrophobic wood treated by TiO_2 and FAS-17

4　Conclusions

In summary, we have demonstrated that the wood treated by TiO_2 and FAS-17 have a remarkable robust superhydrophobic surface against acidic solution of hydrochloric acid, UV light, and high-temperature-humidity condition. The wood surface based on TiO_2 film subsequent modified by FAS-17 showed the WCA still larger than 150° after the treatments containing impregnating in 0. 1 M hydrochloric acid solution for one week, irradiating under UV light for 24 h, or boiling at 150 ℃ for 10 h. Such a functional wood surface may be useful for the development of innovative self-cleaning and durable wooden products for various applications, such as historic structures protection.

Acknowledgments

This work was financially supported by The National Natural Science Foundation of China (grant no. 31270590), China Postdoctoral Science Foundation funded project (2013M540263), and Doctoral Candidate Innovation Research Support Program of Science & Technology Review (kjdb2012006).

References

[1]Toivonen R M. Product quality and value from consumer perspective—An application to wooden products[J]. J. Forest Econ., 2012, 18(2): 157-173.

[2]Hsieh C T, Chang B S, Lin J Y. Improvement of water and oil repellency on wood substrates by using fluorinated silica nanocoating[J]. Appl. Surf. Sci., 2011, 257(18): 7997-8002.

[3]Evans P D, Michell A J, Schmalzl K J. Studies of the degradation and protection of wood surfaces[J]. Wood Sci. Technol., 1992, 26(2): 151-163.

[4]Evans P D, Owen N L, Schmid S, et al. Weathering and photostability of benzoylated wood[J]. Polym. Degrad. Stabil., 2002, 76(2): 291-303.

[5]Hameury S. Moisture buffering capacity of heavy timber structures directly exposed to an indoor climate: a numerical study[J]. Build. Environ., 2005, 40(10): 1400-1412.

[6]Shigo A L, Hillis W E. Heartwood, discolored wood, and microorganisms in living trees[J]. Annu. Rev. Phytopathol., 1973, 11(1): 197-222.

[7]Faix O, Bremer J, Schmidt O, et al. Monitoring of chemical changes in white-rot degraded beech wood by pyrolysis—Gas chromatography and Fourier-transform infrared spectroscopy[J]. J. Anal. Appl. Pyrol., 1991, 21(1-2): 147-162.

[8]Lee B H, Kim H J, Lee J J, et al. Effects of acid rain on coatings for exterior wooden panels[J]. J. Ind. Engl. Chem., 2003, 9(5): 500-507.

[9]Han D, Steckl A J. Superhydrophobic and oleophobic fibers by coaxial electrospinning[J]. Langmuir, 2009, 25(16): 9454-9462.

[10]Gonçalves G, Marques P A A P, Trindade T, et al. Superhydrophobic cellulose nanocomposites[J]. J. Colloid Interf. Sci., 2008, 324(1-2): 42-46.

[11]Kang S M, You I, Cho W K, et al. One-step modification of superhydrophobic surfaces by a mussel-inspired polymer coating[J]. Angew. Chem. Int. Edit., 2010, 49(49): 9401-9404.

[12]Saleema N, Sarkar D K, Gallant D, et al. Chemical nature of superhydrophobic aluminum alloy surfaces produced via a one-step process using fluoroalkyl-silane in a base medium[J]. ACS Appl. Mater. & Inter., 2011, 3(12): 4775-4781.

[13]Liang J, Li D, Wang D, et al. Preparation of stable superhydrophobic film on stainless steel substrate by a combined approach using electrodeposition and fluorinated modification[J]. Appl. Surf. Sci., 2014, 293: 265-270.

[14]Paul B, Martens W N, Frost R L. Surface modification of alumina nanofibres for the selective adsorption of alachlor and imazaquin herbicides[J]. J. Colloid Interf. Sci., 2011, 360(1): 132-138.

[15]Baba K, Hatada R. Synthesis and properties of TiO_2 thin films by plasma source ion implantation[J]. Surf. Coat. Tech., 2001, 136(1-3): 241-243.

[16]Guan K. Relationship between photocatalytic activity, hydrophilicity and self-cleaning effect of TiO_2/SiO_2 films[J]. Surf. Coat. Tech., 2005, 191(2-3): 155-160.

[17]Feng X, Zhai J, Jiang L. The fabrication and switchable superhydrophobicity of TiO_2 nanorod films[J]. Angew. Chem. Int. Edit., 2005, 117(32): 5245-5248.

[18]Kujawa J, Cerneaux S, Kujawski W. Characterization of the surface modification process of Al_2O_3, TiO_2 and ZrO_2 powders by PFAS molecules[J]. Colloid. Surface A, 2014, 447: 14-22.

[19]Chang K C, Chen Y K, Chen H. Fabrication of highly transparent and superhydrophobic silica-based surface by TEOS/PPG hybrid with adjustment of the pH value[J]. Surf. Coat. Tech., 2008, 202(16): 3822-3831.

[20]Sarkar D K, Saleema N. One-step fabrication process of superhydrophobic green coatings[J]. Surf. Coat. Tech., 2010, 204(15): 2483-2486.

[21]Kujawski W, Krajewska S, Kujawski M, et al. Pervaporation properties of fluoroalkylsilane (FAS) grafted ceramic membranes[J]. Desalination, 2007, 205(1-3): 75-86.

[22]Nakajima A, Hashimoto K, Watanabe T, et al. Transparent superhydrophobic thin films with self-cleaning properties [J]. Langmuir, 2000, 16(17): 7044-7047.

[23]Guo M, Kang Z, Li W, et al. A facile approach to fabricate a stable superhydrophobic film with switchable water adhesion on titanium surface[J]. Surf. Coat. Tech., 2014, 239: 227-232.

[24]Wang Y, Wang L, Wang S, et al. From natural lotus leaf to highly hard-flexible diamond-like carbon surface with superhydrophobic and good tribological performance[J]. Surf. Coat. Tech., 2012, 206(8-9): 2258-2264.

[25]Kujawa J, Cerneaux S, Koter S, et al. Highly efficient hydrophobic titania ceramic membranes for water desalination [J]. ACS Appl. Mater. Interfaces, 2014, 6(16): 14223-14230.

[26]Barthlott W, Neinhuis C. Purity of the sacred lotus, or escape from contamination in biological surfaces[J]. Planta, 1997, 202(1): 1-8.

[27]Feng L, Li S, Li Y, et al. Super-hydrophobic surfaces: from natural to artificial[J]. Adv. Mater., 2002, 14(24): 1857-1860.

[28]Andersson S, Serimaa R, Paakkari T, et al. Crystallinity of wood and the size of cellulose crystallites in Norway spruce (Picea abies)[J]. J. Wood Sci., 2003, 49(6): 531-537.

[29]Fu Y, Li G, Yu H, et al. Hydrophobic modification of wood via surface-initiated ARGET ATRP of MMA[J]. Appl. Surf. Sci., 2012, 258(7): 2529-2533.

[30] Zhou W, Sun F, Pan K, et al. Well-ordered large-pore mesoporous anatase TiO_2 with remarkably high thermal stability and improved crystallinity: preparation, characterization, and photocatalytic performance[J]. Adv. Funct. Mater., 2011, 21(10): 1922-1930.

[31]Herrera R, Erdocia X, Llano-Ponte R, et al. Characterization of hydrothermally treated wood in relation to changes on its chemical composition and physical properties[J]. J. Anal. Appl. Pyrol., 2014, 107: 256-266.

[32] Li X, Lei B, Lin Z, et al. The utilization of bamboo charcoal enhances wood plastic composites with excellent mechanical and thermal properties[J]. Mater Design, 2014, 53: 419-424.

[33]Essabir H, Hilali E, Elgharad A, et al. Mechanical and thermal properties of bio-composites based on polypropylene reinforced with Nut-shells of Argan particles[J]. Mater Design, 2013, 49: 442-448.

[34]Yang H, Yan R, Chen H, et al. Characteristics of hemicellulose, cellulose and lignin pyrolysis[J]. Fuel, 2007, 86 (12-13): 1781-1788.

[35]Yang H, Yan R, Chen H, et al. In-depth investigation of biomass pyrolysis based on three major components: hemicellulose, cellulose and lignin[J]. Energ Fuel, 2006, 20(1): 388-393.

[36]Ouajai S, Shanks R A. Composition, structure and thermal degradation of hemp cellulose after chemical treatments [J]. Polym. Degrad. Stabil., 2005, 89(2): 327-335.

[37]Strohm H, Sgraja M, Bertling J, et al. Preparation of TiO_2-polymer hybrid microcapsules[J]. J. Mater. Sci., 2003, 38(8): 1605-1609.

[38]Kijlstra J, Reihs K, Klamt A. Roughness and topology of ultra-hydrophobic surfaces[J]. Colloid. Surface A, 2002, 206(1-3): 521-529.

[39] Milne A J B, Amirfazli A. The Cassie equation: How it is meant to be used[J]. Adv. Colloid Interfac., 2012, 170 (1-2): 48-55.

[40]Kujawa J, Rozicka A, Cerneaux S, et al. The influence of surface modification on the physicochemical properties of ceramic membranes[J]. Colloid. Surface A, 2014, 443: 567-575.

[41]Wang S, Liu C, Liu G, et al. Fabrication of superhydrophobic wood surface by a sol-gel process[J]. Appl. Surf. Sci., 2011, 258(2): 806-810.

[42]Yang H, Deng Y. Preparation and physical properties of superhydrophobic papers[J]. J Colloid Interface Sci, 2008, 325(2): 588-593.

[43] Razmjou A, Arifin E, Dong G, et al. Superhydrophobic modification of TiO_2 nanocomposite PVDF membranes for applications in membrane distillation[J]. J Membrane Sci, 2012, 415: 850-863.

[44]Schondelmaier D, Cramm S, Klingeler R, et al. Orientation and self-assembly of hydrophobic fluoroalkylsilanes[J]. Langmuir, 2002, 18(16): 6242-6245.

中文题目：基于氟硅烷改性的二氧化钛薄膜制备耐久、耐酸、耐高温高湿的木材表面

作者：高丽坤，卢芸，詹先旭，李坚，孙庆丰

摘要：以木材为基质，通过两步法，即低温水热合成二氧化钛和氟硅烷的进一步改性，构建超疏水的木材表面。由电子扫描显微镜结构可知，二氧化钛处理后木材表面负载了一层均匀的二氧化钛微球，使木材表面变粗糙，有利于超疏水表面的构建。此外，十七氟葵基三甲氧基硅烷对二氧化钛/木材的修饰对提高拒水性能具有重要作用。制备的超疏水木材表面的水接触角为152.9°。在0.1 M的盐酸溶液中浸泡一周，紫外光照射24 h或150 ℃水煮10 h后水接触角仍大于150°。制备的木材表面展现出多功能性，即超拒水性、耐酸、耐高温高湿性。该木材的功能性薄膜拓展了木材工程的应用，如历史结构保护。

关键词：耐酸；氟硅烷；二氧化钛薄膜；超疏水；耐高温高湿

In Situ Deposition of Graphene Nanosheets on Wood Surface by One-pot Hydrothermal Method for Enhanced UV-resistant Ability*

Caichao Wan, Yue Jiao, Jian Li

Abstract: Graphene nanosheets were successfully in situ deposited on the surface of the wood matrix via a mild fast one-pot hydrothermal method, and the resulting hybrid graphene/wood (GW) were characterized by scanning electron microscope (SEM), energy dispersive X-ray spectroscopy (EDX), X-ray diffraction (XRD), Raman spectroscopy, and thermogravimetric analysis (TGA). According to the results, the wood matrix was evenly coated by dense uninterrupted multilayer graphene membrane structure, which was formed by layer-by-layer self-assembly of graphene nanosheets. Meanwhile, the graphene coating also induced significant improvement in the thermal stability of GW in comparison with that of the original wood (OW). Accelerated weathering tests were employed to measure and determine the UV-resistant ability of OW and GW. After about six hundred hours of experiments, the surface color change of GW was much less than that of OW; besides, the Fourier transform infrared spectroscopy (FTIR) analysis also proved the less significant changes in surface chemical compositions of GW. The results both indicated that the graphene coating effectively protected wood surface from UV damage. Therefore, this class of GW composite might be expected to be served as high-performance wooden building material for outdoor or some particular harsh environments like strong UV radiation regions use.

Keywords: Wood; Graphene; Hydrothermal method; Composites; UV resistance

1 Introduction

Wood, a kind of natural organic polymer material, has played an important role in human activity since before recorded history because of its unique inherent properties, such as strong mechanical properties, good formability, high strength weight ratio, and aesthetic texture[1-3]. According to statistics, roundwood consumption in the world including fuel wood, charcoal, and industrial wood, is more than 3000 million m^3 every year[4], which demonstrates that wood products are indispensable parts of national product. Growth of wood depends on energy from sun. However, when wood is exposed to solar ultraviolet (UV) light, one of its most important compositions named lignin could easily adsorb UV light and react with free radicals, which could induce a series of photolytic, photooxidative, and thermo-oxidative reactions of wood[5-8]. Meanwhile, these chemical reactions happen all the time and lead to many fatal defects like discoloration, fracturing and

* 本文摘自 Applied Surface Science, 2015, 347: 891-897.

clouding, severely lowering utilization value of wood.

For avoiding this class of wood degradation caused by UV light, many effective methods like coating film[9, 10], adding surface additive[11], thermal treatment[12], and chemical modification[13, 14] had been widely proposed. Indeed, chemical modification is regarded as a kind of common approach to reducing the damage from UV radiation by changing the chemical structures of some wood compositions that participate in photochemical reactions[15]. To date, many chemical reagents such as chromium trioxide[16], chromium nitrate[17], and ferric chloride[18] had been tried to modify wood surface. In addition, some nanomaterials like TiO_2, ZnO and ZrO_2, for their superb properties including low cost, non-toxic, chemical stability, and ability to absorb or scatter UV irradiation, have emerged as quite intriguing materials that hold promise as photoprotective agents for wood, which have been demonstrated by some previous researches[19-21].

Graphene is a flat monolayer of hexagonally arrayed sp^2-bonded carbon atoms tightly packed into a two-dimensional honey-comb lattice, which has superb optical, electrical, thermal, mechanical properties, and high surface area as well as attractive transport phenomena (such as the quantum spin hall effect). These numerous alluring properties make itself useful for supercapacitors, electrode materials, biosensors, and microwave absorbing materials[22, 23]. Recently, Nurxat et al. reported that the addition of graphene significantly improved the resistance of the polyurethane top coatings against UV degradation and photocorrosion since graphene could absorb most of incident light[24]. Therefore, adopting fast mild one-pot hydrothermal method, which is generally regarded as an efficient way of synthesizing hybrid materials[25, 26], to in situ deposit graphene nanosheets on the surface of wood matrix might provide a novel effectual approach to protect wood from UV irradiation.

Herein, the hybrid graphene/wood (GW) composite had been fabricated via a simple quick mild one-pot hydrothermal method. The incorporation of graphene nanosheets resulted in the improvement of thermal stability of GW. Under the harsh accelerated weathering tests for about 600 h, GW showed less significant changes in appearance color and surface chemical compositions compared with the original wood (OW), indicating more superior ability of resistance to UV corrosion. In addition, the possible preparation and UV resistance mechanisms of GW were both proposed.

2 Materials and methods

2.1 Materials

Polar wood slices 20 mm (longitudinal)×20 mm (tangential)×10 mm (radial) in size were ultrasonically rinsed with deionized water for 30 min and then dried at 60 ℃ for 24 h in a vacuum to remove all adsorbed water. All the chemicals were supplied by Shanghai Boyle Chemical Co. Ltd. and used as received.

2.2 Preparation of graphene

The preparation of graphene was based on the method reported by Zheng et al[27]. Briefly, graphite (40 g) was first added into a reaction flask containing nitric acid (270 mL) and sulfuric acid (525 mL) with magnetic stirring in an ice bath. Then, the mixture was slowly mixed with potassium chlorate (330 g) to minimize the risk of explosion, and the resulting mixture was allowed to react for 120 h at room temperature. Thereafter, the resultant was filtered and washed with excess deionized water and HCl solution

(5%) to remove the sulfuric ions, and the following graphene oxide (GO) aqueous solution was neutralized with potassium hydroxide solution. After high-speed centrifugation, the GO powder was obtained and then dried in an air-circulating oven at 135 ℃ for 24 h followed by another 24 h at 135 ℃ in a vacuum oven. Finally, the graphene was successfully synthesized by placing the dried GO powder into a muffle furnace preheated to 1050 ℃ and holding it in the furnace for 30 s to thermally exfoliate GO.

2.3 Preparation of GW

The preparation of GW was performed as follows: 10 mg thermal exfoliated graphene was added into 50 mL distilled water. After the mixture was magnetically stirred for about 30 min, the mixed solution was subjected to an ultrasonic processing for 4 h by an ultrasonic cell crusher (JY99-IID, Ningbo Scientz Biotechnology Co. Ltd., China) with an output power of 900 W in an ice/water bath. Afterwards, the graphene aqueous dispersion was transferred into a Teflon-lined stainless-steel autoclave, and the wood specimens were subsequently put into the above solution. The hydrothermal reaction was performed at reaction temperatures of 120 ℃ for 4 h. The resulting products were repeatedly rinsed with deionized water to remove unattached graphene nanosheets, and then dried at 45 ℃ for over 24 h in vacuum to remove most water.

2.4 Characterizations

The surface morphologies and chemical compositions of OW and GW were characterized using scanning electron microscopy (SEM, FEI, and Quanta 200) attached with an energy dispersive X-ray spectrometer (EDX). The X-ray diffraction (XRD) measurements were performed on a XRD instrument (Rigaku, D/MAX 2200). The 2θ scanning range was between 5 and 60° with a scan rate (2θ) of $4° \cdot min^{-1}$. Raman analysis was performed using a Raman spectrometer (Renishaw inVia, Germany) employing a helium/neon laser (633 nm) as the excitation source. The heat stabilities of the samples were investigated by a TG analyzer (TA, Q600) from 25 to 800 ℃ with a heating rate of $10\ ℃ \cdot min^{-1}$ under nitrogen atmosphere. The Fourier Transform Infrared (FTIR) spectra were recorded on a FTIR spectrophotometer (Magna 560, Nicolet, Thermo Electron Corp) with a resolution of $4\ cm^{-1}$.

2.5 Accelerated weathering tests

Properties of UV resistance of GW and OW samples were investigated by accelerated weathering tests that could induce photo-discoloration of the samples carried out on a QUV accelerated weathering tester (Accelerated Weathering Tester, Q-Panel, Cleveland, OH, USA) with an average irradiance of $0.85\ W \cdot m^{-2}$. The samples were first exposed to fluorescent UV radiation at the wavelength of 340 nm and the temperature of 60 ℃ for 2.5 h. After the radiation, the samples were successively subjected to water spray treatment at 25 ℃ for 0.5 h and condensation process at 45 ℃ for 24 h. Besides, the whole exposure time was set within the scope of 0 to 600 h, and the surface color changes were measured at regular intervals (about every 24 h for 0-200 h, 48 h for 200-400 h, and 96 h for 400-600 h). The color changes were characterized by a portable spectrophotometer (NF-333, Nippon Denshoku Company, Japan) equipped with a CIELAB system. For the CIELAB system, three characteristic parameters, namely L^*, a^*, and b^*, were used to describe lightness and chromaticity indices of samples. In detail, L^* referred to lightness varying from 100 (white) to 0 (black), besides, chromaticity indices were determined by a^* and b^*, and positive and negative a^* represented red and green direction, and plus and minus b^* were yellow and blue, respectively. The changes of L^*, a^*, and b^* were calculated using Eq. (1)-(3).

$$\Delta L^* = L_2 - L_1 \tag{1}$$

$$\Delta a^* = a_2 - a_1 \tag{2}$$

$$\Delta b^* = b_2 - b_1 \tag{3}$$

where ΔL^*, Δa^*, and Δb^* represented the differences of initial and final values of L^*, a^* and b^*; L_1, a_1 and b_1 were the initial color parameters; L_2, a_2 and b_2 were the ultimate color parameters after UV irradiation, respectively.

Eq. 4 was adopted to characterize the overall color changes (ΔE^*). The lower the ΔE^* value, the less significant the color change, that is, the ability of resistance to UV radiation was stronger. The accelerated weathering tests were performed in triplicate and were done at least two different times to ensure reproducibility. All parameters were measured at eight locations on each sample, and the average values were calculated.

$$\Delta E^* = \sqrt{(L_2^* - L_1^*)^2 + (a_2^* - a_1^*)^2 + (b_2^* - b_1^*)^2} \tag{4}$$

3 Results and discussion

The surface microscopic morphologies were observed by SEM. As shown in Fig. 1b, compared with the smooth surface of OW (Fig. 1a), the surface of GW was coated with dense uninterrupted multilayer graphene membrane structure. And the membrane structure was formed by layer-by-layer self-assembly of graphene nanosheets by virtue of the energy provided by hydrothermal reaction[28, 29]; besides, the native wood structures of GW were also disappeared. In the magnified image (Fig. 1c), it could be more clearly seen that the original wood surface was completely covered by these graphene films, and there is no doubt that these

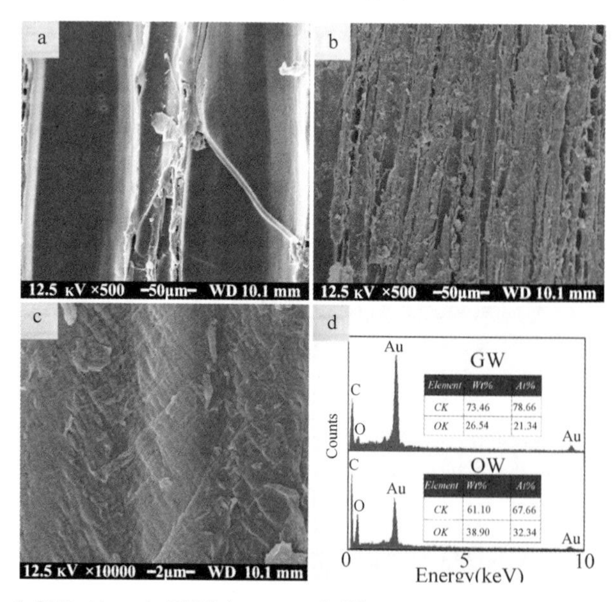

Fig. 1 (a) SEM image of OW. (b, c) SEM images of different magnifications of GW. (d) EDX spectra of OW and GW, and the insets showed the corresponding mass proportions and atomic proportions of C and O elements, respectively

compact accumulated graphene nanosheets would play key roles in protecting the wood surface from UV damages. Furthermore, according to the results of EDX spectra (Fig. 1d), the C/O mass ratio and atom ratio both obviously increased after the hydrothermal process from 1.57 to 2.77, and 2.09 to 3.69, respectively, which further revealed the presence of graphene on the wood matrix surface.

The crystal structures were surveyed by XRD measurements, and the resulting curves were normalized to the same height so that the band maximum occurred at the same depth, to allow comparison of their shapes. The XRD patterns of as-prepared graphene in Fig. 2 showed a broad band at 2θ of ~26°, which could be indexed to the characteristic (002) plane reflection of graphite from the graphene[30, 31]. For OW and GW, the similar diffraction bands at approximately 15°, 22° and 35° were all attributed to the wood characteristics[32], and there was no other characteristic band related to graphene structure in the pattern of GW, possibly due to the relatively low diffraction intensity of graphene characteristic band and the low content of graphene in GW.

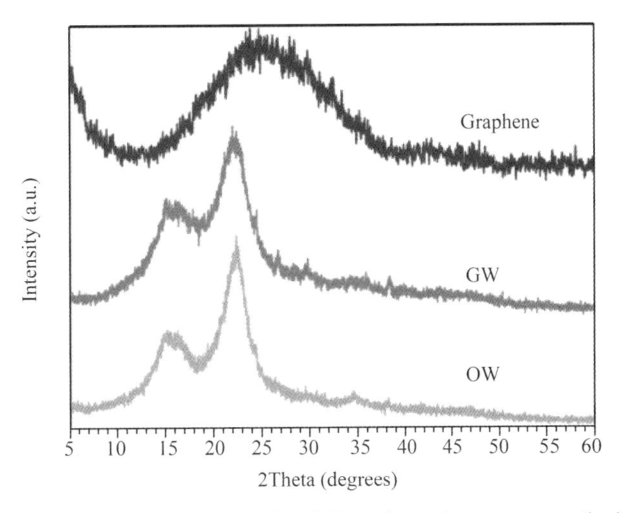

Fig. 2 XRD patterns of OW, GW and graphene, respectively

The Raman spectrum of GW was presented in Fig. 3, which was fitted with pure Gaussian functions. The bands at around 1107 and 2913 cm^{-1} were derived from wood components (mostly cellulose)[33]. The former was dominated by CC and CO stretching motions and some amounts of HCC, HCO bendings, and the latter was primarily corresponding to CH stretching vibrations[34, 35]. Moreover, two characteristic bands of graphene were also determined. The G band as the main spectral feature of graphene was derived from in-plane motion of the carbon atoms and appeared at around 1599 cm^{-1}[36]. Besides, the D band known as the disorder band originated from lattice motion away from the center of the Brillouin zone appeared at around 1338 cm^{-1}[37, 38]. The broadening of the D and G bands with a strong D line indicated localized in plane sp^2 domains and disordered graphitic crystal stacking of the graphene nanosheets[39]. Furthermore, the results also suggested that the graphene nanosheets were successfully deposited on the wood matrix surface.

Fig. 4 presented the TG and DTG curves of OW and GW, and it was apparent that the TG and DTG curves of GW shifted to higher temperature compared with those of OW, which revealed the more superior thermal stability of GW. The slight weight loss before 150 ℃ for both samples was due to the evaporation of adsorbed water[40]. According to the DTG curve, the thermal decomposition process of OW mainly displayed

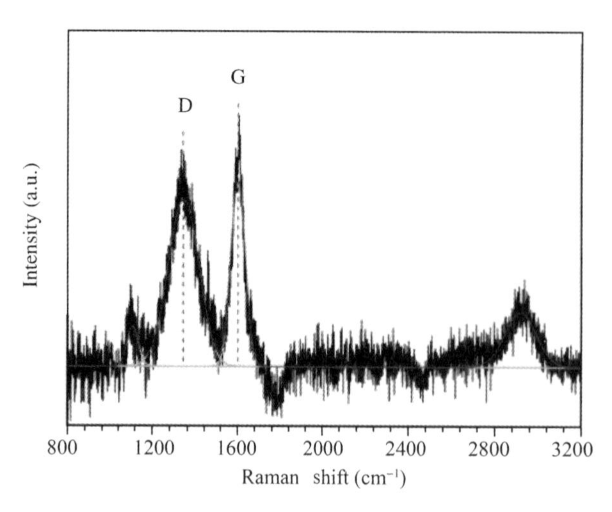

Fig. 3 Raman spectrum of GW with its characteristic D band (~1338 cm^{-1}) and G band (~1599 cm^{-1}) that were characteristics of graphene

two stages as follows: the first stage from 189 ℃ to 320 ℃ was mainly owing to decomposition of hemicellulose (mass loss~26%) with a pronounced shoulder[41]; the second stage from 320 ℃ to 400 ℃ was primarily attributed to cellulose decomposition (mass loss ~ 47%) with a strong exothermic peak (~ 365 ℃)[42]. Besides, the pyrolysis of lignin with superior thermal stability was slow and happened under the whole temperature without obvious characteristic exothermic peak[43]. For GW, it is well-known that graphene has superior thermal stability[44], and the combination of wood matrix with graphene might also be conducive for GW to obtaining improved heat resistance. GW started to sharply decompose at around 225 ℃ posterior to OW (ca. 189 ℃). At maximum degradation rate, the decomposition temperatures of GW and OW are 378 ℃ and 365 ℃, respectively. These increasing trends of thermal decomposition temperature for GW implied that the deposition of graphene on the wood matrix significantly enhanced the thermal stability of the composite.

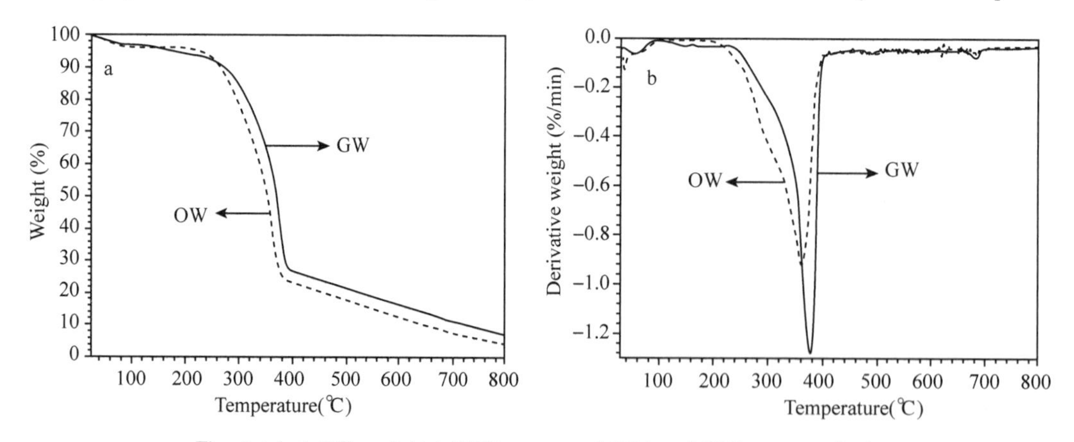

Fig. 4 (a) TG and (b) DTG curves of OW and GW, respectively

Color changes of OW and GW before and after the accelerated weathering tests were represented in Fig. 5. As shown in Fig. 5a, the ΔL^* values of OW and GW both turned negative indicating that the wood samples became black, and exhibited similar change trends with prolonged UV irradiation time, but it was observed that the ΔL^* change of GW was much smaller (ca. 1/4) than that of OW. On the other hand, the similar cases occurring in Δa^* and Δb^* that GW displayed the similar change trends and relatively smaller amplitude of variation than those of OW, could be found (Fig. 5b, c). The higher Δa^* and Δb^* values for OW revealed that the surface color of OW turned a deeper shade of red and dark yellow with increasing UV irradiation time, respectively. Furthermore, the overall color changes (ΔE^*) of GW was also obviously lower (ca. 1/3) than that of OW as shown in Fig. 5d, which suggested that GW suffered less serious damage from UV. Therefore, it could be concluded that the graphene coating effectively prevented discoloration resulted from UV irradiation and slowed down the rate of wood photodegradation.

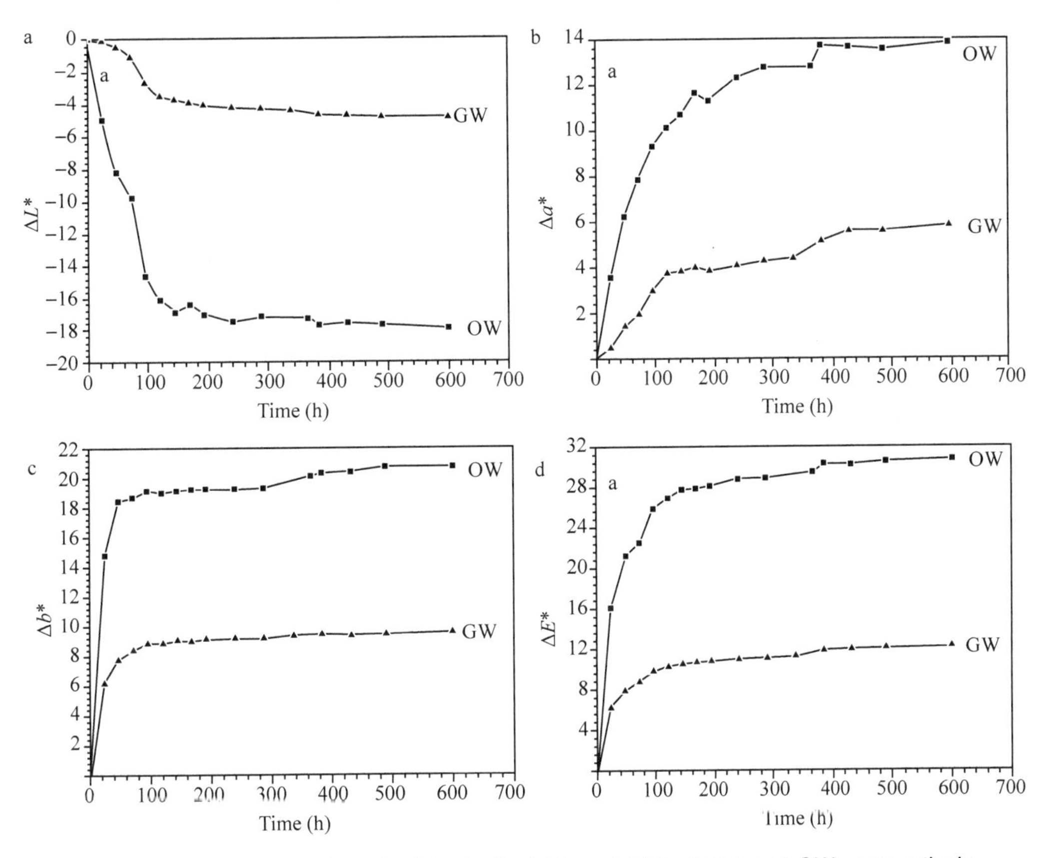

Fig. 5 Change tendencies of ΔL^*, Δa^*, Δb^* and ΔE^* of OW and GW, respectively

For further investigating the changes in surface chemical compositions of OW and GW before and after the tests, FTIR analysis was carried out. As shown in Fig. 6, the adsorption at around 1732 cm^{-1} was assigned as the C = O stretching in non – conjugated ketones, carbonyls and in ester groups, which belonged to characteristic of hemicellulose[45]. Besides, the bands centered at around 1637 and 1512 cm^{-1} were attributed to the side chain carbonyl C=O and aromatic phenyl C=C stretching of lignin[46]. The bands at around 1423, 1382, 1321 and 1267 cm^{-1} were originated from the CH_2 symmetric bending, O–H bending, C–H bending, and C–O symmetric stretching[47], respectively. The absorbances at around 1157 and 1076 cm^{-1} arose from the C–O anti–symmetric stretching and C–O–C pyranose ring skeletal vibration[48]. A small sharp peak at 897 cm^{-1} was characteristic of β–glycosidic linkages between glucose in cellulose, corresponding to the glycosidic

C_1–H deformation with ring vibration contribution[49].

During the weathering process, phenolic hydroxyl groups of lignin interacted with light rapidly to produce phenolic radicals, which successively transformed into *o*–and *p*–quinonoid structures by demethylation or by cleavage of the side chain and formation of carbonyl based chromophoric groups[50]. For OW, the effects of UV exposure on the surface chemical compositions of OW were quite apparent. Indeed, the characteristic aromatic lignin band at around 1512 cm^{-1} dramatically decreased; nevertheless, carbonyl absorption at 1732 cm^{-1} significantly enhanced (Fig. 6a). In Fig. 6b, there was no noticeable decline in adsorption at 1512 cm^{-1} or increase in absorption at 1732 cm^{-1} due to UV irradiation for GW, which indicated that wood coated with graphene nanosheets had the ability to resist to photodegradation.

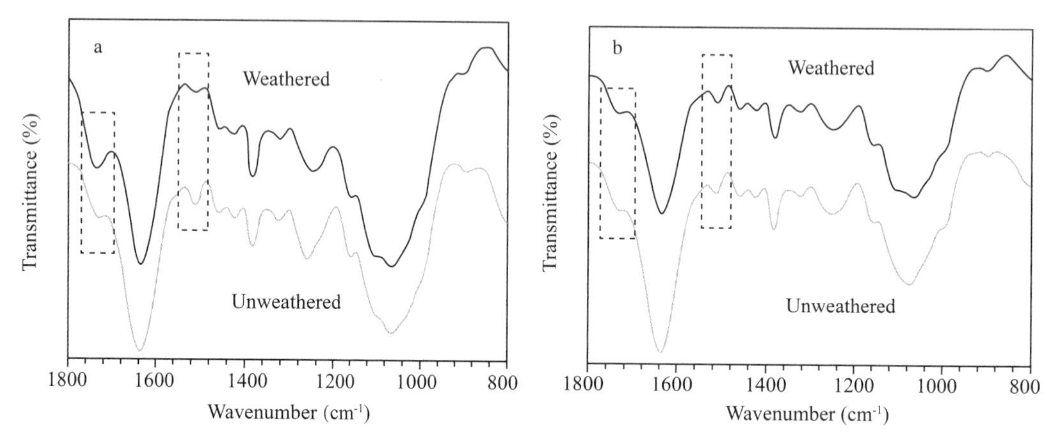

Fig. 6 FTIR spectra of (a) OW and (b) GW before and after the accelerated weathering tests, respectively

Fig. 7 presented a possible schematic illustration for the preparation and UV resistance of GW based on the results mentioned above. Hydrothermal treatment provided not only great chance for the interaction of graphene nanosheets with wood matrix within the narrow space in the autoclave, but also huge power contributing to further combination between wood and graphene nanosheets[51, 52]. During the hydrothermal process, under the high temperature and high pressure, a plenty of graphene nanosheets were layer–by–layer deposited on the surface of wood. The main interactions between the wood matrix and the nanosheets were of physical nature (i. e. Van der Waals' force and strong adsorption ability of graphene)[53]. Eventually, this layer–by–layer self–assembly process would lead to the formation of the dense uninterrupted multilayer graphene membrane structure on wood surface. Undoubtedly, the compact multilayer graphene coating was helpful to absorb and scatter most of UV light[54], which efficiently protected wood from these damages derived from UV irradiation including discoloration, fracturing and clouding, and prolonged service lives of wood products.

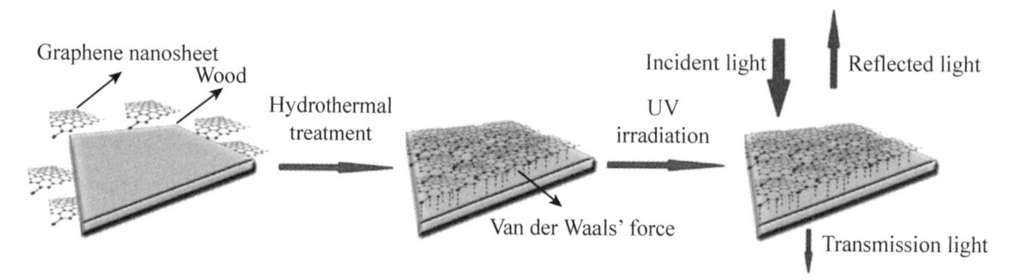

Fig. 7 Schematic illustration for the preparation and ultraviolet resistance of GW

4 Conclusions

Graphene nanosheets had been successfully in situ deposited on the wood surface by the simple mild quick one-pot hydrothermal method. Through the process of layer-by-layer self-assembly of garphene nanosheets during the hydrothermal treatment, the dense uninterrupted multilayer graphene membrane structure was generated on the surface of the wood matrix. And this class of membrane structure was subsequently demonstrated an effective UV photoprotective agent for wood due to the strong UV absorption and scattering abilities, which was beneficial to prolong the service life of wood. Meanwhile, the growth of graphene nanosheets on the wood surface also made contribution to the enhancement of GW thermal stability. Consequently, this kind of GW composite might be expected to find potential applications in novel wood products for outdoor or some special harsh environments like strong UV radiation regions use.

Acknowledgments

This work was financially supported by the National Natural Science Foundation of China (grant no. 31270590).

References

[1] Wegener G, Przyklenk M, Fengel D. Hexafluoropropanol as valuable solvent for lignin in UV and IR spectroscopy [M]. 1983.

[2] Schniewind A P. Concise encyclopedia of wood and wood-based materials[M]. Pergamon Press, 1989.

[3] Pettersen R C. The chemical composition of wood[J]. The Chemistry of Solid Wood, 1984, 207: 57-126.

[4] Shmulsky R, Jones P D. Forest Products and Wood Science An introduction[M]. John Wiley & Sons. 2011.

[5] Ncube E, Meincken M. Surface characteristics of coated soft-and hardwoods due to UV-B ageing[J]. Applied Surface Science, 2010, 256(24): 7504-7509.

[6] Pandey K K. Study of the effect of photo-irradiation on the surface chemistry of wood[J]. Polymer Degradation and Stability, 2005, 90(1): 9-20.

[7] Temiz A, Yildiz U C, Aydin I, et al. Surface roughness and color characteristics of wood treated with preservatives after accelerated weathering test[J]. Applied Surface Science, 2005, 250(1-4): 35-42.

[8] Vanucci C, De Violet P F, Bouas-Laurent H, et al. Photodegradation of lignin: A photophysical and photochemical study of a non-phenolic α-carbonyl β-O-4 lignin model dimer, 3, 4-dimethoxy-α-(2′-methoxypenoxy) acetophenone[J]. Journal of Photochemistry and Photobiology A: Chemistry, 1988, 41(2): 251-265.

[9] Parejo P G, Zayat M, Levy D. Highly efficient UV-absorbing thin-film coatings for protection of organic materials against photodegradation[J]. Journal of Materials Chemistry, 2006, 16(22): 2165-2169.

[10] Denes A R, Young R A. Reduction of weathering degradation of wood through plasma-polymer coating[J]. 1999.

[11] Lee B H, Kim H J. Influence of isocyanate type of acrylated urethane oligomer and of additives on weathering of UV-cured films[J]. Polymer Degradation and Stability, 2006, 91(5): 1025-1035.

[12] Ayadi N, Lejeune F, Charrier F, et al. Color stability of heat-treated wood during artificial weathering[J]. Holz als Roh-und Werkstoff, 2003, 61(3): 221-226.

[13] Xie Y, Krause A, Mai C, et al. Weathering of wood modified with the N-methylol compound 1, 3-dimethylol-4, 5-dihydroxyethyleneurea[J]. Polymer Degradation and Stability, 2005, 89(2): 189-199.

[14] Kiguchi M, Kataoka Y, Matsunaga H, et al. Surface deterioration of wood-flour polypropylene composites by weathering trials[J]. Journal of Wood Science, 2007, 53(3): 234-238.

[15] Hill C A S, Cetin N S, Quinney R F, et al. An investigation of the potential for chemical modification and

subsequent polymeric grafting as a means of protecting wood against photodegradation[J]. Polymer Degradation and Stability, 2001, 72(1): 133-139.

[16] Chang S T, Hon D N S, Feist W C. Photodegradation and photoprotection of wood surfaces[J]. Wood and Fiber Science, 1982, 14(2): 104-117.

[17] Zhang J, Kamdem D P, Temiz A. Weathering of copper-amine treated wood[J]. Applied Surface Science, 2009, 256(3): 842-846.

[18] Pandey K K, Pitman A J. Weathering characteristics of modified rubberwood (Hevea brasiliensis)[J]. Journal of Applied Polymer Science, 2002, 85(3): 622-631.

[19] Wan C, Lu Y, Sun Q, et al. Hydrothermal synthesis of zirconium dioxide coating on the surface of wood with improved UV resistance[J]. Applied Surface Science, 2014, 321: 38-42.

[20] Veronovski N, Verhovšek D, Godnjavec J. The influence of surface-treated nano-TiO_2 (rutile) incorporation in water-based acrylic coatings on wood protection[J]. Wood Science and Technology, 2013, 47(2): 317-328.

[21] Weichelt, F, Emmler, et al. ZnO-Based UV Nanocomposites for Wood Coatings in Outdoor Applications[J]. Macromol Mater Eng, 2010, 295: 130-136.

[22] Geim A K, Novoselov K S. The rise of graphene[M]//Nanoscience and technology: A collection of reviews from nature journals. 2010: 11-19.

[23] Stankovich S, Dikin D A, Dommett G, et al. Graphene-based composite materials[J]. Nature, 2006, 442.

[24] Nuraje N, Khan S I, Misak H, et al. The addition of graphene to polymer coatings for improved weathering[J]. International Scholarly Research Notices, 2013, 2013.

[25] Feng S, Xu R. New materials in hydrothermal synthesis[J]. Accounts of Chemical Research, 2001, 34(3): 239-247.

[26] Finn R C, Zubieta J. A New Class of Organic-Inorganic Hybrid Materials: Hydrothermal Synthesis and Structural Characterization of Bimetallic Organophosphonate Oxide Phases of the Mo/Cu/O/RPO_{32}-Family[J]. Inorganic chemistry, 2001, 40(11): 2466-2467.

[27] Zhang H B, Zheng W G, Yan Q, et al. Electrically conductive polyethylene terephthalate/graphene nanocomposites prepared by melt compounding[J]. Polymer, 2010, 51(5): 1191-1196.

[28] Xu Y, Sheng K, Li C, et al. Self-assembled graphene hydrogel via a one-step hydrothermal process[J]. ACS Nano, 2010, 4(7): 4324-4330.

[29] Chen P, Yang J J, Li S S, et al. Hydrothermal synthesis of macroscopic nitrogen-doped graphene hydrogels for ultrafast supercapacitor[J]. Nano Energy, 2013, 2(2): 249-256.

[30] Ding S, Luan D, Boey F Y C, et al. SnO_2 nanosheets grown on graphene sheets with enhanced lithium storage properties[J]. Chemical Communications, 2011, 47(25): 7155-7157.

[31] Wang G, Yang J, Park J, et al. Facile synthesis and characterization of graphene nanosheets[J]. The Journal of Physical Chemistry C, 2008, 112(22): 8192-8195.

[32] Andersson S, Serimaa R, Paakkari T, et al. Crystallinity of wood and the size of cellulose crystallites in Norway spruce (Picea abies)[J]. Journal of Wood Science, 2003, 49(6): 531-537.

[33] Agarwal U P, Ralph S A. FT-Raman spectroscopy of wood: Identifying contributions of lignin and carbohydrate polymers in the spectrum of black spruce (Picea mariana)[J]. Applied Spectroscopy, 1997, 51(11): 1648-1655.

[34] Agarwal U P. An overview of Raman spectroscopy as applied to lignocellulosic materials[J]. Advances in Lignocellulosics Characterization, 1999: 201-225.

[35] M Szymańska-Chargot, Cybulska J, Zdunek A. Sensing the Structural Differences in Cellulose from Apple and Bacterial Cell Wall Materials by Raman and FT-IR Spectroscopy[J]. Sensors (Basel, Switzerland), 2011, 11(6).

[36] Ferrari A C, Meyer J C, Scardaci V, et al. Raman spectrum of graphene and graphene layers[J]. Physical Review

letters, 2006, 97(18): 187401.

[37] Malard L M, Pimenta M A, Dresselhaus G, et al. Raman spectroscopy in graphene[J]. Physics reports, 2009, 473(5-6): 51-87.

[38] Ni Z, Wang Y, Yu T, et al. Raman spectroscopy and imaging of graphene. Nano Res 1: 273-291[J]. 2008.

[39] Shi Y, Chou S L, Wang J Z, et al. Graphene wrapped $LiFePO_4$/C composites as cathode materials for Li-ion batteries with enhanced rate capability[J]. Journal of Materials Chemistry, 2012, 22(32): 16465-16470.

[40] Ramiah M V. Thermogravimetric and differential thermal analysis of cellulose, hemicellulose, and lignin[J]. Journal of applied polymer science, 1970, 14(5): 1323-1337.

[41] Peng Y, Wu S. The structural and thermal characteristics of wheat straw hemicellulose[J]. Journal of Analytical and Applied Pyrolysis, 2010, 88(2): 134-139.

[42] Shafizadeh F, Fu Y L. Pyrolysis of cellulose[J]. Carbohydrate Research, 1973, 29(1): 113-122.

[43] Yang H, Yan R, Chen H, et al. Characteristics of hemicellulose, cellulose and lignin pyrolysis[J]. Fuel, 2007, 86(12-13): 1781-1788.

[44] Wu Z S, Ren W, Gao L, et al. Synthesis of graphene sheets with high electrical conductivity and good thermal stability by hydrogen arc discharge exfoliation[J]. ACS Nano, 2009, 3(2): 411-417.

[45] Patachia S, Croitoru C, Friedrich C. Effect of UV exposure on the surface chemistry of wood veneers treated with ionic liquids[J]. Applied Surface Science, 2012, 258(18): 6723-6729.

[46] Chen H, Ferrari C, Angiuli M, et al. Qualitative and quantitative analysis of wood samples by Fourier transform infrared spectroscopy and multivariate analysis[J]. Carbohydrate Polymers, 2010, 82(3): 772-778.

[47] Sun X F, Sun R C, Tomkinson J, et al. Degradation of wheat straw lignin and hemicellulosic polymers by a totally chlorine-free method[J]. Polymer Degradation and Stability, 2004, 83(1): 47-57.

[48] Pandey K K. A study of chemical structure of soft and hardwood and wood polymers by FTIR spectroscopy[J]. Journal of Applied Polymer Science, 1999, 71(12): 1969-1975.

[49] Colom X, Carrillo F, Nogués F, et al. Structural analysis of photodegraded wood by means of FTIR spectroscopy[J]. Polymer Degradation and Stability, 2003, 80(3): 543-549.

[50] George B, Suttie E, Merlin A, et al. Photodegradation and photostabilisation of wood-the state of the art[J]. Polymer Degradation and Stability, 2005, 88(2): 268-274.

[51] Byrappa K, Adschiri T. Hydrothermal technology for nanotechnology[J]. Progress in Crystal Growth and Characterization of Materials, 2007, 53(2): 117-166.

[52] Suchanek W L, Riman R E. Hydrothermal synthesis of advanced ceramic powders[C]//Advances in Science and Technology. Trans Tech Publications Ltd, 2006, 45: 184-193.

[53] Song B, Li D, Qi W, et al. Graphene on Au (111): a highly conductive material with excellent adsorption properties for high-resolution bio/nanodetection and identification[J]. Chem. Phys. Chem., 2010, 11(3): 585-589.

[54] Jia L, Wang D H, Huang Y X, et al. Highly durable N-doped graphene/CdS nanocomposites with enhanced photocatalytic hydrogen evolution from water under visible light irradiation[J]. The Journal of Physical Chemistry C, 2011, 115(23): 11466-11473.

中文题目：一锅水热法在木材表面原位沉积石墨烯纳米片以增强抗紫外线能力

作者：万才超，焦月，李坚

摘要：采用温和快速的一锅水热法成功地将石墨烯纳米片原位沉积在木材基体表面，并通过扫描电子显微镜(SEM)、能量色散X射线光谱(EDX)、X射线衍射(XRD)、拉曼光谱和热重分析(TGA)对所得到的混合石墨烯/木材(GW)进行了表征。根据结果显示，木材基体上均匀地涂覆了致密的不间断的多层石墨烯膜结构，该结构是由石墨烯纳米片逐层自组装形成的。同时，与原始木材(OW)相比石墨烯涂层也显著提高了混合石墨烯/木材(GW)的热稳定性。采用加速老化试验来测量和确定原始木材(OW)和混合石墨烯/木材(GW)的抗紫外线能力。经过约600小时的实验，混合石墨烯/木材(GW)的表面颜色变化远小于原始木材(OW)；另外，傅里叶变换红外光谱(FTIR)分析也证明混合石墨烯/木材(GW)的表面化学成分变化较小。这些结果均表明，石墨烯涂层能有效地保护木材表面免受紫外线的破坏。因此，该类混合石墨烯/木材(GW)复合材料有望作为高性能的木质建筑材料，用于户外或一些特殊的恶劣环境，如强紫外线辐射地区使用。

关键词：木材；石墨烯；水热法；复合材料；抗紫外

Superhydrophobic Conductive Wood with Oil Repellency Obtained by Coating with Silver Nanoparticles Modified by Fluoroalkyl Silane*

Likun Gao, Yun Lu, Jian Li, Qingfeng Sun

Abstract: A simple and effective method for preparing superhydrophobic conductive wood surface with super oil repellency is presented. Silver nanoparticles (Ag NPs) were prepared on wood surfaces by the treatment with $AgNO_3$, followed by a reduction treatment with glucose to generate a dual-size surface roughness. Further modification of the surface coated with Ag NPs with a fluoroalkyl silane led to a superhydrophobic surface with water contact angle of 155.2°. This surface is also super repellent toward motor oil with the maximal contact angles around 151.8°. Interestingly, the dense Ag NPs coating on the surface is electrically conductive. The presented multifunctional coating could be a commercialized for various applications, especially for self-cleaning and biomedical electronic devices.

Keywords: Conductive wood surface; Fluoroalkyl silane; Oil repellency; Silver nanoparticles; Superhydrophobic wood surface

1 Introduction

Modified wood with enhanced functionalities could be attractive for high-value-added products, such as wood with surfaces, which are superhydrophobic, oil repellent, and electrical conductive (Hu et al., 2009; Wang et al., 2011; Lu et al., 2014). Such properties can be achieved by combined surface treatment. The literature describes treatments, for example, with TiO_2 (Fujishima et al., 2000; Mahr et al., 2013), Ag (Yuranova et al., 2006; Radetić et al., 2008), ZnO (Li et al., 2015; Weichelt et al., 2010; Yu et al., 2010; Fu et al., 2012), which improve the water repellency if the treatment is followed by coating with low-surface-energy subtances (Wang et al., 2009; Yang et al., 2010). Treatments with epoxy/silica coating (Liu et al. 2014), silica nanoparticles modified with long-chain alkylsilane (Wang et al., 2013), and CeO_2 coating deposited onto the surface (Lu et al., 2014) are also under investigation.

Silver is a well known bactericide since ancient times (Xue et al., 2012). Silver nanoparticles (Ag NPs) were described as antimicrobial materials for biomedical applications (Sondi and Salopek-Sondi 2004; Dubas et al., 2006; Rai et al., 2009). Ag NPs also impart electrical conductivity to the substrates (Chou et al., 2005; Tien et al., 2011). Ag NPs coating on cotton fabric and textiles can be obtained by the treatment with aqueous KOH and $AgNO_3$, which are suited to a subsequent surface hydrophobization (Shateri Khalil-

---

* 本文摘自 Holzforschung, 2016, 70(1): 63-68.

Abad and Yazdanshenas 2010; Xue et al., 2012). The dual-size surface structure modified by fluoroalkyl silane (FAS), which has low surface energy, led to remarkable superhydrophobic surfaces with water contact angles (CAs) larger than 150° (Zhai et al., 2004; Saleema et al., 2011). The modified substrates also show metallic features of the electrical conductivity. However, the literature with this regard is scarce.

The aim of the present study is to enlarge our knowledge concerning the surface modification by means of Ag NPs. Wood will be pretreated with NaOH to make the surface negatively charged, and then Ag NPs will be produced by in situ reduction of $[Ag(NH_3)_2]^+$ with glucose. Then, the treated wood should be modified by fluoroalkyl silane containing long chains with the expectation that a surface will be obtained with superhydrophobicity, super oil repellency and electrical conductivity. Such a material could be well suited for self-cleaning applications and in biomedical electronic devices.

2 Materials and methods

All chemicals were supplied by Shanghai Boyle Chemical Company, Limited (China), and used as received. Each piece of poplar wood (*Populus ussuriensis* Kom) was carefully cut into pieces of 20 mm×20 mm with a thickness of 5 mm. Then the samples were ultrasonically treated in deionized water for 30 min and dried in a vacuum chamber at 103 ℃ for 24 h.

The samples were treated with 1% aqueous NaOH solution at a room temperature (r. t.) for 20 min followed by ultrasonical washing with deionized water for 30 min. Aqueous ammonia (28%) was added by dropping into a solution of aqueous 0.5 M $AgNO_3$ solution under stirring until a transparent colorless $[Ag(NH_3)_2]^+$ solution was formed. The alkali-treated samples were dipped into the $[Ag(NH_3)_2]^+$ solution for 1 h and then transferred into a 0.1 M glucose stock solution. After 5 min, the residual $[Ag(NH_3)_2]^+$ solution was also poured into the glucose solution. The reaction was continued for 15 min. Finally, the prepared samples were removed from the solution, ultrasonically rinsed with deionized water for 30 min, and dried at 60 ℃ for more than 24 h in a vacuum chamber.

Hydrophobization with FAS: A methyl alcohol solution of (heptadecafluoro-1, 1, 2, 2-tetradecyl) trimethoxysilane ($CF_3(CF_2)_7CH_2CH_2Si(OCH_3)_3$, (denoted as FAS-17) was hydrolyzed by addition of a 3 fold-molar excess of water at r. t. The Ag-treated samples were immersed into the solution at r. t. for 24 h. Subsequently, the samples were washed with ethyl alcohol (EtOH) to remove any residual chemicals and were dried at r. t. Then the sampled was dried at 60 ℃ for more than 24 h in a vacuum chamber.

Characterization: The surface morphologies were characterized by field-emission SEM (Quanta 200, FEI, Holland) operating at 12.5 kV in combination with EDS (Genesis, EDAX, Holland). X-ray diffraction (XRD, Bruker D8 Advance, Germany): Cu $K\alpha$ radiation ($\lambda = 1.5418$ Å), step scan mode (step size of 0.02°, and a scan rate of 4° min^{-1}), 40 kV, 40 mA ranging from 5° to 70°. FT-IR spectra: Magna-IR 560 spectrometer (Nicolet, USA), KBr pellet mode, resolution 4 cm^{-1}. Thermogravimetric and Differential Thermal Analysis (TG-DTA): SDT Q600 thermogravimetric analyzer (TA Instruments, USA), 10 mg sample size, N_2 as carrier gas (150 mL · min^{-1}), 10 ℃ min^{-1}, 25 → 700 ℃. Contact angles with water and motor oil (CA_{water} and CA_{oil}): OCA40 instrument (Dataphysics, Germany) at r. t., droplet size 5 μL (applied at five different points), the presented data are averages of these measurement.

The conductivity is evaluated by randomly measuring the resistance between two points in 1 cm distance. Lower resistance means higher conductivity.

3 Results and discussion

The preparation process of the superhydrophobic wood treated with Ag NPs and (heptadecafluoro-1, 1, 2, 2-tetradecyl) trimethoxysilane ($CF_3(CF_2)_7CH_2CH_2Si(OCH_3)_3$ (hereafter abbreviated as FAS-17) is illustrated in Fig. 1. The pristine wood with abundant hydroxyl groups on its surface were pretreated with NaOH, making the wood surface negatively charged. Then, dipping this wood into the silver ammonia complex solution, the complexes of $[Ag(NH_3)_2]^+$ are easily and abundantly absorbed on the surface. Upon reduction with glucose in presence of $[Ag(NH_3)_2]^+$, the silver ions are reduced into silver and absorbed onto the wood surface. In the subsequent modification, the $Si-OCH_3$ groups in FAS-17 are firstly hydrolyzed into Si-OH groups and the hydrolyzed long-chain polymers crosslink with each other through the hydrogen bonds by a self-assemble process. Then, the layer made up of a network of long-chain polymer compounds with Si-O groups overlapping the film of Ag NPs. Finally, under the combined effects of Ag NPs and FAS-17, the wood surface becomes superhydrophobic.

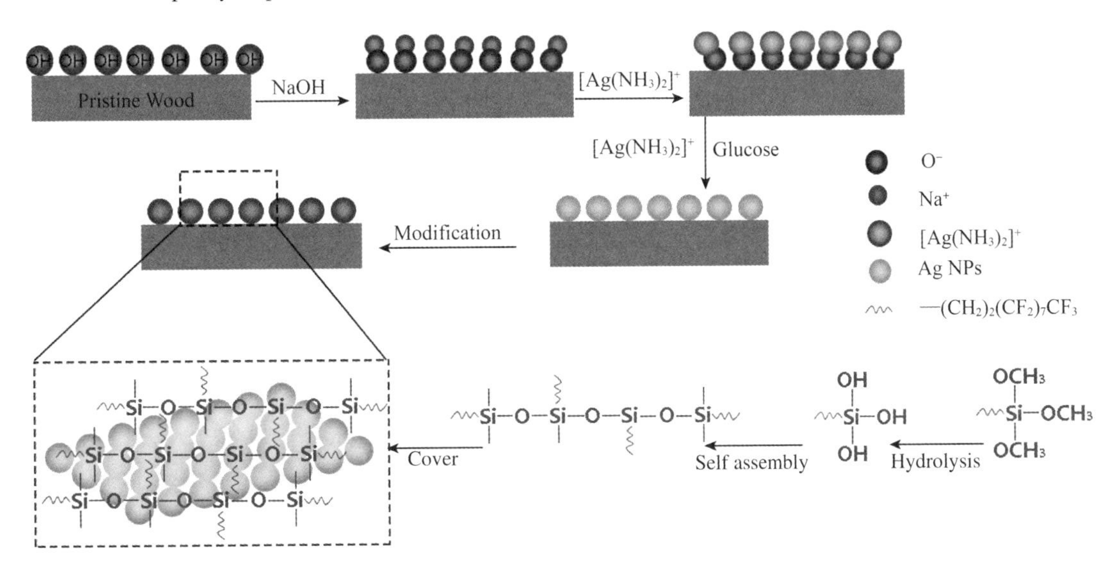

Fig. 1 Schematic illustration of the preparation of a superhydrophobic wood surface obtained by treatment with silver nanoparticles (Ag NPs) and subsequently with (heptadecafluoro-1, 1, 2, 2-tetradecyl) trimethoxysilane ($CF_3(CF_2)_7CH_2CH_2Si(OCH_3)_3$, shortly FAS-17

The SEM images of the surfaces are presented in Fig. 2, where the pristine surface of the poplar wood is smooth and free of any other materials, though some pits are visible (Fig. 2a). The roughness of the surface is increased after coating by dense and even Ag nanoparticles (Fig. 2b), which mask the anatomical details and fill the pits. The magnified SEM image in the insert in Fig. 2b shows an average size of the Ag nanoparticles of ca. 110 nm, which makes the surface rough, i. e. a dual-size surface structure is visible. Such surfaces are potentially superhydrophobic. The elemental compositions of the Ag NPs-treated wood surface was determined based on the EDS spectrum (Fig. 2d), which shows the presence of oxygen, silver, gold, and carbon. The gold originates from the coating layer used for the SEM observation and carbon is from the wood substrate.

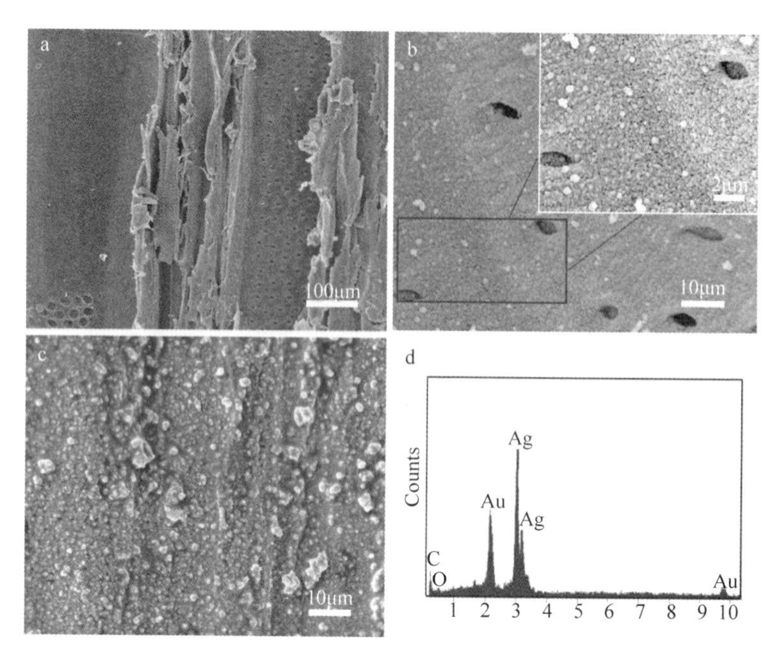

Fig. 2 SEM images of the surfaces of (a) the original wood, (b) the Ag NPs-treated wood, (c) the wood treated by Ag NPs and (heptadecafluoro-1, 1, 2, 2-tetradecyl) trimethoxysilane ($CF_3(CF_2)_7CH_2CH_2Si(OCH_3)_3$, and (d) EDS spectrum of the Ag NPs treated wood

Fig. 3 presents the XRD pattern of the Ag coated wood. Apart from the diffraction peaks at 15° and 22° belonging to characteristic diffraction peaks of cellulose (Andersson et al., 2003), four peaks at 38°, 44°, 64°, and 77° are visible, correspond to (111), (200), (220), and (311) planes of silver crystal (JCPDS cards No. 4-0783), respectively. Peaks of the potential impurities of Ag_2O are not visible. Thus, the coating consists of pour silver crystals.

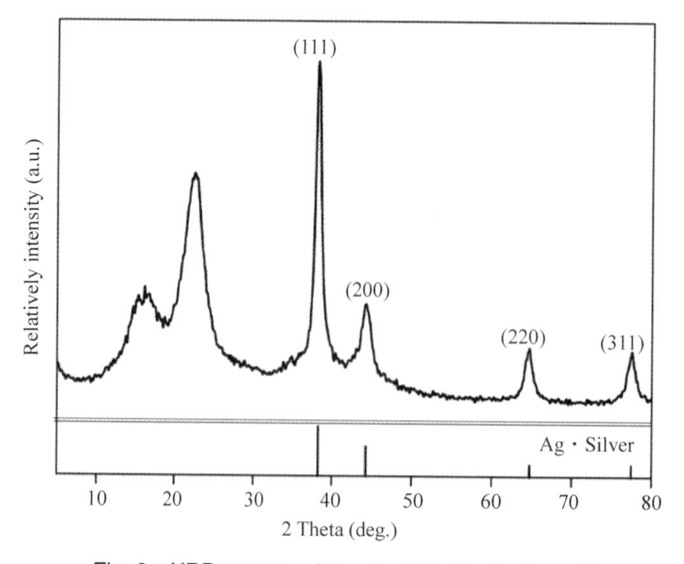

Fig. 3 XRD pattern of the Ag NPs treated wood

Fig. 4 shows FT-IR spectra of the original wood and the wood treated by Ag NPs + FAS-17. The band at 3355 cm^{-1} (stretching vibrations of OH groups) (Fig. 4a) shifted to 3415 cm^{-1} (Fig. 4b) and have a lower intensity indicating that the hydrophilic OH groups are consumed by the coating. Also the other bands are the typical for wood: 2905 cm^{-1} (C-H stretching of the CH_3 and CH_2 groups), 1426 cm^{-1} and 1372 cm^{-1} (CH_2 and CH bending mode), 1163 cm^{-1} and 1058 cm^{-1} (C=O stretching), 1742 cm^{-1} (unconjugated carbonyl group typical of xylan and hemicellulose), 1595 cm^{-1} and 1508 cm^{-1} (C=C stretching of the aromatic ring), and 1462 cm^{-1} (CH_3 bending of lignin) (Lionetto et al., 2012). The expectable new adsorption band at around 745 cm^{-1} (Fig. 4b), stretching and deformation of C-F of the CF_3 and CF_2, is very weak because of the thin coating layer compared to the unmodified bulk. Nevertheless, the FAS-17 molecules seem to be incorporated into the surface (Jin et al., 2014).

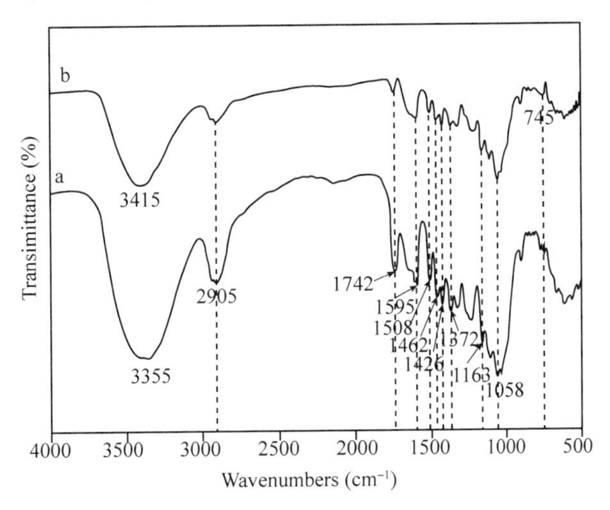

Fig. 4 FT-IR spectra of (a) the original wood, (b) the wood treated by Ag NPs and (heptadecafluoro-1, 1, 2, 2-tetradecyl) trimethoxysilane [$CF_3(CF_2)_7CH_2CH_2Si(OCH_3)_3$]

The TG and DTG curves of the original and the treated samples with Ag NP and Ag NP + FAS-17 are shown in Fig. 5. The weight loss (WL) in the range of 50-80 ℃ is low (loss of physically adsorbed water). In Fig. 5a is visible that the WL takes places in three stages: 1. between 190 - 250 ℃ (thermal depolymerisation of hemicelluloses), 2. between 250 - 380 ℃ (cellulose degradation, accompanied with continuous degradation of lignin), 3. between 380 - 700 ℃ (all wood components degrade leading to aromatization and carbonization). Lignin decomposition occurs slowly and continuously (Ouajai and Shanks 2005). At last, carbon residues (charcoal) with 18.0% yield remained (Fig. 5a). The DTG curves illustrate the above indicated degradation steps with high precision: shoulder peak at about 290 ℃ (hemicelluloses), and the major second decomposition peak at about 377 ℃ (cellulose). The sharp endothermic peaks at 373 ℃ (Ag NPs treated) and 367 ℃ (Ag NPs + FAS-17 treated) are a bit shifted in comparison to that of the untreated wood, probably because of the catalytic effect of Ag accelerating pyrolysis (Song et al. 2007). This effect is low because of the thin modified layer compared to the huge unmodified bulk. The charcoal residue of the treated woods was 21.2% (Ag NP) and 23.8% Ag NP + FAS-17) compared to 18.0% of the untreated reference. The maximum degradation rate of the wood treated by TiO_2 and FAS-17 was the lowest (Figures

5c), which may be due to the decomposition of FAS-17, as an effective protecting barrier. Accordingly, films combined Ag NPs and FAS-17 show slightly higher degradation resistance.

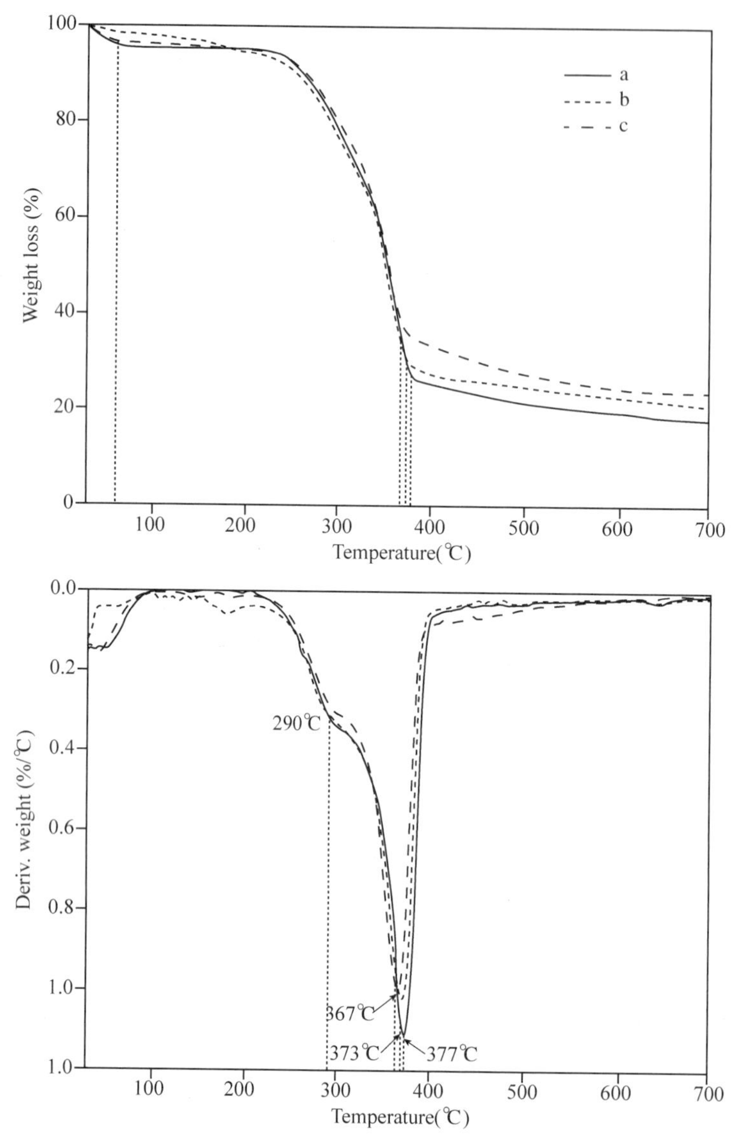

Fig. 5 Thermal analysis (TG/DTG) of (a) the original wood, (b) the Ag NPs-treated wood, and (c) the wood treated by Ag NPs and (heptadecafluoro-1, 1, 2, 2-tetradecyl) trimethoxysilane [$CF_3(CF_2)_7CH_2CH_2Si(OCH_3)_3$]

Fig. 6 presents the photographs of water and oil repellency on the observed surfaces; the data of CA_{water} and CA_{oil} are illustrated as insertions in the corresponding figures. Figures 6a and d show the water and motor oil droplets on the untreated reference. The CAs of the reference are 40.4° (water) and 0° (oil). After treatment with Ag NPs, the CA_{water} is 37.7° (Fig. 6b), i. e. the surface is still hydrophilic, and CA_{oil} is also unchanged (0°) (Fig. 6e). However, wood treated with Ag NPs + FAS-17 became super-liquid-repellent. All the water and oil droplets looked like spheres, with CA_{water} of 155.2° and CA_{oil} of 151.8° (Figures 6c and f). The superhydrophobic coating on the wood surface with the CA_{oil} larger than 150° could be considered as an effective barrier to prevent wood from oil pollution.

Fig. 6 Photographs illustrating the water (a, b, and c) and oil repellencies (d, e, and f) on the wood surfaces. CA_{water}: a) the original wood, b) Ag NPs treated wood, and c) wood treated with Ag NPs and (heptadecafluoro-1, 1, 2, 2-tetradecyl) trimethoxysilane ($CF_3(CF_2)_7CH_2CH_2Si(OCH_3)_3$. CA_{oils}: d) the original wood, e) the Ag NPs-treated wood, and f) the wood treated with Ag NPs and (heptadecafluoro-1, 1, 2, 2-tetradecyl) trimethoxysilane ($CF_3(CF_2)_7CH_2CH_2Si(OCH_3)_3$

The conductivities of the samples were characterized by bulb glow and resistance measurements. When the bulb was connected with the original wood through the wires, no light could be seen due to its insulation with an infinite resistance. In the case of the Ag NPs treated wood, the pure silver crystals impart wood a high conductivity with an electric resistance of 17.2 Ω measured by a multimeter and in the bulb test the shine was very bright. After the additional modification with FAS-17, the brightness of the bulb decreased slightly compared to Ag NPs-treated wood, and the electric resistance was elevated to 41.0 Ω.

4 Conclusions

The wood samples coated with silver nanoparticles (Ag NPs) followed by hydrophobization with FAS-17 have a remarkable super repellency towards water and motor oil, and an excellent electric conductivity. The Ag NPs are roughening the wood surface and the dense silver nanoparticle coating imparts the surface metallic feature with a good conductivity. The multifunctional wood may have various applications, especially for self-cleaning and biomedical electronic devices.

Acknowledgments

This work was financially supported by The National Natural Science Foundation of China (grant no. 31470584), China Postdoctoral Science Foundation funded project (2013M540263), and Doctoral Candidate Innovation Research Support Program of Science & Technology Review (kjdb2012006).

References

[1] Andersson S, Serimaa R, Paakkari T, et al. Crystallinity of wood and the size of cellulose crystallites in Norway spruce (Picea abies)[J]. J. Wood Sci., 2003, 49(6): 531-537.

[2]Chou K S, Huang K C, Lee HH. Fabrication and sintering effect on the morphologies and conductivity of nano-Ag particle films by the spin coating method[J]. Nanotechnology, 2005, 16(6): 779.

[3]Dubas S T, Kumlangdudsana P, Potiyaraj P. Layer-by-layer deposition of antimicrobial silver nanoparticles on textile fibers[J]. Colloid. Surface. A, 2006, 289(1-3): 105-109.

[4]Fu Y, Yu H, Sun Q, et al. Testing of thesuperhydrophobicity of a zinc oxide nanorod array coating on wood surface prepared by hydrothermal treatment[J]. Holzforschung, 2012, 66(6): 739-744.

[5]Fujishima A, Rao T N, Tryk D A. Titanium dioxide photocatalysis[J]. J. Photoch. Photobio., 2000, 1(1): 1-21.

[6]Hu W, Chen S, Li X, et al. In situ synthesis of silver chloride nanoparticles into bacterial cellulose membranes[J]. Mat. Sci. Eng. C-Mater., 2009, 29(4): 1216-1219.

[7]Jin C, Li J, Han S, et al. A durable, superhydrophobic, superoleophobic and corrosion-resistant coating with rose-like ZnO nanoflowers on a bamboo surface[J]. Appl. Surf. Sci., 2014, 320: 322-327.

[8]Li J, Sun Q, Jin C, et al. Comprehensive studies of the hydrothermal growth of ZnO nanocrystals on the surface of bamboo[J]. Ceram. Int., 2015, 41(1): 921-929.

[9]Liu F, Gao Z, Zang D, et al. Mechanical stability of superhydrophobic epoxy/silica coating for better water resistance of wood[J]. Holzforschung, 2015, 69(3): 367-374.

[10]Lionetto F, Del Sole R, Cannoletta D, et al. Monitoring wood degradation during weathering by cellulose crystallinity[J]. Materials, 2012, 5(10): 1910-1922.

[11]Lu Y, Xiao S, Gao R, et al. Improved weathering performance and wettability of wood protected by CeO_2 coating deposited onto the surface[J]. Holzforschung, 2014, 68(3): 345-351.

[12]Mahr M S, Hübert T, Stephan I, et al. Reducing copper leaching from treated wood by sol-gel derived TiO_2 and SiO_2 depositions[J]. Holzforschung, 2013, 67(4): 429-435.

[13]Ouajai S, Shanks R A. Composition, structure and thermal degradation of hemp cellulose after chemical treatments[J]. Polym. Degrad. Stabil., 2005, 89(2): 327-335.

[14]Radetić M, Ilić V, Vodnik V, et al. Antibacterial effect of silver nanoparticles deposited on corona-treated polyester and polyamide fabrics[J]. Polym. Advan. Technol., 2008, 19(12): 1816-1821.

[15]Rai M, Yadav A, Gade A. Silver nanoparticles as a new generation of antimicrobials[J]. Biotechnol. Adv., 2009, 27(1): 76-83.

[16]Saleema N, Sarkar D K, Gallant D, et al. Chemical nature of superhydrophobic aluminum alloy surfaces produced via a one-step process using fluoroalkyl-silane in a base medium[J]. ACS Appl. Mater. Inter., 2011, 3(12): 4775-4781.

[17]Khalil-Abad M S, Yazdanshenas M E. Superhydrophobic antibacterial cotton textiles[J]. J. Colloid. Interf. Sci., 2010, 351(1): 293-298.

[18]Sondi I, Salopek-Sondi B. Silver nanoparticles as antimicrobial agent: A case study on E. coli as a model for Gram-negative bacteria[J]. J. Colloid. Interf. Sci., 2004, 275(1): 177-182.

[19]Song Q, Li Y, Xing J, et al. Thermal stability of composite phase change material microcapsules incorporated with silver nano-particles[J]. Polymer, 2007, 48(11): 3317-3323.

[20]Tien H W, Huang Y L, Yang S Y, et al. The production of graphene nanosheets decorated with silver nanoparticles for use in transparent, conductive films[J]. Carbon, 2011, 49(5): 1550-1560.

[21]Wang X, Chai Y, Liu J. Formation of highly hydrophobic wood surfaces using silica nanoparticles modified with long-chainalkylsilane[J]. Holzforschung, 2013, 67(6): 667-672.

[22]Wang C, Yao T, Wu J, et al. Facile approach in fabricating superhydrophobic andsuperoleophilic surface for water and oil mixture separation[J]. ACS Appl. Mater. Inter., 2009, 1(11): 2613-2617.

[23]Wang S, Liu C, Liu G, et al. Fabrication of superhydrophobic wood surface by a sol-gel process[J]. Appl. Surf. Sci., 2011, 258(2): 806-810.

[24] Weichelt F, Emmler R, Flyunt R, et al. ZnO-based UV nanocomposites for wood coatings in outdoor applications[J]. Macromol. Mater. Eng., 2010, 295(2): 130-136.

[25] Xue C H, Chen J, Yin W, et al. Superhydrophobic conductive textiles with antibacterial property by coating fibers with silver nanoparticles[J]. Appl. Surf. Sci., 2012, 258(7): 2468-2472.

[26] Yang H, Pi P, Cai Z Q, et al. Facile preparation of super-hydrophobic and super-oleophilic silica film on stainless steel mesh via sol-gel process[J]. Appl. Surf. Sci., 2010, 256(13): 4095-4102.

[27] Yu Y, Jiang Z, Wang G, et al. Growth ofZnO nanofilms on wood with improved photostability[J]. Holzforschung, 2010, 64: 385-390.

[28] Yuranova T, Rincon A G, Pulgarin C, et al. Performance and characterization of Ag-cotton and Ag/TiO_2 loaded textiles during the abatement of E. coli[J]. J. Photoch. Photobio. A, 2006, 181(2-3): 363-369.

[29] Zhai L, Cebeci F C, Cohen R E, et al. Stable superhydrophobic coatings from polyelectrolyte multilayers[J]. Nano Lett., 2004, 4(7): 1349-1353.

中文题目：以氟硅烷改性的银纳米薄膜制备超疏水导电木材

作者：高丽坤，卢芸，李坚，孙庆丰

摘要：本文以一种简单有效的方法构筑一种超疏水、导电、拒油的木材。通过硝酸银处理及葡萄糖还原后在木材表面形成一层粗糙的银纳米颗粒构成的薄膜。氟硅烷的进一步改性得到了水接触角为155.2°的超疏水木材表面。该表面同时展现了良好的拒油性，机油液滴的接触角约为151.8°。此外，木材表面致密的银纳米颗粒赋予了木材导电性。该方法制备的木材图层可用于多个商业领域，特别是自清洁和生物医学电子设备。

关键词：导电木材；氟硅烷；拒油；银纳米颗粒；超疏水木材表面

A Robust Superhydrophobic Antibacterial Ag-TiO_2 Composite Film Immobilized on Wood Substrate for Photodegradation of Phenol under Visible-Light Illumination*

Likun Gao, Shaoliang Xiao, Wentao Gan, Xianxu Zhan, Jian Li

Abstract: Ag−TiO_2 heterostructures with Ag nanocrystals and TiO_2 particles well−grown on wood substrate was achieved by a two − step protocol combining hydrothermal synthesis and silver mirror reaction. The investigation of the photocatalytic ability demonstrated that the wood coated with Ag − TiO_2 composite film possess excellent photocatalytic activity, superior to the wood coated with pure TiO_2 particles, for the degradation of phenol under visible − light illumination. The Ag/TiO_2 − coated wood was further modified by (heptadecafluoro − 1, 1, 2, 2 − tetradecyl) trimethoxysilane, which acted as a crucial role in improving the repellency toward water and imparting self − cleaning property to the wood products. The modified wood has potent antibacterial activity toward both Gram−positive and Gram−negative bacterium. The multifunctional film coated on wood surface exhibits a good photodegradation of organic pollutant and a robust superhydrophobicity, leading to an important application in self−cleaning.

Keywords: Ag−TiO_2 composite film; Antibacterial; Photodegradation; Superhydrophobic; Wood

1 Introduction

Recently, organic-inorganic hybrid materials have attracted much attention due to their superior characters, such as electrical, magnetic and optical properties, combining properties of organic and inorganic materials[1]. Furthermore, developing an efficient, environment-friendly, and low-cost method for removal of organic pollutant compounds is essential for the current environment. As a natural organic polymer material, wood is one of the good candidates as host materials of inorganic particles because the excellent electrical, magnetic and optical properties of inorganic materials can be preserved in the polymer matrix. Therefore, the inorganic micro and nano particles/wood composite materials can be considered as a portable catalyst. Meanwhile, wood products with multiple functions, such as superhydrophobicity, self-cleaning, antibacterial activity and photodegradation of organic pollutant, would be greatly appreciated by a more discerning and demanding consumer market for high-value-added products[2, 3]. Multifunctional wood could be fabricated through combined treatments using several materials with one or more specific properties on its surface.

---

* 本文摘自 Ceramics International, 2016, 42: 2170−2179.

As semiconductors materials, the micro-nano structures of metals and oxide metals, such as TiO_2[4], ZnO[5], $CoFe_2O_4$[6], Ag[7], and so forth, have broad applications in the fields of photoelectronics, photocatalysis, and biology, because they have distinct optoelectronic properties. Among them, TiO_2 is widely selected for various applications including self-cleaning surfaces, photocatalysts for air and water purification[8]. Moreover, many organic pollutants, for example, phenols, alcohols and dye, have been proved to be photodegraded by TiO_2 catalyst under UV irradiation[9, 10]. However, there is still a basic problem that the semiconductor TiO_2 with the band gap of 3.2 eV could only be excited by UV light ($\lambda<388$ nm) to inject electrons into conduction band and to leave holes in valence band. Thus, the limitation of the use of TiO_2 under sunlight or visible light in photocatalytic process has become a key factor to expand the advanced application in photocatalysis[11, 12]. Therefore, the composite of TiO_2 with other materials, such as ZnO[13], Ag[14], Cu[15], and Nd[16], have been attempted to efficiently extend the photoresponse from UV to visible-light. It is worth noting that Ag-TiO_2 composite materials have become one of the most attractive candidate materials primarily due to their extraordinary properties and superb photocatalytic performance in practical applications[17]. In addition, silver has been known as a bactericide since ancient times. And Ag nanoparticles have been reported to exhibit antimicrobial and conductive properties[18, 19]. Therefore, the incorporation of Ag nanoparticles into the TiO_2-treated wood matrix may extend their utility in materials and biomedical applications[20].

Based on above consideration, the wood-based products composited with a film of catalysts were proposed. In this paper, the two-step method combining with hydrothermal synthesis and silver mirror reaction was adopted to build a dual-size surface structure on wood surface, and the further modification with low surface energy of (heptadecafluoro-1, 1, 2, 2-tetradecyl) trimethoxysilane was employed to create remarkable superhydrophobic surface. The wood decorated with Ag-TiO_2 composite film after hydrophobization might also be imparted with multiple properties, such as superhydrophobicity, antibacterial actions against both Gram-negative (Escherichia coli) and Gram-positive (Staphylococcus aureus) bacterium, and photodegradation of phenol under visible light.

2 Materials and methods

2.1 Materials

All chemicals supplied by Shanghai Boyle Chemical Company, Limited were of analytical reagent-grade quality and used without further purification. Deionized water was used throughout the study. Wood blocks of 20 mm (R)×20 mm (T)×30 mm (L) were obtained from the sapwood sections of poplar wood (*Populus ussuriensis* Kom), which is one of the most common tree species in the northeast of China. The wood specimens were oven-dried (24 h, 103±2 ℃) to constant weight after ultrasonically rinsing in deionized water for 30 min, and oven-dried weight were determined.

2.2 Preparation of anatase TiO_2 film on the wood surface

Ammonium fluorotitanate (0.4 M) and boric acid (1.2 M) were dissolved in distilled water in a 500 mL glass container at room temperature under vigorous magnetic stirring. Then, a solution of 0.3 M hydrochloric acid was added until the pH reached approximately 3. Then, 75 mL of the adjusted solution was transferred into a 100 mL Teflon container. Wood specimens were subsequently placed into the above reaction solution, separately. The autoclave was sealed and maintained at 90 ℃ for 5 h, then allowed to naturally cool to room temperature. Finally, the prepared samples were removed from the solution, ultrasonically rinsed with deionized water for 3 times, and dried at 45 ℃ for more than 24 h in vacuum. Thus, the TiO_2-treated wood was obtained.

2.3 Coating the TiO_2-treated wood surface with Ag nanoparticles

The Ag nanoparticles were prepared based on a common silver mirror reaction with some modifications[21]. Aqua ammonia (28 wt. %) was added dropwise into a 0.5 M $AgNO_3$ aqueous solution with stirring until a transparent colorless silver ammonia complex solution ($[Ag(NH_3)_2]^+$) was formed. The TiO_2-treated wood samples were dipped into the silver ammonia complex solution for 1 h, the complexes of $[Ag(NH_3)_2]^+$ were easily and abundantly absorbed on the surfaces. And then the silver ammonia complex loaded TiO_2-treated wood was transferred into a 0.1 M glucose stock solution, $[Ag(NH_3)_2]^+$ was in situ reduced into silver seeds on the surfaces. After 5 min, the residual $[Ag(NH_3)_2]^+$ solution was also poured into the glucose solution. The reaction was continued for 15 min. In this process, more and more silver ions were absorbed onto the TiO_2-treated wood surface and reduced into silver, and the silver seeds grew larger into silver nanoparticles, attached to each other forming a compact film on the TiO_2-treated wood. Finally, the prepared samples were removed from the solution, ultrasonically rinsed with deionized water for 30 min, and dried at 45 ℃ for more than 24 h in vacuum. Then the Ag/TiO_2-coated wood was obtained.

2.4 Hydrophobization of wood surfaces with FAS-17

A methyl alcohol solution of (heptadecafluoro-1,1,2,2-tetradecyl) trimethoxysilane ($CF_3(CF_2)_7CH_2CH_2Si(OCH_3)_3$, hereafter denoted as FAS-17) was hydrolyzed by the addition of a 3 fold-molar excess of water at room temperature. Then, 75 mL of the adjusted solution was transferred into a 100 mL Teflon container. The Ag/TiO_2-coated wood samples were subsequently placed into the above reaction solution. The autoclave was sealed and maintained at 75 ℃ for 5 h, then allowed to naturally cool to room temperature. Subsequently, the samples were washed with ethyl alcohol to remove any residual chemicals and allowed to dry in air at room temperature, and they were then dried at 45 ℃ for more than 24 h in vacuum. Thus, a superhydrophobic wood surface was obtained.

2.5 Characterization

The morphology of the wood surfaces was observed through field emission scanning electron microscopy (SEM, Quanta 200, FEI, Holland) operating at 12.5 kV in combination with EDS (Genesis, EDAX, Holland). The transmission electron microscopy (TEM) experiment was performed on a Tecnai G20 electron microscope (FEI, USA) with an acceleration voltage of 200 kV. Carbon-coated copper grids were used as the sample holders. The transmission electron microscopy (TEM) experiment was performed on a Tecnai G20 electron microscope (FEI, USA) with an acceleration voltage of 200 kV. Carbon-coated copper grids were used as the sample holders. X-ray diffraction (XRD, Bruker D8 Advance, Germany) was employed to analyze the crystal structures of all samples applying graphite monochromatic with Cu $K\alpha$ radiation ($\lambda = 1.5418$ Å) in the 2 range from 5° to 80° and a position-sensitive detector using a step size of 0.02° and a scan rate of 4° min^{-1}. FTIR spectra were obtained on KBr tablets and recorded using a Magna-IR 560 spectrometer (Nicolet) with a resolution of 4 cm^{-1} by scanning the region between 4000 and 500 cm^{-1}. Water contact angles

(WCAs) were measured on an OCA40 contact angle system (Dataphysics, Germany) at room temperature. In each measurement, a 5 μL droplet of deionized water was injected onto the surfaces of the wood samples and the contact angles were measured at five different points of each sample. The final values of the contact angles were obtained as an average of five measurements. Optical properties of the material were characterized by UV-vis diffuse reflectance spectroscopy (UV-vis DRS, Beijing Purkinje TU-190, China) equipped with an integrating sphere attachment, which $BaSO_4$ was the reference. Total organic carbon (TOC) in the solution after phenol removal was measured by Multi N/C 2100S/1 Analyzer (Analytik Jena, Germany). To determine the change of phenol concentration and content of total organic carbon in the solution during visible light irradiation, a few milliliters of the solution was taken from the reaction mixture and loaded in Multi N/C 2100S/1 Analyzer after filtration.

2.6 Antibacterial test

Antimicrobial tests were conducted by the bacterial inhibition ring method (agar plate diffusion test/CEN/TC 248 WG 13) and the reduction of bacterial growth test (EN ISO 20743: 2007 Transfer Method)[22]. The antibacterial activity of the hydrophobized Ag/TiO_2-coated wood was evaluated against Escherichia coli (ATCC 25923, Gram-negative bacterium) and Staphylococcus aureus (ATCC 25922, Gram-positive bacterium). The evaluation of the test was as the followings. A mixture of nutrient broth and nutrient agar in 1 L distilled water at pH 7.2 as well as the empty Petri plates were autoclaved. The agar medium was then cast into the Petri plates and cooled in laminar airflow. Approximately 10^5 colony-forming units of each bacterium were inoculated on plates, and then each wood samples was planted onto the agar plates. All the plates were incubated at 37 ℃ for 24 h, following which the zone of inhibition was measured.

2.7 Photodegradation of phenol

The photocatalytic activities of the hydrophobized Ag/TiO_2-coated wood samples were assessed by monitoring the degradation of phenol and photodegradation experiments were conducted in an obturator (Fig. 1). Two pieces of the samples were immersed into 20 mL 67.2 mg · L^{-1} aqueous phenol solution in the dark for 6 h. Then, the solution with the samples was irradiated under a flat-type LED-light (wavelength is in the range of 400-760 nm, light intensity is approximately 3.6 mW · cm^{-2}) as a simulative visible light. In the whole experiment, temperature was detected with temperature detector and controlled with a ventilator. The concentration of phenol was monitored by detecting UV absorbance intensity at 270 nm at given irradiation time intervals. The hydrophobized TiO_2-treated wood were used in the photodegradation of phenol, compared with the hydrophobized Ag/TiO_2-coated wood.

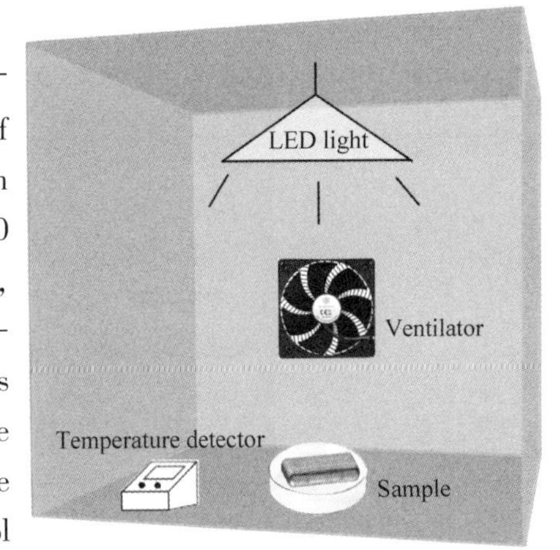

Fig. 1 Schematic diagram of experimental set-up

3 Results and discussion

Fig. 2 shows the SEM images, TEM images and HRTEM images of the wood samples. It can be seen from

Fig. 2a that the original wood shows typical cell wall structure with clean and smooth surface. After hydrothermal synthesis, the wood surface show a covering of TiO_2 microparticles, as shown in Fig. 2b. Then the TiO_2-treated wood surface was coated with a compact and uniform covering of Ag nanoparticles through a silver mirror reaction. TEM image of the Ag/TiO_2-coated wood in Fig. 2e shows that the nanoparticles of Ag are clearly visible, making the wood surface rough, thus generating a dual-size surface structure on the wood substrate. After further hydrophobization with FAS-17 (Fig. 2d), the wood surface exhibits a microrelief similar to that of the plant surface, which mainly combines with wax crystalloids and cuticular folds and provides a water-repellent surface[23]. To further investigate the distribution of Ag and TiO_2 particles, the HRTEM image of the Ag/TiO_2-coated wood is displayed in Fig. 2f. The clear lattice fringe of $d=0.35$ nm matches that of the (101) plane of anatase phase of TiO_2, while the fringe of $d=0.24$ nm corresponds to the (111) plane of cubic phase of Ag[24]. Therefore, Fig. 2 proves that the films made by a two-step method combining with hydrothermal synthesis and silver mirror reaction are satisfactory.

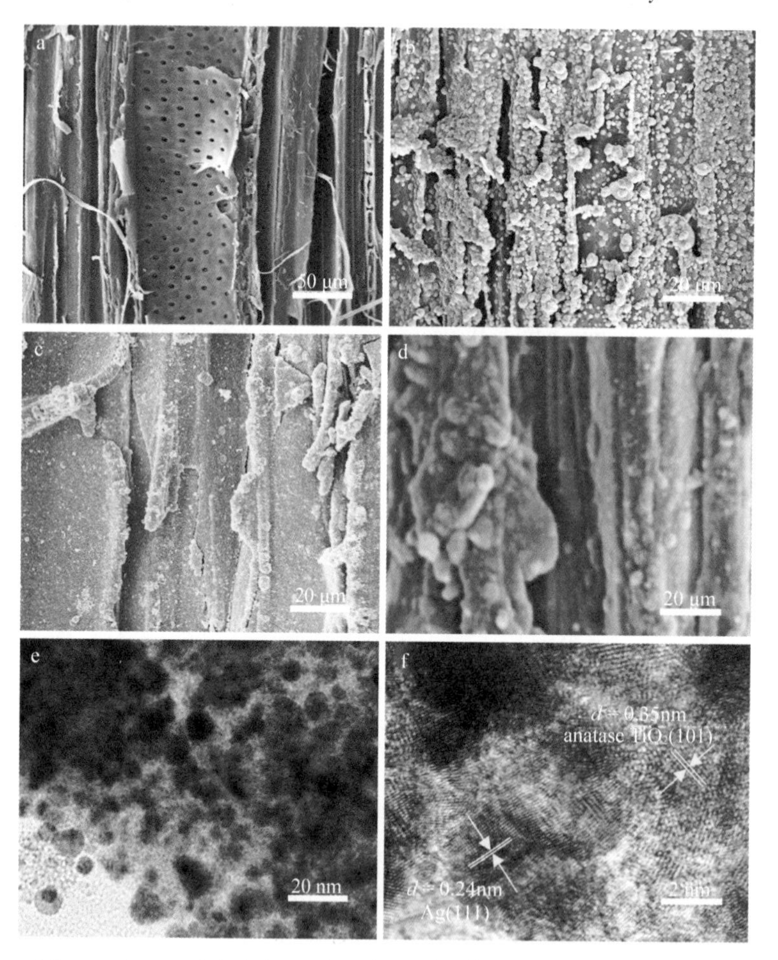

Fig. 2 SEM images of the surfaces of (a) the original wood, (b) the TiO_2-treated wood, (c) the Ag/TiO_2-coated wood, and (d) the hydrophobized Ag/TiO_2-coated wood. (e) TEM image of the Ag/TiO_2-coated wood. (f) HRTEM image of the Ag/TiO_2-coated wood

The surface chemical elemental compositions of the treated wood are determined via energy-dispersive X-ray spectroscopy (EDS), and the results are presented in Fig. 3a and b. Evidence of only fluorine, titanium, gold, and carbon elements could be detected in Fig. 3a, while the additional presence of silver element exists in Fig. 3b. The gold element is originated from the coating layer used for SEM observation, the fluorine element may be originated from the precursor ammonium fluorotitanate, and carbon element was from the wood substrate. No other elements were detected, thus confirming the composition of the microparticles on the TiO_2-treated wood sample to be TiO_2 as well as revealing the loading of metallic Ag on the Ag/TiO_2-coated wood sample. Fig. 2c shows the XRD patterns of the TiO_2-treated wood and the Ag/TiO_2-coated wood. Apparently, the diffraction peaks at 14.8° and 22.5° belonging to the (101) and (002) crystal planes of cellulose in the wood are observed in both the spectrum of the TiO_2-treated wood and the Ag/TiO_2-coated wood[25]. It can be found that the diffraction peaks are well indexed to the standard diffraction pattern of anatase phase TiO_2 and the face-centered cubic (fcc) phase of Ag, indicating that the present synthesis strategy successfully achieves Ag/TiO_2 hierarchical heterostructures with high crystallinity on wood substrate. The curve in Fig. 2c shows that the diffraction peaks of pure anatase TiO_2 particles center at 25.2°, 38.0°, 47.8°, 54.2°, 62.5°, 68.8°, and 74.9°, which agree with (101), (004), (200), (211), (204), (116) and (215) crystal planes of anatase TiO_2(JCPDS file No. 21-1272), respectively[26]. After coated with Ag nanoparticles (Fig. 2c), all of the new diffraction peaks at 38.1°, 44.2°, 64.4°, and 77.4°, except the diffraction peaks of TiO_2, can be perfectly indexed to (111), (200), (220) and (311) reflections of the face-centered cubic (fcc) phase of Ag (JCPDS file No. 04-0783)[21]. The corresponding selected area electron diffraction (SAED) pattern of the

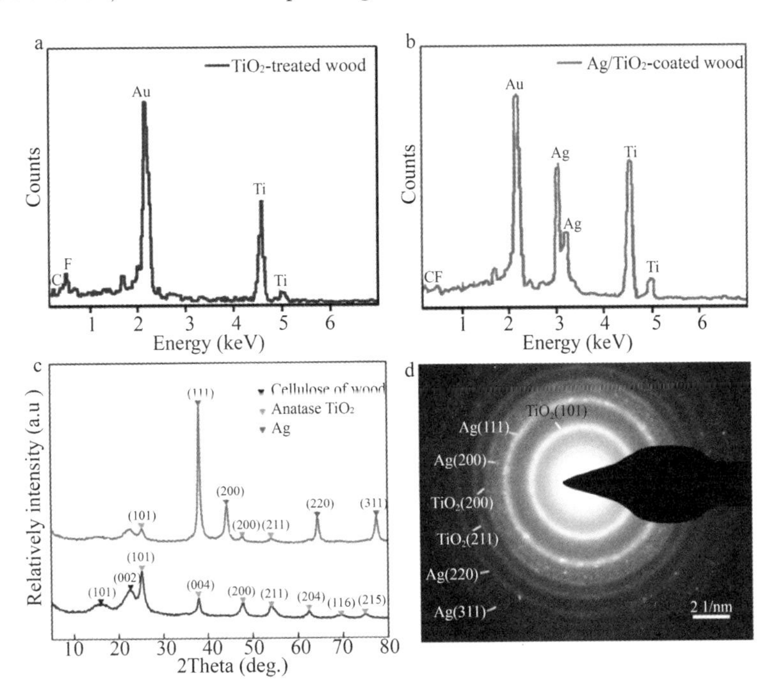

Fig. 3 (a, b) EDS spectra of the TiO_2-treated wood and the Ag/TiO_2-coated wood, (c) XRD patterns of the TiO_2-treated wood and the Ag/TiO_2-coated wood, and (d) the selected electron diffraction (SAED) pattern of the Ag/TiO_2-coated wood

Ag-doped TiO_2/wood is shown in Fig. 3d. The SAED pattern of TiO_2 microparticles in anatase phase and Ag nanoparticles in cubic phase shows that the diffraction patterns and rings are co-existent, and the different crystal planes are identified as (101), (200) and (211) diffraction planes of anatase TiO_2 as well as (111), (200), (220) and (311) diffraction planes of cubic Ag[27], respectively, which is in agreement with XRD pattern results in Fig 3c.

Fig. 4 shows FTIR spectra of the original wood, the Ag/TiO_2-coated wood, and the hydrophobized Ag/TiO_2-coated wood. The band at 3340 cm^{-1} corresponding to the stretching vibrations of OH groups in the wood (Fig. 4a) shifts to larger wavenumbers of 3345 cm^{-1} (Fig. 4b) and 3415 cm^{-1} (Fig. 4c), and the intensity of them comparatively decrease, indicating that the hydrophilic groups (-OH) of the wood decease in both the Ag/TiO_2-coated wood and the hydrophobized Ag/TiO_2-coated wood samples. Typical bands assign to cellulose located at 2903 cm^{-1} for the C-H stretching vibrations of the CH_3 and CH_2 groups, at 1426 cm^{-1} and 1372 cm^{-1} for CH_2 and CH bending mode, and at 1163 cm^{-1} and 1058 cm^{-1} for C=O stretching vibrations, respectively[28]. In addition, a new adsorption peak exists at approximately 745 cm^{-1} in Fig. 4c is attributed the stretching and deformation vibration peak of C-F of the CF_3 and CF_2[29], which illustrates that the FAS-17 molecules incorporates into the surface.

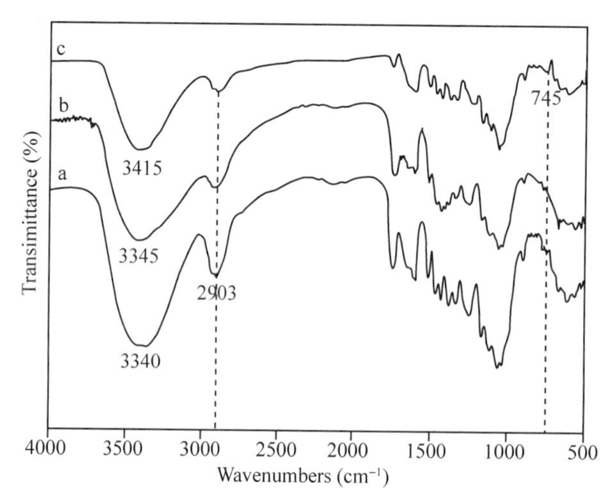

Fig. 4 FTIR spectra of (a) the original wood, (b) the Ag/TiO_2-coated wood, and (c) the hydrophobized Ag/TiO_2-coated wood

The TG and DTG curves of the original wood, the Ag/TiO_2-coated wood and the hydrophobized Ag/TiO_2-coated wood are shown in Fig. 5. The TG curves of both the original and treated woods exhibit a small weight loss at about 50-180 ℃, which corresponds to a mass loss of physically adsorbed water of approximately 5%. After this peak, the DTG curves in Fig. 4b shows two decomposition steps[30]: (1) the first decomposition shoulder peak at about 250-320 ℃, is attributed to thermal depolymerisation of hemicelluloses or pectin; (2) the major second decomposition peak at about 375 ℃, 350 ℃, and 345 ℃, respectively, is attributed to cellulose decomposition. Lignin is the most difficult one to decompose, and its decomposition keeps on along the whole calcining process. The final decomposition process at about 380 - 700 ℃ was attributed to all the wood components degradation gradually leading to the aromatization and carbonization. At last, carbon residues with the weight of 18% are for the original wood remains, as observed from the TG curves (Fig. 5a). Due to the decomposition of cellulose and lignin, the maximum degradation rates of the Ag/TiO_2-coated wood and the hydrophobized Ag/TiO_2-coated wood become lower than that of the original wood. This may be due to the catalysis of Ag - TiO_2 composite film, which generates an accelerated pyrolysis action on wood components. Moreover, the Ag/TiO_2-coated wood and the hydrophobized Ag/TiO_2-coated wood are able to keep more than 24% and 40% of the weight left, respectively. The maximum degradation rate of the hydrophobized Ag/TiO_2-coated wood is the lowest (Fig. 5a), which may be due to the decomposition of FAS-17, as a barrier, effectively protect the wood sample. The results indicate that the film combined with Ag-TiO_2 composite and FAS-17 provide the protection for wood.

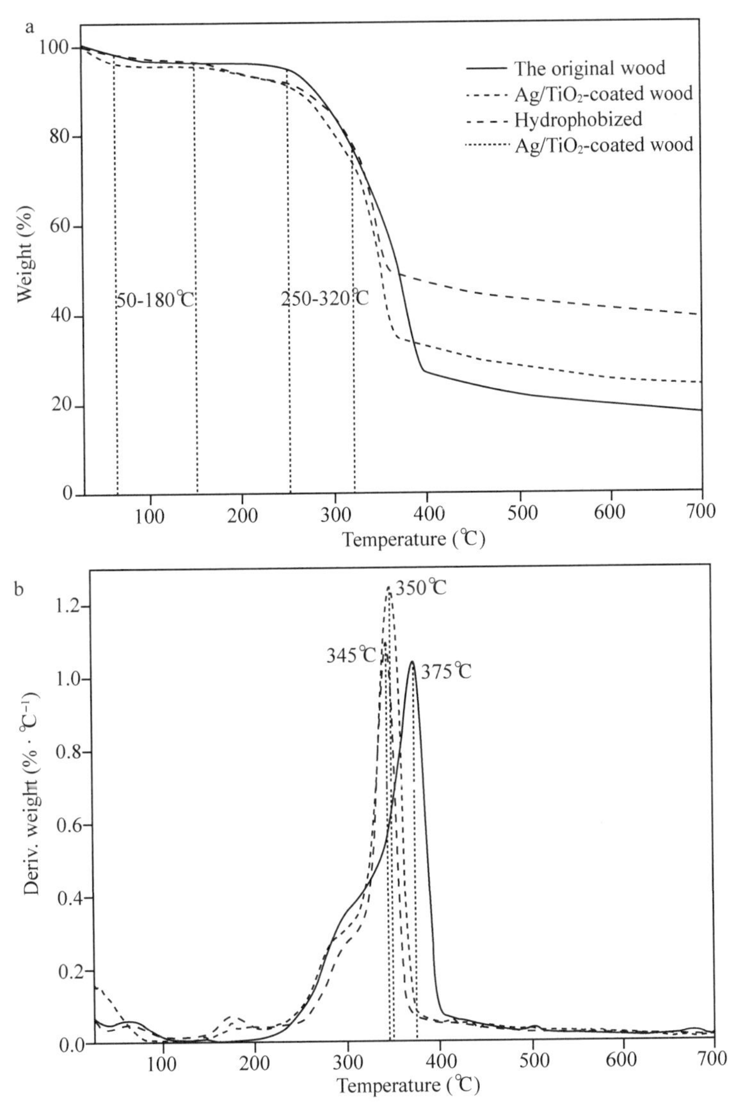

Fig. 5 (a) TG profiles and (b) DTG profiles of the original wood, the Ag/TiO_2-coated wood and the hydrophobized Ag/TiO_2-coated wood

Fig. 6 demonstrates the variation of the water contact angles of the original wood, the TiO_2-treated wood, the Ag/TiO_2-coated wood and the hydrophobized Ag/TiO_2-coated wood. The original wood surface in Fig. 6a presents hydrophilicity with the WCA of 59.7°. As shown in Fig. 6b, the TiO_2-treated wood possesses a superhydrophilic surface with the WCA of 0°. After coated with Ag nanoparticles (Fig. 6c), the wood surface converted into hydrophobic one with the WCA of 121.6°. After further hydrophobization with FAS-17 (Fig. 6d), the hydrophobicity of the wood surface is raised to superhydrophobicity, and the WCA reaches 153.2°. The results are in accordance with the FTIR results that the hydrophilic groups (-OH) of the wood decease in both the Ag/TiO_2-coated wood and the hydrophobized Ag/TiO_2-coated wood, that is, the hydrophobicity of wood surfaces is significantly increased.

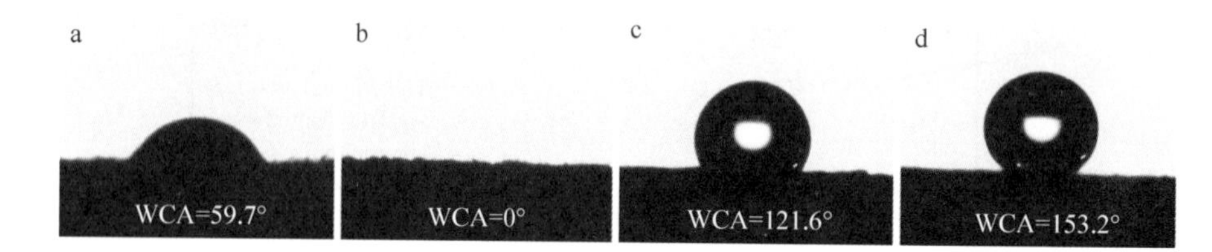

Fig. 6 WCAs of (a) the original wood, (b) the TiO_2-treated wood, (c) the Ag/TiO_2-coated wood, and (d) the hydrophobized Ag/TiO_2-coated wood

Antibacterial activity of wood samples is determined in terms of inhibition zone formed on agar medium. Fig. 7a and b present that the TiO_2-treated wood, which is used as control group, do not show any antibacterial activity for both Escherichia coli and Staphylococcus aureus. The hydrophobized Ag/TiO_2-coated wood placed on the bacteria-inoculated surfaces kill all the bacteria under and around them. It can be observed that the distinct zones of inhibition (clear areas with no bacterial growth) around the wood samples for both Escherichia coli and Staphylococcus aureus (Fig. 7c and d). For Escherichia coli, the width of the inhibition

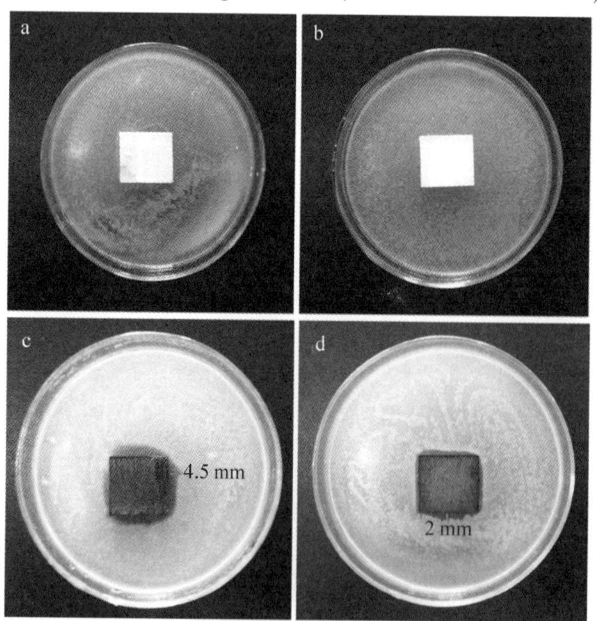

Fig. 7 Antibacterial activity of (a, b) the hydrophobized TiO_2-treated wood in Escherichia coli and Staphylococcus aureus, and (c, d) the hydrophobized Ag/TiO_2-coated wood in Escherichia coli and Staphylococcus aureus, respectively

zone around the hydrophobized Ag/TiO_2-coated wood is approximately 4.5 mm, while for Staphylococcus aureus, the width is approximately 2 mm. The observed zone of inhibition is a result of the leaching of active biocidal species Ag nanoparticles existed in the wood into the surrounding aqueous medium. In other words, the presences of the inhibition zone clearly indicate that the mechanism of the biocidal action of the wood is owing to the effect of Ag nanoparticles.

Fig. 8 ascertains more detailed information concerning the elemental and chemical state of the resulting samples. The fully scanned spectra (Fig. 8a) shows that the Ti, O, and C elements exist on the surface of the pure TiO_2-treated wood sample, while Ti, O, Ag, and C elements exist on the surface of the Ag/TiO_2-coated wood sample. The C element can be ascribed to the wood substrate or the adventitious carbon-based contaminant. From Fig. 8b, the Ti 2p XPS spectra of the TiO_2-treated wood with two peaks at binding energies of 459.0 eV and 464.8 eV, corresponding to the Ti $2p_{3/2}$ and Ti $2p_{1/2}$ peaks, respectively. The gap of 5.8 eV between the two peaks indicates the existence of the Ti^{4+} oxidation state[31]. However, the peak position for Ti 2p in the Ag/TiO_2-coated wood sample shifts to a higher binding energy band than that in the pure TiO_2-treated wood sample. This confirms a lower electron density of the Ti atoms after doped with Ag nanoparticles, and there is a strong interaction between metallic Ag and TiO_2 in the Ag/TiO_2-coated wood. In addition, the Ti $2p_{3/2}$ XPS peak of the Ag/TiO_2-coated wood sample can be fitted into two components, one located at 459.3 eV, attributed to a Ti^{4+} species, and the other located at 458.9 eV, assigned to a Ti^{3+} species (in Fig. 8c), further indicating the strong interaction formed between Ag and TiO_2 species. Certainly, the presence of Ti^{3+} oxide with narrow band gap and energy level located between the valence and the conduction band of TiO_2 may be advantageous to the higher photocatalytic activity of the Ag/TiO_2-coated wood heterostructures driven by visible light. Moreover, the defect sites of TiO_2 surface induced by light, that is, Ti^{3+} species, are necessary for adsorption and photoactivation of oxygen, which are necessary for the photooxidation of pollutants. As shown in Fig. 8d, the Ag 3d XPS spectra with two peaks at binding energies of 367.6 eV and 373.6 eV, corresponding to the Ag $3d_{5/2}$ and Ag $3d_{3/2}$ peaks, respectively. The gap of 6.0 eV between the two peaks is also indicative of metallic Ag, and there is no evidence for the presence of Ag^+, indicating that the Ag^+ ions in the composites are fully reduced to metallic Ag. Additionally, it is obvious that the peaks of Ag 3d shift to the lower position compared with bulk Ag (368.3 eV for Ag $3d_{5/2}$, and 374.3 eV for Ag $3d_{3/2}$)[32], suggesting that the electrons may migrate from the TiO_2 particles to metallic Ag, which reveals that there is a strong interaction between Ag nanoparticles and TiO_2 particles in the interface of heterostructures.

Briefly, it can be demonstrated that Ag nanoparticles are grown on the TiO_2-treated wood matrix, and there is a strong interaction between Ag and TiO_2 in the interface of the Ag/TiO_2-coated wood heterostructures. Furthermore, Ag nanoparticles can act as electron acceptors and be advantageous for separating the photoexcited electron-hole pairs, the heterostructures can inhibit the recombination of excited electrons and holes and then enhance the photocatalytic activity of the products.

In order to investigate the light absorbance of the samples, the UV-vis diffuse reflection spectra of the original wood, the hydrophobized TiO_2-treated wood, the Ag/TiO_2-coated wood, and the hydrophobized Ag/TiO_2-coated wood are depicted in Fig. 9a. As for the original wood and the hydrophobized TiO_2-treated wood, it presents prominent adsorptions below 350 nm wavelength. Whereas both the Ag/TiO_2-coated wood and the hydrophobized Ag/TiO_2-coated wood exhibit a much higher absorption in the region 420-800 nm, indicating

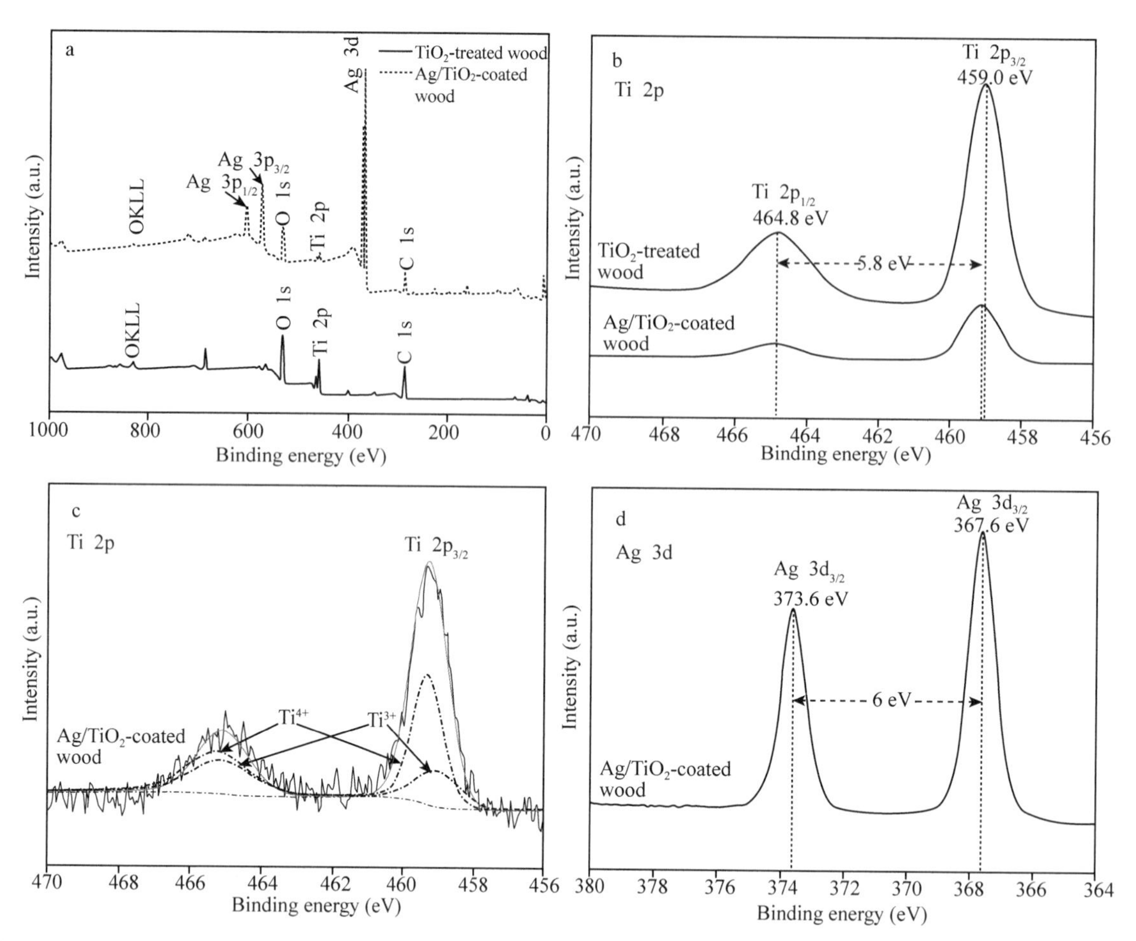

Fig. 8 (a) Survey scan and (b) Ti 2p XPS spectra of TiO_2-treated wood and the Ag/TiO_2-coated wood, (c) peaking-fitting results of Ti 2p XPS spectra of the Ag/TiO_2-coated wood, (d) Ag 3d XPS spectra of the Ag/TiO_2-coated wood

the absorptions of them are significantly red-shifted to the visible due to the loading of the Ag nanoparticles onto the surface of the TiO_2-treated wood. Theoretically, the wide visible absorption should be attributed to the characteristic absorption of surface plasmon resonance originating from the metallic Ag nanoparticles in the Ag/TiO_2-coated wood and the hydrophobized Ag/TiO_2-coated wood. Moreover, it is worth noting that the Ag/TiO_2-coated wood with absorptions in the visible region may implicate highest activity in photocatalysis. The absorptions of the hydrophobized Ag/TiO_2-coated wood in 600-800 nm is slightly decreased, indicating that there is no apparent adverse effect after modification with FAS-17.

Fig. 9 b presents the photodegradation curves of phenol. It can be observed that the original wood has no photocatalytic activity, and the treated wood exhibited photocatalytic activity. The photocatalytic activity of the TiO_2-treated wood is comparatively weaker than the Ag/TiO_2-coated wood and hydrophobized Ag/TiO_2-coated wood under visible-light irradiation. In the findings, the Ag/TiO_2-coated wood exhibited highly efficient photocatalytic activity, moreover, the photocatalytic activity of the hydrophobized Ag/TiO_2-coated wood is slightly reduced. This could be explained by the results in Fig. 8a, which the visible absorption of the

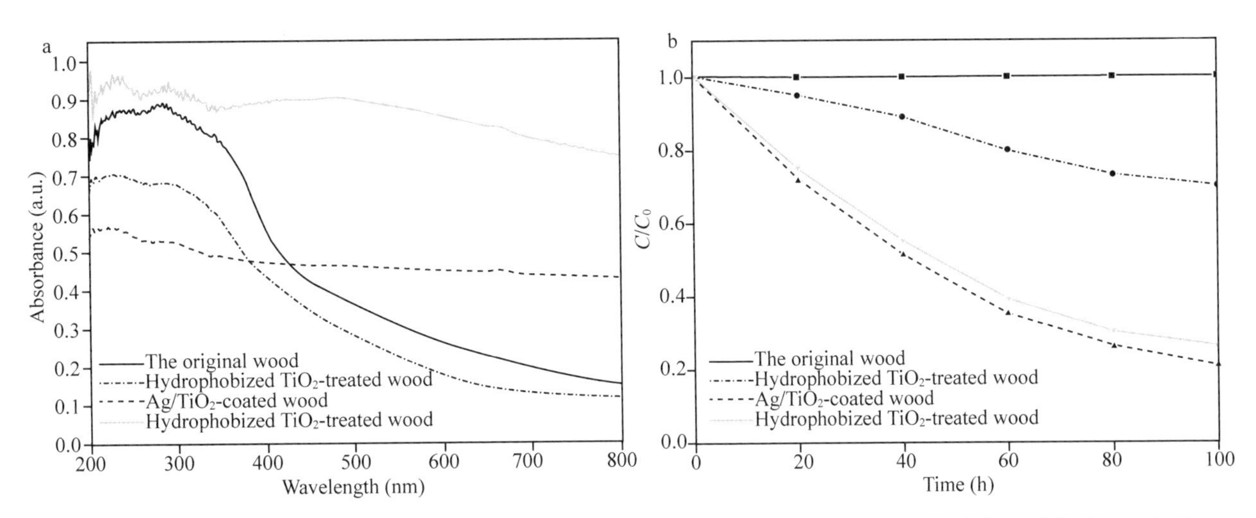

Fig. 9 (a) UV-vis absorption spectra and (b) photodegradation curves of phenol of the original wood, the hydrophobized TiO_2-treated wood, the Ag/TiO_2-coated wood, and the hydrophobized Ag/TiO_2-coated wood under visible-light irradiation, respectively

hydrophobized Ag/TiO_2 - coated wood is slightly decreased. It may be due to the immobilization of Ag nanoparticles and FAS-17 molecules on the TiO_2-treated wood matrix leading to a weak protection under visible-light irradiation. Therefore, it is well worth noting that the hydrophobized Ag/TiO_2-coated wood are a portable photocatalyst. The portable hydrophobized Ag/TiO_2-coated wood catalyst lead to a green application in the degradation of phenol, and the catalyst could be removed out easily from the polluted water after used.

Fig. 10 illustrates that phenol is removed by the hydrophobized TiO_2-treated wood and the hydrophobized Ag/TiO_2-coated wood during visible light irradiation for 100 h comparing the decrease in TOC with starting TOC, and the mineralization of phenol are nearly identical in the pathways. Based on the TOC data, phenol removal by both the hydrophobized TiO_2-treated wood and the hydrophobized Ag/TiO_2-coated wood can be divided into two stages. The first stage is a lag state in the photodegradation system, where TOC reduction is slow. That is probably because phenol photodegradation is launched by branch dissociations, benzene rings

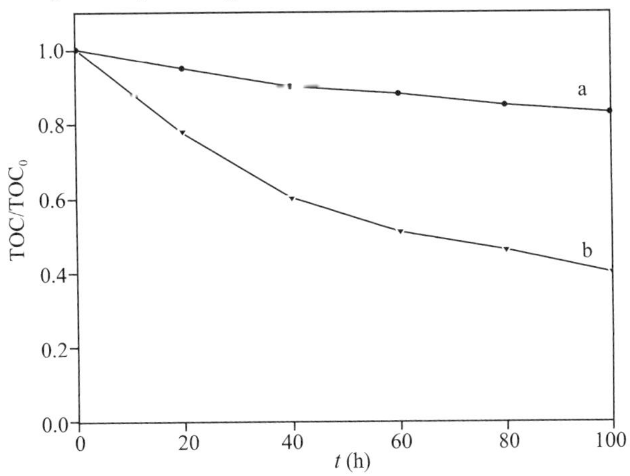

Fig. 10 TOC removal of (a) the hydrophobized TiO_2-treated wood, and (b) the hydrophobized Ag/TiO_2-coated wood during phenol photodegradation under visible light irradiation

remain complete, and the hydroxyl radicals produced in the process react with organic molecules from the sample surfaces or in the solution released from the sample surfaces, which leads to an initially slow photodegradation. Then, a fast degradation in second stage follows. For comparison, it can be observed that for the hydrophobized TiO_2-treated wood (Fig. 10a), after a fast degradation (40 h) follows by a rather slower second stage. Moreover, for the hydrophobized Ag/TiO_2-coated wood (Fig. 10b), TOC removal is greatly accelerated. Thus, combining Ag with TiO_2 further enhances photodegradation for organic pollutants than TiO_2 alone.

4 Conclusions

A robust superhydrophobic antibacterial Ag-TiO_2 composite film immobilized on wood substrate was fabricated successfully by in situ coating wood surfaces with TiO_2 microparticles and Ag nanoparticles followed by hydrophobization with (heptadecafluoro-1, 1, 2, 2-tetradecyl) trimethoxysilane. The hydrophobized Ag/TiO_2-coated wood not only exhibits robust superhydrophobicity and thermal stability, but also is capable of killing both Gram-negative and Gram-positive bacteria due to the bactericidal effects of Ag nanoparticles. Meanwhile, the hydrophobized Ag/TiO_2-coated wood could be used as a portable catalyst for phenol decomposition under visible light. The results indicate that the wood coated with the composite film contained Ag nanoparticles even after hydrophobization exhibits an enhanced photocatalytic activity in the decomposition of phenol driven by the visible light. Therefore, considering the multifunction of robust superhydrophobicity, antibacterial properties and photodecomposition of phenol under visible light, the hydrophobized Ag/TiO_2-coated wood will find potential applications in self-cleaning, antibacterial, photocatalytic and high-value-added wood products.

Acknowledgments

This work was financially supported by the National Natural Science Foundation of China (grant no. 31470584).

References

[1] Zeng J, Liu S, Cai J, et al. TiO_2 immobilized in cellulose matrix for photocatalytic degradation of phenol under weak UV light irradiation[J]. J. Phys. Chem. C, 2010, 114(17): 7806-7811.

[2] Hu W, Chen S, Li X, et al. In situ synthesis of silver chloride nanoparticles into bacterial cellulose membranes[J]. Mat. Sci. Eng. C-Mater., 2009, 29(4): 1216-1219.

[3] Lu Y, Xiao S, Gao R, et al. Improved weathering performance and wettability of wood protected by CeO_2 coating deposited onto the surface[J]. Holzforschung, 2014, 68(3): 345-351.

[4] Šegota S, Ćurković L, Ljubas D, et al. Synthesis, characterization and photocatalytic properties of sol-gel TiO_2 films [J]. Ceram. Int., 2011, 37(4): 1153-1160.

[5] Suwanboon S, Amornpitoksuk P, Muensit N. Dependence of photocatalytic activity on structural and optical properties of nanocrystalline ZnO powders[J]. Ceram. Int., 2011, 37(7): 2247-2253.

[6] Gan W, Gao L, Sun Q, et al. Multifunctional wood materials with magnetic, superhydrophobic and anti-ultraviolet properties[J]. Appl. Surf. Sci., 2015, 332: 565-572.

[7] Fan X, Fan J, Hu X, et al. Preparation and characterization of Ag deposited and Fe doped TiO_2 nanotube arrays for photocatalytic hydrogen production by water splitting[J]. Ceram. Int., 2014, 40(10): 15907-15917.

[8]Fujishima A, Rao T N, Tryk D A. Titanium dioxide photocatalysis[J]. J. Photoch. Photobio. C, 2000, 1(1): 1-21.

[9]Liang W, Li J, Jin Y. Photo-catalytic degradation of gaseous formaldehyde by TiO_2/UV, Ag/TiO_2/UV and Ce/TiO_2/UV [J]. Build. Environ., 2012, 51: 345-350.

[10]Prado Ó J, Veiga M C, Kennes C. Biofiltration of waste gases containing a mixture of formaldehyde and methanol[J]. Appl. Microbio. Biot., 2004, 65(2): 235-242.

[11]Hou X G, Huang M D, Wu X L, et al. Preparation and studies of photocatalytic silver-loaded TiO_2 films by hybrid sol-gel method[J]. Chemical Eng. J., 2009, 146(1): 42-48.

[12] Cozzoli P D, Fanizza E, Comparelli R, et al. Role of metal nanoparticles in TiO_2/Ag nanocomposite-based microheterogeneous photocatalysis[J]. J. Phys. Chem. B, 2004, 108(28): 9623-9630.

[13]Tao R H, Wu J M, Xiao J Z, et al. Conformal growth of ZnO on TiO_2 nanowire array for enhanced photocatalytic activity[J]. Appl. Surf. Sci., 2013, 279: 324-328.

[14]Suwanchawalit C, Wongnawa S, Sriprang P, et al. Enhancement of the photocatalytic performance of Ag-modified TiO_2 photocatalyst under visible light[J]. Ceram. Int., 2012, 38(6): 5201-5207.

[15]Khalid N R, Ahmed E, Hong Z, et al. Cu-doped TiO_2 nanoparticles/graphene composites for efficient visible-light photocatalysis[J]. Ceram. Int., 2013, 39(6): 7107-7113.

[16]Khalid N R, Ahmed E, Hong Z, et al. Graphene modified Nd/TiO_2 photocatalyst for methyl orange degradation under visible light irradiation[J]. Ceram. Int., 2013, 39(4): 3569-3575.

[17] Seery M K, George R, Floris P, et al. Silver doped titanium dioxide nanomaterials for enhanced visible light photocatalysis[J]. J. Photoch. Photobio. A, 2007, 189(2-3): 258-263.

[18]Su W, Wei S S, Hu S Q, et al. Preparation of TiO_2/Ag colloids with ultraviolet resistance and antibacterial property using short chain polyethylene glycol[J]. J. Hazard. Mater., 2009, 172(2-3): 716-720.

[19] Elahifard M R, Rahimnejad S, Haghighi S, et al. Apatite-coated Ag/AgBr/TiO_2 visible-light photocatalyst for destruction of bacteria[J]. J. Am. Chem. Soc., 2007, 129(31): 9552-9553.

[20]Khalil-Abad M S, Yazdanshenas M E. Superhydrophobic antibacterial cotton textiles[J]. J. Colloid Interf. Sci., 2010, 351(1): 293-298.

[21]Gao L, Gan W, Xiao S, et al. Enhancement of photo-catalytic degradation of formaldehyde through loading anatase TiO_2 and silver nanoparticle films on wood substrates[J]. RSC Adv., 2015, 5(65): 52985-52992.

[22]Xue C H, Chen J, Yin W, et al. Superhydrophobic conductive textiles with antibacterial property by coating fibers with silver nanoparticles[J]. Appl. Surf. Sci., 2012, 258(7): 2468-2472.

[23]Gao L, Lu Y, Zhan X, et al. A robust, anti-acid, and high-temperature-humidity-resistant superhydrophobic surface of wood based on a modified TiO_2 film by fluoroalkyl silane[J]. Surf. Coat. Tech., 2015, 262: 33-39.

[24]Wang S, Qian H, Hu Y, et al. Facile one-pot synthesis of uniform TiO_2-Ag hybrid hollow spheres with enhanced photocatalytic activity[J]. Dalton T., 2013, 42(4): 1122-1128.

[25]Andersson S, Serimaa R, Paakkari T, et al. Crystallinity of wood and the size of cellulose crystallites in Norway spruce (Picea abies) [J]. J. Wood Sci., 2003, 49(6): 531-537.

[26]Gao L, Zhan X, Lu Y, et al. pH-dependent structure and wettability of TiO_2-based wood surface[J]. Mater. Lett., 2015, 142: 217-220.

[27]Yuan T, Zhao B, Cai R, et al. Electrospinning based fabrication and performance improvement of film electrodes for lithium-ion batteries composed of TiO_2 hollow fibers[J]. J. Mater. Chem., 2011, 21(38): 15041-15048.

[28]Lionetto F, Del Sole R, Cannoletta D, et al. Monitoring wood degradation during weathering by cellulose crystallinity [J]. Materials, 2012, 5(10): 1910-1922.

[29]Jin C, Li J, Han S, et al. A durable, superhydrophobic, superoleophobic and corrosion-resistant coating with rose-

like ZnO nanoflowers on a bamboo surface[J]. Appl. Surf. Sci., 2014, 320: 322-327.

[30]Ouajai S, Shanks R A. Composition, structure and thermal degradation of hemp cellulose after chemical treatments [J]. Polym. Degrad. Stabil., 2005, 89(2): 327-335.

[31]Sarma B K, Pal A R, Bailung H, et al. A hybrid heterojunction with reverse rectifying characteristics fabricated by magnetron sputtered TiO_x and plasma polymerized aniline structure[J]. J. Phys. D: Appl. Phys., 2012, 45(27): 275401.

[32]Wang C, Yifeng E, Fan L, et al. CdS-Ag nanocomposite arrays: Enhanced electro-chemiluminescence but quenched photoluminescence[J]. J. Mater. Chem., 2009, 19(23): 3841-3846.

中文题目：超疏水、抑菌性木材基银-二氧化钛复合薄膜在可见光照射下光催化降解苯酚

作者：高丽坤，甘文涛，肖少良，詹先旭，李坚

摘要：通过两步法，水热合成和银镜反应，在木材上制备银-二氧化钛异质结构。研究结果表明，相比于单纯的二氧化钛负载的木材，银-二氧化钛复合薄膜负载的木材具有优异的可见光照射光催化降解苯酚的性能。银-二氧化钛复合薄膜负载的木材经过十七氟葵基三甲氧基硅烷的进一步疏水处理，赋予了木材超疏水和自清洁性能。改性的木材对革兰氏阳性菌和革兰氏阴性菌都展现出良好的抑菌性。此木材表明的多功能性薄膜具有良好的光降解有机污染物和超疏水性能，在自清洁领域具有重要应用。

关键词：银-二氧化钛复合薄膜；抑菌；光降解；超疏水；木材

Negative Oxygen Ions Production by Superamphiphobic and Antibacterial TiO_2/Cu_2O Composite Film Anchored on Wooden Substrates*

Likun Gao, Zhe Qiu, Wentao Gan, Xianxu Zhan, Jian Li, Tiangang Qiang

Abstract: According to statistics, early in the 20th century, the proportion of positive and negative air ions on the earth is 1 : 1.2. However, after more than one century, the equilibrium state of the proportion had an obvious change, which the proportion of positive and negative air ions became 1.2 : 1, leading to a surrounding of positive air ions in human living environment. Therefore, it is urgent to adopt effective methods to improve the proportion of negative oxygen ions, which are known as "air vitamin". In this study, negative oxygen ions production by the TiO_2/Cu_2O-treated wood under UV irradiation was first reported. Anatase TiO_2 particles with Cu_2O particles were doped on wooden substrates through a two-step method and further modification is employed to create remarkable superamphiphobic surface. The effect of Cu_2O particles dopant on the negative oxygen ions production of the TiO_2-treated wood was investigated. The results showed that the production of negative oxygen ions was drastically improved by doping with Cu_2O particles under UV irradiation. The wood modified with TiO_2/Cu_2O composite film after hydrophobization is imparted with superamphiphobicity, antibacterial actions against Escherichia coli, and negative oxygen ions production under UV irradiation.

Keywords: Negative oxygen ion; TiO_2/Cu_2O composite film; Superamphiphobic; Antibacterial; Wood

1 Introduction

Nowadays, with the development of social and economic and the improvement of living standards, the human health awareness is also growing. Especially after the mechanism of the action of negative oxygen ions and its effects have been increasingly understood, the negative oxygen ions products will be developed as a kind of functional products[1-3]. As the oxygen molecule captured an electron, the oxygen molecule is defined as negative oxygen ion (O_2^-)[4,5]. The negative oxygen ions are also known as "air vitamin" due to its important meanings to human life activities. Meanwhile, the effects of negative oxygen ions on human health and ecological environment have been validated by domestic and foreign medical experts through clinical practices[6,7]. The researchers have found that negative oxygen ions could combine with bacteria, dust, smoke and some other positively charged particles in the atmosphere, and fall to the ground after gathering into balls, which would lead to the sterilization and the elimination of peculiar smell, that is, a purification of

* 本文摘自 Scientific Reports, 2016, 6: 26055-26064.

atmosphere. And the biological effects of superoxide (O_2^-) are considered in relation to the reactions leading to cell death and to the metabolism of certain endogenous compounds[8]. At the same time, O_2^- production probably take place by surface ionization, which is identified as a viable ionization technique to have the potential to meet the requirements concerning ionization efficiency for the energy range of 10 eV to 1 keV within the limitations imposed by the resources (space, weight, powder, light, ect.) available on a proper surface[5]. Therefore, artificial method to achieve negative oxygen ions include UV irradiation, anion incentive, thermionic emission, corona discharge, charge separation, high - pressure water injection, tourmaline and other natural ores, and so on. In this paper, we adopt anion incentive method, which selected suitable photocatalyst irradiated by light to excite energy, and the excitation energy ionized the oxygen molecule to produce negative ions. At the same time, O_2^- production probably takes place under the natural conditions as a result of gas ionization in the atmosphere.

So far, in most applications of photocatalyst, an *n* - type TiO_2 semiconductor has been used[9-11]. Semiconductor materials are materials whose valence band and conduction band are separated by an energy gap or band gap. When the semiconductor material absorbs photons with energy equal or larger than its band gap, electrons in the valence band can be excited and migrate to the conduction band. For example, when TiO_2 semiconductor materials with the forbidden bandgap of 3. 2 eV are irradiated by light consisting of wavelengths shorter than 415 nm, the electron-hole pairs are produced after the migration of electrons. Then, the electron and hole react with water and oxygen, it would produce hydroxyl radicals (•OH) and superoxide (O_2^-). However, the recombination of the electron and the hole would happen, that is, the superoxide (O_2^-) is reduced. Therefore, in order to suppress the recombination of the electron and the hole and enhance the migration of charges, much attention has been focused on the modification of TiO_2 by adding metal ions or oxides[12-14]. Among the metal ions and oxides, Cu_2O is a nonstoichiometric p - type semiconductor with a direct forbidden bandgap of about 2. 0 eV, whose component elements are inexpensive and abundantly available[15,16]. And doping Cu_2O with TiO_2 forming an n-p heterostructure is an effective way to enhance the photocatalytic activity of TiO_2 photocatalyst.

As a natural material, wood is one of the most versatile and widely used structural materials for indoor and outdoor applications because of its many attractive properties, such as its low density, thermal expansion, renewability, and aesthetics[17]. Especially for indoor applications, such as floorboards and furniture, however, the process technology with the using of urea formaldehyde resin causes serious air pollution leading to human disease[18]. In addition, negative oxygen ions have a lot of advantages for environment including dust extraction, bacteriostasis and deodorization. Therefore, negative - ion wooden materials may have promising prospect in the future development of multifunctional materials.

Based on above considerations, in this paper, we imagine a TiO_2/Cu_2O composite coating on a wooden substrate, and the further modification with low surface energy of (heptadecafluoro - 1, 1, 2, 2 - tetradecyl) trimethoxysilane is employed to create remarkable superamphiphobic surface. The wood decorated with TiO_2/Cu_2O composite film after hydrophobization might also be imparted with superamphiphobicity, antibacterial actions against Escherichia coli, and negative oxygen ions production under UV irradiation. Moreover, the mechanism of the negative oxygen ions production on TiO_2/Cu_2O - treated wood under UV irradiation was discussed in the paper.

2 Materials and methods

2.1 Materials

All chemicals supplied by Shanghai Boyle Chemical Company, Limited were of analytical reagent-grade quality and used without further purification. Deionized water was used throughout the study. Wood blocks of 20 mm (R)×20 mm (T)×30 mm (L) were obtained from the sapwood sections of poplar wood (*Populus ussuriensis* Kom), which is one of the most common tree species in the northeast of China. The wood specimens were oven-dried (24 h, 103±2 ℃) to constant weight after ultrasonically rinsing in deionized water for 30 min, and oven-dried weight were determined.

2.2 Preparation of anatase TiO_2 film on the wood surface

Ammonium fluorotitanate (0.4 M) and boric acid (1.2 M) were dissolved in distilled water at room temperature under vigorous magnetic stirring. Then, a solution of 0.3 M hydrochloric acid was added until the pH reached approximately 3. Then, 75 mL of the adjusted solution was transferred into a 100 mL Teflon container. Wood specimens were subsequently placed into the above reaction solution, separately. The autoclave was sealed and maintained at 90 ℃ for 5 h, then allowed to naturally cool to room temperature. Finally, the prepared samples were removed from the solution, ultrasonically rinsed with deionized water for 3 times, and dried at 45 ℃ for more than 24 h in vacuum. Thus, the TiO_2-treated wood was obtained.

2.3 Coating the TiO_2-treated wood surface with Cu_2O particles

$Cu(CH_3COO)_2 \cdot H_2O$ (0.1 M) was dissolved in deionized water, then D-glucose (0.1 M) and PVP (K-30, 0.3 g) were added under continued stirring. After 1 h, the above solution was transferred into a 50 mL Teflon-lined autoclave, and the TiO_2-treated wood was immersed in the solution. The autoclave was kept in a vacuum oven at 180 ℃ for 2 h. After cooling to room temperature in air, the wood loaded with Cu_2O was removed, washed several times with deionized water and ethanol, and dried at 45 ℃ for more than 24 h in vacuum. Thus, the TiO_2/Cu_2O-treated wood was obtained.

2.4 Hydrophobization of wood surfaces with FAS-17

A methyl alcohol solution of (heptadecafluoro-1,1,2,2-tetradecyl) trimethoxysilane $CF_3(CF_2)_7CH_2CH_2Si(OCH_3)_3$, hereafter denoted as FAS-17) was hydrolyzed by the addition of a 3 fold-molar excess of water at room temperature. Then, 75 mL of the adjusted solution was transferred into a 100 mL Teflon container. The TiO_2/Cu_2O-treated wood samples were subsequently placed into the above reaction solution. The autoclave was sealed and maintained at 75 ℃ for 5 h, then allowed to naturally cool to room temperature. Subsequently, the samples were washed with ethyl alcohol to remove any residual chemicals and allowed to dry in air at room temperature, and they were then dried at 45 ℃ for more than 24 h in vacuum. Thus, a superamphiphobic wood surface was obtained.

2.5 Characterization

The morphology of the wood surfaces was observed through field emission scanning electron microscopy (SEM, Quanta 200, FEI, Holland) operating at 12.5 kV. The transmission electron microscopy (TEM) experiment was performed on a Tecnai G20 electron microscope (FEI, USA) with an acceleration voltage of 200 kV. Carbon-coated copper grids were used as the sample holders. X-ray diffraction (XRD, Bruker D8 Advance, Germany) was employed to analyze the crystal structures of all samples applying graphite monochromatic with Cu $K\alpha$ radiation ($\lambda = 1.5418$ Å) in the 2θ range from 5° to 80° and a position-sensitive detector using a step size of 0.02° and a scan rate of 4° min^{-1}. FTIR spectra were obtained on KBr tablets and recorded using a Magna-IR 560 spectrometer (Nicolet) with a resolution of 4 cm^{-1} by scanning the region between 4000 and 400 cm^{-1}. Water contact angles (WCAs) and oil (hexadecane) contact angles (hereafter defined as OCAs) were measured on an OCA40 contact angle system (Dataphysics, Germany) at room temperature. In each measurement, a 5 μL droplet of deionized water was injected onto the surfaces of the wood samples and the contact angles were measured at five different points of each sample. The final values of the contact angles were obtained as an average of five measurements. Further evidence for the composition of the product was inferred from the X-ray photoelectron spectroscopy (XPS, Thermo ESCALAB 250XI, USA), using an ESCALab MKII X-ray photoelectron spectrometer with Mg-$K\alpha$ X-rays as the excitation source. The flat band potentials of the samples were evaluated with a standard electrochemical workstation (CHI660C) in the three-electrode configuration. The counter electrode was a Pt wire and the reference electrode was saturated calomel electrode (SCE). 1.0 M Na_2SO_4 solution buffered at pH 4.9 was used as the electrolyte. In the electrochemical impedance spectroscopy (EIS) measurement, the fixed frequency of the M-S plots was 10^3 Hz.

2.6 Antibacterial tests

Antimicrobial tests were conducted by the bacterial inhibition ring method (agar plate diffusion test/CEN/TC 248 WG 13) and the reduction of bacterial growth test (EN ISO 20743: 2007 Transfer Method). The antibacterial activity of the hydrophobized TiO_2/Cu_2O-treated wood was evaluated against Escherichia coli (ATCC 25923) and Staphylococcus aureus (ATCC 25922). The evaluation of the test was as the followings. A mixture of nutrient broth and nutrient agar in 1 L distilled water at pH 7.2 as well as the empty Petri plates were autoclaved. The agar medium was then cast into the Petri plates and cooled in laminar airflow. Approximately 10^5 colony-forming units of each bacterium were inoculated on plates, and then each wood samples was planted onto the agar plates. All the plates were incubated at 37 ℃ for 24 h, following which the zone of inhibition was measured.

2.7 Negative oxygen ions production tests under UV irradiation

The experiments were performed in an obturator with the size of 500 mm (L)×300 mm (W)×300 mm (H). A UV light with a wavelength of 365 nm and light intensity of approximately 1.5 mW · cm^{-2} installed in the open central region was used. Negative oxygen ions concentration (D_a) in the obturator was determined by the AIC1000 air anion counter with the resolution ratio of 10 ions · cm^{-3} at 5 min intervals. Initially, the air anion counter must be re-zeroed and keep the numerical value constant for 5 seconds before each test. Then, the UV light was turned on to irradiate the sample. Each set of negative oxygen ions production under UV irradiation lasted for 60 min.

3 Results and discussion

Field-emission scanning electron microscopy (FE-SEM) images in Fig. 1 show microstructural features of the procedure of the TiO_2/Cu_2O composite film deposited on the wood substrates. For the pristine wood (Fig. 1a), the surface is smooth and clean, but there are a few of pits, which indicates that the poplar wood is a type of heterogeneous and porous material. After low-temperature hydrothermal synthesis, the wood surface is covered by TiO_2 microparticles with average diameter of 2.25 μm, as shown in Fig. 1b. Low magnification image in Fig. 1c shows that the TiO_2/Cu_2O composite layer is very dense, and the Cu_2O particles spread all over the top surface of the TiO_2-treated wood. The composite film replicates the grain structures with TiO_2 particles and Cu_2O particles. The transmission electron microscopy (TEM) image of the TiO_2/Cu_2O-treated wood in Fig. 1d shows that the film is formed by rough grains with average diameter of 0.60 μm. It is worth pointing out that the rough grains would significantly increase the water-repellent properties of the surfaces[19].

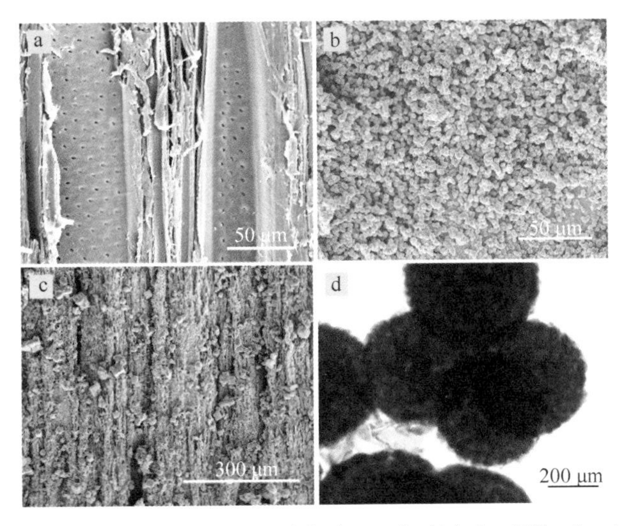

Fig. 1 SEM images of the surfaces of (a) the original wood, (b) the TiO_2-treated wood, and (c) the TiO_2/Cu_2O-treated wood. (d) TEM image of the TiO_2/Cu_2O-treated wood

Fig. 2a shows the XRD patterns of the original wood, the TiO_2-treated wood and the TiO_2/Cu_2O-treated wood. For the original wood, there is no peak except the diffraction peaks at 14.8° and 22.5° belonging to the (101) and (002) crystal planes of cellulose in the wood, which could be observed in both the spectrum of the original wood and the treated wood samples[20]. It can be found that the diffraction peaks are well indexed to the standard diffraction pattern of anatase phase TiO_2 (JCPDS file No. 21-1272) and the cuprite phase of Cu_2O (JCPDS file No. 05-0667)[21], indicating that the present synthesis strategy successfully achieves TiO_2/Cu_2O heterostructures with high crystallinity on wood substrate. After deposition with TiO_2 particles, diffraction peaks at 25.2°, 38.0°, 47.8°, 54.2°, 62.5°, 68.8° and 74.9°, can be perfectly identified to (101), (004), (200), (211), (204), (116) and (215) crystal planes of anatase TiO_2, respectively[22]. The red curve in Fig. 2a shows that all of the new diffraction peaks of the TiO_2/Cu_2O-treated wood center at $2\theta = 36.4°$, 42.3°, 61.6° and 73.8°, except the diffraction peaks of TiO_2, are agree with (111), (200), (220) and (311) planes of pure cuprite Cu_2O[17].

Fig. 2b shows FTIR spectra of the pristine wood and the hydrophobized TiO_2/Cu_2O-treated wood. The band at 3348 cm^{-1} corresponding to the stretching vibrations of OH groups in the wood (black curve) shifts to larger wavenumbers of 3413 cm^{-1} (green curve), and the intensity comparatively decreases, indicating that the

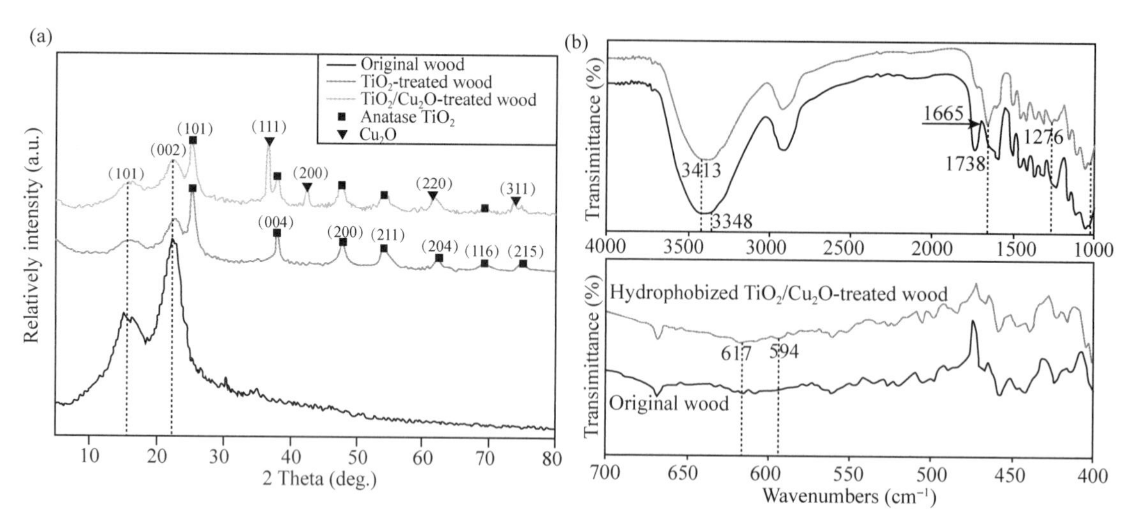

Fig. 2 (a) XRD patterns of the original wood, the TiO_2-treated wood, and the TiO_2/Cu_2O-treated wood. (b) FTIR spectra of the original wood and the hydrophobized TiO_2/Cu_2O-treated wood

hydrophilic groups (−OH) of the wood decreases in the hydrophobized TiO_2/Cu_2O-treated wood[23]. The band at 1738 cm^{-1} is attributed to the C = O stretching vibrations of carboxylic acid and acetyl group in hemicelluloses (black curve), while the band at 1665 cm^{-1} assigned to an conjugated carbonyl in lignin is observed (green curve). In Fig. 2b, the existence of C−F bonds in CF, CF_2 or CF_3 are located at 1276 cm^{-1} and 594 cm^{-1},[24] and the band at 617 cm^{-1} corresponds to stretching vibration of Cu(Ⅰ)−O bond (optically active lattice vibration in the oxide)[25,26]. That is, the as-synthesized TiO_2/Cu_2O heterostructures on the wood surface contain Cu_2O without CuO.

Fig. 3 demonstrates the variation (from region A to region B) of the water contact angles and oil (hexadecane) contact angles of the original wood, the TiO_2-treated wood, TiO_2/Cu_2O-treated wood, and the hydrophobized TiO_2/Cu_2O-treated wood. The original wood surface presents hydrophilicity with the WCA of 58.7° and superoleophilicity with the OCA of 0°. And the TiO_2-treated wood possesses a superamphiphilic surface with both the WCA and the OCA of 0°. After coated with Cu_2O particles, the wood surface converts into hydrophobic one with the WCA of 120.1°, while remains superoleophilic with the OCA of 0°. According to the literatures, higher treatment temperature levels had a likely round-shape of water droplets on the surface, however it is more flat shape on the original sample since the liquid rapidly absorbed and lead to small contact angle[27-29]. Surface of wood becomes smoother with increasing temperatures of heat treatment having better wettability expressed with larger contact angle value. Hemicellulose is greatly affected during the heat treatment and the organic acid is released from the degradation of the hemicellulose, which influences the cross-linking reduction in hydroxyl groups (−OH). Hydrophobicity of wood will also increase with decreasing of the (−OH) groups. Furthermore, it is known that the larger roughness and the lower surface energy of surfaces would lead to the hydrophobicity. After coating with Cu_2O particles, the roughness of the surface became larger to produce a hydrophobic surface.

After further hydrophobization with FAS-17, the hydrophobicity of the wood surface is raised to superamphiphobicity, while the WCA and the OCA reach 158.6° and 154.3°, respectively. The results are in

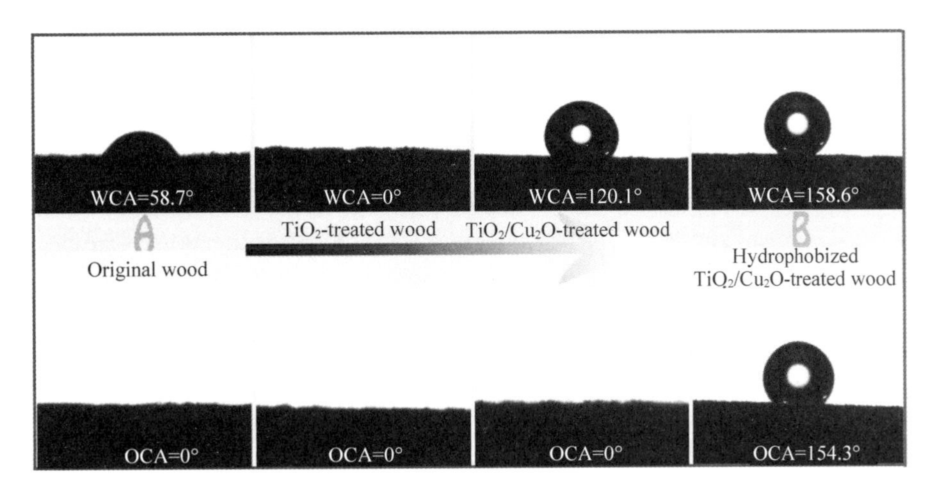

Fig. 3 WCAs and OCAs of the original wood, the TiO_2-treated wood, the TiO_2/Cu_2O-treated wood, and the hydrophobized TiO_2/Cu_2O-treated wood

accordance with the FTIR results that the hydrophilic groups (-OH) of the wood decease in the hydrophobized TiO_2/Cu_2O-treated wood, that is, the liquid-repellent properties of wood surfaces are significantly increased.

Fig. 4 ascertains more detailed information concerning the elemental and chemical state of the TiO_2-treated wood and the TiO_2/Cu_2O-treated wood examined by using XPS. The fully scanned spectra (Fig. 4a) shows that the O, Ti and C elements exist on the surface of the pure TiO_2-treated wood, while the Cu, O, Ti and C elements exist on the surface of the TiO_2/Cu_2O-treated wood. The C element can be ascribed to the wood substrate or the adventitious carbon-based contaminant.

From the peaking-fitting for the Ti 2p (Fig. 4b) in the TiO_2-treated wood and the TiO_2/Cu_2O-treated wood, two peaks at binding energies of 459. 0 eV and 464. 8 eV corresponding to the Ti $2p_{3/2}$ and Ti $2p_{1/2}$ peaks can be observed. And the gap of 5. 8 eV between the two peaks indicates the existence of the Ti^{4+} oxidation state[30]. However, the peak position for Ti 2p in the TiO_2/Cu_2O-treated wood shifts to a higher binding energy band than that in the pure TiO_2-treated wood. This confirms a lower electron density of the Ti atoms after coated with Cu_2O particles, and there is a strong interaction between Cu_2O and TiO_2 in the TiO_2/Cu_2O-treated wood. According to the standard binding energy of Ti $2p_{3/2}$, one located at 459. 5 eV, usually attributed to a Ti^{4+} species, and the other located at 457. 7 eV, assigned to a Ti^{3+} species. The binding energy of Ti $2p_{3/2}$ in the pure TiO_2-treated wood and the TiO_2/Cu_2O-treated wood without UV light irradiation is 459. 0 eV and 459. 2 eV, respectively. The binding energy of Ti $2p_{3/2}$ shown in Fig. 4c is 458. 8 eV, whose peak is much broader than that for the pure TiO_2-treated wood and the TiO_2/Cu_2O-treated wood without UV light irradiation. The spectrum of Ti $2p_{3/2}$ of the TiO_2/Cu_2O-treated wood after UV light irradiation (Fig. 4c) is simulated with Gaussian simulation. The fitting peak at 458. 2 eV is attributed to Ti^{3+}, and that at 459. 0 eV is attributed to Ti^{4+}, respectively[31]. Thus, Ti^{3+} does exist in the TiO_2/Cu_2O-treated wood when it is irradiated under UV light. Certainly, the presence of Ti^{3+} oxide with narrow band gap and energy level located between the valence band and the conduction band of TiO_2, may be advantageous to the higher photocatalytic activity of the TiO_2/Cu_2O heterostructures[32].

As shown in Fig. 4d, it can be found that two main XPS peaks at 952. 6 eV and 932. 9 eV, which can be

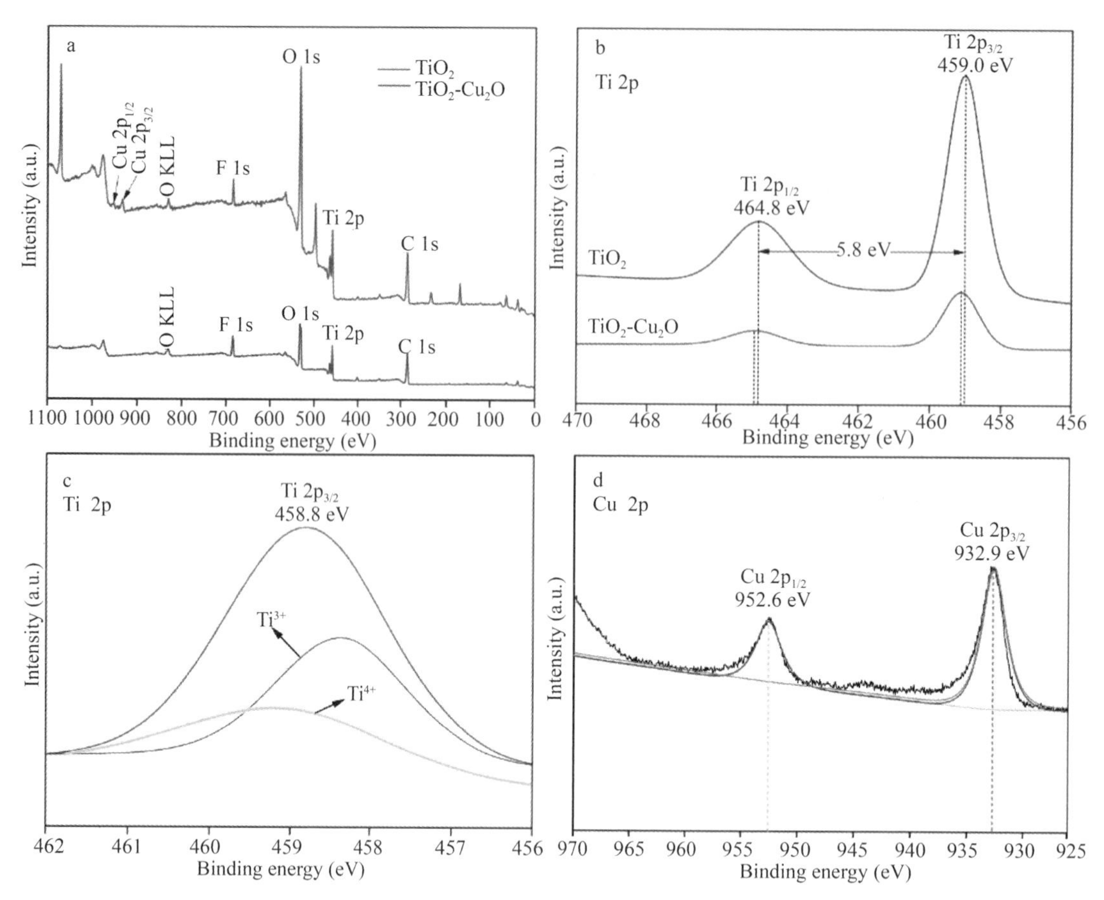

Fig. 4 (a) Survey scan and (b) Ti 2p XPS spectra of the TiO_2-treated wood and the TiO_2/Cu_2O-treated wood, (c) peaking-fitting results of Ti 2p XPS spectra of the TiO_2/Cu_2O-treated wood after UV light irradiation, (d) Cu 2p XPS spectra of the TiO_2/Cu_2O-treated wood

attributed to the double peaks for Cu $2p_{3/2}$ and Cu $2p_{1/2}$ of Cu_2O, respectively, indicating that the oxidation state of Cu is +1[25].

To investigate the band alignment and thus the charge transfer between TiO_2 and Cu_2O on the TiO_2/Cu_2O-treated wood, the Mott-Schottky (M-S) plots of the pure TiO_2-treated wood and the TiO_2/Cu_2O-treated wood (Fig. 5) based on the following equation[33]:

$$C^{-2}=2(V-V_{fb}-k_BT/e)/\varepsilon_0\varepsilon_r eN_A$$

are used to determine their flat band potential V_{fb} and charge carrier density N_A. Here, C is the space-charge capacitance of the semiconductor; ε_0 is the permittivity in vacuum (ε_0 = 8.85 × 10^{-14} F/cm); ε_r is the dielectric constant; V is the applied potential; T is the temperature and k_B is Boltzmann constant (k_B = 1.38× 10^{-23} J/K). From the linear fit of C^{-2} versus V, V_{fb} for the pure TiO_2-treated wood and the TiO_2/Cu_2O-treated wood is obtained to be −0.48 eV and −0.57 eV vs. SCE, respectively. Therefore, it can be observed a negative shift of the flat band potential in Fig. 5b that indicates the sample after coated with Cu_2O particles needs to cross smaller potential barrier and is beneficial to the production of photogenerated electrons and holes.

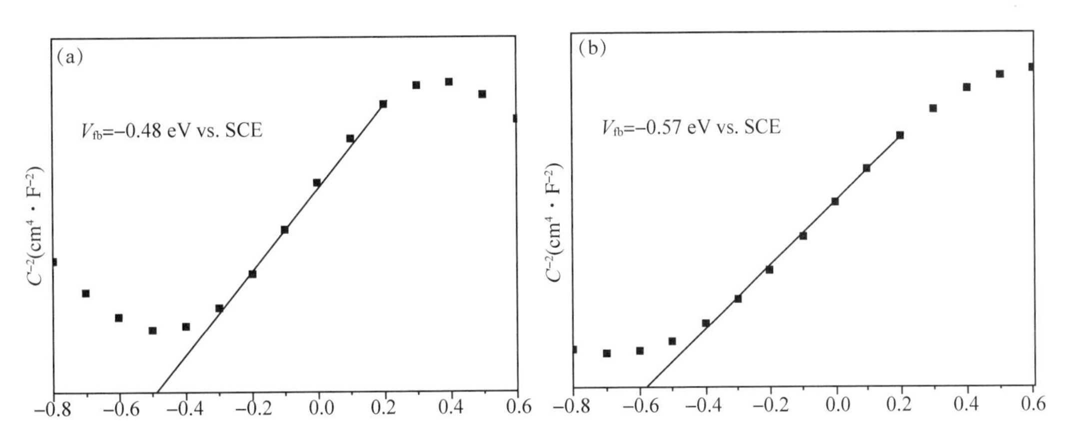

Fig. 5 Mott-Schottky plots of (a) the pure TiO_2-treated wood and (b) the TiO_2/Cu_2O-treated wood

Considering the respective band gaps of 3.2 eV and 2.0 eV for TiO_2 and Cu_2O, the band alignment diagram of the TiO_2/Cu_2O heterostructures is shown in Fig. 6a. When TiO_2 and Cu_2O came into contact, an n-p junction formed at their interface. It is well known that the conduction band of TiO_2 is about -0.2 eV, and the potential of Cu_2O conduction band is -1.4 eV. The photogenerated electrons from the conduction band of Cu_2O were captured by Ti^{4+} ions in TiO_2, and Ti^{4+} ions were further reduced to Ti^{3+} ions. The Ti^{3+} ions have a long lifetime and bear the photogenerated electrons as a form of energy[34]. The electron-transfer process is shown in Eqs. (1) and (2):

$$Cu_2O+h\nu \rightarrow h^+_{vb}+e^-_{cb} \quad (1)$$

$$e^-_{cb}+Ti^{4+} \rightarrow Ti^{3+} \quad (2)$$

While on the TiO_2/Cu_2O heterostructures, the ·OH and H^+ came from the indirect oxidization of the adsorbed water or hydroxyl groups at the positive holes[35]. Meanwhile, the negative oxygen ions (O_2^-) are produced through the reaction between the negative electron and the oxygen molecules. Moreover, the energy stored in the Ti^{3+}, may promote a trapping of electrons leading to produce negative oxygen ions under the natural conditions[34]. Under UV irradiation, the negative oxygen ions production process (Fig. 6b) is shown in Eqs. (3)-(5):

$$TiO_2/Cu_2O+h\nu \rightarrow h^+ +e^- \quad (3)$$

$$h^+ +OH^-/H_2O \rightarrow \cdot OH+H^+ \quad (4)$$

$$e^- +O_2 \rightarrow O_2^- \quad (5)$$

In order to investigate the negative oxygen ions production in the obturator when UV irradiated the samples surfaces, the negative oxygen ions concentrations in the obturator for every 5 minutes are depicted in Fig. 7a. Firstly, the initial negative oxygen ions concentrations in the obturator for the three samples are 0 ions/cm^3. It is obvious that, for both the TiO_2-treated wood and the hydrophobized TiO_2/Cu_2O-treated wood, the negative oxygen ions concentrations are dramatically increased, after 35 or 40 minutes irradiation, the concentrations is decreased but still superior to that of the original wood. The blame for falling lays on that the negative oxygen ions will not increase indefinitely, and the negative oxygen ions concentration in the nature locates between 700 and 4000 ions/cm^3, which may ascribed to the neutralization of the positive and negative ions, and the inhibition effect of the negative ions in the atmosphere to maintain a level when the negative ions

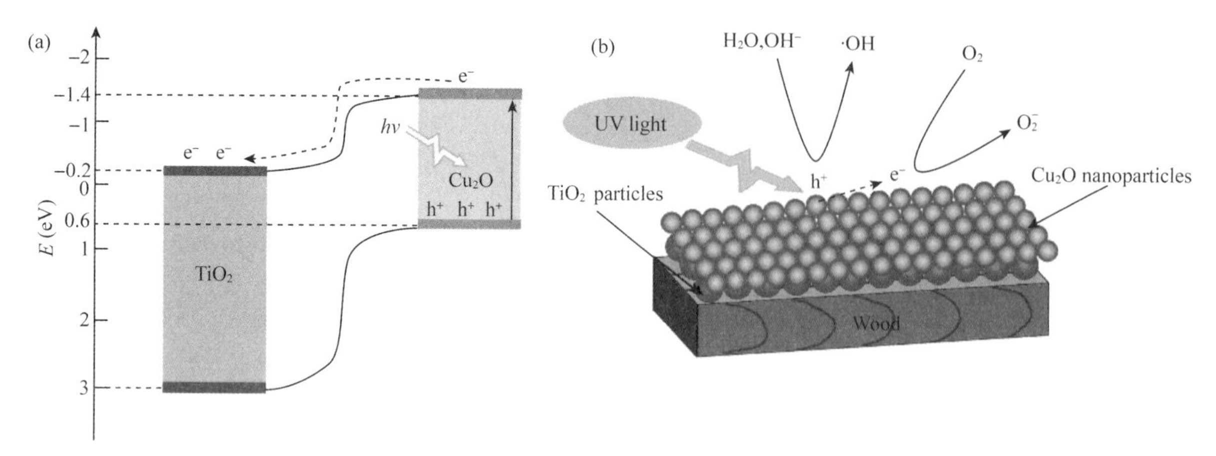

Fig. 6 (a) Illustration of the photocatalysis mechanism at the TiO_2/Cu_2O heterostructures, and (b) possible scheme for negative oxygen ions production on the TiO_2/Cu_2O-treated wood under UV irradiation

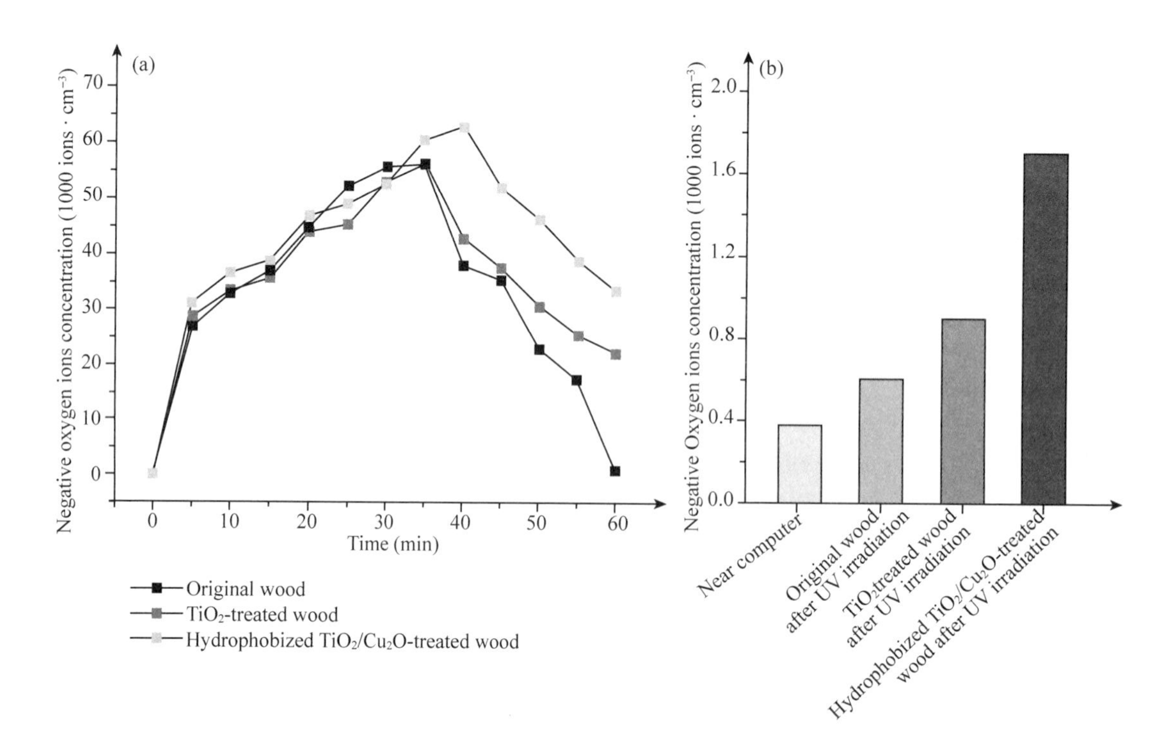

Fig. 7 (a) Negative oxygen ions concentrations in the obturator when UV irradiated the samples surfaces. (b) Negative oxygen ions concentration in the environment near computer and negative oxygen ions concentrations in the obturator after UV irradiated the samples surfaces for 60 minutes

concentration saturates to a certain extent.

Fig. 7b presents the negative oxygen ions concentration in the environment near computer and the negative oxygen ions concentrations in the obturator after UV irradiated the samples surfaces for 60 minutes. The negative oxygen ions concentration in the environment near computer is achieved by calculating the average of the total amount concentration of every 5 minutes in 60 minutes. As for the environment near computer, the negative oxygen ions concentration in the atmosphere is 380 ions · cm^{-3}, which may be the reason that causes physiological barrier and severe illness. And for the original wood and the TiO_2-treated wood after irradiation, the negative oxygen ions concentrations in the obturator are 600 and 900 ions · cm^{-3}, respectively. Whereas for the hydrophobized TiO_2/Cu_2O-treated wood, it exhibits a much higher concentrations of 1700 ions · cm^{-3}, indicating the photocatalytic activity and negative oxygen ions production efficiency of the hydrophobized TiO_2/Cu_2O-treated wood is significantly improved due to the loading of the Cu_2O particles onto the surface of the TiO_2-treated wood. According to the regulation of World Health Organization, when the concentration of negative oxygen ions in the air is more than 1000 - 1500 ions · cm^{-3}, the air is regarded as the fresh air[36]. Thus, the air including negative oxygen ions produced by the hydrophobized TiO_2/Cu_2O-treated wood under UV irradiation is up to the standard of the fresh air.

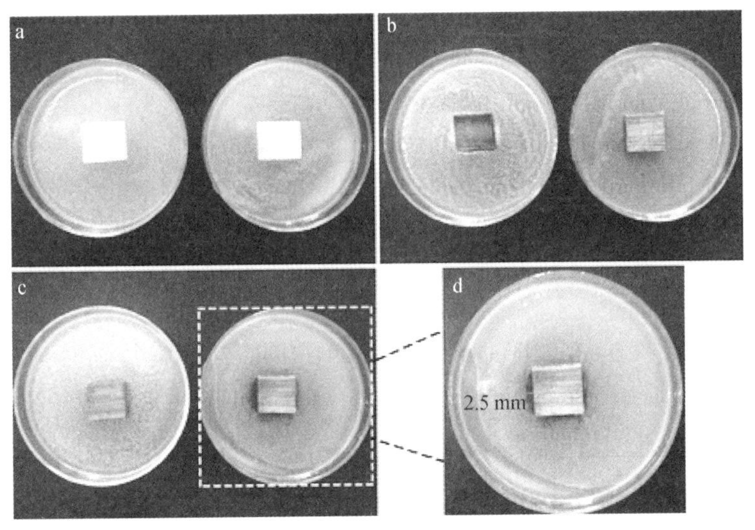

Fig. 8 Antibacterial activity of (a) the TiO_2-treated wood, (b) the Cu_2O-treated wood and (c) the hydrophobized TiO_2/Cu_2O-treated wood in Staphylococcus aureus and Escherichia coli, respectively, and (d) the magnified picture of the hydrophobized Ag/TiO_2-coated wood in Escherichia coli

Antibacterial activities of wood samples are determined in terms of inhibition zone formed on agar medium. Fig. 8a and b present that the TiO_2-treated wood and the Cu_2O-treated wood, which are used as control groups, do not show any antibacterial activity for both Escherichia coli and Staphylococcus aureus. In Fig. 8c, it could be seen that for Escherichia coli, the hydrophobized TiO_2/Cu_2O-treated wood placed on the bacteria-inoculated surfaces kill all the bacteria under and around them, while it has no antibacterial activity for Staphylococcus aureus. It can be observed that the distinct zones of inhibition (clear areas with no bacterial growth) around the wood samples for Escherichia coli (Fig. 8d). For Escherichia coli, the width of the inhibition zone around the hydrophobized TiO_2/Cu_2O-treated wood is approximately 2.5 mm. The observed zone of inhibition is a result of the performance of the negative oxygen ions production of the hydrophobized TiO_2/Cu_2O film anchored on the wood surface into the surrounding aqueous medium, which further proves that the hydrophobized TiO_2/Cu_2O-treated wood could produce negative oxygen ions not only under UV light but also under the natural conditions. In other words, the presences of the inhibition zone clearly indicate that the mechanism of the biocidal action of the wood is owing to the effect of negative oxygen ions.

4 Conclusions

The enhanced photocatalytic properties and excellent multifunctional performances of the hydrophobized TiO_2/Cu_2O-treated wood were probably attributed to the following four points: (a) The anatase TiO_2 particles settled onto the wood surfaces and formed a dense film, which offered large n-p heterojunction interface area when in combination with Cu_2O film. The treatment at high temperature and the rough TiO_2/Cu_2O grains would significantly increase the water-repellent properties of the surfaces; (b) After modification with FAS-17, the amphiphilic wood surface transformed into superamphiphobic one with the WCA of 158.6° and OCA of 154.3°; (c) The step-wise energy band structure facilitated the photogenerated electrons from the conduction band of Cu_2O were captured by Ti^{4+} ions in TiO_2 while promoted Ti^{4+} ions further to reduce to Ti^{3+} ions. The presence of Ti^{3+} ions led to a higher photocatalytic activity and even extended the photo-response from UV to the visible light. Moreover, the energy stored in the Ti^{3+}, may promote a trapping of electrons leading to produce negative oxygen ions under the natural conditions. ; (d) The introduction of Cu_2O significantly improved the photocatalysis efficiency of the TiO_2-treated wood, so that the air including negative oxygen ions produced by the hydrophobized TiO_2/Cu_2O-treated wood after UV irradiation is up to the standard of the fresh air; (e) Compared with the pure TiO_2-treated wood and the pure Cu_2O-treated wood, the hydrophobized TiO_2/Cu_2O-treated wood possesses antibacterial activity for Escherichia coli, which may be due to the biocidal action of the negative oxygen ions.

In brief, the TiO_2/Cu_2O n-p heterostructure was successfully fabricated anchored on the wood surface, and after further modification with FAS-17 the wood surface transformed into superamphiphobic. The results demonstrated the doped Cu_2O particles resulted in a higher photocatalytic activity, which the final produced negative oxygen ions after UV irradiation was up to the standard of the fresh air. Moreover, the presence of Ti^{3+} ions trapped the electrons leading to produce negative oxygen ions under the natural conditions, so that the hydrophobized TiO_2/Cu_2O-treated wood showed a good antibacterial activity for Escherichia coli. Moreover, the excellent superamphiphobicity, antibacterial activity and ability of negative oxygen ions production suggested that such heterostructured film anchored on wood surface is a potential candidate for advanced multifunctional materials, maybe for the extended field of medical or environment purification.

Acknowledgments

This work was financially supported by the National Natural Science Foundation of China (grant no. 31470584).

References

[1]Hawkins L H, Barker T. Air ions and human performance[J]. Ergonomics, 1978, 21(4): 273-278.

[2]Hedge A, Collis M D. Do negative air ions affect human mood and performance? [J]. Ann. Occup. Hyg., 1987, 31(3): 285-290.

[3]Sakoda, A. et al. A comparative study on the characteristics of radioactivities and negative air ions originating from the minerals in some radon hot springs. Appl. Radiat. Isotopes, 65, 50-56 (2007).

[4]Goldstein N I, Goldstein R N, Merzlyak M N. Negative air ions as a source of superoxide[J]. Int. J. Biometeorol., 1992, 36(2): 118-122.

[5]Nagato K, Matsui Y, Miyata T, et al. An analysis of the evolution of negative ions produced by a corona ionizer in air

[J]. Int. J. Mass Spectrom., 2006, 248(3): 142-147.

[6]Sirota T V, Safronova V G, Amelina A G, et al. The effect of negative air ions on the respiratory organs and blood[J]. Biophysics, 2008, 53(5): 457-462.

[7]Tom G, Poole M F, Galla J, et al. The influence of negative air ions on human performance and mood[J]. Hum. Factors, 1981, 23(5): 633-636.

[8]Krueger A P, Smith R F, Go I G. The action of air ions on bacteria: I. Protective and lethal effects on suspensions of Staphylococci in droplets[J]. J. Gen. Physiol., 1957, 41(2): 359-381.

[9]Zeng J, Liu S, Cai J, et al. TiO_2 immobilized in cellulose matrix for photocatalytic degradation of phenol under weak UV light irradiation[J]. J. Phys. Chem. C, 2010, 114(17): 7806-7811.

[10]Zhang X, Hu C, Bai H, et al. Construction of self-supported three-dimensional TiO_2 sheeted networks with enhanced photocatalytic activity[J]. Sci. Rep., 2013, 3(1): 1-6.

[11]Hou H, Shang M, Wang L, et al. Efficient photocatalytic activities of TiO_2 hollow fibers with mixed phases and mesoporous walls[J]. Sci. Rep., 2015, 5(1): 1-9.

[12]Qiu B, Zhou Y, Ma Y, et al. Facile synthesis of the Ti^{3+} self-doped TiO_2-graphene nanosheet composites with enhanced photocatalysis[J]. Sci. Rep., 2015, 5(1): 1-6.

[13]Cheng C, Amini A, Zhu C, et al. Enhanced photocatalytic performance of TiO_2-ZnO hybrid nanostructures[J]. Sci. Rep., 2014, 4(1): 1-5.

[14]Shao X, Lu W, Zhang R, et al. Enhanced photocatalytic activity of TiO_2-C hybrid aerogels for methylene blue degradation[J]. Sci. Rep., 2013, 3(1): 1-9.

[15]Wee S H, Huang P S, Lee J K, et al. Heteroepitaxial Cu_2O thin film solar cell on metallic substrates[J]. Sci. Rep., 2015, 5(1): 1-7.

[16]Kang Z, Yan X, Wang Y, et al. Electronic structure engineering of Cu_2O film/ZnO nanorods array all-oxide pn heterostructure for enhanced photoelectrochemical property and self-powered biosensing application[J]. Sci. Rep., 2015, 5(1): 1-7..

[17]Gao L, Xiao S, Gan W, et al. Durable superamphiphobic wood surfaces from Cu_2O film modified with fluorinated alkyl silane[J]. RSC Adv., 2015, 5(119): 98203-98208.

[18]Gao L, Gan W, Xiao S, et al. Enhancement of photo-catalytic degradation of formaldehyde through loading anatase TiO2 and silver nanoparticle films on wood substrates[J]. RSC Adv., 2015, 5(65): 52985-52992.

[19]Sun Z, Liao T, Liu K, et al. Robust superhydrophobicity of hierarchical ZnO hollow microspheres fabricated by two-step self-assembly[J]. Nano Res., 2013, 6(10): 726-735.

[20]Gao L, Gan W, Xiao S, et al. A robust superhydrophobic antibacterial Ag-TiO_2 composite film immobilized on wood substrate for photodegradation of phenol under visible-light illumination[J]. Ceram. Int., 2016, 42(2): 2170-2179.

[21]Li J, Liu L, Yu Y, et al. Preparation of highly photocatalytic active nano-size TiO_2-Cu_2O particle composites with a novel electrochemical method[J]. Electrochem. Commun., 2004, 6(9): 940-943.

[22]Gao L, Zhan X, Lu Y, et al. pH-dependent structure and wettability of TiO_2-based wood surface[J]. Mater. Lett., 2015, 142: 217-220.

[23]Gao L, Lu Y, Li J, et al. Superhydrophobic conductive wood with oil repellency obtained by coating with silver nanoparticles modified by fluoroalkyl silane[J]. Holzforschung, 2016, 70(1): 63-68.

[24]Brassard J D, Sarkar D K, Perron J. Fluorine based superhydrophobic coatings[J]. Appl. Sci., 2012, 2(2): 453-464.

[25]Zhu H, Du M L, Yu D L, et al. A new strategy for the surface-free-energy-distribution induced selective growth and controlled formation of Cu_2O-Au hierarchical heterostructures with a series of morphological evolutions[J]. J. Mater. Chem. A, 2013, 1(3): 919-929.

[26]Zhang Z, Wang P. Highly stable copper oxide composite as an effective photocathode for water splitting via a facile electrochemical synthesis strategy[J]. J. Mater. Chem., 2012, 22(6): 2456-2464.

[27]Bakar B F A, Hiziroglu S, Tahir P M. Properties of some thermally modified wood species[J]. Mater. Design, 2013, 43: 348-355.

[28]Huang X, Kocaefe D, Kocaefe Y, et al. Structural analysis of heat-treated birch (*Betule papyrifera*) surface during artificial weathering[J]. Appl. Surf. Sci., 2013, 264: 117-127.

[29]Kasemsiri P, Hiziroglu S, Rimdusit S. Characterization of heat treated eastern redcedar (*Juniperus virginiana* L.)[J]. J. Mater. Process. Tech., 2012, 212(6): 1324-1330.

[30]Su C, Liu L, Zhang M, et al. Fabrication of Ag/TiO_2 nanoheterostructures with visible light photocatalytic function via a solvothermal approach[J]. CrystEngComm, 2012, 14(11): 3989-3999.

[31]Zhang Y G, Ma L L, Li J L, et al. In situ Fenton reagent generated from TiO_2/Cu_2O composite film: A new way to utilize TiO2 under visible light irradiation[J]. Environ. Sci. Technol., 2007, 41(17): 6264-6269.

[32]Wang S, Qian H, Hu Y, et al. Facile one-pot synthesis of uniform TiO_2-Ag hybrid hollow spheres with enhanced photocatalytic activity[J]. Dalton T., 2013, 42(4): 1122-1128.

[33]Huang Q, Kang F, Liu H, et al. Highly aligned Cu_2O/CuO/TiO_2 core/shell nanowire arrays as photocathodes for water photoelectrolysis[J]. J. Mater. Chem. A, 2013, 1(7): 2418-2425.

[34]Xiong L, Ouyang M, Yan L, et al. Visible-light energy storage by Ti^{3+} in TiO_2/Cu_2O bilayer film[J]. Chem. Lett., 2009, 38(12): 1154-1155.

[35]Yang L, Luo S, Li Y, et al. High efficient photocatalytic degradation of p-nitrophenol on a unique Cu_2O/TiO_2 pn heterojunction network catalyst[J]. Environ. Sci. Technol., 2010, 44(19): 7641-7646.

[36]Linsheng Z, Duning X, Chucai W. Aeroanion Researches in Evaluation of Forest Recreation Resources[J]. Chinese Journal of Ecology, 1998, 17(6): 56-60.

中文题目：一种可释放负氧离子的超双疏水性抗菌 TiO_2/Cu_2O 复合膜负载的木材

作者：高丽坤，邱哲，甘文涛，詹先旭，李坚，强添刚

摘要：据统计，20世纪初，地球上正负离子的比例为1∶1.2。然而，在一个多世纪后，正负离子的比例发生明显的变化，即1.2∶1，导致人类居住的环境被正离子包围。因此，采取有效的方法提高负离子的比例迫在眉睫。本研究首次报道了二氧化钛/氧化亚铜/木材在紫外光照射下释放负氧离子。以木材为基质材料，采用两步法负载锐钛矿二氧化钛粒子和氧化亚铜粒子，并通过低表面能物质(氟硅烷)的表面修饰构建超双疏的表面。本文研究了氧化亚铜粒子的掺杂对二氧化钛/木材释放负氧离子性能的影响。结果表明，氧化亚铜粒子的掺杂可极大促进紫外光下释放负氧离子。疏水化处理后，木材被赋予了超双疏、抑菌和紫外光下释放负氧离子的性能。

关键词：负氧离子；超双疏；抑菌；木材

Visible-light Activate Ag/WO_3 Films Based on Wood with Enhanced Negative Oxygen Ions Production Properties*

Likun Gao, Wentao Gan, Guoliang Cao, Xianxu Zhan, Tiangang Qiang, Jian Li

Abstract: The Ag/WO_3-wood was fabricated through a hydrothermal method and a silver mirror reaction. The system of visible-light activate Ag/WO_3-wood was used to produce negative oxygen ions, and the effect of Ag nanoparticles on negative oxygen ions production was investigated. From the results of negative oxygen ions production tests, it can be observed that the sample doped with Ag nanoparticles, the concentration of negative oxygen ions is up to 1660 ions · cm^{-3} after 60 minutes visible light irradiation. Moreover, for the Ag/WO_3-wood, even after 60 minutes without irradiation, the concentration of negative oxygen ions could keep more than 1000 ions · cm^{-3}, which is up to the standard of the fresh air. Moreover, due to the porous structure of wood, the wood acted as substrate could promote the nucleation of nanoparticles, prevent the agglomeration of the particles, and thus lead the improvement of photocatalytic properties. And such wood-based functional materials with the property of negative oxygen ions production could be one of the most promising materials in the application of indoor decoration materials, which would meet people's pursuit of healthy, environment-friendly life.

Keywords: WO_3; Ag; Visible light; Negative oxygen ions; Wood

1 Introduction

Nowadays, more and more people pursuit of green and environment-friendly health concept, thus, negative oxygen ions and their effects on environment and human health have increasingly come into the public spotlight[1-3]. Sirota et al. have reported that the inhalation of negative oxygen ions influences the tracheal mucosa and its antioxidant enzymes also expand its action on the blood[4]. Moreover, negative oxygen ions could increase mental alertness and psychomotor performance; provide greater energy and feelings of exhilaration; alleviate irritability, tension and insomnia; and heal burns faster and with fewer scars[5, 6]. Daniels has proved that air ionization could significantly reduce airborne microbials, neutralize odors, and specific volatile organic compounds (VOCs)[7]. The process of air ionization involves the electronically induced formation of small air ions, such as reactive oxygen species (superoxide O_2^-), the diatomic oxygen radical anion, which could react with airborne VOCs[7]. Based on the above positive functions of negative oxygen ions, many attentions have been focused on the approaches of negative oxygen ions production. There

---

* 本文摘自 Applied Surface Science, 2017, 425: 889-895.

may be an increasment in the concentration level of negative oxygen ions in the nature and lives, such as after a rainstorm and waterfalls. And after taking a shower, the negative oxygen ions concentration in the air would increase. But not all the approaches could produce enough negative oxygen ions to reach the level of "fresh air" (according to the regulation of World Health Organization, the air is regarded as the fresh air when the negative oxygen ions concentration in the air is more than 1000 ~ 1500 ions · cm^{-3})[8]. Moreover, among structural materials, wood is one of the most closely related to human's lives. As a natural material, wood is one of the most versatile and widely used structural materials for indoor and outdoor applications such as furniture and floorboards, due to its low density, thermal expansion, renewability, and aesthetics[9]. Thus, wood-based functional materials would meet the needs of people.

Tungsten oxide (WO_3), as an important n-type semiconductor with a narrow gap (~2.8 eV), is widely applied in electrochromic, photochromic, photocatalytic and gas sensing materials[10-13]. In stoichiometric WO_3, the tungsten ions, which could determine the catalytic properties of WO_3 through changing their valence state upon reduction-oxidation processes, have no cation *d* electrons available to transfer to adsorbates, because there are the 6+ valence with the 5d shell empty[14]. When oxygen vacancies appear on the surface of d^0 oxides, the *d*-electron orbitals on adjacent cations would be partially occupied leading to a reduction of surface cations. Meanwhile, the surface cations provide the active sites for much of the chemisorption and catalytic activity that occurred in d^0 transition-metal oxides. In order to achieve the target of improving quantum efficiency of WO_3, noble metals (Pd, Pt, Au) layers have been added to activate the WO_3 films[15, 16]. Ag with a good light absorption capability is widely reported to extend the response of photocatalyst to visible light, for example, the Ag-doped TiO_2/wood reported in our previous work have enhanced the formaldehyde response under visible light[17]. Thus, we supposed that as doping the Ag nanoparticles into WO_3, the performance of the catalysis under visible light could be enhanced.

In this paper, we report a facile route to synthesis Ag/WO_3 films based on wood through a hydrothermal method and a silver mirror reaction. The negative oxygen ions production properties of the obtained Ag/WO_3-wood under visible light are systematically investigated. The mechanism of negative oxygen ions production performance of Ag/WO_3-wood under visible light is explained.

2 Materials and methods

2.1 Materials

All chemicals supplied by Shanghai Boyle Chemical Company, Limited were of analytical reagent-grade quality and used without further purification. Deionized water was used throughout the study. Wood blocks of 20 mm (R)×20 mm (T)×30 mm (L) were obtained from the sapwood sections of poplar wood (*Populus ussuriensis* Kom), which is one of the most common tree species in the northeast of China. The wood specimens were oven-dried (24 h, 103±2 ℃) to constant weight after ultrasonically rinsing in deionized water for 30 min, and oven-dried weight were determined.

2.2 Preparation of the WO_3-wood

Firstly, the 0.97 g of $Na_2WO_4 \cdot 2H_2O$ was dissolved into 50 mL mixed precursor solution of 10 mL absolute ethanol and 40 mL distilled water at room temperature, followed by being acidified to 1.0 of pH value using the H_2SO_4 solution. The mixed solution was conveyed into a stainless steel autoclave. Then the wood was soaked into the reaction solution. The autoclave was sealed and maintained at 110 ℃ for 24 h, and then cooled down to room temperature. Finally, the WO_3 films were formed on the wood surfaces and then washed with distilled water and absolute ethanol for several times, and dried in an oven. Thus the WO_3-wood was finished.

2.3 Doped with Ag nanoparticles

The Ag/WO_3-wood was fabricated through a silver mirror reaction as follows. Aqua ammonia (28 wt. %) was added drop by drop into a 0.5 M $AgNO_3$ aqueous solution with stirring until a transparent colorless $[Ag(NH_3)_2]^+$ solution was formed. The WO_3 sensors based on wooden materials were dipped into the $[Ag(NH_3)_2]^+$ solution for 1 h, then transferred into a 0.1 M glucose stock solution. After 5 min, the residual $[Ag(NH_3)_2]^+$ solution was also poured into the glucose solution. The reaction was continued for 15 min. Finally, the prepared samples were removed from the solution, ultrasonically rinsed with deionized water for 30 min, and dried at 45 ℃ for more than 24 h in vacuum. Thus the Ag/WO_3-wood was finished.

2.4 Characterization

The crystal structure of the as-prepared product was investigated by X-ray diffraction (XRD, Bruker D8 Advance, Germany) with Cu $K\alpha$ radiation of wavelength $\lambda = 1.5418$ Å, using a step scan mode with the step size of 0.02° and a scan rate of 4° min^{-1}, at 40 kV and 40 mA ranging from 5° to 80°. The morphology and microstructure were characterized by field-emission scanning electron microscopy (FE-SEM, Quanta 200, FEI, Holland) operating at 12.5 kV in combination with EDS (Genesis, EDAX, Holland). Transmission electron microscopy (TEM) and high-resolution transmission electron microscopy (HRTEM) were obtained on a Tecnai G20 electron microscope (FEI, USA) with an acceleration voltage of 200 kV. Carbon-coated copper grids were used as the sample holders. FTIR spectra were obtained on KBr tablets and recorded using a Magna-IR 560 spectrometer (Nicolet) with a resolution of 4 cm^{-1} by scanning the region between 4000 and 500 cm^{-1}. Further evidence for the composition of the product was inferred from the X-ray photoelectron spectroscopy (XPS, Thermo ESCALAB 250XI, USA), using an ESCALab MKII X-ray photoelectron spectrometer with Mg-$K\alpha$ X-rays as the excitation source. Specific surface areas of the prepared products were measured by the Brunauer-Emmett-Teller (BET) method based on N_2 adsorption at the liquid nitrogen temperature using a 3H-2000PS2 unit (Beishide Instrument S&T Co., Ltd). Optical properties of the materials were characterized by the UV-vis diffuse reflectance spectroscopy (UV-vis DRS, Beijing Purkinje TU-190, China) equipped with an integrating sphere attachment, which $BaSO_4$ was the reference.

2.5 Negative oxygen ions production tests under visible light

The experiments were performed in an obturator with the size of 500 mm (L)×300 mm (W)×300 mm (H). A visible light with a wavelength of 420 nm was installed in the open central region. Negative oxygen ions concentration in the obturator was determined by the AIC1000 air anion counter with the resolution ratio of 10 ions/cm^3 at 5 min intervals. Initially, the air anion counter must be re-zeroed and keep the numerical value constant for 5 seconds before each test. Then, the visible light was turned on to irradiate the sample. Each set of negative oxygen ions production under visible light lasted for 60 min.

3 Results and discussion

Fig. 1 presents the XRD patterns of the pristine wood, the pure WO_3, the pure Ag/WO_3, the WO_3-wood and the Ag/WO_3-wood. For the pristine wood (Fig. 1a), the diffraction peaks at 14.8° and 22.5° belonging to the (101) and (002) crystal planes of cellulose in the wood[18]. In Fig. 1b–e, it can be found that the diffraction peaks are well indexed to the standard diffraction pattern of the hexagonal WO_3 (JCPDS file No. 75-2187) and the face-centered cubic (fcc) phase of Ag (JCPDS file No. 04-0783), indicating that the present synthesis strategy successfully achieves Ag/WO_3 heterostructures and the high crystallinity on wood substrate[19,20]. As shown in Fig. 1b and d, it can be observed that after treated with WO_3, diffraction peaks at 13.8°, 22.6°, 24.2°, 26.7°, 28.1°, 33.4°, 36.4°, 46.5°, 49.7°, 55.3°, 58.2°, and 63.2°, can be perfectly identified to (100), (001), (110), (101), (200), (111), (201), (002), (220), (202), (400) and (401) crystal planes of hexagonal WO_3, respectively. The curve in Fig. 1e shows that all of the new diffraction peaks of the Ag/WO_3-wood center at $2\theta = 38.2°$, 44.1°, and 64.7°, except the diffraction peaks of WO_3, are agree with (111), (200) and (220) planes of pure Ag. However, in Fig. 1c, the curve of the pure Ag/WO_3, the peaks of Ag nanoparticles could not be seen. Moreover, in Fig. 1d and e, we could not find the peaks of the wood; this is probably ascribed to the thick composite film loaded on the wood leading to a decrease of the peaks intensities of the wood, and meanwhile, the peaks of WO_3 at 13.7° and 22.6° are close to the peaks of the wood at 14.8° and 22.5°.

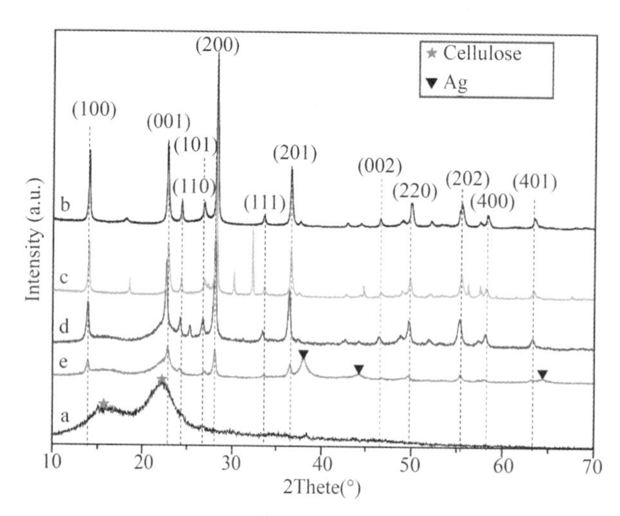

Fig. 1 XRD patterns of (a) the pristine wood, (b) the pure WO_3, (c) the pure Ag/WO_3, (d) the WO_3-wood, and (e) the Ag/WO_3-wood

Fig. 2 shows the HRTEM images of WO_3 and Ag in the Ag/WO_3-wood, and the SAED patterns of the Ag/WO_3-wood. As shown in Fig. 2a, a lattice fringe spacing of 0.383 nm is identified in this image, which corresponds to the (001) lattice plane of hexagonal WO_3[21]. Meanwhile, it is clearly discriminated from 0.236 nm corresponding to the (111) plane of Ag in Fig. 2b[22]. It can be seen that WO_3 and Ag are highly ordered with a high degree of crystallinity (are also confirmed by the XRD spectra). The SAED pattern of WO_3 in hexagonal phase and Ag in cubic phase shows that the diffraction patterns and rings are co-existent, and the different crystal planes are identified as (100), (001), (200), and (201) diffraction planes of hexagonal WO_3 as well as (111) and (200) diffraction planes of cubic Ag, respectively. Additionally, the SAED pattern (Fig. 2c) also displays a polycrystalline nature of hierarchical structure.

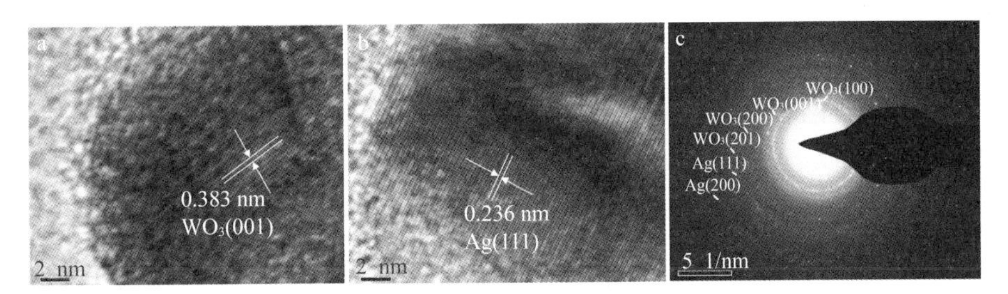

Fig. 2 High resolution TEM (HRTEM) images of (a) WO_3 and (b) Ag in the Ag/WO_3-wood. (c) The selected electron diffraction (SAED) pattern of the Ag/WO_3-wood

The morphologies of the samples analyzed in Fig. 3 are investigated using SEM. In Fig. 3c, the pristine wood surface appears to be smooth and free of any other materials. As shown in Fig. 3a and b, it could be observed that the particles of the pure WO_3 and the pure Ag/WO_3 without wood substrate have agglomerate together and uneven. However, in Fig. 3d and e, the particles are dispersed more evenly on the wood substrate. The reason is ascribed to the effect of wood substrate, which could promote the nucleation of nanoparticles and prevent the agglomeration of the particles due to its porous structure. Fig. 3b is the SEM image of WO_3-wood, which clearly shows that the wood surface is covered by a rough film of WO_3. As shown in Fig. 3c, after Ag nanoparticles doped, it is observed that abundant nanoparticles composed by Ag

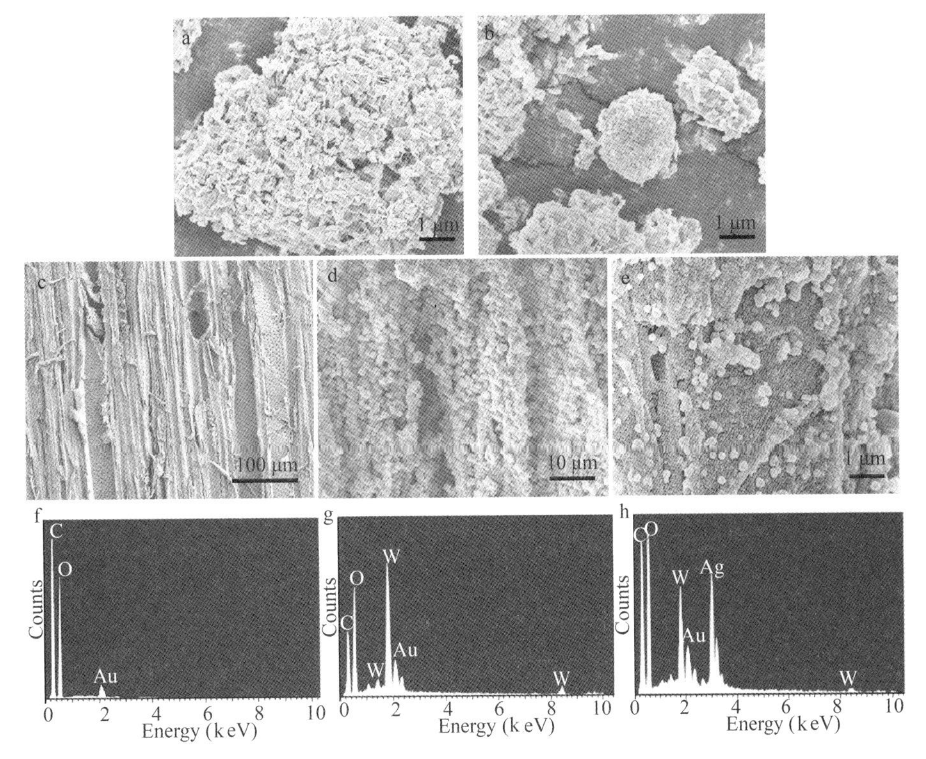

Fig. 3 SEM images of the surfaces of (a) the pure WO_3, (b) the pure Ag/WO_3, (c) the pristine wood, (d) the WO_3-wood, and (e) the Ag/WO_3-wood. EDS spectra of (f) the pristine wood, (g) the WO_3-wood, and (h) the Ag/WO_3-wood

nanoparticles fill the interspaces on the sample surface, and sporadic several microspheres expose on the surface. Fig. 3d shows that the carbon, oxygen and gold elements are detected in the EDS spectrum of the pristine wood. The gold element derives from the conductive layer on wood surface, which is used as the SEM observations. The carbon and oxygen elements resulted from the pristine wood. However, after coating the WO_3 film on wood surface, the W peaks are detected, and the carbon and oxygen are present (Fig. 3g). In Fig. 3h, an added Ag element could be found except the C, O, W and Au. No other obvious elements are detected in the spectra, thus confirming the composition of the particles are WO_3 and Ag, respectively.

Fig. 4 shows FTIR spectra of the pristine wood, the WO_3-wood, and the Ag/WO_3-wood. The bands at 3424 cm^{-1} correspond to the stretching vibrations of OH groups in all the wood samples[9]. Typical bands assign to cellulose located at 2917 cm^{-1} for the C-H stretching vibrations of the CH_3 and CH_2 groups, and at 1032 cm^{-1} for C=O stretching vibrations, respectively. Lignin bands are found at 1597 cm^{-1} for C=C stretching of the aromatic ring. In addition, the new adsorption peaks of both the modified wood samples existed at 559 cm^{-1} and 557 cm^{-1} in Fig. 4b and c are appointed to the intrinsic vibrations of W—O in the tungsten oxide[23].

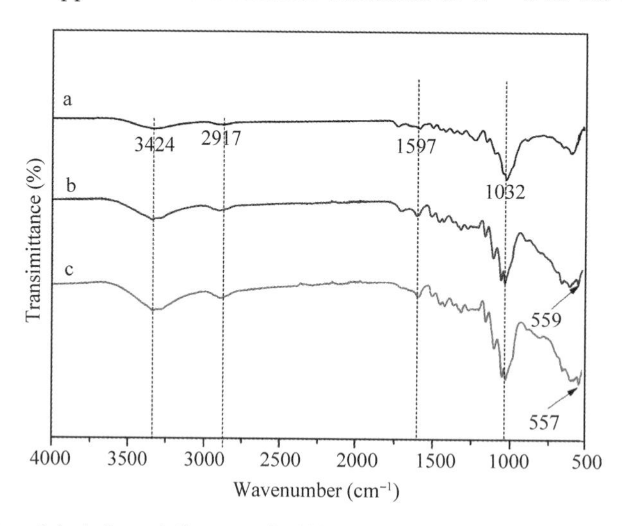

Fig. 4 FTIR spectra of (a) the pristine wood, (b) the WO_3-wood, and (c) the Ag/WO_3-wood

The elemental and chemical characterization of the films is performed by XPS. The fully scanned spectra (Fig. 5a) shows that the W, O, and C elements exist on the surface of the WO_3-wood, as well as the W, O, Ag, and C elements exist on the surface of the Ag/WO_3-wood. The C element can be ascribed to the wood substrate or the adventitious carbon-based contaminant.

The high resolution XPS spectra of the O 1s of the WO_3-wood and the Ag/WO_3-wood are displayed in Fig. 5b, respectively. For both the samples, O 1s regions are dominated by oxygen bound to carbon (such as the C—O in cellulose or C=O—C in lignin)[23] and the crystal lattice oxygen (W-O)[24].

Fig. 5c shows the W 4f and the W $5p_{3/2}$ core level spectra recorded on the WO_3-wood, the results of its fitting analysis. In the spectra, it can be observed that two doublet functions are used for the W 4f component and a singlet for the W $5p_{3/2}$ component near 42.0 eV. One doublet contains its highest intensity peak (W $4f_{7/2}$) located near 36.4 eV, which is generated by photoelectrons emitted from tungsten atoms with an oxidation state of +6; that is, stoichiometric WO_3[14]. After doping with Ag film, it is observed that the

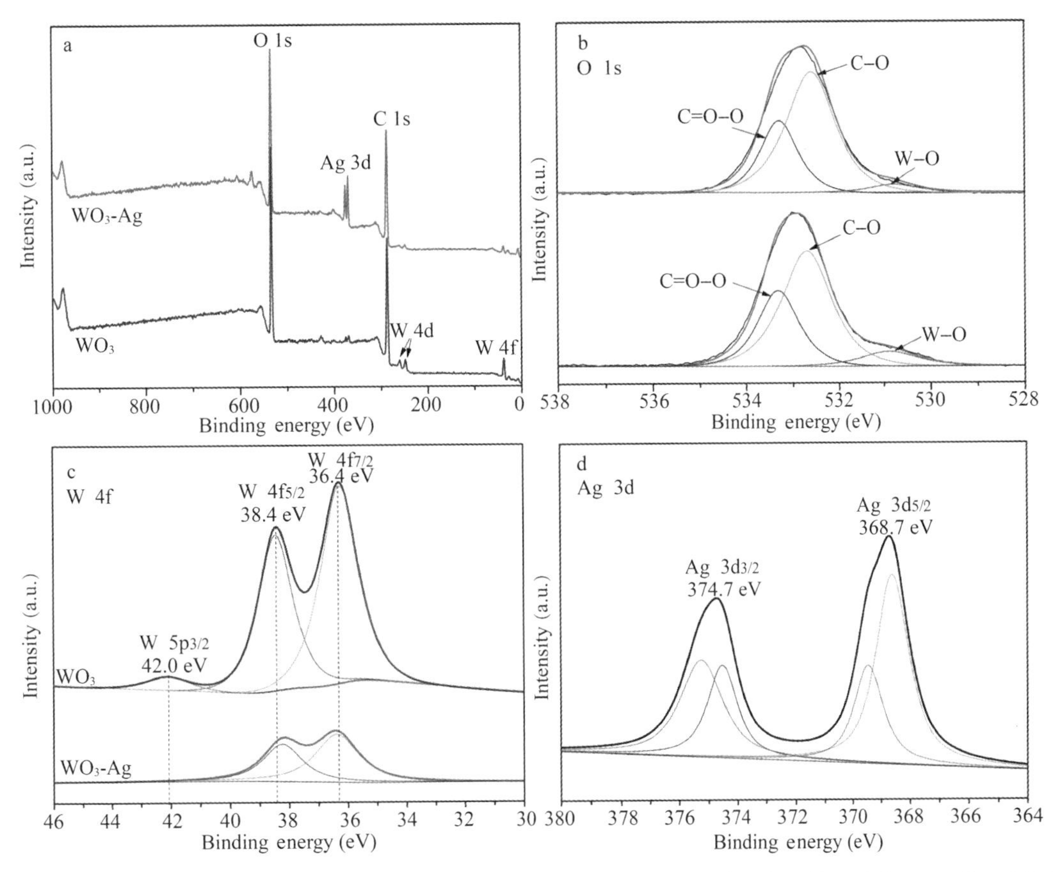

Fig. 5 (a) Survey scans of the WO_3-wood and the Ag/WO_3-wood. (b) Peaking-fitting results of O 1s XPS spectra of the WO_3-wood and the Ag/WO_3-wood, (c) W 4f XPS spectra of the WO_3-wood and the Ag/WO_3-wood, (d) Ag 3d XPS spectra of the Ag/WO_3-wood

intensities of the W doublets and the peak of W-O in the O 1s decrease, which illustrates a lower electron density of the W atoms, and there is a strong interaction between metallic Ag and WO_3 in the Ag/WO_3-wood.

Fig. 5d shows the Ag core level spectrum recorded on the Ag/WO_3-wood. It is made up of two main components associated with 3d doublet at 368.7 eV (Ag $3d_{5/2}$) and 374.7 eV (Ag $3d_{3/2}$), with a spin energy separation of 6.0 eV[25]. The results indicate this is characteristic of metallic silver.

To have an insight into the effect of the wood fibers on the porous structure of the samples, BET analysis was carried out. Fig. 6 shows the N_2 adsorption-desorption isotherms of the pure WO_3, the pure Ag/WO_3, the WO_3-wood and the Ag/WO_3-wood. These curves all exhibit small hysteresis loops, which are attributed to type IV isotherms and the representative of mesoporous materials, indicating the presence of mesopores (2-50 nm)[26, 27]. This result is further confirmed by the corresponding pore-size distribution curves (inset in Fig. 6). Furthermore, the isotherm profile of the Ag/WO_3-wood shows typical H1 type hysteresis loops in the relative pressure range from 0.2 to 0.9 according to the uniform sized spherical-particles aggregates and hysteresis loops close to H3 type from 0.9 to 1.0, indicating the presence of slit-like pores. The pore size distributions of the Ag/WO_3-wood exhibit most broadened pore size ranges among the samples (inset in Fig. 6).

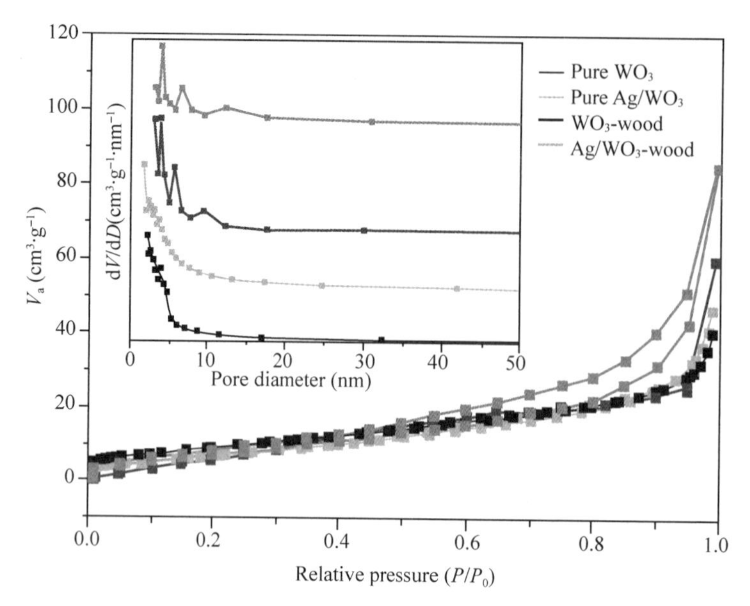

Fig. 6 N_2 adsorption-desorption isotherms of (a) the pure WO_3, (b) the pure Ag/WO_3, (c) the WO_3-wood, and (d) the Ag/WO_3-wood. The inset shows the pore size distributions

The optical absorptions of the pure WO_3 and the pure Ag/WO_3 the WO_3-wood and the Ag/WO_3-wood were measured by using an UV-vis spectrometer. As shown in Fig. 7a, all the samples have photo absorptions from UV light to visible light, whereas the Ag/WO_3-wood after calcination exhibits a highest absorption, indicating more intensive effect to light, especially the visible light. The optical band gap of the samples were estimated from the absorption data by the following equation near the band edge[28, 29]: $(\alpha h v)^n = A\ (hv - E_g)$, where α, v, E_g and A are absorption coefficient, light frequency, band gap and a constant, respectively. The value of n is 0. 5 or 2 for a direct or indirect transition respectively. When an proximate value of E_g was used and the line of $\ln(\alpha h v)$ vs $\ln(hv - E_g)$ was plotted, the value of n was determined to be 0. 5 indicating a direct optical transition which is similar to the previous report on WO_3. The band gap energies of the WO_3 and the WO_3-wood can thus be obtained from the plots of $(\alpha h v)^{1/2}$ versus photon energy E_g, as shown in Fig. 7b. For the WO_3 and the WO_3-wood, the values estimated from the intercept of the tangent to the plot are 2. 65 eV and 2. 5 eV, respectively, which indicates that the photocatalytic property of the sample would be changed. In the study, wood was selected as substrate to prevent the agglomeration of particles, and thus the structures of catalysts were changes in accordance with the results in Fig. 3. Moreover, the cures of the Ag/WO_3 and the Ag/WO_3-wood in Fig. 7b also illustrate that loading silver on the samples could reduce the band gaps of the catalysts, that is, enhances the visible-light absorption.

To test the properties of negative oxygen ions production in the obturator when visible light irradiated the samples surfaces, the negative oxygen ions concentrations in the obturator were recorded every 5 minutes and the concentrations curves are depicted in Fig. 8a. For verifying the effects of the negative oxygen ions production by the samples, the blank experiments were taken by irradiation without samples. Firstly, the initial negative oxygen ions concentrations in the obturator are 0 ions · cm^{-3}. It is obvious that when irradiation without samples, there is scarcely any negative oxygen ions produced. For both the WO_3-wood and the Ag/

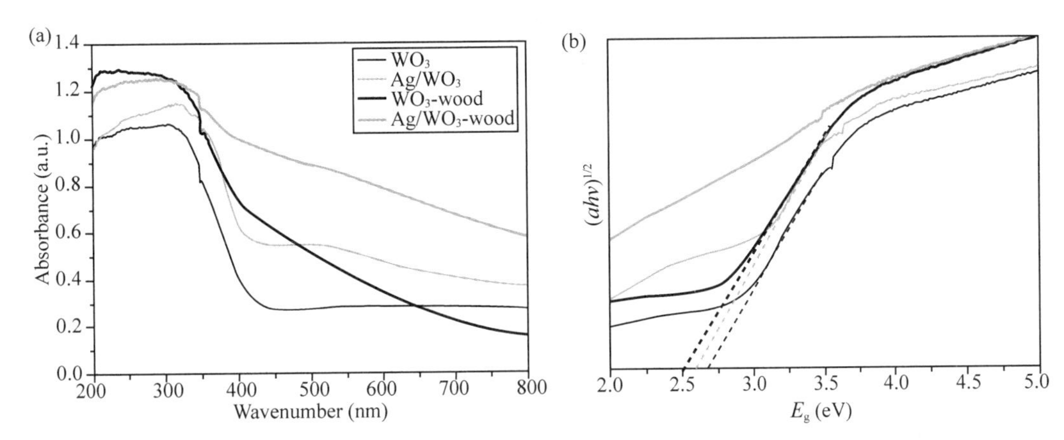

Fig. 7 (a) Typical diffuse reflection spectra of the pure WO_3, the pure Ag/WO_3, the WO_3-wood and the Ag/WO_3-wood. (b) The $\alpha h\nu - E_g$ curve of the pure WO_3, the pure Ag/WO_3, the WO_3-wood and the Ag/WO_3-wood

WO_3-wood, the negative oxygen ions concentrations are dramatically increased, and after 40 minutes irradiation, the concentrations are slightly decreased. Finally, after 60 minutes irradiation, the concentrations of negative oxygen ions production for the WO_3-wood and the Ag/WO_3-wood are 1200 and 1660 ions · cm^{-3}, respectively. However, for the samples without wood substrates, the concentrations of negative oxygen ions production are lower than that for the samples with wood substrates. After 60 minutes irradiation, the concentration of negative oxygen ions production for the pure WO_3 and the pure Ag/WO_3 are 470 and 530 ions · cm^{-3}, respectively. As shown in Fig. 6b, after 60 minutes irradiation, the negative oxygen ions concentrations for all the samples descend gradually, but the concentration for the Ag/WO_3-wood is still superior to those for the pure WO_3, the pure Ag/WO_3 and the WO_3-wood. After 60 minutes without irradiation, the residual negative oxygen ions concentrations for the pure WO_3, the pure Ag/WO_3, the WO_3-

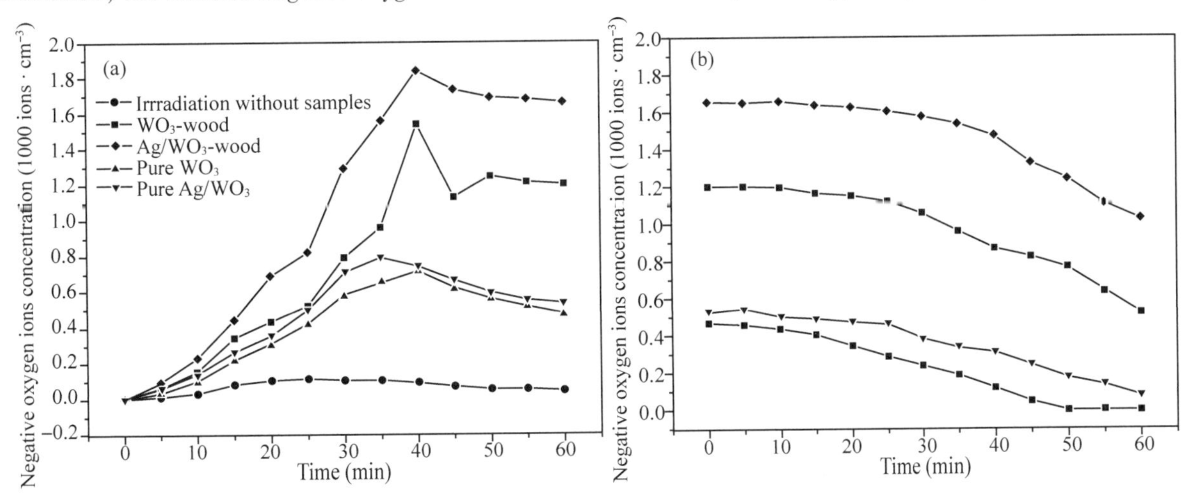

Fig. 8 (a) Negative oxygen ions concentrations in the obturator when visible light irradiated the surfaces of the pure WO_3, the pure Ag/WO_3, the WO_3-wood and the Ag/WO_3-wood, and without samples. (b) The changes of negative oxygen ions concentrations in the obturator after visible light irradiated the surfaces of the pure WO_3, the pure Ag/WO_3, the WO_3-wood and the Ag/WO_3-wood

wood and the Ag/WO_3-wood are 0, 50, 520 and 1030 ions · cm^{-3}, respectively, which indicates negative oxygen ions production efficiency of the Ag/WO_3-wood under visible light is significantly improved due to the loading of the Ag nanoparticles onto the WO_3-wood surface and the effect of the wood substrate that promotes the nucleation of nanoparticles, prevents the agglomeration of the particles, and thus enhances the photocatalytic property. Moreover, the concentration of negative oxygen ions for the Ag/WO_3-wood could keep more than 1000 ions · cm^{-3}, which is up to the standard of the fresh air.

To investigate the recyclability of the Ag/WO_3-wood for negative oxygen ions production, the sample was tested under visible light for 5 cycles. As shown in Fig. 9, after 5 cycles under visible light irradiation, the negative oxygen ions production property of the Ag/WO_3-wood was well-maintained. The concentrations of negative oxygen ions did not exhibit any significant loss after each one hour irradiation, confirming the Ag/WO_3-wood is not photocorroded during the long-time light irradiation. As is well known, the stability of a product is important to its practical application.

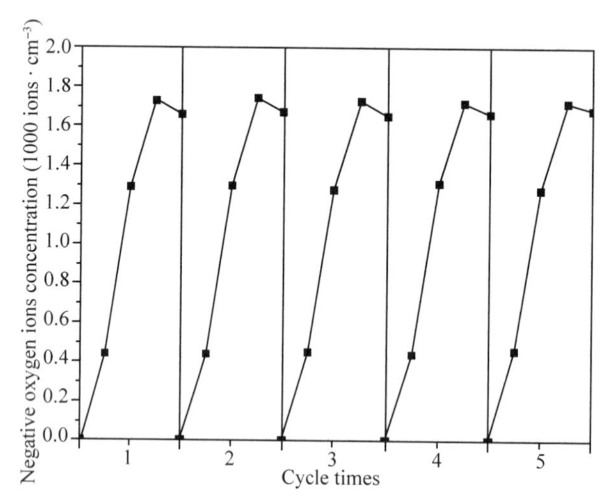

Fig. 9 Cycling runs of negative oxygen ions production in the obturator when visible light irradiated the surfaces the Ag/WO_3-wood

The enhanced negative oxygen ions production property of the Ag/WO_3-wood is ascribed to the effect of Ag nanoparticles loading arising from the following mechanisms. Firstly, when the semiconductor WO_3 contacts the metallic Ag, the charge carriers would redistribute due to the difference of Fermi energy level between WO_3 and Ag[30, 31]. The electrons will migrate from conduction band (CB) of WO_3 to the Ag to achieve the Fermi level equilibration. As a result, the Ag nanoparticles accumulate excess negative charge, while the WO_3 exhibits an excess positive charge, thus the Schottky barrier produces between the semiconductor WO_3 and the metallic Ag, which could trap the photogenerated electrons and inhibit the recombination of excited electrons and holes. The specific schematic diagram of negative oxygen ions production upon the Ag/WO_3 - wood is depicted in Fig. 10. Under visible light irradiation, the photogenerated electrons could react with electron acceptors (O_2) existed in the air, reducing it to superoxide radical anion O_2^-, that is, negative oxygen ions. Moreover, the valence holes photogenerated on the surface of WO_3 could react with H_2O or OH^- molecules to form ·OH radicals[32].

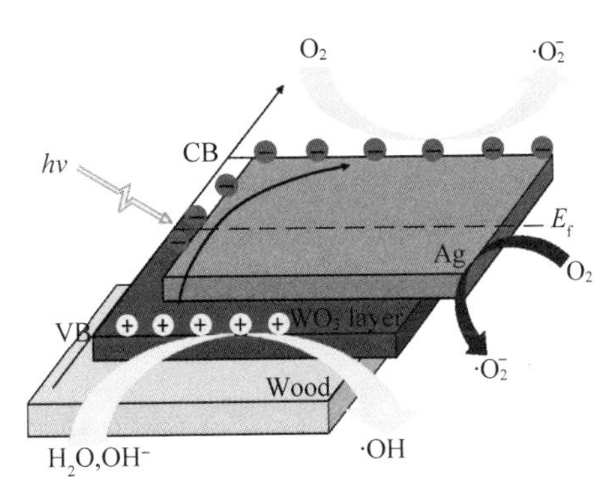

Fig. 10 Schematic illustration of the mechanism of negative oxygen ions production upon the Ag/WO_3-wood

4 Conclusions

In the study, a system of visible light activated Ag/WO_3-wood was built to produce negative oxygen ions. After doped with Ag nanoparticles, the properties of negative oxygen ions production was improved, which was probably attributed to the following reason: (1) the wood selected as substrate could promote the nucleation of nanoparticles, prevent the agglomeration of the particles, and thus lead the improvement of photocatalytic properties. (2) In the visible light irradiation process, there is a strong interaction between metallic Ag and WO_3 in the Ag/WO_3-wood, and the Schottky barrier produced to achieve the Fermi level equilibration. The Ag nanoparticles accumulate excess negative charge and the WO_3 exhibits an excess positive charge, which could trap the photogenerated electrons and inhibit the recombination of excited electrons and holes. Thus the photocatalytic property of the Ag/WO_3-wood was enhanced. Then the photogenerated electrons react with O_2 molecule existed in the air, and produce negative oxygen ions (O_2^-). As a result, after 60 minutes irradiation, the concentration of negative oxygen ions production for the Ag/WO_3-wood is 1660 ions · cm^{-3}. After 60 minutes without irradiation, the residual negative oxygen ions concentration for the Ag/WO_3-wood is 1030 ions · cm^{-3}, which is up to the standard of the fresh air.

Acknowledgments

This work was financially supported by the National Natural Science Foundation of China (grant no. 31470584).

References

[1]Hawkins L H, Barker T. Air ions and human performance[J]. Ergonomics, 1978, 21(4): 273-278.

[2]Krueger A P, Reed E J. Biological impact of small air ions[J]. Science, 1976, 193(4259): 1209-1213.

[3]Kellogg E W. Superoxide involvement in the bactericidal effects of negative air ions on Staphylococcus albus, Nature, 1979, 281: 400-401.

[4]Sirota T V, Novoselov V I, Safronova V G, et al. The effect of inhaled air ions generated by technical ionizers and a bioionizer on rat trachea mucosa and the phagocytic activity of blood cells[J]. IEEE T. Plasma. Sci., 2006, 34(4): 1351-1358.

[5]Tom G, Poole M F, Galla J, et al. The influence of negative air ions on human performance and mood[J]. Hum. Factors, 1981, 23(5): 633-636.

[6]Hedge A, Collis M D. Do negative air ions affect human mood and performance? [J]. Ann. Occup. Hyg., 1987, 31

(3) : 285-290.

[7]Daniels S. L. On the ionization of air for removal of noxious effluvia (Air ionization of indoor environments for control of volatile and particulate contaminants with nonthermal plasmas generated by dielectric-barrier discharge)[J]. IEEE T. Plasma. Sci., 2002, 30(4) : 1471-1481.

[8]Gao L, Qiu Z, Gan W, et al. Negative oxygen ions production by superamphiphobic and antibacterial TiO_2/Cu_2O composite film anchored on wooden substrates[J]. Sci. Rep., 2016, 6(1) : 1-10.

[9]Gao L, Lu Y, Li J, et al. Superhydrophobic conductive wood with oil repellency obtained by coating with silver nanoparticles modified by fluoroalkyl silane[J]. Holzforschung, 2016, 70(1): 63-68.

[10]Memar A, Phan C M, Tade M O. Controlling particle size and photoelectrochemical properties of nanostructured WO_3 with surfactants[J]. Appl. Surf. Sci., 2014, 305: 760-767.

[11]Karuppasamy A. Electrochromism and photocatalysis in dendrite structured Ti: WO_3 thin films grown by sputtering [J]. Appl. Surf. Sci., 2015, 359: 841-846.

[12]Gavrilyuk A I. Aging of the nanosized photochromic WO_3 films and the role of adsorbed water in the photochromism [J]. Appl. Surf. Sci., 2016, 364: 498-504.

[13]Arfaoui A, Touihri S, Mhamdi A, et al. Structural, morphological, gas sensing and photocatalytic characterization of MoO_3 and WO_3 thin films prepared by the thermal vacuum evaporation technique[J]. Appl. Surf. Sci., 2015, 357: 1089-1096.

[14]Bittencourt C, Llobet E, Ivanov P, et al. Ag induced modifications on WO_3 films studied by AFM, Raman and x-ray photoelectron spectroscopy[J]. J. Phys. D: Appl. Phys., 2004, 37(24) : 3383.

[15]Garavand N T, Mahdavi S M. Pt and Pd as catalyst deposited by hydrogen reduction of metal salts on WO_3 films for gasochromic application[J]. Appl. Surf. Sci., 2013, 273: 261-267.

[16]Tahmasebi N, Mahdavi S M. Synthesis and optical properties of Au decorated colloidal tungsten oxide nanoparticles [J]. Appl. Surf. Sci., 2015, 355: 884-890.

[17]Gao L, Gan W, Xiao S, et al. A robust superhydrophobic antibacterial Ag-TiO_2 composite film immobilized on wood substrate for photodegradation of phenol under visible-light illumination[J]. Ceram. Int., 2016, 42(2) : 2170-2179.

[18]Andersson S, Serimaa R, Paakkari T, et al. Crystallinity of wood and the size of cellulose crystallites in Norway spruce (Picea abies)[J]. J. Wood. Sci., 2003, 49(6) : 531-537.

[19]Adhikari S, Sarkar D. High efficient electrochromic WO_3 nanofibers[J]. Electrochim. Acta, 2014, 138: 115-123.

[20] Xue C H, Chen J, Yin W, et al. Superhydrophobic conductive textiles with antibacterial property by coating fibers with silver nanoparticles[J]. Appl. Surf. Sci., 2012, 258(7) : 2468-2472.

[21] Tokunaga T, Kawamoto T, Tanaka K, et al. Growth and structure analysis of tungsten oxide nanorods using environmental TEM[J]. Nanoscale Res. Lett., 2012, 7(1) : 1-7.

[22]Su C, Liu L, Zhang M, et al. Fabrication of Ag/TiO_2 nanoheterostructures with visible light photocatalytic function via a solvothermal approach[J]. Cryst. Eng. Comm, 2012, 14(11) : 3989-3999.

[23] Sheng C, Wang C, Wang H, et al. Self-photodegradation of formaldehyde under visible-light by solid wood modified via nanostructured Fe-doped WO_3 accompanied with superior dimensional stability[J]. J. Hazard. Mater., 2017, 328: 127-139.

[24] Kharade R R, Mali S S, Patil S P, et al. Enhanced electrochromic coloration in Ag nanoparticle decorated WO_3 thin films[J]. Electrochim. Acta, 2013, 102: 358-368.

[25] Zhang L, Chen L, Chen L, et al. A facile synthesis of flower-shaped TiO_2/Ag microspheres and their application in photocatalysts[J]. RSC Adv., 2014, 4(97) : 54463-54468.

[26] Tahir M, Cao C, Butt F K, et al. Large scale production of novel g-C_3N_4 micro strings with high surface area and versatile photodegradation ability[J]. Cryst. Eng. Comm, 2014, 16(9) : 1825-1830.

[27] Gao L, Gan W, Qiu Z, et al. Preparation of heterostructured WO_3/TiO_2 catalysts from wood fibers and its versatile photodegradation abilities[J]. Sci. Rep., 2017, 7(1): 1-13.

[28] Satoh N, Nakashima T, Kamikura K, et al. Quantum size effect in TiO_2 nanoparticles prepared by finely controlled metal assembly on dendrimer templates[J], Nat. Nanotecnol., 2008, 3: 106-111.

[29] Shen Y, Ding D, Yang Y, et al. 6-Fold-symmetrical WO_3 hierarchical nanostructures: Synthesis and photochromic properties[J]. Mater. Res. Bull., 2013, 48(6): 2317-2324.

[30] Ge L, Han C, Liu J, et al. Enhanced visible light photocatalytic activity of novel polymeric g-C_3N_4 loaded with Ag nanoparticles[J]. Appl. Catal. A-Gen., 2011, 409: 215-222.

[31] Sun S, Wang W, Zeng S, et al. Preparation of ordered mesoporous Ag/WO_3 and its highly efficient degradation of acetaldehyde under visible-light irradiation[J]. J. Hazard. Mater., 2010, 178(1-3): 427-433.

[32] Ren J, Wang W, Sun S, et al. Enhanced photocatalytic activity of Bi_2WO_6 loaded with Ag nanoparticles under visible light irradiation[J]. Appl. Catal. B-Environ., 2009, 92(1-2): 50-55.

中文题目：银/三氧化钨/木材可见光下释放负氧离子的性能研究

作者：高丽坤，甘文涛，曹国良，詹先旭，强添刚，李坚

摘要：本研究采用水热合成法和银镜反应成功制备了银/三氧化钨/木材。银/三氧化钨/木材用于可见光下释放负氧离子，并深入探讨了银对负氧离子释放的作用。由负氧离子释放性能测试结果可知，负载银纳米粒子后，样品在可见光照射60 min后负氧离子的释放浓度可达1660个·cm^{-3}。此外，借助于木材的多孔结构，木材作为基质材料，可促使纳米颗粒的成核生长，防止粒子的团聚，从而提高光催化的性能。该木材基功能性材料具有释放负氧离子的性能，是室内装饰材料应用中最有前景的材料之一，将满足人们对健康、环保生活的追求。

关键词：三氧化钨；银；可见光；负氧离子；木材

Cu Thin Films on Wood Surface for Robust Superhydrophobicity by Magnetron Sputtering Treatment with Perfluorocarboxylic Acid*

Wenhui Bao, Ming Zhang, Zhen Jia, Yue Jiao, Liping Cai, Daxin Liang, Jian Li

Abstract: Superhydrophobic wood was fabricated using a magnetron sputtering method, which can be applied to substrates with different shapes for its facile and controllable advantages. Copper was deposited on the cross-section surface of pristine wood followed by a treatment of perfluorocarboxylic acid. The results revealed that the deposition thickness played an important role in varying the surface morphology. In the optimized condition, water contract angle (CA) and sliding angle (SA) can reach 154° and 3.5°, respectively, when the deposition thickness was 50 nm. More importantly, the superhydrophobic surface can resist up to 100 times impingement of sand abrasion, and CA retained to be 151°. This method will have promising applications in wooden artifacts, such as clock frame, art, carvings and decorations, etc.

Keywords: Wood; Mangnetron sputtering; Superhydrophobic; Thin films

1 Introduction

As a type of polymer composite material, wood has the heterogeneous, rough and even porous surfaces, which is the widely used engineering and structural material for various applications due to its numerous attractive properties including esthetic appeal, mechanical strength, and low density. However, wooden materials also suffer some limitations. For example, when wood is exposed to the sunlight and water without any protection, the surfaces of wood materials deteriorate fast. This can lead to the formation of micro-checks and turn into surface cracks due to the repeated swelling and shrinking. To improve the water repellency of wood, the surface coating and bulk treatment are two effective measures to impede hydroxyl groups of wood cell wall polymers to absorb water (Xie et al., 2008). Conventionally, superhydrophobic surfaces exhibit great water-repellent properties with water contact angles greater than 150° (Feng et al., 2002). Therefore, superhydrophobic surfaces on the wood substrate are effective measures to protect exterior wood from deterioration. So far, superhydrophobic coatings have been successfully developed on wood substrates by using various techniques to control the surface micro-nano structure. Specifically, numerous efforts have been made to fabricate superhydrophobic TiO_2/ ZnO surfaces on wood substrates. Wang et al. (2011). synthesized zinc oxide nanorods on wood surface via a wet chemical method, and the product appears to be superhydrophobic with the water contact angle of about 152°. Gao et al. (2015b) reported that a simple hydrothermal process

* 本文摘自 European Journal of Wood and Wood Products, 2017, 77(1): 115-123.

with further hydrophobization was developed for fabricating durable superamphiphobic films of cuprous oxide (Cu_2O) microspheres on a wood substrate. Gao et al. (Gao et al. , 2015a) studied the reversible photocontrol of wood-surface wettability between the superhydrophilicity and superhydrophobicity based on a TiO_2 film modified with octadecyl trichlorosilane (OTS). However, the drawbacks of poor reproducibility, complicated procedures, high manufacturing cost of these methods limit their practical applications. Meanwhile, most superhydrophobic surfaces were obtained on tangential section of wood. No reports regarding the production of superhydrophobic surfaces of cross-section surfaces on wood substrate were found. More importantly, the conventional methods are unable to be adapted for industrial applications. Thus, the newly developed technique of magnetron sputtering can address the previously mentioned drawbacks because it can easily be scaled up in the industries.

As a Physical Vapor Deposition (PVD) method, magnetron sputtering has become the process alternative for the deposition of a wide range of industrially important coatings (Kelly and Arnell, 1999) . Examples include wear-resistant coatings, low friction coatings, corrosion resistant coatings, decorative coatings and coatings with specific optical or electrical properties. Compared with the traditional approaches, such as chemical vapor deposition (Reina et al. , 2009) , sol-gel processing (Pang and Anderson, 2000) , electrospinning (Yao et al. , 2016) , and chemical etching method (And and Shen, 2005) , this method is simpler and easier to control, which works for substrates in any shape and does not need rigorous conditions. Bang et al. (Bang et al. , 2003) demonstrated that ZnO films were deposited on c-plane sapphire substrates with different buffer layer thicknesses, and the roughness could be improved through radio-frequency magnetron sputtering. One-dimensional nanostructural thin films fabricated by magnetron sputtering were conducted by (Khedir et al. , 2010) . Fe and Cu single-layer films, their structures and magnetic properties were investigated by (Kozono et al. , 1987). However, there is no report on the effects of the Cu layer on the wood surface by magnetron sputtering. Importantly, it will create a new prospect avenue for the fundamental research as well as the practical applications in industries (Mukherjee et al. , 2011).

Tree disk cross-cut from a log for making wood artifacts is a promising way to utilize wood effectively (Kang and Lee, 2004) . Kang and Lee (2002) suggested that tree disks could be used for making wooden artifacts, such as clock frame, art, carvings and decor. The cross-section of a tree shows fresh appearance, which would be interested by people in their daily life. Moreover, the annual rings in timber are naturally harmonious, elegant, and beautiful. However, tree disk used in exterior application is susceptible to moisture and liquid water. According to published results, pigmented coating is the only effective approach to prevent wood products from aging, abrasion and rupture (Xie et al. , 2006; Xie et al. , 2008) . In this study, a simple method was proposed to make robust, artificial coatings, providing an efficient way to address the mentioned problems.

Herein, the adjustment of the copper films morphology of the cross-section cell wall on wood substrate was designed. The protruding structure of copper lamellar particles and the recessed cell wall structure further increased the roughness. The surface topography design will retain the wood texture and make full use of its own structure to create more space by creating more trapped air pockets on the surfaces. Copper (Cu) is widely used in industrial applications. Therefore, the Cu film was selected as a rough coating on wood surfaces (Inoue et al. , 1988; Yu et al. , 2007) . Obviously, the coating can be easily applied and repaired. Then, the as-prepared rough surface was modified by the perfluorocarboxylic acid, and rough structures with superhydrophobicity are achieved with excellent chemical stability. In this research, a mechanical resistance experiment was designed to explore the durable properties of the products. The results indicated that the coatings showed excellent mechanical stability. At the same time, the morphology, chemical composition, crystallization structure, and surface roughness properties of superhydrophobic wood were investigated in detail.

2 Material and methods

2.1 Materials

Perfluorocarboxylic acid (97.0%) and ethanol (95.0%) with analytical grade were purchased from the Shanghai Boyle Chemical Company Limited. The Cu target was supplied by the Beijing Guanjinli New Matrial Company Limited. After ultrasonically rinsing in ethanol and deionized water for 10 min, ten samples of radiata pine with a cross-section of 20×20 mm^2 and a length of 10 mm were oven-dried (24 h, 103 ℃) to a constant weight, and their weights were determined.

2.2 Preparation of Cu coating on the wood surface

All Cu films were deposited with a DC magnetron sputtering system (Fig. 1). The target used was a copper (99.999% purity) plate with a diameter of 5 cm. Sputtering was carried out in argon (Ar-99.995% purity), and the sputtering system was equipped with a diffusion pump backed by a rotary pump to pump down the sputtering system to a base pressure of 3×10^{-3}Pa. The thicknesses of the Cu films were checked in situ with a quartz crystal monitor located near the wood substrate during the sputtering process. The wood substrate was pre-conditioned at a room temperature and the sputtering power was 100 W with a fixed target-substrate distance of 6 cm.

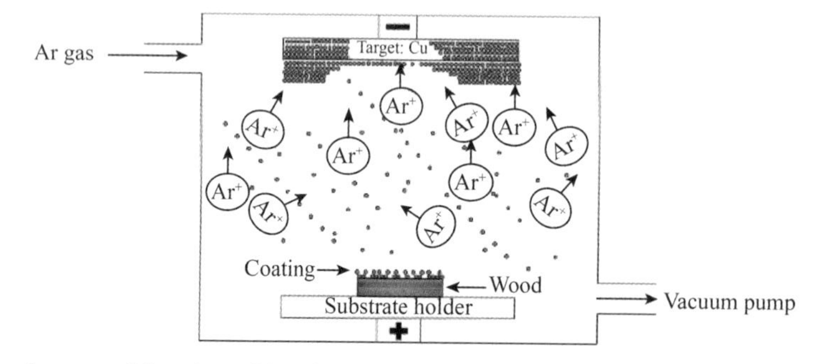

Fig. 1 Schematic diagram of the deposition Cu films on wood substrate by magnetron sputtering equipment

2.3 Modification of Cu coating on the wood surface

The as-prepared Cu coating on the wood surface was immersed into a mixture of ethanol solution of 0.01M perfluorocarboxylic acid for 12 h. Afterwards the sample was rinsed with ethanol for ten minutes and dried naturally in air while waiting for further examination of its properties.

2.4 Characterization

The morphology structure of the as-prepared wood surface was observed using the scanning electron microscopy (SEM, FEIQUANTA200). The chemical composition of the treated and untreated wood was analyzed by the Fourier transform infrared spectroscopy (FTIR, Magna-IR 560, Nicolet). The CA and SA were measured with 5 μL deionized water droplet at room temperature using an optical contact angle meter (Hitachi, CA-A). The surface chemical state and structure of the treated samples were determined using the X-ray powder diffraction (XRD, Philips, PW 1840 diffractometer) operating with Cu-K radiation at a scan rate of 4°/min and accelerating voltage of 40 kV and applied current of 30 mA ranging from 5° to 80°. The roughness (Ra) of the sample was measured using the surface profilometer (2300A-C, Harbin Measuring & Cutting Tool Group Co.), which was equipped with a diamond tip having a radius of 10 μm and being sensitive to vertical movements to an accuracy of ± 0.01 μm. The scanning distance was 5 mm each, and the value of Ra was obtained by averaging the five values at different positions.

2.5 Abrasion test of the as-prepared surfaces

With respect to real application, it is desired that the superhydrophobic surfaces can be survived from harsh weather conditions. To assess the mechanical resistance of the superhydrophobic wood, sand abrasion (Deng et al. 2012) and sandpaper abrasion were used, as shown in Fig. 2. In Fig. 2a, 20 g sand grains with a diameter ranging from 100 to 300 μm impinged on the surface from a height of 30 cm. Fig. 2b shows that the as-prepared sample with 100 g weight on the top was placed face-down on the sandpaper, and the surface was moved in one direction. The changes of the CA and the contact angle hysteresis (CAH) of the superhydrophobic wood surface were measured after the abrasion.

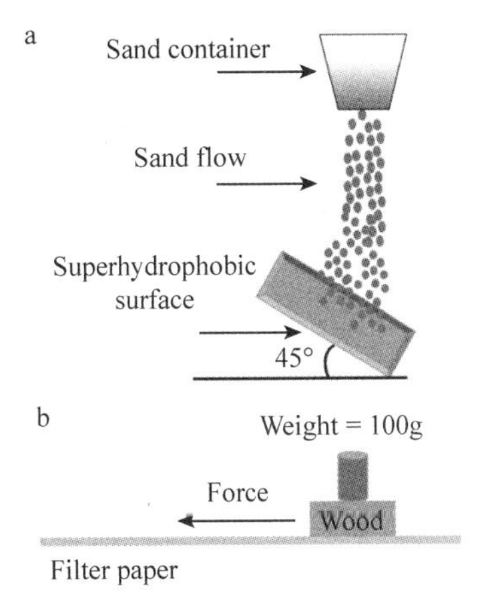

Fig. 2 (a) Schematic drawing of a sand abrasion test. (b) Schematic illustration of the sandpaper abrasion test

3 Results and discussion

3.1 Microstructure of the Cu coating on the wood surface

Cu layers with different thicknesses ranging from 30 to 150 nm were deposited on wood substrates by the DC magnetron sputtering in argon atmosphere. The surface morphology of Cu films was monitored by the scanning electron microscopy (SEM). The cross sections of wood surface exhibited that various inter-cellular fractures and the cell lumina were empty, while the wood fibers showed an irregular and jagged surface. It was observed that the pristine wood was a type of heterogeneous and porous material (Fig. 3a). After the distribution of Cu layers, the lamellar particle size was not uniform when films were deposited with a thickness of 30nm. As shown in Fig. 3b, most cell lumina and cell walls could easily be observed. Fig. 3c and its inset display the SEM images of Cu films over wood cell walls. As can be seen, the irregular shape of lamellar particles completely covered the wood surface with the thickness of approximately 50 nm.

However, as the thickness of Cu films further increased, the films deposited on the 100 nm thick Cu layer. Fig. 3d shows that lamellar particles formed spherical - like particles due to the lamellar particles coalescence. The excessive spherical-like particles may evenly coat the outside surface of wood cell walls and cell lumens (Wang et al., 2012). Higher density of coalescence could cause the particle nucleation occurring faster, resulting in forming a flat and smooth surface of the main layer. Therefore, a further increase in Cu layer thickness over 150nm led to a smooth and crack-free surface because the cell wall topography and the closely packed lamellar structure disappeared as shown in Fig. 3e. The morphology of the Cu films on the wood surface strongly depended on the thickness of the sputtering process. Fig. 3f shows the elements carbon, oxygen, and copper from the spectra of the Cu-coated wood. The carbon and oxygen stemmed from the wood substrate. A signal from Cu present in the spectrum of the Cu-coated wood substrate was found, and no other elements were detected, implying that the nanostructures were primarily caused by Cu (Supplementary Material).

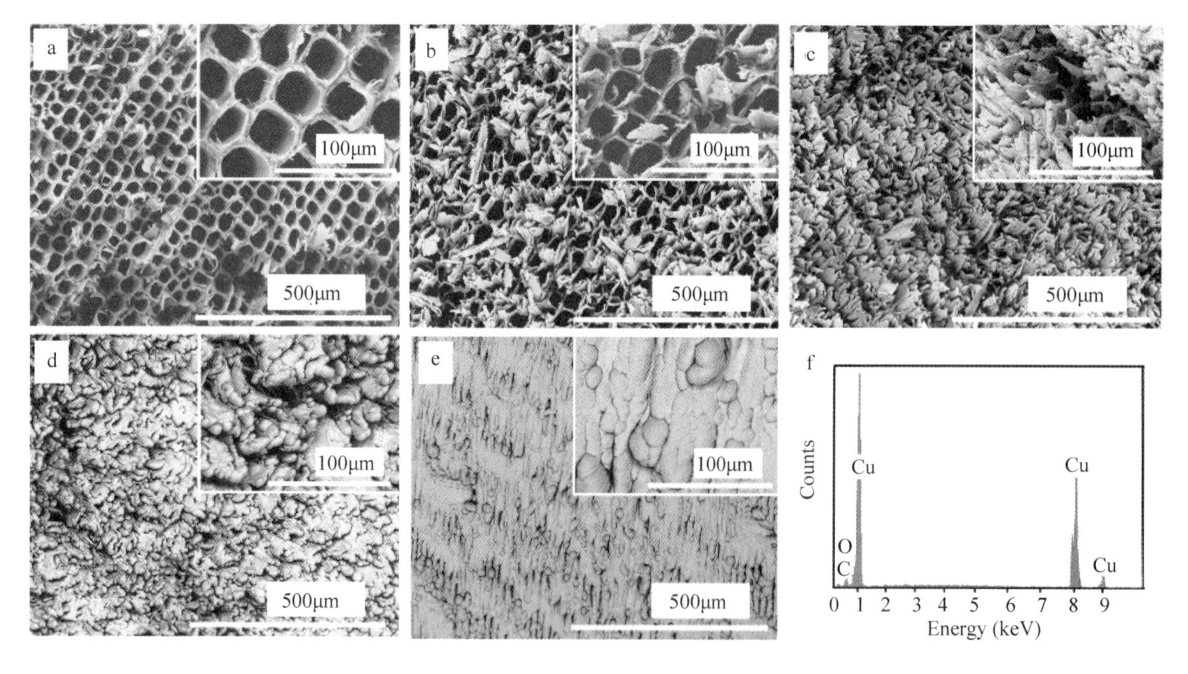

Fig. 3 SEM images of the surfaces of (a) the pristine wood, (b~e) the Cu-treated wood with different deposition thicknesses of 30nm, 50nm, 100nm, 150nm, respectively. (f) EDS spectrum of the Cu-treated wood

3.2 Crystal structure and chemical composition of the as-prepared surface

In Fig. 4b, the diffraction peaks at 14.91° and 23.1° belonged to the (101) and (002) crystal planes of cellulose in the pristine wood. For the Cu-treated wood, all spectra were in agreement with the standard data available for Cu (JCPDS file 4-0836).

Cu has a face centered cubic lattice (Yuan et al. 2013). The XRD spectrum of the Cu-treated wood showed three main peaks at 43.30°, 50.43° and 74.13°, which were attributed to the cubic copper phases with orientation planes (111), (200) and (220), respectively, and no other phases were detected. Therefore, it can be concluded that the as-prepared coating on the wood surface was pure Cu.

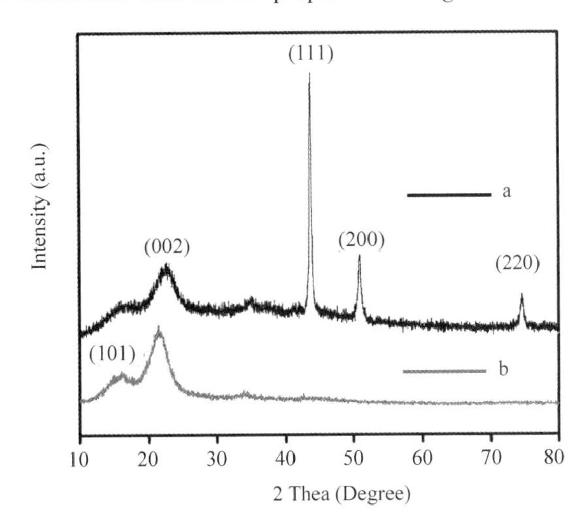

Fig. 4 XRD spectra of the Cu-treated wood (curve a, top) and the pristine wood (curve b, bottom)

In order to obtain a superhydrophobic surface, the perfluorocarboxylic acid was used as a low surface energy material to modify the rough Cu-treated wood (Wang et al. 2008).

In Fig. 5, the FTIR spectra of the pristine wood and the hydrophobized Cu-treated wood. The band at 3330.29 cm^{-1} corresponding to the stretching vibrations of OH groups in the wood shifted to larger wavenumbers of 3338.48 cm^{-1}, and the intensity decreased relatively, indicating that the hydrophilic groups of the wood decreased in the hydrophobized Cu-treated wood. The new band appeared at 1681.32 cm^{-1} and 1602.52 cm^{-1}, corresponding to the C=O stretching vibration of coordinated COO-moieties to Cu^{2+} ions. The presence of the CF, CF_2 and CF_3 bonds located at 1208 cm^{-1} and 610 cm^{-1} revealed that the fluoro-carbon groups were constructed on the Cu-treated wood surface, which might be responsible for the superhydrophobic performance (Gao et al. 2016). Copper nanoplates can be oxidized by the dissolved perfluorocarboxylic acid in ethanol, which provided an acidic environment for the copper oxidization (Liu and Jiang 2011). Cu^{2+} ions were released continuously from the copper nanoplates into the perfluorocarboxylic acid solution according to the reaction equation (1), while copper ions can be captured by coordination with perfluorocarboxylic acid molecules (Xi et al. 2008), forming the copper carboxylate according to the reaction equation (2). It can be deduced that hydrophobic groups were formed on the Cu-treated wood surface after the modification with perfluorocarboxylic acid (Supplementary Material).

$$2Cu+O_2+4H^+ \rightarrow 2Cu^{2+}+2H_2O \tag{1}$$

$$2Cu^{2+}+4CF_3(CF_2)_8COOH \rightarrow Cu_2[CF_3(CF_2)_8COO]_4+4H^+ \tag{2}$$

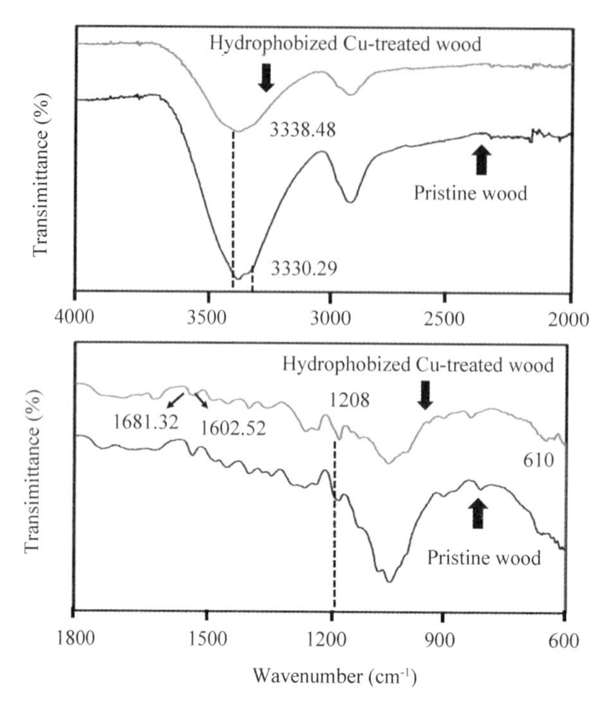

Fig. 5 FTIR spectra of pristine wood and the hydrophobized Cu-treated wood

3.3 Wettability of the superhydrophobic surface

The surface wettability of the as-prepared wood was examined by CA measurements as shown in Fig. 6. The pristine wood was hydrophilic with a CA of only 69° (Fig. 6a). CA of the wood surface modified by the self-assembly of perfluorodecanoic acid monolayer was increased to about 131°, showing the hydrophobic property (Fig. 6b). However, the wood deposited with the 50nm thick Cu layer presented hydrophobicity with a CA of 142°and SA of 15° (Fig. 6c). In Fig. 6d, the CA reached 131° due to the further increase in Cu layer thickness. After the 50nm thick Cu layer deposition followed by the self-assembly of perfluorodecanoic acid monolayer, the wood surface exhibited excellent superhydrophobic property. The water (5μL) displayed a typical spherical shape on the wood surface with a CA of 154±1° and SA of 3.5°. The results indicated that the wood surface wetting property was transformed from hydrophilic to superhydrophobic.

To further understand the relationship between surface roughness and water wetting behavior, Wenzel and Cassie models were proposed. When depositing 50nm thick films on the irregular lamellar, the particles completely covered the wood sample and were closely packed on the wood surface, creating a high surface roughness. In this case, larger amounts of air were trapped in the interspaces or cavities of the superhydrophobic wood surfaces, representing the Cassie state model (Cassie 1944),

$$\cos \theta c = f(\cos\theta + 1) - 1 \quad (3)$$

where f is the fraction of the solid surface in contact with liquid, the fraction of air in contact with liquid at the surface is $1-f$; and θc and θ represent the CAs on rough and smooth surfaces, respectively. The CA of θc was approximately 154° and the CA of θ on the smooth surface modified with perfluorodecanoic acid was 131°. Therefore, the fraction of air in contact with liquid at the surface was calculated as 0.98 by the use of the Cassie equation (Nosonovsky 2007). Once the deposition thickness reached 150 nm, the roughness did not further increase with the particle coalescence caused by higher densities of nucleation. In this case, the water droplet did invade into or fill up the grooves according to the Wenzel model (Lafuma and Quéré 2003).

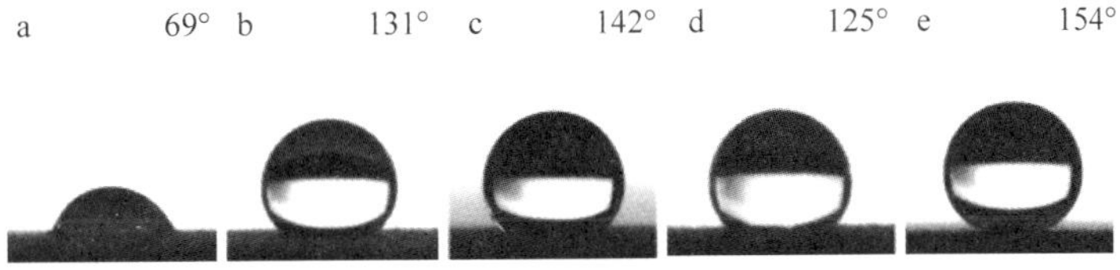

a 69° b 131° c 142° d 125° e 154°

Fig. 6 Contact angle profiles of water droplets on different surfaces: (a) pristine wood; (b) wood surface modified by perfluorodecanoic acid; (c and d) wood surface with 50nm and 150nm thick Cu layer; (e) wood surface with 50nm thick Cu layer treated with perfluorodecanoic acid

3.4 Mechanical, chemical and thermal durability of the superhydrophobic coatings

To assess the mechanical resistance of the superhydrophobic wood, the sand abrasion and sandpaper abrasion test were performed. After 100 times of tests by the sand impacts, a cave formed underneath the Cu-treated wood surface (Fig. 7a). The fracture modes of impact tests are displayed in Fig. 7b. The Cu lamellar particles were fractured after the impact of sand abrasion experiment; the fractured line of Cu-treated wood was jagged and irregular, exposing the porous structure of the wood surface. Fractal-like lamellar particles almost spread all over the wood surface. The as-prepared rough surface completely maintained the submicrometer morphology. This result indicated that the surface retained its superhydrophobicity after 100 sand impact tests (Fig. 7c). By contrast, a small CAH increased from 4.9° to 5.7° (Fig. 7d). Thus, the wood coated with Cu films had a good superhydrophobic property with high water resistance.

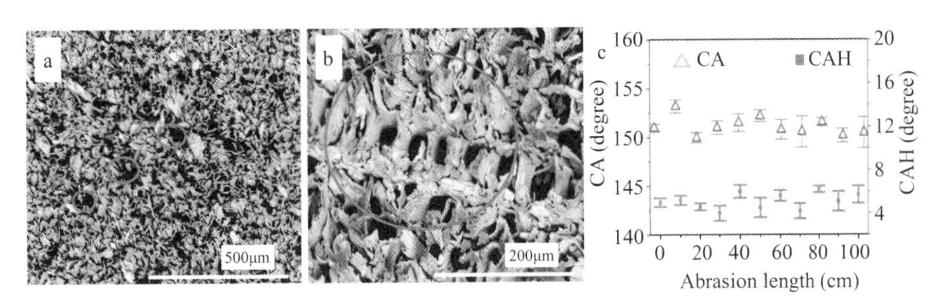

Fig. 7 (a and b) SEM image of the surface topography after 100 impingements. (c) CA and CAH changes depending on the abrasion cycles

As a contrast, superhydrophobic wood surfaces were investigated by SEM examination after the samples were sanded on sanding paper with the abrasion length ranging from 50 to 250 cm as shown in Fig. 8. In Fig. 8a, the irregular shape of lamellar particles covered the wood surface after the abrasion of 50 cm length. When the abrasion length was 100 cm, sporadic scratches on the surface are shown in Fig. 8b. The scratched area gradually expanded with increasing abrasion length (Fig. 8c, d). After an abrasion length of 250 cm length, the initially vertical lamellar structures on the wood surface were broken (Fig. 8e). Fig. 8f shows the changes of CA and CAH with the increase in abrasion length. When the as-prepared sample abrasion length was less than 150 cm, the surface appeared to be superhydrophobic with a CA of about 152°, while the CAH increased to 9.5°. Surface morphology changes generated high surface energy due to the remarkable roughness changes, resulting in a decrease in CA after the abrasion of 200 cm length. After the 250 cm abrasion, the CA on the surface decreased below 110° and CAH increased to over 30°, resulting in the loss of the superhydrophobicity property due to the removal of most surface lamellar structures (Supplementary Material).

Fig. 8 SEM images of surface (a~e) after abrasion on sandpaper (50-250 cm abrasion length). (f) Effect of abrasion on CA and CAH at longer abrasion length

To assess the chemical durability, the superhydrophobic wood was immersed in the conventional solvent for 72 h. It was clearly illustrated that CAs still exhibited its superhydrophobic property, which can be attributed to the insoluble $Cu_2[CF_3(CF_2)_8COO]_4$ in most conventional solvents (Fig. 9a). The thermal stability of the superhydrophobic wood was investigated via the contact angle tests by immersing samples in water at different temperatures for 1 h. The results indicated that the superhydrophobic wood presented high repellency (Fig. 9b), which can resist 283-413 K water immersion for 1 h because the perfluoroalkyl groups formed a mesogen-like structure on the wood surfaces as well as in the bulk, which might prevent the reconstruction of the surface structure. In addition, the superhydrophobic durability of as-prepared samples was evaluated by exposing samples to ambient condition for six month. The resulted CAs still maintained greater than 150°.

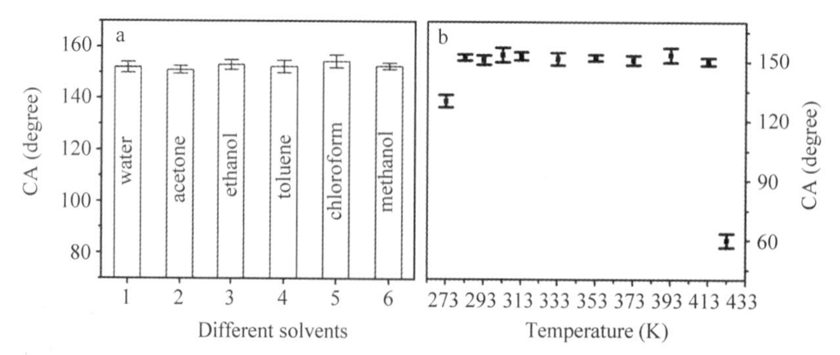

Fig. 9 (a) CAs of the superhydrophobic wood after immersing in different solvents for 72 h. (b) CAs of the superhydrophobic wood after immersing in water for 1 h at different temperatures

4 Conclusions

In summary, in this study, an easy and efficient method for wood surface superhydrophobicity was developed through the DC magnetron sputtering process. The hydrophobic effects of the thickness variation of Cu layer on wood surface were firstly explored. The 50 nm thick Cu films deposited on the wood surface had the roughest surface, and subsequent surface modification with the perfluorodecanoic acid led to a superhydrophobic surface. The results demonstrated that the deposition of Cu films with proper thickness on wood surfaces could be an effective method to achieve the roughest surface. This method did not need rigorous conditions and can also be easily scaled up to create large - area uniform superhydrophobic surface structures. More importantly, wood resources could potentially be further used in various applications. Overall, this paper proposed a method that can deposite Cu films on wood surface by magnetron sputtering.

Acknowledgements

This research was supported by The National Natural Science Foundation of China (grant numbers 31470584, 31400497), the Fundamental Research Funds for the Central Universities (grant numbers 2572016AB64), Heilongjiang Postdoctoral Financial Assistance (grant numbers LBH - Z13001), General Financial Grant from the China Postdoctoral Science Foundation (grant numbers 2014M561311). Overseas Expertise Introduction Project for Discipline Innovation, 111 Project (No. B08016).

Reference

[1] And B Q, Shen Z. Fabrication of Superhydrophobic Surfaces by Dislocation-Selective Chemical Etching on Aluminum, Copper, and Zinc Substrates Langmuir the Acs[J]. J. Sur & Coll., 2005, 21: 9007-9009.

[2] Bang K H, Hwang D K, Jeong M C, et al. Comparative studies on structural and optical properties of ZnO films grown on c-plane sapphire and GaAs (001) by MOCVD[J]. Solid State Commun., 2003, 126: 623-627.

[3] Cassie ABD. Wettability of porous surfaces[J]. Trans Faraday Soc., 1994, 40: 546-551.

[4] Deng X, Mammen L, Butt H J, et al. Candle Soot as a Template for a Transparent Robust Superamphiphobic[J]. Coating Science., 2012, 335: 67-70.

[5] Feng L et al. Super-Hydrophobic Surfaces: From Natural to Artificial[J]. Adv Mater., 2002, 14: 1857-1860.

[6] Gao L, Lu Y, Cao J, Li J, Sun Q. Reversible Photocontrol of Wood-Surface Wettability Between Superhydrophilicity and Superhydrophobicity Based on a TiO_2 Film[J]. J. Wood Chem Technol., 2015, 35: 365-373.

[7] Gao L, Lu Y, Li J, et al. Superhydrophobic conductive wood with oil repellency obtained by coating with silver nanoparticles modified by fluoroalkyl silane[J]. Holzforschung., 2015, 70: 63-68.

[8] Gao L, Xiao S, Gan W, et al. Durable superamphiphobic wood surfaces from Cu_2O film modified with fluorinated alkyl silane[J]. RSC Adv., 2016, 5: 98203-98208.

[9] Hill CAS. Wood modification : chemical, thermal and other processes[J]. John Wiley & Sons, Inc., 2006.

[10] Inoue M, Hashizume K, Tsuchikawa H. The properties of aluminum thin films sputter deposited at elevated temperatures[J]. J Vac Sci Technol A., 1988, 6: 1636-1639.

[11] Jin C, Yao Q, Li J, et al. Fabrication, superhydrophobicity, and microwave absorbing properties of the magnetic γ-Fe_2O_3/bamboo composites[J]. Mater Design., 2015, 85: 205-210.

[12] Kang W, Lee NH. Mathematical modeling to predict drying deformation and stress due to the differential shrinkage within a tree disk[J]. Wood Sci and Tech., 2002, 36: 463-476.

[13] Kang W, Lee NH. Relationship between radial variations in shrinkage and drying defects of tree disks[J]. J wood sci., 2004, 50: 209-216.

[14] Kelly PJ, Arnell RD. Magnetron sputtering: a review of recent developments and applications Vacuum. 2000, 56: 159-172.

[15] Khedir KR, Kannarpady GK, Ishihara H, et al. Morphology control of tungsten nanorods grown by glancing angle RF magnetron sputtering under variable argon pressure and flow rate[J]. Phys Lett A., 2010, 374: 4430-4437.

[16] Kozono Y, Komuro M, Narishige S, et al. Structures and magnetic properties of Fe/Cu multilayered films fabricated by a magnetron sputtering method[J]. J. Appl Phys., 1987, 61: 4311.

[17] Lafuma A, Quéré D. Superhydrophobic states[J]. Nat Mater Nature materials., 2003, 2: 457-460.

[18] Li J, Yu H, Sun Q, et al. Growth of TiO_2 coating on wood surface using controlled hydrothermal method at low temperatures[J]. Appl Surf Sci., 2010, 256: 5046-5050.

[19] Li X, Liu C, Shi T, et al. Preparation of multifunctional Al alloys substrates based on micro/nanostructures and surface modification [J]. Mater Design., 2017, 122: 21-30.

[20] Lin Y, Shen Y, Liu A, et al. Bio-inspiredly fabricating the hierarchical 3D porous structure superhydrophobic surfaces for corrosion prevention[J]. Mater Design., 2016, 103: 300-307.

[21] Liu K, Jiang L. Metallic surfaces with special wettability[J]. Nanoscale., 2011, 3: 825-838.

[22] Mukherjee SK, Joshi L, Barhai PK. A comparative study of nanocrystalline Cu film deposited using anodic vacuum arc and dc magnetron sputtering[J]. Surf Coat Technol., 2011, 205: 4582-4595.

[23] Nosonovsky M. On the range of applicability of the Wenzel and Cassie equations[J]. Langmuir the Acs Journal of Surfaces & Colloids., 2007, 23: 9919-9920.

[24] Pang SC, Anderson MA. Novel Electrode Materials for Thin-Film Ultracapacitors: Comparison of Electrochemical Properties of Sol-Gel-Derived and Electrodeposited Manganese Dioxide[J]. J. Electrochem Soc., 2000, 147: 444-450.

[25] Reina A et al. Large Area, Few-Layer Graphene Films on Arbitrary Substrates by Chemical Vapor Deposition[J]. Nano Lett., 2009, 9: 655-663.

[26] Sun Q, Lu Y, Liu Y. Growth of hydrophobic TiO_2 on wood surface using a hydrothermal method[J]. J. Mater sci., 2011, 46: 7706-7712.

[27] Sun Q et al. Hydrothermal fabrication of rutile TiO_2 submicrospheres on wood surface: An efficient method to prepare UV-protective wood[J]. Mater Chem Phys., 2012, 133: 253-258.

[28] Wang C, Piao C, Lucas C. Synthesis and characterization of superhydrophobic wood surfaces[J]. J. Appl Polym Sci., 2011, 119: 1667-1672.

[29] Wang Q, Zhang B, Qu M, et al. Fabrication of superhydrophobic surfaces on engineering material surfaces with stearic acid[J]. Appl Surf Sci., 2008, 254: 2009-2012.

[30] Wang S, Wang C, Liu C, et al. Fabrication of superhydrophobic spherical-like α-FeOOH films on the wood surface by a hydrothermal method[J]. Colloid Surface A., 2012, 403: 29-34.

[31] Xi J, Feng L, Jiang L. A general approach for fabrication of superhydrophobic and superamphiphobic surfaces[J]. Appl Phys Lett., 2008, 92: 053102.

[32] Xie Y, Krause A, Militz H, Mai C. Coating performance of finishes on wood modified with an N-methylol compound [J]. Prog Org Coat., 2006, 57: 291-300.

[33] Xie Y, Krause A, Militz H, Mai C. Weathering of uncoated and coated wood treated with methylated 1,3-dimethylol-4,5-dihydroxyethyleneurea (mDMDHEU)[J]. Holz als Roh- und Werkstoff., 2008, 66: 455-464.

[34] Yao Q, Wang C, Fan B, et al. One-step solvothermal deposition of ZnO nanorod arrays on a wood surface for robust superamphiphobic performance and superior ultraviolet resistance[J]. Sci Rep., 2016, 6: 35505.

[35] Yu B, Woo P, Erb U. Corrosion behaviour of nanocrystalline copper foil in sodium hydroxide solution[J]. Scripta Mater., 2007, 56: 353-356.

[36] Yuan Z et al. A novel fabrication of a superhydrophobic surface with highly similar hierarchical structure of the lotus

leaf on a copper sheet[J]. Appl Surf Sci., 2018, 285: 205-210.

[37] Zheng J, Bao S, Jin P. TiO_2(R)/VO2(M)/TiO_2(A) multilayer film as smart window: Combination of energy-saving, antifogging and self-cleaning functions[J]. Nano Energy., 2014, 11: 136-145.

[38] Zheng J, Lv Y, Xu S, et al. Nanostructured TiN-based thin films by a novel and facile synthetic route[J]. Mater Design., 2015, 113: 142-148.

[39] Zheng JY, Bao SH, Guo Y, et al. Natural hydrophobicity and reversible wettability conversion of flat anatase TiO_2 thin film[J]. ACS Appl Mater Interfaces., 2017, 6: 1351-1355.

[40] Zhu H et al. Wood-Derived Materials for Green Electronics, Biological Devices, and Energy Applications[J]. Chem Rev., 2016, 116: 9305-9374.

中文题目：木材表面通过磁控溅射制备的 Cu 薄膜后经全氟羧酸改性超疏水性研究

作者：包文慧，张明，贯贞，蔡力平，梁大鑫，李坚

摘要：采用磁控溅射的方法制备超疏水木材，该方法具有简单、可控的优点，可应用于不同形状的基底。铜薄膜沉积在木材的横截面上，再进行全氟羧酸处理。结果表明，溅射沉积厚度对表面形貌的改变起着重要作用。在优化条件下，当沉积厚度在 50 nm 时，接触角为 154°，滑动角可达 3.5°。重要的是，超疏水表面可以抵抗多达 100 次的砂粒磨损冲击，接触角依然保持在 151°。此方法在钟表框架、艺术品、雕刻和装饰品等木制品领域具有广阔的应用前景。

关键词：木材；磁控溅射；超疏水；薄膜

Facile One-pot Synthesis of Wood Based Bismuth Molybdate Nano-eggshells with Efficient Visible-light Photocatalytic Activity*

Yingying Li, Bin Hui, Likun Gao, Fenglong Li, Jian Li

Abstract: Photocatalysis, which harnesses light to degrade pollutants, has been considered as the most green and low-cost approach for environmental remediation. In this work, a novel wood-based photocatalyst was prepared by a simple one-pot way. The as-prepared samples were characterized by X-ray diffraction, scanning electron microscopy, energy dispersive X-ray spectroscopy, UV-Vis diffuse reflectance spectroscopy and X-ray photoelectron spectra, respectively. Results showed that the synthesized bismuth molybdate nano-eggshells were evenly grown on the surface of wood substrate. Additionally, its morphologies were strongly dependent on the initial pH of precursor solution and the presence of wood substrate. The photocatalytic activity of the samples was evaluated based on the decomposition of rhodamine B under visible light irradiation. The sample prepared at pH 6 showed superior performance for the photodegradation of rhodamine B, in comparison to the pure bismuth molybdate powders and the samples prepared at other pH value. The stability, photocatalytic mechanism and the reasons for improving photocatalytic activity were also discussed. This study offers exciting opportunities to achieve the industrialization of photocatalytic wood products.

Keywords: Wood; Photocatalysis; Bismuth molybdate; One-pot synthesis

1 Introduction

As a green technology, photocatalysis has been gathering increasing interest because it can be widely applied to many domains that offer an economic and ecologically safe option to solve energy and pollution problems[1-4]. A representative example is the photosynthesis of green plants, in which trees absorb carbon dioxide from the atmosphere under visible light in the photocatalytic process and store it in the plant tissues[5-8]. However, trees will lose their photocatalytic ability when they are processed into wood products. Nowadays, wooden products are extensively used in daily applications, such as buildings, bridges, furniture and indoor decorations[9]. Furthermore, due to the use of the non-renewable materials, the applications for wood increase significantly with the increasing concerns of population and sustainability. To meet market requirements and compete with other advanced materials, many attempts have been made to create new functions for wood[10-12]. Therefore, the future looks bright for functional wooden products with photocatalytic ability.

---

* 本文摘自 Colloids and Surfaces A, 2018, 556: 284-290.

It is commonly known that wood has three major components which are cellulose, hemicellulose, and lignin. The network of the components and the space of parallel hollow tubes construct the multi-scale porous structure, which offers the penetrability, accessibility and reactivity of wood materials[13, 14]. As a result, the unique composition and subtle hierarchical structures of wood lay the foundation for their further functionalization. In recent years, more and more attention has been forced on semiconductor photocatalytic materials such as TiO_2, ZnO and ZrO_2 due to their potential applications in environmental cleaning, solar energy utilization and hydrogen generation from water splitting[15-18]. Among them, bismuth-based compounds have attracted considerable attention for the strong visible light photocatalytic activities[19, 20]. Bi_2MoO_6 with the layered bismuth oxide family is of special interest due to its dielectric, ion-conductive, luminescent and catalytic properties[21-23]. This was demonstrated in a number of studies that the nontoxic Bi_2MoO_6 could perform as an excellent photocatalyst and solar energy conversion material for pollutant degradation, water splitting and CO_2 reduction, due to its suitable band location and superior spectral properties[24-26].

There are several works investigated on the preparation and properties of titania - wood hybrid materials[27-30], and they found the hybrid photocatalyst shows great photocatalytic activity for organic degradation. However, to the best of our knowledge, there is no work focused on the preparation and photocatalytic properties of wood-based bismuth molybdate (W-BMO). To realize this goal, we propose a simple one-pot way to prepare W-BMO photocatalyst with eggshell structure. The fabricated W-BMO has good visible light responsive photocatalytic activity with the decomposition of rhodamine B (RhB). Its morphology, phase structure, chemical composition, optical properties, stability, photocatalytic mechanism and the reasons of improving photocatalytic activity were investigated.

2 Experimental

2.1 Materials

Wood slices (15 mm×15 mm×0.5 mm) of white poplar (Populus tomentosa Carr.) were used in this experiment. Bismuth nitrate ($Bi(NO_3)_3 \cdot 5H_2O$), sodium molybdate ($Na_2MoO_4 \cdot 2H_2O$), ethanol, ammonia, and rhodamine-B (RhB) was purchased from Shanghai Aladdin Biochemical Technology Co., Ltd. (Shanghai, China). Nitric acid (HNO_3) and sodium hydroxide (NaOH) were purchased from Shanghai Chemical Co. Ltd (Shanghai, China). The ethanol (99.5%) was purchased from Kaitong Chemical Reagent Co., Ltd (Tian jian, China). All of the chemicals were used as received without further purification.

2.2 Fabrication of W-BMO photocatalytic material

Four mmol $Bi(NO_3)_3 \cdot 5H_2O$ was dissolved in 40 mL nitric acid solution (4 mol/L) and stirred for several minutes. A total of 0.484 g of $Na_2MoO_4 \cdot 2H_2O$ was dissolved in 36 mL of deionized water. The white precipitate first appeared and then disappeared when the Na_2MoO_4 solution was dropwise added into the above mixed solution. The different pH value of the precursor suspension was adjusted with 2 mol/L NaOH solution under stirring, while amorphous white precipitate formed during this process. After being stirred for 3 h, the wood slice and the above solution were transferred into a 100 mL stainless steel autoclave. The autoclave was sealed and maintained at 140 ℃ for 12 h, and then cooled down to room temperature. Finally, the resulting samples were removed from the solution, washed with deionized water for several times and dried at 60 ℃ for over 24 h in vacuum. In this experiment, five samples with different pH value of 5, 6, 7, 8 and 9 were prepared, and they were defined as W-BMO-5, W-BMO-6, W-BMO-7, W-BMO-8 and W-BMO-9. It should be mention that the loading amount of bismuth molybdate on the W-BMO-6 was about 20 wt %. Besides, for the purpose of comparison, pure Bi_2MoO_6 was also prepared with the same method except that the wood slice was absent.

2.3 Characterization

The phase structure of the as-prepared products was checked by the X-ray diffraction measurements (XRD, Rigaku, and D/MAX 2200) operated with Cu target radiation (λ = 1.54 Å). The UV-Vis diffuse reflection spectra of the samples were recorded on a UV-Vis spectrophotometer (UV-Vis DRS, TU-1901, China) equipped with an integrated sphere attachment by using $BaSO_4$ as a reference. The surface morphologies and microstructures of as-prepared samples were analyzed by a field-emission scanning electron microscope (JEOL JSM-7500F). The surface components of the samples were determined via energy dispersive X-ray analysis (EDXA). X-ray photoelectron spectra (XPS) measurement was recorded on an ESCALAB 250Xi photoelectron spectrometer using Al $K\alpha$ (1486.6 eV) radiation. Specific surface areas of the samples were measured by the Brunauer-Emmett-Teller (BET) method based on nitrogen adsorption at the liquid nitrogen temperature using a 3H-2000PS2 unit (Beishide Instrument S&T Co., Ltd). Photoluminescence (PL) spectra of the samples were taken with a FluoroMax-4 spectrophotometer.

2.4 Photocatalytic Activity Experiments

The photocatalytic performance of the samples was evaluated by the degradation of rhodamine B (RhB) under visible-light irradiation using a 500 W solar simulator (XES-40S2-CE, SAN-EI Electric) with a 420 nm cutoff filter. The samples were placed 15 cm away from the light source, and the experiments were carried out at room temperature. In each experiment, photocatalyst was added into 50 mL 10 mg · L^{-1} of RhB. Before illumination, the solution was magnetically stirred in the dark for 3 h to establish adsorption/desorption equilibrium between the dye and the catalyst. At given irradiation time intervals, 3 mL of the solution was extracted and centrifuged to remove the potential impurities. The centrifuged solution was monitored with an UV-Vis spectrophotometer in terms of the absorbance at 553 nm during the photodegradation process. An air conditioner was employed to maintain the temperature at 25 ℃.

3 Results and discussion

3.1 Phase structures of samples

The formation of the as-prepared samples was well confirmed by XRD analysis. Fig. 1 presents the XRD patterns of the pristine wood and the W-BMO samples prepared at different pH values. The diffraction peaks centered at 16.1° and 22.2°, which correspond to the (101) and (002) diffraction planes of cellulose in wood substrate. After hydrothermal treatment, the above two peaks were almost the same and some new peaks appeared. As we can see, when the pH values in the reaction system were kept at 5 and 6, peaks centered at 10.8°, 28.0°, 31.9°, 32.5°, 46.2°, 46.6°, 54.9°, 55.4° and 57.7° were observed. These peaks match well with the crystal planes of Bi_2MoO_6(JCPDS 21-0102). Additionally, as the sample prepared at pH 10, its diffraction peaks at 26.9°, 31.0°, 44.1°, 52.7° and 55.1° can be perfectly identified to (111), (200),

(220), (311) and (222) crystal planes of $Bi_{3.64}Mo_{0.36}O_{6.55}$ (JCPDS 43-0446), respectively. As the pH value increased from 5 to 9, the diffraction peaks of Bi_2MoO_6 gradually disappeared whereas $Bi_{3.64}Mo_{0.36}O_{6.55}$ gradually emerged, and no other peaks from possible impurities were detected. The above results indicated that the bismuth molybdate was formed on wood surface through the hydrothermal process.

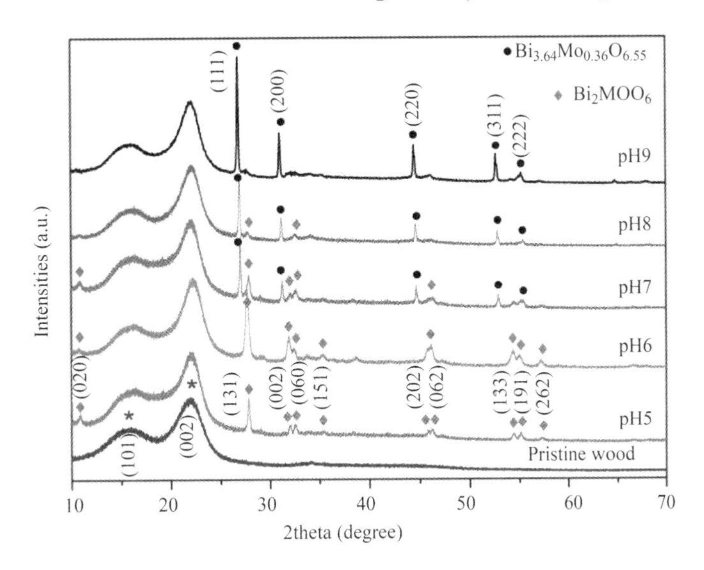

Fig. 1 XRD patterns of the pristine wood and the W-BMO samples prepared at different pH values

3.2 Surface morphologies

SEM experiments were performed in order to observe the micromorphology of the sample surfaces for determining the morphology and spatial distribution of the crystals. Fig. 2 shows the morphology of the pristine wood and the morphological evolution of the W-BMO samples, and the inset at the upper right corner shows the digital photograph of the samples. The surface of pristine wood is shown in Fig. 2a, which was smooth and free of any other materials. In Fig. 2b, after the hydrothermal reaction, the wood structures were rough but still clearly visible. Furthermore, it can be clearly seen from Fig. 2c that the eggshell crystals were evenly distributed on the wood surface. Additionally, when the pH value of the precursor solution increased to 9, the sample consisted of uniform cubic crystals. That is, increasing the pH value changes the shape of the as-obtained samples from microspheres to cubic crystals. Furthermore, pure Bi_2MoO_6 was prepared at the same condition, as shown in Fig. 3, the morphologies of pure Bi_2MoO_6 were very different from the samples that were grown on wood surface. The above results suggested that the surface morphologies of W-BMO samples strongly depend on the initial pH of the precursor solution and the presence of wood substrate. The EDS pattern of the W-BMO-6 demonstrates the existence of C, Bi, Mo and O elements (Fig. 2g). It proved that bismuth molybdate was grown on the wood surface through the hydrothermal process, which is consistent with the above XRD and SEM results.

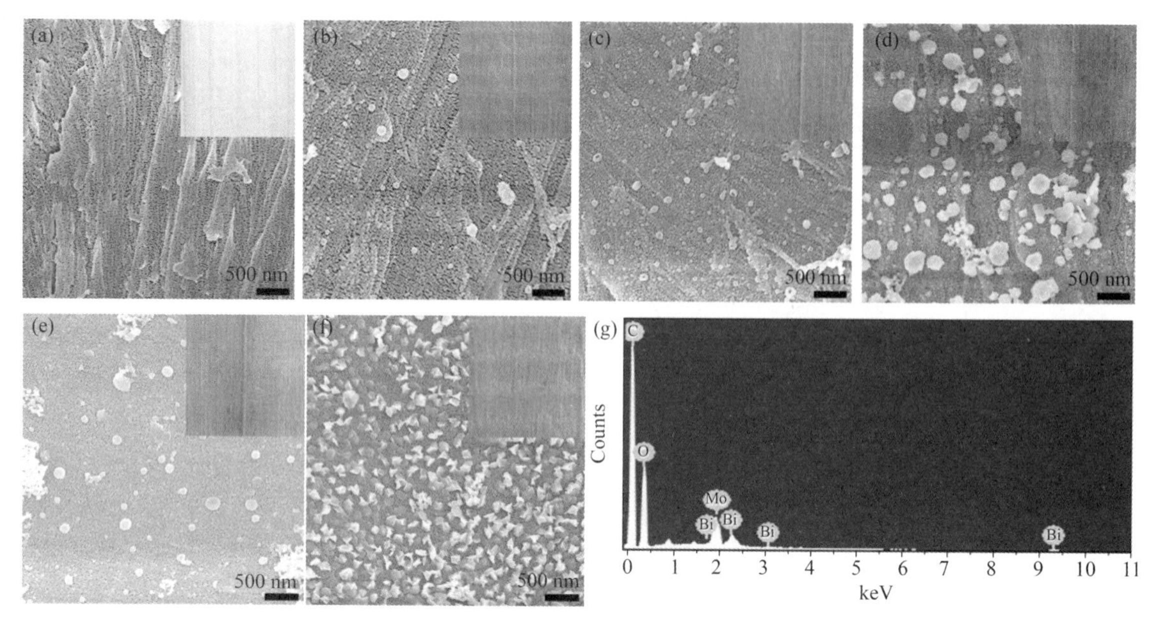

Fig. 2 SEM images of pristine wood (a), W-BMO-5 (b), W-BMO-6 (c), W-BMO-7 (d), W-BMO-8 (e), W-BMO-9 (f) and EDS spectra of W-BMO-6 (g). (inset, the digital photograph of the samples)

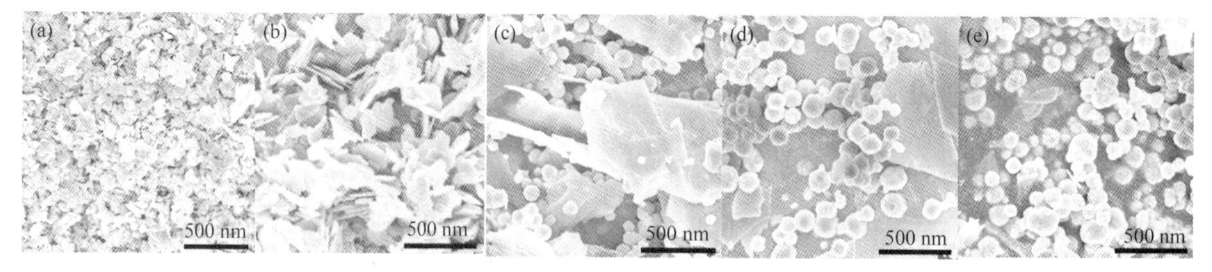

Fig. 3 SEM images of the as-prepared Bi_2MoO_6(a-e: pH 5, pH 6, pH 7, pH 8 and pH 9)

3.3 Optical properties

The UV-Vis diffuse reflectance spectra (DRS) of W-BMO samples and pure Bi_2MoO_6 were measured to evaluate light absorption properties. From Fig. 4, the shape of the UV-Vis DRS spectra was almost the same and exhibited an intense absorption in the visible-light range, which reveals the property of being photoactive under visible light illumination. Furthermore, the absorption intensity of the as-prepared samples changed with different pH values. We can clearly see that, with the exception of W-BMO-8, the absorption intensities of W-BMO samples were all higher than that of pure Bi_2MoO_6 in the visible light region. The DRS results indicated that the visible light absorption of W-BMO was enhanced. Therefore, the visible light response of the W-BMO-6 was significantly improved, and thus it should have enhanced photocatalytic activity compared with the pure Bi_2MoO_6. Furthermore, the estimated band gap of the Bi_2MoO_6 was around 3.17 eV, which is consistent with the literature value[31].

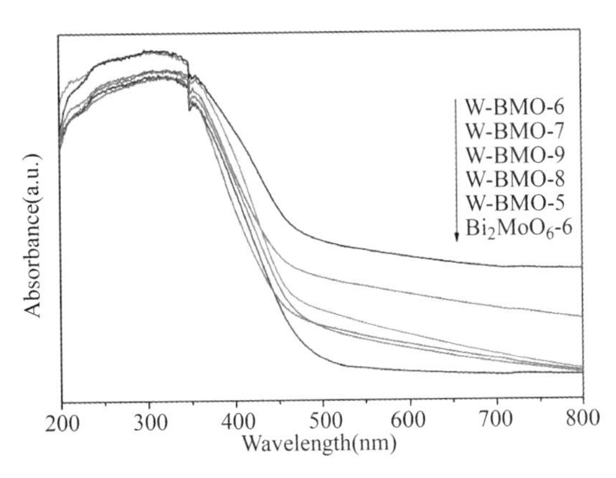

Fig. 4 UV-Vis diffuse reflectance spectra of the as-prepared samples

3.4 Photocatalytic activity

To evaluate the photocatalytic activity of the samples, the typical RhB dye is selected as representative. Before illumination, the dark adsorption over pristine wood, Bi_2MoO_6 and W-BMO samples were performed. As shown in Fig. 5, all the samples could reach adsorption-desorption equilibrium within 180 min, and different quantity of RhB was absorbed. Additionally, the degradation dynamic curves of the RhB dye over different samples are also shown in Fig. 5. C is the concentration of RhB during the photocatalytic process, and C_0 is the initial concentration of RhB before illumination (monitored at 554 nm). Without any catalyst, only a slight decomposition (less than 0.5%) of RhB was detected under visible light irradiation for 60 min. Therefore, the direct photolysis of RhB under visible irradiation was negligible. Besides, the concentration of RhB is almost not changed in the presence of wood, which suggested that the pristine wood lack the ability of photocatalytic degradation. It should be pointed out that the W-BMO-6 had a greater photocatalytic degradation ability than other samples and the photodegradation efficiency of RhB was nearly 99% after 60 min irradiation. On the basis of the above analysis, we concluded that the W-BMO-6 possesses greatest photocatalytic activity among all the

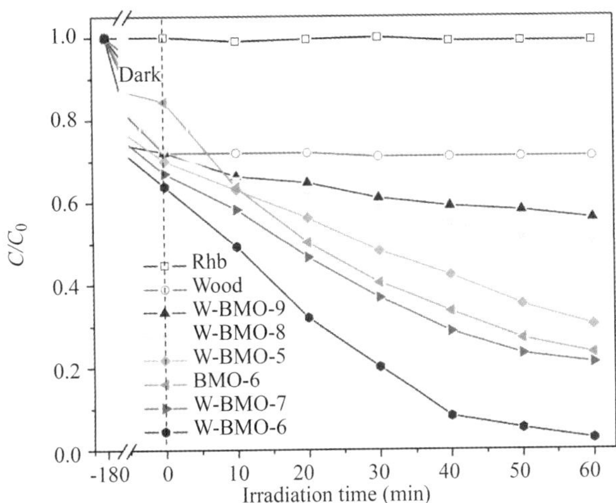

Fig. 5 Photocatalytic degradation efficiencies of RhB under visible-light irradiation in the presence of different photocatalysts

samples. Furthermore, the photocatalytic performance of W-BMO-6 in the first four consecutive cycles was carried out. As can be seen from Fig. 6, the RhB degradation efficiency decreased from 99% (first run) to 90% (fourth run). Besides, the XRD pattern of the freshly prepared sample was similar to the sample after recycling reactions (Fig. 7). The above results confirm that the good stability and durability of W-BMO-6.

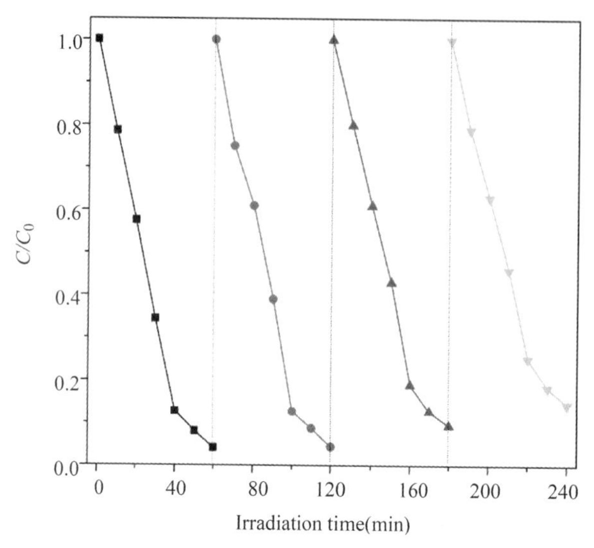

Fig. 6 Cycling runs in photocatalytic degradation of RhB in the presence of W-BMO-6 under visible light irradiation

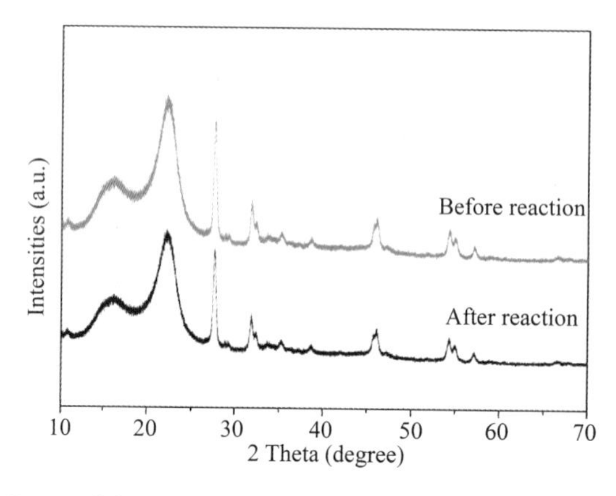

Fig. 7 XRD patterns of the freshly prepared sample and the sample after reactions

3.5 XPS analysis

The XPS analysis was conducted to determine the chemical composition of the W-BMO composites and to identify the chemical state of the element. The fully scanned XPS spectra of the pristine wood and the as-prepared W-BMO-6 samples are shown in Fig. 8a, demonstrating that Bi, Mo, O and C elements existed on the surface of the W-BMO-6 sample. Furthermore, the high resolution XPS spectrum of W-BMO-6 in the Bi 4f region is shown in Fig. 8b. The two peaks located at 158.9 and 164.1 eV could be assigned to the $4f_{7/2}$ and $4f_{5/2}$ orbits, respectively. In Fig. 8c, the Mo $3d_{5/2}$ and Mo $3d_{3/2}$ peaks were observed at 232.1 and 235.3 eV, respectively, indicating that the chemical state of Mo^{6+} was present in Bi_2MoO_6. In the O 1s spectra, the wide asymmetric peaks can be deconvoluted into three component peaks. The peaks were attributed to the Bi-O bond at 529. 5 eV, the Mo-O bond at 530. 2 eV and hydroxyl groups (O-H) at 531. 5 eV[22, 31, 32]. The Bi-O and M-O binding energies in BMO are about 0. 4 and 0. 3 eV higher than those in Bi_2MoO_6, respectively. The similar phenomenon was also observed in the Bi 4f and Mo 3d spectra. This could be attributed to the interaction between wood and Bi_2MoO_6, or the formation of eggshell Bi_2MoO_6.

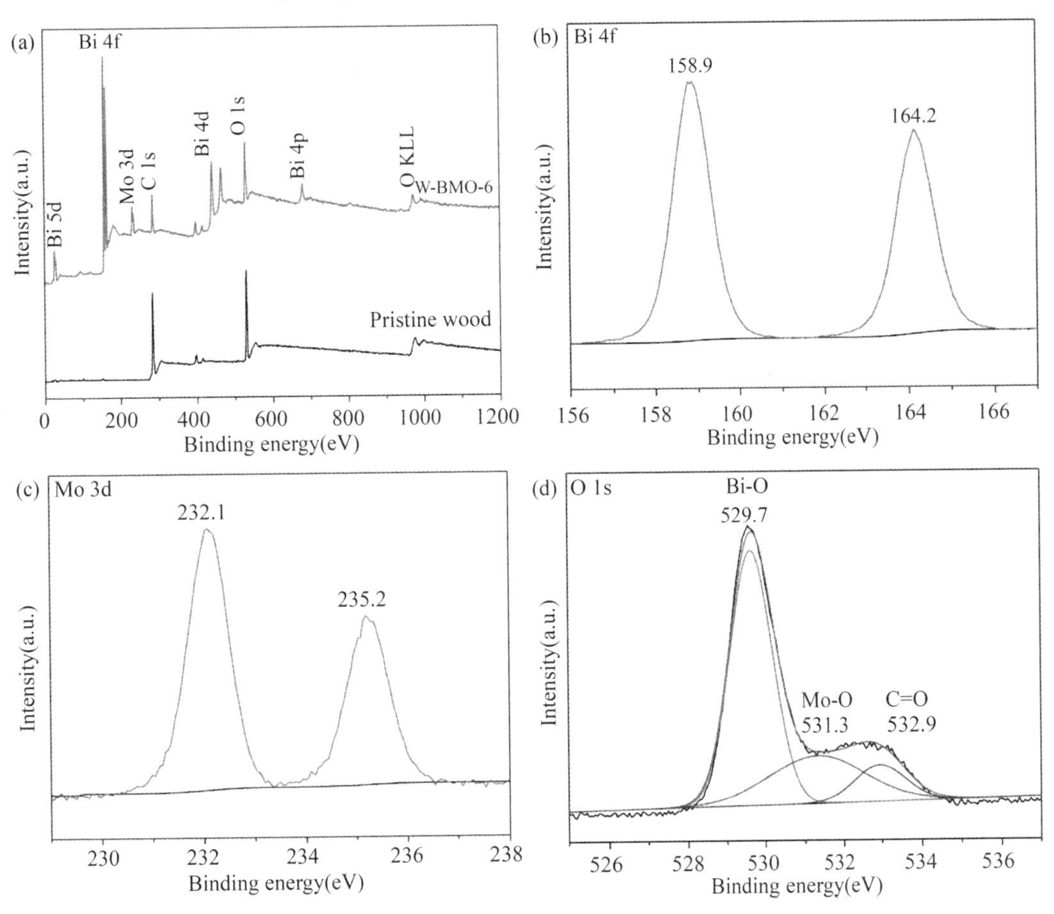

Fig. 8 XPS survey spectra (a) for the W-BMO-6 and Bi_2MoO_6. High-resolution XPS spectra of the Bi 4f (b), Mo 3d (c) and O 1s (d) for W-BMO-6

3. 6 Reasons of photocatalytic activity enhancement

In order to obtain an insight into the main reasons for improving photocatalytic activity of samples, the adsorption activity (surface area) and light-harvesting ability that significantly influence photocatalytic activity are investigated in detail. In Fig. 3, it is clearly seen that the pure bismuth molybdate crystals were severely agglomerated, since the Gibbs free energy of the surface of small-sized nanoparticles is usually very high due to the large surface to volume ratio[33-35]. On the contrary, the eggshell bismuth molybdate crystals of the W-BMO-6 sample were evenly grown on the wood substrate (Fig. 2c). Therefore, the W-BMO-6 sample could supply greater contact area of catalyst to dye than that of pure bismuth molybdate. Furthermore, the specific surface area and the total pore volume of the samples were carried out and listed in Table 1. As we can see,

W-BMO-6 exhibited the largest BET surface area (22.081 $m^2 \cdot g^{-1}$), while the value was 12.446, 19.38 and 20.597 $m^2 \cdot g^{-1}$ for Bi_2MoO_6, W-BMO-5 and W-BMO-7, respectively. Additionally, the W-BMO-6 sample also exhibited the largest pore volume. The above results suggested that the W-BMO-6 sample can supply more surface active sites, leading to the high adsorption of RhB and thus is beneficial to the photocatalytic performance (Fig. 5). In order to investigate the transfer and recombination processes of the photogenerated electron-hole pairs in Bi_2MoO_6 and W-BMO-6, PL spectra were obtained. Fig. 9 shows the PL spectra of pure Bi_2MoO_6 and W-BMO-6 excited by 354 nm. From the figure, it can be observed that the pure Bi_2MoO_6 has a strong emission peak centered at around 468 nm, which is consistent with the previous study[35-37]. Besides, here is a significant decrease in the PL intensity of W-BMO-6 compared to that of pure Bi_2MoO_6. It is generally accepted that a weaker intensity of the peak represents a lower recombination rate of charge carriers[38-41]. The results suggested that the W-BMO-6 sample could effectively restrain the recombination of photoinduced electron-hole pairs. The DRS results indicated that the visible light absorptions of W-BMO were enhanced (Fig. 4). The increasing absorption intensity can be ascribed to the carbonization of wood materials, which is caused by the hydrothermal process. Therefore, the visible light response of the W-BMO-6 was significantly improved, which therefore enhances the utilization of the solar spectrum to increase the photocatalytic activity.

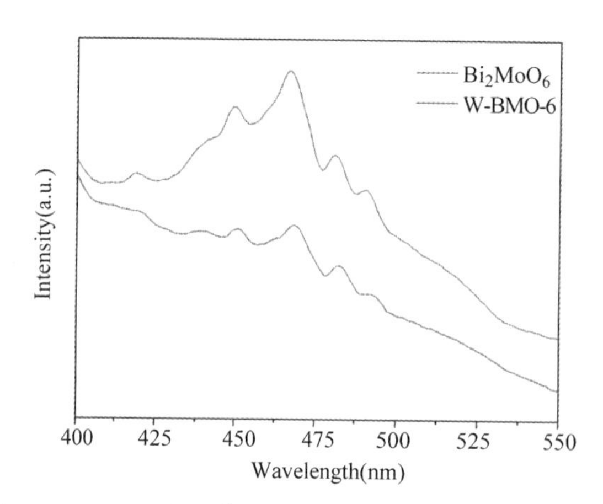

Fig. 9 Photoluminescence spectra of pure Bi_2MoO_6 and W-BMO-6 excited by 354 nm

Table 1 The surface area and total pore volume values of the W-BMO-6 and Bi_2MoO_6

Sample	W-BMO-5	W-BMO-6	W-BMO-7	Bi_2MoO_6
Surface area ($m^2 \cdot g^{-1}$)	19.38	22.081	20.597	12.446
Total pore volume ($cm^3 \cdot g^{-1}$)	0.069	0.084	0.082	0.050

3.7 Photocatalytic mechanism

The wood-based bismuth molybdate photocatalyst was successfully synthesized by one pot reaction. The shape of the photocatalyst crystals could be controlled by tuning of the pH value of the precursor solution. The presence of wood substrate also affected the final product. Photocatalysis investigations revealed that the W-BMO-6 exhibited the greatest photocatalytic performance for the photodegradation of RhB compared with the pure Bi_2MoO_6 and the samples prepared at other pH values. The improved photocatalytic activity could be attributed to the greater light absorption intensities and the superior adsorption capacity of the sample. The as-prepared wood -based photocatalyst is promising for applications in visible-light-driven catalyst, indoor air purification and water pollutant degradation. This work provided new insight into the mass preparation of wood-based functional nanomaterial with high photocatalytic capacity[42].

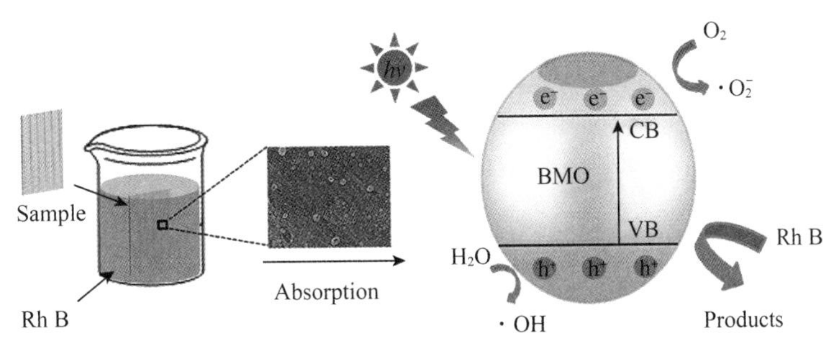

Fig. 10 Schematic diagram of the possible mechanism for degrading organic pollutants over the W-BMO-6 photocatalyst

4 Conclusions

The wood based bismuth molybdate photocatalyst was successfully synthesized by one pot reaction. By tuning of the pH value of the precursor solution, the shape of the photocatalyst crystals could be controlled. The presence of wood substrate also affected the final product. Photocatalysis investigations revealed that the W-BMO-6 exhibited the highest photocatalytic performance for the photodegradation of RhB compared with the pure Bi_2MoO_6 and the samples prepared at other pH value. The improved photocatalytic activity could be attributed to the higher light absorption intensities, the superior adsorption capacity and a lower recombination rate of charge carriers of the sample. The as-prepared wood based photocatalyst is promising for applications in visible-light-driven catalyst, indoor air purification and water pollutant degradation. This work provided new insight into the design and preparation of wood based functional eggshell nanomaterial with high photocatalytic capacity.

Acknowledgements

The authors gratefully acknowledge the financial support from the National Natural Science Foundation of China (31470584), the Overseas Expertise Introduction Project for Discipline Innovation, 111Project (No. B08016) and the Fundamental Research Funds for the Central Universities (Grant 2572017AB08)

Conflict of interest

The authors declared that they have no conflict of interest.

Reference

[1] Malato S, Blanco J, Vidal A, Richter C. Photocatalysis with solar energy at a pilot-plant scale: An overview[J]. Appl. Catal. B-Environ., 2002, 37: 1-15.

[2] Tu W, Zhou Y, Zou. Versatile Graphene-Promoting Photocatalytic Performance of Semiconductors: Basic Principles, Synthesis, Solar Energy Conversion, and Environmental Applications[J]. Adv. Funct. Mater., 2013, 23: 4996-5008.

[3] Lin X, Xu D, Xi Y, et al. Construction of leaf-like g-C_3N_4/Ag/$BiVO_4$ nanoheterostructures with enhanced

photocatalysis performance under visible-light irradiation[J]. Colloids Surf. A, 2017, 513: 117-124.

[4] Liu B, Wang Y, Shang S, et al. One-step synthesis of mulberry-shaped TiO_2-Au nanocomposite and its H_2 evolution property under visible light[J]. Colloids Surf. A, 2018, 553: 203-209.

[5] Arétouyap Z, Bisso D, Njandjock Nouck P, et al. The equatorial rainforest of Central Africa between economic needs and sustainability requirements[J]. J. Environ. Manage., 2018, 206: 20-27.

[6] Schimel D S. Terrestrial ecosystems and the carbon cycle[J]. Global Change Biol., 1995, 1: 77-91.

[7] Gunderson C A, Wullschleger S D. Photosynthetic acclimation in trees to rising atmospheric CO_2: A broader perspective[J]. Photosynth. Res., 1994, 39: 369-388.

[8] Štursová M, Šnajdr J, Cajthaml T, et al. When the forest dies: the response of forest soil fungi to a bark beetle-induced tree dieback[J]. Isme J., 2014, 8: 1920.

[9] Gustavsson L, Sathre R. Variability in energy and carbon dioxide balances of wood and concrete building materials[J]. Build. Environ., 2006, 41: 940-951.

[10] Li Y, Li J. Fabrication of reversible thermoresponsive thin films on wood surfaces with hydrophobic performance[J]. Prog. Org. Coat., 2018, 119: 15-22.

[11] Köklükaya O, Carosio F, Grunlan J, et al. Flame-Retardant Paper from Wood Fibers Functionalized via Layer-by-Layer Assembly[J]. ACS Appl. Mater. Inter., 2015, 7: 23750-23759.

[12] Gao L, Gan W, Cao G, et al. Fabrication of biomass-derived C-doped Bi_2WO_6 templated from wood fibers and its excellent sensing of the gases containing carbonyl groups[J]. J. Colloid. Interf. Sci., 2017, 529: 487-494.

[13] Yang C, Gao Q, Tian W, et al. Superlow load of nanosized MnO on a porous carbon matrix from wood fibre with superior lithium ion storage performance[J]. J. Mater. Chem. A, 2014, 2: 19975-19982.

[14] Ma G, Yang Q, Sun K, et al. Nitrogen-doped porous carbon derived from biomass waste for high-performance supercapacitor[J]. Bioresource Technol., 2015, 197: 137-142.

[15] Yu J, Yu X. Hydrothermal Synthesis and Photocatalytic Activity of Zinc Oxide Hollow Spheres[J]. Environ. Sci. Technol., 2008, 42: 4902-4907.

[16] Wu N. Plasmonic metal-semiconductor photocatalysts and photoelectrochemical cells: A review[J]. Nanoscale, 2018, 10(6): 2679-2696.

[17] Ran J, Jaroniec M, Qiao S Z. Cocatalysts in semiconductor-based photocatalytic CO_2 reduction: Achievements, challenges, and opportunities[J]. Adv. Mater., 2018, 30(7): 1704649.

[18] Yin S, Fan W, Di J, et al. La^{3+} doped BiOBr microsphere with enhanced visible light photocatalytic activity[J]. Colloids Surf. A, 2017, 513: 160-167.

[19] Yan L, Gu Z, Zheng X, et al. Elemental bismuth-graphene heterostructures for photocatalysis from ultraviolet to infrared light[J]. ACS Catal., 2017, 7(10): 7043-7050.

[20] Tian Y, Li W, Zhao C, et al. Fabrication of hollow mesoporousSiO_2-BiOCl@PANI@Pd photocatalysts to improve the photocatalytic performance under visible light[J]. Appl. Catal. B: Environ., 2017, 213: 136-146.

[21] Ren J, Wang W, Shang M, et al. Heterostructured bismuth molybdate composite: Preparation and improved photocatalytic activity under visible-light irradiation[J]. ACS Appl. Mater. Inter., 2011, 3(7): 2529-2533.

[22] Zhou T, Hu J, Li J. Er^{3+} doped bismuth molybdate nanosheets with exposed {010} facets and enhanced photocatalytic performance[J]. Appl. Catal. B: Environ. 2011, 110: 221-230.

[23] Margolis L Y. Present state of ideas on the mechanism of catalytic hydrocarbon oxidation[J]. Catalysis Reviews, 1974, 8(1): 241-267.

[24] Li H, Hou W, Tao X, et al. Conjugated polyene-modified Bi_2MO_6 (MMo or W) for enhancing visible light photocatalytic activity[J]. Applied Catalysis B: Environmental, 2015, 172: 27-36.

[25] Belver C, Adán C, Fernández-García M. Photocatalytic behaviour of Bi_2MO_6 polymetalates for rhodamine B

degradation[J]. Catal. Today., 2009, 143(3-4): 274-281.

[26]Shang M, Wang W, Ren J, et al. Nanoscale Kirkendall effect for the synthesis of Bi_2MoO_6 boxes via a facile solution-phase method[J]. Nanoscale, 2011, 3(4): 1474-1476.

[27]Zhang L, Xu T, Zhao X, et al. Controllable synthesis ofBi_2MoO_6 and effect of morphology and variation in local structure on photocatalytic activities[J]. Appl. Catal. B: Environ., 2010, 98(3-4): 138-146.

[28]Gao L, Gan W, Xiao S, et al. A robust superhydrophobic antibacterial Ag-TiO_2 composite film immobilized on wood substrate for photodegradation of phenol under visible-light illumination[J]. Ceram. Int., 2016, 42(2): 2170-2179.

[29]Kallio T, Alajoki S, Pore V, et al. Antifouling properties of TiO_2: Photocatalytic decomposition and adhesion of fatty and rosin acids, sterols and lipophilic wood extractives[J]. Colloids Surf. A, 2006, 291(1-3): 162-176.

[30]Chen F, Yang X, Wu Q. Antifungal capability of TiO_2 coated film on moist wood[J]. Build. Environ., 2009, 44(5): 1088-1093.

[31]Tian J, Hao P, Wei N, et al. 3DBi_2MoO_6 Nanosheet/TiO_2 nanobelt heterostructure: Enhanced photocatalytic activities and photoelectochemistry performance[J]. ACS Catal., 2015, 5(8): 4530-4536.

[32]Dai Z, Qin F, Zhao H, et al. Time-dependent evolution of the$Bi_{3.64}Mo_{0.36}O_{6.55}$/$Bi_2MoO_6$ heterostructure for enhanced photocatalytic activity via the interfacial hole migration[J]. Nanoscale, 2015, 7(28): 11991-11999.

[33]Chang K, Hai X, Ye J. Transition Metal Disulfides asNoble-Metal-Alternative Co-Catalysts for Solar Hydrogen Production[J]. Adv. Energy Mater., 2016, 6(10): 1502555.

[34]Zhang Z, Fu Q, Xue Y, et al. Theoretical and experimental researches of size-dependent surface thermodynamic properties of nanovaterite[J]. J. Phys. Chem. C, 2016, 120(38): 21652-21658.

[35]Li S, Hu S, Jiang W, et al. Facile synthesis of cerium oxide nanoparticles decorated flower-like bismuth molybdate for enhanced photocatalytic activity toward organic pollutant degradation[J]. J. Colloid. Interf. Sci., 2018, 530: 171-178.

[36]Li S, Hu S, Zhang J, et al. Facile synthesis ofFe_2O_3 nanoparticles anchored on Bi_2MoO_6 microflowers with improved visible light photocatalytic activity[J]. J. Colloid. Interf. Sci., 2017, 497: 93-101.

[37] Li S, Hu S, Jiang W, et al. Hierarchical architectures of bismuth molybdate nanosheets onto nickel titanate nanofibers: Facile synthesis and efficient photocatalytic removal of tetracycline hydrochloride[J]. J. Colloid. Interf. Sci., 2018, 521: 42-49.

[38]Ge L, Han C, Liu J, et al. Enhanced visible light photocatalytic activity of novel polymericg-C_3N_4 loaded with Ag nanoparticles[J]. Appl. Catal. A-Gen., 2011, 409: 215-222.

[39]Li S, Hu S, Jiang W, et al. Facile synthesis of flower-likeAg_3VO_4/Bi_2WO_6 heterojunction with enhanced visible-light photocatalytic activity[J]. J. Colloid. Interf. Sci., 2017, 501: 156-163.

[40]Jiang D, Du X, Chen D, et al. Facile wet chemical method for fabricating p-type BiOBr/n-type nitrogen doped graphene composites: Efficient visible-excited charge separation, and high-performance photoelectrochemical sensing[J]. Carbon, 2016, 102: 10-17.

[41]Li S, Xu K, Hu S, et al. Synthesis of flower-likeAg_2O/BiOCOOH *p-n* heterojunction with enhanced visible light photocatalytic activity[J]. Appl. Surf. Sci., 2017, 397: 95-103.

[42]Barzegar M, Habibi-Yangjeh A, Behboudnia M. Template-free preparation and characterization of nanocrystalline ZnO in aqueous solution of [EMIM][$EtSO_4$] as a low-cost ionic liquid using ultrasonic irradiation and photocatalytic activity [J]. J. Phys. Chem. Solids, 2009, 70(10): 1353-1358.

中文题目：一步法合成具有高效可见光催化活性的木基蛋壳状钼酸铋

作者：李莹莹，惠彬，高丽坤，李凤龙，李坚

摘要：光催化被认为是最环保、最低成本的环境修复方法，可以利用光来降解污染物。这项工作通过简单的一锅法制备了新型的木质光催化剂。所制备的样品分别通过 X 射线衍射、扫描电子显微

镜、X射线光谱、UV可见漫反射光谱和X射线光电子光谱进行表征。结果表明，合成的蛋壳状纳米钼酸铋均匀地生长在木质基材的表面。另外，其形态强烈依赖于前驱液的初始pH值和木材基质的存在与否。基于罗丹明B在可见光照射下的分解情况来评价样品的光催化活性。与纯钼酸铋粉末和在其他pH值下制备的样品相比，在pH为6时制备的样品对罗丹明B的光降解表现出优异的性能。此外，讨论了样品的稳定性、光催化机理以及光催化活性提高的原因。

关键词：木材；光催化；钼酸铋；一锅法制备

02

第二部分

木材仿生与智能响应

大自然给予的启发——木材仿生科学刍议*

李坚，孙庆丰

摘要：本文有针对性地列举了自然界某些生物体所固有的智能行为和独特的自然属性；由自然现象给予的启发，阐述了构建木材仿生科学的理论基础；提出了依据生物学原理和现代技术，赋予木材奇异功能或创生新型复合材料的发展空间。

关键词：木材；仿生科学；超疏水；气凝胶；智能；界面

1 前 言

自然界的生物体经过数十亿年的物竞天择、优胜劣汰，其结构与功能已趋完美，实现了宏观性能和微观结构的有机统一[1]。从大自然给予的启发，向自然界学习，模仿自然界生物体功能的某一方面，构筑相似甚至超越自然生物体功能的新型仿生材料，完成智能操纵过程，进而获得高效、低能耗、环境和谐且快速智能应变的新材料及其新性质，研究和构筑高性能的仿生智能材料是人类发展进程中的一个永恒课题[1]。

木材是一种天然的有机复合材料，具有结构层次分明、构造复杂有序、分级结构鲜明、多孔结构精细等特性，同时具有各向异性、低密度、高弹性、机械性能优良和来源丰富、可再生、清洁等特点，为木材仿生奠定了广阔的空间[2]。木材仿生智能科学与应用技术研究是木材科学发展中的一个具有里程碑意义的研究领域，它使木材从更微观的层次师法自然，利用从生物体获得的启示为木材的功能拓展和高值化开发提供新的研究思路，通过构筑具有仿生结构的智能木材或复合材料，解决木材资源不足和使用中的种种限制，实现木材的自增值性、自修复性、自诊断性、自学习性和环境适应性，使得木材从更高的技术层次上为人类的文明进步服务。

2 仿生学概要

人类很早就有了仿生的思想。《韩非子》曾记载古代工匠用竹木作鸟“成而飞之，三日不下”，这可以认为是人类仿生学的先驱，也是仿生学的萌芽。虽然仿生学的历史可以追溯到许多世纪以前，但一般把 1960 年 9 月由美国空军航空局在俄亥俄州空军基地戴通召开的第一次仿生学会议作为仿生学正式诞生的标志。仿生学一词最早是由美国人斯蒂尔(Jack Ellwood Steele)取自拉丁文“bio”(生命方式)和词尾“nic”(具有……性质的)合成的，他认为：仿生学是研究模仿生物系统方式，或是以具有生物系统特征的方式或类似于生物系统的方式建造技术系统的科学。后来又出现了 biomimetics 一词，意思是模仿生物[3]；近年来 bioinspired 一词逐渐为研究者们所关注，意为受生物启发而研制的材料或进行的过程。路甬祥[4]定义仿生学为：研究生物系统的结构、性状、原理、行为以及相互作用从而为工程技术提供新的设计思想、工作原理和系统构成的技术科学。

---

* 本文摘自中国工程科学，2014，16(4)：4-12.

随着材料学、化学、分子生物学、系统生物学以及纳米技术的发展，仿生学向微纳结构和微纳系统方向发展，实现结构与功能一体化将是仿生材料研究前沿的重要分支。开展仿生结构、功能及结构—功能一体化材料的仿生研究具有重要的科学意义。它将认识自然、模仿自然、在某一方面超越自然有机结合，将结构及功能的协同互补有机结合，并在基础学科和应用技术之间架起了一座桥梁，为新型结构、功能及结构—功能一体化材料的设计、制备和加工提供了新概念、新原理、新方法和新途径。

3 大自然给予的启发

3.1 荷叶的滴水不沾特性

荷叶的滴水不沾特性是自然界中植物表面超疏水性能的典型描述，该特性与荷叶表面的独特结构密切相关。Barthlott 等[5, 6]通过对植物叶片表面的研究表明：植物叶面的超疏水特性和植物叶面粗糙的微米级乳突结构及疏水蜡质材料相关。江雷课题组[7]经研究发现荷叶表面除微米结构外还有纳米结构存在，他们认为荷叶表面的微/纳米多级结构和低表面能的蜡质物使其具有超疏水和自清洁功能，在世界上首次提出了“二元协同纳米界面材料”的新概念。在此启发下，众多科研工作者开展了仿生荷叶滴水不沾特性的研究工作。江雷课题组[8~10]利用静电纺丝技术制备了多孔微球与纳米纤维复合结构的超疏水薄膜材料和具有类荷叶结构的聚苯胺/聚苯乙烯复合膜，该复合膜表现出高导电性和自清洁效应；还对无机金属氧化物如 ZnO、TiO_2、SnO_2 等纳米材料的光响应智能超疏水/超亲水可逆界面进行了分析研究，研究结果为智能性响应界面材料提供了新的研究思路。刘维民课题组在金属材料的表面如铝、铜、钢等仿生构筑具有超疏水性能的功能性薄膜，该薄膜同时具有与基材结合力强、耐酸碱、耐高低温和时效性强等特征。在材料表面仿生荷叶滴水不沾特性构筑功能性薄膜，不仅在理论研究上有重要意义，在实际生产中同样具有重要的应用价值。

3.2 棉花的轻柔飘逸特性

棉花是锦葵科棉花属植物的种子纤维，纤维白色至白中带黄，长为 2~4 cm。棉花纤维是唯一的天然纯净纤维素材料，纤维素含量高达 95%~97%。棉花纤维由直径为 100~200 nm 的纤丝组成，纤丝交错排列在一起，构成细胞壁的网状结构。仿生棉花轻质飘逸特性，可将其用于生物质废弃资源(秸秆、椰壳、甘蔗渣等)高值化开发利用的研究，如分离制备高纯纤维素、轻质高强气凝胶等。Olsson 等[11]以纳米纤维气凝胶为模板，制备得到磁性气凝胶纳米材料。Korhonen 等[12]以纤维素纤维为模板，获得 TiO_2 气凝胶，该气凝胶在传感器、药物释放、载体等领域具有很好的潜在的应用前景。浙江大学高超课题组[13]制备出目前世界上最轻的气凝胶。本课题组以农林业废弃物为原料制备了可漂浮在花瓣上的纤维素基气凝胶，有着优良的弹性和吸油能力，与棉花的轻柔飘逸特性有着类似之处。仿生棉花的轻柔飘逸特性制备功能化气凝胶，探究纤维素气凝胶的形成工艺及机理，调控纤维素气凝胶的孔隙结构，制备疏水亲油性和具相反物性的纤维素气凝胶，为利用可再生的纤维素资源获得高新产品提供理论依据[14]。同时，也为促进生物质产业向高尖端发展提供技术保障。制备的轻质高强纤维素气凝胶在吸附海上泄漏污油、太阳能电池、土壤保水剂、催化剂及载体、气体过滤材料等领域中具有较大的应用价值。

3.3 海鞘的环境响应特性

环境响应型材料是指在外界环境微小变化的刺激下，材料自身的某些物理或化学性质会发生动态且可逆变化的材料，因而也被称为“智能”材料或刺激响应型材料。环境响应型材料广泛存在于自然界中，自然界中的生物都会根据外界环境的改变调节自身的性质和功能。例如，海鞘根据所处环境条件的不同，通过神经控制其体内的色素细胞，快速改变身体的图案和颜色。在海鞘环境响应特性的启发下，人们开始积极探索创造与其相似且具有精巧结构和功能的环境响应型材料，发展用于环境响应型材料的合成技术和理论，这些材料可以对光、温度、pH、电、磁等外界刺激产生(多重)响应，调节自身的形状、相态、表面能、反应速率、渗透速率、亲疏水性、吸附力、识别性能等一些关键性质，广泛地应用于药物传递、生物诊断、组织工程、光学传感、微电机、涂料和纺织材料等领域[15]。Weder 课题组[16]利用环氧乙烷、环氧氯丙烷、纤维素纳米纤维等制备了智能的能在僵硬与松软状态间转换的材料。江雷等[17]研究制备了具有热敏性及 pH 响应性的功能性材料，同时他们还制备了温度、pH 和葡萄糖浓度多响应浸润型表面材料。构筑特定性质的环境响应型智能功能材料不仅能满足实际应用的需求，而且将大幅提升材料的设计空间，赋予材料新的功能，强化其现有性能，突破现有材料应用瓶颈，拓展其应用领域，直接与重大实际应用需求实现对接。

3.4 扇贝的层级结构

扇贝为软体动物门双壳纲翼形亚纲珍珠贝目中的一科，属于贝壳的一种。贝壳的结构是典型的层级结构，一般可分为 3 层：最外一层为角质层，很薄，透明，有光泽，由壳基质构成，不受酸碱的侵蚀，可保护贝壳；中间一层为壳层，又称棱柱层，占贝壳的大部分，由极细的棱柱状的方解石($CaCO_3$，三方晶系)构成；最内一层为壳底，即珍珠质层，富光泽，由小平板状的结构单元累积而成，成层排列，组成成分是多角片型的文石结晶体($CaCO_3$，斜方晶系)[18~20]。对贝壳珍珠层的结构分析表明，其并不是单纯的层片结构，而可以看成两级尺度结构的耦合，是一种天然的无机—有机层级分明的复合材料，它主要由约 95%的 $CaCO_3$ 和 5%的有机基质构成，其抗张强度是普通 $CaCO_3$ 的几千倍，因此贝壳轻质高强的原因与其独特的多尺度、多级次组装结构密切相关[21]。受贝壳轻质高强层级结构影响，Podsiadlo 等[22]利用层层组装技术(LBL)制备了聚乙烯醇/蒙脱土透明层状复合材料，其拉伸强度和杨氏模量较纯聚乙烯醇材料分别提高了近 10 倍和 100 倍。Bonderer 等[23]研究小组利用自下而上的胶体组装技术仿生制备了陶瓷板—壳聚糖层状复合材料，该材料质量仅有钢的 1/2~1/4，却有钢一般的强度。通过观察研究扇贝等贝壳通过自身矿化调控形成高度有序的有机—无机复合结构的形成机理，仿生构筑贝壳类结构的功能材料为不同领域内的新型材料开发和研究提供了重要的发展空间，具有重要的科学意义和应用价值。

3.5 候鸟、海龟的“千里迁徙”和“万里洄游”特性

众所周知，燕子等候鸟每年都在春秋两季分别从南方飞回北方，又从北方飞到南方；一些海龟从栖息的海湾游出几百甚至几千公里后又能回到原来的栖息处。它们是如何辨别方向的？尤其是在茫茫的海洋上？候鸟、海龟的“千里迁徙”和“万里洄游”特性主要是和这些动物利用地球的磁场有关。它们主要依赖地球的磁场来进行定位，候鸟体内的“导航地图”和海龟的“生物罗盘”与地球磁场产生作用，从而使它们能丝毫无误地回到自己的栖息地。物质具有的磁性可用来进行精确定位已被现代科学技术所证实[1]。本课题组目前正以木材、竹材为原料，选择含有某些金属元素的前驱体，采用水热法成功制备了趋磁性木材，相关研究工作近期将予以发表。在现代社会，通过仿生候鸟、海龟的“千里迁徙”和“万里洄游”特性，研究开发了先进的高能加速器、粒子检测器、磁共振成像以及现代通信技术；同时，通过仿生一些动物利用日月星辰导航，也有些动物利用海流、海水成分、地磁场、重力场等进行导航，为研制通信设备和新型导航仪器提供了启示。

3.6 树根的自修复特性

树根在受伤后，经过一段时间，受伤部位可以通过生物体的自身作用而完整愈合。这种现象在许

多植物中都存在，生物愈合过程存在着大量共性：首先，愈合过程是由损伤引起的，在生命机能没有受到致命伤害的情况下，损伤是启动愈合机制的最基本条件；其次，在愈合初期，损伤逐渐被由损伤刺激而产生的增生组织所填充；随后通过机体的输运、化学反应，填充在损伤部位的物质(如薄壁组织、凝块等)发生变化，强度提高，构成与周围组织的有效连接；同时愈合过程需要一定的物质及能量供应，以产生填充损伤的组织，而向损伤处供应物质的输运过程都有液相的参与；最后，生物的愈合是使损伤处的有效连接恢复。受此现象启发，科学家们针对工程、建筑、路面中存在的材料破坏仿生研究了自修复材料。自修复材料是一种智能材料，同时具有感知和激励双重功能。自修复材料可延长产品的使用寿命，并提升产品的安全性。仿生树根自修复特性，杨红等[24]将灌注胶液的液芯光纤埋入玻璃钢复合材料中制成兼有自诊断和自修复功能的智能材料，测得其对拉伸能力的修复达到原始值的1/3，对压缩的修复达到2/3以上。White 等[25]制备了一种具有自动修复裂纹能力的聚合物材料，这种材料嵌有内装修复剂的微胶囊，每个微胶囊约有头发丝宽，这些微胶囊遇到裂纹入侵时破裂，并通过毛细作用释放修复剂到裂纹面，修复剂接触预先埋入环氧基体的催化剂而引发聚合，键合裂纹面，这种损伤诱导的引发聚合使得裂纹修复实现了就地自动控制。Wong 等[26]仿照猪笼草的疏水策略，开发出一种极为光滑的涂层材料，几乎能排斥包括血液、油在内的任何液体，甚至在高压、冰冻等极端环境条件下，仍能保持排斥液体或固体的能力，同时该涂层材料具有自修复功能，即使用刀子刮坏一部分，也能立即自行修复，修复后仍保持疏水性能，该材料在运输燃料和水的管道、医用导管(如导尿管和输血系统)、自动清洁窗、无菌无垢表面等领域具有广泛的应用。自修复材料是一种新型智能材料，在这方面的研究还相对较少，然而从它的功效来看，应具有广阔的应用前景，智能自修复材料对提高产品的安全性和可靠性有着深远的意义。

4 木材仿生科学理论基础

4.1 木材的多尺度分级结构

木材是由各种不同的组织结构、细胞形态、孔隙结构和化学组分构成的，是一类结构层次分明、构造有序的聚合物基天然复合材料，从米级的树干，分米、厘米级的木纤维，毫米级的年轮，微米级的木材细胞，直到纳米级的纤维素分子，具有层次分明、复杂有序的多尺度分级结构。木材单个细胞由薄的初生壁、厚的次生壁和细胞腔组成，细胞腔大而空。其中次生壁又呈多尺度分级结构特点：次生壁是由次生壁外层(S1，厚约为 0.5 μm)、次生壁中层(S2，厚约为 5 μm)和次生壁内层(S3，厚约为 0.1 μm)组成。在光学显微镜下，细胞壁仅能见到宽 0.4~1.0 μm 的丝状结构，称为粗纤丝。如果将粗纤丝再细分下去，在电子显微镜下观察到的细胞壁线形结构，则称为微纤丝。木材细胞壁中微纤丝的宽度为 10~30 nm，微纤丝之间存在着约为 10 nm 的空隙，木质素及半纤维素等物质聚集于此空隙中。其断面约有 40 根纤维素分子链组成的最小丝状结构单元，称为基本纤丝。如果把纤维素分子链的断面看成圆截面，则可以推算其直径约为 0.6 nm。木材细胞壁的组织结构是以纤维素作为骨架的。它的基本组成是一些长短不等的链状纤维素分子，这些纤维素分子链平行排列，有规则地聚集在一起成为微团(即基本纤丝)；由微团组成一种纤丝状的微团系统即微纤丝；由微纤丝组成纤丝；纤丝再聚集成粗纤丝；粗纤丝相互结合形成薄层；许多薄层再聚集成细胞壁。次生壁微纤丝的排列不像初生壁那样无定向，而是整齐地排列成一定方向。各层微纤丝都形成螺旋取向，但是斜度不同。在S1 层，微纤丝有 4~6 个薄层，一般为细胞壁厚度的 10%~22%，微纤丝呈“S”“Z”形交叉缠绕的螺旋线状，并与细胞长轴成 50°~70°。S2 层是次生壁中最厚的一层，在早材管胞的胞壁中，其微纤丝薄层数为 30~40 层，而晚材管胞可达 150 个薄层或以上，一般为细胞壁厚度的 70%~90%；S2 层微纤丝排列与细胞长轴成10°~30°，甚至几乎平行。在S3层，微纤丝有0~6个薄层，一般为细胞壁厚度的2%~8%，微纤丝的排列近似S1层，与细胞长轴成60°~90°，呈比较规则的环状排列。

因此，木材在大自然中形成的精妙细胞结构及其层次分明、排列复杂有序的多级多尺度结构，为木材仿生高性能化材料和制备特殊的多级多尺度结构新型材料奠定了坚实良好的基础。

4.2 木材的分级多孔结构

木材除拥有精妙的多尺度分级结构外，还具有天然形成的精细分级多孔结构。阔叶材中管孔形状多种多样，呈现出不规则的圆形、椭圆形和多边形。在孔径尺寸上从粗到细变化范围很宽，明显呈现出分级特征，且孔径较大的管道和孔径较小的管道形成相间分布结构。针叶材的孔径尺寸则比较均匀，分布较为规则。木材形态各异的管孔形状、尺寸和分布特征，为设计和制备各种分级多孔材料提供了广阔的选材空间。赵广杰[27]按尺度大小把木材中的空隙划分为宏观空隙、微观空隙和介观空隙。宏观空隙是指用肉眼能够看到的空隙，以树脂道、细胞腔为下限空隙，不同树种细胞大小不同，其宽度为50~1 500 μm，长度从0.1~10 mm不等。微观空隙则是以分子链断面数量级为最大起点的空隙，如纤维素分子链的断面数量级的空隙。介观空隙是指三维、两维或一维尺度在纳米量级(1~100 nm)的空隙，也可称为纳米空隙。

形态各异的木材的分级多孔结构为仿生制备新型材料提供了无须加工修饰处理的天然模板，为木材仿生高性能、高附加值功能材料的研究开发提供了无限空间。多级多孔材料在分离提纯、选择性吸附、催化剂装载、光电器件和传感器研制等许多功能领域有重要的研究和应用价值。木材多级多孔特点使得木材本身即可收容其他纳米材料，可使木材实现木材功能化、纳米化、智能化的追求。

4.3 木材的智能性调湿调温功能

木材自身的生物结构和组成物质赋予其某些具有智能性调节作用的性质。木质住宅在暑夏时具有隔热性，在寒冬时具有保温性。木质墙壁可以缓和外部气温变化所引起的室内温度变化。因此，木质住宅具备缓解夏季炎热或冬季寒冷的性能，即“冬暖夏凉”。由于木材组分中含有大量的亲水性基团，又具有极大的比表面积，使木材具有吸湿与解吸性质。当空气中的水蒸气压力大于木材表面水蒸气压力时，木材从空气中吸收水分，称为吸湿；反之，则有一部分水分自木材表面向空气中释放，称为解吸。木材吸湿性的变化取决于木材的构造学特性、木材的化学组成及其所在环境的湿度与温度。在通常情况下，如果室内的木材用量较多，当室内温度提高时，由于木材可以解吸放出水分，室内湿度几乎保持不变；反之，当温度降低时，室内湿度将相应增大，此时木材可以吸收水分，从而仍可保持室内的湿度稳定。

4.4 木材的智能性生物调节功能

木材是一种具有生态学属性的生物质，与人的生命活动息息相关，形成了“木材—人类—环境”的关系。自古以来，适于人类居住的木质环境比较适合人们的生理和心理需要。其内在的奥秘在于木材的视感与人的心理生理学反应遵循并符合1/*f*涨落的潜在规则[28]。自然界存在的事物涨落现象，其能谱密度与频率(*f*)成比例关系，被称为1/*f*涨落。具有1/*f*涨落特征的物体可视后使人感到舒适。木材具有天然生长形成的生物结构、纹理和花纹，还有独特的光泽和颜色，给人们带来视觉上的自然感、亲切感和舒适感。因此，木质结构的房屋、木质家具和木质材料的内装，无一不得到人们的喜爱。其原因是映入人们眼帘的木材(木质材料)具有的1/*f*涨落与人体中的生物节律(节奏)之涨落一致时，人们就产生平静、愉快的心情而有舒适之感。

4.5 木材的智能性调磁功能

木材具有调节“磁气”和减少辐射的智能性功能。尽人皆知，地球是一块大磁石。人类和地球上的全部生物体生活在地球磁场之中，地球提供给人类在地球表面生活所必需的适度的安定性磁力（“磁气”）。动物的感觉器官很敏锐，尤其对微小的磁场变化也有所感知，这正表明其具有与磁力作用不可分离的关系，而磁力感觉是人类生活环境所必需的。空间中的钢筋混凝土或铁金属材料和器具会将地球磁力变弱或屏蔽，易引起生物体各种生物机能的紊乱或使生物体出现异常行为。相反，在木质环境中，因为木材不能屏蔽地球磁力作用，所以生物体可以保持正常、安定的生活节奏。一些研究者已通过对小白鼠的实验对这种影响和作用进行证实。木材对人体不足的“磁气”又具有自然补充的机能，因此可以促进自律神经活动，适宜的“磁气”对减少高血压、风湿症、肾病等多种疾病的发生有一定影响。因此，木结构住宅和室内木材设置较多的微环境空间有利于人居健康。

4.6 木材是天然的气凝胶结构体[29]

首先，木材是天然生长形成的多孔性有限膨胀胶体，是一种天然高分子凝胶材料。依据细胞壁微观形态学，Wardrop 等认为细胞壁由基质物质、构架物质和结壳物质 3 类基本构造物质组成。可塑性的基质形成后立即被纤维素纤丝增强；在后期阶段木质素形成结壳。按照细胞壁个体发育划分为 3 个阶段：a. 基质形成阶段；b. 凝胶被纤丝增强的阶段；c. 结壳作用阶段。木材的基质可认为是一种亲水的凝胶体，主要包括半纤维素和果胶。在最初阶段，细胞壁呈极端可塑性并表现如高度蒙古滞的流体一样，具有较高的膨胀度和塑性变形，在基质形成以后，可塑性的基质立即被纤维素纤丝增强，因而弹性被赋予该系统。Frey-Wysslir 等认为幼嫩细胞壁的最初阶段代表着一种各向同性、没有任何双折射的凝胶组成，此种各向同性物质称为细胞壁的基质。基质、纤丝和覆层有不同的胶态性质。基质是一种所谓的干凝胶，即一种在干燥时硬化并变成半透明的凝胶。构成基质的碳水化合物(果胶、半纤维素等)通过化学提取或酶催化消化，将纤丝游离成气凝胶，易于接近空气的超微结构空间，由于光的折射，致使气凝胶呈白色。这与相关学科气凝胶和干凝胶的原理是一致的。木材细胞壁具备凝胶材料的基本条件和特征。

其次，从木材的组成和结构上看，木材细胞壁中约 50%是纤维素，半纤维素、果胶等占木材质量的 25%以上。纤维素除结晶区与无定形区以外，还包含许多空隙，形成空隙系统，空隙的大小一般为 1~10 nm，最大可达 100 nm，满足作为气凝胶网络纳米结构的基本条件。这与气凝胶材料的结构原理是一致的。此外，一些木材的物理特性具备气凝胶材料的性质。

总之，木材源于自然，拥有大自然赐予的精妙的多尺度分级结构，天然形成的精细分级多孔结构和调湿、调温、生物调节、调磁等多种智能性功能。木材的这些自然属性为木材仿生科学奠定了坚实的理论基础，为仿生构筑高性能木质新材料提供了广阔的发展空间。将木材科学与现代仿生学、材料学、生物学、信息学、能源学及纳米科学等学科互相交叉融合，有效利用木材天然的独特结构和优越的性能，由木材宏观复合向微观复合发展，由木材结构特征复合向功能—结构一体化发展，由木材一元体系向二元甚至多元复合体系扩展，由木材单一传统利用向复合化、智能化、环境化和能动化的研究开发利用发展，进一步研究木材的内在结构和性能，进而抽象并设计出木材仿生材料模型，彻底揭开木材内幕，是木材科学当前的一个关键性课题。木材仿生科学的提出无疑会给传统木材科学增添新的内涵，同时也会给木材科学带来重要进步，具有里程碑式的意义，将极大地延伸木材科学的内涵，使得木材从更高层次上为人类服务。

5 木材仿生功能材料构建研究

5.1 木材仿生构筑超疏水表面

木材作为一种可再生的、多功能的天然资源环境材料，广泛应用于人类生活的方方面面，如木建筑、室内外家具、乐器材和装饰材料等，但木材也存在着吸水膨胀导致尺寸稳定性不佳、易被细菌侵蚀、易被有机物污染等缺陷。由于这些木材固有缺陷的存在，在实际应用中较大程度地限制了木材的使用范围和领域。在木材表面仿生构建超疏水表面后，将木材由亲水性转变为疏水性，实现了相反物性的转换，使得木材不再吸收外界水分，可有效缓解木材变形开裂、霉变、腐朽、降解。目前，在木材表面仿生构建超疏水表面的方法主要有溶胶-凝胶法、水热法、气相沉积法、自组装法、浸渍法、低温等离子体法、液相沉积法等。吴义强等[30]对当前木材表面仿生构建超疏水膜层进行了较为系统的综述，在此不再赘述。李坚课题组[31~33]采用低温水热共溶剂法在木材表面构建了仿生超疏水表面并实现了智能性光控亲疏转换。仿生荷叶滴水不沾特性在木材表面构筑超疏水自清洁表面将极大拓展木材的使用范围和领域。但目前对木材仿生超疏水表面的结构尚缺乏系统的研究数据，仿生超疏水表面的动力学尚未引起关注，同时现有木材表面仿生超疏水表面一般尚处于实验室研究阶段，需要精密的实验设备和昂贵的化学物质，距离规模化生产还有很长的路要走。

5.2 木材仿生构筑异质复合材料

据记载，我国每年所需的70%的天然橡胶和40%以上的合成橡胶需进口，而我国废旧轮胎等物质的循环利用率仅为20%左右，废而不用的废旧轮胎、胶管、胶带、胶鞋等造成了严重的“黑色污染”。木质基-橡胶复合材料能够以小径木、间伐材和加工剩余物与废旧橡胶为原料，通过选择适宜的胶粘剂和热压工艺参数能够制造出木材刨花(木材纤维)-废旧橡胶复合材料，其性能指标达到国家标准。制备这种新型复合材料的关键技术是要通过大量实验确定木材与橡胶的配比及其热压成板时的最佳热压工艺参数。这种复合材料具有良好的防水、防腐、防静电、隔音、隔热和阻尼减震等性能，用途宽泛。

5.3 木材仿生构筑分级多孔氧化物

分级多孔材料在分离提纯、选择性吸附、催化剂装载、光电器件和传感器研制等多个领域具有重要的研究和应用价值。上海交通大学张荻课题组[34~37]以木材为模板，遗传其形态和结构，合成制备木材结构分级多孔 Fe_2O_3、ZnO 和 NiO 材料，获得了 20~100 μm 和 0.1~1 μm 的分级大孔分布，氧化物内有 10~50 nm 的介孔分布。其中，制备的杉木结构 ZnO 具有最高的分形维数，并且孔隙率最高，具有良好的网络连通性；分级多孔 Fe_2O_3 具有优于常规 Fe_2O_3 的气敏性能；分级多孔 ZnO 对 H_2S 具有非常优异的气体选择性。李坚课题组[38]利用杨木木材作为模板制备了具有良好光催化活性的 TiO_2。首先使用溶胶-凝胶法将 TiO_2 溶胶负载于木材表面，通过高温煅烧的方法除掉木材模板即可制备大块的多孔木材结构的新型光催化剂 TiO_2，该光催化剂具有良好的光催化性能和沉降性能。Cao 等[39~41]以白松为模板制备得到多孔 Al_2O_3、ZrO_2、TiO_2 陶瓷。利用木材独特的多层级、多孔结构制备的多孔氧化物材料具有低密度、高比强度、高比表面积、高渗透性、耐高温、抗热冲击强和膨胀系数小等优异性能。此外，木材原料来源广且可再生，制造成本低且可实现复杂形状的原位成形。这些使具有木材结构的多孔材料具有广阔的应用前景。

5.4 木材仿生构筑木陶瓷

以低质材料、废旧木材等木质材料为原料，先经过预切削加工成一定形状，然后用酚醛树脂浸渍，隔氧高温烧结，最后再进行磨削加工制得产品。这种材料具有多孔结构、强重比高、耐磨、耐腐、耐热和吸附性能好等诸多特点，可作为房屋保温和取暖、吸附、抗摩擦及电磁屏蔽材料等。Greil 等[42]用液相浸渗反应法制备了 SiC 基多孔陶瓷；张荻课题组[43~45]仿生木材生态遗传结构制备了一系列的氧化物陶瓷，制备的材料在电、磁、光学、催化等方面有着极大的应用潜力。东北林业大学李淑君等[46~48]采用酚醛树脂浸渍木质材料，经过高温烧结制得木陶瓷产品，并对产品得率、性能、影响因素、微观结构以及烧结过程中的化学变化等进行了系统分析，并探讨了阻燃处理提高产品性能的可能性。木材仿生构筑木陶瓷在加工过程中应注意的技术问题有：a. 在制造过程中要避免木材的变形和开裂；b. 高温烧结时应避免试件的氧化烧失，须采用氮气保护；c. 产品性能与树脂浸渍量、烧结温度和升温速率关系密切，因此采用均匀设计法优化得出相适宜的工艺参数。

5.5 木材仿生构建木材-无机纳米复合材料

21 世纪木材功能性改良将面临巨大的发展机遇与挑战，制备新型多功能化的木质基材料将是木材科学与技术发展的一个重要趋势。木材功能性改良不但要合理利用木材，注重木材基本性质的改善，还要以高新科技为先导，赋予木材诸如超疏水、抗紫外、阻燃等新的功能[49~50]。选择具有不同特性的有机质调控的纳米粒子仿生制备形成的木材-无机纳米复合材料会产生许多新的、奇特的性能。譬如，在木材与纳米 $CaCO_3$ 复合时，利用不同的有机质控制可得到具有疏水、疏油、超疏水(油)的系列功能性材料；通过溶胶-凝胶法制成的 SiO_2、TiO_2 的木材-无机纳米复合材料具有良好的力学强度、阻燃性和尺寸稳定性[51~56]；在自然界中，如柚木等名贵木材，由于无机矿物质以纳米粒子的形式渗入木材基体中进行生物矿化和生理生化作用，形成了天然的木材-无机纳米复合材料，使这类木材在树木生长过程中形成了美丽的材色和肌理以及坚硬的材质和较高的耐久性。木材作为天然有机高分子材料与无机纳米材料复合形成的木质基无机纳米复合材料，不仅具有纳米材料的颗粒体积效应和表面效应等性质，而且将无机物的刚性、尺寸稳定性、热稳定性与木材的韧性、加工性、介电性及独特的环境学特性融为一体，从而产生许多特异的性质。

5.6 木材仿生构筑气凝胶材料

气凝胶是一种用气体代替凝胶中的液体而本质上不改变凝胶本身的网络结构或体积的特殊凝胶，是水凝胶或有机凝胶干燥后的产物，被称为固体烟雾，具有高孔隙率、高比表面积、低热传导系数、低介电常数、低光折射率和低声速等独特的性能。这些独特的性质不仅使得该材料在基础研究中引起人们的兴趣，而且被广泛地用于组织工程、控释系统、血液净化、传感器、农业、水净化、色谱分析、超级高效隔热隔声材料、生物医药、高效可充电电池、超级电容器、催化剂及载体、气体过滤材料和化妆品等领域。Berglund 等[57]用植物纤维素制备出纤维素气凝胶，并仿照木质纤维素结构特性，将木葡聚糖与纤维素复合，利用超临界 CO_2 干燥法组装纤维素-葡聚糖复合气凝胶，其力学强度显著提高。李坚课题组[58,59]利用离子液体和冷冻干燥的方法直接从木粉制得了木质纤维素气凝胶，通过循环冻融工艺可实现气凝胶内部结构、密度及比表面积的调控；邱坚等[60,61]通过超临界干燥技术结合溶胶-凝胶法制备新型木材/SiO_2 气凝胶复合材料，从制备工艺学原理、SiO_2 气凝胶在木材中的分布与界面状态、性能评价以及木材与 SiO_2 气凝胶复合的机理等方面进行了系统的研究。张俐娜等[62,63]以碱/尿素为溶剂，制备得到具有很高机械性能的纤维素气凝胶。木材仿生构建气凝胶是向自然学习的一个重要方面，体现了“师法自然”的科学思想，为发展和构建高值化木质纤维素气凝胶材料提供科学依据和理论指导。

6 木材仿生科学展望

木材仿生科学期望通过模仿具有特殊功能的自然界生物体的结构，充分利用自身独特的天然结构与属性，将其与纳米技术、分子生物学、界面化学、物理模型等相结合，从仿生学的角度出发，以自然界给予的各种现象为启发，制备具有特殊表面润湿性、电磁屏蔽效应、高机械强度的仿生高性能木质基新型材料；引入对热、pH、光或电等刺激有响应的智能元素，通过合理设计材料的组成及结构，制备木质基智能响应材料；发展木材表面仿生多尺度表面微观结构构建方法，探讨材料多尺度微结构对异质材料结合性能的调控机理，制备具有不同物质组成或多尺度微观结构的木质基新型复合材料；基于多尺度界面的仿生结构原理、调控界面分子、纳米及微米多尺度上的多重协同作用，构筑木质基新型微纳结构仿生智能材料。木材仿生科学将更深入地延伸木材科学的内涵，使得科研工作者从更深层次上通过认知、模拟与调控3个步骤揭开木材内幕，同时也为木材科学和其他学科间的交叉融合架起一座桥梁，实现“他山之石，可以攻玉”。

参考文献

[1]王女，赵勇，江雷．受生物启发的多尺度微/纳米结构材料[J]．高等学校化学学报，2011，32(03)：421-428.

[2]李坚．木材科学(第二版)[M]．高等教育出版社，2002.

[3]刘克松，江雷．仿生结构及其功能材料研究进展[J]．科学通报，2009，54(18)：2667-2681.

[4]路甬祥．仿生学的科学意义与前沿：仿生学的意义与发展[J]．科学中国人，2004，000(004)：22-24.

[5]Barthlott W, Neinhuis C. Purity of the sacred lotus, or escape from contamination in biological surfaces[J]. Planta, 1997, 202 (1): 1-8.

[6]Neinhuis C, Barthlott W. Characterization and distribution of water-repellent, self-cleaning plant surfaces[J]. Ann. Bot., 1997, 79(6): 667-677.

[7]Feng L, Li S H, Li Y S, Li H J, Zhu D B. Super-hydrophobic surfaces: From natural to artificial [J]. Adv. Mater., 2002, 14(24): 1857-1860.

[8]Jiang L, Zhao Y, Zhai J. A lotus-leaf-like superhydrophobic surface: A porous microsphere/nanofiber composite film prepared by electrohydrodynamics[J]. Angew. Chem., 2004, 116(33): 4438-4441.

[9]Feng X J, Feng L, Jin M H, Zhai J, Jiang L, Zhu D B. Reversible super-hydrophobicity to super-hydrophilicity transition of aligned ZnO nanorod films[J]. J. Am. Chem. Soc., 2004, 126(1): 62-63.

[10]Feng X J, Zhai J, Jiang L. The fabrication and switch-able superhydrophobicity of TiO_2 nanorod films[J]. Angew. Chem., Int. Ed., 2005, 44(32): 5115-5118.

[11]Olsson R T, Azizi Samir M A S A, Salazar-Alvarez G, Belova L, Strom V, Berglund L A, Ikkala O, Nogues J, Gedde U W. Making flexible magnetic aerogels and stiff magnetic nanopaper using cellulose nanofibrils as templates[J]. Nat. Nanotechnol., 2010, 5(8): 584-588.

[12]Korhonen J T, Hiekkataipale P, Malm J, Karppinen M, Ikkala O, Ras R H A. Inorganic hollow nanotube aerogels by atomic layer deposition onto native nanocellulose templates[J]. ACS Nano, 2011, 5(3): 1967-1974.

[13]Sun H Y, Xu Z, Gao C. Multifunctional, ultra-flyweight, synergistically assembled carbon aerogels[J]. Adv. Mater., 2013, 25(18): 2554-2560.

[14]卢芸，孙庆丰，于海鹏，刘一星．离子液体中的纤维素溶解、再生及材料制备研究进展[J]．有机化学，2010，30(10)：1593-1602.

[15]吕威鹏．若干多功能环境响应型微纳米材料的合成、改性与应用探索[D]．天津大学，2012.

[16]Capadona J R, Shanmuganathan K, Tyler D J, Rowan S J, Weder C. Stimuli-responsive polymer nanocomposites inspired by the sea cucumber dermis[J]. Science, 2008, 319(5868): 1370- 1374.

[17] Xia F, Ge H, Hou Y, Sun T L, Chen L, Zhang G Z, Jiang L. Multiresponsive surfaces change between superhydrophilicity and superhydrophobicity[J]. Adv. Mater., 2007, 19(18): 2520-2524.

[18]崔福斋，郑传林．仿生材料[M]．化学工业出版社，2004.

[19]蔡国斌，万勇，俞书宏．受生物启发模拟合成生物矿物材料及其机理研究进展[J]．无机化学学报，2008，24(5)：673-683.

[20]胡巧玲，李晓东，沈家骢．仿生结构材料的研究进展[J]．材料研究学报，2003，17(4)：337-344.

[21]贾贤．天然生物材料及其仿生工程材料[M]．化学工业出版社，2007.

[22]Podsiadlo P, Kaushik A K, Arruda E M, Waas A M, Shim B S, Xu J, Nandivada H, Pumplin B G, Lahann J, Ramamoorthy A. Ultra-strong and stiff layered polymer nanocomposites[J]. Science, 2007, 318(5847): 80-83.

[23]Bonderer L J, Studart A R, Gauckler L J. Bioinspired design and assembly of platelet reinforced polymer films[J]. Science, 2008, 319(5866): 1069-1073.

[24]杨红，梁大开，陶宝祺，邱浩，曹振新. 光纤智能结构自诊断、自修复的研究[J]. 功能材料，2001，32(4)：419-424.

[25]White S R, Sottos N R, Geubelle P H, Moore J S, Kessler M R, Sriram S R, Brown E N, Viswanathan S. Autonomic healing of polymer composites[J]. Nature, 2001, 409(6822): 794-797.

[26]Wong T S, Kang S H, Tang S K Y, Smythe E J, Hatton B D, Grinthal A, Aizenberg J. Bioinspired self-repairing slippery surfaces with pressure-stable omniphobicity[J]. Nature, 2011, 477(7365): 443-447.

[27]赵广杰. 木材中的纳米尺度、纳米木材及木材-无机纳米复合材料[J]. 北京林业大学学报，2002，24(5/6)：204-207.

[28]李坚，吴玉章，马岩. 功能性木材[M]. 科学出版社，2011.

[29]邱坚，高景然，李坚，刘一星. 基于树木天然生物结构的气凝胶型木材的理论分析[J]. 东北林业大学学报，2008，36(12)：73-75.

[30]杨守禄，吴义强，张新荔. 木材仿生功能性超疏水表面制备的研究进展[C]//第六届中国木材保护大会暨2012中国景观木竹结构与材料产业发展高峰论坛2012橡胶木高效利用专题论坛论文集，2012，187-191.

[31]Li J, Yu H P, Sun Q F, Liu Y X, Cui Y Z, Lu Y. Growth of TiO_2 coating on wood surface using controlled hydrothermal method at low temperatures[J]. Appl. Surf. Sci., 2010, 256(16): 5046-5050.

[32]Sun Q F, Lu Y, Yang D J, Li J, Liu Y X. Preliminary observations of hydrothermal growth of nanomaterials on wood surfaces[J]. Wood Sci. Technol., 2014, 48(1): 51- 58.

[33]Sun Q F, Lu Y, Liu Y X. Growth of hydrophobic TiO_2 on wood surface using a hydrothermal method[J]. J. Mater. Sci., 2011, 46(24): 7706-7712.

[34]Liu Z T, Fan T X, Zhang D, Gong X L, Xu J Q. Hierarchically porous ZnO with high sensitivity and selectivity to H_2S derived from biotemplates[J]. Sens. Actuators B Chem., 2009, 136(2): 499-509.

[35]Liu Z T, Fan T X, Zhang D. Synthesis of biomor-phous nickel oxide from a pinewood template and investigation on a hierarchical porous structure[J]. J. Am. Ceram. Soc., 2006, 89(2): 662-665.

[36]Liu Z T, Fan T X, Gu J J, Zhang D, Gong X L, Guo Q X, Xu J Q. Preparation of porous Fe from biomorphic Fe_2O_3 precursors with wood templates[J]. Mater. Trans., 2007, 48(4): 878-881.

[37]Li X F, Fan T X, Liu Z T, Ding J, Guo Q X, Zhang D. Synthesis and hierarchical pore structure of biomorphic manganese oxide derived from woods[J]. J. Eur. Ceram. Soc., 2006, 26(16): 3657-3664.

[38]Lu Y, Sun Q F, Liu T C, Yang D J, Liu Y X, Li J. Fabrication, characterization and photocatalytic properties of millimeter-long TiO_2 fiber with nanostructures using cellulose fiber as a template[J]. J. Alloys Compd., 2013, 577: 569-574.

[39]Cao J, Rambo C R, Sieber H. Manufacturing of microcellular, biomorphous oxide ceramics from native pine wood [J]. Ceram. Int., 2004, 30(7): 1967-1970.

[40]Cao J, Rambo C R, Sieber H. Preparation of porous Al_2O_3-ceramics by biotemplating of wood[J]. J. Porous Mater., 2004, 11(3): 163-172.

[41]Cao J, Rusina O, Sieber H. Processing of porous TiO_2-ceramics from biological preforms[J]. Ceram. Int., 2004, 30(7): 1971-1974.

[42]Zollfrank C, Kladny R, Sieber H, Greil P. Biomorphous SiOC/C-ceramic composites from chemically modified wood templates[J]. J. Eur. Ceram. Soc., 2004, 24(2): 479-487.

[43]张荻，孙炳合，范同祥. 遗态材料的制备及微观组织分析[J]. 中国科学E辑：工程科学 材料科学，2004，34(7)：721-729.

[44]孙炳合，张荻，范同祥，谢贤清. 木质材料陶瓷化的研究进展[J]. 功能材料，2003，01：20-22，28.

[45]刘兆婷. 木材结构分级多孔氧化物制备、表征及其功能特性研究[D]. 上海交通大学，2008.

[46]李淑君，李坚，刘一星. 木陶瓷的制造(Ⅰ)——实木陶瓷[J]. 东北林业大学学报，2002，30(4)：5-7.

[47]李淑君，李坚，刘一星. 新型炭材料——木陶瓷[J]. 上海建材，2002(4)：19-22.

[48]李淑君. 新型多孔炭材料——木陶瓷的研究[D]. 东北林业大学，2001.

[49]Wang S L, Shi J Y, Liu C Y, Xie C, Wang C Y. Fabrication of a superhydrophobic surface on a wood substrate[J]. Appl. Surf. Sci., 2011, 257(22): 9362-9365.

[50]Wang S L, Wang C Y, Liu C Y, Zhang M, Ma H, Li J. Fabrication of superhydrophobic spherical-like α-FeOOH films on the wood surface by a hydrothermal method[J]. Colloids Surf., A, 2012, 403: 29-34.

[51]袁光明，吴义强，胡云楚. 用无机纳米材料复合改性木材的机理研究进展[J]. 中南林业科技大学学报，2010，30(5)：163-167.

[52]孙庆丰. 外负载无机纳米/木材功能型材料的低温水热共溶剂法可控制备及性能研究[D]. 东北林业大学，2012.

[53]杨星，姜维娜，周晓燕，徐莉. 杨木纤维/无机纳米 Al_2O_3 复合材料的阻燃性能[J]. 林业科技开发，2010，24(2)：58-61.

[54]Sun F B, Yu Y, Jiang Z H, Ren H Q, Liu X E. Nano TiO_2 modification of bamboo and its antibacterial and mildew resistance performance[J]. Spectrosc Spect Anal, 2010, 30 (4): 1056-1060.

[55]符韵林，赵广杰，全寿京. 二氧化硅/木材复合材料的微观结构与物理性能[J]. 复合材料学报，2006，23(4)：52-59.

[56]符韵林，赵广杰. 溶胶-凝胶法在木材/无机纳米复合材料上的应用[J]. 林产工业，2005，32(1)：6-9.

[57]Sehaqui H, Salajkova M, Zhou Q, Berglund L A. Mechanical performance tailoring of tough ultra-high porosity foams prepared from cellulose Ⅰ nanofiber suspensions[J]. Soft Matter, 2010, 6(8): 1824-1832.

[58]Li J, Lu Y, Yang D J, Sun Q F, Liu Y X, Zhao H J. Lignocellulose aerogel from wood-ionic liquid solution (1-allyl-3-methylimidazolium chloride) under freezing and thawing conditions[J]. Biomacromolecules, 2011, 12(5): 1860-1867.

[59]Lu Y, Sun Q F, Yang D J, She X L, Yao X D, Zhu G S, Liu Y X, Zhao H J, Li, J. Fabrication of mesoporous lignocellulose aerogels from wood via cyclic liquid nitrogen freezing-thawing in ionic liquid solution[J]. J. Mater. Chem. A, 2012, 22(27): 13548-13557.

[60]邱坚，李坚. 超临界干燥制备木材-SiO_2 气凝胶复合材料及其纳米结构[J]. 东北林业大学学报，2005，33(3)：3-4.

[61]邱坚. 木材/SiO_2 气凝胶纳米复合材的研究[D]. 东北林业大学，2004.

[62]Cai J, Zhang L. Rapid dissolution of cellulose in LiOH/ urea and NaOH/urea aqueous solutions [J]. Macromol. Biosci., 2005, 5(6): 539-548.

[63]Cai J, Liu S L, Feng J, Kimura S, Wada M, Kuga S, Zhang L N. Cellulose-silica nanocomposite aerogels by in situ formation of silica in cellulose gel[J]. Angew. Chem. Int. Ed., 2012, 124(9): 2118-2121.

英文题目： Inspirations from nature——Preliminary discussion of wood bionics

作者： Li Jian, Sun Qingfeng

摘要： Inherently intelligent behaviors and uniquely natural attributes of some organisms in nature were specifically listed in this paper. Inspired by nature, the theoretical foundation for constructing wood bionics was preliminarily stated. Under the guidance of biology and current technology, wood will be endowed with some untraditional properties and newly innovated wood-based materials will possess a much larger developing space.

关键词： wood; bionics; superhydrophobicity; aerogel; intelligence; interface

光控润湿性转换的抑菌性木材基银钛复合薄膜*

李坚，高丽坤

摘要： 以水热法和银镜法在木材表面制备出 Ag-TiO_2 复合微纳米结构薄膜，并通过有机物氟硅烷修饰使木材表面具有超疏水性。采用场发射扫描电子显微镜(FE-SEM)、X 射线衍射能谱(XRD)、傅立叶变换红外光谱仪(FTIR)和接触角测试等方法对木材表面进行了分析和表征。研究结果显示，经氟硅烷修饰后的 Ag-TiO_2 负载的木材表面具有良好的紫外光驱动润湿性转换的特性，即光照前为超疏水性(152.8°)和亲油性(25°)，光照一段时间后转变为超疏油性(150.2°)和亲水性(26.2°)。这是由于氟硅烷受到紫外光照射后会光致分解破坏一部分的烷基链，并在紫外光的激发下产生亲水基团所致。同时，与单纯 TiO_2 负载的木材相比，Ag-TiO_2 复合薄膜中银纳米颗粒赋予了木材良好的抑菌性能，可提高木材的生物耐久性。以上研究为木材润湿性转换的智能化设计和多功能化设计开辟了新的途径。

关键词： 木材；银钛复合薄膜；光控；润湿性；转换

1 前 言

润湿性是固体表面的特征之一，它是由表面的化学组成和微观几何结构共同决定的，在日常生活、工业及建筑方面具有重要意义[1-2]。当一个固体表面与水或油的接触角大于 150°时，称为超疏水或超疏油表面；反之，当它与水或油的接触角小于 10°时，称为超亲水或超亲油表面[3]。近年来，一类具有超疏水功能的无机微纳米薄膜引起了研究者的广泛关注。将无机微纳米薄膜负载在有机物基质表面，综合有机和无机材料的优异特性(如电性、磁性、光学性质等等)，从而制备的多功能材料在自清洁、油水分离、生物医药等领域有潜在的应用[4]。木材作为一种天然的有机材料，由于其具有低密度、可再生性、环境友好性和外观美丽等优点，被广泛地用作室内和室外材料。而当前解决木材资源不足和使用中的种种限制，实现木材的自增值性和提高木材的高附加值利用，是木材科学发展的首要任务[5]。文中采用简便的水热法和银镜法，将微米级 TiO_2 粒子和纳米级 Ag 粒子沉积到木材表面，构建出微纳米银钛复合薄膜负载的木材，并采用低表面能物质(氟硅烷)进行表面修饰，制备超疏水/亲油木材，紫外光(Ultra violet，UV)照射后产生羟基等亲水/疏油基团，使其表面发生相反物性的转变——超疏油/亲水。并探索出微纳米结构与光控木材表面润湿性转换效应之间的关系，由图 1 可见，光便是使物性发生转变的可逆开关。

2 材料与方法

2.1 原料

试材为 20 mm(长)×20 mm(宽)×5 mm(厚)的杨木(*Populus ussuriensis* Kom)锯材。试验前，用酒精和蒸馏水依次清洗木块后烘干备用；氟钛酸铵；硼酸；硝酸银；氨水；葡萄糖；氟硅烷；甲醇；蒸

* 本文摘自森林与环境学报，2015，35(3)：193-198.

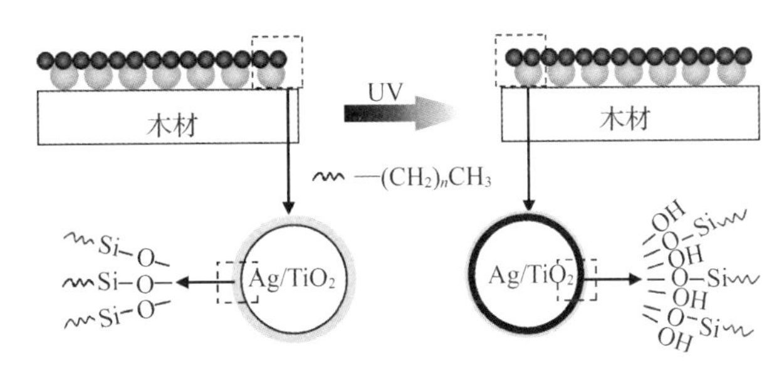

图 1　Ag/TiO_2 负载的木材表面 UV 光控润湿性转换机理图

馏水；十六烷；大肠埃希菌(ATCC 25923)。

2.2　在木材表面制备复合薄膜

取适量的氟钛酸铵和硼酸溶于蒸馏水中，室温下磁力搅拌 30 min；然后，将木材和混合溶液置于聚四氟乙烯反应釜中，在 90 ℃的烘箱内进行水热合成过程，反应 5 h 后取出，自然冷却至室温，用蒸馏水冲洗木材表面，放置烘箱内烘干。即得到二氧化钛(TiO_2)薄膜负载的木材。

依据银镜反应制备银纳米粒子(Ag NPs)[6]。配置一定浓度的硝酸银水溶液，用氨水逐滴加入溶液中直至溶液变透明；将 TiO_2 负载的木材置于透明的银氨溶液中浸泡 1 h；然后，将处理的木材转移到一定浓度的葡萄糖溶液中处理 5 min 后，将剩余的银氨溶液到了葡萄糖溶液中，继续反应 15 min。反应结束后，将木材取出，用蒸馏水冲洗后烘干。即得到 Ag/TiO_2 复合薄膜负载的木材。

将一定比例的氟硅烷和甲醇混合溶液溶于 3 倍体积的蒸馏水中，在室温下搅拌 5 min，将负载 Ag/TiO_2 复合薄膜的木材置于混合溶液中，室温下磁力搅拌 24 h，最后，将样品取出，蒸馏水冲洗后，烘干。即得到氟硅烷改性的 Ag/TiO_2 复合薄膜负载的木材。

为了探明复合微纳米结构对润湿性转换及抑菌性的影响，另制备了单纯微米 TiO_2 负载的木材作为对照组，制备方法参见上述步骤及参考文献[7]。

2.3　样品表面形貌表征及元素组成分析

利用 Quanta 200 型场发射扫描电子显微镜(FE-SEM，荷兰 FEI 公司)和 Tecnai G20 型透射电子显微镜(TEM，美国 FEI 公司)对样品的表面形貌进行表征；采用 D8 Advance 型 X 射线衍射仪(XRD，德国 Bruker 公司)进行物相分析，X 射线源为 Cu 射线，扫描范围 5°~80°，步宽 0.02°，扫描速度 $4°\cdot min^{-1}$；将固体样品打磨成粉末，取约 0.1 mg 的样品粉末与溴化钾粉末混合并充分研磨，随后置于 Magna-IR 560 型傅立叶变换红外光谱仪(FTIR，美国 Nicolet 公司)中，对样品表面化学组分的变化进行表征分析，扫描范围 4000~500 cm^{-1}，分辨率 4 cm^{-1}。

2.4　Ag/TiO_2 负载的木材表面光控润湿性转换测试

Ag/TiO_2 复合薄膜负载的木材表面润湿性转换性能测试，以水滴和油滴(十六烷)与样品表面接触角大小的变化衡量。采用 OCA40 型接触角测定仪(德国 Dataphysics 仪器公司)在室温下进行测试，液滴量为 5 μL，每个样品至少选取 5 个不同点进行测量，取其平均值，其中，水与样品表面的接触角记为 WCA，油与样品表面的接触角记为 OCA。测定样品初始接触角后，以紫外灯(36 W，波长为 400 nm，中国)作为紫外光光源，对样品表面照射一段时间，样品与紫外灯的垂直距离约为 10 cm，分别测试光照后样品表面的水接触角和油接触角。

2.5　Ag/TiO_2 复合薄膜负载的木材抑菌试验

木材样品的抑菌活性采用细菌抑制环法进行表征分析(琼脂扩散试验/CEN/TC 248 WG 13)。样品的抑菌活性选用大肠埃希菌(ATCC 25923)评估[8]。实验步骤如下：采用马铃薯、葡萄糖为营养物质制备细菌培养基。取去皮洗净马铃薯200 g，切成小块，加入1000 mL水，煮沸30 min，滤去马铃薯块，将滤液用蒸馏水补足到1000 mL，倒入500 mL锥形瓶中，加入葡萄糖20 g，琼脂15 g，加热溶化后密封，待用。将上述马铃薯琼脂放于电炉上融化灭菌，制成琼脂培养基，冷却后将样品放入琼脂培养皿内，在恒温恒湿箱中培养24 h。

3 结果与分析

图2(左上)为木材素材的表面形貌，素材经酒精和蒸馏水依次清洗干燥后，可以清楚的看到木材表面无任何杂质。如图2(右上)所示，在木材表面成功构建出微纳米 Ag/TiO_2 复合薄膜。Ag/TiO_2 复合粒子几乎能够填满木材表面的孔隙，分布均匀且致密。氟硅烷改性修饰后，薄膜表面的 Ag/TiO_2 复合粒子仍然清晰可见[图2(左下)]，说明氟硅烷的改性并不影响薄膜的复合结构。图2(右下)为 Ag/TiO_2 复合薄膜负载的木材表面透射电镜图片。均匀分布的球形黑点即为纳米Ag粒子，铺覆在 TiO_2 负载的木材基材表面。

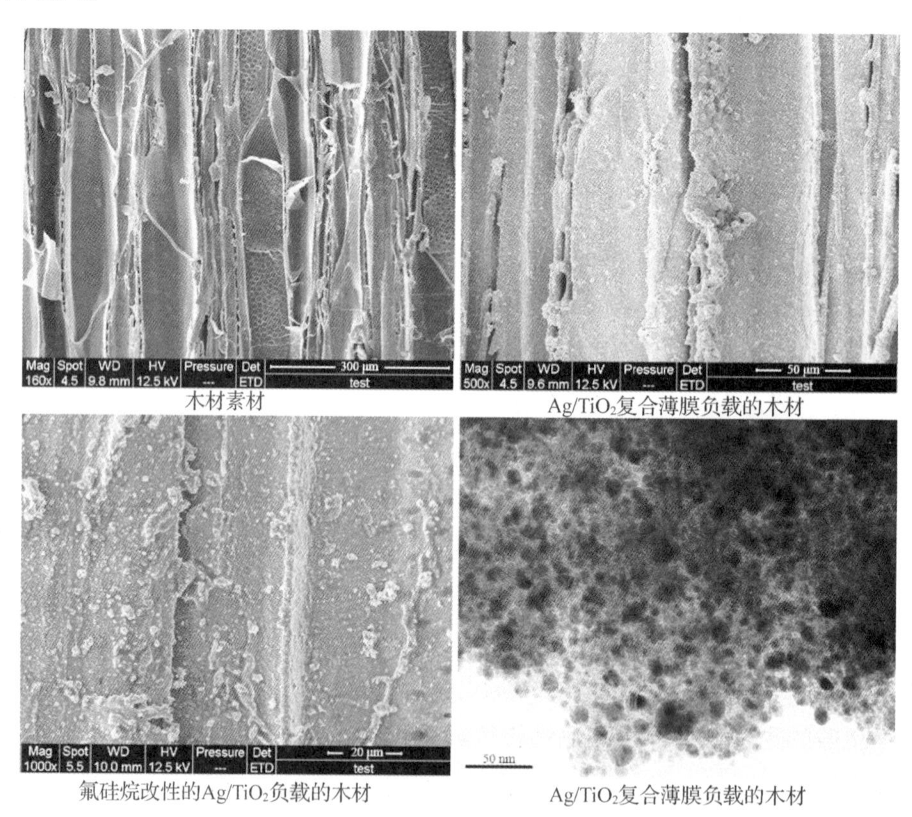

图2 木材表面 FE-SEM 及 TEM

图3为单纯 TiO_2 负载的木材和 Ag/TiO_2 负载的木材的XRD谱图。其中，2个谱图中均可见14.8°和22.5°处的衍射峰，分别对应木材中纤维素(101)和(002)晶面[9-10]。TiO_2 负载的木材的谱线中25.2°、38.0°、47.8°、54.2°、62.5°、68.8°和74.9°处的衍射峰与锐钛矿 TiO_2 的标准衍射图谱(JCPDS file No. 21-1272)相吻合[11]；Ag/TiO_2 负载的木材的谱线中除锐钛矿 TiO_2 的衍射峰可见，还存在38.1°、44.2°、64.4°和77.4°处的衍射峰，分别归属于Ag(JCPDS file No. 04-0783)晶体(111)、

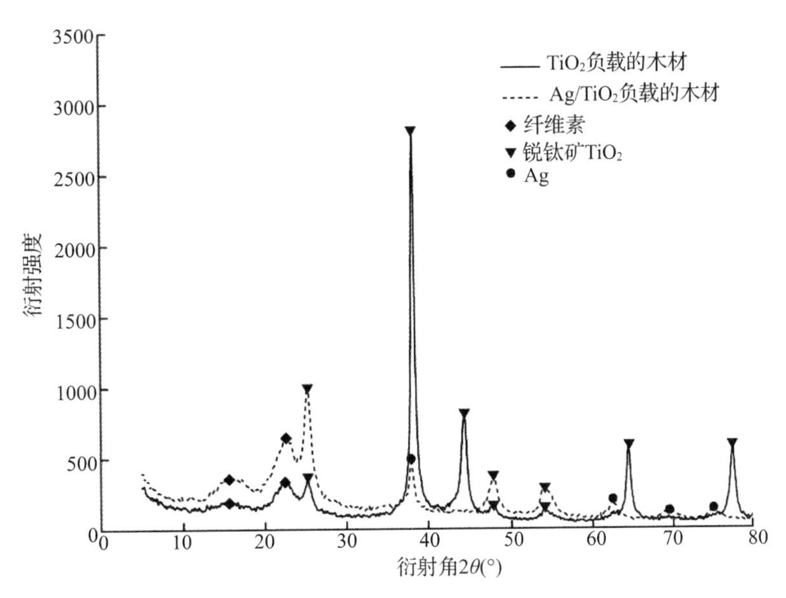

图 3 木材样品的 XRD 图谱

(200)、(220)和(311)面的衍射峰[12]。结果证明，运用该方法以木材为基质，可成功制备出 Ag/TiO_2 微纳米复合薄膜。

图 4 为样品的 FTIR 光谱。从图 4 中可以看出，木材素材中的羟基(OH)伸缩振动峰位于 3340 cm^{-1} 处，而 Ag/TiO_2 负载的木材和氟硅烷改性的 Ag/TiO_2 负载的木材中分别红移至 3345 cm^{-1} 和 3415 cm^{-1} 处，并且峰的强度相对减小，说明 Ag/TiO_2 负载的木材和氟硅烷改性的 Ag/TiO_2 负载的木材化学组分中羟基数目减少，即样品的亲水性降低。纤维素的主要特征峰分别为：2903 cm^{-1} 处 CH_3 和 CH_2 中 C—H 的伸缩振动峰；1426 cm^{-1} 和 1372 cm^{-1} 处 CH_2 和 CH 中 C—H 的弯曲振动峰；1163 cm^{-1} 和 1058 cm^{-1} 处 C ═O 的伸缩振动峰[13]。此外，在氟硅烷改性的 Ag/TiO_2 负载的木材中

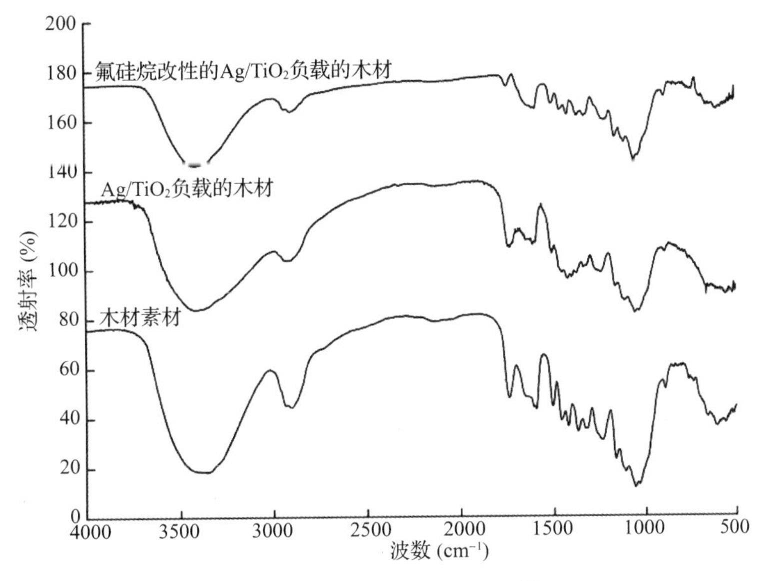

图 4 木材样品的 FTIR 图谱

745 cm^{-1} 左右的峰为氟硅烷分子中 C—F 伸缩振动和变形振动的贡献[14]，表明氟硅烷分子已负载在木材表面。

图 5 为氟硅烷改性的 Ag/TiO_2 负载的木材表面 UV 光照射前后水接触角和油接触角照片。图 5(左上)中 UV 照射前水滴在其表面的接触角为 152.8°，图 5(左下)中油滴在其表面的接触角为 25°，即 UV 光照射前，样品表面为超疏水/亲油性，这是因为微纳米复合薄膜表面所修饰的氟硅烷的长链烷基基团的作用。经过 UV 光照射一段时间后，测得水滴在其表面的接触角为 26.2°[图 5(右上)]，油滴在其表面的接触角为 150.2°[图 5(右下)]，变为超疏油/亲水性。这是由于氟硅烷受到 UV 光照射后会光致分解破坏一部分的烷基链，使得复合薄膜中 TiO_2 裸露在外并被紫外光激发，在表面形成化学吸附水，即产生亲水基团所致[15]。以上实验结果证明，经氟硅烷修饰后的 Ag/TiO_2 负载的木材表面具备 UV 光控润湿性转换能力。同时，此研究为木材润湿性转换的智能化设计提供了新的途径。

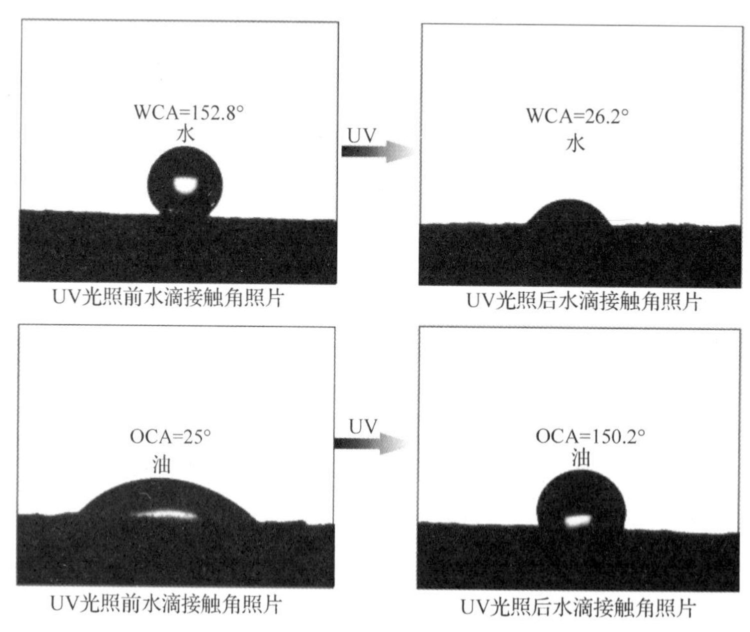

图 5　木材表面 UV 光致润湿性转换的接触角

图 6 为木材样品在琼脂培养皿中的抑菌活性图片。图 6(左)为 TiO_2 负载的木材样品对照组，显而易见地，其对大肠埃希菌无任何的抑菌活性。图 6(右)为 Ag/TiO_2 负载的木材，可以很清楚看见其培养皿中的抑菌圈，该样品在接种细菌的表面可杀死在其下方和周围所有的细菌，抑菌效果良好。该结果表明，木材样品的抑菌生物活性是由于其表面 Ag 纳米粒子具有杀菌作用。

4　结　论

采用水热法和银镜法制备出一种新型氟硅烷改性的 Ag/TiO_2 负载的木材。以木材为基质，负载银钛微纳米复合薄膜，并选用低表面能的氟硅烷进行修饰，实验证明，氟硅烷对复合薄膜的结构不产生影响。UV 光照射一段时间后，木材表面从超疏水/亲油态转变为超疏油/亲水态。因此，木材表面的复合薄膜显示出良好的光控润湿性转换特性。即紫外光便是材料表面发生物性转变的可逆开关。其研究结果为助推木材的智能化设计和利用提供了新的启发。

同时，该无机微纳米复合薄膜赋予木材抑菌活性，这是因为其表面负载具有杀菌活性的 Ag 纳米

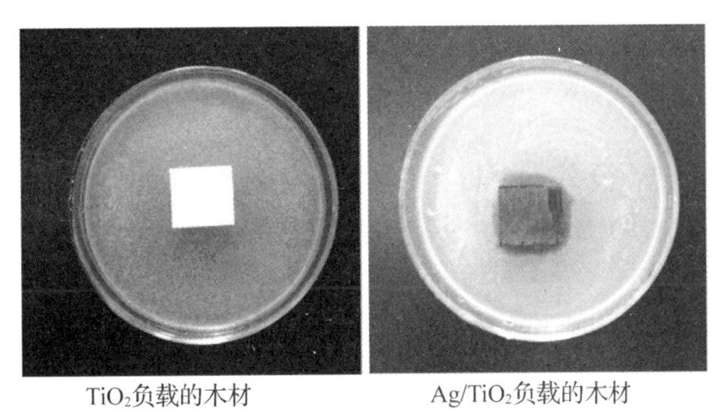

TiO_2负载的木材　　Ag/TiO_2负载的木材

图6　木材样品的抑菌效果

粒子所致。这种多功能化材料的形成，有利于扩大木材的应用范围和实现木材的高附加值利用。

基金项目

国家自然科学基面上项目(31470584)。

参考文献

[1] Sun T, Feng L, Gao X, et al. Bioinspired surfaces with special wettability[J]. Accounts of Chemical Research, 2005, 38(8): 644-652.

[2] 陈洪燕，江雷. 受生物启发特殊浸润表面的设计和制备[J]. 生命科学，2008，20(3)：323-330.

[3] Zhang X, Shi F, Niu J, et al. Superhydrophobic surfaces: from structural control to functional application[J]. Journal of Materials Chemistry, 2008, 18(6): 621-633.

[4] Zeng J, Liu S, Cai J, et al. TiO_2 immobilized in cellulose matrix for photocatalytic degradation of phenol under weak UV light irradiation [J]. The Journal of Physical Chemistry C, 2010, 114(17): 7806-7811.

[5] 李坚，孙庆丰. 大自然给予的启发——木材仿生科学刍议[J]. 中国工程科学，2014，16(4)：4-12.

[6] Jin C, Li J, Han S, et al. Silver mirror reaction as an approach to construct a durable, robust superhydrophobic surface of bamboo timber with high conductivity[J]. Journal of Alloys and Compounds, 2015, 635: 300-306.

[7] Gao L, Lu Y, Zhan X, et al. A robust, anti-acid, and high-temperature-humidity-resistant superhydrophobic surface of wood based on a modified TiO_2 film by fluoroalkyl silane[J]. Surface and Coatings Technology, 2015, 262: 33-39.

[8] Xue CH, Chen J, Yin W, et al. Superhydrophobic conductive textiles with antibacterial property by coating fibers with silver nanoparticles[J]. Applied Surface Science, 2012, 258(7): 2468-2472.

[9] Andersson S, Serimaa R, Paakkari T, et al. Crystallinity of wood and the size of cellulose crystallites in Norway spruce (Picea abies)[J]. Journal of wood science, 2003, 49(6): 531-537.

[10] 李坚. 木材波谱学[M]. 北京：科学出版社，2003.

[11] Gao L, Zhan X, Lu Y, et al. pH-dependent structure and wettability of TiO_2-based wood surface [J]. Materials Letters, 2015, 142: 217-220.

[12] Seery MK, George R, Floris P, et al. Silver doped titanium dioxide nanomaterials for enhanced visible light photocatalysis [J]. Journal of Photochemistry and Photobiology A: Chemistry, 2007, 189(2): 258-263.

[13] 李坚，吴玉章，马岩. 功能性木材[M]. 北京：科学出版社，2011.

[14] Jin C, Li J, Han S, et al. A durable, superhydrophobic, superoleophobic and corrosion-resistant coating with rose-like ZnO nanoflowers on a bamboo surface[J]. Applied Surface Science, 2014, 320: 322-327.

[15] Risse G, Matys S, Böttcher H. Investigation into the photo-induced change in wettability of hydrophobized TiO_2 films [J]. Applied Surface Science, 2008, 254(18): 5994-6001.

英文题目: Photocontrolled Wettability Conversion Properties of Antibacterial Ag-Ti Composite Film Based on Wood Substrate

作者: Li Jian, Gao Likun

摘要: The micro-nano Ag-TiO_2 composite film based on wood surface has been fabricated using hydrothermal synthesis and silver mirror method. Further modification of the treated wood surface with the organic material of fluoroalkyl silane led to a superhydrophobic surface. Field emission scanning electron microscopy (FE-SEM), X-ray diffraction (XRD), Fourier transform infrared spectroscopy (FTIR) and contact angle were used to characterize the wood surfaces. The results indicated that after modified with fluoroalkyl silane, the Ag-TiO_2-coated wood surface possessed the good wettability conversion properties drived by UV light. That is, the modified wood surfaces were originally superhydrophobic (150.2°) and oleophylicity (26.2°), and became superoleophobic (150.2°) and hydrophilicity (26.2°) after UV irradiation for a period of time. That may be due to the photo-decomposition of fluoroalkyl silane leading a destruction of alkyl chain and the production of hydrophilic groups under UV irradiation. Meanwhile, compared with the pure TiO_2-coated wood, the Ag nanoparticles in the Ag-TiO_2 composite film has also impart excellent antibacterial property to wood, which could improve the biological resistance properties of wood. The findings develop a new way for design of smart wood with switchable wettability and multifunctional wood.

关键词: Wood; Ag-Ti composite film; Photocontrol; Wettability; Conversion

pH-dependent Structure and Wettability of TiO_2 -based Wood Surface*

Likun Gao, Xianxu Zhan, Yun Lu, Jian Li, Qingfeng Sun

Abstract: TiO_2 thin films with different wettability were fabricated on the wood surfaces by a simple low-temperature hydrothermal method with the precursor solution pH adjusted by hydrochloric acid/sodium hydroxide. The morphologies of TiO_2 films have been changed from sphere species to filmy ones by adjusting pH of precursor solution. The TiO_2 films synthesized at the precursor solution pH of 1~10 were mainly primary existed in anatase phase without other structure, and the TiO_2-treated wood surfaces presented different wettability with the water contact angles ranged from 9.6° to 132.7°, when the precursor solution pH were controlled to the range of 1~14. Such a wood surface with the desirable wettability shows great potential, as it may be selectively used in the various environments with different humidity.

Keywords: pH-dependent; Thin films; Desirable wettability; Surfaces

1 Introduction

Wood as a natural renewable resource has become one of the most appropriate and versatile raw materials for a variety of uses in our daily lives[1]. Apart from its satisfactory performance, wood also presents some property limitations. For example, the wood dimension changes caused by the unstable moisture may be a serious problem to its durability and use value, which due to the cell wall polymers contain hydroxyl and other oxygen-containing groups that attract moisture through hydrogen bonding[2]. Thus, many researches on developing a variety of wood surfaces with special wettability suitable for various conditions have been constructed by using different methods[3].

The formation of a thin film containing inorganic phases such as TiO_2, ZnO, Ag, and so forth, can lead to novel applications in the areas of self-cleaning, catalysis, electronics, and biomedical engineering[4-6]. Recently, the inorganic-particles-containing films have been studied extensively, and it was found that the properties of the films could be tuned by controlling the preparation conditions, such as solution pH conditions[7]. To create a thin film of inorganic particles with the control over the physicochemical properties becomes one of the challenges in creating various types of functional surfaces. Also, these inorganic-particles-containing films can be readily coated onto the surface of soda-lime glass, paper, wood, textiles, bamboo, and plastic[8-14].

Against this background, the aim of the present work was to realize the desirable wettability of wood surface that was employed to map the precursor solution pH. To achieve this goal, TiO_2 films on the wood

* 本文摘自 Materials Letters, 2015, 142: 217-220.

surfaces were prepared by a simple low-temperature hydrothermal method with the precursor solution pH adjusted by hydrochloric acid/sodium hydroxide.

2 Materials and methods

Polar wood slices (*Populus ussuriensis* Kom) 20 mm × 20 mm × 10 mm were ultrasonically rinsed in distilled water for 30 min and dried in vacuum at 80 ℃ for 24 h.

In TiO_2 preparation, ammonium fluorotitanate (0.4 M) and boric acid (1.2 M) were dissolved in distilled water at room temperature under vigorous magnetic stirring and the prepared solution has a pH value of 6.53 measured by a pH meter. Then, the solutions pH were adjusted approximately from 1 to 6 by 0.1 M hydrochloric acid (HCl) and adjusted approximately from 7 to 14 by 0.1 M sodium hydroxide (NaOH). Then, the adjusted solutions were transferred into the Teflon containers, respectively. Wood specimens were subsequently placed into the above reaction solutions, separately. The autoclaves were sealed and maintained at 90 ℃ for 5 h, then allowed to naturally cool to room temperature. Finally, the prepared samples were removed from the solutions, ultrasonically rinsed with deionized water for 30 min, and dried at 45 ℃ for more than 24 h in vacuum. For comparative studies, blank wood samples were also examined.

The morphology of the wood samples was examined by scanning electron microscope (SEM, Quanta 200, FEI) operating at 12.5 kV. X-ray diffraction (XRD, Bruker D8 Advance, Germany) operating with Cu $K\alpha$ radiation ($\lambda = 1.5418$ Å), using a step scan mode with a scan rate of 4° min^{-1}, 40 kV, 40 mA ranging from 5° to 70°. Water contact angle (WCA) was measured on an OCA40 contact angle system (Dataphysics, Germany) at room temperature. The final value of the CA was obtained as an average of five measurements.

3 Results and discussion

Fig. 1 shows the WCAs of the TiO_2-treated wood surfaces obtained in the different precursor solutions with the pH varied from 1 to 14. The WCAs of the wood surfaces obtained in the precursor solution pH 1~10 are smaller than 90°. At solution pH 11, the WCA shows that hydrophilic films begin to transfer into hydrophobic ones. And the WCAs of the wood surfaces obtained in the solution pH 11~14 are larger than 90°, which was probably due to the low content of TiO_2 on the surfaces with less improvement on the hydrophilic property of the

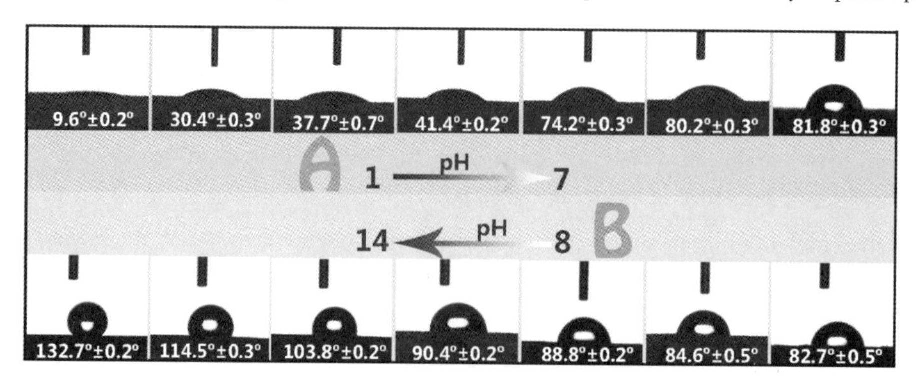

Fig. 1 The WCAs of the TiO_2-treated wood surfaces obtained in the different precursor solutions with pH varied from 1 to 14

surfaces. Obviously, at solution pH 14, the WCA reaches the maximum of 132.7° (standard deviation of 0.2°). The surfaces presented hydrophilicity with the WCAs ranged from 9.6° to 80.2° (standard deviations of 0.2° and 0.3°, respectively), when the precursor solutions were adjusted by HCl. In addition, when the precursor solutions were adjusted by NaOH, it showed that the WCAs of the wood surfaces ranged from 81.8° to 132.7° (standard deviations of 0.3° and 0.2°, respectively).

Fig. 2 displays the SEM images of the control wood and the TiO_2-treated woods obtained in the different precursor solutions with the pH varied from 1 to 14, respectively. In Fig. 2a, the pristine surface of the poplar

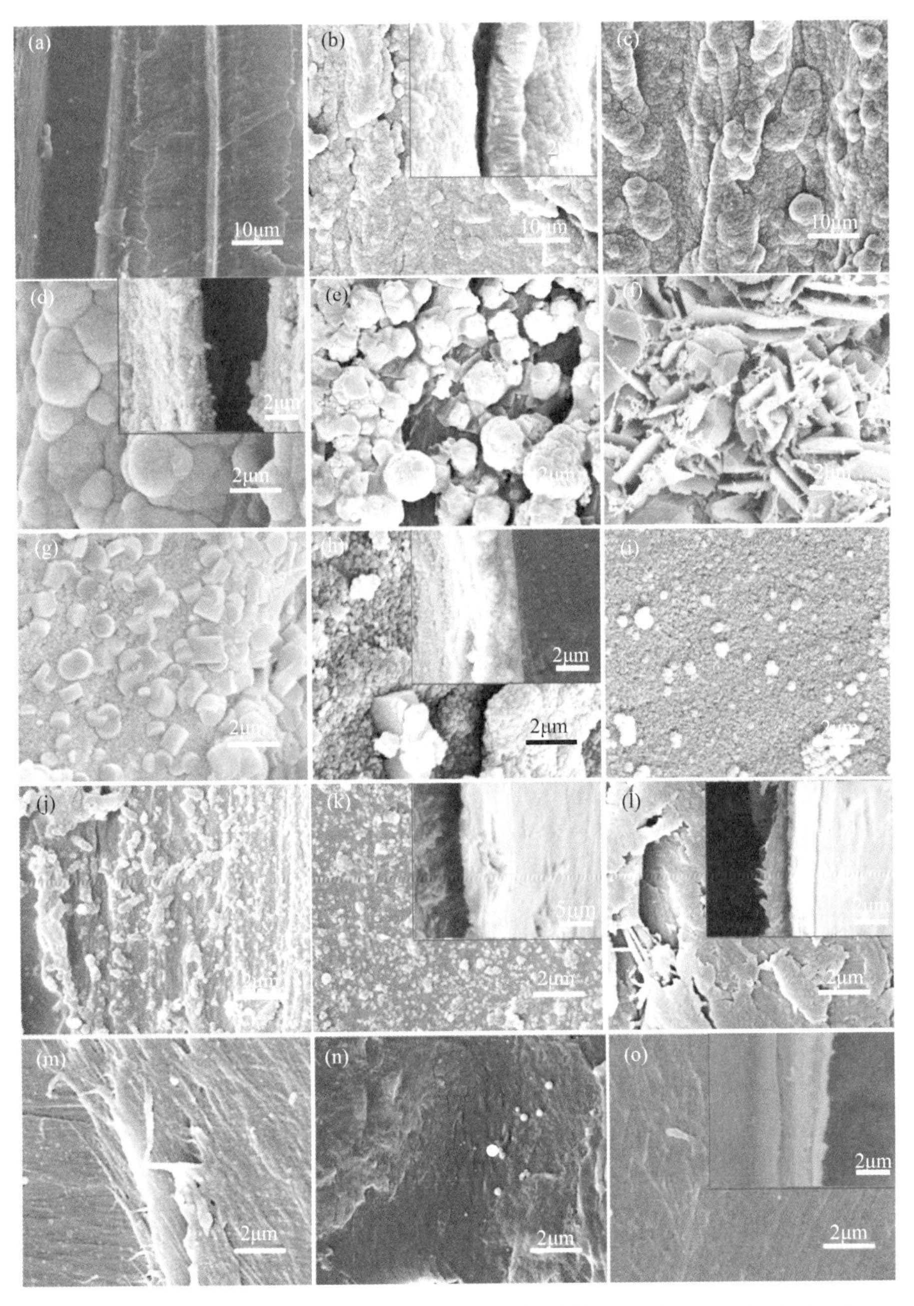

Fig. 2 SEM images of (a) the control wood, and (b–o) the TiO_2-treated woods obtained successively in the precursor solution pH 1~14

wood appeared to be smooth and free of any other materials. As the precursor solution pH increased from 1 to 6, the predominant crystal habit changed from sphere species to cylindrical ones. The SEM images in Fig. 2b-g also exhibited that the range of the precursor solution pH synthesizing dense and even structures of TiO_2 sphere particles was from 1 to 6. The titanium dioxide films were grown to different thicknesses according to the solution pH (SEM images are shown in the insets of Fig. 2). The films were about 1.05 μm thickest and 0.79 μm thick when solution pH were 1 and 3, respectively. When the precursor solution pH were adjusted into 7-10 by NaOH (Fig. 2h-k), the morphologies existed with a preponderance of sphere species. The films were about 0.57 μm thick and 0.21 μm thick when solution pH were 7 and 10, respectively. Finally, the coated surfaces became smoother with the precursor solution pH adjusted into 11-14 (Fig. 2l-o). The pH value of the precursor solution for limiting the formation of TiO_2 sphere particles on the wood surface was judged to be 11. As the insets in Fig. 2l and 2o shown, the films became more and more thin, and finally the film was too thin to observe. Moreover, it could identify a distinct morphology evolution of the TiO_2 at precursor solution pH.

Fig. 3 presents the XRD patterns of the pristine wood and the TiO_2-treated woods obtained in the different precursor solutions with the pH of 1, 6, 11, and 14. Apparently, compared with the control wood, excepting the diffraction peaks at 15° and 22° belonging to the (101) and (002) diffraction planes of cellulose in the wood[15], some new strong diffraction peaks are observed in the TiO_2-treated woods prepared by the precursor solutions with pH of 1 and 6, indicating the formation of new crystal structures on the wood surfaces. The crystallinity of TiO_2 was increased with the reduction of the solution pH down to 1. At solution pH 1, the diffraction peaks at 25.3°, 37.8°, 48.0°, 53.8°, 54.8° and 62.5°, are assigned to the (101), (004), (200), (105), (211) and (204) diffraction planes of anatase TiO_2 (JCPDS card no. 21 - 1272),

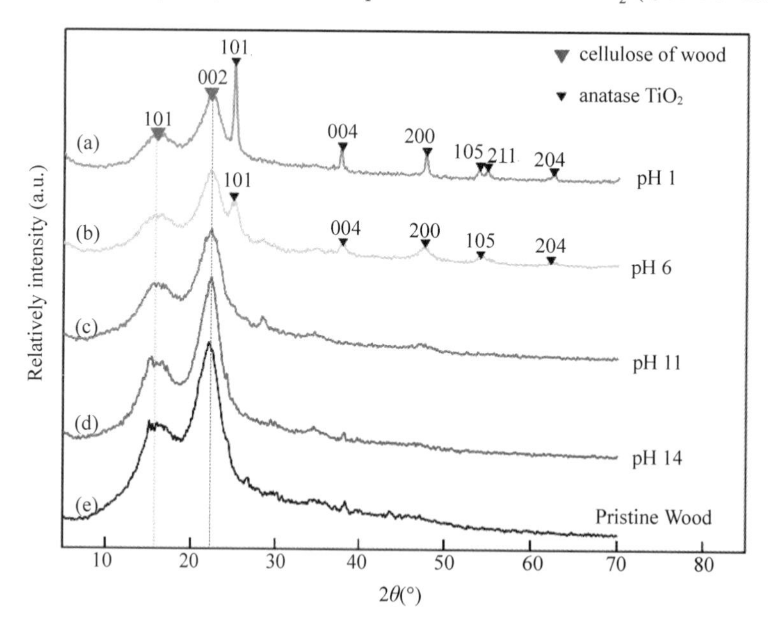

Fig. 3 XRD patterns of the TiO_2-treated woods obtained in the different precursor solutions with (a) pH of 1, (b) pH of 6, (c) pH of 11, (d) pH of 14, and (e) the pristine wood

respectively[11]. While at solution pH 6, the diffraction peaks are same to the treated wood at solution pH 1, but the absence of the diffraction peaks at 54.8° and 62.5°, which may be due to the decrease of TiO_2 crystal[11]. Moreover, the absence of other peaks indicates the purity of the anatase pure TiO_2 coating. However, at the precursor solutions with pH of 11 and 14, there are no differences compared with the control wood, which indicated that there was no anatase TiO_2 crystal settled on the surfaces.

Fig. 4 depicts the effect of the precursor solution pH adjusted by HCl/NaOH. Fig. 4b is the initial precursor solution and the pristine wood with hydroxyl groups on its surface. The initial solution contains numerous $[TiF_6]^{2-}$ ions, H^+ ions, and H_2O molecules. The main reaction equilibrium equations in the mixed solution are as followed[16, 17]: Eq. (1) is the hydrolyzation of $[TiF_6]^{2-}$ ions; Eq. (2) is the reaction between H_3BO_3 and the F^- ions produced in the Eq. (1), the F^- ions are consumed through Eq. (2), leading to the chemical equilibrium in Eq. (1) shifting from left to right; Eq. (3) is the further hydrolyzation of $[TiF_{6-n}(OH)_n]^{2-}$ ions into TiO_2. As shown in Fig. 4a, with the addition of HCl, the increased H^+ ions of the solution accelerated Eq. (2) to move forward. The more F^- ions were consumed, the more $[TiF_{6-n}(OH)_n]^{2-}$ ions were produced with Eq. (1) moved forward. As a result, more TiO_2 particles settled onto the wood surface through the hydrogen bonding in accord with SEM images, which the thick films were composed of dense and even TiO_2 sphere particles. TiO_2 normally contained surface OH groups, namely the surface of TiO_2 was hydroxylated[17], which exposed more hydrophilic groups (OH) and improved the hydrophilicity of the wood surface. Therefore, the water contact angles of the sample surfaces became smaller and smaller when the precursor solutions were adjusted by more and more HCl. However, in Fig. 4c, when adjusting the precursor solution pH by NaOH, the OH^- ions of the solution were increased, which inhibited the hydrolyzation of $[TiF_{6-n}(OH)_n]^{2-}$ ions in Eq. (3), as well as the processes of Eq. (1) and (2). Therefore, the formation of TiO_2 particles on the wood surface was suppressed proved by the films which became thinner as shown in SEM images. Moreover, the hydrophilic groups on the wood surface would be reduced through the reaction between the hydroxyl groups of wood surface and the superfluous OH^- of the solutions. So, it was obvious that the water contact angles of the sample surfaces became larger when adding the solvent containing OH^-. That is, the effective linking degreed of TiO_2 particles and wood surface increased by adding the solvent containing H^+, and decreased by adding the solvent containing OH^-. Therefore, the hydrophilic properties of the treated wood surfaces presented differences when the precursor solutions pH changed.

$$[TiF_6]^{2-} + nH_2O \rightarrow [TiF_{6-n}(OH)_n]^{2-} + nH^+ + nF^- (n=0\sim6) \quad (1)$$

$$H_3BO_3 + 4HF \rightarrow HBF_4 + 3H_2O \quad (2)$$

$$[TiF_{6-n}(OH)_n]^{2-} + (2-n)H_2O \rightarrow TiO_2 + (4-n)H^+ + (6-n)F^- \quad (3)$$

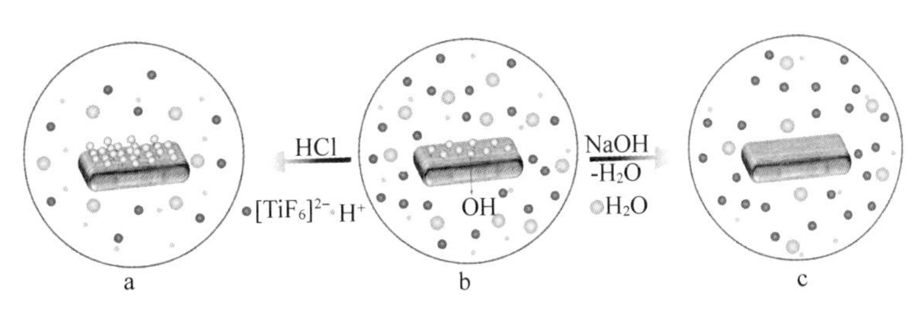

Fig. 4 Proposed mechanism for the fabrication of the TiO_2-treated woods obtained in the different precursor solutions adjusted by HCl/NaOH

4 Conclusions

In the present study, the TiO_2-treated woods with different wettability were prepared by a simple low-temperature hydrothermal method with the precursor solutions adjusted by HCl/NaOH. The morphologies of TiO_2 were changed by adjusting the pH of precursor solutions, and the wood surfaces based on TiO_2 films presented different wettability with the water contact angles ranged from 9.6° to 132.7°, when the precursor solutions pH were controlled to the range of 1~14. Such a wood surface with the desired wettability depended on solution pH may be selectively used in the various environments with different humidity.

Acknowledgments

This work was financially supported by The National Natural Science Foundation of China (grant no. 31270590), China Postdoctoral Science Foundation funded project (2013M540263), and Doctoral Candidate Innovation Research Support Program of Science & Technology Review (kjdb2012006).

References

[1] Ramakrishna S, Mayer J, Wintermantel E, et al. Biomedical applications of polymer-composite materials: A review [J]. Compos. Sci. Technol., 2001, 61(9): 1189-1224.

[2] Evans P D, Wallis A F A, Owen N L. Weathering of chemically modified wood surfaces[J]. Wood Sci. Technol., 2000, 34(2): 151-165.

[3] Evans P D, Michell A J, Schmalzl K J. Studies of the degradation and protection of wood surfaces[J]. Wood Sci. Technol., 1992, 26(2): 151-163.

[4] Athauda T J, Decker D S, Ozer R R. Effect of surface metrology on the wettability of SiO_2 nanoparticle coating[J]. Mater. Lett., 2012, 67(1): 338-341.

[5] Liu K, Cao M, Fujishima A, et al. Bio-inspired titanium dioxide materials with special wettability and their applications[J]. Chem. Rev., 2014, 114(19): 10044-10094.

[6] Sun Z, Liao T, Liu K, et al. Robust superhydrophobicity of hierarchical ZnO hollow microspheres fabricated by two-step self-assembly[J]. Nano Res., 2013, 6(10): 726-735.

[7] Lee D, Omolade D, Cohen R E, et al. pH-dependent structure and properties of TiO_2/SiO_2 nanoparticle multilayer thin films[J]. Chem. Mater., 2007, 19(6): 1427-1433.

[8] Athauda T J, Hari P, Ozer R R. Tuning physical and optical properties of ZnO nanowire arrays grown on cotton fibers [J]. ACS Appl. Mater. Inter., 2013, 5(13): 6237-6246.

[9] Athauda T J, Ozer R R. Investigation of the effect of dual-size coatings on the hydrophobicity of cotton surface[J]. Cellulose, 2012, 19(3): 1031-1040.

[10] Athauda T J, Williams W, Roberts K P, et al. On the surface roughness and hydrophobicity of dual-size double-layer silica nanoparticles[J]. J. Mater. Sci., 2013, 48(18): 6115-6120.

[11] Chae S Y, Park M K, Lee S K, et al. Preparation of size-controlled TiO_2 nanoparticles and derivation of optically transparent photocatalytic films[J]. Chem. Mater., 2003, 15(17): 3326-3331.

[12] Athauda T J, Butt U, Ozer R R. Hydrothermal growth of ZnO nanorods on electrospun polyamide nanofibers[J]. MRS Communications, 2013, 3(1): 51-55.

[13] Jin C, Li J, Wang J, et al. Cross-linked ZnO nanowalls immobilized onto bamboo surface and their use as recyclable photocatalysts[J]. J. Nanomater., 2014, 3: 3.

[14] Li J, Sun Q, Jin C, et al. Comprehensive studies of the hydrothermal growth of ZnO nanocrystals on the surface of bamboo[J]. Ceramics International, 2015, 41(1): 921-929.

[15] Andersson S, Serimaa R, Paakkari T, et al. Crystallinity of wood and the size of cellulose crystallites in Norway spruce (Picea abies)[J]. J. Wood Sci., 2003, 49(6): 531-537.

[16] Tao J, Zhao J, Wang X, et al. Fabrication of titania nanotube arrays on curved surface[J]. Electrochem. Commun., 2008, 10(8): 1161-1163.

[17] Zhang J, Yang C, Chang G, et al. Voltammetric behavior of TiO_2 films on graphite electrodes prepared by liquid phase deposition[J]. Mater. Chem. Phys., 2004, 88(2-3): 398-403.

中文题目：二氧化钛基木材表面随pH变化的结构和润湿性

作者：高丽坤，詹先旭，卢芸，李坚，孙庆丰

摘要：本文采用简单的低温水热法，通过盐酸/氢氧化钠调节前驱体溶液的pH值，制备了不同润湿性的二氧化钛薄膜。通过调节前驱体溶液的pH值，二氧化钛薄膜的形貌由球形变为薄膜状。当前驱体溶液pH值为1~10时，二氧化钛薄膜主要为锐钛矿型，无其他结构。当前驱体pH值为1~14时，二氧化钛处理的木材表面呈现不同的润湿性，接触角由9.6°变化至132.7°。这种可设计的润湿性可选择性地用于不同湿度的各种环境中。

关键词：pH响应；薄膜；可设计的润湿性；表面

Multifunctional Wood Materials with Magnetic, Superhydrophobic and Anti-ultraviolet Properties*

Wentao Gan, Likun Gao, Qingfeng Sun, Chunde Jin, Yun Lu, Jian Li

Abstract: Multifunctional wood materials with magnetic, superhydrophobic and anti-ultraviolet properties were obtained successfully by precipitated $CoFe_2O_4$ nanoparticles on the wood surface and then treated with a layer of octadecyltrichlorosilane (OTS). The as-fabricated wood composites exhibited excellent magnetic property and the water contact angle of the OTS-modified magnetic wood surface reached as high as 150°, revealed the superhydrophobic property. Moreover, accelerated aging tests suggested that the treated wood composites also have an excellent anti-ultraviolet property.

Keywords: $CoFe_2O_4$; Magnetic property; Hydrothermal method; Superhydrophobicity; Anti-ultraviolet property

1 Introduction

Magnetic wood would be a potential for electromagnetic shielding, indoor electromagnetic wave absorber and heavy metal adsorption, which achieved a good harmony of both woody and magnetic characteristics, it was firstly proposed and tested by the Oka group[1-7]. Although the magnetic wood materials have excellent magnetic property, some unfavorable end-product prop-erties of wood materials such as hygroscopic property, dimensional instability and photodegradability have not been fully discussed.

When wood materials were exposed outdoors above ground, a complex combination of solar radiation and other environmental factors like water and temperature contributed to what was described as weathering. Ultraviolet (UV) light irradiation in sun-light was believed to be the most important factor resulted in wood components degradation, among which lignin was the most sensitive one. Under UV light irradiation, the structure of lignin would be destroyed and produced a great amount of free radicals, inducing a reaction with oxygen to produce chromophoric carbonyl and carboxyl groups, and leading to the color change[8-11]. More over, moisture, heat/cold environment were also contributed to the degradation process[12]. In principle, natural wood materials with porous structure were hydrophilic, and water can be easily absorbed and penetrated into the deeper layers of wood by the capillary action of wood cell, which resulting in warping, cupping, face checking and new cracks on the wood surfaces, and accelerating the photodegradation process[13-15]. Thus, there was a great significance to achieve a good harmony of both the superhydrophobicity and anti-ultraviolet property in the wood substrate.

Hybrid wood/inorganic composites were the subject of the intense research activities. These hybrid

* 本文摘自 Applied Surface Science, 2015, 332: 565-572.

materials not only exhibited the inherent properties of wood substrate, such as, the porosity, good strength to weight ratio and esthetic appearance[16], the inorganic properties, in particular anti-ultraviolet property[11] and magnetic properties[17], but also showed the potential synergistic properties, such as superhydrophobicity[18], fire-resistance[19] and biological resistance[20]. Many methods have been applied to the preparation of hybrid wood/inorganic composites[18,21,22]. Among these methods, hydrothermal pro-cess has been recently received much attention. Wang et al. [23] fabricated the superhydrophobic a-FeOOH film on the wood sur-face using the hydrothermal method, and the maximal contact angles of the superhydrophobic wood surface reached as high as 158°. In our previous reports[24-26], the superhydrophobic TiO_2 film with water contact angle about 154° and the CeO_2, ZnO nanoparticles film with excellent anti-ultraviolet property were fabricated on wood substrate via the facile hydrothermal process.

Herein, the magnetic wood was produced successfully by precipitated $CoFe_2O_4$ nanoparticles on the wood surface. Then the surface should be subsequently modified with a layer of octadecyltrichlorosilane (OTS) to obtain superhydrophobicity and UV protection. The modified wood samples were characterized by scanning electron microscopy (SEM), transmission electron microscopy (TEM), atomic force microscopy (AFM), energy dispersive spectroscopy (EDXA), FTIR spectroscopy (FTIR) and X-ray diffraction (XRD). Moreover, the magnetic, hydrophobic andanti-ultraviolet properties of the modified wood composites were also evaluated.

2 Materials and methods

2.1 Materials

Poplar wood slices with the sizes of 15 mm (longitu-dinal)×8 mm (tangential)×1 mm (radial) were ultrasonically washed in deionized water for 30 min and dried at 100 ℃ for 24 h in a vacuum. Cobalt chloride hexahydrate ($CoCl_2 \cdot 6H_2O$), ferrous sulfate heptahydrate ($FeSO_4 \cdot 7H_2O$), potassium nitrate (KNO_3), sodium hydroxide (NaOH) and octadecyltrichlorosilane (OTS, 95%) used in this study were supplied by Shanghai Boyle Chemical Company Limited, and used without further purification.

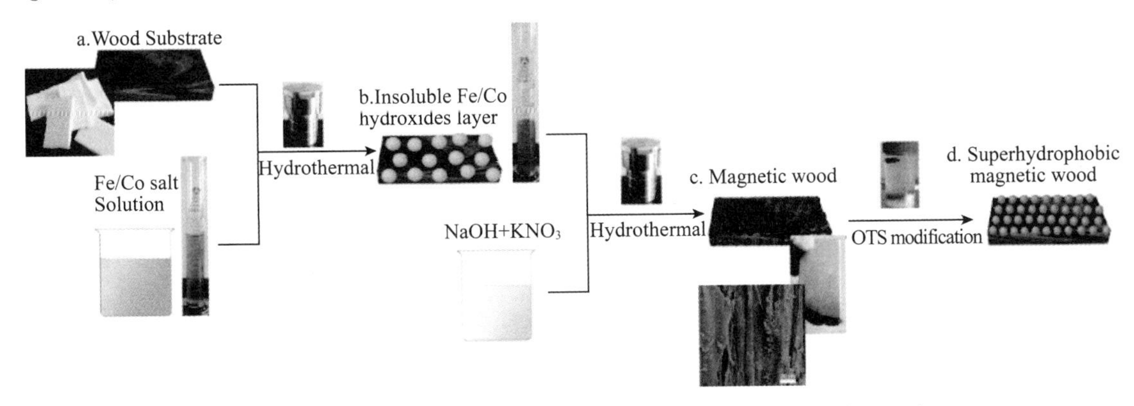

Fig. 1 Preparing process of superhydrophobic and magnetic wood

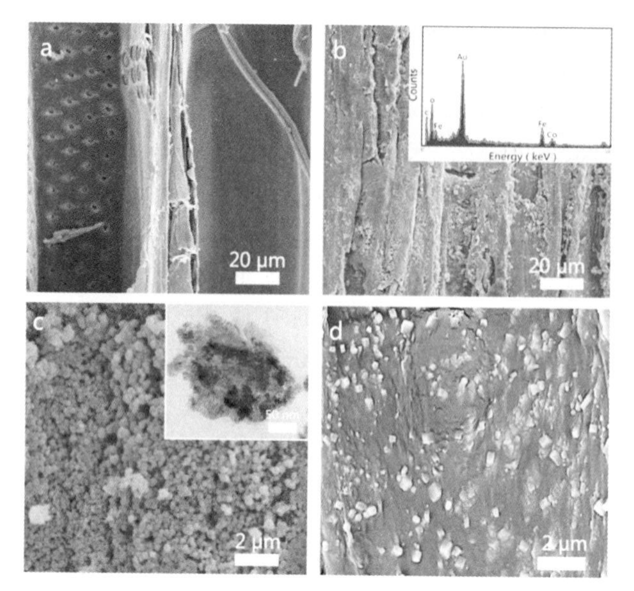

Fig. 2 SEM images of the surfaces of (a) untreated wood surface, and (b, c) magnetic wood at different magnifications; (d) the magnetic wood treated by OTS (Inset: EDXA spectrum of the magnetic wood and TEM image of magnetic nanoparticle)

2.2 Synthesis

The synthesis pathway of superhydrophobic and magnetic wood was described in Fig. 1. The untreated wood materials (step a) were suspended in the 20 mL of 0.2 mol · L^{-1} of freshly pre-pared aqueous solution of $FeSO_4$ and $CoCl_2$ with a molar ration of $[Fe^{2+}]/[Co^{2+}]=2$ via a vacuum process before transferred into a Teflon-lined stainless steel autoclave and heated the system to 90 ℃ for 3 h to thermally precipitate the non-magnetic metal hydroxides on the template (step b). Heating changed the color from transparent to translucent orange[27]. Until the end of the reaction, the supernatant was poured out and 15 mL of 1.32 mol · L^{-1} NaOH solution with KNO_3($[Fe^{2+}]/[NO_3^-]=0.44$) was added into the Teflon-lined stainless-steel autoclave. After an additional hydro-thermal processing for 6 h at 90 ℃, the precipitated precursors were converted into ferrite nanoparticles on the wood surfaces, resulting in high surface roughness with excellent magnetic property (step c). Finally, the prepared wood specimens were removed from the solution and washed by ultrasonically rinsed in deionized water for 30 min, and dried at 50 ℃ for over 24 h in the vacuum cham-ber. Using the above methods, the hydrophilic $CoFe_2O_4$ films were synthesized on the wood surfaces.

The surface modification of the magnetic wood was carried out by a self-assembly of OTS layer. The magnetic woods were immersed into the 20 mL of 5% (*V/V*) OTS ethanol solution at room temperatures under continuous mechanical stirring for 24 h and a layer of OTS molecular covered on magnetic wood surface (step d).

Then the samples were dried at 50 ℃ for over 24 h in the vacuum.

Finally, the superhydrophobic wood surface was obtained.

2.3 Characterizations

The surface morphology of the samples were characterized by the scanning electron microscopy (SEM, FEI, Quanta 200) and transmission electron microscope (TEM, FEI, Tecnai G20). Atomic force microscopy

(AFM) images were obtained using a Multimode Nanoscope Ⅲa controller (Veeco Inc., USA) with a silicon tip oper ated in a tapping mode to characterize the surface morphology and roughness. Crystalline structures of the samples were identified by the X-ray diffraction technique (XRD, Rigaku, D/MAX 2200) oper-ating with Cu Ka radiation (h=1.5418 A°) at a scan rate (2θ) of 4° min^{-1} and the accelerating voltage of 40 kV and the applied cur-rent of 30 mA ranging from 5° to 80°. For FTIR analysis, thin sample disks were made by grinding small portion of the treated wood composites and pressing them with potassium bromide. The FTIR spectra for the wood samples were recorded using FTIR (Magna - IR 560, Nicolet). For magnetic characterization, wood specimens of 4 mm (tangential) 4 mm (radial) 4 mm (longitudinal) were used. The magnetic properties of the composites were measured by a superconducting quantum interference device (MPMS XL-7, Quantum Design Corp.) at a room temperature (300 K). The contact angle analyzer (JC2000C, Beijing Code Tong Technology Co., Ltd) at ambient temperatures with a droplet volume of 5 μL was employed to measure the WCA of the samples. An average of the five mea-surements taken at different positions on each sample was applied to calculate the final WCA angle. DR-UV/Visible spectra for the wood samples were obtained with the instrument TU-190, Beijing Purkinje, China, equipped with an integrating sphere attachment. $BaSO_4$ was the reference.

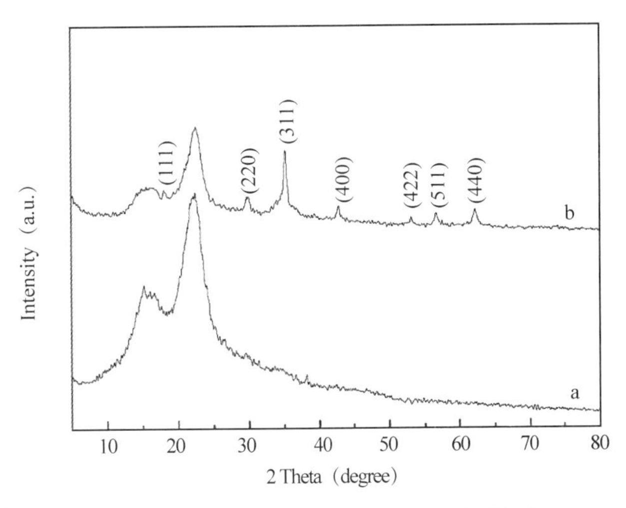

Fig. 3 XRD patterns of (a) the untreated wood, and (b) the magnetic wood

2.3.1 Accelerated aging test

A QUV accelerated weathering tester (Accelerated Weathering Tester, Q-Panl, Cleveland, OH, USA) was used in this experiment, which allowed for water spray and condensation. The samples were fixed in stainless steel holders and then subjected to accelerated weathering through constant exposure to fluorescent UV-light radiation at a wavelength of 340 nm and a temperature of 50 ℃ for 8 h, followed by water spray for 4 h. The average irradiance was set to 0.77 W · m^{-2}, and the spray temperature was fixed at 25 ℃. The exposure time ranged from 0 to 1200 h. The color change of the wood surface before and after the UV irradiation was measured using the portable spectrophotometer (NF-333, Nippon Denshoku Company, Japan) with CIELAB system in accordance with the ISO-2470 standard. CIELAB L^*, a^*, b^*, parameters were measured at six locations on each specimen and average value was calculated. In the CIELAB system, L^* axis represents the lightness, and varies from 100 (white) to 0 (black); a^* and b^* are the chromaticity indices;

$+a^*$ is the red direction; a^* is green; $+b^*$ is yellow; and b^* is blue. The change of L^*, a^* and b^* are calculated according to Eqs. (1)-(3).

$$\Delta a^* = a_2 - a_1 \tag{1}$$

$$\Delta b^* = b_2 - b_1 \tag{2}$$

$$\Delta L^* = L_2 - L_1 \tag{3}$$

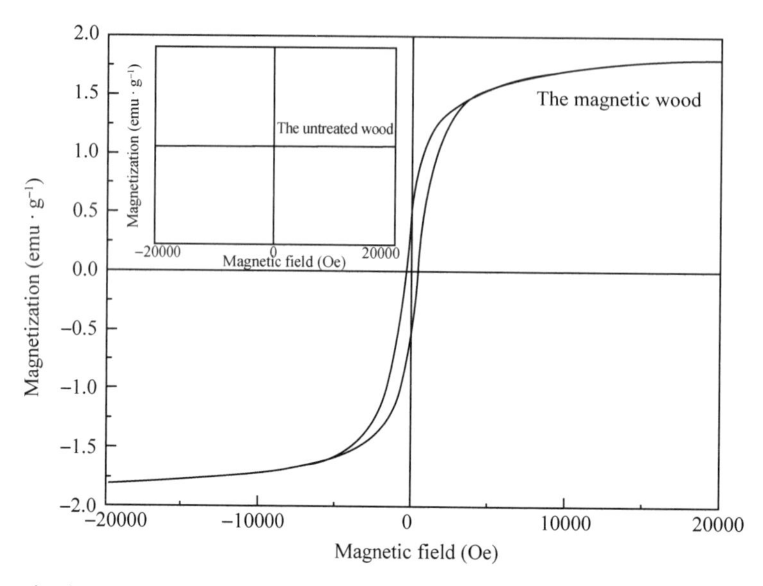

Fig. 4 Magnetization curves at room temperatures of the untreated wood and the magnetic wood

where ΔL^*, Δa^* and Δb^* values were calculated from the difference of the final and initial values (i. e., after and before UV irradiation) of L^*, a^* and b^*, respectively. a_1, b_1 and L_1 are the initial color parameters; a_2, b_2 and L_2 are the final parameters. These values were used to calculate the overall color change ΔE^* as a function of the weathering time, according to the following equation: $CoFe_2O_4$ and OTS. In Fig. 2a, the pristine tracheids of poplar were smooth. However, the $CoFe_2O_4$ nanoparticles formed on wood masks the piths and other details (Fig. 2b and c). The surface chemical elemental compositions of the magnetic wood were determined by EDXA spectrum, and the results were presented as an insert in Fig. 2b. Evidence of only carbon, oxygen, iron, cobalt and gold could be detected in the spectra. The gold originated from the coating layer used for scanning electron microscope observation, carbon element was from the wood substrate and no other elements were a lower ΔE^* value corresponded to a smaller color difference and indicated strong resistance to UV radiation.

$$\Delta E^* = (L_2^* - L_1^*)2 + (a_2 - a_1)^2 + (b_2 - b_1)^2 \tag{4}$$

3 Results and discussion

The SEM images in Fig. 2 illustrate the surface morphologies of the untreated wood, magnetic wood and the wood treated by detected. Accordingly, $CoFe_2O_4$ nanoparticles were coated effectively on the wood surface. The TEM image shows the average size of the $CoFe_2O_4$ nanoparticles was 200 nm which peeled off from the magnetic wood samples by the ultrasonic treatment (Fig. 2c, Inset). The wood surface treated with $CoFe_2O_4$ and OTS was presented in Fig. 2d. Obviously, a thin layer of the wax crystalloids and nanoscale protuberances uniformly covered the wood surface. The presence of such a rough binary structure on the wood surface was the main requirement for superhydrophobicity[28-30].

Fig. 3 shows the XRD patterns of the untreated wood and mag-netic wood. In Fig. 3a, the diffraction peaks at 15° and 22° represent the characteristic diffraction peaks of the wood. In Fig. 3b, additional diffraction peaks represent the new crystal structures of the magnetic wood. The diffraction peaks at 18°, 30°, 35°, 43°, 53°, 56°and 62° could be assigned to the diffractions of the (1 1 1), (2 2 0), (3 1 1), (4 0 0), (4 2 2), (5 1 1) and (4 4 0) planes of $CoFe_2O_4$(PDF No. 22-1086), respectively.

Fig. 4 illustrates the magnetic characterization of the untreated wood and the magnetic wood. The untreated wood shows non-magnetic characterizes, with maximum applied field up to 20 kOe (Fig. 4, Inset). Interestingly, the magnetic wood exhibits a polar property and shows a typical ferromagnetic behavior, the satura-tion magnetization (M_S) and coercivity (H_C) of the magnetic wood were 1.81 emu g^{-1} and 451 Oe, respectively. The results reveal the wood after $CoFe_2O_4$ treated possessed excellent magnetic property compared with the untreated wood. An important question that still remains was whether a superhydrophobic wood surface can be created by coating with a rough $CoFe_2O_4$ nanoparticles film and self-assembly modification of OTS leading to a low surface energy. In Fig. 5 top view and three-dimensional surface plots (300 nm 300 nm) AFM representations were presented for the untreated wood, the magnetic wood and the wood treated by $CoFe_2O_4$ and OTS, respectively. Compared with the untreated wood surface (Fig. 5a), the magnetic wood surface (Fig. 5b) exhibited a fine microstructure and more complex surface texture which was consistent with a rough topography. The surface of the wood treated by $CoFe_2O_4$ and OTS (Fig. 5c) was composed of high "mountains" and deep "valleys". The difference of the surface roughness can be reflected in the rms roughness values (rms = root mean square = the standard deviation of the Z value, Z being the total height range analyzed) of the three surface, which were 26.9 nm (the untreated wood), 49.9 nm (the magnetic wood) and 96.1 nm (the wood treated by $CoFe_2O_4$ and OTS), respectively. Therefore, it can be concluded that the modified surface topology was responsible for the film generated hydrophobic characteristics, due to a change in surface roughness.

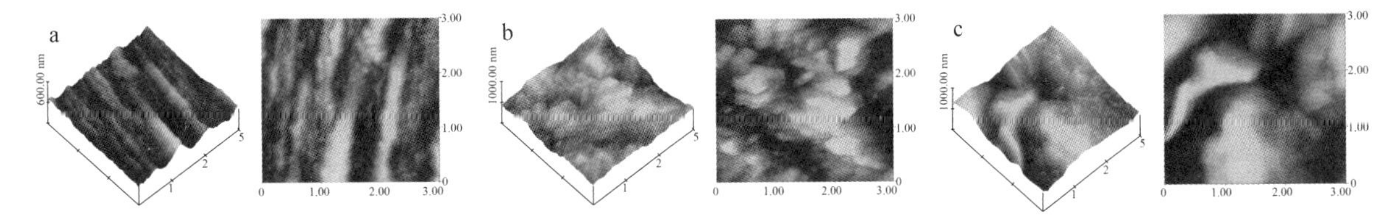

Fig. 5 AFM images of (a) the untreated wood, (b) the magnetic wood, and (c) the wood treated by $CoFe_2O_4$ and OTS

Fig. 6 shows the FTIR spectra of the untreated wood, magnetic wood and the magnetic wood treated by OTS. For magnetic wood sample, the absorption band at 3423 cm^{-1} assigned to the stretching vibration of OH groups became comparatively narrow after modification, probably due to interaction between the OH groups of wood surface and the deposited $CoFe_2O_4$ nanoparticles[14]. The bands at 425 cm^{-1} were attributed to Fe-O stretching vibrations at the octahedral site[31]. With the addition of OTS, the new adsorption bands at 2919 cm^{-1} (C H asymmetric stretching vibrations) and 2850 cm^{-1}(C-H asymmetric stretching vibrations), were for the long-chain alkyl group of the coating surface. The bands around 1106 and 906 cm^{-1} were corresponded to Si-O-Si vibration bands. The characteristic band near 436 cm^{-1}, which arises from the stretching vibration of Fe-O bonds, has a shifted to high wavenumbers. The phenomenon can be explained according to the formation of Fe-O-Si bonds where Fe-OH groups on the surface of the $CoFe_2O_4$ nanoparticles were replaced by $Fe-O-Si(O)_2R$ as shown in Fig. 9[32].

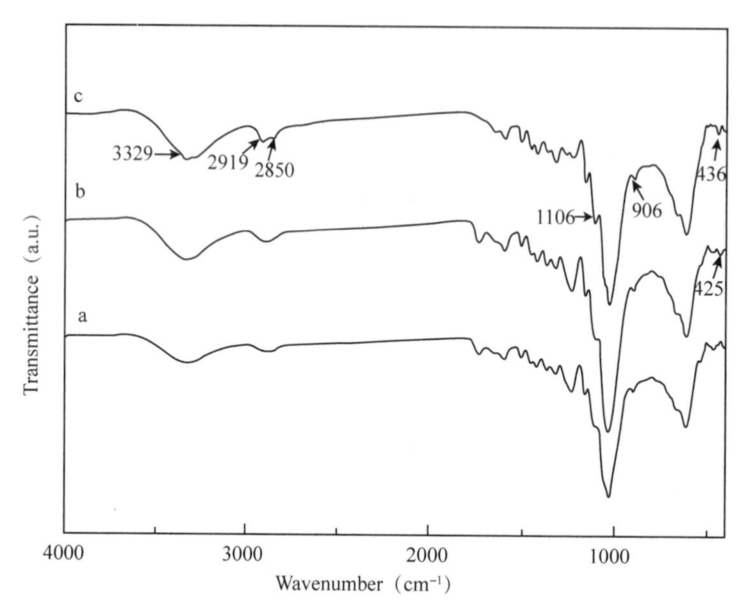

Fig. 6 FTIR spectra of (a) the untreated wood, (b) the magnetic wood, and (c) the magnetic wood and hydrolyzed octadecyltrichlorosilane (OTS)

Superhydrophobic surface with a water contact angle (CA) higher than 150°, has a great potential for research and practical applications[28-30,33]. Fig. 7 shows the wettability of the wood sample by static CA measurements. CA of the untreated wood and the magnetic wood were 40° and 7°, respectively, which demonstrated the hydrophilic property obviously. For the wood surface only treated with OTS can achieved a maximum water contact angle of about 100° and showed the hydrophobic property to a certain extent. However, after the combined treatment by $CoFe_2O_4$ and OTS, CA was found to be 150°, indicated an excellent hydrophobic property clearly.

To further understand the superhydrophobic property of the treated wood surface, the Cassie's equation, which was generally applicable to heterogeneous roughness and low-surface energy surfaces that exhibit superhydrophobicity, was employed[34]:

$$\cos\theta_c = f(\cos\theta + 1) - 1 \tag{5}$$

where f was the fraction of the solid surface in contact with liquid, the fraction of air in contact with liquid at the surface was $1/f$, θ_c and & represented the water contact angle on rough and smooth surfaces, respectively. Here, the Cassie's equation can be used to calculate the fraction of the air pockets at the surface. In this paper, the water contact angle θ_c on the magnetic wood surface treated with OTS was approximately 150°, and the water contact angle on the smooth surface modified with OTS was 100°. Therefore, by the use of Cassie equation, the fraction of air in contact with liquid at the surface was calculated as 0.84.

In addition, the environmental stability and durability of the superhydrophobic wood surface have been investigated. The sample can be maintained at least for three months of storage in air, the water contact of wood surface has no negligible change, indicated its good stability in air.

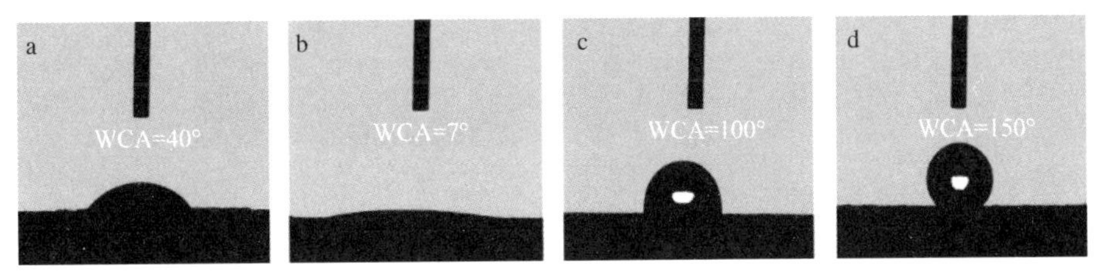

Fig. 7 Contact angels (CA) of 5-μL water droplet on the surface of (a) the untreated wood, (b) the wood sample only treated by $CoFe_2O_4$ nanoparticles, (c) the wood sample only treated with octadecyltrichlorosilane (OTS), and (d) the wood sample treated with $CoFe_2O_4$ nanoparticles and OTS

Fig. 8 shows the excellent magnetic and superhydrophobic properties of the wood samples. The superhydrophobic and magnetic wood was floating on the surface of the water, and could be easily attracted by using the external magnet (Fig. 8a). The magnetic wood samples were placed on the top of a desk, could be attracted and firmly adsorbed by the external magnet (Fig. 8b). The water droplet would drop down from the wood surface by driving the wooden platelet (Fig. 8c).

Fig. 9 presents a schematic illustration of the superhydrophobic and magnetic wood. Since there were large surface-to-volume atomic ratio, high surface activity and amount of dangling bonds on nanoparticle surface, the atoms on the surface were apt to absorb ions or molecules in solution. For $CoFe_2O_4$ nanoparticles dispersed in aqueous solution, the bare atoms of Fe on the particle surface would adsorb OH^-, so that there was OH^- rich surface[32]. The untreated wood sample (Fig. 9a) was immersed into the pre-cursor solution through a hydrothermal process and the plentiful OH groups between the $CoFe_2O_4$ nanoparticles and wood surface reacted with each other, and a $CoFe_2O_4$ layer arose on the sur face, resulting in a increased surface roughness (Fig. 9b). When the OTS was introduced, the Si-Cl groups of the OTS molecule firstly hydrolyzed into Si-OH groups, the -OH on the surface of $CoFe_2O_4$ can react with Si-OH (Fig. 9c). With the continuous dehydration reaction, Si-O groups with the superhydrophobic long chain alkyl groups were formed onto the magnetic wood surface by chemical bond. Finally, under the combined effects of $CoFe_2O_4$ nanoparticles and OTS, the surface became superhydrophobicity[33,35].

The color changes before and after UV irradiations were mea-sured in accelerated aging tests to evaluate the anti-ultraviolet property of the magnetic wood. The experimental results were presented in Fig. 10. In Fig. 10a, the change of Δa^* value of the untreated wood sample under UV irradiation indicated that the sur-face color of the untreated wood sample became gradually redder with increasing UV irradiation time. The Δa^* values of the magnetic wood showed the similar changing tendency compared with the untreated wood (only 1/10 of the untreated wood), which showed the stronger UV-resistant ability of the magnetic wood compared with that of the untreated wood. In Fig. 10b, the change of Δb^* values indicated that the color of the untreated wood turned to dark yellow, while that of the magnetic wood turned to slight blue with increasing UV irradiation time. In Fig. 10c, ΔL^* values of the untreated wood, magnetic wood became negative with the time,

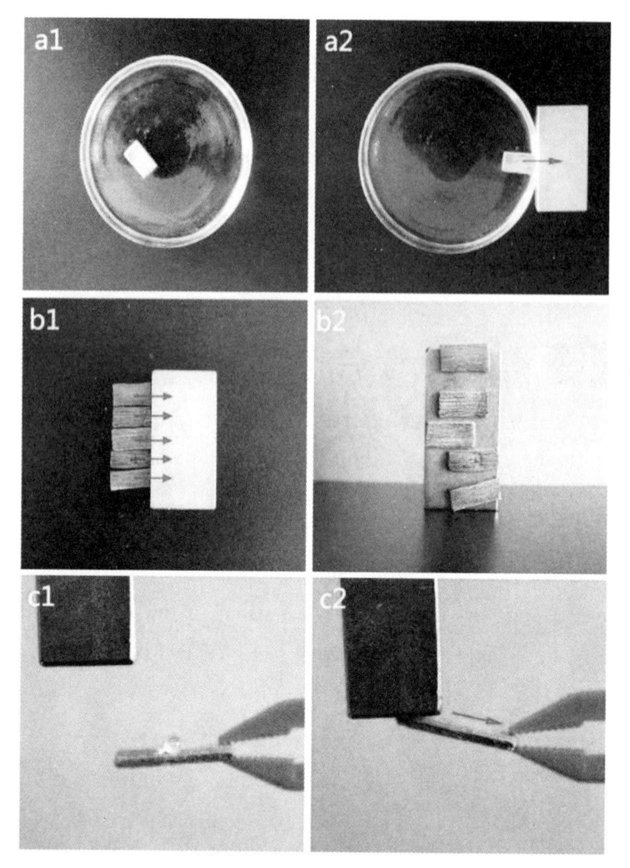

Fig. 8　The superhydrophobic and magnetic wood samples, under the influence of an external magnet. (a1 and a2) The floating specimen moved toward the magnet. (b1 and b2) Actuation of magnetic wood platelet. (c1 and c2) The water droplet would drop down from the wood surface by driving the magnetic wood

revealed the lightness turned to black. However, the tendency of ΔL^* change of the magnetic wood had slight variation in comparison with the untreated wood, suggested a superior UV-resistant ability of the magnetic wood was obtained. In Fig. 10d, it could be seen that the total color change (ΔE^*) of the magnetic wood had a slighter change compared with the untreated wood with increasing UV irradiation time, and the ΔE^* values of the magnetic wood was only 1/4 of the untreated wood. This further confirmed that the magnetic wood have

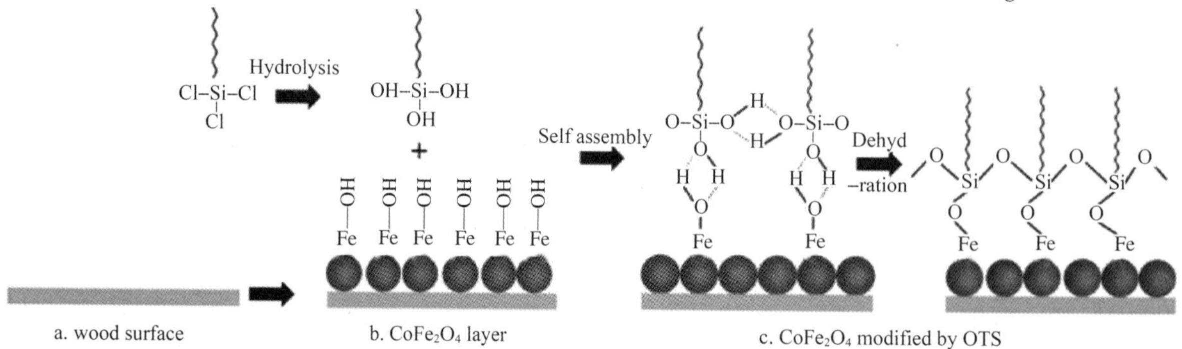

Fig. 9　Schematic illustration of the preparing process of superhydrophobic and magnetic wood surface

excellent UV-resistant ability.

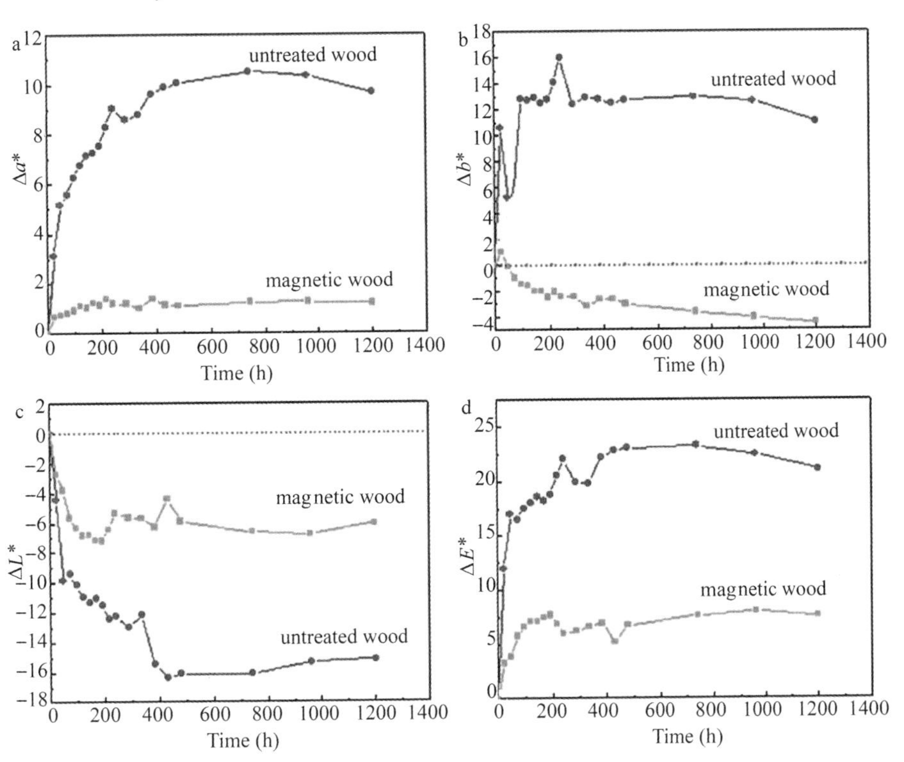

Fig. 10 Change tendency of Δa^*, Δb^*, ΔL^* and ΔE^* of the untreated wood and the magnetic wood

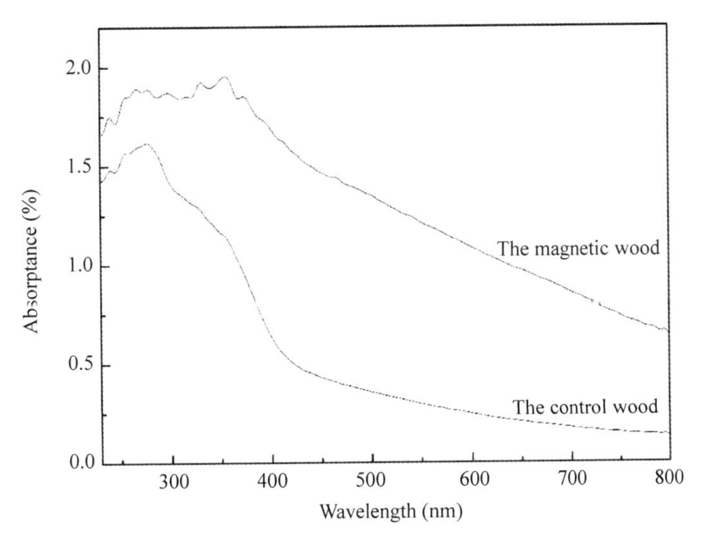

Fig. 11 UV-vis diffuse reflection spectra of the untreated wood and the magnetic wood

The UV-vis diffuse reflection spectra of the untreated wood and magnetic wood were shown in Fig. 11. Zhu et al. [36] reported that the $CoFe_2O_4$ samples have the UV absorption at the region of 200-800 nm. This was consistent with our experimental results in Fig. 11. Magnetic wood clearly exhibited a much higher absorption at the 200-800 nm wavelength range compared with the untreated wood, indicated the magnetic wood has higher

UV resistance than the untreated wood. Hence, better anti-ultraviolet property was achieved by modified with $CoFe_2O_4$ compared to the untreated wood samples. We attributed this to the $CoFe_2O_4$ nanoparticles film hindered the UV light from reaching the wood surface and thus prevented photodegradation of wood components[37].

4 Conclusions

Multifunctional wood materials with magnetic, superhydrophobic and anti-ultraviolet properties can be prepared through the combined effects of $CoFe_2O_4$ nanoparticles and octadecyltrichlorosilane (OTS). The saturation magnetization (M_S) and coercivity (H_C) of the magnetic wood were 1.8 emu · g^{-1} and 450 Oe, respectively. The contact angel (CA) of the treated wood was around 150°. Moreover, the modified wood sample also showed a superior anti-ultraviolet performance. The presented approach may provide future direction for the multifunctional modification of wood.

Acknowledgments

This work was financially supported by The National Natural Science Foundation of China (grant no. 31470584) and Doctoral Candidate Innovation Research Support Program of Science & Technology Review (kjdb2012006).

References

[1]Oka H, Fujita H. Experimental study on magnetic and heating characteristics of magnetic wood[J]. J. Appl. Phys., 1999, 85(8): 5732-5734.

[2]Oka H, Hojo A, Seki K, et al. Wood construction and magnetic characteristics of impregnated type magnetic wood[J]. J. Magn. Magn. Mater., 2002, 239(1-3): 617-619.

[3]Oka H, Narita K, Osada H, et al. Experimental results on indoor electromagnetic wave absorber using magnetic wood [J]. J. Appl. Phys., 2002, 91(10): 7008-7010.

[4]Oka H, Hamano H, Chiba S. Experimental study on actuation functions of coating-type magnetic wood[J]. J. Magn. Magn. Mater., 2004, 272: E1693-E1694.

[5]Oka H, Tokuta H, Namizaki Y, et al. Effects of humidity on the magnetic and woody characteristics of powder-type magnetic wood[J]. J. Magn. Magn. Mater., 2004, 272: 1515-1517.

[6]Oka H, Kataoka Y, Osada H, et al. Experimental study on electromagnetic wave absorbing control of coating-type magnetic wood using a grooving process[J]. J. Magn. Magne. Mater., 2007, 310(2): E1028-E1029.

[7]Oka H, Uchidate S, Sekino N, et al. Electromagnetic wave absorption characteristics of half carbonized powder-type magnetic wood[J]. IEEE Trans. Magn., 2011, 47(10): 3078-3080.

[8]Aloui F, Ahajji A, Irmouli Y, et al. Inorganic UV absorbers for the photostabilisation of wood-clearcoating systems: Comparison with organic UV absorbers[J]. Appl. Surf. Sci., 2007, 253(8): 3737-3745.

[9]Jirous-Rajkovic V, Bogner A, Radovan D. The efficiency of various treatments in protecting wood surfaces against weathering[J]. Surf. Coat. Int., 2004, 87(1): 15-19.

[10]Patachia S, Croitoru C, Friedrich C. Effect of UV exposure on the surface chemistry of wood veneers treated with ionic liquids[J]. Appl. Surf. Sci., 2012, 258(18): 6723-6729.

[11]Salla J, Pandey K K, Srinivas K. Improvement of UV resistance of wood surfaces by using ZnO nanoparticles[J]. Polym. Degrad. Stab., 2012, 97(4): 592-596.

[12]Hayoz P, Peter W, Rogez D. A new innovative stabilization method for the protection of natural wood[J]. Prog. Org. Coat., 2003, 48(2-4): 297-309.

[13]Einchhorn S, Dufresne A, Aranguren M, et al. Review: current international research into cellulose nanofibres and composites[J]. J. Mater. Sci., 2010, 45: 1-33.

[14]Lu Y, Xiao S L, Gao R N, et al. Improved weathering performance and wettability of wood protected by CeO_2 coating deposited onto the surface[J]. Holzforschung, 2014, 68(3): 345-351.

[15]Hakkou M, Pétrissans M, Zoulalian A, et al. Investigation of wood wettability changes during heat treatment on the basis of chemical analysis[J]. Polym. Degrad. Stab., 2005, 89(1): 1-5.

[16]Saka S, Ueno T. Several SiO_2 wood-inorganic composites and their fire-resisting properties[J]. Wood Sci. Technol., 1997, 31(6): 457-466.

[17]Merk V, Chanana M, Gierlinger N, et al. Hybrid wood materials with magnetic anisotropy dictated by the hierarchical cell structure[J]. ACS Appl. Mater. Interf., 2014, 6(12): 9760-9767.

[18]Wang C Y, Piao C, Lucas C. Synthesis and characterization of superhydrophobic wood surfaces[J]. J. Appl. Polym. Sci., 2011, 119(3): 1667-1672.

[19]Miyafuji H, Saka S. Fire-resisting properties in several TiO_2 wood-inorganic composites and their topochemistry[J]. Wood Sci. Technol., 1997, 31(6): 449-455.

[20]Furuno T, Imamura Y. Combinations of wood and silicate Part 6. Biological resistances of wood-mineral composites using water glass-boron compound system[J]. Wood Sci. Technol., 1998, 32(3): 161-170.

[21]Hübert T, Unger B, Bücker M. Sol-gel derived TiO_2 wood composites[J]. J. Sol-Gel Sci. Technol., 2010, 53(2): 384-389.

[22]Yu Y, Jiang Z H, Wang G, et al. Surface functionalization of bamboo with nanostructured ZnO[J]. Wood Sci. Technol., 2012, 46(4): 781-790.

[23]Wang S L, Wang C Y, Liu C Y, et al. Fabrication of superhydrophobic spherical-like α-FeOOH films on the wood surface by a hydrothermal method[J]. Colloids Surf. A, 2012, 403: 29-34.

[24]Sun Q F, Lu Y, Liu Y X. Growth of hydrophobic TiO_2 on wood surface using a hydrothermal method[J]. J. Mater. Sci., 2011, 46(24): 7706-7712.

[25]Sun Q F, Lu Y, Zhang H M, et al. Improved UV resistance in wood through the hydrothermal growth of highly ordered ZnO nanorod arrays[J]. J. Mater. Sci., 2012, 47(10): 4457-4462.

[26]Sun Q F, Lu Y, Yang D J, et al. Preliminary observations of hydrothermal growth of nanomaterials on wood surfaces [J]. Wood Sci. Technol., 2014, 48(1): 51-58.

[27]Olsson R T, Samir M A, Salazar-Alvarez G, et al. Making flexible magnetic aerogels and stiff magnetic nanopaper using cellulose nanofibrils as templates[J]. Nat. Nanotechnol., 2010, 5(8): 584-588.

[28]Li N, Xia T, Heng L P, et al. Superhydrophobic Zr-based metallic glass surface with high adhesive force[J]. Appl. Phys. Lett., 2013, 102(25): 251603.

[29]Xia T, Li N, Wu Y, et al. Patterned superhydrophobic surface based on Pd-based metallic glass[J]. Appl. Phys. Lett., 2012, 101(8): 081601.

[30]Lu B P, Li N. Versatile aluminum alloy surface with various wettability[J]. Appl. Surf. Sci., 2014, 326: 168-173.

[31]Köseoğlu Y, Alan F, Tan M, et al. Low temperature hydrothermal synthesis and characterization of Mn doped cobalt ferrite nanoparticles[J]. Ceram. Int., 2012, 38(5): 3625-3634.

[32]Ma M, Zhang Y, Yu W, et al. Preparation and characterization of magnetite nanoparticles coated by amino silane [J]. Colloids Surf. A, 2003, 212(2-3): 219-226.

[33]De Palma R, Peeters S, Van Bael M J, et al. Silane ligand exchange to make hydrophobic superparamagnetic nanoparticles water-dispersible[J]. Chem. Mater., 2007, 19(7): 1821-1831.

[34]Gao L K, Lu Y, Zhan X X, et al. A robust, anti-acid, and high-temperature-humidity-resistant superhydrophobic

surface of wood based on a modified TiO_2 film by fluoroalkyl silane[J]. Surf. Coat. Technol., 2015, 262: 33-39.

[35] Degen P, Shukla A, Boetcher U, et al. Self-assembled ultra-thin coatings of octadecyltrichlorosilane (OTS) formed at the surface of iron oxide nanoparticles[J]. Colloid Polym. Sci., 2008, 286(2): 159-168.

[36] Zhu Z R, Li X Y, Zhao Q D, et al. Surface photovoltage properties and photocatalytic activities of nanocrystalline $CoFe_2O_4$ particles with porous superstructure fabricated by a modified chemical coprecipitation method[J]. J. Nanopart. Res., 2011, 13(5): 2147-2155.

[37] Donath S, Militz H, Mai C. Weathering of silane treated wood[J]. Holz. Roh. Werkst., 2007, 65(1): 35.

中文题目：具有磁性、超疏水和抗紫外线性能的多功能木质材料

作者：甘文涛，高丽坤，孙庆丰，金春德，卢芸，李坚

摘要：通过在木材表面沉淀 $CoFe_2O_4$ 纳米颗粒，然后用一层十八烷基三氯硅烷(OTS)处理，成功获得了具有磁性，超疏水性和抗紫外线性能的多功能木质材料。制成的木质复合材料具有优异的磁性，OTS 改性磁性木质表面的水接触角高达 150°，具有超疏水性。此外，加速老化测试表明，经处理的木质复合材料也具有出色的抗紫外线性能。

关键词：$CoFe_2O_4$；磁性能；水热法；超疏水性；抗紫外线性能

Growth of $CoFe_2O_4$ Particles on Wood Template Using Controlled Hydrothermal Method at Low Temperature*

Wentao Gan, Ying Liu, Likun Gao, Xianxu Zhan, Jian Li

Abstract: Magnetic wood would be a potential for electromagnetic shielding, indoor electromagnetic wave absorber and special decoration. This work focuses on the development of a method to make magnetic wood/$CoFe_2O_4$ composites using a hydrothermal process. A particular emphasis is devoted to the understanding of the role of the chemical parameters involved in the hydrothermal technique, and of the hydrothermal treatment on the structures and properties of the materials obtained. The magnetic wood/$CoFe_2O_4$ composites could be obtained at 90 ℃ within 8h. The crystallization of the spinel ferrites that precipitated on the wood template was promoted by the increase in reaction temperature. The saturation magnetization of the magnetic wood composites increases with holding time rapidly, attaining a maximum value of 2.59 emu · g^{-1} at 8h. Moreover, it has been found that the concentration of KNO_3 and NaOH have a critical influence on the resultant structure, morphology and magnetic property of the magnetic wood composites.

Key words: $CoFe_2O_4$; Hydrothermal method; Magnetic materials; Wood modification

1 Introduction

Magnetic nanoparticles are widely used in applications such as magnetic fluids, catalysis, biotechnology, data storage and environmental remediation[1-4]. The mixing of magnetic nanoparticles with polymers to form magnetically responsive composite materials has been practiced for decades, which was opening pathways for engineering flexible composites that exhibit advantageous optical, mechanical and magnetic properties[5-9]. In this context, wood was a fiber composite with three-dimensional architecture made of cellulose microfibrils embedded into a matrix of hemicelluloses and lignin. The network of cellulose fibrils, hemicelluloses, lignin and the space of parallel hollow tubes defines the porous structure, which also determines the penetrability, accessibility and reactivity of wood component[10, 11]. Therefore, the wood was also an ideal template to combine with magnetic nanoparticles to produce inexpensive, lightweight and hierarchical materials.

The first feasible method for the synthesis of the magnetic wood composites was introduced by Oka in 1991. Oka et al. fabricated three types of magnetic wood by impregnating wood fiberboards (impregnated type), coating (coating type) and hot-pressing (powder type) with commercial magnetic materials, such as magnetite Fe_3O_4 and Mn-Zn ferrite powder[12-15]. Recently, Merk et al. precipitated magnetic iron oxide particles via the coprecipitation of ferric and ferrous ions in the presence of wood to produce hierarchical cell structure with a unique magnetic anisotropy[16]. Similar studies also confirmed that the $CoFe_2O_4$ nanoparticles

* 本文出自 Ceramics International, 2016, 41: 14876-14885.

with lager magnetocrystalline anisotropy and reasonable magnetization could be easily grown on the wood template by a hydrothermal process[17-19]. However, the greatest hurdle to the broader use of wood/magnetic composites was the absence of the structure-property relationships.

Therefore, the present research was aimed at probing the influence of the morphology and crystallization of the magnetic $CoFe_2O_4$ particles on the magnetic properties of wood/$CoFe_2O_4$ composites. The magnetic wood/$CoFe_2O_4$ composites were prepared by precipitated $CoFe_2O_4$ nanoparticles on the wood matrix using a hydrothermal method with characterizations of scanning electron microscope (SEM), energy dispersive spectroscopy (EDXA), X-ray diffraction (XRD) and Fourier transform infrared spectroscopy (FTIR). With the regulation of the reaction conditions, the magnetic property of the magnetic wood could be varied. Thus, controlled properties of magnetic wood could be achieved for different requirements.

2 Materials and methods

2.1 Materials

The wood samples [20 mm (R)×20 mm (T)×20 mm (L), 4 mm (R)×4 mm (T)×4 mm (L)] obtained from the sapwood portions of poplar wood (*Populus ussuriensis Kom.*). All reagents were supplied by Shanghai Boyle Chemical Company Limited and were used without any further purification. Cobalt chloride hexahydrate ($CoCl_2 \cdot 6H_2O$) and ferrous sulfate heptahydrate ($FeSO_4 \cdot 7H_2O$) were used to achieve a Fe/Co mole ratio of 2. Potassium nitrate (KNO_3) was used as a mild oxidizing agent. The intermediate compound was precipitated by contacting the metals and nitrate solution with suitable amounts of sodium hydroxide (NaOH).

2.2 Grown process of $CoFe_2O_4$ particles on wood template

The synthetic conditions used for the growth of $CoFe_2O_4$ on the wood template were showed in Table 1. A typical synthesis pathway of magnetic wood was described as following. The untreated wood was suspended in the 40 mL of 0.165 mol · L^{-1} of freshly prepared aqueous solution of $FeSO_4$ and $CoCl_2$ via a vacuum process before transferred into a Teflon-lined stainless steel autoclave and heated the system to 90 ℃ for 3 h. Heating changed the color from transparent to translucent orange. At the end of the reaction, the supernatant was poured out and 35 mL of 1.32 M NaOH solution with 0.25 M KNO_3 was added into the Teflon-lined stainless-steel autoclave. After an additional hydrothermal processing for 8 h at 90 ℃, the precipitated precursors were converted into cobalt ferrite particles on the wood templates, resulting in excellent magnetic property. Finally, the prepared wood specimens were removed from the solution, ultrasonically rinsed with deionized water for 10 min, and dried at 50 ℃ for over 24 h in vacuum.

2.3 Characterization

The morphology and elemental compositions of the wood samples were examined by scanning electron microscopy (SEM, FEI, and Quanta 200) combined with energy dispersive X-ray spectroscopy (EDXA, Genesis). The crystalline structure of these composites were identified via X-ray diffraction (XRD, Rigaku, and D/MAX 2200) operating with Cu $K\alpha$ radiation (λ = 1.5418 A) at a scan rate (2θ) of 4° · min^{-1}, an accelerating voltage of 40 kV, and an applied current of 30 mA ranging from 5° to 80°. The surface chemical compositions of the samples were determined via the Fourier transformation infrared spectroscopy (FTIR, Nicolet, Magna-IR 560). For magnetic characterization, wood specimens of 4 mm (R)×4 mm (T)×

4 mm (L) were used. The magnetization measurements were carried out at room temperature by using a superconducting quantum interference device (MPMS XL-7, Quantum Design Corp).

3 Results and discussion

Fig. 1 displays the EDXA spectra of the untreated wood and the magnetic wood samples. As shown in Fig. 1a, carbon, oxygen and gold could be detected in the EDXA spectrum of the untreated wood. The gold resulted from the conductive layer on the surface of the samples used for the SEM observation. The carbon and oxygen originated from the wood substrate. After hydrothermal process, the strong peak corresponding to iron and cobalt could be detected in the EDXA spectrum (Fig. 2b). No elements other than C, O, Au, Fe and Co could be detected in the spectra.

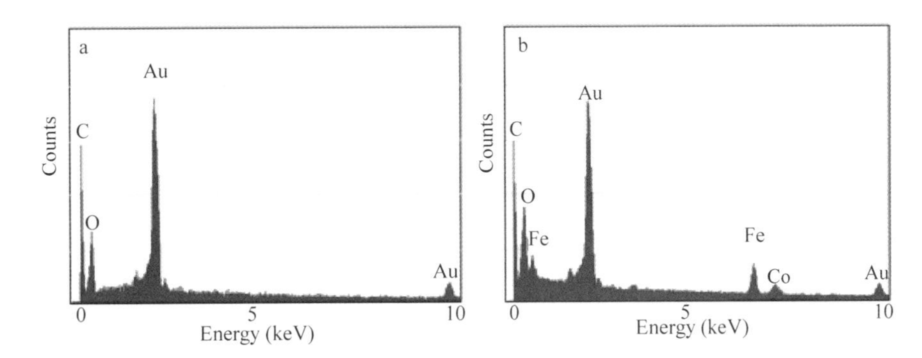

Fig. 1 The EDXA spectra of (a) untreated wood and (b) magnetic wood, respectively

Fig. 2 exhibits the FTIR spectra of the untreated wood and the magnetic wood composites. The main absorption bands in the FTIR spectra are located at ~ 3334 cm^{-1}, ~ 2892 cm^{-1}, ~ 1600 cm^{-1}, ~ 1380 cm^{-1}, and 1031 cm^{-1}, corresponding to the O—H, C—H, C=O, C—H and C—O stretching vibrations, respectively, and are attributed to the wood template. The absorption peaks at ~ 435 cm^{-1} are mainly attributed to the Fe—O stretching vibration, and is assigned to $CoFe_2O_4$ particles[20, 21]. It is also found that the characteristic absorption bands of the O—H bond of the magnetic wood shift to low wavenumbers of 3295 cm^{-1} compared with that of untreated wood (in 3334 cm^{-1}). The phenomenon can be explained according to a strong interaction between the hydroxyl groups of wood and $CoFe_2O_4$ particles through hydrogen bonds, which leads to weakening of bond force constant for O—H bonds, so that the absorption bands shift to low wavenumber. This strong interaction led to the immobilization of inorganic particles on the wood template[22-24].

In principle, two basic steps are involved in the generation of the magnetic wood materials during the hydrothermal reaction, as shown in Fig. 3. Initially, the non-magnetic metal hydroxides precipitated into the 3D network of the wood matrix after the first hydrothermal process. In alkaline aqueous solution, the precipitated precursors were converted into cobalt ferrite nanoparticles on the wood cell walls. A possible mechanism of the formation of the $CoFe_2O_4$ nanostructured materials can be expressed by the reactions presented in Eqs. (1) and (2)[25, 26]:

$$Fe^{2+} + Co^{2+} + 4OH^- \rightarrow Fe(OH)_2 + Co(OH)_2 \quad (1)$$

$$8Fe(OH)_2 + 4Co(OH)_2 + NO_{3-} \rightarrow 4CoFe_2O_4 + NH_3 + 10H_2O + OH^- \quad (2)$$

For $CoFe_2O_4$ nanoparticles dispersed in alkaline solution, the bare atoms of Fe would adsorb OH^- due to the high surface activity and large surface-to-volume atomic ratio, so that there is OH-rich surface[27]. Besides, wood also features numerous hydroxyl groups that confer the hydrophilic nature of the cell wall. When the wood templates were immersed into the precursor solution, the hydrophilic nature makes the Fe-OH trapped inside the wood samples by interaction with the wood hydroxyl.

The above characterizations proved that magnetic $CoFe_2O_4$ bonded to the wood through the interaction of hydrogen groups during the hydrothermal process. To further understand the structure-property relationship of the magnetic wood materials, a systemic investigation was carried out according to Table 1.

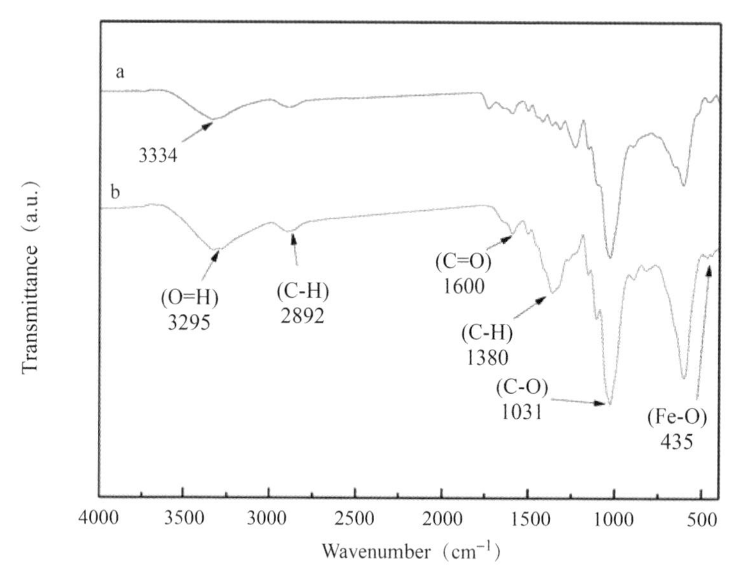

Fig. 2 FTIR spectra of (a) untreated wood and (b) magnetic wood, respectively

Table 1 lists the different synthetic conditions adopted to grow $CoFe_2O_4$ on the wood. The influence of experimental parameters such as temperature, reaction time and the concentration of reactants on the magnetic property, morphology and crystallization of the magnetic wood were characterized by the magnetization curves, SEM images and XRD patterns, respectively.

3.1 Effect of growth temperature

To study the effect of growth temperature on the morphology of magnetic wood composites, the $CoFe_2O_4$ nanostructured materials were hydrothermally grown on the wood surface at various temperatures ranging from 50 ℃ to 130 ℃ in increments of 20 ℃. The other parameters were kept constant (reaction time 8h, $[OH^-]$ = 1.32 M, $[NO^{3-}]$ = 0.25 M). The corresponding SEM images $CoFe_2O_4$ nanostructured materials grown at each temperature shown in Fig. 4. Fig. 4a shows an SEM image of the untreated wood surface, which is smooth. Fig. 4b-f displays the SEM photographs of the $CoFe_2O_4$ particles grown on the wood surface at various temperatures from 50 ℃ to 130 ℃ in increment of 20 ℃. As expected, higher reaction temperatures favored the cobalt ferrite formation on the wood template. When the temperature was 50 ℃ (Fig. 4b), no $CoFe_2O_4$ particles were obtained on the wood template. This indicates that the activation barrier for the growth of $CoFe_2O_4$ cannot be completely overcome in this temperature. A thin film of $CoFe_2O_4$ nanoparticles was obtained

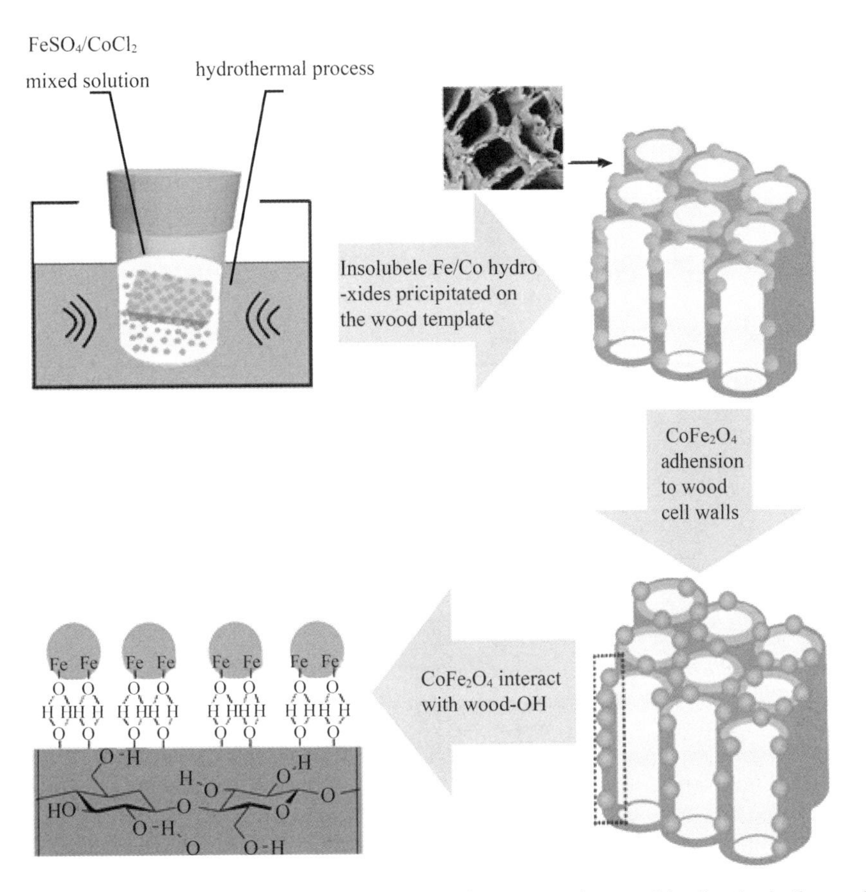

Fig. 3 Schematic of experimental set-up for preparing magnetic wood by the hydrothermal method

on the wood surface at 70 ℃. When the temperature was increased to 90 ℃, the amount of $CoFe_2O_4$ particles that grown on the wood surface was obviously increased. Moreover, if the temperature was higher than 90 ℃, the size of the $CoFe_2O_4$ particles that grown on the wood surface was observed to be increasing linearly with reaction temperature, the maximum diameter of $CoFe_2O_4$ particles about 1 μm were observed at 130 ℃. In general, very small primary particles precipitated on the wood template at the lower temperature, with the temperature increased, the large particles are formed on the wood template by a coalescence of crystallites[28]. Thus, temperature has a significant influence on the morphology of the magnetic wood composites.

Table 1 Synthetic condition for growing $CoFe_2O_4$ on the wood template

Samples	Temperature(℃)	Time(h)	$FeSO_4/CoCl_2$(mol · L^{-1})	KNO_3(mol · L^{-1})	NaOH(mol · L^{-1})	M_S(emu · g^{-1})	H_C(Oe)
S1	50	8	0.165	0.25	1.32	0.11	85.4
S2	70	8	0.165	0.25	1.32	0.67	98.7
S3	90	8	0.165	0.25	1.32	2.59	135.5
S4	110	8	0.165	0.25	1.32	3.41	150.0
S5	130	8	0.165	0.25	1.32	4.53	100.9

续表

Samples	Temperature(℃)	Time(h)	$FeSO_4/CoCl_2$(mol · L^{-1})	KNO_3(mol · L^{-1})	NaOH(mol · L^{-1})	M_S(emu · g^{-1})	H_C(Oe)
S6	90	3	0. 165	0. 25	1. 32	0. 17	160. 8
S7	90	5	0. 165	0. 25	1. 32	1. 63	155. 7
S8	90	12	0. 165	0. 25	1. 32	2. 24	120. 1
S9	90	8	0. 165	0	1. 32	1. 59	175. 5
S10	90	8	0. 165	0. 125	1. 32	1. 97	150. 7
S11	90	8	0. 165	0. 25	0. 33	1. 84	125. 5
S12	90	8	0. 165	0. 25	0. 66	2. 33	100. 2

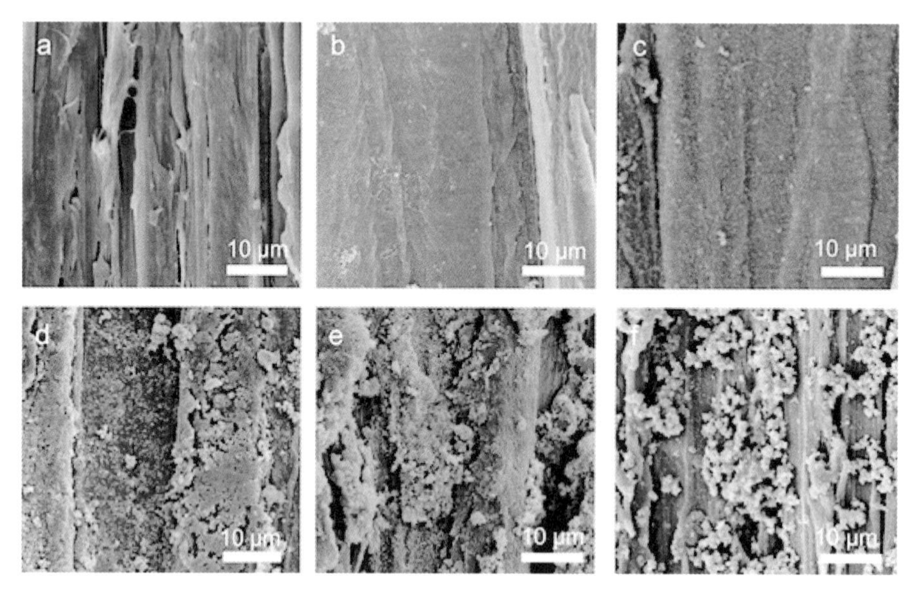

Fig. 4 (a) SEM image of the untreated wood surface, (b-f) SEM photographs of $CoFe_2O_4$ particles grown on the wood surface by a hydrothermal process with temperatures in range from 50 ℃ to 130 ℃ with a temperature gap of 20 ℃, respectively

The XRD patterns of the samples synthesized at different reaction temperatures were showed in Fig. 5. The strong diffraction peaks at 14. 8° and 22. 6° correspond to the crystalline region of the cellulose in wood[29, 30]. Besides, the diffraction peaks at 18°, 30°, 35°, 43°, 53°, 56° and 62° could be assigned to the diffractions of the (111), (220), (311), (400), (422), (511) and (440) planes of $CoFe_2O_4$(PDF No. 22-1086), respectively. However, the XRD patterns show that only broadened (220) and (311) peaks appear as samples were prepared at 50 ℃ and 70 ℃, more peaks of $CoFe_2O_4$ appear and become sharper with the increase of reaction temperature. The others correspond to the crystalline region of cellulose have no obviously change. These demonstrate that spinel crystallization process of $CoFe_2O_4$ nanoparticles that precipitated on the wood template has appeared even at 90 ℃.

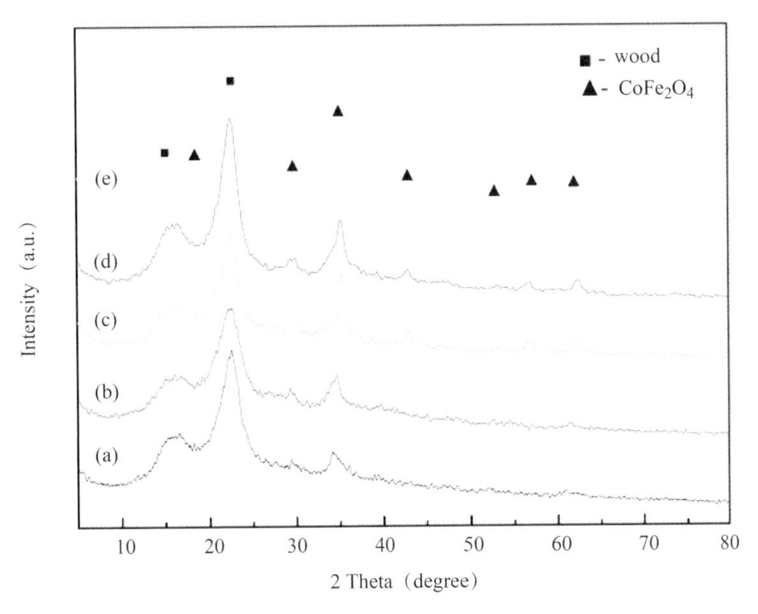

Fig. 5 X-ray diffraction patterns of the hydrothermally synthesized samples at (a) 50 ℃, (b) 70 ℃, (c) 90 ℃, (d) 110 ℃ and (e) 130 ℃, respectively

Fig. 6 illustrates the magnetic characterization of the samples synthesized at different reaction temperatures. As our easily report, the untreated wood shows non-magnetic characterizes. But the magnetic wood exhibits a polar property and shows a typical ferromagnetic behavior[18, 19]. The saturation magnetization (M_S) of the samples synthesized at 50 ℃, 70 ℃, 90 ℃, 110 ℃ and 130 ℃ were 0.11 emu · g^{-1}, 0.67 emu · g^{-1}, 2.59 emu · g^{-1}, 3.41 emu · g^{-1} and 4.53 emu · g^{-1}, respectively (shown in Table 1). Inset shows the

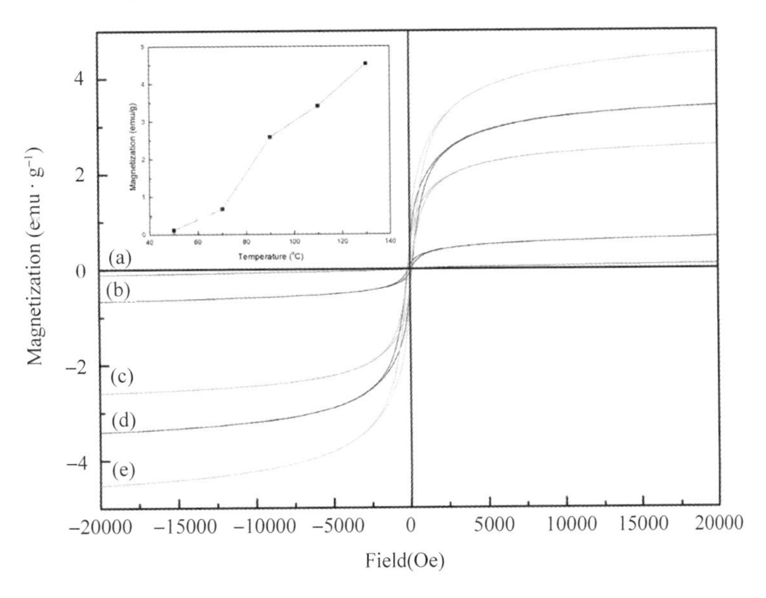

Fig. 6 Magnetization curves at room temperatures of the untreated wood and the hydrothermally synthesized samples at (a) 50 ℃, (b) 70 ℃, (c) 90 ℃, (d) 110 ℃ and (e) 130 ℃, respectively. Inset: the saturation magnetization (M_S) as a function of temperature for magnetic wood composites

saturation magnetization (M_S) as a function of reaction temperature for magnetic wood composites. The data shows the saturation magnetization increases with reaction temperature rapidly, attaining a maximum value of 4.53 emu · g^{-1} at 130 ℃. This increase at higher temperature could be attributed to the larger size of particles and the better crystallization of $CoFe_2O_4$ that precipitated on the wood template, which could be observed in SEM images and XRD analysis. And the coercivity (H_c) of the magnetic wood samples synthesized at 50 ℃, 70 ℃, 90 ℃, 110 ℃ and 130 ℃ were 85.4 Oe, 98.7 Oe, 135.5 Oe, 150 Oe and 100.5 Oe, respectively (shown in Table 1). The coercivity increases was attained a maximum value of 150 Oe at 110 ℃ and then decrease with the reaction temperature. This phenomenon could be explained by the coercivity increases with the size of the $CoFe_2O_4$ particles and was expected to decrease as the particle size increases beyond the critical single-domain diameter[31, 32].

3.2 Effect of reaction time

To study the effect of reaction time on the morphology of the magnetic wood composites, the $CoFe_2O_4$ nanostructured materials were hydrothermally grown on the wood template for 3 h, 5 h, 8 h and 12 h. The other parameters were kept constant (growth temperature 90 ℃, $[OH^-]$ = 1.32 M, $[NO_3^-]$ = 0.25 M). Fig. 7 displays the SEM images of $CoFe_2O_4$ nanostructured materials grown for (a) 3 h, (b) 5 h, (c) 8 h, and (d) 12h. Fig. 7a shows only sparse small particles were grown on the wood surface for 3 h. A densely layer of $CoFe_2O_4$ nanoparticles was covered on the wood surface when the hydrothermal time was 5 h. Increasing the synthesis time to 8 h leads to the agglomeration of particles on the wood template (Fig. 7c). When the reaction time was prolonged to 12 h, the morphology of $CoFe_2O_4$ nanostructured materials seems to be no obvious change (Fig. 7d), suggesting the magnetic $CoFe_2O_4$ particles that grown on the wood template could be saturated.

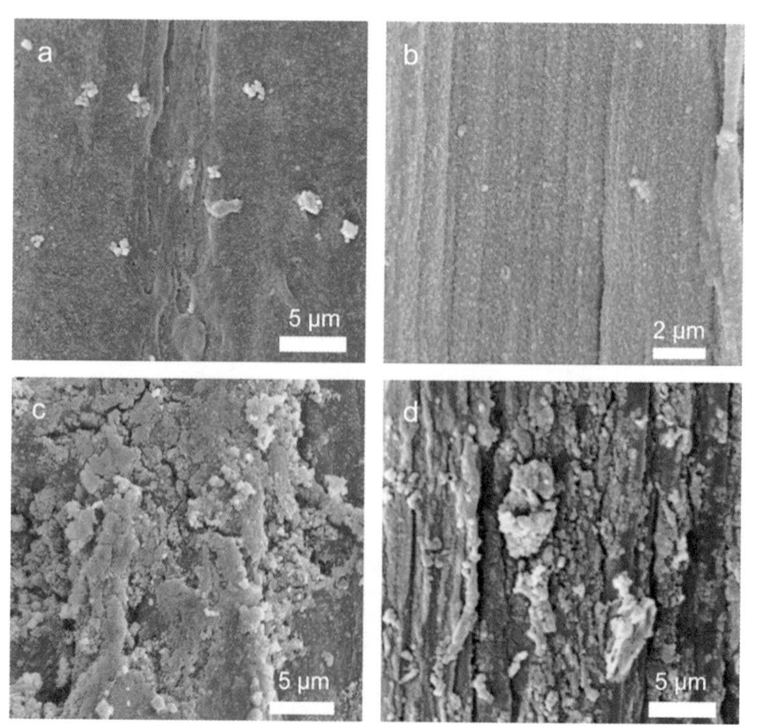

Fig. 7 SEM image of the hydrothermally synthesized samples at 90 ℃ for (a) 3 h, (b) 5 h, (c) 8 h and (d) 12 h, respectively

Fig. 8 shows the XRD patterns of the samples formed by the hydrothermal process and held at 90 ℃ for different times. An amorphous phase of magnetic particles were precipitated on the wood template when the holding time was 3 h, more peaks of $CoFe_2O_4$ appear and become sharper rapidly from 5 h to 8 h. The crystallization of $CoFe_2O_4$ that precipitated on the wood template for 12 h seem to show no obvious change compared with that for 8 h. The results mean that the reaction times at 8 h and above suitable to the growth of the $CoFe_2O_4$ crystallite on wood template, in agreement with SEM observation.

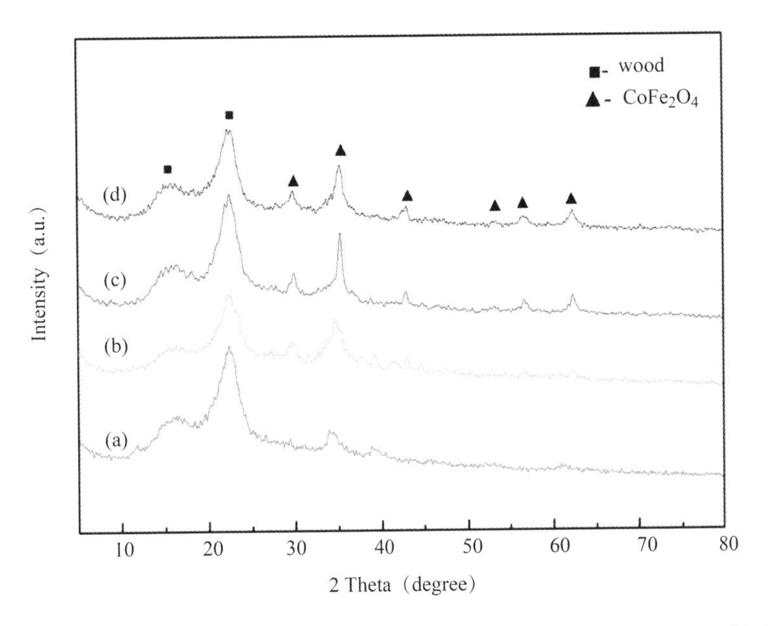

Fig. 8 X-ray diffraction patterns of the hydrothermally synthesized samples at 90 ℃ for (a) 3h, (b) 5h, (c) 8h and (d) 12h, respectively

Fig. 9 illustrates the magnetic characterization of the samples synthesized at different reaction temperatures. The saturation magnetization (M_S) of the samples synthesized for 3 h, 5 h, 8 h, 12 h were 0.17 emu · g^{-1}, 1.63 emu · g^{-1}, 2.59 emu · g^{-1} and 2.24 emu · g^{-1}, respectively (seen in Table 1). Inset shows the saturation magnetization as a function of reaction time for magnetic wood composites at room temperature. The data shows the saturation magnetization increases with reaction time rapidly, attaining a maximum value of 2.59 emu · g^{-1} at 8 h, and then slightly decrease with the reaction time. This phenomenon proved the reaction time has a significant influence on the saturation magnetization of the magnetic wood composites in initial formation process, and this effect becomes insignificant when the reaction time is more than 8 h. As for the coercivity (H_c) of the samples synthesized for 3 h, 5 h, 8 h and 12 h were 160.8 Oe, 155.7 Oe, 135.5 Oe and 128.1 Oe, respectively (shown in Table 1). The decrease of coercivity could be attributed to some agglomeration of the magnetic particles that precipitated on the wood template, which strengthen the exchange interaction between the particles and reduces the coercivity of the magnetic wood[33, 34].

3.3 Effect of the concentration of KNO_3

To elucidate the effect of the concentrations of KNO_3 used on the growth of the $CoFe_2O_4$ nanostructred materials, the KNO_3 concentration was varied from 0 M to 0.25 M while the other growth parameters were kept

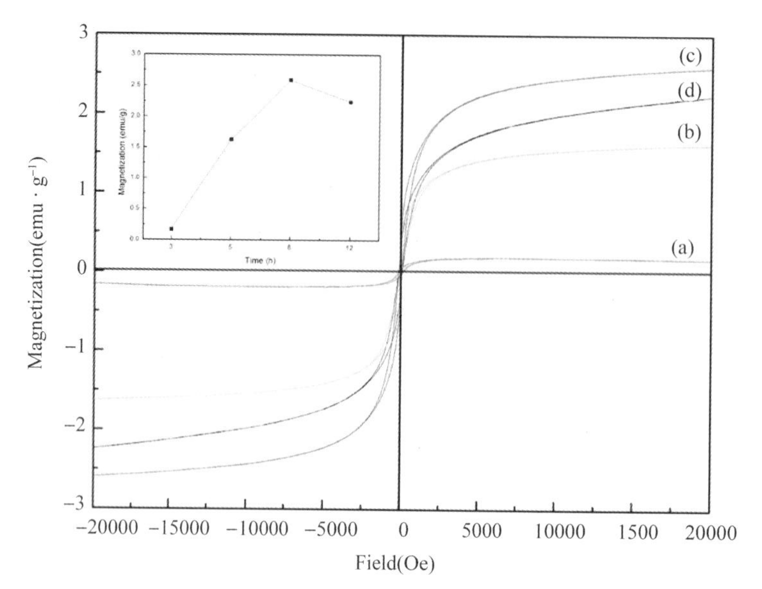

Fig. 9 Magnetization curves at room temperatures of the hydrothermally synthesized samples for (a) 3 h, (b) 5 h, (c) 8 h and (d) 12 h, respectively. Inset: the saturation magnetization (M_S) as a function of reaction time for magnetic wood composites

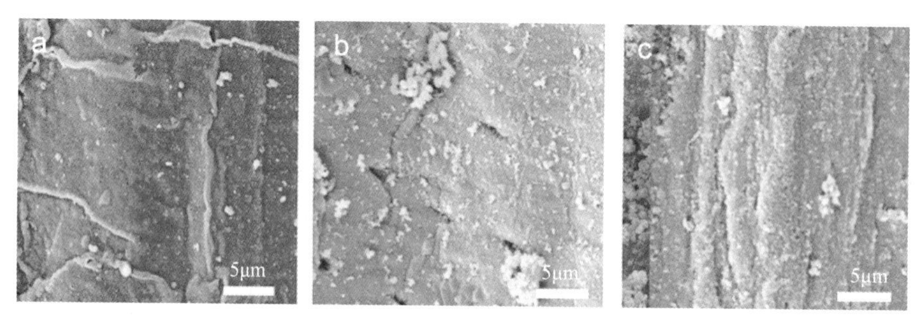

Fig. 10 SEM image of hydrothermally synthesized samples at 90 ℃ for 8 h from aqueous solutions of (a) $[NO_3^-]$=0 M, (b) $[NO_3^-]$=0. 125 M and (c) $[NO_3^-]$=0. 25 M, respectively

constant (growth temperature 90 ℃, growth time 8 h, $[OH^-]$ = 1. 32 M). SEM images and XRD patterns of $CoFe_2O_4$ nanostructured materials grown at different concentrations of KNO_3 ware shown in Fig. 10 and Fig. 11, respectively. In Fig. 11, when the concentration of KNO_3 was lower than 0. 125 M, the sample does not reveal a spinel phase of $CoFe_2O_4$; impurity phases such as $Fe(OH)_2$, FeOOH, $Co(OH)_2$ and $Fe_{0.67}Co_{0.33}OOH$ were observed in Fig. 11(a) and (b). However, a spinel phase of $CoFe_2O_4$ was formed on the wood template when the concentration of KNO_3 was 0. 25 M. Thus, the concentration of KNO_3 has a significant influence on the crystal structure of cobalt ferrite and it required for the complete conversion of the hydroxides to ferrite on wood template was nearly 0. 25 M. Moreover, the XRD data suggest that the spinel $CoFe_2O_4$ particles were produced by oxidation of the metal hydroxyl, which was good in agreement with the reaction Eqs. (1) and (2). The SEM observations clearly showed the aggregation of primary particles that precipitated on wood template to give large and well crystals in Fig. 10.

Fig. 12 illustrates the magnetic characterization of the samples synthesized at different KNO_3 concentration. The

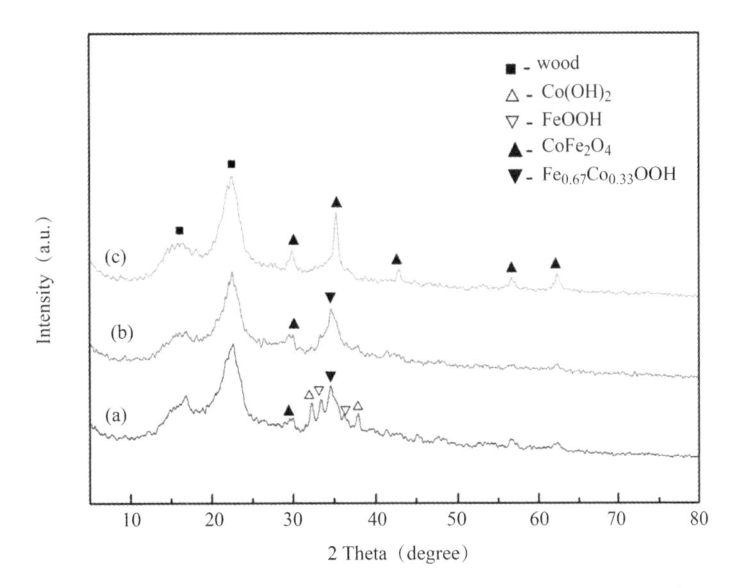

Fig. 11 X-ray diffraction patterns of the hydrothermally synthesized samples at 90 ℃ for 8 h from aqueous solutions of (a) $[NO_3^-]=0$ M, (b) $[NO_3^-]=0.125$ M and (c) $[NO_3^-]=0.25$ M, respectively

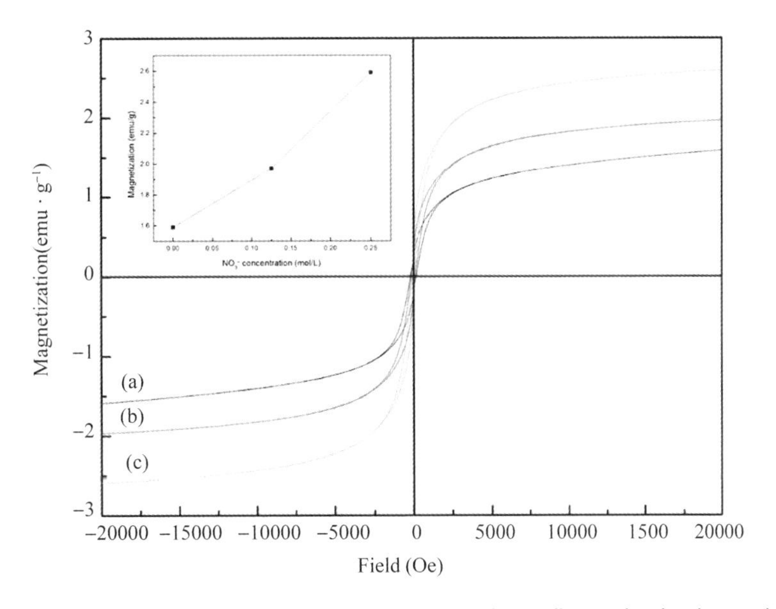

Fig. 12 Magnetization curves at room temperatures of the hydrothermally synthesized samples at 90 ℃ for 8 h from aqueous solutions of (a) $[NO_3^-]=0$ M, (b) $[NO_3^-]=0.125$ M and (c) $[NO_3^-]=0.25$ M, respectively. Inset: the saturation magnetization (M_S) as a function of KNO_3 concentration for magnetic wood composites

saturation magnetization (M_S) of the samples synthesized from aqueous solutions of $[NO_3^-]=0$ M, 0.125 M and 0.25 M were 1.59 emu · g^{-1}, 1.97 emu · g^{-1} and 2.59 emu · g^{-1}, respectively. Inset shows the saturation magnetization (M_S) as a function of KNO_3 concentration for magnetic wood composites at room temperature. The data shows the saturation magnetization increases with KNO_3 concentration rapidly. Besides the coercivity of the magnetic wood samples were also showed in Table 1. The differences in saturation magnetization and coercivity are probably due to cation inversion, vacancies and the non-stoichiometric composition typical of water-based synthesis of spinel nanoparticles from hydroxides[7, 32]. Note that the hydroxides are non-magnetic property and should not contribute to the magnetic properties of the magnetic wood materials.

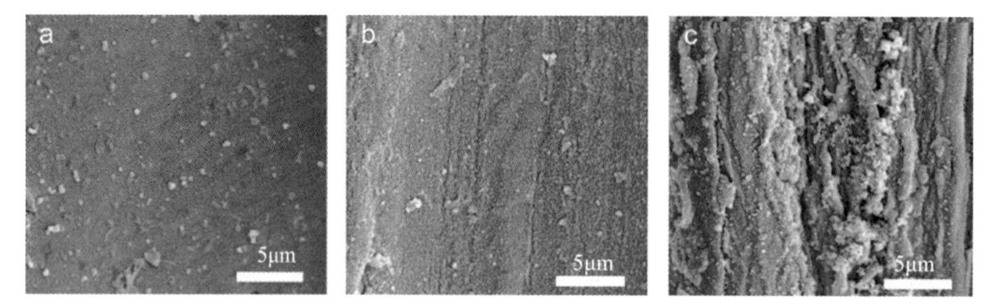

Fig. 13 SEM image of the hydrothermally synthesized samples at 90 ℃ for 8 h from aqueous solutions of (a) $[OH^-]=0.33$ M, (b) $[OH^-]=0.66$ M and (c) $[OH^-]=1.32$ M, respectively

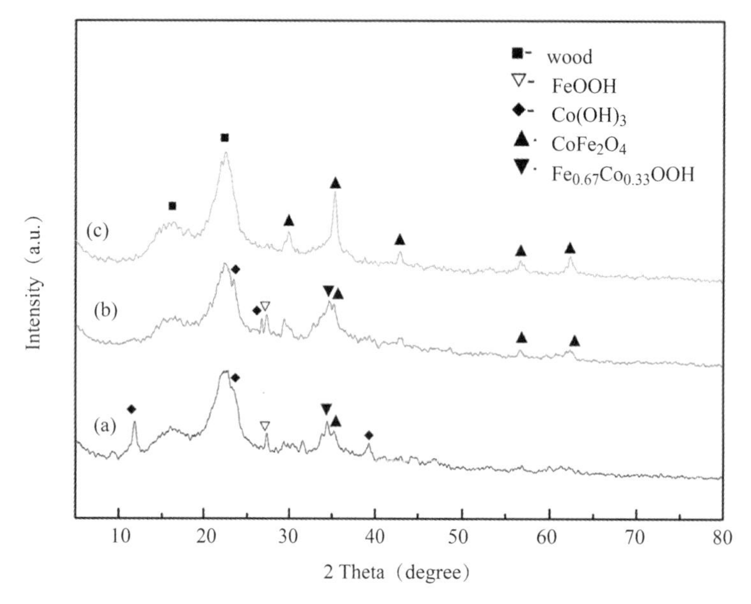

Fig. 14 X-ray diffraction patterns of the hydrothermally synthesized samples at 90 ℃ for 8 h from aqueous solutions of (a) $[OH^-]=0.33$ M, (b) $[OH^-]=0.66$ M and (c) $[OH^-]=1.32$ M, respectively

3.4 Effect of the concentration of NaOH

To elucidate the effect of the concentrations of NaOH used on the growth of the $CoFe_2O_4$ nanostructred materials, the NaOH concentration was varied from 0.33 M to 1.32 M while the other growth parameters were kept constant (growth temperature 90 ℃, reaction time 8 h, $[NO_3^-]=0.25$ M). SEM images and XRD patterns of $CoFe_2O_4$ nanostructured materials grown at different concentrations of NaOH were showed in Fig. 13 and Fig. 14, respectively. In Fig. 14, when the OH^- concentration was less than the 0.66 M, the reaction was not completed within 8 h and also the intermediate hydroxide phase such as $Fe(OH)_3$, FeOOH, and $Fe_{0.67}Co_{0.33}OOH$ were detected in the XRD analysis. In turn, a well defined crystals of cobalt ferrite were grown on the wood surface at OH^- concentrations was 1.32 M. The SEM observations clearly showed the aggregation of primary particles to give large crystals on the wood surface with the increased of NaOH concentration in Fig. 13.

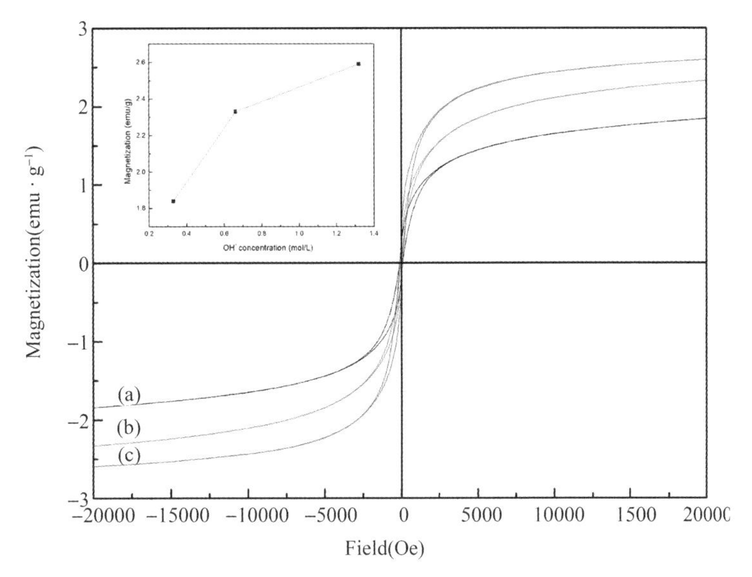

Fig. 15 Magnetization curves at room temperatures of the hydrothermally synthesized samples at 90 ℃ for 8 h from aqueous solutions of (a) [OH^-] = 0.33 M, (b) [OH^-] = 0.66 M and (c) [OH^-] = 1.32 M, respectively. Inset: the saturation magnetization (M_S) as a function of NaOH concentration for magnetic wood composites

Fig. 15 illustrates the magnetic characterization of the samples synthesized at different NaOH concentration. The saturation magnetization (M_S) of the samples synthesized from aqueous solutions of [OH^-] = 0.33 M, [OH^-] = 0.66 M and [OH^-] = 1.32 M were 1.84 emu · g^{-1}, 2.22 emu · g^{-1} and 2.59 emu · g^{-1}, respectively. Inset shows the saturation magnetization (M_S) as a function of NaOH concentration for magnetic wood composites at room temperature. Under our experimental conditions, pure $CoFe_2O_4$ spinel phase that precipitated on the wood template was obtained only in the tests using 1.32 M of NaOH. For other amounts of NaOH, the ferrite formation was not completed and the intermediate hydroxide phase was present. Thus, the saturation magnetization increases with NaOH concentration. Besides, we also concluded that the concentration of NaOH also plays an important role in controlling the coercivity. The maximum coercivity of 135.5 Oe at room temperature was observed for the magnetic wood prepared by using 1.32 M of NaOH as shown in Table 1. The possible reason for the increases of the saturation magnetization and coercivity with the concentration of NaOH is the better crystallization of $CoFe_2O_4$ that precipitated on the wood template.

4 Conclusions

Magnetic wood composites have been synthesized employing the hydromel method. A particular focus was given to the investigation of the fundamental role of the growth temperature, reaction time, the concentration of reactants on the final crystalline and morphology of the materials obtained. It is clearly showed that KNO_3 and NaOH enables the $CoFe_2O_4$ particles with a higher purity were grown on the wood template. The morphology was strongly influenced by the hydrothermal temperature. Higher reaction temperatures favored the magnetic wood with lager particles size and stronger saturation magnetization formation. Moreover, the reaction time also has a significant influence on the saturation magnetization and the morphology of the magnetic wood composites in initial formation process, and this effect becomes insignificant when the reaction time is more than 8 h. The results given in this paper can be considered as a study for the development of methods to prepare magnetic wood with the desirable structure and magnetic properties, which is crucial for fully utilizing the abundant wood resources.

Acknowledgments

This work was financially supported by The National Natural Science Foundation of China (grant no. 31470584).

References

[1]Lu A H, Salabas E E, Schüth F. Magnetic nanoparticles: synthesis, protection, functionalization, and application[J]. Angew. Chem. Int. Ed., 2007, 46(8): 1222-1244.

[2]Maaz K, Mumtaz A, Hasanain S K, et al. Synthesis and magnetic properties of cobalt ferrite ($CoFe_2O_4$) nanoparticles prepared by wet chemical route[J]. J. Magn. Magn. Mater., 2007, 308(2): 289-295.

[3]Pathak T K, Vasoya N H, Lakhani V K, et al. Structural and magnetic phase evolution study on needle-shaped nanoparticles of magnesium ferrite[J]. Ceram. Int., 2010, 36(1): 275-281.

[4]Zhang S, Yang D, Jing D. Enhanced photodynamic therapy of mixed phase TiO_2(B)/anatase nanofibers for killing of HeLa cells[J]. Nano Res., 2014, 7(11): 1659-1669.

[5]Balazs A C, Emrick T, Russell T P. Nanoparticle polymer composites: where two small worlds meet[J]. Science, 2006, 314(5802): 1107-1110.

[6]Way A E, Hsu L, Shanmuganathan K, et al. pH-responsive cellulose nanocrystal gels and nanocomposites[J]. ACS Macro Lett., 2012, 1(8): 1001-1006.

[7]Olsson R T, Samir M A, Salazar-Alvarez G, et al. Making flexible magnetic aerogels and stiff magnetic nanopaper using cellulose nanofibrils as templates[J]. Nat. Nanotechnol., 2010, 5(8): 584-588.

[8]Liu S, Zhou J, Zhang L. In situ synthesis of plate-like Fe_2O_3 nanoparticles in porous cellulose films with obvious magnetic anisotropy[J]. Cellulose, 2011, 18(3): 663-673.

[9]Liu S, Li R, Zhou J, et al. Effects of external factors on the arrangement of plate-liked Fe_2O_3 nanoparticles in cellulose scaffolds[J]. Carbohyd. Polym., 2012, , 87(1): 830-838.

[10]Ugolev B N. Wood as a natural smart material[J]. Wood Sci. Technol., 2014, 48(3): 553-568.

[11]Rana R, Langenfeld-Heyser R, Finkeldey R, et al. FTIR spectroscopy, chemical and histochemical characterisation of wood and lignin of five tropical timber wood species of the family of Dipterocarpaceae[J]. Wood Sci. Technol., 2010, 44(2): 225-242.

[12]Oka H, Fujita H. Experimental study on magnetic and heating characteristics of magnetic wood[J]. J. Appl. Phys., 1999, 85(8): 5732-5734.

[13]Oka H, Hojo A, Seki K, et al. Wood construction and magnetic characteristics of impregnated type magnetic wood[J]. J. Magn. Magn. Mater., 2002, 239(1-3): 617-619.

[14]Oka H, Hamano H, Chiba S. Experimental study on actuation functions of coating-type magnetic wood[J]. J. Magn. Magn. Mater., 2004, 272: E1693-E1694.

[15]Oka H, Uchidate S, Sekino N, et al. Electromagnetic wave absorption characteristics of half carbonized powder-type magnetic wood[J]. IEEE T. Magn., 2011, 47(10): 3078-3080.

[16] Merk V, Chanan M a, Gierlinger N, et al. Hybrid wood materials with magnetic anisotropy dictated by the hierarchical cell structure[J]. ACS Appl. Mater. Interfaces, 2014, 6(12): 9760-9767.

[17] Trey S, Olsson R T, Ström V, et al. Controlled deposition of magnetic particles within the 3-D template of wood: making use of the natural hierarchical structure of wood[J]. RSC Adv., 2014, 4(67): 35678-35685.

[18] Gan W, Gao L, Sun Q, et al. Multifunctional wood materials with magnetic, superhydrophobic and anti-ultraviolet properties[J]. Appl. Surf. Sci., 2015, 332: 565-572.

[19] Gan W, Gao L, Zhan X, et al. Hydrothermal synthesis of magnetic wood composites and improved wood properties by precipitation with $CoFe_2O_4$/hydroxyapatite[J]. RSC Adv., 2015, 5(57): 45919-45927.

[20] Köseoğlu Y, Alan F, Tan M, et al. Low temperature hydrothermal synthesis and characterization of Mn doped cobalt ferrite nanoparticles[J]. Ceram. Int., 2012, 38(5): 3625-3634.

[21] Waldron R D. Infrared spectra of ferrites[J]. Phys. Rev., 1995, 99(6): 1727.

[22] Li J, Sun Q, Jin C, et al. Comprehensive studies of the hydrothermal growth of ZnO nanocrystals on the surface of bamboo[J]. Ceram. Int., 2015, 41(1): 921-929.

[23] Wang B, Feng M, Zhan H. Improvement of wood properties by impregnation with TiO_2 via ultrasonic-assisted sol-gel process[J]. RSC Adv., 2014, 4(99): 56355-56360.

[24] Li J, Yu H, Sun Q, et al. Growth of TiO_2 coating on wood surface using controlled hydrothermal method at low temperatures[J]. Appl. Surf. Sci., 2010, 256(16): 5046-5050.

[25] Sugimoto T, Matijević E. Formation of uniform spherical magnetite particles by crystallization from ferrous hydroxide gels[J]. J. Colloid Interf. Sci., 1980, 74(1): 227-243.

[26] Tamura H, Matijevic E. Precipitation of cobalt ferrites[J]. J. Colloid Interf. Sci., 1982, 90(1): 100-109.

[27] Ma M, Zhang Y, Yu W, et al. Preparation and characterization of magnetite nanoparticles coated by amino silane [J]. Colloid Surface A., 2003, 212(2-3): 219-226.

[28] Chinnasamy C N, Senoue M, Jeyadevan B, et al. Synthesis of size-controlled cobalt ferrite particles with high coercivity and squareness ratio[J]. J. Colloid Interf. Sci., 2003, 263(1): 80-83.

[29] Lu Y, Sun Q, Yang D, et al. Fabrication of mesoporous lignocellulose aerogels from wood via cyclic liquid nitrogen freezing-thawing in ionic liquid solution[J]. J. Mater. Chem., 2012, 22(27): 13548-13557.

[30] Li J, Lu Y, Yang D, et al. Lignocellulose aerogel from wood-ionic liquid solution (1-allyl-3-methylimidazolium chloride) under freezing and thawing conditions[J]. Biomacromolecules, 2011, 12(5): 1860-1867.

[31] Maaz K, Mumtaz A, Hasanain S K, et al. Synthesis and magnetic properties of cobalt ferrite ($CoFe_2O_4$) nanoparticles prepared by wet chemical route[J]. J. Magn. Magn. Mater., 2007, 308(2): 289-295.

[32] Salazar-Alvarez G, Olsson R T, Sort J, et al. Enhanced Coercivity in Co-Rich Near-Stoichiometric $Co_xFe_{3-x}O_{4+\delta}$ Nanoparticles Prepared in Large Batches[J]. Chem. Mater., 2007, 19(20): 4957-4963.

[33] Schrefl T, Schmidts H F, Fidler J, et al. Nucleation of reversed domains at grain boundaries[J]. J. Appl. Phys., 1993, 73(10): 6510-6512.

[34] Sugimoto T, Yamaguchi G. Contact recrystallization of silver halide microcrystals in solution[J]. J. Cryst. Growth, 1976, 34(2): 253-262.

中文题目： 可控低温水热法在木材模板上生长 $CoFe_2O_4$ 颗粒

作者： 甘文涛，刘莹，高丽坤，詹先旭，李坚

摘要： 磁性木材有望成为电磁屏蔽、室内电磁波吸收器和特殊装饰材料。这项工作的重点是开发一种利用水热法制备磁性木材/$CoFe_2O_4$ 复合材料的方法。特别强调理解水热技术中所涉及的化学参数的作用，以及水热处理对所获得材料的结构和性质的影响。在 90 ℃温度下 8 h 内可获得磁性木材/

钴铁氧化合物。反应温度的升高促进了尖晶石铁氧体在木材模板上的结晶。磁性木质复合材料的饱和磁化强度随着保温时间的延长而迅速增加，在 8 h 达到最大值 2. 59 emu · g^{-1}。此外，硝酸钾和氢氧化钠的浓度对磁性木质复合材料的结构、形貌和磁性能有重要影响。

关键词：$CoFe_2O_4$；水热法；磁性材料；木材改性

Hydrothermal Synthesis of Magnetic Wood Composites and Improved Wood Properties by Precipitation with $CoFe_2O_4$ /Hydroxyapatite*

Wentao Gan, Likun Gao, Xianxu Zhan, Jian Li

Abstract: Magnetic wood would be a potential for electromagnetic shielding, indoor electromagnetic wave absorber and heavy metal adsorption. In this study, magnetic wood with improved thermal stability and mechanical property as well as UV resistant ability were prepared by modification with magnetic $CoFe_2O_4$ and hydroxylapatite (HAP) via a hydrothermal process. The functional groups and morphology of the modified wood were examined by Fourier transform infrared spectroscopy (FTIR), X-ray diffraction (XRD) and scanning electron microscopy (SEM) with energy-dispersive X-ray analysis (EDXA). The results indicated that the magnetic $CoFe_2O_4$ nanoparticles precipitated on the wood substrate through the alkali treatment, and then the hydroxyapatite was grown on the wood surface by the electrostatic interactions between calcium cations and the negatively charged OH^- anions. Moreover, the magnetic, thermal, mechanical performances and UV-resistance of the $CoFe_2O_4$/HAP-modified wood composites were also evaluated.

Keywords: Sol-gel process; Inorganic composites; Nanoparticles; Hydroxyapatite; Surface

1 Introduction

Wood was an excellent natural composite material for interior design, architecture and decoration because the hierarchical structure, porosity and adapted mechanical properties. [1-3] The dominating functions of wood in the tree/from trees resulted in an excellent light-weight engineering material and provided a natural porosity scaffold for further functionalization.

Hybrid wood/inorganic composites were attractive for us in applications required multifunctional properties, and their mechanical, optical and magnetic properties, among others, have been extensively explored. [4-6] Numerous approaches based on sol-gel process, vacuum impregnation, brush painting and hydrothermal method have been developed to synthesize hybrid wood/inorganic composites, [7-11] however most methods were concentrated on a single-layer of inorganic particles. Saka and co-workers have reported that the synthesis of wood/SiO_2 composites based on a sol-gel method, and the fire-resistance and photostability of wood composites were improved. [4, 12] Sun and co-workers have precipitated a layer of TiO_2 nanoparticles on the wood surface by a hydrothermal method, the water resistance and dimensional stability of wood/TiO_2 composites were fully discussed. [13] Recently, some studies have been paid much attention to improving the

---

* 本文出自 RSC Adavances, 2015, 5: 45919-45927.

performance of wood using the magnetic nanoparticles.[5, 14, 15]

Magnetic materials were receiving increasing interest in recent years, because of their possibly applications in the fields of biomedicine, data storage, catalyzer and microwave absorbing materials.[16-19] Using a templating approach, a minor weight fraction of magnetic nanoparticles was simply mixed with wood polymer matrix. Oka et al. fabricated three kind of magnetic wood by loading, coatings and impregnating wood fiber board with commercial magnetic fluids. On the basis of the approach, wood could be functionalized with enhanced properties in terms of magnetic property, heating characteristics, actuation and electromagnetic wave absorption.[20-23] Although the wood materials exhibited the excellent magnetic property, some unfavorable end-product properties such as combustibility and photodegradability have not been discussed.

In this study, magnetic wood composites were prepared by precipitating $CoFe_2O_4$ and hydroxyapatite (HAP) on the wood surface. A combination of wood with HAP was expected to offer such striking features as high mechanical property, good incombustibility and photostability.[24, 25] The $CoFe_2O_4$/HAP-modified wood composites not only possessed magnetic property, but also enhanced the thermal, mechanical and anti-ultraviolet properties.

2 Experimental

2.1 Materials

Poplar wood samples used have cube shapes with $a = 20$ mm were ultrasonically rinsed in deionized water for 30 min and vacuum-dried at 103 ℃ for 24 h. Cobalt chloride hexahydrate ($CoCl_2 \cdot 6H_2O$), ferrous sulfate heptahydrate ($FeSO_4 \cdot 7H_2O$), potassium nitrate (KNO_3), sodium hydroxide (NaOH), calcium nitrate [$Ca(NO_3)_2 \cdot 4H_2O$], diammonium hydrogen phosphate [$(NH_4)_4HPO_4$] and ammonia solution ($NH_3 \cdot H_2O$, 25%) used in this study were supplied by Shanghai Boyle Chemical Company Limited., and used without further purification.

2.2 Preparation of the magnetic $CoFe_2O_4$-modified wood composites

The wood samples were immersed in the 40 mL of a 0.2 mol of freshly prepared aqueous solution of $FeSO_4$ and $CoCl_2$ ($[Fe^{2+}]/[Co^{2+}] = 2$). Then the system transferred into a Teflon-lined stainless-steel autoclave and heated to 90 ℃ for 3 h. At the end of the reaction, the supernatant was poured out and 30 mL of a1.32 mol L^{-1} NaOH solution with KNO_3($[Fe^{2+}]/[NO_3^-] = 0.44$) was added to the Teflon-lined stainless-steel autoclave. To complete the reaction, the system was kept for an additional hydrothermal processing for 6 h at 90 ℃. Finally, the magnetic wood specimens were removed from the solution, ultrasonically rinsed with deionized water for 10 min, and dried at 50 ℃ for 48 h in vacuum.

2.3 Preparation of the $CoFe_2O_4$/HAP-modified wood composites

100mL $Ca(NO_3)_2$(33.7mmol) and $(NH_4)_2HPO_4$(20mmol) solution was adjusted to pH = 11 by the drop-wise addition of 25% ammonia solution, then the $CoFe_2O_4$-modified wood composites and the above mixture were transferred into a Teflon-lined stainless-steel autoclave and heated to 90 ℃. After 3 h, the mixture was cooled to room temperature and aged overnight. The as synthesized samples were removed and ultrasonically rinsed with deionized water for 10 min, and then dried at 50 ℃ for 48 h in vacuum.

2.4 Characterizations

The morphologies of the untreated wood, $CoFe_2O_4$ – modified wood composites and $CoFe_2O_4$/HAP – modified wood composites were characterized using scanning electron microscopy (SEM, FEI, and Quanta 200). The surface chemical compositions of the samples were determined via energy-dispersive X-ray analysis (EDXA). The crystalline structure of these composites were identified via X-ray diffraction (XRD, Rigaku, and D/MAX 2200) operating with Cu $K\alpha$ radiation (λ = 1.5418 A) at a scan rate (2θ) of 4° · min^{-1}, an accelerating voltage of 40 kV, and an applied current of 30 mA ranging from 5° to 80°. The samples used for XRD analysis were prepared by grinding the dried wood samples and pressing them into concavity slide. The surface chemical compositions of the samples were determined via the Fourier transformation infrared spectroscopy (FTIR, Nicolet, Magna-IR 560). For FTIR analysis, thin sample disks were made by grinding small portion of the dried wood samples and pressing them with potassium bromide. DR-UV/Visible spectra were obtained with the instrument TU-190, Beijing Purkinje, China, equipped with an integrating sphere attachment. $BaSO_4$ was the reference.

For magnetic characterization, wood specimens used have cube shapes with a = 4 mm, the magnetic properties of the composites were measured by a superconducting quantum interference device (MPMS XL-7, Quantum Design Corp.) at room temperature (300 K).

The thermal performances of the composites were examined using a thermal analyzer (TGA, SDT Q600) in the temperature range from room temperature up to 800 ℃ at a heating rate of 10 ℃ · min^{-1}, with a flow of dried air (100 mL · min^{-1}).

For mechanical characterization, wood specimens of 20 (R)×20 (T)×30 (L)/mm were used (7-10 specimens per type), and their deformation behavior under compression in longitudinal direction was recorded using a universal testing machine (Instron 1185) at a crosshead speed of 5 mm · min^{-1}.

The weathering was performed with a QUV Accelerated Weathering Tester (Atlas, Chicago, IL, USA), which allowed water spray and condensation. The samples were fixed in stainless steel holders and under irradiation of fluorescent UV light at wavelength of 340 nm with a temperature of 50 ℃ for 8 h, followed by a spray of water for 4 h. The average irradiance was set to 0.77 W · m^{-2}, and the temperature of the spray was set to 25 ℃. The exposure time ranged from 0 to 600 h. And the color change of the wood surface after UV irradiation were measured using the portable spectrophotometer (NF-333, Nippon Denshoku Company, Japan) with CIELAB system in accordance with the ISO-2470 standard. CIELAB L^*, a^*, b^*, parameters were measured at six locations on each specimen and average value was calculated. In the CIELAB system, L^* axis represents the lightness, and varies from 100 (white) to 0 (black); a^* and b^* are the chromaticity indices; $+a^*$ is the red direction; $-a^*$ is green; $+b^*$ is yellow; and $-b^*$ is blue. The change of L^*, a^* and b^* are calculated according to Eqn. (1)-(3).

$$\Delta a^* = a_2 - a_1 \quad (1)$$

$$\Delta b^* = b_2 - b_1 \quad (2)$$

$$\Delta L^* = L_2 - L_1 \quad (3)$$

Where Δa^*, Δb^*, and ΔL^* were the differences in the initial and final values of a^*, b^*, and L^*, respectively; a_1, b_1, and L_1 were the initial color parameters; and a_2, b_2, and L_2 were the color parameters after UV irradiation. The overall color changes (ΔE^*) were used to evaluate the total color change using Eqn. (4).

$$\Delta E^* = \sqrt{(L_2^* - L_1^*)^2 + (a_2^* - a_1^*)^2 + (b_2^* - b_1^*)^2} \tag{4}$$

A lower ΔE^* value corresponds to a smaller color difference and indicates strong resistance to UV radiation.

3 Results and discussion

3.1 Structural characterization and morphology of the $CoFe_2O_4$/HAP-modified wood composites

The morphology and element distribution of the wood samples before and after hydrothermal treatment were analyzed with the SEM image and EDXA spectra. It is known that wood possesses porous structure with plenty of interconnected pores. Fig. 1a shows the typical SEM images of the poplar wood surface and pits (namely the smallest holes of all pores) were found on the walls of tracheids. At present, there is a tendency to use hydrothermal method to improve the compatibility between the hydrophilic wood substrates and inorganic nanoparticles. Because hydrothermal solution could break the van der Waals force between cellulose molecules and dissolve part of the lignin and hemicelluloses in wood, which causes more cellulose molecules exposed to hydrothermal solution and make wood surface rough (Fig. 1b, marked oval region). [13, 26-28] Therefore, magnetic nanoparticles and HAP can be easily precipitated on the wood substrate. Fig. 1b shows the SEM images of the magnetic $CoFe_2O_4$ nanoparticles deposited on the wood surface at different magnification. At the low magnification, the magnetic particles were particularly visible in the luminous vessels of the poplar wood. Higher magnification revealed a densely packed layer of magnetic nanoparticles covered the wood surface. In the composited with HAP (Fig. 1c), a new kind of inorganic particles with the sizes around 20 μm had obviously been deposited on the wood surface, indicating the new layer of HAP particles was grown on the wood surface.

The EDXA spectra revealed that carbon and oxygen were the only elements presented at the untreated wood surface (Fig. 1a, Insert). The additional strong signal of iron and cobalt element presented in the spectra of magnetic $CoFe_2O_4$-modified wood composites (Fig. 1b, Insert). After HAP modification, the strong peak of calcium and phosphorus were also showed in treated wood (Fig. 1c, Insert), confirming the HAP and $CoFe_2O_4$ precipitated on the wood surface effectively.

To investigate the functional group composition of the wood samples, FTIR measurement performed before and after the hydrothermal treatment (Fig. 2). In Fig. 2, the band at 2910, 1640, 1458 and 1420 cm^{-1} correspond to —CH_3 asymmetric, O—H bending vibration, C—H deformation and C—O stretching vibration, respectively. Most of the observed peaks of the untreated wood represent major cell wall components such as cellulose (1163, 897 cm^{-1}), hemicelluloses (1740, 1118, 1056 cm^{-1}) and lignin (1594, 1507, 1230 cm^{-1}). [29, 30] For the $CoFe_2O_4$/HAP-modified wood composites, the bands at 1740, 1590, 1510, 1230, 1118 and 1056 cm^{-1} were disappeared, indicating the degradation of the wood component of hemicelluloses and lignin after hydrothermal treatment. The absorption band assigned to the O—H stretch vibration of hydroxyl groups weakened at 3423 cm^{-1}, revealing a decrease in the relative content of hydroxyl groups on the treated wood composites. The results were consistent with that observed by Wan et al. [24] It was

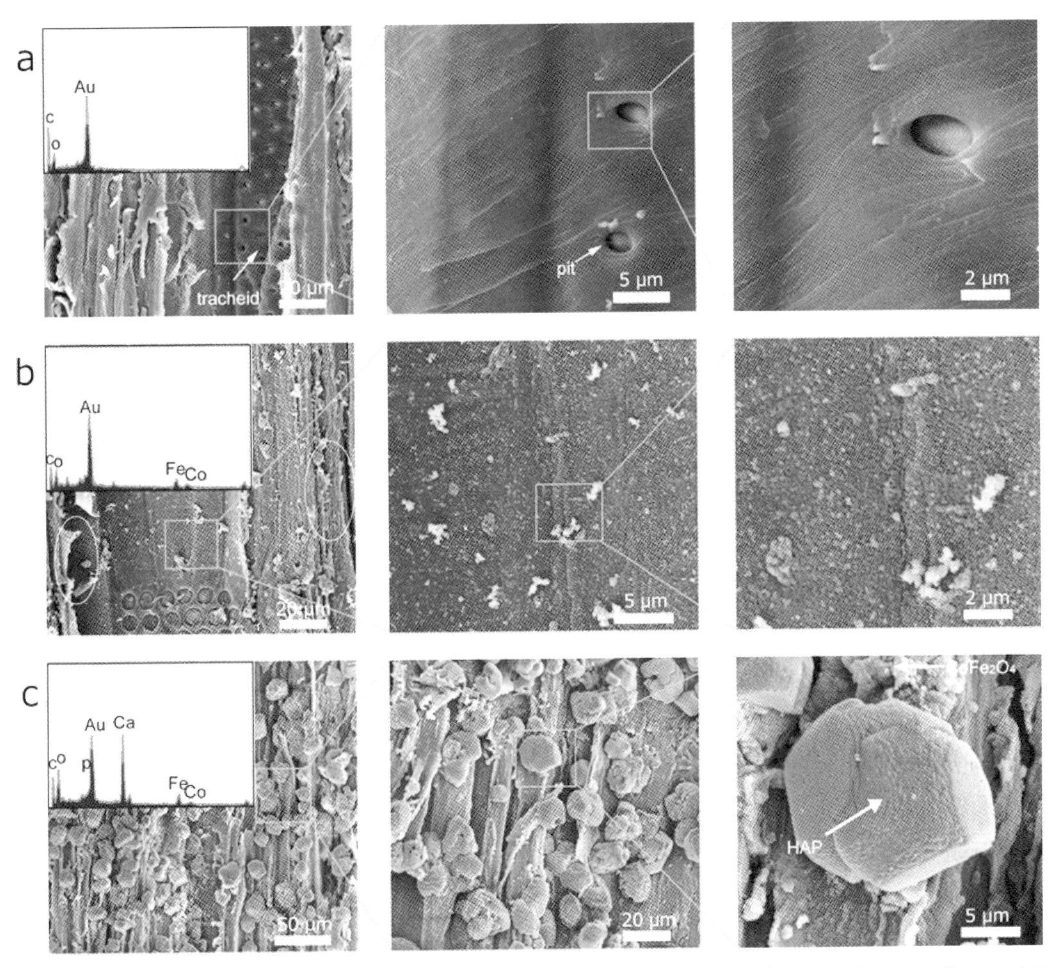

Fig. 1 SEM images of (a) untreated wood, (b) magnetic $CoFe_2O_4$-modified wood composites and (c) magnetic $CoFe_2O_4$/HAP-modified wood composites at different magnification. The insert are the corresponding EDAX spectra of wood surface

believed that calcium ions was adsorbed onto the surface of the wood by ionic interaction between calcium ions and the negatively charged OH^- ions.[31, 32] And the incorporated calcium ions can bind phosphate ions and hydroxyl ions to form the hydroxyapatite. The adsorption bands attributed to the asymmetric stretching (ν_3) and bending vibrations (ν_4) of the PO_{43-} ions were clear at 1042 and around 562-604 cm^{-1}, respectively.[33-36] The characteristic band at 453 cm^{-1} was arises from the stretching vibrations of Fe-O bonds at octahedral site.[37,38] These results not only strongly confirmed the $CoFe_2O_4$ and HAP particles precipitated on the wood template, but also indicated that no obvious interaction existed between $CoFe_2O_4$ and HAP. In order to further verify the presence of HAP and $CoFe_2O_4$, XRD was employed to gain evidence for the $CoFe_2O_4$/HAP-modified wood composites.

Fig. 3 shows the XRD patterns for the untreated wood, magnetic $CoFe_2O_4$-modified wood composites, as well as magnetic $CoFe_2O_4$/HAP-modified wood composites. The main peaks at $2\theta = 15°$ and $22°$, which were characteristic of wood,[28, 39] were observed in the $CoFe_2O_4$-modified wood composites and $CoFe_2O_4$/HAP-

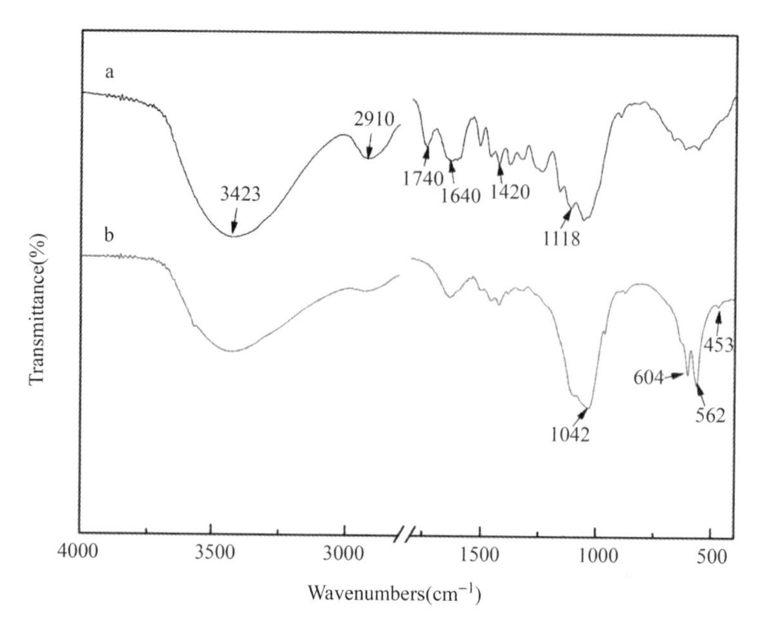

Fig. 2 FTIR spectra of (a) untreated wood and (b) $CoFe_2O_4$/HAP-modified wood composites

modified wood composites. The other peaks at $2\theta = 18°$, $30°$, $35°$, $43°$, $53°$, $56°$ and $62°$, which were assigned to the diffraction of the (111), (220), (311), (400), (422), (511) and (440) planes of $CoFe_2O_4$ (PDF No. 22 - 1086), were also observed in the magnetic $CoFe_2O_4$/HAP - modified wood composites. Additionally, extra peaks located at 26°, 32°, 39°, 46°, 49° and 53° noted. These peaks were characteristic peaks of HAP crystals, [40-42] suggesting that HAP crystals obtained on wood template via the hydrothermal process. Furthermore, this also revealed that the phase change of $CoFe_2O_4$ did not take place in the following treatment, which was in good agreement with the result by FTIR analysis.

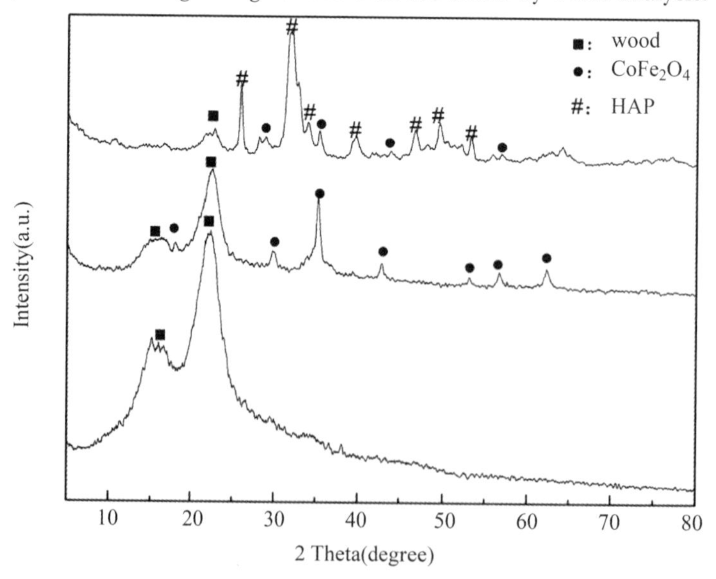

Fig. 3 XRD patterns for the untreated wood, magnetic $CoFe_2O_4$-modified wood composites and $CoFe_2O_4$/HAP-modified wood composites

3.2 Incorporation mechanism of $CoFe_2O_4$ and HAP particles on the wood template by the hydrothermal treatment

Based on the above analysis, a schematic presentation of the formation mechanism of $CoFe_2O_4$ and HAP particles on the wood template showed in Fig. 4. The main component of wood matrix was cellulose, which also features numerous hydroxyl groups that confer the hydrophilic nature of the cell wall. The hydrophilic wood substrate was immersed in a aqueous $FeSO_4/CoCl_2$ solution at room temperature before heating the system to 90 ℃ to thermally precipitated the non-magnetic metal hydroxides on the template easily.[43] Then the precipitated precursors were converted into ferrite crystal nanoparticles on immersion in $NaOH/KNO_3$ solution at 90 ℃. The KNO_3 was used as an oxidizing agent and the reactions of the spinel formation can be written as follows:[44, 45]

$$Fe^{2+} + Co^{2+} + 4OH^- \longrightarrow Fe(OH)_2 + Co(OH)_2 \quad (5)$$

$$8Fe(OH)_2 + 4Co(OH)_2 + NO_{3-} \longrightarrow 4CoFe_2O_4 + NH_3 + 10H_2O + OH^- \quad (6)$$

For $CoFe_2O_4$ nanoparticles dispersed in alkali solution, the bare atoms of Fe on the particle surface would adsorb OH^-, so that there is OH^--rich surface.[46] Based on the electronic structure nature of Ca^{2+} cation, it is speculated that the ion-polar interaction occur in the formation of HAP crystals in the wood surface.[31,47] The OH^- anions were electronegative and can adsorbed cations, which could provide nucleation sites for HAP on the wood surface through the electrostatic interactions with Ca^{2+}. A divalent Ca^{2+} cation may attach itself to two adjacent hydroxyl ions, which can donate two pairs of electrons to the cation. Once the nuclei formed, HAP could spontaneously grow by consuming the Ca^{2+}, OH^- and PO_4^{3-} form the surrounding solution and finally the $CoFe_2O_4$/HAP-modified wood composites were obtained.[26]

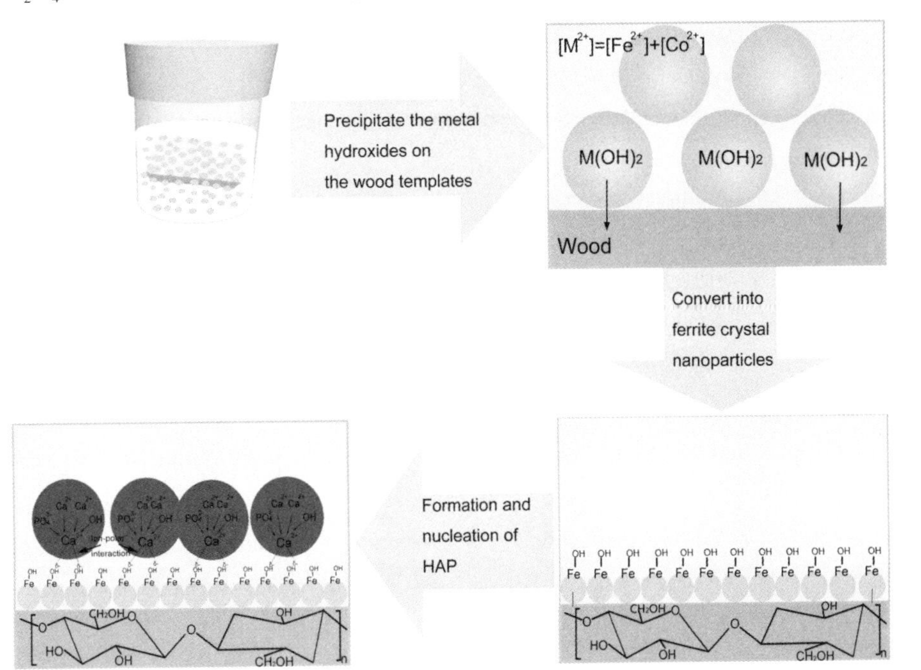

Fig. 4 Schematic of experimental set-up for preparing $CoFe_2O_4$/HAP-modified wood composites

3.3 Magnetic, Thermal, Mechanical properties and UV-resistance of the $CoFe_2O_4$/HAP-modified wood composites

Magnetic property of the wood samples showed in Fig. 5. In Fig. 5a, both of the $CoFe_2O_4$-modified wood composites and $CoFe_2O_4$/HAP-modified wood composites showed the typical hysteresis loops at the room temperature. The magnetization saturation values (M_S) of the $CoFe_2O_4$-modified wood composites and $CoFe_2O_4$/HAP-modified wood composites were 2.0 emu · g^{-1} and 1.84 emu · g^{-1}, respectively. The coercivity (H_C) of the $CoFe_2O_4$-modified wood composites and $CoFe_2O_4$/HAP-modified wood composites were 355.3 and 301.5 Oe, respectively. The reduced M_S and H_C can be explained by considering the diamagnetic contribution of the HAP surrounding the $CoFe_2O_4$ on the wood template[48, 49] In Fig. 5b, the magnetization can easily be probed by moving and lifting up a magnetic wood block under the influence of a permanent magnet. Moreover, the magnetic wooden sticks and platelet also can actuate by the magnetic field.

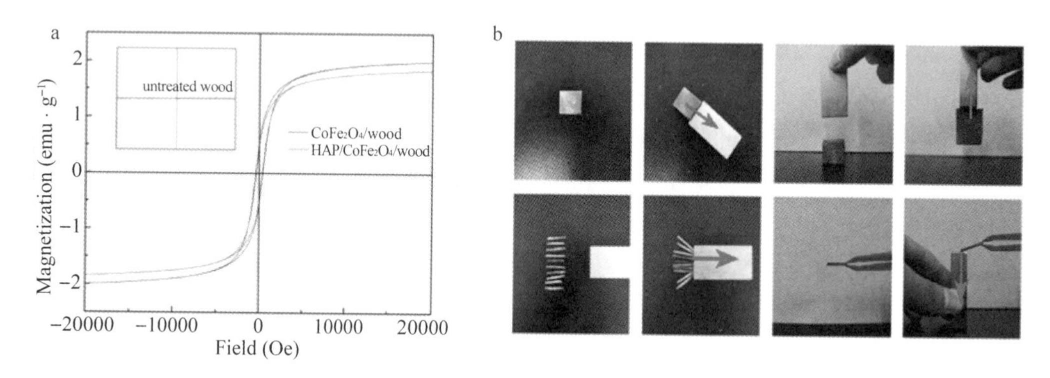

Fig. 5 (a) Magnetic hysteresis curves of the untreated wood (blue line), $CoFe_2O_4$-modified wood composites (black line) and $CoFe_2O_4$/HAP-modified wood composites (red line), at room temperature (300 K). (b) $CoFe_2O_4$/HAP-modified wood composites under the influence of an external magnet

The thermal analyses performed to examine the thermal stability of the functional magnetic wood composites. Curves representing thermal degradation (TG and DTG) for wood samples displayed in Fig. 6a and b. A small weight loss of 8%~10% around 100 ℃ was mainly attributed to the evaporation of adsorbed water. In Fig. 6a, the untreated wood showed two obvious weight loss steps with elevating temperature. The first weight loss by flaming was found in the temperature ranged from 300 ~ 350 ℃. Subsequently, another characteristic decrease in its weight by glowing was observes at 350 ~ 450 ℃. In comparison with the untreated wood, the TG curves of the $CoFe_2O_4$-modified wood composites shifted to lower temperature during the first abrupt decrease in flaming. The decreased decomposition temperature could be ascribed to the catalytic property of the precipitated inorganic nanoparticles. [50, 51] However, the initial decomposition temperature for this decrease shifted to the higher temperature after a combination of HAP particles on the wood surface. The initial decomposition temperature of the $CoFe_2O_4$-modified wood was around 170 ℃, which increased to 215 ℃, it was 45 ℃ higher than that of the $CoFe_2O_4$-modified wood. In addition, between 300 ℃ and 470 ℃ more residues resulted after flaming. The results can be explained by the incombustible HAP particles coated on $CoFe_2O_4$-modified wood surface. Above 470 ℃, the TG curves of these composites show an abrupt decrease by glowing.

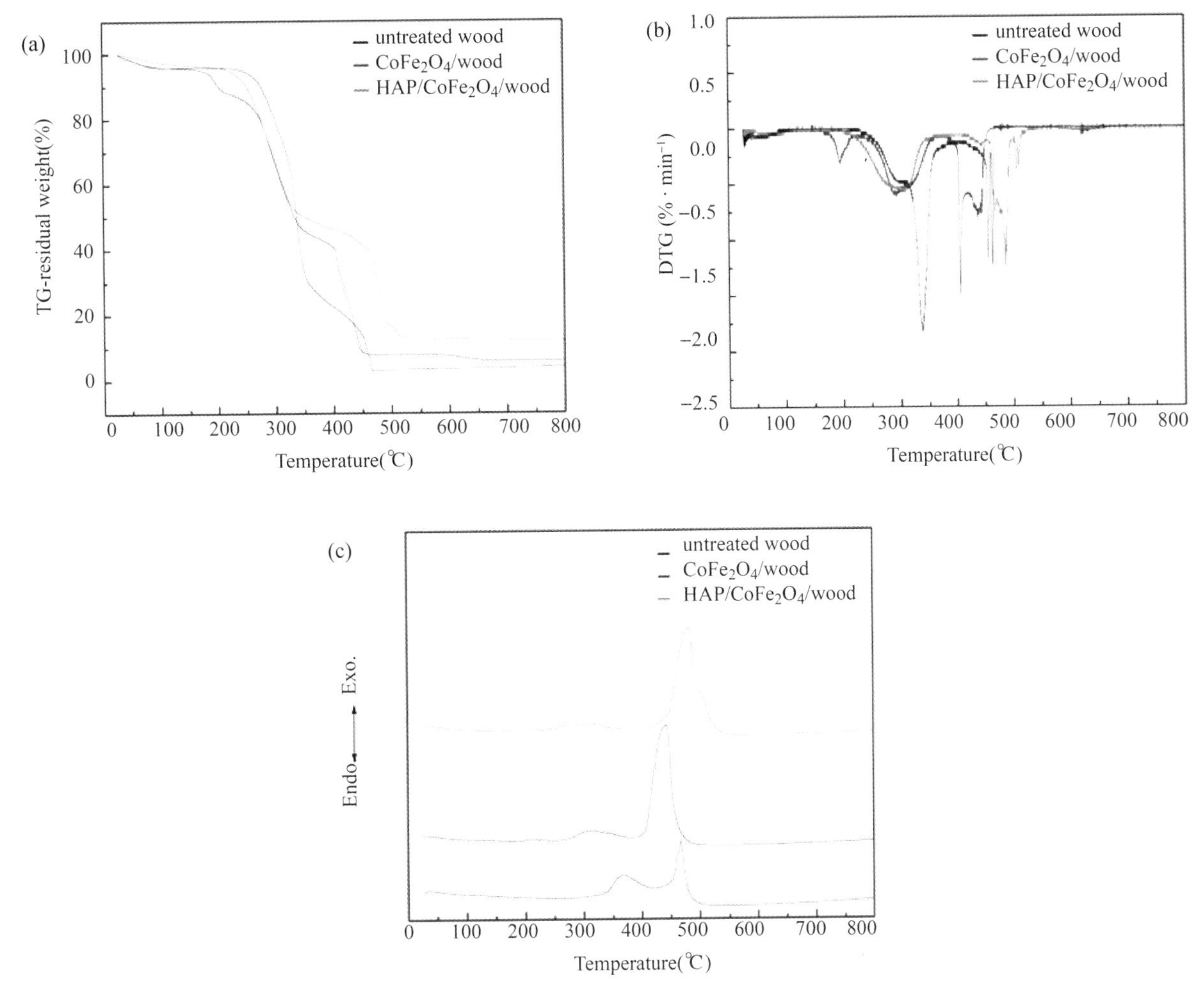

Fig. 6 (a) TG, (b) DTG and (c) DTA curves showing the thermal behaviors of the untreated wood (black line), magnetic $CoFe_2O_4$-modified wood composites (blue line) and magnetic $CoFe_2O_4$/HAP-modified wood composites (red line)

For the DTG curves in Fig. 6b, the obvious peaks of the $CoFe_2O_4$-modified wood composites and $CoFe_2O_4$/HAP-modified wood composites weakened compared to that of untreated wood, indicating a lower rate of weight loss of the modified wood. It can assume that the wood components were likely to shielded by the fire-retardant inorganic coating, which may hamper the wood components from being accessible to oxygen and reduced the rate of combustion.[7]

Fig. 6c shows the results of the differential thermal analysis (DTA) for the wood composites. The prominent exothermic peaks for flaming and glowing observed, respectively, at 360 and 470 ℃ in the DTA of the untreated wood. In the modified wood composites, the exothermic peak for flaming was weakened, shifting its peak to the lower temperature compared with the untreated wood, and only a broad exothermic peak hill was observed at about 300 ℃. These results indicated the enhance fire resistance to flaming in the modified composites.[52] On the DTA curves of the modified composites at the higher temperature (>400 ℃) prominent peaks were observed that were due to glowing. In addition, the peak temperature shifted to the higher temperature after the HAP precipitated on the wood surface. This shift corresponded to that of the temperature of the second abrupt weight decrease in the TG curves, as shown in Fig. 6a. Based on these results, it was obvious that the $CoFe_2O_4$/HAP-modified wood composites can raise the glowing temperature and have a great fire resistance effect.

The mechanical behavior of the wood composites under compression in the longitudinal direction was characterized and the average values of each type were showed in Fig. 7. The modified wood shows a deformation behavior generally similar to that of untreated wood. However, an elevated stress threshold was observed for the modified wood composites compared with untreated wood. This indicated that the yield strength of wood materials in compression have been improved after modification with $CoFe_2O_4$ and HAP. An increase in yield strength of the modified wood composites can be due to condensation processes in the lignin and hemicelluloses, as molecules degrade and can form new chemical bonds.[27]

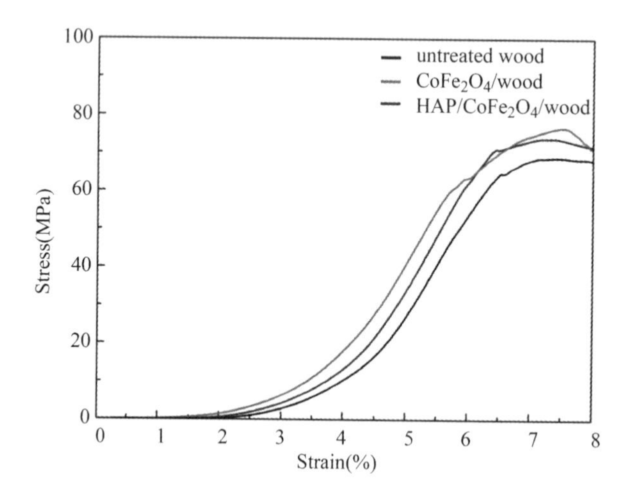

Fig. 7 Stress-strain curves of the untreated wood, $CoFe_2O_4$-modified wood composites and $CoFe_2O_4$/HAP-modified wood composites under compression in the longitudinal direction

The accelerated aging tests performed and the color changes before and after UV irradiations measured to evaluate the UV resistance of the $CoFe_2O_4$/HAP - modified wood composites. The experimental results presented in Fig. 8. The change in the Δa^* values of the untreated wood under UV irradiation indicated that the color turned a deeper shade of red with increasing UV irradiation. The Δa^* values of the $CoFe_2O_4$-modified wood composites and $CoFe_2O_4$/HAP-modified wood composites showed a similar trend, however, the Δa^* change in $CoFe_2O_4$/HAP-modified wood composites were much smaller (approximately 1/4) than that of the untreated wood (Fig. 8a). The Δb^* values of the untreated wood, $CoFe_2O_4$-modified wood composites and the $CoFe_2O_4$/HAP-modified wood composites indicated that their colors turned dark yellow, slight blue and slight yellow, respectively, with prolonged UV irradiation time (Fig. 8b). The ΔL^* values of the untreated wood, $CoFe_2O_4$-modified wood composites and $CoFe_2O_4$/HAP-modified wood composites became negative with time, indicating the lightness turns to black. Besides, the ΔL^* values of the $CoFe_2O_4$/HAP-modified wood composites were much smaller (only 1/6 and 1/8) than that of the $CoFe_2O_4$-modified wood composites and untreated wood, respectively, which meant much stronger UV-resistant ability of the $CoFe_2O_4$/HAP-modified wood composites (Fig. 8c). The total color change (ΔE^*) of thetes occurred dramatic variation with increasing UV irradiation time, while only a slight change occurred for the $CoFe_2O_4$/HAP-modified wood composites (Fig. 8d). Based on the above results the appearance of the untreated wood would seriously destroyed by UV irradiation without the employed protective measure. Although the $CoFe_2O_4$-modified wood composites showed UV-resistant ability to some extent, damage to the appearance could not be avoided. The $CoFe_2O_4$/HAP-modified wood composites exhibited an excellent ability to resist UV irradiation, and thus prevented the wood surface from damage.

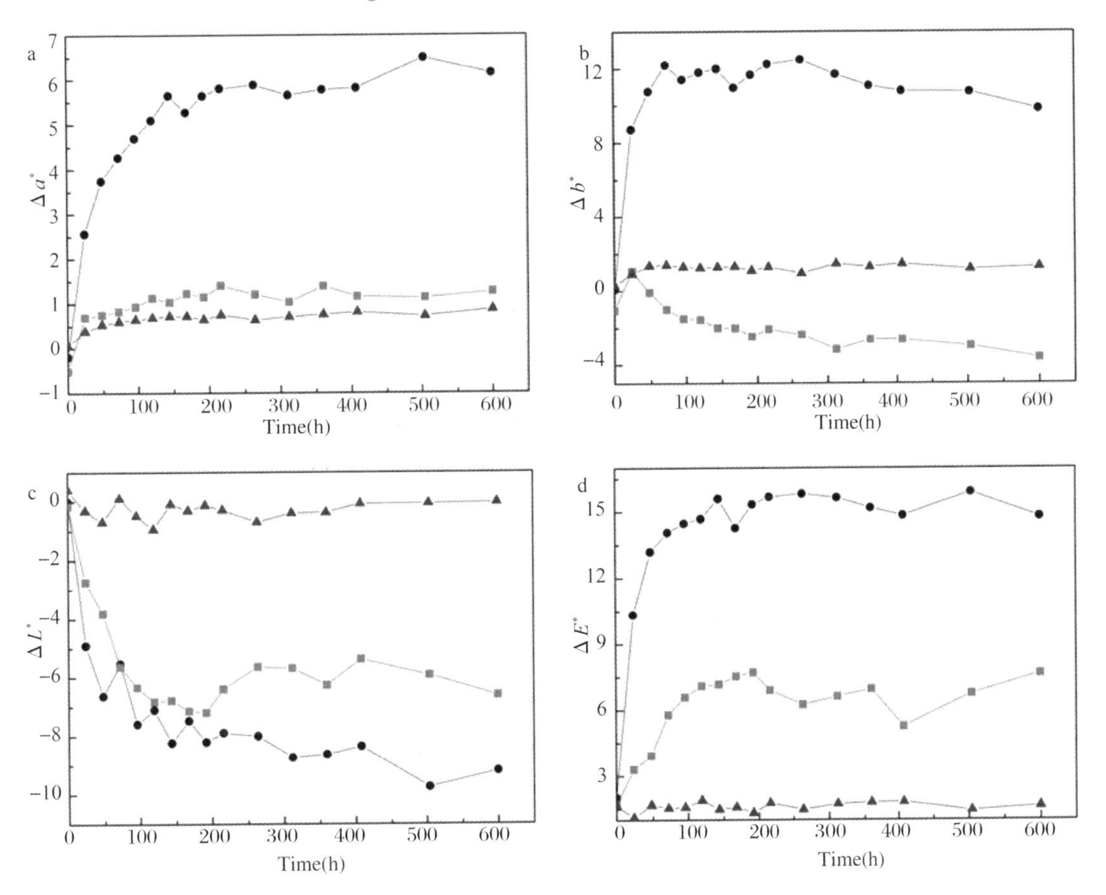

Fig. 8 Tendencies of the CIELAB (a) Δa^*, (b) Δb^*, (c) ΔL^* and (d) ΔE^* color changes of the untreated wood (●), $CoFe_2O_4$-modified wood composites (■) and $CoFe_2O_4$/HAP-modified wood composites (▲)

The UV-Vis diffuse reflection spectra of the untreated wood and $CoFe_2O_4$/HAP-modified wood composites shown in Fig. 9. Previously, Hu et al. [53] reported that the absorption edge of HAP may appear at 230 nm. Zhu et al. [54] reported that the $CoFe_2O_4$ samples have the UV absorption at the region of 200-700 nm. There were consistent with our experimental results. In Fig. 9, UV-Vis peak characteristic of HAP appears at around 210 nm and the $CoFe_2O_4$/HAP-modified wood composites has a much higher absorption in the region of 200~700 nm than the untreated wood, indicating that $CoFe_2O_4$/HAP-modified wood composites has better UV absorption. Hence, better anti-ultraviolet property achieved by modified with $CoFe_2O_4$/HAP compared to the untreated wood sample. We attribute this to the absorption of $CoFe_2O_4$ and HAP particles,

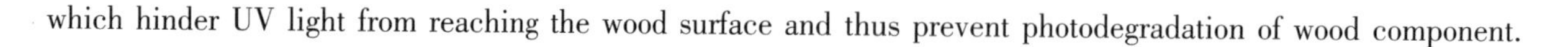

which hinder UV light from reaching the wood surface and thus prevent photodegradation of wood component.

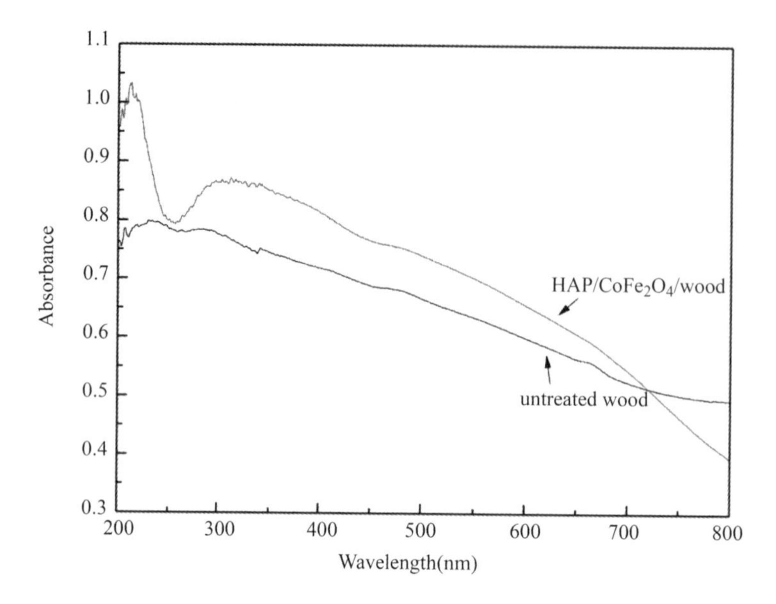

Fig. 9 UV-Vis diffuse reflection spectra of the untreated wood and $CoFe_2O_4$/HAP-modified wood composites

4 Conclusions

In this study, magnetic $CoFe_2O_4$ and HAP particles precipitated on the wood surface via a hydrothermal process, resulting into a functional wood/inorganic composites with excellent magnetic, thermal, mechanical and UV-resistance. The alkali treatment in preparation of magnetic $CoFe_2O_4$-modified wood material promoted the HAP formation. Owing to the excellent magnetic properties, some complex movements such as lift, drag, alignment and actuation can perform under the influence of magnetic field, and the improved thermal, mechanical properties and UV-resistance expanded the application of the modified magnetic wood composites in the field of special decoration.

Acknowledgments

This work was financially supported by The National Natural Science Foundation of China (grant no. 31470584).

References

[1]Ugolev B N. Wood as a natural smart material[J]. Wood Sci. Technol., 2014, 48(3): 553-568.

[2]Fratzl P, Weinkamer R. Nature's hierarchical materials[J]. Prog. Mater. Sci., 2007, 52(8): 1263-1334.

[3]Persson P V, Hafren J, Fogden A, et al. Silica nanocasts of wood fibers: a study of cell-wall accessibility and structure [J]. Biomacromolecules, 2004, 5(3): 1097-1101.

[4]Miyafuji H, Kokaji H, Saka S. Photostable wood-inorganic composites prepared by the sol-gel process with UV absorbent[J]. J. Wood Sci., 2004, 50(2): 130-135.

[5]Merk V, Chanana M, Gierlinger N, et al. Hybrid wood materials with magnetic anisotropy dictated by the hierarchical cell structure[J]. ACS Appl. Mater. Interfaces, 2014, 6(12): 9760-9767.

[6]Wang X Q, Liu J L, Chai Y B. Thermal, mechanical, and moisture absorption properties of wood-TiO_2 composites

prepared by a sol-gel process[J]. BioResources, 2012, 7(1): 893-901.

[7]Wang B L, Feng M, Zhan H B. Improvement of wood properties by impregnation with TiO_2 via ultrasonic-assisted sol-gel process[J]. RSC Adv., 2014, 4(99): 56355-56360.

[8]Devi R R, Gogoi K, Konwar B K, et al. Synergistic effect of nano TiO_2 and nanoclay on mechanical, flame retardancy, UV stability, and antibacterial properties of wood polymer composites[J]. Polym. Bull., 2013, 70(4): 1397-1413.

[9]Donath S, Militz H, Mai C. Weathering of silane treated wood[J]. Holz Roh werkst, 2007, 65(1): 35-42.

[10]Noodam J, Sirisathitkul C, Matan N, et al. Magnetic properties of NdFeB-coated rubberwood composites[J]. Int. J. Miner., Metall. Mater., 2013, 20(1): 65-70.

[11]Wang S L, Wang C Y, Liu C Y, et al. Fabrication of superhydrophobic spherical-like α-FeOOH films on the wood surface by a hydrothermal method[J]. Colloids Surf., A, 2012, 403: 29-34.

[12]Saka S, Ueno T. Several SiO_2 wood-inorganic composites and their fire-resisting properties[J]. Wood Sci. Technol., 1997, 31(6): 457-466.

[13]Sun Q F, Yu H P, Liu Y X, et al. Improvement of water resistance and dimensional stability of wood through titanium dioxide coating[J]. Holzforschung, 2010, 64(6): 757-761.

[14]Trey S, Olsson R T, Ström V, et al. Controlled deposition of magnetic particles within the 3-D template of wood: making use of the natural hierarchical structure of wood[J]. RSC Adv., 2014, 4(67): 35678-35685.

[15]Gan W T, Gao L K, Sun Q F, et al. Multifunctional wood materials with magnetic, superhydrophobic and anti-ultraviolet properties[J]. Appl. Surf. Sci., 2015, 332: 565-572.

[16]Lu A H, Salabas E L, Schuth F. Magnetic nanoparticles: synthesis, protection, functionalization, and application [J]. Angew. Chem. Int. Ed., 2007, 46(8): 1222-1244.

[17]Sun S H, Zeng H, Robinson D B, et al. Monodisperse MFe_2O_4(M= Fe, Co, Mn) nanoparticles[J]. J. Am. Chem. Soc., 2004, 126(1): 273-279.

[18]Hyeon T. Chemical synthesis of magnetic nanoparticles[J]. Chem. Commun., 2003, 8: 927-934.

[19]Gu H W, Xu K M, Xu C J, et al. Biofunctional magnetic nanoparticles for protein separation and pathogen detection [J]. Chem. Commun., 2006, 9: 941-949.

[20]Oka H, Fujita H. Experimental study on magnetic and heating characteristics of magnetic wood[J]. J. Appl. Phys., 1999, 85(8): 5732-5734.

[21]Oka H, Hojo A, Takashiba T. Wood construction and magnetic characteristics of impregnated type magnetic wood[J]. J. Magn. Magn. Mater., 2002, 239(1-3): 617-619.

[22]Oka H, Hamano H, Chiba S. Experimental study on actuation functions of coating-type magnetic wood[J]. J. Magn. Magn. Mater., 2004, 272: E1693-E1694.

[23]Oka H, Uchidate S, Sekino N, et al. Electromagnetic wave absorption characteristics of half carbonized powder-type magnetic wood[J]. IEEE Trans. Magn., 2011, 47(10): 3078-3080.

[24]Wan Y Z, Hong L, Jia S R, et al. Synthesis and characterization of hydroxyapatite-bacterial cellulose nanocomposites [J]. Compos. Sci. Technol., 2006, 66(11-12): 1825-1832.

[25]Nishikawa H. Surface changes and radical formation on hydroxyapatite by UV irradiation for inducing photocatalytic activation[J]. J. Mol. Catal. A-Chem., 2003, 206(1-2): 331-338.

[26]Wang N, Cai C J, Cai D Q, et al. Hydrothermal fabrication of hydroxyapatite on the PEG-grafted surface of wood from Chinese Glossy Privet[J]. Appl. Surf. Sci., 2012, 259: 643-649.

[27]Sundqvist B, Karlsson O, Westermark U. Determination of formic-acid and acetic acid concentrations formed during hydrothermal treatment of birch wood and its relation to colour, strength and hardness[J]. Wood Sci. Technol., 2006, 40(7): 549-561.

[28]Lu Y, Sun Q F, Liu T C, et al. Fabrication, characterization and photocatalytic properties of millimeter-long TiO_2

fiber with nanostructures using cellulose fiber as a template[J]. J. Alloys Compd., 2013, 577: 569-574.

[29]Gierlinger N, Goswami L, Schmidt M, et al. In situ FTIR microscopic study on enzymatic treatment of poplar wood cross-sections[J]. Biomacromolecules, 2008, 9: 2194-2201.

[30]Rana R, Langenfeld-Heyser R, Finkeldey R, et al. FTIR spectroscopy, chemical and histochemical characterisation of wood and lignin of five tropical timber wood species of the family of Dipterocarpaceae[J]. Wood Sci. Technol., 2010, 44(2): 225-242.

[31]Shi S K, Chen S Y, Zhang X, et al. Biomimetic mineralization synthesis of calcium-deficient carbonate-containing hydroxyapatite in a three-dimensional network of bacterial cellulose[J]. J. Chem. Technol. Biotechnol., 2009, 84(2): 285-290.

[32]He J H., Kunitake T, Nakao A. Facile in situ synthesis of noble metal nanoparticles in porous cellulose fibers[J]. Chem. Mater., 2003, 15(23): 4401-4406.

[33]Wan Y Z, Huang Y, Yuan C D, et al. Biomimetic synthesis of hydroxyapatite/bacterial cellulose nanocomposites for biomedical applications[J]. Mater. Sci. Eng. C, 2007, 27(4): 855-864.

[34]Borum L, Wilson O C. Surface modification of hydroxyapatite. Part II. Silica[J]. Biomaterials, 2003, 24(21): 3681-3688.

[35]Pan H H, Liu X Y, Tang R K, et al. Mystery of the transformation from amorphous calcium phosphate to hydroxyapatite[J]. Chem. Commun., 2010, 46(39): 7415-7417.

[36]Mohandes F, Salavati-Niasari M. Freeze-drying synthesis, characterization and in vitro bioactivity of chitosan/graphene oxide/hydroxyapatite nanocomposite[J]. RSC Adv., 2014, 4(49): 25993-26001.

[37]Koseoglu Y, Alan F, Tan M, et al. Low temperature hydrothermal synthesis and characterization of Mn doped cobalt ferrite nanoparticles[J]. Ceram. Int., 2012, 38(5): 3625-3634.

[38]Waldron R D. Infrared spectra of ferrites[J]. Phys. Rev., 1955, 99(6): 1727-1735.

[39]Li J, Lu Y, Yang D J, et al. Lignocellulose aerogel from wood-ionic liquid solution (1-allyl-3-methylimidazolium chloride) under freezing and thawing conditions[J]. Biomacromolecules, 2011, 12(5): 1860-1867.

[40]Wang D P, Duan X, Zhang J Y, et al. Fabrication of superparamagnetic hydroxyapatite with highly ordered three-dimensional pores[J]. J. Mater. Sci., 2009, 44(15): 4020-4025.

[41]Deng J, Mo L P, Zhao F Y, et al. Sulfonic acid supported on hydroxyapatite-encapsulated-γ-Fe_2O_3 nanocrystallites as a magnetically separable catalyst for one-pot reductive amination of carbonyl compounds[J]. Green Chem., 2011, 13(9): 2576-2584.

[42]Zhou Y Y, Qi P W, Zhao Z H, et al. Fabrication and characterization of fibrous HAP/PVP/PEO composites prepared by sol-electrospinning[J]. RSC Adv., 2014, 4(32): 16731-16738.

[43]Olsson R T, Samir M A S A, Salazar-Alvarez G, et al. Making flexible magnetic aerogels and stiff magnetic nanopaper using cellulose nanofibrils as templates[J]. Nat. nanotechnol., 2010, 5(8): 584-588.

[44]Sugimoto T, Matijevic E. Formation of uniform spherical magnetite particles by crystallization from ferrous hydroxide gels[J]. J. Colloid Interface Sci., 1980, 74(1): 227-243.

[45]Tamura H, Matijevic E. Precipitation of cobalt ferrites[J]. J. Colloid Interface Sci., 1982, 90(1): 100-109.

[46]Ma M, Zhang Y, Yu W, et al. Preparation and characterization of magnetite nanoparticles coated by amino silane [J]. Colloid. Surf., A, 2003, 212(2-3): 219-226.

[47]Nge T T, Sugiyama J. Surface functional group dependent apatite formation on bacterial cellulose microfibrils network in a simulated body fluid[J]. J. Biomed. Mater. Res., Part A, 2007, 81(1): 124-134.

[48]Zhao X L, Zhao H L, Yuan H H, et al. Multifunctional superparamagnetic Fe_3O_4@ SiO_2 core/shell nanoparticles: design and application for cell imaging[J]. J. Biomed. Nanotechnol., 2014, 10(2): 262-270.

[49]Yang Z P, Gong X Y, Zhang C J. Recyclable Fe_3O_4/hydroxyapatite composite nanoparticles for photocatalytic

applications[J]. Chem. Eng. J., 2010, 165(1): 117-121.

[50]Ma Z Y, Li F S, Bai H P. Effect of Fe_2O_3 in Fe_2O_3/AP composite particles on thermal decomposition of AP and on burning rate of the composite propellant[J]. Propell. Explos. Pyrot., 2006, 31(6): 447-451.

[51]Liu S L, Ke D N, Zeng J., et al. Construction of inorganic nanoparticles by micro-nano-porous structure of cellulose matrix[J]. Cellulose, 2011, 18: 945-956.

[52]Miyafuji H, Saka S. Na_2O-SiO_2 wood-inorganic composites prepared by the sol-gel process and their fire-resistant properties[J]. J. Wood Sci., 2001, 47(6): 483-489.

[53]Hu A M, Li M, Chang C K, et al. Preparation and characterization of a titanium-substituted hydroxyapatite photocatalyst[J]. J. Mol. Catal. A-Chem., 2007, 267(1-2): 79-85.

[54]Zhu Z R, Li X Y, Zhao Q D, et al. Surface photovoltage properties and photocatalytic activities of nanocrystalline $CoFe_2O_4$ particles with porous superstructure fabricated by a modified chemical coprecipitation method[J]. J. Nanopart. Res., 2011, 13(5): 2147-2155.

中文题目：$CoFe_2O_4$/羟基磷灰石沉淀法水热合成磁性木复合材料并改善木材性能

作者：甘文涛，高丽坤，詹先旭，李坚

摘要：磁性木材在电磁屏蔽、室内电磁波吸收和重金属吸附方面具有潜在的用途。本研究通过水热法制备了具有改善的热稳定性、力学性能和抗紫外性能的磁性木质复合材料。用傅里叶变换红外光谱、X射线衍射和扫描电子显微镜结合能量色散X射线分析(EDXA)研究了改性木材的官能团和形态。结果表明，磁性纳米二氧化钴在碱处理过程中沉淀在木材表面，然后通过钙离子与带负电荷的羟基磷灰石之间的静电相互作用在木材表面生成羟基磷灰石阴离子。此外，还对复合材料的磁性、热性能、力学性能和抗紫外线性能进行了评价。

关键词：溶胶-凝胶过程；无机复合材料；纳米颗粒；羟基磷灰石；表面

Mechanical Stable Superhydrophobic Epoxy/Silica Coating for Better Water Resistance of Wood*

Feng Liu, Zhengxin Gao, Deli Zang, Chengyu Wang, Jian Li

Abstract: A three-step procedure has been developed for superhydrophobic coating on wood based on epoxy/silica materials in combination with hydrophobization. Firstly, the epoxy resin is adhered to wood by immersing the samples into an epoxy resin acetone solution, then amino-functionalized silica particles are anchored by the epoxide groups, and finally, the created surface is modified by octadecyltrichlorosilane (OTS). The superhydrophobic wood surface is not only water repellent, as shown by contact angle tests, but also decreases essentially the wood's water absorption as determined by a 120-day water immersion test. The good mechanical stability of the coating was confirmed by a sand collision method.

Keywords: Coating; Mechanical Stability; Superhydrophobic; Water resistance; Wood

1 Introduction

The excellent properties of wood in outdoor applications are detracted by swelling and shrinking and the attack of microorganisms, while all these effects are moisture dependent. To keep the wood in optimal condition as long as possible, a proper protection is necessary (Denes et al., 1999; Schultz et al., 2007; Donath et al., 2007; De Vetter et al., 2009; Hsieh et al., 2011). One of the protection methods is coating of the wood surface.

Superhydrophobic surfaces with water contact angle (CA) higher than 150° and sliding angles less than 10° have attracted considerable attention because they are not only water repellent but are also easy to clean and have antifouling characteristics (Chapman et al., 2009; Chunder et al., 2009; Dorrer and Ruhe, 2009; Fu et al., 2012; Wang et al., 2013; She et al., 2013; Zhang et al., 2013; Lu et al., 2014; Wang et al., 2014). Best known are the natural superhydrophobic species such as lotus leaves, rose petals, butterfly's wing and water strider's legs (Sun et al., 2005; Feng et al., 2008). Superhydrophobic treatment of wood has a great potential in terms of better water and dust repellency, dimensional stability, and a longer service life (Mohammed-Ziegler et al., 2008), and research in this context was very intense in the last years in China (Wang et al., 2013; She et al., 2013; Zhang et al., 2013).

Coatings obtained by a combination of suitable hierarchical micro/nano-structures superimposed with materials having low surface free energy seem to be the key for obtaining superhydrophobic wood surfaces (Li et al., 2007; Lee and McCarthy, 2007; Koch et al., 2009; Feng et al., 2013). To this purpose, a number of synthetic strategies have been reported, including sol-gel processing (Su et al., 2011; Liu et al., 2014), anodic oxidation (Wu et al., 2009), electrodeposition (Yu et al., 2013). Various materials have

---

* 本文摘自 Holzforschung, 2015, 69(3): 565-572.

been tested, including polymers (Hurst et al., 2012; Zhou et al., 2012), metals (Rao et al., 2011; Feng et al., 2012), oxides (Li et al., 2011; Fu et al., 2012; Lu et al., 2014; Wang et al., 2014), fabric (Zhu et al., 2012), and composites (Wang et al., 2011a). Silica nanoparticles modified with long-chain alkylsilane are also useful for increasing hydrophobicity (Wang et al., 2013).

Superhydrophobic wood surfaces are also investigated in our laboratories. Wang et al. (Wang et al., 2011c) obtained such surfaces by a wet chemical method followed by the surface modification with stearic acid through synthetized zinc oxide lamellar particles. Moreover, superhydrophobic silica films with water CA of about 164° were obtained on wood surfaces by means of a sol-gel process and a self-assembly of 1H, 1H, 2H, 2H-perfluoroalkyltriethoxysilanes (POTS) monolayer (Wang et al., 2011b). The desired effects were achieved by an effective and straightforward hydrothermal reaction process of ferric sulfate urea superimposed by a self-assembly of octadecyltrichlorosilane (OTS) monolayer (Wang et al., 2012).

In most cases, the superhydrophobic wood surface was not suited for practical applications because of the limited mechanical stability of the layers created (Xiu et al., 2010). In general, the mechanical stability of superhydrophobic wood surfaces is neglected in the literature, though this is important for a long term self-cleaning and water repellency applications in the praxis. If the layer is destroyed irreversibly, the decreasing CAs are indicative for deteriorating water repellency properties (Deng et al., 2011).

Accordingly, the mechanical stability of the developed superhydrophobic wood surfaces was also in focus in the present study. The wood samples should be prepared in a three-step process: 1. Coating the wood with epoxy resin. 2. Anchoring silica particles, and 3. Coating with a layer having low surface energy. In step one, the epoxy resin will be adhered to wood by immersing the samples into an epoxy resin acetone solution. In step two, the silica particles will be anchored through the strong bond between epoxide group of the resin and amino group of amino-functionalized silica particles. Finally, the prepared wood surfaces will be modified by the OTS reagent.

2 Materials and methods

2.1 Materials

Larch lumbers, grown in Harbin, were obtained from the sapwood of larch [*Larix gmelinii* (Ruprecht) Kuzeneva]. Tetraethoxysilane (TEOS) (98.0%), EtOH (99.7%), ammonia (28.0%) acetone (99.0%) were purchased from Tianjin Kaitong Chemical Reagent Co., Ltd. Bisphenol A epoxy resin (E-54) was provided by Wuxi Resin Plant. 3-Triethoxysilylpropylamine (APS) was purchased from Aladdin Chemistry Co., Ltd. Octadecyltrichlorosilane (OTS) used for surface hydrophobic modification was purchased from New Jersey. Deionized water was self-made. All of the chemicals were used as received without further purification.

2.2 Synthesis

The silica particles were produced following the method of Stöber et al. (1968). Briefly, 20 mL of TEOS was added dropwise to a flask containing 10 mL of ammonia solution (28.0%, catalyst) under magnetic stirring, 20 mL deionized water and 180 mL of EtOH. The stirring was continued for 2 h, and then the solution was placed 12 h at room temperature (r. t.). The particles were separated by centrifugation and the supernatant was discarded. The particles were then washed by EtOH three times. The white power was dried at

60 ℃ in an oven overnight.

Then 4. 0 g silica particles were redispersed into 100 mL anhyd. EtOH, and APS hydrolysate was added dropwise to the silica suspension with vigorous stirring. APS hydrolysate was prepared by mixing the 5 mL APS, 5 mL deionized water, and 25 mL anhyd. EtOH with a magnetic stirrer for 1. 5 h at r. t. The silica suspension was stirred at 60 ℃ for 12 h. The silica particles were then separated by centrifugation and washed with EtOH three times. The amino-functionalized silica particles were dried at 60 ℃ for 12 h. Fig. 1 shows the grafting of APS molecule onto the silica particles surface.

Each piece of larch wood was cut into blocks of 10×3×0. 15 cm^3. The wood samples were ultrasonically washed with acetone, EtOH and deionized water for 15 min, respectively. Then, the wooden substrates dried at 60 ℃ in an oven overnight.

SiO_2 particle　　amino-functionalized silica particle

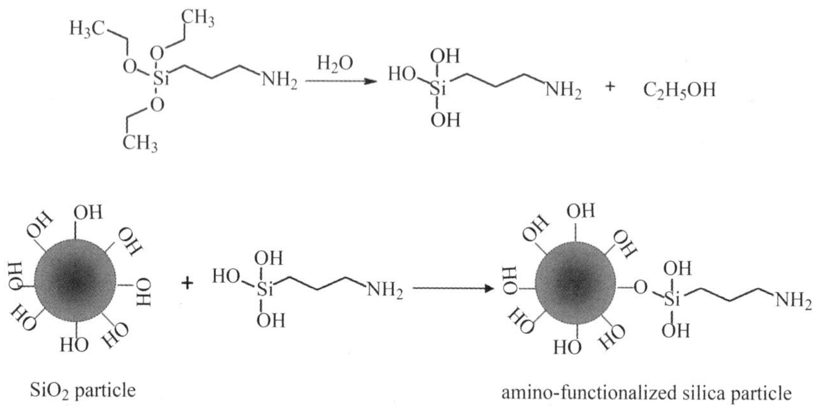

Fig. 1　Amino-functionalization of silica particle

As detailed in Fig. 2, the wood sample was first immersed in the epoxy resin acetone solution (4%, m/v) at r. t. for 1. 5 h, and then dried with nitrogen. Subsequently, the silica particles were fixed in the surface of epoxidized wood sample, by immersing them in a 0. 5% (wt/vol) amino-functionalized silica aqueous solution for 1. 5 h. Ultimately, the surface was washed by deionized water for three times, blown to dry by nitrogen and dried in a vacuum drying oven at 80 ℃ for 12 h.

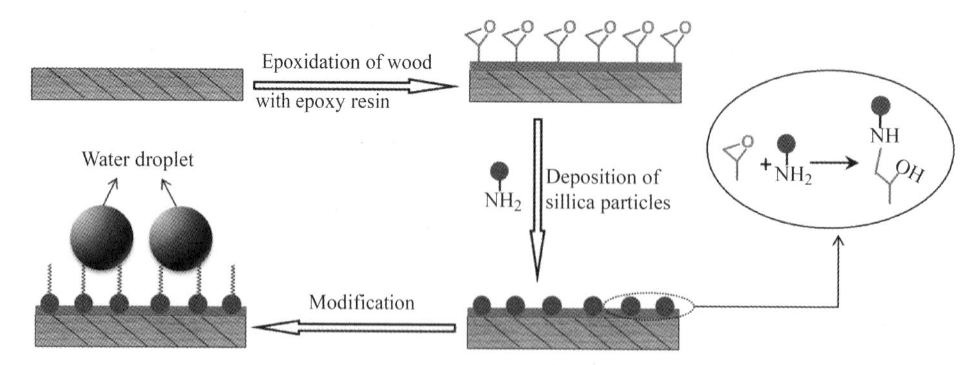

Fig. 2　The schematic diagram of the fabricating process for superhydrophobic wood surface

The last step was completed with a self-assembly OTS monolayer. First, the OTS EtOH solution was prepared by magnetic stirring the mixture of the 100 mL anhyd. EtOH, 2 mL OTS, 0. 25 mL H_2O, 0. 05 mL glacial acetic acid at r. t. for 4 h. Then, the sample was immersed into the OTS EtOH solution. The modification was maintained at 60 ℃ for 2 h. The resulting surface was washed several times with anhyd. EtOH, and dried in a drying oven.

2.3 Characterizations

The surface morphologies were observed by scanning electron microscopy (SEM, FEI QUANTA200) operating at 12.5 kV. Contact angle (CA) and CA hysteresis measurements were performed with an optical CA meter (Hitachi, CA-A) equipped with a CCD camera for image capture at r. t., and the average value of five measurements taken at different places of the sample was calculated.

The wood samples were immersed into the deionized water at r. t. and weighed and measured after 1, 2, 4, 8, 10, 20, 40, 60, 80, 100 and 120 days of immersion (im.). The weight percentage gain *WPG* was calculated as $\%WPG = (W_{afterim.} - W_{beforeim.}) / W_{beforeim.}$ (Sun et al., 2010).

The sand collision test for determination of the mechanical stability is illustrated in Fig. 3 (Deng et al., 2011). 50 g commercial sand grain (diameter: 100 μm to 300 μm) are impacted on a superhydrophobic wood surface from a height of 30 cm, while the wood sample was fixed at 45° to the horizontal surface. This process of sand impact lasted about 25 sec. The sand collision test is inspired by Standard ASTM D 968-05 and Deng et al. (2011). It simulated the damage produced by the impact of solid particles that fall under gravity. The changes of the CA and CA hysteresis of the superhydrophobic surface were measured after collision test.

In addition, the sand collision test was also performed on two kinds of superhydrophobic wood surfaces prepared by Wang et al. (2011b and c) for comparison. The ZnO-wood surface was prepared by immersing the sample into 150 mL solution containing 0.33 g of zinc acetate dehydrate and 2 mL of triethylamine for 24 h at r. t. and subsequent stearic acid modification (Wang et al., 2011c). The SiO_2-wood surface was obtained through immersing the wood sample into the mixture solution of 90 mL EtOH, 10 mL TEOS, 10 mL deionized water and 10 mL NH_4OH for 6 h at r. t. and subsequent self-assembly of POTS monolayer (Wang et al., 2011b).

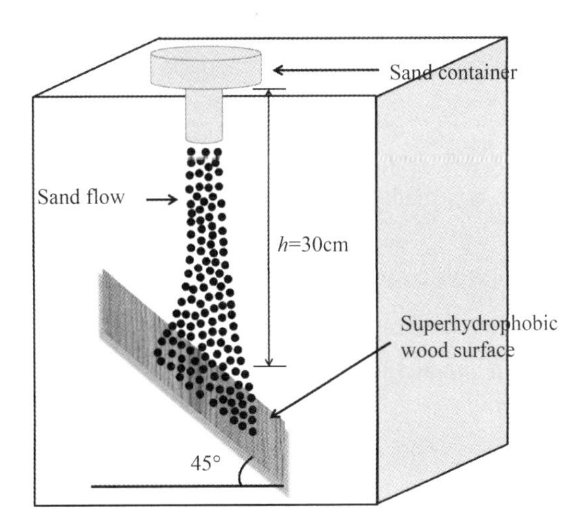

Fig. 3 Sketch of the collision test invoked to evaluate mechanical stability of the superhydrophobic wood surface against sand collision

3 Results and discussion

3.1 Microstructure of the epoxy/silica coating

Fig. 4 shows SEM images of an untreated wood surface, the one covered with epoxy/silica coating at low and high magnification, respectively. Fig. 4a, b clearly reveals that untreated larch wood is a heterogeneous material with a smooth surface. As shown in Fig. 4c, d, a comparatively homogenous layer of epoxy/silica coating was present on the surface after modification, in which epoxy resin acts as an adhesive for anchoring the silica particles. The adhesive forces between wood surface and epoxy resin are strong. Fig. 4e demonstrates that a great deal of silica particles were imbedded in the epoxy resin on wood surface in a random distribution with diameters ranging from 300 to 500 nm, which significantly roughen the wood surface. After coating with a layer of low surface energy OTS film, a larger amount of air is trapped into the interspaces of the wood surface, and the water droplets are mainly in contact with the trapped air (the CA of a liquid with air is assumed to be 180°).

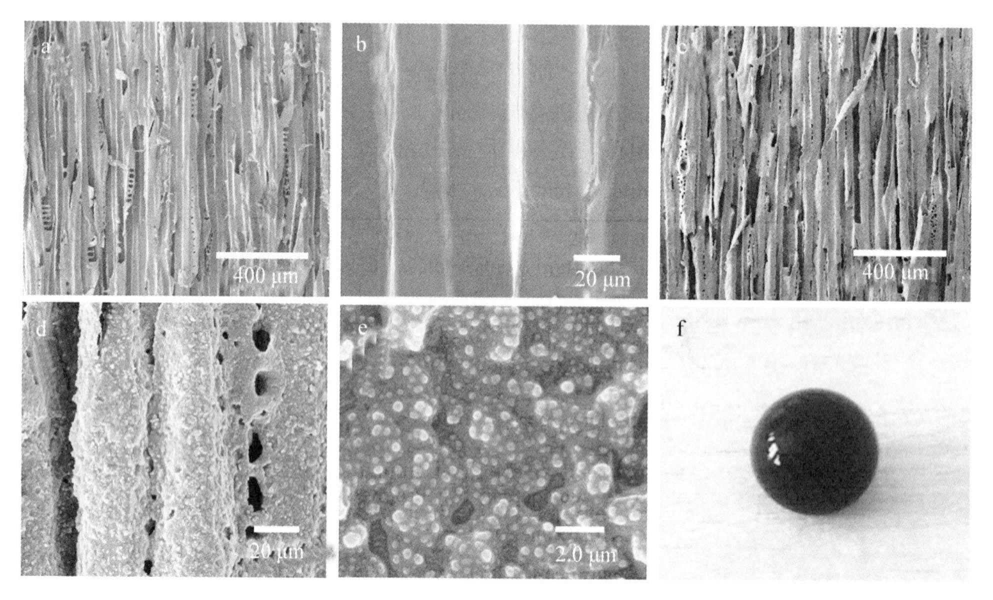

Fig. 4 SEM images of untreated larch wood surface at (a) low and (d) high magnification and the surface of superhydrophobic wood at (c) low and (d, e) high magnification. (f) Photographs of the static water droplets (dyed with methylene blue) on superhydrophobic wood surface

3.2 Wettability of the superhydrophobic surface

The wettability was evaluated by CA measurement on the wood surface. If the water CA is <90°, the surface is hydrophilic, and above 90°, hydrophobic. If the water CA is >150°, the surface is superhydrophobic (Wang et al., 2013). The untreated wood is hydrophilic with CA < 90°. The modification with OTS monolayer can achieve a maximal CA about 120°, which is equivalent to a hydrophobic material, but which does not mean superhydrophobicity. On the wood surface covered with epoxy/silica coating without OTS modification, the water droplets can spread out quickly and have CAs nearly 0°. However, after the epoxy/silica coating modified by OTS, the wood surface is highly hydrophobic. By combining the rough structure and the long - chain hydrophobic alkyls, the CA increased to 154°, and the sliding angle was 7°. The superhydrophobicity is the combined effect of the rough surface with the trapped air in the pores.

Generally, Young equation describes well the wettability of an absolutely smooth surface (Liu et al., 2013), Fig. 5a:

$$\cos\theta = \frac{\sigma_{sv} - \sigma_{sl}}{\sigma_{lv}} \tag{1}$$

Where σ_{sv}, σ_{sl}, *and* σ_{lv} are the interfacial tensions of the solid-vapor, solid-liquid, and liquid-vapor phases, respectively. In reality, surfaces are not absolutely smooth, thus the Wenzel equation (Eq. 2) is better suited with a CA θ_w of liquids on a rough homogeneous surface (Fig. 5b):

$$\cos\theta_w = r\left(\frac{\sigma_{sv} - \sigma_{sl}}{\sigma_{lv}}\right) = r\cos\theta \tag{2}$$

Where r is the roughness factor defined as the ratio of the actual surface area of the rough surface to the geometric projected area, and it is always larger than 1. For an in-depth consideration of wettability (Fig. 5c), the Cassie equation was also employed:

$$\cos\theta_c = f(\cos\theta + 1) - 1 \tag{3}$$

Where f is the fraction of the solid surface in contact with liquid, and the fraction of air in contact with liquid is $1-f$, θ and θ_c represent the CA on smooth and rough surfaces, respectively. Eq. 3 can be employed to calculate the fraction of the air pockets at the superhydrophobic wood surface (Liu et al., 2013). Here, the value of water CA θ_c on the superhydrophobic surface is 154°, and the water CA θ on the smooth wood surface (modified by OTS) is 120°. Therefore, the fraction of air is calculated by Eq. 3 to be 0.8.

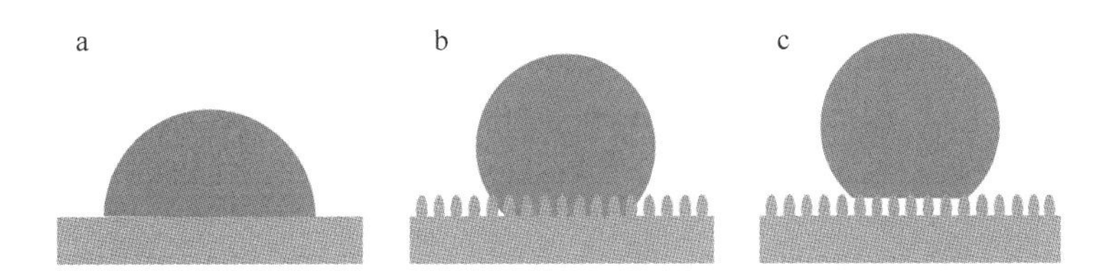

Fig. 5 Effect of surface structure on the wetting behavior of solid substrates: (a) droplet on flat substrate, Young model; (b) Wenzel model; (c) Cassie model

3.3 Long-time water immersion test

Fig. 6 shows the weight percentage gain (*WPG*) at the initial 20 days, after which *WPG* reaches a steady state. The WPG_{max} of wood treated with epoxy/silica samples only reached 5.7% after 120-day water immersion, whereas the untreated wood absorbed up to 155.1% water, i.e. the water absorption was reduced by a factor of 27. The superhydrophobic treatment does not only reduce the water uptake rate but also the moisture content at the equilibrium.

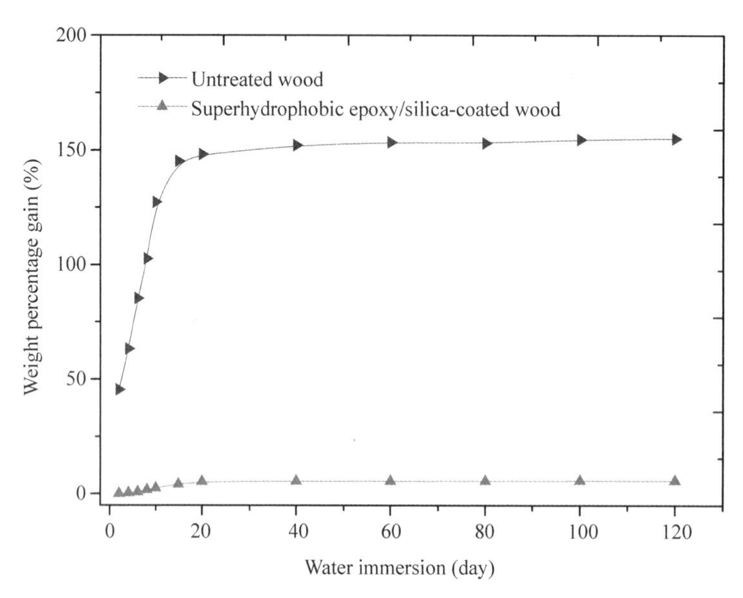

Fig. 6 The weight percentage gain of untreated wood and superhydrophobic epoxy/silica-coated wood during 120-days of water immersion experiment

3.4 Mechanical stability by sand collision test

Abrasion deteriorates the hydrophobicity and the self-cleaning effect (Liu et al., 2013) and this can be tested by sand collision. Fig. 7a shows the CAs of the prepared wood surfaces with increasing concentration of epoxy resin. The CA = 146° on the surface prepared with 0.5% epoxy resin, up to maximum contact angles CA = 159° for 1% epoxy resin. After that, the CAs decrease with increasing epoxy resin concentration but the CA remains above 150° up to 4% concentration before it is lowered again gradually. Obviously, epoxy concentration of 1% is a good choice for good water repellency.

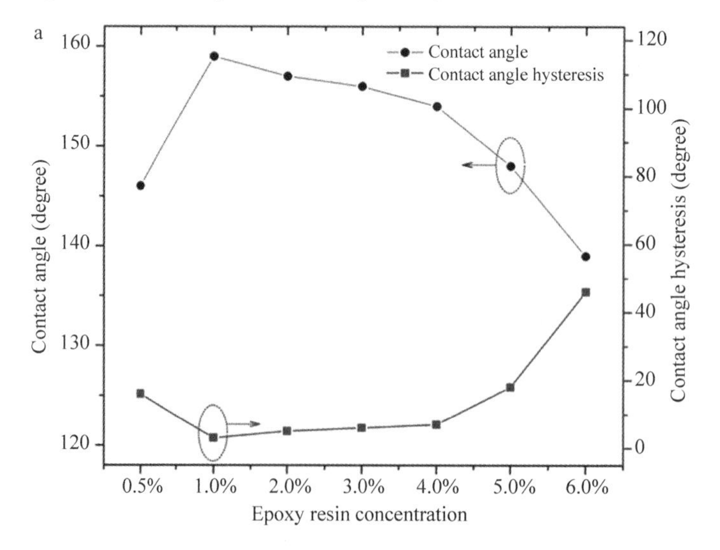

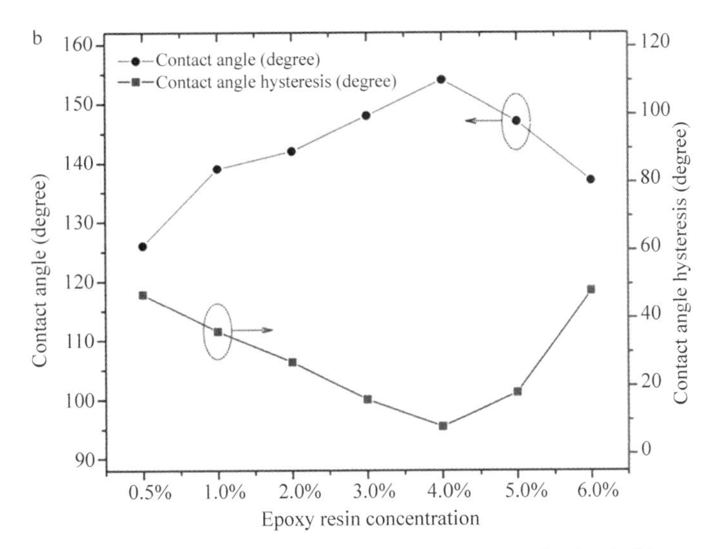

Fig. 7 Water Contact angles and contact angle hysteresis on superhydrophobic wood surface coated with different epoxy resin concentrations (a) before collision test and (b) after collision test

However, in the sand collision test, the CAs dropped below 150° except in cases when the epoxy resin concentration was 4% (Fig. 7b). Obviously, the amount of available epoxy groups influence the anchoring strengh of the micro-silica particles. Fig. 8a, b and Fig. 8e, f show the changes of the SEM images of surfaces obtained by 1 and 4% epoxy treatment before and after sand collision test. Deep grooves are visible between the

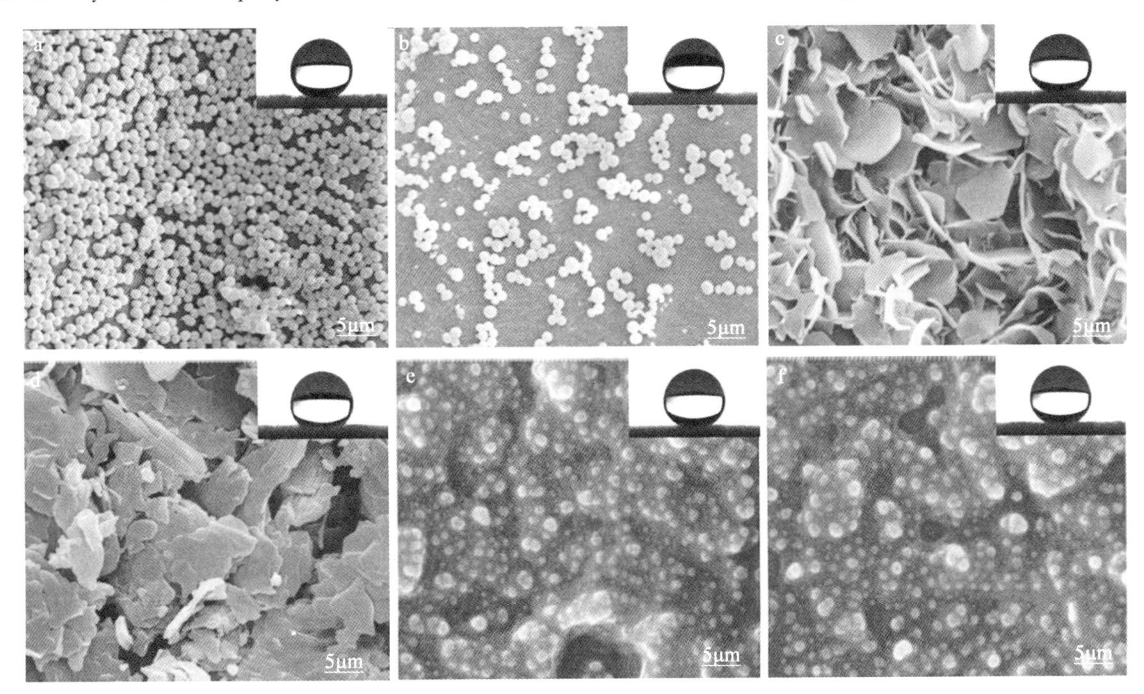

Fig. 8 SEM images of superhydrophobic wood surfaces prepared by epoxy/silica coating (a) before and (b) after collision test at the epoxy resin concentration of 1.0%; SEM images of superhydrophobic wood surfaces prepared by lamellar ZnO coating (c) before and (d) after collision test; SEM images of superhydrophobic wood surfaces prepared by epoxy/silica coating (e) before and (f) after collision test at the epoxy resin concentration of 4.0%

micro-particles. Nevertheless, at an epoxy concentration of 4% (Fig. 8e, f) the grooves are less pronounced, thus at this concentration is the mechanical stability of the surface better. For comparison, the sand collision test was also performed on the superhydrophobic surfaces described earlier (Wang et al., 2011b, c) (Table 1). As shown, these samples lost a lot of their hydrophobicity upon sand collision, i. e. the CA decreased from above 150° to below 140° (ZnO-wood and SiO_2-wood coatings) and the increase of the CA hysteresis was more than 60°. The water droplets adhered to the surface even in deep angles, i. e. the self-cleaning effect on the ZnO - wood and SiO_2 - wood surface has vanished after abrasion. The surface of ZnO- wood, the microstructure changed dramatically after sand collision test (Fig. 8c, d). The uniformly distributed lamellar ZnO particles are destroyed because of the brittleness of this inorganic oxide (Liu et al., 2013). The lamellar ZnO structures are dashed into pieces and are no longer erected (Fig. 8d).

The CA and CA hysteresis of the superhydrophobic epoxy/silica coatings are almost unchanged after sand collision test (Fig. 8f, Table 1) and the spherical silica particles are still imbedded in the epoxy resin in a random distribution.

Table 1 Contact angles (CA) and CA hysteresis angles before and after sand collision test.

Wood surfaces with	Lite-rature	Contact angles (°)		CA hysteresis angles (°)	
		before	after	before	after
Epoxy/silica	Present	154±1.5	153±1.5	7	8±1.5
ZnO	1	151	135±5	~5	>60
SiO_2	2	164	124±3	~3	>6

1: Wang et al. 2011c; 2: Wang et al. 2011b.

In reality, just because the superhydrophobic wood is fabricated in the laboratories does not mean it can be applied in practice. When superhydrophobic wood is exposed to the atmosphere during practical applications, it is easy to be influenced by dirt, impurities and especially hard thing impact, resulting in deteriorating water repellency properties. However, up to date, there is no a uniform and standard method to evaluating the mechanical stability of superhydrophobic surface. Therefore, the sand collision test, inspired by Standard ASTM D 968-05 and Deng et al. (2011), is designed in this paper.

In this paper, the as - prepared superhydrophobic epoxy/silica coating on wood surface is not unbreakable. However, based on the results from the sand collision test, which compared with several other superhydrophobic woods, the superhydrophobic epoxy/silica coating is undoubtedly successful and its mechanical stability obviously has a good improvement.

4 Conclusions

A superhydrophobic epoxy/silica coating on wooden substrates shows a super repellency toward water on the surface and the water absorption capacity of the wood body is radically lowered in water immersion tests. The simple and facile immersion method increases the commercial feasibility. The enhancement of the water-repellency can be attributed to the fact that the water droplets partially sit on OTS-epoxy/silica coating and that there are air pockets underneath the droplets. Promising is also the mechanical stability of the described superhydrophobic surfaces, which can be attributed to the strong adhesive power of epoxy resins, which anchor the silica particles. The commercial application of the method presented depends mainly on the economical viability of the process steps with the expensive chemicals.

Acknowledgments

This research was supported by the Fundamental Research Funds for the Central Universities (DL12EB05-01).

References

[1] Chapman G M, Bai H, Li C, et al. Micro-nanoscale binary structured silver films fabricated by electrochemical deposition[J]. Mater Chem. Phys., 2009, 114: 120-124.

[2] Chunder A, Etcheverry K, Londe G, et al. Conformal switchable superhydrophobic/hydrophilic surfaces for microscale flow control[J]. Colloids Surf. A, 2009, 333: 187-193.

[3] De Vetter L, Stevens M, Van Acker J. Fungal decay resistance and durability of organosilicon-treated wood[J]. Int. Biodeter Biodegr, 2009, 63: 130-134.

[4] Denes A R, Tshabalala M A, Rowell R, et al. Hexamethyldisiloxane-plasma coating of wood surfaces for creating water repellent characteristics[J]. Holzforschung, 1999, 53: 318-326.

[5] Deng X, Mammen L, Zhao Y, et al. Transparent, Thermally Stable and Mechanically Robust Superhydrophobic Surfaces Made from Porous Silica Capsules[J]. Adv. Mater, 2011, 23: 2962-2965.

[6] Donath S, Militz H, Mai C. Creating water-repellent effects on wood by treatment with silanes[J]. Holzforschung, 2006, 60: 40-46.

[7] Dorrer C, Ruhe J. Some thoughts on superhydrophobic wetting[J]. Soft Matter, 2009, 5: 51-61.

[8] Feng L, Liu Y, Zhang H, et al. Superhydrophobic alumina surface with high adhesive force and long-term stability [J]. Colloids Surf. A, 2012, 410: 66-71.

[9] Feng L, Zhang H, Wang Z, et al. Superhydrophobic aluminum alloy surface: Fabrication, structure, and corrosion resistance[J]. Colloids Surf A, 2014, 441: 319-325.

[10] Feng L, Zhang Y, Xi J, et al. Petal Effect: A Superhydrophobic State with High Adhesive Force[J]. Langmuir, 2008, 24: 4114-4119.

[11] Fu Y, Yu H, Sun Q, et al. Testing of the superhydrophobicity of a zinc oxide nanorod array coating on wood surface prepared by hydrothermal treatment[J]. Holzforschung, 2012, 66: 739-744.

[12] Hsieh C T, Chang B S, Lin J Y. Improvement of water and oil repellency on wood substrates by using fluorinated silica nanocoating[J]. Appl. Surf. Sci., 2011, 257: 7997-8002.

[13] Hurst S M, Farshchian B, Choi J, et al. A universally applicable method for fabricating superhydrophobic polymer surfaces[J]. Colloids Surf. A, 2012, 407: 85-90.

[14] Koch K, Bhushan B, Barthlott W. Multifunctional surface structures of plants: An inspiration for biomimetics[J]. Prog. Mater Sci., 2009, 54: 137-178.

[15] Lee J A, McCarthy T J. Polymer Surface Modification Topography Effects Leading to Extreme Wettability Behavior [J]. Macromolecules, 2007, 40: 3965-3969.

[16] Li J, Liu X, Ye Y, et al. Gecko-inspired synthesis of superhydrophobic ZnO surfaces with high water adhesion[J]. Colloids Surf. A, 2011, 384: 109-114.

[17] Li X M, Reinhoudt D, Crego-Calama M. What do we need for a superhydrophobic surface? A review on the recent progress in the preparation of superhydrophobic surfaces. Chem[J]. Soc. Rev., 2007, 36: 1350-1368.

[18] Liu F, Ma M, Zang D, et al. Fabrication of superhydrophobic/superoleophilic cotton for application in the field of water/oil separation[J]. Carbohyd Polym., 2014, 103: 480-487.

[19] Liu F, Wang S, Zhang M, et al. Improvement of mechanical robustness of the superhydrophobic wood surface by

coating PVA/SiO_2 composite polymer[J]. Appl. Surf. Sci., 2013, 280: 686-692.

[20]Lu Y, Xiao S, Gao R, et al. Improved weathering performance and wettability of wood protected by CeO_2 coating deposited onto the surface[J]. Holzforschung, 2014, 68: 345-351.

[21]Mohammed-Ziegler I, Tánczos I, Hórvölgyi Z, et al. Water-repellent acylated and silylated wood samples and their surface analytical characterization[J]. Colloids Surf. A, 2008, 319: 204-212.

[22]Rao A V, Latthe S S, Mahadik S A, et al. Mechanically stable and corrosion resistant superhydrophobic sol-gel coatings on copper substrate[J]. Appl. Surf. Sci., 2011, 257: 5772-5776.

[23]Schultz T P, Nicholas D D, Ingram L. Laboratory and outdoor water repellency and dimensional stability of southern pine sapwood treated with a waterborne water repellent made from resin acids[J]. Holzforschung, 2007, 61: 317-322.

[24]She Z, Li Q, Wang Z, et al. Researching the Fabrication of Anticorrosion Superhydrophobic Surface on Magnesium Alloy and its Mechanical Stability and Durability[J]. Chem. Eng. J, 2013, 228: 415-424.

[25]Standard ASTM D 968-05. Standard test methods for abrasion resistance of organin coatings by falling abrasive. 2005.

[26]Stöber W, Fink A, Bohn E. Controlled growth of monodisperse silica spheres in the micron size range[J]. J. Colloid Interf Sci., 1968, 26: 62-69.

[27]Su D, Huang C, Hu Y, et al. Preparation of superhydrophobic surface with a novel sol-gel system[J]. Appl. Surf. Sci., 2011, 258: 928-934.

[28]Sun Q, Yu H, Liu Y, et al. Improvement of water resistance and dimensional stability of wood through titanium dioxide coating[J]. Holzforschung, 2010, 64: 757-761.

[29]Sun T, Feng L, Gao X, et al. Bioinspired Surfaces with Special Wettability[J]. Accounts Chem. Res., 2005, 38: 644-652.

[30]Wang C, Piao C, Lucas C. Synthesis and characterization of superhydrophobic wood surfaces[J]. J., Appl., Polym., Sci., 2011, 119: 1667-1672.

[31]Wang S, Liu C, Liu G, et al. Fabrication of superhydrophobic wood surface by a sol-gel process[J]. Appl. Surf. Sci., 2011, 258: 806-810.

[32]Wang S, Shi J, Liu C, et al. Fabrication of a superhydrophobic surface on a wood substrate[J]. Appl. Surf. Sci., 2011, 257: 9362-9365.

[33]Wang S, Wang C, Liu C, et al. Fabrication of superhydrophobic spherical-like α-FeOOH films on the wood surface by a hydrothermal method[J]. Colloids Surf. A, 2012, 403: 29-34.

[34] Wang X, Chai Y, Liu J. Formation of highly hydrophobic wood surfaces using silica nanoparticles modified with long-chain alkylsilane[J]. Holzforschung, 2013: 67: 667-672.

[35]Wang N, Fu Y, Liu Y, et al. Synthesis of aluminum hydroxide thin coating and its influence on the thermomechanical and fire-resistant properties of wood[J]. Holzforschung, 2014, 68: 7

[36]Wu W, Wang X, Wang D, et al. Alumina nanowire forests via unconventional anodization and super-repellency plus low adhesion to diverse liquids[J]. Chem. Commun, 2009, 1043-1045.

[37]Xiu Y, Liu Y, Hess D W, et al. Mechanically robust superhydrophobicity on hierarchically structured Si surfaces[J]. Nanotechnology, 2010, 21: 155705.

[38]Yu Q, Zeng Z, Zhao W, et al. Fabrication of adhesive superhydrophobic Ni-Cu-P alloy coatings with high mechanical strength by one step electrodeposition[J]. Colloids Surf. A, 2013, 427: 1-6.

[39]Zhang X, Geng T, Guo Y, et al. Facile fabrication of stable superhydrophobic SiO_2/polystyrene coating and separation of liquids with different surface tension[J]. Chem. Eng. J., 2013, 231: 414-419.

[40]Zhou X, Zhang Z, Xu X, et al. Facile fabrication of recoverable and stable superhydrophobic polyaniline films[J]. Colloids Surf. A, 2012, 412: 129-134.

[41]Zhu X, Zhang Z, Yang J, et al. Facile fabrication of a superhydrophobic fabric with mechanical stability and easy-

repairability[J]. J. Colloid Interf. Sci., 2012, 380: 182-186.

中文题目：机械稳定性的超疏水环氧/硅树脂涂层，提高木材的耐水性

作者：刘峰，高正新，臧德利，王成毓，李坚

摘要：研究了以环氧树脂/二氧化硅材料为基材的木材超疏水涂层与疏水相结合的三步法。首先，将样品浸入环氧树脂丙酮溶液中，将环氧树脂黏在木材上，其次用环氧基团锚定氨基功能化的二氧化硅颗粒，最后，用十八烷基三氯硅烷(OTS)修饰生成的表面。超疏水表面不仅具有接触角(CA)试验所示的拒水性，而且120天水浸试验所示的木材吸水率也基本下降。用冲砂法验证了涂层良好的力学稳定性。

关键词：涂层；机械稳定性；超疏水；耐水；木材

特殊润湿性油水分离材料的研究进展*

李　坚，张　明，强添刚

摘要：近几年，石油和有机化学品泄漏对人们的生存环境乃至生态系统造成了极为严重的损害，油水分离手段及其材料的研发已成为一个全球性的，需要众人为之努力的重要任务。受到大自然的启发，许多具有特殊润湿性的材料得以合成与发展，尤其是超疏水—超亲油、超亲水—超疏油材料已成功用于选择性油水分离，并显示出诱人的应用前景。简要介绍了自然界的特殊浸润现象以及微—纳多级结构的制备方法；综述了以金属网、海绵、无机粉体为基材的新型油水分离材料；提出了该特殊材料存在的问题与拟解决办法；重点介绍并总结了以生物质材料如木粉、秸秆粉、滤纸、棉花、棉织物等作为基材，合成特殊润湿性油水分离生物质材料的优势与发展现状，并对该领域的研究趋势进行了展望。

关键词：油水分离；特殊润湿性；超疏水/超亲油性；超亲水/超疏油性；生物质材料

据统计，每年世界上高达1000万吨的油类通过食品、医药、纺织、皮革、机械加工、交通运输、石油开采、石油化工等途径流入海洋。油类产品多轻于河流与海洋，会在水表面形成油膜，油膜阻碍氧气进入水体，使水体缺氧，造成水生生物大量死亡，而且油被冲到海滩，也会对海滩上的其他生物造成严重的影响。而环境治理、油类回收以及进一步水的循环利用，均须对含油污水进行有效地分离处理，这就迫使人类亟需开发大量优良的吸油材料。根据油污在水中的存在形态，可分为4种[1-2]：浮油(粒径大于150 μm)、分散油(粒径在20~150 μm)、称为乳化油(粒径小于20 μm)、溶解油(粒径小于几微米)。传统的含油污水的处理方法有絮凝法，浮选法，刮渣法，破乳法和重力分离法等。但这些方法存在添加过多、化学药剂造成二次污染、分离效率不高、能耗高、费用高昂等缺点[2]。

自然界的生物体经过数十亿的物竞天择，其结构与功能已经趋于完美。因此，向大自然学习，模仿自然界生物体的功能，构建类似甚至超越自然界生物体功能的，具备高效、低能耗、环境友好以及新性能的仿生材料，将是人类发展进程中的永恒课题。1997年，BARTHLOTT et al.[3]与NEINHUIS et al.[4]对多种植物表面微观结构进行了观察与研究，结果显示荷叶表面存在许多乳头状的凸起结构，而该结构进一步被大量纳米级蜡状晶体所覆盖，如图1(b)所示[5]。2002年，FENG et al.[6]在荷叶的乳突结构表面进一步发现了大量密集的纳米柱状结构，其简图如图1(c)所示[5]。由此得出结论：荷叶效应的根本原因在于其表面微/纳二元分级结构与低表面能物质的有机结合。如今，随着纳米技术与仿生科学的蓬勃发展，通过在材料表面构建微/纳二元分级结构，并进一步以低表面能物质进行修饰来制备超疏水性材料已成为可能。目前，常用技术有溶胶—凝胶技术[7]、化学气相沉积技术[8]、辐射接枝法[9]、电化学沉积技术[10]、刻蚀技术[11]、模板法[12]、相分离技术[13]、熔融—冷却凝固成型技术[14]、层层自组装技术[15]、静电纺丝技术[16]、聚合物成膜法[17]等。已经挖掘的应用领域包括自清洁表面[18]、太阳能电池[19]、微流体器件[20]、防污[21]、防雾[22]、抗结冰[23]、减阻[24]、减低细

* 本文摘自森林与环境学报，2016，36(3)：257-265.

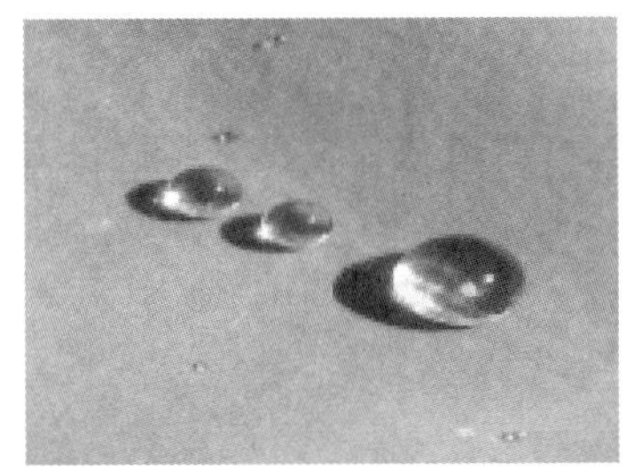

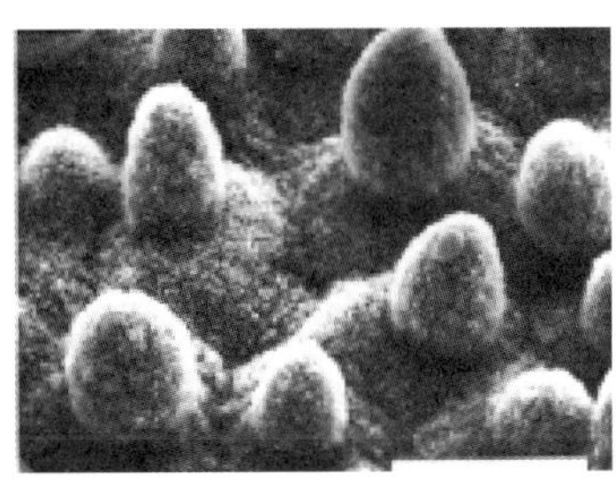

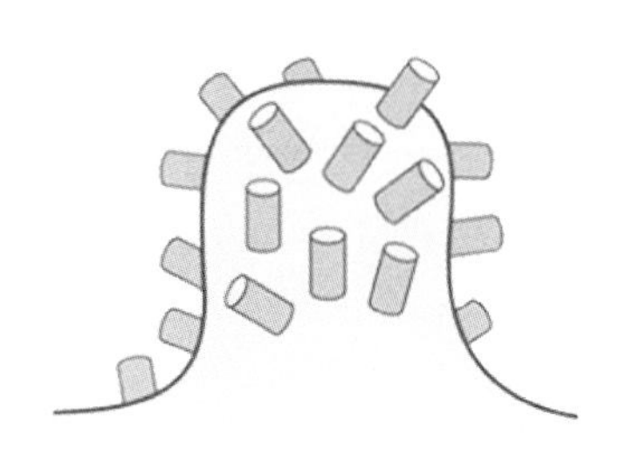

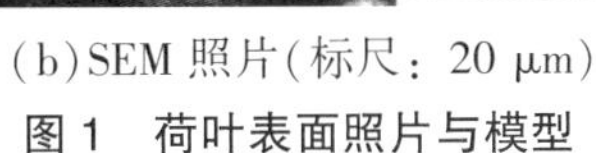

(a)普通照片　(b)SEM 照片(标尺：20 μm)　(c)微结构模型

图 1　荷叶表面照片与模型

菌粘附[25]、水收集[26]、油水分离[27]。其中，油水分离领域是具有特殊浸润性的表面的一个重要应用，也是本文主要的研究内容。

1　特殊润湿性油水分离材料

随着超疏水性材料合成技术的不断完善，人们已不满于仅仅追求功能单一的超疏水性材料，如何开发智能超疏水性材料以及多功能超疏水性材料，成为世界各国专家与学者们的主要研究课题。如图 2 所示，固体表面的特殊润湿性能包括以下 8 种：超疏水性(Super-Hydrophobicity)、超亲水性(Super-Hydrophilicity)、超疏油性(Super-Oileophilicity)、超亲油性(Super-Oileophilicity)、超双疏性(Super-Amphiphobicity)、超双亲性(Super-Amphiphilicity)、超疏水—超亲油性(Superhydrophobicity-Superoileophilicity)、超亲水—超疏油性(Superhydrophilicity-Superoileophilicity)[28]。另外，存在 2 种特殊情况：超疏水—超亲水性在特定情况下的相互转换，以及超疏油—超亲油性在特定情况下的相互转换，即智能开关(Smart Switch)。以上 10 中材料，其特殊润湿性能得以成功表达的关键，在于材料表面微/纳二元分级结构与功能材料修饰的协同作用。目前，前 8 中特殊润湿性材料的制备技术渐渐趋于成熟。其中，超疏水—超亲油性(近 100%除水)或超亲水—超疏油性(近 100%除油)功能的新材料的开发，更是为油水分离(Water-Oil Separation)材料界带来新鲜的血液。而该特殊材料也因其无论在分离速度还是分离精度方面均优于传统分离手段而备受期待，应作为未来油水分离材料研究的主要内容。

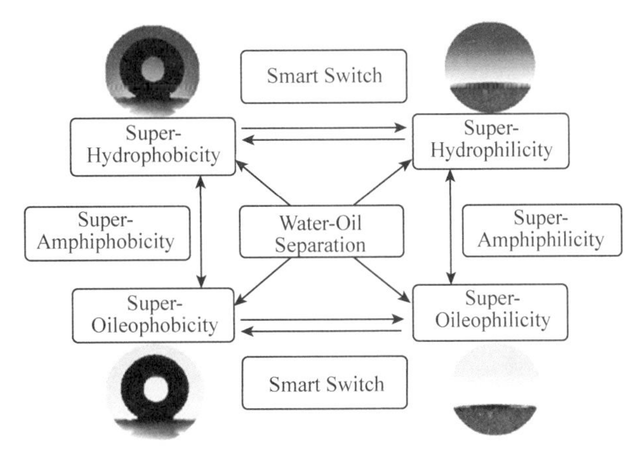

图 2　二元界面材料体系下的特殊浸润性多元组合图

1.1 金属网

1.1.1 超疏水—超亲油性

2004 年，FENG et al.[29]通过喷雾干燥法制备了一种超疏水—超亲油不锈钢网，并成功地被应用于油水混合物的分离。通过简单的喷雾干燥法，WANG et al.[30]在金属网表面成功构筑了微米级粗糙结构，进一步以聚四氟乙烯作为前驱体，在不锈钢网面上修饰了一层疏水亲油性薄膜，从而形成超疏水超亲油的金属网。同时，LUO et al.[31]在不同的金属表面成功构建了微—纳多级结构，进一步负载一层疏水涂层，从而获得具有超疏水—超亲油效果的基材表面。DENG et al.[32]通过调节低密度聚乙烯在二甲苯溶剂中的浓度，再将不锈钢网浸渍于上述溶液中，成功地使该金属网获得可调控的粗糙疏水亲油表面，并应用于除油实验，如图 3 所示。LEE et al.[33]将多孔的碳纳米管以化学气相沉积技术为手段，成功接枝于不锈钢网表面，经过化学修饰后，制得了一种超疏水—超亲油性的三维多孔碳纳米管不锈钢网，而该不锈钢网具有较好的油水分离功能。

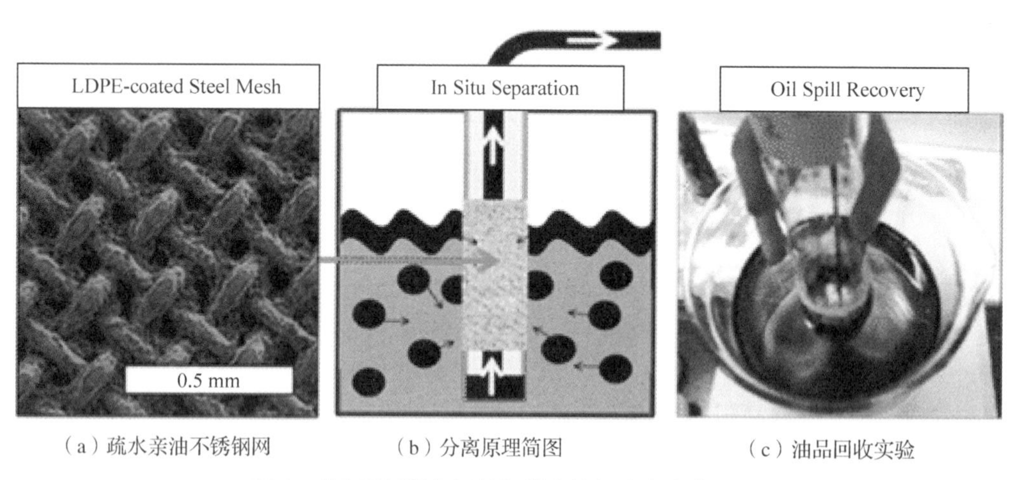

（a）疏水亲油不锈钢网　（b）分离原理简图　（c）油品回收实验

图 3　处理后不锈钢网的扫描电镜与油水分离图像

1.1.2 超亲水—超疏油性

YANG et al.[34]引入粗糙度使疏油亲水性扩大到超疏油超亲水性，成功地制备出超疏油—超亲水的聚二甲基二烯丙基氯化铵—全氟辛酸钠二氧化硅涂层，聚二甲基二烯丙基氯化铵—全氟辛酸钠同时含有氟元素和亲水基团，纳米二氧化硅则进一步增加了粗糙度，致使油滴的接触角高达 155°±1°，而水滴的接触角几乎为 0°。将这种涂料涂覆于不锈钢网表面，即得到超疏油—超亲水网，这种网膜高

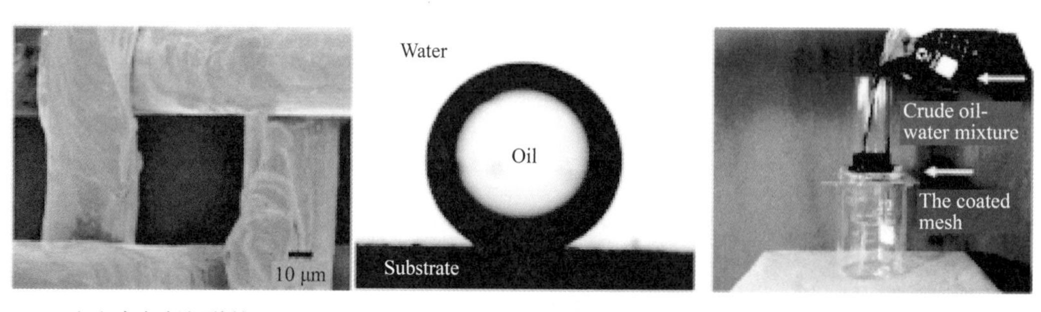

（a）亲水疏油不锈钢网　（b）水下与油滴接触角　（c）油水分离实验

图 4　处理后不锈钢网的形貌、润湿性及油水分离图像

效地实现了油水分离。XUE et al.[35]采用水热法制备出沸石涂层不锈钢网，该不锈钢网在空气中呈现出超亲水性，而在水下油滴接触角超过150°，如图4所示。因此，只要将不锈钢网用水完全润湿，即可在重力驱动下对多种类型的油水混合物体系进行分离。相比于超疏水—超亲油的油水分离金属网，超亲水—超疏油与超亲水—水下超疏油的金属网更易清洗、重复性与稳定性更强。遗憾的是，目前有关此类材料的报道和研究相对较少，需要得到科研人员的更多关注。

1.2 海绵

RUAN et al.[36]先通过浸渍的方法在三聚氰胺甲醛树脂海绵表面负载一层多巴胺薄膜，再利用全氟硅烷进行疏水改性，制得了一种吸附性较强且具备阻燃性的超疏水性海绵。CALCAGNILE et al.[37]将聚四氟乙烯颗粒和超顺磁性的氧化铁负载于聚氨酯海绵表面，再对其进行聚四氟乙烯改性，制得具有磁性的超疏水—超亲油聚氨酯海绵。ZHU et al.[38]先将海绵在浓硫酸、氧化铬的混合液中进行浸渍刻蚀，之后再用五水硫酸铜、甲醛和乙二胺四乙酸二钠盐的混合液浸泡，最后，先后以硝酸银和正十二烷酸进行处理，即制得具有超疏水—超亲油特性的海绵。NGUYEN et al.[39]采用浸渍涂覆法将石墨烯负载于海绵表面，再以聚二甲基硅氧烷进行疏水处理，制得超疏水—超亲油海绵，如图5所示。处理后，该海绵表面与水的接触角高达162°，与油的接触角接近0°。另外，研究表明，即使水中只有一滴油，该海绵也可从水中将油滴回收。

（a）处理前

（b）处理中

（c）处理后

图5 处理前后海绵扫描电镜图像

1.3 无机粉体及高分子材料

从墨西哥湾石油事件中得到启发，澳大利亚研究员ARBATAN et al.[40]以硬脂酸和碳酸钙粉体为原料，制得超疏水亲油碳酸钙，而该碳酸钙可将油水混合液中油液完全吸收干净，如图6所示。SU et al.[41]使用溶胶—凝胶法制备二氧化硅溶胶，并以聚二甲基硅氧烷和聚氨酯进行疏水改性，进一步将处理溶胶负载于多孔陶瓷管表面，成功制得超疏水—超亲油性多孔陶瓷管。当油水混合液流过管路时，油品透过管壁汇集流出，水则沿管路方向直接排出。XU et al.[42]报道了一种超疏水超亲油的磁

（a）处理前

（b）处理中

（c）处理后

图6 处理后碳酸钙分离油水混合液图像

性颗粒，这些颗粒能够自发包裹住水面的油滴，在外界磁场下，能够将水面的油滴移除，然后将包裹油滴的粉末移至乙醇中，能将油滴和颗粒分别回收利用。SUN et al.[43]利用石墨烯自身的蜂窝结构，制得疏水亲油性石墨烯，并检测了该产品对油水混合液的分离效果：在水中滴加少量油液，5 s 内该石墨烯可将油渍吸收干净。另外，各国研究人员还将碳纳米管棉球[44]，纳米多孔高分子材料[45]，独立式锰氧化物纳米线[46]等材料进行处理，取得了突出的研究成果，成功应用于油水分离。

1.4 生物质

生物质是自然界中丰富且可再生利用的资源，全球每年可产生 135 亿 t 生物质。其中相当部分是以棉料、废纸、麦草、秸秆、木屑等形式存在，且往往被以有机肥料或垃圾的方式处理掉，不但形成资源浪费，而且严重污染环境[47]。因此，设计生物质材料的回收利用工艺，进而提高其利用率与附加值便成为各领域专家学者们的主要研究任务，同时也是未来生物质产业的发展方向。

目前，生物质材料主要作为燃料能源开发，涉及气化、液化、热解、固化和直接燃烧等技术[48]；转化高附加值化学品，例如葡萄糖、木糖、苯丙烷单体及二聚体，气态小分子如 CH_4 和 CO，液态小分子如有机酸、醛、醇，重要基础平台化合物糠醛、乙酰丙酸、木糖醇、乙醇等[49]；通过高温与助剂共同作用下生物质材料的动态塑化改性，生物质原材料的基因技术材质改良，研制开发生物质材料塑性加工的专用设备，木质纤维—热塑性聚合物复合材料技术等手段制备木塑材料[50]。迄今为止，鲜少将生物质材赋予超疏水—超亲油或超亲水—超疏油性能并应用于油品回收领域的相关报道。

1.4.1 生物质基粉体

ZANG et al.[51]将木粉分别通过氢氧化钠、过氧化氢溶液预处理，然后制备疏水性 SiO_2 微球，将两者加入聚苯乙烯的四氢呋喃溶液中，充分混合、干燥后，制得超疏水—超亲油性木粉。该木粉与水的接触角可达 153°，与油接触角为 0°，可快速将正己烷、汽油、柴油、原油、机器润滑油从其水的混合物中分离出来，如图 7 所示。此外，ZANG et al.[52]将废弃的玉米秸秆磨成粉，以氢氧化钠、过氧化氢溶液进行预处理，然后以氢氧化钠与硝酸锌为原料合成氧化锌纳米颗粒，再将秸秆粉与氧化锌颗粒加入含有十二烷基磺酸钠、十六烷基三甲氧基硅烷的乙醇溶液中进一步改性，最终获得超疏水—超亲油性秸秆粉。该粉体与水的接触角可达 155°，与油接触角为 0°，不但可快速将柴油、汽油、原油、豆油、正己烷、辛烷、甲苯、氯仿从其水的混合物中分离出来，酸碱稳定性良好，而且可多次重复使用。

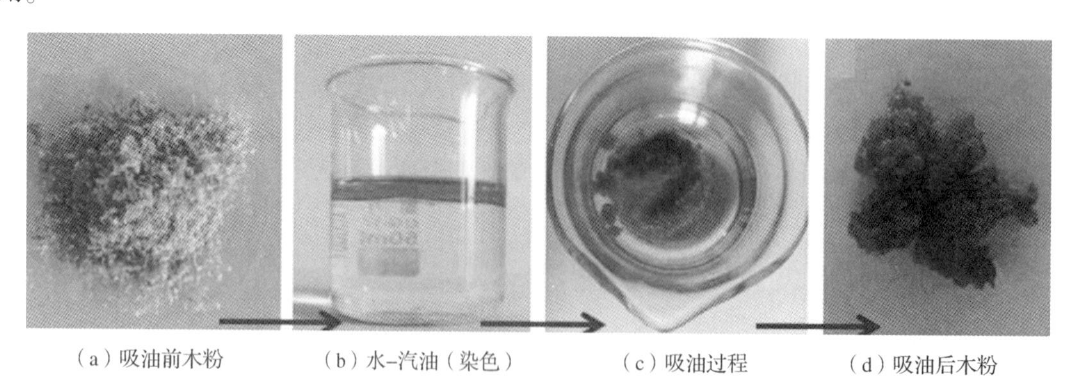

(a) 吸油前木粉　(b) 水–汽油（染色）　(c) 吸油过程　(d) 吸油后木粉

图 7 超疏水—超亲油性木粉处理水与汽油混合物过程图像

1.4.2 滤纸

黄相璇[53]首先通过阳离子改性法将二乙醇胺与双酚 A 酚醛环氧树脂中的环氧基团进行开环反应

制备成水性环氧树脂乳液，并应用于机油滤纸。而后通过苯甲酸和马来酸酐共同改性滤纸，并在此基础上通过自由基聚合引入甲基丙烯酸十二氟庚酯单体来合成具有较低表面能且亲油的滤纸。ZHANG et al.[54]以三甲基氯硅烷合成疏水性 SiO_2 纳米微球，并掺杂于四氢呋喃中，再加入端乙烯基聚二甲基硅氧烷，负载滤纸表面，制得超疏水—超亲油滤纸，如图 8 所示。ZHANG et al.[55]采用溶胶—凝胶法合成 SiO_2 纳米颗粒，再以十八烷基三氯硅烷(OTS)疏水改性，浸渍于四氢呋喃/聚苯乙烯溶液，最后负载于滤纸表面，制得超疏水—超亲油滤纸。另外，GAO et al.[56]还将滤纸以环氧树脂预处理，再将制得的 TiO_2 纳米颗粒以 KH550 氨基化通过化学键和作用负载于滤纸表面，最后通过 OTS 降低其表面能，制得超疏水—超亲油滤纸。该滤纸经过油水分离测试后，油品的回收率可达 98.2%，除水效果则为 100%。

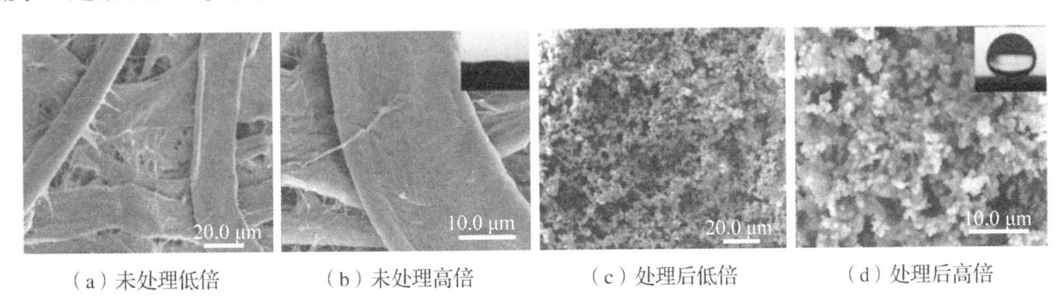

（a）未处理低倍　（b）未处理高倍　（c）处理后低倍　（d）处理后高倍

图 8　处理前后滤纸扫描电镜图像

1.4.3　棉花与棉织物

邓晓庆等[57]采用溶胶—凝胶法制备了 SiO_2 水溶胶，并用三甲基甲氧基硅烷进行疏水处理，通过一步浸泡法将疏水性 SiO_2 修饰到棉花上制备出超疏水—超亲油棉花。改性棉花对柴油展示了高达 8.3 $g \cdot g^{-1}$ 的饱和吸附量，在 5 次重复使用中，吸附能力和接触角均可维持在 4.0 $g \cdot g^{-1}$ 和大于 150°水平。LIU et al.[58]预先采用碱刻蚀法处理棉花，再通过溶胶—凝胶法将 SiO_2 负载于棉纤维表面，最后使用十八烷基三氯硅烷进行疏水改性，制得用于油水分离的超疏水—超亲油棉花。

ZHANG et al.[59]以正硅酸乙酯为前驱体，氨水为催化剂合成 SiO_2 纳米微球，再以 OTS 进行疏水改性，与四氢呋喃/聚苯乙烯溶液超声复合，最后负载于棉织物表面，制得超疏水—超亲油棉织物，如图 9 所示。为了充分说明得到的产品可以用于油水分离，做了油酸、正己烷、正庚烷、甲苯等含水混合物的分离试验，效果十分显著。另外，ZHANG et al.[60]将棉织物环氧化预处理，将制得的 ZnO 纳米颗粒氨基化预处理，再通过化学键和作用增强棉织物与纳米颗粒间的结合力，最后通过全氟硅烷降低其表面能，制得超疏水—超亲油棉织物。该棉织物不但拥有突出的油水分离效果，还具备一定的阻燃性、化学机械稳定性、抗紫外性等。ZHOU et al.[61]通过聚苯胺与氟烷基硅烷的协同作用，以化

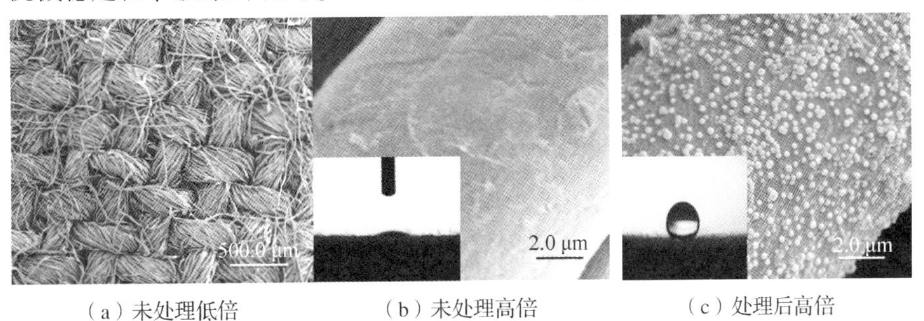

（a）未处理低倍　（b）未处理高倍　（c）处理后高倍

图 9　处理前后棉织物扫描电镜图像

学气相沉积技术为手段，赋予棉织物表面超疏水—超亲油特性。该产品对油水混合物的分离效率高达97.8%，具备较好的耐酸碱性。

2 存在问题与拟解决方案

近年来，世界各国的专家学者们在油水分离材料的合成与应用领域已经取得了一定的研究成果，但该领域仍存在较多问题。首先，目前常见的油水分离材料一般采用无机金属氧化物或者聚合物制得，无机金属氧化物对饮用水具有一定的毒性，而聚合物在油性液体中容易溶胀或者溶解。因此，目前首要任务是开发不被油性液体溶解的油水分离材料，如丙烯酸烷基酯型、苯乙烯—二乙烯苯型、橡胶型、丙烯酸叔丁酯和二乙烯基共聚物等。其次是研究不同表面张力液体、不同结构固体表面对固液接触时表现的润湿性、吸附能力的影响，探索不同固体材料应用于分离或吸附不同极性和表面张力液体的可行性。

现如今，大部分研究所选用的实验基材主要是金属网、海绵、多孔陶瓷、无机纳米颗粒等，而这些材料在实际应用过程中均存在较多问题，例如在金属网表面构筑粗糙结构的难度较大，实验条件苛刻；海绵因具有极强的吸水性，须选择低表面能含氟有机化合物处理，严重污染环境，更不利于批量生产；多孔陶瓷、无机纳米颗粒的结构不稳定，容易脱落到油品或者水中，不但造成材料浪费，而且还会对人类及其他生物的健康产生威胁。因此，建议研究者们可以将超疏水—超亲油性或超亲水—超疏油性材料的基材转向密度小、孔隙度、比表面积大、易降解、原料易得、价格低廉的生物质材料。

此外，针对油品回收，生物质材料存在以下自身优势：密度小——油品常漂浮于湖泊、河流、海水之上；孔隙度、比表面积大——吸附、分离油品；易降解——便于后续处理；原料易得，产量大——可大量处理海上油污。如选用生物质材料作为基材，开发出新型油水分离生物质材料，不但回收利用了即将废弃的生物质材料，而且使生物质材料与高分子材料、无机纳米材料、油水分离材料等得到较好的复合，充分发挥各自的优点，并进一步提升了生物质材料的附加值。另外，由于可以选用的生物质材料种类很多，如木粉、秸秆粉、棉花、滤纸、棉织物等，合成的特种油水分离型生物质材料也多种多样，其应用前景极为可观。

3 总结与展望

综上，无论是制备超疏水—超亲油性材料还是超亲水—超疏油性材料，成功的关键是在材料表面合成适当的粗糙结构，并接枝合适的界面改性剂。根据不同的使用情况与使用要求：除了可以选用金属网、海绵、无机粉体、高分子材料作基材以外，生物质材料如木粉、秸秆粉、棉花、棉织物、滤纸等也可作为制备油水分离材料的首选原料；常用合成技术有溶胶—凝胶技术、化学气相沉积技术、熔融—冷却凝固成型技术、电化学沉积技术、刻蚀技术、模板法、相分离技术、水热法、层层自组装技术、静电纺丝技术、喷涂技术等；适当地添加性质性能不同的无机纳米材料来提升材料的各方面性能，如铁氧化物、钴氧化物、镍氧化物、铜氧化物、纳米银、二氧化硅、氧化锌、二氧化钛等；可以选用的界面改性剂(低表面能材料以及油水分离改性材料)有硬脂酸类、具有长碳链硅氧烷类、含氟聚合物、有机硅聚合物、苯乙烯—二乙烯苯型、石蜡、橡胶型、丙烯酸烷基酯型、丙烯酸叔丁酯和二乙烯基共聚物等。

特殊浸润性油水分离材料，无论在处理海洋溢油方面，还是在处理工业油水分离领域均展现了良好的应用前景。尤其是生物质材料在该领域的应用，更是给人们带来了新希望。但由于受开发成本及技术水平等限制，相关产品的商业化程度远远不够，后续应着重加强对以下问题的研究：根据不同使用环境，选用合适的基材，制备稳定性强、具有自修复性的特殊润湿性油水分离材料；基于浸润性原理的基础研究，开发经济有效的仿生微观结构，优化制备工艺及后处理过程；提高人们对生物质材料应用于油水分离领域的认识与重视程度；改善不同材料表面的低表面能修饰方法，用于指导油水分离相关产品的开发；建立低成本高效油水分离体系，尤其是特殊条件下油水混合物中微量油或者微量水的去除方法。

参考文献

[1]钮劲涛. 纤维素基高吸油树脂的制备及应用研究[D]. 阜新：辽宁工程技术大学，2011.

[2]王枢，褚良银，陈文梅，等. 油水分离膜的研究新进展[J]. 油田化学，2003，20(4)：387-390.

[3] Barthlott W，Neinhuis C. Purity of the sacred lotus，or escape from contamination in biological surfaces[J]. Planta，1997，202(1)：1-8.

[4] Neinhuis C，Barthlott W. Characterization and distribution of water-repellent，self-cleaning plant surfaces[J]. Ann Bot.，1997，79(6)：667-677.

[5] 姬鹏婷. 超亲油超疏水油水分离纺织品的制备与研究[D]. 西安：陕西科技大学，2014.

[6]Feng L，Li S，Li Y，et al. Super-hydrophobic surfaces：from natural to artificial[J]. Adv. Mater，2002，14(24)：1857-1860.

[7] Xu L H，Zhuang W，Xu B，et al. Fabrication of superhydrophobic cotton fabrics by silica hydrosol and hydrophobization[J]. Appl . Surf. Sci.，2011，257(13)：5491-5498.

[8]符开伟，郑振荣，李岳，等. 化学气相沉积制备自清洁涤纶织物[J]. 产业用纺织品，2012(6)：37-40.

[9]吴景霞. 超疏水纺织品的辐射方法制备及其服用性能研究[D]. 上海：中国科学院研究生院，2015.

[10] Zhang X，Shi F，Yu H，et al. Polyelectrolyte multilayer as matrix for electrochemical deposition of gold clusters：Toward super-hydrophobic surface[J]. J. Am. Chem. Soc.，2004，126(10)：3064-3075.

[11]张平. 基于纤维表面化学刻蚀及疏水化改性制备超疏水纺织品的研究[D]. 西安：陕西科技大学，2014.

[12] Wang X D，Graugnard E，King J S，et al. Large-scale fabrication of ordered nanobowl arrays[J]. Nano. Lett.，2004，4(11)：2223-2226.

[13]杨艳丽. 基于相分离法的棉织物超疏水整理研究[D]. 无锡：江南大学，2014.

[14]Öner D，Mccarthy T J. Ultrahydrophobic surfaces. Effects of topography length scales on wettability[J]. Langmuir，2000，16(20)：7777-7782.

[15] Zhang M，Wang S L，Wang C Y，et al. A facile method to fabricate superhydrophobic cotton fabrics[J]. Appl. Surf. Sci.，2012，261：561-566.

[16] Cakir M，Kartal I，Yildiz Z. The preparation of UV-cured superhydrophobic cotton fabric surfaces by electrospinning method[J]. Text Res. J.，2014，84(14)：1528-1538.

[17] Wang C Y，Zhang M，Xu Y，et al. One step synthesis of unique silica particles for the fabrication of bionic and stably superhydrophobic coatings on wood surface[J]. Adv. Powder Technol.，2014，25(2)：530-535.

[18] NystrÖm D，Lindqvist J，Östmark E，et al. Superhydrophobic and self-cleaning bio-fiber surfaces via ATRP and subsequent postfunctionalization[J]. ACS Appl. Mater Interfaces，2009，1(4)：816-823.

[19] Park Y，Im H，Im M，et al. Self-cleaning effect of highly water-repellent microshell structures for solar cell applications[J]. J. Mater Chem.，2011，21(3)：633-636.

[20] Elsharkawy M，Schutzius T M，Megaridis C M. Inkjet patterned superhydrophobic paper for open-air surface microfluidic devices[J]. Lab. Chip.，2014，14(6)：1168-1175.

[21] Banerjee I，Pangule R C，KANE R S. Antifouling coatings：recent developments in the design of surfaces that prevent fouling by proteins，bacteria，and marine organisms[J]. Adv. Mater，2011，23(6)：690-718.

[22] Zhang L B，Li Y，Sun J Q，et al. Mechanically stable antireflection and antifogging coatings fabricated by the layer-by-layer deposition process and postcalcination[J]. Langmuir，2008，24(19)：10851-10857.

[23]阎映弟，罗能镇，相咸高，等. 防覆冰涂层构建机理及制备[J]. 化学进展，2014，26(1)：214-222.

[24] Bixler G D, Bhushan B. Fluid drag reduction with shark-skin riblet inspired microstructured surfaces[J]. Adv Funct Mater, 2013, 23(36): 4507-4528.

[25]李坚, 高丽坤. 光控润湿性转换的抑菌性木材基银钛复合薄膜[J]. 森林与环境学报, 2015, 35(3): 193-198.

[26] Lee A, Moon M W, Lim H, et al. Water harvest via dewing[J]. Langmuir, 2012, 28(27): 10183-10191.

[27]Zhang M, Wang C Y, Wang S L, et al. Fabrication of superhydrophobic cotton textiles for water-oil separation based on drop-coating route[J]. Carbohyd. Polym., 2013, 97(1): 59-64.

[28]卢晟. 油水分离滤纸及吸油聚合物制备及其性能研究[D]. 上海: 上海交通大学, 2013.

[29] Feng L, Zhang Z Y, Mai Z H, et al. A super-hydrophobic and super-oleophilic coating mesh film for the separation of oil and water[J]. Angew. Chem. Int. Edit, 2004, 43(15): 2012-2014.

[30] Wang L F, Zhao Y, Wang J M, et al. Ultra-fast spreading on superhydrophilic fibrous mesh with nanochannels[J]. Appl. Surf. Sci., 2009, 255(9): 4944-4949.

[31] Luo Z Z, Zhang Z, Hu L T, et al. Stable bionic superhydrophobic coating surface fabricated by a conventional curing process[J]. Adv. Mater, 2008, 20(5): 970-974.

[32] Deng D, Prendergast D P, Macfarlane J, et al. Hydrophobic meshes for oil spill recovery devices[J]. ACS Appl. Mater Inter., 2013, 5(3): 774-781.

[33] Lee C H, Johnson N, Drelich J, et al. The performance of superhydrophobic and superoleophilic carbon nanotube meshes in water-oil filtration[J]. Carbon, 2011, 49(2): 669-676.

[34] Yang J, Zhang Z Z, Xu X H, et al. Superhydrophilic-superoleophobic coatings[J]. J. Mater Chem., 2012, 22(7): 2834-2837.

[35] Xue Z X, Wang S T, Lin L, et al. A Novel Superhydrophilic and underwater superoleophobic hydrogel-coated mesh for oil/water separation[J]. Adv. Mater, 2011, 23(37): 4270-4273.

[36] Ruan C P, Ai K L, Li X B, et al. A superhydrophobic sponge with excellent absorbency and flame retardancy[J]. Angew. Chem. Int. Edi., 2014, 53(22): 5556-5560.

[37] Calcagnile P, FragouliD, Bayer I S, et al. Magnetically driven floating foams for the removal of oil contaminants from water[J]. ACS Nano., 2012, 6(6): 5413-5419.

[38] Zhu Q, Pan Q M, Liu F T. Facile removal and collection of oils from water surfaces through superhydrophobic and superoleophilic sponges[J]. J. Phys. Chem. C, 2011, 115(35): 17464-17470.

[39] Nguyen D D, Tai N H, Lee S B, et al. Superhydrophobic and superoleophilic properties of graphene-based sponges fabricated using a facile dip coating method[J]. Energy Environ. Sci., 2012, 5(7): 7 908-7 912.

[40] ArbatanT, Fang X Y, Shen W. Superhydrophobic and oleophilic calcium carbonate powder as a selective oil sorbent with potential use in oil spill clean-ups[J]. Chem. Eng. J., 2011, 166(2): 787-791.

[41] Su C H, Xu Y Q, Zhang W, et al. Porous ceramic membrane with superhydrophobic and superoleophilic surface for reclaiming oil from oily water[J]. Appl. Surf. Sci., 2012, 258(7): 2319-2323.

[42] Xu L P, Wu X W, Meng J X, et al. Papilla-like magnetic particles with hierarchical structure for oil removal from water[J]. Chem. Commun., 2013, 49(78): 8752-8754.

[43] Sun H Y, Xu Z, Gao C. Multifunctional, ultra-flyweight, synergistically assembled carbon aerogels[J]. Adv. Mater, 2013, 25(18): 2554-2560.

[44] Gui X C, Wei J Q, Wang K L, et al. Carbon nanotube sponges[J]. Adv. Mater, 2010, 22(5): 617-621.

[45] Zhang Y L, Wei S, Liu F J, et al. Superhydrophobic nanoporous polymers as efficient adsorbents for organic compounds[J]. Nano. Today, 2009, 4(2): 135-142.

[46] Yuan J K, Liu X G, Akbulut O, et al. Superwetting nanowire membranes for selective absorption[J]. Nat Nanotech, 2008, 3(6): 332-336.

[47]鲍甫成. 发展生物质材料与生物质材料科学[J]. 林产工业，2008，35(4)：3-7.

[48] 杨丰科，程伟，王艳，等. 生物质热裂解生产生物燃料的研究进展[J]. 氨基酸和生物资源，2008，30(4)：37-41.

[49]林鹿，何北海，孙润仓，等. 木质生物质转化高附加值化学品[J]. 化学进展，2007，19(7/8)：1 206-1 216.

[50] 王清文，欧荣贤. 生物质材料的塑性加工研究进展[C]//第二届中国林业学术大会——S11 木材及生物质资源高效增值利用与木材安全论文集. 南宁：中国林学会，2009.

[51] Zang D L, Liu F, Zhang M, et al. Novel superhydrophobic and superoleophilic sawdust as a selective oil sorbent for oil spill cleanup[J]. Chem. Eng. Res. Des., 2015, 102: 34-41.

[52] Zang D L, Zhang M, Liu F, et al. Superhydrophobic/superoleophilic corn straw fibers as effective oil sorbents for the recovery of spilled oil[J]. J. Chem. Technol. Biot., 2015, doi: 10. 1002/jctb. 4834.

[53]黄相璇. 超疏水/超亲油水性环氧树脂乳液涂层的制备及在油水分离滤纸中的应用研究[D]. 广州：华南理工大学，2012.

[54] Zhang X, Guo Y G, Zhang P Y, et al. Superhydrophobic and superoleophilic nanoparticle film: synthesis and reversible wettability switching behavior[J]. ACS Appl. Mater Inter., 2012, 4(3): 1742-1746.

[55] Zhang M, Wang C Y, Wang S L, et al. Fabrication of coral-like superhydrophobic coating on filter paper for water-oil separation[J]. Appl. Surf. Sci., 2012, 261: 764-769.

[56] Gao Z X, Zhai X L, Liu F, et al. Fabrication of TiO_2/EP super-hydrophobic thin film on filter paper surface[J]. Carbohyd. Polym., 2015, 128: 24-31.

[57] 邓晓庆，商静芬，王侦，等. 超疏水棉花的制备及用于油水分离[J]. 应用化学，2013，30(7)：826-833.

[58] Liu F, Ma M L, Zang D L, et al. Fabrication of superhydrophobic/superoleophilic cotton for application in the field of water/oil separation[J]. Carbohyd. Polym., 2014, 103: 480-487.

[59] Zhang M, Li J, Zang D L, et al. Preparation and characterization of cotton fabric with potential use in UV resistance and oil reclaim[J]. Carbohyd. Polym., 2016, 137: 264-270.

[60] Zhang M, Zang D L, Shi J Y, et al. Superhydrophobic cotton textile with robust composite film and flame retardancy [J]. RSC Adv., 2015, 5(83): 67780-67786.

[61] Zhou X Y, Zhang Z Z, Xu X H, et al. Robust and durable superhydrophobic cotton fabrics for oil/water separation [J]. ACS Appl. Mater Inter., 2013, 5(15): 7208-7214.

英文题目： Research Progress of the Materials with Special Wettability for Oil-water Separation

作者： Li Jian, Zhang Ming, Qiang Tiangang

摘要： In recent years, oil and organic spillages have caused serious damages to environment and ecosystems. It is a worldwide and important mission for all the people to fabricate the suitable separating materials and to find the effective method for oil-water separation. Inspired by the nature, more and more materials with special wetting behavior have got great development. Most importantly, super-hydrophobicity/super-oleophylicity and super-hydrophilicity/super-lipophobicity materials have been successfully applied in to water-oil separation, showing the great potential in practical applications. In this paper, the special wetting phenomena in nature and different fabricating method for micro-nano multi-structures were introduced; the novel water-oil materials based on metal net, sponge and inorganic powder were reviewed; the problem and the solution were presented; the advantage and the development status of water-oil biomaterials such as wood powder, straw powder, filter paper, cotton and cotton fabrics were stressed in details. Then the research trend of this field has been prospected.

关键词： Oil-water separation; Special wetting behavior; Super-hydrophobicity/Super-oleophylicity; Super-hydrophilicity/super-lipophobicity; Bio-based materials

Removal of Oils from Water Surface Via Useful Recyclable $CoFe_2O_4$ /Sawdust Composites under Magnetic Field*

Wentao Gan, Likun Gao, Wenbo Zhang, Jian Li, Liping Cai, Xianxu Zhan

Abstract: In this study, the poplar sawdust was used as a recyclable, low cost, environmentally friend and highly selective oil absorbent material. The sawdust with high hydrophobicity, superoleophilicity and magnetic property was obtained by precipitating $CoFe_2O_4$ on sawdust surface and chemically modifying the low surface energy polysiloxane layers. These composites were able to absorb lubrication oil up to 11.5 times of the sawdust weight while completely repelling water. In addition, the oil-absorption composites were able to collect oils conveniently from water surface with the help of an external magnet due to its exceptional magnetic property. Furthermore, the sawdust could be cleared up and re-used for water-oil separation at least 10 times. This work provides a potential platform for developing a water-protection strategy using natural waste resources.

Keywords: Sawdust absorbent; Oil removal; Magnetic $CoFe_2O_4$ particles; Recyclability

1 Introduction

Interest has recently arisen in the development of a facile method for fast removing oils from water surface, driven by the problem of oil contaminants causing severe environmental pollution and health problems[1-3]. Especially an increasing attention is being devoted to the utilizing of natural waste resources to replace chemicals in the sustainable society. Treatments for oil spillage included burning[4], using chemical dispersants[5] and absorbent materials[6], mechanically collecting[7], etc. Among these methods, the approach of using adsorbent material is attractive for some applications because of its easy collection and ready availability[8, 9]. For examples, using superamphiphobic filter paper[10], carbon nanotubes[11], polymer films[12], metal nanoparticles[13], highly porous materials[14], etc., were reported to efficiently separate oils from water. However, the high costs and recycling operations limited the application.

Recently, high hydrophobicity, superoleophilicity and magnetism were introduced to synthetic absorbent in order to reduce the costs and improve the efficiency. It seems that the magnetic absorbent, possessing the capability to remove oil from water surface and being convenient for magnetic separation, are promising materials for oil-water separation[15]. Thanikaivelan et al. proposed a stable magnetic oil-absorbent of collagen and superparamagnetic iron oxide nanoparticles. These nanocomposite have selective oil absorption and magnetic property, allowing it to be used in oil removal applications[16]. Palchoudhury et al. prepared a kind of polyvinylpyrrolidone-coated iron oxide nanoparticles for fast and selective adsorption of oil from water

* 本文摘自 Materials & Design, 2016, 98: 194-200.

surface[17]. As a solid waste product obtained from mechanical wood processing, wood sawdust has great-quantities but largely underutilized presently and is available for conversion into value - added products. Furthermore, the porous appearances and abundant hydroxyl groups of the wood sawdust provide an essential capability for further modification[18, 19].

In this study, using the poplar sawdust, a recyclable, low cost, environmentally friend and highly selective oil absorbent material was developed. These magnetic $CoFe_2O_4$/sawdust composites could selectively absorb lubricating oils and organic solvents while completely repelling water. Moreover, the oil - absorbed composites were able to be quickly cleared up by a magnet bar and the absorbed oils were easily desorbed in the ethanol solution, which provided a possibility to be re-used for the oil-removal from water.

2 Experimental section

2.1 Materials

Poplar sawdust used for this experiment was obtained from a local saw mill in Heilongjiang Province, China. The sawdust was first washed with distilled water, dried, cut, and sieved through100-mesh screen. The cobalt chloride hexahydrate ($CoCl_2 \cdot 6H_2O$), ferrous sulfate heptahydrate ($FeSO_4 \cdot 7H_2O$), potassium nitrate (KNO_3), sodium hydroxide (NaOH) and vinyl triethoxysilaneused used in this study were supplied by the Shanghai Boyle Chemical Company Limited., and used without further purification.

2.2 Synthesis

Fig. 1 shows the process for synthesizing highly hydrophobic $CoFe_2O_4$/sawdust composites. 5 g wood sawdust was added into a mixture of 80 mL freshly prepared aqueous solution of 0.2 mol · L^{-1} $FeSO_4$ and $CoCl_2$ with a molar ration of $[Fe^{2+}]/[Co^{2+}] = 2$ before being transferred into a Teflon-lined stainless steel autoclave and the system was heated to 90 ℃ for 3 h. At the end of the reaction, the supernatant was poured out and 65 mL of 1.32 M NaOH solution with 0.25 M KNO_3 was added into the Teflon-lined stainless-steel autoclave. After the additional hydrothermal processing for 8 h at 90 ℃, the precipitated precursors were converted into ferrite crystal nanoparticles, and the magnetic $CoFe_2O_4$/sawdust composites were collected by an external magnet, rinsed with deionized water for 3 times, and dried at 90 ℃ for over 24 h in vacuum. Finally, the obtained powder was treated with the 2 wt% vinyl triethoxysilane ethanol solution containing 95 mL anhydrous ethanol, 5 mL H_2O and 1 mL glacial acetic acid at room temperature for 3 h. After the filtration, these composites were dried at 110 ℃ for another 24 h.

2.3 Adsorption and recyclability study

The oil absorbent ability of the composites was determined by weight measurements. The weights of composites before and after the oil absorption were measured as m_1 and m_2 by an electronic balance with an accuracy of 0.1 mg, respectively. The oil absorbent capacity q of the composites was calculated using the formula $q=(m_2-m_1)/m_1$. Three replicates were utilized for each experiment and the mean values were used for the analyses.

The removal of the lubrication oil from water surface was carried out according to the procedure illustrated in Fig. 2. Firstly, the 0.1 g treated wood sawdust were scattered on the surface of the mixture (6 mL oil on the surface of 20 mL water). The oil was absorbed by the composites quickly. After 1 min, the oil-absorbed composites were collected from the water surface by a magnet bar. The adsorbed oil was removed from the wood sawdust by vigorously stirring in ethanol solution for 5 min. After being dried at 110 ℃ for 12 h, the sawdust was re-used for the removal of oil from water surface. The recyclability was investigated by the measurements of the water contact angles, magnetic property and oil absorbent ability.

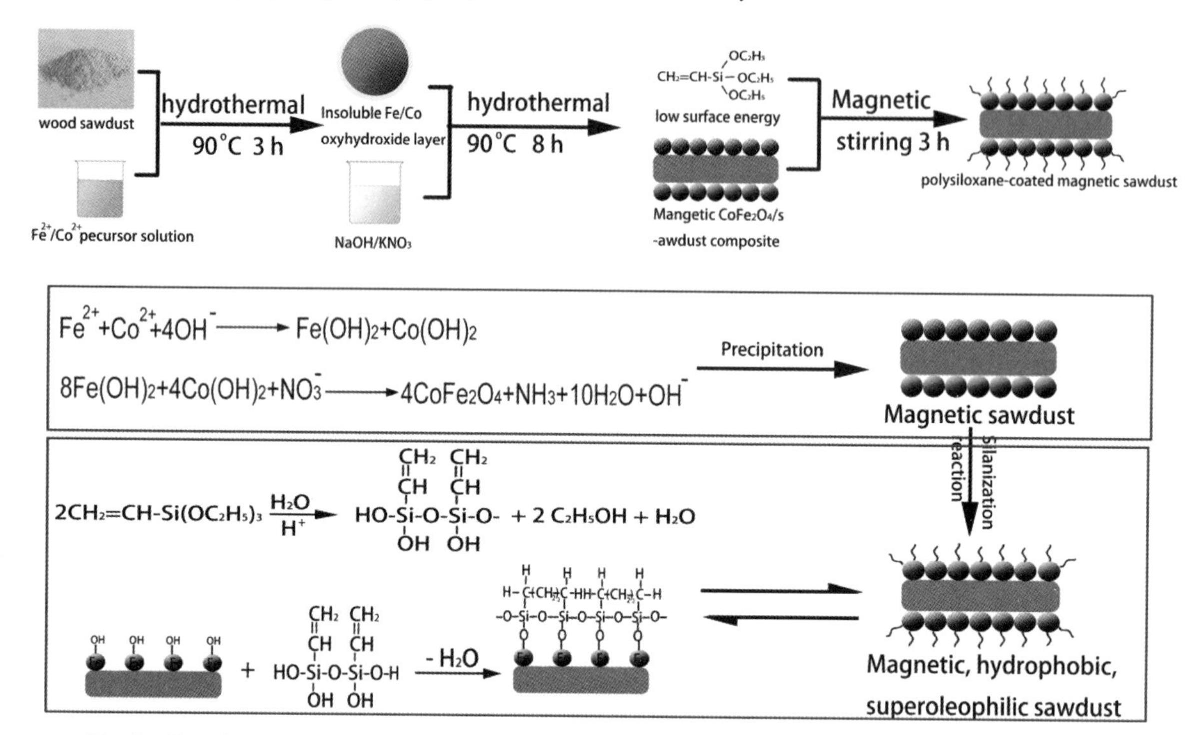

Fig. 1 Experimental strategy for the fabrication of highly hydrophobic $CoFe_2O_4$/sawdust composites

2.4 Characterization

The morphology of the specimen was examined by the scanning electron microscopy (SEM, FEI, and Quanta 200) and transmission electron microscopy (TEM, FEI, and Tecnai G20). The crystalline structure of these composites were identified using the X-ray diffraction (XRD, Rigaku, and D/MAX 2200) operating with Cu $K\alpha$ radiation ($\lambda = 1.5418$ A) at a scan rate (2θ) of 4° · min^{-1}, an accelerating voltage of 40 kV, and an applied current of 30 mA ranging from 5° to 80°. The X-ray photoelectron spectroscope (XPS) was recorded on Thermo ESCALAB 250XI. The deconvolution of the overlapping peaks was performed using a mixed Gaussian-Lorentzian fit program. The surface chemical compositions of the samples were determined via the Fourier transformation infrared spectroscopy (FTIR, Nicolet, Magna-IR 560). The magnetization measurements were carried out at room temperature using a superconducting quantum interference device (MPMS XL-7, Quantum Design Corp.). The water contact angles were measured by the commercial instrument (Analyzer JC2000C, Beijing code tong technology co., LTD) at ambient temperatures. Before the measurements, the sawdust was placed on a slide and pressed into a flat film. A distilled water droplet and an oil droplet of 5 μL were employed as the indicators. The average of the five measurements taken at different locations on each sample was used to calculate the final WCA angle.

3 Results and discussion

3.1 Preparation of highly hydrophobic $CoFe_2O_4$/sawdust composites

Fig. 1 shows the reactive scheme of the highly hydrophobic $CoFe_2O_4$/sawdust composites. The hydrophilic sawdust was immersed in an aqueous $FeSO_4/CoCl_2$ solution at room temperature before the system was heated to 90 ℃ to thermally precipitate the insoluble Co/Fe oxyhydroxide complexes on sawdust surface[20]. Then the precipitated precursors were converted into ferrite crystal nanoparticles upon the immersion in $NaOH/KNO_3$ solution at 90 ℃ for 8 h. After that, the silane polymer was coated onto the surface of $CoFe_2O_4$/sawdust composites by the silanization reaction. Firstly, the organosilane was hydrolyzed, and ethyoxyl groups were replaced by hydroxyl groups to form reactive silanol groups when placed into an aqueous acid solution that acted as a catalyst. Then the reactive silanol groups were condensed with other silanol groups to produce siloxane bonds (Si−O−Si) and formed a silane polymer through the condensation reaction. Secondly, the bare atoms of Fe on the particle surface were adsorbed OH^- to form an OH^--rich surface when the $CoFe_2O_4$ nanoparticles dispersed in an aqueous solution[21]. Finally, the silane polymer was associated with the magnetic nanoparticles and formed a covalent bond with OH groups. With the continuous dehydration and polymerization reaction of vinyl-silanes, the silane polymer was coated on the magnetic sawdust surface by chemical bonds[22, 23].

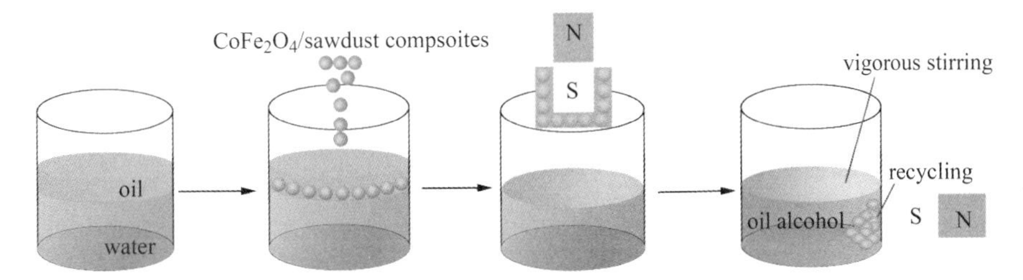

Fig. 2 Illustration for the removal of oil from water surface through highly hydrophobic $CoFe_2O_4$/sawdust composites under external magnetic field

3.2 SEM images of the sawdust

Fig. 3 shows the SEM images of the wood sawdust before and after the chemical modification. As well known, wood possesses porous structures with plenty of interconnected pores[24]. As shown on the SEM images (Fig. 3a) of the untreated sawdust, the sawdust surface was clean and smooth. There were some pits (namely the smallest holes of all pores) inside the sawdust, indicating that the sawdust also possesses the minute structure of porous wood. After the hydrothermal treatment, a layer of magnetic nanoparticles was obviously precipitated on the sawdust surface, resulting in a rough morphology (Fig. 3b and c), which could be responsible for the performance of hydrophobicity[25]. The inset in Fig. 3c illustrates that the diameters of magnetic nanoparticles peeled off from the treated sawdust by the high-power ultrasonic oscillator (1200 W) were ranged from 35 to 101 nm.

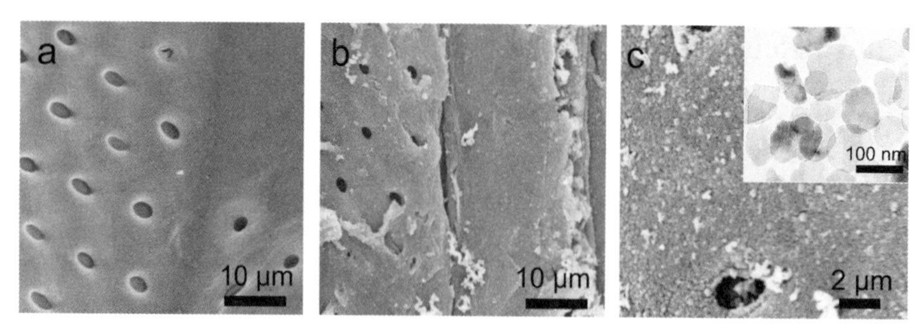

Fig. 3 SEM images of (a) pristine sawdust, (b) and (c) as-prepared sawdust at different magnifications

3.3 Chemical component analysis

Fig. 4 presents the XRD pattern of magnetic $CoFe_2O_4$/sawdust composites. The diffraction peaks at 14.8° and 22.6° represented the crystalline region of the cellulose[26]. The additional peaks were for the new crystal structures of the modified sawdust. The diffraction peaks at 18°, 30°, 35°, 43°, 53°, 57°, 62° were assigned to the diffractions of the (111), (220), (311), (400), (422), (511) and (440) planes of $CoFe_2O_4$ (JCPDS No. 22-1086), respectively. The absence of other peaks indicated the purity of the products.

The composition of the modified sawdust was examined accurately using XPS. Fig. 5a shows that the characteristic signals for C, O, Fe, Co and Si elements were clearly observed in the XPS survey spectrum of the silane-coated magnetic sawdust. Figs. 5b, c and d illustrate the high-resolution XPS spectra of 2p Co, Fe and Si, respectively. A $Co2p_{3/2}$ signal appears at 781.0 eV with a satellite peak at 786.1 eV; the peak at 796.8 eV was ascribed to the $Co2p_{1/2}$ level. The Fe2p level with binding energies of 711.7 eV and 725.2 eV was assigned to $Fe2p_{3/2}$ and $Fe2p_{1/2}$, respectively. No obvious broadening was observed for the XPS peaks, indicating the pure $CoFe_2O_4$ phase of the product[27]. In Fig. 5d, the Si2p spectra was attributed to the Si-O-Si bond[28], suggesting the formation of silane layers on the sawdust surface after the vinyltriethoxysilane modification.

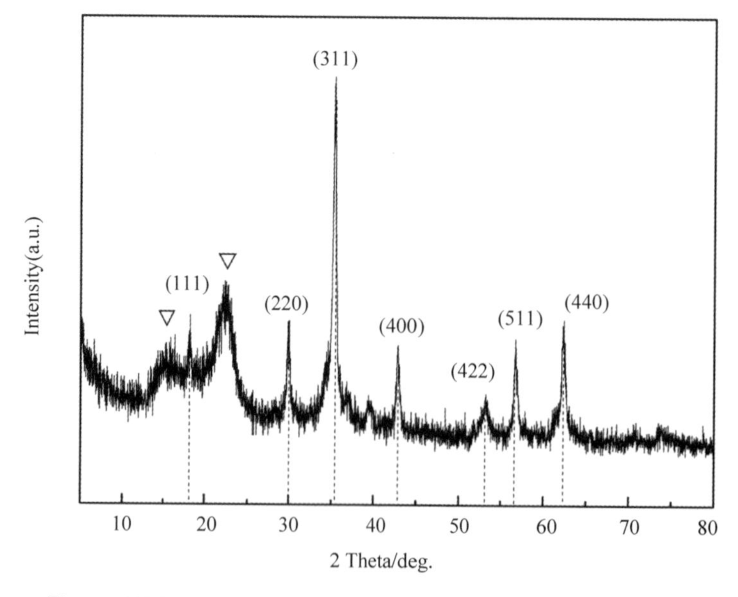

Fig. 4 XRD pattern of magnetic $CoFe_2O_4$/sawdust composites

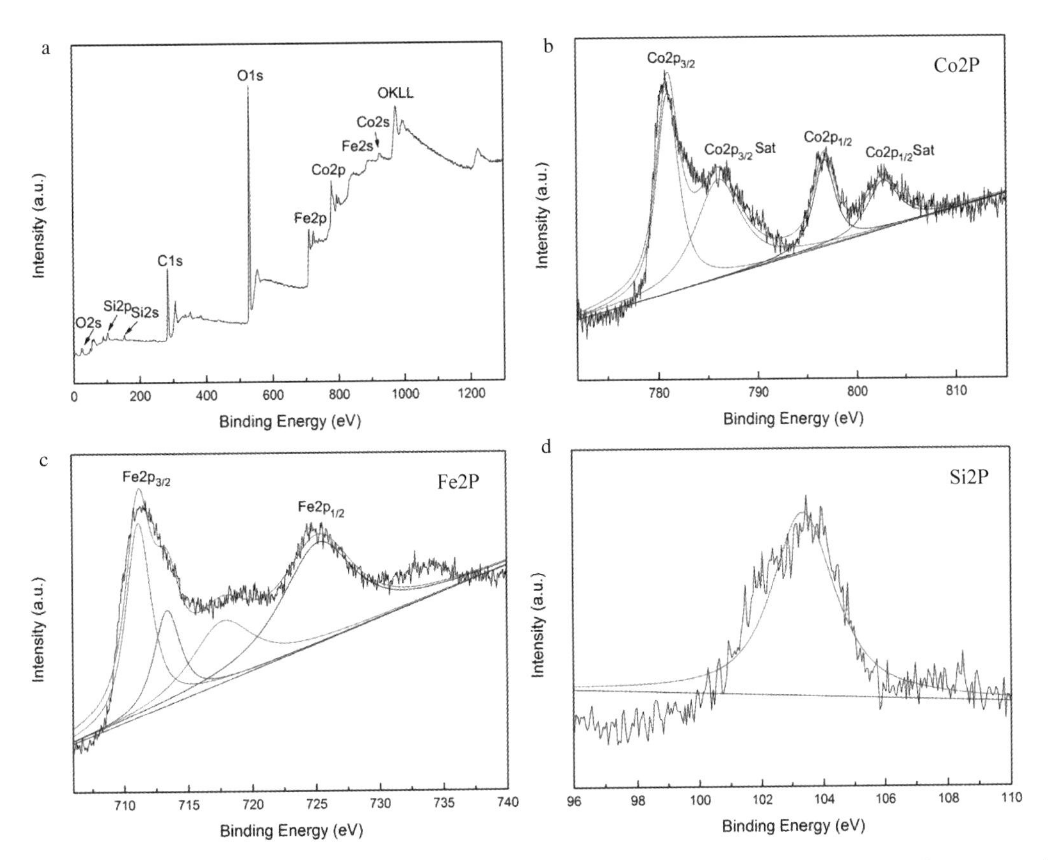

Fig. 5 XPS spectra of (a) survey spectrum, (b) Co2p, (c) Fe2p and (d) Si2p in silane-coated magnetic sawdust, respectively

In order to confirm the coating of the magnetic sawdust surface by the silanization reaction, the FTIR spectrum of the pristine sawdust and silane-coated magnetic sawdust were investigated. In Fig. 6, the bands at 3422, 2927, 1631, and 1430 cm^{-1} corresponded to O—H stretch vibration, $—CH_3$ asymmetric, O—H bending vibration and C—O stretching vibration, respectively. The most observed peaks of the untreated sawdust represented major cell wall components, such as cellulose (890 cm^{-1}), hemicelluloses (1739, 1124, 1020 cm^{-1}) and lignin (1240 cm^{-1})[29]. After the hydrothermal treatment, two absorption bands at around 622 and 567 cm^{-1} were corresponded to the stretching vibrations of Fe—O bonds at the tetrahedron site, indicating the presence of magnetic nanoparticles. Moreover, the adsorption band was observed at 459 cm^{-1}, which corresponded to the shifting of the stretching vibrations of Fe—O bonds at octahedron site of bulk magnetite (at 375 cm^{-1}) to a higher wave number. It could be caused by the silica coating that was absorbed on the particles surface by Fe—O—Si bonds[30]. The adsorption of silica polymer onto the surface of magnetic nanoparticles was characterized by the absorptions at 1113 and 1050 cm^{-1} assigned to the Si—O—Si groups[31]. Furthermore, the absorption band at 2899 cm^{-1} confirmed the presence of $—CH_2$ groups. No absorption bands of the vinyl bands (3060 and 1599 cm^{-1}) appeared in the spectra, revealing probability appears the reaction of polymerization when the vinyltriethoxysilane exposure to air, leading to the formation of $(—CH_2CH_2—)$ compounds[23].

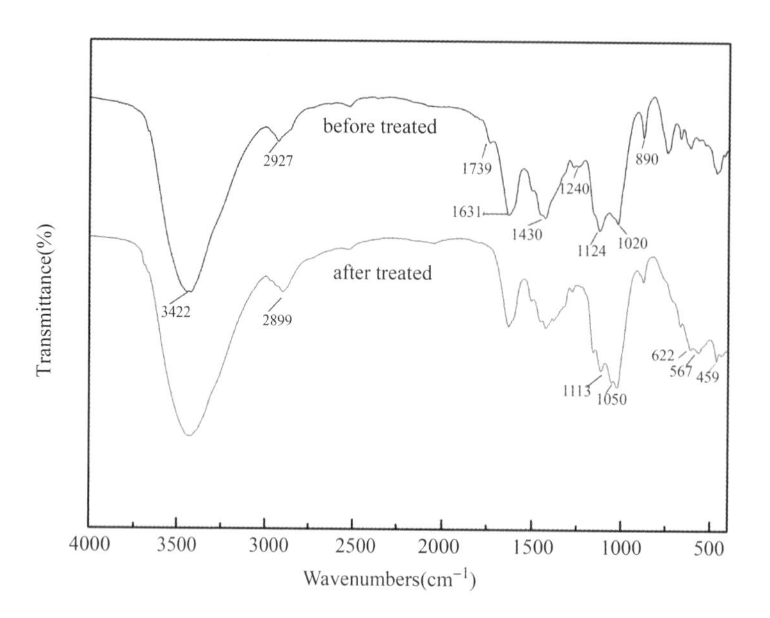

Fig. 6 FTIR spectrum of the pristine sawdust and silane-coated magnetic sawdust

3.4 Surface wettability of hydrophobic/superoleophilic sawdust

The wettability of the wood sawdust after the polysiloxane modification was evaluated by examining water contact angle through a sessile drop on the specimen. As presented in Fig. 7a, the measured water contact angle on pristine sawdust surface was 0°, which could be explained by the abundant hydroxyl groups existed on sawdust surface. However, the water contact angle of the silane-coated magnetic sawdust was 151° (Fig. 7b), higher than that of the pristine sawdust, confirming the high hydrophobicity of the sawdust after polysiloxane modification. In contrast, a drop of oil quickly diffused into silane-coated magnetic sawdust, indicating superoleophilic property of the treated sawdust (Fig. 7c). It was revealed that the highly hydrophobic and superoleophilic properties of the silane-coated sawdust were obtained by arising the low-surface-energy polysiloxane and the roughness structure.

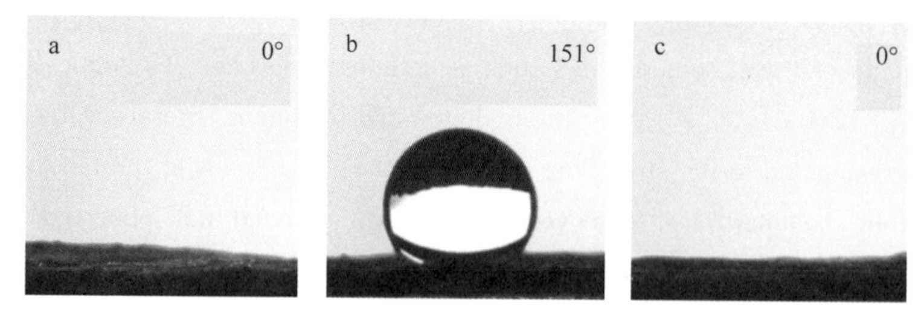

Fig. 7 Optical images of (a) water droplet on untreated wood sawdust, (b) water droplet and (c) oil droplet on as-prepared hydrophobic/superoleophilic sawdust

3.5 Magnetic properties

The magnetic characterization of the pristine sawdust and treated sawdust were conducted using the superconducting quantum interference device at room temperature, with the maximum applied field up to 20 kOe (Fig. 8). The hysteresis curve of the pristine sawdust was a straight line close to zero, suggesting the diamagnetic property of wood sawdust. However, the modified sawdust exhibited a polar property and showed a typical ferromagnetic behavior. The saturation magnetization (M_S) and coercive force (h_c) of the modified sawdust obtained at room temperature was 6.1emu · g^{-1} and 250.9 Oe, respectively. However, M_S and H_c of the as-prepared sawdust were much smaller than that of $CoFe_2O_4$ (M_S = 68 emu · g^{-1}, H_c = 750 Oe at 300 k)[32]. Taking the surface origin[33] and the finite size effect[34] into account, the decline in M_S could be attributed to the presence of the nonmagnetic sawdust substrate. On the one hand, the sawdust with partial diamagnetic substitutions made a random canting of magnetic particles surface spins thus reduced the saturation magnetization[35]. On the other hand, the presence of the hierarchical porous sawdust substrate with hydroxyl groups provided magnetic nanoparticles with a limited nucleating location[36], which effectively reduced the agglomeration of particles and reduced the saturation magnetization[37]. In addition, H_c was expected to decrease as the particle size increased beyond the critical single-domain diameter[38]. Thus, the increase of the particles size on the sawdust surface (TEM images in Fig. 3) indicated that H_c of the composite was decreased due to the performance-related structural effects.

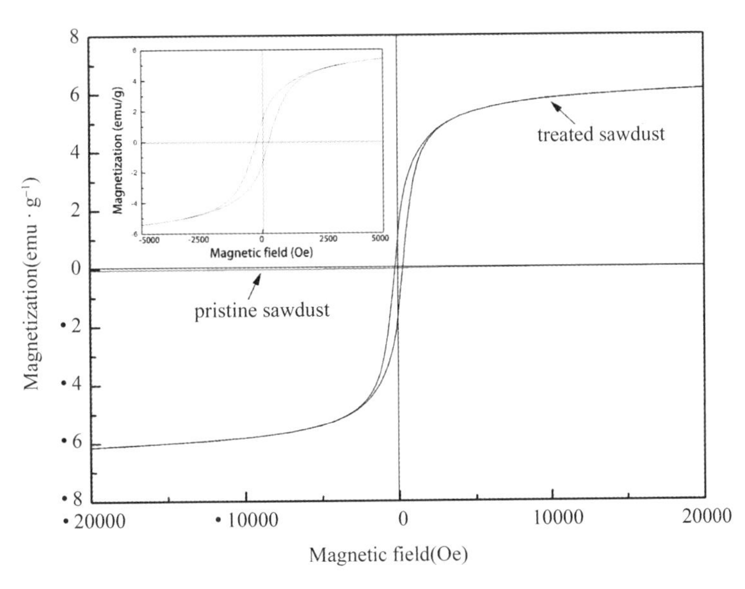

Fig. 8 Hysteresis loop of the pristine sawdust and silane-coated magnetic sawdust at room temperature

3.6 Stability analysis and application in water-oil separation

As a kind of hydrophobic/superoleophilic materials, the chemical stability and environmental durability of the as-prepared sawdust play a significant role in practical applications. The contact angle tests and magnetic measurements were carried out using an aqueous solution with a pH variation from 1 to 14 for 12 h to inspect the resistance of the as-prepared sawdust to acid and alkali. Fig. 9a reveals the relationship between the water contact angle and the saturation magnetization (M_S) of as-obtained samples and aqueous solutions with different pH values. The as-prepared sawdust showed that the water contact angles were larger than 145° after the floating on the aqueous solutions with pH values ranging from 1 to 14 for 12 h. Only a slight decrease in contact angle were observed for the solution with pH 1 and pH 14, illustrating that the hydrophobic silane cloth owned good acid and alkali resistance property. In addition, the saturation magnetization of the as-prepared sawdust were stable between 4. 5 and 6. 1 emu · g^{-1}, suggesting that the as-prepared sawdust could be recycled easily by a extern magnetic field even used in an extreme acid/alkali condition. Moreover, exposing samples in an air environment for 60 days at room temperature, the as-prepared sawdust remained high hydrophobicity with water contact angle higher than 149° and excellent magnetic property with M_S values higher than 5. 8 emu · g^{-1}(Fig. 9b), revealing the greatly environmental durability. The remarkable chemical stability and environmental durability of the as-prepared sawdust were attributed to the protective character of the silane cloth employed to adhere abundant magnetic nanoparticles to sawdust surfaces.

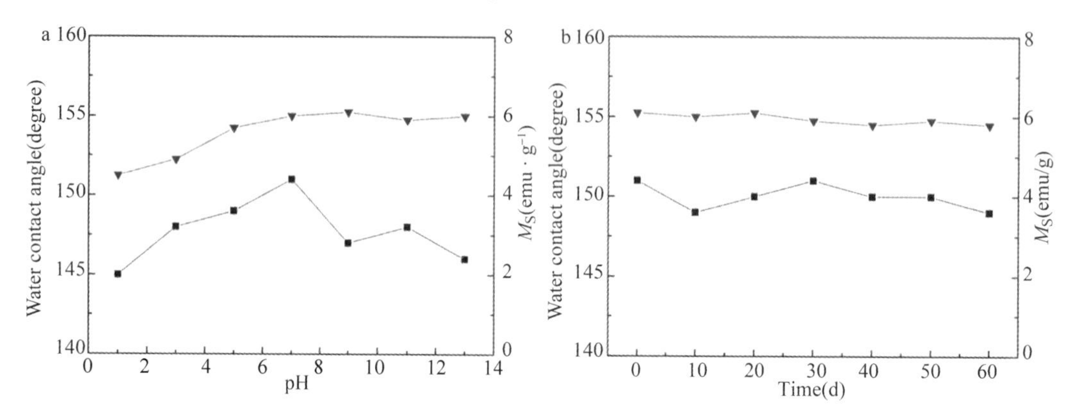

Fig. 9 Variation of water contact angle and M_S of as-prepared sawdust with (a) different pH of aqueous solution and (b) days of storage in air environment

The highly hydrophobic and superoleophilic sawdust exhibited the selective oil adsorption capacity (Supply information Movie. 1). Fig. 10 presents the process of the as-prepared sawdust as the selective oil absorbent for the separation of water and lubrication oil mixture. As expected, the as-prepared sawdust quickly absorbed the red oil while completely repelling water. More importantly, the oil-absorbed sawdust was moved easily with a magnet bar over the water surface and then was quickly collected. The maximal lubrication oil absorption capacity of the as-prepared sawdust was up to 11. 5 times of their weight in 1 min, whereas that of the pristine sawdust was 2. 1 (Supply information Fig. S1), confirming the importance of the superoleophilic and magnetic coatings for the oil absorption. A comparison of the maximum oil adsorption capacity of this product with those of some other absorbents reported in literature was given in Table 1. It was observed that this absorbent has higher adsorption in comparison with other oil absorbents. The oil absorption capacity of the as-prepared absorbent for the methanol, ethanol, hexane, octane, diesel oil and crude oil were also studied (Supply information Fig. S1). These results indicated that the sawdust composites could be used to remove organic contaminants by applying an external magnetic field.

Fig. 10 Removal of lubrication oil from water surface by the as-prepared wood sawdust under magnetic field. The lubricating oil was labeled by oil red Ⅲ dye for clarity

3.7 Regeneration study

The reuse cycles in oil-water mixture have a significant influence on the operation cost. Fig. 11 shows the relationship between the number of reuse cycles and the oil adsorption capacity, magnetic properties, water contact angles of the regenerated materials. It was observed that the adsorption capacity of the adsorbent was not obviously changed within the ten cycles. In the tenth cycle, the adsorption capacity of the sawdust was still 9.5 $g \cdot g^{-1}$, which was higher than other oil adsorbents[37-43]. Moreover, only slight changes in water contact angles and M_S were observed in each cycle. As presented, the water contact angles and M_S remained above 142° and 4.3 $emu \cdot g^{-1}$, respectively. These results indicated that the as-prepared adsorbent possessed excellent chemical stability and reusability. Combined with the advantages of low cost, easy fabrication, environmental friendliness, highly hydrophobic $CoFe_2O_4$/sawdust composites have potential applications for removing spilled oil from water surface.

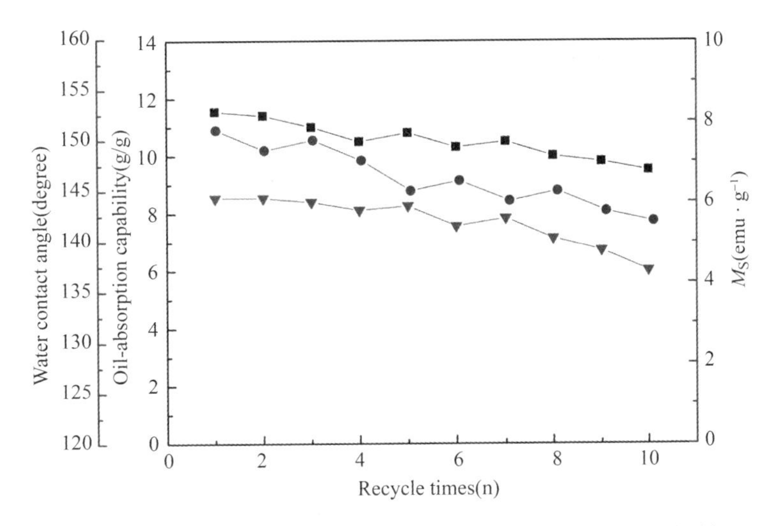

Fig. 11 Oil-absorption capability, water contact angles and M_S of highly hydrophobic sawdust composites after water-oil separation for ten cycles

4 Conclusions

In summary, the present study provided an alternative method for the fast removal of oil from water surface via the highly hydrophobic/superoleophilic magnetic sawdust under an external magnetic field. These sawdust composites were prepared by the combination of wood sawdust and magnetic $CoFe_2O_4$ nanoparticles, and then the modification of the low surface energy polysiloxane layers. The as-prepared sawdust with appropriate magnetic property not only exhibited an oil absorbent capacity up to 11.5 times of their own weight, good chemical stability in corrosive media and environmental durability, but also showed the excellent reusability. The oil absorbent capability was not obviously changed after 10 cycle times, which was of importance in practical applications. This approach provided further routes for improving the utility value of natural waste resources.

Acknowledgments

This work was financially supported by The National Natural Science Foundation of China (grant no. 31470584).

References

[1]Crone T J, Inagaki, M. Magnitude of the 2010 Gulf of Mexico oil leak[J]. Science, 2010, 330(6004): 634-634.

[2]Toyoda M, Inagaki, M. Sorption and recovery of heavy oils by using exfoliated graphite[J]. Spill Sci. Technol. Bull., 2003, 8(5-6): 467-474.

[3]Dalton T, Jin D. Extent and frequency of vessel oil spills in US marine protected areas[J]. Mar. Pollut. Bull., 2010, 60(11): 1939-1945.

[4]Bellino P W, Rangwala A S, Flynn M R. A study of in situ burning of crude oil in an ice channel[J]. Proc. Combust. Inst., 2013, 34: 2539-2546.

[5]Aeppli C, Carmichael C A, Nelson R K, et al. Oil weathering after the Deepwater Horizon disaster led to the formation of oxygenated residues[J]. Environ. Sci. technol., 2012, 46(16): 8799-8807.

[6]Guix M, Orozco J, García M, et al. Superhydrophobic alkanethiol-coated microsubmarines for effective removal of oil [J]. ACS Nano, 2012, 6(5): 4445-4451.

[7]Broje V, Keller A A. Improved mechanical oil spill recovery using an optimized geometry for the skimmer surface[J]. Environ. Sci. Technol., 2006, 40(24): 7914-7918.

[8]Bayat A, Aghamiri S F, Moheb A, et al. Oil spill cleanup from sea water by sorbent materials[J]. Chem. Eng. Technol., 2005, 28(12): 1525-1528.

[9]Korhonen J T, Kettunen M, Ras R H A, et al. Hydrophobic nanocellulose aerogels as floating, sustainable, reusable, and recyclable oil absorbents[J]. ACS Appl. Mater. Interfaces, 2011, 3(6): 1813-1816.

[10]Li L, Breedveld V, Hess, D W. Design and fabrication of superamphiphobic paper surfaces[J]. ACS Appl. Mater. Interfaces, 2013, 5(11): 5381-5386.

[11]Dong X C, Chen J, Ma Y W, et al. Superhydrophobic and superoleophilic hybrid foam of graphene and carbon nanotube for selective removal of oils or organic solvents from the surface of water[J]. Chem. Commun., 2012, 48(36): 10660-10662.

[12]Cao L L, Price T P, Weiss M, et al. Super water-and oil-repellent surfaces on intrinsically hydrophilic and oleophilic porous silicon films[J]. Langmuir, 2008, 24(8): 1640-1643.

[13]Zhu Q, Tao F, Pan Q M. Fast and selective removal of oils from water surface via highly hydrophobic core-shell Fe_2O_3 @C nanoparticles under magnetic field[J]. ACS Appl. Mater. Interfaces, 2010, 2(11): 3141-3146.

[14]Adebajo M O, Frost R L, Kloprogge J T, et al. Porous materials for oil spill cleanup: a review of synthesis and absorbing properties[J]. J. Porous Mater, 2003, 10(3): 159-170.

[15]Chu Y, Pan Q M. Three-dimensionally macroporous Fe/C nanocomposites as highly selective oil-absorption materials [J]. ACS Appl. Mater. Interfaces, 2012, 4(5): 2420-2425.

[16]Thanikaivelan P, Narayanan N T, Pradhan B K, et al. Collagen based magnetic nanocomposites for oil removal applications[J]. Sci Rep-UK, 2012, 2: 230.

[17]Palchoudhury S, Lead J R. A facile and cost-effective method for separation of oil-water mixtures using polymer-coated iron oxide nanoparticles[J]. Environ. Sci. Technol., 2014, 48(24): 14558-14563.

[18]Hui B, Li Y Y, Huang Q T, et al. Fabrication of smart coatings based on wood substrates with photoresponsive behavior and hydrophobic performance[J]. Mater. Des., 2015, 84: 277-284.

[19]Jin C D, Yao Q F, Li J P, et al. Fabrication, superhydrophobicity, and microwave absorbing properties of the magnetic γ-Fe_2O_3/bamboo composites[J]. Mater. Des., 2015, 85: 205-210.

[20]Gan W T, Gao L K, Zhan X X, et al. Hydrothermal synthesis of magnetic wood composites and improved wood properties by precipitation with $CoFe_2O_4$/hydroxyapatite[J]. RSC Adv., 2015, 5(57): 45919-45927.

[21]Ma M, Zhang Y, Yu W, et al. Preparation and characterization of magnetite nanoparticles coated by amino silane

[J]. Colloids Surface., A, 2003, 212(2-3): 219-226.

[22] Yamaura M, Camilo R L, Sampaio L C, et al. Preparation and characterization of (3-aminopropyl) triethoxysilane-coated magnetite nanoparticles[J]. J. Magn. Mater., 2004, 279(2-3): 210-217.

[23] Flis J, Kanoza M. Electrochemical and surface analytical study of vinyl-triethoxy silane films on iron after exposure to air[J]. Electrochim. Acta, 2006, 51(11): 2338-2345.

[24] Ugolev B N. Wood as a natural smart material[J]. Wood Sci. Technol., 2014, 48(3): 553-568.

[25] Feng L, Li S H, Zhu, D B. Super-hydrophobic surfaces: from natural to artificial[J]. Adv. Mater., 2002, 14 (24): 1857-1860.

[26] Li J, Lu Y, Yang D, et al. Lignocellulose aerogel from wood-ionic liquid solution (1-allyl-3-methylimidazolium chloride) under freezing and thawing conditions[J]. Biomacromolecules, 2011, 12(5): 1860-1867.

[27] Yamashita T, Hayes P. Analysis of XPS spectra of Fe^{2+} and Fe^{3+} ions in oxide materials[J]. Appl. Surf. Sci., 2008, 254(8): 2441-2449.

[28] Kim H M, Uenoyama M, Kokubo T. Biomimetic apatite formation on polyethylene photografted with vinyltrimethoxysilane and hydrolyzed[J]. Biomaterials, 2001, 22(18): 2489-2494.

[29] Gierlinger N, Goswami L, Schmidt M, et al. In situ FT-IR microscopic study on enzymatic treatment of poplar wood cross-sections[J]. Biomacromolecules, 2008, 9(8): 2194-2201.

[30] Bruni S, Cariati F, Solinas S. IR and NMR study of nanoparticle-support interactions in a Fe_2O_3-SiO_2 nanocomposite prepared by a sol-gel method[J]. Nanostruct. Mater., 1999, 11(5): 573-586.

[31] Li G S, Li P L, Inomata H. Characterization of the dispersion process for $NiFe_2O_4$ nanocrystals in a silica matrix with infrared spectroscopy and electron paramagnetic resonance[J]. J. Mol. Struct., 2001, 560(1-3): 87-93.

[32] Maaz K, Mumtaz A, Hasanain S K, et al. Synthesis and magnetic properties of cobalt ferrite ($CoFe_2O_4$) nanoparticles prepared by wet chemical route[J]. J. Magn. Magn. Mater., 2007, 308(2): 289-295.

[33] Lu H M, Zheng W T, Jiang Q. Saturation magnetization of ferromagnetic and ferrimagnetic nanocrystals at room temperature[J]. J. Phys. D Appl. Phys., 2007, 40(2): 320.

[34] Berkowitz A E, Schuele W J, Flanders P J. Influence of crystallite size on the magnetic properties of acicular γ-Fe_2O_3 particles[J]. J. Appl. Phys., 1968, 39(2): 1261-1263.

[35] Coey J M D. Noncollinear spin arrangement in ultrafine ferrimagnetic crystallites[J]. Phys. Rev. Lett., 1971, 27 (17): 1140.

[36] Gan W T, Liu Y, Gao L K, et al. Growth of $CoFe_2O_4$ particles on wood template using controlled hydrothermal method at low temperature[J]. Ceram. Int., 2015, 41(10): 14876-14885.

[37] Kodama R H. Magnetic nanoparticles[J]. J. Magn. Magn. Mater., 1999, 200(1-3). 359-372.

[38] Chinnasamy C N, Senoue M, Jeyadevan B, et al. Synthesis of size-controlled cobalt ferrite particles with high coercivity and squareness ratio[J]. J. Colloid Interface Sci., 2003, 263(1): 80-83.

[39] Kumar A, Sharma G, Naushad M, et al. SPION/β-cyclodextrin core-shell nanostructures for oil spill remediation and organic pollutant removal from waste water[J]. Chem. Eng. J., 2015, 280: 175-187.

[40] Banerjee S S, Joshi M V, Jayaram R V. Treatment of oil spill by sorption technique using fatty acid grafted sawdust [J]. Chemosphere, 2006, 64(6): 1026-1031.

[41] Wang D, Silbaugh T, Pfeffer R, et al. Removal of emulsified oil from water by inverse fluidization of hydrophobic aerogels[J]. Powder Technol., 2010, 203(2): 298-309.

[42] Li J, Luo M, Zhao C J, et al. Oil removal from water with yellow horn shell residues treated by ionic liquid[J]. Bioresour. Technol., 2013, 128: 673-678.

[43] Ali N, El-Harbawi M, Jabal A A, et al. Characteristics and oil sorption effectiveness of kapok fibre, sugarcane bagasse and rice husks: oil removal suitability matrix[J]. Environ. Technol., 2012, 33(4): 481-486.

中文题目：在磁场作用下通过有用的可循环 $CoFe_2O_4$/木粉复合材料从水面去除油

作者：甘文涛，高丽坤，张文博，李坚，蔡力平，詹先旭

摘要：在本研究中，杨树木屑被用作可回收、低成本、环境友好和高选择性的吸油材料。通过在木粉表面沉淀 $CoFe_2O_4$ 并对低表面能聚硅氧烷层进行化学改性，获得了具有高疏水性、超亲油性和磁性的木粉。这些复合材料能够吸收高达木粉重量 11.5 倍的润滑油，同时完全排斥水。此外，由于其特殊的磁性，吸油复合材料能够在外部磁体的帮助下方便地从水面收集油。此外，木粉可以被清理并至少 10 次重新用于水—油分离。这项工作为利用自然废弃资源处理水污染提供了一条可行途径。

关键词：木屑吸附剂；除油；磁性 $CoFe_2O_4$ 颗粒；可回收性

Fabrication of Microwave Absorbing $CoFe_2O_4$ Coatings with Robust Superhydrophobicity on Natural Wood Surfaces*

Wentao Gan, Likun Gao, Wenbo Zhang, Jian Li, Xianxu Zhan

Abstract: A superhydrophobic wood surface with microwave absorption property was prepared based on the formation of $CoFe_2O_4$ nanoparticles and subsequent hydrophobization using fluorinated alkylsilane (FAS). Meanwhile, sticky epoxy resin was worked as a caking agent by adhering abundant of $CoFe_2O_4$ nanoparticles to wood surface. The as prepared superhydrophobic coatings on wood maintain stable super hydrophobicity after suffering a significant abrasion. Moreover, the complex permeability and permittivity of the coated wood composites were measured in the frequency range of 2-18 GHz by vector network analysis. The microwave absorption properties were elucidated by the traditional coaxial line method. The results show that the as-prepared wood composites have excellent microwave absorption properties at the frequency of 16 GHz, and the minimum reflection loss can reach -12.3 dB. The approach presented may provide further routes for designing outdoor wood wave absorbers with a specified absorption frequency.

Keywords: Wood; $CoFe_2O_4$ nanoparticles; Superhydrophobicity; Microwave absorption

1 Introduction

With the rapid development of electronic industry, the gigahertz electromagnetic wave results in a serious problem of electromagnetic pollution, which will do harm to human health and decrease the sensitivity of electronic equipment, causing severe fault of date or accidents[1-3]. An effective solution for the electromagnetic interference problem is using the electromagnetic wave absorbing materials. Nano materials[4,5] and composite materials[6,7] are frequently studied for eliminating and preventing the electromagnetic radiation.

Wood has been used as an important material in the building and construction industries for thousands of years because of the remarkable features such as good mechanical properties, light weight, aesthetic appearance, excellent processability and sustainability[8-10]. Thus, the electromagnetic wave absorbing materials based on natural wood are necessary to be designed and studied extensively. Oka et al. have proposed a kind of powdertype magnetic wood to achieve a specific matching frequency for an electromagnetic wave absorber[11]. However, wood materials have shortcomings due to its organic constitution, such as the poor fire resistance, slow destruction by fungi, insects and microorganisms, and easy to warp, split and decay from water absorption or UV exposure[12]. Therefore, wood protective treatments have been necessary as long as it has been in use. Nowadays, functionalizing wood by inorganic particles application has been proved to be an effective method to improve the wood performance and prolong its service life[13]. For example, Saka and co-

* 本文摘自 Ceramics International, 2016: 13199-13206.

workers fabricated TiO_2/wood composites, which showed the TiO_2 gels deposited within the wood could improve the properties of wood in dimensional stability and fire resistance[14]. Wang et al. prepared a superhydrophobic spherical-like α-FeOOH film on wood surface, yielding a superhydrophobic wood material with reduced water uptake as well as increased dimensional stability[15]. Salla et al. found that the wood materials coated with ZnO nanoparticles exhibited enhanced UV resistance[16].

Magnetic nanoparticles have been developed by researchers from a wide range of disciplines, such as catalysis, biomedicine, data storage, magnetic recording and microwave absorption fields[17-19]. Recently, many studies also found that wood materials could be the effective scaffolds for magnetic functionalization, and the magnetic wood materials not only possess wood structure but also exhibit excellent magnetic properties due to the remarkable physical and chemical properties of magnetic nanoparticles[20,21].

In this study, an electromagnetic wave absorbing materials based on natural wood were developed through combining the wood materials with superhydrophobic $CoFe_2O_4$ coatings. Epoxy resin provides a robust connection between the wood and $CoFe_2O_4$ nanoparticles. We created an ethanol-based magnetic suspension that can be painted onto wood surfaces to create a superhydrophobic and electromagnetic wave absorbing coating. Moreover, the morphology and structure of the wood composite were also investigated.

2 Experimental section

2.1 Materials

Wood cubes of Cathay poplar (Populus cathayana Rehd) with dimensions 20×20×20 mm^3 (longitudinal× radial×tangential) were ultrasonically rinsed in deionized water for 30 min and then dried at 103 ℃ for 48 h. Cobalt chloride hexahydrate ($CoCl_2 \cdot 6H_2O$), ferrous sulfate heptahydrate ($FeSO_4 \cdot 7H_2O$), sodium hydroxide (NaOH), potassium nitrate (KNO_3), 1H, 1H, 2H, 2H - perfluorodecyltriethoxysilane ($C_{16}F_{17}H_{19}O_3Si$, FAS), absolute ethyl alcohol and ethyl acetate used in this study were supplied by Shanghai Boyle Chemical Company Limited and used without further purification. Epoxy resin and curing agent 651 were purchased from Nantong Synthetic Material Co., Ltd (China).

2.2 Preparation of magnetic $CoFe_2O_4$ nanoparticles

As shown in Fig. 1, in a typical synthesis, $CoCl_2 \cdot 6H_2O$ (0.05 M) and $FeSO_4 \cdot 7H_2O$ (0.1 M) were dissolved in water at room temperature. Then the homogeneous suspension was transferred into a Teflon-lined stainless-steel autoclave (100 mL) and heated to 110 ℃ for 3 h to produce the nonmagnetic metal hydroxides. At the end of the reaction, the supernatant was poured out and 60 mL of a 1.32 mol · L^{-1} NaOH solution with KNO_3($[Fe^{2+}]/[NO^{3-}] = 0.44$) was added to the Teflon-lined stainless-steel autoclave. The KNO_3 was used as an oxidizing agent. After a reaction at 110 ℃ for 10 h, the solution was cooled to room temperature. Then the resultant product was washed with deionized water and absolute ethanol for several times, and finally dried at 60 ℃ for 12 h. The formation of the $CoFe_2O_4$ was shown in Fig. 1. 2.2 Preparation of magnetic $CoFe_2O_4$ nanoparticles.

2.3 Preparation of the coated wood composites

$CoFe_2O_4$ nanoparticles (1 g) were dispersed in absolute ethyl alcohol (10 mL) of FAS with the volumetric concentration of TO for 24 la at room temperature. Subsequently, the samples were washed, and then dried at 60 ℃ for 12 h to obtain hydrophobic $CoFe_2O_4$ nanoparticles. The FAS was coated onto the surface of $CoFe_2O_4$ by a silanization reaction. Firstly, the FAS was hydrolyzed, and ethyoxyl groups were replaced by hydroxyl groups. Then the reactive silane groups were condensed with others to produce siloxane bonds (Si-O-Si) and formed a silane polymer through the condensation reaction. Subsequently, the silane polymer was as—sociated with the magnetic nanoparticles and formed a covalent bond with OH groups on the bare Fe atoms[22]. With the continuous dehydration reaction, the silane polymer was formed on the $CoFe_2O_4$ surface (Fig. 1).

Epoxy resin (20 g) was mixed with ethyl acetate (30 mL) under magnetic stirring for 20 min to form the Solution A, and the corresponding curing agent (40 g) was dissolved in ethyl acetate (30 mL) to form the Solution B. The Solution A and B were mixed with magnetic stirring for 30 min to obtain a primer coating solution of epoxy resin. The wood samples were dipped into the solution for 3 min, and then dried in air for 20 min This procedure was repeated for five times. Prior to the coating treatment, 1 g FAS-modified $CoFe_2O_4$ nanoparticles were placed into 5 mL ethyl al cohol, and the solution was mechanically stirred for 30 min. The epoxy resin precoated wood samples were dipped into the as—prepared coating solution for 4 min, and dried in air for 20 min This procedure was repeated for 3 times to allow full deposition of the $CoFe_2O_4$, nanoparticles on the wood surface. Finally, the treated samples were dried at 80 ℃ for 12 h.

2.4 Characterization

The surface morphology of the samples was characterized by the scanning electron microscopy (SEM, FEI, Quanta 200). Atomic force microscopy (AFM) images were obtained using a Multimode Nanoscope Ⅲa controller (Veeco Inc., USA) with a silicon tip operated in a tapping mode to characterize the surface roughness. Crystalline structures of the samples were identified by the X-ray diffraction technique (XRD, Rigaku, D/MAX 2200) operating with Cu $K\alpha$ radiation ($\lambda_{1/4}$ = 1.5418 Å) at a scan rate (2θ) of 4° min1 and the accelerating voltage of 40 kV and the applied current of 30 mA ranging from 5° to 80°. The FTIR spectra for the wood samples were recorded using FTIR (Magna-IR 560, Nicolet). The X-ray photoelectron spectroscope (XPS) was recorded on Thermo ESCALAB 250XI. The deconvolution of the overlapping peaks was performed using a mixed Gaussian-Lorentzian fit program. For magnetic characterization, wood specimens of 4×4×4 mm^3 (longitudinal×radial×tangential) were used. The magnetic properties of the composites were measured by a superconducting quantum interference device (MPMS XL-7, Quantum Design Corp.) at a room temperature (300 K). The contact angle analyzer (JC2000C, Beijing code tong technology co., LTD.) at ambient temperatures with a droplet volume of 5 μL was employed to measure the WCA of the samples. An average of the five measurements taken at different positions on each sample was applied to calculate the final WCA angle. For EM absorption measurement, the ring wood samples (Φ out1/4 7.00 mm, Φ in 1/4 3.04 mm, thickness 1/4 2 mm) were prepared for measuring the material constants. The complex permittivity and permeability values were measured with coaxial wire method in the frequency range of 2-18 GHz with the network analyzer (N5244A PNA-X).

The sandpaper abrasion test was performed according to the reported methods[23]. As schematically depicted in Fig. 6a, under a load of 200 g weight (5 kPa), the sample surface was rubbed against the sandpaper (1500 mesh), and moved for 25 cm along the ruler with a speed of 5 cm · s^{-1}. The process was defined as one-abrasion cycle, and ten cycles were conducted. CAs was measured after each abrasion cycle. Changes in surface morphology of the coated wood were examined by scanning electron microscopy.

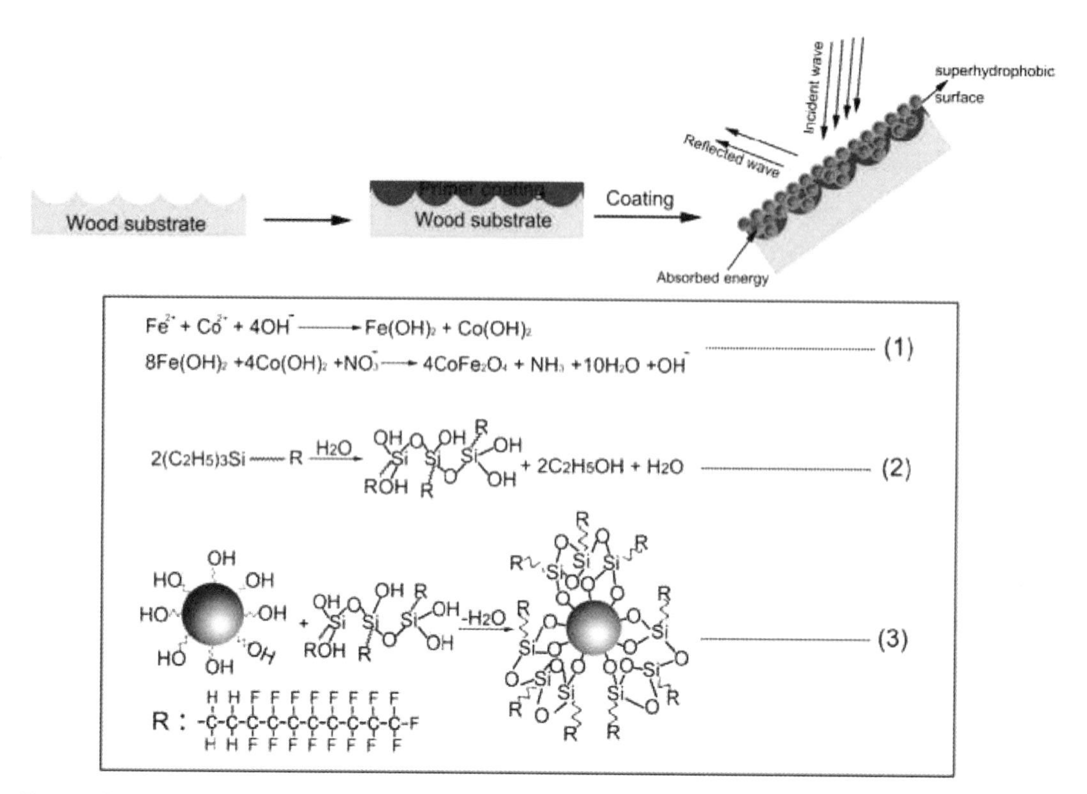

Fig. 1 Sketch of the procedure to prepare superhydrophobic $CoFe_2O_4$ coatings on the wood surface

3 Results and discussion

Fig. 2a shows a typical SEM image of the pristine wood surface, on which the micro-grooved structures with minor features such as pits and tracheid can be clearly observed. Upon treatment with FAS/ $CoFe_2O_4$/ epoxy resin nanocomposites, a particulate morphology can be found on the surface (Fig. 2b). The high-magnification image (Fig. 2c) indicates dual-scale roughness patterns consisting of FAS-modified $CoFe_2O_4$ nanoparticles superimposed on random microstructures formed by the aggregation of magnetic nanoparticles. The surface roughness was also examined by the atomic force microscopy, and it can be reflected in the rms roughness values (rms1/4 root mean square1/4 the standard deviation of the Z value, Z being the total height range analyzed) of 84.02 nm (Fig. S1). Moreover, cross-sectional SEM image of the coated wood (Fig. S2) shows that the $CoFe_2O_4$ coating with a thickness of 150 mm is sufficient to cover the wood substrate. Fig. S3 shows the SEM image of the as-prepared wood and corresponding elemental maps of C, Fe and Co, it can confirm that the $CoFe_2O_4$ nanoparticles were coated on wood surface homogeneously.

The XRD patterns of the samples are exhibited in Fig. 3. Apparently, the diffraction peaks at 16° and 22°, corresponding to the (101) and (002) crystal planes of cellulose, are observed in both the spectrum of the pristine wood and the coated wood[24]. The peaks at 2θ values of 30°, 35°, 37°, 43°, 53°, 56° and 62° attributed to the (220), (311), (222), (400), (422), (511) and (440) planes of cobalt ferrite crystallite (JCPDS file no. 22-1086), respectively[25]. Compared with the XRD patterns of the pristine wood and as prepared $CoFe_2O_4$, all the diffraction peaks of wood and $CoFe_2O_4$ exist in the coated wood composites,

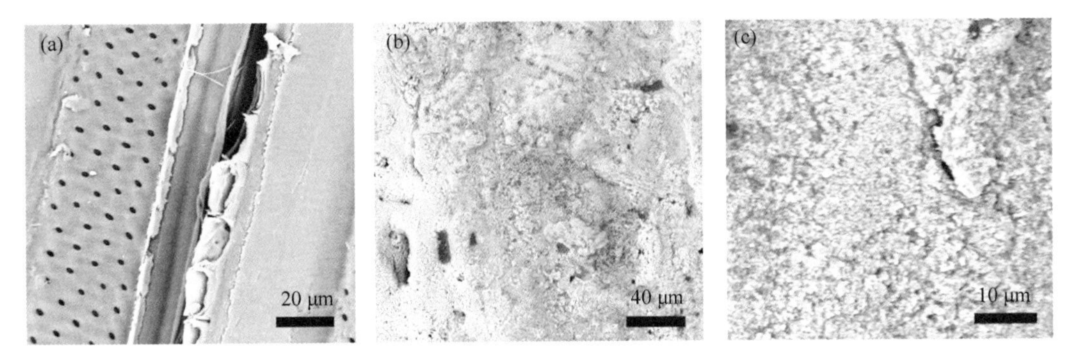

Fig. 2 SEM images of (a) the pristine wood, (b) and (c) the coated wood at different magnification

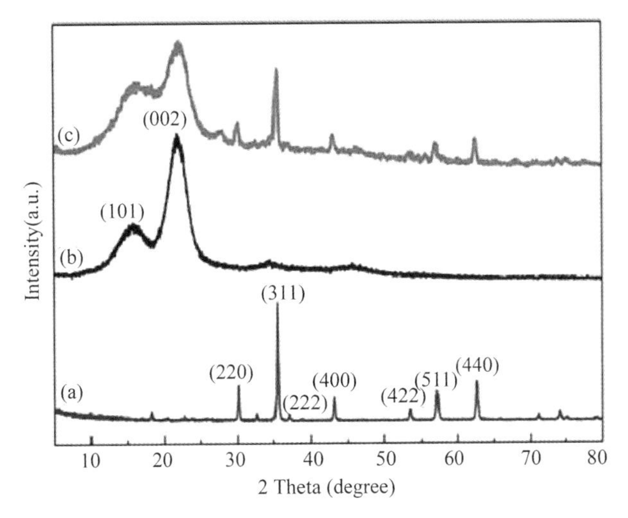

Fig 3 XRD patterns of (a) the as-prepared $CoFe_2O_4$ nanoparticles, (b) the pristine wood and (c) the coated wood composites

revealing that the $CoFe_2O_4$ covered on the wood surface effectively.

The chemical nature of the wood composites before and after the coating treatment has been studied by FTIR and XPS. From the FTIR spectra (Fig. 4) of the pristine wood, it is saw that the broad and intense absorption peak at around 3376 cm^{-1} corresponds to the O—H stretching vibrations of wood compounds (cellulose and lignin) and absorbed water. The peak at 2904 cm^{-1} is attributed to the C—H stretching vibrations of aliphatic acids. The peak around 1740 cm^{-1} is due to the carbonyl (C1/4 O) stretching vibration of the carboxyl groups of hemicelluloses and lignin in wood. The peak at 1597 cm^{-1} shows the C1/4 C stretching vibration of the aromatic or benzene ring in lignin. The vibration at 1426 cm^{-1} could be due to C—H groups in the plane deformation vibrations of methyl, methylene and methoxyl groups. The bands in the range 1300-1000 cm^{-1} could be assigned to the C—O stretching vibration of hemicelluloses and lignin[26]. For the coating treatment, new peaks appear at 1244 cm^{-1} and 1161 cm^{-1}, corresponding to the C—F stretching vibrations[27]. Peak at 1035 cm^{-1}, characteristic of Si-O-Si asymmetric vibrations, is also observed. Besides, the peaks at 587 cm^{-1} and 826 cm^{-1} are attributed to the Fe-O-Si bond vibration of FAS-modified $CoFe_2O_4$ and epoxy groups in the epoxy resin, respectively.

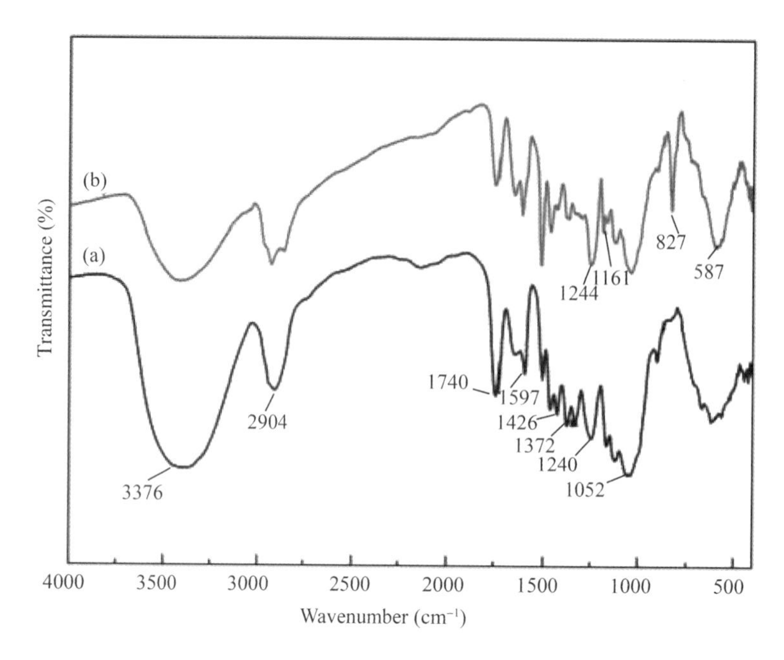

Fig. 4 FTIR spectra of (a) the pristine wood and (b) the coated wood

The composition of the coated wood was examined accurately using XPS. Fig. 5a shows that the characteristic signals for C, O, Fe, Co, N, F and Si elements are clearly observed in the XPS survey spectrum of the coated wood composites. The peaks of Si2p and F1s belong to the layer of fluorinated alkylsilane ($F_3CF_2CF_2CF_2CF_2CF_2CF_2CF_2CH_2CH_2CSi-$) could be found at 154. 3 eV and 686. 6 eV, respectively. Besides, the peak of N1 s at 398. 2 eV is attributed to the functional groups in curing agent 651. Fig. 5b-d illustrate the high-resolution XPS spectra of 2p Co, Fe and Si, respectively. The peaks at 781. 5 eV is from $Co2p_{3/2}$ with a satellite peak at 785. 7 eV, while the peak at 796. 8 eV is ascribed to the $Co2p_{1/2}$ level, with a satellite peak at 805. 0 eV. The Fe2p level with binding energies of 711. 8 and 725. 2 eV is assigned to $Fe2p_{3/2}$ and $Fe2p_{1/2}$, respectively. No obvious broadening is observed for the XPS peaks, revealing the pure $CoFe_2O_4$ phase of the product[30]. In Fig. 5d, the Si2p spectra is attributed to the Si-O-Si bond[31]. These suggest that the FAS-modified $CoFe_2O_4$ coating is successfully covered on the wood surface.

The wettability of wood surface before and after painting was evaluated by exploring contact angle through a sessile drop on specimen. An apparent water contact angle of 0° is observed on the pristine wood surface as seen in Fig. 6a, which attribute to the subsistence of abundant hydroxyl groups in basic wood materials as proved in FTIR analysis. The wood surface also exhibits hydrophilicity with a water contact angle of 69° after covering with epoxy resin coating (Fig. 6b). Further treatment with FAS-modified $CoFe_2O_4$ nanoparticles made the sample turn into superhydrophobicity with a high CA value of 158° (Fig. 6c). The superhydrophobicity of the as-prepared wood can be attributed to the $CoFe_2O_4$ aggregations that increased the surface roughness and FAS decreased the surface energy of wood surface by the long chain fluoroalkyl.

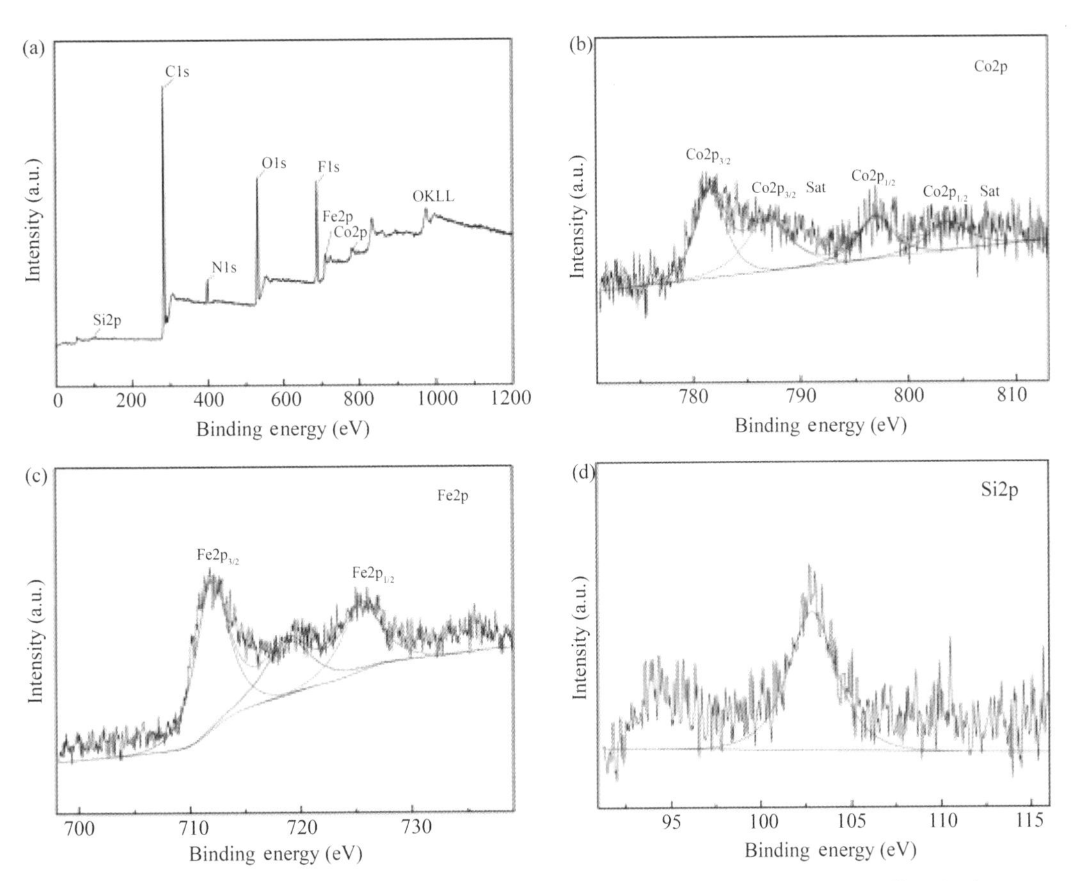

Fig. 5 XPS spectra of (a) survey spectrum, (b) Co2p, (c) Fe2p and (d) Si2p in the coated wood composites, respectively

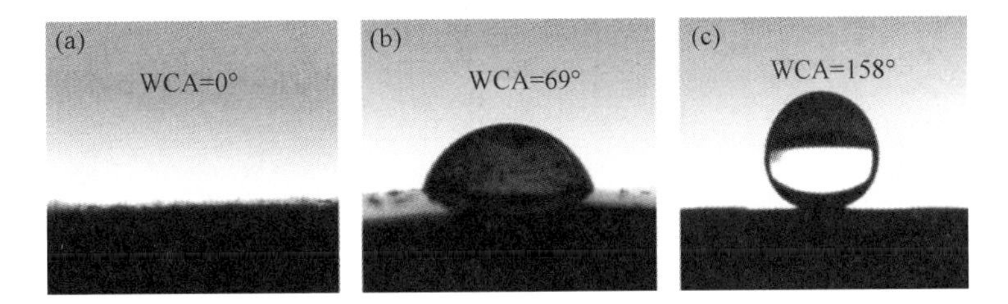

Fig. 6 Photographs of water droplets on (a) the pristine wood, (b) the epoxy resin pre-coated wood, and (c) the hydrophobic $CoFe_2O_4$ coated wood

For most of textured surfaces, the big drawback is that they are readily abraded, because the surface roughness at the micro-or nanoscale is usually mechanically weak[32]. Thus, assessing the durability of superhydrophobic wood surface has an important significance. The sandpaper abrasion tests were carried out to examine the abrasion durability of the fabricated superhydrophobic wood surface. With a pressure of 5 kPa exerted, the coated wood was rubbed against the sandpaper (1500 mesh) and moved for 25 cm along the ruler (Fig. 7a). After ten abrasion cycles, it is observed that the surface microstructures with micro-and nanoscale roughness patterns are still retained upon being abraded repeatedly, but some scratches present in wood surface

(Fig. 7b). The water CAs after each abrasion cycle is shown in Fig. 7c. It can be noted that the coated wood surface still sustained the high hydrophobicity with CAs around 148°, after being scratched repeatedly. The stable hydrophobicity can be attributed to the adequate thickness of the magnetic FAS/ $CoFe_2O_4$ film. The surface texture can sustain the abrasion damage by regenerating the surface patterns, because a removal of the top layer of the hydrophobic $CoFe_2O_4$ nanoparticles will expose a new rough surface, which is similar to the previous study[29].

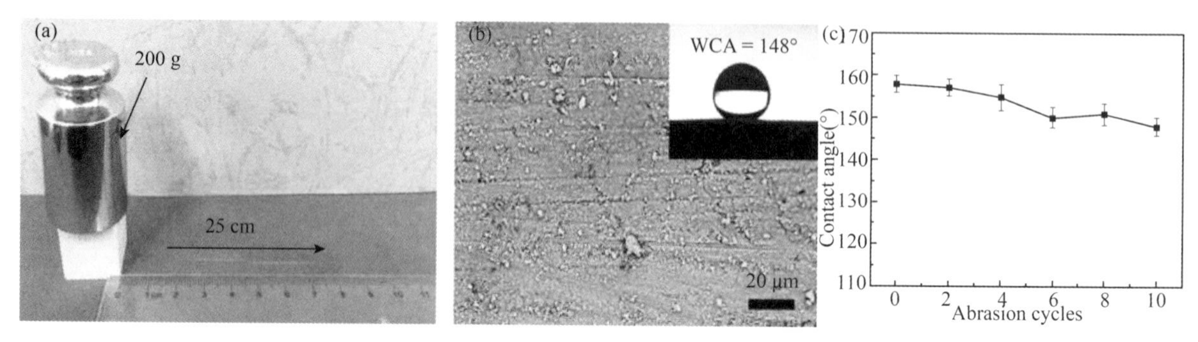

Fig. 7 (a) Photograph depicting the sandpaper abrasion test; (b) SEM images of the coated wood surface after 10 cycles of abrasion; (c) CAs as a function of mechanical abrasion cycles for the coated wood

The magnetic property is important to investigate the electromagnetic wave absorption properties; the magnetic properties of the pristine wood and the coated wood were measured at room temperature. As observed in Fig. 8, the hysteresis curve of the pristine wood is a straight line close to zero, suggesting the nonmagnetic property of wood materials. However, the coated wood exhibits a polar property and shows a typical ferromagnetic behavior. The saturation magnetization (M_S) and coercive force (H_c) of the treated wood composites obtained at room temperature is 20.5 emu · g^{-1} and 2006.4 Oe, respectively. The difference in magnetic properties of these two composites may be attributed to the introduction of magnetic $CoFe_2O_4$.

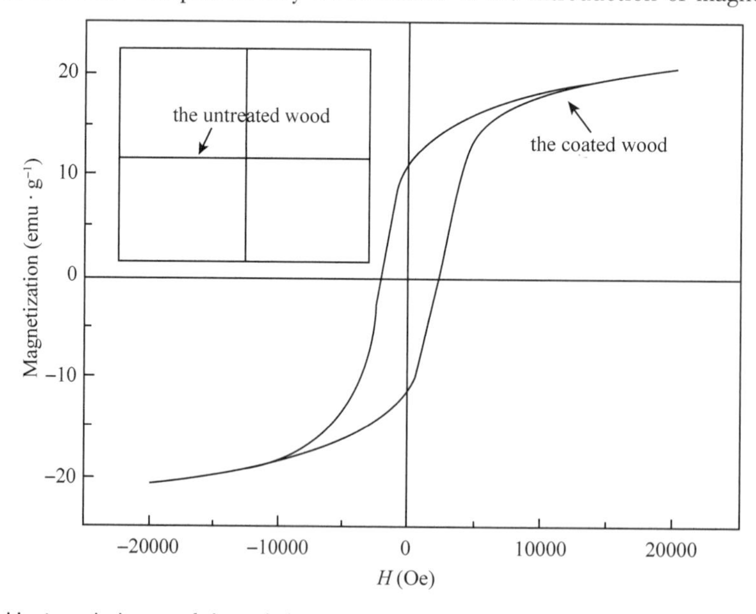

Fig 8 Hysteresis loops of the pristine wood and the coated wood at room temperature

The complex permittivity and permeability of the pristine wood and the coated wood are showed in Fig. 9. The real permittivity (ε') and real permeability (μ') represent the storage ability of electromagnetic energy, and the imaginary permittivity (ε'') and imaginary permeability (μ'') are connected with the energy dissipation and magnetic loss, respectively[33]. It is observed that the values of ε' and ε'' for the coated wood are larger than that of pristine wood. However, the ε'' of the coated wood shows one major peak at around $f1/4$ 16 GHz. The corresponding real part of permittivity (ε') decreases in the range of 15.5-16.5 GHz, then goes up again. The resonance peak in the curves of complex permittivity could be interpreted because of atomic and electronic polarization[34,35]. From Fig. 9c and d, it is also known that the relative permeability of the pristine wood and the coated wood are almost unity, and the real part and the imaginary part are nearly 1 and 0, respectively. However, the coated wood shows a negative μ'' appear in 16 GHz, which signifies the magnetic energy radiate out without any absorption[36].

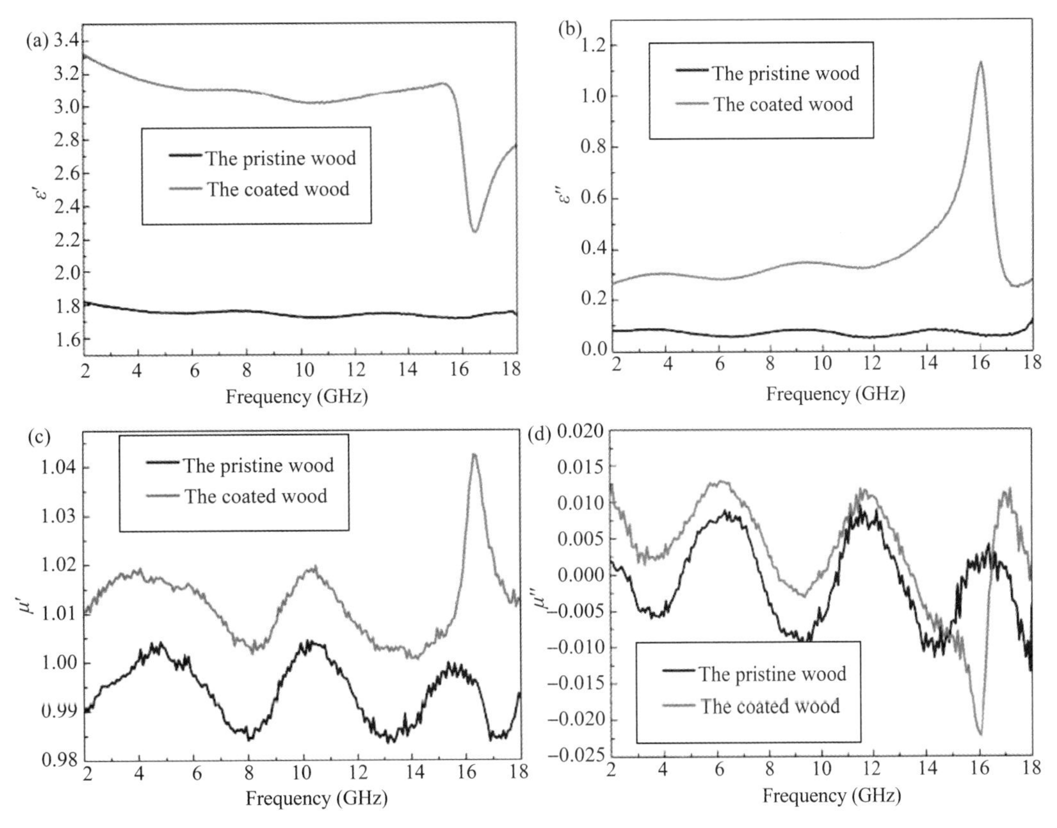

Fig. 9 Frequency dependence of (a) real and (b) imaginary parts of relative complex permittivity; (c) real and (d) imaginary parts of relative complex permeability of samples

From the values of dielectric loss and magnetic loss (shown in Fig. 10), it can be concluded that the dielectric loss and magnetic loss of the pristine wood are 0, indicating the electromagnetic wave can be broke through the wood materials without any energy loss. Moreover, the dielectric loss values of the coated wood are higher than their magnetic loss, so it can be deduced that the as-prepared wood is a dielectric loss absorbent, and the maximum value of dielectric loss is 0.44 (16 GHz), which are in accordance with the reflection loss peak (shown in Fig. 11a). The causes for dielectric loss could be explained by the electronic dipole polarization of the $CoFe_2O_4$ layer. Besides, the magnetic loss of magnetic materials mainly originates from domain wall resonance, eddy current effect, natural resonance, and hysteresis[37]. In this study, the eddy current effect and natural resonance may be responsible for the attenuation of electromagnetic wave over 2–18 GHz frequency range, because the hysteresis loss in negligible in the weak field and the domain wall resonance loss occurs at MHz frequency. The eddy current loss is defined in Eq. (1):

$$\mu^{n} \approx 2\pi\mu_0(\mu')^2\sigma d^2 f/3 \tag{1}$$

where μ_0(H · m^{-1}) is the electric permeability and s (S · cm^{-1}) is the electric conductivity in vacuum. If the reflection loss results from the eddy current effect, the values of $C_0[C_0=\mu''(\mu')^{-2}f^{-1}]$ are constant when the frequency is changing. As shown in Fig. S4, the values of C_0 for the as-prepared wood are almost constant in the range of 2–18 GHz, revealing the presence of the eddy current loss. Moreover, the natural resonance can be evaluated by Eqs. (2) and (3):

$$2\pi f_r = rH_a \tag{2}$$

$$H_a = 4|K_1|/3\mu_0 M_3 \tag{3}$$

where r is the gyromagnetic ratio and Ha is the anisotropy energy. $|K_1|$ and Ms are the anisotropy coefficient and the saturation magnetization, respectively. The resonance frequency depends on the effective anisotropy field, which is associated with coercivity values of the materials[38]. As shown in Fig. 8, the coercivity value of the as-prepared wood composites is higher than the pristine wood, which is also beneficial to electromagnetic wave absorption.

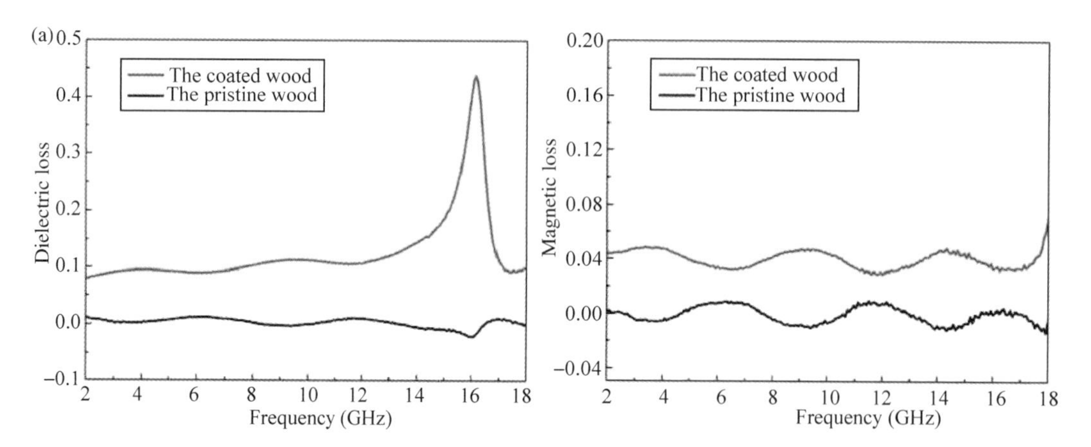

Fig. 10 Frequency dependence of (a) dielectric loss and (b) magnetic loss of samples

To study the microwave absorption property, the reflection loss (RL) of the electromagnetic radiation was calculated. On the basis of the measured data of permittivity and permeability, reflection loss (RL) usually can be calculated by the following expression[39]:

$$R = 20\log\left|\frac{Z_{in}-1}{Z_{in}+1}\right| \tag{4}$$

where Z_{in} is the input characteristic impedance, which can be expressed as[40]:

$$Z_{in} = \sqrt{\frac{\mu_r}{\varepsilon_r}}\tanh\left[j\left(\frac{2f_\pi d}{c}\right)\sqrt{\mu_r\varepsilon_r}\right] \tag{5}$$

where ε_r and μ_r are the complex permittivity and permeability of the absorber, respectively, f is the frequency, d is the thickness of the absorbent and c is the velocity of light in free space.

As observed in Fig. 11, the coated wood composite shows enhanced wave absorption properties. Fig. 11a shows the theoretical reflection loss (RLs) of the pristine wood and the coated wood at a thickness of 3.5 mm in the range of 2–18 GHz. It is observed that the reflection loss of the coated wood composites is much lower than that of the pristine wood. The minimum reflection loss of the coated wood reaches −12.3 dB at 16.1 GHz. Fig. 11b, shows the three-dimensional presentations of calculated theoretical RLs of the coated wood at various thicknesses (1−5 mm) in the frequency range of 2−18 GHz. This indicates that the microwave absorbing ability of the coated wood at different frequencies can be adjusted by controlling the thickness of the absorbents.

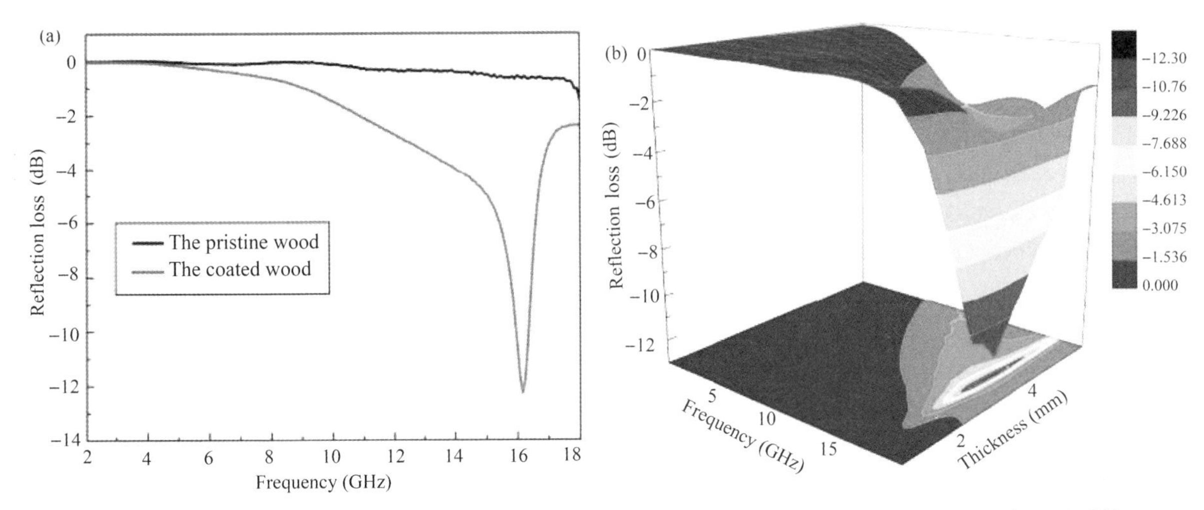

Fig. 11 (a) Reflection coefficient of the products with a thickness of 3.5 mm in the range of 2–48 GHz, (b) Three—dimensional presentations of the reflection loss of the coated wood

On the one hand, good absorbing materials should have a good impedance matching with free space and low surface reflectivity of the electromagnetic wave. On the other hand, it should have high absorptive capacity[41]. In this study, the coated wood composites can reach this condition. Because the diameter of $CoFe_2O_4$ nanoparticles is much less than the incident wavelength, so it can greatly reduce the reflection from the wood surface, which makes better impedance matching. More important, compare with the pristine wood, when the electromagnetic wave penetrates the coated wood, the electron of the coating layer will be polarized to cause dielectric loss, and thus lead to better wave absorbing performance.

4 Conclusions

Superhydrophobic wood composites with microwave absorption property were prepared by a simple process: using the combination of FAS-modified $CoFe_2O_4$ nanoparticles and epoxy resin to obtain textured wood surface. The resulting superhydrophobic coating on wood showed remarkable tolerance against sandpaper abrasion while retaining their hydrophobicity. Moreover, the as prepared wood also exhibited excellent microwave absorption property at the frequency of 16 GHz, and the minimum reflection loss can reach −12.3 dB. The main microwave absorbing mechanism of the coated wood is the dielectric loss. The approach presented in this work may provide further routes for designing outdoor wood wave absorbers with a specified absorption frequency.

Acknowledgments

This work was financially supported by The National Natural Science Foundation of China (grant no. 31470584).

Appendix A. Supplementary material

Supplementary data associated with this article can be found in the online version at http://dx. doi. org/10. 10J6/j. cei-anoint. 2016. OS. J12.

References

[1] Wang G S, Wu Y, Wei Y Z, et al. Fabrication of reduced graphene oxide (RGO)/Co_3O_4 nanohybrid particles and a RGO/Co_3O_4/poly (vinylidene fluoride) composite with enhanced wave-absorption properties[J]. ChemPlusChem, 2014, 79(3): 375.

[2] Xu H L, Bi H, Yang R B. Enhanced microwave absorption property of bowl-like Fe_3O_4 hollow spheres/reduced graphene oxide composites[J]. J. Appl. Phys., 2012, 111(7): 07A522.

[3] Bai X, Zhai Y H, Zhang Y. Green approach to prepare graphene-based composites with high microwave absorption capacity[J]. J. Phys. Chem. C, 2011, 115(23): 11673-11677.

[4] Shah A, Wang Y H, Huang H, et al. Microwave absorption and flexural properties of Fe nanoparticle/carbon fiber/epoxy resin composite plates[J]. Compos. Struct., 2015, 131: 1132-1141.

[5] Qiang C W, Xu J C, Zhang Z Q, et al. Magnetic properties and microwave absorption properties of carbon fibers coated by Fe_3O_4 nanoparticles[J]. J. Alloys Compd., 2010, 506(1): 93-97.

[6] Xie S, Yang Y, Hou G Y, et al. Development of layer structured wave absorbing mineral wool boards for indoor electromagnetic radiation protection[J]. J. Build. Eng., 2016, 5: 79-85.

[7] Li B Y, Duan Y P, Zhang Y F, et al. Electromagnetic wave absorption properties of cement-based composites filled with porous materials[J]. Mater. Des., 2011, 32(5): 3017-3020.

[8] Ermeydan M A, Cabane E, Hass P, et al. Fully biodegradable modification of wood for improvement of dimensional stability and water absorption properties by poly (ε-caprolactone) grafting into the cell walls[J]. Green Chem., 2014, 16(6): 3313-3321.

[9] Ugolev B N. Wood as a natural smart material[J]. Wood Sci. Technol., 2014, 48(3): 553-568.

[10] W. Gan, Y. Liu, L. Gao, et al. Growth of $CoFe_2O_4$ particles on wood template using controlled hydrothermal method at low temperature[J]. Ceram. Int., 2015, 41(10): 14876-14885.

[11] Oka H, Uchidate S, Sekino N, et al. Electromagnetic wave absorption characteristics of half carbonized powder-type magnetic wood[J]. IEEE Trans. Magn., 2011, 47(10): 3078-3080.

[12] Keplinger T, Cabane E, Chanana M, et al. Aversatile strategy for grafting polymers to wood cell walls[J]. Acta Biomater., 2015, 11: 256-263.

[13] Mahltig B, Swaboda C, Roessler A, et al. Functionalising wood by nanosol application[J]. J. Mater. Chem., 2008, 18(27): 3180-3192.

[14] Miyafuji H, Saka S. Fire-resisting properties in several TiO_2 wood-inorganic composites and their topochemistry[J]. Wood SCI. Technol., 1997, 31(6): 449-455.

[15] Wang S L, Wang C Y, Liu C Y, et al. Fabrication of superhydrophobic spherical-like α-FeOOH films on the wood surface by a hydrothermal method[J]. Colloid Surf. A-Physicochem. Eng. Asp., 2012, 403: 29-34.

[16] Salla J, Pandey K K, Srinivas K. Improvement of UV resistance of wood surfaces by using ZnO nanoparticles[J].

Polym. Degrad. Stabil., 2012, 97(4): 592-596.

[17] Kodama R H. Magnetic nanoparticles[J]. J. Magn. Magn. Mater., 1999, 200(1-3): 359-372.

[18] Laurent D, Forge D, Port M, et al. Magnetic iron oxide nanoparticles: synthesis, stabilization, vectorization, physicochemical characterizations, and biological applications[J]. Chem. Rev., 2008, 108(6): 2064-2110.

[19] Lu A H, Salabas E L, Schüth F. Magnetic nanoparticles: synthesis, protection, functionalization, and application [J]. Angew. Chem. -Int. Edit., 2007, 46(8): 1222-1244.

[20] Merk V, Chanana M, Gierlinger N, et al. Hybrid wood materials with magnetic anisotropy dictated by the hierarchical cell structure[J]. ACS Appl. Mater. Interfaces, 2014, 6(12): 9760-9767.

[21] Gan W T, Gao L K, Liu Y, et al. The magnetic, mechanical, thermal properties and UV resistance of $CoFe_2O_4$/SiO_2-coated film on wood[J]. J. Wood Chem. Technol., 2016, 36(2): 94-104.

[22] Ma M, Zhang Y, Yu W, et al. Preparation and char-acterization of magnetite nanoparticles coated by amino silane [J]. Colloid Surf. A-Physicochem. Eng. Asp., 2003, 212(2-3): 219-226.

[23] Lu Y, Sathasivam S, Song J, et al. Robust self-cleaning surfaces that function when exposed to either air or oil[J]. Science, 2015, 347(6226): 1132-1135.

[24] Lu Y, Sun Q F, Yang D J, et al. Fabrication of mesoporous lignocellulose aerogels from wood via cyclic liquid nitrogen freezing-thawing in ionic liquid solution[J]. J. Mater. Chem. A, 2012, 22(27): 13548-13557.

[25] Gan W T, Gao L K, Sun Q F, et al. Multifunctional wood materials with magnetic, superhydrophobic and anti-ultraviolet properties[J]. Appl. Surf. Sci., 2015, 332: 565-572.

[26] Gierlinger N, Goswami L, Schmidt M, et al. In situ FTIR microscopic study on enzymatic treatment of poplar wood cross-sections[J]. Biomacromolecules, 2008, 9(8): 2194-2201.

[27] Wang H X, Fang J, Cheng T, et al. One-step coating of fluoro-containing silica nanoparticles for universal generation of surface superhydrophobicity[J]. Chem. Commun., 2008 (7): 877-879.

[28] Bruni S, Cariati F, Casu M, et al. IR and NMR study of nanoparticle-support interactions in a Fe_2O_3-SiO_2 nanocomposite prepared by a Sol-gel method, Nanostruct[J]. Nanostruct. Mater., 1999, 11(5): 573-586.

[29] Tu K K, Wang X Q, Kong L Z, et al. Fabrication of robust, damage-tolerant superhydrophobic coatings on naturally micro-grooved wood surfaces[J]. RSC Adv., 2016, 6(1): 701-707.

[30] Yamashita T, Hayes P. Analysis of XPS spectra of Fe^{2+} and Fe^{3+} ions in oxide materials[J]. Appl. Surf. Sci., 2008, 254(8): 2441-2449.

[31] Li J P, Lu Y, Wu Z X, et al. Durable, self-cleaning and superhydrophobic bamboo timber surfaces based on TiO_2 films combined with fluoroalkylsilane[J]. Ceram. Int., 2016, 42(8): 9621-9629.

[32] Zimmermann J, Reifler F A, Fortunato G, et al. A simple, one-step approach to durable and robust superhydrophobic textiles[J]. Adv. Funct. Mater., 2008, 18(22): 3662-3669.

[33] Zhou W C, Hu X J, Bai X X, et al. Synthesis and electro- magnetic, microwave absorbing properties of core-shell Fe_3O_4-poly (3, 4-ethylenedioxythiophene) microspheres[J]. ACS Appl. Mater. Interfaces, 2011, 3(10): 3839-3845.

[34] Zhu W M, Wang L, Zhao R, et al. Electromagnetic and micro- wave-absorbing properties of magnetic nickel ferrite nanocrystals[J]. Nanoscale, 2011, 3(7): 2862-2864.

[35] Liu X G, Jiang J J, Geng D Y, et al. Dual nonlinear dielectric resonance and strong natural resonance in Ni/ZnO nanocapsules[J]. Appl. Phys. Lett., 2009, 94(5): 053119.

[36] Zhang X J, Wang G S, Cao W Q, et al. Enhanced microwave absorption property of reduced graphene oxide (RGO)-$MnFe_2O_4$ nanocomposites and polyvinylidene fluoride[J]. ACS Appl. Mater. Interfaces, 2014, 6(10): 7471-7478.

[37] Du Y C, Liu W W, Qiang R, et al. Shell thickness-dependent microwave absorption of core-shell Fe_3O_4@C composites[J]. ACS Appl. Mater. Interfaces, 2014, 6(15): 12997-13006.

[38] Pan Y F, Wang G S, Yue Y H. Fabrication of Fe_3O_4@ SiO_2@ RGO nanocomposites and their excellent absorption

properties with low filler content[J]. RSC Adv., 2015, 5(88): 71718-71723.

[39]Wang G S, Nie L Z, Yu S H. Tunable wave absorption properties of $\beta-MnO_2$ nanorods and their application in dielectric composites[J]. RSC Adv., 2012, 2(15): 6216-6221.

[40]Cao M S, Shi X L, Fang X Y, et al. Microwave absorption properties and mechanism of cagelike ZnO/SiO_2 nanocomposites[J]. Appl. Phys. Lett., 2007, 91(20): 203110.

[41]Duan Y, Yang Y, He M, et al. Absorbing properties of α-manganese dioxide/carbon black double-layer composites [J]. J. Phys. D-Appl. Phys., 2008, 41(12): 125403.

中文题目：在天然木材表面上制备具有强超疏水性的吸收微波的 $CoFe_2O_4$ 涂层

作者：甘文涛，高丽坤，张文博，李坚，詹先旭

摘要：基于 $CoFe_2O_4$ 纳米颗粒的形成以及随后使用氟化烷基硅烷(FAS)的疏水化作用，制备了具有微波吸收特性的超疏水木材表面。同时，通过将大量的 $CoFe_2O_4$ 纳米颗粒黏附到木材表面，黏性环氧树脂被用作粘结剂。木材上制备的超疏水性涂料在遭受严重磨损后仍保持稳定的超疏水性。此外，通过矢量网络分析在12~18GHz的频率范围内测量了涂层木材复合材料的复磁导率和介电常数。通过传统的同轴线方法阐明了微波吸收特性。结果表明，所制备的木质复合材料在16 GHz频率下具有优异的微波吸收性能，最小反射损耗可达-12.3 dB。本方法可以为设计具有吸收特定频率的户外吸波木材提供可行的途径。

关键词：木材；$CoFe_2O_4$ 纳米粒子；超疏水性；微波吸收

The Magnetic, Mechanical, Thermal Properties and UV Resistance of $CoFe_2O_4/SiO_2$-Coated Film on Wood*

Wentao Gan, Likun Gao, Ying Liu, Xianxu Zhan, Jian Li

Abstract: Magnetic wood has potential for electromagnetic shielding, as an indoor electromagnetic wave absorber, and heavy metal adsorption. In this article, magnetic wood materials were prepared by precipitating magnetic $CoFe_2O_4$ nanoparticles and SiO_2 on a wood surface via a hydrothermal process. The morphology, crystalline phase and chemical structure of the magnetic wood composites were characterized by scanning electron microscope (SEM), transmission electron microscope (TEM), X-ray diffraction (XRD), Raman spectroscopy, and Fourier transform infrared spectroscopy (FTIR). The as-fabricated wood composites exhibited excellent magnetic and anti-ultraviolet properties. Thermal analysis showed that the incorporation of SiO_2 gel retarded the thermal decomposition of magnetic wood matrix and improved the thermal stability of wood. Moreover, the mechanical performance of the modified wood materials was also evaluated.

Keywords: Composite materials; Magnetic nanoparticles; Thin films; Wood modification

1 Introduction

Magnetic wood was first introduced and developed by the Oka group in 1991; they achieved a good harmony of both woody and magnetic functions through combining the recycled magnetic materials and waste wood. The magnetic wood has potential for electromagnetic shielding, as an indoor electromagnetic wave absorber, and for heating and magnetic attraction materials due to the excellent magnetic property.[1-4] However, some unfavorable end-product properties of wood materials, such as combustibility and photodegradability, have not been fully discussed.

As a natural biological material, wood was sensitive to fire and sunlight, which led to thermal instability and weathering performance, but natural wood materials provided an excellent scaffold for further functionalization because of the porous structure and hydrophilic property. Nowadays, inorganic modification of wood to form hybrid wood/inorganic composites is one of the promising methods for improvement of wood properties. By incorporation of inorganic components into wood matrix, wood could be functionalized with enhanced properties in terms of UV resistance, fire resistance, biodegradation resistance, water repellence, and magnetic property, while simultaneously retaining the desirable materials properties of wood itself[5-9]. Numerous approaches, such as the sol-gel process, vacuum impregnation, dip coating, and hydrothermal method have been developed to synthesize hybrid wood/inorganic composites; however, most studies concentrated on compositing a single type of inorganic particles with wood template.[10-12] Studies by Saka et

* 本文摘自 Journal of Wood Chemistry and Technology, 2016, 36: 94-104.

al. reported the synthesis of wood/SiO_2 composites based on a sol-gel method, and the fire resistance and photostability of wood composites have been greatly improved.[13, 14] Similar studies by Sun et al. showed that the layer of TiO_2, ZnO, and CeO_2 nanoparticles could be grown on the wood surface by a hydrothermal process, and the modified wood composites exhibited excellent dimensional stability and UV-resistance.[15-17] Moreover, Merk et al. produced a kind of hybrid wood materials with magnetic anisotropy via the incorporation of magnetic iron oxides inside the wood materials.[18]

Herein, we selected the magnetic $CoFe_2O_4$ as inorganic nanoparticles because of their moderate saturation magnetization, high coercivity, and excellent physical and chemical stability,[19, 20] and prepared magnetic wood composites by precipitating $CoFe_2O_4$ nanoparticles on the wood template via the hydrothermal process, and then treated this with a layer of SiO_2. The $CoFe_2O_4$/SiO_2-modified wood composites not only featured magnetic property, but also enhanced thermal property and UV-resistance. In addition, the mechanical property of $CoFe_2O_4$/SiO_2-modified wood composites was also evaluated.

2 Materials and method

2.1 Materials

Poplar wood 20 mm (longitudinal) × 20 mm (tangential) × 20 mm (radial) in size was ultrasonically rinsed in deionized water for 30 min and then vacuum-dried at 103 ℃ for 24 h. Cobalt chloride hexahydrate ($CoCl_2 \cdot 6H_2O$), ferrous sulfate heptahydrate ($FeSO_4 \cdot 7H_2O$), potassium nitrate (KNO_3), sodium hydroxide (NaOH), tetraethyl orthosilicate (TEOS, 98%), ethanol (C_2H_5OH, 99.8%) and ammonia solution ($NH_3 \cdot H_2O$, 25%) used in this study were supplied by Shanghai Boyle Chemical Company Limited., and used without further purification.

2.2 Preparation of the $CoFe_2O_4$/SiO_2-modified wood

The experimental procedure for the preparation of the magnetic $CoFe_2O_4$/SiO_2-modified wood composites is displayed in Fig. 1. The wood samples were immersed in the 40 mL of 0.2 mol of freshly prepared aqueous solution of $FeSO_4$ and $CoCl_2$($Fe^{2+}/Co^{2+}=2$). Then, the system was transferred into a Teflon-lined stainless-steel autoclave and heated to 90 ℃ for 3 h. The non-magnetic metal hydroxides were precipitated on the wood template. At the end of the reaction, the supernatant was poured out and 30 mL of 1.32 mol/L NaOH solution with KNO_3 ($[Fe^{2+}] \cdot [NO_3^-]^{-1} = 0.44$) was added to the Teflon-lined stainless-steel autoclave. The precipitated precursors were converted into ferrite crystal nanoparticles via an additional hydrothermal processing for 6 h at 90 ℃. KNO_3 was added to act as a mild oxidizing agent.[21, 22] After hydrothermal reaction, the magnetic wood samples were sufficiently rinsed in deionized water and added to a beaker with ethanol (280 mL), deionized water (70 mL), and ammonia solution (5.0 mL, 28 wt%) under continuous mechanical stirring for 15 min. Afterward, 4.0 mL of TEOS was added dropwise in 10 min. It was a necessary that all of the $Si-OCH_2CH_3$ groups underwent hydrolysis, causing the formation of Si-OH groups. The Si-OH groups were attached onto the magnetic $CoFe_2O_4$-modified wood surface through a self-assembly process. Finally, SiO_2 was deposited on a wood template through the hydrothermal process at 90 ℃ for 10 h via the continuous dehydration reaction. At the end of reaction, the treated wood specimens were removed from the solution, ultrasonically rinsed with deionized water for 30 min, and dried at 50 ℃ for 24 h in vacuum.

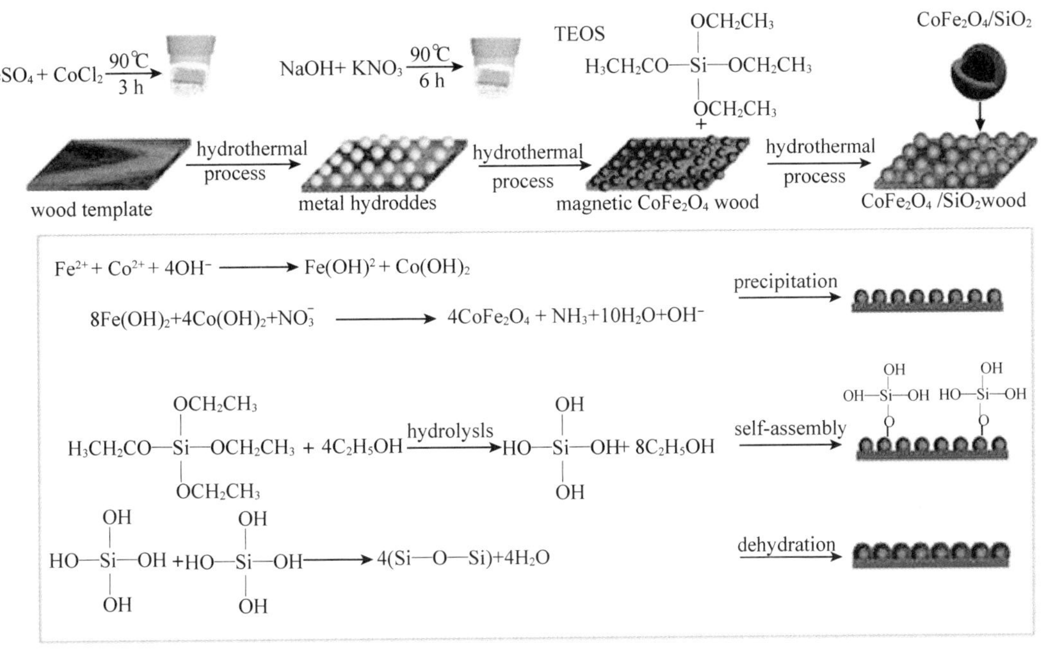

Fig. 1 Schematic diagrams of the procedure for the preparation of magnetic $CoFe_2O_4/SiO_2$-modified wood composites

2.3 Characterizations

The morphologies of the $CoFe_2O_4/SiO_2$-modified wood composites were characterized using scanning electron microscopy (SEM, FEI, and Quanta 200) and transmission electron microscopy (TEM, FEI, and Tecnai G20). The surface chemical compositions of the samples were determined via energy-dispersive X-ray analysis (EDXA). The crystalline structure of these composites were identified via X-ray diffraction (XRD, Rigaku, and D/MAX 2200) operating with Cu $K\alpha$ radiation ($\lambda = 1.5418$ Å) at a scan rate (2θ) of $4° \cdot min^{-1}$, an accelerating voltage of 40 kV, and an applied current of 30 mA ranging from 5° to 80°. All Raman spectra were collected with a laser confocal Micro-Raman spectroscopy (Horiba Jobin Yvon) ($\lambda = 532$ nm). The laser power was set to low values (1%) to avoid sample degradation. The surface chemical compositions of the samples were determined via Fourier transformation infrared spectroscopy (FTIR, Nicolet, Magna-IR 560). For FTIR analysis, thin sample disks were made by grinding a small portion of the $CoFe_2O_4/SiO_2$-modified wood composites and pressing them with potassium bromide. The magnetic nanoparticles used for the TEM observation and Raman measurement were from the sediment of hydrothermal solution in preparation.

For magnetic characterization, wood specimens of 4 mm (tangential) × 4 mm (radial) × 4 mm (longitudinal) were used, and the magnetic properties of the composites were measured by a superconducting quantum interference device (MPMS XL-7, Quantum Design Corp.) at room temperature (300 K).

For TGA analysis, small amounts of the wood powder about-5 mg-were prepared by grinding the wood block using a high-speed universal grinder. The thermal performances of the composites were examined using a thermal analyzer (TGA, SDT Q600) in the temperature range from room temperature up to 600℃ at a heating rate of 20 ℃ · min^{-1} under nitrogen atmosphere.

For mechanical characterization, wood specimens (4–5 specimens per type) of 20 mm (tangential)×20 mm (radial)×30 mm (longitudinal) with approximately 13 wt% moisture content were used; the temperature and relative humidity of the room for the wood mechanically tested were 23 ℃ and 51%, respectively. Their deformation behavior under compression in a longitudinal direction was recorded using a universal testing machine (Instron 1185) at a crosshead speed of 5 mm · min^{-1}.

The wood samples of 20 mm (tangential)×5 mm (radial)×70 mm (longitudinal) were subjected to weathering using a QUV Accelerated Weathering Tester (Atlas, Chicago, IL, USA), which allowed water spray and condensation. The samples were fixed in stainless steel holders under irradiation of fluorescent UV light at wavelength of 340 nm with a temperature of 50 ℃ for 8 h, followed by water spray for 4 h. The average irradiance was set to 0.77 W · m^{-2}, and the spray temperature was fixed 25 ℃. The exposure time ranged from 0 to 600 h. The color change of the wood surface after UV irradiation was measured using a portable spectrophotometer (NF-333, Nippon Denshoku Company, Japan) with CIELAB system in accordance with the ISO-2470 standard. CIELAB L^*, a^*, b^*, parameters were measured at six locations on each specimen and average value was calculated. In the CIELAB system, L^* axis represents the lightness, and varies from 100 (white) to 0 (black); a^* and b^* are the chromaticity indices; $+a^*$ is the red direction; $-a^*$ is green; $+b^*$ is yellow; and $-b^*$ is blue. The change of L^*, a^* and b^* are calculated according to Eqs. (1) –(3).

$$\Delta a^* = a_2 - a_1 \tag{1}$$

$$\Delta b^* = b_2 - b_1 \tag{2}$$

$$\Delta L^* = L_2 - L_1 \tag{3}$$

Where Δa^*, Δb^*, and ΔL^* were the differences in the initial and final values of a^*, b^*, and L^*, respectively; a_1, b_1, and L_1 were the initial color parameters; and a_2, b_2, and L_2 were the color parameters after UV irradiation. The color differences (ΔE^*) were calculated by Eqs. (4):

$$\Delta E^* = [(\Delta L^*)^2 + (\Delta a^*)^2 + (\Delta b^*)^2]^{1/2} \tag{4}$$

3 Results and discussion

3.1 Structural characterization and morphology of the $CoFe_2O_4/SiO_2$-modified wood composites

XRD patterns of the untreated wood and $CoFe_2O_4$-modified wood composites were presented in Fig. 2. In Fig. 2a, the diffraction peaks at 15° and 22° represented the crystalline region of the cellulose.[23, 24] In Fig. 2b of the $CoFe_2O_4$-modified wood, additional diffraction peaks at $2\theta = 30°$, 35°, 43°, 56°, and 62° could be assigned to the diffraction of the (220), (311), (400), (511), and (440) planes of $CoFe_2O_4$ (PDF No. 22-1086), respectively.

To further confirm the formation of $CoFe_2O_4$ nanoparticles, Raman spectra of the sediment in hydrothermal process were measured at room temperature, as presented in Fig. 3. In the Raman spectra, although most peaks of $CoFe_2O_4$ were close to γ-Fe_2O_3 peaks, γ-Fe_2O_3 exhibits much stronger peaks at 1378 and 1576 cm^{-1} than observed for $CoFe_2O_4$.[25, 26] As shown in Fig. 3, very weak peaks were observed in the range of 800 to 2000 cm^{-1}, suggesting that the formation of γ-Fe_2O_3 was very unlikely. Raman peaks observed at 192, 305, 470 and 670 cm^{-1} were characteristic of $CoFe_2O_4$.[27] An important question that still remains was whether a silica layer was formed on the $CoFe_2O_4$-modified wood surface upon the addition of TEOS to a mixture containing the wood composites and aqueous ammonia.

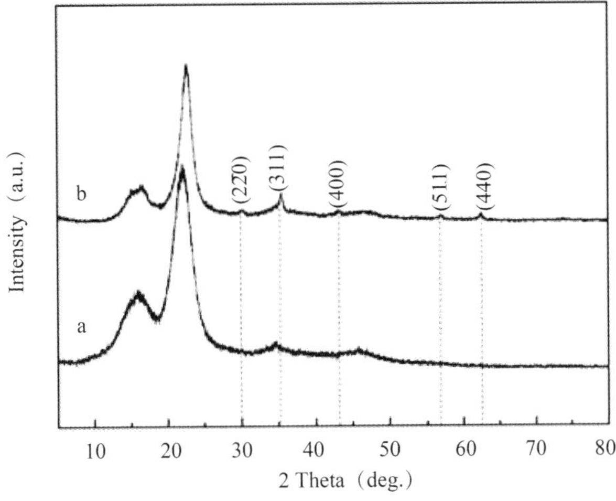

Fig. 2 XRD pattern of (a) untreated wood and (b) $CoFe_2O_4$-modified wood composites

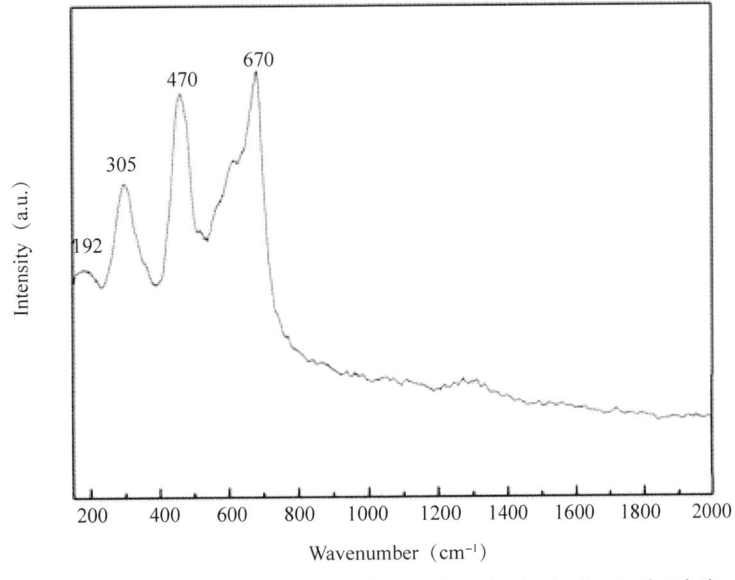

Fig. 3 Roman spectrum of the sediment into the hydrothermal solution

SEM and EDXA were therefore made for the $CoFe_2O_4/SiO_2$-modified wood composites, and the results are showed in Fig. 4, along with the results of untreated wood for comparison (Fig. 4a). In the wood composited with SiO_2, the inorganic nanoparticles were found to be deposited on the cell lumens at low magnification (Fig. 4b). Higher magnifications (Fig. 4c) revealed densely packed layer of inorganic nanoparticles precipitated on the wood template and some nanoparticles aggregates of several microns. The TEM image, as an insertion in Fig. 4c, revealed the sizes of $CoFe_2O_4/SiO_2$ nanoparticles, in which core-shell structure was 367 nm. EDXA spectrum presented in Fig. 4d shows the chemical elements of the treated wood: carbon, oxygen, gold, cobalt, iron and silicon. The peaks for carbon and oxygen elements were from the wood template, gold originated from the coating layer used during the SEM observation, whereas additional peaks for cobalt, iron and silicon elements confirmed the presence of $CoFe_2O_4/SiO_2$ on the wood template.

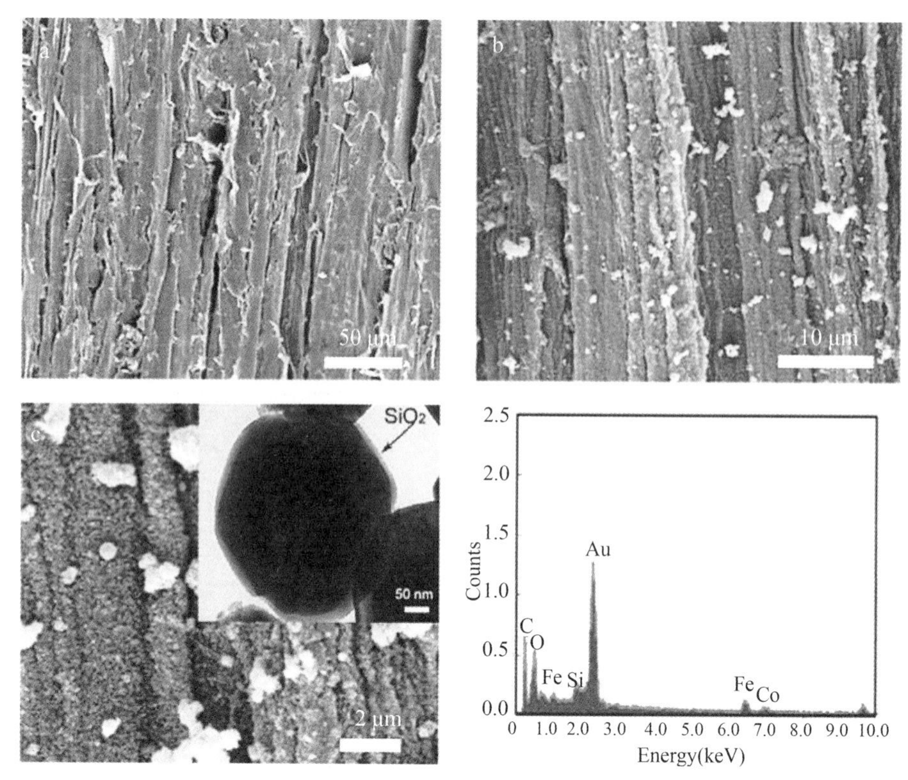

Fig. 4 SEM images of (a) untreated wood; (b, c) $CoFe_2O_4/SiO_2$-modified wood composites at different magnifications; (d) EDXA spectrum of the $CoFe_2O_4/SiO_2$-modified wood composites

To investigate the functional group composition of the wood samples, FTIR measurements were performed before and after the hydrothermal treatment (Fig. 5). In Fig. 5, the bands at 3425 cm^{-1}, 2920 cm^{-1}, 1637, 1458, and 1420 cm^{-1} correspond to O—H stretching vibration, —CH_3 asymmetric, O—H bending vibration, C—H deformation, and C—O stretching vibration, respectively. Most of the observed peaks of the untreated wood represent major cell wall components such as cellulose (1160, 897 cm^{-1}), hemicelluloses (1740, 1056 cm^{-1}) and lignin (1507, 1240 cm^{-1}). [28, 29] For the $CoFe_2O_4/SiO_2$-modified wood composites, the bands at 1740 and 1240 cm^{-1} disappeared, indicating the degradation of hemicelluloses and lignin after alkali treatment. The new bands at 1110, 805, and 471 cm^{-1} could be attributed to the vibration modes of SiO_2, [30, 31] and the other characteristic band near 570 cm^{-1} arises from the stretch vibration of Fe—O bonds. [32, 33] These results strongly confirmed the $CoFe_2O_4/SiO_2$ nanoparticles were precipitated on the wood template.

3.2 Magnetic, Thermal, Mechanical properties and UV-resistance of the $CoFe_2O_4/SiO_2$-modified wood composites

The room-temperature magnetization hysteresis curves of the wood composites were measured with superconducting quantum interference device, as shown in Fig. 6. The hysteresis curve of the untreated wood is approximation of a straight line close to zero that indicated that the wood was a non-magnetic material, but the $CoFe_2O_4$-modified wood composites and $CoFe_2O_4/SiO_2$-modified wood composites were exhibiting a polar property and showing a typical ferromagnetic behavior. The magnetization saturation values (M_S) of the

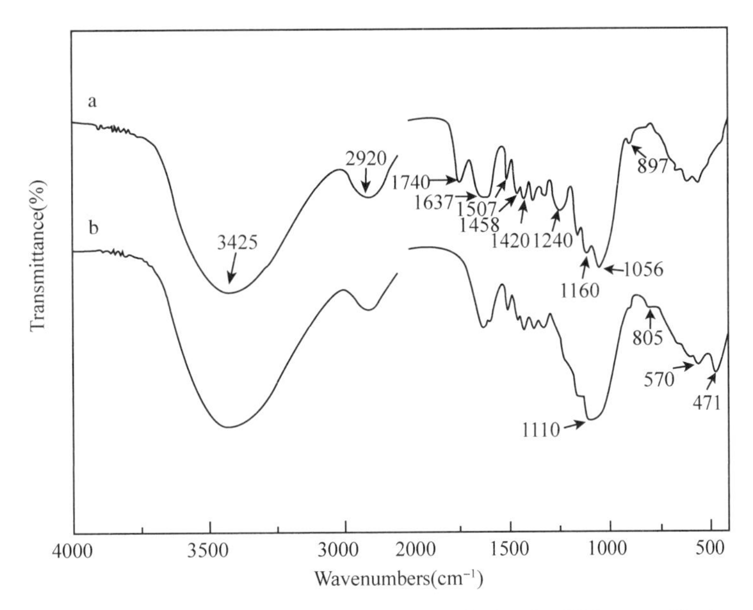

Fig. 5 FTIR spectra of (a) the untreated wood and (b) $CoFe_2O_4/SiO_2$-modified wood composites

$CoFe_2O_4$-modified wood composites and $CoFe_2O_4/SiO_2$-modified wood composites were 1.8 and 1.5 emu/g, respectively. The coercivity (Hc) of the $CoFe_2O_4$-modified wood composites and $CoFe_2O_4$ SiO_2-modified wood composites were 351 and 301 Oe, respectively. The reduced Ms could be explained by considering the diamagnetic contribution of the SiO_2 surrounding the $CoFe_2O_4$ on the wood surface, which will weaken the magnetic moment, whereas the low H_c may result from the size of $CoFe_2O_4$ nanoparticles embedded into SiO_2.[34, 35] The insertion revealed the magnetization can easily be probed by moving and lifting up a magnetic wood block using a permanent magnet.

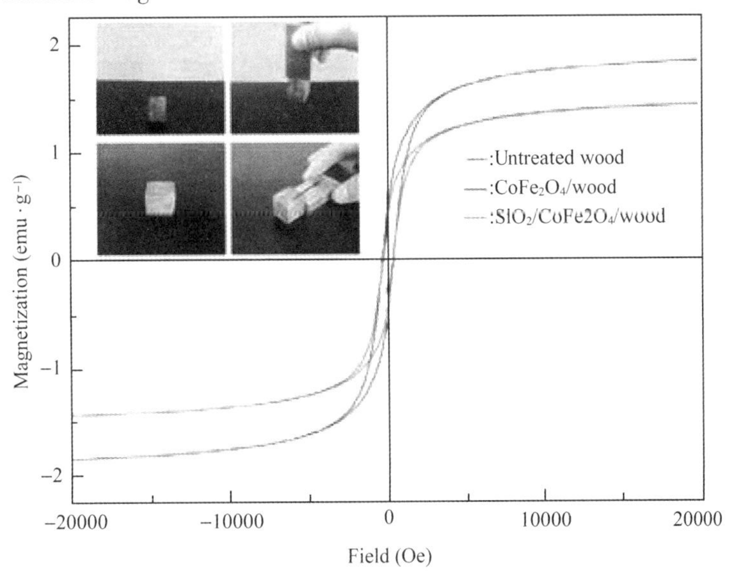

Fig. 6 Magnetic hysteresis curves of the untreated wood (black line), $CoFe_2O_4$-modified wood composites (blue line), and $CoFe_2O_4/SiO_2$-modified wood composites (red line). Inset: magnetic $CoFe_2O_4/SiO_2$-modified wood sample under the influence of an external magnet

The mechanical behavior of wood composites under compression in the longitudinal direction was characterized (Fig. 7). The stress-strain curve of the untreated wood showed a liner elastic region at the beginning of compression, followed by a yield point, indicating the onset of cell wall buckling and collapse, after which wood continued to deform under compression with a steadily decreasing stress.[36] The modified wood composites showed a deformation behavior generally similar to that of untreated wood. However, an elevated stress threshold required to initiate cell wall collapse was observed for the modified wood. In addition, after yield point the composites showed an initially abrupt and then steady decrease in compressive stress. This revealed that the wood brittleness and yield strength in compression were improved by the inorganic modification.

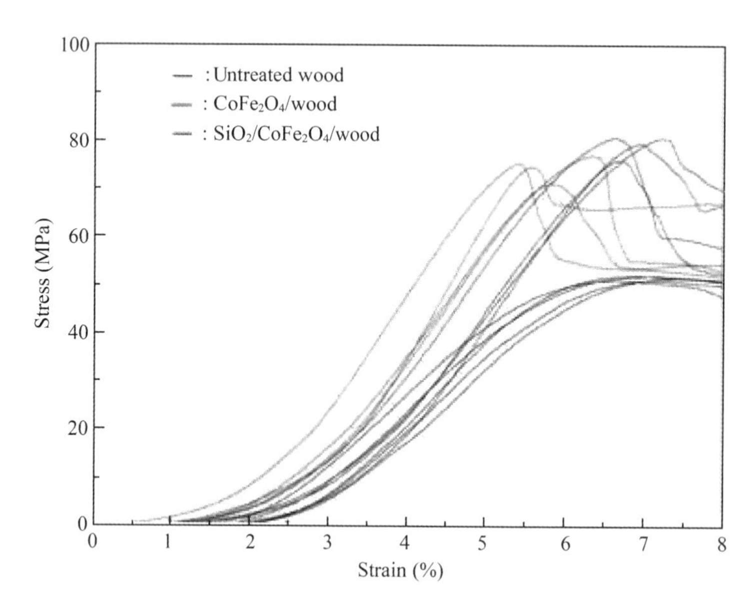

Fig. 7 Stress-strain curves of the untreated wood, $CoFe_2O_4$-modified wood composites, and $CoFe_2O_4/SiO_2$-modified wood composites under compression in the longitudinal direction

Thermal analyses were performed to examine the thermal stability of the prepared modified wood samples. Curves representing thermal degradation (TG and DTG) for tested wood specimens were displayed in Fig. 8. In Fig. 8a, for the TG curves, the initial small mass loss from room temperature to 100 ℃ was mainly attributed to the evaporation of adsorbed water. An abrupt weight loss was observed between 250 ℃ and 400 ℃. These distinct weight losses were due to oxidation and pyrolysis of wood components.[37] As opposed to the untreated control, the $CoFe_2O_4$-modified wood composites displayed similar TG curves up to 200 ℃, after which the temperatures at which the thermal degradation took place were decreased, from 221 ℃ for the untreated wood to 210 ℃ for the $CoFe_2O_4$-modified wood composites. The decreased decomposition temperature could be ascribed to the catalytic property of the precipitated inorganic nanoparticles.[38, 39] As for $CoFe_2O_4/SiO_2$-modified wood composites, the temperatures at which the thermal degradation took place were increased, from 221 ℃ for the untreated wood to 243 ℃ for the $CoFe_2O_4/SiO_2$-modified wood composites, indicating the high thermal stability after SiO_2 modification. It can be assumed that the wood components were likely to be shielded by the fire-retardant SiO_2 coating and the combustion was delayed, thus enhancing the thermal stability of wood at some extent. Similar results were found for the thermal stability of wood samples that has been impregnated or coated with silicon nano sols. [40]

For the DTG curves in Fig. 8b, two obvious peaks of the untreated wood were found between 280 ℃ and 360 ℃. The pyrolysis of hemicelluloses in the untreated wood presented a shoulder at 280 ℃ in the cellulose pyrolysis peak. After $CoFe_2O_4$ and SiO_2 modification, this peak was missing in the DTG curves, which suggested that most of the hemicelluloses were already degraded in the treated wood, which was in good agreement with the result by FTIR analysis. Moreover, compared to that of untreated wood, the peak of the modified wood weakened, indicating the decrease of combustion rate.

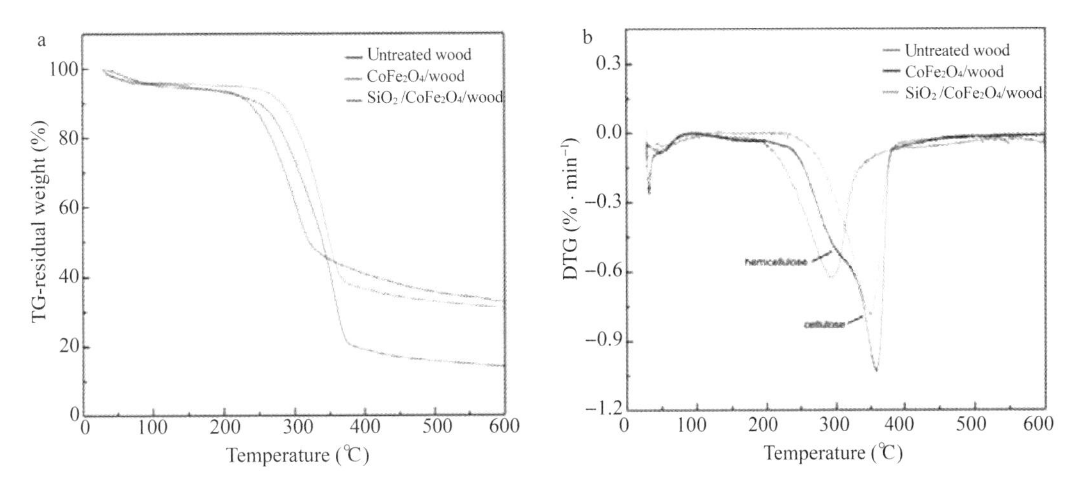

Fig. 8 Thermal behavior (a) TG and (b) DTG curves of the untreated wood, $CoFe_2O_4$-modified wood composites, and $CoFeO_4/SiO_2$-modified wood composites

The accelerated aging tests were performed and the color changes before and after UV irradiations were measured to evaluate the UV resistance of the modified wood composites. The experimental results were presented in Fig. 9. In Fig. 9a, a positive Δa^* values of the untreated wood indicated that the color of the untreated wood sample became gradually redder with increasing UV irradiation. The Δa^* values of the $CoFe_2O_4$-modified wood composites and $CoFe_2O_4$/ SiO_2-modified wood composites showed a similar trend; however, the Δa^* change in $CoFe_2O_4$/ SiO_2-modified wood composites were much smaller than that of the $CoFe_2O_4$-modified wood and the untreated wood. In Fig. 9b, the changes in Δb^* values indicated the untreated wood was subjected to a strong yellowing, while there is almost no change of the yellow index of $CoFe_2O_4/SiO_2$-modified wood, and a slight coloration in the blue direction of the $CoFe_2O_4$-modified wood. In Fig. 9c, ΔL^* values of all samples became negative with time, indicating the color of the samples was darkening. But the ΔL^* values of the $CoFe_2O_4/SiO_2$-modified wood composites were much smaller (only 1/3 and 1/5) than that of the $CoFe_2O_4$-modified wood composites and untreated wood, respectively, which meant much stronger UV-resistant ability of the $CoFe_2O_4/SiO_2$-modified wood composites. The total color change (ΔE^*) was showed in Fig. 9d. It could be seen that ΔE^* values of all three were different with increasing UV irradiation time, while only a slight change occurred for the $CoFe_2O_4/SiO_2$-modified wood composites. Based on the above results the natural color of the untreated wood would be changed by UV irradiation without the employed protective measure. Although the $CoFe_2O_4$-modified wood composites showed UV-resistant ability to some extent, damage to the appearance could not be avoided. The $CoFe_2O_4/SiO_2$-modified wood exhibited an excellent ability to resist UV irradiation, and thus prevented the wood materials from damage. The UV-resistance of $CoFe_2O_4$-modified wood could be attributed to the UV absorption of the $CoFe_2O_4$ nanoparticles, [41] which hinder UV light from reaching the wood surface and thus prevent photodegradation of wood component, while $CoFe_2O_4$ nanoparticles coated with silica shows the enhanced optical properties, especially the higher diffuse reflectance compared with that of the uncoated nanoparticles, [42] so the excellent anti-ultraviolet property was achieved by modified with $CoFe_2O_4/SiO_2$.

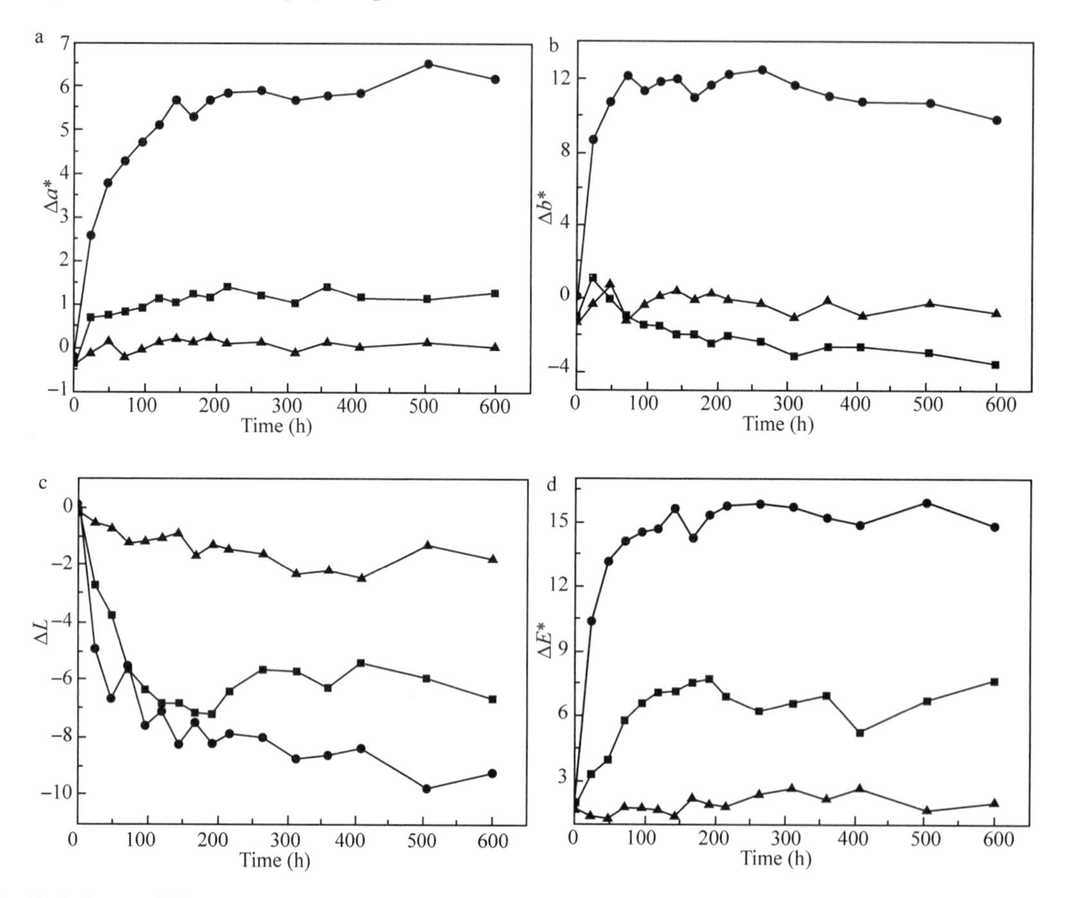

Fig. 9 Variation in CIELAB parameters ΔL^*, Δa^*, Δb^*, and ΔE^* at different irradiation times for the untreated wood (●), $CoFe_2O_4$-modified wood composites (■), and $CoFe_2O_4/SiO_2$-modified wood composites (▲)

4 Conclusions

Magnetic wood composites were prepared by precipitating $CoFe_2O_4$ nanoparticles on the wood surface and then treating with a layer of SiO_2 via a hydrothermal process. The $CoFe_2O_4$ nanoparticles were deposited on the wood template by converted metal hydroxides into ferrite crystal nanoparticles via the hydrothermal process and the other layer of SiO_2 was covered on $CoFe_2O_4$-modified wood surface via the hydrolysis of TEOS. The $CoFe_2O_4/SiO_2$-modified wood composites not only exhibited excellent magnetic property, but some simple movements such as lift, drag, and actuation can be performed under the influence of the magnetic field, but also possessed improved thermal property and UV-resistance. Moreover, the wood brittleness and yield strength in compression were also improved after the inorganic modification.

Acknowledgments

This work was financially supported by The National Natural Science Foundation of China (grant no. 31470584).

References

[1]Oka H, Fujita H. Experimental study on magnetic and heating characteristics of magnetic wood[J]. J. Appl. Phys., 1999, 85(8): 5732-5734.

[2]Oka H, Hojo A, Takashiba T. Wood construction and magnetic characteristics of impregnated type magnetic wood[J]. J. Magn. Magn. Mater., 2002, 239(1-3): 617-619.

[3]Oka H, Hamano H, Chiba S. Experimental study on actuation functions of coating-type magnetic wood[J]. J. Magn. Magn. Mater., 2004, 272: E1693-E1694.

[4]Oka H, Uchidate S, Sekino N, et al. Electromagnetic wave absorption characteristics of half carbonized powder-type magnetic wood[J]. IEEE Trans. Magn., 2011, 47(10): 3078-3080.

[5]Devi R R, Gogoi K, Konwar B K, et al. Synergistic effect of nano TiO_2 and nanoclay on mechanical, flame retardancy, UV stability, and antibacterial properties of wood polymer composites[J]. Polym. Bull., 2013, 70(4): 1397-1413.

[6]Akaki T, Maehara H, Tooyama M. Development of wood and wood ash-based hydroxyapatite composites and their fire-retarding properties[J]. J. Wood Sci., 2012, 58(6): 532-537.

[7]Furuno T, Imamura Y. Combinations of wood and silicate Part 6. Biological resistances of wood-mineral composites using water glass-boron compound system[J]. Wood Sci. Technol., 1998, 32(3): 161-170.

[8]Hübert T, Unger B, Bücker M. Sol-gel derived TiO_2 wood composites[J]. J. Sol-gel Sci. Technol., 2010, 53(2): 384-389.

[9]Gan W T, Gao L K, Sun Q F, et al. Multifunctional wood materials with magnetic, superhydrophobic and anti-ultraviolet properties[J]. Appl. Surf. Sci., 2015, 332: 565-572.

[10]Wang B, Feng M, Zhan H. Improvement of wood properties by impregnation with TiO_2 via ultrasonic-assisted sol-gel process[J]. RSC Adv., 2014, 4(99): 56355-56360.

[11]Donath S, Militz H, Mai C. Weathering of silane treated wood[J]. Holz Roh werkst, 2007, 65(1): 35-42.

[12]Sun Q F, Yu H P, Liu Y X, et al. Prolonging the combustion duration of wood by TiO_2 coating synthesized using cosolvent-controlled hydrothermal method[J]. J. Mater. Sci., 2010, 45(24): 6661-6667.

[13]Miyafuji H, Kokaji H, Saka S. Photostable wood-inorganic composites prepared by the sol-gel process with UV absorbent[J]. J. Wood Sci., 2004, 50(2): 130-135.

[14]Saka S, Ueno T. Several SiO_2 wood-inorganic composites and their fire-resisting properties[J]. Wood Sci. Technol., 1997, 31(6): 457-466.

[15]Sun Q F, Yu H P, Liu Y X, et al. Improvement of water resistance and dimensional stability of wood through titanium dioxide coating[J]. Holzforschung, 2010, 64(6): 757-761.

[16]Sun Q F, Lu Y, Zhang H M, et al. Improved UV resistance in wood through the hydrothermal growth of highly ordered ZnO nanorod arrays[J]. J. Wood Sci., 2012, 47(10): 4457-4462.

[17]Lu Y, Xiao S L, Gao R N, et al. Improved weathering performance and wettability of wood protected by CeO_2 coating deposited onto the surface[J]. Holzforschung, 2014, 68(3): 345-351.

[18]Merk V, Chanana M, Gierlinger N, et al. Hybrid wood materials with magnetic anisotropy dictated by the hierarchical cell structure[J]. ACS Appl. Mater. Inter., 2014, 6(12): 9760-9767.

[19]Ren X Z, Shi J H, Tong L Z, et al. Magnetic and luminescence properties of the porous $CoFe_2O_4$@Y_2O_3: Eu^{3+} nanocomposite with higher coercivity[J]. J. Nanopart. Res., 2013, 15(6): 1-10.

[20]Zhao L X, Zhang H J, Zhou L, et al. Synthesis and characterization of 1D Co/$CoFe_2O_4$ composites with tunable morphologies[J]. Chem. Commun., 2008, (30): 3570-3572.

[21]Tamura H, Matijevic E. Precipitation of cobalt ferrites[J]. J. Colloid. Interf. Sci., 1982, 90(1): 100-109.

[22]Sugimoto T, Matijević E. Formation of uniform spherical magnetite particles by crystallization from ferrous hydroxide gels[J]. J. Colloid. Interf. Sci., 1980, 74(1): 227-243.

[23]Lu Y, Sun Q F, Yang D J, et al. Fabrication of mesoporous lignocellulose aerogels from wood via cyclic liquid nitrogen freezing-thawing in ionic liquid solution[J]. J. Mater. Chem., 2012, 22(27): 13548-13557.

[24]Li J, Lu Y, Yang D J, et al. Lignocellulose aerogel from wood-ionic liquid solution (1-allyl-3-methylimidazolium chloride) under freezing and thawing conditions[J]. Biomacromolecules, 2011, 12(5): 1860-1867.

[25]Tang D P, Yuan R, Chai Y Q, et al. Magnetic-core/porous-shell $CoFe_2O_4$/SiO_2 composite nanoparticles as immobilized affinity supports for clinical immunoassays[J]. Adv. Funct. Mater., 2007, 17(6): 976-982.

[26]Wang Z W, Downs R T, Pischedda V, et al. High-pressure x-ray diffraction and Raman spectroscopic studies of the tetragonal spinel $CoFe_2O_4$[J]. Phys. Rev. B, 2003, 68(9): 094101.

[27]Qu Y Q, Yang H B, Yang N, et al. The effect of reaction temperature on the particle size, structure and magnetic properties of coprecipitated $CoFe_2O_4$ nanoparticles[J]. Mater. Lett., 2006, 60(29-30): 3548-3552.

[28]Rana R, Langenfeld-Heyser R, Finkeldey R, et al. FTIR spectroscopy, chemical and histochemical characterisation of wood and lignin of five tropical timber wood species of the family of Dipterocarpaceae[J]. Wood Sci. Technol., 2010, 44(2): 225-242.

[29]Gierlinger N, Goswami L, Schmidt M, et al. In situ FT-IR microscopic study on enzymatic treatment of poplar wood cross-sections[J]. Biomacromolecules, 2008, 9(8): 2194-2201.

[30]Cannas C, Musinu A, Ardu A, et al. $CoFe_2O_4$ and $CoFe_2O_4$/SiO_2 core/shell nanoparticles: Magnetic and spectroscopic study[J]. Chem. Mat., 2010, 22(11): 3353-3361.

[31]Andersson J, Areva S, Spliethoff B, et al. Sol-gel synthesis of a multifunctional, hierarchically porous silica/apatite composite[J]. Biomaterials, 2005, 26(34): 6827-6835.

[32]Waldron R D. Infrared spectra of ferrites[J]. Phys. Rev., 1955, 99(6): 1727-1735.

[33]Koseoglu Y, Alan F, Tan M, et al. Low temperature hydrothermal synthesis and characterization of Mn doped cobalt ferrite nanoparticles[J]. Ceram. Int., 2012, 38(5): 3625-3634.

[34]Zhao X L, Zhao H L, Yuan H H, et al. Multifunctional superparamagnetic Fe_3O_4@SiO_2 core/shell nanoparticles: Design and application for cell imaging[J]. J. Biomed. Nanotechnol., 2014, 10(2): 262-270.

[35]Yang Z P, Gong X Y, Zhang C J. Recyclable Fe_3O_4/hydroxyapatite composite nanoparticles for photocatalytic applications[J]. Chem. Eng. J., 2010, 165(1): 117-121.

[36]Wang X Q, Liu J L, Chai Y B. Thermal, mechanical, and moisture absorption properties of wood-TiO_2 composites prepared by a sol-gel process[J]. BioResources, 2012, 7(1): 0893-0901.

[37]Rosa M F, Chiou B S, Medeiros E S, et al. Effect of fiber treatments on tensile and thermal properties of starch/ethylene vinyl alcohol copolymers/coir biocomposites[J]. Bioresour. Technol., 2009, 100(21): 5196-5202.

[38]Ma Z Y, Li F S, Bai H P. Effect of Fe_2O_3 in Fe_2O_3/AP composite particles on thermal decomposition of AP and on burning rate of the composite propellant[J]. Propellants, Explos., Pyrotech., 2006, 31(6): 447-451.

[39]Liu S L, Ke D N, Zeng J, et al. Construction of inorganic nanoparticles by micro-nano-porous structure of cellulose matrix[J]. Cellulose, 2011, 18(4): 945-956.

[40]Miyafuji H, Saka S. Na_2O-SiO_2 wood-inorganic composites prepared by the sol-gel process and their fire-resistant properties[J]. J. Wood Sci., 2001, 47(6): 483-489.

[41] Zhu Z R, Li X Y, Zhao Q D, et al. Surface photovoltage properties and photocatalytic activities of nanocrystalline $CoFe_2O_4$ particles with porous superstructure fabricated by a modified chemical coprecipitation method[J]. J. Nanopart. Res., 2011, 13(5): 2147-2155.

[42] Girgis E, Wahsh M M S, Othman A G M, et al. Synthesis, magnetic and optical properties of core/shell $Co_{1-x}Zn_xFe_2O_4/SiO_2$ nanoparticles[J]. Nanoscale Res. Lett., 2011, 6: 460.

中文标题：木材上的 $CoFe_2O_4/SiO_2$ 涂层薄膜的磁性、机械性、热性能和抗紫外线性

作者：甘文涛，高丽坤，刘莹，詹先旭，李坚

摘要：磁性木材将是一种潜在的电磁屏蔽、室内电磁波吸收和重金属吸附的材料。本文通过水热法将磁性 $CoFe_2O_4$ 纳米颗粒和 SiO_2 沉淀在木材表面，制备了磁性木材材料。通过扫描电子显微镜(SEM)、透射电子显微镜(TEM)、X 射线衍射(XRD)、拉曼光谱和傅里叶变换红外光谱(FTIR)对磁性木质复合材料的形态、晶相和化学结构进行了表征。所制备的木质复合材料表现出优异的磁性和抗紫外线性能。热分析表明，SiO_2 凝胶的加入延缓了磁性木材基体的热分解，提高了木材的热稳定性。此外，还对改性木质材料的机械性能进行了评估。

关键词：复合材料；磁性纳米颗粒；薄膜；木材改性

Preparation of Thiol-functionalized Magnetic Sawdust Composites As the Adsorbent to Remove Heavy Metal Ions*

Wentao Gan, Likun Gao, Xianxu Zhan, Jian Li

Abstract: Thiol-functionalized magnetic sawdust, synthesized by precipitating the γ-Fe_2O_3 nanoparticles on sawdust surface and then modifying the 3-mercaptopropyltrimethoxysilane layers, has been investigated as an-environmental-friendly and recyclable adsorbent for heavy metal ions. The process of modifying was confirmed by Scanning electron microscopy, Fourier transform infrared spectroscopy, X-ray photoelectron spectroscope and X-ray diffraction. Compared with the nonmagnetic sawdust, the thiol-functionalized magnetic sawdust possesses high saturation magnetization (7.28 emu · g^{-1}) and can be easier and faster to separated from water under an external magnetic field. The adsorption equilibrium was reached within 20 min and the adsorption kinetics was elucidated by pseudo-second-order model. The adsorption isotherms of Cu^{2+}, Pb^{2+} and Cd^{2+} fitted well with Langmuir model, exhibiting adsorption capacity of 5.49 mg · g^{-1}, 12.5 mg · g^{-1} and 3.80 mg · g^{-1}, respectively. Competitive adsorption among the three metal ions showed a preferential adsorption of Pb^{2+}>Cu^{2+}>Cd^{2+}. In addition, the magnetic sawdust adsorbent exhibited excellent acid-alkali stability and the metal-loaded adsorbent was able to regenerate in acid solution without significant adsorption capacity loss.

Keywords: Sawdust; Magnetic γ-Fe_2O_3; Thiol functionalization; Heavy metal; Adsorption

1 Introduction

Heavy metal ions such as cadmium, lead and copper, often found in industrial wastewater, which are non-biodegradable materials in the environment and can be accumulated in human bodies throughout the food chain, causing significant physiological disorders.[1-3] Therefore, it is necessary to remove heavy metal ions from the wastewater before their discharge. A variety of methods and techniques have been developed for the removal of toxic heavy metal ions from aqueous solution, such as chemical precipitation, adsorption, ion exchange, and membrane separation, etc..[4-6] Considering the efficiency and operation of the available methods, adsorption is regarded as one of the best available control technique for the removal of low concentrations of heavy metals from wastewater. However, some effective adsorbents such as activated carbon and ion exchange resins are not suitable for practical application owing to the high capital costs.

The application of low-cost adsorbents has been investigated as a replacement for costly conventional technologies of removing heavy metal ions from solution. Plant wastes have increasingly received more attention because they are abundant in nature, biodegradable, environmental-friendly, and renewable.[7] A lot of adsorption experiments have been concentrated on plant wastes such as waste tea,[8] orange peel,[9] chestnut

---

* 本文摘自 RSC Advances, 2016, 6: 37600-37609.

shell,[10] sugarcane bagasse,[11] Cinnamomum camphora leaves,[12] neem bark,[13] sawdust,[14] etc. These plant wastes are mostly composed of cellulose, hemicellulose and lignin, both with a capacity for binding metal cations. However, the adsorption capacity of these untreated plant wastes is usually limited by the low quantity of functional groups. The direct use of the untreated materials as absorbents not only can increases chemical oxygen demand (COD) of water due to the release of soluble organic compounds such as lignin, pectin and tannin into the solution,[15] but also requires an additional separation step to remove such absorbents from solution. Therefore, the design and exploration of recyclable biosorbents based on plant wastes are still necessary.

Some studies have proved that the chemical modification could be used to enhance the adsorption capacity of plant waste through introducing functional groups and decrease the COD of water by the release of the organic compounds in the chemical pretreatment process.[16] Functionalized silica layer with organic functional groups are commonly used since the silica provides many advantages such as good adsorption, cation exchange capacity, easy to impregnate medium to create sever modified silica surface with high mechanic strength and thermal stability.[17] Among the types examined, those with thiol-functionalised groups have been found to be efficient for the removal of heavy metal ions.[18] Moreover, Merk et al. has confirmed that the wood materials can act as efficient substrates to the nucleation and growth of magnetic particles.[19] Many researches also have proved that the wood/magnetic particles composites with appropriate magnetic property could be collected by the permanent magnet.[20, 21] As an abundant by-product, sawdust with the porous structure and a hydrophilic nature, obtained from mechanical wood processing, which is a renewable and low cost waste materials, is an excellent scaffold for further magnetic particles modification. Furthermore, wood sawdust is mainly composed of cellulose (45%-50%) and lignin (23%-30%), both has a capacity for binding metal cations because of the hydroxyl, phenolic and carboxylic groups present in their structure.[22]

In this study, poplar sawdust was chemically modified by magnetic γ-Fe_2O_3 nanoparticles and 3-mercaptopropytrimethoxysilane (MPTMS), used as a kind of recyclable adsorbent to remove heavy metal ions from water. Thiol-functionalized magnetic sawdust (TF-MS) were characterized using scanning electron microscope (SEM), X-ray diffraction (XRD), X-ray photoelectron spectroscope (XPS), Fourier transform infrared spectroscopy (FTIR), thermogravimetric analyses (TGA) and magnetization measurement. Also the adsorption behaviors including kinetic, isotherm, competitive adsorption and stability of TF-MS were evaluated.

2 Materials and methods

2.1 Materials

Poplar sawdust used for this experiment was obtained from a local saw mill in Heilongjiang Province, China. The sawdust was first washed with distilled water, dried, cut, and sieved through 100-mesh screen. The ferricchloride hexahydrate ($FeCl_3 \cdot 6H_2O$), ferrous chloride tetrahydrate ($FeCl_2 \cdot 4H_2O$), ammonia solution (25%), anhydrous ethanol (≥99.9%), 3-mercaptopropyltrimethoxysilane (MPTMS, ≥95%), glacial acetic acid (≥99.5%) used in this study were supplied by Shanghai Boyle Chemical Company Limited., and used without further purification.

2.2 Synthesis of magnetic sawdust composites (MS)

TF-MS was prepared as shown in Fig. 1. Firstly, 10 g wood sawdust, 5.40 g $FeCl_3 \cdot 6H_2O$ (0.02mol) and 1.98 g $FeCl_2 \cdot 4H_2O$ (0.01mol) were dissolved in 200 mL distilled water, the Fe^{3+}/Fe^{2+} ratio was 2. The pH value of the iron ions precursors were adjusted to ca. 10 by the dropwise addition of 10% ammonia solution. After stirring for 15 min, the system was transferred into a Teflon-lined stainless-steel autoclave, and heated at 90 ℃ to precipitate the magnetic Fe_3O_4 on the sawdust substrate. Then the oxidation reaction was appeared when the Fe_3O_4 nanoparticles exposure to air, leading to form γ-Fe_2O_3. Finally, the prepared specimens were removed, rinsed with distilled water for three times, and dried at 50 ℃ for over 24 h in vacuum.

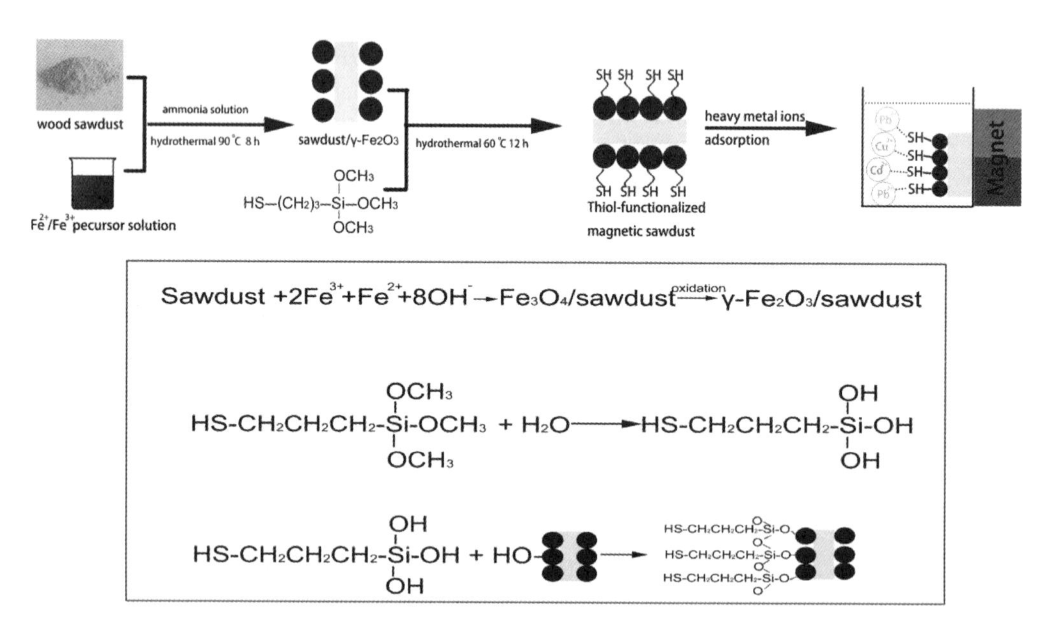

Fig. 1 The scheme of the preparation of TF-MS

2.3 Surface modification of MS

The 3-mercaptopropyltrimethoxysilane ethanol solution was prepared by stirring the mixture of the 100 mL anhydrous ethanol, 5 mL 3-mercaptopropyltrimethoxysilane (MPTMS), 2 mL H_2O, 5 mL glacial acetic acid at room temperature. 10 g magnetic sawdust was immersed into the 3-mercaptopropyltrimethoxysilane ethanol solution. The modification was maintained at 60 ℃ for 12 h. In a simplified scheme (Fig. 1), the silanization reaction was described as following: Firstly, the organosilane was placed into an aqueous solution of an acid that acts as a catalyst. It is hydrolyzed, the methoxy groups ($-OCH_3$) are replaced by hydroxyl groups to form reactive silanol groups. Then, the hydrolyzed silane condensed with the hydroxyl groups on the surface of magnetic sawdust forming a covalent bond with OH groups. With the continuous dehydration reaction, the -SH groups were successfully formed onto the surface of magnetic sawdust. The reaction product was collected with a magnet, washed repeatedly with ethanol and distilled water, dried in a vacuum oven at 50 ℃.

2.4 Characterization

The morphology of the samples was examined by the scanning electron microscopy (SEM, FEI, and Quanta 200). The crystalline structure of these composites were identified using the X-ray diffraction (XRD, Rigaku, and D/MAX 2200) operating with Cu $K\alpha$ radiation ($\lambda = 1.5418$ Å) at a scan rate (2θ) of $4° \ min^{-1}$, an accelerating voltage of 40 kV, and an applied current of 30 mA ranging from 5° to 80°. The X-ray photoelectron spectroscope (XPS) was recorded on the Thermo ESCALAB 250XI. The deconvolution of the overlapping peaks was performed using a mixed Gaussian-Lorentzian fit program. The surface chemical compositions of the samples were determined via the Fourier transformation infrared spectroscopy (FTIR, Nicolet, Magna-IR 560). Thermogravimetric analyses (TGA) were examined using a thermal analyzer (TGA, SDT Q600) in the temperature range from room temperature up to 800 ℃ at a heating rate of $10 \ ℃ \cdot min^{-1}$, with a flow of dried air ($100 \ mL \cdot min^{-1}$). The magnetization measurements were carried out at room temperature using a superconducting quantum interference device (MPMS XL-7, Quantum Design Corp.). The potassium dichromate method was used to determine the chemical oxygen demand (COD).

2.5 Batch adsorption experiments

Batch experiments were conducted in duplicates in 250 mL glass conical flasks under ultrasonication for several minutes at 25 ℃. An amount of 0.1 g adsorbent and 100 mL heavy metal solution was used for every treatment unless otherwise stated. Separation of the adsorbent was completed in 1 min by applying an external magnet. The concentration of heavy metal was determined by the Flame Atomic Absorption Spectrometer (TAS-990AFG). The equilibrium sorption capacity, q_e, was calculated according to eqn. (1):

$$q_e = \frac{(C_0 - C_e)V}{m} \quad (1)$$

where C_0 ($mg \cdot L^{-1}$) is the initial concentration of metal ion, C_e ($mg \cdot L^{-1}$) is the equilibrium concentration in solution, V (L) is the total volume of solution, and m (g) is the sorbent mass.

To survey competitive adsorption among Cu^{2+}, Pb^{2+} and Cd^{2+}, 0.1 g adsorbent was added into 100 mL multi-metal solution with the initial pH 6. For bi-metal solution, the concentration was $5 \ mg \cdot L^{-1}$ for each and for tri-metal solution, $3.33 \ mg \cdot L^{-1}$ for each metal.

All the experiments were replicated three times, and the mean values were used in our analyses. If the standard error (S.E.) were greater than 0.05, the test was repeated to control for errors.

2.6 Stability and regeneration of adsorbent

Stability of TF-MS under acidic and alkaline conditions was evaluated by dispersing 0.1 g adsorbent in 100 mL different concentration of HCl or NaOH solution. After shaking for 6 h at 25 ℃, and then the treated adsorbent was washed to neutrality for reuse. The adsorption by the acid or alkali treated TF-MS was investigated in aqueous solution containing Pb^{2+} ions ($15 \ mg \cdot L^{-1}$) at 25 ℃ for 30 min. By determining the magnetic property and adsorption ability of the treated adsorbent, stability could be inferred.

Regeneration studied was conducted by dispersing 0.1 g Pb^{2+} loaded adsorbent in 100 mL of 1 M HCl solution for 30 min. Then the adsorbent was collected and washed for reused.

3 Results and discussion

3.1 Characterization of adsorbent.

Fig. 2 shows the representative SEM images of the untreated sawdust and TF-MS. Fig. 2a shows the typical SEM image of the poplar sawdust surface, the smooth surface and some pits (namely the smallest holes of all pores) found on the walls of tracheids. After chemical treatment, the SEM images of TF-MS (Fig. 2b and c) reveal that many tiny γ-Fe_2O_3 nanoparticles are precipitated on the surface of the sawdust. There is no detected bare surface area on the surface of sawdust, indicating the high efficiency of our synthesis method.

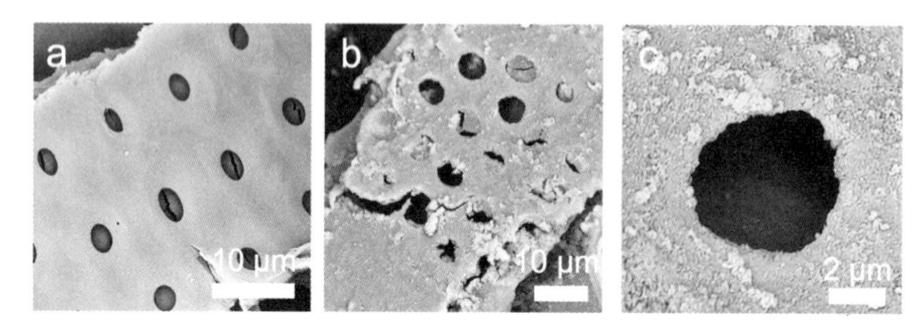

Fig. 2 SEM images of (a) the untreated sawdust, (b, c) TF-MS at different magnifications

The FTIR spectra of untreated sawdust, MS and TF-MS are shown in Fig. 3. In both spectra, the broad and intense absorption peak at 3420 cm^{-1} attributed to the O—H stretching vibrations of cellulose, hemicellulose, lignin and absorbed water. The peak observed at 2901 cm^{-1} corresponded to the C—H stretching vibrations of methyl, methylene and methoxy groups. The peak at 1739 cm^{-1} in the untreated sawdust spectrum shows the carbonyl (C=O) stretching vibration of the carboxyl groups of hemicelluloses and lignin in the sawdust. The peak around 1594 cm^{-1} is due to the aromatic stretching vibration of lignin. The peaks around 1424 cm^{-1} could be attributed to the C—H deformation vibrations of methyl, methylene and methoxy groups. The peaks in the range 1300 - 1000 cm^{-1} could be assigned to the C—O stretching vibration of

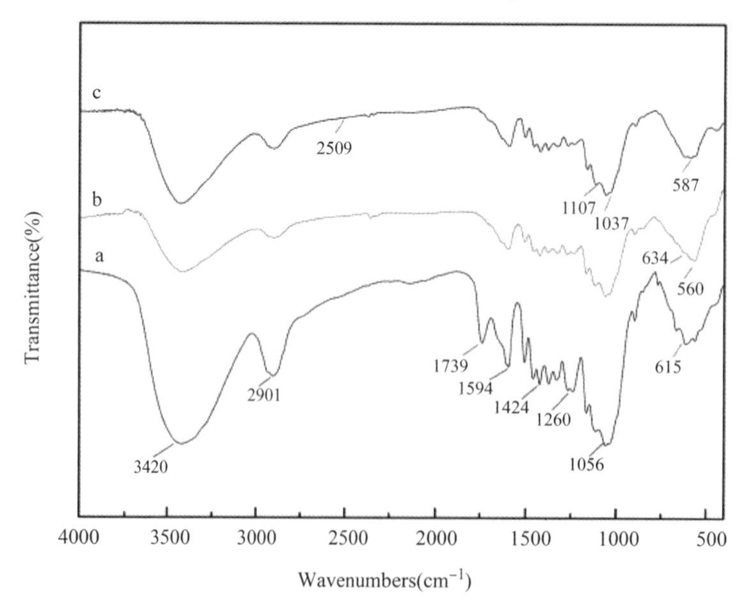

Fig. 3 FTIR spectra: (a) the untreated sawdust, (b) MS and (c) TF-MS

hemicelluloses and lignin.[23] For magnetic sawdust, the peaks at 560 cm^{-1} and 634 cm^{-1} attributed to the Fe—O bond vibration of γ-Fe_2O_3.[24] The FTIR spectrum of TF—MS shows the strong adsorption characteristics of Fe—O—Si bands at 587 cm^{-1}.[25] There is also presented the characteristic bands of Si—O—Si bonds (ν_{as} Si—O—Si at 1106 cm^{-1}, ν_{as} Si—OH at 1037 cm^{-1}). S—H stretches were found at 2509 cm^{-1}, which are typically very weak and cannot be detected in the spectra of thin film.[26] The results illustrated that the sawdust was functionalized with magnetic particles and 3-mercaptopropyltrimethoxysilane in the synthetic process.

To further confirm the successful modifying the magnetic sawdust with mercaptopropyl silica coating, the XPS spectra of the TF-MS was measured. Fig. 4a shows that the characteristic signals for C, O, Fe, S and Si elements are clearly observed in the XPS survey spectrum of TF-MS. The peaks of Si2s, Si2p and S2s belonged to the layer of mercaptopropyl silica ($HSCH_2CH_2CH_2Si$—) could be found at 102.2 eV, 154.3eV and 165.3 eV, respectively.[27] These suggested that the silane coating was successfully modified on the surface of MS. Fig. 4b illustrates the high-resolution XPS spectra of 2p Fe. The Fe2p level with binding energies of 711.7 eV and 724.2 eV closely correspond to $Fe2p_{3/2}$ and $Fe2p_{1/2}$ spin-orbit peaks of γ-Fe_2O_3.[28] The satellite peak at 719.2 eV also proves the presence of Fe^{3+}.

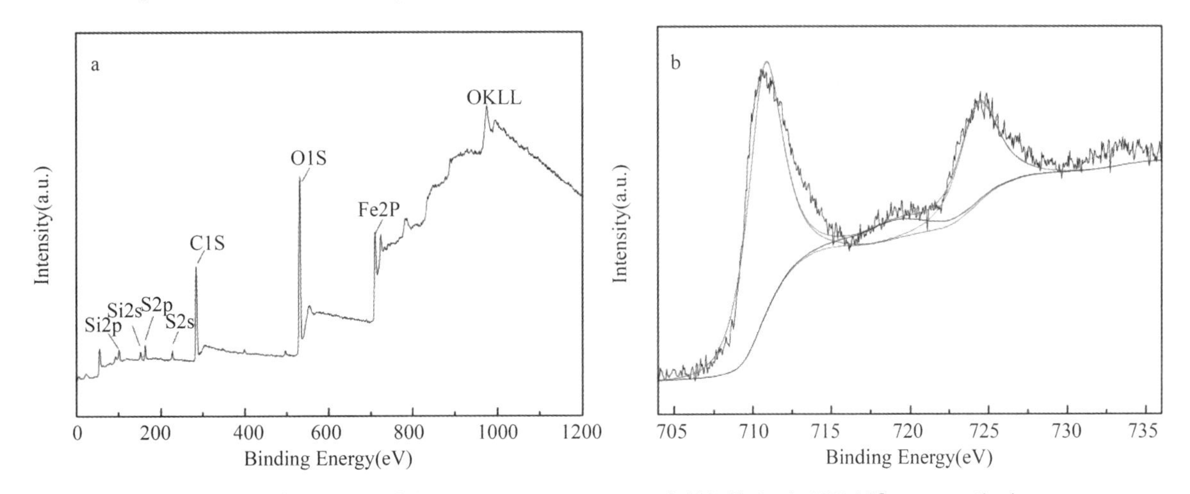

Fig. 4 XPS spectra of (a) survey spectrum and (b) Fe2p in TF-MS, respectively

For further clarifying the crystal structure and phase purity of the produces, the XRD patterns of the untreated sawdust, MS and TF-MS were measured, as shown in Fig. 5. The cellulose characteristic peaks could be seen at around 15.5° and 22.0° for both the untreated sawdust, MS and TF-MS. The additional diffraction peaks of MS and TF-MS at around 30.2°, 35.6°, 43.1°, 53.4°, 57.2° and 62.7° are corresponding to the diffractions of the (220), (311), (400), (422), (511) and (440) planes of γ-Fe_2O_3 (JCPDS no. 39-1346), respectively. The absence of the other peaks of TF-MS suggested that the 3-mercaptopropyltrimethoxysilane modification does not result in the phase change of γ-Fe_2O_3.

The introduction of magnetic γ-Fe_2O_3 and mercaptopropyl silica into sawdust is also confirmed by the results of TGA, as shown in Fig. 6. A small weight loss of about 5 wt% around 100 ℃ is due to the evaporation of adsorbed water. For the untreated sawdust, the obvious weight loss (about 60 wt%) in the temperature range from 250-320 ℃ is attributed to the oxidation and pyrolysis of cellulose, and the further weight loss (about 30 wt%) in the temperature range from 320-420 ℃ corresponded to the degradation of lignin. For the

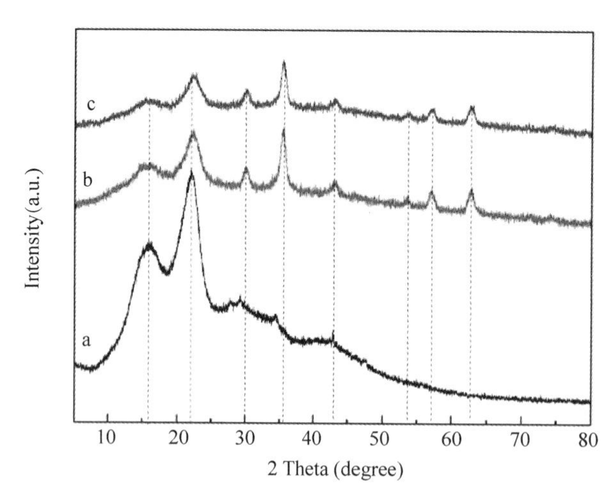

Fig. 5 XRD patterns of (a) the untreated sawdust, (b) MS and (c) TF-MS, respectively

MS, there is still a weight loss of about 40 wt% and 30 wt% in the temperature range from 250-320 ℃ and 320-420 ℃, respectively, indicating the degradation of sawdust components. However, more residues resulted after flaming, indicated some incombustible iron oxides (about 17 wt%) were precipitated on the sawdust substrate. As compared with MS, the residues of TF - MS increased weight for approximately 2 wt%, corresponding to the further modification of 3-mercaptopropyltrimethoxysilane.

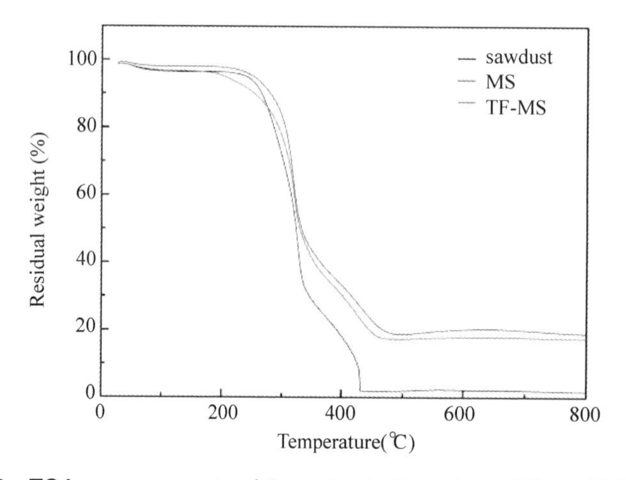

Fig. 6 TGA measurements of the untreated sawdust, MS and TF-MS

Magnetic adsorbent has been attracted increasing attentions because they could be easily separated under a magnetic field. Such magnetic separation is essential to improve the operation efficiency and reduced the cost during wastewater treatment. [29] Fig. 7 shows the magnetic hysteresis loop of the untreated sawdust, MS and TF-MS at room temperature. The untreated sawdust exhibit diamagnetic behavior at room temperature and the magnetization saturation value for sawdust obtained at room temperature is 0 emu · g^{-1}. However, MS and TF-MS show typical superparamagnetism at room temperature with no coercivity and remanence. The saturation magnetization values for MS and TF-MS are 11. 58 and 7. 28 emu · g^{-1} at 2 T, respectively. It can be clearly seen that the saturation magnetization value is decreased after modification by MPTMS, which can be attributed to the formation of the nonmagnetic thiol-functionalized layer. Besides, the TF-MS can be collected by a magnet as shown in inset of Fig. 7, indicating that it can be easily recycled by an external magnetic field.

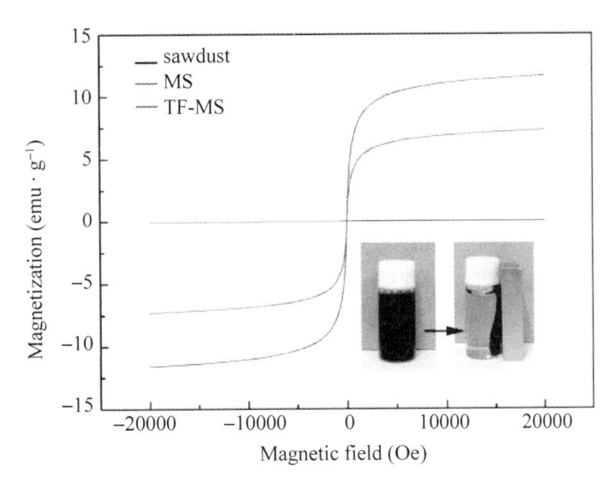

Fig. 7 Room-temperature hysteresis loops of the untreated sawdust, MS and TF-MS. The inset gives the photograph of TF-MS dispersed in water (left) and their response to a magnet (right)

3.2 Effect of chemical pretreatment on heavy metal ions adsorption

The adsorption capacity of heavy metal ions on the untreated sawdust, magnetic sawdust and TF-MS were examined and Pb^{+} as a representative ion (Table 1). The results clearly show that TF-MS more effectively adsorbed Pb^{2+} ion than the untreated sawdust performed. This could be the superior ion exchange capacity of TF-MS compared to untreated sawdust and MS because of the increasing number of thiol groups on sawdust after grafting of 3-mercaptopropyltrimethoxysilane on sawdust. The chemical modification of sawdust made TF-MS to absorb more Pb^{2+} ions without an increase in COD, which was due to the release of the organic compounds such as lignin, pectin and tannin in the chemical pretreatment process. [15]

Table 1 Effect of chemical pretreatment on heavy metal ions adsorption

Adsorbent	COD ($mg\ O_2 \cdot L^{-1}$)	Pb^{2+} adsorption
Untreated sawdust	125	2.73
MS	55	4.45
TF-MS	46	9.62

3.3 Effect of pH

The adsorption capacity of an adsorbent for metal ions not only depends on the chemical and physical properties of the adsorbent, but also on the competitive adsorption of coexisting matters in aqueous solution and the hydrolysis capacity of the metal ions. [30] Cu^{2+}, Pb^{2+} and Cd^{2+} removal by TF-MS are measured in batch experiments with various pH values from 2 to 8, as shown in Fig. 8. The removal efficiency of Pb^{2+} is higher than that of Cu^{2+} and Cd^{2+} in the pH range from 2-7 and both of them present increasing trend with the rise of pH values. The minimum adsorption for Cu^{2+}, Pb^{2+} and Cd^{2+} at pH 2.0 may be due to the fact that the competition of metal ions with H^{+} for combination with thiol groups and the strengthening protonation of the thiol groups on sawdust surface ($pK_a = 9.65$) in lower acid solution. Besides, the metal ions have the tendency to hydrate to form $M(OH)_2$ that has smaller effective size and higher mobility than metal ions under high pH solution, thus resulting in increased adsorption of adsorbent with the increase of the pH value.[31] Therefore, the pH value of this study is fixed at 6.0 due to its positive effect.

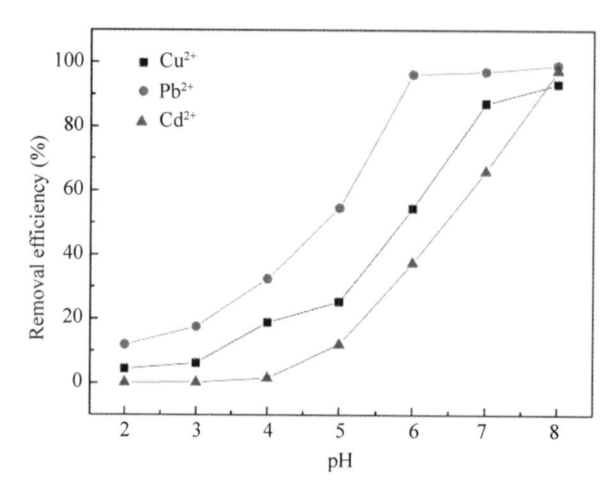

Fig. 8 Effect of pH on the removal of Cu^{2+}, Pb^{2+} and Cd^{2+} by the TF-MS at initial metal ions concentration of 10 mg · L^{-1}, adsorbent dose 1 g · L^{-1} and temperature 25 ℃ for 30 min, pH was adjusted by 0.1 mol · L^{-1} HCl and NaOH

3.4 Effect of contact time

The effect of contact time on adsorption of Cu^{2+}, Pb^{2+} and Cd^{2+} was studied. Fig. 9 showed that the adsorption rate was fast and achieved adsorption equilibrium within 20 min. It is possible that the surface coverage is relatively low in the early stage, so the heavy metal ions occupy the active surface sites rapidly. As the adsorption progress, the surface of TF-MS is occupied by metal ions gradually, and the rate uptake becomes slower and reaching saturation adsorption in the latter stage.

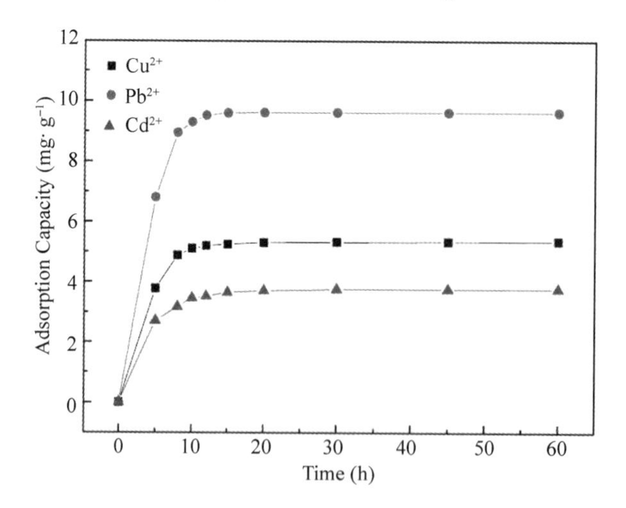

Fig. 9 Effect of contact time on adsorption of Cu^{2+}, Pb^{2+} and Cd^{2+} (initial concentration: 10 mg · L^{-1}, adsorbent dose: 1 g · L^{-1}, pH 6.0)

In order to show the most suitable kinetic model for Cu^{2+}, Pb^{2+} and Cd^{2+} removal, the pseudo-first-order [Eqn. (2)], pseudo-second-order [Eqn. (3)], Elovich [Eqn. (4)] and intra-particles diffusion [Eqn. (5)] kinetic models are used to fit our experimental data. The pseudo-first-order, pseudo-second-order, Elovich and intra-particles diffusion rate equations are expressed as follows:[32, 33]

$$\ln(q_e-q_t)=\ln q_e-k_1 t \tag{2}$$

$$\frac{t}{q_t}=\frac{1}{k_2 q_e^2}+\frac{t}{q_e} \tag{3}$$

$$q_t=\frac{\ln(\alpha\beta)}{\beta}+\frac{\ln(t)}{\beta} \tag{4}$$

$$q_t=k_{\text{intra}}t^{1/2}+c \tag{5}$$

where q_e and q_t are the amounts of the metal ions adsorbed ($mg \cdot g^{-1}$) at equilibrium and at time t (min), respectively. k_1 (min^{-1}) is the constant of pseudo first-order rate; k_2 ($g \cdot mg^{-1} \cdot min^{-1}$) is the constant of pseudo-second-order rate. α and β are the initial adsorption rate ($mg \cdot g^{-1} \cdot min^{-1}$) and the desorption constant ($mg \cdot g^{-1} \cdot min^{-1}$), respectively. k_{intra} is the intra-particles diffusion rate constant ($mg \cdot g^{-1} \cdot min^{-1}$), which may be taken as a rate factor, i.e., percent metal ions adsorbed per unit time, c is a constant related with the boundary layer thickness ($mg \cdot g^{-1}$).

The parameters and the correlation coefficients (R^2) for the four models are shown in Table 2. Taking the adsorption of Cu^{2+} by TF-MS for example, the R^2 value of the pseudo-second-order model is 0.9991, which is higher than that of pseudo-first-order (0.9635), Elovich (0.9058) and intra-particles diffusion model (0.9814), revealing that the better fit to pseudo-second-order model. Besides, the adsorption capacity (q_e, cal.) calculated from the pseudo-second-order model is 5.46 $mg \cdot g^{-1}$, which is much closer to the experimental data (q_e, exp=5.35 $mg \cdot g^{-1}$) than the other three kinetic models. Similar phenomenon can be obtained for the adsorption of Pb^{2+} and Cd^{2+}. Therefore, the obtained data are in well agreement with the pseudo-second-order kinetic model, indicating the adsorption rate of the heavy metal ions is controlled by chemical process. The order of adsorption rates was $Pb^{2+} > Cu^{2+} > Cd^{2+}$.

Table 2 Kinetic parameters of different models for heavy metal ions adsorption onto TF-MS

Kinetic models and parameters	Cu^{2+} C_0($mg \cdot L^{-1}$) 10.0	Pb^{2+} C_0($mg \cdot L^{-1}$) 10.0	Cd^{2+} C_0($mg \cdot L^{-1}$) 10.0
q_e(exp.) ($mg \cdot L^{-1}$)	5.35	9.62	3.77
Lagergren pseudo-first-order			
q_e(calc.) ($mg \cdot L^{-1}$)	4.28	12.10	3.17
k_1	0.252	0.37	0.211
R^2	0.9635	0.9754	0.9831
Pseudo-second-order equation			
q_e(calc.) ($mg \cdot L^{-1}$)	5.46	10.0	4.0
k_2($g \cdot mg^{-1} \cdot min^{-1}$)	0.189	0.125	0.178
H ($mg \cdot g^{-1} \cdot min^{-1}$)	5.643	12.5	2.857
R^2	0.9991	0.9988	0.9993

(续)

Kinetic models and parameters	Cu^{2+} C_0(mg · L^{-1}) 10.0	Pb^{2+} C_0(mg · L^{-1}) 10.0	Cd^{2+} C_0(mg · L^{-1}) 10.0
Elovich equation			
q_e(calc.) (mg · L^{-1})	6.20	11.40	4.05
α(mg · g^{-1} · min^{-2})	3.402	5.747	3.218
β(g · mg^{-1} · min^{-1})	0.596	0.314	1.033
R^2	0.9058	0.8956	0.9745
Intra-particle diffusion equation			
q_e(calc.) (mg · L^{-1})	6.09	11.07	4.09
k_{intra}(mg · g^{-1} · $min^{1/2}$)	1.574	2.87	1.05
C	0.111	0.17	0.10
R^2	0.9814	0.9821	0.9790

Recently, Abdel-Ghani et al. suggested that 2 h was required to reach equilibrium using chemically treated wood sawdust to adsorb 25 mg · L^{-1} Pb^{2+}.[34] Cheng et al. pointed out that 30 min were needed for 20 mg · L^{-1} strontium ions adsorption equilibrium using Fe_3O_4 modified sawdust.[35] Hence, it is remarked that the TF-MS reached equilibrium within 20 min compared with other chemical modified sawdust. It is strongly believed that the thiol-groups are mainly responsible for the rapid adsorption equilibrium. The fast adsorption equilibrium provides the advantages for water treatment of highly effective and roboticized design.

3.5 Adsorption isotherm

The adsorption isotherms of Cu^{2+}, Pb^{2+} and Cd^{2+} with corresponding Langmuir plots were showed in Fig. 10. The Langmuir adsorption model assumes that the adsorption occurs on monolayer and takes place at specific homogenous sites.[2, 32] It can be expressed as:

$$\frac{C_e}{q_e}=\frac{1}{K_L q_m}+\frac{C_e}{q_m} \tag{6}$$

where C_e is the equilibrium metal ions concentration (mg · L^{-1}) and q_e represents the amount of metal ions adsorbed by per unit the absorbent (mg · L^{-1}); q_m and K_L are the maximum adsorption capacity (mg · g^{-1}) and the Langmuir adsorption equilibrium constant (L · mg^{-1}), respectively. They can be obtained from the slop and intercept of linear plot of C_e/q_e *vs.* Ce and all the parameters are listed in Table 3.

The Freundlich model is applied for multilayer sorption and non-ideal sorption on heterogeneous surface sites, and can be represented in linear form as follows:

$$\ln q_e=\ln K_F+\frac{1}{n}+\ln C_e \tag{7}$$

where K_F and n are the Freundlich constants related to the adsorption capacity (mg · g^{-1}) and the heterogeneity factor (mg^{-1}), respectively. These constants are evaluated from slop and intercept of the linear plots of C_e $q_e vs^{-1}$ C_e. All the parameters are listed in Table 3.

Table 3 Langmuir and Freundlich isotherm constants for the adsorption of Cu^{2+}, Pb^{2+} and Cd^{2+} ions onto TF-MS

	Langmuir model			Freundlich model		
	q_m(mg · g^{-1})	K_L(L · mg^{-1})	r^2	K_F	n	r^2
Cu^{2+}	5.49	10.71	0.9816	4.19	4.23	0.8949
Pb^{2+}	12.50	8	0.9808	6.75	2.22	0.9029
Cd^{2+}	3.80	13.85	0.9938	3.00	5.95	0.9821

Taking Cu^{2+} for example, from Table 3, the correlation coefficients r^2 of Langmuir equation is 0.9816, which is higher than that of Freundlich equation (0.8949). Therefore, the Langmuir isotherm model correlates better than the Freundlich isotherm model with the experimental data, suggesting the adsorbed Cu^{2+} by TF-MS is a monolayer adsorption. Similar phenomenon can be obtained for the adsorption of Pb^{2+} and Cd^{2+}. According to the Langmuir equation, the adsorption capacity of Cu^{2+}, Pb^{2+} and Cd^{2+} on TF-MS were calculated to be 5.49 mg · g^{-1}, 12.5 mg · g^{-1} and 3.80 mg · g^{-1}, respectively. A comparison of the maximum capacity, q_{max}, of TF-MS with those of some other biomass materials reported in literature is given in Table S2. Comparison of heavy metal ions adsorption on TF-MS with the literature data indicates that the adsorption capacity of TF-MS is greater than other biomass materials. Moreover, the maximum adsorption capacity ranked in the order of $Pb^{2+}>Cu^{2+}>Cd^{2+}$, which is the same with the calculated q_e order as expected in the kinetic studies and the one deduced from the study of the impact of pH. The accordant conclusions coming from the adsorption kinetic, isotherms and pH study indicated a differential binding capacity $Pb^{2+}>Cu^{2+}>Cd^{2+}$.

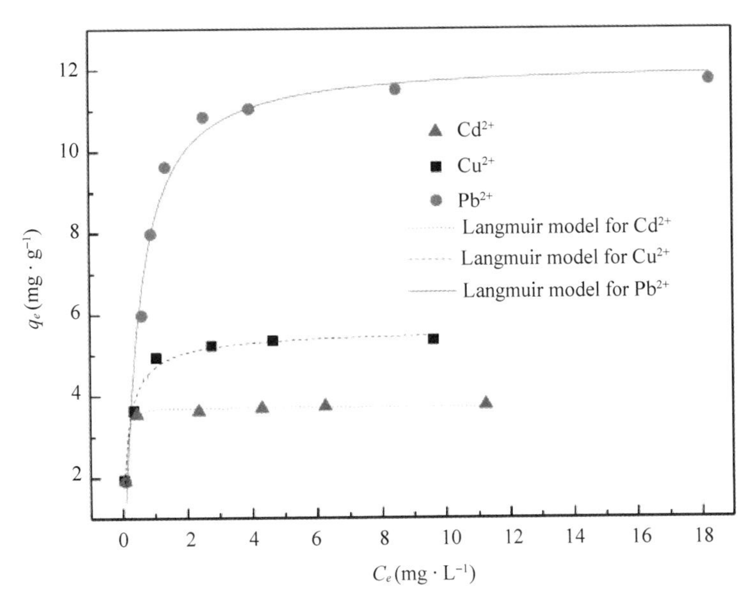

Fig. 10 Experimental metal adsorption isotherms and modeled results using Langmuir equation (initial concentration ranging rbm 2 to 30 mg · L^{-1}, adsorbent dose: 1 g · L^{-1}, pH 6.0, contact time: 30 min)

3.6 Competitive adsorption

The competitive adsorption among Cu^{2+}, Pb^{2+} and Cd^{2+} was investigated, as shown in Fig. 11. For Cu^{2+}

and Cd^{2+} bi-metal solution, the efficiency of Cu^{2+} and Cd^{2+} is 71.1% and 45.7%, respectively. In Pb^{2+} and Cd^{2+} mixed solution, the removal of Pb^{2+} is triple of Cd^{2+}. And in Pb^{2+} and Cu^{2+} mixed solution, the removal of Pb^{2+} is double of Cu^{2+}. Obviously, in the presence of Pb^{2+}, adsorption of Cu^{2+} and Cd^{2+} is strongly restricted. For the tri-metal solution with equal concentration of Cu^{2+}, Pb^{2+} and Cd^{2+}, the corresponding removal efficiency is 42.5%, 90.6%, and 29.1%, respectively. These indicated that the TF-MS have a preferential adsorption of $Pb^{2+} > Cu^{2+} > Cd^{2+}$. It could be explained by Pearson's theory that soft ligands like the thiol group on the surface of sawdust are soft base, which can form a highly polarizable donor centres, thereby have the capable of interacting and strong affinity with low-lying orbitals of soft acid, and Pb is considered to be softer than Cu and Cd.[31, 36]

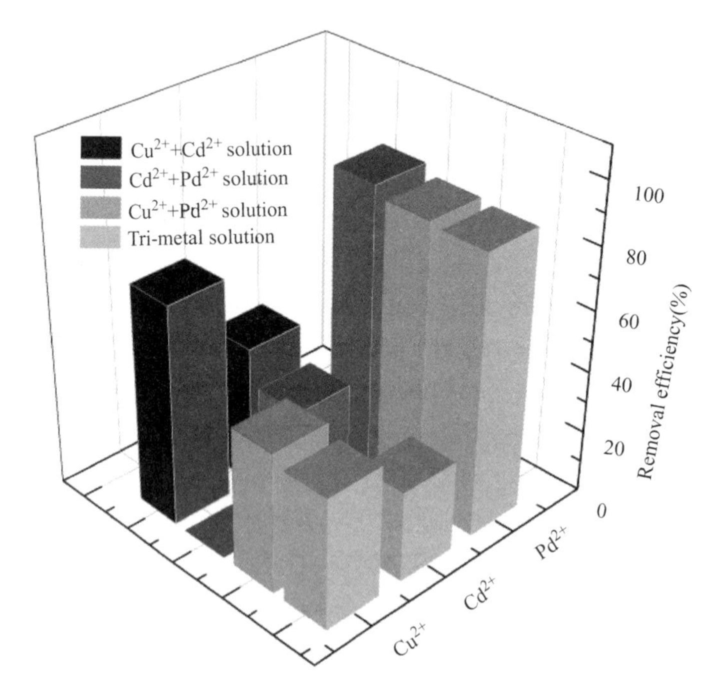

Fig. 11 Competitive adsorption of multi-metal solution (adsorbent dose: 1 g · L^{-1}, pH: 6.0, contact time: 30 min)

3.7 Stability and regeneration

Stability in different conditions is important for the practical application of the adsorbent. As for the degree of acid and alkali resistance, the results are showed in Table 4. The magnetization values of TF-MS after 2 M HCl treatment is still maintained at 4.41 emu · g^{-1}, which is only decreased 2.87 emu · g^{-1} compared with the untreated TF-MS. This value is smaller than that reported in many studies,[37, 38] indicating the good magnetization stability of TF-MS. The magnetization values of TF-MS increased slightly after alkali treatment. It is possibly due to the chemical reaction between the silane layer and NaOH, which reduced the content of nonmagnetic silane in TF-MS and increased the magnetic values. Moreover, the magnetization values of all treated TF-MS are strong enough for separation in 1 min. As for the removal efficiency of treated TF-MS, it can observe that the removal efficiency of Pb^{2+} decreased after acid treatment, which is caused by the protonation of TF-MS adsorbent in the process of acid treatment. Although the treated adsorbent was washed repeatedly to reach neutral for reuse, some H^+ occupied the binding sites on the adsorbent surface was irrevocable, which led to the decrease of removal efficiency. On the contrary, the removal efficiency increased after alkali treatment. It is believed that the adsorbent was deprotonated and some hydroxyl groups might form on the TF-MS surface to chelate metal ions, resulting in the subsequent increase of removal efficiency. Therefore, it can be expected that TF-MS can be used in wastewater even under extreme condition.

Table 4 Effect of different concentration HCl and NaOH treatment on magnetization and removal efficiency.

	HCl			NaOH				Untreated
Concentration ($mol \cdot L^{-1}$)	0.1	1	2	0.1	1	2	5	
Magnetization ($emu \cdot g^{-1}$)	7.07	6.57	4.41	8.08	8.12	9.09	10.58	7.28
Removal(%)	56.5	49.1	43.3	74.5	75.9	77.3	78.9	73.6

The regeneration of TF-MS is surveyed by eluted solution of 1 M HCl under sonication for 30 min. Fig. 12 shows the removal efficiency of Pb^{2+} over four successive adsorption-desorption cycles. It is observed that approximate 96% removal efficiency is reached in the first cycle. Even though the efficiency decreased with the increasing of cycle, over 87% efficiency was obtained in the further adsorption-desorption cycle.

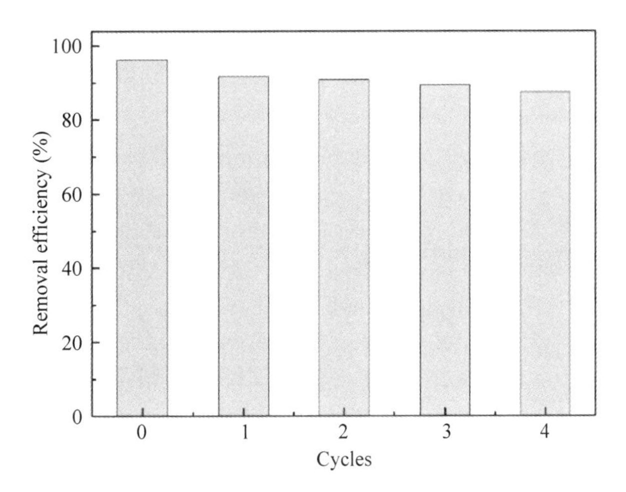

Fig. 12 Pb^{2+} adsorption on regenerated TF-MS adsorbent with four adsorption-regeneration cycles (initial concentration: 10 $mg \cdot L^{-1}$, adsorbent dose: 1 $g \cdot L^{-1}$, pH 6.0)

4 Conclusions

Thiol-functionalized magnetic sawdust adsorbent was prepared for removal of heavy metals. Such prepared adsorbent possessed the superparamagnetic character, which make it an effective and convenient adsorbent for heavy metal removal. The adsorption of Cu^{2+}, Pb^{2+} and Cd^{2+} was well modeled by pseudo-second-order model and Langmuir adsorption isotherm, and the adsorbent presented a preferential binding capacity of $Pb^{2+} > Cu^{2+} > Cd^{2+}$. Besides, the adsorbent possess excellent stability in strong acid and alkaline condition, and could be used repeatedly by effective regeneration using acid solution. Coupled with the advantages of fast, easy separation, environmental friendliness and appropriate adsorption capacity for heavy metal ions, the approach presented may provide further routes for the development of efficient and recyclable biosorbent.

Acknowledgments

This work was financially supported by The National Natural Science Foundation of China (grant no. 31470584).

References

[1] Yantasee W, Warner C L, Sangvanich T, et al. Removal of heavy metals from aqueous systems with thiol functionalized superparamagnetic nanoparticles[J]. Environ. Sci. Technol., 2007, 41(14): 5114-5119.

[2] Hao Y M, Man C, Hu Z B. Effective removal of Cu (II) ions from aqueous solution by amino-functionalized magnetic nanoparticles[J]. J. Hazard. Mater., 2010, 184(1-3): 392-399.

[3] Bailey S E, Olin T J, Adrian D D. A review of potentially low-cost sorbents for heavy metals[J]. Water Res., 1999, 33(11): 2469-2479.

[4] Liu X W, Hu Q Y, Fang Z, et al. Magnetic chitosan nanocomposites: a useful recyclable tool for heavy metal ion removal[J]. Langmuir, 2009, 25(1): 3-8.

[5] Wang S G, Wang K L, Dai C, et al. Adsorption of Pb^{2+} on amino-functionalized core-shell magnetic mesoporous SBA-15 silica composite[J]. Chem. Eng. J., 2015, 262: 897-903.

[6] Ozmen M, Can K, Arslan G, et al. Adsorption of Cu(II) from aqueous solution by using modified Fe_3O_4 magnetic nanoparticles[J]. Desalination, 2010, 254(1-3): 162-169.

[7] Ngah W S W, Hanafiah M A K M. Removal of heavy metal ions from wastewater by chemically modified plant wastes as adsorbents: a review[J]. Bioresour. Technol., 2008, 99(10): 3935-3948.

[8] Tee T W, Khan A R M. Removal of lead, cadmium and zinc by waste tea leaves[J]. Environ. Technol. Lett., 1988, 9(11): 1223-1232.

[9] Liang X, Guo X Y, Feng N C, et al. Application of orange peel xanthate for the adsorption of Pb^{2+} from aqueous solutions[J]. J. Hazard. Mater., 2009, 170(1): 425-429.

[10] Yao Z Y, Qi J H, Wang L H. Equilibrium, kinetic and thermodynamic studies on the biosorption of Cu(II) onto chestnut shell[J]. J. Hazard. Mater., 2010, 174(1-3): 137-143.

[11] Karnitz O, Gurgel L V A, De Melo J C P, et al. Adsorption of heavy metal ion from aqueous single metal solution by chemically modified sugarcane bagasse[J]. Bioresour. Technol., 2007, 98(6): 1291-1297.

[12] Chen H, Zhao J, Dai G L, et al. Adsorption characteristics of Pb(II) from aqueous solution onto a natural biosorbent, fallen Cinnamomum camphora leaves[J]. Desalination, 2010, 262(1-3): 174-182.

[13] Bhattacharya A K, Mandal S N, Das S K. Adsorption of Zn(II) from aqueous solution by using different adsorbents [J]. Chem. Eng. J., 2006, 123(1-2): 43-51.

[14] Sciban M, Klasnja M, Skrbic B. Modified hardwood sawdust as adsorbent of heavy metal ions from water[J]. Wood Sci. Technol., 2006, 40(3): 217-227.

[15] Feng N C, Guo X Y, Liang S. Adsorption study of copper(II) by chemically modified orange peel[J]. J. Hazard. Mater., 2009, 164(2-3): 1286-1292.

[16] Bai R S, Abraham T E. Studies on enhancement of Cr(VI) biosorption by chemically modified biomass of Rhizopus nigricans[J]. Water Res., 2002, 36(5): 1224-1236.

[17] Najafi M, Rostamian R, Rafati A A. Chemically modified silica gel with thiol group as an adsorbent for retention of some toxic soft metal ions from water and industrial effluent[J]. Chem. Eng. J., 2011, 168(1): 426-432.

[18] Rostamian R, Najafi M, Rafati A A. Synthesis and characterization of thiol-functionalized silica nano hollow sphere as a novel adsorbent for removal of poisonous heavy metal ions from water: Kinetics, isotherms and error analysis[J]. Chem. Eng. J., 2011, 171(3): 1004-1011.

[19] Merk V, Chanana M, Gierlinger N, et al. Hybrid wood materials with magnetic anisotropy dictated by the hierarchical

cell structure[J]. ACS Appl. Mater. Interfaces, 2014, 6(12): 9760-9767.

[20]Trey S, Olsson R T, Strom V, et al. Controlled deposition of magnetic particles within the 3-D template of wood: making use of the natural hierarchical structure of wood[J]. RSC Adv., 2014, 4(67): 35678-35685.

[21]Gan W T, Gao L K, Zhan X X, et al. Hydrothermal synthesis of magnetic wood composites and improved wood properties by precipitation with $CoFe_2O_4$/hydroxyapatite[J]. RSC Adv., 2015, 5(57): 45919-45927.

[22]Sciban M, Radetic B, Kevresan D, et al. Adsorption of heavy metals from electroplating wastewater by wood sawdust [J]. Bioresour. Technol., 2007, 98(2): 402-409.

[23]Gierlinger N, Goswami L, Schmidt M, et al. In situ FT-IR microscopic study on enzymatic treatment of poplar wood cross-sections[J]. Biomacromolecules, 2008, 9(8): 2194-2201.

[24]Waldron R D. Infrared spectra of ferrites[J]. Phys. Rev., 1955, 99(6): 1727-1735.

[25]Hakami O, Zhang Y, Banks C J. Thiol-functionalised mesoporous silica-coated magnetite nanoparticles for high efficiency removal and recovery of Hg from water[J]. Water Res., 2012, 46(12): 3913-3922.

[26]Walcarius A, Delacote C. Mercury(II) binding to thiol-functionalized mesoporous silicas: critical effect of pH and sorbent properties on capacity and selectivity[J]. Anal. Chim. Acta, 2005, 547(1): 3-13.

[27]Li S Z, Yue X L, Jing Y M, et al. Fabrication of zonal thiol-functionalized silica nanofibers for removal of heavy metal ions from wastewater[J]. Colloids Surf., A, 2011, 380(1-3): 229-233.

[28]Wan C C, Li J. Facile synthesis of well-dispersed superparamagnetic gamma-Fe_2O_3 nanoparticles encapsulated in three-dimensional architectures of cellulose aerogels and their applications for Cr(VI) removal from contaminated water[J]. ACS Sustain. Chem. Eng., 2015, 3(9): 2142-2152.

[29]Hua M, Zhang S J, Pan B C, et al. Heavy metal removal from water/wastewater by nanosized metal oxides: a review [J]. J. Hazard. Mater., 2012, 211: 317-331.

[30]Zhang C, Sui J H, Li J, et al. Efficient removal of heavy metal ions by thiol-functionalized superparamagnetic carbon nanotubes[J]. Chem. Eng. J., 2012, 210: 45-52.

[31]Li G L, Zhao Z S, Liu J Y, et al. Effective heavy metal removal from aqueous systems by thiol functionalized magnetic mesoporous silica[J]. J. Hazard. Mater., 2011, 192(1): 277-283.

[32]Tan Y Q, Chen M, Hao Y M. High efficient removal of Pb(II) by amino-functionalized Fe_3O_4 magnetic nano-particles[J]. Chem. Eng. J., 2012, 191: 104-111.

[33]Ozmen M, Can K, Akin I, et al. Surface modification of glass beads with glutaraldehyde: Characterization and their adsorption property for metal ions[J]. J. Hazard. Mater., 2009, 171(1-3): 594-600.

[34]Abdel-Ghani N T, El-Chaghaby G A, Helal F S. Simultaneous removal of aluminum, iron, copper, zinc, and lead from aqueous solution using raw and chemically treated African beech wood sawdust[J]. Desalin. Water Treat., 2013, 51(16-18): 3558-3575.

[35]Cheng Z H, Gao Z X, Ma W, et al. Preparation of magnetic Fe_3O_4 particles modified sawdust as the adsorbent to remove strontium ions[J]. Chem. Eng. J., 2012, 209: 451-457.

[36]Pearson R G. Hard and soft acids and bases[J]. J. Am. Chem. Soc., 1963, 85(22): 3533-3539.

[37]Pang Y, Zeng G M, Tang L, et al. PEI-grafted magnetic porous powder for highly effective adsorption of heavy metal ions[J]. Desalination, 2011, 281: 278-284.

[38]Wang J H, Zheng S R, Shao Y, et al. Amino-functionalized Fe_3O_4@SiO_2 core-shell magnetic nanomaterial as a novel adsorbent for aqueous heavy metals removal[J]. J. Colloid Interface Sci., 2010, 349(1): 293-299.

中文题目：硫醇功能化的磁性锯末复合材料的制备，作为去除重金属离子的吸附剂

作者：甘文涛，高丽坤，詹先旭，李坚

摘要：通过在锯末表面沉淀 γ-Fe_2O_3 纳米颗粒，然后利用3-巯基丙基三甲氧基硅烷层进行改性，

合成了硫醇功能化的磁性锯末，并对其作为一种环境友好和可回收的重金属离子吸附剂进行了研究。修饰过程通过扫描电子显微镜、傅里叶变换红外光谱、X 射线光电子能谱和 X 射线衍射进行了检测。与非磁性锯末相比，硫醇功能化的磁性锯末具有较高的饱和磁化率(7.28 emu · g^{-1})，在外部磁场下可以更容易、更快地与水分离。吸附平衡在20分钟内达到，吸附动力学用准二阶动力学模型进行了阐释。Cu^{2+}、Pb^{2+}和Cd^{2+}的吸附等温线与 Langmuir 模型拟合良好，显示出的吸附能力分别为 5.49 mg · g^{-1}、12.5 mg · g^{-a1} 和 3.80 mg · g^{-1}。三种金属离子之间的竞争性吸附显示出对 Pb^{2+}> Cu^{2+}> Cd^{2+}的优先吸附。此外，磁性锯末吸附剂表现出良好的酸碱稳定性，负载金属的吸附剂能够在酸性溶液中再生而没有明显的吸附能力损失。

关键词：锯末；磁性 γ-Fe_2O_3；硫醇功能化；重金属；吸附

Fabrication of Biomass-derived C-doped Bi_2WO_6 Templated from Wood Fibers and Its Excellent Sensing of the Gases Containing Carbonyl Groups*

Likun Gao, Wentao Gan, Guoliang Cao, Xianxu Zhan, Tiangang Qiang, Jian Li

Abstract: In this work, wood fibers (*Populus ussuriensis* Kom) were used as biotemplate to synthesize Bi_2WO_6. In corporation of biomass-derived C, the sensitivity to the gases containing carbonyl groups of gas sensors is improved. The experimental results showed that the C-doped Bi_2WO_6 sensors presented high sensitivities to acetone, acetic acid and ethyl acetate molecules at low temperature (370 ℃). The gas-sensitivities of C-doped Bi_2WO_6 sensors on exposure to 1000 ppm of acetone, acetic acid and ethyl acetate gas are 4.4, 3.0 and 2.4 times higher than that exposure to 1000 ppm of ethanol, respectively. That is because the biomass-derived C can adsorb more oxygen molecules and the carbonyl groups in the gases can react with more oxygen molecules and release more electrons leading a resistance decrease of the sensors due to the electronic structure of carbonyl groups. The reported strategy opened up a new way for the development of gas sensors.

Keywords: Biomass-derived C; Bi_2WO_6; Wood fibers; Gas sensor; Carbonyl groups

1 Introduction

There has been a considerably increased interest in detecting and controlling the harmful gases. Recently, many semiconductors, such as SnO_2, ZnO, TiO_2, etc., owing to their highly sensitive to the changes in different chemical environment, have been used widely in gas sensor systems for the detection of hazardous gases[1-6]. Bismuth tungstate (Bi_2WO_6) among the most attractive and important *Aurivillius* family, has been found to possess interesting physical properties, such as ferroelectric piezoelectricity, pyroelectricity, catalytic behavior and a non-linear dielectric susceptibility[7]. *Aurivillius* family is defined as structurally related oxides with general formula $Bi_2A_{n-1}B_nO_{3n+3}$ (A=Ca, Sr, Ba, Pb, Bi, Na, K and B=Ti, Nb, Ta, Mo, W, Fe)[8, 9]. Bi_2WO_6 is the simplest members of the family with $n=1$, whose layer structure is composed by alternating $(Bi_2O_2)_n^{2n+}$ layers and perovskite-like $(WO_4)_n^{2n+}$ layers[10-14]. Bi_2WO_6 has gotten more and more concerns in recent years due to its prospect in photoluminescence, microwave, optical fibers, catalysts, magnetic devices, and gas sensors.

The morphological parameters of semiconductor materials such as surface area, total pore volume, particle and pore size distribution, crystallinity, thermal stability, phase composition, etc., play important roles in their performances[15]. Many different synthetic approaches were adopted to enhance the particular properties

---

* 本文摘自 Colloids and Surfaces A: Physicochemical and Engineering Aspects, 2017, 529: 487-494.

through fabricating varied surface morphology of semiconductor. It is well known that there is a key step to fabricate a gas sensor, that is, calcination. And some calcination processes need to be conducted under the protection of inert or reductive gases, such as H_2, Ar and NH_3, to protect the materials structures. However, for example, if H_2 was selected as reductive gas in the calcination, there would be several problems such as the potential risks of explosion and the high costs of pure gases[16]. Among the reductive gases, CO could be easily obtained by oxidizing the carbon during calcination in air.

It is well known that the physical/chemical properties of nanostructured materials are closely interrelated to their size, shape and structure. Much interest has been focused on designing and fabricating novel nano-and micro-structured materials. The most-applied method for obtaining morphology controllable inorganic materials with structural specialty, complexity, and related unique properties is templating of biomass materials via surface precipitation of suitable inorganic molecular precursors and following with a calcination to remove the templates[17-20]. Colón *et al.* have reported that TiO_2 with the carbon doped showed excellent photocatalytic properties due to the enhanced high surface area[21]. This can be attributed not only to the C doping in the structure, but also to the excellent physical properties such as the porous structure of C with a high surface area and large pore volume available for the mesoporous structures[22]. Among the biotemplates, wood is a natural composite material mainly composed of cellulose, hemicellulose and lignin[23]. The tracheidal cells or vessels extended from micrometers to several meters in the wood make wood a perfect candidate as a template for generating inorganic materials with hierarchical structures. All those works thus inspired us to fabricate an *Aurivillius* type Bi_2WO_6 sensor biotemplated from wood fibers substrate. The wood fibers would not only act as templates to fabricate high-surface-area inorganic materials, but also produce CO gas to protect the materials during calcination process.

In this work, we design and fabricate biomass-derived C-doped Bi_2WO_6 templated from wood fibers via a facile hydrothermal process and a calcination process. The gas sensing properties of the biomass-derived C-doped Bi_2WO_6 templated from wood fibers to different molecules including ethanol, acetone, acetic acid and ethyl acetate have been investigated. The biomass-derived C-doped Bi_2WO_6 from wood fibers showed high sensing properties of the acetone, acetic acid and ethyl acetate molecules. The reaction mechanism of biomass-derived C-doped Bi_2WO_6 based gas sensor was also investigated in the paper.

2 Materials and methods

2.1 Materials

All chemicals supplied by Shanghai Boyle Chemical Company, Limited were of analytical reagent-grade quality and used without further purification. Deionized water was used throughout the study. Wood fibers were obtained from the sapwood sections of poplar wood (*Populus ussuriensis* Kom), which is one of the most common tree species in the northeast of China. The wood fibers were oven-dried (24 h, 103±2 ℃) to constant weight after ultrasonically rinsing in deionized water for 30 min, and oven-dried weight were determined.

2.2 Synthesis of biomass-derived C-doped Bi_2WO_6 templated from wood fibers

Biomass-derived C-doped Bi_2WO_6 was synthesized through a hydrothermal method. Firstly, $Bi(NO_3)_3 \cdot 5H_2O$ (2.425 g) was completely dissolved in ethylene glycol (EG, 40 mL) under magnetic stirring. Then,

$Na_2WO_4 \cdot 2H_2O$ (0.8247 g) was added into the EG solution with continuous stirring and a transparent mixture was obtained. The whole solution and 5 g wood fibers were transferred into a Teflon-lined stainless autoclave (45 mL) and heated at 160 ℃ for 12 h. After cooling down to room temperature naturally, the samples were taken out and washed by ethanol for several times. The samples were dried at 80 ℃ for 24 h and subsequently annealed at 500 ℃ for 3 h in air. Finally, the biomass-derived C-doped Bi_2WO_6 were obtained and the C content in the production was 58.9%. For comparison, the pure Bi_2WO_6 was prepared using the same method without wood fibers.

2.3 Characterization

The morphology and microstructure were characterized by field-emission scanning electron microscopy (FE-SEM, JSM-7500F, JEOL, Japan) operating at 12.5 kV in combination with EDS (EDS Inca X-Max, Oxford Instruments, England). The crystal structure of the as-prepared product was investigated by X-ray diffraction (XRD, Bruker D8 Advance, Germany) with Cu *Kα* radiation of wavelength $\lambda = 1.5418$ Å, using a step scan mode with the step size of 0.02° and a scan rate of 4° min^{-1}, at 40 kV and 40 mA ranging from 5° to 80°. Further evidence for the composition of the product was inferred from the X-ray photoelectron spectroscopy (XPS, Thermo ESCALAB 250XI, USA), using an ESCALab MKII X-ray photoelectron spectrometer with Mg-*Kα* X-rays as the excitation source. Thermogravimetric and Differential Thermal Analysis (TG-DTA) spectra were performed using a PE-TGA7 thermogravimetric analyzer (Perkin Elmer Company) and a DTA/9050311 high temperature differential analyzer. 10 mg of the samples were taken and measured in air, and then treated in 150 mL/min of dry pure N_2 with temperatures at the rate of 10 C · min^{-1} ranging from 20 ℃ to 800 ℃. Specific surface areas of the prepared products were measured by the Brunauer-Emmett-Teller (BET) method based on N_2 adsorption at the liquid nitrogen temperature using a 3H-2000PS2 unit (Beishide Instrument S&T Co., Ltd).

2.4 Gas sensor fabrication and measurements

Schematic diagram of the testing system is displayed in Fig. 1a. It shows the working principle of the sensors. The sensor export voltage was tested by using a common electrical circuit with voltage of 5 V. Gas sensing of the samples was performed in a WS-30A static gas-sensing system (Zhengzhou Wei-Sheng Electronics Technology Co., Ltd., Henan, P. R. China). The sensors were fabricated as follows. Firstly, the samples were mixed with ethanol in an agate mortar to form the paste. Fig. 1b shows a diagram of the sensor. A Ni-Cr heating wire was placed inside the component as a resistor to adjust the whole working temperature of the gas sensors. Finally, the ceramic chip was pasted to the conductive pedestal and welded together with wire. All the sensors were aged at 300 ℃ for 3 days to improve their stability. The measurement was processed by a static process in a test chamber (320 mm×320 mm×250 mm). The desired gases concentration was obtained by evaporating the certain volume of liquids through a heater in the testing chamber. Then, the gas sensing measurement was set out. The sensitivity, S, is determined as the ratio, $R_a/R_g \times 100\%$, where R_a is the resistance in air and R_g is the resistance in the tested gas atmosphere. The response time is defined as the time taken for the sensor to reach 90% of the saturation value after the sensor is exposed to gases, and the recovery time is defined as the time taken for the sensor to decrease to 10% of the saturation value after the removal of gases.

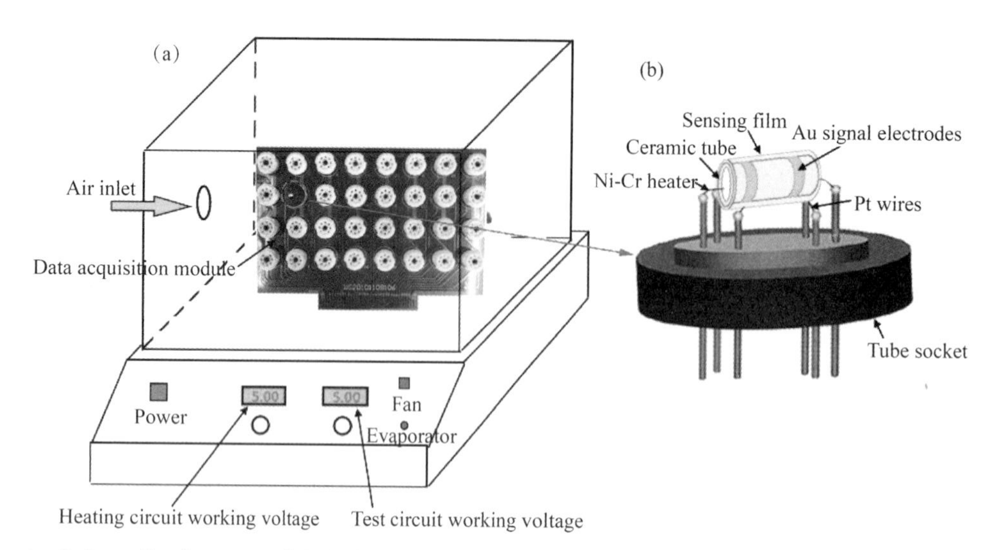

Fig. 1 Schematic diagrams of (a) the testing system for gas sensing properties and (b) the sensor

3 Results and discussion

The XRD patterns of the pure Bi_2WO_6 and the C-doped Bi_2WO_6 before and after calcination are shown in Fig. 2. It is shown that before calcination there are two diffraction peaks at 14.8° and 22.5° belonging to the (101) and (002) crystal planes of cellulose in the wood (Fig. 2b)[24, 25]. And in Fig. 2a and 2b, the diffraction peaks at 28.4°, 33.0°, 47.2°, 55.4° and 76.1° can be perfectly identified to (113), (020), (220), (208) and (333) crystal planes of orthorhombic bismuth tungstate (lattice constants a = 5.464(8) Å, b = 5.432(5) Å and c = 16.44(1) Å, JCPDS No. 73-1126), respectively[26, 27]. As it is known, both starting materials might undergo phase and morphological changes during the sensor fabrication process at high temperatures so that the pure Bi_2WO_6 and the C-doped Bi_2WO_6 after calcination are investigated accordingly,

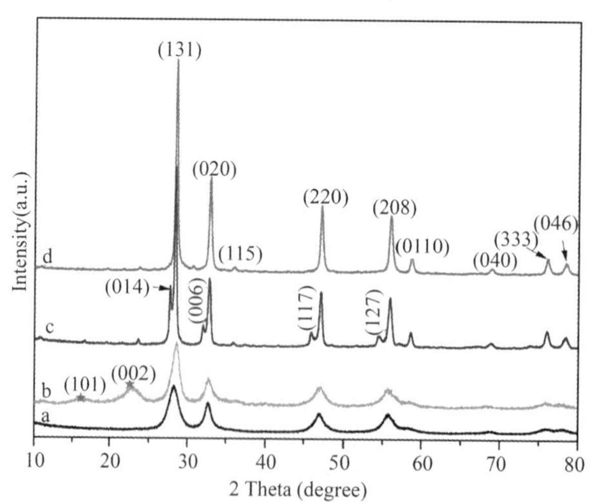

Fig. 2 XRD patterns of the pure Bi_2WO_6(a) before and (c) after calcination, and the C-doped Bi_2WO_6(b) before and (d) after calcination, respectively

as shown in Fig. 2c and 2d. The new diffraction peaks of the C-doped Bi_2WO_6 after calcination (Fig. 2d) center at 2θ = 36.1°, 58.7°, 69.0° and 78.5°, are also agree with (115), (0110), (040) and (046) planes of pure orthorhombic Bi_2WO_6, whereas there are four more peaks of the pure Bi_2WO_6 after calcination (Fig. 2c) center at 2θ=27.2°, 32.6°, 45.2° and 54.2° attributing to (014), (006), (117) and (127) planes of pure orthorhombic Bi_2WO_6, No characteristic peaks of the other impurities are detected. The results are in line with the Aurivillius structure type, indicating the Bi_2WO_6 materials remain unchanged after calcination at 500 ℃. Moreover, in Fig. 2c and 2d, the sharper diffraction peaks of the Bi_2WO_6 suggest that calcination at 500 ℃ for 3 h is sufficient to crystallize the pure Bi_2WO_6 nanostructures.

X-ray photoelectron spectroscopy surface measurements are performed on the C-doped Bi_2WO_6 after calcination to determine the elemental composition and oxidation state. Fig. 3 provides the high resolution spectra of Bi 4f, W 4f, O 1s and C 1s. In Fig. 3a, the peaks of Bi $4f_{7/2}$ and Bi $4f_{5/2}$ located at 158.66 and 163.95 eV are attributed to the trivalent oxidation state of bismuth[28, 29]. The binding energies of W $4f_{7/2}$ and W $4f_{5/2}$ peaks (Fig. 3b) are 34.93 and 37.09 eV, respectively[30]. The energy gap between the two peaks of W $4f_{7/2}$ and W $4f_{5/2}$ is 2.16 belonging to the tungsten in the W^{6+} valance state. The O 1s can be divided into three peaks (Fig. 3c), positioned at 529.69 eV, 532.30 eV and 534.91 eV, which are assigned to crystal lattice oxygen $[Bi_2O_2]^{2+}$ and $[WO_4]^{2-}$ layers of Bi_2WO_6 and the adsorbed oxygen in the form of hydrated species OH^- on the surface, respectively[31]. All these peaks indicate that the sample is Bi_2WO_6. Fig. 3d represents the C 1s XPS spectra of the C-doped Bi_2WO_6 after calcination as well as its fitting results. The

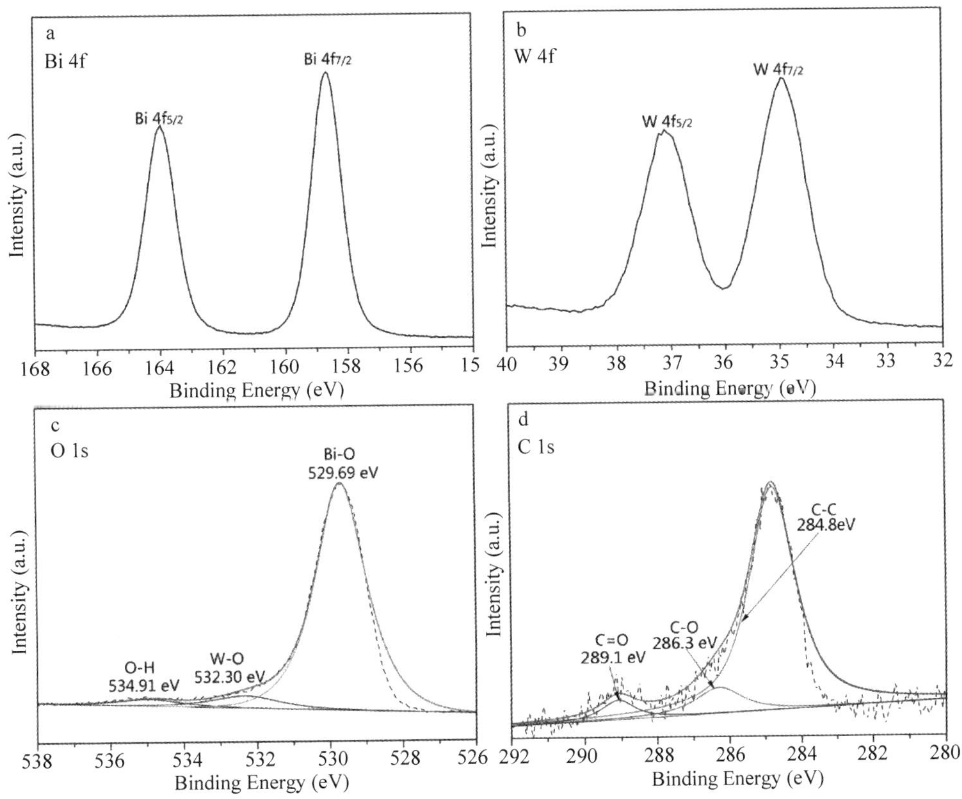

Fig. 3 XPS spectra of the C-doped Bi_2WO_6 after calcination: (a) Bi 4f spectra, (b) W 4f spectra, (c) O 1s spectra, and (d) C 1s spectra

spectra are deconvoluted into three separated Gaussian distributions. The peak around 284. 8 eV is assigned to nonoxygenated ring C, and the peak observed at 286. 3 eV corresponds to C—O bonds[32]. The other peak around 289. 1 eV is attributed to C=O bonds, which may be typical for adsorbed carbon on the samples[33]. Moreover, the C peaks indicates that there is C existed in the sample after calcination.

SEM images of the pure Bi_2WO_6, the pure Bi_2WO_6 after calcination, the C-doped Bi_2WO_6 and the C-doped Bi_2WO_6 after calcination are presented in Fig. 4. The pure Bi_2WO_6 (Fig. 4a) presents structure of agglomeration, while the C-doped Bi_2WO_6(Fig. 4c) has a uniformly dispersed morphology. Moreover, the C-doped Bi_2WO_6 possesses a nanoplate-like appearance with a thickness of approximately 40 nm. Interestingly, after calcination the particles of Bi_2WO_6 are converted into irregular shaped particles. It can be seen that the structure of the C-doped Bi_2WO_6 after calcination is also more even than that of the pure Bi_2WO_6 after calcination in the case of the presence of wood fibers.

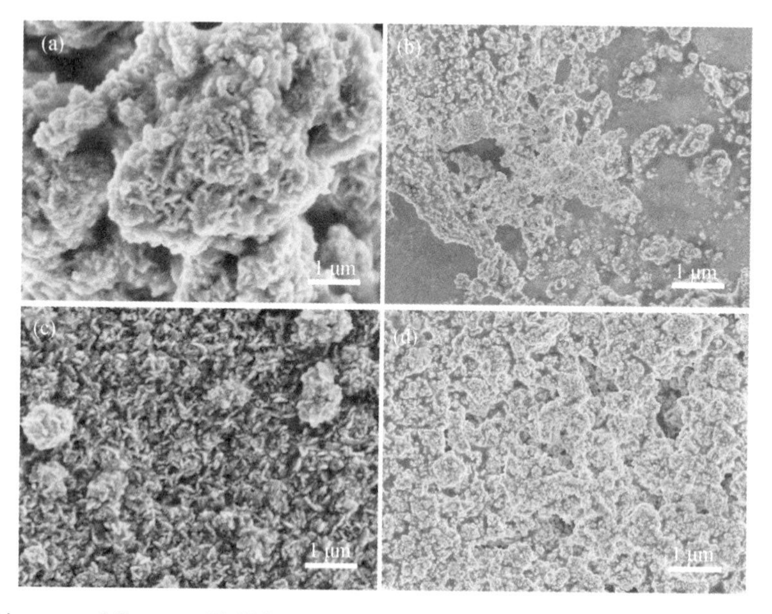

Fig. 4 SEM images of the pure Bi_2WO_6(a) before and (b) after calcination, the C-doped Bi_2WO_6 (c) before and (d) after calcination

From thermal analysis of samples we obtain information about the evolution of the C-doped Bi_2WO_6 system during calcination. Fig. 5 shows the thermogravimetric diagrams of the C-doped Bi_2WO_6. As is well known, cellulose is the main component with the weight of about 50% in wood, and the content of C in the cellulose is about 40%-50%. According to the results of thermogravimetry, the decomposition at about 190-250 ℃ is attributed to thermal depolymerisation of hemicellulose, the decomposition at about 250-380 ℃ is attributed to cellulose decomposition, and the decomposition at about 380-700 ℃ is attributed to oxidative degradation of the charred residue leading to aromatization and carbonization[34]. Thus, the C content is related to calcination temperature. The pure Bi_2WO_6 obtained through hydrothermal synthesis in the absence of wood fibers leads to a total weight loss of about 12. 2% after calcination at 500 ℃. However, the weight loss of the C-doped Bi_2WO_6 after calcination at 500 ℃ is 58. 1%, including the losses of carbon and Bi_2WO_6. Thus, it can be calculated that the C content of the C-doped Bi_2WO_6 after calcination at 500 ℃ is 13. 0%, which

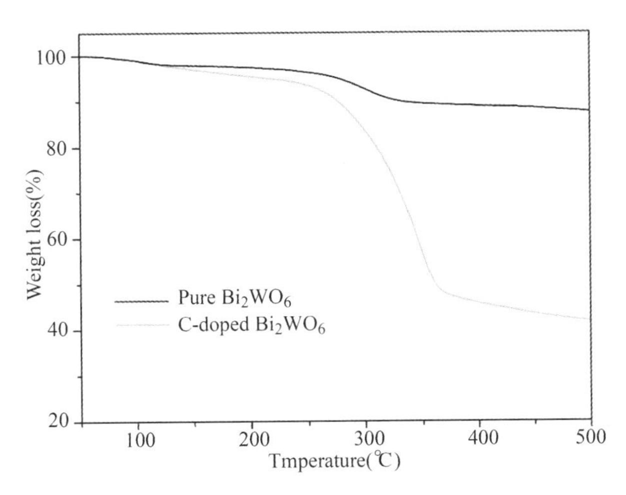

Fig. 5 TG profiles of the pure Bi_2WO_6 and the C-doped Bi_2WO_6

further proves the result in Fig. 3.

In order to investigate the porous structures of the samples and the effect of the wood fibers on the porous structure of the samples, BET analysis was carried out. Fig. 6 shows the N_2 adsorption-desorption isotherms of the pure Bi_2WO_6 after calcination and the C-doped Bi_2WO_6 after calcination. The nitrogen adsorption and desorption isotherms can be categorized as type IV isotherm with a hysteresis loop observed in the relative (P/P_0) range of 0.4-1.0, which implies the presence of mesopores in the size of 2-50 nm[35]. This result is also confirmed by the corresponding pore-size distribution (inset in Fig. 6). Furthermore, the isotherm profile of the C-doped Bi_2WO_6 after calcination shows typical H3 type hysteresis loops, indicating the presence of slit-like pores. The pore size distribution of the C-doped Bi_2WO_6 after calcination exhibits a broadened pore size range (inset in Fig. 6).

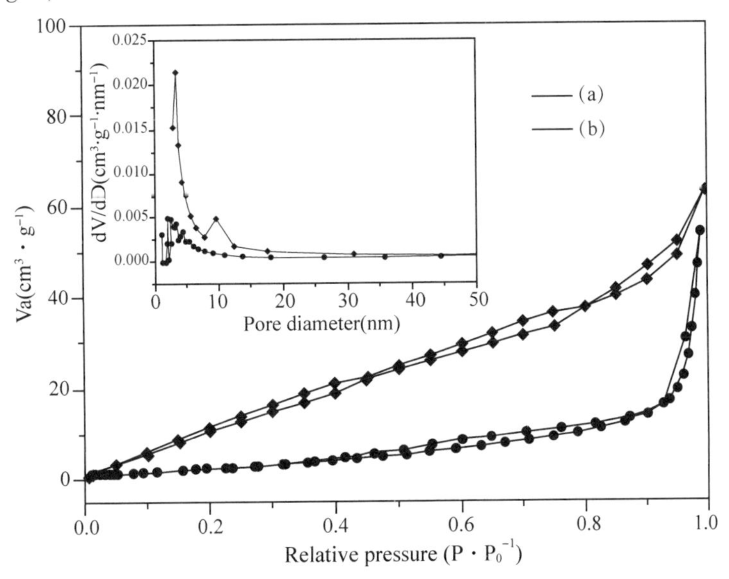

Fig. 6 N_2 adsorption-desorption isotherms of (a) the pure Bi_2WO_6 after calcination and (b) the C-doped Bi_2WO_6 after calcination. The inset shows the pore size distributions

Structure parameters of the pure Bi_2WO_6 after calcination and the C-doped Bi_2WO_6 after calcination are presented in Table 1. The C-doped Bi_2WO_6 after calcination exhibits higher surface area (52.77 $m^2 \cdot g^{-1}$) and larger pore volume (0.94 $cm^3 \cdot g^{-1}$). These large values of the physical parameters together with a sharp capillary condensation step clearly reflect the presence of the well-ordered mesoporosity in the sample. That is, the method used in the study seems to produce a certain heterogeneous system, in terms of the surface properties (surface area, pore size distribution, etc.) of obtained the C-doped Bi_2WO_6 after calcination. Thus, because of its large surface area, the C-doped Bi_2WO_6 after calcination provides more reaction sites for the adsorption of reactant molecules and increases the response speed of gas, so the gas sensing property of the C-doped Bi_2WO_6 after calcination is enhanced.

Table 1 The structure parameters of the pure Bi_2WO_6 after calcination and the C-doped Bi_2WO_6 after calcination

Sample	BET surface area ($m^2 \cdot g^{-1}$)	Pore size (nm)	Pores volume ($cm^3 \cdot g^{-1}$)
Pure Bi_2WO_6 after calcination	22.84	15.94	0.091
C-doped Bi_2WO_6 after calcination	52.77	18.25	0.094

As is known, the morphology, size and surface structure of nanostructures can dramatically change its optical, electrical, and physical properties. Thus, it is expected that such a uniform hierarchical C-doped Bi_2WO_6 after calcination with high surface area might be a potential candidate for gas sensing applications. To determine the optimum working temperature for C-doped Bi_2WO_6 based sensors, the gas sensing properties were investigated. The responses of sensors to 50 ppm ethanol, acetone, acetic acid and ethyl acetate gases in dry air were examined vs. temperature (Fig. 7a). It is obviously seen that gas responses to all the four kinds of gases increase with increasing temperature, and the gas response to acetone gas is higher than those to other gases at all temperatures. The responses show a maximum at 370 ℃, and afterwards the responses decrease again. That is, the optimum operating temperature of 370 ℃ has therefore been performed in further experiments.

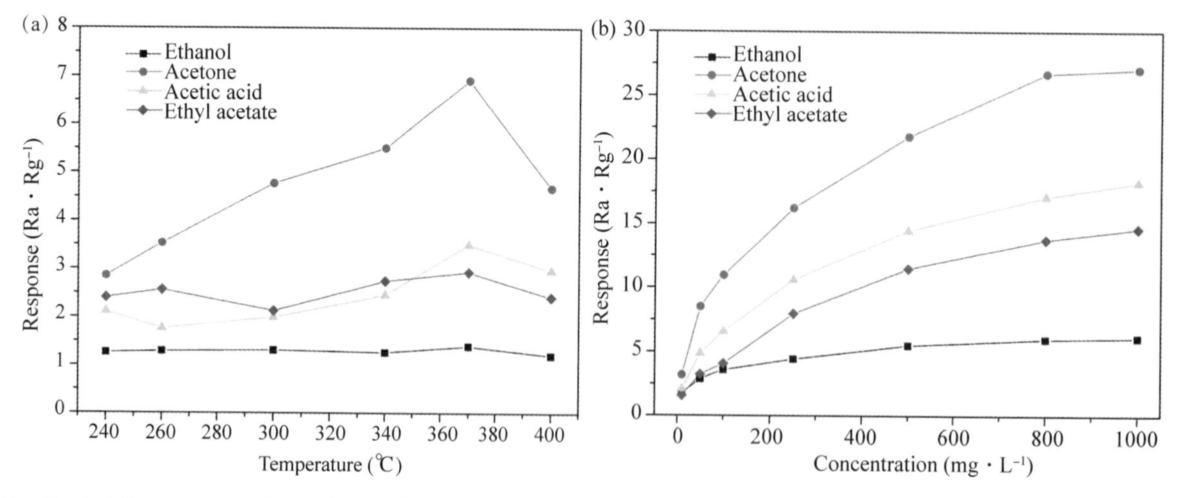

Fig. 7 (a) Temperature dependence of gas response of sensors [the concentration of each gases was 50 ppm, relative humidity (RH), 10%]. (b) Responses vs. different gases concentrations for the gas sensor based on the C-doped Bi_2WO_6 after calcination when operated at 370 ℃ [relative humidity (RH), 10%]

Sensitivity is an important factor of chemical sensors, a higher sensitivity can usually allow for a low detection limit. The sensitivity of C-doped Bi_2WO_6 based gas sensor to 10, 50, 100, 250, 500, 800 and 1000 ppm of gases are compared in Fig. 7b, showing the sensitivity of C-doped Bi_2WO_6 gas sensor is improved with increasing gas concentration. The responses of sensor reach up to 27.06, 18.23 and 14.63 on exposure to 1000 ppm acetone, acetic acid and ethyl acetate gas compared with 6.10 on exposure to 1000 ppm ethanol, respectively. In other words, the gas sensitivities of C-doped Bi_2WO_6 gas sensor on exposure to 1000 ppm of acetone, acetic acid and ethyl acetate gas are 4.4, 3.0 and 2.4 times higher than that exposure to 1000 ppm of ethanol, respectively. The selectivity of the sensing of different organic gases over C-doped Bi_2WO_6 based gas sensor decreases in the following order: acetone > acetic acid > ethyl acetate > ethanol.

Fig. 8a shows the response-recovery curves of the gas sensor based on the C-doped Bi_2WO_6 after calcination to ethanol, acetone, acetic acid and ethyl acetate gases with concentration increased to 10, 50, 100, 250, 500, 800 and 1000 ppm at 370 ℃. The gas responses of the gas sensor based on the C-doped Bi_2WO_6 after calcination to acetone, acetic acid and ethyl acetate gases are better than that to ethanol, which can be ascribed to the selectivity to the gases containing carbonyl groups. In addition, the sensor exhibited satisfactory sensing properties at low temperature (370 ℃).

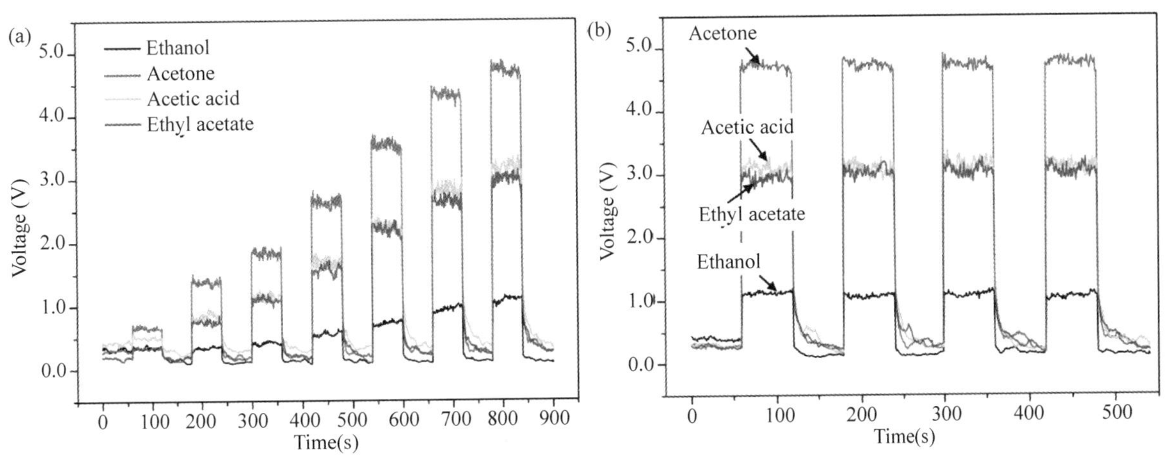

Fig. 8 (a) Response-recovery curve of the gas sensor based on the C-doped Bi_2WO_6 after calcination when operated at 370 ℃ (RH 10%). (b) The repeatability of the gas sensor based on the C-doped Bi_2WO_6 for 1000 ppm ethanol, acetone, acetic acid and ethyl acetate gases when operated at 370 ℃ (RH 10%)

For illustrating the mechanism of the reaction mechanism of C-doped Bi_2WO_6 based gas sensor, the structure characters of Bi_2WO_6 is presented in advance. Bi_2WO_6 is the simplest members of *Aurivillius* family with perovskite-layer structure. The crystal structure is consisted of $[Bi_2O_2]^{2+}$ layers and a perovskite layer $[WO_4]^{2-}$ (Fig. 9a) [36]. The Bi 6s orbitals hybridized with O 2p orbitals and formed valence band, while W 5d was regarded as conduction band. And according to the result in Fig. 3d, most of C exists in the C-doped Bi_2WO_6 as nonoxygenated ring C, which evenly disperses in the sample as shown in the Fig. 9b.

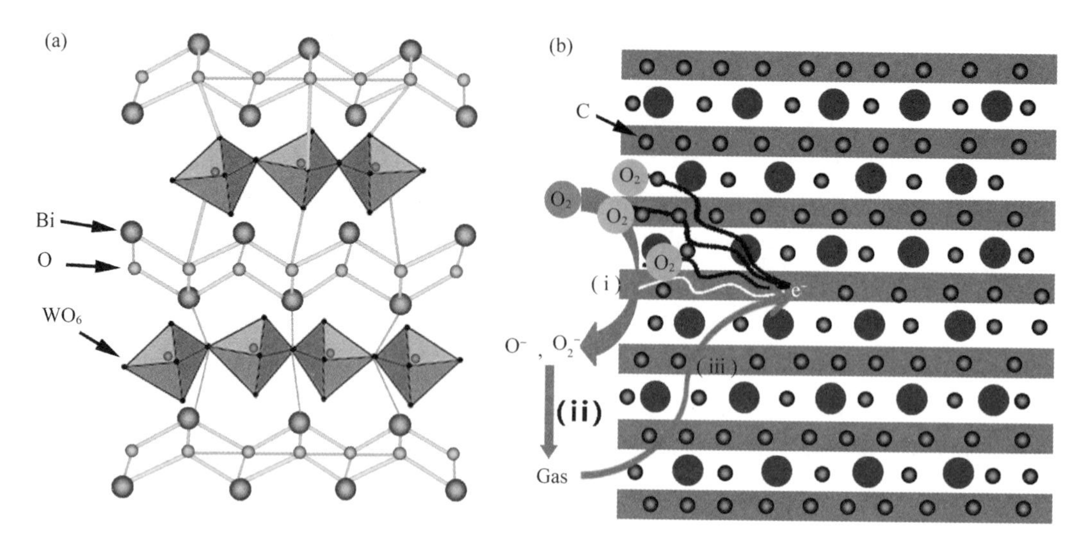

Fig. 9 (a) The crystal structure of Bi_2WO_6 showing the layer structure. (b) Schematic diagram of the proposed reaction mechanism of the C-doped Bi_2WO_6 based gas sensor in the gases containing carbonyl groups

Moreover, due to its layer structure, the oxidation and reduction sites exist on the surface and edges of the ultrathin layer unit, which is benefited to provide a more favorable condition for the migration and separation of the electron-hole pairs[37]. In other words, the holes generated in the layer structure need a shorter distance to reach the surface of the ultrathin layer unit, and then they can be trapped by interlayer water molecules absorbed on the surface. This rapid hole-trapping process allows electrons to freely and sufficiently migrate in the ultrathin layer unit until reaching the unit edges.

For semiconductor-based sensor, the change of resistance is mainly caused by the adsorption and desorption of gas molecules on the surface of the sensing structure[38]. The experimental observations described above can be explained by the surface-depletion model as shown in Fig. 9b. When the as-fabricated sensor is exposed to air, molecular oxygen can adsorb on the surface of Bi_2WO_6 nanostructure and form O_2^-, O_2^- and O^- by capturing electrons from the conduction band. Moreover, the C species in the sample would absorb more molecular oxygen due to its absorption property and action as a charge carrier[32]. When the sensor is exposed to ethanol, acetone, acetic acid or ethyl acetate gases, which are reductive gases, the gases would react with adsorbed oxygen species on the Bi_2WO_6 surface to form CO_2 and H_2O. This leads to an increasing carrier concentration of the sample and decreasing resistance of the sensor.

According to the schematic diagram in Fig. 9b, the acetone gas is selected as an example to illustrate the reaction mechanism. First, acetone molecules (C_3H_6O) would be oxidized by the surface absorbed oxygen species along with the production of electrons. The produced electrons would be released to the sample leading to an increase of carrier concentration. The electronic states of the adsorbed oxygen species on the surface of C-doped Bi_2WO_6 based sensors change as follows:

$$O_2 + e^- \rightarrow 2O_2^- \quad (1)$$

$$2O_2^- + e^- \rightarrow 2O^- \quad (2)$$

$$C_3H_6O + 8O^- \rightarrow 3CO_2 + 3H_2O + 8e^- \quad (3)$$

Thus, the above reactions can release electrons, and then the released electrons go back into the conduction band of the samples, which causes a resistance decrease of the sensors.

From a viewpoint of electronic structure, the organic gases contain hydroxyl group (OH) or carbonyl group (C=O), the oxygen atoms of which possess electron-withdrawing properties. These groups can thus be responsible for the acceleration of Reaction (3). It is remarkable that among the organic gases, the sensitivity order is acetone>acetic acid>ethyl acetate>ethanol. This order coincides with the order of electron-withdrawing strength, which is C=O>OH. And based on the above results, it can be concluded that the C-doped Bi_2WO_6 based gas sensor shows high sensitivity to the gases contained carbonyl groups.

4 Conclusions

In summary, uniform hierarchical C-doped Bi_2WO_6 templated from wood fibers was successfully synthesized via a hydrothermal process and a calcination process. The as-prepared sample with high surface area of 52.77 m^2/g and large pore volume of 0.94 $cm^3 \cdot g^{-1}$ is achieved when there is the existence of wood fibers in the synthesis. The samples were used to fabricate gas sensors, and the responses to acetone, acetic acid, ethyl acetate and ethanol gases were investigated. The results show that the sensors exhibit higher response sensitivity to acetone, acetic acid and ethyl acetate gases than ethanol gas. The sensitivities of C-doped Bi_2WO_6 based sensors on exposure to 1000 ppm of acetone, acetic acid and ethyl acetate gas are 4.4, 3.0 and 2.4 times higher than that exposure to 1000 ppm of ethanol, respectively. The C-doped Bi_2WO_6 based sensors exhibit higher sensitivities to the gases containing carbonyl groups because of the following reasons: (a) the mesoporous structure has a high surface area and large pore volume for the gases to adsorb and entering into and come out of the mesoporous structure freely. (b) The existence of C species in the sample would absorb more molecular oxygen due to its absorption property and action as a charge carrier. (c) The electron-withdrawing strength of C=O is higher than that of OH, and the carbonyl groups can react with more oxygen molecules and release more electrons leading a resistance decrease of the sensors. That is, the gases molecules containing carbonyl groups could not only be adsorbed onto the surface of the sample, but also fully interact with O^-, resulting in the high gas response.

Acknowledgments

This work was financially supported by the National Natural Science Foundation of China (grant no. 31470584).

References

[1] Wang L, Lou Z, Zhang T, et al. Facile synthesis of hierarchical SnO_2 semiconductor microspheres for gas sensor application[J]. Sensor. Actuat. B-Chem., 2011, 155(1): 285-289.

[2] Van Hieu N. Comparative study of gas sensor performance of SnO_2 nanowires and their hierarchical nanostructures[J]. Sensor. Actuat. B-Chem., 2010, 150(1): 112-119.

[3] An D, Li Y, Lian X, et al. Synthesis of porous ZnO structure for gas sensor and photocatalytic applications[J]. Colloid. Surface. A., 2014, 447: 81-87.

[4] Han P G, Wong H, Poon M C. Sensitivity and stability of porous polycrystalline silicon gas sensor[J]. Colloid. Surface. A., 2001, 179(2-3): 171-175.

[5] Pang Z, Yang Z, Chen Y, et al. A room temperature ammonia gas sensor based on cellulose/TiO_2/PANI composite nanofibers[J]. Colloid. Surface. A., 2016, 494: 248-255.

[6]Lin S, Li D, Wu J, et al. A selective room temperature formaldehyde gas sensor using TiO_2 nanotube arrays[J]. Sensor. Actuat. B-Chem., 2011, 156(2): 505-509.

[7]Zheng K, Zhou Y, Gu L, et al. Humidity sensors based on Aurivillius type Bi_2MO_6(M= W, Mo) oxide films[J]. Sensor. Actuat. B-Chem., 2010, 148(1): 240-246.

[8]Zhang Z, Wang W, Jiang D, et al. Synthesis of dumbbell-like Bi_2WO_6@ $CaWO_4$ composite photocatalyst and application in water treatment[J]. Appl. Surf. Sci., 2014, 292: 948-953.

[9]Xu Q C, Wellia D V, Ng Y H, et al. Synthesis of porous and visible-light absorbing Bi_2WO_6/TiO_2 heterojunction films with improved photoelectrochemical and photocatalytic performances[J]. J. Phys. Chem. C, 2011, 115(15): 7419-7428.

[10]Natarajan T S, Bajaj H C, Tayade R J. Synthesis of homogeneous sphere-like Bi_2WO_6 nanostructure by silica protected calcination with high visible-light-driven photocatalytic activity under direct sunlight[J]. CrystEngComm, 2015, 17(5): 1037-1049.

[11]Liu Y, Cai R, Fang T, et al. Low temperature synthesis of Bi_2WO_6 and its photocatalytic activities[J]. Mater. Res. Bull., 2015, 66: 96-100.

[12]Wang X, Tian P, Lin Y, et al. Hierarchical nanostructures assembled from ultrathin Bi_2WO_6 nanoflakes and their visible-light induced photocatalytic property[J]. J. Alloy. Compd., 2015, 620: 228-232.

[13]Ge M, Liu L. Sunlight-induced photocatalytic performance of Bi_2WO_6 hierarchical microspheres synthesized via a relatively green hydrothermal route[J]. Mat. Sci. Semicon. Proc., 2014, 25: 258-263.

[14]Chen S H, Yin Z, Luo S L, et al. Photoreactive mesoporous carbon/Bi_2WO_6 composites: synthesis and reactivity[J]. Appl. Surf. Sci., 2012, 259: 7-12.

[15]Kim S J, Hwang I S, Kang Y C, et al. Design of selective gas sensors using additive-loaded In_2O_3 hollow spheres prepared by combinatorial hydrothermal reactions[J]. Sensors, 2011, 11(11): 10603-10614.

[16]Saıdi M Y, Barker J, Huang H, et al. Performance characteristics of lithium vanadium phosphate as a cathode material for lithium-ion batteries[J]. J. Power Sources, 2003, 119: 266-272.

[17]Zhou M, Zang D, Zhai X, et al. Preparation of biomorphic porous zinc oxide by wood template method[J]. Ceram. Int., 2016, 42(9): 10704-10710.

[18]Zhou H, Fan T, Zhang D. Biotemplated materials for sustainable energy and environment: current status and challenges[J]. ChemSusChem, 2011, 4(10): 1344-1387.

[19]Caruso F. Nanoengineering of particle surfaces[J]. Advanced materials, 2001, 13(1): 11-22.

[20]Yao F, Yang Q, Yin C, et al. Biomimetic Bi_2WO_6 with hierarchical structures from butterfly wings for visible light absorption[J]. Mater. Lett., 2012, 77: 21-24.

[21]Colón G, Hidalgo M C, Navıo J A. A novel preparation of high surface area TiO_2 nanoparticles from alkoxide precursor and using active carbon as additive[J]. Catal. Today, 2002, 76(2-4): 91-101.

[22]Song M Y, Park H Y, Yang D S, et al. Seaweed-derived heteroatom-doped highly porous carbon as an electrocatalyst for the oxygen reduction reaction[J]. ChemSusChem, 2014, 7(6): 1755-1763.

[23]Dong A, Wang Y, Tang Y, et al. Zeolitic tissue through wood cell templating[J]. Adv. Mater., 2002, 14(12): 926-929.

[24]Gao L, Qiu Z, Gan W, et al. Negative oxygen ions production by superamphiphobic and antibacterial TiO_2/Cu_2O composite film anchored on wooden substrates[J]. Sci. Rep., 2016, 6(1): 1-10.

[25]Gao L, Lu Y, Li J, et al. Superhydrophobic conductive wood with oil repellency obtained by coating with silver nanoparticles modified by fluoroalkyl silane[J]. Holzforschung, 2016, 70(1): 63-68.

[26]Lou Z, Deng J, Wang L, et al. Curling-like Bi_2WO_6 microdiscs with lamellar structure for enhanced gas-sensing properties[J]. Sensor. Actuat. B, 2013, 182: 217-222.

[27]Zhou Y, Vuille K, Heel A, et al. Studies on Nanostructured Bi_2WO_6: Convenient Hydrothermal and TiO_2-Coating

Pathways[J]. Z. Anorg. Allg. Chem., 2010, 635: 1848-1855.

[28]He J, Wang W, Long F, et al. Hydrothermal synthesis of hierarchical rose-like Bi_2WO_6 microspheres with high photocatalytic activities under visible-light irradiation[J]. Mat. Sci. Eng. B, 2012, 177(12): 967-974.

[29]Li X, Huang R, Hu Y, et al. A templated method to Bi_2WO_6 hollow microspheres and their conversion to double-shell Bi_2O_3/Bi_2WO_6 hollow microspheres with improved photocatalytic performance[J]. Inorg. Chem., 2012, 51(11): 6245-6250.

[30]Wu L, Bi J, Li Z, et al. Rapid preparation of Bi_2WO_6 photocatalyst with nanosheet morphology via microwave-assisted solvothermal synthesis[J]. Catal. Today, 2008, 131(1-4): 15-20.

[31]Liu Y, Wei B, Xu L, et al. Generation of oxygen vacancy and OH radicals: A comparative study of Bi_2WO_6 and Bi_2WO_{6-x} nanoplates[J]. Chem. Cat. Chem, 2015, 7(24): 4076-4084.

[32]Yin C, Zhu S, Chen Z, et al. One step fabrication of C-doped $BiVO_4$ with hierarchical structures for a high-performance photocatalyst under visible light irradiation[J]. J. Mater. Chem. A, 2013, 1(29): 8367-8378.

[33]Gao L, Gan W, Xiao S, et al. Enhancement of photo-catalytic degradation of formaldehyde through loading anatase TiO_2 and silver nanoparticle films on wood substrates[J]. RSC Adv., 2015, 5(65): 52985-52992.

[34] Ouajai S, Shanks R A. Composition, structure and thermal degradation of hemp cellulose after chemical treatments [J]. Polym. Degrad. Stabil., 2005, 89(2): 327-335.

[35] Tahir M, Cao C, Butt F K, et al. Large scale production of novel g-C_3N_4 micro strings with high surface area and versatile photodegradation ability[J]. Cryst. Eng. Comm, 2014, 16(9): 1825-1830.

[36]McDowell N A, Knight K S, Lightfoot P. Unusual high-ttemperature structural behaviour in ferroelectric Bi_2WO_6[J]. Chem. Eur. J., 2006, 12(5): 1493-1499.

[37]Liu G, Wang L, Yang H G, et al. Titania-based photocatalysts—crystal growth, doping and heterostructuring[J]. J. Mater. Chem., 2010, 20(5): 831-843.

[38]Wang D, Zhen Y, Xue G, et al. Synthesis of mesoporous Bi_2WO_6 architectures and their gas sensitivity to ethanol [J]. J. Mater. Chem. C, 2013, 1(26): 4153-4162.

中文题目：生物质衍生碳掺杂 Bi_2WO_6 的制备及气敏性能研究

作者：高丽坤，甘文涛，曹国良，詹先旭，强添刚，李坚

摘要：本研究以木粉为模板合成钨酸铋，并掺杂生物质衍生碳以提高气敏元件对含羰基气体的灵敏度。实验结果表明，碳掺杂的钨酸铋气敏元件在低温（370 ℃）下对丙酮、乙酸和乙酸乙酯分子具有较高的灵敏度。碳掺杂的钨酸铋气敏元件在 1000 ppm 的丙酮、乙酸和乙酸乙酯气体中的灵敏度分别比在 1000 ppm 的乙醇中高 4.4 倍、3.0 倍和 2.4 倍。这是因为生物质衍生的碳能够吸附更多的氧分子，而气体中的羰基能够与更多的氧分子发生反应，释放出更多的电子，从而导致气敏元件的电阻下降。该策略为气敏元件的发展开辟了一种新方法。

关键词：生物质衍生碳；钨酸铋；木粉；气敏元件；羰基

Bioinspired C/TiO_2 Photocatalyst for Rhodamine B Degradation under Visible Light Irradiation*

Likun Gao, Jian Li, Wentao Gan

Abstract: *Papilio paris* butterfly wings were replicated by a sol-gel method and a calcination process, which could take advantage of the spatial features of the wing to enhance their photocatalytic properties. Hierarchical structures of *P. paris* - carbon - TiO_2 (PP - C - TiO_2) were confirmed by SEM observations. By applying the Brunauer-Emmett-Teller method, it was concluded that in the presence of wings the product shows higher surface area with respect to the pure TiO_2 made in the absence of the wings. The higher surface area is also beneficial for the improvement of photocatalytic property. Furthermore, the conduction and valence bands of the PP-C-TiO_2 are more negative than the corresponding bands of pure TiO_2, allowing the electrons to migrate from the valence band to the conduction band upon absorbing visible light. That is, the presence of C originating from wings in the PP - C - TiO_2 could extend the photoresponsiveness of the products to visible light. This strategy provides a simple method to fabricate a high-performance photocatalyst, which enables the simultaneous control of the morphology and carbon element doping.

Keywords: Bioinspired; Butterfly wings; C/TiO_2; Photocatalyst; Visible light

1 Introduction

The most commonly used photocatalysts such as TiO_2 with a band gap of 3.2 eV show relatively high reactivity and chemical stability under UV light. However, this part of the UV light spectrum ($\lambda < 387$ nm) accounts for only 4% of the solar energy[1]. This greatly limits the application of TiO_2 photocatalyst in practice. To use solar irradiation or interior lighting efficiently, many efforts have been made to seek a photocatalyst with high reactivity under visible light. One approach has been to dope nonmetal elements (e.g., C, N and P) into TiO_2, which could improve the optical response of TiO_2 under visible light excitation[2-5]. Irie et al. have reported a carbon-doped anatase TiO_2 powders generated by oxidizing commercial TiC powders, and the products showed photocatalytic activities for IPA decomposition under visible light (400-530 nm) irradiation[6]. This occurred because carbon occupied the oxygen sites and the substitution caused the absorbance edge of TiO_2 to shift to the higher wavelength region.

Moreover, researchers have found that the hierarchical structure is a significant feature contributing to high photocatalytic performance[7, 8] because it can provide more active sites and enhance the capture efficiency of incident light[9]. In fact, natural materials possess an astonishing variety of hierarchical structures, which are not easily synthesized artificially. Song et al. fabricated an artificial N-doped ZnO photocatalyst by copying

* 本文摘自 Frontiers of Agricultural Science and Engineering, 2017, 4(4): 459-464.

the elaborate architecture of green leaves through a two-steps infiltration and the N contained in the leaves self-doped into the products[10]. The artificial N-doped ZnO showed superior photocatalytic activity in the visible light region and excellent methylene blue degradation under solar energy irradiation. In addition to green leaves, there are many biotemplate materials in nature, such as bamboo, butterfly wings, cotton fibers, kelp, seaweed and wood[11-14]. Among these natural biotemplates, butterfly wings with uniform architecture often display special properties, such as high absorption range to visible light[13]. Furthermore, there are abundant biogenic C elements preserved in the wings.

Thus, we designed a simple method by using *Papilio paris* butterfly wings as biotemplates, coating with TiO_2 films and calcination in air to fabricate a *P. paris*-carbon-TiO_2(PP-C-TiO_2) photocatalyst. In addition, the rhodamine B (RhB) photodegradation of the product was compared with that obtained with a measured amount of pure TiO_2.

2 Materials and methods

2.1 Materials

A *P. paris* butterfly sample was borrowed from Shanghai Natural Wild Insect Kingdom Co., Ltd. (Shanghai, China). *P. paris* butterfly wings were pretreated and washed with anhydrous ethanol, and dried in air for 12 h. All chemicals supplied by Shanghai Boyle Chemical Company (Shanghai, China) were of analytical reagent-grade quality and used without further purification. Deionized water was used throughout the study.

2.2 Synthesis

Moderate diethanol amine was added to a solution of a mixture of butyl titanate and anhydrous ethanol (molar ratio 17 : 1) with magnetic stirring. Subsequently, adequate 85% ($w \cdot w^{-1}$) ethanol aqueous solution was added dropwise to the solution with continuous stirring. The solution was magnetically stirred for >3 h until the Tyndall phenomenon could be seen and a TiO_2 sol was obtained. The wings were carefully immersed into the TiO_2 sol for 20 h. The treated wings were then washed with deionized water and dried overnight at room temperature. The synthesis process is described in Fig. 1. The treated wings were calcined in air at 550 ℃ for 3 h with a constant heating rate of 1 ℃ min^{-1} to crystallize the TiO_2 and control the carbon content from the organic templates. The carbon content in the sample was about 0.6%. For comparison, pure TiO_2 was prepared through the same sol-gel method and calcination process without the wings.

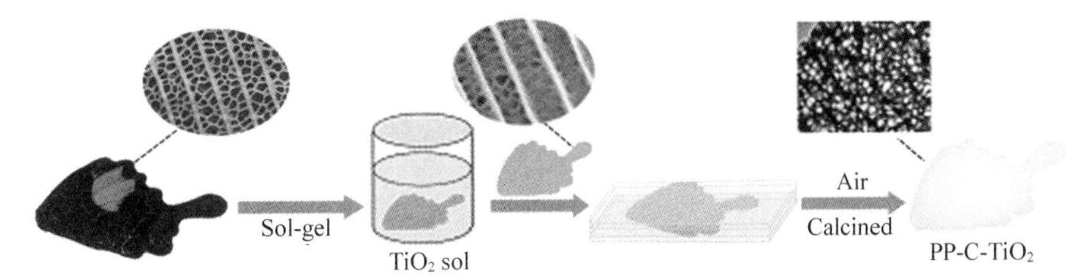

Fig. 1 Synthesis of TiO_2-replicated *Papilio paris* wings (PP-C-TiO_2)

2.3 Characterizations

The morphologies of the samples were characterized by field-emission scanning electron microscopy (FE-SEM, JSM-7500F, JEOL, Tokyo, Japan) operating at 12.5 kV in combination with energy dispersive spectroscopy (X-Max, Oxford Instruments, Abingdon, Oxfordshire, UK). The crystal structure of the as-prepared product was investigated by X-ray diffraction (D8 Advance, Bruker, Billerica, MA, USA) with Cu $K\alpha$ radiation of wavelength $\lambda = 1.5418$ Å, using a step scan mode with the step size of 0.02° and a scan rate of 4° · min^{-1}, at 40 kV and 40 mA ranging from 5°-80°. Further evidence for the composition of the product was inferred from the specific surface area of the prepared products measured by the Brunauer-Emmett-Teller (BET) method based on N_2 adsorption at the temperature of liquid nitrogen using a 3H-2000PS2 unit (Beishide Instrument ST Co., Ltd, Beijing, China). Optical properties were characterized by the UV-vis diffuse reflectance spectroscopy (TU-190, Beijing Purkinje General Instrument Co., Ltd., Beijing, China) equipped with an integrating sphere attachment, using $BaSO_4$ as the reference. Thermogravimetric analysis was carried on using a simultaneous thermal analyzer (SDT Q600, TA Instruments, New Castle, DE, USA) in a temperature range of 25-700 ℃ under a dynamic N_2 atmosphere.

2.4 Photocatalytic test

For photocatalytic tests, a measured quantity of sample was dissolved in 100 mL aqueous solutions of RhB in glass beakers. The concentration of RhB was 10 mg in 1 L of H_2O. At first, the solution was stirred continuously in the dark for 60 min to establish adsorption-desorption equilibrium among the photocatalysts and dye solution, then this solution was illuminated with visible light. A 500 W xenon lamp with the wavelength range of 425 nm was used as light source and the intensity of the incident visible light on the solution was 110 W/m^2. The glass beaker was then placed in front of the lamp with continuous magnetic stirring. Five mL of solution was collected and centrifuged. The UV absorption measurements were then used to observe the photodegradation at specific time intervals. The absorption peaks for RhB were observed at 553 nm. Stability of the material was measured in the solution with the above steps repeated four times.

3 Results and discussion

SEM images of the original butterfly wings, TiO_2-treated wings and PP-C-TiO_2 are presented in Fig. 2. The wings had a well-organized porous framework and a hierarchical architecture assembled in ridges and pillars (Fig. 2a). After treatment with TiO_2 sol, the pores were all filled (Fig. 2b). Through the calcination in air, the PP-C-TiO_2 faithfully replicated the well-organized porous hierarchical architecture of the natural wings (Fig. 2c).

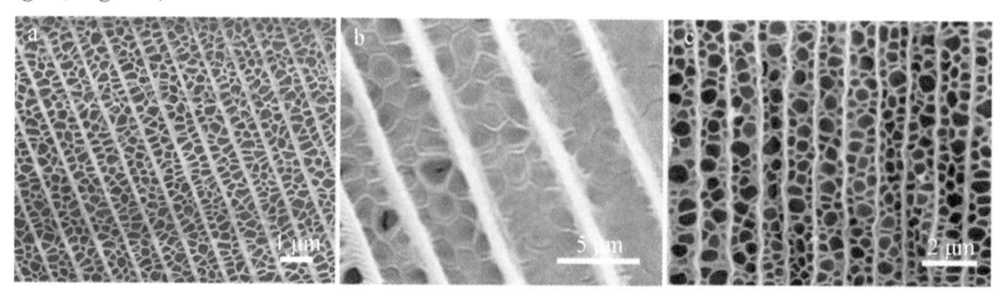

Fig. 2 SEM images of (a) original *Papilio paris* butterfly wings, (b) TiO_2-treated *P. paris* wings and (c) TiO_2-replicated wings (PP-C-TiO_2)

The distributions of C, O and Ti in the PP−C−TiO_2 are presented by element mapping in Fig. 3. The C present came from the wing exoskeleton, whereas the O and the Ti were mainly from the added TiO_2. Element mapping also indicated that Ti and O were uniformly distributed after calcination in air.

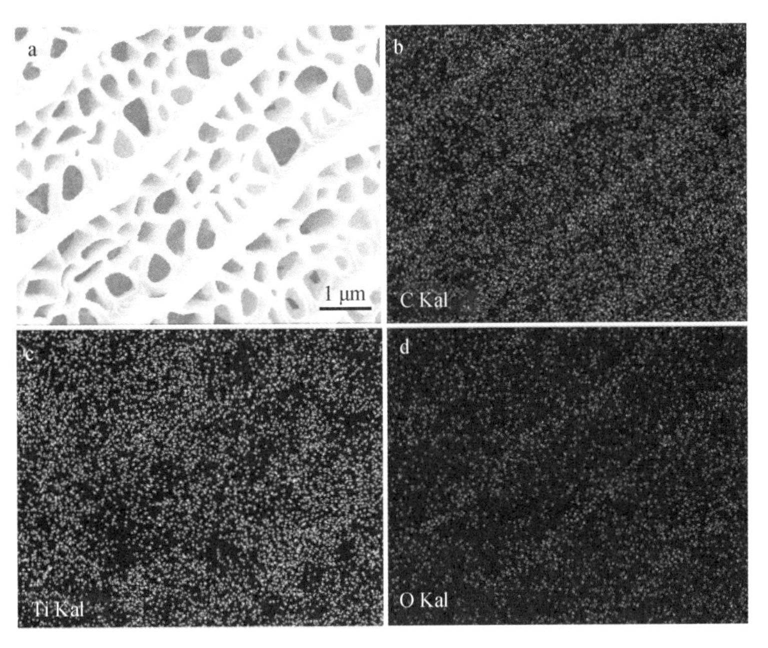

Fig. 3 SEM images and element mappings of the TiO_2-replicated *Papilio paris* wings (PP−C−TiO_2)

Fig. 4 shows the X−ray diffraction patterns of the TiO_2−treated wings, the pure TiO_2 and the PP−C−TiO_2. The diffraction peaks at 28. 4° (Fig. 4a and b) belong to the (210) crystal planes of TiO_2. The diffraction peaks of the pure TiO_2 are well indexed to the standard diffraction pattern of anatase phase TiO_2 (JCPDS file no. 21−1272)[15, 16]. After calcination in air, the diffraction peaks of the PP−C−TiO_2(Fig. 4c) were the same as the pure TiO_2(Fig. 4b). The curves in Fig. 4b and c show the peaks center at 2q = 25. 2°, 38. 0°, 47. 8°, 54. 2°, 55. 3°, 62. 5°, 68. 8°, 70. 5° and 74. 9°, which agrees with the (101), (004), (200), (105), (211), (204), (116), (220) and (215) crystal planes of anatase TiO_2.

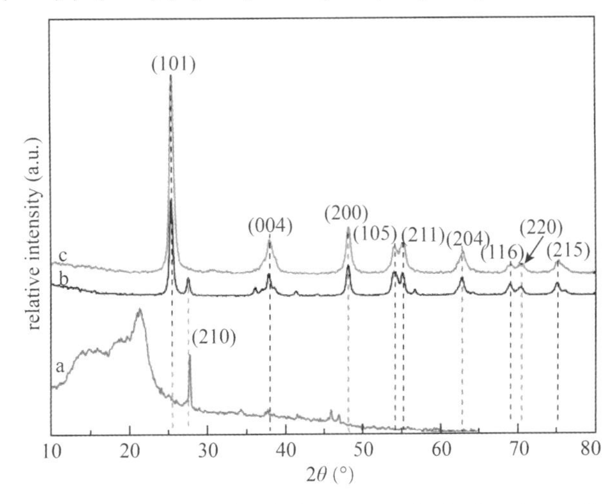

Fig. 4 X-ray diffraction patterns of (a) TiO_2-treated butterfly wings, (b) pure TiO_2 and (c) TiO_2-replicated *Papilio paris* wings (PP−C−TiO_2)

BET analysis was conducted in order to understand the effect of the wings on the porous structure of the samples. Fig. 5 shows the N_2 adsorption-desorption isotherms of the pure TiO_2 and the PP-C-TiO_2. It could be seen the curve for the PP-C-TiO_2 exhibits a hysteresis loop, which is attributed to type IV isotherms and the representative of mesoporous materials, indicating the presence of mesopores (2-50 nm)[17, 18], while the curve for the pure d TiO_2 shows a type III isotherm, which indicates that the absorption property of the material is weak.

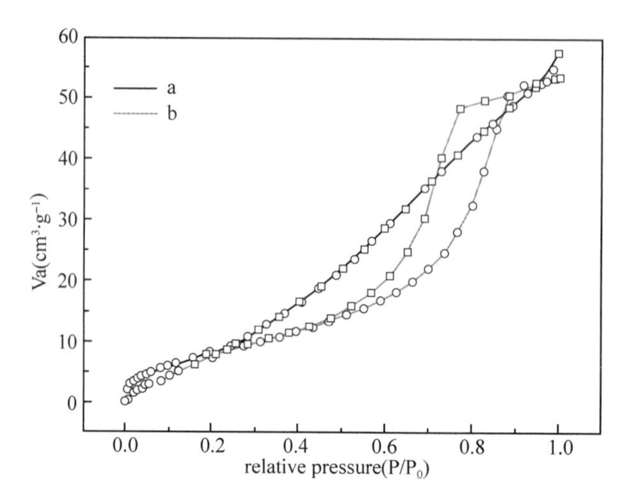

Fig. 5 N_2 adsorption-desorption isotherms of (a) pure TiO_2 and (b) TiO_2-replicated *Papilio paris* wings (PP-C-TiO_2)

BET surface areas, pore sizes and pore volumes of the pure TiO_2 and the PP-C-TiO_2 are presented in Table 1. From these data, it is clear that the preparation of TiO_2 in the presence of the wings after calcination produced a significantly higher (up to 1.7 times) surface area than the pure TiO_2. The pure TiO_2 had a relatively low surface area of 29.7 m^2 g^{-1}. The sample prepared in the presence of the wings, S_{BET} is higher than 50 m^2 g^{-1}. That is, the method used seems to have produced a certain heterogeneous system with respect to the wings, in terms of the surface properties (e. g., surface area and pore size distribution) of the PP-C-TiO_2. Thus, because of its large surface area, the PP-C-TiO_2 provides more photocatalytic reaction sites for the adsorption of reactant molecules and increases the efficiency of the electron-hole separation, so the photocatalytic activity of the PP-C-TiO_2 is enhanced.

Table 1 Structural parameters of pure TiO_2 and TiO_2-replicated *Papilio paris* wings (PP-C-TiO_2)

Sample	BET surface area($m^2 \cdot g^{-1}$)	Pore size(nm)	Pores volume($cm^3 \cdot g^{-1}$)
pure TiO_2	29.7	11.20	0.08
PP-C-TiO_2	51.9	9.05	0.12

In order to investigate the light absorbance of the samples, the UV-vis diffuse reflection spectra of the pure TiO_2 and the PP-C-TiO_2 were obtained (Fig. 6a). For the pure TiO_2, there was prominent adsorption below 350 nm, whereas for the PP-C-TiO_2 there was a much higher absorption in the region 380-500 nm, indicating the absorption of PP-C-TiO_2 is significantly red-shifted to visible wavelengths, due to the presence of C. Moreover, it is worth noting that the PP-C-TiO_2 heterostructures with absorptions in the visible region may indicate a greater potential for photocatalysis. To calculate valence band position, the optical band gap was determined by the following Tauc equation[19]:

$$(\alpha h\nu)^n = A(h\nu - E_g)$$

Where, A = constant, $h\nu$ = light energy, E_g = optical band gap energy, α = measure absorption coefficient, and n = 0.5 for indirect band gap. Given that TiO_2 has an indirect band gap, the y axis of the Tauc plot is $(\alpha h\nu)^{1/2}$ for TiO_2[20]. In Fig. 6b, the extrapolation of the Tauc plot to x intercepts gives optical band gaps of 3.02 eV and 2.23 eV for the pure TiO_2 and the PP-C-TiO_2, respectively. Therefore, the conduction band and valence band of the PP-C-TiO_2 are more negative than the corresponding bands of the pure TiO_2. It is possible for the electrons to migrate from the valence band to the conduction band upon absorbing visible light, which could lead to the visible light activity of the PP-C-TiO_2.

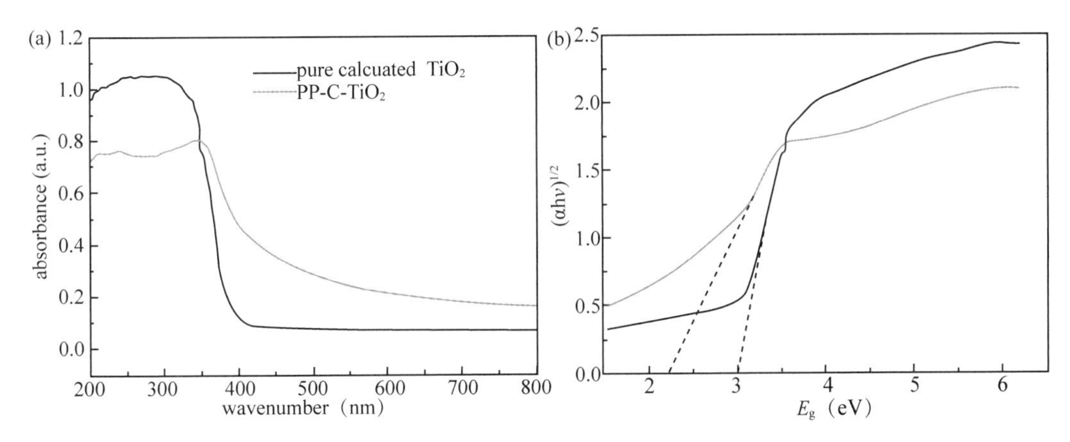

Fig. 6 (a) UV-vis absorption spectra of the pure TiO_2 and TiO_2-replicated *Papilio paris* wings (PP-C-TiO_2), and (b) the evaluation of the optical band gap using the Tauc plot

To test the photodegradation abilities of the samples, the photocatalytic degradation of RhB was measured. Fig. 7a shows the relationships between concentration ratio (C/C_0) and time for RhB degradation with 50 mg pure TiO_2, 50 mg PP-C-TiO_2 and irradiation without photocatalysts. The PP-C-TiO_2 took 90 min to degrade phenol. However, the pure TiO_2 could not degrade RhB. Fig. 7b shows the first order rate constant k (min^{-1}) of the pure TiO_2 and the PP-C-TiO_2 for RhB, which was calculated by the following first order equation[21, 22]:

$$\ln(C_0/C) = kt$$

Where, C_0 is the initial concentration of the dye in solution and C is the concentration of dye at time t. This shows the k value of 0.02614 min^{-1} for RhB in the case of the pure TiO_2 as compared to the value of 0.00039 min^{-1} in the case of the PP-C-TiO_2. This indicates that the PP-C-TiO_2 possesses photodegradation abilities.

PP-C-TiO_2 as a photocatalyst can easily be recycled by a simple filtration. After four cycles for the photodegradation of RhB, the catalyst did not exhibit any significant loss of activity (Fig. 8), confirming that the PP-C-TiO_2 is not photocorroded during the photocatalytic oxidation of the dye pollutant. Therefore, this stable photocatalyst show a great potential for practical applications.

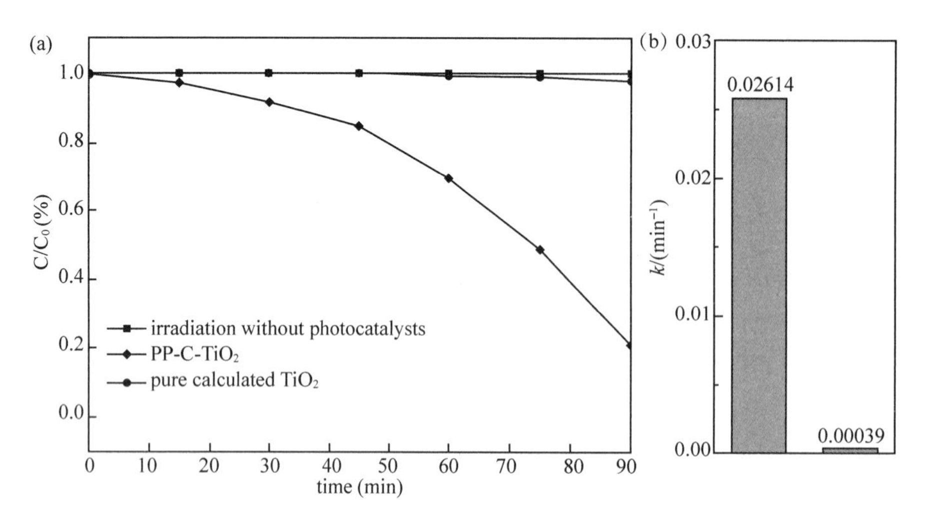

Fig. 7 (a) Concentration ratio (C/C_0) of photocatalytic rhodamine B with pure TiO_2, TiO_2-replicated *Papilio paris* wings (PP-C-TiO_2) and irradiation without photocatalysts. (b) First order rate constant k (min^{-1}) of pure TiO_2 and PP-C-TiO_2 for RhB

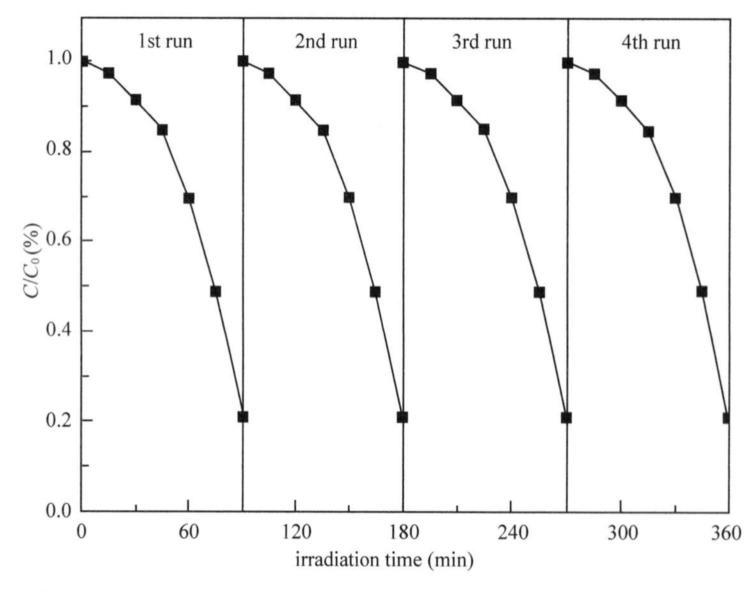

Fig. 8 Comparison of repeated cycles of photocatalytic degradation of rhodamine B in the presence of 50 mg TiO_2-replicated *Papilio paris* wings (PP-C-TiO_2) under visible irradiation

4 Conclusions

A sol-gel method and a calcination process were combined to fabricate heterostructured C/TiO_2 photocatalysts from *P. paris* butterfly wings. The elaborate architecture of the wings was copied by immersing the wings into the TiO_2 sol, followed by calcination in air to remove the wings. Moreover, the C contained in the wings was self-doped into the products. These *P. paris*-C-TiO_2 exhibited high potential for application as a visible light photocatalyst for degradation of organic pollutants. This is ascribed to the hierarchical structures and the higher surface area that provide more reactive sites for photocatalysis. Also, the C dopant decreases the band gap of TiO_2 and shifts the optical absorption to the visible light region. This work provides a pathway for constructing high-performance materials and efficiently utilizing of solar energy.

Acknowledgments

This work was supported by the National Natural Science Foundation of China (grant no. 31470584) and the Fundamental Research Funds for the Central Universities (grant no. 2572017AB08).

References

[1] Vernardou D, Drosos H, Spanakis E, et al. Electrochemical and photocatalytic properties of WO_3 coatings grown at low temperatures[J]. J. Mater. Chem., 2011, 21(2): 513-517

[2] Asahi R, Morikawa T, Ohwaki T, et al. Visible-light photocatalysis in nitrogen-doped titanium oxides[J]. Science, 2001, 293(5528): 269-271

[3] Khan S U M, Al-Shahry M, Ingler W B. Efficient photochemical water splitting by a chemically modified n-TiO_2[J]. Science, 2002, 297(5590): 2243-2245

[4] Lin L, Zheng R Y, Xie J L, et al. Synthesis and characterization of phosphor and nitrogen co-doped titania[J]. Appl. Catal. B-Environ., 2007, 76(1): 196-202

[5] Ohno T, Tsubota T, Nishijima K, et al. Degradation of methylene blue on carbonate species-doped TiO_2 photocatalysts under visible light[J]. Chem. Lett., 2004, 33(6): 750-751

[6] Irie H, Watanabe Y, Hashimoto K. Carbon-doped anatase TiO_2 powders as a visible-light sensitive photocatalyst[J]. Chem. Lett., 2003, 32(8): 772-773

[7] Parlett C M A, Wilson K, Lee A F. Hierarchical porous materials: catalytic applications[J]. Chem. Soc. Rev., 2013, 42(9): 3876-3893

[8] Cho K, Na K, Kim J, et al. Zeolite synthesis using hierarchical structure-directing surfactants: retaining porous structure of initial synthesis gel and precursors[J]. Chem. Mater., 2012, 24(14): 2733-2738

[9] Yin C, Zhu S, Chen Z, et al. One step fabrication of C-doped $BiVO_4$ with hierarchical structures for a high-performance photocatalyst under visible light irradiation[J]. J. Mater. Chem. A, 2013, 1(29): 8367-8378

[10] Song F, Su H, Han J, et al. Fabrication and good ethanol sensing of biomorphic SnO_2 with architecture hierarchy of butterfly wings[J]. Nanotechnology, 2009, 20(49): 495502

[11] Shi N, Li X, Fan T, et al. Biogenic NI-codoped TiO_2 photocatalyst derived from kelp for efficient dye degradation [J]. Energ. Environ. Sci., 2011, 4(1): 172-180

[12] Song M Y, Park H Y, Yang D S, et al. Seaweed-Derived Heteroatom-Doped Highly Porous Carbon as an Electrocatalyst for the Oxygen Reduction Reaction[J]. ChemSusChem, 2014, 7(6): 1755-1763

[13] Li Y, Meng Q, Ma J, et al. Bioinspired carbon/SnO_2 composite anodes prepared from a photonic hierarchical structure for lithium batteries[J]. ACS Appl. Mater. Inter., 2015, 7(21): 11146-11154

[14] Peng W, Hu X, Zhang D. Bioinspired fabrication of magneto-optic hierarchical architecture by hydrothermal process from butterfly wing[J]. J. Magn. Magn. Mater., 2011, 323(15): 2064-2069

[15] Gao L, Gan W, Xiao S, et al. Enhancement of photo-catalytic degradation of formaldehyde through loading anatase TiO_2 and silver nanoparticle films on wood substrates[J]. RSC Adv., 2015, 5(65): 52985-52992

[16] Gao L, Qiu Z, Gan W, et al. Negative oxygen ions production by superamphiphobic and antibacterial TiO_2/Cu_2O composite film anchored on wooden substrates[J]. Sci. Rep., 2016, 6: 26055-26064

[17] Gao L, Gan W, Qiu Z, et al. Preparation of heterostructured WO_3/TiO_2 catalysts from wood fibers and its versatile photodegradation abilities[J]. Sci. Rep., 2017, 7: 1102-1114

[18] Tahir M, Cao C, Butt F K, et al. Large scale production of novel gC_3N_4 micro strings with high surface area and

versatile photodegradation ability[J]. CrystEngComm, 2014, 16(9): 1825-1830

[19]Mor G K, Varghese O K, Paulose M, et al. Transparent highly ordered TiO_2 nanotube arrays via anodization of titanium thin films[J]. Adv. Funct. Mater., 2005, 15(8): 1291-1296

[20]Satoh N, Nakashima T, Kamikura K, et al. Quantum size effect in TiO_2 nanoparticles prepared by finely controlled metal assembly on dendrimer templates[J]. Nat. Nanotechnol., 2008, 3(2): 106-111

[21]Han C, Wang Y, Lei Y, et al. In situ synthesis of graphitic-C_3N_4 nanosheet hybridized N-doped TiO_2 nanofibers for efficient photocatalytic H_2 production and degradation[J]. Nano Res., 2015, 8(4): 1199-1209

[22]Zhao S, Chen S, Yu H, et al. g-C_3N_4/TiO_2 hybrid photocatalyst with wide absorption wavelength range and effective photogenerated charge separation[J]. Sep. Purif. Technol., 2012, 99: 50-54

中文题目：仿生碳/二氧化钛光催化剂用于可见光照射下罗丹明B的降解

作者：李坚，高丽坤，甘文涛

摘要：本研究采用溶胶-凝胶法和煅烧处理复制了巴黎凤尾蝶的蝶翅结构，利用蝶翅的多孔特性提高其光催化性能。通过扫描电子显微镜观察到蝶翅-碳-二氧化钛(PP-C-TiO_2)的层次结构。通过比表面积测试可以得出，相比于没有蝶翅的单纯二氧化钛，有蝶翅的样品具有较高的比表面积，从而有利于提高光催化性能。此外，PP-C-TiO_2 的导带和价带比单纯的二氧化钛更负，使得电子在吸收可见光后从价带跃迁至导带。也就是说，蝶翅的碳的存在可将 PP-C-TiO_2 的光响应范围拓展至可见光区域。该策略提供了一种制备高催化活性光催化的简单方法，能够同时调控形貌和碳的掺杂含量。

关键词：仿生；蝶翅；碳/二氧化钛；光催化；可见光

Fabrication of Smart Wood with Reversible Thermoresponsive Performance*

Yingying Li, Bin Hui, Guoliang Li, Jian Li

Abstract: The smart thermoresponsive wood was developed by depositing thermochromic materials on wood surfaces. The as-fabricated wood composites can reversibly change their colors with the changing temperature. The scanning electron microscope (SEM) showed that the wood surface was made up of even spherical particles. The attenuated total reflectance Fourier transform infrared spectroscopy (ATR-FTIR) spectra demonstrated that the thermochromic materials were successfully settled onto wood surfaces, and the cross-cut test results further proved that all samples can meet general application in the furniture industry. As the concentrations of thermochromic materials increased from 0 to 4.0%, the total color changes (ΔE^*) of samples remarkably increased from 1.81 to 39.56, exhibiting a good thermoresponsive property. TG results showed that the sample had the excellent thermal stability below the temperature of 216 ℃. Furthermore, the accelerated UV-aging tests demonstrated that the thermoresponsive property of the samples was highly efficient when it was exposed under the UV irradiation for 100 h.

Keywords: Thermoresponse; Wood surfaces; Reversibility; Temperature

1 Introduction

Wood has been extensively used in buildings, decorations, industries and other daily lives due to its unique performances of the lightweight, esthetic, and amendment for indoor environment[1]. The demands for wood increase significantly with the increasing concerns of population and sustainability due to the use of the non-renewable materials[2]. To meet market requirements and compete with other advanced materials, many attempts have been made to create new functions for wood[3,4], such as high-temperature heat treatment[5], retardant treatment[6,7], preservative treatment[8,9] and superhydrophobic treatment[10,11]. Among them, the interest has recently arisen in the development of smart stimuli-responsive materials because of its potential applications as sensors, smart switches or energy storage and conversion. Thus, some smart wood materials with photoresponse[12,13], magnetic response[14,15] and humidity response[16,17] have been explored.

Due to the fast response to temperature, reversible thermochromic materials have become one of the most attracting smart materials[18-20]. Owing to the unique properties, reversible thermochromic materials are widely used in many fields such as temperature sensor, thermal relays and intelligent coating[21-24]. According to the switching characteristics, the color of the thermochromic wood materials with antidromic reversible thermochromism can be repeatedly changed. Furthermore, the previously mentioned materials are colored at

* 本文摘自 Journal of Materials Science, 2017, 52: 7688-7697.

room temperature and turned colourless with an increase in temperature[25,26]. However, one of the major limitations for the materials is that the naturally beautiful wood texture suffers a great loss. To resolve this problem, an appropriate recipe of the orthodromic reversible thermochromic materials, polyvinyl alcohol (PVA) and dextrin (DT) was tested. Here, PVA is an environmental-friendly polyhydroxy polymer with prominent adhesion to wood and has the good water solubility and excellent film-forming property. DT was added into the PVA solution to improve the hardness of films[27]. This film has both advantages of non-poison, non-pollution and film-forming property, which is better than the conventional paint.

With naturally porous structures, wood has hollow vessel, pits, and resin ducts[28,29]. The porous structures could provide enough loci for the insertion of thermochromic materials. Herein, the orthodromic reversible thermoresponsive wood was designed using a simple drop-coating method. The as-fabricated wood can intelligently change colors between the color of the pristine wood and the dying color according to temperature. That is, the material is colored when temperature increases and turns back to the natural wood color when cooled to room temperature. The smart thermoresponsive wood could be a promising material in the applications of graphical display, information storage and temperature indicators.

2 Experimental

2.1 Materials

Obtained from Harbin, China, Larix gmelinii wood samples, with a moisture content of 7% and a size of 50 mm × 25 mm × 5 mm, were ultrasonically treated by distilled water, acetone, and ethanol, respectively. The ethanol (99.7%), acetone (99.5%), the polyvinyl alcohol (PVA) (alcoholysis 98.0%~99.0%, model 1750±50) and dextrin (DT) (maltodextrin, 98.0%) were purchased from Tian jian Kaitong Chemical Reagent Co., Ltd. Thermochromic materials namely, 2'-chloro-6'-(diethylamino)-3'TR-2 (white powder), were manufactured by Chong Yu Technology Co., Ltd. All of the chemicals were used as received without further purification.

2.2 Fabrication of smart thermoresponsive wood

The main experimental strategy for the fabrication of smart thermoresponsive wood is shown in Fig. 1. Firstly 2 g PVA and 2 g DT were dissolved in 100 mL deionized water. The mixture solution was vigorously stirred at 75 ℃ for 2 h. Secondly the thermochromic materials were added into the solutions and stirred for 2 h at 45 ℃, and then ultrasonically dispersed for 10 min. The percentage of thermochromic materials was pre-set at 0, 1.5%, 2.0%, 2.5%, 3%, 3.5% and 4.0%. Thirdly the previous-mentioned solutions deposited on the wood surfaces using a drop-coating method and placed at room temperature for 24 h to be naturally dried. Finally, the smart thermoresponsive wood surfaces were obtained with a thickness of 12 μm.

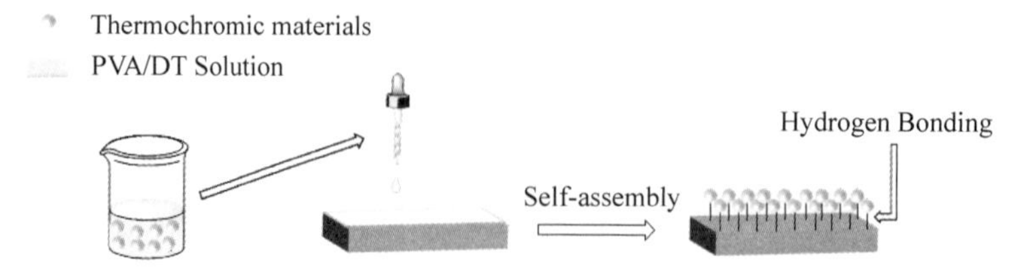

Fig. 1 Experimental strategy for the fabrication of smart thermoresponsive wood

2.3 Characterizations

The digital photos of the smart wood were obtained using a Nikon D7000 digital camera. A scanning electron microscopy (SEM, FEI Quanta200) was used to investigate the surface morphologies. The samples were sprayed by gold to ensure the conductivity. The attenuated total reflectance of the Fourier transform infrared spectroscopy (ATR-FTIR) was recorded by the Nicolet Nexus 670 FTIR instrument in the range of 400–4000 cm^{-1} with a resolution of 4 cm^{-1}. The thermal stabilities of the samples were investigated using a thermal gravity (TG) analyzer (TA, Q600), and the experiments were performed in the temperature range 25–450 ℃ at a heating rate of 20 ℃/min under the nitrogen atmosphere.

2.4 Color tests

The samples were placed into a DHG-9023A conditioning chamber (Jiangsu Jun Instrument Technology Co., Ltd.). The temperature sensors were calibrated to an accuracy of 0.1 ℃, and the temperature was monitored at 200 seconds interval and increased in an interval of 1 ℃. The digital camera was placed in a fixed position to record the whole process over a temperature range of 25–65 ℃, and these records were imported to a computer. The Commission Internationale de L'Eclairage (CIE) L^*, a^*, and b^* values were measured by using the Adobe Photoshop CS6 software installed in the computer[12]. The average values were measured at five different locations.

2.5 Lacquer adhesion tests

Adhesion of the thermochromic materials to wood substrates was measured according to the cross-cut test (ISO 2409)[30-32]. In the cross-cut test, two sets of six cuts with 1-mm spacing were made perpendicular to each other, the cutting tool was used to make the cross-cut pattern at 90° angles through the coating, thus making a lattice of 25 small blocks, and then a standardized tape was stuck on the lattice and pulled off at a take-off angle of 60°[33-35]. The number of removed blocks was the indication of adhesion.

2.6 Accelerated UV-ageing test

Using an UV lamp with a wavelength of 340 nm, the UV accelerating aging treatment was carried out in an oven equipped in order to measure the fatigue resistance of thermoresponsive wood in practical applications. The samples were fixed in stainless steel holders and rotated around, and then placed into a ZN-PUV aging test chamber (Shanghai Maijie Experimental Equipment Co., Ltd.), which underwent UV irradiation for 240 h at 60 ℃ in the oven. The radiation power was 4 kW with the irradiating distance of 500 mm. Because thermoresponsive wood was colored up to 40 ℃, the samples can keep colored all the time under the above experimental conditions. That is, the accelerated UV-ageing test was designed to identify the fatigue resistance of the samples under the continuous coloring condition.

3 Results and discussion

3.1 ATR-FTIR analyses

The FTIR of the thermochromic material is shown in Fig. 2. The absorption peak at 3362 cm^{-1} was attributed to the N-H stretching vibration of the secondary amine. The peak at 812 cm^{-1} was ascribed to the skeleton vibration of the 1,3,5- triazine ring, and other two absorption peaks at 1691 and 1342 cm^{-1} were assigned to C ═N and C—N, respectively[36,37]. The previously-mentioned absorption peaks were the characteristic peaks of the MF resin. In addition, the absorption peaks at 1462 and 1549 cm^{-1} were due to the skeleton vibration stretching of the benzene ring, and the multiple strong peaks at 2848 and 2916 cm^{-1} corresponded to the aliphatic C—H stretching vibrations of methylene and methyl groups. The absorption peaks at 1242 and 1022 cm^{-1} were ascribed to the stretching vibration of C—O—C, and the peak at around 1762 cm^{-1} was attributed to the vibration of the closed ring of lactone carbonyl and the peak at 719 $cm^{-1}$was related to the stretching vibration of-Cl[38].

The FTIR spectra of un-treated wood, PVA/DT coated wood and thermoresponsive wood are presented in Fig. 3. Compared with Fig. 3a, the new absorption peaks appeared in Fig. 3b. The peaks at 1234 and 1591 cm^{-1} corresponded to the stretching vibration of C—O and C ═C, respectively. The absorption peak at 630 cm^{-1} was attributed to the C—CO—C in-plane bending vibration. As shown in Fig. 3c, compared with the un-treated wood, some new peaks appeared in the spectra of the thermoresponsive wood, which were resulted from the thermochromic material. The peak at 2923 cm^{-1} corresponded to the aliphatic C—H stretching vibrations of methylene groups. The absorption peaks at 1691 cm^{-1} and 670 cm^{-1} were owing to the Ar—CO stretching vibration and Ar—H bending vibration, respectively. Moreover, the absorption peak at 1641 cm^{-1} was attributed to the bending vibrations of N—H (intermolecular hydrogen bond in amide). These observations indicated that a new coating which consisted of PVA, DT and thermochromic materials was coated on wood surfaces.

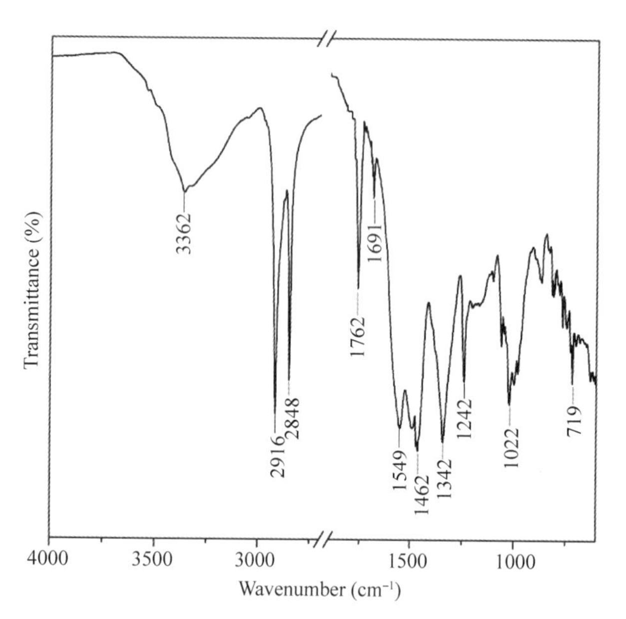

Fig. 2 FTIR spectrum of thermochromic materials

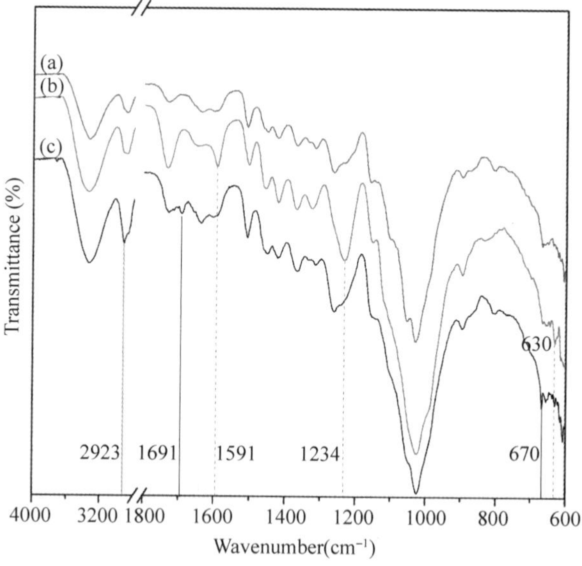

Fig. 3 ATR-FTIR spectra of (a) the un-treated wood, (b) PVA/DT coated wood, (c) smart thermoresponsive wood

3.2 Surface morphologies

The size of the thermochromic materials were analyzed by the Nano measurer. According to the diagram of the particle immersions, as illustrated in Fig. 4, the size of 0.5-1.5 μm made the maximum proportion and the largest size of thermochromic material particles was about 4 μm. Fig. 5a shows the SEM images of untreated Larch wood, the average diameter of tracheid string was approximately 25 μm, and the average pore size of pit was approximately 6 μm. It was observed clearly that the thermochromic material particles were much smaller than the wood diameters and wood pits. Thus, the thermochromic materials (spherical particles) can be easily inserted into these microstructure of wood surfaces, thereby forming a homogeneous and smooth surface on the smart thermoresponsive wood (Fig. 5b).

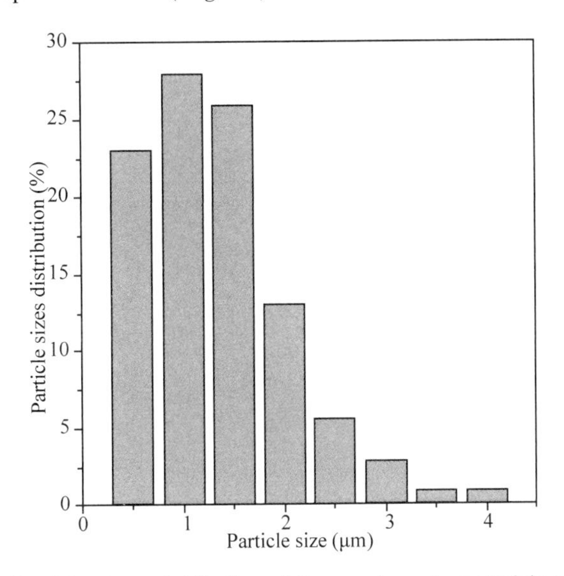

Fig. 4 Particle sizes distribution of thermochromic materials particles

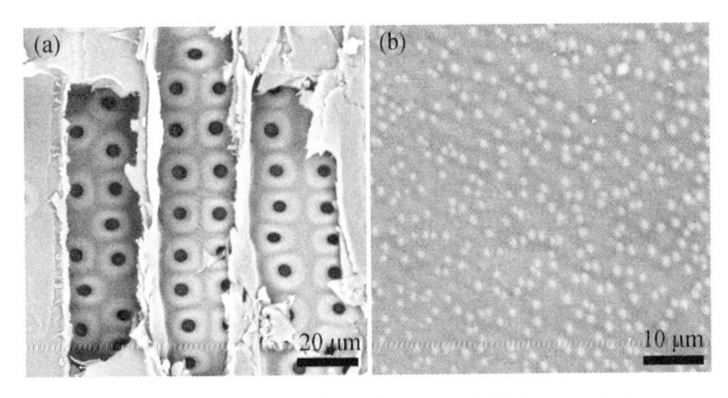

Fig. 5 SEM images of the (a) un-treated wood surface and (b) smart thermoresponsive wood surface

3.3 Thermoresponsive properties

Color changes of the thermoresponsive materials were related to the increasing number of chromophores. Fig. 6 shows the color parameters ΔE^* (the total color changes), L (lightness index), a (red-green index) and b (yellow-blue index) of the thermoresponsive wood with different thermochromic material concentrations before and after heating. The ΔE^* values of the samples were calculated using the following equations:

$$\Delta L^* = L_2 - L_1 \quad (1)$$

$$\Delta a^* = a_2 - a_1 \quad (2)$$

$$\Delta b^* = b_2 - b_1 \quad (3)$$

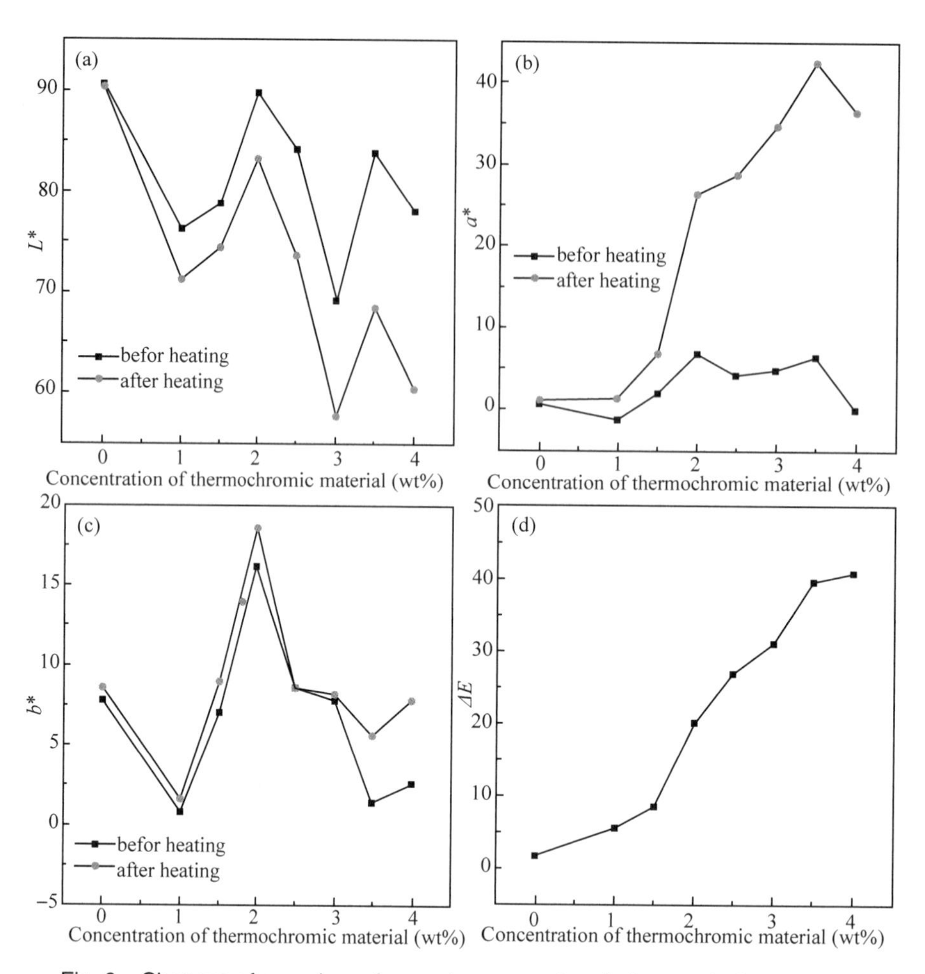

Fig. 6 Changes of sample surface color parameters before and after heating

$$\Delta E^* = \sqrt{(\Delta L^*)^2 + (\Delta a^*)^2 + (\Delta b^*)^2} \tag{4}$$

When the concentrations of thermochromic materials increased from 0 to 4.0%, the lightness index L^* of the samples decreased from −0.2 to −17.2, indicating that the surface color turned darker. The change in a^* value remarkably increased from 0.6 to 36.2, implying the surface color turned a deeper shade of red. Because the main color change of the sample was red, the change of b^* value was irregular. The ΔE^* values occurred dramatic variation from 1.81 to 39.56 with the increase in concentration from 0 to 3.5%, exhibiting a much superior thermoresponsive property. However, when the concentration increased from 3.5% to 4.0%, the L^*, a^*, b^*, and ΔE^* values changed slightly. Based on the above results, the surface color of the samples was close to saturation when concentration was set at 3.5%, and it was not necessary to continue the increase in concentration.

Fig. 7 shows the color-changed process of the sample with a thermochromic material concentration of 3.5%. Under room temperature, the thermoresponsive wood showed the natural color of wood. With the temperature increasing from 25 ℃ to 40 ℃, the thermoresponsive wood displayed excellent thermochromic process. After being cooled to the room temperature (25 ℃), the wood turned back to the natural color of wood, indicating a good orthodromic reversible behaviour. Furthermore, the natural wood texture was completely unaffected because the formed film was transparent. Thus, the method for fabricating the orthodromic reversible thermoresponsive wood is feasible.

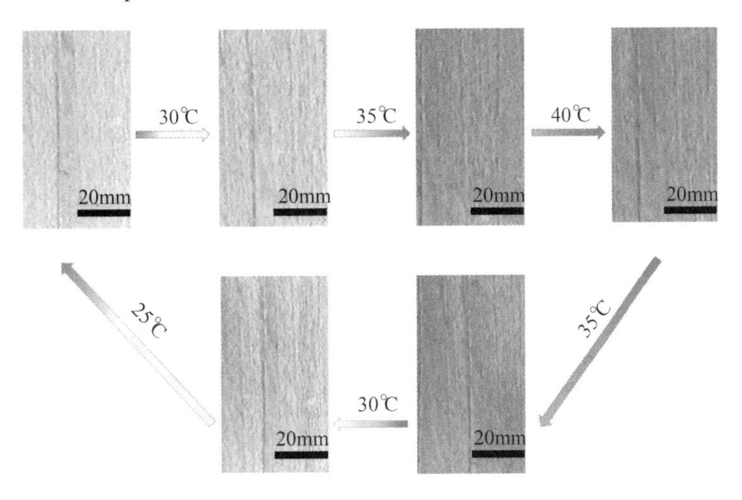

Fig. 7 Color-changed process of thermoresponsive wood

3.4 Lacquer adhesion of coating to wood

According to GB/T 9286, eqv ISO 2409—2013, the adhesion quality of the thermochromic materials to wood substrates is ranked by different numbers ranging from 0 to 5[39]. As shown in Figs. 8a, b and c, the lacquer adhesion of the samples with thermochromic material concentrations of 0%, 2.0% and 3.0% were classified as grade 0 because the cut edges were smooth for all groups but a number of small flakes. When the concentration of thermochromic materials reached 4.0%, the adhesion quality was slightly decreased due to the increasing of the film thickness. According to Fig. 8d, the surface of sample shows that small flakes were detached along edges but there was no area affected at intersections of cuts, leading to the lacquer adhesion classified as grade 1. This is the second highest adhesion grade, which can completely meet general applications in the furniture industry. (Lacquer adhesion Grade 1-2)[40].

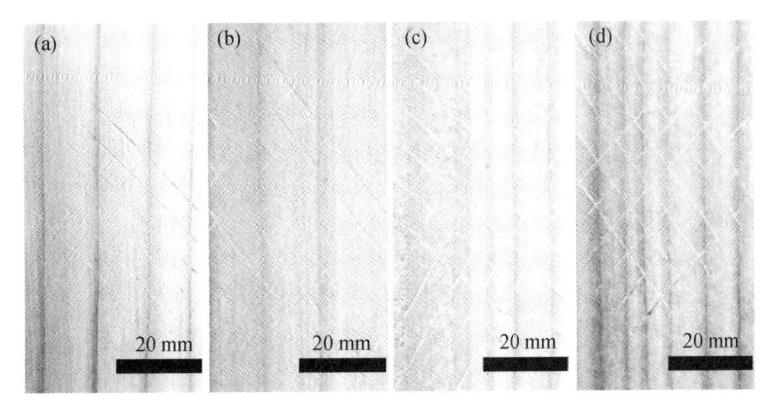

Fig. 8 Appearance of cut area of the cross-cut test for the samples with thermochromic materials concentrations of (a) 0%, (b) 2.0%, (c) 3.0%, and (d) 4.0%

3.5 Thermal gravimetric analysis

To evaluate the thermal stability of the developed smart wood, the thermal behaviors of the un-treated wood, PVA/DT coated wood, thermochromic materials and thermoresponsive wood were examined using the thermal gravimetric analysis. In Fig. 9a, the un-treated wood exhibited a small weight loss from 50 ℃ to 150 ℃, which was due to moisture and highly volatile matters. The main pyrolysis process came up in a range from 150 ℃ to 380 ℃. The decomposition of hemicellulose and cellulose took place in the ranges of 200–380 ℃ and 250–380 ℃, respectively, while the decomposition of lignin occurred in the range from 150 ℃ to 400 ℃. In Fig. 9b, no sharp distinction was found between the PVA/DT-coated wood curve and the un-treated wood curve, demonstrating that the PVA/DT had no effect on the thermal stability of wood samples. Fig. 9c presents a thermogravimetric analysis trace of the thermochromic materials. Up to 220 ℃, only 6% mass loss was observed, which was probably related to the loss of adsorbed and bound water. The degradation of thermochromic materials took place in a single process with the beginning at 220 ℃ and the maximum degradation rate was located at 315 ℃. During this stage, the breakage of the microcapsules and the evaporation of the major core materials took place, and the thermochromic materials had completely lost its thermochromic properties. Based on the above analysis, it can be verified that the thermochromic materials possessed excellent thermal stability before 220 ℃ and can keep stable below the temperature of 315 ℃. In Fig. 9d, in the range of 100 ℃ and 216 ℃, the thermoresponsive wood had slight weight loss, while there were continuous weight losses of the un-treated wood and PVA/DT coated wood, indicating that the thermochromic materials can absorb energy from the surroundings to protect wood substrate against pyrolysis. Therefore, it was concluded that the thermoresponsive wood possessed a better heat-resistant performance.

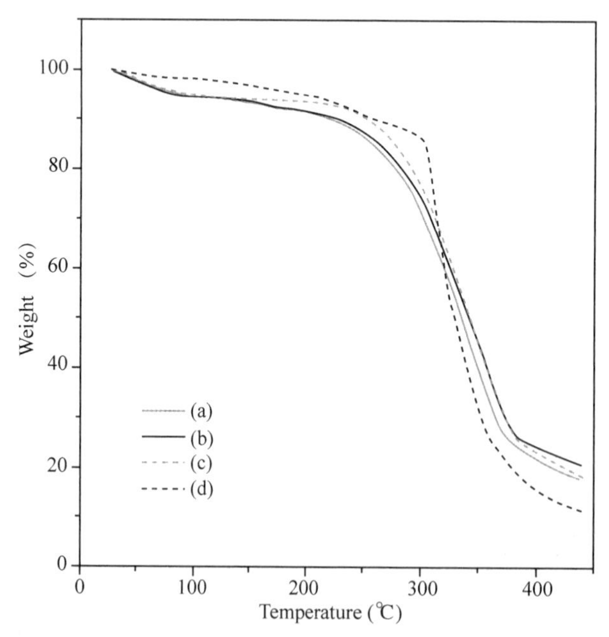

Fig. 9 TGA trace of (a) un-treated wood, (b) PVA/DT coated wood, (c) thermochromic materials and (d) thermoresponsive wood

3.6 Accelerated UV aging tests

Aging tests under UV irradiation were performed to simulate the photo-oxidative aging of samples in practical applications. Colorimetric measurements (CIE Lab) were carried out to show the color alteration of the smart thermoresponsive wood. ΔE^* values of the samples were calculated with the same method as the color tests. Fig. 10 presents the ΔE^* values of the samples with thermochromic material concentrations of 0%, 1.0%, 3.0% and 4%. It was observed that the ΔE^* value of Fig. 10d was much higher than those of Figs. 10a, b, and c, indicating that the sample with higher thermochromic material concentrations possessed a better fatigue resistance. Observed from the whole variant trend of Fig. 10, with the increase in UV radiation time form 0 to 100 h, the ΔE^* values decreased slowly. When the radiation time continuously increased from 100 to 150 h, the ΔE^* values rapidly decreased, implying that the thermoresponsive property was weakened by the increase of the aging time. In other words, this thermoresponsive wood could be extensively used as indoor materials. However, the stability of the sample under long-time UV irradiation still requires more exploration for the applications as outdoor materials.

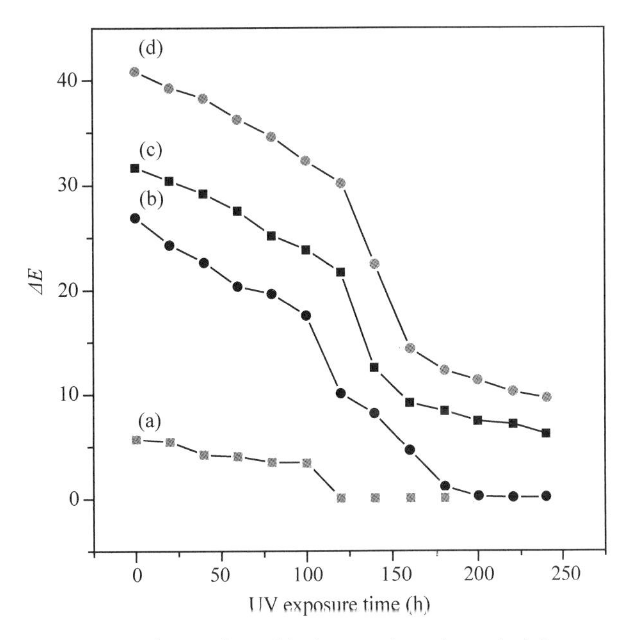

Fig. 10 UV-resistant aging tests of samples with thermochromic materials concentrations of (a) 0%, (b) 2.0%, (c) 3.0%, and (d) 4.0%

4 Conclusions

The smart wood with orthodromic reversible thermoresponsive property was successfully fabricated through a simple drop-coating method. The lacquer adhesion of the samples with thermochromic material concentrations of 0%, 2.0% and 3.0% reached grade 0, while the sample with the concentration of 4.0% was grade 1. When the concentrations of thermochromic materials increased from 0 to 3.5%, the ΔE^* values remarkably increased from 1.81 to 39.56. The as-prepared wood possessed good thermoresponsive properties. The smart wood was colored when the temperature increased and turned back to the natural color of wood after being cooled to room temperature. In addition, the natural wood texture was completely unaffected by the transparent film through the visual inspection. The TG results showed that the initial un-treated wood decomposition temperature was approximately 150 ℃, and the sample possessed excellent thermal stability below a temperature of 216 ℃. The thermoresponsive property of the sample under the continuous activated condition was durable after the UV irradiation for 100 h. The thermoresponsive wood could broaden the path for the intelligent applications of wood materials. It could be applied as temperature indicators, information storage and expected to be used as a type of novel environmentally friendly anti-counterfeit measures.

Acknowledgements

The authors gratefully acknowledge the financial support from the National Natural Science Foundation of China (31470584).

Conflict of interest

The authors declared that they have no conflict of interest.

References

[1] Hui B, Li J. Low-temperature synthesis of hierarchical flower-like hexagonal molybdenum trioxide films on wood surfaces and their light-driven molecular responses[J]. J. Mater. Sci., 2016, 51(24): 10926-10934.

[2] Raunikar R, Buongiorno J, Turner J A, et al. Global outlook for wood and forests with the bioenergy demand implied by scenarios of the Intergovernmental Panel on Climate Change[J]. For Policy Econ. 2010, 12(1): 48-56.

[3] Hui B, Li G, Zhao X, et al. h-MoO_3 microrods grown on wood substrates through a low-temperature hydrothermal route and their optical properties[J]. J. Mater. Sci. Mater. Electron., 2017, 28(4): 3264-3271.

[4] Sun Q, Lu Y, Liu Y. Growth of hydrophobic TiO_2 on wood surface using a hydrothermal method[J]. J. Mater. Sci., 2011, 46(24): 7706-7712.

[5] Bekhta P, Niemz P. Effect of high temperature on the change in color, dimensional stability and mechanical properties of spruce wood[J]. 2003.

[6] Taghiyari H R. Fire-retarding properties of nano-silver in solid woods[J]. Wood Sci. Technol., 2012, 46(5): 939-952.

[7] Koklukaya O, Carosio F, Grunlan J C, et al. Flame-retardant paper from wood fibers functionalized via layer-by-layer assembly[J]. ACS Appl. Mater. Interfaces, 2015, 7(42): 23750-23759.

[8] Aydin I, Colakoglu G. Effects of surface inactivation, high temperature drying and preservative treatment on surface roughness and colour of alder and beech wood[J]. Appl. Surf. Sci., 2005, 252(2): 430-440.

[9] Muin M, Adachi A, Inoue M, et al. Feasibility of supercritical carbon dioxide as a carrier solvent for preservative treatment of wood-based composites[J]. J. Wood Sci., 2003, 49(1): 0065-0072.

[10] Tsioptsias C, Panayiotou C. Thermal stability and hydrophobicity enhancement of wood through impregnation with aqueous solutions and supercritical carbon dioxide[J]. J. Mater. Sci., 2011, 46(16): 5406-5411.

[11] Fang Q, Chen B, Lin Y, et al. Aromatic and hydrophobic surfaces of wood-derived biochar enhance perchlorate adsorption via hydrogen bonding to oxygen-containing organic groups[J]. Environ. Sci. Technol., 2014, 48(1): 279-288.

[12] Hui B, Li Y, Huang Q, et al. Fabrication of smart coatings based on wood substrates with photoresponsive behavior and hydrophobic performance[J]. Mater. Des., 2015, 84: 277-284.

[13] Pandey K K. A note on the influence of extractives on the photo-discoloration and photo-degradation of wood[J]. Polym. Degrad. Stab., 2005, 87(2): 375-379.

[14] Merk V, Chanana M, Gierlinger N, et al. Hybrid wood materials with magnetic anisotropy dictated by the hierarchical cell structure[J]. ACS Appl. Mater. Interfaces, 2014, 6(12): 9760-9767.

[15]Hui B, Li G, Han G, et al. Fabrication of magnetic response composite based on wood veneers by a simple in situ synthesis method[J]. Wood Sci. Technol., 2015, 49(4): 755-767.

[16]Husson J M, Dubois F, Sauvat N. Elastic response in wood under moisture content variations: analytic development [J]. Mech Time Depend Mater, 2010, 14(2): 203-217.

[17]Thygesen L G, Tang Engelund E, Hoffmeyer P. Water sorption in wood and modified wood at high values of relative humidity. Part I: results for untreated, acetylated, and furfurylated Norway spruce[J]. 2010.

[18]Perruchas S, Le Goff X F, Maron S, et al. Mechanochromic and thermochromic luminescence of a copper iodide cluster[J]. J. Am. Chem. Soc., 2010, 132(32): 10967-10969.

[19]Hu L, Lyu S, Fu F, et al. Preparation and properties of multifunctional thermochromic energy-storage wood materials [J]. J. Mater. Sci., 51(5): 2716-2726.

[20]Bräunlich I, Lienemann S, Mair C, et al. Tuning the spin-crossover temperature of polynuclear iron (II)-triazole complexes in solution by water and preparation of thermochromic fibers[J]. J. Mater. Sci., 2015, 50(6): 2355-2364.

[21]Choi J O, Lee H S, Ko K H. Oxidation potential control of VO_2 thin films by metal oxide co-sputtering[J]. J. Mater. Sci., 2014, 49(14): 5087-5092.

[22]Sun Y, Xiao X, Xu G, et al. Anisotropic vanadium dioxide sculptured thin films with superior thermochromic properties[J]. Sci. Rep., 2013, 3(1): 1-10.

[23]Giuliano M R, Advani S G, Prasad A K. Thermal analysis and management of lithium-titanate batteries[J]. J. Power Sources, 2011, 196(15): 6517-6524.

[24]Lee J, Stein I Y, Kessler S S, et al. Aligned carbon nanotube film enables thermally induced state transformations in layered polymeric materials[J]. ACS Appl. Mater. Interfaces, 2015, 7(16): 8900-8905.

[25] Lee S, Lee J, Lee M, et al. Construction and Molecular Understanding of an Unprecedented, Reversibly Thermochromic Bis-Polydiacetylene[J]. Adv. Funct. Mater., 2014.

[26]Pairas G N, Tsoungas P G. H-Bond: The Chemistry-Biology H-Bridge[J]. Chemistry Select, 2016, 1(15): 4520.

[27]Cano A, Fortunati E, Cháfer M, et al. Effect of cellulose nanocrystals on the properties of pea starch-poly (vinyl alcohol) blend films[J]. J. Mater. Sci., 2015, 50(21): 6979-6992.

[28]Zhu H, Jia Z, Chen Y, et al. Tin anode for sodium-ion batteries using natural wood fiber as a mechanical buffer and electrolyte reservoir[J]. Nano Lett., 2013, 13(7): 3093-3100.

[29]YongFeng L, XiaoYing D, ZeGuang L, et al. Effect of polymer in situ synthesized from methyl methacrylate and styrene on the morphology, thermal behavior, and durability of wood[J]. J. Appl. Polym. Sci., 2013, 128(1): 13-20.

[30]Andreatta F, Lanzutti A, Paussa L, et al. Addition of phosphates or copper nitrate in a fluotitanate conversion coating containing a silane coupling agent for aluminium alloy AA6014[J]. Prog. Org. Coat., 2014, 77(12): 2107-2115.

[31]Hočevar M, Berginc M, Topič M, et al. Sponge-like TiO_2 layers for dye-sensitized solar cells[J]. J Solgel Sci Technol, 2010, 53(3): 647-654.

[32]Fateh R, Ismail A A, Dillert R, et al. Highly active crystalline mesoporous TiO_2 films coated onto polycarbonate substrates for self-cleaning applications[J]. J. Phys. Chem. C, 2011, 115(21): 10405-10411.

[33]Gao X, Tang J, Zuo Y, et al. The electroplated palladium-copper alloy film on 316L stainless steel and its corrosion resistance in mixture of acetic and formic acids[J]. Corros. Sci., 2009, 51(8): 1822-1827.

[34]Kieckow F, Kwietniewski C, Tentardini E K, et al. XPS and ion scattering studies on compound formation and interfacial mixing in TiN/Ti nanolayers on plasma nitrided tool steel[J]. Surf. Coat. Technol., 2006, 201(6): 3066-3073.

[35]Kim K, Kang T H, Ihm K, et al. Initial stage of nitridation on Si(100) surface using low-energy nitrogen ion implantation[J]. Surf. Sci., 2006, 600(17): 3496-3501.

[36]Luo W, Yang W, Jiang S, et al. Microencapsulation of decabromodiphenyl ether by in situ polymerization: preparation and characterization[J]. Polym. Degrad. Stab., 2007, 92(7): 1359-1364.

[37] Zuo J D, Liu S M, Sheng Q. Synthesis and application in polypropylene of a novel of phosphorus-containing intumescent flame retardant[J]. Molecules, 2010, 15(11): 7593-7602.

[38] Hu Y, Yang Y, Ning Y, et al. Facile preparation of artemisia argyi oil-loaded antibacterial microcapsules by hydroxyapatite-stabilized Pickering emulsion templating[J]. Colloids Surf. B, 2013, 112: 96-102.

[39] Zhao H, Huang Z, Cui J. A new method for electroless Ni-P plating on AZ31 magnesium alloy[J]. Surf. Coat. Technol., 2007, 202(1): 133-139.

[40] Liu H H, Xu L, Wang Y X, et al. Prevalence and progression of myopic retinopathy in Chinese adults: The Beijing Eye Study[J]. Ophthalmology, 2010, 117(9): 1763-1768.

中文题目：具有可逆温度响应性智能木材的制备

作者：李莹莹，惠彬，李国梁，李坚

摘要：通过在木材表面自组装温敏材料制备了温度响应木材。样品可以随着温度的变化可逆地改变其颜色。扫描电子显微镜显示，木材表面由均匀的球形颗粒组成。红外光谱图表明，温敏材料已成功沉积在木材表面上，漆膜附着力测试结果进一步证明，所有样品都可以满足家具行业的一般应用。随着温敏材料的浓度从0%增加到4.0%，样品的总颜色变化(ΔE^*)从1.81显著增加到39.56，表现出良好的温度响应性能。热重结果表明，该样品在216 ℃以下具有优异的热稳定性。此外，紫外线老化测试表明，样品在暴露于紫外线照射下100 h后仍然具有很好的温度响应性能。

关键词：温度响应；木材表面；可逆性；温度

Luminescent and Transparent Wood Composites Fabricated by PMMA and γ-Fe_2O_3@YVO_4: Eu^{3+} Nanoparticles Impregnation*

Wentao Gan, Shaoliang Xiao, Likun Gao, Runan Gao, Jian Li, Xianxu Zhan

Abstract: Natural wood is functionalized using the index matching poly (methyl methacrylate) (PMMA) and luminescent γ-Fe_2O_3@ YVO_4: Eu^{3+} nanoparticles to form a novel type of luminescent and transparent wood composite. First, the delignified wood template was obtained from natural wood though a lignin removal process, which can be used as support for transparent polymer and phosphor nanoparticles. Then, the functionalization occurs in the lumen of wood, which benefits from PMMA that fills the cell lumen and enhances cellulose nanofiber interaction, leading to wood composites with excellent thermal property, dimensional stability, and mechanical properties. More importantly, this wood composite displays a high optical transmittance in a broad wavelength range between 350 and 800 nm, magnetic responsiveness, and brightly colored photoluminescence under UV excitation at 254 nm. The unique properties and green nature of the luminescent wood composite have great potential in applications including green LED lighting equipment, luminescent magnetic switches, and anti-counterfeiting facilities.

Keywords: Luminescent wood; Transparent; Magnetic, γ-Fe_2O_3@ YVO_4: Eu^{3+}

1 Introduction

Wood is a sustainable and renewable material with an remarkable mechanical properties and sophisticated hierarchical structure, which has been widely used in daily life in various applications, including homes, furniture, artwork, heating and decoration.[1-3] However, the biocomposite nature of wood results in numerous harmful effects, including easy absorption of water and moisture, cracking, mildew, and biological and light degradation, which strongly reduce the durability of wood in service.[4-6] Therefore, many modified approaches at the molecular or nanoscale level or surface of the wood have been proposed, in order to improve the shortcomings of the wood.[7-10] With the development of materials science, wood materials have been increasingly considered as a biobased template to endow with some novel properties.[11] For example, Merk et al. reported on magnetic wood composites based on in-depth penetration of ferric and ferrous ions in wood matrices,[12] which have some potential in the field of electromagnetic wave absorption, heating, and magnetic actuation.[13-15] Wu et al. fabricated a superhydrophobic inorganic nanoparticles film on wood surface, yielding a durable superhydrophobic wood material with excellent dimensional stability and self-

* 本文摘自 ACS Sustainable Chem. Eng. 2017, 5: 3855-3862.

cleaning.[16] Moreover, Hu et al. fabricated a unique transparent and haze wood materials by epoxy resin impregnation.[17] Firstly, colored lignin were removed from the wood. The porous and hierarchical structure was well preserved after epoxy resin impregnation and curing.[18] Further reserch conclusively showed that the transparent wood composites can be utilized for a range of optoelectronics and energy efficient building materials.[19,20]

Nowadays, considerable interest in the field of luminescent materials has been focus of nanomaterials doped with lanthanide ions, these nanomaterials exhibit attractive chemical features such as blinking, low toxicity, resistance to photochemical degradation and photobleaching.[21-23] To the best of our knowledge, most of the available preparation involving the sol-gel method, hydrothermal process, the Pechini method, and solvothermal synthesis give excellent opportunities to obtain lanthanide-doped nanomaterials with various particles size, shapes, and doping levels,[24-26] and these parameters greatly influence the chemical activity and luminescent properties of such materials.

To expand the application of wood materials in the optical field, it is advantageous to impregnate some luminescent nanomaterials to the wood template, leading to the excellently luminescent wood composites. In this study, a novel approach, based on the removal of light absorbed lignin from the wood template and then impregnation of PMMA and γ-Fe_2O_3 @ YVO_4: Eu^{3+} nanoparticles, was proposed to synthesize highly transparent, luminescent, and magnetic wood composite. The resulting nanoscale structure of the wood template is analyzed, and effects on optical, magnetic, thermal, and mechanical properties are investigated. The luminescent wood with high transparency will offer diversified application toward the preparation of new types of green LED lighting equipment, luminescent magnetic switches, and anticounterfeiting facilities.

2 Experimental section

2.1 Materials

Wood slices (20 mm×10 mm×0.5 mm) of Cathay poplar (Populus cathayana Rehd) were used in this experiment. Ferric trichloride hexahydrate ($FeCl_3 \cdot 6H_2O$), ferrous chloride tetrahydrate ($FeCl_2 \cdot 4H_2O$), ammonia solution ($NH_3 \cdot H_2O$, 25%), sodium chlorite ($NaClO_2$), glacial acetic acid, hydrogen peroxide (30% solution), sodium acetate (NaAc), ammonium metavanadate (NH_4VO_3), citric acid monohydrate, and ethanol in this study were supplied by Shanghai Boyle Chemical Company Limited. Methyl methacrylate (MMA), 2, 2'-azobis-(2-methylpropionitrile) (AIBN, Sigma-Aldrich), yttrium oxide (Y_2O_3, 99.9%), and europium oxide (Eu_2O_3, 99.9%) were purchased from Shanghai Aladdin Biochemical Technology Co., Ltd (China). All chemicals were of analytic-reagent grade.

2.2 Synthesis of γ-Fe_2O_3@YVO_4: Eu^{3+} Nanoparticles

Fig. 1 shows the reactive scheme for synthesizing the luminescent and transparent wood. The γ-Fe_2O_3@YVO_4: Eu^{3+} nanoparticles were synthesized according to solvothermal method. Firstly, 2 mmol Y_2O_3, 0.1 mmol of Eu_2O_3, 24 mmol of citric acid, and 4 mmol of NH_4VO_3 were dissolved in dilute nitric acid with stirring at 80 ℃ for several minutes, and then dried at 120 ℃ for 24 h. After that, the samples was calcined at 1000 ℃ for 4 h to obtain the phosphor samples. Then, 0.1 g of phosphor particles, 0.25 g of CTAB,

0.92 g of $FeCl_3$ and 0.34 g of $FeCl_2$ were introduced into 50 mL distilled water with continuous stirring. The pH value of the mixture was adjusted to ca. 10 through the addition of 25% ammonia solution. After stirring for 30 min, the system was transferred into a Teflon-lined stainless-steel autoclave and heated at 110 ℃ for 8 h. Finally, the powder samples was dried at 60 ℃ for 12 h.

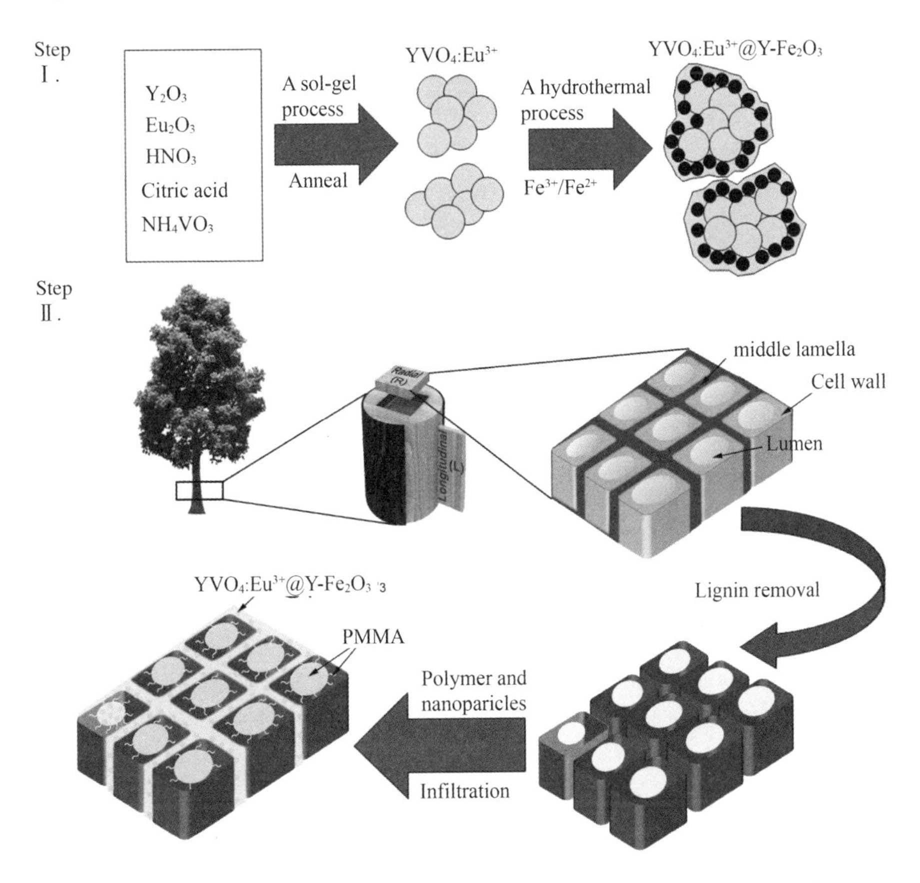

Fig. 1 Experimental strategy for the fabrication of luminescent wood

2.3 Synthesis of luminescent wood (LW)

The delignified wood template was obtained thought a lignin removal process that the natural wood was treated with a $NaClO_2$(2 wt%) and glacial acetic acid solution (pH=4.6) at 80 ℃ for 12 h. The treated wood templates were rinsed with distilled water, following by placement in an aqueous solution of H_2O_2($5\ mol \cdot L^{-1}$) and boiling for 4 h; then they were dried at 45 ℃ for 24 h. Then, the polymer solution was prepared by mixing the methyl methacrylate (MMA) and 2, 2'-azobis (2-methylpropionitrile) with a weight ratio of 1000 to 3, which was prepolymerized at 75 ℃ for 20 min. The prepolymerized methyl methacrylate solution was cooled down to room temperature in ice-water bath. After that, the prepolymerized polymer solution and γ-Fe_2O_3@ YVO_4: Eu^{3+} nanoparticles were uniformly dispersed and mixed together with the weight ratios of 1000 : 1, 1000 : 2, 1000 : 5, and 1000 : 10 for luminescent wood containing different particle concentrations. The delignified wood template was infiltrated in the mixture under vacuum for 30 min, and this process was repeated three times to ensure the full infiltration. Finally, the infiltrated wood was sandwiched between two glass slides and kept static at 50 ℃ for 6 h. The luminescent and transparent wood composites was peeled off from the glass slides after the methyl methacrylate was completely solidified. In order to determine the concentration of Eu^{3+} in LW, corresponding ICP-MS analysis results of as-prepared wood samples were added in Table S1.

2.4 Characterization

The surface morphology of the as-prepared samples was analyzed using scanning electron microscopy (SEM, HITACHI TM3030) and transmission electron microscopy (TEM, Tecnai G20). The surface components of the samples were determined via energy dispersive X-ray analysis (EDXA) and Fourier transformation infrared spectroscopy (Magna-IR 560). The phase structure of the as-prepared products was checked by the X-ray diffraction measurements (XRD, Rigaku, and D/MAX 2200) operated with Cu target radiation (λ = 1.54 Å). Inductively coupled plasma mass spectrometry (ICP-MS) was carried out on an Agilent 725. The thermal performances of the wood composites were measured by a thermal analyzer (TGA, SDT Q600) at a heating rate of 10 ℃ · min^{-1} under under nitrogen atmosphere. The transmittance of wood composites was analyzed with a U-4100 UV-vis spectrometer. The excitation and emission spectra of LW were recorded at room temperature using a FLsp920 spectrofluorometer. When the excitation wavelength (300 nm) was selected, the quantum yield values were calculated automatically. Magnetic properties of wood samples with the size of 4 mm×4 mm×0.5 mm were determined using a superconducting quantum interference device (MPMS XL-7, Quantum Design Corp.) at room temperature. The mechanical properties were performed using an Instron 1185 testing machine. The wood was selected without joints or fasteners with a dimension of about 50 mm×10 mm×3 mm. The dimensional stability was evaluated by placed the wood samples (20 mm×20 mm× 0.5 mm) in chambers containing distilled water at the room temperature (25 ℃). Volume change of the wood samples can be calculated according to eq 1:

$$VC(\%) = \frac{V_1 - V_0}{V_0} i \times 100\% \qquad (1)$$

Where V_0 and V_1 are the volumes of wood before and after impregnation, respectively.

3 Results and discussion

For confirmation of particles size and distribution of YVO_4: Eu^{3+} and γ-Fe_2O_3@ YVO_4: Eu^{3+}, the TEM measurement was applied. Fig. 2 shows the TEM images of YVO_4: Eu^{3+} and γ-Fe_2O_3 @ YVO_4: Eu^{3+} NPs. Fig. S1 demonstrates the histograms of diameter of YVO_4: Eu^{3+} NPs and γ-Fe_2O_3@ YVO_4: Eu^{3+} NPs. As revealed in Fig. 2a, the YVO_4: Eu^{3+} NPs are uniform and spherical in shape, with an average diameter of~ 15 nm (Fig. S1a). Fig. 2b and c shows the TEM images of γ-Fe_2O_3 @ YVO_4: Eu^{3+} NPs under different magnifications. The TEM images indicate that the YVO_4: Eu^{3+} NPs with uniform size are assembled around the γ-Fe_2O_3 NPs. The diameter of the γ-Fe_2O_3@ YVO_4: Eu^{3+} NPs is mainly centered at 95 nm (Fig. S1b). The energy dispersive spectrum elemental analysis data of γ-Fe_2O_3@ YVO_4: Eu^{3+} NPs shown in Table S2 confirms the presence of iron, oxygen, yttrium, vanadium, and europium in the as-prepared nanocomposites.

To determine the spatial distribution of PMMA and γ-Fe_2O_3@ Y_2O_3: Eu^{3+} NPs after polymer impregnation at the cell and tissue levels, a series of SEM pictures of top-view and cross-sectional SEM images of the wood samples was taken. In Fig. 3a and b, after PMMA impregnation, the original tracheid with some pits of wood

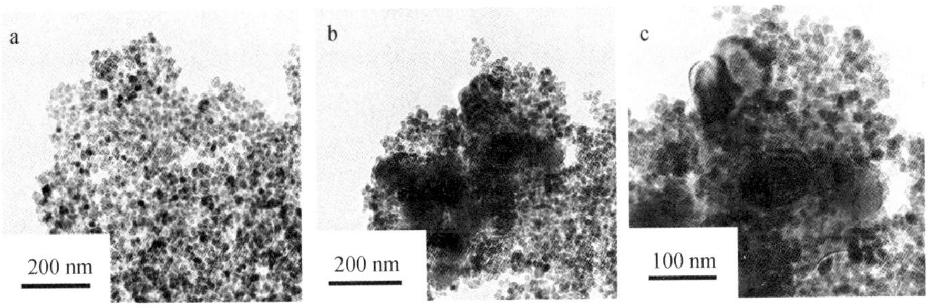

Fig. 2 TEM micrographs of (a) YVO_4: Eu^{3+} and (b, c) γ-Fe_2O_3@YVO_4: Eu^{3+} NPs

surface were observed to be covered by a continuous smooth polymer film. Upon addition of γ-Fe_2O_3@ YVO_4: Eu^{3+} NPs, some particles are widely distributed on the wood surface (Fig. 3c). the energy dispersive spectrum presents in Fig. S2 shows the chemical elements of LW: carbon, oxygen, iron, europium, yttrium, and gold (from the coating for SEM imaging). Moreover, the corresponding elemental maps of C, Fe, Y and Eu (Fig. S2) also confirm that the γ-Fe_2O_3@ YVO_4: Eu^{3+} NPs are deposited on the wood template. A cross-sectional SEM photograph of the untreated wood is shown in Fig. 3d, in which the porous structure is well presented. During the experiment process, it is highly expected that polymerization takes place in the cell lumen, because the porous structure in wood will result in light scattering in the visible range, and leads to the opacity of wood.[27] Fig. 3e and f shows that almost all of cell lumen are filled with the polymer, while there are some nanoparticles located on the cell lumen after addition of γ-Fe_2O_3@ YVO_4: Eu^{3+} NPs.

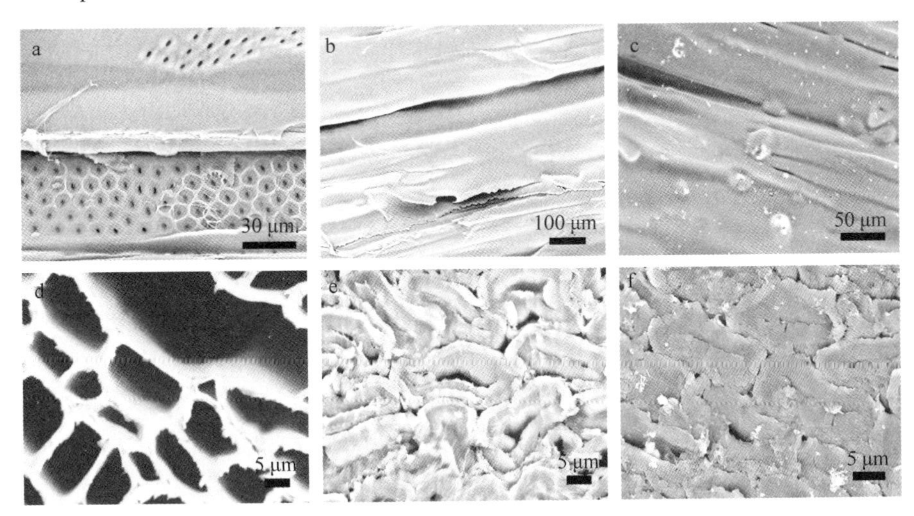

Fig. 3 Top-view SEM image of (a) natural wood, (b) transparent wood, and (c) luminescent wood with a γ-Fe_2O_3@YVO_4: Eu^{3+} NP concentration of 0. 1 wt%, respectively. Cross-sectional SEM image of (d) natural wood, (e) transparent wood, and (f) luminescent wood with a γ-Fe_2O_3@YVO_4: Eu^{3+} NP concentration of 0. 1 wt%, respectively

To determine the crystal structure of the samples obtained from different stages of the preparation, XRD was performed (Fig. 4). The XRD pattern, Fig. 4a of YVO_4: Eu^{3+} shows the main XRD diffraction peaks at angles $2\theta = 25°$, 33. 5°, and 50°, which are assigned to the diffraction of the (200), (112) and (312) planes of YVO_4 (JCPDS 17-0341), suggesting the successful crystallization of YVO_4: Eu^{3+}.[28] Fig. 4b shows the XRD patterns of γ-Fe_2O_3@ YVO_4: Eu^{3+}. The XRD diffraction peaks at $2\theta = 18°$, 30°, 35°, 43°, 53°, 57°, and 62° of γ-Fe_2O_3 can be indexed according to the standard data of maghemite (JCPDS 39-1346),[29] all diffraction peaks belonging to YVO_4: Eu^{3+} are present, revealing the γ-Fe_2O_3@ YVO_4: Eu^{3+} NPs was successful synthesized in the experiments. In Fig. 4c, two major diffraction peaks are observed for natural wood at $2\theta = 16°$ and 22°, which are characteristic of cellulose.[3,30] After polymer impregnation (Fig. 4d), two broad peaks are observed for the transparent wood at $2\theta = 15°$ and 30°, which are ascribed to PMMA.[31] Upon further addition of γ-Fe_2O_3@ YVO_4: Eu^{3+} NPs (Fig. 4e), the main diffraction peaks at $2\theta = 25°$, 33°, 35°, 40°, 50°, 53°, 62°, and 63° of γ-Fe_2O_3@ YVO_4: Eu^{3+} exist in LW; all diffraction peaks belonging to cellulose and PMMA are present, indicating that the γ-Fe_2O_3@ YVO_4: Eu^{3+} and PMMA are immobilized into the wood template successfully.

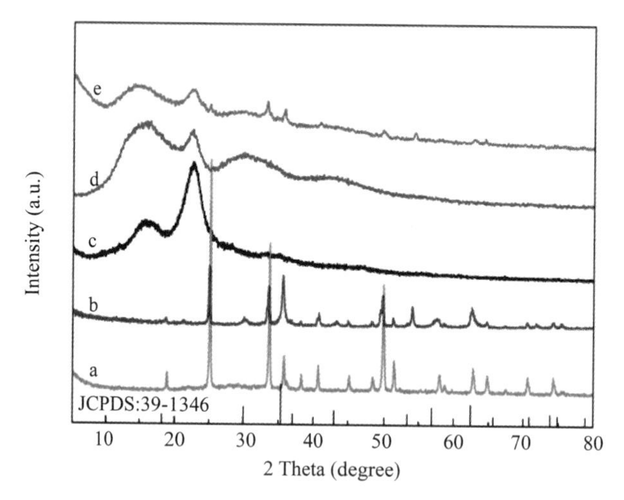

Fig. 4 XRD patterns of (a) YVO_4: Eu^{3+}, (b) γ-Fe_2O_3@YVO_4: Eu^{3+}, (c) natural wood, (d) transparent wood, and (e) luminescent wood with a γ-Fe_2O_3@YVO_4: Eu^{3+} NP concentration of 0.2 wt%

Fig. 5 shows the FTIR spectra of natural wood, transparent wood, and LW with 0.2 wt% γ-Fe_2O_3@ YVO_4: Eu^{3+} NP loading. The natural wood (Fig. 5a) presents characteristic absorption peaks such as O—H stretching vibrations at 3435 cm^{-1}, C—H stretching vibration at 2925 cm^{-1}, acetyl groups of hemicelluloses at 1742 cm^{-1}, C=O stretching vibrations at 1632 cm^{-1}, aromatic stretching vibrations of lignin at 1584 cm^{-1}, aliphatic acid group vibration at 1250 cm^{-1}, and aromatic C—H in-plane deformation assigned to lignin at 1060 cm^{-1}, which are in agreement with earlier research.[32] After lignin removal and polymer infiltration, the absorption peaks at 1584 and 1060 cm^{-1} disappeared, indicating light-absorbing lignin was degraded during the preparation period. Moreover, the FTIR spectrum of the transparent wood also possesses the characteristic peaks of PMMA (2996 and 2950 cm^{-1} for C—H, 1733 cm^{-1} for C=O, and 1210 and 1169 cm^{-1} for C—O, as presented in Fig. 5b).[33] However, when the γ-Fe_2O_3@ YVO_4: Eu^{3+} NPs were added to the polymer (Fig. 5c), two absorption peaks at 835 and 450 cm^{-1} appeared, which corresponded to the characteristic absorption of V-O and Y(Eu)-O bonds, respectively.[34] This further proves that YVO_4: Eu^{3+} has impregnated into the wood template. Furthermore, the stretching vibrations of the Fe-O can be clearly seen around at 558 cm^{-1},[35] which confirms that the γ-Fe_2O_3 NPs are deposited on the wood composites successfully.

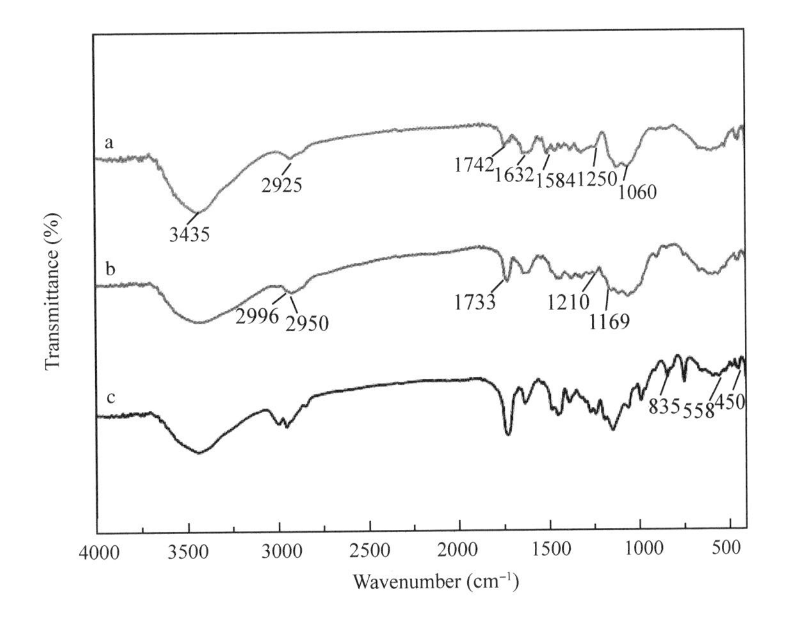

Fig. 5 FTIR spectra of (a) natural wood, (b) transparent wood, and (c) luminescent wood with a γ-Fe_2O_3@YVO_4: Eu^{3+} NP concentration of 0.2 wt%

The results of the TG and DTA measurements of the samples are shown in Fig. 6. In Fig. 6a, PMMA undergoes three weight loss stages. The first step, started at around 160 ℃, would be caused by the scissions of head-to-head linkages (H−H). The second step, started at around 275 ℃, was attributed to the scissions at unsaturated ends including a hemolytic scission β to the vinyl group. The last step, started at around 340 ℃, could be explained by the random scission within the polymer chain. [36] However, the natural wood samples undergo two degradation steps. The first step started at around 100 ℃ could be caused by the evaporation of water, and the second step started at around 260 ℃ would be ascribed to the degradation of wood component such as cellulose, hemicellulose, and lignin. [37] Similar degradation behaviors could be observed in LW. The first degradation step in the range of 100−260 ℃ due to the evaporation of water and the scissions of head-to-head linkages (H—H) in PMMA. The second degradation steps in the range of 260−420 ℃ could be caused by the degradation of wood component and PMMA. Compared with the TG curve of PMMA, more residue of LW remained, which can be explained by the inorganic γ-Fe_2O_3@ YVO_4: Eu^{3+} NPs, and the remaining carbon from the delignified wood template. And compared with the TG curve of natural wood, less residue was observed in LW, which may be attributed to the impregnation of complete combustion of PMMA in wood template. In Fig. 6b, PMMA shows three main exothermic peaks at around 228, 294, and 377 ℃ in the DTA curve, corresponding to the three weight looses stages of PMMA in Fig. 6a. As for the natural wood, two prominent exothermic peaks for flaming and glowing are observed at 305 and 367 ℃, respectively. [38] Moreover, compared with the DTA results of PMMA and natural wood, the exothermic peaks of LW are weakened and shifted to the high temperature, revealing the good thermostability of LW.

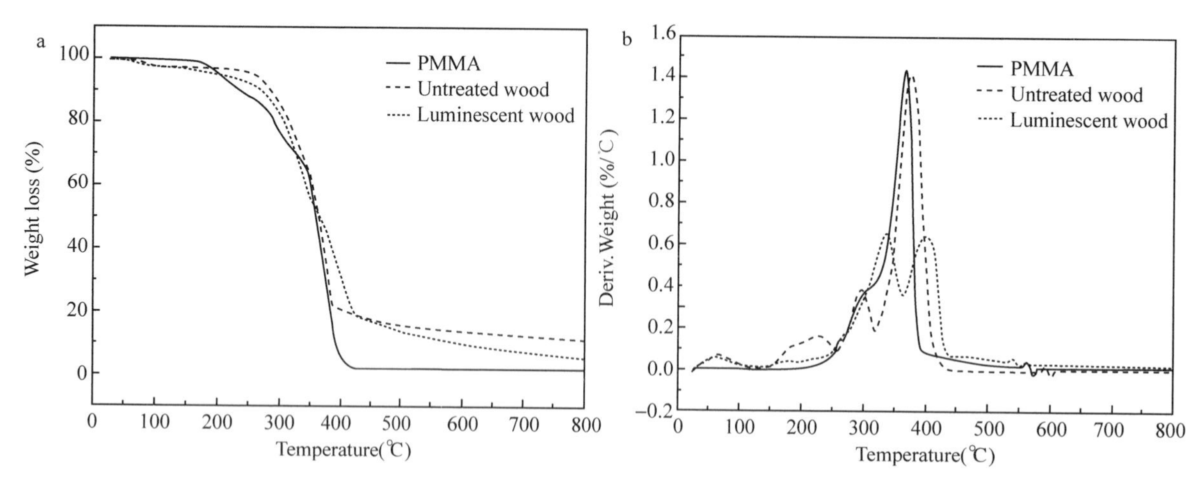

Fig. 6 (a) TG and (b) DTA curves of the PMMA, natural wood, and luminescent wood with a γ-Fe_2O_3@YVO_4: Eu^{3+} NP concentration of 0.5 wt%

The other noticeable characteristic of LW is optical properties. The transmittance of the wood samples obtained from different stages of processing is presented in Fig. 7. Natural wood shows negligible transmittance due to the light absorption of lignin and light scattering in porous wood structure. After lignin removal, light scattering still in porous wood structure, thus the transmittance of delignified wood is only 4.9%. Until polymer impregnation (without γ-Fe_2O_3@ YVO_4: Eu^{3+} NPs), a highly transparent wood composite with the transmittance of 86.1 % was achieved. As the concentration of γ-Fe_2O_3@ YVO_4: Eu^{3+} NPs increased, the transmittance of LW was decreased in 350-800 nm wavelength region, while it is generally higher than that of the natural wood (Fig. 7). This is mainly caused by the optical absorption of as-prepared nanoparticles.[39] Additionally, the larger amount of γ-Fe_2O_3@ YVO_4: Eu^{3+} NPs that fills into the wood template can lead to the agglomeration of particles, which also will increase the light reflection and further decrease the transmittance of wood composites. However, the transmittance of LW that possesses γ-Fe_2O_3@ YVO_4: Eu^{3+} NPs of 0.1, 0.2 and 0.5 wt% is 80.6 %, 73.2 % and 40.8 %, respectively. The highest transmittance of the LW is 80.6 %, which is good enough to be applied in transparent devices.

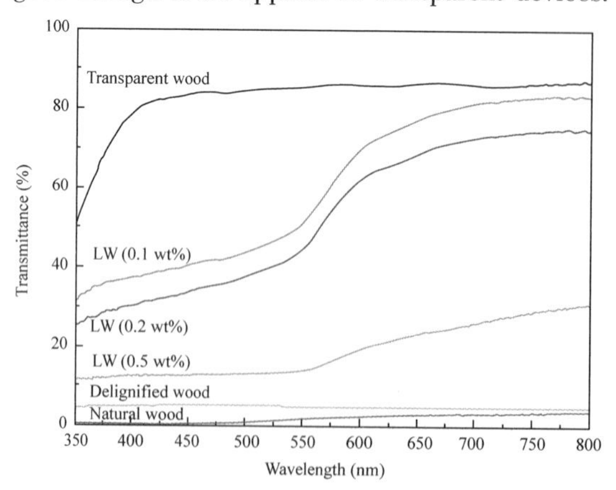

Fig. 7 Transmittance of the wood samples obtained from different stages of processing

The corresponding photoluminescence spectra of LW with different concentrations of γ-Fe_2O_3@ YVO_4: Eu^{3+} NPs are shown in Fig. 8. Fig. S3 demonstrates the photoluminescence spectra of YVO_4: Eu^{3+} NPs and γ-Fe_2O_3@ YVO_4: Eu^{3+} NPs. For the excitation spectrum of LW, there is a broad band extending from 250 to 350 nm with a maximum value at 300 nm in Fig. 8a. The band corresponds to charge transfer bands of Eu-O and V-O and VO_{43-} absorption bands.[40,41] Both the LW, YVO_4: Eu^{3+}, and γ-Fe_2O_3@ YVO_4: Eu^{3+} NPs show similar excitation spectra in the range of 250–350 nm, but two main peaks at 397 and 467 nm disappear in the excitation spectra of LW and γ-Fe_2O_3@ YVO_4: Eu^{3+} NPs (Fig. S3), which correspond to the electron transitions of VO_{43-} groups to Eu^{3+} from the $^7F_0$-$^5L_6$ and $^7F_0$-$^5L_2$, respectively. This phenomenon may be caused by the light absorption of γ-Fe_2O_3 NPs impregnated into the composites.[42] Moreover, the excitation intensity was increased at first and then decreased with adding more γ-Fe_2O_3@ YVO_4: Eu^{3+} NPs. The highest intensity took place when the weight ratio of γ-Fe_2O_3@ YVO_4: Eu^{3+} NPs to PMMA was 0. 5 wt%.

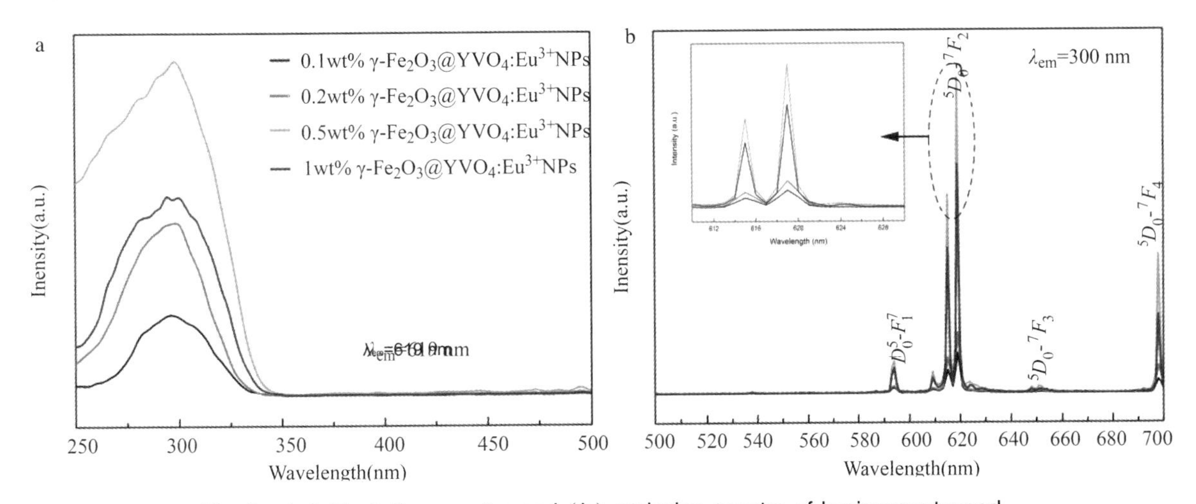

Fig. 8 (a) Excitation spectra and (b) emission spectra of luminescent wood

For the emission spectra of LW, as shown in Fig. 8b, three peaks at 594, 650, and 698 nm are caused by $^5D_0$-$^7F_1$, $^5D_0$-$^7F_3$, and $^5D_0$-$^7F_4$ transitions of Eu^{3+} ions in YVO_4, respectively. However, the main emission peak at 619 nm corresponds to the $^5D_0$-$^7F_2$ transition of Eu^{3+} ions, revealing the 3p electrons of Eu^{3+} ions are occupied as a site without an inversion symmetry.[43] As reported by Reisfeld et al.,[44] when the ratio $I(D_0$-$F_2)/I(D_0$-$F_1)$ is inferior to 1, indicating the Eu^{3+} site is totally symmetry. In our case, the ratio $I(D_0$-$F_2)/I(D_0$-$F_1)$ is around 6, suggesting the Eu^{3+} site is low symmetry. The inset of Fig. 8b is plots of emission intensity at 619 nm variation with various (Eu^{3+} ions) γ-Fe_2O_3@ YVO_4: Eu^{3+} NPs doping concentration. A comparison among LW with different NPs doping concentration suggests that LW with the γ-Fe_2O_3@ YVO_4: Eu^{3+} NPs concentration of 0. 5 wt% has the strongest emission intensity. Higher Eu^{3+} concentrations in LW result in quenching the photoluminescence intensity. Magnetic γ-Fe_2O_3 NPs also make some side effects on emission intensity. Therefore, the optimum doping concentration of γ-Fe_2O_3@ YVO_4: Eu^{3+} NPs is 0. 5 wt%.

Meanwhile, the effect on quantum yields via impregnation of different amounts of γ-Fe_2O_3@ YVO_4: Eu^{3+} NPs into the wood composites was studies, and the results are shown in Table 1. The quantum yield of γ-Fe_2O_3@ YVO_4: Eu^{3+} NPs is found much smaller than YVO_4: Eu^{3+} NPs, which may be caused by the quenching of the luminescence by magnetic γ-Fe_2O_3 NPs. Furthermore, compared with the bulk γ-Fe_2O_3@YVO_4: Eu^{3+} NPs, there is a decline on the quantum yields of LW. A explanation may be that the wood template is nonluminescent material, so the PL efficiency of the LW will decrease with its percentage of increase. Thus, a low quantum yield of LW is obtained. For LW with different concentrations of γ-Fe_2O_3@YVO_4: Eu^{3+} NPs, with the increasing amount of Eu^{3+}, more energy transfer takes place between vanadates groups to europium ions, contributing to a improvement of the quantum yield. More significantly, however, when the concentration of γ-Fe_2O_3@YVO_4: Eu^{3+} NPs reached 1 wt%, an additional quenching path occur from energy transfer between neighboring Eu^{3+} which results in a drop of the quantum yield.[45] All in all, the highest quantum yield of LW is obtained for the γ-Fe_2O_3@YVO_4: Eu^{3+} concentration of 0.5 wt%.

Table 1 Quantum yields of as-Prepared samples

Sample	Quantum yield (%)
YVO_4: Eu^{3+}	74.44
γ-Fe_2O_3@YVO_4: Eu^{3+}	1.15
LW-0.1wt%	0.10
LW-0.2wt%	0.11
LW-0.5wt%	0.64
LW-1wt%	0.44

Not only does LW possess luminescent property, but it is also a kind of superparamagnetic material, which provides potential for magneto-optical applications. Fig. 9 is magnetization curves of LW with different concentrations of γ-Fe_2O_3@YVO_4: Eu^{3+} NPs at room temperature. The inset shows the magnetization curves of γ-Fe_2O_3 NPs and γ-Fe_2O_3@YVO_4: Eu^{3+} NPs. The saturation magnetizations M_s of as-prepared γ-Fe_2O_3 NPs and γ-Fe_2O_3@YVO_4: Eu^{3+} NPs are 68.02 and 27.77 emu · g^{-1}, respectively. The large drop in M_s of the γ-Fe_2O_3@YVO_4: Eu^{3+} NPs is caused by the presence of the nonmagnetic YVO_4: Eu^{3+} NPs. Similar results are observed in LW. The natural wood is observed to be a nonmagnetic material as expected. Interestingly, all of the LW samples show polar properties and exhibit a typical superparamagnetic behavior at room temperature due to the immeasurable coercivity and remanence. Generally, the Langevin equation can use to define the magnetic properties of composites in a superparamagnetic system:[46]

$$\frac{M}{M_S}=\coth x-\frac{1}{x} \tag{2}$$

Where M (emu · g^{-1}) is the magnetization, and M_S(emu · g^{-1}) is the saturation magnetization. Usually, $x=\alpha H$. Where, H (Oe) is the magnetic field. α is a function of the electron spin magnetic moment m (μ_B, μ_B is the Bohr magneton) of the individual molecules, which can be expressed as in eq 3:

$$\alpha=\frac{m}{k_B T} \tag{3}$$

Where, k_B is Boltzmann's constant and T is the absolute temperature. According to Fig. 9, the magnetization (M) of LW increases with the increasing magnetic field (H) until it reaches the saturation value. The M_S values of LW which possesses γ-Fe_2O_3@YVO_4: Eu^{3+} NPs of 0.1, 0.2, and 0.5 wt% are 0.26, 1.19 and 2.10 emu · g^{-1}, respectively. As expected, the Ms values of LW with the γ-Fe_2O_3@YVO_4:

Eu^{3+} NP concentration of 1 wt% is 3. 35 emu · g^{-1}, which is greater than that of the LW with lower γ-Fe_2O_3 @ YVO4: Eu^{3+} NP concentration (Fig. S4). The results suggest a higher concentration of γ-Fe_2O_3@ YVO_4: Eu^{3+} NPs in wood template will better maintain the magnetic properties.

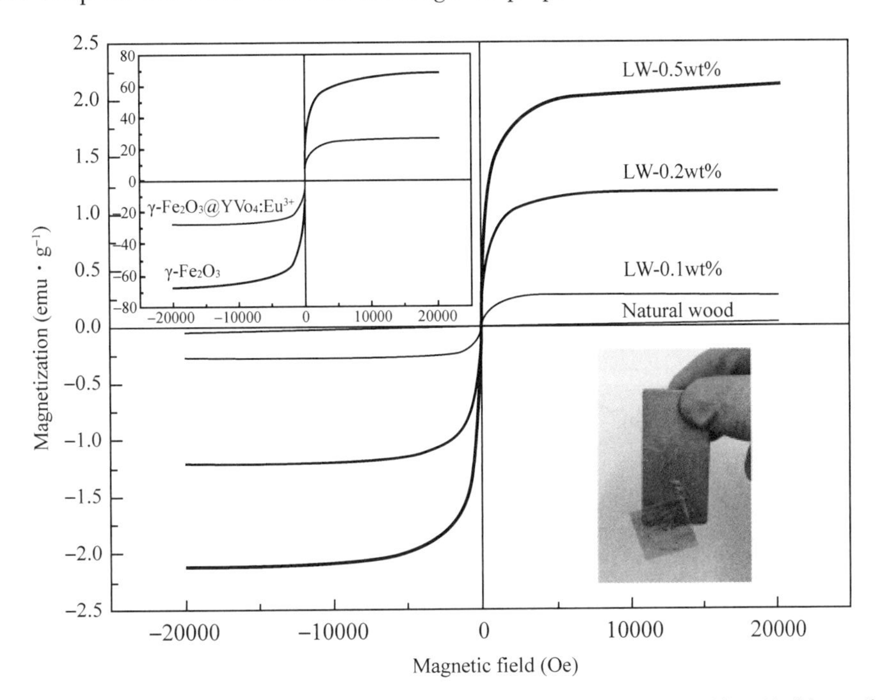

Fig. 9 Magnetization curves of luminescent wood at room temperature. (inset) Magnetization curves of as-prepared γ-Fe_2O_3 NPs and γ-Fe_2O_3 @ YVO_4: Eu^{3+} NPs. The digital photograph shows that the luminescent wood with a γ-Fe_2O_3@YVO_4: Eu^{3+} NP concentration of 0. 2 wt% is easily lifted with a magnet

The best fit to eq 2 is obtained using a nonlinear fitting of magnetization (M) and magnetic field (H) by Polymer software. The α parameter of for the γ-Fe_2O_3 NPs, γ-Fe_2O_3@ YVO_4: Eu^{3+} NPs, and LW with the γ-Fe_2O_3@ YVO_4: Eu^{3+} NP concentrations of 0. 1, 0. 2, and 0. 5 wt%, determined from the fit, are 4.75×10^{-3}, 4.57×10^{-3}, 4.50×10^{-3}, 4.49×10^{-3}, and 4.69×10^{-3} T^{-1}, respectively. The fitting correlation coefficient R^2 of all samples are 0. 9967, 0. 9970, 0. 9988, 0. 9986, and 0. 9981, respectively. According to the eq 3, the magnetic moment (m) was calculated from α, and the values of the γ-Fe_2O_3 NPs, γ-Fe_2O_3 @ YVO_4: Eu^{3+} NPs, and LW with the γ-Fe_2O_3@ YVO_4: Eu^{3+} NP concentrations of 0. 1, 0. 2, and 0. 5 wt% are 2. 04, 1. 96, 1. 94, 1. 93, and 2. 02μ_B, respectively. The calculated magnetic moment (m) does not change much between the five samples, suggesting the fabrication process has little effect on the magnetic moment of γ-Fe_2O_3 NPs.

For practical applications, it is important to evaluate the mechanical properties and dimensional stability of LW. The as-prepared wood samples under the stretching process in the longitudinal direction are shown in Fig. 10a. Natural wood possesses the fracture strength and modulus of 38. 32 MPa and 5. 04 GPa, respectively, because of the hierarchical structure of wood and the strong interactions among wood components such as cellulose and lignin. [47] Compared with the natural wood, the delignified wood shows reduced mechanical properties with a fracture strength and modulus down to 17.78 and 0.66 GPa, respectively. The low strength and modulus of the delignified wood are caused by the lack of strong load transfer mechanisms between the cellulose nanofibers after lignin removal.[18] However, LW shows improved mechanical properties with a fracture strength and modulus up to 45.92 MPa and 2.66 GPa, respectively, which can be explained enhanced cellulose nanofibers interaction after PMMA infiltration. The results of the dimensional stability studies of LW are presented in Fig. 10b. The gain in volume during storage in water of LW is reduced significantly compared with the natural wood. The maximum volume gain of natural wood reached 40.4% after 60 days of soaking, whereas LW displays a 15.8% volume gain, indicating that the better dimensional stability of LW was obtained compared with the natural wood. We attribute this to the blocking effect of the filling PMMA, which are impregnating the cell lumen as presented in Fig. 3f, thus resulting in reduced water absorption of LW.

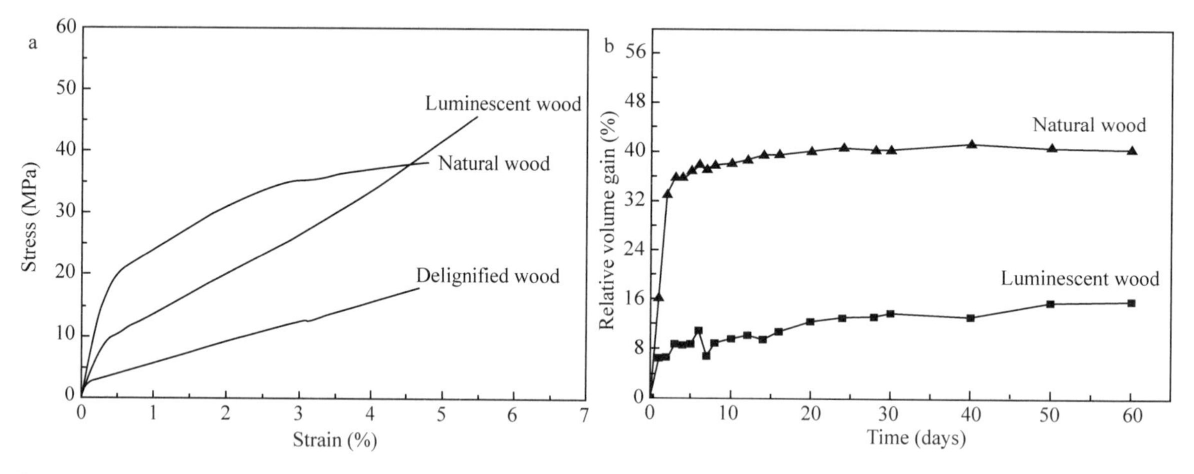

Fig. 10 (a) Experimental stress-strain curves of natural wood, delignified wood, and luminescent wood with the a γ-Fe_2O_3@YVO_4: Eu^{3+} NP concentration of 0.5 wt%. (b) Volume changes of natural wood and luminescent wood with a γ-Fe_2O_3@YVO_4: Eu^{3+} NP concentration of 0.5 wt% during the water immersion test

Fig. 11 shows photographs of the wood samples. In Fig. 11a, the wood slice with yellow color is optically opaque. After lignin removal, the wood slice is still opaque, but the color of wood is white (Fig. 11b). Until the polymer infiltration, the wood slice is optically clear (Fig. 11c). With the addition of γ-Fe_2O_3@ YVO_4: Eu^{3+} NPs, a highly transparent and brown wood slice was obtained (Fig. 11d). In Fig. 11e, it can be obviously seen that LW emit red luminescence, under UV excitation at 254 nm.

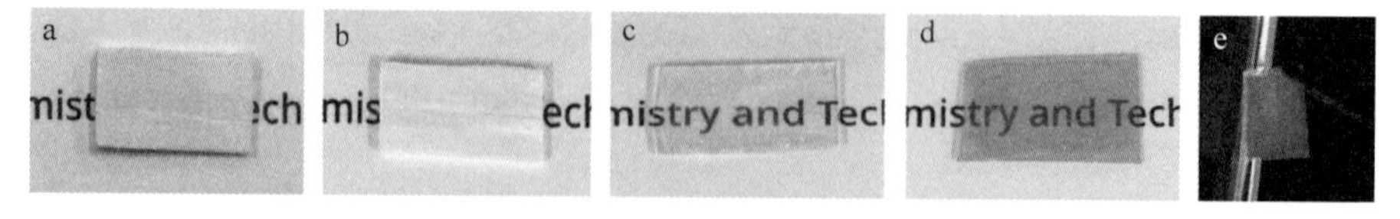

Fig. 11 Photographs of (a) natural wood, (b) delignified wood, (c) transparent wood, (d) luminescent wood with a γ-Fe_2O_3 @ YVO_4: Eu^{3+} NP concentration of 0.1 wt%, and (e) luminescent wood with the γ-Fe_2O_3@YVO_4: Eu^{3+} NP concentration of 0.1 wt% under UV light excitation at 254 nm

4 Conclusions

In summary, we have demonstrated a facial method for incorporating luminescent γ-Fe_2O_3@ YVO_4: Eu^{3+} NPs and index matching poly (methyl methacrylate) (PMMA) into a delignified wood template. The resultant luminescent wood glows a bright red color under a single wavelength excitation. Moreover, high optical transmittance of 80. 6 % and saturation magnetization of 0. 26 emu · g^{-1} were achieved at a luminescent wood with 0. 1 wt% γ-Fe_2O_3@ YVO_4: Eu^{3+} NP loading. Optical transmittance decreased as the γ-Fe_2O_3@ YVO_4: Eu^{3+} NP mass fraction increased, whereas saturation magnetization increased for the same changes. Furthermore, the luminescent wood possessed higher thermal properties, dimensional stability, and mechanical properties than the natural wood. With the advantages of high transparency, unique luminescence, moderate magnetism, good thermal properties, dimensional stability, and excellent mechanical properties, these luminescent wood samples exhibit great promise as green LED lighting equipment, luminescent magnetic switches, and anticounterfeiting facilities.

Associated content

Supporting Information

The Supporting Information is available free of charge on the ACS Publications website at DOI: 10. 1021/acssuschemeng. 6b029 Histograms of diameter of nanocomposites, EDS elemental analysis, ICP-MS measurements, photoluminescence spectra, and magnetization curve (PDF)

Acknowledgments

This work was financially supported by The National Natural Science Foundation of China (grant no. 31470584).

References

[1]Zhu H L, Luo W, Ciesielski P N, et al. Wood-derived materials for green electronics, biological devices, and energy applications[J]. Chem. Rev., 2016, 116(16): 9305-9374.

[2]Cabane E, Keplinger T, Merk V, et al. Renewable and functional wood materials by grafting polymerization within cell walls[J]. ChemSusChem, 2014, 7(4): 1020-1025.

[3]Li J, Lu Y, Yang D J, et al. Lignocellulose aerogel from wood-ionic liquid solution (1-allyl-3-methylimidazolium chloride) under freezing and thawing conditions[J]. Biomacromolecules, 2011, 12(5): 1860-1867.

[4]Ugolev B N. Wood as a natural smart material[J]. Wood Sci. Technol., 2014, 48(3): 553-568.

[5]Merk V, Chanana M, Keplinger T, et al. Hybrid wood materials with improved fire retardance by bio-inspired mineralisation on the nano- and submicron level[J]. Green Chem., 2015, 17(3): 1423-1428.

[6]Keplinger T, Cabane E, Chanana M, et al. A versatile strategy for grafting polymers to wood cell walls[J]. Acta Biomater., 2014, 11: 256-263.

[7]Persson P V, Hafren J, Fogden A, et al. Silica nanocasts of wood fibers: A study of cell-wall accessibility and structure[J]. Biomacromolecules, 2004, 5(3): 1097-1101.

[8]Ermeydan M A, Cabane E, Hass P, et al. Fully biodegradable modification of wood for improvement of dimensional stability and water absorption properties by poly(ε-caprolactone) grafting into the cell walls[J]. Green Chem., 2014, 16(6): 3313-3321.

[9]Sun Q F, Lu Y, Yang D J, et al. Preliminary observations of hydrothermal growth of nanomaterials on wood surfaces

[J]. Wood Sci. Technol., 2014, 48(1): 51-58.

[10]Gan W T, Gao L K, Sun Q F, et al. Multifunctional wood materials with magnetic, superhydrophobic and anti-ultraviolet properties[J]. Appl. Surf. Sci., 2015, 332: 565-572.

[11]Paris O, Fritz-Popovski G, Van Opdenbosch D, et al. Recent progress in the replication of hierarchical biological tissues[J]. Adv. Funct. Mater., 2013, 23(36): 4408-4422.

[12]Merk V, Chanana M, Gierlinger N, et al. Hybrid wood materials with magnetic anisotropy dictated by the hierarchical cell structure[J]. ACS Appl. Mater. Interfaces, 2014, 6(12): 9760-9767.

[13] Gan W T, Gao L K, Zhang W B, et al. Fabrication of microwave absorbing $CoFe_2O_4$ coatings with robust superhydrophobicity on natural wood surfaces[J]. Ceram. Int., 2016, 42(11): 13199-13206.

[14]Oka H, Fujita H. Experimental study on magnetic and heating characteristics of magnetic wood[J]. J. Appl. Phys., 1999, 85(8): 5732-5734.

[15]Oka H, Narita K, Seki K. Experimental results on indoor electromagnetic wave absorber using magnetic wood[J]. J. Appl. Phys., 2002, 91(10): 7008-7010.

[16]Wu Y Q, Jia S S, Qing Y, et al. A versatile and efficient method to fabricate durablesuperhydrophobic surfaces on wood, lignocellulosic fiber, glass, and metal substrates[J]. J. Mater. Chem. A, 2016, 4(37): 14111-14121.

[17]Zhu M W, Song J W, Li T, et al. Highly anisotropic, highly transparent wood composites[J]. Adv. Mater., 2016, 28(26): 5181-5187.

[18]Li Y Y, Fu Q L, Yu S, et al. Optically transparent wood from a nanoporous cellulosic template: Combining functional and structural performance[J]. Biomacromolecules, 2016, 17(4): 1358-1364.

[19] Zhu M W, Li T, Davis C S, et al. Transparent and haze wood composites for highly efficient broadband light management in solar cells[J]. Nano Energy, 2016, 26: 332-339.

[20] Li T, Zhu M W, Yang Z, et al. Wood composite as an energy efficient building material: guided sunlight transmittance and effective thermal insulation[J]. Adv. Energy Mater., 2016, 6(22): 1601122.

[21] Liu C H, Chen D P. Controlled synthesis of hexagon shaped lanthanide-doped LaF_3 nanoplates with multicolor upconversion fluorescence[J]. J. Mater. Chem., 2007, 17(37): 3875-3880.

[22]Pinto R J B, Carlos L D, Marques P A A P, et al. An overview of luminescent bio-based composites[J]. J. Appl. Polym. Sci., 2014, 131(22): 547-557.

[23]Wu X C, Tao Y R, Song C Y, et al. Morphological control and luminescent properties of YVO_4: Eu nanocrystals[J]. J. Phys. Chem. B, 2006, 110(32): 15791-15796.

[24] Rzepka A, Ryba-Romanowski W, Diduszko R, et al. Growth and characterization of Nd, Yb-yttrium oxide nanopowders obtained by sol-gel method[J]. Cryst. Res. Technol., 2007, 42(12): 1314-1319.

[25] Ninjbadgar T, Garnweitner G, Borger A, et al. Synthesis of luminescent ZrO_2: Eu^{3+} nanoparticles and their holographic sub-micrometer patterning in polymer composites[J]. Adv. Funct. Mater., 2009, 19(11): 1819-1825.

[26]Wiglusz R J, Grzyb T, Lis S, et al. Hydrothermal preparation and photoluminescent properties of $MgAl_2O_4$: Eu^{3+} spinel nanocrystals[J]. J. Lumin., 2010, 130(3): 434-441.

[27]Fink S. Transparent wood-a new approach in the functional study of wood structure[J]. Holzforschung, 1992, 46(5): 403-408.

[28]Liu D M, Tong L Z, Shi J H, et al. Luminescent and magnetic properties of YVO_4: Ln^{3+}@Fe_3O_4(Ln^{3+} = Eu^{3+} or Dy^{3+}) nanocomposites[J]. J. Alloys. Compd., 2012, 512(1): 361-365.

[29]Wan C C, Li J. Facile synthesis of well-dispersed superparamagnetic γ-Fe_2O_3 nanoparticles encapsulated in three-dimensional architectures of cellulose aerogels and their applications for cr(vi) removal from contaminated water[J]. ACS Sustainable Chem. Eng., 2015, 3(9): 2142-2152.

[30] Lu Y, Sun Q F, Yang D J, et al. Fabrication of mesoporous lignocellulose aerogels from wood via cyclic liquid

nitrogen freezing-thawing in ionic liquid solution[J]. J. Mater. Chem., 2012, 22(27): 13548-13557.

[31]Baskaran R, Selvasekarapandian S, Kuwata N, et al. Conductivity and thermal studies of blend polymer electrolytes based on PVAc-PMMA[J]. Solid State Ionics, 2006, 177(26-32): 2679-2682.

[32]Gierlinger N, Goswami L, Schmidt M, et al. In situ FT-IR microscopic study on enzymatic treatment of poplar wood cross-sections[J]. Biomacromolecules, 2008, 9(8): 2194-2201.

[33]Jiang S H, Gui Z, Bao C L, et al. Preparation of functionalized graphene by simultaneous reduction and surface modification and its polymethyl methacrylate composites through latex technology and melt blending[J]. Chem. Eng. J., 2013, 226: 326-335.

[34]Yu M, Lin J, Fang J. Silica spheres coated with YVO_4: Eu^{3+} layers via Sol-Gel process: a simple method to obtain spherical core-shell phosphors[J]. Chem. Mater., 2005, 17(7): 1783-1791.

[35]Koseoglu Y, Alan F, Tan M, et al. Low temperature hydrothermal synthesis and characterization of Mn doped cobalt ferrite nanoparticles[J]. Ceram. Int., 2012, 38(5): 3625-3634.

[36]Ferriol M, Gentilhomme A, Cochez M, et al. Thermal degradation of poly(methyl methacrylate) (PMMA): modelling of DTG and TG curves[J]. Polym. Degrad. Stab., 2003, 79(2): 271-281.

[37]Gan W T, Gao L K, Zhan X X, et al. Hydrothermal synthesis of magnetic wood composites and improved wood properties by precipitation with $CoFe_2O_4$/hydroxyapatite[J]. RSC Adv., 2015, 5(57): 45919-45927.

[38]Miyafuji H, Saka S. Na_2O-SiO_2 wood-inorganic composites prepared by the sol-gel process and their fire-resistant properties[J]. J. Wood Sci., 2001, 47(6): 483-489.

[39]Schlegel A, Alvarado S F, Wachter P. Optical properties of magnetite (Fe_3O_4)[J]. J. Phys. C: Solid State Phys., 1980, 12(6): 1157-1164.

[40]Xu W, Wang Y, Bai X, et al. Controllable synthesis and size-dependent luminescent properties of YVO_4: Eu^{3+} nanospheres and microspheres[J]. J. Phys. Chem. C., 2010, 114(33): 14018-14024.

[41]Zhu H L, Zuo D T. Highly enhanced photoluminescence from YVO_4: Eu^{3+}@ YPO_4 core/shell heteronanostructures [J]. J. Phys. Chem. C, 2009, 113(24): 10402-10406.

[42]Tong L Z, Liu D M, Shi J H, et al. Magnetic and luminescent properties of Fe_3O_4@ Y_2O_3: Eu^{3+} nanocomposites[J]. J. Mater. Sci., 2012, 47(1): 132-137.

[43]Bao A, Lai H, Yang Y M, et al. Luminescent properties of YVO_4: Eu/SiO_2 core-shell composite particles[J]. J. Nanopart. Res., 2009, 12(2): 635-643.

[44]Reisfeld R, Zigansky E, Gaft M. Europium probe for estimation of site symmetry in glass films, glasses and crystals [J]. Mol. Phys., 2004, 102(11-12): 1319-1330.

[45]Duee N, Ambard C, Pereira F, et al. New synthesis strategies for luminescent YVO_4: Eu and $EuVO_4$ nanoparticles with H_2O_2 selective sensing properties[J]. Chem. Mater., 2015, 27(15): 5198-5205.

[46] Gu H B, Huang Y D, Zhang X, et al. Magnetoresistive polyaniline-magnetite nanocomposites with negative dielectrical properties[J]. Polymer, 2012, 53(3): 801-809.

[47]Bekhta P, Niemz P. Effect of high temperature on the change in color, dimensional stability and mechanical properties of spruce wood[J]. Holzforschung, 2003, 57(5): 539-546.

中文题目：聚甲基丙烯酸甲酯和 γ-Fe_2O_3@ YVO_4：Eu^{3+}纳米粒子浸渍制备发光透明木质复合材料

作者：甘文涛，肖少良，高丽坤，高汝楠，李坚，詹先旭

摘要：利用折射率匹配的聚甲基丙烯酸甲酯(PMMA)和发光 γ-Fe_2O_3@ YVO_4：Eu^{3+}纳米粒子对天然木材进行功能化，形成一种新型的发光透明木材复合材料。首先，通过木质素去除过程从天然木材获得脱木质素的木材模板，其可用作透明聚合物和磷光体纳米颗粒的载体。然后，功能化发生在木材

的内腔中，这得益于填充细胞内腔并增强纤维素纳米纤维相互作用的聚甲基丙烯酸甲酯，导致木材复合材料具有优异的热性能、尺寸稳定性和机械性能。更重要的是，这种木材复合材料在350和800 nm之间的宽波长范围内显示出高透光率、磁响应性和在254 nm的紫外激发下的亮色光致发光。发光木质复合材料的独特性能和绿色特性在绿色发光二极管照明设备、发光磁开关和防伪设施等方面具有巨大的应用潜力。

关键词：发光木材；透明；磁性；γ-Fe_2O_3@YVO_4：Eu^{3+}

Magnetic Wood As An Effective Induction Heating Material: Magnetocaloric Effect and Thermal Insulation*

Wentao Gan, Likun Gao, Shaoliang Xiao, Runan Gao,
Wenbo Zhang, Jian Li, Xianxu Zhan

Abstract: An effective induction heating material composed of wood and magnetic Fe_3O_4 particles is prepared via a simple hydrothermal method. Fe_3O_4 particles deposit on the porous wood substrate provide excellent magnetic properties and high magnetothermal conversion resulting in rapid increase in the temperature at the wood substrate. Benefiting from the effective magnetocaloric effect of magnetic particles, the temperature of magnetic wood exhibits a noticeable rises (from 25.9 to 70.1 ℃ in 10 min) at low-frequency magnetic field. For simulating the real environment, the indoor temperature rise of magnetic wooden model also reveals the excellent heating performance of magnetic wood. Besides, the magnetic wood also serves as a thermal insulator to slow down the transport of thermal energy from the high temperature area to its surroundings due to its low thermal conductivity, which benefit the rational use of thermal energy in daily life. In view of highly scalable, simple operation, good durability, excellent magnetic properties, effective magnetocaloric effect, and thermal insulation, the novel magnetic wood composite demonstrated here is an attractive induction heating material for application in building, decoration, and massage furniture.

Keywords: Fe_3O_4 particles; Induction heating material; Magnetic properties; Wood

1 Introduction

The problems of resource depletion caused by the consumption of a large number of fossil fuels have attracted wide attentions from all over the world.[1-4] The conventional heating supply needs to consume large number of nonrenewable resources such as coal and natural gas. Therefore, the renewable and environment-friendly materials are becoming a new focus of research as an alternative to the nonrenewable materials for environment and energy applications. Wood is an oldest biomass material, which has been exploited and utilized by humans for thousands of years, because of its renewable characteristic, low weight-to-strength ratio, beautiful texture, high mechanical strength, and low thermal conductivity.[5-7] Based on the 3D porous architecture consists of cellulose, hemicellulose, and lignin of natural wood, many approaches on the modification and functionalization of wood have been presented to improve its quality and raise its added value.[8,9] For example, Li and co-workers proposed a new kind of wood-based nanocomposites with

* 本文摘自 Advanced Materials Interfaces, 2017, 4(22): 1700777.

satisfactory durability and high specific strength via a vacuum impregnation of organic inorganic polymer into wood porous structure.[10] Hu and co-workers fabricated a transparent wood composite with high optical transparency and broadband optical haze through a delignification and epoxy resin impregnating process, which can be used as a heat-shielding window instead of glass for energy efficient building material application.[11] The transparent wood can be used to increase the sunlight usage and reduce the heat losing due to the cooperation of guided sunlight transmittance and effective thermal insulation.[12,13] Moreover, Oka et al. demonstrated the magnetic wood by loading and coating wood fiber boards with magnetic powder.[14,15] Further study by Merk et al. showed that the incorporation of magnetic nanoparticles inside the wood is also an effective way to prepare magnetic wood composite.[16] The authors suggested that the magnetic wood with hierarchical structures is a kind of engineered material, but did not further examine the application field of magnetic wood in terms of magnetic properties.

Recently, considerable researches about magnetic nanoparticles have been concentrated on the magnetocaloric effect. The conversion of magnetic energy into thermal energy by magnetic nanoparticles has the potential for biomedical sciences such as cancer therapy, drug release, and magnetic hyperthermia.[17-19] The conversion efficiencies of magnetic nanoparticles are mainly depended on the size, composition, and concentration of magnetic nanoparticles, and the applied magnetic field and frequency.[20] Generally, under the high frequency of magnetic field, the high conversion efficiency of magnetic nanoparticles was obtained, but the high frequency of magnetic field (300 kHz to 1MHz) poses great harm to the human body due to the potential for nonspecific heating of human tissues via eddy currents at these frequencies.[21,22] Thus, it is of great significance to study the induction heating of magnetic materials at low-frequency magnetic field.

In this work, we report on the fabrication of magnetic Fe_3O_4/wood composites with the excellent magnetic properties for potential induction heating application at low-frequency magnetic field. Magnetic Fe_3O_4 particles were immersed into wood substrate via a vacuum impregnation and hydrothermal processes. The induction heating experiments showed that the magnetic wood with the Fe_3O_4 particles concentration of 25 wt% had a 44 ℃ increase in temperature at an excitation frequency of 35 kHz, with a magnetic field of 0.06 mT. More importantly, the magnetic wood possesses excellent thermal insulation with a thermal conductivity of 0.17 $W \cdot (m \cdot K)^{-1}$ along the radial direction of wood. In addition, the induction heating performance of a small model house was characterized for simulating the real environment. The results showed that the magnetic wood was a potential candidate for application as green and new induction heating material in building, decoration, and massage furniture.

2 Results and discussion

2.1 Synthesis and Characterization of magnetic wood

The synthesis process of magnetic wood by a hydrothermal process is investigated herein, as shown in Figure1a. The wood block obtained from tree trunk comprised of many channels, which are perpendicular to the plane direction of wood. In order to achieve the magnetic wood with good magnetic properties, it is highly expected that Fe_3O_4 nanoparticles have immersed into the cell lumen of wood through the vertical channels from the wood surface. However, magnetic nanoparticles with low stability are usually deposited in aqueous solution due to the Waals attraction and the gravity of particles,[23] which made it difficult for Fe_3O_4 nanoparticles to immerse into wood. Thus, in this experiment, the precursor solution was prepared by the mixture of Fe_3O_4 nanoparticles and sodium oleate. The sodium oleate is used as surfactant to improve the dispersion of Fe_3O_4 in aqueous solution (Fig. S1, Supporting Information), so that Fe_3O_4 nanoparticles could immerse into the wood substrate via the impregnation and hydrothermal process. Subsequently, wood samples were put in a gently sealed vessel with the precursor solution. The resultant solution in the sealed glass bottle gradually varied into vapor along with the heated process at 130 ℃, and the pressure in the sealed space was significantly increased, which promoted further penetration of Fe_3O_4 nanoparticles. Furthermore, the sealed condition slowed the evaporation of water, and provided Fe_3O_4 nanoparticles with enough time to precipitate on the wood surface and inside the wood substrate. Finally, the solution penetration on the surface and inner of the wood ended until the complete vaporization of the solution, and Fe_3O_4 particles were deposited on the wood substrate successfully.

As shown in Fig. 1b, the magnetic wooden model with black color can be prepared via the hydrothermal process, where the yellow colored wood is coated by black Fe_3O_4 particles, and the original wood structure is well preserved. Although the color of wood has been changed after hydrothermal treatment, we can beautify the surface color of wood products through some mature technologies, such as painting and veneering processes. Benefiting from the magnetic particles deposited on the wood substrate, the magnetic energy could convert into the thermal energy via magnetic wood when we placed the magnetic wooden products in an alternating magnetic field (Fig. 1c).

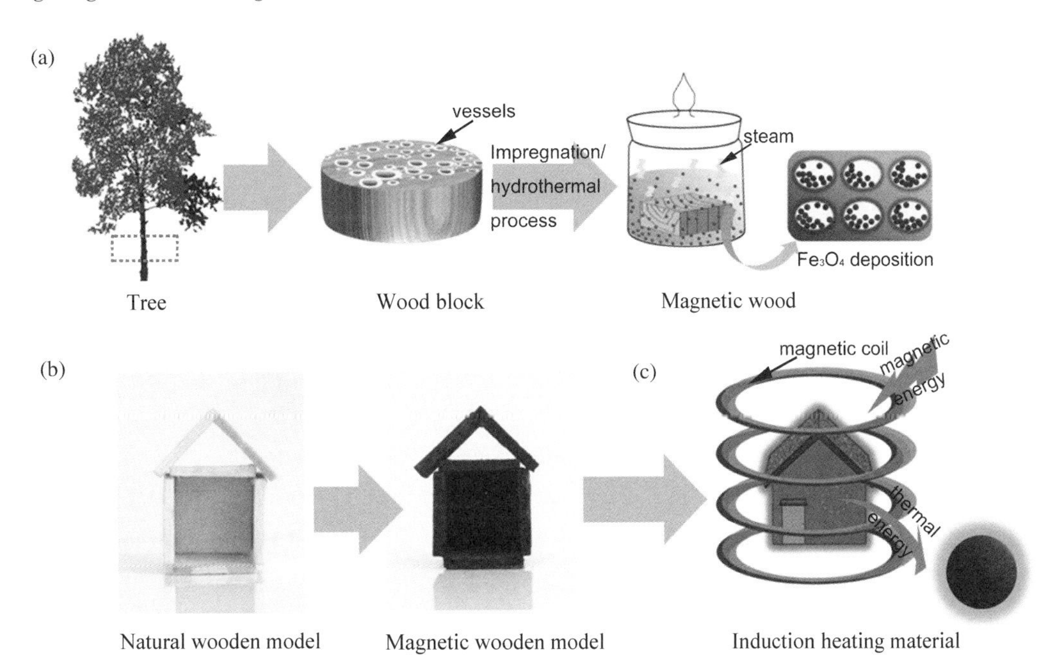

Fig. 1 (a) Synthetic route of magnetic wood. (b) Photographs of model house before and after hydrothermal treatment, and (c) Schematic of the magnetocaloric effect of magnetic wooden model

To confirm the particles size and morphology of Fe_3O_4 nanoparticles, teansmission electron micoscopy (TEM) measurement was applied. As shown in Fig. S2a (Supporting Information), it is clear that these sphere-like nanoparticles are well dispersed. The histogram of diameter of Fe_3O_4 nanoparticles (Fig. S2a,

Supporting Information, inset) reveals the diameter of the nanoparticles is centered at 10 nm. The high-resolution TEM (HRTTEM) image shown in Fig. S2b (Supporting Information) exhibits that the lattice spacing between two planes is 0.29 nm, corresponding to the characteristic of (220) spinel planes. Furthermore, the corresponding selected area electron diffraction (SAED) pattern shows spots/rings that are consistent with the magnetite (Fe_3O_4) structure (Fig. S2b, Supporting Information, inset).[24]

A series of scanning electron microscopy (SEM) pictures of cross section and radial section of wood sample are presented in Fig. 2. Before treatment, the innate porous architecture of natural wood is well viewed from the cross section (Fig. 2a). In Fig. 2b, it can be clearly observed that a layer of magnetic particles is deposited on the cell wall of natural wood after treatment. However, a small number of magnetic particles are permeated into the cell lumen, and the original pores of wood are well maintained (Fig. 2b, marked oval region). Fig. 2c displays the SEM image of MW-2 and it can be seen that the deposition of Fe_3O_4 particles on wood substrate is increasingly favored at high Fe_3O_4 concentration. Although a large number of Fe_3O_4 particles are permeated into the cell lumen, some original pores of wood are still maintained (Fig. 2c, marked oval region). Further increasing Fe_3O_4 concentration, Fe_3O_4 particles block the pores of wood and completely cover the wood surface (Fig. 2d). Energy-dispersive X-ray analysis (EDXA) spectrum presented in Fig. S3 (Supporting Information) shows the major chemical composition of magnetic wood: carbon, oxygen, sodium, iron, and gold (from the coating for SEM observation). Moreover, the mappings of Fe in the cross section of magnetic wood also reveal that Fe_3O_4 particles are dispersed throughout the whole wood surface (Fig. 2, inset). Fig. 2e-h shows the radial-sectional SEM images of the wood samples before and after treatment. As shown in Fig. 2e, some pits are found on the wall of tracheid, which is the typical morphology of poplar wood. After treatment, some sparse Fe_3O_4 particles are particularly visible in the vessels of wood (Fig. 2f). A dense layer of Fe_3O_4 particles is deposited on the wood surface with the increase of Fe_3O_4 concentration (Fig. 2g). Further increasing the mass of Fe_3O_4 to 6 g leads to the uniform distribution of magnetic particles on the wood substrate, and completely covers the original surface structure of wood. Moreover, Table 1 shows that the Fe_3O_4 particles loading rate of MW-1, MW-2, and MW-3 is 9.4, 17.2, and 25.0 %, respectively, confirming that the deposition of Fe_3O_4 particles on the wood substrate is increased with the increasing concentration of Fe_3O_4.

X-ray diffraction (XRD) results were used to determine the crystal structure of the synthesized samples. As shown in Fig. 3a, the characteristic peaks of natural wood can be observed at around 15° and 22.6°, corresponding to (101) and (002) planes of cellulose.[25,26] The peaks of Fe_3O_4 at around 30.1°, 35.4°, 43.1°, 53.4°, 56.9°, and 62.5° are in good agreement with the JCPDS card of magnetite (No. 19-0629). After treatment, all diffraction peaks belonging to Fe_3O_4 and wood are present in magnetic wood, indicating the magnetic Fe_3O_4 particles are deposited on the wood substrate successfully.

Fig. 3b shows the Fourier transform infrared (FTIR) spectra of natural wood and magnetic wood. In the natural wood spectrum, bands at 3414 and 2908 cm^{-1}, attributing to O—H stretching vibrations and C—H stretching vibrations, respectively.[27] The bands at 1736 and 1245 cm^{-1} correspond to acetyl groups in hemicellulose, and the aromatic stretching vibrations of lignin at 1599 and 1507 cm^{-1}.[28] The peaks around 1425 and 1376 cm^{-1} are due to the C—H deformation vibrations expected from cellulose.[29] The peak at 1057 cm^{-1} is originated from the C—O stretching vibrations.[29] Apart from the characteristic peaks derived

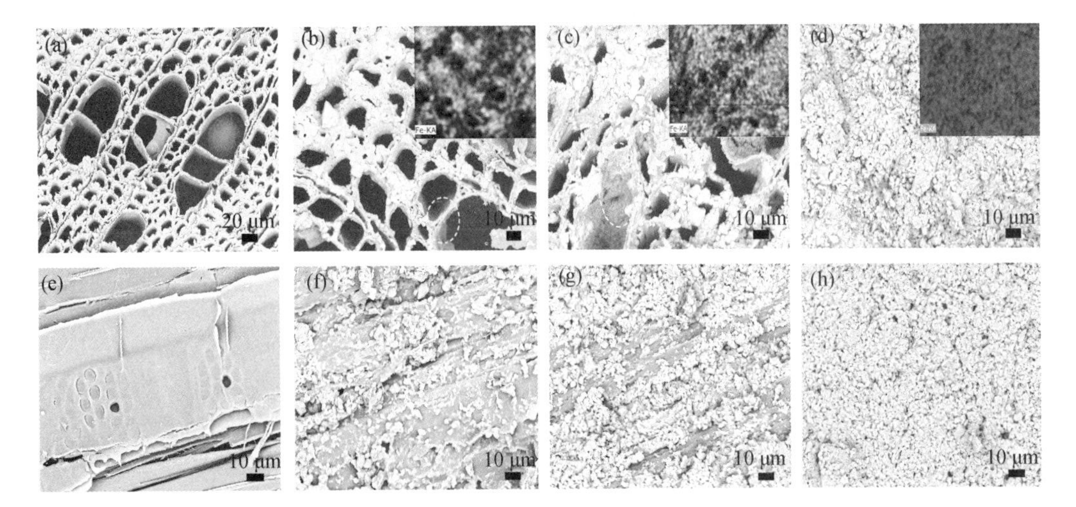

Fig. 2 (a)-(d) Cross-sectional SEM images of natural wood, MW-1, MW-2, and MW-3, respectively. Inset: Fe element mapping of the cross section of magnetic wood. (e)-(h) Radial sectional SEM images of natural wood, MW-1, MW-2, and MW-3, respectively

from wood, the FTIR spectrum of magnetic wood presents two peaks located at 2922 and 2852 cm^{-1} relating to C—H stretching vibrations in the —CH_2 and —CH_3 groups of sodium oleate.[30] The new peak at 586 cm^{-1} relates to stretching vibration of Fe-O,[31] indicating that the Fe_3O_4 particles have deposited on the wood substrate.

Table 1 The weight changes of wood before and after treatment

Sample	W_n(g)	W_m(g)	R_m(%)
MW-1	1.6576	1.8300	9.4
MW-2	1.6018	1.9350	17.2
MW-3	1.6236	2.1656	25.0

The introduction of magnetic Fe_3O_4 particles into wood substrate is also confirmed by the results of thermogravimetric analysis (TGA), as shown in Fig. 3c. Two-stage weight losses are observed in the natural wood. The first stage at around 100 ℃ is due to the elimination of moisture. The major weight loss of the natural wood in the range of 250-390 ℃ is caused by the pyrolysis of wood components (cellulose, hemicellulose and lignin). The terminal degradation temperature of the natural wood is 800 ℃, with 7.20 wt% residue remaining, suggesting the wood has been converted to charcoal under the experiment conditions. However, TG curves of the magnetic wood with different particles loadings are different from that of the natural wood. Two obvious weight losses in the temperature range of 220-340 ℃ and 340-450 ℃ are caused by the pyrolysis of wood components and sodium oleate,[32,33] respectively. More importantly, the additional decomposition step started at the temperature around 650 ℃ corresponding to the reduction reaction between the Fe_3O_4 and the char obtained from the pyrolysis of wood.[34] The reaction occurring between Fe_3O_4 and the residual char leads to the oxidation of char to carbon dioxide and results in the decrease of residues. Moreover, the weight of residues in the magnetic wood with different Fe_3O_4 particles loading is 21.67, 30.34, and 33.05 wt%, respectively,

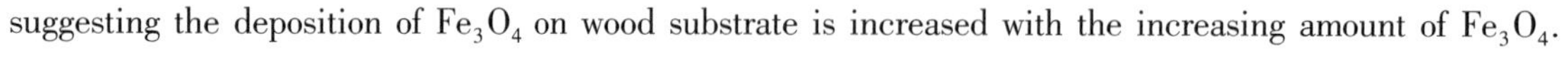
suggesting the deposition of Fe_3O_4 on wood substrate is increased with the increasing amount of Fe_3O_4.

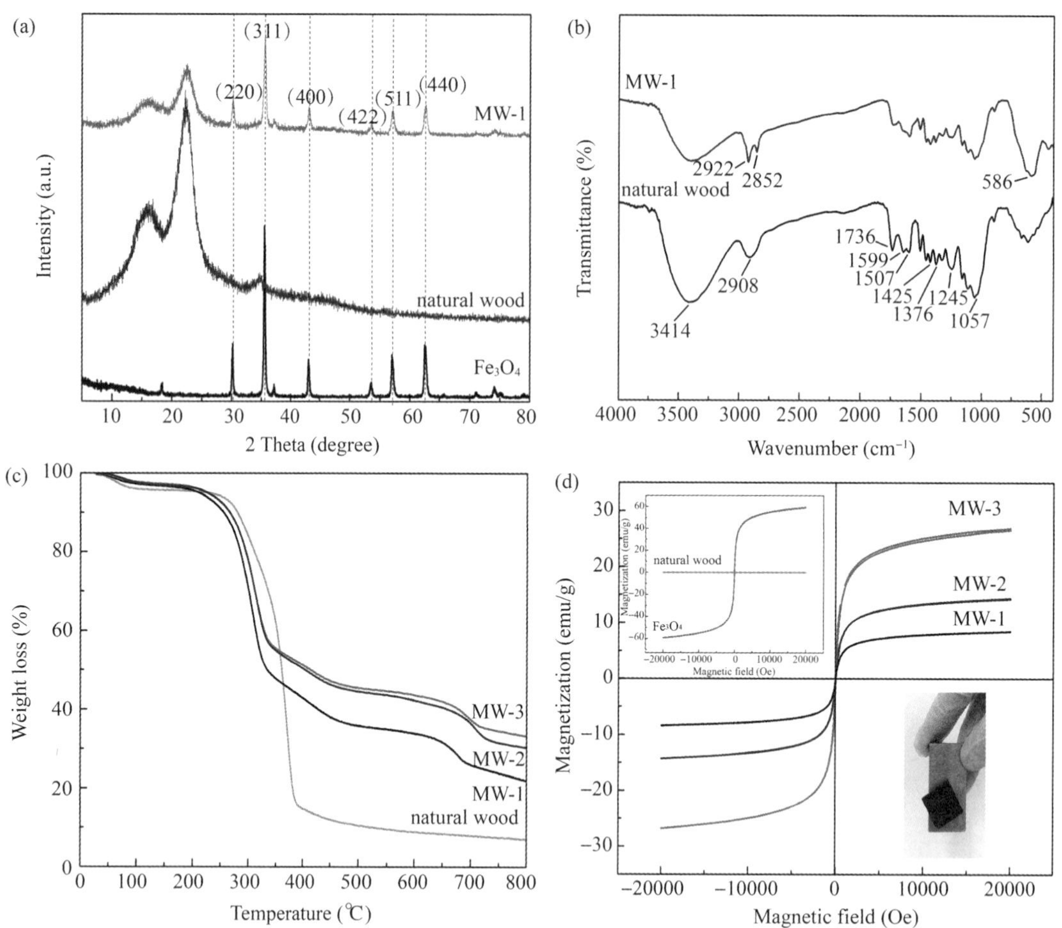

Fig. 3 (a) XRD patterns of Fe_3O_4 nanoparticles, natural wood, and MW-1, respectively. (b) FTIR spectra of natural wood and MW-1, respectively. (c) TG curves of natural wood, MW-1, MW-2, and MW-3, respectively. (d) Magnetization curves of magnetic wood at room temperature, inset shows the magnetization curves of Fe_3O_4 and natural wood, respectively. The digital image exhibits that MW-1 is easily lifted with a bar magnet

The noticeable advantage of magnetic wood is magnetic properties. Fig. 3d shows the magnetization curves of magnetic wood with different concentration of Fe_3O_4 at room temperature. The inset shows the magnetization curves of natural wood and Fe_3O_4 nanoparticles. The saturation magnetization (M_S) of Fe_3O_4 nanoparticles is 58.8 emu · g^{-1}, and natural wood is observed to be a typical nonmagnetic material as expected. Interestingly, all of the magnetic wood samples show a typical superparamagnetic behavior, because the coercivity and remanence of magnetic wood are vanishingly low. The M_S values of Fe_3O_4 particles containing magnetic wood of 9.4, 17.2, and 25.0 wt% are 8.4, 14.3, and 26.5 emu · g^{-1}, respectively. The results reveal that a higher concentration of magnetic Fe_3O_4 particles in wood will obtain better magnetic properties. The saturation magnetization of magnetic wood is lower than that of the pure Fe_3O_4 nanoparticles, which could be explained by the presence of nonmagnetic wood substrate. Nevertheless, a comparison of the saturation magnetization of this product with those of other magnetic composites reported in literature is given in Table 2. It is obvious that the saturation magnetization of MW-3 is higher than the others at room temperature, which is important to magnetocaloric effect of magnetic wood.

Table 2 Saturation magnetization comparison of various magnetic wood

Magnetic wood	Saturation magnetization [emu · g^{-1}]	Reference
Fe_3O_4/wood	26.5	This study
Fe_3O_4/wood	13.0	Dong et al.[44]
FA-treated magnetic wood	10.5	Dong et al.[44]
γ-Fe_2O_3/bamboo	6.1	Jin et al.[45]
Magnetic cellulose sponge	12.8	Peng et al.[46]
Fe_2O_3/C foams	17.2	Chen and Pan[47]
Carbon aerogel	19.9	Yu et al.[48]
Transparent magnetic nanopaper	14.1	Li et al.[49]
γFe_2O_3@YVO_4: Eu^{3+}/wood	3.3	Rana et al.[28]

In general, the heating of magnetic nanoparticles is accomplished through rotating the magnetic moment of each nanoparticles against the energy barrier,[41] so it is necessary to know the change of magnetic moment of magnetic materials before and after treatment. The magnetic properties of a superparamagnetic material can be determined on the basis of the Langevin equation:[42]

$$M/M_S = \coth x - 1/x \tag{1}$$

Where M and M_S are the magnetization (emu · g^{-1}) in magnetic field H (Oe) and saturation magnetization, respectively. $x = \alpha H$, where α is a function of the electron spin magnetic moment m of each molecules (μ_B, μ_B is the Bohr magneton)

$$\alpha = m/kT \tag{2}$$

Where k and T are the Boltzmann constant and absolute temperature, respectively.

To best fit the Equation (1), the nonlinear fitting of magnetization (M) and magnetic field (H) by Polymath software was applied. The values of M_S for pure Fe_3O_4 nanoparticles and magnetic wood with Fe_3O_4 loading of 9.4, 17.2, and 25.0 wt%, determined from the fit, are 53.4, 7.7, 13.1, and 24.3 emu · g^{-1}, respectively. More importantly, the parameter α for the pure Fe_3O_4 nanoparticles and magnetic wood with the Fe_3O_4 particle concentration of 9.4, 17.2, and 25.0 wt% are 3.88×10^{-3}, 3.21×10^{-3}, 3.32×10^{-3}, and 3.62×10^{-3}, respectively. The fitting correlation coefficient R^2 of all the samples are 0.9951, 0.9944, 0.9959, and 0.9908, respectively. Thus, the magnetic moment m was calculated according to the Equation (2), and the value of pure Fe_3O_4 nanoparticles and magnetic wood with the Fe_3O_4 concentration of 9.4, 17.2, and 25.0 wt% are 1.68, 1.40, 1.45, and 1.57 μ_B, respectively. The magnetic moment does not change much between pure Fe_3O_4 nanoparticles and magnetic wood samples, indicating the hydrothermal treatment and the wood substrate have little effect on the magnetic moment of Fe_3O_4 particles.

2.2 Magnetocaloric effect of magnetic wood

We then evaluated the magnetocaloric effect of magnetic wood via measuring the temperature increase in response to an induction heating setup. The current of copper coil was 5 A, and the frequency of the alternating magnetic field was fixed at 35 kHz, which was much lower than that of most of previous studies.[17-22,37]

The temperature of the magnetic wood was monitored with an IR camera, as shown in Fig. 4a. After 10 min of excitation, it can be seen that the color of IR image in natural wood is cool blue, and the temperature of natural wood was 25 ℃, close to room temperature, revealing the magnetic energy would not convert into thermal energy in natural wood at the alternating magnetic field with low frequency. However, the color of IR images in magnetic wood are hot red, the temperature of MW-1, MW-2, and MW-3 reaches 46. 3 ℃ (22. 3 ℃ increase), 60. 9 ℃ (36. 1 ℃ increase) and 70. 1 ℃ (44. 2 ℃ increase), respectively, indicating the magnetic energy is effectively converted into thermal energy in magnetic wood.

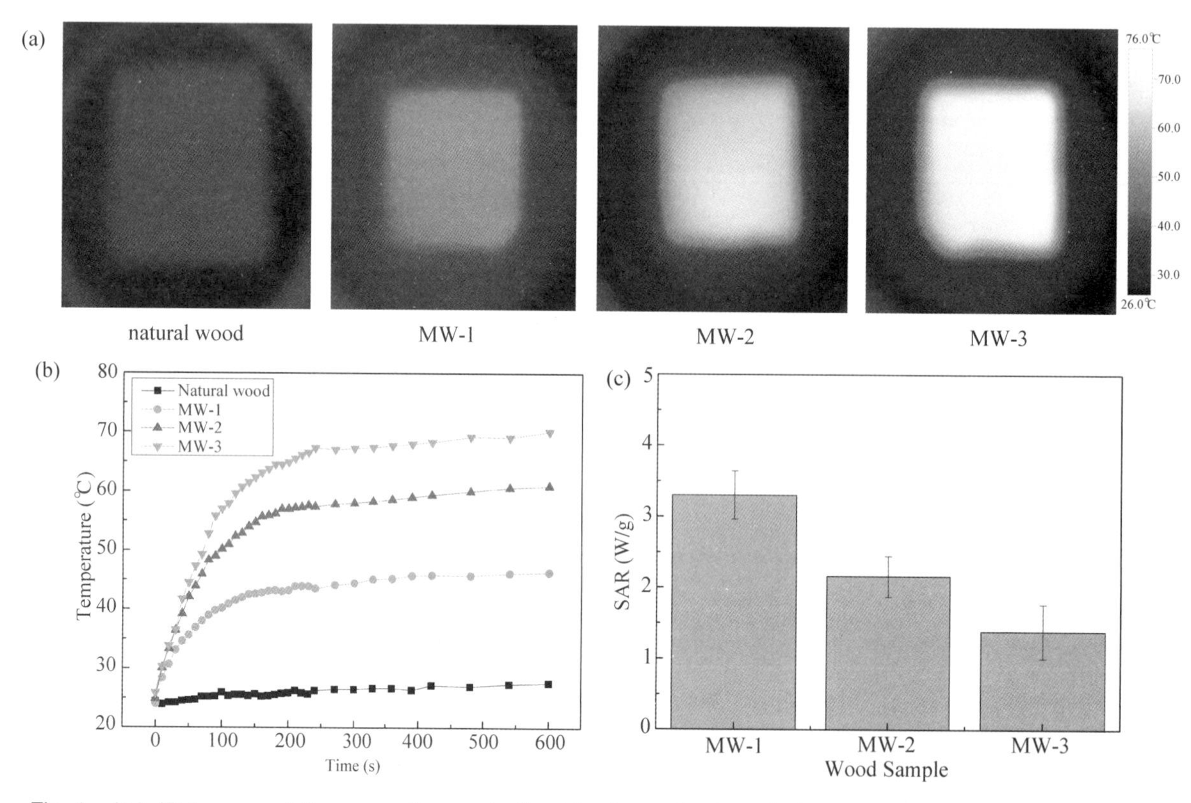

Fig. 4 (a) IR images of the magnetic wood after 10 min application of alternating magnetic field (35 kHz, 0. 06 mT). (b) Time-dependent temperature curves of natural wood and magnetic wood samples under alternating magnetic field with the frequency of 35 kHz. (c) SAR values of the magnetic wood

In order to study the change regulation of temperature in magnetic wood under alternating magnetic field at low frequency, Fig. 4b shows the experimental hyperthermia curves for wood samples. As for natural wood, the heating profile is almost a straight line close to 25 ℃, revealing the natural wood has no magnetocaloric property at the low-frequency magnetic field. However, the temperature of magnetic wood is increased rapidly, and then reached to the stable state as heating time continued. As expected, the heating curves of magnetic wood are found to be affected by the concentration of magnetic Fe_3O_4 particles that were deposited on the wood substrate. Comparing all self-heating curves of the magnetic wood, we find the temperature of magnetic wood increases with the increase of the Fe_3O_4 concentration, and the MW-3 shows the highest rise in temperature. Besides, the rank of initial calefactive velocity is MW-3>MW-2>MW-1. On the basis of the initial calefactive velocity of the magnetic wood, the theoretical specific adsorption rate (SAR) values of magnetic wood are calculated, as shown in Fig. 4c. The SAR value of MW-1, MW-2, and MW-3 are 3.30, 2.16, and 1.38 $W \cdot g^{-1}$, respectively. Compared with the SAR values of magnetic nanoparticles in aqueous solution,[39,44] the low SAR values of magnetic wood are obtained, which can be explained by the effect of aggregation of Fe_3O_4 nanoparticles, as shown in Fig. 2. As described by Néel relaxation, for the nanoparticle and magnetic field such that $\omega\tau<1$, the decreasing relaxation time will decrease the SAR value of magnetic nanoparticles.[35,40] In this study, the diameter of Fe_3O_4 is 10 nm and the frequency of magnetic field is 35 kHz, so $\omega\tau<1$. However, the diameter of Fe_3O_4 that deposited on the wood substrate is mainly centered at 5 μm (Fig. S4, Supporting Information), which is much larger than the results of TEM measurement, revealing the aggregation of Fe_3O_4 nanoparticles. The aggregation of Fe_3O_4 nanoparticles that deposited on the wood substrate reduced the interparticle distance and increased the interparticle interaction, thus the Néel time is decreased, which lead to the decrease of SAR values of magnetic wood. It is worth noting that SAR value of MW-1 is higher than those of other magnetic wood, indicating the highest heating efficiency of MW-1.

For most of hybrid organic/inorganic composites, it is important to assess their durability. In this study, the ultrasonic treatments and sandpaper abrasion tests were carried out to examine the durability of magnetic wood. As shown in Fig. S5 (Supporting Information), the weight change of wood samples after ultrasonic treatment and abrasion test is 0.1094 and 0.1082 g, respectively. As for the magnetocaloric effect, compared with the hyperthermia curves of MW-1 and MW-2 without treatments, the temperature of MW-1 after ultrasonic treatment and MW-2 after abrasion test was slightly decreased, but the highest temperature of MW-1 after ultrasonic treatment and MW-2 after abrasion test still reached 45.8 and 59.5 ℃, respectively, indicating the good durability of magnetic wood.

2.3 Thermal insulation of magnetic wood

Besides the requirement for generating heat and durability, another noticeable property of magnetic wood material is the outstanding thermal insulation. As can been seen in Fig. 5a, the heat flow is usually moved along the radial and longitudinal direction of wood vessel, but the radial heat travelling pathway produces an large phonon scattering effect, and the wood components (mainly cellulose and hemicellulose) also provide the high phonon resistance when the heat flow is transferred along the radial and longitudinal direction of wood. Beyond that, the magnetic wood is also a thermal insulator with air pockets in the original structure due to the poorest conductor of air compared to that of the solid and liquid.[13] It can be expected that the magnetic wood provides a structural barrier between wooden products and the outdoor environment, and the transport of magnetothermal energy through the magnetic wooden shell will be slowed down. Thus, we can monitor the temperature changes of magnetic wood in time, and make rational use of thermal energy in our daily life. As shown in Fig. 5b, the thermal conductivities of natural wood, MW-1, MW-2, and MW-3 at radial direction are 0.1457, 0.1684, 0.1727, and 0.1735 $W \cdot mK^{-1}$, respectively. Interestingly, the thermal conductivity of wood is increased slightly after Fe_3O_4 particles deposited on the wood substrate. As reported by Tabkeda,[47] the pure Fe_3O_4 shows a high thermal conductivity of around 5 $W \cdot m \cdot K^{-1}$, which is much higher than that of the magnetic wood. The resulting lower thermal conductivity of magnetic wood is likely caused by the alignment of wood cells and the chemical components of wood cell walls (cellulose and hemicellulose), which provided a high phonon resistance across the wood vessels and the multiple interface phonon scattering effect. Moreover, we find that the cement also has a much higher thermal conductivity of 1.2 W · m · K^{-1},[48] proving the magnetic wood more effective in reducing heat conduction.

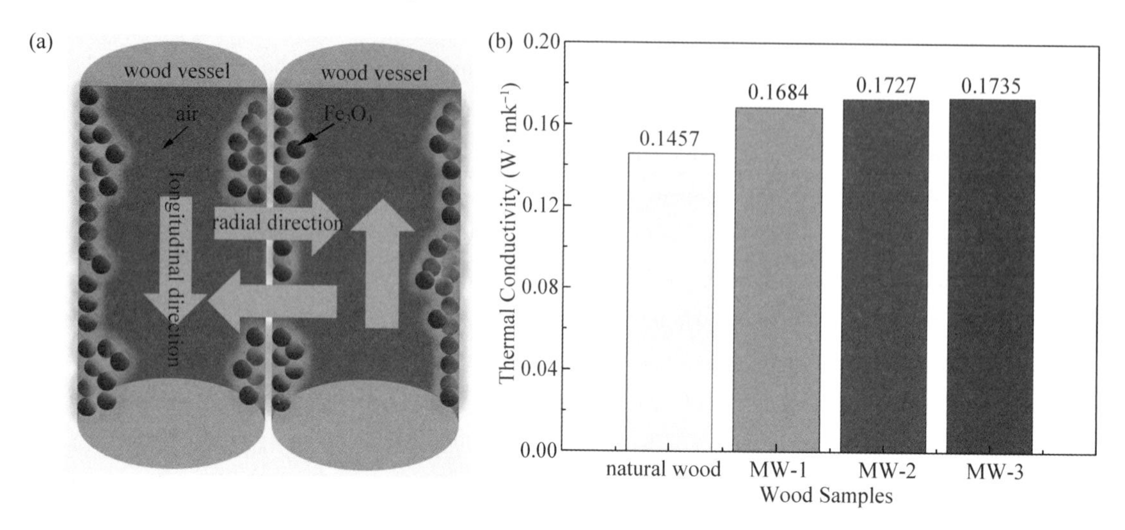

Fig. 5 (a) An illustration of the radial and longitudinal heat transport in magnetic wood. (b) The measured thermal conductivities of natural wood and magnetic wood at radial direction

2.4 Simulation test

For simulating the real environment, the studies of the thermoregulation effect of natural wood and magnetic wood (MW-1) under alternating magnetic field with the frequency of 35 kHz were investigated with the model house (Fig. 6). After continuous induction heating for 5 min, the temperature of magnetic wooden shell rapidly increases from room temperature (24.9 ℃) to around 43.6 ℃, whereas the temperature of natural wood remained at room temperature. The large temperature rise of magnetic wooden shell (ΔT = 18.7 ℃) compared to the relatively constant temperature of natural wooden shell upon induction heating demonstrates the effective conversion between magnetic energy and thermal energy of magnetic wood under low-frequency magnetic field. Not only that, we also examined the indoor temperature change of magnetic wooden model in the alternating magnetic field. As shown in Fig. 6c, it can be observed that the indoor temperature increases with the increase of exciting time, and the indoor temperature of magnetic wooden model is rising from room temperature (24.9 ℃) to 33.4 ℃ in 5 min, indicating the excellent heating performance of magnetic wood. According to IEEE Standards (IEEE C95.1 "Standard for Safety Levels with Respect to Human Exposure to Radio Frequency Electromagnetic Fields, 3 kHz to 300 GHz."),[43] the alternating magnetic field with a frequency of 35 kHz used in this study did not exceed the external field intensity limits for public, thus the magnetic wood could be used as an attractive induction heating material.

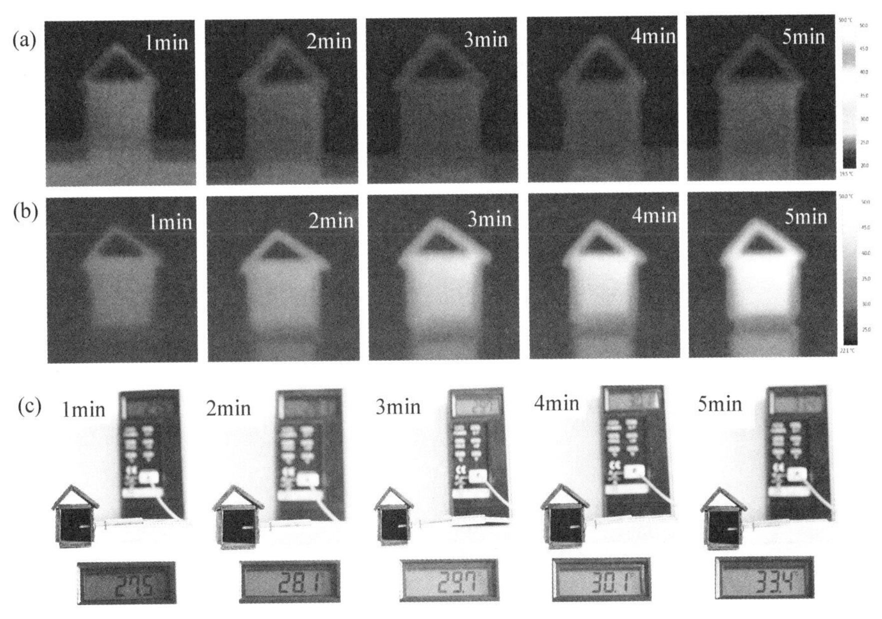

Fig. 6 IR images of (a) natural wooden model and (b) magnetic wooden model in order to show the temperature change after induction heating process. (c) Digital images of magnetic wooden model in order to exhibit the indoor temperature change after continuous induction heating

3 Conclusions

In summary, we have demonstrated a scalable and facile method for depositing magnetic Fe_3O_4 particles on the wood substrate, for magnetocaloric effect application. Owing to its natural vessel structure, excellent magnetic properties, effective magnetocaloric effect, and thermal insulation, wood form an excellent candidate as induction heating material. The excellent magnetic properties of magnetic wood result in a larger temperature rise under the frequency of magnetic field of 35 kHz, and the temperature of magnetic wood increasing with the increase of the concentration of Fe_3O_4 particles. However, the high Fe_3O_4 concentration leads to the aggregation of magnetic particles that deposited on the wood substrate, thus the heating efficiency of magnetic wood was found to be decreased with the increase of Fe_3O_4 concentration. Besides, simulation tests reveal the excellent heating performance of the magnetic wooden model. With the advantages of high efficiency, simply fabrication, and renewable characteristic, the novel magnetic wood is highly attractive for induction heating application in buildings, decoration, and massage furniture.

4 Experimental section

4.1 Materials

Wood samples (20 mm× 20 mm× 10 mm) of Cathay poplar (Populus cathayana Rehd) were used in this experiment. Each sample was ultrasonically rinsed in distilled water for 30 min, and vacuum dried at 80 ℃ for 48 h before use. Sodium oleate (analytic reagent) and Fe_3O_4 nanoparticles (purity quotient of 99%) were purchased from Shanghai Aladdin Biochemical Technology Co. , Ltd (China).

4.2 Synthesis

Fe_3O_4 nanoparticles and sodium oleate with a weight ratio of 4 : 1 were added into 50 mL of distilled water. After stirring at 80 ℃ for 60 min, the precursor solution was impregnated into wood sample via a vacuum impregnation: 0.09 MPa for 10 min subsequent released to air pressure, this process was repeated 10 times. Then the mixture was transferred into a Teflon-lined stainless-steel autoclave and heated to 110 ℃ for 8 h. After that, the solution and wood samples were placed in a gently sealed glass bottle and heated at 130 ℃ for 5 h to evaporate the solution. Finally, the magnetic Fe_3O_4 particles were successfully deposited on the wood substrate at the driven power of pressure and gravity. For wood samples containing different concentration of magnetic particles, the additive amount of Fe_3O_4 nanoparticles is 2, 4 and 6 g, and the corresponding magnetic wood sample is named as MW-1, MW-2, and MW-3, respectively. The Fe_3O_4 particles loading rate is calculated as follows:

$$R_m(\%) = (W_m - W_n)/W_m \times 100 \tag{3}$$

Where R_m is Fe_3O_4 particles loading rate, W_m is the weight of magnetic wood, and W_n means the weight of natural wood before treatment. Table 1 shows the whole parameters of wood samples.

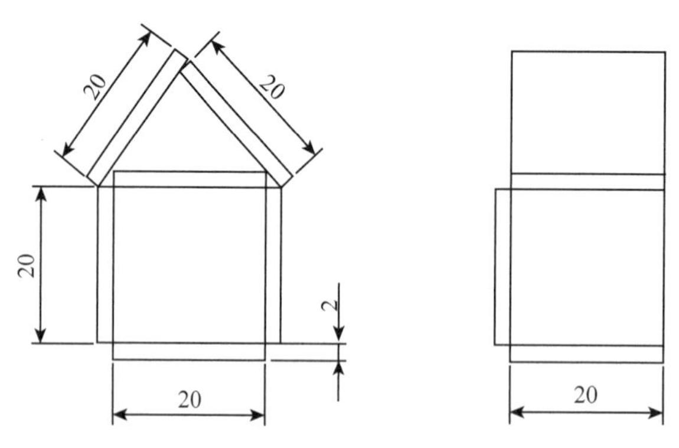

Scheme 1 Schematic description of a model house

4.3 Characterization

The morphology of the samples was characterized by SEM (HITACHI TM3030) and TEM (Tecnai G2 20) equipped with an EDXA. The surface components of the wood samples were analysis using FTIR (Magna-IR 560). XRD tests (Rigaku D/MAX 2200) operated with Cu target radiation (λ = 1.54 Å) at scan rate (2θ) of 4° · min^{-1}. The thermal performances of the wood composites were evaluated via the TGA (Q500) at a heating rate of 10 ℃ · min^{-1} under nitrogen atmosphere. Magnetic properties of wood samples with the dimension of 4 mm × 4 mm× 4 mm were determined using a superconducting quantum interference device (MPMS XL-7) at room temperature. The thermal conductivities of wood samples along the radial direction with the dimensions of Φ 12 ×1 mm^3 were measured up to 80 ℃ using the steady-state heat flow method (Longwin TIM LW-9389).

4.4 Measurement of magnetocaloric effect

A constant alternating magnetic field was generated with a homemade magnetothermal equipment consisting of a 5 mm copper coil and a resonant RLC circuit. During the experiment, wood samples are stimulated under a magnetic field frequency of 35 kHz, with a magnetic field of 0.06 mT and a current of 5

A. The change of temperature was recorded with an infrared thermal imager (Testo 869) in real time. To avoid thermal fluctuation, the temperature of the coil was maintained at room temperature with a water cooling circuit. The SAR of magnetic wood was calculated as follows[22]

$$SAR = \frac{C}{m}\frac{dT}{dt} \tag{4}$$

where C is the specific heat capacity of wood [$C = 1.7\ J \cdot (g \cdot K)^{-1}$], m is the mass of magnetic nanoparticle, and dT/dt is the initial slope ($t = 20$ s) of the graph of the change in temperature versus time.

4.5 Simulation test

The indoor temperature of magnetic wooden house was characterized by using a system that simulated the real environment. The model house was made of wood slices (MW-1) with the size of $20 \times 20 \times 2\ mm^3$, and epoxy resin was used for adhesion of wood, as shown in Scheme 1. After continuous induction heating under the alternating magnetic field, a thermoelectric couples (3 mm in diameter and 10 mm in length) were employed as temperature probes to measure the indoor temperature, which was monitored by a meter.

Supporting Information

Supporting Information is available from the Wiley Online Library or from the author.

Acknowledgements

This work was financially supported by The National Natural Science Foundation of China (grant no. 31470584) and Fundamental Research Funds for the Central Universities (grant no. 2572017AB10).

References

[1] Chu S, Cui Y, Liu N. The path towards sustainable energy[J]. Nat. Mater., 2016, 16(1): 16-22.

[2] Hook M, Tang X. Depletion of fossil fuels and anthropogenic climate change-A review[J]. Energy Policy, 2013, 52: 797-809.

[3] Chow J, Kopp R J, Portney P R. Energy resources and global development[J]. Science, 2003, 302(5650): 1528-1531.

[4] Mohamed A R, Lee K T. Energy for sustainable development in Malaysia: Energy policy and alternative energy[J]. Energy Policy, 2006, 34(15): 2388-2397.

[5] Merk V, Chanana M, Keplinger T, et al. Hybrid wood materials with improved fire retardance by bio-inspired mineralisation on the nano- and submicron level[J]. Green Chem., 2015, 17(3): 1423-1428.

[6] Wu Y Q, Jia S S, Qing Y, et al. A versatile and efficient method to fabricate durablesuperhydrophobic surfaces on wood, lignocellulosic fiber, glass, and metal substrates[J]. J. Mater. Chem. A, 2016, 4(37): 14111-14121.

[7] Burgert I, Cabane E, Zollfrank C, et al. Bio-inspired functional wood-based materials-hybrids and replicates[J]. Int. Mater. Rev., 2015, 60(8): 431-450.

[8] Zhao G L, Yu Z L. Recent research and development advances of wood science and technology in China: impacts of funding support from National Natural Science Foundation of China[J]. Wood Sci. Technol., 2016, 50(1): 193-215.

[9] Liu K K, Jiang Q, Tadepalli S, et al. Wood graphene oxide composite for highly efficient solar steam generation and desalination[J]. ACS Appl. Mater. Inter., 2017, 9(8): 7675-7681.

[10] Dong X Y, Zhuo X, Wei J, et al. wood-based nanocomposite derived by in situ formation of organic-inorganic hybrid polymer within wood via a sol-gel method[J]. ACS Appl. Mater. Inter., 2017, 9(10): 9070-9078.

[11] Zhu M W, Song J W, Li T, et al. Highly anisotropic, highly transparent wood composites[J]. Adv. Mater., 2016, 28(26): 5181-5187.

[12]Yu Z Y, Yao Y J, Yao J N, et al. Transparent wood containing Cs_xWO_3 nanoparticles for heat-shielding window applications[J]. J. Mater. Chem. A, 2017, 5(13): 6019-6024.

[13]Li T, Zhu M, Yang Z, et al. Wood composite as an energy efficient building material: Guided sunlight transmittance and effective thermal insulation[J]. Adv. Energy Mater., 2016, 6(22): 1601122.

[14]Oka H, Uchidate S, Sekino N, et al. Electromagnetic wave absorption characteristics of half carbonized powder-type magnetic wood[J]. IEEE Trans. Magn., 2011, 47(10): 3078-3080.

[15]Oka H, Fujita H. Experimental study on magnetic and heating characteristics of magnetic wood[J]. J. Appl. Phys., 1999, 85(8): 5732-5734.

[16]Merk V, Chanana M, Gierlinger N, et al. Hybrid wood materials with magnetic anisotropy dictated by the hierarchical cell structure[J]. ACS Appl. Mater. Interfaces, 2014, 6(12): 9760-9767.

[17]Lee J H, Jang J T, Choi J S, et al. Exchange-coupled magnetic nanoparticles for efficient heat induction[J]. Nat. Nanotechnol., 2011, 6(7): 418-422.

[18]Hervault A, Thanh N T K. Magnetic nanoparticle-based therapeutic agents for thermo-chemotherapy treatment of cancer[J]. Nanoscale, 2014, 6(20): 11553-11573.

[19]Wabler M, Zhu W L, Hedayati M, et al. Magnetic resonance imaging contrast of iron oxide nanoparticles developed for hyperthermia is dominated by iron content[J]. Int. J. Hyperthermia., 2014, 30(3): 192-200.

[20]Presa P D L, Luengo Y, Multigner M, et al. Study of heating efficiency as a function of concentration, size, and applied field in gamma-Fe_2O_3 nanoparticles[J]. J. Phys. Chem. C, 2012, 116(48): 25602-25610.

[21]Kozissnik B, Bohorquez A C, Dobson J, et al. Magnetic fluid hyperthermia: Advances, challenges, and opportunity [J]. Int. J. Hyperthermia, 2013, 29(8): 706-714.

[22]Guardia P, Di Corato R, Lartigue L, et al. Water-soluble iron oxide nanocubes with high values of specific absorption rate for cancer cell hyperthermia treatment[J]. ACS Nano, 2012, 6(4): 3080-3091.

[23]Connolly J, St Pierre T G. Proposed biosensors based on time-dependent properties of magnetic fluids[J]. J. Magn. Magn. Mater., 2001, 225(1-2): 156-160.

[24]Daou T J, Pourroy G, Begin-Colin S, et al. Hydrothermal synthesis of monodisperse magnetite nanoparticles[J]. Chem. Mater., 2006, 18(18): 4399-4404.

[25]Lu Y, Sun Q F, Yang D J, et al. Fabrication of mesoporous lignocellulose aerogels from wood via cyclic liquid nitrogen freezing-thawing in ionic liquid solution[J]. J. Mater. Chem., 2012, 22(27): 13548-13557.

[26]Li J, Lu Y, Yang D J, et al. Lignocellulose aerogel from wood-ionic liquid solution (1-Allyl-3-methylimidazolium chloride) under freezing and thawing conditions[J]. Biomacromolecules, 2011, 12(5): 1860-1867.

[27]Gan W T, Xiao S L, Gao L K, et al. Luminescent and transparent wood composites fabricated by poly-(methyl methacrylate) and gamma-Fe_2O_3@YVO_4: Eu^{3+} nanoparticle impregnation[J]. ACS Sustain. Chem. Eng., 2017, 5(5): 3855-3862.

[28]Rana R, Langenfeld-Heyser R, Finkeldey R, et al. FTIR spectroscopy, chemical and histochemical characterisation of wood and lignin of five tropical timber wood species of the family of Dipterocarpaceae[J]. Wood Sci. Technol., 2010, 44(2): 225-242.

[29]Gierlinger N, Goswami L, Schmidt M, et al. In situ FT-IR microscopic study on enzymatic treatment of poplar wood cross-sections[J]. Biomacromolecules, 2008, 9(8): 2194-2201.

[30]Araujo-Neto R P, Silva-Freitas E L, Carvalho J F, et al. Monodisperse sodium oleate coated magnetite high susceptibility nanoparticles for hyperthermia applications[J]. J. Magn. Magn. Mater., 2014, 364: 72-79.

[31]Ma M, Zhang Y, Yu W, et al. Preparation and characterization of magnetite nanoparticles coated by amino silane[J]. Colloid. Surf., A, 2003, 212(2-3): 219-226.

[32]Saka S, Ueno T. Several SiO_2 wood-inorganic composites and their fire-resisting properties[J]. Wood Sci. Technol.,

1997, 31(6): 457-466.

[33]Hu B, Zhao C H, Jin X Y, et al. Antagonistic effect in pickering emulsion stabilized by mixtures of hydroxyapatite nanoparticles and sodium oleate[J]. Colloid Surf., A, 2015, 484: 278-287.

[34]Wan C C, Li J. Facile synthesis of well-dispersed superparamagnetic gamma-Fe_2O_3 nanoparticles encapsulated in three-dimensional architectures of cellulose aerogels and their applications for Cr(VI) removal from contaminated water[J]. ACS Sustain. Chem. Eng., 2015, 3(9): 2142-2152.

[35]Deatsch A D, Evans B A. Heating efficiency in magnetic nanoparticle hyperthermia[J]. J. Magn. Magn. Mater., 2014, 354: 163-172.

[36]Gu H B, Huang Y D, Zhang X, et al. Magnetoresistive polyaniline-magnetite nanocomposites with negative dielectrical properties[J]. Polymer, 2012, 53(3): 801-809.

[37]Kolen'ko Y V, Banobre-Lopez M, Rodrigue-zabreu C, et al. Large-scale synthesis of colloidal Fe_3O_4 nanoparticles exhibiting high heating efficiency in magnetic hyperthermia[J]. J. Phys. Chem. C, 2014, 118(16): 8691-8701.

[38]Espinosa A, Corato R D, Kolosnjaj-tabi J, et al. Duality of iron oxide nanoparticles in cancer therapy: amplification of heating efficiency by magnetic hyperthermia and photothermal bimodal treatment[J]. ACS Nano, 2016, 10(2): 2436-2446.

[39]Dong J Y, Zink J I. Taking the temperature of the interiors of magnetically heated nanoparticles[J]. ACS Nano, 2014, 8(5): 5199-5207.

[40]Di C R, Espinosa A, Lartigue L, et al. Magnetic hyperthermia efficiency in the cellular environment for different nanoparticle designs[J]. Biomaterials, 2014, 35(24): 6400-6411.

[41]Takeda M, Onishi T, Nakakubo S, et al. Physical properties of iron-oxide scales on Si-containing steels at high temperature[J]. Mater. Trans., 2009, 50(9): 2242-2246.

[42]Uysal H, Demirboga R, Şahin R, et al. The effects of different cement dosages, slumps, andpumice aggregate ratios on the thermal conductivity and density of concrete[J]. Cem. Concr. Res., 2004, 34(5): 845-848.

[43]Lin J C. A new IEEE standard for safety levels with respect to human exposure to radio-frequency radiation[J]. IEEE Antennas Propag Mag, 2006, 48(1): 157-159.

[44]Dong Y M, Yan Y T, Zhang Y, et al. Combined treatment for conversion of fast-growing poplar wood to magnetic wood with high dimensional stability[J]. Wood Sci. Technol., 2016, 50(3): 503-517.

[45]Jin C D, Yao Q F, Li J P, et al. Fabrication, superhydrophobicity, and microwave absorbing properties of the magnetic gamma-Fe_2O_3/bamboo composites[J]. Mater. Des., 2015, 85: 205-210.

[46]Peng H L, Wang H, Wu J N, et al. Preparation of superhydrophobic magnetic cellulose sponge for removing oil from water[J]. Ind. Eng. Chem. Res., 2016, 55(3): 832-838.

[47]Chen N, Pan Q M. Versatile fabrication of ultralight magnetic foams and application for oil-water separation[J]. ACS nano, 2013, 7(8): 6875-6883.

[48]Yu M, Li J, Wang L J. Preparation and characterization of magnetic carbon aerogel from pyrolysis of sodium carboxymethyl cellulose aerogel crosslinked by iron trichloride[J]. J. Porous Mat., 2016, 23(4): 997-1003.

[49]Li Y Y, Zhu H L, Gu H B, et al. Strong transparent magnetic nanopaper prepared by immobilization of Fe_3O_4 nanoparticles in a nanofibrillated cellulose network[J]. J. Mater. Chem. A, 2013, 1(48): 15278-15283.

中文题目：磁性木材作为有效感应加热材料：磁热效应和绝热

作者：甘文涛，高丽坤，肖少良，高汝楠，张文博，李坚

摘要：通过简单的水热法制备了一种由木材和磁性 Fe_2O_3 颗粒组成的高效感应加热材料。沉积在多孔木材基底上的 Fe_2O_3 颗粒提供了优异的磁性和高的磁热转换，导致木材基底的温度快速升高。得益于磁性粒子的有效磁热效应，磁性木材的温度在低频磁场下呈现出显著上升(10min 内从 25.9 ℃上

升到 70.1 ℃)。为了模拟真实环境，磁性木模型的室内温升也揭示了磁性木优异的加热性能。此外，磁性木材还具有隔热作用，由于其导热系数低，可以减缓热能从高温区域向周围环境的传输，有利于日常生活中热能的合理利用。考虑到高度可扩展、操作简单、良好的耐久性、优异的磁性能、有效的磁热效应和绝热性，这里展示的新型磁性木材复合材料是一种用于建筑、装饰和按摩家具的有吸引力的感应加热材料。

关键词：Fe_3O_4 粒子；感应加热材料；磁性；木材

Transparent Magnetic Wood Composites Based on Immobilizing Fe_3O_4 Nanoparticles into A Delignified Wood Template*

Wentao Gan, Likun Gao, Shaoliang Xiao, Wenbo Zhang, Xianxu Zhan, Jian Li

Abstract: Wood is one of the key renewable resources due to its excellent structure and physical properties. Functionalized wood and wood-based materials not only possess important engineering applications but also great potential in some new technology fields, such as electronics, optics and energy. Endowing these functional wood-based materials with magnetic properties, has an important significance for exploring lightweight building materials or electronic devices. In this study, we report the fabrication of a transparent magnetic wood (TMW) base on filling the index-matching methyl methacrylate and magnetic Fe_3O_4 nanoparticles into the delignified wood template. The presence of the polymer and Fe_3O_4 nanoparticles within the wood structure is monitored by scanning electron microscopy, energy-dispersive X-ray analysis, and Fourier transformation infrared spectroscopy. The resulting TMW possesses moderate transparency and magnetic properties combining with outstanding mechanical performance. Moreover, the influence of the concentration of Fe_3O_4 nanoparticles on the final optical, magnetic and mechanical properties of TMW is also discussed. This work provides a potential strategy to develop wood-based materials for magneto-optical application.

Keywords: Transparent wood; Magnetic wood; Fe_3O_4; Magneto-optical application

1 Introduction

Nowadays, the materials derived from biological resources are ready to be explored for applications in new technology areas, such as biomedicine, electronics and energy[1-5]. Wood is a fundamental engineering material, which has remarkable mechanical properties, processability, and lightweight due to the unique structures from its natural growth[6,7]. The cell wall of natural wood has a three-dimensional porous structure that consists of three main components including cellulose, hemicelluloses and lignin[8,9]. Although cellulose and hemicelluloses are optically colorless, lignin has an extremely complex components with optical opaqueness, so wood is not optical transparence. Researches in the field of wood modification have proved that the chemicals have penetrated into the cell wall structure in order to improve the wood properties or added functional properties[10-14]. The recent observation of transparent wood composites with excellent optical and mechanical properties was prepared by removing the light-absorbing lignin and filling index-matching polymer to achieve high optical transparency[15-18]. However, less effort has been made in the preparation of transparent

* 本文摘自 Journal of Materials Science, 2017, 52: 3321-3329.

magnetic materials based on wood templates, despite their importance for various engineering applications.

Optically transparent magnetic materials have attracted increasing attention in materials science community because of their potential applications including in the information storage, magneto-optical switches, bioanalytical applications, optical isolators, modulators, etc[19-22]. Designs of transparent magnetic materials usually focus on embedding the magnetic nanoparticles such as Fe, Co, Fe_3O_4, and γ-Fe_2O_3 in various host matrices. For example, a transparent magnetic Fe_3O_4-SiO_2 material was prepared through a sol-gel method with tetraethoxysilane and Fe_3O_4 ferrofluids as precursors[23]. Li et al. obtained a transparent magnetic nanopaper based on immobilizing the magnetic nanoparticles into NFC matrices[24]. Song et al. fabricated a kind of flexible, optically homogeneous magnetic composites by covalently bonding liquid crystals to the siloxane backbones and then linking them to dopamine-functionalized ferrite nanoparticles[25]. The magnetic and optical properties of the composites mainly depend on the size and loading of the magnetic nanoparticles. The diameters of magnetic nanoparticles are much less than the wavelength of visible light, so they can greatly reduce the reflection from surface, which is important to guarantee transparency and benefit to magnetic properties.

Based on these concepts, we report the fabrication of a transparent magnetic wood (TMW) base on filling the index matching methyl methacrylate (MMA) and magnetic Fe_3O_4 nanoparticles into a delignified wood template. The morphology, magnetic characters, and mechanical properties of the as-prepared samples were characterized. Not only the original structure of the wood is preserved, but using an impregnation approach also can able to control the final concentration of magnetic nanoparticles and further control the optical, magnetic, and mechanical properties of these composites.

2 Experimental section

2.1 Materials

Wood slices of Cathay poplar (Populus cathayana Rehd) were used in this study. The size of samples is 20 mm×20 mm×0.5 mm. Sodium chlorite ($NaClO_2$), glacial acetic acid, hydrogen peroxide (30%) were supplied by Shanghai Boyle Chemical Co., Ltd, and used without further purification. Methyl methacrylate (MMA), 2, 2'-Azobis (2-methylpropionitrile) (AIBN, Sigma-Aldrich), and Fe_3O_4 nanoparticles (NPs) were purchased from Shanghai Aladdin Biochemical Technology Co., Ltd (China). The morphologies and sizes of the Fe_3O_4 are shown in Fig. S1.

2.2 Synthesis

Sketch of the procedure to prepare the transparent magnetic wood (TMW) is showed in Fig. 1. Firstly, the wood slices were treated with a glacial acetic acid solution (pH 4.6) of $NaClO_2$ (2 wt%) to remove the lignin at 80 ℃ for 12 h, following by rinsing in distilled water. Then, the treated wood were placed in an aqueous solution of H_2O_2 (5 mol · L^{-1}) and kept boiling for 4 h. After that, the samples were rinsed with distilled water, and dried at 45 ℃ for 24 h in vacuum. Secondly, the infiltrated polymer was prepared by mixing the methyl methacrylate (MMA) and 2, 2′-Azobis (2-methylpropionitrile) with a weight ratio of 1000 to 3, which was prepolymerized at 70 ℃ for 20 min. Then, the prepolymerized polymer was cooled down to room temperature in ice-water bath. After that, the polymer and Fe_3O_4 NPs were uniformly mixed with weight ratios of 1000 : 1, 1000 : 2, and 1000 : 5 to form dark impregnation solutions. The lignin-removed wood was infiltrated in the above solutions under vacuum for 30 min, and this process was repeated three times to ensure the full infiltration. Finally, the infiltrated wood samples were sandwiched between two glass slides and dried at 50 ℃ for 6 h. The TMW samples were obtained after being peeled off from the glass slides.

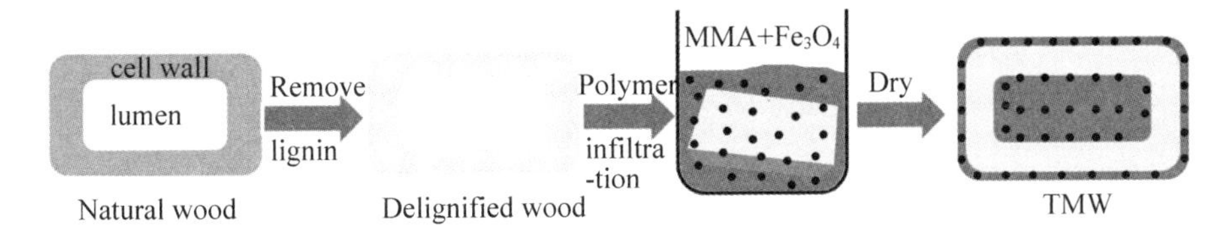

Fig. 1 Sketch of the procedure to prepare the transparent magnetic wood composites

2.3 Characterization

The surface and cross section of wood samples were observed with a scanning electron microscopy (SEM, HITACHI TM3030). The surface chemical compositions of the samples were determined via energy-dispersive X-ray analysis (EDXA) and Fourier transformation infrared spectroscopy (FTIR, Nicolet, Magna-IR 560). The crystalline structure of these composites were identified via X-ray diffraction (XRD, Rigaku, and D/MAX 2200) operating with Cu $K\alpha$ radiation (λ = 1.5418 A) at a scan rate (2θ) of $4° \cdot min^{-1}$, an accelerating voltage of 40 kV, and an applied current of 30 mA ranging from 5° to 80°. The transmittances of the transparent magnetic wood samples were analyzed with a U-4100 UV-Vis Spectrometer (Hitachi Limited, Japan). The magnetic properties of the composites with the size of 4 mm×4 mm×0.5 mm were measured by a superconducting quantum interference device (MPMS XL-7, Quantum Design Corp.) at room temperature (300 K). For mechanical characterization, wood specimens of 50 mm × 10 mm ×3 mm with approximately 13 wt% moisture content were used (3-5 specimens were tested for each type), their tensile behavior under the stretching process in the longitudinal direction was recorded using a universal testing machine (Instron 1185) with a speed of $2 mm \cdot min^{-1}$, and the temperature and the relative humidity of the room for the wood mechanical testing were 25 ℃ and 60%, respectively.

3 Results and discussion

To determine the spatial distribution of Fe_3O_4 NPs after the polymer infiltration at the cell levels, a series of SEM pictures of cube cross sections was performed. The honeycomb-like structure and porous architecture of the untreated wood are well presented in Fig. 2a. Fig. 2b shows that no noticeable destructions in cell walls are observed after the lignin removing process, which indicates the wood structure is well preserved. After the polymer infiltration, wood slices are completely immersed in the reaction solution, and it is expected that polymerization took place in the cell lumina. Fig. 2c-2f reveal that the polymer remains in the cell lumina and completely fills into the micropores of the wood. However, there are some Fe_3O_4 NPs anchored on the polymer. In Fig. 2d, the nanoparticles are widely spaced. With an increasing concentration of Fe_3O_4 NPs, the nanoparticles are more densely distributed as shown in Fig. 2d-2f. EDXA spectrum presented in Fig. 2e shows the major chemical compositions of the TMW: carbon, oxygen, iron, and gold (from the coating for SEM imaging). Furthermore, the inset in Fig. 2f is the mapping of Fe in the cross-section of TMW. These results indicate that the Fe_3O_4 NPs are dispersed throughout the whole wood. Fig. 2g-l display the top-view SEM images of the wood before and after the treatment. After the MMA infiltration, the original micro-grooved structures with minor features (e. g. , pits, tracheids) of wood surface (Fig. 2g, h) are covered by a smooth polymer film (Fig. 2i). As expected, less nanoparticles are deposited on the wood surface for lower content of Fe_3O_4 NPs (Fig. 2j); uniformly dispersed nanoparticles are covered on the wood surface when the Fe_3O_4 content increases to 0. 2 wt%. Increasing Fe_3O_4 content to 0. 5 wt% leads to the agglomeration of nanoparticles on the wood template. The insets in Fig. 2g-l show the digital images of wood samples obtained from different stages of the preparation. Wood is originally of yellow color due to the light-absorbing lignin. After lignin removal, the resulting wood becomes white, while it is not optically transparent due to the light scattering at the interface between the cell wall and air[26]. The transparent wood has been further developed by infiltrating the index matching polymer into the lignin-removed wood. The clear appearance of the letter beneath the transparent wood sample illustrates the high transparency. Additionally, the TMW samples with lower contents of Fe_3O_4 NPs display higher transparency. As the concentration of Fe_3O_4 is increased from 0. 1 to 0. 5 wt%, the letters beneath the wood samples are more obscured, suggesting that the transparence of the magnetic wood can be tailored by controlling the contents of Fe_3O_4 NPs.

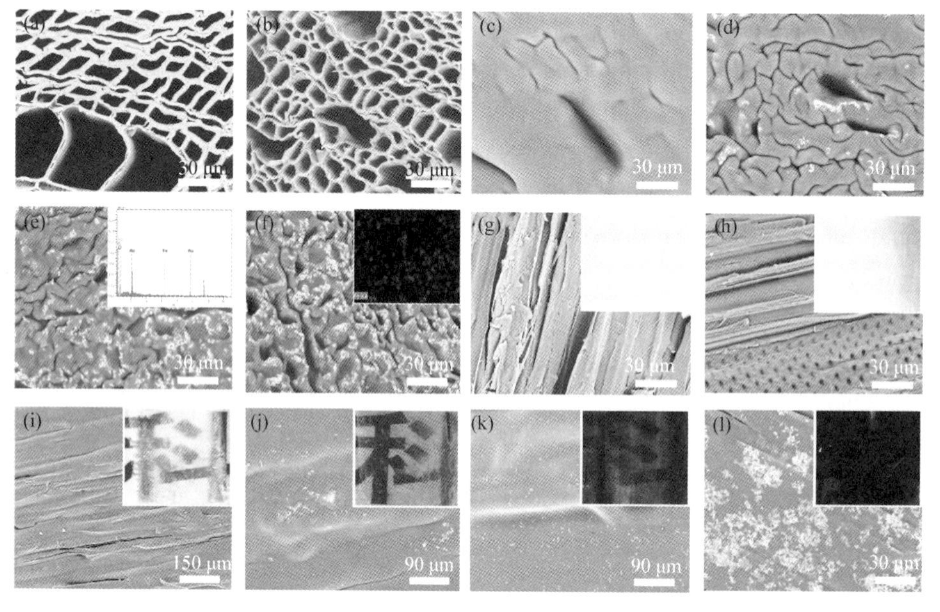

Fig. 2 (a)-(b) Cross-sectional SEM images of the wood before and after lignin removal without polymer infiltration, respectively. (c)-(f) Cross-sectional SEM photographs of the wood samples possesses 0, 0. 1, 0. 2 wt% and 0. 5 wt% Fe_3O_4 NPs, respectively. Inset: EDXA spectrum and Fe element mapping of TMW cross section, respectively. (g)-(h) Top-view SEM images of the wood before and after lignin removal without polymer infiltration, respectively; i-l Top-view SEM photographs of the wood samples possesses 0, 0. 1, 0. 2, and 0. 5 wt% Fe_3O_4 NPs, respectively. Inset the digital images of wood samples show the transparency

To determine the crystal structure of the wood samples before and after treatment, XRD was performed (Fig. 3). The XRD pattern, Fig. 3a, of Fe_3O_4 nanoparticles shows intense peaks at angles $2\theta = 30°$, $35°$, $37°$, $43°$, $53°$, $56°$, and $62°$, which indicate the crystalline nature of magnetite (JCPDS: 19-0629). Two major diffraction peaks were observed for the natural wood at approximately $16°$ and $22°$, which attributed to the (101) and (002) crystal planes of cellulose, respectively (Fig. 3b)[27,28]. After PMMA impregnation (Fig. 3c), two broad peaks at angles $2\theta = 15°$ and $30°$, which are ascribed to PMMA[29]. Further addition the Fe_3O_4 NPs (Fig. 3d), the diffraction peaks at $2\theta = 30°$, $35°$, $43°$, $56°$, and $62°$ of Fe_3O_4 are existed in the TMW, all diffraction peaks belonging to cellulose and PMMA are present, revealing that the Fe_3O_4 NPs and PMMA were immobilized into the wood template successfully.

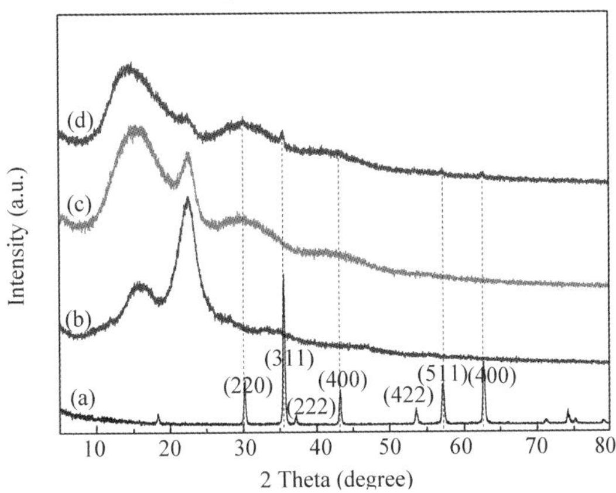

Fig. 3 XRD patterns for a Fe_3O_4 nanoparticles, b natural wood, c transparent wood, and d TMW possesses 0.2 wt% Fe_3O_4 NPs

Fig. 4 shows the FTIR spectra of the natural wood, the wood after lignin removal, transparent wood and TMW possesses 0.2 wt% Fe_3O_4 NPs. The natural wood (Fig. 4a) presents characteristic absorption peaks including 3430 cm^{-1}(O-H stretching vibrations), 2928 cm^{-1}(C-H stretching vibration), 1731 cm^{-1}(acetyl groups in hemicelluloses), 1642 cm^{-1}(C=O stretching vibrations), 1510 cm^{-1}(the aromatic stretching vibrations of lignin), 1459 cm^{-1}(C=O symmetric stretching vibrations), 1260 cm^{-1}(aliphatic acid group vibration), 1165 cm^{-1}(C—O—C asymmetric stretching vibration) and 1055 cm^{-1}(aromatic C—H in-plane deformation assigned to lignin), which are in agreement with earlier researches[30,31]. After lignin removal, the bands at around 1510 cm^{-1} and 1055 cm^{-1} are disappeared, further proving the degradation of light-absorbing lignin (Fig. 4b). After polymer infiltration, the FTIR spectrum of the transparent wood without Fe_3O_4 NPs possesses not only the characteristic peaks of the delignified wood, but also the peaks of PMMA (2997 and 2949 cm^{-1} for C—H, 1731 cm^{-1} for C=O, and 1204 and 1168 cm^{-1} for C—O, as presented in Fig. 4c)[32,33]. However, when the Fe_3O_4 NPs were added to the polymer (Fig. 4d), the stretching vibrations of the Fe-O at the tetrahedral site and octahedral site can be clearly seen around at 560 and 433 cm^{-1}, respectively[34]. These results further confirm that the polymerization of MMA to form PMMA in the wood template and the Fe_3O_4 NPs are deposited on the wood composites successfully.

The transmittance of the wood samples before and after the treatment is presented in Fig. 5, and the thickness of all the samples are around 0.5 mm. Natural wood shows almost negligible transmittance because of the strong lignin absorption. After lignin removal and polymer infiltration (without Fe_3O_4 NPs), the transparent wood composites show dramatically high transmittance, and the transmittance is up to 90.4%. As the loading of Fe_3O_4 NPs increases, the transmittance decreases in the 350-800 nm wavelength region, but it is generally higher than that of the natural wood samples (Fig. 5a). The dependence of optical transmittance on

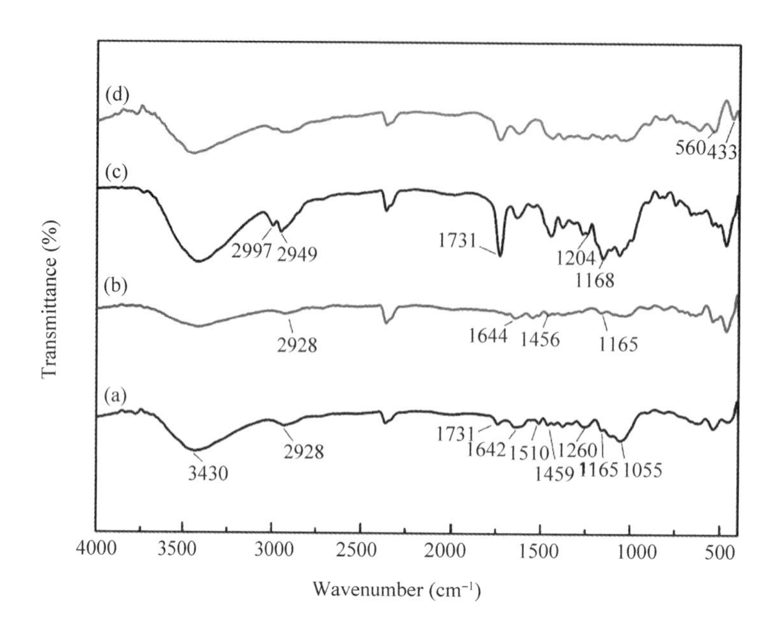

Fig. 4 FTIR spectra of a natural wood, b delignified wood, c transparent wood, and d TMW possesses 0. 2 wt% Fe_3O_4 NPs

the concentration of Fe_3O_4 NPs is shown in Fig. 5b. When the concentration of Fe_3O_4 NPs increases from 0. 1 to 0. 5 wt%, the transmittance decreases from 63. 8 to 19. 5%. This is mainly caused by the optical absorption of Fe_3O_4 NPs[35]. Besides, the larger amount of Fe_3O_4 NPs that deposited on the wood template leads to the agglomeration of magnetic nanoparticles (Fig. 2i), which can also increase the light reflection from the surface and further decreases the transmittance of the wood composites. However, the highest transmittance of the TMW is 63. 8 %, which is enough to be applied in transparent magnetic devices.

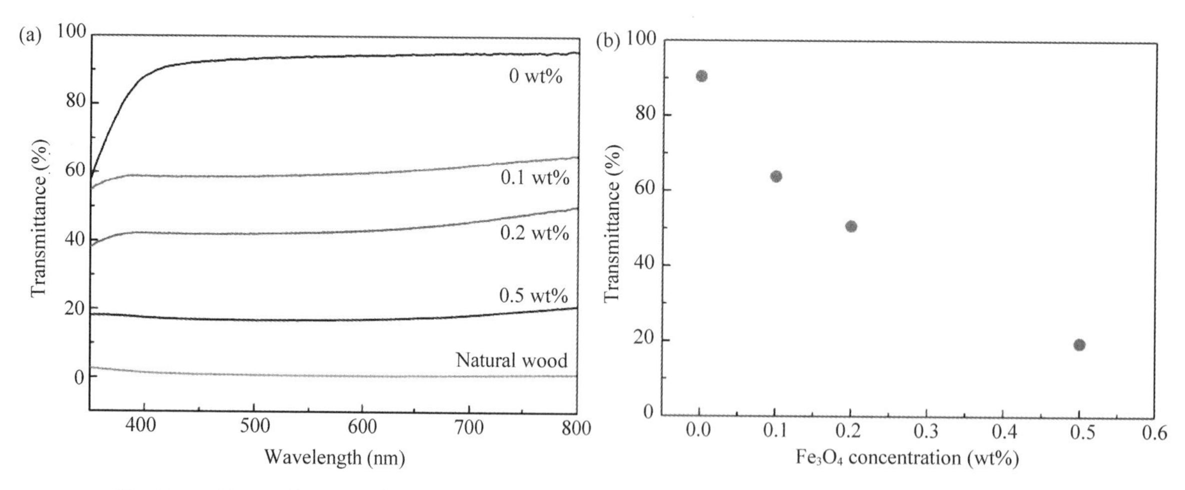

Fig. 5 a Transmittance of the wood samples before and after treatment. b the dependence of optical transmittance on the concentration of Fe_3O_4 NPs

The excellent magnetic properties of the transparent magnetic wood are show in Fig. 6. The entire wood sample with a Fe_3O_4 NPs loading of 0. 2 wt% is easily lifted with a magnet (Fig. 6a). Additionally, the TMW shows a good conductivity when the conductive ink is written on it (Fig. 6b). More importantly, the TMW with a written circuit is an excellent magnetic switch as presented in Fig. 6c and d. The TMW switch is actuated when the magnet is closed, illuminating an LED connected to the switch with a nonmagnetic Cu wire. A video of the switch actuation to illuminate the LED is provided (Supporting Information Movie 1). Magnetization curves of all the wood samples measured at room temperature are shown in Fig. 6e, the hysteresis curve of the natural wood is a straight line close to zero, demonstrating the nonmagnetic property of wood materials. However, all of the treated wood shows almost immeasurable coercivity and remanence, suggesting that these composites are superparamagnetic at room temperature. The magnetic properties in a superparamagnetic system can be defined according to the Langevin Equation[36]:

$$\frac{M}{M_S}=\coth x-\frac{1}{x} \tag{1}$$

Where M_S ($emu \cdot g^{-1}$) is the saturation magnetization, M is the magnetization in a magnetic field H (Oe), and $x=\alpha H$. α is a function of the electron spin magnetic moment m (JT^{-1} or μ_B, μ_B is the Bohr magneton) of the individual molecules as expressed in Eq. (2):

$$\alpha=\frac{m}{k_B T} \tag{2}$$

Where k_B is Boltzmann's constant, and T is the absolute temperature. According to Fig. 6e, the magnetization of all the TMW increases with the increasing magnetic field until it reaches the saturation magnetization. M_S of the TMW with Fe_3O_4 NPs loading of 0.1, 0.2 and 0.5 wt% are 0.35, 0.87 and 1.58 $emu \cdot g^{-1}$, respectively. The results indicate a higher loading of magnetic nanoparticles in wood template would better maintain the saturation magnetization.

The best fit to Eq. (1) is obtained by a nonlinear fitting of magnetization (M) and magnetic field (H) using Polymer software. The values of M_s for the TMW with Fe_3O_4 NPs loading of 0.1, 0.2 and 0.5 wt%, determined from the fit, are 0.33, 0.79, and 1.46 $emu \cdot g^{-1}$, respectively. The fitting correlation coefficient, R^2, of the TMW with Fe_3O_4 NPs loading of 0.1, 0.2 and 0.5 wt% are 0.9968, 0.9934, and 0.9959, respectively. However, the α parameter of the TMW with Fe_3O_4 NPs loading of 0.1, 0.2, and 0.5 wt% are 3.69×10^{-3}, 3.46×10^{-3}, and 3.41×10^{-3} T^{-1}, respectively. The magnetic moment m can be calculated according to the Eq. (2) and the values of the TMW with Fe_3O_4 NPs loading of 0.1, 0.2, and 0.5 wt% is 1.60, 1.51, and 1.48 μ_B, respectively. The calculated μ does not change much between the three samples, revealing that wood template has little effect on the magnetic moment of the Fe_3O_4 NPs.

Mechanical properties of the natural wood and TMW with different contents of Fe_3O_4 NPs under the stretching process in the longitudinal direction are shown in Fig. 7. The natural wood possesses the fracture strength of 28.5 MPa, due to the hierarchical structure and the strong interactions among cellulose, hemicelluloses and lignin[37]. Compared with the natural wood, TMW still possesses a higher strength and ductility, which make the TMW highly desirable for structural materials application (Fig. 7a). However, TMW with Fe_3O_4 loading of 0.1, 0.2, and 0.5 wt% possesses tensile strength of 40.6, 36.2, and 28.7 MPa, respectively. The elastic modulus of TMW with a Fe_3O_4 loading of 0.1, 0.2, and 0.5 wt% is 0.79, 0.66, and 0.57 GPa, respectively (Fig. 7b). As reported by Coto et al.[38], when the inorganic nanoparticles are added to an PMMA matrix without coupling, the mechanical properties of the matrix is

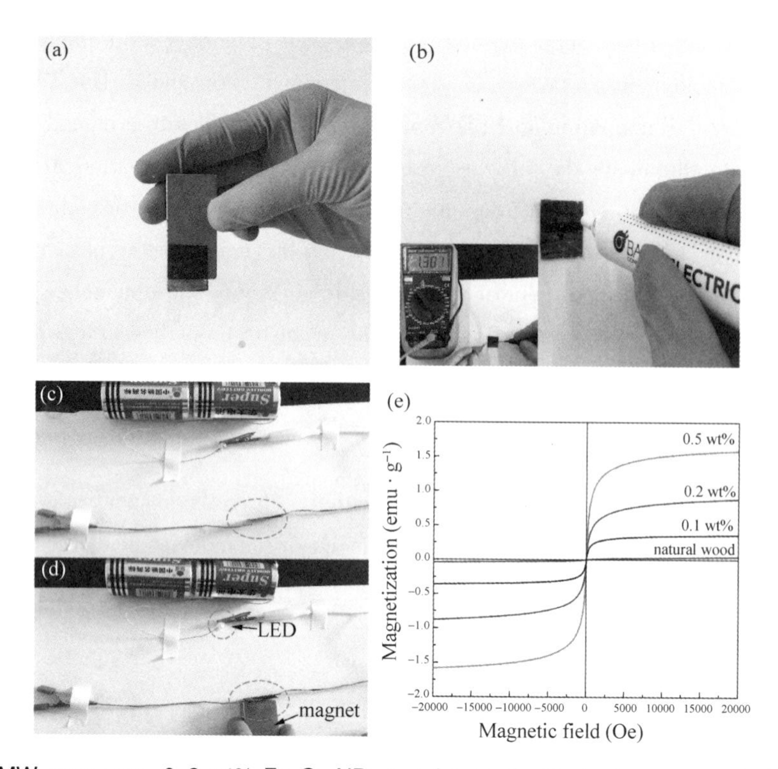

Fig. 6 (a) A TMW possesses 0. 2 wt% Fe_3O_4 NPs can be easily lifted with a magnet. (b) The magnetic wood with CNT ink written on it, the inset illustrating the conductivity of the ink on wood, and the TMW contains 0. 2 wt% Fe_3O_4 NPs. (c) and (d) The magnetic switch: there is a gap between conductive TMW, and when the magnet was placed close to the conductive magnetic wood, the wood moved to contact with Cu wire and lighted the LED. (e) The magnetization curves of the wood samples at room temperature

reduced. In this experiment, increasing the Fe_3O_4 NPs loading leads to the agglomeration of magnetic nanoparticles on the wood template (as shown in Fig. 2), and tensile stress in the wood composites was presumably concentrated in the cracks in the aggregates; as a result, the mechanical properties was decreased.

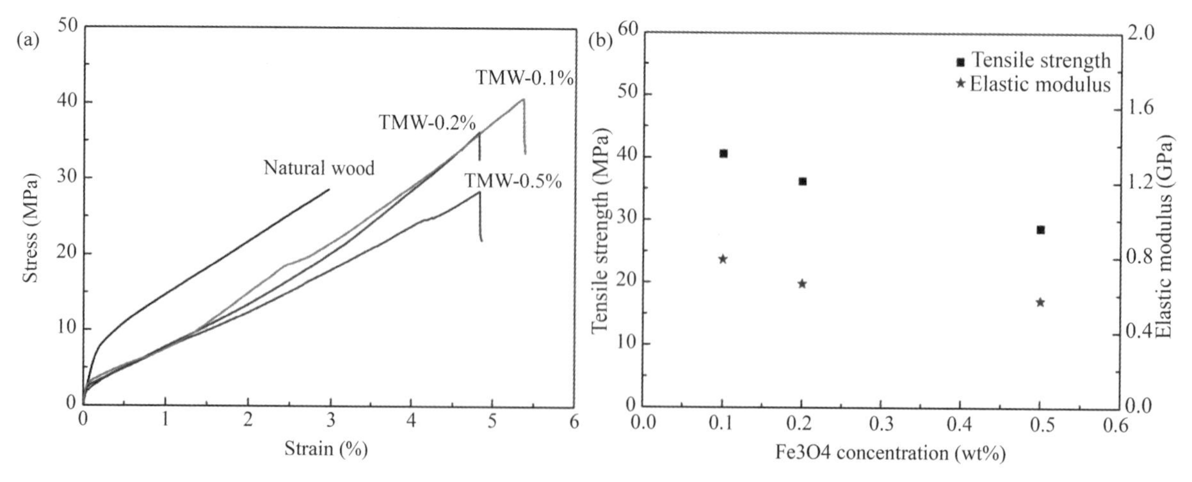

Fig. 7 (a) Experimental stress-strain curves of nature wood and TMW with different contents of Fe_3O_4 NPs, and (b) average tensile strength and elastic modulus of TMW with different contents of Fe_3O_4 NPs

4 Conclusions

TMW was fabricated by infiltration of index MMA and magnetic Fe_3O_4 nanoparticles into a delignified wood template. The saturation magnetization of TMW increases with the increasing contents of Fe_3O_4 nanoparticles, while the transmittance and mechanical properties decreases. The obtained TMW is transparent in visible light with a transmittance of 63% and possesses a saturation magnetization of 0.35 emu · g^{-1}. Combining the excellent tensile strength, simple manufacturing procedure, green, and low cost makes this material an excellent candidate for light-transmitting magnetic buildings and magneto-optical devices.

Acknowledgments

This work was financially supported by The National Natural Science Foundation of China (Grant No. 31470584).

References

[1]Zhu H L, Luo W, Ciesielski P N, et al. Wood-derived materials for green electronics, biological devices, and energy applications[J]. Chem. Rev., 2016, 116(16): 9305-9374.

[2] Li D H, Lv C X, Liu L, et al. Egg-box structure in cobalt alginate: A new approach to multifunctional hierarchical mesoporous N-doped carbon nanofibers for efficient catalysis and energy storage[J]. ACS Cent. Sci., 2015, 1(5): 261-269.

[3] Zou Y H, Chen S, Yang X F, et al. Suppressing Fe-Li antisite defects in $LiFePO_4$/carbon hybrid microtube to enhance the lithium-ion storage[J]. Adv. Energy Mater., 2016, 6(24): 80.

[4] Ma N, Jia Y, Yang X F, et al. Seaweed biomass derived (Ni, Co)/CNT nanoaerogels: efficient bifunctional electrocatalysts for oxygen evolution and reduction reactions[J]. J. Mater. Chem. A, 2016, 4(17): 6376-6384.

[5] Liu L, Yang X F, Ma N, et al. Scalable and cost-effective synthesis of highly efficient Fe_2N-based oxygen reduction catalyst derived from seaweed biomass[J]. Small, 2016, 12(10): 1295-1301.

[6] Cabane E, Keplinger T, Merk V, et al. Renewable and functional wood materials by grafting polymerization within cell walls[J]. ChemSusChem, 2014, 7(4): 1020-1025.

[7] Ugolev B N. Wood as a natural smart material[J]. Wood Sci. Technol., 2014, 48(3): 553-568.

[8] Ermeydan M A, Cabane E, Hass P, et al. Fully biodegradable modification of wood for improvement of dimensional stability and water absorption properties by poly (ε-caprolactone) grafting into the cell walls[J]. Green Chem., 2014, 16(6): 3313-3321.

[9] Persson P V, Hafrén J, Fogden A, et al. Silica nanocasts of wood fibers: a study of cell-wall accessibility and structure[J]. Biomacromolecules, 2004, 5(3): 1097-1101.

[10]Mahltig B, Swaboda C, Roessler A, et al. Functionalising wood by nanosol application[J]. J. Mater. Chem., 2008, 18(27): 3180-3192.

[11] Merk V, Chanana M, Gierlinger N, et al. Hybrid wood materials with magnetic anisotropy dictated by the hierarchical cell structure[J]. ACS Appl. Mater. Interfaces, 2014, 6(12): 9760-9767.

[12] Gan W T, Liu Y, Gao L K, et al. Growth of $CoFe_2O_4$ particles on wood template using controlled hydrothermal method at low temperature[J]. Ceram. Int., 2015, 41(10): 14876-14885.

[13] Sun Q F, Lu Y, Liu Y X. Growth of hydrophobic TiO_2 on wood surface using a hydrothermal method[J]. J. Mater. Sci., 2011, 46(24): 7706-7712.

[14] Sun Q F, Lu Y, Zhang H M, et al. Improved UV resistance in wood through the hydrothermal growth of highly ordered ZnO nanorod arrays[J]. J. Mater. Sci., 2012, 47(10): 4457-4462.

[15] Li Y Y, Fu Q L, Yu S, et al. Optically transparent wood from a nanoporous cellulosic template: combining functional

and structural performance[J]. Biomacromolecules, 2016, 17(4): 1358-1364.

[16] Zhu M W, Song J W, Li T, et al. Highly anisotropic, highly transparent wood composites[J]. Adv. Mater., 2016, 28(26): 5181-5187.

[17] Zhu M W, Li T, Davis C S, et al. Transparent and haze wood composites for highly efficient broadband light management in solar cells[J]. Nano Energy, 2016, 26: 332-339.

[18] Li T, Zhu M W, Yang Z, et al. Wood composite as an energy efficient building material: guided sunlight transmittance and effective thermal insulation[J]. Adv. Energy Mater., 2016, 6(22): 1601122.

[19] Lopez-Santiago A, Gangopadhyay P, Thomas J, et al. Faraday rotation in magnetite-polymethylmethacrylate core-shell nanocomposites with high optical quality[J]. Appl. Phys. Lett., 2009, 95(14): 143302.

[20] Patoka P, Skeren T, Hilgendorff M, et al. Transmission of light through magnetic nanocavities[J]. Small, 2011, 7(21): 3096-3100.

[21] Gach P C, Sims C E, Allbritton N L. Transparent magnetic photoresists for bioanalytical applications[J]. Biomaterials, 2010, 31(33): 8810-8817.

[22] Louzguine-Luzgin D V, Hitosugi T, Chen N, et al. Investigation of transparent magnetic material formed by selective oxidation of a metallic glass[J]. Thin Solid Films, 2013, 531: 471-475.

[23] Thomas S, Sakthikumar D, Joy P A, et al. Optically transparent magnetic nanocomposites based on encapsulated Fe_3O_4 nanoparticles in a sol-gel silica network[J]. Nanotechnology, 2006, 17(22): 5565-5572.

[24] Li Y Y, Zhu H L, Gu H B, et al. Strong transparent magnetic nanopaper prepared by immobilization of Fe_3O_4 nanoparticles in a nanofibrillated cellulose network[J]. J. Mater. Chem. A, 2013, 1(48): 15278-15283.

[25] Song H M, Kim J C, Hong J H, et al. Magnetic and transparent composites by linking liquid crystals to ferrite nanoparticles through covalent networks[J]. Adv. Funct. Mater., 2007, 17: 2070-2076.

[26] Fink S. Transparent wood-a new approach in the functional study of wood structure[J]. Holzforschung, 1992, 46(5): 403-408.

[27] Li J, Lu Y, Yang D J, et al. Lignocellulose aerogel from wood-ionic liquid solution (1-allyl-3-methylimidazolium chloride) under freezing and thawing conditions[J]. Biomacromolecules, 2011, 12(5): 1860-1867.

[28] Lu Y, Sun Q F, Yang D J, et al. Fabrication of mesoporous lignocellulose aerogels from wood via cyclic liquid nitrogen freezing-thawing in ionic liquid solution[J]. J. Mater. Chem., 2012, 22(27): 13548-13557.

[29] Baskaran R, Selvasekarapandian S, Kuwata N, et al. Conductivity and thermal studies of blend polymer electrolytes based on PVAc-PMMA[J]. Solid State Ionics, 2006, 177(26-32): 2679-2682.

[30] Gierlinger N, Goswami L, Schmidt M, et al. In situ FT-IR microscopic study on enzymatic treatment of poplar wood cross-sections[J]. Biomacromolecules, 2008, 9(8): 2194-2201.

[31] Rana R, Langenfeld-Heyser R, Finkeldey R, et al. FTIR spectroscopy, chemical and histochemical characterisation of wood and lignin of five tropical timber wood species of the family of dipterocarpaceae[J]. Wood Sci. Technol., 2010, 44: 225-242.

[32] Jiang S H, Gui Z, Bao C L, et al. Preparation of functionalized graphene by simultaneous reduction and surface modification and its polymethyl methacrylate composites through latex technology and melt blending[J]. Chem. Eng. J., 2013, 226: 326-335.

[33] Kavale M S, Mahadik D B, Parale V G, et al. Optically transparent, superhydrophobic methyltrimethoxysilane based silica coatings without silylating reagent[J]. Appl. Surf. Sci., 2011, 258(1): 158-162.

[34] Koseoglu Y, Alan F, Tan M, et al. Low temperature hydrothermal synthesis and characterization of Mn doped cobalt ferrite nanoparticles[J]. Ceram. Int., 2012, 38(5): 3625-3634.

[35] Schlegel A, Alvarado S F, Wachter P. Optical properties of magnetite (Fe_3O_4)[J]. J. Phys. C: Solid State Phys., 1979, 12(6): 1157-1164.

[36] Gu H B, Huang Y D, Zhang X, et al. Magnetoresistive polyaniline - magnetite nanocomposites with negative dielectrical properties[J]. Polymer, 2012, 53(3): 801-809.

[37] Bekhta P, Niemz P. Effect of high temperature on the change in color, dimensional stability and mechanical properties of spruce wood[J]. Holzforschung, 2003, 57(5): 539-546.

[38] Goto K, Tamura J, Shinzato S, et al. Bioactive bone cements containing nano-sized titania particles for use as bone substitutes[J]. Biomaterials, 2005, 26(33): 6496-6505.

中文题目：基于将 Fe_2O_3 纳米粒子固定在脱木质素木质模板中的透明磁性木质复合材料

作者：甘文涛，高丽坤，肖少良，张文博，詹先旭，李坚

摘要：木材因其优异的结构和物理性能而成为重要的可再生资源之一。功能化木材和木质材料不仅具有重要的工程应用，而且在电子、光学、能源等新技术领域也具有巨大的潜力。赋予这些功能性木质材料以磁性对重量轻的建筑材料或电子器件具有重要意义。在这项研究中，我们在去木质模板中填充了甲基丙烯酸甲酯和磁性 Fe_3O_4 纳米粒子，从而实现了一种新的磁性制备了一种新的磁性木材(TMW)。通过扫描电子显微镜、能量色散 X 光分析和傅里叶变换红外光谱法监测木材结构中的聚合物和 Fe_3O_4 纳米粒子的存在。由此产生的 TMW 具有适度的透明度和磁性，并结合出色的机械性能。此外，还讨论了 Fe_3O_4 纳米粒子浓度对薄膜最终光学、磁性和力学性能的影响。这项工作探索了木材在磁光学领域的应用前景。

关键词：透明木材；磁性木材；Fe_3O_4；磁光应用

Removal of Cu^{2+} Ions from Aqueous Solution by Amino-functionalized Magnetic Sawdust Composites*

Wentao Gan, Likun Gao, Xianxu Zhan, Jian Li

Abstract: In this study, amino-functionalized magnetic γ-Fe_2O_3/sawdust composites (MSC-NH_2) were investigated as biological absorption materials for removing Cu^{2+} ions from aqueous solution. These composites were fabricated by precipitated γ-Fe_2O_3 nanoparticles on sawdust substrate, then functionalized with 1, 6-hexanediamine. Characterization of MSC-NH_2 was investigated by means of SEM, TEM, XRD, FTIR, BET, MPMS and XPS analysis to discuss the uptake mechanism. As a result, the amino groups are grafted upon the sawdust surfaces. The MSC-NH_2 could be effectively used to remove Cu^{2+} from aqueous solution and be separated conveniently from the solution with the help of an external magnet. A batch experiments show that the adsorption equilibrium is achieved in 150 min and the adsorption capacity is 7.55 mg · g^{-1} at pH 6 and room temperature. The isotherm analysis indicates that the sorption data could be represented by Langmuir isotherm models. The kinetics is evaluated utilizing the Lagergren pseudo-first-order, pseudo-second-order, Elovich and intra-particle diffusion models. Thermodynamic parameters reveal the spontaneous, endothermic and chemical nature of adsorption.

Keywords: Magnetic sawdust; Amino-functionalized; Adsorption; Copper

1 Introduction

Copper is one of the undesirable heavy metals that can be accumulated in human bodies throughout the food chain and cause several physiological disorders (Zhou et al. 2004; Witek-Krowiak et al. 2013). The main industrial sources of copper pollution are paint, electroplating, pigment industries, fertilizer and wood manufacturing (Hao et al. 2010). According to Chinese standards (GB 8978-1996), the acceptable limit of Cu^{2+} for industrial effluents to be discharged to surface water is 0.5 mg · L^{-1}. Thus, it is necessary to remove Cu^{2+} ions from industrial effluents before their discharge. A wide range of physical and chemical processes is available for the removal of Cu^{2+} from wastewater, such as ion exchange (Fonseca et al. 2005), chemical precipitation (Espana et al. 2006), absorption (Zheng et al. 2009), electrodialysis (Ogutveren et al. 1997), etc.

Natural wastes have been investigated as absorbent materials for a long time, because they are inexpensive (Ngah et al. 2008). Most of the adsorption research focus on plant wastes such as poplar wood sawdust (Li et al. 2007), barley straw (Pehlivan et al. 2012), barks (Seki et al. 1997), rubber (Hanafiah et al. 2006a), tree fern (Ho et al. 2004), pineapple peel fibre (Hu et al. 2011), leaf powder (Hanafiah et al. 2006b) and sugarcane bagasse (Karnitz et al. 2007). However, the absorption capacity of natural

* 本文摘自 Wood Science and Technology, 2017, 51: 207-225.

resources is relatively low. The direct use of the untreated plant wastes as adsorbents can bring several problems such as high biological chemical demand (BOD), chemical oxygen demand (COD) and total organic carbon (TOC) due to the release of soluble organic compounds such as lignin, pectin and tannin into the solution contained in the plant wastes (Feng et al. 2009). Besides, the use of plant wastes as absorbents also requires an additional separation step to remove such absorbents from solution, which involves further expense. Therefore, the design and exploration of novel adsorbents based on plant wastes are still necessary.

Nowadays, the mixing of magnetic nanoparticles with lignocellulosic materials to form magnetic adsorbents has exhibited many excellent properties. On the one hand, the lignocellulosic substrate can effectively prevent the aggregation of magnetic nanoparticles. On the other hand, the lignocellulosic substrate also can increase the mechanical strength of absorbents and improve the applicability of magnetic nanoparticles in real wastewater treatment. Furthermore, after adsorption, the magnetic absorbents could be separated effectively from the wastewater by applying an external magnetic field and avoided secondary pollution (Reddy et al. 2013). For example, Song et al. prepared an easily separable amine-functionalize magnetic corn stalk composites and used for toxic hexavalent chromium absorption from water. Its optimum Cr(Ⅵ) adsorption capacity can reach as high as 231.1 $mg \cdot g^{-1}$ at 45 ℃ (Song et al. 2015). Tian et al. found that wheat straw mixed with Fe_3O_4 nanoparticles exhibited enhanced arsenic absorption property from water (Tian et al., 2011).

As a forestry and solid waste, sawdust with the porous structure and a lignocellulosic nature, obtained from mechanical wood processing, which is a renewable and low-cost resource, is an excellent scaffold for further magnetic particles modification. In this study, sawdust is being investigated as a magnetic adsorbent for removing Cu^{2+} from aqueous solution. Firstly, wood sawdust was loaded with γ-Fe_2O_3 nanoparticles via a hydrothermal process, and then in order to improve the cation exchange capacity, the 1, 6- hexanediamine with a large number of amine groups was modified on the magnetic γ-Fe_2O_3/sawdust composites surface. The physicochemical characteristics of the amino-functionalized magnetic sawdust composites (MSC-NH_2) were characterized by scanning electron microscope (SEM), energy dispersive spectroscopy (EDXA), X-ray diffraction (XRD), Fourier transform infrared spectroscopy (FTIR), X-ray photoelectron spectroscopy (XPS), Brunauer-Emmett-Teller (BET) principle and superconducting quantum interference device (MPMS). Furthermore, to fully understand the sorption behavior, various factors affecting the absorption behavior such as temperature, pH, contact time, initial concentration of Cu^{2+} ions and reusability were also examined in bath experiments.

2 Experimental

2.1 Materials

Poplar sawdust used for this experiment was obtained from a local saw mill in Heilongjiang Province, China. The sawdust was first washed with distilled water, dried, cut, and sieved through 100-mesh screen. All chemicals including $FeCl_3$, $FeCl_2$, KCl, KNO_3, K_2SO_4, $CuSO_4 \cdot 5H_2O$, ammonium hydroxide, ethylene glycol and 1, 6-hexanediamine used in this study were supplied by Shanghai Boyle Chemical Company Limited, and used without further purification.

2.2 Synthesis of magnetic sawdust composites

$MSC-NH_2$ were prepared as shown in Fig. 1; 10 g wood sawdust, 1.98 g $FeCl_2 \cdot 4H_2O$ (0.01mol) and 5.40 g $FeCl_3 \cdot 6H_2O$ (0.02 mol) were dissolved in 200 mL distilled water, the Fe^{3+}/Fe^{2+} ratio was 2. Then, the pH value of the iron ion precursor was adjusted to ca. 10 by the dropwise addition of 10% ammonia solution. After stirring for 15 min, the system was transferred into a Teflon-lined stainless-steel autoclave and heated at 90 ℃ for 8 h. Finally, the prepared specimens were removed, rinsed with distilled water for three times, and dried at 50 ℃ for over 24 h in vacuum.

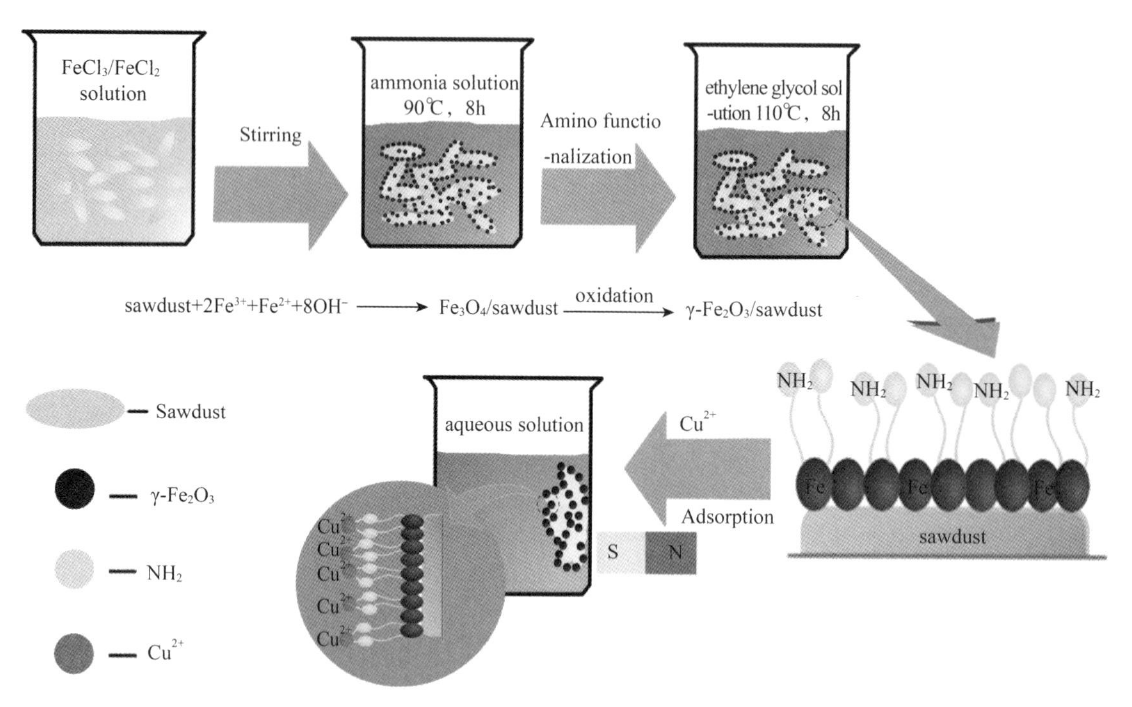

Fig. 1 The scheme of the preparation of $MSC-NH_2$ and adsorption of Cu^{2+} ions

2.3 Surface coating of γ-Fe_2O_3/sawdust by 1, 6-hexanediamine

1,6-hexanediamine (6.5 g) was dissolved in ethylene glycol (30 mL) to form a clear solution at 50 ℃ under vigorous stirring for 15 min. Then the magnetic sawdust was added to the mixture and placed into a Teflon-lined stainless-steel autoclave, thereafter the autoclave was heated at 110 ℃ for 8 h. The reaction product was collected with a magnet, washed repeatedly with ethanol and distilled water, dried in a vacuum oven at 50 ℃.

2.4 Adsorption and desorption studies

The adsorption of Cu^{2+} ions by the $MSC-NH_2$ was investigated in aqueous solution at 25 ℃. The stock solution of Cu^{2+} ions were prepared by dissolving an appropriate quantity of $CuSO_4$ in double distilled water to give the final concentration range from 1 $mg \cdot L^{-1}$ to 12 $mg \cdot L^{-1}$. The final pH was adjusted to a certain pH value using 0.1 M NaOH or 0.1 M HCl. In general, 0.1 g of $MSC-NH_2$ was added to 100 mL copper-containing solution and mixed by ultrasonication for several minutes. Then, the absorbent was separated from the mixture with a permanent hand-held magnet. The concentration of Cu^{2+} ions in the solution was measured by Flame Atomic Absorption Spectrometer (TAS-990AFG). Unless otherwise specified, the absorption experiments were performed in aqueous solution at pH 6. The equilibrium sorption capacity, q_e, was calculated according to Eq. (1):

$$q_e = \frac{(C_0 - C_e)V}{m} \tag{1}$$

where C_0 ($mg \cdot L^{-1}$) is the initial concentration of metal ion, C_e ($mg \cdot L^{-1}$) is the equilibrium concentration in solution, V (L) is the total volume of solution, and m (g) is the sorbent mass.

Desorption of Cu^{2+} was performed by mixing 0.1 g of Cu^{2+}-adsorbed MSC-NH_2 in to 100 mL of HCl solution (0.1 $mol \cdot L^{-1}$) and sonicating for 60 min. Then, the MSC-NH_2 was separated from the solution by an external magnet, washed with double distilled water three times and dried for reuse. The consecutive adsorption-desorption process was carried out five times.

All the experiments were replicated three times, and the mean values were used in our analyses. If the standard error (SE) were greater than 0.05, the test was repeated to control for errors.

2.5 Characterization

The morphology and elemental compositions of the wood samples were examined by the scanning electron microscopy (SEM, FEI, and Quanta 200) combined with energy dispersive X-ray spectroscopy (EDXA, Genesis) and transmission electron microscopy (TEM, FEI, and Tecnai G20). The crystalline structure of these composites was identified using the X-ray diffraction (XRD, Rigaku, and D/MAX 2200) operating with Cu $K\alpha$ radiation ($\lambda = 1.5418$Å) at a scan rate (2θ) of $4° \ min^{-1}$, an accelerating voltage of 40 kV, and an applied current of 30 mA ranging from 5° to 80°. The surface chemical compositions of the samples were determined via the Fourier transformation infrared spectroscopy (FTIR, Nicolet, Magna-IR 560). The X-ray photoelectron spectroscope (XPS) was recorded on the Thermo ESCALAB 250XI. The deconvolution of the overlapping peaks was performed using a mixed Gaussian-Lorentzian fit program. The magnetization measurements were carried out at room temperature using a superconducting quantum interference device (MPMS XL-7, Quantum Design Corp.). The Brunauer-Emmett-Teller (BET) nitrogen absorption was carried out with a Micromeritics JW-BK132F instrument (Beijing JWGB Sci. & Tech. co., Ltd) to get the pore size distribution of as-prepared sawdust. Samples were degassed at 60 ℃ for 5 h, followed by analyzing at −196 ℃ N_2 physisorption. The data was collected at a relative pressure between 0.05 and 0.3. The zeta potential of sawdust suspension was determined by the Zetasizer NANO ZS potential analyzer (Malvern Instrument Ltd.). The potassium dichromate method was used to determine the chemical oxygen demand (COD).

3 Results and discussion

3.1 Characterization of the adsorbent

The SEM images in Fig. 2 illustrate the surface morphologies of the original and treated sawdust. Fig. 2a shows the typical SEM image of the poplar sawdust surface, the pits and some sawdust particles found on the walls of tracheids. However, after chemical modification the micrographs reveal the presence of magnetic nanoparticles on the sawdust surface (2b and 2c). The TEM images in Fig. S1 shows that the particles with the sizes ranging from 6 to 22 nm are clearly visible. EDXA spectrum present in Fig. 2c shows the chemical elements of $MSC\text{-}NH_2$: carbon, oxygen, nitrogen, iron and gold (from the coating for SEM imaging). Moreover, the SEM image of the $MSC\text{-}NH_2$ and corresponding elemental maps of C, Fe and N (Fig. S2) can confirm that the amine functionalized magnetic $\gamma\text{-}Fe_2O_3$ nanoparticles are coated on wood surface homogeneously.

Fig. 2d shows the XRD patterns of $\gamma\text{-}Fe_2O_3$ nanoparticles, wood sawdust and $MSC\text{-}NH_2$. From the XRD pattern of $\gamma\text{-}Fe_2O_3$ nanoparticles, it can be seen that all the diffraction peaks are readily indexed to the face-centered cubic structure of maghemite (JCPDS NO. 39–1346). The positions at 15° and 23° of wood sawdust represented the crystalline region of the cellulose (Li et al., 2011). Compared with the XRD patterns of $\gamma\text{-}Fe_2O_3$ nanoparticles and wood sawdust, all the diffraction peaks of $\gamma\text{-}Fe_2O_3$ and sawdust exist in the $MSC\text{-}NH_2$, indicating the $\gamma\text{-}Fe_2O_3$ nanoparticles coated on the sawdust surface effectively.

To provide direct proof for the amine functionalization, FTIR spectroscopy was also used to characterize the amine-functionalized magnetic sawdust composites (Fig. 2e). In both spectra, the broad and intense absorption peak at 3420 cm^{-1} is attributed to the O—H stretching vibrations of cellulose, hemicelluloses, lignin and absorbed water. The peak observes at 2917 cm^{-1} corresponds to the C—H stretching vibrations of methyl and methylene groups. The peaks at 1731 cm^{-1} and 1594 cm^{-1} in the untreated sawdust spectrum show the C=O stretching vibration of the carboxyl groups and the aromatic stretching vibration of lignin, respectively. The peak around 1426 cm^{-1} can be attributed to the C—H deformation vibrations of methyl, methylene and methoxy groups. Moreover, the bands in the range 1300–1000 cm^{-1} can be assigned to the C—O stretching vibration of carboxylic groups in hemicelluloses and lignin in the untreated sawdust (Gierlinger et al., 2008; Rana et al., 2010). After magnetic nanoparticles modification, the peak at 573 cm^{-1} attributes to the Fe—O bond vibration of $\gamma\text{-}Fe_2O_3$ is observed obviously (Wang et al., 2006). For FTIR spectrum of $MSC\text{-}NH_2$, the broad band with the range from 3000 to 3600 cm^{-1} is due to the O—H and overlapped N-H stretching vibration. The new bands observe at 1625, 1480, and 878 cm^{-1} from the amine-functionalized magnetic wood composites match well with that from free 1, 6-hexadiamine, indicating the existence of the —NH_2 group in the amine-functionalized materials (Wang et al., 2006; Liu et al., 2012). The results from FTIR reveal that these composites are functionalized with magnetic nanoparticles and amino groups in the synthetic process.

The magnetic properties of $MSC\text{-}NH_2$ were studied using the superconducting quantum interference device (MPMS). The magnetic hysteresis loops measured at room temperature are illustrated in Fig. 2f. The M-H curve shows that the diamagnetic property of untreated sawdust and the magnetization saturation of untreated sawdust obtained at room temperature is 0 emu · g^{-1}. However, $MSC\text{-}NH_2$ exhibits a polar property and shows a typical paramagnetism. The saturation magnetization (M_S) of $MSC\text{-}NH_2$ obtained at room temperature is 7.0 emu · g^{-1}, which are much smaller than that of the bulk $\gamma\text{-}Fe_2O_3$(M_S = 76 emu · g^{-1}). The above differences can be explained by the sawdust substrate with partial diamagnetic substitutions not only make a random canting of magnetic particles surface spins but also effectively reduce the agglomeration of magnetic particles, thus reduce the saturation magnetization of the composites (Lu et al., 2007; Berkowitz et al., 1968). The inset indicates that the as-synthesized adsorbents disperse in water are easily collected by external magnetic field, which can reduce the separation of such absorbents from solution. Such facile separation is essential to improve the operation efficiency and reduce the cost during wastewater treatment. These results suggest that the

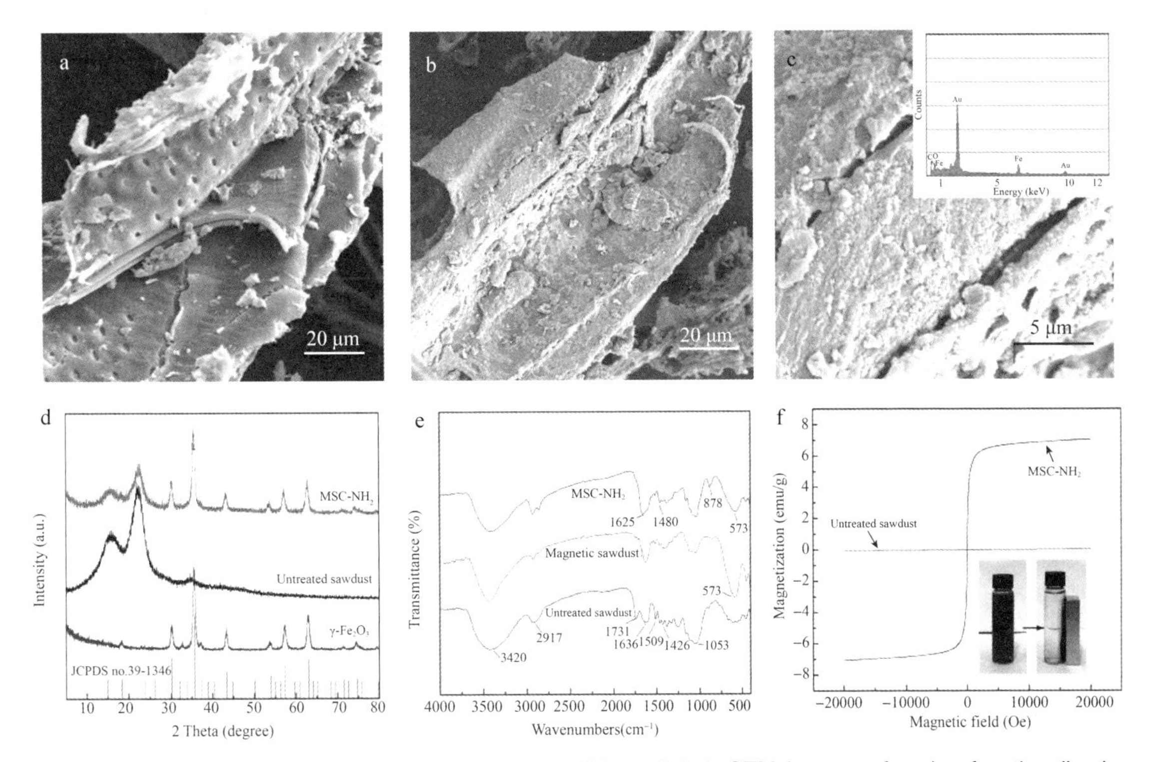

Fig. 2 (a) SEM images of untreated sawdust; (b) and (c) SEM images of amine functionalized magnetic sawdust; The inset in panel c shows the EDXA spectrum of the $MSC\text{-}NH_2$; (d) XRD patterns of as-prepared samples; (e) FTIR spectra of untreated sawdust, magnetic sawdust and $MSC\text{-}NH_2$, respectively; (f) Room-temperature hysteresis loops of untreated sawdust and $MSC\text{-}NH_2$

$MSC\text{-}NH_2$ with good magnetic property can be used as recyclable adsorbents.

The textural characteristics of the untreated sawdust, $\gamma\text{-}Fe_2O_3$-modified sawdust and $MSC\text{-}NH_2$ calculated from nitrogen adsorption-desorption isotherms are listed in Table 1. As shown in Table 1, the BET surface area and total volume of untreated sawdust are 4.289 $m^2 \cdot g^{-1}$ and 0.0212 $cm^3 \cdot g^{-1}$, respectively. BET surface area and total pore volume are increase almost 5 times after modification with $\gamma\text{-}Fe_2O_3$ nanoparticles; this indicates that the modified sawdust can provide a high level of contact area and sufficient active sites for Cu^{2+} removal. However, after grafting with amine groups on magnetic sawdust surface, BET surface area and total pore volume decrease, because grafting of amino groups onto the pore surface produces contraction of the mesopore (Wang et al., 2015). Furthermore, the average pore diameters and micropore volume in the order $\gamma\text{-}Fe_2O_3$-modified sawdust>$MSC\text{-}NH_2$>untreated sawdust.

Table 1 BET parameters of as-prepared sawdust

Sample	BET surface area ($m^2 \cdot g^{-1}$)	Average pore diameter (nm)	Total pore volume ($cm^3 \cdot g^{-1}$)	Micropore volume ($cm^3 \cdot g^{-1}$)
Untreated sawdust	4.289	5.673	0.0212	0.00149
Magnetic sawdust	22.357	12.339	0.113	0.00788
$MSC\text{-}NH_2$	4.327	6.561	0.0268	0.00158

3.2 Effect of chemical pretreatment on Cu^{2+} adsorption

The sorption capacity of Cu^{2+} on the untreated sawdust, magnetic sawdust and MSC-NH_2 was examined, and the results are shown in Table 2. The results exhibit that magnetic sawdust more effectively adsorb Cu^{2+} ion than the untreated sawdust performed, which could be caused by the large BET surface area of magnetic sawdust. However, there is a remarkable difference in removal efficiency of Cu^{2+} ions between magnetic sawdust and MSC-NH_2, and MSC-NH_2 exhibits larger adsorption capacity than magnetic sawdust. Considering the difference between magnetic sawdust and MSC-NH_2, the lager adsorption capacity of MSC-NH_2 attributes to the amino groups modified on the surface of magnetic sawdust. Moreover, the chemical modification of sawdust makes MSC-NH_2 to absorb more Cu^{2+} ions without an increase in COD, which is due to the release of the organic compounds such as lignin, pectin and tannin in the chemical pretreatment process.

Table 2 Effect of chemical pretreatment on Cu^{2+} adsorption

Adsorbent	COD ($mg\ O_2 \cdot L^{-1}$)	Cu^{2+} sorption capacity
Untreated sawdust	184	3.12
Magnetic sawdust	55	3.81
MSC-NH_2	56	7.52

Conditions: pH: 6.0; contact time: 150 min; $C_0 = 10\ mg \cdot L^{-1}$; adsorbent concentration: $1\ g \cdot L^{-1}$.

3.3 Effect of contact time and adsorption kinetics

To examine the influence of time on Cu^{2+} adsorption, kinetics experiments were carried out by adding 0.1 g new prepared MSC-NH_2 to 100 mL solution containing 2.0, 5.0 and 7.0 $mg \cdot L^{-1}$ Cu^{2+} at pH 6.0 and 25 ℃. The results shown in Fig. 3 reveal that the adsorption equilibriums are reached within 150 min. The initial Cu^{2+} concentration do not have a significant effect on the time to reach equilibrium.

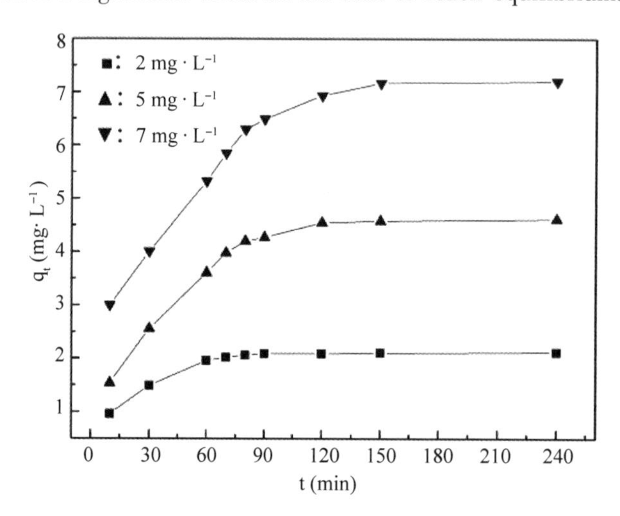

Fig. 3 Effect of contact time and initial metal concentration on sorption capacity of Cu^{2+} on MSC-NH_2 (pH 6.0, MSC-NH_2 dosage $1\ g \cdot L^{-1}$, temperature 25 ℃)

The experimental data of adsorption for Cu^{2+} onto MSC-NH_2 are analyzed using Lagergren pseudo-first order, pseudo-second order, Elovich and intra-particle diffusion kinetic models. The uniformity between the experimental data and model-predicted values is expressed by the correlation coefficients R^2.

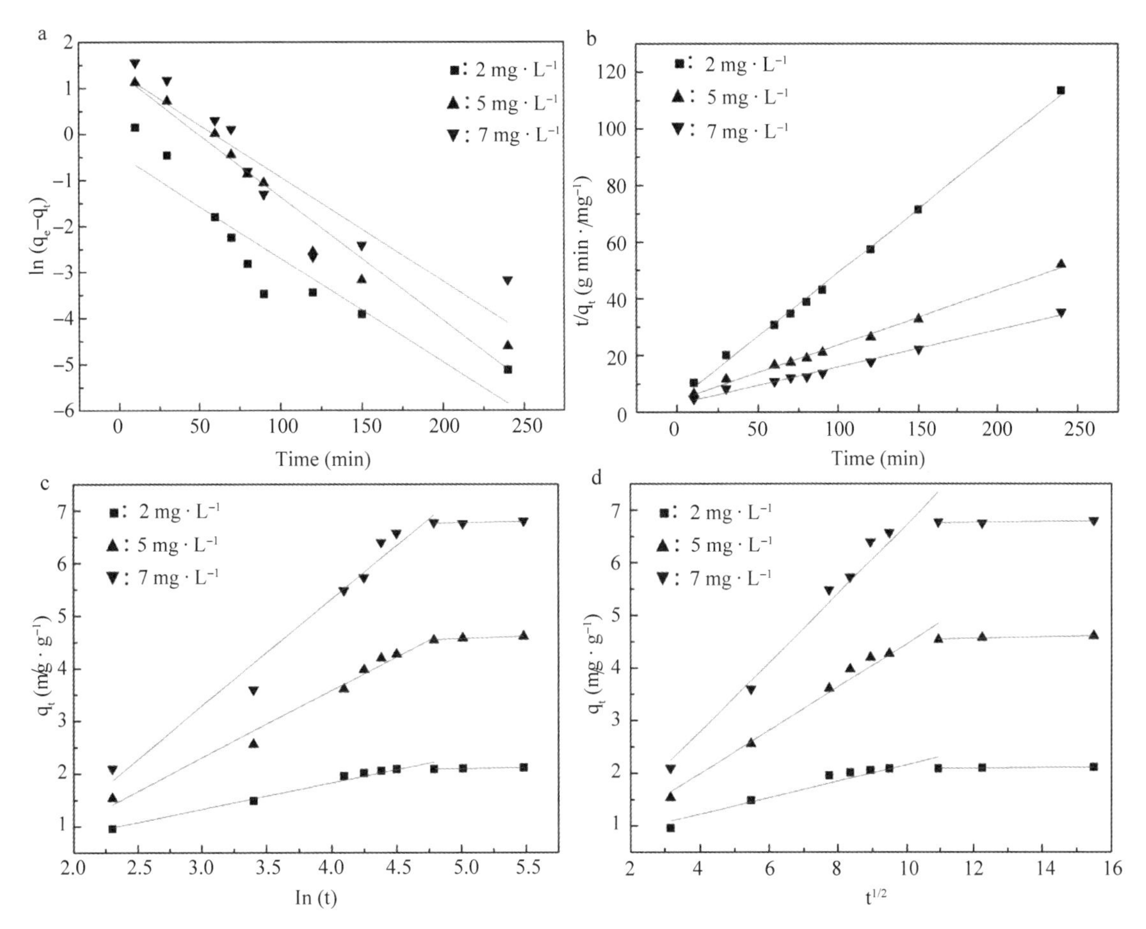

Fig. 4 The fitting of different kinetic models for Cu^{2+} adsorption onto MSC-NH_2 at different initial concentration (2 mg · L^{-1}, 5 mg · L^{-1}, and 7 mg · L^{-1}) at 25 ℃ (a lagergrenpseudo-first order, b Pseudo-second order, c Elovich, d Intra-particle diffusion)

The Lagergren pseudo-first order model is expressed as

$$\ln(q_e - q_t) = \ln q_e - k_1 t \tag{2}$$

where q_e and q_t are the amounts of the metal ions adsorbed (mg · g^{-1}) at equilibrium and at time t(min), respectively, and k_1 (min^{-1}) is the constant of pseudo first-order rate. Fig. 4a shows the fitting result. The plots of $\ln(q_e - q_t)$ *vs.* t gave straight lines under different initial concentrations. The calculated k_1, q_e and R^2 are presented in Table 3.

Table 3 Kinetic parameters of different models for Cu^{2+} ions adsorption onto MSC-NH_2

Kinetic models and parameters	Cu^{2+} C_0(mg · L^{-1})		
	2.0	5.0	7.0
q_e(exp.) (mg · L^{-1})	2.12	4.62	6.84
Lagergren pseudo-first-order			
q_e(calc.) (mg · L^{-1})	0.64	3.78	3.82

(续)

Kinetic models and parameters	Cu^{2+} C_0(mg·L^{-1})		
	2.0	5.0	7.0
k_1	0.0221	0.0268	0.0226
R^2	0.841	0.895	0.814
Pseudo-second-order equation			
q_e(calc.) (mg·L^{-1})	2.91	4.76	7.35
k_2(g·mg^{-1}·min^{-1})	0.010	0.011	0.013
H (mg·g^{-1}·min^{-1})	0.086	0.23	0.71
R^2	0.9975	0.9940	0.9902
Elovich equation			
q_e(calc.) (mg·L^{-1})	3.62	2.98	3.28
α(mg·g^{-1}·min^{-2})	0.355	0.237	0.509
β(g mg^{-1}·min^{-1})	1.036	0.776	0.492
R^2	0.9655	0.9800	0.9682
Intra-particle diffusion equation			
q_e(calc.) (mg·L^{-1})	2.51	5.38	8.19
k_{intra}(mg·g^{-1}·$min^{1/2}$)	0.156	0.410	0.655
C	0.602	0.383	0.194
R^2	0.8852	0.9691	0.9603

The linearized form of pseudo-second order kinetic rate equation is:

$$\frac{t}{q_t}=\frac{1}{k_2 q2_e}+\frac{t}{q_e} \tag{3}$$

where q_e and q_t are the amounts of the metal ions adsorbed (mg·g^{-1}) at equilibrium and at time t (min), respectively and k_2(g·mg^{-1}·min^{-1}) is the constant of pseudo-second-order rate. Fig. 4b shows the fitting result. The plots of t/q_t *vs.* t give straight lines under different initial concentrations. The calculated q_e, k_2 and R^2 are presented in Table 3.

The Elovich kinetic equation is described as Eq. (4), which can be used to interpret the kinetics of chemisorpotion on heterogeneous sorbents.

$$q_t=\frac{\ln(\alpha\beta)}{\beta}+\frac{\ln(t)}{\beta} \tag{4}$$

where α and β are the initial adsorption rate (mg·g^{-1}·min^{-1}) and the desorption constant (mg·g^{-1}·min^{-1}), respectively, and q_t is the amounts of the metal ions adsorbed at time t (min). As shown in Fig. 4c, before reaching the adsorption equilibrium, the plots show a linear relationship. Elovich constants are calculated from the slope and intercept of these graphs and are presented in Table 3.

The intra-particle diffusion model is expressed as Eq. (5), which can be used to explore the possibility of intra-particle diffusion of species from the solution to the solid phase.

$$q_t=k_{intra}t\frac{1}{2}+c \tag{5}$$

where k_{intra} is the intra-particles diffusion rate constant (mg·g^{-1}·min^{-1}), which may be taken as a rate factor, i.e., percent Cu^{2+} adsorbed per unit time, c is a constant related with the boundary layer thickness

($mg \cdot g^{-1}$). Generally, a straight linear plot of q_t *vs.* $t^{1/2}$ reveals that the sorption process is controlled by intra-particle diffusion only. The multi-liner plots represent that two or more steps influence the sorption process.

As shown in Fig. 4d, the plot of q_t *vs.* $t^{1/2}$ for Cu^{2+} exhibit two stages, the first straight portion representing macropore diffusion and the second depicting micropore diffusion (Tan et al., 2012). All calculated parameters k_{intra} and c are listed in Table 3.

From the parameters listed in Table 3, it is obvious that the adsorption kinetic follows the pseudo-second kinetic model better than the Lagergren pseudo-first order, Elovich and intra-particle diffusion kinetic models for its higher correlation coefficients (R^2). Moreover, the theoretical vales of q_e from the pseudo-second order also agrees well with the experimental values, suggesting that the adsorption process is controlled by chemical adsorption (Wang et al., 2015).

3.4 Effect of pH on Cu^{2+} ions adsorption

The pH value of the aqueous solution is an important factor in adsorption process. The effect of solution pH on the adsorption of Cu^{2+} ions by MSC-NH_2 at 25 ℃ for 150 min with an initial Cu^{2+} ion concentration of 2 $mg \cdot L^{-1}$ is showed in Fig. 5. As presented in Fig. 5a, the adsorption capacity of Cu^{2+} was low, dropping off to nearly zero when the pH is about 2. A growth is observed at higher pH levels the removal efficiency increased from 20.5% to 97.9% when the initial pH varies from 4 to 6. The variation of Cu^{2+} adsorption with solution pH could be explained by amine complexation adsorption mechanism, as shown in Eqs. (6) and (7).

$$MSC\text{-}NH_2 + H^+ = MSC\text{-}NH_{3+} \quad (6)$$

$$MSC\text{-}NH_2 + Cu^{2+} = MSC\text{-}NH_2Cu^{2+} \quad (7)$$

Eq. (6) indicates the protonation reactions of NH_2 to form NH_{3+} at low pH, and Eq. (7) reveals that Cu^{2+} ions form complexes with the amino groups on the surface of MSC-NH_2. As more NH_2 groups are protonated to form NH_{3+} at lower pH values, the adsorption sites on MSC-NH_2 surface for Cu^{2+} adsorption through Eq. (7) are passivated and hence metal adsorption is suppressed. Conversely, the reaction in Eq. (6) proceeded to the left with increasing solution pH, and more NH_2 sites are exposed on the surface of MSC-NH_2 for Cu^{2+} ion adsorption, thus increasing the adsorption capacity. It is supported that the isoelectric point (pI) of MSC-NH_2 is 5.4, indicating that the amino-functionalized magnetic sawdust are positively charged at $pH<5.4$ (Fig. 5b).

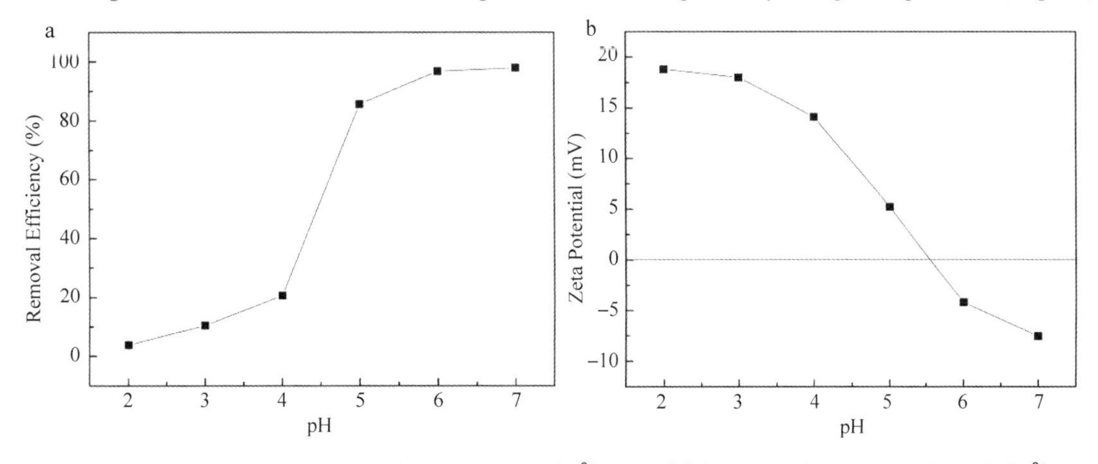

Fig. 5 (a) Effect of pH on the adsorption of Cu^{2+} by MSC-NH_2. Initial concentration of Cu^{2+}: 2 $mg \cdot L^{-1}$; MSC-NH_2: 0.1 g; solution volume: 100 mL. (b) Zeta potentials of MSC-NH_2

Earlier speciation data also reveals that the Cu^{2+} precipitated as $Cu(OH)_2$ when the concentration of Cu^{2+} is larger than 12 mg · L^{-1} at pH 6.0, according to solubility product constant of $Cu(OH)_2$ [$\log(K_{sp})$ = 19.66] (Hao et al., 2010). When values are over 6.0, various neutral hydrolysis species are presented. In order to ensure that the heavy metal exist in their ionic states during adsorption, in this study, the largest concentration of Cu^{2+} was 12 mg · L^{-1} and the pH of the solution was controlled to below 6.

3.5 Effect of initial Cu^{2+} concentration and adsorption isotherms

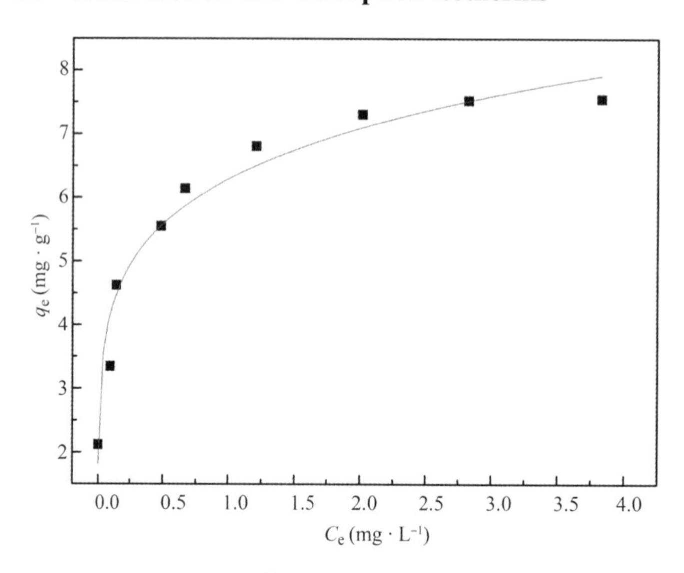

Fig. 6 Adsorption isotherms of Cu^{2+}. Conditions: pH: 6.0; contact time: 150 min; Temperature: 25 ℃; Adsorbent conceOntration: 1 g · L^{-1}

Fig. 6 presents the adsorption isotherms of Cu^{2+} on MSC-NH_2. The Langmuir and Freundlich isotherm models were selected in this study. The Langmuir isotherm is based on monolayer adsorption onto the surface with a finite number of active sites of the adsorbent and is depicted in linear form:

$$\frac{C_e}{q_e}=\frac{1}{K_L q_m}+\frac{C_e}{q_m} \tag{8}$$

where C_e is the equilibrium Cu^{2+} concentration (mg · L^{-1}) and q_e represents the amount of Cu^{2+} adsorbed into unit mass of absorbent (mg · g^{-1}); q_m and K_L are the maximum adsorption capacity (mg · g^{-1}) and the Langmuir adsorption equilibrium constant (L · mg^{-1}), respectively. They can be obtained from the slop and intercept of linear plot of C_e/q_e *vs.* C_e and all the parameters are listed in Table 4. The dimensionless constant separation factor R_L also is determined to predict the type of adsorption process, which can be represented as follows:

$$R_L=\frac{1}{1+K_L C_0} \tag{9}$$

Here C_0 is the initial Cu^{2+} concentration (mg · L^{-1}). The value of R_L indicates the type of isotherm to be linear ($R_L=1$), favorable ($0<R_L<1$) or unfavorable ($R_L=0$). In this study, values of R_L ranging from 0.403 to 0.784 suggested the favorable adsorption between MSC-NH_2 and Cu^{2+} ions.

The Freundlich model was applied for multilayer sorption and non-ideal sorption on heterogeneous surfaces, and can be represented in linear form as follows:

$$\ln q_e = \ln K_F + \frac{1}{n}\ln C_e \tag{10}$$

where K_F and n are the Freundlich constants related to the adsorption capacity ($mg \cdot g^{-1}$) and the heterogeneity factor (mg^{-1}), respectively. These constants are evaluated from slop and intercept of the linear plots of $C_e/q_e vs. C_e$, respectively. All the parameters are listed in Table 4.

Table 4 Langmuir and Freundlich constants for the adsorption of Cu^{2+} ions onto MSC-NH_2

Langmuir			
q_m($mg \cdot g^{-1}$)	K_L($L \cdot mg^{-1}$)	R_L range	r_L^2
7.72	0.13	0.403-0.784	0.9984
Freundlich			
n	K_F($L\ g^{-1}$)	r_F^2	
5.95	6.39	0.9163	

From Table 4, the high coefficient of determination ($r^2 = 0.9984$) shows that the linear Langmuir equation gives a good fit to the adsorption isotherm. The Freundlich model is not able to describe the relationship between the amount of adsorbed Cu^{2+} and its equilibrium concentration in the solution. According to the Langmuir equation, the adsorption capacity of Cu^{2+} on MSC-NH_2 was calculated to be 7.72 $mg \cdot g^{-1}$. In addition, a comparison of the Cu^{2+} adsorption capacity of MSC-NH_2 with those of some other adsorbents reported in literature is given in Table 5. It is obvious that the adsorption capacity of MSC-NH_2 is higher than most of the other adsorbents under similar conditions. Therefore, the studies above indicate that MSC-NH_2 prepared in this study could be efficient for Cu^{2+} removal in aqueous solutions.

Table 5 Adsorption properties comparison of various adsorbents for Cu^{2+} removal

Adsorbent	q_{max}($mg \cdot L^{-1}$)	References
Sawdust	1.5	(Larous et al., 2005)
Low-rank Turkish coal	1.6	(Karabulut et al., 2000)
Expanded perlite (EP)	1.95	(Ghassabzadeh et al., 2010)
Shells of rice	2.95	(Aydın et al., 2008)
Modified oak sawdust	3.5	(Argun et al., 2007)
Corn stalk	3.74	(Šćiban et al., 2008)
Sawdust	5.76	(Šćiban et al., 2007)
Soybean straw	5.4	(Šćiban et al., 2008)
H_3PO_4-activated rubber wood sawdust	5.7	(Kalavathy et al. 2005)
Peat	6.2	(Hanzlik et al. 2004)
Oak sawdust	7.04	(Sciban et al. 2006)
Bentonite	9.72	(Olu-Owolabi et al. 2010)
MSC-NH_2	7.5	Present study

3.6 Effect of contact temperature and adsorption thermodynamics

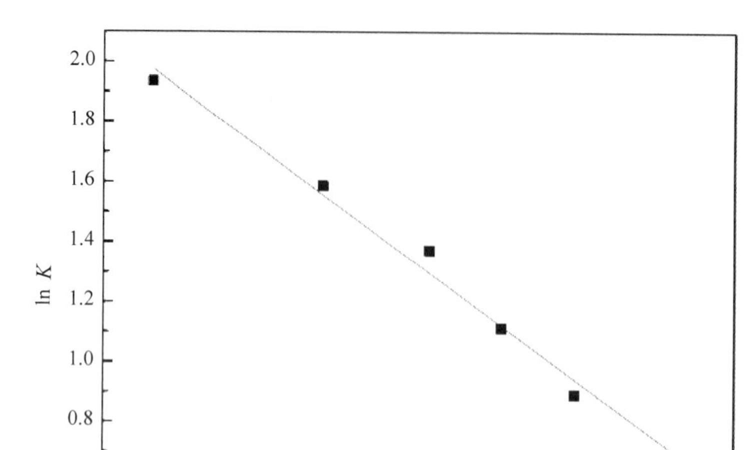

Fig. 7 Plot of ln *K* vs. 1/T to predict the thermodynamic parameters for the adsorption of Cu^{2+} ions onto MSC-NH_2

The influence of temperature on Cu^{2+} ions adsorption onto MSC-NH_2 was carried out at a temperature ranging from 298 to 313 K. The energy and entropy were considered to determine what process would take place spontaneously. The change in free energy ΔG^0, enthalpy ΔH^0 and entropy ΔS^0 associated with the adsorption process were obtained by the following equations:

$$\Delta G^0 = -RT\ln K \quad (11)$$

$$\ln K = -\frac{\Delta H^0}{RT} + \frac{\Delta S^0}{R} \quad (12)$$

where K (q_e/C_e) is the equilibrium constant, R is the gas constant [8.314 J · (mol K)$^{-1}$], T the absolute temperature (K). ΔG^0, ΔH^0 and ΔS^0arecalculated from the slope and intercept of the plot of ln K *vs.* $1/T$ (Fig. 7), and all values were collected and are shown in Table 6. The negative ΔG^0 values indicated that the adsorption is spontaneous, and the decreasing of ΔG^0 as temperature raised confirms that the adsorption is more favorable at high temperatures. The positive ΔH^0 values suggests the endothermic nature of adsorption, which is also supported by the increase in value of Cu^{2+} uptake with the raised temperature. The positive ΔS^0 values can be used to describe the increasing randomness at the MSC-NH_2-solution interface during the adsorption.

Table 6 Thermodynamic parameters of Cu^{2+} adsorption onto MSC-NH_2 at different temperatures

T (K)	ln K (L · mg^{-1})	ΔG^0(kJ · mol^{-1})	ΔH^0(kJ · mol^{-1})	ΔS^0[J · (mol K)$^{-1}$]
298	0.67	-16.62	67.23	224.1
301	0.89	-22.39		
303	1.11	-28.07		
305	1.37	-34.80		
308	1.58	-40.61		
313	1.93	-50.34		

4 Desorption and regeneration studies

Taking the practical application into account, the reuse performance of the $MSC\text{-}NH_2$ adsorbent is very important. Desorption and regeneration studied were carried out using 100 mL 0.1 mol · L^{-1} HCl as eluted solution and was repeated 5 times using the same adsorbent. The adsorbent was washed with distilled water before each adsorption. Table 7 shows $MSC\text{-}NH_2$ can be used repeatedly without significant loss in their adsorption capacity. After five cycles, the removal efficiency of Cu^{2+} decreases from 97.0% to 78.4%. The removal efficiency of the recycled $MSC\text{-}NH_2$ exhibits a loss of 20.9 % in the five cycles. The loss of adsorption capacity may be attributed to the detachment of the amino groups from the magnetic sawdust surface during the recycling processes, as proved by the element content analysis that the nitrogen element content is decreased from 9.49 wt% of the first cycle to 3.72 wt% of the fifth cycle (Table. S1). Further investigations are necessary to develop strategies to improve the regeneration performance of the synthetic magnetic adsorbent.

Table 7 Five cycles of Cu^{2+} adsorption-desorption with 0.1 mol · L^{-1} HCl as the desorbing agent

Cycles	Amount of Cu^{2+} before adsorption (mg · L^{-1})	Amount of Cu^{2+} after adsorption (mg · L^{-1})	Adsorption (%)	Amount of Cu^{2+} desorbed (mg · L^{-1})	Recovery (%)
1	5	0.15	97.0	4.75	97.9
2	5	0.23	95.4	4.55	95.4
3	5	0.46	90.8	4.09	90.1
4	5	0.71	85.8	3.70	86.3
5	5	1.19	76.1	2.66	70.1

5 Uptake mechanism

XPS analysis of the $MSC\text{-}NH_2$ before and after adsorption of Cu^{2+} was carried out for understanding of the adsorption mechanism. As shown in Fig. S3, the elemental composition of $MSC\text{-}NH_2$ is determined and peaks are successfully assigned to the corresponding carbon, oxygen, nitrogen, iron and copper atoms. The Fe 2p peaks are located at 711.2eV and 724.8 eV, which indicated the presence of $\gamma\text{-}Fe_2O_3$ in the composite, further confirming the magnetic property of $MSC\text{-}NH_2$. In Fig. S3 (c), before Cu^{2+} adsorption, the main peak in the N 1s spectrum at BE of about 399.2 eV could be attributed to the nitrogen atom of free-NH_2. After Cu^{2+} adsorption, two peaks appear in the N 1s spectrum [Fig. S3 (d)]. The new peak at a BE of 402.0 eV indicates a decrease in electron density for the nitrogen atoms, which might explained by the formation of NH_2Cu^{2+} complexes, a lone pair of electrons in the nitrogen atom was donated to the shared bond between N and Cu^{2+}, thus resulting in a higher BE peak observed (Jin et al., 2011). Similar results were obtained for the Cu $2p_{3/2}$ spectrum, considering that the binding energy of Cu^{2+} located in the ionic binding is normally at 935 eV (Sheng et al., 2004). Given that Cu^{2+} with lower electron densities can gain electrons from the amino group, copper with lower binding energy was observed [Fig. S3 (j)]. In short, strong complexation occurred in the adsorption of Cu^{2+}. In addition, the carbon atoms in the C 1s spectrum are considered in the forms of C—C (284.6 eV) and C—O (285.5 eV) [Fig. S3 (e)]. The oxygen atoms in the O 1s spectrum are in the forms of O—C (532. 1 eV) and O—Fe (530. 0 eV) [Fig. S3 (g)]. No significant changes were observed in the C 1s and O 1s spectra before and after copper ion adsorption. As mentioned above, the XPS results support an adsorption mechanism of covalent bonds formed between Cu^{2+} and amino groups.

6 Conclusions

In this study, a novel magnetic sawdust adsorbent for the removal of Cu^{2+} ions from aqueous solution was synthesized by precipitating γ-Fe_2O_3 particles on natural poplar sawdust substrate and then grafting the amino groups on the composite surfaces. The presence of γ-Fe_2O_3 particles and amino groups on the product were supported by the examinations of the scanning electron microscopy, x-ray diffraction and FTIR spectroscopy. Compared with the untreated sawdust, the adsorption of MSC-NH_2 increased 2. 5 times for Cu^{2+} ions. This was due to the increase in amino groups imparted onto sawdust. More importantly, the Cu^{2+}-loaded MSC-NH_2could be quickly collected by magnetic separation and regenerated easily by acid treatment. Four adsorption kinetic models were applied to fit the experimental data and the pseudo-second-order kinetic was proved to be the best kinetic model, suggesting that the adsorption was chemical adsorption process. Besides, the adsorption process also followed the Langmuir adsorption isotherm model. Thermodynamic parameters indicated that the adsorption was spontaneous and endothermic in nature. Coupled with the advantages of low cost, easy separation, environmental friendliness and appropriate adsorption capacity for Cu^{2+} solution, the amino-functional magnetic sawdust could be regarded as a potential candidate for effective Cu^{2+} ion adsorbent.

Acknowledgments

This work was financially supported by The National Natural Science Foundation of China (grant no. 31470584).

References

[1]Argun M E, Dursun S, Ozdemir C, et al. Heavy metal adsorption by modified oak sawdust: thermodynamics and kinetics[J]. J. Hazard. Mater., 2007, 141(1): 77-85.

[2]Aydın H, Bulut Y, Yerlikaya C. Removal of copper (II) from aqueous solution by adsorption onto low-cost adsorbents [J]. J. Environ. Manage., 2008, 87(1): 37-45.

[3]Berkowitz A E, Schuele W J, Flanders P J. Influence of crystallite size on the magnetic properties of acicular γ-Fe_2O_3 particles[J]. J. Appl. Phys., 1968, 39(2): 1261.

[4]Espana J S, Pamo E L, Pastor E S, et al. The removal of dissolved metals by hydroxysulphate precipitates during oxidation and neutralization of acid mine waters[J]. Aquat. Geochem., 2006, 12(3): 269-298.

[5]Feng N C, Guo X Y, Liang S. Adsorption study of copper(II) by chemically modified orange peel[J]. J. Hazard. Mater., 2009, 164(2-3): 1286-1292.

[6]da Fonseca M G, de Oliveora M M, Arakaki L N H, et al. Natural vermiculite as an exchanger support for heavy cations in aqueous solution[J]. J. Colloid Interface Sci., 2005, 285(1): 50-55.

[7]Ghassabzadeh H, Mohadespour A, Torab-Mostaedi M, et al. Adsorption of Ag, Cu and Hg from aqueous solutions using expanded perlite[J]. J. Hazard. Mater., 2010, 177(1-3): 950-955.

[8]Gierlinger N, Goswami L, Schmidt M, et al. In situ FT-IR microscopic study on enzymatic treatment of poplar wood cross-sections[J]. Biomacromolecules, 2008, 9(8): 2194-2201.

[9]Hanafiah M A K M, Shafiei S, Harun M K, Yahya M Z A. Kinetic and thermodynamic study of Cd^{2+} adsorption onto rubber tree (hevea brasiliensis) leaf powder[J]. Mater. Sci. Forum, 2006, 517: 217-221.

[10]Hanafiah M A K, Ibrahim S C, Yahaya M Z A. Equilibrium adsorption study of lead ions onto sodium hydroxide modified lalang (imperata cylindrica) leaf powder[J]. J. Appl. Sci. Res., 2006, 2: 1169-1174.

[11]Hanzlik P, Jehlicka J, Weishauptova Z, et al. Adsorption of copper, cadmium and silver from aqueous solutions onto natural carbonaceous materials[J]. Plant, Soil Environ., 2004, 50(6): 257-264.

[12]Hao Y M, Man C, Hu Z B. Effective removal of Cu(II) ions from aqueous solution by amino-functionalized magnetic nanoparticles[J]. J. Hazard. Mater., 2010, 184(1-3): 392-399.

[13]Ho Y S, Chiu W T, Hsu C S, et al. Sorption of lead ions from aqueous solution using tree fern as a sorbent[J]. Hydrometallurgy, 2004, 73(1-2): 55-61.

[14]Hu X Y, Zhao M M, Song G S, et al. Modification of pineapple peel fibre with succinic anhydride for Cu^{2+}, Cd^{2+} and Pb^{2+} removal from aqueous solutions[J]. Environ. Technol., 2011, 32(7): 739-746.

[15]Jin X L, Yu C, Li Y F, et al. Preparation of novel nano-adsorbent based on organic-inorganic hybrid and their adsorption for heavy metals and organic pollutants presented in water environment[J]. J. Hazard. Mater., 2011, 186(2-3): 1672-1680.

[16]Kalavathy M H, Karthikeyan T, Rajgopal S, et al. Kinetic and isotherm studies of Cu(II) adsorption onto H_3PO_4-activated rubber wood sawdust[J]. J. Colloid Interface Sci., 2005, 292(2): 354-362.

[17]Karabulut S, Karabakan A, Denizli A, et al. Batch removal of copper(II) and zinc(II) from aqueous solutions with low-rank turkish coals[J]. Sep. Purif. Technol., 2000, 18(3): 177-184.

[18]Karnitz O, Gurgel L V A, de Melo J C P, et al. Adsorption of heavy metal ion from aqueous single metal solution by chemically modified sugarcane bagasse[J]. Bioresour. Technol., 2007, 98(6): 1291-1297.

[19]Larous S, Meniai A H, Lehocine M B. Experimental study of the removal of copper from aqueous solutions by adsorption using sawdust[J]. Desalination, 2005, 185(1-3): 483-490.

[20]Li J, Lu Y, Yang D J, et al. Lignocellulose aerogel from wood-ionic liquid solution (1-allyl-3-methylimidazolium chloride) under freezing and thawing conditions[J]. Biomacromolecules, 2011, 12(5): 1860-1867.

[21]Li Q, Zhai J P, Zhang W Y, et al. Kinetic studies of adsorption of Pb(II), Cr(III) and Cu(II) from aqueous solution by sawdust and modified peanut husk[J]. J. Hazard. Mater., 2007, 141(1): 163-167.

[22]Liu Z, Wang H S, Liu C, et al. Magnetic cellulose-chitosan hydrogels prepared from ionic liquids as reusable adsorbent for removal of heavy metal ions[J]. Chem. Commun., 2012, 48(59): 7350-7352.

[23]Lu H M, Zheng W T, Jiang Q. Saturation magnetization of ferromagnetic and ferrimagnetic nanocrystals at room temperature[J]. J. Phys. D: Appl. Phys., 2007, 40(2): 320-325.

[24]Ngah W S W, Hanafiah M A K M. Removal of heavy metal ions from wastewater by chemically modified plant wastes as adsorbents: A review[J]. Bioresour. Technol., 2008, 99(10): 3935-3948.

[25]Ogutveren U B, Koparal A S, Ozel E. Electrodialysis for the removal of copper ions from wastewater[J]. J. Environ. Sci. Health, Part A: Toxic/Hazard. Subst. Environ. Eng., 1997, 32(3): 749-761.

[26]Olu-Owolabi B I, Popoola D B, Unuabonah E I. Removal of Cu^{2+} and Cd^{2+} from aqueous solution by bentonite clay modified with binary mixture of goethite and humic acid[J]. Water, Air, Soil Pollut., 2010, 211(1-4): 459-474.

[27]Pehlivan E, Altun T, Parlayici S. Modified barley straw as a potential biosorbent for removal of copper ions from aqueous solution[J]. Food Chem., 2012, 135(4): 2229-2234.

[28]Rana R, Langenfeld-Heyser R, Finkeldey R, et al. FTIR spectroscopy, chemical and histochemical characterisation of wood and lignin of five tropical timber wood species of the family of dipterocarpaceae[J]. Wood Sci. Technol., 2010, 44(2): 225-242.

[29]Reddy D H K, Lee S M. Application of magnetic chitosan composites for the removal of toxic metal and dyes from aqueous solutions[J]. Adv. Colloid Interface Sci., 2013, 201: 68-93.

[30]Sciban M, Klasnja M, Skrbic B. Modified hardwood sawdust as adsorbent of heavy metal ions from water[J]. Wood

Sci. Technol., 2006, 40(3): 217-227

[31]Sciban M, Klasnja M, Skrbic B. Adsorption of copper ions from water by modified agricultural by-products[J]. Desalination, 2008, 229(1-3): 170-180.

[32]Sciban M, Radetic B, Kevresan D, et al. Adsorption of heavy metals from electroplating wastewater by wood sawdust [J]. Bioresour. Technol., 2007, 98(2): 402-409.

[33]Seki K, Saito N, Aoyama M. Removal of heavy metal ions from solutions by coniferous barks[J]. Wood Sci. Technol., 1997, 31(6): 441-447.

[34]Sheng P X, Ting Y P, Chen J P, et al. Sorption of lead, copper, cadmium, zinc, and nickel by marine algal biomass: Characterization of biosorptive capacity and investigation of mechanisms[J]. J. Colloid Interface Sci., 2004, 275(1): 131-141.

[35]Song W, Gao B Y, Zhang T G, et al. High-capacity adsorption of dissolved hexavalent chromium using amine-functionalized magnetic corn stalk composites[J]. Bioresour. Technol., 2015, 190: 550-557.

[36]Tan Y Q, Chen M, Hao Y M. High efficient removal of Pb(II) by amino-functionalized Fe_3O_4 magnetic nano-particles[J]. Chem. Eng. J., 2012, 191: 104-111

[37]Tian Y, Wu M, Lin X B, et al. Synthesis of magnetic wheat straw for arsenic adsorption[J]. J. Hazard. Mater., 2011, 193: 10-16.

[38]Wang L Y, Bao J, Wang L, et al. One-pot synthesis and bioapplication of amine-functionalized magnetite nanoparticles and hollow nanospheres[J]. Chem. -Eur. J., 2006, 12(24): 6341-6347

[39]Wang S G, Wang K K, Dai C, et al. Adsorption of Pb^{2+} on amino-functionalized core-shell magnetic mesoporous SBA-15 silica composite[J]. Chem. Eng. J., 2015, 262: 897-903.

[40]Witek-Krowiak A. Application of beech sawdust for removal of heavy metals from water: Biosorption and desorption studies[J]. Eur. J. Wood Wood Prod., 2013, 71(2): 227-236.

[41]Zheng J C, Feng H M, Lam M H W, et al. Removal of Cu(II) in aqueous media by biosorption using water hyacinth roots as a biosorbent material[J]. J. Hazard. Mater., 2009, 171(1-3): 780-785.

[42]Zhou D, Zhang L N, Zhou J P, et al. Cellulose/chitin beads for adsorption of heavy metals in aqueous solution[J]. Water Res., 2004, 38(11): 2643-2650.

中文题目：氨基功能化磁性木屑复合材料去除水溶液中 Cu^{2+} 离子

作者：甘文涛，高丽坤，詹先旭，李坚

摘要：本文研究了氨基功能化磁性氧化铁/锯末复合材料作为生物吸附材料去除 Cu^{2+} 水溶液中的离子。这些复合材料是通过在锯屑基底上沉淀 γ-Fe_3O_4 纳米颗粒，然后用 1，6-己二胺官能化来制备的。通过扫描电镜、透射电镜、XRD、红外光谱、BET、MPMS 和 XPS 等手段对其进行表征，探讨其吸收机理。结果发现，氨基被接枝到锯屑表面。MSC-NH_2 能有效去除 Cu^{2+} 并在外部磁铁的帮助下方便地从溶液中分离出来。批量实验表明，吸附平衡在 150 min 内达到，在 pH=6 和室温下吸附量为 7.55 mg/g。等温线分析表明，吸附数据可用朗缪尔等温线模型表示。动力学利用拉格尔格伦拟一级反应、二级、埃洛维奇和粒子内扩散模型进行评估。热力学参数揭示了该吸附过程是自发、吸热的化学吸附过程。

关键词：磁性木屑；氨基功能化；吸附作用；铜离子

Magnetic Property, Thermal Stability, UV-Resistance and Moisture Absorption Behavior of Magnetic Wood Composites*

Wentao Gan, Ying Liu, Likun Gao, Xianxu Zhan, Jian Li

Abstract: Magnetic wood composites with improved anti-ultraviolet property and dimensional stability were prepared by modification with iron oxides (Fe_3O_4/ γ-Fe_2O_3) via a facile one-step hydrothermal method using ferric trichloride and ferrous chloride as the precursors at 90 ℃. The morphology, crystalline phase and chemical structure of the wood composites before and after hydrothermal process were characterized by scanning electron microscope, transmission electron microscope, X-ray diffraction, Raman spectroscopy and Fourier transformation infrared spectroscopy (FTIR). The results indicated that the magnetic nanoparticles precipitated on the wood substrate by adhesion to wood surface in the form of aggregates. A possible hydrothermal fabrication mechanism was proposed. After hydrothermal treatment, the magnetic wood composites exhibited appropriate magnetic and anti-ultraviolet properties. Thermal analysis showed that the incorporation of magnetic nanoparticles retarded the thermal decomposition of wood matrix and improved the thermal stability of wood. Moreover, the dimensional stability of the modified wood materials was also evaluated.

Keywords: Magnetic Wood; Iron oxides nanoparticles; Hydrothermal method; Wood modification

1 Introduction

Magnetic nanoparticles have recently attracted significant interest due to their unusual optical, electronic, and magnetic properties, which often differ from the bulk products. These properties have wide applications in the fields of magnetic storage, target-drug delivery, magnetic record media, lithium-ion batteries, catalysis, and so on[1-5].

The mixing of polymer and nanoparticles is opening pathways for engineering flexible composites that exhibit advantageous electrical, optical, or mechanical properties[6]. Nowadays, magnetically responsive cellulose materials are specific subset of smart materials due to the unique physical properties with a wide range of potential applications, such as military use, magnetic filters, sensors, and low frequency magnetic shielding[7-9]. As a kind of the cellulose-based materials, wood have been used for many applications due to the excellent material properties, aesthetic appearance, environmental aspects and superior mechanical performance[10]. The dominating functions of wood resulting in an excellent lightweight engineering material and providing a natural scaffold for further functionalization.

---

* 本文摘自 Polymer Composites, 2015, 38: 1646-1654.

Magnetic wood, which was proposed by H. Oka in 1991, magnetically attracts in spite of the fact that material itself was essentially wood[11]. In addition to strong magnetic property, it is also possesses a wood texture, low specific gravity, and is easy to process. Various synthetic routes have been employed to produce the magnetic wood, such as impregnating methods, powder methods and coating methods[12]. The authors examined the magnetic property, heating characteristics, actuation function and indoor electromagnetic wave absorbing characteristics of the magnetic wood[13-15], but the effects in terms of spatial distribution, size, aggregation state and surface properties of the incorporated particles have not been fully discussed. Recently, some studies have been prepared a new kind of magnetic wood composites by in situ precipitation of the magnetic particles in the wood structure. The morphology, crystal structures and magnetic character of the modified wood samples were studied[16-18]. However, as a natural biological materials wood was sensitive to sunlight, fire and humidity changes, which led to weathering, thermal, and dimensional instability. Few studies have been concentrated on improving the inherent disadvantages of wood such as weathering, thermal, and dimensional instability using the magnetic particles modification.

Hydrothermal solution synthesis method has been proven to a versatile route for preparation of magnetic nanoparticles due to the convenience and simplicity in the fabrication[19,20]. Simultaneously, Barata et al.[21] showed that cellulosic fibres act as efficient hydrophilic substrates to the nucleation and growth of inorganic particles in aqueous medium. Similar studies[22-24] were conducted to demonstrate that wood surface acted a substrate containing plentiful hydroxyl groups, which induced the growth of deposited inorganic particles, such as SiO_2 and TiO_2. In our previous researches[25-27], hydrothermal method has been confirmed as to an efficient and facile method for depositing inorganic nanoparticles onto the wood surface.

Herein, we prepared a kind of magnetic wood by deposited the magnetic nanoparticles (Fe_3O_4/ γ-Fe_2O_3) on the wood template via a one-step hydrothermal method. The morphology, crystalline phase and chemical structure of the wood composites before and after hydrothermal process were characterized and a possible formation mechanism of the magnetic wood composite was also proposed. The magnetic, thermal performances and UV-resistance as well as dimensional stability of the magnetic wood composites were subsequently evaluated.

2 Materials and methods

2.1 Materials

Poplar wood samples [20 mm (L) × 20 mm (T) × 20 mm (R), 20 mm (L) × 20 mm (T) × 5 mm (R), 70 mm (L) ×20 mm (T) ×5 mm (R), and 4 mm (L) × 4 mm (T) × 4 mm (R)] were ultrasonically rinsed in deionized water for 30 min and dried at 100 ℃ for 24 h in vacuum. Ferric trichloride hexahydrate ($FeCl_3 \cdot 6H_2O$), ferrous chloride tetrahydrate ($FeCl_2 \cdot 4H_2O$), and ammonia solution ($NH_3 \cdot H_2O$, 25%) were supplied by Shanghai Boyle Chemical Company Limited and used without further purification.

2.2 Preparation of the magnetic wood composites

The Teflon containers with 100 mL were used in our experiment and the wood samples with different size were separately prepared. In a typical synthetic procedure, 1.8 g $FeCl_3 \cdot 6H_2O$ and 1.8 g $FeCl_2 \cdot 4H_2O$ were dissolved in 100 mL deionized water with vigorous stirring. The pH value of the iron ion precursor was adjusted to ca. 10 by the dropwise addition of 25% ammonia solution. After stirring for 30 min, two wood samples with the size of 20 mm (L) × 20 mm (T) × 20 mm (R) were immersed in the obtained solution, then the system was transferred into a Teflon-lined stainless-steel autoclave and heated at 90 ℃ for 8 h. Finally, the prepared wood specimens were removed and ultrasonically rinsed with deionized water for 30 min, and then dried at 50 ℃ for 48 h in vacuum. The untreated wood specimens were employed as contrast.

2.3 Characterization

The morphology of the samples was observed by scanning electron microscopy (SEM, FEI, Quanta 200) and transmission electron microscopy (TEM, FEI, and Tecnai G20). Surface chemical composition of specimens was determined by energy disperse X-ray analysis (EDXA) that connects the SEM. Crystalline structure of the samples was identified by the X-ray diffraction technique (XRD, Rigaku, and D/MAX 2200) operated with Cu $K\alpha$ radiation (λ = 1.5418 Å) at a scan rate (2θ) of 4° min^{-1} and an accelerated voltage of 40 kV, with the applied current of 30 mA ranging from 5° to 80°. All Raman spectra were collected with a laser confocal Micro-Raman spectroscopy (Horiba Jobin Yvon) (λ = 532 nm). The laser power was set to low values (1 %) to avoid sample degradation. The magnetic nanoparticles used for the Raman measurement and TEM observation were from the sediment of hydrothermal solution in preparation. The surface chemical compositions of the samples were determined via the Fourier transformation infrared spectroscopy (FTIR, Nicolet, Magna-IR 560). For FTIR analysis, thin sample disks were made by grinding small portion of the magnetic wood composites and pressing them with potassium bromide. DR-UV/Visible spectra were obtained with the instrument TU - 190, Beijing Purkinje, China, equipped with an integrating sphere attachment. $BaSO_4$ was the reference.

For magnetic characterization, wood specimens used with the size of 4 mm (L) × 4 mm (T) × 4 mm (R), the magnetic properties of the composites were measured by a superconducting quantum interference device (MPMS XL-7, Quantum Design Corp.) at room temperature (300 K).

For TGA analysis, small amounts of the wood powder about 5 mg was prepared by grinding the wood block using the high-speed universal grinder. The thermal performances of the composites were examined using a thermal analyzer (TGA, SDT Q600) in the temperature range from room temperature up to 600 ℃ at a heating rate of 20 ℃ · min^{-1} under nitrogen atmosphere.

The wood samples with the size of 70 mm (L) ×20 mm (T) ×5 mm (R) were subjected to weathering using a QUV Accelerated Weathering Tester (Atlas, Chicago, IL), which allowed water spray and condensation. The samples were fixed in stainless steel holders and under irradiation of fluorescent UV light at wavelength of 340 nm with a temperature of 50 ℃ for 8 h, followed by water spray for 4 h. The average irradiance was set to 0.77 W · m^{-2}, and the spray temperature was fixed 25 ℃. The exposure time ranged from 0 to 1200 h. And the color change of the wood surface after UV irradiation were measured using the portable spectrophotometer (NF-333, Nippon Denshoku Company, Japan) with CIELAB system in accordance with the ISO-2470 standard. CIELAB L^*, a^*, b^*, parameters were measured at six locations on each specimen and average value was calculated. In the CIELAB system, L^* axis represents the lightness, and varies from 100 (white) to 0 (black); a^* and b^* are the chromaticity indices; $+a^*$ is the red direction; $-a^*$ is green; $+b^*$ is yellow; and $-b^*$ is blue. The change of L^*, a^* and b^* are calculated according to Eqs. (1)-(3).

$$\Delta a^* = a_2 - a_1 \tag{1}$$

$$\Delta b^* = b_2 - b_1 \tag{2}$$

$$\Delta L^* = L_2 - L_1 \tag{3}$$

Where Δa^*, Δb^*, and ΔL^* were the differences in the initial and final values of a^*, b^*, and L^*, respectively; a_1, b_1, and L_1 were the initial color parameters; and a_2, b_2, and L_2 were the color parameters after UV irradiation. The overall color changes (ΔE^*) were used to evaluate the total color change using Eq. 4.

$$\Delta E^* = \sqrt{(L^*{}_2 - L^*{}_1)2 + (a^*{}_2 - a^*{}_1)2 + (b^*{}_2 - b^*{}_1)2} \tag{4}$$

A lower ΔE^* value corresponds to a smaller color difference and indicates strong resistance to UV radiation.

The moisture absorption behavior and dimensional stability was evaluated by placed the wood specimens [20 mm (L) × 20 mm (T) × 20 mm (R)] in chambers containing distilled water of 20 ± 2 ℃ with a stainless steel mesh cover in accordance to the Chinese national standard GB 1934. 1-2009 "Testing methods for physical and mechanical properties of wood". The specimens were weighed and measured after 1, 2, 3, 4, 5, 6, 7, 8, 9, 10, 12, 14, 16, 18, 20, 24, 28, 30, 40, 50, 60, 70, 80, and 90 days of immersion.

Weight percentage changes and volume percentage changes were calculated to evaluate the water absorption and dimensional stability of wood.

$$\text{Weight change: WC}(\%) = \frac{W_1 - W_0}{W_0} \times 100\% \tag{5}$$

Where W_1 is weight of the specimen after treatment and W_0 is weight of the specimen before treatment.

$$\text{Volume change: VC}(\%) = \frac{V_1 - V_0}{V_0} \times 100\% \tag{6}$$

Where V_1 is volume of the specimen after treatment and V_0 is volume of the specimen before treatment.

3 Results and discussion

3.1 Structural characterization and morphology of the magnetic wood composites

Fig. 1 shows the surface morphology of the untreated wood and magnetic wood. In Fig. 1a, surface of the poplar specimen were clear and some pits were found on the walls of tracheids. A thin layer of magnetic particles visible formed on wood surface masks the piths and other details after hydrothermal process at 90 ℃ (Fig. 1b). High magnification SEM image revealed the magnetic nanoparticles were densely distributed on the whole wood surface area (Fig. 1c). The TEM images as an insertion in Fig. 1c shows the average size of the iron oxides particles that precipitated on the wood surface was 10 nm. EDXA spectrum presented in Fig. 1d shows the chemical elements of the magnetic wood: carbon, oxygen, iron and gold. Carbon and oxygen were originated from the wood substrate, and gold was from the coating for SEM imaging. Accordingly, ferric oxides nanoparticles precipitated effectively on the wood template.

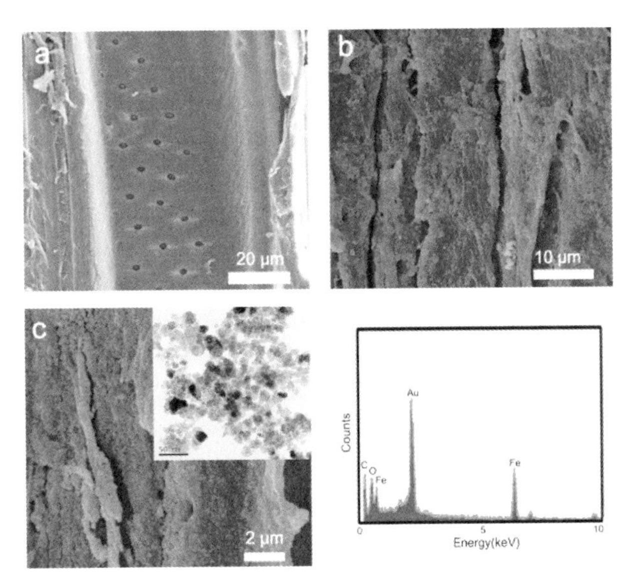

Fig. 1 SEM images of (a) untreated wood, (b) and (c) magnetic wood at different magnification; (d) EDXA spectrum of the magnetic wood(Color Fig. can be viewed in the online issue, which is available at wileyonlinelibrary. com.)

The XRD patterns of the wood sample were showed in Fig. 2. According to the Fig. 2a, there were characteristic peaks at 14.8° and 22.6° for the untreated wood, which were believed to represent the crystalline region of the cellulose in wood[28,29]. Apparently, in Fig. 2b, except for the diffraction peaks at 14.8° and 22.6°, new others diffraction peaks of at 2θ values of 18°, 30°, 35° could be assigned to the diffractions of the (220), (311) and (400) planes of magnetite Fe_3O_4 (PDF No. 22 - 1086), respectively. However, the signal diffraction peaks of iron oxide to wood was too low, due to the relative low concentration of the magnetic nanoparticles in the wood materials, and the similar spinel structures of Fe_3O_4 and γ-Fe_2O_3, XRD phase analysis might not be effective at differentiating between the two phase.

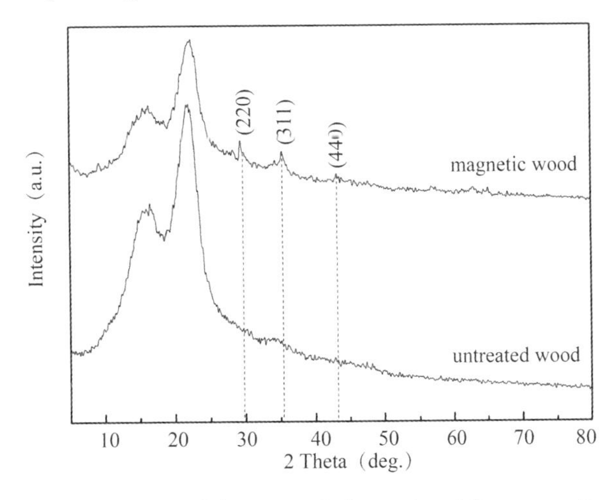

Fig. 2 XRD patterns of the untreated wood and the magnetic wood

To identify the composition of ferromagnetic phases in the wood matrix, Raman spectroscopic mapping was presented in Fig. 3. The Raman spectra demonstrated the predominance ferromagnetic phases deposited on the wood surface were magnetite Fe_3O_4 and maghemite γ-Fe_2O_3. The cubic spinel structured magnetite Fe_3O_4 nanoparticles comprise two magnetic sublattices with Fe^{2+} and Fe^{3+} occupy either tetrahedral or octahedral sites[30]. In Fig. 3, the crystal structure of magnetite Fe_3O_4 corresponded to five Raman active phonon modes, $3T_{2g}$, E_g and A_{1g}[31]. The stronger peak at 664 cm^{-1} in this study can be assigned to A_{1g}, the band around 501 and 321 cm^{-1} were corresponded to T_{2g} and E_g. Weak T_{2g} modes of magnetite were detected in 455 and 470 cm^{-1}. As compared with reported spectra[32], the scattering bands at 355, 501, 680, 702, 1325, and 1598 cm^{-1} can be ascribed to maghemite. These results strongly confirmed magnetite Fe_3O_4 and maghemite γ-Fe_2O_3 nanoparticles were precipitated on the wood template.

Fig. 4 shows the FTIR spectra of the untreated wood and magnetic wood. In Fig. 4a, the absorption band assigned to the O—H stretching vibration of hydroxyl groups was observed at 3340 cm^{-1} and the bands at 2901 cm^{-1} correspond to -CH_3 asymmetric. Most of the observed peaks of the untreated wood represent major cell wall components such as cellulose (1153 cm^{-1}), hemicelluloses (1736 cm^{-1}) and lignin (1601, 1505, 1228 cm^{-1})[33,34]. For the magnetic wood, the bands at 1736 cm^{-1} and 1505 cm^{-1} were disappeared, indicating the degradation of the wood component of hemicelluloses and lignin after hydrothermal treatment. The absorption band assigned to the O—H stretching vibration of hydroxyl groups was became comparatively narrow and shifted to 3250 cm^{-1}, indicating a interaction between the OH groups of the wood surface and the inorganic nanoparticles through hydrogen bonds[35-37]. In addition, the characteristic band at 580 cm^{-1} was arises from the stretching vibrations of Fe-O bonds[38].

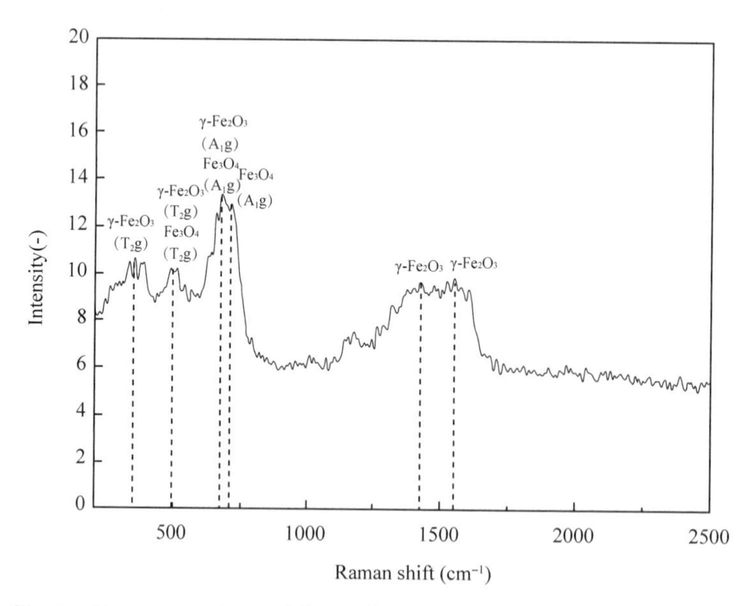

Fig. 3 Raman spectrum of the sediment into the hydrothermal solution
(Color Fig. can be viewed in the online issue, which is available at wileyonlinelibrary. com.)

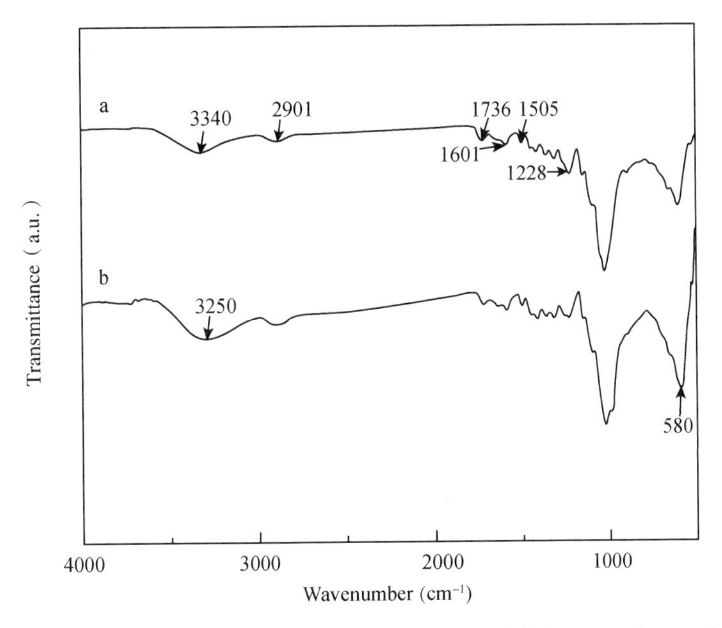

Fig. 4 FTIR spectra of (a) untreated wood and (b) magnetic wood

3.2 Incorporation mechanism of magnetic nanoparticles on the wood template by the hydrothermal treatment

Based on the above investigation of the experimental for the magnetic wood, a possible formation mechanism could be described as follows (Fig. 5): the precipitation reaction, which took place as the results of mixing iron ion aqueous solution and ammonia, might refer to the Eq. (4).

$$Fe^{2+} + 2Fe^{3+} + 8OH^{-} = Fe_3O_4 + 4H_2O \quad (7)$$

However, magnetite (Fe_3O_4) was not very stable and was sensitive to oxidation. Magnetite was transformed into maghemite (γ-Fe_2O_3) in the presence of oxygen[39].

Since there are large surface-to-volume atomic ratio and high surface activity on nanoparticle surface, the atoms on the surface are apt to adsorb ions or molecules in solution. For the magnetic nanoparticles dispersed in an aqueous solution, the bare atoms of Fe on the particle surface would adsorb OH^-, and formed an OH-rich surface[40]. Wood also features numerous hydroxyl groups that confer the hydrophilic nature of the cell wall. When the wood templates were immersed into the precursor solution, the hydrophilic nature makes the Fe-OH trapped inside the wood samples by interaction with the wood hydroxyl. Subsequently, owing to the magnetic adsorption force, increasing magnetic nanoparticles can be attracted to wood surface and finally the magnetic wood was obtained.

3.3 Magnetic, Thermal properties, UV-resistance and Dimension stability of the magnetic wood composites

Magnetic properties of the wood samples were showed in Fig. 6. In Fig. 6a, the hysteresis curve of the untreated wood approximated a straight line close to zero, indicating the nonmagnetic property of wood. However, the magnetic wood was exhibited a polar property and showed a typical ferromagnetic behavior. The magnetic intensity of the magnetic wood increased rapidly with the increase of the magnetic field firstly. Then the curves were evolved to a loop structure with the change of the direction of magnetic field and

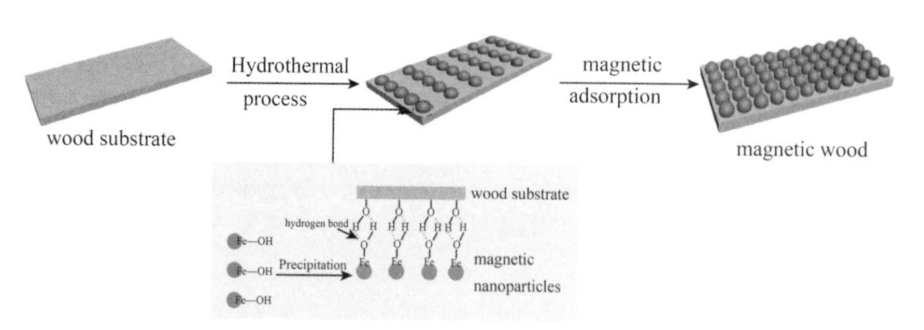

Fig. 5 Schematic illustrations for fabricating the magnetic wood by the hydrothermal method (Color Fig. can be viewed in the online issue, which is available at wileyonlinelibrary. com.)

exhibited a symmetric disposition along the axis. Magnetic intensity of the magnetic wood composites up to a saturation condition when the magnetic field was 7 KOe. The saturation magnetization (M_s) of the magnetic wood samples was 2 emu · g^{-1}, no coercivity and remanence, suggesting that these composites have excellent superparamagnetic property at room temperature. The M_s value of the magnetic wood was the same as that reported in literature[16,18], but it was much smaller than the values of corresponding bulk Fe_3O_4(63 emu · g^{-1}) and γ-Fe_2O_3(50 emu · g^{-1}) nanoparticles synthesized by coprecipitation method[41,42]. The reduced M_s could be explained by considering the small amount of magnetic nanoparticles (25%, seen in TG curves) that precipitated on the wood template and the nonmagnetic contribution of the wood substrate. The camera images of the magnetic wood under the influence of bar magnet were showed in Fig. 6b. The obvious magnetic property of the magnetic wood can be seen, when a magnetic field was applied, the magnetic wood could be lifted, dragged, alignment and attracted to the bar magnet.

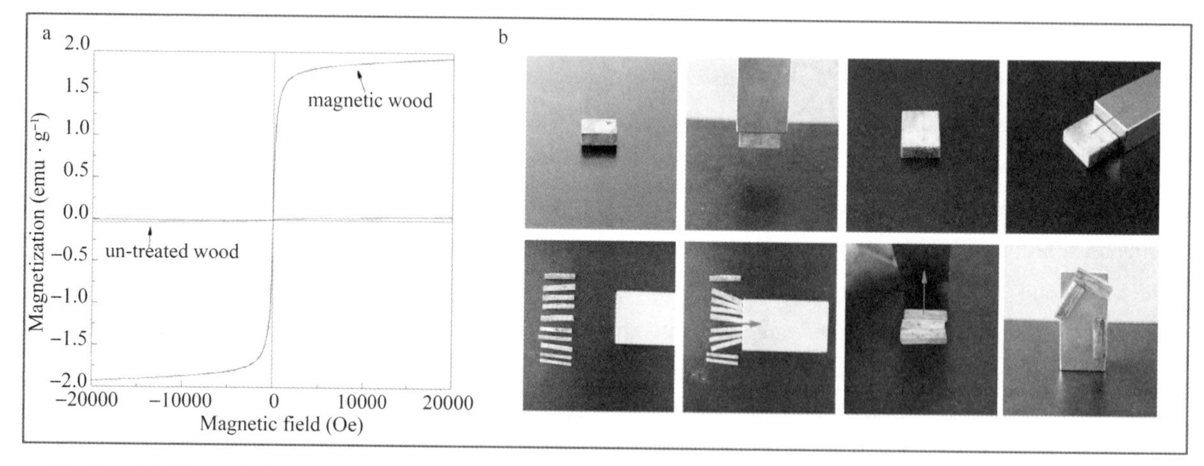

Fig. 6 (a) Magnetic hysteresis curves of the untreated wood and magnetic wood at room temperature (300 K). (b) Magnetic wood samples under the influence of an external magnet(Color Fig. can be viewed in the online issue, which is available at wileyonlinelibrary. com.)

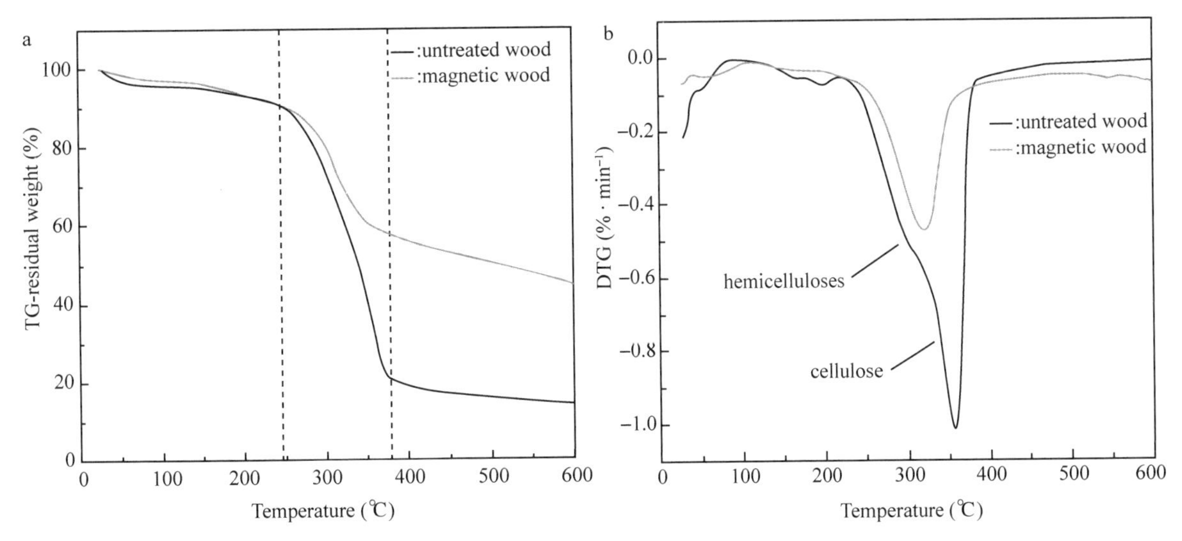

Fig. 7 Thermal behavior (a) TG and (b) DTG curves of the wood samples
(Color Fig. can be viewed in the online issue, which is available at wileyonlinelibrary. com.)

Thermal analyses were performed to examine the thermal stability of the magnetic wood samples. Curves represented thermal degradation (TG and DTG) for tested wood samples were displayed in Fig. 7. On the TG curve of untreated wood in Fig. 7a, the initial small mass loss from room temperature to 100 ℃ was mainly attributed to the evaporation of adsorbed water. An abrupt remarkable weight loss was found between 250 to 370 ℃, which were caused by the oxidation and pyrolysis of wood components. As opposed to the untreated wood, the magnetic wood samples displayed the magnetic wood composites displayed similar TG curves up to 200 ℃, after which a lower rate of weight loss was observed and eventually a higher amount of residues remained. Moreover, the temperature at which the wood components degradation took place increased, from 221 ℃ for the untreated wood to 240 ℃ for the magnetic wood composites, indicating a good thermal stability of the magnetic wood composites.

For the DTG curves in Fig. 7b, two obvious peaks of the untreated wood were found between 280 and 360 ℃. A remarkable peak at 360 ℃ was caused by the oxidation and pyrolysis of cellulose. The pyrolysis of hemicelluloses in the untreated wood, presented a shoulder at 280 ℃ in the cellulose pyrolysis peak[43]. After hydrothermal treatment, the peak at 280 ℃ was missing in the DTG curves suggested that most of the hemicelluloses were already degraded in the magnetic wood, which was in good agreement with the result by FTIR analysis. Moreover, compared to that of untreated wood, the peak of the magnetic wood weakened, indicating the decrease of combustion rate after magnetic particles modification. It can be assumed that the wood components were likely to be shielded by the uninflammable inorganic coating. So the wood components were which may hampered from being accessible to oxygen and the complete combustion was delayed, thus enhancing the thermal stability of wood[22].

To evaluate UV irradiation resistant ability of the magnetic wood materials, color changes were measured before and after accelerated aging. The experimental results were showed in Fig. 8. In Fig. 8a, Δa^* values of the untreated wood turned to red, indicating the color of the untreated wood sample became gradually redder with increasing UV irradiation time. The Δa^* values of the magnetic wood shows the similar changing tendency compared with the untreated wood. However, the Δa^* values of the magnetic wood were much smaller (only 2/5) than that of the untreated wood, which meant stronger UV-resistant ability of the magnetic wood. In Fig. 8b, the changes in Δb^* values indicated that the untreated wood and magnetic wood turned to strong yellowing and slight bluing with increasing UV irradiation time, respectively. In Fig. 8c, ΔL^* values of the untreated wood and magnetic wood became negative with time, revealing the samples was darkening. The total color change (ΔE^*) was shown in Fig. 8d. It could be seen that ΔE^* values of the untreated wood was significant, whereas that for magnetic wood was slight. These results showed that magnetic wood exhibited an UV resistance at some extent and prevented wood surface damage.

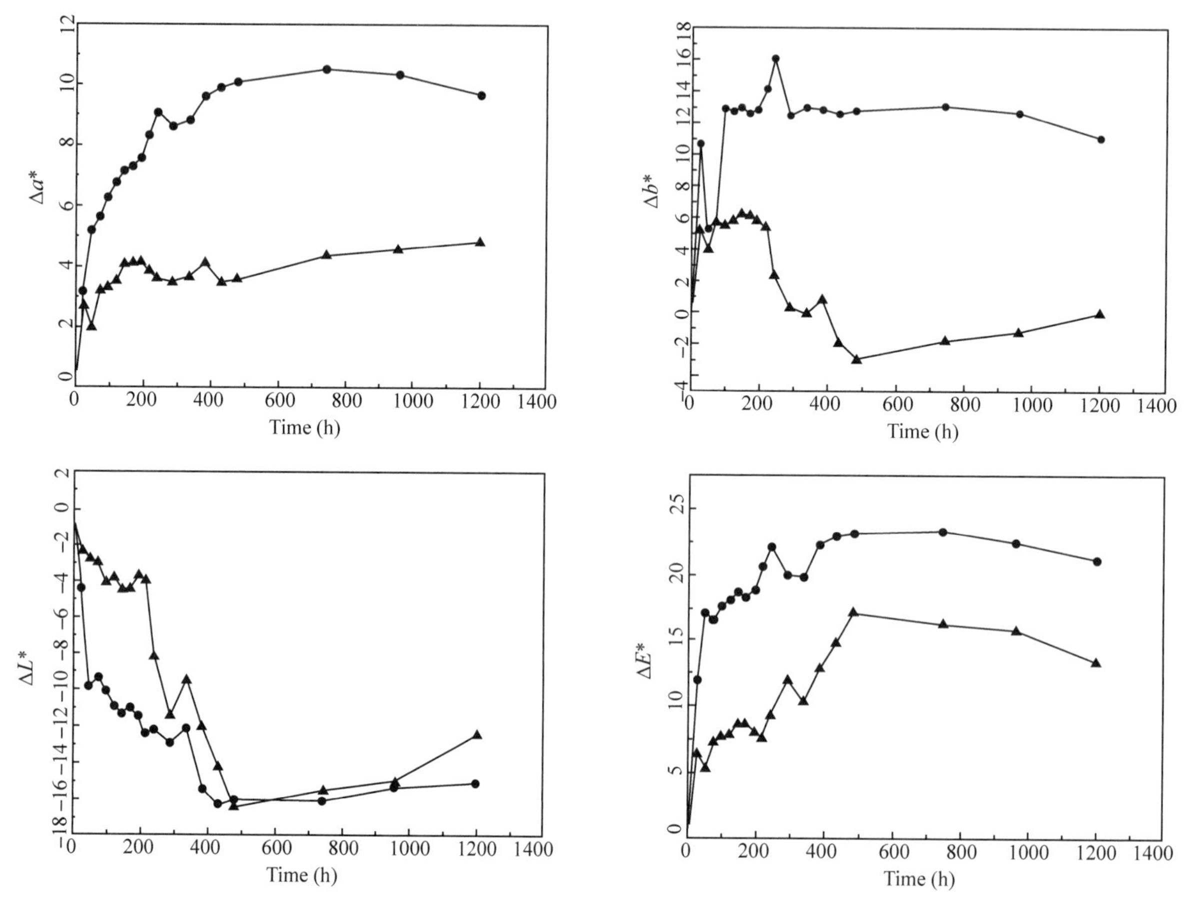

Fig. 8 Change tendency of Δa^*, Δb^*, ΔL^* and ΔE^* of the untreated wood (●) and magnetic wood (▲), respectively (Color Fig. can be viewed in the online issue, which is available at wileyonlinelibrary. com.)

The UV-Vis diffuse reflection spectra of the untreated wood and magnetic wood were showed in Fig. 9. Previously, Yu et al. [44] reported that the Fe_3O_4 nanoparticles have the UV absorption at the region of 200–700 nm. This was consistent with our experimental results. In Fig. 9, the magnetic wood has higher absorption in the region of 200–700 nm than the untreated wood, indicating that magnetic wood composites has better UV absorption. The better anti-ultraviolet properties were achieved by modifying with magnetic nanoparticles compared to the untreated wood. We attribute this to the UV absorption by the Fe_3O_4 nanoparticles, which hinder the UV light from reaching the wood surface and thus prevent photodegradation of wood component.

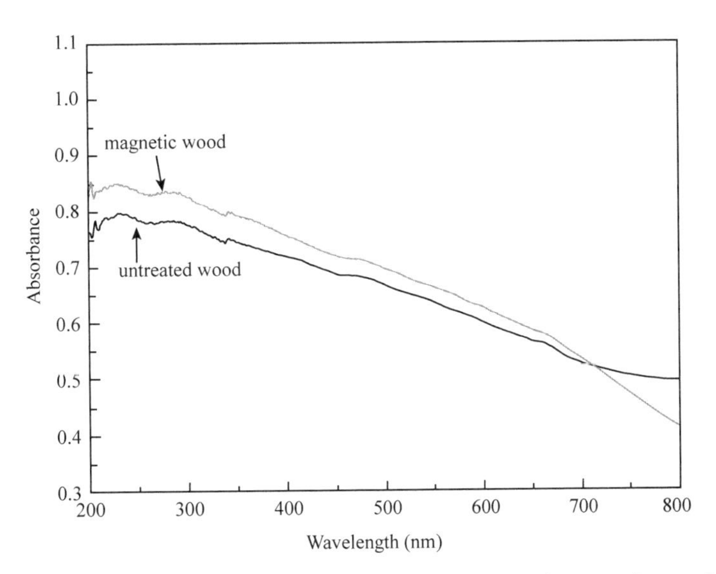

Fig. 9 UV-Vis diffuse reflection spectra of the untreated wood and magnetic wood, respectively (Color Fig. can be viewed in the online issue, which is available at wileyonlinelibrary. com.)

The results of the hygroscopicity and dimensional stability studies of magnetic wood were presented in Fig. 10. In Fig. 10a, the moisture uptake of the magnetic wood was reduced significantly compared with the untreated wood. The maximum weight percentage change of magnetic wood specimens reached 150% after 90 days of soaking, whereas that for untreated wood specimens can absorb up to 230% water. In Fig. 10b, the gains in volume during storage in water of poplar sapwood were contrasted with the magnetic wood. Storage in water results in saturation values of bulking between about 18%-20% of untreated wood. While, magnetic wood displayed a 10%-12% volume gains. This indicated that better dimensional stability was achieved by modified with iron oxides compared to the untreated wood sample. We attribute this to the blocking effect of the precipitated inorganic nanoparticles [35] and the decomposition of the hygroscopic hemicelluloses in the wood, which were converted to small organic molecules with poor hygroscopicity and improved the dimensional stability of wood materials[45].

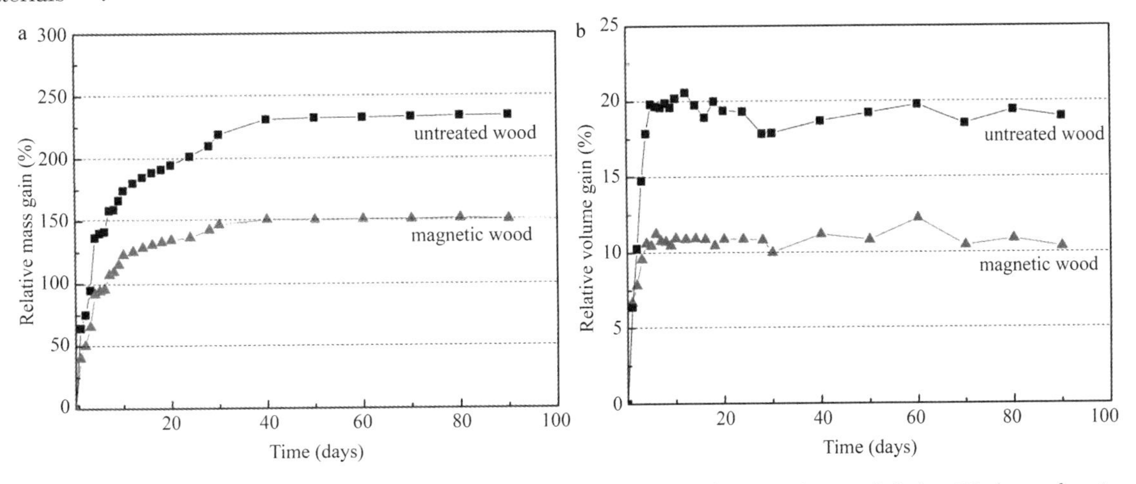

Fig. 10 (a) Mass and (b) volume changes of untreated wood and magnetic wood during 90 days of water immersion test(Color Fig. can be viewed in the online issue, which is available at wileyonlinelibrary. com.)

4 Conclusions

Magnetic wood materials were prepared by deposited iron oxide nanoparticles (magnetite Fe_3O_4 and maghemite γ-Fe_2O_3) on the wood template via a facile one-step hydrothermal method. The magnetic nanoparticles were attached to the wood surface through the interaction with the hydroxyl groups of wood leading to formation of hydrogen bonds, contributing to the depositing of magnetic nanoparticles on the wood substrate. Moreover, the magnetic wood exhibited appropriate magnetic property compared to the untreated wood materials that can be easily lifted, dragged, alignment and attracted by the external magnetic field, and it also possessed improved UV-resistance, thermal and dimensional stability. The approach presented may provide further routes for the functional applications of wood.

Acknowledgments

This work was financially supported by The National Natural Science Foundation of China (grant no. 31470584).

References

[1]Wang H L, Cui LF, Yang Y A, et al. Mn_3O_4-graphene hybrid as a high-capacity anode material for lithium-ion batteries[J]. J. Am. Chem. Soc., 2010, 132(40): 13978-13980.

[2] Lv G, He F, Wang X M, et al. Novel nanocomposite of nano Fe_3O_4 and polylactide nanofibers for application in drug uptake and induction of cell death of leukemia cancer cells[J]. Langmuir, 2008, 24(5): 2151-2156.

[3] Zeng J F, Jing L H, Hou Y, et al. Anchoring group effects of surface ligand on magnetic properties of Fe_3O_4 nanoparticles: Towards high performance MRI contrast agents[J]. Adv. Mater., 2014, 26(17): 2694-2698.

[4] Zhou G M, Wang D W, Li F, et al. Graphene-wrapped Fe_3O_4 anode material with improved reversible capacity and cyclic stability for lithium-ion batteries[J]. Chem. Mater., 2010, 22(18): 5306-5313.

[5] Sun X L, Guo S J, Liu Y, et al. Dumbbell-like PtPd-Fe_3O_4 nanoparticles for enhanced electrochemical detection of H_2O_2[J]. Nano Lett., 2012, 12(9): 4859-4863.

[6] Balazs A C, Emrick T, Russell T P. Nanoparticle polymer composites: where two small worlds meet[J]. Science, 2006, 314(5802): 1107-1110.

[7] Olsson R T, Samir M A, Salazar-Alvarez G, et al. Making flexible magnetic aerogels and stiff magnetic nanopaper using cellulose nanofibrils as templates[J]. Nat. Nanotechnol., 2010, 5(8): 584-588.

[8] Liu S L, Luo X G, Zhou J P. Magnetic responsive cellulose nanocomposites and their applications[M]. Cellulose-Medical, Pharmaceutical and Electronic Applications., 2013.

[9] Liu S L, Zhang L, Zhou J P, et al. Structure and properties of cellulose/Fe_2O_3 nanocomposite fibers spun via an effective pathway[J]. J. Phys. Chem. C, 2008, 112(12): 4538-4544.

[10] Rowell R M. Handbook of wood chemistry and wood composites[M]. CRC press, 2012.

[11] Oka H, Hamano H, Chiba S. Experimental study on actuation functions of coating-type magnetic wood[J]. J. Magn. Magn. Mater., 2004, 272: E1693-E1694.

[12] Oka H, Hojo A, Osada H, et al. Manufacturing methods and magnetic characteristics of magnetic wood[J]. J. Magn. Magn. Mater., 2004, 272: 2332-2334.

[13] Oka H, Fujita H. Experimental study on magnetic and heating characteristics of magneticwood[J]. J. Appl. Phys., 1999, 85(8): 5732-5734.

[14] Oka H, Hojo A, Takashiba T. Wood construction and magnetic characteristics of impregnated type magnetic wood [J]. J. Magn. Magn. Mater., 2002, 239(1-3): 617-619.

[15] Oka H, Narita K, Seki K. Experimental results on indoor electromagnetic wave absorber using magnetic wood[J]. J. Appl. Phys., 2002, 91(10): 7008-7010.

[16] Merk V, Chanana M, Gierlinger N, et al. Hybrid wood materials with magnetic anisotropy dictated by the hierarchical cell structure[J]. ACS Appl. Mater. Interfaces, 2014, 6(12): 9760-9767.

[17] Trey S, Olsson R T, Strom V, et al. Controlled deposition of magnetic particles within the 3-D template of wood: making use of the natural hierarchical structure of wood[J]. RSC Adv., 2014, 4(67): 35678-35685.

[18] Gan W T, Gao L K, Sun Q F, et al. Multifunctional wood materials with magnetic, superhydrophobic and anti-ultraviolet properties[J]. Appl. Surf. Sci., 2015, 332: 565-572.

[19] Li X Y, Si Z J, Lei Y Q, et al. Direct hydrothermal synthesis of single-crystalline triangular Fe_3O_4 nanoprisms[J]. CrystEngComm, 2010, 12(7): 2060-2063.

[20] Sun X H, Zheng C M, Zhang F X, et al. Size-controlled synthesis of magnetite (Fe_3O_4) nanoparticles coated with glucose and gluconic acid from a single Fe (III) precursor by a sucrose bifunctional hydrothermal method[J]. J. Phys. Chem. C, 2009, 113(36): 16002-16008.

[21] Barata M A B, Neves M C, Neto C P, et al. Growth of $BiVO_4$ particles in cellulosic fibres by in situ reaction[J]. Dyes Pigm., 2005, 65(2): 125-127.

[22] Saka S, Ueno T. Several SiO_2 wood-inorganic composites and their fire-resisting properties[J]. Wood Sci. Technol., 1997, 31(6): 457-466.

[23] Tshabalala M A, Sung L P. Wood surface modification by in-situ sol-gel deposition of hybrid inorganic-organic thin films[J]. J. Coat. Technol. Res., 2007, 4: 483-490.

[24] Tshabalala M A, Kingshott P, VanLandingham M R, et al. Surface chemistry and moisture sorption properties of wood coated with multifunctional alkoxysilanes by sol-gel process[J]. J. Appl. Polym. Sci., 2003, 88(12): 2828-2841.

[25] Sun Q F, Lu Y, Yang D J, et al. Preliminary observations of hydrothermal growth of nanomaterials on wood surfaces [J]. Wood Sci. Technol., 2014, 48(1): 51-58.

[26] Li J, Yu H, Sun Q, et al. Growth of TiO_2 coating on wood surface using controlled hydrothermal method at low temperatures[J]. Appl. Surf. Sci., 2010, 256(16): 5046-5050.

[27] Gan W T, Gao L K, Zhan X X, et al. Hydrothermal synthesis of magnetic wood composites and improved wood properties by precipitation with $CoFe_2O_4$/hydroxyapatite[J]. RSC Adv., 2015, 5(57): 45919-45927.

[28] Lu Y, Sun Q F, Yang D J, et al. Fabrication of mesoporous lignocellulose aerogels from wood via cyclic liquid nitrogen freezing-thawing in ionic liquid solution[J]. J. Mater. Chem., 2012, 22(27): 13548-13557.

[29] Li J, Lu Y, Yang D J, et al. Lignocellulose aerogel from wood-ionic liquid solution (1-allyl-3-methylimidazolium chloride) under freezing and thawing conditions[J]. Biomacromolecules, 2011, 12(5): 1860-1867.

[30] Long J W, Logan M S, Rhodes C P, et al. Nanocrystalline iron oxide aerogels as mesoporous magnetic architectures [J]. J. Am. Chem. Soc., 2004, 126(51): 16879-16889.

[31] Faria D L A, Venancio Silva S, de Oliveira M T. Raman microspectroscopy of some iron oxides and oxyhydroxides [J]. J. Raman Spectrosc., 1997, 28: 873-878.

[32] Jubb A M, Allen H C. Vibrational spectroscopic characterization of hematite, maghemite, andmagnetite thin films produced by vapor deposition[J]. ACS Appl. Mater. Interfaces, 2010, 2(10): 2804-2812.

[33] Gierlinger N, Goswami L, Schmidt M, et al. In situ FTIR microscopic study on enzymatic treatment of poplar wood cross-sections[J]. Biomacromolecules, 2008, 9(8): 2194-2201.

[34] Rana R, Langenfeld-Heyser R, Finkeldey R, et al. FTIR spectroscopy, chemical and histochemical characterisation of wood and lignin of five tropical timber wood species of the family of dipterocarpaceae[J]. Wood Sci. Technol., 2010, 44(2): 225-242.

[35] Wang B L, Feng M, Zhan H B. Improvement of wood properties by impregnation with TiO_2 via ultrasonic-assisted

sol-gel process[J]. RSC Adv., 2014, 4(99): 56355-56360.

[36] Wan C C, Lu Y, Sun Q F, et al. Hydrothermal synthesis of zirconium dioxide coating on the surface of wood with improved UV resistance[J]. Appl. Surf. Sci., 2014, 321: 38-42.

[37] Li J P, Sun Q F, Jin C D, et al. Comprehensive studies of the hydrothermal growth of ZnO nanocrystals on the surface of bamboo[J]. Ceram. Int., 2015, 41(1): 921-929.

[38] Koseoglu Y, Alan F, Tan M, et al. Low temperature hydrothermal synthesis and characterization of Mn doped cobalt ferrite nanoparticles[J]. Ceram. Int., 2012, 38(5): 3625-3634.

[39] Laurent S, Forge D, Port M, et al. Magnetic iron oxide nanoparticles: synthesis, stabilization, vectorization, physicochemical characterizations, and biological applications[J]. Chem. Rev., 2008, 108(6): 2064-2110.

[40] Ma M, Zhang Y, Yu W, et al. Preparation and characterization of magnetite nanoparticles coated by amino silane [J]. Colloids Surf., A, 2003, 212(2-3): 219-226.

[41] Tao K, Dou H J, Sun K. Interfacial coprecipitation to prepare magnetite nanoparticles: concentration and temperature dependence[J]. Colloids Surf. A, 2008, 320(1-3): 115-122.

[42] Sreeja V, Joy P A. Microwave-hydrothermal synthesis of γ-Fe_2O_3 nanoparticles and their magnetic properties[J]. Mater. Res. Bull., 2007, 42(8): 1570-1576.

[43] Rosa M F, Chiou B S, Medeiros E S, et al. Effect of fiber treatments on tensile and thermal properties of starch/ethylene vinyl alcohol copolymers/coir biocomposites[J]. Bioresour. Technol., 2009, 100(21): 5196-5202.

[44] Yu C M, Guo J W, Gu H Y. Direct electrochemical behavior of hemoglobin at surface of Au@ Fe_3O_4 magnetic nanoparticles[J]. Microchim. Acta, 2009, 166(3-4): 215-220.

[45] Rowell R M, Ibach R E, McSweeny J, et al. Understanding decay resistance, dimensional stability and strength changes in heat-treated and acetylated wood[J]. Wood Mater. Sci. Eng., 2009, 4(1-2): 14-22.

中文题目：磁性木质复合材料的磁性、热稳定性、抗紫外老化性和吸湿性

作者：甘文涛，刘莹，高丽坤，詹先旭，李坚

摘要：以三氯化铁和氯化亚铁为前驱体，在90℃条件下，通过一步水热法，用氧化铁(Fe_3O_4/γ-Fe_2O_3)改性制备了具有更好抗紫外线性能和尺寸稳定性的磁性木材复合材料，采用扫描电镜、透射电镜、X射线衍射、拉曼光谱和傅里叶变换红外光谱(FTIR)对水热处理前后木材复合材料的晶相和化学结构进行了表征。结果表明，磁性纳米颗粒以聚集体形式沉积在木材表面，并提出了一种水热制备机理。水热处理后，磁性木材复合材料表现出适中的磁性和抗紫外线性能。热分析表明，磁性纳米颗粒的加入延缓了木材基体的热分解，提高了木材的热稳定性。此外，还对改性木材的尺寸稳定性进行了评估。

关键词：磁性木材；氧化铁纳米粒子；水热法；木材改性

Fabrication of Reversible Thermoresponsive Thin Films on Wood Surfaces with Hydrophobic Performance*

Yingying Li, Jian Li

Abstract: The smart reversible thermoresponsive wood with hydrophobic performance was fabricated by depositing modified thermoresponsive coatings on wood surfaces. The modified composite films were prepared by entrapping thermochromic materials (TM) supported on the 3-Aminopropyltriethoxysilane (AEPT) into polyvinyl alcohol solution (PVA). All the samples possessed superior reversible thermoresponsive property, and the thermochromic response rate of TM-APTES/PVA modified wood was higher than the other samples. The attenuated total reflectance Fourier transform infrared spectroscopy (ATR-FTIR) spectra demonstrated that the TM-APTES/PVA composite polymer films have been successfully settled onto wood surfaces. The scanning electron microscope (SEM) and pull-off test results proved that the modified wood possessed better dispersion stability of TM particles and stronger interfacial adhesion than TM/PVA coated wood. Furthermore, the surface properties of wood materials changed from hydrophilic to hydrophobic after chemical modification.

Keywords: Thermoresponsive coating; Hydrophobicity; Wood surfaces; 3-Aminopropyltriethoxysilane

1 Introduction

As a natural and renewable biopolymer, wood has been extensively used in buildings, decorations, industries and other daily lives in virtue of its unique performances of the high ratio of strength to weight, esthetic, and amendment for indoor environment[1-3]. Whereas the wood supplies increase rapidly, many attempts have been made to create new functions for wood to meet market requirements and compete with other advanced materials[4-7]. Among them, the interest has recently arisen in the development of smart stimuli-responsive materials due to its potential applications as sensors, smart switches and energy storage[8-10].

The reversible thermochromic materials have become one of the most advanced materials in smart materials because of its fast response to temperature[11,12]. Due to the unique properties, reversible thermochromic materials are widely used in many fields such as molecular switch[13], temperature sensor[14,15], thermal relays[16] and intelligent coating[17,18]. As a novel type of smart material, the orthodromic reversible thermoresponsive wood was developed by coating wood substrates with transparent films. The evident advantage of the orthodromic thermoresponsive wood is that its color can be repeatedly changed with the changing temperature without destroying the naturally beautiful wood texture. However, the hydrophilicity of wood may limit their application[19].

* 本文摘自 Progress in Organic Coatings, 2018, 119: 15-22.

The main substances of wood cell wall are cellulose, hemicellulose and lignin, making up approximately 97~99% of the total weight[20]. The hydroxyl and other oxygen-containing groups of wood cell wall polymers can attract and retain moisture through hydrogen bonding[21]. The moisture absorption of wood may cause dimensional changes of the composites and cause a serious problem to its durability and use value of the interfacial adhesion[22-24]. The chemical modification may make wood materials more dimensionally stable by reducing water sorption[25].

Binding of a functional group to wood surface via a covalent bond is a reliable method of modification of the wood substrate. Silane coupling agent is a chemical that functions at the interface to create a chemical bridge between the reinforcement and matrix[26,27]. That is, using its organofunctional group (Y) and hydrolysable groups (X) to form a stable link. Here, the schematic of organosilane bonded onto wood substrate is shown in Fig. 1. Three hydrolysable groups in silane coupling agent are expected to hydrolyze forming the Si-O-Si network structure, resulting into the preparation of structure[28-30]. At the same time, the organofunctional groups are chosen for reactivity or compatibility with the polymer, resulting into good contact with other phase[31-33]. As a kind of aminosilanes, 3-Aminopropyltriethoxysilane (APTES) may form the strong intra- and intermolecular hydrogen bonds with its amino group, meanwhile, possessing a strong affinity towards the hydroxyl groups of wood surface[34,35]. Although the interaction among thermochromic materials and wood surface can be reinforced by covalent bonds, the thermochromic materials may be destroyed by the tensile stress initiated from coating shrinkage during drying. Therefore, PVA was added to relieve the tensile stress and also serve as a transparent and environmental coating material[33,36].

In this paper, we prepared a novel reversible thermoresponsive wood with hydrophobic performance by a sample drop-coating method. The thermoresponsive properties, surface morphologies, wettability and bond strength of samples were investigated.

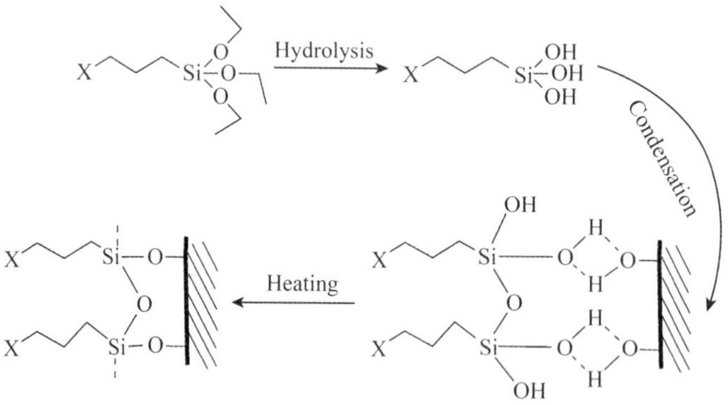

Fig. 1 The interaction reaction between organosilane and wood substrate

2 Experimental

2.1 Materials

Wood samples (radial section) of 50 mm × 25 mm × 5 mm were obtained from Harbin, China, Manchurian ash (Fraxinus mandschurica Rupr.). The wood specimens were oven-dried (24 h, 103± 2 ℃) to constant weight after ultrasonically rinsing in deionized water, acetone and ethanol for 10 min, respectively. The ethanol (99.7%), acetone (99.5%) and polyvinyl alcohol (PVA) (alcoholysis 98.0%–99.0%, model 1750 ± 50) were purchased from Tian jian Kaitong Chemical Reagent Co., Ltd. Thermochromic materials (TM) namely, 2'-chloro-6'-(diethylamino)-3'TR-2 (white powder), were manufactured by Chong Yu Technology Co., Ltd. 3-Aminopropyltriethoxysilane (APTES) was supplied by Sa en Chemical Technology (Shanghai) Co., Ltd. All of the chemicals were used as received without further purification.

2.2 Fabrication of APTES /PVA coated wood

Firstly the PVA solution was prepared by dissolving 4 g PVA into 100 mL deionized water at 75 ℃ for 2h with magnetically stirring. Secondly 1.5 mL APTES was added into 8.5 mL alcohol solution, and then slowly added into 20 mL PVA solution under continuous stirring. Thirdly six separate experiments were set at 40 ℃ with different hydrolyzing time of 20 min, 25 min, 30 min, 35 min, 40 min and 45 min. Next, 1.0 mL above suspension was dropped onto the wooden substrates. Finally, the treated samples were placed in a 110 ℃ oven for 10 min to be dried.

2.3 Fabrication of thermoresponsive wood

To prepare coating solution, 0.35 g TM and 1.5 mL APTES were added into 8.5 mL alcohol solution stirred for 10 min at room temperature, and then slowly added into 20 mL PVA solution under magnetically stirring at 40 ℃ for 30 min. Next, 1.0 mL coating solution was dropped onto the wood surface. Then, the treated samples were dried in oven at 110 ℃ for 1h. Finally, the TM-APTES/PVA modified wood was obtained. Furthermore, a named TM/PVA coated wood was also prepared for the purpose of comparison. The same experimental procedures were used except that the 1.5 mL APTES was absent.

2.4 Characterizations

The digital photos of the samples were obtained using a Nikon D7000 digital camera. A scanning electron microscopy (SEM, FEI Quanta200) was used to investigate the surface morphologies, the samples were glued onto a specific holder and were sprayed by gold to ensure the conductivity. The attenuated total reflectance of the Fourier transform infrared spectroscopy (ATR-FTIR) was recorded by the Nicolet Nexus 670 FTIR instrument in the range of 400–4000 cm^{-1} with a resolution of 4 cm^{-1}. An OCA 40 contact angle system (Dataphysics, Germany) was used to measure the water contact angles (WCAs) by injected a 5 μl droplet of deionized water onto the surfaces of the samples at room temperature. The WCAs were measured at five different points for each sample and the average value was reported as the final contact angle value.

2.5 Color tests

The samples were placed into a DHG-9023A conditioning chamber (Jiangsu Jun Instrument Technology Co., Ltd.). The temperature sensor was calibrated to an accuracy of 0.1 ℃, and the temperature was monitored at 200 seconds interval and increased in an interval of 1 ℃. A digital camera was placed in a fixed position to record the whole process over a temperature range of 25–65 ℃, and these records were imported to a computer. The Commission Internationale de L'Eclairage (CIE) L^*, a^*, and b^* values were measured by using the Adobe Photoshop CS6 software installed in the computer[19].

2.6 Response time tests

Response time test was measured according to the analysis of images captured by the camera. With the increase in temperature, the ΔE^* values increased from initial value to maximum value (Fig. 2a). As temperature continues to rise, ΔE^* values remain unchanged (Fig. 2b). Finally the ΔE^* values decreased from maximum value to initial value with the decreasing temperature (Fig. 2c). The response time Δt_1 was identified as the coloration time. Correspondingly, the response time Δt_2 was the bleaching time.

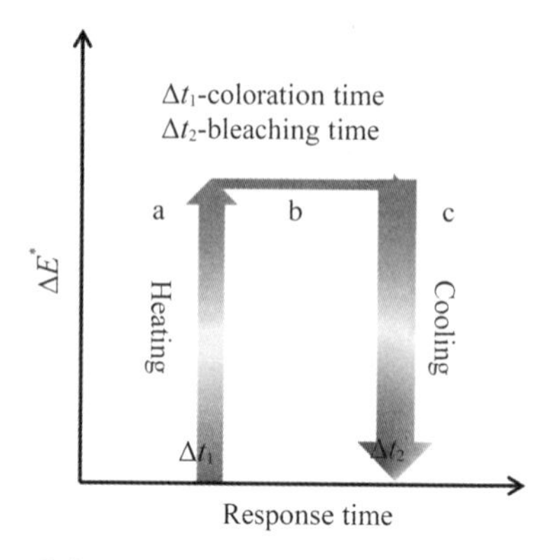

Fig. 2 Schematic illustration of the response time

2.7 Adhesion test

The bond strength between the coating and wood substrate was measured according to ISO 4624 with an automatic adhesion tester (PosiTest AT-A, DeFelsko)[37]. The schematic diagram of the coating-pull-off test is shown in Fig. 3. The dollies of 20 mm diameter were degreased by ethyl alcohol and then attached to the prepared samples by a two-component epoxy based cyanoacrylate adhesive. After adhesive curing, a testing apparatus was attached to the loading fixture and was strained at 0.7 MPa · s^{-1} until rupture occurs and the maximum pressure was marked as the adhesion strength[38,39]. For each test, three replicate samples were employed, and the arithmetic average value was quoted.

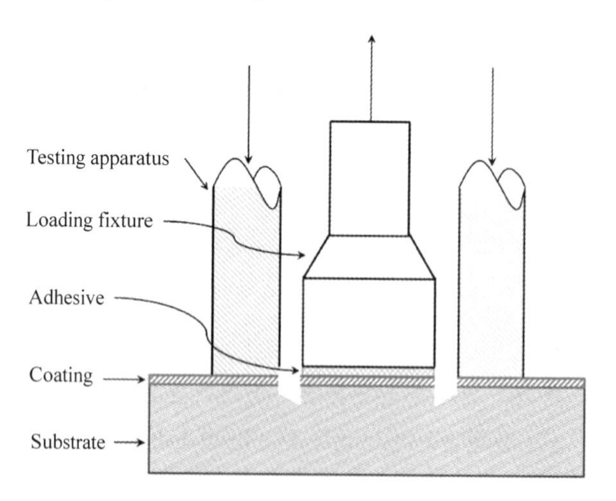

Fig. 3 Schematic illustration of pull-off test

3 Results and discussion

3.1 Thermoresponsive properties

Table 1 shows the color parameters L (lightness index), a (red-green index), b (yellow-blue index) and ΔE^* (the total color changes) of the thermoresponsive wood before and after heating. The color changes of the as-prepared samples were related to the increasing number of chromophores. The ΔE^* values were calculated using the following equations:

$$\Delta L^* = L_2 - L_1 \tag{1}$$

$$\Delta a^* = a_2 - a_1 \tag{2}$$

$$\Delta b^* = b_2 - b_1 \tag{3}$$

$$\Delta E^* = \sqrt{(\Delta L^*)^2 + (\Delta a^*)^2 + (\Delta b^*)^2} \tag{4}$$

The L_1, a_1 and b_1 are the color parameters before heating; L_2, a_2 and b_2 are the color parameters after heating. The color parameters were measured at five different points on each sample. The final value for L, a and b were obtained as an average of these five measurements.

As shown in Table1, the ΔE^* value of pristine wood was 0.2, verifying that the heating processes can be performed concurrently with minimal interference to the results. The lightness index L^* of the thermoresponsive wood showed a decreasing trend, implying the surface color turned darker. The change in a^* value remarkably increased, indicating that the surface color turned a deeper shade of red. In addition, the change of b^* value was irregular due to the main color change of sample was red. Above all, the total color changes of the TM-APTES/PVA modified wood and TM/PVA coated wood reached up to 41.0 and 38.2, respectively. All the samples possessed superior thermoresponsive property.

Fig. 4 shows the color-changed process of sample with a thermochromic material concentration of 3.5%. Under room temperature, the thermoresponsive wood showed the natural color of wood. With the temperature increasing from 25 ℃ to 40 ℃, the thermoresponsive wood displayed excellent thermochromic process. After being cooled to the room temperature (25 ℃), the wood turned back to the natural color of wood, indicating a good reversible behaviour. Furthermore, the natural wood texture was completely unaffected because the formed film was transparent.

Table 1 The color parameters of the pristine wood and thermoresponsive wood before and after heating

Sample	L_1	L_2	a_1	a_2	b_1	b_2	ΔE
pristine wood	74.8	74.9	7.2	7.4	14.8	14.8	0.2
TM/PVA coated wood	74.0	62.2	5.8	41.4	12.6	19.6	38.2
TM-APTES/PVA coated wood	77.6	64	7.6	45.0	11.0	1.0	41.0

3.2 Response time

Response time including coloration time and bleaching time is an important factor in the practical applications. The Fig. 5 shows the response time of the thermoresponsive wood. The average coloration time of the TM/PVA coated wood was 16.3 s, which was about four times bigger than the TM-APTES/PVA modified wood (4.4s). Besides this, as shown in Fig. 5b, the average bleaching time of the TM/PVA coated wood

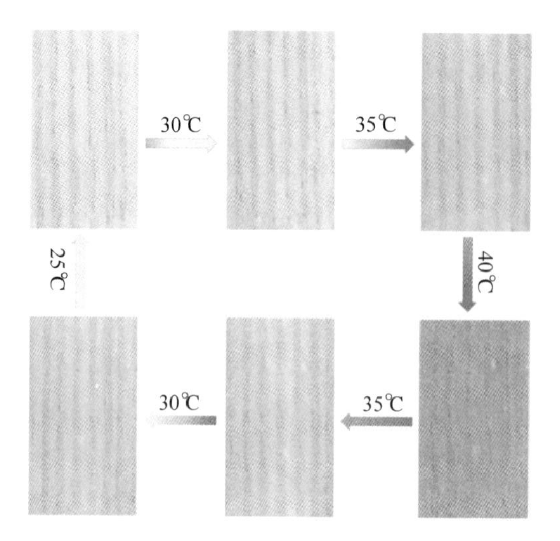

Fig. 4 Color-changed process of thermoresponsive wood

(26. 8 s) was more than tripled the TM-APTES/PVA modified wood (7. 4s). In general, the response time of the TM/PVA coated wood was bigger than the TM-APTES/PVA modified wood, implying that the TM-APTES/PVA modified wood has a higher thermochromic response rate. The reason may be that the dispersion stability of the TM particles in TM-APTES/PVA modified wood was better, thus resulting better thermal absorptivity and diffusivity.

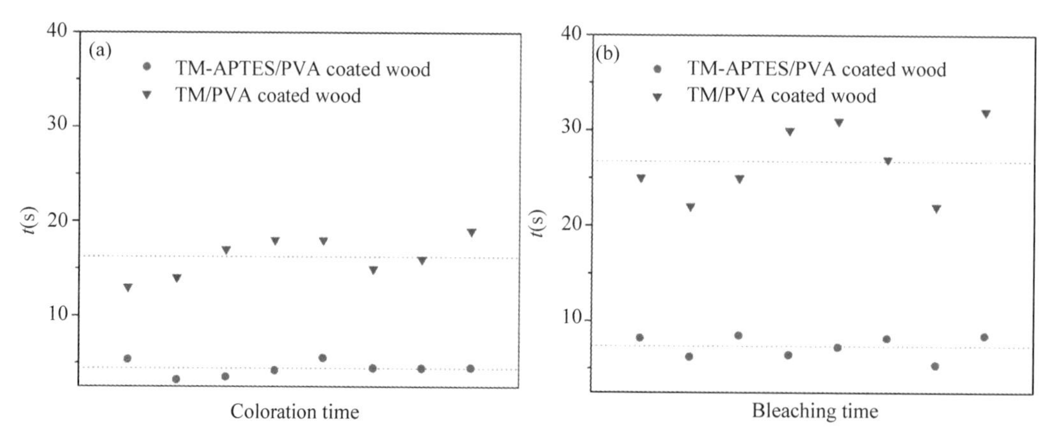

Fig. 5 The (a) coloration time and (b) bleaching time of the TM/PVA coated wood and TM-APTES/PVA coated wood

3. 3 ATR-FTIR analyses

To effectively couple the wood substrate and polymer matrices, obtain the best adhesive strength between the films and wood surfaces. The condensation supposed to be minimized to leave the silanols free for being adsorbed to the hydroxyl groups of wood substrate during the hydrolysis process of APTES. It is critical in choosing a synthetic strategy for surface functionalization. Herein, six sets of experiments were carried out to identify the best hydrolyzing time.

The FTIR spectra of APTES/PVA coated wood with different hydrolyzing time are shown in Fig. 6. The absorption peak at 2840 cm^{-1} corresponded to the stretching vibration of SiO-CH_3, and its absorption intensity firstly decreases and then increases with the increasing hydrolyzing time[40]. It is worth noting that the absorption peak of SiO—$CH_3$was almost disappeared at the hydrolyzing time of 30min, implying that the amount of hydrolyzed silane has reached a saturation point. In addition, the absorption peaks at 1026 cm^{-1} was attributed to the stretching vibration of Si-O-Si, which indicated the condensation reaction of the hydrolyzed silane. The absorption peak of Si-O-Si became strongest at the hydrolyzing time of 30 min, which confirming that the APTES reached a maximum condensation reaction after the maximum hydrolysis reaction. To further confirm the best adhesive strength between the films and wood surfaces. Pull-off adhesion test of the APTES/PVA coated wood was performed and the results were listed in Table 2. The results suggested that the bond strength of APTES/PVA coated wood with 30 min hydrolyzing time is stronger than the other samples. Therefore, the hydrolyzing time adopted in this paper is 30 min.

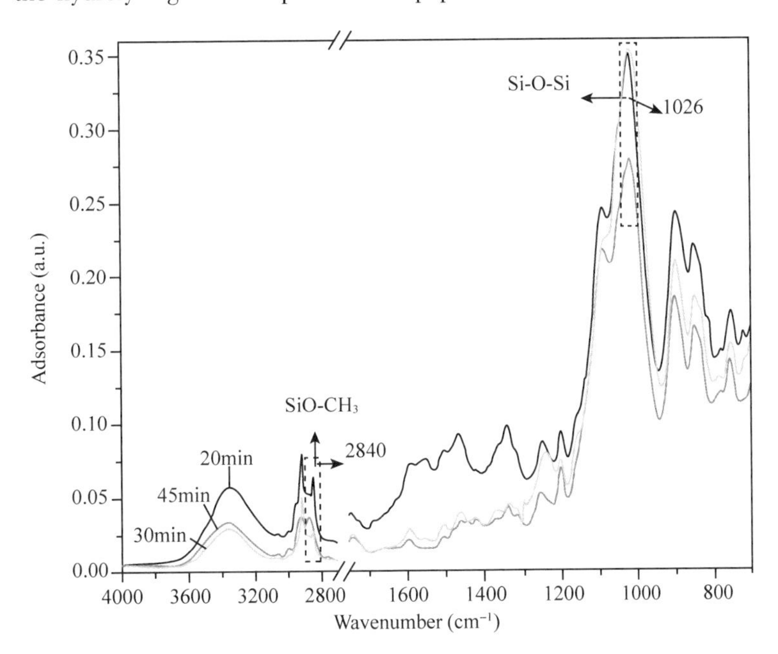

Fig. 6 ATR-FTIR spectra of (a) pristine wood and the APTES/PVA coated wood with different hydrolyzing time

Table 2 Adhesion test results of APTES/PVA coated wood.

Hydrolyzing time (min)	Adhesion (MPa)
20	2.7
25	2.9
30	3.8
35	3.6
40	3.0
45	2.9

Fig. 7 shows the ATR-FTIR spectra of pristine wood and TM-APTES/PVA modified wood. Compared with Fig. 7a, some new peaks appeared in TM-APTES/PVA modified wood. The absorption peak at 3301 cm^{-1}

was attributed to the N-H stretching vibration of the secondary amine. The peak at 1415 cm^{-1} was ascribed to the stretching vibration of the C-N. Another two absorption peaks at 1085 and 1021 cm^{-1} were owing to the Si-O-Si and Si-O-C vibrations, respectively[41]. These observations confirming that the TM-APTES/PVA composite polymer coating have been successfully settled onto wood surfaces.

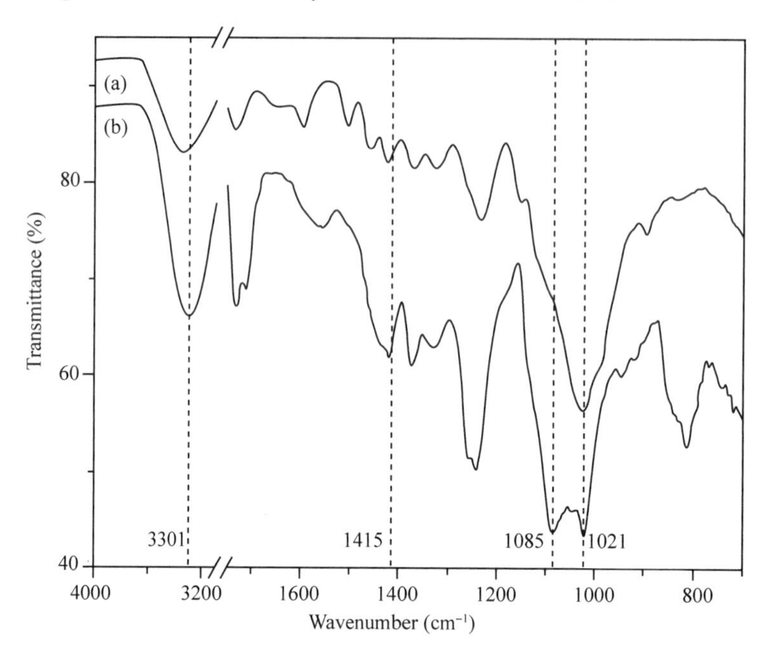

Fig. 7 ATR-FTIR spectra of (a) pristine wood and (b) TM-APTES/PVA wood

3.4 Surface morphologies

The SEM images of the TM/PVA coated wood and the TM-APTES/PVA modified wood were shown in Fig. 8. It was observed clearly that the TM particles dispersed unevenly in TM/PVA hybrid film (Fig. 8a). However, the TM particles were regularly distributed in the surface of TM-APTES/PVA modified wood (Fig. 8b). More interestingly, the glossy spherical particles were arranged in a fish-scale shape closely linked with each other and there was no agglomeration occurred. The possible formation mechanism of the TM-APTES/PVA modified wood is shown in Fig. 9. The strategy is to activate the APTES by hydrolyzing the alkoxy groups off, thereby forming the reactive silanol groups, as similar with the hydrolysis process of organosilane (Fig. 1). Then the reactive silanol react with the hydroxyl groups meanwhile condense with themselves forming macromolecular networks. Moreover, the bifunctional groups of APTES respectively react with the TM particles and PVA polymer or wood surface, forming a chemical bridge between them. The differences of surface morphologies between the TM/PVA coated wood and TM-APTES/PVA modified wood indicated that the APTES were successfully created a chemical bridge between the particles at each interface.

3.5 Pull-off adhesion test

Pull-off adhesion tests were performed in order to evaluate which sample has stronger adhesion strength of the coating to wood substrate. All kinds of failure modes are illustrated in Fig. 10 to clarify the different case. (1) Epoxy failure happens between the epoxy and the coating, which signals a poor bonding between the epoxy and coating surface. (2) Cohesive failure occurs in a polymer layer interpreted as cohesion strength of

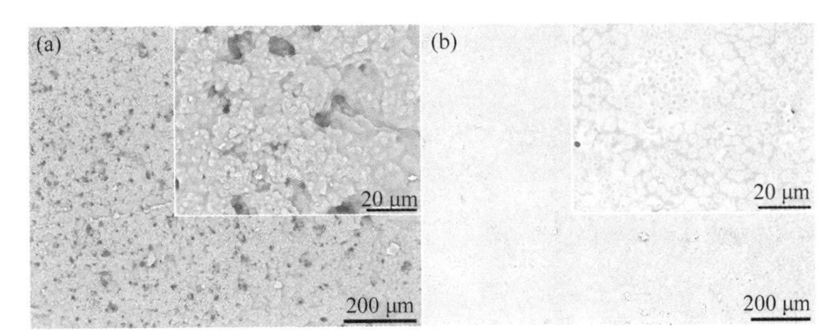

Fig. 8 SEM images of the (a) TM/PVA coated wood and (b) TM-APTES/PVA coated wood at low and high magnifications

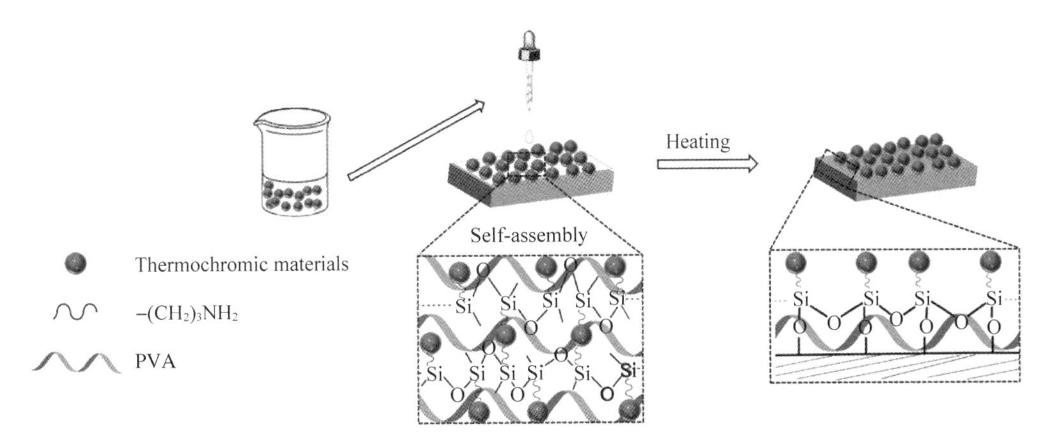

Fig. 9 Schematic of the mechanism of the TM-APTES/PVA coated wood

coating. Cohesion strength measures how strong the coating itself holds together. (3) Adhesive failure happens at the coating-substrate interface can be interpreted as adhesion strength of coating. (4) Substrate failure happens when wood substrate is torn[42].

The visual appearance of the samples after pull-off adhesion test is presented in Fig. 11. As can be seen in Fig. 10a, there is no evidence of adhesive failure and coating is still covered the substrate but with different thickness. Most parts of area have cohesive failure mode. It is evident that cohesion strength of TM/PVA coating is 1.21 MPa and it is necessary to improve the cohesion strength of coating itself. Fig. 10b shows the epoxy, adhesion and substrate failures of the TM/PVA and TM-APTES/PVA coated wood after pull-off adhesion test. The bond strength is 1.57 MPa and 3.01 MPa, respectively. Most parts of area had adhesive failure mode and the rest of the area had epoxy failure and substrate failure. The epoxy failure occurred because of wood is naturally porous thus the interface between epoxy and the coating surface is not flat enough. Therefore, the data cannot entirely represent the adhesion between the coating and the wood surface, and they only show strength of either the adhesive layer or the wood part. However, the experimental phenomena indicated that the bond strength of TM-APTES/PVA modified wood is stronger than TM/PVA coated wood. This conclusion was considered reasonable because it is well known that the covalent bond is stronger than the hydrogen bond (between polar molecules). Therefore, the APTES improved the interfacial adhesion when one end of the molecule is tethered to the wood surface and the functional group at the other end reacts with the polymer phase.

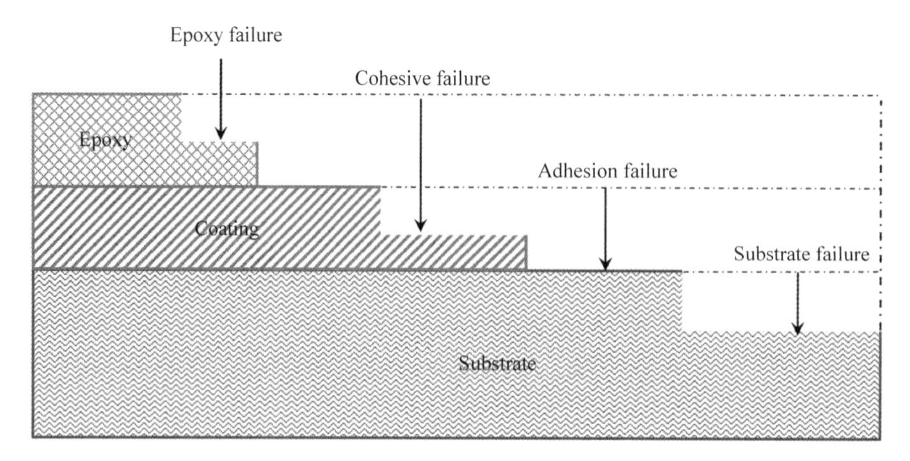

Fig. 10 Schematic illustration of all kinds of failure modes

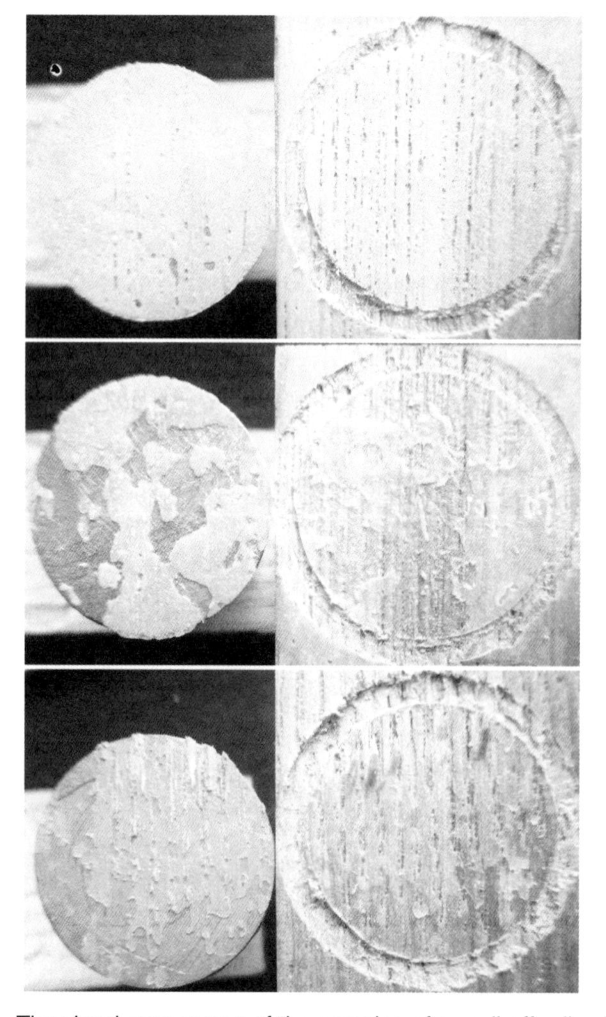

Fig. 11 The visual appearance of the samples after pull-off adhesion test

3.6 *Wettability analyses*

The wettability of the sample was evaluated by measuring the water contact angle (WAC) of liquid on the surface. Fig. 12 demonstrates the contact angles of the pristine wood, PVA coated wood, TM/PVA coated wood and TM-APTES/PVA modified wood. The pristine wood surface in Fig. 12a presents hydrophilicity (the contact angle of water droplet on a surface is lesser than 90°) with the WCA of 83.1°, and the WCA became 53.0° after 15 s. After coated with PVA film, the sample surface in Fig. 12b also presents hydrophilicity with the WCA of 84.5°. As shown in Fig. 12c, the TM/PVA coated wood possesses a hydrophobicity surface with the WCA of 98° but the contact angle became 85° after 15 s, which was attributable to the hydroxyl groups from the PVA polymer. However, it's worth noting that the wettability of the TM-APTES/PVA modified wood was reduced remarkably. Fig. 12d shows a hydrophobic surface with the WCA of 123° and the contact angle retained at 118° after 15 s. As illustrated in Fig. 9, the APTES reacted with the hydroxyl groups of wood substrate and condensed with themselves on the surface of film forming a macromolecular network. This series of reactions reduced the hydroxyl of the wood substrate and the hydrophobic macromolecular network was formed on the film surface consequently blocking these hygroscopic hydroxyl sites. As a result, the surface properties of wood changed from hydrophilic to hydrophobic.

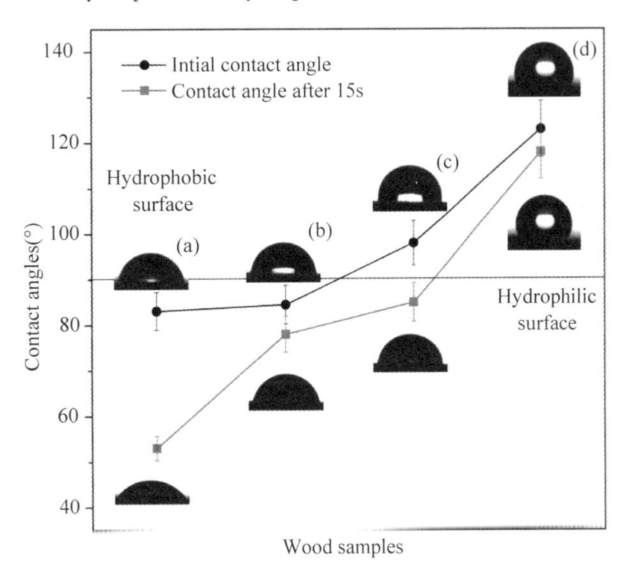

Fig. 12 Contact angles of droplets of water on the as-prepared wood surface

4 Conclusions

In this study, reversible thermoresponsive wood with hydrophobic performance was successfully prepared by a drop-coating process. The as-fabricated samples can reversibly change their colors with the changing temperature. The ΔE^* values of the TM/PVA and TM-APTES/PVA coated wood remarkably reached up to 38.2 and 41.0, respectively, exhibiting a superior thermoresponsive property, and the TM-APTES/PVA modified wood had a higher thermochromic response rate among them. Equally important, the natural wood texture was completely unaffected by the transparent film through the visual inspection. In this work, AEPT plays a key role in improving the performances of modified wood samples. The SEM showed that the glossy spherical TM particles were linked with each other arranged in a fish-scale shape on the surface of TM-APTES/PVA modified wood, however, dispersed unevenly in TM/PVA hybrid film. Furthermore, the TM-APTES/PVA composite polymer coating has been successfully settled onto wood surface and the pull-off test proved that it has stronger adhesion strength than the other samples. Furthermore, the pristine wood presented hydrophilicity with the WCA of 83.1° and the WCA became 53.0° after 15 s. the TM-APTES/PVA modified wood showed a hydrophobicity surface with the WCA of 123 and the contact angle retained at 118° after 15 s. The result demonstrated that the TM-APTES/PVA film formed a hydrophobic macromolecular network on wood surface. Hence, the surface properties of wood materials changed from hydrophilic to hydrophobic.

Acknowledgements

The authors gratefully acknowledge the financial support from the National Natural Science Foundation of China (31470584).

Conflict of interest

The authors declared that they have no conflict of interest.

Reference

[1] Hingston J A, Collins C D, Murphy R J, et al. Leaching of chromated copper arsenate wood preservatives: A review [J]. Environ. Pollut., 2001, 111(1): 53-66.

[2] Raj R G, Kokta B V, Maldas D, et al. Use of wood fibers in thermoplastics. VII. The effect of coupling agents in polyethylene-wood fiber composites[J]. J. Appl. Polym. Sci., 1989, 37(4): 1089-1103.

[3] Lozhechnikova A, Bellanger H, Michen B, et al. Surfactant-free carnauba wax dispersion and its use for layer-by-layer assembled protective surface coatings on wood[J]. Appl. Surf. Sci., 2017, 396: 1273-1281.

[4] Canadell J G, Schulze E D. Global potential of biospheric carbon management for climate mitigation[J]. Nat. Commun., 2014, 5(1): 1-12.

[5] Rowell R M, Tillman A M, Zhengtian L. Dimensional stabilization of flakeboard by chemical modification[J]. Wood Sci. Technol., 1986, 20(1): 83-95.

[6] Kaboorani A, Auclair N, Riedl B, et al. Physical and morphological properties of UV-cured cellulose nanocrystal (CNC) based nanocomposite coatings for wood furniture[J]. Prog. Org. Coat., 2016, 93: 17-22.

[7] Kilpeläinen I, Xie H, King A, et al. Dissolution of wood in ionic liquids[J]. J. Agric. Food Chem., 2007, 55(22): 9142-9148.

[8] Ng K K, Shakiba M, Huynh E, et al. Stimuli-responsive photoacoustic nanoswitch for in vivo sensing applications[J]. ACS nano, 2014, 8(8): 8363-8373.

[9] Gitsov I, Fréchet J M J. Stimuli-responsive hybrid macromolecules: Novel amphiphilic star copolymers with dendritic groups at the periphery[J]. J. Am. Chem. Soc., 1996, 118(15): 3785-3786.

[10] Sokolovskaya E, Rahmani S, Misra A C, et al. Dual-stimuli-responsive microparticles[J]. ACS Appl. Mater. Interfaces, 2015, 7(18): 9744-9751.

[11] Yan D, Lu J, Ma J, et al. Reversibly thermochromic, fluorescent ultrathin films with a supramolecular architecture [J]. Angew. Chem., 2011, 123(3): 746-749.

[12] Park I S, Park H J, Jeong W, et al. Low temperature thermochromic polydiacetylenes: Design, colorimetric properties, and nanofiber formation[J]. Macromolecules, 2016, 49(4): 1270-1278.

[13] Gilat S L, Kawai S H, Lehn J M. Light-triggered molecular devices: Photochemical switching of optical and electrochemical properties in molecular wire type diarylethene species[J]. Chem. Eur. J., 1995, 1(5): 275-284.

[14] Gou M L, Guo G, Zhang J, et al. Time-temperature chromatic sensor based on polydiacetylene (PDA) vesicle and

amphiphilic copolymer[J]. Sens. Actuators B Chem., 2010, 150(1): 406-411.

[15]Chen X, Yoon J. A thermally reversible temperature sensor based on polydiacetylene: Synthesis and thermochromic properties[J]. Dyes Pigm., 2011, 89(3): 194-198.

[16]Lu Y, Zhou S, Gu G, et al. Preparation of transparent, hard thermochromic polysiloxane/tungsten-doped vanadium dioxide nanocomposite coatings at ambient temperature[J]. Thin Solid Films, 2013, 534: 231-237.

[17]Kim H, Kim Y, Kim K S, et al. Flexible thermochromic window based on hybridized VO_2/graphene[J]. ACS nano, 2013, 7(7): 5769-5776.

[18]Pucci A, Ruggeri G, Bronco S, et al. Colour responsive smart polymers and biopolymers films through nanodispersion of organic chromophores and metal particles[J]. Prog. Org. Coat., 2011, 72(1-2): 21-25.

[19]Cataldi A, Corcione C E, Frigione M, et al. Photocurable resin/nanocellulose composite coatings for wood protection [J]. Prog. Org. Coat., 2017, 106: 128-136.

[20] Pasangulapati V, Ramachandriya K D, Kumar A, et al. Effects of cellulose, hemicellulose and lignin on thermochemical conversion characteristics of the selected biomass[J]. Bioresour. Technol., 2012, 114: 663-669.

[21] Puglia D, Biagiotti J, Kenny J M. A review on natural fibre-based composites—Part II: Application of natural reinforcements in composite materials for automotive industry[J]. J. Nat. Fibers, 2005, 1(3): 23-65.

[22]Afra E, Mohammadnejad S, Saraeyan A. Cellulose nanofibils as coating material and its effects on paper properties [J]. Prog. Org. Coat., 2016, 101: 455-460.

[23]Kamdem D P, Pizzi A, Jermannaud A. Durability of heat-treated wood[J]. Holz Roh-Werkst, 2002, 60(1): 1-6.

[24]Bekhta P, Niemz P. Effect of high temperature on the change in color, dimensional stability and mechanical properties of spruce wood[J]. 2003.

[25]Wang J Y, Cooper P A. Effect of oil type, temperature and time on moisture properties of hot oil-treated wood[J]. Holz Roh-Werkst, 2005, 63(6): 417-422.

[26]Lorenzo V, Acebo C, Ramis X, et al. Mechanical characterization of sol-gel epoxy-silylated hyperbranched poly (ethyleneimine) coatings by means of Depth Sensing Indentation methods[J]. Prog. Org. Coat., 2016, 92: 16-22.

[27]Jiang D, Xing L, Liu L, et al. Interfacially reinforced unsaturated polyester composites by chemically grafting different functional POSS onto carbon fibers[J]. J. Mater. Chem. A, 2014, 2(43): 18293-18303.

[28]Ishida H, Miller J D. Cyclization of methacrylate-functional silane on particulate clay[J]. J. Polym. Sci. B Polym. Phys., 1985, 23(11): 2227-2242.

[29]Li X, Tabil L G, Panigrahi S. Chemical treatments of natural fiber for use in natural fiber-reinforced composites: A review[J]. J. Polym. Environ., 2007, 15(1): 25-33.

[30]Yu L, Dean K, Li L. Polymer blends and composites from renewable resources[J]. Prog. Polym. Sci., 2006, 31 (6): 576-602.

[31]Yong V, Hahn H T. Dispersant optimization using design of experiments for SiC/vinyl ester nanocomposites[J]. Nanotechnology, 2005, 16(4): 354.

[32] Holubová B, Cílová Z Z, Kučerová I, et al. Weatherability of hybrid organic-inorganic silica protective coatings on glass[J]. Prog. Org. Coat., 2015, 88: 172-180.

[33]Xie Y, Hill C A S, Xiao Z, et al. Silane coupling agents used for natural fiber/polymer composites: A review[J]. Compos. Part A Appl. Sci. Manuf., 2010, 41(7): 806-819.

[34]Horne J C, Huang Y, Liu G Y, et al. Correspondence between Layer Morphology and Intralayer Excitation Transport Dynamics in Zirconium-Phosphonate Monolayers[J]. J. Am. Chem. Soc., 1999, 121(18): 4419-4426.

[35]Zhu M, Lerum M Z, Chen W. How to prepare reproducible, homogeneous, and hydrolytically stable aminosilane-derived layers on silica[J]. Langmuir, 2012, 28(1): 416-423.

[36]Fuad M Y A, Ismail Z, Ishak Z A M, et al. Application of rice husk ash as fillers in polypropylene: Effect of titanate,

zirconate and silane coupling agents[J]. Eur. Polym. J., 1995, 31(9): 885-893.

[37]Yuan X, Yue Z F, Liu Z Q, et al. Comparison of the failure mechanisms of silicone-epoxy hybrid coatings on type A3 mild steel and 2024 Al-alloy[J]. Prog. Org. Coat., 2016, 90: 101-113.

[38]Ghanbari A, Attar M M. Surface free energy characterization and adhesion performance of mild steel treated based on zirconium conversion coating: a comparative study[J]. Surf. Coat. Technol., 2014, 246: 26-33.

[39]Papaj E A, Mills D J, Jamali S S. Effect of hardener variation on protective properties of polyurethane coating[J]. Prog. Org. Coat., 2014, 77(12): 2086-2090.

[40]Ehsani M, Borsi H, Gockenbach E, et al. Modified silicone rubber for use as high voltage outdoor insulators[J]. Adv. Polym. Tech., 2005, 24(1): 51-61.

[41] Li S M, Jia N, Zhu J F, et al. Synthesis of cellulose-calcium silicate nanocomposites in ethanol/water mixed solvents and their characterization[J]. Carbohyd. Polym., 2010, 80(1): 270-275.

[42]Zhang S, Wang Y S, Zeng X T, et al. Evaluation of adhesion strength and toughness of fluoridated hydroxyapatite coatings[J]. Thin Solid Films, 2008, 516(16): 5162-5167.

中文题目：木基质可逆温度响应疏水薄膜的制备

作者：李莹莹，李坚

摘要：通过在木材表面沉积改性的温敏涂层制造出了具有疏水性能的智能可逆温度响应木材。通过将负载在3-氨基丙基三乙氧基硅烷(AEPT)上的热致变色材料(TM)加入聚乙烯醇溶液(PVA)中来制备改性复合膜。所有样品均具有优异的可逆温度响应性能，TM-APTES/PVA改性木材的温致变色响应率高于其他样品。红外光谱(ATR-FTIR)表明，TM APTES/PVA复合聚合物薄膜已成功沉积在木材表面上。扫描电子显微镜(SEM)和漆膜剥离试验结果证明，改性木材比TM/PVA涂层木材具有更好的TM颗粒分散稳定性和更强的漆膜附着力。此外，经过化学改性后，木质材料的表面性质从亲水性变为疏水性。

关键词：温度响应涂层；疏水；木材表面；3-氨基丙基三乙氧基硅烷

Inorganic-organic Hybrid Wood in Response to Visible Light*

Yingying Li, Bin Hui, Miao Lv, Jian Li, Guoliang Li

Abstract: The inorganic-organic hybrid photoresponsive wood was fabricated by incorporating the phosphomolybdic acid (PMA)/polyvinylpyrrolidone (PVP) composites into the multi-scale hierarchical structure of wood using a simple pressure impregnation method. The as-prepared PMA/PVP coated wood (PPW) can change their colors in response to visible light. Scanning electron microscopy (SEM) and energy dispersive X-ray spectroscopy (EDS) demonstrated that the PMA/PVP composites coated on the wood surfaces successfully and the wood structures were still clearly visible. The atomic force microscopy (AFM) images confirmed that the surface topography of the samples was changed after visible light irradiation, which was ascribed to the formation of heteropoly blues. The analysis of the color parameters proved that as-prepared PPW possessed the excellent photoresponsive property. The ultraviolet-visible spectra (UV-vis) and X-ray photoelectron spectroscopy (XPS) were conducted in order to explore the photoresponsive mechanism, which confirmed the photoreduction reaction occurred, producing the heteropoly blues during the process of visible light irradiation.

Keywords: Photoresponsive wood; Inorganic-organic hybrid; Surface; Visible light

1 Introduction

Wood is a natural and renewable material with many attractive properties, such as desirable amendment for indoor environment, low thermal expansion, good formability and aesthetic appearance, which has been extensively applied in decorations, construction industries, and daily lives fields[1-6]. In addition to these traditional applications, wood materials with enhanced functionalities have great potential as high value-added products[7-9]. As we know, wood has three main components including cellulose, hemicellulose and lignin. The network of these components and the space of parallel hollow tubes builds the sophisticated hierarchical structure, which also determines the penetrability and accessibility of wood component[10-12]. Therefore, wood materials offer exciting opportunities for further functionalization. In recent years, many attempts have been made to create new functions for wood to meet market requirements and compete with other advanced materials, including decay resistance, fire resistance, hydrophobic property, magnetic property and thermochromic property, while simultaneously retaining the desirable properties of wood itself[7,13-16].

Stimuli-responsive materials are smart materials which can show noticeable changes in their properties with environmental stimulus variations. They have attracted much attention for their potential applications as

* 本文摘自 Journal of Materials Science, 2018, 53: 3889-3898.

sensors, smart switches or energy storage and conversion[17-21]. Photoresponsive wood as a novel type of stimuli-responsive material was recently developed. Up to now, photoresponsive wood can be divided into two major types. The organic type was fabricated by depositing organic photochromic material coating on wood surface by a simple drop-coating method. Although the raw photochromic materials are a little expensive, they have reversible color change and rapid response rate[22]. As for the inorganic type, MoO_3 or WO_3 films were coated on wood substrate through a hydrothermal method. The inorganic photoresponsive wood materials have excellent thermostability and extensive sources, but they were in response to UV light[23,24]. It is worthwhile mentioning that the artificial UV light sources are expensive and consume large quantities of electrical power[25]. Although solar energy is the most stable and readily available renewable resource, the wavelength of most commercial solar energy is usually in the visible light region[26]. Therefore, it is necessary to extend the UV light response of photoresponsive wood to the visible light region for high density data storage and solar energy applications. Phosphomolybdic acid (PMA) is a unique class of nanosized transition metal oxide clusters, which has been intensively studied due to its intriguing structure and attractive properties[27,28]. One of the most interesting properties of PMA is that the material can exhibit reversible multi-electron redox processes, retaining its composition and structure[29-31]. On the basis of these properties, PMA has been widely used in catalysis, biomedicine, sensing, and photoelectronic materials[32-34]. As a kind of water-soluble polymer, polyvinylpyrrolidone (PVP) attracts attention due to its outstanding physical and chemical properties, making it a kind good of material as coating, additive, structure-directing agent and stabilizing agent[35,36]. It should be note that the PMA can respond to visible light when combined with PVP.

In this paper, the inorganic-organic hybrid photoresponsive wood was proposed by combining the PMA nanoparticles with PVP polymeric networks on the wood surface. The surface morphologies and chemical components were characterized with scanning electron microscopy (SEM), atomic force microscopy (AFM), energy dispersive X-ray spectrometry (EDS) and attenuated total reflectance of the Fourier transform infrared spectroscopy (ATR-FTIR). The photoresponse mechanism was studied via UV-vis diffuse reflectance spectroscopy (UV-vis) and X-ray photoelectron spectra (XPS).

2 Materials and methods

2.1 Materials

Obtained from Harbin, China, birch veneer samples, with a size of 20 mm×20 mm×0.6 mm. The acetone (99.5%) and ethanol (99.7%) were purchased from Kaitong Chemical Reagent Co., Ltd (Tian jian, China). Polyvinylpyrrolidone [$(C_6H_9NO)_n$, K29-32, $M_w \approx 58000$] was purchased from Aladdin Bio-chem Technology Co. Ltd (Shanghai, China). Phosphomolybdic acid (HPMo, $H_3[P(Mo_3O_{10})_4]$) was purchased from Sinopharm Chemical Reagent Co., Ltd (Shanghai, China). All of the chemicals were used as received without further purification.

2.2 Fabrication of inorganic-organic hybrid photoresponsive wood

In order to remove the impurities formed during raw material processing, the wood samples were ultrasonically treated by distilled water, acetone, and ethanol, respectively, and then dried in an air-circulating oven at 35 ℃ for 30 min. The coating solution was prepared by dissolving 0.07 g PMA and 0.05 g

PVP into 20 mL ethanol under vigorous stirring at room temperature, respectively. After stirring for 1h, the wood samples were added into the coating solution for 15 min, and then the wood samples was moved to a vacuum drying oven at 80 ℃ and 0.06 MPa for 15 min. Finally, the inorganic-organic hybrid photoresponsive wood were obtained and then stored under N_2 in dark until analysis. All the experiments and samples were sheltered from light.

2.3 Characterization

The attenuated total reflectance of the Fourier transform infrared spectroscopy (ATR-FTIR) was carried out at room temperature with a Nicolet Nexus 670 FTIR instrument in the range of 400—4000 cm^{-1} with a resolution of 4 cm^{-1}. The surface morphology and elemental compositions of the sample surfaces were analyzed by scanning electron microscope (SEM) coupled to energy dispersive X-ray spectrometry (EDS). X-ray photoelectron spectra (XPS) was recorded on an ESCALAB 250Xi photoelectron spectrometer using Al $K\alpha$ (1486.6 eV) radiation. The UV-vis diffuse reflection spectra of the samples were obtained from a UV-vis spectrophotometer (UV-vis DRS, TU-1901, China) equipped with an integrating sphere attachment using $BaSO_4$ as the baseline correction. The Atomic force microscopy (AFM) measurements were performed using a Nanoscope V Multimode 8 scanning probe microscope. All experiments were carried out with the same AFM probe under ambient conditions (temperature of 25 ℃, relative humidity of 25%).

2.4 Photochromic text

Photochromic experiments were performed using a 500W Xe lamp as the light source, and a digital camera was used to record the whole process. The distance between the lamp and samples was 20 cm, and the samples were exposed to air during the process of visible light irradiation. Colorimetric analysis of the samples is performed in the Commission Internationale de L'Eclairage L a b space (CIELAB), which is designed to accurately approximate human vision. The color parameters L^* (lightness index), a^* (red-green index) and b^* (yellow-blue index) values were measured by using the Adobe Photoshop CS6 software installed in the computer[37]. The total color change ΔE^* values were calculated using the following equations:

$$\Delta E^* = \sqrt{(L_2-L_1)^2+(a_2-a_1)^2+(b_2-b_1)^2} \quad (1)$$

The L_1, a_1 and b_1 are the color parameters before visible light irradiation; L_2, a_2 and b_2 are the color parameters under irradiation with the visible light. The color parameters were measured at five different points on each sample, and the final value for L, a and b were obtained as an average of these five measurements.

3 Results and discussion

3.1 ATR-FTIR analyses

Fig. 1 presented the ATR-FTIR transmission spectra of pristine wood and PPW before and after visible light irradiation. As shown in Fig. 1b, several new peaks appeared in the spectrum, compared with pristine wood. The peaks at 1658 and 1290 cm^{-1} were ascribed to the stretching vibration of C=O and C—N, respectively. The previously mentioned absorption peaks were attributed to the organic groups of PVP. Moreover, the absorption peaks at 1060, 963, 875 and 799 cm^{-1} were the characteristic peaks of the Keggin structure. These peaks were ascribed to ν(P-O), ν(Mo=Ot), ν(Mo-Oc-Mo) and ν(Mo-Oe-Mo), where Ot, Oc, and Oe refer to the terminal, corner and edge oxygens, respectively, which matches well with earlier reports[38-40]. The broad blunt vibration band around 3300 cm^{-1} may be ascribed to N—H and O—H, which was due to the interaction between the oxygen atom of PMA and the active hydrogen of the amidic groups. Significantly, it can be seen from the curves c that the irradiation resulted in a decrease in the intensity of the absorption band at 3330 cm^{-1}, indicating that the vibrations of N—H bond are disturbed during the irradiation, which may be caused by the hydrogen bond interaction between PMA and PVP. Furthermore, the vibration bands associated to PMA and PVP were all preserved, verifying that the basic structure of Keggin and the organic polymeric matrix were not destroyed after visible light irradiation.

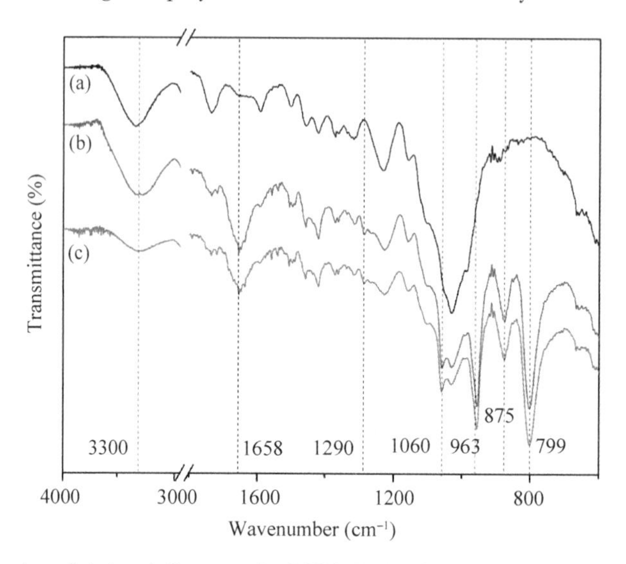

Fig. 1 ATR-FTIR spectra of (a) pristine wood, PPW (b) before and (c) after visible light irradiation

3.2 SEM observations and EDS analysis of samples

Fig. 2 shows the SEM images and EDS spectra of the pristine wood and the PPW. The main microstructures like pits and vessel members of pristine wood were clearly detected by SEM (Fig. 2a). In Fig. 2b, it was observed that the PMA/PVP composites were evenly coated on wood surfaces. The wood structures were rough but still clearly visible, which illustrated that the as-prepared PPW still possessed the subtle multiscale hierarchical structure of wood.

The surface elemental composition of the pristine wood and the PPW were analyzed using the EDS spectra, as shown in Fig. 1c and d, respectively. The carbon, oxygen and platinum elements were found in the pristine wood. The platinum element is originated from the coating layer used for electric conduction during the SEM observation. In Fig. 2d, the strong peaks around 2.0 keV and 2.3 keV were detected, which confirmed the presence of phosphorus and molybdenum on the wood surface.

3.3 AFM measurements

The wood samples were further studied by AFM to obtain the details of the change of surface morphology. Tapping mode AFM images depicted the surface topography of the samples before and after visible light irradiation. AFM image of PPW before visible light irradiation is shown in Fig. 3a, the compact regular rolling mountain peaks with similar shape and the uniform size were observed over the surface. The structure of linear topography feature derived from the structure of lignocellulose. When the samples were exposed to visible

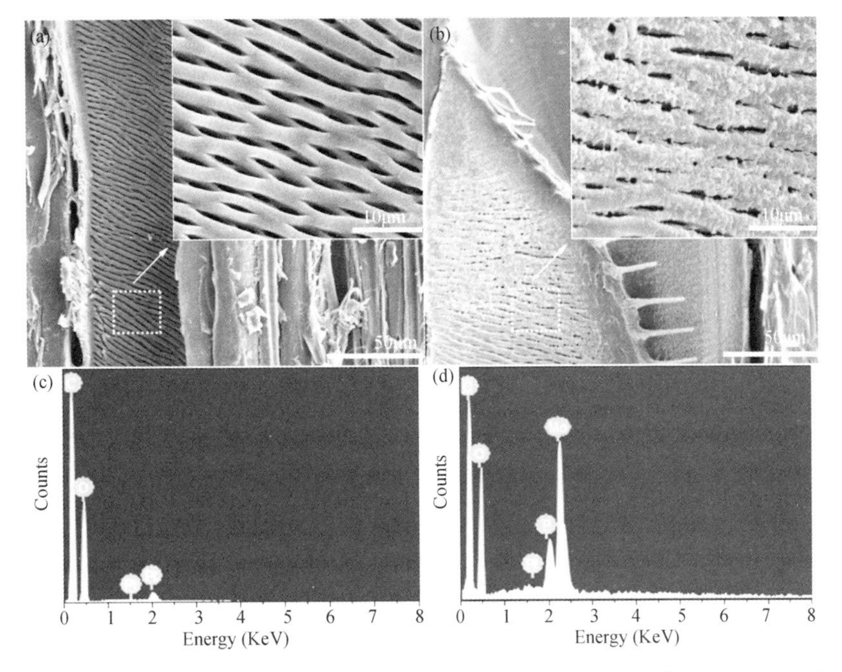

Fig. 2 Low- and high-magnification SEM images of (a) the pristine wood and (b) the PPW. EDS spectra of (c) the pristine wood and (d) the PPW

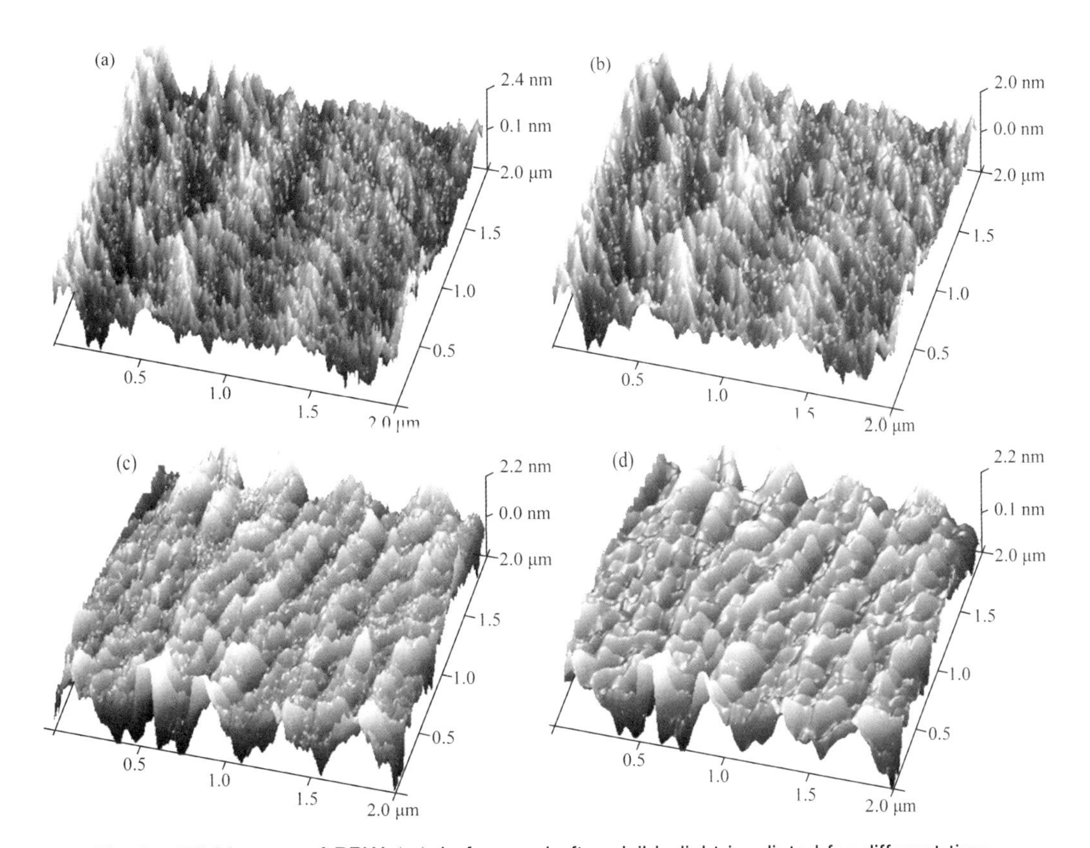

Fig. 3 AFM images of PPW (a) before and after visible light irradiated for different time

light for 2 min (Fig. 3b), the shape of the peaks was similar, but the height was lower than the peaks in Fig. 3a. After irradiated for 20 min, the surface topography changed from the peak to the hill structure (Fig. 3c). The surfaces of the sample tend to be smooth and the size of the composite particles became larger, which might be attributed to the formation of heteropoly blues. As shown in Fig. 3d, when the sample was exposed to visible light for 50 min, the shape and the size of the hill structure was almost constant, but the surface topography was more uniform and smooth. Beside this, compare with Fig. 3b, the height of the particles has increased, which might be corresponding to a certain degree of heteropoly blues particles during the irradiation.

3.4 Photochromic behaviors

Fig. 4 shows the color parameters of the pristine wood and photoresponsive wood with different visible light irradiation time. After visible light irradiation, the color parameters of the pristine wood almost unchanged, indicating that the pristine wood lack the photochromic capacity. For PPW, when the samples were exposed under visible light for 80 min, the lightness index L^* of the samples remarkably decreased from 69.8 to 23.2, implying the surface color turned much darker. With increasing irradiation time, a^* value first decreased from 4.0 to −3.0 and then increased gradually, which demonstrated that the surface color turned gray green. In addition, the change of b^* value decreased from 28.6 to −2.0, verifying that the sample had yellow surface at first, and then turned dark blue. Furthermore, the ΔE^* value dramatically reached 56.0, exhibiting a superior photoresponsive property. Fig. 5a shows the color-changed process of the sample. Before visible light irradiation, the samples were pale yellow in color. With the irradiation time increasing to 80 min, the color of the photoresponsive wood turned dark blue, which was consistent with the analytic results of color parameters. Furthermore, the coloring and bleaching process of the sample irradiated for 5 min was shown in

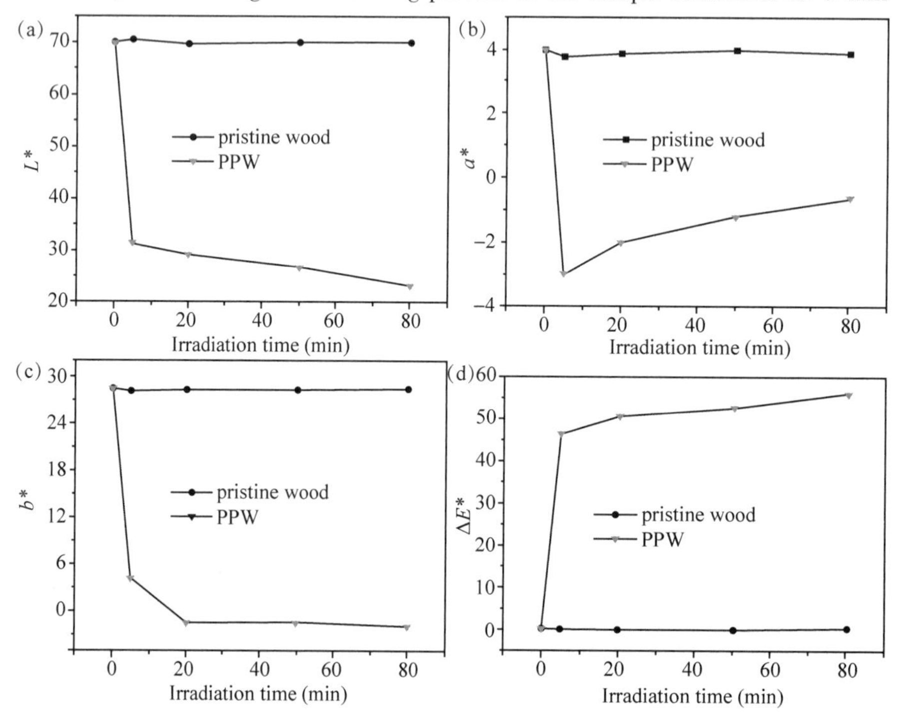

Fig. 4 Color parameters with different visible light irradiation time

Fig. 6. As we can see, the colored sample gradually faded in the air after turned off the light source. It is worth noting that the faded samples will turn dark blue when exposed to visible light again, and this process can be repeated multiple times.

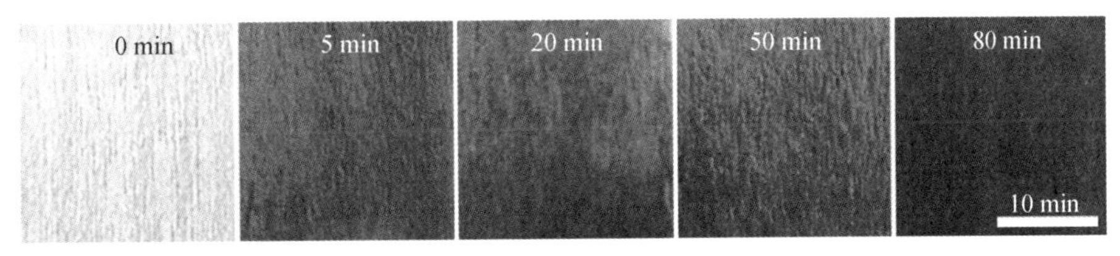

Fig. 5 Color-changed process of PPW irradiated for different time

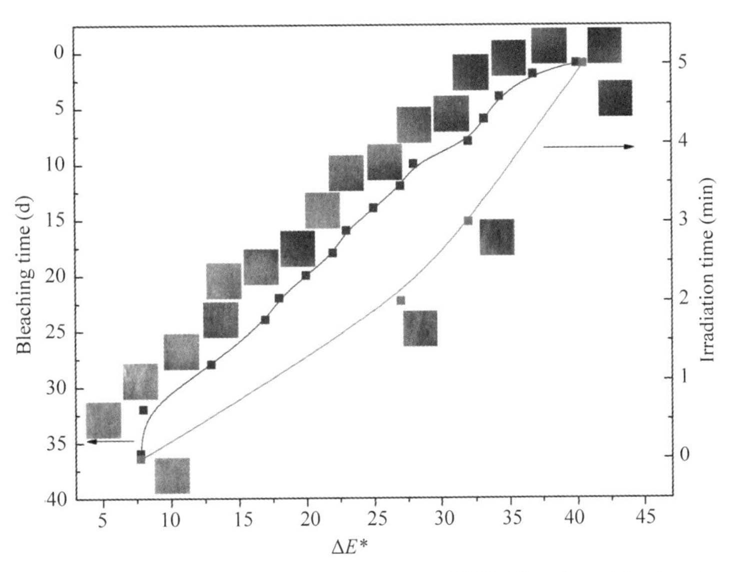

Fig. 6 Coloring and bleaching process of PPW irradiated for 5 min

3.5 UV-vis absorption spectra

In order to explain the photoresponsive mechanism, the UV-vis absorption spectra of the pristine wood and the PPW with different visible light irradiation time were carried out in Fig. 7. It was observed that the maximum absorption peak of the pristine wood at 289 nm was attributed to $\pi \rightarrow \pi$ transition of aromatic C—C bonds[41,42]. For PPW before visible light irradiation, the peak at 289 nm red-shifted to 369 nm, which was due to the Mo—O bonds on the wood surface. Furthermore, the low-energy tail of the absorption shifted to the blue light region (400-500 nm). After visible light irradiation, two broad absorption bands appeared, which were attributed to metal-to-metal d-d transition at about 530 nm and intervalence charge transfer (IVCT) ($Mo^{6+} \rightarrow Mo^{5+}$) at about 711 nm, respectively. These results indicated that the photoreduction reaction occurred and the heteropoly blues were formed as the sample irradiated with visible light. Besides, the intensity of the two absorption bands continuously increase with increasing irradiation time (inset), which suggested that the as-prepared samples are promising for high density data storage.

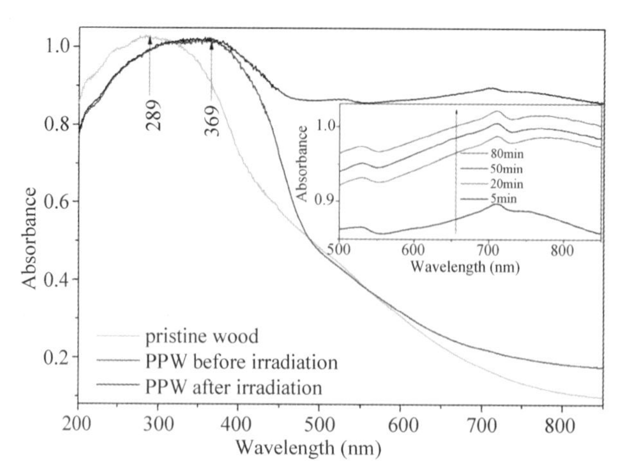

Fig. 7 UV-vis spectra of the pristine wood and PPW before and after visible light irradiation (inset, UV-vis spectra of the PPW with different visible light irradiation time)

3.6 XPS analyses

X-ray photoelectron spectroscopy (XPS) experiments were further performed to characterize the Mo oxidation state. The measured spectra were deconvoluted using Gaussian functions. Fig. 8 shows the XPS spectra of the Mo 3d energy level of PPW before and after exposure to visible light irradiation. For the samples before irradiation, the spectrum exhibits two contributions, $3d_{5/2}$ and $3d_{3/2}$ (resulting from the spin-orbit splitting), located at 232.7 eV and 235.8 eV, respectively, which were due to Mo^{6+}. It can be seen that the two peaks became broad and asymmetric after the sample was exposed to visible light (Fig. 8b). After deconvolution, binding energy of the Mo3d was separated into four peaks. The first predominant doublet at 235.8 eV and 232.7eV were characteristic of Mo^{6+}. In addition, the peaks at 231.6 eV and 234.6 eV were assigned to Mo^{5+}($3d_{5/2}$) and Mo^{5+}($3d_{3/2}$), respectively. As discussed above, the results further indicated that the Mo^{5+} ions have been produced during the visible light irradiation process, indicating the photoreduction reaction occurred. Based on the above-mentioned results, the possible photo-reduction/oxidation process of the sample is shown in Fig. 9.

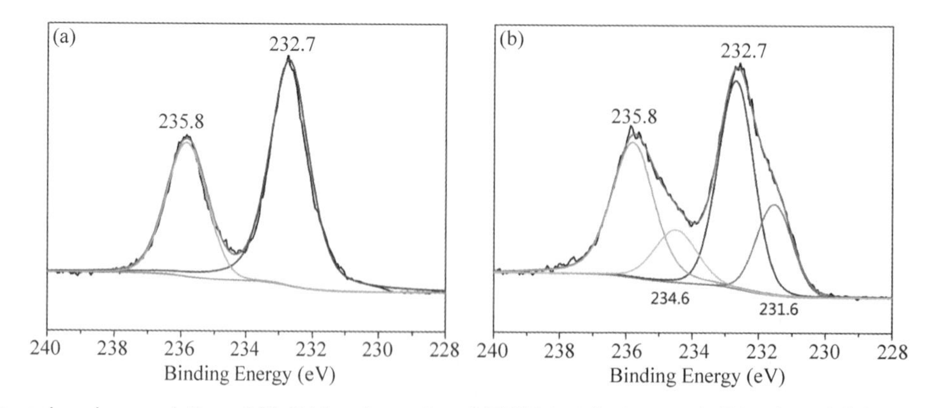

Fig. 8 Gaussian deconvolution of Mo3d level spectra of PPW (a) before and (b) after visible light irradiation

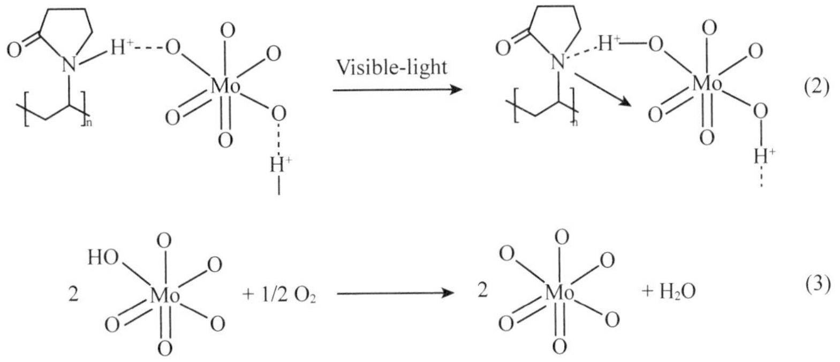

Fig. 9 The possible photo-reduction/oxidation process

4 Conclusions

The inorganic-organic hybrid photoresponsive wood was successful prepared using the combination of PMA nanoparticles and PVP polymeric networks. The EDS spectra and SEM images show that the PMA/PVP composites were successful coated on wood surfaces and the multiscale hierarchical structures of wood were still clearly visible. Besides this, the as-prepared PPW can change their colors in response to visible light. After visible light irradiation, the structure of PMA and the organic polymeric matrix were all preserved. In addition, the surface topography of the samples changed after visible light irradiation, which was attributed to the formation of heteropoly blues. Furthermore, the investigation of the color parameters demonstrated that the PPW possessed excellent photoresponsive property and the photoresponsive mechanism was analyzed. The UV-vis and XPS results indicated that the photoreduction reaction occurred and part of Mo^{6+} was converted to Mo^{5+}, demonstrating the heteropoly blues formed during the visible light irradiation process. Due to the property of responding to visible light, the inorganic-organic hybrid photoresponsive wood could be applied in smart home, molecular switch, high density data storage and solar energy applications.

Acknowledgements

The authors gratefully acknowledge the financial support from the National Natural Science Foundation of China (31470584) and the Special Fund for Forest Scientific Research in the Public Welfare (No. 201504602-5).

Conflict of interest

The authors declared that they have no conflict of interest.

Reference

[1] Steinbüchel A. Non-biodegradable biopolymers from renewable resources: Perspectives and impacts[J]. Curr. Opin. Biotechnol., 2005, 16(6): 607-613.

[2] Wensing M, Uhde E, Salthammer T. Plastics additives in the indoor environment—flame retardants and plasticizers [J]. Sci. Total Environ., 2005, 339(1-3): 19-40.

[3] Nishiyama Y. Structure and properties of the cellulose microfibril[J]. J. Wood Sci., 2009, 55(4): 241-249.

[4] Regazzi A, Dumont P J J, Harthong B, et al. Effectiveness of thermo-compression for manufacturing native starch bulk materials[J]. J. Mater. Sci., 2016, 51(11): 5146-5159.

[5] Lei Y, Liu S, Li J, et al. Effect of hot-water extraction on alkaline pulping of bagasse[J]. Biotechnol. Adv., 2010,

28(5): 609-612.

[6]Gustavsson L, Joelsson A. Life cycle primary energy analysis of residential buildings[J]. Energy Build., 2010, 42(2): 210-220.

[7]Hu L, Lyu S, Fu F, et al. Preparation and properties of multifunctional thermochromic energy-storage wood materials[J]. J. Mater. Sci., 2016, 51(5): 2716-2726.

[8]Anjum M A R, Lee J S. Sulfur and nitrogen dual-doped molybdenum phosphide nanocrystallites as an active and stable hydrogen evolution reaction electrocatalyst in acidic and alkaline media[J]. ACS Catal., 2017, 7(4): 3030-3038.

[9]Gallezot P. Conversion of biomass to selected chemical products[J]. Chem. Soc. Rev., 2012, 41(4): 1538-1558.

[10]Pérez J, Munoz-Dorado J, De la Rubia T, et al. Biodegradation and biological treatments of cellulose, hemicellulose and lignin: an overview[J]. Int. Microbiol., 2002, 5(2): 53-63.

[11] Yang H, Yan R, Chen H, et al. In-depth investigation of biomass pyrolysis based on three major components: hemicellulose, cellulose and lignin[J]. Energy Fuels, 2006, 20(1): 388-393.

[12] Yang H, Yan R, Chen H, et al. Characteristics of hemicellulose, cellulose and lignin pyrolysis[J]. Fuel, 2007, 86(12-13): 1781-1788.

[13]Chen Y, Wang H, Yao Q, et al. Biomimetic taro leaf-like films decorated on wood surfaces using soft lithography for superparamagnetic and superhydrophobic performance[J]. J. Mater. Sci., 2017, 52(12): 7428-7438.

[14]Fang Q, Chen B, Lin Y, et al. Aromatic and hydrophobic surfaces of wood-derived biochar enhance perchlorate adsorption via hydrogen bonding to oxygen-containing organic groups[J]. Environ. Sci. Technol., 2014, 48(1): 279-288.

[15]Ringman R, Pilgård A, Brischke C, et al. Mode of action of brown rot decay resistance in modified wood: a review[J]. Holzforschung, 2014, 68(2): 239-246.

[16]Mohan D, Kumar H, Sarswat A, et al. Cadmium and lead remediation using magnetic oak wood and oak bark fast pyrolysis bio-chars[J]. Chem. Eng. J., 2014, 236: 513-528.

[17]Alkemade R, Van Oorschot M, Miles L, et al. $GLOBIO_3$: a framework to investigate options for reducing global terrestrial biodiversity loss[J]. Ecosystems (N. Y., Print), 2009, 12(3): 374-390.

[18]Zuo Y, Zhao J, Gao Y, et al. Controllable synthesis of P (NIPAM-co-MPTMS)/PAA-Au composite materials with tunable LSPR performance[J]. J. Mater. Sci., 2017, 52(16): 9584-9601.

[19]Liu J, Lu Y. Stimuli-responsive disassembly of nanoparticle aggregates for light-up colorimetric sensing[J]. J. Am. Chem. Soc., 2005, 127(36): 12677-12683.

[20]Paris J L, Colilla M, Izquierdo-Barba I, et al. Tuning mesoporous silica dissolution in physiological environments: A review[J]. J. Mater. Sci., 2017, 52(15): 8761-8771.

[21]Nunes S P, Behzad A R, Hooghan B, et al. Switchable pH-responsive polymeric membranes prepared via block copolymer micelle assembly[J]. ACS nano, 2011, 5(5): 3516-3522.

[22]Hui B, Li G, Li J, et al. Hydrothermal deposition and photoresponsive properties of WO_3 thin films on wood surfaces using ethanol as an assistant agent[J]. J Taiwan Inst Chem Eng, 2016, 64: 336-342.

[23]Hui B, Li Y, Huang Q, et al. Fabrication of smart coatings based on wood substrates with photoresponsive behavior and hydrophobic performance[J]. Mater. Des., 2015, 84: 277-284.

[24]Hui B, Wu D, Huang Q, et al. Photoresponsive and wetting performances of sheet-like nanostructures of tungsten trioxide thin films grown on wood surfaces[J]. RSC Adv., 2015, 5(90): 73566-73574.

[25]Zhao W, Chen C, Li X, et al. Photodegradation of sulforhodamine-B dye in platinized titania dispersions under visible light irradiation: influence of platinum as a functional co-catalyst[J]. J. Phys. Chem. B, 2002, 106(19): 5022-5028.

[26] Laurier K G M, Vermoortele F, Ameloot R, et al. Iron (III)-based metal-organic frameworks as visible light photocatalysts[J]. J. Am. Chem. Soc., 2013, 135(39): 14488-14491.

[27]Chilivery R, Rana R K. Microcapsule structure with a tunable textured surface via the assembly of polyoxomolybdate

clusters: a bioinspired strategy and enhanced activities in alkene oxidation[J]. ACS Appl. Mater. Interfaces, 2017, 9(3): 3161-3167.

[28]Jameel U, Zhu M, Chen X, et al. Recent progress of synthesis and applications in polyoxometalate and nanogold hybrid materials[J]. J. Mater. Sci., 2016, 51(5): 2181-2198.

[29]Jia X, Shen L, Yao M, et al. Highly efficient low-bandgap polymer solar cells with solution-processed and annealing-free phosphomolybdic acid as hole-transport layers[J]. ACS Appl. Mater. Interfaces, 2015, 7(9): 5367-5372.

[30]Uma T, Nogami M. Proton-conducting glass electrolyte[J]. Anal. Chem., 2008, 80(2): 506-508.

[31]Li K, Wang J, Zou Y, et al. Surfactant-assisted sol-gel synthesis of zirconia supported phosphotungstates or Ti-substituted phosphotungstates for catalytic oxidation of cyclohexene[J]. APPL CATAL A-GEN, 2014, 482: 84-91.

[32]Sang B, Li Z, Li X, et al. Graphene-based flame retardants: A review[J]. J. Mater. Sci., 2016, 51(18): 8271-8295.

[33]Othman A M, Rizk N M H, El-Shahawi M S. Polymer membrane sensors for sildenafil citrate (Viagra) determination in pharmaceutical preparations[J]. Anal. Chim. Acta, 2004, 515(2): 303-309.

[34]Xiao Y, Zheng L, Cao M. Hybridization and pore engineering for achieving high-performance lithium storage of carbide as anode material[J]. Nano Energy, 2015, 12: 152-160.

[35] Yin B, Wang J, Jia H, et al. Enhanced mechanical properties and thermal conductivity of styrene-butadiene rubber reinforced with polyvinylpyrrolidone-modified graphene oxide[J]. J. Mater. Sci., 2016, 51(12): 5724-5737.

[36] O'Haire T, Russell S J, Carr C M. Centrifugal melt spinning of polyvinylpyrrolidone (PVP)/triacontene copolymer fibres[J]. J. Mater. Sci., 2016, 51(16): 7512-7522.

[37]Li Y, Hui B, Li G, et al. Fabrication of smart wood with reversible thermoresponsive performance[J]. J. Mater. Sci., 2017, 52(13).

[38]Ohisa S, Kagami S, Pu Y J, et al. A solution-processed heteropoly acid containing MoO_3 units as a hole-injection material for highly stable organic light-emitting devices[J]. ACS Appl. Mater. Interfaces, 2016, 8(32): 20946-20954.

[39]Zhang B, Asakura H, Yan N. Atomically dispersed rhodium on self-assembled phosphotungstic acid: structural features and catalytic CO oxidation properties[J]. Ind. Eng. Chem. Res., 2017, 56(13): 3578-3587.

[40]Yang H, Song T, Liu L, et al. Polyaniline/polyoxometalate hybrid nanofibers as cathode for lithium ion batteries with improved lithium storage capacity[J]. J. Phys. Chem. C, 2013, 117(34): 17376-17381.

[41]Li J, Yu H, Sun Q, et al. Growth of TiO_2 coating on wood surface using controlled hydrothermal method at low temperatures[J]. Appl. Surf. Sci., 2010, 256(16): 5046-5050.

[42]Singh B, Fang Y, Cowie B C C, et al. NEXAFS and XPS characterisation of carbon functional groups of fresh and aged biochars[J]. Org. Geochem., 2014, 77: 1-10.

中文题目：无机-有机杂化光响应木材

作者：李莹莹，惠彬，于森，李坚，李国梁

摘要：以木材的多尺度层次结构为模板，通过使用简单的压力浸渍方法将磷钼酸(PMA)/聚乙烯吡咯烷酮(PVP)复合其中，制成了无机-有机杂化光敏木材。制备的PMA / PVP涂层木材(PPW)可以根据可见光改变颜色。扫描电子显微镜(SEM)和X射线光谱表明，PMA/PVP复合材料已成功涂覆在木材表面，并且木材结构仍然清晰可见。原子力显微镜图像证实可见光照射后样品的表面形貌发生了变化，这归因于杂多蓝的形成。颜色参数的分析证明，制备的PPW具有优异的光响应性。为了探索光响应机理，进行了紫外可见光谱和X射线光电子能谱分析，证实了光还原反应的发生，并且在可见光照射过程中产生了杂多蓝。

关键词：光响应木材；有机无机杂化；表面；可见光

木质仿生智能响应材料的研究进展*

李坚*，李莹莹

摘要： 智能响应材料需具备三个基本要素，即感知、驱动和控制，在全球新材料研究领域中，仿生智能响应材料是目前世界各国技术战略发展中的竞争热点。木材是一种天然且可再生的生物质材料，具有良好的结构和功能特性。作为人类使用最早的材料，木材具有轻质、美观、生物调节等优良特性，是绿色环境人体健康的贡献者。木材的纤维素、半纤维素和木质素构成了木材精妙的微结构同时提供了许多官能团，为木材仿生智能材料的合成奠定了优良的基础。此综述简要介绍了木质仿生智能响应材料的研究进展，综述了 pH、气体、光、机械力、湿度、温度和双重/多重刺激响应木质材料的制备、性能与潜在应用；重点介绍并总结了以木质材料为基材的仿生智能响应材料的发展现状。

关键词： 木质材料；仿生；智能；刺激-响应

变色龙是一种典型的智能响应物种，其颜色变化取决于周围环境因素如光线、温度以及情绪（惊吓、开心或沮丧），在不同环境刺激下，变色龙身体颜色可变为绿色、黄色、米色或深棕色等[1]；向日葵具有一种茎干细胞生长素，因惧光多聚集在茎干背光的部分，导致背光面茎干生长速度更快，产生“向阳”现象[2]；光合作用是绿色植物和藻类在阳光刺激作用下将二氧化碳和水转化为碳水化合物和氧气的智能过程，等等。智能响应材料是指在受到光、热、湿度和力等物理刺激，以及 pH 值、离子、葡萄糖和酶等化学刺激下，其微观分子结构及分子构象发生转变，从而导致材料自身的结构、物理和化学性能等发生相应变化的材料[3-4]，是药物载体[5]、传感元件[6]、分子开关[7]、建筑工程材料[8]等领域的研究热点。

木材具有机械性能好、热膨胀低、强重比高、美观、可持续性强等优异特点，自古以来就被人类所使用。然而，随着生活水平的提高，消费者对多功能绿色环保材料的关注与日俱增。传统意义上的木质产品已经不能满足大众需求。木材是由纤维素、半纤维素和木质素组成的天然复合材料，具有多尺度各向异性取向的孔道结构，同时也具有渗透性和大量的活性官能团。基于此，木材的层次结构和反应活性为其与功能性材料的结合提供了夯实的基础。为了满足消费者市场要求并与其他先进材料进行竞争，相关研究人员在木材功能化方面已经做出了许多成果[9-12]。通过仿生手段进行木材智能响应功能化使木材具有更加良好的性能具有重要意义和应用前景。本文的主要目的是综述近年来木质仿生智能响应材料的研究进展，介绍 pH、气体、光、机械力、湿度、温度和双重/多重刺激响应木质材料的制备、性能与潜在应用。

1 pH 智能响应木质功能材料

纤维素是一种葡萄糖大分子多糖，骨架上具有大量规则排列的可反应性羟基，易于进行化学修饰[13]，其生产原料来源于木材、棉花、麻和秸秆等[14]。木质纤维素来源于木材，是一种很有发展前

* 本文摘自 Journal of Forest and Environment，2019，39：337-343.

途的材料，具有广泛的应用前景。木质纤维素的三维微孔结构使其具备优异的理化性质。学者们制备的表面化学和形态学可控的智能响应纤维素材料为木材添加了功能性。材料的 pH 响应性源于 pH 敏感基团，这些基团本质上是一些弱酸或弱碱基团，响应的本质即是弱酸或弱碱的解离平衡发生改变。纤维素中含有大量活性官能团，可用作构建 pH 响应体系。

CHINGA-CARRASCO *et al.*[15] 从辐射松纸浆中提取纳米纤维素，对纳米纤维素进行羧甲基化和高碘酸盐氧化预处理，制备了一种含有大量醛基和羧基的高度纤维化材料。样品因可离子化官能团的三维微孔结构材料具有膨胀能力而具备 pH 响应特性。此纳米纤维素凝胶在中性和碱性条件下的溶胀度明显高于酸性环境（pH=3），可应用于慢性伤口管理、控制和生物膜中抗菌成分的智能释放。WANG *et al.* 采用原子转移自由基聚合法（ATRP）合成具有 pH 响应功能的乙基纤维素-g-聚甲基丙烯酸（2-N，N-二乙胺基）乙酯共聚物（EC-g-PDEAEMA），此胶束具有可逆的 pH 响应性。在相同侧链长度下，临界胶束浓度随着接枝密度的增加而降低。pH 在 6~6.9 时，随着胶束水动力半径减小，脱质子作用导致胶束壳内侧 PDEAEMA 链塌陷。实验发现，利发霉素在 pH 为 6.6 的缓冲液中的累积释放量高于 pH 为 7.4 时，证明负载利发霉素接枝共聚物胶束在药物传递系统中具有可控的 pH 响应释放功能[16]。TAN *et al* 利用羧甲基纤维素（CMC）直接还原氯金酸（$HAuCl_4$）制备了金纳米粒子（AuNPs）。将半胱胺盐酸盐（CA）与 CMC 静电复合物通过配体交换修饰 AuNPs，得到 Au-CA/CMC 的分散体系。此 Au-CA/CMC 分散体系在不同 pH 值下具有明显的可逆 pH 响应行为。如图 1 所示，Au-CA/CMC 溶液的初始 pH 值为 6.4 时溶液呈亮红宝石色，其颜色随溶液的 pH 值的改变而变化。此产品可作为金属纳米颗粒的还原剂和响应稳定剂，具有广泛药用价值[17]。

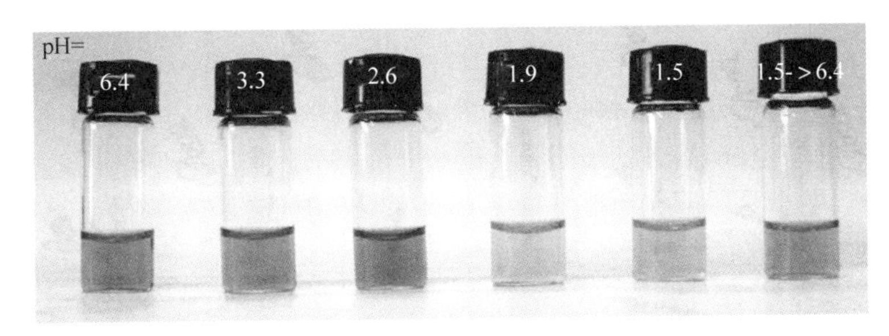

图 1　在不同 pH 值下的 Au-CA/CMC 分散液[17]

2　气体智能响应木质功能材料

每种智能响应材料都有缺点，这限制了它们的适用性。例如，pH 智能响应材料存在盐积累问题，这可能会削弱 pH 反应聚合物的可切换性。电智能响应材料存在体积过大则能耗较高的问题。但是在气体响应聚合物中，大多数刺激气体在大体积操作中易于添加或去除，因此，气体智能响应聚合物在工业应用中具有重要的意义[18]。

GAO *et al.* 以木纤维为生物模板加以 Bi_2WO_6 合成了一种气体响应材料，配合生物源 C 提高了气体传感器对含羰基气体的敏感性。掺杂 C 的 Bi_2WO_6 传感器对丙酮、乙酸和乙酸乙酯分子的低温敏感性分别是未掺 C 的 Bi_2WO_6 传感器敏感性的 4.4 倍、3.0 倍和 2.4 倍。由于生物源 C 可以吸附更多氧分子与气体中的羰基进行反应，从而释放更多的电子，导致传感器的电阻降低。该研究为气体传感器的开发开辟了一条新的途径[19]。GARCIA-VALDEZ *et al.* 采用表面引发的硝基氧介导聚合（SI-NMP）将三种二氧化碳响应聚合物接枝到纤维素纳米晶体（CNC）表面。如图 2 所示，使用二氧化碳（CO_2）作为触发器样品能够可逆的从疏水状态切换为亲水状态。此研究提出的策略是一种通用的平台技术，合成步骤并不复杂，且很容易扩展。可以方便的将具备可逆切换表面特性的 CNC 融入各种复合材料中[20]。

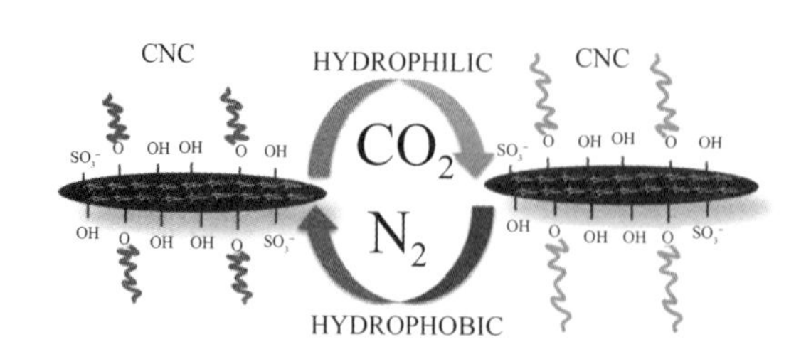

图2 具有 CO_2 响应性能的纤维素纳米晶体[20]

3 光智能响应木质功能材料

3.1 可见光响应

GAO *et al.* 在木材表面原位涂覆了锐钛矿二氧化钛（TiO_2）微粒，在可见光的驱动下，可降解甲醛和苯酚气体。添加 Ag 纳米颗粒可加强其可见光催化性质，研究后续对样品表面进行了疏水处理，成功地制备了一种超强的疏水抗菌复合膜。此产品在可见光下具有超强的超疏水性、抗菌性和光分解苯酚的多功能特性，在自清洁、抗菌、光催化、高附加值木材产品等方面具有潜在的应用前景[21,22]。LI *et al.* 以木材单板的多尺寸多孔结构为天然生物基模板，利用水热法制备了木质基磷钼酸可见光光催化剂。在可见光照射下对罗丹明 B 具有优越的光降解能力[23]。如图 3 所示，LI *et al.* 将磷钼酸包裹入壳聚糖/聚乙烯基吡咯烷酮混合物中，并将这种混合物固定在木材表面从而制备了一种光响应木材。在可见光照射下，随着辐照时间从 0 增加到 90 min，样品的总色差值从 0.7 显著增加到 42.5，证明样品具有良好的光响应性能。样品具有优越的灵敏度、准确的响应能力和环保性能，在传感器、智能家居、太阳能转换等领域具有广阔的应用前景[24]。

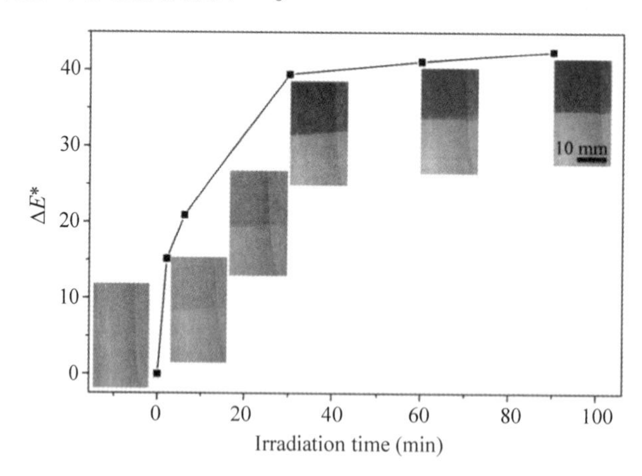

图3 光响应木材的光致变色过程[24]

3.2 紫外光响应

GAO *et al.* 首次制备并报道了在木材表面负载锐钛矿 TiO_2 和 Cu_2O 粒子，得到具有光催化能力的样品。在紫外辐照下，样品产生了负氧离子并且对大肠杆菌有抗菌作用[25]。采用低温水热法，HUI *et al.* 在木材表面生长了具有光智能响应的层状花型三氧化钼（MoO_3）薄膜，样品对紫外光具有智能响应性。

如图 4 紫外可见吸收光谱所示，该产品在 365 nm 处对紫外光的刺激反应良好，颜色变化显著[26]。如图 5 所示，GAO *et al.* 采用低温水热法用十八烷基三氯硅烷（OTS）对二氧化钛木质基材进行改性，在紫外光下，样品表面的水接触角约为 0°。当放置在黑暗中，样品表面的水接触角大约为 152°[27]。

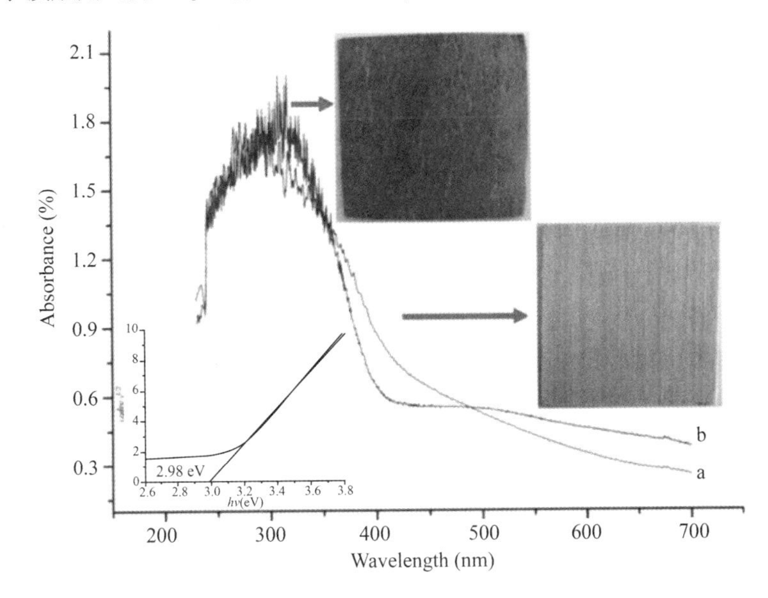

图 4 素材（a）和三氧化钼涂层木材（b）的紫外可见吸收光谱[26]

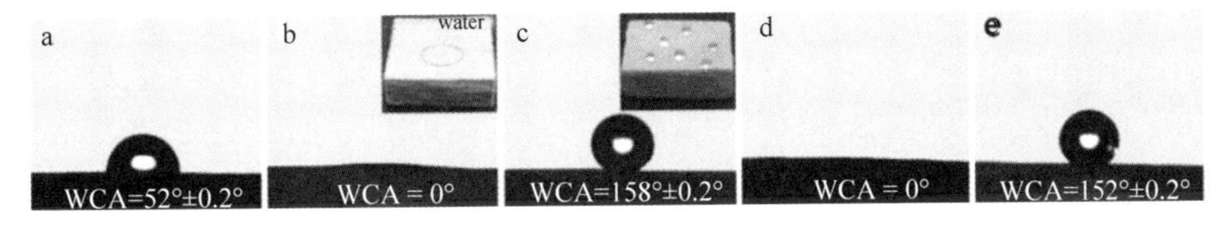

图 5 不同样品的水接触角[27]

4 机械力智能响应木质功能材料

IMATO *et al.* 采用一种可断裂可逆重组的动态共价力学团对 CNC 表面进行改性，研究表明，由于分子迁移率的限制，活化表面修饰的机械载体表现出对机械应力高度敏感性。在纳米复合材料保持自愈能力的同时，改性的 CNC 可以有效增强含有类似动态共价键的自愈聚合物。此研究结果对于设计具有损伤自报告和自愈合等功能特性的复合材料具有广泛的应用价值[28]。如图 6 所示，YANG *et al.* 以肼与醛基碳纳米管交联为基础，制备了化学交联的 CNC 气凝胶。与以往物理交联 CNC 气凝胶相比，化学交联 CNC 气凝胶具有更好的力学性能和变形恢复能力，尤其是在水中。甚至在压缩和释放 20 次循环之后，气凝胶在压缩 85%后依旧可以恢复 80%以上。此方法证明，CNC 气凝胶既可以用作高吸附剂和油/水分离，也可以作为绝缘或减震材料[29]。

5 湿度智能响应木质功能材料

KOTRESH *et al.* 利用一种经济有效的生物高分子材料 CMC，合成了一种室温湿度传感装置。以 100 Hz 的频率测试了复合材料在室温下的湿度感应响应，实验结果表明，相对湿度（RH）从 25%变化

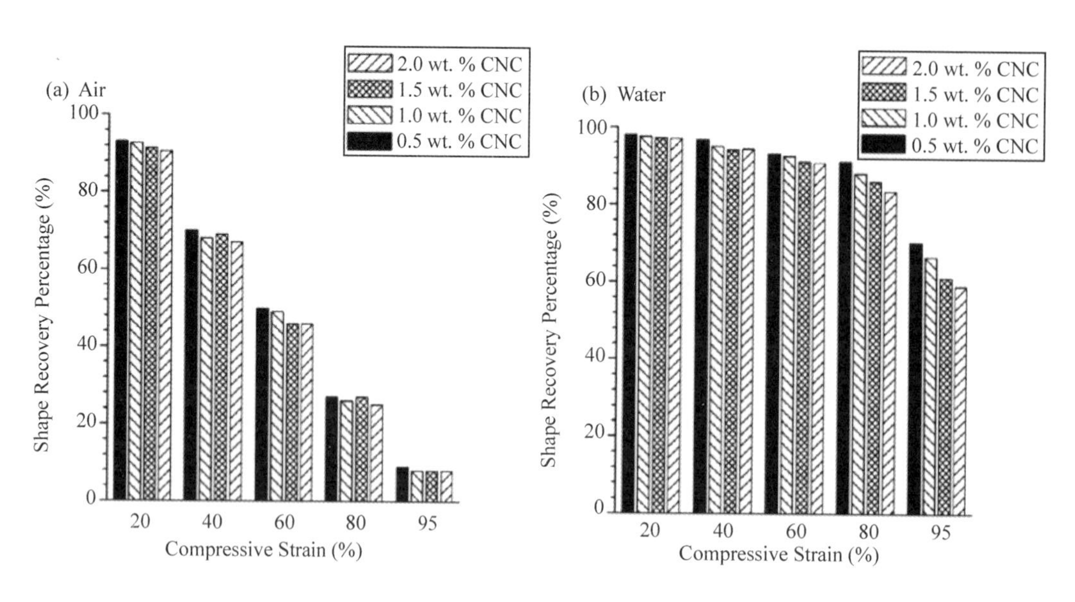

图 6 不同样品在(a)空气中和(b)水中的形状恢复率[29]

到 75%时，复合膜的阻抗变化约为 3 个数量级。又通过 1 个月的稳定性研究，验证了该材料作为湿度传感器的实际可行性。这些结果表明，该复合材料具有较好的加工性能和性价比，有望成为一种比金属氧化物传感器更高效的湿度传感器[30]。YAO *et al.* 利用聚乙二醇(PEG)和纤维素纳米晶制备了湿度智能响应光子结构，在缓慢干燥过程中得到螺旋结构均匀的固体薄膜。如图 7 所示，柔性具有智能响应性的手性向列相 CNC/PEG 复合薄膜进行可逆膨胀脱水时，随着其相对湿度在 50%~100%增加或减少，其结构颜色在绿色和透明之间呈现可逆、均匀的变化。该复合材料还具有良好的力学性能和耐热性能，可以用于低成本智能响应光子材料的制备。可将其应用于比色传感器、光学活性元件、油墨和装饰涂料等领域[31]。ZHANG *et al.* 以纤维素为原料，用硬脂酰基团对其进行修饰，制造出了湿度智能响应的独立膜。由于薄膜表面水分子的吸湿、解吸作用，此样品可以在局部湿度梯度内以做有节奏的弯曲运动，可以可逆地折叠和展开。通过双层制作得到的薄膜表现出一边为亲水另一边为疏水。如图 6 所示，只要此样品保持在 55 ℃和 5.9% RH 条件下，就保持为平面。若在冷却到 22 ℃的同时增加 RH 至 35%，样品便开始卷曲，这种现象是薄膜的凝固和轻微收缩造成的[32]。

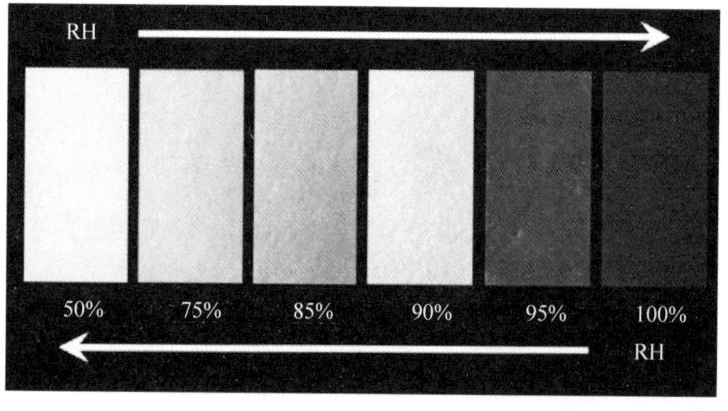

图 7 样品在不同相对湿度下的结构色[31]

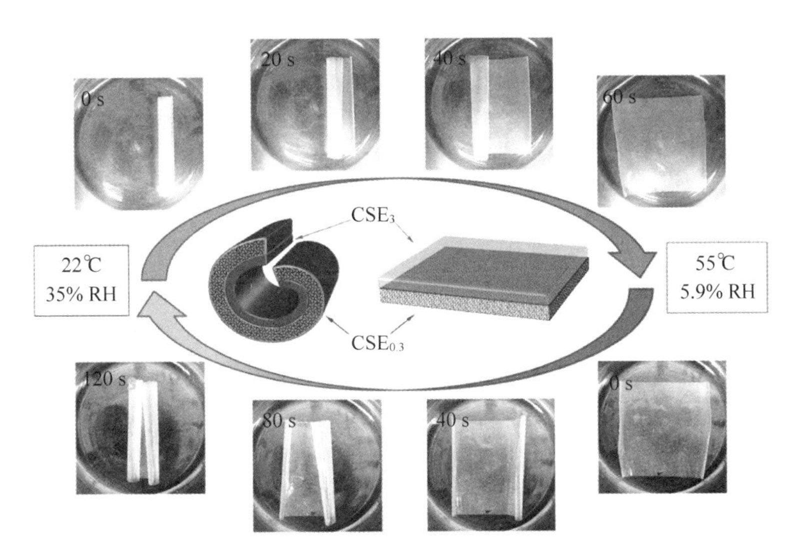

图 8 双层膜在不用温湿度条件下的照片[32]

6 温度智能响应木质功能材料

在智能响应材料中，温度智能响应型材料是一种能够感应外界温度的变化进而发生预定响应的智能响应型材料，因不依赖于其它化学助剂在众多环境刺激响应性聚合物中脱颖而出[33,34]。温度响应是日常生活中最常接触的一种响应，温度变化不仅在自然界存在，靠人工也很容易实现，所以对温度响应材料的研究具有非常重要的现实意义。温度响应性能赋予了材料多种属性与功能，在信息存储[35]、太阳能电池[36]、能量储存与转换[37]、传感器[38]、生物医学[39]等领域有广阔的应用前景。

LI *et al.* 将负载在3-氨丙基三乙氧基硅烷上的温敏变色材料接枝到聚乙烯中，并将此复合膜锚定在木材表面得到温度智能响应木材。如图9所示，所有制备的样品具有优异的正向可逆温度响应的特性，并具有较快的变色响应速率，表现出良好的温度变色响应特性。此研究对促进木质材料产品结构调整，和推动行业进步和技术升级具有重要的理论和现实意义。具有良好的实际应用潜力[40]。如图10所示，KEPLINGER *et al.* 利用木材的微观孔道结构为模板，首先利用甲基丙烯酸酯化对木材横截面进行预处理，利用酸酐与木材固有的OH基团反应，在木材细胞腔内原位形成聚异丙基丙烯酰水凝胶，赋予其温度智能响应的能力。这种混合材料为木材资源的全新利用开辟了道路[41]。

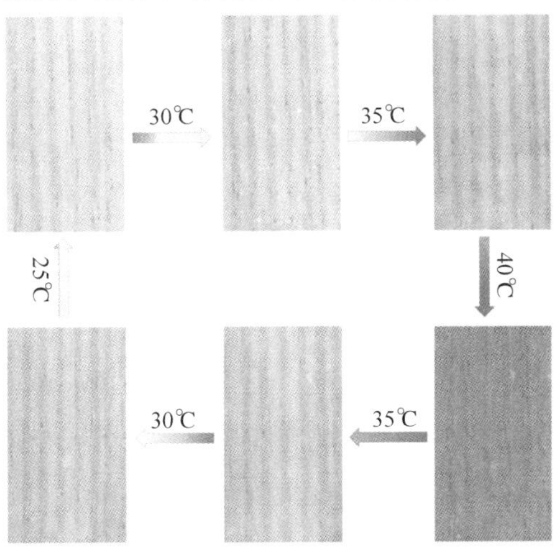

图 9 温度响应木材的变色过程[40]

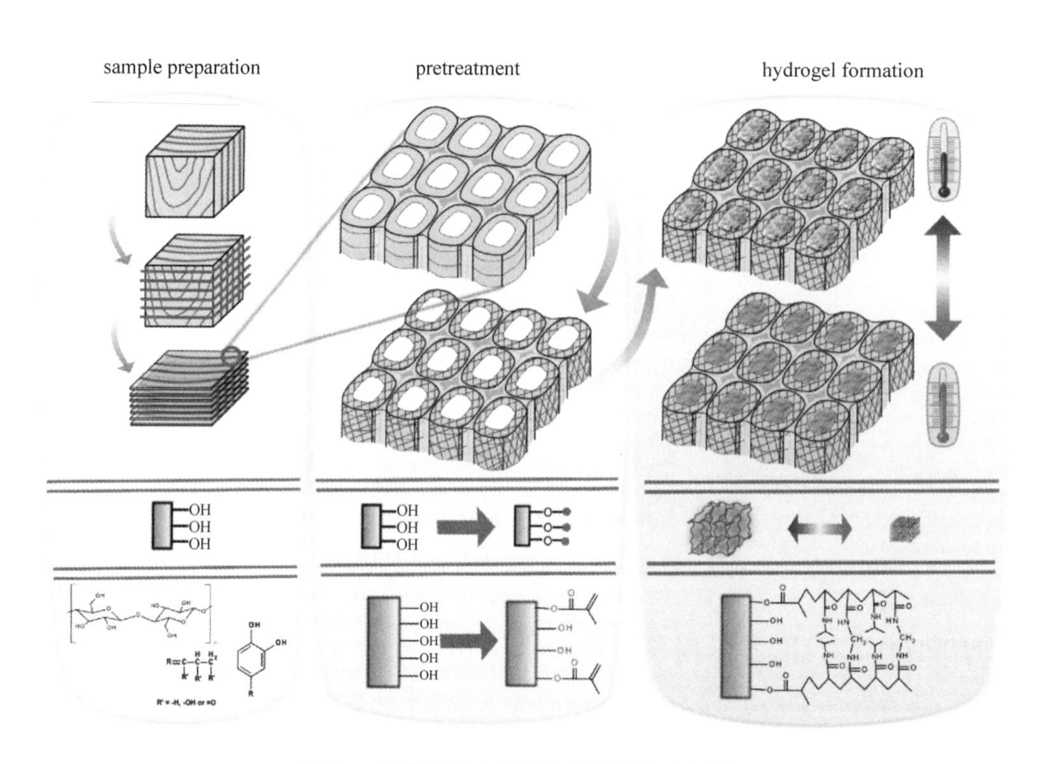

图 10 温度响应水凝胶木材的制备过程[41]

7 双重/多重智能响应木质功能材料

双重/多重智能响应功能材料具有广阔的应用前景，如药物传递、传感器和仿生机械等[42]。例如，LIU *et al.* 将乙酸纤维素(CAA)溶液与盐酸二胱氨酸(CYS)溶液混合制备了一种新型的纤维素水凝胶。载满小分子的纤维素水凝胶在 pH 和氧化还原反应中表现出可逆的溶胶—凝胶转变，从而表现出不同的释放特性。此研究制备的 pH/氧化还原双响应水凝胶在靶向给药和智能传感器方面具有广阔的应用前景[43]。

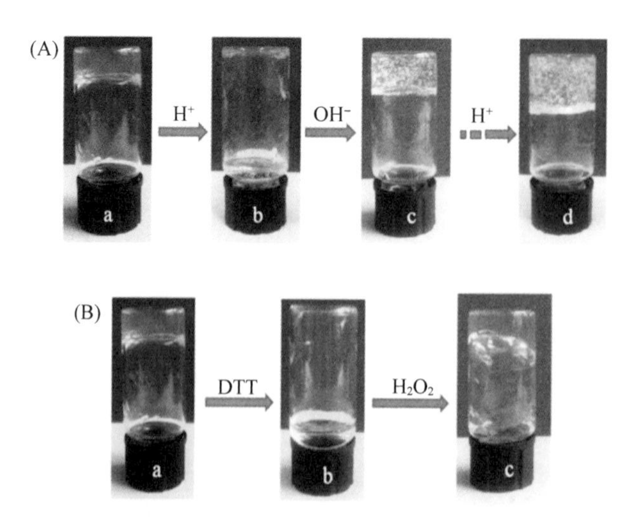

图 11 纤维素水凝胶的 (a) pH 响应性和 (b) 氧化还原响应性[43]

8 展 望

目前，具有智能响应性能的仿生智能木质材料的研究尚处在早期阶段。这种材料不但在木材领域而且在先进材料领域均展现了良好的生命力。虽然学者们已经制备了多种刺激响应的木质仿生智能材料，但受开发成本及技术水平等限制，目前有些研究成果仅可在实验室实现，相关产品的商业化程度远远不够。一部分研究的内容不够清晰透彻，部分关键技术尚未解决并没有形成完整的技术体系，距离规模化生产还有很长的路要走。要解决这些问题还需进行大量的研究工作。要充分利用木材自身独特的天然结构与属性，将其与纳米技术、分子生物学、界面化学、物理模型等多学科交叉融合，制备出具有奇异性能的仿生木质基新型材料，实现木材的高效高值化利用。

参考文献

[1] Isapour G, Lattuada M. Bioinspired stimuli-responsive color-changing systems[J]. Adv. Mater., 2018, 30(19): 1707069.

[2] Yang H, Leow W R, Wang T, et al. 3D printed photoresponsive devices based on shape memory composites[J]. Adv. Mater., 2017, 29(33): 1701627.

[3] Guragain S, Bastakoti B P, Malgras V, et al. Multi-stimuli-responsive polymeric materials[J]. Chem. Eur. J., 2015, 21(38): 13164-13174.

[4] Stuart M A C, Huck W T S, Genzer J, et al. Emerging applications of stimuli-responsive polymer materials[J]. Nat. Mater., 2010, 9(2): 101-113.

[5] Mura S, Nicolas J, Couvreur P. Stimuli-responsive nanocarriers for drug delivery[J]. Nat. Mater., 2013, 12(11): 991-1003.

[6] Richter A, Paschew G, Klatt S, et al. Review on hydrogel-based pH sensors and microsensors[J]. Sensors, 2008, 8(1): 561-581.

[7] Motoyama K, Li H, Koike T, et al. Photo-and electro-chromic organometallics with dithienylethene (DTE) linker, L2CpM-DTE-MCpL2: Dually stimuli-responsive molecular switch[J]. Dalton Trans., 2011, 40(40): 10643-10657.

[8] Li Y, Hui B, Li G, et al. Fabrication of smart wood with reversible thermoresponsive performance[J]. J. Mater. Sci., 2017, 52(13).

[9] Burgert I, Cabane E, Zollfrank C, et al. Bio-inspired functional wood-based materials-hybrids and replicates[J]. Int. Mater. Rev., 2015, 60(8): 431-450.

[10] Cabane E, Keplinger T, Merk V, et al. Renewable and functional wood materials by grafting polymerization within cell walls[J]. ChemSusChem, 2014, 7(4): 1020-1025.

[11] Olakanmi E O, Strydom M J. Critical materials and processing challenges affecting the interface and functional performance of wood polymer composites (WPCs)[J]. Mater. Chem. Phys., 2016, 171: 290-302.

[12] Fink S. Transparent wood-a new approach in the functional study of wood structure[J]. Holzforschung, 1992, 46(5): 403-408.

[13] Yang H, Yan R, Chen H, et al. Characteristics of hemicellulose, cellulose and lignin pyrolysis[J]. Fuel, 2007, 86(12-13): 1781-1788.

[14] Klemm D, Heublein B, Fink H P, et al. Cellulose: fascinating biopolymer and sustainable raw material[J]. Angew. Chem. Int. Ed., 2005, 44(22): 3358-3393.

[15] Chinga-Carrasco G, Syverud K. Pretreatment-dependent surface chemistry of wood nanocellulose for pH-sensitive hydrogels[J]. J Biomater Appl, 2014, 29(3): 423-432.

[16] Wang D, Tan J, Kang H, et al. Synthesis, self-assembly and drug release behaviors of pH-responsive copolymers ethyl cellulose-graft-PDEAEMA through ATRP[J]. Carbohydr. Polym., 2011, 84(1): 195-202.

[17] Tan J, Liu R, Wang W, et al. Controllable aggregation and reversible pH sensitivity of AuNPs regulated by carboxymethyl cellulose[J]. Langmuir, 2010, 26(3): 2093-2098.

[18] Zhang Q, Lei L, Zhu S. Gas-Responsive Polymers[J]. ACS Macro Lett., 2017, 6(5): 515-522.

[19] Gao L, Gan W, Cao G, et al. Fabrication of biomass-derived C-doped Bi_2WO_6 templated from wood fibers and its excellent sensing of the gases containing carbonyl groups[J]. Colloids Surf. A Physicochem. Eng. Asp., 2017, 529: 487-494.

[20] Garcia-Valdez O, Brescacin T, Arredondo J, et al. Grafting CO_2-responsive polymers from cellulose nanocrystals via nitroxide-mediated polymerisation[J]. Polym. Chem., 2017, 8(28): 4124-4131.

[21] Gao L, Gan W, Xiao S, et al. A robust superhydrophobic antibacterial Ag-TiO_2 composite film immobilized on wood substrate for photodegradation of phenol under visible-light illumination[J]. Ceram. Int., 2016, 42(2): 2170-2179.

[22] Gao L, Zhan X, Lu Y, et al. pH-dependent structure and wettability of TiO_2-based wood surface[J]. Mater. Lett., 2015, 142: 217-220.

[23] Li Y, Hui B, Gao L, et al. Facile one-pot synthesis of wood based bismuth molybdate nano-eggshells with efficient visible-light photocatalytic activity[J]. Colloids Surf. A Physicochem. Eng. Asp., 2018, 556: 284-290.

[24] Li Y, Gao L, Li J. Photoresponsive wood-based composite fabricated by a simple drop-coating procedure[J]. Wood Sci. Technol., 2019, 53(1): 211-226.

[25] Gao L, Qiu Z, Gan W, et al. Negative oxygen ions production by superamphiphobic and antibacterial TiO_2/Cu_2O composite film anchored on wooden substrates[J]. Sci. Rep., 2016, 6(1): 1-10.

[26] Hui B, Li J. Low-temperature synthesis of hierarchical flower-like hexagonal molybdenum trioxide films on wood surfaces and their light-driven molecular responses[J]. J. Mater. Sci., 2016, 51(24): 10926-10934.

[27] Gao L, Lu Y, Cao J, et al. Reversible photocontrol of wood-surface wettability between superhydrophilicity and superhydrophobicity based on a TiO_2 film[J]. J. Wood Chem. Technol., 2015, 35(5): 365-373.

[28] Imato K, Natterodt J C, Sapkota J, et al. Dynamic covalent diarylbibenzofuranone-modified nanocellulose: mechanochromic behaviour and application in self-healing polymer composites[J]. Polym. Chem., 2017, 8(13): 2115-2122.

[29] Yang X, Cranston E D. Chemically cross-linked cellulose nanocrystal aerogels with shape recovery and superabsorbent properties[J]. Chem. Mater., 2014, 26(20): 6016-6025.

[30] Kotresh S, Ravikiran Y T, Prakash H G R, et al. Humidity sensing performance of spin coated polyaniline-carboxymethyl cellulose composite at room temperature[J]. Cellulose, 2016, 23(5): 3177-3186.

[31] Yao K, Meng Q, Bulone V, et al. Flexible and responsive chiral nematic cellulose nanocrystal/poly (ethylene glycol) composite films with uniform and tunable structural color[J]. Adv. Mater., 2017, 29(28): 1701323.

[32] Zhang K, Geissler A, Standhardt M, et al. Moisture-responsive films of cellulose stearoyl esters showing reversible shape transitions[J]. Sci. Rep., 2015, 5(1): 1-13.

[33] Nakayama M, Okano T, Miyazaki T, et al. Molecular design of biodegradable polymeric micelles for temperature-responsive drug release[J]. J Control Release, 2006, 115(1): 46-56.

[34] Aoki T, Kawashima M, Katono H, et al. Temperature-responsive interpenetrating polymer networks constructed with poly (acrylic acid) and poly (N, N-dimethylacrylamide)[J]. Macromolecules, 1994, 27(4): 947-952.

[35] Wan P, Chen X. Stimuli-Responsive Supramolecular Interfaces for Controllable Bioelectrocatalysis [J]. ChemElectroChem, 2014, 1(10): 1602-1612.

[36] Zhou Y, Cai Y, Hu X, et al. Temperature-responsive hydrogel with ultra-large solar modulation and high luminous transmission for “smart window” applications[J]. J. Mater. Chem. A, 2014, 2(33): 13550-13555.

[37] Huang Y, Zhu M, Huang Y, et al. Multifunctional energy storage and conversion devices[J]. Adv. Mater., 2016, 28(38): 8344-8364.

[38] Yin L, He C, Huang C, et al. A dual pH and temperature responsive polymeric fluorescent sensor and its imaging application in living cells[J]. ChemComm, 2012, 48(37): 4486-4488.

[39]Gulzar A, Gai S, Yang P, et al. Stimuli responsive drug delivery application of polymer and silica in biomedicine[J]. J. Mater. Chem. B, 2015, 3(44): 8599-8622.

[40]Li Y, Li J. Fabrication of reversible thermoresponsive thin films on wood surfaces with hydrophobic performance[J]. Prog. Org. Coat., 2018, 119: 15-22.

[41]Keplinger T, Cabane E, Berg J K, et al. Smart Hierarchical Bio-Based Materials by Formation of Stimuli-Responsive Hydrogels inside the Microporous Structure of Wood[J]. Adv. Mater. Interfaces, 2016, 3(16): 1600233.

[42]Liu P, Mai C, Zhang K. Formation of uniform multi-stimuli-responsive and multiblock hydrogels from dialdehyde cellulose[J]. ACS Sustain. Chem. Eng., 2017, 5(6): 5313-5319.

[43]Liu H, Rong L, Wang B, et al. Facile fabrication of redox/pH dual stimuli responsive cellulose hydrogel[J]. Carbohydr. Polym., 2017, 176: 299-306.

[44]Berglund L A, Burgert I. Bioinspired wood nanotechnology for functional materials[J]. Adv. Mater., 2018, 30(19): 1704285.

[45]Ugolev B N. Wood as a natural smart material[J]. Wood science and technology, 2014, 48(3): 553-568.

[46]李坚. 大自然的启发——木材仿生与智能响应[J]. 科技导报, 2016, 34(19): 1-1.

[47]Merk V, Chanana M, Gierlinger N, et al. Hybrid wood materials with magnetic anisotropy dictated by the hierarchical cell structure[J]. ACS Appl. Mater. Interfaces, 2014, 6(12): 9760-9767.

英文题目: Research Progress of Wood Materials with Stimuli-responsive Properties

作者: LI Jian, LI Yingying

摘要: Intelligent responsive materials have three basic elements: perception, actuation and control, bio-inspired intelligent responsive materials have become one of the most exciting fields in the research of new materials. Wood is the oldest material used by human beings for construction after stone. As a natural and renewable biopolymer, wood has been extensively used in buildings, decorations, industries and other daily lives in virtue of its unique performances of the high ratio of strength to weight, esthetic, and amendment for indoor environment. It is commonly known that wood has three major components which are cellulose, hemicellulose, and lignin. The network of the components and the space of parallel hollow tubes construct the multi-scale porous structure, which offers the penetrability, accessibility and reactivity of wood materials. As a result, the unique composition and subtle hierarchical structures of wood lay the foundation for their further functionalization. In this paper, the research progress of wood materials with stimuli-responsive properties was introduced, the preparation, properties and potential applications of wood materials in responses to pH, gas, light, mechanical forces, humidity, temperature and double/multiple stimuli were detailed explained. This article focuses on the development of bio-inspired intelligent responsive materials based on wood materials.

关键词: Wood; Bio-inspired; Intelligent materials; Stimulus-response

Photoresponsive Wood-based Composite Fabricated by A Simple Drop-coating Procedure*

Yingying Li, Likun Gao, Jian Li

Abstract: We report a photoresponsive wood prepared by entrapping phosphomolybdic acid (PMA) into chitin/polyvinyl pyrrolidone (CS/PVP) blend based on wood substrate. Well dispersed mixed matrix membrane was observed by SEM and the change of surface morphology was measured by AFM. Colorimetric parameters of the samples under visible light irradiation were investigated. With the irradiation time increasing from 0 to 90 min, total color change of the sample remarkably increased from 0.7 to 42.5, verifying an excellent photoresponsive property. The interaction between the PMA particles and CS/PVP blend based on wood substrate was investigated by using FTIR. Furthermore, the results of UV-Vis and XPS spectra revealed that the photoreduction reaction occurred and the heteropoly blues were formed during the process of irradiation. The coating adhesion and thermal stability results demonstrated that the coating was firmly adhered to wood substrate and the as-prepared sample possessed a better heat-resistant performance than the pristine wood.

Keywords: Photoresponsive wood; Phosphomolybdic acid; Chitin/polyvinyl pyrrolidone; Mixed matrix membrane

1 Introduction

Wood is the oldest material used by human beings for construction after stone. As a natural and renewable composite material, wood have sophisticated hierarchical structure and many attractive properties, rendering them ideal candidates in various fields like buildings, papermaking and decorations (Kaya, 2018; Liu et al., 2016; Schrettl et al., 2016; Vnučec et al., 2015). The demands for wood increase significantly with the increasing concerns of population and sustainability due to the use of the non-renewable materials, so many attempts have been made to create new functions for wood to meet market requirements. Wood has three major components including cellulose, hemicellulose, and lignin. The network of the components and the space of parallel hollow tubes construct the multi-scale porous structure, which offers the penetrability, accessibility and reactivity of wood materials (Jiang et al., 2017; Koivuranta et al., 2017). As a result, the unique composition and subtle hierarchical structures of wood materials lay the foundation for its further functionalization. By incorporation of other components into the wood matrix, wood could be functionalized with enhanced properties, including fire resistance (Pečenko et al., 2016), corrosion resistance (Shi et al., 2017), hydrophobicity (Tang et al., 2017) and magnetism (Hui et al., 2015).

* 本文摘自 Wood Science and Technology, 2019, 53: 211-226.

Photochromic material is a kind of smart stimuli-responsive materials which have attracted many attentions because of their potential applications in optical storage media, multi-photon devices and optical sensor (Barile et al., 2016; Chan et al., 2014; Cong et al., 2014; Yu et al., 2015). As a class of nanosized transition metal oxide clusters, phosphomolybdic acid (PMA) has a typical classical Keggin structure with one P atom in the center surrounded by four oxygen atoms to form a tetrahedron. The P atom is caged by 12 octahedral MO6-units linked to one another by the neighboring oxygen atoms (Zhang et al., 2016). Another 12 oxygen atoms complete the structure, each of which is double-bonded with an additional Mo atom. As a result of the unique structure, PMA furnishes favorable accessibility of electron transfer from empty d orbitals for metal-oxygen π bonding (Kim and Shanmugam 2013). Meanwhile, PMA can undergo a multi-electron redox process photo-chemically without any structural changes (Magalad et al., 2010). Therefore, the excellent physicochemical properties of PMA make it suitable for use as photochromic material.

For certain applications, fast response time and good photochromic property are required. However, in some cases PMA showed poorer photochromic responses. In order to improve the photoresponsive performance, organic proton donor was used to combine with PMA (Ai et al., 2008; Jiang et al., 2008; Zhang et al., 2004). At the same time, to realize practical applications in various types of wood materials, PMA must be easily processed into coatings or any other forms. Many studies were introducing POM particles into polymer substrate, such as polyacrylamide and polyether chains (Aminabhavi and Naik 2002; Jing et al., 2014; Sun et al., 2015). Chitin is the second most abundant polysaccharide after cellulose. It commonly exists in the cell walls of fungi, insects, and the exoskeletons of crustaceans, which is low-cost and easily available (Cano et al., 2017). Chitosan (CS) is derivative of the chitin, its non-toxic, bioadhesive and antifungal biocompatible properties provide potential for many applications (Chen et al., 2017; Galhoum et al., 2015; Pan et al., 2016; Tirino et al., 2014). It should be pointed out that the largely free amino and hydroxyl group in chitosan can act as electron donors and interact with PMA. However, sometimes there are agglomeration occurred when PMA combine with CS and wood substrate. Therefore, a CS/PVP network matrix was formed to improve the dispersion stability of the coating. At the meantime, the choice of CS/PVP network composite not only due to its ability to act as proton reservoir, but also because of its good film-forming ability, which could bonded with wood materials.

In this paper, we fabricate a novel transparent photochromic CS/PVP-PMA mixed matrix membrane based on wood substrate using a sample and facile drop-coating method. The samples were characterized with ATR-FTIR, SEM, AFM, UV-Vis and XPS. The photoresponsive property, coating adhesion and thermal stability of samples were also investigated. The as-prepared samples are expected to have novel optical properties and applied to anti-counterfeiting, optical information storage and optical sensor.

2 Experimental

2.1 Materials

Obtained from Harbin, China, Larch samples, with a moisture content of 6.5% and a size of 20 mm× 50 mm×5 mm. The acetone (99.5%) and ethanol (99.7%) were purchased from Kaitong Chemical Reagent Co., Ltd (Tian jian, China). The polyvinyl alcohol (PVA) (alcoholysis 87.0%–89.0%, model 1788), chitosan (viscosity 100-200 mPa s) with 95% of deacetylation and polyvinyl pyrrolidone (PVP) (K29-32,

Mw≈58, 000) was purchased from Aladdin Industrial Corporation (Shanghai, China). Phosphomolybdic acid (PMA, $H_3PO_4 \cdot 12MoO_3$) was purchased from Sinopharm Chemical Reagent Co., Ltd (Shanghai, China). All of the chemicals were used as received without further purification.

2.2 Fabrication of photoresponsive wood

Prior to film deposition, the wood samples were ultrasonically treated by distilled water, acetone, and ethanol, respectively, and then dried in an air-circulating oven at 35 ℃ for 12h. The main experimental strategy for the fabrication of photoresponsive wood is shown in Fig. 1. The CS solution was prepared by dissolving 0.6 g CS into 100 mL deionized water at 60 ℃ for 90 min with continuous stirring. To prepare CS/PVP blend, 0.2 g PVP was dispersed in 70 mL of ethyl alcohol and added into the CS solution, which was then stirred for 60 min at room temperature. After that, 0.3 g PMA was dispersed in 30 mL of deionized water then added into the mixture solution and stirred for 30 min. Then the homogeneous solution was deposited on the wood surfaces using a drop-coating method and placed at room temperature for 24 h to be naturally dried. Finally, the CS/PVP-PMA composite wood was obtained. Besides this, a named PVA/PVP-PMA coated wood was also prepared for the purpose of comparison. The same experimental procedures were used except that the 0.4 g CS was replaced by 4 g PVA. It should be noted that all the experiments were sheltered from light and the as-prepared sample were stored under N_2 in dark until analysis.

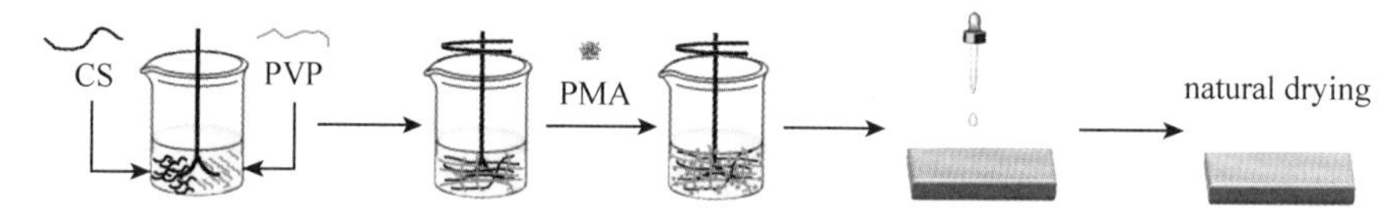

Fig. 1 Experimental strategy for the fabrication of CS/PVP-PMA composite wood

2.3 Characterization

The attenuated total reflectance of the Fourier transform infrared spectroscopy (ATR-FTIR) was carried out at room temperature with a Nicolet Nexus 670 FTIR instrument in the range of 400–4000 cm^{-1} with a resolution of 4 cm^{-1}. The morphologies of the sample surfaces were analyzed by scanning electron microscope (SEM). The X-ray photoelectron spectra (XPS) of the samples were recorded on an ESCALAB 250Xi photoelectron spectrometer using Al $K\alpha$ (1486.6 eV) radiation. The UV-Vis spectra were obtained from a UV-Vis spectrophotometer (UV-Vis DRS, TU-1901, China). The pyrolysis experiments were carried out in a thermal gravity (TG) analyzer (TA, Q600) under the nitrogen atmosphere at a heating rate of 10 ℃/min. The Atomic force microscopy (AFM) measurements were performed using a Nanoscope V Multimode 8 scanning probe microscope. The samples were focus on one position during the whole visible light radiation process while the light source was placed over samples at a distance of 20 cm. All experiments were carried out with the same AFM probe under ambient conditions (temperature of 25 ℃, relative humidity of 25%).

2.4 Photoresponsive text

Photoresponsive experiments were performed using a solar simulator (XES-40S2-CE, SAN-EI Electric) as the light source. The UV light and Vis light were transformed each other by controlling the UV filter (300–380 nm) and VIS filter (380–780 nm), respectively. In this work here, the results were obtained using the VIS irradiation. A high resolution digital camera was used to record the whole process of the as-prepared samples under irradiation. The samples were placed 20 cm away from the light source and exposed to air during the whole process. According to the Commission Internationale de L'Eclairage $L^* a^* b^*$ space, colorimetric analysis of the samples was performed, which is designed to accurately approximate human vision. Adobe Photoshop CS6 software installed in the computer were used to measure the color parameters L^*, a^* and b^* values, where L^* is the lightness index ($-L^*$ = dark, $+L^*$ = light), a^* is red-green index ($-a^*$ = green, $+a^*$ = red) and b^* is yellow-blue index ($-b^*$ = blue, $+b^*$ = yellow), respectively. The total color change ΔE^* values were calculated using the following equations:

$$\Delta E^* = \sqrt{(L_2^* - L_1^*)^2 + (a_2^* - a_1^*)^2 + (b_2^* - b_1^*)^2} \tag{1}$$

The L_1^*, a_1^* and b_1^* are the color parameters before irradiation; L_2^*, a_2^* and b_2^* are the color parameters under irradiation. The color parameters were measured at five different points on each sample, and the final value for L^*, a^* and b^* was obtained as an average of these five measurements.

2.5 Adhesion test

For adhesion measurement, wood samples were cut into a size of 50 mm × 50 mm × 10 mm. The pull-off text was performed using an automatic adhesion tester (PosiTest AT-A, DeFelsko) according to ISO 4624. Before testing, the coating around the edges of the dolly was cut through with a 20 mm cutting tool. Aluminium dollies of 20 mm diameter were degreased by ethyl alcohol and then glued to the samples by a two-component epoxy-based adhesive. After adhesive curing, a uniaxial tensile load at pull rate of 0.70 MPa/s was applied to the dolly by the tester until rupture occurs. The highest pulling force was marked as the adhesion strength. Ten replicate samples were applied for every test and the average values were then reported.

3 Results and discussion

3.1 ATR-FTIR analyses

Fig. 2 shows the FTIR spectra of pristine wood, CS coated wood, PMA powder and CS/PVP-PMA composite wood. Fig. 2a shows the major peaks associated with the pristine wood. After coated with CS coating (Fig. 2b), several peaks appeared in the spectrum compared with that of pristine wood. The peaks at 1653 cm^{-1} and 1558 cm^{-1} are assigned to amide I stretching of C=O and NH_2 bending vibration of amine II, respectively. The CO-NH deformation vibration of amide III appears at 1338 cm^{-1}. The previously mentioned absorption peaks were the characteristic peaks of CS. In order to understand the interaction between PMA particles, CS/PVP blend and wood substrate, the spectrum of pure PMA powder was acquired (Fig. 2c). The characteristic peaks of PMA particles appeared at 1060 cm^{-1} (ν_{as} P-O), 957 cm^{-1} (ν_{as} Mo=O_d), 889 cm^{-1} (ν_{as} Mo-O_b-Mo) and 780 cm^{-1} (ν_{as} Mo-O_c-Mo), which are in agreement with previously research (Zhang et al., 2016). As shown in CS/PVP-PMA composite wood (Fig. 2d), the peak at 1651 cm^{-1} is caused by the C=O stretching vibration of PVP. In addition, the NH_2 bending vibration at 1558 cm^{-1} and CO-NH deformation vibration at 1338 cm^{-1} almost disappeared after the formation of the film, which is caused by the strong interaction between PMA particles and CS/PVP blend. The results suggested that the PMA particles have been incorporated into the CS/PVP blend. It is worth mentioning that when PMA particles are added, the peaks at 1060, 957 and 889 cm^{-1} shifted to 1055, 953 and 879 cm^{-1}, respectively. Besides this, the band of Mo-Oc-Mo undergoes a blue shift from 780 cm^{-1} to 801 cm^{-1}. This may be ascribed to the electron transfer interactions between the CS/PVP blend and PMA particles (see Fig. 3). Furthermore, the bridged oxygen of two octahedra sharing a corner (ν_{as} Mo-O_c-Mo) vibration shows a higher peak shift than other octahedral oxygen. This reflects the occurrence of partial protonation, resulting in electrostatic interaction between the membrane and the wood substrate[17,18]. These observations indicated that the electrostatic interaction contributes more than the electron transfer interaction because of the large amount of oxygen containing functional groups on the wood surface.

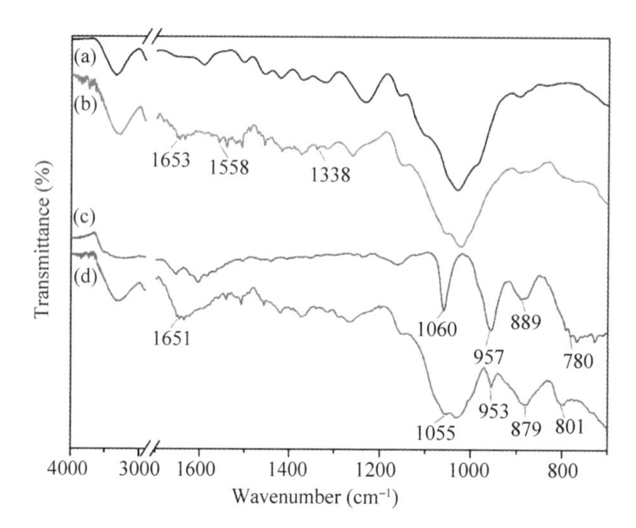

Fig. 2 FTIR spectra of (a) pristine wood, (b) CS coated wood, (c) PMA powder and (d) CS/PVP-PMA composite wood

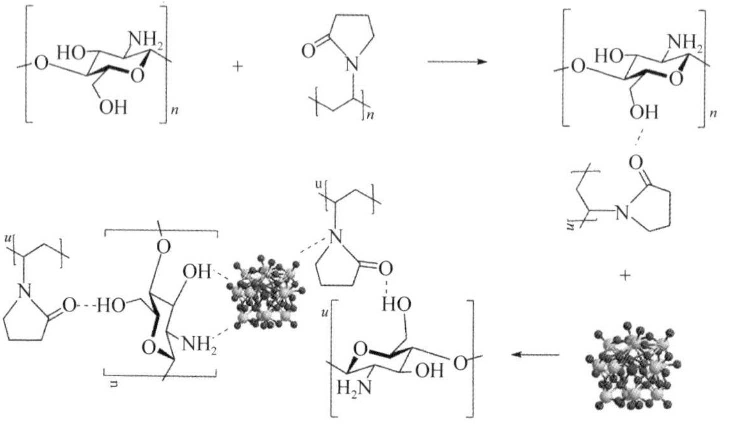

Fig. 3 Schematics of the interaction between CS/PVP blend and PMA particles

3.2 Surface morphologies

To observe the micromorphology of the sample surfaces for determining the internal interactions and the spatial distribution of the composites, low and high magnification SEM pictures were performed. In Fig. 4a, the pristine surface of the Larch wood, which contains some pits and tracheid string appear to be free of deposits of any other materials. After coated with CS, the wood surface was observed to be covered by an extraordinary smooth and homogeneous polymer film. As for CS/PVP-PMA composite wood (Fig. 4c, b), interconnected networks have been successfully constructed and dispersed throughout the wood surface, furthermore, the PMA particles were wrapped by the CS/PVP blend. The low and high magnification SEM pictures of PMA/PVP-PVA composite wood were presented in Fig. 4e, f, respectively. The surfaces showed that the agglomeration of PMA particles occurred with uneven distribution. These observed results revealed that CS and PVP can build a good network structure, and gives a direct revelation that PMA particles are embedded and evenly dispersed within CS/PVP blend. The excellent dispersion of PMA particles in the blend matrix may be directly correlated with its effectiveness in improving the photoresponsive properties of the samples.

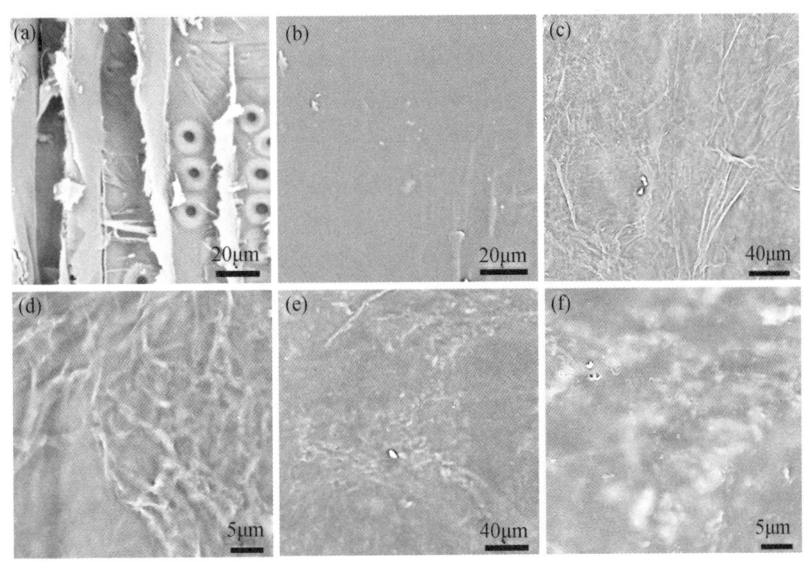

Fig. 4 SEM images of (a) pristine wood, (b) CS coated wood, (c, d) CS/PVP-PMA composite wood and (e, f) PVA/PVP-PMA coated wood

3.3 Photoresponsive behaviors

The photochromic properties were vital features in this study and directly influence the applications of the samples. Here, colorimetric analysis is designed to provide a quantitative representation of the perceived color difference between a pair of samples. The as-prepared samples were exposed to the solar simulator (adjusted to AM 1.5G at 600 W/m^2) for 30 min. The color parameters of the pristine wood (Sample A), PVA/PVP-PMA coated wood (Sample B), and CS/PVP-PMA composite wood (Sample C) were shown in Fig. 5. As we can see, all the samples owns high L^* and b^* value before irradiation, implying that the pristine wood had a light yellow surface and the transparent film remained the natural beauty of wood texture. As for the photoresponsive wood, the lightness factor L^* were reduced after irradiation, which confirmed that the surface color turned darker. Moreover, the decrease of a^* and b^* value suggested that the surface color turned green and blue, respectively. On the basis of these results, we concluded that the photoresponsive wood had light yellow surface at first, and then turned dark blue after irradiation.

Furthermore, the total color change (ΔE^*) of the samples were shown in Fig. 5d. After irradiation, the ΔE^* value of the pristine wood almost unchanged, indicating that the pristine wood had no response to irradiation. However, as for photoresponsive wood, the ΔE^* values occurred dramatic variation after irradiation. The ΔE^* values of the PVA/PVP-PMA coated wood and CS/PVP-PMA composite wood were 24.8 and 39.8, respectively. The results exhibited that the CS/PVP-PMA composite wood had a better photochromic performance. Fig. 6 shows the color-changed process of the CS/PVP-PMA composite wood. Before visible light irradiation, the sample had a light yellow surface. With the irradiation time increasing to 90 min, the color of the sample turned dark blue, coinciding with the analyzed results from color parameters.

For certain applications, fast response is required. Different from the photochromic text, here, the intensity of light is adjusted to AM 1.5 G at 20 W · m^{-2} and the samples were exposed to the irradiation for 40 s. The total color change (ΔE^*) and color-changed process of the samples were shown in Fig. 7. As we can see, after irradiation, the PVA/PVP-PMA coated wood was nearly the same with the sample before irradiation and the ΔE^* value was 1.2. On the contrary, the color change of CS/PVP-PMA composite wood is pretty obvious by using macroscopic observation and the ΔE^* value was reach to 5.2. The observation indicated that CS/PVP-PMA composite wood had a higher response speed than PVA/PVP-PMA coated wood. The previously mentioned results give a direct revelation that the CS/PVP-PMA composite wood has a prominent photoresponsive property.

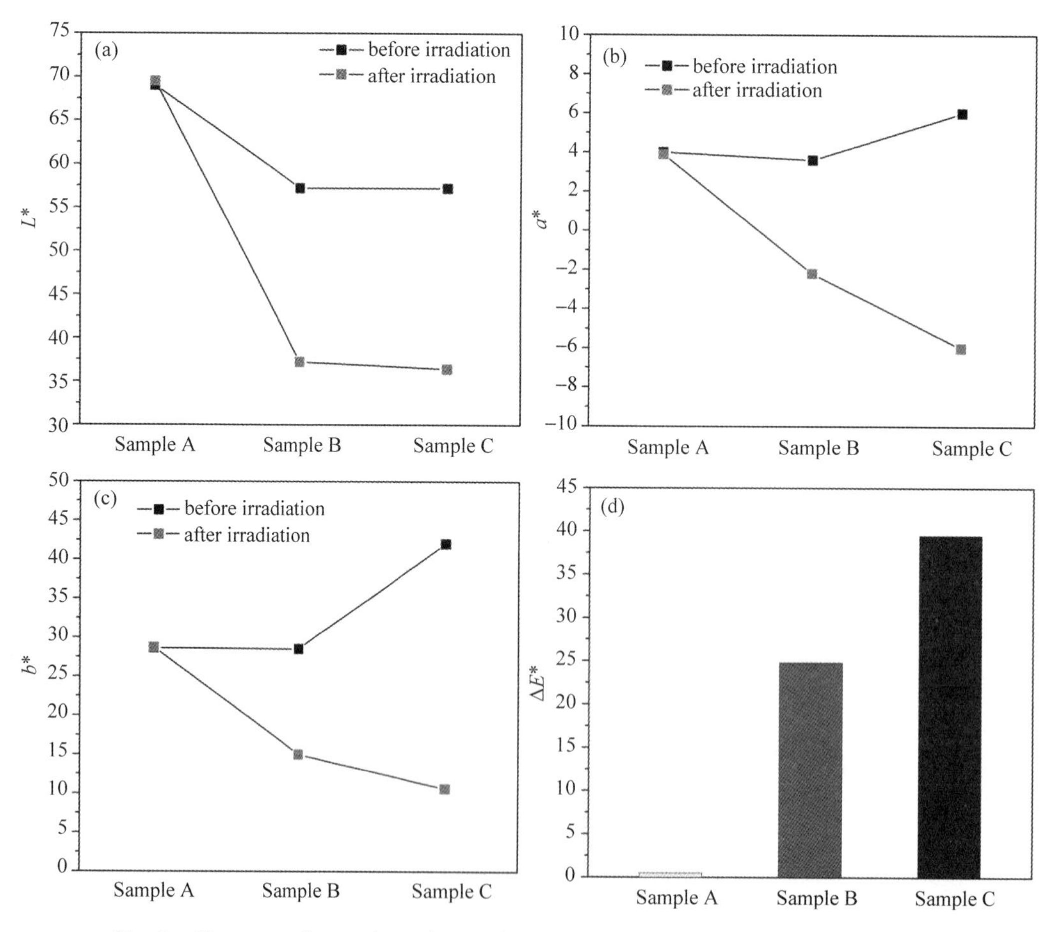

Fig. 5 Changes of sample surface color parameters before and after irradiation

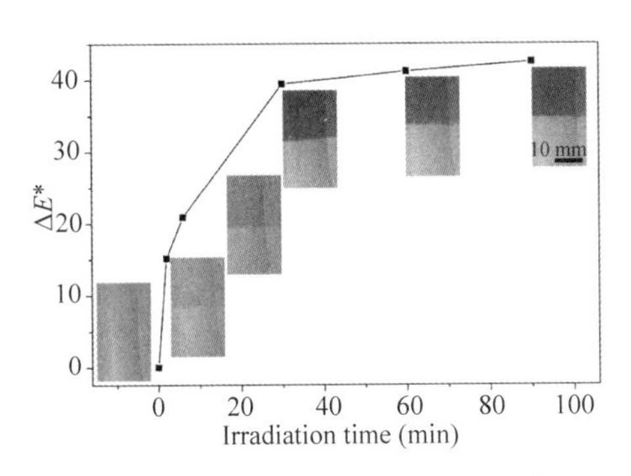

Fig. 6 Color-changed process of CS/PVP-PMA composite wood

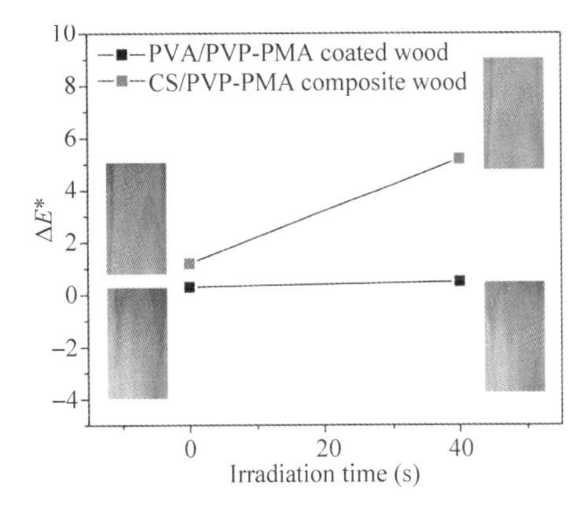

Fig. 7 Coloring process of samples irradiated for 40 s

3.4 Atomic force microscopy

To obtain information involving the change of surface morphology, tapping mode 3D AFM images of the CS/PVP-PMA composite wood surfaces before and after irradiation are shown in Fig. 8. Sample before irradiation is shown in Fig. 3a, resembled sharp rolling mountain peaks were observed over the surface. After irradiation, the peaks became obtuse and smooth, indicating the formation of heteropoly blues. A review of the literature on this topic shows a discrepancy. In the literature, when the PMA hybrid film was exposed under visible light, the surface of the hybrid film became much more rough and the sizes of PMA composite particles became much larger, besides this, the agglomeration of composite particles became more pronounced (Wang et al., 2014). The differences in the morphology may occur because of that the PMA particles were surrounded by CS/PVP blend, so no obvious changes were observed in this text.

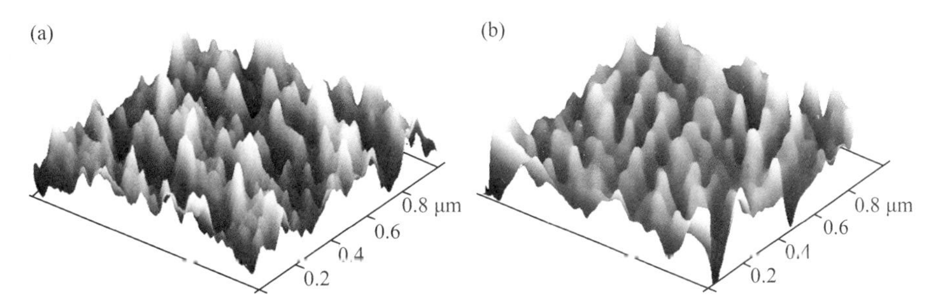

Fig. 8 AFM images of CS/PVP-PMA composite wood (a) before and (b) after irradiation

3.5 Optical properties

In order to further understand the changes of the samples during the irradiation. Coloration process of the CS/PVP-PMA composite wood surfaces with different irradiation time was demonstrated by UV-Vis absorption spectra. As shown in Fig. 9, the UV-Vis absorption spectrum of the sample before irradiation exhibited the characteristic absorption peak of PMA at around 345 nm, which belongs to the charge transfer transition from oxygen to molybdenum. After irradiated for 2 min, a shift is observed in the peak absorbance from 345 nm to 350 nm, which may result of the interaction between the PMA and the CS/PVP blend. Furthermore, one broad absorption band appeared at 711 nm, which attributed to the intervalence charge transfer (IVCT) of

$Mo^{6+} \rightarrow Mo^{5+}$. With increasing irradiation time, the intensity of the IVCT band enhanced and another weak band around 523 nm appeared, which attributed to the d-d transition. The appearance of the absorption peaks verified the existence of the electron transfer. And the electron transfer leaded to the formation of heteropoly blue (a mixed-valence Mo^{V}/Mo^{VI} Complex), giving rise to a dark blue-colored, long-lived charge-separate state.

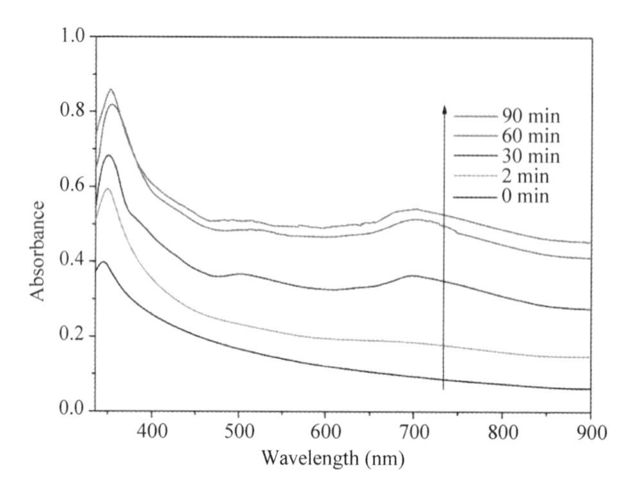

Fig. 9 UV-Vis spectra of the CS/PVP-PMA composite wood with different irradiation time

3.6 XPS analyses

X-ray photoelectron spectroscopy (XPS) experiments were further performed to determine the surface chemical composition and electronic structure. The XPS spectra of the Mo 3d energy level of the samples before and after irradiation are shown in Fig. 10. The spectrum clearly evidences the presence of two chemical environments for molybdenum atom (Fig. 10a). The first predominant doublet, present at 232.9 eV and 236.0 eV, is characteristic of Mo^{6+}. The second one, with a lower intensity, is located at lower binding energies, 231.7 eV and 234.6 eV, correspond to Mo^{5+}. The appearance of characterized signal of Mo^{5+} before irradiation might be resulted from its highly sensitive to light, which may be caused by the excitation of X-ray inspection. As shown in Fig. 10b, the peaks became broad and asymmetric after light exposure. It is noteworthy that the characteristic of Mo^{5+} peak value is remarkable increased, which indicated that the Mo^{5+} ions have produced during the irradiation process. As discussed above, the results further proved that the photoreduction reaction occurred, and the reaction is shown in Fig. 10c (Jiang et al., 2014; Nagul et al., 2015).

3.7 Adhesion of the coating to wood

In order to evaluate the adhesion strength of coating to wood substrate, pull-off adhesion tests were performed. In the tests, some samples were damaged in the wood part, and the pulling force was marked as the cohesion strength of wood substrate, which indicated that the adhesion between the coating and the wood surface is bigger than the cohesion strength of some wood substrate. Therefore, we chose wood sample with good quality and the test results of adhesion between the coating and the wood substrate were listed in Table 1. The average value of coating adhesion was 2.44 MPa. Several reviews describing the adhesion strength of coating to wood substrate, in the literature, adhesion strength of 2.32 MPa can completely meet general

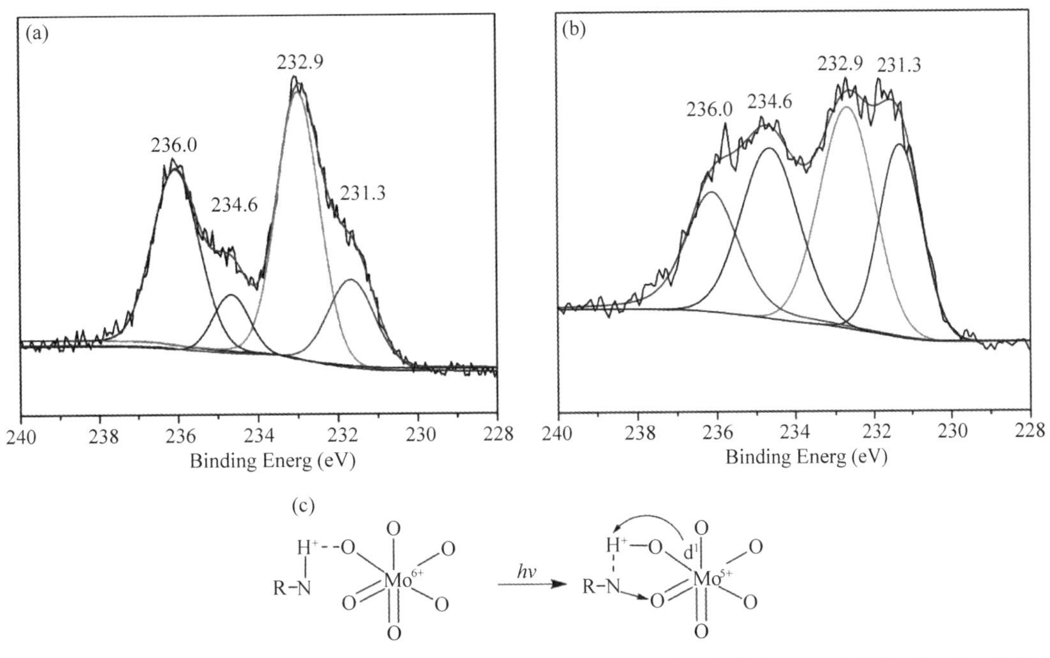

Fig. 10 Mo3d level spectra of the CS/PVP-PMA composite wood (a) before and (b) after irradiation

applications in the furniture industry (Hui et al., 2015). The results revealed that the coatings were firmly adhered to the wood surface, which can be attributed to two reasons. First, the irregular porous structure of wood substrate is suitable for mechanical anchorage, so the coating deposited on wood surface can adhere firmly due to a twist lock effect. The second reason is that CS and PVP themselves was a kind of excellent adhesive, which can firmly grab wood surfaces.

Table 1 Adhesion test results

	Adhesion (MPa)
1	2.42
2	2.55
3	2.31
4	2.75
5	2.42
6	2.51
7	2.35
8	2.46
9	2.33
10	2.28

3.8 Thermal gravimetric analysis

Considering the application, the thermal stability of the samples was evaluated using the thermal gravimetric analysis. Fig. 11 shows the mass loss and the derivative of mass loss curves obtained during the pyrolysis of the pristine wood and CS/PVP-PMA composite wood. In Fig. 11a, the pristine wood exhibited a 6% mass loss below 105 ℃, which can be attributed to the evaporation of moisture, adsorbed water and highly volatile matters. The main degradation process occurs as a two-step process and in a range from 200 ℃ to 390 ℃. In the first step, the slight shoulder in the DTG curve takes place at around 290 ℃ is related to the degradation of hemicellulose. The main degradation of cellulose occurs in a range from approximately 320 ℃ to 350 ℃, while lignin is decomposed in both two pyrolysis process without characteristic peaks. According to Fig. 11b, no sharp distinction was found between the CS/PVP-PMA composite wood curve and the pristine wood curve, demonstrating that the coating had no bad effect on the thermal stability of wood samples. On the contrary, up to 200 ℃, only 5% mass loss was observed, and the shoulder in the DTG curve occurred at higher temperatures of 305 ℃ and 376 ℃, respectively. The results concluded that the CS/PVP-PMA composite wood possessed a better heat-resistant performance than the pristine wood.

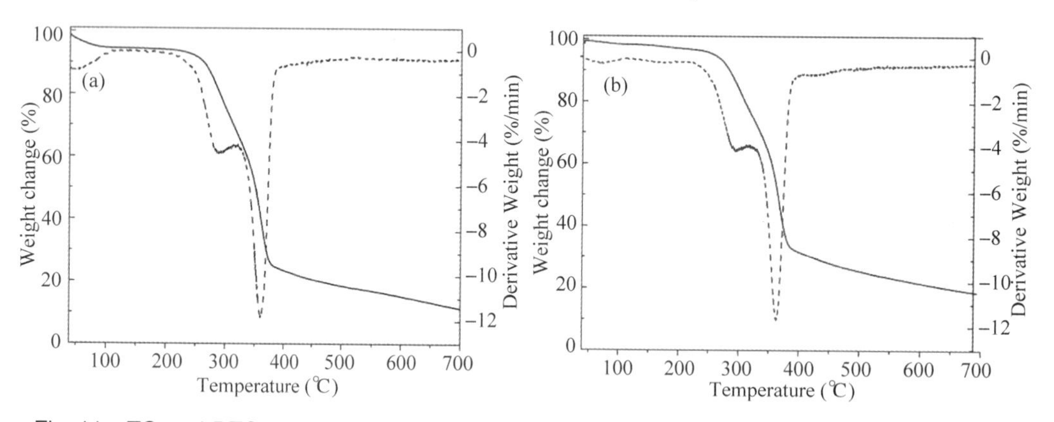

Fig. 11 TG and DTG graph of the (a) pristine wood and (b) CS/PVP-PMA composite wood

4 Conclusions

In this paper, a novel transparent photochromic CS/PVP-PMA mixed matrix membrane based on wood substrate was successfully developed through a simple and facile drop-coating method. The results of colorimetric analysis demonstrated that the CS/PVP-PMA composite wood has remarkable photoresponsive behavior. The PMA particles were embedded and evenly dispersed within the CS/PVP network structure. The good dispersion of the CS/PVP blend on wood surface established the basis for its excellent photoresponsive properties. The electron transfer interaction and the electrostatic interaction between PMA, CS/PVP mixed blend and wood substrate were investigated by FTIR spectra. During the process of irradiation, the photoreduction reaction occurred and the heteropoly blues were formed, so the samples turned to dark blue from colorless at the same time. The results of adhesion and TG texts proved that the CS/PVP-PMA mixed matrix membrane was firmly adhered to wood surface and the CS/PVP-PMA composite wood possessed a better heat-resistant performance than the pristine wood, respectively. The as-prepared photoresponsive wood with novel optical properties was cost-effective synthesis, and promising for applications such as optical information storage, solar converter, a type of environmentally friendly anti-counterfeit measures and smart home.

Acknowledgements

The authors gratefully acknowledge the financial support from the National Natural Science Foundation of

China (31470584) and the Overseas Expertise Introduction Project for Discipline Innovation, 111Project (No. B08016).

Conflict of interest

The authors declared that they have no conflict of interest.

References

[1] Ai L M, Feng W, Chen J, et al. Evaluation of microstructure and photochromic behavior of polyvinyl alcohol nanocomposite films containing polyoxometalates[J]. Mater. Chem. Phys., 2008, 109(1): 131-136.

[2] Aminabhavi T M, Naik H G. Synthesis of graft copolymeric membranes of poly (vinyl alcohol) and polyacrylamide for the pervaporation separation of water/acetic acid mixtures[J]. J. Appl. Polym. Sci., 2002, 83(2): 244-258.

[3] Barile C J, Slotcavage D J, McGehee M D. Polymer-nanoparticle electrochromic materials that selectively modulate visible and near-infrared light[J]. Chem. Mater., 2016, 28(5): 1439-1445.

[4] Cano L, Pollet E, Avérous L, et al. Effect of TiO_2 nanoparticles on the properties of thermoplastic chitosan-based nano-biocomposites obtained by mechanical kneading[J]. Compos. Part A Appl. Sci. Manuf., 2017, 93: 33-40.

[5] Chan J C H, Lam W H, Yam V W W. A highly efficient silole-containing dithienylethene with excellent thermal stability and fatigue resistance: a promising candidate for optical memory storage materials[J]. J. Am. Chem. Soc., 2014, 136 (49): 16994-16997.

[6] Yu-Han C, Lin C Y, Pei-Ling Y, et al. Antifungal agents from heartwood extract of Taiwania cryptomerioides against brown root rot fungus Phellinus noxius[J]. Wood Sci. Technol., 2017, 51(3): 639-651.

[7] Cong S, Tian Y, Li Q, et al. Single-crystalline tungsten oxide quantum dots for fast pseudocapacitor and electrochromic applications[J]. Adv. Mater., 2014, 26(25): 4260-4267.

[8] Galhoum A A, Mahfouz M G, Atia A A, et al. Amino acid functionalized chitosan magnetic nanobased particles for uranyl sorption[J]. Ind. Eng. Chem. Res., 2015, 54(49): 12374-12385.

[9] Hui B, Li G, Han G, et al. Fabrication of magnetic response composite based on wood veneers by a simple in situ synthesis method[J]. Wood Sci. Technol., 2015, 49(4): 755-767.

[10] Hui B, Li Y, Huang Q, et al. Fabrication of smart coatings based on wood substrates with photoresponsive behavior and hydrophobic performance[J]. Mater. Des., 2015, 84: 277-284.

[11] Jiang J, Wang J, Zhang X, et al. Assessing multi-scale deconstruction of wood cell wall subjected to mechanical milling for enhancing enzymatic hydrolysis[J]. Ind Crops Prod, 2017, 109: 498-508.

[12] Jiang M, Jiao T, Liu M. Photochromic Langmuir-Blodgett films based on polyoxomolybdate and gemini amphiphiles [J]. New J. Chem., 2008, 32(6): 959-965

[13] Jiang M, Zhu D, Cai J, et al. Electrocatalytic hydrogen evolution and oxygen reduction on polyoxotungstates/graphene nanocomposite multilayers[J]. J. Phys. Chem. C, 2014, 118(26): 14371-14378.

[14] Jing X, Zou D, Meng Q, et al. Fabrication and visible-light photochromism of novel hybrid inorganic-organic film based on polyoxometalates and ethyl cellulose[J]. Inorg Chem Commun, 2014, 46: 149-154.

[15] Kaya M. Evaluation of a novel woody waste obtained from tea tree sawdust as an adsorbent for dye removal[J]. Wood Sci. Technol., 2018, 52(1): 245-260.

[16] Kim Y, Shanmugam S. Polyoxometalate-reduced graphene oxide hybrid catalyst: Synthesis, structure, and electrochemical properties[J]. ACS Appl. Mater. Interfaces, 2013, 5(22): 12197-12204.

[17] Koivuranta E, Hietala M, Ämmälä A, et al. Improved durability of lignocellulose-polypropylene composites manufactured using twin-screw extrusion[J]. Compos. Part A Appl. Sci. Manuf., 2017, 101: 265-272.

[18] Liu L, Qian M, Song P, et al. Fabrication of green lignin-based flame retardants for enhancing the thermal and fire retardancy properties of polypropylene/wood composites[J]. ACS Sustain. Chem. Eng., 2016, 4(4): 2422-2431.

[19] Magalad V T, Gokavi G S, Raju K V S N, et al. Mixed matrix blend membranes of poly (vinyl alcohol)-poly (vinyl pyrrolidone) loaded with phosphomolybdic acid used in pervaporation dehydration of ethanol[J]. J. Membr. Sci., 2010, 354 (1-2): 150-161.

[20] Nagul E A, McKelvie I D, Kolev S D. The use of on-line UV photoreduction in the flow analysis determination of dissolved reactive phosphate in natural waters[J]. Talanta, 2015, 133: 155-161.

[21] Pan Q, Lv Y, Williams G R, et al. Lactobionic acid and carboxymethyl chitosan functionalized graphene oxide nanocomposites as targeted anticancer drug delivery systems[J]. Carbohydr. Polym., 2016, 151: 812-820.

[22] Pečenko R, Svensson S, Hozjan T. Model evaluation of heat and mass transfer in wood exposed to fire[J]. Wood Sci. Technol., 2016, 50(4): 727-737.

[23] Schrettl S, Schulte B, Frauenrath H. Templating for hierarchical structure control in carbon materials[J]. Nanoscale, 2016, 8(45): 18828-18848.

[24] Shi C, Tang Z, Wang L, et al. Preparation and characterization of conductive and corrosion-resistant wood-based composite by electroless Ni-W-P plating on birch veneer[J]. Wood Sci. Technol., 2017, 51(3): 685-698.

[25] Sun H, Gao N, Ren J, et al. Polyoxometalate-based rewritable paper[J]. Chem. Mater., 2015, 27(22): 7573-7576.

[26] Tang Z, Xie L, Hess D W, et al. Fabrication of amphiphobic softwood and hardwood by treatment with non-fluorinated chemicals[J]. Wood Sci. Technol., 2017, 51(1): 97-113.

[27] Tirino P, Laurino R, Maglio G, et al. Synthesis of chitosan-PEO hydrogels via mesylation and regioselective Cu (I)-catalyzed cycloaddition[J]. Carbohydr. Polym., 2014, 112: 736-745.

[28] Vnučec D, Goršek A, Kutnar A, et al. Thermal modification of soy proteins in the vacuum chamber and wood adhesion[J]. Wood Sci. Technol., 2015, 49(2): 225-239.

[29] Wang X, Dong Q, Meng Q, et al. Visible-light photochromic nanocomposite thin films based on polyvinylpyrrolidone and polyoxometalates supported on clay minerals[J]. Appl. Surf. Sci., 2014, 316: 637-642.

[30] Yu J, Cui Y, Wu C D, et al. Two-photon responsive metal-organic framework[J]. J. Am. Chem. Soc., 2015, 137 (12): 4026-4029.

[31] Zhang B, Asakura H, Zhang J, et al. Stabilizing a platinum1 single-atom catalyst on supported phosphomolybdic acid without compromising hydrogenation activity[J]. Angew. Chem., 2016, 128(29): 8459-8463.

中文题目：通过简单的滴涂工艺制备光响应木基复合材料

作者：李莹莹，高丽坤，李坚

摘要：本文报道了一种以磷钼酸(PMA)为包埋剂，将其包埋在甲壳素/聚乙烯吡咯烷酮(CS/PVP)共混物中制备的光敏木材。用扫描电镜观察了分散良好的混合基质膜，用原子力显微镜观察了表面形貌的变化。研究了样品在可见光照射下的比色参数。随着辐照时间的延长，样品的总显色率从0.7显著提高到42.5，证明了该样品具有优良的光响应性能。采用傅里叶变换红外光谱研究了PMA粒子与CS/PVP共混物之间的相互作用。紫外-可见光谱和XPS光谱结果显示，在辐照过程中发生了光还原反应，并形成了杂多蓝。涂层的附着力和热稳定性结果表明，涂层与木材基体粘附牢固，制备的样品比原始木材具有更好的耐热性能。

关键词：光响应木材；磷钼酸；甲壳素/聚乙烯吡咯烷酮；复合膜

Energy Saving Wood Composite with Temperature Regulatory Ability and Thermoresponsive Performance*

Yingying Li, Runan Gao, Jian Li

Abstract: Energy saving materials are economically and environmentally beneficial as they effectively reduce energy consumption and greenhouse gas emissions. In this study, energy saving wood composite was prepared by injecting a methyl methacrylate dispersion containing a mixture of VO_2 (M) @ SiO_2 core-shell nanocomposites and thermochromic microcapsules into wood templates. APTS was added to stabilize the suspension. The good compatibility between the injected substances and the wood template benefited from the PMMA and APTS that filled the cell lumen and bonded the substances, leading to energy saving wood composite with excellent dimensional stability, optical properties, and mechanical performance. More importantly, the energy saving wood composite exhibited outstanding temperature regulatory ability by shielding needless infrared radiation, resulting in thermal comfort. The unique properties of the energy saving wood composite could potentially be applied in building, furniture, thermal insulation, energy conversion, energy storage, and conservation.

Keywords: Wood composite; Temperature regulation; Energy saving; Phase transition

1 Introduction

Anthropogenic climate change (ACC) has the potential to alter Earth's temperature, weather, and sea level, leading to increased risk of floods, diseases, and extinction among vulnerable species[1-3]. Burning fossil fuels, excess use of energy, and other human activities are the proximate cause of ACC. Therefore, considerable research efforts have been devoted to increase energy efficiency, reduce energy usage, and decrease carbon greenhouse gas emissions[4-6]. Recently, particular attention has been focus on the potential for energy savings in the building environment as this sector uses more than 48% of the primary energy used in most countries[7-9]. This has encouraged many studies on the exploitation and application of energy conservation products and services to decrease the requiring energy to provide a comfortable environment for both work and life[10,11]. In this regard, energy saving walls and windows are the first line of defense between the indoor and outdoor environments[12,13]. Electroresponsive materials, thermoresponsive materials and photoresponsive materials are the three most widely used materials as energy saving coatings[14,15]. Among them, thermoresponsive materials have attracted considerable attention because they can intelligently response to temperature without further help of chemical additions[16,17]. At the same time, extra energy is unnecessary for thermoresponsive materials to operation because of their transmitting/reflecting irradiation switch[18-21].

* 本文摘自 European Polymer Journal, 2019, 118: 163-169.

Monoclinic vanadium dioxide (VO_2) with a monoclinic structure (M) has expanded the applications to the energy saving windows because it exhibits a reversible structural transition to the tetragonal rutile phase (VO_2(R)) at about 68 ℃[22]. That is, VO_2(M) is a semiconductor with reasonable infrared transmission below the transformation temperature that changes to the VO_2(R) with high infrared reflection above its transformation temperature[23,24]. Therefore, VO_2 coated thermochromic windows have received scientific and technical interest for their solar-heat shielding ability. Although VO_2 thermochromic smart windows have drawn a lot of attention, some issues have arisen because of the glass substrate. Glass is a fragile material with poor elasticity, and shattered glass pieces could lead to severe safety issues[25,26]. Due to high thermal conductivity of the glass windows, thirty percent of the energy used in providing indoor environment is lost. In contrary, wood as a natural and renewable composite material can withstand higher impact and afford better heat preservation compared with ordinary glass[27-31]. Moreover, wood is an ideal bio-based template to endow novel properties due to its sophisticated hierarchical structure and unique physiochemical properties[32-36].

The current approach for preparing transparent wood is by infilling a suitable polymer into the delignified wood temple. However, removing the lignin damages and weakens the wood structure; thus, the wood slices might break into pieces after the delignification step[25]. To solve this problem, once prepared a wood temple by removing only the chromophoric structures of the lignin; however, the temple turned yellow again after a period of time. In the present study, wood temple was obtained by removing part of the lignin but retaining sufficient lignin in the wood temple to confer structural support, and then the chromophoric structures of the retained lignin were removed. The as-prepared transparent wood temple had a high transmittance in the visible light region. In the present study, a facile method for fabricating energy saving wood composite (EW) was demonstrated by injecting a VO_2 (M) @ SiO_2/thermochromic microcapsule-containing modified methyl methacrylate dispersion into an as-prepared template. The optical properties, weather ability, and mechanical performances of EW were investigated. In addition, the temperature regulatory ability of the EW was studied using a model house system. The results show that the EW is a potential candidate energy saving material due to excellent energy storage and solar heat control properties.

2 Experimental

2.1 Materials

Balsa wood (250 mm×250 mm×15 mm) was purchased from Hangzhou Synhong Chemical Co., Ltd. (Zhejiang, China), which reprocessed into radial section wood slices with a size of 50 mm × 50 mm × 5 mm. Vanadium pentoxide (V_2O_5), oxalic acid, tetraethyl orthosilicate (TEOS), sodium hydroxide (NaOH), sodium sulfite (Na_2SO_3), diethylenetriaminepentaacetic acid (DTPA), 3-aminopropyltriethoxysilane (APTS), methyl methylbenzoate (MMA) and 2, 2'-azobis (2-methylpropionitrile) were purchased from Shanghai Aladdin Biochemical Technology Co., Ltd. (Shanghai, China). Thermochromic microcapsules (TM), white powder, which consisted of 3, 3-bis (1-butyl-2-methyl-1H-indol-3-yl) -1 (3H) -isobenzofuranone, dibehenyl phosphate, myristyl alcohol, and melamine formaldehyde resin. Ethanol (99.5%), acetone, hydrogen peroxide (H_2O_2, 30 wt%), oxalic acid, sodium silicate and magnesium sulfate were purchased from Kaitong Chemical Reagent Co., Ltd (Tian jian, China). All chemicals were used as received without further purification.

2.2 Preparation of the VO_2@SiO_2 nanoparticles

The VO_2 nanoparticles were synthesized by the hydrothermal method. In a typical procedure, 0.25 g V_2O_5 was added to 80 mL of aqueous oxalic acid solutions (0.15 M concentration) to form a tawny suspension. The suspension then placed to a 100 mL stainless-steel autoclave reactor with filling rate about 80% (volume) under electric magnetic agitation for 180min. The reactor was sealed and maintained at 240 ℃ for 24 h, and then cooled down to 25 ℃. The VO_2 nanoparticles were centrifuged, collected, rinsed with water and ethanol, and dried. After this, 35 mL ethanol solution containing 0.3 g VO_2 nanoparticles were obtained, and 3% TEOS deionized water solution was added to above ethanol solution. Immediately, a 7 mL solution of ammonia was added to enhance hydrolysis of the TEOS. The hydrolysis process was held at 50 ℃ with magnetic stirring for 2 h[37]. When the reaction was complete, the sample was centrifuged, collected, rinsed with water and ethanol, and dried at 60 ℃ in a vacuum. The obtained VO_2(M)@SiO_2 nanoparticles were stored under N_2.

2.3 Synthesis of the wood template

The pristine wood was treated with a boiled mixed aqueous solution of NaOH (2.5 M) and Na_2SO_3(0.4 M) for 1.5 h, and washed thoroughly with deionized water. Then the above wood samples were immersed in a mixed solution of deionized water, sodium silicate (3.0 wt%), NaOH (3.0 wt%), and H_2O_2(4.0 wt%) at 70 ℃ for 3 h. The wood templates were rinsed with deionized water and preserved in deionized water for later use.

2.4 Preparation of the energy saving wood composite (EW)

The wood template was treated with ethanol and acetone through a vacuuming procedure prior to polymer infiltration. Fig. 1 schematically depicts the experimental procedure for the EW. The prepolymerized solution was obtained by mixing MMA and 0.4 wt% 2, 2'-azobis(2-methylpropionitrile) at 80 ℃ for 30 min, and the reaction was stopped by transferring the container to ice water. The 0.5 wt% VO_2(M)@SiO_2 and 0.5 wt% TM particles were uniformly dispersed in a prepolymerized solution, and an appropriate amount of APTS was added and stirred for 1 h to stabilize the suspension. This solution was infiltrated into the wood template under through a vacuuming procedure. Finally, the as-prepared sample was pre-cured at 110 ℃ for 5 min, and then dried at 70 ℃ for 4h.

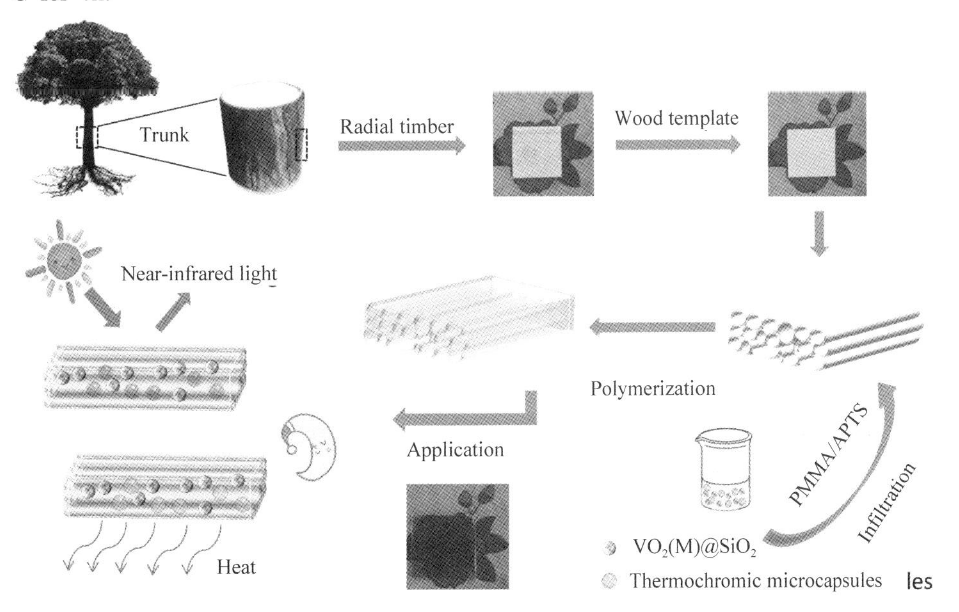

Fig. 1 A schematic of the fabrication of energy saving wood composite

2.5 Characterization

A scanning electron microscope (SEM; Quanta200; FEI, Hillsboro, OR, USA) was used to analyze surface morphology, and the samples were glued onto a specific holder and sprayed with gold to ensure conductivity. The phase structure of the as-prepared sample was checked by X-ray diffraction (XRD: Rigaku, and D/MAX 2200) measurements operated with Cu target radiation (λ = 1.54 Å). The attenuated total reflectance Fourier transform-infrared spectra (ATR-FTIR) were recorded by the Nicolet Nexus 670 FTIR instrument in the range of 400-4000 cm^{-1} with a resolution of 4 cm^{-1}. A WDW-300 electronic universal testing machine was used to carry out the stress-strain measurements of the samples. Transmittance of the wood composites was analyzed with a UV-Vis spectrophotometer (UV-vis DRS: TU-1901, China) equipped with an integrated sphere attachment. Haze was measured with a haze meter (CS-700) according to ASTM D1003. The differential scanning calorimetry (DSC) experiments were carried out on a Q100 DSC (TA Instruments, New Castle, DE, USA) under a nitrogen atmosphere at a rate of 2 ℃/min. The temperature range utilized was from 20 ℃ to 90 ℃.

3 Results and discussion

3.1 XRD analyses

XRD was performed to determine the crystalline structure of the samples. Fig. 2 (a) is the XRD pattern and SEM picture of the VO_2(M)@SiO_2 particles. The XRD patterns of the pristine wood, transparent wood, and EW are shown in Fig. 2 (b). The diffraction peaks of wood cellulose were located at 16.1° and 22.2°. The two broad peaks in transparent wood were observed at 2θ = 15° and 30°, which were corresponded to PMMA. The characteristic diffraction peaks of VO_2(M) were located at 28.0°, 34.2°, 44.7°, 53.1°, and 73.7° in EW which is consistent with literature values (JCPDS 43-1051). In addition, the peaks of PMMA and cellulose were all preserved. These results confirm that the VO_2(M)@SiO_2 was successfully immobilized in the wood composites.

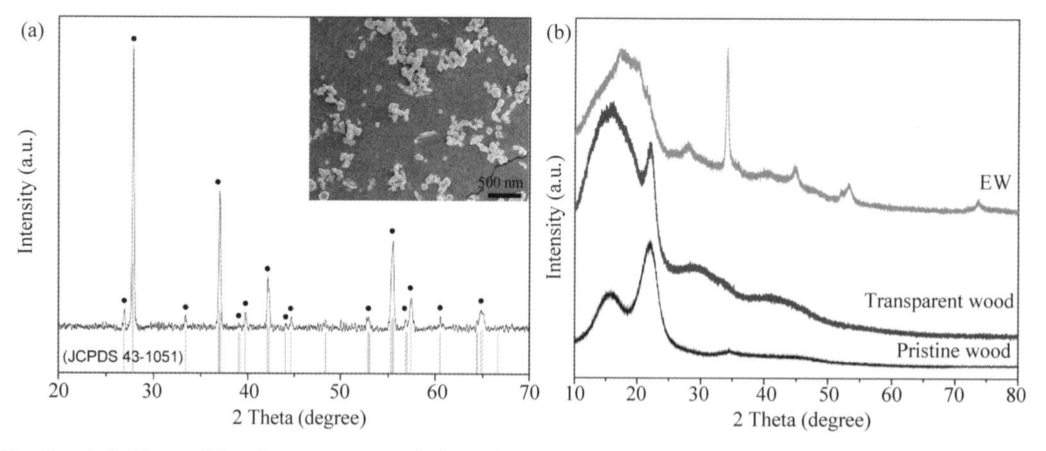

Fig. 2 (a) X-ray diffraction patterns and Scanning electron microscopic picture of the VO_2(M)@SiO_2 particles, and (b) X-ray diffraction patterns of pristine wood, transparent wood, and energy saving wood composite (EW)

3.2 FT-IR spectra

FT-IR spectra of pristine wood are shown in Fig. 3, wood template, transparent wood and EW. Characteristic cellulose peaks appeared at 2918 cm^{-1} was ascribed to -CH. The -CH bending vibration appeared at 1375 cm^{-1}, and the stretching vibration of C−O appeared at 1035 cm^{-1}, which results from cellulose and hemicellulose. The aromatic skeletal vibration plus C=O stretching at 1591 cm^{-1}, the aromatic skeletal vibration at 1504 cm^{-1}, the C−O deformation of the syringyl ring at 1235 cm^{-1}, and the aromatic C−H in-plane at 1108 cm^{-1} were characteristic lignin peaks. These results are in agreement with earlier studies[38,39]. The intensities of the peaks at 1591, 1504, and 1235 cm^{-1} decreased in wood template. However no intensity decrease occurred at 1108 cm^{-1} in the spectrum, which means that part of the lignin was removed while part was preserved. After MMA infiltration, several peaks appeared in curve c. The peaks at 2968 cm^{-1} and 1748 cm^{-1} were ascribed to C−H and C=O stretching vibrations, respectively. Those peaks at 1208 and 1166 cm^{-1} were assigned to C−O−C stretching vibrations. The absorption peak at 1453 cm^{-1} in EW (Fig. 3d) was due to the skeletal vibration stretching of the benzene ring, and the peaks centered at 1359 and 710 cm^{-1} belonged to the stretching vibrations of the 1, 3, 5-triazine ring. The above mentioned absorption peaks were the characteristic peaks of the MF resin, which is the TM shell. Peaks at 1251, 1128 and 1042 cm^{-1} corresponded to Si−C, Si−O−C and Si−O−Si, respectively, demonstrating that the APTS chemical bond was formed. In addition, new peaks at 980 and 530 cm^{-1} were assigned to V=O and V−O−V bending vibrations, respectively, which further confirmed successful addition of the VO_2 (M) @ SiO_2 nanoparticles.

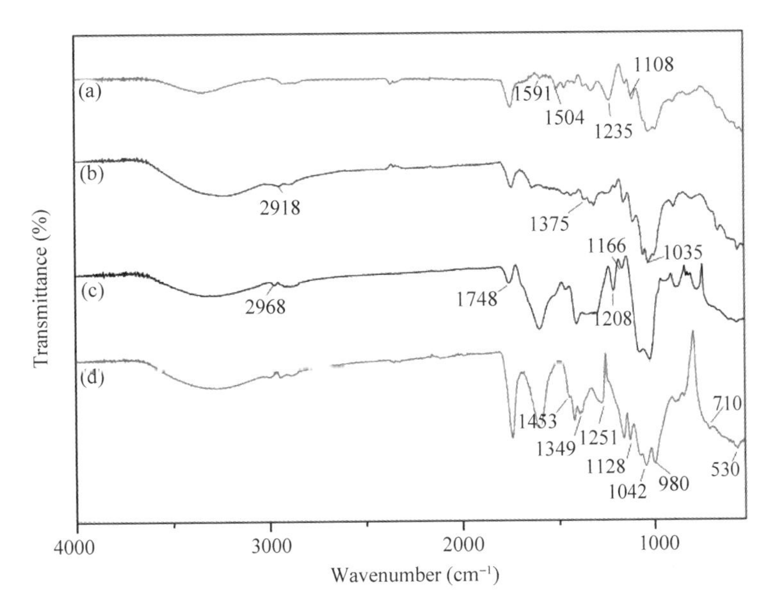

Fig. 3 Fourier-transform infrared spectra of (a) pristine wood, (b) wood template, (c) transparent wood, and (d) energy saving transparent wood (EW)

3.3 Surface morphologies

SEM pictures of the samples were obtained to observe the distribution of the infusion and the micromorphology of the samples. The honeycomb-like microstructures and porous architecture of the pristine wood are shown in Fig. 4a, which provide space for the later polymerization reaction of the injected substances. As shown in Fig. 4b, polymerization occurred in the cell lumen and resulted in highly transparent wood composites. However, when the VO_2(M)@SiO_2 and TM particles were added (Fig. 1c), the cell lumen and injected substances bonded poorly, and the interface between them appeared detached. The reasons might be that the compatibility between them was insufficient and the substances in the cell lumen were too heavy for surface tension to hold. Therefore, APTS was used to tether to the wood cell lumen and the substances. As shown in Fig. 4d, the injected substances were firmly held after adding APTS, and the interface between the cell wall appeared well-integrated, demonstrating improved compatibility between the injected substances and the wood template.

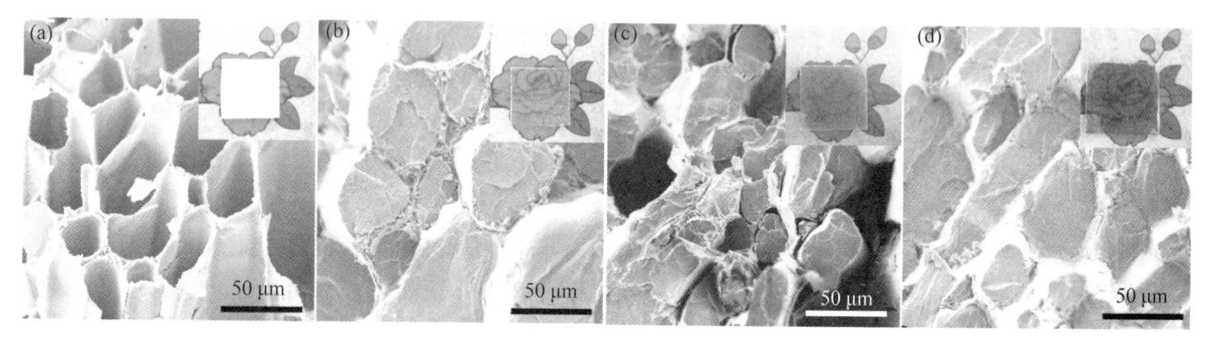

Fig. 4 Scanning electron microscopic images of (a) pristine wood, (b) transparent wood, (c) energy saving wood without APTS, and (d) energy saving transparent wood (EW). (Insets are photographs of the samples)

3.4 Optical properties

Transmittance and haze of the samples are shown in Fig. 5. Pristine wood showed inappreciable transmittance because of its light dispersion and absorption. Transmittance of the transparent wood reached 85 ± 1% in the visible wavelength range. Transmittance dropped to 69 ± 1% after added the VO_2(M)@SiO_2 and TM particles. The main reason is that the particles agglomerated and causing optical heterogeneity and resulting in lower optical transmittance. Interestingly, the EW exhibited higher transmittance than energy saving wood without APTS. A possible reason is that the substances were evenly distributed due to the APTS. The transmittance of EW was 78 ± 1%, which was sufficient to be applied to transparent devices. The inset shows the EW haze, which covered the entire visible wavelength (88 ± 1%). This property can be used for materials with indoor daylighting and privacy requirements. These results suggest that the EW could meet the requirement for transparent building applications due to its advantageous optical transparency and high haze.

3.5 Thermoresponsive performance

In order to investigate the color characteristics of the samples, CIE $L^* a^* b^*$ values of the samples are measured. Among them, L^* is the lightness; a^* = red (+), green(−); b^* = yellow (+), blue (−). ΔE represents the total color difference, and the higher ΔE value, the more significant the color effect is. As shown in Fig. 6a, the color of the sample is almost colorless, and after heating, the appearance color of sample showed bigger redness and yellowness, this led to the highest ΔE value (48.52), producing obvious color effects. In addition, the phase temperature of the EW was characterized by DSC analysis. As shown in

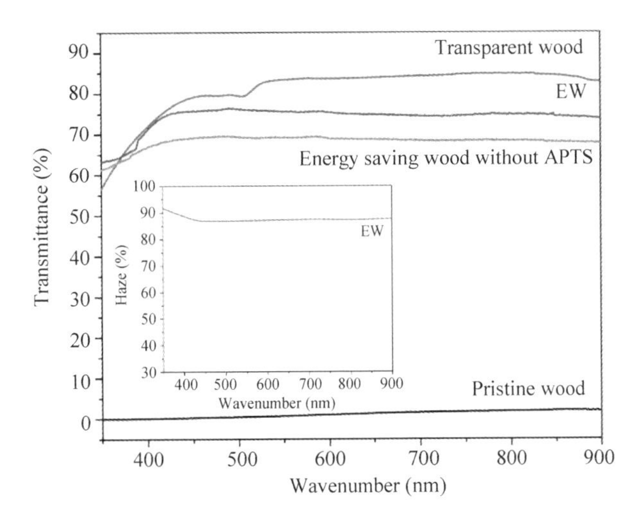

Fig. 5 Transmittance of pristine wood, energy saving wood without APTS, energy saving transparent wood (EW), and transparent wood. The inset is the EW haze

Fig. 6b, endothermic and exothermic peaks were apparent during the heating and cooling cycles. The peaks at about 64 ℃ in the heating process and at about 52 ℃ in the cooling process were ascribed to $VO_2(M)@SiO_2$, which is in agreement with previously reported data[39]. These results provide definitive evidence for the formation of $VO_2(M)@SiO_2$ due to the noticeable first-order structural transition of VO_2(M-R). The TM curve revealed one endothermic peak at 34.5 ℃ during the cooling process and two exothermic peaks at 44.9 ℃ and 59.0 ℃ during the heating process. Notably, the phase transition temperature of the $VO_2(M)@SiO_2$ nanocrystals can be tuned with dopants. In addition, many choices are available for TM at different phase transition temperatures. That is, the sample can be adjusted to many desired phase temperatures, which lays the foundation for EW in different applications.

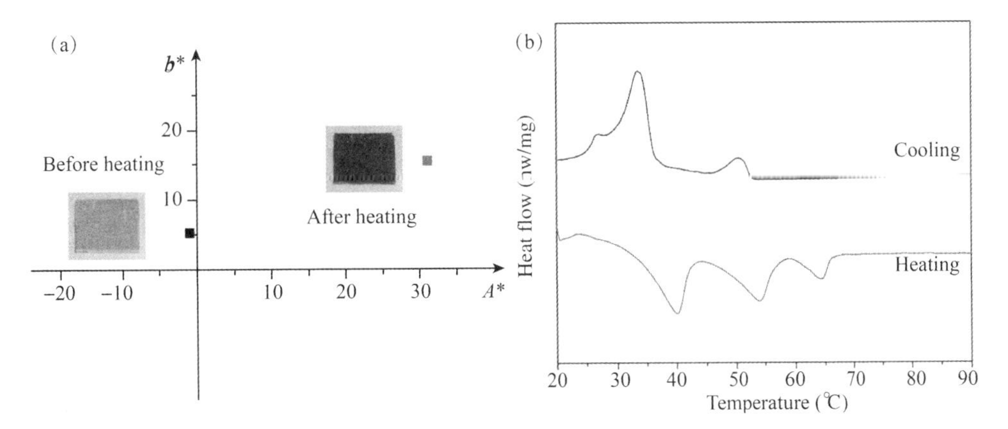

Fig. 6 The a^* and b^* values (a) and the differential scanning calorimetry curves (b) of energy saving transparent wood (EW)

3.6 Model house test

A contrast test of the thermoregulatory ability of the EW and glass was conducted with the model houses to verify the heat shielding and energy storage performance of the sample in a practical application. As shown in

Fig. 7a, the model house was made of 250 mm×250 mm×15 mm wood planks, and the roofs of the two rooms were made of the EW and reference glass, respectively. The rooms were sealed during the testing process. Infrared lights (BR125, IR 250 W; Philips, Best, The Netherlands) were used to simulate natural light, while two thermoelectric couples were employed to monitor the temperature changes inside the houses. As shown in Fig. 7b, after continuous radiation for 15 min, temperature of the EW roof house was about 33 ℃ lower than that of glass roof house, suggesting that a large proportion of infrared was hindered by the EW roof. The infrared-heat shielding ability was owing to the VO_2(M) nanoparticles, the SiO_2 shell serves as a capping layer, enhances the mechanical stability of the VO_2(M), and protects the particles against oxidation. In addition, owing to the thermochromic microcapsule, after turning off the light and cooling the house to 19 ℃ for 30 min, the temperature of EM roof house varied from 33 ℃ to 28 ℃, while the temperature of the glass roof house decreased from 64 ℃ to 20 ℃. These results indicate that the EW is an ideal material for smart thermoregulation, as it maintained a stable temperature range during the heating and cooling process.

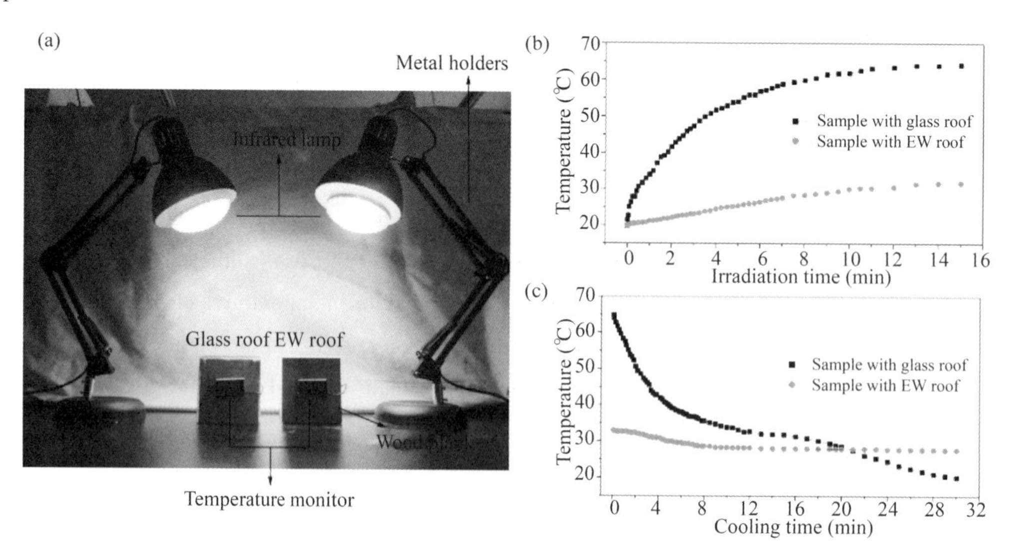

Fig. 7 (a) Illustration of the testing system, the temperature of the samples with (b) different irradiation times and (c) different cooling times

3.7 Mechanical properties and dimensional stability

Fig. 8 presents the stress-strain curves of pristine wood, delignified wood, wood template and EW. The fracture strength and modulus of the pristine wood were 31.86 MPa and 5.26 GPa, respectively, due to the sophisticated microstructure of wood and the binding interaction of lignin. Therefore, the delignified wood after lignin treatment had a very fragile structure with lower fracture strength and modulus of 2.55 MPa and 0.48 GPa while the wood template had an improved fracture strength of 14.83 MPa and modulus of 1.97 GPa. The EW showed superior fracture strength and modulus up to 50.05 MPa and 2.54 GPa due to the favorable synergy between PMMA, APTS, and wood template. The EW possessed a strain of 6%, which was much higher than that of pristine wood. The high fracture strength, modulus, and ductility of EW are highly desirable for practical applications. A water-resistant property is important for long-life applications. However, wood products applications are usually limited due to their hygroscopic nature. In this study, the pristine wood and EW were immersed in water, and the volume and weight changes of the samples are presented in Fig. 9. After being immersed in water for 30 days, the volume of wood increased by 49.8%, whereas EW displays a 17.1% volume gain; and the weight growth rate of pristine wood and EW were about 415.3% and 5.8%, respectively. These results show that the dimensional stability of EW in water was much better than that of pristine wood. The excellent dimensional stability of the EW was because of the blocking effect of the polymer substance.

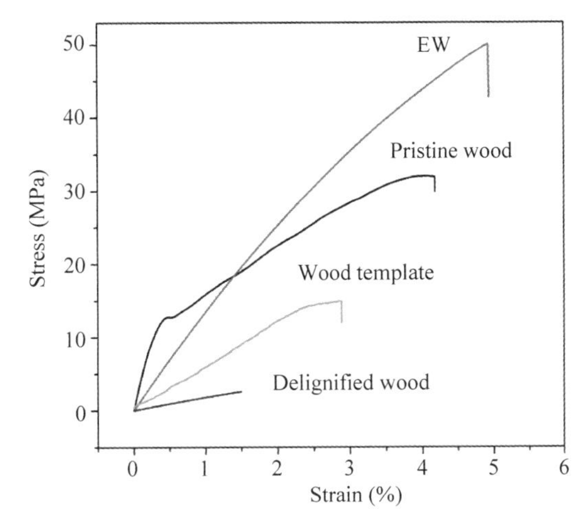

Fig. 8 Stress-strain curves of delignified wood, wood template, pristine wood, and energy saving transparent wood (EW)

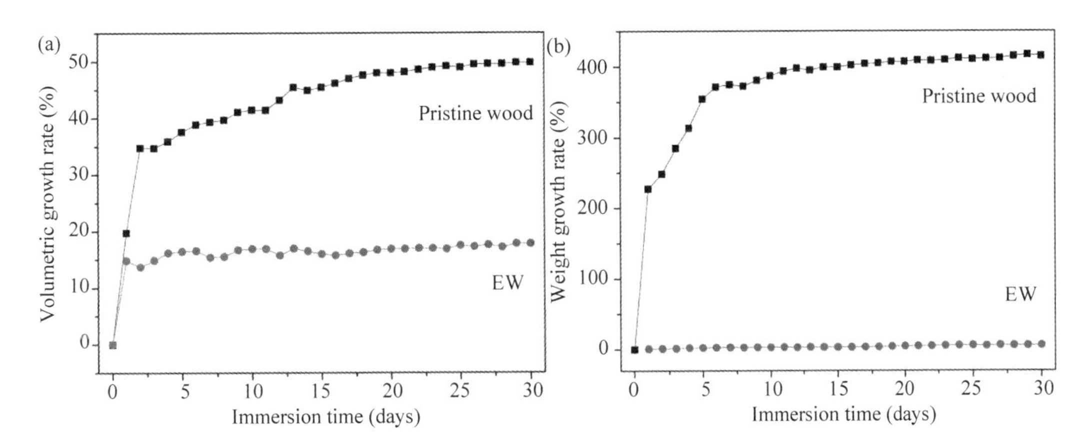

Fig. 9 Volume growth rate (a) and weight growth rate (b) of pristine wood and energy saving transparent wood (EW) during the water immersion test

4 Conclusions

The EW was obtained by dispersing VO_2@SiO_2/TM particles in a modified MMA solution, followed by infiltration into a wood template. The XRD and FT-IR results confirm that the VO_2(M)@SiO_2 and TM particles were successfully immobilized in the wood template. The SEM photographs show that the injected substances were firmly held and the interface between the cell wall appeared well-integrated. The high visible light transmittance and high haze of the EW will be useful for applications with requirements of daylighting and privacy. The EW also exhibited excellent mechanical properties with a fracture strength of 50.05 MPa, modulus of 2.54 GPa, and strain of 6%, which are much higher than those for pristine wood. In addition, the superior dimensional stability of the EW was due to the blocking effect of the polymer substance. The real house simulation test proved that the EW had excellent infrared-heat shielding ability and energy storage performance. All results indicate that the EW could be an ideal candidate for energy saving material, and could have potential applications in the fields of building, furniture, thermal insulation, energy conversion, energy storage and conservation.

Acknowledgements

This work was supported by the financial support from the Fundamental Research Funds for the Central Universities (Grant 2572018AB16), the Overseas Expertise Introduction Project for Discipline Innovation, 111Project (No. B08016), and the National Natural Science Foundation of China (31470584).

Conflict of interest

The authors declared that they have no conflict of interest.

Reference

[1] O'Beirne M D, Werne J P, Hecky R E, et al. Anthropogenic climate change has altered primary productivity in Lake Superior[J]. Nat. Commun., 2017, 8(1): 1-8.

[2] Noernberg B, Borchardt E, Luinstra G A, et al. Wood plastic composites from poly (propylene carbonate) and poplar wood flour-Mechanical, thermal and morphological properties[J]. Eur. Polym. J., 2014, 51: 167-176.

[3] Brimelow J C, Burrows W R, Hanesiak J M. The changing hail threat over North America in response to anthropogenic climate change[J]. Nat Clim Chang, 2017, 7(7): 516-522.

[4] Patz J A, Campbell-Lendrum D, Holloway T, et al. Impact of regional climate change on human health[J]. Nature, 2005, 438(7066): 310-317.

[5] Benniston A C, Harriman A, Li P, et al. Temperature-Induced Switching of the Mechanism for Intramolecular Energy Transfer in a 2, 2': 6', 2"-Terpyridine-Based Ru (II)- Os (II) Trinuclear Array[J]. J. Am. Chem. Soc., 2005, 127(8): 2553-2564.

[6] Yang Y, Chen X. Investigation on energy absorption efficiency of each layer in ballistic armour panel for applications in hybrid design[J]. Compos. Struct., 2017, 164: 1-9.

[7] Sahiner N, Demirci S. PEI-based hydrogels with different morphology and sizes: Bulkgel, microgel, and cryogel for catalytic energy and environmental catalytic applications[J]. Eur. Polym. J., 2016, 76: 156-169.

[8] Cao X, Dai X, Liu J. Building energy-consumption status worldwide and the state-of-the-art technologies for zero-energy buildings during the past decade[J]. Energy Build., 2016, 128: 198-213.

[9] Allouhi A, El Fouih Y, Kousksou T, et al. Energy consumption and efficiency in buildings: Current status and future trends[J]. J. Clean. Prod., 2015, 109: 118-130.

[10] Powell M J, Quesada-Cabrera R, Taylor A, et al. Intelligent multifunctional $VO_2/SiO_2/TiO_2$ coatings for self-cleaning, energy-saving window panels[J]. Chem. Mater., 2016, 28(5): 1369-1376.

[11] Chan K Y, Jia B, Lin H, et al. A critical review on multifunctional composites as structural capacitors for energy storage[J]. Compos. Struct., 2018, 188: 126-142.

[12] Zhao Z, Li Z, Xia Q, et al. Fast synthesis of temperature-sensitive PNIPAAm hydrogels by microwave irradiation[J]. Eur. Polym. J., 2008, 44(4): 1217-1224.

[13]Pacheco R, Ordóñez J, Martínez G. Energy efficient design of building: A review[J]. Renew. Sust. Energ. Rev., 2012, 16(6): 3559-3573.

[14]Gong C, Wei Y, Chen M, et al. Double imprinted photoresponsive polymer for simultaneous detection of phthalate esters in plastics[J]. Eur. Polym. J., 2018, 108: 295-303.

[15] Li J, Guo Q, Lu Y, et al. Polyindole vertical nanowire array based electrochromic-supercapacitor difunctional device for energy storage and utilization[J]. Eur. Polym. J., 2019, 113: 29-35.

[16] Yuan T, Vazquez M, Goldner A N, et al. Versatile thermochromic supramolecular materials based on competing charge transfer interactions[J]. Adv. Funct. Mater., 2016, 26(47): 8604-8612.

[17]Liu X, Padilla W J. Thermochromic infrared metamaterials[J]. Adv. Mater., 2016, 28(5): 871-875.

[18]Ren Y, Palstra T T M, Khomskii D I, et al. Temperature-induced magnetization reversal in a YVO_3 single crystal[J]. Nature, 1998, 396(6710): 441-444.

[19] Chen Y, Zeng X, Zhu J, et al. High performance and enhanced durability of thermochromic films using VO_2@ ZnO core-shell nanoparticles[J]. ACS Appl. Mater. Interfaces, 2017, 9(33): 27784-27791.

[20] Zheng Z, Zhang L, Ling Y, et al. Triblock copolymers containing UCST polypeptide and poly (propylene glycol): synthesis, thermoresponsive properties, and modification of PVA hydrogel[J]. Eur. Polym. J., 2019, 115: 244-250.

[21] Lacroix F, Allheily V, Diener K, et al. Thermomechanical behavior of aeronautic structural carbon epoxy composite submitted to a laser irradiation[J]. Compos. Struct., 2016, 143: 220-229.

[22] Chen K, Liu N, Zhang M, et al. Oxidative desulfurization of dibenzothiophene over monoclinic VO_2 phase-transition catalysts[J]. Appl. Catal. B, 2017, 212: 32-40.

[23] Kamalisarvestani M, Saidur R, Mekhilef S, et al. Performance, materials and coating technologies of thermochromic thin films on smart windows[J]. Renew. Sust. Energ. Rev., 2013, 26: 353-364.

[24] Kim M W, Bae T S, Hong W K. Formation of a core-shell-like vanadium dioxide nanobeam via reduction and surface oxidation and its metal-insulator phase transition behavior[J]. Appl. Surf. Sci., 2018, 455: 1185-1191.

[25] Zhu M, Song J, Li T, et al. Highly anisotropic, highly transparent wood composites[J]. Adv. Mater., 2016, 28(26): 5181-5187.

[26] Saiter A, Saiter J M, Grenet J. Cooperative rearranging regions in polymeric materials: Relationship with the fragility of glass-forming liquids[J]. Eur. Polym. J., 2006, 42(1): 213-219.

[27]Dang B, Chen Y, Yang N, et al. Effect of carbon fiber addition on the electromagnetic shielding properties of carbon fiber/polyacrylamide/wood based fiberboards[J]. Nanotechnology, 2018, 29(19): 195605.

[28]Bergamonti L, Graiff C, Tegoni M, et al. Facile preparation of functionalized poly (amidoamine) s with biocidal activity on wood substrates[J]. Eur. Polym. J., 2019, 116: 232-241.

[29]Chen Y, Dang B, Jin C, et al. Processing lignocellulose-based composites into an ultrastrong structural material[J]. ACS Nano, 2018, 13(1): 371-376.

[30]Lu Y, Ye G, She X, et al. Sustainable route for molecularly thin cellulose nanoribbons and derived nitrogen-doped carbon electrocatalysts[J]. ACS Sustain. Chem. Eng., 2017, 5(10): 8729-8737.

[31]Shi S Q, Gardner D J. Dynamic adhesive wettability of wood[J]. Wood Fiber Sci., 2001, 33(1): 58-68.

[32]Sheng C, Wang C, Wang H, et al. Self-photodegradation of formaldehyde under visible-light by solid wood modified via nanostructured Fe-doped WO_3 accompanied with superior dimensional stability[J]. J. Hazard. Mater., 2017, 328: 127-139.

[33]Wang H, Yao Q, Wang C, et al. Hydrothermal synthesis of nanooctahedra $MnFe_2O_4$ onto the wood surface with soft magnetism, fire resistance and electromagnetic wave absorption[J]. Nanomaterials, 2017, 7(6): 118.

[34]Chen Y, Wang H, Yao Q, et al. Biomimetic taro leaf-like films decorated on wood surfaces using soft lithography for superparamagnetic and superhydrophobic performance[J]. J. Mater. Sci., 2017, 52(12): 7428-7438.

[35]Yuan T Q, Sun S N, Xu F, et al. Characterization of lignin structures and lignin-carbohydrate complex (LCC) linkages by quantitative 13C and 2D HSQC NMR spectroscopy[J]. J. Agric. Food Chem., 2011, 59(19): 10604-10614.

[36]Baishya P, Nath D, Begum P, et al. Effects of wheat gluten protein on the properties of starch based sustainable wood polymer nanocomposites[J]. Eur. Polym. J., 2018, 100: 137-145.

[37] Kobayashi Y, Katakami H, Mine E, et al. Silica coating of silver nanoparticles using a modified Stöber method[J]. J. Colloid Interface Sci., 2005, 283(2): 392-396.

[38]Li Y, Hui B, Li G, et al. Fabrication of smart wood with reversible thermoresponsive performance[J]. J. Mater. Sci., 2017, 52(13).

[39] Jiang S, Gui Z, Bao C, et al. Preparation of functionalized graphene by simultaneous reduction and surface modification and its polymethyl methacrylate composites through latex technology and melt blending[J]. Chem. Eng. J., 2013, 226: 326-335.

中文题目：具有温度调节能力和热响应性能的节能木材复合材料

作者：李莹莹，高汝南，李坚

摘要：节能材料在经济和环境方面都是有益的，因为它们有效地减少了能源消耗和温室气体排放。在这项研究中，通过将含有 VO_2(M)@SiO_2 核壳纳米复合材料和温致变色微胶囊的混合物分散在甲基丙烯酸甲酯溶液并注入木材模板中，从而制备了节能木材复合材料。加入 APTS 以稳定悬浮液。注入的物质与木材模板之间的良好相容性得益于 PMMA 和 APTS 填充了细胞腔并粘结了物质，从而导致了节能木质复合材料，具有出色的尺寸稳定性、光学性能和机械性能。更重要的是，节能型木质复合材料通过屏蔽不必要的红外辐射而表现出出色的温度调节能力。节能木质复合材料的独特性能可潜在地应用于建筑、家具、隔热、能量转换、能量存储和节约中。

关键词：木质复合材料；温度调节；节能；相变

Structurally Colored Wood Composite with Reflective Heat Insulation and Hydrophobicity*

Yingying Li, Likun Gao, Yingtao Liu, Jian Li

Abstract: Structural colors are environmentally beneficial as they originate from the physical structure of the material, and they cannot be imitated by chemical dyes and are free from photo-bleaching. In this study, structurally colored wood was prepared by building a structurally colored film based on a wood substrate. The prepared samples were characterized by X-ray diffraction (XRD), scanning electron microscopy (SEM), and Fourier transform-infrared (FT-IR). The angel-dependent optical effects were confirmed by comparing the CIE chromatic coordinates over a range of viewing angles. Through octadecyltrichlorosilane (OTS) treatment, the sample surface was made hydrophobic and had a water contact angle of 139°, which broadens the range of applications for the material. Thermal gravimetric analysis results showed that the thermal stability of the modified structurally colored wood (MCW) was much better than that of pristine wood. The reflectance spectra and the model room test results demonstrated that the MCW possesses the reflective heat insulation ability. The unique and promising properties of the MCW could potentially be applied in buildings, furniture, and for energy conservation.

Keywords: Structural colors; Wood composite; Energy saving; Hydrophobic performance

1 Introduction

Nature has been the inspiration for many important human inventions. The unique, brilliant colors in the natural world have attracted a significant amount of research interest[1-3]. To the best of our knowledge, an organism's color is produced either through pigments, bioluminescence, or structurally[4-6]. The observed colors in a pigment are caused by their absorption of certain spectral wavelengths. Bioluminescence originates from chemical reactions in certain organisms that contain photophores[7,8]. Structural colors are ubiquitous in nature, they result from periodic self-organized microstructures, which interact with light to produce brilliant colors in a complex manner[9-11]. Structural colors have many properties that differ from those of pigmentary colors and bioluminescence[12,13]. The structural colors are environmentally friendly because they originate from the physical structure of the material. Furthermore, structural colors cannot be imitated by chemical dyes or pigments and are free from photo-bleaching; hence, they are usually more durable than conventional chemical pigments and dyes[14,15]. Therefore, structurally colored materials are potential candidates to replace dyes and pigments in many applications[16]. As a kind of structurally colored material, mica-titania exhibits a pearl-shine effect due to the angle-dependent optical effects that arise from alternating transparent layers with

* 本文摘自 Journal of Wood Chemistry and Technology, 2019, 39(6): 454-463.

different refractive indices[17,18]. Mica-titania pigments are typically produced by the deposition of TiO_2 on mica from an aqueous suspension, followed by a calcination process[19,20]. In general, the higher refractive index of rutile allows for a better color effect to be achieved when TiO_2 is coated on mica[21]. Therefore, rutile mica-titania has attracted intensive attention in the field of fillers. Furthermore, rutile mica-titania has also attracted significant attention in the fields of decoration, energy conservation, and as a durable coating because of its excellent reflective heat insulation, weather resistance, and thermal stability.

Wood has been used by humans since prehistoric times due to their attractive features, such as good mechanical properties, low thermal expansion, low weight, aesthetics, excellent formability, and sustainability[22,23]. It is commonly known that wood is a natural composite composed of cellulose, hemicellulose, and lignin[24]. A network of these three basic components and hollow tubes construct the multi-scale porous structure, which also determines the penetrability, accessibility, and reactivity of wood-based materials[25-27]. As a result, its unique composition and subtle hierarchical structures make wood an ideal template to combine with functional particles. This results in the possibility for it to meet future market requirement and compete with other advanced materials[28].

Several studies have reported wood-based materials with additional thermochromic or photochromic properties, but the color of the samples all originate from chemical dyes[29,30]. To the best of our knowledge, there have been no reports on structurally colored solid wood materials. In this paper, we demonstrate a simple and facile process for fabricating structurally colored wood. The phase composition, morphology, optical properties, thermal stability, and hydrophobic performance were investigated. The structurally colored wood displayed not only a beautiful decorative function but was also an ideal candidate for use as an insulative energy saving material. The modified structurally colored wood (MCW) can be economically beneficial because they could reduce the need for air-conditioning by reflecting excess solar heat, thus reducing energy consumption and greenhouse gas emissions.

2 Experimental

2.1 Materials

For this study, mica was used as the substrate and was supplied by Sanbao Pearl Luster Mica Tech CO. LTD, China. Hydrochloric acid (HCl), stannic chloride pentahydrate ($SnCl_4 \cdot 5H_2O$), sodium hydroxide (NaOH), titanium tetrachloride ($TiCl_4$), and polyvinyl alcohol (PVA) (alcoholysis 98.0%–99.0%, model 1750±50) were purchased from Kaitong Chemical Reagent Co. Ltd. (Tian jian, China). Octadecyltrichlorosilane (OTS) (95.0%) was used for modification of the surface hydrophobicity and was purchased from Shanghai Aladdin Biochemical Technology Co. Ltd. (Shanghai, China). Larch, Birch, and Manchurian ash were obtained from Harbin, China, which reprocessed into wood slices (radial section) with a size of 250 mm×250 mm×15 mm. The wood samples were oven-dried (24 h, 103±2 ℃) to constant weight after ultrasonically rinsing in deionized water. All chemicals were used as received without further purification.

2.2 Synthesis of Rutile Titania-Mica

The preparation of the titania-mica pigment was carried out in the following manner. The 10 wt% mica was uniformly dispersed in distilled water. The solution was then maintained at 70 ℃ and the pH value was adjusted to 2.0 with diluted hydrochloric acid under stirring. Then, $SnCl_4$ solution was added in a drop wise manner while the pH was held constant through the simultaneous addition of NaOH. Next, a 1.5 M $TiCl_4$ solution was introduced into the above suspension at a constant speed of 1.0 mL · min^{-1}. The pH value of the slurry was kept constant by the simultaneous addition of an NaOH solution. The samples were collected by centrifugation, washed with deionized water several times, and dried at 70 ℃ for more than 24 h in vacuum environment. Finally, the titania-mica pigment was obtained by calcination in a muffle furnace at 850 ℃ for 1 h.

2.3 Synthesis of the structurally colored wood

The wood samples were ultrasonically rinsed in deionized water, acetone, and ethanol for 10 min each, and then oven-dried (24 h, 103±2 ℃) to a constant weight after. A 4 wt% PVA solution was prepared at 75 ℃ for 2 h with magnetic stirring. Next, 0.3 wt% of titania-mica pigments was added to the above solution and stirred for 15 min. Next, a 1.0 mL solution was dropped onto the wood surface and then transferred to a vacuum environment for 10 min in order to make sure the solution fully contacted with the wood substance. Finally, the treated samples were placed in a 30 ℃ oven until completely dry. Unless explicitly stated, the substances of the samples were Larch wood.

2.4 Preparation of modified structurally colored wood (MCW).

As shown in Fig. 1, the formation of the hydrophobic structurally colored wood was accomplished by building a self-assembled OTS monolayer. First, the as-prepared samples were submerged into 25 mL of a 1.0% OTS/chloroform solution at room temperature for 1 h, followed by rinsing with chloroform. The prepared samples were then dried at room temperature for 3 h to obtain the hydrophobic structurally colored wood. Unless explicitly stated, the substances of the samples were Larch wood.

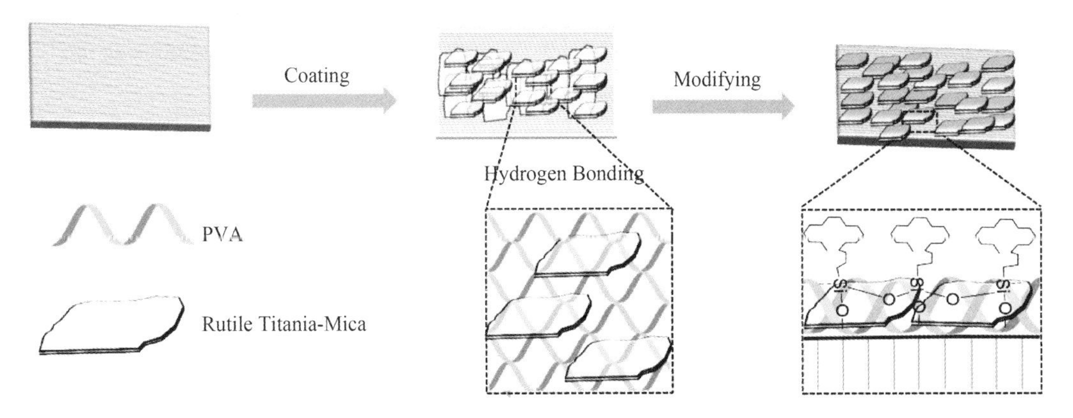

Fig. 1 Schematic illustration of the fabrication of the modified structurally colored wood (MCW)

2.5 Characterization

A scanning electron microscope (SEM) (Quanta200; FEI, Hillsboro, OR, USA) was used to analyze the surface morphology. For this, the samples were glued onto a specific holder and coated with gold to ensure conductivity. The phase of the as-prepared sample was checked by X-ray diffraction (XRD) (Rigaku, and D/MAX 2200) measurements operated using Cu target radiation (λ = 1.54 Å). Attenuated total reflectance

Fourier transform-infrared spectra (FT-IR) (Nicolet Nexus 670 FT-IR) were recorded in the range of 400-4000 cm^{-1} with a resolution of 4 cm^{-1}. Reflectance spectra of the samples were analyzed with a UV-Vis spectrophotometer (UV-Vis DRS) (TU-1901, China) equipped with an integrating sphere. The thermal stabilities of the samples were investigated using a thermal gravimetric analyzer (TGA) (TA, Q600), and the experiments were performed in the temperature range of 25-700 ℃ at a heating rate of 10 ℃ · min^{-1} under a nitrogen atmosphere. An OCA 40 contact angle system (Dataphysics, Germany) was used to measure the water contact angles (WCAs). For this, a 5 μl droplet of deionized water was injected onto the sample surface at room temperature.

2.6 Color test

The samples at different observation angles were obtained by using a digital camera, and these records were imported to a computer. The Commission Internationale de L'Eclairage (CIE) L^*, a^*, and b^* values were measured by using the Adobe Photoshop CS6 software installed in the computer[29]. The total color difference is represented by ΔE, which was calculated using the following equations:

$$\Delta E^* = \sqrt{(L_2-L_1)^2+(a_2-a_1)^2+(b_2-b_1)^2} \tag{1}$$

The L_1, a_1 and b_1 are the color parameters of untreated wood; L_2, a_2 and b_2 are the color parameters of treated samples at different observation angles. The color parameters were measured at five different points on each sample, and the final value for color parameters were obtained as an average of these five measurements[30].

2.7 Environment simulation test

A model house room was constructed using 250 mm×250 mm×15 mm wood planks; the roofs of the two rooms were made of MCW and pristine wood, respectively. The rooms were sealed during the testing process. A solar simulator (XES-40S2-CE, Lamp L150SS, Japan) with a wavelength range of 300-1100 nm was used to simulate natural light. Two thermoelectric couples were employed to monitor the temperature change inside the two respective rooms.

3 Results and discussion

3.1 XRD analyses

XRD was carried out to determine the crystalline structure of the samples. The XRD patterns of rutile titania-mica, pristine wood, and MSW are shown in Fig. 2. In Fig. 2a, we see that the XRD patterns exhibit strong diffraction peaks at 27°, 36° and 55°, indicating the presence of TiO_2 in the rutile phase (JCPDS No. 21-1276). The characteristic diffraction peaks of $KMg_3(AlSi_3O_{10})F_2$ were also observed. The results prove that rutile titania-mica was successfully synthesized and is consistent with values in literature[31,32]. In Fig. 2b, the diffraction peaks centered at 16.1° and 22.2° corresponded to the (101) and (002) diffraction planes of cellulose in wood. As shown in Fig. 2c, characteristic peaks belonging to wood, TiO_2, and $KMg_3(AlSi_3O_{10})F_2$ also appeared. This confirmed that the rutile titania-mica was successfully anchored onto the surface of wood substance.

3.2 FT-IR spectra

Fig. 3 shows the FT-IR spectra of pristine wood and MCM. As shown in Fig. 3a, the broad peak at

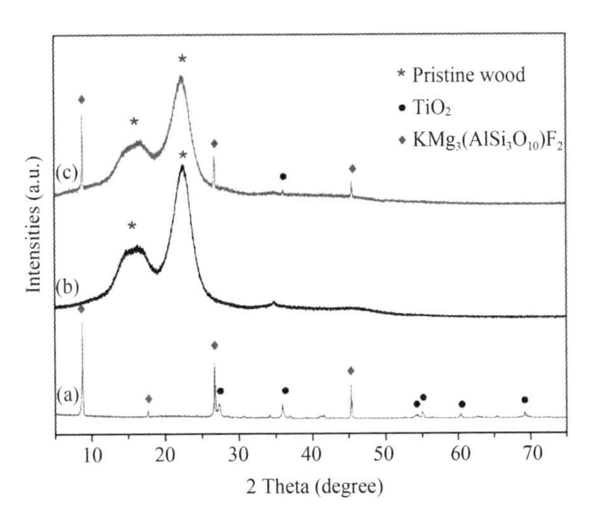

Fig. 2 X-ray diffraction patterns of (a) rutile titania-mica, (b) pristine wood, and modified structurally colored wood (MCW)

3332 cm^{-1} was assigned to the OH group. The peaks at 1371 cm^{-1} and 1024 cm^{-1} were ascribed to the -CH bending vibration and the C－O stretching vibration, respectively, which correspond to cellulose and hemicellulose. In Fig. 3b we see that the intensity of the peak corresponding to the OH group decreased dramatically, which indicates a reduction in hydrophilicity of the MCW. The character peaks associated to cellulose and hemicellulose were preserved, verifying that the basic structure of the wood was un-affected by the modification. The peaks at 792 cm^{-1} and 686 cm^{-1} were assigned to Si-O-Al and confirmed the successful immobilization of the titania-mica particles, which is consistent with the XRD results. The peaks at 1464 cm^{-1} and 1051 cm^{-1} are ascribed to the methylene deformation vibration and Si-O-Si, respectively. The results indicated the condensation reaction of the silane, which demonstrate that the OTS film was formed. The results confirm that the hydrophobic coating was successfully applied onto the wood surface.

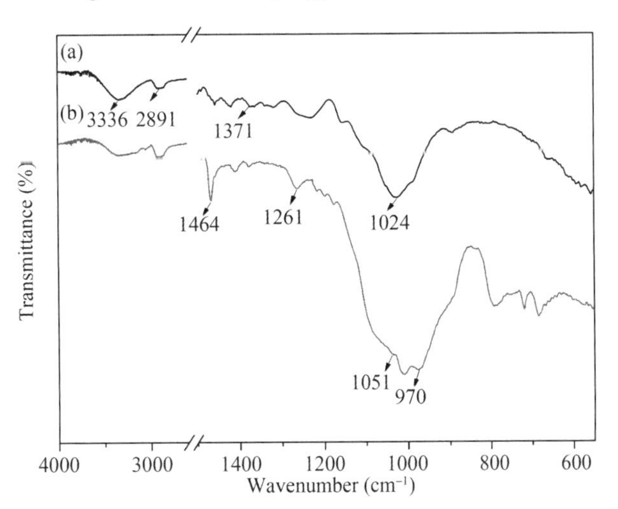

Fig. 3 Fourier-transform infrared spectra of (a) pristine wood, and (b) modified structurally colored wood (MCW)

3.3 Surface morphologies

The micromorphology of the samples was observed by SEM. The morphology of the broad, smooth, and flaky rutile titania-mica is shown in Fig. 4a. In Fig. 4b, we see that the TiO_2 deposited homogenously onto the mica surface. Fig. 4c shows the multilevel microstructure and multi-scale porous architecture of the pristine Larch wood, the wood cell lumens were very trimly and free of other foreign material. From Fig. 4d, it can be clearly seen that the rutile titania-mica particles are dispersed on the surface of the wood substrate while lying flat. The results suggest that the rutile titania-mica particles were successfully attached to the wood surface and are consistent with the XRD results. With the OTS treatment, molecular layers were formed by self-assembly of long-chain hydrophobic alkyls (Fig. e and f), forming a hydrophobic film on the surface of the sample.

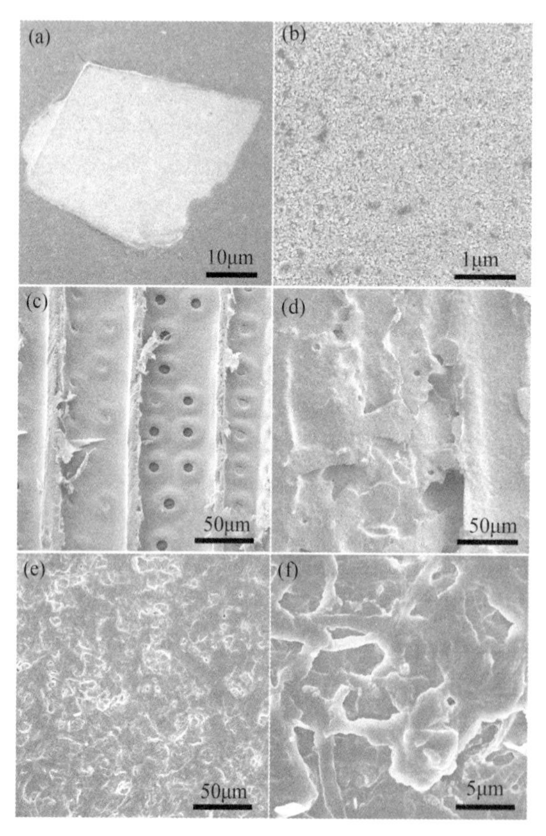

Fig. 4 Scanning electron microscopy images of rutile titania-mica at (a) low and (b) high magnification, pristine wood (c), structurally colored wood (d), and the modified structurally colored wood (MCW) at (e) low and (f) high magnification

3.4 Optical properties

In order to investigate the color characteristics of the samples, the CIE $L^* a^* b^*$ *color space* values of the samples were measured; where L^* is the lightness; a^* = red (+), green (-); and b^* = yellow (+), blue (-). The total color difference is represented by ΔE. It is worth mention that a higher ΔE value indicates a more significant colors effect[33,34]. The L^*, a^*, b^*, and ΔE values of the MSW samples with different substrates at an observed angle of 10° are given at Table 1. When the sample was made of Larch wood, the green and blue had the maximum values, as shown in the Table 1. This led to the highest ΔE value and produced obvious color effects. The results prove that the type of wood species used for the substrate has a direct impact on the color effect of the samples. Among substrates made of Larch, Birch, and Manchurian, it was observed that Larch has the most significant color effect due to its darker texture. In addition, color characters of the untreated Larch wood and modified structurally colored wood (MCW) at different observed angle were also given at Table 2. In Fig. 5 we can see the angel-dependent optical effects by comparing the CIE chromatic coordinates. The color of pristine wood is yellow, and the apparent color of MCW varies from blue to red purple on changing the observation angles.

Table 1 Color characters of the samples at an observed angle of 10°

Wood species	L^*	a^*	b^*	ΔE
Larch	12.0	-14.0	-61.0	63.7
Birch	6	-1.3	-4.3	7.5
Manchurian ash	24.7	-5.0	-23.7	34.5

Table 2 Color characters of the untreated Larch wood and modified structurally colored wood (MCW) at different observed angle

Larch sample	L^*	a^*	b^*
Untreated wood	65.0	17.7	36.7
MCW at10°	77.0	3.6	-24.3
MCW at110°	68.3	18.0	-16

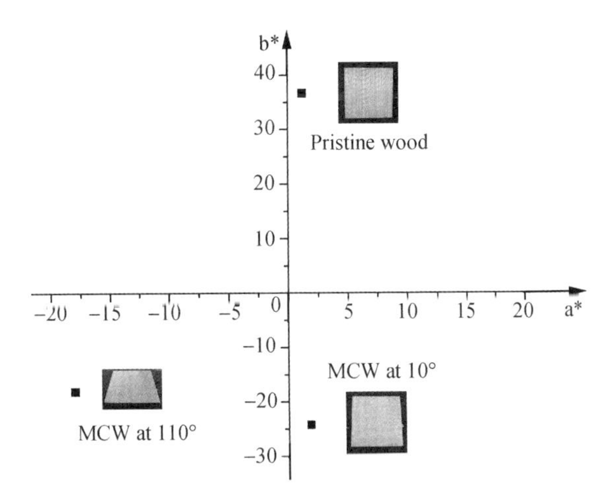

Fig. 5 The a^* and b^* values of pristine wood and modified structurally colored wood (MCW) at different observation angles

3.5 Wettability analyses

The wettability of the samples was evaluated by measuring the water contact angle on the sample surface (the sample exhibits hydrophilicity when the water contact angle is less than 90°). The water contact angle on pristine wood, structurally colored wood, and MSW are shown in Fig. 6. The pristine wood surface in Fig. 7a demonstrates hydrophilicity with a water contact angle of 83. 5°. And the water contact angle subsequently reduces to 43. 0° after 15 s, which proved the hydrophillic nature of pristine wood. As shown in Fig. 6b, the structurally colored wood possesses a hydrophobic surface with a water contact angle of 100°, the contact angle reduces to 89° after 15 s, which can be mainly ascribed to the hydroxyl groups from PVA and the wood substrate. During OTS treatment, the hydroxyl groups produced by OTS hydrolysis react with the surface hydroxyl groups of PVA and the wood substrate; in this manner, long-chain hydrophobic alkyls are introduced on the sample surface. This result in a remarkable reduction in the wettability of the sample, and the MCW shows a hydrophobic surface with a water contact angle of 139°, and the contact angle was maintained at 134° after 15 s (Fig. 7c). The results suggested that the surface property of wood sample is successfully changed from hydrophilic to hydrophobic by the OTS treatment.

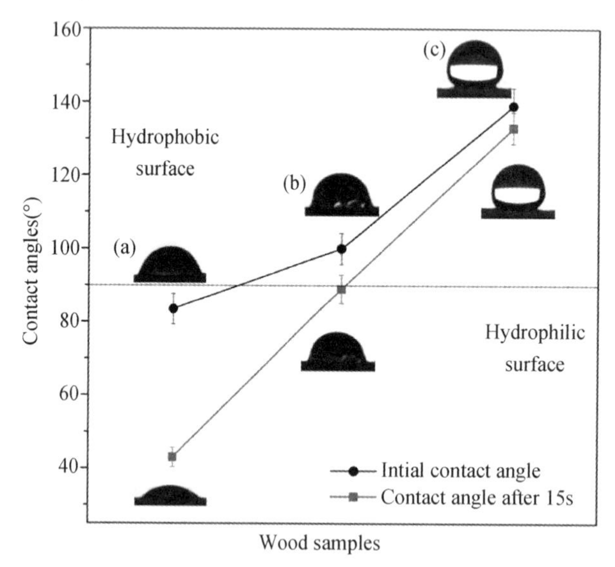

Fig. 6 Contact angles of droplets of water on the surface of (a) pristine wood (b) structurally colored wood, and (c) modified structurally colored wood (MCW)

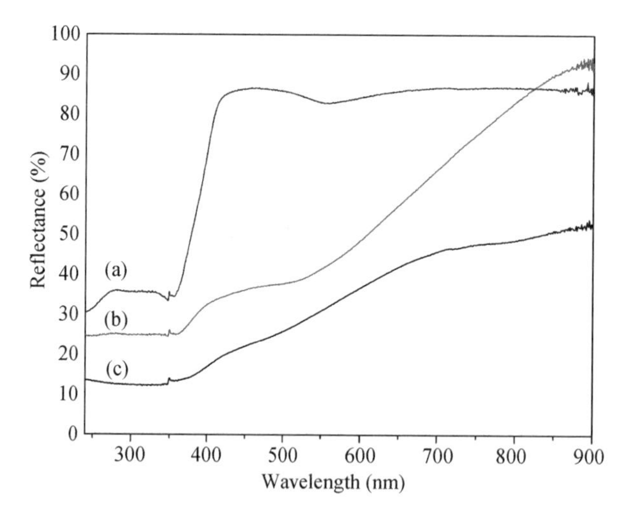

Fig. 7 UV-Vis diffuse reflectance spectra of the (a) rutile titania-mica, (b) modified structurally colored wood (MCW), and (c) pristine wood

3.6 Reflective heat insulative ability

Fig. 7 shows the reflectance spectra of the mica titanium pigments, pristine wood, and MCW. From Fig. 7a, we can see that the rutile titania-mica shows a high reflectance across the whole visible region. In addition, the reflectance of pristine wood increased in the region of 400–900 nm and reached 50% at 900 nm (Fig. 7c). As shown in Fig. 7b, the reflectance of MSW was generally higher than that of pristine wood and reached 91% at 900 nm, which indicated the radiant barrier ability of MSW. Furthermore, in order to verify the heat shielding performance of the samples in practical application, a comparative experiment between MCW and the pristine wood was carried out with an environment simulation test. As shown in Fig. 8, after continuous radiation for 15 min, the temperature of the model room with the MCW roof was about 10 ℃ lower than that of the model room with the pristine wood roof. This demonstrates that a part of the solar radiation was blocked by the MCW roof. The results indicate that MCW is an ideal material for smart thermoregulation because of its reflective insulating ability.

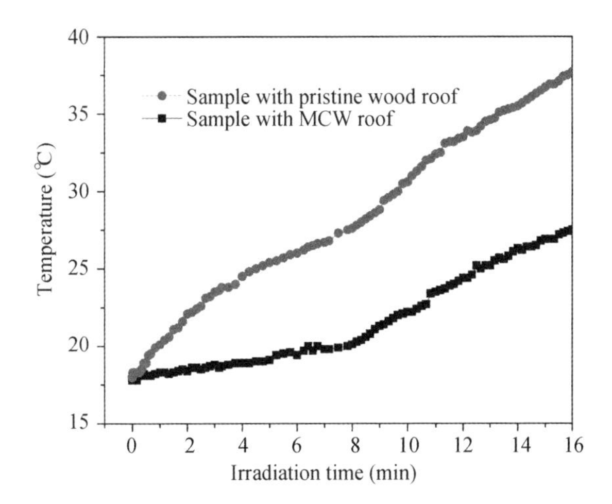

Fig. 8 The temperature of the model rooms with different irradiation times

3.7 Thermal gravimetric analysis

The thermal stability of the samples was evaluated using thermal gravimetric analysis. Fig. 9 shows the TG curves of rutile titania-mica, MCW, and pristine wood. The thermogravimetric trace in Fig. 9a shows that almost no degradation of the rutile titania-mica occurs, which demonstrates that the rutile titania-mica has excellent thermal stability. As seen in Fig. 9c, from 40 to 100 ℃, the pristine wood exhibited a small weight loss, which was due to a loss of moisture and highly volatile matter. The main pyrolysis process occurred in the range of 200-380 ℃, which can be attributed to the decomposition of lignin, hemicellulose, and cellulose[35]. At 700 ℃, a 89% loss in total mass was observed. After modification of the pristine wood with the composite coating (Fig. 9b), the MCW displays similar thermal decomposition behavior to the pristine wood with a lower mass loss (total mass loss for the MCW is ~78%). We see that the protective coating on the MCW results in it having a lower total mass loss than pristine wood during the pyrolysis process. This suggests that the thermal stability of MCW is much better than that of pristine wood.

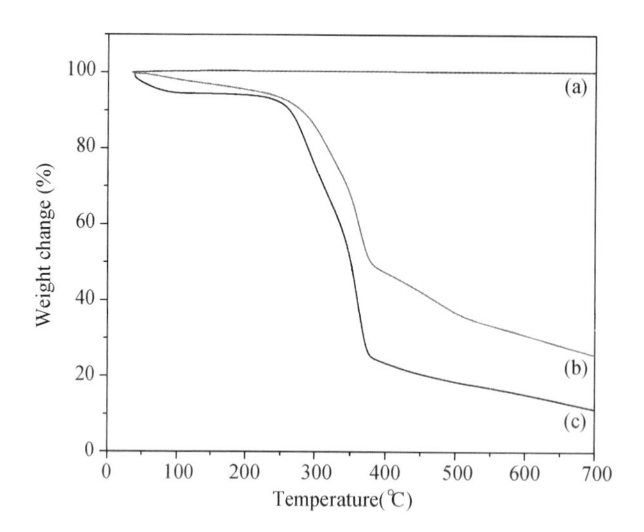

Fig. 9 TG curves of (a) rutile titania-mica, (b) modified structurally colored wood, and (c) pristine wood

4 Conclusions

Structurally colored wood was fabricated by building a structurally colored film on a wood-based substrate. The XRD, FT-IR, and SEM results confirmed that rutile titania-mica was immobilized on the wood substance. The angel-dependent optical effects of the samples were determined by comparing the CIE chromatic coordinates. The CIE $L^* a^* b^*$ values indicated that the substrate made from Larch had the most significant color effect. The wettability analysis confirmed that a hydrophobic film was successful formed on the wood surface and had a water contact angle of 139°. The reflectance spectra and environmental simulation test results demonstrated that MCW has good reflective insulation properties. Finally, the thermal gravimetric analysis showed that the thermal stability of the MCW was much better than that of pristine wood.

Acknowledgements

The authors gratefully acknowledge the financial support from the Fundamental Research Funds for the Central Universities (Grant 2572018AB16), the National Natural Science Foundation of China (31470584), the Overseas Expertise Introduction Project for Discipline Innovation, 111Project (No. B08016), and the Fundamental Research Funds for the Central Universities (Grant 2572017AB08).

Conflict of interest

The authors declared that they have no conflict of interest.

Reference

[1]Kinoshita S, Yoshioka S, Miyazaki J. Physics of structural colors[J]. Rep. Prog. Phys., 2008, 71(7): 076401.

[2] Ziobro G C. Origin and nature of kraft colour: 2 The role of bleaching in the formation of the extraction stage effluent colour[J]. J. Wood Chem. Technol., 1990, 10(2): 151-168.

[3]Diao Y Y, Liu X Y, Toh G W, et al. Multiple structural coloring of silk-fibroin photonic crystals and humidity-responsive color sensing[J]. Adv. Funct. Mater., 2013, 23(43): 5373-5380.

[4]Wei P, Zhou M, Pan L, et al. Suitability of Printing Materials for Heat-Induced Inkless Eco-Printing[J]. J. Wood Chem. Technol, 2016, 36(2): 129-139.

[5]Dumanli A G, Savin T. Recent advances in the biomimicry of structural colours[J]. Chem. Soc. Rev., 2016, 45(24): 6698-6724.

[6]Liu H, Yang S, Ni Y. Dye stability in the presence of hydrogen peroxide and its implication for using dye in the HYP manufacturing process[J]. J. Wood Chem. Technol, 2009, 29(1): 1-10.

[7]Yang Y, Shao Q, Deng R, et al. In vitro and in vivo uncaging and bioluminescence imaging by using photocaged upconversion nanoparticles[J]. Angew. Chem., 2012, 124(13): 3179-3183.

[8] Hu W, Matsumura M. Structure and thickness dependence of p-n heterojunction solar cells based on copper phthalocyanine and perylene pigments[J]. J. Phys. D, 2004, 37(10): 1434.

[9]Sato O, Kubo S, Gu Z Z. Structural color films with lotus effects, superhydrophilicity, and tunable stop-bands[J]. Acc. Chem. Res., 2009, 42(1): 1-10.

[10]Carmo J F, Miranda I, Quilhó T, et al. Copaifera langsdorffii bark as a source of chemicals: Structural and chemical characterization[J]. J. Wood Chem. Technol, 2016, 36(5): 305-317.

[11]Park S Y, Hong C Y, Kim S H, et al. Photodegradation of Natural Wood Veneer and Studies on Its Color Stabilization for Automotive Interior Materials[J]. J. Wood Chem. Technol., 2018, 38(4): 301-312.

[12] Zhang S Y, Xu Q, Wang Z J, et al. Preparation and characterization of anodic alumina films with high-saturation structural colors in a mixed organic electrolyte[J]. Surf. Coat. Technol., 2018, 346: 48-52.

[15] Kim H, Ge J, Kim J, et al. Structural colour printing using a magnetically tunable and lithographically fixable photonic crystal[J]. Nat. Photonics, 2009, 3(9): 534-540.

[16]Takeoka Y. Stimuli-responsive opals: colloidal crystals and colloidal amorphous arrays for use in functional structurally colored materials[J]. J. Mater. Chem. C, 2013, 1(38): 6059-6074.

[17]Du H, Liu C, Sun J, et al. An investigation of angle-dependent optical properties of multi-layer structure pigments formed by metal-oxide-coated mica[J]. Powder Technol., 2008, 185(3): 291-296.

[18] Bai Y, Ling Y, Bai X. Breakdown characteristics of titanium dioxide-silicone-fluorophlogopite nanocomposite coating [J]. Surf. Coat. Technol., 2011, 205(16): 4073-4078.

[19]Su C Y, Tang H Z, Zhu G D, et al. The optical properties and sunscreen application of spherical h-BN-TiO_2/mica composite powder[J]. Ceram. Int., 2014, 40(3): 4691-4696.

[20]Cavalcante P M T, Dondi M, Guarini G, et al. Ceramic application of mica titania pearlescent pigments[J]. Dyes Pigm., 2007, 74(1): 1-8

[21]Gao Q, Wu X, Fan Y, et al. Low temperature synthesis and characterization of rutile TiO_2-coated mica-titania pigments[J]. Dyes Pigm., 2012, 95(3): 534-539.

[22]Kumar V, Tyagi L, Sinha S. Wood flour-reinforced plastic composites: A review[J]. Rev. Chem. Eng., 2011, 27(5-6): 253-64.

[23]Chen Y, Dang B, Wang H, et al. Hydrothermal deposition of $CoFe_2O_4$ with a micro nano binary structure onto a wood surface with related magnetic property and microwave absorption[J]. J. Wood Chem. Technol., 2019, 39(1): 31-42.

[24]Yuan T Q, Sun S N, Xu F, et al. Characterization of lignin structures and lignin-carbohydrate complex (LCC) linkages by quantitative ^{13}C and 2D HSQC NMR spectroscopy[J]. J. Agric. Food Chem., 2011, 59(19): 10604-10614.

[25]Huang K, Luo J, Cao R, et al. Enhanced xylooligosaccharides yields and enzymatic hydrolyzability of cellulose using acetic acid catalysis of poplar sawdust[J]. J. Wood Chem. Technol., 2018, 38(5): 371-384.

[26]Chen Y, Wang H, Yao Q, et al. Biomimetic taro leaf-like films decorated on wood surfaces using soft lithography for superparamagnetic and superhydrophobic performance[J]. J. Mater. Sci., 2017, 52(12): 7428-7438.

[27]Tomak E D, Gonultas O. The wood preservative potentials of valonia, chestnut, tara and sulphited oak tannins[J]. J. Wood Chem. Technol., 2018, 38(3): 183-197.

[28]Wang H, Yao Q, Wang C, et al. Hydrothermal synthesis of nanooctahedra $MnFe_2O_4$ onto the wood surface with soft

magnetism, fire resistance and electromagnetic wave absorption[J]. Nanomater., 2017, 7(6): 118.

[29] Li Y, Hui B, Li G, et al. Fabrication of smart wood with reversible thermoresponsive performance[J]. J. Mater. Sci., 2017, 52(13): 7688-7697.

[30] Li Y, Li J. Fabrication of reversible thermoresponsive thin films on wood surfaces with hydrophobic performance[J]. Prog. Org. Coat., 2018, 119: 15-22.

[31] Miljević B, van der Bergh J M, Vučetić S, et al. Molybdenum doped TiO_2 nanocomposite coatings: Visible light driven photocatalytic self-cleaning of mineral substrates[J]. Ceram. Int., 2017, 43(11): 8214-8221.

[32] Topuz B B, Gündüz G, Mavis B, et al. Synthesis and characterization of copper phthalocyanine and tetracarboxamide copper phthalocyanine deposited mica-titania pigments[J]. Dyes Pigm., 2013, 96(1): 31-37.

[33] Topuz B B, Gündüz G, Mavis B, et al. The effect of tin dioxide (SnO_2) on the anatase-rutile phase transformation of titania (TiO_2) in mica-titania pigments and their use in paint[J]. Dyes Pigm., 2011, 90(2): 123-128.

[34] Jiménez-Aguilar D M, Ortega-Regules A E, Lozada-Ramírez J D, et al. Color and chemical stability of spray-dried blueberry extract using mesquite gum as wall material[J]. J Food Compost Anal, 2011, 24(6): 889-894.

[35] Li Y, Gao L, Li J. Photoresponsive wood-based composite fabricated by a simple drop-coating procedure[J]. Wood Sci. Technol., 2019, 53(1): 211-226.

中文题目：具有反射隔热和疏水性能的结构色木质复合材料

作者：李莹莹，高丽坤，刘迎涛，李坚

摘要：结构色源于材料的物理结构，因此对环境有益，并且不能被化学染料模仿，并且不会发生光漂白。在这项研究中，通过在木质基材上构建结构色膜来制备结构色木材。通过 X 射线衍射(XRD)，扫描电子显微镜(SEM)和傅立叶变换红外光谱(FT-IR)对制备的样品进行表征。通过比较不同视角下的 CIE 色坐标，可以确认与角度相关的光学效果。通过十八烷基三氯硅烷(OTS)处理后，样品表面改为疏水性，并且水接触角为 139°，从而扩大了该材料的应用范围。热重分析结果表明，改性结构色木材(MCW)的热稳定性比原始木材好得多。反射光谱和模型室测试结果表明，MCW 具有反射隔热能力，可潜在地应用于建筑物、家具和节能中。

关键词：结构色；木基质复合材料；节能；疏水性

Natural Phenolic Compound-iron Complexes: Sustainable Solar Absorbers for Wood-based Solar Steam Generation Devices*

He Gao, Mingming Yang, Ben Dang, Xiongfei Luo,
Shouxin Liu, Shujun Li, Zhijun Chen, Jian Li

Abstract: Wood-based solar steam generation devices (W-SSGDs) show great promise for desalination and wastewater treatment since they are cheap and sustainable. The fabrication of green, sustainable and efficient solar-to-thermal materials for use in W-SSGDs, however, remains a challenge. Here, we have developed coordination complexes between Fe^{3+} and naturally occurring phenolic compounds as solar-to-thermal materials. The as-prepared solar-to-thermal material prepared by coordinating Fe^{3+} with catechin showed wide optical absorbance and efficient conversion efficiency, and was stable under different pH conditions. The good photothermal properties of this as-prepared solar-to-thermal material allowed us to construct a high performance W-SSGD that had a steam generation efficiency of 54. 32% and an evaporation rate as high as 0. 9204 $kg \cdot m^{-2} \cdot h^{-1}$.

Keywords: Solar steam generation; Natural phenolic compound; Iron ions; Coordination; Solar-to-thermal conversion

1 Introduction

Solar steam generation techniques show promise for desalination and wastewater treatment.[1-9] Wood-based solar steam generation devices (W-SSGDs), in particular, show great potential for water distillation because they are sustainable, easily prepared and green.[10] W-SSGDs typically have two components, a solar-to-thermal layer and a wood matrix, which serve as thermal insulation and water transportation layer, respectively.[11] Recently, many efficient materials, including carbonous materials, such as carbon nanotubes and graphene, and plasma metals, have been developed as the solar-to-thermal layer for W-SSGDs.[12-24] For example, Li et al.[25] reported an efficient (80% under one-sun illumination) and effective (four orders salinity decrement) solar desalination device. A foldable graphene oxide film, was served as efficient solar absorbers (>94%), vapor channels, and thermal insulators by a scalable process. Chen et al.[26] reported the use of carbon nanotube (CNT)-modified flexible wood membrane (F-Wood/CNTs) is demonstrated as a flexible, portable, recyclable, and efficient solar steam generation device for low-cost and scalable solar steam generation applications. Solar steam generation device based on the F-Wood/CNTs

---

* 本文摘自 RSC Advances，2020，10：1152–1158.

membrane demonstrates a high efficiency of 81% at 10 kW · cm^{-2}, representing one of the highest values ever reported. Nevertheless, the identification of new solar-to-thermal materials, with higher photothermal conversion efficiency, sustainability and easy preparation, for use in W-SSGDs remains an important goal. Recently, biomass-derived catechol-containing compounds have been reported to show efficient photothermal conversion when they are coordinated with metal cations.[27-36] The fabrication of these light-to-thermal materials was convenient, green and cheap. Inspired by this, we now report the successful preparation of new solar-to-thermal materials (termed PCF-n, where n denotes different biomass-derived phenolic compounds), which are formed by coordination between natural phenolic substances and Fe^{3+}, for use in W-SSGDs. As-prepared PCF-1 (coordination between catechin and Fe^{3+}), particularly, showed good thermal stability and efficient photothermal conversion. As-prepared PCF-1 was coated onto the surface of basswood to prepare a W-SSGD (Fig. 1). The as-prepared W-SSGD showed a high steam generation efficiency (54%) under one sun irradiation, with an evaporation rate as high as 0.92 kg · m^{-2} · h^{-1}, which was attributed to the highly efficient photothermal layer. PCF-1 is thus a promising sustainable solar-to-thermal material that can be used for the construction of a high-performance W-SSGD in a convenient and green manner.

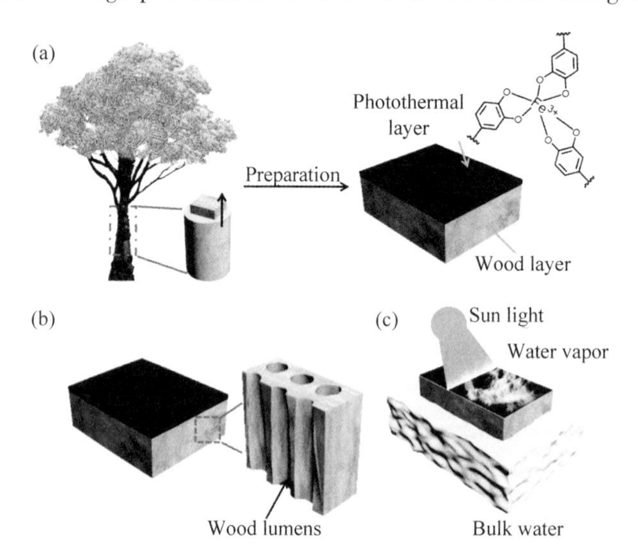

Fig. 1 Schematic illustration of (a) Preparation of W-SSGD; (b) Magnified structure of wood; (c) Solar desalination of sea water

2 Experimental Section

2.1 Materials

The basswood was purchased from Alibaba (Hangzhou, China). Natural phenolics were purchased from Sigma (Shanghai Warehouse, China). All other reagents and solvents were purchased from Merck Life Science Co., Ltd. (Shanghai, China) or Aladdin Bio-Chem Technology Co., Ltd. (Shanghai, China).

2.2 Characterization

SEM images were captured using an FEI Sirion 200 scanning electron microscope (Philips Research, Eindhoven, Netherlands). Light absorbance of the photothermal materials was measured using a Cary 5000

UV-vis-NIR spectrophotometer (Agilent Technologies, Santa Clara, CA, USA) over the spectral range 200–1200 nm. An integrating sphere was used to collect the reflected light. A CEL-S500 xenon lamp (Aulight Co., Ltd., Beijing, China), which simulates solar radiation, was used as the light source. Temperatures were measured using a DTM-180A digital thermometer (Zhaohui Instruments Co., Ltd., Hengshui, China). All photographs were taken using a Huawei P20 mobile phone (Huawei Technologies Co., Ltd., Shenzheng, China).

2.3 Preparation of PCF-n

Natural phenolic compound (catechin, cyanidin, chlorogenic acid and tannic acid, 1 g) was stirred with water until completely dissolved. $FeCl_3 \cdot 6H_2O$ (10% $w \cdot w^{-1}$) was then added to the solution and the mixture was stirred for 30 min at room temperature. PCF-1, 3, 4, 5 were obtained by centrifugation and washed three times with deionized water. For PCF-2, quercetin (1 g) was stirred with ethanol until completely dissolved. $FeCl_3 \cdot 6H_2O$ (10% $w \cdot w^{-1}$) was then added to the solution and the mixture was stirred for 30 min at room temperature. PCF-2 were obtained by centrifugation and washed three times with deionized water.

2.4 Preparation of W-SSGD

The W-SSGD was prepared by coating the surface of a bulk basswood sample (4 cm × 2 cm × 2 cm) with PCF-1 (0.5 g). Specifically, the dispersion of PCF-1 (0.5 g) in water (10 mL) was brushed on the wood surface. After that, the wood was dried at the room temperature for 24 h.

2.5 Solar steam generation efficiency calculation

Typically, steam generation efficiency can be calculated using equation (1):

$$\eta = \frac{h_{LV}}{C_{\text{opt}} q_i} \tag{1}$$

where $\dot{m}$ is the evaporation rate, h_{LV} is the total enthalpy, including both sensible heat and heat used in the phase change of liquid water to steam, C_{opt} is the optical concentration and q_i is the normal solar irradiation (100 $mW \cdot cm^{-2}$).

3 Results and Discussion

PCF-n (n = 1-5) were prepared by coordinating different biomass-derived phenolic compounds, catechin (1), quercetin (2), cyanidin (3), chlorogenic acid (4) and tannic acid (5), with Fe^{3+} (Fig. S1†). A possible structure for the coordination complexes is shown in Fig. 2a. As-prepared PCF-n (n = 1-5) were black powders (Fig. 2b-f) and could be easily produced on a large scale. The UV-Vis-NIR spectra of PCF-n (n = 1-5) were investigated over the range 400-2500 nm, which encompasses the majority of the sun's output. All of the PCF samples showed absorbance over this range (Fig. 2g), with PCF-1 and PCF-2, formed by coordination between catechol and Fe^{3+} and quercetin and Fe^{3+}, respectively, showing the highest absorbance. These two complexes thus have the greatest potential for absorbing and converting solar energy. The phenolic molecules themselves, on the other hand, showed significant absorbance only in the UV-Vis region of the spectrum (wavelength<700 nm), confirming that coordination of biomass-derived phenolic compounds with Fe^{3+} significantly enhances their absorption in the NIR region and thus increases their potential as solar-to-thermal materials. The reason for enhancement of NIR region might be attributed to the characteristic ligand (phenolic compounds)-to-metal (iron) charge transfer (LMCT) caused by coordination. The LMCT enabled the molecule to have low energy gap, which eventually red shifted the absorbance wavelength.[37] PCF-1 and PCF-2 might have strongest LMCT, compared to other complexes, which triggered their nice performance in the NIR absorbance. Following this proposed mechanism, stronger NIR absorbance of PCF-1 and PCF-2 might be caused by their more intensive LMCT. Since PCF-1 and PCF-2 showed the most efficient solar absorbance, they were selected for further investigation. X-ray photoelectron spectroscopic (XPS) analysis showed that as-prepared PCF-1 and PCF-2 contain C, O and Fe atoms (Fig. S2†), indicating successful coordination between the phenolic compounds and Fe^{3+}.

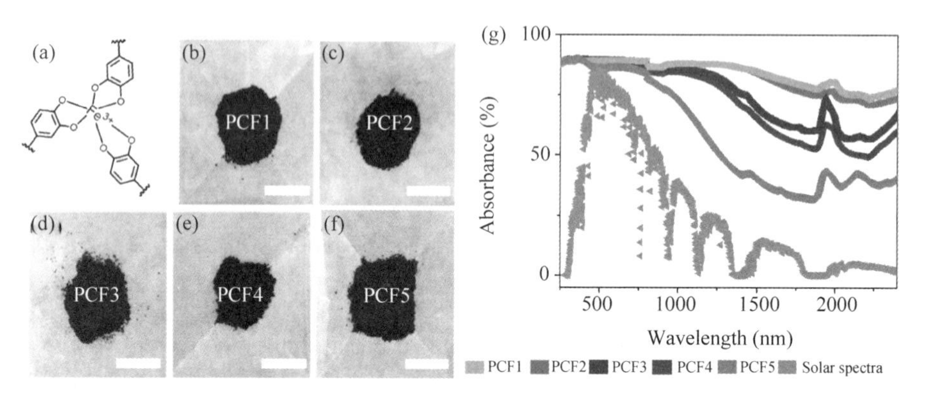

Fig. 2 (a) Possible structure of coordination complexes PCF-n(n=1–5); (b-f) Images of PCF-n(n=1–5) powders, scale bar=1 cm; g) Absorbance spectra of PCF-n(n=1–5) and solar irradiation spectrum

The photothermal properties of PCF–1 and PCF-2 were investigated next, using a xenon lamp to simulate solar radiation. Under standard one sun irradiation (100 mW · cm^{-2}), the surface temperatures of PCF–1 and PCF-2 increased from ~24 ℃ to ~58 ℃ and from ~23 ℃ to ~58 ℃, respectively, over 10 min (Fig. 3a and h), suggesting good photothermal conversion. To evaluate stability, the photothermal effects of PCF-1 and PCF-2 were first studied under different pH conditions. The photothermal properties of PCF-1 were very similar under acidic (pH=4. 2), neutral (pH=7) and basic (pH=9. 3) conditions, indicating that coordination between Fe^{3+} and catechin is stable in aqueous solution over a range of pH values (Fig. 3a and b-g). Additionally, PCF-1 and the samples after acidic/basic treatment did not show the obvious temperature change aer 14 min (Fig. S3†). Even after 1 h, the value did not obviously change, which demonstrated that PCF-1 had nice photothermal stability (PCF-1 without treatment, 55. 2 ℃; PCF-1 after basic treatment, 57. 9 ℃; PCF-1 after acidic treatment 57. 1 ℃) (Fig. S3†). The photothermal efficiency of PCF-2 showed a noticeable decrease under acidic or basic conditions (Fig. 3h and 3i-3n), demonstrating that PCF-2 is less stable in acidic or basic environments. Additionally, the photothermal effect of PCF-1 and PCF-2 was found to be reversible, with both complexes maintaining a good photothermal effect after 15 cycles (Fig. S4 and Fig. S5†).

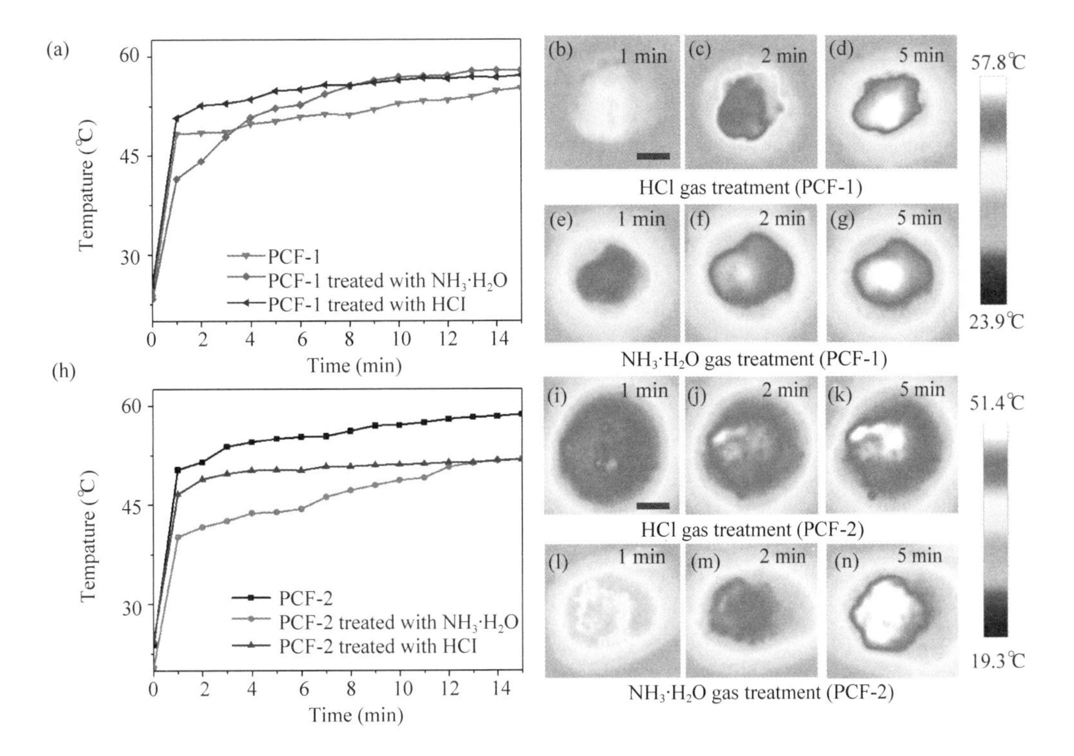

Fig. 3 (a) Temperature change of PCF-1 upon light irradiation at different pH; (b-g) IR images of PCF-1 upon light irradiation; (h) Temperature change of PCF-2 upon light irradiation at different pH; (i-n) IR images of PCF-2 upon light irradiation

Since PCF-1 demonstrated the best photothermal properties, it was subsequently coated onto basswood to construct a W-SSGD for solar steam generation (Fig. S6†). The scanning electron microscope (SEM) images of as-employed wood channels are shown in Fig. 4a-4f. As-prepared W-SSGD was also characterized by SEM (Fig. S6†). SEM results also showed that the PCF-1 still adhered firmly to the wood after acidic (pH=4.2) or alkaline treatment (pH=9.3) and the morphology did not change obviously (Fig. S7†). These strong adhesion for PCF-1 to wood might be attributed to the dopamine-like catechol moieties. The catechol could form efficient hydrogen bond with the hydroxyl moieties of wood surface. Natural basswood has an overall porosity of ~70%. The mesoporous microstructure of basswood consists of lumens (with average size ~50 μm, Fig. 4a-f) that are surrounded by fiber tracheids. A slice of wood 2 cm × 2 cm × 0.2 cm was used to demonstrate fluidic transport across the lumens. The water transportation speed for the basswood reached 0.18 $kg \cdot kg^{-1} \cdot h^{-1}$ (Fig. 4g). The device was then placed in a container of similar size to minimize the effect of evaporation of water surrounding the W-SSGD. The temperature change at the water evaporation surface under one standard sun irradiation was monitored using an infrared camera. The surface temperature of the W-SSGD increased rapidly from ~16 ℃ to ~37 ℃ after simulated solar radiation for 4 min (Fig. 4h-4k). The fact that the temperature almost reached equilibrium after 4 min of irradiation suggests that the W-SSGD has a rapid response to solar radiation. The photothermal stability of the W-SSGD in water was also measured over 15 cycles (Fig. S8†). The temperature increase did not noticeably alter, suggesting that the W-SSGD has good photothermal stability. The W-SSGD constructed using PCF-1 achieved an evaporation rate of 1.25 $kg \cdot m^{-2} \cdot h^{-1}$. The spontaneous evaporation in dark field was 0. 34 kg · m^{-2} · h^{-1}. Based on this, the W-SSGD exhibited an efficiency of 54. 32% under one sun irradiation (Fig. 4i). The stability of the W-SSGD was investigated over 50 cycles. The reading for each cycle was taken after 30 min, when the performance had stabilized (Fig. S 9†). The results showed that W-SSGD was very stable as a solar steam generation device.

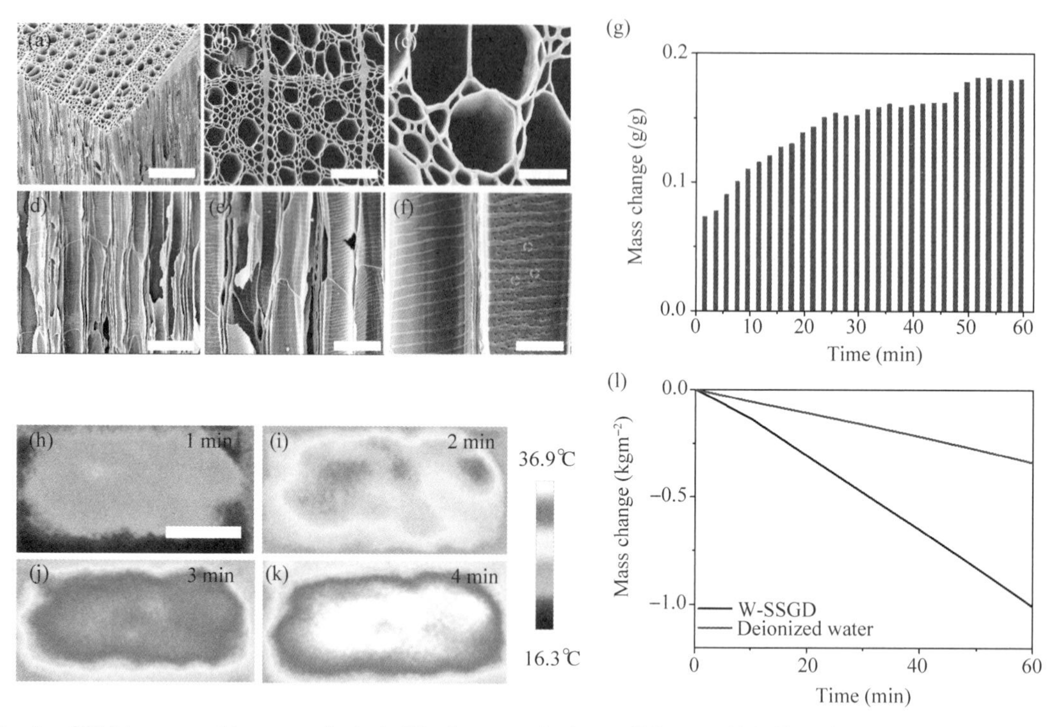

Fig. 4 SEM images of basswood: (a) 3D view, scale bar=400 μm; (b) Top-down view, scale bar=100 μm; (c) Magnified top-down view, scale bar=50 μm; (d) Cross-sectional view, scale bar=150 μm; (e) Magnified cross-sectional view, scale bar=100 μm; (f) Pit structure across the lumens, scale bar=50 μm; (g) Weight changes of basswood in water; (h-k) Temperature changes of W-SSGD upon light irradiation (100 mW cm^{-2}); (l) Mass of water evaporated by W-SSGD under one sun irradiation compared with evaporation of deionized water as the control

Encouraged by this result, the W-SSGD was tested using a real seawater sample from the South China Sea. Before that, PCF-1 was investigated for its stability in sea water. Immersing PCF-1 in the seawater for 24 h did not change its UV-vis spectra (Fig. S10†), indicating its nice stability. The water purification rate by the W-SSGD could be tuned by the intensity of the irradiation. Light intensities of 50 mW · cm^{-2} and 100 mW · cm^{-2} enabled average water purification rates of ~0. 71 kg · m^{-2} · h^{-1} and ~1. 03 kg · m^{-2} · h^{-1}, respectively (Fig. 5a). The water purification rate was as high as 1. 18 kg · m^{-2} · h^{-1} when the irradiation intensity was 150 mW · cm^{-2}. For seawater desalination, the salt deposition should influence the water evaporation. However, Fig. 5 a shows that the mass change rate has no obvious change. That was attributed to the inherent advantages of bimodal porous structure of wood.[38] Taking advantage of the inherent bimodal porous and interconnected microstructures of the balsa wood, rapid capillary transport through the microchannels and efficient transport between the micro-and macrochannels through ray cells and pits in the bimodal evaporator can lead to quick replenishment of surface vaporized brine to ensure fast and continuous clean water vapor generation. The quality of the collected purified water was measured by inductively coupled plasma spectroscopy. The concentration of all four primary ions in seawater (Na^+, Mg^{2+}, K^+ and Ca^{2+}) was reduced by two orders of magnitude after solar desalination (Fig. 5b), demonstrating the effectiveness of solar desalination based on W-SSGD.

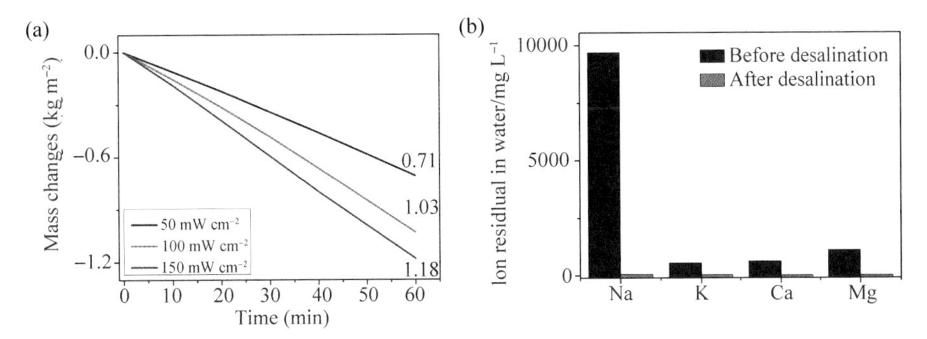

Fig. 5 (a) Mass of water lost from the W-SSGD under different intensity light irradiation; (b) Measured concentrations of four primary ions in seawater sample before and after desalination

4 Conclusions

In summary, we have demonstrated an efficient and stable solar-to-thermal material, PCF-1, which was used to construct a W-SSGD. Under ambient sun conditions, the as-prepared W-SSGD device exhibited an efficiency of~54%, with an evaporation rate of 0.92 kg · m^{-2} · h^{-1} and a water-vapor interface temperature of~ 37 ℃. Since our strategy to prepare such solar-to-thermal materials is easy, cheap and green, this approach toward solar-to-thermal materials is promising for cost-effective real-world application. Moreover, the efficiency might be further enhanced by using other naturally occurring phenolic compounds.

Acknowledgements

This work was supported by the National Key Research and Development Program of China (2018YFD0600302), the Special Project for Double First-Class-Cultivation of Innovative Talents (000/41113102), Hei Longjiang Province Postdoc Funding (LBH-Z18005) and the Young Elite Scientists Sponsorship Program by CAST (2018QNRC001).

References

[1] Wu X, Robson M E, Phelps J L, et al. A flexible photothermal cotton-CuS nanocageagarose aerogel towards portable solar steam generation[J], Nano energy, 2019, 56: 708-715.

[2] Wu X, Gao T, Han C, et al. A photothermal reservoir for highly efficient solar steam generation without bulk water [J], Sci. Bull., 2019, 64: 1625-1633.

[3] Ammar H E, Swellam W S, Mohamed K A A, et al. Thin film technology for solar steam generation: A new dawn[J], Solar Energy, 2019, 177(1): 561-575.

[4] Luo X, Wu D X, Huang C L, et al. Skeleton double layer structure for high solar steam generation[J], Energy, 2019, 183: 1032-1039.

[5] Jin H, Lin G, Bai L, et al. Steam generation in a nanoparticle-based solar receiver[J], Nano Energy, 2016, 28: 397-406.

[6] Liu S, Huang C, Luo X, et al. High-performance solar steam generation of a paper-based carbon particle system[J],

Appl. Therm. Eng., 2018, 142: 566-572.

[7] Luo X, Wu D X, Huang C L, et al. Skeleton double layer structure for high solar steam generation[J], Energy, 2019, 183: 1032-1039.

[8] Shannon M A, Bohn P W, Elimelech M, et al. Science and technology for water purification in the coming decades [J], Nature, 2008, 452: 310-310.

[9] Vivar M, Skryabin I, Everett V, et al. A concept for a hybrid solar water purification and photovoltaic system[J], Sol. Energy Mater. Sol. Cells, 2010, 94: 1772-1782.

[10] Jia C, Li Y J, Yang Z, et al. Rich mesostructures derived from natural wood for solar steam generation[J], Joule, 2017, 1(3, 15): 588-599.

[11] Ghafurian M M, Niazmand H, Ebrahimnia-Bajestan E, et al. Wood surface treatment techniques for enhanced solar steam generation[J], Renewable Energy, 2020, 146: 2308-2315.

[12] Sajadi S M, Farokhnia N, Irajizad P, et al. Flexible artificially-networked structure for ambient/high pressure solar steam generation[J], J. Mater. Chem. A, 2016, 4: 4700-4705.

[13] Zeng Y, Wang K, Yao J, et al. Hollow carbon beads for significant water evaporation enhancement[J], Chem. Eng. Sci., 2014, 116: 704-709.

[14] Deng Z, Miao L, Liu P F, et al. Extremely high water-production created by a nanoink-stained PVA evaporator with embossment structure[J], Nano Energy, 2019, 55: 368-376.

[15] Wan Y, Wang C, Song X, et al. A facile nanocomposite strategy to fabricate a rGO-MWCNT photothermal layer for efficient water evaporation[J], J. Mater. Chem. A, 2018, 6: 963-971.

[16] Bai B L, Yang X H, Tian R, et al. High-efficiency solar steam generation based on blue brick-graphene inverted cone evaporator[J], Applied Thermal Engineering, 2019: 114379.

[17] Yang J, Pang Y, Huang W, et al. Functionalized graphene enables highly efficient solar thermal steam generation [J], ACS Nano, 2017, 11(6): 5510-5518.

[18] Wang X Z, He Y R, Liu X, et al. Enhanced direct steam generation via a bio-inspired solar heating method using carbon nanotube films[J], Powder Technology, 2017, 321: 276-285.

[19] Hou B F, Zhu Z Q, Liu X H, et al. Functionalized carbon materials for efficient solar steam and electricity generation [J], Mater. Chem. Phys., 2019, 222: 159-164.

[20] Zhang C, Gong L, Xiang L, et al. Deposition and adhesion of polydopamine on the surfaces of varying wettability[J], ASC Appl. Mater. Interfaces, 2017, 9: 30943-30950.

[21] Wang X, He Y, Liu X, et al. Solar steam generation through bio-inspired interface heating of broadband-absorbing plasmonic membranes[J], Appl. Energy, 2017, 195: 414-425.

[22] Liu C, Huang J, Hsiung C E, et al. Fratalocchi, High-Performance Large-Scale Solar Steam Generation with Nanolayers of Reusable Biomimetic Nanoparticles[J], Adv. Sustainable Syst., 2017, 1: 1600013.

[23] Chang C, Yang C, Liu Y, et al. Efficient solar-thermal energy harvest driven by interfacial plasmonic heating-assisted evaporation[J], ACS Appl. Mater. Interfaces, 2016, 8: 23412-23418.

[24] Kiriarachchi H D, Awad F S, Hassan A A, et al. Plasmonic chemically modified cotton nanocomposite fibers for efficient solar water desalination and wastewater treatment[J], Nanoscale, 2018, 10: 18531-18539.

[25] Li X, Xu W, Tang M, et al. Graphene oxide-based efficient and scalable solar desalination under one sun with a confined 2D water path[J], Proc. Natl. Acad. Sci. U. S. A., 2016, 113: 13953-13958.

[26] Chen C, Li Y, Song J, et al. Highly flexible and efficient solar steam generation device[J], Adv. Mater., 2017, 29: 1701756.

[27] Luo X F, Ma C H, Chen Z J, et al. Biomass-derived solar-to-thermal materials: promising energy absorbers to convert light to mechanical motion[J], J. Mater. Chem. A, 2019, 7: 4002-4008.

[28] Wang C, Bai J, Liu Y, et al. Polydopamine coated selenide molybdenum: A new photothermal nanocarrier for highly effective chemo-photothermal synergistic therapy[J], ACS Biomater. Sci. Eng., 2016, 2: 2011-2017.

[29] Rossella F, Soldano C, Bellani V, et al. Metal-filled nanotubes as a novel class of photothermal nanomaterials[J], Adv. Mater., 2012, 24: 2453-2458.

[30] Wu X, Jiang Q, Ghim D, et al. Localized heating with a photothermal polydopamine coating facilitates a novel membrane distillation process[J], J. Mater. Chem. A, 2018, 6: 18799-18807.

[31] Lei W, Ren K, Chen T, et al. Polydopamine nanocoating for effective photothermal killing of bacteria and fungus upon near-infrared irradiation[J], Adv. Mater. Interfaces, 2016, 3: 1600767.

[32] Pérez-Mitta G, Tuninetti J S, Knoll W, et al. Polydopamine meets solid-state nanopores: A bioinspired integrative surface chemistry approach to tailor the functional properties of nanofluidic diodes[J], J. Am. Chem. Soc., 2015, 137: 6011-6017.

[33] Zheng R, Wang S, Tian Y, et al. Polydopamine-coated magnetic composite particles with an enhanced photothermal effect[J], ACS Appl. Mater. Interfaces, 2015, 7: 15876-15884.

[34] Lin L S, Cong Z X, Cao J B, et al. Multifunctional Fe_3O_4@polydopamine core-shell nanocomposites for intracellular mRNA detection and imaging-guided photothermal therapy[J], ACS Nano, 2014, 8: 3876-3883.

[35] Kang S M, Park S, Kim D, et al. Simultaneous reduction and surface functionalization of graphene oxide by mussel-inspired chemistry[J], Adv. Funct. Mater., 2011, 21: 108-112.

[36] Wu X, Chen G Y, Zhang W, et al. A plant-transpiration-process-inspired strategy foe highly efficient solar evaporation[J], Adv. Sustain. Syst., 2017, 1(6): 1700046.

[37] Rahim M A, ornmalm M, Bertleff-Zieschang N, et al. Multiligand Metal-Phenolic Assembly from Green Tea Infusions[J], ACS Appl. Mater. Interfaces, 2017, 10: 7632-7639.

[38] He S, Chen C, Kuang Y, et al. Natureinspired salt resistant bimodal porous solar evaporator for efficient and stable water desalination[J], Energy Environ. Sci., 2019, 12: 1558-1567.

中文题目：天然酚醛化合物-铁配合物：木质太阳能蒸汽发生装置的可持续太阳能吸收剂

作者：高鹤，杨明明，党奔，罗雄飞，刘守新，李淑君，陈志俊，李坚

摘要：木质太阳能蒸汽发生装置(W-SSGDs)具有价格低廉，可持续发展的特点，有望用于海水淡化和废水处理。然而，制造用于 W-SSGDs 的绿色，可持续和高效的太阳热能材料仍然是一个挑战。在这里，我们开发了 Fe^{3+} 与作为太阳热材料的天然酚类化合物之间的配位配合物。Fe^{3+} 与儿茶素配位制得的太阳热材料具有很宽的光吸收率和有效的转化效率，并且在不同的 pH 条件下稳定。这种制备的太阳能热材料具有良好的光热性能，使我们能够构建出高性能的 W-SSGD，其蒸汽产生效率为 54.32%，蒸发速率高达 0.9204 $kg \cdot m^{-2} \cdot h^{-1}$。

关键词：产生太阳蒸汽；天然酚类化合物；铁离子；配位；太阳热转换

03

第三部分
生物质基气凝胶

新型木质纤维素气凝胶的制备、表征及疏水吸油性能研究*

万才超，卢芸，孙庆丰，李坚

摘要：为制备一种新型的木质纤维素气凝胶，采用化学预处理、溶解再生与冷冻干燥相结合的方法对废弃的麦秸杆进行提纯、溶解、置换和干燥，并采用绿色、无毒、低廉的氢氧化钠/聚乙二醇溶液作为纤维素溶剂。采用扫描电子显微镜(SEM)、BET比表面积分析、X射线衍射仪(XRD)、傅里叶变换红外吸收光谱(FTIR)和热重分析(TGA)对制备的新型木质纤维素气凝胶的微观形貌、比表面积与孔径分布、晶型结构、化学结构及热稳定性进行表征。结果表明，制备的新型木质纤维素气凝胶具有连续、层叠的三维网状结构，比表面积为99.17 $m^2 \cdot g^{-1}$，总孔容为0.45 $cm^3 \cdot g^{-1}$；纤维素气凝胶的晶型由纤维素Ⅰ型转变为纤维素Ⅱ型，结晶度为72.3%，相对于原料提高了23.4%，热稳定性也略微升高；并利用三甲基氯硅烷(TMCS)进行疏水改性，制备出了具有疏水性能的纤维素气凝胶。文中提供了一种新的制备木质纤维素气凝胶的有效溶剂，且具有高吸附性能、高承重能力、高结晶度的纤维素气凝胶是一种具有较大应用潜力的新型功能材料。

关键词：木质纤维素气凝胶；疏水；吸油；纤维素溶剂；溶解再生

中国是世界农业大国，农作物秸秆的年产量高达8亿t以上，居世界各国秸秆总产量之首，但利用率极低。据统计，2008年，国内秸秆可收集利用总量中，直接燃用量21000万t，占32.26%；饲料用量17660万t，占27.13%；废弃和焚烧量10922万t，占16.78%；直接还田量9200万t，占14.13%；工业生产利用量4300万t，占6.61%；食用菌培养利用量1300万t，占2.00%；新能源开发利用量720万t，占1.11%。数据表明，直接燃用、废弃和焚烧量近50%，而新能源开发利用量仅占1.11%。因此，利用农作物秸秆开发新能源、新材料，实现秸秆的高值化利用，仍存在很大的发展空间[1]。

纤维素是自然界中最常见的有机聚合物，可利用的纤维素资源大约有1900亿t[2]。纤维素是构成植物细胞壁的主要成分之一，因此，广泛存在于各种天然植物中。农作物秸秆中，纤维素含量约为40%；木材中，纤维素的含量约为45%；此外，棉花中纤维素的含量高达98%。因此，在石油资源逐渐枯竭、环境污染越来越严重的当下，大力发展绿色环保、性能优良的天然纤维素材料，提高纤维素资源的综合利用，被认为是有效的解决途径之一。纤维素气凝胶作为新生的第3代气凝胶，兼具了硅气凝胶[3]和聚合物基气凝胶[4,5]的高孔隙率、高比表面积、低声速、低光折射率、低介电常数和低热传导系数等优良性质[6-9]，同时融入了自身的优异性能，如良好的生物相容性、可降解性和亲水性[10,11]，在制药、处理原油泄漏、制备绿色的隔音材料等领域[12,13]具有很大的应用前景(图1)。但是，在纤维素气凝胶的制备过程中，由于纤维素结晶区的存在，使得纤维素很难溶于一般的简单溶剂，这一点也严重地限制了纤维素的推广与应用[14]。

20世纪，虽然很多的新溶剂体系被发现，如液氨/硫氰酸铵(NH_3/NH_4SCN)[15]，金属硫氰酸盐[$Ca(SCN)_2$，NaSCN，KSCN，LiSCN][16,17]，氯化锂/N-N-二甲基乙酰胺(LiCl/DMAC)[18]和二甲基亚砜(DMSO)[19]等，但是由于挥发性、毒性和高成本等原因，大部分的溶剂体系仍局限于实验室范围

* 本文摘自科技导报，2014，32(4)：79-85.

内应用。2008 年，Yan 等[20]报道了一个新型、绿色和低成本的纤维素溶解体系，即氢氧化钠/聚乙二醇(NaOH/PEG)体系，氢氧化钠与聚乙二醇以质量比 9：1 混合，通过搅拌、冻融等方式可以获得均一的纤维素水溶液，其纤维素的最大溶解度可达 13%。2010 年，Zhang 等[21]通过氢氧化钠/聚乙二醇(NaOH/PEG)体系处理棉绒，制备了一种新型的纤维素纤维，证明了氢氧化钠/聚乙二醇水溶液具有优异纤维素溶解性能。

本研究以废弃的麦秸秆为原料，制备纤维素气凝胶，并探索其疏水吸油性能，旨在实现农、林作物废弃物的高值化利用。本研究先采用化学方法预处理麦秸秆以脱除秸秆细胞壁内的半纤维素、木质素及少量的抽提物，再利用氢氧化钠/聚乙二醇(NaOH/PEG)体系在室温下溶解纤维素，并通过搅拌、冻融等方式获得溶解度为 2%的均一的纤维素溶液。同时，通过溶解再生法和冷冻干燥法，制备出超轻的麦秸秆纤维素气凝胶。并使用扫描电子显微镜(SEM)、BET 比表面积分析、X 射线衍射仪(XRD)、傅里叶变换红外光谱(FTIR)和热重分析(TGA)这些常规的检测方法，对制备的麦秸秆纤维素气凝胶的微观形貌、比表面积与孔径分布、晶型结构、化学结构和热稳定性进行分析表征。

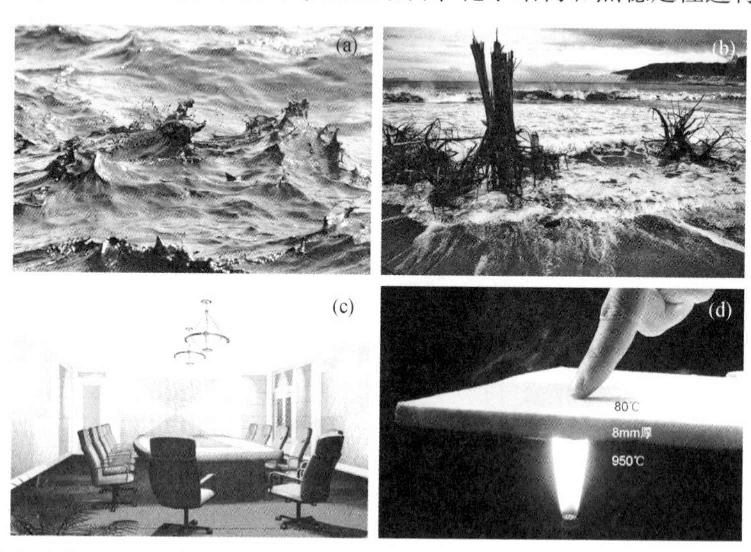

图 1　纤维素气凝胶在(a)原油泄漏，核(b)废水，(c)隔音材料，(d)隔热保温材料上的潜在应用

1　材料与方法

1.1　原料

麦秸秆(wheat straw)购自江苏连云港美味农家绿色农产品店，在室温下干燥并粉碎，过 60 目筛，作为纤维素气凝胶的原料。试验中所用氢氧化钠、聚乙二醇(DP=4000)、盐酸、叔丁醇、乙醇、苯、亚氯酸钠、醋酸和氢氧化钾均为分析纯，购自天津科密欧化学试剂开发中心。

1.2　纤维素的化学提纯

麦秸秆粉末中纤维素的提纯过程如图 2 所示。分析天平准确称取 2 g 麦秸秆粉末，置于索氏抽提器中，利用体积比为 2：1 的苯/乙醇溶液在 90 ℃下抽提 6 h，脱除抽提物。然后，利用亚氯酸钠在酸性条件下(pH 值 4~5，冰醋酸调节)脱除木质素，得到棕纤维素。最后，利用氢氧化钾处理棕纤维素，脱除半纤维素制成纯化纤维素[22]。

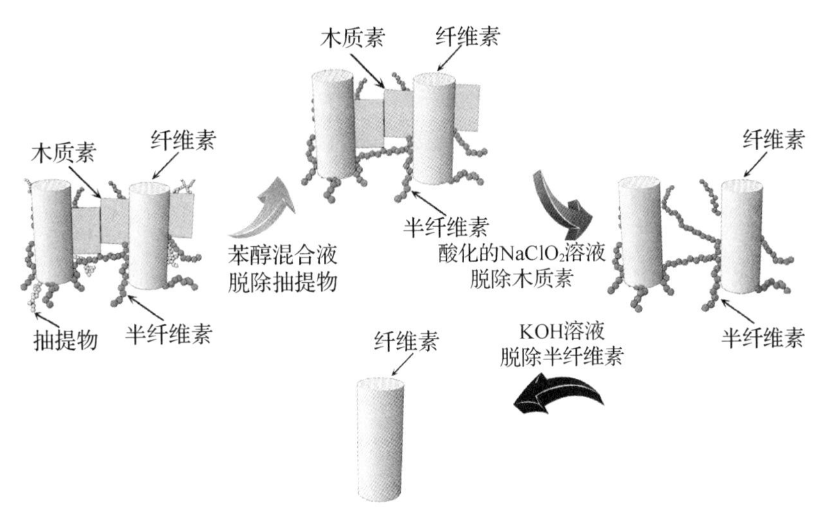

图2 麦秸秆纤维素化学提纯过程示意图

1.3 纤维素气凝胶的制备

纤维素气凝胶的制备过程如图3所示。首先将经过干燥处理的纤维素分散在质量比为9∶1的氢氧化钠/聚乙二醇溶液中，室温下自由搅拌5 h，溶解度为2%。其次，将得到的混合溶液冷却至-10 ℃，并在该温度下冰冻12 h，得到冰冻的固体。冰冻完成后取出，在室温下解冻并剧烈搅拌30 min，得到均一的纤维素水溶液。将纤维素水溶液在-10 ℃下再次冰冻12 h后，依次使用1%的盐酸、蒸馏水和叔丁醇进行置换，得到圆柱状的白色水凝胶。最后，将水凝胶在-55 ℃下冷冻后于25 Pa下冷冻干燥。干燥后得到多孔性的纤维素气凝胶。

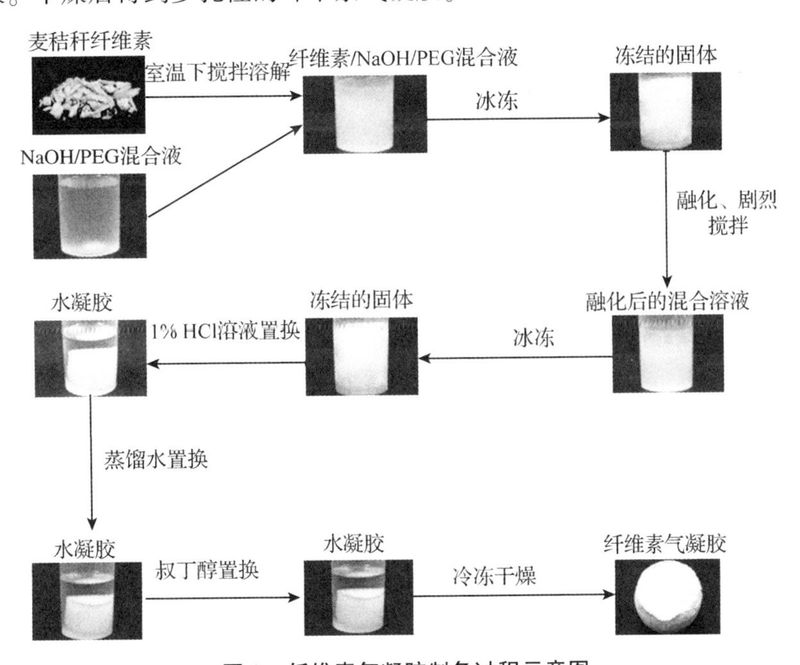

图3 纤维素气凝胶制备过程示意图

1.4 纤维素气凝胶的吸油性能测试与疏水改性

将纤维素气凝胶切成规格为直径30 mm×厚度10 mm的块状圆柱体，称重后置入废弃的机油中浸泡 48 h。浸泡完全后，将固体取出，用滤纸轻轻刮掉表面上残留的机油，称重。取另一块纤维素气凝胶，在干燥器内利用三甲基氯硅烷(TMCS)处理 24 h，制备出具有疏水性能的纤维素气凝胶[23,24]。

1.5 表征

制备出的纤维素气凝胶的微观形貌结构采用美国 FEI 公司的 Quanta200 型扫描电镜(SEM)观察，高真空模式，工作电压 12.5 kV，束斑 5.0。

原材料和纤维素气凝胶的晶型结构和结晶度采用日本理学的 D/MAX220 型 X 射线衍射仪(XRD)进行表征，测试采用铜靶，射线波长为 0.154 nm，扫描角度范围 2θ 为 5°~40°，扫描速度为 4°·min^{-1}，步距 0.02°，管电压为40kV，管电流为 30 mA。结晶度采用 Segal 法经验公式[25]计算。

原材料和纤维素气凝胶的化学组分采用美国 Nicolet Magna 560(Nicolet, Thermo Electron Corp)型傅里叶变换红外光谱仪(FTIR)进行表征，所有的表征样品都研磨成粉，溴化钾压片。波长范围为 500~4000 cm^{-1}，分辨率为 4 cm^{-1}。

原材料和纤维素气凝胶的热稳定性采用美国 TA 公司 SDT Q600 型同步热分析仪进行表征。温度范围为 22~700 ℃，升温速率为 20 ℃·min^{-1}；载气为高纯氮气，气体流量为 100 mL·min^{-1}。

原材料和纤维素气凝胶的 N_2 吸、脱附测量采用美国 Tristar Ⅱ 3020 吸附仪进行，测试前，样品在真空条件下 100 ℃预先脱气 6 h。BET 比表面积通过 the Brunauer-Emmett-Teller (BET)方法计算得知。孔容(Vt)和孔径(D)由等温线吸附分支采用 the Barrett-Joyner-Halenda (BJH)模型计算所得，其中孔体积用相对压力 p/p_0=0.99 处的吸附量计算得到。

2 结果与分析

如图 4 所示，以废弃的麦秸秆为原材料，通过抽提—溶解—冰冻—解冻—冰冻—置换—冷冻干燥的过程，制备出了新型的木质纤维素气凝胶，其密度为 0.057 g·cm^{-3}。同时，这种以氢氧化钠/聚乙二醇体系为溶剂制备的纤维素气凝胶，具有良好的承重能力且不易破碎，1.22 g 该样品可以承受 25 N 以上的压力。

图 4 麦秸秆纤维素及纤维素气凝胶

(a)化学提纯的麦秸秆纤维素 (b)纤维素气凝胶 (c)受压状态下的纤维素气凝胶

为了考查纤维素气凝胶的吸油性能，本研究采用了一个简易的吸油试验(图 5)。试验结果表明，吸油后样品的质量达到了吸油前的 14.21 倍，这一数值与市售的吸附剂相近。同时，吸油前后，样品的尺寸变化不大，且无明显的破损、塌陷现象。此外，通过三甲基氯硅烷处理获得的具有疏水性能的纤维素气凝胶对水、罗丹明 B、甲基橙、咖啡、牛奶等物质均具有较明显的疏水作用(图 6)。因此，这种基于氢氧化钠/聚乙二醇体系制备的纤维素气凝胶，由于自身良好的疏水吸油性能、尺寸稳定性、抗压且不易破坏，使其在处理原油泄漏、实现油水分离领域具有较好的应用前景。

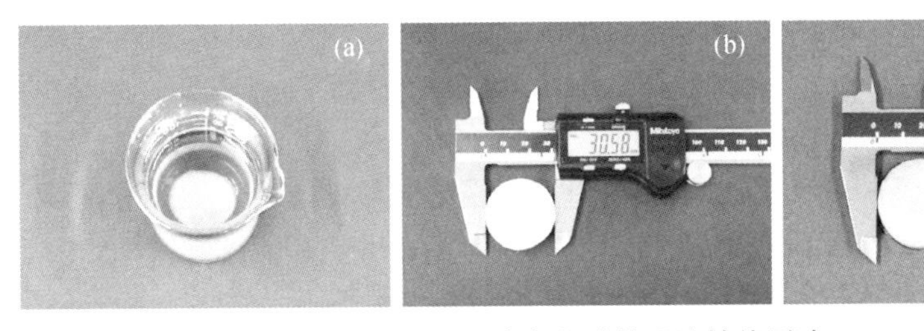

图 5　纤维素气凝胶的吸油性能测试

(a)吸油状态　(b)吸油前尺寸测量　(c)吸油后尺寸测量

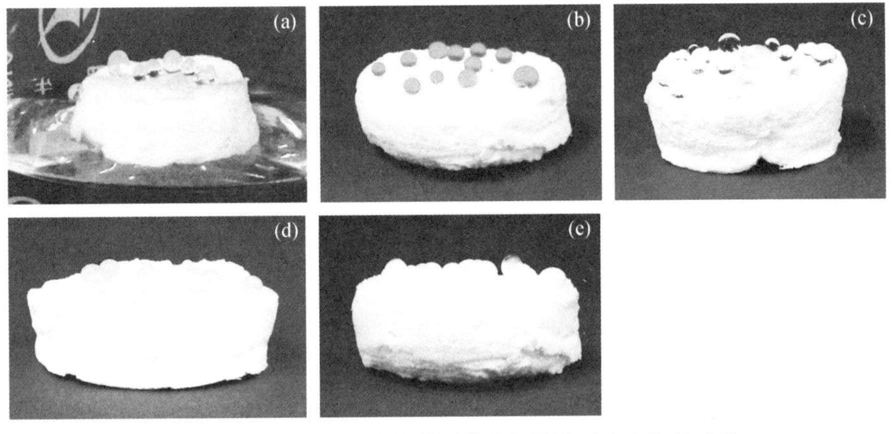

图 6　三甲基氯硅烷改性的纤维素气凝胶的疏水性能试验

(a)水　(b)罗丹明 B　(c)甲基橙　(d)咖啡　(e)牛奶

2.1　纤维素气凝胶的微观形貌结构

图 7 为采用冷冻干燥所得到的纤维素气凝胶样品的 SEM 照片。如图 7(a)所示，可以观察到连续的、致密的、多孔性的三维网状结构。在冷冻干燥过程中，样品的微观结构在原始位置上冰冻，而孔隙结构中的叔丁醇逐渐升华，最终被空气替代[26]。整个过程处于真空、低温(25 Pa，−55 ℃)状态，叔丁醇升华时产生的表面张力极小，有效地保留了纤维素气凝胶的三维网状结构，同时也减缓了样品的皱缩[27]。此外，网络中孔的大小分布略显不均，这可能是由于依次采用多种溶剂置换时产生的表面张力不均，导致有些网络结构的坍陷，造成孔洞的大小不一。放大倍数较高的图 7(b)中可以观察到相互层叠、交织的三维网状结构，表明可能存在较高的比表面积和发达的孔隙结构，有利于获得优良的吸油性能。

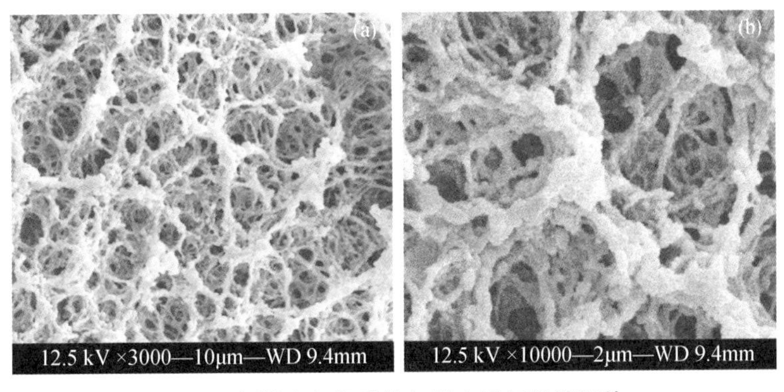

图 7　纤维素气凝胶的扫描电子显微镜照片

(a)3000×　(b)放大倍数为 10000×

2.2 纤维素气凝胶和原料的孔径分布

液氮温度下的 N_2 吸附分析为多孔材料的微观结构的探究提供了许多有价值的信息。经冷冻干燥得到的纤维素气凝胶的比表面积为 99.17 $m^2 \cdot g^{-1}$，总孔容为 0.45 $cm^3 \cdot g^{-1}$。木质纤维素气凝胶的 N_2 吸附-脱附等温线如图 8 所示。根据 IUPAC 气体吸附等温线的分类标准[28]，图 8 的吸附等温线符合Ⅱ型吸附曲线，表明孔的直径大于 10 nm。使用 BJH 方法确定材料的孔径分布，分析结果表明，材料的平均孔隙直径为 16.47 nm，属于介孔材料。

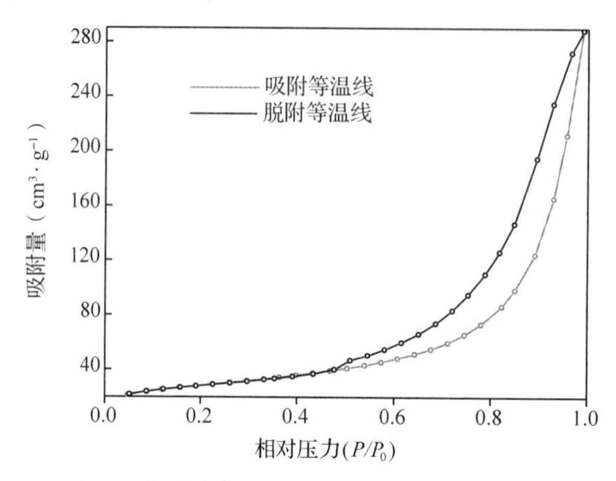

图 8　纤维素气凝胶的 N_2 吸附-脱附等温线

2.3 纤维素气凝胶和原料的晶型与化学组成

图 9(a)显示了麦秸秆、化学提纯的麦秸秆纤维素与纤维素气凝胶的 XRD 衍射谱图。其中麦秸秆和化学提纯的纤维素呈现典型的天然纤维素Ⅰ型结构，在 22.18°附近出现的峰对应此结构的典型晶面(002)；此外，在 15.74°附近出现的峰是此结构的典型晶面(101)和($10\bar{1}$)的复合。而经过溶解再生法得到的纤维素气凝胶的晶型结构明显发生改变，由纤维素Ⅰ型结构转变为纤维素Ⅱ型结构；纤维素Ⅱ型结构的典型晶面(101)、($10\bar{1}$)和(002)的峰位在 12.34°、20.22°和 21.96°附近出现[20]。根据 Segal 法经验公式，麦秸秆的结晶度为 48.9%，而提纯的纤维素的结晶度升高至 67.1%，这是因为无定形的半纤维素、木质素和少量抽提物的脱除，引起样品结晶度的相对升高。此外，纤维素气凝胶的结晶度为 72.3%，表明氢氧化钠/聚乙二醇体系的溶解处理及随后的置换处理并没有降低样品的结晶度反而使样品的结晶度略微升高。

麦秸秆、化学提纯的麦秸秆纤维素与纤维素气凝胶的 FTIR 光谱如图 9(b)所示。由图 9(b)可以看出，麦秸秆、提纯的纤维素与纤维素气凝胶的红外谱图在 3313 cm^{-1} 附近均出现强烈的吸收峰，代表聚合物上 O—H 伸缩振动。此外，相对于麦秸秆和纤维素气凝胶，提纯的纤维素的红外谱图在 3313 cm^{-1} 附近的吸收峰的强度明显较高，这是因为前期的氢氧化钾脱半纤维素处理破坏了大量的分子内氢键所引起的；同时，纤维素气凝胶的红外谱图在此处的吸收峰的强度较低，这可能是因为纤维素上的部分羟基与聚乙二醇上的羟基之间形成了分子间氢键。此外，在 2883、1430、1353 和 893 cm^{-1} 附近的吸收峰，分别代表 C—H 伸缩振动、CH_2 剪式振动、C—H 弯曲振动和 β-D 葡萄糖苷键，均为纤维素的特征吸收峰[29]。其中，麦秸秆的红外谱图在 2883 cm^{-1} 附近分成两个强度稍低的吸收峰，这是因为木质素的存在所引起的。同时，麦秸秆的红外谱图在 1727 cm^{-1} 和 1636 cm^{-1} 附近具有明显的吸收峰，分别代表木质素侧链羰基的吸收谱带和半纤维素的 C═O 伸缩振动；而提纯的纤维素与纤维素气凝胶的红外谱图在这两个峰位处的吸收峰的强度明显减弱，甚至消失，表明亚氯酸钠和氢氧化钾处理有效地脱除了原组分中木质素和半纤维素[30]。此外，麦秸秆和提纯的纤维素在 1430 cm^{-1} 处均具有较明显的吸收峰，代表纤维素Ⅰ型结构；而纤维素气凝胶的红外谱图在此处的吸收峰减弱并向低波数方向移动，预示纤维素Ⅱ型结构的形成[21]。

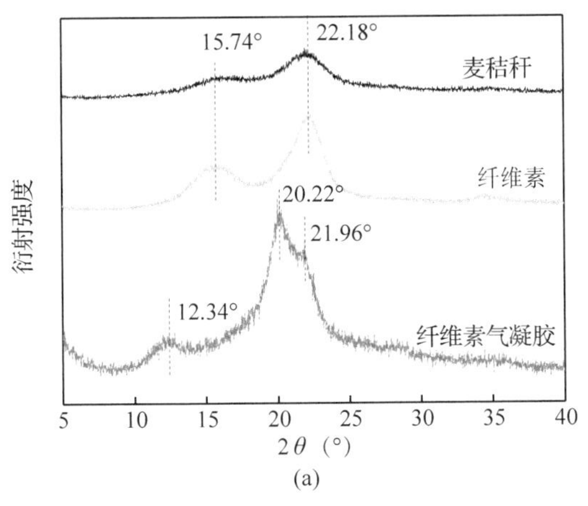

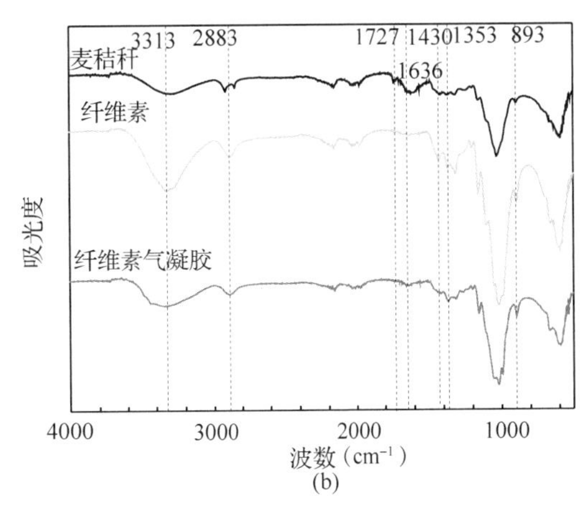

图9 麦秸秆、纤维素与纤维素气凝胶的X射线衍射谱图(a)和FTIR谱图(b)

2.4 纤维素气凝胶和原料的热稳定性

麦秸秆、化学提纯的纤维素与纤维素气凝胶的TG与DTG曲线如图10所示。在所有的TG曲线中，150 ℃之前的微小失重均是由于样品内部水分的蒸发造成的。由麦秸秆的TG和DTG曲线可以得出，麦秸秆的第一阶段和第二阶段降解分别发生在266 ℃和325 ℃附近，降解率分别为13%和33%。对于提纯的纤维素和纤维素气凝胶，均只存在一步降解；从图10(b)中可以得出，两者的最大峰值位置在350 ℃附近。由于半纤维素的降解主要发生在200~300 ℃，且半纤维素的降解速率比纤维素的降解速率慢[31]，因此麦秸秆在热解过程中形成了两个降解阶段[32]。此外，提纯的纤维素和纤维素气凝胶的DTG曲线在350 ℃附近形成的最大峰值与纤维素的降解有关。如图10所示，麦秸秆的降解速率比提纯的纤维素和纤维素气凝胶明显慢，且麦秸秆的剧烈降解阶段的起始温度也相对较低，表明前期的亚氯酸钠和氢氧化钾处理一定程度上提高了提纯的纤维素和纤维素气凝胶的热稳定性。同时，提纯的纤维素和纤维素气凝胶的TG和DTG曲线相似，说明氢氧化钠/聚乙二醇体系的溶解处理及随后的置换处理并没有降低样品的热稳定性。此外，由图10(a)中可以看出，麦秸秆在700 ℃处的残渣含量最高，约为15%，这是因为麦秸秆中木质素的降解发生在整个过程中，在700 ℃时木质素仍有残渣剩余[33]。而纤维素气凝胶在此温度下的残渣含量比纤维素的残渣含量高，这可能是由于在氢氧化钠/聚乙二醇溶液处理过程中引入了聚乙二醇分子了。

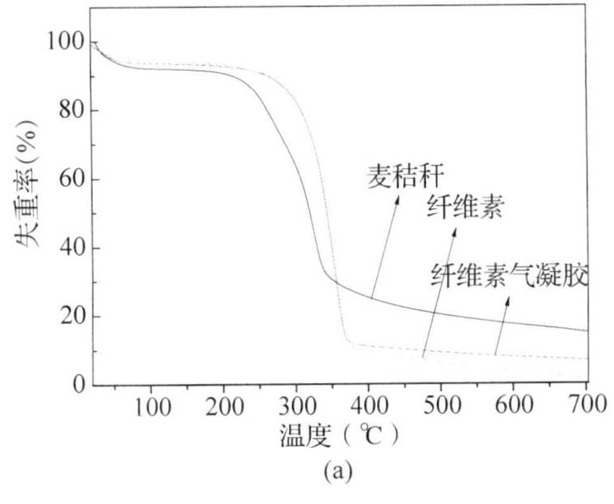

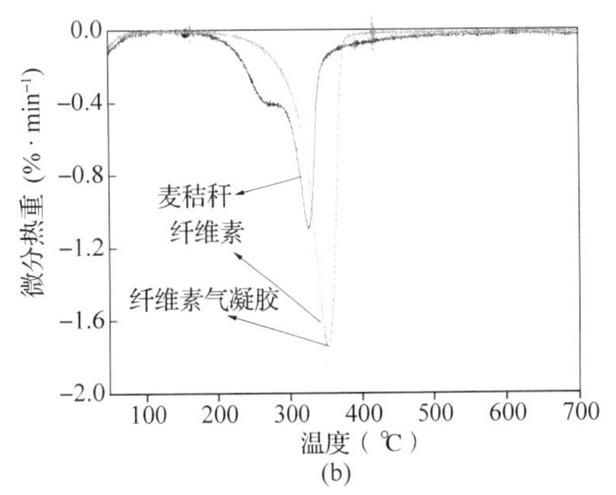

图10 麦秸秆、纤维素与纤维素气凝胶的TG(a)和DTG(b)曲线

3 结论

以化学处理提纯的麦秸秆纤维素为原料，通过溶解再生法制备出多孔性的纤维素气凝胶。本文中制备出的纤维素气凝胶具有优良的成型、承重和吸附能力，在扫描电子显微镜下可以观察到连续、层叠的三维网状结构，比表面积为 99.17 $m^2 \cdot g^{-1}$，总孔容为 0.45 $cm^3 \cdot g^{-1}$。且经过三甲基氯硅烷改性后，具有良好的疏水性能。纤维素气凝胶的晶型从天然的纤维素Ⅰ型结构转变为纤维素Ⅱ型结构，结晶度为 72.3%，相对于原料提高了 23.4%；同时，纤维素气凝胶的热稳定性也略微升高。证明了绿色、无毒、低廉的氢氧化钠/聚乙二醇溶液不仅可以作为良好的纤维素溶剂，而且在处理过程中并未降低样品的结晶度和热稳定性。因此，试验制备的新型纤维素气凝胶可以作为绿色功能材料，在处理核废水、原油泄漏等领域具有良好的应用前景。除了材料本身具有的功能特性，该材料有望成为先进材料功能化的新平台。

参考文献

[1] Bi Y Y. Study on straw resources evaluation and utilization in China[D]. Beijing: Chinese Academy of Agricultural Sciences, 2010.

[2] Browne R, Saxena I. Cellulose: Molecular and structural biology: selected articles on the synthesis, structure and applications of cellulsoe[M]. Dordrecht: Springer, 2007.

[3] Schaefer D W, Keefer K D. Structure of random porous materials: silica aerogel[J]. Physical review letters, 1986, 56(20): 2199-2022.

[4] Lee J K, Gould G L. Polydicyclopentadiene based aerogel: a new insulation material[J]. Journal of Sol-Gel Science and Technology, 2007, 44(1): 29-40.

[5] Biesmans G, Randall D, Francais E, et al. Polyurethane-based organic aerogels' thermal performance[J]. Journal of Non-Crystalline Solids, 1998, 225: 36-40.

[6] Tamon H, Ishizaka H, Mikami M, et al. Porous structure of organic and carbon aerogels synthesized by sol-gel polycondensation of resorcinol with formaldehyde[J]. Carbon, 1997, 35(6): 791-796.

[7] Gronauer M, Fricke J. Acoustic properties of microporous SiO_2-aerogel[J]. Acta Acustica United with Acustica, 1986, 59(3): 177-181.

[8] Hrubesh L W, Pekala R W. Thermal properties of organic and inorganic aerogels[J]. Journal of Materials Research, 1994, 9(3): 731-738.

[9] Fischer F, Rigacci A, Pirard R, et al. Cellulose-based aerogels[J]. Polymer, 2006, 47(22): 7636-7645.

[10] Ding B, Cai J, Huang J, et al. Facile preparation of robust and biocompatible chitin aerogels[J]. Journal of Materials Chemistry, 2012, 22(12): 5801-5809.

[11] Pinkert A, Marsh K N, Pang S, et al. Ionic liquids and their interaction with cellulose[J]. Chemical Reviews, 2009, 109(12): 6712-6728.

[12] Schwertfeger F, Zimmermann A, Krempel H . Use of inorganic aerogels in pharmacy: US, US6280744 B1[P]. 2001.

[13] Korhonen J T, Kettunen M, Ras R H A, et al. Hydrophobic nanocellulose aerogels as floating, sustainable, reusable, and recyclable oil absorbents[J]. ACS applied materials & interfaces, 2011, 3(6): 1813-1816.

[14] Lindman B, Karlström G, Stigsson L. On the mechanism of dissolution of cellulose[J]. Journal of molecular liquids, 2010, 156(1): 76-81.

[15] Cuculo J A, Smith C B, Sangwatanaroj U, et al. A study on the mechanism of dissolution of the cellulose/NH3/NH4SCN system. I[J]. Journal of Polymer Science Part A: Polymer Chemistry, 1994, 32(2): 229-239.

[16] Hattori M, Koga T, Shimaya Y, et al. Aqueous calcium thiocyanate solution as a cellulose solvent. Structure and interactions with cellulose[J]. Polymer Journal, 1998, 30(1): 43-48.

[17] Frey M W, Li L, Xiao M, et al. Dissolution of cellulose in ethylene diamine/salt solvent systems[J]. Cellulose, 2006, 13(2): 147-155.

[18] Matsumoto T, Tatsumi D, Tamai N, et al. Solution properties of celluloses from different biological origins in LiCl · DMAc[J]. Cellulose, 2001, 8(4): 275-282.

[19] Johnson D C, Nicholson M D, Haigh F C. Dimethyl sulfoxide/paraformaldehyde: a nondegrading solvent for cellulose[J]. 1975.

[20] Yan L, Gao Z. Dissolving of cellulose in PEG/NaOH aqueous solution[J]. Cellulose, 2008, 15(6): 789-796.

[21] Zhang S, Li F X, Yu J Y. Structure and properties of novel cellulose fibres produced from NaOH/PEG-treated cotton linters[J]. Iranian Polymer Journal, 2010, 19(12): 949-957.

[22] Lu Yun, Sun Qingfeng, Li Jian. Preparation and Characterization of Nanofiber Films and Foams Based on Ultrasonic Nanofibrillated Cellulose from Wood[J]. Science & Technology Review, 2013, 31(15): 17-22.

[23] Schwertfeger F, Frank D, Schmidt M. Hydrophobic waterglass based aerogels without solvent exchange or supercritical drying[J]. Journal of Non-Crystalline Solids, 1998, 225: 24-29.

[24] Shi F, Wang L, Liu J. Synthesis and characterization of silica aerogels by a novel fast ambient pressure drying process [J]. Materials Letters, 2006, 60(29-30): 3718-3722.

[25] Weimer P J, Hackney J M, French A D. Effects of chemical treatments and heating on the crystallinity of celluloses and their implications for evaluating the effect of crystallinity on cellulose biodegradation[J]. Biotechnology and bioengineering, 1995, 48(2): 169-178.

[26] Mellor J D. Fundamentals of freeze-drying[M]. Academic Press Inc. (London) Ltd., 1978.

[27] Inoue T, Osatake H. A new drying method of biological specimens for scanning electron microscopy: The t-butyl alcohol freeze-drying method[J]. Arch Histol Cytol, 1988, 51(1): 53-59.

[28] Hu X, Hu K, Zeng L, et al. Hydrogels prepared from pineapple peel cellulose using ionic liquid and their characterization and primary sodium salicylate release study[J]. Carbohydrate Polymers, 2010, 82(1): 62-68.

[29] Yang H, Yan R, Chen H, et al. Characteristics of hemicellulose, cellulose and lignin pyrolysis[J]. Fuel, 2007, 86 (12): 1781-1788.

[30] Li J, Lu Y, Yang D, et al. Lignocellulose aerogel from wood-ionic liquid solution (1-allyl-3-methylimidazolium chloride) under freezing and thawing conditions[J]. Biomacromolecules, 2011, 12(5): 1860-1867.

[31] Ren J, Sun R, Liu C, et al. Acetylation of wheat straw hemicelluloses in ionic liquid using iodine as a catalyst[J]. Carbohydrate Polymers, 2007, 70(4): 406-414.

[32] Rodrigues Filho G, Monteiro DS, Meireles CdS, et al. Synthesis and characterization of cellulose acetate produced from recycled newspaper[J]. Carbohydrate Polymers, 2008, 73(1): 74-82.

[33] Tejado A, Pena C, Labidi J, et al. Physico-chemical characterization of lignins from different sources for use in phenol-formaldehyde resin synthesis[J]. Bioresource Technology, 2007, 98(8): 1655-1663.

英文题目： Preparation and Characterization of the Novel Lignocellulose Aerogel with Hydrophobicity and Oil Absorption Property

作者： WAN Caichao, LU Yun, SUN Qingfeng, LI Jian

摘要： In order to obtain the novel lignocellulose aerogel, the raw material, namely waste wheat straw, was purified, dissolved, replaced and dried in sequence via corresponding chemical pretreatment, dissolution and regeneration as well as freeze drying, furthermore, a green, non-toxic and inexpensive NaOH/PEG aqueous solution was chose to dissolve cellulose. Morphological feature, pore size distribution, crystal form, chemical construction and thermostability of the novel lignocellulose aerogel are analyzed by the Scanning Electron Microscopy (SEM), the BET measurement, the X-Ray Diffraction (XRD), the Fourier Transform Infrared Spectroscopy (FTIR) and the Thermogravimetric Analysis (TGA). It is shown that the obtained novel lignocellulose aerogel has continuous and tiered 3D network structure, moreover, its specific surface area reaches 99.17 $m^2 \cdot g^{-1}$, and total pore volume reaches 0.45 $cm^3 \cdot g^{-1}$. Crystal form of the novel lignocellulose aerogel is transformed from the cellulose Ⅰ crystalline structure to cellulose Ⅱ crystalline structure, and the crystallinity reaches 72.3%, increased by 23.4% as compared with the raw material straw. Meanwhile, the thermostability is slightly improved. Moreover, trimethylchlorosilane (TMCS) was used to hydrophobically modify the lignocellulose aerogel. In this article, a new and effective solvent fabricating the lignocellulose aerogel is offered, and the novel lignocellulose aerogel having superior adsorptive property, excellent weight capacity and high crystallinity has great application potential as a new-style functional material.

关键词： Lignocellulose aerogel; Hydrophobicity; Oil absorption property; Cellulose solvent; Dissolution and regeneration

Embedding ZnO Nanorods into Porous Cellulose Aerogels via A Facile One-step Low-Temperature Hydrothermal Method*

Caichao Wan, Jian Li

Abstract: A facile effective one-step low-temperature hydrothermal approach was employed to in situ embed ZnO nanorods into the porous cellulose aerogels. Besides, the preparation of cellulose aerogels is based on a green NaOH/polyethylene glycol solution. The rod-like ZnO has average diameter of about 348 nm and length of about 1.49 μm, and display wurtzite phase. Meanwhile, the scanning electron microscope and transmission electron microscopy observations confirm that the nanorods are tightly anchored to the aerogels matrixes, and exhibit good dispersion without dramatic agglomeration, indicating that the cellulose aerogels are a class of idea green matrix materials to support nanoparticles. Moreover, the method might also be extended to fabricate other multifunctional cellulose-based nanocomposites.

Keywords: Zinc oxide; Cellulose aerogels; Hydrothermal method; Nanocomposites; Polymers

1 Introduction

Nanoscale zinc oxide (ZnO) is a class of promisin and high-profile inorganic nanomaterials. The lack of a center of symmetry in wurtzite, combined with large electromechanical coupling, results in strong piezoelectric and pyroelectric properties and the consequent use of ZnO in mechanical actuators and piezoelectric sensors[1]. Moreover, a direct wide band gap (3.37 eV) and large exciton binding energy (60 meV) of this material ensure that it is a promising candidate for stable room temperature luminescent and lasing devices[2]. Besides, ZnO is also a strong antibacterial material[3]. Currently, ZnO nanostructures (nanorod, nanowire, nanotube, nanobelt, nanopropellers, etc.) could be fabricated by various methods such as catalytic growth via thermal evaporation, cyclic feeding chemical vapor deposition, vapor-liquid-solid method, template-based growth, and hydrothermal process[4-7]. Among these methods, undoubtedly, hydrothermal method is widely used because of its easy and economical process.

Recently, ZnO nanostructures combined with various organic polymers (e.g., nylon fibers, cotton fabrics, thermoplastic polyurethane, and aromatic polyamide fiber) have been the subject of increasing interest[8,9]. The resulting nanocomposites are expected to display synergistically improved properties by combining the attractive functionalities of both components. Cellulose aerogels are a class of important cellulose products, whose raw materials (namely cellulose) are widely available from a wide range of native biomass resources. The dissolution of cellulose is a necessary process for the preparation of cellulose aerogels. However, chemical processing of cellulose is rather difficult, due to the complexity of biopolymeric network, the partially crystalline structure and the extended non-covalent interactions among molecules. To date, a series of

* 本文摘自 Materials & Design, 2015, 83: 620-625.

effective cellulose solvents have been successively found[10]; hereinto, a green low-cost solvent named NaOH/PEG solution is worthy of attention, which is recently reported to have a strong dissolution capacity of cellulose[11]. Based on our previous research[12], the aerogels prepared from the cellulose-NaOH/PEG solution are composed of abundant micro/nano-scale pores, and this unique architectural feature endows themselves with low density (54 mg · cm^{-3}), large specific surface area (204 m^2 · g^{-1}), and high pore volume (0.99 m^3 · g^{-1}). In addition, the cellulose aerogels also have plentiful oxygen-containing functional groups on the surface, which could serve as nano-reacting sites where inorganic nanoparticles could be anchored tightly to avoid particles agglomeration and achieve controlled growth[13]. Therefore, exterior and interior decoration of cellulose aerogels with nano-ZnO will lead to a new class of nanocomposites that inherit numerous merits from the both aerogels and ZnO, which might be useful for adsorbents, UV protection, construction of superhydrophobic surface, etc.

In this study, a facile one-step low-temperature hydrothermal method was employed to in situ embed ZnO nanorods into the porous cellulose aerogels, and a green NaOH/PEG solution was used to fabricate the cellulose hydrogels. We utilized transmission electron microscopy (TEM) and scanning electron microscope (SEM) to characterize the microstructures of the hybrid ZnO nanorods/cellulose (ZRC) aerogels. The surface chemical compositions and crystal structures were investigated by energy dispersive X-ray (EDX) analysis, X-ray photoelectron spectroscopy (XPS), associated selected-area electron diffraction (SAED), and X-ray diffraction (XRD). Additionally, the pore structural features were measured by nitrogen adsorption measurements. Needless to say, the research on incorporation of ZnO nanorods into pure cellulose aerogels matrixes via a hydrothermal treatment is still very rare, which is the subject of this study.

2 Materials and methods

2.1 Materials

Native wheat straw was grinded and then screened through a sixty mesh sieve, and the resulting powder was subsequently collected and dried in a vacuum oven at 60 ℃ for 24 h before used. Chemicals including benzene, absolute ethyl alcohol, sodium chlorite ($NaClO_2$), glacial acetic acid (CH_3COOH), potassium hydroxide (KOH), hydrochloric acid (HCl, 37%), sodium hydroxide (NaOH), polyethylene glycol (PEG) 4000, zinc nitrate hexahydrate [$(Zn(NO_3)_2 \cdot 6H_2O]$, ammonium hydroxide ($NH_3 \cdot H_2O$, 25%), and tert-butyl alcohol are all of analytical reagent grade and purchased from Shanghai Boyle Chemical Co. Ltd. (China).

2.2 Purification of cellulose from wheat straw

The purification of the cellulose from wheat straw was performed following the procedures of benzene/ethanol extraction, bleaching process, alkali treatment and acid hydrolysis process[14-16]. Briefly, the dried wheat straw (2 g) was first treated by a mixed aqueous solution of 2 : 1 V/V% benzene/ethanol at 90 ℃ for 6 h via a soxhlet extractor, keeping the liquid boiling briskly so that siphoning from the extractor was no less than four times per hour. Then, the treated sample was air-dried and treated with 65 mL 1% $NaClO_2$ solution (pH = 4–5 adjusted by glacial acetic acid) at 75 ℃ for 5 h. Thereafter, the resulting residue was collected by filtration and subsequently rinsed with deionised water three times, and then immediately treated with 50 mL 2% NaOH at 90 ℃ for 2 h. The next step was to collect and wash the product again before treating it with 30 mL 1% HCl at 80 ℃ for 2 h. The acid treatment could further hydrolyze the residual hemicelluloses and pectin by breaking down the polysaccharides to simple sugars and hence release the cellulose fibres, and neutralise residual alkali ions from NaOH and remove mineral traces like Na element[17,18]. Finally, the residue was collected and washed again and dried at 60 ℃ for 24 h, and the following relatively pure cellulose was obtained.

2.3 Preparation of cellulose hydrogel

The purified cellulose was added to a 10% aqueous solution of NaOH/PEG-4000 (9 : 1 wt/wt) with magnetic stirring for 5 h to form a homogeneous solution. Then, the cellulose solution was frozen for 12 h at −15 ℃ and subsequently thawed at ambient temperature with vigorous stirring for 30 min. After being frozen for 5 h at −15 ℃ again, the frozen cake was immersed in the 1% HCl solution for 6 h, and this process was repeated until the formation of white hydrogel. Finally, the hydrogel was rinsed with a great deal of distilled water to remove superfluous hydrogen ions and chlorine anions, and the following clean cellulose hydrogel was obtained[12,19].

2.4 Preparation of ZRC aerogel

The precursory Zn^{2+} solution was obtained by dissolving one grams of zinc nitrate hexahydrate in 100 mL distilled water with magnetic stirring and then adjusting the pH of the solution to 9.5 by dropwise adding ammonium hydroxide solution (25%), and the mixed solution was sequentially stirred for 1 h at room temperature. After that, the aforementioned solution and the cellulose hydrogel were transferred to a Teflon lined steel autoclave and maintained at 120 ℃ for 6 h. After the hydrothermal treatment, the autoclave was allowed to cool to room temperature, and the following product was successively rinsed several times with distilled water and tert butyl alcohol, respectively. Finally, the product was subjected to a freezing dry process at −30 ℃ in a vacuum for 48 h to remove all liquids, and the following ZRC aerogel was obtained. For comparison, the pure cellulose (PC) aerogels were prepared following the aforementioned method and also subjected to the hydrothermal treatment using distilled water instead of the precursory Zn^{2+} solution.

2.5 Characterizations

XPS spectra were recorded using a Thermo Escalab 250Xi XPS spectrometer (Germany). Deconvolution of the overlapping peaks was performed using a mixed Gaussian-Lorentzian fitting program (Origin 8.5, Originlab Corporation). XRD spectroscopy was carried out on a XRD instrument (Rigaku, D/MAX 2200) operating with Cu *Kα* radiation ($\lambda = 1.5418$Å) at a scan rate (2θ) of 4° min^{-1} and the accelerating voltage of 40 kV and the applied current of 30 mA ranging from 5° to 80°. TEM observation and SAED were performed with a FEI, Tecnai G2 F20 TEM with an accelerating voltage of 100 kV. SEM observation was performed with a Hitachi S4800 SEM equipped with an EDX detector for element analysis. Nitrogen adsorption measurements were implemented at −196 ℃ using an accelerated surface area and porosimetry system (3H-2000PS2 unit, Beishide Instrument S&T Co., Ltd).

3 Results and discussion

ZnO nanostructure was fabricated and embedded into the porous cellulose aerogels by the facile one-step low-temperature hydrothermal method. Some involved chemical reactions are summarized in the following equations:

$$NH_3 \cdot H_2O \leftrightarrow NH_4^+ + OH^- \tag{1}$$

$$Zn^{2+} + 2OH^- \leftrightarrow Zn(OH)_2 \downarrow \tag{2}$$

$$Zn(OH)_2 + 4NH_3 \cdot H_2O \leftrightarrow Zn(NH_3)_4^{2+} + 4H_2O + 2OH^- \tag{3}$$

$$Zn(OH)_2 \leftrightarrow ZnO + H_2O \tag{4}$$

When ammonium hydroxide was firstly introduced into the solution, $Zn(OH)_2$ sediment was formed in solution due to the attraction of Zn^{2+} and OH^-, as shown in Equation (1) and (2). The resulting $Zn(OH)_2$ was subsequently disappeared under stirring on account that Zn^{2+} ions were combined with NH_{4+} ions and transformed into stable zinc ammine, as shown in Equation (3). Moreover, ammonium hydroxide plays a two-fold role in the reaction process, which could not only offer OH^- as the source of O in ZnO, but also form a complex with zinc ions as a buffering mechanism to slowly release Zn^{2+}[20]. Furthermore, the high temperature and high pressure in the autoclave would promote and eventually lead to the formation and growth of ZnO [Equation (4)].

The chemical states of the compositional elements in ZRC aerogels are surveyed by XPS analysis. In Fig. 1a, it can be seen that several peaks are found corresponding to Zn 2p, Zn LMN, Zn 3s, Zn 3p, Zn 3d, C 1s, and O 1s, indicating the presence of zinc, oxygen and carbon in ZRC aerogels. From the Zn 2p core-level XPS spectrum (Fig. 1b), the doublet spectral lines of Zn 2p are observed at the binding energy of 1021.6 eV (Zn 2p3/2) and 1044.7 eV (Zn 2p1/2) with a spin-orbit splitting of 23.1 eV, which reveals that the zinc element is in the oxidation state Ⅱ, corresponding to Zn^{2+}(ZnO)[21]. From the O 1s spectrum (Fig. 1c), the two oxygen species with significant different intensity are present in the near-surface region. The peak at around 530.0 eV is attributed to oxygen in the ZnO crystal lattice, and the another peak at around 532.9 eV is assigned to chemisorbed oxygen resulted from plentiful surface hydroxyls[22]. Consequently, the results confirm that ZnO has been successfully fabricated and combined with cellulose aerogels by the hydrothermal method.

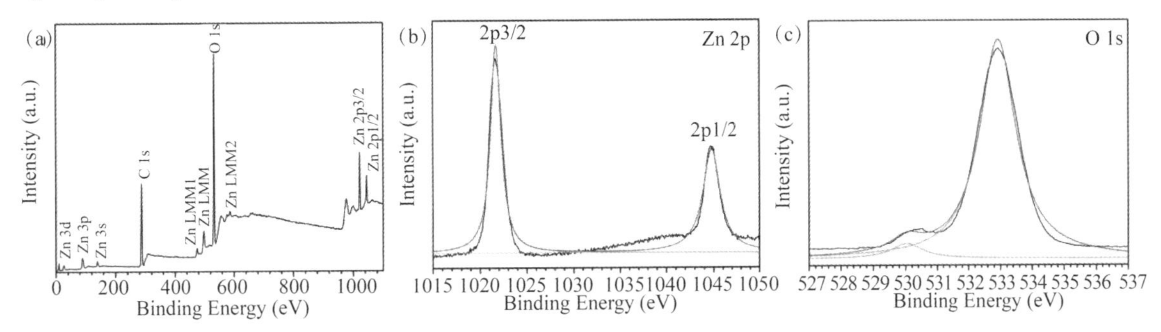

Fig. 1 (a) XPS survey spectrum, high resolution scan of (b) Zn 2p, and (c) O 1s of ZRC aerogels

The XRD patterns of PC aerogels and ZRC aerogels are shown in Fig. 2. Apparently, PC aerogels exhibit typical cellulose Ⅱ crystal structure with characteristic peaks at around 12.4°, 20.2°, and 22.1° corresponding to ($\bar{1}$10), (110), and (200) planes, respectively. Whereas for ZRC aerogels, apart from the cellulose characteristic peaks, the diffraction peaks at around 31.9°, 34.6°, 36.4°, 47.6°, 56.7°, 62.9°, 66.4°, 68.0°, 69.2°, 72.7°, and 77.1° have been keenly indexed as hexagonal wurtzite phase of ZnO (JPCDS file No. 36-1451)[23], which are assigned to the (100), (002), (101), (102), (110), (103), (200), (112), (201), (004), and (202) planes, further indicating that the wurtzite ZnO nano-crystalline has been successfully incorporated into the porous cellulose aerogels. In addition, unlike some previous reports[24,25] that exhibit overwhelming (002) peak resulted from the preferred orientation growth of the ZnO crystals along the *c*-axis, the peak intensity ratio of (100), (002), and (101) is 56 : 49 : 100 in this study, which is approximately consistent with the values from the aforementioned JCPDS card (57 : 44 : 100). The result reveals no apparent preferred orientation growth. Moreover, the preferred growth of the (*hkl*) planes has been expressed in terms of the texture coefficient $T_{c(hkl)}$[26,27]. Quantitative information concerning the preferential crystallite orientation can been obtained from the texture coefficient, T_c, defined as:

$$T_{c(hkl)} = (I_{(hkl)}/I_{r(hkl)})/[1/n\sum(I_{(hkl)}/I_{r(hkl)}) \tag{5}$$

where $T_{c(hkl)}$ is the texture coefficient, I_{hkl} are the XRD intensities obtained from ZRC aerogels, and n is the number of considered diffraction peaks. $I_{r(hkl)}$ are the intensities of the XRD reference (JCPDS card 36-1451). If $T_{c(hkl)} \approx 1$ for all the (*hkl*) planes considered, then the ZnO nanorods are with a randomly oriented crystallite similar to the JCPDS reference, while values higher than 1 indicate the abundance of grains in a given (*hkl*) direction. Values 0<$T_{c(hkl)}$<1 indicate the lack of grains oriented in that direction. As $T_{c(hkl)}$ increases, then the greater is the preferential growth of the crystallites in the direction perpendicular to the (*hkl*) plane. Since six diffraction peaks were used [(100), (002), (101), (102), (110), (103)], the maximum value $T_{c(hkl)}$ possible is 6. Table 1 presents the I_{hkl} and $I_{r(hkl)}$ of the aforementioned six diffraction peaks. The calculated result shows that the $T_{c(002)}$ is 1.06≈1, which reveals that the ZnO nanorods in ZRC aerogels are with a randomly oriented crystallite. The subsequent SEM and TEM observations also confirm the randomly orientation growth of ZnO nanorods. Besides, the strong interaction between particle-substrate and the different plane growth sites (i.e., cross-linked cellulose chains surface) might be responsible for this randomly orientation[28].

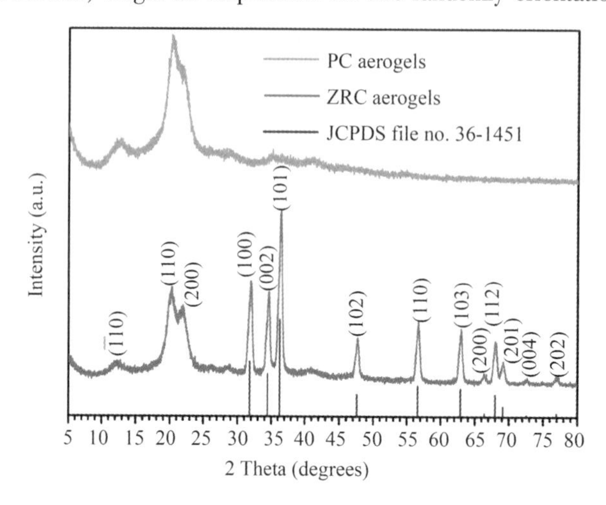

Fig. 2 XRD patterns of PC aerogels and ZRC aerogels, respectively

Table 1 The I_{hkl} and $I_{r(hkl)}$ of the peaks of (100), (002), (101), (102), (110), and (103).

	31.8°, (100) *	34.6°, (002)	36.4°, (101)	47.7°, (102)	56.7°, (110)	62.9°, (103)
I_{hkl}	55.5	49.4	100	24.2	35.4	32.1
$I_{r(hkl)}$	57	44	100	23	32	29

* The 2θ and corresponding crystal plane.

The pore structure characteristics were investigated by nitrogen adsorption measurements. In Fig. 3a, the both isotherms are of type Ⅳ isotherm according to the IUPAC classification, typical for mesoporous structures. Besides, the both hysteresis loops do not exhibit limiting adsorptions at high pressures, and belong to type H3 hysteresis loops which are indicative of slit-shaped pores[29]. Fig. 3b presents the pore diameter distributions, which were calculated from the data of the adsorption branch of the isotherm using Barrett-Joyner-Halenda (BJH) method. As shown, the pore diameter distribution diagrams of PC aerogels and ZRC aerogels are extremely similar, which have the pore diameters of 3-90 nm and exhibit two peaks at around 3.8 nm and 9.6 nm. The specific surface area was calculated over a relative pressure range of 0.05 – 0.30 from multipoint Brunauer-Emmett-Teller (BET) method. The nitrogen adsorption volume at the relative pressure (P/P_0) of 0.994 was used to determine the pore volume. PC aerogels have the specific surface area and pore diameter of around 120 $m^2 \cdot g^{-1}$ and 0.62 $m^3 \cdot g^{-1}$. The results are relatively lower compared with our previous report (204 $m^2 \cdot g^{-1}$ and 0.99 $m^3 \cdot g^{-1}$)[12], which is possibly attributed to the collapse of partial pores caused by the hydrothermal treatment (high temperature and high pressure). Compared with PC aerogels, the specific surface area and pore volume of ZRC aerogels are lower (ca. 115 $m^2 \cdot g^{-1}$ and 0.54 $m^3 \cdot g^{-1}$) notwithstanding that the incorporation of ZnO is potentially helpful to reduce the collapse and shrinkage of pore during the hydrothermal and drying processes due to the strong interaction between the ZnO and the aerogels, and thus improve the aforementioned pore characteristic parameters. However, the presence of alkali ions in high concentration would further exacerbate the collapse of pore structure, reducing the specific surface area and pore volume of ZRC aerogels.

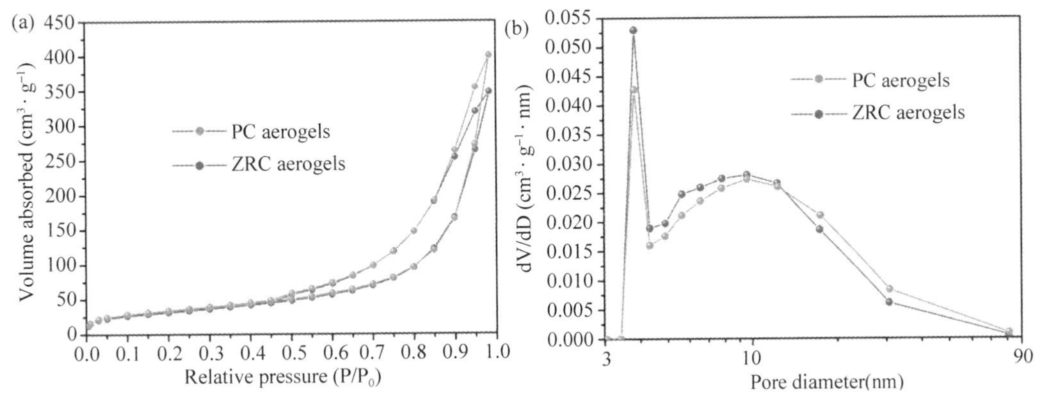

Fig. 3 (a) Nitrogen adsorption-desorption isotherms and (b) pore diameter distributions of PC aerogels and ZRC aerogels, respectively

From the SEM image in Fig. 4a, b, ZRC aerogels maintain abundant porous structure similar to that of PC aerogels after the hydrothermal treatment, and the pore diameters are around several dozens of nanometers, consistent with the results of nitrogen adsorption tests. Besides, the plentiful ZnO nanorods are generated and tightly anchored to the aerogels matrixes, which exhibit good dispersion without significant aggregation. From the insets, there is no obvious difference in appearance between PC aerogels and ZRC aerogels, and the both samples keep well-defined shape indicating the good molding ability. The bulk density is calculated by dividing the sample weight by the sample volume. The results show that the hybrid aerogels have higher density of 76.6 $mg \cdot cm^{-3}$ than that of PC aerogels (60.7 $mg \cdot cm^{-3}$). Fig. 4c shows the EDX spectra of PC aerogels and ZRC aerogels. Apart from the common C and O elements from the cellulose, a sharp peak assigned to Zn element could be seen, which reveals the presence of ZnO. Moreover, the enhancement in the intensity of oxygen peak of ZRC aerogels is another evidence for the presence of ZnO. According to the SEM image, the diameter and length distributions of the ZnO nanorods in ZRC aerogels are plotted and presented in Fig. 4d. The histograms show Gaussian-like distributions, and the mean diameter (d) and standard deviation (σ) of diameter and length are estimated to be 348 ± 61 nm and 1.49±0.28 μm, respectively, which are similar to some previous reports based on similar hydrothermal process[30,31].

The TEM observation further confirms the results of SEM image. As shown in Fig. 5a, the rod-like ZnO nanostructures are tightly adhered to the porous aerogels matrixes without dramatic agglomeration, further demonstrating the strong interaction between the aerogels and the ZnO nanorods. Moreover, Fig. 5b shows the SAED pattern, in which five main diffraction rings correspond to the (100), (101), (102), (103) and (112) planes of polycrystalline ZnO with a hexagonal structure. Besides, the diffraction rings are discontinuous and consist of rather sharp spots, indicating that the nanorods are well crystallized[32].

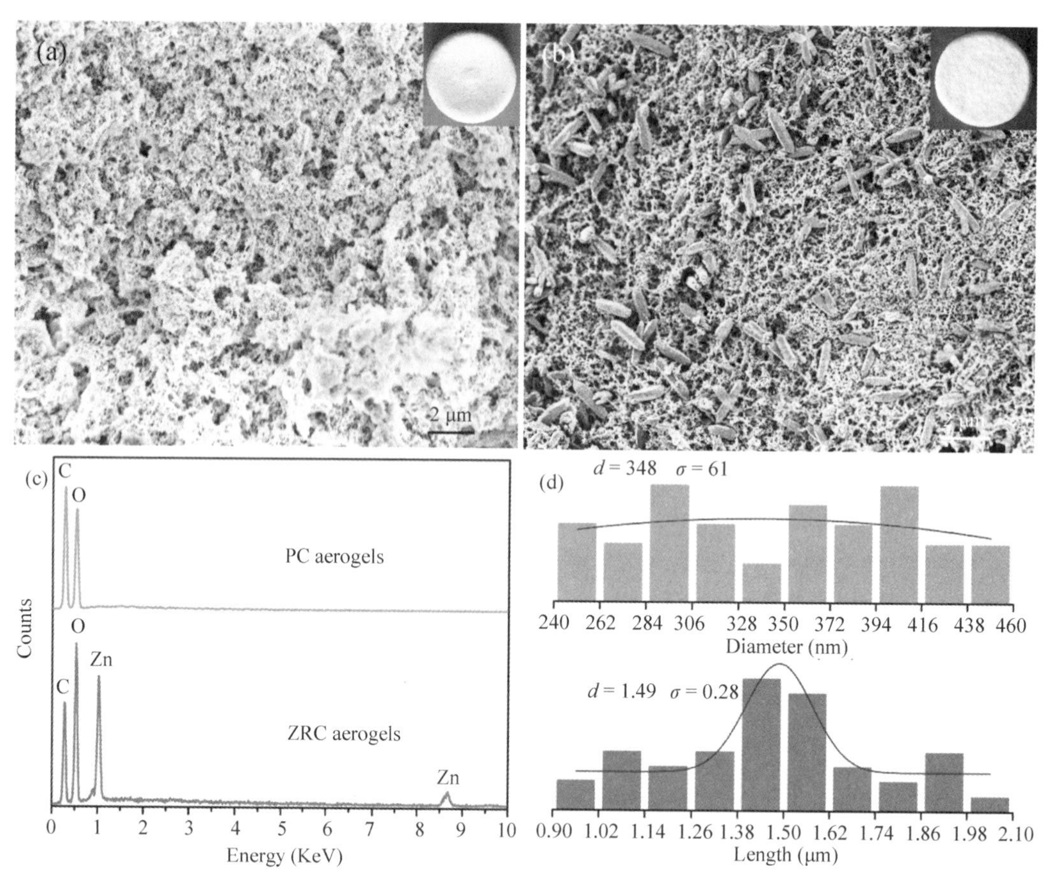

Fig. 4 (a, b) SEM images of PC aerogels and ZRC aerogels, respectively. The insets show the corresponding macroscopic images. (c) EDX spectra of PC aerogels and ZRC aerogels. (d) Diameter and length distributions of the ZnO nanorods in ZRC aerogels

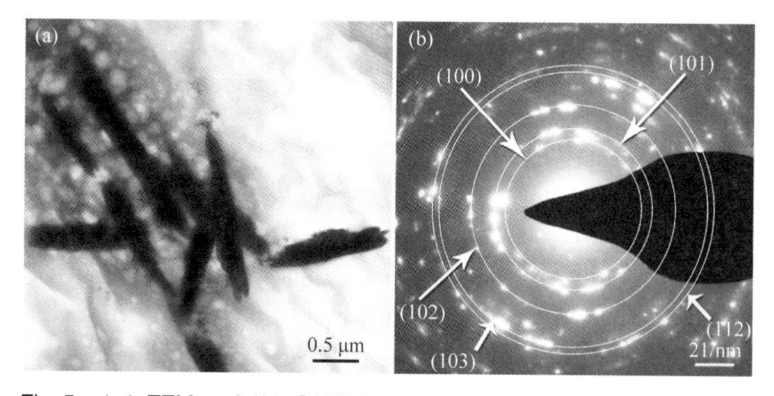

Fig. 5 (a) TEM and (b) SAED images of ZRC aerogels, respectively

4 Conclusions

ZnO nanorods were successfully prepared and embedded into the nanoporous cellulose aerogels matrixes by the mild simple one-pot low-temperature hydrothermal treatment, which exhibit good dispersion without obvious agglomeration. Furthermore, the rod-like ZnO has the average diameter of about 348 nm and length of about 1.49 μm, and exhibits hexagonal wurtzite phase. Thus, this class of novel ZRC aerogels composites might be useful for some photocatalytic, piezoelectric, fluorescent, and antimicrobial applications. Meanwhile, the method might also be extended to fabricate other versatile cellulose-based nanocomposites.

Acknowledgments

This study was supported by the National Natural Science Foundation of China (grant no. 31270590 and 31470584).

References

[1] Wang Z L. Zinc oxide nanostructures: growth, properties and applications[J]. Journal of physics: condensed matter, 2004, 16(25): R829.

[2] Geburt S, Lorke M, da Rosa A L, et al. Intense intrashell luminescence of Eu-doped single ZnO nanowires at room temperature by implantation created Eu-Oi complexes[J]. Nano letters, 2014, 14(8): 4523-4528.

[3] Yang S, Zhang Y, Yu J, et al. Antibacterial and mechanical properties of honeycomb ceramic materials incorporated with silver and zinc[J]. Materials & Design, 2014, 59: 461-465.

[4] Özgür Ü, Alivov Y I, Liu C, et al. A comprehensive review of ZnO materials and devices[J]. Journal of applied physics, 2005, 98(4): 11.

[5] Nguyen T, Tuan N T, Cuong N D, et al. Near-infrared emission from ZnO nanorods grown by thermal evaporation[J]. Journal of luminescence, 2014, 156: 199-204.

[6] Kim B J, Kim M W, Jang J S, et al. Real time observation of ZnO nanostructure formation via the solid-vapor and solid-solid-vapor mechanisms[J]. Nanoscale, 2014, 6(12): 6984-6990.

[7] Tien H N, Hur S H. A highly sensitive UV sensor composed of 2D NiO nanosheets and 1D ZnO nanorods fabricated by a hydrothermal process[J]. Sensors and Actuators A: Physical, 2014, 207: 20-24.

[8] Matei A, Cernica I, Cadar O, et al. Synthesis and characterization of ZnO-polymer nanocomposites[J]. International Journal of Material Forming, 2008, 1(1): 767-770.

[9] Chang B P, Akil H M, Affendy M G, et al. Comparative study of wear performance of particulate and fiber-reinforced nano-ZnO/ultra-high molecular weight polyethylene hybrid composites using response surface methodology[J]. Materials & Design, 2014, 63: 805-819.

[10] Medronho B, Lindman B. Competing forces during cellulose dissolution: From solvents to mechanisms[J]. Current Opinion in Colloid & Interface Science, 2014, 19(1): 32-40.

[11] Zhang S H, Li F X, Yu J Y. Structure and properties of novel cellulose fibres produced from NaOH/PEG-treated cotton linters[J]. Iranian polymer journal, 2010; 19: 949-957.

[12] Wan C, Lu Y, Jiao Y, et al. Preparation of mechanically strong and lightweight cellulose aerogels from cellulose-NaOH/PEG solution[J]. Journal of Sol-Gel Science and Technology, 2015, 74(1): 256-259.

[13] Tingaut P, Zimmermann T, Sèbe G. Cellulose nanocrystals and microfibrillated cellulose as building blocks for the design of hierarchical functional materials[J]. Journal of Materials Chemistry, 2012, 22(38): 20105-20111.

[14] Li J, Wan C, Lu Y, et al. Fabrication of cellulose aerogel from wheat straw with strong absorptive capacity[J]. Frontiers of Agricultural Science and Engineering, 2014, 1(1): 46-52.

[15] Wan C, Lu Y, Cao J, et al. Preparation, characterization and oil adsorption properties of cellulose aerogels from four kinds of plant materials via a NAOH/PEG aqueous solution[J]. Fibers and Polymers, 2015, 16(2): 302-307.

[16] Zuluaga R, Putaux J L, Cruz J, et al. Cellulose microfibrils from banana rachis: Effect of alkaline treatments on structural and morphological features[J]. Carbohydrate Polymers, 2009, 76(1): 51-59.

[17] Alemdar A, Sain M. Isolation and characterization of nanofibers from agricultural residues-Wheat straw and soy hulls

[J]. Bioresource technology, 2008, 99(6): 1664-1671.

[18] Johar N, Ahmad I, Dufresne A. Extraction, preparation and characterization of cellulose fibres and nanocrystals from rice husk[J]. Industrial Crops and Products, 2012, 37(1): 93-99.

[19] Wan C, Jiao Y, Sun Q, et al. Preparation, characterization, and antibacterial properties of silver nanoparticles embedded into cellulose aerogels[J]. Polymer Composites, 2016, 37(4): 1137-1142.

[20] Wen X, Wu W, Ding Y, et al. Seedless synthesis of patterned ZnO nanowire arrays on metal thin films (Au, Ag, Cu, Sn) and their application for flexible electromechanical sensing[J]. Journal of Materials Chemistry, 2012, 22(19): 9469-9476.

[21] Marrani A G, Caprioli F, Boccia A, et al. Electrochemically deposited ZnO films: an XPS study on the evolution of their surface hydroxide and defect composition upon thermal annealing[J]. Journal of Solid State Electrochemistry, 2014, 18(2): 505-513.

[22] Snigurenko D, Jakiela R, Guziewicz E, et al. XPS study of arsenic doped ZnO grown by Atomic Layer Deposition [J]. Journal of alloys and compounds, 2014, 582: 594-597.

[23] Ristić M, Musić S, Ivanda M, et al. Sol-gel synthesis and characterization of nanocrystalline ZnO powders[J]. Journal of Alloys and Compounds, 2005, 397(1-2): L1-L4.

[24] Zhang X, Liu Y, Kang Z. 3D branched ZnO nanowire arrays decorated with plasmonic Au nanoparticles for high-performance photoelectrochemical water splitting[J]. ACS applied materials & interfaces, 2014, 6(6): 4480-4489.

[25] Tseng Y K, Huang C J, Cheng H M, et al. Characterization and field-emission properties of needle-like zinc oxide nanowires grown vertically on conductive zinc oxide films[J]. Advanced functional materials, 2003, 13(10): 811-814.

[26] Barrett C S, Massalski T B. Structure of Metals, McGraw-Hill[J]. New York, 1966.

[27] Romero R, Leinen D, Dalchiele E A, et al. The effects of zinc acetate and zinc chloride precursors on the preferred crystalline orientation of ZnO and Al-doped ZnO thin films obtained by spray pyrolysis[J]. Thin Solid Films, 2006, 515(4): 1942-1949.

[28] Znaidi L, Illia G S, Benyahia S, et al. Oriented ZnO thin films synthesis by sol-gel process for laser application[J]. Thin solid films, 2003, 428(1-2): 257-262.

[29] Wei T Y, Kuo C Y, Hsu Y J, et al. Tin oxide nanocrystals embedded in silica aerogel: Photoluminescence and photocatalysis[J]. Microporous and Mesoporous Materials, 2008, 112(1-3): 580-588.

[30] Lu C H, Yeh C H. Influence of hydrothermal conditions on the morphology and particle size of zinc oxide powder[J]. Ceramics International, 2000, 26(4): 351-357.

[31] Ni Y, Wei X, Hong J, et al. Hydrothermal preparation and optical properties of ZnO nanorods[J]. Materials Science and Engineering: B, 2005, 121(1-2): 42-47.

[32] Zhang H, Ma X, Xu J, et al. Arrays of ZnO nanowires fabricated by a simple chemical solution route[J]. Nanotechnology, 2003, 14(4): 423.

中文题目：通过简单的一步低温水热法将氧化锌纳米棒嵌入多孔纤维素气凝胶中

作者：万才超，李坚

摘要：本文采用了一种简便有效的一步式低温水热法将氧化锌纳米棒原位嵌入多孔纤维素气凝胶中。此外，纤维素气凝胶的制备是基于绿色氢氧化钠/聚乙二醇溶液。棒状氧化锌的平均直径约为348 nm，长度约为1.49 μm，呈纤锌矿相。同时，扫描电镜和透射电镜观察证实，纳米棒紧紧地固定在气凝胶基质上，并表现出良好的分散性，没有剧烈的团聚现象，表明纤维素气凝胶是一类理想的支撑纳米颗粒的绿色基质材料。此外，该方法还可以扩展到制造其他多功能纤维素基纳米复合材料。

关键词：氧化锌；纤维素气凝胶；水热法；纳米复合材料；聚合物

Synthesis of Well-dispersed Magnetic $CoFe_2O_4$ Nanoparticles in Cellulose Aerogels via A Facile Oxidative Co-precipitation Method*

Caichao Wan, Jian Li

Abstract: With the increasing emphasis on green chemistry, it is becoming more important to develop environmentally friendly matrix materials for the synthesis of nanocomposites. Cellulose aerogels with hierarchical micro/nano-scale three-dimensional network beneficial to control and guide the growth of nanoparticles, are suitable as a class of ideal green nanoparticles hosts to fabricate multifunctional nanocomposites. Herein, a facile oxidative co-precipitation method was carried out to disperse $CoFe_2O_4$ nanoparticles in the cellulose aerogels matrixes, and the cellulose aerogels were prepared from the native wheat straw based on a green NaOH/polyethylene glycol solution. The mean diameter of the well-dispersed $CoFe_2O_4$ nanoparticles in the hybrid aerogels is 98.5 nm. Besides, the hybrid aerogels exhibit strong magnetic responsiveness, which could be flexibly actuated by a small magnet. And this feature also makes this class of magnetic aerogels possibly useful as recyclable adsorbents and some magnetic devices. Meanwhile, the mild green preparation method could also be extended to fabricate other miscellaneous cellulose-based nanocomposites.

Keywords: Cellulose aerogels; $CoFe_2O_4$; Magnetic properties; Template synthesis; Nanocomposites

1 Introduction

Spinel nanocrystals, which are extensively considered as one of the most important inorganic nanomaterials due to their superb electrical, optical, magnetic, and catalytic properties, have typical AB_2O_4 structure in which A and B exhibit tetrahedral and octahedral cation sites, respectively, and O displays the oxygen anion site (Ching, Mo, Tanaka & Yoshiya, 2001; Walsh et al., 2007). Moreover, spinel ferrite nanoparticles with the general chemical formula MFe_2O_4 (e.g., M = Mg, Mn, Co, Ni, or Zn) have been the subject of much scientific interest in recent years because of their multitudinous potential applications ranging from fundamental research to industrial use (Beal et al., 2011; Mathew & Juang, 2007; Zeng & Sun, 2008). Among the spinel ferrites compounds, $CoFe_2O_4$, one of the well-known hard magnetic materials, has high magnetocrystalline anisotropy, coercivity, and moderate saturation magnetization, which is a promising candidate suitable for microwave devices, contrast enhancers in magnetic resonance imaging (Hutlova, Niznansky, Rehspringer, Estournès & Kurmoo, 2003; Manova et al., 2004), etc.

* 本文摘自 Carbohydrate Polymers, 2015, 134: 144-150.

Along with the increasing research interest in organic-inorganic nanocomposites often presenting the best properties of each of the components in a synergic way, dispersing nanoparticle ferrites (like $CoFe_2O_4$) in some organic polymer matrixes to produce magnetically responsive multifunctional nanocomposites has attracted much more attention. From some previous literatures (Fan, Luo, Sun, Li, Lu & Qiu, 2012; Ghosh, Sheridon & Fischer, 2008; Müller-Schulte & Schmitz-Rode, 2006), some organic polymer matrixes like thermosensitive *N*-isopropylacrylamide, chitosan, and polyethylene have been employed to combine with various magnetic nanoparticles to fabricate high-performance composites with all sorts of functions. These composites could be used as contactless controllable drug carriers, recyclable novel adsorbing or separating materials, special catalysts, etc. Furthermore, as a kind of green organic matrix materials, cellulose aerogels prepared from cellulose isolated from native biomass resources have not only biodegradability and good compatibility with the environment, but also high chemical stability, large surface area, and high porosity (Guilminot, Gavillon, Chatenet, Berthon-Fabry, Rigacci & Budtova, 2008; Liebner, Potthast, Rosenau, Haimer & Wendland, 2007; Sehaqui, Zhou & Berglund, 2011). Besides, their micro/nano-scale pore structure and plentiful surface hydroxide radicals are beneficial to control and guide growth of nanoparticles (especially for these various ferrites) (Cai, Kimura, Wada & Kuga, 2008; Li, Wang, Wang & Yu, 2014). Therefore, cellulose aerogels are widely regarded as ideal environmentally friendly nanoparticles hosts. However, notwithstanding numerous reports of organic-inorganic magnetic nanocomposites using nano-ferrites as filling agents, the researches on green magnetic materials involving cellulose products like aerogels, foams and films are still rare (Liu, Yan, Tao, Yu & Liu, 2012; Olsson et al., 2010), especially for cellulose aerogels.

Therefore, herein we report the synthesis of green magnetic nanocomposites using cellulose aerogels as templates to combine with magnetic $CoFe_2O_4$ nanoparticles via a facile oxidative co-precipitation approach. Meanwhile, a green NaOH/polyethylene glycol (PEG) solution was used to fabricate cellulose hydrogels (the precursor of cellulose aerogels). The hybrid $CoFe_2O_4$@ cellulose aerogels ($CoFe_2O_4$@ CA) were characterized by scanning electron microscope (SEM) attached with energy dispersive spectrometer (EDS), transmission electron microscopy (TEM) associated with selected area electron diffraction (SAED), X-ray photoelectron spectroscopy (XPS), X-ray diffraction (XRD), Fourier transform infrared spectroscopy (FTIR), and thermogravimetric analysis (TGA). Furthermore, the magnetic behaviors were investigated by measuring the magnetic hysteresis loop by superconducting quantum interference device (SQUID) magnetometer.

2 Materials and methods

2.1 Materials

Native wheat straw was grinded and then screened through a sixty mesh sieve, and the resulting powder was subsequently collected and dried in a vacuum oven at 60 ℃ for 24 h before used. All chemical reagents with analytical grade are supplied by Shanghai Boyle Chemical Co. Ltd. (China) and used without further purification.

2.2 Preparation of $CoFe_2O_4$@CA

The isolation of cellulose component from the native wheat straw powder and the subsequent synthesis of cellulose hydrogel based on the green NaOH/PEG-4000 solution could be referred to our previous reports (Wan & Li, 2015; Wan, Lu, Jiao, Cao, Li & Sun, 2014; Wan, Lu, Jiao, Jin, Sun & Li, 2015). The obtained hydrogel with the size of about 20 mm (diameter) × 10 mm (height) was immersed in 40 mL aqueous solutions of $FeSO_4$ and $CoCl_2$ with a molar ratio of [Fe]/[Co] = 2 for 24 h to ensure saturation adsorption and homogeneous distribution, and the concentration of $CoCl_2$ salt was 0.1 mol · L^{-1}. Then, the system was heated to 90 ℃ for 3 h, and the hydrogel filled with Fe^{2+} and Co^{2+} was subsequently transferred into 50 mL aqueous solutions of NaOH (2 mol · L^{-1}) with KNO_3([Fe^{2+}]/[NO_3^-] = 0.44) at 90 ℃. The system was kept for an additional 6 h at 90 ℃ to complete the formation of $CoFe_2O_4$ in the cellulose aerogel matrix, and all reactions were performed in air. After that, the $CoFe_2O_4$@ cellulose hydrogel ($CoFe_2O_4$@ CH) was washed several times with a large amount of distilled water and tert-butyl alcohol in sequence, and then underwent a freeze drying process to remove all liquids, and the following $CoFe_2O_4$ @ CA was obtained. Additionally, the pure cellulose aerogel (PCA) was prepared following the aforementioned processes without the immersion and heating treatments.

2.3 Characterizations

SEM observation and EDS analysis were performed using a FEI, Quanta 250 SEM operating at 15 – 20 kV. TEM observation and SAED were performed with a FEI, Tecnai G2 F20 TEM at 200 kV. XPS spectra were recorded using a Thermo Escalab 250Xi XPS spectrometer. Deconvolution of the overlapping peaks was performed using a mixed Gaussian-Lorentzian fit program. XRD was carried out on a Bruker D8 Advance TXS XRD instrument operating with Cu Kα radiation (λ = 1.5405 Å) at a scan rate (2θ) of 4° · min^{-1} and the accelerating voltage of 40 kV and the applied current of 40 mA ranging from 5° to 90°. FTIR spectra were recorded on a Nicolet Nexus 670 FTIR instrument in the range of 400 – 4000 cm^{-1} with a resolution of 4 cm^{-1}. Thermal stabilities were determined using a TG analyzer (TA, Q600) with a heating rate of 10 ℃ · min^{-1} in a N_2 environment from room temperature to 800 ℃. Magnetic properties were measured on a MPMS XL-7 SQUID magnetometer at 300 K within a field range of ± 2.0 T.

3 Result and discussion

3.1 Crystal structures and chemical components

As shown in Fig. 1a, $CoFe_2O_4$@ CA exhibits clear spinel diffraction pattern, which presents obvious diffraction peaks at around 18.4°, 30.2°, 35.5°, 37.2°, 43.1°, 53.4°, 57.0°, 62.6°, 70.9°, 73.9°, 75.0°, 79.0°, and 86.7°, corresponding to the (111), (220), (311), (222), (400), (422), (511), (440), (620), (533), (622), (444) and (642) planes, respectively. The results are consistent with the standard JCPDS card no. 22 – 1086 for $CoFe_2O_4$. In addition, similar to PCA, the three XRD diffraction peaks located at around 12.1°, 20.0° and 22.0° are assigned to the ($\bar{1}$10), (110) and (200) planes of crystalline cellulose Ⅱ of the aerogels matrixes (Zhao, Kwak, Zhang, Brown, Arey & Holladay, 2007), revealing that $CoFe_2O_4$ nanoparticles have been successfully synthesized and dispersed in the cellulose aerogels. The average crystallite size of the $CoFe_2O_4$ nanoparticles in $CoFe_2O_4$ @ CA was calculated from the full width at half maximum of the strongest diffraction peak (311) using Scherrer equation (Burton, Ong, Rea & Chan, 2009):

$$d = K\lambda / (\beta \cos\theta) \tag{1}$$

where λ is the X-ray wavelength, K is a constant taken as 0.89, θ is the angle of Bragg diffraction and β is the full width at half maximum. The average crystallite size of the $CoFe_2O_4$ nanoparticles is approximatively 34.9 nm.

For FTIR analysis, in addition to some cellulose characteristic peaks (Lan, Liu, Yue, Sun & Kennedy, 2011) as labeled in Fig. 1b and Table 1, the FTIR spectrum of $CoFe_2O_4$@CA also displays a new peak at around 561 cm^{-1} due to the stretching of Fe-O bond (Ayyappan, Mahadevan, Chandramohan, Srinivasan, Philip & Raj, 2010), which further confirms the incorporation of the $CoFe_2O_4$ nanoparticles into the aerogels matrixes.

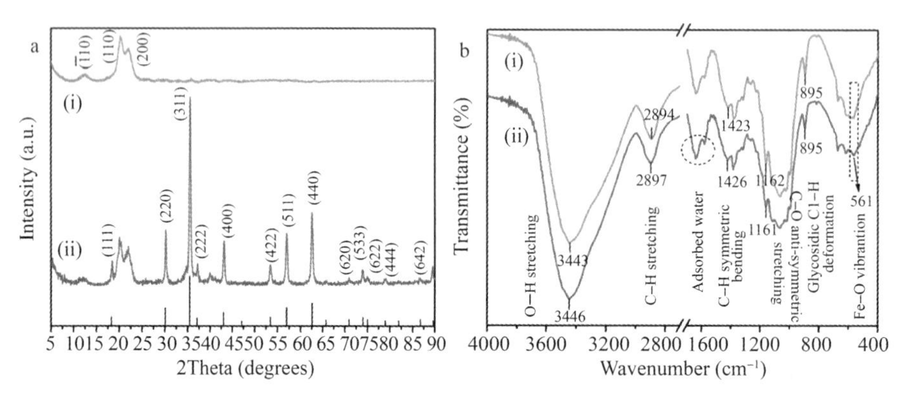

Fig. 1 (a) XRD patterns and (b) FTIR spectra of PCA (i) and $CoFe_2O_4$@CA (ii), respectively. The bottom line in a is the standard JCPDS card no. 22-1086 for $CoFe_2O_4$

Table 1 FTIR absorption frequencies (cm^{-1}) of the main signals of $CoFe_2O_4$@CA and their assignments

Absorption band (cm^{-1})	Assignment
3446	O—H stretching
2897	C—H stretching
1635	Adsorbed water
1426	C—H symmetric bending
1161	C—O anti-symmetric stretching
895	Glycosidic C_1—H deformation
561	Fe—O stretching

In Fig. 2a, the characteristic signals for C, O, Fe, and Co elements are clearly observed in the XPS survey spectrum of $CoFe_2O_4$@CA. In detail, there are three types of oxygen species in the O1s spectrum (Fig. 2b), where the fitting peaks at around 529.8, 531.2, and 533.0 eV are attributed to the metal-oxygen bond, oxygen in OH groups, and multiplicity of the adsorbed water (Lu, Yang, Bian, Chao & Wang, 2014), respectively. The presence of the peak at 531.2 eV also suggests that the surface of $CoFe_2O_4$@CA is partly hydroxylated. In Fig. 2c, the Co2p XPS spectrum is well fitted into four peaks situated at

780.9, 785.5, 796.8, and 803.0 eV from $Co2p_{3/2}$ and $Co2p_{1/2}$ with their satellite peaks, which are consistent with the presence of Co^{2+} in the high-spin state (Yuan et al., 2010). Moreover, the Fe2p spectrum shows the main peaks of $Fe2p_{3/2}$ and $Fe2p_{1/2}$ at around 711.8 and 724.8 eV accompanied by the corresponding satellite lines visible at 717.3 and 732.8 eV, only indicative of the presence of Fe^{3+} cations (Yang, Mao, Li, Liu & Tong, 2013).

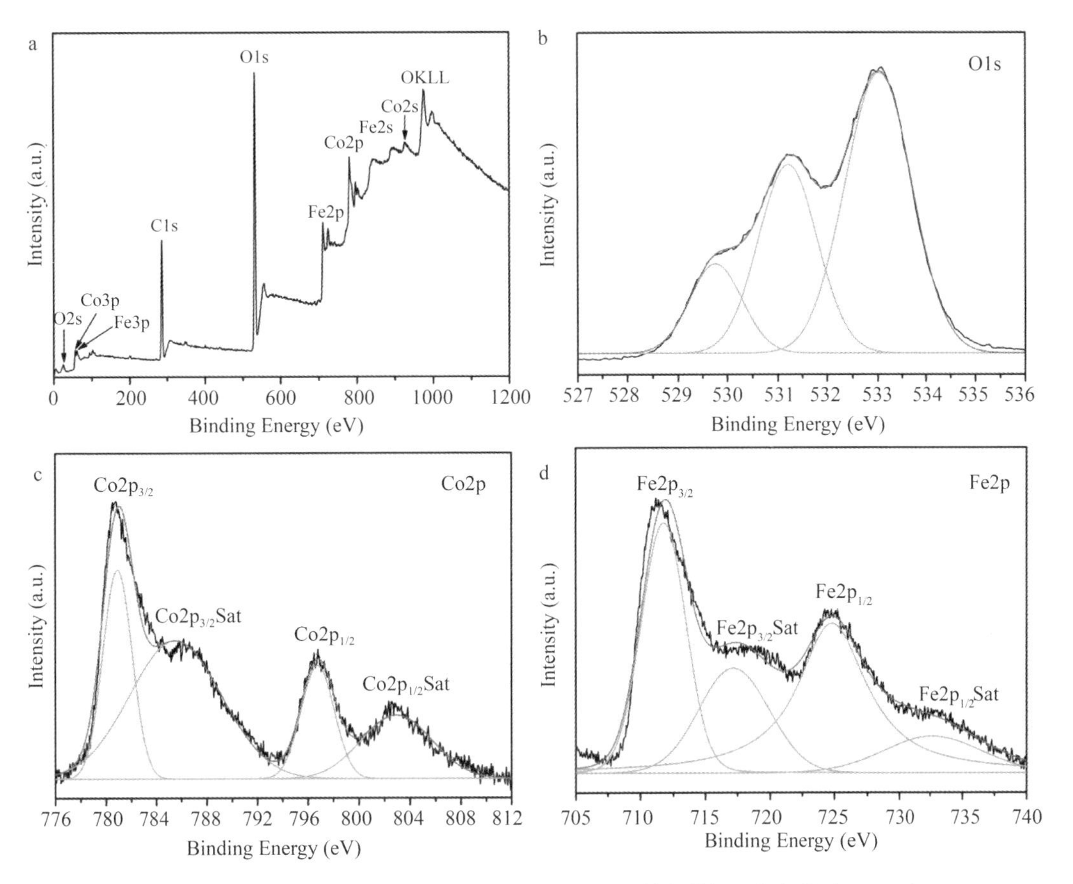

Fig. 2 XPS spectra of (a) survey spectrum, (b) O1s, (c) Co2p, and (d) Fe2p in $CoFe_2O_4$@CA, respectively

3.2 Microscopic morphologies

Fig. 3 shows the SEM images of PCA and $CoFe_2O_4$@ CA, respectively. In Fig. 3a, b, compared with the neat surface of PCA, the plentiful $CoFe_2O_4$ nanoparticles as indicated by the white points are deposited in the aerogels matrixes, which exhibit good dispersion without obvious agglomeration. Meanwhile, comparing the corresponding EDS spectra, apart from the common C and O elements, the new strong Fe and Co peaks are also detected for the hybrid aerogels, which further confirms the presence of $CoFe_2O_4$ nanoparticles in $CoFe_2O_4$@ CA. Besides, the significant improvement in the intensity of oxygen peak is another proof. Fig. 3c presents the higher-magnification SEM images, and it is observed that these nanoparticles are tightly anchored to the three-dimensional network of the aerogels, which demonstrates that the aerogels could be served as idea matrixes to support well-dispersed nanoparticles.

Fig. 3 (a) SEM image of PCA. (b and c) SEM images of different magnifications of $CoFe_2O_4$@CA. Insets of a and b present the corresponding EDS spectra

The TEM observation also confirms the good dispersion (Fig. 4a), which shows that the particles are uniformly anchored to the aerogels matrixes without any free particles isolated from the matrixes, and the such strong combination might contribute to preventing the agglomeration of these nanoparticles to large particles in the pore structures. The diameter histogram of these particles as measured from the TEM micrograph using an image analyzer, displays Gaussian-like distribution, and their mean diameter (d) and standard deviation (σ) are estimated to be 98.5 and 29.4 nm. Apparently, the aforementioned calculated size value of the crystallites (ca. 34.9 nm) is smaller than the particle dimensions observed, indicating that the particles are multidomain (Nypelö, Rodriguez-Abreu, Rivas, Dickey & Rojas, 2014). High-resolution TEM image as shown in Fig. 4b exhibits a planar space of lattice spacing of about 0.48 nm, which corresponds to the (111) lattice plane (Meng, Chen & Jiao, 2008). The SAED image in Fig. 4c shows a series of strong diffraction rings originated from the random orientations of the nanoparticles, indicating the highly crystalline of the $CoFe_2O_4$.

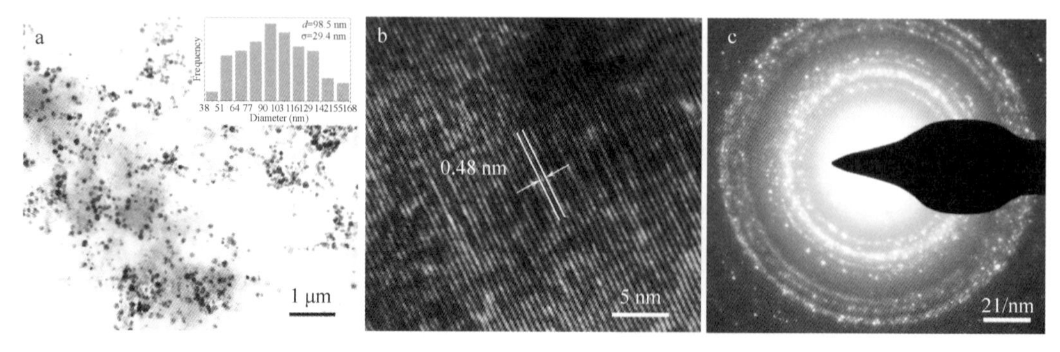

Fig. 4 (a) TEM, (b) high-resolution TEM, and (c) SAED images of $CoFe_2O_4$@CA. Inset of a presents the size distribution of the $CoFe_2O_4$ nanoparticles in $CoFe_2O_4$@CA

3.3 Thermal stabilities and magnetic properties

The strong exothermic peaks at around 365.1 ℃ and 334.4 ℃ for PCA and $CoFe_2O_4$@ CA are ascribed to the cellulose pyrolysis (Wan, Lu, Jin, Sun & Li, 2015; Yang, Yan, Chen, Lee & Zheng, 2007) (Fig. 5a). The degradation of the thermal stability of $CoFe_2O_4$@ CA might be ascribed to the hot alkali solution treatment leading to the damage of cellulose crystal structure. In addition, for $CoFe_2O_4$@ CA, the small weight loss at 525-595 ℃ is probably attributed to the reaction between the $CoFe_2O_4$ nanoparticles and the carbon residue obtained from the pyrolysis of cellulose (Wu, Xiao, Li, Lei, Zhang & Wang, 2012). The residual masses above 800 ℃ are 2.7% for PCA and 19.2% for $CoFe_2O_4$@CA. Ignoring the influence of the aforementioned reaction occurring at 525-595 ℃, the $CoFe_2O_4$ content in the composites could be roughly estimated to be 16.5%.

From the typical hysteresis loop in Fig. 5b, the curves indicate hysteresis ferromagnetism in the range of around ± 2 kOe for $CoFe_2O_4$@CA, while exceeding this range the specific magnetization increases with increasing field and tends to saturate. $CoFe_2O_4$@CA exhibits a specific saturation magnetization (M_s) of 8.6 emu · g^{-1}. Taking the $CoFe_2O_4$ proportion in $CoFe_2O_4$@CA into consideration, the M_s value could be roughly calculated as 52.1 emu · g^{-1}, much less than that of bulk cobalt ferrite (80.8 emu · g^{-1} at 300k) (Grigorova et al., 1998). The decline in M_s is possibly explained by the relatively smaller particle sizes due to the reduction of particle agglomeration resulted from the presence of the hierarchical porous aerogels matrixes, hence inducing the enhancement of surface effects. In detail, when the size of magnetic particles reduces to the nanoscale, various surface defects will become appreciable, resulting in a magnetically disordered layer at the particle surface. Nevertheless, the surface layer makes no contribution to the magnetization, and becomes more remarkable by reducing the particle size, consequently leading to the decline of M_s value (Wang, Yang, Xian & Jiang, 2012). Moreover, the inversion degree is possibly another reason for the lower M_s value. Usually, the decrease in the particle dimension results in modification of the inversion degree and consequently in a variation of the magnetic properties (Cannas et al., 2006). The coercive force (H_c) of $CoFe_2O_4$@CA is around 0.75 kOe. Considering the $CoFe_2O_4$ proportion, the H_c value was calculated to be as high as 4.5 kOe, close to the theoretical H_c value of bulk $CoFe_2O_4$ (5.4 kOe at 300 k), which is attributed to the reduction of the interactions between the particles due to the porous aerogels matrixes serving as inert barriers that space the particles sufficiently apart, and hence leading to a relatively high H_c (Morales, Munoz-Aguado, Garcıa-Palacios, Lázaro & Serna, 1998). In addition, it has been confirmed that the size of the magnetic crystallites exceeding the single domain critical size will induce the decrease of the coercive force owing to the appearance of multidomain grains (Yan et al., 1999), which is possibly responsive for the slight difference in H_c between $CoFe_2O_4$@CA and bulk $CoFe_2O_4$.

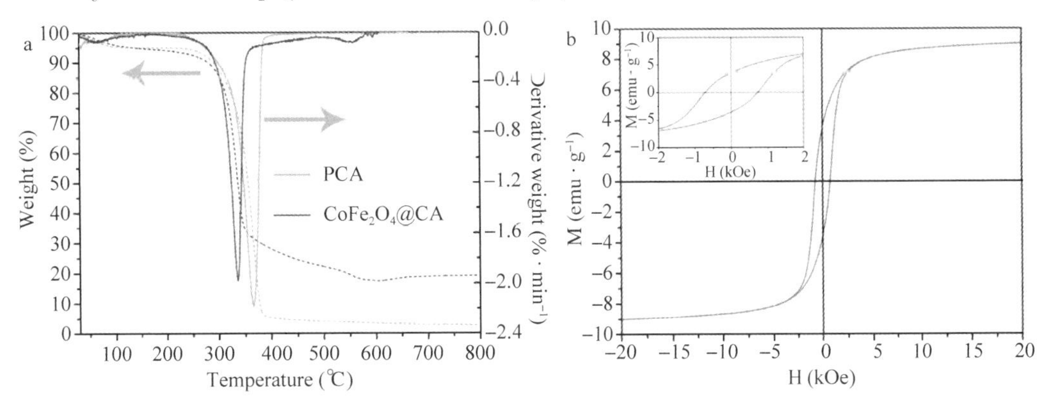

Fig. 5 (a) TG (dotted line) and DTG (solid line) curves of PCA and $CoFe_2O_4$@CA. (b) Hysteresis loop of $CoFe_2O_4$@CA at 300 k; the insert presents the initial magnetization curves as a function of applied magnetic field

3.4 Magnetically responsive phenomena

Fig. 6 displays some interesting magnetically responsive phenomena of the $CoFe_2O_4$ nanoparticles and the hybrid hydrogels or aerogels. The nanoparticles are completely gathered near the magnet, and easily separated from water under an external magnetic field (Fig. 6a); besides, the glass bottle containing the nanoparticles dispersion moves quickly towards the magnet due to the strong magnetic responsibility (see Supplementary Video 1). The hybrid hydrogels also exhibit strong magnetic responsiveness, which could be flexibly actuated by the magnet and float to the surface overcoming gravity and water resistance. Moreover, the hydrogels in water also move with quickness and lightness following the movement of the magnet (see Supplementary Video 2), which are expected to be used as easily recyclable adsorption materials for water purification and oil spills. Fig. 6c shows the uplifted magnetic aerogels ($m \approx 0.84$ g, $\rho \approx 146.13$ mg · cm^{-3}) by a magnet. The aerogels could also be attracted and firmly held on the glass plates under an external magnetic field, even though there are still far distances between the aerogels and the magnet (5.3 and 11.7 mm) (Fig. 6d, e; see

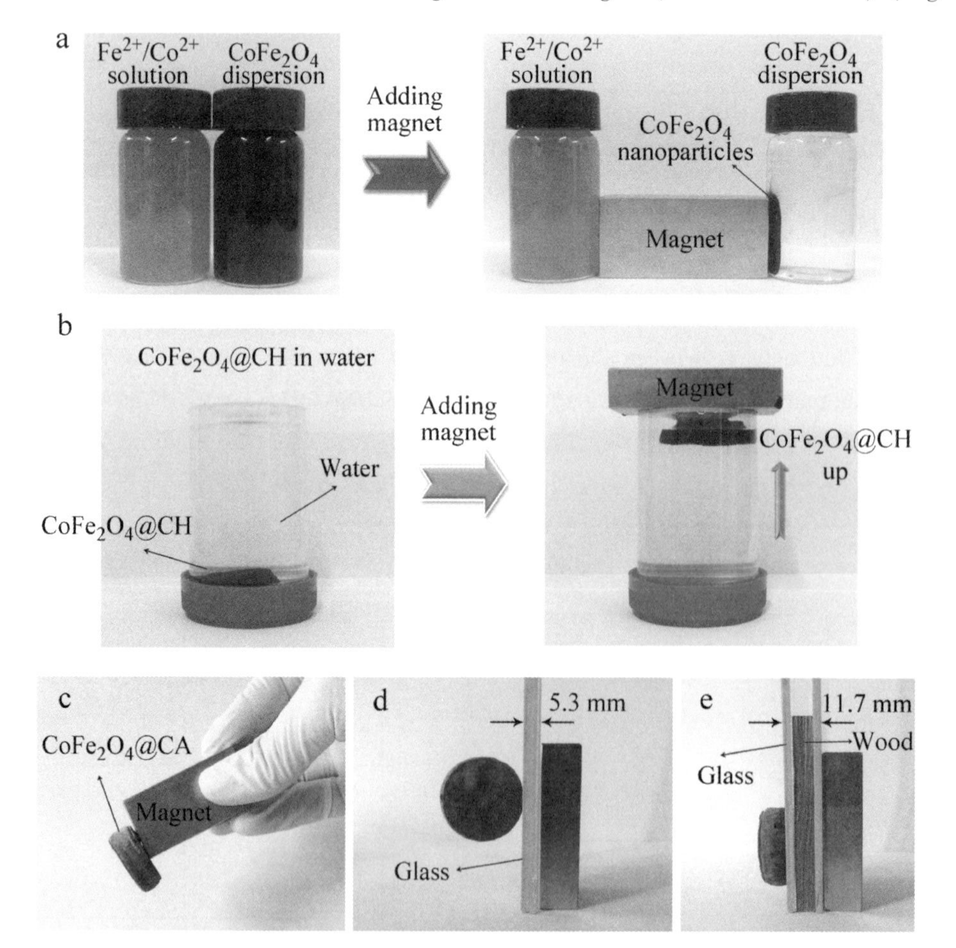

Fig. 6 Magnetic responsiveness of the $CoFe_2O_4$ nanoparticles, $CoFe_2O_4$@CH, and $CoFe_2O_4$@CA. (a) Separation of the $CoFe_2O_4$ nanoparticles from the aqueous solution under an external magnetic field. (b) $CoFe_2O_4$@CH floating from bottom to top of the container under an external magnetic field. (c-e) $CoFe_2O_4$@CA could be magnetically attracted under an external magnetic field

also Supplementary Video 3, 4), indicative of strong magnetic responsibility.

4 Conclusions

We have fabricated a class of green, lightweight and strongly magnetically responsive $CoFe_2O_4$@ CA hybrids via a facile oxidative co-precipitation method. The magnetic $CoFe_2O_4$ nanoparticles with a mean diameter of 98.5 nm are homogeneously dispersed in the three-dimensional networks of the cellulose aerogels, and the loading content of $CoFe_2O_4$ is around 16.5%. Moreover, the $CoFe_2O_4$ nanoparticles and the hybrid hydrogels or aerogels all exhibit strong magnetic responsiveness, which might find numerous potential applications as recyclable adsorption materials, magneto-dependent sensors, etc. In addition, the green convenient and effective preparation method is expected to be extended to the fabrication of other cellulose-based functional materials.

Supporting Information

Please see the supporting information (Video 1-4) through the following links: https://mega.co.nz/#!7MZWUCwS!Hf5O8k2ShFthL02HQNHeEEZMFJ5cn4Y6Wq-3nBUdLEM

Acknowledgments

This study was supported by the National Natural Science Foundation of China (grant no. 31270590 and 31470584).

References

[1] Ayyappan S, Mahadevan S, Chandramohan P, et al. Influence of Co^{2+} ion concentration on the size, magnetic properties, and purity of $CoFe_2O_4$ spinel ferrite nanoparticles[J]. The Journal of Physical Chemistry C, 2010, 114(14): 6334-6341.

[2] Beal J H L, Prabakar S, Gaston N, et al. Synthesis and comparison of the magnetic properties of iron sulfide spinel and iron oxide spinel nanocrystals[J]. Chemistry of Materials, 2011, 23(10): 2514-2517.

[3] Burton A W, Ong K, Rea T, et al. On the estimation of average crystallite size of zeolites from the Scherrer equation: A critical evaluation of its application to zeolites with one-dimensional pore systems[J]. Microporous and Mesoporous Materials, 2009, 117(1-2): 75-90.

[4] Cai J, Kimura S, Wada M, et al. Nanoporous cellulose as metal nanoparticles support[J]. Biomacromolecules, 2009, 10(1): 87-94.

[5] Cannas C, Musinu A, Piccaluga G, et al. Magnetic properties of cobalt ferrite-silica nanocomposites prepared by a sol-gel autocombustion technique[J]. The Journal of chemical physics, 2006, 125(16): 164714.

[6] Ching W Y, Mo S D, Tanaka I, et al. Prediction of spinel structure and properties of single and double nitrides[J]. Physical Review B, 2001, 63(6): 064102.

[7] Fan L, Luo C, Sun M, et al. Preparation of novel magnetic chitosan/graphene oxide composite as effective adsorbents toward methylene blue[J]. Bioresource technology, 2012, 114: 703-706.

[8] Ghosh A, Sheridon N K, Fischer P. Voltage-Controllable Magnetic Composite Based on Multifunctional Polyethylene Microparticles[J]. Small, 2008, 4(11): 1956-1958.

[9] Grigorova M, Blythe H J, Blaskov V, et al. Magnetic properties and Mössbauer spectra of nanosized $CoFe_2O_4$ powders [J]. Journal of magnetism and magnetic materials, 1998, 183(1-2): 163-172.

[10] Guilminot E, Gavillon R, Chatenet M, et al. New nanostructured carbons based on porous cellulose: Elaboration,

pyrolysis and use as platinum nanoparticles substrate for oxygen reduction electrocatalysis[J]. Journal of Power Sources, 2008, 185(2): 717-726.

[11] Hutlova A, Niznansky D, Rehspringer J L, et al. High coercive field for nanoparticles of $CoFe_2O_4$ in amorphous silica sol-gel[J]. Advanced Materials, 2003, 15(19): 1622-1625.

[12] Lan W, Liu C F, Yue F X, et al. Ultrasound-assisted dissolution of cellulose in ionic liquid[J]. Carbohydrate polymers, 2011, 86(2): 672-677.

[13] Li F, Wang W, Wang X, et al. Changes of structure and property of alkali soluble hydroxyethyl celluloses (HECs) and their regenerated films with the molar substitution[J]. Carbohydrate polymers, 2014, 114: 206-212.

[14] Liebner F, Potthast A, Rosenau T, et al. Ultralight-weight cellulose aerogels from NBnMO-stabilized Lyocell dopes [J]. Research Letters in Materials Science, 2007, 73724.

[15] Liu S, Yan Q, Tao D, et al. Highly flexible magnetic composite aerogels prepared by using cellulose nanofibril networks as templates[J]. Carbohydrate polymers, 2012, 89(2): 551-557.

[16] Lu X, Yang L, Bian X, et al. Rapid, Microwave-Assisted, and One-Pot Synthesis of Magnetic Palladium-$CoFe_2O_4$-Graphene Composite Nanosheets and Their Applications as Recyclable Catalysts [J]. Particle & Particle Systems Characterization, 2014, 31(2): 245-251.

[17] Müller-Schulte D, Schmitz-Rode T. Thermosensitive magnetic polymer particles as contactless controllable drug carriers[J]. Journal of Magnetism and Magnetic Materials, 2006, 302(1): 267-271.

[18] Manova E, Kunev B, Paneva D, et al. Mechano-synthesis, characterization, and magnetic properties of nanoparticles of cobalt ferrite, $CoFe_2O_4$[J]. Chemistry of materials, 2004, 16(26): 5689-5696.

[19] Mathew D S, Juang R S. An overview of the structure and magnetism of spinel ferrite nanoparticles and their synthesis in microemulsions[J]. Chemical engineering journal, 2007, 129(1-3): 51-65.

[20] Meng Y, Chen D, Jiao X. Synthesis and characterization of $CoFe_2O_4$ hollow spheres[J]. European Journal of Inorganic Chemistry, 2008(25), 4019-4023.

[21] Morales M P, Munoz-Aguado M J, Garcıa-Palacios J L, et al. Coercivity enhancement in γ-Fe_2O_3 particles dispersed at low-volume fraction[J]. Journal of magnetism and magnetic materials, 1998, 183(1-2): 232-240.

[22] Nypelö T, Rodriguez-Abreu C, Rivas J, et al. Magneto-responsive hybrid materials based on cellulose nanocrystals [J]. Cellulose, 2014, 21(4): 2557-2566.

[23] Olsson R T, Samir M A S A, Salazar-Alvarez G, et al. Making flexible magnetic aerogels and stiff magnetic nanopaper using cellulose nanofibrils as templates[J]. Nature nanotechnology, 2010, 5(8): 584-588.

[24] Sehaqui H, Zhou Q, Berglund L A. High-porosity aerogels of high specific surface area prepared from nanofibrillated cellulose (NFC)[J]. Composites science and technology, 2011, 71(13): 1593-1599.

[25] Walsh A, Wei S H, Yan Y, et al. Structural, magnetic, and electronic properties of the Co-Fe-Al oxide spinel system: Density-functional theory calculations[J]. Physical review B, 2007, 76(16): 165119.

[26] Wan C, Li J. Embedding ZnO nanorods into porous cellulose aerogels via a facile one-step low-temperature hydrothermal method[J]. Materials & Design, 2015, 83: 620-625.

[27] Wan C, Lu Y, Jiao Y, et al. Cellulose aerogels from cellulose-NaOH/PEG solution and comparison with different cellulose contents[J]. Materials Science and Technology, 2015, 31(9): 1096-1102.

[28] Wan C, Lu Y, Jiao Y, et al. Fabrication of hydrophobic, electrically conductive and flame-resistant carbon aerogels by pyrolysis of regenerated cellulose aerogels[J]. Carbohydrate polymers, 2015, 118: 115-118.

[29] Wan C, Lu Y, Jin C, et al. A facile low-temperature hydrothermal method to prepare anatase titania/cellulose aerogels with strong photocatalytic activities for rhodamine B and methyl orange degradations[J]. Journal of Nanomaterials, 2015, 1-8.

[30] WP W, Yang H, Xian T, et al. XPS and magnetic properties of $CoFe_2O_4$ nanoparticles synthesized by a

polyacrylamide gel route[J]. Materials Transactions, 2012, 53(9): 1586-1589.

[31] Wu L, Xiao Q, Li Z, et al. $CoFe_2O_4$/C composite fibers as anode materials for lithium-ion batteries with stable and high electrochemical performance[J]. Solid State Ionics, 2012, 215: 24-28.

[32] Yan C H, Xu Z G, Cheng F X, et al. Nanophased $CoFe_2O_4$ prepared by combustion method[J]. Solid State Communications, 1999, 111(5): 287-291.

[33] Yang H, Mao Y, Li M, et al. Electrochemical synthesis of $CoFe_2O_4$ porous nanosheets for visible light driven photoelectrochemical applications[J]. New Journal of Chemistry, 2013, 37(10): 2965-2968.

[34] Yang H, Yan R, Chen H, et al. Characteristics of hemicellulose, cellulose and lignin pyrolysis[J]. Fuel, 2007, 86(12-13): 1781-1788.

[35] Yuan H L, Wang Y Q, Zhou S M, et al. Low-temperature preparation of superparamagnetic $CoFe_2O_4$ microspheres with high saturation magnetization[J]. Nanoscale research letters, 2010, 5(11): 1817-1821.

[36] Zeng H, Sun S. Syntheses, properties, and potential applications of multicomponent magnetic nanoparticles[J]. Advanced Functional Materials, 2008, 18(3): 391-400.

[37] Zhao H, Kwak J H, Zhang Z C, et al. Studying cellulose fiber structure by SEM, XRD, NMR and acid hydrolysis[J]. Carbohydrate polymers, 2007, 68(2): 235-241.

中文题目：利用简单的氧化共沉淀法在纤维素气凝胶中合成分散良好的磁性铁酸钴纳米粒子

作者：万才超，李坚

摘要：随着对绿色化学的日益重视，开发环境友好型基质材料合成纳米复合材料变得越来越重要。纤维素气凝胶具有层次分明的微/纳米尺度三维网络，有利于控制和引导纳米粒子的生长，适合作为一类理想的绿色纳米粒子宿主来制造多功能纳米复合材料。在此，采用简便的氧化共沉淀法将 $CoFe_2O_4$ 纳米颗粒分散在纤维素气凝胶基质中，以原生小麦秸秆为原料，以绿色的 NaOH/聚乙二醇溶液为基础制备纤维素气凝胶。混合气凝胶中分散良好的 $CoFe_2O_4$ 纳米颗粒的平均直径为 98.5 nm。此外，混合气凝胶还表现出很强的磁性响应性，可以用小磁铁灵活驱动。而这一特点也使得该类磁性气凝胶可能作为可回收的吸附剂和一些磁性器件。同时，该温和的绿色制备方法还可以推广到其他杂色纤维素基纳米复合材料的制造中。

关键词：纤维素气凝胶；$CoFe_2O_4$；磁性；模板合成；纳米复合材料

Facile Synthesis of Well-dispersed Superparamagnetic γ-Fe_2O_3 Nanoparticles Encapsulated in Three-dimensional Architectures of Cellulose Aerogels and Their Applications for Cr(Ⅵ) Removal from Contaminated Water*

Caichao Wan, Jian Li

Abstract: With the increasing emphasis on green chemistry, cellulose aerogels which consist of abundant three-dimensional (3D) architectures, have been considered as a class of idea green matrix materials to encapsulate various nanoparticles for synthesis of miscellaneous functional materials. Herein, a facile template synthesis combined with chemical co-precipitation was implemented to prepare hybrid γ-Fe_2O_3@ cellulose aerogels (γ-Fe_2O_3@CA). The γ-Fe_2O_3 nanoparticles are well dispersed and immobilized in the micro/nano-scale pore structure of the aerogels, and exhibit superparamagnetic behavior. The particle sizes, pore characteristic parameters, magnetic property, and mechanical strength of the synthetic γ-Fe_2O_3@CA could be flexibly tailored by adjusting the concentrations of the initial reactants. In addition, γ-Fe_2O_3@CA exhibits rapid adsorption rate and excellent adsorption ability to remove Cr(Ⅵ) heavy metal ions. Moreover, combining with the advantages of environmental benefits, facile convenient preparation method, high specific surface area and strong mechanical strength, and strong magnetic responsiveness, this class of green γ-Fe_2O_3@CA is more favorable and suitable for Cr(Ⅵ) removal from contaminated water, and also useful in many other applications.

Keywords: Cellulose aerogels; Superparamagnetic; γ-Fe_2O_3; Composites; Chromium; Adsorption

1 Introduction

Iron oxides in the nano range, have been extensively considered as a fascinating class of materials from the viewpoints of both theory and practical applications, which exhibit tremendous potential in catalytic materials, adsorbents for water purification, gas sensors, ion exchangers, magnetic recording devices, magnetic resonance imaging, bioseparation, etc.[1-5] As one of the most important forms of iron oxides, the cubic spinel structured maghemite (γ-Fe_2O_3) with oxygen atoms forming an fcc close-packed structure, represents non-toxicity, biocompatibility, thermal and chemical stability, favorable hysteric properties, and

* 本文摘自 ACS Sustainable Chemistry & Engineering, 2015, 3: 2142-2152.

excellent magnetic properties. In addition, γ-Fe_2O_3 could exhibit unusual superparamagnetic behavior when the particles occur below a critical and material-dependent particle size (the thermal energy exceeds the magnetic anisotropy energy). Besides, superparamagnetic particles do not retain any permanent magnetization after removal of an applied magnetic field, which contributes to their stability and dispersion. Therefore, these characteristics make γ-Fe_2O_3 frequently combined with various matrix materials to fabricate magnetically responsive miscellaneous γ-Fe_2O_3/polymer composites, which usually inherit advantages from both individual components. These polymers include polyacrylate (PEA), polyaniline (PANI), polypyrrole (PPY), polystyrene (PS), poly (3,4-ethylenedioxythiophene) (PEDOT), poly (N-isopropylacrylamide) (PNIPAAm), poly(methyl methacrylate) (PMMA), poly(vinyl chloride) (PVC), etc.[6-10]

With the increasing emphasis on green chemistry, it is becoming more important to develop environmentally friendly matrix materials for the synthesis of γ-Fe_2O_3/polymer composites.[11,12] Cellulose aerogels are getting an increasing interest as a class of natural biodegradable template, which consist of cross-linked three-dimensional (3D) architectures formed by self-assembly of cellulose chains as result of hydrogen-bond interaction. This unique structural feature endows themselves low density, large specific surface area, and high porosity. Indeed, the ordered interior pores of cellulose aerogels are appropriately used as reaction sites, which could provide confined spaces (micro-nano scale) for insertion of nanoparticles.[13,14] Meanwhile, the strong interactions would occur between abundant surface hydroxyl groups of cellulose and γ-Fe_2O_3 nanoparticles, which are beneficial to immobilization of the particles. The special advantage of this approach is the sharp decline of particle aggregation, and individual, randomly distributed nanoparticles are formed. Apart from the aforementioned merits, this kind of so-called template-based synthesis method generally has low cost, easily available experimental facilities, and mild technical process, and allows flexibility to design desired morphology, structure and dimension.[15] Therefore, the template-based synthesis method is suitable for fabricating nanomaterials. To date, numerous researches have employed template-based synthesis method to synthesize miscellaneous iron oxides (e.g., α-Fe_2O_3, γ-Fe_2O_3, Fe_3O_4).[16-22] The involved templates include carbon nanotube, hydrochar, ionic liquid [bmim][Cl], iron-based metal organic framework, mesoporous silica, polyethylene glycol (PEG), surfactant micelles (cetyltrimethylammonium bromide), etc. However, the reports using the porous bio-template (namely cellulose aerogels) to prepare various iron oxides are still rare.[23] Therefore, it might be interesting to synthesize this class of novel biomaterials using cellulose aerogels as template to support magnetic iron oxides, and investigate their potential applications. In addition, cellulose aerogels are generally prepared following the procedures of cellulose dissolution-regeneration-freeze drying or supercritical drying. Nevertheless, owing to strong inter- and intra-molecular hydrogen bonds, high degree of polymerization and high degree of crystallinity, the raw material (namely cellulose) is difficult to be dissolved or processed in some common water or organic solvents. So far, some effective green cellulose solvents with less harmful to people's health and environment have been successively found, such as ionic liquid, *N*methylmorpholine-*N*-oxide (NMMO), and NaOH/PEG solution.[24-26]

Herein, a facile chemical co-precipitation approach was reported to synthesize the well-dispersed superparamagnetic γ-Fe_2O_3 nanoparticles encapsulated in the 3D architectures of cellulose aerogels. Meanwhile, a green low-cost NaOH/PEG aqueous solution was carried out to fabricate cellulose hydrogels (the precursor of cellulose aerogels), and the waste wheat straw was used as the cellulose source. The nanoparticle sizes, pore characteristic parameters, magnetic properties, and mechanical properties of the synthetic γ-Fe_2O_3 @ cellulose aerogels (γ-Fe_2O_3 @ CA) could be flexibly tailored by adjusting the concentrations of the initial reactants. Meanwhile, a possible mechanism schematic for the preparation of the composites is proposed. As an example of potential applications, this class of green γ-Fe_2O_3 @ CA nanocomposites were used as environmentally friendly adsorbents in heavy metal waste water treatment [hexavalent chromium, Cr(Ⅵ)]. In addition, unlike some previous reports[18~20,22] with a final removal treatment of template, the template (i. e., cellulose aerogels) was retained in this paper, considering its numerous good features (e. g., large specific area and pore volume) and importance in the dispersion of nanoparticles which plays an improtant role in adsorption property. Moreover, we expect that the hybrid γ-Fe_2O_3@ CA will display a synergistic effect of obtaining a strong ability to adsorb the heavy metal ions.

2 Materials and methods

2.1 Materials

Native wheat straw was grinded and then screened through a sixty mesh sieve, and the resulting powder was subsequently collected and dried in a vacuum oven at 60 ℃ for 24 h before used. All chemical reagents with analytical grade are purchased from Tianjin Kemiou Chemical Reagent Co., Ltd. (China) and used without further purification.

2.2 Preparation of Hybrid γ-Fe_2O_3@CA

The synthesis pathway of superhydrophobic and magnetic wood was described in Fig. 1. The untreated wood materials (step a) were suspended in the 20 mL of 0.2 mol · L^{-1} of freshly pre- pared aqueous solution of $FeSO_4$ and $CoCl_2$ with a molar ration of $[Fe^{2+}]/[Co^{2+}]=2$via a vacuum process before transferred into a Teflon-lined stainless steel autoclave and heated the system to 90 ℃ for 3 h to thermally precipitate the non-magnetic metal hydroxides on the template (step b). Heating changed the color from transparent to translucent orange[27]. Until the end of the reaction, the supernatant was poured out and 15 mL of 1.32 mol · L^{-1} NaOH solution with KNO_3 ($[Fe^{2+}]/[NO_3^-]=0.44$) was added into the Teflon-lined stainless-steel autoclave. After an additional hydro- thermal processing for 6 h at 90 ℃, the precipitated precursors were converted into ferrite nanoparticles on the wood surfaces, resulting in high surface roughness with excellent magnetic property (step c). Finally, the prepared wood specimens were removed from the solution and washed by ultrasonically rinsed in deionized water for 30 min, and dried at 50 ℃ for over 24 h in the vacuum chamber. Using the above methods, the hydrophilic $CoFe_2O_4$ films were synthesized on the wood surfaces.

The surface modification of the magnetic wood was carried out by a self-assembly of OTS layer. The magnetic woods were immersed into the 20 mL of 5% (V/V) OTS ethanol solution at room temperatures under continuous mechanical stirring for 24 h and a layer of OTS molecular covered on magnetic wood surface (step d).

Then the samples were dried at 50 ℃ for over 24 h in the vacuum.

Finally, the superhydrophobic wood surface was obtained.

2.3 Characterizations

Transmission electron microscopy (TEM) and high-resolution TEM (HRTEM) observations, and selected area electron diffraction (SAED) were performed with a FEI, Tecnai G2 F20 TEM with a field-emission gun operating at 200 kV. The samples were suspended in ethanol and were prepared by being drop-cast onto a carbon-coated 200-mesh copper grid and subsequently dried at room temperature.

N_2 adsorption-desorption measurements were implemented at −196 ℃ using an accelerated surface area and porosimetry system (3H-2000PS2 unit, Beishide Instrument S&T Co., Ltd.). Prior to the measurements, all of the samples were outgassed at 90 ℃ for 10 h for the removal of any moisture or adsorbed contaminants. The specific surface area was calculated over a relative pressure range of 0.05-0.30 from multipoint Brunauer-Emmett-Teller (BET) method. The nitrogen adsorption volume at the relative pressure (P/P_0) of 0.994 was used to determine the pore volume. The pore diameter distributions were calculated from the data of the adsorption branch of the isotherm using Barrett-Joyner-Halenda (BJH) method.

X-ray photoelectron spectroscopy (XPS) was carried out using a Thermo Escalab 250Xi XPS spectrometer equipped with a dual X-ray source using Al-Kα. Deconvolution of the overlapping peaks was performed using a mixed Gaussian-Lorentzian fitting program (Origin 8.5, Originlab Corporation). X-ray diffraction (XRD) spectroscopy was implemented on a Bruker D8 Advance TXS XRD instrument with Cu Kα (target) radiation (λ = 1.5418 Å) at a scan rate (2θ) of 4° · min^{-1} and a scan range from 5 to 90°. Fourier transform infrared spectra (FTIR) were recorded by a Nicolet Nexus 670 FTIR instrument in the range of 400-4000 cm^{-1} with a resolution of 4 cm^{-1}. All of the samples were ground into a powder and then blended with KBr before pressing the mixture into ultra-thin pellets. Thermal stabilities were determined using a thermal gravity (TG) analyzer (TA, Q600) from room temperature to 800 ℃ at a heating rate of 10 ℃ · min^{-1} under a nitrogen atmosphere. A Quantum Design Magnetometer (MPMS XL-7) using a superconducting quantum interference device (SQUID) sensor was used to make measurements of the magnetic properties of the composites in applied magnetic fields over the range from −20000 to +20000 Oe at 298 K.

Compression tests were performed with a universal testing machine (Suns, UTM4304X), and the compressing velocity was set to 2 mm · min^{-1}. Before the tests, the aerogels were tailored to the same sizes of around 30 mm in diameter and 10 mm in length. The modulus was calculated from the initial linear region of the stress-strain curves in low strain (<6%). The energy absorption by the samples was taken as the area below the stress-strain curve between 0% and 90% strain.

2.4 Cr(Ⅵ) Adsorption Experiments

Solutions with different concentrations of Cr(Ⅵ) were prepared using $K_2Cr_2O_7$ as the source of heavy metal ions, respectively. The pH value was adjusted to 3 using HCl (0.1 M). The as-prepared PCA and γ-Fe_2O_3@CA were used as adsorbents. For the adsorption kinetic study, 100 mg of adsorbent was added to 50 mL solutions with a Cr(Ⅵ) initial concentration of 10 mg · L^{-1}. After a specified time, the adsorbent was separated by centrifugation from the Cr(Ⅵ) solutions, and the supernatant was collected and analyzed using an inductively coupled plasma-mass spectrometer (ICP-MS; Agilent 7700, Agilent Technologies, Santa Clara, California) to measure the concentration of Cr(Ⅵ) in the remaining solution. For the adsorption isothermal study, 50 mg of adsorbent was added to 25 mL of solution with different concentrations under stirring for 2 h at room temperature. After the centrifugal separation, the concentration of Cr(Ⅵ) in the supernatant was also tested. Note that all the adsorption experiments were carried out at room temperature.

3 Result and discussion

3.1 Characterizations of γ-Fe_2O_3@CA by TEM, SAED, XPS, XRD, and FTIR

Micro/nano-scale 3D architectures of the aerogels are suitable reacting sites, which allow guest molecules to penetrate into their inner spaces (Fig. 1a). Meanwhile, strong electrostatic interactions would occur between the incorporated Fe^{2+}/Fe^{3+} ions and cellulose macromolecules, on account that the electron-rich oxygen atoms of polar hydroxyl and ether groups of cellulose are expected to interact with the electropositive transition metal cations. This interaction would result in the immobilization of cations in the aerogels matrixes, contributing to uniform distribution and stability.[28-30] The subsequent dipping process accompanied by heating results in the transformation of cations and the ultimate in-situ formation of γ-Fe_2O_3 nanoparticles, and the nanoparticles are stabilized at the previous ionized selective sites of the cellulose surface without drastic aggregation as shown in Fig. 1b.

The HRTEM image of γ-Fe_2O_3@ CA reveals the presence of γ-Fe_2O_3 as shown in Fig. 1c; i. e., lattice fringes with spacings of around 0. 48 nm and 0. 29 nm agree well with the (111) and (220) lattice spacings.[31] Besides, the corresponding SAED pattern exhibits homogeneous ring patterns typical of nanocrystalline materials, and is consistent with the lattice spacing of cubic γ-Fe_2O_3 (the inset in Fig. 1c). XPS was used to precisely determine the oxidation state of the synthesized particles. The Fe 2p peaks at binding energies of 711. 2 and 724. 7 eV (Fig. 1d), closely correspond to the Fe $2p_{3/2}$ and Fe $2p_{1/2}$ spin-orbit peaks of γ-Fe_2O_3.[32] The satellite peak at around 719. 3 eV is another evidence of Fe^{3+}. For further clarifying the crystal structure and phase purity of the synthesized particles, XRD analysis was carried out. For comparision, the XRD patterns were both normalized to the maximum line intensity. As shown in Fig. 1e, the cellulose characteristic peaks could be seen at around 12. 3°, 20. 2°, and 22. 0° for both PCA and γ-Fe_2O_3@ CA, corresponding to (101), $(10\bar{1})$, and (002) planes. Besides, the peaks of γ-Fe_2O_3 @ CA at around 30. 2°, 35. 6°, 43. 4°, 53. 7°, 57. 4°, and 62. 9° are consistent with the JCPDS file of γ-Fe_2O_3 (No. 39-1346). Moreover, the *d*-spacings also match well with the JCPDS file (see Table S1 and Fig. S1), indicating the formation of γ-Fe_2O_3 with a cubic spinel crystalline structure. Meanwhile, there is no indication of any other iron oxide phases in the XRD pattern of γ-Fe_2O_3@ CA. Fig. 1f present the FTIR spectra of PCA and γ-Fe_2O_3@ CA. Apart from the characteristic peaks derived from cellulose (for details, see supplementary information), the FTIR spectrum of γ-Fe_2O_3 @ CA presents two new peaks located at 546 and 634 cm^{-1} related to stretching vibration of Fe-O (Fig. 1f),[33] which further prove the generation of γ-Fe_2O_3. Meanwhile, the band at 3039-3710 cm^{-1} assigned to the stretching vibrations of hydroxyl groups, shifts from 3444 cm^{-1} to lower wavenumber 3423 cm^{-1}, suggesting the strong interaction between the hydroxyl groups of cellulose aerogels and the encapsulated γ-Fe_2O_3. From the above, the γ-Fe_2O_3 nanoparticles were successfully fabricated and encapsulated in the 3D architectures of cellulose aerogels with good dispersion via the aforementioned facile approach.

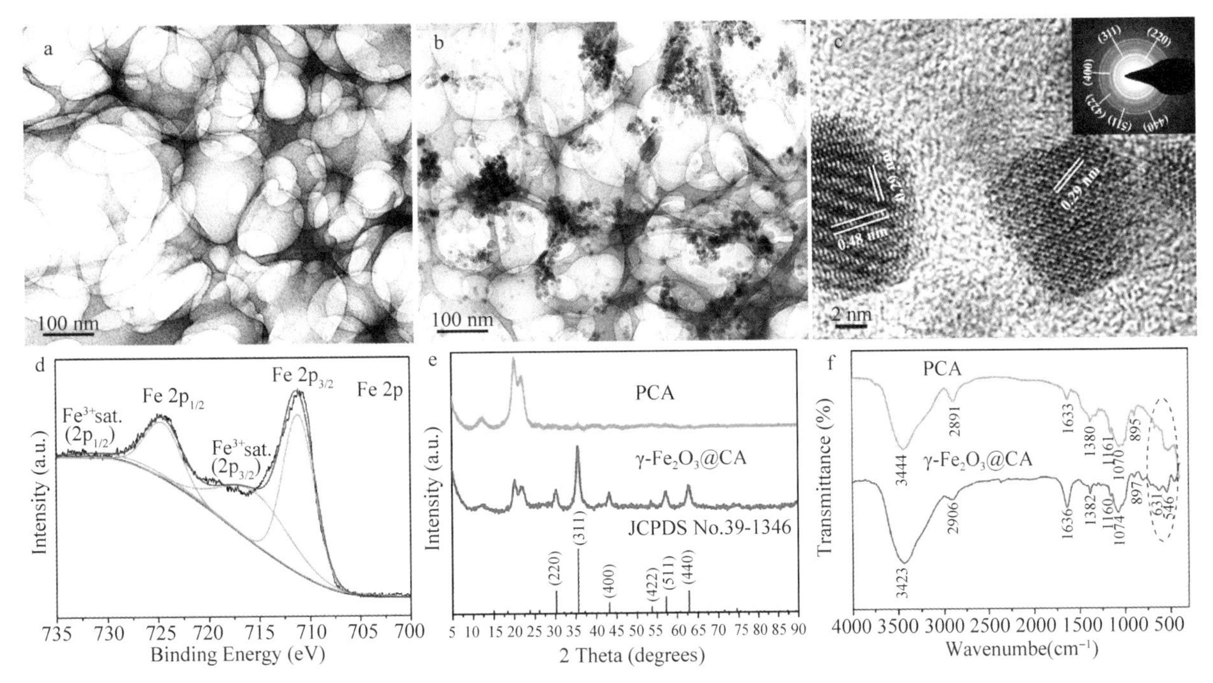

Fig. 1 (a) TEM image of PCA. (b) TEM image, (c) HRTEM image, and (d) XPS spectrum of γ-Fe_2O_3@CA. The inset in c shows the SAED pattern of γ-Fe_2O_3@CA. (e) XRD patterns of PCA and γ-Fe_2O_3@CA, respectively. The positions and intensities of the diffraction lines of γ-Fe_2O_3(i. e., JCPDSNo. 39-1346) are shown at the bottom. (f) FTIR spectra of PCA and γ-Fe_2O_3@CA, respectively

3.2 Tunability of γ-Fe_2O_3 Nanoparticles Sizes by Concentrations of Initial Reactants

The sizes of nanoparticles could be tailored by altering the total concentration of the $FeCl_3/FeSO_4$ salts (with a fixed molar ratio of 2) from 0.01 mol · L^{-1} to 0.05 mol · L^{-1} and 0.1 mol · L^{-1}, and the resulting composites are marked as γ-Fe_2O_3@ CA-001, γ-Fe_2O_3@ CA-005, and γ-Fe_2O_3@ CA-01, respectively. As shown in Fig. 2a-c, the faceted shaped nanoparticles exhibit favorable dispersion, while their sizes are obviously varied with the concentrations of initial reactants. Their size distributions are presented in Fig 2d-e. It could be seen that the average size (d) increases from 8.1 nm to 9.4 nm and eventually to 11.7 nm when the concentration of ferrous ions gradually increases from 0.01 mol · L^{-1} to 0.05 mol · L^{-1} and eventually to 0.1 mol · L^{-1}. This upward tendency could be ascribed to the nucleation stage of the nanoparticles. When the concentration of consistent species reaches the supersaturation, a nucleation stage would occur; i. e., the nuclei starts to grow by flocculation or aggregation of nuclei into larger particles until the aforementioned final size is obtained. Therefore, the increase of reactant concentration produces positive effect to reach the sufficient number of nanoparticles for the nucleation and growth stages. Besides, the formed greater number of small crystals are prone to aggregation owing to the strong interaction and high reactant concentration, which results in higher sizes. Meanwhile, the accompanying Ostwald ripening-small crystals would shrink while large ones would grow-then lead to a broad size distribution. [34]

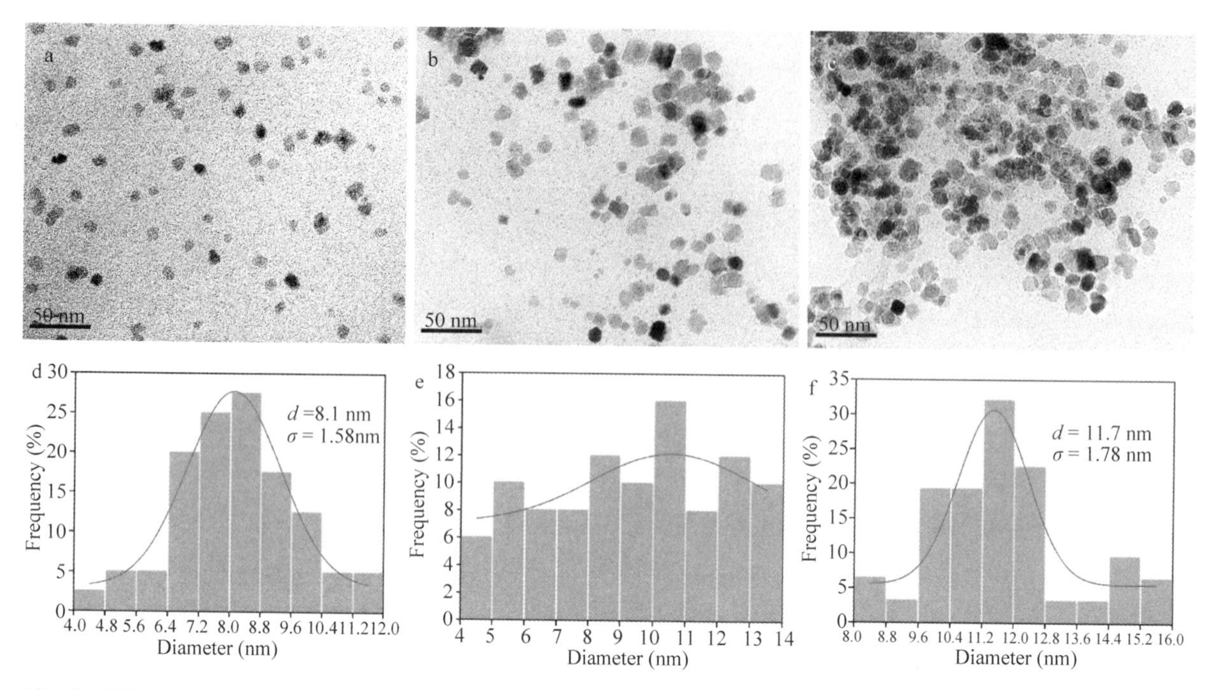

Fig. 2 TEM images of (a) γ-Fe_2O_3@CA-001, (b) γ-Fe_2O_3@CA-005, and (c) γ-Fe_2O_3@CA-01. (d), (e), and (f) are the corresponding particle diameter distributions of the composites of (a), (b), and (c)

3.3 Pore Structure Characteristics, and Thermal, Mechanical and Magnetic Properties of γ-Fe_2O_3@CA

N_2 adsorption-desorption isotherms provide much useful information about the pore structures of this class of hybrid γ-Fe_2O_3@ CA (Fig. 3a). These isotherms could all be classified to type Ⅳ according to IUPAC classification,[35] and the adsorption-desorption loops belong to the type H_3,[36] which reveals the existence of slit-shaped pores. The adsorption uptakes in the P/P_0 range between 0 and 0.6 increase slightly for all the samples, indicating the existence of few micropores. Meanwhile, nitrogen molecules are gradually adsorbed on the internal surface of porous structures from single to multilayer in this period. Thereafter, the adsorption amounts increase quickly in the range of 0.6-1, which generates obvious hysteresis loops, and still do not reach a plateau near the P/P_0 of 1.0, revealing the presence of mesopores and macropores. The results are in good agreement with the observation from TEM images. Moreover, the pore size distributions as shown in Fig 3b also prove the existence of mesopores and macropores, which exhibit that the pore sizes are within the scope of 3-140 nm. Besides, the two maximum peaks at around 3.8 nm and 16.1 nm could also be clearly distinguished. BET and BJH analysis gives specific surface areas and pore volumes of 136.8 $m^2 \cdot g^{-1}$ and 0.88 $cm^3 \cdot g^{-1}$ for PCA, 140.8 $m^2 \cdot g^{-1}$ and 0.92 $cm^3 \cdot g^{-1}$ for γ-Fe_2O_3@ CA-001, 156.3 $m^2 \cdot g^{-1}$ and 0.97 $cm^3 \cdot g^{-1}$ for γ-Fe_2O_3@ CA-005, and 173.5 $m^2 \cdot g^{-1}$ and 1.02 $cm^3 \cdot g^{-1}$ for γ-Fe_2O_3@ CA-01, as shown in Table 1, respectively. It could be seen that the insertion of γ-Fe_2O_3 improves the pore characteristic parameters of the composites compared with those of PCA. Besides, the specific surface area and pore volume of the composites are effectively controlled by the content of incorporated γ-Fe_2O_3 nanoparticles. The values apparently increase with the increasing of the γ-Fe_2O_3 nanoparticles content. These enhancement phenomena might be on account that the insertion of higher content of nanoparticles more effectively strengthens the 3D skeleton structure of the aerogels, leading to more resistant to shrinkage and collapse during the drying treatment.

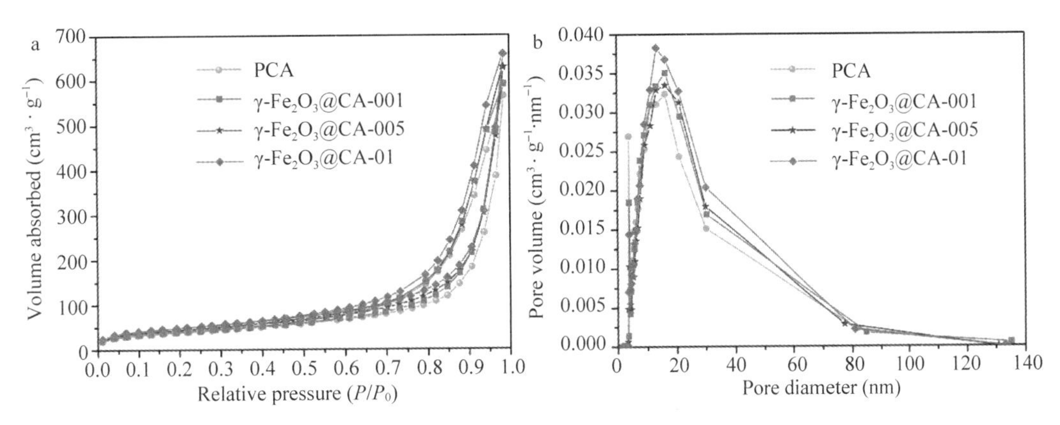

Fig. 3 (a) N_2 adsorption-desorption isotherms and (b) pore size distributions of PCA, γ-Fe_2O_3@CA-001, γ-Fe_2O_3@CA-005, and γ-Fe_2O_3@CA-01, respectively

Considering the importance of thermal stability in the applications of nanocomposites, the thermal behaviors of the synthetic composites and PCA were studied by TG technique. TG and derivative thermal gravity (DTG) traces of these materials exhibit three major weight loss peaks at around 40–120 ℃, 217–400 ℃, and 585–648 ℃ (Fig. 4a, b). The two former peaks are attributed to the evaporation of absorbed water and the decomposition of cellulose[37], respectively. Moreover, for the investigation of the weight loss at 585–648 ℃, γ-Fe_2O_3@CA-01 was selected as an example to undergo the thermal treatments at various temperatures (the details about the processes are given in Supporting information), and the resulting products were tested by XRD to identify the compositions. As shown in Fig. S2, after the thermal treatment at 610 ℃, γ-Fe_2O_3@CA-01 is reduced to Fe (JCPDS No. 65-4899, denoted with triangles) possibly due to the reaction occurring between the γ-Fe_2O_3 nanoparticles and the residual char obtained from the pyrolysis of cellulose. Meanwhile, this reaction also results in the oxidation of residual char to carbon dioxide, leading to weakening or disappearance of graphite characteristic peaks of residual char. Therefore, the reaction between the γ-Fe_2O_3 and the residual char is responsible for this weight loss. As the processing temperature rises from 610 ℃ to 660 ℃ and eventually to 800 ℃, the XRD peaks of γ-Fe_2O_3 gradually weaken and disappear, while the new peaks corresponding to $CFe_{15.1}$ (JCPDS No. 52-0512) are generated at 800 ℃. In addition, regardless of the impurities with low content (i. e., $CFe_{15.1}$ and carbon residue), we suppose that the products at 800 ℃ only consist of Fe. Thus, the contents of γ-Fe_2O_3 could be roughly calculated to be 14.1% in γ-Fe_2O_3@CA-001, 26.2% in γ-Fe_2O_3@CA-005, and 34.1% in γ-Fe_2O_3@CA-01, respectively.

In addition, the well-known inferior mechanical properties of this class of porous aerogels, which result from the defects such as dangling ends and loops in their structures,[38] are the main challenges that require urgent attention and immediate improvement. However, the combination with various fillers such as $CoFe_2O_4$, clay, graphene oxide, silica, and polypyrrole has been reported to make contribution to significantly improving the mechanical properties and fragility of aerogels.[39-43] Therefore, we expect that the incorporation of γ-Fe_2O_3 will also be helpful to enhance the mechanical strength of the hybrid aerogels. The stress-strain behaviors were measured to determine and compare the mechanical properties of PCA and γ-Fe_2O_3@ CA. As shown in Fig. 4c, the compression stress-strain curve could be roughly divided into four stages.[44] First, the linear elastic behavior resulted from elastic cell wall bending takes place in low strain (<6%); second, the curves gradually transform from linear to non-linear in higher strain; third, a horizontal plateau region appears after reaching yield stress, and cell collapse typically occurs; final, the loose porous 3D network structure starts to touch resulting in considerable stiffening. The composites have higher compressibility than that of PCA. The elasticity modulus value of PCA is 1.58 Mpa, and the value increase from 2.09 MPa for γ-Fe_2O_3@ CA-001 to 2.68 MPa for γ-Fe_2O_3@ CA-005 and eventually to 3.19 MPa for γ-Fe_2O_3@ CA-01. Moreover, the energy absorption in compression is calculated to be 240.8 kJ · m^{-3} for PCA, 338.9 kJ · m^{-3} for γ-Fe_2O_3@ CA-001, 487.8 kJ · m^{-3} for γ-Fe_2O_3@ CA-005, and 658.0 kJ · m^{-3} for γ-Fe_2O_3@ CA-01. The higher values reveal the stronger resistance to compression deformation. These results indicate that the incorporation of γ-Fe_2O_3 nanoparticles contributes to apparent improvement in the compression mechanical properties of the composites. Meanwhile, the higher particles loading contents possibly induce further enhancement in the mechanical properties from γ-Fe_2O_3@ CA-001 to γ-Fe_2O_3@ CA-01.

The room-temperature hysteresis cycles of the composites show absence of hysteresis and coercivity (Fig. 4d), which is characteristic of superparamagnetic behavior caused by the small particle size of γ-Fe_2O_3 smaller than the related critical single-domain size.[45] The curves all increase rapidly with increasing applied magnetic field, which exhibit saturation magnetization (M_s) values of 1.6 emu · g^{-1} for γ-Fe_2O_3@ CA-001, 9.9 emu · g^{-1} for γ-Fe_2O_3@ CA-005, and 18.0 for γ-Fe_2O_3@ CA-01. Apparently, the M_s values significantly increase as the total concentration of the $FeCl_3$/$FeSO_4$ salts increases, which demonstrates the tailorability of the magnetic property. Additionally, taking the nanoparticles contents in the samples (14.1%, 26.2%, and 34.1%) into consideration, the corresponding calculated M_s values are 11.3 emu · g^{-1}, 37.8 emu · g^{-1} and 52.8 emu · g^{-1}, which are much smaller than the theoretical M_s of 76 emu · g^{-1} for bulk γ-Fe_2O_3. The above differences might be explained by the small particle size effect.[46] In detail, bulk γ-Fe_2O_3 has a spinel-type collinear ferrimagnetic spin structure. Noncollinear spin arrangement at or near the surface of the particle due to a surface effect, is more pronounced in small particle sizes, which significantly affects the magnetic properties. Meanwhie, the cation disorder and magnetically disordered surface layer around the particles could also be responsible for the lower saturation magnetization. Besides, the presence of amorphous impurities (e.g., carbon residue) and carbide impurities (e.g. $CFe_{15.1}$) also can reduce the total magnetization. Additionally, these magnetic materials reach the saturation magnetization at relatively low applied fields (ca. 5 kOe), which might be useful in some special occasions, where a strong magnetic signal is required at small applied magnetic fields (e.g. magnatic resonance imaging, magnetic separation, drug delivery, etc).[47]

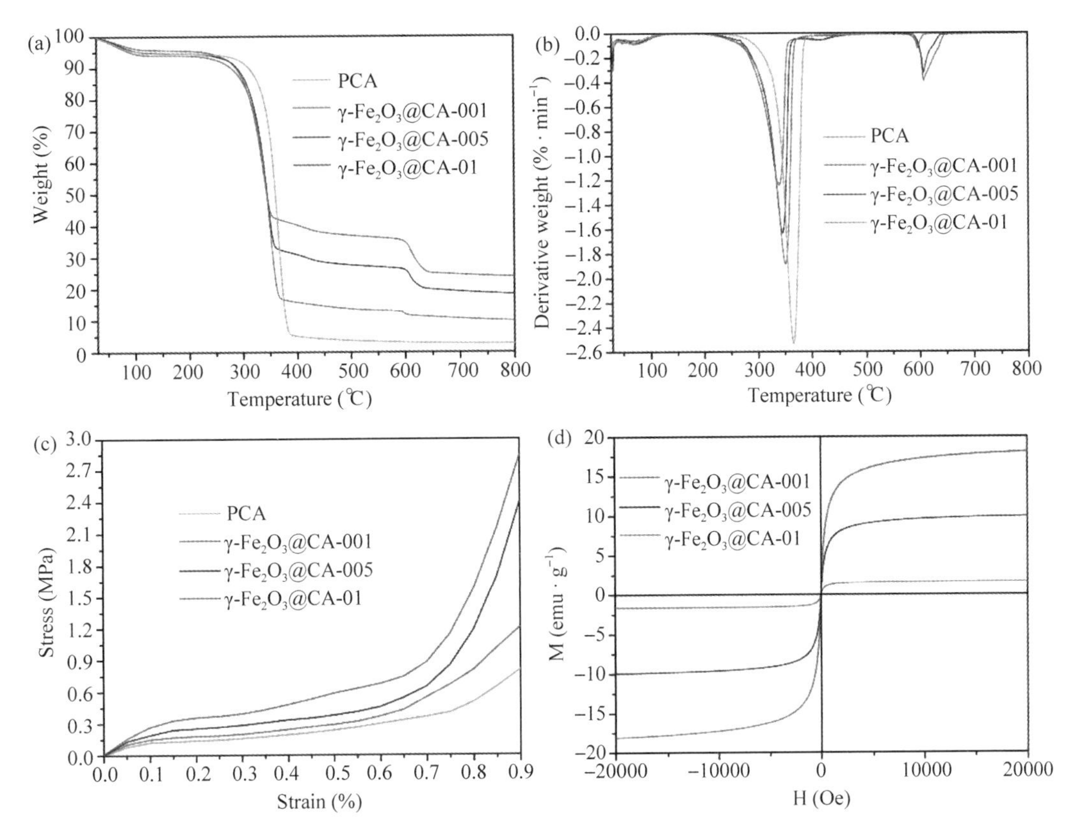

Fig. 4 (a) TG curves, (b) DTG curves, and (c) compression stress-strain curves of PCA, γ-Fe_2O_3@CA-001, γ-Fe_2O_3@CA-005, and γ-Fe_2O_3@CA-01, respectively. (d) Hysteresis cycles of γ-Fe_2O_3@CA-001, γ-Fe_2O_3@CA-005, and γ-Fe_2O_3@CA-01

Table 1 Characteristics of PCA and γ-Fe_2O_3@CA with the different precursor concentrations

Sample	γ-Fe_2O_3 fraction (wt%)	Particle size (nm)	Specific surface area($m^2 \cdot g^{-1}$)	Pore volume ($cm^3 \cdot g^{-1}$)	Elasticity modulus (MPa)	Energy absorption ($kJ \cdot m^{-3}$)	Saturation magnetization ($emu \cdot g^{-1}$)
PCA	—	—	136.8	0.88	1.58	240.8	—
γ-Fe_2O_3 @CA-001	14.1%	8.1±1.58	140.8	0.92	2.09	338.9	1.6
γ-Fe_2O_3 @CA-005	26.2%	9.4±2.76	156.3	0.97	2.68	487.8	9.9
γ-Fe_2O_3 @CA-01	34.1%	11.7±1.78	173.5	1.02	3.19	658.0	18.0

3.4 Macroscopic Phenomena of Magnetic Responsiveness of γ-Fe_2O_3 Nanoparticles, γ-Fe_2O_3@ Cellulose Hydrogels, and γ-Fe_2O_3@ CA

The synthetic γ-Fe_2O_3 nanoparticles and the related hybrid products including hydrogels and aerogels are demonstrated to be magnetically responsive materials and actuators as shown in Fig. 5. It could be seen that the nanoparticles could be easily recycled from the black suspension by magnet attraction (see also Supplementary

Video 1). The hydrogels are sensitive to the applied magnetic field, which could flexibly float in the water following the magnet movement (see also Supplementary Video 2). In addition, the dried aerogels from the different reactants concentrations (γ-Fe_2O_3@CA-001, -005, and -01) also exhibit favorable magnetic responsiveness. They could tightly adhere to the magnet surface, and maintain well-defined shapes (see also Supplementary Video 3). Therefore, this class of magnetically responsive aerogels with high mechanical strength and large specific surface might find some potential applications like recyclable green magnetically driven adsorbents for water purification, and some biodegradable electromagnetic devices.

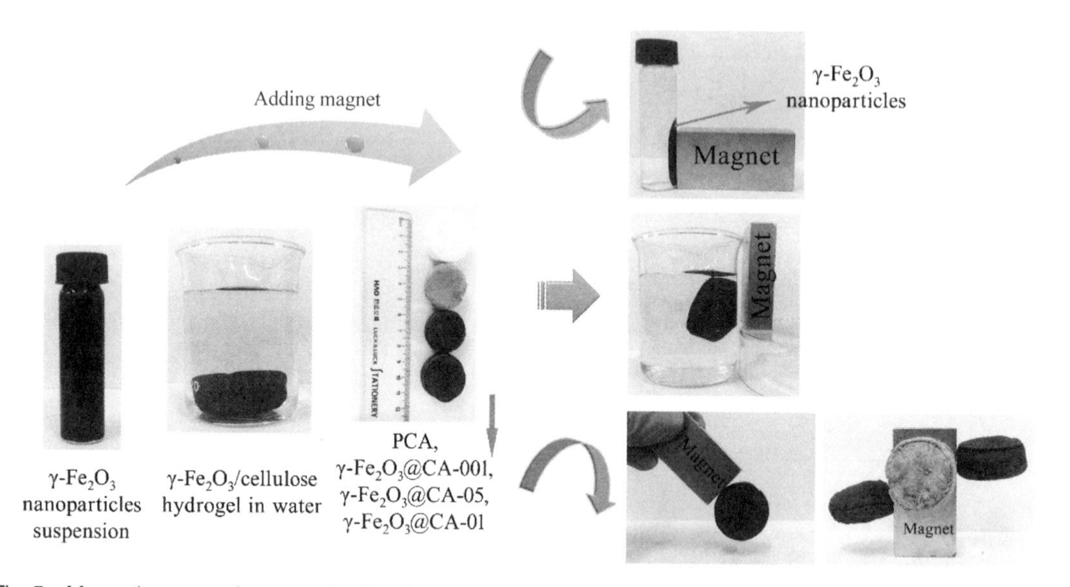

Fig. 5 Magnetic responsiveness of γ-Fe_2O_3 nanoparticles, γ-Fe_2O_3@cellulose hydrogels, and γ-Fe_2O_3@CA

3.5 Schematic Illustration for Preparation of γ-Fe_2O_3@CA

The schematic illustration for the preparation of γ-Fe_2O_3@CA is presented in Fig. 6. As previously mentioned, the metal ions are tightly anchored to the surface of cellulose chains by strong electrostatic (i. e., ion-dipole) interactions. The subsequent immersion in NaOH solution leads to the formation of γ-Fe_2O_3 nanoparticles, which are stabilized at the previous ionized selective sites resulting in a good dispersion. Moreover, the drying method (i. e., freeze drying) causes limited shrinkage, which contributes to maintaining the original 3D skeleton structure of the hydrogels. The insertion of γ-Fe_2O_3 further strengthens the 3D skeleton, and enhances resistance to shrinkage and collapse.[48]

3.6 Heavy Metal Ion Cr(Ⅵ) Adsorption Properties of γ-Fe_2O_3@CA

For the investigation of the potential applications of γ-Fe_2O_3@CA, we used this class of nanocomposites as environmentally friendly adsorbents in heavy metal waste water treatment. Cr(Ⅵ) is a common heavy metal contaminant in water resources, which has been designated as one of the top-priority toxic pollutants by the U. S. EPA due to its carcinogenesis and strong toxicity to humans and animals. Cr(Ⅵ) is usually released from various industrial operations, such as chromate production, mining and metallurgy, petroleum refining, electroplating, leather tanning, etc.[49] Unreasonably treated or even untreated Cr(Ⅵ)-containing wastewater is arbitrarily discharged, resulting in the serious Cr(Ⅵ) contamination. Undoubtedly, effective removal of Cr

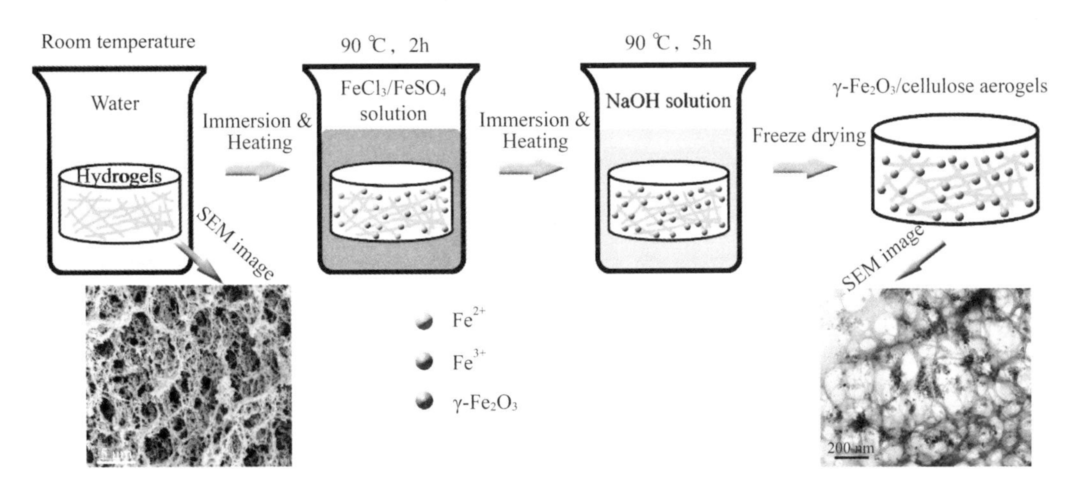

Fig. 6 Schematic illustration for the preparation of γ-Fe_2O_3@CA

(Ⅵ) from water is of great significance. So far, all sorts of adsorbents including natural materials (e. g., clay, rice husk ash, activated alumina, fuller's earth, fly ash, saw dust) and synthetic materials (e. g., active carbon, mesoporous silica, polyaniline coated ethyl cellulose) have been carried out to treat with Cr (Ⅵ)-containing wastewater.[50-54] However, compared with these adsorbents, γ-Fe_2O_3@ CA has not only good environmental benefits and strong selectiveness in Cr(Ⅵ) removal,[55] but also excellent magnetic responsibility contributing to achieving a given level of magnetic separation with much less energy. Besides, due to the large specific surface area and 3D hierarchical porous structure of the aerogels template as well as good dispersion of nanoparticles, we expect that the hybrid γ-Fe_2O_3@ CA will display a synergistic effect of obtaining a strong ability to adsorb Cr(Ⅵ).

The adsorption kinetics of Cr(Ⅵ) on PCA, γ-Fe_2O_3@ CA-001, γ-Fe_2O_3@ CA-005, and γ-Fe_2O_3@ CA-01 are presented in Fig. 7a. Interestingly, the four kinds of materials all have rapid Cr(Ⅵ) adsorption capacity, and reach adsorption equilibrium within several minutes. Moreover, γ-Fe_2O_3@ CA-01 displays the Cr(IV) removal efficiency of 91.8%, much higher than those of γ-Fe_2O_3@ CA-005 (42.7%), γ-Fe_2O_3@ CA-001 (12.3%), and PCA (6.7%) under the same conditions, indicating the potential stronger adsorption property for Cr(Ⅵ) removal. In Fig. 7b, the adsorption kinetics data is well fitted by the pseudo-second-order rate kinetic model expressed as Eq. (1).[56]

$$\frac{t}{Q_t}=\frac{t}{Q_e}+\frac{1}{K_2Q_e^2} \tag{1}$$

where Q_e and Q_t ($mg \cdot g^{-1}$) is the amount of adsorbed heavy metal at equilibrium and at time t, respectively. K_2 is the adsorption rate constant ($g \cdot mg^{-1} \cdot min^{-1}$). The correlation coefficients (R^2) for PCA, γ-Fe_2O_3@ CA-001, γ-Fe_2O_3@ CA-005 and γ-Fe_2O_3@ CA-01 are 0.9988, 0.9992, 0.9999 and 0.9999, indicating that the Cr(Ⅵ) adsorption follows the pseudo-second-order kinetic model. According to the slope and intercept, the K_2 values are calculated to be 0.3433, 0.3800, 0.6035 and 0.7364 $g \cdot mg^{-1} \cdot min^{-1}$ for PCA, γ-Fe_2O_3@ CA-001, γ-Fe_2O_3@ CA-005, and γ-Fe_2O_3@ CA-01, respectively. Apparently, the K_2 values increase with the increase of γ-Fe_2O_3 content, which reveals that the higher γ-Fe_2O_3 proportion

contributes to the improvement in Cr(Ⅵ) adsorption rate.

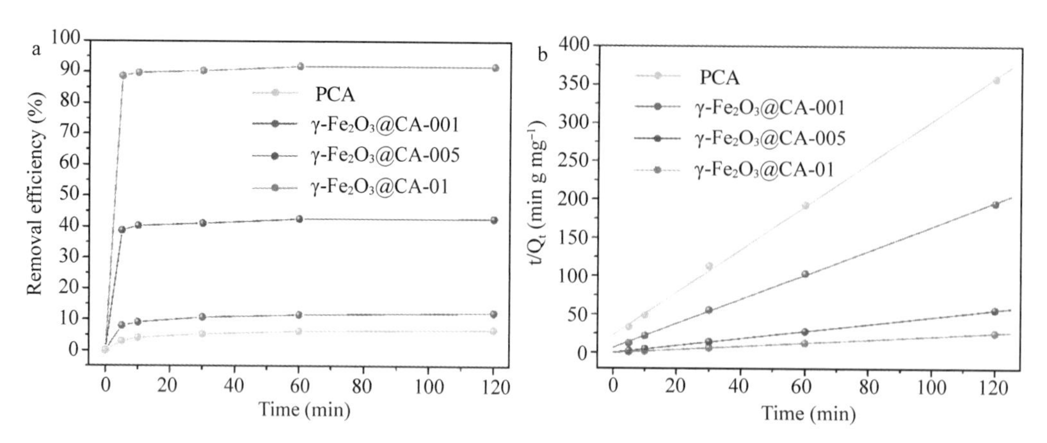

Fig. 7 (a) Adsorption kinetics of Cr(Ⅵ) and (b) pseudo-second order model for adsorption of Cr(Ⅵ) on PCA, γ-Fe_2O_3@CA-001, γ-Fe_2O_3@CA-005, and γ-Fe_2O_3@CA-01, respectively

Additionally, the adsorption mechanisms of γ-Fe_2O_3@ CA for Cr(Ⅵ) is believed to primarily involve electrostatic attraction. [55,57] It has been proposed that cellulose and iron-oxide surfaces are protonated at low pH, so that the surface charge is positive. Meanwhile, when pH value is lower than 6.8, $HCrO^{4-}$ is the dominant form of hexavalent chromium in solution. Therefore, the strong electrostatic attraction would occur between the positively charged γ-Fe_2O_3@ CA and the negatively charged Cr(Ⅵ) ($HCrO^{4-}$) species, allowing the approach of the Cr(Ⅵ) ions to the adsorbent surfaces. Moreover, the above phenomenon [i. e. , the positive correlation between the γ-Fe_2O_3 proportion and the Cr(Ⅵ) adsorption rate] is possibly attributed to more adsorption sites on the surface of the absorbent when γ-Fe_2O_3@ CA with higher γ-Fe_2O_3 proportion was used.

Fig. 8a shows the adsorption isotherms of Cr(Ⅵ) on PCA, γ-Fe_2O_3@ CA-001, γ-Fe_2O_3@ CA-005, and γ-Fe_2O_3@ CA-01, which are obtained with different initial concentrations ranging from 10 to 100 mg · L^{-1}. Apparently, for all the samples, the isotherms rapidly increase in the lower concentration region, and subsequently gradually flatten out in the high concentration region. The Langmuir adsorption model is adopted to fit the equilibrium adsorption data and estimate the maximum adsorption capacity. The linearized form of the Langmuir equation isotherm is expressed as Eq. (2): [58]

$$\frac{C_e}{Q_e}=\frac{1}{K_L Q_m}+\frac{C_e}{Q_m} \tag{2}$$

where Q_e is the amount of Cr(Ⅵ) adsorbed per unit weight of the adsorbent at equilibrium (mg · g^{-1}), C_e is the equilibrium concentration of Cr(Ⅵ) (mg/L), K_L is the Langmuir adsorption equilibrium constant related to the energy, and Q_m is the maximum adsorption capacity (mg · g^{-1}). As shown in Fig. 8b, the experimental data fits the Langmuir adsorption model well, with R^2 of 0.9992 for PCA, 0.9993 for γ-Fe_2O_3 @ CA-001, 0.9989 for γ-Fe_2O_3@ CA-005, and 0.9997 for γ-Fe_2O_3@ CA-01. In addition, the maximum adsorption capacities (Q_m) of these samples calculated from the Langmuir isotherm model are listed in Table 2. For comparison, some literature values of Q_m of other iron oxide-based adsorbents for Cr(Ⅵ) adsorption are also given.[59-63] Among the four kinds of adsorbents in this study, PCA has the lowest Q_m of 0.64 mg · g^{-1}. The similar phenomenon also has been observed by Dupont and Guillon.[64] They conducted studies on Cr(Ⅵ) adsorption with pure cellulose at acidic pH, and the results also display no significant adsorption onto cellulose. However, the Q_m of PCA is still close to that of commercial α-Fe_2O_3 (0.68 mg · g^{-1}). With the incorporation of γ-Fe_2O_3, the Q_m values significantly increase from 1.14 mg · g^{-1} for γ-Fe_2O_3@CA-001 to 3.94 mg · g^{-1} for γ-Fe_2O_3@CA-005 and eventually to 10.2 mg · g^{-1} for γ-Fe_2O_3@CA-01. Obviously, the higher content of γ-Fe_2O_3 will lead to stronger Cr(Ⅵ) adsorption capacity; besides, the results also reveal that the Cr(Ⅵ) adsorption capacity is controllable by adjusting the content of γ-Fe_2O_3 (namely the concentrations of the initial reactants). Moreover, the adsorption capacity of γ-Fe_2O_3@CA-01 is much higher than or comparable with those of other iron oxide-based adsorbents listed in Table 2, indicating the superior Cr(Ⅵ) adsorption capability. Meanwhile, considering the advantages of environmental benefits, facile convenient preparation method, high specific surface area and strong mechanical strength, easy recycle and separation by magnet attraction, and tunability of Cr(Ⅵ) adsorption capability, this class of green γ-Fe_2O_3@CA is more favorable and suitable for Cr(Ⅵ) removal from contaminated water.

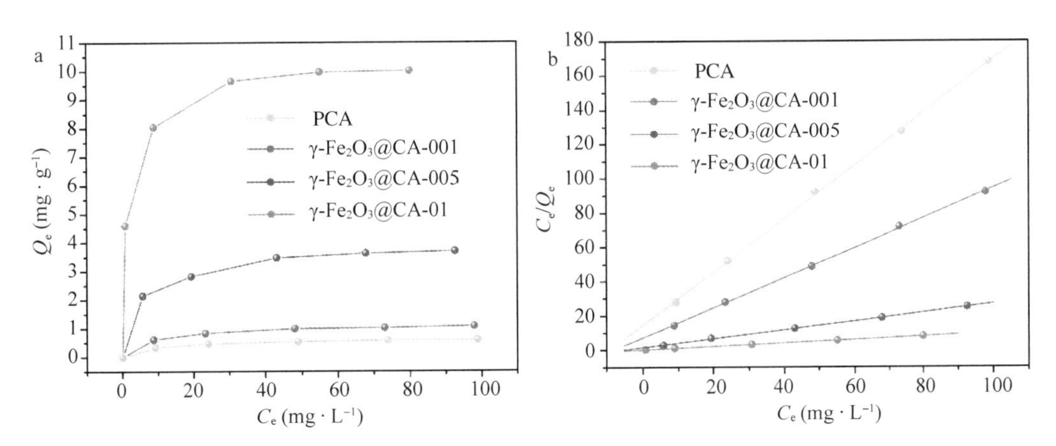

Fig. 8 (a) Adsorption isotherms of Cr(Ⅵ) and (b) Langmuir isotherms plots of Cr(Ⅵ) adsorption on PCA, γ-Fe_2O_3@CA-001, γ-Fe_2O_3@CA-005, and γ-Fe_2O_3@CA-01, respectively.

Table 2 Comparison of maximum adsorption capacities of PCA, γ-Fe_2O_3@CA-001, γ-Fe_2O_3@CA-005, and γ-Fe_2O_3@CA-01 for removal of Cr(Ⅵ) with those of other iron oxide-based adsorbents

Adsorbent	Q_m (mg · g^{-1})	pH	References
PCA	0.64	3	Present study
γ-Fe_2O_3@CA-001	1.14	3	Present study
γ-Fe_2O_3@CA-005	3.94	3	Present study
γ-Fe_2O_3@CA-01	10.2	3	Present study
Commercial α-Fe_2O_3	0.68	3	(59)
Fe_3O_4@n-SiO_2 nanoparticles	3.78	2	(60)
Iron oxide nanoparticles embedded in orange peel pith	5.37	1	(61)
3D flowerlike α-Fe_2O_3	5.4	3	(59)
Fe_3O_4(PAA coated and amino functionalized)	11.2	2	(62)
Diatomite-supported magnetite nanoparticles	11.4	2-2.5	(63)

4 Conclusions

We demonstrate that the 3D architectures of cellulose aerogels based on the green NaOH/PEG solvent, could be served as suitable template for encapsulating the well-dispersed superparamagnetic γ-Fe_2O_3 nanoparticles via a facile chemical co-precipitation approach. The particle sizes, pore characteristic parameters, magnetic property, and mechanical strength of the synthetic γ-Fe_2O_3@CA could be flexibly tailored by adjusting the total concentration of the $FeCl_3/FeSO_4$ salts. Moreover, the composites exhibit superior magnetic responsiveness, which are easily actuated by an applied magnetic field. Meanwhile, the hybrid γ-Fe_2O_3@CA also displays rapid adsorption rate and excellent adsorption ability to remove Cr(Ⅵ) heavy metal ions. Therefore, this class of γ-Fe_2O_3@CA is expected to be useful as a kind of novel environmentally friendly magnetically actuated adsorbent for Cr(Ⅵ) removal from contaminated water.

Acknowledgments

This study was supported by the National Natural Science Foundation of China (grant no. 31270590 and 31470584).

References

[1] Laurent S, Forge D, Port M, et al. Magnetic iron oxide nanoparticles: synthesis, stabilization, vectorization, physicochemical characterizations, and biological applications[J]. Chemical reviews, 2008, 108(6): 2064-2110.

[2] Pouran S R, Raman A A A, Daud W M A W. Review on the application of modified iron oxides as heterogeneous catalysts in Fenton reactions[J]. Journal of Cleaner Production, 2014, 64: 24-35.

[3] Urbanova V, Magro M, Gedanken A, et al. Nanocrystalline iron oxides, composites, and related materials as a platform for electrochemical, magnetic, and chemical biosensors[J]. Chemistry of Materials, 2014, 26(23): 6653-6673.

[4] Gallo J, Kamaly N, Lavdas I, et al. CXCR4-targeted and MMP-responsive iron oxide nanoparticles for enhanced magnetic resonance imaging[J]. Angewandte Chemie International Edition, 2014, 53(36): 9550-9554.

[5] Dhadge V L, Morgado P I, Freitas F, et al. An extracellular polymer at the interface of magnetic bioseparations[J]. Journal of the Royal Society Interface, 2014, 11(100): 20140743.

[6] Lin Y S, Wu S H, Hung Y, et al. Multifunctional composite nanoparticles: magnetic, luminescent, and mesoporous [J]. Chemistry of Materials, 2006, 18(22): 5170-5172.

[7] Rubio-Retama J, Zafeiropoulos N E, Frick B, et al. Investigation of the Relationship between Hydrogen Bonds and Macroscopic Properties in Hybrid Core-Shell γ-Fe_2O_3-P (NIPAM-AAS) Microgels[J]. Langmuir, 2010, 26(10): 7101-7106.

[8] Yanik J, Uddin M A, Ikeuchi K, et al. The catalytic effect of Red Mud on the degradation of poly (vinyl chloride) containing polymer mixture into fuel oil[J]. Polymer degradation and stability, 2001, 73(2): 335-346.

[9] Xiaotun Y, Lingge X, Choon N S, et al. Magnetic and electrical properties of polypyrrole-coated γ-Fe_2O_3 nanocomposite particles[J]. Nanotechnology, 2003, 14(6): 624.

[10] Wang Y, Teng X, Wang J S, et al. Solvent-free atom transfer radical polymerization in the synthesis of Fe_2O_3@ polystyrene core-shell nanoparticles[J]. Nano letters, 2003, 3(6): 789-793.

[11] Yang J, Zhang H, Yu M, et al. High-content, well-dispersed γ-Fe_2O_3 nanoparticles encapsulated in macroporous silica with superior arsenic removal performance[J]. Advanced Functional Materials, 2014, 24(10): 1354-1363.

[12] Behrens S. Preparation of functional magnetic nanocomposites and hybrid materials: recent progress and future directions[J]. Nanoscale, 2011, 3(3): 877-892.

[13] Guo Y, Wang X, Sun R. Cellulose-based self-assembled nanoparticles for antitumor drug delivery[J]. Journal of

Controlled Release, 2013, 1(172): e85.

[14] Cai J, Kimura S, Wada M, et al. Nanoporous cellulose as metal nanoparticles support[J]. Biomacromolecules, 2009, 10(1): 87-94.

[15] Cao G, Liu D. Template-based synthesis of nanorod, nanowire, and nanotube arrays[J]. Advances in colloid and interface science, 2008, 136(1-2): 45-64.

[16] Sun Z, Yuan H, Liu Z, et al. A highly efficient chemical sensor material for H_2S: α-Fe_2O_3 nanotubes fabricated using carbon nanotube templates[J]. Advanced Materials, 2005, 17(24): 2993-2997.

[17] Zhu X, Liu Y, Zhou C, et al. Novel and high-performance magnetic carbon composite prepared from waste hydrochar for dye removal[J]. ACS Sustainable Chemistry & Engineering, 2014, 2(4): 969-977.

[18] Lian J, Duan X, Ma J, et al. Hematite (α-Fe_2O_3) with various morphologies: ionic liquid-assisted synthesis, formation mechanism, and properties[J]. ACS nano, 2009, 3(11): 3749-3761.

[19] Xu X, Cao R, Jeong S, et al. Spindle-like mesoporous α-Fe_2O_3 anode material prepared from MOF template for high-rate lithium batteries[J]. Nano letters, 2012, 12(9): 4988-4991.

[20] Manukyan K V, Chen Y S, Rouvimov S, et al. Ultrasmall α-Fe_2O_3 superparamagnetic nanoparticles with high magnetization prepared by template-assisted combustion process[J]. The Journal of Physical Chemistry C, 2014, 118(29): 16264-16271.

[21] Luo X, Guo B, Luo J, et al. Recovery of lithium from wastewater using development of Li ion-imprinted polymers [J]. ACS Sustainable Chemistry & Engineering, 2015, 3(3): 460-467.

[22] Liu X M, Fu S Y, Xiao H M, et al. Preparation and characterization of shuttle-like α-Fe_2O_3 nanoparticles by supermolecular template[J]. Journal of Solid State Chemistry, 2005, 178(9): 2798-2803.

[23] Xiong R, Lu C, Wang Y, et al. Nanofibrillated cellulose as the support and reductant for the facile synthesis of Fe_3O_4/Ag nanocomposites with catalytic and antibacterial activity[J]. Journal of Materials Chemistry A, 2013, 1(47): 14910-14918.

[24] Moon R J, Martini A, Nairn J, et al. Cellulose nanomaterials review: structure, properties and nanocomposites[J]. Chemical Society Reviews, 2011, 40(7): 3941-3994.

[25] Zhang H, Wu J, Zhang J, et al. 1-Allyl-3-methylimidazolium chloride room temperature ionic liquid: a new and powerful nonderivatizing solvent for cellulose[J]. Macromolecules, 2005, 38(20): 8272-8277.

[26] Fink H P, Weigel P, Purz H J, et al. Structure formation of regenerated cellulose materials from NMMO-solutions [J]. Progress in Polymer Science, 2001, 26(9): 1473-1524.

[27] Wan C, Lu Y, Jiao Y, et al. Cellulose aerogels from cellulose-NaOH/PEG solution and comparison with different cellulose contents[J]. Materials Science and Technology, 2015, 31(9): 1096-1102.

[28] He J, Kunitake T, Nakao A. Facile in situ synthesis of noble metal nanoparticles in porous cellulose fibers[J]. Chemistry of Materials, 2003, 15(23): 4401-4406.

[29] Liu S, Yan Q, Tao D, et al. Highly flexible magnetic composite aerogels prepared by using cellulose nanofibril networks as templates[J]. Carbohydrate polymers, 2012, 89(2): 551-557.

[30] Tingaut P, Zimmermann T, Sèbe G. Cellulose nanocrystals and microfibrillated cellulose as building blocks for the design of hierarchical functional materials[J]. Journal of Materials Chemistry, 2012, 22(38): 20105-20111.

[31] Stagi L, De Toro J A, Ardu A, et al. Surface effects under visible irradiation and heat treatment on the phase stability of γ-Fe_2O_3 nanoparticles and γ-Fe_2O_3-SiO_2 core-shell nanostructures[J]. The Journal of Physical Chemistry C, 2014, 118(5): 2857-2866.

[32] Liu S, Zhou J, Zhang L. In situ synthesis of plate-like Fe_2O_3 nanoparticles in porous cellulose films with obvious magnetic anisotropy[J]. Cellulose, 2011, 18(3): 663-673.

[33] Zhua H Y, Jianga R, Xiaob L, et al. A novel magnetically separable-Fe_2O_3/crosslinked chitosan adsorbent:

Preparation, characterization and adsorption application for removal of hazardous azo dye[J]. Journal of Hazardous Materials, 2010, 179: 251-257.

[34] Yin Y, Alivisatos A P. Colloidal nanocrystal synthesis and the organic-inorganic interface[J]. Nature, 2005, 437(7059): 664-670.

[35] Rudaz C, Courson R, Bonnet L, et al. Aeropectin: fully biomass-based mechanically strong and thermal superinsulating aerogel[J]. Biomacromolecules, 2014, 15(6): 2188-2195.

[36] Yang H, Zhu W, Sun, S, Guo X. Preparation of monolithic titania aerogels with high surface area by a sol-gel process combined surface modification[J]. RSC Adv. 2014, 4, 32934-32940.

[37] Jiao Y, Wan C, Li J. Room-temperature embedment of anatase titania nanoparticles into porous cellulose aerogels[J]. Applied Physics A, 2015, 120(1): 341-347.

[38] Sescousse R, Gavillon R, Budtova T. Aerocellulose from cellulose-ionic liquid solutions: preparation, properties and comparison with cellulose-NaOH and cellulose-NMMO routes[J]. Carbohydrate polymers, 2011, 83(4): 1766-1774.

[39] Olsson R T, Samir M A S A, Salazar-Alvarez G, et al. Making flexible magnetic aerogels and stiff magnetic nanopaper using cellulose nanofibrils as templates[J]. Nature nanotechnology, 2010, 5(8): 584-588.

[40] Gawryla M D, van den Berg O, Weder C, et al. Clay aerogel/cellulose whisker nanocomposites: a nanoscale wattle and daub[J]. Journal of Materials Chemistry, 2009, 19(15): 2118-2124.

[41] Ouyang W, Sun J, Memon J, et al. Scalable preparation of three-dimensional porous structures of reduced graphene oxide/cellulose composites and their application in supercapacitors[J]. Carbon, 2013, 62: 501-509.

[42] Cai J, Liu S, Feng J, et al. Cellulose-silica nanocomposite aerogels by in situ formation of silica in cellulose gel[J]. Angewandte Chemie, 2012, 124(9): 2118-2121.

[43] Shi Z, Gao H, Feng J, et al. In situ synthesis of robust conductive cellulose/polypyrrole composite aerogels and their potential application in nerve regeneration[J]. Angewandte Chemie International Edition, 2014, 53(21): 5380-5384.

[44] Sehaqui H, Salajková M, Zhou Q, et al. Mechanical performance tailoring of tough ultra-high porosity foams prepared from cellulose I nanofiber suspensions[J]. Soft Matter, 2010, 6(8): 1824-1832.

[45] Gaudisson T, Artus M, Acevedo U, et al. On the microstructural and magnetic properties of fine-grained $CoFe_2O_4$ ceramics produced by combining polyol process and spark plasma sintering[J]. Journal of magnetism and magnetic materials, 2014, 370: 87-95.

[46] Shafi K V P M, Ulman A, Dyal A, et al. Magnetic enhancement of γ-Fe_2O_3 nanoparticles by sonochemical coating[J]. Chemistry of materials, 2002, 14(4): 1778-1787.

[47] Gutfleisch O, Willard M A, Brück E, et al. Magnetic materials and devices for the 21st century: stronger, lighter, and more energy efficient[J]. Advanced materials, 2011, 23(7): 821-842.

[48] Zhang J, Cao Y, Feng J, et al. Graphene-oxide-sheet-induced gelation of cellulose and promoted mechanical properties of composite aerogels[J]. The Journal of Physical Chemistry C, 2012, 116(14): 8063-8068.

[49] Jiang W, Pelaez M, Dionysiou D D, et al. Chromium (Ⅵ) removal by maghemite nanoparticles[J]. Chemical Engineering Journal, 2013, 222: 527-533.

[50] Unuabonah E I, Günter C, Weber J, et al. Hybrid clay: a new highly efficient adsorbent for water treatment[J]. ACSSustainable Chemistry & Engineering, 2013, 1(8): 966-973.

[51] Bhattacharya A K, Naiya T K, Mandal S N, et al. Adsorption, kinetics and equilibrium studies on removal of Cr (Ⅵ) from aqueous solutions using different low-cost adsorbents[J]. Chemical engineering journal, 2008, 137(3): 529-541.

[52] Li J, Wang L, Qi T, et al. Different N-containing functional groups modified mesoporous adsorbents for Cr (Ⅵ) sequestration: Synthesis, characterization and comparison[J]. Microporous and Mesoporous Materials, 2008, 110(2-3): 442-450.

[53] Khezami L, Capart R. Removal of chromium (Ⅵ) from aqueous solution by activated carbons: kinetic and

equilibrium studies[J]. Journal of hazardous materials, 2005, 123(1-3): 223-231.

[54] Qiu B, Xu C, Sun D, et al. Polyaniline coated ethyl cellulose with improved hexavalent chromium removal[J]. ACS Sustainable Chemistry & Engineering, 2014, 2(8): 2070-2080.

[55] Wang P, Lo I M C. Synthesis of mesoporous magnetic γ-Fe_2O_3 and its application to Cr (VI) removal from contaminated water[J]. Water research, 2009, 43(15): 3727-3734.

[56] Sağ Y, Aktay Y. Kinetic studies on sorption of Cr (Ⅵ) and Cu (Ⅱ) ions by chitin, chitosan and Rhizopus arrhizus [J]. Biochemical Engineering Journal, 2002, 12(2): 143-153.

[57] Miretzky P, Cirelli A F. Cr (Ⅵ) and Cr (Ⅲ) removal from aqueous solution by raw and modified lignocellulosic materials: a review[J]. Journal of hazardous materials, 2010, 180(1-3): 1-19.

[58] Jia Z, Wang Q, Ren D, et al. Fabrication of one-dimensional mesoporous α-Fe_2O_3 nanostructure via self-sacrificial template and its enhanced Cr (Ⅵ) adsorption capacity[J]. Applied Surface Science, 2013, 264: 255-260.

[59] Zhong L S, Hu J S, Liang H P, et al. Self-Assembled 3D flowerlike iron oxide nanostructures and their application in water treatment[J]. Advanced materials, 2006, 18(18): 2426-2431.

[60] Srivastava V, Sharma Y C. Synthesis and characterization of Fe_3O_4@ n-SiO_2 nanoparticles from an agrowaste material and its application for the removal of Cr (Ⅵ) from aqueous solutions[J]. Water, Air, & Soil Pollution, 2014, 225(1): 1-16.

[61] López-Téllez G, Barrera-Díaz C E, Balderas-Hernández P, et al. Removal of hexavalent chromium in aquatic solutions by iron nanoparticles embedded in orange peel pith[J]. Chemical Engineering Journal, 2011, 173(2): 480-485.

[62] Huang S H, Chen D H. Rapid removal of heavy metal cations and anions from aqueous solutions by an amino-functionalized magnetic nano-adsorbent[J]. Journal of hazardous materials, 2009, 163(1): 174-179.

[63] Yuan P, Liu D, Fan M, et al. Removal of hexavalent chromium [Cr (Ⅵ)] from aqueous solutions by the diatomite-supported/unsupported magnetite nanoparticles[J]. Journal of hazardous materials, 2010, 173(1-3): 614-621.

[64] Dupont L, Guillon E. Removal of hexavalent chromium with a lignocellulosic substrate extracted from wheat bran[J]. Environmental science & technology, 2003, 37(18): 4235-4241.

中文题目: 纤维素气凝胶三维结构中包覆的超顺磁性γ-Fe_2O_3纳米颗粒的简易合成及其在去除污染水中的六价铬方面的应用研究

作者: 万才超，李坚

摘要: 随着人们对绿色化学的日益重视，有丰富的三维(3D)结构组成的纤维素气凝胶被认为是一类想法的绿色基质材料，可以封装各种纳米颗粒以合成各种功能材料。在此，采用简易模板合成与化学共沉淀相结合的方法制备了混合型γ-Fe_2O_3@纤维素气凝胶(γ-Fe_2O_3@CA)。γ-Fe_2O_3纳米颗粒在气凝胶的微/纳米级孔隙结构中得到良好的分散和固定，并表现出超顺磁性行为。通过调整初始反应物的浓度，可以灵活定制合成γ-Fe_2O_3@CA的粒径、孔隙特征参数、磁性能和机械强度。此外，γ-Fe_2O_3@CA对Cr(Ⅵ)重金属离子表现出快速的吸附速率和优异的吸附能力。此外，结合环保效益、制备方法简单方便、比表面积高和机械强度强、磁性响应性强等优点，该类绿色γ-Fe_2O_3@CA更有利于和适用于去除污染水体中的Cr(Ⅵ)，并且在其他许多应用领域也很有用。

关键词: 纤维素气凝胶；超顺磁性；γ-Fe_2O_3；复合材料；铬；吸附性

Room-temperature Embedment of Anatase Titania Nanoparticles into Porous Cellulose Aerogels*

Yue Jiao, Caichao Wan, Jian Li

Abstract: In this paper, a facile easy method for room-temperature embedment of anatase titania (TiO_2) nanoparticles into porous cellulose aerogels was reported. The obtained anatase TiO_2/cellulose (ATC) aerogels were characterized by scanning electron microscopy (SEM), energy-dispersive X-ray spectrometer (EDXS), transmission electron microscopy (TEM), X-ray photoelectron spectroscopy (XPS), X-ray diffraction (XRD), nitrogen adsorption measurements, and thermogravimetric analysis (TGA). The results showed that high-purity anatase TiO_2 nanoparticles with sizes of 3.69 ± 0.77 nm were evenly dispersed in the cellulose aerogels, which leaded to the significant improvement in specific surface area and pore volume of ATC aerogels. Meanwhile, the hybrid ATC aerogels also had a high loading content of TiO_2 (ca. 17.7%). Furthermore, through a simple photocatalytic degradation test of indigo carmine dye under UV light, ATC aerogels exhibited superior photocatalytic activity and shape stability, which might be useful in some fields like governance of water pollution, and chemical leaks.

Keywords: Cellulose aerogels; Anatase titania; Nanocomposites; Porous materials; Photocatalysis

1 Introduction

Organic/inorganic hybrid materials have emerged as quite intriguing materials, due to their combining organic and inorganic materials' superb characters, which hold great potential in many applications such as adsorbents, catalysts, fuel cells and sensors[1-4]. Among various natural organic polymers, undoubtedly, cellulose is high-profile; this kind of unbranched polymer of β-1, 4-linked glucopyranose is acclaimed, and has been extensively applied in every walk of life such as papermaking, adhesive, architectural coating, and biomedicine. However, it is difficult for cellulose to be dissolved, hydrolyzed, and processed in some common aqueous or organic solvents, owing to its some inherent structure features like strong inter and intra-molecular hydrogen bonds, high degree of polymerization, and high degrees of crystallinity[5]. In recent decades, some effective cellulose solvents were successively reported[6-9], contributing to further exploiting and fabricating novel cellulose products (like films, foams, and aerogels). As one of the most promising cellulose products, cellulose aerogels consist of cross-linked three-dimensional (3D) network. The unique structural characteristic endows themselves with low density, high porosity, and large specific surface area. As a result, cellulose aerogels may find applications in multitudinous fields such as adsorbing materials, catalyst supports, super-thermal and sound insulators, and electronic devices[10-12]. In particular, cellulose aerogels

* 本文摘自 Applied Physics A, 2015, 120: 341-347.

are also extremely promising matrix materials in the domain of nanocomposites synthesis. Their hierarchical micro-nano porous structures are beneficial to achieve controlled growth of nanoparticles; besides, abundant surface hydroxyl groups of cellulose are also suitable binders for immobilization of nanoparticles in the matrix[13]. In recent years, plentiful investigations have been undertaken extensively for organic/inorganic hybrid materials using cellulose aerogels as matrixes[14-16]. The synthetic composites exhibit numerous attractive performances including photocatalytic ability, antibacterial property, high mechanical strength, and magnetic property.

Titania (TiO_2) nanoparticles are one of the most noticeable inorganic nanoparticles at present, which could be served as solar cells, photocatalysts for air and water purification, high permittivity dielectric layers for electronic devices, sensors for gas and biomolecules, and biocompatible coatings for biomaterial petroleum processing,[17]. In nature, TiO_2 mainly occurs in four distinct crystallographic phases: anatase (tetragonal, space group $I4_1/amd$), rutile (tetragonal, space group $P4_2/mnm$), brookite (orthorhombic, space group P*bca*), and TiO_2(B) (monoclinic, space group $C2/m$)[18]. Meanwhile, anatase TiO_2 phase is wildly believed to have more superior photocatalytic activity within these four kinds of polymorphs[19], which is more dominant in some applications such as photodecomposition and solar energy conversion. However, owing to some intractable problems like difficultly recycling TiO_2 powders from treated aqueous solutions, nano-TiO_2 is still limited to small-scale applications. Consequently, nanocomposites that combine nano-TiO_2 with some special substrates with large specific surface area (like cellulose aerogels), have attracted increasing attention[20,21]. Taking examples of cellulose aerogels, embedding anatase TiO_2 nanoparticles into nanoporous cellulose aerogels might contribute to not only guiding growth of the nanoparticles, but also inhibiting spontaneous agglomeration of nanoparticles caused by high surface energy and chemical activity of nanoparticles. Besides, anatase TiO_2/cellulose (ATC) aerogels composites probably inherit various functions from the individual components such as photocatalytic property, bacteriostatic activity, and strong adsorption capacity. This multifunctional feature makes ATC aerogels applied to a wider range of occasions.

In general, anatase TiO_2 phase could be prepared by multiple methods such as sol-gel technique, calcination, hydrothermal method, chemical vapor deposition, and thermal hydrolysis[22-26]. In continuation of environmentally benign and low energy consumption methods for synthesis of nanoparticles, herein, we reported a mild and effective method to fabricate anatase TiO_2 nanoparticles that were embedded in porous cellulose aerogels at room temperature by simply immersing cellulose hydrogels in an anatase TiO_2 nanosol. Meanwhile, a green cost-effective NaOH/polyethylene glycol (PEG) aqueous solution was carried out to fabricate cellulose hydrogels (the precursor of cellulose aerogels). Moreover, an easy photocatalytic degradation experiment for ATC aerogels was perform under UV light to roughly demonstrate the photochemical utilization potentiality.

2 Materials and methods

2.1 Materials

All reagents were commercially available and of reagent grade. Cellulose was isolated from waste wheat straw by chemical pretreatment, and the pretreatment process could refer to our previous reports[27]. The purified cellulose was dried at 60 ℃ for 24 h before used. Deionized water was used for all experiments.

2.2 Preparation of anatase TiO_2 nanosol

Anatase TiO_2 nanosol was fabricated according to the method reported by Wu et al.[28]. Briefly, absolute ethyl alcohol (20 mL) and tetrabutyl titanate (5 mL) were firstly mixed homogeneously with magnetic stirring for 30 min at room temperature, and then the obtained mixed solution was dropwise added into 0.04 M HNO_3 solution (200 mL) at room temperature with continuous magnetic stirring. After that, the mixture was kept stirring for 48 h to ensure complete hydrolysis, nucleation and growth of TiO_2 crystallites. Finally, the mixed solution was stirred for another 48 h, and the following pale blue anatase TiO_2 nanosol was obtained. During this 96-h stirring, it could be seen that the previous emulsion was significantly transformed to a transparent solution. Moreover, the sol could be stably kept at room temperature for several months without sedimentation or delamination.

2.3 Preparation of anatase TiO_2/cellulose aerogel

The desired amount of cellulose was added to a mixed aqueous solution of NaOH/PEG-4000 at a mass ratio of 9/1 with magnetic stirring for 5 h at room temperature to form a homogeneous cellulose solution with 2% concentration. Then, the cellulose solution was frozen for 12 h at −15 ℃, and subsequently thawed at ambient temperature with vigorous stirring for 30 min. After being frozen again for 5 h at −15 ℃, the solution was placed into a 1% HCl coagulating bath for 6 h, repeating this impregnation process until the formation of a white hydrogel. Thereafter, the hydrogel was repeatedly rinsed with a good deal of distilled water, and the obtained clean cellulose hydrogel was immediately immersed in the above-mentioned anatase TiO_2 nanosol for 3 h at room temperature. After the immersion treatment, the obtained ATC hydrogel was heated at 60 ℃ for 5 min in a thermostatic water bath, and subsequently cured at 100 ℃ for 5 min to complete the formation of nano-TiO_2 in the pore structure of cellulose hydrogel. Finally, ATC hydrogel was washed several times by distilled water and tert-butyl alcohol in sequence to remove unattached TiO_2 particles and other residual chemicals, and then subjected to a freeze-drying process for 48 h at −30 ℃ in vacuum, and the following ATC aerogel was fabricated. In addition, the pure cellulose (PC) aerogel was prepared by direct tert-butyl alcohol freeze-drying treatment of cellulose hydrogel without the above-mentioned immersion process. The anatase TiO_2 particles were obtained by the drying treatment of the nanosol in a vacuum oven at 60 ℃ for 48 h.

2.4 Indigo carmine dye photocatalytic degradation

The photocatalytic activity of ATC aerogels was roughly evaluated by observing the changes in color during the photocatalytic degradation process of indigo carmine dye. The indigo carmine dye (50 mL, 5×10^{-5} mol · L^{-1}) and ATC aerogels sample with sizes of about 30 mm (diameter) × 15 mm (height) (ca. 0.79 g) were put into a clean glass dish, and then stirred in the dark for 30 min to achieve adsorption equilibrium. Thereafter, the dish was placed in a closed chamber with a UV source (mercury lamp, 120 W, and 365 nm). The distance between the UV source and the container was around 10 cm. The dishes were exposed to the mercury lamp for about 1 h, monitoring the color changes during the entire photocatalytic degradation process.

2.5 Characterizations

Microstructures and surface chemical compositions were determined by scanning electron microscopy (SEM, FEI, and Quanta 200) equipped with an energy-dispersive X-ray spectrometer (EDXS). Transmission electron microscopy (TEM) observations were carried out on a FEI, Tecnai G2 F20 TEM. X-ray photoelectron spectra (XPS) were recorded in the range of 0–1400 eV using Thermo Escalab 250Xi XPS spectrometer (Germany). Crystal structures were characterized via X-ray diffraction (XRD, Rigaku, and D/MAX 2200) operating with Cu Kα radiation ($\lambda = 1.5418$ Å) at a scan rate (2θ) of 4° · min^{-1} ranging from 5° to 70°. Nitrogen adsorption measurements were performed with an accelerated surface area and porosimetry system (3H-2000PS2 unit, Beishide Instrument S&T Co., Ltd). Brunauer-Emmett-Teller (BET) analysis was carried out for a relative vapor pressure of 0.05–0.3. Barrett-Joyner-Halenda (BJH) analysis was performed for desorption branch. Thermogravimetric analysis (TGA) was performed with a synchronous thermal analyzer (the United States, SDT-Q600) from room temperature to 800 ℃ at a heating rate of 10 ℃ · min^{-1} under a nitrogen atmosphere.

3 Result and discussion

The SEM image in Fig. 1a demonstrated that ATC aerogels maintained interconnected 3D network after the immersion and heating processes. Moreover, the 3D architecture was possibly generated by self-assembly of cellulose chains due to large-scale hydrogen bond connection as the result of the plentiful surface hydroxide radicals[29]. However, it is hard to distinguish the synthetic nanoparticles from the complicated network structure. Even higher magnification SEM image (50000 times) still could not clearly identify the nanoparticles (Fig. 1b), possibly due to their extremely small particle size, and the following TEM observation would prove that. In addition, compared with PC aerogels, apart from the common C and O elements, the EDX spectrum of ATC aerogels exhibited new strong peak assigned to Ti element (inset in Fig. 1b), which indicated the presence of plentiful titanium compound. Furthermore, the dramatic decline of the atomic ratio of C/O from 1.12 to 0.83 was another potential evidence of the presence of Titanium oxide.

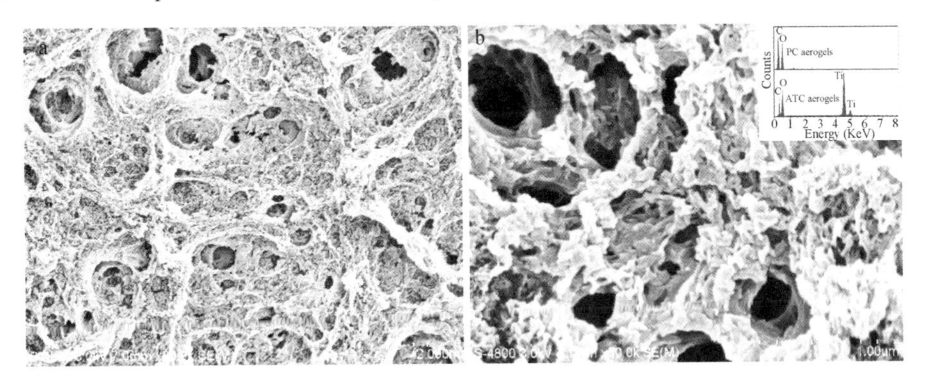

Fig. 1 (a, b) SEM images of different magnification of ATC aerogels. The inset in Fig. 1b exhibited the corresponding EDX spectra of PC aerogels and ATC aerogels

It is of note that the properties of nanocomposites are tightly associated with the diameters and dispersion of nanoparticles in matrixes[30,31]. According to the TEM observation (Fig. 2a, b), the formed nanoparticles were homogeneously dispersed and immobilized in cellulose aerogels matrixes as indicated by the dark spots, and no severe agglomeration phenomena occurred, which indicated that the 3D architecture of porous cellulose aerogels might be a suitable template for the synthesis of nanoparticles. Meanwhile, this class of matrix materials might also be expanded to fabricate other versatile oxide nanoparticles/cellulose aerogels composites. High-resolution transmission electron microscopy image demonstrated the presence of anatase TiO_2

as shown in Fig. 2c; i. e. , lattice fringes with a spacing of 0. 351 nm and 0. 237 nm agreed well with the (101) and (004) planes of anatase TiO_2[32]. Moreover, the particle size distribution that was calculated based on the TEM images, was shown in Fig. 2d. The results showed that the diameter histogram exhibited Gaussian-like distributions, and the synthetic nanoparticles were polydisperse. Their mean diameter (d) and standard deviation (σ) were estimated to be 3. 69 nm and 0. 77 nm, respectively.

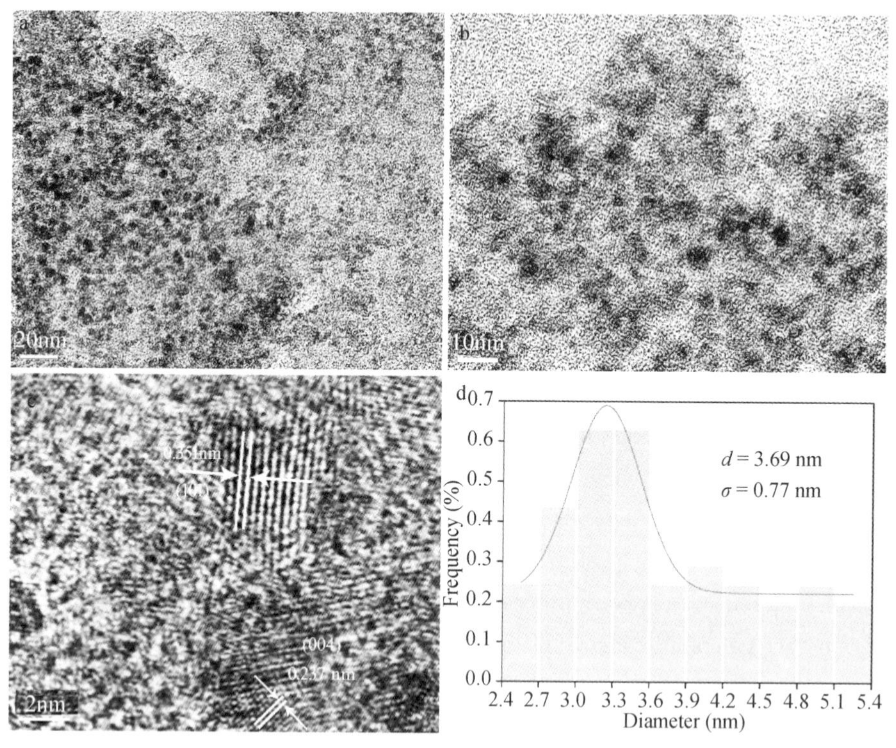

Fig. 2 (a, b) TEM images of different magnification and (c) HRTEM image of ATC aerogels. (d) Size distribution of the TiO_2 nanoparticles in the cellulose aerogels

Fig. 3a showed survey scan XPS spectrum of ATC aerogels, and the spectrum clearly revealed that the main elements on the sample surface were Ti, O, and C. Fig. 3b presented the Ti ($2p_{3/2}$, $2p_{1/2}$) spectra. The binding energies of Ti $2p_{3/2}$ and Ti $2p_{1/2}$ were 458. 6 and 464. 4 eV, respectively, indicating the typical presence of Ti^{4+}(TiO_2)[33,34]. Fig. 3c displayed the O1s spectra. The O 1s peak at 530. 0 eV was corresponding to lattice oxygen of Ti^{4+}—O, while the higher binding energy of 532. 8 eV was assigned to hydroxyl groups (O—H)[35], respectively. Fig. 4d showed the C 1s spectra. Three peaks were observed after multi-peak Gaussian fitting. The peak located at 284. 8 eV was ascribed to carbon atoms in C—C, C ═C, and C—H bonds. The peak at 286. 4 eV was assigned to the C—O bond, and the peak at 287. 8 eV was attributed to the C ═O bond[36]. Therefore, the XPS results further confirmed the formation of TiO_2 via the aforementioned preparation approach.

To determine the crystal phase of the formed TiO_2 particles, XRD measurements were carried out. Deconvolution of the overlapping peaks was performed using a mixed Gaussian-Lorentzian fit program. As shown in Fig. 4, the pattern of the synthetic TiO_2 nanoparticles showed the presence of peaks (2θ = 25. 14°,

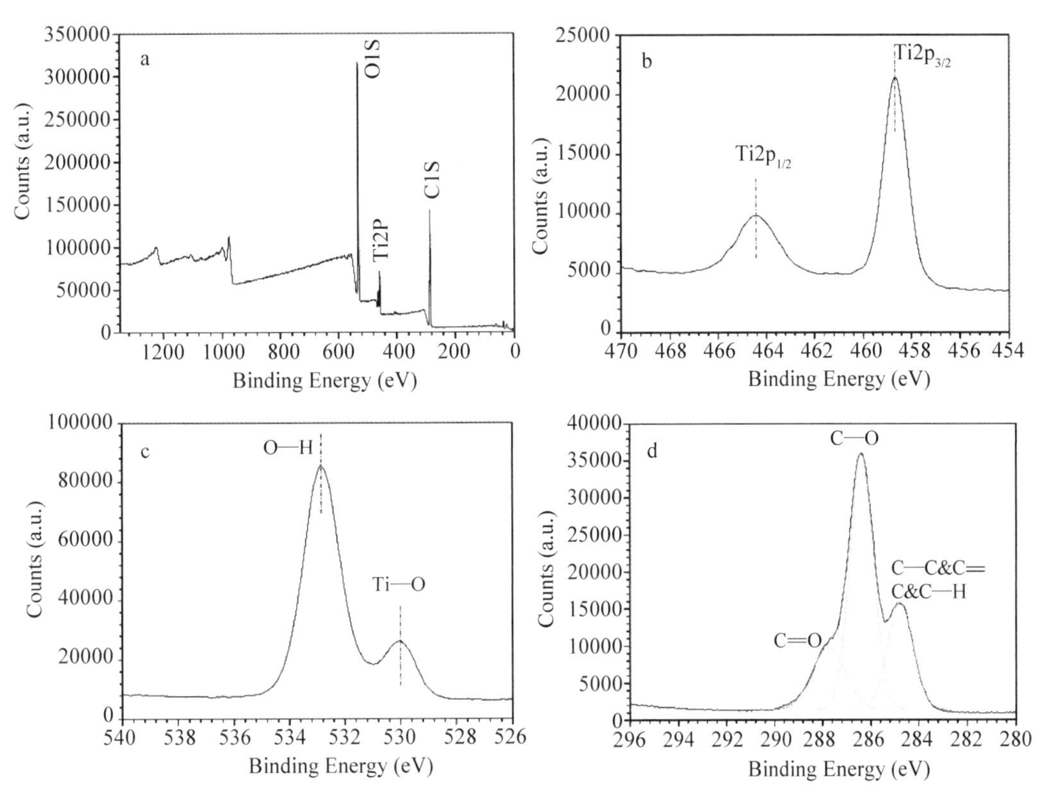

Fig. 3 (a) Survey scan, (b) Ti 2p, (c) O 1s, and (d) C 1s XPS spectra of ATC aerogels, respectively

37.64°, 47.78°, 54.14°, 55.62°, 62.76°, and 68.96°) assigned to anatase-type TiO_2 (JCPDS, 21-1272)[37], and no excess peaks related to rutile or brookite TiO_2 were detected, implying that only high-purity anatase TiO_2 nanoparticles were formed via the mild room-temperature synthesis method. Compared with the TiO_2 nanoparticles, the new diffraction peaks (2θ = 12.56°, 20.20°, and 21.68°) as shown in the pattern of ATC aerogels were originated from the (101), (10$\bar{1}$) and (002) planes of cellulose Ⅱ crystal structure, corresponding to the characteristics of the cellulose aerogels matrixes[38]. In addition, some peaks belonging to anatase TiO_2 could also be clearly distinguished at around 25.58°, 37.80°, 47.97°, 54.15°, 55.65°, 63.18°and 68.86°, demonstrating that the anatase TiO_2 nanoparticles were successfully embedded into the porous cellulose aerogels matrixes.

Nitrogen adsorption measurements were used to estimate specific surface area and porosity characteristics of PC aerogels and ATC aerogels, and the corresponding sorption isotherms were presented in Fig. 5. According to the IUPAC classification, the both sorption isotherms were of type Ⅳ[39]. At low relative pressure, the adsorption uptakes of the samples increased slowly, and nitrogen molecules was gradually adsorbed on the internal surface of porous structures from single to multilayer. In addition, the samples showed the isotherms with clear upward deviations (P/P_0 at 0.6-0.9) which was caused by capillary condensation, suggesting the characteristic of mesoporous structure. Furthermore, the larger amount of adsorption (without apparent limitation) occurring at relative pressures above 0.9, demonstrated the presence of macropores. The hysteresis loops for the both samples were of type H_3 without obvious adsorption limits at P/P_0 close to 1, possibly revealed the existence of slit-shaped pores[40]. Specific surface area and pore volume are two important

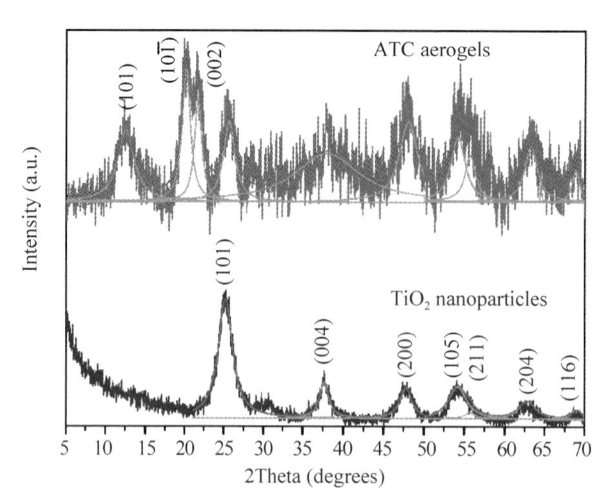

Fig. 4 XRD patterns of ATC aerogels and the TiO_2 nanoparticles, respectively

structural characteristics of aerogels, and high values are desirable for applications such as functional carriers and adsorbing materials. The BET analysis of PC aerogels and ATC aerogels gave specific surface areas of around 185 and 201 $m^2 \cdot g^{-1}$, respectively. The mesopore analysis by BJH method gave pore volumes of approximately 1.0 and 1.3 $cm^3 \cdot g^{-1}$ for PC aerogels and ATC aerogels, respectively. From these data, it was observed that specific surface area and pore volume of ATC aerogels were 8.6% and 30.0% higher than those of PC aerogels, which indicated that the incorporation of TiO_2 particles into porous cellulose aerogels dramatically improved the pore characteristics.

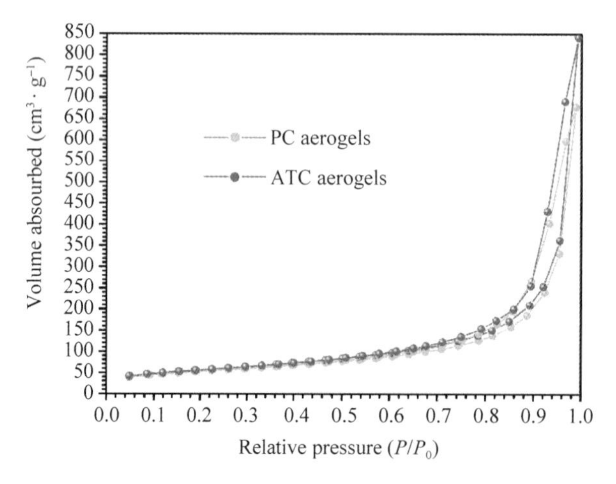

Fig. 5 Nitrogen adsorption and desorption isotherms of PC aerogels and ATC aerogels, respectively

Fig. 6 showed the TG and DTG curves of ATC aerogels, which were recorded under a nitrogen atmosphere. For comparative purposes, the TG and DTG curve of PC aerogels were also displayed, and exhibited only a strong exothermic peak at around 364 ℃ (Fig. 6b) corresponding to the thermal degradation of cellulose[41]. For the DTG curve of ATC aerogels, two exothermic peaks appeared during the whole pyrolysis process. The peak at around 297 ℃ was primarily due to decomposition and oxidation of residual organic substances during the preparation process[42]. In addition, the peak that was induced by cellulose pyrolysis occurred at around 353 ℃ for ATC aerogels, 11 ℃ lower than that of PC aerogels. The degradation in thermal stability of the composites might be due to the catalytic character of TiO_2, the loosening of molecular chains in crystalline regions of cellulose resulted from infusion of TiO_2 particles during the impregnation process, and the damage of cellulose chains caused by a small amount of HNO_3 in the sol[43,44]. In addition, the residual char yields above 800 ℃ were 4.3% for PC aerogels and 22.0% for ATC aerogels, respectively. Owing to the high decomposition temperature of TiO_2 (> 1000 ℃), the content of TiO_2 was almost not changed during the pyrolysis, indicating that the TiO_2 loading content in cellulose aerogels was approximately 17.7%.

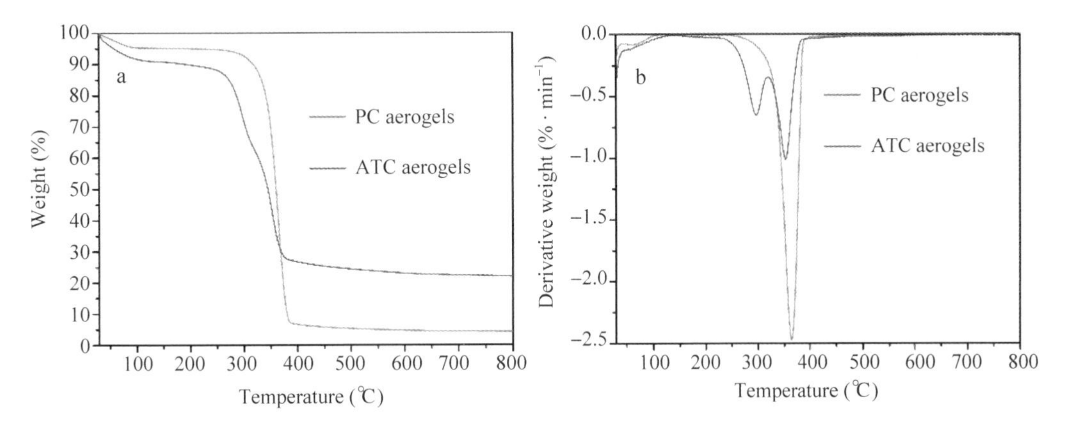

Fig. 6 (a) TG and (b) DTG curves of PC aerogels and ATC aerogels, respectively

To investigate the potential photocatalytic ability of ATC aerogels, a simple photocatalytic degradation test of indigo carmine dye was implemented under UV light for approximately 1 h (Fig. 7). For eliminating interference from adsorption effect of the sample, before the irradiation process, the aerogels sample was immersed in the indigo carmine solution, and stirred for about 0.5 h to reach adsorption equilibrium. The color of the aerogels was transformed from white to nattier blue, however, the color of the solution still kept deep blue, which indicated that the absorption action of the porous aerogels didn't significantly affect the solution color. In addition, during the UV illumination process, apparently, the drastic color changes did occur to the indigo carmine solution, which was gradually transformed from dark blue to transparent colorless with the prolongation of time. The results revealed some promising photocatalytic applications of ATC aerogels like treatment of organic dye wastewater. Meanwhile, the aerogels sample maintained the well-defined shape without any collapse throughout the entire UV radiation process, which provided convenience for recovery processing of the aerogels.

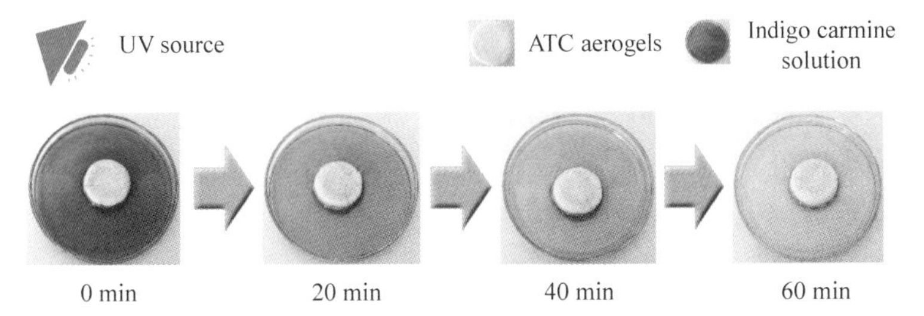

Fig. 7 Photocatalytic activity of ATC aerogels for indigo carmine dye degradation

4 Conclusions

In this study, a simple facile room-temperature impregnation treatment of the cellulose hydrogels (the precursor of cellulose aerogels) in an anatase TiO_2 nanosol was carried out to complete the embedment of anatase titania into the porous structures. Meanwhile, the preparation of the cellulose hydrogels was based on a green cost-effective and low-pollution cellulose solvent named NaOH/PEG solution. After the freeze-drying process, the resulting hybrid ATC aerogels exhibited that the TiO_2 nanoparticles with diameters of 3.69 ± 0.77 nm were inserted into the cross-linked 3D networks. Moreover, the hybrid aerogels had improved specific surface area of around 201 $m^2 \cdot g^{-1}$ and pore volume of approximately 1.3 $cm^3 \cdot g^{-1}$, 8.6% and 30.0% higher than those of PC aerogels (185 $m^2 \cdot g^{-1}$ and 1.0 $cm^3 \cdot g^{-1}$). But the additional handlings including impregnation and subsequent heating treatments lowered the thermal stability of ATC aerogels, because of the shift of the characteristic exothermic peaks to lower temperatures. In addition, ATC aerogels also showed a high photocatalytic activity for the indigo carmine dye degradation under UV light, and maintained the well-defined shape throughout the entire radiation process, which indicated some potential applications in water purification, air cleaning, and chemical leaks.

Acknowledgments

This work was financially supported by the National Natural Science Foundation of China (grant no. 31270590 and 31470584).

References

[1] Chujo Y. Organic-inorganic hybrid materials[J]. Current Opinion in Solid State and Materials Science, 1996, 1(6): 806-811.

[2] Kumar P, Guliants V V. Periodic mesoporous organic-inorganic hybrid materials: Applications in membrane separations and adsorption[J]. Microporous and Mesoporous Materials, 2010, 132(1-2): 1-14.

[3] Dufaud V, Davis M E. Design of heterogeneous catalysts via multiple active site positioning in organic-inorganic hybrid materials[J]. Journal of the American Chemical Society, 2003, 125(31): 9403-9413.

[4] Steele B C H, Heinzel A. Materials for fuel-cell technologies[M]//Materials for sustainable energy: a collection of peer-reviewed research and review articles from nature publishing group. 2011: 224-231.

[5] Swatloski R P, Spear S K, Holbrey J D, et al. Dissolution of cellose with ionic liquids[J]. Journal of the American chemical society, 2002, 124(18): 4974-4975.

[6] Dawsey T R, McCormick C L. The lithium chloride/dimethylacetamide solvent for cellulose: a literature review[J]. Journal of Macromolecular Science-Reviews in Macromolecular Chemistry and Physics, 1990, 30(3-4): 405-440.

[7] Frey M W, Cuculo J A, Khan S A. Rheology and gelation of cellulose/ammonia/ammonium thiocyanate solutions[J]. Journal of polymer science part B: Polymer physics, 1996, 34(14): 2375-2381.

[8] Qi H, Chang C, Zhang L. Effects of temperature and molecular weight on dissolution of cellulose in NaOH/urea aqueous solution[J]. Cellulose, 2008, 15(6): 779-787.

[9] Yan L, Gao Z. Dissolving of cellulose in PEG/NaOH aqueous solution[J]. Cellulose, 2008, 15(6): 789-796.

[10] Korhonen J T, Kettunen M, Ras R H A, et al. Hydrophobic nanocellulose aerogels as floating, sustainable, reusable, and recyclable oil absorbents[J]. ACS applied materials & interfaces, 2011, 3(6): 1813-1816.

[11] Siqueira G, Bras J, Dufresne A. Cellulosic bionanocomposites: a review of preparation, properties and applications

[J]. Polymers, 2010, 2(4): 728-765.

[12] Zheng G, Cui Y, Karabulut E, et al. Nanostructured paper for flexible energy and electronic devices[J]. MRS bulletin, 2013, 38(4): 320-325.

[13] Cai J, Kimura S, Wada M, et al. Nanoporous cellulose as metal nanoparticles support[J]. Biomacromolecules, 2009, 10(1): 87-94.

[14] Zhang J, Cao Y, Feng J, et al. Graphene-oxide-sheet-induced gelation of cellulose and promoted mechanical properties of composite aerogels[J]. The Journal of Physical Chemistry C, 2012, 116(14): 8063-8068.

[15] Wan C, Jiao Y, Sun Q, et al. Preparation, characterization, and antibacterial properties of silver nanoparticles embedded into cellulose aerogels[J]. Polymer Composites, 2014.

[16] Olsson R T, Samir M A S A, Salazar-Alvarez G, et al. Making flexible magnetic aerogels and stiff magnetic nanopaper using cellulose nanofibrils as templates[J]. Nature nanotechnology, 2010, 5(8): 584-588.

[17] Diebold U. Structure and properties of TiO_2 surfaces: a brief review[J]. Applied physics A, 2003, 76(5): 681-687.

[18] Dambournet D, Belharouak I, Amine K. Tailored Preparation Methods of TiO_2 Anatase, Rutile, Brookite: Mechanism of Formation and Electrochemical Properties[J]. Chemistry of Materials, 2010, 22(3): 1173-1179.

[19] Peng T, Zhao D, Dai K, et al. Synthesis of titanium dioxide nanoparticles with mesoporous anatase wall and high photocatalytic activity[J]. The journal of physical chemistry B, 2005, 109(11): 4947-4952.

[20] Woan K, Pyrgiotakis G, Sigmund W. Photocatalytic carbon-nanotube-TiO_2 composites[J]. Advanced Materials, 2009, 21(21): 2233-2239.

[21] Khataee A, Mansoori G A. Nanostructured titanium dioxide materials: Properties, preparation and applications[M]. World scientific, 2011.

[22] Bischoff B L, Anderson M A. Peptization process in the sol-gel preparation of porous anatase (TiO_2)[J]. Chemistry of materials, 1995, 7(10): 1772-1778.

[23] Pillai S C, Periyat P, George R, et al. Synthesis of high-temperature stable anatase TiO_2 photocatalyst[J]. The Journal of Physical Chemistry C, 2007, 111(4): 1605-1611.

[24] Nian J N, Teng H. Hydrothermal synthesis of single-crystalline anatase TiO_2 nanorods with nanotubes as the precursor [J]. The Journal of Physical Chemistry B, 2006, 110(9): 4193-4198.

[25] Boschloo G K, Goossens A, Schoonman J. Photoelectrochemical study of thin anatase TiO_2 films prepared by metallorganic chemical vapor deposition[J]. Journal of the Electrochemical Society, 1997, 144(4): 1311.

[26] Sun J, Gao L, Zhang Q. Synthesizing and comparing the photocatalytic properties of high surface area rutile and anatase titania nanoparticles[J]. Journal of the American Ceramic Society, 2003, 86(10): 1677-1682.

[27] Li J, Wan C, Lu Y, et al. Fabrication of cellulose aerogel from wheat straw with strong absorptive capacity[J]. Frontiers of Agricultural Science and Engineering, 2014, 1(1): 46-52.

[28] Wu D, Long M, Zhou J, et al. Synthesis and characterization of self-cleaning cotton fabrics modified by TiO_2 through a facile approach[J]. Surface and Coatings Technology, 2009, 203(24): 3728-3733.

[29] Tingaut P, Zimmermann T, Sèbe G. Cellulose nanocrystals and microfibrillated cellulose as building blocks for the design of hierarchical functional materials[J]. Journal of Materials Chemistry, 2012, 22(38): 20105-20111.

[30] Kashiwagi T, Du F, Winey K I, et al. Flammability properties of polymer nanocomposites with single-walled carbonnanotubes: effects of nanotube dispersion and concentration[J]. Polymer, 2005, 46(2): 471-481.

[31] Jordan J, Jacob K I, Tannenbaum R, et al. Experimental trends in polymer nanocomposites-a review[J]. Materials science and engineering: A, 2005, 393(1-2): 1-11.

[32] Liu B, Khare A, Aydil E S. Synthesis of single-crystalline anatase nanorods and nanoflakes on transparent conducting substrates[J]. Chemical Communications, 2012, 48(68): 8565-8567.

[33] Jiang Y, Yang D, Zhang L, et al. Preparation of Protamine-Titania Microcapsules Through Synergy Between Layer-by-Layer Assembly and Biomimetic Mineralization[J]. Advanced Functional Materials, 2009, 19(1): 150-156.

[34] Erdem B, Hunsicker R A, Simmons G W, et al. XPS and FTIR surface characterization of TiO_2 particles used in polymer encapsulation[J]. Langmuir, 2001, 17(9): 2664-2669.

[35] Gong W J, Tao H W, Zi G L, et al. Visible light photodegradation of dyes over mesoporous titania prepared by using chrome azurol S as template[J]. Research on Chemical Intermediates, 2009, 35(6): 751-760.

[36] How G , Pandikumar A , Huang N M , et al. Highly Exposed {001 Facets of Titanium Dioxide Modified with Reduced Graphene Oxide for Dopamine Sensing[J]. Scientific Reports, 2014, 4(1): 5044.

[37] Wen P, Itoh H, Tang W, et al. Single nanocrystals of anatase-type TiO_2 prepared from layered titanate nanosheets: Formation mechanism and characterization of surface properties[J]. Langmuir, 2007, 23(23): 11782-11790.

[38] Nishiyama Y, Langan P, Chanzy H. Crystal structure and hydrogen-bonding system in cellulose Iβ from synchrotron X-ray and neutron fiber diffraction[J]. Journal of the American Chemical Society, 2002, 124(31): 9074-9082.

[39] Sehaqui H, Zhou Q, Berglund L A. High-porosity aerogels of high specific surface area prepared from nanofibrillated cellulose (NFC)[J]. Composites science and technology, 2011, 71(13): 1593-1599.

[40] Wei T Y, Kuo C Y, Hsu Y J, et al. Tin oxide nanocrystals embedded in silica aerogel: Photoluminescence and photocatalysis[J]. Microporous and Mesoporous Materials, 2008, 112(1-3): 580-588.

[41] Baker R R. Thermal decomposition of cellulose[J]. Journal of Thermal Analysis and Calorimetry, 1975, 8(1): 163-173.

[42] Tang A, Deng Y, Jin J, et al. $ZnFe_2O_4$-TiO_2 Nanoparticles within Mesoporous MCM-41[J]. The Scientific World Journal, 2012(24): 480527.

[43] Yu Q, Wu P, Xu P, et al. Synthesis of cellulose/titanium dioxide hybrids in supercritical carbon dioxide[J]. Green Chemistry, 2008, 10(10): 1061-1067.

[44] H Wang, Zhong W , Xu P , et al. Polyimide/silica/titania nanohybrids via a novel non-hydrolytic sol-gel route-ScienceDirect[J]. Composites Part A: Applied Science and Manufacturing, 2005, 36(7): 909-914.

中文题目：天然芦苇纳米纤化纤维素制备超疏水超轻气凝胶的研究

作者：焦月，万才超，强添刚，李坚

摘要：芦苇是一种可广泛利用的水生植物资源，其应用一般局限于造纸、动物饲料等传统领域。此外，每年大部分芦苇被浪费或直接焚烧，造成严重的空气污染(如大气雾霾)。因此，进一步开发芦苇高价值应用的新形式是值得研究的。本论文以天然芦苇为原料，通过化学提纯、超声波处理、冷冻干燥等一系列操作简便的方法，制备了纳米纤维化纤维素气凝胶(NFC)这一超轻质吸附剂。用扫描电子显微镜、能谱仪、傅立叶变换红外光谱、X 射线衍射和热重分析等手段对超低密度(4.9 mg · cm^{-3})气凝胶进行了表征。为了获得良好的疏水性，用甲基三氯硅烷对 NFC 气凝胶进行疏水处理。接触角高达 151°~155°的超疏水 NFC 气凝胶对各种有机溶剂和废油有很好的吸附效率(53~93 g · g^{-1})。更重要的是，气凝胶还表现出良好的吸附回收性能，在 5 次循环后仍能保持 80%以上的初始吸附效率。

关键词：芦苇；纳米原纤维化纤维素；气凝胶；超疏水性；吸附剂

Cellulose Aerogels from Cellulose-NaOH/PEG Solution and Comparison with Different Cellulose Contents*

Caichao Wan, Yun Lu, Yue Jiao, Jun Cao, Jian Li, Qingfeng Sun

Abstract: A green mild NaOH/polyethylene glycol solution was employed to dissolve the cellulose extracted from waste wheat straw. Subsequently combined with the freezing-thawing treatment, regeneration process and freeze drying, the cellulose aerogels with different cellulose contents were fabricated at mass ratios of 1/100, 3/100, 5/100 and 7/100 cellulose to the solution. Moreover, the influences of cellulose contents on the morphologies, crystal structures, crystallinity indexes, thermal stabilities and pore characteristic parameters of the aerogels were investigated by scanning electron microscope, X-ray diffraction, thermal gravity analysis and nitrogen adsorption. Meanwhile, the oil adsorption properties of the aerogels with different cellulose contents were also surveyed; besides, for improving lipophilicity, the hydrophobic modifications of trimethylchlorosilane for the aerogels were carried out before the adsorption tests.

Keywords: Cellulose aerogels; NaOH/PEG solution; Oil adsorption; Freeze-drying; Polymers

1 Introduction

Aerogels composed of cross-linked three-dimensional (3D) network are being paid increasing attention from scientific communities to industrial communities due to their extraordinary structure, thermal, optical, mechanical and electrical properties,[1-5] which hold great potentials in the fields of adsorbents, catalysts, thermal insulation materials, optical elements, engineering materials, and supercapacitors.[6-12] Aerogels based on native cellulose have not only excellent performances of traditional inorganic or polymer-based aerogels, but also new alluring properties such as biodegradability, biocompatibility and flexibility.[13,14] Generally, the preparation of cellulose aerogels needs be via a cellulose dissolution process. However, cellulose is difficult to process in solution or as a melt, owing to its strong inter- and intra-molecular hydrogen bonds, high degree of polymerization (DP) and high degrees of crystallinity, which interrupt the dissolution of cellulose solid into solution.[15,16] In the past century, some cellulose solvents such as ammonium thiocyanate,[17] calcium and sodium thiocyanate,[18] lithium chloride/N, N-dimethylacetamide (LiCl/DMAc)[19] and liquid ammonia/ammonium thiocyanate (NH_3/NH_4SCN)[20] were successively reported, but most of them are still limited in a laboratory scale or bring some serious environmental problems. Recently, some novel green cellulose solvents have been developed like ionic liquid,[21,22] Nmethylmorpholine-N-oxide (NMMO)[23] and water-based solvent systems such as NaOH/polyethylene glycol (PEG) aqueous solution,[24,25] which have laid a solid foundation for broadening cellulose industrial applications. Especially,

* 本文摘自 Materials Science and Technology, 2015, 31(9): 1096-1102.

in the NaOH/PEG solvent system, PEG as a typical environmental benign molecule with repeat unit of —(CH_2—CH_2—O)— has plentiful oxygen atoms that could sever as the hydrogen bonding acceptor to connect with the hydroxyl groups of split cellulose chains as the result of NaOH interaction to form new hydrogen bonds for preventing the regeneration of cellulose through the inter- and intra-chains association, eventually stabilizing the cellulose solution.[24,26]

Wheat straw, one of the most common crop residues, contains abundant cellulose component (ca. 30%–50%), which is generally directly burned or partly used as direct fuel, cattle food and fertilizer. Utilizing the cellulose extracted from wheat straw to fabricate novel high-performance functional materials might be beneficial to high value utilization of these crop residues.

Therefore, herein, we adopted cellulose isolated from waste wheat straw to prepare cellulose aerogels via a mild, low-cost and green NaOH/PEG aqueous solution. Apart from the traditional preparation processes of regeneration cellulose aerogels including dissolution, regeneration and freeze drying (or supercritical drying), a special freezing-thawing process was also employed in this study to increase the degree of crosslinking of cellulose chains and to promote the formation of 3D architecture, and hydrogels obtained by this method have been reported to have a high degree of swelling, good elastic properties, and excellent mechanical strengths as well as superior pore characteristics.[27-29] The as-prepared aerogels were characterized by scanning electron microscope (SEM), X-ray diffraction (XRD), thermal gravity analysis (TGA) and nitrogen adsorption. Meanwhile, the effects of cellulose contents on the aerogels' morphologies, structures and properties were also investigated. Before oil adsorption tests, the aerogels with different cellulose contents all underwent hydrophobic modification by trimethylchlorosilane (TMCS) to improve the oil adsorption properties.

2 Materials and methods

2.1 Materials

Sixty-mesh powder of wheat straw after grinding and sieving was collected and dried in a vacuum at 60 ℃ for 24 h before used. All chemical reagents were supplied by Tianjin Kemiou chemical reagent Co. Ltd., and used without further purification.

2.2 Isolation of the cellulose from wheat straw

The isolation of the cellulose from wheat straw was following the procedures of benzene/ethanol extraction, bleaching process, alkali treatment and acid hydrolysis process.[30,31] Briefly, the dried wheat straw (2 g) was first treated by a mixed aqueous solution of 2/1 V/V% benzene/ethanol at 90 ℃ for 6 h via a soxhlet extractor keeping the liquid boiling briskly so that siphoning from the extractor was no less than four times per hour. Then, the treated sample was air dried and treated with 65 mL 1% $NaClO_2$ solution (pH=4–5 adjusted by glacial acetic acid) at 75 ℃ for 5 h. Whereafter, the resulting residue was collected by filtration and subsequently rinsed with deionized water for three times, and then immediately treated with 50 mL 2% NaOH at 90 ℃ for 2 h. The next step was to collect and wash the product again before treating it with 30 mL 1% HCl at 80 ℃ for 2 h. The acid treatment could further hydrolyze the residual hemicelluloses and pectin by breaking down the polysaccharides to simple sugars and hence release the cellulose fibers, and neutralize residual alkali ions from NaOH and remove mineral traces like Na element.[32,33] Finally, the residue was collected and washed again and dried at 60 ℃ for 24 h, and the following relatively pure cellulose was obtained.

2.3 Preparation of the cellulose aerogels

Sub-samples of the purified cellulose were added to a 10% aqueous solution of NaOH/PEG-4000 (9 : 1 wt/wt) with magnetic stirring for 5 h to form a homogeneous solution. This was repeated at mass ratios of 1/100, 3/100, 5/100, and 7/100 cellulose to the NaOH/PEG solution, which were marked as S1, S3, S5 and S7, respectively. Then, these cellulose solutions were frozen for 12 h at −15 ℃ and subsequently thawed at ambient temperature with vigorous stirring for 30 min. After being frozen for 5 h at −15 ℃ again, the products were successively regenerated following the procedures as follows: ① the frozen cakes were placed in the 1% HCl solution for 6 h, and this process was repeated until the formation of amber-like hydrogels; ② the resulting hydrogels were immersed in water for 6 h, and this process was repeated for three times to remove superfluous hydrogen ions and chlorine anions; ③ the hydrogels were immersed in tertiary butanol for 6 h, and this process was repeated for three times to remove water from the previous step. Finally, the resultant cellulose hydrogels were freeze-dried at −30 ℃ for 48 h in an approximate vacuum (25 Pa). After depressurization, the cellulose aerogels were obtained and characterized.

2.4 Characterizations

The micromorphology was observed by SEM (Quanta 200, FEI). The XRD patterns were measured with a XRD instrument (D/max 2200, Rigaku) using Ni-filtered Cu Kα radiation ($\lambda = 1.5406$ Å) at 40 kV and 30 mA. Scattered radiation was detected ranging from 5° to 40° at a scan rate of $4° \cdot min^{-1}$. TGA experiments were performed by a synchronous thermal analyzer (the United States, SDT-Q600) from room temperature to 700 ℃ at a rate of 10 ℃ $\cdot\ min^{-1}$ under a nitrogen atmosphere. Nitrogen adsorption was carried out on a Micromeritics Tristar Ⅱ 3020 surface area analyzer. The characteristic parameters such as specific surface area and pore size distributions were calculated by the Brunauer-Emmett-Teller (BET) and Barrett-Joyner-Halenda (BJH) methods. A contact angle analyzer (JC2000C) was employed to measure the water contact angle (WCA) of the samples.

2.5 Estimation of cellulose molecular weight

The viscosity-average molecular weight ($M\eta$) of purified cellulose can be estimated from an average intrinsic viscosity value ($[\eta]$). The viscosity measurements were performed on dissolved cellulose with LiOH/urea aqueous solution as solvent by an Ubbelohde viscometer at 25±0.1 ℃, and the corresponding $M\eta$ was calculated by the following equation[34]:

$$[\eta] = 3.72\times10-2M_{\eta}^{0.77}(mL \cdot g^{-1}) \quad (1)$$

3 Result and discussion

The purified cellulose was obtained from the waste straw after the benzene/ethanol extraction, bleaching process, alkali treatment and acid hydrolysis process, and had a high yield of about 56.04%. Moreover, according to the viscosity measurements (Fig. S1), the intrinsic viscosity of the as-prepared cellulose was 265.1 $mL \cdot g^{-1}$, and the corresponding viscosity-average molecular weight was 1.0×10^{5} $g \cdot mol^{-1}$. The cellulose aerogels were also successfully prepared following the procedure of dissolution, freezing-thawing process, regeneration and freeze drying (Fig. 1). As some literatures reported, the freezing-thawing process was employed to increase the degree of crosslinking of cellulose chains and to promote the formation of 3D architecture for improving commonly poor mechanical strengths and enhancing the pore properties of cellulose aerogels. Fig. 2 showed the digital photographs and bulk density distributions of the as-prepared cellulose aerogels. All cellulose aerogels with different cellulose contents showed no obvious differences in appearance, and maintained well-defined shapes after dissolution, regeneration, freezing-thawing process and freeze drying (Fig. 2a), which indicated the superior forming ability. The bulk density was calculated by dividing the weight by the sample volume measured with a micrometer. As shown in Fig. 2b, the bulk densities of the aerogels were low ranging from 0. 044 to 0. 150 $g \cdot cm^{-1}$.

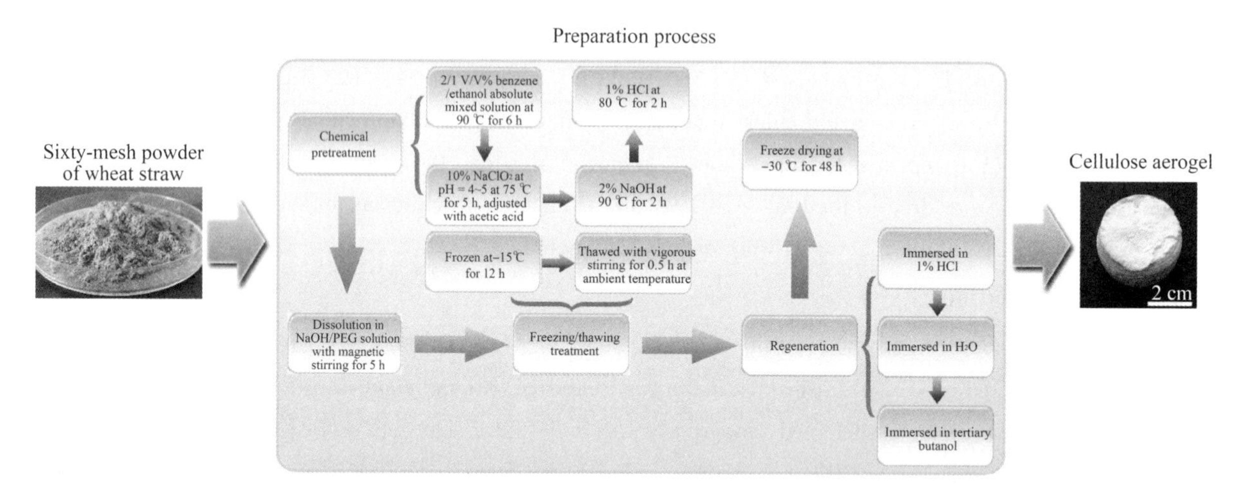

Fig. 1 Schematic representation of the cellulose aerogel preparation from wheat straw

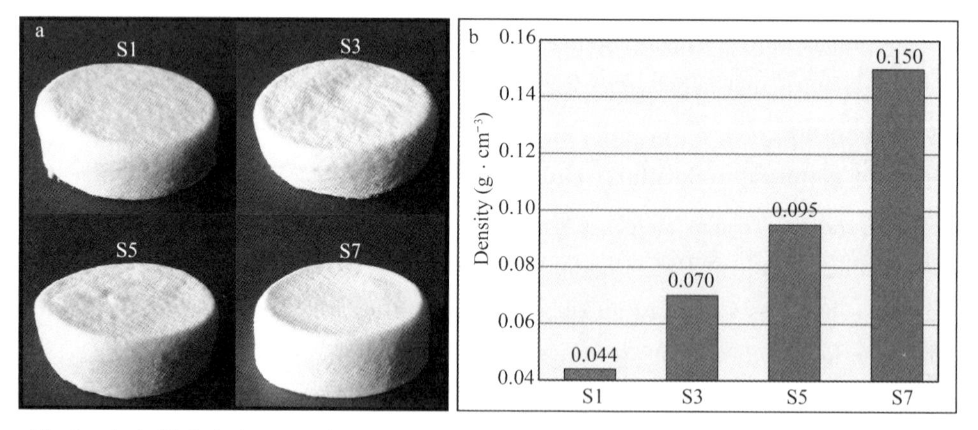

Fig. 2 (a) Digital photographs and (b) bulk density distribution of the cellulose aerogels

SEM images of the aerogels from S1 to S7 were represented in Fig. 3. It could be seen that S1 and S3 showed typical 3D network. Meanwhile, the pore structures of S1 and S3 was interconnected and hierarchical with diameters in the range of several decade nanometers to several microns, and the synergistic effects of mesopores and macropores might contribute to enhancing adsorption property. [35] While increasing the cellulose contents from 3/100 to 7/100, the 3D network gradually became unconspicuous, whereas incompletely dissolved fibers gradually increased. Especially, the sample S7 contained plenty of incompletely dissolved fibers. These SEM images results revealed that the cellulose dissolving capacity of the NaOH/PEG aqueous solution might be approximately the mass ratios of 3/100-5/100 cellulose to the solution.

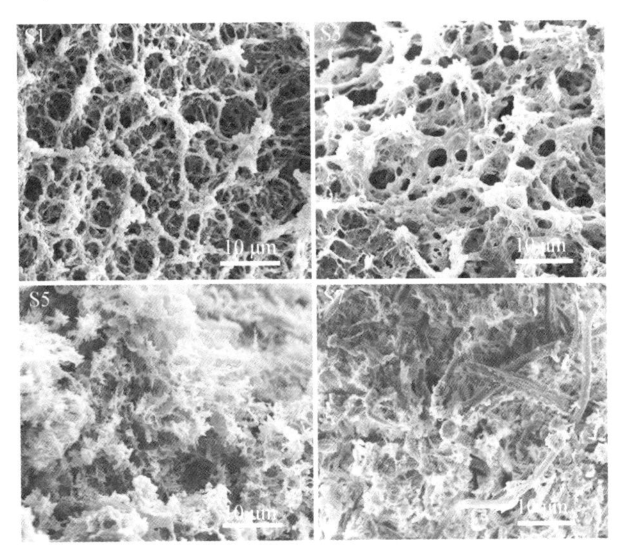

Fig. 3 SEM images of the cellulose aerogels with different cellulose contents

As shown in Fig. 4a, the wheat straw and the purified cellulose displayed typical cellulose Ⅰ crystal structure with parallel up arrangement of cellulose chains,[36] which exhibited (101), (10) and (002) peaks at around 16.05°, 17.28° and 22.15°, and 16.11°, 17.33° and 22.14°, respectively. Nevertheless, the characteristic peaks of the aerogels changed a lot. For S1, the peaks appeared at 11.89°, 19.89° and 21.63°, corresponding to (101), (10) and (002) planes of cellulose Ⅱ with antiparallel configurations of cellulose chains, suggesting a transformation of cellulose Ⅰ to cellulose Ⅱ.[37] Similarly, the (101), (10-1) and (002) peaks appeared at 12.09°, 20.01° and 21.72° for S3, 12.30°, 20.14° and 21.86° for S5, and 12.55°, 20.25° and 21.96° for S7, respectively. Apparently, the characteristic peaks gradually moved to the high degrees from S1 to S7 (as shown in Fig. S2), which indicated the mixed crystal structures of cellulose Ⅰ and cellulose Ⅱ as well as the presence of undissolved cellulose with cellulose Ⅰ,[38] and the undissolved fractions might increase with the increase of cellulose contents.

Crystallinity index was obtained as the ratio of the area arising from the crystalline phase to the total area, and the results were shown in Fig. 4b. Compared with the wheat straw, the crystallinity index of the purified cellulose increased significantly by 20.4% from 41.5% to 61.9%, due to the removal of amorphous substances like hemicellulose and lignin. Nevertheless, the crystallinity indexes of the aerogels decreased compared with that of the cellulose owing to the damage of crystalline texture by alkali ions during dissolution.[39,40] Meanwhile, it was observed that the crystallinity index increased with the increase of cellulose contents, which suggested the possible presence of incompletely dissolved and undamaged cellulose, and the proportion of cellulose with relatively intact structures might increase as the cellulose contents increased while the cellulose content exceeded the cellulose solubility limit of NaOH/PEG solvent system (ca. 3%-5%).

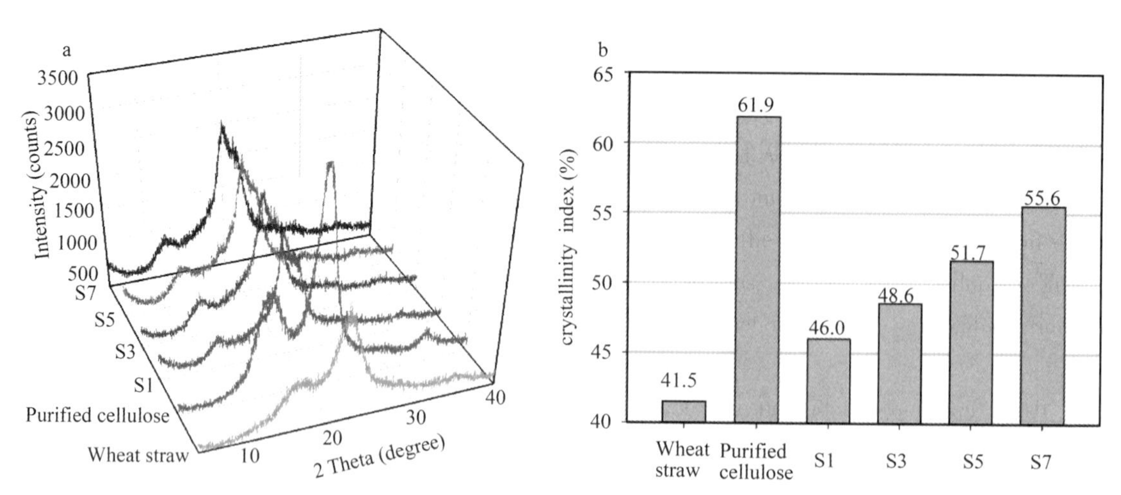

Fig. 4 (a) XRD patterns and (b) crystallinity index distribution of the cellulose aerogels, the purified cellulose and the wheat straw

The pyrolysis characteristics, both TG and DTG curves of the cellulose aerogels, the purified cellulose and the wheat straw, were shown in Fig. 5. The slight weight loss for all samples before 150 ℃ was attributed to the evaporation of absorbed water. For the wheat straw, the pyrolysis process could be divided into three stages[41,42]: ① hemicellulose started its decomposition easily corresponding to the first decomposition shoulder peak at around 205-300 ℃; ② cellulose pyrolysis occurred at a higher temperature range (300-400 ℃) with the maximum weight loss rate attained at 330 ℃; ③ lignin was the most difficult one to decompose, which displayed slow pyrolysis process under the whole temperature range from ambient to 700 ℃. For the other samples, there was only a severe decomposition peak of cellulose with the maximum weight loss rate ranging from 345 ℃ to 367 ℃.

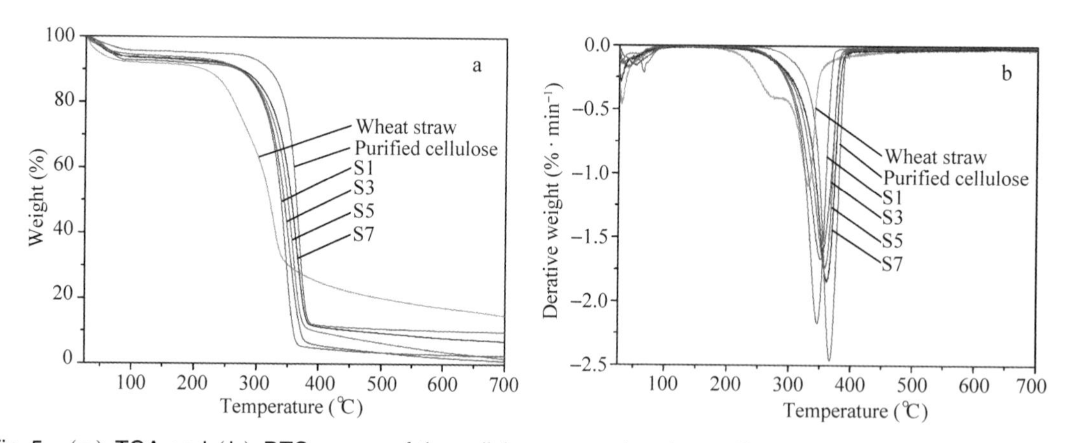

Fig. 5 (a) TGA and (b) DTG curves of the cellulose aerogels, the purified cellulose and the wheat straw

Comparing these TG and DTG curves, at 50% weight loss, the decomposition temperature occurred at 320 ℃ for the wheat straw, 338 ℃ for S1, 343 ℃ for S3, 349 ℃ for S5, 355 ℃ for S7 and 365 ℃ for the cellulose. Moreover, the wheat straw reached maximum weight loss rate at the earliest at 330 ℃, and following S1 (346 ℃), S3 (350 ℃), S5 (359 ℃), S7 (361 ℃) and the cellulose (366 ℃) got the maximum weight loss rate later. According to these trends, a significant positive correlation between thermal stability and crystallinity index was not hard to found; meanwhile, it could also be seen that the higher cellulose contents contributed to the thermal stability enhancement of the as-prepared aerogels.

Before the oil adsorption tests, the cellulose aerogels were subjected to hydrophobic modification to improve the oil adsorption properties by TMCS via a mild chemical vapor deposition method; [43] namely, the samples and 300 μL TMCS were placed in a sealed container at room temperature for 24 h, and the volatile TMCS would react with the hydroxide radicals on the surface of the aerogels leading to a hydrophobic coating on the surface.

After the hydrophobic modifications, the pore features of TMCS modified cellulose aerogels were investigated by nitrogen adsorption measurements to evaluate the effect of cellulose contents on the pore structures of aerogels and the potential adsorption properties. As shown in Fig. 6a, the TMCS treated aerogels all exhibited type-Ⅳ adsorption isotherms which indicated the presence of mesopores, and the obvious hysteresis loops between adsorption and desorption isotherms were generally resulted from the capillary condensation occurring in the mesopores. [44] Moreover, from the pore diameter distributions curves (Fig. 6b), it could be seen that the pore diameters ranged from 1 nm to 60 nm, which suggested that the cellulose aerogels were all mainly composed of mesopores (2–50 nm).

Some pore characteristic parameters including specific surface area, pore volume and average pore size were calculated by BET and BJH methods based on the adsorption-desorption isotherms, and the detail results were illuminated in Table 1. As shown, S1 with developed 3D network observed by SEM had the largest specific surface area (143.2 $m^2 \cdot g^{-1}$), pore volume (0.62 $cm^3 \cdot g^{-1}$) and average pore width (17.5 nm); besides, with the increase of cellulose contents from 1% to 7%, these pore characteristic parameters significantly decreased. Especially for S7 with plentiful incompletely dissolved fibers, the surface area (36.5 $m^2 \cdot g^{-1}$) and pore volume (0.15 $cm^3 \cdot g^{-1}$) were far lower (approximately 1/4) than those of S1, indicating potential inferior adsorption capacity. On the contrary, the samples with relatively higher surface area and pore volume such as S1 and S3 might have more superior adsorption performances.

To further explore the differences of oil adsorption properties of the aerogels with different cellulose contents, the simple adsorption experiments were implemented by immersing the TMCS treated samples in motor oil for 6 h (Fig. 7b). Moreover, the oil adsorption property was estimated by the oil adsorption ratio calculated by dividing the weight after the test by the original dry weight.

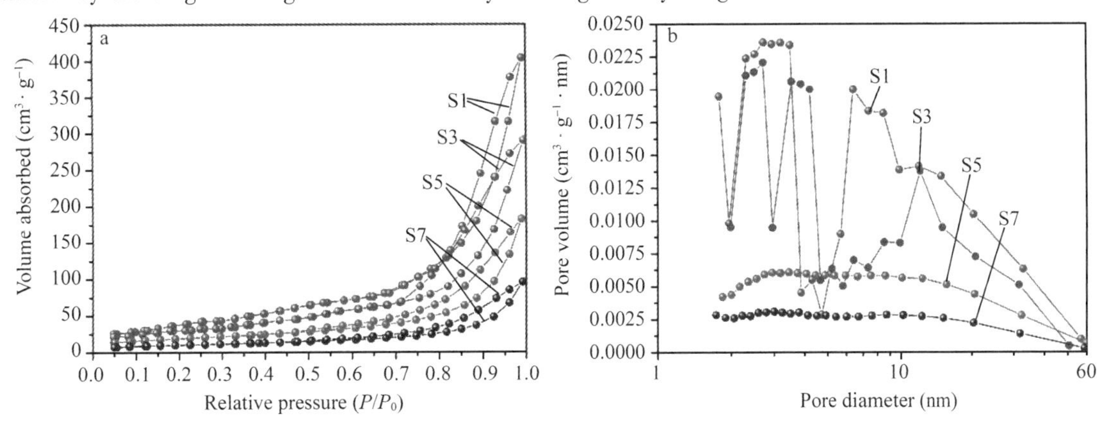

Fig. 6 (a) N_2 adsorption-desorption isotherms and (b) pore diameter distributions of the TMCS treated cellulose aerogels, respectively

As shown in Fig. 7a, the TMCS treated aerogels exhibited strong hydrophobic properties with WCA of 136°, which could stably hold water drop on the surface. Subsequently, these hydrophobic aerogels were immersed in motor oil, and the resulting aerogels all maintained original shapes without significant collapse and size variation after the oil adsorption tests (Fig. 7c), which revealed superior stabilization of the modified aerogels in motor oil. Fig. 7d showed the oil adsorption ratio distributions of the hydrophobic aerogels, and it was observed that the cellulose contents had great influence on the oil adsorption property. The oil adsorption ratios were about 19.2, 17.7, 13.6 and 6.9 times for S1, S3, S5 and S9, respectively. There was a negative correlation between oil adsorption ratio and cellulose contents, and the aerogels (S1 and S3) with better cellulose dissolution effects displayed higher oil adsorption, which were far higher (approximately 2.5–2.8 times) than that of S7 containing plentiful incompletely dissolved fibers; meanwhile, the oil adsorption ratio distributions results were also consistent with the results of nitrogen adsorption analysis, namely, the samples with higher surface area and pore volume gained stronger oil adsorption properties. Therefore, it could be concluded that it might be easier to obtain stronger oil adsorption property within the mass ratio of 3/100 cellulose to the NaOH/PEG solution.

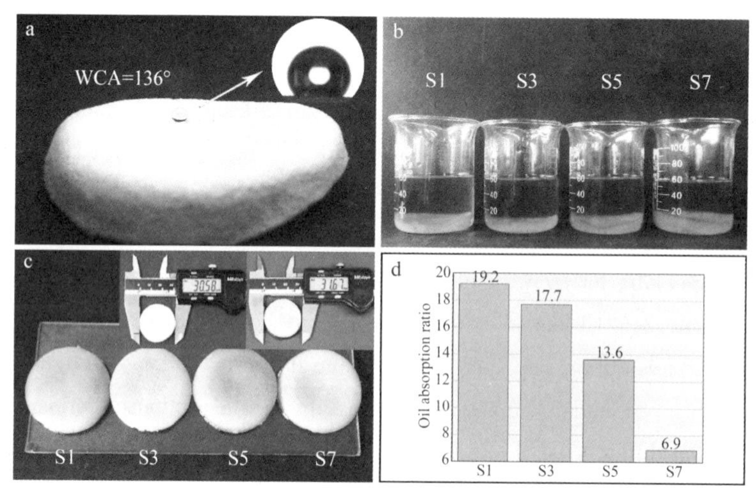

Fig. 7 (a) A water drop stably standing on the TMCS treated aerogels surface, the inset showed the water contact angle measurements; (b) oil adsorption tests of the hydrophobic cellulose aerogels in motor oil; (c) the resulting aerogels after the oil adsorption tests, the insets showed the dimensional change before and after the tests; (d) oil adsorption ratio distributions

Table 1 Pore characteristic parameters of the cellulose aerogels, respectively.

Sample	Specific surface area ($m^2 \cdot g^{-1}$)	Pore volume ($cm^3 \cdot g^{-1}$)	Average pore width (nm)
S1	143.2	0.62	17.5
S3	105.3	0.45	17.1
S5	68.3	0.28	16.5
S7	36.5	0.15	16.3

4 Conclusions

The cellulose aerogels based on a green NaOH/PEG solution system were successfully prepared by dissolution, freezing-thawing treatment, regeneration and freeze drying. Within the weight ratios of cellulose to the solvent of 1% – 3%, the obtained aerogels exhibited typical 3D network, and then the 3D network gradually disappeared as the cellulose contents increased from 3% to 7%. Moreover, with the increase of cellulose contents, the XRD characteristic peaks of the aerogels gradually moved to higher degrees indicating the mixed crystal structures of cellulose Ⅰ and cellulose Ⅱ as well as the presence of undissolved cellulose with cellulose Ⅰ, and the crystallinity index obviously increased from 46.0% (S1) to 55.6% (S7), which might contribute to the improvement of thermal stability. Besides, the higher cellulose contents leaded to the decline of pore characteristic parameters including specific surface area, pore volume and average pore width possibly owing to the increase of undissolved proportion. Furthermore, the highly hydrophobic aerogels treated by TMCS also displayed distinguishing oil adsorption ratios (approximately 6–20 times of their dry weights), and the aerogels with higher pore characteristic parameters and excellent dissolution effect gained stronger oil adsorption properties. Meanwhile, based on these results, for this cellulose purified from the waste wheat straw, the solubility limit of the NaOH/PEG solution system might be around 3% – 5% without other improvement programs at room temperature.

Acknowledgments

This work was financially supported by the National Natural Science Foundation of China (grant no. 31270590), a project funded by the China Postdoctoral Science Foundation (2013M540263), and the Doctoral Candidate Innovation Research Support Program of Science & Technology Review (kjdb2012006).

References

[1] Tamon H, Ishizaka H, Mikami M, et al. Porous structure of organic and carbon aerogels synthesized by sol-gel polycondensation of resorcinol with formaldehyde[J]. Carbon, 1997, 35(6): 791–796.

[2] Lu X, Caps R, Fricke J, et al. Correlation between structure and thermal conductivity of organic aerogels[J]. Journal of Non-Crystalline Solids, 1995, 188(3): 226–234.

[3] Emmerling A, Petricevic R, Beck A, et al. Relationship between optical transparency and nanostructural features of silica aerogels[J]. Journal of non-crystalline solids, 1995, 185(3): 240–248.

[4] Hrubesh L W, Poco J F. Thin aerogel films for optical, thermal, acoustic and electronic applications[J]. Journal of non-crystalline solids, 1995, 188(1–2): 46–53.

[5] Sachithanadam M, Joshi S C. High strain recovery with improved mechanical properties of gelatin-silica aerogel composites post-binding treatment[J]. Journal of Materials Science, 2014, 49(1): 163–179.

[6] Meena A K, Mishra G K, Rai P K, et al. Removal of heavy metal ions from aqueous solutions using carbon aerogel as an adsorbent[J]. Journal of hazardous materials, 2005, 122(1–2): 161–170.

[7] Pajonk G M. Aerogel catalysts[J]. Applied Catalysis, 1991, 72(2): 217–266.

[8] Smith D M, Maskara A, Boes U. Aerogel-based thermal insulation[J]. Journal of non-crystalline solids, 1998, 225: 254–259.

[9] Rao A V, Pajonk G M. Effect of methyltrimethoxysilane as a co-precursor on the optical properties of silica aerogels [J]. Journal of Non-Crystalline Solids, 2001, 285(1–3): 202–209.

[10] Saliger R, Fischer U, Herta C, et al. High surface area carbon aerogels for supercapacitors[J]. Journal of Non-Crystalline Solids, 1998, 225: 81-85.

[11] Kim S J, Hwang S W, Hyun S H. Preparation of carbon aerogel electrodes for supercapacitor and their electrochemical characteristics[J]. Journal of materials science, 2005, 40(3): 725-731.

[12] Kim G S, Hyun S H. Effect of mixing on thermal and mechanical properties of aerogel-PVB composites[J]. Journal ofmaterials science, 2003, 38(9): 1961-1966.

[13] Tsioptsias C, Stefopoulos A, Kokkinomalis I, et al. Development of micro-and nano-porous composite materials by processing cellulose with ionic liquids and supercritical CO_2[J]. Green Chemistry, 2008, 10(9): 965-971.

[14] Aulin C, Netrval J, Wågberg L, et al. Aerogels from nanofibrillated cellulose with tunable oleophobicity[J]. Soft Matter, 2010, 6(14): 3298-3305.

[15] Ding Z D, Chi Z, Gu W X, et al. Theoretical and experimental investigation on dissolution and regeneration of cellulose in ionic liquid[J]. Carbohydrate polymers, 2012, 89(1): 7-16.

[16] Trygg J, Fardim P. Enhancement of cellulose dissolution in water-based solvent via ethanol-hydrochloric acid pretreatment[J]. Cellulose, 2011, 18(4): 987-994.

[17] Yang K S, Theil M H, Chen Y S, et al. Formation and characterization of the fibres and films from mesophase solutions of cellulose in ammonia/ammonium thiocyanate solvent[J]. Polymer, 1992, 33(1): 170-174.

[18] Kuga S. The porous structure of cellulose gel regenerated from calcium thiocyanate solution[J]. Journal of colloid and interface science, 1980, 77(2): 413-417.

[19] Ass B A P, Belgacem M N, Frollini E. Mercerized linters cellulose: characterization and acetylation in N, N-dimethylacetamide/lithium chloride[J]. Carbohydrate Polymers, 2006, 63(1): 19-29.

[20] Frey M W, Theil M H. Calculated phase diagrams for cellulose/ammonia/ammonium thiocyanate solutions in comparison to experimental results[J]. Cellulose, 2004, 11(1): 53-63.

[21] Olsson C, Hedlund A, Idström A, et al. Effect of methylimidazole on cellulose/ionic liquid solutions and regenerated material therefrom[J]. Journal of Materials Science, 2014, 49(9): 3423-3433.

[22] Lu Y, Sun Q, Yang D, et al. Fabrication of mesoporous lignocellulose aerogels from wood via cyclic liquid nitrogen freezing-thawing in ionic liquid solution[J]. Journal of Materials Chemistry, 2012, 22(27): 13548-13557.

[23] Liu X, Chen Q, Pan H. Rheological behavior of chitosan derivative/cellulose polyblends from N-methylmorpholine N-oxide/H_2O solution[J]. Journal of materials science, 2007, 42(16): 6510-6514.

[24] Yan L, Gao Z. Dissolving of cellulose in PEG/NaOH aqueous solution[J]. Cellulose, 2008, 15(6): 789-796.

[25] Han D, Yan L. Preparation of all-cellulose composite by selective dissolving of cellulose surface in PEG/NaOH aqueous solution[J]. Carbohydrate Polymers, 2010, 79(3): 614-619.

[26] Chen J, Spear S K, Huddleston J G, et al. Polyethylene glycol and solutions of polyethylene glycol as green reaction media[J]. Green Chemistry, 2005, 7(2): 64-82.

[27] Resendiz-Hernandez P J, Rodriguez-Fernandez O S, Garcia-Cerda L A. Synthesis of poly (vinyl alcohol)-magnetite ferrogel obtained by freezing-thawing technique[J]. Journal of Magnetism and Magnetic Materials, 2008, 320(14): e373-e376.

[28] Wang Y, Chang C, Zhang L. Effects of freezing/thawing cycles and cellulose nanowhiskers on structure and properties of biocompatible starch/PVA sponges[J]. Macromolecular Materials and Engineering, 2010, 295(2): 137-145.

[29] Chang C, Lue A, Zhang L. Effects of crosslinking methods on structure and properties of cellulose/PVA hydrogels [J]. Macromolecular chemistry and physics, 2008, 209(12): 1266-1273.

[30] Li J, Wan C, Lu Y, et al. Fabrication of cellulose aerogel from wheat straw with strong absorptive capacity[J]. Frontiers of Agricultural Science and Engineering, 2014, 1(1): 46-52.

[31] Zuluaga R, Putaux J L, Cruz J, et al. Cellulose microfibrils from banana rachis: Effect of alkaline treatments on

structural and morphological features[J]. Carbohydrate Polymers, 2009, 76(1): 51-59.

[32] Alemdar A, Sain M. Isolation and characterization of nanofibers from agricultural residues-Wheat straw and soy hulls [J]. Bioresource technology, 2008, 99(6): 1664-1671.

[33] Johar N, Ahmad I, Dufresne A. Extraction, preparation and characterization of cellulose fibres and nanocrystals from rice husk[J]. Industrial Crops and Products, 2012, 37(1): 93-99.

[34] Cai J, Liu Y, Zhang L. Dilute solution properties of cellulose in LiOH/urea aqueous system[J]. Journal of Polymer Science Part B: Polymer Physics, 2006, 44(21): 3093-3101.

[35] Zhang W, Qu Z, Li X, et al. Comparison of dynamic adsorption/desorption characteristics of toluene on different porous materials[J]. Journal of Environmental Sciences, 2012, 24(3): 520-528.

[36] Oh S Y, Yoo D I, Shin Y, et al. Crystalline structure analysis of cellulose treated with sodium hydroxide and carbon dioxide by means of X-ray diffraction and FTIR spectroscopy[J]. Carbohydrate research, 2005, 340(15): 2376-2391.

[37] Gupta P K, Uniyal V, Naithani S. Polymorphic transformation of cellulose Ⅰ to cellulose Ⅱ by alkali pretreatment and urea as an additive[J]. Carbohydrate polymers, 2013, 94(2): 843-849.

[38] Cheng G, Varanasi P, Li C, et al. Transition of cellulose crystalline structure and surface morphology of biomass as a function of ionic liquid pretreatment and its relation to enzymatic hydrolysis[J]. Biomacromolecules, 2011, 12(4): 933-941.

[39] Wang Y, Zhao Y, Deng Y. Effect of enzymatic treatment on cotton fiber dissolution in NaOH/urea solution at cold temperature[J]. Carbohydrate Polymers, 2008, 72(1): 178-184.

[40] Sghaier A E O B, Chaabouni Y, Msahli S, et al. Morphological and crystalline characterization of NaOH and NaOCl treated Agave americana L. fiber[J]. Industrial Crops and Products, 2012, 36(1): 257-266.

[41] Yang H, Yan R, Chen H, et al. Characteristics of hemicellulose, cellulose and lignin pyrolysis[J]. Fuel, 2007, 86 (12-13): 1781-1788.

[42] Orfão J J M, Antunes F J A, Figueiredo J L. Pyrolysis kinetics of lignocellulosic materials-three independent reactionsmodel[J]. Fuel, 1999, 78(3): 349-358.

[43] Song J, Rojas O J. Paper chemistry: approaching super-hydrophobicity from cellulosic materials: a review[J]. Nordic pulp & paper research journal, 2013, 28(2): 216-238.

[44] Tao Y, Kanoh H, Kaneko K. Uniform mesopore-donated zeolite Y using carbon aerogel templating[J]. The Journal of Physical Chemistry B, 2003, 107(40): 10974-10976.

中文题目：使用纤维素-氢氧化钠/聚乙二醇溶液制备纤维素气凝胶以及与不同纤维素含量的比较

作者：万才超，卢芸，焦月，曹军，孙庆丰，李坚

摘要：使用绿色的温和的 NaOH/聚乙二醇溶液溶解从废麦草中提取的纤维素。随后结合冻融处理，再生过程和冷冻干燥，分别以质量比为 1∶100、3∶100、5∶100 和 7∶100 制备具有不同纤维素含量的纤维素气凝胶。此外，通过扫描电子显微镜，X 射线衍射、热重分析和氮吸附研究了纤维素含量对气凝胶的形貌、晶体结构、结晶度指标、热稳定性和孔特征参数的影响。同时，考察了不同纤维素含量的气凝胶的吸油性能。此外，为了提高亲脂性，在吸附试验之前对气凝胶进行了三甲基氯硅烷的疏水改性。

关键词：纤维素气凝胶；氢氧化钠/聚乙二醇溶液；油吸附；冷冻干燥；聚合物

Preparation, Characterization and Oil Adsorption Properties of Cellulose Aerogels from Four Kinds of Plant Materials via A NaOH/PEG Aqueous Solution *

Caichao Wan, Yun Lu, Yue Jiao, Jun Cao, Jian Li, Qingfeng Sun

Abstract: In this paper, cellulose aerogels based on four kinds of plant materials including wheat straw, bamboo fiber, filter paper and cotton were prepared by alternative chemical pretreatment, dissolution in a green NaOH/PEG solution, freeze-thaw treatment, regeneration and freeze drying, respectively, and were subsequently characterized by scanning electron microscopy (SEM), nitrogen adsorption measurement, X-ray diffraction (XRD), thermogravimetric analysis (TGA). The differences in morphology, pore feature, crystalline structure and thermal property for the four kinds of aerogels were investigated. Meanwhile, the oil adsorption capacities differences of the aerogels were also studied after the aerogels were subjected to hydrophobic modifications by methyltrichlorosilane (MTCS); besides, the effect of the hydrophobic modifications was explored by Fourier transform infrared spectroscopy (FTIR).

Keywords: Cellulose aerogels; Adsorption properties; Hydrophobic modification; Porous materials; Polymers

1 Introduction

Cellulose, the one of the most abundant biopolymers on earth, widely exists in wood, cotton, straw and other plant-based materials, and works as the dominant reinforcing phase in plant structures[1,2]. Despite well-known biodegradability and renewability, cellulose also has numerous charming properties such as high strength and stiffness, low density and superior biocompatibility[3-6], which has continuously attracted much more attentions. In general, it is difficult to dissolve cellulose in some common water or organic solvents due to its strong inter- and intra-molecular hydrogen bonds, high degree of polymerization and high degree of crystallinity[7,8], which imposes huge restrictions on application of cellulose. In recent decades, as some novel green cellulose solvents have been successively found like ionic liquid[9,10], NH_3/NH_4SCN[11] and NaOH/PEG aqueous solution[12,13], the exploitation and development of cellulose functional materials especially for cellulose aerogels with numerous excellent performances, have gradually become focus topics again.

Cellulose aerogels have not only low density, high specific surface area and high porosity similar to traditional inorganic or polymer-based aerogels, but also flexibility and biodegradability, which hold great potential in the fields of adsorbents, thermal (electrical or sound) insulation materials and tissue

* 本文摘自 Fibers and Polymers, 2015, 16(2): 302-307.

engineering[14-18]. Owing to the unique characteristic features of cellulose, the preparation of regeneration cellulose aerogels commonly needs to firstly dissolve cellulose, and following regeneration and freeze drying (or supercritical drying). Actually, the cellulose isolated from different cellulosic materials has subtle differences in structure and property, such as content of crystalline, paracrystalline and amorphous regions, crystallite dimensions and molecular rigidity[19], which might have significant effects on the macro-performance of the prepared cellulose materials.

Therefore, in this paper, four kinds of plant materials including wheat straw, bamboo fiber, filter paper and cotton were employed to fabricate cellulose aerogels by alternative chemical pretreatment, dissolution in a green NaOH/PEG solution, freeze-thaw treatment, regeneration and freeze drying. The morphologies, structures and properties of obtained aerogels were characterized and compared by scanning electron microscopy (SEM), nitrogen adsorption measurements, X-ray diffraction (XRD), and thermogravimetric analysis (TGA). Moreover, the oil adsorption capacities differences of the aerogels were also investigated after the aerogels underwent hydrophobic modifications of methyltrichlorosilane (MTCS); meanwhile, Fourier transform infrared spectroscopy (FTIR) was carried out to explore the modification effect.

2 Materials and methods

2.1 Materials

Four kinds of plant materials including wheat straw, bamboo fiber, filter paper and cotton were selected as the precursors of the cellulose aerogels. Especially, the cellulose from the wheat straw was firstly isolated through a referred approach: typically, the wheat straw was first treated by a mixed solution of benzene/absolute ethanol (2 : 1 v/v) in a Soxhlet extractor at 90 ℃ for 6 h. Then, the treated sample was air-dried and then treated with a 10 wt% $NaClO_2$ solution (pH=4.5 adjusted by glacial acetic acid) at 75 ℃ for 5 h. The next step was to collect the sample by filtration and wash it three times with deionized water, and then immediately treat it with 2 wt% NaOH at 90 ℃ for 2 h. The product was again collected by filtration and washed three times with deionized water before treating it with 1 wt% HCl at 80 ℃ for 2 h. Finally, the purified cellulose was collected by filtration and washed three times with deionized water and dried at 60 ℃ for 24 h. Other materials involving bamboo fiber, filter paper and cotton were used directly without further purification due to the ultra-high cellulose content. All other chemicals were purchased from Tianjin Kermel chemical Co. Ltd. and used as received.

2.2 Preparation of the cellulose aerogels

The four kinds of cellulose with the 2% weight were added to 10 wt% aqueous solutions of NaOH/PEG (4 g of NaOH and 0.4 g of PEG-4000 were added into 40 mL of distilled water) with magnetic stirring for 5 h to form homogeneous solutions, respectively. These solutions were subsequently frozen for 12 h at −15 ℃ to form solid frozen masses, and then thawed at ambient temperature with vigorous stirring for 30 min. After being frozen again for 3 h at −15 ℃, the frozen solutions were immediately placed into the coagulation bath of 1 v% hydrochloric acid for six hours, repeating this process until the formation of the amber-like hydrogels. Afterwards, the cellulose hydrogels were successively rinsed with deionized water and tertiary butyl alcohol for removing the residual Cl^- anions and water molecules, respectively. Finally, the cellulose hydrogels filled with tertiary butyl alcohol underwent freeze-drying at −35 ℃ for 48 h in vacuum, and the following cellulose aerogels were obtained. The cellulose aerogels from wheat straw, bamboo fiber, filter paper and cotton were referred to as Aerogel W, Aerogel B, Aerogel F and Aerogel C, respectively.

2.3 Hydrophobic modification of the cellulose aerogels

The cellulose aerogels surfaces were modified by MTCS via a chemical vapor deposition (CVD) method in order to get hydrophobicity. The aerogels were placed in a sealed desiccator with a container filled with 300 μL MTCS precursor liquid. There was no direct contact between the liquid and the aerogels. The desiccator was put at ambient temperature for 24 h. The precursor vaporized and reacted with the hydroxide radical groups of cellulose surface, resulting in the formation of hydrophobicity.

2.4 Oil adsorption tests of the aerogels

The oil adsorption tests were implemented by immersing the MTCS treated aerogels (ca. 0.5 g) in the 50 mL waste motor oil for 6 h at room temperature. After the impregnation, the wet aerogels were slowly lifted up, and the excess oil was allowed to drain for 30 s. Residual oil on the surface of the samples was removed with filter paper. The weights of the samples were measured before and after the tests. The absorption efficiency can be referred to as weight gain defined as the weight of absorbed substances per unit weight of original cellulose aerogel.

2.5 Characterizations

Morphology was characterized by SEM (FEI, Quanta 200). The pore characteristics were determined by N_2 adsorption-desorption measurements at −196 ℃ using an accelerated surface area and porosimetry system (3H-2000PS2 unit, Beishide Instrument S&T Co., Ltd), the specific surface area and the pore-size distributions were estimated by the Brunauer-Emmet-Teller (BET) and the Barrett-Joyner-Halenda (BJH) methods. Crystalline structures were identified by XRD (Rigaku, D/MAX 2200) operating with Cu Kα radiation (λ = 1.5418 Å) at a scan rate (2θ) of 4° · min^{-1} and the accelerating voltage of 40 kV and the applied current of 30 mA ranging from 5° to 40°. The thermal stability was determined using a TG analyzer (TA Q600) with a heating rate of 10 ℃ · min^{-1} in a N_2 environment. FTIR spectra were recorded on a FTIR spectrophotometer (Magna 560, Nicolet, Thermo Electron Corp) in the range of 650 − 4000 cm^{-1} with a resolution of 4 cm^{-1}.

2.6 Results and discussion

Fig. 1 showed the digital photographs and the SEM images of Aerogel W, Aerogel B, Aerogel F, and Aerogel C, respectively. Apparently, the as-prepared cellulose aerogels from four kinds of plant materials all maintained well-defined shape without any collapse, indicating superior forming ability. In addition, apart from Aerogel C mainly composed of plentiful undissolved fibers, other aerogels all displayed typical and dense three-dimensional (3D) network, indicating that the NaOH/PEG aqueous solution was a direct solvent of cellulose. Meanwhile, the inferior dissolution effect of cotton might be on account of the high degree of crystallinity, which would be verified by other subsequent characterization means.

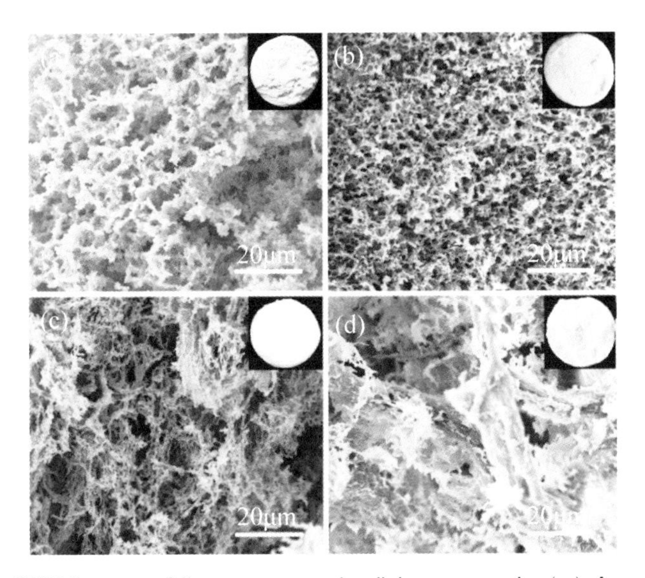

Fig. 1 SEM images of the as-prepared cellulose aerogels. (a) Aerogel W, (b) Aerogel B, (c) Aerogel F, and (d) Aerogel C. The insets showed the digital photographs of the aerogels

The pore features were investigated by N_2 adsorption-desorption measurements, and the resulting N_2 adsorption-desorption isotherms and pore size distributions were presented in Fig. 2. It was observed that all aerogels exhibited type-Ⅳ adsorption isotherms[20] (Fig. 2a), which indicated the presence of mesopores, and the significant hysteresis loops were generally attributed to the capillary condensation occurring in the mesopores[21]. Moreover, as shown in Fig. 2b, the aerogels had pore diameters of about 1 - 200 nm, indicating that the materials possibly contained micropores (<2 nm), mesopores (2-50 nm) and macropores (>50 nm), and synergistic effects of hierarchical and multi-scale pore structures might be beneficial to adsorb adsorbates with various dimensions. Among the four aerogels, Aerogel W had relatively narrower pore size distribution (ca. 1-60 nm), primarily made up of mesopores.

Table 1 represented the detailed experimental results of the specific surface area, pore volume, and bulk density of the aerogels. Nanoporous Aerogel W, Aerogel B and Aerogel F with developed network structures exhibited much larger specific surface areas (99.17 $m^2 \cdot g^{-1}$, 152.5 $m^2 \cdot g^{-1}$ and 137.11 $m^2 \cdot g^{-1}$) and pore volumes (0.45 $cm^3 \cdot g^{-1}$, 0.51 $cm^3 \cdot g^{-1}$ and 0.45 $cm^3 \cdot g^{-1}$), compared with those of Aerogel C (63.3 $m^2 \cdot g^{-1}$ and 0.22 $cm^3 \cdot g^{-1}$), which suggested potential stronger adsorption properties. Moreover, the aerogels all had low bulk densities calculated by dividing the weight by the volume (53.37-91.96 $mg \cdot cm^{-3}$); besides, it could be also observed that the specific surface area was sensitive to the bulk density, that was, the samples with higher bulk densities conversely displayed relatively lower specific surface area. It was not hard to comprehend the subtle correlation between the bulk density and the specific surface area; in general, for a certain quality of porous solids, the solids with more micro-nano pore structures might have higher volume and specific surface area leading to lower bulk density.

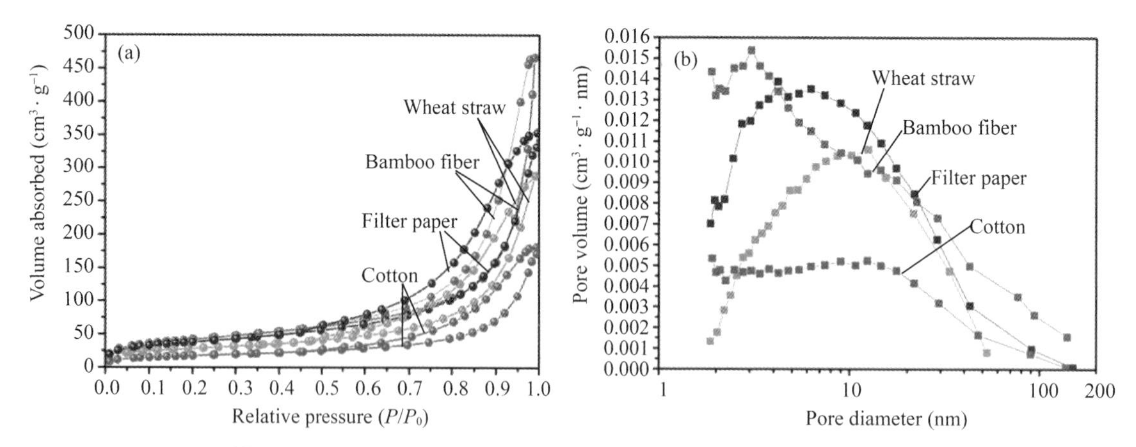

Fig. 2 (a) N_2 adsorption-desorption isotherms and (b) pore size distributions of Aerogel W, Aerogel B, Aerogel F and Aerogel C, respectively

Table 1 Pore properties of Aerogel W, Aerogel B, Aerogel F and Aerogel C

Sample	Specific surface area ($m^2 \cdot g^{-1}$)	Pore volume ($cm^3 \cdot g^{-1}$)	Bulk density ($mg \cdot cm^{-3}$)
Aerogel W	99. 17	0. 45	60. 74
Aerogel B	152. 50	0. 51	53. 37
Aerogel F	137. 11	0. 45	57. 83
Aerogel C	63. 30	0. 22	91. 96

Fig. 3a showed the XRD patterns of the aerogels and the corresponding raw materials. It was observed that the raw materials including wheat straw, bamboo fiber, filter paper and cotton all exhibited typical cellulose I crystalline structure, owing to the strong diffraction peaks at around 14. 6°, 16. 3° and 22. 5° corresponding to (101), ($10\bar{1}$) and (002) planes[22]. Whereas, as the result of NaOH treatment during dissolution, obvious transformation from cellulose Ⅰ to cellulose Ⅱ for the aerogels could be found due to the generation of new peaks at around 12. 0°, 19. 9° and 21. 8°[23], respectively.

NaOH treatment resulted in the splitting and formation of new intra- and intermolecular hydrogen bonds, and rearranged or transformed the crystalline structure, which had significant influences on crystallinity index[24-26]. Crystallinity index was estimated as the ratio of the area arising from the crystalline phase to the total area. As shown in Fig. 3b, compared with the raw materials, the crystallinity indexes of the corresponding aerogels significantly decreased by 10. 1%-19. 3%. Meanwhile, it could be seen that cotton had relatively higher crystallinity index of 72. 1%; besides, it is well-known that cotton fiber also has the highest molecular weight among native cellulosic materials[27,28], leading to the undesirable dissolution effect in this solvent system.

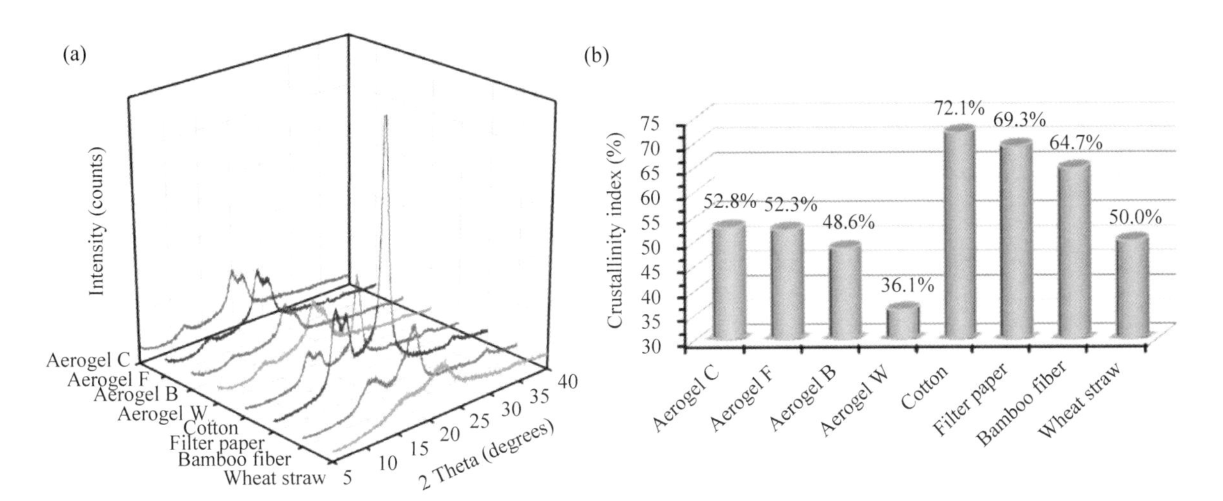

Fig. 3 (a) XRD patterns and (b) crystallinity index distributions of wheat straw, bamboo fiber, filter paper, cotton, Aerogel W, Aerogel B, Aerogel F and Aerogel C, respectively

TG and DTG curves of Aerogel W, Aerogel B, Aerogel F and Aerogel C were presented in Fig. 4, respectively. The weight loss below 150 ℃ was principally attributed to the evaporation of adsorbed water for all aerogels samples[29]; besides, the aerogels all appeared one severe decomposition stage with strong pyrolysis peaks occurring at 337 - 353 ℃, which was related to pyrolysis of cellulose. Moreover, comparing the pyrolysis processes of the four kinds of aerogels, Aerogel W displayed inferior thermal stability, which started to decompose at 207 ℃ and reached the maximum degradation rate at 337 ℃ at the earliest; on the contrary, Aerogel C exhibited more superior thermal stability with relatively higher degradation onset (230 ℃) and decomposition temperature at maximum degradation rate (353 ℃). Furthermore, Aerogel F and Aerogel B began to degrade at around 225 ℃ and 222 ℃, and realized the pyrolysis peaks at 354 ℃ and 347 ℃, respectively. According to these decomposition tendencies, it was not hard to find a positive correlation between thermal stability and crystallinity index, and higher crystallinity index might be beneficial to get better thermal property, which was in accordance with some previous reports[30,31].

Before the oil adsorption experiments of the aerogels from different raw materials, the mild hydrophobic modifications of MTCS were carried out via a CVD method. The formation of the hydrophobic coating of the cellulose aerogels was confirmed by FTIR analysis (Fig. 5). Apart from the characteristic vibrations of cellulose structures like O—H stretching (ca. 3372 cm^{-1}), C—H stretching (ca. 2899 cm^{-1}), O—H bending (ca. 1371 cm^{-1}), and C—H bending (ca. 1314 cm^{-1})[32], the FTIR spectrum of the MTCS treated cellulose aerogel displayed two new strong absorption peaks at around 775 cm^{-1} and 1277 cm^{-1}, which were possibly attributed to the stretching vibrations of the Si—C bonds and to —CH_3 deformation vibrations of the siloxane compounds[33], respectively. Furthermore, the intensities of the absorption bands centered at around 1022 cm^{-1} significantly enhanced, which might be on account that the typical absorption peaks of the Si-O-Si bonds of MTCS in the region of 1000-1130 cm^{-1} appeared to be overlapped by the cellulose bands due to C—O bending modes[34]. Therefore, based on the FTIR results mentioned above, it might be concluded that the MTCS molecules had been successfully grafted on the surface of cellulose.

Fig. 6 showed the hydrophobic performance and the oil adsorption tests of the MTCS treated cellulose

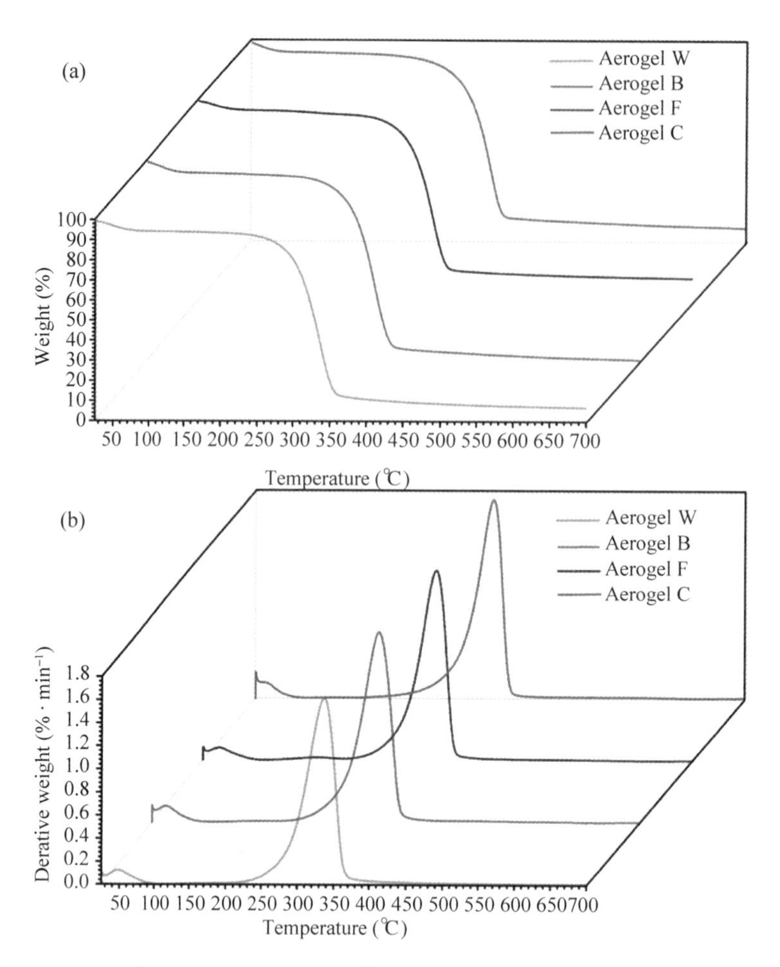

Fig. 4 (a) TG and (b) DTG curves of Aerogel W, Aerogel B, Aerogel F and Aerogel C, respectively

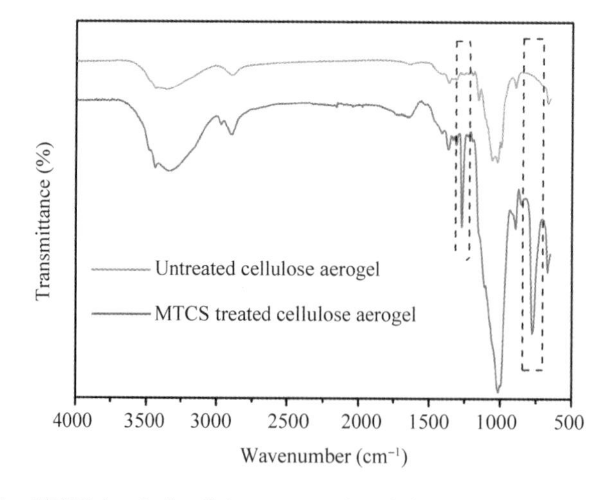

Fig. 5 FTIR spectra of the MTCS treated cellulose aerogel and the untreated cellulose aerogel, respectively

aerogels. As shown, a MTCS treated aerogel floating on water due to water repellent and low density, and several water drops could stably stand on the surface of the aerogel (Fig. 6a) indicating the formation of hydrophobicity. Fig. 6b presented the oil adsorption properties distributions of the modified aerogels, and approximately 13.5–20.6 times of their dry weights of waste motor oil were absorbed comparable with some commercially available adsorbents such as activated carbon ($<5\ g \cdot g^{-1}$), raw cotton fiber (ca. $30\ g \cdot g^{-1}$), polyurethane foams ($40\ g \cdot g^{-1}$) and nonwoven polypropylene ($15\ g \cdot g^{-1}$)[35-37], which was mainly because of physical absorption. Aerogel B, Aerogel F and Aerogel W with large specific surface areas and pore volumes adsorbed more oil (approximate 1.3–1.6 times) than that of Aerogel C, indicating stronger oil adsorption capacity.

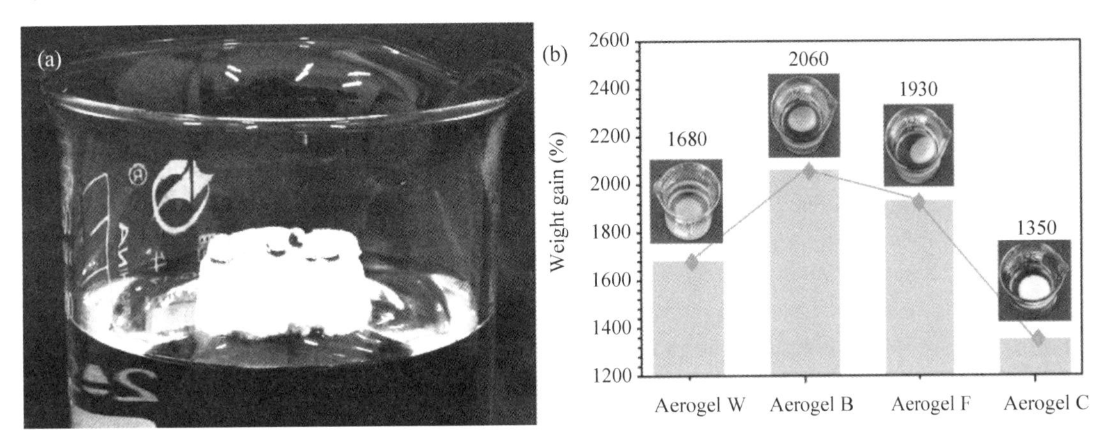

Fig. 6 (a) A MTCS treated cellulose aerogel with several water drops on the surface floating on the water. (b) Oil adsorption properties of Aerogel W, Aerogel B, Aerogel F and Aerogel C that had been modified by MTCS

3 Conclusions

We successfully fabricated cellulose aerogels from four kinds of plant materials including wheat straw, bamboo fiber, filter paper and cotton via alternative chemical pretreatment, dissolution in NaOH/PEG solution, freeze-thaw treatment, regeneration and freeze drying. Owing to high crystallinity index and molecular weight of cotton, Aerogel C displayed inferior dissolution effect leading to low specific surface area and pore volume; instead, Aerogel W, Aerogel B and Aerogel F with typical 3D network had more superior pore characteristics. Moreover, the aerogels all exhibited cellulose Ⅱ crystalline structure, whose crystallinity indexes significantly decreased compared with the corresponding raw materials due to NaOH treatment during dissolution; meanwhile, higher crystallinity index might be beneficial to get better thermal property. Furthermore, the MTCS modified cellulose aerogels could adsorb 13.5–20.6 times of their dry weights of waste motor oil, and the aerogels with larger specific surface areas and pore volumes displayed higher oil adsorption volumes.

Acknowledgments

This work was financially supported by the National Natural Science Foundation of China (grant no. 31270590), a project funded by the China Postdoctoral Science Foundation (2013M540263), and the Doctoral Candidate Innovation Research Support Program of Science & Technology Review (kjdb2012006).

References

[1] Pei A, Malho J M, Ruokolainen J, et al. Strong nanocomposite reinforcement effects in polyurethane elastomer with low volume fraction of cellulose nanocrystals[J]. Macromolecules, 2011, 44(11): 4422-4427.

[2]Fan J, Li Y. Maximizing the yield of nanocrystalline cellulose from cotton pulp fiber[J]. Carbohydrate polymers, 2012, 88(4): 1184-1188.

[3] Wu Q, Henriksson M, Liu X, et al. A high strength nanocomposite based on microcrystalline cellulose and polyurethane[J]. Biomacromolecules, 2007, 8(12): 3687-3692.

[4] Reddy N, Yang Y. Structure and properties of high quality natural cellulose fibers from cornstalks[J]. Polymer, 2005, 46(15): 5494-5500.

[5] Siró I, Plackett D. Microfibrillated cellulose and new nanocomposite materials: a review[J]. Cellulose, 2010, 17(3): 459-494.

[6] Shenqi W, Yaoting Y, Tao C, et al. Cellulose amphiphilic adsorbent for the removal of low density lipoprotein[J]. Artificial Cells, Blood Substitutes, and Biotechnology, 2002, 30(4): 285-292.

[7] Dadi A P, Varanasi S, Schall C A. Enhancement of cellulose saccharification kinetics using an ionic liquid pretreatment step[J]. Biotechnology and bioengineering, 2006, 95(5): 904-910.

[8] Lee S Y, Mohan D J, Kang I A, et al. Nanocellulose reinforced PVA composite films: effects of acid treatment and filler loading[J]. Fibers and Polymers, 2009, 10(1): 77-82.

[9] Zhang H, Wu J, Zhang J, et al. 1-Allyl-3-methylimidazolium chloride room temperature ionic liquid: a new and powerful nonderivatizing solvent for cellulose[J]. Macromolecules, 2005, 38(20): 8272-8277.

[10] Lu Y, Sun Q, Yang D, et al. Fabrication of mesoporous lignocellulose aerogels from wood via cyclic liquid nitrogen freezing-thawing in ionic liquid solution[J]. Journal of Materials Chemistry, 2012, 22(27): 13548-13557.

[11] Cuculo J A, Smith C B, Sangwatanaroj U, et al. A study on the mechanism of dissolution of the cellulose/NH_3/NH_4SCN system. I[J]. Journal of Polymer Science Part A: Polymer Chemistry, 1994, 32(2): 229-239.

[12] Yan L, Gao Z. Dissolving of cellulose in PEG/NaOH aqueous solution[J]. Cellulose, 2008, 15(6): 789-796.

[13] Zhang S H, Li F X, Yu J Y. Structure and properties of novel cellulose fibres produced from NaOH/PEG-treated cotton linters[J]. Iranian polymer journal, 2010, 12: 949-957.

[14] Liebner F, Haimer E, Wendland M, et al. er, P. Miethe, T. Heinze, A. Potthast and T. Rosenau[J]. Macromol. Biosci, 2010, 10: 349-352.

[15] Korhonen J T, Kettunen M, Ras R H A, et al. Hydrophobic nanocellulose aerogels as floating, sustainable, reusable, and recyclable oil absorbents[J]. ACS applied materials & interfaces, 2011, 3(6): 1813-1816.

[16] Lee J K, Gould G L. Polydicyclopentadiene based aerogel: a new insulation material[J]. Journal of Sol-Gel Science and Technology, 2007, 44(1): 29-40.

[17] Shi J, Lu L, Guo W, et al. An environment-friendly thermal insulation material from cellulose and plasma modification[J]. Journal of Applied Polymer Science, 2013, 130(5): 3652-3658.

[18] Aulin C, Netrval J, Wågberg L, et al. Aerogels from nanofibrillated cellulose with tunable oleophobicity[J]. Soft Matter, 2010, 6(14): 3298-3305.

[19] Focher B, Palma M T, Canetti M, et al. Structural differences between non-wood plant celluloses: evidence from solid state NMR, vibrational spectroscopy and X-ray diffractometry[J]. Industrial Crops and Products, 2001, 13(3): 193-208.

[20] Liu H, Sha W, Cooper A T, et al. Preparation and characterization of a novel silica aerogel as adsorbent for toxic organic compounds[J]. Colloids and Surfaces A: Physicochemical and Engineering Aspects, 2009, 347(1-3): 38-44.

[21] Kierlik E, Monson P A, Rosinberg M L, et al. Capillary condensation in disordered porous materials: Hysteresis versus equilibrium behavior[J]. Physical review letters, 2001, 87(5): 055701.

[22] Oh S Y, Yoo D I, Shin Y. HC Ki m, HY Ki m, YS Chung, WH Park, JH Youk[J]. Carbohyd. Res, 2005, 340:

2376.

[23] Okano T, Sarko A. Mercerization of cellulose. II. Alkali-cellulose intermediates and a possible mercerization mechanism[J]. Journal of Applied Polymer Science, 1985, 30(1): 325-332.

[24] Ouajai S, Shanks R A. Composition, structure and thermal degradation of hemp cellulose after chemical treatments [J]. Polymer degradation and stability, 2005, 89(2): 327-335.

[25] Mittal A, Katahira R, Himmel M E, et al. Effects of alkaline or liquid-ammonia treatment on crystalline cellulose: changes in crystalline structure and effects on enzymatic digestibility[J]. Biotechnology for biofuels, 2011, 4(1): 1-16.

[26] Borysiak S, Doczekalska B. X-ray diffraction study of pine wood treated with NaOH[J]. Fibres Text. East. Eur., 2005, 13(5): 87-89.

[27] Goring D A I, Timell T E. Molecular weight of native celluloses[J]. Tappi, 1962, 45(6): 454-460.

[28] Timpa J D. Application of universal calibration in gel permeation chromatography for molecular weight determinations of plant cell wall polymers: cotton fiber[J]. Journal of Agricultural and Food Chemistry, 1991, 39(2): 270-275.

[29] Li J, Lu Y, Yang D, et al. Lignocellulose aerogel from wood-ionic liquid solution (1-allyl-3-methylimidazolium chloride) under freezing and thawing conditions[J]. Biomacromolecules, 2011, 12(5): 1860-1867.

[30] Nada A M A, Hassan M L. Thermal behavior of cellulose and some cellulose derivatives[J]. Polymer Degradation and Stability, 2000, 67(1): 111-115.

[31] Alemdar A, Sain M. Biocomposites from wheat straw nanofibers: Morphology, thermal and mechanical properties[J]. Composites Science and Technology, 2008, 68(2): 557-565.

[32] Lan W, Liu C F, Yue F X, et al. Ultrasound-assisted dissolution of cellulose in ionic liquid[J]. Carbohydrate polymers, 2011, 86(2): 672-677.

[33] Shirgholami M A. Shateri Khalil-Abad M, Khajavi R, et al. Fabrication of superhydrophobic polymethylsilsesquioxane nanostructures on cotton textiles by a solution–immersion process[J]. J. Colloid Interface Sci. 2011, 359: 530-535.

[34] Li S, Zhang S, Wang X. Fabrication of superhydrophobic cellulose-based materials through a solution-immersion process[J]. Langmuir, 2008, 24(10): 5585-5590.

[35] Deschamps G, Caruel H. M., Borredon, C., et al. "Oil removal from water by selective sorption on hydrophobic cotton fibers. 1. Study of sorption properties and comparison with other cotton fiber-based sorbents"[J]. Environ. Sci. Technol, 2003, 37: 1013-1015.

[36] Li H, Liu L, Yang F. Hydrophobic modification of polyurethane foam for oil spill cleanup[J]. Marine Pollution Bulletin, 2012, 64(8): 1648-1653.

[37] Radetic M, Ilic V, Radojevic D, et al. Efficiency of recycled wool–based nonwoven material for the removal of oils from water[J]. Chemosphere, 2008, 70(3): 525-530.

中文题目: 用氢氧化钠/聚乙二醇水溶液制备四种植物材料纤维素气凝胶，并对其进行表征和吸附性能研究

作者: 万才超，卢芸，焦月，曹军，孙庆丰，李坚

摘要: 本文以麦秸、竹纤维、滤纸和棉花四种植物材料为基础，分别采用化学预处理、在NaOH/PEG溶液中溶解、冻融处理、再生和冷冻干燥等方法制备了纤维素气凝胶，并通过扫描电镜(SEM)、氮吸附测量、X射线衍射(XRD)、热重分析(TGA)对其进行了表征。研究了四种气凝胶在形貌、孔隙特征、结晶结构和热性能方面的差异。同时，还研究了气凝胶经甲基三氯硅烷(MTCS)疏水改性后的吸油能力差异，并通过傅立叶变换红外光谱(FTIR)探讨了疏水改性的效果。

关键词: 纤维素气凝胶；吸附性能；疏水改性；多孔材料；聚合物

Ultra-light and Hydrophobic Nanofibrillated Cellulose Aerogels from Coconut Shell with Ultra-strong Adsorption Properties*

Caichao Wan, Yun Lu, Yue Jiao, Chunde Jin, Qingfeng Sun, Jian Li

Abstract: Ultra-light aerogels based on nanofibrillated cellulose (NFC) isolated from coconut shell were successfully prepared via a mild fast method including chemical pretreatment, ultrasonic isolation, solvent-exchange and tert-butanol freeze drying. The as-prepared NFC aerogels with complex three-dimensional (3D) fibrillar network had low bulk density of 0.84 $mg \cdot cm^{-3}$, whose specific surface area and pore volume were 9.1 $m^2 \cdot g^{-1}$ and 0.025 $cm^3 \cdot g^{-1}$, and maintained cellulose Ⅰ crystal structure and showed more superior thermal stability than that of the coconut shell raw materials. After the hydrophobic modification by Methyltrichlorosilane (MTCS), the NFC aerogels exhibited high water repellent property and ultra-strong oil adsorption capacity (542 times of the original dry weight of diesel oil) as well as superior oil/water separation performance. Moreover, the absorption capabilities of the MTCS treated NFC aerogels can be as high as 296–669 times their own weights for various organic solvents and oil. Thus, this class of high-performance adsorbing materials might be useful for dealing with chemical leaks and oil spills.

Keywords: Cellulose and other wood products; Nanostructured polymers; Surfaces and interfaces; Adsorption; Functionalization of polymers

1 Introduction

Cellulose aerogels, the green biodegradable nanoporous materials, combine excellent properties of traditional silica aerogels and hydrocarbon copolymer or polymer-based aerogels such as low density, high specific surface area and high porosity,[1-4] and unique features from native cellulose like hydrophilia and biocompatibility.[5,6] These intriguing characteristics make cellulose aerogels potential substitutes for some petrochemicals in catalysts,[7] adsorbents,[8,9] fuel cells[10,11] and thermal or electrical insulation materials.[12-14]

Cellulose aerogels are generally divided into regeneration cellulose aerogels and nanofibrillated cellulose (NFC) aerogels. In preparation of regeneration cellulose aerogels, a procedure of dissolution, regeneration and freeze drying or supercritical drying is frequently-used;[15,16] besides, cellulose is difficult to be hydrolyzed or processed in common aqueous or organic solvents due to its strong inter and intra-molecular hydrogen bonds, high degree of polymerization, and high degrees of crystallinity,[17] thus, some special cellulose solvents like ammonium thiocyanate,[18] calcium and sodium thiocyanate,[19] lithium chloride/N, N-

* 本文摘自 Journal of Applied Polymer Science, 2015, 132(24): 42037.

dimethylacetamide (LiCl/DMAc),[20] ionic liquid,[21] and NaOH/PEG aqueous solution[22] are used during the above-mentioned dissolution process. However, among the cellulose solvents systems, numerous systems might cause serious environmental pollution. In comparison, the aerogels originated from NFC, whose preparation process mainly depends on isolation of NFC and freeze-drying treatment, show considerable advantages from the perspective of environmental protection since no harmful solvents are required during NFC processing.[23] Furthermore, some time-consuming necessary courses such as dissolution and gelation related to regeneration cellulose aerogels are also not involved in preparation of NFC aerogels; instead, some mild fast means such as ultrasonic processing[24] used for isolation of NFC are employed. Meanwhile, for NFC aerogels, the particular three-dimensional (3D) network formed by self-assembly of NFCs with diameters of a few nanometers to several decade nanometers also creates a large number of extraordinary performances in density, adsorption and conductivity.[25-28]

In this study, a mild method including chemical pretreatment, ultrasonic processing, solvent-exchange, and tert-butanol freeze drying for the preparation of the ultra-light NFC aerogels from NFC hydrocolloidal dispersions separated from coconut shell was described. The as-prepared aerogels were characterized by scanning electron microscope (SEM), N_2 adsorption measurements, X-ray diffraction (XRD) and thermal gravity analysis (TGA). Moreover, the NFC aerogels underwent hydrophobic modification by methyltrichlorosilane (MTCS) for improving hydrophobicity and lipophilicity, and the interaction between the aerogels and the modifying agent was investigated by Fourier transform infrared spectroscopy (FTIR), and the modified samples were subsequently subjected to oil adsorption and oil/water separation tests. Meanwhile, the adsorption properties of the modified aerogels for various organic solvents and oil were also investigated.

2 Materials and methods

2.1 Materials

Sixty-mesh powder of coconut shell after grinding and sieving was collected and dried in a vacuum at 60 ℃ for 24 h before used. All chemicals were supplied by Tianjin Kemiou Chemical Reagent Co. Ltd. and used without further purification.

2.2 Preparation of the NFC aerogel

The purified cellulose from the coconut shell powder was referred to our previous report[29] via a chemical pretreatment process. Then, a certain amount of purified cellulose was uniformly dispersed in distilled water with magnetic stirring to form 0.5 wt% aqueous dispersion, and the resulting mixture was subsequently underwent an ultrasonic treatment for 60 min with an output power of 300 W in an ice/water bath. The ultrasonic process was performed at 60 kHz with a sonifier (JY99-IID, Scientz Technology, China) with a 1-cm^2-diameter titanium horn under a 50% duty cycle (i. e., a repeating cycle of 1-s ultrasonic treatment and 1-s shutdown). After the ultrasonication, the suspension was subjected to dialysis tubing cellulose membrane (avg. flat width = 76 mm, molecular weight cut-off = 14000, Sigma-Aldrich), and underwent solvent-exchange with t-BuOH by immersing the dialysis tubing in t-BuOH for 12 h. The obtained concentrated suspension with high viscosity and gel-like appearance was freeze-dried at -35 ℃ at 25 Pa for 48 h, and following ultra-light NFC aerogel was fabricated.

2.3 Hydrophobic modification of the NFC aerogel

The as-prepared NFC aerogel was hydrophobicly modified using MTCS through a chemical vapor deposition method. A 5 mL beaker containing MTCS (300 μL) and NFC aerogels samples were placed in a sealed desiccator, avoiding direct contact between the samples and the modifying agent. Within the desiccator, the samples and the volatile MTCS were allowed to react for 24 h at ambient temperature. This process successfully fabricated hydrophobic aerogel.

2.4 Characterization

SEM observations were carried out on a FEI, Quanta 200 SEM. N_2 adsorption measurements were implemented by an accelerated surface area and porosimetry system (3H-2000PS2 unit, Beishide Instrument S&T Co., Ltd). XRD patterns were run on a X-ray diffractometer (Rigaku, D/MAX 2200) operating with Cu Kα radiation (λ = 1.5418 Å) at a scan rate (2θ) of 4° · min^{-1}. Thermal stabilities were characterized by TG analyzer (TA, Q600) from 25 ℃ to 800 ℃ with a heating rate of 10 ℃ · min^{-1} under nitrogen atmosphere. Surface chemical compositions were investigated by a FTIR spectrophotometer (Magna 560, Nicolet, Thermo Electron Corp) in the range of 650–4000 cm^{-1} with a resolution of 4 cm^{-1}. A contact angle analyzer (JC2000C) with a droplet volume of 5 μL was employed to measure the water contact angle (WCA) of the samples.

3 Result and discussion

As shown in Fig. 1a, the NFC aerogels with complex 3D fibrillar network mainly consisted of interlaced tangled NFC aggregations (the slender filiform texture), and thin membranes possibly formed by two-dimensional self-assembly of NFCs. In a high magnification (Fig. 1b), it was observed that the slender NFCs were cross-linked and had large length-diameter ratio, whose diameters were within the scope of 30–180 nm, and the mean diameter of these NFCs was ~80.6 nm. Additionally, the NFC aerogels have ultralow bulk density of 0.84 mg · cm^{-3} determined by the weighed mass of aerogels divided by the measured volume, which is comparable to those of ultralight single-walled carbon nanotubes aerogel (2.7 mg · cm^{-3}),[30] N-doped graphene aerogels (2.1 mg · cm^{-3}),[31] three-dimensional hierarchical boron nitride foam (1.6 mg · cm^{-3}),[32] and periodic microlattice-constructed Ni foams (0.9 mg · cm^{-3}),[33] and is slightly larger than those of ultra-flyweight all-carbon aerogels (0.16 mg · cm^{-3}),[34] the lowest density of ultralight materials ever reported. Furthermore, the NFC aerogels could stably stand on the leaf (inset in Fig. 1a) further indicating the lightweight characteristic.

Fig. 1 (a) Low-magnification and (b) high-magnification SEM images of the NFC aerogels. Inset of (a) presented the digital photograph of the lightweight NFC aerogels, and inset of (b) presented the diameter distribution of the NFCs

According to the N_2 adsorption measurements, the NFC aerogels showed Ⅳ-type adsorption isotherm (Fig. 2a), and the obvious hysteresis loop between the adsorption and desorption isotherms was formed by capillary condensation[35], indicating the presence of mesopore in the NFC aerogels. Besides, the pore diameter distribution curve in Fig. 2b showed that the aerogels had pore sizes of 1 – 60 nm, which further confirmed that the aerogels belonged to mesoporous material. The Barrett-Joyner-Halenda and Brunauer-Emmett-Teller methods were carried out to calculate the pore characteristic parameters, and the results showed that the samples had specific surface area and pore volume of 9.1 $m^2 \cdot g^{-1}$ and 0.025 $cm^3 \cdot g^{-1}$, respectively.

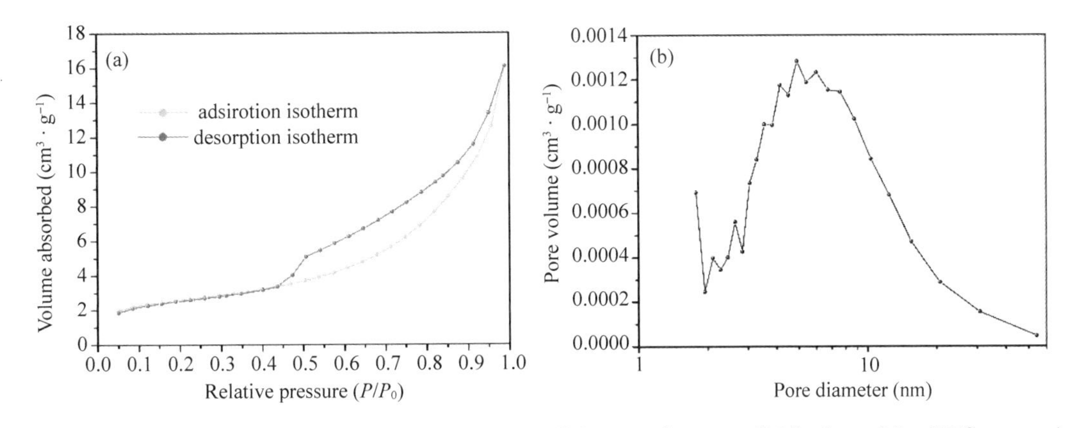

Fig. 2 (a) N_2 adsorption-desorption isotherms and (b) pore diameter distribution of the NFC aerogels

Fig. 3 presented the XRD patterns of the coconut shell and the NFC aerogels. It was observed that the both samples exhibited typical cellulose Ⅰ crystal structure with characteristic peaks at around 15.1°, 16.8° and 22.2° corresponding to the (101), (10$\bar{1}$) and (002) planes,[36] which revealed that the chemical pretreatment and the ultrasonic processing didn't change the crystal form. The inset in Fig. 3 showed the crystallinity index distribution calculated by the Segal method.[37] Compared with the coconut shell, the crystallinity index of the NFC aerogels exhibited a significant improvement by 19.2% due to the removal of some amorphous substances like hemicellulose and lignin during chemical pretreatment,[38] and the improvement might also contribute to enhancing thermal stability.[39]

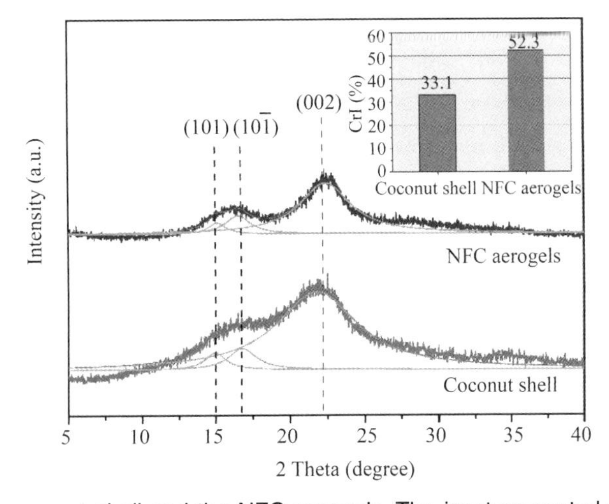

Fig. 3 XRD patterns of the coconut shell and the NFC aerogels. The inset presented crystallinity index distributions

Fig. 4 represented the TG and DTG plots of the coconut shell and the NFC aerogels. The slight mass loss for the both samples before 150 ℃ was due to the evaporation of retained moisture. For the coconut shell pyrolysis, in agreement with previous findings, the DTG curves displayed two main regions.[40,41] The first region (200 - 310 ℃) was mainly associated with the hemicellulose decomposition which exhibited a pronounced shoulder; the second region (310 - 400 ℃) was related to the attainment of the maximum, principally because of cellulose degradation, followed by a rapid decay and a long tail. Moreover, lignin was the most difficult one to decompose among the three main components, whose decomposition happened slowly under the whole temperature ranged from ambient to 800 ℃, and the wide range of temperatures also caused absence of its characteristic peak. For the aerogels, owing to the massive removal of amorphous substances such as hemicellulose and lignin, there was only a severe decomposition stage of cellulose. For the both samples, at degradation onset, 50% weight loss, and maximum degradation rate, compared with the coconut shell (200 ℃, 345 ℃ and 349 ℃), the aerogels all happened at higher temperature (235 ℃, 353 ℃ and 357 ℃), indicating the improvement of the aerogels thermal stability.

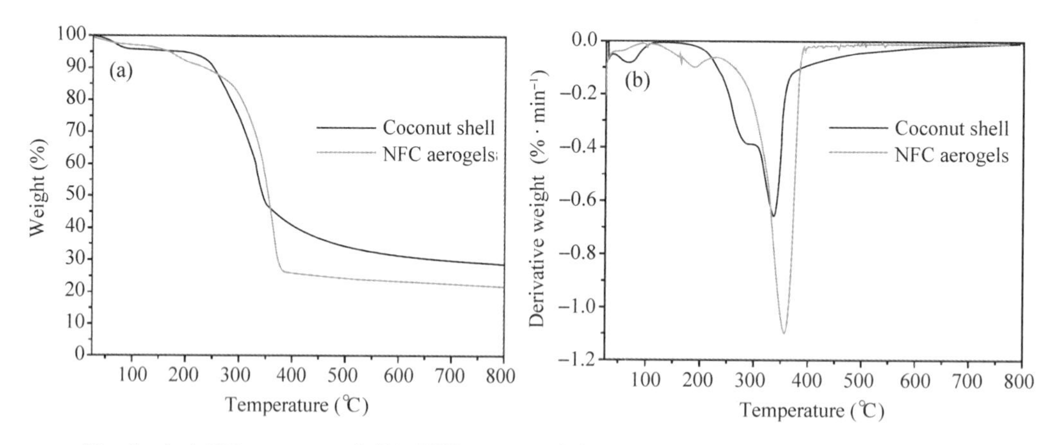

Fig. 4 (a) TG curves and (b) DTG curves of the coconut shell and the NFC aerogels

The modification by MTCS contributed to the formation of a low-energy surface of the cellulose aerogels due to the replacement of surface hydrophilic hydroxyl groups on the cellulose with plentiful hydrophobic groups, leading to the generation of hydrophobicity.[42] Fig. 5a presented the FTIR spectra of the NFC aerogels and the MTCS treated aerogels. Evidently, the peaks in the two spectra were rather similar; detailedly, the broad adsorption band centered at around 3325 cm^{-1} was attributed to the O—H stretching. The peak at 2895 cm^{-1} was assigned to the C—H stretching, nevertheless, the NFC aerogels exhibited two peaks at this region (2915 cm^{-1} and 2854 cm^{-1}) corresponding to alkane C—H asymmetric and symmetric stretching vibrations, respectively. Moreover, the peaks at around 1640, 1419, 1371 and 1316 cm^{-1} were derived from absorbed water, the CH_2 symmetric bending, the O—H bending, and the C—H bending[43], respectively. Apart from the similar peaks, the MTCS treated NFC aerogels displayed two new strong peaks at around 1273 cm^{-1} and 776 cm^{-1} attributed to the stretching vibrations of the Si-C bonds and to -CH_3 deformation vibrations of the siloxane compounds[44], confirming the formation of the hydrophobic coating of MTCS on the NFC aerogels. As shown in Fig. 5b, the MTCS treated NFC aerogels showed strong hydrophobic performance with a WCA of 139°, and could stably hold water drops on the surface; moreover, the modified aerogels could persistently stay on the oil layer and did not sink into the water layer under gravity after being put in the oil/water mixture (Fig. 5c), which further suggested the generation of hydrophobicity. Subsequently, the modified aerogels were subjected to oil adsorption tests by immersing the samples in diesel oil for 5 min (Fig. 5d) and weighting the samples before and after the immersion. The result indicated that the aerogels exhibited rapid adsorption rate and amazing oil adsorption property (approximately 542 times of the original dry weight) without any collapse in oil.

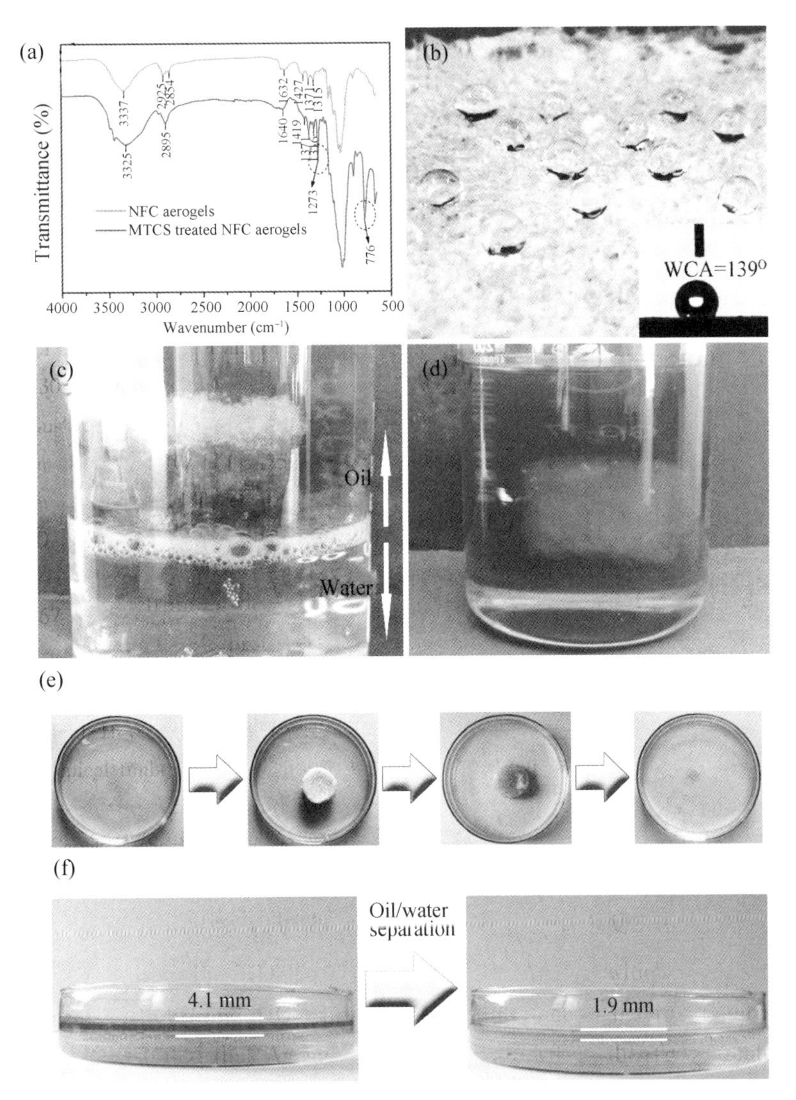

Fig. 5 (a) FTIR spectra of the NFC aerogels and the MTCS treated aerogels. (b) Water drops stably standing on the MTCS treated aerogel surface, the inset presented the water contact angle measurement. (c) MTCS treated aerogel persistently staying in oil layer instead of water layer due to the water repellency property. (d) Oil adsorption and (e) oil/water separation tests of the modified aerogel. (f) Oil layer thickness change before and after the oil/water separation test

Considering the excellent surface hydrophobicity and ultra-strong oil adsorption ability, the MTCS treated NFC aerogels might be ideal candidates for highly efficient separation/extraction of specific substances for oil spills. As shown in Fig. 5e, when a small piece of the aerogel was forced to the oil/water mixture, it could largely adsorb diesel oil instead of water. The whole process lasted 10 min, and the significant color difference before and after the test could be clearly distinguished; meanwhile, the oil layer thickness change from 4. 1 mm to 1. 9 mm (Fig. 5f) further demonstrated the superior oil/water separation performance of the aerogels.

The adsorption properties of the hydrophobic aerogels for some typical organic solvents and oil like industrial alcohol (methanol, ethanol, etc.), aromatic compounds (benzene), and commercial petroleum product (motor oil), which are regard as common pollutants in daily life or industry, were investigated (Fig. 6a). The absorption efficiency can be referred to as weight gain (wt%) defined as the weight of absorbed substances per unit weight of original NFC aerogel. The aerogels displayed ultra-strong absorption capacities for all of these organic liquids and oil, which could adsorb the liquids up to 296 times to 669 times their own weights (Fig. 6b), and the absorption efficiency is comparable with that of other high-performance adsorbing materials such as carbon nanofiber aerogels (106–312 times),[45] ultra-flyweight all-carbon aerogels (215–743 times),[34] and carbon nanotube/graphene oxide aerogels (215–913 times),[46] thereby indicating a facile, effective and promising route for dealing with chemical leaks and oil spills. Nevertheless, owing to the inferior flexibility, the aerogels had limited recyclability, and the defect would be improved in future work.

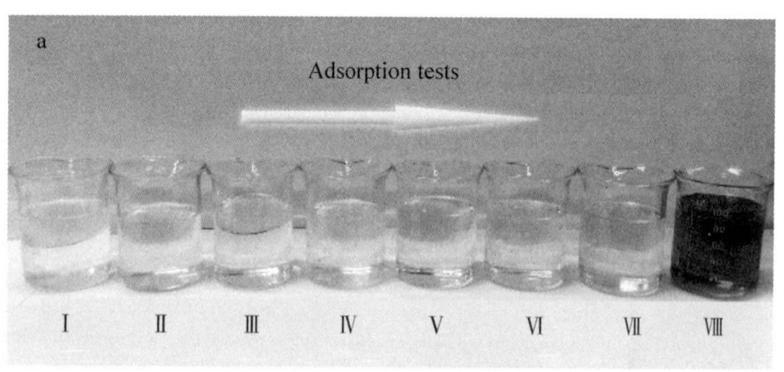

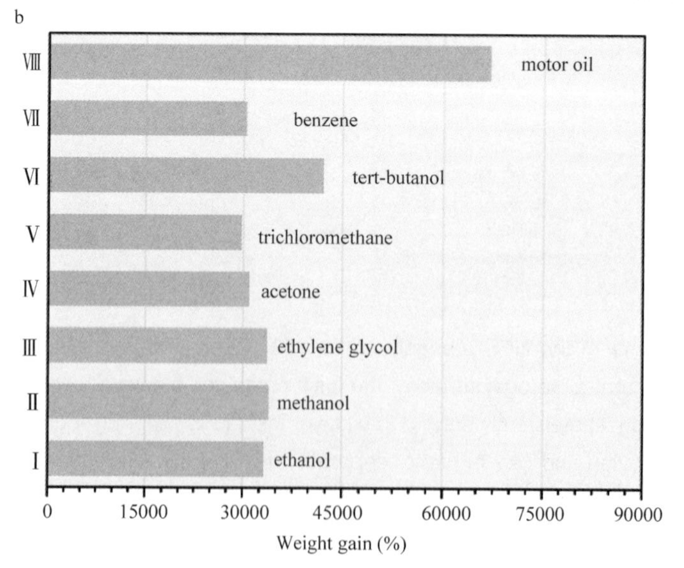

Fig. 6 (a) Adsorption tests and (b) absorption efficiency of the MTCS treated NFC aerogels for various organic liquids and oil

4 Conclusions

We had prepared NFC aerogels with ultra-low bulk density (ca. 0.84 mg · cm^{-3}) using coconut shell as the raw material following a mild quick chemical pretreatment-ultrasonic processing-solvent-exchange-tert-butanol freeze drying process. The as-prepared aerogels with complex three-dimensional fibrillar network showed cellulose Ⅰ crystal structure and improved thermal stability compared with the coconut shell, and had specific surface area and pore volume of 9.1 m^2 · g^{-1} and 0.025 cm^3 · g^{-1}, respectively. Moreover, the MTCS modified NFC aerogels with high hydrophobic property (WCA = 139°) could adsorb 542 times the original dry weight of diesel oil, and exhibited superior oil/water separation performance. Moreover, the hydrophobic aerogels also displayed strong adsorption capacity for various organic solvents and oil (296–669 times the pristine weight), thereby indicating a facile and useful route for dealing with chemical leaks and oil spills.

Acknowledgments

This work was financially supported by the National Natural Science Foundation of China (grant no. 31270590), a project funded by the China Postdoctoral Science Foundation (2013M540263), and the Doctoral Candidate Innovation Research Support Program of Science & Technology Review (kjdb2012006).

References

[1] Liebner F, Haimer E, Potthast A, et al. Cellulosic aerogels as ultra-lightweight materials. Part 2: Synthesis and properties 2nd ICC 2007, Tokyo, Japan, October 25-29, 2007[J]. Holzforschung, 2009.

[2] Yang W, Wu D, Fu R. Porous structure and liquid-phase adsorption properties of activated carbon aerogels[J]. Journal of Applied Polymer Science, 2007, 106(4): 2775-2779.

[3] Mihranyan A. Cellulose from cladophorales green algae: From environmental problem to high-tech composite materials [J]. Journal of Applied Polymer Science, 2011, 119(4): 2449-2460.

[4] Sehaqui H, Salajková M, Zhou Q, et al. Mechanical performance tailoring of tough ultra-high porosity foams prepared from cellulose I nanofiber suspensions[J]. Soft Matter, 2010, 6(8): 1824-1832.

[5] Tsioptsias C, Stefopoulos A, Kokkinomalis I, et al. Development of micro-and nano-porous composite materials by processing cellulose with ionic liquids and supercritical CO_2[J]. Green Chemistry, 2008, 10(9): 965-971.

[6] Heath L, Thielemans W. Cellulose nanowhisker aerogels[J]. Green Chemistry, 2010, 12(8): 1448-1453.

[7] Xiong R, Lu C, Wang Y, et al. Nanofibrillated cellulose as the support and reductant for the facile synthesis of Fe_3O_4/Ag nanocomposites with catalytic and antibacterial activity[J]. Journal of Materials Chemistry A, 2013, 1(47): 14910-14918.

[8] Gebald C, Wurzbacher J A, Tingaut P, et al. Amine-based nanofibrillated cellulose as adsorbent for CO_2 capture from air[J]. Environmental science & technology, 2011, 45(20): 9101-9108.

[9] Isobe N, Chen X, Kim U J, et al. TEMPO-oxidized cellulose hydrogel as a high-capacity and reusable heavy metal ion adsorbent[J]. Journal of hazardous materials, 2013, 260: 195-201.

[10] Guilminot E, Fischer F, Chatenet M, et al. Use of cellulose-based carbon aerogels as catalyst support for PEM fuel cell electrodes: Electrochemical characterization[J]. Journal of Power Sources, 2007, 166(1): 104-111.

[11] Osman M H, Shah A A, Walsh F C. Recent progress and continuing challenges in bio-fuel cells. Part I: Enzymatic cells[J]. Biosensors and Bioelectronics, 2011, 26(7): 3087-3102.

[12] Tingaut P, Zimmermann T, Sèbe G. Cellulose nanocrystals and microfibrillated cellulose as building blocks for the

design of hierarchical functional materials[J]. Journal of Materials Chemistry, 2012, 22(38): 20105-20111.

[13] Shi J, Lu L, Guo W, et al. Heat insulation performance, mechanics and hydrophobic modification of cellulose-SiO_2 composite aerogels[J]. Carbohydrate polymers, 2013, 98(1): 282-289.

[14] Shi J, Lu L, Guo W, et al. An environment-friendly thermal insulation material from cellulose and plasma modification[J]. Journal of Applied Polymer Science, 2013, 130(5): 3652-3658.

[15] Cai J, Kimura S, Wada M, et al. Cellulose aerogels from aqueous alkali hydroxide-urea solution[J]. ChemSusChem: Chemistry & Sustainability Energy & Materials, 2008, 1(1-2): 149-154.

[16] Sescousse R, Gavillon R, Budtova T. Aerocellulose from cellulose-ionic liquid solutions: Preparation, properties and comparison with cellulose-NaOH and cellulose-NMMO routes[J]. Carbohydrate Polymers, 2011, 83(4): 1766-1774.

[17] Zhang X, Liu X, Zheng W, et al. Regenerated cellulose/graphene nanocomposite films prepared in DMAC/LiCl solution[J]. Carbohydrate Polymers, 2012, 88(1): 26-30.

[18] Liu C, Cuculo J A, Smith B. Coagulation studies for cellulose in the ammonia/ammonium thiocyanate (NH_3/NH_4SCN) direct solvent system[J]. Journal of Polymer Science Part B: Polymer Physics, 1989, 27(12): 2493-2511.

[19] Hattori M, Saito M, Shimaya Y. J. Polym. Sci., Polym. Lett. Ed. J. Polym. Sci., Polym. Lett. Ed. 17, 479, 1979[J]. Polymer journal, 1998, 30(1): 49-55.

[20] Dupont A L. Cellulose in lithium chloride/N, N-dimethylacetamide, optimisation of a dissolution method using paper substrates and stability of the solutions[J]. Polymer, 2003, 44(15): 4117-4126.

[21] Li J, Lu Y, Yang D, et al. Lignocellulose aerogel from wood-ionic liquid solution (1-allyl-3-methylimidazolium chloride) under freezing and thawing conditions[J]. Biomacromolecules, 2011, 12(5): 1860-1867.

[22] Wan C, Lu Y, Jiao Y, et al. Cellulose aerogels from cellulose-NaOH/PEG solution and comparison with different cellulose contents[J]. Materials Science and Technology, 2015, 31(9): 1096-1102.

[23] Dong H, Snyder J F, Tran D T, et al. Hydrogel, aerogel and film of cellulose nanofibrils functionalized with silver nanoparticles[J]. Carbohydrate polymers, 2013, 95(2): 760-767.

[24] Saito T, Kuramae R, Wohlert J, et al. An ultrastrong nanofibrillar biomaterial: the strength of single cellulose nanofibrils revealed via sonication-induced fragmentation[J]. Biomacromolecules, 2013, 14(1): 248-253.

[25] Valo H, Arola S, Laaksonen P, et al. Drug release from nanoparticles embedded in four different nanofibrillar cellulose aerogels[J]. European Journal of Pharmaceutical Sciences, 2013, 50(1): 69-77.

[26] Carlsson D O, Nyström G, Zhou Q, et al. Electroactive nanofibrillated cellulose aerogel composites with tunable structural and electrochemical properties[J]. Journal of Materials Chemistry, 2012, 22(36): 19014-19024.

[27] Cervin N T, Aulin C, Larsson P T, et al. Ultra porous nanocellulose aerogels as separation medium for mixtures of oil/water liquids[J]. Cellulose, 2012, 19(2): 401-410.

[28] Aulin C, Netrval J, Wågberg L, et al. Aerogels from nanofibrillated cellulose with tunable oleophobicity[J]. Soft Matter, 2010, 6(14): 3298-3305.

[29] Li J, Wan C, Lu Y, et al. Fabrication of cellulose aerogel from wheat straw with strong absorptive capacity[J]. Frontiers of Agricultural Science and Engineering, 2014, 1(1): 46-52.

[30] Jung S M, Jung H Y, Dresselhaus M S, et al. A facile route for 3D aerogels from nanostructured 1D and 2D materials[J]. Rep, 2013, 2.

[31] Zhao Y, Hu C, Hu Y, et al. A versatile, ultralight, nitrogen-doped graphene framework[J]. Angewandte Chemie International Edition, 2012, 51(45): 11371-11375.

[32] Yin J, Li X, Zhou J, et al. Ultralight three-dimensional boron nitride foam with ultralow permittivity and superelasticity[J]. Nano letters, 2013, 13(7): 3232-3236.

[33] Schaedler T A, Jacobsen A J, Torrents A, et al. Ultralight metallic microlattices[J]. Science, 2011, 334(6058): 962-965.

[34] Sun H, Xu Z, Gao C. Multifunctional, ultra-flyweight, synergistically assembled carbon aerogels[J]. Advanced materials, 2013, 25(18): 2554-2560.

[35] Wan C, Lu Y, Jiao Y, et al. Fabrication of hydrophobic, electrically conductive and flame-resistant carbon aerogels by pyrolysis of regenerated cellulose aerogels[J]. Carbohydrate polymers, 2015, 118: 115-118.

[36] Oh S Y, Yoo D I, Shin Y, et al. Crystalline structure analysis of cellulose treated with sodium hydroxide and carbon dioxide by means of X-ray diffraction and FTIR spectroscopy[J]. Carbohydrate research, 2005, 340(15): 2376-2391.

[37] Segal L, Creely J J, Martin Jr A E, et al. An empirical method for estimating the degree of crystallinity of native celluloseusing the X-ray diffractometer[J]. Textile research journal, 1959, 29(10): 786-794.

[38] Fahma F, Iwamoto S, Hori N, et al. Effect of pre-acid-hydrolysis treatment on morphology and properties of cellulose nanowhiskers from coconut husk[J]. Cellulose, 2011, 18(2): 443-450.

[39] Rodrigues Filho G, de Assunção R M N, Vieira J G, et al. Characterization of methylcellulose produced from sugar cane bagasse cellulose: Crystallinity and thermal properties[J]. Polymer degradation and stability, 2007, 92(2): 205-210.

[40] Grønli M G, Várhegyi G, Di Blasi C. Thermogravimetric analysis and devolatilization kinetics of wood[J]. Industrial & Engineering Chemistry Research, 2002, 41(17): 4201-4208.

[41] Yang H, Yan R, Chen H, et al. Characteristics of hemicellulose, cellulose and lignin pyrolysis[J]. Fuel, 2007, 86(12-13): 1781-1788.

[42] Shewale P M, Rao A V, Rao A P. Effect of different trimethyl silylating agents on the hydrophobic and physical properties of silica aerogels[J]. Applied Surface Science, 2008, 254(21): 6902-6907.

[43] Lan W, Liu C F, Yue F X, et al. Ultrasound-assisted dissolution of cellulose in ionic liquid[J]. Carbohydrate polymers, 2011, 86(2): 672-677.

[44] Shirgholami M A. Shateri Khalil-Abad M, Khajavi R, et al. Fabrication of superhydrophobic polymethylsilsesquioxane nanostructures on cotton textiles by a solution-immersion process[J]. J Colloid Interface Sci, 2011, 359: 530-535.

[45] Wu Z Y, Li C, Liang H W, et al. Ultralight, flexible, and fire-resistant carbon nanofiber aerogels from bacterial cellulose[J]. Angewandte Chemie, 2013, 125(10): 2997-3001.

[46] Pan H, Zhu S, Mao L. Graphene nanoarchitectonics: approaching the excellent properties of graphene from microscale to macroscale[J]. Journal of Inorganic and Organometallic Polymers and Materials, 2015, 25(2): 179-188.

中文题目：具有超强吸附性能的椰壳超轻疏水性纳米纤维素气凝胶

作者：万才超，卢芸，焦月，金春德，孙庆丰，李坚

摘要：通过温和快速法成功制备了基于从椰子壳中分离出的纳米纤维素(NFC)的超轻气凝胶，该方法包括化学预处理、超声分离、溶剂交换和叔丁醇冷冻干燥。所制备的NFC气凝胶具有复杂的三维纤维网络，其体积密度低至0.84 mg·cm^{-3}(比表面积59.1 m^2·g^{-1}，孔隙体积50.025 cm^3·g^{-1})，保持了纤维素Ⅰ的晶体结构，并表现出比椰壳原料更优越的热稳定性。经甲基三氯硅烷(MTCS)疏水改性后，NFC气凝胶表现出较高的憎水性能，超强的吸油能力(是柴油原干重的542倍)，油水分离性能优越。此外，MTCS处理的NFC气凝胶对各种有机溶剂和油品的吸附能力高达其自重的296~669倍。因此，该类高性能吸附材料或可用于处理化学品泄漏和溢油。

关键词：吸附；纤维素和其他木制品；聚合物的功能化；纳米结构聚合物；表面和界面

Fabrication of Hydrophobic, Electrically Conductive and Flame-resistant Carbon Aerogels by Pyrolysis of Regenerated Cellulose Aerogels*

Caichao Wan, Yun Lu, Yue Jiao, Chunde Jin, Qingfeng Sun, Jian Li

Abstract: In this paper, we reported miscellaneous carbon aerogels prepared by pyrolysis of regenerated cellulose aerogels that were fabricated by dissolution in a mild NaOH/PEG solution, freeze-thaw treatment, regeneration, and freeze drying. The as-prepared carbon aerogels were subsequently characterized by scanning electron microscopy (SEM), energy-dispersive X-ray spectroscopy (EDX), nitrogen adsorption measurements, X-ray diffraction (XRD), Raman spectroscopy, and water contact angle (WCA) tests. The results showed that the carbon aerogels with pore diameters of 1-60 nm maintained interconnected three-dimensional (3D) network after the pyrolysis, and showed type-IV adsorption isotherm. The pyrolysis process leaded to the decomposition of oxygen-containing functional groups, the destruction of cellulose crystalline structure, and the formation of highly disordered amorphous graphite. Moreover, the carbon aerogels also had strong hydrophobicity, electrical conductivity and flame retardance, which held great potential in the fields of waterproof, electronic devices and fireproofing.

Keywords: Carbon; Aerogels; Pyrolysis; Regenerated cellulose; Porous materials

1 Introduction

Carbon aerogels with large specific surface area, high porosity and high electrical conductivity have been widely considered as promising high-performance materials for adsorbents, catalyst supports, artificial muscles, electrodes for electrical double-layer capacitors, and gas sensors (Li, Wang, Huang, Gamboa & Sebastian, 2006; Moreno-Castilla & Maldonado-Hódar, 2005; Waghuley, Yenorkar, Yawale & Yawale, 2008). Traditionally, carbon aerogels are synthesized by carbonization of organic aerogels, which are prepared by sol-gel procedure from polycondensation of different organic monomers (Moreno-Castilla et al., 2005). Typically, resorcinol-formaldehyde system is the most extensively used raw materials for fabrication of carbon aerogels, and the resulting carbon aerogels have tunable surface properties related to the synthesis and processing conditions, which can design various materials with unique properties. Apart from the above-mentioned numerous merits, some defects of the resorcinol-formaldehyde organic aerogels, such as high density (100~800 mg · cm^{-1}) (Fu et al., 2003; Wu, Fu, Dresselhaus & Dresselhaus, 2006), fragility, toxicity and environmental pollution, dramatically hamper the development and industry application of this

* 本文摘自 Carbohydrate Polymers, 2015, 118: 115-118.

class of carbon aerogels. Recently, some novel carbon aerogels using natural polymers like glucose and bacterial cellulose as precursors have been prepared by pyrolysis, and show superior flexibility, adsorption properties, fire resistance, and electric properties (Chen et al., 2013; Liang et al., 2012; Liang et al., 2012; Wu, Li, Liang, Chen & Yu, 2013; Wu et al., 2014). Nevertheless, regenerated cellulose aerogels, composed of cross-linked three-dimensional (3D) network similar to bacterial cellulose, might also be ideal precursors for carbon aerogels.

Thereby, in this paper, versatile carbon aerogels were fabricated by pyrolysis of regenerated cellulose aerogels that were prepared following the procedures of dissolving cellulose in a green NaOH/PEG solution, freeze-thaw treatment, regeneration, and freeze drying. The as-prepared carbon aerogels were characterized by scanning electron microscopy (SEM), energy-dispersive X-ray spectroscopy (EDX), nitrogen adsorption measurements, X-ray diffraction (XRD), Raman spectrum, and water contact angle (WCA) tests. Moreover, the resulting carbon aerogels showed strong hydrophobicity, electrical conductivity and flame retardance.

2 Materials and methods

2.1 Materials

Sixty-mesh powder of wheat straw after grinding and sieving was collected and dried in a vacuum at 60 ℃ for 24 h before used. All chemicals were supplied by Shanghai Boyle Chemical Co. Ltd. and used without further purification.

2.2 Fabrication of carbon aerogel via a pyrolysis method

The purified cellulose from the wheat straw powder was referred to our previous report via a chemical pretreatment process (Li, Wan, Lu & Sun, 2014); meanwhile, the α-cellulose content of the obtained cellulose was determined by a modification of the TAPPI Method No. T 203 08-61 (Antrim, Chan, Crary Jr & Harris, 1980). The preparation of homogenous cellulose solution could be referred to the method based on the green NaOH/PEG solution (Yan & Gao, 2007). Briefly, the purified cellulose was mixed with a NaOH/PEG-4000 aqueous solution (9 : 1 by weight) with magnetic stirring for 6 h to form a 2 wt% cellulose solution. After being frozen at −15 ℃ for 12 h, the solution was thawed out at ambient temperature under vigorous stirring for 30 min, and the following homogenous cellulose solution was obtained. After being frozen at −15 ℃ for 3 h again, the frozen cellulose solution underwent a regeneration process by solvent exchange with 1% hydrochloric acid, distilled water and tert butyl alcohol in sequence until the formation of hydrogel without residual chlorine anions. Afterwards, the hydrogel was subjected to a freeze drying process at −35 ℃ for 48 h in a vacuum, and the following cellulose aerogel was pyrolyzed at 1000 ℃ under argon atmosphere to generate black carbon aerogel.

2.3 Characterizations

Micromorphology and surface chemical compositions were determined by SEM attached with EDX (FEI, Quanta 200). Nitrogen adsorption measurements were run on an accelerated surface area and porosimetry system (3H-2000PS2 unit, Beishide Instrument S&T Co. Ltd.), and pore characteristic parameters were calculated based on the Brunauer-Emmet-Teller (BET) and Barrett-Joyner-Halenda (BJH) methods. Crystal structures were identified by XRD (Rigaku, D/MAX 2200). Raman spectra were performed using a Raman spectrometer (Renishaw, inVia). The wetting properties were characterized by a WCA analyzer (JC2000C).

3 Result and discussion

The as-prepared cellulose after the chemical pretreatment had high α-cellulose content of 90.03% because of the massive removal of hemicellulose and lignin, indicating high purity for the cellulose samples. Moreover, the cellulose aerogels and the carbon aerogels from the cellulose-NaOH/PEG solution were also successfully fabricated. SEM images of the cellulose aerogels and the corresponding carbon aerogels were shown in Fig. 1. Typical 3D network structure of the cellulose aerogels could be clearly found, whose pore sizes ranged from approximately dozens of nanometers to several microns. After the pyrolysis process, the cross-linked nanoporous 3D network was maintained for the carbon aerogels, whereas the network became relatively tenuous and uneven. Moreover, compared with the cellulose aerogels, the oxygen peak in the EDX spectrum of carbon aerogels was almost entirely disappeared (inset in Fig. 1A), and the carbon peak improved sharply, which indicated the severe decomposition of oxygen-containing functional groups during the pyrolysis contributing to the formation of hydrophobicity.

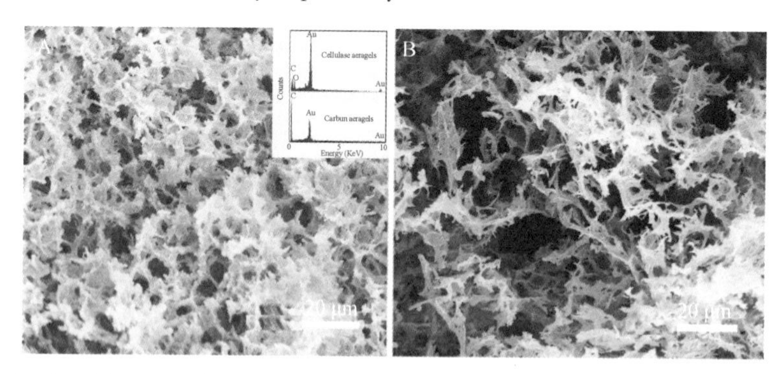

Fig. 1 SEM images of the cellulose aerogels (A) and the carbon aerogels (B). The inset showed the corresponding EDX spectra

Detailed pore features of the carbon aerogels were investigated by nitrogen adsorption measurements. As shown in Fig. 2A, the nanoporous carbon aerogels with specific surface area (113 $m^2 \cdot g^{-1}$), average pore size (7.2 nm) and total pore volume (0.21 $cm^3 \cdot g^{-1}$) exhibited type-IV adsorption isotherm according to the IUPAC classification, and the hysteresis loop was generally attributed to the capillary condensation occurring in the mesopores. Moreover, it was observed in Fig. 2B that the pore diameters of the carbon aerogels were within the scope of 1–60 nm, which suggested that the aerogels were primarily composed of mesopores (2–50 nm).

The XRD pattern of the wheat straw exhibited typical cellulose I crystalline structure (Fig. 2C) due to the characteristic peaks at around 15.1°, 16.9° and 21.9° corresponding to the (101), ($10\bar{1}$) and (002) planes. As the result of NaOH treatment, the cellulose aerogels showed strong peaks at 12.0°, 19.8° and 21.6°, indicating the generation of cellulose Ⅱ. After the pyrolysis at 1000 ℃, the cellulose characteristic peaks were completely disappeared, revealing that the cellulose crystalline structure was destroyed, and two new broad peaks centered at around 22.0° and 43.3° were corresponding to the (002) and (100) planes of graphite, which demonstrated that amorphous carbon was formed (Wu et al., 2013). From the Raman spectra as shown in Fig. 2D, the characteristic peaks of the wheat straw and the cellulose aerogels at around 1101 cm^{-1}, 1378 cm^{-1} and 1460 cm^{-1}, which were primarily originated from cellulose components (Agarwal & Ralph, 1997), were all disappeared in the Raman spectrum of the carbon aerogels; instead, the graphitic D- and G bands at 1346 cm^{-1} and 1594 cm^{-1} related to graphitic sp^2-bonding were generated. Especially, the shift of the G-peak from a value of 1581 cm^{-1} to 1594 cm^{-1} revealed the presences of nanocrystalline (highly disordered) graphite and aromatic cluster (Sevilla & Fuertes, 2009). Meanwhile, the broad D-peak also suggested disorder in the graphite structure.

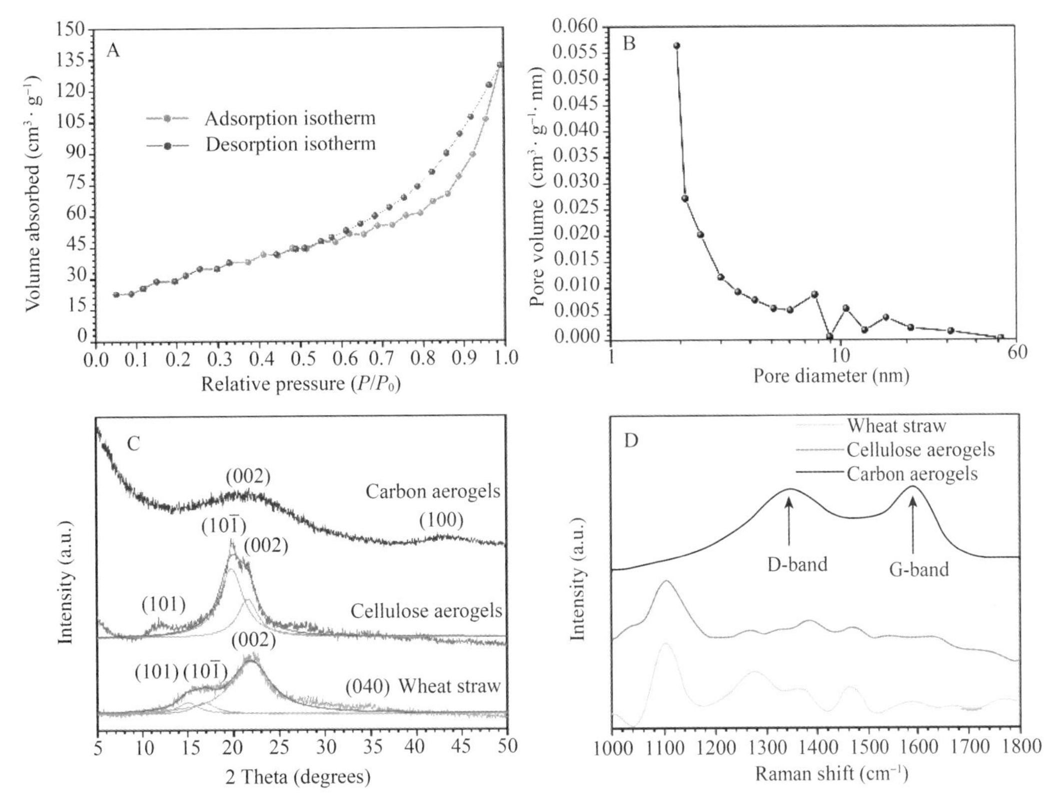

Fig. 2 (A) N_2 adsorption-desorption isotherms and (B) pore size distribution of the carbon aerogels. (C) XRD patterns and (D) Raman spectra of the wheat straw, the cellulose aerogels and the carbon aerogels

Some unique macro-properties of the carbon aerogels were presented in Fig. 3. Fig. 3A showed that carbon aerogels with a WCA of 139° could stably hold several water drops on the surface, indicating the formation of strong hydrophobicity, which might be used as high-performance separating or adsorbing materials for cleanup of oil spillage and chemical leakage. Moreover, it was observed in Fig. 3B that the cellulose aerogels had inferior conductivity when connected to a closed circuit system, and could not support current flow leading to no luminescence of the bulb. On the contrary, the carbon aerogels were good electrical conductors, which allowed current flow to pass through and lighted the bulb, and the brightness of the bulb was comparable with that of the bulb in the circuit with aluminum mash, which may find applications in electronic devices. The carbon aerogels were also superior fire retardants, which did not support any burning and release obvious smoke all the time when exposed to a flame of an alcohol burner (Fig. 3C). Thus, as previously reported carbon aerogels based on bacterial cellulose with similar strong hydrophobic properties, electrical conductivity, fire resistance, and adsorption properties (Liang et al., 2012; Wu et al., 2014), this class of carbon aerogels in this study generated from the regenerated cellulose aerogels might also be expected to be exploited more excellent performances. Furthermore, the detailed adsorption, thermal, and electrical properties of the carbon aerogels would be intensively studied in our future research.

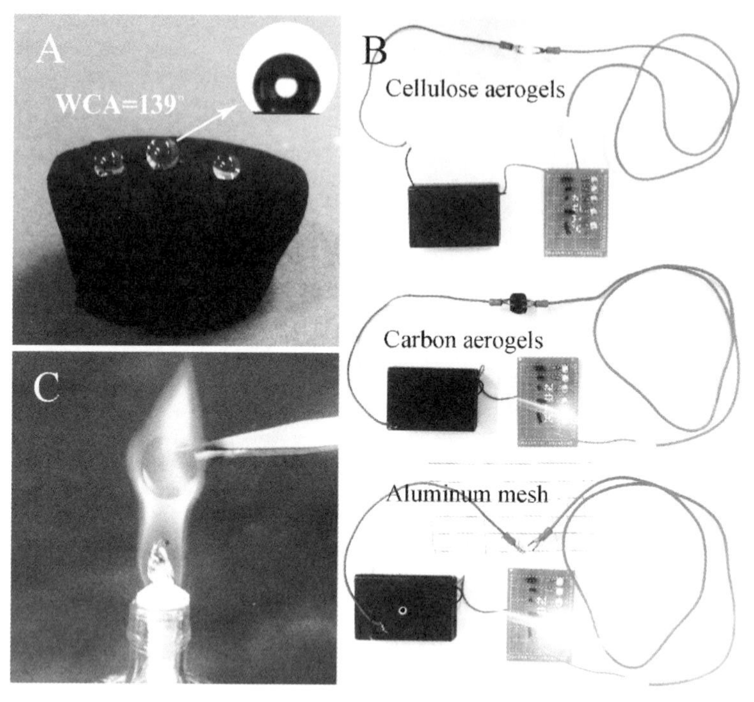

Fig. 3 (A) Hydrophobic carbon aerogels with several water drops on the surface, and the inset showed the water contact angle measurement. (B) Electrical conductivity tests of the cellulose aerogels, the carbon aerogels and the aluminum mesh. (C) Digital photograph of the carbon aerogels in a hot flame of an alcohol burner

4 Conclusions

In conclusion, we successfully fabricated miscellaneous carbon aerogels by pyrolysis of regenerated cellulose aerogels. After pyrolysis, the nanoporous carbon aerogels maintained cross-linked 3D network, and exhibited type-IV adsorption isotherm, which was mainly composed of mesopores. Moreover, the pyrolysis process resulted in the decomposition of oxygen-containing functional groups, the destruction of cellulose crystalline structure, and the formation of highly disordered amorphous graphite. Meanwhile, the carbon aerogels also displayed strong hydrophobicity, electrical conductivity and flame retardance, which might be useful for waterproof materials, electronic devices and flame retardants.

Acknowledgments

We appreciate the financial support of the China Postdoctoral Science Foundation (2013M540263), the National Natural Science Foundation of China (grant no. 31270590), and the Doctoral Candidate Innovation

Research Support Program of Science & Technology Review (kjdb2012006).

References

[1] Agarwal U P, Ralph S A. FT-Raman spectroscopy of wood: identifying contributions of lignin and carbohydrate polymers in the spectrum of black spruce (Picea mariana)[J]. Applied spectroscopy, 1997, 51(11): 1648-1655.

[2] Antrim R L, Chan Y C, Crary Jr J R, et al. Method for obtaining a purified cellulose product from corn hulls: U. S. Patent 4, 239, 906[P]. 1980-12-16.

[3] Chen L F, Huang Z H, Liang H W, et al. Flexible all-solid-state high-power supercapacitor fabricated with nitrogen-doped carbon nanofiber electrode material derived from bacterial cellulose[J]. Energy & Environmental Science, 2013, 6(11): 3331-3338.

[4] Fu R, Zheng B, Liu J, et al. The fabrication and characterization of carbon aerogels by gelation and supercritical drying in isopropanol[J]. Advanced Functional Materials, 2003, 13(7): 558-562.

[5] Li J, Wan C, Lu Y, et al. Fabrication of cellulose aerogel from wheat straw with strong absorptive capacity[J]. Frontiers of Agricultural Science and Engineering, 2014, 1(1): 46-52.

[6] Li J, Wang X, Huang Q, et al. Studies on preparation and performances of carbon aerogel electrodes for the application of supercapacitor[J]. Journal of Power Sources, 2006, 158(1): 784-788.

[7] Liang H W, Guan Q F, Chen L F, et al. Macroscopic-scale template synthesis of robust carbonaceous nanofiber hydrogels and aerogels and their applications[J]. Angewandte Chemie International Edition, 2012, 51(21): 5101-5105.

[8] Liang H W, Guan Q F, Zhu Z, et al. Highly conductive and stretchable conductors fabricated from bacterial cellulose [J]. NPG Asia Materials, 2012, 4(6): e19-e19.

[9] Moreno-Castilla C, Maldonado-Hódar F J. Carbon aerogels for catalysis applications: An overview[J]. Carbon, 2005, 43(3): 455-465.

[10] Sevilla M, Fuertes A B. The production of carbon materials by hydrothermal carbonization of cellulose[J]. Carbon, 2009, 47(9): 2281-2289.

[11] Waghuley S A, Yenorkar S M, Yawale S S, et al. Application of chemically synthesized conducting polymer-polypyrrole as a carbon dioxide gas sensor[J]. Sensors and Actuators B: Chemical, 2008, 128(2): 366-373.

[12] Wu D, Fu R, Dresselhaus M S, et al. Fabrication and nano-structure control of carbon aerogels via a microemulsion-templated sol-gel polymerization method[J]. Carbon, 2006, 44(4): 675-681.

[13] Wu Z Y, Li C, Liang H W, et al. Ultralight, flexible, and fire-resistant carbon nanofiber aerogels from bacterial cellulose[J]. Angewandte Chemie, 2013, 125(10): 2997-3001.

[14] Wu Z Y, Li C, Liang H W, et al. Carbon nanofiber aerogels for emergent cleanup of oil spillage and chemical leakage under harsh conditions[J]. Scientific reports, 2014, 4(1): 1-6.

[15] Yan L, Gao Z. Dissolving of cellulose in PEG/NaOH aqueous solution[J]. Cellulose, 2008, 15(6): 789-796.

中文题目：通过热解再生纤维素气凝胶制备疏水，导电和阻燃的碳气凝胶

作者：万才超，卢芸，焦月，金春德，孙庆丰，李坚

摘要：在本文中，我们报道了通过热解再生纤维素气凝胶制备的杂碳气凝胶，这些纤维素气凝胶通过在温和的 NaOH/PEG 溶液中溶解、冻融处理、再生和冷冻干燥来制造。通过扫描电子显微镜(SEM)、能量分散 X 射线光谱(EDX)、氮气吸附测量、X 射线衍射(XRD)、拉曼光谱和水接触角(WCA)测试对制备的碳气凝胶进行了表征。结果表明，孔径为 1~60 nm 的碳气凝胶在热解后保持了相互连接的三维网络，并呈现出Ⅳ型吸附等温线。在热解过程中，含氧官能团被分解，纤维素结晶结

构被破坏，形成了高度无序的非晶态石墨。此外，碳气凝胶还具有较强的疏水性、导电性和阻燃性，在防水、电子器件和防火等领域具有很大的潜力。

关键词：碳；气凝胶；热解；再生纤维素；多孔材料

Thermally Induced Gel from Cellulose-NaOH/PEG Solution: Preparation, Characterization and Mechanical Properties*

Caichao Wan, Yun Lu, Chunde Jin, Qingfeng Sun, Jian Li

Abstract: In this paper, we reported a thermally induced gel with strong mechanical properties prepared from cellulose-NaOH/PEG aqueous solution following the procedures of dissolution, heating and freeze drying. The as-prepared gel showed undeveloped networks composed of cross-linked fiber aggregations tightly coated with plenty of $NaOH \cdot H_2O$ and PEG aggregated fine particles, which leaded to the significant enhancement of thermal stability and the disappearance of the original cellulose crystalline structures. Furthermore, the elastic modulus, yield stress and toughness of the mechanical strongly gel were measured to be up to 3210 kPa, 325 kPa and 38.9 $kJ \cdot m^{-3}$, respectively, comparable to those of cross-linked polymer gel materials with strong mechanical strength like the microfibrillated cellulose aerogels and the three-dimensional architectures of graphene hydrogels.

Keywords: Cellulose; Gel; Mechanical properties; Polymers

1 Introduction

Natural renewable resources have attracted much attention from scientific community to industrial community as the petrochemical resources are increasingly depleted and the environmental pollutions become even more violent. Cellulose, the almost inexhaustible natural polymer, has numerous alluring properties like renewability, biodegradability, biocompatibility and high strength and stiffness[1-3], which has been widely considered as a substitute for petrochemically derived compounds in many cases[4,5]. Nevertheless, owing to some unique characteristic features such as strong inter- and intra-molecular hydrogen bonding, stiff molecules and close chain packing via the numerous hydrogen bonds, it is hard to dissolve cellulose in most common aqueous or organic solvents, which greatly hinders its application. To date, there are only limited numbers of solvent systems for cellulose, such as NH_3/NH_4SCN[6], ionic liquid[7] and NaOH/PEG aqueous solution[8].

In cellulose/NaOH/PEG aqueous solution, strong intermolecular hydrogen bonds of cellulose are broken by NaOH to create a significant ion-pair interaction; meanwhile, the oxygen atoms in the PEG chain with repeat unit of $—(CH_2—CH_2—O)—$ serve as hydrogen bonding donor and receptor between PEG and cellulose to form new intermolecular interactions, which impedes self-association of hydroxyls on the cellulose chains and leads to the dissolution of cellulose in the aqueous solution. Actually, the covering layer of sodium

* 本文摘自 Applied Physics A, 2015, 119: 45-48.

ions, water and PEG molecules surrounding the cellulose chain is extremely sensitive to temperature. While rising temperature, the covering layer would be broken due to the rapid thermal escape motion of solvent molecules, and the resulting exposed hydroxyl groups of the cellulose will subsequently randomly associate and tangle with each other, which lays a solid foundation for the formation of gel networks[9-11].

Thereby, in this study, a novel, mild and fast method for the preparation of a thermally induced gel from cellulose-NaOH/PEG aqueous solution following the procedures of dissolution, heating and freeze drying was described (Fig. 1). The as-prepared cellulose gel was subsequently characterized by scanning electron microscope (SEM), energy dispersive X-ray analysis (EDXA), X-ray diffraction (XRD), thermogravimetric analysis (TGA) and compression mechanical property tests. Moreover, the preparation of the novel gel material might be useful to provide a new thought for cellulose application.

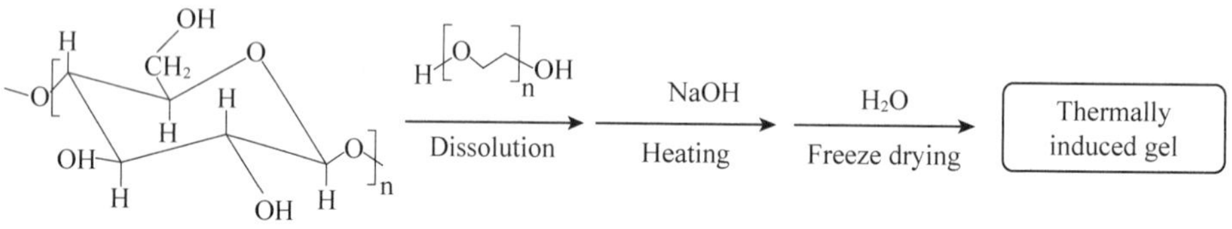

Fig. 1 Procedures for the preparation of thermally induced gel from cellulose-NaOH/PEG solution

2 Materials and methods

2.1 Materials

Filter paper was provided by the Hangzhou special paper Co., Ltd. and dried at 50 ℃ for 24 h to remove all absorbed water serving as the cellulosic resource. Other all chemical reagents including NaOH and PEG-4000 were purchased from Tianjin Kermel chemical Co. Ltd., and used without further purification.

2.2 Preparation of the thermally induced gel

The aqueous solution of NaOH/PEG/H_2O (9 : 1 : 90 by weight) was used as the solvent of cellulose. The mixture of cellulose and NaOH/PEG aqueous solution was magnetic stirring at room temperature for about 6 h to form a 2 wt% homogenous cellulose solution. Then, the solution was placed into a 95 ℃ constant temperature water bath for 5 h to gradually form an amber-like hydrogel. After being frozen at −15 ℃ for about 3 h, the mixture was subjected to a freeze drying process at −35 ℃ for about 48 h in vacuum, and the following gel material was successfully fabricated.

2.3 Characterizations

The as-prepared gel material was observed using FEI Quanta 200 SEM attached with EDXA spectrometer unit. The crystalline structures were determined by XRD (Rigaku, D/MAX 2200) generated at 40 kV and 30 mA using Cu Kα radiation (λ = 0.1542 nm). The diffraction angle ranged from 5° to 70°. Thermal stabilities were investigated by TG analyzer (TA, Q600) from room temperature to 800 ℃ with a heating rate of 10 ℃ · min^{-1} under nitrogen environment. The compression tests were performed in an electronic universal testing machine (Suns, UTM4304X) with compressing velocity of 2 mm · min^{-1}.

3 Result and discussion

The SEM images of the thermally induced gel were shown in Fig. 2. As shown, the gel with well-defined shape (inset in Fig. 2a) was composed of cross-linked fiber aggregations tightly coated with plenty of fine particles that coincided to form a network extended to several microns. In higher magnification (Fig. 2b), it was observed that the particles with diameters of a few hundred nanometers to several microns were ellipsoidal and connected to each other, which were subsequently demonstrated to be mainly sodium compounds according to the EDXA spectra (inset in Fig. 2b). Moreover, compared with the filter paper, the carbon peak of the gel weakened significantly, and the oxygen peak slightly enhanced, which might be due to the covering layer densely coating on the cellulose surface; meanwhile, EDXA studies exhibited that the atomic ratio of Na/O of the gel was 0.84 : 1, which was close to 1 : 1, and this was consistent with NaOH, indicating the possible presence of NaOH in the covering layer.

Fig. 2 (a) Low-magnification and (b) high-magnification SEM images of the thermally induced gel. Inset of (a) presented the digital photograph of the gel, and inset of (b) presented the EDXA spectra of the gel and the filter paper

As shown in Fig. 3a, the filter paper exhibited strong peaks at around 14.6°, 16.1° and 22.3° corresponding to the (101), ($10\bar{1}$) and (002) planes, which belonged to typical cellulose Ⅰ crystalline structure[12]. Nevertheless, for the thermally induced gel, the characteristic peaks of cellulose were completely disappeared, possibly owing to the cellulose degradation[13,14] and the shield of the covering layer; instead, some new peaks in the XRD pattern of the gel could be well-indexed to PEG-4000 and $NaOH \cdot H_2O$ (JCPDS no. 30-1194), which revealed that the covering layer on the surface of the gel principally consisted of PEG-4000 and $NaOH \cdot H_2O$.

Fig. 3b, c presented the TG and the DTG curves of the filter paper and the thermally induced gel, respectively. In both TG curves, the small weight losses below 150 ℃ primarily resulted from evaporation of adsorbed water. The decomposition behavior of the filter paper was mainly related to cellulose pyrolysis with maximum degradation rate at 366 ℃, followed by a rapid decay and a long tail, and displayed a low char yield (ca. 6%). Whereas for the gel, on account that the pyrolysis of PEG approximately takes place ranging from 120 ℃ to 200 ℃ under N_2[15], thus the slightly high mass loss of the gel before 200 ℃ (ca. 27%) was attributed to not only evaporation of adsorbed water below 150 ℃ which was originated from strong hygroscopicity of NaOH and PEG on the gel surface, but also thermal decomposition of PEG molecules. While sequentially rising temperature from 200 ℃ to 800 ℃, the gel weight almost maintained constant with slight mass loss of 5%, and showed a high char yield of 68%, which indicated strong thermal stability. The high char yield of the gel might be owing to the covering layer mainly composed of $NaOH \cdot H_2O$ and PEG on the surface of the cross-linked fiber skeleton structures. Due to high fusion point (ca. 318 ℃) and boiling point (ca. 1390 ℃) of NaOH, the $NaOH \cdot H_2O$ aggregations might gradually loss the water of crystallization and melt during the temperature-rise period; whereas, the liquid NaOH might still adhere to the surface of the cross-linked fiber skeleton structures, possibly contributing to effectively protecting the cellulose from thermal decomposition.

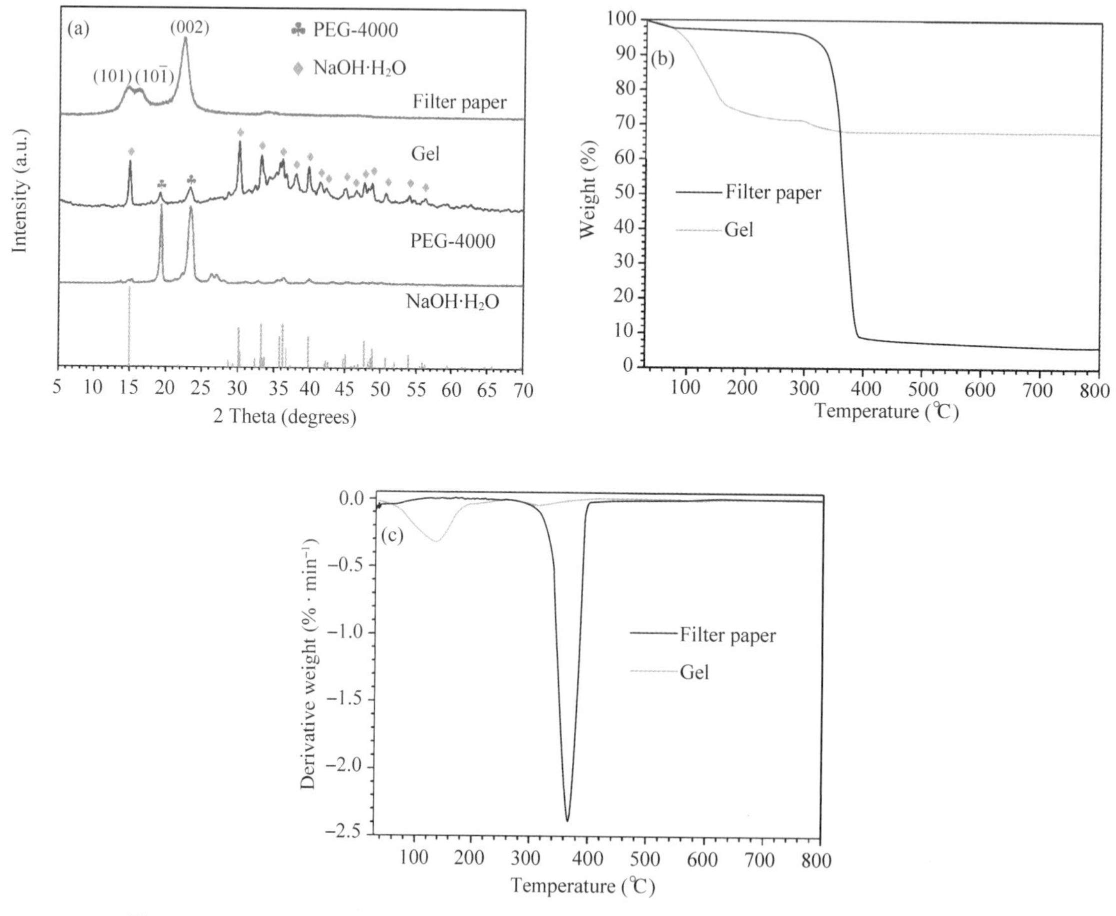

Fig. 3 (a) XRD patterns of the filter paper, the thermally induced gel, NaOH and PEG-4000. (b) TG and (c) DTG curves of the filter paper and the thermally induced gel

The mechanical properties of the thermally induced gel were also high. The compression stress-strain curve of the gel was presented in Fig. 4. The curve showed an initial linear region (< 10%), and a plateau with gradually increasing slope until very high strains up to 40%, and a densification region with inelastic hardening from 40% to 60%. The elastic modulus, yield stress and toughness of the gel were calculated to be about 3210 kPa, 325 kPa and 38.9 $kJ \cdot m^{-3}$, respectively, which were comparable to those of cross-linked polymer gel materials with high mechanical strength[16-18]. Moreover, the inset showed the gel supporting a mass of 5 kg without any collapse, further indicating the strong mechanical strength.

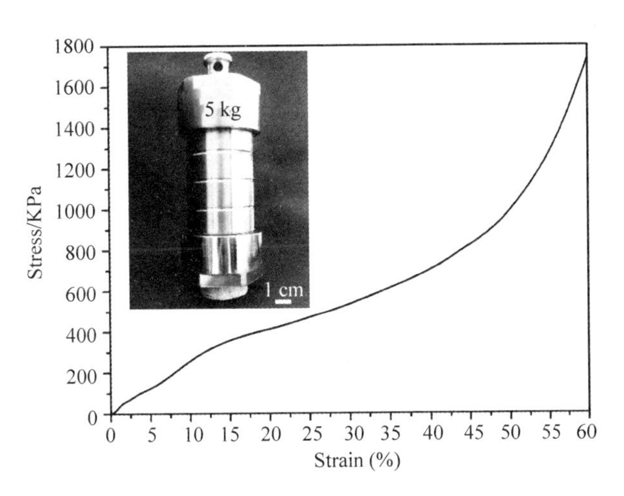

Fig. 4 Compression stress-strain curve of the thermally induced gel. The inset presented the gel with a load bearing of 5 kg

4 Conclusions

An easy, quick and mild method was reported to fabricate a thermally induced gel with strong mechanical strength based on cellulose-NaOH/PEG aqueous solution. The gel consisted of cross-linked fiber aggregations with dense covering layer of NaOH · H_2O and PEG on the surface, and the covering layer made contribution to the improvement of thermal stability and resulted in the disappearance of cellulose crystalline structure. Moreover, the mechanically strong gel also had high elastic modulus of 3210 kPa, yield stress of 325 kPa and toughness of 38.9 kJ · m^{-3}. Consequently, the method for the preparation of the novel gel material with strong mechanical strength might be helpful to broaden the potential applications of cellulose productions.

Acknowledgments

We appreciate the financial support of the China Postdoctoral Science Foundation (2013M540263), the National Natural Science Foundation of China (grant no. 31270590), and the Doctoral Candidate Innovation Research Support Program of Science & Technology Review (kjdb2012006).

References

[1] Hult E L, Iversen T, Sugiyama J. Characterization of the supermolecular structure of cellulose in wood pulp fibres[J]. Cellulose, 2003, 10(2): 103-110.

[2] Zografi G, Kontny M J, Yang A Y S, et al. Surface area and water vapor sorption of macrocrystalline cellulose[J]. International journal of pharmaceutics, 1984, 18(1-2): 99-116.

[3] Brandt A, Gräsvik J, Hallett J P, et al. Deconstruction of lignocellulosic biomass with ionic liquids[J]. Green chemistry, 2013, 15(3): 550-583.

[4] Zhu S, Wu Y, Chen Q, et al. Dissolution of cellulose with ionic liquids and its application: a mini-review[J]. Green Chemistry, 2006, 8(4): 325-327.

[5] Orts W J, Shey J, Imam S H, et al. Application of cellulose microfibrils in polymer nanocomposites[J]. Journal of

Polymers and the Environment, 2005, 13(4): 301-306.

[6] Cuculo J A, Smith C B, Sangwatanaroj U, et al. A study on the mechanism of dissolution of the cellulose/NH_3/NH_4SCN system. [J]. Journal of Polymer Science Part A: Polymer Chemistry, 1994, 32(2): 229-239.

[7] Zhang H, Wu J, Zhang J, et al. 1-Allyl-3-methylimidazolium chloride room temperature ionic liquid: a new and powerful nonderivatizing solvent for cellulose[J]. Macromolecules, 2005, 38(20): 8272-8277.

[8] Yan L, Gao Z. Dissolving of cellulose in PEG/NaOH aqueous solution[J]. Cellulose, 2008, 15(6): 789-796.

[9] Weng L, Zhang L, Ruan D, et al. Thermal gelation of cellulose in a NaOH/thiourea aqueous solution[J]. Langmuir, 2004, 20(6): 2086-2093.

[10] Kadokawa J I , Murakami M A , Kaneko Y . A facile preparation of gel materials from a solution of cellulose in ionic liquid[J]. Carbohydrate Research, 2008, 343(4): 769-772.

[11] Cai J, Zhang L. Unique gelation behavior of cellulose in NaOH/urea aqueous solution[J]. Biomacromolecules, 2006, 7(1): 183-189.

[12]Oh S Y, Yoo D I, Shin Y, et al. Crystalline structure analysis of cellulose treated with sodium hydroxide and carbon dioxide by means of X-ray diffraction and FTIR spectroscopy[J]. Carbohydrate Research, 2005, 340(15): 2376-2391.

[13] Van Loon L R, Glaus M A. Experimental and theoretical studies on alkaline degradation of cellulose and its impact on the sorption of radionuclides[J]. Paul Scherrer Institut, 1998.

[14]Knill C J, Kennedy J F. Degradation of cellulose under alkaline conditions[J]. Carbohydrate Polymers, 2003, 51(3): 281-300.

[15]Chen M, Wang Y, Song L, et al. Urchin-like ZnO microspheres synthesized by thermal decomposition of hydrozincite as a copper catalyst promoter for the Rochow reaction[J]. RSC Advances, 2012, 2(10): 4164-4168.

[16] Chen W, Yan L. In situ self-assembly of mild chemical reduction graphene for three-dimensional architectures[J]. Nanoscale, 2011, 3(8): 3132-3137.

[17] Grieshaber S E, Farran A J E, Lin-Gibson S, et al. Synthesis and characterization of elastin-mimetic hybrid polymers with multiblock, alternating molecular architecture and elastomeric properties[J]. Macromolecules, 2009, 42(7): 2532-2541.

[18] Sehaqui H , M Salajková, Zhou Q , et al. Mechanical performance tailoring of tough ultra-high porosity foams prepared from cellulose I nanofiber suspensions[J]. Soft Matter, 2010, 6(8): 1824-1832.

中文题目：纤维素/NaOH/PEG溶液的热诱导凝胶：制备、表征和机械性能

作者：万才超，卢芸，焦月，金春德，孙庆丰，李坚

摘要：本文报道了以纤维素/NaOH/PEG水溶液为原料，经过溶解、加热、冷冻干燥等程序制备的具有较强力学性能的热诱导凝胶。凝胶呈现出由交联纤维聚集而成的不发达网络，紧紧包覆着大量NaOH·H_2O和PEG聚集的细小颗粒，从而热稳定性显著增强，并且原有纤维素结晶结构消失。此外，测得机械强度高的凝胶的弹性模量、屈服应力和韧性分别高达3210 kPa、325 kPa和389 kJ m^{-3}，且与微纤维化纤维素气凝胶和石墨烯水凝胶三维结构等机械强度高的交联高分子凝胶材料相当。

关键词：纤维素；凝胶；机械性能；聚合物

A Facile Low-temperature Hydrothermal Method to Prepare Anatase Titania/Cellulose Aerogels with Strong Photocatalytic Activities for Rhodamine B and Methyl Orange Degradations*

Caichao Wan, Yun Lu, Chunde Jin, Qingfeng Sun, Jian Li

Abstract: In this paper, a facile low-temperature hydrothermal method for in-situ preparation of anatase titania (TiO_2) homogeneously dispersed in cellulose aerogels substrates was described. The formed anatase TiO_2 aggregations composed of a mass of evenly dispersed TiO_2 nanoparticles with sizes of 2 – 5 nm were embedded in the interconnected three-dimensional (3D) architecture of the cellulose aerogels matrixes without large-scale reunion phenomenon; meanwhile, the obtained anatase titania/cellulose (ATC) aerogels also had a high loading amount of TiO_2 (ca. 35.7%). Furthermore, compared with commercially available Degussa P25, ATC aerogels displayed comparable photocatalytic activities for Rhodamine B and methyl orange degradations under UV radiation, which might be useful in the fields of catalysts, wastewater treatment and organic pollutant degradation. Meanwhile, the photocatalytic reaction behaviors of ATC aerogels under UV irradiation were also illuminated.

Keywords: Titania; Cellulose aerogels; Hydrothermal synthesis; Photocatalysis; Nanocomposites

1 Introduction

Titanium dioxide (TiO_2), one of the most important semiconductor materials, has been extensively applied as photocatalyst for decontamination of water polluted with organic pollutants due to its high optical reactivity, chemically and biologically inert, nontoxicity and low cost[1-3]. TiO_2 crystals mainly exist in two stable polymorphic forms: anatase (tetragonal, space group $I4_1/amd$) and rutile (tetragonal, space group $P4_2/mnm$)[4]. Generally, it is accepted that anatase TiO_2 is more efficient as photocatalyst, and rutile TiO_2 is preferable for blocking UV irradiation, notwithstanding that some catalysts with mixed phases exhibit significantly higher catalytic activities like Degussa P25[5,6].

In fact, nano-scale particles with large specific surface area and high surface energy are very easy to reunite leading to conspicuous performance degradation. The confinement of nanoparticles can be achieved by using some special matrixes with unique structures as hosts like multi-walled carbon nanotubes[7-9], graphene[10], fibers[11], and some traditional inorganic or synthetic polymer-based aerogels[12,13], which

* 本文摘自 Journal of Nanomaterials, 2015, 2015: 717016.

could provide a means of particle dispersion and act as protection of metallic nanoparticles against air-oxidation. Especially, owing to large specific surface area, high porosity, multi-scale micro-nano structure, stable chemical property, renewability, and biodegradability, cellulose aerogels are widely considered as good supports for nanomaterials, such as TiO_2, ZnO, Ag and ZrO_2[14-17]. Especially, TiO_2/cellulose aerogels hybrids are useful for photocatalysts, UV shields, antibacterial materials and synthesis of transition-metal carbide materials[18-21]. Furthermore, synthesis of anatase TiO_2/cellulose (ATC) composites generally depends on high-temperature calcination treatment using cellulose or cellulose derivatives as templates at 400–700 ℃[22-24], so as to get anatase TiO_2 crystals and avoid excessive crystal transformation from anatase to rutile while exceeding the critical temperature.

Herein, a facile low-temperature hydrothermal method was employed to prepare ATC aerogels instead of the common calcination process. Moreover, distinguishing with some previous literatures about TiO_2/cellulose composites with improved activity, in this study, a green NaOH/polyethylene glycol (PEG) aqueous solution was used to fabricate the cellulose aerogels, and the resulting environmentally friendly nanoporous aerogels were selected as hosts to support the nano-TiO_2. The obtained ATC aerogels were characterized by scanning electron microscopy (SEM), transmission electron microscopy (TEM), energy dispersive X-ray spectroscopy (EDX), X-ray photoelectron spectroscopy (XPS), X-ray diffraction (XRD), and thermogravimetry (TG). Meanwhile, the photocatalytic activities of ATC aerogels for two typical organic pollutants including Rhodamine B (RhB) and methyl orange (MO) degradations were also investigated and compared with those of P25. Based on which, photocatalytic reaction behaviors of ATC aerogels under UV irradiation were illuminated.

2 Materials and methods

2.1 Materials

Sixty-mesh powder of wheat straw after grinding and sieving was collected and dried in a vacuum at 60 ℃ for 24 h before used. All chemicals were supplied by Tianjin Kemiou Chemical Reagent Co. Ltd. and used as received.

2.2 Cellulose hydrogel preparation

The purified cellulose from the wheat straw powder was referred to our previous report via a chemical pretreatment process[25]. The obtained dried cellulose was mixed with an aqueous solution of NaOH/PEG-4000 (9 : 1 by weight) with magnetic stirring at room temperature for about 6 h to form a 2 wt% homogenous cellulose solution. Then, the cellulose solution underwent a freeze-thaw process by freezing the solution at −15 ℃ for 12 h and subsequently thawing it out at room temperature for 0.5 h under vigorously stirring. After being frozen again for 3 h at −15 ℃, the solution was regenerated by placing it into a coagulation bath of 1 v% hydrochloric acid for 6 h, repeating this process until the formation of the amber-like hydrogel. Afterwards, the obtained hydrogel was rinsed repeatedly with distilled water and anhydrous ethyl alcohol (EtOH) to remove superfluous hydrions and chlorine anions.

2.3 ATC aerogel preparation

The low-temperature hydrothermal preparation of ATC aerogel was described in Scheme 1. Tetrabutyl orthotita (TBOT) (5mL) and EtOH (200mL) were mixed with magnetic stirring for 30 min, and then transferred into a Teflon-lined stainless-steel autoclave. The cellulose hydrogel was subsequently placed into the above reaction solution. After the autoclave was sealed and heated to 110 ℃ for 4 h, 100 mL deionized water (pH=6.5) was added and the solution was re-heated to 70 ℃ for 4 h. After reaction, the prepared sample was removed from the solution, and ultrasonically rinsed with deionized water and tert butyl alcohol in sequence for 30 min. Finally, the sample was subjected to a freeze-drying process at −35 ℃ for 48 h, and the following ATC aerogel was successfully fabricated. Moreover, the pure TiO_2 particles powders were prepared following the above hydrothermal process without adding the cellulose hydrogel, and the mixed solution after reaction was filtered, washed by distilled water and dried at 60 ℃ for 24 h to get dried TiO_2 particles powders. The pure cellulose aerogel was obtained by the direct tert butyl alcohol freeze-drying process of the cellulose hydrogel.

Scheme 1 was inserted here.

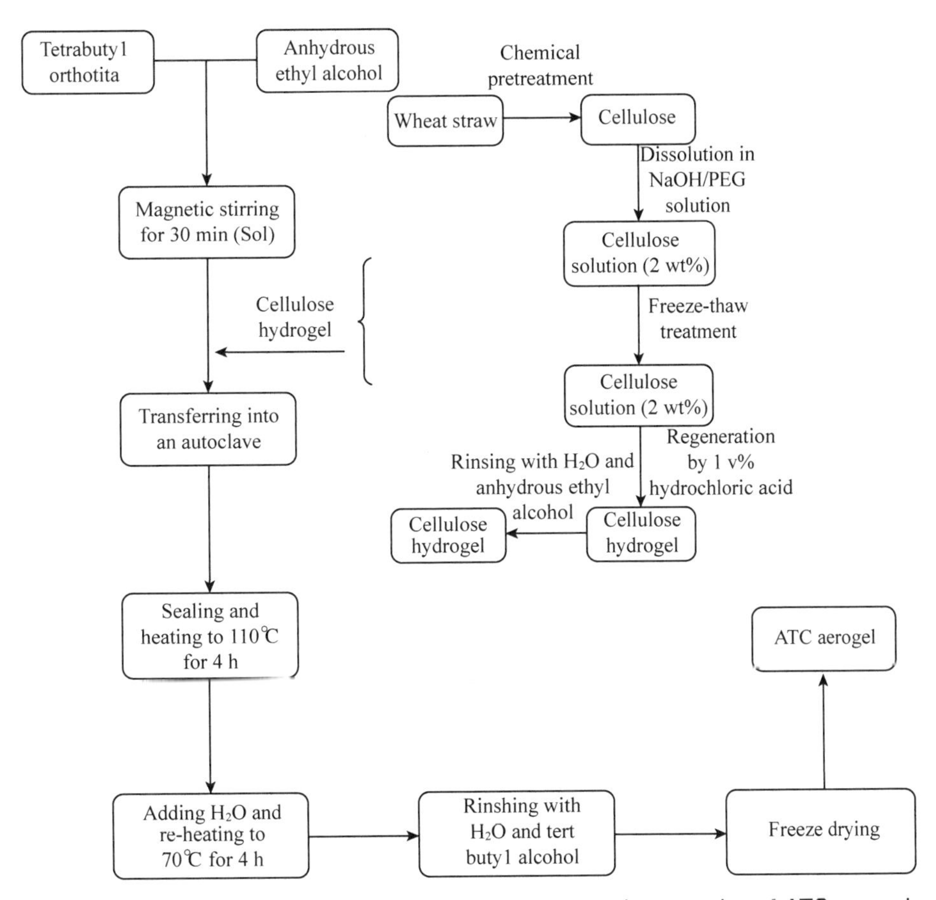

Scheme 1 Diagram of the low-temperature hydrothermal preparation of ATC aerogel

2.4 Characterizations

Microstructures and surface chemical compositions were evaluated by SEM (FEI, Quanta 200) equipped with EDX spectrometer operating at 15–20 kV on gold-sputtered samples. TEM observation was carried out on a FEI, Tecnai G2 F20 TEM. Crystalline structures were identified by XRD instrument (Rigaku, D/MAX

2200) operating with Cu Kα radiation (λ = 1.5418 Å) at a scan rate (2θ) of 4° · min^{-1} and the accelerating voltage of 40 kV and the applied current of 30 mA ranging from 10° to 80°. XPS spectra were recorded using a Thermo Escalab 250Xi XPS spectrometer (Germany). Deconvolution of the overlapping peaks was performed using a mixed Gaussian-Lorentzian fit program. Thermal stabilities were determined using a TG analyzer (TA Q600) with a heating rate of 10 ℃ · min^{-1} in a N_2 environment.

2.5 Photocatalytic activities measurements

The photocatalytic activities of ATC aerogels were evaluated by measuring the decomposition of RhB and MO, while pure TiO_2 nanoparticles (commercial Degussa P25) were used for comparison, and the amount of P25 and the amount of ATC aerogels were controlled by the same titania loading determined by the following TG tests. A 100 W mercury lamp with main wavelength of 365 nm was purchased from Beijing BrightStars Science and Technology Corp and selected as UV light source. For typical photocatalytic experiments of the hybrid aerogels, ATC aerogels samples with sizes of about 20 mm (diameter) × 10 mm (height) were immersed into the circular disks with 30 mL RhB or MO aqueous solutions (50 mg · L^{-1}), respectively. Before irradiation, the solutions with ATC aerogels were magnetically stirred in the dark for 30 min to achieve adsorption equilibrium. Then, the dishes were exposed to the mercury lamp for about 1 h, and the distances between the lamps and the dishes were approximately 10 cm. Samples were taken at every time interval (20 min) and centrifuged for removing photocatalysts. Concentration of the filtrate was analyzed by a TU-1901 UV-vis spectrophotometer (Beijing Purkinje, China) at 554 nm for RhB or 464 nm for MO. The efficiency was calculated by Eq. (1):

$$Y = 100 \times [(C - C_0)/C_0] \tag{1}$$

where C_0 and C are the initial dye concentration and the concentration at time t, respectively. Similarly, the photocatalytic experiments of P25 for RhB and MO were following the above procedures.

3 Results and discussion

Fig. 1a showed the SEM image of ATC aerogels. As shown, porous and interconnected three-dimensional (3D) architecture with pore sizes of a few hundred nanometers to several micrometers, originated from the original cellulose hydrogels architecture, could be clearly identified. In addition, it was observed that numerous white ellipsoidal substances surrounding the porous skeleton structure were generated, which might be derived from titanium compounds. To further investigate the chemical components of the newly generated white ellipsoidal substances, the area in the green box with some evident aggregations of particles (Fig. 1a) was tested by EDX, and the results demonstrated that apart from C and O elements from the cellulose substrates as well as Au element from the coating layer used for SEM observation, new strong peaks corresponding to Ti element were detected (inset in Fig. 1a), confirming that the white ellipsoidal substances primarily consisted of titanium compounds. Moreover, from the higher magnification SEM image (Fig. 1b), it was clear that the ellipsoidal substances containing Ti element tightly adhered to the cellulose matrixes, indicating potential good interface bonding.

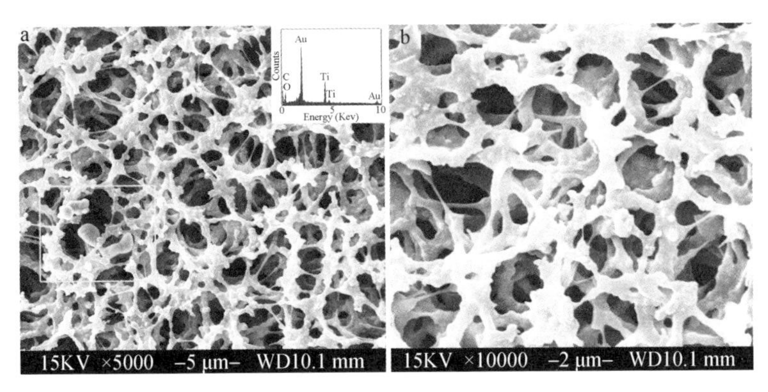

Fig. 1 (a) Low-magnification and (b) high-magnification SEM images of ATC aerogels, respectively. The green box was the area of EDX analysis, and the corresponding EDX spectrum was shown in Fig. 1 as an inset

In order to further explore the composition and structure of the ellipsoidal substances as well as the interaction between the ellipsoidal substances and the cellulose aerogels matrixes, the samples were observed by TEM. The TEM images of different magnifications of ATC aerogels were represented in Fig. 2. Apparently, the cellulose aerogels matrixes were loaded with a mass of evenly dispersed nanoparticles as indicated by the dark spots in the TEM image with lower magnification (Fig. 2a). These abundant particles might be responsible for the formation of the ellipsoidal substances. The higher magnification TEM image (Fig. 2b) showed that the size of nanoparticles was found to range between 2–5 nm, and no large-scale aggregation was observed, indicating that the combination of nanoparticles and cellulose aerogels effectively hindered the reunion phenomenon.

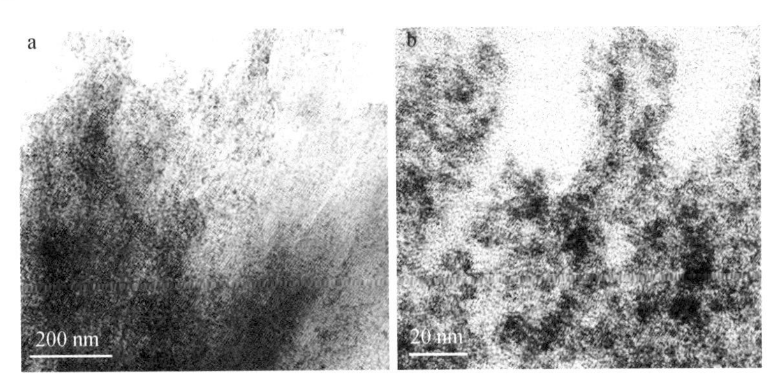

Fig. 2 TEM images of different magnifications of ATC aerogels, respectively

XPS measurements were performed to elucidate the surface chemical composition and the oxidation state of ATC aerogels. In the high-resolution XPS spectra of Ti 2p (Fig. 3a), the binding energies of Ti $2p_{3/2}$ and $2p_{1/2}$ were centered at 457.2 eV and 462.5 eV, respectively, which revealed that the titanium elements were in the oxidation state IV, corresponding to Ti^{4+}(TiO_2)[26,27]. Meanwhile, it also suggested that the ellipsoidal substances observed by SEM were composed of TiO_2 particles; besides, the nanoparticles in the TEM images might be exactly the generated TiO_2 particles. The O 1s spectra of ATC aerogels displayed the contributions of three components (Fig. 3b) including TiO_2(529.8 eV), hydroxyl groups (O—H) (531.2 eV) and C=O groups (532.1 eV)[28]. Moreover, in the C 1s spectra (Fig. 3c), major peak at 284.7 eV was related to C—C, C=C and C—H bonds, and the peak at around 287.7 eV suggested the presence of Ti—O—C bond[29].

The XRD patterns of the TiO_2 particles and ATC aerogels were shown in Fig. 3d. For the TiO_2 particles, a series of characteristic peaks, corresponding to the (103), (004), (112), (200), (105), (211), (213), (204), (116), (220), (107), (215) and (301) planes, were observed, which were regarded as an attributive indicator of anatase phase TiO_2 crystallites (JCPDS file no. 21-1272)[30]. Similarly, ATC aerogels also exhibited typical strong peaks assigned to anatase phase as well as cellulose characteristic peaks at 11.9°, 19.8° and 22.0° belonging to (101), (10$\bar{1}$) and (002) planes[31]. Especially, no peaks of rutile TiO_2 phase (JCPDS file no. 21-1276) were detected, indicating the high anatase phase purity of ATC aerogels. Meanwhile, it also revealed that the anatase TiO_2 was successfully fabricated and combined with cellulose aerogels by the facile low-temperature hydrothermal method. Moreover, it was obvious that the characteristic peaks of the anatase TiO_2 were broad, which was an indication of the small size of the TiO_2 particles formed in the ATC aerogels. The crystallite size as calculated by the Debye-Scherrer equation[32] was 4.3 nm, which was consistent with the TEM results from Fig. 2b.

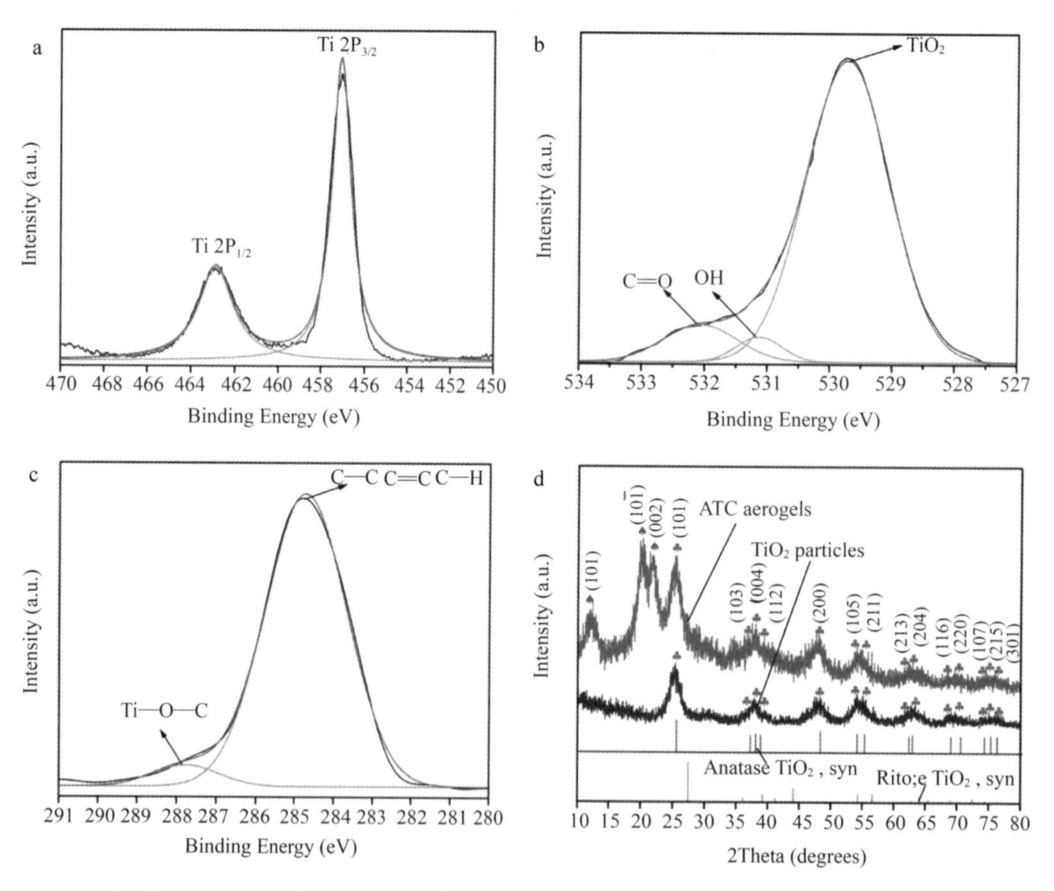

Fig. 3 High-resolution XPS spectra of ATC aerogels in the (a) Ti 2p, (b) O 1s, and (c) C 1s regions. (d) XRD patterns of the TiO_2 particles and ATC aerogels

Fig. 4 showed the TG and DTG curves of ATC aerogels and the pure cellulose aerogels, respectively. The

DTG curve for pure cellulose aerogels showed one major exothermic peak at around 362 ℃, which represented a typical thermal decomposition behavior of cellulose in an inert atmosphere, as previously reported[33]. In comparison, the maximum loss of weight of ATC aerogels shifted to a temperature of ~323 ℃, 39 ℃ lower than that of the pure cellulose aerogels according to the DTG curves. The potential reasons for the shift might be the dehydroxylation and the removal of residual organics from the titania[34], the catalytic character of TiO_2[35], and the loosening of molecular chains in crystalline regions of cellulose as the result of infusion of TiO_2 particles[36]. Moreover, ATC aerogels had high residual ash (ca. 44.1%), far higher than that of the pure cellulose aerogels samples (ca. 8.4%), which indicated high loading amount of TiO_2 in the hybrid ATC aerogels samples (ca. 35.7%).

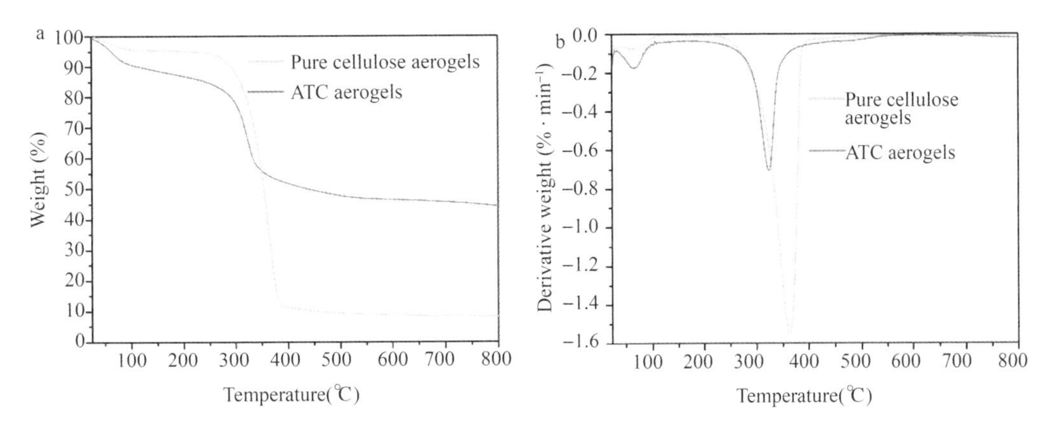

Fig. 4 TG and DTG curves of ATC aerogels and the pure cellulose aerogels, respectively

RhB and MO, the two kinds of common organic pollutants, were employed to evaluate the photocatalytic activities of ATC aerogels. For the sake of comprehending the photocatalytic property, the commercially available TiO_2 P25 (Degussa, Germany) was used as a photocatalytic reference. As shown in Fig. 5a, the concentrations of the RhB and the MO solutions changed little for the wheat straw, indicating an inferior photocatalytic performance under UV irradiation. On the contrary, both of P25 and ATC aerogels showed strong photodegradation abilities for RhB and MO due to the drastic changes of concentration before and after the experiments. Compared with P25, ATC aerogels exhibited similar concentration change tendency for RhB and slightly faster rate of photocatalytic degradation for MO, respectively, which proved that the photocatalytic activities of ATC aerogels were comparable with those of P25 only taking account of these two kinds of organic pollutants. Moreover, the strong photocatalytic activities of ATC aerogels might be due to good dispersion, potential higher surface area and smaller crystallite size of the formed anatase TiO_2[37,38]. Furthermore, Fig. 5b showed the macrographs of RhB and MO degradations by ATC aerogels, and the significant color differences before and after the UV radiation further indicated the superior photocatalytic activities of ATC aerogels.

Based on the results mentioned above, photocatalytic reaction behaviors of ATC aerogels under UV irradiation for RhB and MO degradations were illuminated in Fig. 6. When the surface of ATC aerogels was illuminated by UV light with energy equal to or larger than the band gap energy of anatase TiO_2 nanocrystals, the electrons were excited in the valance band (VB) to the conduction band (CB), leading to the formation of a positive hole (h^+) in the VB and an electron (e^-) in the CB[39-41]. The generated electron-hole pairs

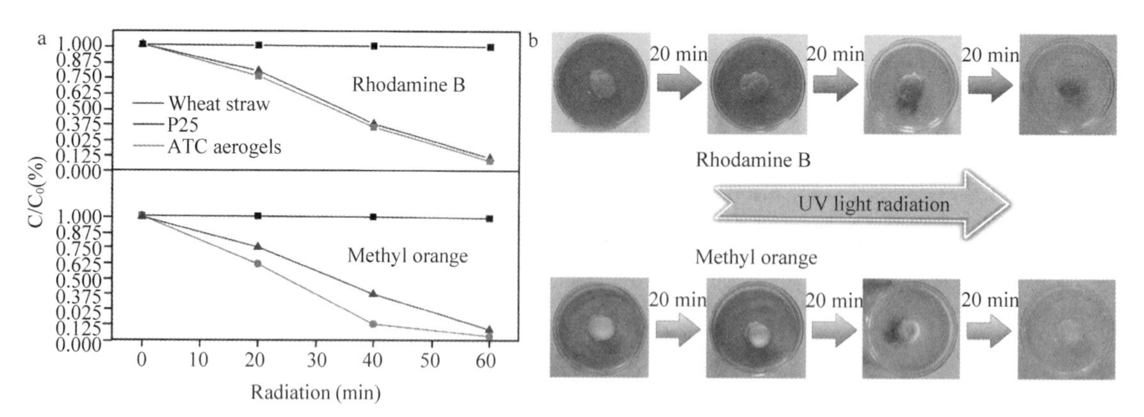

Fig. 5 (a) Photocatalytic activities of the wheat straw, P25 and ATC aerogels for RhB and MO degradations. (b) Macrographs of RhB and MO degradations by ATC aerogels under UV radiation

immediately interacted with surface adsorbed molecular oxygen to yield superoxide radical anions ($\cdot O_2^-$), and with water to produce the highly reactive HO · radicals, respectively. These radicals groups repeatedly attacked RhB and MO molecules, eventually resulting in their degradations into CO_2 and H_2O[42,43], which were responsible for the photocatalytic activity.

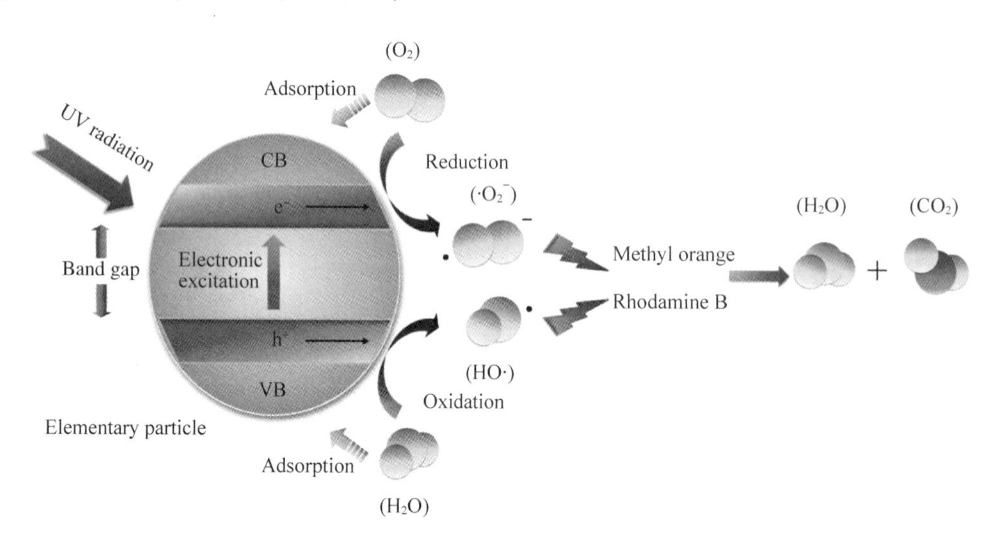

Fig. 6 Schematic representation of the photocatalytic mechanism of ATC aerogels

4 Conclusions

In conclusion, ATC aerogels were successfully fabricated via a mild simple low-temperature hydrothermal method. The as-prepared ATC aerogels were composed of cross-linked 3D architecture originated from the original cellulose hydrogels surrounding by plentiful ellipsoidal anatase TiO_2 particles aggregations, and these ellipsoidal aggregations consisted of a mass of evenly dispersed TiO_2 nanoparticles with sizes of 2–5 nm. Meanwhile, a high loading amount of TiO_2 in the hybrid ATC aerogels samples (ca. 35.7%) was also obtained according to the TG results. Moreover, the ATC aerogels exhibited comparable photocatalytic activities for RhB and MO degradations with Degussa P25 under UV radiation, which might be served as novel green catalysts for water decontamination and organic pollutants decomposition.

Acknowledgments

We appreciate the financial support of the China Postdoctoral Science Foundation (2013M540263), the National Natural Science Foundation of China (grant no. 31270590), and the Doctoral Candidate Innovation Research Support Program of Science & Technology Review (kjdb2012006).

References

[1] Chen X, Mao S S. Titanium dioxide nanomaterials: synthesis, properties, modifications, and applications[J]. Chemical reviews, 2007, 107(7): 2891-2959.

[2] Gelover S, Mondragón P, Jiménez A. Titanium dioxide sol-gel deposited over glass and its application as a photocatalyst for water decontamination[J]. Journal of Photochemistry and Photobiology A: Chemistry, 2004, 165(1-3): 241-246.

[3] Pozzo R L, Baltanas M A, Cassano A E. Supported titanium oxide as photocatalyst in water decontamination: state of the art[J]. Catalysis Today, 1997, 39(3): 219-231.

[4] Sun Q, Lu Y, Zhang H, et al. Hydrothermal fabrication of rutile TiO_2 submicrospheres on wood surface: An efficient method to prepare UV-protective wood[J]. Materials Chemistry and Physics, 2012, 133(1): 253-258.

[5] Sun J, Gao L, Zhang Q. Synthesizing and comparing the photocatalytic properties of high surface area rutile and anatase titania nanoparticles[J]. Journal of the American Ceramic Society, 2003, 86(10): 1677-1682.

[6] Hurum D C, Agrios A G, Gray K A, et al. Explaining the enhanced photocatalytic activity of Degussa P25 mixed-phase TiO_2 using EPR[J]. The Journal of Physical Chemistry B, 2003, 107(19): 4545-4549.

[7] Xia X H, Jia Z J, Yu Y, et al. Preparation of multi-walled carbon nanotube supported TiO_2 and its photocatalytic activity in the reduction of CO_2 with H_2O[J]. Carbon, 2007, 45(4): 717-721.

[8] Yu Y, Jimmy C Y, Chan C Y, et al. Enhancement of adsorption and photocatalytic activity of TiO_2 by using carbon nanotubes for the treatment of azo dye[J]. Applied Catalysis B: Environmental, 2005, 61(1-2): 1-11.

[9] Yu Y, Jimmy C Y, Yu J G, et al. Enhancement of photocatalytic activity of mesoporous TiO_2 by using carbon nanotubes[J]. Applied Catalysis A: General, 2005, 289(2): 186-196.

[10] Wu Z S, Ren W, Wen L, et al. Graphene anchored with Co_3O_4 nanoparticles as anode of lithium ion batteries with enhanced reversible capacity and cyclic performance[J]. ACS nano, 2010, 4(6): 3187-3194.

[11] Zheng X, Guo D, Shao Y, et al. Photochemical modification of an optical fiber tip with a silver nanoparticle film: a SERS chemical sensor[J]. Langmuir, 2008, 24(8): 4394-4398.

[12] Miller J M, Dunn B, Tran T D, et al. Deposition of ruthenium nanoparticles on carbon aerogels for high energy density supercapacitor electrodes[J]. Journal of the Electrochemical Society, 1997, 144(12): L309.

[13] Saquing C D, Kang D, Aindow M, et al. Investigation of the supercritical deposition of platinum nanoparticles into carbon aerogels[J]. Microporous and Mesoporous Materials, 2005, 80(1-3): 11-23.

[14] Cai J, Kimura S, Wada M, et al. Nanoporous cellulose as metal nanoparticles support[J]. Biomacromolecules, 2009, 10(1): 87-94.

[15] Moon R J, Martini A, Nairn J, et al. Cellulose nanomaterials review: structure, properties and nanocomposites[J]. Chemical Society Reviews, 2011, 40(7): 3941-3994.

[16] Tingaut P, Zimmermann T, Sèbe G. Cellulose nanocrystals and microfibrillated cellulose as building blocks for the design of hierarchical functional materials[J]. Journal of Materials Chemistry, 2012, 22(38): 20105-20111.

[17] Luong N D, Lee Y, Nam J D. Highly-loaded silver nanoparticles in ultrafine cellulose acetate nanofibrillar aerogel

[J]. European Polymer Journal, 2008, 44(10): 3116-3121.

[18] Lu Y, Sun Q F, Li J, et al. Fabrication, characterization and photocatalytic activity of TiO_2/cellulose composite aerogel[C]//Key Engineering Materials. Trans Tech Publications Ltd, 2014, 609: 542-546.

[19] Hamann T W, Martinson A B F, Elam J W, et al. Atomic layer deposition of TiO_2 on aerogel templates: new photoanodes for dye-sensitized solar cells[J]. The Journal of Physical Chemistry C, 2008, 112(27): 10303-10307.

[20] Kusabe M, Kozuka H, Abe S, et al. Sol-gel preparation and properties of hydroxypropylcellulose-titania hybrid thin films[J]. Journal of sol-gel science and technology, 2007, 44(2): 111-118.

[21] Shin H, Jeong D K, Lee J, et al. Formation of TiO_2 and ZrO_2 nanotubes using atomic layer deposition with ultraprecise control of the wall thickness[J]. Advanced Materials, 2004, 16(14): 1197-1200.

[22] Melone L, Altomare L, Alfieri I, et al. Ceramic aerogels from TEMPO-oxidized cellulose nanofibre templates: Synthesis, characterization, and photocatalytic properties[J]. Journal of Photochemistry and Photobiology A: Chemistry, 2013, 261: 53-60.

[23] Venkataramanan N S, Matsui K, Kawanami H, et al. Green synthesis of titania nanowire composites on natural cellulose fibers[J]. Green Chemistry, 2007, 9(1): 18-19.

[24] Nelson K, Deng Y. Enhanced light scattering from hollow polycrystalline TiO_2 particles in a cellulose matrix[J]. Langmuir, 2008, 24(3): 975-982.

[25] Li J, Wan C, Lu Y, et al. Fabrication of cellulose aerogel from wheat straw with strong absorptive capacity[J]. Frontiers of Agricultural Science and Engineering, 2014, 1(1): 46-52.

[26] Wang B, Karthikeyan R, Lu X Y, et al. High photocatalytic activity of immobilized TiO_2 nanorods on carbonized cotton fibers[J]. Journal of hazardous materials, 2013, 263: 659-669.

[27] Erdem B, Hunsicker R A, Simmons G W, et al. XPS and FTIR surface characterization of TiO_2 particles used in polymer encapsulation[J]. Langmuir, 2001, 17(9): 2664-2669.

[28] Hu C, Duo S, Liu T, et al. Low temperature facile synthesis of anatase TiO_2 coated multiwalled carbon nanotube nanocomposites[J]. Materials Letters, 2010, 64(22): 2472-2474.

[29] Gong W J, Tao H W, Zi G L, et al. Visible light photodegradation of dyes over mesoporous titania prepared by using chrome azurol S as template[J]. Research on Chemical Intermediates, 2009, 35(6): 751-760.

[30] Ye J, Liu W, Cai J, et al. Nanoporous anatase TiO_2 mesocrystals: additive-free synthesis, remarkable crystalline-phase stability, and improved lithium insertion behavior. [J]. Journal of the American Chemical Society, 2010, 133(4): 933-940.

[31] Mansikkamäki P, Lahtinen M, Rissanen K. Structural changes of cellulose crystallites induced by mercerisation in different solvent systems; determined by powder X-ray diffraction method[J]. Cellulose, 2005, 12(3): 233-242.

[32] Trentler T J, Denler T E, Bertone J F, et al. Synthesis of TiO_2 nanocrystals by nonhydrolytic solution-based reactions [J]. Journal of the American Chemical Society, 1999, 121(7): 1613-1614.

[33] Shafizadeh F, Bradbury A G W. Thermal degradation of cellulose in air and nitrogen at low temperatures[J]. Journal of applied polymer science, 1979, 23(5): 1431-1442.

[34] Schnitzler D C, Zarbin A J G. Organic/inorganic hybrid materials formed from TiO_2 nanoparticles and polyaniline[J]. Journal of the Brazilian Chemical Society, 2004, 15: 378-384.

[35] Wang H, Zhong W, Xu P, et al. Polyimide/silica/titania nanohybrids via a novel non-hydrolytic sol-gel route[J]. Composites Part A: Applied Science and Manufacturing, 2005, 36(7): 909-914.

[36] Yu Q, Wu P, Xu P, et al. Synthesis of cellulose/titanium dioxide hybrids in supercritical carbon dioxide[J]. Green Chemistry, 2008, 10(10): 1061-1067.

[37] Sun D, Yang J, Wang X. Bacterial cellulose/TiO_2 hybrid nanofibers prepared by the surface hydrolysis method with molecular precision[J]. Nanoscale, 2010, 2(2): 287-292.

[38] Guo W L, Yang Z X, Wang X K. Sonochemical Deposition And Characterization Of Nanophasic TiO_2 On Silica Particles[J]. Materials Research Innovations, 2013, 10(1): 63-68.

[39] Thiruvenkatachari R, Vigneswaran S, Moon I S. A review on UV/TiO_2 photocatalytic oxidation process (Journal Review)[J]. Korean Journal of Chemical Engineering, 2008, 25(1): 64-72.

[40] Laoufi N A, Tassalit D, Bentahar F. The degradation of phenol in water solution by TiO_2 photocatalysis in a helical reactor[J]. Global NEST Journal, 2008, 10(3): 404-418.

[41] Sirimahachai U, Ndiege N, Chandrasekharan R, et al. Nanosized TiO_2 particles decorated on SiO_2 spheres (TiO_2/SiO_2): synthesis and photocatalytic activities[J]. Journal of sol-gel science and technology, 2010, 56(1): 53-60.

[42] Rani M, Gupta N, Pal B. Superior photodecomposition of pyrene by metal ion-loaded TiO_2 catalyst under UV light irradiation[J]. Environmental Science and Pollution Research, 2012, 19(6): 2305-2312.

[43] Răileanu M, Crişan M, Niţoi I, et al. TiO_2-based nanomaterials with photocatalytic properties for the advanced degradation of xenobiotic compounds from water. A literature survey[J]. Water, Air, & Soil Pollution, 2013, 224(6): 1-45.

中文题目：一种简单的低温水热法制备具有强光催化活性的罗丹明B和甲基橙降解的锐钛矿型二氧化钛/纤维素气凝胶

作者：万才超，卢芸，金春德，孙庆丰，李坚

摘要：本文介绍了一种简便的低温水热方法，用于原位制备均匀分散在纤维素气凝胶基质中的锐钛矿型二氧化钛(TiO_2)的方法。所形成的锐钛矿型TiO_2聚集体由大量均匀分散的TiO_2纳米颗粒组成，其大小为2~3 nm，被嵌入纤维素气凝胶基质的互连三维(3D)结构中，而没有大规模的团聚现象；同时，所得到的锐钛型二氧化钛/纤维素(ATC)气凝胶也具有较高的TiO_2负载量(约35.7%)。此外，与市售的Degussa P25相比，ATC气凝胶在紫外光下对罗丹明B和甲基橙的降解表现出了相当的光催化活性，这可能在催化剂、废水处理和有机污染物降解领域有用。同时，还阐明了ATC气凝胶在紫外光照射下的光催化反应行为。

关键词：二氧化钛；纤维素气凝胶；水热合成；光催化；纳米复合材料

Preparation of Mechanically Strong and Lightweight Cellulose Aerogels from Cellulose-NaOH/PEG Solution*

Caichao Wan, Yun Lu, Yue Jiao, Jun Cao, Qingfeng Sun, Jian Li

Abstract: Novel mechanically strong and lightweight cellulose aerogels were successfully prepared by the procedures in four steps: (i) dissolving bamboo fiber in a mild NaOH/PEG solution; (ii) freeze-thaw treatment; (iii) regeneration; (iv) freeze drying. The aerogels with dense interconnected and hierarchical pore structures had high specific surface area of 152 $m^2 \cdot g^{-1}$, large pore volume of 0.51 $cm^3 \cdot g^{-1}$, high porosity as high as 97% and low density of 0.054 $g \cdot cm^{-3}$, and showed cellulose Ⅱ crystal structure. Moreover, the aerogels exhibited strong resistance to compression load with high Young' modulus of 1.85 MPa, yield stress of 83.57 kPa, and toughness of 52.34 $kJ \cdot m^{-3}$.

Keywords: Cellulose aerogels; Freeze drying; Mechanical properties; Porous materials; Polymers

1 Introduction

Aerogels with their large specific surface area, high porosity and low density have been extensively considered as potential candidates for multifarious advanced applications[1]. However, some drawbacks including fragility, hydrophilicity, and demand of supercritical drying in production hamper commercialization of aerogels[2,3], especially for fragility. The last two issues could be conceivably dealt with by some special modification and alternative means, nevertheless, the inherent fragility problems widely occurring in some inorganic and thermoset polymer aerogels impose severe restrictions on the handling and long-term use. Recently, green cellulose aerogels combined traditional good qualities with some new properties from cellulose such as biocompatibility have attracted increasing attention[4,5]. Meanwhile, native cellulose aerogels are not found to significantly suffer from the fragility problems, and usually show excellent flexibility according to some literatures[6-8]. Notwithstanding some hybrid cellulose aerogels incorporated with reinforcing agents exhibit improved mechanical properties[9,10], the reports aiming at mechanically strong pure cellulose aerogels are not abundant. Therefore, in this paper, a kind of mechanically strong and lightweight native cellulose aerogels had been successfully prepared by dissolving cellulose in a green cellulose solvent named NaOH/PEG solution, followed by freeze-thaw treatment, regeneration and freeze drying. The micromorphology, structure and properties of the products were characterized by scanning electron microscope (SEM), nitrogen adsorption measurements, X-ray diffraction (XRD), and universal testing machine.

* 本文摘自 Journal of Sol-Gel Science and Technology, 2015, 74: 256-259.

2 Materials and methods

2.1 Materials

Bamboo fiber was supplied by Beijing Murun Technology Development Co. Ltd. and further completely cleaned and dried at 60 ℃ for 24 h. All chemicals were purchased from Tianjin Kemiou Chemical Reagent Co. Ltd. and used as received.

2.2 Preparation of cellulose aerogel

Dried bamboo fiber was added to 10% aqueous solution of NaOH/PEG-4000 (9 : 1 wt/wt) with magnetic stirring at room temperature for 6 h to form 2wt% homogeneous cellulose solution. Then, the cellulose solution was frozen at −15 ℃ for 12 h, and subsequently thawed at room temperature with vigorous stirring for 30 min. After being frozen again at −15 ℃ for 5 h, the frozen cake was successively regenerated in 1% HCl solution, distilled water and tertiary butanol until the formation of an amber-like hydrogel. Finally, the cellulose aerogel was successfully prepared after the forty-eight hours of freeze drying at −30 ℃ of the hydrogel.

2.3 Characterizations

SEM observations of cellulose aerogels morphology were performed using a FEI, Quanta 200 SEM at the acceleration voltage of 10−15 kV. Nitrogen adsorption measurements were carried out by an accelerated surface area and porosimetry system (3H-2000PS2 unit, Beishide Instrument S & T Co. Ltd.). Specific surface area and pore characteristic parameters were calculated by the Brunauer-Emmett-Teller and Barrett-Joyner-Halenda methods. Crystalline structures were identified by XRD (Rigaku, D/MAX 2200) operating with Cu Kα radiation ($\lambda = 1.5418$ Å) at a scan rate (2θ) of $4° \cdot min^{-1}$ and the accelerating voltage of 40 kV and the applied current of 30 mA ranging from 5° to 40°. Compression tests were performed in a universal testing machine (Suns, UTM4304X) with a compressing velocity of $2\ mm \cdot min^{-1}$.

3 Result and discussion

Fig. 1 showed the SEM images of the cellulose aerogels. As can be seen in the image at a low magnification (Fig. 1a), the aerogels exhibited homogeneous three-dimensional (3D) network

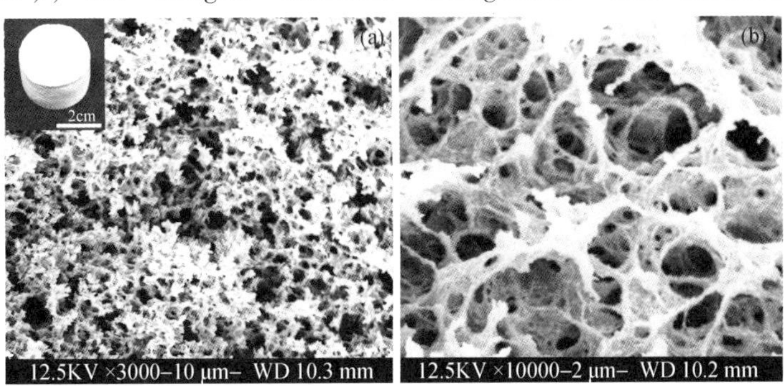

Fig. 1 (a) Low-magnification and (b) high-magnification SEM images of the cellulose aerogels. The inset showed the digital photograph of the dried cellulose aerogels sample

structure. Moreover, the higher magnification image (Fig. 1b) showed dense interconnected and hierarchical pore structures with pores of different sizes from micro- to nano-scale. The inset in Fig. 1a showed the macrograph of the aerogels, and the aerogels sample maintained well-defined form without significant collapse, indicating the superior molding ability. The bulk density of the aerogels was calculated by dividing the weight by the sample volume measured with a micrometer, and the value was low (ca. 0.054 g · cm^{-3}).

Fig. 2 presented N_2 adsorption-desorption isotherms and pore diameter distribution of the cellulose aerogels. As shown in Fig. 2a, the typical IV adsorption isotherm could be identified according to the IUPAC classification, involving adsorption on mesoporous adsorbents with strong adsorbate-adsorbent interaction[11]. Furthermore, the as-prepared aerogels had high specific surface area of 152 m^2 · g^{-1} and large pore volume of 0.51 cm^3 · g^{-1} with porosity as high as 97%, which was comparable to some porous cellulose aerogels from other approaches[12,13], and high values were desirable for applications like catalyst carrier, supercapacitor, fuel cell and drug release. In Fig. 2b, the aerogels were mainly made up of micropores (<2 nm) and mesopores (2-50 nm), and the calculated average pore diameter was 13.37 nm according to the BJH method.

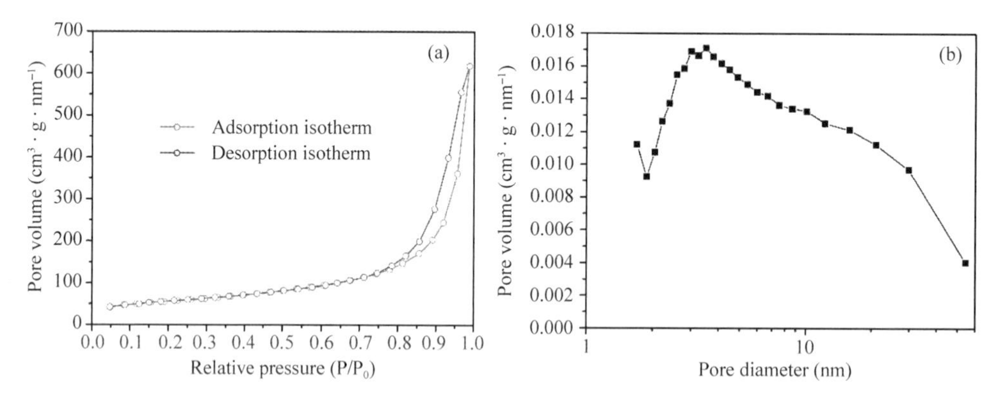

Fig. 2 (a) N_2 adsorption-desorption isotherms and (b) pore diameter distribution of the cellulose aerogels

The XRD patterns of the cellulose aerogels and the bamboo fiber were shown in Fig. 3. The bamboo fiber exhibited peaks at around 14.9°, 16.4°, 22.1° and 34.7°, corresponding to the (101), (10$\bar{1}$), (002) and (030) planes of cellulose Ⅰ crystal structure. For the aerogels, the existence of diffraction peaks at around 12.0°, 20.0° and 21.6° revealed the transformation of cellulose Ⅰ to cellulose Ⅱ[14].

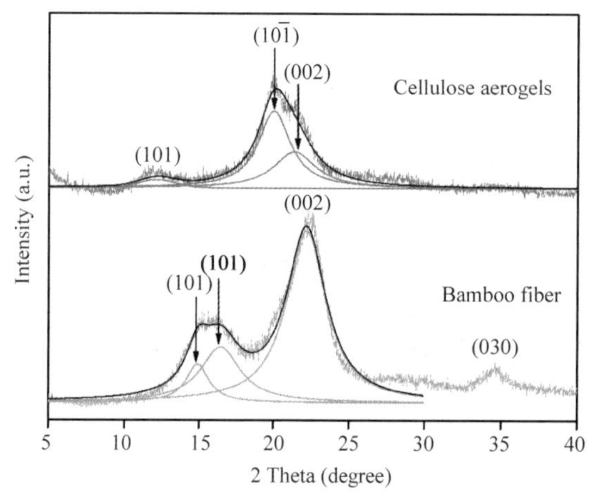

Fig. 3 XRD patterns of the cellulose aerogels and the bamboo fiber

The compression mechanical properties of the aerogels were investigated by universal testing machine, and the tests were performed in triplicate and were done at least two different times to ensure reproducibility. Fig. 4 showed the compression stress-strain curve and the digital photograph of the aerogels. It was observed that the aerogels sample could bear 2.5 kg load without obvious collapse and deformation (inset a in Fig. 4), indicating promising compression resistance. Moreover, the compression stress-strain curve could be divided into four stages[15]. First, the linear elastic behavior in nature (inset b in Fig. 4) attributed to elastic cell wall bending occurred in low strain (<6%); second, the curve gradually transformed from linear to non-linear in higher strain, and the material would collapse in this region; third, a horizontal plateau region appeared after reaching yield stress, and the resulting plastic hinges would result in cell collapse; final, the loose porous 3D network structure started to touch leading to considerable stiffening. The aerogels had high Young' modulus of 1.85 MPa, yield stress of 83.57 kPa, and toughness of 52.34 kJ · m^{-3}, which were calculated based on the curve. Especially, the cross-linked and hierarchical micro-nano pore structures could effectively resist compression load and relieve deformation.

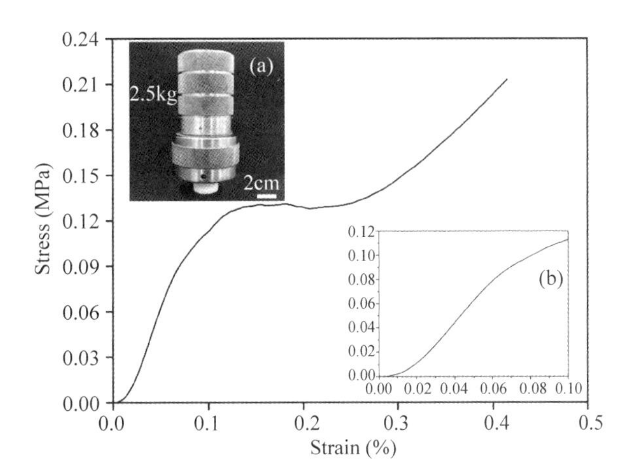

Fig. 4 Compression stress-strain curve of the cellulose aerogels. Inset a: the aerogels sample with a load bearing of 5 kg. Inset b: expanded low-strain range

4 Conclusions

In conclusion, we successfully fabricated a class of mechanically strong and lightweight nanoporous cellulose aerogels with cellulose Ⅱ crystal structure via dissolution in a mild NaOH/PEG solution, freeze-thaw treatment, regeneration and freeze drying. The aerogels with hierarchical pore structures had high specific surface area (152 m^2 · g^{-1}), pore volume (0.51 cm^3 · g^{-1}), and porosity (97%) as well as low density (0.054 g · cm^{-3}); meanwhile, the material also exhibited strong resistant to compression deformation.

Acknowledgments

This work was financially supported by the National Natural Science Foundation of China (grant no. 31270590), a project funded by the China Postdoctoral Science Foundation (2013M540263), and the Doctoral Candidate Innovation Research Support Program of Science & Technology Review (kjdb2012006).

References

[1] Hrubesh L W. Aerogel applications[J]. Journal of Non-Crystalline Solids, 1998, 225: 335-342.

[2] Wei T Y, Lu S Y, Chang Y C. Transparent, hydrophobic composite aerogels with high mechanical strength and low high-temperature thermal conductivities[J]. The Journal of Physical Chemistry B, 2008, 112(38): 11881-11886.

[3] Cai J, Liu S, Feng J, et al. Cellulose-silica nanocomposite aerogels by in situ formation of silica in cellulose gel[J]. Angewandte Chemie, 2012, 124(9): 2118-2121.

[4] Chin S F, Romainor A N B, Pang S C. Fabrication of hydrophobic and magnetic cellulose aerogel with high oil absorption capacity[J]. Materials Letters, 2014, 115: 241-243.

[5] Fischer F, Rigacci A, Pirard R, et al. Cellulose-based aerogels[J]. Polymer, 2006, 47(22): 7636-7645.

[6] Olsson R T, Samir M A S A, Salazar-Alvarez G, et al. Making flexible magnetic aerogels and stiff magnetic nanopaper using cellulose nanofibrils as templates[J]. Nature nanotechnology, 2010, 5(8): 584-588.

[7] Pääkkö M, Vapaavuori J, Silvennoinen R, et al. Long and entangled native cellulose I nanofibers allow flexible aerogels and hierarchically porous templates for functionalities[J]. Soft Matter, 2008, 4(12): 2492-2499.

[8] Liu S, Yan Q, Tao D, et al. Highly flexible magnetic composite aerogels prepared by using cellulose nanofibril networks as templates[J]. Carbohydrate polymers, 2012, 89(2): 551-557.

[9] Zhang J, Cao Y, Feng J, et al. Graphene-oxide-sheet-induced gelation of cellulose and promoted mechanical properties of composite aerogels[J]. The Journal of Physical Chemistry C, 2012, 116(14): 8063-8068.

[10] Gawryla M D, van den Berg O, Weder C, et al. Clay aerogel/cellulose whisker nanocomposites: a nanoscale wattle and daub[J]. Journal of Materials Chemistry, 2009, 19(15): 2118-2124.

[11] Sehaqui H, Zhou Q, Berglund L A. High-porosity aerogels of high specific surface area prepared from nanofibrillated cellulose (NFC)[J]. Composites science and technology, 2011, 71(13): 1593-1599.

[12] Deng M, Zhou Q, Du A, et al. Preparation of nanoporous cellulose foams from cellulose-ionic liquid solutions[J]. Materials Letters, 2009, 63(21): 1851-1854.

[13] Tsioptsias C, Stefopoulos A, Kokkinomalis I, et al. Development of micro-and nano-porous composite materials by processing cellulose with ionic liquids and supercritical CO_2[J]. Green Chemistry, 2008, 10(9): 965-971.

[14] Oh S Y, Yoo D I, Shin Y, et al. Crystalline structure analysis of cellulose treated with sodium hydroxide and carbon dioxide by means of X-ray diffraction and FTIR spectroscopy[J]. Carbohydrate research, 2005, 340(15): 2376-2391.

[15] Sehaqui H, Salajková M, Zhou Q, et al. Mechanical performance tailoring of tough ultra-high porosity foams prepared from cellulose I nanofiber suspensions[J]. Soft Matter, 2010, 6(8): 1824-1832.

中文题目：用纤维素-NaOH/PEG溶液制备机械强度高、重量轻的纤维素气凝胶

作者：万才超，卢芸，焦月，曹军，孙庆丰，李坚

摘要：通过以下四个步骤成功制备了新型机械强度高、重量轻的纤维素气凝胶。①将竹纤维溶解在温和的NaOH/PEG溶液中；②冻融处理；③再生；④冷冻干燥。实验结果表明，该气凝胶具有致密的相互连接和层次分明的孔隙结构，比表面积高达204 $m^2 \cdot g^{-1}$，大孔体积为0.99 $cm^3 \cdot g^{-1}$，高孔隙率高达97%，低密度为0.054 $g \cdot cm^{-3}$，呈现纤维素Ⅱ型晶体结构。此外，该气凝胶还表现出较强的抗压能力，杨氏模量高达1.85 MPa，屈服应力为83.57 kPa，韧性为52.34 kJ m^{-3}。

关键词：纤维素气凝胶；冷冻干燥；力学性能；多孔材料

Fabrication and Characterization of Nanofibrillated Cellulose and Its Aerogels from Natural Pine Needles*

Shaoliang Xiao[1], Runan Gao[1], Yun Lu[2*], Jian Li[1*], Qingfeng Sun[1*]

Abstract: To obtain the nanofibriled cellulose from natural pine needles, a combination of chemical pretreatments and subsequently ultrasonic treatments was employed for removing the hemicelluloses and lignins and splitting the bundled cellulose into pine needle nanofibres. Using SEM and diameter distribution method, it was confirmed that the obtained pine needle nanofibres had a narrow diameter from 30-70 nm. The crystalline type of the pine needle nanofibres was the cellulose Ⅰ type. The crystallinity reached 66.19%, which was increased by 7.61% as compared with the raw material pine needles. The TGA and DTG results showed that the degradation temperature of the nanofibres was increased to approximately 267 ℃ and 352 ℃ compared with 221 ℃ and 343 ℃ of the raw material fibres, respectively. Furthermore, the highly flexible and ultralight pine needle nanofibres aerogels were prepared from the aqueous pine needle nanofibres solution using the freezing-drying technique. Aerogels were studied by SEM observation and nitrogen gas adsorption. The mechanical properties were measured in compression for aerogels. This study provides a new opportunity to fabricate novel nanomaterials from waste biomass materials, which is crucial for the fully utilising of abundant biomass resources.

Keywords: Aerogels; Nanofibrillated cellulose; Pine needles; Ultrasonic; Hydrophobic

1 Introduction

It is well-known that cellulose, which accounts for approximately 40% of plant biomass, is the most abundant biopolymer in the world. Interest has recently arisen in the utilization of cellulose due to its unique characteristics, such as biodegradability, biocompatibility, renewability, and sustainability (Ifuku & Saimoto, 2012). Especially an increasing attention is being devoted to the nanometer-sized single cellulose fibres, generally defined as fibres with a diameter below 100 nm (Li & Xia, 2004; Sun et al., 2003), because of their intrinsic properties, such as the nanoscale dimensions, high surface area, unique morphology, low density and high mechanical strength (Habibi et al., 2010). A great number of applications of nanometer-sized single cellulose fibres, such as the controlled drug release formulations (Valo et al., 2013), food presentation (Klemm et al., 2011), tissue engineering (Bodin et al., 2007), cosmetics (Klemm et al., 2006), biodegradable packaging materials (Khan et al., 2010), reinforcement components in flexible display panels and oxygen-barrier layers (Shinoda et al., 2012), were undertaken. Therefore, the development of effective methods for extracting nanofibres from biomass has

* 本文摘自 Carbohydrate Polymers, 2015, 119: 202-209.

received an arising interest. The existing approaches include mechanical , chemical and biological treatments (Chen et al. , 2009), TEMPO-mediated oxidation on the surface of microfibrils and a subsequent mild mechanical treatment, electrospinning methods (Zhang & Yu, 2014), and ultrasound treatment (Huang et al. , 2003; Kim et al. , 2006). Among these methods, ultrasound for materials synthesis has been sufficiently investigated and is considered as one of the most powerful tools in nanostructured materials synthesis (Bang & Suslick, 2010; Zeiger & Suslick, 2011). The chemical effect of ultrasonication is caused by acoustic cavitation, which generates localised hot spots with very high temperatures (>5000 K), pressures (>20 MPa), and heating/cooling rates (>1010K/s) (Suslick, 1998). Such extreme environments provide a unique platform to break the strong cellulose interfibrillar hydrogen bonding, allowing nanofibres to be gradually disintegrated (Cheng et al. , 2010; Tischer et al. , 2010; Zhao et al. , 2007). Furthermore, this isolation technology may be universally applicable to all the biomass resources consisting of nanofibres and other embedding matrixes (Lu et al. , 2013). Accordingly, this study investigates the extraction of natural pine needle nanofibres from natural pine needles under pulsed ultrasonication.

Pine needles are a renewable natural bioresource and abundant across all China. With several advantages including rapid growth, renewability, relatively high strength, and good flexibility, pine needles can be harvested all the year round and one ton needles can be collected in two acres of pine generally (Thakur & Singha, 2011). Pine needle is composed of approximately 30% cellulose, 23% lignin and others substrates. The previous studies on cellulose nanofibres that were extracted from the pine needles by a Chemical-mechanical technique were reported to examine their potential applications.

Plant cell walls usually consist of rigid cellulosic microfibrils embedded in the soft hemicelluloses and lignin matrix. Cellulose is the fibrillar component of plant cells, which are, however, tightly hooked to one another by multiple hydrogen bonds. Moreover, cellulose nanofibres are embedded in matrix substances such as hemicellulose and lignin in cell walls. Thus it is difficult to split cellulose fibrils only by mechanical treatments (Saito et al. , 2007). Many researchers reported that the matrix substances were removed using chemical methods before the fibrillation process and kept the material in the water-swollen state. As a result, the narrow cellulose nanofibres can be obtained. Herein, a successful process of cellulose fibre nano-fibrillation from pine needles using the facile pulsed ultrasonication combined with chemical pretreatments was reported. The process could yield high-crystallinity and high-quantity pine needle nanofibres with a narrow width ranging from 30-70 nm (Abe et al. , 2007; Alemdar & Sain, 2008b; Lu et al. , 2013). This high-intensity ultrasonication treatment can separate nanofibres from natural nanofibre-embedding matrixes after the removal of the matrix substances. More interestingly, ultralight and highly flexible cellulose nanoporous aerogels were also prepared from narrow diameter distribution nanofibres using vacuum freezing-drying technique. Additionally, Cervin et al. reported a type of hydrophobic and oleophilic NFC aerogels for oil/water separation *via* a vapor phase deposition method, which is considerably simple, feasible and facile. (Cervin et al. , 2012). In the present work, the hydrophilic and the hydrophobic NFC aerogels from natural pine needles were both fabricated and their own features were also characterized. The hydrophobic NFC aerogel from natural pine needles showed superior oil/water separation.

2 Materials and methods

2.1 Raw materials

The pine needle powder with a 60 mesh was used for the analysis of cellulose nanofibres. 2 g dried powder was extracted with benzene / ethanol (2 : 1, v : v) in soxhlet 90 ℃ for 6 h. The dewaxed samples were dried in an oven for 24 h at 50 ℃ before use. Potassium hydroxide (KOH), hydrochloric acid (HCl), ethanol, benzene, acetic acid, sodium chlorite, trimethylchlorosilane and other chemicals used were of the analytical or reagent grade.

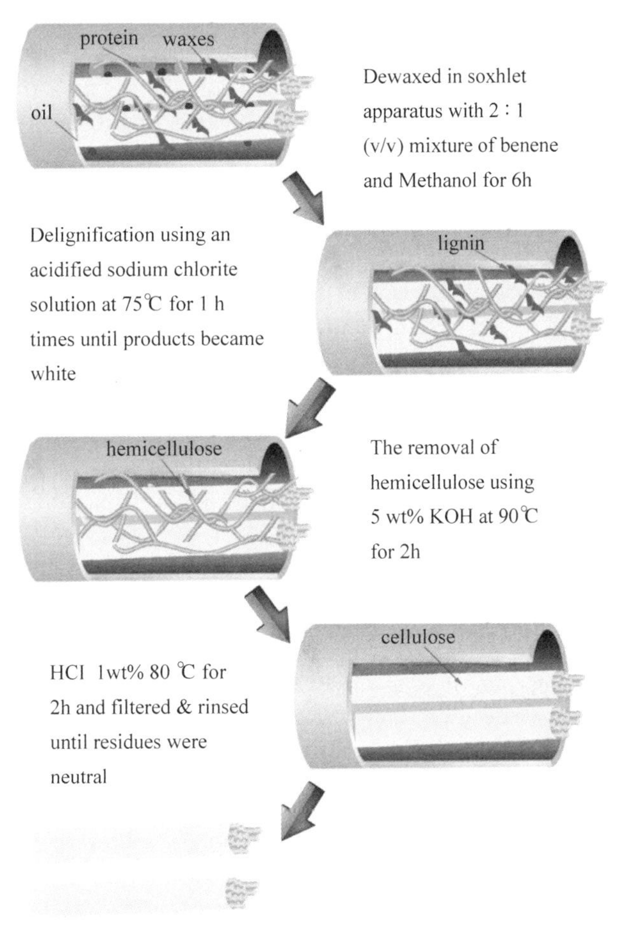

Fig. 1 The isolation process of a series of chemical treatment

2.2 Preparation of pine needle nanofibres

The isolation process consist of a series of chemical treatments in Fig. 1 and high-intensity ultrasonication according to the flowchart shown in Fig. 2. First, acidified sodium chlorite treatment was carried out at 75 ℃ for 1 h, which was repeated five times until the product became white. Following the method of Abe & Yano (2009), the major lignin was removed. Next, an alkaline treatment with potassium hydroxide (KOH) was conducted for removing the hemicelluloses, residual starch and pectin. Then the samples were treated using the hydrochloric acid (HCl) at 80 ℃ for 2 h. After the series of chemical treatments, the samples were filtered and rinsed with distilled water until the residues were neutralized. The samples were kept in a water-swollen state during the whole chemical process to avoid generating strong hydrogen bonding among nanofibres after the matrix was removed.

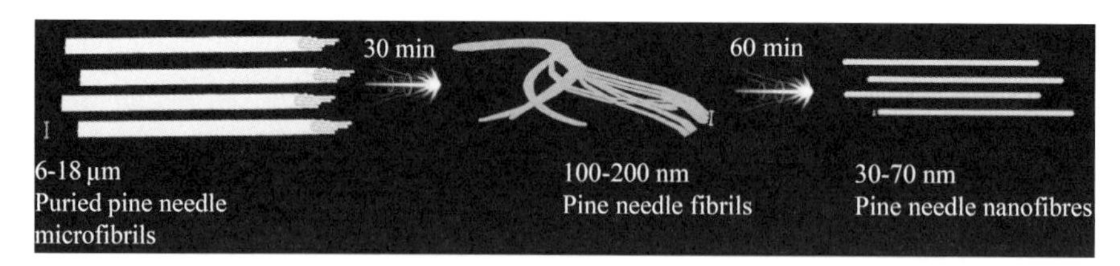

Fig. 2 The individualization of pine needle nanofibers from pine needle powers with a typical laboratory-scale ultrasonication apparatus under neutral condition

The purified pine needle microfibrils with a diameter ranging from 6-18 μm were split to thinner pine needle fibrils after 30 min ultrasonic treatment along the axes of the fibres. All these pine needle fibrils are further separated to 30-70 nm width nanofibres.

After the chemical pretreatment, to facilitate nanofibrillation, the purified cellulose fibres were dispersed in distilled water such that approximately 450 mL of the water slurry contained 0. 5 wt% samples. Sonication was performed at 60 kHz with a Sonifiers ® Cell Disruptor/Homogeniser (JY99-IID, Scientz Technology, China) with a 1-cm^2-diameter titanium horn under a 50% duty cycle (i. e. , a repeating cycle of 0. 5-s ultrasonic treatment and 0. 5-s shutdown) to reduce temperature variation. The sonication then was conducted for 1 h with an output power of 1500 W to isolate the fibres. The ultrasonic treatment was carried out in an ice/water bath, and the ice was maintained throughout the entire ultrasonication time.

2. 3 Preparation of pine needle nanofibres aerogels

The pine needle nanofibres suspensions were poured into moulds and then placed in a refrigerator. Next, the frozen samples were freeze-dried using a Scientz-10N freeze-dryer (BT6K-ES, Virtis) to sublime the materials directly from the solid phase to the gas phase. The cold trap temperature was below −55 ℃ and the vacuum pressure was below 25 μPa during the freeze-drying process. The freeze-dried pine needle nanofibre aerogels were used for the characterization.

2. 4 Hydrophobic coating of pine needle nanofibres aerogels

The hydrophilic pine needle nanofibres aerogels were obtained by the treatment of trimethylchlorosilane through the vapor phase deposition. The aerogels were placed in big glass bottle. A small glass vial containing trimethylchlorosilane was added into the glass bottle. Within the desiccator, the sample and trimethylchlorosilane allowed to react for 48 h at ambient temperature. This process successfully fabricated hydrophobic aerogels.

2. 5 Characterization

Surface morphology of specimens was characterized by the field emission scanning electron microscopy (SEM, FEI, Sirion 200) operating at 20. 0 kV. The chemical analysis was accomplished using the energy dispersive X-ray spectroscopy (EDXA) in connection with SEM and by FTIR spectroscopy (MagnaIR 560, Nicolet, KBr method). X-ray diffraction (XRD: Bruker D8 Advance, Germany) operating with Cu-K αradiation (λ = 1. 5418 Å) at a scan rate of 4 °min^{-1}, 40 kV, 40 mA ranging from 5 ° to 60 °. The thermal stability was characterized by TG analyzer (TA, Q600) from 25 ℃ to 800 ℃ with a heating rate of 10 ℃ · min^{-1} under nitrogen atmosphere. Water contact angle (WCA) was measured on an OCA40 contact angle system (Dataphysics, Germany) at room temperature. The final value of the CA was obtained as an average of five measurements. The specific surface area using Brunauer-Emmett-Teller (BET) analysis and nitrogen adsorption/desorption isotherms of the aerogels were determined by Brunauer-Emmett-Teller (BET) analysis *via* an accelerated surface area and porosimetry system (3H-2000PS2 unit, Beishide Instrument S & T Co. Ltd.). Approximately 0. 1 g of aerogel sample was first degassed in instrument at 105 ℃ for 4 h prior to the analysis followed by N_2 adsorption at -196 ℃. The BET surface area and the BJH pore size distribution were calculated. The compressive strength of the aerogel samples with a cylindrical shape was evaluated by an electronic universal mechanical testing machine (RGT-20A, Qingdao, China), which is equipped with a load cell of 500 N load cell at a strain rate of 10% min^{-1} in a conditioned room at 24 ℃ and 64% relative humidity.

3 Results and discussion

3.1 Examination of pine needle nanofibres

Fig. 3a shows the microstructure of the cross-section of the original materials. The microstructure of the pine needle was of an outer epidermis and the cellular structure. The epidermis was rich in cellulose (Hornsby et al., 1997). Intercellular material consisted of lignin, hemicelluloses and extractives (waxes, oil, pectin etc), which were the cementing materials around the fibre-bundles. Fig. 3b shows the structure of the pine needle fibres after the chemical treatment. The chemical-purified cellulose with cleaner surface suggested that the hemicelluloses, lignin and extractives were partially removed after a series of chemical pretreatments in Fig. 1. The pine needle fibres were separated into individual micro-sized fibres with the mean diameter of 12. 43 μm (Fig. 2e). In fact, these microsized fibres were reportedly composed of strong hydrogen bonding nanofibres, and individualized nanofibre bundles with a width of 30-70 nm can be seen in Fig. 3f. Fig. 3c shows SEM images of the sample after ultrasonic treatment, yielding plentiful slender fibrils. It is clear that from the image the average diameter of the fibres is about 50. 91 nm (Fig. 3f). The insert graph of Fig. 3a shows the nanofibres without any chemical treatment which was directly by treated by the ultrasonication. Furthermore, the morphology structure of aerogels before and after vapor phase deposition are shown in Fig. S1. The NFCs apparently were interconnected with each other and formed a typical porous three dimensional network (3D) structure, which suggested the NFC aerogel would be a potential substrate for 3D functional materials (Cervin et al., 2013; Zhang et al., 2014). The elemental compositions in the samples were detected by an energy dispersive X-ray spectroscopy (EDXA). The EDXA spectra (as shown in Fig. 2d, insets in Fig. 3e and f) show the peaks for C, N, O, K, Ca, Mg, and P according to their binding energies, respectively. The main components contain C (48. 77 wt%), O (26. 87 wt%) and N (18. 54 wt%) in the raw material. The inorganic elements in the raw material contain K (1. 13 wt%), Ca (1. 44 wt%), Mg (0. 31 wt%), and P (2. 95 wt%). Additionally, the peaks of N, K, Ca, Mg and P disappeared in the EDXA spectra of the samples (insets in Fig. 3e and f), indicating that these substances had been completely removed during the chemical pretreatment process.

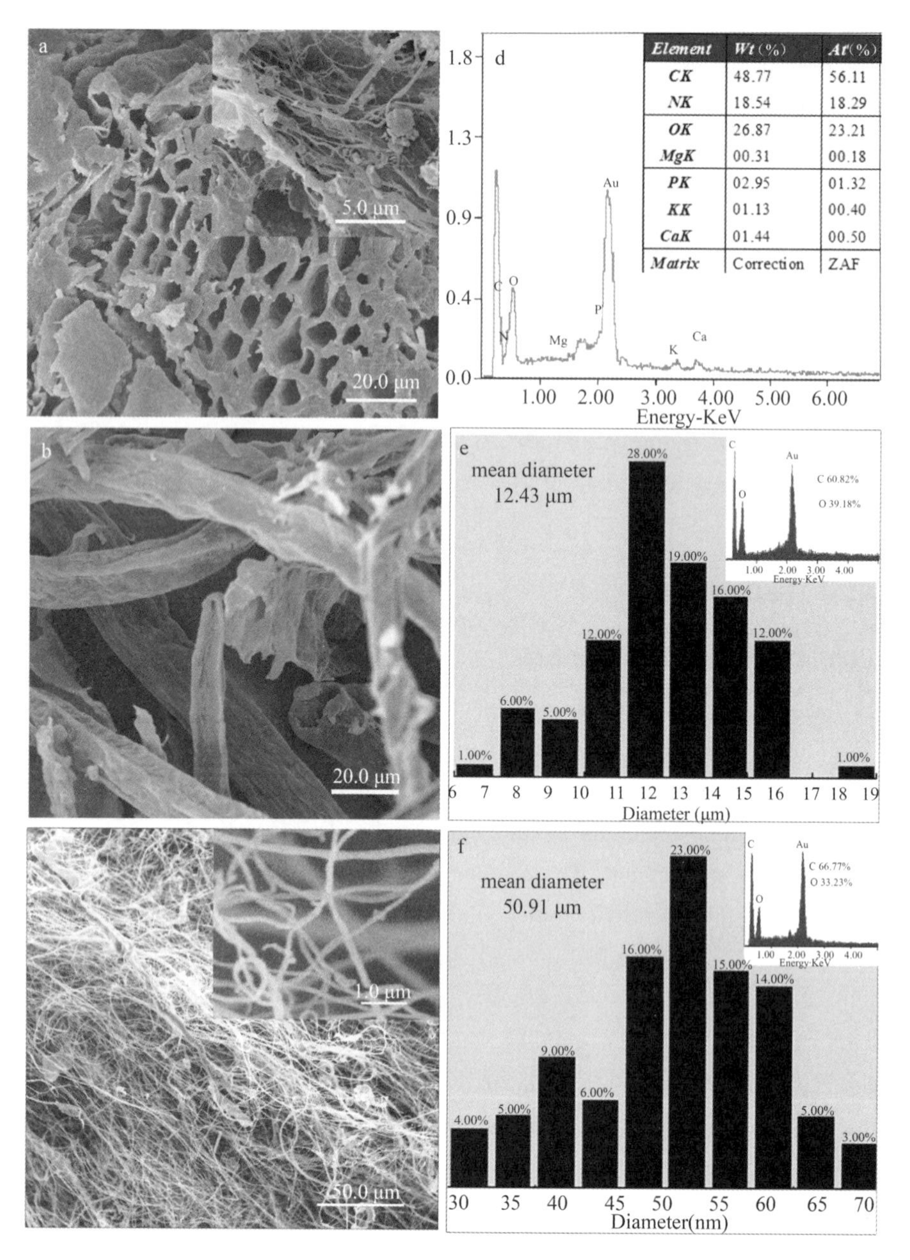

Fig. 3 SEM micrographs of (a) the raw materials, the insets of SEM graphs are the raw materials nanofibres using directly ultrasonic treatment without any chemical pretreatment. (b) the chemical-purified cellulose. (d) the energy dispersive X-ray spectroscopy (EDXA) of the raw materials, (e) and (f) are the corresponding diameter distributions of the fibres of (b), (c) and insets are corresponding the EDS spectra, respectively

Their compact agglomeration of pine needle nanofibres showed that cellulose chains had an intermolecular hydrogen bonding and a strong hydrophilic interaction in between the cellulosic chains. It may be explained by the effect of acoustic cavitation of high frequency (20-25 kHz) ultrasound in the formation, expansion, and implosion of microbubbles in aqueous solution. This violent collapse caused direct particle-shock wave interactions and was the primary pathway to split pine needle cellulose fibres along the axial direction. Thus, the sonification impact broke the relatively weak pine needle interfibrillar hydrogen bonding and the Van der Waals force, gradually disintegrating the micro-scale pine needle fibres into nanofibres. (Lu et al., 2013; Zeiger & Suslick, 2011; Zhao et al., 2007) The ultrasonic treatment method was an efficient pathway to prepare ultrafine nanofibrils from pine needle microfibres.

3.2 Characterizations of the samples

Cellulose crystallinity in the samples played a key role in the determination of the mechanical and thermal properties. To obtain the crystallinity of the samples and analyse the effect of chemical pretreatment and ultrasonic treatment on the reorganization of pine needle nanofibres, the XRD examination was carried out. Fig. 4 shows the diffractogrames of (a) the raw materials, (b) chemical-purified cellulose fibres, and (c) nanofibres. All samples exhibited sharp diffraction peaks around $2\theta = 14.8°$ (101), 16.5° (101) and 22.5° (002) (Kim et al., 2006), which are supposed to represent the typical cellulose-Ⅰ structure (Nishiyama et al., 2003). This suggested that the crystal structure of cellulose was not changed during the preparation. Moreover, Fig. 4a shows the XRD pattern of the raw materials. The diffraction peaks at $2\theta = 14.75$ (200), 24.37 (004), 29.86 (214), 31.14 (124), 35.45 (040), and 38.10 (305) can be assigned to the phase of $Mg(NH_4)_2H_2(PO_4)_2 \cdot 4H_2O$ (JCPDF 16-0353), $C_2CaO_4 \cdot H_2O$ (JCPDF 20-0231) and $K_2CaP_4O_{12}$ (JCPDF 29-0992). These mineral crystals were all removed from the raw materials by the potassium hydroxide (KOH) and the hydrochloric acid (HCl) treatment. The crystallinity of each sample was also calculated according the Segal method and is listed in Table. 1. The crystallinity values were estimated as 58.58%, 71.14%, 66.19% for (a) the raw materials, (b) chemical-purified cellulose fibres, and (c) nanofibres, respectively. The increase in crystallinity after chemical pretreatment was due to the removal of the most hemicelluloses which exist in the amorphous regions and was reported by several authors (Alemdar & Sain, 2008b; Li et al., 2009). This leaded to the realignment of cellulose molecules. However, after ultrasonic treatment, the crystallinity of nanofibres was lower than that of the chemical-purified cellulose fibres, which implied that the ultrasonic treatment had an effect on the destruction of the crystal regions in the

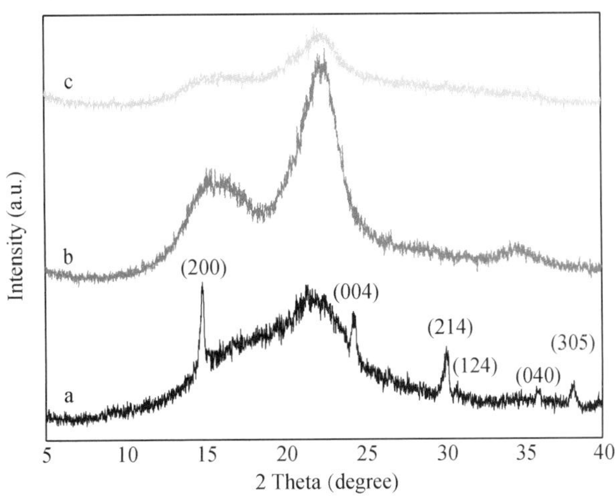

Fig. 4 X-ray diffraction patterns of (a) the raw materials, (b) chemical-purified cellulose fibres, and (c) pine needle nanofibres

cellulose nanofibres. It was probably explained that the energy release during the cavitation processes made the aligned microfibril unparallel and transferred to the hydrogen junction zones along the microfibril direction (Chen et al., 2013). The increase in the number of crystallinity regions increased the rigidity of cellulose, leading higher tensile strength of the fibres. Therefore it was more effective for providing better reinforcement for composite materials using these nanofibres (Alemdar & Sain, 2008b; Bhatnagar & Sain, 2005; Rong et al., 2001).

Table 1 The crystallinity of the pine needle cellulose sample

Sample	I_{002}	I_{am}	relative crystallinity(%)
The raw materials	458	324	58.58
Chemical-purified cellulose	742	301	71.14
Pine needle nanofibres	276	141	66.19

FT-IR spectroscopy is a nondestructive method for examining the physico-chemical properties of lignocellulosic materials. The FT-IR spectra of (a) the raw materials, (b) purified pine needle fibres, (c) pine needle nanofibres (d) the trimethylchlorosilane treated NFC aerogel are shown in Fig. 5. Furthermore, the characteristic absorption peaks from the above samples were summarized in Table 2 (Esteves et al., 2013; Kahar, 2013; Pandey, 1999; Liu et al., 2009; Yang et al., 2007). The dominant peaks in the region between around 3325-3340 cm^{-1} and 2889-2915 cm^{-1} were due to stretching vibrations of O—H and C—H (Oh et al., 2005). The peak at 1509 in the spectrum of (a) the raw materials, which was attributed to Aromatic C—O stretching mode for lignin and guaiacol ring of lignin, disappeared completely in (b) purified pine needle fibres and (c) pine needle nanofibres. This indicated that the lignin was well removed from the newly prepared pine needle nanofibres by the $NaClO_2$ treatment. The prominent peak at 1727 cm^{-1} in the (a) the raw materials was attributed to either the acetyl and uronic ester groups of the hemicelluloses or the ester linkage of carboxylic group of the ferulic and p-coumeric acids of the

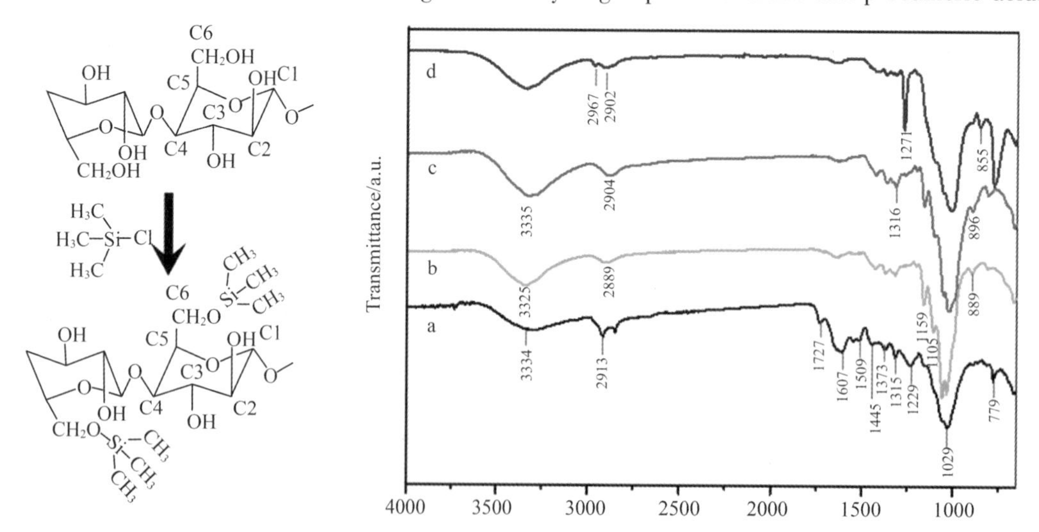

Fig. 5 FT-IR spectra of (a) the raw materials, (b) chemical-purified cellulose fibres, (c) pine needle nanofibres, (d) the trimethylchlorosilane treated NFC aerogel

lignin and/or hemicelluloses. This peak disappeared completely in the chemically treated pine needles because of the removal of the most hemicelluloses and lignin from the pine needles by applying the chemical extraction (Sain & Panthapulakkal, 2006). Interestingly, no difference between the spectrum of cellulose nanofibres obtained under different ultrasonic output powers and chemical-purified cellulose fibres was found. This result suggested that the molecular structures of cellulose remained unchanged in the case of ultrasonic treatment. The increase of the band at 889-896 cm^{-1} in the chemically treated pine needles illustrates the typical structure of cellulose. Furthermore, after the surface modification, (d) the trimethylchlorosilane treated NFC aerogel showed visible absorption peaks at 2967 cm^{-1}, 1271cm^{-1} and 855 cm^{-1}, which correspond to —CH_3 terminal groups. The adsorption peaks of the O—H bond weakened as the trimethylchlorosilane modified NFC aerogel became hydrophobic with fewer adsorbed water molecules. With this treatment, the —OH groups were replaced with —$OSi(CH_3)_3$ groups from trimethylchlorosilane to make the aerogels hydrophobic. (Wei, Lu & Chang, 2008)

Table 2 Frequencies (cm^{-1}) of the main signals of (a) the raw materials, (b) purified pine needle fibres, (c) pine needle nanofibres and the assignments

Absorption band (cm^{-1})	Assignment
3325-3340 (m)	O—H stretch (hydrogen-bonded)
2889-2915 (m)	C—H stretching
1727 (m)	Alkyl esther from cell wall hemicellulose C═O; strong carbonyl groups in branched hemicellulose
1509 (m)	Aromatic C—O stretching mode for lignin; guayacyl ring of lignin; lignocellulose
1445 (m)	O—CH_3
1373 (m)	C—H stretch of cellulose
1315-1316 (m)	CH_2 wagging
1229 (m)	C—O—C stretching
1159 (m)	Antisymmetric stretching C—O—C glycoside; C—O—C β-1, 4 glycosil linkage of cellulose.
1105 (m)	C—O vibration of crystalline cellulose; glucose ring stretch from cellulose
1029-1035 (m)	C—O vibration of cellulose
889-896 (m)	Amorphous cellulose vibration; glucose ring stretch
700-900 (m)	C—H
9671271855 (m)	—CH_3

Investigating the thermal properties of reinforcing materials is important in order to identify their applicability for biocomposite processing at high temperatures (Alemdar & Sain, 2008a). The thermogravimetry (TGA) and derivative thermogravimetry (DTG) results obtained from (a) the raw material, (b) chemical-purified cellulose and (c) pine needle nanofibres are shown in Fig. 6. It was clearly that a small weight loss was found in the range of 25-120 ℃ due to the evaporation of the humidity of the materials or low molecular weight compounds remaining from the isolation procedures in all TGA and DTG curves (Kumar et al., 2014; Soares et al., 1995; Xiao et al., 2001). The onset of weight loss of the main step of degradation lain at 221 ℃, 267 ℃ and 312 ℃ for (a) the raw material, (b) chemical-purified cellulose and (c) pine needle nanofibres, respectively in TGA. In terms of thermal degradation, the three samples in the order of the easiest to the most difficult to degrade were a>c>b. The rate of degradation reached its peak at

343 ℃, 363 ℃ and 352 ℃ for (a) the raw material, (b) chemical-purified cellulose and (c) pine needle nanofibres, respectively in DTG. The thermal stability of (b) chemical-purified cellulose and (c) pine needle nanofibres is higher than that of (a) the raw material from the result of data analyses. It can be concluded that the higher temperature of thermal decomposition and lesser obtained residual mass of the fibres was contributed to the removal of hemicelluloses (Hemicellulose appears a random, amorphous structure, rich of branches, which are very easy to be removed from the main stem and to degrade to volatiles evolving out at low temperatures.) and lignin (Lignin is full of aromatic rings with various branches.) from the fibres and higher crystallinity of the cellulose after chemi-mechanical treatment (Nguyen et al., 1981; Yang et al., 2007). Furthermore, thermal degradation of pine needle nanofibres occurred at a lower temperature within broader ranges of temperature and exhibited lower thermal stability than chemical-purified cellulose due to their nano-sizes, greater number of free ends in the chain of pine needle nanofibres, and may show drastic reduction in the molecular weight (Kumar, Negi, Choudhary & Bhardwaj, 2014). Therefore, the nanofibres exhibited enhanced thermal properties, making them promising candidates for the application in thermoplastic composites.

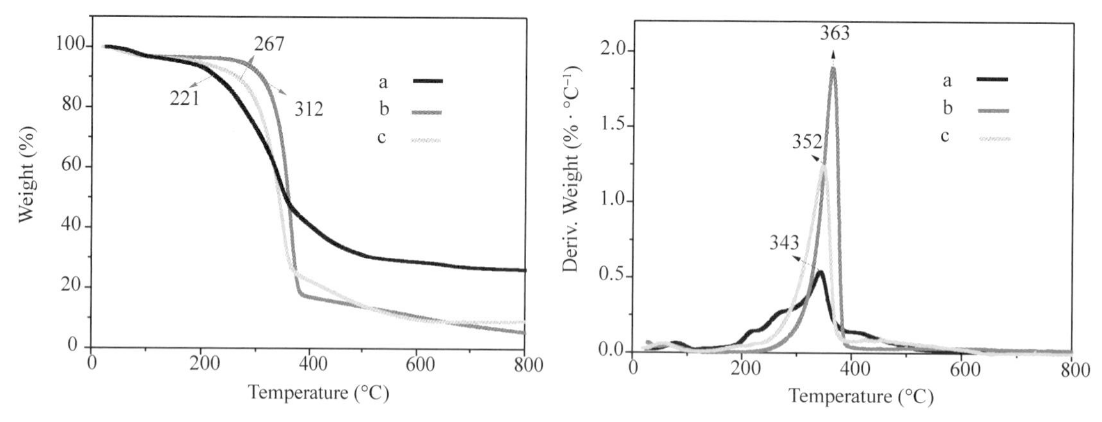

Fig. 6 TGA and DTG thermograms of (a) the raw material, (b) chemical-purified cellulose fibres and (c) pine needle nanofibres, under nitrogen atmosphere

3.3 Fabrication of nanofibre ultralight flexible aerogels and hydrophobic nanofibre aerogels

The condensed cellulose nanofibrillated hydrogels were collected. After freeze-drying the hydrogels, highly flexible nanofibre aerogels were fabricated. Fig. 7a shows nanofibre highly flexible aerogels, which can be repeatedly bent without destroying their structural integrity. Fig. 7b exhibits ultralight nanofibre aerogels, which had a white appearance and a bulk density of 3.12 mg · cm^{-3}. Thus, the separated nanofibrils can be easily stored and transported. These findings also verified that the highly flexible and low-density nanofibre flexible aerogels containing pine needle nanofibres were successfully prepared. Such nanofibre aerogels are expected to be used in various fundamental and applied research fields, such as nanofibre templates for hollow inorganic tubes, water purification, catalysis, sensing, separation, filtration, tissue engineering, and packaging materials etc. (Deng et al., 2009).

The wettability of the uncoated and the hydrophobic aerogel were evaluated by static WCA measurements (Fig. 8.). WCA of the pristine aerogel was 5° according to (a). The WCA of the samples treated with

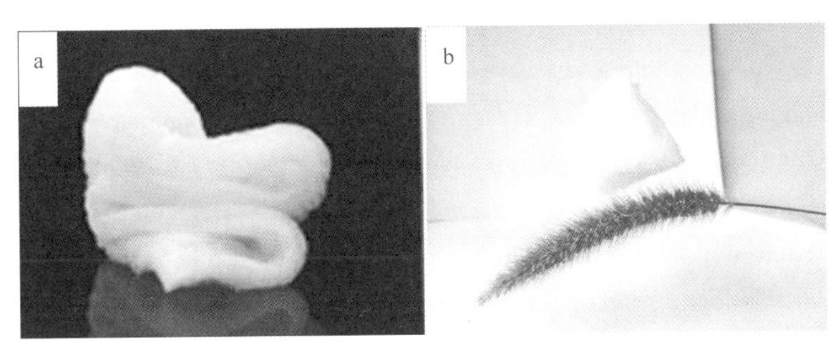

Fig. 7 (a) The nanofibre highly flexible aerogels were bent and folded without destroying the ristructural integrity, (b) the nanofibre ultralight aerogels were placed on dog tail grass

trimethylchlorosilane increased to 135° according to (b), which is an unmistakable sign for hydrophobicity. The modification by trimethylchlorosilane contributed to the formation of a low-energy surface of the cellulose aerogel due to the replacement of surface hydroxyl groups on the cellulose with plentiful —Si—$(CH_3)_3$ groups. Moreover, it showed that (c) photographs of the coated aerogel was pressed into water and (d) floated on the water surface. The treated aerogel had hydrophobic for many other types of polar solvent such as the static (e) milk, (h) tea water, (f) drop, (i) RhB and (g) methyl orange in Fig. 8. Especially, it can rest on water droplet and at the same time absorb oil [(h) cooking oil and drop]. The trimethylchlorosilane treated NFC aerogel might be an ideal candidate for highly efficient water-oil separation substances for oil spills (Cervin et al., 2012; Xue et al., 2014).

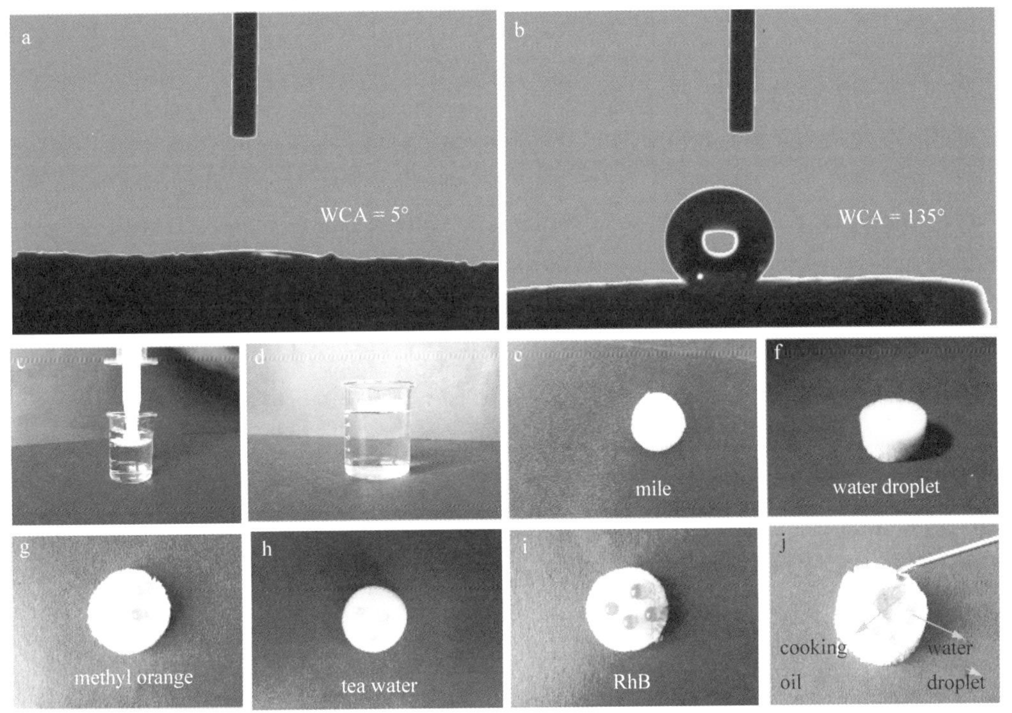

Fig. 8 Water contact angle (a) on the surface of the uncoated aerogel and (b) on the surface of the coated aerogel; (c) photographs of the coated aerogel was been pressed into water and (d) floated on the water surface; photographs of the static (e) milk, (f) water droplet, (g) methyl orange, (h) tea water, (i) RhB, and (j) cooking oil on the coated aerogel

Fig. S2 shows the nitrogen adsorption-desorption isotherms, BJH pore size distribution, the specific surface areas, total pore volume and average pore diameter of NFC aerogels. In Fig. S2, NFC aerogels exhibit type IV adsorption isotherms according to the IUPAC classification, which indicate the presence of mesopores (Gregg et al., 1967; Rouquerol et al., 1994). The desorption cycles of the isotherms exhibited a hysteresis loop which is generally attributed to the capillary condensation occurring in the mesopores. The pore diameter distributions was evaluated from the desorption branch of nitrogen isotherms by using the Barrett-Joyner-Halenda (BJH) method. The pore sizes of NFC aerogels was within the range 1 nm to 60 nm, and mainly consisted of micropores (<2 nm) and mesopores (2-50 nm). The specific surface areas, average pore diameter and total pore volume of NFC aerogel were 20.0927 $m^2 \cdot g^{-1}$, 25.4465 nm, and 0.07762 $m^3 \cdot g^{-1}$, respectively.

Mechanical property of NFC aerogels are summarized in Table S1. The NFC aerogels exhibit ductile behavior, can be compressed to large strains (>70%, see Fig. S3) and have low compression strength (37kPa). As shown in Fig. S4, the modified NFC aerogel could stably float on the surface of the water. Subsequently, the modified aerogel could persistently stay on the oil layer under gravity after being put in the oil/water mixture. The oil absorbed cellulose aerogel is removed easily by simply squeezing it. The aerogels combine both hydrophobic and oleophilic properties and prove to be very efficient in removing oil from a water surface with an excellent selectivity and recyclability. Moreover, this aerogel could absorbed 52 times oil than that of its own weight without any collapse in oil (see in detail Movie S1).

4 Conclusions

In the present work, cellulose nanofibres were extracted from pine needles by chemical-mechanical treatments. Experimental results showed that the diameters of the produced pine needle nanofibres were within the range of 30-70 nm. Chemical analysis and FTIR measurements of the fibres revealed that the partial hemicelluloses and lignin were due to the success of the chemical treatment applied. The crystallinity of the cellulose was increased by 7.61% for the pine needle nanofibres. The nanofibres exhibited enhanced thermal properties, thus making them promising candidates for the use in thermoplastic composites. Furthermore, highly flexible and ultralight pine needle nanofibres and hydrophobic coating of pine needle nanofibres aerogels were manufactured.

Acknowledgments

This work was financially supported by The National Natural Science Foundation of China (grant no. 31270590), China Postdoctoral Science Foundation funded project (2013M540263), and Doctoral Candidate Innovation Research Support Program of Science & Technology Review (kjdb2012006).

References

[1] Abe K, Iwamoto S, & Yano H. Obtaining cellulose nanofibers with a uniform width of 15 nm from wood[J]. Biomacromolecules, 2007, 8(10), 3276-3278.

[2] Abe K, & Yano H. Comparison of the characteristics of cellulose microfibril aggregates of wood, rice straw and potato tuber. Cellulose, 2009, 16(6), 1017-1023.

[3] Alemdar A, & Sain M. Biocomposites from wheat straw nanofibers: Morphology, thermal and mechanical properties

[J]. Compos. Sci. Technol., 2008, 68(2), 557-565.

[4]Alemdar A, & Sain M. Isolation and characterization of nanofibers from agricultural residues-Wheat straw and soy hulls [J]. Bioresour. Technol., 2008, 99(6), 1664-1671.

[5]Bang J H, & Suslick K S. Applications of ultrasound to the synthesis of nanostructured materials. Adv Mater. [J], 2010, 22(10), 1039-1059.

[6]Bhatnagar A, & Sain M. Processing of cellulose nanofiber-reinforced composites[J]. J. Reinf. Plast. Comp., 2005, 24(12), 1259-1268.

[7]Bodin A, Ahrenstedt L, Fink H, etal. Modification of nanocellulose with a xyloglucan-RGD conjugate enhances adhesion and proliferation of endothelial cells: implications for tissue engineering[J]. Biomacromolecules, 2007, 8(12), 3697-3704.

[8]Cervin N T, Aulin C, Larsson P T, etal. Ultra porous nanocellulose aerogels as separation medium for mixtures of oil/water liquids[J]. Cellulose, 2012, 19(2), 401-410.

[9]Chen Y, Liu C, Chang P R, etal. Bionanocomposites based on pea starch and cellulose nanowhiskers hydrolyzed from pea hull fibre: effect of hydrolysis time[J]. Carbohydr. Polym., 2009, 76(4), 607-615.

[10]Cheng Q, Wang S, & Han Q. Novel process for isolating fibrils from cellulose fibers by high-intensity ultrasonication. II. Fibril characterization[J]. J. Appl. Polym. Sci., 2010, 115(5), 2756-2762.

[11]Deng M, Zhou Q, Du A, etal. Preparation of nanoporous cellulose foams from cellulose-ionic liquid solutions[J]. Mater. Lett., 2009, 63(21), 1851-1854.

[12]Esteves B, Velez Marques A, Domingos I, etal. Chemical changes of heat treated pine and eucalypt wood monitored by FTIR[J]. Maderas-cienc Tecnol., 2013.

[13]Habibi Y, Lucia L A, & Rojas O J. Cellulose nanocrystals: chemistry, self-assembly, and applications[J]. Chem. Rev., 2010, 110(6), 3479-3500.

[14]Hornsby P, Hinrichsen E, & Tarverdi K. Preparation and properties of polypropylene composites reinforced with wheat and flax straw fibres: part I fibre characterization[J]. J. Mater. Sci., 1997, 32(2), 443-449.

[15]Huang ZM, Zhang YZ, Kotaki M, et al. A review on polymer nanofibers by electrospinning and their applications in nanocomposites[J]. Compos. Sci. Technol., 2003, 63(15), 2223-2253.

[16]Ifuku S, & Saimoto H. Chitin nanofibers: preparations, modifications, and applications[J]. Nanoscale, 2012, 4(11), 3308-3318.

[17]Kahar P. Synergistic Effects of Pretreatment Process on Enzymatic Digestion of Rice Straw for Efficient Ethanol Fermentation. Environmental Biotechnology-New Approaches and Prospective Applications, M. Petre, ed. (InTech), 2013.

[18]Khan R A, Salmieri S, Dussault D, etal. Production and properties of nanocellulose-reinforced methylcellulose-based biodegradable films[J]. J. Agric. Food. Chem., 2010, 58(13), 7878-7885.

[19]Kim C-W, Kim D-S, Kang S-Y, etal. Structural studies of electrospun cellulose nanofibers[J]. Polymer, 2006, 47(14), 5097-5107.

[20]Klemm D, Kramer F, Moritz S, etal. Nanocelluloses: a new family of nature-based materials[J]. Angew. Chem. Int. Ed. Engl., 2011, 50(24), 5438-5466.

[21]Klemm D, Schumann D, Kramer F, etal. Nanocelluloses as innovative polymers in research and application[J]. Polysaccharides Ii (pp. 49-96): Springer, 2006.

[22]Kumar A, Negi Y S, Choudhary V, et al. Characterization of cellulose nanocrystals produced by acid-hydrolysis from sugarcane bagasse as agro-waste[J]. J. Mater. Phy. Chem., 2014, 2(1), 1-8.

[23]Li D, & Xia Y. Electrospinning of nanofibers: reinventing the wheel[J]. Adv. Mater., 2004, 16(14), 1151-1170.

[24]Li R, Fei J, Cai Y, etal. Cellulose whiskers extracted from mulberry: a novel biomass production[J]. Carbohydr.

Polym. , 2009, 76(1), 94-99.

[25]Lu Y, Sun Q, She X, et al. Fabrication and characterisation of α-chitin nanofibers and highly transparent chitin films by pulsed ultrasonication[J]. Carbohydr. Polym. , 2013, 98(2), 1497-1504.

[26]Nguyen T, Zavarin E, & Barrall E M. Thermal analysis of lignocellulosic materials: Part I. Unmodified Materials. Journal of Macromolecular Science—Reviews in Macromolecular Chemistry, 1981, 20(1), 1-65.

[27]Nishiyama Y, Sugiyama J, Chanzy H, et al. Crystal structure and hydrogen bonding system in cellulose Iα from synchrotron X-ray and neutron fiber diffraction[J]. J. Am. Chem. Soc. , 2003, 125(47), 14300-14306.

[28]Oh S Y, Yoo D I, Shin Y, et al. Crystalline structure analysis of cellulose treated with sodium hydroxide and carbon dioxide by means of X-ray diffraction and FTIR spectroscopy[J]. Carbohydr. Res. , 2005, 340(15), 2376-2391.

[29]Pandey K. A study of chemical structure of soft and hardwood and wood polymers by FTIR spectroscopy[J]. J. Appl. Polym. Sci. , 1999, 71(12), 1969-1975.

[30]Rong M Z, Zhang M Q, Liu Y, et al. The effect of fiber treatment on the mechanical properties of unidirectional sisal-reinforced epoxy composites[J]. Compos. Sci. Technol. , 2001, 61(10), 1437-1447.

[31]Sain M, & Panthapulakkal S. Bioprocess preparation of wheat straw fibers and their characterization[J]. Ind. Crops. Prod. , 2006, 23(1), 1-8.

[32]Saito T, Kimura S, Nishiyama Y, et al. Cellulose nanofibers prepared by TEMPO-mediated oxidation of native cellulose[J]. Biomacromolecules, 2007, 8(8), 2485-2491.

[33]Shinoda R, Saito T, Okita Y, et al. Relationship between length and degree of polymerization of TEMPO-oxidized cellulose nanofibrils[J]. Biomacromolecules, 2012, 13(3), 842-849.

[34]Soares S, Camino G, & Levchik S. Comparative study of the thermal decomposition of pure cellulose and pulp paper [J]. Polym. Degrad. Stab. , 1995, 49(2), 275-283.

[35]Sun Y, Mayers B, Herricks T, et al. Polyol synthesis of uniform silver nanowires: a plausible growth mechanism and the supporting evidence[J]. Nano Lett. , 2003, 3(7), 955-960.

[36]Suslick K S. Sonochemistry. Kirk-Othmer Encyclopedia of Chemical Technology, 1998.

[37]Thakur V, & Singha A. Physicochemical and mechanical behavior of cellulosic pine needle-based biocomposites[J]. Int. J. Polym. Anal. Ch. , 2011, 16(6), 390-398.

[38]Tischer P C F, Sierakowski M R, Westfahl Jr H, et al. Nanostructural reorganization of bacterial cellulose by ultrasonic treatment[J]. Biomacromolecules, 2010, 11(5), 1217-1224.

[39]Valo H, Arola S, Laaksonen P, et al. Drug release from nanoparticles embedded in four different nanofibrillar cellulose aerogels[J]. Eur. J. Pharm. Sci. , 2013, 50(1), 69-77.

[40]Xiao B, Sun X, & Sun R. Chemical, structural, and thermal characterizations of alkali-soluble lignins and hemicelluloses, and cellulose from maize stems, rye straw, and rice straw[J]. Polym. Degrad. Stab. , 2001, 74(2), 307-319.

[41]Xue Z, Cao Y, Liu N, et al. Special wettable materials for oil/water separation[J]. J. Mater. Chem. A Mater. , 2014, 2(8), 2445-2460.

[42]Yang H, Yan R, Chen H, et al. Characteristics of hemicellulose, cellulose and lignin pyrolysis[J]. Fuel, 2007, 86 (12), 1781-1788.

[43]Zeiger B W, & Suslick K S. Sonofragmentation of molecular crystals[J]. J. Am. Chem. Soc. , 2011, 133(37), 14530-14533.

[44]Zhao HP, Feng XQ, & Gao H. Ultrasonic technique for extracting nanofibers from nature materials[J]. Appl. Phys. Lett. , 2007, 90(7), 073112.

[45]Zhang C-L, & Yu S-H. Nanoparticles meet electrospinning: recent advances and future prospects[J]. Chem. Soc. Rev. , 2014.

[46] Liu H, Sha W, Cooper A T, et al. Preparation and characterization of a novel silica aerogel as adsorbent for toxic organic compounds[J]. ColloidsSurf. APhysicochem. Eng. Asp., 2009, 347(1), 38-44.

[47] Gregg S J, Sing K S W, & Salzberg H. Adsorption surface area and porosity[J]. J. Electrochem. Soc., 1967, 114 (11), 279C-279C.

[48] Rouquerol J, Avnir D, Fairbridge C, et al. Recommendations for the characterization of porous solids (Technical Report)[J]. Pure. Appl. Chem., 1994, 66(8), 1739-1758.

[49] Cervin N T, Andersson L A, Ng J B S, et al. Lightweight and strong cellulose materials made from aqueous foams stabilized by nanofibrillated cellulose[J]. Biomacromolecules, 2013, 14(2), 503-511.

[50] Zhang Z, Sèbe G, Rentsch D, Z, et al. Ultralightweight and flexible silylated nanocellulose sponges for the selective removal of oil from Water[J]. Chem. Mater., 2014.

[51] Chen P, Yu H, Liu Y, et al. Concentration effects on the isolation and dynamic rheological behavior of cellulose nanofibers via ultrasonic processing[J]. Cellulose, 2013, 20(1), 149-157.

[52] Wei TY, Lu S-Y, & Chang Y-C. Transparent, hydrophobic composite aerogels with high mechanical strength and low high-temperature thermal conductivities[J]. J. Phys. Chem. B., 2008, 112(38), 11881-11886.

中文题目：松针纳米纤维素及其气凝胶的制备与表征

作者：肖少良，高汝楠，卢芸，李坚，孙庆丰

摘要：为了实现农林废弃物的高效利用，本工作利用化学预处理和超声破碎相结合的方法从松针中制备出了纳米纤维素，制备出的松针纳米纤维素具有高长径比和高结晶度。进一步地通过冷冻干燥和化学气相沉积法制备了油水分离纤维素气凝胶并探讨其在油水分离领域的应用。该项工作为实现农林废弃物的高值化利用提供了新策略。

关键词：纳米纤维素；松针；超声破碎；疏水

Incorporation of Graphene Nanosheets into Cellulose Aerogels: Enhanced Mechanical, Thermal, and Oil Adsorption Properties*

Caichao Wan, Jian Li

Abstract: In this paper, novel graphene/cellulose (GC) aerogels were prepared based on a green NaOH/PEG solution. Scanning electron microscope (SEM) observation indicates that the three-dimensional network skeleton structure of cellulose aerogels is tightly covered by the compact sheet structure. X-ray diffraction (XRD) and Raman spectroscopy demonstrate that the graphene nanosheets have been successfully synthesized and embedded in the cellulose aerogels. The incorporation of graphene nanosheets gives rise to the significant improvement in the specific surface area and pore volume, thermal stability, mechanical strength, and oil adsorption efficiency of GC aerogels. Therefore, the green hybrid GC aerogels have more advantages over the pure cellulose aerogels in treating oil-containing wastewater or oil spills under the harsh environment.

Keywords: Graphene; Cellulose aerogels; Mechanical properties; Thermal stabilities; Oil adsorption.

1 Introduction

Cellulose aerogels are a class of important cellulose products, and have attracted much attention since the first report by Kistler in 1931s[1]. Their porous three-dimensional (3D) network structure endows themselves numerous merits like large specific surface area, high porosity, and low density[2-4]. Compared with inorganic aerogels, cellulose aerogels are gaining more interest due to their environment friendliness, biodegradability, and biocompatibility. As a result, cellulose aerogels have many potential applications such as catalyst carriers, adsorbents, biomedicines, tissue engineering, fuel cells and sound or heat insulation materials[5-8]. Cellulose aerogels are generally prepared by the cellulose dissolution, regeneration and dry process. However, the poor dissolution of cellulose in common aqueous or organic solution, which is attributed to the complexity of bio-polymeric network, the partially crystalline structure and the extended non-covalent interactions among molecules[9], hastens the development of cellulose solvents. Especially, among the various cellulose solvents, several green, nontoxic, and nonvolatile solvent systems are high-profile such as ionic liquid, *N*-methylmorpholine-*N*-oxide (NMMO), and some alkali-based systems like NaOH/PEG solution[10-13].

For their more widespread utilization, the critical weakness related to the inferior mechanical properties of cellulose aerogels, which is possibly derived from the defects such as dangling ends and loops in their structures[14], are the main challenges requiring urgent improvement. Besides, some author also acclaim that

* 本文摘自 Applied Physics A, 2016, 122: 105.

the fragility is ascribed to the weak hydrogen bonding interaction and the quick rearrangement of cellulose chains[15]. Pre-gelation processing of cellulose solution is an interesting approach to enhancing the mechanical properties of cellulose aerogels, which is conducted by storing cellulose solution at pre-gelation temperatures leading to the formation of network units with stronger crosslinking structure[15,16]. However, the pre-gelation process is usually extraordinarily time-consuming, and the improvement in mechanical properties is also rather limited. In contrast, incorporating with reinforcement additives is relatively wider technique to improve the mechanical properties of cellulose aerogels. The reported fillers include cellulose whisker[17], clay[18], silica[19], organic polymers [e.g., polystyrene, poly(methyl methacrylate), and polypyrrole][20,21], metal nanoparticles or their related compounds (e.g., Ag and $CoFe_2O_4$)[22,23], etc. However, the exploration on the appropriate reinforcement additives still needs to be continued, which is attributed to not only the demand for improving mechanical strength of cellulose aerogels, but also various fantastic performances of the composites inherited from both aerogels and fillers.

Graphene is composed of sp2 bonded carbon atoms arranged in hexagonal pattern in a 2D plane, which has high aspect ratio, supernormal mechanical property, and intriguing transport phenomena. Therefore, graphene is wildly considered to be useful in different fields such as composites, optical devices, ultrasensitive sensors, and genie widget[24]. Moreover, graphene have been extensively reported as a kind of good filler for various polymer matrixes, and the resulting products exhibit many striking features in structure, strength and macro-micro performances[25,26]. Therefore, incorporation of graphene nanosheets into cellulose aerogels might effectively improve the inferior mechanical properties of cellulose aerogels, and introduce new properties.

Herein, we show a facile, simple and green preparation method of graphene/cellulose (GC) aerogels by incorporating graphene nanosheets into the cellulose aerogels matrix using the NaOH/PEG aqueous solution as the processing solvent. The resulting GC aerogels were subsequently characterized by scanning electron microscope (SEM), energy dispersive X-ray spectroscopy (EDX), N_2 adsorption measurements, X-ray diffraction (XRD), Raman spectroscopy, and thermogravimetric analysis (TGA). The mechanical properties were investigated using a mechanical tester. As a kind of green porous materials, the application potential of GC aerogels was also studied as eco-friendly oil adsorbent. The oil adsorption efficiency of GC aerogels was measured and compared with that of the pure cellulose (PC) aerogels.

2 Materials and methods

2.1 Materials

Waste coconut shell was grinded and then screened through a sixty-mesh sieve, and the resulting powder was subsequently collected and dried in a vacuum oven at 60 ℃ for 24 h before used. All chemical reagents with analytical grade were supplied by Shanghai Boyle Chemical Co. Ltd. (China) and used without further purification.

2.2 Preparation of graphene

Typically, graphene was synthesized from natural graphite powder by a modified Hummers method[27,28]. Briefly, graphite powder was first oxidized by concentrated nitric acid and sulfuric acid (1 : 2, v/v). The reaction was conducted in an ice bath, and potassium chlorate was added slowly. Then the mixture was heated to 65 ℃, and kept at this temperature for 120 h to complete the oxidation reaction. After being thoroughly washing and filtering by distilled water (until pH = 7), the graphite oxide was obtained. Thereafter, a mild two-hour ultrasonication treatment was carried out to exfoliate the graphite oxide in a mixture of ethanol and water. The resultant was subsequently reduced to graphene by hydrazine hydrate at 100 ℃ for 24 h. After being centrifuged, washed, and vacuum-dried, the graphene powder was obtained.

2.3 Preparation of GC aerogel

The cellulose was isolated from the coconut shell powder by a facile chemical treatment method, and the purified cellulose was then dissolved in the NaOH/PEG solution to form a 2 wt% homogeneous cellulose aqueous solution. The detailed process could be referred to our previous reports[29-31]. The graphene aqueous dispersion (0.2 mg · mL^{-1}) obtained by four-hour ultrasonication (500 W) was added in the aforementioned homogeneous cellulose solution at a 99.5 : 0.5 mass ratio of cellulose to graphene. Then, the mixture was magnetically stirred for 30 min, and immediately frozen at −15 ℃ for 3 h. After that, the frozen cake was subjected to the regeneration treatment by immersing it in a 1% hydrochloric acid for 6 h at room temperature, repeating this process until the formation of a cylindrical GC hydrogel. Finally, GC hydrogel was successively rinsed with distilled water and tert butyl alcohol several times to remove residual ions and water, and then underwent a freeze drying treatment at −35 ℃ for 48 h in an approximate vacuum (25 Pa), and the following GC aerogel was obtained. Moreover, PC aerogel was prepared following the aforementioned process without adding the graphene.

2.4 Characterizations

SEM images were obtained by a FEI Quanta 200 SEM coupled with EDX. N_2 adsorption measurements were carried out on a Micromeritics Tristar Ⅱ 3020 surface area analyzer. The specific surface area and pore size distribution were calculated by the Brunauer-Emmett-Teller (BET) and Barrett-Joyner-Halenda (BJH) methods. XRD technique was performed on a XRD instrument (D/max 2200, Rigaku) using Ni-filtered Cu Kα radiation (λ = 1.5406 Å) at 40 kV and 30 mA. Scattered radiation was detected in the range of $2\theta = 5°-70°$ at a scan rate of 4° · min^{-1}. Raman spectra were recorded using a Raman spectrometer (Renishaw inVia, Germany) with a helium/neon laser (785 nm) as the excitation source. TGA was implemented on a TG analyzer (TA, Q600) under N_2 atmosphere at a heating rate of 10 ℃ · min^{-1}. The compression tests were performed on a mechanical tester (CMT6503, MTS) with a compressing velocity of 2 mm · min^{-1}. The Young's modulus was calculated as the slope of the linear region of the stress-strain curve corresponding with 2%-10% strain. The energy absorption was taken as the area below the stress-strain curve between 0% and 70% strain.

3 Result and discussion

Fig. 1 shows the SEM images of PC aerogels and GC aerogels, respectively. It is observed in Fig. 1a that the nanoporous PC aerogels have dense and interconnected 3D network structure, which is formed by spontaneous physical cross-linking of cellulose chains due to many surface hydroxyl groups that can easily form hydrogen bonding linked networks[32]. Meanwhile, the developed 3D network also demonstrates the good dissolution capacity of NaOH/PEG solvent. Whereas for GC aerogels, it is obvious that the 3D architecture is almost completely disappeared, and covered tightly with the dense sheet structure (Fig. 1b). Moreover, from the EDX spectra (the inset in Fig. 1), the C, O and Au elements are detected. The Au element originates from the coating layer used for electric conduction during the SEM observation. Besides, the mass ratio of C/O of GC aerogels is approximately 2.77, much higher than that of PC aerogels (ca. 1.44). The remarkable improvement in carbon proportion reveals the potential presence of graphene.

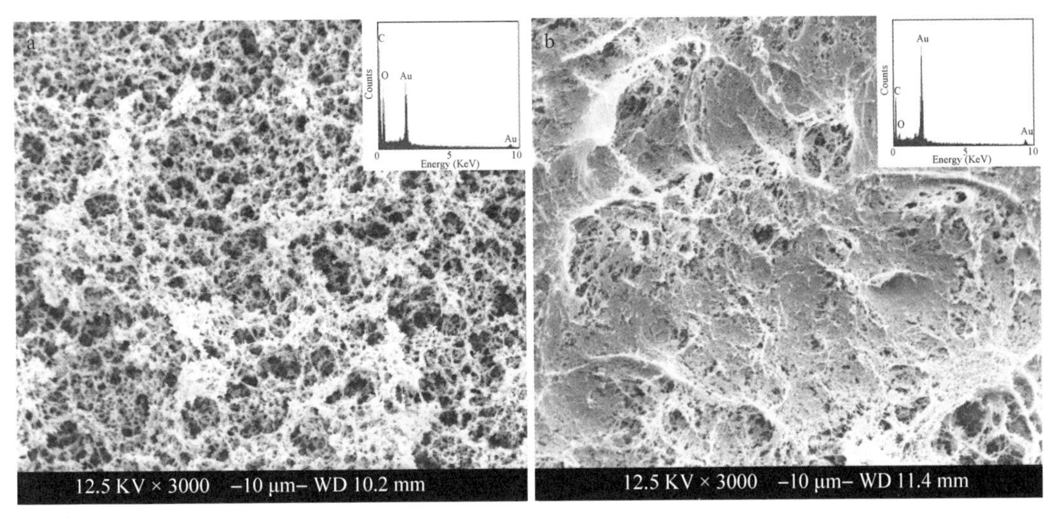

Fig. 1 SEM images of (a) PC aerogels and (b) GC aerogels, respectively. The insets show the corresponding EDX spectra

Fig. 2a presents the N_2 adsorption-desorption isotherms of PC aerogels and GC aerogels, respectively. Apparently, the both aerogels samples exhibit similar change tendency belonging to type-Ⅳ adsorption isotherms according to the IUPAC classification[33], which indicates the presence of mesopores (2-50 nm). Moreover, the curves rise slowly at low relative pressure, which reveals the presence of little or no micropore (<2 nm)[34]. At higher relative pressure (P/P_0 at 0.6−0.9), the obvious hysteresis loops could be found, which is caused by capillary condensation, suggesting the characteristic of mesoporous structure[35]. Furthermore, the large amount of adsorption (without apparent limitation) occurring at relative pressures above 0.9, demonstrates the presence of macropores (>50 nm)[36]. Moreover, it could be seen in Fig. 2b that the pore diameters are within the scope of 3−110 nm with several peaks concentrated at 3−20 nm, which further demonstrates that both PC aerogels and GC aerogels are primarily composed of mesopores and macropores. The BET calculation gives the BET surface area for PC aerogels equal to 156 $m^2 \cdot g^{-1}$; and the BJH method gives the pore volume equal to 0.95 $cm^3 \cdot g^{-1}$, respectively. For GC aerogels, after being embedded with graphene nanosheets with high specific surface area, GC aerogels show larger BET surface area (205 $m^2 \cdot g^{-1}$) and total pore volume (1.11 $cm^3 \cdot g^{-1}$). The results suggest that the incorporation of graphene makes contribution to the significant enhancement in the pore characteristic parameters.

The XRD patterns of as-prepared graphene, PC aerogels, and GC aerogels are shown in Fig. 3a, respectively. The graphene presents a broad diffraction peak at around 25.0° with an obvious disappearance of the characteristic peaks (generally at $2\theta = 12°$ and $2\theta = 26.5°$). It can be attributed to the exfoliation of

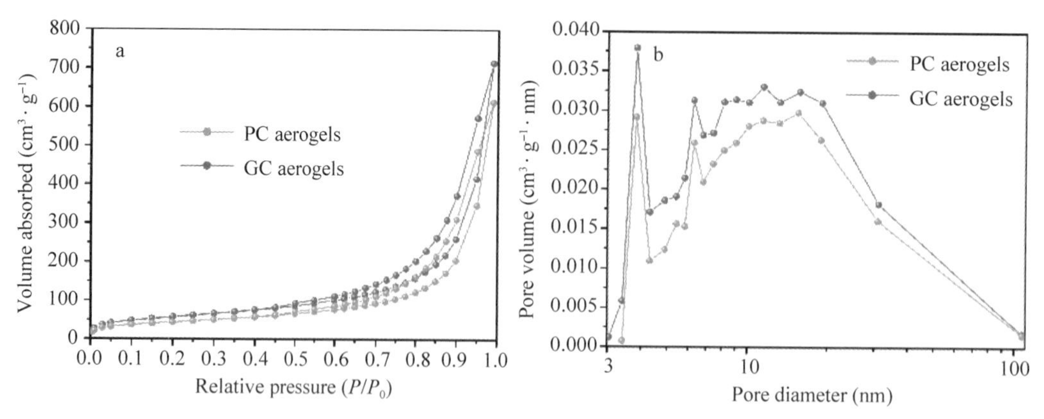

Fig. 2 (a) N_2 adsorption-desorption isotherms and (b) pore diameter distributions of PC aerogels and GC aerogels, respectively

graphene into a monolayer or few layers and long range disorder in graphene[37-39]. Moreover, PC aerogels and GC aerogels exhibit similar XRD patterns, in which the all diffraction peaks at around 11.7°, 20.1° and 21.9° could be indexed to typical cellulose Ⅱ crystalline structure corresponding to ($\bar{1}10$), (110) and (200) planes[40]. In addition, the XRD pattern of GC aerogels doesn't show any characteristic peaks derived from graphene, possibly attributed to the disorder, low content, and good dispersion of graphene[41,42].

Fig. 3b presents the Raman spectra of PC aerogels and GC aerogels, and the fast Fourier transform (FFT) filtering algorithm was used to create smooth spectra. As shown, the peaks at 897 cm^{-1}, 1103 cm^{-1}, 1377 cm^{-1} and 1459 cm^{-1} in the Raman spectrum of PC aerogels are originated from HCC and HCO bending at C6, CC and CO stretching, HCC, HCO and HOC bending, and HCH and HOC bending of cellulose structure[43], respectively. For GC aerogels, apart from the characteristic peaks from cellulose (899 cm^{-1} and 1095 cm^{-1}), two strong peaks at around 1340 cm^{-1} and 1600 cm^{-1} are attributed to the D-band (K-point phonons of A_{1g} symmetry) and G-band (E_{2g} phonons of Csp^2 atoms) of graphene[44], which demonstrates that the graphene nanosheets have been successfully embedded into the cellulose aerogels. Furthermore, the broadening of D and G bands with a strong D line reveals disordered graphitic crystal stacking of the graphene nanosheets[45], in accordance with the aforementioned XRD results.

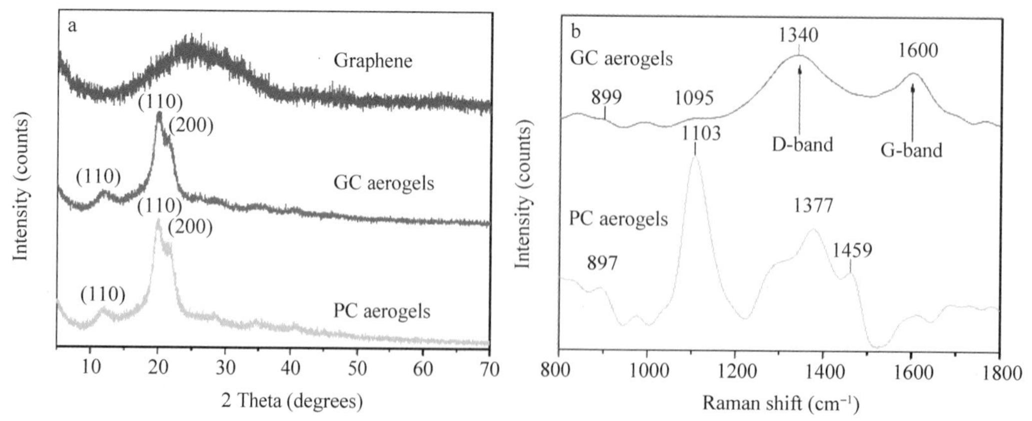

Fig. 3 (a) XRD patterns of as-prepared graphene, PC aerogels and GC aerogels, respectively. (b) Raman spectra of PC aerogels and GC aerogels, respectively

In view of the importance of thermal stabilities of nanocomposites in practical applications, the thermal stabilities of PC aerogels and GC aerogels were investigated by TGA and derivative thermogravimetry (DTG). As shown in Fig. 4, for the both samples, the small weight loss below 150 ℃ is attributed to the evaporation of adsorbed water; besides, the only one exothermic peak is centered at 335-370 ℃, which is related to cellulose pyrolysis[46]. Compared with PC aerogels, GC aerogels apparently show more superior thermal stability. In detail, PC aerogels start to decompose at 216.8 ℃, whereas GC aerogels begin at 253.7 ℃; at maximum degradation rate, the decomposition temperature occurs at 337.9 ℃ for PC aerogels and 366.7 ℃ for GC aerogels. These rising trend of decomposition temperature implies that the incorporation of graphene with high heat resistance in the aerogel is helpful to significantly increase the heat resistance.

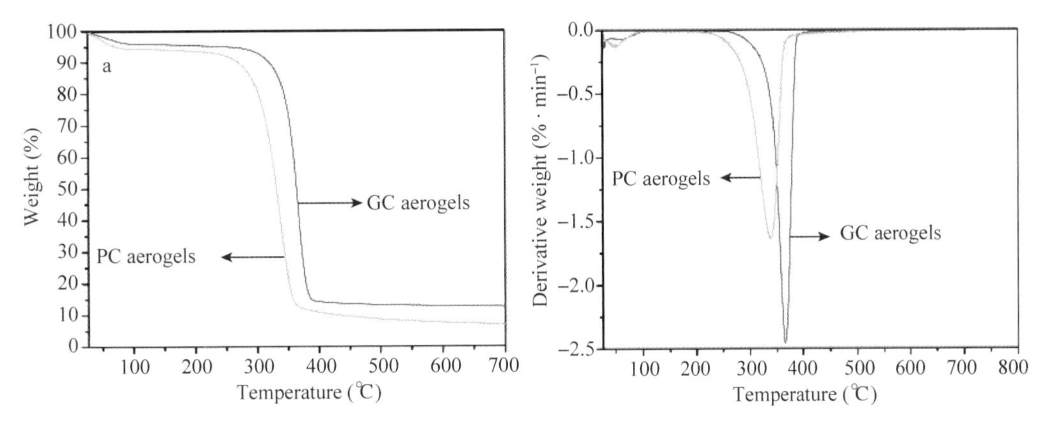

Fig. 4 (a) TG and (b) DTG curves of PC aerogels and GC aerogels, respectively

The stress-strain curves of PC aerogels and GC aerogels are presented in Fig. 5, respectively. The reinforcing effect of graphene is rather significant. Compared with PC aerogels, the Young's modulus and energy absorption of GC aerogels increase by 168% from 508 kPa to 1361 kPa and 77% from 104 kJ · m^{-3} to 184 kJ · m^{-3}, respectively. The significant improvement further confirms that the graphene is a good reinforcing additive for the cellulose aerogels. In addition, for the curves, the compression stress-strain behavior could roughly be divided into four regions[47]. First, at low strains (< 6%), the stress-strain behavior is linear elastic in nature primarily due to elastic cell wall bending. Second, a stark transformation from linear behavior to nonlinear behavior could be found. Third, in higher strain, a horizontal plateau region appears after reaching yield stress, and the collapse behavior would happen due to the plastic hinges resulted from plastic yielding of the cell wall[48]. Final, considerable stiffening will occur on account that the loose porous 3D network structure starts to touch. Actually, the two-dimension (2D) nanopaper-like structure of graphene has stronger load bearing capacity than the 3D skeleton of cellulose aerogels[14]. Besides, the high aspect ratio of graphene also plays an important role in the enhancement in the resistance to pressure.

Frequently occurred oil spills have brought great hazard to natural environment and human health. As a class of green biomaterials with large specific surface area and pore volume, cellulose aerogels are frequently considered as promising adsorbents for the treatment of oil-containing wastewater or oil spills. Therefore, it has some significance to research the oil adsorption property of PC aerogels and GC aerogels. An easily-operated oil adsorption test was designed, i. e., immersing the samples (ca. 30 mm in diameter and 15 mm in length.) into

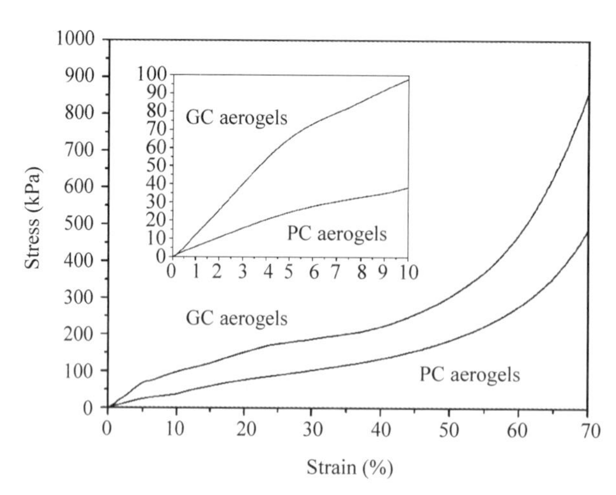

Fig. 5 Stress-strain curves of PC aerogels and GC aerogels, respectively

100 mL beakers with 80 mL waste machine oil for 6 h. The weights of the samples were measured before and after the tests. The oil adsorption property is estimated by the oil adsorption ratio calculated by dividing the weight after the test by the original dry weight. After the six-hour adsorption, as shown in Fig. 6, the aerogels both naturally sank below the liquid surface indicating that a large number of oil molecules enter into the pores of the aerogels. The absorption efficiency of GC aerogels is estimated to be 20.4 $g \cdot g^{-1}$, which is 33.3% higher than that of PC aerogels (15.3 $g \cdot g^{-1}$). Besides, the aerogels maintain the original sizes and shape before and after the tests, and no obvious cracks and collapse can be seen, which suggest the favorable shape stability of the aerogels in the oil. In addition, apart from the superior oil adsorption ability, the stronger mechanical strength of GC aerogels is also considered as more attractive alternative for the fabrication of green oil adsorbent, which is beneficial to resist various potential damages such as impact and compression. Therefore, GC aerogels draw more advantages than PC aerogels to develop green high-performance adsorbents.

Fig. 6 PC aerogels and GC aerogels at the beginning (a) and ending (b) of the oil adsorption tests

4 Conclusions

A mild, easy and environmentally friendly method is proposed to fabricate GC aerogels nanocomposites using the green NaOH/PEG aqueous solution as the processing solution. The stacked self-assembly sheet structure is tightly covered on the 3D network skeleton of cellulose aerogels, leading to the significant improvement in the specific surface area, pore volume, thermal stability, mechanical properties, and oil adsorption efficiency of GC aerogels. Considering the stronger mechanical strength and thermal property as well as oil-absorbing ability, GC aerogels are more suitable for the treatment of oil spills under the harsh environment comparing to PC aerogels.

Acknowledgments

This study was supported by the National Natural Science Foundation of China (grant no. 31270590 and 31470584).

References

[1] Kistler S S. Coherent expanded aerogels and jellies[J]. Nature, 1931, 127(3211): 741-741.

[2] Cai J, Kimura S, Wada M, et al. Cellulose aerogels from aqueous alkali hydroxide-urea solution[J]. ChemSusChem: Chemistry & Sustainability Energy & Materials, 2008, 1(1-2): 149-154.

[3] Sehaqui H, Zhou Q, Berglund L A. High-porosity aerogels of high specific surface area prepared from nanofibrillated cellulose (NFC)[J]. Composites science and technology, 2011, 71(13): 1593-1599.

[4] Liebner F, Potthast A, Rosenau T, et al. Cellulose aerogels: Highly porous, ultra-lightweight materials[J]. 2008.

[5] Fischer F, Rigacci A, Pirard R, et al. Cellulose-based aerogels[J]. Polymer, 2006, 47(22): 7636-7645.

[6] Isobe N, Chen X, Kim U J, et al. TEMPO-oxidized cellulose hydrogel as a high-capacity and reusable heavy metal ion adsorbent[J]. Journal of hazardous materials, 2013, 260: 195-201.

[7] Pröbstle H, Schmitt C, Fricke J. Button cell supercapacitors with monolithic carbon aerogels[J]. Journal of Power Sources, 2002, 105(2): 189-194.

[8] Arenas J P, Crocker M J. Recent trends in porous sound-absorbing materials[J]. Sound & vibration, 2010, 44(7): 12-18.

[9] Medronho B, Lindman B. Competing forces during cellulose dissolution: From solvents to mechanisms[J]. Current Opinion in Colloid & Interface Science, 2014.

[10] Zhang H, Wu J, Zhang J, et al. 1-Allyl-3-methylimidazolium chloride room temperature ionic liquid: a new and powerful nonderivatizing solvent for cellulose[J]. Macromolecules, 2005, 38(20): 8272-8277.

[11] Rosenau T, Potthast A, Adorjan I, et al. Cellulose solutions in N-methylmorpholine-N-oxide (NMMO)-degradation processes and stabilizers[J]. Cellulose, 2002, 9(3): 283-291.

[12] Zhang S, Li F X, Yu J Y. Structure and Properties of Novel Cellulose Fibres Produced From NaOH/PEG-treated Cotton Linters[J]. Iranian Polymer Journal, 2010, 19(12): 949-957.

[13] Medronho B, Lindman B. Competing forces during cellulose dissolution: From solvents to mechanisms[J]. Current Opinion in Colloid & Interface Science, 2014, 19(1): 32-40.

[14] Zhang J, Cao Y, Feng J, et al. Graphene-oxide-sheet-induced gelation of cellulose and promoted mechanical properties of composite aerogels[J]. The Journal of Physical Chemistry C, 2012, 116(14): 8063-8068.

[15] Liang S, Zhang L, Li Y, et al. Fabrication and properties of cellulose hydrated membrane with unique structure[J]. Macromolecular Chemistry and Physics, 2007, 208(6): 594-602.

[16] Wu J, Liang S, Dai H, et al. Structure and properties of cellulose/chitin blended hydrogel membranes fabricated via a solution pre-gelation technique[J]. Carbohydrate Polymers, 2010, 79(3): 677-684.

[17] Azizi Samir M A S, Alloin F, Dufresne A. Review of recent research into cellulosic whiskers, their properties and their application in nanocomposite field[J]. Biomacromolecules, 2005, 6(2): 612-626.

[18] Liang W, M Sánchez-Soto, Maspoch M L. Polymer/clay aerogel composites with flame retardant agents: Mechanical, thermal and fire behavior[J]. Materials & Design (1980-2015), 2013, 52(24): 609-614.

[19] Cai J, Liu S, Feng J, et al. Cellulose-silica nanocomposite aerogels by in situ formation of silica in cellulose gel[J]. Angewandte Chemie, 2012, 124(9): 2118-2121.

[20] Isobe N, Sekine M, Kimura S, et al. Anomalous reinforcing effects in cellulose gel-based polymeric nanocomposites [J]. Cellulose, 2011, 18(2): 327-333.

[21] Carlsson D O, Nyström G, Zhou Q, et al. Electroactive nanofibrillated cellulose aerogel composites with tunable structural and electrochemical properties[J]. Journal of Materials Chemistry, 2012, 22(36): 19014-19024.

[22] Bober P, Liu J, Mikkonen K S, et al. Biocomposites of nanofibrillated cellulose, polypyrrole, and silver nanoparticles with electroconductive and antimicrobial properties[J]. Biomacromolecules, 2014, 15(10): 3655-3663.

[23] Olsson R T, Samir M A S A, Salazar-Alvarez G, et al. Making flexible magnetic aerogels and stiff magnetic nanopaper using cellulose nanofibrils as templates[J]. Nature nanotechnology, 2010, 5(8): 584-588.

[24] Bai S. Graphene-Based Inorganic Nanocomposites[J]. Progress in Chemistry, 2010.

[25] Zhang H, Lv X, Li Y, et al. P25-graphene composite as a high performance photocatalyst[J]. ACS nano, 2010, 4(1): 380-386.

[26] Kuilla T, Bhadra S, Yao D, et al. Recent advances in graphene based polymer composites[J]. Progress in polymer science, 2010, 35(11): 1350-1375.

[27] Hummers Jr W S, Offeman R E. Preparation of graphitic oxide[J]. Journal of the american chemical society, 1958, 80(6): 1339-1339.

[28] Ding Y, Jiang Y, Xu F, et al. Preparation of nano-structured $LiFePO_4$/graphene composites by co-precipitation method[J]. Electrochemistry Communications, 2010, 12(1): 10-13.

[29] Wan C, Li J. Facile synthesis of well-dispersed superparamagnetic γ-Fe_2O_3 nanoparticles encapsulated in three-dimensional architectures of cellulose aerogels and their applications for Cr (VI) removal from contaminated water[J]. ACS Sustainable Chemistry & Engineering, 2015, 3(9): 2142-2152.

[30] Wan C, Li J. Synthesis of well-dispersed magnetic $CoFe_2O_4$ nanoparticles in cellulose aerogels via a facile oxidative co-precipitation method[J]. Carbohydrate polymers, 2015, 134: 144-150.

[31] Wan C, Li J. Embedding ZnO nanorods into porous cellulose aerogels via a facile one-step low-temperature hydrothermal method[J]. Materials & Design, 2015, 83: 620-625.

[32] Simard M, Su D, Wuest J D. Use of hydrogen bonds to control molecular aggregation. Self-assembly of three-dimensional networks with large chambers[J]. Journal of the American Chemical Society, 1991, 113(12): 4696-4698.

[33] Bag S, Trikalitis P N, Chupas P J, et al. Porous semiconducting gels and aerogels from chalcogenide clusters[J]. Science, 2007, 317(5837): 490-493.

[34] Cranston R W, Inkley F A. 17 the determination of pore structures from nitrogen adsorption isotherms[M]//Advances in catalysis. Academic Press, 1957, 9: 143-154.

[35] Liang Y, Wu D, Fu R. Carbon Microfibers with Hierarchical Porous Structure from Electrospun Fiber-Like Natural Biopolymer[J]. Rep, 2013, 3: 1119.

[36] Estella J, Echeverría J C, Laguna M, et al. Effects of aging and drying conditions on the structural and textural properties of silica gels[J]. Microporous and mesoporous materials, 2007, 102(1-3): 274-282.

[37] Li Y, Gao W, Ci L, et al. Catalytic performance of Pt nanoparticles on reduced graphene oxide for methanol electro-

oxidation[J]. Carbon, 2010, 48(4): 1124-1130.

[38] Guo H L, Wang X F, Qian Q Y, et al. A green approach to the synthesis of graphene nanosheets[J]. ACS nano, 2009, 3(9): 2653-2659.

[39] Hajian M, Reisi M R, Koohmareh G A, et al. Preparation and characterization of polyvinylbutyral/graphene nanocomposite[J]. Journal of Polymer Research, 2012, 19(10): 1-7.

[40] Dinand E, Vignon M, Chanzy H, et al. Mercerization of primary wall cellulose and its implication for the conversion of cellulose I→ cellulose Ⅱ[J]. Cellulose, 2002, 9(1): 7-18.

[41] Yuan X. Enhanced interfacial interaction for effective reinforcement of poly (vinyl alcohol) nanocomposites at low loading of graphene[J]. Polymer bulletin, 2011, 67(9): 1785-1797.

[42] Liang J, Huang Y, Zhang L, et al. Molecular-level dispersion of graphene into poly (vinyl alcohol) and effective reinforcement of their nanocomposites[J]. Advanced Functional Materials, 2009, 19(14): 2297-2302.

[43] Agarwal U P, Atalla R H . Using Raman Spectroscopy to Identify Chromophores in Lignin-Lignocellulosics[J]. ACS Symposium Series, 1999, 742: 250-264.

[44] Shi Y, Chou S L, Wang J Z, et al. Graphene wrapped $LiFePO_4$/C composites as cathode materials for Li-ion batteries with enhanced rate capability[J]. Journal of Materials Chemistry, 2012, 22(32): 16465-16470.

[45] Evers S, Nazar L F. Graphene-enveloped sulfur in a one pot reaction: a cathode with good coulombic efficiency and high practical sulfur content[J]. Chemical Communications, 2012, 48(9): 1233-1235.

[46] Shafizadeh F, Fu Y L. Pyrolysis of cellulose[J]. Carbohydrate Research, 1973, 29(1): 113-122.

[47] Wan C, Lu Y, Jin C, et al. Thermally induced gel from cellulose/NaOH/PEG solution: preparation, characterization and mechanical properties[J]. Applied Physics A, 2015, 119(1): 45-48.

[48] Sehaqui H, Salajková M, Zhou Q, et al. Mechanical performance tailoring of tough ultra-high porosity foams prepared from cellulose I nanofiber suspensions[J]. Soft Matter, 2010, 6(8): 1824-1832.

中文题目：在纤维素气凝胶中加入石墨烯纳米片：增强机械、热和油的吸附性能

作者：万才超，李坚

摘要：本文以一种绿色的NaOH/PEG溶液为基体，制备了新型石墨烯/纤维素气凝胶。扫描电镜观察表明，纤维素气凝胶的三维网状骨架结构被紧凑的片状结构紧紧覆盖。X射线衍射和拉曼光谱分析表明，石墨烯纳米片已成功合成并嵌入纤维素气凝胶中。石墨烯纳米片的加入显著提高了气凝胶的比表面积、孔容、热稳定性、机械强度和吸油效率。因此，与纯纤维素气凝胶相比，绿色复合气凝胶在处理恶劣环境下的含油废水或溢油时具有更大的优势。

关键词：石墨烯；纤维素气凝胶；机械性能；热稳定性；吸油性能

Cellulose Aerogels Functionalized with Polypyrrole and Silver Nanoparticles: In-situ Synthesis, Characterization and Antibacterial Activity*

Caichao Wan, Jian Li

Abstract: Green porous and lightweight cellulose aerogels have been considered as promising candidates to substitute some petrochemical host materials to support various nanomaterials. In this work, waste wheat straw was collected as feedstock to fabricate cellulose hydrogels, and a green inexpensive NaOH/polyethylene glycol solution was used as cellulose solvent. Prior to freeze-drying treatment, the cellulose hydrogels were integrated with polypyrrole and silver nanoparticles by easily-operated in-situ oxidative polymerization of pyrrole using silver ions as oxidizing agent. The tri-component hybrid aerogels were characterized by scanning electron microscope, transmission electron microscope, energy dispersive X-ray spectroscopy, selected area electron diffraction, X-ray photoelectron spectroscopy, and X-ray diffraction. Moreover, the antibacterial activity of the hybrid aerogels against *E. coli* (Gram-negative), *S. aureus* (Gram-positive) and *L. monocytogenes* (intracellular bacteria) was qualitatively and quantitatively investigated by parallel streak method and determination of minimal inhibitory concentration, respectively. This work provides an example of combining cellulose aerogels with nanomaterials, and helps to develop novel forms of cellulose-based functional materials.

Keywords: Cellulose aerogels; Polypyrrole; Silver nanoparticles; Antibacterial activity; Composites

1 Introduction

Bacteria are everywhere: in soil, in water, in air, and in the bodies of every person and animal. Most of them are harmless and even highly helpful. For example, they can break down dead plants and animals to promote carbon cycle, help green plants get nitrogen (biological nitrogen fixation), and keep our digestive tract working smoothly. On the other hand, similar to viruses, bacteria can cause hundreds of illnesses. Some plants, animals and even microbes themselves have been known to have a great variety of mechanisms to keep microbes at bay (Gennaro, Zanetti, Benincasa, Podda & Miani, 2002; Maróti, Kereszt, Kondorosi & Mergaert, 2011; Zasloff, 2002). However, bacterial infection has become a serious problem in human society. Thus it is urgent to develop multifarious green novel antibacterial agents. In recent decades, antimicrobial polymers show good prospects for development because of their good chemical stability and biocompatibility, lack of diffusion through skin, non-volatility, low toxicity, long residence time, and

---

* 本文摘自 Carbohydrate Polymers, 2016, 146: 362-367.

biological activity (Kenawy, Worley & Broughton, 2007; Tew, Scott, Klein & DeGrado, 2009; Timofeeva & Kleshcheva, 2011). Polypyrrole (PPy), which is formed from a number of connected pyrrole ring structures, is one of the typical antimicrobial polymers and has been demonstrated to have good antibacterial activity against *E. coli* (Varesano et al., 2013). The antibacterial mechanism of PPy is possibly attributed to leakage of cytoplasm since the positive charges in PPy chains interact with bacteria cell wall. The preparation of PPy is implemented by oxidation of pyrrole, generally dependent on chemical or electrochemical approaches (Chen, Xu, Wang, Qian & Sun, 2015; Efimov, 1997; Joo et al., 2000). Especially, the chemical method has numerous merits such as low cost, low requirements for equipment and high output, and is more conducive to large-scale production. Silver nitrate ($AgNO_3$) is one of the effective oxidants for the chemical oxidative polymerization of pyrrole. Also, the polymerization process induces reduction of Ag^+ to metallic silver on the PPy surface (Bober et al., 2014). Therefore, this method is usually considered as the most facile efficient route to obtain PPy/Ag composites. Nanoscale silver has long been known to have strong bacteriostatic and antibacterial effects as well as a broad spectrum of antimicrobial activity (Durán, Marcato, De Souza, Alves & Esposito, 2007; Marambio-Jones & Hoek, 2010). Thus PPy/Ag hybrids are expected to display synergistically improved antibacterial activity by combining the antibacterial effects of both components. However, considering some restrictions such as inferior processability of PPy, spontaneous self-agglomeration of nanomaterials and recycling problems, it needs to seek for green host materials to support and immobilize PPy and nano-Ag. In our previous reports, cellulose aerogels have been repeatedly used as host materials to support various nanomaterials including γ-Fe_2O_3 (Wan & Li, 2015b), cobalt ferrite (Wan & Li, 2015c), rod-like ZnO (Wan & Li, 2015a), and anatase titania (Jiao, Wan & Li, 2015; Wan, Lu, Jin, Sun & Li, 2015). The large specific area of cellulose aerogels makes contribution to acquiring high loading content of nanomaterials. Also, the abundant hydroxyl groups on the cellulose surface are beneficial to fix nanoparticles by hydrogen-bond or electrostatic interactions (Cai, Kimura, Wada & Kuga, 2008). In addition, other advantages include biodegradability, widely available feedstock and low density (Sehaqui, Zhou & Berglund, 2011). Thus cellulose aerogels have been regarded as extremely promising candidates to substitute some petrochemical host materials like styrene-methyl acrylate copolymer (Borthakur, Sharma & Dolui, 2011) and poly (tetrafluoroethylene) (PTFE) (Shi, Zhou, Qing, Dai & Lu, 2012). Furthermore, we believe that the tri-component hybrid aerogels can not only retain the excellent properties of PPy and nano-Ag, but also inherit chemical, structural and mechanical features of the substrates.

In the present work, cellulose aerogels were functionalized with PPy and Ag nanoparticles by in-situ oxidative polymerization of pyrrole using $AgNO_3$ as oxidizing agent. Besides, waste wheat straw was used as the raw materials of cellulose aerogels, and a low-cost aqueous solution of NaOH/polyethylene glycol (PEG) was used as cellulose solvent. The tri-component hybrid aerogels (coded as PPy/Ag/CA) were characterized by transmission electron microscopy (TEM), scanning electron microscope (SEM), energy dispersive X-ray spectroscopy (EDX), selected area electron diffraction (SAED), X-ray photoelectron spectroscopy (XPS) and X-ray diffraction (XRD). Moreover, the antibacterial activity of PPy/Ag/CA against gram-negative *E. coli*, gram-positive *S. aureus* and intracellular bacteria *L. monocytogenes* was evaluated. To the best of our knowledge, there are still rare reports about the tri-component porous composites containing cellulose, PPy and nano-Ag.

2 Materials and methods

2.1 Materials

Native wheat straw was grinded and then screened through a sixty mesh sieve, and the resulting powder was subsequently collected and dried in a vacuum oven at 60 ℃ for 24 h before use. *E. coli* (ATCC 25922), *S. aureus* (ATCC 6538) and *L. monocytogenes* (NICPBP 54002) were purchased from Guangzhou Jiake Microbiological Research Center, China. All other chemicals were purchased from Kemiou Chemical Reagent Co. Ltd. (Tianjin, China) and used as received.

2.2 Preparation of PPy/Ag/CA

The cellulose hydrogels were prepared in three steps: (1) chemically purifying cellulose from the wheat straw; (2) dissolving the purified cellulose in the NaOH/PEG-4000 aqueous solution; (3) regeneration of cellulose solution in 1% HCl solution. The detailed processes can be referred to our previous reports (Wan, Lu, Cao, Sun & Li, 2015; Wan, Lu, Jiao, Jin, Sun & Li, 2015). Regarding the preparation of PPy/Ag/CA, pyrrole (0.145 g) was firstly mixed with 50 mL distilled water with magnetic stirring for 30 min. Thereafter, the cellulose hydrogel was immersed in the aqueous pyrrole solution at room temperature for 12 h. After that, the mixed system was mixed with 180 mL $AgNO_3$ solution (0.03 mol · L^{-1}), and then kept at room temperature for a week. During this process, the color of hydrogels was gradually changed from white to gray and eventually to black. The resultant hybrid hydrogel was repeatedly rinsed with a large amount of distilled water and tert-butyl alcohol. Finally, the hybrid hydrogel was subjected to a freeze-drying process to remove most liquids.

2.3 Characterizations

TEM and high-resolution TEM (HRTEM) observations and SAED were performed with a FEI, Tecnai G2 F20 TEM. SEM observation was performed with a Hitachi S4800 SEM equipped with an EDX detector for element analysis. XPS was carried out using a Thermo Escalab 250Xi XPS spectrometer. Deconvolution of the overlapping peaks was performed using a mixed Gaussian-Lorentzian fitting program (Origin 8.5, Originlab Corporation). XRD spectroscopy was implemented on a Bruker D8 Advance TXS XRD instrument.

2.4 Antibacterial activity studies

Antibacterial activity of PPy/Ag/CA has been investigated against *E. coli* as the model Gram-negative bacteria, *S. aureus* as the model Gram-positive bacteria, and *L. monocytogenes* as the model intracellular bacteria. Qualitative and quantitative antibacterial experiments on PPy/Ag/CA were carried out by two methods [i.e., parallel streak method and determination of minimal inhibitory concentration (MIC)], respectively.

2.4.1 Parallel streak method

The antibacterial activity of PPy/Ag/CA was evaluated qualitatively according to AATCC Test Method 147-2004 (Parallel Streak Method). Using a 4 mm inoculating loop, one loopful of the diluted inoculum was transferred to the surface of sterile agar plate by making five parallel streaks. The test specimen was gently pressed transversely across the agar surface, and the Petri plate was subsequently incubated for 24 h at 37 ℃. Incubated plate was examined for interruption of growth along the streaks of inoculum beneath the specimen and for a clean zone of inhibition beyond its edge. The average width of an inhibition zone along a streak on either side of the test specimen was calculated using the following equation:

$$W=\frac{T-D}{2} \tag{1}$$

where W is the width of clear zone of inhibition, T is the width of specimen and clear zone, and D is the width of test specimen.

2.4.2 Determination of MIC

Quantitative analysis of the antibacterial activity of PPy/Ag/CA was implemented by determination of MIC. The MIC value is defined as the lowest concentration (in $\mu g \cdot mL^{-1}$) of the antibacterial agent that prevents visible growth of a microorganism compared with the growth in the control plate. The determination of MIC was performed by the agar dilution method. In a typical process, the PPy/Ag/CA specimen was grinded and dispersed in sterile phosphate buffer solution. The different concentrations of suspension solutions were added at twofold concentrations (0.125 to 128 $\mu g \cdot ml^{-1}$) to sterilize Mueller-Hinton agar (tempered to 50 ℃) prior to pouring plates. After solidification, 10 μl (ca. 10^4 cfu) of each prepared cultures was spotted (five spots per plate) on the agar surfaces. All test plates were incubated at 37 ℃ for 24 h. Inoculated agar plates without added PPy/Ag/CA served as positive controls. All experiments were performed in triplicates.

3 Results and discussion

3.1 Schematic diagram for the preparation of PPy/Ag/CA

A schematic diagram for the preparation of PPy/Ag/CA is presented in Fig. 1. Cellulose was chemically purified from the wheat straw and subsequently dissolved in the NaOH/PEG solution. After the regeneration of cellulose solution in hydrochloric acid, the cellulose hydrogels were generated. The hydrogels were immersed in the aqueous pyrrole solution. During the immersion, the hydroxyl groups of cellulose might interact with the imine groups of pyrrole by hydrogen-bond interaction, leading to firm immobilization of the monomers on the surfaces of cellulose chains. After the addition of $AgNO_3$, the oxidative polymerization of pyrrole individually took place at the original binding sites on the cellulose surface. The oxidative polymerization also resulted in the reduction of silver ions to zero-valent Ag nanoparticles. The formed Ag nanoparticles are expected to be tightly anchored to the surfaces of the resultant PPy molecules. Therefore, we believe that the method can acquire a favorable dispersion of PPy and nano-Ag, which will be confirmed by the SEM and TEM observations. The mole ratio of silver ions to pyrrole was set as 2.5 in this work. The value has been confirmed to contribute to acquiring strong antimicrobial activity (Bober et al., 2014; Omastová et al., 2013).

3.2 Morphology observations and elemental analysis

The morphology of PPy/Ag/CA was investigated by TEM and SEM observations. According to the SEM observation (see Fig. S1 in the Supporting Information), it can be seen that the surface of aerogels was densely coated with numerous strip-shaped, flake-like and granular substances, possibly originated from PPy and Ag. According to the EDX analysis (see the inset of Fig. 2a), several elements including C (52.45 At%), O (41.32 At%), N (4.89 At%) and Ag (1.34 At%) were detected. The Au element was derived from the coating layer used for electric conduction during the SEM observation. Based on the TEM observation

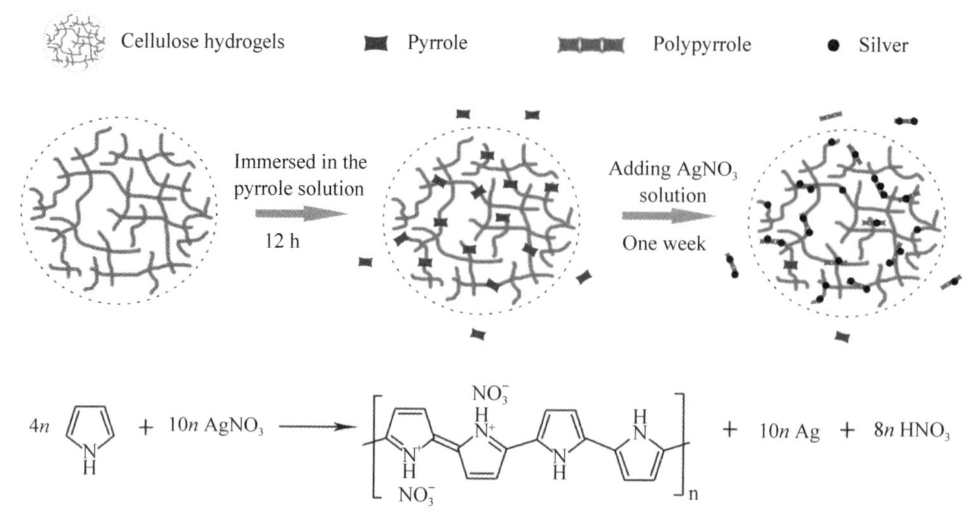

Fig. 1 Schematic diagram for the preparation of PPy/Ag/CA

in Fig. 2a, it is observed that a plenty of nanomaterials were homogeneously dispersed in the three-dimensional (3D) network of cellulose aerogels. No obvious self-agglomeration occurred. Additionally, there are no any nanomaterials separated from the substrates, which suggests that the generated PPy and nano-Ag were tightly immobilized on the surface of cellulose. These observations demonstrate that the porous cellulose aerogels with hierarchical micro/nano-scaled 3D network can serve as ideal host materials to integrate with Ag nanoparticles and PPy. It is well-known that good dispersion plays a crucial role in properties of nanomaterials (Ji, Chen, Wai & Fulton, 1999). The higher-magnification TEM image (Fig. 2b) shows that the spherical Ag nanoparticles were adhered to the surfaces of flake-like and strip-shaped PPy, indicating good interface bonding.

The HRTEM image of PPy/Ag/CA (Fig. 2c) shows that the selected region are highly crystalline with lattice fringes measured to be 0.235 nm which is in conformity to (111) plane of face-centered cubic Ag. The SAED pattern is presented in Fig. 1d, in which the symmetric hexagonal spots are indexed to ($\bar{1}1\bar{1}$), ($\bar{2}20$), ($\bar{1}11$), (002) and ($\bar{1}13$) reflections. Besides, the spot pattern obtained for Ag nanoparticles reveals the single crystalline natures.

3.3 Crystal structure and chemical compositions

For the investigation of the crystal structure of PPy/Ag/CA, the XRD analysis was carried out. For comparison, the XRD pattern of the pure cellulose aerogels (PCA) was also measured. As shown in Fig. 3, the three diffraction peaks of PCA at around 12.2°, 20.2° and 21.9° are assigned to (101), ($10\bar{1}$) and (002) planes of cellulose Ⅱ crystal structure (Nishiyama, Langan & Chanzy, 2002), respectively. Apart from the above three cellulose characteristic peaks, another five peaks of PPy/Ag/CA at 38.0°, 44.2°, 64.3°, 77.3° and 81.5° belong to face-centered cubic silver, because the corresponding d-spacings match well with the JCPDS file (No. 04-0783) (see Table S1 and Fig. S2). Moreover, the average crystallite size of the Ag nanoparticles in PPy/Ag/CA was calculated from the full width at half maximum of the strongest Ag (111) diffraction line using Scherrer's equation (Holzwarth & Gibson, 2011):

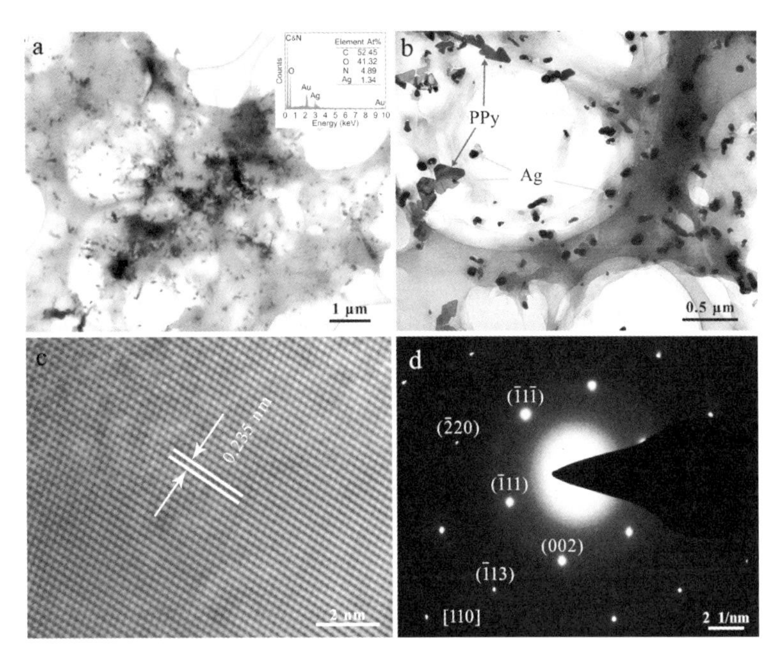

Fig. 2 (a) Low-magnification TEM image, (b) high-magnification TEM image, (c) HRTEM image and (d) SAED pattern of PPy/Ag/CA, respectively. The inset in (a) is the corresponding EDX spectrum

$$d = K\lambda / (\beta\cos\theta) \tag{2}$$

where λ is the X-ray wavelength, K is a constant taken as 0.89, θ is the angle of Bragg diffraction and β is the full width at half maximum. The average crystallite size of the Ag nanoparticles is approximatively 15.8 nm, consistent with the particle sizes (13-28 nm) observed from the TEM images (Fig. 2b).

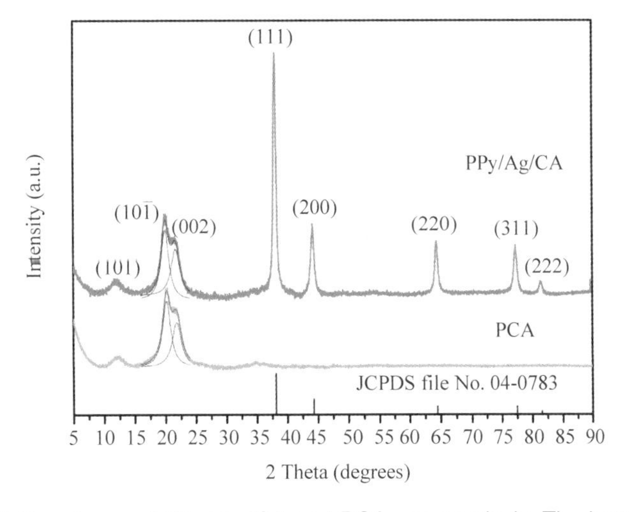

Fig. 3 XRD patterns of PPy/Ag/CA and PCA, respectively. The bottom line is the standard JCPDS file No. 04-0783 for face-centered cubic silver

The chemical states of the compositions in PPy/Ag/CA were characterized by XPS analysis. According to the XPS survey spectrum in Fig. 4a, it can be seen that several peaks are found corresponding to O 1s, N 1s, Ag 3d and C 1s, indicating the presence of oxygen, carbon, nitrogen and silver in PPy/Ag/CA. The Ag 3d high resolution XPS spectrum is presented in Fig. 4b, in which the two significant XPS signals at around 374. 2 eV and 368. 2 eV with 6. 0 eV separation are assigned to Ag $3d_{3/2}$ and Ag $3d_{5/2}$ binding energies of Ag^0 (Wei, Li, Yang, Pan, Yan & Yu, 2010), respectively. The results demonstrate that the silver ions were successfully reduced to zero-valent silver nanoparticles. Regarding the C1s high resolution XPS spectrum (Fig. 4c), the four fitting peaks at 288. 1 eV, 286. 8 eV, 286. 1 eV and 284. 8 eV are ascribed to O—C—O, C—OH, C—N, and C—C or C—H (Babu, Dhandapani, Maruthamuthu & Kulandainathan, 2012), respectively. The N 1s spectrum in Fig. 4d can be principally deconvoluted into three Gaussian peaks centered at 401. 5 eV, 399. 8 eV and 398. 2 eV, respectively. The former two peaks are attributed to—N^+—and—NH—species (áL Tan, 1998; Feng, Sun, Hou & Zhu, 2007), respectively. The peak at 398. 2 eV is derived from the uncharged deprotonated imine (=N—) nitrogen atoms or imine defects (Wei, Li, Yang, Pan, Yan & Yu, 2010). Moreover, an N^+/N ratio of 0. 26 for PPy/Ag/CA is in close agreement with the doping level generally reported for PPy in the literature (Menon, Lei & Martin, 1996).

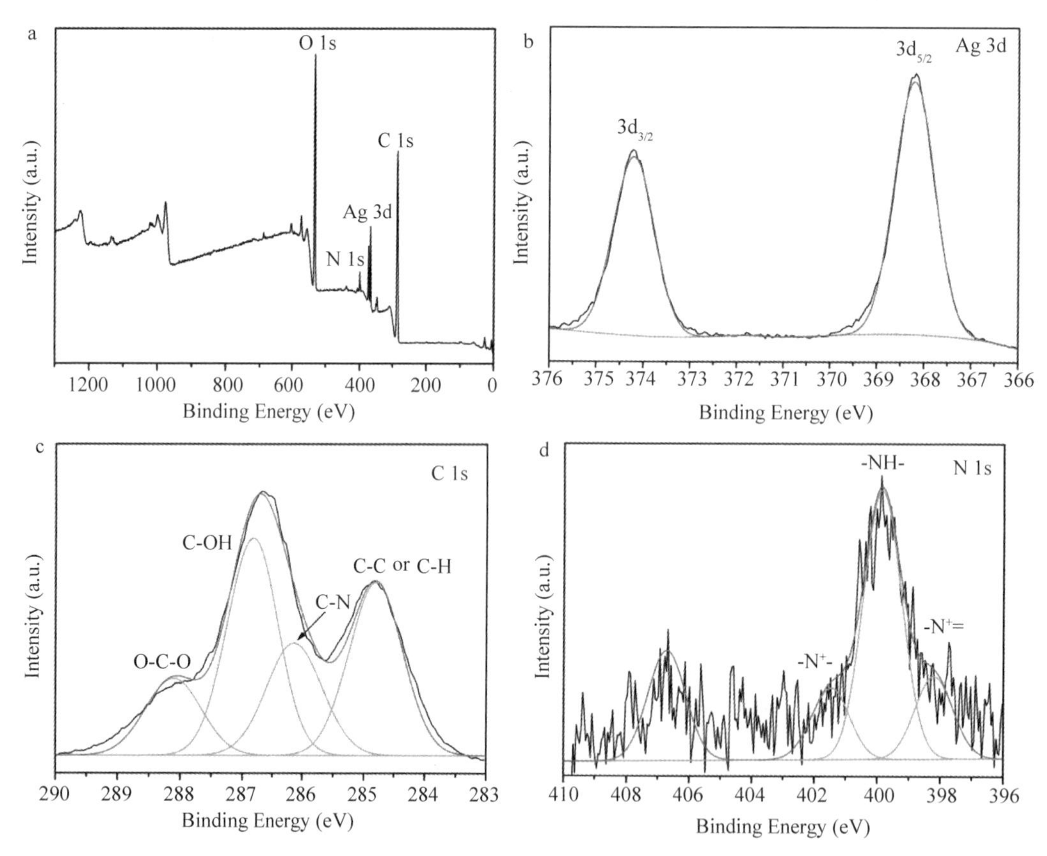

Fig. 4 (a) XPS survey spectrum and high resolution scan of Ag 3d (b), C 1s (c) and N 1s (d) of PPy/Ag/CA, respectively

3. 4 Antibacterial activity

The antibacterial activity of PPy/Ag/CA was qualitatively evaluated according to AATCC Test Method 147-2004 (Parallel Streak Method) against the three representative bacterial strains, i. e. , gram-negative *E. coli*, gram-positive *S. aureus* and intracellular bacteria *L. monocytogenes*. The test results are shown in

Fig. 5. The color differences among these PPy/Ag/CA samples are attributed to the use of flash lamp when taking photos. It is observed that PPy/Ag/CA successfully inhibited the growth of *E. coli*, *S. aureus* and *L. monocytogenes*. There are no bacterial streaks near the samples, indicating that PPy/Ag/CA has the ability to kill the cultures while in contact with them. The clear inhibition zones can be observed on both sides of the samples (4.5 mm for *E. coli*, 2.0 mm for *S. aureus* and 2.4 mm for *L. monocytogenes*). The formation of inhibition zones is possibly attributed to the ionization of Ag and its ability to be released from the aerogels and diffuse to the bacterial cell membranes (Monteiro, Gorup, Takamiya, Ruvollo-Filho, de Camargo & Barbosa, 2009; Radheshkumar & Münstedt, 2006; Sharma, Yngard & Lin, 2009). It is well-known that Ag^{+} has antibacterial capability because it can disrupt plasma membrane permeability of bacterial cells (Chen, Wu & Zeng, 2005). In addition, we previous work (Wan, Jiao, Sun & Li, 2014) has confirmed that the cellulose aerogels didn't exhibit obviously antibacterial function. Therefore, it might be concluded that the insertion of PPy and nano-Ag endows the cellulose aerogels with high efficiency to kill Gram-negative and Gram-positive bacteria and intracellular bacteria at direct contact and to inhibit their growth in the proximity of the aerogels.

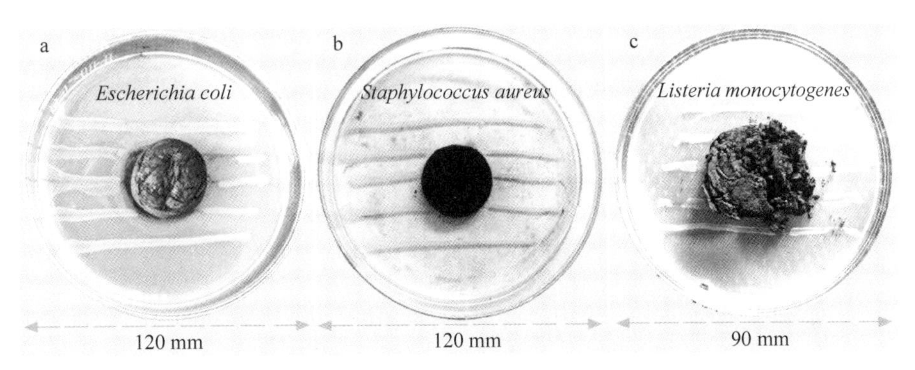

Fig. 5 Antibacterial activity of PPy/Ag/CA against *E. coli* (a), *S. aureus* (b) and *L. monocytogenes* (c), respectively

In order to quantitatively analyze the antibacterial activity of PPy/Ag/CA against the aforementioned strains, the MIC was determined by the agar dilution method. For the agar dilution, solutions with defined numbers of bacterial cells were spotted directly onto the nutrient agar plates that have incorporated different concentrations of antibacterial agent (Wiegand, Hilpert & Hancock, 2008). After the incubation, the presence of bacterial colonies on the plates indicates growth of the organism. Therefore, the MIC is defined as the lowest concentration of the antibacterial agent that prevents visible growth of a microorganism under defined conditions. The results presented in Table 1 show the MIC values for *E. coli* of 16 $\mu g \cdot mL^{-1}$, for *S. aureus* of 32 $\mu g \cdot mL^{-1}$, and for *L. monocytogenes* of 32 $\mu g \cdot mL^{-1}$, respectively. Visible growth was observed when the lower concentrations were used, suggesting no inhibitory effects on the growth of the microorganisms under these circumstances. In addition, we compare our data with the data from some other recently reported cellulose-based antibacterial agents, as shown in Table 2. Obviously, the MIC values of PPy/Ag/CA are much lower than those of other antibacterial agents, such as cellulose nanocrystals/Ag nanospheres (Xiong et al., 2013), lysozyme-conjugated nanocellulose (Jebali et al., 2013), allicin-conjugated nanocellulose (Jebali et al., 2013), *azadirachta indica* (Fabry, Okemo & Ansorg, 1998), *thymus serpyllum* (Alzoreky

& Nakahara, 2003), and *cinnamomum cassia* (Alzoreky & Nakahara, 2003). The results further verify the strong antibacterial activity of PPy/Ag/CA.

Table 1 Determination of MIC of PPy/Ag/CA for the studied strains

Concentration ($\mu g \cdot mL^{-1}$)	0.125	0.25	0.5	1	2	4	8	16	32	64	128
E. coli	+	+	+	+	+	+	+	–	–	–	–
S. aureus	+	+	+	+	+	+	+	+	–	–	–
L. monocytogenes	+	+	+	+	+	+	+	+	–	–	–

Table 2 Comparison of MIC ($\mu g \cdot ml^{-1}$) of PPy/Ag/CA with other recently reported cellulose-based antibacterial agents

Antibacterial agents	*E. coli*	*S. aureus*	*L. monocytogenes*
PPy/Ag/CA (this work)	16	32	32
Cellulose nanocrystals/Ag nanospheres	66	66	—
Lysozyme-conjugated nanocellulose	1000	1000	—
Allicin-conjugated nanocellulose	1000	1000	—
Azadirachta indica	2000	1000	—
Thymus serpyllum	>2640	1320	1320
Cinnamomum cassia	2640	1320	2640

4 Conclusions

The cellulose hydrogels were fabricated from the wheat straw and subsequently integrated with PPy and nano-Ag to prepare tri-component hybrid aerogels. The EDX, HRTEM, SAED, XRD and XPS analysis demonstrates that the PPy and Ag nanoparticles were successfully synthesized by the in-situ oxidative polymerization of pyrrole using the silver ions as oxidizing agent. Moreover, the SEM and TEM observations indicate that the well-dispersed PPy was tightly adhered to the substrates, suggesting that the aerogels can serve as green host materials to immobilize nanomaterials. The Ag nanoparticles were adhered to the surfaces of flake-like and strip-shaped PPy, indicating good interface bonding. In addition, the qualitative and quantitative antibacterial assays suggest that the cellulose aerogels functionalized with PPy and nano-Ag have the ability to kill Gram-negative bacteria (*E. coli*), Gram-positive bacteria (*S. aureus*) and intracellular bacteria (*L. monocytogenes*) at direct contact and to inhibit their growth in the proximity of the aerogels. In summary, the cellulose aerogels can provide a potential platform to combine with a large variety of nanomaterials to exploit new forms of advanced functional materials.

Acknowledgements

This study was supported by the National Natural Science Foundation of China (grant nos. 31270590 and 31470584).

References

[1] áL Tan K. Palladium-containing polyaniline and polypyrrole microparticles[J]. Journal of Materials Chemistry, 1998,

8(8): 1743-1748.

[2] Alzoreky N S, Nakahara K. Antibacterial activity of extracts from some edible plants commonly consumed in Asia[J]. International journal of food microbiology, 2003, 80(3): 223-230.

[3] Babu K F, Dhandapani P, Maruthamuthu S, et al. One pot synthesis of polypyrrole silver nanocomposite on cotton fabrics for multifunctional property[J]. Carbohydrate polymers, 2012, 90(4): 1557-1563.

[4] Bober P, Liu J, Mikkonen K S, et al. Biocomposites of nanofibrillated cellulose, polypyrrole, and silver nanoparticles with electroconductive and antimicrobial properties[J]. Biomacromolecules, 2014, 15(10): 3655-3663.

[5] Borthakur L J, Sharma S, Dolui S K. Studies on Ag/Polypyrrole composite deposited on the surface of styrene-methyl acrylate copolymer microparticles and their electrical and electrochemical properties[J]. Journal of Materials Science: Materials in Electronics, 2011, 22(8): 949-958.

[6] Cai J, Kimura S, Wada M, et al. Nanoporous cellulose as metal nanoparticles support[J]. Biomacromolecules, 2009, 10(1): 87-94.

[7] Chen J, Xu J, Wang K, et al. Highly thermostable, flexible, and conductive films prepared from cellulose, graphite, and polypyrrole nanoparticles[J]. ACS applied materials & interfaces, 2015, 7(28): 15641-15648.

[8] Chen S, Wu G, Zeng H. Preparation of high antimicrobial activity thiourea chitosan-Ag^+ complex[J]. Carbohydrate Polymers, 2005, 60(1): 33-38.

[9] Durán N, Marcato P D, De Souza G I H, et al. Antibacterial effect of silver nanoparticles produced by fungal process on textile fabrics and their effluent treatment[J]. Journal of biomedical nanotechnology, 2007, 3(2): 203-208.

[10] Tat'yana V V, Efimov O N. Polypyrrole: a conducting polymer; its synthesis, properties and applications[J]. Russian chemical reviews, 1997, 66(5): 443.

[11] Fabry W, Okemo P O, Ansorg R. Antibacterial activity of East African medicinal plants[J]. Journal of ethnopharmacology, 1998, 60(1): 79-84.

[12] Feng X, Sun Z, Hou W, et al. Synthesis of functional polypyrrole/prussian blue and polypyrrole/Ag composite microtubes by using a reactive template[J]. Nanotechnology, 2007, 18(19): 195603.

[13] Gennaro R, Zanetti M, Benincasa M, et al. Pro-rich antimicrobial peptides from animals: structure, biological functions and mechanism of action[J]. Current pharmaceutical design, 2002, 8(9): 763-778.

[14] Holzwarth U, Gibson N. The Scherrer equation versus the 'Debye-Scherrer equation'[J]. Nature nanotechnology, 2011, 6(9): 534-534.

[15] Jebali A, Hekmatimoghaddam S, Behzadi A, et al. Antimicrobial activity of nanocellulose conjugated with allicin and lysozyme[J]. Cellulose, 2013, 20(6): 2897-2907.

[16] Ji M, Chen X, Wai C M, et al. Synthesizing and dispersing silver nanoparticles in a water-in-supercritical carbon dioxide microemulsion[J]. Journal of the American Chemical Society, 1999, 121(11): 2631.

[17] Jiao Y, Wan C, Li J. Room-temperature embedment of anatase titania nanoparticles into porous cellulose aerogels [J]. Applied Physics A, 2015, 120(1): 341-347.

[18] Joo J, Lee J K, Lee S Y, et al. Physical characterization of electrochemically and chemically synthesized polypyrroles [J]. Macromolecules, 2000, 33(14): 5131-5136.

[19] Kenawy E R, Worley S D, Broughton R. The chemistry and applications of antimicrobial polymers: a state-of-the-art review[J]. Biomacromolecules, 2007, 8(5): 1359-1384.

[20] Maróti G, Kereszt A, Kondorosi É, et al. Natural roles of antimicrobial peptides in microbes, plants and animals[J]. Research in microbiology, 2011, 162(4): 363-374.

[21] Marambio-Jones C, Hoek E M V. A review of the antibacterial effects of silver nanomaterials and potential implications for human health and the environment[J]. Journal of nanoparticle research, 2010, 12(5): 1531-1551.

[22] Menon V P, Lei J, Martin C R. Investigation of molecular and supermolecular structure in template-synthesized

polypyrrole tubules and fibrils[J]. Chemistry of materials, 1996, 8(9): 2382-2390.

[23] Monteiro D R, Gorup L F, Takamiya A S, et al. The growing importance of materials that prevent microbial adhesion: antimicrobial effect of medical devices containing silver[J]. International journal of antimicrobial agents, 2009, 34(2): 103-110.

[24] Nishiyama Y, Langan P, Chanzy H. Crystal structure and hydrogen-bonding system in cellulose Iβ from synchrotron X-ray and neutron fiber diffraction[J]. Journal of the American Chemical Society, 2002, 124(31): 9074-9082.

[25] Omastová M, Mosnáčková K, Fedorko P, et al. Polypyrrole/silver composites prepared by single-step synthesis[J]. Synthetic Metals, 2013, 166: 57-62.

[26] Radheshkumar C, Münstedt H. Antimicrobial polymers from polypropylene/silver composites-Ag^+ release measured by anode stripping voltammetry[J]. Reactive and Functional Polymers, 2006, 66(7): 780-788.

[27] Sehaqui H, Zhou Q, Berglund L A. High-porosity aerogels of high specific surface area prepared from nanofibrillated cellulose (NFC)[J]. Composites science and technology, 2011, 71(13): 1593-1599.

[28] Sharma V K, Yngard R A, Lin Y. Silver nanoparticles: green synthesis and their antimicrobial activities[J]. Advances in colloid and interface science, 2009, 145(1-2): 83-96.

[29] Shi Z, Zhou H, Qing X, et al. Facile fabrication and characterization of poly (tetrafluoroethylene)@polypyrrole/nano-silver composite membranes with conducting and antibacterial property[J]. Applied Surface Science, 2012, 258(17): 6359-6365.

[30] Tew G N, Scott R W, Klein M L, et al. De novo design of antimicrobial polymers, foldamers, and small molecules: from discovery to practical applications[J]. Accounts of chemical research, 2010, 43(1): 30-39.

[31] Timofeeva L, Kleshcheva N. Antimicrobial polymers: mechanism of action, factors of activity, and applications[J]. Applied microbiology and biotechnology, 2011, 89(3): 475-492.

[32] Varesano A, Vineis C, Aluigi A, et al. Antibacterial efficacy of polypyrrole in textile applications[J]. Fibers and Polymers, 2013, 14(1): 36-42.

[33] Wan C, Jiao Y, Sun Q, et al. Preparation, characterization, and antibacterial properties of silver nanoparticles embedded into cellulose aerogels[J]. Polymer Composites, 2014.

[34] Wan C, Li J. Embedding ZnO nanorods into porous cellulose aerogels via a facile one-step low-temperature hydrothermal method[J]. Materials & Design, 2015, 83: 620-625.

[35] Wan C, Li J. Facile synthesis of well-dispersed superparamagnetic γ-Fe_2O_3 nanoparticles encapsulated in three-dimensional architectures of cellulose aerogels and their applications for Cr (VI) removal from contaminated water[J]. ACS Sustainable Chemistry & Engineering, 2015, 3(9): 2142-2152.

[36] Wan C, Li J. Synthesis of well-dispersed magnetic $CoFe_2O_4$ nanoparticles in cellulose aerogels via a facile oxidative coprecipitation method[J]. Carbohydrate polymers, 2015, 134: 144-150.

[37] Wan C, Lu Y, Cao J, et al. Preparation, characterization and oil adsorption properties of cellulose aerogels from four kinds of plant materials via a NAOH/PEG aqueous solution[J]. Fibers and Polymers, 2015, 16(2): 302-307.

[38] Wan, CC, Lu, et al. Fabrication of hydrophobic, electrically conductive and flame-resistant carbon aerogels by pyrolysis of regenerated cellulose aerogels[J]. Carbohyd Polym, 2015.

[39] Wan C, Lu Y, Jin C, et al. A facile low-temperature hydrothermal method to prepare anatase titania/cellulose aerogels with strong photocatalytic activities for rhodamine B and methyl orange degradations[J]. Journal of Nanomaterials, 2015, 2015: 4.

[40] Wei Y, Li L, Yang X, et al. One-step UV-induced synthesis of polypyrrole/Ag nanocomposites at the water/ionic liquid interface[J]. Nanoscale research letters, 2010, 5(2): 433-437.

[41] Wiegand I, Hilpert K, Hancock R E W. Agar and broth dilution methods to determine the minimal inhibitory concentration (MIC) of antimicrobial substances[J]. Nature protocols, 2008, 3(2): 163-175.

[42] Xiong R, Lu C, Zhang W, et al. Facile synthesis of tunable silver nanostructures for antibacterial application using cellulose nanocrystals[J]. Carbohydrate polymers, 2013, 95(1): 214-219.

[43] Zasloff M. Antimicrobial peptides of multicellular organisms[J]. Nature, 2002, 415(6870): 389-395.

中文题目: 聚吡咯和纳米银功能化纤维素气凝胶的原位合成、表征和抗菌活性

作者: 万才超,李坚

摘要: 绿色多孔轻质纤维素气凝胶被认为是替代某些石化基质材料来支撑各种纳米材料的理想材料。本研究以废弃麦秸为原料制备纤维素水凝胶,并以廉价的氢氧化钠/聚乙二醇溶液作为纤维素溶剂。在冻干处理前,以银离子为氧化剂,通过操作简便的吡咯原位氧化聚合将纤维素水凝胶与聚吡咯和银纳米粒子结合。通过扫描电镜、透射电镜、X射线能谱、选区电子衍射、X射线光电子能谱和X射线衍射等手段对三组分复合气凝胶进行了表征。采用平行条纹法和最低抑菌浓度法分别对混合气凝胶对大肠杆菌(革兰氏阴性)、金黄色葡萄球菌(革兰氏阳性)和单核细胞增生李斯特菌(胞内细菌)的抑菌活性进行了定性和定量研究。这项工作为纤维素气凝胶与纳米材料的结合提供了一个范例,有助于开发新型的纤维素基功能材料。

关键词: 纤维素气凝胶;聚吡咯;纳米银;抗菌活性;复合材料

Graphene Oxide/Cellulose Aerogels Nanocomposite: Preparation, Pyrolysis, and Application for Electromagnetic Interference Shielding*

Caichao Wan, Jian Li

Abstract: Hybrid aerogels consisting of graphene oxide (GO) and cellulose were prepared via a solution mixing-regeneration-freeze drying process. The presence of GO affected the micromorphology of the hybrid aerogels, and a self-assembly behavior of cellulose was observed after the incorporation of GO. Moreover, there is no remarkable modification in the crystallinity index and thermal stability after the insertion of GO. After the reduction of GO in the hybrid aerogels by L-ascorbic acid and the subsequent pyrolysis of the aerogels, the resultant displays some interesting characteristics, including good electromagnetic interference (EMI) shielding capacity (SE_{total} = 58.4 dB), high electrical conductivity (19.1 $S \cdot m^{-1}$), hydrophobicity, and fire resistance, which provide an opportunity for some advanced applications such as EMI protection, electrochemical devices, water-proofing agents, and fire retardants. Moreover, this work possibly helps to facilitate the development of both cellulose and GO-based materials and expand their application scope.

Keywords: Graphene oxide; Cellulose aerogels; Pyrolysis; Electromagnetic interference shielding; Multifunction; Nanocomposite

1 Introduction

Graphene, a single-atom-thick sheet of honeycomb carbon lattice, has attracted huge interests from both theoretical and experimental studies since its first discovery in 2004 (Allen, Tung & Kaner, 2009; Geim & Novoselov, 2007). Long-range π-conjugation in graphene yields extraordinary thermal, mechanical, and electrical properties, which qualify it for the reinforcement of polymer matrixes. As the precursor of graphene, graphene oxide (GO) is also considered as a good nanofiller due to its outstanding properties similar to graphene. Because of abundant oxygen-containing functional groups (e.g., hydroxyl, epoxy, ether, diol, and ketone groups) on its basal planes and edges, GO has strong interaction with polar substances, which allows it to be readily dispersed in polar polymers or solvents (e.g., water) (Dreyer, Park, Bielawski & Ruoff, 2010; Zhu et al., 2010). To date, extensive efforts have been devoted to investigating the GO or graphene-based bi-component or tri-component hybrids and their applications. Recently, Liang et al. (Liang et al., 2009a) successfully prepared the first poly(vinyl alcohol) nanocomposite using graphene oxide as a mechanical reinforcement material. The nanocomposite exhibited a 76% increase in tensile strength and a 62%

* 本文摘自 Carbohydrate Polymers, 2016, 150: 172-179.

improvement of Young's modulus by the addition of only 0.7% GO. More recently, Yan et al. (Yan et al., 2014) synthesized a promising electrode material based on aniline tetramers and GO, which presented good cycling stability and high specific capacitance. Barras et al. (Barras et al., 2013) reported a tri-component gold nanoparticle/molybdenum cluster/GO nanocomposite with high photocatalytic activity for the degradation of rhodamine B under visible light irradiation. Apart from the aforementioned applications like high-strength materials, electrode materials, and photocatalytic degradation, GO or graphene-based hybrids also hold great potential in electromagnetic interference (EMI) shielding due to their high electrical conductivity and high surface area and aspect ratio. The extensive use of commercial, military and scientific electronic devices and communication instruments creates serious concern for electromagnetic pollution. The electromagnetic waves not only interfere with the normal operation of equipment leading to loss of energy, time and money, but also jeopardize even human health (Zamanian & Hardiman, 2005). Therefore, there is a strong desire to develop novel lightweight and environmentally friendly EMI shielding materials to protect human and electronic equipment from these undesired signals. To date, numerous polymers such as epoxy (Liang et al., 2009b), phenolic resin (Singh et al., 2012), polyaniline (Chang et al., 2012), poly(methyl methacrylate) (Zhang, Yan, Zheng, He & Yu, 2011), polystyrene (Yan, Ren, Pang, Fu, Yang & Li, 2012), polyetherimide (Ling et al., 2013), polyurethane (Hsiao et al., 2013), poly(3, 4-ethylenedioxythiophene) (Liu, Huang & Sun, 2013), and polyvinylidene fluoride (Eswaraiah, Sankaranarayanan & Ramaprabhu, 2011) have been reported to combine with graphene or GO to fabricate various EMI shielding materials.

With the increasing emphasis on green chemistry, some natural polymers such as cellulose and chitin have been paid increasing attention (Kumar, 2000; Moon, Martini, Nairn, Simonsen & Youngblood, 2011), which are considered as promising candidates to substitute some petroleum-based products. Cellulose is a linear natural polymer of anhydroglucose units linked at the one and four carbon atoms by β-glycosidic bond. Cellulose is the most abundant native biopolymer on earth. As one of the typical cellulose products, cellulose aerogels have attracted increasing attention, and some typical merits (including low density, large specific surface area and high porosity) have been verified by many recent reports (Aaltonen & Jauhiainen, 2009; Jin, Nishiyama, Wada & Kuga, 2004; Li, Lu, Yang, Sun, Liu & Zhao, 2011; Liebner, Potthast, Rosenau, Haimer & Wendland, 2008; Ratke, 2011; Sescousse, Gavillon & Budtova, 2011). Therefore, cellulose aerogels are suitable to serve as green matrix materials (Cai, Kimura, Wada & Kuga, 2008; Olsson et al., 2010; Pääkkö et al., 2008). Besides, our previous works also have demonstrated that the cellulose aerogels could serve as good hosts to support well-dispersed nanoscale silver (Wan, Jiao, Sun & Li, 2016), titanium dioxide (Jiao, Wan & Li; Wan, Lu, Jin, Sun & Li, 2015), rod-like zinc oxide (Wan & Li, 2015a), cobalt ferrite (Wan & Li, 2015c), and γ-Fe_2O_3 (Wan & Li, 2015b). In addition, through a pyrolysis treatment under inert gas protection, the cellulose aerogels will be transformed to carbon aerogels, and some new advantages which are different from the cellulose aerogels will appear, such as high electrical conductivity, waterproof ability, and flame resistance (Ishida, Kim, Kuga, Nishiyama & Brown, 2004; Wu, Li, Liang, Chen & Yu, 2013). Also, it has been confirmed that pyrolysis treatment has little influence on the micromorphology of cellulose aerogels (Wan, Lu, Jiao, Jin, Sun & Li, 2015). Undoubtedly, the transformation of cellulose aerogels to carbon aerogels is beneficial to obtain more favorable EMI shielding ability due to increased electrical conductivity. In addition, compared with some conventional metal-based EMI shielding materials, the carbon aerogels integrated with graphene are lightweight and resistant to corrosion, and offer processing advantages.

Taking these considerations into account, herein, lightweight cellulose-based nanocomposite was prepared by incorporating GO into the cellulose aerogels using a low-cost NaOH/polyethylene glycol (PEG) solution (Zhang, Li & Yu, 2010) as the processing solvent. We investigated the influences of the incorporation of GO on the micromorphology, crystal structures, and thermal stabilities of GO/cellulose aerogels (GO/CA) nanocomposite by scanning electron microscope (SEM), transmission electron microscopy (TEM), Raman spectroscopy, X-ray diffraction (XRD), and thermogravimetric analysis (TGA). For the investigation of the potential application in EMI shielding, the GO sheets in the precursor GO/cellulose hydrogels (GO/CH) were reduced by L-ascorbic acid (L-AA) for improving electrical conductivity. After the freeze drying treatment, the resultant graphene nanosheets/cellulose aerogels underwent a pyrolysis process under nitrogen protection. The obtained graphene nanosheets/carbon aerogels (GN/CNA) composite displays good EMI shielding ability. Furthermore, some features including favorable electrical conductivity, hydrophobicity, and flame resistance also appeared.

2 Materials and methods

2.1 Materials

Bamboo fiber with cellulose content of about 72% was supplied by Zhejiang Mingtong Textile Technology Co., Ltd. and further washed several times with distilled water and dried at 60 ℃ for 24 h. Owing to the high cellulose content, the bamboo fiber was directly used as the cellulosic source. GO was purchased from Nanjing XFNANO Materials Tech Co. Ltd., which was synthesized by a modified Hummers method. Chemicals including hydrochloric acid (HCl, 37%), sodium hydroxide (NaOH), PEG-4000, tert-butyl alcohol and L-AA are all of analytical reagent grade and purchased from Shanghai Boyle Chemical Co. Ltd. (China).

2.2 Preparation of GO/CA and GN/CNA

The clean dried bamboo fiber was mixed with a 10% aqueous solution of NaOH/PEG-4000 (9 : 1 wt/wt) with magnetic stirring at room temperature for 5 h to form a 1% homogeneous cellulose solution. Then, the cellulose solution was frozen at −15 ℃ for 12 h, and subsequently thawed at room temperature under vigorous stirring for 30 min. The GO sheets in the water ($2\ mg \cdot mL^{-1}$) were dispersed with a sonifier (Scientz Technology, JY99-IID 900 W/20 kHz) for 3 h. The GO aqueous dispersion was added into the aforementioned thawed cellulose solution according to a 1: 100 mass ratio of GO to cellulose, and the mixed solution (labeled as GO/cellulose-1) was subsequently magnetically stirred for 30 min. After the mixture was frozen at -15 ℃ for 5 h, and the frozen cake immediately underwent a regeneration process by immersing it in a 1% hydrochloric acid for 6 h at room temperature, repeating the immersion process until the formation of a cylindrical hydrogel (namely GO/CH, labeled as GO/CH-1). Finally, the hydrogel was successively rinsed with distilled water and tert-butyl alcohol several times to remove residual ions and water, and then underwent a freeze drying treatment at −35 ℃ and a pressure of 25 Pa for 48 h, and the following GO/CA nanocomposite (marked as GO/CA-1) was obtained. Repeating the above processes using the GO/cellulose mixed solution with a higher GO proportion (i. e., 5 : 100 mass ratio of GO to cellulose), the resulting GO/cellulose mixed solution,

GO/CH, and GO/CA were marked as GO/cellulose-5, GO/CH-5 and GO/CA-5, respectively. Moreover, the pure cellulose aerogel (PCA) was prepared for comparison following the aforementioned processes without adding the GO.

For the preparation of GN/CNA, the clean GO/CH samples were immersed in 100 mL L-AA aqueous solution (5 mg · mL^{-1}) to reduce GO component for 24 h under vigorous stirring. After being rinsed and freeze-dried, the graphene nanosheets/cellulose aerogels were transferred into a tubular furnace for pyrolysis under a flow of nitrogen. The samples were heated to 500 ℃ at a heating rate of 5 ℃ · min^{-1}, and this temperature was maintained for 1 h; then, the temperature raised to 1000 ℃ at 5 ℃ · min^{-1} and held at this temperature for 2 h to allow for complete pyrolysis. Thereafter, the temperature decreased to 500 ℃ at 5 ℃ · min^{-1}, and finally decreased naturally to the room temperature, and the following GN/CNA was obtained. The GN/CNA samples obtained from GO/CH-1 and GO/CH-5 are coded as GN/CNA-1 and GN/CNA-5, respectively. For comparison, PCA was also pyrolyzed following the above process, and the resultant is labeled as PCNA.

2.3 Characterizations

Morphologies were characterized by SEM (Hitachi, S4800) and TEM (FEI, Tecnai G2 F20). Crystalline structures were identified by XRD (Rigaku, D/MAX 2200) operating with Cu Kα radiation (λ = 1.5418 Å) at a scan rate (2θ) of 4° · min^{-1} and the accelerating voltage of 40 kV and the applied current of 30 mA ranging from 8° to 40°. Raman analysis was performed using a Raman spectrometer (Renishaw inVia, Germany) employing a helium/neon laser (633 nm) as the excitation source. Thermal stabilities were determined using a TG analyzer (TA, Q600) with a heating rate of 10 ℃ · min^{-1} in a nitrogen environment. EMI shielding effectiveness was measured using a PNA-X network analyzer (N5244a) at the frequency range of 8.2-12.4 GHz (X-band). The measured samples were prepared by uniformly mixing 20% of GN/CNA (grinded into powder) with 80% of paraffin wax. The mixture was then pressed into a toroidal shaped mould (7.0 mm in external diameter, 3.0 mm in inner diameter and 2.0 mm in thickness). For the measurements of conductivity (σ), the samples were grinded into fine powder and then pressed into a homemade transparent PC tube with an inner diameter of 8 mm. The bottom of the tube is closed by a stationary brass piston, after the introduction of powder sample (ca. 2 cm^3), the upper side is closed by a stainless steel plunger which could be freely moved up and down in the tube. Then a weighed load was put on the upper piston to allow for good electrical contacts between powders. Electrical resistance (R) of the pressed powder was measured at room temperature using an Elecall EM90A digital multimeter, and the conductivity (σ) is calculated according to $\sigma = (1/R)(L/\pi r^2)$, where L is the length of the sample in the tube, and r is the radius of the tube.

3 Results and discussion

3.1 Schematic diagram of preparation of GO/CA

In this work, we fabricated a class of GO/CA nanocomposite by a series of operations including the dissolution of cellulose, freezing-thawing treatment of cellulose solution, mixing the cellulose solution with the

GO dispersion, freezing the mixed solution, regeneration, and freeze drying (Fig. 1). With the introduction of hydrochloric acid during the regeneration, the hydroxyl ions from NaOH were neutralized, and thus the original splitting cellulose molecules caused by NaOH tended to reconnect and entangle with each other to regenerate inter- and intramolecular hydrogen bonds of cellulose. The self-assembly process led to the formation of GO/CH. The final freeze-drying treatment worked by freezing the material and then reducing the surrounding pressure to allow the frozen liquids in the material to sublimate directly from the solid phase to the gas phase. Compared with air drying and oven drying, the freeze-drying method generally cause less shrinkage and collapse to porous structure (Inoue & Osatake, 1988).

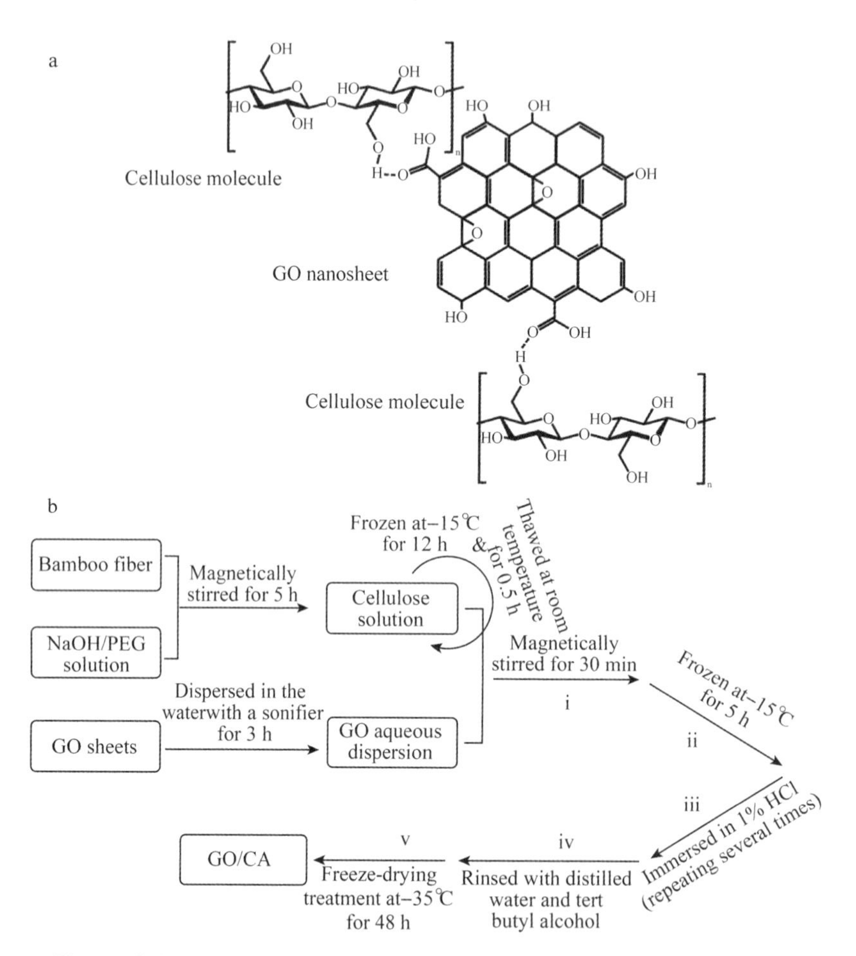

Fig. 1 Schematic illustration of hydrogen bond interactions between GO sheets and cellulose molecules (a) and fabrication of GO/CA nanocomposite (b)

3.2 Optical observation of GO in cellulose solution or cellulose solvent

Fig. 2 shows the digital pictures of the GO aqueous dispersion and the mixture of GO dispersion and cellulose solution (or cellulose solvent), respectively. It is found that the GO dispersion could remain stable for a month at least (Fig. 2b), due to the strong electric repulsion resulted from the abundant oxygen-containing groups on the basal planes and edges of GO sheets (Li, Mueller, Gilje, Kaner & Wallace,

2008). Interestingly, the color of GO dispersion became more and more black with the extending of time, which reveals that the deoxygenation of GO probably occurred (Fan et al., 2008). Fig. 2c and e present the mixture of GO dispersion and cellulose solvent (i. e., NaOH/PEG solution) with different proportions of GO. Apparently, GO sheets tended to aggregate because the inclusion of NaOH destroyed the electrostatic repulsion balance of the GO lattices (Feng, Zhang, Shen, Yoshino & Feng, 2012). However, the mixed solutions of GO dispersion and cellulose solution (Fig. 2a and d) were stable for several days. With the time prolonging, for the mixture with the lower cellulose proportion (5 : 100 mass ratio of GO to cellulose), a phase separation had already taken place (Fig. 2d). However, the mixed solution with the higher cellulose proportion (1 : 100 mass ratio of GO to cellulose) was still homogeneous after a month as shown in Fig. 2a. These results indicate that the presence of cellulose slowed down the aggregation of GO at high NaOH concentration. The homogeneous GO/cellulose mixed solution is important to fabricate the hybrid aerogels.

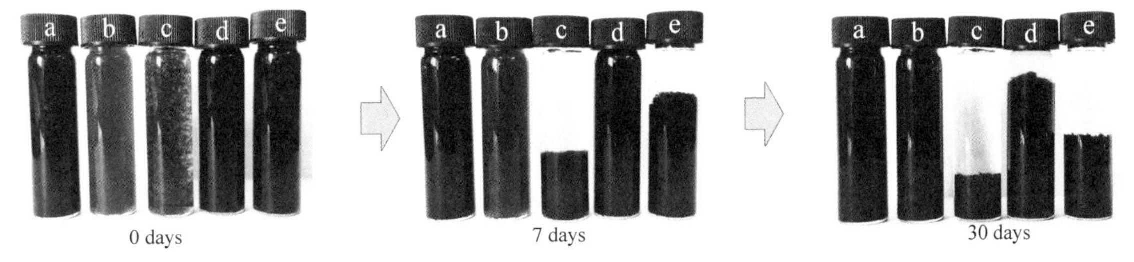

Fig. 2 Digital pictures of GO/cellulose-1 (a), GO aqueous dispersion (b), cellulose solvent with 1 ‱ GO (c), GO/cellulose-5 (d), and cellulose solvent with 5 ‱ GO (e)

3.3 Morphology, crystal structure and thermal stability of GO/CA

The TEM observation in Fig. 3a shows large (micron and submicron scale), transparent and thin sheets with wrinkled and scrolled structure, which is the typical features of GO (Chen, Song, Tang, Lu & Xue, 2012). The porous structure below the thin sheets might be originated from the cellulose aerogels. Fig. 3b shows the Raman spectrum of GO/CA-1, in which two characteristic peaks at 1344 and 1591 cm^{-1} that correspond to the D and G bands can be found (Bora & Dolui, 2012; Kudin et al., 2008). The D band results from the structural imperfections created by the hydroxyl and epoxide groups on the carbon basal plane, and the G band assigns to the first-order scattering of the E_{2g} phonon from sp^2 carbon (graphite lattice). Besides, the intensity ratio (I_D : I_G) is found to be 1.32, higher than some previous reports (Guo et al., 2012; Yadav, Rhee, Jung & Park, 2013) about GO-based composites. The possible reason is attributed to the deoxygenation of GO under alkaline condition, which imports structural defects and disorders (Bo et al., 2014; Bose, Kuila, Uddin, Kim, Lau & Lee, 2010). In addition, the fitting bands at 1100 and 2899 cm^{-1} are originated from the cellulose, corresponding to the COC stretching symmetric and CH stretching vibrations (Agarwal, Reiner & Ralph, 2010), respectively.

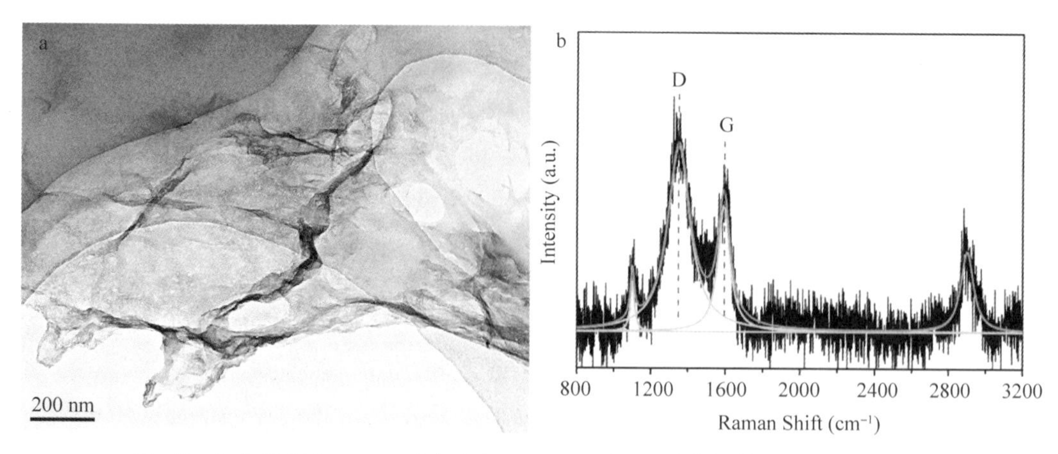

Fig. 3 (a) TEM image and (b) Raman spectrum of GO/CA-1, respectively

Fig. 4 presents the SEM images of PCA, GO/CA-1, and GO/CA-5, respectively. Apparently, the network-like structure can be observed for PCA in Fig. 4a. After the incorporation of GO, there are some changes in morphology. For GO/CA-1 (Fig. 4b), the porous structure is still maintained, but the pore structure became less homogeneous. We can find obvious aggregation phenomena. Similarly to GO/CA-5, the aggregation phenomena can also be seen in Fig. 4c, in which the abundant sheet structures with the sizes of several microns were formed. Higher-magnification SEM images further confirm the aggregation phenomena for GO/CA. PCA exhibits numerous individual fibril structures with the diameters of 10-20 nm randomly entangling with each other (Fig. 4d). While for GO/CA, apart from the stronger aggregation and entanglement leading to the denser networks, the fibril structures are more inclined to assemble within a two-dimensional (2D) plane with the increasing GO content (Fig. 4e and f). These results reveal that the presence of GO could affect the micromorphology of the hybrid aerogels even at such low GO contents. The detail mechanism might be still unknown, but the self-assembly behavior of cellulose structure is worth to be deeply explored in our future research. Regarding the bulk density of the aerogels which was calculated by dividing the sample mass by its volume, the value was calculated to be about 52.1 $mg \cdot cm^{-3}$ for PCA, 55.3 $mg \cdot cm^{-3}$ for GO/CA-1, and 56.9 $mg \cdot cm^{-3}$ for GO/CA-5, respectively.

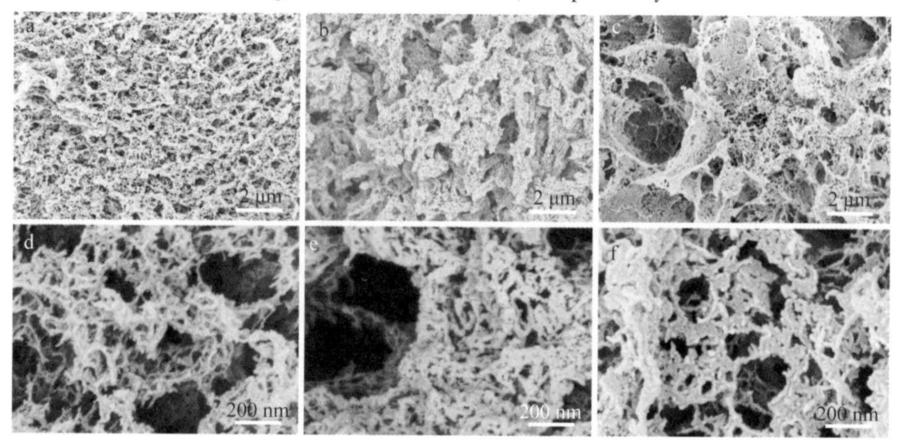

Fig. 4 (a-c) SEM images and (d-f) the corresponding higher-magnification SEM images of PCA, GO/CA-1, and GO/CA-5, respectively

The XRD patterns of GO/CA with different GO proportions are shown in Fig. S1. In the case of PCA, four characteristic peaks centered at 12.1°, 20.2°, 21.8° and 34.8° are observed, corresponding to the (101), $(10\bar{1})$, (002) and (040) planes of the typical profile of cellulose Ⅱ (Nishiyama, Langan & Chanzy, 2002). In the case of GO/CA, the peak at around 12° is possibly attributed either to cellulose or to GO or to both. The intensities of the aforementioned diffraction peaks changed little after the incorporation of GO. We calculated the relative crystallinity index (C_I) as the ratio of the area arising from the crystalline phase to the total area. The results show that the C_I of PCA equals 64.2%. After the incorporation of GO, the C_I of GO/CA-1 and GO/CA-5 was calculated to be 67.5% and 58.6%, respectively. These differences are within the errors. Therefore, the presence of GO might have little effects on the crystallization behaviors of cellulose in this work.

We adopted TG and derivative thermogravimetry (DTG) to compare the thermal stability of GO/CA and PCA, and the results are given in Fig. S2. The slight weight loss for all samples below 150 ℃ results from the evaporation of adsorbed water. The strong peaks centered at 350-400 ℃ are attributed to the pyrolysis of cellulose composition (Xiao, Sun & Sun, 2001). There is no remarkable mass loss related to the decomposition of the oxygen-containing groups on the GO surface, possibly due to the deoxygenation and low proportion of GO in the composite. The temperature at the highest degradation rate is 363 ℃ for PCA, 367 ℃ for GO/CA-1, and 353 ℃ for GO/CA-5, respectively. These differences are within the errors, indicating that the incorporation of GO didn't significantly modify the thermal stability of the aerogels.

3.4 EMI shielding ability of GN/CNA

For the investigation of the potential application in EMI shielding, the GO component was reduced to graphene by L-AA for improving electrical conductivity. After the freeze drying, the graphene nanosheets/cellulose aerogels were obtained and subsequently subjected to a pyrolysis process to fabricate the carbon aerogels containing the graphene nanosheets (i.e., GN/CNA).

A vector network analyzer is an instrument that measures the network parameters of electrical networks. The analyzer is commonly used to measure S-parameters and characterize two-port networks. The EMI shielding property is evaluated by shielding effectiveness expressed in decibels (dB) over the frequency range of 8.2-12.4 GHz. A higher decibel level suggests less energy transmitted through shielding material. The total shielding effectiveness (SE_{total}) can be expressed as (Liu et al., 2014):

$$SE_{total}(\text{dB}) = 10\log\frac{P_i}{P} = SE_A + SE_R + SE_M \quad (1)$$

where P_i and P_t are the incident and transmitted electromagnetic power, respectively. SE_R and SE_A are the shielding effectiveness due to reflection and absorption, respectively. SE_M is multiple reflection effectiveness inside the material, which can be negligible when $SE_{total}>10$ dB. Besides, SE_R and SE_A can be described as (Cao et al., 2012):

$$SE_R = -10\log(1-R) \quad (2)$$

$$SE_A = -10\log[T/(1-R)] \quad (3)$$

where R and T are reflected power and transmitted power, respectively. The scattering parameters (S_{ij} denotes the power transmitted from port i to port j) of the vector network analyzer adopting two-port network test method are related to R and T, respectively, that is, $R=|S_{11}|^2=|S_{22}|^2$ *and* $T=|S_{12}|^2=|S_{21}|^2$.

Fig. 5a demonstrates that EMI SE_{total} presents typical graphene-loading dependent performance. The SE_{total} values of GN/CNA-1 and GN/CNA-5 are approximately 39.2 and 58.4 dB, which are 15.0% and 71.3% higher than that of PCNA (34.1 dB), respectively. All the shielding effectiveness reaches the commercially achievable level(> 20 dB). It is not hard to find that the increase of graphene content leads to the increase of SE_{total} value. In addition, regardless of the error between coaxial and waveguide methods, GN/CNA-5 obviously presents more superior shielding effectiveness, as compared to some previously reported carbon-based EMI shielding materials with equal or greater thickness (≥ 2 mm), such as carbon fiber/silica (12.4 dB) (Cao, Song, Hou, Wen & Yuan, 2010), multiwalled carbon nanotubes/portland cement (28 dB) (Singh et al., 2013), biomorphic porous carbon (40 dB) (Liu et al., 2012), exfoliated few-layer graphene (12 dB) (Song et al., 2013), and highly ordered porous carbon (50 dB) (Song et al., 2014a), as shown in Table 1. For comparing the SE_{total} value of GN/CNA-5 with that of other shielding materials with less thickness (1 mm) in Table 1 [including Graphene/ethylene-vinyl acetate copolymers resins (Song et al., 2014b) and multiwalled carbon nanotubes/polymethyl methacrylate (Yuen et al., 2008)], we adopted some empirical equations to calculate and predict the theoretical SE_{total} value of GN/CNA-5 in the case that the thickness is 1 mm. The SE_{total} value depends on the relative value of the sample thickness (t) compared to the skin depth (δ), which in meters is defined by (Zhang, Ni, Fu & Kurashik, 2007):

$$\delta = \frac{1}{\sqrt{\pi f \mu \sigma}} \tag{4}$$

where f is the frequency (in Hz), μ is the magnetic permeability by $\mu = \mu_0 \mu_r$, $\mu_0 = 4\pi \times 10^{-7}$ H · m^{-1} is the absolute permeability of free space (air), and σ is the electrical conductivity in S · m^{-1}. Since GN/CNA-5 can be considered as a non-magnetic substance, we have $\mu_r = 1$. According to the report by Li et al. (Li et al., 2010), when $t/\delta \geq 1.3$, SE_{total} can be expressed as follows:

$$SE_{total} = 20\log\left(\frac{Z_0 \delta \sigma}{2\sqrt{2}}\right) + 8.68\frac{t}{\delta} \tag{5}$$

where $Z_0 = 120\pi$, and t is the specimen thickness in meters. When $t/\delta \leq 1.3$, SE_{total} will result in following formula

$$SE_{total} = 20\log\left(1 + \frac{Z_0 t \sigma}{2}\right) \tag{6}$$

In this work, the frequency (f) is $8.2\times 10^9 - 12.4 \times 10^9$ Hz, and thus the skin depth (δ) can be calculated to be $1.0\text{-}1.3 \times 10^{-3}$ m ($\sigma = 19.1$ S · m^{-1}, appearing in the next section). So $t/\delta = 0.8\text{-}1$ ($t = 1$ mm), and SE_{total} can be calculated by the equation (6). The calculated results show that the SE_{total} value is around 13.2 dB, which is lower than that of the EMI shielding materials with a thickness of 1 mm as shown in Table 1. However, considering the error between the SE_{total} theoretical value (ca. 25 dB) and measured value (58.4 dB) of GN/CNA-5 when the thickness is 2 mm, we believe that the measured SE_{total} value of GN/CNA-5 with a thickness of 1 mm is probably much higher than the above theoretical value.

Moreover, the improvement in SE_{total} with the increase of graphene content might be ascribed to the high conductivity and large number density of graphene nanosheets, contributing to constructing denser conducting networks and providing more plentiful mobile charge carriers and conductive channels, which help a lot in increasing the conductivity of the composite (confirmed in the next section).[56,59] Besides, according to the research by Wang et al. (Wang et al., 2011), the residual defects and groups in the graphene cannot only prompt energy transition from contiguous states to Fermi level (Belavin, Okotrub & Bulusheva, 2002; Watts, Hsu, Barnes & Chambers, 2003), but also can introduce defect polarization relaxation (Che, Peng, Duan, Chen & Liang, 2004) and groups' electronic dipole relaxation (Paredes, Villar-Rodil, Martinez-Alonso & Tascon, 2008), which are all in favor of electromagnetic wave penetration and absorption.

The EMI shielding mechanism was analyzed by comparing the contribution of reflection (SE_R) and absorption (SE_A) to the overall EMI shielding effectiveness (Fig. 5b and c). Apparently, SE_A exhibits the much higher values comparing to those of SE_R for all samples, which implies that the EMI shielding mechanism is absorption dominant. Undoubtedly, an absorption-dominant shielding mechanism is beneficial to alleviate secondary radiation, which is considered as more attractive alternative for the fabrication of electromagnetic radiation protection products (e.g., coating and fabric). Additionally, the overall growth trends of SE_{total} over whole frequency have been theoretically predicted and experimentally confirmed for the absorption-dominant materials (Zhang, Zheng, Yan, Jiang & Yu, 2012).

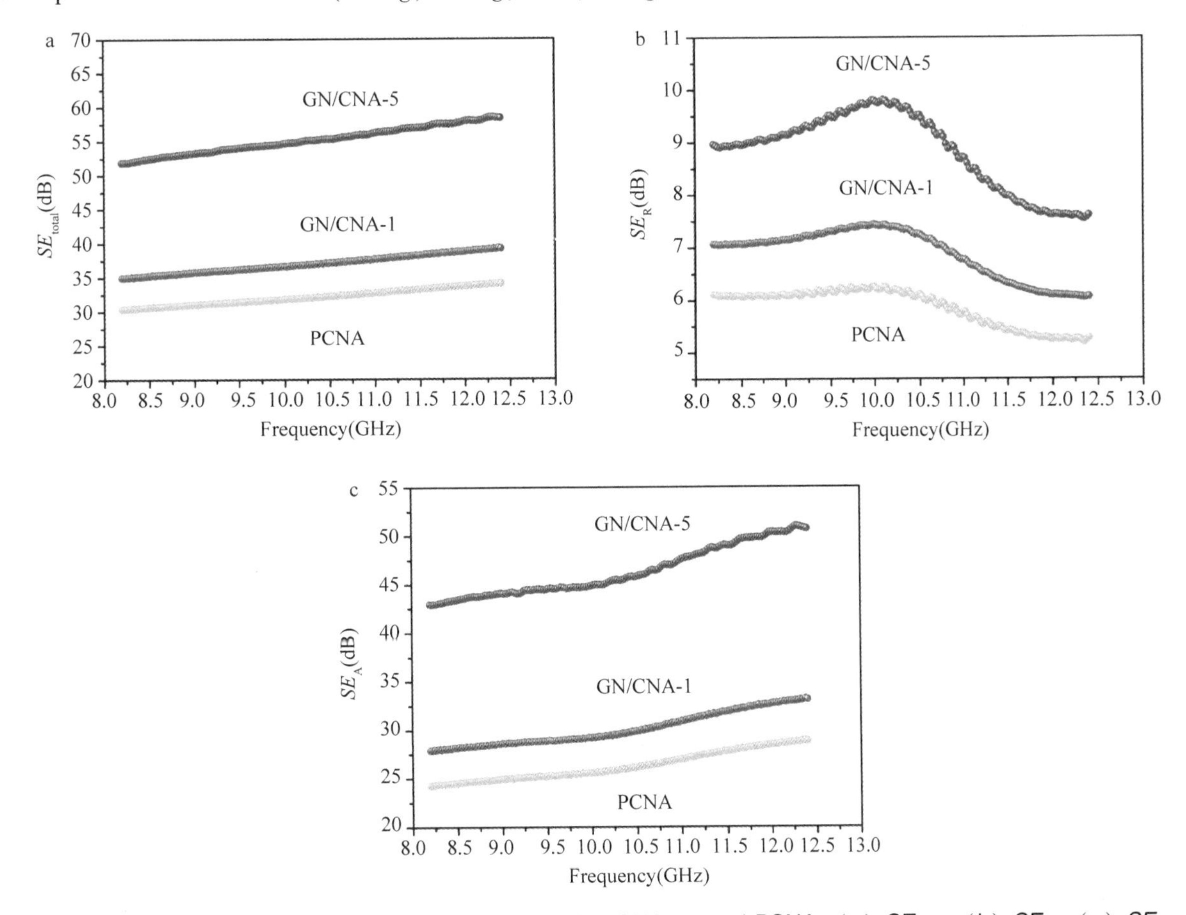

Fig. 5 EMI shielding effectiveness of GN/CNA-1, GN/CNA-5, and PCNA: (a) SE_{total}, (b) SE_R, (c) SE_A

Table 1 Typical carbon-based materials for EMI shielding

Materials	Method	Best SE_{total}(dB)	Thickness (mm)	Frequency range (GHz)	Refs
Carbon fiber/silica	Waveguide	12.4	2.5	8.2–12.4	Cao, Song, Hou, Wen & Yuan, 2010
Graphene/EVA[a] resins	Waveguide	27	1	8.2–12.4	Song et al., 2014b
MWCNT[b]/portland cement	Waveguide	28	2	8.2–12.4	Singh et al., 2013
Biomorphic porous carbon	Waveguide	40	4	8.2–12.4	Liu et al., 2012
MWCNT/PMMA[c]	Waveguide	58.7	1	2-18	Yuen et al., 2008
Exfoliated few-layer graphene	Coaxial(85% wax)	12	2.5	8.2-12.4	Song et al., 2013
Highly ordered porous carbon	Coaxial(80% wax)	50	2	2-18	Song et al., 2014a
GN/CNA-5	Coaxial(80% wax)	58.4	2	8.2–12.4	This work

[a]ethylene-vinyl acetate copolymers; [b]multiwalled carbon nanotubes; [c]polymethyl methacrylate.

3.5 Electrical conductivity, hydrophobicity, and flame resistance of GN/CNA

Apart from the good EMI shielding ability, some other interesting features also appeared. The electrical conductivity of GN/CNA-1 was calculated as 8.2 S · m^{-1}. With the increase of graphene content, the conductivity of GN/CNA-5 increases to 19.1 S · m^{-1}, which is around three times as high as that of PCNA (ca. 5.9 S · m^{-1}). Therefore, the results reveal that the incorporation of graphene sheets dramatically enhances the electrical conductivity of the composite. The electrical conductivity of GN/CNA is also comparable with that of some previously reported carbon-based materials measured by the similar method, such as carbon xerogels (5-40 S · m^{-1}) (Alegre et al., 2011) and carbon nanofibers (2-388 S · m^{-1}) (Sebastián, Suelves, Moliner & Lázaro, 2010). The high conductivity of GN/CNA assures that it can be used as a flexible lead to connect components in circuits. As shown in Fig. S3a, the battery could light a blue light-emitting-diode (LED) by using GN/CNA as a connecting lead. In addition, the good hydrophobicity can also be seen for GN/CNA due to the destruction of the original hydrophilic groups (like hydroxyl groups) of cellulose during the pyrolysis. GN/CNA could support a spherical water droplet (colored with methyl orange for clarity) on its surface (Fig. S3b). Besides, GN/CNA also exhibits obvious mirror-reflection phenomenon on its surface when immersed in water (Fig. S3c), further confirming the formation of hydrophobicity. The phenomenon is ascribed to the formation of an interface between the entrapped air in the hybrid 3D aerogels and the surrounding water (Yao, Yu, Xu, Chen & Fu, 2015). Fig. S3d displays the burning process of GN/CNA in an alcohol lamp. Apparently, GN/CNA shows excellent flame resistance. The sample not only maintained its shape during the whole burning process, but also didn't support any burning and release obvious smoke all the time.

4 Conclusions

The GO/CA nanocomposite was fabricated by a solution mixing-regeneration-freeze drying process. The presence of cellulose helps to slow down the sedimentation of GO at the high NaOH concentration. Moreover, because of the incorporation of GO, the micromorphology of the hybrid aerogels changed, and we observed the self-assembly behavior of cellulose, which is worth to be further studied. In addition, there is no remarkable modification in the crystallinity index and thermal stability of the hybrid aerogels after the insertion of GO. After the in-situ reduction of GO composition and the pyrolysis of graphene nanosheets/cellulose aerogels, the resulting GN/CNA displays good EMI shielding ability with the best SE_{total} of 58.4 dB. Moreover, the absorption-dominant shielding mechanism is beneficial to alleviate secondary radiation, which is considered as more attractive alternative for the fabrication of electromagnetic radiation protection products. Some characteristics caused by the pyrolysis (i. e., good electrical conductivity, hydrophobicity, and flame retardancy) make it of interest to use this class of miscellaneous nanocomposite as advanced electrochemical devices and waterproof and flame-resistant materials.

Acknowledgment

This study was supported by the National Natural Science Foundation of China (grant nos. 31270590 and 31470584).

References

[1] Aaltonen O, Jauhiainen O. The preparation of lignocellulosic aerogels from ionic liquid solutions[J]. Carbohydrate Polymers, 2009, 75(1): 125-129.

[2] Agarwal U P, Reiner R S, Ralph S A. Cellulose I crystallinity determination using FT-Raman spectroscopy: univariate and multivariate methods[J]. Cellulose, 2010, 17(4): 721-733.

[3] Alegre C, Calvillo L, Moliner R, et al. Pt and PtRu electrocatalysts supported on carbon xerogels for direct methanol fuel cells[J]. Journal of Power Sources, 2011, 196(9): 4226-4235.

[4] Allen M J, Tung V C, Kaner R B. Honeycomb carbon: a review of graphene[J]. Chemical reviews, 2010, 110(1): 132-145.

[5] Barras A, Das M R, Devarapalli R R, et al. One-pot synthesis of gold nanoparticle/molybdenum cluster/graphene oxide nanocomposite and its photocatalytic activity[J]. Applied Catalysis B: Environmental, 2013, 130: 270-276.

[6] Belavin V V, Okotrub A V, Bulusheva L G. A study of the influence of structural imperfection on the electronic structure of carbon nanotubes by X-ray spectroscopy and quantum-chemical methods[J]. Physics of the Solid State, 2002, 44(4): 663-665.

[7] Bo Z, Shuai X, Mao S, et al. Green preparation of reduced graphene oxide for sensing and energy storage applications[J]. Scientific reports, 2014, 4(1): 1-8.

[8] Bora C, Dolui S K. Fabrication of polypyrrole/graphene oxide nanocomposites by liquid/liquid interfacial polymerization and evaluation of their optical, electrical and electrochemical properties[J]. Polymer, 2012, 53(4): 923-932.

[9] Bose S, Kuila T, Uddin M E, et al. In-situ synthesis and characterization of electrically conductive polypyrrole/graphene nanocomposites[J]. Polymer, 2010, 51(25): 5921-5928.

[10] Cai J, Kimura S, Wada M, et al. Nanoporous cellulose as metal nanoparticles support[J]. Biomacromolecules, 2009, 10(1): 87-94.

[11] Cao M S, Song W L, Hou Z L, et al. The effects of temperature and frequency on the dielectric properties, electromagnetic interference shielding and microwave-absorption of short carbon fiber/silica composites[J]. Carbon, 2010, 48(3): 788-796.

[12] Cao M S, Yang J, Song W L, et al. Ferroferric oxide/multiwalled carbon nanotube vs polyaniline/ferroferric oxide/multiwalled carbon nanotube multiheterostructures for highly effective microwave absorption[J]. ACS applied materials & interfaces, 2012, 4(12): 6949-6956.

[13] Chang C H, Huang T C, Peng C W, et al. Novel anticorrosion coatings prepared from polyaniline/graphene

composites[J]. Carbon, 2012, 50(14): 5044-5051.

[14]Che R C, Peng L M, Duan X F, et al. Microwave absorption enhancement and complex permittivity and permeability of Fe encapsulated within carbon nanotubes[J]. Advanced Materials, 2004, 16(5): 401-405.

[15]Chen Y, Song B, Tang X, et al. One-step synthesis of hollow porous Fe_3O_4 beads-reduced graphene oxide composites with superior battery performance[J]. Journal of Materials Chemistry, 2012, 22(34): 17656-17662.

[16]Dreyer D R, Park S, Bielawski C W, et al. The chemistry of graphene oxide[J]. Chemical society reviews, 2010, 39(1): 228-240.

[17]Eswaraiah V, Sankaranarayanan V, Ramaprabhu S. Functionalized graphene-VDF foam composites for EMI shielding [J]. Macromolecular Materials and Engineering, 2011, 296(10): 894-898.

[18]Fan X, Peng W, Li Y, et al. Deoxygenation of exfoliated graphite oxide under alkaline conditions: a green route to graphene preparation[J]. Advanced Materials, 2008, 20(23): 4490-4493.

[19]Geim A K, Novoselov K S. The rise of graphene[J]. Nature Materials, 2007, 6: 183-191.

[20]Guo Y, Sun X, Liu Y, et al. One pot preparation of reduced graphene oxide (RGO) or Au (Ag) nanoparticle-RGO hybrids using chitosan as a reducing and stabilizing agent and their use in methanol electrooxidation[J]. Carbon, 2012, 50(7): 2513-2523.

[21]Hsiao S T, Ma CC M, Tien H W, et al. Using a non-covalent modification to prepare a high electromagnetic interference shielding performance graphene nanosheet/water-borne polyurethane composite[J]. Carbon, 2013, 60: 57-66.

[22]INOUÉ T, OSATAKE H. A new drying method of biological specimens for scanning electron microscopy: the t-butyl alcohol freeze-dryingmethod[J]. Archives of histology and cytology, 1988, 51(1): 53-59.

[23]Ishida O, Kim D Y, Kuga S, et al. Microfibrillar carbon from native cellulose[J]. Cellulose, 2004, 11(3): 475-480.

[24]Jiao Y, Wan C, Li J. Room-temperature embedment ofanatase titania nanoparticles into porous cellulose aerogels[J]. Applied Physics A, 2015, 120(1): 341-347.

[25]Jin H, Nishiyama Y, Wada M, et al. Nanofibrillar cellulose aerogels[J]. Colloids and Surfaces A: Physicochemical and Engineering Aspects, 2004, 240(1-3): 63-67.

[26]Kudin K N, Ozbas B, Schniepp H C, et al. Raman spectra of graphite oxide and functionalized graphene sheets[J]. Nano letters, 2008, 8(1): 36-41.

[27]Kumar M N V R. A review of chitin and chitosanapplications[J]. Reactive and functional polymers, 2000, 46(1): 1-27.

[28]Li Y, Chen C, Li J T, et al. Enhanced dielectric constant for efficient electromagnetic shielding based on carbon-nanotube-added styrene acrylic emulsion based composite[J]. Nanoscale research letters, 2010, 5(7): 1170-1176.

[29] Li D, Müller M B, Gilje S, et al. Processable aqueous dispersions of graphene nanosheets [J]. Nature nanotechnology, 2008, 3(2): 101-105.

[30]Li J, Lu Y, Yang D, et al. Lignocellulose aerogel from wood-ionic liquid solution (1-allyl-3-methylimidazolium chloride) under freezing and thawingconditions[J]. Biomacromolecules, 2011, 12(5): 1860-1867.

[31]Liang J, Huang Y, Zhang L, et al. Molecular-level dispersion ofgraphene into poly (vinyl alcohol) and effective reinforcement of their nanocomposites[J]. Advanced Functional Materials, 2009, 19(14): 2297-2302.

[32]Liang J, Wang Y, Huang Y, et al. Electromagnetic interference shielding ofgraphene/epoxy composites[J]. Carbon, 2009, 47(3): 922-925.

[33]Liebner F, Potthast A, Rosenau T, et al. Cellulose aerogels: Highly porous, ultra-lightweight materials[J]. 2008, 62(2), 129-135.

[34]Ling J, Zhai W, Feng W, et al. Facile preparation of lightweight microcellular polyetherimide/graphene composite foams for electromagnetic interference shielding[J]. ACS applied materials & interfaces, 2013, 5(7): 2677-2684.

[35]Liu P B, Huang Y, Sun X. Excellent electromagnetic absorption properties of poly (3, 4-ethylenedioxythiophene)-reduced graphene oxide-Co_3O_4 composites prepared by a hydrothermal method[J]. ACS applied materials & interfaces, 2013, 5 (23): 12355-12360.

[36]Liu Q, Gu J, Zhang W, et al. Biomorphic porous graphitic carbon for electromagnetic interference shielding[J]. Journal of Materials Chemistry, 2012, 22(39): 21183-21188.

[37]Liu X, Yin X, Kong L, et al. Fabrication and electromagnetic interference shielding effectiveness of carbon nanotube reinforced carbon fiber/pyrolytic carbon composites[J]. Carbon, 2014, 68: 501-510.

[38]Moon R J, Martini A, Nairn J, et al. Cellulose nanomaterials review: structure, properties and nanocomposites[J]. Chemical Society Reviews, 2011, 40(7): 3941-3994.

[39]Nishiyama Y, Langan P, Chanzy H. Crystal structure and hydrogen-bonding system in cellulose Iβ from synchrotron X-ray and neutron fiber diffraction[J]. Journal of the American Chemical Society, 2002, 124(31): 9074-9082.

[40]Olsson R T, Samir M A S A, Salazar-Alvarez G, et al. Making flexible magnetic aerogels and stiff magneticnanopaper using cellulose nanofibrils as templates[J]. Nature nanotechnology, 2010, 5(8): 584-588.

[41]Paakko M, Vapaavuori J, Silvennoinen R, et al. Long and entangled native cellulose I nanofibers allow flexible aerogels and hierarchically porous templates for functionalities[J]. Soft Matter, 2008, 4(12): 2492-2499.

[42]Paredes J I, Villar-Rodil S, Martínez-Alonso A, et al. Graphene oxide dispersions in organic solvents[J]. Langmuir, 2008, 24(19): 10560-10564.

[43]Ratke L. Monoliths and fibrous cellulose aerogels[M]//Aerogels handbook. Springer, New York, NY, 2011: 173-190.

[44]Sebastian D, Suelves I, Moliner R, et al. The effect of the functionalization of carbon nanofibers on their electronic conductivity[J]. Carbon, 2010, 48(15): 4421-4431.

[45]Sescousse R, Gavillon R, Budtova T. Aerocellulose from cellulose-ionic liquid solutions: preparation, properties and comparison with cellulose-NaOH and cellulose-NMMO routes[J]. Carbohydrate polymers, 2011, 83(4): 1766-1774.

[46]Singh A P, Garg P, Alam F, et al. Phenolic resin-based composite sheets filled with mixtures of reduced graphene oxide, γ-Fe_2O_3 and carbon fibers for excellent electromagnetic interference shielding in the X-band[J]. Carbon, 2012, 50(10): 3868-3875.

[47]Singh A P, Gupta B K, Mishra M, et al. Multiwalled carbon nanotube/cement composites with exceptional electromagnetic interference shielding properties[J]. Carbon, 2013, 56: 86-96.

[48]Song W L, Cao M S, Fan L Z, et al. Highly ordered porous carbon/wax composites for effective electromagnetic attenuation andshielding[J]. Carbon, 2014, 77: 130-142.

[49]Song W L, Cao M S, Lu M M, et al. Flexible graphene/polymer composite films in sandwich structures for effective electromagnetic interference shielding[J]. Carbon, 2014, 66: 67-76.

[50]Song W L, Cao M S, Lu M M, et al. Alignment of graphene sheets in wax composites for electromagnetic interference shielding improvement[J]. Nanotechnology, 2013, 24(11): 1-10.

[51]Wan C, Jiao Y, Sun Q, et al. Preparation, characterization, and antibacterial properties of silver nanoparticles embedded into celluloseaerogels[J]. Polymer Composites, 2016, 37(4), 1137-1142.

[52]Wan C, Li J. Embedding ZnO nanorods into porous cellulose aerogels via a facile one-step low-temperature hydrothermal method[J]. Materials & Design, 2015, 83: 620-625.

[53]Wan C, Li J. Facile synthesis of well-dispersedsuperparamagnetic γ-Fe_2O_3 nanoparticles encapsulated in three-dimensional architectures of cellulose aerogels and their applications for Cr (VI) removal from contaminated water[J]. ACS Sustainable Chemistry & Engineering, 2015, 3(9): 2142-2152.

[54]Wan C, Li J. Synthesis of well-dispersed magnetic $CoFe_2O_4$ nanoparticles in cellulose aerogels via a facile oxidative co-precipitation method[J]. Carbohydrate polymers, 2015, 134: 144-150.

[55] Wan, CC, Lu, et al. Fabrication of hydrophobic, electrically conductive and flame-resistant carbon aerogels by pyrolysis of regenerated celluloseaerogels[J]. Carbohyd Polym, 2015, 118: 115-118.

[56] Wan C, Lu Y, Jin C, et al. A facile low-temperature hydrothermal method to prepareanatase titania/cellulose aerogels with strong photocatalytic activities for rhodamine B and methyl orange degradations[J]. Journal of Nanomaterials, 2015, 1-8.

[57] Wang C, Han X, Xu P, et al. The electromagnetic property of chemically reduced graphene oxide and its application as microwave absorbing material[J]. Applied Physics Letters, 2011, 98(7): 1-3.

[58] Watts P C P, Hsu W K, Barnes A, et al. High permittivity from defectivemultiwalled carbon nanotubes in the X-band [J]. Advanced Materials, 2003, 15(7-8): 600-603.

[59] Wu Z Y, Li C, Liang H W, et al. Ultralight, flexible, and fire-resistant carbon nanofiber aerogels from bacterial cellulose[J]. Angewandte Chemie, 2013, 125(10): 2997-3001.

[60] Xiao B, Sun X F, Sun R C. Chemical, structural, and thermal characterizations of alkali-solublelignins and hemicelluloses, and cellulose from maize stems, rye straw, and rice straw[J]. Polymer degradation and stability, 2001, 74(2): 307-319.

[61] Yadav M, Rhee K Y, Jung I H, et al. Eco-friendly synthesis, characterization and properties of a sodium carboxymethyl cellulose/graphene oxide nanocomposite film[J]. Cellulose, 2013, 20(2): 687-698.

[62] Yan D X, Ren P G, Pang H, et al. Efficient electromagnetic interference shielding of lightweight graphene/polystyrene composite[J]. Journal of Materials Chemistry, 2012, 22(36): 18772-18774.

[63] Yan J, Yang L, Cui M, et al. Aniline Tetramer-Graphene Oxide Composites for High Performance Supercapacitors [J]. Advanced Energy Materials, 2014, 4(18), 1-7.

[64] Yao X, Yu W, Xu X, et al. Amphiphilic, ultralight, and multifunctional graphene/nanofibrillated cellulose aerogel achieved by cation-induced gelation and chemical reduction[J]. Nanoscale, 2015, 7(9): 3959-3964.

[65] Yuen S M, Ma CC M, Chuang C Y, et al. Effect of processing method on the shielding effectiveness of electromagnetic interference of MWCNT/PMMA composites[J]. Composites Science and Technology, 2008, 68(3-4): 963-968.

[66] Zamanian A, Hardiman C. Electromagnetic radiation and human health: A review of sources and effects[J]. High Frequency Electronics, 2005, 4(3): 16-26.

[67] Zhang C S, Ni QQ, Fu S Y, et al. Electromagnetic interference shielding effect of nanocomposites with carbon nanotube and shape memory polymer[J]. Composites Science and Technology, 2007, 67(14): 2973-2980.

[68] Zhang H B, Yan Q, Zheng W G, et al. Tough graphene-polymer microcellular foams for electromagnetic interference shielding[J]. ACS applied materials & interfaces, 2011, 3(3): 918-924.

[69] Zhang H B, Zheng W G, Yan Q, et al. The effect of surface chemistry of graphene on rheological and electrical properties of polymethylmethacrylate composites[J]. Carbon, 2012, 50(14): 5117-5125.

[70] Zhang S H, Li F X, Yu J Y. Structure and properties of novel cellulosefibres produced from NaOH/PEG-treated cotton linters[J]. Iranian Polymer Journal, 2010, 19(12): 949-957.

[71] Zhu Y, Murali S, Cai W, et al. Graphene and graphene oxide: synthesis, properties, and applications [J]. Advanced materials, 2010, 22(35): 3906-3924.

中文题目：氧化石墨烯/纤维素气凝胶纳米复合材料的制备、热解及电磁屏蔽的应用

作者：万才超，李坚

摘要：采用溶液混合—再生—冷冻干燥的工艺制备由氧化石墨烯(GO)和纤维素组成的复合气凝胶。GO的存在影响了复合气凝胶的微观形貌，在纤维素中加入GO后的出现自组装行为。此外，GO的加入对结晶度和热稳定性没有明显的影响。用L-抗坏血酸还原复合气凝胶中的GO并随后对气凝胶进行热解，所得物显示一些有趣的特性，包括良好的电磁（EMI）屏蔽能力(SE_{total}=58.4 dB)，高导电

率(19.1 S · m^{-1}),疏水性,防火性,为EMI保护,电化学装置,防水剂和阻燃剂等一些先进应用提供了机会。此外,这项工作可能有助于促进纤维素和氧化石墨烯基材料的开发,扩大其应用范围。

关键词: 氧化石墨烯;纤维素气凝胶;热解;电磁屏蔽;多功能;纳米复合材料

Preparation, Characterization and Antibacterial Properties of Silver Nanoparticles Embedded into Cellulose Aerogels*

Caichao Wan, Yue Jiao, Jian Li, Qingfeng Sun

Abstract: In this paper, highly-loaded silver (Ag) nanoparticles with mean diameters of about 7.83 nm were synthesized by reducing the Ag ions by $NaBH_4$ with strong reducibility, and homogeneously embedded into cellulose aerogels without obvious reunion. The as-prepared nano-Ag/cellulose (NAC) aerogels maintained nanoporous and multi-scale morphology similar to the pure cellulose aerogels, and showed strong antibacterial activities for both *Escherichia coli* (Gram-negative) and *Staphylococcus aureus* (Gram-positive). Meanwhile, after the incorporation of Ag nanoparticles, NAC aerogels also displayed more superior thermal stability. Thus, the novel NAC aerogels might be expected to be used as various biomedical applications, especially green heat-resistant high-performance antibacterial materials.

Keywords: Antibacterial; Silver; Cellulose aerogels; Nanocomposites; Polymers

1 Introduction

Cellulose aerogels, derived from spontaneous physical cross-linking of cellulose chains due to many hydroxyl groups on the surface that can easily form hydrogen bonding linked networks, display multitudinous alluring properties such as high porosity, large specific surface area, low density, biodegradability, good formability and biocompatibility[1-5]. Moreover, otherwise than the corresponding primary structure namely cellulose molecule: an unbranched polymer of β-1, 4-linked glucopyranose, nanoporous cellulose aerogels with high specific surface area endow the surface of micro-fiber with a great deal of hydroxyls and ether bonds[6], form the main controlling site of the template to guide the synthesis of nanoparticles[7], and show multi-scale micro-nano pore structure leading to numerous superior macro-performances including adsorption, catalysis and low conduction[8-10]. Thereby, cellulose aerogels may be considered as ideal hydrophilic matrixes to support nanoparticles for synthesizing high-performance multifunctional nanocomposites.

Nanoscale metals particles generally exhibit significant differences in micro and macro properties compared with their corresponding bulk metals due to the nanometer size effect, and thus may find various applications in biomedicine, catalysts, photoelectric devices, sensors and magnetic materials[11-15]. Among various metal nanoparticles, nano-sized silver (Ag) metal and its compounds are well-known to have strong bacteriostatic and bactericidal effects as well as broad-spectrum antimicrobial activities[16,17], which hold great

* 本文摘自 Polymer Composites, 2016, 37(4): 1137-1142.

potential application in every walk of life. Actually, clothing, respirators, household water filters, contraconceptives, antibacterial sprays, cosmetics, detergent, dietary supplements, cutting boards, sox, shoes, cell phones, laptop keyboards, and children's toys are among the retail products that purportedly exploit the antimicrobial properties of Ag nanomaterials[18]. Furthermore, unlike the Ag ions as antimicrobial agents with limited use due to low solubility and rapid precipitation of Ag ions in biological and environmental media containing Cl^-[19], Ag nanoparticles have relatively less usage limitation and are easily available via various approaches such as chemical reduction of a silver salt, irradiation, and biological reduction[20-22]. In addition, owing to large surface area and high chemical activity as well as the effects of endless Brownian movement and Van der Waals force, Ag nanoparticles are liable to aggregate leading to performance degradation[23,24]. Evenly dispersing nano-Ag particles in nanoporous cellulose aerogels substrates with micro-nano three dimensional (3D) architectures might effectively hinder reunion of nanoparticles. Furthermore, the large specific surface area and abundant surface oxygen-containing groups and binding sites of cellulose aerogels make contribution to fabricating highly-loaded nano-Ag composites.

Therefore, in the current paper, novel nano-Ag/cellulose (NAC) aerogels hybrids were successfully prepared based on a green cellulose solvent (NaOH/PEG aqueous solution) via an in-situ direct metallization method. The as-prepared NAC aerogels were subsequently characterized by scanning electron microscope (SEM), transmission electron microscopy (TEM), energy dispersive X-ray spectroscopy (EDX), X-ray diffraction (XRD) and thermogravimetric analysis (TGA). Meanwhile, the antibacterial activities of NAC aerogels for both *Escherichia coli* (Gram-negative) and *Staphylococcus aureus* (Gram-positive) were also investigated. Moreover, a schematic diagram of the embedment of nano-Ag in nanoporous cellulose aerogels substrates was also proposed based on the experimental results.

2 Materials and Methods

2.1 Materials

Natural bamboo fiber was supplied by Beijing Murun technology development Co. Ltd., and further completely rinsed with distilled water and then dried at 60 ℃ for 24 h, which was used as cellulosic source. *Escherichia coli* (ATCC25922) and *Staphylococcus aureus* (ATCC6538) were purchased from Guangzhou Jiake Microbiological Research Center, China. All other chemicals were purchased from Tianjin Kemiou chemical reagent Co. Ltd. and used as received.

2.2 Preparation of cellulose hydrogel

A certain amount of dried bamboo fibers were added to 10% aqueous solution of NaOH/PEG-4000 (9 : 1 wt/wt) with magnetic stirring at room temperature for 5 h to form a 2% homogeneous cellulose solution, and then the solution was subjected to a freeze-thaw process by freezing the solution at −15 ℃ for 12 h and subsequently thawing out it at room temperature for 0.5 h under vigorously stirring. After being frozen again at −15 ℃ for 3 h, the frozen solution immediately underwent a regeneration process by immerging it in a 1 v% HCl solution for about 6 h, repeating this process until the formation of an amber-like cellulose hydrogel. After that, the hydrogel was repeatedly rinsed with distilled water to remove residual hydrogen ions and chlorine anions.

2.3 Preparation of NAC aerogel

Preparation of NAC aerogel was implemented via a mild in-situ direct metallization method, and the specific steps were as follows: the as-prepared hydrogel was first immersed in 0.1 M of the aqueous $AgNO_3$ for 3 h, followed by rinsing with distilled water for ca. 30 s. Then, the Ag^+-saturated cellulose hydrogel was reduced in 0.1 M of the aqueous $NaBH_4$ for 10 min, and subsequently washed by large amount of distilled water and tertiary butanol in sequence for 10 min to remove the excess chemicals. Finally, the sample was freezing-dried at −35 ℃ for 48 h, and the following NAC aerogel was successfully obtained. In addition, the pure cellulose aerogel was prepared by direct tertiary butanol freeze drying of the cellulose hydrogel without the impregnation and the reduction processes.

2.4 Antimicrobial activity studies

The disc diffusion method was carried out to evaluate the antimicrobial activities of NAC aerogels against *Escherichia coli* (ATCC25922) and *Staphylococcus aureus* (ATCC6538) that were selected as the model Gram-negative and Gram-positive bacteria, respectively, and performed in Luria-Bertani (LB) medium solid agar Petri dish[25]; meanwhile, for the sake of comparison, the pure cellulose aerogels were selected as reference. After being sterilized by autoclaving for 15 min at 120 ℃, NAC aerogels and the cellulose aerogels were both placed on *E. coli*-cultured agar plates and *S. aureus*-cultured agar plates which were then incubated for 24 h at 37 ℃. After the incubation, the inhibition zones were generated around the samples with antibacterial property, and their diameters in millimeters were monitored and measured. All antibacterial activity tests were performed in triplicate and were done at least two different times to ensure reproducibility.

2.5 Characterizations

The morphologies were observed by a FEI, Quanta 200 SEM. TEM observation and EDX analysis were carried out on a FEI, Tecnai G2 F20 TEM equipped with an EDX spectrometer. XRD spectroscopy was performed by a XRD instrument (Rigaku, D/MAX 2200) operating with Cu Kα radiation ($\lambda = 1.5418$ Å) at a scan rate (2θ) of 4° · min^{-1} and the accelerating voltage of 40 kV and the applied current of 30 mA ranging from 5° to 90°. The thermal stabilities were determined using a TG analyzer (TA Q600) with a heating rate of 10 ℃ · min^{-1} in a N_2 environment.

3 Results and Discussion

Fig. 1 showed the SEM images of the pure cellulose aerogels and NAC aerogels, respectively. As shown, the cellulose aerogels consisted of hierarchical and cross-linking 3D network with pores diameters ranging from several decades nanometers to a few microns. Meanwhile, NAC aerogels also displayed similar nanoporous and multi-scale morphology, which maintained the 3D network skeletal structures without obvious collapse and significant pores sizes differences after the Ag ions were reduced to form Ag metal particles, demonstrating that the metallization process of $NaBH_4$ had less effect on the original structure of the aerogels substrates. Especially, the small nanoscale pores were useful for anchoring and stabilizing the nanoparticles, and supplied channels for the facile migration of the reactants and by-products during the reduction reactions[26].

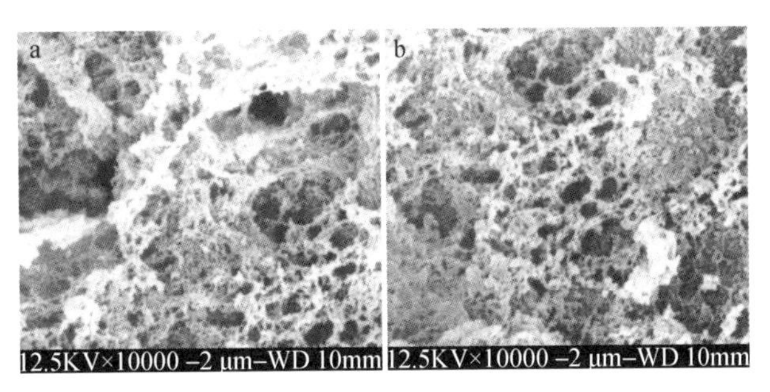

Fig. 1 SEM images of (a) the pure cellulose aerogels and (b) NAC aerogels, respectively

It can be seen in the TEM images of NAC aerogels (Fig. 2a, b) that the formed plentiful Ag nanoparticles were tightly bound to and homogeneously dispersed in cellulose aerogels matrixes without apparent reunion phenomenon. Moreover, from the EDX spectrum of NAC aerogels (Fig. 2c), apart from C and O elements from cellulose and Cu element from the stray signal originating with copper TEM grids as well as Na element from $NaBH_4$, the strong Ag peaks were generated, and the high peak intensity revealed that the composite was loaded with plenty of Ag nanoparticles. Additionally, according to the TEM images (Fig. 2b), the particle diameter histogram of NAC aerogels was plotted and presented in Fig. 2d. The diameter histogram showed Gaussian-like distributions, and their mean diameter (d) and standard deviation (σ) were estimated to be 7.83 and 2.99 nm, respectively.

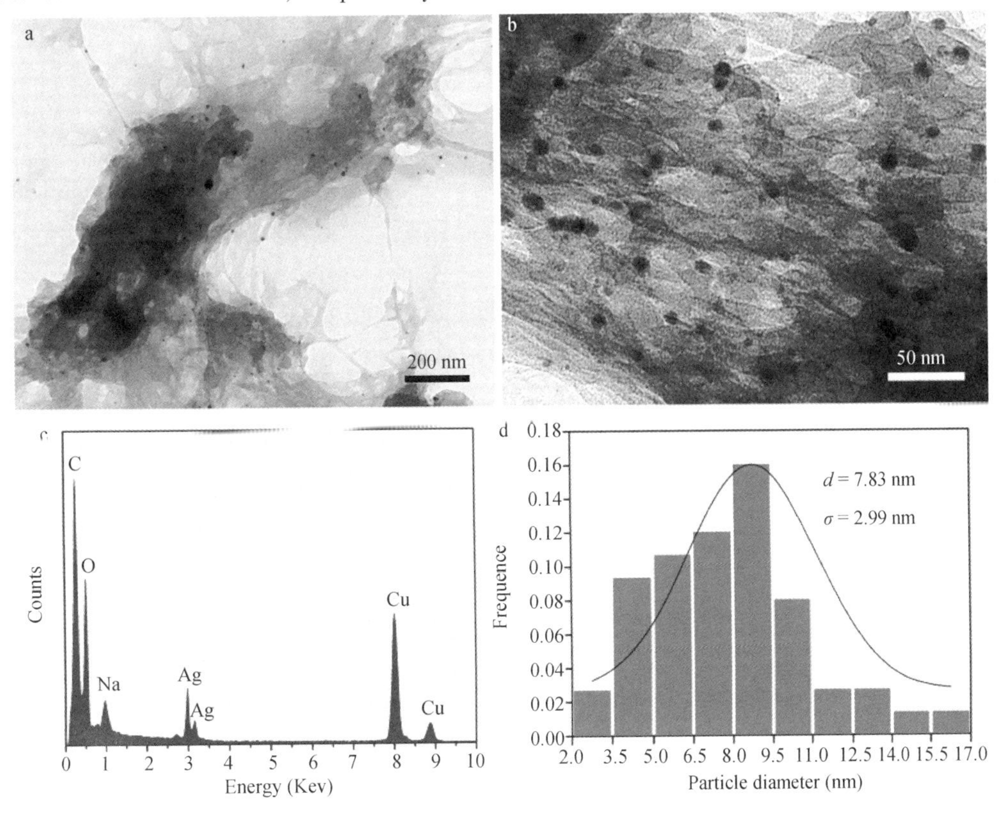

Fig. 2 TEM images (a and b), EDX spectrum (c) and particle diameter histogram (d) of NAC aerogels, respectively

The XRD patterns of the bamboo fiber and NAC aerogels were shown in Fig. 3. Apparently, the bamboo fiber exhibited typical cellulose I crystalline structure with characteristic peaks at around 14.6°, 16.4° and 22.2° corresponding to ($1\bar{1}0$), (110) and (200) planes[27], respectively. For NAC aerogels, in addition to the peaks related to transformational cellulose Ⅱ crystalline structure at around 12.0°, 19.9° and 21.6° as the result of NaOH during dissolution[28,29], the newly generated diffraction peaks at around 38.1°, 44.4°, 64.5°, 77.5° and 81.5° were corresponding to the reflections of the (111), (200), (220), (311) and (222) crystalline planes of the metallic face-centered cubic (fcc) Ag, respectively, indicating that the Ag nanoparticles had been successfully formed and embedded into cellulose aerogels.

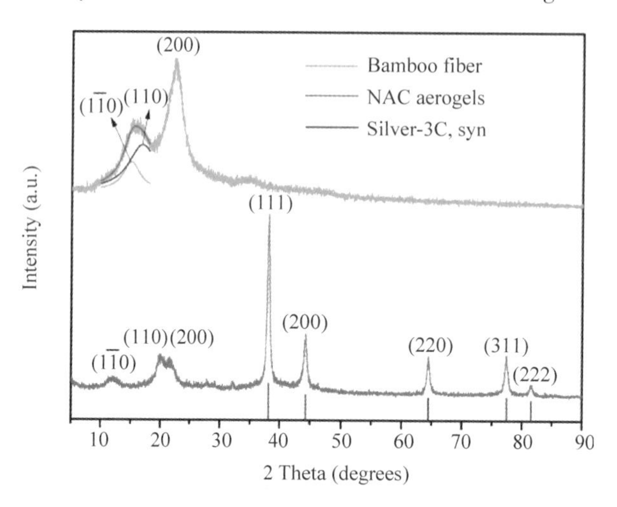

Fig. 3 XRD patterns of the bamboo fiber and NAC aerogels, respectively

In consideration of the importance of thermal stability in many applications of nanocomposites, thermal decompositions of NAC aerogels and the corresponding raw materials namely bamboo fiber were examined by TGA, and the results were illuminated in Fig. 4. The small weight losses below 150 ℃ for both samples were resulted from evaporation of adsorbed water[30]. For the bamboo fiber, there was only a dramatic decomposition stage assigned to cellulose with maximum peak centered at 337 ℃[31]. Compared with the bamboo fiber, NAC aerogels showed more superior thermal stability after being incorporated with Ag nanoparticles that have a high melting point of approximately 962 ℃, which reached the maximum degradation rate at higher temperature (ca. 351 ℃). Furthermore, the presence of Ag nanoparticles also caused increase in remaining char at 800 ℃ for NAC aerogels (ca. 17.6%) far higher than that of the bamboo fiber (ca. 2.2%), further suggesting the high Ag loading content in cellulose aerogels matrixes.

The antibacterial activities of NAC aerogels for *E. coli* and *S. aureus* were measured by the disc diffusion method, and the results were shown in Fig. 5. It was clearly observed that the inhibition zones around NAC aerogels were formed, and the growth inhibition rings of *E. coli* and *S. aureus* were about 4.3 and 4.1 mm, respectively, which suggested the formation of antimicrobial property. On the contrary, there was no inhibition zone for the pure cellulose aerogels as control, and the result demonstrated that the antibacterial activities were only originated from the embedded Ag nanoparticles instead of the individual cellulose aerogels.

Based on the results mentioned above, a schematic representation of the preparation of NAC aerogels by

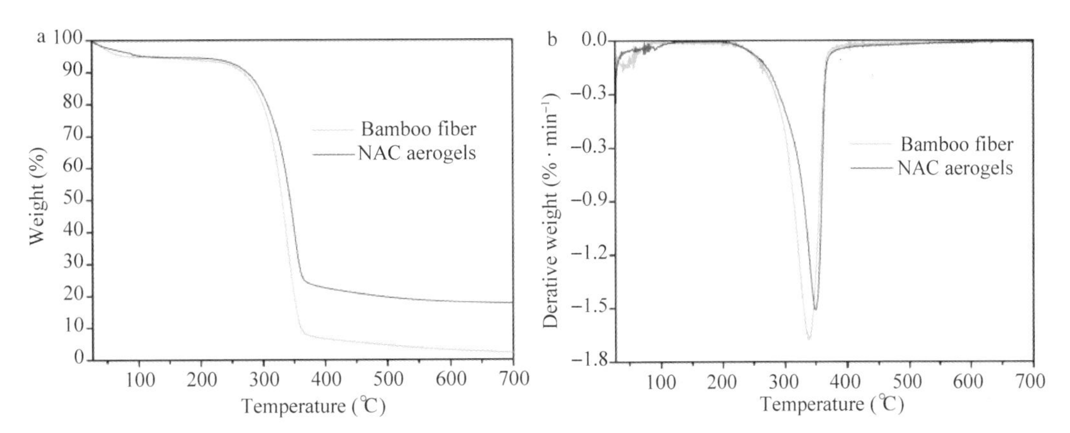

Fig. 4 (a) TG and (b) DTG curves of the bamboo fiber and NAC aerogels, respectively

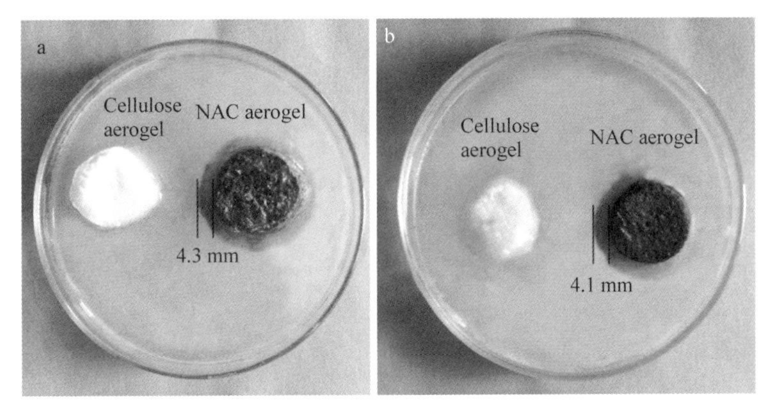

Fig. 5 Antimicrobial activities of the cellulose aerogels and NAC aerogels against (a) *Escherichia coli* and (b) *Staphylococcus aureus*, respectively

the in-situ metallization method was proposed and illuminated in Fig. 6. As shown, after dissolution, freeze—thaw process and regeneration, the resulting cellulose hydrogel was composed of physical cross-linking of cellulose chains, and contained abundant oxygen-containing groups such as hydroxyl and ether groups on the surface. After the hydrogel was immerged in $AgNO_3$ aqueous solution, electron-rich oxygen atoms of oxygen-containing groups of cellulose attracted and combined with the Ag ions due to strong electrostatic interactions like ion—dipole interaction leading to immobilization of Ag ions[32-34], and thereby the Ag ions were kept stable and uniformly distributed on the cellulose chains surfaces. When the cellulose hydrogel with Ag ions was dipped in the aqueous solution of $NaBH_4$ with strong reducibility, the Ag ions were quickly reduced to zero-valent Ag nanoparticles. Meanwhile, owing to being tightly anchored by the oxygen-containing groups, the generated nano-Ag was also stabilized at the original ionized selective sites of cellulose chains surfaces, which impeded the aggregation of nanoparticles resulting in excellent dispersion effect[35].

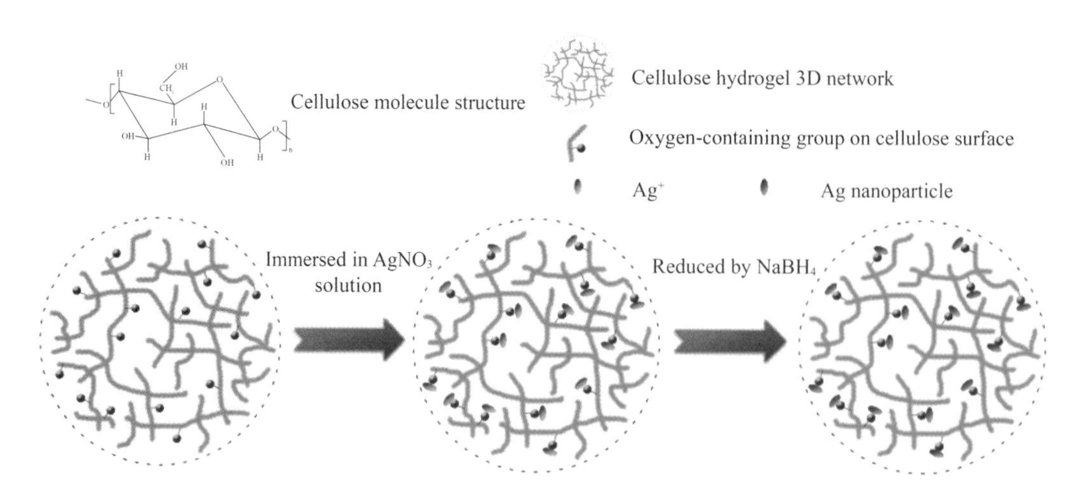

Fig. 6 Schematic representation of the preparation of NAC aerogels by the in-situ metallization method

4 Conclusions

The novel NAC aerogels were successfully prepared by the in-situ direct metallization method. The formed Ag nanoparticles with mean diameters of about 7. 83 nm were tightly bound to and homogeneously dispersed in cellulose aerogels matrixes without apparent reunion phenomenon due to immobilization of Ag ions by the electron-rich oxygen atoms of cellulose under the effect of electrostatic interactions. The as-prepared NAC aerogels maintained nanoporous and multi-scale morphology similar to the pure cellulose aerogels, showed more superior thermal stability than that of the bamboo fiber raw material, and exhibited strong antibacterial activities for *E. coli* and *S. aureus*, which might be useful for various biomedical applications, especially green heat-resistant high-performance antibacterial materials.

Acknowledgements

We appreciate the financial support of the China Postdoctoral Science Foundation (2013M540263) and the National Natural Science Foundation of China (grant no. 31270590).

References

[1]Sehaqui H, Zhou Q, Berglund L A. High-porosity aerogels of high specific surface area prepared from nanofibrillated cellulose (NFC)[J]. Composites science and technology, 2011, 71(13): 1593-1599.

[2]Olsson R T, Samir M A S A, Salazar-Alvarez G, et al. Making flexible magnetic aerogels and stiff magneticnanopaper using cellulose nanofibrils as templates[J]. Nature nanotechnology, 2010, 5(8): 584-588.

[3]Liebner F, Potthast A, Rosenau T, et al. Ultralight-weight cellulose aerogels from NBnMO-stabilized Lyocell dopes [J]. Research Letters in Materials Science, 2007, 2007.

[4]Fischer F, Rigacci A, Pirard R, et al. Cellulose-based aerogels[J]. Polymer, 2006, 47(22): 7636-7645.

[5]Miyamoto T, Takahashi S, Ito H, et al. Tissue biocompatibility of cellulose and its derivatives[J]. Journal of biomedical materials research, 1989, 23(1): 125-133.

[6]Hult E L, Iversen T, Sugiyama J. Characterization of the supermolecular structure of cellulose in wood pulp fibres[J]. Cellulose, 2003, 10(2): 103-110.

[7]He J, Kunitake T, Watanabe T. Porous and nonporous Ag nanostructures fabricated using cellulose fiber as a template

[J]. Chemical communications, 2005 (6): 795-796.

[8]Hu W, Chen S, Li X, et al. In situ synthesis of silver chloride nanoparticles into bacterial cellulosemembranes[J]. Materials Science and Engineering: C, 2009, 29(4): 1216-1219.

[9]Ras R, Ikkala O, Korhonen J T, et al. Hydrophobic nanocellulose aerogels as floating, sustainable, reusable, and recyclable oil absorbents. [J]. Acs Applied Materials & Interfaces, 2011, 3(6): 1813.

[10]Guilminot E, Fischer F, Chatenet M, et al. Use of cellulose-based carbon aerogels as catalyst support for PEM fuel cell electrodes: Electrochemical characterization[J]. Journal of Power Sources, 2007, 166(1): 104-111.

[11]DellaRocca J, Liu D, Lin W. Nanoscale metal-organic frameworks for biomedical imaging and drug delivery[J]. Accounts of chemical research, 2011, 44(10): 957-968.

[12]Jain P K, Huang X, El-Sayed I H, et al. Noble metals on the nanoscale: optical and photothermal properties and some applications in imaging, sensing, biology, and medicine[J]. Accounts of chemical research, 2008, 41(12): 1578-1586.

[13]Bell A T. The impact ofnanoscience on heterogeneous catalysis[J]. Science, 2003, 299(5613): 1688-1691.

[14]Leslie-Pelecky D L, Rieke R D. Magnetic properties of nanostructured materials[J]. Chemistry of materials, 1996, 8 (8): 1770-1783.

[15]Rieter W J, Taylor K M L, Lin W. Surface modification and functionalization of nanoscale metal-organic frameworks for controlled release and luminescence sensing[J]. Journal of the American Chemical Society, 2007, 129(32): 9852-9853.

[16]Atiyeh B S, Costagliola M, Hayek S N, et al. Effect of silver on burn wound infection control and healing: review of the literature[J]. Burns, 2007, 33(2): 139-148.

[17]Maneerung T, Tokura S, Rujiravanit R. Impregnation of silver nanoparticles into bacterial cellulose for antimicrobial wound dressing[J]. Carbohydrate polymers, 2008, 72(1): 43-51.

[18]Marambio-Jones C, Hoek E M V. A review of the antibacterial effects of silver nanomaterials and potential implications for human health and the environment[J]. Journal of nanoparticle research, 2010, 12(5): 1531-1551.

[19] Lesniak W, Bielinska A U, Sun K, et al. Silver/dendrimer nanocomposites as biomarkers: fabrication, characterization, in vitro toxicity, and intracellular detection[J]. Nano letters, 2005, 5(11): 2123-2130.

[20]Martínez-Castañón G A, Niño-Martínez N, Loyola-Rodríguez J P, et al. Synthesis of silver particles with different sizes and morphologies[J]. Materials Letters, 2009, 63(15): 1266-1268.

[21]Sharma V K, Yngard R A, Lin Y. Silver nanoparticles: green synthesis and their antimicrobial activities[J]. Advances in colloid and interface science, 2009, 145(1-2): 83-96.

[22]Saifuddin N, Wong C W, Yasumira A A. Rapid biosynthesis of silver nanoparticles using culture supernatant of bacteria with microwave irradiation[J]. E-journal of Chemistry, 2009, 6(1): 61-70.

[23]Hakim L F, Portman J L, Casper M D, et al. Aggregation behavior of nanoparticles in fluidizedbeds[J]. Powder Technology, 2005, 160(3): 149-160.

[24]Badawy A M E, Luxton T P, Silva R G, et al. Impact of environmental conditions (pH, ionic strength, and electrolyte type) on the surface charge and aggregation of silver nanoparticles suspensions[J]. Environmental science & technology, 2010, 44(4): 1260-1266.

[25]Drew W L, Barry A L, O'Toole R, et al. Reliability of the Kirby-Bauer disc diffusion method for detecting methicillin-resistant strains of Staphylococcusaureus[J]. Applied microbiology, 1972, 24(2): 240-247.

[26]Luong N D, Lee Y, Nam J D. Highly-loaded silver nanoparticles in ultrafine cellulose acetate nanofibrillar aerogel [J]. European Polymer Journal, 2008, 44(10): 3116-3121.

[27]Cai J, Kimura S, Wada M, et al. Nanoporous cellulose as metal nanoparticles support[J]. Biomacromolecules, 2009, 10(1): 87-94.

[28]Oh S Y, Yoo D I. Y, Shin, et al. Crystalline structure analysis of cellulose treated with sodium hydroxide and carbon dioxide by means of X-ray diffraction and FTIR spectroscopy[J], Carbohydr. Res, 2005, 340(15): 2376-2391.

[29] Okano T, Sarko A. Mercerization of cellulose. II. Alkali-cellulose intermediates and a possible mercerization mechanism[J]. Journal of Applied Polymer Science, 1985, 30(1): 325-332.

[30] Li J, Lu Y, Yang D, et al. Lignocellulose aerogel from wood-ionic liquid solution (1-allyl-3-methylimidazolium chloride) under freezing and thawingconditions[J]. Biomacromolecules, 2011, 12(5): 1860-1867.

[31] Yang H, Yan R, Chen H, et al. Characteristics of Hemicellulose, Cellulose and LigninPyrolysis[J]. Fuel, 2007, 86(12-13): 1781-1788.

[32] He J, Kunitake T, Nakao A. Facile In Situ Synthesis of Noble Metal Nanoparticles in Porous Cellulose Fibers[J]. Chemistry of Materials, 2003, 15(23): 4401-4406.

[33] Kesting R E. Semipermeable membranes of cellulose acetate for desalination in the process of reverse osmosis. I. Lyotropic swelling of secondary cellulose acetate[J]. Journal of Applied Polymer Science, 1965, 9(2): 663-688.

[34] Sharp K A, Honig B. Electrostatic interactions in macromolecules: theory and applications[J]. Annual review of biophysics and biophysical chemistry, 1990, 19(1): 301-332.

[35] Mlambo M, Mpelane S, Mdluli P S, et al. Unique flexible silver dendrites thin films fabricated on cellulose dialysis cassettes[J]. Journal of Materials Science, 2013, 48(18): 6418-6425.

中文题目：纳米银嵌入纤维素气凝胶的制备、表征及抗菌性能研究

作者：万才超，焦月，孙庆丰，李坚

摘要：本文采用强还原性 $NaBH_4$ 还原 Ag 离子，制备了平均粒径约为 7.83nm 的高负载 Ag 纳米粒子，且均匀的嵌入在纤维素气凝胶中，没有明显团聚现象。制备的纳米 Ag/纤维素(NAC)气凝胶保持了与纯纤维素气凝胶相似的纳米多孔和多尺度形态，并且对大肠杆菌(革兰氏阴性)和金黄色葡萄球菌(革兰氏阳性)均有较强的抗菌活性。同时，在加入银纳米粒子后，NAC 气凝胶也表现出更优越的热稳定性。因此，新型 NAC 气凝胶有望应用在多种生物医学方面，尤其是绿色耐热高性能抗菌材料。

关键词：抗菌；银；纤维素气凝胶；纳米复合材料；聚合物

Influence of Pre-gelation Temperature on Mechanical Properties of Cellulose Aerogels Based on A Green NaOH/PEG Solution—A Comparative Study*

Caichao Wan, Yue Jiao, Jian Li

Abstract: Cellulose aerogels with porous three - dimensional network structure often suffer from weak mechanical strength, which imposes huge restrictions on their applications. In this work, an easily-operated pre-gelation treatment was introduced to the conventional fabrication technique of regenerated cellulose aerogels (i. e., dissolution, regeneration and drying treatment) to improve their structure uniformity, crystallinity, and mechanical strength. These properties were found to be closely related to the pre-gelation temperature, and a relatively lower temperature (50 ℃) made the greatest contribution. Especially for the mechanical strength, the elasticity modulus and energy absorption of the aerogels treated at 50 ℃ reach the maximums of 1.03 MPa and 476.2 $kJ \cdot m^{-3}$, which are 5.4 times and 94.3% higher than those of the untreated aerogels. As an example of potential application, the surface of the aerogels was decorated with multiwalled carbon nanotubes. The obtained composite displays a good electromagnetic interference shielding ability.

Keywords: Cellulose aerogels; Pre-gelation; Mechanical properties; MWCNTs; Electromagnetic interference shielding; Polymers

1 Introduction

Eco-friendly regenerated cellulose aerogels (RCA) have attracted intensive attention because of their unique structure characteristics, such as hierarchical porous structure, an ocean of surface hydroxyl groups, high specific surface area and porosity, and ultra-low density[1,2]. Therefore, RCA has been extensively used in numerous frontier fields, such as green host materials to support various nanoparticles[3-5], high-performance adsorbents or separation materials to treat with organic wastewater[6,7], and special super-insulating materials for heat, sound and electric insulation[8]. However, such a promising biomaterial still suffers from lack of mechanical strength due to the weak hydrogen-bonding interaction between cellulose chains and the loose mesh-like structure. At present, the conventional preparation method of RCA is based on cellulose dissolution in direct solvents, regeneration in non-solvents, and special drying processes to prevent pore collapse. For cellulose dissolution, a multitude of cellulose solvents have been successively reported in the past several decades[9-11]. Hereinto, NaOH/polyethylene glycol (PEG) aqueous solution is a kind of green

* 本文摘自 Colloid and Polymer Science, 2016, 294(8): 1281-1287.

and cost-effective cellulose solvent, and has been confirmed to have the ability to effectively dissolve cellulose[12]. Regarding the mechanism of dissolution, although it is still a matter of continuing debate, the widely accepted opinion considers that the cellulose dissolution results from the solvent ability to eliminate the inter- and intramolecular hydrogen bonds of cellulose and prevent the self-association of the splitting cellulose chains[13]. The formed inclusion complex structure in the aqueous solution (i. e., the solvent molecules or their hydrates coating on the surface of the splitting cellulose chains) is relatively stable at low temperatures.

In the subsequent regeneration process, the quick rearrangement of cellulose chains and the phase disengagement take place after the introduction of coagulant (e. g., hydrochloric acid), decreasing the structure homogeneity and inducing the generation of structure defects. Thus the weak mechanical strength is obtained for the aerogels.

Pre-gelation treatment of the cellulose solution has been confirmed as an effective pathway to improve the mechanical strength of regenerated cellulose membrane based on a NaOH/thiourea solvent system[14]. The thermal irreversible gelation greatly restricts the mobility of cellulose chains by increasing their physical crosslinking, and thus reduces the speed of rearrangement during the regeneration, eventually leading to a denser, more homogeneous, and mechanically stronger frame structure. Therefore, it is interesting and meaningful to investigate whether the pre-gelation treatment can also enhance the mechanical property of RCA based on the NaOH/PEG solution; meanwhile, a comparative study of different pre-gelation temperatures is worth being carried out.

In this work, a mild pre-gelation treatment combined with a traditional technique (i. e., dissolution, regeneration and freeze drying) was introduced to fabricate RCA and enhance the mechanical strength. The mechanical properties were characterized with stress-strain curves measured by a universal testing machine. The morphology and crystal structure were studied by scanning electron microscopy (SEM) and X-ray diffraction (XRD) to clarify the influences of pre-gelation temperature. In addition, multiwalled carbon nanotubes (MWCNTs) were deposited on the surface of RCA, and the potential application of the composite (coded as MWCNT/RCA) was studied in electromagnetic interference (EMI) shielding.

2 Materials and methods

2.1 Materials

Bamboo fiber was supplied by Zhejiang Mingtong Textile Technology Co., Ltd. and further washed serval times with distilled water and dried at 60 ℃ for 24 h. The bamboo fiber was directly used as the cellulose feedstock considering its high cellulose content. MWCNTs were provided by Shanghai Aladdin Industrial Inc. (China). The diameters vary from 20 to 40 nm, and the lengths are in the range of 1-2 μm. Prior to use, MWCNTs were purified further in concentrated 3 mol · L^{-1} nitric acid for 24 h by magnetic stirring. All other chemicals were purchased from Kemiou Chemical Reagent Co. Ltd. (Tianjin, China) and used as received.

2.2 Preparation of RCA via a solution pre-gelation technique

Based on our previous work[15] on the influences of cellulose concentration on the morphology and properties of cellulose aerogels, in this work, cellulose concentration was set as 1% to fabricate cellulose aerogels. This concentration has been verified to contribute to obtaining developed three-dimensional (3D) network and superior properties. Cellulose solution was fabricated by mixing the bamboo fiber with the NaOH/PEG-4000 solution (9 : 1 wt/wt) with magnetic stirring for 5 h. The cellulose solutions were subsequently heated at pre-gelation temperatures of 50, 70 and 90 ℃, respectively, for 12 h to obtain cellulose physical gel. Thereafter, the gel blocks were frozen at −15 ℃ for 5 h and then immersed in a coagulating bath (1% hydrochloric acid) to induce the cellulose rearrangement. The immersion process generally continued for three days, and the coagulant was changed every other 6 h. After the generation of cellulose hydrogels, the products were rinsed with a large amount of distilled water and tert-butyl alcohol to remove residual impurities. Finally, the hydrogels underwent a freeze drying treatment at −35 ℃ and a pressure of 25 Pa for 48 h, and the following RCA was obtained. The RCA samples fabricated at the pre-gelation temperatures of 50, 70 and 90 ℃ were labeled as RCA-50, RCA-70 and RCA-90, respectively. Besides, the RCA sample prepared from the traditional path without the pre-gelation treatment was coded as TRCA.

2.3 Depositing MWCNTs on the surface of RCA

A simple method was carried out to deposit purified MWCNTs on the surface of RCA. The purified MWCNTs (0.10 g) were firstly dispersed in 100 mL dimethyl formamide (DMF) by a three-hour ultrasonic treatment (900 W, 20 kHz), and then powder-like RCA (0.30 g) was immersed in the aforementioned dispersion for 1 h. The DMF was expelled from the mixture by distillation under reduced pressure as shown in Fig. 1. The resultant product was washed with a large amount of distilled water and subsequently collected by centrifugation. The washing-centrifuging process was repeated several times to remove residual DMF. Finally, the collected product was dried at 50 ℃ for 48 h. The resulting composite was coded as MWCNT/RCA.

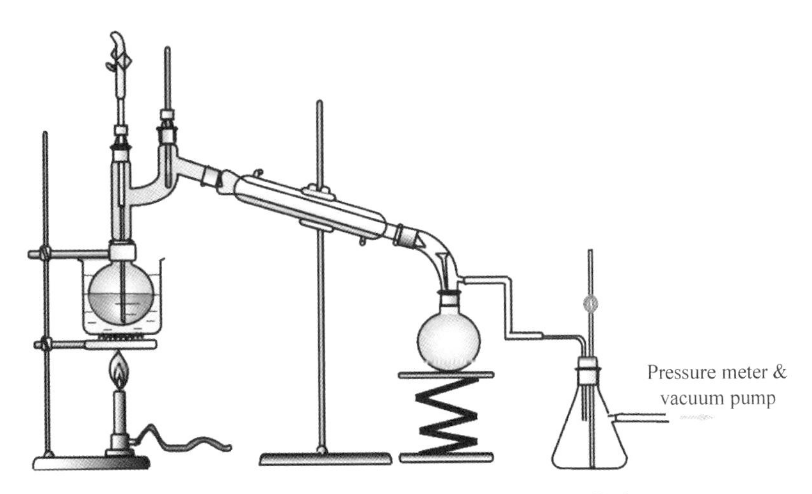

Fig. 1 Schematic diagram of reduced pressure distillation apparatus

2.4 Characterizations

SEM observation was performed with a Hitachi S4800 SEM. XRD spectroscopy was implemented on a Rigaku D/MAX 2200 XRD instrument operating with Cu Kα radiation (λ = 1.5418 Å) at a scan rate (2θ) of 4° · min^{-1} ranging from 5° to 40°. N_2 adsorption-desorption measurements were implemented at −196 ℃ using an accelerated surface area and porosimetry system (3H−2000PS2 unit, Beishide Instrument S&T Co., Ltd.). Compression tests were conducted using an Instron 3382 universal testing machine at a test speed of

1 mm · min^{-1}. Load-displacement data was recorded and used to calculate the mechanical performance parameters and draw the stress-strain curves. EMI shielding effectiveness (EMI SE) was measured on a PNA-X network analyzer (N5244a) with the coaxial method at the frequency range of 8.2–12.4 GHz (X-band). For preparing the test specimen, MWCNT/RCA was mixed with paraffin wax (mass ratio is 50/50) and then pressed into a toroidal shaped mould (Φ_{in} : 3.0 mm, Φ_{out} : 7.0 mm, H : 2.0 mm).

3 Results and discussion

For the investigation of the morphology of RCA and the pre-gelation temperature dependence, SEM observations were done (Fig. 2). As shown, the differences can be clearly identified notwithstanding that all the samples display 3D network structure. On the basis of the traditional technique, a typical intertangled mesh-like structure was generated for TRCA (Fig. 2a). With the introduction of pre-gelation treatment, the network structure obviously became more homogeneous at the relatively lower pre-gelation temperatures of 50 ℃ (Fig. 2b). Further increasing the pre-gelation temperature from 50 ℃ to 70 ℃ and eventually to 90 ℃, the notable changes took place. It can be seen in Fig. 2c and d that the 3D network gradually becomes inhomogeneous, and massive agglomeration appears.

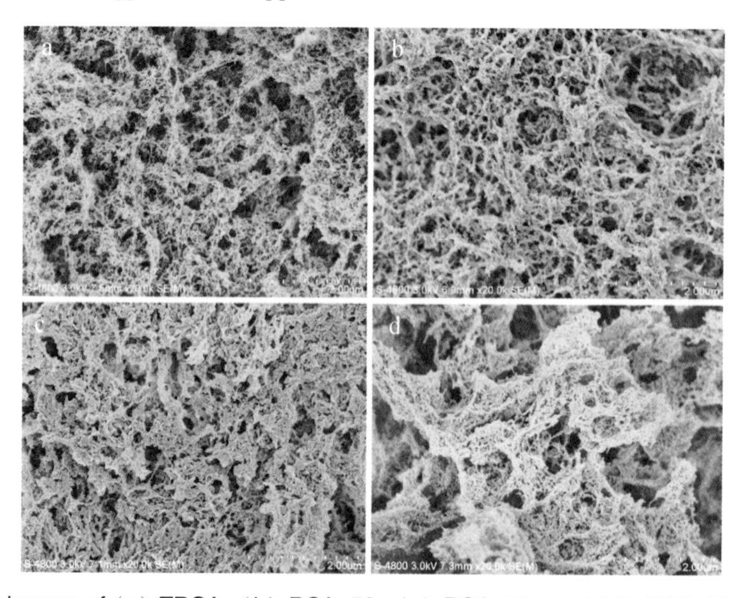

Fig. 2 SEM images of (a) TRCA, (b) RCA-50, (c) RCA-70, and (d) RCA-90, respectively

The nitrogen adsorption-desorption measurements were carried out to further determine the pore size distributions of TRCA, RCA-50, RCA-70 and RCA-90. The pore size distributions were estimated by the Barrett-Joyner-Halenda (BJH) method. As shown in Fig. 3, RCA-90 clearly shows a sawtooth curve with several peaks at 3.4, 5.4, 6.3, 8.9, 15.4 and 30.1 nm, which reveals the heterogeneous pore structure. In contrast to RCA-90, the pore size distribution curve of RCA-70 is relatively gentler with two peaks at 3.0 and 6.2 nm. For TRCA and RCA-50, it can be seen that their pore size distribution curves are smoother, and almost no apparent peaks appear, suggesting that TRCA and RCA-50 have more homogeneous pore structure. These results are consistent with the above SEM observations.

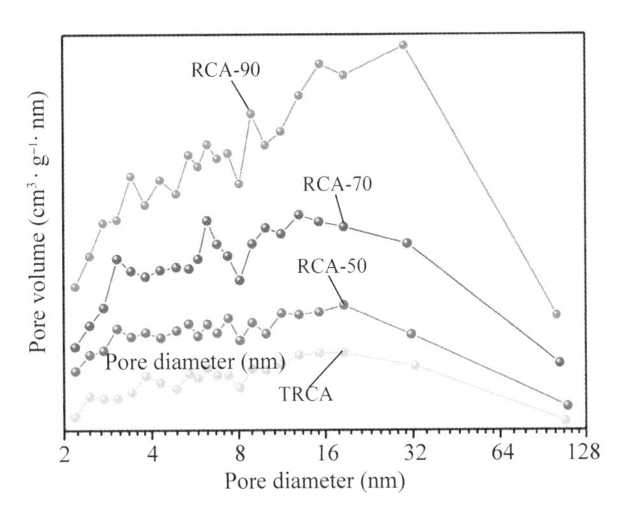

Fig. 3 Pore size distributions of TRCA, RCA-50, RCA-70 and RCA-90, respectively

The possible reason for the improved homogeneity at 50 ℃ is the strengthening of physical cross-linking between cellulose chains, which largely limits the quick rearrangement and the irregular agglomeration of cellulose molecules in the coagulant and thus reduces the probability of formation of structural defects. The relatively ideal pre-gelation temperature (50 ℃) induces the slow and strong entanglements of cellulose chains due to the higher energy of hydrogen bonding at the lower temperature, which promotes the formation of homogenous and dense structure. With the increase of pre-gelation temperature (70 and 90 ℃), the inclusion complex structure becomes unstable and easier to be destroyed. Besides, the self-aggregation rate of cellulose chains sharply rises and thus increases the probability of defects formation and leads to the relatively non-uniform structure[14, 16, 17].

Fig. 4a shows the XRD patterns of TRCA, RCA-50, RCA-70 and RCA-90. Apparently, these patterns are extremely similar. All the aerogels show typical cellulose Ⅱ crystal structure, and exhibit similar diffraction peaks at around 12.2°, 20.1°, 21.9° and 34.6°, corresponding to the (101), ($10\bar{1}$), (002) and (040) planes[18]. However, the differences in the crystallinity index of the aerogels reveal the effects of pre-gelation treatment on the cellulose crystallization behavior. As shown in Fig 4b, the small increment in crystallinity index from 68.3% (TRCA) to 70.5% (RCA-50) indicates the positive contribution of low-temperature pre-gelation treatment to the cellulose crystallization. In addition, the crystallinity index decreases from 70.5% to 65.6% and eventually to 61.7% with the increasing pre-gelation temperature from 50 ℃ to 70 ℃ and eventually to 90 ℃, suggesting that the higher pre-gelation temperature disturbs crystallization of the cellulose molecules and induces relative unordered crystallization behavior[19].

The reinforcing effects of the pre-gelation treatment on the mechanical properties of RCA were measured by a universal testing machine. The compression stress-strain curves are presented in Fig. 5a. It is clear that the treatment significantly strengthens the mechanical properties of RCA. The elasticity modulus values were calculated to be 1.03 MPa for RCA-50, 0.80 MPa for RCA-70, and 0.45 MPa for RCA-90, which are 5.4, 4.0, and 1.8 times higher than that of TRCA (0.16 MPa), respectively. Meanwhile, the energy absorption values also substantially increase by 94.3%, 31.4%, and 36.3%, from 245 kJ · m^{-3} (TRCA) to

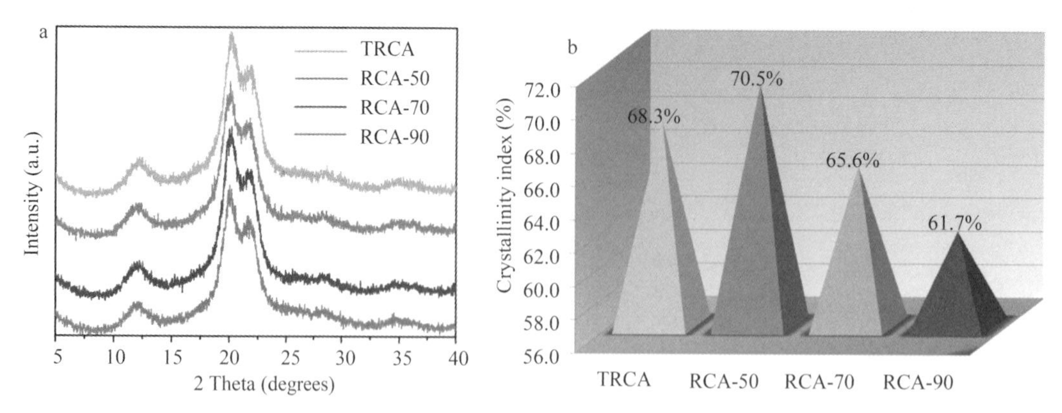

Fig. 4 (a) XRD patterns and (b) crystallinity index distributions of TRCA, RCA-50, RCA-70 and RCA-90, respectively

476 kJ · m^{-3}(RCA-50), 322 kJ · m^{-3}(RCA-70) and 334 kJ · m^{-3}(RCA-90), respectively. These rising tendencies reveal the good reinforcing ability of the pre-gelation treatment. In addition, the pre-gelation temperature also plays an important role in the reinforcing effects. Comparing the mechanical properties of the treated samples, the obvious advantages can be seen for RCA-50, whose elasticity modulus and energy absorption values are 28.8% and 47.8% higher than those of RCA-70, and 128.9% and 42.5% higher than those of RCA-90, respectively. The results indicate that the pre-gelation treatment at 50 ℃ is more beneficial to obtain stronger mechanical strength. The possible reason is ascribed to the more homogeneous network structure.

For comparison, we adopted the well-known cellulose solvent namely NaOH/urea aqueous solution to fabricate cellulose aerogels (coded as URCA). The cellulose solution (1%) was prepared in terms of the work reported by Zhou and Zhang[20], and the subsequent regeneration and drying treatment were done using the method of this paper. As shown in Fig. 5b, URCA has the elasticity modulus and energy absorption of only 0.019 MPa and 121 kJ · m^{-3}, much lower than those of the aerogels based on the NaOH/PEG solvent system. Moreover, it's also interesting to compare these RCA samples with a class of gel reported in our previous work[21], which was prepared by the pre-gelation treatment of cellulose solution at 95 ℃ for 5 h and subsequent direct freeze drying without involving the regeneration process. The gel has the high elasticity modulus and energy absorption values of 3.21 MPa and 389 kJ · m^{-3}. For comparing their mechanical properties, the energy absorption value was taken as the area below the stress-strain curve between 0% and 60% strain. Regardless of the influences of feedstock and cellulose solution concentration, the elasticity modulus and energy absorption of the gel are higher than those of RCA-50 (1.03 MPa and 77.7 kJ · m^{-3}). However, the strong hygroscopicity of the gel attributed to the plentiful NaOH and PEG on its surface imposes huge restrictions on its applications.

With the swift development of broadcast, television and microwave technology, rapidly increasing power of radio-frequency equipment results in serious EMI pollution, which has posed a threat to the health of human beings[22]. For minimizing the threat, it is worth to develop high-performance and eco-friendly EMI shielding materials. Carbon nanotubes (CNTs) possess huge potential for utilization in EMI shielding[23, 24]. Depositing CNTs on the surface of cellulose aerogels possibly leads to the formation of a desirable EMI shielding

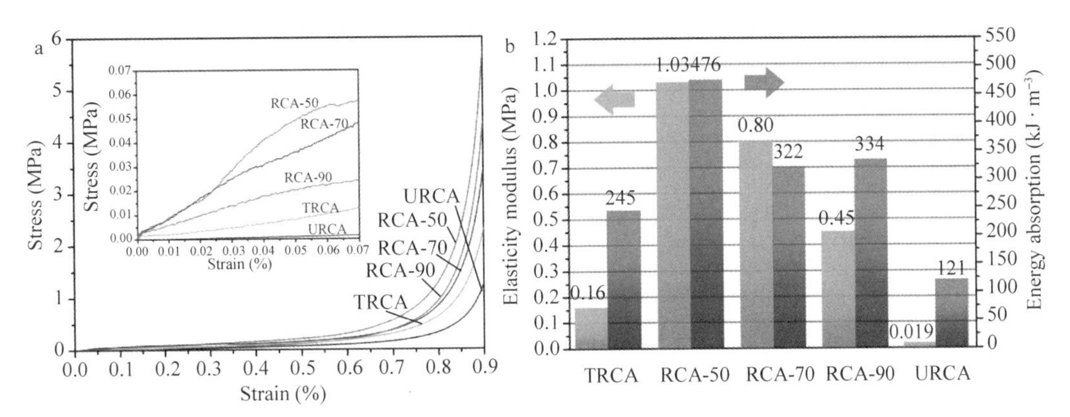

Fig. 5 (a) Compression stress-strain curves and (b) elasticity modulus and energy absorption values of TRCA, RCA-50, RCA-70, RCA-90 and URCA, respectively. The inset in (a) shows the expanded low-strain range

material. We chose RCA - 50 as an example to combine with the purified MWCNTs. Regarding the EMI shielding ability, EMI SE was measured with the coaxial method. A higher EMI SE expressed in decibels (dB) suggests less energy transmitted through shielding material. The total shielding effectiveness (SE_{total}) can be divided into absorption loss (SE_A), reflection loss (SE_R), and multiple reflection loss (SE_M), which can be expressed as the following equations[25]:

$$SE_{total}(\mathrm{dB}) = 10\log\frac{P_i}{P_t} = SE_A + SE_R + SE_M \quad (1)$$

$$SE_R = -10\log(1 - |S_{11}|^2) \quad (2)$$

$$SE_A = -10\log[|S_{12}|^2/(1 - |S_{11}|^2)] \quad (3)$$

where P_i and P_t are the incident power and the transmitted power. When $SE_{total} > 10$ dB, SE_M can be negligible. $|S_{12}|$ and $|S_{11}|$ are the scattering parameters (S-parameters) of the two-port vector network analyzer system.

Fig. 6 presents the variation of the SE_{total}, SE_A and SE_R values over the frequency range of 8.2 - 12.4 GHz for MWCNT/RCA. The SE_{total} is approximately 18.7-19.1 dB, close to the commercially achievable level (20 dB). Regardless of the error between coaxial and waveguide methods, the value is also higher than that of some previously reported cellulose-based EMI shielding materials[26-31] as listed in Table 1, such as polyaniline/$BaTiO_3$/cotton fabrics (16.8 dB), nafion-MWCNT/cotton fabrics (11.5 dB) and butanetetracarboxylic acid/MWCNT/cotton (5 dB). The good EMI shielding ability is primarily ascribed to the extraordinary electromagnetic characteristics of CNTs. In addition, comparing the contribution of

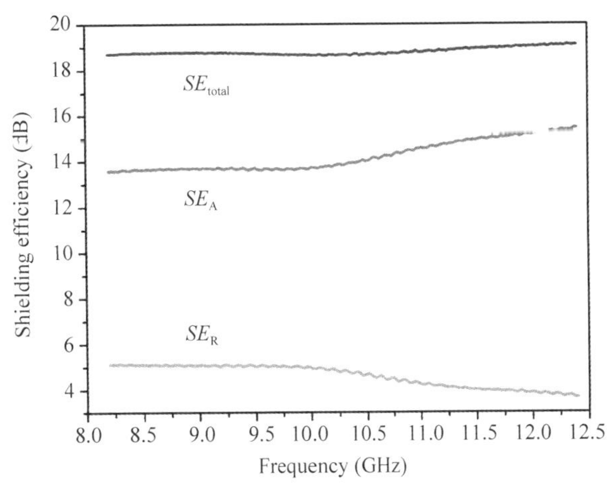

Fig. 6 SE_{total}, SE_A and SE_R of MWCNT/RCA as a function of frequency

reflection (SE_R) and absorption (SE_A) to the SE_{total}, the SE_A is 13.6–15.4 dB, much higher than the SE_R value of 3.7–5.1 dB, which suggests that the EMI shielding mechanism is absorption dominant. An absorption-dominant shielding mechanism helps a lot to reduce secondary radiation, and thus is regarded to be an alluring alternative for the development of electromagnetic radiation protection products[32].

Table 1 Cellulose-based EMI shielding materials

Materials	Method	Best SE_{total} (dB)	Frequency range (GHz)	Refs.
MWCNT/RCA	Coaxial (40% wax)	19.1	8.2–12.4	This work
Polyaniline/$BaTiO_3$/cotton fabrics	Waveguide	16.8	12.4–18.0	[26]
Nafion-MWCNT/cotton fabrics	Waveguide	11.5	8–12	[27]
Butanetetracarboxylic acid/MWCNT/cotton	Waveguide	5	8.2–12.4	[28]
Polyaniline/bacterial cellulose	Waveguide	5	8.2–12.4	[29]
Polyaniline/cotton fabrics	Waveguide	1.6	8–12	[30]
Bacterial cellulose/magnetite	Waveguide	1.5	8.2–12.4	[31]

4 Conclusions

A facile pre-gelation treatment was introduced to improve the mechanical property of RCA based on an inexpensive and eco-friendly NaOH/PEG solution. The relative lower pre-gelation temperature (50 ℃) is confirmed to be more beneficial to obtain more compact and uniform 3D network structure and higher crystallinity. The higher temperatures (70 ℃ and 90 ℃) result in the performance deterioration. The temperature at 50 ℃ makes the greatest contribution to the reinforcement of mechanical strength. RCA treated at 50 ℃ has the highest elasticity modulus of 1.03 MPa and energy absorption of 476 $kJ \cdot m^{-3}$. Thus this work provides a simple effective method to enhance the mechanical properties of RCA, which is helpful to ameliorate their application limitations. In addition, RCA decorated with MWCNTs is confirmed to be suitable for microwave absorption, and can act as a potential candidate for EMI shielding applications.

Acknowledgements

This study was supported by the National Natural Science Foundation of China (grant nos. 31270590 and 31470584).

References

[1] Fischer F, Rigacci A, Pirard R, et al. Cellulose-based aerogels[J]. Polymer, 2006, 47(22): 7636–7645.

[2] Gavillon R, Budtova T. Aerocellulose: new highly porous cellulose prepared from cellulose-NaOH aqueous solutions [J]. Biomacromolecules, 2008, 9(1): 269–277.

[3] Wan C, Li J. Facile Synthesis of Well-Dispersed Superparamagnetic γ-Fe_2O_3 Nanoparticles Encapsulated in Three Dimensional Architectures of Cellulose Aerogels and Their Applications for Cr(VI) Removal from Contaminated Water[J]. Acs Sustainable Chemistry & Engineering, 2015: 150723104428008.

[4] Wan C, Li J. Synthesis of well-dispersed magnetic $CoFe_2O_4$ nanoparticles in cellulose aerogels via a facile oxidative co-precipitation method[J]. Carbohydrate polymers, 2015, 134: 144–150.

[5] Shi Z, Huang J, Liu C, et al. Three-dimensional nanoporous cellulose gels as a flexible reinforcement matrix for

polymer nanocomposites[J]. ACS applied materials & interfaces, 2015, 7(41): 22990-22998.

[6] Isobe N, Chen X, Kim U J, et al. TEMPO-oxidized cellulose hydrogel as a high-capacity and reusable heavy metal ion adsorbent[J]. Journal of hazardous materials, 2013, 260: 195-201.

[7] Korhonen J T, Kettunen M, Ras R H A, et al. Hydrophobic nanocellulose aerogels as floating, sustainable, reusable, and recyclable oil absorbents[J]. ACS applied materials & interfaces, 2011, 3(6): 1813-1816.

[8] Bendahou D, Bendahou A, Seantier B, et al. Nano-fibrillated cellulose-zeolites based new hybrid composites aerogels with super thermal insulating properties[J]. Industrial Crops and Products, 2015, 65: 374-382.

[9] Moon R J, Martini A, Nairn J, et al. Cellulose nanomaterials review: structure, properties and nanocomposites[J]. Chemical Society Reviews, 2011, 40(7): 3941-3994.

[10] Huber T, Müssig J, Curnow O, et al. A critical review of all-cellulose composites[J]. Journal of Materials Science, 2012, 47(3): 1171-1186.

[11] Sathitsuksanoh N, George A, Zhang Y H P. New lignocellulose pretreatments using cellulose solvents: a review[J]. Journal of Chemical Technology & Biotechnology, 2013, 88(2): 169-180.

[12] Yan L, Gao Z. Dissolving of cellulose in PEG/NaOH aqueous solution[J]. Cellulose, 2008, 15(6): 789-796.

[13] Medronho B, Lindman B. Competing forces during cellulose dissolution: From solvents to mechanisms[J]. Current Opinion in Colloid & Interface Science, 2014, 19(1): 32-40.

[14] Liang S, Zhang L, Li Y, et al. Fabrication and properties of cellulose hydrated membrane with unique structure[J]. Macromolecular Chemistry and Physics, 2007, 208(6): 594-602.

[15] Wan C, Lu Y, Jiao Y, et al. Cellulose aerogels from cellulose-NaOH/PEG solution and comparison with different cellulose contents[J]. Materials Science and Technology, 2015, 31(9): 1096-1102.

[16] Wu J, Liang S, Dai H, et al. Structure and properties of cellulose/chitin blended hydrogel membranes fabricated via a solution pre-gelation technique[J]. Carbohydrate Polymers, 2010, 79(3): 677-684.

[17] Surapolchai W, Schiraldi D A. The effects of physical and chemical interactions in the formation of cellulose aerogels [J]. Polymer bulletin, 2010, 65(9): 951-960.

[18] Sèbe G. Ham-Pichavant F. dr; Ibarboure E.; Koffi ALC; Tingaut P[J]. Supramolecular Structure Characterization of Cellulose II Nanowhiskers Produced by Acid Hydrolysis of Cellulose I Substrates. Biomacromolecules, 2012, 13: 570-578.

[19] Cai J, Kimura S, Wada M, et al. Cellulose aerogels from aqueous alkali hydroxide - urea solution [J]. ChemSusChem: Chemistry & Sustainability Energy & Materials, 2008, 1(1-2): 149-154.

[20] Zhou J, Zhang L. Solubility of cellulose in NaOH/urea aqueous solution[J]. Polymer journal, 2000, 32(10): 866-870.

[21] Wan C, Lu Y, Jin C, et al. Thermally induced gel from cellulose/NaOH/PEG solution: preparation, characterization and mechanical properties[J]. Applied Physics A, 2015, 119(1): 45-48.

[22] Thomassin J M, Vuluga D, Alexandre M, et al. A convenient route for the dispersion of carbon nanotubes in polymers: Application to the preparation of electromagnetic interference (EMI) absorbers[J]. Polymer, 2012, 53(1): 169-174.

[23] Yang Y, Gupta M C, Dudley K L, et al. Novel carbon nanotube- polystyrene foam composites for electromagnetic interference shielding[J]. Nano letters, 2005, 5(11): 2131-2134.

[24] Zhang C S, Ni Q Q, Fu S Y, et al. Electromagnetic interference shielding effect of nanocomposites with carbon nanotube and shape memory polymer[J]. Composites Science and Technology, 2007, 67(14): 2973-2980.

[25] Liu X, Yin X, Kong L, et al. Fabrication and electromagnetic interference shielding effectiveness of carbon nanotube reinforced carbon fiber/pyrolytic carbon composites[J]. Carbon, 2014, 68: 501-510.

[26] Saini P, Choudhary V, Vijayan N, et al. Improved electromagnetic interference shielding response of poly(aniline)-coated fabrics containing dielectric and magnetic nanoparticles[J]. The Journal of Physical Chemistry C, 2012, 116(24):

13403-13412.

[27] Zou L, Yao L, Ma Y, et al. Comparison of polyelectrolyte and sodium dodecyl benzene sulfonate as dispersants for multiwalled carbon nanotubes on cotton fabrics for electromagnetic interference shielding[J]. Journal of Applied Polymer Science, 2014, 131(15).

[28] Alimohammadi F, Gashti M P, Shamei A. Functional cellulose fibers via polycarboxylic acid/carbon nanotube composite coating[J]. Journal of Coatings Technology and Research, 2013, 10(1): 123-132.

[29] Marins J A, Soares B G, Fraga M, et al. Self-supported bacterial cellulose polyaniline conducting membrane as electromagnetic interference shielding material: effect of the oxidizing agent[J]. Cellulose, 2014, 21(3): 1409-1418.

[30] Muthukumar N, Thilagavathi G. Development and characterization of electrically conductive polyaniline coated fabrics [J]. Indian Journal of Chemical Technology, 2012, 19: 434-441.

[31] Marins J A, Soares B G, Barud H S, et al. Flexible magnetic membranes based on bacterial cellulose and its evaluation as electromagnetic interference shielding material[J]. Materials Science and Engineering: C, 2013, 33(7): 3994-4001.

[32] Chung D D L. Corrosion control of steel-reinforced concrete[J]. Journal of Materials Engineering and Performance, 2000, 9(5): 585-588.

中文题目：预凝胶温度对绿色 NaOH/PEG 纤维素气凝胶力学性能影响的比较

作者：万才超，焦月，李坚

摘要：具有多孔三维网络结构的纤维素气凝胶往往机械强度较弱，这极大地限制了其应用。在本研究中，将一种易于操作的预凝胶处理方法引入再生纤维素气凝胶的常规制备技术中(即溶解、再生和干燥处理)，以改善其结构均匀性、结晶度和机械强度。这些性质与预凝胶温度密切相关，在相对较低的温度(50 ℃)贡献最大，特别是在机械强度方面，经 50 ℃处理的气凝胶的弹性模量和吸能达到最大值 1.03 MPa 和 476.2 $kJ \cdot m^{-3}$，分别比未经处理的气凝胶提高了 5.4 倍和 94.3%。作为潜在应用的一个范例，气凝胶的表面被多壁碳纳米管修饰，所得复合材料显示出良好的电磁屏蔽能力。

关键词：纤维素气凝胶；预凝胶；机械性能；多壁碳纳米管；电磁屏蔽；聚合物

Synthesis of Carbon Fiber Aerogel from Natural Bamboo Fiber and Its Application As A Green High-efficiency and Recyclable Adsorbent*

Yue Jiao, Caichao Wan, Jian Li

Abstract: We report an eco-friendly and inexpensive pyrolysis route to produce lightweight and superhydrophobic carbon fiber aerogel (CFA) by using cheap natural bamboo fiber as raw material. When used as an adsorbent, CFA can effectively and rapidly adsorb organic liquid from wastewater and didn't allow any water adsorption. CFA can adsorb an extensive range of organic liquids. The adsorption efficiency can reach up to 23–51 times the weight of original CFA. In addition, CFA shows excellent recyclability, and remained high adsorption efficiency even after five cycles through burning, squeezing, or extracting. In view of the lightweight feature, good adsorption selectivity, substantial adsorption efficiency, excellent adsorption recyclability, and easily-operated cost-effective preparation technique, consequently, we believe that such natural bamboo fiber-derived CFA holds potential for environmental remediation. More importantly, this work provides a good example to expand high-value applications of low-cost biomass resources.

Keywords: Bamboo fiber; Carbon fiber aerogel; Superhydrophobic; Recyclable; Green adsorbent

1 Introduction

The increasing seriousness of environmental pollution arising from oil spills, chemical leaks and industrial organic wastewater discharge has become an important issue demanding prompt solution. Extensive researches have been carried out to develop novel high-performance absorbent materials. For instance, Hrubesh et al.[1-3] fabricated different types of hydrophobic silica aerogels, which effectively removed organic liquids from aqueous solutions and mixtures. Zheng et al.[4] reported hybrid aerogels composed of polyvinyl alcohol and cellulose nanofibril. After being treated with methyltrichlorosilane, the superhydrophobic and superoleophilic aerogels not only exhibited excellent absorption performance for various types of oils or organic solvents (44–96 $g \cdot g^{-1}$), but also exhibited a good scavenging capability for heavy metal ions. Zhai et al.[5] synthesized superhydrophobic poly(vinyl alcohol)/cellulose nanofibril aerogel microspheres, whose crude oil absorption and organic solvent uptake capacity can reach up to 116 and 140 times of their own weight, respectively. In recent decades, with the increasing emphasis on green chemistry, natural sustainable and biodegradable resources have attracted huge research interest. Several investigations have been undertaken for removal of organic pollutants from wastewater by using different biomass materials or their derivatives. For instance,

* 本文摘自 Materials & Design, 2016, 107: 26–32.

Rashwan and Girgis[6] reported a kind of activated carbons derived from pyrolysis of rice straw. The activated carbons exhibited high adsorption capacities of 139 mg · g^{-1} for methylene blue, 154 mg · g^{-1} for congo red, 59 mg · g^{-1} for phenol, and 149 mg · g^{-1} for *p*-nitrophenol. Recently, Srinivasan and Viraraghavan[7] used walnut shell media to treat with various oils. It was found that adsorbed oil could be recovered from walnut shell media by applying pressure. The oil sorption capacities of 0.3-0.74 g · g^{-1} were obtained. More recently, Venkatanarasimhan and Raghavachari[8] synthesized eco-friendly nanocomposite materials based on mild epoxidation of natural rubber and magnetite nanoparticles. The magnetically recoverable materials could adsorb 7 g of petrol per gram without any mass loss. In addition, some other similar bioadsorbents based on biomass materials (e. g., pine bark[9], avocado kernel seeds[10], passion fruit peel[11], bamboo[12], and orange peel[13]) have been widely reported.

As one of the main components of plants, cellulose is a kind of typical hydrophilic carbohydrate polymer generated from repeating β-D-glucopyranose molecules that are covalently linked through acetal functions between the equatorial OH group of C4 and the C1 carbon atom (β-1, 4-glucan), as shown in Fig. 1. However, the strong hydrophilicity of cellulose is not beneficial to adsorb hydrophobic liquids like most organic solvents and oils. Therefore, prior to the use as adsorbents for organic liquids, these natural hydrophilic bioresources generally need to be subjected to hydrophobic treatment (e. g., acetylation[14], esterification[15] and silanization[16]) for grafting oleophilic groups. Also, some nanomaterial coatings (e. g., TiO_2) have been verified to facilitate construction of a low-energy and hydrophobic surface[17]. Recently, an easily-operated pyrolysis process is highly regarded. The process makes contribution to the transformation from hydrophilic cellulose-based materials to hydrophobic carbon products by destroying hydrophilic groups[18, 19]. Especially, the micromorphology changed little before and after the pyrolysis[20]. Compared with the aforementioned traditional chemical modifications, the pyrolysis technique is relatively easier to operate and more environmentally friendly; and no poisonous chemicals are needed. Although several reports[19, 21] have adopted pyrolysis technique to fabricate carbon-based adsorbents using cellulose-based materials as raw materials (like bacterial cellulose), it is still worth to develop cheaper and widely available natural biomass materials as feedstocks to fabricate novel high-performance carbon products.

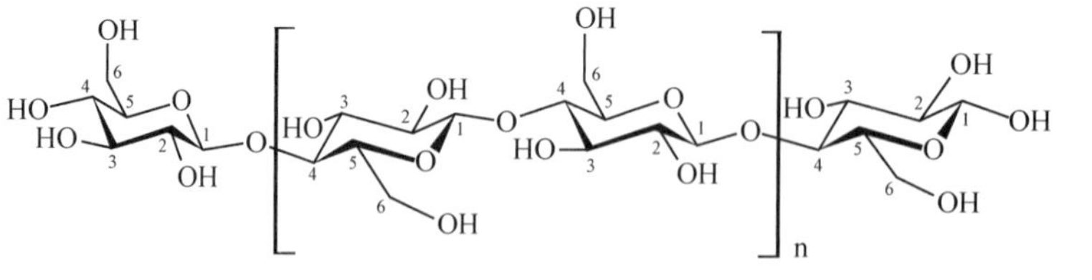

Fig. 1 Chemical structure of cellulose

Natural bamboo fiber, as a type of cellulose fibers, comes from an abundant and renewable resource (namely bamboo) at low cost, which ensures a continuous fiber supply and a significant material cost saving. Some researches[22, 23] show that the bamboo fiber is alike bast fibers in chemical composition; that is, cellulose constitutes the major portion (73.83%), and other components include hemicellulose (12.49%), lignin (10.15%), pectin, tannin, pigment, etc. Moreover, owing to loose structure and existence of disordered non-cellulose substances, bamboo fiber has a larger moisture regain capability than that of cotton, ramie and flax fibers. Also, bamboo fiber has high water - retention rate. Therefore, these characteristics indicate that bamboo fiber is a favorable potential adsorbent material. Nevertheless, the weak organic liquids sorption property, poor hydrophobicity, and unsatisfactory recyclability impose great restrictions on its application in adsorption and separation of organic pollutants from wastewater. Therefore, in the present work, we report a simple synthesis of carbon fiber aerogel (CFA) from inexpensive sustainable and degradable natural bamboo fiber via an eco - friendly pyrolysis technique. The changes in morphology, chemical components, crystal structure, and water wettability were investigated before and after the pyrolysis treatment by using scanning electron microscopy (SEM), energy-dispersive X-ray (EDX) analysis, Fourier transform infrared (FTIR) spectroscopy, X-ray diffraction (XRD), and water contact angle (WCA) tests. When used as adsorbents, the adsorption efficiency of CFA towards various common organic solvents and oil was measured. In addition, the adsorption selectivity and recyclability were also evaluated.

2 Materials and method

2.1 Materials

Natural bamboo fiber was rinsed several times by deionized water, and then dried in vacuum at 60 ℃ for 24 h. All other chemicals were of analytical grade and used without further purification.

2.2 Preparation of carbon fiber aerogel (CFA)

A piece of dried bamboo fiber was put into a cylindrical quartz container, and then the container was placed into a tubular furnace. Prior to the pyrolysis treatment, the air in the tubular furnace and sample were completely removed by vacuum pumping followed by ventilating nitrogen gas for 15 min. After that, the sample was heated to 500 ℃ at a heating rate of 5 ℃ · min^{-1}, and this temperature was maintained for 1 h; then, the temperature was raised to 1000 ℃ at 5 ℃ · min^{-1} and remained for 2 h to allow for complete pyrolysis. Thereafter, the temperature decreased to 500 ℃ at 5 ℃ · min^{-1}, and finally decreased naturally to the room temperature, and the following CFA was obtained. The whole pyrolysis process was conducted under the protection of nitrogen.

2.3 Adsorption property of CFA for oil and organic solvents

A typical adsorption test contains following steps: CFA was firstly weighed before the test, and subsequently placed in contact with an organic solvent or oil until the aerogel was completely filled with the liquid (about 3 min for organic solvents and 10 min for oil), and finally taken out for weight measurement. To avoid the effects of evaporation of adsorbates, especially for those with low boiling points and strong volatility, the weight measurement must be done quickly. The adsorption property was characterized by adsorption efficiency, which was calculated by dividing the wet weight of the samples after the adsorptions by the initial weight. The adsorption recyclability was assessed by repeating the aforementioned adsorption process for five times and monitoring the changes of adsorption efficiency. The adsorbates were removed by three methods, i. e., combusting for cheap flammable and toxic liquid pollutants, squeezing for low-cost and low-viscosity solvents, and extracting for expensive and high - viscosity solvents (residual extraction agent in CFA was removed by burning).

2.4 Characterizations

The morphology was observed by SEM (HITACHI, TM3030) operating at 15.0 kV. Elemental analysis was conducted by an EDX spectrometer. FTIR spectra were recorded on a FTIR instrument (Nicolet 6700, Thermo Fisher Scientific., USA) in the range of 400 – 4000 cm^{-1} with a resolution of 4 cm^{-1}. The crystal structures were characterized by XRD (Rigaku D/MAX 2200) operating with Cu Ka radiation ($\lambda = 1.5418$ Å) at a scan rate (2θ) of 4 min^{-1} ranging from 5° to 40°. WCA was measured on a Powereach JC2000C contact angle analyzer. The Brunauer–Emmet–Teller (BET) surface area and pore property of CFA were determined from N_2 adsorption–desorption experiments at 196 ℃ using an accelerated surface area and porosimetry system (3H – 2000PS2 unit, Beishide Instrument S&T Co., Ltd). Meanwhile, the pore volume and pore – size distribution were estimated by the Barrett–Joyner–Halenda (BJH) method.

3 Results and discussion

3.1 Characterizations of CFA by SEM, EDX, XRD and FTIR

The SEM image in Fig. 2a shows the microstructure of the natural bamboo fiber. As shown, the long fibers are intertwined leading to the formation of three–dimensional (3D) networks. The 3D porous frame provides congenital condition for adsorption[24]. The average size (d) and standard deviation (σ) of the bamboo fiber are calculated as 15.33 and 2.27 μm, respectively, according to the diameter frequency histogram (Fig. 2b). Additionally, the main elements of the bamboo fiber are shown in Fig. 2c, where the C and O elements were detected by EDX and account for approximately 35.5% and 64.5%, respectively. The Au element is originated from the coating layer used for electric conduction during the SEM observation. In contrast, CFA fabricated from the pyrolysis of bamboo fiber is clearly composed of thinner fibers (Fig. 2d). Meanwhile, it can be found that the 3D fibers network maintains well and becomes denser after the pyrolysis. Furthermore, the fibers are relatively thinner, and the corresponding diameter frequency histogram (Fig. 2e) suggests that the average size and standard deviation of the fibers decrease to 6.02 and 0.84 μm, respectively. In addition, the EDX spectrum of CFA (Fig. 2f) shows a remarkable improvement in the mass ratio of carbon to oxygen (from 1.8 to 8.2), which demonstrates that the most oxygen–containing groups had been damaged. We roughly calculated the bulk density of CFA by dividing the sample mass by the sample volume, and the result is as low as 39.6 $mg \cdot cm^{-3}$.

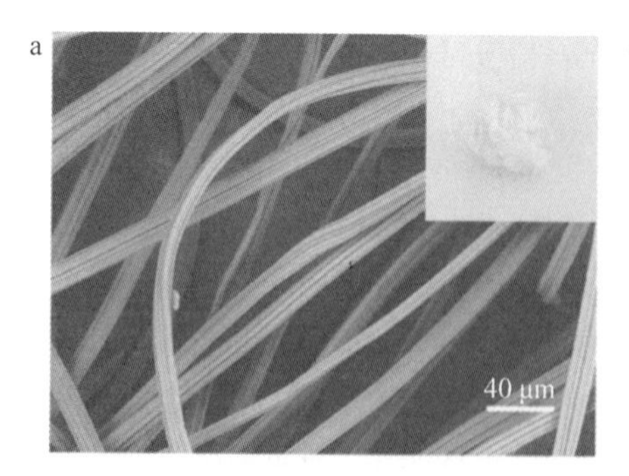

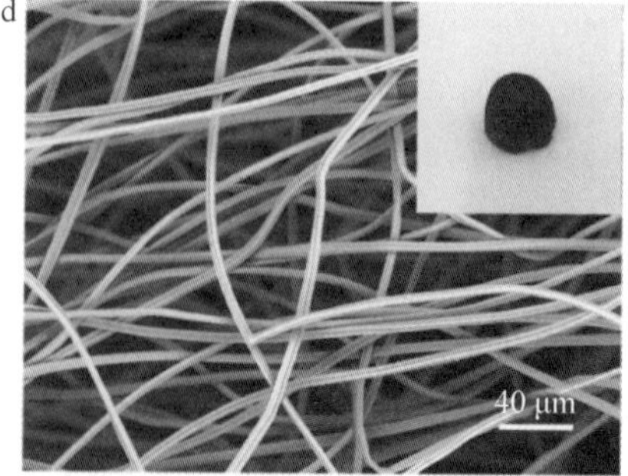

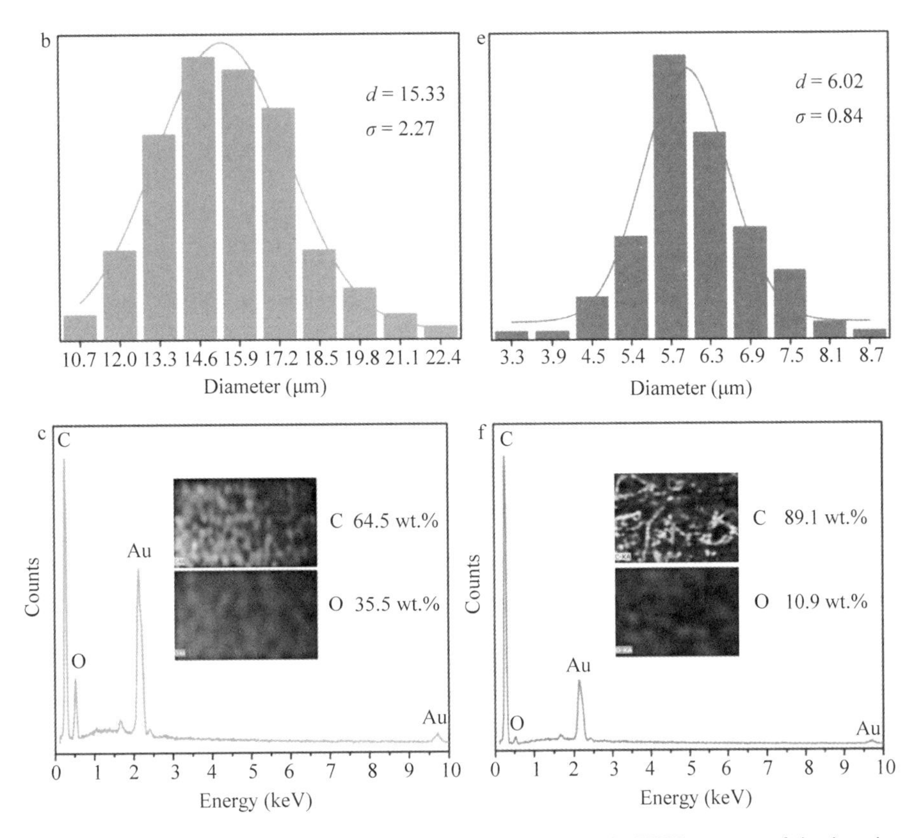

Fig. 2 (a, d) SEM images, (b, e) diameter distributions and (c, f) EDX spectra of the bamboo fiber and CFA, respectively. The insets in (a, d) present the macroscopic images, and the insets in (c, f) give the corresponding elemental maps and contents

The XRD pattern of the natural bamboo fiber (Fig. 3a) suggests typical cellulose Ⅰ crystal structure with parallel up arrangement of cellulose chains. The sharp peak at around 22.1° is derived from the (002) plane, and the broad peak at around 15.8° is due to the mixture of peaks originated from (101) and (10$\bar{1}$) planes[25]. After the pyrolysis, the cellulose peaks disappeared absolutely; instead, a broad peak centered at

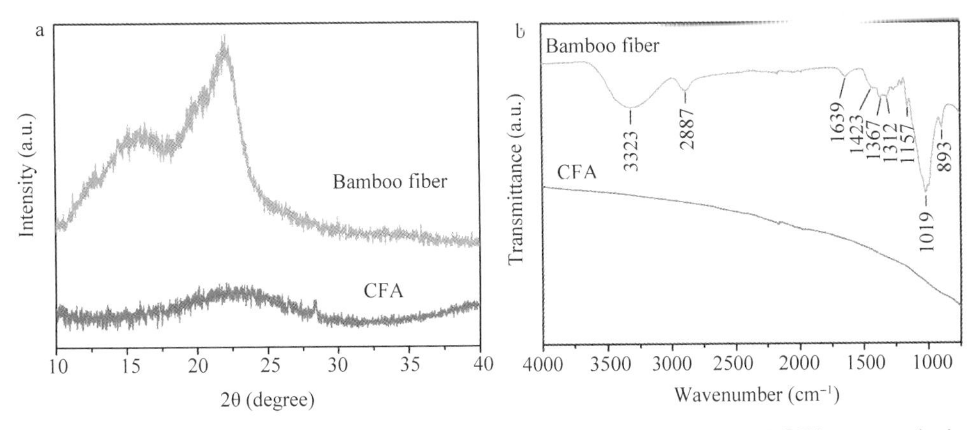

Fig. 3 (a) XRD patterns and (b) FTIR spectra of the bamboo fiber and CFA, respectively

22.7° was generated, corresponding to the (002) plane of graphite[26]. These results indicate that the crystalline structure of the bamboo fiber was destroyed during the pyrolysis process, and amorphous carbon was generated[27]. FTIR analysis further confirms the removal of hydrophilic oxygen-containing functional groups of the bamboo fiber. As shown in Fig. 3b and Table 1, the bamboo fiber shows several peaks of hydrophilic groups[28], such as C=O, C-O and -OH. Nevertheless, after the pyrolysis, the resultant CFA presents no apparent characteristic signals related to the hydrophilic groups, which reveals the formation of hydrophobicity.

Table 1 Frequencies (cm^{-1}) of the main signals of the bamboo fiber

Absorption band (cm^{-1})	Assignment
3323	Valence vibration of hydrogen bonded OH-groups
2887	CH_2 valence vibration
1639	C=O stretch
1423	CH_2 scissoring
1367	C-H stretch in CH_3
1312	CH_2 rocking vibration
1157	C-O-C asymmetric valence vibration
1019	C-O-C pyranose ring skeletal vibration
893	Anomere C-groups, C_1-H eformation, ring valence vibration

3.2 Specific surface area and pore diameter distribution of CFA

N_2 adsorption-desorption isotherms can provide much useful information about pore structures. As shown in Fig. 4a, the adsorption isotherm of CFA can be classified to type Ⅳ according to IUPAC classification. A clear hysteresis loop between the adsorption isotherm and desorption isotherm is generally attributed to the capillary condensation occurring in the mesopores[29]. The adsorption-desorption loop belongs to the H_2-type hysteresis loop[30], which reveals the existence of inkbottle shaped pores. The BET specific surface area and pore volume of CFA are around 26.2 $m^2 \cdot g^{-1}$ and 0.095 $cm^3 \cdot g^{-1}$, respectively. The pore size distribution curve obtained using BJH method is displayed in Fig. 4b. Two regions can be confirmed: (1) mesopores (2-50 nm) with three peaks at 2.1, 2.7 and 3.7 nm; and (2) macropores (50-170 nm), indicating that CFA is mainly composed of mesopores and macropores. The average pore diameter of CFA was calculated to be 14.0 nm.

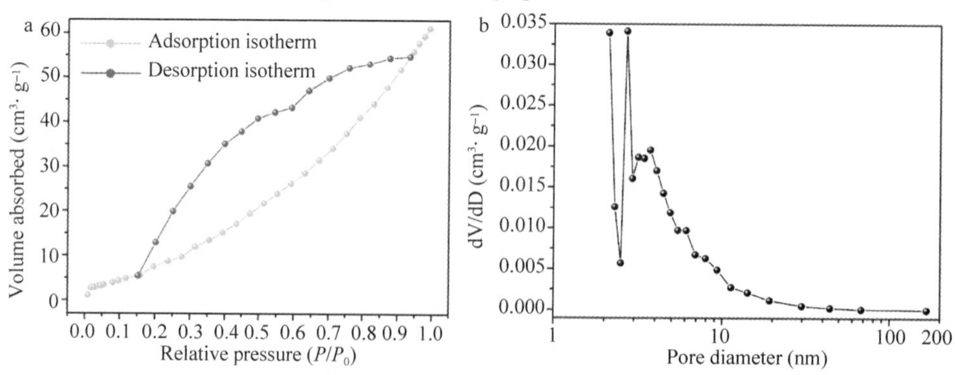

Fig. 4 (a) N_2 adsorption-desorption isotherms and (b) pore-size distribution of CFA, respectively

3.3 Water wettability and adsorption selectivity of CFA

It is extremely attractive to develop selective materials that do not adsorb water but are able to rapidly adsorb a large amount of oleophilic solvents. The materials with good adsorption selectivity hold great potential as oil – water separation materials or high – performance adsorbents for treatment of organic wastewater. In Fig. 5a, when water came into contact with the raw material of bamboo fiber, the bamboo fiber was quickly submerged into the water, which indicates strong hydrophilicity and poor buoyancy characteristics, and thus the raw material cannot be applied in removal of organic pollutants from water. On the contrary, the pyrolysis product (i. e., CFA) could easily float on the water surface (Fig. 5b). Based on the research by Marmur and Ras[31], the favorable floatation of CFA is mainly resulted from its water repellency, whereas the low density has little influence. Moreover, when water droplet came into the surface of CFA, it could stably stand and maintain complete spherical shape (Fig. 5c), which also indicates excellent hydrophobicity. Fig. 5d presents the WCA test of CFA. The WCA reaches up to 151°, suggesting that CFA is superhydrophobic. This result can be well explained by the aforementioned results of EDX, XRD and FTIR analyses. Fig. 5e and 5f present the adsorption processes of heptane stained with Sudan Ⅲ at the bottom of water or on the water surface by using CFA. As shown, when CFA was dipped into the water and touched the heptane (dyed with Sudan Ⅲ), it did not allow water adsorption, but it adsorbed the heptane rapidly and completely in eight seconds regardless of the location of the heptane (see also Video S1 and S2 in the Supporting Information). Therefore, these results confirm that CFA can effectively remove organic solvent from water. Considering the strong water repellency, good adsorption selectivity and lightweight feature, CFA is undoubtedly a promising candidate for treatment of wastewater containing organic pollutants.

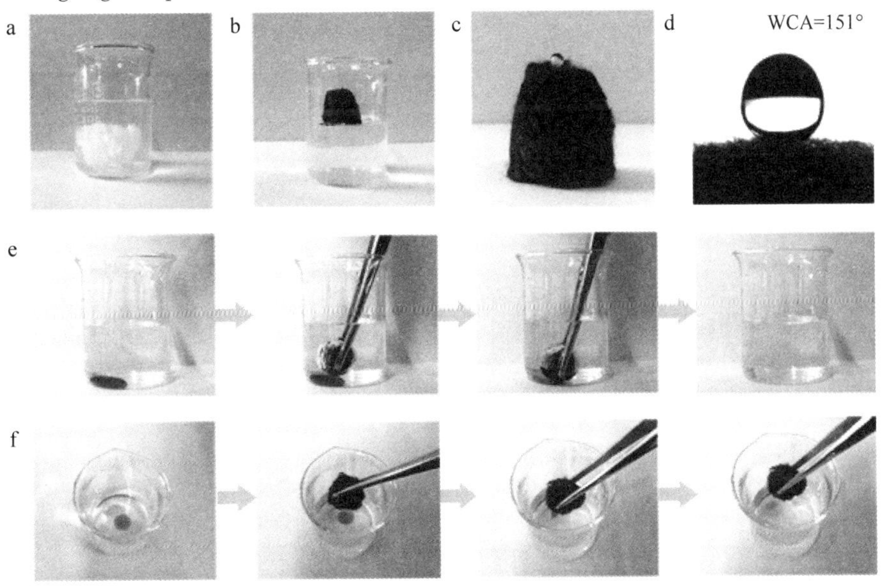

Fig. 5 (a) The Bamboo fiber submerged in the water. (b) CFA floated on the water. (c) A water droplet stably stood on the surface of CFA. (d) WCA measurement of CFA. (e) Adsorption process of heptane stained with Sudan Ⅲ at the bottom of water by using CFA. (f) Adsorption process of a layer of heptane stained with Sudan Ⅲ on the water surface by using CFA

3.4 Adsorption property of CFA for various organic solvents and oil

Some organic solvents (e.g., benzene, chloroform, dimethylformamide, ethanol, ethyl acetate, oleic acid and tetrachloromethane toluene) and oils are common pollutants in wastewater from industry as well as in our daily life. Besides the aforementioned investigation on the adsorption selectivity of CFA, it makes sense to quantitatively survey the adsorption efficiency of CFAs for various organic solvents and oil. The results in Fig. 6 show that CFA exhibits a high adsorption efficiency for all of these organic liquids, and the adsorption efficiency of CFA reaches up to 23–51 $g \cdot g^{-1}$. Regardless of errors due to types of organic adsorbates, the value is comparable to or even higher than that of other adsorbents in some literatures, as summarized in Table 2, including activated carbons (<1 $g \cdot g^{-1}$)[32], carbon/expanded vermiculite (1–4 $g \cdot g^{-1}$)[33], nanowire membranes (4–20 $g \cdot g^{-1}$)[34], conjugated microporous polymers (5–25 $g \cdot g^{-1}$)[35], magnetic exfoliated graphite (30–50 $g \cdot g^{-1}$)[36], spongy graphene (20–86 $g \cdot g^{-1}$)[37], and boron-doped multiwalled carbon nanotubes (25–125 $g \cdot g^{-1}$)[38]. Furthermore, taking the easily-operated cost-effective and eco-friendly preparation process into consideration, CFA is more potential.

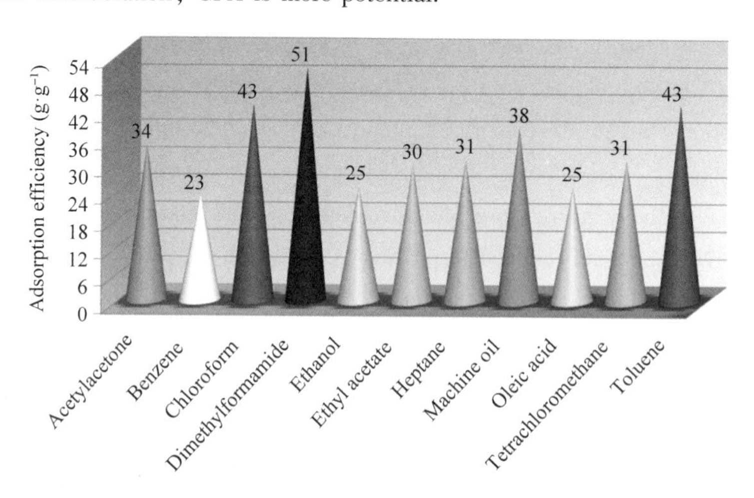

Fig. 6 Adsorption efficiency of CFA for various organic solvents and machine oil

Table 2 Comparison of adsorption efficiency of CFA with other recently reported adsorbents

Adsorbents	Preparation method	Adsorption efficiency ($g \cdot g^{-1}$)	Ref.
Activated carbons	Activation method	<1	[32]
Carbon/expanded vermiculite	Chemical vapor deposition	1–4	[22]
Nanowire membranes	Hydrothermal method and vapor deposition technique	4–20	[34]
Conjugated microporous polymers	Homocoupling polymerization	5–25	[35]
Magnetic exfoliated graphite	Sol-gel reaction	30–50	[36]
Spongy graphene	Reducing and moulding	20–86	[37]
Boron-doped multiwalled carbon nanotubes	Aerosol-assisted chemical vapor deposition	25–125	[38]
CFA	Pyrolysis	23–51	This work

3.5 Adsorption recyclability of CFA and recoverability of organic liquids

Besides strong adsorption selectivity and substantial adsorption efficiency, adsorption recyclability is also of essential importance for adsorbent materials. High adsorption recyclability makes contribution to saving cost, reducing energy consumption of preparation, reducing generation of garbage, etc. For the realization of adsorption recyclability, it needs to separate adsorbed liquids from the adsorbent before the next adsorption process. In the present work, we adopted three simple methods to recycle or remove adsorbates, i. e., combusting for cheap flammable and toxic liquid pollutants, squeezing for low-cost and low-viscosity solvents, and extracting for expensive and high-viscosity solvents (residual extraction agent in CFA was removed by burning). As shown in Fig. 7, the toxic benzene was removed from CFA by burning; inexpensive and low-viscosity ethanol could be easily separated from CFA by pressuring the sample; high-viscosity machine oil was separated from CFA via an extracting process by using ethanol, and the oil can be readily recovered by fractional distillation.

All kinds of adsorption-desorption processes were repeated for five times, and the changes in the adsorbed mass and remnant mass of the adsorbents are presented in Fig. 8. Direct burning in air was applied to recycle CFA; on this occasion the adsorbed pollutions can be efficiently used for heating. As shown in Fig. 8a, the weight of CFA decreased little from 0.135 g to 0.130 g after the five cycles. The adsorbed mass for benzene decreased little from 3.056 g to 2.963 g after the five cycles (see also Video S3 in the Supporting Information). Squeezing was adopted as another simple method for recycling CFA (Fig. 8b). As mentioned above, this method is more suitable for those pollutants with low viscosity or being nontoxic (like ethanol). In the first cycle, 28.7 times weight gain was achieved for ethanol, but the remnant mass increased to 669 mg (about five times of their original mass) after squeezing because of incomplete removal of ethanol. From the

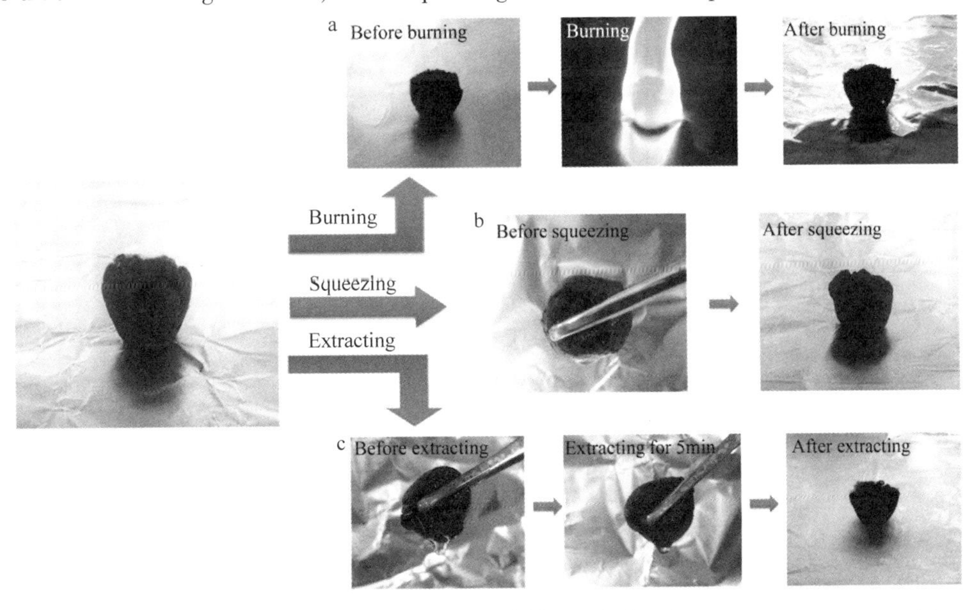

Fig. 7 Separation processes of adsorbed liquids from CFA. (a) Burning process was applied to separate benzene. (b) Squeezing process was applied to separate ethanol. (c) Extracting process was applied to separate machine oil by using ethanol

second cycle onwards, the adsorption efficiency became stable and decreased to ~5.0 g · g^{-1}. This decrease is not only attributed to the high remnant mass, but also the damage of porous structure caused by the repeated compression (the adsorbed mass decreased from 37 to 33 g after the fiver cycles). As for the cyclic adsorption-extracting test, particularly for those organic liquids with high viscosity (like oils) or high cost, the waste machine oil was used as an example. We adopted ethanol to extract the machine oil from CFA, and the oil can be separated from the mixture by fractionation due to the difference of boiling point. In addition, the complete removal of residual ethanol from CFA was achieved by burning. It can be seen in Fig. 8c that the adsorption efficiency did not change a lot after the five cycles. Consequently, the above three easily-operated methods including burning, squeezing and extracting, or a combination of them can be applied to recycle CFA dependent on type of organic liquids.

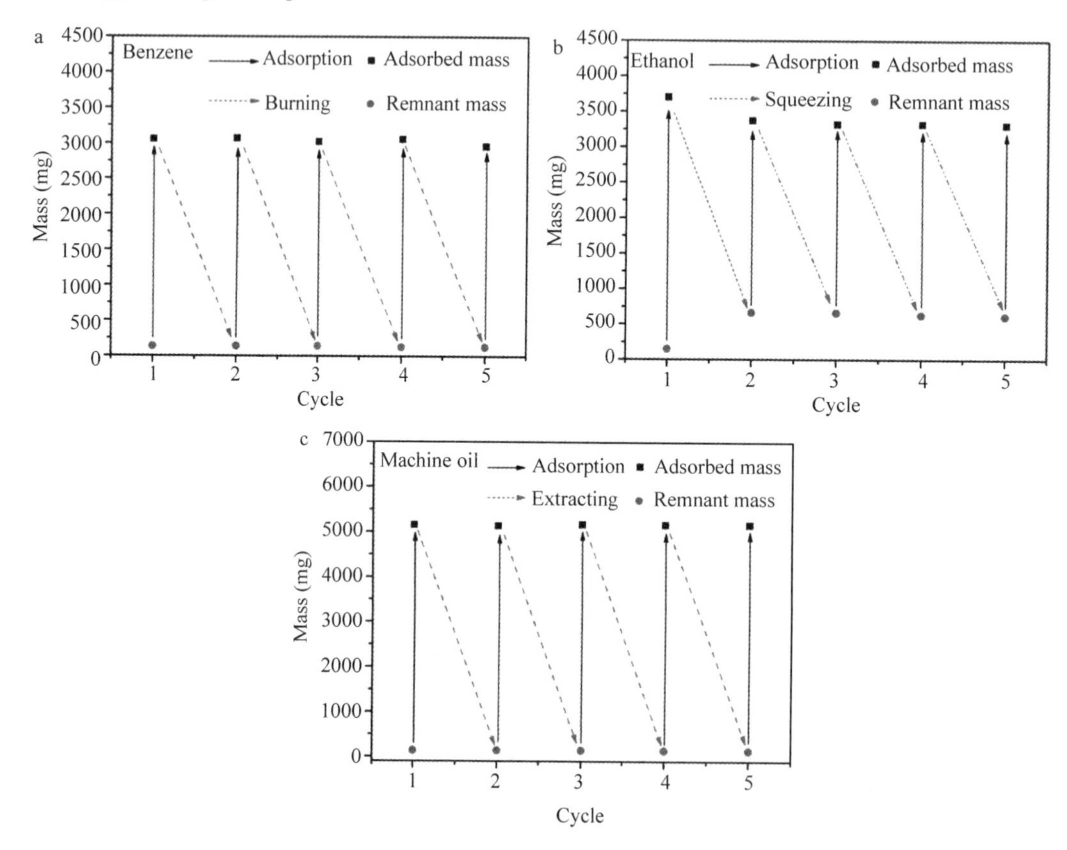

Fig. 8 Adsorption recyclability tests of CFA. (a) Burning was used to recycle CFA for adsorption of benzene. (b) Squeezing was applied to recycle CFA for adsorption of ethanol. (c) Extracting was applied to recycle CFA for adsorption of machine oil

4 Conclusions

We successfully fabricated superhydrophobic lightweight CFA with satisfactory adsorption efficiency and excellent recyclability via a simple and cheap pyrolysis method using extensively available and low-cost natural material (namely bamboo fiber) as feedstock. CFA with strong water repellency possesses good adsorption selectivity. It can quickly and effectively adsorbed organic pollutant from wastewater, but didn't allow water adsorption. Moreover, CFA can uptake various organic liquids at 23 to 51 times its own weight. CFA can also be recycled and reused through three simple pathways, i. e. , burning, squeezing and extracting, or a combination of them. Therefore, considering the lightweight feature, good adsorption selectivity, substantial adsorption efficiency and excellent adsorption recyclability, CFA can serve as a promising candidate for treatment of oil spillage and for recovering organic liquids from wastewater. More importantly, this work provides a good example to expand high-value applications of low-cost biomass resources.

Supporting Information

Please see the supporting information (Video 1-3).

Acknowledgments

This study was supported by the National Natural Science Foundation of China (grant nos. 31270590 and 31470584).

References

[1]Hrubesh L W, Coronado P R, Satcher Jr J H. Solvent removal from water with hydrophobic aerogels[J]. Journal of Non-Crystalline Solids, 2001, 285(1-3): 328-332.

[2]Reynolds J G, Coronado P R, Hrubesh L W. Hydrophobic aerogels for oil-spill clean up-synthesis and characterization [J]. Journal of Non-Crystalline Solids, 2001, 292(1-3): 127-137.

[3]Reynolds J G, Coronado P R, Hrubesh L W. Hydrophobic aerogels for oil-spill cleanup? Intrinsic absorbing properties [J]. Energy Sources, 2001, 23(9): 831-843.

[4]Zheng Q, Cai Z, Gong S. Green synthesis of polyvinyl alcohol (PVA)-cellulose nanofibril (CNF) hybrid aerogels and their use as superabsorbents[J]. Journal of materials chemistry A, 2014, 2(9): 3110-3118.

[5]Zhai T, Zheng Q, Cai Z, et al. Synthesis of polyvinyl alcohol/cellulose nanofibril hybrid aerogel microspheres and their use as oil/solvent superabsorbents[J]. Carbohydrate polymers, 2016, 148: 300-308.

[6]Rashwan W E, Girgis B S. Adsorption capacities of activated carbons derived from rice straw and water hyacinth in the removal of organic pollutants from water[J]. Adsorption Science & Technology, 2004, 22(3): 181-194.

[7]Srinivasan A, Viraraghavan T. Removal of oil by walnut shell media[J]. Bioresource technology, 2008, 99(17): 8217-8220.

[8]Venkatanarasimhan S, Raghavachari D. Epoxidized natural rubber-magnetite nanocomposites for oil spill recovery[J]. Journal of Materials Chemistry A, 2013, 1(3): 868-876.

[9]Björklund K, Li L. Evaluation of low-cost materials for sorption of hydrophobic organic pollutants in stormwater[J]. Journal of environmental management, 2015, 159: 106-114.

[10]Elizalde-González M P, Mattusch J, Peláez-Cid A A, et al. Characterization of adsorbent materials prepared from avocado kernel seeds: Natural, activated and carbonized forms[J]. Journal of Analytical and Applied Pyrolysis, 2007, 78(1): 185-193.

[11]Pavan F A, Mazzocato A C, Gushikem Y. Removal of methylene bluc dye from aqueous solutions by adsorption using yellow passion fruit peel as adsorbent[J]. Bioresource technology, 2008, 99(8): 3162-3165.

[12]Guo J Z, Li B, Liu L, et al. Removal of methylene blue from aqueous solutions by chemically modified bamboo[J]. Chemosphere, 2014, 111: 225-231.

[13]Namasivayam C, Muniasamy N, Gayatri K, et al. Removal of dyes from aqueous solutions by cellulosic waste orange peel[J]. Bioresource Technology, 1996, 57(1): 37-43.

[14]Tserki V, Zafeiropoulos N E, Simon F, et al. A study of the effect of acetylation and propionylation surface treatments

on natural fibres[J]. Composites Part A: applied science and manufacturing, 2005, 36(8): 1110-1118.

[15]Sidik S M, Jalil A A, Triwahyono S, et al. Modified oil palm leaves adsorbent with enhanced hydrophobicity for crude oil removal[J]. Chemical Engineering Journal, 2012, 203: 9-18.

[16]Likon M, Saarela J. The conversion of paper mill sludge into absorbent for oil spill sanitation-the life cycle assessment [C]//Macromolecular Symposia. Weinheim: WILEY-VCH Verlag, 2012, 320(1): 50-56.

[17]Korhonen J T, Kettunen M, Ras R H A, et al. Hydrophobic nanocellulose aerogels as floating, sustainable, reusable, and recyclable oil absorbents[J]. ACS applied materials & interfaces, 2011, 3(6): 1813-1816.

[18]Wan, CC, Jiao, et al. Fabrication of hydrophobic, electrically conductive and flame-resistant carbon aerogels by pyrolysis of regenerated celluloseaerogels[J]. CARBOHYD POLYM, 2015, 2015, 118: 115-118.

[19]Wu Z Y, Li C, Liang H W, et al. Ultralight, Flexible, and Fire-Resistant Carbon Nanofiber Aerogels from Bacterial Cellulose[J]. Angewandte Chemie International Edition, 2013, 52(10): 2925-2929.

[20]Bi H, Yin Z, Cao X, et al. Carbon fiber aerogel made from raw cotton: a novel, efficient and recyclable sorbent for oils and organic solvents[J]. Advanced Materials, 2013, 25(41): 5916-5921.

[21]Savova D, Apak E, Ekinci E, et al. Biomass conversion to carbon adsorbents and gas[J]. Biomass and Bioenergy, 2001, 21(2): 133-142.

[22]Li L J, Wang Y P, Wang G, et al. Evaluation of properties of natural bamboo fiber for application in summer textiles [J]. Journal of Fiber Bioengineering and Informatics, 2010, 3(2): 94-99.

[23]Khalil H P S A, Bhat I U H, Jawaid M, et al. Bamboo fibre reinforced biocomposites: A review[J]. Materials & Design, 2012, 42: 353-368.

[24]Maji T K, Ohba M, Kitagawa S. Transformation from a 2D stacked layer to 3D interpenetrated framework by changing the spacer functionality: Synthesis, structure, adsorption, and magnetic properties[J]. Inorganic chemistry, 2005, 44(25): 9225-9231.

[25]Nishiyama Y, Langan P, Chanzy H. Crystal structure and hydrogen-bonding system in cellulose Iβ from synchrotron X-ray and neutron fiber diffraction[J]. Journal of the American Chemical Society, 2002, 124(31): 9074-9082.

[26] Mao Y, Duan H, Xu B, et al. Lithium storage in nitrogen-rich mesoporous carbon materials [J]. Energy & Environmental Science, 2012, 5(7): 7950-7955.

[27]Qiu W, Xia J, He S, et al. Facile synthesis of hollow MoS_2 microspheres/amorphous carbon composites and their lithium storage properties[J]. Electrochimica Acta, 2014, 117: 145-152.

[28]Abidi N, Cabrales L, Haigler C H. Changes in the cell wall and cellulose content of developing cotton fibers investigated by FTIR spectroscopy[J]. Carbohydrate Polymers, 2014, 100: 9-16.

[29]Lu Y, Sun Q, Yang D, et al. Fabrication ofmesoporous lignocellulose aerogels from wood via cyclic liquid nitrogen freezing-thawing in ionic liquid solution[J]. Journal of Materials Chemistry, 2012, 22(27): 13548-13557.

[30]Leofanti G, Padovan M, Tozzola G, et al. Surface area and pore texture of catalysts[J]. Catalysis today, 1998, 41 (1-3): 207-219.

[31]Marmur A, Ras R H A. The porous nano-fibers raft: analysis of load-carrying mechanism and capacity[J]. Soft Matter, 2011, 7(16): 7382-7385.

[32]Lillo-Ródenas M A, Cazorla-Amorós D, Linares-Solano A. Behaviour of activated carbons with different pore size distributions and surface oxygen groups for benzene and toluene adsorption at low concentrations[J]. Carbon, 2005, 43(8): 1758-1767.

[33]Moura F C C, Lago R M. Catalytic growth of carbon nanotubes and nanofibers on vermiculite to produce floatable hydrophobic "nanosponges" for oil spill remediation[J]. Applied Catalysis B: Environmental, 2009, 90(3-4): 436-440.

[34] Yuan J, Liu X, Akbulut O, et al. Superwetting nanowire membranes for selective absorption [J]. Nature nanotechnology, 2008, 3(6): 332-336.

[35]Li A, Sun H X, Tan D Z, et al. Superhydrophobic conjugated microporous polymers for separation and adsorption [J]. Energy & Environmental Science, 2011, 4(6): 2062-2065.

[36]Wang G, Sun Q, Zhang Y, et al. Sorption and regeneration of magnetic exfoliated graphite as a new sorbent for oilpollution[J]. Desalination, 2010, 263(1-3): 183-188.

[37]Bi H, Xie X, Yin K, et al. Spongy graphene as a highly efficient and recyclable sorbent for oils and organic solvents [J]. Advanced Functional Materials, 2012, 22(21): 4421-4425.

[38]Hashim D P, Narayanan N T, Romo-Herrera J M, et al. Covalently bonded three-dimensional carbon nanotube solids via boron induced nanojunctions[J]. Scientific reports, 2012, 2(1): 1-8.

中文题目：天然竹纤维合成碳纤维气凝胶及其作为绿色高效可回收吸附剂的应用

作者：焦月，万才超，李坚

摘要：报道了一种以廉价的天然竹纤维为原料制备轻质超疏水碳纤维气凝胶(CFA)的低成本、环保的热解工艺。当用作吸附剂时，CFA 能有效、快速地吸附废水中的有机液体，不允许任何水的吸附。CFA 可以吸附多种有机液体。吸附效率可达原 CFA 重量的 23~51 倍。此外，CFA 表现出优异的可回收性，即使在燃烧、挤压或提取五次循环后仍保持较高的吸附效率。鉴于其轻质、吸附选择性好、吸附效率高、吸附回收性能好、制备工艺简单、成本低等特点，我们认为这种天然竹纤维衍生 CFA 在环境修复方面具有潜在的应用前景。更重要的是，这项工作为拓展低成本生物质资源的高价值应用提供了一个很好的范例。

关键词：竹纤维；碳纤维气凝胶；超疏水；可回收；绿色吸附剂

Synthesis of Superhydrophobic Ultralight Aaerogels from Nanofibrillated Cellulose Isolated from Natural Reed for High-performance Adsorbents*

Yue Jiao, Caichao Wan, Tiangang Qiang, Jian Li

Abstract: Reed is one of the widely available aquatic plant resources, and its applications are generally limited to some traditional areas like papermaking and animals' fodder. Besides, most of reed is wasted or directly burned every year causing serious air pollution (like atmospheric haze). Therefore, it is worth to further develop new forms of high-value applications of reed. Herein, the natural reed was collected to fabricate ultralight adsorbents namely nanofibrillated cellulose (NFC) aerogels via an easily-operated method, which includes chemical purification, ultrasonication, and freeze drying. The NFC aerogels with the ultra-low density of 4.9 mg · cm^{-3} were characterized by scanning electron microscopy, energy dispersive X-ray spectrometer, Fourier transform infrared spectroscopy, X-ray diffraction, and thermogravimetric analysis. For acquiring good hydrophobicity, the NFC aerogels were subjected to a hydrophobic treatment by methyltrichlorosilane. The superhydrophobic NFC aerogels with contact angles of as high as 151° – 155° have excellent adsorption efficiency (53–93 g · g^{-1}) for various organic solvents and waste oil. More importantly, the aerogels also exhibit favorable adsorption recyclability, which can maintain more than 80% of the initial adsorption efficiency after the five cycles.

Keywords: Nanofibrillated cellulose; Aerogels; Ultralight; Superhydrophobic; Recyclable adsorbents

1 Introduction

With the rapid consumption of oil resource day by day, it is necessary to use green, economic, and renewable resources to exploit new materials to substitute for traditional petrochemical products. Cellulose is one of the most abundant reproducible natural polymers on earth. Especially, the individual cellulose nanofibrils have gained increasing attentions. Beside their nano-size dimensions, cellulose nanofibrils have other advantages like outstanding mechanical properties, low density, biocompatibility, and high reactivity[1, 2]. As a result, cellulose nanofibrils can be useful in various fields, such as reinforcement components in flexible display panels[3], biodegradable packaging materials[4], and oxygen-barrier layers[5]. In recent decades, several techniques have been proposed to extract highly purified nanofibrils from various cellulosic resources, such as high pressure homogenization[6], high-speed shearing[7], and TEMPO-mediated oxidation combined with mechanical treatment[8]. All these methods can obtain different types of nanofibrils,

* 本文摘自 Applied Physics A, 2016, 122(7): 1-10.

primarily depending on both raw materials and disintegration process. Among the existent approaches, ultrasonic technique has been extensively examined for materials synthesis and considered as one of the most powerful tools to isolate cellulose nanofibrils[9, 10]. Such powerful ultrasonic environments provide a unique platform to break the firm hydrogen bonding between nanofibrils, allowing individual nanofibrils to be gradually separated. Apart from the remarkable merits, it is also worth to mention that the ultrasonication causes cellulose degradation to some extent. To date, the effect of ultrasonication in degrading polysaccharide linkages has been well described[11-13].

Reed is a kind of annual herb plant which is distributed in temperate and tropical zone. It mainly grows in irrigation canals, river near swamps, etc. Reed is composed of about 49.4% cellulose, 31.5% hemicellulose, 8.8% lignin, and other substrates[14]. The applications of reed are generally limited to some traditional areas like papermaking and animals' fodder. Besides, most of reed is wasted or burned every year. Therefore, it is important to further develop new forms of high-value application of this resource. The cell walls of plant usually consist of rigid cellulose microfibrils embedding in the soft hemicellulose and lignin matrix[15, 16]. Cellulose microfibrils consist of numerous cellulose nanofibrils which are tightly hooked to one another by multiple hydrogen bonds. Thus it is difficult to split cellulose nanofibrils only by mechanical treatments. Many researchers[17, 18] reported that the matrix substances can be removed using chemical methods before the fibrillation process. As a result, the narrow cellulose nanofibrils can be obtained by the combination of chemical treatment and ultrasonication. More interestingly, a class of aerogels can be prepared from the aqueous or alcohol dispersion of cellulose nanofibrils using freeze-drying technique. Paakko et al[19] firstly reported cellulose I nanofiber aerogels with a specific surface area of 66 $m^2 \cdot g^{-1}$, which was fabricated by freeze-drying treatment of NFC dispersion. Sehaqui et al.[20] then prepared nanofibrillated cellulose (NFC) aerogels with high specific surface area (153-284 $m^2 \cdot g^{-1}$), high porosity (93%-99%), and low density (14-105 $mg \cdot cm^{-3}$) via the multiple-stage solvent-exchange technique followed by freeze drying. Also, the NFC aerogels were used as templates to combine with hydrophobic silanes[21], Fe_3O_4/Ag[22], polypyrrole[23], and N-(2-aminoethyl)-3-aminopropylmethyldimethoxysilane[24] for some advanced applications including oil-water separation, catalyst, antibacterial agent, CO_2 capture from air, and electrochemical devices. It is well-known that silica aerogels[25, 26] and carbon aerogels (typically like carbon nanofiber aerogels[27, 28], all-carbon aerogels containing graphene sheets and carbon nanotubes[29], and carbon aerogels from the carbonization of organic aerogels[30]) have been extensively used as absorbent materials to treat with organic pollutants. However, there are not plentiful reports focusing on the adsorption property of biodegradable NFC aerogels for toxic organic liquids and their adsorption recyclability. Therefore, it is desirable to carry out these researches.

In this study, we used waste natural reed as raw material to fabricate ultralight NFC aerogels following the procedures of chemical purification, ultrasonication, and freeze drying. Compared with the commercially available cellulosic feedstocks such as wood pulp and dissolving pulp, reed is cheaper and widely available. Moreover, the utilization of reed is consistent with our objective of achieving high-value utilization of low-value bioresources. The effects of the chemical purification and ultrasonication on the morphology, chemical composition, crystal structure, and thermal stability were investigated. After the modification by methyltrichlorosilane (MTCS), the aerogels acquired superhydrophobicity with contact angles of as high as

151−155°. In addition, their adsorption efficiency for various common organic solvents and waste oil can reach up to 53 − 93 g · g^{-1}. For measuring their adsorption recyclability, we put the aerogels into a thin and lightweight nonwoven cloth bag considering the poor mechanical property of the aerogels, and the bag filled with the samples were dipped into various organic liquids. Two simple methods (i. e. , squeezing-vacuum drying and air drying) were used to separate the liquids from the bag, dependent on the type of pollutants. The results show that the adsorbent can maintain more than 80% of the initial adsorption efficiency after the five cycles. These make the aerogels alternative biodegradable adsorbents to treat with organic pollutants.

2 Materials and methods

2.1 Materials

The reed powder was sieved through a 60 mesh screen and used as the raw material. All the chemicals including benzene, ethanol, sodium chlorite, potassium hydroxide, hydrochloric acid (37%), and tert-butyl alcohol were analytically pure, supplied by Tianjin Kermel Chemical Reagent Co. Ltd. (China), and used without further purification.

2.2 Separation of NFCs from reed

The separation process involves chemical purification and ultrasonication[31, 32]. For the chemical purification, the dried reed powder (2 g) was firstly extracted with benzene/ethanol (2 : 1, v : v) in soxhlet at 90 ℃ for 24 h to remove pectin and waxes, and then dried in an oven for 24 h at 50 ℃. Secondly, lignin was removed by mixing the resulting sample with acidified sodium chlorite solution at 75℃ for an hour. This process was repeated five times. Thirdly, hemicellulose was removed by an alkaline treatment (2% potassium hydroxide solution) at 90 ℃ for 2 h. Finally, the samples were treated with the hydrochloric acid (1%) at 80 ℃ for 2 h, and subsequently filtered and rinsed with a large amount of distilled water. The purified cellulose was dried at 50 ℃ to constant weight. The yield of cellulose is around 22.5%, which was roughly calculated by dividing the product dry weight by the weight of the dried reed powder.

For the ultrasonication process, the purified cellulose was firstly dispersed in distilled water to form a 0.05% aqueous suspension (300 mL), and the suspension was then placed into an ultrasonic generator. Sonication was performed at 20 − 25 kHz with a sonifier cell disruptor (JY99 − IID, Scientz Technology, China) with a 1−cm^2-diameter titanium horn under a 50% duty cycle (i. e. , a repeating cycle of 0.5 s ultrasonic treatment and 0.5 s shutdown). The sonication was conducted for 1 h with an output power of 900 W in an ice/water bath. After that, the NFC suspension was obtained.

2.3 Preparation of NFC aerogels

The NFC suspension was subjected to a solvent exchange with tertiary butanol by dipping the dialysis tubing containing the suspension into tertiary butanol until the volume of the suspension was reduced to approximately one fifth. The concentrated NFC suspension was collected and poured into moulds and then freeze-dried to sublimate tertiary butanol directly from solid phase to gas phase. The cold trap temperature and pressure are around −55 ℃ and 25 Pa during the whole freeze-drying process.

2.4 Hydrophobic treatment of NFC aerogels

The hydrophobic NFC aerogels were obtained by the treatment of MTCS through the vapour phase deposition. In a typical process, the aerogels were placed in a glass desiccator, and a small glass vial containing 300 μL MTCS was then added into the desiccator. The desiccator was sealed, and the samples and the reagent were allowed to react at room temperature for 24 h.

2.5 Characterizations

The morphology was characterized by scanning electron microscopy (SEM, FEI Sirion 3030) operating at 15.0 kV. Elemental analysis was conducted by an energy-dispersive X-ray (EDX) spectrometer. Fourier transform infrared (FTIR) spectra were recorded on a FTIR instrument (Nicolet 6700, Thermo Fisher Scientific., USA) in the range of 400-4000 cm^{-1} with a resolution of 4 cm^{-1}. The crystal structures were characterized via X-ray diffraction (XRD, Rigaku D/MAX 2200) operating with Cu Ka radiation ($\lambda = 1.5418$ Å) at a scan rate (2θ) of 4 min^{-1} ranging from 5° to 40°. Thermogravimetric analysis (TGA) was performed with a synchronous thermal analyser (TA, Q600) from room temperature to 800 ℃ at a heating rate of 10 ℃ · min^{-1} under a nitrogen atmosphere. The surface wettability was evaluated by contact angle measurement, using a Powereach JC2000C contact angle analyser. N_2 adsorption-desorption measurements were implemented at -196 ℃ using an accelerated surface area and porosimetry system (3H-2000PS2 unit, Beishide Instrument S&T Co. Ltd.). The specific surface area was calculated over a relative pressure range of 0.05-0.30 from the multipoint Brunauer-Emmett-Teller (BET) method. The nitrogen adsorption volume at the relative pressure (P/P0) of 0.99 was used to determine the pore volume. The pore diameter distributions were calculated from the data of the adsorption branch of the isotherm using the Barrett-Joyner-Halenda (BJH) method.

2.6 Adsorption tests for various organic liquids

The adsorption tests were conducted by dipping the MTCS-treated NFC aerogels into various organic liquids for 1 min. The sample weight was measured before and after the tests. The adsorption property was characterized by adsorption efficiency, which was calculated by dividing the wet weight of the sample after the adsorption tests by the initial weight.

2.7 Cyclic adsorption tests

We firstly put the MTCS-treated NFC aerogels (ca. 0.05 g) into a thin and lightweight nonwoven cloth bag (6 cm × 8 cm, 0.25-0.35 g). Then the bag was sealed and dipped into methylbenzene (or ethyl alcohol) for 1 min. The adsorbed methylbenzene (or ethyl alcohol) was separated from the adsorbent by squeezing-vacuum drying (or air drying). The adsorption-desorption process was repeated five times. The weight of the bag was measured before and after each period.

3 Result and discussion

3.1 Morphology observation and elemental analysis

The SEM images of the reed, purified cellulose, and NFC aerogels are presented in Fig. 1a-c. As shown in Fig. 1a, the microstructure of reed outer epidermis is provided, and a plenty of pits can be clearly seen. Besides, according to the EDX analysis (Fig. 1d), the elements including C, O, Si, Cl, K and Na were detected for the reed, and the corresponding proportions are 54.64%, 25.33%, 17.49%, 0.91%, 0.88%, and 0.75%, respectively. The Au element is originated from the coating layer used for electric conduction during the SEM observation. Through the removal of non-cellulosic substances (such as waxes, pectin, hemicellulose, and lignin) by the purification, it can be seen that the macrofibrils composed of a bundle of microfibrils are disintegrated[33], and the numerous microfibrils were successfully isolated (Fig. 1b). In addition, from the EDX spectrum in Fig. 1e, there are only three elements here (i. e., C, O and Si), which indicates that the most mineral substances had been removed by the chemical purification.

As mentioned above, cellulose microfibrils are composed of numerous tightly connected cellulose nanofibrils which are hard to be split. Ultrasonication can provide local hot spots with high temperature and pressure as well as rapid heating/cooling rate, which can effectively destroy the intermolecular hydrogen bonding between the linear polymers. Although ultrasonication is a powerful tool to isolate cellulose nanofibrils, ultrasonication causes cellulose degradation. Sonication of cellulose in aqueous suspension leads to the depolymerization of cellulose. It is found that changes in the molecular structure of cellulose are attributed to the hydrolysis (chain cleavage)[11]. Fig. 1c shows the morphology of NFC aerogels formed by the self-assemble of abundant NFCs. The cross-linked 3D network can be identified. Moreover, the inset of Fig. 1c confirms that the nano-scale NFCs are successfully separated. We drew the frequency distribution histogram of NFCs diameters (see Fig. S1 in the Supporting Information), and the plot exhibits a Gaussian-like distribution. The diameter of NFCs ranges from 10 to 37 nm, similar to that of some previously reported cellulose nanofibers obtained by ultrasonic method in literatures[34] (10–40 nm),[35] (30–100 nm), and[36] (2–50 nm). In addition, it can be found that the NFCs in our paper have narrower diameter distribution. Moreover, the average size (d) and standard deviation (σ) are calculated as 22.67 nm and 5.79 nm, respectively.

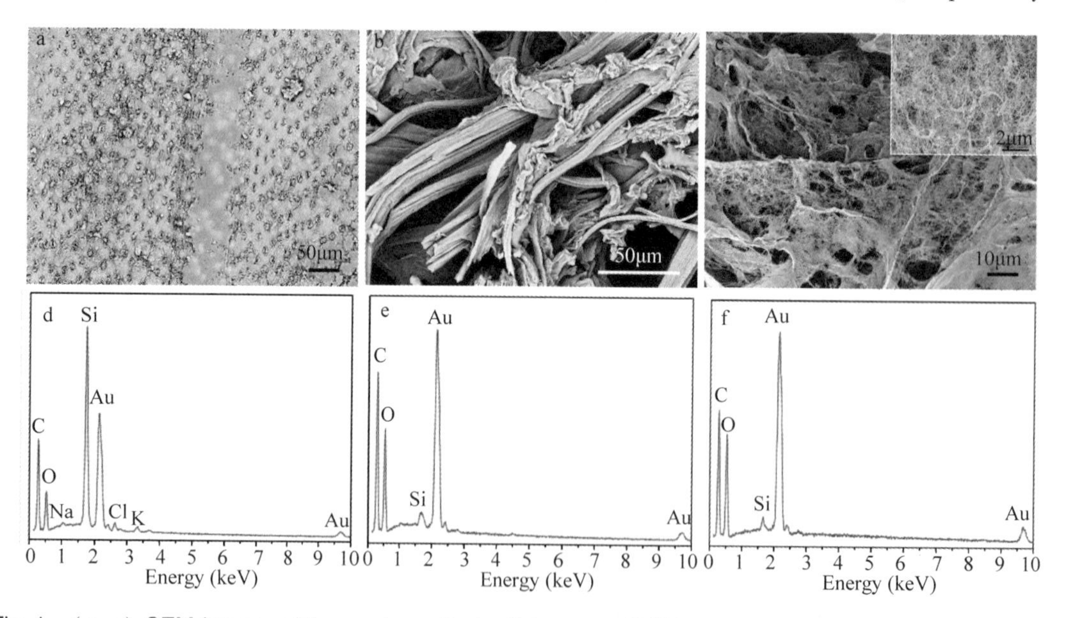

Fig. 1 (a–c) SEM images of the reed, purified cellulose, and NFC aerogels, respectively. Inset of (c) is the corresponding enlarged image. (d–f) present the EDX spectra of the reed, purified cellulose, and NFC aerogels, respectively

3.2 Chemical composition, crystal structure, and thermal stability

FTIR was carried out to clarify the differences in chemical compositions. The FTIR spectra of reed, purified cellulose, and NFC aerogels are presented in Fig. 2, and their main signals are illustrated in Table 1. For the reed, the absorption at 3329 cm^{-1} is attributed to the O—H stretching, and the bands at 2917 and 2850 cm^{-1} are attributed to C—H asymmetric and symmetric stretching vibrations[37], respectively. The bands at 1421, 1369, 1315, and 1234 cm^{-1} originate from CH_2 scissoring, C—H stretch in CH_3, CH_2 rocking vibration, and C—O—C stretching. The C—O—C pyranose ring skeletal vibration gives a prominent band at 1029 cm^{-1}. A small sharp peak at 891 cm^{-1} corresponds to the glycosidic C_1—H deformation with ring vibration contribution, which is characteristic of β-glycosidic linkages between glucose in cellulose[38]. The band at 803 cm^{-1} is due to pyran vibration. Compared with the spectrum of the reed, there are small differences in the spectra of purified cellulose and NFC aerogels. The peak at 1507 cm^{-1} attributed to aromatic skeletal vibrations of lignin[39] disappears completely after chemical purification. The outstanding peak at 1735 cm^{-1} also disappears, which is derived from either the acetyl and uronic ester groups of the hemicellulose or the ester linkage of the carboxylic groups of the ferulic and p-coumeric acids of lignin and/or hemicellulose[40]. Therefore, these results suggest that hemicellulose and lignin were effectively removed by the chemical purification. In addition, comparing the purified cellulose and the NFC aerogels, there is no remarkable difference in some bands (typically at 1421 and 897 cm^{-1}) sensitive to crystal form[41], which reveals that the ultrasonication didn't change the cellulose crystal form.

Table 1 Frequencies (cm^{-1}) of the main signals of the reed, purified cellulose, and NFC aerogels, respectively

Absorption band (cm^{-1})	Assignment
3329	Valence vibration of hydrogen bonded OH-groups
2917-2850	CH_2 valence vibration
1735	C=O valence vibration of acetyl or COOH-groups
1507	Aromatic skeletal vibrations;
1421	CH_2 scissoring
1369	C-H stretch in CH_3
1315	CH_2 rocking vibration
1234	C-O-C stretching
1108	Ring asymmetric valence vibration
1029	C O C pyranose ring skeletal vibration
891	Anomere C-groups, C_1-H eformation, ring valence vibration
803	Pyran vibration

For further investigating the changes in crystal structure, the XRD patterns of the reed, purified cellulose, and NFC aerogels are compared. In Fig. 3, we can see that all samples exhibit a sharp peak at around 22.3° from the (002) plane and a broad peak at around 16.5° due to the mixture of peaks from (101) and ($10\bar{1}$), corresponding to typical cellulose Ⅰ crystal structure with parallel up arrangement of cellulose chains[42]. It indicates that the chemical purification and ultrasonication did not change the crystal form of cellulose. However, these treatments significantly affect the peak intensity, suggesting the probable variations in crystallinity. The crystallinity index (CrI) was calculated as the ratio of the area arising from the crystalline phase to the total area, and the fitting of XRD patterns are provided in Fig. S2 (see Supporting Information).

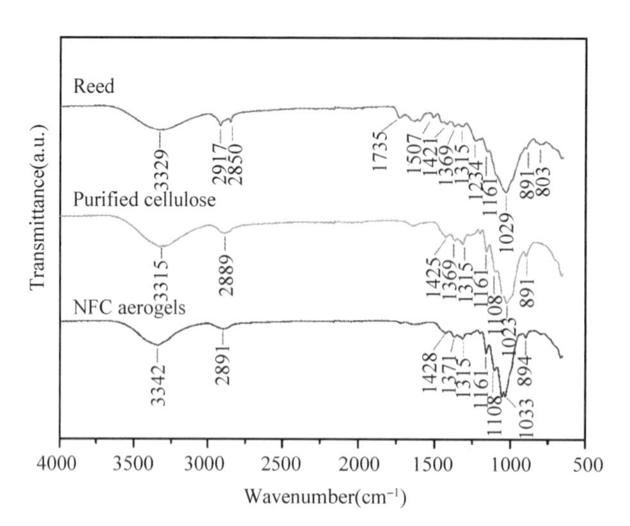

Fig. 2 FTIR spectra of the reed, purified cellulose, and NFC aerogels, respectively

Compared to the reed, the crystallinity index of the purified cellulose increases by 8. 81% (from 58. 61% to 67. 42%), due to the removal of hemicellulose and lignin which exist in the amorphous substances. Nevertheless, the crystallinity index of NFC aerogels (62. 88%) was 4. 54% lower than that of the cellulose. Li et al. [12] attributed the decrease in crystallinity index to the non - selective effect of ultrasonication, meaning that it can remove both amorphous and crystalline cellulose.

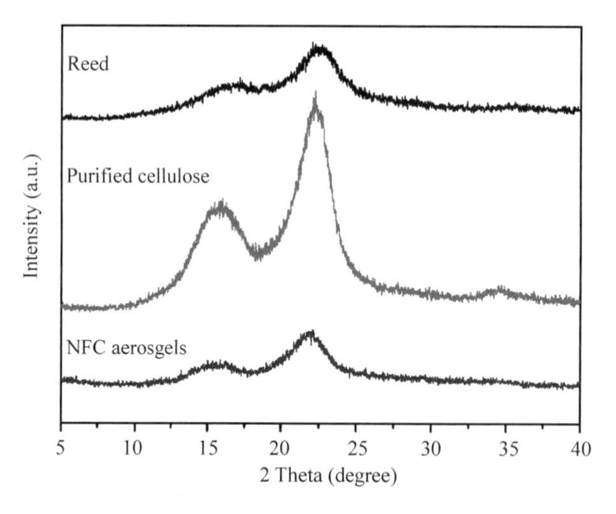

Fig. 3 X-ray diffraction patterns of the reed, purified cellulose, and NFC aerogels, respectively

The thermal stability of the reed, purified cellulose, and NFC aerogels was studied by TGA and derivative thermogravimetry (DTG) in Fig. 4. The reed shows two main exothermic peaks. The former (200–310 ℃) is mainly associated with the hemicellulose decomposition, which exhibits a pronounced shoulder[43]. The latter (310 - 400 ℃) is principally assigned to cellulose degradation[44]. Owing to the wide decomposition temperature and slow decomposition rate of lignin, no characteristic peak related to lignin can be seen in the whole pyrolysis process[45]. In addition, it is worth to mention that cellulose decomposition is expressed by two competitive degradation reactions, one ascribed to the formation of tars (mainly levoglucosan) and char, the other to the light gases[46]. The formation of levoglucosan from cellulose pyrolysis has been proposed as the cleavage of the 1, 4-glycosidic linkage in the cellulose polymer followed by intramolecular rearrangement of the monomer units[47]. Recent literatures[48] confirm that the thermal decomposition of levoglucosan is extended over a wider temperature range (200–800 ℃) according to the interaction of hemicellulose or lignin upon the pyrolysis of cellulose. Comparing the reed and the purified cellulose, it is not hard to see that the purified cellulose displays more superior thermal stability. The strongest peak of the cellulose is located at 361.7 ℃, 15.3 ℃ higher than that of the reed (346.4 ℃). The removal of amorphous substances contributes to this improvement. In addition, the fast heating rate and the different thermal conductivity of substances extracted from the reed possibly create a more remarkable heat transfer problem between the sample and the instrument[49], as compared to the purified cellulose. Thus the maximum degradation rate of the reed might shift to a higher temperature. In contrast, the NFC aerogels have a poorer thermal stability, whose strongest peak is centered at 337.9 ℃ much lower than the others. It is well-known that crystallinity plays an important role in thermal properties of cellulose products[50]. Therefore, this decline can be explained by the damage of crystal structure by the ultrasonication.

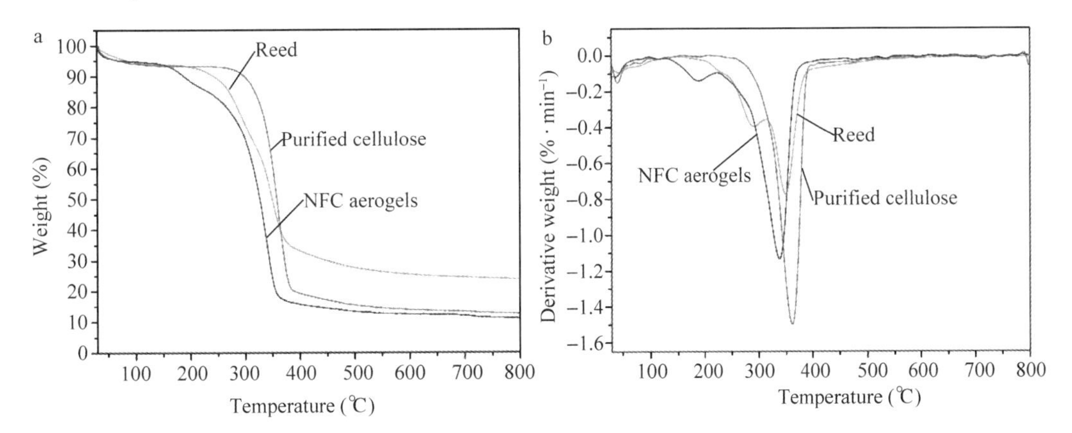

Fig. 4 (a) TGA and (b) DTG of the reed, purified cellulose, and NFC aerogels, respectively

3.3 Hydrophobic modification and adsorption property

After a series of treatments including chemical purification, ultrasonication, and freeze drying, we can obtain the ultralight NFC aerogels with an ultra-low bulk density of 4.9 mg · cm^{-3}. Table 2 summarizes some lightweight cellulose aerogels. The density of the NFC aerogels reported in this paper is much higher than that reported by Chen et al. (0.2 mg · cm^{-3})[51], but comparable or slightly lower than others[19, 20, 52-55]. Nowadays, lightweight materials have attracted great interests from numerous advanced areas, especially for adsorbing materials. Herein, as an example of potential application, the NFC aerogels were adopted as adsorbents to adsorb various organic solvents and waste oil. Prior to the adsorption, the aerogels were subjected to a facile hydrophobic modification by MTCS for enhancing the interfacial compatibility between hydrophilic cellulose and hydrophobic organic liquids. The hydrophobicity was evaluated by the contact angle measurement. It is obvious that the MTCS-treated aerogels have good hydrophobic property for various liquids (Fig. 5a), from up to down, including Coca-Cola, methyl orange, indigo carmine, rhodamine b, and milk. Fig. 5b presents the corresponding images of contact angle tests. The untreated NFC aerogels show strong hydrophilicity, and cannot support water drop on the surface. After treated with MTCS, the aerogels acquired a high hydrophobicity, and the contact angle values reach up to 151°–155° for these liquids. The FTIR spectrum in Fig. 5c further confirms the formation of hydrophobicity for the treated aerogels. The two new strong peaks at around 1273 and 776 cm^{-1} are attributed to the stretching vibrations of the Si–C bonds and the —CH_3 deformation vibrations[56]. Moreover, the possible reaction mechanism between the cellulose and the reagent is given in Fig. 5c.

Table 2 Bulk density of some lightweight cellulose aerogels

Materials	Feedstock	Method	Bulk density ($mg \cdot cm^{-3}$)	Reference
NFC aerogels	Dissolving pulp	Freeze drying	7–30	[52]
NFC aerogels	Hardwood	Freeze drying	9.8	[53]
NFC aerogels	Cellulose fibers	Supercritical drying	10–60	[54]
NFC aerogels	Wood pulp	Freeze drying	14–105	[20]
Bacterial cellulose aerogels	*Gluconacetobacter xylinum*	Supercritical drying	8	[55]
Cellulose I nanofibers aerogels	Softwood cellulose pulp	Freeze drying	20	[19]
Cellulose I nanofibers aerogels	Wood fibers	Freeze drying	0.2	[51]
NFC aerogels	Reed	Freeze drying	5	This work

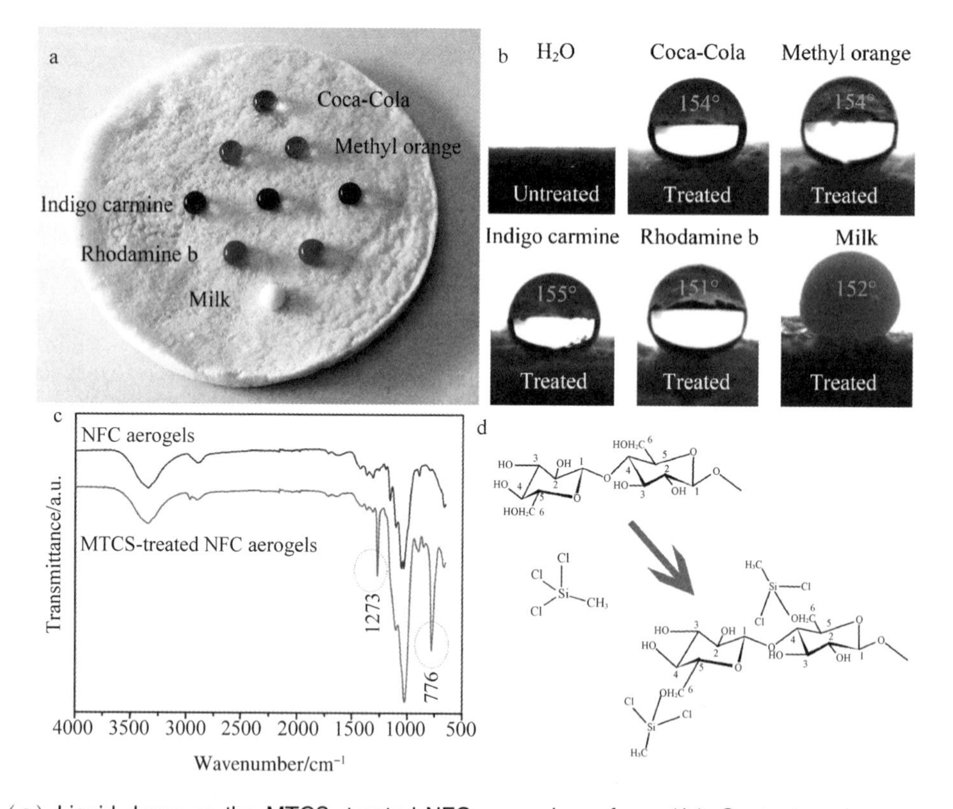

Fig. 5 (a) Liquid drops on the MTCS-treated NFC aerogels surface. (b) Contact angle measurements and (c) FTIR spectra of the untreated and MTCS-treated NFC aerogels. (d) Possible reaction mechanism between cellulose and MTCS

N_2 adsorption-desorption isotherms can provide much useful information about the pore structures of the MTCS-treated NFC aerogels. As shown in Fig. 6a, the aerogels display an Ⅳ-type adsorption isotherm according to the IUPAC classification, and the obvious hysteresis loop between the adsorption and desorption isotherms was originated from capillary condensation. Moreover, the pore size distribution shown in Fig. 6b exh ibits that the pore sizes are within the scope of 1−106 nm, indicating the existence of micropores (<2 nm), mesopores (2−50 nm) and macropores (>50 nm). BET and BJH analysis gives specific surface area and pore volume of 55.2 $m^2 \cdot g^{-1}$ and 0.18 $m^3 \cdot g^{-1}$ for the aerogels. The value of specific surface area is much lower than that reported by Sehaqui et al. [20] (249 $m^2 \cdot g^{-1}$), but comparable to or even slightly higher than that in literatures[19] (66 $m^2 \cdot g^{-1}$), [21] (42 $m^2 \cdot$ g−1), [53] (18.4 $m^2 \cdot g^{-1}$), and[52] (15 $m^2 \cdot g^{-1}$). Moreover, the aerogels have a very high porosity, ca. 99.7%, where the porosity Φ is defined as $\Phi = 1 - (\rho/\rho s)$, where ρ and ρs (1.5 $g \cdot cm^{-3}$)[21] are the bulk densities of the aerogel and the crystalline cellulose, respectively.

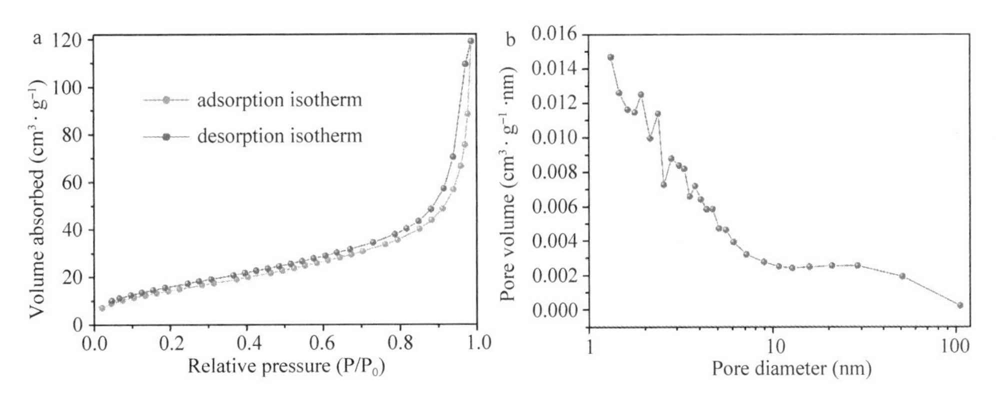

Fig. 6 (a) N_2 adsorption-desorption isotherms and (b) pore size distribution of the MTCS-treated NFC aerogels

The wastewater of industry includes various organic pollutants like acetone, benzene, ethyl acetate, oleic acid, toluene and trichloromethane, and waste oil. It is meaningful to investigate the adsorption property of the MTCS-treated NFC aerogels for these organic pollutants. The bulk density of all the aerogels used to perform adsorption tests is around 5 $mg \cdot cm^{-3}$. The adsorption results in Fig. 7 show that the adsorption efficiency of the MTCS-treated aerogels reaches up to 53−93 $g \cdot g^{-1}$. The adsorption efficiency for methylbenzene was taken as an example to make a comparison with other cellulose-based adsorbents reported elsewhere. The methylbenzene adsorption efficiency of the aerogels (ca. 92 $g \cdot g^{-1}$) in this work is slightly higher than that of the silylated nanocellulose sponges (ca. 62 $g \cdot g^{-1}$)[57] and the cellulose nanofibrils aerogels modified by styreneacrylic monomers (ca. 30 $g \cdot g^{-1}$)[58]. Recently, Jiang and Hseih[59] achieved greater adsorption efficiency of methylbenzene (ca. 250 $g \cdot g^{-1}$) by vapor depositing triethoxyl (octyl) silane onto cellulose nanofibrils aerogels.

Besides substantial adsorption efficiency, adsorption recyclability is also of essential importance for ideal adsorbing materials. Good adsorption recyclability is beneficial to save costs, reduce energy consumption of preparation, and reduce generation of garbage. Considering some characteristics of the NFC aerogels adverse to the cyclic adsorption (e.g., poor mechanical property and strong shrinkage), the MTCS-treated NFC aerogels were put into a thin and lightweight nonwoven cloth bag. The bag was sealed and subjected to the cyclic

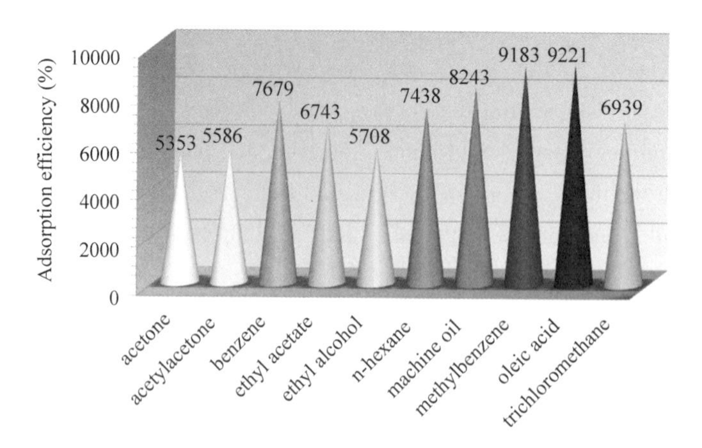

Fig. 7 Absorption efficiency of the MTCS-treated NFC aerogels for various organic solvents and waste machine oil

adsorption tests. In this way, we needn't worry about the shrinkage, operational and recycling problems. We adopted two methods to recycle the adsorbent, i. e., squeezing-vacuum drying for toxic expensive liquids and air drying for cheap volatile and innoxious liquids. The adsorption-desorption process was repeated five times. As shown in Fig. 8a, the poisonous methylbenzene was used as an example of squeezing-vacuum drying experiment. In the first cycle, 5.4 g of methylbenzene was adsorbed. But the mass of the adsorbed methylbenzene decreased to 4.7 g in the second cycle. From the second cycle onwards, the adsorption efficiency became stable, i. e., the weight increment kept constant (4.7–4.8 g). It can be seen that the adsorption efficiency maintained 87% of the initial adsorption efficiency after the five cycles. As for the cyclic sorption-air drying test, the drying process was conducted in the fume hood at room temperature, and the next adsorption process didn't begin until the sample weight maintained constant. It can be observed in Fig. 8b that the adsorbent still can adsorb 4.1 g ethyl alcohol after the five cycles, which is around 80% of the first adsorption weight. Therefore, the two simple methods can be applied for recycling the adsorbent. Also, these results indicate that the adsorbent (i. e., the bag filled with the aerogels) has favorable adsorption recyclability.

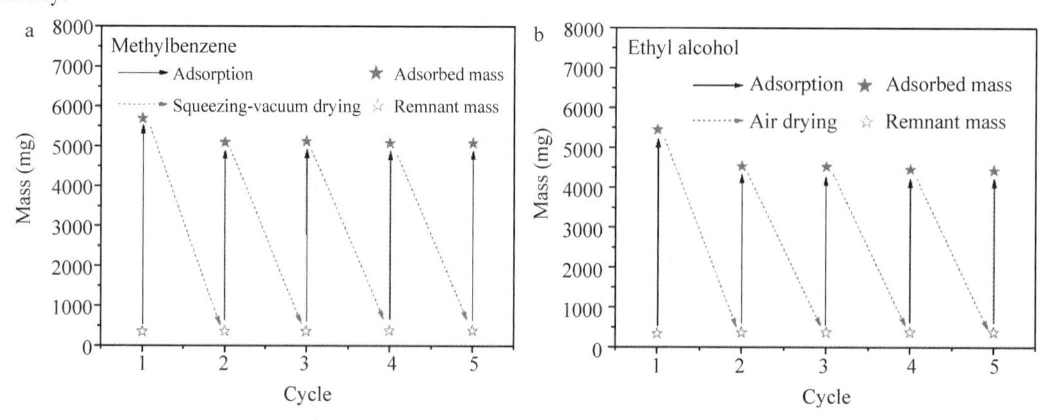

Fig. 8 Cyclic adsorption tests of the nonwoven cloth bag filled with the MTCS-treated NFC aerogels. (a) Squeezing-vacuum drying was used to recycle the adsorbent for adsorption of methylbenzene. (b) Air drying was applied to recycle the adsorbent for adsorption of ethyl alcohol

4 Conclusions

We successfully used the waste natural reed to fabricate ultralight NFC aerogels via chemical purification, ultrasonication, and freeze drying. The FTIR analysis indicates the removal of non-cellulosic substances by the chemical purification. The SEM observation confirms that the NFCs with an average size of 22.67 nm were isolated. The XRD analysis demonstrates that the purification and ultrasonication didn't change the cellulose crystal form (cellulose I). The NFC aerogels have a poorer thermal stability as a result of the ultrasonication, as compared to that of the reed and the purified cellulose. Through the hydrophobic modification by MTCS, the contact angles are as high as 151−155°. The adsorption efficiency of the hydrophobic aerogels reaches up to 53−93 $g \cdot g^{-1}$ for various organic solvents and waste machine oil. For eliminating the effects of the poor mechanical property and strong shrinkage of the aerogels, we put the aerogels into a nonwoven cloth bag, and the bag was sealed and subjected to the cyclic adsorption tests. The two simple methods (i.e., squeezing-vacuum drying and air drying) can be applied for recycling the adsorbent dependent on the type of pollutants. The adsorbent can maintain more than 80% of the initial adsorption efficiency after the five cycles. In summary, this work reports low-density superhydrophobic and high-performance adsorbents, and provides an example to recycle waste bioresources and expands their high-value applications.

Acknowledgments

This study was supported by the National Natural Science Foundation of China (grant nos. 31270590 and 31470584).

References

[1] Kahil A, Bhat H, Ireana A H, et al. Green composites from sustainable cellulose nanofibrils[J]. Carbohydrate Polymers, 2011, 87: 963−979.

[2] Kaushik A, Singh M, Verma G. Green nanocomposites based on thermoplastic starch and steam exploded cellulose nanofibrils from wheat straw[J]. Carbohydrate Polymers, 2010, 82(2): 337−345.

[3] Hu L, Zheng G, Yao J, et al. Transparent and conductive paper from nanocellulose fibers[J]. Energy & Environmental Science, 2013, 6(2): 513−518.

[4] Siró I, Plackett D. Microfibrillated cellulose and new nanocomposite materials: a review[J]. Cellulose, 2010, 17(3): 459−494.

[5] Liu A, Walther A, Ikkala O, et al. Clay nanopaper with tough cellulose nanofiber matrix for fire retardancy and gas barrier functions. [J]. Biomacromolecules, 2011, 12(3): 633−641.

[6] Lee S Y, Chun S J, Kang I A, et al. Preparation of cellulosenanofibrils by high-pressure homogenizer and cellulose-based composite films[J]. Journal of Industrial and Engineering Chemistry, 2009, 15(1): 50−55.

[7] Zhao J, Zhang W, Zhang X, et al. Extraction of cellulosenanofibrils from dry softwood pulp using high shear homogenization[J]. Carbohydrate Polymers, 2013, 97(2): 695−702.

[8] Isogai T, Saito T, Isogai A. Wood cellulose nanofibrils prepared by TEMPO electro-mediated oxidation[J]. Cellulose, 2011, 18(2): 421−431.

[9] Chen W, Yu H, Liu Y, et al. Individualization of cellulosenanofibers from wood using high-intensity ultrasonication combined with chemical pretreatments[J]. Carbohydrate Polymers, 2011, 83(4): 1804−1811.

[10] Wan C, Lu Y, Jiao Y, et al. Ultralight and hydrophobic nanofibrillated cellulose aerogels from coconut shell with

ultrastrong adsorption properties[J]. Journal of Applied Polymer Science, 2015, 132(24).

[11]Stefanovic B, Rosenau T, Potthast A. Effect of sonochemical treatments on the integrity and oxidation state of cellulose [J]. Carbohydrate polymers, 2013, 92(1): 921-927.

[12]Li W, Yue J, Liu S. Preparation of nanocrystalline cellulose via ultrasound and its reinforcement capability for poly (vinyl alcohol) composites[J]. Ultrasonics sonochemistry, 2012, 19(3): 479-485.

[13]Grönroos A, Pirkonen P, Ruppert O. Ultrasonic depolymerization of aqueous carboxymethylcellulose[J]. Ultrasonics Sonochemistry, 2004, 11(1): 9-12.

[14]Tutt M, Kikas T, Kahr H, et al. Using steam explosion pretreatment method for bioethanol production from floodplain meadow hay[J]. Agronomy Research, 2014, 12(2): 417-424.

[15]Somerville C, Bauer S, Brininstool G, et al. Toward a systems approach to understanding plant cell walls[J]. Science, 2004, 306(5705): 2206-2211.

[16]Cosgrove, Daniel J. Assembly and enlargement of the primary cell wall in plants. [J]. Annu Rev Cell Dev Biol, 1997, 13(1): 171-201.

[17] Nm A, Cw B, Bd C, et al. Features of promising technologies for pretreatment of lignocellulosic biomass - ScienceDirect[J]. Bioresource Technology, 2005, 96(6): 673-686.

[18] Zhu S, Wu Y, Yu Z, et al. Comparison of Three Microwave/Chemical Pretreatment Processes for Enzymatic Hydrolysis of Rice Straw[J]. Biosystems Engineering, 2006, 93(3): 279-283.

[19]Pääkkö M, Vapaavuori J, Silvennoinen R, et al. Long and entangled native cellulose I nanofibers allow flexible aerogels and hierarchically porous templates for functionalities[J]. Soft Matter, 2008, 4(12): 2492-2499.

[20]Sehaqui H, Zhou Q, Berglund L A. High-porosity aerogels of high specific surface area prepared from nanofibrillated cellulose (NFC)[J]. Composites science and technology, 2011, 71(13): 1593-1599.

[21]Cervin N T, Aulin C, Larsson P T, et al. Ultra porous nanocellulose aerogels as separation medium for mixtures of oil/water liquids[J]. Cellulose, 2012, 19(2): 401-410.

[22]Xiong R, Lu C, Wang Y, et al. Nanofibrillated cellulose as the support and reductant for the facile synthesis of Fe_3O_4/Ag nanocomposites with catalytic and antibacterial activity[J]. Journal of Materials Chemistry A, 2013, 1(47): 14910-14918.

[23] Carlsson D O, Nyström G, Zhou Q, et al. Electroactive nanofibrillated cellulose aerogel composites with tunable structural and electrochemical properties[J]. Journal of Materials Chemistry, 2012, 22(36): 19014-19024.

[24]Gebald C, Wurzbacher J A, Tingaut P, et al. Amine-based nanofibrillated cellulose as adsorbent for CO_2 capture from air[J]. Environmental science & technology, 2011, 45(20): 9101-9108.

[25]Liu H, Sha W, Cooper A T, et al. Preparation and characterization of a novel silica aerogel as adsorbent for toxic organic compounds[J]. Colloids and Surfaces A: Physicochemical and Engineering Aspects, 2009, 347(1-3): 38-44.

[26]Rao A V, Hegde N D, Hirashima H. Absorption and desorption of organic liquids in elastic superhydrophobic silica aerogels[J]. Journal of colloid and interface science, 2007, 305(1): 124-132.

[27]Wu Z Y, Li C, Liang H W, et al. Carbonnanofiber aerogels for emergent cleanup of oil spillage and chemical leakage under harsh conditions[J]. Scientific reports, 2014, 4(1): 1-6.

[28] Liang H W, Guan Q F, Chen L F, et al. Macroscopic-scale template synthesis of robust carbonaceousnanofiber hydrogels and aerogels and their applications[J]. Angewandte Chemie International Edition, 2012, 51(21): 5101-5105.

[29]Sun H, Xu Z, Gao C. Multifunctional, ultra-flyweight, synergistically assembled carbon aerogels[J]. Advanced materials, 2013, 25(18): 2554-2560.

[30]Carrasco-Marin F, Fairén-Jiménez D, Moreno-Castilla C. Carbon aerogels from gallic acid-resorcinol mixtures as adsorbents of benzene, toluene and xylenes from dry and wet air under dynamic conditions[J]. Carbon, 2009, 47(2): 463-469.

[31]Wan C, Li J. Synthesis of well-dispersed magnetic $CoFe_2O_4$ nanoparticles in cellulose aerogels via a facile oxidative co-precipitation method[J]. Carbohydrate polymers, 2015, 134: 144-150.

[32]Y. Jiao, C. Wan, J. Li, Appl. Phys. A-Mater 120, 341 (2015).

[33]Boufi S, Gandini A. Triticale crop residue: a cheap material for high performance nanofibrillated cellulose[J]. RSC advances, 2015, 5(5): 3141-3151.

[34]Chen W, Yu H, Liu Y, et al. Isolation and characterization of cellulose nanofibers from four plant cellulose fibers using a chemical-ultrasonic process[J]. Cellulose, 2011, 18(2): 433-442.

[35]Wang H, Li D, Zhang R. Preparation ofultralong cellulose nanofibers and optically transparent nanopapers derived from waste corrugated paper pulp[J]. BioResources, 2013, 8(1): 1374-1384.

[36]Qua E H, Hornsby P R, Sharma H SS, et al. Preparation and characterization of poly (vinyl alcohol) nanocomposites made from cellulose nanofibers[J]. Journal of Applied Polymer Science, 2009, 113(4): 2238-2247.

[37]Gao Z, Ma M, Zhai X, et al. Improvement of chemical stability and durability of superhydrophobic wood surface via a film of TiO_2 coated $CaCO_3$ micro-/nano-composite particles[J]. Rsc Advances, 2015, 5(79): 63978-63984.

[38]Sun R, Hughes S. Fractional extraction andphysico-chemical characterization of hemicelluloses and cellulose from sugar beet pulp[J]. Carbohydrate polymers, 1998, 36(4): 293-299.

[39]Faix O, Beinhoff O. FTIR spectra of milled wood lignins and lignin polymer models (DHP's) with enhanced resolution obtained by deconvolution[J]. Journal of wood chemistry and technology, 1988, 8(4): 505-522.

[40]Lu Y, Sun Q, Yang D, et al. Fabrication ofmesoporous lignocellulose aerogels from wood via cyclic liquid nitrogen freezing-thawing in ionic liquid solution[J]. Journal of Materials Chemistry, 2012, 22(27): 13548-13557.

[41]Abidi N, Cabrales L, Haigler C H. Changes in the cell wall and cellulose content of developing cotton fibers investigated by FTIR spectroscopy[J]. Carbohydrate Polymers, 2014, 100: 9-16.

[42]Wan C, Lu Y, Sun Q, et al. Hydrothermal synthesis of zirconium dioxide coating on the surface of wood with improved UVresistance[J]. Applied surface science, 2014, 321: 38-42.

[43]Wan C, Jiao Y, Li J. In situ deposition ofgraphene nanosheets on wood surface by one-pot hydrothermal method for enhanced UV-resistant ability[J]. Applied Surface Science, 2015, 347: 891-897.

[44]Yang H, Yan R, Chen H, et al. Characteristics of hemicellulose, cellulose and ligninpyrolysis[J]. Fuel, 2007, 86 (12-13): 1781-1788.

[45]Alves A, Schwanninger M, Pereira H, et al. Analytical pyrolysis as a direct method to determine the lignin content in wood: Part 1: Comparison of pyrolysis lignin with Klason lignin[J]. Journal of analytical and applied pyrolysis, 2006, 76(1-2): 209-213.

[46]Capart R, Khezami L, Burnham A K. Assessment of various kinetic models for the pyrolysis of a microgranular cellulose[J]. Thermochimica Acta, 2004, 417(1): 79-89.

[47]Li S, Lyons-Hart J, Banyasz J, et al. Real-time evolved gas analysis by FTIR method: an experimental study of cellulose pyrolysis[J]. Fuel, 2001, 80(12): 1809-1817.

[48]Wang S, Guo X, Wang K, et al. Influence of the interaction of components on the pyrolysis behavior of biomass[J]. Journal of Analytical and Applied Pyrolysis, 2011, 91(1): 183-189.

[49]Alvarez V A, Vázquez A. Thermal degradation of cellulose derivatives/starch blends and sisal fibre biocomposites [J]. Polymer degradation and stability, 2004, 84(1): 13-21.

[50]Lu P, Hsieh Y L. Preparation and properties of cellulosenanocrystals: rods, spheres, and network[J]. Carbohydrate polymers, 2010, 82(2): 329-336.

[51]Chen W, Yu H, Li Q, et al. Ultralight and highly flexible aerogels with long cellulose I nanofibers[J]. Soft matter, 2011, 7(21): 10360-10368.

[52]Aulin C, Netrval J, Wågberg L, et al. Aerogels from nanofibrillated cellulose with tunable oleophobicity[J]. Soft

Matter, 2010, 6(14): 3298-3305.

[53] Silva T C F, Habibi Y, Colodette J L, et al. A fundamental investigation of the microarchitecture and mechanical properties of tempo-oxidized nanofibrillated cellulose (NFC)-based aerogels[J]. Cellulose, 2012, 19(6): 1945-1956.

[54] Hoepfner S, Ratke L, Milow B. Synthesis and characterisation of nanofibrillar cellulose aerogels[J]. Cellulose, 2008, 15(1): 121-129.

[55] Liebner F, Haimer E, Wendland M, et al. er, P. Miethe, T. Heinze, A. Potthast and T. Rosenau[J]. Macromol. Biosci, 2010, 10: 349-352.

[56] Shateri-Khalilabad M, Yazdanshenas M E. One-pot sonochemical synthesis of superhydrophobic organic-inorganic hybrid coatings on cotton cellulose[J]. Cellulose, 2013, 20(6): 3039-3051.

[57] Zhang Z, Sèbe G, Rentsch D, et al. Ultralightweight and flexible silylated nanocellulose sponges for the selective removal of oil from water[J]. Chemistry of Materials, 2014, 26(8): 2659-2668.

[58] Mulyadi A, Zhang Z, Deng Y. Fluorine-free oil absorbents made from cellulose nanofibril aerogels[J]. ACS applied materials & interfaces, 2016, 8(4): 2732-2740.

[59] Jiang F, Hsieh Y L. Amphiphilic superabsorbent cellulose nanofibril aerogels[J]. Journal of Materials Chemistry A, 2014, 2(18): 6337-6342.

中文题目：天然芦苇纳米纤化纤维素制备超疏水超轻气凝胶的研究

作者：焦月，万才超，强添刚，李坚

摘要：芦苇是一种可广泛利用的水生植物资源，其应用一般局限于造纸、动物饲料等传统领域。此外，每年大部分芦苇被浪费或直接焚烧，造成严重的空气污染(如大气雾霾)。因此，进一步开发芦苇高价值应用的新形式是值得研究的。本论文以天然芦苇为原料，通过化学提纯、超声波处理、冷冻干燥等一系列操作简便的方法，制备了纳米纤维化纤维素气凝胶(NFC)这一超轻质吸附剂。用扫描电子显微镜、能谱仪、傅立叶变换红外光谱、X 射线衍射和热重分析等手段对超低密度($4.9\ mg\cdot cm^{-3}$)气凝胶进行了表征。为了获得良好的疏水性，用甲基三氯硅烷对 NFC 气凝胶进行疏水处理。接触角高达 151°~155°的超疏水 NFC 气凝胶对各种有机溶剂和废油有很好的吸附效率($53\sim93\ g\cdot g^{-1}$)。更重要的是，气凝胶还表现出良好的吸附回收性能，在 5 次循环后仍能保持 80%以上的初始吸附效率。

关键词：芦苇；纳米原纤维化纤维素；气凝胶；超疏水性；吸附剂

Cellulose-derived Carbon Aerogels Supported Goethite (α-FeOOH) Nanoneedles and Nanoflowers for Electromagnetic Interference Shielding*

Caichao Wan, Yue Jiao, Tiangang Qiang, Jian Li

Abstract: We describe a rapid and facile chemical precipitation method to grow goethite (α-FeOOH) nanoneedles and nanoflowers on the carbon aerogels which was obtained from the pyrolysis of cellulose aerogels. When evaluated as electromagnetic interference (EMI) shielding materials, the α-FeOOH/cellulose-derived carbon aerogels composite displays the highest SE_{total} value of 34.0 dB at the Fe^{3+}/Fe^{2+} concentration of 0.01 M, which is about 4.8 times higher than that of the individual α-FeOOH (5.9 dB). When the higher or lower Fe^{3+}/Fe^{2+} concentrations were used, the EMI shielding performance deterioration occurred. The integration of α-FeOOH with the carbon aerogels transforms the reflection-dominant mechanism for α-FeOOH into the adsorption-dominant mechanism for the composite. The adsorption-dominant mechanism undoubtedly makes contribution to alleviating secondary radiation, which is regarded as more attractive alternative for developing electromagnetic radiation protection products.

Keywords: Cellulose; Carbon aerogels; Goethite; Electromagnetic interference shielding; Composites

1 Introduction

Widespread utilization of commercial, military and scientific electronic devices and communication instruments has created serious concern for electromagnetic pollution. Electromagnetic waves not only interfere with normal operation of equipment causing loss of energy, time and money, but also jeopardize even human health (Ahlbom & Feychting, 2003). Therefore, there is a strong desire to develop high-performance electromagnetic interference (EMI) shielding materials to protect human and electronic equipment from these undesired signals. Because of outstanding electrical, anticorrosive and antioxidative properties and excellent thermal stability, carbon materials have been extensively utilized in EMI shielding, such as carbon nanofibers (Yang, Lozano, Lomeli, Foltz & Jones, 2005), carbon nanotubes (Yang, Gupta, Dudley & Lawrence, 2005), carbon foam (Fang, Li, Sun, Zhang & Zhang, 2007) and graphene (Liang et al., 2009). Some literatures have demonstrated that integrating carbon materials with iron oxides makes contribution to acquiring more superior EMI shielding property, as compared to individual carbon materials or iron oxides. Singh et al. (Singh et al., 2012) fabricated composite sheets consisting of phenolic resin filled with a mixture of reduced graphene oxide, γ-Fe_2O_3 and carbon fibers. They found that the maximum shielding effectiveness strongly

* 本文摘自 Carbohydrate Polymers, 2017, 156: 427-434.

depended on volume fraction of γ-Fe_2O_3 in reduced graphene oxide matrix. Chen et al. (Chen et al., 2011) reported porous Fe_3O_4/carbon core/shell nanorods with a maximum reflection loss of −27.9 dB, much higher than that of Fe_3O_4/wax composite (less than −10 dB). Among a great variety of iron oxides, iron oxyhydroxides (FeOOH), such as α-, β- and γ-type, have proven to be efficient for various applications like magnetic recording media materials (Wang et al., 2004). Goethite (α-FeOOH), which exists primarily in soil, sediment and iron ore, is a kind of abundant mineral with interesting magnetic properties (Forsyth, Hedley & Johnson, 1968). Compared to bulk material, one-dimensional (1D) nanoscale α-FeOOH (e.g., nanorods) has superior magnetic properties because it bears a small longitudinal remnant magnetic moment and has a negative anisotropy of magnetic susceptibility (Lemaire, Davidson, Panine & Jolivet, 2004; Xiao, Zhang & Fu, 2010). Therefore, the incorporation of α-FeOOH into carbon-based materials may be useful to improve the EMI shielding property due to effective complementarity between dielectric loss and magnetic loss (Chen et al., 2011).

Among various carbon materials, porous carbon aerogels have received considerable attention in numerous fields like catalysis (Moreno-Castilla & Maldonado-Hódar, 2005), hydrogen storage (Kabbour, Baumann, Satcher, Saulnier & Ahn, 2006), and supercapacitors (Saliger, Fischer, Herta & Fricke, 1998; Zhang & Zhao, 2009). Carbon aerogels are traditionally obtained from the carbonization of organic aerogels (Li, Reichenauer & Fricke, 2002; Tamon, Ishizaka, Mikami & Okazaki, 1997), which are prepared from the sol-gel polycondensation of certain organic monomers (e.g., resorcinol and formaldehyde). In contrast to the toxic organic monomers, a precursor derived from green renewable resources (e.g., cellulose) holds more advantages, such as environmental friendliness and harmlessness. Cellulose aerogels, a class of porous biodegradable materials, are well-known to have cross-linked three-dimensional (3D) network, low density, and large specific surface area (Jin, Nishiyama, Wada & Kuga, 2004; Liebner, Potthast, Rosenau, Haimer & Wendland, 2008; Sehaqui, Zhou & Berglund, 2011). Some literatures (Hao et al., 2014; Wu, Li, Liang, Chen & Yu, 2013) and our previous work (Wan, Lu, Jiao, Jin, Sun & Li, 2015) have demonstrated that cellulose aerogels can be pyrolyzed into carbon aerogels under inert gas protection. The pyrolysis not only does not damage the network structure, but also leads to the formation of new properties like high electrical conductivity. In addition, the large specific area and porous feature of cellulose-derived carbon aerogels (CDCA) is beneficial to acquire a high loading content of guest nanomaterials, which is expected to enhance the contribution of magnetic loss of α-FeOOH and thus strengthen the EMI shielding property of α-FeOOH/CDCA. So far, only several literatures reported EMI shielding products consisting of α-FeOOH (Wu, Wu & Wang, 2015; Xiao, Zhang & Fu, 2010); to the best of our knowledge, there is no report on the α-FeOOH/CDCA composite. We now wish to report this composite and its EMI shielding property.

In this work, nanoneedles and nanoflowers of α-FeOOH were in-situ grown on CDCA substrate via an easily-operated chemical precipitation method. CDCA was obtained by pyrolyzing cellulose aerogels prepared based on a green cheap NaOH/polyethylene glycol solvent (Yan & Gao, 2008; Zhang, Li & Yu, 2010). Prior to the deposition of α-FeOOH, CDCA was subjected to an acid (HNO_3) treatment for introduction of oxygen-containing groups on its surface to interact with the iron ions. These functional groups can act as nucleation sites for α-FeOOH. The as-prepared α-FeOOH/CDCA was characterized by scanning electron microscopy (SEM), transmission electron microscopy (TEM), selected area electron diffraction (SAED), energy dispersive X-ray spectroscopy (EDX), X-ray photoelectron spectroscopy (XPS), and X-ray diffraction (XRD). The influences of the total concentration of $FeCl_3/FeSO_4$ salts on the EMI shielding property of α-FeOOH/CDCA were studied.

2 Materials and methods

2.1 Materials

Bamboo fiber was supplied by Zhejiang Mingtong Textile Technology Co., Ltd. (China) and further washed serval times with distilled water and dried at 60 ℃ for 24 h. The bamboo fiber was directly used as the cellulose feedstock considering its high cellulose content. Chemicals including nitric acid (HNO_3, 65%), sodium hydroxide (NaOH), ferric chloride hexahydrate ($FeCl_3 \cdot 6H_2O$), and ferrous sulfate heptahydrate ($FeSO_4 \cdot 7H_2O$) are all of analytical reagent grade and purchased from Shanghai Boyle Chemical Co. Ltd. (China).

2.2 Preparation of cellulose-derived carbon aerogels (CDCA)

The cellulose aerogels (starting materials) were prepared in three steps: (1) dissolving bamboo fiber in NaOH/polyethylene glycol solution; (2) regeneration of cellulose solution in 1% HCl solution; (3) freeze drying. The detailed processes can be referred to our previous reports (Wan & Li, 2015a, c, 2016). The preparation of CDCA was carried out by transferring the cellulose aerogels into a tubular furnace for pyrolysis under the protection of nitrogen (Xu et al., 2015). In detail, the cellulose aerogels were firstly heated to 500 ℃ at a heating rate of 5 ℃ $\cdot$ min^{-1}, and this temperature was maintained for 1 h; then the sample was heated to 1000 ℃ at 5 ℃ $\cdot$ min^{-1} and held at this temperature for 2 h to allow for complete pyrolysis. Thereafter, the temperature decreased to 500 ℃ at 5 ℃ $\cdot$ min^{-1} and finally decreased naturally to the room temperature.

2.3 Preparation of α-FeOOH/CDCA

Prior to the deposition of α-FeOOH, CDCA was treated with HNO_3(65%) for 3 h at room temperature to introduce various oxygen-containing groups on its surface. The synthesis of α-FeOOH was implemented referring to the literature method with some modifications (ZÁVIŠOVÁ, TOMAŠOVIČOVÁ, KOVÁČ, KONERACKÁ, KOPČANSKÝ & VÁVRA, 2010). In a typical process, acid-treated CDCA (0.5 g) was firstly immersed in a freshly prepared mixed solution of $FeCl_3$ and $FeSO_4$(50 mL) with a constant molar ratio of $[Fe^{3+}]/[Fe^{2+}]=2$ for 30 min. Then the mixture was heated to 90 ℃ and held at this temperature for 2 h. After that, the mixed system was placed into 0.2 M NaOH solution, and the mixture was kept for 12 h at 90 ℃. Finally, the resultant was washed several times with distilled water and dried at 60 ℃ for 12 h, and the α-FeOOH/CDCA composite was obtained.

2.4 Characterization

SEM observations were performed with a Hitachi S4800 SEM equipped with an EDX detector for element analysis. TEM observations and SAED were performed with a FEI, Tecnai G2 F20 TEM with a field-emission gun operating at 200 kV. XPS analysis was carried out using a Thermo Escalab 250Xi XPS spectrometer equipped with a dual X-ray source using Al-Kα. Deconvolution of the overlapping peaks was performed using a mixed Gaussian - Lorentzian fitting program (Origin 8.5, Originlab Corporation). XRD analysis was implemented on a Bruker D8 Advance TXS XRD instrument with Cu Kα (target) radiation ($\lambda = 1.5418$ Å) at a scan rate (2θ) of 4° · min^{-1} and a scan range from 5° to 80°.

2.5 Measurements of EMI shielding effectiveness

EMI shielding effectiveness was measured using a PNA-X network analyzer (N5244a) at the frequency range of 8.2-12.4 GHz (X-band). The measured sample was prepared by mixing 40% of α-FeOOH/CDCA (grinded into powder) or α-FeOOH with 60% of paraffin wax. The mixture was then pressed into a toroidal shaped mould (7.0 mm in external diameter, 3.0 mm in inner diameter and 2.0 mm in thickness).

2.6 Measurements of electrical conductivity

For the measurements of electrical conductivity (σ), the samples were firstly grinded into fine powder and then pressed into a transparent PC tube with an inner diameter of 8 mm. The bottom of the tube is closed by a stationary brass piston. After the introduction of powder sample (ca. 2 cm^3), the upper side is closed by a stainless steel plunger which could be freely moved up and down in the tube. Then a weighed load was put on the upper piston to allow for good electrical contacts between powders. Electrical resistance (R) of the pressed powder was measured at room temperature using an Elecall EM90A digital multimeter. Therefore, the electrical conductivity (σ) can be calculated by $\sigma = (1/R)(L/\pi r^2)$, where L is the length of the sample in the tube, and r is the radius of the tube.

3 Results and discussion

3.1 Schematic diagram for preparation of α-FeOOH/CDCA

In the present work, we would like to synthesize a kind of hybrid aerogels (i.e., α-FeOOH/CDCA) composed of cellulose - derived carbon and α-FeOOH nanoneedles and nanoflowers for EMI shielding applications. A schematic diagram for the preparation of α-FeOOH/CDCA is presented in Fig. 1. For acquiring good electrical conductivity which is one of the most important parameters for EMI shielding, the cellulose aerogels were firstly pyrolyzed into carbon aerogels (i.e., CDCA). The oxygen-containing groups of cellulose were destroyed during this process. Then, a HNO_3 treatment was used to introduce new oxygen-containing groups on the surface of CDCA for anchoring and interacting with the iron ions. During the immersion of acid-treated CDCA in the solution of ferrous and ferric ions and the subsequent heating process, the strong electrostatic interactions (e.g., ion-dipole interaction) occurred between the electron-rich oxygen atoms of polar hydroxyl and carboxyl groups of acid-treated CDCA and the electropositive transition metal cations (He, Kunitake & Nakao, 2003). These interactions result in the immobilization of cations on CDCA. After dipping CDCA with attached Fe^{2+}/Fe^{3+} into the hot alkali solution, nanonuclei were quickly generated at the original ionized selective sites of CDCA surface. Numerous unattached and isolated particles were continuously deposited on the small raised areas of the nanonuclei, resulting in the generation of nanoneedles along a specific direction. Regarding the generation of nanoflowers, these thermodynamically unstable nanonuclei firstly gathered together, and then the random growth of these nanonuclei led to the formation of nanoflowers (Li et al., 2011; Penn & Banfield, 1998).

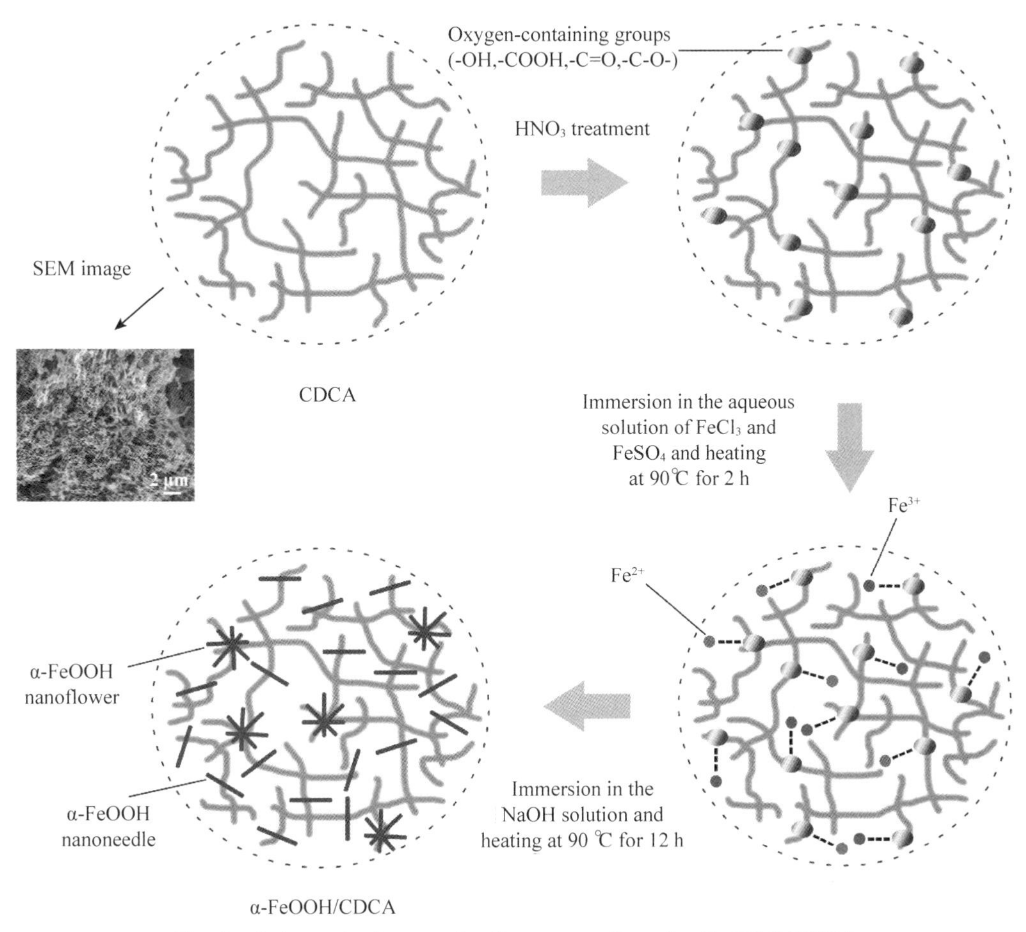

Fig. 1 Schematic diagram for the preparation of α-FeOOH/CDCA

3.2 Morphology observations and elemental analysis

Morphological characteristics of CDCA before and after combining with α-FeOOH were studied by SEM observations. Fig. 2a shows obviously that CDCA is a porous material, and a plenty of sheet-like structures with widths of tens of micrometers and thicknesses of several hundred nanometers (see Fig. 2b) were tightly intertwined with each other. Moreover, as shown in Fig. 2c, the magnified image of the yellow-squared marker region in Fig. 1a shows that these micron-scale thin sheets consist of abundant pore structures with sizes ranging from several decades nanometers to hundreds of nanometers. For α-FeOOH/CDCA, it is found in Fig. 2d and e that the surface of the sheet-like structure of CDCA was coated with a mass of needle-like α-FeOOH, and some of α-FeOOH nanoneedles self-assembled into nanoflowers. The self-assembly mechanism can be explained by the oriented attachment crystal growth model reported by Penn and Banfield (Penn & Banfield, 1998). As mentioned above, nanonuclei were firstly formed at the original reaction stage. The freshly formed nanonuclei were thermodynamically unstable due to their high surface energy, and thus tended to aggregate together to minimize the interfacial energy, leading to the formation of many agglomerates. As the reaction proceeded, the continuous deposition of new particles on the small raised areas of the agglomerates resulted in the generation of short nanoneedles along a specific direction. As the reaction went on, each nanoneedle continued to grow along its axis, and the following nanoflower-like structures were formed. EDX analysis was used to obtain additional information about the elemental compositions. For CDCA, C, O and Au elements were detected (Fig. 2f). The Au element originates from the coating layer used for electric conduction during SEM observation. The low oxygen content of 11.5% is attributed to the pyrolysis damaging the oxygen-containing groups of cellulose. In contrast to CDCA, some new signals assigned to Fe element were detected for α-FeOOH/CDCA. In addition, the incorporation of α-FeOOH results in a significant improvement in the mass ratio of oxygen to carbon from 0.13 (CDCA) to 0.25 (α-FeOOH/CDCA).

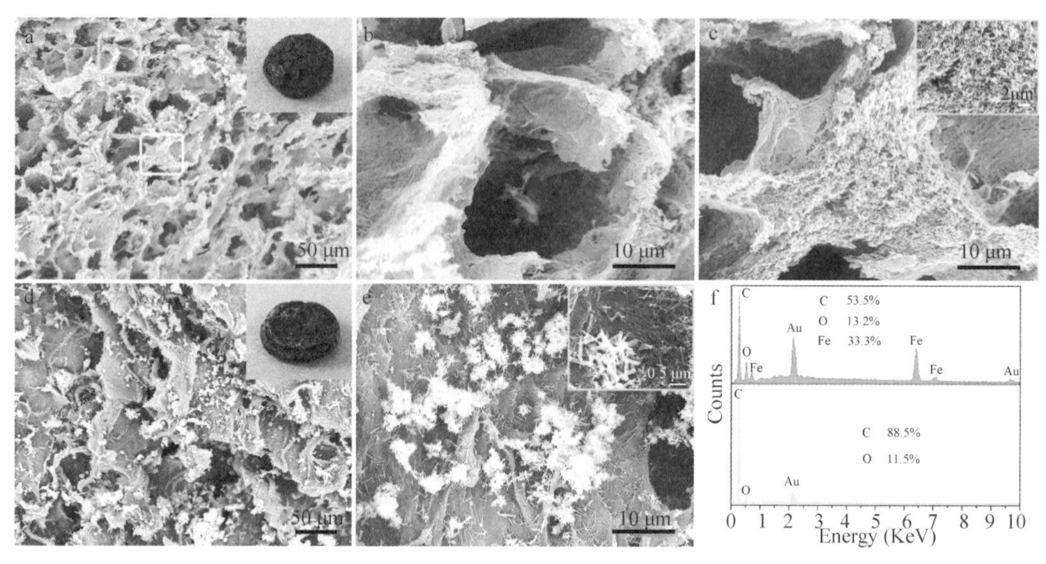

Fig. 2 (a) SEM image of CDCA and the inset shows its digital image. Magnified images of (b) the green-squared marker region and (c) the yellow-squared marker region in (a). The inset in (c) shows the corresponding magnified image. (d–e) SEM images of α-FeOOH/CDCA under different magnifications, and the inset in (d) shows its digital image, and the inset in (e) presents the corresponding magnified image. (f) EDX spectra of CDCA and α-FeOOH/CDCA

TEM analysis provides further insight into the morphology and size. As shown in Fig. 3a, a large number of α-FeOOH nanoneedles were accumulated on the surface of CDCA, consistent with the SEM observations. The nanoneedles have diameters of 38–115 nm and lengths of 410–960 nm (Fig. 3b). The corresponding SAED pattern exhibits a polycrystalline feature (Fig. 3c) due to the random orientation of different α-FeOOH nanocrystals (Li et al., 2011). In addition, the diffraction rings can be indexed to the (121), (130), (120), (141) and (231) planes of α-FeOOH.

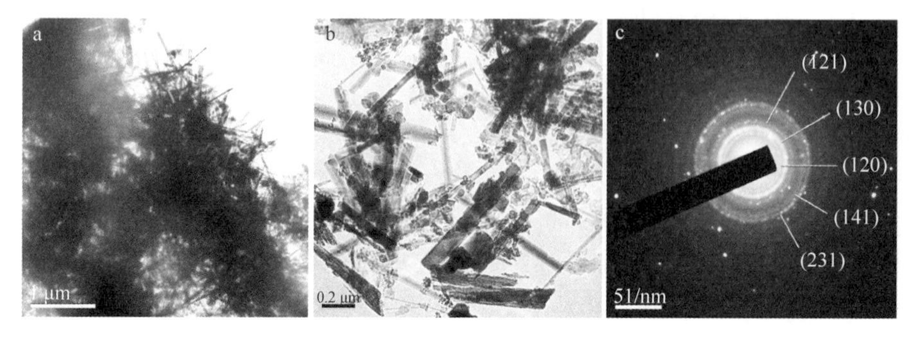

Fig. 3 (a) Low- and (b) high-magnification TEM images and (c) SAED pattern of α-FeOOH/CDCA, respectively

3.3 Crystal structure and chemical compositions

For clarifying the crystal structure and phase purity of the synthesized α-FeOOH in α-FeOOH/CDCA, XRD analysis was carried out. As shown in Fig. 4, CDCA shows two broad diffraction peaks centered at 22.0° and 43.8°, which are ascribed to the (002) and (100) planes of amorphous carbon originated from the pyrolysis of cellulose (Chang & Chen, 2011). However, the two peaks related to CDCA cannot be identified in the XRD pattern of α-FeOOH/CDCA, possibly due to the plentiful nanoneedles and nanoflowers coating on CDCA substrate. Moreover, a series of new sharp peaks can be observed and well indexed to the standard data of α-FeOOH (JCPDS card no. 29 - 0713), indicating that α-FeOOH were successfully synthesized and integrated with CDCA. In addition, there is no indication of any other iron oxide phases in the XRD pattern of α-FeOOH/CDCA, suggesting a high purity of α-FeOOH in the composite.

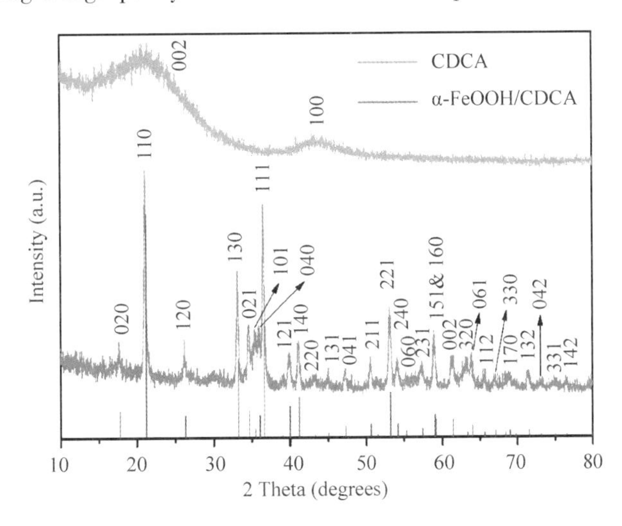

Fig. 4 XRD patterns of CDCA and α-FeOOH/CDCA, respectively. The bottom line is the standard JCPDS card no. 29-0713 for α-FeOOH

XPS analysis was used to investigate the valence states of Fe ion and the chemical bonding states of α-FeOOH/CDCA. Fig. 5a shows the XPS survey spectrum of α-FeOOH/CDCA. The signals from C, O and Fe elements were detected, agreeing well with the results of EDX analysis. Generally, Fe 2p states were used to distinguish the difference between Fe^{2+} and Fe^{3+} (Descostes, Mercier, Thromat, Beaucaire & Gautier-Soyer, 2000). The Fe 2p peaks at binding energies of 711.6 and 725.2 eV (Fig. 5b), corresponding to the Fe $2p_{3/2}$ and Fe $2p_{1/2}$ spin-orbit peaks (Zeng, Ren, Zheng, Wu & Cui, 2012), respectively, are in good agreement with the binding energy values of Fe^{3+} in α-FeOOH. The satellite peak of Fe $2p_{3/2}$ centered at 719.7 eV is another evidence of Fe^{3+} (Wan & Li, 2015b). The absence of Fe(Ⅱ) component and associated shake-up features implies that the Fe^{2+} turned to α-FeOOH crystals during the reaction. The XPS signals of O 1s can be deconvoluted into four components (Fig. 5c), which reveal the presence of different oxygen - containing functional groups. The peaks at 533.1 and 531.8 eV are assigned to the O-C=O and C-O of CDCA (Zhang, Li & Chen, 2010), indicating that the oxygen-containing groups were successfully introduced on the surface of CDCA by the HNO_3 treatment. In addition, the peaks at 530.6 and 529.9 eV are related to the hydroxide (O-H) and oxide (O^{2-}) components of α-FeOOH (Harvey & Linton, 1981), respectively. The C 1s spectrum further confirms the presence of oxygen-containing groups. As shown in Fig. 5d, three fitting peaks located at 284. 8, 285. 8, and 289. 3 eV can be identified, which correspond to the C-H/C-C, C-O, and O-C=O (Peng, Luo, Qiu & Yuan, 2013), respectively.

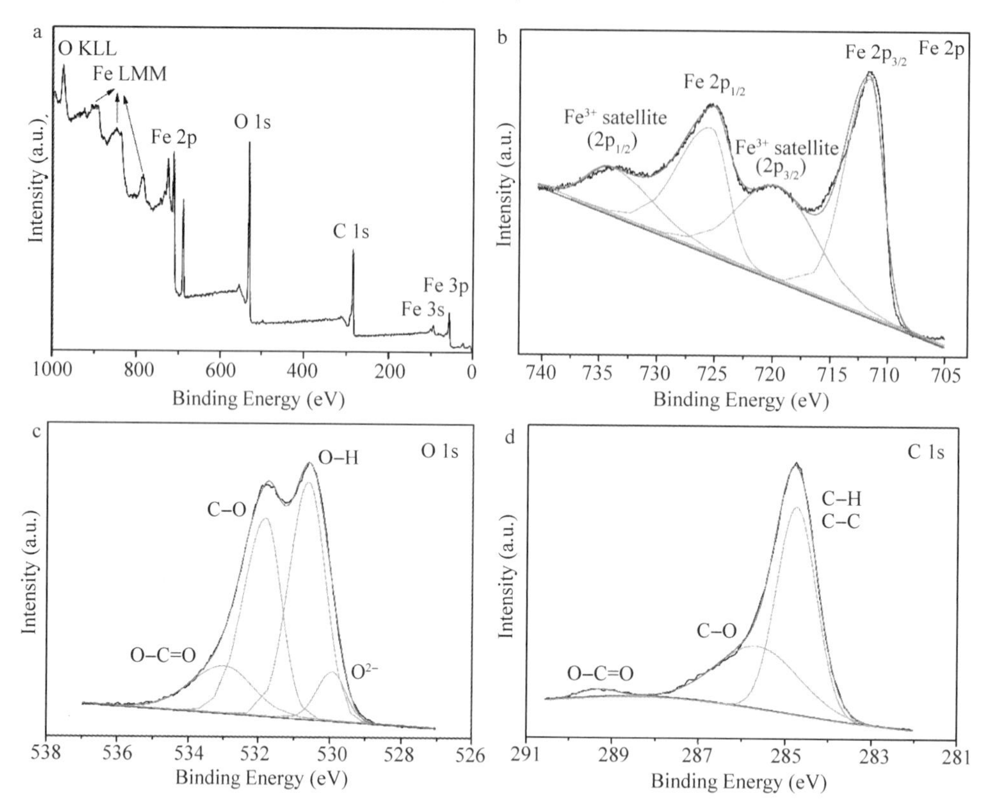

Fig. 5 XPS spectra of α-FeOOH/CDCA: (a) wide scan survey spectrum; the core-level XPS signals of (b) Fe 2p, (c) O 1s, and (d) C 1s

3. 4 EMI shielding properties

For an EMI shielding material, when an electromagnetic wave (I) impacts on the material, two waves are created on the surface: a reflected wave (R) and a transmitted wave into the material ($I-R$). Inside the material, a fraction of the wave ($I-R$) may be dissipated as heat (A) until it reaches the second surface of the material. At this point, two new waves appear: one that is transmitted through the surface (T) and a new wave that is reflected into the material. This process is repeated successively until it meets the criteria stated in equation 1:

$$I=R+A+T \tag{1}$$

Both in the reflection and transmission processes, waves generated at each step may cause constructive and destructive interferences depending on the sample thickness and frequency. The reflection process on each plane of the material is what is called multiple reflections. Therefore, the total shielding effectiveness, SE_{total}, of a material can be quantified as the sum of three contributions: reflection (SE_R), absorption (SE_A) and multiple reflections (SE_M).

$$SE_{total}(\text{dB}) = SE_A + SE_R + SE_M \tag{2}$$

SE_M can be negligible when $SE_{total} > 10$ dB. Besides, SE_R and SE_A can be described as (Al-Saleh & Sundararaj, 2012):

$$SE_R = -10\log(1-R) \quad (3)$$

$$SE_A = -10\log[T/(1-R)] \quad (4)$$

In addition, the scattering parameters (S_{ij} denotes the power transmitted from port i to port j) of the vector network analyzer adopting two-port network test method are related to R and T, respectively, that is, $R = |S_{11}|^2 = |S_{22}|^2$ and $T = |S_{12}|^2 = |S_{21}|^2$.

For the investigation of the influences of the total concentration of $FeCl_3/FeSO_4$ salts on the EMI shielding property of α-FeOOH/CDCA, we used different concentrations of Fe^{3+}/Fe^{2+} (including 0.06, 0.03, 0.01, and 0.003 M) to fabricate α-FeOOH/CDCA, and the resultant composites are coded as Fe@C-6, Fe@C-3, Fe@C-1, and Fe@C-03, respectively. As shown in Fig. 6 and Table 1, the electrically insulating α-FeOOH has the lowest SE_{total} of 5.9 dB. After the α-FeOOH was integrated with CDCA, the SE_{total} value significantly increases to 17.4 dB for Fe@C-6, 29.5 dB for Fe@C-3, 34.0 dB for Fe@C-1, and 32.2 dB for Fe@C-03. These EMI shielding efficiencies are close to or even higher than the commercially achievable level (>20 dB) (Song et al., 2014). In addition, it is obvious that the SE_{total} value gradually increases with the decreasing concentration of Fe^{3+}/Fe^{2+} from 0.06 to 0.01 M, which is possibly associated with the increasing electrical conductivity (from 0.67 to 1.93 $S \cdot m^{-1}$) along with the increasing proportion of electrically conductive carbon in the composite. In addition, when the concentration decreases from 0.01 to 0.003 M, it is clear that the value of electrical conductivity increases from 1.93 to 2.10 $S \cdot m^{-1}$, while the SE_{total} value decreases from 34.0 to 32.2 dB. The more superior EMI shielding property of Fe@C-1 might be ascribed to the more effective complementarity between the dielectric loss and the magnetic loss (Chen et al., 2011). The EMI shielding effectiveness is closely related to the dielectric properties of the material dependent on ionic, electronic, orientational and space charge polarization. The contribution to the space charge polarization appears due to the heterogeneity of the material. The presence of insulating magnetic α-FeOOH in the conducting matrix results in the formation of more interfaces and a heterogeneous system due to some space charge accumulating at the interface that contributes to the higher microwave absorption in the composite (Singh et al., 2012). Flower-like α-FeOOH is known to have larger specific surface area than that of commercial bulk α-FeOOH (Li et al., 2011), which can provide plentiful interfaces. Furthermore, the incident electromagnetic waves undergo reflection and scattering many times between the micro/nano-scaled surfaces of flower-like α-FeOOH, and the large surface area is beneficial for the adsorption and dissipation of electromagnetic waves in this process. Also, the waves may encounter the local interfacial polarization, and then produce an energy level transition or complete attenuation. Therefore, the interface may act as a polarization center, resulting in the improvement of the EMI shielding effectiveness (Guo et al., 2009).

The EMI shielding mechanism was estimated by comparing the contribution of reflection (SE_R) and absorption (SE_A) to the overall EMI shielding effectiveness (Park et al., 2009). It is seen in Fig. 6a that α-FeOOH has a SE_R of 3.5 dB, 45.8% higher than its SE_A(2.4 dB), which indicates that the EMI shielding mechanism is reflection-dominant. The reflection-dominant mechanism is common for iron oxides (Kim, Kim & Kim, 2012; Kumar, Dhakate, Saini & Mathur, 2013). All the composites (i.e., α-FeOOH/CDCA) clearly show adsorption-dominant mechanism (Fig. 6b-e). As listed in Table 2, the SE_A values of α-

FeOOH/CDCA (14.3–29.1 dB) are 3.6–4.9 times higher than the corresponding SE_R values (3.1–4.9 dB). Also, the overall growth trends of SE_{total} over whole frequency have been theoretically predicted and experimentally confirmed for the absorption-dominant materials (Zhang, Zheng, Yan, Jiang & Yu, 2012). An absorption - dominant shielding mechanism is beneficial to alleviate secondary radiation, which is considered as more attractive alternative for the fabrication of electromagnetic radiation protection products (like coating). According to electromagnetic theory, SE_R and SE_A can be expressed as:

$$SE_R = 10\log(\sigma_{ac}/16\omega\varepsilon_0\mu') \tag{5}$$

$$SE_A = 8.68t\sqrt{\sigma\omega\mu'/2} \tag{6}$$

where σ_{ac} is the alternative conductivity, ω is the frequency, t is the thickness, ε_0 is the vacuum permittivity, and μ' is the real permeability. It can be seen that SE_R and SE_A are the functions of electrical conductivity, magnetic permeability, thickness, and measurement frequency for an EMI shielding material. Moreover, there is a positive correlation between electrical conductivity and SE_R (or SE_A) according to these formulas. Obviously, the increase of electrical conductivity results in the improvements in the SE_R and SE_A of Fe@C-6, Fe@C-3, and Fe@C-1, consistent with the above equations. However, in contrast to Fe@C-1, the lower SE_A of Fe@C-03 is possibly related to the lower permeability. In conclusion, these results reveal that the Fe^{3+}/Fe^{2+} concentration of 0.01 M is beneficial to fabricate α-FeOOH/CDCA with higher EMI shielding properties.

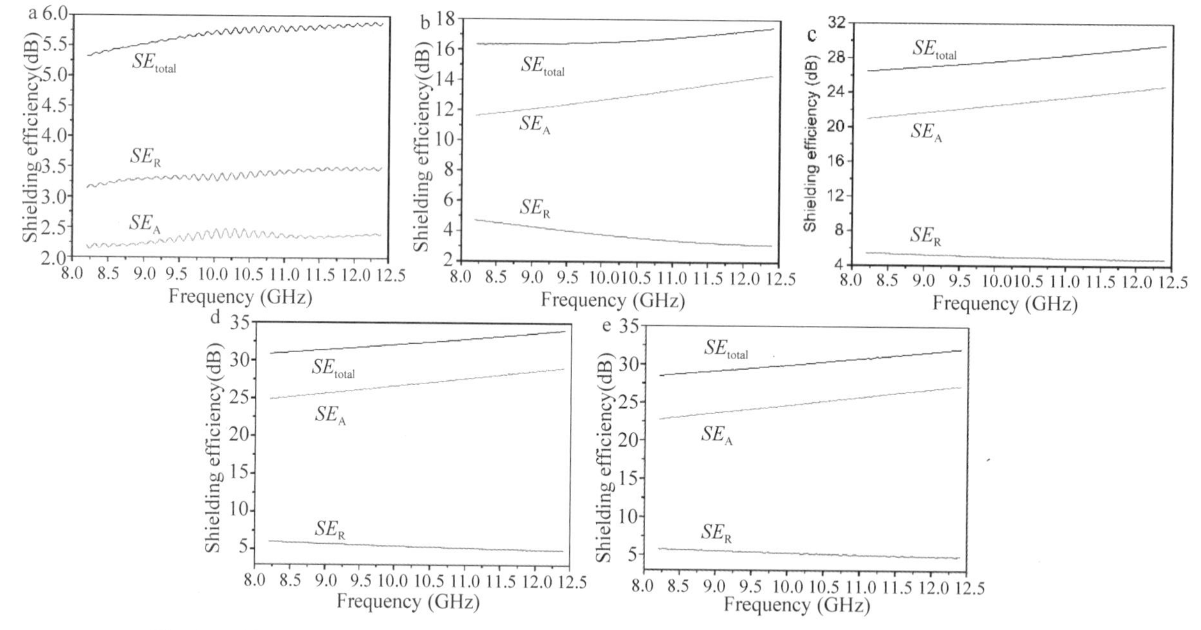

Fig. 6 SE_{total}, SE_R, and SE_A values of (a) α-FeOOH, (b) Fe@C-6, (c) Fe@C-3, (d) Fe@C-1, and (e) Fe@C-03 as a function of frequency, respectively

Table 1 Electrical conductivity, SE_{total}, SE_R, and SE_A values of α-FeOOH, Fe@C-6, Fe@C-3, Fe@C-1 and Fe@C-03, respectively

Samples	Concentration of Fe^{3+}/Fe^{2+} ($mol \cdot L^{-1}$)	Electrical conductivity ($S \cdot m^{-1}$)	SE_{total} at 12.4 GHz (dB)	SE_R at 12.4 GHz (dB)	SE_A at 12.4 GHz (dB)
α-FeOOH	0.06	Insulation	5.9	3.5	2.4
Fe@C-6	0.06	0.67	17.4	3.1	14.3
Fe@C-3	0.03	1.31	29.5	4.7	24.8
Fe@C-1	0.01	1.93	34.0	4.9	29.1
Fe@C-03	0.03	2.10	32.2	4.9	27.3

4 Conclusions

The porous carbon aerogels prepared from the pyrolysis of cellulose aerogels were used as substrates to support α-FeOOH nanoneedles and nanoflowers. Prior to the deposition of α-FeOOH, the carbon aerogels underwent an acid treatment for the introduction of oxygen-containing groups to anchor and interact with the iron ions. The nanoneedles have diameters of 38 - 115 nm and lengths of 410 - 960 nm. The formation of nanoflowers might be due to the random growth of the aggregation of nanonuclei. As an example of potential applications, the α-FeOOH/CDCA composite can serve as EMI shielding materials. It is found that the Fe^{3+}/Fe^{2+} concentration of 0.01 M is beneficial to acquire more superior EMI shielding properties. Moreover, the presence of α-FeOOH in the carbon substrate might contribute to the space charge polarization causing the improvement of microwave absorption, and the large surface area of nanoflowers is beneficial for the adsorption and dissipation of electromagnetic waves. Additionally, the absorption-dominant mechanism of α-FeOOH/CDCA is helpful to alleviate secondary radiation, which is regarded as an attractive alternative for the fabrication of electromagnetic radiation protection products.

Acknowledgment

This study was supported by the National Natural Science Foundation of China (grant nos. 31270590 and 31470584) and the Fundamental Research Funds for the Central Universities (grant nos. 2572016AB22).

References

[1] Ahlbom A, Feychting M. Electromagnetic radiation: Environmental pollution and health[J]. British medical bulletin, 2003, 68(1): 157-165.

[2] Al-Saleh M H, Sundararaj U. Electromagnetic interference shielding mechanisms of CNT/polymer composites[J]. Carbon, 2009, 47(7): 1738-1746.

[3] Al-Saleh M H, Sundararaj U. X-band EMI shielding mechanisms and shielding effectiveness of high structure carbon black/polypropylene composites[J]. Journal of Physics D Applied Physics, 2013, 46(3): 035304.

[4] Chang K, Chen W. Single-layer MoS_2/graphene dispersed in amorphous carbon: towards high electrochemical performances in rechargeable lithium ion batteries[J]. Journal of Materials Chemistry, 2011, 21(43): 17175-17184.

[5] Chen Y J, Xiao G, Wang T S, et al. Porous Fe_3O_4/Carbon Core/Shell Nanorods: Synthesis and Electromagnetic Properties[J]. The Journal of Physical Chemistry C, 2011, 115(28): 10061-10064.

[6] Chen Y J, Xiao G, Wang T S, et al. Porous Fe_3O_4/carbon core/shell nanorods: synthesis and electromagnetic

properties[J]. The Journal of Physical Chemistry C, 2011, 115(28): 13603-13608.

[7]Descostes M, Mercier F, Thromat N, et al. Use of XPS in the determination of chemical environment and oxidation state of iron and sulfur samples: constitution of a data basis in binding energies for Fe and S reference compounds and applications to the evidence of surface species of an oxidized pyr[J]. Applied Surface Science, 2000, 165(4): 288-302.

[8]Fang Z, Li C, Sun J, et al. The electromagnetic characteristics of carbonfoams[J]. Carbon, 2007, 45(15): 2873-2879.

[9]Forsyth J B, Hedley I G, Johnson C E. The magnetic structure and hyperfine field of goethite (α-FeOOH)[J]. Journal of Physics C: Solid State Physics, 1968, 1(1): 179.

[10]Guo X, Deng Y, Gu D, et al. Synthesis and microwave absorption of uniform hematite nanoparticles and their core-shell mesoporous silica nanocomposites[J]. Journal of Materials Chemistry, 2009, 19(37): 6706-6712.

[11]Hao P, Zhao Z, Tian J, et al. Hierarchical porous carbon aerogel derived from bagasse for high performance supercapacitor electrode[J]. Nanoscale, 2014, 6(20): 12120-12129.

[12]Harvey D T, Linton R W. Chemical characterization of hydrous ferric oxides by X-ray photoelectronspectroscopy[J]. Analytical Chemistry, 1981, 53(11): 1684-1688.

[13]He J, Kunitake T, Nakao A. Facile in situ synthesis of noble metal nanoparticles in porous cellulose fibers[J]. Chemistry of Materials, 2003, 15(23): 4401-4406.

[14]Jin H, Nishiyama Y, Wada M, et al. Nanofibrillar cellulose aerogels[J]. Colloids & Surfaces A Physicochemical & Engineering Aspects, 2004, 240(1-3): 63-67.

[15]Kabbour H, Baumann T F, Satcher J H, et al. Toward new candidates for hydrogen storage: high-surface-area carbon aerogels[J]. Chemistry of Materials, 2006, 18(26): 6085-6087.

[16]Kim H R, Kim B S, Kim I S. Fabrication and EMI shielding effectiveness of Ag-decorated highly porous poly (vinyl alcohol)/Fe_2O_3 nanofibrous composites[J]. Materials Chemistry and Physics, 2012, 135(2-3): 1024-1029.

[17]Kumar R, Dhakate S R, Saini P, et al. Improved electromagnetic interference shielding effectiveness of light weight carbon foam by ferrocene accumulation[J]. RSC advances, 2013, 3(13): 4145-4151.

[18]Lemaire B J, Davidson P, Panine P, et al. Magnetic-Field-Induced Nematic-Columnar Phase Transition in Aqueous Suspensions of Goethite (α-FeOOH) Nanorods[J]. Physical review letters, 2004, 93(26): 267801.

[19]Li H, Li W, Zhang Y, et al. Chrysanthemum-like α-FeOOH microspheres produced by a simple green method and their outstanding ability in heavy metal ion removal[J]. Journal of Materials Chemistry, 2011, 21(22): 7878-7881.

[20]Li J, Wan C. Cellulose aerogels decorated with multi-walled carbon nanotubes: preparation, characterization, and application for electromagnetic interferenceshielding[J]. Frontiers of Agricultural Science and Engineering, 2016, 2(4): 341-346.

[21]Li W, Reichenauer G, Fricke J. Carbon aerogels derived from cresol-resorcinol-formaldehyde for supercapacitors[J]. Carbon, 2002, 40(15): 2955-2959.

[22]Liang J, Wang Y, Huang Y, et al. Electromagnetic interference shielding of graphene/epoxy composites[J]. Carbon, 2009, 47(3): 922-925.

[23]Liebner F, Potthast A, Rosenau T, et al. Cellulose aerogels: Highly porous, ultra-lightweight materials[J]. Holzforschung, 2008, 62(2): 129-135.

[24]Moreno-Castilla C, Maldonado-Hódar F J. Carbon aerogels for catalysis applications: An overview[J]. Carbon, 2005, 43(3): 455-465.

[25]Park J G, Louis J, Cheng Q, et al. Electromagnetic interference shielding properties of carbon nanotube buckypaper composites[J]. Nanotechnology, 2009, 20(41): 415702.

[26]Peng F, Luo T, Qiu L, et al. An easy method to synthesize graphene oxide-FeOOH composites and their potential application in water purification[J]. Materials Research Bulletin, 2013, 48(6): 2180-2185.

[27] Penn R L, Banfield J F. Imperfect oriented attachment: dislocation generation in defect-free nanocrystals[J]. Science, 1998, 281(5379): 969-971.

[28] Saliger R, Fischer U, Herta C, et al. High surface area carbon aerogels for supercapacitors[J]. Journal of Non-Crystalline Solids, 1998, 225: 81-85.

[29] Sehaqui H, Zhou Q, Berglund L A. High-porosity aerogels of high specific surface area prepared from nanofibrillated cellulose (NFC)[J]. Composites science and technology, 2011, 71(13): 1593-1599.

[30] Singh A P, Garg P, Alam F, et al. Phenolic resin-based composite sheets filled with mixtures of reduced graphene oxide, γ-Fe_2O_3 and carbon fibers for excellent electromagnetic interference shielding in the X-band[J]. Carbon, 2012, 50(10): 3868-3875.

[31] Song W L, Fan L Z, Cao M S, et al. Facile fabrication of ultrathingraphene papers for effective electromagnetic shielding[J]. Journal of Materials Chemistry C, 2014, 2(25): 5057-5064.

[32] Tamon H, Ishizaka H, Mikami M, et al. Porous structure of organic and carbon aerogels synthesized by sol-gel polycondensation of resorcinol with formaldehyde[J]. Carbon, 1997, 35(6): 791-796.

[33] Wan C, Li J. Embedding ZnO nanorods into porous cellulose aerogels via a facile one-step low-temperature hydrothermal method[J]. Materials & Design, 2015, 83: 620-625.

[34] Wan C, Li J. Facile synthesis of well-dispersedsuperparamagnetic γ-Fe_2O_3 nanoparticles encapsulated in three-dimensional architectures of cellulose aerogels and their applications for Cr (VI) removal from contaminated water[J]. ACS Sustainable Chemistry & Engineering, 2015, 3(9): 2142-2152.

[35] Wan C, Li J. Synthesis of well-dispersed magnetic $CoFe_2O_4$ nanoparticles in cellulose aerogels via a facile oxidative co-precipitation method[J]. Carbohydrate polymers, 2015, 134: 144-150.

[36] Wan C, Li J. Cellulose aerogels functionalized withpolypyrrole and silver nanoparticles: In-situ synthesis, characterization and antibacterial activity[J]. Carbohydrate polymers, 2016, 146: 362-367.

[37] Wan C, Lu Y, Jiao Y, et al. Fabrication of hydrophobic, electrically conductive and flame-resistant carbon aerogels by pyrolysis of regenerated celluloseaerogels[J]. Carbohydrate polymers, 2015, 118: 115-118.

[38] Wang X, Chen X, Gao L, et al. Synthesis of β-FeOOH and α-Fe_2O_3 nanorods and electrochemical properties of β-FeOOH[J]. Journal of Materials Chemistry, 2004, 14(5): 905-907.

[39] Wu H, Wu G, Wang L. Peculiar porous α-Fe_2O_3, γ-Fe_2O_3 and Fe_3O_4 nanospheres: facile synthesis and electromagnetic properties[J]. Powder Technology, 2015, 269: 443-451.

[40] Wu Z Y, Li C, Liang H W, et al. Innenrücktitelbild: Ultralight, Flexible, and Fire-Resistant Carbon Nanofiber Aerogels from Bacterial Cellulose (Angew. Chem. 10/2013)[J]. Angewandte Chemie, 2013, 125(10): 3113-3113

[41] Xiao H M, Zhang W D, Fu S Y. One-step synthesis, electromagnetic and microwave absorbing properties of α-FeOOH/polypyrrole nanocomposites[J]. Composites science and technology, 2010, 70(6): 909-915.

[42] Xu X, Zhou J, Nagaraju D H, et al. Flexible, highly graphitized carbon aerogels based on bacterial cellulose/lignin: Catalyst-free synthesis and its application in energy storage devices[J]. Advanced Functional Materials, 2015, 25(21): 3193-3202.

[43] Yan L, Gao Z. Dissolving of cellulose in PEG/NaOH aqueous solution[J]. Cellulose, 2008, 15(6): 789-796.

[44] Yang S, Lozano K, Lomeli A, et al. Electromagnetic interference shielding effectiveness of carbon nanofiber/LCP composites[J]. Composites Part A: applied science and manufacturing, 2005, 36(5): 691-697.

[45] Yang Y, Gupta M C, Dudley K L, et al. Novel carbon nanotube-polystyrene foam composites for electromagnetic interference shielding[J]. Nano letters, 2005, 5(11): 2131-2134.

[46] Závišová V, Tomašovičová N, KOVÁČ J, et al. Synthesis and characterisation of rod-like magnetic nanoparticles[J]. Olomouc, Czech Republic, 2010, 10: 12-14.

[47] Zeng L, Ren W, Zheng J, et al. Synthesis of water-soluble FeOOH nanospindles and their performance for magnetic

resonance imaging[J]. Applied surface science, 2012, 258(7): 2570-2575.

[48]Zhang H B, Zheng W G, Yan Q, et al. The effect of surface chemistry of graphene on rheological and electrical properties of polymethylmethacrylate composites[J]. Carbon, 2012, 50(14): 5117-5125.

[49]Zhang LL, Zhao X S. Carbon-based materials as supercapacitor electrodes[J]. Chemical Society Reviews, 2009, 38(9): 2520-2531.

[50]Zhang S, Li F X, Yu J Y. Structure and Properties of Novel Cellulose Fibres Produced From NaOH/PEG-treated Cotton Linters[J]. Iranian Polymer Journal, 2010, 19(12): 949-957.

[51]Zhang S, Li X, Chen J P. An XPS study for mechanisms of arsenate adsorption onto a magnetite-doped activated carbonfiber[J]. Journal of colloid and interface science, 2010, 343(1): 232-238.

中文题目：用于电磁屏蔽的纤维素衍生碳气凝胶负载针铁矿(α-FeOOH)纳米针和纳米花

作者：万才超，焦月，强添刚，李坚

摘要：我们描述了一种快速简便的化学沉淀法，在纤维素气凝胶热解得到的碳气凝胶上生长出针铁矿(α-FeOOH)纳米针和纳米花。当作为电磁屏蔽材料进行评估时，α-FeOOH /纤维素衍生碳气凝胶复合材料在 Fe^{3+}/Fe^{2+} 浓度为 0.01 μm 时 SE_{total} 值最高，达到 34.0 dB，是 α-FeOOH (5.9 dB)的 4.8 倍。当 Fe^{3+}/Fe^{2+} 浓度升高或降低时，电磁屏蔽性能下降。α-FeOOH 与碳气凝胶的结合将 α-FeOOH 的反射主导机制转变为吸附主导机制。吸附主导机制无疑有助于缓解二次辐射，被认为是开发电磁辐射防护产品更有吸引力的选择。

关键词：纤维素；碳气凝胶；针铁矿；电磁屏蔽；复合材料

Synthesis and Electromagnetic Interference Shielding of Cellulose-derived Carbon Aerogels Functionalized with α-Fe_2O_3 and Polypyrrole*

Caichao Wan, Jian Li

Abstract: Eco-friendly cellulose-derived carbon aerogels (CDCA) were employed as porous substrate to integrate with α-Fe_2O_3 and polypyrrole (PPy) via pyrolysis and vapor-phase deposition. The SEM and TEM observations present that the wrinkled PPy sheets and the α-Fe_2O_3 nanoparticles were well dispersed in CDCA. The strong interactions (such as hydrogen bonding) between the substrate and the nanomaterials were demonstrated by the FTIR and XPS analysis. When utilized as electromagnetic interference (EMI) shielding materials, the α-Fe_2O_3/PPy/CDCA (FPCA) composite has the highest total shielding effectiveness (SE_{total}) of 39.4 dB, about 2.0, 2.9, and 1.3 times that of the acid-treated CDCA (19.3 dB), PPy (13.6 dB), and α-Fe_2O_3/CDCA (29.3 dB), respectively. Moreover, the shielding effectiveness due to absorption accounts for 78.2%-84.2% of SE_{total} for FPCA, indicative of the absorption-dominant shielding mechanism contributing to alleviating secondary radiation. These features make the composite a useful alternative candidate for EMI shielding.

Keywords: Cellulose; Carbon aerogels; α-Fe_2O_3; Polypyrrole; Electromagnetic interference shielding; Composites

1 Introduction

Nanostructured organic-inorganic composites have attracted increasing attention from various fields like adsorbents (Schumacher, Gonzalez, Pérez-Mendoza, Wright & Seaton, 2006; Wan & Li, 2015b), supercapacitors (Gómez-Romero et al., 2003), and catalysts (Wight & Davis, 2002), due to their combining organic and inorganic materials' superb characters (Kango, Kalia, Celli, Njuguna, Habibi & Kumar, 2013; Kickelbick, 2003; Sanchez, Belleville, Popall & Nicole, 2011). Cellulose is the most abundant renewable biopolymers on earth, which is generated from repeating β-D glucopyranose molecules that are covalently linked through acetal functions between the equatorial OH group of C4 and the C1 carbon atom (β-1,4-glucan) (Klemm, Heublein, Fink & Bohn, 2005). With the increasing emphasis on green chemistry and sustainable development in recent decade, cellulose products have been paid more and more attention. Cellulose aerogels, one of important cellulose products, have large specific surface area (Sehaqui, Zhou & Berglund, 2011), high porosity (Liebner, Potthast, Rosenau, Haimer & Wendland, 2008), and

---

* 本文摘自 Carbohydrate Polymers, 2017, 161: 158-165.

plentiful surface hydroxyl groups, which provide convenience to integrate with various organic and inorganic guest substances for the development of novel functional composites (Cai et al., 2014; Olsson et al., 2010; Pääkkö et al., 2008). Cellulose aerogels can be transformed into carbon product (called as cellulose-derived carbon aerogels, CDCA) via a pyrolysis treatment in oxygen-limited condition, and its porous structure cannot be remarkably changed after the pyrolysis (Wan, Lu, Jiao, Jin, Sun & Li, 2015; Wu, Li, Liang, Chen & Yu, 2013). As a result, CDCA can also serve as porous substrate to support functional guest materials. Especially, CDCA-based composites inherit not only the good electrical conductivity and porous structure of CDCA, but also new fascinating functions dependent on the inserted guest substances.

Carbon aerogels are traditionally obtained from the carbonization of organic aerogels (Moreno-Castilla & Maldonado-Hódar, 2005), which are prepared from the sol-gel polycondensation of certain organic monomers, such as resorcinol-formaldehyde (Al-Muhtaseb & Ritter, 2003), phenol-resorcinol-formaldehyde (Wu, Fu, Sun & Yu, 2005), phenolic-furfural (Wu & Fu, 2006), and melamine-formaldehyde (Friedel & Greulich-Weber, 2006). For the synthesis of carbon aerogels, it is clear that cellulose aerogels serving as precursor cause less pollution because no toxic chemicals are needed and generated, as compared to the above organic monomers. Recently, some novel carbon aerogels constructed by the self-assembly of carbon materials (e.g., graphene (Zhang et al., 2011), graphene oxide (Mi, Huang, Xie, Wang, Liu & Gao, 2012), carbon nanotube (Bryning, Milkie, Islam, Hough, Kikkawa & Yodh, 2007), or their mixtures [Sui, Meng, Zhang, Ma & Cao, 2012; Sun, Xu & Gao, 2013)] have attracted huge attention. Nevertheless, these carbon aerogels might be more applicable to some advanced fields such as high-performance optoelectronic devices (Li, Yang, Sun, He, Xu & Ding, 2014), considering their high cost and complicated preparation technics.

To date, the applications of CDCA and CDCA-based composites mainly focus on the fields of environmental remediation (like adsorbents) (Wu, Li, Liang, Chen & Yu, 2013) and energy storage systems [like supercapacitors (Hao et al., 2014), Zn-air battery (Liang, Wu, Chen, Li & Yu, 2015], and lithium ion batteries (Wang et al., 2013)). It is worth to further extend the application scopes of CDCA. Nowadays, widespread utilization of commercial, military and scientific electronic devices and communication instruments has created serious concern for electromagnetic pollution (Balmori, 2009; Leitgeb, Schröttner & Böhm, 2005). Owing to good electrical, anticorrosive and antioxidative properties and ideal thermal stability, CDCA holds some potential in electromagnetic interference (EMI) shielding (Chung, 2001). Besides carbon materials, conductive polymers [like polypyrrole (PPy) and polyaniline (PANI)] (Wang & Jing, 2005), metals (like silver, copper, iron, aluminum, and nickel) (Chung, 2000), and metallic derivatives (like metallic oxide) (Shen, Zhai, Tao, Ling & Zheng, 2013; Xiao, Zhang & Fu, 2010) can also be used as EMI shielding materials. Thus it is meaningful to integrate these various types of materials to develop new-type high-performance EMI shielding products.

In this work, we report an easily-operated cost-effective process (including pyrolysis and vapor phase polymerization) to functionalize CDCA with α-Fe_2O_3 and PPy. The α-Fe_2O_3/PPy/CDCA composite (coded as FPCA) was characterized by scanning electron microscopy (SEM), energy dispersive X-ray spectroscopy (EDX), transmission electron microscopy (TEM), high-resolution TEM (HRTEM), selected area electron diffraction (SAED), X-ray photoelectron spectroscopy (XPS), X-ray diffraction (XRD), and Fourier transform infrared spectroscopy (FTIR). For the investigation of the reinforcement effects of α-Fe_2O_3 and PPy, the EMI shielding effectiveness of FPCA was tested and compared to that of the α-Fe_2O_3/CDCA (coded as FCA), PPy and acid-treated CDCA.

2 Materials and methods

2.1 Materials

All chemicals including iron (Ⅲ) chloride hexahydrate ($FeCl_3 \cdot 6H_2O$), iron (Ⅲ) nitrate nonahydrate [$Fe(NO_3)_3 \cdot 9H_2O$], p-toluenesulfonic acid (p-TSA), hydrochloric acid (HCl), and pyrrole were supplied by Kemiou Chemical Reagent Co., Ltd. (Tianjin, China).

2.2 Preparation of cellulose-derived carbon aerogels (CDCA)

The preparation processes of cellulose aerogels can be referred to our previous works (Wan & Li, 2015a, 2016a; Wan, Lu, Jiao, Cao, Sun & Li, 2015). CDCA was obtained by transferring the cellulose aerogels into a tubular furnace for pyrolysis under the protection of nitrogen (Wan & Li, 2016b). In a typical process, the cellulose aerogels was firstly heated to 500 ℃ at a heating rate of 5 ℃ $\cdot$ min^{-1}, and this temperature was maintained for 1 h; then the sample was heated to 1000 ℃ at 5 ℃ $\cdot$ min^{-1} and held at this temperature for 2 h to allow for complete pyrolysis. Thereafter, the temperature decreased to 500 ℃ at a cooling rate of 5 ℃ $\cdot$ min^{-1} and finally decreased naturally to room temperature.

2.3 Preparation of α-Fe_2O_3/PPy/CDCA (FPCA) composite

Prior to the deposition of α-Fe_2O_3 and PPy, CDCA was treated with HNO_3 (65%) for 24 h at room temperature for the introduction of various oxygen-containing groups to fasten the iron ions and pyrrole monomers. The α-Fe_2O_3 was firstly deposited on CDCA substrate via a simple impregnation-pyrolysis method (Han, Li, Li, Lei, Sun & Lu, 2013). In a typical process, the solution of $Fe(NO_3)_3 \cdot 9H_2O$ (50 mL, 0.94 M) was firstly mixed with 500 μL HCl (37%). Subsequently, the acid-treated CDCA (0.3 g) was immersed in the above solution for 3 h and then dried at 50 ℃ for 3 h. After that, the sample was transferred into a tubular furnace and heated to 400 ℃ with a heating rate of 2 ℃ $\cdot$ min^{-1} under a nitrogen atmosphere, and this temperature was maintained for 4 h. Finally, the temperature decreased naturally to room temperature, and the following FCA was obtained. This impregnation-pyrolysis process was repeated one more time in order to increase the α-Fe_2O_3 loading.

The incorporation of PPy into FCA was conducted by vapor-phase polymerization (Winther-Jensen, Chen, West & Wallace, 2004). In a typical process, 0.15 g FCA was firstly immersed in 1 mL of solution containing 0.3 M $FeCl_3 \cdot 6H_2O$ as the oxidant and 0.033 M p-TSA as the dopant. After the immersion, the sample was placed into a glass desiccator, and then a vial containing 2 mL pyrrole was also placed into the desiccator. After sealing, the desiccator was left at room temperature for 12 h, allowing the polymerization of vapor-phase pyrrole on the surface of FCA. Finally, the resultant was washed and dried at 50 ℃, and the following FPCA was prepared. In addition, the PPy powder was also prepared by the aforementioned vapor-phase polymerization method without being combined with FCA.

2.4 Characterizations

SEM observations were performed with a Hitachi S4800 SEM equipped with an EDX detector for element analysis. TEM and HRTEM observations and SAED were performed with a FEI, Tecnai G2 F20 TEM with a field-emission gun operating at 200 kV. XPS analysis was implemented using a Thermo Escalab 250Xi XPS spectrometer equipped with a dual X-ray source using Al-Kα. Deconvolution of the overlapping peaks was performed using a mixed Gaussian-Lorentzian fitting program (Origin 9.0, Originlab Corporation). XRD analysis was performed on a Bruker D8 Advance TXS XRD instrument with Cu Kα (target) radiation (λ = 1.5418 Å) at a scan rate (2θ) of 4° min^{-1} and a scan range from 5 to 80°. FTIR spectra were recorded by a Nicolet Nexus 670 FTIR instrument in the range of 650-4000 cm^{-1} with a resolution of 4 cm^{-1}.

2.5 Measurements of EMI shielding property

The EMI shielding property was investigated by using a PNA-X network analyzer (N5244a) at the frequency range of 8-13 GHz (X-band). The measured samples were prepared by uniformly mixing 50% of FPCA, FCA, PPy or CDCA (grinded into powder) with 50% of paraffin wax. The mixture was then pressed into a toroidal shaped mould (7.0 · mm in external diameter, 3.0 mm in inner diameter, and 2.0 mm in thickness).

3 Results and discussion

3.1 Characterizations of FPCA by SEM, TEM, SAED, XPS, XRD, and FTIR

The morphology of CDCA, FCA and FPCA was studied by SEM observations. As shown in Fig. 1a, CDCA exhibits cross-linked three-dimensional (3D) network structure, and this micro/nano-scale porous structure provides convenience for the insertion of guest substances. Fig. 1b displays the SEM images of FCA. As shown, the 3D network was still maintained, while we cannot clearly identify the synthetic α-Fe_2O_3 nanoparticles from the network, which is possibly due to their extremely small sizes (the following TEM observations would prove that). The SEM image of FPCA is presented in Fig. 1c, where a plenty of sheet-like PPy was generated and

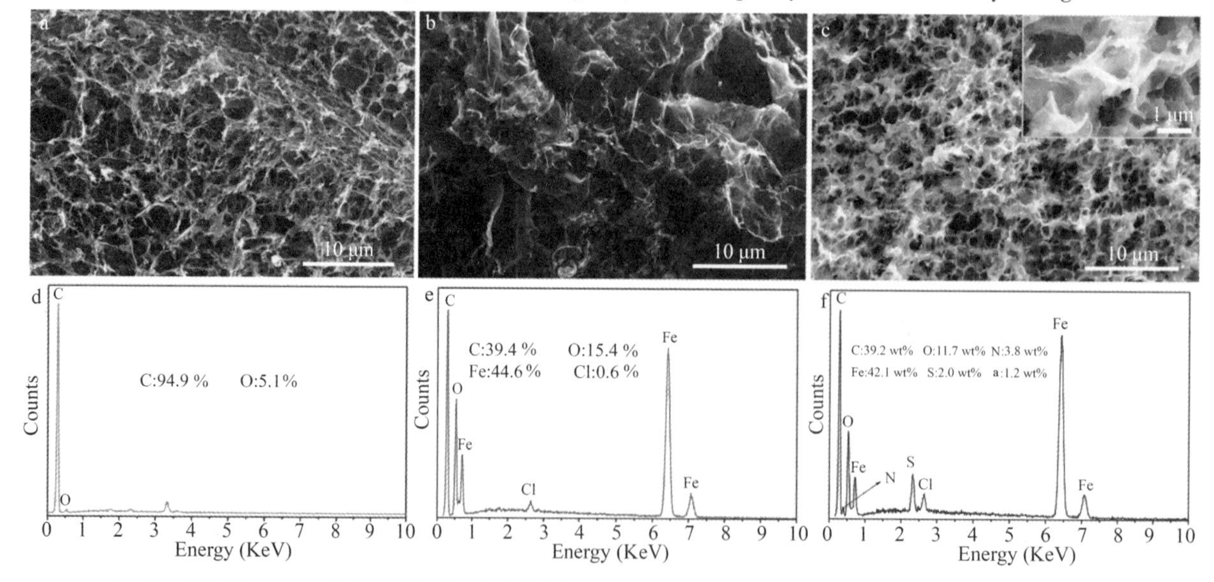

Fig. 1 SEM images of (a) CDCA, (b) FCA, and (c) FPCA. The inset in *c* presents the higher-magnification SEM image of FPCA. EDX spectra of (d) CDCA, (e) FCA, and (f) FPCA

homogeneously deposited on the surface of the substrate. According to the higher-magnification SEM image (the inset in Fig. 1c), these PPy sheets with wrinkles and ripples were dozens of nanometers in thickness and tightly intertwined with each other. CDCA is composed of C and O elements based on the EDX analysis (Fig. 1d). After the incorporation of α-Fe_2O_3, the strong Fe peaks appeared at 0.72, 6.43 and 7.06 keV (Fig. 1e). Also, the C/O ratio decreases from 18.6 for CDCA to 2.6 for FCA, which is attributed to the HNO_3 treatment and the presence of α-Fe_2O_3 in FCA. For FPCA, apart from the strong signals related to Fe element, a small peak corresponding to N element was also detected (Fig. 1f), which is derived from the deposited PPy. In addition, compared to FCA, the C/O ratio of FPCA increases to 3.4 possibly because of the existence of PPy.

The TEM image of CDCA is presented in Fig. 2a, where the cross-linked porous 3D network can be observed. Fig. 2b shows the HRTEM image of CDCA. There is no obvious long-range ordered structure, revealing its amorphous nature (Ci et al., 2001). Moreover, the SAED pattern of CDCA shows dispersing diffraction rings (Fig. 2c), which is another demonstration of the amorphous structure (Han, Tay, Shakerzadeh & Ostrikov, 2009). The TEM image of FPCA confirms the generation of well-dispersed flake-like PPy nanostructures (Fig. 2d), with the widths of a few hundred nanometers. The higher-magnification TEM images, as shown in Fig. 2e and f, exhibit that some nanoparticles were tightly attached to the surface of PPy sheets, indicative of good interface bonding between them. We drew the distribution histogram of the diameters of these nanoparticles (Fig. 2g). The plot exhibits a Gaussian-like distribution, and the average size (d) and standard deviation (σ) were calculated to be 8.9 nm and 3.1 nm, respectively. In addition, the HRTEM image of FPCA displays several lattice fringes with spacings of 0.188, 0.221, and 0.249 nm (Fig. 2h), well agreeing with the (024), (113), and (110) lattice spacings of α-Fe_2O_3, respectively. The result reveals that these nanoparticles consist of α-Fe_2O_3. Moreover, the SAED pattern of FPCA (Fig. 2i) shows ring patterns typical of nanocrystalline materials, and is consistent with the lattice spacings of α-Fe_2O_3.

The crystallographic structure of CDCA, FCA and FPCA was analyzed by XRD. As shown in Fig. 3a, CDCA shows a broad diffraction peak centered at 21.1°, which is ascribed to the (002) plane of amorphous carbon originated from the pyrolysis of cellulose. In addition, a series of new peaks (2θ = 24.3°, 33.3°, 35.8°, 41.0°, 49.6°, 54.2°, 57.7°, 62.6°, 64.1°, 72.1° and 75.7°) can be observed for FCA and well indexed to the standard data of α-Fe_2O_3 (JCPDS card No. 33-0664), further indicating that α-Fe_2O_3 was successfully synthesized and integrated with CDCA. FPCA also shows the characteristic peaks of α-Fe_2O_3. Nevertheless, there is no obvious peak related to the synthetic PPy in the XRD pattern of FPCA, which is possibly attributed to the high intensity of α-Fe_2O_3 peaks shielding the characteristic signals of PPy.

The FTIR spectra of CDCA before and after the HNO_3 treatment are displayed in Fig. 3b. There are no any bands related to cellulose in the spectrum of CDCA, indicating the destruction of cellulose by the pyrolysis. This result is in agreement with the XRD analysis. After the HNO_3 treatment, CDCA displays new vibrational features at 3589 and 1651 cm^{-1} corresponding to the O-H and C=O stretching (Schwanninger, Rodrigues, Pereira & Hinterstoisser, 2004), respectively, which indicates that the oxygen-containing groups were successfully introduced onto the surface of CDCA. After the deposition of α-Fe_2O_3 and PPy, there are several new bands in the spectrum of FPCA. The two broad bands centered at 3355 and 1614 cm^{-1} are mainly attributed to the adsorbed water (Kalutskaya & Gusev, 1980). In addition, the band related to the O-H

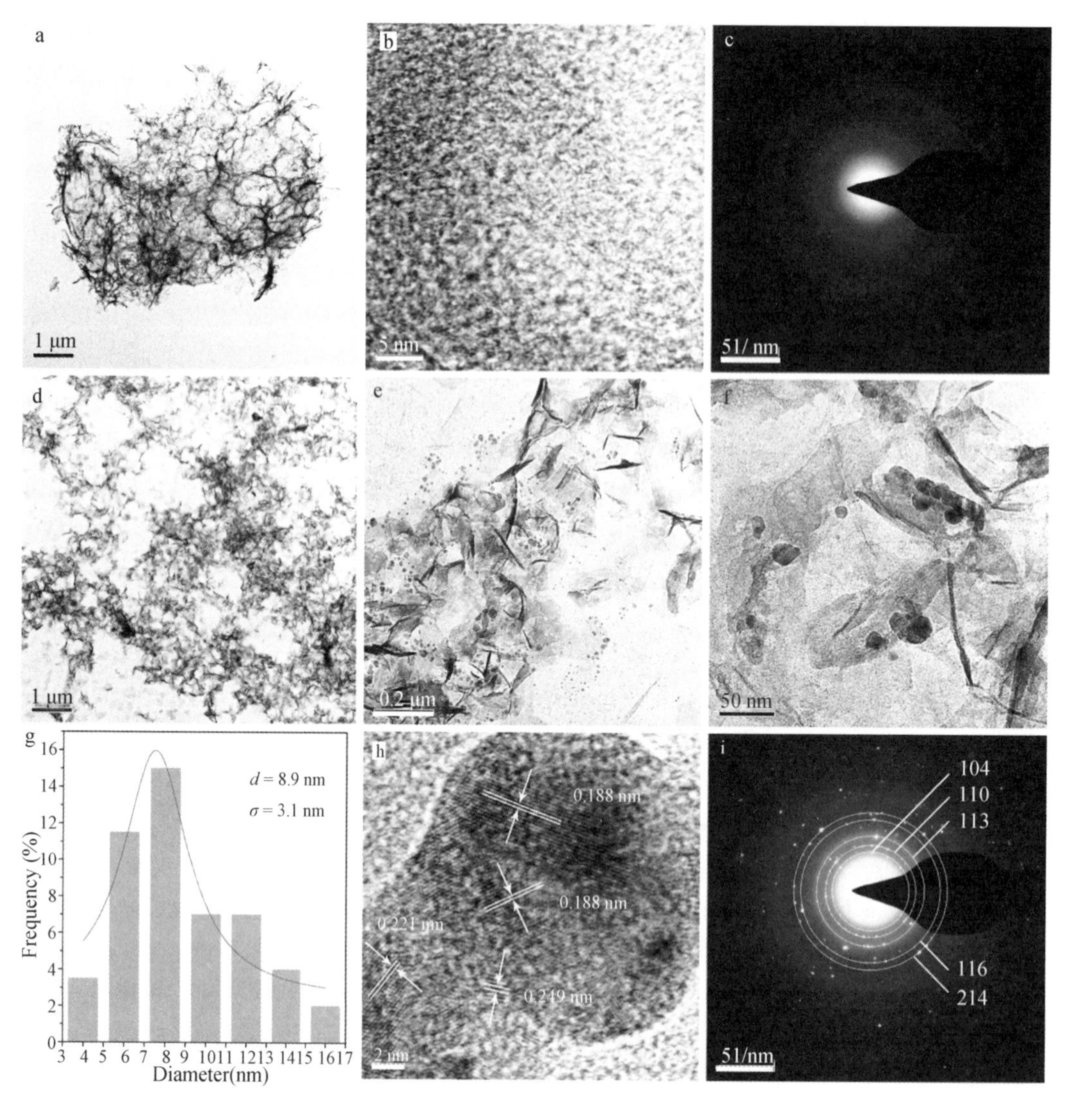

Fig. 2 (a) TEM and (b) HRTEM images and (c) SAED pattern of CDCA, respectively. (d–f) TEM images of FPCA under different magnifications. (g) Diameter distribution of α-Fe_2O_3 in FPCA. (h) HRTEM image and (i) SAED pattern of FPCA, respectively

stretching remarkably shifts to the lower wavenumber as compared to that of the acid-treated CDCA, due to the strong interactions between the hydroxyl groups of CDCA and the nitrogen lone pairs ($\ddot{N}$) of PPy (or the oxygen components of α-Fe_2O_3) (Wang, Zhu, Yang, Zhou, Sun & Tang, 2012). The bands at 1182 and 1036 cm^{-1} belong to the N–C stretching vibration and the C–H in-plane vibration of PPy ring (Xu, Sun & Gao, 2011), respectively. The bands at 810 and 1005 cm^{-1} confirm the presence of polymerized pyrrole (Selvaraj, Alagar & Kumar, 2007). Moreover, we cannot identify the characteristic bands of PPy at around 1540 and 1450 cm^{-1} in the spectrum of FPCA (Xu et al., 2013), which were shielded by the broad band at 1614 cm^{-1}. In addition, the band at 674 cm^{-1} is assigned to the stretching vibration of Fe–O (Zhua, Jianga, Xiaob & Lib, 2010).

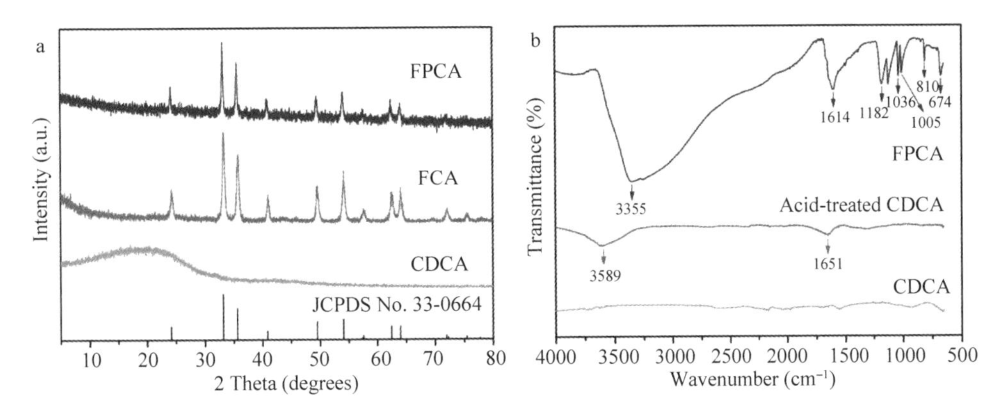

Fig. 3 (a) XRD patterns of CDCA, FCA, and FPCA, respectively. The bottom line is the standard JCPDS card No. 33-0664 for α-Fe_2O_3. (b) FTIR spectra of CDCA, acid-treated CDCA, and FPCA, respectively

A comparison of O 1s XPS core level spectra for the acid-treated CDCA and FPCA is shown in Fig. 4a and b. The acid-treated CDCA exhibits peaks with binding energy at 532. 5 eV and 530. 4 eV, corresponding to the ether (C—OC) and hydroxyl (C—OH) groups (Wang, Bian, Zhou, Tang & Tang, 2013), respectively. This result further indicates that the HNO_3 treatment introduced oxygen-containing groups onto the surface of CDCA. After the deposition of α-Fe_2O_3 and PPy, the peak corresponding to the hydroxyl groups becomes less pronounced. Furthermore, O 1s core level shifts to higher binding energy due to both the shielding effects of α-Fe_2O_3 and PPy and the strong interactions between CDCA and PPy or α-Fe_2O_3, such as the hydrogen bonding between the nitrogen lone pairs ($\ddot{N}$) of PPy and the -OH groups of the underlying CDCA substrate (Ding, Qian, Yu & An, 2010).

Generally, Fe 2p states are used to distinguish the difference between Fe^{2+} and Fe^{3+}. The Fe 2p peaks at binding energies of 711. 3 and 724. 7 eV (Fig. 4c), corresponding to the Fe $2p_{3/2}$ and Fe $2p_{1/2}$ spin-orbit peaks, respectively, are in good agreement with the binding energy values of Fe^{3+} in α-Fe_2O_3 (Descostes, Mercier, Thromat, Beaucaire & Gautier-Soyer, 2000). The satellite peak for Fe $2p_{3/2}$ at 719. 5 eV is another evidence of Fe^{3+} (Liu, Zhou & Zhang, 2011). Regarding the N 1s spectrum (Fig. 4d), the two fitting peaks at 401. 0 and 399. 9 eV are assigned to the positively charged nitrogen (N^+) and the neutral amine-like nitrogen (—NH—) in the pyrrole structure (Chu, Peng, Kilmartin, Bowmaker, Cooney & Travas-Sejdic, 2008), respectively, which further demonstrate the existence of PPy.

3. 2 Schematic illustration for the preparation of FPCA

According to the discussions above, the deposition mechanism of α-Fe_2O_3 and PPy on the surface of CDCA substrate may be speculated as in Fig. 5. CDCA was obtained from the pyrolysis of cellulose aerogels, and maintained the cross-linked 3D network before and after the pyrolysis. However, the pyrolysis damaged the oxygen-containing groups of cellulose. Therefore, CDCA was subjected to the HNO_3 treatment for the introduction of new oxygen-containing groups to immobilize the iron ions and pyrrole monomers. The deposition of α-Fe_2O_3 was carried out by the impregnation-pyrolysis method. Through the impregnation of the acid-treated CDCA in the solution of $Fe(NO_3)_3$ and HCl, the ferric ions were anchored to the surface of carbon skeleton via the strong interactions between the oxygen-containing groups of CDCA and the ferric ions. After the pyrolysis, α-Fe_2O_3 nanoparticles were formed at the original binding sites on the surface of CDCA. The

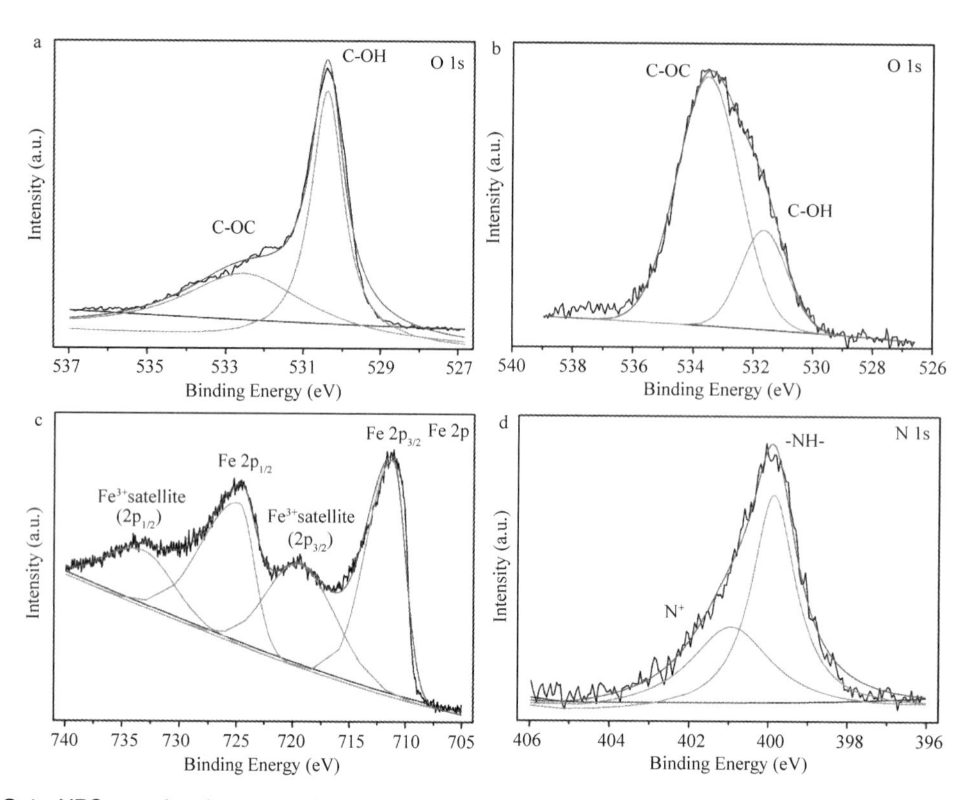

Fig. 4 O 1s XPS core level spectra of (a) the acid-treated CDCA and (b) FPCA. (c) Fe 2p and (d) N 1s XPS core level spectra of FPCA, respectively

deposition of PPy was implemented by the vapor-phase polymerization method. The ferric ions (oxidant) and p-TSA (dopant) were firstly attached to the surface of FCA by the immersion process. Thereafter, the vapor-phase pyrrole monomers were reacted with the oxidant and dopant, resulting in the polymerization of pyrrole on the surface of FCA, and thus PPy was successfully deposited onto the substrate.

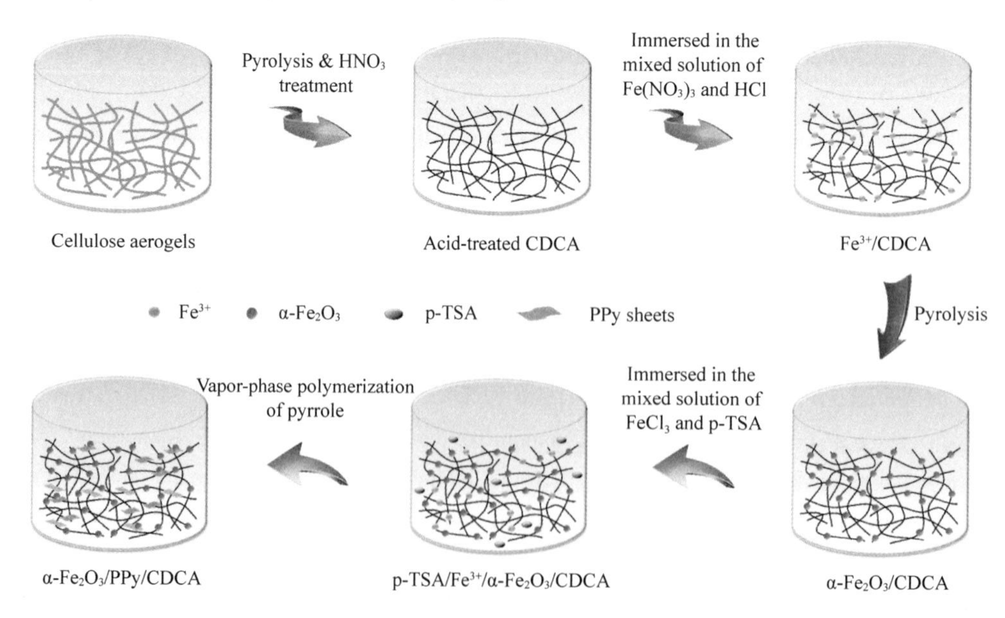

Fig. 5 Schematic illustration of the synthetic procedures for FPCA

3.3 EMI shielding property

A vector network analyzer is an instrument that measures the network parameters of electrical networks. The analyzer is commonly used to measure S-parameters and characterize two-port networks. The EMI shielding property is evaluated by shielding effectiveness expressed in decibels (dB) over the frequency range of 8-13 GHz. A higher decibel level represents less energy transmitted through shielding material. The total shielding effectiveness (SE_{total}) can be expressed as (Liu et al., 2014):

$$SE_{total}(\text{dB}) = 10\log\frac{P_i}{P_t} = SE_A + SE_R + SE_M \quad (1)$$

where P_i and P_t are the incident and transmitted electromagnetic power, respectively. SE_R and SE_A are the shielding effectiveness due to reflection and absorption, respectively. SE_M is multiple reflection effectiveness inside the material, which can be negligible when $SE_{total} > 10$ dB. Besides, SE_R and SE_A can be described as (Al-Saleh & Sundararaj, 2009):

$$SE_R = -10\log(1 - R) \quad (2)$$

$$SE_A = -10\log[T/(1 - R)] \quad (3)$$

where R and T are reflected power and transmitted power, respectively. The scattering parameters (S_{ij} denotes the power transmitted from port i to port j) of the vector network analyzer adopting two-port network test method are related to R and T, that is, $R = |S_{11}|^2 = |S_{22}|^2$ and $T = |S_{12}|^2 = |S_{21}|^2$ (Al-Saleh & Sundararaj, 2012).

For the investigation of the reinforcement effects of α-Fe_2O_3 and Ppy, the EMI shielding property of FPCA was compared to that of FCA, Ppy (synthesized by the vapor-phase polymerization method) and acid-treated CDCA. As shown in Fig. 6a and b, the SE_{total} values of the acid-treated CDCA and Ppy can reach up to 19.3 and 13.6 dB, respectively, which are close to the commercially achievable level (>20 dB). In contrast to the acid-treated CDCA, FCA has a higher SE_{total} value of 29.3 dB (Fig. 6c), indicative of favorable reinforcement effects of α-Fe_2O_3. For FPCA, the SE_{total} value further increases to 39.4 dB (Fig. 6d), which is around 2.0, 2.9 and 1.3 times that of the acid-treated CDCA, Ppy and FCA, respectively. The remarkably increasing trends reflect that the existence of α-Fe_2O_3 and Ppy effectively facilitates the improvement of the EMI shielding property of the composites. In addition, regardless of the error between coaxial and waveguide methods, the SE_{total} value of FPCA is comparable to or even slightly higher than that of some previously reported carbon-based EMI shielding materials with same or greater thickness ($\geq$ 2 mm), such as exfoliated few-layer graphene (12 dB) (Song et al., 2013), carbon fiber/silica (12.4 dB) (Cao, Song, Hou, Wen & Yuan, 2010), multiwalled carbon nanotubes/ortland cement (28 dB) (Singh, Gupta, Mishra, Chandra, Mathur & Dhawan, 2013), biomorphic porous carbon (40 dB) (Liu, Gu, Zhang, Miyamoto, Chen & Zhang, 2012), and highly ordered porous carbon (50 dB) (Song et al., 2014). EMI shielding mechanism was estimated by comparing the contribution of reflection (SE_R) and absorption (SE_A) to the overall EMI SE_{total}. Obviously, all the samples display absorption-dominant mechanism because their SE_A values are much higher than the corresponding SE_R values. Also, the overall growth trends of SE_{total} value for all the samples over whole frequency have been predicted theoretically and confirmed experimentally for absorption-dominant materials (Zhang, Zheng, Yan, Jiang & Yu, 2012). An absorption-dominant shielding mechanism is undoubtedly beneficial to alleviate secondary radiation, which is considered as an attractive alternative for the

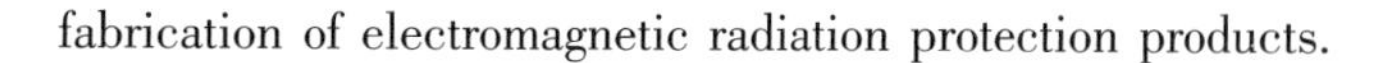

fabrication of electromagnetic radiation protection products.

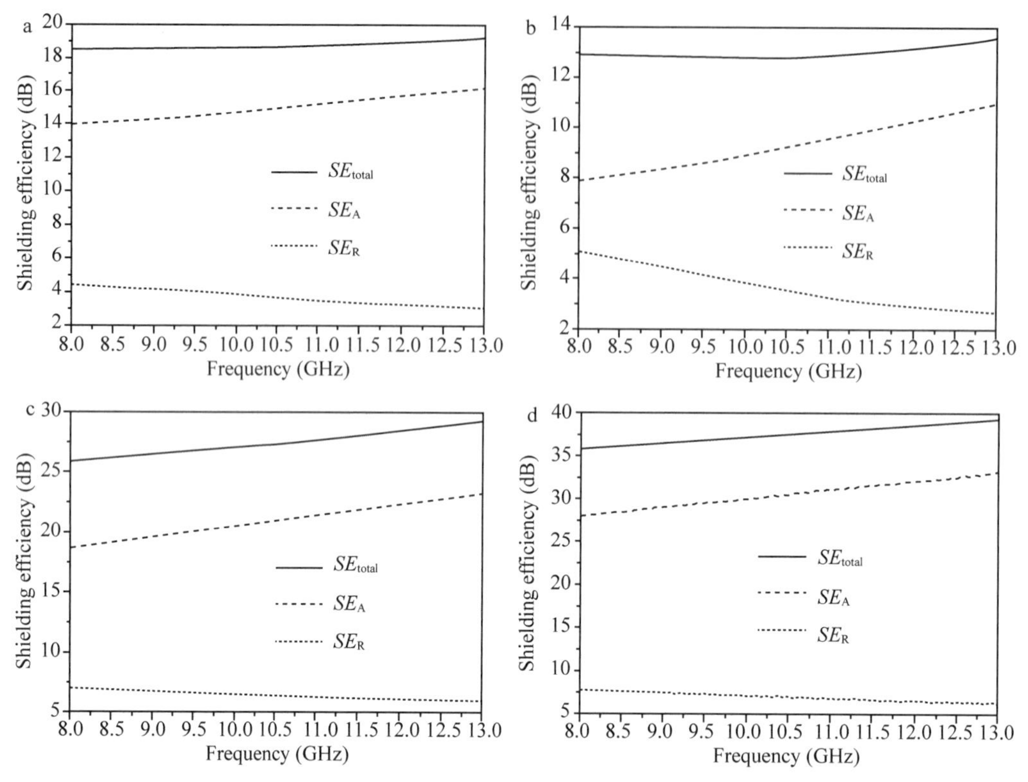

Fig. 6 SE_{total}, SE_A, and SE_R values of (a) the acid-treated CDCA, (b) Ppy, (c) FCA, and (d) FPCA as a function of frequency, respectively

Table 1 Some carbon-based materials for EMI shielding applications

Materials	Method	Best SE_{total} (dB)	Thickness (mm)	Frequency range (GHz)	Refs.
Exfoliated few-layer graphene	Coaxial (85% wax)	12	2.5	8.2-12.4	Song et al., 2013
Carbon fiber/silica	Waveguide	12.4	2.5	8.2-12.4	Cao et al., 2010
Multiwalled carbon nanotubes/portland cement	Waveguide	28	2	8.2-12.4	Singh et al., 2013
Biomorphic porous carbon	Waveguide	40	4	8.2-12.4	Liu et al., 2012
Highly ordered porous carbon	Coaxial (80% wax)	50	2	2-18	Song et al., 2014
Acid-treated CDCA	Coaxial (50% wax)	19.3	2	8-13	This work
PPy	Coaxial (50% wax)	13.6	2	8-13	This work
FCA	Coaxial (50% wax)	29.3	2	8-13	This work
FPCA	Coaxial (50% wax)	39.4	2	8-13	This work

4 Conclusions

A simple low-cost process including pyrolysis and vapor phase polymerization was employed to combine CDCA substrate with α-Fe_2O_3 and PPy. CDCA was fabricated from the pyrolysis of cellulose aerogels, and underwent the treatment of HNO_3 to introduce new oxygen-containing groups onto the surface of CDCA to immobilize the ferric ions and pyrrole monomers. The SEM and TEM observations show that the PPy sheets and α-Fe_2O_3 nanoparticles were evenly deposited onto CDCA. The strong interactions between the oxygen-containing groups of CDCA and the nitrogen lone pairs ($\ddot{N}$) of PPy (or the oxygen components of α-Fe_2O_3) were confirmed by FTIR and XPS analysis. When evaluated as EMI shielding materials, FPCA has the highest SE_{total} value of 39.4 dB, as compared to 19.3 dB of the acid-treated CDCA, 13.6 dB of the as-prepared PPy, and 29.3 dB of FCA, which indicates the good reinforcement effects of α-Fe_2O_3 and PPy. FPCA has the absorption-dominant EMI shielding mechanism, contributing to alleviating secondary radiation, which is regarded as a fascinating alternative for the development of electromagnetic radiation protection products. Moreover, this eco-friendly porous CDCA substrate is expected to integrate with more types of functional substances to develop novel functional composites.

Acknowledgments

This study was supported by the National Natural Science Foundation of China (grant nos. 31270590 and 31470584) and the Fundamental Research Funds for the Central Universities (grant nos. 2572016AB22).

References

[1] Al-Saleh M H, Sundararaj U. Electromagnetic interference shielding mechanisms of CNT/polymer composites[J]. Carbon, 2009, 47(7): 1738-1746.

[2] Al-Saleh M H, Sundararaj U. X-band EMI shielding mechanisms and shielding effectiveness of high structure carbon black/polypropylene composites[J]. Journal of Physics D Applied Physics, 2013, 46(3): 035304.

[3] Al-Muhtaseb S A, Ritter J A. Preparation and properties of resorcinol-formaldehyde organic and carbon gels[J]. Advanced materials, 2003, 15(2): 101-114.

[4] Balmori A. Electromagnetic pollution from phone masts. Effects on wildlife[J]. Pathophysiology, 2009, 16(2-3): 191-199.

[5] Bryning M B, Milkie D E, Islam M F, et al. Carbon nanotube aerogels[J]. Advanced materials, 2007, 19(5): 661-664.

[6] Cai H, Sharma S, Liu W, et al. Aerogel microspheres from natural cellulose nanofibrils and their application as cell culture scaffold[J]. Biomacromolecules, 2014, 15(7): 2540-2547.

[7] Cao M S, Song W L, Hou Z L, et al. The effects of temperature and frequency on the dielectric properties, electromagnetic interference shielding and microwave-absorption of short carbon fiber/silica composites[J]. Carbon, 2010, 48(3): 788-796.

[8] Chu S Y, Peng H, Kilmartin P A, et al. Effect of deposition current density on the linear actuation behaviour of PPy (CF_3SO_3) films[J]. Current Applied Physics, 2008, 8(3-4): 324-327.

[9] DD L Chung. Materials for electromagnetic interferenceshielding[J]. Journal of Materials Engineering & Performance, 2000, 9(3): 350-354.

[10] Chung D. Electromagnetic interference shielding effectiveness of carbon materials[J]. Carbon, 2001, 39(2): 279-

285.

[11]Ci L, Wei B, Xu C, et al. Crystallization behavior of the amorphous carbon nanotubes prepared by the CVD method [J]. Journal of Crystal Growth, 2001, 233(4): 823-828.

[12]Descostes M, Mercier F, Thromat N, et al. Use of XPS in the determination of chemical environment and oxidation state of iron and sulfur samples: constitution of a data basis in binding energies for Fe and S reference compounds and applications to the evidence of surface species of an oxidized pyr[J]. Applied Surface Science, 2000, 165(4): 288-302.

[13]Ding C, Qian X, Yu G, et al. Dopant effect and characterization of polypyrrole-cellulose composites prepared by in situ polymerization process[J]. Cellulose, 2010, 17(6): 1067-1077.

[14]Friedel B, Greulich-Weber S. Preparation of Monodisperse, Submicrometer Carbon Spheres by Pyrolysis of Melamine-Formaldehyde Resin[J]. Small, 2010, 2(7): 859-863.

[15] Gómez-Romero P, Chojak M, Cuentas-Gallegos K, et al. Hybrid organic-inorganic nanocomposite materials for application in solid state electrochemical supercapacitors[J]. Electrochemistry Communications, 2003, 5(2): 149-153.

[16]Han F, Li D, Li W C, et al. Nanoengineered polypyrrole-coated Fe2O3@ C multifunctional composites with an improved cycle stability as lithium-ion anodes[J]. Advanced Functional Materials, 2013, 23(13): 1692-1700.

[17]Han Z J, Tay B K, Shakerzadeh M, et al. Superhydrophobic amorphous carbon/carbon nanotube nanocomposites[J]. Applied Physics Letters, 2009, 94(22): 223106.

[18] Hao P, Zhao Z, Tian J, et al. Hierarchical porous carbon aerogel derived from bagasse for high performance supercapacitor electrode[J]. Nanoscale, 2014, 6(20): 12120-12129.

[19]Kalutskaya E P, Gusev S S. An infrared spectroscopic investigation of the hydration of cellulose[J]. Polymer Science USSR, 1980, 22(3): 550-556.

[20]Kango S, Kalia S, Celli A, et al. Surface modification of inorganic nanoparticles for development of organic-inorganic nanocomposites—A review[J]. Progress in Polymer Science, 2013, 38(8): 1232-1261.

[21]Kickelbick G. Concepts for the incorporation of inorganic building blocks into organic polymers on a nanoscale[J]. Progress in polymer science, 2003, 28(1): 83-114.

[22]Klemm D, Heublein B, Fink H P, et al. Cellulose: fascinating biopolymer and sustainable raw material[J]. Angewandte chemie international edition, 2005, 44(22): 3358-3393.

[23]Leitgeb N, Schröttner J, Böhm M. Does "electromagnetic pollution" cause illness? [J]. Wiener Medizinische Wochenschrift, 2005, 155(9): 237-241.

[24]Li X, Yang S, Sun J, et al. Tungsten oxide nanowire-reducedgraphene oxide aerogel for high-efficiency visible light photocatalysis[J]. Carbon, 2014, 78: 38-48.

[25]Liang H W, Wu Z Y, Chen L F, et al. Bacterial cellulose derived nitrogen-doped carbonnanofiber aerogel: an efficient metal-free oxygen reduction electrocatalyst for zinc-air battery[J]. Nano Energy, 2015, 11: 366-376.

[26] Liebner F, Potthast A, Rosenau T, et al. Cellulose aerogels: Highly porous, ultra-lightweight materials[J]. Holzforschung, 2008, 62(2): 129-135.

[27]Liu Q, Gu J, Zhang W, et al. Biomorphic porous graphitic carbon for electromagnetic interference shielding[J]. Journal of Materials Chemistry, 2012, 22(39): 21183-21188.

[28]Liu S L, Jin P, et al. In situ synthesis of plate-like Fe_2O_3 nanoparticles in porous cellulose films with obvious magnetic anisotropy[J]. Cellulose, 2011, 18(3): 663-673.

[29]Liu X, Yin X, Kong L, et al. Fabrication and electromagnetic interference shielding effectiveness of carbon nanotube reinforced carbon fiber/pyrolytic carbon composites[J]. Carbon, 2014, 68: 501-510.

[30]Mi X, Huang G, Xie W, et al. Preparation of graphene oxide aerogel and its adsorption for Cu^{2+} ions[J]. Carbon, 2012, 50(13): 4856-4864.

[31]Moreno-Castilla C, Maldonado-Hódar F J. Carbon aerogels for catalysis applications: An overview[J]. Carbon, 2005,

43(3): 455-465.

[32]Olsson R T, Samir M A S A, Salazar-Alvarez G, et al. Making flexible magnetic aerogels and stiff magneticnanopaper using cellulose nanofibrils as templates[J]. Nature nanotechnology, 2010, 5(8): 584-588.

[33]Pääkkö M, Vapaavuori J, Silvennoinen R, et al. Long and entangled native cellulose I nanofibers allow flexible aerogels and hierarchically porous templates for functionalities[J]. Soft Matter, 2008, 4(12): 2492-2499.

[34]Sanchez, Clément, Belleville P, Popall M , et al. Applications of advanced hybrid organic-inorganic nanomaterials: from laboratory to market[J]. Chemical Society Reviews, 2011, 40(2): 696-753.

[35] Schumacher C, Gonzalez J, Pérez-Mendoza M, et al. Design of hybrid organic/inorganic adsorbents based on periodicmesoporous silica[J]. Industrial & engineering chemistry research, 2006, 45(16): 5586-5597.

[36]Schwanninger M, Rodrigues J C, Pereira H, et al. Effects of short-time vibratory ball milling on the shape of FT-IR spectra of wood and cellulose[J]. Vibrational spectroscopy, 2004, 36(1): 23-40.

[37]Sehaqui H, Zhou Q, Berglund L A. High-porosity aerogels of high specific surface area prepared from nanofibrillated cellulose (NFC)[J]. Composites science and technology, 2011, 71(13): 1593-1599.

[38] Selvaraj V, Alagar M, Kumar K S. Synthesis and characterization of metal nanoparticles-decorated PPY-CNT composite and their electrocatalytic oxidation of formic acid and formaldehyde for fuel cell applications[J]. Applied Catalysis B Environmental, 2007, 75(1-2): 129-138.

[39]Shen B, Zhai W, Tao M, et al. Lightweight, multifunctional polyetherimide/graphene@ Fe_3O_4 composite foams for shielding of electromagnetic pollution. [J]. Applied Materials & Interfaces, 2013, 5(21): 11383-11391.

[40] Singh A P, Gupta B K, Mishra M, et al. Multiwalled carbon nanotube/cement composites with exceptional electromagnetic interference shielding properties[J]. Carbon, 2013, 56: 86-96.

[41]Song W L, Cao M S, Fan L Z, et al. Highly ordered porous carbon/wax composites for effective electromagnetic attenuation andshielding[J]. Carbon, 2014, 77: 130-142.

[42]Song W L, Cao M S, Lu MM, et al. Alignment of graphene sheets in wax composites for electromagnetic interference shielding improvement[J]. Nanotechnology, 2013, 24(11): 115708.

[43] Sui Z, Meng Q, Zhang X, et al. Green synthesis of carbon nanotube-graphene hybrid aerogels and their use as versatile agents for water purification[J]. Journal of Materials Chemistry, 2012, 22(18): 8767-8771.

[44]Sun H, Xu Z, Gao C. Multifunctional, ultra-flyweight, synergistically assembled carbon aerogels[J]. Advanced materials, 2013, 25(18): 2554-2560.

[45]Wan CC. EmbeddingZnO nanorods into porous cellulose aerogels via a facile one-step low-temperature hydrothermal method[J]. Mater Design, 2015, 2015, 83: 620-625.

[46] Wan C, Li J. Facile synthesis of well-dispersedsuperparamagnetic γ-Fe_2O_3 nanoparticles encapsulated in three-dimensional architectures of cellulose aerogels and their applications for Cr (VI) removal from contaminated water[J]. ACS Sustainable Chemistry & Engineering, 2015, 3(9): 2142-2152.

[47] Wan C, Li J. Cellulose aerogels functionalized withpolypyrrole and silver nanoparticles: In-situ synthesis, characterization and antibacterial activity[J]. Carbohydrate polymers, 2016, 146: 362-367.

[48] Wan C, Li J. Graphene oxide/cellulose aerogels nanocomposite: Preparation, pyrolysis, and application for electromagnetic interference shielding[J]. Carbohydrate polymers, 2016, 150: 172-179.

[49]Wan C, Lu Y, Jiao Y, et al. Cellulose aerogels from cellulose-NaOH/PEG solution and comparison with different cellulose contents[J]. Materials Science and Technology, 2015, 31(9): 1096-1102.

[50]Wan C, Lu Y, Jiao Y, et al. Fabrication of hydrophobic, electrically conductive and flame-resistant carbon aerogels by pyrolysis of regenerated celluloseaerogels[J]. Carbohydrate polymers, 2015, 118: 115-118.

[51]Wang B, Li X, Luo B, et al. Pyrolyzed bacterial cellulose: a versatile support for lithium ion battery anode materials [J]. Small, 2013, 9(14): 2399-2404.

[52]Wang H, Bian L, Zhou P, et al. Core-sheath structured bacterial cellulose/polypyrrole nanocomposites with excellent conductivity as supercapacitors[J]. Journal of Materials Chemistry A, 2013, 1(3): 578-584.

[53]Wang H, Zhu E, Yang J, et al. Bacterial cellulosenanofiber-supported polyaniline nanocomposites with flake-shaped morphology as supercapacitor electrodes[J]. The Journal of Physical Chemistry C, 2012, 116(24): 13013-13019.

[54]Wang Y, Jing X. Intrinsically conducting polymers for electromagnetic interferenceshielding[J]. Polymers for advanced technologies, 2005, 16(4): 344-351.

[55]Wight A P, Davis M E. Design and preparation of organic-inorganic hybridcatalysts[J]. Chemical reviews, 2002, 102(10): 3589-3614.

[56]Winther-Jensen B, Chen J, West K, et al. Vapor Phase Polymerization of Pyrrole and Thiophene Using Iron(Ⅲ) Sulfonates as Oxidizing Agents[J]. Macromolecules, 2004, 37(16): 5930-5935.

[57]Wu D, Fu R. Synthesis of organic and carbon aerogels from phenol-furfural by two-steppolymerization[J]. Microporous and mesoporous materials, 2006, 96(1-3): 115-120.

[58]Wu D, Fu R, Sun Z, et al. Low-density organic and carbon aerogels from the sol-gel polymerization of phenol withformaldehyde[J]. Journal of Non-Crystalline Solids, 2005, 351(10-11): 915-921.

[59]Wu Z Y, Li C, Liang H W, et al. Ultralight, flexible, and fire-resistant carbon nanofiber aerogels from bacterial cellulose[J]. Angewandte Chemie, 2013, 125(10): 2997-3001.

[60]Xiao H M, Zhang W D, Fu S Y. One-step synthesis, electromagnetic and microwave absorbing properties of α-FeOOH/polypyrrole nanocomposites[J]. Composites science and technology, 2010, 70(6): 909-915.

[61]Xu C, Sun J, Gao L. Synthesis of novel hierarchical graphene/polypyrrole nanosheet composites and their superior electrochemical performance[J]. Journal of Materials Chemistry, 2011, 21(30): 11253-11258.

[62]Xu J, Zhu L, Bai Z, et al. Conductive polypyrrole-bacterial cellulose nanocomposite membranes as flexible supercapacitor electrode[J]. Organic Electronics, 2013, 14(12): 3331-3338.

[63]Zhang H B, Zheng W G, Yan Q, et al. The effect of surface chemistry of graphene on rheological and electrical properties of polymethylmethacrylate composites[J]. Carbon, 2012, 50(14): 5117-5125.

[64]Zhang X, Sui Z, Xu B, et al. Mechanically strong and highly conductive graphene aerogel and its use as electrodes for electrochemical power sources[J]. Journal of Materials Chemistry, 2011, 21(18): 6494-6497.

[65]Zhua H Y, Jianga R, Xiaob L, et al. A novel magnetically separable-Fe2O3/crosslinked chitosan adsorbent: Preparation, characterization and adsorption application for removal of hazardous azo dye[J]. Journal of Hazardous Materials, 2010, 179: 251-257.

中文题目：α-Fe_2O_3 和聚吡咯功能化纤维素衍生碳气凝胶的合成及其电磁屏蔽性能

作者：万才超，李坚

摘要：以环保纤维素衍生碳气凝胶(CDCA)为多孔基质，通过热解和气相聚合与 α-Fe_2O_3 和聚吡咯(PPy)相结合。扫描电镜(SEM)和透射电镜(TEM)观察表明，皱缩的 PPy 片和 α-Fe_2O_3 纳米粒子在 CDCA 中分散良好。FTIR 和 XPS 分析表明基体与纳米材料之间存在很强的相互作用(氢键)。α-Fe_2O_3/PPy/CDCA(FPCA)复合材料作为电磁(EMI)屏蔽材料的总屏蔽效果(SE_{total})最高为 39.4 dB，分别是酸处理 CDCA(19.3 dB)、PPy (13.6 dB)和 α-Fe_2O_3/CDCA (29.3 dB)的 2.0、2.9 和 1.3 倍。此外，由于吸收导致的屏蔽效率占 FPCA 的 SE_{total} 的 78.2%~84.2%，这表明以吸收为主的屏蔽机制有助于减轻二次辐射，这些特点使该复合材料成为电磁屏蔽的极具优势的备选材料。

关键词：纤维素；碳气凝胶；α-Fe_2O_3；聚吡咯；电磁屏蔽；复合材料

Facile Fabrication of Nanofibrillated Chitin/Ag_2O Heterostructured Aerogels with High Iodine Capture Efficiency*

Runan Gao, Yun Lu, Shaoliang Xiao, Jian Li

Abstract: Nanofibrillated chitin/Ag_2O aerogels were fabricated for radioiodine removal. Chitin was first fabricated into nanofibers with abundant acetyl amino groups (—$NHCOCH_3$) on the surface. Then, highly porous chitin nanofiber (ChNF) aerogels were obtained via freeze-drying. The ChNF aerogels exhibited a low bulk density of 2.19 mg · cm^{-3} and a high specific surface area of 179.71 m^2 · g^{-1}. Ag_2O nanoparticles were evenly anchored on the surfaces of ChNF scaffolds via strong interactions with —$NHCOCH_3$ groups, subsequently yielding Ag_2O@ ChNF heterostructured aerogels. The composites were used as efficient absorbents to remove radioiodine anions from water and capture a high amount of I_2 vapor in the forms of AgI and iodine molecules. The adsorption capacity of the composite monoliths can reach up to 2.81 mmol · g^{-1} of I^- anions. The high adsorbability of the composite monolithic aerogel signifies its potential applications in radioactive waste disposal.

Keywords: Chitin nanofibers; Ag_2O nanoparticles; Aerogel; Radioactive; Adsorption

1 Introduction

Aerogels are low-density solid materials and are characterized by a highly accessible mesoporous network that is composed of three-dimensionally interconnected particles[1]. Aerogels integrate unusual properties, such as low density, high specific surface area, low heat conductivity, and high transparency, that are individually present in other materials. Given these outstanding characteristics, aerogels have been utilized as Cherenkov detectors, catalyst support media, and absorbers[2,3]. Organic aerogels appeared as a compelling alternative to inorganic aerogels after Pekala et al.[4] conducted fundamental research on phenolic-resin-based aerogels. Resorcinol/formaldehyde aerogel is the first developed and the most investigated organic aerogel[5]. Many other reactant systems, such as melamine-formaldehyde[6], resorcinol-furfural[7], cresol-formaldehyde[8], phenol-furfural[9], polyurethane[10], polyisocyanate, and polyolefin[11,12], have been developed in later works by other researchers. Recently, cellulose or chitin have received increased attention as aerogel building blocks because of their outstanding high stiffness, low density, high aspect ratio, and large specific surface area[13-17]. Numerous reports on nanofabrillated-cellulose-based aerogels have been widely published[18]. Similar studies on chitin aerogels, however, have just come into force. Yun Lu et al. successfully

* 本文摘自 Scientific Reports, 2017, 7: 4303.

fabricated chitin nanofibers for assembly into spongy foam[19]. Bo Duan et al. reported the use of chitin sponge materials for oil-water separation[20]. Chitin-based aerogels have also been developed and reported as efficient base catalysts[21].

Ag_2O has received intensive research attention because of its high reactivity with I to form insoluble AgI. This Ag_2O characteristic can be applied in radioactive iodine waste disposal. Among a variety of harmful radioactive isotopes derived from the fission of uranium 235, I[129] and I[131] are major factors that increase morbidity rate. Directly immobilizing radioactive iodine with Ag_2O, however, is unfeasible as removal capacity and dynamics depend on the specific surface area of Ag_2O. Although nanosized Ag_2O particles with large specific surface areas have high capacity for radioiodine capture, the separation and recovery of the nanoparticles remain problematic. A leading research study has anchored Ag_2O nanoparticles on titanate-based nano-adsorbents and indicated that the key countermeasure to prevent agglomeration is the even dispersion of fine inorganic nanoparticle granules on a carrier with a high specific area[22].

In the present work, we introduced porous ChNF aerogels as supports for Ag_2O-nanoparticle decoration to develop a high-efficiency, sustainable absorbent. One-dimensional (1D) nanofibers were isolated from chitin, an environmentally friendly bio-resource with superior mechanical properties, such as strength and flexibility, which are comparable with those of collagen. These properties make ChNF a favorable scaffold for inorganic nanoparticle decoration. Specific acetyl amino groups (—$NHCOCH_3$) on ChNF scaffolds facilitate the intensive chelation of Ag^+ on the ChNF surface through the donation of a lone electron pair by acetyl amino groups (—$NHCOCH_3$)[23]. Hydroxyl groups on fibers also provide a unique platform for surface modification. After preparation into ChNF aerogels, the 3D network of entangled nanofibers provides a large specific area for the even dispersion of fine Ag_2O nanoparticles without agglomeration. When used as adsorbents, numerous mesopores allow the access of radioactive ions in water to Ag_2O nanoparticles, thus guaranteeing the high adsorbability of radioactive iodine anions.

To develop a high-efficiency iodine absorbent, we prepared Ag_2O-anchored ChNF (Ag_2O@ ChNF) hetero-structured aerogel via a facile process. First, we prepared ChNF nanofiber aerogels through chemical pretreatment combined with high-intensity ultrasonication and freeze-drying. We then isolated purified chitin into nanofibers with individual diameters of approximately 50 to 300 nm. We replaced water with tert-butanol to induce the crosslinking of chitin nanofibers to form gels. Finally, we obtained monolithic ChNF aerogels through freeze-drying. This monolithic material showed low bulk density and high specific surface area. Ag_2O nanoparticles were anchored on chitin nanofibers via simple chemical reactions in solution. The size of Ag_2O nanoparticles could be controlled by adjusting the initial concentration of ammoniated silver solution (Tollen's reagent). Ultrasonication treatment was also introduced to prevent the agglomeration of nanoparticles. The Ag_2O-loaded monoliths efficiently remove I^- from water and capture a significant amount of I_2vapor for effective fixation in the forms of AgI and iodine molecules.

2 Results and Discussion

Acid pretreatment partially deacetylated ChNFs. The deacetylation degree (DD) of ChNFs was 32.8%, as evaluated from ^{13}C solid-state NMR (see Supplementary Fig. S1). Chemical pretreatments slackened tightly bonded chitin fibril bundles in addition to eliminating proteins, mineral salts, and lipids from chitin samples. The microstructure of the chemically purified chitin is presented in Fig. 1(a). After pretreatment, interlaced chitin fibril bundles are clearly visible in the scanning electron microscopy (SEM) image. Subsequently, to obtain ChNFs high-intensity ultrasonic treatment was applied. During ultrasonication, an active zone with a high concentration of cavities was created. The sonication probe of the ultrasonicator worked as an energy transfer device that vibrated the liquid to form hollow bubbles, or cavities[24-26]. Cavitation gradually disassembled fibril bundles into nanofibers because the intensive collapse of cavities on the surfaces of the fibril bundles effectively split fibers along the axial direction. Thus, the final nanofibers were obtained. Although ultrasonication treatment can avoid the use of chemicals and has a smaller impact on the environment, it can broaden the diameter distribution of nanofibers. Fig. 1(b) presents the morphology of ChNFs after 30 min of ultrasonication. The Fig. shows long, straight individual nanofibers with individual diameters of 50 to 300 nm. The mean diameter of ChNFs was 153 nm. More detailed information about the morphology and diameter distribution of ChNFs is presented in Supplementary Fig. S2. The individual nanofibers were then assembled into ultralight aerogels. After replacing the water in the sonication suspension with tert-butanol, the viscosity of the suspension increased due to the entanglement of ChNFs and a pulpy gel body was obtained (see Supplementary Fig. S3). After freeze-drying, monolithic ChNF aerogels were obtained, as shown in Fig. 1(c). The ChNF aerogels that were prepared in the present study had a very low density of approximately 2.19 $mg \cdot cm^{-3}$. This monolithic material did not obviously shrink during freeze-drying. Moreover, its shape corresponded with the shape of the cylindrical plastic molds, indicating the absence of structural collapse. Fig. 1(d) shows the microstructure of the ChNF aerogels. The three-dimensionally interconnected nanofiber skeleton of nanofibrillated chitin is clearly visible in the figure.

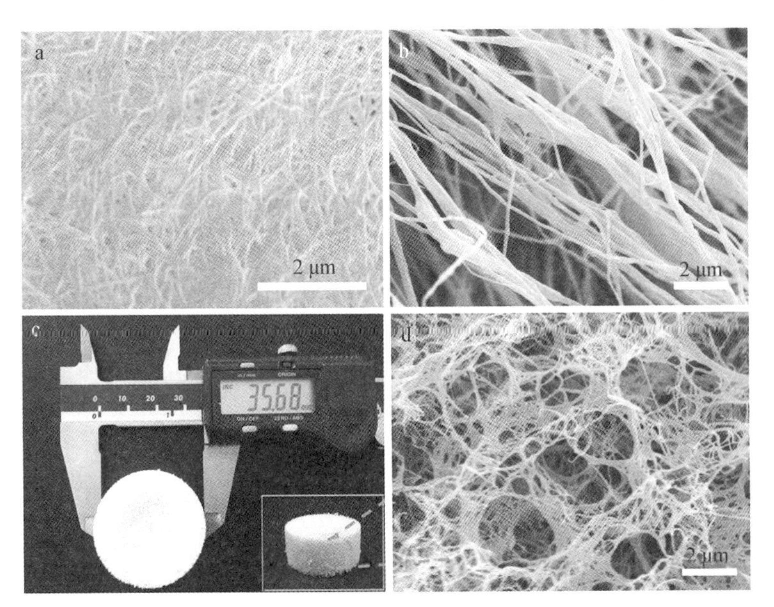

Fig. 1 SEM images of (a) chitin after chemical pretreatment and (b) ChNFs after 30 min of ultrasonication. (c) Macrogragh of the ChNF aerogel. (d) Microscopic structure of the ChNF aerogel

N_2 adsorption-desorption under liquid nitrogen flow provides valuable structural information about porous materials. ChNF aerogels have a large specific surface area of 179.71 $m^2 \cdot g^{-1}$ and high porosity of

99.85%. The average pore diameter and the total pore volume of ChNF aerogels were 16.37 nm and 0.76 $cm^3 \cdot g^{-1}$, respectively. The porosity of aerogels was calculated by:

$$P_a = \left(1 - \frac{\rho_a}{\rho_c}\right) \times 100$$

where ρ_a is the density of the ChNF aerogel and ρ_c is the density of bulk chitin ($1.425\ g \cdot cm^{-3}$)[27].

Fig. 2(a) and (b) show the nitrogen adsorption-desorption isotherms and Barrett-Joyner-Halenda (BJH) pore size distribution of ChNF aerogels. According to IUPAC classification, ChNF aerogels exhibited type IV adsorption isotherms. The adsorption branch nearly overlapped with the desorption branch when P/P_0 was less than 0.4. The presence of mesopores was demonstrated when a type H3 hysteresis loop appeared when P/P_0 exceeded 0.4[28]. The main pore diameter distribution of ChNF aerogels was 2-20 nm, as shown in Fig. 2(b), which further confirmed the mesoporous structure of the samples.

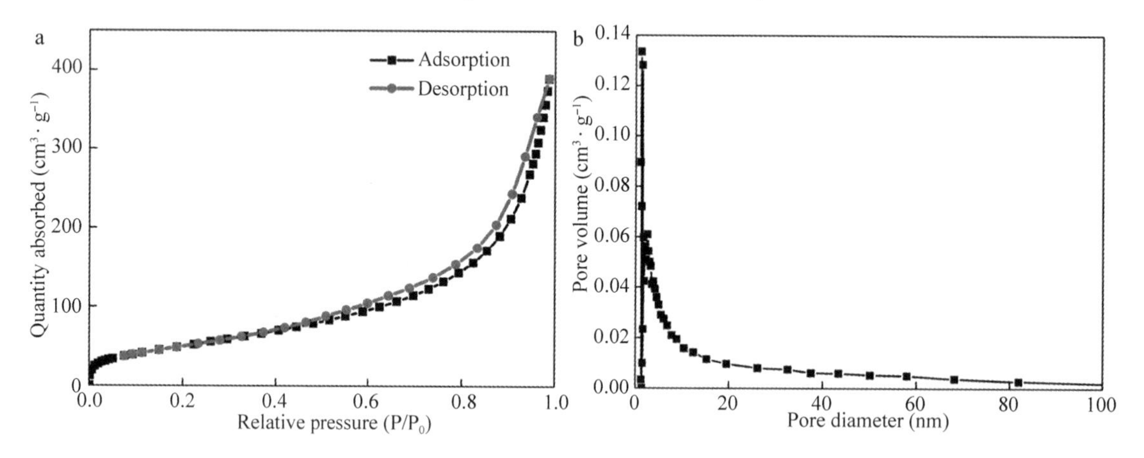

Fig. 2 Adsorption-desorption isotherms (a) and pore size distribution (b) of ChNF aerogels

Ag_2O@ChNFs with different Ag_2O loading concentrations were investigated via N_2 adsorption-desorption. The bulk density, specific surface area, and pore volume of pristine and modified aerogels are listed in Table 1. Specific surface area and pore volume gradually decreased as Ag_2O concentration increased. This behavior indicated that porosity decreases in accordance with the thickening of the ChNF scaffold after anchoring inorganic nanoparticles.

Table 1 Bulk densities, specific surface areas, and porosities of pristine and Ag_2O-anchored aerogels

Ag_2O(%)	$\rho_{acrogel}$($mg \cdot cm^{-3}$)	BET specific surface area ($m^2 \cdot g^{-1}$)	Pore volume($cm^3 \cdot g^{-1}$)
0	2.19	171.91	0.76
31	4.20	131.96	0.66
56	4.46	92.58	0.33
122	5.31	74.31	0.27

Ag_2O@ChNF aerogels were obtained after immersing ChNF networks in ammoniated silver solution followed by freeze-drying. Prior to freezing, the samples were treated with ultrasonication again to ensure homogeneous dispersion and to prevent the agglomeration of fine nano-granules. Ag_2O nanoparticles formed and anchored on the ChNF network through following reactions:

X-ray diffraction (XRD) and FTIR analyses provided idence for the formation of Ag_2O nanoparticles via the above reactions. XRD data are presented in Fig. 3(a). In this figure, the spectral line of pristine monolithic ChNF materials is at the bottom and the spectra of aerogels with different loading concentrations are listed upward in succession. The diffraction peaks of monolithic ChNF aerogels appeared at 9.4°, 12.8°, 19.3°, 20.5°, 23.5°, and 26.3°, which corresponded to the planes of (020), (101), (110), (120), (130), and (013), respectively. The crystallinity of ChNFs was 84.1% and was estimated by crystalline index (CrI%) expressed as $CrI_{110} = (I_{110} - I_{am}) \times 100/I_{110}$. Characteristic Ag_2O peaks appeared after the precipitation of precursors on ChNF scaffolds. The peaks at 32.8° and 38.1° corresponded to the (111) and (200) planes of Ag_2O. Ag_2O nanoparticles covered the ChNF network as the concentration of starting ammoniated silver solution increased. This process was indicated in the XRD pattern as the gradual decay or even the overlapping of characteristic chitin peaks.

The interaction between Ag_2O nanoparticles and ChNF scaffolds was further investigated via FTIR analysis, as shown in Fig. 3(b). For ChNF, the peak at the frequency of 1680 cm^{-1} was characterized as C=O stretching, which proved the presence of acetyl amino groups. After loading, a negative shift to 1660 cm^{-1} occurred. In addition, a characteristic peak at 1558 cm^{-1} that was assigned to C-N stretching and N-H deformation broadened and decreased. This pattern indicated a strong interaction between acetyl amino groups and Ag^+. The adsorption peak at 3260 cm^{-1} was attributed to the stretching vibrations of intermolecular hydrogen bonds in chitin, where C(6)OH groups are hydrogen-bonded to N-H in adjacent molecular chains. After Ag_2O deposition, a negative shift to 3244 cm^{-1} occurred and the peak was broadened. This result demonstrated that hydrogen bonds formed between the hydroxyl groups of chitin and Ag_2O nanoparticles, which increased the attraction of Ag^+ to ChNF scaffolds.

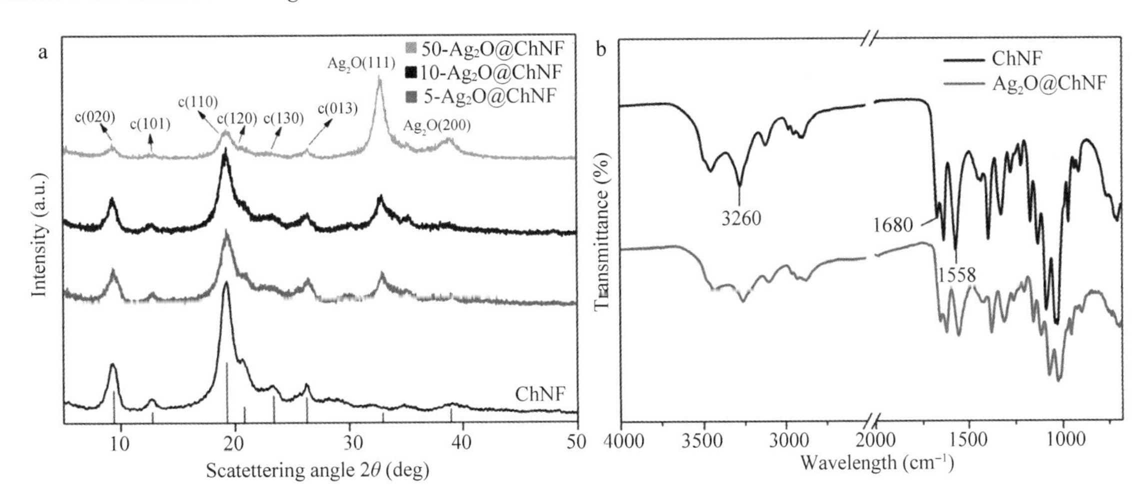

Fig. 3 (a) XRD spectrogram of ChNF and Ag_2O@ChNF aerogels with different loading concentrations. (b) FTIR spectrogram of ChNF and Ag_2O@ChNF

Characterization via transmission electron microscopy (TEM) provided additional structural information about Ag_2O@ ChNF aerogels. Fig. 4(a-c) shows the TEM images of Ag_2O@ ChNF samples with different loading concentrations. In 50-Ag_2O@ ChNF, the sample with the highest loading concentration (the labeling method of samples is detailed in the experimental section), nanoparticles were densely and evenly anchored on

ChNF networks with negligible agglomeration. By contrast, smaller nanoparticles were sparsely distributed and scattered on 10 - Ag_2O @ ChNF and 5 - Ag_2O @ ChNF. The loading content in the final product could be controlled to ~122% (50-Ag_2O@ ChNF), ~56% (10-Ag_2O@ ChNF), and ~31% (5-Ag_2O@ ChNF). Furthermore, the size of Ag_2O nanoparticles could be controlled by the initial concentration of Tollen's reagent. Particle sizes were then measured and recorded (Supplementary Fig. S4). As the concentration of Tollen's reagent increased, the mean diameter of Ag_2O nanoparticles increased from 8 nm in 5 - Ag_2O @ ChNF, to 13 nm in 10-Ag_2O@ ChNF, and finally to 16 nm in 50-Ag_2O@ ChNF. These results were in a good agreement with results that were calculated using the Scherrer equation (See Supplementary Table S2). The HRTEM image of Ag_2O@ ChNF is presented in Fig. 4(d). Nanoparticles were evenly distributed on ChNF scaffolds without agglomeration. High surface energies make nanoparticles more likely to aggregate. Ultrasonication with the appropriate handling time, however, can effectively alleviate aggregation. Before the standing process for precursor precipitation, the samples were treated with ultrasonication for 5 min. During ultrasonication, the explosion of cavities exerted shear force on nanoparticles. Shear force breaks down Van der Waals attraction and prevents agglomeration. The inset in Fig. 4(d) shows that the Ag_2O phase has a face-centered cubic structure. This particle formed a five-fold twinning feature that is common to the cubic structural phase. Fig. 4(e) provides information about lattice fringes of Ag_2O nanoparticles. The lattice spacing of plane (002) was 0.24 nm and that of plane (111) was 0.27 nm. The SAED image showed a series of intermittent diffraction rings that originated from the random orientations of the Ag_2O nanoparticles.

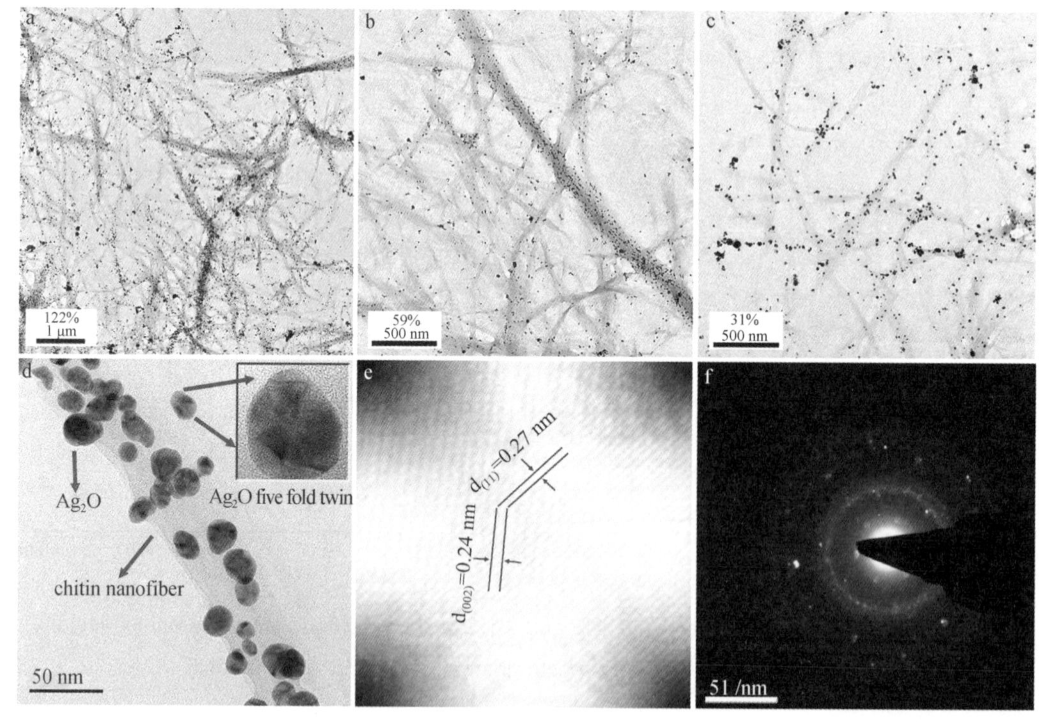

Fig. 4 TEM images of (a) 50-Ag_2O@ChNF, (b) 10-Ag_2O@ChNF, and (c) 5-Ag_2O@ChNF. (d) HRTEM image of 50-Ag_2O@ChNF, inset: the five-fold twinning particle of Ag_2O. (e) Inverse FFT (IFFT) image-scaled lattice fringe of Ag_2O (111) plane and (002) plane. (f) SEAD image of Ag_2O

Iodine adsorption. As shown in Fig. 5(a), below 20 ppm NaI, all adsorbents immobilized more than 80% of I^-, whereas 50-Ag_2O@ChNF removed 100% of I^-. The kinetic isotherms of I^- adsorption at 20 ppm I^- are presented in Fig. 5(b). During the first 7 h of adsorption at NaI concentrations of 20 ppm or less, 50-Ag_2O@ChNF, 10-Ag_2O@ChNF, and 5-Ag_2O@ChNF reached saturated sorption capacity, which is equivalent to adsorption quantity at 48 h. The adsorption capacities of the samples showed the following order: 50-Ag_2O@ChNF, ~2.81 mmol · g^{-1}; 10-Ag_2O@ChNF, ~2.72 mmol · g^{-1}; and 5-Ag_2O@ChNF ~2.40 mmol · g^{-1}. Ag_2O-free ChNF aerogels were also used to adsorb iodine anions as a control. Pristine aerogels removed 13.8% iodine anions in solution with an adsorption capacity of 0.41 mmol · g^{-1}. Amine groups at deacetylated C2 positions in ChNFs may be attracted to and capture I^- via electrostatic interaction. Capture capacity, however, was not proportional to Ag_2O content. This disproportionate relationship may be attributed to change in Ag_2O particle size, specific surface area, and pore volume. When loading content increased from 31% to 122%, the specific surface area decreased from 131.96 m^2 · g^{-1} to 74.31 m^2 · g^{-1} and pore volume decreased from 0.66 cm^3 · g^{-1} to 0.27 cm^3 · g^{-1}. Larger Ag_2O particles thickened the pore walls of ChNF aerogels and decreased specific surface area and pore volume. These effects collectively decreased the adsorption capacity of aerogels with high Ag_2O content below those of aerogels of low Ag_2O content. The absorbability of Ag_2O@ChNF aerogels remains considerably higher than those of Hg- and Cu-based adsorbents: the maximum adsorption capacity of cinnabar is only 19 μmol · g^{-1} and that of Cu_2S is 48 μmol · g^{-1}[29,30]. The iodide adsorption capacity of several sorption materials are listed in Supplementary Table S1.

The high-efficiency adsorption capacity and fast adsorption kinetics of Ag_2O@ChNF are dependent on the specific area of Ag_2O particles. Nanosized Ag_2O particles with large specific area are more active and have more opportunities for exposure to I^- than Ag_2O agglomerates or large Ag_2O particles. In addition, the large specific area and high porosity of ChNF aerogels provide fine Ag_2O nanoparticles with ample space for uniform dispersion to avoid particle aggregation, thus accelerating adsorption.

We investigated the selectivity of Ag_2O@ChNF aerogels for capturing I^- when I^- coexisted with Cl^-. As shown in Fig. 5(c), I^- adsorption decreased at high Cl^- concentrations. At a Cl^- and I^- molar ratio of 100, 88% I^- was removed. When Cl^- and I^- molar ratio increased to 1000, 48% I^- was removed. Although competitive adsorption occurred when Cl^- coexisted with I^-, adsorption capacity still reached 1.59 mmol · g^{-1}(Cl^- : I^- = 1000), which is higher than those reported for Bi-based adsorbents (see Supplementary Table S1).

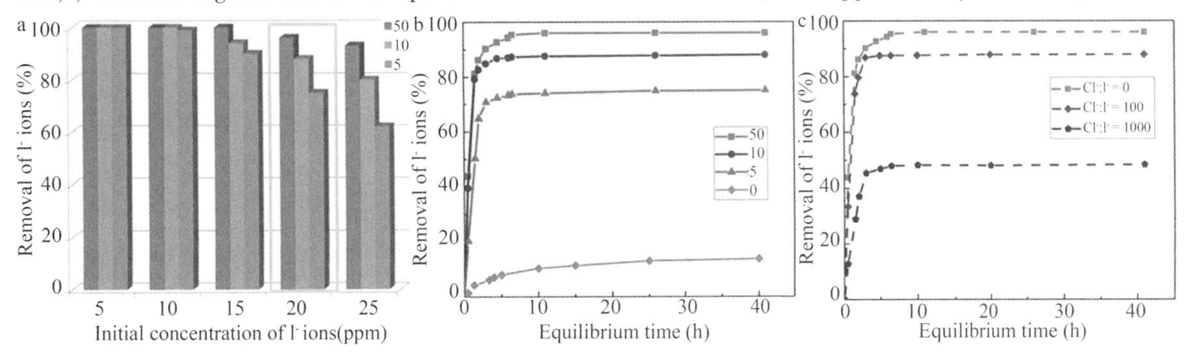

Fig. 5 (a) Removal of I^- anions at different concentrations by 50-Ag_2O@ChNF, 10-Ag_2O@ChNF, and 5-Ag_2O@ChNF; (b) The kinetic adsorption curve of I^- anions by Ag_2O@ChNF composites and ChNF aerogel; (c) I^- adsorption kinetics with 20 ppm I^- anions with different concentration of Cl^- as a competitive anion

During reaction with NaI solution, the formation of AgI@ ChNF composite caused the color of Ag_2O@ ChNF aerogels to change to yellowish-white. Fig. 6 shows the XRD patterns of Ag_2O@ ChNF samples with different loading concentrations after the adsorption test. The diffraction peak of hexagonal β-AgI appeared at 22.3°, 23.7°, 25.4°, 39.2°, 42.6°, and 46.3°, which corresponded to the (100), (002), (101), (110), (103), and (112) planes of β-AgI. By contrast, the previous diffraction peaks of Ag_2O disappeared. The TEM images of AgI nanoparticles formed via I^- adsorption were collected (Supplementary Fig. S5).

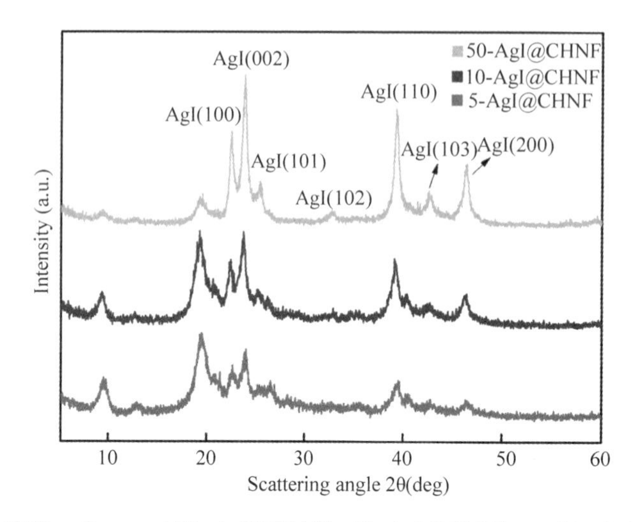

Fig. 6 XRD patterns of 50-AgI@ChNF, 10-AgI@ChNF, and 5-AgI@ChNF

To investigate the effects of ultrasonication on the agglomeration of nanoparticles, we did not treat some samples with ultrasonication prior to freeze-drying. AgI was formed as disordered agglomerates without a specific orientation. This result further proved the effectiveness of ultrasonication on the prevention of nanoparticle agglomeration. Agglomeration prevented AgI from detaching from the ChNF network. The cross-linked 3D ChNF network locked agglomerates in mesopores, preventing their leakage into the solution. During the capture test, Ag_2O chemically transformed to AgI. On a macroscopic level, the color change of the aerogel indicated this chemical phase transformation. As show in Fig. 7(a), after exposure to iodine vapor, the color of Ag_2O@ ChNF turned from brownish-black to light yellow. Further investigation revealed that I^- anchored on the ChNF scaffolds not only in the form of AgI but also as molecules. This result was verified by thermoanalysis. The thermogravimetric (TG) analysis of pristine ChNF aerogel before and after the capture test is presented in Fig. 7(b). A broad mass loss step is observed from 155 ℃ to 190 ℃ (I_2 bp = 184 ℃), which was designated as the decomposition of I_2. Moreover, a weight loss peak also appeared in a similar temperature range in the DTG spectra line of AgI@ ChNF, as shown in Fig. 7(c). Therefore, AgI@ ChNF aerogels can capture a portion of iodine vapor in the form of iodine molecules. In DTG analysis, as shown in Fig. 7(c), a major weight loss peak at ~380 ℃ appeared in the spectral lines of pristine ChNF and AgI@ ChNF aerogels. This major weight loss peak indicated the decomposition of the chitin scaffold. An inapparent peak that appeared at ~318 ℃ represented the loss of AgI. The above results were further confirmed by the X-ray photoelectron spectroscopy (XPS) spectra of Ag 3d, as shown in Fig. 7(d). However, given that the shifts in the binding energies of chemical phase transformation in XPS line shapes for Ag are inconspicuous, Ag MNN Auger analysis was performed to help identify the chemical states of composite materials[31]. Before I_2 adsorption, the binding energy of the Ag $3d_{5/2}$ photoemission peak was 368. 3 eV. Ag MNN bimodal peaks were observed in the spectrum and the binding energy of Ag $M_5N_{45}N_{45}$ was 355. 2 eV, which confirmed that Ag_2O particles anchored on the surfaces of the ChNF scaffold. After the adsorption test, a positive shift of binding energy to 0. 4 eV was observed, the binding energy of Ag $3d_{5/2}$ in samples turned to 368. 7 eV, and the doublet peak in the Auger spectrum disappeared. Moreover, the binding energy of Ag $M_5N_{45}N_{45}$ was 350. 1 eV. These results collectively provide evidence for the conversion of Ag_2O to AgI. The XPS spectrum of I 3d was also investigated to verify the results of TG analysis. The I $3d_{5/2}$ binding energy of 50−Ag_2O@ ChNF at the adsorption time of 10 min was measured at 619. 6 eV. After 30 min, the binding energy of I $3d_{5/2}$ was 620 eV, which corresponded to the binding energy of I_2. The analytical results mentioned above indicated that Ag_2O@ ChNF rapidly and efficiently fixed I_2 in the form of AgI. Moreover, by prolonging the duration of adsorption, molecular I_2 can also be anchored inside the aerogels. Therefore, the high efficiency of Ag_2O @ ChNF heterostructured aerogel indicates its potential application in radioactive iodine adsorption.

In conclusion, a green and facile fabrication process was developed for the preparation of nanofibrillated Chitin/Ag_2O heterostructured aerogels. ChNFs were first dissociated from purified chitin via a simple ultrasonication treatment. The ChNF building blocks provided abundant acetyl amino groups as active sides for interaction with Ag_2O. Host materials were then obtained after the assembly of ChNFs into pristine monolithic aerogels. ChNF aerogel had an ultra−low density of 2. 19 mg · cm^{-3}. The final products were obtained after immersion in Tollen's reagent followed by freeze−drying. Ag_2O@ ChNF aerogels have characteristics that are advantageous for radioactive iodine capture. The physical entanglement of ChNFs provided the composited aerogels with a highly mesoporous structure and a high specific surface area. Fine nanosized Ag_2O particles conferred the high efficiency of this material efficient for iodine immobilization. The adsorption capacity of Ag_2O @ ChNF aerogels for I^- anions reached up to 2. 81 mmol · g^{-1}. The results of the vapor adsorption test revealed that this monolithic composite material rapidly captured iodine vapor in the form of AgI and as molecular iodine. Therefore, this material, which was derived from green bio−resources, is an innovative alternative for radioactive iodine capture and has considerable potential applications in radioactive waste disposal.

3 Methods

3. 1 Chemical purification of chitin.

To eliminate lipids and pigments from the samples, 2 g carapace powders was submerged in a 2 : 1 (v : v) mixture of ethyl acetate and ethanol (95%) and purified using a Soxhlet apparatus at 90 ℃ for 6 h under reflux. The samples were rinsed with distilled water and then treated with 5% KOH under strong agitation for 48 h to remove protein. The samples were cooled to room temperature, thoroughly washed with distilled water, and treated with 7% hydrochloric acid under magnetic stirring at room temperature for an additional 48 h to remove mineral salts. The treated samples were then washed with distilled water and filtered. Next, the samples were treated with an acidified sodium chlorite solution (1. 5% $NaClO_2$, buffered with acetic acid to pH 4. 0) at 80 ℃ for 1 h. This process was repeated 6 times. After the above treatments, the purified chitin was immersed in distilled water to maintain moisture and to prevent strong hydrogen bonding between fibers during drying.

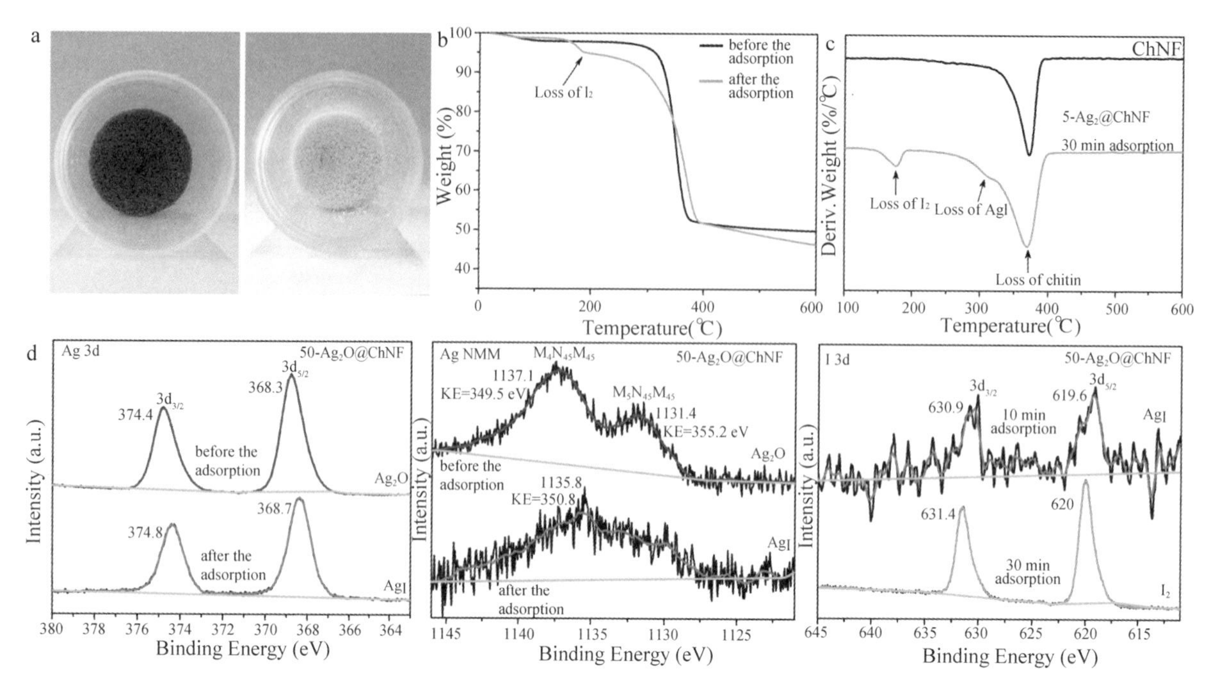

Fig. 7 (a) Image of the aerogel before and after the I_2 adsorption. (b) TG analysis of ChNF aerogel before and after I_2 adsorption. (c) DTG analysis of pristine ChNF and AgI @ ChNF of 5-Ag_2O @ ChNF, 30 min adsorption. (d) Ag 3d photoelectron spectra and Ag MNN Auger spectra of 50-Ag_2O@ChNF before and after I_2 capture. I 3d photoelectron spectra of 50-Ag_2O@ChNF with adsorption time of 10 and 30 min

3.2 Fabrication of chitin nanofibers (ChNFs).

A total of 0.5% purified wet chitin was dispersed in 400 mL distilled water. The sample was sonicated at 20 kHz at the work cycle of 1 s with a YJ99-IIDN sonicator (NingboScientz Technology Co., ltd) with a 1 cm^2 diameter titanium horn and an output power of 1300 W. Sonication was performed in an ice-water bath. The ChNF solution was obtained after sonication.

3.3 Synthesis of ChNF aerogels.

The ChNF suspension was loaded into dialysis bags, which were then subjected to solvent-exchange with *tert*-butanol. After this process, the low-viscosity ChNF suspension first turned into a diaphanous, pulpy substance, and finally to a gel body. The obtained gel was divided equally and poured into molds. The samples were then freeze-dried at −55 ℃ for 12 h at 25 μpa in vacuo, finally yielding ChNF aerogels.

3.4 Synthesis of Ag_2O@ ChNF aerogels.

ChNF aerogels were immersed in freshly prepared Tollen's reagent at a molar ratio of [Ag]/[NH_3] = 1 : 2. Samples were then treated via ultrasonication for 5 min to ensure that particles were homogeneously dispersed in the ChNF network. Samples were prepared with 5×10^{-2}, 1×10^{-2}, or 5×10^{-3} mol/L $Ag(NH_3)_2OH$; transferred to NaOH solution (pH = 13); and allowed to stand in open containers for 12 h at room temperature for the full reaction. The colors of samples gradually changed from white to brownish black as nanoparticles formed. Nanoparticle-loaded ChNF networks were then washed with distilled water thrice to remove heteroions. Composited networks were frozen at −30 ℃ and freeze−dried to obtain monolithic aerogels with fine Ag_2O fine granules anchored on ChNF networks. Final products prepared with 5×10^{-2}, 1×10^{-2}, or 5×10^{-3} mol/L Ag (NH_3)$_2$OH were denoted as 50 − Ag_2O @ ChNF, 10 − Ag_2O @ ChNF, or 5 − Ag_2O @ ChNF, respectively.

3.5 I^-adsorption.

For safety concerns, we used the nonradioactive I isotope to monitor the behavior of [131] I^- and [129] I^-. Ag_2O @ ChNF adsorbents with different loading concentration were added to a series of aqueous NaI solutions at concentrations of 5 to 25 ppm. Per milligram absorbents was assigned to 15 mL of NaI solution. An appropriate amount of reaction solution was sampled and centrifuged at specific time intervals. The supernatant was used to measure the adsorption spectra of I^- via UV − Vis measurement. The kinetic isotherms of I^- adsorption by Ag_2O @ ChNF samples were determined at 20 ppm with a piece of pristine aerogel as the control. The complete color conversion of the adsorbents indicated the completion of the reaction. Adsorbents were recovered and dried for further characterization.

3.6 Selective uptake of I^- anions by Ag_2O@ChNF aerogels.

The selective uptake of I^- was investigated in the presence of high concentrations of Cl^- anions. Per milligram absorbents (50−Ag_2O@ ChNF) was assigned to a 15−mL mixture solution of NaI and NaCl, where NaI concentration was 20 ppm and the molar ratios of Cl^- to I^- were set to 100 : 1 and 1000 : 1, respectively. The solution was shaken for 45 h. At specific time intervals, an appropriate amount of solution was sampled and centrifuged. The supernatant was subjected to UV−vis measurement.

3.7 Capture of iodine.

A 500 mg sample of I_2 was placed in a 50−mL beaker. Then, a piece of fritted glassware with a piece of Ag_2O@ ChNF gel was placed on the beaker. The apparatus was heated to 75 ℃ to evaporate iodine. During this process, a distinct color distribution appeared on the aerogels. The complete color change of the aerogel sample indicated that the reaction was completed. A piece of ChNF aerogel was also used to capture iodine vapor. Materials after I^- adsorption and I_2 vapor capture were denoted as AgI@ ChNF.

3.8 Characterization.

The [13] C NMR experiments were performed on a 500 MHz nuclear magnetic resonance spectorometer (Bruker advance Ⅲ HD 500 MHz). The bulk density of ChNF aerogels was determined on the basis of the physical dimensions and weights of the samples. The microstructures of ChNF and monolithic aerogels were observed via SEM (Sirion 200, FEI). Samples were coated with gold in advance to improve conductivity. Adsorption−desorption isotherm measurements were obtained with a JW−BK132F specific surface area and pore size analyzer. Specific surface area was calculated in accordance with the Brunauer−Emmett−Teller method. Pore size distribution was estimated via BJH method. The TEM analysis of the samples was performed on a FEI, Tecnai G2 F20 TEM at 200 kV. The appropriate amount of Ag_2O@ ChNF was dispersed in alcohol and dispersed with an ultrasonic cleaner. One microliter of suspension was taken out by pipette and dropped on the support film and air dried for TEM observation. The XRD patterns of the Ag_2O@ ChNF samples before and after adsorption of I^- were measured with D/max 2200, Rigaku powder XRD instrument using Ni−

filtered Cu Kα radiation ($\lambda = 1.5406$ Å). Data were collected over a 2θ from 5° to 60° at a scanning rate of 4° · min^{-1}. FTIR spectra were recorded on a Nicolet Nexus 670 FTIR instrument in the range of 500-4000 cm^{-1} at a resolution of 4 cm^{-1}. I^- concentration was measured and calculated by a Shimadzu UV-2501 PC spectrometer. Thermogravimetric analysis (TGA) was performed in a Shimadzu TGA-50 thermal analyzer by heating each sample from room temperature to 800 ℃ with a ramping rate of 5 ℃ · min^{-1} under nitrogen flow. The composition of the composite samples was investigated via X-ray photoelectron spectroscopy using a Thermo Escalab 250 Xi XPS spectrometer. The C1s peak of adventitious carbon at 285 eV was used as a binding energy reference.

References

[1] Gudrun R. Kirk-Othmer Encyclopedia of Chemical Technology[M]. John Wiley & Sons, Inc., 2008.

[2] Hrubesh L. Aerogel applications[J]. J. Non-Cryst. Solids, 1998, 225: 335-342.

[3] Hüsing N, Schubert U. Aerogels—airy materials: chemistry, structure, and properties[J]. Angew. Chem. Int. Edit., 1998, 37: 22-45.

[4] Katsoulidis A, He J, Kanatzidis M, Functional monolithic polymeric organic framework aerogel as reducing and hostingmedia for Ag nanoparticles and application in capturing of iodine vapors[J]. Chem. Mater., 2012, 24: 1937-1943.

[5] Pekala R. Organic aerogels from the polycondensation of resorcinol with formaldehyde[J]. J. Mater. Sci., 1989, 24: 3221-3227.

[6] Nguyen M, Dao L. Effects of processing variable on melamine-formaldehyde aerogel formation[J]. J Non-Cryst Solids, 1998, 225: 51-57.

[7] Wu D, Fu R, Zhang S, et al. The preparation of carbon aerogels based upon the gelation ofresorcinol-furfural in isopropanol with organic base catalyst[J]. J Non-Cryst Solids, 2004, 336: 26-31.

[8] Li W, Reichenauer G, Fricke J. Carbon aerogels derived from cresol-resorcinol-formaldehyde for supercapacitors[J]. Carbon, 2002, 40: 2955-2959.

[9] Wu D, Fu R. Synthesis of organic and carbon aerogels from phenol-furfural by two-step polymerization[J]. Micropor. Mesopor. Mat., 2006, 96: 115-120 (2006).

[10] Biesmans G, Mertens A, Duffours L, et al. Polyurethane based organic aerogels and their transformationinto carbon aerogels[J]. J Non-Cryst Solids, 1998, 225: 64-68.

[11] Lee J, Gould G. [P]. 2010.

[12] Leventis N. Multifunctional Polyurea Aerogels from Isocyanates and Water. A Structure-Property Case Study[J]. Chem. Mater., 2010, 22: 6692-6710.

[13] MushiN, Kochumalayil J, Cervin N, et al. Nanostructurally Controlled Hydrogel Based on Small-Diameter Native Chitin Nanofibers: Preparation, Structure, and Properties[J]. ChemSusChem, 2016, 225: 64-68.

[14] Kurita K. Chemistry and application of chitin and chitosan[J]. Polym. Degrad. Stabil., 1998, 59: 117-120.

[15] Kurita K. Chitinand chitosan: functional biopolymers from marine crustaceans[J]. Mar Biotechnol, 2006, 8: 203-226.

[16] Pillai C, Paul W, Sharma C. Chitin and chitosan polymers: Chemistry, solubility and fiber formation[J]. Prog Polym. Sci., 2009, 34: 001.

[17] Kobayashi Y, Saito T, Isogai A. Aerogels with 3D Ordered Nanofiber Skeletons of Liquid-Crystalline Nanocellulose Derivatives as Tough and Transparent Insulators[J]. Angew. Chem. Int. Edit., 2014, 126: 10562-10565.

[18] Xiao S, Gao R, Lu Y, et al. Fabrication and characterization of nanofibrillated cellulose and its aerogels from naturalpine needles[J]. Carbohyd. Polym., 2015, 119: 202-209.

[19] Lu Y. Xiao S, Gao R, et al. Fabrication and characterisation of α-chitin nanofibers and highly transparent chitin films by pulsed ultrasonication[J]. Carbohyd. Polym., 2013, 98: 1497-1504.

[20] Duan, B, Gao H, He M, et al. Hydrophobic Modification on Surface of Chitin Sponges for Highly Effective Separation of Oil[J]. ACS Appl. Mater. Inter., 2014, 6: 19933-19942.

[21] Tsutsumi Y, Koga H, Qi Z, et al. Nanofibrillar chitin aerogels as renewable base catalysts[J]. Biomacromolecules, 2014, 15: 4314-4319.

[22] Bo A. Removal of radioactive iodine from water using Ag_2O grafted titanate nanolamina as efficient adsorbent[J]. J. Hazard. Mater., 2013, 246: 199-205.

[23] Noishiki Y. Alkali-induced conversion of β-chitin to α-chitin[J]. Biomacromolecules, 2003, 4: 896-899.

[24] Mandzy N, Grulke E, Druffel T. Breakage of TiO_2 agglomerates in electrostatically stabilized aqueous dispersions[J]. Powder. Technol., 2005, 160: 121-126.

[25] Kusters K., Pratsinis S, Thoma S, et al. Energy - size reduction laws for ultrasonic fragmentation[J]. Powder Technol, 1994, 80: 253-263.

[26] Jiang J, Oberdörster G, Biswas P. Characterization of size, surface charge, and agglomeration state of nanoparticle dispersionsfor toxicological studies[J]. J. Nanopart. Res., 2009, 11: 77-89.

[27] Ding B. Facile preparation of robust and biocompatible chitin aerogels[J]. J. Mater. Chem., 2012, 22: 5801-5809.

[28] Tompsett G, Krogh L, Griffin D, et al. Hysteresis and scanning behavior of mesoporous molecular sieves[J]. Langmuir, 2005, 21: 8214-8225.

[29] Sazarashi M. Adsorption of I Ions on Cinnabar for I129 Waste Management: Radiochimica Acta[J]. Radiochim Acta, 1994, 65: 195-198.

[30] Lefèvre G, Bessière J, Ehrhardt J, et al. Immobilization of iodide on copper(I) sulfide minerals[J]. J Environ Radioactiv, 2003, 4: 70-73.

[31] Bera S, Gangopadhyay P, Nair K, et al. Electron spectroscopic analysis of silver nanoparticles in a sodaglass matrix [J]. J Electron Spectrosc, 2006, 152: 91-95.

中文题目：高效制备具有放射性碘捕捉功能的纳米纤丝化甲壳素/氧化银异质复合气凝胶材料

作者：高汝楠，卢芸，肖少良，李坚

摘要：本文报道了用纳米纤丝化甲壳素/氧化银气凝胶捕捉放射性碘元素的工作。首先制备了表面富有乙酰氨基的甲壳素纳米纤丝，其次通过冷冻干燥组装了甲壳素纳米纤丝气凝胶材料。该材料具有超低密度(2.19 $mg \cdot cm^{-3}$)和超高比表面积(179.71 $m^2 \cdot g^{-1}$)。通过银离子和乙酰氨基之间的强相互作用，最终制备出了 Ag_2O@ ChNF 异质复合气凝胶材料。该复合材料可以高效吸附水中的放射性碘离子，还可以快速捕捉碘蒸气。气凝胶材料对碘离子的吸附量高达 2.81 $mmol \cdot g^{-1}$，在放射性废物处理领域展示出了强劲的应用潜力。

关键词：甲壳素纳米纤维，氧化银纳米颗粒，气凝胶，放射性，吸附

Synthesis of $MnFe_2O_4$/Cellulose Aerogel Nanocomposite with Strong Magnetic Responsiveness*

Jian Li, Yue Jiao, Caichao Wan

Abstract: Cellulose aerogel, with abundant three-dimensional architecture, has been considered as a class of ideal eco-friendly matrix materials to encapsulate various nanoparticles for synthesis of miscellaneous functional materials. In the present paper, hexagonal single-crystalline $MnFe_2O_4$ was fabricated and inserted into the cellulose aerogel using an in-situ chemical precipitation method. The as-prepared $MnFe_2O_4$ nanoparticles were well dispersed and immobilized in the micro/nanoscale pore structure of the aerogel, and exhibited superparamagnetic behavior. In addition, the nanocomposite was easily actuated under the effect of an external magnetic field, revealing its strong magnetic responsiveness. Combined with the advantages of environmental benefits, facile synthesis method, strong magnetic responsiveness, and unique structural feature, this class of $MnFe_2O_4$/cellulose aerogel nanocomposite has possible uses for applications such as magnetically actuated adsorbents.

Keywords: Cellulose aerogel; $MnFe_2O_4$; Magnetic responsiveness; Nanocomposite

1 Introduction

Aerogel based on cellulose, the most abundant and renewable natural polymer, is considered as one of the most promising cellulose products. Cellulose aerogel consists of a cross-linked three-dimensional (3D) network. The unique structural characteristic endows itself with low density, high porosity and large specific surface area[1-3]. As a result, cellulose aerogel finds applications in multitudinous fields such as adsorbing materials[4], catalyst supports[5], and super-thermal and sound insulators[6]. In recent years, cellulose aerogel has gained increasing attention for magnetic devices. Magnetic cellulose aerogel is a kind of cellulose aerogel which shows magnetic responsiveness under the action of an additional magnetic field. In general, magnetic nanoparticles (e.g., α-FeOOH, $CoFe_2O_4$, Fe_2O_3, Fe_3O_4 and $MnFe_2O_4$) are loaded in the aerogel in order to endow magnetic properties, and the resultant magnetic aerogel has diamagnetic, paramagnetic, ferromagnetic, ferrimagnetic or antiferromagnetic characteristics[7,8], mainly dependent on the inserted magnetic nanoparticles. Chin et al. prepared magnetic cellulose aerogel by in-situ incorporation of magnetic Fe_3O_4 nanoparticles into the cellulose aerogel[9]. The hybrid aerogel can be easily recovered from water by applying an external magnetic field. Liu et al. obtained magnetic $CoFe_2O_4$/cellulose hybrid aerogel by using cellulose aerogel as a template, and the formed magnetic composite aerogel was lightweight, flexible, and highly porous[10]. Wan and Li prepared superparamagnetic γ-Fe_2O_3 nanoparticles encapsulated in 3D

* 本文摘自 Frontiers of Agricultural Science and Engineering, 2017, 4(1): 116-120.

architecture of cellulose aerogel for the application of removing Cr^{6+} ions from contaminated water[11]. Based on these reports, it was found that the magnetic composite aerogel not only had strong magnetic responsiveness, but also the unique physicochemical properties of cellulose aerogel. These multifarious functions contribute to expanding the scope of its potential application.

$MnFe_2O_4$ is a typical soft ferrite with a small magnetic anisotropy constant of about 10^3 $J \cdot m^{-3}$[12]. $MnFe_2O_4$ nanoparticles have attracted considerable attention because of their potential as contrast enhancement agents in magnetic resonance imaging technology[13,14]. It is well-known that nanoparticles readily agglomerate due to high surface energy, which leads to performance deterioration. Therefore, integration of magnetic nanoparticles, such as $MnFe_2O_4$, within a porous substrate, such as cellulose aerogel, might not only provide magnetically actuated aerogel, but also effectively reduce the particle agglomeration.

In the present study, hexagonal single-crystalline $MnFe_2O_4$ was fabricated and inserted into the cellulose aerogel using an in-situ chemical precipitation method. The morphology, crystal structure, chemical components and thermodynamic stability of the as-prepared $MnFe_2O_4$/cellulose aerogel (coded as $MnFe_2O_4$/CA) nanocomposite were investigated by scanning electron microscopy (SEM), energy-dispersive X-ray (EDX), transmission electron microscopy (TEM), selected area electron diffraction (SAED), X-ray diffraction (XRD), thermogravimetric (TG) analysis and X-ray photoelectron spectroscopy (XPS). The magnetic properties of the nanocomposite was also tested by superconducting quantum interference device.

2 Materials and methods

2.1 Materials

Chemical reagents including $FeSO_4 \cdot 7H_2O$, KNO_3, $MnCl_2 \cdot 4H_2O$, and NaOH were of analytical grade and purchased from Tianjin Kemiou Chemical Reagent Co., Ltd. (Tianjin, China) and used without further purification.

2.2 Preparation of $MnFe_2O_4$/CA

The preparation process of cellulose hydrogel (the precursor of cellulose aerogel) was described in previous reports[15,16]. The prepared cellulose hydrogel was immersed in a freshly prepared aqueous solution of $FeSO_4 \cdot 7H_2O$ and $MnCl_2 \cdot 4H_2O$ (40 mL) with the stoichiometric ratio of Mn : Fe of 1 : 2. After 1 h of immersion, the mixture was heated to 90 ℃ and held at this temperature for 2 h. KNO_3 and NaOH were dissolved in 40 mL of distilled water. This solution was heated to 90 ℃, and quickly added to the suspension of metal ions/cellulose hydrogel. The molar ratios of $[Mn^{2+}]$: $[OH^-]$ and $[Mn^{2+}]$: $[KNO_3]$ were 1 : 2 and 1 : 3, respectively. The mixed system was kept for an additional 6 h at 90 ℃. All reactions were performed in air. The resulting product was washed repeatedly with distilled water and tert-butyl alcohol to remove residual chemicals, and then underwent a tert-butyl alcohol freeze-drying treatment at −35 ℃ for 24 h. In addition, the pure cellulose aerogel was prepared by the tert-butyl alcohol freeze-drying treatment of the cellulose hydrogel.

2.3 Characterization

The morphology was observed by SEM (FEI, Quanta 200) equipped with an EDX spectrometer for elemental analysis. TEM and high-resolution TEM (HRTEM) observations and SAED were performed with a

FEI, Tecnai G2 F20 TEM with a field-emission gun operating at 200 kV. XRD was implemented on a Bruker D8 Advance TXS XRD instrument. XPS was carried out using a Thermo Escalab 250Xi XPS spectrometer equipped with a dual X-ray source using Al Kα. Thermal stabilities were determined using a TG analyzer (TA, Q600) from room temperature to 800 ℃ at a heating rate of 10 ℃ · min^{-1} under a nitrogen atmosphere. A magnetometer (MPMS XL-7, Quantum Design) using a superconducting quantum interference device sensor was used to make a measurement of the magnetic properties of the nanocomposite in applied magnetic fields over the range from −20000 to +20000 Oe at 298 K.

3 Results and discussion

The SEM image demonstrates that $MnFe_2O_4$/CA maintained a 3D interconnected network after the incorporation of $MnFe_2O_4$(Fig. 1a). However, it is difficult to distinguish the $MnFe_2O_4$ nanoparticles from the complicated network structure. Even the higher-magnification SEM image still cannot identify the nanoparticles (Fig. 1b), possibly due to their extremely small particle size. The following TEM observation confirmed this. Compared with the pure cellulose aerogel, apart from the common C and O, the EDX spectrum of $MnFe_2O_4$/CA exhibited new peaks assigned to S, K, Fe and Mn (Fig. 1c). The Fe and Mn originated from the synthetic $MnFe_2O_4$, while the S and K are attributed to the residual chemicals. Further insight into the microstructure of the $MnFe_2O_4$ in the cellulose aerogel was gained by using TEM and HRTEM. It can be seen in Fig. 1d that the nanoparticles (the dark spots of hexagon) with the sizes ranging from 70 to 140 nm were homogeneously dispersed and immobilized in the cellulose aerogel matrix, which indicates that the cellulose aerogel with 3D architecture is a suitable template for the synthesis of nanoparticles. The SAED pattern was obtained on an individual $MnFe_2O_4$ nanoparticle, as shown in Fig. 1e. It clearly demonstrates that the hexagonal $MnFe_2O_4$ was essentially a single crystalline structure. Fig. 1f presents the HRTEM image of the hexagonal $MnFe_2O_4$. The lattice spacing between two adjacent fringes was 0.26 nm, corresponding to the (311) plane of $MnFe_2O_4$.

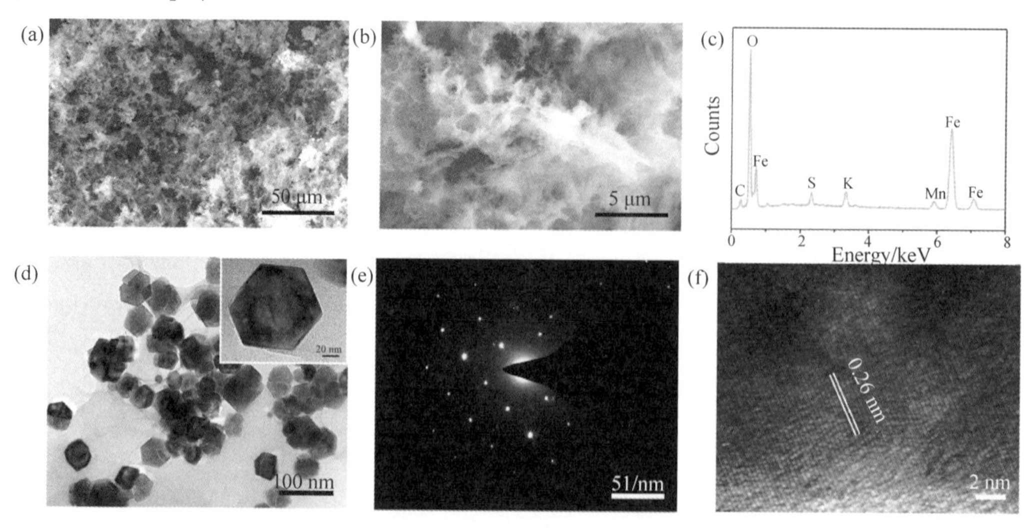

Fig. 1 SEM images (a, b), EDX pattern (c), TEM image (d), SAED pattern (e), and HRTEM image (f) of $MnFe_2O_4$/CA, respectively. The inset in (d) shows the shape of single $MnFe_2O_4$ nanoparticle

For clarifying the crystal structure and phase purity of the synthesized particles, XRD analysis was conducted. As shown in Fig. 2, the characteristic peaks of cellulose can be observed at 12.3°, 20.2° and 22.0° for both the cellulose aerogel and $MnFe_2O_4$/CA, corresponding to ($\bar{1}10$), (110) and (200) planes[17], respectively. Also, the peaks of $MnFe_2O_4$/CA at 29.7°, 34.9°, 42.5°, 52.7°, 56.2° and 61.6° are related to (220), (311), (400), (422), (511) and (440), respectively, according to the Joint Committee on Powder Diffraction Standards file of $MnFe_2O_4$ (No. 10-0319). This analysis indicates the formation of $MnFe_2O_4$ with a spinel ferrite crystalline structure. In addition, there is no indication of any other phases in the XRD pattern of $MnFe_2O_4$/CA.

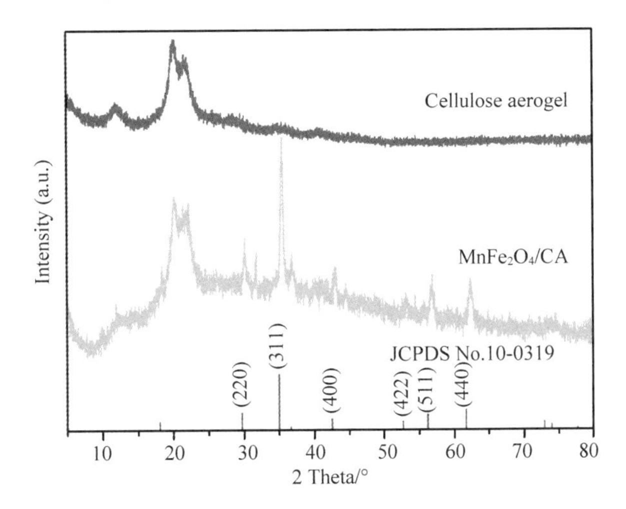

Fig. 2 XRD patterns of the cellulose aerogel and $MnFe_2O_4$/CA

The chemical states of elements in $MnFe_2O_4$/CA were investigated by XPS. The nanocomposite is composed of C, O, Mn and Fe, consistent with the results of EDX analysis. The photoelectron lines at around 284, 531, 642 and 711 eV (Fig. 3a) are attributed to C 1s, O 1s, Mn 2p and Fe 2p, respectively. Three different C 1s signals are observed at 284.8, 286.4 and 287.8 eV (Fig. 3b). The main peak at 284.6 eV is ascribed to the carbon atoms in C—C, C=C, and C—H bonds[18]. The other peaks at 286.4 eV, is assigned to the C—O, and the peak at 287.8 eV is attributed to the C=O bond[19]. The signals at 710.9 and 724.7 eV are related to the Fe $2p_{3/2}$ and Fe $2p_{1/2}$ (Fig. 3c), respectively, confirming the presence of Fe^{3+}[20]. The Mn $2p_{3/2}$ (Fig. 3d) signal appears at 641.6 eV, and the peak at 653.1 eV is ascribed to the Mn $2p_{1/2}$ signal, providing clear evidence for Mn^{2+} chemical state on the nanocomposite surface[21].

The loading content of $MnFe_2O_4$ in $MnFe_2O_4$/CA was measured by TG technique. As shown in Fig. 4a, the residual char yields above 800 ℃ were 21.7% for the cellulose aerogel and 32.2% for $MnFe_2O_4$/CA, respectively. Thus, the content of $MnFe_2O_4$ can be roughly calculated as 10.5% in $MnFe_2O_4$/CA. In addition, it is seen in Fig. 4b that the room-temperature hysteresis curve of the nanocomposite shows the absence of hysteresis and coercivity, which is characteristic of superparamagnetic behavior[22]. The curve increases rapidly with increasing applied magnetic field, and exhibits saturation magnetization values of 7.7 emu · g^{-1}.

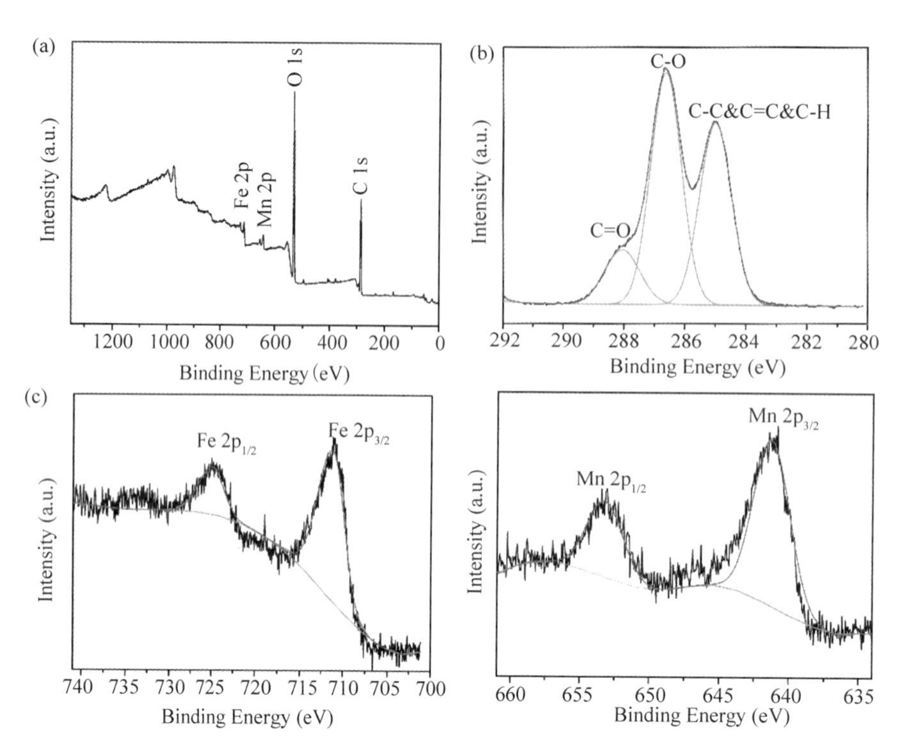

Fig. 3 Survey scan (a), O 1s (b), Fe 2p (c), and Mn 2p (d) XPS spectra of $MnFe_2O_4$/CA, respectively

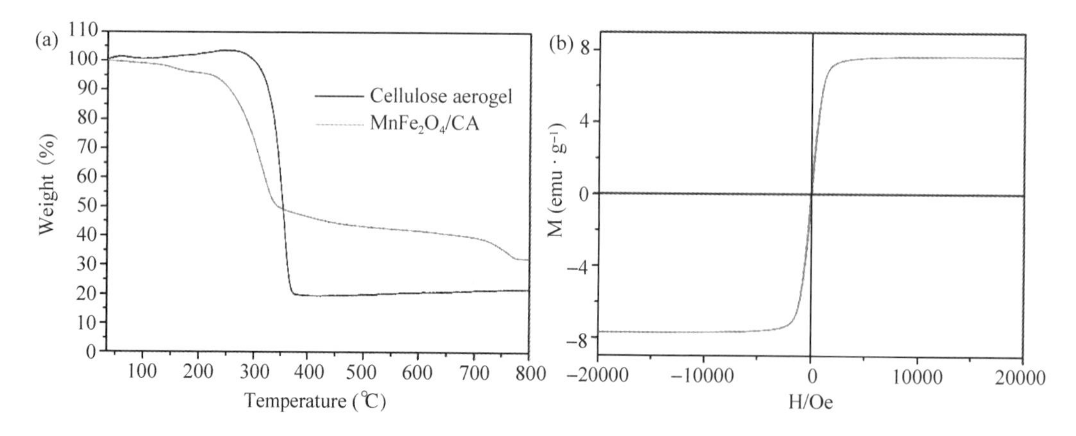

Fig. 4 TG curves (a) of the cellulose aerogel and $MnFe_2O_4$/CA. Hysteresis curve (b) of $MnFe_2O_4$/CA, respectively

The synthetic hybrid aerogel (i. e., $MnFe_2O_4$/CA) was verified to be a magnetically responsive material and actuator, as shown in Fig. 5. It can be tightly adhered to the surface of a magnet and maintain well-defined shape. Therefore, this class of aerogel with favorable shape stability and strong magnetic responsiveness may be used as a recyclable eco-friendly magnetically driven adsorbent for water purification, or a biodegradable electromagnetic device.

Fig. 5 Magnetic responsiveness of $MnFe_2O_4$/CA

4 Conclusions

We demonstrate that the 3D architecture of cellulose aerogel can be used as a suitable eco-friendly template to encapsulate the superparamagnetic $MnFe_2O_4$ nanoparticles via a facile in-situ chemical precipitation approach. The hexagonal $MnFe_2O_4$ nanoparticles have the sizes of 70 to 140 nm and were homogeneously immobilized in the cellulose aerogel. Moreover, the composite aerogel tightly adhered to the magnet surface, indicative of superior magnetic responsiveness. Therefore, this class of $MnFe_2O_4$/CA nanocomposite is expected to be useful as a kind of environmentally-friendly magnetically actuated adsorbent for the treatment of contaminated water.

Acknowledgments

This study was supported by the National Natural Science Foundation of China (grant nos. 31270590 and 31470584).

References

[1] Sehaqui H, Zhou Q, Berglund L A. High-porosity aerogels of high specific surface area prepared from nanofibrillated cellulose (NFC)[J]. Composites science and technology, 2011, 71(13): 1593-1599.

[2] Innerlohinger J, Weber H K, Kraft G. Aerocellulose: Aerogels and Aerogel-like Materials made from Cellulose[J]. Macromolecular Symposia, 2010, 244(1): 126-135.

[3] Pääkkö M, Vapaavuori J, Silvennoinen R, et al. Long and entangled native cellulose I nanofibers allow flexible aerogels and hierarchically porous templates for functionalities[J]. Soft Matter, 2008, 4(12): 2492-2499.

[4] Nguyen S T, Feng J, Shao K N, et al. Advanced thermal insulation and absorption properties of recycled cellulose aerogels[J]. Colloids & Surfaces A Physicochemical & Engineering Aspects, 2014, 445(6): 128-134.

[5] Xiong R, Lu C, Wang Y, et al. Nanofibrillated cellulose as the support and reductant for the facile synthesis of Fe_3O_4/Ag nanocomposites with catalytic and antibacterial activity[J]. Journal of Materials Chemistry A, 2013, 1(47): 14910-14918.

[6] Shi J, Lu L, Guo W, et al. Heat insulation performance, mechanics and hydrophobic modification of cellulose-SiO_2 composite aerogels[J]. Carbohydrate Polymers, 2013, 98(1): 282-289.

[7] Leslie-Pelecky D L, Rieke R D. Magnetic properties of nanostructured materials[J]. Chemistry of materials, 1996, 8(8): 1770-1783.

[8] Woo K, Hong J, Choi S, et al. Easy Synthesis and Magnetic Properties of Iron Oxide Nanoparticles[J]. Chemistry of Materials, 2004, 16(14): 2814-2818.

[9] Chin S F, Romainor A N B, Pang S C. Fabrication of hydrophobic and magnetic cellulose aerogel with high oil

absorption capacity[J]. Materials Letters, 2014, 115: 241-243.

[10]Liu S, Yan Q, Tao D, et al. Highly flexible magnetic composite aerogels prepared by using cellulose nanofibril networks as templates[J]. Carbohydrate polymers, 2012, 89(2): 551-557.

[11]Wan C, Li J. Facile synthesis of well-dispersed superparamagnetic γ-Fe_2O_3 nanoparticles encapsulated in three-dimensional architectures of cellulose aerogels and their applications for Cr (VI) removal from contaminated water[J]. ACS Sustainable Chemistry & Engineering, 2015, 3(9): 2142-2152.

[12]Song Q, Zhang Z J. Controlled synthesis and magnetic properties of bimagnetic spinel ferrite $CoFe_2O_4$ and $MnFe_2O_4$ nanocrystals with core-shell architecture[J]. Journal of the American Chemical Society, 2012, 134(24): 10182-10190.

[13]Lee N, Hyeon T. Designed synthesis of uniformly sized iron oxide nanoparticles for efficient magnetic resonance imaging contrast agents[J]. Chemical Society Reviews, 2012, 41(7): 2575-2589.

[14]Lee J, Yang J, Ko H, et al. Multifunctional magnetic gold nanocomposites: human epithelial cancer detection via magnetic resonance imaging and localized synchronous therapy[J]. Advanced Functional Materials, 2008, 18(2): 258-264.

[15]Li J, Wan C. Cellulose aerogels decorated with multi-walled carbon nanotubes: preparation, characterization, and application for electromagnetic interference shielding[J]. Frontiers of Agricultural Science and Engineering, 2016, 2(4): 341-346.

[16]LI J, WAN C, LU Y, et al. Fabrication of cellulose aerogel from wheat straw with strong absorptive capacity[J]. Frontiers of Agricultural Science and Engineering, 2014, 1(1): 46-52.

[17]Cai J, Kimura S, Wada M, et al. Cellulose aerogels from aqueous alkali hydroxide-urea solution[J]. ChemSusChem: Chemistry & Sustainability Energy & Materials, 2008, 1(1-2): 149-154.

[18]Gong W J, Tao H W, Zi G L, et al. Visible light photodegradation of dyes over mesoporous titania prepared by using chrome azurol S as template[J]. Research on Chemical Intermediates, 2009, 35(6): 751-760.

[19]Bose S, Kuila T, Uddin M E, et al. In-situ synthesis and characterization of electrically conductive polypyrrole/graphene nanocomposites[J]. Polymer, 2010, 51(25): 5921-5928.

[20]Yao Y, Cai Y, Lu F, et al. Magnetic recoverable $MnFe_2O_4$ and $MnFe_2O_4$-graphene hybrid as heterogeneous catalysts of peroxymonosulfate activation for efficient degradation of aqueous organic pollutants[J]. Journal of hazardous materials, 2014, 270: 61-70.

[21]Fu Y, Xiong P, Chen H, et al. High Photocatalytic Activity of Magnetically Separable Manganese Ferrite-Graphene Heteroarchitectures[J]. Industrial & Engineering Chemistry Research, 2012, 51(2): 725-731.

[22]Esmaeili A, Ghobadianpour S. Vancomycin loaded superparamagnetic $MnFe_2O_4$ nanoparticles coated with PEGylated chitosan to enhance antibacterial activity[J]. International journal of pharmaceutics, 2016, 501(1-2): 326-330.

中文题目：强磁响应性 $MnFe_2O_4$/纤维素气凝胶纳米复合材料的制备

作者：李坚，焦月，万才超

摘要：纤维素气凝胶具有丰富的三维结构，被认为是一类理想的生态友好型基质材料，可以包裹各种纳米颗粒，用于合成各种功能材料。本文采用原位化学沉淀法制备了六方单晶 $MnFe_2O_4$，并将其插入纤维素气凝胶中。所制备的 $MnFe_2O_4$ 纳米粒子在气凝胶的微/纳米孔结构中分散和固定得很好，表现出超磁行为。此外，纳米复合材料很容易在外加磁场的作用下被激活，表现出很强的磁响应性。这种 $MnFe_2O_4$/纤维素气凝胶纳米复合材料具有环境友好、合成方法简便、磁响应性强、结构独特等优点，在磁驱动吸附剂等领域具有潜在的应用前景。

关键词：纤维素气凝胶；$MnFe_2O_4$；磁性响应性；纳米复合材料

负载 $Ni(OH)_2/NiOOH$ 微球的碳气凝胶的水热合成与储能应用*

李坚，焦月，万才超

摘要：为发展高性能的超级电容器电极材料和拓展碳气凝胶的应用领域，采用温和、快速的水热合成工艺将 $Ni(OH)_2/NiOOH$ 微球（尺寸为 2~4 μm）嵌插入纤维素衍生的碳气凝胶（CDCA）的三维碳骨架结构中。通过扫描电子显微镜（SEM）可以观察到，该三维的 $Ni(OH)_2/NiOOH$ 微球是由众多的二维片状结构（宽度为 300~800 nm）沿不同的方向聚集、堆叠而形成的。将 $Ni(OH)_2/NiOOH/CDCA$ 纳米复合材料与乙炔黑（导电剂）和聚四氟乙烯（黏结剂）混合，然后负载在泡沫镍上，制成薄片状的超级电容器电极材料。在三电极体系下，该电极的循环伏安（CV）曲线所围面积明显大于泡沫镍支撑的 $Ni(OH)_2/NiOOH$ 电极的 CV 曲线所围的面积。此外，该电极的面电容值可达 639 $mF \cdot cm^{-2}$（电流密度设定为 1 $mA \cdot cm^{-2}$），且在经过 5 000 次连续充放电循环测试后（电流密度设定为 20 $mA \cdot cm^{-2}$），该电极的电容值仍保持初始电容值的 92.37%。

关键词：纤维素；碳气凝胶；氢氧化镍；微球；水热合成；超级电容器

超级电容器具有快速的充放电速率、高功率密度、长循环寿命和较宽的工作温度等特性，是一种重要的新型储能装置[1-6]。超级电容器按储能机理一般可分为两大类：双电层电容器和赝电容器。碳材料是主要的双电层电容器电极材料[7]。在各种碳材料中，碳气凝胶由于具有高孔隙率、高电导率、低密度、高比表面积等优良特性[8]，因此被科研工作者认为是发展高性能的超级电容器电极的一种理想原料。在传统的制备方法中，碳气凝胶是由某些有机单体（如间苯二酚-甲醛体系等）经溶胶-凝胶缩聚反应制备得到有机气凝胶后再经碳化制备而成的[9]。最近，一些文献[10-12]报道了一些特殊的碳材料（包括碳纳米管、氧化石墨烯或它们的混合物）也可通过自组装形成碳气凝胶。实际上，一些纤维素产品（如：纤维素气凝胶）也可作为热解制备碳气凝胶的前驱物[13]。相比于其他类型的碳气凝胶，这种纤维素衍生的碳气凝胶（cellulose-derived carbon aerogels，CDCA）不仅拥有传统碳气凝胶的各种常见优点，而且还具有良好的经济和环境优势。然而，具有双电层电容特性的 CDCA 通常展现出较低的电容和能量密度，因此需要将其与理论上电容和能量密度更高的赝电容器电极材料相结合来提升 CDCA 在储能领域的应用价值。

对于赝电容器来说，能量存贮是利用表面快速、可逆的氧化还原反应来实现的[14,15]。相比于双电层电容器电极材料，赝电容器电极材料虽然展现出了更高的电容和能量密度，但其具有倍率性能差、功率密度低和工作寿命较短等缺点[16]，而这些缺陷正是双电层电容器电极材料的优势所在。因此，导电聚合物和金属氧化物等赝电容器电极材料也常被用作辅助材料与双电层电容器电极材料相结合，从而提高混合电极材料的整体能量密度和循环稳定性[17,18]。

采用简单易行的水热合成方法，向 CDCA 的多孔结构中嵌插 $Ni(OH)_2/NiOOH$ 微球。制得的 $Ni(OH)_2/NiOOH/CDCA$ 纳米复合材料与乙炔黑（导电剂）和聚四氟乙烯（粘结剂）混合，然后涂抹在泡

* 本文摘自森林与环境学报，2018，38（3）：257-264.

沫镍基质上进行压片、干燥，即可制得超级电容器电极。采用三电极体系评估该超级电容器电极的电化学性能。

1　材料与方法

1.1　材料与试剂

试验主要采用的化学试剂包括：浓硝酸(65%)、六水合氯化镍($NiCl_2 \cdot 6H_2O$)、尿素等，购自天津科密欧化学试剂有限公司，均为分析纯等级且使用前未经任何处理。

1.2　制备 $Ni(OH)_2$/NiOOH/CDCA 纳米复合材料

CDCA 是由纤维素气凝胶在惰性气体保护下经高温热解所得到的。纤维素气凝胶的制备过程主要包括“纤维素溶解于氢氧化钠/聚乙二醇溶剂体系”“纤维素溶液的冻融处理”“纤维素溶液的再生处理”和“纤维素凝胶的冷冻干燥”4 个步骤，详细制备过程按参考文献[19-21]进行。纤维素气凝胶的热解工艺过程如下：将纤维素气凝胶放置在管式炉中，以 5 ℃ · min^{-1} 的加热速率加热到 500 ℃，并在该温度下保持 1 h；以 5 ℃ · min^{-1} 的加热速率加热到 1 000 ℃，在该温度下保持 2 h 使其充分热解；以 5 ℃ · min^{-1} 的速度将温度从 1000 ℃下降到 500 ℃；当温度降至 500 ℃时程序立即终止。整个热解过程均在氮气环境下进行，热解结束后即可得到 CDCA。

在室温下采用浓硝酸处理 CDCA 24 h，在其表面引入多种含氧基团。随后，根据文献[22]的方法，采用水热法合成 $Ni(OH)_2$/NiOOH/CDCA，具体合成过程可简述如下：①将 96 mg $NiCl_2 \cdot 6H_2O$ 和 150 mg 尿素与 60 mL 蒸馏水相混合，随后磁力搅拌 5 min 得到均匀的混合溶液；②将浓硝酸处理后的 CDCA 浸渍在上述溶液中 3 h；③将混合物转移至带有聚四氟乙烯内衬的高压反应釜中，填充度控制在 80%左右。高压反应釜密封后，随即在 180 ℃下高温反应 2 h；④反应结束后，待反应釜自然冷却至室温。采用蒸馏水和乙醇对产物进行充分洗涤，随后将产物置于温度为 60 ℃的鼓风干燥箱中干燥 24 h。

1.3　表征方法

样品的微观形貌和表面元素通过装配有能量弥散 X 射线(Energy dispersive X-ray，EDX)探测器的扫描电子显微镜(Scanning electron microscope，SEM，FEIQuanta200)测定。样品的结晶结构通过 X 射线衍射(X-ray diffraction，XRD)仪来表征。XRD 谱图是使用 Bruker D8 Advance TXS XRD 仪在 Cu Kα(靶)辐射(λ=1.541 8 Å)和扫描速率为 4° · min^{-1} 下进行的，扫描范围为 5°~90°。

1.4　电极材料的制备和电化学测试方法

以乙醇为介质，按 80 : 10 : 10 的质量比均匀混合 $Ni(OH)_2$/NiOOH/CDCA、乙炔黑和聚四氟乙烯。将得到的浆料涂抹在泡沫镍基质上并在 10 MPa 下进行压片，随后在 80 ℃下干燥 12 h，即可得到薄片状的电极材料。该电极材料暴露的几何面积为 1 cm^2，其他面积采用环氧树脂胶进行胶封。电极上活性材料的面密度控制在约 2.0 mg · cm^{-2}。

所有的电化学测试均在室温下进行，使用 CS350 电化学工作站(武汉科斯特仪器有限公司)进行测试。电化学测试采用三电极体系。上述制备的电极材料为工作电极，Hg/HgO 电极用作参比电极，铂片用作对电极。电解液为 6 mol · L^{-1} 的氢氧化钾溶液。电化学测试项目包括：循环伏安(Cyclic voltammetry，CV)测试、恒流充放电(Galvanostatic charge discharge，GCD)测试和循环稳定性测试。CV 测试的电势范围为 0~0.5 V，扫描速率从 5 mV · s^{-1} 逐渐增加至 100 mV · s^{-1}。GCD 测试也是在 0~0.5 V 的电势范围内进行的，电流密度从 1 mA · cm^{-2} 逐渐增加至 20 mA · cm^{-2}。循环稳定性测试也是在上述电势范围内进行的，电流密度设定为 20 mA · cm^{-2}，循环充放电过程重复 5000 次，检验电容值的变化情况。

2 结果与分析

2.1 $Ni(OH)_2/NiOOH/CDCA$ 纳米复合材料合成机理

图 1 展示了 $Ni(OH)_2/NiOOH/CDCA$ 纳米复合材料的合成机理示意图，详细过程可描述如下：①经过高温热解过程，纤维素气凝胶转化为 CDCA，但在此过程中纤维素气凝胶上的含氧官能团遭受到了破坏。因此，为了在 CDCA 的碳骨架结构上引入新的含氧基团（为原位生成 $Ni(OH)_2/NiOOH$ 微球提供反应位点），采用浓硝酸处理 CDCA；②将浓硝酸处理后的 CDCA 浸渍在 $NiCl_2 \cdot 6H_2O$/尿素混合溶液中，CDCA 上新引入的含氧官能团带负电因此可以与溶液中带正电的 $[Ni(NH_3)_6]^{2+}$ 发生静电吸引，从而将 $[Ni(NH_3)_6]^{2+}$ 固定在 CDCA 表面上；③采用水热合成工艺在 CDCA 中形成 $Ni(OH)_2/NiOOH$ 微球。

至于水热反应过程中 $Ni(OH)_2/NiOOH$ 微球的形成，可能的反应机理简述如下：①在水热反应的初始阶段，$[Ni(NH_3)_6]^{2+}$ 发生分解释放 NH_3，从而为众多 $Ni(OH)_2/NiOOH$ 核的形成提供 OH^-。众所周知，新生成的纳米晶体具有高表面能，因此并不稳定且趋向于形成更加稳定的几何形态来降低表面能[23]；②对于类似水镁石的晶体[即 $M(OH)_2$ 或 MOOH，M = Mg、Ca、Ni、Co 等]具有 CdI_2 类型的层状结构[24]。对于这种晶体结构，层与层之间是由范德华力相连（作用力较弱），而层内具有离子键性质的共价键（键力较强）。因此，类似水镁石的晶体倾向于形成二维的薄片状结构，因为这种片状的

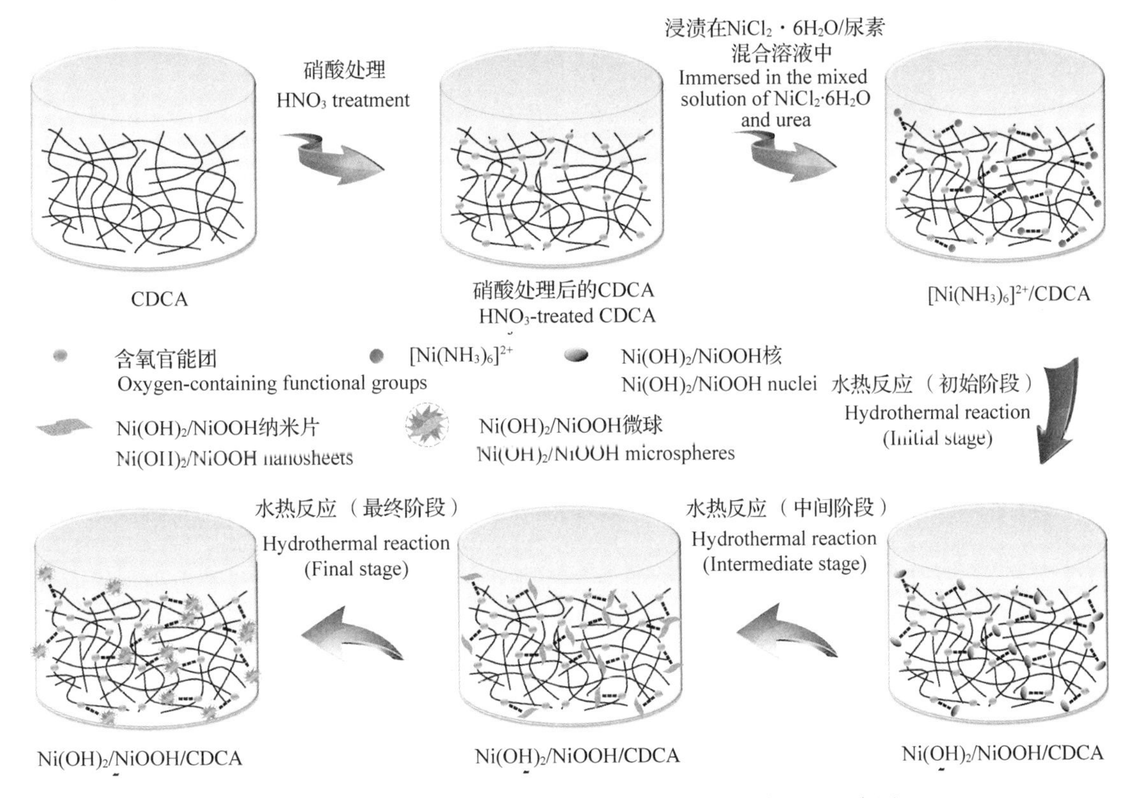

图 1 $Ni(OH)_2/NiOOH/CDCA$ 纳米复合材料的制备机理示意图

Fig. 1 Schematic diagram for the preparationof $Ni(OH)_2/NiOOH/CDCA$

$M(OH)_2$ 或 MOOH 具有低的体系能量[25]；③随着水热反应的继续进行，这些附着在 CDCA 基质上的纳米片和它们周围未附着上的纳米片在水热作用下逐渐发生聚集。这种不同方向的聚集和堆叠最终会导致球状的 $Ni(OH)_2/NiOOH$ 的形成。

2.2 $Ni(OH)_2/NiOOH/CDCA$ 纳米复合材料的微观形貌和元素成分

采用 SEM 来观察 CDCA 和 $Ni(OH)_2/NiOOH/CDCA$ 的微观形貌特征如图 2 所示。由图 2(a)可知，经过高温热解处理后，CDCA 依然保持着多孔、交联的三维网络结构。当 CDCA 用于制备超级电容器电极时，这些孔隙结构可以为电解液离子的移动和运输提供便捷通道[26-27]。经过水热处理后，可以观察到 $Ni(OH)_2/NiOOH/CDCA$ 的 SEM 图像中出现了许多 2~4 μm 的 $Ni(OH)_2/NiOOH$ 微球，这些微球与 CDCA 基质牢牢结合，并未掉落至基质之外[图 2(b)]。图 2c 和 d 展示了这些微球的高倍率 SEM 图像，单个 $Ni(OH)_2/NiOOH$ 微球是由大量宽度为 300~800 nm 的二维薄片沿不同的方向聚集、堆叠而成的。

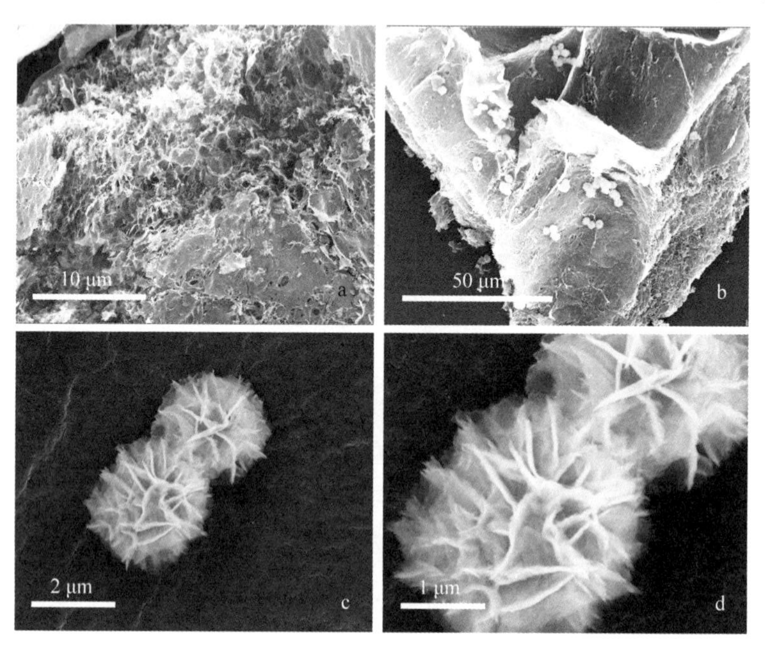

注：(a)CDCA(10 000×)；(b)$Ni(OH)_2/NiOOH/CDCA$(2 000×)；(c，d)$Ni(OH)_2/NiOOH$ 微球(40000×和 80000×)。

Note: (a) CDCA (10 000×), (b) $Ni(OH)_2/NiOOH/CDCA$ (2 000×) and (c, d) $Ni(OH)_2/NiOOH$ microspheres (40000× and 80000×).

图 2 CDCA、$Ni(OH)_2/NiOOH/CDCA$ 和 $Ni(OH)_2/NiOOH$ 微球的 SEM 图像

Fig. 2 SEM images of CDCA, $Ni(OH)_2/NiOOH/CDCA$ and $Ni(OH)_2/NiOOH$ microspheres

EDX 分析用来对材料微区成分的元素种类与含量进行探究，通常配合 SEM 和透射电子显微镜共同使用。$Ni(OH)_2/NiOOH/CDCA$ 的 EDX 图谱如图 3 所示。从图中可以得知，$Ni(OH)_2/NiOOH/CDCA$ 的主要元素包括 C 元素、O 元素和 Ni 元素。在只考虑这 3 种元素的情况下，它们的质量百分率分别为 54.15%、19.92%和 25.93%。值得一提的是，H 元素只有一层电子，而 EDX 需要激发走内层一个电子，外层电子补进内层的时候才能发射出 X 射线从而得到数据。而 H 元素只有一层电子，没有外层电子跃迁所以没有 X 射线，因此 EDX 谱图中没有 H 元素的信号。另外一个原因是，H 的 1s 电子很容易转移，在大多数情况下会转移到其他的原子附近，所以检测起来就更难了。$Ni(OH)_2/NiOOH/CDCA$ 的 EDX 图谱如图 3 所示，图 3 中的金元素是测试前的“喷金”过程所引入的，目的是增加样品表面的导电性，便于后续的 SEM 观察。

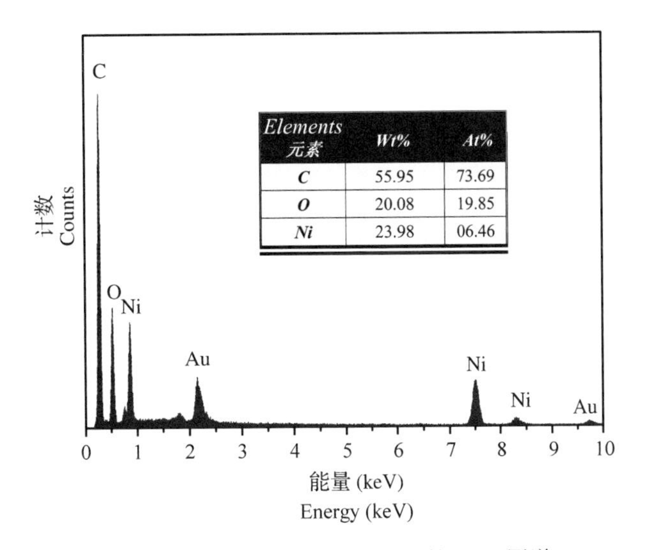

图 3 $Ni(OH)_2$/NiOOH/CDCA 的 EDX 图谱

Fig. 3 EDX pattern of $Ni(OH)_2$/NiOOH/CDCA

2.3 $Ni(OH)_2$/NiOOH/CDCA 纳米复合材料的晶体结构

采用 XRD 来分析 CDCA 和 $Ni(OH)_2$/NiOOH/CDCA 的晶体结构结果如图 4 所示。CDCA 在 22.7°和 44.2°处出现两个宽峰，这主要是来源于纤维素热解所形成的无定形碳的(002)和(100)晶面。此外，随着 $Ni(OH)_2$/NiOOH 微球的插入，一些新的 XRD 峰出现在 19.2°、33.2°、38.6°、52.1°和 59.4°处，对应于 β-$Ni(OH)_2$ 的(001)、(100)、(101)、(102)和(110)晶面[参考 β-$Ni(OH)_2$ 的标准 JCPDS 卡片 no. 14-0117]；此外，出现在 12.7°、25.5°和 43.1°处的 XRD 峰对应于 γ-NiOOH 的(003)、(006)和(105)晶面(参考 γ-NiOOH 的标准 JCPDS 卡片 no. 06-0075)。根据这些分析结果可以推断，SEM 图像中观察到的微球主要是由 β-$Ni(OH)_2$ 和 γ-NiOOH 两种晶体所构成的。

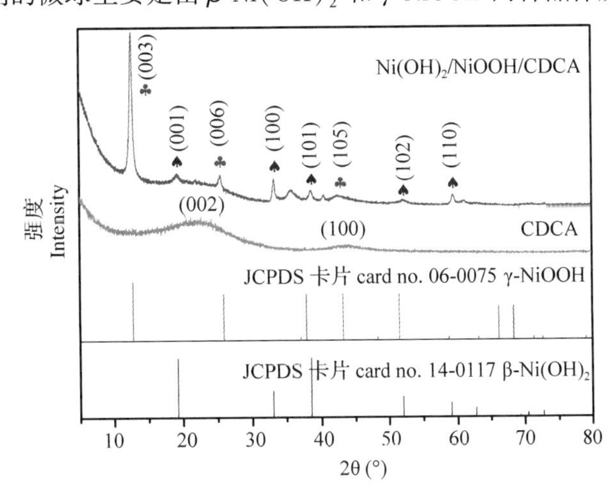

注：底部的直线为 γ-NiOOH(紫色线)和 β-$Ni(OH)_2$(蓝色线)的标准 JCPDS 卡片(nos. 06-0075 和 14-0117)。

Note: the bottom lines are the standard JCPDS cards no. 06-0075 for γ-NiOOH and no. 14-0117 for β-$Ni(OH)_2$, respectively.

图 4 CDCA 和 $Ni(OH)_2$/NiOOH/CDCA 的 XRD 谱图

Fig. 4 XRD patterns of CDCA and $Ni(OH)_2$/NiOOH/CDCA

2.4 $Ni(OH)_2$/NiOOH/CDCA 纳米复合材料的电化学性能

将 $Ni(OH)_2$/NiOOH/CDCA 与乙炔黑和聚四氟乙烯在乙醇介质中均匀混合，涂抹在泡沫镍上进行压片、干燥，即可得到片状的超级电容器电极。这种泡沫镍支撑的 $Ni(OH)_2$/NiOOH/CDCA 超级电容器电极材料的电化学测试是在三电极体系下进行的，电解液采用浓度为 6 $mol \cdot L^{-1}$ 的 KOH，测试项目包括 CV、GCD 和循环稳定性测试。泡沫镍支撑的 $Ni(OH)_2$/NiOOH 电极和泡沫镍支撑的 $Ni(OH)_2$/NiOOH/CDCA 电极的 CV 曲线如图 5(a)所示，扫描速率设定为 5 $mV \cdot s^{-1}$。两种电极的 CV 曲线均呈现出了明显的氧化还原峰，表明了它们主要是通过氧化还原反应来实现电荷存储的，即 $Ni(OH)_2$ 和 NiOOH 之间的可逆氧化还原转化[28-29]，反应式为 $Ni(OH)_2+OH^- \leftrightarrow NiOOH+H_2O+e^-$。从这些变化可以看出，$Ni(OH)_2$/NiOOH 微球的存在向复合电极材料中引入了法拉第赝电容特性。此外，由电化学理论可知，CV 曲线所围面积大小直接反映了电极的电荷存储能力。通过比较图 5a 中两种电极的 CV 曲线，可以发现泡沫镍支撑的 $Ni(OH)_2$/NiOOH/CDCA 电极的 CV 曲线所围面积明显大于泡沫镍支撑的 $Ni(OH)_2$/NiOOH 电极的 CV 曲线所围面积，这表明了经过与 CDCA 结合后所得到的三元复合材料[即 $Ni(OH)_2$/NiOOH/CDCA]的电荷存储能力更强。泡沫镍支撑的 $Ni(OH)_2$/NiOOH/CDCA 电极在不同扫描速率下的 CV 曲线如图 5(b)所示。随着扫描速率从 5 $mV \cdot s^{-1}$ 增加至 100 $mV \cdot s^{-1}$，CV 曲线中的氧化峰逐渐向正方向移动，而还原峰则向负方向移动，该现象主要归因于与随着扫描速率的增加而随之增加的电极电阻[30]。

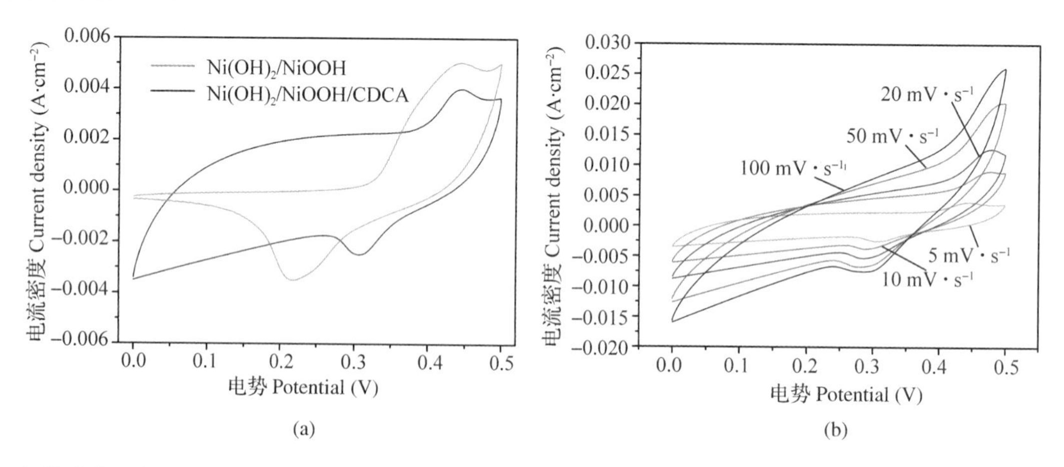

注：(a)扫描速率设定为 5 $mV \cdot s^{-1}$ 时，泡沫镍支撑的 $Ni(OH)_2$/NiOOH 或 $Ni(OH)_2$/NiOOH/CDCA 电极的 CV 曲线；(b)泡沫镍支撑的 $Ni(OH)_2$/NiOOH/CDCA 电极在不同扫描速率时的 CV 曲线。Note: (a) CV curves of Ni foam-supported CDCA or $Ni(OH)_2$/NiOOH/CDCA electrodes at a scan rate of 5 $mV \cdot s^{-1}$; (b) CV curves of Ni foam-supported $Ni(OH)_2$/NiOOH/CDCA electrode at different scan rates.

图 5 泡沫镍支撑的 $Ni(OH)_2$/NiOOH 或 $Ni(OH)_2$/NiOOH/CDCA 电极的 CV 曲线

Fig. 5 CV curves of Ni foam-supported CDCA or $Ni(OH)_2$/NiOOH/CDCA electrodes

泡沫镍支撑的 $Ni(OH)_2$/NiOOH/CDCA 电极在不同的电流密度下的 GCD 曲线和面电容值如图 6 所示。由 GCD 曲线也能观察到 $Ni(OH)_2$/NiOOH 微球的引入所带来的法拉第赝电容特性，即 GCD 曲线出现了明显的曲率，并未呈现出双电层电容器的对称直线特性[图 6(a)]。此外，GCD 曲线的放电部分和充电部分呈近似对称状，这也表明了复合材料的电容中包含有双电层电容和法拉第赝电容。面电容计算公式如下：

$$C_A = (I \times \Delta t)/(\Delta V \times S)$$

式中：C_A 为面电容($F \cdot cm^{-2}$)；I 为充放电电流(A)；Δt 为放电时间(s)；ΔV 为电势窗口(V)；S 为工作电极的暴露面积(cm^2)。由图6(b)可知，该电极材料的最高面电容值可达639 $mF \cdot cm^{-2}$，对应的电流密度为1 $mA \cdot cm^{-2}$。该面电容值高于许多超级电容器电极的最高面电容值，如：Fe_3O_4@SnO_2 核壳纳米棒薄膜(7 $mF \cdot cm^{-2}$)[31]、石墨烯-纤维素纸薄膜(81 $mF \cdot cm^{-2}$)[32]、碳修饰的 MnO_2 纳米片阵列(143 $mF \cdot cm^{-2}$)[33]和 α-Fe_2O_3@NiO 异质结构(557 $mF \cdot cm^{-2}$)[34]。由于电极上活性材料的面密约为2.0 $mg \cdot cm^{-2}$，因此该电极的质量比电容值可计算为319.5 $F \cdot g^{-1}$。此外，当电流密度从1 $mA \cdot cm^{-2}$ 增大至20 $mA \cdot cm^{-2}$ 时，该复合电极的电容值仍能保持64 $mF \cdot cm^{-2}$(32 $F \cdot g^{-1}$)。

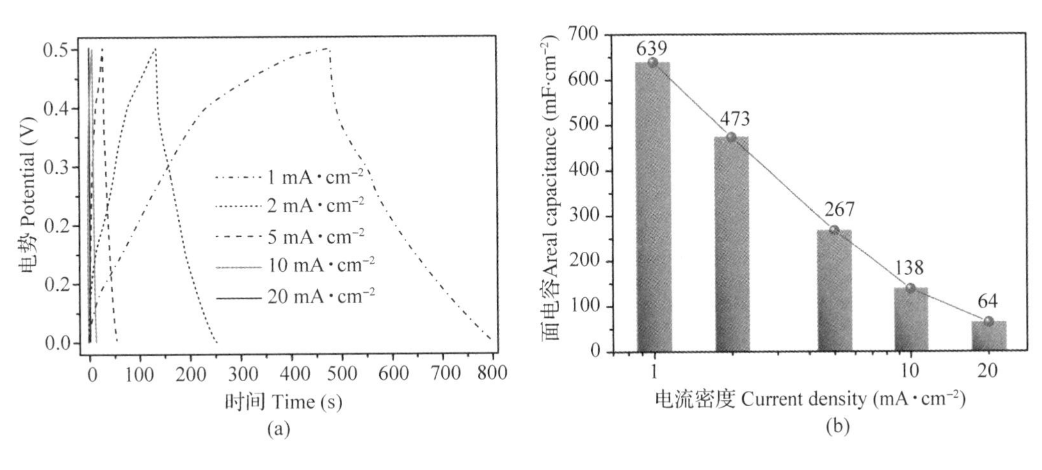

注：(a)GCD 曲线；(b)面电容值。

Note: (a) GCD curves; (b) areal capacitances.

图6 泡沫镍支撑的 $Ni(OH)_2/NiOOH/CDCA$ 电极 GCD 曲线和面电容值

Fig. 6 GCD curves and areal capacitances of Ni foam-supported $Ni(OH)_2/NiOOH/CDCA$electrode

泡沫镍支撑的 $Ni(OH)_2/NiOOH/CDCA$ 电极的长期使用稳定通过重复5000次充放电测试来探究，电流密度设定为20 $mA \cdot cm^{-2}$，泡沫镍支撑的 $Ni(OH)_2/NiOOH/CDCA$ 电极的循环稳定性测试结果如图7所示。经过5000次GCD试验，电容的下降率仅为7.63%；同时，在循环试验前后，该电极的CV曲线的变化也非常微小。这些现象均证实了该复合电极优异的循环稳定性。

对于这些良好的电化学性质，一些可能的原因可归纳如下：①多孔的CDCA基质和泡沫镍在吸附电解液后可以作为电解液的存储容器，从而促进电解液离子的转移；②高导电的泡沫镍和CDCA有利于电子的传导；③具有微纳米多尺度结构的 $Ni(OH)_2/NiOOH$ 微球有助于增加电化学反应过程中的活性位点；④$Ni(OH)_2$ 和 NiOOH 均属于高电化学活性的电极材料，具有极高的理论电容[32,35]。

3 结论

采用温和、快速的水热合成法在CDCA基质的三维碳骨架结构中生长 $Ni(OH)_2/NiOOH$ 微球，该 $Ni(OH)_2/NiOOH$ 微球的尺寸为2~4 μm，且由大量的二维片状结构(宽度为300~800 nm)聚集、堆叠而成；泡沫镍支撑的 $Ni(OH)_2/NiOOH/CDCA$ 电极呈现出高的面电容值(639 $mF \cdot cm^{-2}$)和优异的循环稳定性，即经过5 000次连续充放电测试后，仍保留了92.37%的初始电容值。这些良好的电化学特性与 $Ni(OH)_2/NiOOH$ 微球的独特结构和赝电容特性紧密相关。此外，CDCA和泡沫镍的多孔结构和高电导率也促进了电解液离子和电子的转移。

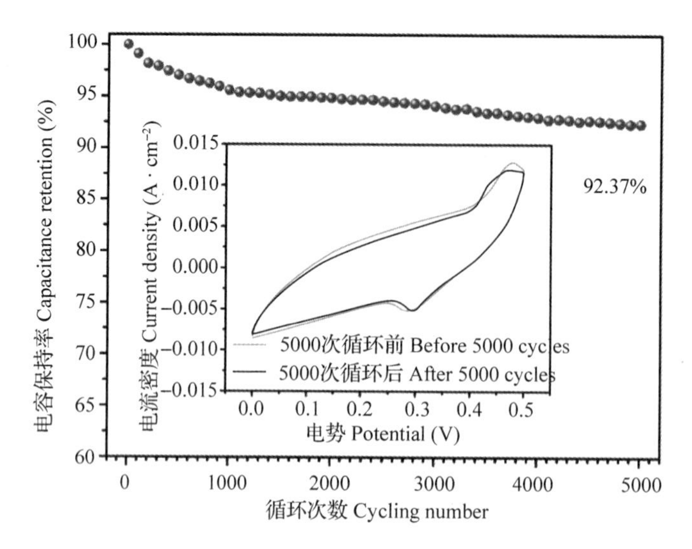

注：循环稳定性测试的电流密度为 20 mA · cm^{-2}，插图中为 5000 次充放电测试前后的该电极的 CV 曲线图。
Note: the current density of cycling stability tests was set as 20 mA · cm^{-2}, and the inset shows the CV curves before and after 5 000 cycles.

图 7　泡沫镍支撑的 $Ni(OH)_2$/NiOOH/CDCA 电极的循环稳定性测试

Fig. 7　Cyclingstability tests of Ni foam-supported $Ni(OH)_2$/NiOOH/CDCAelectrode

参考文献

[1]LARCHER D, TARASCON J M. Towards greener and more sustainable batteries for electrical energy storage[J]. Nature Chemistry, 2015, 7(1): 19-29.

[2] STERNBERG A, BARDOW A. Power-to-What? -Environmental assessment of energy storage systems[J]. Energy & Environmental Science, 2015, 8(2): 389-400.

[3] LUO X, WANG J, DOONER M, et al. Overview of current development in electrical energy storage technologies and the application potential in power system operation[J]. Applied Energy, 2015, 137: 511-536.

[4] YU Z, TETARD L, ZHAI L, et al. Supercapacitor electrode materials: nanostructures from 0 to 3 dimensions[J]. Energy & Environmental Science, 2015, 8(3): 702-730.

[5] ZHANG L, HU X, WANG Z, et al. A review of supercapacitor modeling, estimation, and applications: A control/management perspective[J]. Renewable and Sustainable Energy Reviews, 2017, 81: 1868-1878.

[6] GONZÁLEZ A, GOIKOLEA E, BARRENA J A, et al. Review on supercapacitors: technologies and materials[J]. Renewable and Sustainable Energy Reviews, 2016, 58: 1189-1206.

[7] ZHANG L L, ZHAO X S. Carbon-based materials as supercapacitor electrodes[J]. Chemical Society Reviews, 2009, 38(9): 2520-2531.

[8] ANTONIETTI M, FECHLER N, FELLINGER T P. Carbon aerogels and monoliths: control of porosity and nanoarchitecture via sol-gel routes[J]. Chemistry of Materials, 2013, 26(1): 196-210.

[9] AL-MUHTASEB S A, RITTER J A. Preparation and properties of resorcinol-formaldehyde organic and carbon gels [J]. Advanced Materials, 2003, 15(2): 101-114.

[10] BRYNING M B, MILKIE D E, ISLAM M F, et al. Carbon nanotube aerogels[J]. Advanced Materials, 2007, 19 (5): 661-664.

[11] HU H, ZHAO Z, WAN W, et al. Ultralight and highly compressible graphene aerogels[J]. Advanced Materials,

2013, 25(15): 2219-2223.

[12] NARDECCHIA S, CARRIAZO D, FERRER M L, et al. Three dimensional macroporous architectures and aerogels built of carbon nanotubes and/or graphene: synthesis and applications[J]. Chemical Society Reviews, 2013, 42(2): 794-830.

[13]卿彦, 吴义强, 罗莎, 等. 生物质纳米纤维功能化应用研究进展[J]. 功能材料, 2016, 47(5): 43-49.

[14] SNOOK G A, KAO P, BEST A S. Conducting-polymer-based supercapacitor devices and electrodes[J]. Journal of Power Sources, 2011, 196(1): 1-12.

[15] FARAJI S, ANI F N. Microwave-assisted synthesis of metal oxide/hydroxide composite electrodes for high power supercapacitors-a review[J]. Journal of Power Sources, 2014, 263: 338-360.

[16] WANG G, ZHANG L, ZHANG J. A review of electrode materials for electrochemical supercapacitors[J]. Chemical Society Reviews, 2012, 41(2): 797-828.

[17] YAN J, FAN Z, SUN W, et al. Advanced asymmetric supercapacitors based on Ni(OH)$_2$/graphene and porous graphene electrodes with high energy density[J]. Advanced Functional Materials, 2012, 22(12): 2632-2641.

[18] WANG R, XU C, LEE J M. High performance asymmetric supercapacitors: new NiOOH nanosheet/graphene hydrogels and pure graphene hydrogels[J]. Nano Energy, 2016, 19: 210-221.

[19]万才超, 卢芸, 孙庆丰, 等. 新型木质纤维素气凝胶的制备、表征及疏水吸油性能[J]. 科技导报, 2014, 32(z1): 79-85.

[20]WAN C, LI J. Facile synthesis of well-dispersed superparamagnetic γ-Fe_2O_3 nanoparticles encapsulated in three-dimensional architectures of cellulose aerogels and their applications for Cr (VI) removal from contaminated water[J]. ACS Sustainable Chemistry & Engineering, 2015, 3(9): 2142-2152.

[21] WAN C, LU Y, JIAO Y, et al. Preparation of mechanically strong and lightweight cellulose aerogels from cellulose-NaOH/PEG solution[J]. Journal of Sol-Gel Science and Technology, 2015, 74(1): 256-259.

[22] JI J, ZHANG L L, JI H, et al. Nanoporous Ni(OH)$_2$ thin film on 3D ultrathin-graphite foam for asymmetric supercapacitor[J]. ACS Nano, 2013, 7(7): 6237-6243.

[23]魏遥驰. 纳米材料表面能密度的尺寸及温度效应[D]. 北京: 中国科学院大学, 2015.

[24] YAN X, TONG X, WANG J, et al. Rational synthesis of hierarchically porous NiO hollow spheres and their supercapacitor application[J]. Materials Letters, 2013, 95: 1-4.

[25] DUAN G, CAI W, LUO Y, et al. A hierarchically structured Ni(OH)$_2$ monolayer hollow-sphere array and its tunable optical properties over a large region[J]. Advanced functional materials, 2007, 17(4): 644-650.

[26] SALIGER R, FISCHER U, HERTA C, et al. High surface area carbon aerogels for supercapacitors[J]. Journal of Non-Crystalline Solids, 1998, 225: 81-85.

[27] WANG D W, LI F, LIU M, et al. 3D aperiodic hierarchical porous graphitic carbon material for high-rate electrochemical capacitive energy storage[J]. Angewandte Chemie, 2008, 120(2): 379-382.

[28] CHEN P, CHEN H, QIU J, et al. Inkjet printing of single-walled carbon nanotube/RuO_2 nanowire supercapacitors on cloth fabrics and flexible substrates[J]. Nano Research, 2010, 3(8): 594-603.

[29] HUANG J, XU P, CAO D, et al. Asymmetric supercapacitors based on β-Ni(OH)$_2$ nanosheets and activated carbon with high energy density[J]. Journal of Power Sources, 2014, 246: 371-376.

[30] PATIL U M, GURAV K V, FULARI V J, et al. Characterization of honeycomb-like "β-Ni(OH)$_2$" thin films synthesized by chemical bath deposition method and their supercapacitor application[J]. Journal of Power Sources, 2009, 188(1): 338-342.

[31] LI R, REN X, ZHANG F, et al. Synthesis of Fe_3O_4@SnO_2 core-shell nanorod film and its application as a thin-film supercapacitor electrode[J]. Chemical Communications, 2012, 48(41): 5010-5012.

[32] WENG Z, SU Y, WANG D W, et al. Graphene-cellulose paper flexible supercapacitors[J]. Advanced Energy Materials, 2011, 1(5): 917-922.

[33] HUANG Y, LI Y, HU Z, et al. A carbon modified MnO_2 nanosheet array as a stable high-capacitance supercapacitor electrode[J]. Journal of Materials Chemistry A, 2013, 1(34): 9809-9813.

[34] JIAO Y, LIU Y, YIN B, et al. Hybrid α-Fe_2O_3 @ NiO heterostructures for flexible and high performance supercapacitor electrodes and visible light driven photocatalysts[J]. Nano Energy, 2014, 10: 90-98.

[35]XIA L, LI X, WU Y, et al. Wood-derived carbons with hierarchical porous structures and monolithic shapes prepared by biological-template and self-assembly strategies[J]. ACS Sustainable Chemistry & Engineering, 2015, 3(8): 1724-1731.

英文题目: Hydrothermal Synthesis and Energy Storage Applications of Carbon Aerogels-supported $Ni(OH)_2$/NiOOH Microspheres

作者: LI Jian, JIAO Yue, WAN Caichao

摘要: For developing high-performance supercapacitor electrodes and extending the application area of carbon aerogels, a facile and fast hydrothermal approach was used to embed $Ni(OH)_2$/NiOOH microspheres with a diameter of 2-4 μm into the three-dimensional carbon skeleton structure of CDCA. According to the scanning electron microscope (SEM) observations, the microsphere was formed by the different-direction aggregation and self-assembly of two-dimensional $Ni(OH)_2$/NiOOH nanosheets with a width of 300-800 nm. The as-prepared $Ni(OH)_2$/NiOOH/CDCA composite was mixed with acetylene black (conductive agent) and polytetrafluoroethylene (bonding agent) and then loaded onto a Ni foam. The mixture was pressed into a supercapacitor electrode. In a three-electrode mode, the area enclosed by the CV curve of this electrode is larger than that of the Ni foam-supported $Ni(OH)_2$/NiOOH without carbon aerogels. Moreover, the electrode displays a noticeable capacitance of 639 $mF \cdot cm^{-2}$ at 1 $mA \cdot cm^{-2}$. Also, the electrode presented an excellent cycling stability with a 92.37% of capacitance retention after 5000 cycles.

关键词: Cellulose; Carbon aerogels; Nickel hydroxide; Microspheres; Hydrothermal Synthesis; Supercapacitors

Facile Hydrothermal Synthesis of Fe_3O_4 @ Cellulose Aerogel Nanocomposite and Its Application in Fenton-like Degradation of Rhodamine B*

Yue Jiao, Caichao Wan, Wenhui Bao, He Gao, Daxin Liang, Jian Li

Abstract: A magnetic cellulose aerogel-supported Fe_3O_4 nanoparticles composite was designed as a highly efficient and eco-friendly catalyst for Fenton-like degradation of RhB. The composite (coded as Fe_3O_4@ CA) was formed by embedding well-dispersed Fe_3O_4 nanoparticles into the 3D structure of cellulose aerogels by virtue of a facile and cheap hydrothermal method. Comparative studies indicate that the Rhodamine B (RhB) decolorization ratio is much higher in co-presence of Fe_3O_4 and H_2O_2 than that in presence of Fe_3O_4 or H_2O_2 only, revealing that the Fe_3O_4 @ CA-catalyzed Fenton-like reaction governed the RhB decolorization process. It was also found that almost 100% RhB removal was achieved in the Fenton-like system. Moreover, the composite exhibited higher catalytic activity than that of the individual Fe_3O_4 particles. In addition, the Fe_3O_4@ CA catalyst retained ~97% of its ability to degrade RhB after the six successive degradation experiments, suggesting its excellent reusability. All these merits indicate that the green and low-cost catalyst with strong magnetic responsiveness possesses good potential for H_2O_2-driven Fenton-like treatment of organic dyestuff wastewater.

Keywords: Cellulose aerogels; Fe_3O_4; Hydrothermal method; Fenton-like degradation; Catalysts

1 Introduction

The relationship of industrial activity and environmental pollution is a serious topic and matter of great concern in modern times (Pandey & Ramontja, 2016). Wastewater discharge from industrial units is a large problem for conventional treatment plants in the entire world (Pandey, 2017). The release of the wastewater in natural environment is not only hazardous to aquatic life but also in many cases mutagenic to humans (Pandey, 2016). Amongst various hazardous substances in wastewater (such as oil, heavy metal ions and radioactive substances), dyes are a typical class of contaminants and is extensively utilized for dyeing, printing and several other coloring purposes in industries. Dyestuff wastewater, which is directly discharged into ecosystem without pre-treatments, affects the aquatic flora and fauna and causes environmental diseases (Kant, 2012). In addition, dyes accumulate in sediment and soil, leading to disruption of ecological balance. Also, dyes pollutants can directly pollute ground water systems as they can leach from the soil (Namasivayam & Sumithra, 2005). All these pollution modes are dangerous to human health. For instance, azo dyes (like 4-aminobiphenyl and 4-chloro-2-methylbenzenamine) have been banned by the German

* 本文摘自 Carbohydrate Polymers, 2018, 189: 371-378.

government for dyeing body-contact products since they produce carcinogenic amines during biodegradation (O'Neill et al., 1999). Some disperse dyes with anthraquinone or azo structures were found to cause allergic contact eczema (Hatch & Maibach, 1985). In addition, Clarke and Anliker who surveyed about 3000 colorants demonstrated that only 2% of dyes had an LC_{50} to fish of <1.0 mg dm^{-3} and over 96% of dyes had an LC_{50} above 10 mg dm^{-3} (Clarke & Anliker, 1984). Their work indicates that many kinds of dyes exhibit toxicity to mammals and aquatic organisms. However, it is not easy to treat dyestuff wastewater since the recalcitrant organic dyes molecules have strong resistance to aerobic digestion and are stable to light, heat and oxidizing agents.

Over the past few decades, several physical, chemical and biological methods have been developed to remove dyes released into aquatic environment, such as adsorption (Yagub, Sen, Afroze & Ang, 2014), biodegradation (Kalme, Parshetti, Jadhav & Govindwar, 2007), flocculation (Lee, Robinson & Chong, 2014; Renault, Sancey, Badot & Crini, 2009), reverse osmosis (Uzal, Yilmaz, & Yetis, 2010), thermal wet air oxidation (Kim & Ihm, 2011) and ultrafiltration (Alventosa-deLara, Barredo-Damas, Alcaina-Miranda & Iborra-Clar, 2012). However, there are still some typical defects for these approaches described above. For example, biodegradation is not suitable to treat industrial chemicals due to biomass poisoning. Thermal wet air oxidation of effluents with >100 g · L^{-1} of chemical oxygen demand generates high concentrations of toxic byproducts like dioxins and furans (Tuppurainen, Asikainen, Ruokojärvi & Ruuskanen, 2003). Physico-chemical methods (such as adsorption, flocculation and reverse osmosis) generally require additional post-treatments to prevent secondary disposal and contamination. More importantly, the price of some high-performance adsorbents or membranes is exorbitant; furthermore, some preparation technologies are complex and eco-unfriendly and thus unfit for extending to commercial scale. In addition to this, there are still some new adsorbents that are only selective for specific dyes but there is a still need to develop a material capable of decontaminating all types of pollutants (Qureshi et al., 2017). Recent investigations have verified that advanced oxidation processes (AOPs) possess favorable decolorizing ability for reactive dyes (Antonopoulou, Evgenidou, Lambropoulou & Konstantinou, 2014; Oturan & Aaron, 2014). AOPs operate at near-ambient temperature and pressure and can generate strongly oxidizing hydroxyl radicals (OH·) that aggressively and almost indiscriminately attack all types of organic pollutants in wastewater, leading to complete decomposition of organic dyes pollutants into non-toxic products with low molecular weight (such as CO_2 and H_2O). Especially, one of the green and effective AOPs called Fenton reaction has attracted widespread attention. Fenton reaction operates at acidic pH in the presence of H_2O_2 and ferrous ions yielding hydroxyl radicals with powerful oxidation capacity (Babuponnusami & Muthukumar, 2014; Bagal & Gogate, 2014; Laine & Cheng, 2007). Fenton reaction can be divided into two categories. One is the standard Fenton reaction which utilizes soluble Fe(Ⅱ) as the catalyst, while the other is referred to as Fenton-like process that includes the utilization of Fe(Ⅲ), e.g., FeOOH (Pinto et al., 2012) and iron chelates (Li, Bachas & Bhattacharyya, 2005). In addition, some other metal ions [like Co(Ⅱ), Cr(Ⅲ), Cu(Ⅱ), Mn(Ⅱ) and Ni(Ⅱ)] were also found to catalyze a similar Fenton-like reaction (Yoshikawa & Miyahara, 2003). Compared with the above-mentioned dyes removal methods, Fenton or Fenton-like reaction can provide a more eco-friendly, inexpensive, mild and low-energy process to treat dyestuff wastewater since the reaction merely involves H_2O_2 and Fenton or Fenton-like reagents (like iron oxides with rich reserves, low cost and environmental friendliness). Furthermore, Fenton or Fenton-like reaction directly decomposes dyes molecules during the treatment process without post-processing. By contrast, for adsorption and flocculation, secondary treatments are still required after collecting or separating dyes molecules.

In consideration of widespread availability, low toxicity, nonvolatility, high stability and low cost, inverse spinel magnetite (Fe_3O_4) is considered as a good Fenton-like catalyst (Costa, Moura, Ardisson, Fabris & Lago, 2008; Xu & Wang, 2012). Also, high catalytic effects of Fe_3O_4 in Fenton-like system have been experimentally demonstrated (Munoz, de Pedro, Casas & Rodriguez, 2015; Rahim Pouran, Abdul Raman & Wan Daud, 2014). Research has shown that Fe (Ⅱ) ions play a role in the mechanism and ·OH is the main active species for degradation of organic contaminants (Garrido-Ramírez, Theng & Mora, 2010; Wang, Zheng, Zhang & Wang, 2016). In detail, under acidic or neutral condition, the mechanism of H_2O_2 activation by Fe_3O_4 may refer to the formation of a complex between $\equiv Fe^{III}$ (the three horizontal lines stand for the sites on the surface of substrates where Fe_3O_4 are fasten) and H_2O_2, which is defined as $\equiv Fe^{III}H_2O_2$ (eq 1). The $\equiv Fe^{III}H_2O_2$ are subsequently converted to $\equiv Fe^{II}$ species and $\cdot HO_2$ (eq 2). The generated $\cdot HO_2$ further reacts with $\equiv Fe^{III}$, resulting in the generation of $\equiv Fe^{II}$ species (eq 3). All the produced $\equiv Fe^{II}$ species (eqs 2, 3) react with H_2O_2 to produce $\cdot OH$ radicals (eq 4), which readily degrade and mineralize organic contaminants (eq 5).

$$\equiv Fe^{III} + H_2O_2 \rightarrow \equiv Fe^{III}H_2O_2 \quad (1)$$

$$\equiv Fe^{III}H_2O_2 \rightarrow \equiv Fe^{II} + \cdot HO_2 + H^+ \quad (2)$$

$$\equiv Fe^{III} + \cdot HO_2 \rightarrow \equiv Fe^{II} + O_2 + H^+ \quad (3)$$

$$\equiv Fe^{II} + H_2O_2 \rightarrow \equiv Fe^{III} + \cdot OH + OH^- \quad (4)$$

$$\text{organic contaminants} + \cdot OH \rightarrow \cdots CO_2 + H_2O \quad (5)$$

However, Fe_3O_4 nanoparticles are easy to gather into large particles due to strong anisotropic dipolar interactions specifically in aqueous phase, which causes the reduction of dispersibility and activity of the nanoparticles (Zubir, Yacou, Motuzas, Zhang & Diniz da Costa, 2014). Therefore, it is necessary to immobilize Fe_3O_4 nanoparticles onto some special substrates to preserve their unique properties. For instance, Gu et al. (Gu, Zhu, Guo, Huang, Lou & Yuan, 2013) combined Fe_3O_4 nanoparticles with sewage sludge derived porous carbon, and the composite was applied for the adsorption and degradation of 1-diazo-2-naphthol-4-sulfonic acid in the presence of H_2O_2 with performance superior to that of pure Fe_3O_4 magnetic nanoparticles. Deng et al. (Deng, Wen & Wang, 2012) decomposed Acid Orange Ⅱ by a Fenton-like catalyst (Fe_3O_4-multi-walled carbon nanotubes), and the hybrid also displayed a higher activity than nanometer-size Fe_3O_4. Hua et al. (Hua et al., 2014) adopted graphene oxide to support Fe_3O_4 nanoparticles, which exhibited good stability and reusability to degrade bisphenol A by Fenton-like reaction. In addition, apart from these carbon-based substrates, some other types of substances were also utilized to integrate with Fe_3O_4 to develop various Fenton-like catalysts, such as SiO_2 (Huang, Su, Yang & Lu, 2013), natural maifanite (Zhao, Weng, Cui, Zhang, Xu & Liu, 2016), chitosan hollow fibers (Seyed Dorraji et al., 2015) and polyphenol (Wang, Fang & Megharaj, 2014). Especially, cellulose products are considered extensively as a class of green and cost-effective host materials because of their widely available feedstock, biodegradability and unique structural characteristics such as highly porous structure and large pore volume and surface area (Huang & Gu, 2011). Compared to one-dimensional (like fibers) or two-dimensional (like films and membrane) cellulose products, three-dimensional (3D) cellulose materials (like foams and aerogels) with these aforementioned more superior structural characteristics are beneficial to acquire higher loading content of nanoparticles (Olsson et al., 2010; Tian, Peng, Wu, Li, Deng & Liu, 2017; Wan, Jiao, Qiang & Li, 2017; Wan & Li, 2015a, b). In particular, strong interactions are expected to occur between the hydroxyl groups of cellulose and Fe_3O_4 nanoparticles, which may be the crucial factor to immobilize the nanoparticles (Cai, Kimura, Wada & Kuga, 2009). The existence of cellulose aerogels also introduces a series of new functions to the Fe_3O_4-based composite.

In this paper, magnetic Fe_3O_4 nanoparticles were synthesized and fasten to the 3D framework of cellulose aerogels by a mild, simple and low-cost hydrothermal method. Rhodamine B (RhB), a very common cationic dye with dark red color, was selected as the targeted dye contaminant considering its extensive utilizations in numerous fields (such as dyeing, tracer dye (Asraf-Sni & Gitis, 2011) and fluorescence correlation spectroscopy [LaRochelle, Cobb, Steinauer, Rhoades & Schepartz, 2015)] and health impacts [such as carcinogenicity (Umeda, 1956)]. The as-prepared Fe_3O_4@ cellulose aerogel (labeled as Fe_3O_4@CA) nanocomposite was used as an environmentally friendly catalyst to decompose RhB by Fenton-like reaction. Moreover, the mechanism of H_2O_2 activation by Fe_3O_4 was demonstrated by utilizing three kinds of different ingredients [i. e., (Ⅰ) H_2O_2, (Ⅱ) Fe_3O_4@CA and (Ⅲ) H_2O_2+ Fe_3O_4@CA] to treat water containing RhB. In addition, the catalytic activity of the individual Fe_3O_4 particles in the presence of H_2O_2 was tested and compared to that of Fe_3O_4@CA, for studying the contributions of cellulose aerogels in the Fenton-like reaction. The reusability of Fe_3O_4@CA was also studied by repeating the Fenton-like degradation process of RhB six.

2 Materials and method

2.1 Chemicals

Bamboo fiber was used as the cellulose source. The viscosity average molecular weight M_η of bamboo fiber is about 1.9×10^5 g mol^{-1} according to the supplier (Zhejiang Mingtong Textile Technology Co., Ltd., China), which was determined using viscometer and calculated from the Mark-Houwink equation $[\eta]=kM_\eta^\alpha$. Reagents, including ferric chloride ($FeCl_3\cdot6H_2O$), ethylene glycol, sodium hydroxide (NaOH), sodium acetate (CH_3COONa) and polyethylene glycol-4000 (PEG-4000), were purchased from Tianjin Kemiou Chemical Reagent Co., Ltd. (China) and used without further purification. RhB and 30% (w/w) H_2O_2 were provided by Sinopharm Chemical Reagent Co., Ltd. (Shanghai, China).

2.2 Synthesis of Fe_3O_4@CA

The cellulose hydrogels were prepared as per previous reports (Wan & Li, 2016; Wan, Lu, Cao, Sun & Li, 2015), which can be briefly described as follows: (1) the bamboo fiber was firstly added to a 10% aqueous solution of NaOH/PEG (namely 4 g NaOH and 0.4 g PEG-4000 were added into 40 ml distilled water) with magnetic stirring for 5 h at room temperature to form a homogeneous 2% cellulose solution; (2) the cellulose solution was then frozen at −15 ℃ for 12 h and thawed subsequently at room temperature under vigorous stirring for 30 min; (3) after being frozen at −15 ℃ for 5 h again, the frozen cake was immediately immersed in a 1% HCl solution for 6 h, and this immersion process was repeated several times until the formation of a cylindrical hydrogel; (4) after being repeatedly rinsed with distilled water for removing impurities, cellulose hydrogels were obtained. The Fe_3O_4@CA was synthesized via a hydrothermal method (Deng, Li, Peng, Wang, Chen & Li, 2005). In a typical process, the cellulose hydrogels (ca. 5 g) were immersed in a mixed solution of $FeCl_3 \cdot 6H_2O$ (1.35 g, 5 mmol) and ethylene glycol (40 mL). After 10 min of ultrasonic dispersion, the mixture was placed for 24 h at room temperature. After the impregnation, CH_3COONa (3.6 g) and PEG-4000 (1.0 g) were added into the above-mentioned mixture with gentle stirring for 10 min at room temperature and then transferred into a Teflon-lined stainless-steel autoclave. The autoclave was sealed and heated to 200 ℃ for 8 h. The resulting product was immersed in distilled water for 3 h and this process was repeated five times to remove residual chemicals. Finally, the rinsed product was subjected to a tert-butyl alcohol freeze-drying treatment at −35 ℃ for 24 h. In addition, for the preparation of pure cellulose aerogels (coded as PCA), the cellulose hydrogels were firstly immersed in tert-butyl alcohol to replace water with tert-butyl alcohol and subsequently freeze-dried at −30 ℃ for 48 h in an approximate vacuum (25 Pa). After depressurization, PCA were obtained. Individual Fe_3O_4 particles were synthesized by the same condition as Fe_3O_4@ CA and except without combining with the cellulose aerogels.

2.3 Characterizations

Morphology was characterized by scanning electron microscopy (SEM, Hitachi S4800) equipped with an energy-dispersive X-ray (EDX) spectrometer for elemental analysis. A patch of sample was adhered to a specimen stage by conductive adhesive tapes. Before the observation, the sample was sputtered with gold for the purpose of electrical conduction. Transmission electron microscopy (TEM) and high-resolution TEM (HRTEM) observations and selected area electron diffraction (SAED) were performed with a FEI, Tecnai G2 F20 TEM with a field-emission gun operating at 200 kV. The sample was suspended in ethanol and was prepared by being drop-cast onto a carbon-coated 200-mesh copper grid and subsequently dried at room temperature. The crystal structure was characterized by X-ray diffraction (XRD), which was implemented on a Bruker D8 Advance TXS XRD instrument with Cu Kα (target) radiation (λ = 1.5418 Å) at a scan rate (2θ) of $4° \cdot min^{-1}$ and a scan range of 5° to 80°. We grinded the sample into a fine powder and subsequently put the powder into a groove on a glass slide. After being compacted, the slide with the powder was used for XRD experiments. Chemical valences were characterized by X-ray photoelectron spectroscopy (XPS), which was executed by a Thermo Escalab 250Xi XPS spectrometer equipped with a dual X-ray source using Al Kα. The preparation method of XPS specimens is similar to that of XRD specimens. Fourier transform infrared spectra (FTIR) were recorded by a Nicolet Nexus 670 FTIR instrument in the range of 600−4000 cm^{-1} with a resolution of 4 cm^{-1}. The dry sample was ground into a powder and then blended with KBr before pressing the mixture into an ultra-thin pellet. Thermogravimetric (TG) analysis was performed with a synchronous thermal analyzer (SDTQ600, USA) under a nitrogen atmosphere from room temperature to 800 ℃ at a heating rate of 10 ℃ $\cdot min^{-1}$. A Quantum Design Magnetometer (MPMS XL−7) using a superconducting quantum interference device (SQUID) sensor was used to make measurements of the magnetic properties in applied magnetic fields over the range from −20000 to +20000 Oe at 298 K.

2.4 Fenton-like degradation experiments

The Fenton-like reaction activities of Fe_3O_4@ CA and Fe_3O_4 particles were evaluated by monitoring the degradation rate of RhB. All experiments were carried out in a 120 mL plastic cup exposed to air at 298 K in dark. The RhB solution (10 mg $\cdot L^{-1}$, 100 mL) was initially adjusted to pH 3.0 by using 0.1 M HCl. The additive amount of H_2O_2 is 1 mL (~9.9 mmol). The dosage of Fe_3O_4@ CA catalyst was 0.3 g, and the dosage of Fe_3O_4 particles catalyst was equal to the Fe_3O_4 loading of Fe_3O_4@ CA determined by the TG tests. The Fenton-like reaction process lasted for 64 h. The supernatant was collected at every time interval

(8 h) and used for the concentration measurements. The concentration of RhB was quantified by a TU-1901 UV-vis spectrophotometer (Beijing Purkinje, China) and the corresponding UV-Vis absorption spectra were recorded. The degradation rates (η) of RhB were calculated according to the following equation:

$$\eta=\frac{A_0-A_t}{A_0}\times 100\% \quad (6)$$

where A_0 and A_t are the initial absorbance and the absorbance at timet, respectively.

3 Results and discussion

3.1 Schematic diagram for the preparation of Fe_3O_4@CA

The schematic illustration for the preparation of Fe_3O_4@ CA is presented in Fig. 1. The bamboo fiber was firstly dissolved in the aqueous solution of NaOH/PEG leading to the formation of cellulose solution. The regeneration process was conducted by immersing the frozen solution in HCl solution at room temperature, and this process was repeated several times until the formation of a cylindrical cellulose hydrogel. The cellulose hydrogel was then soaked in the solution of $FeCl_3\cdot 6H_2O$, CH_3COONa, PEG and ethylene glycol. The mixture was hydrothermally treated in a Teflon-lined stainless-steel autoclave to insert Fe_3O_4 nanoparticles into the cellulose hydrogels. The resultant product finally underwent a freeze-drying process to obtain Fe_3O_4@ CA.

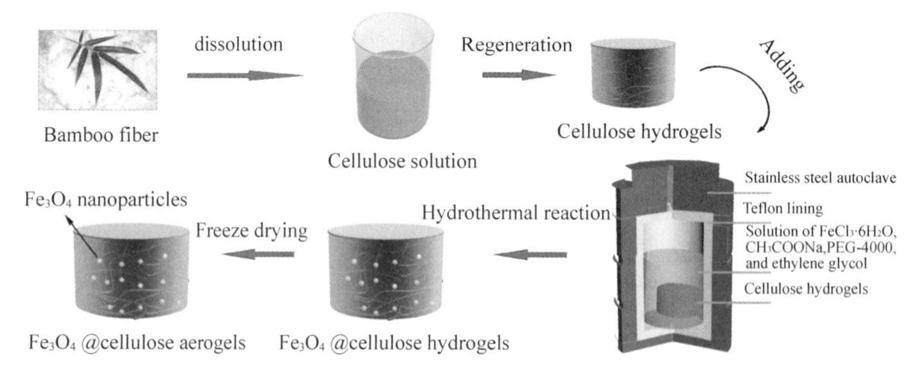

Fig. 1 Schematic diagram for the preparation of Fe_3O_4@CA

3.2 Morphological observations and composition analysis

The SEM image in Fig. 2a indicates that Fe_3O_4@ CA maintained an interconnected 3D network after the immersion and hydrothermal synthesis processes. However, it is hard to distinguish the synthetic Fe_3O_4 nanoparticles from the complicated network structure even when the magnification ratio increased to 20000 (Fig. 2b), possibly due to the small sizes of the nanoparticles. In addition, the EDX spectrum of Fe_3O_4@ CA exhibits strong peak assigned to Fe element (accounting for 22.84%) in Fig. 2c, indicating the plentiful presence of iron oxide.

The TEM images (Fig. 3a and b) show that the nanoparticles were homogeneously dispersed and immobilized in the 3D architecture of porous cellulose aerogels as indicated by the dark spots. No severe agglomeration phenomena occurred. The higher-magnification TEM image (Fig. 3c) clearly displays that the nanoparticles are quasi-spherical and well-dispersed. Undoubtedly, the good dispersion is beneficial to give full play to its nanometer effects. The size distribution of nanoparticles, which was calculated based on the TEM

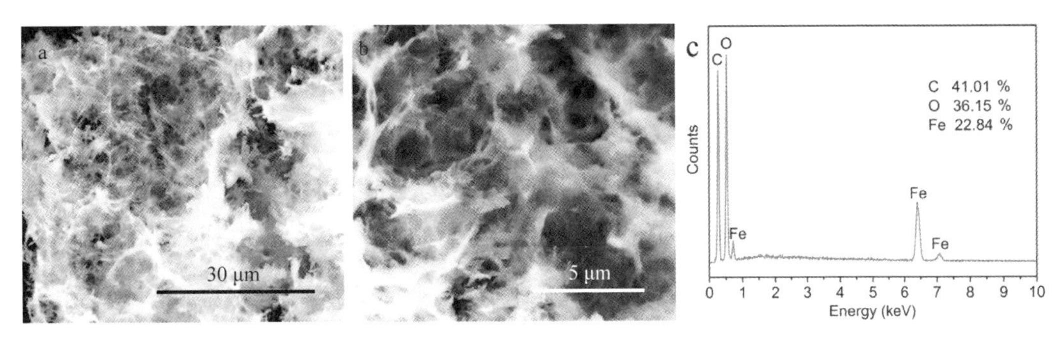

Fig. 2 (a and 3b) Different magnifications of SEM images and (c) EDX spectrum of Fe_3O_4@CA, respectively

images, is shown in Fig. 3d. The mean diameter (d) and standard deviation (σ) were estimated to be 11.85 and 2.02 nm, respectively. The HRTEM image demonstrates the presence of Fe_3O_4 in Fe_3O_4@ CA, as shown in Fig. 3e, i. e., the lattice fringes with a spacing of 0.48 nm agree well with the (111) plane of Fe_3O_4. Besides, the SAED pattern was performed on an individual Fe_3O_4 nanoparticle, as shown in Fig. 3f. The homogeneous ring patterns are consistent with the lattice spacings of Fe_3O_4. In addition, we hypothesize that there may be some interactions between hydroxyl groups of cellulose and Fe_3O_4 nanoparticles according to previous reports (Liu, Zhang, Zhou & Wu, 2008; Luo, Liu, Zhou & Zhang, 2009; Zeng, Liu, Cai & Zhang, 2010), which are expected to be helpful to immobilize Fe_3O_4 nanoparticles in cellulose aerogels. As shown in Fig. S1, a broad FTIR peak at 3000 – 3600 cm^{-1}, which is related to the stretching vibration of hydroxyl groups of cellulose, shifted from 3339 cm^{-1} for PCA to the lower wavenumber (ca. 3295 cm^{-1}) for Fe_3O_4@ CA.

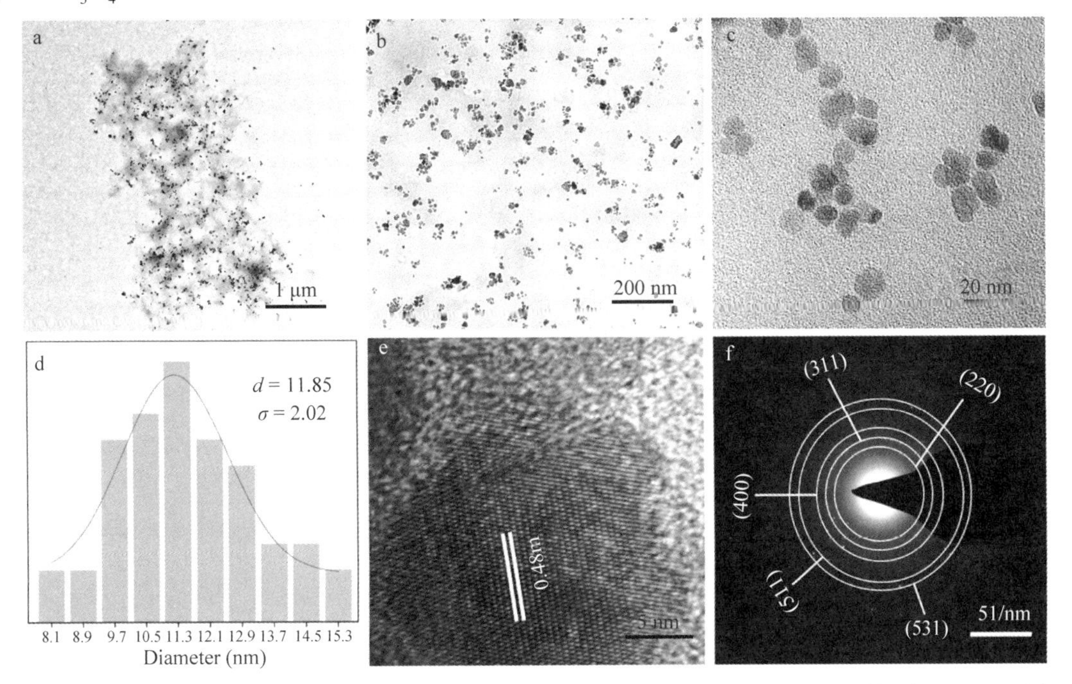

Fig. 3 (a–c) Different magnifications of TEM images and (d) size distribution of the Fe_3O_4 nanoparticles in Fe_3O_4@CA. (e) HRTEM image and (f) SAED pattern of Fe_3O_4@CA, respectively

This shift is related to aforementioned interactions (Zeng, Liu, Cai & Zhang, 2010). In addition, the peak at 2893 cm^{-1} is derived from C–H stretching vibration. The peaks at around 1642, 1417, 1370 and 1313 cm^{-1} are attributed to absorbed water, CH_2 symmetric bending, O–H bending and C–H bending (Lan, Liu, Yue, Sun & Kennedy, 2011), respectively. The peak at 1158 cm^{-1} belongs to C–O anti-symmetric stretching (Liu, Zhang, Li, Yue, & Sun, 2010). A shoulder band at 1110 cm^{-1} arises from C–OH skeletal vibration. The sharp peak at 893 cm^{-1} originates from glycosidic C–H deformation with ring vibration contribution, corresponding to the characteristic of β-glycosidic linkages between glucose in cellulose (Lan et al., 2011).

Fig. 4 presents the morphology of the individual Fe_3O_4 particles synthesized by the same hydrothermal method without combining with the cellulose aerogels. It is observed that the micron-sized blocky structure was generated (Fig. 4a). Based on the magnified SEM image in Fig. 4b, it can be found that the blocky structure was formed by the agglomeration of a great number of Fe_3O_4 nanoparticles. These images suggest that severe agglomeration of Fe_3O_4 nanoparticles occurred during the hydrothermal process because of the absence of cellulose aerogels. On the other hand, this phenomenon also demonstrates that the presence of cellulose aerogels helps to achieve the good dispersion of nano-Fe_3O_4 in nanometer scale.

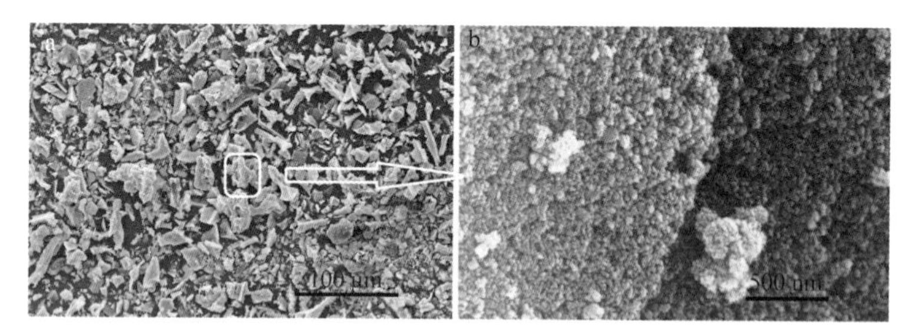

Fig. 4 (a) SEM image of the individual Fe_3O_4 particles synthesized by the same hydrothermal method without combining with the cellulose aerogels. (b) Magnified image of the green-squared marker region in (a)

The crystal structures of PCA and Fe_3O_4@CA were studied by XRD. As shown in Fig. 5a, the three peaks of PCA at 11.8°, 20.3° and 21.9° correspond to the ($\bar{1}$10), (110) and (200) planes of cellulose Ⅱ crystal structure (JCPDS card no. 03-0192) (Bose & Das, 2013). After the insertion of Fe_3O_4, the XRD pattern of the composite (Fe_3O_4@CA) verifies the co-existence of cellulose and Fe_3O_4. The new signals at 29.9°, 35.4°, 43.2°, 57.0° and 62.5° are related to the (220), (311), (400), (511) and (440) planes of Fe_3O_4 (JCPDS card no. 19-0629), further indicative of the successful synthesis of Fe_3O_4 in the cellulose aerogels. The average crystallite size of Fe_3O_4 nanoparticles in Fe_3O_4@CA was calculated from the full width at half maximum of the strongest diffraction peak (311) by Scherrer equation (Burton, Ong, Rea & Chan, 2009):

$$d = K\lambda/(\beta\cos\theta) \quad (7)$$

where λ is the X-ray wavelength, K is a constant taken as 0.89, θ is the angle of Bragg diffraction and β is the full width at half maximum. Therefore, the average crystallite size of Fe_3O_4 nanoparticles was calculated to be approximatively 10.08nm, which is consistent with the result of TEM observations (i.e., 11.85 nm). XPS analysis was executed to investigate the chemical valence of Fe element in Fe_3O_4@CA. Its high-resolution Fe 2p XPS spectrum is displayed in Fig. 5b, where two peaks at 711.2 and 725.0 eV correspond to

Fe $2p_{3/2}$ and Fe $2p_{1/2}$ peaks of Fe_3O_4(Yamashita & Hayes, 2008) , respectively. The broad spin-orbit split Fe 2p peaks are due to the small chemical shift difference between Fe^{2+} and Fe^{3+} in Fe_3O_4(Han, Ma, Sun, Lei & Lu, 2014). In addition, we cannot identify a shake-up satellite peak generally situated at ~719 eV (the fingerprint of the electronic structure of Fe_2O_3) (Grosvenor, Kobe, Biesinger & McIntyre, 2004), demonstrating that Fe_3O_4 rather than Fe_2O_3 was generated. In addition, the C 1s spectrum can be fitted into three peaks using Gaussian method in Fig. 5c. The three peaks located at 288.0, 286.5 and 284.8 eV can be ascribed to the O—C=O, C—OH, and C—C or C—H (Hu, Duo, Liu, Li & Zhang, 2010), respectively, which are primarily derived from the cellulose component.

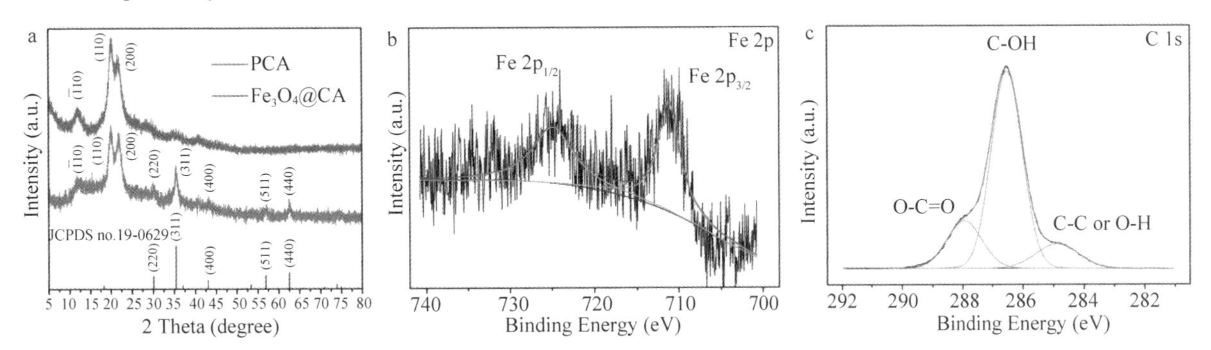

Fig. 5 (a) XRD pattern of PCA and Fe_3O_4@CA, and the bottom line is the standard JCPDS card no. 19-0629 for Fe_3O_4. High-resolution XPS spectra of (b) Fe 2p and (c) C 1s of Fe_3O_4@CA, respectively

3.3 Thermal stability and magnetic behaviors

The thermal stability of PCA and Fe_3O_4@ CA was investigated by TG and DTG. As shown in Fig. S2a, for both samples, the weight loss below 150 ℃ is because of the evaporation of adsorbed water. The strong peak at 300-400 ℃ is attributed to the pyrolysis of cellulose (Wu, Yao, Hu, Yang, Qing & Wu, 2014). In addition, the peak centered at 569.9 ℃ (see Fig. S2b) may be due to the redox reaction between the pyrolysis products of cellulose and the iron oxide based on the previous research, and the zero-valent iron and carbon are the primary product at 800 ℃ (Wan & Li, 2015a). The residue amounts of PCA and Fe_3O_4@ CA are 2.7% and 27.3% at 800 ℃, respectively. Therefore, the amount of zero-valent iron can be calculated to be 24.6 by subtracting the residue amount Fe_3O_4@ CA from the residue amount of PCA. According to the proportion of iron element in Fe_3O_4, the amount of Fe_3O_4 can be calculated to be $24.6\times232\div168\approx34.0\%$.

In Fig. S3a, Fe_3O_4@ CA shows the characteristic of superparamagnetic behavior, and no remanence and coercivity were observed, which may be attributed to the small particle size of Fe_3O_4 smaller than the related critical single-domain size (Gaudisson et al., 2014). The curve increases rapidly with increasing applied magnetic field and tends to saturate. Its saturation magnetization (M_s) value is about 6.7 emu g^{-1}. Taking the nanoparticles content in the sample (i.e., 34.0%) into consideration, the corresponding calculated M_s value can reach up to 19.7 emu g^{-1}. However, the value is still lower than that of the correspondent bulk Fe_3O_4(92 emu · g^{-1}) (Wang, Sun, Sun & Chen, 2003). This decline may be explained by the relatively smaller particle sizes due to the reduction of particles agglomeration resulted from the presence of hierarchical porous aerogels matrixes, hence inducing the enhancement of surface effects. In detail, when the size of magnetic particles reduces to the nanoscale, various surface defects become appreciable, resulting in a magnetically disordered layer at the particle surface. While the surface layer makes no contribution to the magnetization and becomes more remarkable with the decrease of particle size, consequently leading to the decline of M_s value (Wang, Yang, Xian, &Jiang, 2012). In Fig. S3b, it is seen that Fe_3O_4@ CA is easily adhered to the magnet surface by magnetic attraction (see also Supplementary Video 1). Furthermore, even when there is a 2 mm thickness of wood slice between the magnet and Fe_3O_4@ CA, Fe_3O_4@ CA can be still tightly adhered to the magnet surface (see also Supplementary Video 2). All these phenomena indicate that Fe_3O_4@ CA has strong magnetic responsiveness.

3.4 Catalytic activity in the Fenton-like system

RhB was chosen as the model organic pollutant to evaluate the catalytic activity of Fe_3O_4@ CA in the Fenton-like system. As shown in Fig. 6a, the self-degradation of RhB solution without the addition of catalyst and H_2O_2 is almost negligible. It is seen that the color of RhB solution also changed little (Fig. 6b), and only about 1% of RhB molecules were degraded after 64 h. In addition, we observed that when there was only Fe_3O_4@ CA in the solution of RhB, the removal rate of RhB was about 11.4% after 64 h. This result may be because of the adsorption ability of Fe_3O_4@ CA due to the porous structure. Furthermore, the degradation of RhB is 29% in the presence of H_2O_2 only. In contrast, a significant reduction in the RhB concentration (100% removal) was achieved by applying both Fe_3O_4@ CA and H_2O_2 to treat the RhB solution. The oxidation potentials of ·OOH and O_2^-· are much lower than that of ·OH (Hu et al., 2011; Lin & Gurol, 1998). As compared to ·OH in the Fenton-like reaction, ·OOH and O_2·$^-$ of H_2O_2 thus did much less contribution to the oxidation of RhB. In addition, the 100% RhB removal proportion verifies that the decolorization of RhB was governed by Fe_3O_4@ CA-catalyzed Fenton-like reaction. For demonstrating the contributions of the matrix (namely cellulose aerogels) in the Fenton-like reaction, we also measured the catalytic activity of the individual Fe_3O_4 particles in the presence of H_2O_2. It is clear that the degradation velocity of RhB solution using Fe_3O_4 particles is slower than that using Fe_3O_4 @ CA. Also, after the Fenton-like degradation experiment, there was still 7% of RhB remaining in the solution. These results indicate that the activity of Fe_3O_4@ CA is superior to that of Fe_3O_4 particles. Apart from good removal effects, fast removal rate is also pivotal. The pseudo-second-order reaction kinetics was applied to simulate the rate of the degradation of RhB as shown in Fig. S4. The model can be expressed as:

$$\frac{t}{Q_t}=\frac{t}{Q_e}+\frac{1}{K_2Q_e^2} \tag{8}$$

where Q_e and Q_t($mg \cdot g^{-1}$) is the amount of degraded RhB at equilibrium and at time t, respectively. K_2 is the kinetic rate constant ($g \cdot mg^{-1} \cdot min^{-1}$). The fitted rate constants for Fe_3O_4 and Fe_3O_4@ CA were calculated as 8.7×10^{-5} and 1.0×10^{-4} $g \cdot mg^{-1} \cdot min^{-1}$, respectively. The higher K_2 value for Fe_3O_4@ CA reflects the higher Fenton-like degradation activity. The activity enhancement for Fe_3O_4@ CA is possibly attributed to the homogeneous dispersion of Fe_3O_4 nanoparticles in the cellulose aerogels, which increases the active sites for Fenton-like reactions (Deng, Wen & Wang, 2012). Moreover, the smaller sizes of Fe_3O_4 nanoparticles in Fe_3O_4@ CA (see Fig. 3a-c), as compared to the individual Fe_3O_4 particles that were fabricated without the existence of cellulose aerogels (see Fig. 4), may also contribute to the faster RhB removal rate. In addition, regarding the Fenton-like reaction mechanism, as mentioned above, Fe^{III} and Fe^{II} can react with H_2O_2 to produce ·OH radical and the oxidation ability of radicals caused the degradation of RhB molecules into low molecular weight (such as CO_2 and H_2O) (Fig. 6c).

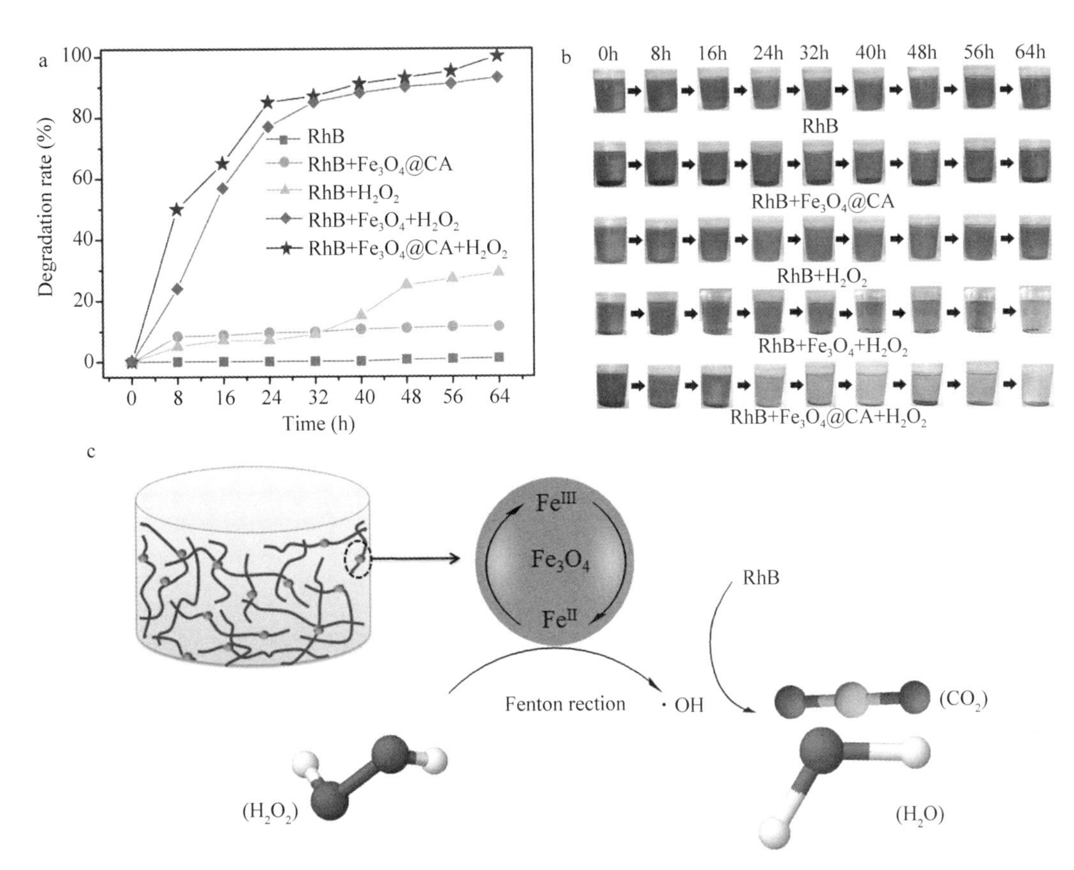

Fig. 6 (a) Degradation kinetics and (b) color changes of RhB solution treated with four kinds of different ingredients, i. e. , (Ⅰ) Fe_3O_4@CA, (Ⅱ) H_2O_2, (Ⅲ) Fe_3O_4+ H_2O_2 and (Ⅲ) Fe_3O_4@CA + H_2O_2. (c) Mechanism diagram of RhB degradation by Fenton-like reaction

The reusability is also one of the most important parameters to assess the practical application potential of catalysts. For the tests of reusability, Fe_3O_4@ CA was removed from the solution by an external magnetic field at the end of the each degradation process of RhB solution, and the supernatant was then collected from the solution for concentration determination. The isolated Fe_3O_4@ CA was washed with distilled water to remove residual chemicals, and was then used for next degradation experiment. Fig. 7a shows the color changes of RhB solutions before and after each Fenton-like reaction process. We can observe that all the RhB solutions were almost totally discolored after the reaction. The degradation rates of RhB solutions are presented in Fig. 7b and the corresponding UV-Vis absorption spectra were provided in Fig. S5 in the Supporting Information. The Fenton-like reaction system can maintain 97% - 100% of RhB degradation rates for the six degradation processes, indicative of an excellent reusability of Fe_3O_4@ CA.

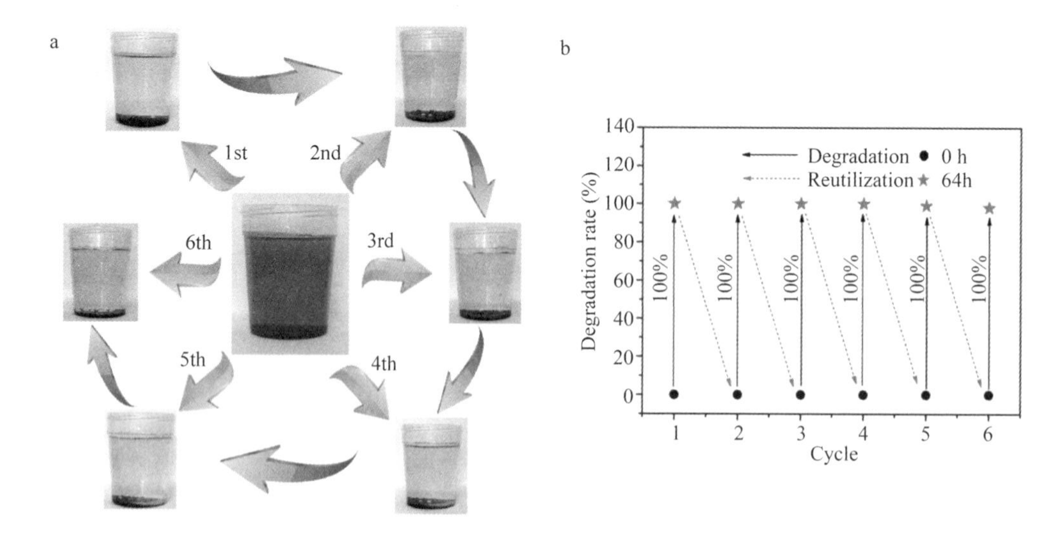

Fig. 7 (a) Color changes and (b) degradation rates of RhB solution using Fe_3O_4@CA before and after each Fenton-like degradation process

4 Conclusions

In summary, the cellulose aerogels were decorated with Fe_3O_4 nanoparticles by using a facile hydrothermal synthesis method. Moreover, the analysis of experiment results indicates that Fe_3O_4@CA was an efficient Fenton-like catalyst. Due to the excellent catalysis of Fe_3O_4@CA, the degradation rate of RhB can reach up to 100% after 64 hours' reaction in the Fenton-like system. Moreover, Fe_3O_4@CA has higher catalytic activity than that of the individual Fe_3O_4 particles. Additionally, Fe_3O_4@CA still remained high catalytic activity for the Fenton-like degradation of RhB after the six successive runs, reflecting its favorable reusability which is a crucial factor for practical application. Moreover, the strong magnetic responsiveness of Fe_3O_4@CA provides convenience for the separation and recycle of Fe_3O_4@CA from the waste water. Consequently, the Fe_3O_4@CA catalyst in this study holds good prospect to be utilized as an eco-friendly, cheap and efficient catalyst for environmental remediation (especially for treatment of organic dyestuff wastewater).

Supporting Information

Please see the supporting information (Video 1 and 2) through the following links:
https://mega.nz/#!jAxlHabT!5aFRL65B5kdHoecmmfECwqqP3_1NgirOqD-yaq8CnMg

Acknowledgements

This study was supported by the National Natural Science Foundation of China (grant no. 31270590 and 31470584) and the Fundamental Research Funds for the Central Universities (grant no. 2572016AB22).

References

[1] Alventosa-Delara E, Barredo-Damas S, Alcaina-Miranda M I, et al. Ultrafiltration technology with a ceramic membrane for reactive dye removal: Optimization of membrane performance[J]. Journal of Hazardous Materials, 2012, 209-210(none): 492-500.

[2] Antonopoulou M, Evgenidou E, D Lambropoulou, et al. A review on advanced oxidation processes for the removal of taste and odor compounds from aqueous media[J]. Water Research, 2014, 53(apr. 15): 215-234.

[3] Asraf-Snir M, Gitis V. Tracer studies with fluorescent-dyed microorganisms—A new method for determination of residence time in chlorination reactors[J]. Chemical Engineering Journal, 2011, 166(2): 579-585.

[4] Babuponnusami A, Muthukumar K. A review on Fenton and improvements to the Fenton process for wastewater treatment[J]. Journal of Environmental Chemical Engineering, 2014, 2(1): 557-572.

[5] Bagal M V, Gogate P R. Wastewater treatment using hybrid treatment schemes based on cavitation and Fenton chemistry: a review[J]. Ultrasonics Sonochemistry, 2014, 21(1): 1-14.

[6] Bose S, Das C. Preparation and characterization of low cost tubular ceramic support membranes using sawdust as a pore-former[J]. Materials Letters, 2013, 110: 152-155.

[7] Burton A W, Ong K, Rea T, et al. On the estimation of average crystallite size of zeolites from the Scherrer equation: A critical evaluation of its application to zeolites with one-dimensional pore systems[J]. Microporous and Mesoporous Materials, 2009, 117(1-2): 75-90.

[8] Cai J, Kimura S, Wada M, et al. Nanoporous cellulose as metal nanoparticles support. [J]. Biomacromolecules, 2009, 10(1): 87-94.

[9] Clarke E A, Anliker R. Safety in use of organic colorants: health and safety aspects[J]. Review of Progress in Coloration and Related Topics, 1984, 14(1): 84-89.

[10] Costa R CC, Moura F C C, Ardisson J D, et al. Highly active heterogeneous Fenton-like systems based on FeO/Fe_3O_4 composites prepared by controlled reduction of iron oxides[J]. Applied Catalysis B: Environmental, 2008, 83(1-2): 131-139.

[11] Deng H, Li X, Peng Q, et al. Monodisperse magnetic single-crystal ferrite microspheres[J]. Angewandte Chemie, 2005, 117(18): 2842-2845.

[12] Deng J, Wen X, Wang Q. Solvothermal in situ synthesis of Fe_3O_4-multi-walled carbon nanotubes with enhanced heterogeneous Fenton-like activity[J]. Materials Research Bulletin, 2012, 47(11): 3369-3376.

[13] Garrido-Ramírez E G, Theng B K G, Mora M L. Clays and oxide minerals as catalysts and nanocatalysts in Fenton-like reactions—a review[J]. Applied Clay Science, 2010, 47(3-4): 182-192.

[14] Gaudisson T, Artus M, Acevedo U, et al. On the microstructural and magnetic properties of fine-grained $CoFe_2O_4$ ceramics produced by combining polyol process and spark plasma sintering[J]. Journal of Magnetism & Magnetic Materials, 2014, 370(dec.): 87-95.

[15] Gijs M A M. Magnetic bead handling on-chip: new opportunities for analytical applications[J]. Microfluidics and nanofluidics, 2004, 1(1): 22-40.

[16] Grosvenor A P, Kobe B A, Biesinger M C, et al. Investigation of multiplet splitting of Fe 2p XPS spectra and bonding in iron compounds[J]. Surface and Interface Analysis: An International Journal devoted to the development and application of techniques for the analysis of surfaces, interfaces and thin films, 2004, 36(12): 1564-1574.

[17] Gu L, Zhu N, Guo H, et al. Adsorption and Fenton-like degradation of naphthalene dye intermediate on sewage sludge derived porous carbon[J]. Journal of Hazardous Materials, 2013, 246: 145-153.

[18] Han F, Ma L, Sun Q, et al. Rationally designed carbon-coated Fe_3O_4 coaxial nanotubes with hierarchical porosity as high-rate anodes for lithium ion batteries[J]. Nano Research, 2014, 7(11): 1706-1717.

[19] Hatch K L, Maibach H I. Textile dye dermatitis: a review[J]. Journal of the American Academy of Dermatology, 1985, 12(6): 1079-1092.

[20] Hu C, Duo S, Liu T, et al. Low temperature facile synthesis of anatase TiO_2 coated multiwalled carbon nanotube nanocomposites[J]. Materials Letters, 2010, 64(22): 2472-2474.

[21] Hu X, Liu B, Deng Y, et al. Adsorption and heterogeneous Fenton degradation of 17α-methyltestosterone on nano Fe_3O_4/MWCNTs in aqueous solution[J]. Applied Catalysis B: Environmental, 2011, 107(3-4): 274-283.

[22] Hua Z, Ma W, Bai X, et al. Heterogeneous Fenton degradation of bisphenol A catalyzed by efficient adsorptive Fe_3O_4/GO nanocomposites[J]. Environ Sci Pollut Res Int, 2014, 21(12): 7737-7745.

[23]Huang Y H, Su CC, Yang Y P, et al. Degradation of aniline catalyzed by heterogeneous Fenton-like reaction using iron oxide/SiO_2[J]. Environmental Progress & Sustainable Energy, 2013, 32(2): 187-192.

[24]Huang J, Gu Y. Self-assembly of various guest substrates in natural cellulose substances to functional nanostructured materials[J]. Current opinion in colloid & interface science, 2011, 16(6): 470-481.

[25]Kant R. Textile dyeing industry an environmental hazard[J]. Natural Science, 2012, 4(1): 22-26.

[26]Kim K H, Ihm S K. Heterogeneous catalytic wet air oxidation of refractory organic pollutants in industrial wastewaters: a review[J]. Journal of Hazardous Materials, 2011, 186(1): 16-34.

[27] Laine D F, Cheng I F. The destruction of organic pollutants under mild reaction conditions: A review [J]. Microchemical Journal, 2007, 85(2): 183-193.

[28]Lan W, Liu C F, Yue F X, et al. Ultrasound-assisted dissolution of cellulose in ionic liquid[J]. Carbohydrate polymers, 2011, 86(2): 672-677.

[29]Larochelle J R, Cobb G B, Steinauer A, et al. Fluorescence Correlation Spectroscopy Reveals Highly Efficient Cytosolic Delivery of Certain Penta-Arg Proteins and Stapled Peptides[J]. Journal of the American Chemical Society, 2015, 137 (7): 2536-2541.

[30]Lee C S, Robinson J, Chong M F. A review on application of flocculants in wastewater treatment[J]. Process Safety and Environmental Protection, 2014, 92(6): 489-508.

[31] Li Y C, Bachas L G, Bhattacharyya D. Kinetics studies of trichlorophenol destruction by chelate-based Fenton reaction[J]. Environmental Engineering Science, 2005, 22(6): 756-771.

[32] Lin SS, Gurol M D. Catalytic decomposition of hydrogen peroxide on iron oxide: kinetics, mechanism, and implications[J]. Environmental science & technology, 1998, 32(10): 1417-1423.

[33]Liu C F, Zhang A P, Li W Y, et al. Succinoylation of cellulose catalyzed with iodine in ionic liquid[J]. Industrial Crops and Products, 2010, 31(2): 363-369.

[34]Liu S, Zhang L, Zhou J, et al. Structure and Properties of Cellulose/Fe_2O_3 Nanocomposite Fibers Spun via an Effective Pathway[J]. The Journal of Physical Chemistry C, 2008, 112(12): 4538-4544.

[35]Luo X, Liu S, Zhou J, et al. In situ synthesis of Fe_3O_4/cellulose microspheres with magnetic-induced protein delivery [J]. Journal of Materials Chemistry, 2009, 19(21): 3538-3545.

[36] Munoz M, Pedro Z D, Casas J A, et al. Preparation of magnetite-based catalysts and their application in heterogeneous Fenton oxidation - A review[J]. Applied Catalysis B Environmental, 2015, 176-177: 249-265.

[37]Namasivayam C, Sumithra S. Removal of direct red 12B and methylene blue from water by adsorption onto Fe (Ⅲ)/Cr (Ⅲ) hydroxide, an industrial solid waste[J]. Journal of environmental management, 2005, 74(3): 207-215.

[38]Olsson R T, Samir M, Salazar-Alvarez G, et al. Making flexible magnetic aerogels and stiff magnetic nanopaper using cellulose nanofibrils as templates[J]. Nature Nanotechnology, 2010, 5(8): 584-588.

[39]O'Neill C, Hawkes F R, Hawkes D L, et al. Colour in textile effluents-sources, measurement, discharge consents and simulation: a review [J]. Journal of Chemical Technology & Biotechnology: International Research in Process, Environmental & Clean Technology, 1999, 74(11): 1009-1018.

[40]Oturan M A, Aaron J J. Advanced oxidation processes in water/wastewater treatment: principles and applications. A review[J]. Critical Reviews in Environmental Science and Technology, 2014, 44(23): 2577-2641.

[41]Pandey S. Highly sensitive and selective chemiresistor gas/vapor sensors based on polyaniline nanocomposite: A comprehensive review[J]. Journal of Science: Advanced Materials and Devices, 2016, 1(4): 431-453.

[42]Pandey S, Ramontja J. Natural bentonite clay and its composites for dye removal: current state and future potential [J]. American Journal of Chemistry and Applications, 2016, 3(2): 8-19.

[43] Pandey S. A comprehensive review on recent developments in bentonite-based materials used as adsorbents for wastewater treatment[J]. Journal of Molecular Liquids, 2017, 241: 1091-1113.

[44]Pinto I S X, Pacheco P H VV, Coelho J V, et al. Nanostructured δ-FeOOH: an efficient Fenton-like catalyst for the oxidation of organics in water[J]. Applied Catalysis B: Environmental, 2012, 119: 175-182.

[45] Qureshi U A, Khatri Z, Ahmed F, et al. Highly efficient and robust electrospun nanofibers for selective removal of acid dye[J]. Journal of Molecular Liquids, 2017, 244: 478-488.

[46] Pouran S R, Raman A, Wan M. Review on the application of modified iron oxides as heterogeneous catalysts in Fenton reactions[J]. Journal of Cleaner Production, 2014, 64(2): 24-35.

[47] Renault F, Sancey B, Badot P M, et al. Chitosan for coagulation/flocculation processes-an eco-friendly approach[J]. European Polymer Journal, 2009, 45(5): 1337-1348.

[48] Dorraji M S S, Mirmohseni A, Carraro M, et al. Fenton-like catalytic activity of wet-spun chitosan hollow fibers loaded with Fe3O4 nanoparticles: batch and continuous flow investigations[J]. Journal of Molecular Catalysis A: Chemical, 2015, 398: 353-357.

[49] Tian J, Peng D, Wu X, et al. Electrodeposition of Ag nanoparticles on conductive polyaniline/cellulose aerogels with increased synergistic effect for energy storage[J]. Carbohydrate polymers, 2017, 156: 19-25.

[50] Tuppurainen K, Asikainen A, Ruokojaervi P, et al. Perspectives on the formation of polychlorinated dibenzo-p-dioxins and dibenzofurans during municipal solid waste (MSW) incineration and other combustion processes[J]. Cheminform, 2003, 34 (48): 652-658.

[51] Umeda M. Experimental study of xanthene dyes as carcinogenic agents[J]. Gann, 1956, 47(1): 51-78.

[52] Uzal N, Yilmaz L, Yetis U. Nanofiltration and Reverse Osmosis for Reuse of Indigo Dye Rinsing Waters[J]. Separation Science and Technology, 2010, 45(3): 331-338.

[53] Wan C, Jiao Y, Qiang T, et al. Cellulose-derived carbon aerogels supported goethite (α-FeOOH) nanoneedles and nanoflowers for electromagnetic interference shielding[J]. Carbohydrate polymers, 2017, 156: 427-434.

[54] Wan C, Li J. Facile synthesis of well-dispersedsuperparamagnetic γ-Fe_2O_3 nanoparticles encapsulated in three-dimensional architectures of cellulose aerogels and their applications for Cr (Ⅵ) removal from contaminated water[J]. ACS Sustainable Chemistry & Engineering, 2015, 3(9): 2142-2152.

[55] Wan C, Li J. Synthesis of well-dispersed magnetic $CoFe_2O_4$ nanoparticles in cellulose aerogels via a facile oxidative co-precipitation method[J]. Carbohydrate polymers, 2015, 134: 144-150.

[56] Wan C, Li J. Cellulose aerogels functionalized withpolypyrrole and silver nanoparticles: In-situ synthesis, characterization and antibacterial activity[J]. Carbohydrate polymers, 2016, 146: 362-367.

[57] Wan C, Lu Y, Cao J, et al. Preparation, characterization and oil adsorption properties of cellulose aerogels from four kinds of plant materials via aNaOH/PEG aqueous solution[J]. Fibers and Polymers, 2015, 16(2): 302-307.

[58] Wang J, Sun J, Sun Q, et al. One-step hydrothermal process to prepare highly crystalline Fe_3O_4 nanoparticles with improved magnetic properties[J]. Materials research bulletin, 2003, 38(7): 1113-1118.

[59] Wang W, Yang H, Xian T, et al. XPS and magnetic properties of $CoFe_2O_4$ nanoparticles synthesized by a polyacrylamide gel route[J]. Materials Transactions, 2012, 53(9): 1586-1589.

[60] Wang N, Tong Z, Zhang G, et al. A review on Fenton-like processes for organic wastewater treatment[J]. Journal of Environmental Chemical Engineering, 2016, 4(1): 762-787.

[61] Wang Z, Fang C, Megharaj M. Characterization of iron-polyphenol nanoparticles synthesized by three plant extracts and their fenton oxidation of azo dye[J]. ACS Sustainable Chemistry & Engineering, 2014, 2(4): 1022-1025.

[62] Wu Y, Yao C, Hu Y, et al. Flame retardancy and thermal degradation behavior of red gum wood treated with hydrate magnesium chloride[J]. Journal of Industrial & Engineering Chemistry, 2014, 20(5): 3536-3542.

[63] Xu L, Wang J. Magnetic nanoscaled Fe_3O_4/CeO_2 composite as an efficient Fenton-like heterogeneous catalyst for degradation of 4-chlorophenol[J]. Environmental science & technology, 2012, 46(18): 10145-10153.

[64] Yagub M T, Sen T K, Afroze S, et al. Dye and its removal from aqueous solution by adsorption: a review[J]. Advances in colloid and interface science, 2014, 209: 172-184.

[65] Yamashita T, Hayes P. Analysis of XPS spectra of Fe^{2+} and Fe^{3+} ions in oxide materials[J]. Applied Surface Science, 2008, 254(8): 2441-2449.

[66] Strlic M, Kolar J, Selih V S, et al. A comparative study of several transition metals in Fenton-like reaction systems at

circum-neutral pH[J]. Acta Chimica Slovenica, 2003, 50(4): 619-632.

[67]Zeng J, Liu S, Cai J, et al. TiO_2 Immobilized in Cellulose Matrix for Photocatalytic Degradation of Phenol under Weak UV Light Irradiation[J]. Journal of Physical Chemistry C, 2010, 114(17): 7806-7811.

[68]Zhao H, Weng L, Cui W W, et al. In situ anchor of magnetic Fe_3O_4 nanoparticles onto natural maifanite as efficient heterogeneous Fenton-like catalyst[J]. Frontiers of Materials Science, 2016, 10(3): 300-309.

[69]Zubir N A, Yacou C, Motuzas J, et al. Structural and functional investigation of graphene oxide-Fe_3O_4 nanocomposites for the heterogeneous Fenton-like reaction[J]. Scientific reports, 2014, 4(1): 1-8.

中文题目： 水热法制备 Fe_3O_4@纤维素气凝胶纳米复合材料及其在类 Fenton 降解罗丹明 B 中的应用

作者： 焦月，万才超，包文慧，高鹤，梁大鑫，李坚

摘要： 设计了一种磁性纤维素气凝胶负载纳米 Fe_3O_4 复合催化剂，用于类 Fenton 降解罗丹明 B (RhB)。这种复合材料(记为 Fe_3O_4@CA)是通过一种简单、廉价的水热方法将分散良好的 Fe_3O_4 纳米颗粒嵌入纤维素气凝胶的 3D 结构中而形成的。对比研究表明，Fe_3O_4 和 H_2O_2 共同存在时，RhB 的脱色率明显高于单独 Fe_3O_4 和 H_2O_2 时的脱色率，说明 Fe_3O_4@CA 催化的类 Fenton 反应控制了 RhB 的脱色过程。研究还发现，在类 Fenton 系统中，RhB 的去除率几乎达到 100%。此外，复合粒子的催化活性高于单独的 Fe_3O_4 粒子。此外，Fe_3O_4@CA 催化剂经过连续 6 次降解实验，对 RhB 的降解能力保持在 97%左右，具有良好的重复使用性能。这些优点表明，这种具有较强磁响应性的绿色、低成本催化剂在 H_2O_2 驱动的类 Fenton 处理有机染料废水方面具有良好的应用前景。

关键词： 纤维素气凝胶；水热法；类芬顿降解；催化剂

Hydrothermal Synthesis of SnO_2-ZnO Aggregates in Cellulose Aerogels for Rhodamine B Degradation*

Yue Jiao, Caichao Wan, Jian Li

Abstract: Cellulose aerogels have been considered as ideal green matrixes to support various nanoparticles due to their hierarchical porous nanostructures and an ocean of surface hydroxyl groups. Herein, an easily-operated low-energy hydrothermal method was implemented to synthesize SnO_2-ZnO aggregates in cellulose aerogels. The well-dispersed aggregates with the mean diameter of 45.4 nm are tightly adhered on the three-dimensional framework of the aerogels, due to the strong interactions between the hydroxyl groups of cellulose and the oxygen atoms of SnO_2 or ZnO. Furthermore, the tri-component composite shows good photocatalytic activity for degradation of rhodamine B under ultraviolet irradiation, which is possibly helpful to deal with dyestuff wastewater as a kind of low-pollution cost-effective photocatalyst.

Keywords: Cellulose aerogels; Nanocomposites; Photocatalysts; Hydrothermal method

1 Introduction

Aerogels are materials prepared by replacing the liquid solvent in a gel with air without substantially altering the network structure or the volume of the gel body[1]. The unique highly porous characteristic endows them with numerous extraordinary performances, such as low density, high specific surface area, and large open pores[2]. Nowadays, aerogels have been paid increasing attention and regarded as promising candidates for a variety of advanced applications, like supercapacitors, oil absorbents, super-insulating materials, and molecular sensors[3].

Aerogels based on cellulose (the most abundant natural polymer) have shown many advantages over traditional aerogels (such as inorganic and synthetic polymer-based aerogels), which not only possess all above merits but also avoid the consumption of nonrenewable resources and have bright environmental advantages. Moreover, the features of cellulose aerogels, like hierarchical micro/nano-scale three-dimensional (3D) network and an ocean of surface hydroxyls, make them suitable candidates as green matrixes to support various nanoparticles, helping to reduce agglomeration and control particle growth[4, 5]. Moreover, the generating composites create a multitude of new properties such as photocatalysis, antibacterial activity, preferential adsorption, and energy conversion[6], primarily dependent on the characteristics of the inserted nanoparticles. Our previous studies[7-10] have confirmed that the cellulose aerogels can serve as ideal hosts to support various nanostructures like γ-Fe_2O_3, $CoFe_2O_4$, rod-like ZnO, and anatase TiO_2. Therefore, it is interesting and meaningful to further develop this new class of composites and expand their functions.

In the present work, SnO_2-ZnO aggregates were synthesized and incorporated into the porous cellulose

* 本文摘自 Cellulose Chemistry and Technology, 2018, 52: 141.

aerogels via a facile and cost-effective hydrothermal method. As an example of potential application, the tri-component hybrid, i. e. , SnO_2-ZnO@ cellulose aerogels (coded as SZ@ CA), was utilized as an eco-friendly photocatalyst for degradation of rhodamine B (RhB) under ultraviolet (UV) irradiation.

2 Materials and methods

2.1 Materials

All chemical reagents, including zinc nitrate hexahydrate [$Zn(NO_3)_2 \cdot 6H_2O$], stannic chloride pentahydrate ($SnCl_4 \cdot 5H_2O$), cetyltrimethyl ammonium bromide (CTAB), sodium hydroxide (NaOH), absolute ethanol, and tert-butyl alcohol, were purchased from Tianjin Kemiou Chemical Reagent Co. , Ltd. (China) and used without further purification.

2.2 Hydrothermal synthesis of SZ@ CA

The preparation of cellulose hydrogel (i. e. , the precursor of cellulose aerogel) can be referred to some previous literatures[7, 11]. The synthesis of SZ @ CA followed a hydrothermal method[12]. Firstly, in the preparation of ZnO rods, 50 mL aqueous solution of $Zn(NO_3)_2 \cdot 6H_2O$ (0.13 M) was mixed with 50 mL aqueous solution of NaOH (1.3 M) with magnetic stirring for 30 min, and then the mixture was heated in a water bath at 50 ℃ for 90 min. After the heating, the mixed solution was allowed to naturally cool to room temperature, and the product was separated by centrifugation and then rinsed with a large amount of absolute ethanol and finally dried at 60 ℃ for 24 h in a vacuum. The resulting ZnO rods were used as the starting zinc source. Secondly, for the hydrothermal treatment, CTAB (2.19 g), $SnCl_4 \cdot 5H_2O$ (1.58 g), ZnO rods (0.1 g), and NaOH aqueous solution (1.8 g, 70 mL) were mixed with magnetic stirring for 30 min. Thereafter, the mixed solution and the cellulose hydrogel were transferred into a 100 mL Teflon-lined stainless-steel autoclave and maintained at 160 ℃ for 12 h. Thirdly, the resulting SZ@ CA was repeatedly rinsed with distilled water and tert-butyl alcohol in sequence for the removal of residual impurities. After that, the hybrid was subjected to a freeze drying process at −35 ℃ for 48 h in an approximate vacuum (25 Pa) to remove most liquid.

2.3 Characterizations

Transmission electron microscopy (TEM) observations and selected area electron diffraction (SAED) were performed with a FEI, Tecnai G2 F20 TEM equipped with an energy dispersive X-ray (EDX) detector for elemental analysis. The samples were suspended in ethanol and were prepared by being drop-cast onto a carbon-coated 200-mesh copper grid and subsequently dried at room temperature.

2.4 Photocatalytic activity measurements

SZ@ CA sample (0.2 g) was grinded into powder and added into an aqueous RhB solution (20 mg · L^{-1}, 50 mL), and the mixture was magnetically stirred in the dark for 30 min to ensure the establishment of an adsorption-desorption equilibrium on the catalyst surface. Thereafter, a dish containing the mixture was placed in a closed chamber with a UV source (mercury lamp, 300 W). The dish was exposed to the mercury lamp for 2 h. The RhB solution was collected at regular time intervals. The mixed solution was centrifuged at 8000 rpm for 5 min to remove the catalyst, and the supernatant was collected. The concentration of RhB was tested by an UV-vis-NIR spectrophotometer (U-4100, Hitachi) at its maximum absorption wavelength of 554 nm.

3 Results and discussion

Fig. 1a shows the TEM image of SZ@ CA. Obviously, these SnO_2-ZnO aggregates were uniformly adhered on the cellulose aerogels matrixes, and no free aggregates separated from the matrixes were observed, which confirm that the cellulose aerogels can act as ideal host materials. Increasing the magnification of TEM image, it can be seen in Fig. 1b that the aggregates are approximately spherical and composed of several smaller particles. The good interface combination between the cellulose aerogels and the SnO_2-ZnO aggregates can be clearly seen (Fig. 1c). The aggregates are inserted and tightly immobilized in the 3D framework of the aerogels, possibly due to the interactions between the hydroxyl groups of cellulose and the oxygen atoms of SnO_2 or ZnO. Randomly collecting the aggregates in Fig. 1b and accurately measuring their diameters using Adobe Photoshop CS5 software, the diameter distribution was plotted in Fig. 1d. We can see that the diameter histogram exhibits Gaussian-like distribution, and the diameters of the aggregates range from 26 to 68 nm. The mean diameter (d) and standard deviation (σ) were calculated to be around 45.4 and 8.56 nm, respectively.

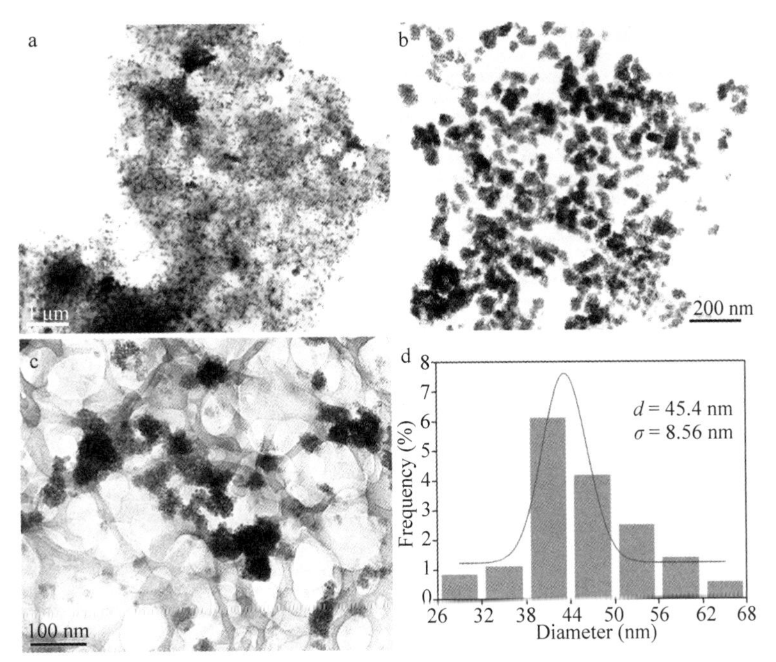

Fig. 1 (a-c) TEM images of SZ@CA under different levels of magnification. (d) Diameter distribution of SnO_2-ZnO aggregates in SZ@CA

The EDX spectrum in Fig. 2a displays the existence of element of C, O, Sn, and Zn, which are derived from the cellulose aerogels and SnO_2-ZnO aggregates. The EDX analysis shows that the atomic ratio of Sn to Zn is around 3 : 1. Moreover, the signals of Cu and partial C are generated from the sample holder. In Fig. 2b, the high resolution TEM (HRTEM) image reveals the simultaneous presence of crystalline ZnO and SnO_2 phases. The measured two lattice fringes with lattice spacings of 0.336 and 0.280 nm agree well with the d-spacings of the (110) plane of the tetragonal rutile SnO_2 and the (100) plane of the hexagonal wurtzite ZnO[13], respectively. The corresponding SAED pattern in Fig. 2c shows several concentric rings, indicating that the aggregates have a polycrystalline structure. These concentric rings can be assigned to diffraction from (110) and (101) planes of ZnO and (110), (101), (211) and (112) planes of SnO_2[14]. Therefore, these aforementioned characterizations confirm that the secondary particles (namely the aggregates) consist of ZnO and SnO_2.

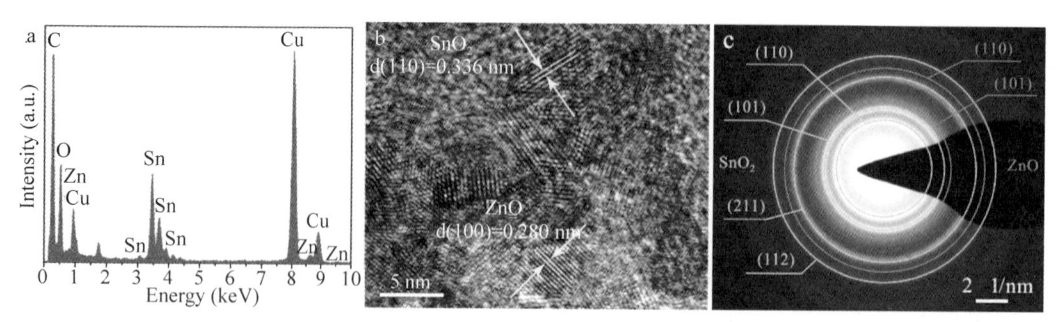

Fig. 2 (a) EDX spectrum, (b) HRTEM image, and (c) SAED pattern of SZ@CA, respectively

As an example of potential applications, SZ@ CA was used as a green biodegradable photocatalyst for the degradation of RhB under UV irradiation. Fig. 3a shows the degradation rates of RhB over no catalyst and SZ@ CA (C_0 and C are the equilibrium concentration of RhB before and after UV irradiation, respectively). The negligible noncatalytic degradation of RhB shown by the blank test was detected after the exposure to UV light for 2 h. By contrast, under identical conditions with exposure to UV light, SZ @ CA displays good photocatalytic activity. The RhB molecules were gradually degraded into colorless organic compounds or small molecules (e. g. , CO_2), leading to color changes of the RhB solution from purple to approximatively transparent colorless (Fig. 3b). The previous literature has verified that the cellulose composition has no obvious photocatalytic activity for degradation of dye molecules[15]. Thus these results reveal that the SnO_2-ZnO aggregates play important roles in the photocatalytic activity of the hybrid. The unique porous nanostructures of the aerogels matrixes help to maintain good dispersion of the aggregates, which is crucial for photocatalytic property.

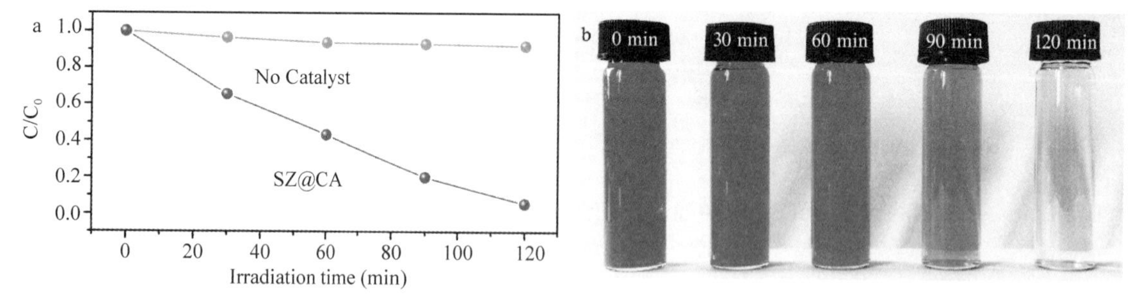

Fig. 3 (a) Photocatalytic degradation rates of RhB under UV irradiation by no catalyst and SZ@CA. (b) Color changes of RhB solution during the photocatalytic degradation process

4 Conclusions

A mild, simple, and low-cost hydrothermal method was employed to synthesize the SnO_2-ZnO aggregates in the porous cellulose aerogels. The approximately spherical aggregates have the mean diameter of 45. 4 nm, and display favorable dispersion in the aerogels matrixes without significant agglomeration. Moreover, the composite displays good photocatalytic activity for the degradation of RhB under UV irradiation, which is possibly useful as a kind of novel environmentally friendly photocatalyst for treatment of dyestuff wastewater.

Acknowledgements

This study was supported by the National Natural Science Foundation of China (grant no. 31270590 and 31470584).

References

[1] Shi J, Lu L, Guo W, et al. An environment-friendly thermal insulation material from cellulose and plasma modification [J]. Journal of Applied Polymer Science, 2013, 130(5): 3652-3658.

[2] Kistler S S. Coherent expanded aerogels and jellies[J]. Nature, 1931, 127(3211): 741-741.

[3] Hrubesh L W. Aerogel Applications[J]. Journal of Non-Crystalline Solids, 1998, 225(1): 335-342.

[4] Tingaut P, Zimmermann T, Sèbe G. Cellulose nanocrystals and microfibrillated cellulose as building blocks for the design of hierarchical functional materials[J]. Journal of Materials Chemistry, 2012, 22(38): 20105-20111.

[5] Cai J, Kimura S, Wada M, et al. Nanoporous cellulose as metal nanoparticles support[J]. Biomacromolecules, 2009, 10(1): 87-94.

[6] Siró I, Plackett D. Microfibrillated cellulose and new nanocomposite materials: a review[J]. Cellulose, 2010, 17(3): 459-494.

[7] Wan C, Li J. Facile synthesis of well-dispersedsuperparamagnetic γ-Fe_2O_3 nanoparticles encapsulated in three-dimensional architectures of cellulose aerogels and their applications for Cr (VI) removal from contaminated water[J]. ACS Sustainable Chemistry & Engineering, 2015, 3(9): 2142-2152.

[8] Wan C, Li J. Embedding ZnO nanorods into porous cellulose aerogels via a facile one-step low-temperature hydrothermal method[J]. Materials & Design, 2015, 83: 620-625.

[9] Wan C, Li J. Synthesis of well-dispersed magnetic $CoFe_2O_4$ nanoparticles in cellulose aerogels via a facile oxidative co-precipitation method[J]. Carbohydrate polymers, 2015, 134: 144-150.

[10] Jiao Y, Wan C, Li J. Room-temperature embedment ofanatase titania nanoparticles into porous cellulose aerogels[J]. Applied Physics A, 2015, 120(1): 341-347.

[11] Wang W, Li F, Yu J, et al. Structure and properties of novel cellulose-based fibers spun from aqueous NaOH solvent under various drawing conditions[J]. Cellulose, 2015, 22(2): 1333-1345.

[12] Wang WW, Zhu Y J, Yang L X. ZnO-SnO_2 hollow spheres and hierarchical nanosheets: hydrothermal preparation, formation mechanism, and photocatalytic properties[J]. Advanced Functional Materials, 2007, 17(1): 59-64.

[13] Tian W, Zhai T, Zhang C, et al. Low-cost fully transparent ultraviolet photodetectors based on electrospun ZnO-SnO_2 heterojunction nanofibers[J]. Advanced Materials, 2013, 25(33): 4625-4630.

[14] Feng N, Qiao L, Hu D, et al. Synthesis, characterization, and lithium-storage of ZnO-SnO_2 hierarchical architectures [J]. RSC Advances, 2013, 3(21): 7758-7764.

[15] Zeng J, Liu S, Cai J, et al. TiO_2 immobilized in cellulose matrix for photocatalytic degradation of phenol under weak UV light irradiation[J]. The Journal of Physical Chemistry C, 2010, 114(17): 7806-7811.

中文题目：纤维素气凝胶/SnO_2-ZnO 聚集体的水热合成及其光催化降解罗丹明 B 性能研究

作者：焦月，万才超，李坚

摘要：纤维素气凝胶由于具有层次化的多孔纳米结构和大量的表面羟基，被认为是支撑各种纳米颗粒的理想绿色基质。本文采用一种易于操作的低能水热法在纤维素气凝胶中合成 SnO_2-ZnO 聚集体。由于纤维素中的羟基与 SnO_2 或 ZnO 中的氧原子之间存在较强的相互作用，使得平均粒径为 45.4 nm 的团聚体紧密地附着在气凝胶的三维骨架上。此外，该三组分复合材料在紫外光照射下对罗丹明 B 的降解表现出良好的光催化活性，有望作为一种低污染、低成本的光催化剂应用于染料废水的处理。

关键词：纤维素气凝胶；纳米复合材料；光催化剂；水热法

A Scalable Top-to-bottom Design on Low Tortuosity of Anisotropic Carbon Aerogels for Fast and Reusable Passive Capillary Absorption and Separation of Organic Leakages*

Caichao Wan, Yue Jiao, Song Wei, Xianjun Li,
Wenyan Tian, Yiqiang Wu, Jian Li

Abstract: Creation of sustainable, cost-effective and scalable absorbents with ideal absorption properties is a worldwide challenge because many high-performance absorbents are still restricted in laboratory scope due to several critical defects (like complex and eco-unfriendly synthesis process, high cost, and difficulty in large-scale production). Herein, a facile and scalable top-to-bottom design is proposed to create a kind of novel anisotropic carbon aerogels with low tortuosity of stacked laminated structure, derived from the hierarchical cellular channels of balsa wood. In virtue of this unique structure and favorable oleophilicity, a fast passive capillary absorption with low flow resistance is achieved (as demonstrated by the theoretical modeling). As a result, the anisotropic carbon aerogels have quite sensitive selectivity to separate organic pollutants from water, broad-spectrum and high absorption capacity for different organic liquids (13277 − 31597 mg · g^{-1}), and superior recyclability (98.7% absorption capacity retention after 5 cycles). Combining these outstanding performances with cheap preparation strategy as well as good environmental friendliness, this work provides a kind of potential scalable materials for efficient reusable absorption and separation of organic leakages.

Keywords: Carbon aerogels; Anisotropic structure; Capillary absorption; Top-to-bottom design; Absorption selectivity; Absorption recyclability; Absorbents

1 Introduction

Serious environmental issues, which are resulted from organic leakages or arbitrary discharge of organic wastewater, have motivated great efforts to develop cost-effective and high-performance absorbents for separation of organic pollutants from water.[1-3] To date, some traditional absorbing materials (such as activated carbon, diatomite, zeolites, clay, molecular sieve and biomass wastes) usually suffer from weak selectivity (namely simultaneously absorbing non- or low-polar organics and polar water), negligible reusability and low absorption capacity. Furthermore, many previously reported high-performance absorbents are still restricted in laboratory scope due to several critical defects (like complex and eco-unfriendly synthesis process, high cost, and difficulty in large-scale production).[4-6] As a result, it is of great importance to develop new, eco-friendly, easily available and cheap porous absorbents meeting all these practical

---

* 本文摘自 ACS Applied Materials Interfaces, 2019, 11, 47846-47857.

performance demands.

Aerogels are highly porous sponge-like materials, whose sizes of pores and solid skeleton are both in the nanometer range. Aerogels have been widely used for the separation of organic liquids-water mixture and absorption of organic liquids.[6-8] Commonly, aerogels (such as silica and biomass aerogels) have non-obvious selectivity; therefore, a hydrophobic and oleophilic coating (achieved by grafting chlorosilane, fluorosilane, and so on) needs to be created before use.[9,10] By comparison, carbon aerogels are a special class of aerogels, which are originated from the pyrolysis of organic precursors. Therefore, carbon aerogels inherently have high hydrophobicity owing to the removal of hydrophilic oxygen-containing components in the precursors.[11] As a result, carbon aerogels have the ability to treat organic liquids-water mixture without any further modifications. Traditionally, various petroleum-based systems (like resorcinol-formaldehyde,[12] cresol-formaldehyde,[13] cresol-resorcinol-formaldehyde,[14] phenol-furfural[15] and phenolic resole-methylolated melamine,[16] as listed in Table S1) are studied most frequently for the preparation of carbon aerogels. Apparently, these systems cannot well meet the increasing requirement for green chemistry. Consequently, some greener carbon aerogels (such as using cellulose acetate aerogels,[17] sodium carboxymethyl cellulose aerogels[18] and regenerated cellulose aerogels[19] as precursors) are emerged. Nevertheless, in the preparation of these precursors, the slow sol-gel process and solvent replacement as well as the high energy consumption of nanocrystallization (like high-frequency ultrasonic treatment,[20] high-pressure homogenization[21] and steam explosion[22]) are always involved. Furthermore, this bottom-to-top way is prone to form an isotropic structure with high tortuosity of cross-linked network.[8,11,23] Nevertheless, recent studies demonstrate that low tortuosity of porous structure can enable a fast absorption with low flow resistance, which is beneficial to acquire a high absorption volume and rate.[24,25] In addition, it is still rare for the study on low tortuosity of anisotropic carbon aerogels and their application for cleanup of organic pollutants.

In this study, we demonstrate a facile and scalable top-to-bottom design on low tortuosity of anisotropic carbon aerogels for the efficient reusable absorption and separation of organic pollutants. This design includes two mild and rapid procedures, namely the removal of lignin and pyrolysis, as illustrated in Fig. 1a. Balsa wood, the lightest wood on the earth, is used as the starting material to build carbon aerogels due to its low density (~ 130 mg · cm^{-3}, Table S2), high porosity ($\sim 91.3\%$) and thin cell wall. The purpose of delignification is to induce structure transformation from hierarchical cellular channels to a low tortuosity of stacked laminated structure, extending the interior volume and reducing the diffusion resistance of absorbed liquids.[26] In addition, the aim of the pyrolysis is to produce a drastic composition variation from hydrophilic matters to strongly hydrophobic carbon. Relying on the low tortuosity of anisotropic structure, strong hydrophobicity, high scalability, low cost, and good environmental friendliness, the delignified wood-based carbon aerogels are expected to act as promising candidate for the efficient and reusable separation of organic liquids/water and absorption of organic liquids (Fig. 1b).

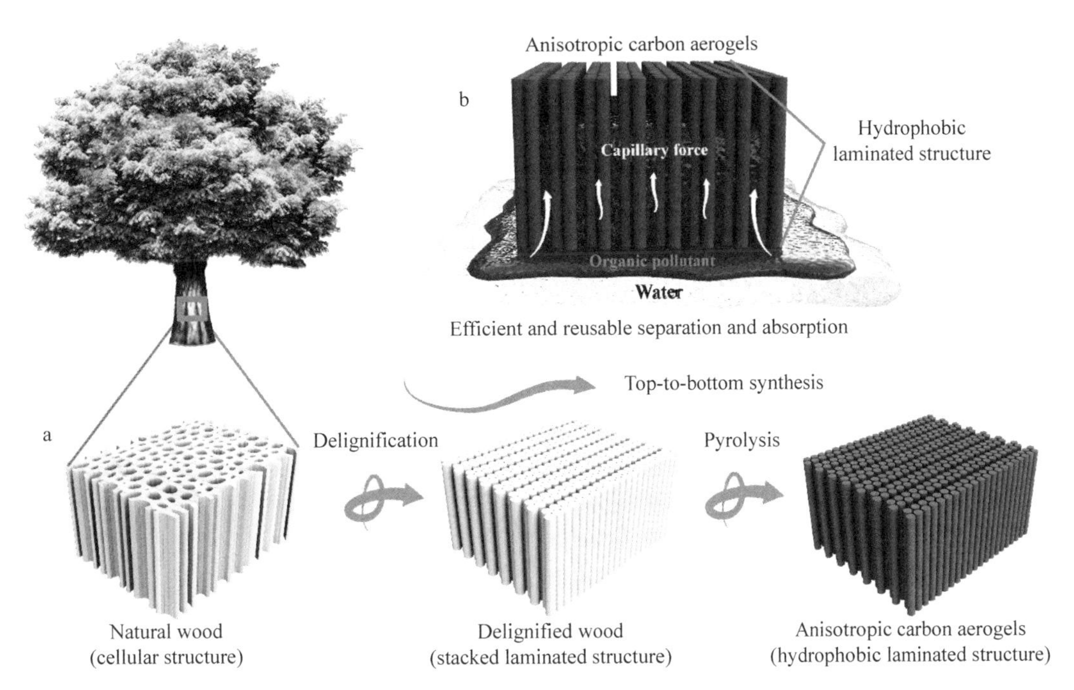

Fig. 1 Schematic illustrations for (a) scalable top-to-bottom design on the anisotropic carbon aerogels and (b) capillary force-assisted organic pollutant cleanup by the carbon aerogels

2 Results and discussion

Changes of the macro- and micro-structure before and after these top-to-bottom treatments are presented in Fig. 2a. By comparing their appearances in Fig. 2a, e and i, the yellow natural wood becomes white due to the delignification (lignin has a complex structure with dark color and is the primary source of wood color) and further turns into black after the pyrolysis. The density decreases from 130 to 72 mg · cm^{-3} after the removal of lignin and further declines to 43 mg · cm^{-3} due to the evolution from polymers to carbon (see Table S2). Such a low density of 43 mg · cm^{-3} is far lower than or comparable to that of various carbon aerogels prepared through the bottom-to-top way, like resorcinol-formaldehyde system (550 mg · cm^{-3}), [12] phenol-melamine-formaldehyde system (300 mg · cm^{-3}), [27] phenol furfural system (180 mg · cm^{-3}), [15] sodium carboxymethyl cellulose system (52 mg · cm^{-3}) [18] and cellulose microfibril system (10 mg · cm^{-3}), [28] as listed in Table S1.

The cross-section scanning electron microscope (SEM) images (XY plane, Fig. 2b and c) of the natural wood shows three-dimensional (3D) porous channels with lumina of dozens of micrometers in diameter. From the radial-section SEM image (XZ plane, Fig. 2d), all these channels are straight aligned along the tree-growth direction. After removing the lignin composition, we can find a clear evolution from the irregular hexagonal lumina to the sheet-like architecture consisting of parallelly arrayed fibers (Fig. 2f and g). Besides, the wall is well maintained (XZ plane), as seen in Fig. 2h, indicative of the retention of channels. Through the subsequent pyrolysis, these sheets (XY plane) become thinner (Fig. 2j and k). In addition, from the SEM image of the carbon aerogels in YZ plane (Fig. 2l and S1), a clear and low tortuosity of anisotropic stacked laminated structure with interlayer spacing of dozens to hundreds of micrometers is well formed. This architecture has the ability to serve as a spontaneously absorbed reservoir for the migration and storage of organic liquids.[29] The transmission electron microscope (TEM) image of the carbon aerogels is shown in Fig. 2m, where the thin sheet-like structure can also be identified, consistent with the above SEM observations. Fig. 2n shows the corresponding high-resolution TEM (HRTEM) image. There is no obvious long-range ordered structure, revealing the amorphous nature of carbon component in the aerogels. Moreover, the selected area electron diffraction (SAED) pattern presents dispersing diffraction rings (Fig. 2o), which is another proof of the amorphous structure.

Fig. 2 Structural characterizations of the natural wood, delignified wood and delignified wood-based carbon aerogels: (a) digital photo of the natural wood; (b–d) SEM images of the natural wood in the XY plane (b, c) and XZ plane (d); (e) digital photo of the delignified wood; (f–h) SEM images of the delignified wood in the XY plane (f, g) and XZ plane (h); (i) digital photo of the delignified wood-based carbon aerogels; (j–l) SEM images of the delignified wood-based carbon aerogels in the XY plane (j, k) and YZ plane (l); (m) TEM and (n) HRTEM images and (o) SAED pattern of the carbon aerogels

The Fourier transformed infrared (FTIR) analysis indicates that the delignification treatment causes a significant decline in the intensity of lignin characteristic bands (Fig. 3a), e. g., 1732, 1592 and 1237 cm^{-1} assigned to the C=O stretching vibration, aromatic skeletal vibrations and stretching vibration of Ar—O,[30] respectively, as summarized in Table S3. The effect of pyrolysis is also verified by FTIR spectra, where the main absorption bands of functional groups (like C=O, C—O, C—H and O—H) disappear in the FTIR spectra of carbon aerogels, revealing the formation of hydrophobicity. The corresponding energy dispersive X-ray patterns (Fig. S2) also reflect the dramatic decrease of oxygen content from 41.25% (natural wood) to 5.96% (carbon aerogels). The result is further proved by X-ray diffraction (XRD) patterns (Fig. 3b), in which the three peaks at $2\theta = 15.1°$, $16.5°$ and $22.9°$ corresponding to the (101), (101-) and (002) planes of cellulose I completely vanish after the pyrolysis.[31] Instead, two broad peaks centred at $2\theta = 23.4°$ and $44.6°$ are generated and assigned to the (002) and (100) planes of graphite.[32] In addition, the increase of crystallinity index from 76.0% to 81.9% (calculated based on the Segel' method[33]) also reflects the removal of amorphous lignin. The thermogravimetric (TG) and derivative thermogravimetric (DTG) curves of the carbon aerogels and the feedstock (nature wood) are measured and plotted in Fig. S3. As shown in Fig. S3a, for the nature wood, the small initial drop occurring before 150 ℃ is due to the evaporation of retained moisture. The DTG curves in Fig. S3b display two main regions: the first region occurs between 200 ℃ and 315 ℃, and a pronounced shoulder related to the hemicellulose decomposition appears; the second region (315-393 ℃) is attributed to the cellulose degradation.[8] Moreover, the lignin degrades slowly in the whole temperature range, leading to no obvious characteristic peaks. By contrast, there are no any exothermic peaks in the DTG curve of the carbon aerogels, and the weight loss during the while pyrolysis is also slight. These features indicate the almost complete removal of the oxygen-containing components in the carbon aerogels.

Nitrogen adsorption-desorption isotherms are used to study the pore structure characteristic. As presented in Fig. 3c, the adsorption uptake of the carbon aerogels in the P/P_0 range of 0 and 0.6 increases slightly, indicating the existence of few micropores.[34] On the contrary, in the range of 0.6-1, the adsorption amount increases quickly and still does not reach a plateau near the P/P_0 of 1.0, revealing the presence of mesopores and macropores. Besides, the adsorption-desorption loop belongs to the type H3, suggesting the existence of slit-shaped pores with parallel plate structure.[35] The result agrees well with the SEM observation in Fig. 2l. Moreover, the pore diameter distribution reveals that the pore sizes are within the scope of 1-200 nm (Fig. 3d), demonstrating the existence of mesopores and macropores. The uptrend of pore volume with the increase of pore size suggests the existence of abundant macropores, beneficial to absorb organic liquids. In addition, the carbon aerogels possess the maximum surface area and pore volume ($37.1\ m^2 \cdot g^{-1}$ and $0.148\ cm^3 \cdot g^{-1}$), 18.7 and 24.7 times of those of the natural wood (Table S2). Also, the carbon aerogels reach the highest porosity of 97.9% and largest average pore diameter of 33.5 nm among these four kinds of materials (Table S2). According to the geometry model for tortuosity of flow path in porous media proposed by Yu and Li' research,[36] the averaged tortuosity (γ_{av}) as a function of porosity (δ) is given by:

$$\gamma_{av} = \frac{1}{2}\left[1 + \frac{1}{2}\sqrt{1-\delta} + \sqrt{1-\delta} \cdot \frac{\sqrt{\left(\frac{1}{\sqrt{1-\delta}} - 1\right)^2 + \frac{1}{4}}}{1 - \sqrt{1-\delta}}\right] \tag{1}$$

As a result, the γ_{av} value of the carbon aerogels is calculated as 1.038, lower than that of the natural wood (1.085), delignified wood (1.060) and wood-derived carbon (1.049) (Table S2). The lower γ_{av} value indicates the lower tortuosity, contributing to a faster absorption with lower flow resistance. These above superiorities of the anisotropic carbon aerogels are expected to acquire more excellent absorption and separation functions.

As expected, the hydrophilic wood is transformed into the hydrophobic and oleophilic carbon aerogels

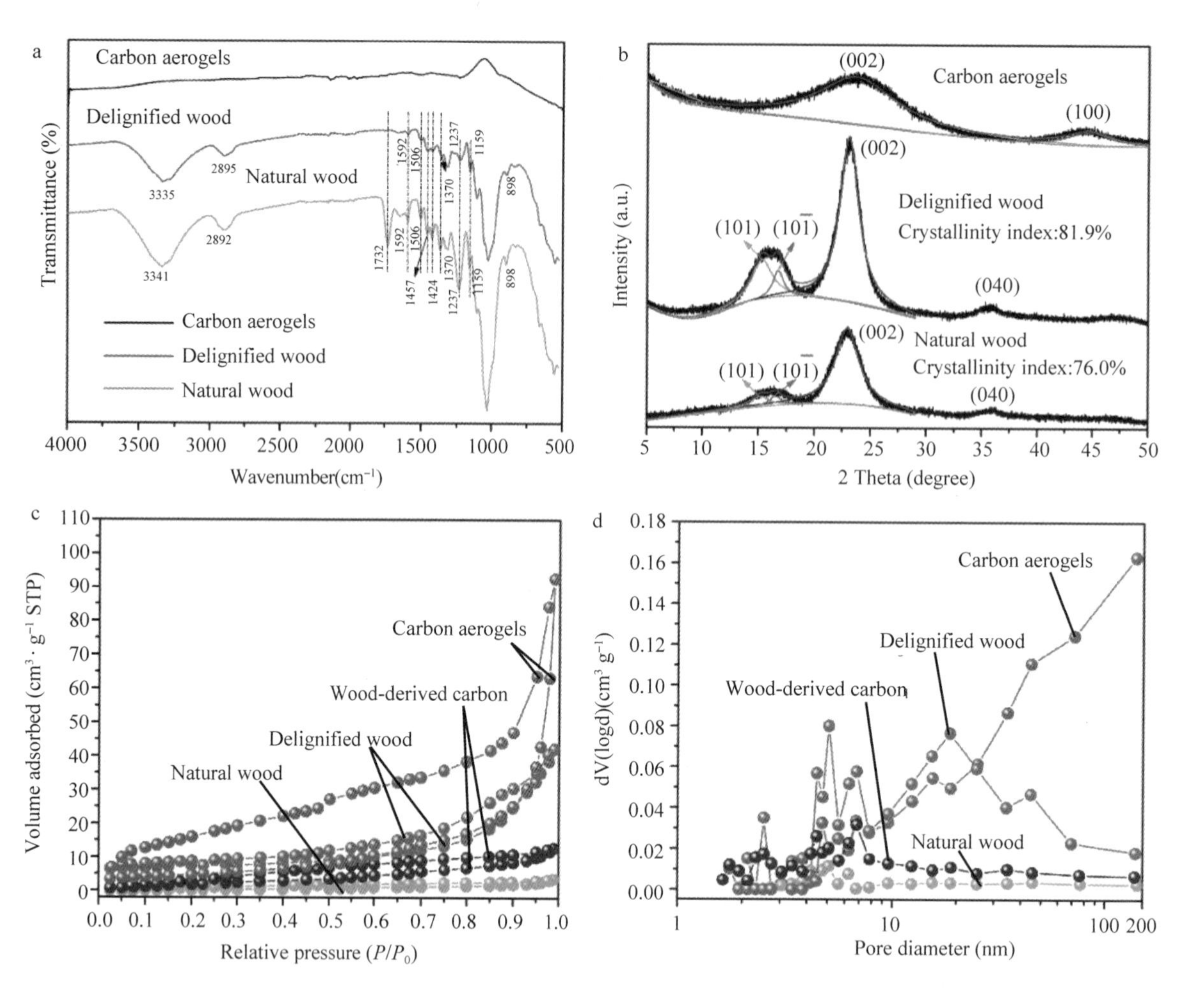

Fig. 3 Chemical compositions, crystal structure and pore structure of the natural wood, delignified wood, wood-derived carbon and delignified wood-based carbon aerogels: (a) FTIR spectra, (b) XRD patterns, (c) nitrogen adsorption-desorption isotherms and (d) pore diameter distributions

after the pyrolysis. The water contact angle image suggests the hydrophilicity of the natural wood (Fig. S4). By contrast, water droplets keep their spherical shapes when placed onto the carbon aerogel (Fig. 4a), and the water contact angle increases from 122.6° to 134.9° when the pyrolysis temperature increases from 700 ℃ to 1300 ℃ (Fig. S5). Therefore, the carbon aerogels pyrolyzed at 1300 ℃ acquire the best hydrophobic property, and this property holds potential in liquid-liquid separation or water treatment technology. On the contrary, the n-octane contact angle of the carbon aerogels approaches zero (Fig. 4c), indicating the good oleophilicity.

Because of the unique structure and wettability as well as low density, the carbon aerogels are regarded as ideal candidate for highly efficient separation/extraction of organic substances from water. As presented in Fig. 4d and e, when a small piece of carbon aerogel is forced to the liquids, the carbon aerogels rapidly absorb the organic pollutants on the surface or at the bottom of water (i.e., n-octane and tetrachloromethane dyed with Sudan Ⅳ) completely within several seconds. The Videos for the overall absorption processes are available in Video S1 and S2. Also, the effective separation of organics from water by the aerogels is also demonstrated by the UV-vis absorption spectra (Fig. S6). These results well display the excellent absorption selectivity of the anisotropic carbon aerogels. In addition, from Fig. 4d, we can find that the carbon aerogel is floating on the water surface after collecting all pollutant, thereby revealing an efficient fast route for the cleanup of organic leakages.

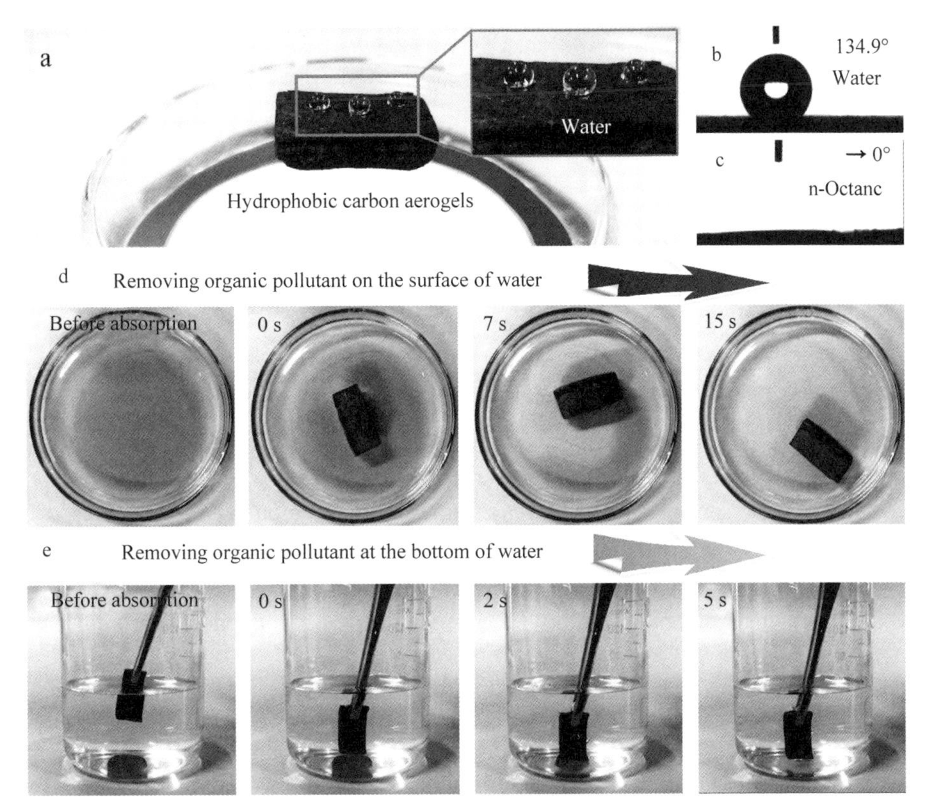

Fig. 4 Hydrophobicity and absorptive selectivity of the anisotropic carbon aerogels: (a) water droplets (as spheres) on the surface of the aerogel; (b) water contact angle of 134.9° and (c) octane contact angle of 0°; snapshots of the aerogels during absorption of (d) n-octane on the surface of water and (e) tetrachloromethane at the bottom of water

For further demonstrating the absorptive selectivity and separation function of the anisotropic carbon aerogels, we assemble a separating setup (Fig. 5), namely using a peristaltic pump to transfer the mixture of water and n-octane/ tetrachloromethane to a fractional column filled with the carbon aerogels. The separating effects are monitored. As shown in Fig. 5a, in the case of overwater organic pollutant, the mixture of water (35 mL, dyed with indigo blue) and n-octane (15 mL, dyed with Sudan Ⅳ), can be fastly and well separated by the carbon aerogels within 50 s. After the separating, it is clear that there is only blue water in the filter liquor and the volume change of water is almost negligible (see Fig. 5b), indicating the good absorptive selectivity and separation function of the anisotropic carbon aerogels. In the case of underwater organic pollutant (tetrachloromethane dyed with Sudan Ⅳ acts as the pollutant), the organic liquid is firstly pumped, and a similar and excellent separation effect is achieved (Fig. 5c and d). The results further verify the outstanding separating function of carbon aerogels towards both the overwater and underwater organic pollutants.

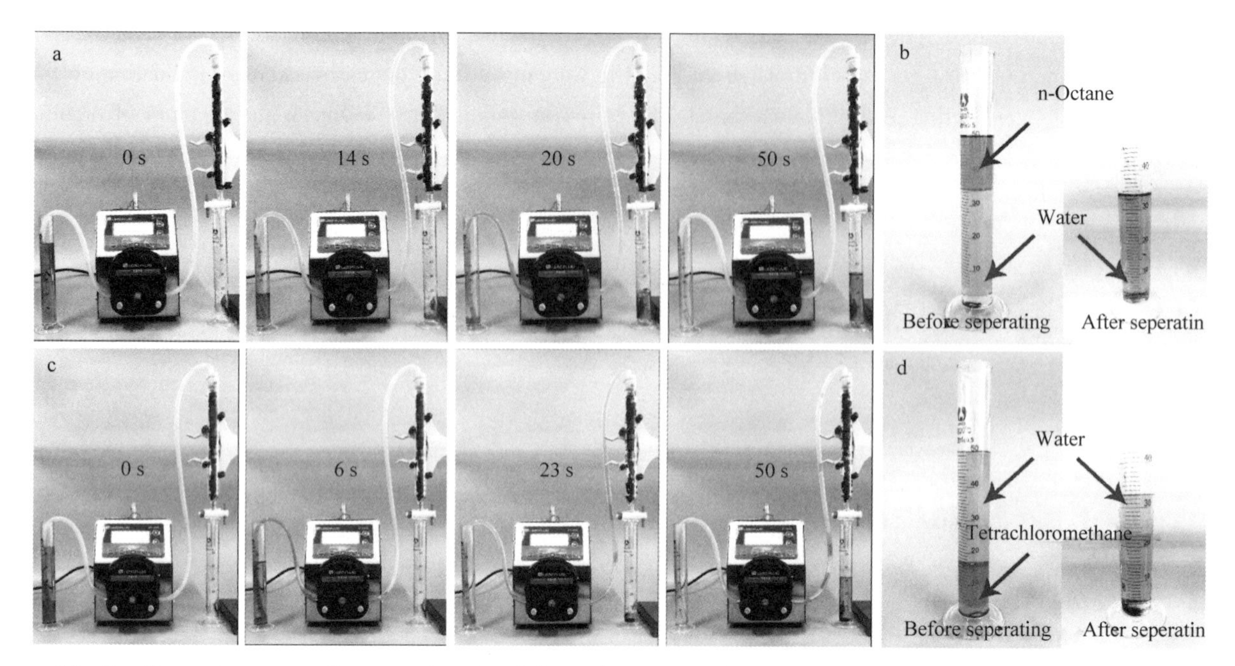

Fig. 5 Absorption separation experiments of the anisotropic carbon aerogels: (a) separation of n-octane (dyed with Sudan Ⅳ) and water (dyed with indigo blue) (overwater organic pollutant); (b) separating effect of overwater organic pollutant; (c) separation of tetrachloromethane (dyed with Sudan Ⅳ) and water (dyed with indigo blue) (underwater organic pollutants); (d) separating effect of underwater organic pollutant

The absorption capacity ($mg \cdot g^{-1}$) of the anisotropic carbon aerogels is defined as the weight of absorbed substances per unit weight of carbon aerogels. A variety of organic solvents or mixtures are studied, for instance, hydrocarbons, aromatic compounds, commercial organic solvents as well as organic mixtures (like petroleum), which easily cause pollutants in daily life because of chemical leakages or arbitrary discharge of organic wastewater. The anisotropic carbon aerogels exhibit high absorption capacities of 13277–31597 $mg \cdot g^{-1}$ for all of these organic liquids (Fig. 6a), meanwhile the absorption capacities per unit volume of carbon aerogels are in the range of 571–1359 $mg \cdot cm^{-3}$, which are obtained by dividing the absorbate mass by the volume of the carbon aerogels. Clearly, the absorption data are distinctly larger than those of petroleum-based carbon aerogels and can also compete with that of some novel biomass-based absorbents prepared by bottom-to-top strategy (see Table 1).[37-44] Apart from the merit of high absorption capacity, this simpler, faster and greener top-to-bottom synthesis way is undoubtedly more suitable for large-scale production and application. Besides, the absorption capacity of the carbon aerogels towards tetrachloromethane is 21.0, 4.9 and 7.6 times higher than that of the natural wood (1439 $mg \cdot g^{-1}$), delignified wood (5361 $mg \cdot g^{-1}$) and wood-derived carbon (3669 $mg \cdot g^{-1}$), as shown in Fig. 6b. Numerous merits of the unique laminated structure (including larger surface area and pore volume, lower density and average tortuosity and higher porosity) play crucial roles for the higher absorption capacity (the detailed mechanism is available in the next part). In addition, the result also confirms the necessity of delignification to acquire this low-tortuosity and low-flow-resistance structure.

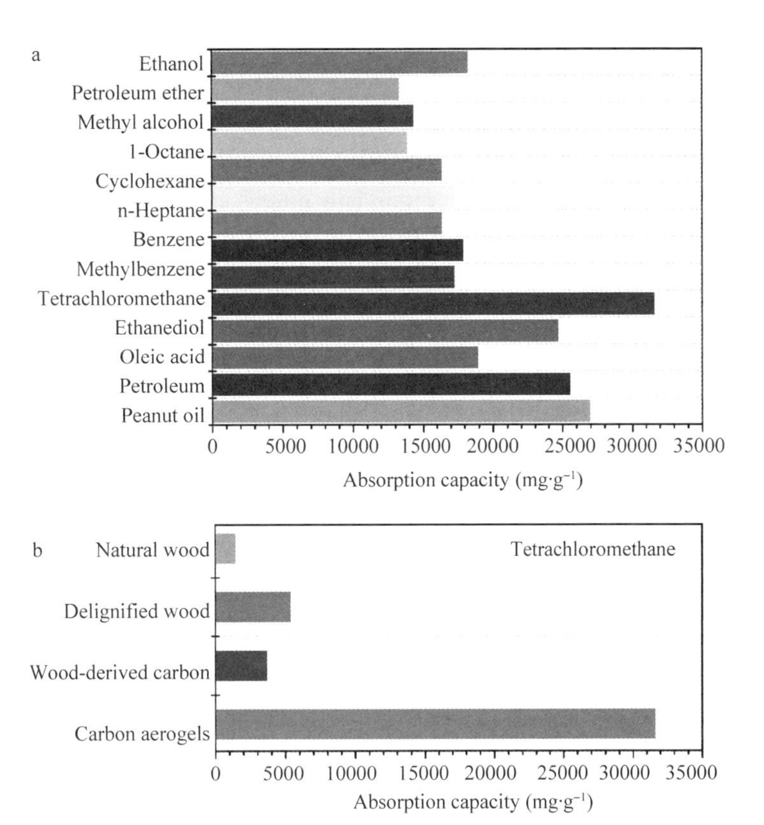

Fig. 6 Absorption capacity of the anisotropic carbon aerogels: (a) absorption capacity of the aerogels for different organic solvents or mixtures; (b) comparison of the absorption capacity of the aerogels, natural wood, delignified wood and wood-derived carbon

Table 1 Comparison of the absorption capacity, bulk density, porosity and averaged tortuosity of the anisotropic carbon aerogels with those of congeneric petroleum-based or biomass-based carbon aerogels

Samples	Absorbates	Absorption capacity ($mg \cdot g^{-1}$)	Absorption capacity in this work ($mg \cdot g^{-1}$)	Bulk density ($mg \cdot cm^{-3}$)	Porosity	Averaged tortuosity	Refs
Cellulose-based carbon aerogels	Tetrachloromethane	~25000	31597	20	98%	1.037	37
Resorcinol-formaldehyde-based carbon aerogels	Methylbenzene	1180	17260	-	-	-	38
Resorcinol-gallic acid-formaldehyde-based carbon aerogels	Methylbenzene	425	17260	640	~68.6%	1.233	39
Carbon-silica aerogels	Benzene	580	17887	-	-	-	40
Silica aerogel-activated carbon	Benzene	875	17887	220	-	-	41
Bacterial cellulose/polyimide-based carbon aerogels	Cyclohexane	~14400	16357	-	-	-	42
Popcorn-based carbon aerogels	Cyclohexane	~8420	16357	95	95.3%	1.059	43
Formaldehyde-melamine-based carbon aerogels	n-Heptane	~5750	16348	95	87.7%	1.106	44
Delignified wood-based carbon aerogels	n-Heptane *et al.*	-	13277-31597	43	97.9%	1.038	This work

Absorption kinetics and isotherm are the important methods to characterize the absorption rate and capacity, respectively. The absorption kinetics of peanut oil and ethanediol on the anisotropic carbon aerogels is presented in Fig. 7a. Interestingly, for the both liquids, the anisotropic carbon aerogels display ultrafast absorption rates and reach absorption equilibrium within only dozens of seconds, in accordance with the results in Fig. 4 and 5. Furthermore, the aerogels achieve the high absorption capacities of 26917 mg · g^{-1} for peanut oil and 24685 mg · g^{-1} for ethanediol. In Fig. 7b, it is clear that these absorption kinetics data can be well fitted by the pseudo-second-order rate kinetic model [the correlation coefficient (R^2) values reach up to 0.9999], which is expressed as Equation (1):[45]

$$\frac{t}{Q_t}=\frac{t}{Q_e}+\frac{1}{K_2Q_e^2} \tag{2}$$

where Q_e and Q_t(g · g^{-1}) is the absorbed amount at equilibrium and at time t, respectively, and K_2 is the absorption rate constant (g · g^{-1} · s^{-1}). A higher K_2 represents the faster absorption rate. After the calculation based on the slope and intercept, the anisotropic carbon aerogels deliver a K_2 value of 11526 g · g^{-1} · s^{-1} for ethanediol, higher than that for peanut oil (5567 g · g^{-1} · s^{-1}). The result reflects the faster absorption rate when absorbing ethanediol because its viscosity is lower than that of peanut oil. The corresponding oil absorption isotherm is shown in Fig. 7c, which is obtained from different initial oil contents ranging from 75 to 1050 g · L^{-1}. It is found that the isotherm rapidly increases in the lower concentration region, and gradually flattens out in the high concentration scale. The equilibrium absorption data can be fitted by using the Langmuir absorption model. The linearized form of the Langmuir equation isotherm is expressed as Equation (2):[46]

$$\frac{C_e}{Q_e}=\frac{1}{K_LQ_m}+\frac{C_e}{Q_m} \tag{3}$$

where C_e is the equilibrium concentration of oil (g · L^{-1}), Q_e is the absorbed oil amount at equilibrium (g · g^{-1}), K_L is the Langmuir absorption equilibrium constant and Q_m is the maximum oil absorption capacity (g · g^{-1}). From Fig. 7d, it is obvious that the experimental data fit the Langmuir absorption model well, with a high R^2 value of 0.9997. Moreover, based on the Equation 2, the Q_m value of the anisotropic carbon aerogels can be calculated to be 27027 mg · g^{-1}.

Recyclability is also one of the most key parameters for organic liquids/water absorption and separation applications. Considering that the mechanism is mainly physical absorption of organic molecules (stored in pores of absorbents), it is easy to remove these molecules from pores by means of various simple methods dependent on their nature. For flammable absorbates (taking ethanol and n-octane as examples), the regeneration of carbon aerogels is achieved by direct combustion. In this case, these absorbed substances can be used for heating. As shown in Fig. 8a and b, after the five absorption/combustion cycles, the absorption capacity declines to 17111 mg · g^{-1} for ethanol (93.8% of initial value) and 14254 mg · g^{-1} for n-octane (96.2% of initial value). The slight performance loss makes the carbon aerogels recyclable for many times. Drying is utilized as an alternative method to recycle volatile organic liquids (e.g., tetrachloromethane). After absorption, the sample is dried to release the absorbed liquids and a device is employed to collect the vapor of liquids. The property loss is quite small (31597-32026 mg · g^{-1}) after the five absorption/drying cycles (Fig. 8c) since the capillary force from the liquid evaporation causes a negligible influence on the laminated structure (as verified by the SEM image after the cycles, Fig. S7, ESI). For

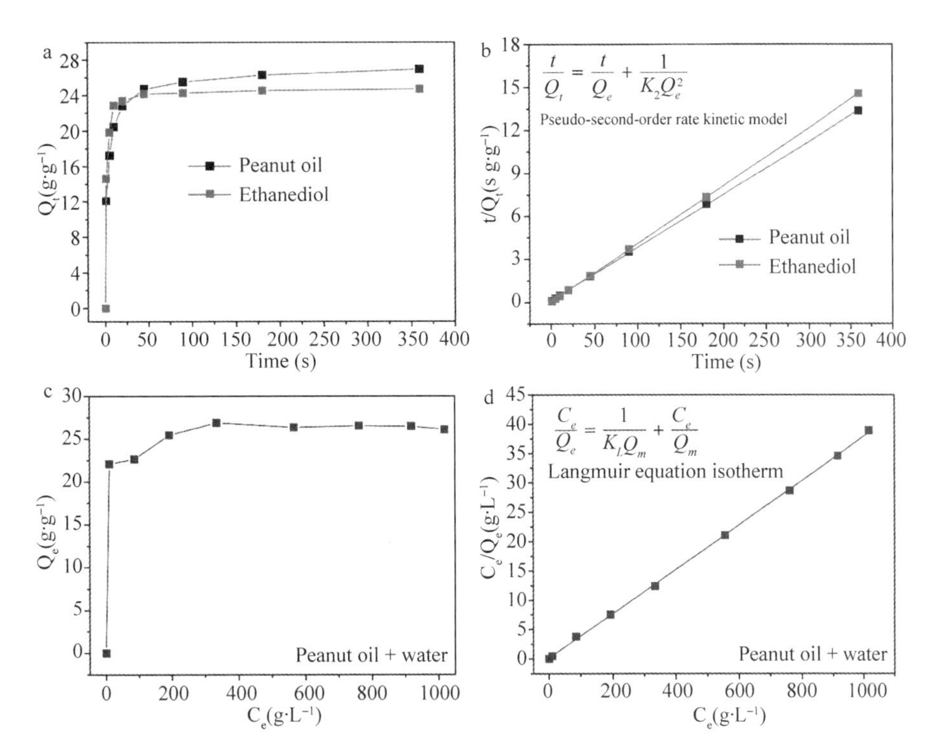

Fig. 7 Absorption kinetics and isotherm on the anisotropic carbon aerogels: (a) absorption kinetics of peanut oil and ethanediol, (b) fitting the experimental data to the pseudo-second-order model, (c) absorption isotherm of water-oil mixture, and (d) fitting the experimental data to the Langmuir equation

uninflammable and high-viscosity organics (like oleic acid), a combination method of extraction-combustion is applied for recycling. An 86.2% of absorption capacity retention is gained after the five cycles (see Fig. 8d). These results clearly verify the outstanding recyclability of the carbon aerogels when used as absorbents.

Besides obvious advantages over the natural wood and delignified wood, the carbon aerogels also show a higher absorption capacity and faster absorption rate than those of the wood-derived carbon with similar wettability (by comparing Fig. 4d and e, 6b and S8), revealing the superiority of this low tortuosity of anisotropic stacked laminated structure. These advantages can be roughly predicted and explained by an analysis of absorption mechanism.

The mechanism of organic liquids absorbed onto the anisotropic carbon aerogels is generally determined by many factors, related to the physicochemical properties of liquids and aerogels and the structure of aerogels. Form the previous study and theory of fluid mechanics,[25] the absorption process of organic liquids onto the carbon aerogels can be simplified to the capillary rise of a liquid in a flat-plate microchannel with a layer spacing of d (Fig. S9). The dynamics of the capillary rise for a column of liquid of height h (velocity $\dot{h}$ and acceleration $\ddot{h}$ are expressed as dh/dt and d^2h/dt^2, respectively) is determined by the equation 3:

$$\sum F = m\ddot{h} = 2(d + l)\sigma\cos\theta + dl(P_B - P_T) - dl\rho gh - 2(d + l)\tau\dot{h} \quad (4)$$

where m is the liquid mass ($m = Ah\rho$, A and ρ are the projection area of the capillary and liquid density), l is the length of flat-plate microchannel, "$2(d+l)\sigma\cos\theta$" is the capillary force (σ and θ are the coefficient

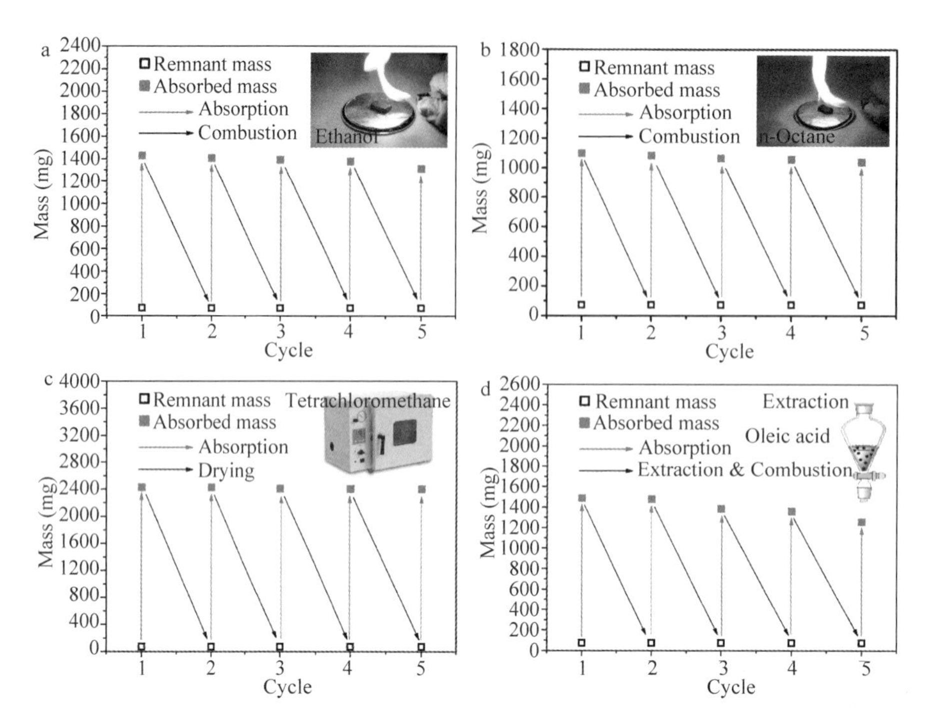

Fig. 8 Absorption recyclability of the anisotropic carbon aerogels: (a, b) recyclability of the aerogels for absorption of ethanol and n-octane (flammable liquids) using combustion method; (c) recyclability of the aerogels for absorption of tetrachloromethane (volatile liquid) using drying method; (d) recyclability of the aerogels for absorption of oleic acid (high-viscosity liquid) using combination method of extraction-combustion. Insets show the corresponding removing methods of organic liquids from the aerogels

of surface tension and dynamic contact angle), "$dl(P_B - P_T)$" is the dynamic pressure force (P_B and P_T are the dynamic pressure at bottom and top), "$dl\rho gh$" is the weight (g is the gravity acceleration) and "$2(d+l)\tau h$" is the viscous force (τ is the shear stress at the wall).

According to the theory of fluid mechanics, when liquids pass through a parallel plate, the shear stress τ can be expressed as:[47]

$$\tau = 6\eta Q/d^2 l \tag{5}$$

where η is the dynamic viscosity and Q is the volume of passed liquids per unit time ($Q = \dot{h}dl$). Therefore, $\tau = 6\eta\dot{h}/d$. Based on Bernoulli's equation, we can obtain:

$$\frac{1}{2}\rho h + P_B = P_T \tag{6}$$

Therefore, equation 3 becomes a function of h, $\dot{h}$ and $\ddot{h}$:

$$h\ddot{h} = \frac{2(d+l)}{\rho dl}\left(\sigma\cos\theta - \frac{6h\dot{h}\eta}{d}\right) - \left(gh + \frac{\dot{h}}{2}\right) \tag{7}$$

Given an enough long microchannel, the maximum height h^* to which the liquid column can rise is gotten by substituting $\dot{h}\to 0$ and $\ddot{h}\to 0$ into equation 7, yielding:

$$h^* = \frac{2(d+l)}{\rho dlg}\sigma\cos\theta \tag{8}$$

In the initial stage (inertial regime), $h \to 0$, the motion is mainly dominated by the capillary force so that equation 6 can be turned into:

$$h\ddot{h} = \frac{2(d+l)}{\rho dl}\sigma \cos \theta - gh \tag{9}$$

As the $\dot{h}$ increases, it reaches its maximum $\dot{h}^*$ when $\ddot{h} \to 0$ and the capillary motion is balanced by the capillary force and the dynamic pressure:

$$\frac{2(d+l)}{\rho dl}\sigma \cos \theta = \frac{1}{2}\dot{h}^* \tag{10}$$

By using equation 7 in equation 9, the maximum velocity is $\dot{h}^* = \sqrt{2gh^*}$. As the liquid column is further rising up, the viscous force dominates over the dynamic pressure and weight, so that equation 6 becomes:

$$\frac{2(d+l)}{\rho dl}\left(\sigma \cos \theta - \frac{6h\dot{h}\eta}{d}\right) = \frac{2(d+l)}{\rho dl}\left(\sigma \cos \theta - \frac{3\eta dh2}{d\ dt}\right) = 0 \tag{11}$$

The capillary filling now occurs in the viscous regime. By using equation 7 and 10, the rising height h' during initial regime can approximatively be predicted as:

$$h' = \frac{\sigma \cos \theta d}{6\dot{h}\eta} \approx \frac{\sigma \cos \theta d}{6\sqrt{2gh^*}\,\eta} = \frac{\sqrt{\rho\sigma \cos \theta \dfrac{d3l}{d+l}}}{12\eta} \tag{12}$$

The capillary motion in the initial regime is negligible since the initial regime progresses very fast and $h' \ll h^*$. Besides, since $d \ll h^*$, the capillary motion is mainly governed by equation 10 in the viscous regime in the process of absorption. Thus, the time-dependent capillary motion of fluid in a flat-plate microchannel can be expressed as:

$$h = \sqrt{\frac{dt\sigma \cos \theta}{3\eta}} \tag{13}$$

$$\dot{h} = \frac{1}{2}\sqrt{\frac{d\sigma \cos \theta}{3\eta t}} \tag{14}$$

which meet the Washburn's equation ($h \sim \sqrt{t}$). Setting the porosity and tortuosity of samples as δ and γ, the oil absorption volume V per unit area is expressed as:

$$V = h\frac{\delta}{\gamma} = \frac{\delta}{\gamma}\sqrt{\frac{dt\sigma \cos\theta}{3\eta}} \tag{15}$$

In addition, the absorption rate at instant t ($\dot{V}$) and the time-dependent average absorption rate $\dot{V}_A$ from $t=0$ to $t=\Delta t$ can be expressed as:

$$\dot{V} = \dot{h}\frac{\delta}{\gamma} = \frac{\delta}{2\gamma}\sqrt{\frac{d\sigma \cos \theta}{3\eta t}} \tag{16}$$

$$\dot{V}_A = \frac{\Delta V}{\Delta t} = \frac{\delta}{\gamma}\sqrt{\frac{d\sigma \cos \theta}{3\eta \Delta t}} \tag{17}$$

Based on equation 14 and 16, it is clear that the key absorption property (namely oil absorption volume V and absorption rate $\dot{V}$) are directly associated with the structure characteristics of samples (i. e., porosity δ, tortuosity γ and layer spacing d). The larger porosity and layer spacing along with lower tortuosity contribute to acquiring more excellent absorption property. As compared to other three kinds of materials (namely the natural wood, delignified wood and wood-derived carbon), the anisotropic carbon aerogels have more advantages in the higher porosity, lower contact angle of organic liquids θ and tortuosity, and larger layer spacing owing to the removal of components (the raised average pore diameter is a good proof, see Table S2), which are responsible for the larger absorption capacity and faster absorption rate of the anisotropic carbon aerogels.

The breakthrough performance for oil absorption on the anisotropic carbon aerogels is studied. The schematic diagram of fixed-bed column test setup is presented in Fig. S10. The breakthrough curves under different flow rates (1, 2 and 5 mL · min^{-1}) are shown in Fig. S11. It is found that the breakthrough time decreases from 350 to 80 min for residual oil as the flow rate increases from 1 to 5 mL · min^{-1}. This tendency may be ascribed to the decreased contact time inside the column at the higher flow rates. As a result, the distribution and diffusion of the liquids become weaker among the absorbent particles, leading to a shorter time to reach the saturation and lower the separating efficiency.[48] The Yoon-Nelson model $ln\frac{C_t}{C_0-C_t}=K_{YN}t-\tau K_{YN}$ is applied to fit these fixed-bed column test data, and the Yoon-Nelson constant (K_{YN}) and τ (the time required for 50% absorbate breakthrough) are determined from the plot of $\ln(C_t/C_0-C_t)$ versus time (where C_0 and C_t are the inlet and effluent oil concentrations).[48] As seen in Fig. S11, good fitting results with R^2 that recorded 0.9908, 0.9889 and 0.9926 are obtained under different flow rates (1, 2 and 5 mL · min^{-1}), respectively. Other calculated parameters related to the flow rates are summarized in Table S4. It is clear that there is a good agreement between the experimental points and model predicted values.

3 Conclusions

In conclusion, we report a kind of balsa wood-derived anisotropic carbon aerogels with low tortuosity of stacked laminated structure for the efficient and reusable absorption and separation of organic pollutants from water. The low tortuosity and anisotropy endow the aerogels with low flow-resistance for passive capillary absorption. Also, experimental results and theoretical modelling illustrate that the stacked laminated structure with larger interlayer spacing and porosity and better oleophilicity is beneficial for larger absorption capacity and faster absorption rate, as compared to the three precursors. The superiorities enable efficient separation of organic pollutants from water, high and broad-spectrum absorption capacity of 13277-31597 mg · g^{-1}, and excellent recyclability with 86.2%-98.7% of absorption capacity retention after 5 cycles. More importantly, in contrast to common bottom-to-top way for synthesis of aerogels, this scalable top-to-bottom design excludes the slow and high-energy-consumption synthesis process and takes full advantage of the structural anisotropy of the precursor. Consequently, the superior properties, facile and rapid preparation way and scalability of the anisotropic carbon aerogels provide a potential feasible approach for cleanup of organic spills.

3.1 Experimental section

Materials. Balsa wood was offered by a wood-working factory in China (Linwei Wood Industry, China). These wood slices were rinsed ultrasonically with absolute ethyl alcohol and distilled water for 30 min and dried at 60 ℃ for 24 h. Chemicals including sodium hydroxide (NaOH), sodium sulfite (Na_2SO_3), hydrogen peroxide (H_2O_2), ammonium hydroxide ($NH_3 \cdot H_2O$) and tert-butyl alcohol were purchased from Shanghai Boyle Chemical Co. Ltd. and used without further purification.

Delignification of wood. Delignification of wood was carried out according to the following processes: (1) the clean and dried balsa wood was immersed in a mixed solution of NaOH (1.25 mol · L^{-1}) and Na_2SO_3 (0.2 mol · L^{-1}) at 100 ℃ for 5 h; (2) the resultant was washed with plenty of distilled water to remove residual chemicals and then added into a 3 v% solution of H_2O_2 (pH = 9 ~ 10, adjusted by $NH_3 \cdot H_2O$) at 85 ℃ for 5 h; (3) the resultant was repeatedly rinsed by distilled water for the removal of impurities and underwent a tert-butyl alcohol freeze-drying process to remove most liquids.

Pyrolysis of delignified wood. The delignified wood was transferred into a tubular furnace for pyrolysis under an argon flow. The delignified wood sample was firstly heated to 500 ℃ at a heating rate of 2 ℃ · min^{-1} and the temperature was kept for 1 h. After this, the sample was further heated to 700, 1000 or 1300 ℃ at 5 ℃ min^{-1} and these temperatures were held for 2 h to allow complete pyrolysis. Then, the sample was cooled to 500 ℃ at 5 ℃ · min^{-1} and finally cooled to room temperature naturally.

Characterizations. SEM observations were performed with a Hitachi S4800 SEM equipped with an EDX detector. TEM and HRTEM observations and SAED were performed with a FEI, Tecnai G2 F20 TEM with a field-emission gun operating at 200 kV. Thermal stability was investigated by a TA Q600TG analyzer from room temperature to 700 ℃ with a heating rate of 10 ℃ · min^{-1} under nitrogen atmosphere. FTIR were recorded on a Nicolet Nexus 670 FTIR instrument in the range of 500 – 4000 cm^{-1} with a resolution of 4 cm^{-1}. XRD analysis was conducted on a Bruker D8 Advance TXS XRD instrument. Pore structure and specific surface area were analysed by N_2 adsorption-desorption tests by an accelerated surface area and porosimeter system (3H- 2000PS2 unit, Beishide Instrument S&T Co., Ltd). The porosity δ was calculated by the equation $\delta(\%) = (1 - \rho_b/\rho_s) \times 100\%$, where ρ_b is the bulk density (calculated by dividing the weight by the bulk volume) and ρ_s is the skeletal density, respectively. According to the previous studies, the ρ_s values of wood, microfibril cellulose and carbon fibers are 1.49, 1.48 and 2.04 g · cm^{-3},[49,50] respectively. UV-vis absorption spectra were recorded by using a UV-vis spectrophotometer (UV-6000PC, Shanghai Metash Instruments Co., Ltd.). Contact angle tests were conducted on a contact angle analyser (JC2000C, Shanghai Zhongchen Digital Technic Apparatus Co., Ltd.).

3.2 Absorption tests

Absorption capacity tests. In a typical experimental process, the samples were immersed into different organic liquids for three minutes (the weight increment is negligible after the three minutes of absorption, i.e., reaching saturation). The absorption capacity is defined as A, which is calculated according to Equation (18):

$$A(mg \cdot g^{-1}) = \frac{(M_{\text{saturated}} - M_{\text{dry}})}{M_{\text{dry}}} \tag{18}$$

where M_{dry} and $M_{\text{saturated}}$ are the weights before and after absorption of different organic liquids, respectively.

Absorption selectivity tests. Absorption selectivity tests were carried out by using the carbon aerogels to absorb organic liquids on the surface of water and at the bottom of water. n-Octane (a liquid whose density is lower than that of water, 0.703 g · mL^{-1}) was stained with Sudan Ⅳ for clear presentation and the mixture was dropped onto the surface of water. Besides, tetrachloromethane (a liquid whose density is higher than that of water, 1.595 g · mL^{-1}) was also stained with Sudan Ⅳ and the mixture was placed at the bottom of water. The separating time and effect were monitored.

Absorption separation tests. Absorption separation tests were studied by assembling a separating setup, i. e. , using a peristaltic pump to transfer the mixture of water and n-octane/ tetrachloromethane to a fractional column filled with the carbon aerogels (the mass is ~ 1 g). The volumes of water and n - octane/ tetrachloromethane are 35 mL and 15 mL, respectively. For clear presentation, the water was stained with indigo blue, and the organic liquids were dyed with Sudan Ⅳ. The separating time and effect were monitored.

Absorption recyclability tests. Absorption recyclability was studied by repeating the absorption process for five times and monitoring changes in the absorption capacity. Absorbates were removed by three approaches based on the characteristic of organic liquids, namely combustion for flammable liquids, drying for volatile liquids (105 ℃, 10 min) and extracting-combustion for high-viscosity liquids (dipping samples filled with high-viscosity liquids in the ethanol extraction agent and then removing the residual extraction agent in samples by burning).

Absorption kinetics and isotherm determination. The research objects of absorption kinetics are the commercially available peanut oil and ethanediol, while the research object of absorption isotherm is the mixture of peanut oil and water under different mixing proportions. The pH value of these liquids is close to 7, and the carbon aerogels are used as absorbents. For the absorption kinetic study, ~60 mg of absorbent was added to 30 mL of peanut oil or ethanediol. After a specified time, the absorbent was taken out and weighed, and the absorption capacity was calculated based on the above Equation (17). For the absorption isothermal study, ~60 mg of absorbent was added to the mixtures of 20 g water and different masses of peanut oil under stirring for 3 min at room temperature. The residual oil content is determined by the Municipal Construction Standard of the People's Republic of China (no. CJ/T 51-2004). Briefly, the oil-water mixture is firstly mixed with 5 mL H_2SO_4(50 v%), and the resultant is then extracted by using 25 mL petroleum ether for three times. The extracting solution is dehydrated for 0.5-2 h by using anhydrous Na_2SO_4. The mixture is filtered and the filter liquor is transferred to an evaporating dish. The evaporating dish is dried in a thermostatic waterbath (65 ℃) and further dried in a drying oven (65 ℃) for 1 h. After cooling, the weight of evaporating dish is weighed. The residual oil content is defined as B, which is calculated based on Equation (19):

$$B(mgL^{-1}) = \frac{M_2 - M_1}{V} \qquad (19)$$

where M_1 is the weight of evaporating dish, M_2 is the total weight of oil and evaporating dish, and V is the volume of mixture liquid.

Fixed-bed column tests. Fixed-bed column tests were implemented by using a glass column (2 cm internal diameter and 12 cm height) packed with the carbon aerogels between two supporting layers of glass wool. The function of glass wool is to prevent the carbon aerogels from floating. The schematic diagram of the fixed-bed column test setup is presented in Fig. S10. The influence of flow rate on the uptake of residual oil from the oil-water mixture (M_{oil} : M_{water} = 1 : 1, stabilized by $Ca(OH)_2$) by the carbon aerogels was explored using three flow rates of 1, 2 and 5 mL · min^{-1}. The bed height of column and pH value of liquids were set as 10 cm and 7, respectively. The oil-water mixture was pumped through the fixed bed of carbon aerogels, and the filter liquor leaving the column was collected and analyzed at regular intervals based on the above standard no. CJ/T 51-2004.

Supporting Information

The supporting information includes: SEM images of carbon aerogels in the YZ plane. EDX patterns of natural wood and carbon aerogels. TG and DTG curves of natural wood and carbon aerogels. Water contact angle of natural wood. Effects of pyrolysis temperature on the hydrophobicity of carbon aerogels. UV－vis absorption spectra of n－octane dyed with Sudan Ⅳ before and after the absorption. Video S1－S2 exhibiting carbon aerogels absorbing overwater and underwater organic pollutants. SEM image of carbon aerogels after five absorption/drying cycles. Snapshots of wood－derived carbon during absorption of n－octane (dyed with Sudan Ⅳ) on the surface of water. Schematic diagram for capillary imbibition motion of organic liquids in the laminated structure of carbon aerogels. Schematic diagram of fixed－bed column test setup. Experimental and theoretical breakthrough curves of carbon aerogels. Table S1～S4.

Conflicts of interest

There are no conflicts to declare.

Acknowledgements

This study was supported by the National Natural Science Foundation of China (grant nos. 31901249, 31890771 and 31530009), the Scientific Research Foundation of Hunan Provincial Education Department (grant no. 18B180) and the the Youth Scientific Research Foundation of Central South University of Forestry and Technology (grant no. QJ2018002A).

References

[1]Sayyad Amin J, Vared Abkenar M, Zendehboudi S. Natural sorbent for oil spill cleanup from water surface: environmental implication[J]. Industrial & Engineering Chemistry Research, 2015, 54(43): 10615－10621.

[2]Pintor A, Ferreira C, Pereira J C, et al. Use of cork powder and granules for the adsorption of pollutants: A review [J]. Water Research, 2012, 46(10): 3152－3166.

[3]Doshi B, Sillanpää M, Kalliola S. A review of bio-based materials for oil spill treatment[J]. Water Research, 2018, 135: 262－277.

[4]Bi H, Xie X, Yin K, et al. Spongy graphene as a highly efficient and recyclable sorbent for oils and organic solvents [J]. Advanced Functional Materials, 2012, 22(21): 4421－4425.

[5]Zhu H, Chen D, Li N, et al. Graphene foam with switchable oil wettability for oil and organic solvents recovery[J]. Advanced Functional Materials, 2015, 25(4): 597－605.

[6]Yang Si, Qiuxia Fu, Xueqin Wang, et al. Superelastic and superhydrophobic nanofiber-assembled cellular aerogels for effective separation of oil/water emulsions. [J]. Acs Nano, 2015, 9(4): 3791－3799.

[7]Bi H, Huang X, Wu X, et al. Carbonmicrobelt aerogel prepared by waste paper: an efficient and recyclable sorbent for oils and organic solvents[J]. Small, 2014, 10(17): 3544－3550.

[8]Wan C, Jiao Y, Wei S, et al. Functionalnanocomposites from sustainable regenerated cellulose aerogels: A review[J]. Chemical Engineering Journal, 2019, 359: 459－475.

[9]Jiang F, Hsieh Y L. Amphiphilic superabsorbent cellulose nanofibril aerogels[J]. Journal of Materials Chemistry A, 2014, 2(18): 6337－6342.

[10]Wang J, Zhao D, Shang K, et al. Ultrasoft gelatin aerogels for oil contaminant removal[J]. Journal of Materials Chemistry A, 2016, 4(24): 9381－9389.

[11]Wu Z Y, Li C, Liang H W, et al. Innenrücktitelbild: Ultralight, Flexible, and Fire-Resistant Carbon Nanofiber Aerogels from Bacterial Cellulose (Angew. Chem. 10/2013)[J]. Angewandte Chemie, 2013, 125(10): 3113－3113.

[12]Pekala R W, Farmer J C, Alviso C T, et al. Carbon aerogels for electrochemical applications[J]. Journal of non-crystalline solids, 1998, 225: 74－80.

[13]Li W C, Lu A H, Guo S C. Characterization of the microstructures of organic and carbon aerogels based upon mixed cresol-formaldehyde[J]. Carbon, 2001, 39(13): 1989－1994.

[14] Li W, Reichenauer G, Fricke J. Carbon aerogels derived from cresol-resorcinol-formaldehyde for supercapacitors[J]. Carbon, 2002, 40(15): 2955-2959.

[15] Wu D, Fu R. Synthesis of organic and carbon aerogels from phenol-furfural by two-steppolymerization [J]. Microporous and mesoporous materials, 2006, 96(1-3): 115-120.

[16] Zhang R, Lv Y G, Zhan L, et al. Monolithic carbon aerogels from sol-gel polymerization of phenolic resoles and methylolated melamine[J]. Carbon, 2003, 41(8): 1660-1663.

[17] Grzyb B, Hildenbrand C, Berthon-Fabry S, et al. Functionalisation and chemical characterisation of cellulose-derived carbon aerogels[J]. Carbon, 2010, 48(8): 2297-2307.

[18] Yu M, Li J, Wang L. KOH-activated carbon aerogels derived from sodiumcarboxymethyl cellulose for high-performance supercapacitors and dye adsorption[J]. Chemical engineering journal, 2017, 310: 300-306.

[19] Yang X, Fei B, Ma J, et al. Porous nanoplatelets wrapped carbon aerogels by pyrolysis of regenerated bamboo cellulose aerogels as supercapacitor electrodes[J]. Carbohydrate polymers, 2018, 180: 385-392.

[20] Rajala S, Siponkoski T, Sarlin E, et al. Cellulose nanofibril film as a piezoelectric sensor material[J]. ACS applied materials & interfaces, 2016, 8(24): 15607-15614.

[21] Yao C, Hernandez A, Yu Y, et al. Triboelectric Nanogenerators and Power-Boards from Cellulose Nanofibrils and Recycled Materials[J]. Nano Energy, 2016: 103-108.

[22] Fortunati E, Luzi F, Jiménez A, et al. Revalorization of sunflower stalks as novel sources of cellulose nanofibrils and nanocrystals and their effect on wheat gluten bionanocomposite properties[J]. Carbohydrate polymers, 2016, 149: 357-368.

[23] Hu H, Zhao Z, Gogotsi Y, et al. Compressible carbon nanotube-graphene hybrid aerogels with superhydrophobicity and superoleophilicity for oil sorption[J]. Environmental Science & Technology Letters, 2014, 1(3): 214-220.

[24] Chen C, Zhang Y, Li Y, et al. All-wood, low tortuosity, aqueous, biodegradable supercapacitors with ultra-high capacitance[J]. Energy Environ Sci, 2017, 10(2): 538-545.

[25] Kuang Y, Chen C, Chen G, et al. Bioinspired Solar-Heated Carbon Absorbent for Efficient Cleanup of Highly Viscous Crude Oil[J]. Advanced Functional Materials, 2019, 29(16): 1900162.

[26] Song J, Chen C, Yang Z, et al. Highly compressible, anisotropic aerogel with aligned cellulosenanofibers[J]. ACS nano, 2018, 12(1): 140-147.

[27] Long D, Zhang J, Yang J, et al. Chemical state of nitrogen in carbon aerogels issued from phenol-melamine-formaldehydegels[J]. Carbon, 2008, 46(9): 1259-1262.

[28] Meng Y, Young T M, Liu P, et al. Ultralight carbon aerogel from nanocellulose as a highly selective oil absorption material[J]. Cellulose, 2015, 22(1): 435-447.

[29] Wan C, Jiao Y, Liang D, et al. A geologic architecture system-inspired micro-/nano-heterostructure design for high-performance energy storage[J]. Advanced Energy Materials, 2018, 8(33): 1802388.

[30] Jing Z, Wang X, Hu J, et al. Thermal degradation of softwood lignin and hardwood lignin by TG-FTIR andPy-GC/MS[J]. Polymer Degradation & Stability, 2014, 108: 133-138.

[31] Majdanac L D, Poleti D, Teodorovic M J. Determination of the crystallinity of cellulose samples by x-ray diffraction [J]. Acta Polymerica, 2010, 42(8): 351-357.

[32] He X, Ling P, Yu M, et al. Rice husk-derived porous carbons with high capacitance by $ZnCl_2$ activation for supercapacitors[J]. Electrochimica Acta, 2013, 105: 635-641.

[33] Segal L, Creely J J, Martin Jr A E, et al. An empirical method for estimating the degree of crystallinity of native cellulose using the X-ray diffractometer[J]. Textile research journal, 1959, 29(10): 786-794.

[34] He X, Ling P, Qiu J, et al. Efficient preparation of biomass-based mesoporous carbons for supercapacitors with both high energy density and high power density[J]. Journal of Power Sources, 2013, 240(oct. 15): 109-113.

[35] Durá G, Budarin V L, Castro-Osma J A, et al. Importance of Micropore-Mesopore Interfaces in Carbon Dioxide Capture by Carbon-Based Materials[J]. Angewandte Chemie, 2016, 128(32): 9319-9323.

[36] Bo-Ming Y, Jian-Hua L. A geometry model for tortuosity of flow path in porous media[J]. Chinese Physics Letters, 2004, 21(8): 1569-1571.

[37] Wang H, Gong Y, Wang Y. Cellulose-based hydrophobic carbon aerogels as versatile and superior adsorbents for

sewage treatment[J]. Rsc Advances, 2014, 4(86): 45753-45759.

[38]Maldonado-Hódar F J, Moreno-Castilla C, Carrasco-Marín F, et al. Reversible toluene adsorption on monolithic carbon aerogels[J]. Journal of Hazardous Materials, 2007, 148(3): 548-552.

[39]F Carrasco-Marín, D Fairén-Jiménez, Moreno-Castilla C . Carbon aerogels from gallic acid-resorcinol mixtures as adsorbents of benzene, toluene and xylenes from dry and wet air under dynamic conditions[J]. Carbon, 2009, 47(2): 463-469.

[40]Dou B, Li J, Wang Y, et al. Adsorption and desorption performance of benzene over hierarchically structured carbon-silica aerogelcomposites[J]. Journal of hazardous materials, 2011, 196: 194-200.

[41]Mohammadi A, Moghaddas J. Synthesis, adsorption and regeneration of nanoporous silica aerogel and silica aerogel-activated carbon composites[J]. Chemical Engineering Research & Design, 2015, 94: 475-484.

[42]Lai F, Miao Y E, Zuo L, et al. Carbon aerogels derived from bacterial cellulose/polyimide composites as versatile adsorbents and supercapacitor electrodes[J]. ChemNanoMat, 2016, 2(3): 212-219.

[43]Dai J, Zhang R, Ge W, et al. 3D macroscopic superhydrophobic magnetic porous carbon aerogel converted from biorenewable popcorn for selective oil-water separation[J]. Materials & Design, 2018, 139: 122-131.

[44]Yu Y, Zhen T, Ngai T, et al. Nitrogen-rich and fire-resistant carbon aerogels for the removal of oil contaminants from water. [J]. Acs Appl Mater Interfaces, 2014, 6(9): 6351-6360.

[45]Feng J, Nguyen S T, Fan Z, et al. Advanced fabrication and oil absorption properties of super-hydrophobic recycled cellulose aerogels[J]. Chemical Engineering Journal, 2015, 270: 168-175.

[46] Wan C, Li J. Facile synthesis of well-dispersedsuperparamagnetic γ-Fe_2O_3 nanoparticles encapsulated in three-dimensional architectures of cellulose aerogels and their applications for Cr (Ⅵ) removal from contaminated water[J]. ACS Sustainable Chemistry & Engineering, 2015, 3(9): 2142-2152.

[47]Miao Y, Ajami N E, Huang T S, et al. Enhancer-associated long non-coding RNA LEENE regulates endothelial nitric oxide synthase and endothelial function[J]. Nature Communications, 2018, 9(1): 292.

[48]El-Sayed M, Ramzi M, Hosny R, et al. Breakthrough curves of oil adsorption on novel amorphous carbon thin film [J]. Water Science and Technology, 2016, 73(10): 2361-2369.

[49]Meng Y, Young T M, Liu P, et al. Ultralight carbon aerogel from nanocellulose as a highly selective oil absorption material[J]. Cellulose, 2015, 22(1): 435-447.

[50]Szczurek A, Jurewicz K, Amaral-Labat G, et al. Structure and electrochemical capacitance of carbon cryogels derived from phenol-formaldehyde resins[J]. Carbon, 2010, 48(13): 3874-3883.

中文题目：自上而下设计低弯度各向异性碳气凝胶用于快速和可重复使用的毛细管吸收和有机泄漏分离

作者：万才超，焦月，魏松，李贤军，田文燕，吴义强，李坚

摘要：由于许多高性能吸附剂存在一些关键缺陷(如合成过程复杂且不环保、成本高、难以大规模生产)，因此，创造具有理想吸附剂性能的可持续、经济、可扩展的吸附剂是一项全球性的挑战。本文提出了一种简单的、可扩展的、从上到下的设计方法，从轻木的层次化细胞通道中衍生出一种具有低弯曲度的叠层结构的新型各向异性碳气凝胶。利用这种独特的结构和良好的亲油性，实现了低流动阻力的快速被动毛细管吸收(理论模型证明)。结果表明，各向异性碳气凝胶对水中有机污染物的分离具有较好的选择性，对不同有机液体具有广谱和高吸收率(13277 ~ 31597 $mg \cdot g^{-1}$)，且具有良好的可循环性(5 次循环后吸收率保持在 98.7%)。将这些优异的性能与廉价的制备策略以及良好的环境友好性相结合，为有机泄漏物的高效可重复利用吸收和分离提供了一种潜在的可扩展材料。

关键词：碳气凝胶；各向异性结构；毛细吸收；自顶向下的设计；吸附选择性；吸收再循环能力；吸附剂

Self-stacked Multilayer FeOCl Supported on A Cellulose-derived Carbon Aerogel: A New and High-performance Anode Material for Supercapacitors*

Caichao Wan, Yue Jiao, Wenhui Bao, He Gao, Yiqiang Wu, Jian Li

Abstract: To build high-energy density of asymmetric supercapacitors (ASC), current studies are always directed towards cathode materials; however, anode materials are paid much less attention. Here we for the first time demonstrate that orthorhombic FeOCl with a self-stacked laminated structure suits to be a high-performance anode material for supercapacitors since its unique laminated structure can provide abundant active sites for migration and intercalation reactions of electrolyte ions. By introducing a highly conductive and porous cellulose-derived carbon aerogel (CDCA) matrix, the mechanical stability and charge-storage kinetics of FeOCl are significantly heightened. FeOCl@ CDCA delivers an ultra-high areal specific capacitance of 1618 mF · cm^{-2} (647 F · g^{-1}) at 2 mA · cm^{-2} and outstanding cycle stability with no more than 10% capacitance loss after 10000 cycles in 1 M Na_2SO_4 between −1 and 0 V vs. Ag/AgCl. An ASC operating at 0−1.8 V was fabricated using a FeOCl@ CDCA anode and a cheap MnO_2 cathode. The ASC displays a highly competitive energy/power density (289 μW · h cm^{-2} at 1.8 mW · cm^{-2}) and excellent rate capability and cycle stability. These findings may open a new pathway to the design of high-energy density of energy-storage systems using FeOCl-based anodes.

1 Introduction

Supercapacitors are always considered as one of the most promising candidates for next-generation energy storage systems due to their distinct merits like fast charging/discharging rate, high power density and long cycle life.[1-3] Compared to another electrochemical energy storage device (namely batteries) with a bigger market at present, supercapacitors are capable to deliver hundred to many thousand times higher power in the same volume, but they cannot store the same amount of charge (usually 3−30 times lower).[2] Therefore, the major challenge for supercapacitors is to increase energy density and concurrently maintain high power density and cycle stability. Electrochemical activity of electrodes plays a key role in determining the property of supercapacitors. Electrode materials can be classified into two categories based on working potential window, i.e., cathode materials (>0 V vs. SCE) and anode materials (<0 V vs. SCE). Up to now, cathode materials have been paid more attention and already achieved brilliant progress,[4-9] while the studies on performance enhancements of anode materials are distinctly fewer. In addition, based on the difference in charge storage

---

* 本文摘自 Journal Materials Chemistry A, 2019, 7: 9556−9564.

mechanisms, electrode materials can also be divided into pseudocapacitive materials (e. g. , transition-metal oxides or hydroxides and conjugated polymers) and electrical double-layer capacitor (EDLC) materials (like carbon materials).[1,10] Carbon materials have already worked as anode materials for the assembly of asymmetric supercapacitors (ASCs) due to their wider working potential window, higher electrical conductivity and more outstanding cycle stability when compared to those of pseudocapacitive anode materials (like Bi_2O_3,[11] V_2O_5,[12] MoO_2[13] and FeO_x[14]); unfortunately, they often suffer from lower capacitance and energy density, which are exactly the advantages of pseudocapacitive materials. Consequently, an effective integration of these two classes of active substances is believed to helpfully develop novel high-performance anode materials.[15,16]

Experimental researches have demonstrated that iron oxides with variable oxidation states are promising candidates as anodes in ASCs.[14,17-19] However, iron oxychloride (FeOCl) has not been paid enough attention, especially for supercapacitors. From the viewpoint of crystalline structure, orthorhombic FeOCl can be regarded that the "OH" of γ-FeOOH is substituted with "Cl". Therefore, FeOCl has the similar structure as MOOH-typed compounds with weak interaction between layers (i. e. , Van der Waals' force) and strong binding within the layered planes.[20] The so-called "van der Waals layer" of FeOCl, as illustrated in Fig. 1a, suits for intercalation reactions by transferring charges between intercalated compounds and the FeOCl matrix.[21] By calculation, we find that the interlayer spacing of FeOCl (d_{010} = 0. 783 nm, Fig. S1) is enough for the entrance and migration of many electrolyte ions (like Na^+ with a hydrated ionic radius of 0. 358 nm).[22] Along with the movement of Na^+, the intercalated Na^+ participates in the reduction reaction of FeOCl as an anode. Obviously, the laminated structure of FeOCl is able to provide abundant channels and active sites for the intercalation reactions and is thus expected to display an ideal electrochemical activity.

In this work, we for the first time report the application of FeOCl as a novel anode material for supercapacitors. Its self-stacked laminated structure is in favour of intercalation reactions of electrolyte ions. After the integration of a conductive matrix (i. e. , cellulose-derived carbon aerogel, CDCA), the mechanical stability and ionic and charge-storage kinetics of FeOCl are heightened. As a result, FeOCl@CDCA displays an ultra-high areal specific capacitance of 1618 $mF \cdot cm^{-2}$ (647 $F \cdot g^{-1}$) at 2 $mA \cdot cm^{-2}$ and outstanding cycle stability (ca. 91% after 10000 cycles). We assembled an asymmetrical supercapacitor (ASC) using a FeOCl@CDCA anode and an inexpensive MnO_2 cathode, which possesses highly competitive energy/power density (289 $\mu W \cdot h \cdot cm^{-2}$ at 1. 8 $mW \cdot cm^{-2}$) amongest current state-of-the-art ASCs.

2 Results and discussion

The FeOCl, which was synthesized via a simple chemical-vapor-transport technique (details are available in SI), exhibits a 2D sheet-like structure with a size of several microns and the sheets stack into a laminated structure (as shown in Fig. 1c and d). To address the issue of poor electrical conductivity, the FeOCl was combined with a conductive CDCA matrix by ball-milling (Fig. 1b) and the laminated structure of FeOCl was well maintained before and after the ball-milling (green frames, Fig. 1e). From the higher-magnification SEM image in Fig. 1f, it is clear that the feature structure of CDCA adheres to the surface of FeOCl sheets. In addition, the Raman spectrum also demonstrates the presence of CDCA with typical D and G-bands (Fig. S2), reflecting a good mixture between the two components. The G-band of FeOCl@CDCA slightly upshifts as compared to that of CDCA due to charge-transfer effects between them,[23] which is another evidence for their co-presence. Moreover, from the TEM image of FeOCl@CDCA, an obvious lamellar structure can be identified (as directed by green circles).

From X-ray diffraction (XRD) patterns (Fig. 1f), the FeOCl and FeOCl@CDCA both show the characteristic peaks of orthorhombic FeOCl (JCPDS no. 24-1005), while the ball-milling results in the broadening and decrease in the peak intensity, indicative of the grain refining of FeOCl by the ball-milling. X-ray photoelectron spectroscopy (XPS) was adopted to identify the elemental compositions and chemical state of Fe in the composite. As shown in Fig. 1i, the signals of O (KLL), Fe 2S, Fe 2P, Fe (LVV), O 1s, C 1s, Cl 2s and Cl 2p were detected. In addition, for the Fe 2p core-level XPS spectrum (Fig. 1j), the Fe 2p peaks at 711.6 eV and 725.4 eV correspond to Fe $2p_{3/2}$ and Fe $2p_{1/2}$ spin-orbit peaks of FeOCl, suggesting that the Fe element appears as Fe(Ⅲ) state.[24] In addition, the Fe $2p_{3/2}$ peak is accompanied by a shake-up satellite situated at 719.3 eV, further representative of Fe (Ⅲ). It is well-known that surface area and pore size of active substances play important roles in electrochemical reactions.[25,26] CDCA is a typical 3D porous material with a large surface area (157 $m^2 \cdot g^{-1}$, its isotherms and morphology are available in Fig. S3 and S4) and has been widely used to support active substances for multifarious applications (such as catalysts,[27] sensors[28] and selective adsorption[28,29]). The surface area of self-stacked multilayer FeOCl is about 40 $m^2 \cdot g^{-1}$. The incorporation of CDCA results in a remarkable increase by 40% to ~56 $m^2 \cdot g^{-1}$ for FeOCl@CDCA. The larger surface area is able to provide more reaction sites and porous CDCA can serve as a reservoir to store electrolyte and facilitate ion transfer. The isotherms of FeOCl@CDCA show a hysteresis loop (Fig. 1k), which reflects the existence of mesopores (2-50 nm).

The adsorption does not reach a plateau near the P/P_0 of 1.0, revealing the presence of macropores (>50 nm). The speculations are proved by the pore size distribution (1.8-260 nm, calculated by the Barret-Joyner-Halenda method). Such a pore size distribution helps for electrolyte penetration, allowing reaction species to quickly access the electrode.[30]

To evaluate electrochemical performances of FeOCl@CDCA as a supercapacitor electrode, its cyclic voltammetry (CV) and galvanostatic chargedischarge (GCD) curves, cycle sta bility, and electrochemical impedance spectra (EIS) were measured and compared to those of CDCA and FeOCl electrodes in a conventional three-electrode system, as shown in Fig. 2a. Fig. 2c compares the CV curves of CDCA, FeOCl and FeOCl@CDCA electrodes at the scan rate of 5 $mV \cdot s^{-1}$. The contribution of Ni foam to the capacitance is negligible (Fig. S5). Besides, the area enclosed within the CV curve of pseudocapacitive FeOCl (Fig. S6), corresponding to energy storage capability, is much larger than that of CDCA with an EDLC characteristic (Fig. S7). Moreover, the combination of two materials results in a larger area of CV curve for FeOCl@CDCA composite, reflecting its stronger charge storage ability. The CV curves of FeOCl@CDCA at different scan rates are presented in Fig. 2d, where a pair of redox peaks are identified (see also Fig. 2c) and represent a redox insertion reaction which is derived from the 3D absorption of electroactive species into the electrode material. Besides, the XPS spectrum of FeOCl@CDCA electrode after electrochemical tests suggests the transformation of Fe (Ⅲ) to Fe (Ⅱ) (Fig. S8). Therefore, the possible charge storage process of FeOCl in the Na_2SO_4 electrolyte is described as:[31]

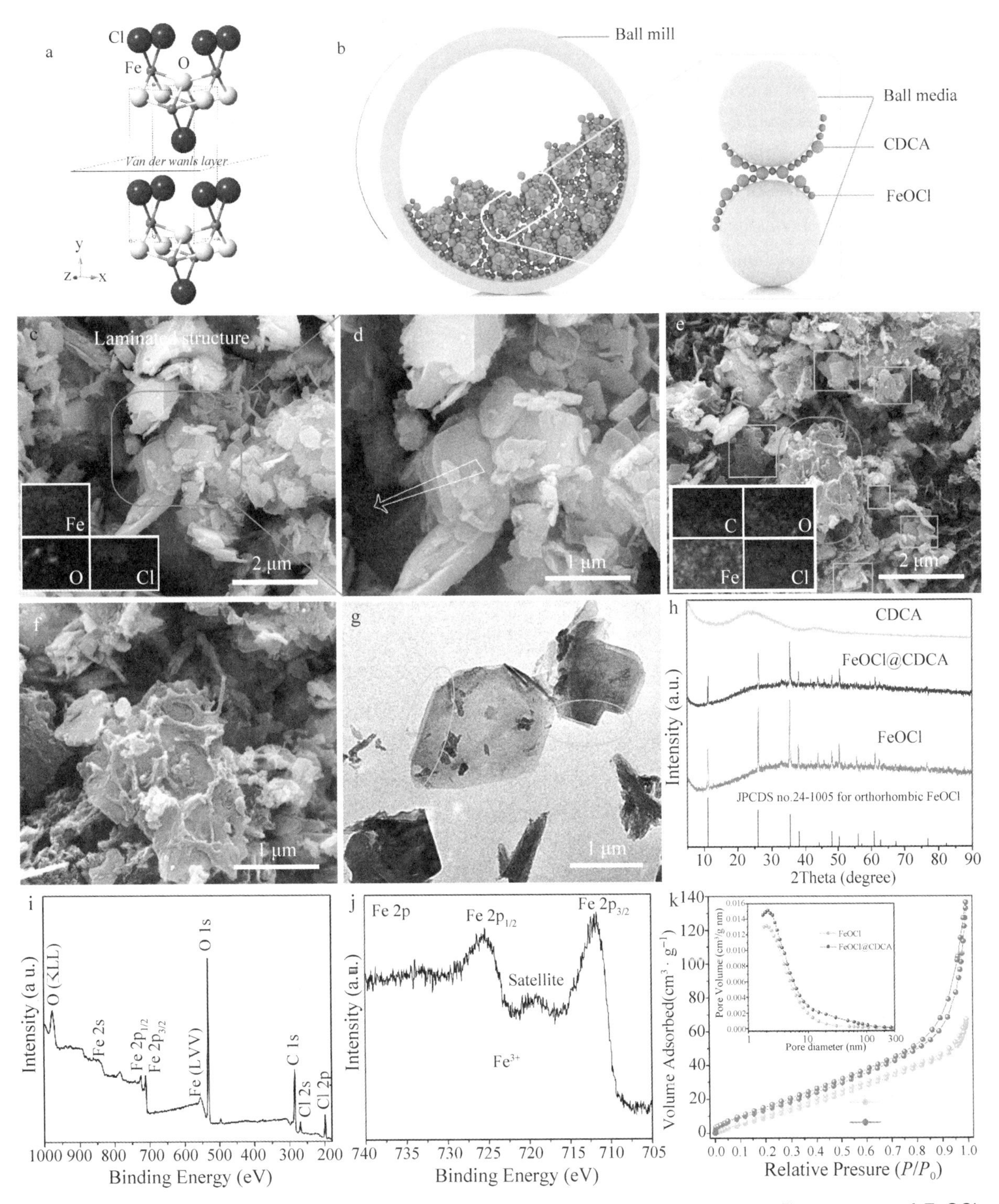

Fig. 1 Micromorphology, synthesis and characterizations of FeOCl@CDCA: (a) crystalline structure of FeOCl; (b) schematic diagram of the ball-milling process; SEM images of (c, d) the FeOCl and (e, f) FeOCl@CDCA (*d* and *f* are the enlarged images of areas within the red frames of *c* and *e*), and the insets of *c* and *e* show the element mappings; (g) TEM images of FeOCl@CDCA; (h) XRD pattern; (i) XPS survey spectrum and (j) Fe 2p core-level XPS spectrum of FeOCl@CDCA; (k) N_2 adsorption-desorption isotherms, and the inset shows the pore size distribution

$$Fe(Ⅲ)ClO+Na^{+}+e^{-}\leftrightarrow NaFe(Ⅱ)ClO \quad (1)$$

To strengthen the validation of the proposed Na^{+}-intercalation mechanism, the CV curve of FeOCl@CDCA was also measured in 1 M $MgSO_4$ electrolyte. As the hydrated ionic diameter of Mg^{2+}(0.856 nm)[22] is larger than that of the interlayer distance of FeOCl (0.783 nm), a notably decreased capacitance measured in the $MgSO_4$ electrolyte will help to demonstrate the intercalation mechanism. As expected, the area enclosed within the CV curve measured in the solution of $MgSO_4$ is much smaller than that measured in the solution of Na_2SO_4(Fig. S9a), indicating a sharp decline in capacitance when tested in $MgSO_4$ electrolyte possibly due to the reduction of intercalation reaction area.

Charge-storage kinetics of electrode materials towards Na^{+} is evaluated by CV analysis. The analysis is performed regarding the behavior of peak currents by assuming that the current (i) obeys a power-law relationship with the scan rate (ν):[32]

$$i=a\nu^{b} \quad (2)$$

where both a and b are adjustable parameters. The b-value can be determined by the slope of log(v)-log(i) plots. In particular, the b-value of 0.5 represents a total diffusion-controlled behavior, whereas 1.0 reveals a capacitive process. Fig. 2e shows the log(v)-log(i) plots for CDCA, FeOCl and FeOCl@CDCA electrodes. The b value of CDCA is calculated to be 0.935 for anodic currents, indicating that the majority of currents at the peak potential is capacitive. The result demonstrates that the capacitive behavior dominates the reaction of CDCA with fast kinetics. By contrast, the FeOCl has the b value of 0.579 (close to 0.5), suggesting that the kinetic behavior of FeOCl is mainly controlled by diffusion process. Moreover, it is clear that the integration of CDCA increases the b value of FeOCl@CDCA to 0.622, reflecing the positive contribution of CDCA to the charge-storage kinetics of FeOCl@CDCA.

To study the capacity contribution, the surface and insertion capacities of FeOCl@CDCA at different scan rates are calculated. The current response i at a fixed potential V is described as the sum of two contributions from surface and intercalation capacities, as discussed below:[33]

$$i(V)=k_1\nu+k_2\nu^{1/2} \quad (3)$$

$$i(V)/\nu^{1/2}=k_1\nu^{1/2}+k_2 \quad (4)$$

where ν represents the scan rate, while $k_1\nu$ and $k_2\nu^{1/2}$ denote the current contributions from the surface capacity (surface pseudocapacity and double-layer capacity) and insertion processes (insertion pseudocapacity), respectively. By determining k_1 and k_2, we can distinguish the fractions of current arising from the surface capacity and insertion processes at a specific potential. At a scan rate of 5 mV·s^{-1}(Fig. 2f), ~9% of the total current, i.e., the capacity, is capacitive in nature. Similarly, the contribution ratios between the two different processes at other scan rates are calculated. The quantified results show that the capacitive contribution gradually improves with the increasing scan rate. The reason may be the fact that higher scan rates restrict the ion diffusion process, which prevents the ions from accessing the laminated structure. Therefore, at high scan rates, the contribution of surface capacity of the electrode becomes more remarkable. In addition, at scan rates of 5-100 mV·s^{-1}, the fraction of insertion capacity is higher (68%-91%), demonstrating that the kinetic behavior of FeOCl@CDCA is mainly controlled by diffusion process. The result is consistent with that of b-value analysis.

For further investigating ionic transport process, apparent diffusion coefficients of the FeOCl and FeOCl@

CDCA electrodes are calculated based on the Randles-Sevick equation:[34]

$$I_{\mathrm{P}} = 0.4463n \times FAC\left(\frac{NFvD}{RT}\right) 1/2 = [269000 \times n^{3/2}AD^{1/2}C]v^{1/2} \tag{5}$$

where I_{p} is the peak current, n is the number of electrons transferred per molecule during the electrochemical reactions (1 in this case), A is the exposed area of the electrode ($1\ \mathrm{cm}^2$), C is the concentration of sodium ions in the electrolyte (1 M in the present system), D is the ionic diffusion coefficient

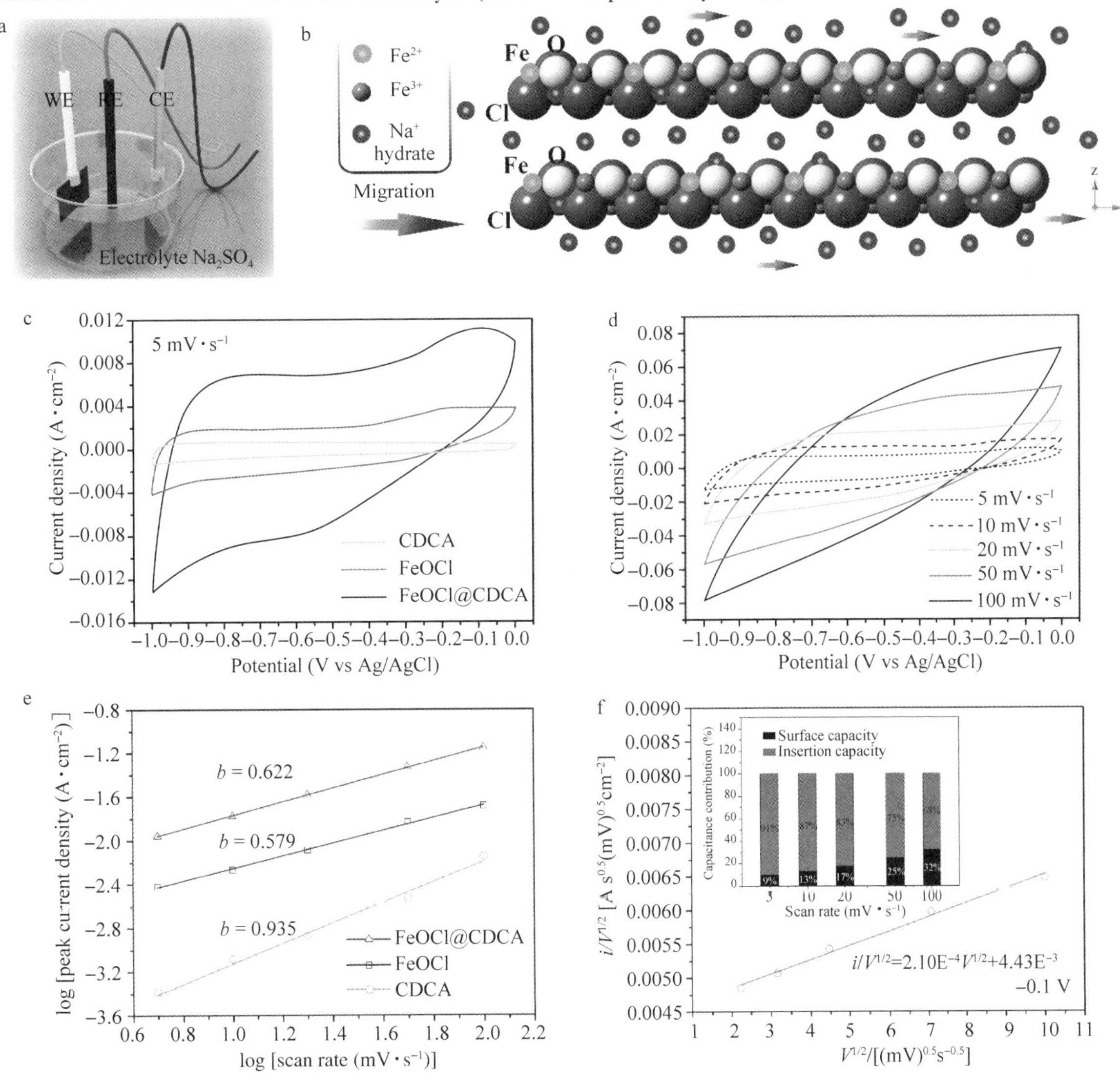

Fig. 2 Kinetic analysis of the electrochemical behavior of FeOCl@CDCA electrode: schematic diagrams of (a) the electrochemical testing apparatus using a conventional three-electrode system and (b) the plausible mechanism of electrochemical reactions; (c) CV curves at the scan rate of $5\ \mathrm{mV \cdot s^{-1}}$; (d) CV curves of FeOCl@CDCA at scan rates of $5\text{-}100\ \mathrm{mV \cdot s^{-1}}$; (e) determination of the b value using the relationship between the peak current and the scan rate according to the voltammograms in d; (f) $i(\mathrm{V})/v^{1/2}$ vs. $v^{1/2}$ plot using the anodic current at a potential of -0.15 V, and the inset shows surface and insertion capacity contributions at different scan rates

(including the diffusion of both sodium and hydroxyl ions, and v is the scan rate. From Fig. S10, the calculated ionic diffusion coefficient of FeOCl@ CDCA is 8.21×10^{-7} $cm^2 \cdot s^{-1}$, which is almost 11.9 times as high as that of the FeOCl (6.92×10^{-8} $cm^2 \cdot s^{-1}$). These results indicate that FeOCl@ CDCA exhibits a faster ionic diffusion rate, resulting in enhanced storage capacity.

GCD is a reliable method for determining capacitances of supercapacitors. As seen in Fig. 3a, the non-linear shape reflects the pseudocapacitance feature of FeOCl @ CDCA, consistent with the result of CV analysis. The good symmetry between the charge and discharge curves demonstrates the good reversibility and coulombic efficiency (namely the ratio of discharging and charging capacitances). Based on these GCD curves, FeOCl@ CDCA achieves the highest areal specific capacitance of about 1618 $mF \cdot cm^{-2}$(647 $F \cdot g^{-1}$) at a discharge current of 2 $mA \cdot cm^{-2}$(0.8 $A \cdot g^{-1}$) (Fig. 3b), which is around 1.3 and 14.7 times higher than that of the FeOCl and CDCA electrodes (ca. 696 and 103 $mF \cdot cm^{-2}$. Details on GCD plots of CDCA and FeOCl electrodes are available in Fig. S11). When the current density notably increases from 2 to 50 $mA \cdot cm^{-2}$, FeOCl@ CDCA remains a high capacitance retention of 54% (875 $mF \cdot cm^{-2}$/350 $F \cdot g^{-1}$), indicating its excellent rate capability. Such high capacitances of FeOCl@ CDCA are more superior to numerous results reported for anodes, e. g. , γ-Fe_2O_3/graphene (76 $F \cdot g^{-1}$),[35] mesoporous MoO_2 (146 $F \cdot g^{-1}$),[36] polypyrrole@ V_2O_5(308 $F \cdot g^{-1}$),[12] Ti-Fe_2O_3@ PEDOT (311.6 $F \cdot g^{-1}$),[37] highly functionalized activated carbons (525 $F \cdot g^{-1}$)[38] and iron nanosheets/graphene (717 $F \cdot g^{-1}$)[39], as listed in Table S1. We also measured the GCD curve of FeOCl@ CDCA in the solution of $MgSO_4$ at 5 $mA \cdot cm^{-2}$. A capacitance value of 759 $mA \cdot cm^{-2}$ is obtained (Fig. S9b), lower than half of that measured in the $NaSO_4$ solution (1552 $mA \cdot cm^{-2}$), which further demonstrates the intercalation mechanism and reduced intercalation reaction sites in $MgSO_4$ solution. Moreover, it is known that the crystal structure of FeOCl is similar to that of FeOOH. By comparing their capacitances, the areal specific capacitance of FeOCl@ CDCA electrode (1.62 $F \cdot cm^{-2}$, 0.8 $A \cdot g^{-1}$) is comparable to or even higher than that of FeOOH-based electrodes, such as FeOOH nanoparticles[40](1.71 $F \cdot cm^{-2}$, 1 $A \cdot g^{-1}$), nano-sized columned β-FeOOH[41](1.16 $F \cdot cm^{-2}$, 0.5 $A \cdot g^{-1}$) and FeOOH nanorod/carbon tube network (0.79 $F \cdot cm^{-2}$, 0.5 $A \cdot g^{-1}$)[42]. Because of the high mass loading (2.5 $mg \cdot cm^{-2}$), the corresponding gravimetric specific capacitance of FeOCl@ CDCA is 647 $F \cdot g^{-1}$, lower than that of FeOOH nanoparticles (1066 $F \cdot g^{-1}$, 1.6 $mg \cdot cm^{-2}$), but higher that that of nano-sized columned β-FeOOH (116 $F \cdot g^{-1}$, 10 $mg \cdot cm^{-2}$) and FeOOH nanorod/carbon tube network (396 $F \cdot g^{-1}$, 2 $mg \cdot cm^{-2}$). In addition, it is known that FeOOH is thermally unstable and is quantitatively converted to the topotactically related γ-Fe_2O_3 when the isostructural γ-FeOOH is treated with organic bases at temperatures of 120-140 ℃.[43] Also, according to the report by Tang et al., α-Fe_2O_3 nanorods are prepared by calcining α-FeOOH at 250 ℃ for 2 h[44]. By contrast, the thermal stability of FeOCl is more superior because FeOCl is obtained by calcining the mixture of α-Fe_2O_3 and $FeCl_3$ at 380 ℃. In consequence, FeOCl also has own unique advantages when acting as supercapacitor electrodes, as compared with FeOOH.

The capacitance contribution from CDCA and FeOCl components to the total capacitance of FeOCl@ CDCA can be estimated based on the above b-value analysis (Fig. 2e). CDCA has a capacitance behavior-dominant charge-storage kinetics while the kinetic behavior of FeOCl is mainly controlled by diffusion process. Therefore, the capacitance contribution of CDCA and FeOCl compositions to the total capacitance of FeOCl @ CDCA can be distinguished roughly as surface capacitance and intercalation capacitance, respectively. By virtue of the capacitance contribution analysis, the fraction of insertion capacitance of FeOCl@ CDCA is much higher (68%-91%) than that of surface capacitance (9%-32%) at scan rates of 5-100 $mV \cdot s^{-1}$. Thus, we can roughly speculate that capacitance contribution from the FeOCl should account for more than 60% while the contribution from CDCA should account for less than 40%. Besides, the increase of scan rates will increase the contribution proportion from CDCA and decrease the contribution proportion from the FeOCl.

Also, FeOCl@ CDCA shows excellent cycle stability with no more than 10% capacitance loss (ca. 9%) after 10000 cycles (Fig. 3c), lower than that of the FeOCl (ca. 25% loss) (Fig. S12). The dominant cycle behavior of FeOCl@ CDCA is attributed to contributions of CDCA to the stability of FeOCl, possibly including two points: (1) 3D interconnected CDCA might act as buffers to alleviate strains of FeOCl in the process of intercalation and de-intercalation of ions;[45] (2) good mixture between CDCA and FeOCl effectively prevent the aggregation of FeOCl. As presented in the inset of Fig. 3c, the SEM image of FeOCl@ CDCA after 10000 cycles demonstrates that the laminated structure is well maintained and the laminated FeOCl surrounded with CDCA does not contact with others, indicating its good shape stability and dispersion. Similar results can be found for other composites composed of carbon materials and metallic oxides/hydroxides.[46,47] Moreover, a high coulombic efficiency of 96%-100% is achieved during the cycles.

EIS was measured to quantify the resistance at the electrode/electrolyte interface and analyze causes for performance differences of CDCA, FeOCl and FeOCl@ CDCA (Fig. 3d). The equivalent circuit diagram used for the fitting of EIS data is shown in the inset of Fig. 3d. By comparing Nyquist plots, we can see that integrating the FeOCl with CDCA decreases the diameter of semicircles (an indicator for charge transfer resistance, R_c) from 1 Ω (FeOCl) to 0.4 Ω (FeOCl@ CDCA), because the highly conductive CDCA constructs dense conducting networks and provides plentiful mobile charge carriers and conductive channels. This result agrees well with that of electrical conductivity measurements showing a significant increase from outrange ($<10^{-5}$ $S \cdot cm^{-1}$) for the FeOCl to 0.26 $S \cdot cm^{-1}$ for FeOCl@ CDCA. This value is close to that of some conjugated polymers (0.01-400 $S \cdot cm^{-1}$).[48] In addition, the projected length of the 45° Warburg region on the real impedance axis (Z') reflects the ion penetration process. Obviously, the length of FeOCl@ CDCA is shorter than that of CDCA and FeOCl, indicative of a shorter ion-diffusion path. Therefore, the EIS suggests that FeOCl@ CDCA has better ionic/electronic transport ability. In the low-frequency region, FeOCl @ CDCA shows a more vertical line along the imaginary axis as compared to that of the FeOCl, demonstrating its more ideally capacitive behavior because of the introduction of non-Faradic charge storage process.

Based on these analyses, the more superior electrochemical properties of FeOCl@ CDCA are possibly attributed to synergistic functions of CDCA and FeOCl, which can be summarized as follows: (1) the large surface area of FeOCl consisting of laminated microsheets has the ability to provide abundant space and sites for the electrochemical reactions; (2) FeOCl has a high theoretical capacitance value of ~899 $F \cdot g^{-1}$ (the calculation is available in SI) and the laminated structure of FeOCl supported on the CDCA shows an excellent mechanical stability, helpful for achieving long service life; (3) the porous CDCA with a large surface area can act as an electrolyte storage and shorten diffusion path of ions between ions and laminated FeOCl (Fig. 3e). The highly conductive CDCA is able to supply abundant mobile charge carriers and conductive channels during the reactions.

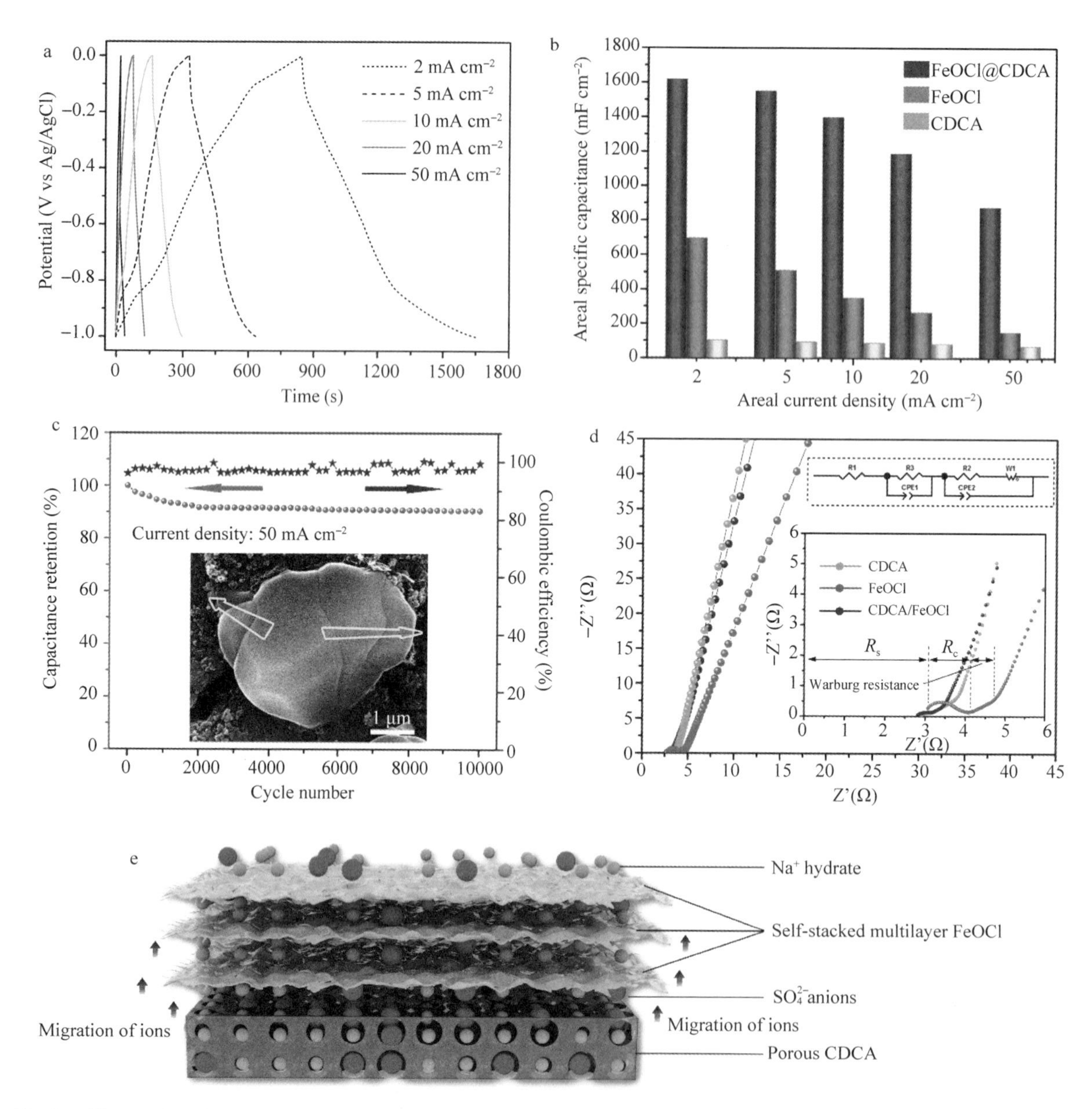

Fig. 3 Electrochemical properties of FeOCl@CDCA electrode: (a) GCD curves of FeOCl@CDCA at current densities of 2–50 mA · cm^{-2}; (b) rate performance; (c) cycle stability and coulombic efficiency of FeOCl@CDCA at 50 mA · cm^{-2} and the inset shows its SEM image after 10000 cycles; (d) Nyquist plots and insets show the enlarged image at the high-frequency region and the equivalent circuit used for EIS data fitting; (e) schematic diagrams of functions of CDCA and FeOCl components on the electrochemical reactions

An ASC was assembled to evaluate the potential of FeOCl@CDCA for practical applications. The schematic diagram (Fig. 4a) depicts the fabrication of ASC with FeOCl@CDCA as an anode and a common cheap Ni foam-supported MnO_2 as a cathode. The MnO_2 cathode shows a moderate areal specific capacitance of 318 mF · cm^{-2} (ca. 127 F · g^{-1}) at 2 mA · cm^{-2} and good rate capability (the detailed electrochemical evaluation is available in Fig. S13). Considering that FeOCl@CDCA was tested within a stable potential window from −1 to 0 V (vs. Ag/AgCl) while that of the MnO_2 electrode was measured within the window of 0~0.8 V (vs. Ag/AgCl), such two electrodes combination as ASC may afford a device with 1.8 V operation voltages in 1 M Na_2SO_4 as electrolyte. As illustrated in Fig. 4b, the stable electrochemical voltage of the ASC can be extended from 0.8 to 1.8 V. In addition, the CV curves of the ASC with a non-rectangular shape at different scan rates in a 1.8 V potential window (Fig. 4c) reveals that the capacitance primarily comes from redox reactions.

The GCD curves of the ASC at different current densities were tested. As seen in Fig. 4d, both charge and discharge curves keep a good symmetry at such a high operation voltage of 1.8 V, verifying its outstanding reversibility and good coulombic efficiency. In addition, the ASC delivers an areal specific capacitance of 641 and 372 mF · cm^{-2} (128 and 74 F · g^{-1}) at a current density of 2 and 50 mA · cm^{-2}, respectively. A superior rate capability with ~58% of the capacitance is retained in that current density range (Fig. 4e). From the Ragone plot (Fig. 4f), the ASC exhibits a maximum areal energy density of 289 μW · h · cm^{-2} at a power density of 1.8 mW · cm^{-2} and an areal energy density of 168 μW · h · cm^{-2} at a maximum power density of 45 mW · cm^{-2}. The attained energy densities of our ASC are superior to those of many recently reported ASCs with a pseudocapacitive storage mechanism (see Fig. 4f and Table S2).[49-54] Moreover, the ASC remains at 88.7% of the initial capacitance after 10000 cycles (Fig. 4g), reflecting its good cycle stability. In addition, based on previous researches, possible reasons for the decrease of the capacitance with cycle number is possibly summarized into three points: (1) A small quantity of structural alternation/disintegration around the active sites, by virtue of repeated ingression and depletion of ions in the electrode materials during the Faradaic reactions;[55] (2) Dissolution/detachment of active materials (like a small portion of Fe dissolution) leading to reduction of electrochemical active sites;[56,57] (3) Consumption of electrolyte caused by tiny minority of irreversible side-reactions and interface reactions.[58,59] Also, the device delivers a high coulombic efficiency (>97%) during the cycles, agreeing well with the speculation from the GCD analysis. The inset of Fig. 4g presents the Nyquist plots of the ASC before and after the cycles. Obviously, the solution resistance (R_s) did not change much (~3 Ω) before and after tests. The corresponding charge transfer resistances R_c are 1.0 and 2.4 Ω, respectively. The slightly enlarged charge transfer resistance after the cycle tests is possibly related to the destruction and dissolution of a part of electroactive materials suffering from plentiful harsh redox reactions. This slight increment further indicates that the ASC has favorable cycle stability. Furthermore, these low R_c and R_s values are expected to enhance its power density. These outstanding electrochemical performances of the MnO_2//FeOCl@ CDCA ASC demonstrate that FeOCl@ CDCA suits to be a cost-effective and high-performance anode material for supercapacitors.

3 Conclusions

In summary, we demonstrate that the self-stacked and multilayer FeOCl supported on CDCA is suitable as a novel and high-performance anode which delivers an ultra-high areal specific capacitance of 1618 mF · cm^{-2} (647 F · g^{-1}) at 2 mA · cm^{-2} and outstanding cycle stability with no more than 10% capacitance loss (ca. 9%) after 10000 cycles. The laminated structure of the hierarchical FeOCl with a superior mechanical stability (strengthened by CDCA) provides abundant channels and active sites for the migration and intercalation reactions of Na^+, which plays a key role in the high electrochemical activity. The assembled

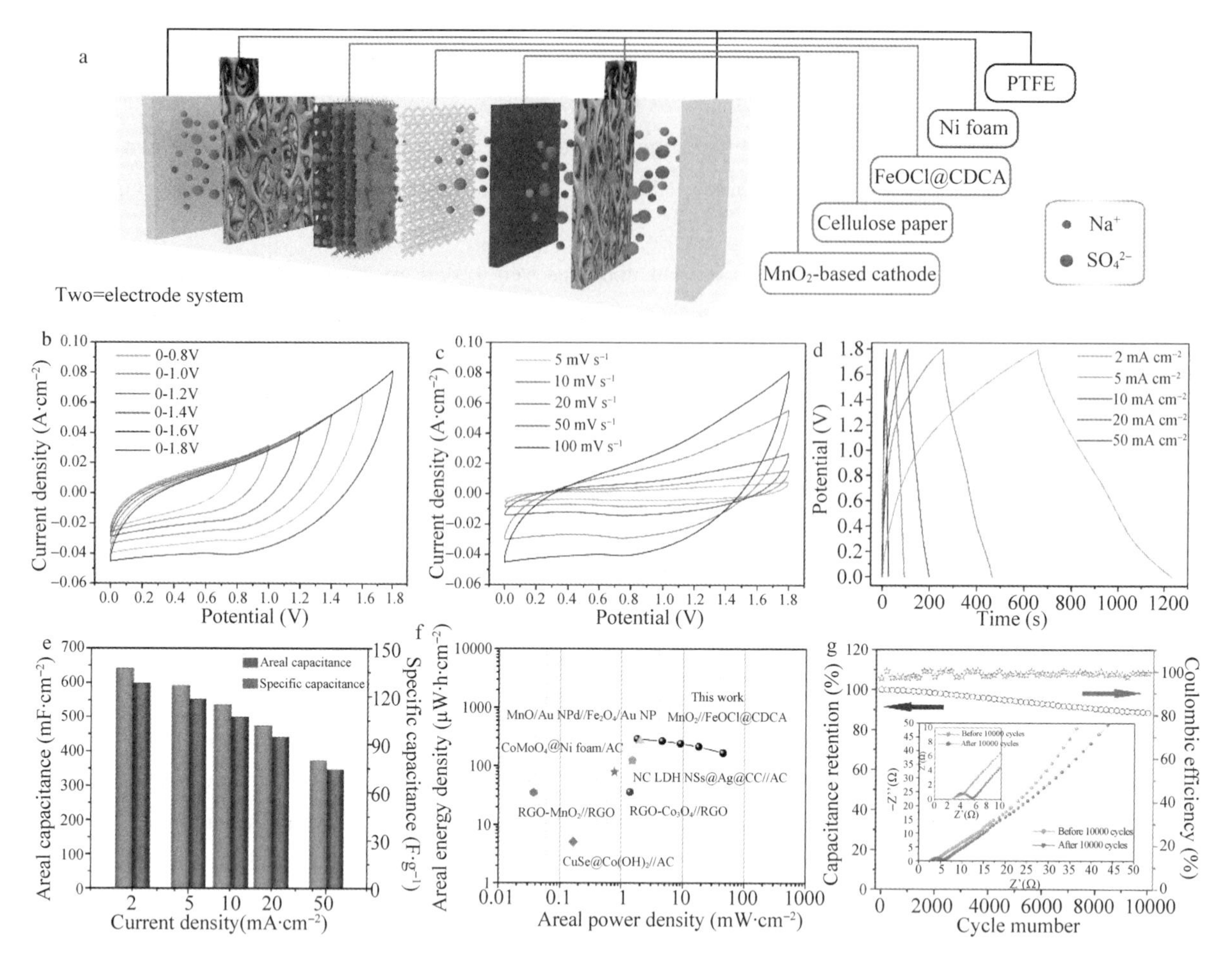

Fig. 4 Electrochemical properties of the MnO_2//FeOCl @ CDCA ASC: (a) schematic diagram of the assembled ASC device in a two-electrode system; (b) CV curves in different potential windows at the scan rate of 100 mV · s^{-1}; (c) CV curves at scan rates of 5–100 mV · s^{-1}; (d) GCD curves at current densities of 2–50 mA · cm^{-2}; (e) rate performance; (f) Ragone plot of the ASC compared with data of some other ASCs in the literature; (g) cycle stability and coulombic efficiency at the current density of 50 mA · cm^{-2}, and the inset shows the Nyquist plots before and after the cycles

MnO_2//FeOCl@ CDCA ASC shows a highly competitive energy/power density (289 μW · h · cm^{-2} at 1. 8 mW · cm^{-2}) and cycle stability among many ever-reported ASCs storing energy by Faradaic reactions. More importantly, the cost-effective and easily synthesized FeOCl@ CDCA anode with outstanding electrochemical properties should be applicable to assembly of diversified high-performance energy-storage devices (e. g. , Li/Na-ion batteries) that are not limited to supercapacitors.

Supporting Information

Supporting Information is available from the Wiley Online Library or from the author.

Acknowledgements

This study was supported by the National Natural Science Foundation of China (grant nos. 31530009 and

31470584), the Youth Scientific Research Foundation, Central South University of Forestry and Technology (grant no. QJ2018002A), and the Fundamental Research Funds for the Central Universities (grant no. 2572018AB09).

References

[1] Wang G, Zhang L, Zhang J. A review of electrode materials for electrochemical supercapacitors[J]. Chem. Soc. Rev., 2012, 41(2): 797-828.

[2] González A, Goikolea E, Barrena J. A, Mysyk R. Review on supercapacitors: Technologies and materials[J]. Renewable Sustainable Energy Rev., 2016, 58: 1189-1206.

[3] Winter M, Brodd R J. What are batteries, fuel cells, and supercapacitors? [J]. Chem. Rev. 2004, 104(10): 4245-4269.

[4] Dubal D. P, Ayyad O, Ruiz V, Gomez-Romero P. Hybrid energy storage: the merging of battery and supercapacitor chemistries [J]. Chem. Soc. Rev., 2015, 44(7): 1777-1790.

[5] Kim H, Cho M Y, Kim M H, Park K Y, Gwon H, Lee Y, Roh K C, Kang K. A Novel high-energy hybrid supercapacitor with an anatase TiO_2-reduced graphene oxide anode and an activated carbon cathode[J], Adv. Energy Mater., 2013, 3(11): 1500-1506.

[6] Chen W, Xia C, Alshareef H N. One-step electrodeposited nickel cobalt sulfide nanosheet arrays for high-performance asymmetric supercapacitors[J]. ACS Nano, 2014, 8(9): 9531-9541.

[7] Zuo W H, Xie C Y, Xu P, Li Y Y, Liu J P. A novel phase-transformation activation process toward Ni-Mn-O nanoprism arrays for 2. 4 V ultrahigh-voltage aqueous supercapacitors[J]. Adv. Mater., 2017, 29(36): 1703463.

[8] Wang X G, Li Q C, Zhang L, Hu Z L, Yu L H, Jiang T, Lu C, Yan C L, Sun J Y, Liu Z F. Caging Nb_2O_5 nanowires in PECVD-derived graphene capsules toward bendable sodium-Ion hybrid supercapacitors[J]. Adv. Mater., 2018, 30 (26): 1800963.

[9] Yu G H, Xie X, Pan L J, Bao Z N, Cui Y. Hybrid nanostructured materials for high-performance electrochemical capacitors[J]. Nano Energy, 2013, 2(2): 213-234.

[10] Simon P, Gogotsi Y. Materials for electrochemical capacitors[J]. Nat. Mater., 2008, 7(11): 845-854.

[11] hinde N M, Xia Q X, Yun J M, Mane R S, Kim K H. Polycrystalline and mesoporous 3-D Bi_2O_3 nanostructured negatrodes for high-energy and power-asymmetric supercapacitors: superfast room-temperature direct wet chemical Growth[J]. ACS Appl. Mater. Interfaces, 2018, 10 (13): 11037-11047.

[12] Qu Q T, Zhu Y S, Gao X W, Wu Y P. Core-shell structure of polypyrrole grown on V_2O_5 nanoribbon as high performance anode material for supercapacitors[J]. Adv. Energy Mater., 2012, 2(8): 950-955.

[13] Lu X F, Huang Z X, Tong Y X, Li G R. Asymmetric supercapacitors with high energy density based on helical hierarchical porous $NaxMnO_2$ and MoO_2[J]. Chem. Sci., 2016, 7(1): 510-517.

[14] Wang H W, Xu Z J, Yi H, Wei H G, Guo Z H, Wang X F. One-step preparation of single-crystalline Fe_2O_3 particles/graphene composite hydrogels as high-performance anode materials for supercapacitors[J]. Nano Energy, 2014, 7: 86-96.

[15] Jiang H, Ma J, Li C Z. Mesoporous carbon incorporated metal oxide nanomaterials as supercapacitor electrodes [J]. Adv. Mater., 2012, 24(30): 4197-4202.

[16] Zhi M J, Xiang C C, Li J T, Li M, Wu N Q. Nanostructured carbon-metal oxide composite electrodes for supercapacitors: a review[J]. Nanoscale 2013, 5(1): 72-88.

[17] Qu Q T, Yang S B, Feng X L. 2D Sandwich-like sheets of iron oxide grown on graphene as high energy anode material for supercapacitors[J]. Adv. Mater., 2011, 23(46): 5574-5580.

[18] Guan C, Liu J L, Wang Y D, Mao L, Fan Z X, Shen Z X, Zhang H, Wang J. Iron oxide-decorated carbon for supercapacitor anodes with ultrahigh energy density and outstanding cycling stability[J]. ACS Nano 2015, 9(5): 5198-5207.

[19] Yang P H, Ding Y, Lin Z Y, Chen Z W, Li Y Z, Qiang P F, Ebrahimi M, Mai W J, Wong C P, Wang Z L. Low-cost high-performance solid-state asymmetric supercapacitors based on MnO_2 nanowires and Fe_2O_3 nanotubes[J]. Nano Lett., 2014, 14(2): 731-736.

[20] Duan G, Cai W, Luo Y, Sun F. A hierarchically structured $Ni(OH)_2$ monolayer hollow-sphere array and its tunable optical properties over a large region[J]. Adv. Funct. Mater., 2007, 17(4): 644-650.

[21] Jarrige I, Cai Y Q, Shieh S R, Ishii H, Hiraoka N, Karna S, Li W H. Charge transfer in FeOCl intercalation compounds and its pressure dependence: An x-ray spectroscopic study[J]. Phys. Rev. B, 2010, 82(16): 165121.

[22] Nightingale E R, Phenomenological theory of ion solvation-effective radii of hydrated ions[J]. J. Phys. Chem., 1959, 63 (9): 1381-1387.

[23] Fagan S B, Souza A G, Lima J O G, Mendes J, Ferreira O P, Mazali I O, Alve O L, Dresselhaus M S. 1, 2-dichlorobenzene interacting with carbon nanotubes[J]. Nano Lett., 2004, 4(7): 1285-1288.

[24] Zhang J, Liu G D, Liu S J. 2D/2D FeOCl/graphite oxide heterojunction with enhanced catalytic performance as a photo-Fenton catalyst[J]. New J. Chem., 2018, 42(9): 6896-6902.

[25] Zhang L L, Zhao X S. Carbon-based materials as supercapacitor electrodes[J]. Chem. Soc. Rev., 2009, 38(9): 2520-2531.

[26] Chmiola J, Yushin G, Gogotsi Y, Portet C, Simon P, Taberna P L. Anomalous increase in carbon capacitance at pore sizes less than 1 nanometer[J]. Science, 2006, 313(5794): 1760-1763.

[27] Liang H W, Wu Z Y, Chen L F, Li C, Yu S H, Bacterial cellulose derived nitrogen-doped carbon nanofiber aerogel: An efficient metal-free oxygen reduction electrocatalyst for zinc-air battery[J]. Nano Energy 2015, 11: 366-376.

[28] Wu Z Y, Li C, Liang H W, Chen J F, Yu S H. Ultralight, flexible, and fire-resistant carbon nanofiber aerogels from bacterial cellulose[J]. Angew. Chem., 2013, 52(10): 2925-2929.

[29] Bi H C, Yin Z Y, Cao X H, Xie X, Tan C L, Huang X, Chen B, Chen F T, Yang Q L, Bu X Y, Lu X H, Sun L T, Zhang H. Carbon fiber aerogel made from raw cotton: a novel, efficient and recyclable sorbent for oils and organic solvents [J]. Adv. Mater., 2013, 25(41): 5916-5921.

[30] Chmiola J, Yushin G, Dash R. Gogotsi Y. Effect of pore size and surface area of carbide derived carbons on specific capacitance[J]. J. Power Sources, 2006, 158(1): 765-772.

[31] Jin W H, Cao G T, Sun J Y. Hybrid supercapacitor based on MnO_2 and columned FeOOH using Li_2SO_4 electrolyte solution[J]. J. Power Sources, 2008, 1751(1): 686-691.

[32] Wang J, Polleux J, Lim J, Dunn B. Pseudocapacitive contributions to electrochemical energy storage in TiO_2 (anatase) nanoparticles[J]. J. Phys. Chem. C, 2007, 111(40): 14925-14931.

[33] Kim H S, Cook J B, Tolbert S H, Dunn B. The Development of Pseudocapacitive Properties in Nanosized-MoO_2[J]. J. Electrochem. Soc., 2015, 162(5): A5083-A5090.

[34] Wang X P, Niu C J, Meng J S, Hu P, Xu X M, Wei X J, Zhou L, Zhao K N, Luo W, Yan M Y, Mai L Q. Novel $K_3V_2(PO_4)_3$/C Bundled Nanowires as Superior Sodium-Ion Battery Electrode with Ultrahigh Cycling Stability[J]. Adv. Energy Mater., 2015, 5(17): 1500716.

[35] Chen H C, Wang C C, Lu S Y, gamma-Fe_2O_3/graphene nanocomposites as a stable high performance anode material for neutral aqueous supercapacitors[J]. J. Mater. Chem. A, 2014, 2(40): 16955-16962.

[36] Li X Y, Shao J, Li J, Zhang L, Qu Q T, Zheng H H Ordered mesoporous MoO_2 as a high-performance anode material for aqueous supercapacitors[J]. J. Power Sources, 2013, 237: 80-83.

[37] Zeng Y X, Han Y, Zhao Y T, Zeng Y, Yu M H, Liu Y J, Tang H L, Tong Y X, Lu X H. Advanced Ti-doped Fe_2O_3@ PEDOT core/shell anode for high-energy asymmetric supercapacitors [J]. Adv. Energy Mater., 2015, 5

(12): 1402176.

[38] Li Z, Xu Z W, Wang HL, Ding J, Zahiri B, Holt C M B, Tan X H, Mitlin D. Colossal pseudocapacitance in a high functionality-high surface area carbon anode doubles the energy of an asymmetric supercapacitor[J]. Energy Environ. Sci., 2014, 7(5): 1708-1718.

[39] Long C L, Wei T, Yan J, Jiang L L, Fan Z J. Supercapacitors based on graphene-supported iron nanosheets as negative electrode materials[J]. ACS Nano, 2013, 7(12): 11325-11332.

[40] Owusu K A, Qu L B, Li J T, Wang Z Y, Zhao K N, Yang C, Hercule K M, Lin C, Shi C W, Wei Q L, Zhou L, Mai L Q. Low-crystalline iron oxide hydroxide nanoparticle anode for high-performance supercapacitors[J]. Nat. Commun., 2017, 8: 14264.

[41] Jin W H, Cao G T, Sun J Y. Hybrid supercapacitor based on MnO_2 and columned FeOOH using Li_2SO_4 electrolyte solution[J]. J. Power Sources, 2008, 175(1): 686-691.

[42] Li J F, Chen D D, Wu Q S, Wang X, Zhang Y, Zhang Q W. FeOOH nanorod arrays aligned on eggplant derived super long carbon tube networks as negative electrodes for supercapacitors[J]. New J. Chem., 2018, 42(6): 4513-4519.

[43] Desiraju G R, Rao M, A mild transformation of gamma-feoom to gamma-Fe_2O_3 using organic-reagents[J]. Mater. Res. Bull., 1982, 17(4): 443-449.

[44] Tang B, Wang G L, Zhuo L H, Ge J C, Cui L J. Facile route to alpha-FeOOH and alpha-Fe_2O_3 nanorods and magnetic property of alpha-Fe_2O_3 nanorods[J]. Inorg. Chem., 2006, 45(13): 5196-5200.

[45] Hao F B, Zhang Z W, Yin L W. Co_3O_4/carbon aerogel hybrids as anode materials for lithium-ion batteries with enhanced electrochemical properties[J]. ACS Appl. Mater. Interfaces, 2013, 5(17): 8337-8344.

[46] Chaudhari N K, Chaudhari S, Yu J S. Cube-like alpha-Fe_2O_3 supported on ordered multimodal porous carbon as high-performance electrode material for supercapacitors[J]. ChemSusChem, 2014, 7(11): 3102-3111.

[47] Cheng X P, Gui X C, Lin Z Q, Zheng Y J, Liu M, Zhan R Z, Zhu Y, Tang Z K. Three-dimensional alpha-Fe_2O_3/carbon nanotube sponges as flexible supercapacitor electrodes[J]. Journal of Materials Chemistry A, 2015, 3(42): 20927-20934.

[48] Snook G A, Kao P, Best A S. Conducting-polymer-based supercapacitor devices and electrodes[J]. J. Power Sources, 2011, 196(1): 1-12.

[49] Gong J F, Tian Y Z, Yang Z Y, Wang Q J, Hong X H, Ding Q P. High-performance flexible all-solid-state asymmetric supercapacitors based on vertically aligned CuSe@ $Co(OH)_2$ nanosheet arrays[J]. J. Phys. Chem. C, 2018, 122(4): 2002-2011.

[50] Afriyanti S, Ce Yao F, Xu W, Pooi See L. Large areal mass, flexible and free-standing reduced graphene oxide/manganese dioxide paper for asymmetric supercapacitor device[J]. Adv. Mater., 2013, 25(20): 2809-2815.

[51] Ghosh D, Lim J, Narayan R, Kim S O. High energy density all solid state asymmetric pseudocapacitors based on free standing reduced graphene oxide-Co_3O_4 composite aerogel electrodes[J]. ACS Appl. Mater. Interfaces, 2016, 8(34): 22253-22260.

[52] Sekhar S C, Nagaraju G, Yu J S. Conductive silver nanowires-fenced carbon cloth fibers-supported layered double hydroxide nanosheets as a flexible and binder-free electrode for high-performance asymmetric supercapacitors[J]. Nano Energy, 2017, 36: 58-67.

[53] Veerasubramani G K, Krishnamoorthy K, Kim S J. Improved electrochemical performances of binder-free CoMoO4 nanoplate arrays@ Ni foam electrode using redox additive electrolyte[J]. J. Power Sources, 2016, 306: 378-386.

[54] Ko Y, Kwon M, Bae W K, Lee B, Lee S W, Cho J. Flexible supercapacitor electrodes based on real metal-like cellulose papers[J]. Nat. Commun., 2017, 8: 536.

[55] Shaikh S M F, Rahman G, Mane R S, Joo O S. Bismuth oxide nanoplates-based efficient DSSCs: Influence of ZnO surface passivation layer[J]. Electrochim. Acta, 2013, 111: 593-600.

[56] Quan H Y, Cheng B C, Xiao Y H, Lei S J. One-pot synthesis of alpha-Fe_2O_3 nanoplates-reduced graphene oxide composites for supercapacitor application[J]. Chem. Eng. J., 2016, 286: 165-173.

[57] Wang H W, Xu Z J, Yi H, Wei HG, Guo Z H, Wang X F. One-step preparation of single-crystalline Fe_2O_3 particles/graphene composite hydrogels as high-performance anode materials for supercapacitors[J]. Nano Energy, 2014, 7: 86-96.

[58] Chaudhari N K, Chaudhari S, Yu JS, Cube-like alpha-Fe_2O_3 supported on ordered multimodal porous carbon as high-performance electrode material for supercapacitors[J]. ChemSusChem, 2014, 7(11): 3102-3111.

[59] Wang Y G, Xia Y Y, Hybrid aqueous energy storage cells using activated carbon and lithium-intercalated compounds I. The C/$LiMn_2O_4$ system[J]. J. Electrochem. Soc., 2006, 153(2): A450-A454.

中文题目：纤维素基碳气凝胶支撑的自堆叠多层 FeOCl：一种新型高性能超级电容器负极材料

作者：万才超，焦月，包文慧，高鹤，吴义强，李坚

摘要：为了构建高能密度不对称超级电容器(ASCs)，目前的研究主要集中在阴极材料方面。然而，负极材料的研究较少。本研究首次证明具有自叠层结构的正交 FeOCl 适合作为高性能的超级电容器负极材料，其独特的叠层结构可以为电解质离子的迁移和插层反应提供丰富的活性位点。通过引入高导电性和多孔纤维素基碳气凝胶(CDCA)基体，FeOCl 的力学稳定性和电荷存储动力学得到了显著提高。FeOCl@ CDCA 在 2 mA · cm^{-2} 下提供了超高的面积比电容 1618 mF · cm^{-2}(647 F · g^{-1})和出色的循环稳定性，在 1 M Na_2SO_4 中 1 和 0 V 之间的 Ag/AgCl 经 10000 次循环后，电容损失不超过 10%。采用 FeOCl@ CDCA 阳极和廉价的 MnO_2 阴极制备了 0 ~ 1.8 V 的 ASC。ASC 显示了高度竞争的能量/功率密度(1.8 mW · cm^{-2} 时为 289 mW · h · cm^{-2})和出色的速率能力和循环稳定性。这些发现为利用 FeOCl 基阳极设计高能量密度储能系统开辟了新的途径。

04

第四部分

纳米纤维素及其他

Effects of Water-borne Rosin on the Fixation and Decay Resistance of Copper-based Preservative Treated Wood *

Nguyen Thi Thanh Hien, Jian Li, Shujun Li

Abstract: A rosin sizing agent designed to impregnate wood and immobilize copper in wood cells for protection against decay was investigated. Poplar (Populus ussuriensis) wood samples were impregnated with combinations of 3% $CuSO_4$ solution and 1.0%, 2.0%, or 4.0% rosin sizing agent. The decay resistance of treated wood blocks was measured by a soil-block culture method. After a 12-week decay test, the weight losses of untreated control blocks were 70.45% by Trametes versicolor and 61.84% by Gloeophyllum trabeum. The wood decay resistance was also slightly improved by the treatment with only the rosin sizing agent. However, after being treated with the rosin sizing agent and $CuSO_4$, the wood had great decay resistance. The average weight losses of the samples degraded by fungi were less than 4%. Notably, the leached wood blocks had a weight loss of less than 3%. After leaching, the copper content in the leachates was analyzed by atomic absorption spectroscopy (AAS). Results showed that the amount of copper ions released from the samples treated with the copper-rosin solutions was half those from the samples treated with copper alone. Scanning electron microscopy coupled with energy dispersive X-ray analysis (SEM-EDX) proved that the copper element was still in the cell lumens of leached wood blocks, which is consistent with the results of AAS analysis. It signifies that the rosin sizing agent is very helpful to fix the copper preservative in wood.

Keywords: Rosin; Fixation; Decay resistance; Copper sulfate; Wood preservative

1 Introduction

Copper compounds were known to have wood protection capacities and have been used for this purpose for more than 200 years. The efficacy of copper sulfate against wood decay due to fungi, insects, and marine borers was established in wood products from the 1970s and 1980s (Freeman and McIntyre, 2008; Ngoc, 2006). However, copper itself cannot ensure sufficient protection against wood destroying organisms because it is easily lost from treated wood (Ruddick, 2000). In order to overcome this problem, copper was usually combined with other compounds such as sodium fluoride (NaF), sodium hydroxide (NaOH), arsenic (As), chromium (Cr), borate, *etc.* Among these compounds, chromate copper arsenate (CCA) has been used extensively for wood preservation for the longest. Nevertheless, recognition of the risks to human's health and potential environmental damage has prompted changes in the types of preservatives used commercially in recent years. Particularly, CCA was completely banned in the European Union and limited to nonresidential uses in the United States (Townsend and Solo-Gabriele, 2006). This has stimulated the wood preserving industry to create new wood preservatives to minimize the environmental impact of treated wood.

* 本文出自 Bioresources, 2012, 7: 3573-3584.

Hence, current research has been focused on the development of more environmentally friendly, alternative wood preservatives to CCA. For example, arsenic/chromium-free alternatives based on copper compounds, such as copper azole (CA) and ammoniacal copper quaternary (ACQ) have been introduced as alternative chemicals (Nicholas and Schultz, 1995; Nicholas and Schultz, 1997); and they have become a predominant choice worldwide in today's wood preservation systems. Subsequently, water-borne micronized copper formulations such as micronized copper azole, micronized copper quat also have come into use (McIntyre and Freeman, 2008). However, the cost of these biocides is much higher in comparison to the CCA.

Several different methods have been researched to decrease copper leaching from wood, such as combined wood impregnation processes including impregnation with a copper-based, chromium-free wood preservative and subsequent impregnation with a hydrophobic product (Treu et al., 2011), incorporating additives into the preservative formulations to limit copper migration (MitsuhashiGonzalez, 2007), and combining copper with other co-biocides like quaternary ammonium compounds, azoles, octanoic acid, amine, and boron to form an insoluble complex Chen, 2011; Humar et al., 2005; Humar et al., 2007; Zhang and Kamdem, 2000). Some natural resources, such as the industrial waste enzymatic-hydrolyzed okara have been used to enhance the fixation of antifungal salts in wood structures (Ahn et al., 2010; Kim et al., 2011), and tannins were also combined with copper and/or boron salts to formulate a new wood preservative system that showed good efficacy against fungus decay (Laks et al., 1998; Tondi et al., 2012). Other studies used soy protein products or commercial extracts as raw materials in their preservative formulations for the fixation of copper (Sen et al., 2009; Yang et al., 2006). However, such renewable resources are difficult to implement as fixatives in newly developed preservative systems because of their high cost and rarity.

Rosin, which comes from softwood, is abundant, natural, and renewable. It has a good hydrophobic and wood affinity. Over the years, its main widespread application has been in the paper industry as a sizing agent (Yao and Zheng, 2000). The investigations to use rosin for wood preservation and different chemical mechanisms between copper, rosin, and wood constituents has been carried out by the research group of Pizzi (Pizzi, 1993a). The copper-rosin soaps obtained when dissolved in a solvent (ethanol) have also been impregnated into wood (Pizzi, 1993b) and the copper-rosin soaps were shown to be extremely efficient towards both fungi (unsterile soil bed test) and termites (field test). In another study, the use of non-solvent rosin-copper formulations to impregnate wood has also been proposed (Roussel, et al., 2000) and treated wood blocks have shown good performance when leached, but a double impregnation system was required. In addition, earlier investigations showed that a rosin sizing agent can improve the moisture absorbing ability of wood and also help improve wood decay resistance (Li, et al., 2011; Li, et al., 2009). Therefore, this study aims to investigate the effect of rosin size on copper fixation to develop new formulations for wood preservation and determine the efficacy of copper-rosin preservatives against fungal decay.

2 Experimental

2.1 Materials

The anion rosin scattered emulsion sizing agent (R) was an industrial product and was supplied by Guangxi Wuzhou Arakawa Chemical Industries Co., Ltd. In this study, it was used to treat wood at three cove salts to protect wood against fungal decay in only one concentration of 3%. All of the other chemical reagents used in this work were provided by Tianjin Kermel Chemical Reagent Co. , Ltd. and were all pure grade reagents.

2.2 Preparation of Test Samples

The wood samples were taken from Poplar trees (*Populus ussuriensis* Komo) at 15 years of age and were selected according to GB 1929-91. Wood specimens were cut from untreated poplar sapwood into block dimensions 20 mm×20 mm×20 mm. Defect-free cubes were selected for the tests. The weight differences of the chosen blocks could not exceed 0.5 g. Feeder strips were also prepared from poplar sapwood. One feeder strip [22 mm×22 mm×3 mm (longitudinal direction)] was needed for each cube in a culture bottle.

2.3 Impregnation Method

Before treatment, all blocks were oven-dried at 103 ℃ to a constant weight (the nearest 0.01 g) and recorded as W_1. After drying, the blocks for each retention group were placed in a suitable beaker and weighed down to prevent eventual floating during treatment. The process was performed in a small-scale impregnation container under a vacuum of 0.01 MPa for 30 minutes followed by injection of the preservative mixture and then brought back to atmospheric pressure. After the blocks were completely saturated, they were individually removed from the solution and lightly blotted with absorbent paper to remove the excess preservative solution and immediately weighed to the nearest 0.01 g to ascertain the mass after impregnation (W_2). The theory retention of each block was calculated using the following formula:

$$\text{Theory retention, kg} \cdot \text{m}^{-3} = \frac{GC}{V} \times 10 \qquad (1)$$

where: $G = W_2 - W_1$, the weight in grams of the treating solution absorbed by the block;

C = Grams of preservative in 100 grams of treating solution;

V = Volume of the block in cubic centimeters.

After calculating retention, the treated samples were air-dried for 48 hours, oven-dried at 103 ℃ overnight, and then weighed to determine the dry weights of the wood blocks after treatment. The difference between the dry weights before and after treatment is the actual retention of each block. And the percentage of actual retention to the theory retention was regarded as treatability of each preservative formulation.

2.4 Leaching

A leaching test was performed according to AWPA E11 (American Wood Preservers' Association 2007 - Method of Determining the Leachability of Wood Preservatives).

Twelve blocks of a given retention group were placed equally on stainless steel in 1000 mL beakers. They were weighed down in each beaker and completely submerged with 50 mL of distilled water for each block. Then the beakers containing the blocks covered with water were placed into a vacuum desiccator. A vacuum process of 100 mm of mercury or less was adopted for one-half hour or until air bubbles ceased to escape from the submerged blocks. Then the vacuum was broken to allow water impregnation of the blocks and the weights were removed from the blocks. After 6, 24, and 48 hours and thereafter at 48-hour intervals for 14 days, the leachates were removed from the beaker and kept for copper analysis. The amount of leachate was replaced by an equal amount of fresh distilled water.

2.5 Analysis of Copper

In order to measure the amount of copper ions leached from the treated wood blocks, the leachates were analyzed according to AWPA Standard A11-93 by using an atomic absorption spectroscopy (AAS) analyzer.

2.6 Decay Test

Wood blocks were tested to evaluate their resistance against biological attack according to Chinese standard LY/T 1283-1998. The white-rot fungi *Trametes versicolor* and brown-rot fungi *Gloeophyllum trabeum* were used as test fungi. Soil culture bottles with feeder strips on the soil surface were inoculated with fungus cultured on potato dextrose agar. After the feeder strips were covered with fungal mycelia, sterilized wood blocks were placed onto the feeder strip. The soil-block culture was incubated in a temperature and humidity-controlled chamber at 28 ℃ and 75% RH for 12 weeks. After exposure to the fungi, the blocks were removed from the decay chambers, gently cleaned to remove the mycelium, dried at 103 ℃ until constant weight was obtained, and weighted to determine weight loss.

2.7 Microscopic Observation by SEM-EDX

After the decay test of wood blocks was completed, the wood blocks were sliced into thin samples using a razor blade. The samples were mounted on a metal stub and were sputter-coated with a thin layer (approximately 20 nm thick) of gold. The specimens were then observed with a scanning electron microscope (SEM, FEI Quanta 200; USA). Random observations were made on different structures to identify the existence of copper in the anatomical structure of the specimens. The element composition was determined by regional analysis using an energy dispersive X-ray spectrometer (EDX) combined with the scanning electron microscope.

3 Results and discussion

3.1 Retention Results

Retention levels of poplar wood samples treated with copper-rosin solutions (as kilograms per cubic meter) and the actual percent retention of preservative formulations in wood blocks are recorded in Table 1. Total uptake of treating solutions in poplar wood, including both rosin alone and in combination with copper, were relatively uniform. The actual retentions of the copper-rosin preservatives were very close to theory retentions, namely above 85%. Results indicated that the concentration of the solutions considered using the impregnation method described did not influence the penetration of the preservative complexes into the wood blocks.

There were slight differences in the treatability of the three rosin formulations (Table 1). The actual percent retention of preservative solution-containing rosin only or containing copper-rosin decreased from 98.22% to 85.33% and from 98.38% to 89.66%, respectively, with the increase in concentration of rosin from 1.0% to 4.0% in the impregnation solution. An explanation for this would be that the increase of rosin concentration, which leads to increase in uptake, also leads to an increased amount of rosin in the outer part of the wood sample due to increasing filtration effect. After impregnation, the outer part of the wood sample is cleaned carefully and therefore a relatively larger amount is removed from the wood surface at higher concentrations of rosin, which might be partially responsible for decreasing the actual retention of preservatives. However, this decrease was not important and the best retention was obtained with 1% rosin size and 3% added copper sulfate.

Table 1 Retention levels and treatability of wood samples treated with solutions

Abbreviation	Solution and Concentrations	Theory Retention ($kg \cdot m^{-3}$)	Actual Retention ($kg \cdot m^{-3}$)	Treatability[a] (%)
1	1% Rosin + 3% $CuSO_4$	33.95 (0.93)[b]	33.48 (1.66)	98.38 (2.32)
2	2% Rosin + 3% $CuSO_4$	41.61 (1.69)	39.30 (2.07)	94.41 (2.17)
3	4% Rosin + 3% $CuSO_4$	58.32 (1.68)	52.29 (1.97)	89.66 (2.52)
4	3% $CuSO_4$	26.06 (1.26)	25.05 (2.14)	96.01 (4.49)
5	1% Rosin	7.9 (0.32)	7.75 (0.52)	98.22 (6.34)
6	2% Rosin	15.83 (0.94)	15.32 (1.32)	96.91 (7.34)
7	4%Rosin	31.72 (0.39)	27.06 (1.34)	85.33 (6.23)

[a] Treatability refers to the percentage of actual retention to the theory retention.

[b] All results are means of 24 samples. Standard deviations are in brackets.

3.2 Leaching

The analyzed results of copper ions released from blocks treated with the copper-rosin solutions and those treated with the copper sulfate solution alone taken at different time intervals is presented in Fig. 1. A significant reduction of copper ions leaching from wood samples treated with the copper-rosin solutions was observed. For all samples, the unfixed copper rapidly leached from wood during the first stages of the leaching process, and decreased significantly over time. However, the leaching of copper occurred much more slowly when wood samples were treated with rosin-copper solutions. This is probably attributable to the presence of rosin. After being penetrated into wood blocks, the rosins present in the cell lumen were either interacted with copper to form a insoluble copper resinate compound (Pizzi, 1993a) or in the form of an adhesive film to cover the copper crystals, which could be proved by SEM-EDX analysis. During the leaching process, the rosin acted as a barrier that slowed copper release from deep inside the samples.

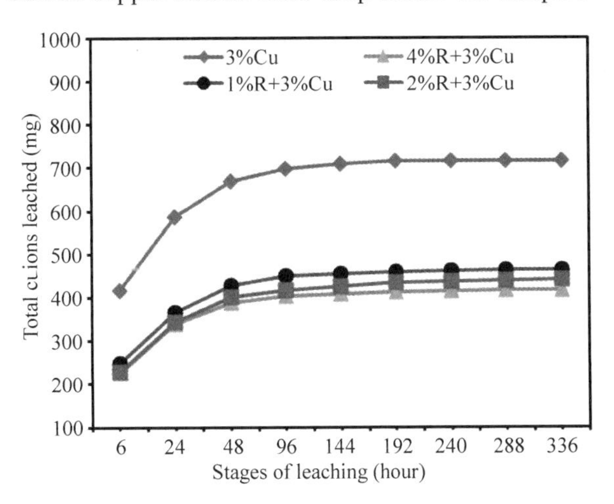

Fig. 1 Copper ions released from treated wood specimens at different time intervals; Cu: copper sulfate ($CuSO_4$), R: rosin sizing agent

There was a large amount of copper ions that leached out from wood samples treated with copper sulfate alone. After 9 leaching cycles, 715.49 mg of copper was leached out from the samples, which represented 69.3% of the copper impregnated in the wood blocks, which is comparable with the results reported by

(Mourant et al., 2009) and (Humar et al., 2005). However, due to the application of rosins in the present investigations, copper ion leaching was effectively reduced, particularly in the first stage of the process (Fig. 1). The total amount of copper ions release from the samples treated with the copper-rosin solutions was 2 times less than those from the samples treated with only copper sulfate. The treatments with copper-rosin showed that content of Cu ions leaching slightly decreases with the increase of rosin concentration in the impregnation mixture. This would be explained that when wood samples were impregnated with increasing concentration of rosin, amount of rosin on the surface of treated wood was probably more formed, which resulted in the reduction of the copper ion diffusion from wood during leaching of the samples.

The results suggested that addition of rosin had a significant effect on the fixation of copper in wood but the concentrations used in this work did not show a difference in the size of the effect.

3.3 Decay Resistance

The results from the decay test are shown in Table 2. The weight losses of the control wood blocks against *Trametes versicolor* and *Gloeophyllum trabeum* were 70.45% and 61.84%, respectively. The unleached wood blocks treated with copper alone had approximately 4% or less weight loss for both test fungi. This result was in agreement with that reported. However, a severe weight loss (approximately 40%) was found for the leached wood samples treated with only copper.

As shown in Table 2, the samples impregnated with only rosin sizing agents and leached had weight losses in the range of 48%-55%, which was much lower than those of the untreated control samples. And no distinct differences for decay capacity could be observed between poplar samples treated with any of the 3 concentrations of rosin (1.0%, 2.0%, or 4.0%). The differences between the weight losses after decay of the leached and unleached samples were not so pronounced in the rosin size samples as for the copper samples. This signifies that the rosin sizing agent itself also has poor performance against fungal wood decay because of its water repellency and inherent decay resistance rather than general toxicity (Eberhardt et al., 1994). This result was in accordance with that reported in previous research (Li et al., 2011).

However, the samples treated with copper-rosin formulations showed good decay resistance against both Trametes versicolor and Gloeophyllum trabeum. The average weight loss of the samples degraded by fungi was in a range of 1.24% to 3.46% after being incubated for 12 weeks. Most of the leached wood blocks treated with copper-rosin formulations showed less than approximately 3% weight loss and were not entirely covered by mycelium of both test fungi. In some cases, the unleached specimens presented a slightly higher average mass loss than the leached ones, which means that mass losses of unleached specimens were not only the result of fungal decay, but the result of leaching, too (Humar et al., 2007). When unleached specimens are exposed to fungi, moisture content of the wood increases and the unfixed copper will diffuse from specimens into soil culture resulting in mass losses, which are not a result of fungal action but of copper sulfate (Goodell et al., 1995). And when specimens are leached prior to fungal exposure, the unfixed copper is removed from wood during leaching procedure, thus detected mass losses are a result of fungal decay only.

No significant differences in performance against wood decaying fungi were found between copper-rosin formulations. All wood samples containing copper-rosin gave a better performance against fungal decay than those only containing copper after leaching. Therefore, the use of rosin size as fixed agents may reduce environmental impact of wood treated with copper-based preservatives.

Table 2 Weight loss (%) of samples exposed to white rot fungus *Trametes versicolor* and *Gloeophyllum trabeum*

Abbreviation	Solution and concentrations	Weight loss (%)			
		Gloeophyllum trabeum		*Trametes versicolor*	
		Unleached	Leached	Unleached	Leached
1	1% Rosin + 3% $CuSO_4$	3.14 (0.61)[a]	2.24 (0.86)	3.6 (0.63)	1.93 (0.47)
2	2% Rosin + 3% $CuSO_4$	3.16 (0.75)	1.24 (0.47)	3.26 (1.04)	2.11 (0.76)
3	4% Rosin + 3% $CuSO_4$	3.43 (0.73)	1.46 (0.57)	3.34 (1.06)	2.31 (0.68)
4	3% $CuSO_4$	3.04 (0.62)	34.14 (3.03)	2.66 (0.85)	39.12(3.23)
5	1% Rosin	48.46 (2.62)	51.04 (3.54)	51.32 (0.85)	54.98 (2.57)
6	2% Rosin	49.22 (3.32)	52.94 (5.06)	52.79 (1.44)	55.92 (2.05)
7	4%Rosin	49.25 (4.16)	51.45 (1.96)	51.25 (3.26)	54.45 (3.34)
8	Control	61.84 (5.68)	—	70.45 (4.94)	—

[a] All results are means of 6 samples. Standard deviations are in brackets.

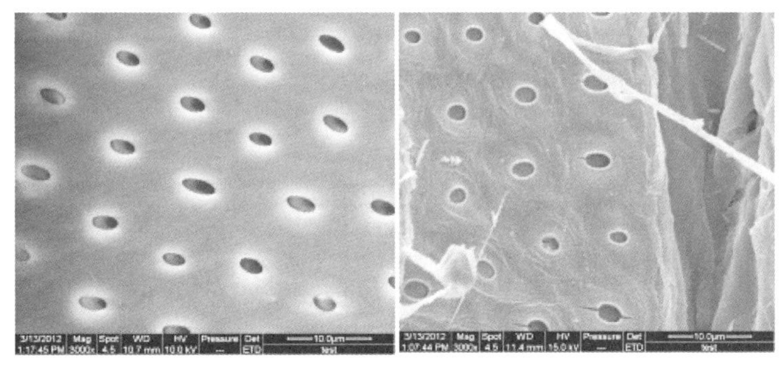

Fig. 2 Scanning electron microscopic images with magnification 10 μm of tangential section of control wood block before (left) and after (right) exposed to fungus

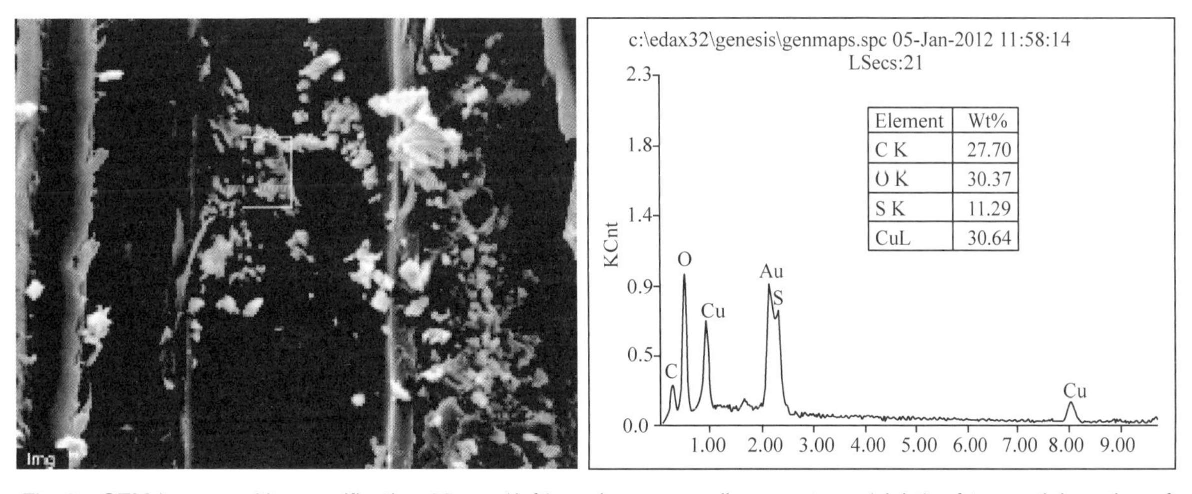

Fig. 3 SEM images with magnification 20 μm (left) and corresponding spectrum (right) of tangential section of unleached wood blocks treated with copper alone

3.4 Microscopic Observation and SEM-EDX Analysis

To confirm the effectiveness of copper fixation by the rosin sizing agent, SEM observation and energy-dispersive X-ray spectroscopy analysis (EDX) were used to identify the presence of copper in the copper-rosin treated and leached wood samples. Fig. 2 shows the SEM images of the control wood sample before and after the fungal exposure. It can be clearly seen that surface of wood cell wall of the control sample before the fungal exposure was extremely smooth (Fig. 2 left). After the fungal exposure, wood cell walls have been completely destroyed by the fungi (Fig. 2 right). When the wood blocks treated with copper sulfate were observed, various crystal particles were found in the cell lumens (Fig. 3 left). The spot analysis using SEM-EDX proved that these particles contained Cu and S originating from copper sulfate (Fig. 3 right).

In the microscopic observation of the wood blocks treated with copper-rosin formulations, various spherical agglomerates were easily detected in the cell lumen (Fig. 4a and b left). The spectrum obtained from the spot analysis confirmed that these agglomerates contained the element Cu (Fig. 4a and b right).

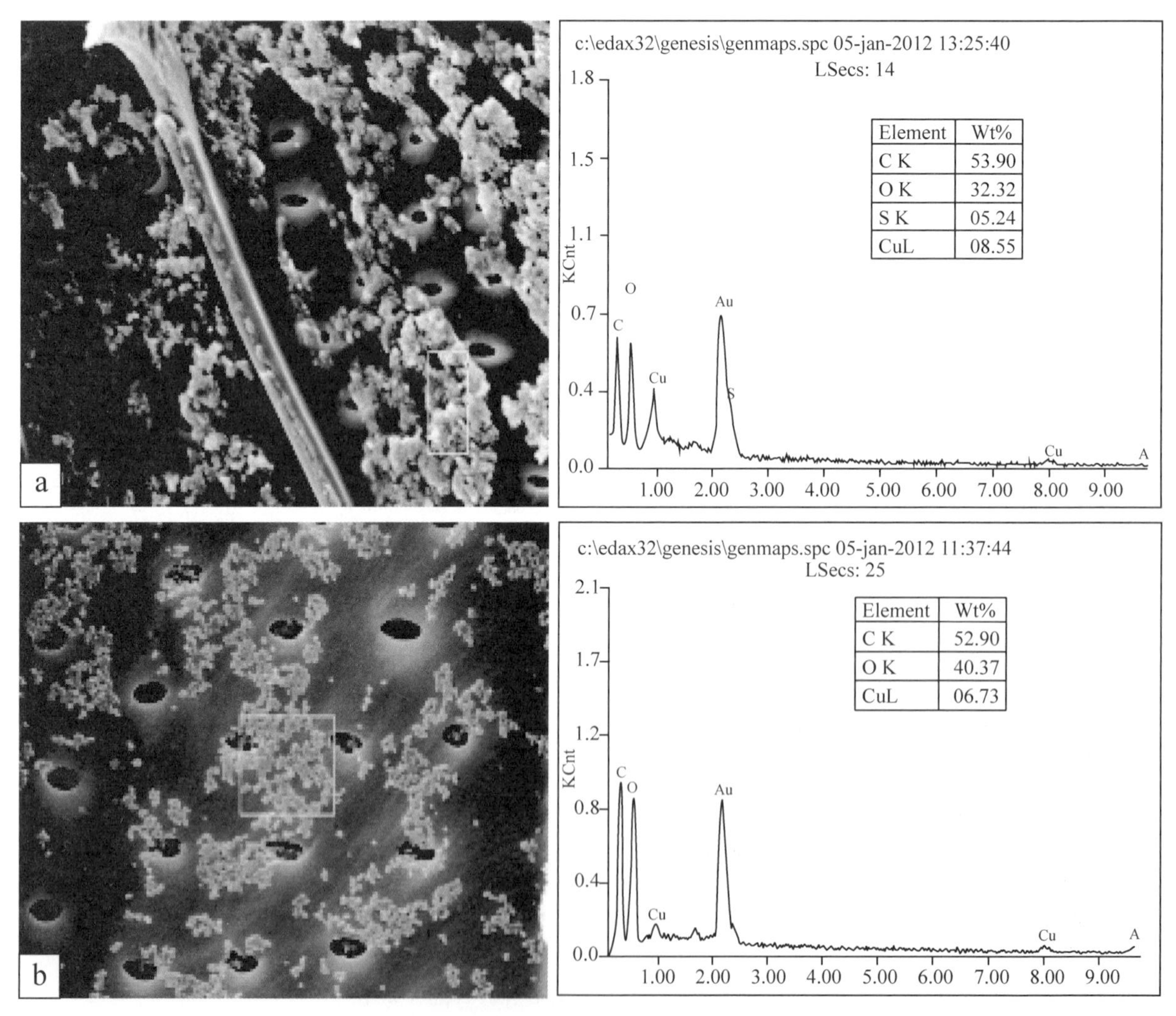

Fig. 4 SEM images (left side) and corresponding spectrum (right side) of tangential section of wood blocks treated with 2%Rosin+3%$CuSO_4$: (a) unleached with magnification 20 μm and (b) leached with magnification 10 μm

Unlike the crystals in Fig. 3, these agglomerates were tightly adhered to the wood cell wall. They had a lower Cu content and much higher C content in comparison to that observed in the crystal particles. This signifies rosin interacted with copper and formed an adhesive film to cover the copper crystals. Therefore, Cu was fixed into the wood blocks. The SEM-EDX analysis results suggested that the presence of the preservative complexes containing Cu contributed to the good decay resistance of the leached wood blocks treated with the mixture of rosin size and copper sulfate.

4 Conclusions

1. This study evaluated effect of rosin size on copper fixation and decay resistance of wood treated with copper sulfate and rosin sizing agent, separately or in combination, against white rot fungi *Trametes versicolor* and brown rot fungi *Gloeophyllum trabeum*. The samples impregnated with copper sulfate and the rosin sizing agent were more effective against fungal wood decay than those impregnated with only copper after leaching. Most of the leached wood samples treated with copper - rosin formulations showed less than approximately 3% weight loss. The rosin size agents themselves also showed poor performance against wood decay fungi.

2. The result of AAS analysis showed that rosin size had a certain effect on fixation of copper. The amount of copper ions released from the samples treated with the copper-rosin solutions was half those from the samples treated with copper alone.

3. The SEM observation and EDX analysis of the wood blocks treated with copper-rosin formulations confirmed that the preservative complexes containing Cu existed in the cell lumens of leached and decayed wood blocks.

4. This study may help in developing a new approach for using rosin size to reduce the hazard of the copper preservative leaching into the environment and lead to wood treated with water - borne copper preservatives being more widely used.

Acknowledgments

The authors are grateful for the support of the Vietnamese Government, the National Natural Science Foundation of China (31070487), and Foundation for University Young Core Teachers of Heilongjiang Province of China (1154G49).

References

[1] Ahn S H, Oh S C, Choi I G, Han G S, Jeong H S, Kim K W, Yoon Y H, Yang I. Environmentally friendly wood preservatives formulated with enzymatic-hydrolyzed okara, copper and/or boron salts[J]. J. Hazard. Mater., 2010, 178(1-3): 604-611.

[2] Chen G. Laboratory evaluation of borate: amine: copper derivatives in wood for fungal decay protection[J]. Wood Fiber Sci., 2011, 43(3): 271-279.

[3] Eberhardt T L, Han J S, Micales J A, Young R A. Decay resistance in conifer seed cones: Role of resin acids as inhibitors of decomposition by white rot fungi[J]. Holzforschung, 1994, 48(4): 278-284.

[4] Freeman M H, McIntyre C R, A comprehensive review of copper-based wood preservatives: With a focus on new micronized or dispersed copper systems[J]. For. Prod. J., 2008, 58(11): 6-27.

[5]Goodell B, Liu J, Slahor J. Evaluation of diffusible preservatives using an accelerated field simulator[J]. For. Prod. J., 1995, 45(6): 74-76.

[6]Humar M, Kalan P, Šentjurc M, Pohleven F. Influence of carboxylic acids on fixation of copper in wood impregnated with copper amine based preservatives[J]. Wood Sci. Technol., 2005, 39(8): 685-693.

[7]Humar M, Žlindra D, Pohleven F. Improvement of fungicidal properties and copper fixation of copper-ethanolamine wood preservatives using octanoic acid and boron compounds[J]. Holz Roh- Werkst., 2007, 65(1): 17-21.

[8]Kim H Y, Jeong H S, Min B C, Ahn S H, Oh S C, Yoon Y H, Choi I G, Yang I. Antigungal efficacy of environmentally friendly wood preservatives formulated with enzymatic-hydrolyzed okara, copper, or boron salts[J]. Environ. Toxicol. Chem., 2011, 30(6): 1297-1305.

[9]Laks P E, McKaig P A, Hemingway R W. Flavonoid biocides: wood preservatives based on condensed tannins[J]. Holzforschung, 1998, 42(5): 299-306

[10]Li S, Thanh-Hien N T, Han S, Li J. Application of rosin in wood preservation[J]. Chem Ind Forest Prod., 2011, 31(5): 117-121.

[11]Li S, Wang X, Li J. Effect of two water borne rosin on wood protection[C]. Transaction of China pulp and paper, 2009, 24: 200-203.

[12]McIntyre C R, Freeman M H. A comprehensive review of copper-based wood preservatives[J]. For. Prod. J., 2008, 58(11): 6-27.

[13]MitsuhashiGonzalez J M. Limiting copper loss from treated wood in or near aquatic environments[D]. Oregon State University, Corvallis, Oregon, 2007.

[14]Mourant D, Yang D Q, Lu X, Riedl B, Roy C. Copper and boron fixation in wood by pyrolytic resins[J]. Bioresour. Technol., 2009, 100(4): 1442-1449.

[15]Ngoc N T B. Wood preservation (in Vietnamese). Agriculture Press, Ha Noi., 2006.

[16]Nicholas D D, Schultz T P. Biocides that have potential as wood preservatives: An overview[C]. Wood Preservation in the '90s and Beyond, Forest Prod. Soc. Proc. No. 7308, Madison Wisconsin, 1995, 169-173.

[17]Nicholas D D, Schultz T P. Comparative performance of several ammoniacal copper preservative systems: Prepared for the 28th Annual Meeting, Whistler. B. C., Canada. International Research Group on Wood Protection, Stockholm Sweden, 1997.

[18]Pizzi A. A new approach to nontoxic, wide-spectrum, ground-contact wood preservatives. 1. Approach and reaction mechanisms[J]. Holzforschung, 1993a, 47(3): 253-260.

[19]Pizzi A. A new approach to nontoxic, wide-spectrum, ground-contact wood preservatives. 2. Accelerated and long-term field tests[J]. Holzforschung, 1993b, 47(4): 343-348.

[20]Roussel C, Haluk J P, Pizzi A, Thévenon M F. Copper based wood preservative: A new approach using fixation with resin acids of rosin[C]. International Research Group on Wood Protectiont, Stockholm, Sweden, 2019.

[21]Ruddick J N R. The use of chemicals to prevent the degradation of wood[M]. Uhlig's corrosion handbook, R. W. Revie, ed., John Wiley & Sons, Inc., Hoboken, New Jersey, Canada, 2000, 503-512.

[22]Sen S, Tascioglu C, Tırak K. Fixation, leachability, and decay resistance of wood treated with some commercial extracts and wood preservative salts[J]. Int. Biodeterior. Biodegrad., 2009, 63(2): 135-141.

[23]Tondi G, Weiland S, Lemenager N, Petutschnigg A, Pizzi A, Thevenon M F. Efficacy of tannin in fixing boron in wood: fungal and termite resistance[J]. BioResources, 2012, 7(1): 1238-1252.

[24] Townsend T, Solo-Gabriele H. Environmental impacts of treated wood [M]. CRC, Boca Raton, Florida, USA., 2006.

[25]Treu A, Larnøy E, Militz H. Process related copper leaching during a combined wood preservation process[J]. Eur. J. Wood Prod., 2011, 69(2): 263-269.

[26] Yang I, Kuo M, Myers D J. Soy protein combined with copper and boron compounds for providing effective wood preservation[J]. JAOCS, 73(3): 239-245.

[27] Yao X, Zheng L. Development potential of rosin sizing agent[J]. Chemical Technology Market, 2000, 10: 21.

[28] Zhang J, Kamdem D P. Interaction of copper-amine with southern pine: Retention and migration[J]. Wood Fiber Sci., 2000, 32(3), 332-339.

Article submitted: April 6, 2012;

中文题目：水载型松香对铜基防腐剂固定和木材耐腐性能的影响

作者：阮氏清贤，李坚，李淑君

摘要：本研究以松香施胶剂浸渍木材和固定铜以提高木材耐腐性能。用3% $CuSO_4$ 溶液与1.0%、2.0%、4.0%松香施胶剂混合浸渍大青杨木块，采用土壤-木块培养法测定了处理后木块的耐腐性能。经过12周的腐朽实验后，未经处理的木块以彩绒革盖菌或密粘褶菌腐朽的质量损失分别为70.45%和61.84%。仅用松香施胶剂处理后，木材的耐腐性略有提高。经松香施胶剂和 $CuSO_4$ 共同处理后，赋予了木材很好的耐腐性能。处理木块平均质量损失不足4%。值得注意的是，流失就可以导致木块质量损失接近3%。用原子吸收光谱法(AAS)对流失液中的铜含量进行分析，结果表明，用铜-松香混合溶液处理后的铜离子流失量仅为单独用铜处理后流失量的一半。扫描电子显微镜结合能量色散X射线分析(SEM-EDX)证实，铜元素仍然存在于流失处理后木块的细胞腔中，这也与原子吸收光谱分析的结果一致。这说明松香施胶剂对铜防腐剂在木材中的固定有很大的帮助。

关键词：松香；固定；耐腐性；硫酸铜；木材防腐

The Combined Effects of Copper Sulfate and Rosin Sizing Agent Treatment on Some Physical and Mechanical Properties of Poplar Wood*

Thi Thanh Hien Nguyen, Shujun Li, Jian Li

Abstract: This aim of this study was to determine the effect of rosin upon some physical and mechanical properties of poplar wood treated with the mixtures of 3% $CuSO_4$ and 1.0%, 2.0%, or 4.0% rosin sizing agent. Rosin-copper treatments decreased the moisture absorption, water absorption and swelling properties of wood, whilst increasing water repellent efficiency and anti-swelling efficiency to approximately 40% after 30-day immersion in water. In general, rosin-copper treatments increased the compression strength parallel to grain and Brinell hardness compared to control, but the MOR and MOE were lower than that of control.
Keywords: Rosin-copper; Moisture absorption; Water absorption; Compression strength; MOR; MOE; Hardness strength

1 Introduction

Wood has successfully been utilized as building material for thousands of years due to its availability, ease of use, and great insulating and strength properties. However, wood also has some negative aspects, most notably its susceptibility to deterioration by microorganisms and insects. Consequently, wood must be treated with preservatives to extend its service life. Copper compounds have been used in wood preservative formulations for more than 200 years[1]. Among them, chromate copper arsenate (CCA) has high resistance to leaching and very good performance in service. Nevertheless, this conventional wood preservative has been banned for some applications due to its mammalian toxicity and its adverse effect on the environment[2]. Therefore, the present investigations were undertaken to develop fungicides that would not only effectively protect wood against biological corrosion, but would also be characterized by lower leaching of the biologically active substance.

Rosin is a product obtained from pines and some other plants, mostly conifers. One of the major components of rosin is abietic acid, a partially unsaturated compound with three fused six-membered rings and one carboxyl group, giving it good hydrophobic properties. Over the years, rosin has been the most widely used in the paper industry as a sizing agent[3]. The new approach and reaction mechanisms between carboxylic acid groups of resin acids of rosin and copper have been investigated by Pizzi[4], and wood blocks treated with solution of rosin-copper soap using benzene or ethanol as solvents have shown to be efficient against both

* 本文摘自 Construction and Building Materials, 2013, 40: 33-39.

fungal and termite in field tests[5]. Furthermore, our earlier investigation showed that rosin sizing agent had a certain effect on fixation of copper in wood and after being treated with the mixture of rosin sizing agent and $CuSO_4$, wood blocks still had great decay resistance, even after leaching[6]. However, besides concern for the efficacy against degradation and low impact on the environment, good physical and mechanical properties are required for treated wood.

Many researchers have shown that the preservatives, especially waterborne preservatives, have a negative impact on mechanical properties of the wood. For example, Yildiz et al. [7] evaluated the impact of various preservatives, including ammoniacal copper quat (ACQ), Wolmanit CX-8 (copper HDO), Thanalit E 3491 (copper tebuconazole) and CCA on modulus of rupture (MOR) and modulus of elasticity (MOE) of yellow pine. Some of these chemicals can reduce the mechanical properties of wood. Samples treated with CCA exhibited a reduction in MOR of 12% and almost 10% in MOE compared with untreated controls[7]. Mourant et al. [8] also reported that the MOR of jack pine samples was generally negatively affected by the combined PF-pyrolytic oil resin treatment and copper chloride or copper chloride-sodium borate mixture treatment, however, the MOE of treated jack pine samples did not differ statistically from that of the untreated wood; Simsek et al. [9] reported that compression strength parallel to grain (CSPG) and MOR of wood samples treated with borates were lower compared to untreated control samples. However, Cao et al. [10] indicated that copper-ethanolamine treatment increase dimensional stability of wood at high atmospheric temperature in long-term application; Tomak et al. [11] studied the effect of oil treatments on several properties of wood treated with boron compounds and found that oil heat treatment decreased water absorption by approximately 20% and increased water repellent efficiency by 80% - 90% after 2 weeks immersion in water. Furthermore, a comprehensive review of the effects of waterborne wood preservatives on the properties of treated wood reported that the important factors affecting properties of the treated wood include the type and uptake of chemicals, wood species, pH of the solution, preand post-treatment conditions such as drying temperature or treatment parameters including vacuum and pressure levels[12]. Therefore, the aim of this research work was to evaluate the effects of new wood preservatives (rosin-copper) on some physical and mechanical properties of poplar wood, such as the moisture absorption and water absorption characteristics, compression strength, Brinell hardness, modulus of rupture (MOR), and modulus of elasticity (MOE). They are the primary criteria for the selection and designing of wood for different applications.

2 Materials and methods

2.1 Preparation of test specimens and treating solutions

Poplar wood (Populus ussuriensis Komo) was selected according to GB 1929-91[13] (same as ISO 3129). Wood samples having four different configurations were prepared from untreated poplar sapwood. The specimen dimensions were 20 mm×20 mm×20 mm (for the moisture absorption and water absorption tests), 30 mm×20 mm×20 mm (for compression strength), 300 mm×20 mm×20 mm (for bending measurement), and 70 mm × 50 mm × 50 mm for hardness strength measurements (longitudinal, tangential, and radial, respectively). Defect-free specimens were selected for the tests. Samples were divided into eight groups: one untreated group served as a control, and the other seven groups were treated with copper sulfate and rosin sizing agent, separately or in combination. Each treatment group had a total of 30 samples.

The anionic rosin emulsion sizing agent was an industrial product and was supplied by Guangxi Wuzhou Arakawa Chemical Industries Co. , Ltd. In this study, rosin was used at three concentration levels (1.0%, 2.0% and 4.0%). Cupric sulfate anhydrous ($CuSO_4$) was used as preservative salts to protect wood against decay fungi at only one concentration of 3%. The other chemical reagents used in this work were provided by Tianjin Kermel Chemical Reagent Co. , Ltd. and were all pure grade reagents.

2.2 Impregnation procedures

Wood samples were oven dried at 103 ℃ for 24 h before being weighed. After drying, the samples were treated using a full-cell pressure process at 0.1 MPa vacuum for 60 min followed by 0.3 MPa pressure for 2 h. Following this, the samples remained in the solutions for 60 minutes at atmospheric pressure. The wood samples were then removed from the treatment solution, wiped lightly to remove the rest of the solution from the wood surface, and weighed (nearest 0.01 g) to determine gross retentions for each treating solution.

All treated samples were subsequently stored at ambient laboratory temperature for air drying 4 weeks prior to testing.

2.3 Microscopic observation

After treatment, small samples (1 cm ×1 cm ×1 cm) were cut from the untreated control and treated wood blocks using a razor blade. The samples were mounted on a metal stub with adhesive, and then placed under vacuum and sputter-coated with a thin layer (approximately 20 nm thick) of gold. The samples were then observed with a scanning electron microscope equipped with an energy dispersive X-ray analyzer (SEM, FEI Quanta 200; USA) at an accelerating voltage of 20 kV. Random observations were made on different areas of the wood to detect the presence of copper in the anatomical structure of the samples.

2.4 Measurement of moisture absorption

The treated samples (20 mm×20 mm×20 mm) were oven-dried at (103±2)℃ for 24 h and weighed to determine the over-dry weight. After drying, the wood samples were placed in a temperature and humidity-controlled chamber at 20℃ and 65% RH for approximately 4 weeks. Thirty samples were used for each treatment group. After stabilization, each sample was weighed using a four-place analytical balance. The equilibrium moisture content (*EMC* in%) was calculated using the following equation:

$$EMC(\%)=[(G-G_0)/G_0]\times 100 \qquad (1)$$

where, G is the equilibrium weight of sample after moisture absorption (g), G_0 was oven-dry weight of wood sample (g).

2.5 Measurement of water absorption and dimensional stability

Oven-dry wood samples (20 mm×20 mm×20 mm) were placed into a 1000 mL beaker filled with distilled water. The samples were placed separately and horizontally in the beakers with stainless steel mesh over the specimens. Thirty replicates were used for each treatment group. The specimens were removed from the beaker after 6 h, 1 day, 2 days, 4 days, 8 days, 12 days and 20 days, and thereafter at 10-day intervals and excess water was removed by dabbing with a tissue. The samples were weighed and again placed in the beaker and the water replaced. The test was continued for a total of 30 days. The radial, tangential, and volumetric swelling at both the water swollen and oven-dried states were measured using a digital micrometer (±0.01 mm). The swelling difference between treated and control samples was used to calculated the anti-swelling efficiency (ASE) according to the following formula:

$$ASE(\%) = \frac{S_u - S}{S_u} \times 100 \tag{2}$$

where S_u is volumetric swelling of untreated wood (%) and S was that of treated wood (%).

$$S(\%) = [(V_w - V_d)/V_d] \times 100 \tag{3}$$

where V_w was volume of the wood sample after saturation with water (mm^3) and V_d was the volume of initial dry wood sample (mm^3).

The swelling coefficient in the radial and tangential direction was also calculated as:

$$S_\alpha(\%) = [(\alpha_w - \alpha_d)/\alpha_d] \times 100 \tag{4}$$

where α_d was the single direction dimensional (radial or tangential) of the initial dry sample (mm) and α_w was the single direction dimension after saturation with water (mm).

Water absorption (WA) was calculated according to the following formula:

$$WA(\%) = [(W_2 - W_1)/W_1] \times 100 \tag{5}$$

where W_2 was the wet weight of the sample after wetting with water (g) and W_1 was the initial dry weight of the sample (g).

Water repellency efficiency (WRE) was calculated according to the following formula:

$$WRE(\%) = [(W_{ac} - W_{at})/W_{ac}] \times 100 \tag{6}$$

where W_{at} is the water absorption rates of treated wood samples (%), and W_{ac} was the water absorption rate of untreated control (%).

2.6 Measurement of mechanical properties

After air-drying, the mechanical properties of all treated samples and untreated control samples were tested according to Chinese standards by using an Instron Universal Testing machine: Compression strength parallel to grain (CSPG) (GB/T 1935 same ISO 3787), modulus of rupture (MOR) (GB/T 1936.1[15] same as ISO 3133), modulus of elasticity in bending (MOE) (GB/T 1936.2[16] same as ISO 3349), and Brinell hardness [(parallel, perpendicular to grain) (BH)] (GB/T 1941[17] same as ISO 3350). Each of these tests was performed on 30 samples per treatment group.

At the end of experiments, moisture contents (W) of samples were measured according to GB 1931-2009[18] (same as ISO 3130) and the moisture contents of sample where moisture content deviated from 12% what then were used to correct strength values (transformed to 12% moisture content) using the following strength conversion equation:

$$\sigma_{12} = \sigma_w \times [1 + \alpha(W - 12)] \tag{7}$$

where σ_{12} = strength at 12% moisture content ($N \cdot mm^{-2}$), σ_w = strength at moisture content deviated from 12% ($N \cdot mm^{-2}$), α = constant value showing relationship between strength and moisture content (α = 0.05, 0.04, 0.015, and 0.03 for CSPG, MOR, MOE, and BH, respectively), W = moisture content during test (%).

2.7 Statistical analysis

In order to determine the effects of wood preservatives on physical and mechanical properties of wood, one-way ANOVA tests were conducted and homogeneous groups were determined by using SPSS 13.0

statistical software package.

3 Results and discussion

3.1 Moisture absorption

The resultant equilibrium moisture content (EMC) of wood samples impregnated with copper sulfate and rosin sizing agent, separately or in combination are provided in Table 1. In comparison to untreated samples, all treatment affected EMC of wood. Absorption capacity was a little lower in wood samples treated with copper sulfate alone. This may be explained in two ways: by weight gain due to copper sulfate treatment and by changes in hygroscopicity due to copper some of the absorption sites for moisture[19].

However, moisture absorption in samples treated with rosin alone or in combination with copper decreased significantly. EMC value of untreated samples was 10.86%, while the one of samples treated with 1%, 2%, or 4% rosin were 9.72%, 9.66% and 9.49%, respectively. Samples impregnated with 3%$CuSO_4$ plus 1%, 2%, or 4% rosin had EMC values of 10.06%, 9.96% and 9.59%, respectively, suggesting that EMC decreased with increasing rosin concentration. This was probably due to the hydrophobic property of rosin. Increase rosin concentration leads to increased uptake as well as the filling of the lumen with rosin, which increasing water repellency, thus reducing the moisture absorbing capacity of the wood sample[20,21]. However, one-way ANOVA revealed that this difference was not significant (Table 1). The results also suggested that addition of rosin can contribute to increase dimensional stability of wood. Cao et al.[10,19] also found that copper-ethanolamine treatment improved the dimensional stability of wood.

Table 1 Equilibrium moisture content (EMC), water absorption (WA), water repellent efficiency (WRE) and Anti-swelling efficiency (ASE) values of wood treated with rosin alone and in combination with copper after saturation with water

Treatments	EMC (%)	WA (%)	WRE (%)	ASE (%)		
				Radial	Tangential	Volumetric
Control	10.86 (0.19)^e	225.40 (7.97)^g	-	-	-	-
3%Cu	10.43 (0.23)^e	210.50 (5.85)^f	6.61	8.61	6.72	6.98
1%R	9.72 (0.11)^{ab}	202.84 (4.88)^e	9.54	13.16	11.65	17.36
2%R	9.66 (0.17)^{ab}	166.87 (4.85)^d	25.97	14.92	14.61	22.52
4%R	9.49 (0.17)^a	157.06 (6.52)^c	30.32	28.15	17.12	23.13
1%R+3%Cu	10.06 (0.19)^c	153.12 (6.02)^c	32.07	35.32	20.63	29.66
2%R+3%Cu	9.96 (0.21)^{bc}	140.90 (9.17)^b	37.49	38.01	25.56	34.73
4%R+3%Cu	9.59 (0.17)^{ab}	133.96 (7.05)^a	40.57	37.45	25.25	34.97

Note: Standard deviations are in brackets; Cu: anhydrous copper sulfate and R: rosin sizing agent; Means within a column followed by the same letter are not significantly different at 5% level of significance using the one-way ANOVA test.

3.2 Water absorption and water repellent efficiency

Relative water absorption (WA) levels of wood samples treated with copper sulfate and rosin sizing agent, separately or in combination measured at different time intervals are presented in Fig. 1 Water uptake for both control and treated samples increased rapidly from the start of soaking to approximately 8 soaking days, after which the rate of increase decreased significantly. These results may be due to WA into available empty pores in wood at the beginning of soaking and the reduction of those wood spaces over time. The increased rate of water uptake for treated wood was much lower than that for controls, throughout the soaking time. Total water absorption level of untreated control samples after 30 days immersion in water were 225.4% (Table 1). However, rosin application reduced the WA level of treated wood, particularly in samples treated with rosin-copper solutions had WA levels in the range of 134.0% - 153.1%. This was much lower than those of the control and copper only treated samples (210.5%). This improvement could be attributed to the formation of rosin deposits or insoluble copper resinate compounds in the cell lumen[4-6], which created a physical and mechanical barrier that hindered water movement. Furthermore, micrographs SEM of the treated poplar wood (Fig. 2b-d) clearly showed that the rosin and copper resinate compound filled the vessels as well as lumens, and other void spaces in the wood structure. This reduced penetration of water into the wood blocks. Indeed, the WA of the copper treated samples was slightly lower than that of the control samples. This can be explained in the same way as for the moisture absorption properties: some of adsorption sites (i. e. hydroxyl group) in the interior of wood cell wall were occupied by copper, thus reducing the water absorbing capacity of wood sample. Such chemical treatment have been reported to exhibit reduced water absorption compared to untreated material[22].

Water repellent efficiency (WRE) values of the treated samples after 30 days of water immersion are shown in Table 1. Samples with lower water uptakes produced higher WRE. Samples treated with rosin alone or in combination with copper had reduced the WA and improved WRE. The analysis of variance (ANOVA) revealed that there were significant differences in the WA values and the water repellent efficiency (WRE) of samples treated with three rosin formulations. WA values decreased with increasing rosin concentration in the rosin alone or plus copper treated specimens. The highest increment of WRE was observed in sample treated with 4% rosin and 3% added copper sulfate.

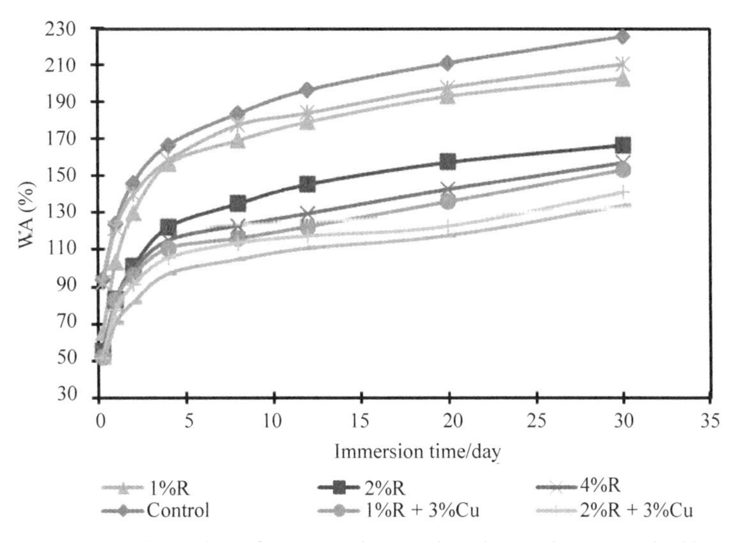

Fig. 1 Changes in water absorption of untreated control and samples treated with rosin alone or in combination with copper during water immersion (%)

3.3 Swelling properties and anti-swelling efficiency (ASE)

The results of swelling (S) in the radial, tangential directions, and volumetric of the untreated control and treated poplar wood samples after a 30 days water immersion are presented in Fig. 3. Untreated samples exhibited higher swelling coefficient values compared to the treated wood samples. Swelling in the radial, tangential, and volumetric for the controls were 6.6%, 8.7% and 16.3%, respectively. The lowest swelling values were obtained for wood samples treated with 4% rosin plus 3% $CuSO_4$, with average of 4.2%, 6.5% and 10.6% in the radial, tangential, and volumetric, respectively. Overall, all treatments in this study improved the swelling coefficient of wood.

The anti-swelling efficiency (ASE) values based on average values of swelling coefficients are also summarized in Table 1. All treated wood samples expressed significant improvement in anti-swelling efficiency (ASE) compared to untreated controls. A value of 6.98% of ASE for the wood samples impregnated with copper sulfate after 30 days soaking in water was observed. It is possible that there was a small amount of copper cross-linking with wood components inside cell wall, which resulted in reduced water absorption and improved in anti-swelling efficiency. However, wood samples treated rosin alone or in combination with copper, also demonstrated significant improvement in ASE compared to untreated samples. This may reflect the

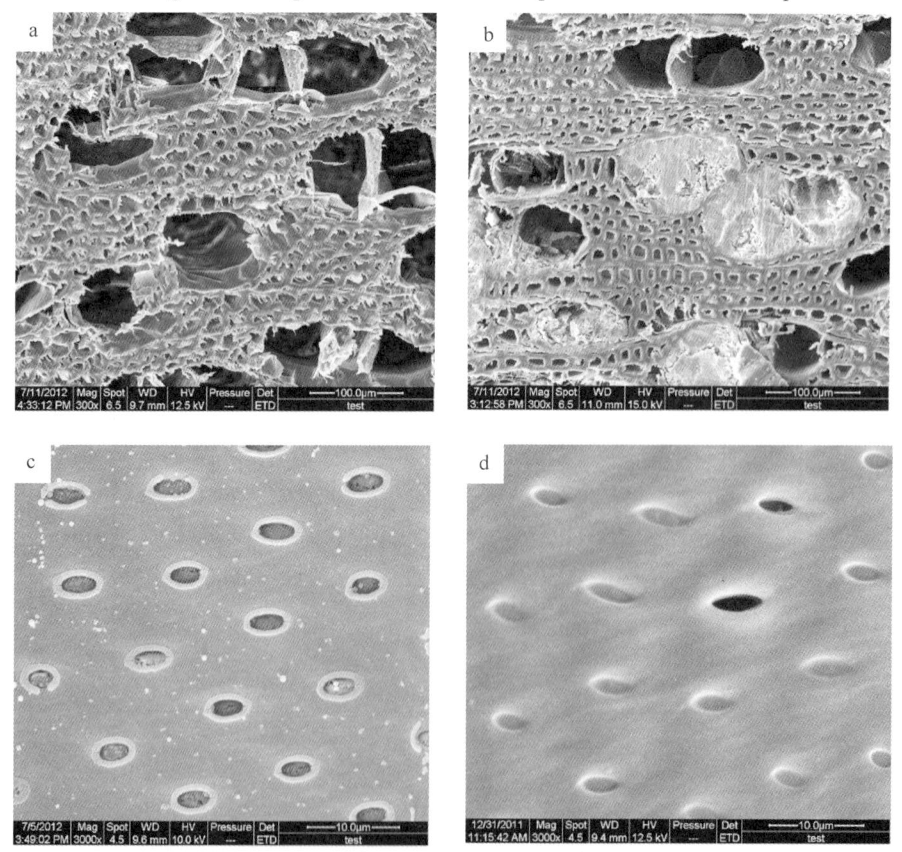

Fig. 2 SEM images of poplar samples treated with rosin alone and in combination with copper; cross-section of untreated control (a) and wood block treated with 2%rosin+3%$CuSO_4$(b); tangential section of wood block treated with 1%rosin (c) and 4%rosin+3%$CuSO_4$(d)

fact that rosin has good hydrophobic properties, but may also interacts with copper to form an insoluble copper resinate compound[4] that blocks water molecule movement inside the wood cell wall. Rosin-copper treated poplar wood had effective water resistance that could contribute to increase dimensional stability of wood. Result showed that the performance of rosin-copper based preservatives were comparable to other new preservative systems[23,24].

3.4 Compression strength

The compression strength parallel to grain (CSPG) values of wood samples are given in Table 2. There were no significant differences in CSPG values between untreated control and samples treated with copper sulfate, or rosins alone. However, there was a significant different between untreated controls and samples treated with the rosin-copper solutions. Rosin concentration had no significant impact on CSPG of wood samples treated with rosin alone or in combination with copper. The highest compression strength value was 44.41 MPa in the 1%rosin + 3%$CuSO_4$ treated poplar samples, while the lowest CSPG value was 36.82 MPa in wood treated with 2%rosin.

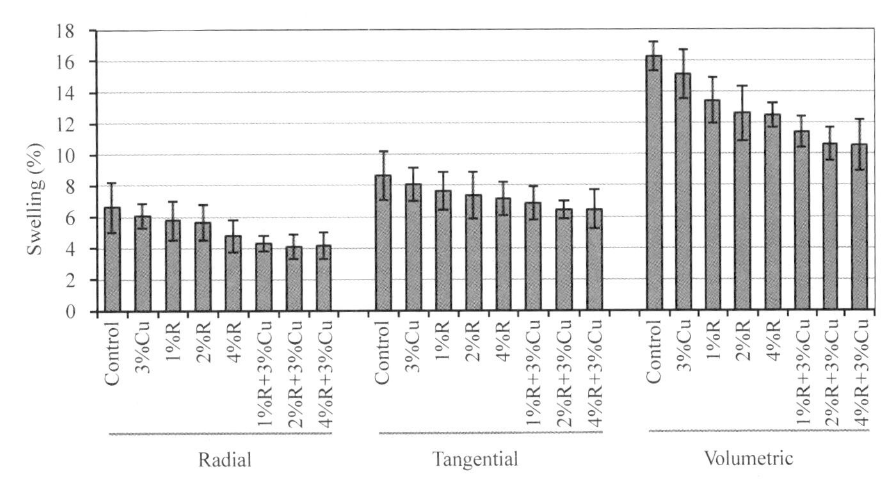

Fig. 3 Swelling (%) of untreated control and samples treated with rosin alone or in combination with copper after saturation with water (Cu: anhydrous copper sulfate and R: rosin sizing agent)

Table 2 Compression strength parallel to grain of poplar wood treated with rosin alone and in combination with copper

Treatments	Retention ($kg \cdot m^{-3}$)		Compression strength, MPa, at 12% MC		Change (%)
	Mean	Std	Mean	Std	
1.0%R	8.08	0.12	37.03[b]	3.46	-3.74
2.0%R	16.7	0.63	36.82[b]	4.62	-4.30
4.0%R	32.92	0.83	37.17[b]	4.42	-3.38
1.0%R + 3.0%Cu	33.73	0.8	44.41[a]	5.12	15.42
2.0%R + 3.0%Cu	40.9	1.75	43.62[a]	4.52	13.37
4.0%R + 3.0%Cu	61.15	1.15	44.24[a]	3.46	14.99
3.0%Cu	25.46	0.35	37.07[b]	3.04	-3.65
Control	-	-	38.47[b]	4.62	-

Note: Standard deviations are in brackets: Cu: anhydrous copper sulfate and R: rosin sizing agent; means within a column followed by the same letter are not significantly different at 5% level of significance using the one-way ANOVA test.

CSPG of wood treated with rosin or copper sulfate alone decreased slightly compared with the untreated control samples (Table 2). These results are in good agreement with those reported by Yildiz et al. [7] and Winandy[12]. However, the CSPG of wood treated with rosin-copper solutions increased around 13% -16% compared with the untreated control samples. This could be attributed to the rosin reacting with copper to form an insoluble copper resinate compound[4]. It was reported that the gross wood probably fails in compression due to the buckling of relatively thin cell walls because of a long-column type of instability[18]. The SEM micrographs (Fig. 2) showed that the copper resinate compound filled the vessels as well as the lumens and coated the cell walls, which greatly increasing their lateral stability. This has also been demonstrated by Yildiz et al. [25] in a study of on mechanical properties of wood-polymer composites.

3.5 Bending test

The modulus of rupture (MOR) and (MOE) values of poplar wood samples are given in Table 3. One-way ANOVA revealed that there were significant differences for MOR values between untreated controls and samples treated with copper sulfate alone. There were no significant differences between untreated controls and samples treated with the rosin alone or rosin-copper solutions nor there were significant differences between copper sulfate and rosins alone or in combination with copper. Table 3 also shows that there were significant differences for MOE values between untreated control and all treated wood samples. However, there was no significant differences between samples treated with copper sulfate and samples treated with rosin alone or in combination with copper.

All wood samples treated with copper sulfate and rosin sizing agent, separately or in combination, have lower MOR and MOE values compared to untreated controls. However, the concentration of rosin used did not affect MOR or MOE. Winandy[12] reported that waterborne preservative treatments generally reduced mechanical properties of wood more than oil-type preservative treatments since waterborne preservative chemicals physically react with the wood cell wall material. Strength losses caused by a waterborne preservative are related directly to the severity of the fixation/precipitation reaction[7]. Our results showed that reduction in MOR of 8% and almost 9% in MOE was observed. Comparing results obtained here across several studies with other waterborne preservative systems shows that wood treated with this new rosin-copper system had equivalent or superior properties. Foe example, yellow pine wood treated with CCA exhibited a reduction in MOR of 12% and almost 10% in MOE[7]. Su[26] found some reductions in MOR and MOE of poplar wood treated with some copper based preservatives. LeVan et al. [27] also reported that the MOR of southern pinewood treated some fire-retardant chemicals were reduced from 10% to 20%.

3.6 Brinell Hardness

The Brinell hardness values perpendicular to the transverse, tangential and radial surfaces of the untreated control and treated poplar wood samples were given in the Table 4. There were no significant differences for Brinell hardness between the rosin alone and rosin-copper treated samples. However, statistically significant differences for Brinell hardness perpendicular to the transverse surface were found between untreated controls as copper sulfate and the rosin alone or the rosin-copper treated samples. There was no significant differences for Brinell hardness perpendicular to the tangential and radial surfaces values between untreated control as well as copper sulfate and the rosin alone or the rosin-copper treated samples.

Table 3 Average values of MOR and MOE of treated poplar wood samples

Treatments	Retention ($kg \cdot m^{-3}$)		MOR at 12% MC (MPa)		MOE at 12% MC (MPa)		Change (%)	
	Mean	Std	Mean	Std	Mean	Std	MOR	MOE
1. 0%R	7. 51	0. 59	86. 85^{ab}	6. 56	9030. 81^{b}	370. 08	-5. 97	-8. 4
2. 0%R	15. 92	0. 56	86. 24^{ab}	4. 52	9015. 64^{b}	557. 28	-6. 63	-8. 56
4. 0%R	32. 51	0. 8	85. 25^{b}	4. 43	9078. 40^{b}	490. 59	-7. 7	-7. 92
1. 0%R + 3. 0%Cu	32. 09	0. 98	86. 99^{ab}	5. 43	9087. 54^{b}	725. 61	-5. 82	-7. 83
2. 0%R + 3. 0%Cu	40. 99	1. 85	87. 35^{ab}	10. 24	9360. 97^{b}	633. 94	-5. 43	-5. 05
4. 0%R + 3. 0%Cu	60. 79	2. 7	87. 16^{ab}	7. 8	9099. 42^{b}	432. 93	-5. 64	-7. 71
3. 0%Cu	25. 7	0. 59	84. 99^{b}	10. 73	9057. 74^{b}	871. 5	-7. 98	-8. 13
Control	0	0	92. 37^{a}	10. 17	9859. 32^{a}	939. 79		

Note: Cu: anhydrous copper sulfate and R: rosin sizing agent; Means within a column followed by the same letter are not significantly different at 5% level of significance using the one-way ANOVA test.

Table 4 Average value of hardness of treated poplar wood samples

Treatments	Retention ($kg \cdot m^{-3}$)	Average hardness at 12% MC (kN)			Change (%)		
		Longitudinal	Tangential	Radial	Longitudinal	Tangential	Radial
1. 0%R	8. 19 (0. 28)	3. 35 (0. 41)^{a}	2. 13 (0. 36)^{a}	2. 12 (0. 24)^{a}	38. 38	-0. 16	9. 81
2. 0%R	16. 46 (0. 27)	3. 22 (0. 46)^{a}	2. 03 (0. 37)^{ab}	2. 10 (0. 33)^{a}	32. 96	-4. 78	8. 84
4. 0%R	31. 81 (0. 86)	3. 35 (0. 51)^{a}	2. 04 (0. 29)^{ab}	2. 11 (0. 35)^{a}	38. 28	-4. 06	9. 24
1. 0%R + 3. 0%Cu	32. 66 (2. 18)	3. 27 (0. 31)^{a}	2. 01 (0. 28)^{ab}	2. 03 (0. 36)^{ab}	34. 77	-5. 49	5. 11
2. 0%R + 3. 0%Cu	40. 06 (2. 17)	3. 21 (0. 36)^{a}	2. 06 (0. 35)^{ab}	2. 02 (0. 27)^{ab}	32. 39	-3. 13	4. 49
4. 0%R + 3. 0%Cu	57. 56 (1. 19)	3. 28 (0. 42)^{a}	1. 99 (0. 34)^{ab}	2. 03 (0. 36)^{ab}	35. 23	-6. 66	5. 22
3. 0%Cu	25. 54 (0. 84)	2. 77 (0. 23)^{b}	1. 80 (0. 32)^{b}	1. 77 (0. 40)^{b}	14. 27	-15. 45	-8. 25
Control	0. 00	2. 42 (0. 31)^{b}	2. 13 (0. 44)^{a}	1. 93 (0. 36)^{ab}	-	-	-

Note: Standard deviations are in brackets; Cu: anhydrous copper sulfate and R: rosin sizing agent; Means within a column followed by the same letter are not significantly different at 5% level of significance.

Treatments with rosin alone or rosin-copper did not adversely affect Brinell hardness perpendicular to the transverse and radial surface of wood, but they had a negative effect on Brinell hardness perpendicular to the tangential surface (Table 4). The Brinell hardness values perpendicular to the transverse and radial surfaces of wood samples treated with rosin alone or in combination with copper increased compared to untreated control. The percentage increased in Brinell hardness on transverse surface ranging from 32% to 39% and from 4% to 10% on radial surface. The surface hardness of a material is known to increase with the density[28]. The SEM in Fig. 2 revealed that the rosin and insoluble copper resinate compound were found in the vessels and the cell lumens and had apparently stiffened the thin cell - walls sufficiently to prevent buckling under a compressive load. This probably accounts for the relatively large increases in the Brinell hardness perpendicular to the longitudinal as well as the CSPG of wood[25]. On the other hand, the Brinell hardness perpendicular to the tangential direction in samples treated with rosin alone or rosin-copper were slightly lower than one of the untreated controls. This suggests that the rosins did not penetrate sufficiently into the cell walls. This lack of hardening resulted in slightly decreased Brinell hardness perpendicular to the tangential surface[8, 29].

4 Conclusions

The effects of treatment with mixture rosin-copper on the concentration level of the rosin on the physical and mechanical properties of poplar wood were examined.

Combinations of rosin sizing agent and copper sulfate obviously decreased moisture absorption, water absorption as well as swelling and improved the WRE and ASE of wood. All rosin-copper formulations used in this study increased the CSPG, the BH on transverse and radial surfaces compared to control, as well as wood treated with copper sulfate alone, but MOR, MOE and BH values on tangential surface were slightly lower than those for untreated control.

The use of rosin sizing agent combined with copper sulfate to impregnate wood increased fixation of copper in wood, while also improving dimensional stability as well as mechanical properties.

Acknowledgements

The authors are grateful for the support of the Vietnamese Government, the National Natural Science Foundation of China (31070487), and Foundation for University Young Core Teachers of Heilongjiang Province of China (1154G49). Using the one-way A NOVA test.

References

[1] Ngoc NTB. Wood preservation (in Vietnamese)[M]. Ha Noi: Agriculture Press, 2006.

[2] Townsend T, Solo-Gabriele H. Environmental impacts of treated wood[M]. CRC, Boca Raton, Florida, USA. 2006.

[3] Yao X, Zheng L. Development potential of rosin sizing agent[M]. Chem Technol Mark, 2000, 10: 21.

[4] Pizzi A. A new approach to nontoxic, wide-spectrum, ground-contact wood preservatives. 1. Approach and reaction mechanisms[J]. Holzforsch, 1993a, 47(3): 253-60.

[5] Pizzi A. A new approach to nontoxic, wide-spectrum, ground-contact wood preservatives. 2. Accelerated and long-term field tests[J]. Holzforsch, 1993b, 47(4): 343-8.

[6] Nguyen TTH, Li J, Li S. Effects of water-borne rosin on the fixation and decay resistance of copper-based preservative treated wood[J]. Bioresour, 2012, 7(3): 3573-84.

[7] Yildiz UC, Temiz A, Gezer ED, Yildiz S. Effects of wood preservatives on mechanical properties of yellow pine (Pinus sylvestris L.) wood[J]. Build Environ, 2004, 39: 1071-5.

[8] Mourant D, Yang D Q, Riedl B, Roy C. Mechanical properties of wood treated with PF-pyrolytic oil resins[J]. Holz Roh Werkst, 2008, 66: 163-71.

[9] Simsek H, Baysal E, Peker H. Some mechanical properties and decay resistance of wood impregnated with environmentally-friendly borates[J]. Constr Build Mater, 2010, 24: 2279-84.

[10] Cao J, Xie M, Zhao G. Tensile stress relaxation of copper-ethanolamine (Cu-EA) treated wood[J]. Wood Sci Technol, 2006, 40: 417-26.

[11] Tomak ED, Viitanen H, Yildiz UC, Hughes M. The combined effects of boron and oil heat treatment on the properties of beech and Scots pine wood. Part 2: Water absorption, compression strength, color changes, and decay resistance[J]. J. Mater. Sci., 2011, 46: 608-15.

[12] Winandy JE. Effects of waterborne preservative treatment on mechanical properties: a review[C]. In: Proceedings, vol. 91. Woodstock (MD): American Wood Preservers' Association, 1995: 17-33.

[13] GB 1929. Wood-sampling methods and general requirements for physical and mechanical tests[S]. China, 2009.

[14]GB/T 1935. Wood-determination of ultimate stress in compression parallel to grain[S]. China, 2009.

[15]GB/T 1936. 1. Wood-determination of ultimate strength in static bending[S]. China, 2009.

[16]GB/T 1936. 2. Wood-determination of modulus of elasticity in static bending[S]. China, 2009.

[17]GB/T 1941. Wood-determination of static hardness[S]. China, 2009.

[18]GB 1931. Wood-determination of moisture content for physical and mechanical tests[S]. China, 2009.

[19]Cao J, Kamdem D P. Moisture adsorption characteristics of copper-ethanolamine (Cu-EA) treated Southern yellow pine (Pinus spp.)[J]. Holzforsch, 2004, 58: 32-38.

[20]Eberhardt T L, HanJ S, Micales J A, Young R A. Decay resistance in conifer seed cones: role of resin acids as inhibitors of decomposition by white rot fungi[J]. Holzforsch, 2019, 48 (4): 278-284.

[21] Mansouri H, Pizzi A, Leban J M, Delmotte L, Lindgren O, Vaziri M. Causes for the improved water resistance in pine wood linear welded joints[J]. J. Adhesion Sci. Technol, 2011, 25(16): 1987-1995.

[22] Islam M S, Hamdan S, Rusop M, Rahman M R, Ahmed A S, Idrus M A M M. stability and water repellent efficiency measurement of chemially modified tropical light hard wood[J]. Bioresource, 2012, 7(1): 1221-1231.

[23] Toussaint-Dauvergne E, Soulounganga P, Gerardin P, Loubinoux B. Glycerol/glyoxal: a new boron fixation system for wood preservation and dimensional stabilization[J]. Holzforsch, 2000, 54: 123-126.

[24] Baysal E, Ozaki S K, Yalinkilic M K. Dimensional stabilization of wood treated with furfuDimensionalryl alcohol catalysed by borates[J]. Wood Sci. Technol, 2004, 38: 405-415.

[25] Yildiz U C, Yildiz S, Gezer E D. Mechanical properties and decay resistance of wood-polymer composites prepared from fast growing species in Turkey[J]. Bioresour Technol, 2005, 96: 1003-1011.

[26] Su W. Functional synthesis of tree extractives and the application of Wood preservation[D]. Harbin (China): Northeast Forestry University, 2008.

[27] LeVan S L, Winandy J E. Effects of fire-retardant treatments on wood strength: a review[J]. Wood Fiber Sci., 1990, 22(1): 113-131.

[28] Shukla S R, Kamdem D P. Effect of copper based preservatives treatment of the properties of southern pine LVL[J]. Constr Build Mater, 2012, 34: 593-601.

[29] Pizzi A. Phenolic Resin Adhesives[M]. In: Pizzi A, Mittal KL, editors. Handbook of Adhesive Technology. second ed. New York: Marcel Dekker, Inc. 2003: 541-71.

中文题目：硫酸铜与松香施胶剂复合处理对杨木物理和力学性能的影响

作者：阮氏清贤，李淑君，李坚

摘要：本文研究了以3%的 $CuSO_4$ 分别与1.0%、2.0%、4.0%松香施胶剂混合处理对人青杨木材物埋和力学性能的影响。松香-铜处理降低了木材的吸湿、吸水和润胀性能，同时提高了木材的防水性能和抗胀率，处理材在水中浸泡30天后的抗胀率达40%左右。总体来说，松香-铜处理提高了木材的顺纹抗压强度和布氏硬度，但是，MOR和MOE均低于对照组。

关键词：松香-铜；吸湿性；吸水率；抗压强度；MOR；MOE；硬度强度

Enhanced Thermal and Mechanical Properties of PVA Composites Formed with Filamentous Nanocellulose Fibrils*

Wei Li, Qiong Wu, Xin Zhao, Zhanhua Huang, Jun Cao, Jian Li, Shouxin Liu

Abstract: Long filamentous nanocellulose fibrils (NCFs) were prepared from chemical-thermomechanical pulps (CTMP) using ultrasonication. Their contribution to enhancements in thermal stability and mechanical properties of poly(vinyl alcohol) films were investigated. The unique chemical pretreatment and mechanical effects of CTMP loosen and unfoldfibers during the pulping process, which enables further chemical purification and subsequent ultrasound treatment for formation of NCFs. The NCFs exhibited higher crystallinity (72.9%) compared with that of CTMP (61.5%), and had diameters ranging from 50 to 120 nm. A NCF content of 6 *wt%* was found to yield the best thermal stability, light transmittance, and mechanical properties in the PVA/NCF composites. The composites also exhibited a visible light transmittance of 73.7%, and the tensile strength and Young's modulus were significantly improved, with values 2.8 and 2.4 times larger, respectively, than that of neat PVA.

Keywords: Nanocellulose fibrils; CTMP; Ultrasonication; Poly(vinyl alcohol)

1 Introduction

Natural fiber-reinforced polymer composites are a potential solution for the environmental burden presented by the use of petroleum-based, non-biodegradable polymeric materials (Pandey, Misra, Mohanty, Drzal, & Singh, 2005; Svagan, Samir, & Berglund, 2008; Shih & Huang, 2011). Nanocellulose fibrils (NCFs) are ideal candidates for reinforced composites owing to their abundance, renewability, biodegradability, exceptional mechanical properties (high specific strength and modulus), low thermal expansion, environmental benefits, and low cost (Habibi, Lucia, & Rojas, 2010; Wegner & Jones, 2006; Zhang, Zhang, Lu, & Deng, 2012; Bhatnagar & Sain, 2005; Abdul Khalil, Bhat, & Ireana Yusra, 2012).

The isolation of NCFs from cellulose fibers using simple, low cost, and environmentally friendly methods is a great challenge (Phong, Gabr, Okubo, Chuong, & Fujii, 2013; Wang & Cheng, 2009). Several mechanical processes have been used to extract nanofibers from cellulosic materials, such as pulping beating (Nakagaito & Yano, 2004; Chakraborty, Sain, & Kortschot, 2005), high pressure homogenizing (Bruce, Hobson, Farrent, & Hepworth, 2005; Stenstad, Andresen, Tanem, & Stenius, 2008; Leitner, Hinterstoisser, Wastyn, Keches, & Gindl, 2007), and cryogenic crushing (Alemdar & Sain, 2008; Wang & Sain, 2007).

---

* 本文摘自 Carbohydrate Polymers, 2014, 113: 403-410.

Recently, ultrasonic techniques have been used to isolate cellulose nanofibers (Chen, Yu, & Liu, 2011a; Cheng, Wang, & Han, 2010; Cheng, Wang, & Rials, 2009). They apply sound energy to disrupt physical and chemical systems, and ultrasonication- induced cavitation for degrading polysaccharide linkages, such as chitosan (Liu, Du, & Kennedy, 2007; Kasaai, Arul, & Charlet, 2008), dextran (Cote & Willet, 1999; Portenlanger & Heusinger, 1997), xyloglucan (Vodeniǎarová, Drímaloví, Hromádková, Malovíková, & Ebringerová, 2006), and carboxymethylcellulose (Gronroos, Pirkonen, & Ruppert, 2004; Aliyu & Hepher, 2000) has been well documented. Moreover, the violent collapse during cavitation cre- ates micro-jets and shock waves on the surface of purified cellulose fibers, causing erosion and splitting along the axial direction. Sonification can break the relatively weak nonbonding interactions, such as van der Waals forces, at the interfaces between the microfibers (Suslick, Choe, Cichowlas, & Grinstaff, 1991; Filson & Dawson-Andoh, 2009; Zhao, Feng, & Gao, 2007). Thus, ultrasonication is well suited for isolating NCFs having relatively long networks (Cheng et al., 2009; Tischer, Sierakowski, Westfahl, & Tischer, 2010).

High-intensity ultrasonication, combined with chemical pre- treatments, has been used to prepare NCFs from several raw materials, such as bamboo, wood, microcrystalline cellulose (MCC) and alkaline peroxide mechanical pulp (APMP) (Chen et al., 2011a, 2011b; Li, Yue, & Liu, 2012; Li, Zhao, & Liu, 2013). Pulps made using these hybrid processes are known as chemical-thermomechanical pulps (CTMP). The process reportedly produces pulps with high yields and has an environmental impact approaching that of mechanical pulping. The chemical pretreatment conditions are much less vigorous (lower temperatures, shorter times, less extreme pHs) than in a chemical pulping process, because the goal is to loosen and split the fibers for easier mechanical treatment. The mechanical process l eads to externalfibrillation of the fibers by gradually peeling off the cell walls (P and S1 layers) and exposing the S2 layer, and also causes internal fibrillation that loosens the fiber wall. Compared with the above-mentioned cellulosic materials, CTMP seems an excellent original material for NCF production. Meanwhile, it also leads to a relatively high crystallinity (Konn, Holmbom, & Nickull, 2002; Law & Daud, 2000).

Polyvinyl alcohol (PVA) has a wide variety of applications because of its high dielectric strength, high elasticity, hydrophilic characteristics, and its ability to form good films via solution casting. It has a carbon chain backbone with hydroxyl groups that can act as a source of hydrogen bonding to enhance the formation of polymer complexes (Sedlarık, Saha, Kuritka, & Saha, 2006). Many researchers have dispersed cellulose fibers in the hydrophilic polymers of poly(vinyl alcohol), resulting in strong reinforcement effects (Paralikar, Simonsen, & Lombardi, 2008).

Here, using CTMP as a starting material, high intensity ultrasonication was used to prepare long, filamentous NCFs with uniform diameters. The chemical composition, morphology, crystalline structure, and thermal behavior of the NCFs and their intermediate products were characterized. The mechanical and optical properties, and the improved thermal stability in particular, of PVA composites prepared with different quantities of NCFs are also discussed in detail.

2 Materials and methods

2.1 Preparation of NCFs

CTMP with a brightness of 40% ISO was obtained from pine wood chips and was kindly supplied by a pulp factory in the Jilin province of China. CTMP (2 g) was mixed with an acidified $NaClO_2$ solution at 75 ℃ for 1 h to remove lignin content; this process was repeated five times. The resulting holocellulose fibers (Ho-CFs) were treated with 4-*wt*% NaOH at 90 ℃ for 4 h to remove hemicelluloses and residual lignin, and the residues were filtered and rinsed with distilled water until pH = 7 was obtained. The solid product was alkali-treated cellulose fibers (Al-CFs). The Al-CFs were soaked in distilled water at approximately 0. 1% (*w/w*) solid content. An Al-CFs solution (150 mL) was placed in a 20-25-kHz ultrasonic generator (JY98-ⅢD, Ningbo Scientz Biotechnology Co. , Ltd. , Ningbo, China), equipped with a 2. 5-cm-diameter cylindrical titanium alloy probe tip. The ultrasonication was conducted in an ice water bath for 30 min at an output power of 1200 W, producing a suspension of NCFs. The NCFs were separated from large bundles by centrifuga- tion, and were freeze dried and stored at 5 ℃.

2. 2 Preparation of NCF-reinforced PVA films

An aqueous PVA solution (10 *wt*%) and an aqueous NCF water suspension (approximately 0. 35 *wt*%) were mixed, stirred manually, and then dispersed via ultrasonic treatment (500 W, Tianjin Automatic Science Co. , Ltd, Tianjin, China) for approximately 5 min at a power level of 85%. The mixtures were degassed in a desiccator at room temperature until films were formed; the films were then dried at 70 ℃ for 4 h, resulting in thicknesses of approximately 0. 1 mm. Composite films with different NCF contents (0, 2, 6, 10 and 14 *wt*%) were prepared (denoted here as PVA, PVA/NCFs-2, PVA/NCFs-6, PVA/NCFs-10 and PVA/NCFs-14, respectively).

2. 3 Chemical composition

The chemical compositions of CTMP, Ho-CFs, Al-CFs, and NCFs were determined in accordance with the standards of the Technical Association of Pulp and Paper Industry (TAPPI). The holocellulose (cellulose + hemicelluloses) content was determined as described in TAPPI T19 m - 54. An acidified sodium chlorite solution method (75 ℃ for 1 h, 3 times) was used to obtain the holocellulose (Sun, Lawther, & Banks, 1995). The a-cellulose content of the fibers was then determined by further NaOH [17. 5%, (25±0. 2) ℃] treatment of the fibers to remove the hemicelluloses (T203os-61, a-cellulose in pulp) (Sun, Sun, Zhao, & Sun, 2004; Teixeira et al. , 2010). The difference between the holocellulose and a-cellulose values yielded the hemicellulose content in the fibers. Three samples of each material were tested, and averaged values were obtained. The degree of polymerization (DP) of CTMP, Ho-CFs, and NCFs was measured using the viscosity method with an aqueous 1. 0 M bis (ethylenediamine) - copper (Ⅱ) hydroxide solvent, as described by Iwamoto (Iwamoto, Nakagaito, Yano, & Nogi, 2005; Iwamoto, Nakagaito, & Yano, 2007).

2. 4 Characterizations

Microstructural analysis was performed using a scanning electron microscope (SEM, FEI QUANTA200 (FEI, Hillsboro, OR, USA). Prior to SEM imaging, the sample surfaces were coated with a thin layer of gold using a BAL - TEC SCD 005 sputter coater (Leica, Wetzlar, Germany) to provide electrical conductivity. Fourier transform infrared spectroscopy (FT-IR) was performed with a Magana-IR560E. S. P (Nicolet, USA) spectrometer to identify functional groups. X-ray diffraction XRD measurements were acquired on a D/max-r B X-ray diffractometer (Rigaku Corp, Tokyo, Japan) using Cu Ka radiation. The samples were scanned over a 20 range varying over 10° - 30°. The crystallinity index of cellulose materials was calculated from the height of the 200 peak (I200, 2θ=22.6°) and the intensity minimum between the peaks at 200 and 110 (Iam, 2θ = 18°) using the Segal method (CI (%) = (1 - Iam/I200) × 100). I200 represents both a crystalline and an amorphous material, Thermogravimetric analysis (TGA) was performed using a TGA-Q50 (TA, USA) instrument. Temperature programs for dynamic tests were run over the range 25 - 600 ℃ at a heating rate of 10 ℃/min. The mechanical properties of the PVA/NCF composite films were analyzed using a RGT-20A instrument. To characterize each type of film, three samples were fabricated with lengths of (150±2) mm, widths of (15±0.3) mm, and thicknesses of (0.1±0.05) mm. The optical transmittance was measured over 300-800 nm using a UV-vis spectrophotometer (TU-1900, Beijing Purkinje General Instrument Co., Ltd., Beijing, China).

3 Result and discussion

3.1 Chemical composition and SEM analysis

Changes in chemical composition of the fibers from different pretreatment processes are presented in Table 1. CTMP contained a higher percentage of hemicelluloses (18.6%) and lignin (Klason lignin 21.3% and acid-insoluble lignin 2.8%), and the lowest percentage of α-cellulose (56.8%). When the fibers were subjected to $NaClO_2$ treatment, the lignin content of the Ho- CFs decreased to 2.4%, whereas the α-cellulose and hemicellulos content increased to 69.4 and 25.4%, respectively. After NaOH treatment, the α-cellulose content of the Al-CFs increased to 84.3%, resulting in highly purified cellulose fibers. The cellulose fiber structure changed as a function of the removal of the primary components (Fig. 1). The unique chemical-mechanical effects of CTMP on raw cellulose fibers were because of a special pulping process in which caustics were used to soften wood chips before mechanical treatment. After mechanical treatment, the cohesive forces of the fiber decreased and the fibers become looser and softer. The S1- layer was peeled off into flakes [Fig. 1 (a)] to enhance the swelling and the absorption capability, making it more easily permeable to $NaClO_2$ and NaOH solutions. This also decreases the treatment time and the amounts of chemical reagents. However, there are no detailed changes of the CTMP surface that can be observed in Fig. 1b; it exhibited some wrinkling, while other local areas exhibited relative smoothness [Fig. 1 (c) and (d)]. The fiber structure was also maintained after $NaClO_2$ treatment removed most of the lignin. After NaOH treatment was performed to remove most hemicelluloses and residual lignin, fibrillation on the cellulose surface can be clearly observed [Fig. 2 (e) and (f)]. These changes were favorable to the subsequent high intensity ultrasonic treatment that can concentrate ultrasonic stress for efficient defibrillation.

Table 1 Chemical compositions and DP of CTMP, Ho-CFs, Al-CFs and NCFs

Material	α-Cellulose	Hemicelluloses(%)	Klasonlignin(%)	Acid-insoluble lignin (%)	DP
CTMP	56.8±1.4	18.6±0.8	21.3±1.5	2.8±0.2	1535±14
Ho-CFs	69.4±1.7	25.4±1.3	2.4±0.1		1210±12
Al-CFs	84.3±2.5	13.2±0.5			1016±10
NCFs	86.5±2.2	11.6±0.3			865±8

Long filamentous NCFs are shown in Fig. 1(g) and h. It is interesting to note that they formed a network structure. When the dispersingaqueous medium was removed by freeze-drying, the NCFs were leftinterlaced to each other. About 80% had diameters ranging from 50 to 120 nm [Fig. 3(e)]. Moreover, as listed in Table 1, the DP of the NCFs was 865, a decrease of 43.6% relative to that of CTMP. This revealed that more long-chain polysaccharides of cellulose were maintained compared with traditional acid hydrolysis methods during chemical and ultrasonic processing (Bondeson, Mathew, & Oksman, 2006). The scission of cellulose polymers results from solvodynamic shear created by ultrasonic cavitation. A liquid jet propagates toward the cellulose surface at a velocity of several hundred meters per second, and makes violent contact, resulting in significant morphology changes (Caruso et al., 2009). The long web-like and filamentous NCFs could be used in a variety of applications, such as reinforced nanocomposites, advanced functional materials, and membrane filters.

3.2 FT-IR and XRD analysis of CTMP, Ho-CFs, Al-CFs and NCFs

FT-IR spectra of CTMP, Ho-CFs, Al-CFs and NCFs are plotted in Fig. 2(a). The 1508 and 1452 cm^{-1} peaks of CTMP are attributed tothe aromatic ring C C stretching vibrations and the C H deformation vibrations of lignin, respectively (Sain & Panthapulakkal, 2006; Sun, Tomkinson, Wang, & Xiao, 2000). The intensity of these peaks almost disappeared for Ho-CFs, because of lignin removal. The peaks at 1270 $cm^{-1}$for CTMP and Ho-CFs are attributed to the C-O of guaiacyl, while the peaks at 1604 $cm^{-1}$for CTMP and Ho-CF are from the aromatic rings of lignin (Lebo & Lonsky, 1990; Imsgard, Falkehag, & Kringstad, 1971). The two peaks disappeared for Al-CFs and NCFs, which is attributed to the removal lignin. These results confirmed the data summarized in Table 1. After ultrasonic treatment, the spectrum of the NCFs was almost the same as that of the Al-CFs. This fact indicated that more hemicelluloses and lignin were removed during the treatment with $NaClO_2$and NaOH, and that the original molecular structure of cellulose was maintained even after these components were removed and after the ultrasonic treatments.

Fig. 2(b) plots X-ray diffraction profiles and lists the crystallinity of the CTMP, Ho-CFs, Al-CFs, and NCFs. All samples had diffraction peaks at $2\theta = 16.5°$ and $22.5°$, and had the typical cellulose Ⅰ pattern (Nishiyama, Sugiyama, Chanzy, & Langan, 2003). The crystallinity of CTMP was 61.5%. After $NaClO_2$ treatment, the crystallinity of Ho-CFs increased to 68.9% because of lignin removal. The crystallinity of Al-CFs increased to 72.0% because of hemicellulose removal from amorphous regions. After ultrasonication, the crystallinity of NCFs increased to 72.9%; and from the NCF XRD profiles, it can be-observed that the intensities of the amorphous and crystalline areas all decreased. Therefore, both amorphous and crystalline cellulose were removed. The increased crystallinity may owed to a slower decline in the proportion of crystalline areas (Li et al., 2012, 2013). The effect of ultrasonication in heterogeneous systems is primarily the consequence of cavitation. In the CF-water system, both the crystalline and amorphous regions were subjected to intense collisions (Cintas & Luche, 1999). It can be concluded that on the CF surface, the intense physical stresses cause hydrogen bond break-age between the CF microfibers that are then defibrillated to NCFs. The high crystallinity of NCFs could provide better strength in composite materials (Sakurada, Nukushina, & Ito, 1962).

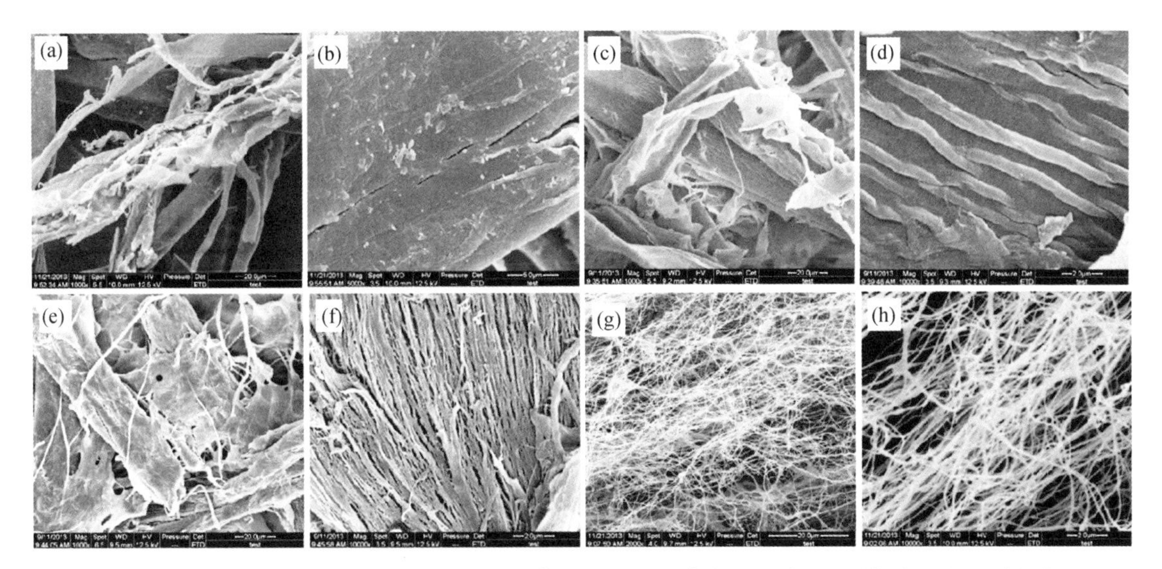

Fig. 1 SEM images of [(a) and (b)] raw CTMP; [(c) and (d)] fibers after acidified sodium chlorite treatment (Ho-CFs); [(e) and (f)] fibers after alkali-treated (Al-CFs); and [(g) and (h)] NCFs

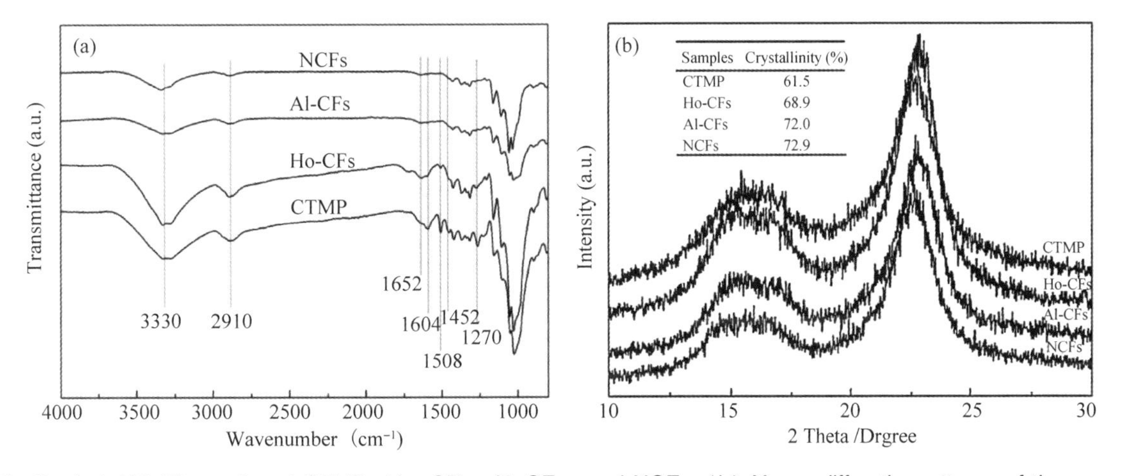

Fig. 2 (a) FT-IR spectra of CTMP, Ho-CFs, Al-CFs, and NCFs. (b) X-ray diffractionpatterns of the same

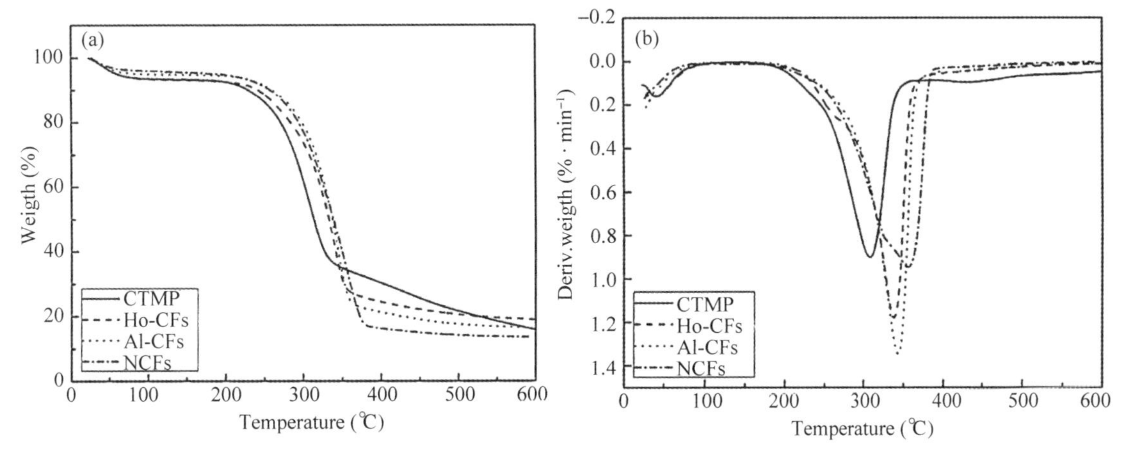

Fig. 3 TGA (a) and DTG (b) curves for CTMP, Ho-CFs, Al-CFs and NCFs

3.3 TG and DTG analysis

Thermogravimetric (TG) and derivative TG (DTG) curves for CTMP, Ho-CFs, Al-CFs, and NCFs (Fig. 3) showed that, for all samples, a small weight loss corresponding to the evaporation of absorbed water was observed at low temperatures (<110 ℃). The primary degradation of CTMP occurred at 201.1 ℃, and the maximum degradation was observed at 308.2 ℃ because of cellulose pyrolysis. Owing to the removal of lignin, the initial degradation and the main degradation peak of Ho-CFs increased to 208.4 ℃ and 338.1 ℃, respectively. After most of the hemicelluloses were removed in the NaOH treatment, the maximum degradation temperature for the Al-CFs was shifted to a higher temperature (342.9 ℃). These results show that removal of hemicelluloses could improve the thermal stability of cellulose fibers. For NCFs, the maximum degradation was observed at 355.4 ℃. The starting and maximum degradation temperatures for NCFs were all significantly higher than those of CTMP, which indicated that the thermal stability of the NCFs increased. The excellent thermal properties are suitable for thermoplastics, for which high processing temperatures are generally needed (Glasser, Taib, Jain, & Kander, 1999).

3.4 Optical properties and SEM analysis of PVA/NCF composites

Fig. 4a shows a photograph of PVA/NCF composite films with 0, 2, 6, 10, and 14 *wt%* NCFs. The optical transmittances of the composites were largely dependent on the dispersion of the NCFs in the PVA matrix. The letters in the background could be clearly seen through the films, which indicated transparency. How-ever, the films became increasingly opaque with increasing NCF contents. When the transmittances of pure PVA and the PVA/NCF Fig. 5. TGA (a) and DTG (b) curves for the PVA and PVA/NCF composite. Composites are plotted as shown in Fig. 4b, the transmittance decreased with increasing NCF

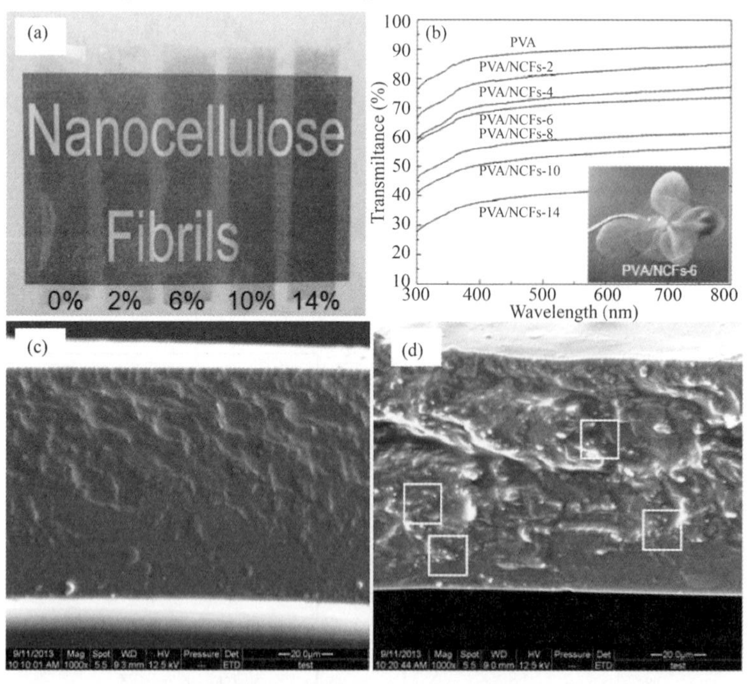

Fig. 4 (a) optical photo, (b) transmittance property curves for PVA/NCF composites, (c) SEM image of PVA, and (d) SEM image of PVA/NCFs-6 composites

content. When the NCF content was 2, 4 and 6 *wt%*, the films show transparent, and the transmittancewas 85%, 77% and 73%, respectively. When the NCFs content was increased to 8 *wt%*, the transmittance decreased from 91% to 61%. Along with the increasing of NCF content was 10%, the transmittance decreased significantly to 56%, owing to the poor dispersion of the NCFs. As shown in Fig. 4(c), the surface of the PVA/NCFs-6composites was smooth. A SEM cross-sectional image of the fracture surface of a PVA/NCF-6 composite [Fig. 4(d)] shows that the dispersion of the NCFs in the PVA was uniform.

3.5 Thermal properties of the PVA/NCF composites

Fig. 5 shows TGA and DTG curves for pure PVA and for PVA/NCF composites. According to the DTG curves, the degradation of pure PVA can be divided into three processes. These take place in the temperature ranges 50-181.7 ℃, 181.7-342.7 ℃ and 342.7-500 ℃. The first degradation process in the PVA/NCF composite began at a slightly higher temperature than that in pure PVA. The most important result, however, was that the temperature of the second degradation process in PVA/NCF was higher than that of pure PVA, which indicates an increased PVA/NCF thermal stability. Combined with results in Table 2, it can be seen that with the introduction of NCFs in PVA, there are higher maxi-mum decomposition temperatures in the first process 0.4-9.1 ℃, and the maximum decomposition temperatures of second process increased 4.9-16.9 ℃. The latter indicated that adding a small portion of NCF dramatically improved the thermal stability of PVA/NCF.

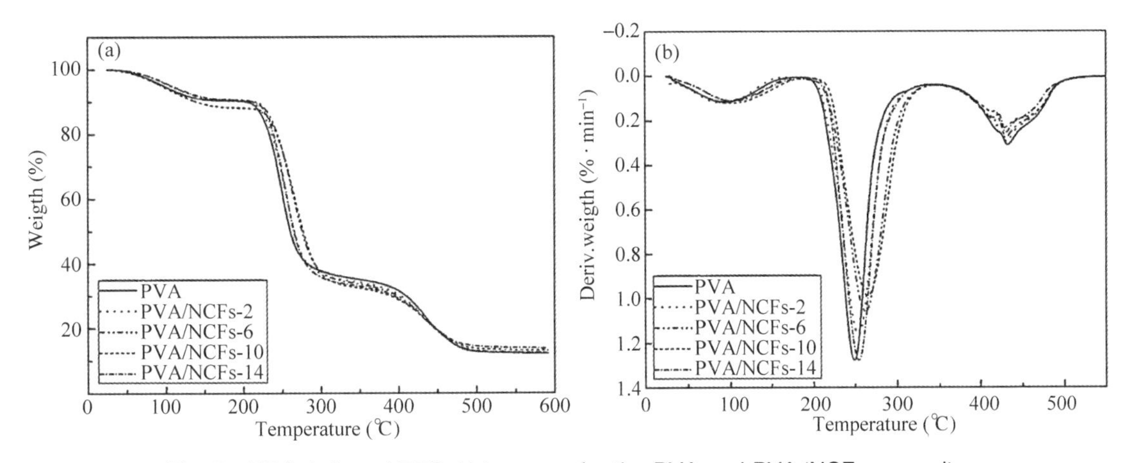

Fig. 5 TGA (a) and DTG (b) curves for the PVA and PVA/NCF composite

Table 2 Onset temperature and T_{max} (degradation temperature) of pure PVA and PVA/NCF composites with NCF weight contents of 2%, 6%, 10%, and 14%

Sample	First process	Second process		Third process	
	T_{max}(℃)	Onset temperature (℃)	T_{max}(℃)	Onset temperature (℃)	T_{max}(℃)
PVA	93.4	181.7	248.7	342.7	432.7
PVA \ NCFs-2	93.8	182.2	253.6	343.9	431.7
PVA \ NCFs-6	102.5	184.6	262.5	346.2	430.6
PVA \ NCFs-10	98.3	189.9	265.6	346.7	428.1
PVA \ NCFs-14	98.6	184.7	253.9	339.5	430.6

3.6 Mechanical properties of PVA/NCF composites

Fig. 6 shows the results of tensile strength and Young's modulus tests on neat PVA and PVA/NCF composites with 2, 4, 6, 8, 10, and 14 *wt%* NCF. The mechanical properties increased with the increasing of NCF content from 2 to 6 *wt%*. When the NCF content was 6%, the tensile strength and Young's modulus of the PVA composites reached maximum values, which were 2.8 and 2.4 times larger than that of neat PVA, respectively. This indicated a strong reinforcement of the PVA matrix with addition of the NCFs. For NCF content greater than 6 *wt%*, the tensile strength and Young's mod-ulus exhibited a declining trend; this may be because of decreased NCF dispersion, making the PVA composites more fragile. Thus, the aggregation of NCFs in PVA caused local concentrations of stress, and tensile failure occurred more readily. Combined with the thermal and transmittance results, these mechanical results lead to the conclusion that a NCF content of 6 wt% is optimal for PVA composites.

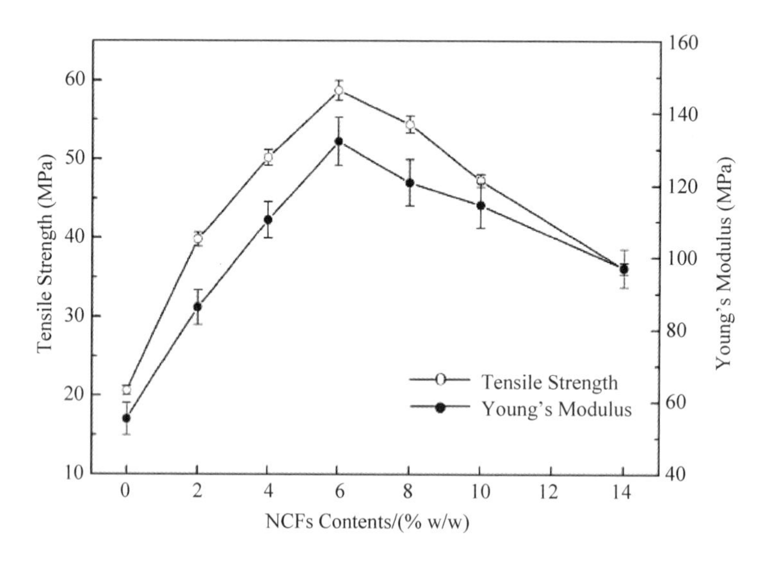

Fig. 6 Mechanical property curves for PVA/NCF composites

4 Conclusions

We have demonstrated a facile preparation of long, filamentous NCFs with 50–120 nm diameters from CTMP by using high intensity ultrasonication. The NCFs were mainly cellulose because the cellulose Ⅰ crystal structure remained following the removal of lignin and parts of the hemicelluloses during the chemical process. The resulting NCFs exhibited higher crystallinity (72.9%) compared with that of CTMP (61.5%). The main thermal degradation temperature of the NCFs (355.4 ℃) was higher than that of CTMP (308.2 ℃) owing to removal of the lignin and parts of the hemicelluloses. Because of the excellent thermal stability of the NCFs, the maximum temperature of the second degradation process for PVA/NCF was increased by 13.8 ℃ for a NCF content of 6 *wt%*. Thus, the thermal stability of PVA/NCF increased significantly. The tensile strength and Young's modulus of the PVA/NCF composites were also significantly improved, with values 2.8 and 2.4 times larger than that of neat PVA, respectively. Finally, the composites also retained excellent transparency.

Acknowledgments

This project was supported financially by the Research Fund for the Fundamental Research Funds for the Central Universities (DL13BBX14, 2572014EB01), the National Natural Science Foundation of China (No. 31170545) and National Key Technology R&D Program (2012BAD32B04).

References

[1] Abdul K H, Bhat A, Ireana Y A Green composites from sustainable cellulose nanofibrils: A review[J]. Carbohyd Polym, 2012, 87(2): 963-979.

[2] Alemdar A, Sain M. Isolation and characterization of nanofibers from agricultural residues-Wheat straw and soy hulls [J]. Bioresour Technol, 2008, 99(6): 1664-1671.

[3] Aliyu M, Hepher M. Effects of ultrasound energy on degradation of cellulose material[J]. Ultrason Sonochem, 2000, 7(4): 265-268.

[4] Bhatnagar A, Sain M. Processing of Cellulose Nanofiber-reinforced Composites[J]. J. Reinf Plast Comp., 2005, 24 (12): 1259-1268.

[5] Bondeson D, Mathew A, Oksman K. Optimization of the isolation of nanocrystals from microcrystalline celluloseby acid hydrolysis[J]. Cellulose, 2006, 13(2): 171.

[6] Bruce D, Hobson R, Farrent J. High-performance composites from low-cost plant primary cell walls[J]. Compos. Part A. Appl. Sci. Manuf., 2005, 36(11): 1486-1493.

[7] Caruso M, Davis D, Shen Q L. Mechanically-Induced Chemical Changes in Polymeric Materials[J]. Chem Rev, 2009, 109(11): 5755-5798.

[8] Chakraborty A, Sain M, Kortschot M Cellulose microfibrils: A novel method of preparation using high shear refining and cryocrushing[J]. Holzforschung, 2005, 59(1): 102-107.

[9] Cheng Q Z, Wang S Q, Han Q Y. Novel process for isolating fibrils from cellulose fibers by high-intensity ultrasonication. II. Fibril characterization[J]. J. Appl. Polym Sci., 2010, 115(5): 2756-2762.

[10] Cheng Q Z, Wang S Q, Rials T. Poly(vinyl alcohol) nanocomposites reinforced with cellulose fibrils isolated by high intensity ultrasonication[J]. Compos. Part A. Appl. Sci. Manuf, 2009, 40(2): 218-224.

[11] Chen W S, Yu H P, Liu Y X. Preparation of millimeter-long cellulose I nanofibers with diameters of 30-80nm from bamboo fibers[J]. Carbohydr Polym, 2011, 86(2): 453-461.

[12] Chen W S, Yu H P, Liu Y X. Individualization of cellulose nanofibers from wood using high-intensity ultrasonication combined with chemical pretreatments[J]. Carbohydr Polym, 2011, 83(4): 1804-1811.

[13] Cintas P, Luche, J. L. Green chemistry: The sonochemical approach[J]. Green Chem, 1999, 1: 115-125.

[14] Cote G L, Willet J L. Thermomechanical depolymerization of dextran[J]. Carbohyd Polym, 1999, 39: 119-126.

[15] Filson P, Dawsonandoh B. Sono-chemical preparation of cellulose nanocrystals from lignocellulose derived materials [J]. Bioresour Technol, 2009, 100(7): 2259-2264.

[16] Glasser W, Taib R, Jain R. Fiber-reinforced cellulosic thermoplastic composites[J]. J. Appl. Polym Sci., 1999, 73(7): 1329-1340.

[17] Grönroos A, Pirkonen P, Ruppert O. Ultrasonic depolymerization of aqueous carboxymethylcellulose[J]. Ultrason Sonochem, 2004, 11(1): 9-12.

[18] Habibi Y, Lucia L, Rojas O. Cellulose Nanocrystals: Chemistry, Self-Assembly, and Applications[J]. Chem Rev., 2010, 110(6): 3479-3500.

[19] Imsgard F, Falkehag S, Kringstad K. On possible chromophoric structures in spruce wood[J]. Tappi Tech Ass Pulp Pap Indus, 1971, 54(10): 1680-1684.

[20] Iwamoto S, Nakagaito A, Yano H. Optically transparent composites reinforced with plant fiber-based nanofibers[J]. Appl. Phys. A., 2005, 81(6): 1109-1112.

[21] Iwamoto S, Nakagaito A, Yano H. Nano-fibrillation of pulp fibers for the processing of transparent nanocomposites [J]. Appl. Phys. A., 2007, 89(2): 461-466.

[22] Kasaai M, Arul J, Charlet G. Fragmentation of chitosan by ultrasonic irradiation[J]. Ultrason Sonochem, 2008, 15 (6): 1001-1008.

[23] Konn J, Holmbom B, Nickull O. Chemical Reactions in Chemimechanical Pulping: Material Balances of Wood Components in a CTMP Process[J]. J. Pulp. Pap. Sci., 2002, 28: 395-399.

[24] Law K, Daud W. CMP and CTMP of a fast-growing tropical wood: Acacia mangium[J]. Tappi J, 2000, 83(7).

[25] Lebo J S, Lonsky W, Mcdonough T. The occurrence and light induced formation of ortho-quinonoid lignin structures in white spruce refiner mechanical pulp[J]. J. Pulp. Pap. Sci., 1988, 16(5): J139-J143.

[26] Leitner J, Hinterstoisser B, Wastyn M. Sugar beet cellulose nanofibril-reinforced composites[J]. Cellulose, 2007, 14(5): 419-425.

[27] Liu H, Du Y, Kennedy J. Hydration energy of the 1, 4-bonds of chitosan and their breakdown by ultrasonic treatment [J]. Carbohydr Polym, 2007, 68(3): 598-600.

[28] Li W, Yue J Q, Liu S X. Preparation of nanocrystalline cellulose via ultrasound and its reinforcement capability for poly (vinyl alcohol) composites[J]. Ultrason Sonochem, 2012, 19(3): 479-485.

[29] Li W Zhao X, Liu S X. Preparation of entangled nanocellulose fibers from APMP and its magnetic functional property as matrix[J]. Carbohydr Polym, 2013, 94(1): 278-285.

[30] Nakagaito A, Yano H. The effect of morphological changes from pulp fiber towards nano-scale fibrillated cellulose on the mechanical properties of high-strength plant fiber based composites[J]. Appl. Phys. A., 2004, 78(4): 547-552.

[31] Nishiyama Y, Sugiyama J, Chanzy H. Crystal structure and hydrogen bonding system in cellulose Ia from synchrotron X-ray and neutron fiber diffraction[J]. J. Am. Chem. Soc., 2003, 125(47): 14300-14306.

[32] Pandey J, Kumar A, Misra M. Recent advances in biodegradable nanocomposites[J]. J. Nanosci Nanotechno, 2005, 5(4): 497-526.

[33] Paralikar S, Simonsen J, Lombardi J. Poly (vinyl alcohol)/cellulose nanocrystal barrier membranes[J]. J. Membr Sci., 2008, 320(1-2): 248-258.

[34] Phong N, Gabr M, Okubo K. Enhancement of mechanical properties of carbon fabric/epoxy composites using micro/nano-sized bamboo fibrils[J]. Mater. Des, 2013, 47: 624-632.

[35] Portenlänger G, Heusinger H. The influence of frequency on the mechanical and radical effects for the ultrasonic degradation of dextranes[J]. Ultrason Sonochem, 1997, 4(2): 127-130.

[36] Sain M, Panthapulakkal S. Bioprocess preparation of wheat straw fibers and their characterization[J]. Ind. Crop. Prod., 2006, 23(1): 1-8.

[37] Sakurada I, Nukushina Y, Ito Taisuke. Experimental determination of the elastic modulus of crystalline regions in oriented polymers[J]. J. Polym. Sci., 1962, 57(165): 651-660.

[38] Sedlarik V, Saha N, Kuritka I. Characterization of polymeric biocomposite based on poly (vinyl alcohol) and poly (vinyl pyrrolidone)[J]. Polym Compos, 2006, 27(2): 147-152.

[39] Shih Y, Huang C. Polylactic acid (PLA)/banana fiber (BF) biodegradable green composites[J]. J. Polym Res., 2011, 18(6): 2335-2340.

[40] Stenstad P, Andresen M, Tanem B. Chemical surface modifications of microfibrillated cellulose[J]. Cellulose, 2008, 15(1): 35-45.

[41] Sun J, Sun X f, Zhao H. Isolation and characterization of cellulose from sugarcane bagasse[J]. Polym Degrad Stab., 2004, 84(2): 331-339.

[42] Sun R C, Lawther J, Banks W. Influence of alkaline pre-treatments on the cell wall components of wheat straw[J]. Ind. Crop Prod., 1995, 4(2): 127-145.

[43] Sun R C, Tomkinson J, Wang Y X. Physico-chemical and structural characterization of hemicelluloses from wheat straw by alkaline peroxide extraction[J]. Polymer, 2000, 41(7): 2647-2656.

[44] Suslick K, Choe S, Cichowlas A. Sonochemical synthesis of amorphous iron[J]. Nature, 1991, 353(6343): 414-416.

[45] Svagan A, Samir M, Berglund L. Biomimetic foams of high mechanical performance based on nanostructured cell walls reinforced by native cellulose nanofibrils[J]. Adv. Mater, 2008, 20(7): 1263-1269.

[46] Morais T E, Corrêa A, Manzoli A. Cellulose nanofibers from white and naturally colored cotton fibers[J]. Cellulose, 2010, 17(3): 595-606.

[47] Tischer P, Sierakowski M, Westfahl J H. Nanostructural reorganization of bacterial cellulose by ultrasonic treatment [J]. Biomacromolecules, 2010, 11(5): 1217-1224.

[48] Vodeniăarová M, Drímalová G, Hromádková Z. Xyloglucan degradation using different radiation sources: A comparative study[J]. Ultrason Sonochem, 2006, 13(2): 157-164.

[49] Wang B, Sain M. Dispersion of soybean stock-based nanofiber in a plastic matrix[J]. Polym Int., 2007, 56(4): 538-546.

[50] Wang S Q, Cheng Q Z. A novel process to isolate fibrils from cellulose fibers by high-intensity ultrasonication, Part 1: Process optimization[J]. J Appl Polym Sci., 2009, 113(2): 1270-1275.

[51] Wegner T, Jones P. Advancing cellulose-based nanotechnology[J]. Cellulose, 2006, 13(2): 115-118.

[52] Zhang W, Zhang Y, Lu Can H. Aerogels from crosslinked cellulose nano/micro-fibrils and their fast shape recovery property in water[J]. J. Mater Chem, 2012, 22(23): 11642-11650.

[53] Zhao H P, Feng X Q, Gao H J. Ultrasonic technique for extracting nanofibers from nature materials[J]. Appl. Phys. Lett., 2007, 90(7): 73112.

中文题目：纳米纤维素纤丝增强聚乙烯醇复合材料的热和力学性能研究

作者：李伟，吴琼，赵鑫，黄占华，曹军，李坚，刘守新

摘要：以化学热磨机械浆为原料，采用超声法制备了长丝状纳米纤维素纤丝。研究了纳米纤丝在聚乙烯醇复合薄膜中的热稳定性和机械性能。在制浆过程中，化学热磨机械浆独特的化学前处理和机械作用使纤维松弛和解离，从而使化学处理能够更深入纤维中进而有助于后续的纯化和超声处理。与化学热磨机械浆61.5%的结晶度相比，所制备的纳米纤丝的结晶度为72.9%，直径范围为50~120 nm。当纳米纤丝含量为6%时，聚乙烯醇/纳米纤丝复合薄膜表现出最佳的热稳定性、透光率和力学性能。复合材料的光透过率为73.7%，其拉伸强度和杨氏模量均有显著提高，分别是纯聚乙烯醇的2.8和2.4倍。

关键词：纳米纤维素纤丝；化学热磨机械浆；超声；聚乙烯醇

Antioxidant Properties of Pyroligneous Acid Obtained by Thermochemical Conversion of *Schisandra chinensis* Baill*

Chunhui Ma, Wei Li, Yuangang Zu, Lei Yang, Jian Li

Abstract: Sustainable development of renewable resources is a major challenge globally. Biomass is an important renewable energy source and an alternative to fossil fuels. Pyrolysis of biomass is a promising method for simultaneous production of biochar, bio-oil, pyroligneous acid (PA), and gaseous fuels. The purpose of this study was to investigate the pyrolysis process and products yields of *Schisandra chinensis* fruits with different pyrolysis powers. The obtained PA was extracted with organic solvents, including ethyl formate, dichloromethane, methanol and tetrahydrofuran. The antioxidant activities, including the free radical scavenging activity and ferric reducing power, of the PA extracts were investigated. The synthetic antioxidants butylated hydroxyanisole and butylated hydroxytoluene were used as positive controls. A dichloromethane extract of PA showed excellent antioxidant properties compared to the other extracts. The chemical compositions of the PA extracts were determined by GC-MS, and further proved that the dichloromethane extract had the best antioxidant characteristics among the extracts tested.

Keywords: *Schisandra chinensis* Baill.; Pyrolysis; Pyroligneous acid; Antioxidant activity; GC-MS

1 Introduction

Biomass is an important renewable resource, and its use for energy production provides economic security that is less environmentally damaging than that of other energy sources. Biomass can be converted by thermochemical processes, including pyrolysis, combustion, gasification, and liquefaction. Pyrolysis is an efficient method for utilization of biomass, especially for agricultural countries with large quantities of available biomass by-products[1,2]. The products of biomass pyrolysis include char, gas, tar and pyroligneous acid (PA).

Also known as wood vinegar or pyroligneous liquor, PA is a complex mixture of water, alcohols, organic acids, phenolics, aldehydes, ketones, esters, furan and pyran derivatives, hydrocarbons, and nitrogen compounds[3,4]. The composition and yield of PA depends on the species the biomass was derived from, and pyrolysis conditions. PA is a reddish-brown, acidic, water-soluble wood distillate, which is used in pesticides, refined food additives[5], and smoke flavoring[6]. PA is useful for soil improvement, and especially for the control of fungal and termite infestations[7,8]. PA also exhibits excellent antioxidant activity. PA from bamboo has superoxide anion scavenging activity and antioxidant activity[9]. Antioxidant and free radical scavenging activities have been reported for PAs from *Rhizophora apiculata*[10], walnut shells[11], and hickory

* 本文摘自 Molecules, 2014, 19: 20821-20838.

shells[12]. PA has potential as a natural antioxidant because it is rich in phenolic compounds and can be used as a food antioxidant[11].

Antioxidants have become important because of their role in health and their effects on cardiovascular disease, atherosclerosis, cancer, and aging. Many antioxidant compounds from plant sources have been identified as free radical scavengers[13,14]. In recent years, the search for naturally occurring antioxidants for use in food and medicine has intensified in order to replace synthetic antioxidants[15,16], which are being restricted because of their side effects (e. g., carcinogenicity)[17]. A number of synthetic antioxidants, such as 2-*tert*-butyl-4-methoxyphenol and 3-*tert*-butyl-4-methoxyphenol (butylated hydroxyanisole, BHA), and *tert*-butylhydroquinone are added to foodstuffs, but their use has come under scrutiny because of toxicity issues[18]. Therefore, attention has been directed towards the discovery of natural antioxidants from plant sources. Crude extracts of plant materials rich in polyphenols are increasingly of interest to the food industry because of their capacity to retard oxidative degradation of lipids and thereby improve the quality and nutritional value of food[13].

The composition and yield of PA depends on the pyrolysis process conditions and the biomass raw materials. Extensive studies of the chemical compositions and applications of PAs from oak, sakura, green tea, bamboo, eucalyptus, mangrove, rosemary and waste biomass have been conducted[7,10,19]. However, the themochemical conversion of *Schisandra chinensis* Baill. (*S. chinensis*) and chemical composition and antioxidant activity of *S. chinensis* PA have not been studied in detail.

Dried fruit of *S. chinensis* are important in herbal medicines and as a food additive in China[20]. They are used extensively in Korea and Japan as a tonic, sedative and astringent agent to treat various diseases[21-23]. Modern pharmacological research has shown that *S. chinensis* has antioxidant, antitumor[24], anti-hepatotoxic, detoxificant, anticarcinogenic[25] and anti-inflammatory[26] activity, and can act on the central nervous system. In addition, it can reduce fatigue and increase endurance, which can contribute to improved physical performance in sports[27]. In China, it is also used as a flavor agent and food additive when stewing fish and meat and making soup, tea, yogurt and porridge[28-31]. The chemical composition of *S. chinensis* fruits includes essential oil terpenoids[32], polysaccharides[33], anthocyanins[34], organic acids, vitamins, tannins[35], and biphenyl cyclooctene lignans[36] was shown in Table 1.

Table 1 Chemical composition of *S. chinensis* fruits

Chemical Composition	Content (%)	Reference
Essential oil	1.2-3.0	[32]
Polysaccharides	1.2-2.2	[33]
Anthocyanins	2.0-3.2	[34]
Terpenoids	<1.5	[35]
Organic acids (citric, malic, fumaric and tartaric acid)	<1.0	[35]
Vitamins C and E	<0.5	[35]
Tannins	<1.5	[35]
Biphenyl cyclooctene lignans and derivatives	7.2-19.2	[36]

This study aimed to investigate the pyrolysis of *S. chinensis*, which can be used to obtained the natural antioxidant *S. chinensis* PA and to reduce environmental pollution from fossil fuels. Another objective of this study was to evaluate the phenolic content, radical scavenging activity, and reducing power of *S. chinensis* PA extracts in ethyl formate, dichloromethane, methanol, and tetrahydrofuran. The extracts were obtained after pyrolysis under different heating conditions. The results could be used for development of antioxidants and preservatives from *S. chinensis* PA.

2 Results and Discussion

2.1 Influence of Moisture Content and Molding Temperature on the Briquetting Effect

Dried residues with different moisture contents (3.5%, 4.0%, 5.0%, 5.5%, 6.5%, 8.0%, 9.5%, 10.5%, 11.5%, 13.0%, 14.0%, 15.5%, 16.0%, 17.5%) were put into the hopper, respectively, and then the electric heating tube was opened, the molding temperature was kept at 200 ℃, and the extrusion machine was started, driving the spiral propellers to push the dried residues into the molding sleeve. The briquetting effect of molding rods is shown in Table 2. When the residual moisture was less than 5% or more than 15%, it cannot be briquetted. While the residual moisture was 5%-8% or 13%-15%, the molding effect was poor, although it can be briquetted. However, when the residual moisture was 8%-13%, the briquetted effect of molding rods was good, so the ideal residual moisture for the briquette production process was 8%-13%.

Table 2 Effect of moisture content on raw residue briquetting

Moisture (%)	<5%	5%-8%	8%-13%	13%-15%	>15%
Can be Briquetted	No	Almost not	Yes	Almost not	No

Dried residue with 10.0% moisture content was put into the hopper, and then the electric heating tube was opened, and the molding temperature were kept at different temperatures (50 ℃, 100 ℃, 150 ℃, 200 ℃, 250 ℃, 300 ℃, 350 ℃ and 400 ℃), respectively, and the extrusion machine was started, driving the spiral propellers to push the dried residues into the molding sleeve. The briquetting effect of molding rods is shown in Table 3. When the molding temperature was less than 150 ℃ or more than 400 ℃, it cannot be briquetted. When the molding temperature was 250-400 ℃ the briquetting effect is poor, and there were some cracks on the molding rods' surface. When the molding temperature was 200 ℃, the briquetting effect of the molding rods was good, there were no cracks and they were not carbonized on surface, and the molding rods output speed was the most fast of all.

Table 3 Effect of temperature on raw residue briquetting

Briquetting Temperature (℃)	50	100	150	200	250	300	350	400
Can be briquetted	No	No	Yes	Yes	Yes	Yes	Almost not	No
Cracks on surface	—	—	Yes	No	Yes	Yes	A lot	—
Carbonized on surface	—	—	No	No	A little	Partial	Totally	Seriously
Briquetting rate ($kg \cdot min^{-1}$)	—	—	2.1	2.0	1.8	1.8	1.7	—

2.2 Pyrolysis Curves and Yields of Pyrolysis Products

The pyrolysis curves shows the pyrolysis temperature against pyrolysis time. The yields of the pyrolysis products, including bio-gas, bio-PA, bio-oil, and bio-char, are shown in Fig. 1. The pyrolysis curves (Fig. 1a) showed that with a higher heating power, the temperature increased more rapidly and the pyrolysis rate was higher.

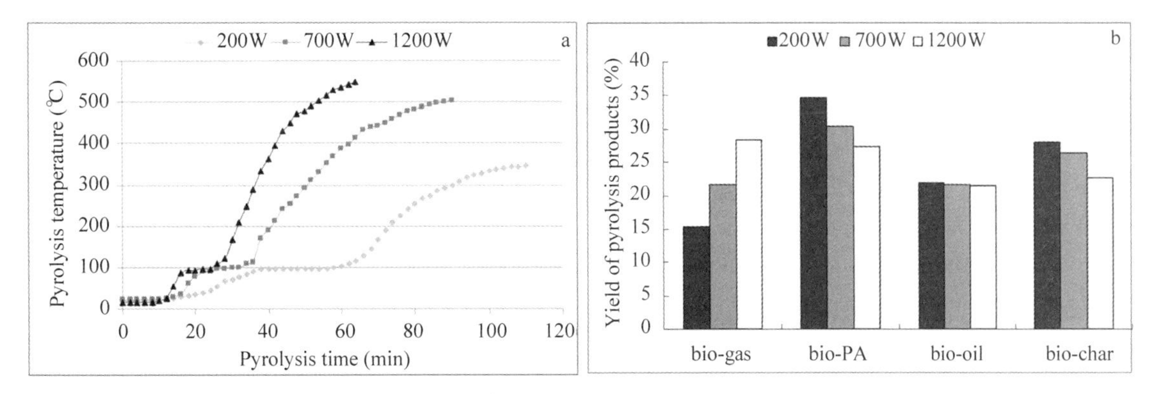

Fig. 1 Pyrolysis curves (a) and the yields of pyrolysis products (b) generated with different heating powers

The maximum final temperature of the pyrolysis reaction was close to 600 ℃ (1200 W). Depending on the thermal environment and the final temperature, pyrolysis will yield mainly biochar at low temperatures (<450 ℃) when the heating rate is quite slow, and mainly gases at high temperatures (>800 ℃) with rapid heating rates. At an intermediate temperature and under relatively high heating rates, the main product is bio-oil. From Fig. 1b, the yields of bio-oil pyrolysis with different powers were similar, and the yields of bio-oil were 21.5% (200 W), 21.4% (700 W), and 21.4% (1200 W), respectively, of the total pyrolysis products. However, when the pyrolysis power increased, the bio-gas yield increased and the bio-PA and bio-char yields decreased, the PA yield with 200 W pyrolysis was 34.5% of the total pyrolysis products, more than that with pyrolysis at higher power 30.4% (700 W), and 27.5% (1200 W), respectively. The char yield with 200 W pyrolysis was 28.4% of the total pyrolysis products, more than that with pyrolysis at higher power 26.1% (700 W), and 22.5% (1200 W), respectively. Thus, to obtain high yields of bio-PA and bio-char, a low power pyrolysis (200 W) should be used. The mass difference from input and output products was determined by the product mass of non-condensable gas. The bio-gas yields were 15.6% (200 W), 22.1% (700 W), and 28.6% (1200 W) of the total pyrolysis products, respectively. Therefore, the higher the pyrolysis power, the faster the pyrolysis rate was, and the larger the amount of non-condensable gas produced was.

2.3 Separation of PA from Bio-Oil

The condensate collected from the condenser was a mixture of PA and bio-oil, and separation by cooling was a simple way to separate the bio-PA (the upper layer) from the bio-oil (the lower layer). To investigate the effect of the separation temperature on separation time, five 200 mL aliquots of condensate were accurately measured and then poured into a separatory funnel, cooled at different temperatures (-5-20 ℃). The volume of PA and bio-oil were recorded every hour, and the separation time was recorded at the time at which the volumes stopped changing. The separation curve for PA and bio-oil is shown in Fig. 2.

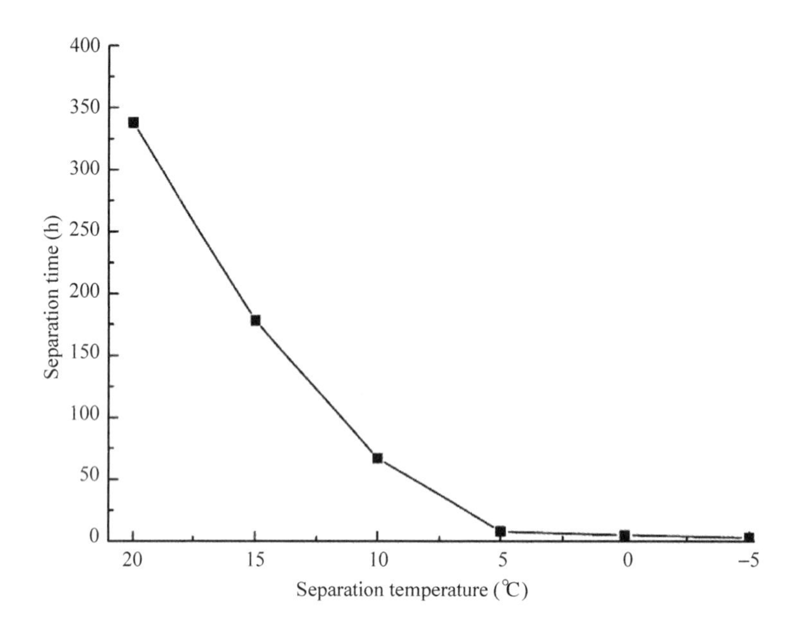

Fig. 2 Effect of temperature on time required to separate bio-PA and bio-oil

Lower cooling temperatures gave better separation of the two phases than higher temperatures and this reduced the separation time. Separation of the condensate at 20 ℃ required nearly 350 h for complete separation. The separation rate was higher below 5 ℃, and the condensate was completely separated within 12 h. If the condensate was separated at a temperature below 0 ℃, it would need refrigeration, which would increase energy consumption. Therefore, 5 ℃ was selected as the operation temperature to separate bio-PA from bio-oil. After stratification in the closed separatory funnel, the bio-oil (the lower layer) was removed and the antioxidant capacities of the bio-PA was determined.

2.4 Determination of Total Phenolics Content

The Folin-Ciocalteu assay is a fast and simple method to rapidly determine the content of phenolics in samples. Most of the work dealing with the content of phenolics in natural products uses gallic acid as a standard, and the content of phenolics in this work is expressed as gallic acid equivalents to facilitate comparison to earlier studies[37]. As shown in Table 4, the total phenolic content of PA-200 (pyrolysis at 200 W) was higher than that of PA-700 (pyrolysis at 700 W) and PA-1200 (pyrolysis at 1200 W), because the phenolic compounds were not destroyed (*i.e.*, pyrolyzed) under the low temperature conditions. Furthermore, some of the water-soluble phenolic compounds were extracted into water under the lower temperature conditions, because of slow evaporation of moisture from the residue. Consequently, the total phenolic content of PA-200 was the highest, and PA-200 was selected for the antioxidant activity tests. As shown in Table 4, in all the PA extracts, the total phenolic contents of DMEP and MEP were higher than those of EFEP and TFEP. This means the phenolic compounds in PA were more likely to be soluble in dichloromethane and methanol.

Table 4 Total phenolic content of PA extracts

Heating Power (W)	Symbols	Gallic Acid Equivalents ($mg \cdot g^{-1}$)			
		EFEP [a]	DMEP [b]	MEP [c]	TFEP [d]
200	PA-200	2.36±0.05	3.91±0.13	3.79±0.14	2.14±0.10
700	PA-700	2.35±0.12	3.79±0.11	3.47±0.15	1.99±0.06
1200	PA-1200	2.30±0.13	2.81±0.14	2.69±0.10	1.58±0.08

a EFEP: ethyl formate extract phase of pyroligneous acid; b DMEP: dichloromethane extract phase of pyroligneous acid; c MEP: methanol extract phase of pyroligneous acid; d TFEP: tetrahydrofuran extract phase of pyroligneous acid.

2.5 Ferric Reducing Power

The antioxidant potentials of the PA extracts were estimated from their ability to reduce TPTZ-Fe(Ⅲ) to TPTZ-Fe(Ⅱ)[38]. A higher absorbance indicated a higher ferric reducing power. The reducing antioxidant power of each sample is expressed as Trolox equivalents, and the reducing antioxidant power of solvent was deducted as a blank. Fig. 3 shows the reduction capacities of the various extracts from samples subjected to

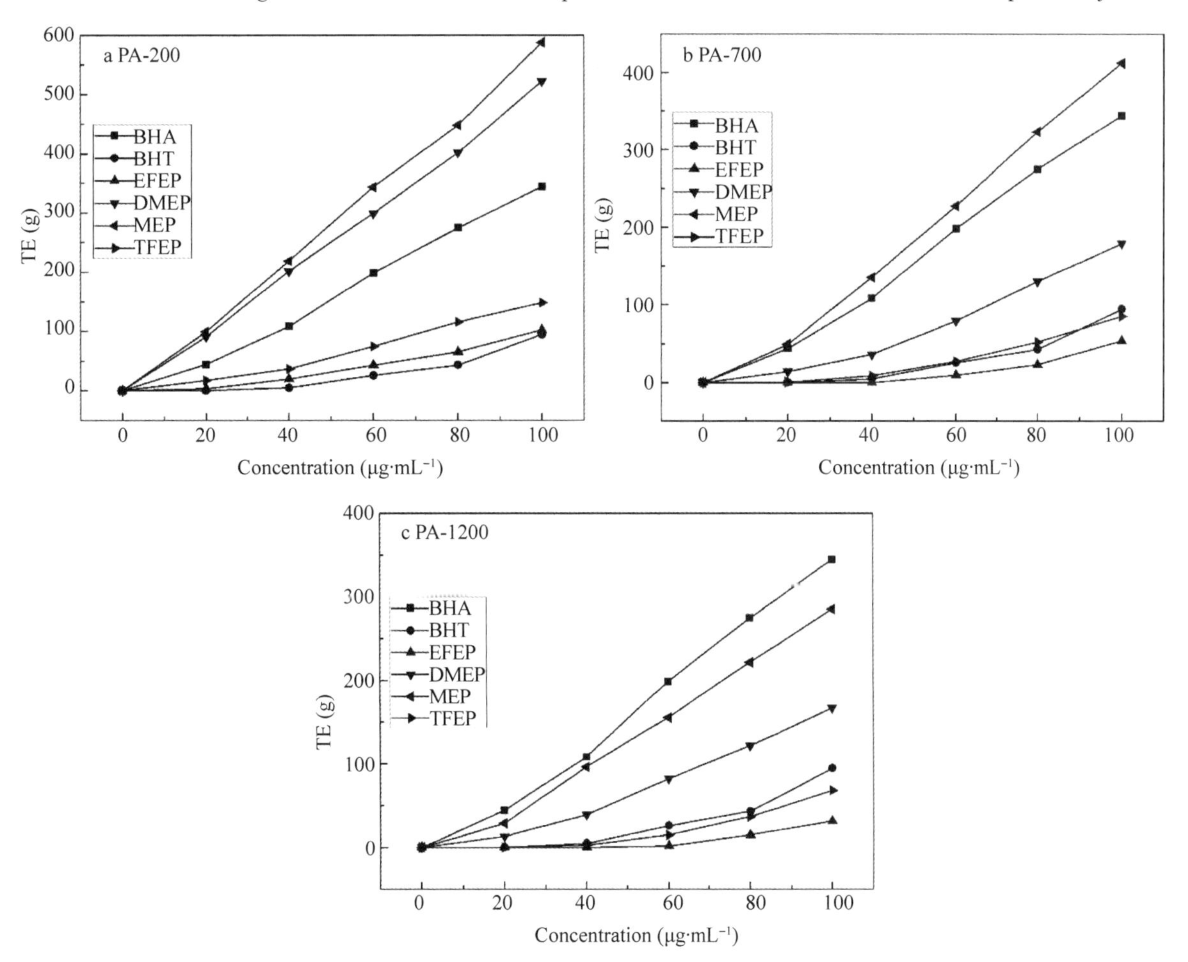

Fig. 3 Reducing antioxidant capacities of PA extracts after (a) pyrolysis at 200 W; (b) pyrolysis at 700 W; (c) pyrolysis at 1200 W

different heating powers during pyrolysis. Fig. 3a (PA−200) shows that the reduction capacities of MEP and DMEP at 100 μg · mL^{-1} were 587. 55±22. 50 and 522. 05±32. 20 (TE) · g^{-1}, respectively. These results were higher than those of EFEP (102. 35±4. 20 (TE) · g^{-1}) and TFEP (148. 05±8. 43 (TE) · g^{-1}) at the same concentration. The reduction capacities of BHA and BHT at 100 μg · mL^{-1} were 344. 05±11. 11 (TE) · g^{-1} and 94. 35±2. 34 (TE) · g^{-1}, respectively. Fig. 3b (PA−700) and Fig. 3c (PA−1200) show the same trends. The reducing power order of PA−200 was MEP>DMEP>BHA>TFEP>EFEP>BHT, while the order of PA−700 was MEP>BHA>DMEP>BHT>TFEP>EFEP, and the order of PA−1200 was BHA>MEP>DMEP>BHT>TFEP>EFEP.

2. 6 DPPH Free Radical Scavenging Activity

DPPH is a stable free radical, and can accept an electron or hydrogen free radical to reach a steady state[39]. Consequently, DPPH has been widely used to determine the free−radical scavenging abilities of various samples. When the scavenging rate of DPPH is 50%, the corresponding value of sample concentration is SC50. So, a lower SC50 value indicates the stronger antioxidant activity. As shown in Fig. 4, an obvious increase in SC% value occurred when the concentration of PA increased.

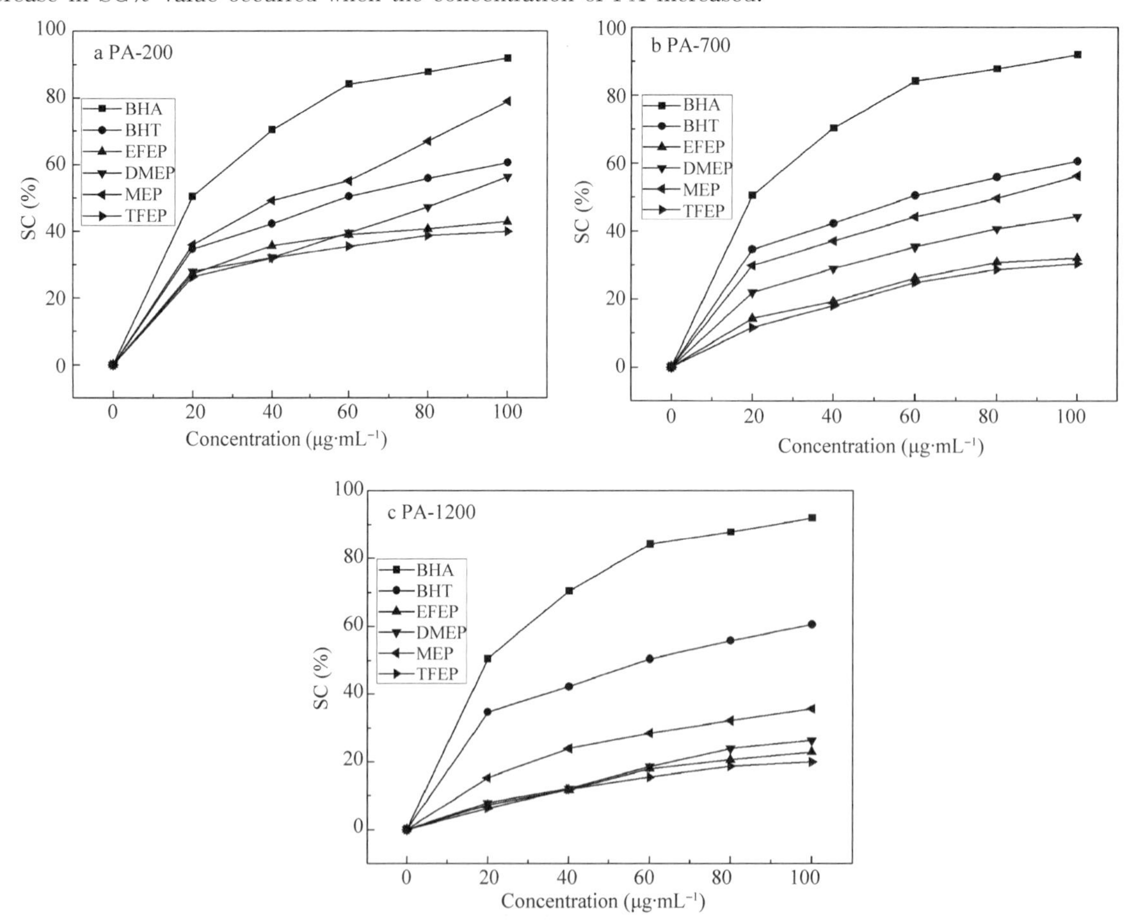

Fig. 4 DPPH scavenging activities of PA extracts after (a) pyrolysis at 200 W; (b) pyrolysis at 700 W; (c) pyrolysis at 1200 W

Thus, the DPPH scavenging activities decreased order was MEP > DMEP > EFEP > TFEP. The DPPH scavenging activity of PA-200 (Fig. 4a) was higher than that of PA-700 (Fig. 4b) and PA-1200 (Fig. 4c), as the lower heating power enhance the residence time of liquid pyrolysis products, increasing the content of water soluble bioactive small molecules[10]. In detail, the yields of pyrolysis products, such as the condenser and bio-char depend on the heating rate of pyrolysis. In other words, the lower heating power and the slower pyrolysis rate, the pyrolysis reaction is more completely, and the yield of the condenser liquid is higher. And vice the higher heating power and faster pyrolysis, enhance the solid pyrolysis products thus increasing bio-char formation[2]. When the pyrolysis power was 200 W, the SC50 values of MEP and DMEP were (41±1.2) and (92±2.3) $\mu g \cdot mL^{-1}$, respectively, and were lower than the SC50 values of EFEP and TFEP. The SC50 values of BHA and BHT were (19±0.23) and (60±0.30) $\mu g \cdot mL^{-1}$, respectively. The DPPH scavenging activities of PA-200 were in the order BHA>MEP>BHT>DMEP>EFEP>TFEP. While for PA-700 and PA-1200, the order was BHA>BHT>MEP>DMEP>EFEP>TFEP.

2.7 Chemical Composition of PA Extracts

The compositions of the *S. chinensis* PA extracts were analyzed by GC-MS. Results were accepted when a match >90% was obtained. The relative contents of the compounds were determined using the normalization method. Among the pyrolysis powers tested, 200 W pyrolysis gave the highest yield of PA, and PA-200 had the highest total phenolic content. Therefore, we used PA-200 for the antioxidant test, and extracted it with four different organic solvents. The GC-MS results from this study are shown in Table 5.

Table 5 Volatile compounds in PA-200 extracts

NO.	Retention time (min)	Compounds	CAS Number	Molecular Formula	RA(%)			
					EFEP [b]	DMEP [c]	MEP [d]	TFEP [e]
1	5.498		000583-58-4	C_7H_9N	—	—	—	9.07
2	6.106		000616-02-4	$C_5H_4O_3$	16.43	7.20	17.06	14.35
3	6.489		002758-18-1	C_6H_8O	4.89	6.87	—	2.82
4	7.078		001864-94-4	$C_7H_6O_2$	19.88	20.89	17.80	18.39
5	8.065		000765-70-8	$C_6H_8O_2$	2.50	—	—	—
6	8.686		001528-21-8	C_6H_{10}	1.70	3.15	—	1.79
7	9.261		000106-44-5	C_7H_8O	10.41	16.24	22.84	10.35
8	9.429		171741-07-4	$C_{11}H_{24}FO_2P$	9.57	12.51	9.50	8.85
9	9.755		000932-52-5	$C_4H_5N_3O_2$	9.98	12.61	9.29	9.07
10	9.929		000826-36-8	$C_9H_{17}NO$	2.41	3.74	1.87	2.29
11	9.996		006094-02-6	C_7H_{14}	0.71	—	2.38	—
12	10.243		002314-78-5	$C_6H_9NO_2$	1.21	—	2.59	—
13	10.294		000626-64-2	C_5H_5NO	1.66	—	2.83	—
14	10.510		000123-56-8	$C_4H_5NO_2$	2.03	7.71	—	2.87
15	10.698		001121-89-7	$C_5H_7NO_2$	3.60	—	6.04	3.73
16	10.869		000118-70-7	$C_4H_7N_5$	2.23	2.65	—	3.68
17	10.926		003437-29-4	$C_6H_9NO_2$	1.56	—	—	3.27

(continued)

NO.	Retention time (min)	Compounds	CAS Number	Molecular Formula	RA%			
					EFEP [b]	DMEP [c]	MEP [d]	TFEP [e]
18	11.261		000093-51-6	$C_5H_{10}O_2$	1.24	—	—	1.95
19	11.490		001122-43-6	C_7H_9NO	—	—	—	0.54
20	11.655		100009-81-8	$C_6H_8O_4$	5.01	3.51	—	2.09
21	13.578		000123-31-9	$C_6H_6O_2$	—	—	5.75	—
22	33.801		000119-47-1	$C_{23}H_{32}O_2$	2.98	2.92	2.05	4.89

a. RA: relative area of total peak area (removed the blank solvent); b. EFEP: ethyl formate extract phase of pyroligneous acid; c. DMEP: dichloromethane extract phase of pyroligneous acid; d. MEP: methanol extract phase of pyroligneous acid; e. TFEP: tetrahydrofuran extract phase of pyroligneous acid; f. Compound No. 22: Butylated hydroxytoluene (BHT) is added in the solvent to keep the stability of solvent, is not the composition of PA.

According to previous studies, the thermal degradation characteristics of wood materials or plants are strongly influenced by their chemical composition (cellulose, hemicellulose and lignin)[40-42]. Biomass undergoing pyrolytic reactions at high temperatures forms different compounds. Saccharides, including cellulose, hemicellulose and pectin, are thermally degraded into ketones, alcohols, and furan and pyran derivatives. However, lignin is converted into phenol, guaiacol, syringol, pyrocatechol, and their derivatives, which are dissolved in the bio-oil layer[39]. After cryogenic separation, smaller molecules partition into the water layer (*i.e.*, the PA layer) and larger molecules in the bio-oil layer. The gas chromatograms and the main constituents of MEP are shown in Fig. 5.

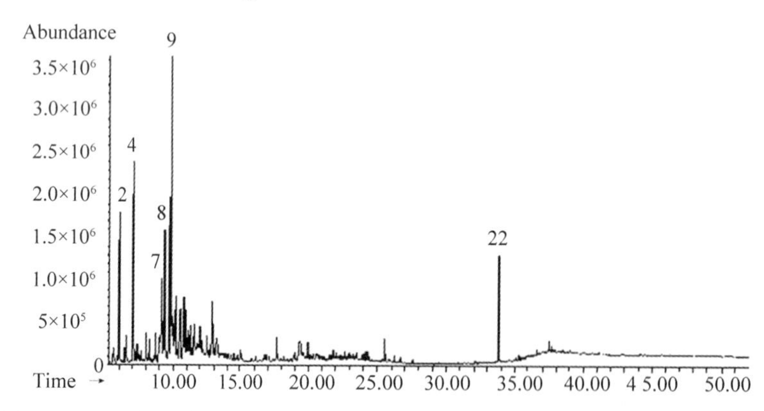

Fig. 5 Total ion chromatogram and the main constituents of PA-200 MEP

The compound numbers in Fig. 5 correspond to those in Table 5. The main constituents of MEP were *p*-cresol (22.84%), formic acid phenyl ester (17.80%), and furandione (17.06%). The main constituents of DMEP were *p*-cresol (16.24%), formic acid phenyl ester (20.89%), and furandione (7.20%). However, which in EFEP and TFEP was slightly lower, and the antioxidant activity of MEP and DMEP was higher than that of EFEP and TFEP. Therefore, in combination with the results of total phenolics content, ferric reducing power, and free radical scavenging activity, the phenolic hydroxyl and polyenes played an important role in improving the oxidative resistance of PA after pyrolysis. Twelve compounds from DMEP and MEP were identified, and for EFEP and TFEP, 19 and 17 compounds were identified, respectively. BHT

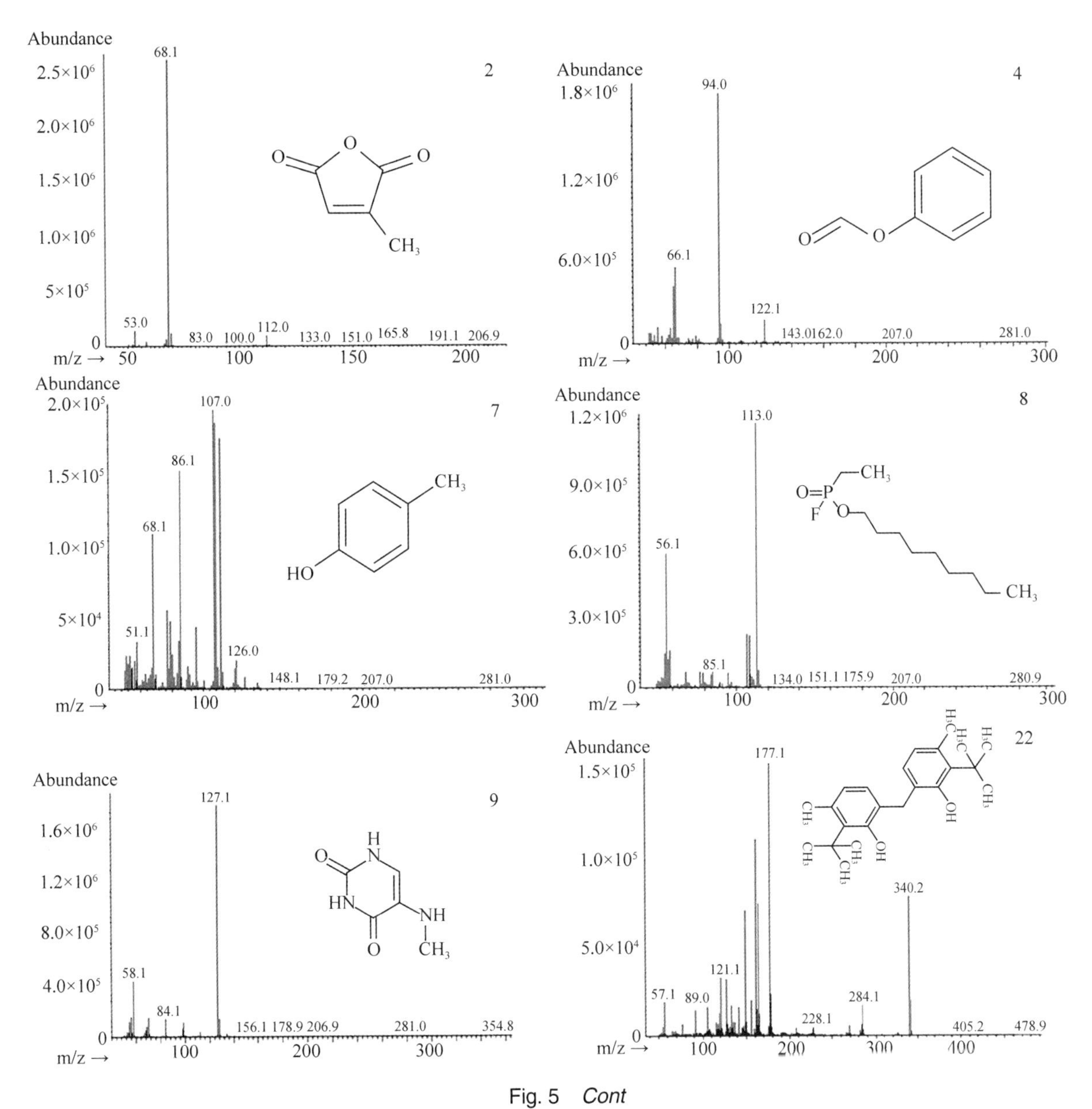

Fig. 5 *Cont*

(No. 22 compounds), which was added to the solvent to maintain its stability, was present in small amounts (<5.0%).

These results indicate that compounds containing the phenolic hydroxyl group are functional components responsible for the reduction of oxidants and scavenging of free radicals. This conclusion is supported by similar results obtained in other studies[15,39].

3 Experimental Section

3.1 Material

3.1.1 Chemical Reagents

Folin－Ciocalteu's reagent, 1, 1－diphenyl－1－picrylhydrazyl (DPPH, 95%), gallic acid, 2, 4, 6－tripyridyl－S－trizine (TPTZ), 6－hydroxy－2, 5, 7, 8－tetramethylchromane－2－carboxylic acid (Trolox), butylated hydroxyanisole and butylated hydroxytoluene (BHT) were purchased from Sigma－Aldrich (St. Louis, MO, USA). All other chemicals of analytical grade and were obtained from Sinopharm Chemical Reagent Co., Ltd (Beijing, China). Reverse osmosis Milli-Q water (Millipore, Billerica, MA, USA) was used for all solutions and dilutions.

3.1.2 Raw Materials

S. chinensis fruit were purchased from San Keshu Trading (Heilongjiang, China) and identified by Professor Shao－quan Nie from the Key Laboratory of Forest Plant Ecology, Northeast Forestry University (Harbin, China). *S. chinensis* fruit were crushed and refluxed twice with an ethanol-water (80: 20, *v/v*) solution at 90 ℃ for 2 h to obtain the active compounds, such as polysaccharides, anthocyanins, terpenoids, organic acids, vitamins, tannins, biphenyl cyclooctene lignans and derivatives, *etc*. And then the extracts were concentrated under the vacuum conditions, and the residues were stored under －4 ℃ until used for thermochemicalconversion.

3.2 Methods

3.2.1 Molding Method

The dried residues prepared in Section 3.1.2 were dried and then compact molding in briquetting equipment. In the process of compact molding, there is continuous feeding and continuous discharge mode with the improved briquetting equipment. The dried residual were first put into the hopper, and then the electric heating tube and the extrusion machine were started, driving the spiral propellers put the dried residual into the molding sleeve. Finally, the molding rods were export from the exit of molding sleeve. The schematic diagram of briquetting equipment was shown in Fig. 6.

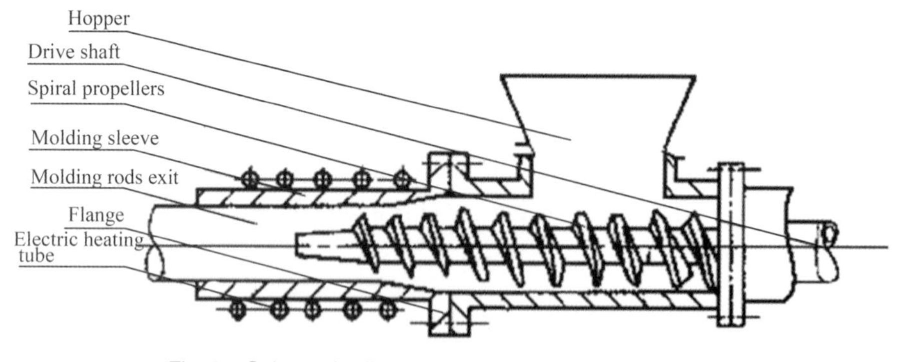

Fig. 6 Schematic diagram of the briquetting equipment

3.2.2 Pyrolysis Method

The molding rods prepared in Section 3.2.1 was heated for pyrolysis in a system consisting of a pyrolysis kettle and condensers (Fig. 7). The molding rods were placed in the pyrolysis kettle, and the liquefier and pyrolysis kettle were started. The pyrolysis rate was controlled by adjusting the heating power, and the condensate was collected and the residual tar was purified. Then, the electric heating tube was opened. The end point of the pyrolytic reaction was determined by weighing the transducer. At the end point, the final weight increase of the condensate was less than 2.0%, and the valve was turned off. Condensate collected from the condenser contained PA in the upper layer and bio-oil in the lower layer. The raw PA obtained had a clear reddish-brown color similar to black tea. The solid residue remaining in the pyrolysis kettle was biochar. According to the law of conservation of mass, the mass difference from input and output products was determined by the product mass of non-condensable gas.

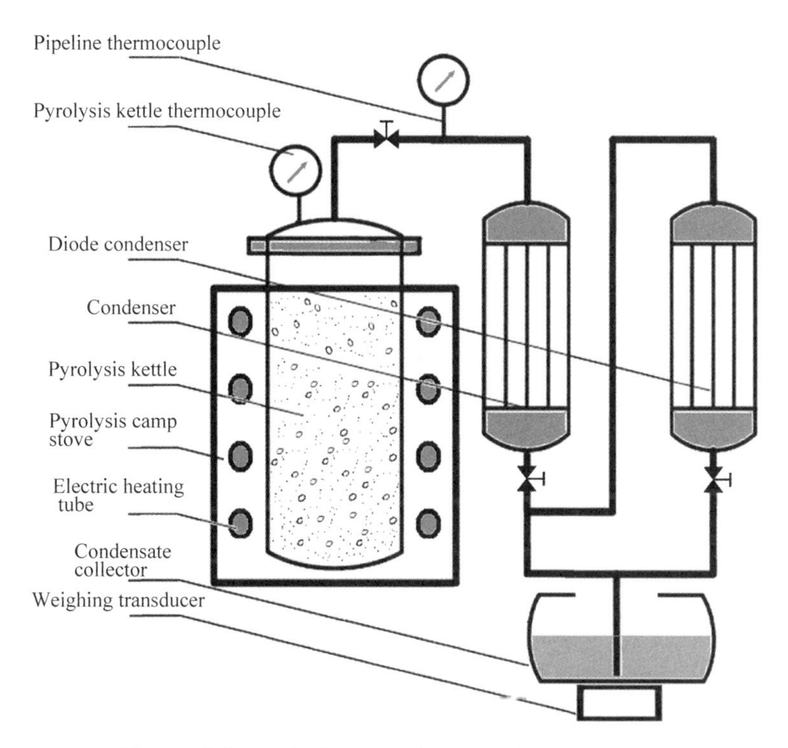

Fig. 7 Schematic diagram of the pyrolysis equipment

3.2.3 Preparation Method of PA Extracts

Three PA pyrolysis products were obtained, including PA-200 (heating power 200 W, highest pyrolysis temperature 310 ℃), PA-700 (heating power 700 W, highest pyrolysis temperature 440 ℃), and PA-1200 (heating power 1200 W, highest pyrolysis temperature 530 ℃). Each was obtained using 200 g of *S. chinensis* fruit. The ending of the pyrolysis process was that no more yellow gas was generated and no more liquid was collected. After pyrolysis, the PAs were dried over anhydrous sodium sulfate. They were then extracted at room temperature by mixing (100 rpm) for 0.5 h with each of the following solvents: ethyl formate, dichloromethane, methanol, and tetrahydrofuran. The sodium sulfate was removed by filtration, and the organic extracts were used for antioxidant tests. While these solvents were selected for this study, they are not recommended for use in food and medicine industry, due to the residues of toxicity solvent. Especially formate, dichloromethane, and tetrahydrofuran, bacause the large consumption of organic solvent was not conducive to environmental protection. The volume of each organic solvent used was half the volume of PA. The ethyl formate extract phase (EFEP), dichloromethane extract phase (DMEP), methanol extract phase (MEP) and tetrahydrofuran extract phase (TFEP) were thick brown liquids and were stored at 4–8 ℃ in the dark.

3.2.4 Folin Ciocalteu Assay

A colorimetric assay based on a published procedure with slight modifications[10,43] was used for estimating the total phenolic content of PA. Each sample (1 mL) was pipetted into a tube, and 50% Folin Ciocalteu's reagent (1 mL) and 10% sodium carbonate solution (1 mL) were added. The solutions were vortex mixed for 30 s, and then left to stand at room temperature for 2 h. Absorbance measurements were recorded at 765 nm, using gallic acid to construct calibration curves. The results are reported as mean values expressed as milligrams of gallic acid equivalents per gram of sample.

3.2.5 Ferric Reducing Antioxidant Power (FRAP)

The reducing antioxidant capacity of PA and its extracts were examined using a modified FRAP assay[4]. The FRAP reagent was prepared from 300 mmol · L^{-1} acetate buffer (pH 3.6), 20 mmol · L^{-1} ferric chloride, and 10 mmol · L^{-1} 2,4,6-tripyridyl-s-triazine diluted in 40 mmol/L hydrochloric acid. The three solutions were mixed in a ratio of 25 : 2.5 : 2.5 (*v/v/v*). The FRAP assay was performed using reagents preheated to 38 ℃. Before analysis, the initial absorbance of 3 mL of the reagent and 3 mL of acetate buffer used as a blank were measured at 593 nm. The samples (100 μL) were transferred into test tubes containing the reagent. The mixtures were shaken thoroughly and he absorbance values at 593 nm were recorded after 90 min. A higher absorbance indicated a higher ferric reducing power. The reducing antioxidant power of the sample is expressed as Trolox equivalent (TE · g^{-1}), which is the ratio between the slope of the sample's regression line and that of Trolox.

3.2.6 DPPH Free Radical Scavenging Activity

The DPPH radical scavenging activities of PA and its extracts were examined using the method of Lee *et al.*[44] and compared with those of the synthetic antioxidants BHA and BHT. Briefly, a 1.0 mL sample was mixed with 2.0 mL of a methanolic solution of DPPH (35 mg · L^{-1}) in a lightproof container. The mixture was shaken (100 rpm) and allowed to stand at room temperature for 30 min. Then, the absorbance was measured at 517 nm (UV-2550 spectrophotometer, Shimadzu, Kyoto, Japan) against methanol as a blank. Lower absorbance values for the reaction mixture indicated higher free radical scavenging activity. The percentage of DPPH discoloration of the samples was calculated according to the formula:

$$SC\% = (A_0 - A)/A_0 \times 100$$

where A_0 is the absorbance of the control reaction, which contained all reagents except the test compound; and A is the absorbance of the sample. The sample concentration providing 50% inhibition ($SC50$) was calculated from a graph of inhibition percentage against sample concentration.

3.2.7 GC-MS Analysis

GC-MS analysis was carried out on an Agilent 7890A GC system (Agilent Technologies, Palo Alto, CA, USA) fitted with a DB-17MS capillary column (30 mm×0.25 mm, film thickness 0.25 μm) and equipped with an Agilent 7693 auto sampler, and 5975C inert XL EI/CI mass selective detector with triple-axis detector. A sample volume of 2 μL was injected manually in splitless mode. The carrier gas was helium (He) at a flow rate of 1.0 mL · min^{-1}. The oven temperature was set at 60 ℃ for the initial 5 min, then increased to 120 ℃ at a rate of 10 ℃ · min^{-1} and held for 5 min, increased to 200 ℃ at a rate of 10 ℃ · min^{-1} and held for 5 min, and then increased to 280 ℃ at a rate of 10 ℃ · min^{-1} and held for 15 min. The injector and detector temperatures were 280 ℃ and 230 ℃, respectively. The pressure and flow rate of the injector were 15.0 psi and 25.0 mL · min^{-1}, respectively. The MS was operated at 70 eV, and the scan range (TIC) was 50-500 m · z^{-1}.

4 Conclusions

The pyrolysis process and products of *S. chinensis* fruit were studied. The yield of PA was higher with a lower pyrolysis heating power than with a higher pyrolysis heating power. Based on the total phenolic contents of the PA extracts, PA-200 was selected to test for antioxidant activities, including free radical scavenging activity and ferric reducing power. MEP and DMEP of PA-200 showed superior characteristics compared to the other extracts. Phenols and polyenes were the major components in MEP and DMEP, followed by ketones and furan derivatives. The results indicated that phenols were the functional components from *S. chinensis* fruit responsible for the reduction of oxidants and scavenging of free radicals in PA. The organic solvents in the PA extracts were vacuum evaporated, and then dissolved in ethanol, used in food, medicine, cosmetics and pesticides industry, respectively. Moreover, the further analysis of PA extracts, such as toxicity tests are necessary.

Acknowledgments

The authors thank the Fundamental Research Funds for the Central Universities (Grant No. 2572014EY01) and Hei Long Jiang postdoctoral foundation for financial support.

Author Contributions

LY and YZ conceived and designed the experiments; CM and WL Performed the experiments and analyzed the data; YZ and JL contributed reagents, materials, and analysis tools; CM wrote the paper. All authors read and approved the final manuscript.

Conflicts of Interest

The authors declare no conflict of interest.

References

[1] Minkova V, Marinov S P, Zanzi R, et al. Thermochemical treatment of biomass in a flow of steam or in a mixture of steam and carbon dioxide[J]. Fuel Process Technol, 2000, 62: 45-52.

[2] Schroder E. Experiments on the pyrolysis of large beechwood particles in fixed beds[J]. J. Anal. Appl. Pyrolysis, 2004, 71: 669-694.

[3]Guerrero A, Ruiz M P, Alzueta M U, et al. Pyrolysis of eucalyptus at different heating rates: Studies of char characterization and oxidative reactivity[J]. J. Anal. Appl. Pyrolysis, 2005, 74: 307-314.

[4]Guillén M D, Manzanos M J. Study of the volatile composition of an aqueous oak smoke preparation[J]. Food Chem., 2002, 79: 283-292.

[5]Bubonja-Sonje M, Giacometti J, Abram M. Antioxidant and antilisterial activity of olive oil, cocoa and rosemary extract polyphenols[J]. Food Chem, 2011, 127: 1821-1827.

[6] Mohan D, Pittman C U, Steele P H. Pyrolysis of wood/biomass for bio-oil: A critical review[J]. Energy Fuel., 2006, 20: 848-889.

[7]Nakai T, Kartel S N, Hata T, et al. Chemical characterization of pyrolysis liquids of wood-based composites and evaluation of their bio-efficiency[J]. Build Environ, 2007, 42: 1236-1241.

[8]Ratanapisit J, Apiraksakul S, Rerngnarong A, et al. Preliminary evaluation of production and characterization of wood vinegar from rubber wood[J]. Songklanakarin J. Sci. Technol, 2009, 31: 343-349.

[9] Chang Y, Zhao S, Ni W, et al. Research of the antioxidative properties of bamboo vinegar[J]. J. East China Univ. Sci. Technol (Nat. Sci. Ed.), 2004, 30: 640-643.

[10]Loo A Y, Jain K, Darah I. Antioxidant and radical scavenging activities of the pyroligneous acid from a mangrove plant, Rhizophora apiculate[J]. Food Chem., 2007, 107: 300-307.

[11]Wei Q, Ma X, Zhao Z, et al. Antioxidant activities and chemical profiles of pyroligneous acids from walnut shell[J]. J. Anal. Appl. Pyrolysis, 2010, 88: 149-154.

[12]Cai K, He Y. Antioxidant activities of the pyroligneous acid in living Caenorhabditis elegans[J]. Adv. Mater Res., 2011, 236-238: 2564-2569.

[13] Amarowicz R, Estrella I, Hernández T, et al. Free radical-scavenging capacity, antioxidant activity, and phenolic composition of green lentil (Lens culinaris)[J]. Food Chem., 2010, 121: 705-711.

[14]Lu Q, Liu W, Yang L, et al. Investigation of the effects of different organosolv pulping methods on antioxidant capacity and extraction efficiency of lignin[J]. Food Chem., 2012, 131: 313-317.

[15]Yang L, Huang, J M, Zu Y G, et al. Preparation and radical scavenging activities of polymeric procyanidins nanoparticles by a supercritical antisolvent (SAS) process[J]. Food Chem., 2011, 128: 1152-1159.

[16]Wang H, Zu G, Yang L, et al. Effects of heat and ultraviolet radiation on the oxidative stability of pine nut oil supplemented with carnosic acid[J]. J. Agric Food Chem., 2011, 59: 13018-13025.

[17] Ahmeda A, Hossain M A, Ismail Z. Antioxidant properties of the isolated flavonoids from the medicinal plant Phyllanthus niruri[J]. Asian J. Food Agro-Ind, 2009, 2: 373-381.

[18] Valentao P, Fernandes E, Carvalho F, et al. Antioxidative properties of cardoon (Cynara cardunculus L.) infusion against superoxide radical hydroxyl radical, and hypochlorous acid[J]. J. Agric Food Chem., 2002, 50: 4989-4993.

[19]Ma C H, Song K G, Yu J H, et al. Pyrolysis process and antioxidant activity of pyroligneous acid from Rosmarinus officinalis leaves[J]. J. Anal. Appl. Pyrolysis, 2013, 104: 38-47.

[20]Pharmacopoeia of the People's Republic of China; Peoples Medicinal Publishing House: Beijing, China, 2010; Volume I, pp. 61-62.

[21] Park J Y, Shin H K, Lee Y J, et al. The mechanism of vasorelaxation induced by Schisandra chinensis extract in rat thoracic aorta[J]. J. Ethnopharmacology, 2009, 121: 69-73.

[22] Chang G T, Kang S K, Kim J H, et al. Inhibitory effect of the Korean herbal medicine, Dae-Jo-Whan, on platelet-activating factor-induced platelet aggregation[J]. J. Ethnopharmacol, 2005, 102: 430-439.

[23] Chen C G. Basic Theory of Traditional Chinese Medicine; Shanghai of TCM Press: Shanghai, China, 2002.

[24] Deng XX, Chen X H, Yin R, et al. Determination of deoxyschizandrin in rat plasma by LC-MS[J]. J Pharma Biomed Anal., 2008, 46: 121-126.

[25] Fu M, Sun Z H, Zong M, et al. Deoxyschisandrin modulates synchronized Ca^{2+} oscillations and spontaneous synaptic transmission of cultured hippocampal neurons[J]. Acta. Pharmacol, 2008, 29: 891-898.

[26] Guo L Y, Hung T M, Bae K H, et al. Anti-inflammatory effects ofschisandrin isolated from the fruit of Schisandra chinensis Baill[J]. Eur. J. Pharmacol, 2008, 591: 293-299.

[27] Peng J Y, Fan G R, Qu L P, et al. Application of preparative high-speed counter-current chromatography for isolation and separation of schizandrin and gomisin A from Schisandra chinensis[J]. J. Chromatogr A., 2005, 1082: 203-207.

[28] Zhao J, Meng Z, Li G, et al. Study on processing of coagulation Schizandra chinensis yogurt[J]. Food Sci Technol, 2009, 34: 49-51.

[29] Liang X, Wen J. Progress on Schisandra chinensis (Turcz.) Baill[J]. Health Food Drug., 2009, 11: 70-73.

[30] Guan X, Huo F, Yu W, et al. The exploitation for the extractive of fructus schisandrae chinensis in natural sapid substance[J]. China Food Addit, 2007, 18: 110-113.

[31] Sun C, Wang S, Ding X, et al. Application and development prospects of the fruits of Schisandra chinensis in food industry[J]. Food Mach, 2003, 19: 9-10.

[32] Ma C H, Yang L, Zu Y G, et al. Optimization of conditions of solvent-free microwave extraction and study on antioxidant capacity of essential oil from Schisandra chinensis (Turcz.) Baill[J]. Food Chem, 2012, 4: 2532-2539.

[33] Chen Y, Tang J B, Wang X K, et al. An immunostimulatory polysaccharide (SCP-IIa) from the fruit of Schisandra chinensis (Turcz.) Baill[J]. Int. J. Boil Macromol, 2012, 50: 844-848.

[34] Ma C, Yang L, Yang F, et al. Content and colour stability of anthocyanins isolated from Schisandra chinensis (Turcz.) Baill fruits[J]. Int. J. Mol. Sci., 2012, 13: 14294-14310.

[35] Hancke J L, Burgos R A, Ahumada F. Review Schisandra chinensis (Turcz.) Baill[J]. Fitoterapia, 1999, 70: 451-471.

[36] Ma C H, Liu T T, Yang L, et al. Preparation of high purity biphenyl cyclooctene lignans from Schisandra extract by ion exchange resin catalytictransformation combined with macroporous resin separation[J]. J. Chromatogr B., 2011, 879: 3444-3451.

[37] Jerez M, Pinelo M, Sineiro J, et al. Influence of extraction conditions on phenolicyields from pine bark: Assessment of procyanidins polymerization degree by thiolysis[J]. Food Chem., 2004, 94: 406-414.

[38] Siddhuraju P. Antioxidant activity of polyphenolic compounds extracted from defatted raw and dry heated Tamarindus indica seed coat[J]. LWT-Food Sci. Technol, 2007, 40: 982-990.

[39] Soares J R, Dins T C, Cunha A P, et al. Antioxidant activity of some extracts of Thymus zygis[J]. Free Radic Res., 1997, 26: 469-478.

[40] Mansaray K G, Ghaly A E. Physical and thermochemical properties of rice husk[J]. Energy Sources, 1997, 19: 989-1004.

[41] Baker R R. Kinetic parameters from the nonisothermal decomposition of a multicomponent solid[J]. Thermochim Acta, 1978, 23: 201-212.

[42] Bradbury A G W, Sakai Y, Shafizadeh F. A kinetic model for pyrolysis of cellulose[J]. J. Appl Polym Sci., 1979, 23: 3271-3280.

[43] Liu X P, Jia J, Yang L, et al. Evaluation of antioxidant activities of aqueous extracts and fractionation from different parts of Elsholtzia ciliate[J]. Molecules, 2012, 17: 5430-5441.

[44] Zu S C, Yang L, Huang J M, et al. Micronization of taxifolin by supercritical antisolvent process and evaluation of radical scavenging activity[J]. Int. J. Mol. Sci., 2012, 13: 8869-8881.

Sample Availability: Samples of *S. chinensis* fruits are available from the authors.

中文题目：热解北五味子果实制备醋液的抗氧化特性分析

作者：马春慧，李伟，祖元刚，杨磊，李坚

摘要：可再生资源的可持续发展是全球面临的一项重大挑战。生物质是一种重要的可再生能源，是化石燃料的替代品。热解同时获得生物炭、生物焦油、木醋酸(PA)和气态燃料。本研究的目的是探讨不同功率热解五味子果实的热解过程及产物收率，所得木醋液分别用甲酸乙酯、二氯甲烷、噻吩和四氢呋喃等有机溶剂萃取后的抗氧化特性分析。包括自由基清除能力和铁还原能力，并与合成抗氧化剂丁基化羟基茴香醚和丁基化羟甲苯作为阳性对照。醋液的二氯甲烷萃取物与其他萃取物相比具有优良的抗氧化性能。用 GC-MS 法测定了醋液的化学组成，进一步证实二氯甲烷萃取物的抗氧化性最好。

关键词：北五味子果实；热解；木醋酸；抗氧化；GC-MS

Ultrasound-Assisted Extraction of Arabinogalactan and Dihydroquercetin Simultaneously from Larix Gmelinii As A Pretreatment for Pulping and Papermaking*

Chunhui Ma, Lei Yang, Wei Li, Jinquan Yue, Jian Li, Yuangang Zu

Abstract: An ultrasound-assisted extraction (UAE) method using ethanol was applied for extracting arabinogalactan (AG) and dihydroquercetin (DHQ) simultaneously from larch wood, as a pretreatment for pulping and papermaking. The extraction parameters were optimized by a Box-Behnken experimental design with the yields of AG and DHQ as the response values. Under optimum conditions (three extractions, each using 40% ethanol, for 50 min, 200 W ultrasound power and 1 : 18 solid-liquid ratio), the yields of AG and DHQ were 183.4 and 36.76 mg · g^{-1}, respectively. After UAE pretreated, the wood chips were used for Kraft pulping (KP) and high boiling solvent pulping (HBSP). The pulping yield after pretreatment was higher than that of untreated (the pulping yields of untreated HBSP and KP were 42.37% and 39.60%, and the pulping yields of HBSP and KP after UAE-pretreated were 44.23% and 41.50% respectively), as indicated by a lower kappa number (77.91 and 27.30 for untreated HBSP and KP; 77.01 and 26.83 for UAE-pretreated HBSP and KP). Furthermore, the characteristics of paper produced from pretreated wood chips were superior to those from the untreated chips: the basis weight was lower (85.67 and 82.48 g · cm^{-1} for paper from untreated KP and HBSP; 79.94 and 80.25 g · cm^{-1} for paper from UAE-pretreated KP and HBSP), and the tensile-strengths, tearing strengths, bursting strengths, and folding strengths were higher than these of paper after UAE-pretreated, respectively.

Keywords: Flavonoids; Taxifolin

1 Introduction

The larch is a conifer of the Larix genus, from the Pinaceae family[1], which is native to the cooler temperate climate of the northern hemisphere, especially in the boreal forests of Russia and Canada[2]. Larix gmelinii (syn. L. dahurica) is a species, unique to the northeastern forests of China, a temperate coniferous broad-leaved mixed forest area, accounting for the largest proportion of forest trees[3]. The chemical composition of Larix gmelinii comprises cellulose, lignins, hemicellulose (arabinogalactan), ash and extractable compounds (resins and dihydroquercetin). It is also an important source of pulp wood for papermaking.

However, arabinogalactan (AG) and dihydroquercetin (DHQ) cannot be ignored in the pulping

* 本文摘自 Applied Surface Science, 2015, 332: 565-572.

process. A large amount of alkali is consumed and the temperature has to rise slowly, because of AG in the raw material, such as the Kraft pulping process. Because the mechanism of lignin removal relies on the degradation of hemicelluloses, the degradation of AG occurs at the heating stage under alkaline conditions[4]. If AG could be extracted before the pulping operation, this would not only reduce the amount of alkali used, but also shorten the heating time. Meanwhile, DHQ, as well as tannins and polyphenols, is also present in larch wood. The higher degree of condensation of the phenolic hydroxyl group causes difficulty in the pulping process[5]. If the compounds containing phenolic hydroxyl groups could be extracted before pulping operation, this could facilitate the bleaching process for pulping.

AG and DHQ are active compounds in larch wood and have usually been wasted in the pulping process, but they also have many pharmacological effects.

AG is a biopolymer consisting of arabinose and galactose monosaccharides, which has a galactan core connected to D-galactopyranose by b-1, 3-O-glucoside links[6]. AG is present in the larch woody tissue at a level of 15%-30%[7]. Larch AG is a highly branched polysaccharide consisting of b-D-galactopyranose, a-L-arabinofuranose and b-L-arabinopyranose residues[8]. AG was generally recognized as safe by the FDA, USA in 1974. It is well accepted by consumers as a dietary supplement[9] and food ingredient, because of its water-soluble and non- viscous properties[10]. Moreover, many pharmacological effects of AG have been reported, such as anti-inflammatory[6], gastro-protective[6], membranetropic[11], and immune-modulating activities[12]. Additionally, AG can be converted into more valuable products through a sugar platform[8]. Therefore, AG has been recognized as a multi-purpose natural product with great economic potential and environmental value, which has attracted increasing attention by researchers. AG is commonly extracted with water and precipitated using the ethanol method, which needs a higher-volume fraction of ethanol solution to precipitate AG thoroughly[13].

DHQ also known as taxifolin [2-(3,4-dihydroxyphenyl)-2,3-dihydro-3,5,7- trihydroxy-4H-benzopyran-4-one] and vitamin P[14], and the structure is shown in Fig. 1(a). DHQ have many biological activities, including the inhibition or activation of a variety of enzymes, resulting in different physiological effects; the protection of cells against oxidative stress, attributed to its antioxidant

properties[15]; and also anti-radiation[16], anti-viral[17], anti-tumor[18] and scavenging free radical[19] properties because of its rich content of phenolic hydroxyl groups. The antioxidant properties of DHQ can be comparable or superior to many synthetic or natural antioxidants. Moreover, it is not toxic to the fetus, teratogenic, mutagenic or allergenic. DHQ is commonly extracted from wood using a polar solvent, such as methanol or ethanol, or by Soxhlet, reflux, ultrasonic or accelerated solvent extraction[20]. Supercritical fluid extraction (CO_2) can be applied, but a modifier such as methanol is then required[21]. The use of enzymatic water extraction of DHQ from wood material has been proposed by Wang et al.[22], but the increasing cost of enzyme disposal has created a growing demand for better and cheaper extraction methods for DHQ.

Recently, much more attention has been given to the application of ultrasound-assisted extraction (UAE) in the fields of analytical chemistry and sample preparation (i. e. digestion, extraction and dissolution)[23]. Compared withconventional solvent extraction, the use of ultrasound makes the extraction of valuable compounds more efficient by using shorter time frames and lower extraction temperatures[24]. The possible benefits of ultrasound in extraction are the intensification of mass transfer, cell disruption, improved

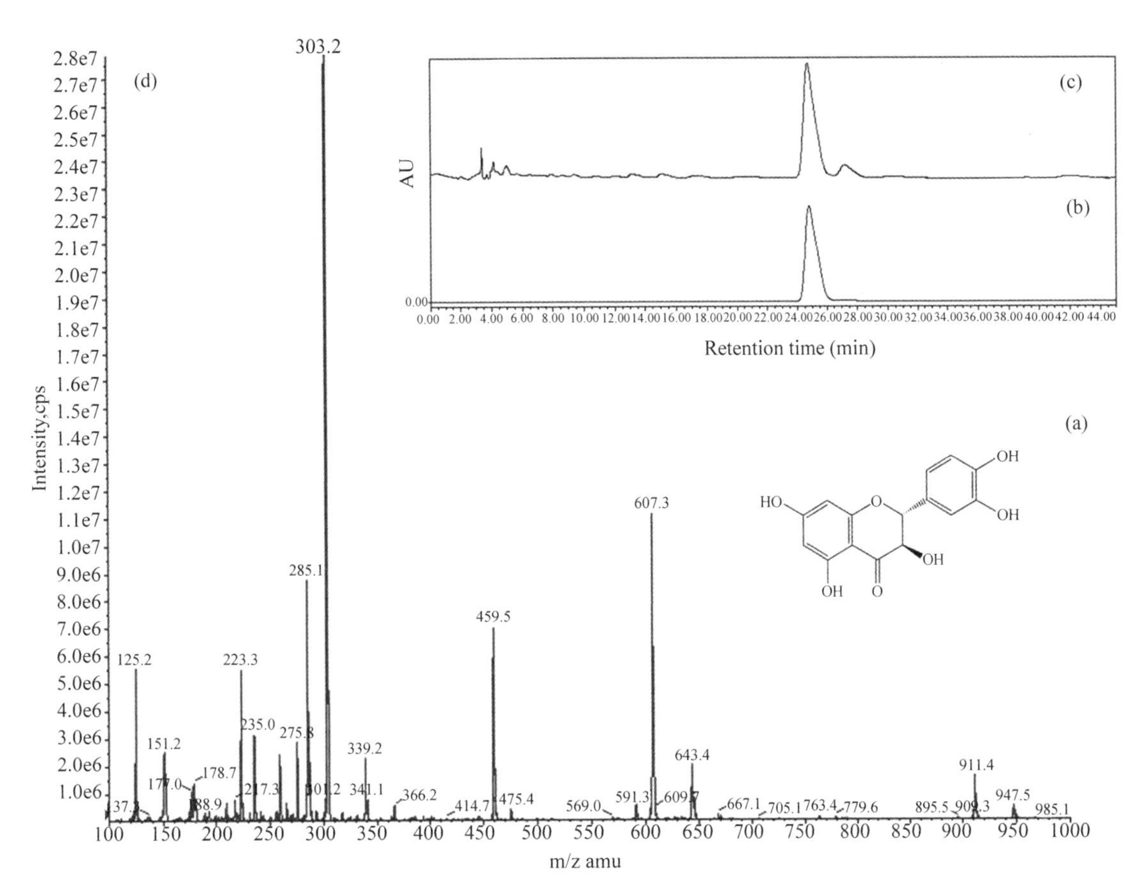

Fig. 1 HPLC-UV and LC-ESI-MS analysis of DHQ. (a) Chemical structure of DHQ. (b) HPLC-UV chromatogram of DHQ standard. (c) HPLC-UV chromatogram of Larch wood sample. (d) LC-ESI-MS chromatogram of DHQ

penetration and capillary effects[25]. Ultrasound is currently used to extract such pharmacologically active compounds as polysaccharides[13, 25], flavonoids[21, 22] and isoflavonoids[24], alkaloids[26], lignans[27], anthraquinone[28], carnosic and rosmarinic acids[29], steroids and triterpenoids[30] from plant materials. However, there are few reports of applying UAE for extracting two or more kinds of natural products simultaneously from wood or plant materials.

The primary aim of the present study is to optimize the UAE conditions for isolating AG and DHQ simultaneously from larch wood using a lower-volume fraction of ethanol solution as a pretreatment for the pulping and papermaking processes. AG will then be precipitated using a higher-volume fraction of ethanol solution so that AG and DHQ can be separated. The wood chips pretreated by UAE will be used as the raw material for pulping. Any changes in the pulping characteristics and physical properties of the paper will be determined. Compared with the conventional extraction methods, the invocation of this study is not only obtained the active compounds, AG and DHQ in one step, also improve the quality of pulping and papermaking which made from the pretreatment larch wood after extracting AG and DHQ.

2 Materials and Methods

2.1 Wood materials

Larix gmelinii wood chips (average size 2562066 mm) were purchased from Greater Khingan Mountains (Heilongjiang, China) and identified by Academician Jian Li from the College of Material Science and Engineering, Northeast Forestry University, Harbin, China. The moisture content of the wood chips was 5.9% after drying in a shaded and ventilated place, with no further grinding processing before use.

For determining the composition, 100.0 g wood flour after crushing was precisely weighed. Larch wood contains cellulose (52.4%±0.6%), lignin (28.2%±0.3%), benzene alcohol extractives (5.2%±0.1%), arabinogalactan (22.3%±0.5% using Soxhlet extraction 12 h with water) and dihydroquercetin (6.3%±0.2% using Soxhlet extraction 12 h with 40% ethanol).

2.2 Chemical materials

Dihydroquercetin (DHQ) standards (1257117-20607065, 98% purity) were purchased from the National Institute for the Control of Pharmaceutical and Biological Products (Beijing, China). Deionized water was purified using a Milli-Q Water Purification system (Millipore, Billerica, MA, USA). Acetonitrile and acetic acid of HPLC grade were purchased from J&K Chemical Ltd. (shanghai, China). The rest of the solvents and chemicals used in this study were of analytical grade and purchased from Beijing Chemical Reagents Co. (Beijing, China). All solutions prepared for HPLC were filtered through 0.45-mm membranes (Guang Fu Chemical Reagents Co., Tianjin, China) before use.

2.3 HPLC-UV quantitative analysis method of DHQ

The HPLC-UV system consisted of a Waters 717 automatic sample handling system composed of an HPLC system equipped with 1525 Bin pump, 717 automatic column temperature control box and 2487 UV-detector (Waters, Milford, MA, USA). Chromatographic separation was performed on a HiQ sil-C18 reversed-phase column (4.6 mm 6250 mm, 5 mm, KYA TECH Corp., Tokyo, Japan) for the determination of DHQ.

For HPLC-UV analysis, acetonitrile-water-acetic acid (18 : 82 : 0.1, *v/v/v*) was used as the mobile phase at a flow rate of 1.0 mL · min^{-1} with a 10 mL injection volume and 25 ℃ column temperature. The absorbance was measured at a wavelength of 294 nm for the detection of DHQ with a run time of 30 min. The retention time of DHQ is 24 min. The corresponding calibration curve for DHQ is *Y*53.17556107 *X* + 2.59936104 (r50.9999). A high degree of linearity was found for DHQ over the range 0.0312-0.5000 mg/mL. The HPLC-UV chroma- tograms of the DHQ standard and a larch wood sample are shown in Fig. 1(b) and 1(c).

2.4 LC-ESI-MS qualitative analysis method of DHQ

The HPLC-ESI-MS system consisted of an Agilent 1100 series HPLC system equipped with a G1312A Bin pump, a G1379A Degasser (Agilent, San Jose, CA, USA) and a G1316A automatic column temperature control box. Chromatographic separation was performed on a HiQ sil-C18 reversed-phase column (4.6 mm 6250 mm, 5 mm, KYA TECH). An API3000 Triple tandem quadrupole mass spectrometer with a TurboIon-Spray interface from Applied Biosystems (Foster City, CA, USA) was operated in the positive electrospray ionization (ESI+) source mode. All mass spectra were acquired in multiple reaction monitoring transitions.

For HPLC-ESI-MS analysis, acetonitrile-water-acetic acid (18 : 82 : 0.1, *v/v/v*) was used at a flow rate of 1.0 mL · min^{-1} with a run time of 65 min. The injection volume was 10 mL and the column temperature was maintained at 25 ℃. The ion source was operated at a temperature of 250 ℃. The nebulizing gas, curtain gas and collision gas were set at 12, 10 and 6 a. u., respectively. The ion spray voltage was 5500 V. The entrance potential and focusing potential were set at 10 and 400 V, respectively. Analyst software (version 1.4, Ab Sciex, Framingham, MA, USA) installed on a Dell computer was used for data acquisition and processing. The LC-ESI-MS chromatogram of DHQ is shown in Fig. 1(d).

2.5 UAE method for extracting AG and DHQ as pretreatment

For the ultrasonic-assisted extraction (UAE) experiments, an ultrasonic bath was used as an ultrasonic source. The bath (KQ-250DB, Kunshan Ultrasonic Co. Ltd., China) was a open rectangular container (23.5613.3610.2 cm), to which 50 kHz transducers were annealed at the bottom. The bath power rating is 250 W, and the power is divided into 5 file, are 50 W, 100 W, 150 W, 200 W and 250 W, respectively.

The extraction experiment was performed by adding 30.0 g of wood materials into different extraction solvent in a 500 mL glass flask. The flask was then partially immersed into the ultrasonic bath, which contains 2.5 L of water. The water in the ultrasonic bath is circulated and regulated at the desired temperature to maintain the water temperature at a constant value (25 ℃) and prevent it from being influenced by the ultrasonic exposure. After ultrasonic procedure, the extracts are then cooled down to the room temperature, and filtrated through a 0.45 mm filter prior to HPLC analysis.

2.6 Effect of the volume fraction of ethanol for extracting AG and DHQ as pretreatment method

Polysaccharides are commonly extracted with water and precipitated using a higher-volume fraction of ethanol solution. To extract AG and DHQ from larch wood simultaneously, the volume fraction of ethanol as the extraction solvent was investigated. First, AG and DHQ were extracted simultaneously using UAE with a lower-volume fraction of ethanol solution, and then AG was precipitated using a higher-volume fraction of ethanol solution. The precipitated effect was shown in Table 1 (the original data was in File S1). Finally, AG and DHQ were separated using centrifugation at 10000 r/min.

2.7 Single-factor test method

To study the effect of the solid-liquid ratio, soaking time, ultrasound extraction time and ultrasound power on the extraction yields of AG and DHQ, the single- factor test was used three times and the average value was recorded.

2.8 Optimization test by Response Surface Methodology (RSM)

To further investigate the interaction between the factors in the process of UAE, we optimized the operating conditions using Response Surface Methodology (RSM) and the Box-Behnken software for data processing. The bounds of the factors were 40-60 min for extraction time, 150-250 W for ultrasound power and 1 : 16-1 : 20 for the solid-liquid ratio. As the volume fraction of ethanol and soaking time are not parameters in the process of UAE, it cannot be optimized by the software. The volume fraction of ethanol was 40% and the soaking time was 8.0 h. Specific protocols for experimental conditions and response values are shown in Table 2 (the original data, File S1).

Table 1 Effect of ethanol volume fraction on precipitation yield of AG

Volume fraction of ethanol	Rotated speed ($r \cdot min^{-1}$)	precipitation yield of AG ($mg \cdot g^{-1}$)	Precipitation phenomenon of AG
20%	10000	6. 28±0. 34	No obvious precipitation
30%	10000	11. 86±0. 24	No obvious precipitation
40%	10000	25. 37±0. 75	No obvious precipitation
50%	10000	40. 35±0. 41	Milky turbidity
60%	10000	58. 48±0. 79	Milky turbidity
70%	10000	141. 60±1. 47	White flocculent precipitation
80%	10000	168. 79±1. 74	White flocculent precipitation
90%	10000	172. 87±1. 76	White flocculent precipitation

2. 9 Effect of the extraction cycles on the yields and extraction efficiency of AG and DHQ

The number of extraction cycles is a crucial factor in the extraction process; the more extraction cycles, the higher the dissolution efficiency of the active ingredient. However, there is difficulty in the consumption and recycling of extraction solvent as the number of extraction cycles increases. Four extraction cycles were used and the total extraction efficiency from them was defined as 100%.

Table 2 Experimental design matrix to screen important variables for extraction yields of AG and DHQ

Run	Factor *A* Extraction time (min)	Factor *B* Ultrasound power (W)	Factor *C* Solid-liquid ratio ($g \cdot mL^{-1}$)	Response Y_1 Extraction yield of DHQ ($mg \cdot g^{-1}$)	Response Y_2 Extraction yield of AG ($mg \cdot g^{-1}$)
1	40	250	1 : 18	36. 1	163. 4
2	50	150	1 : 16	34. 2	154. 3
3	50	250	1 : 20	36. 9	183. 2
4	60	200	1 : 20	37. 1	185. 4
5	40	200	1 : 20	36. 3	169. 5
6	60	250	1 : 18	36. 6	179. 7
7	40	200	1 : 16	36. 2	161. 7
8	50	200	1 : 18	36. 7	183. 4
9	50	200	1 : 18	36. 8	183. 2
10	50	150	1 : 20	35. 6	158. 3
11	60	150	1 : 18	35. 1	160. 2
12	40	150	1 : 18	34. 5	156. 8
13	50	200	1 : 18	36. 7	183. 8
14	50	200	1 : 18	36. 9	183. 5
15	50	200	1 : 18	36. 7	183. 1
16	60	200	1 : 16	36. 0	163. 3
17	50	250	1 : 16	36. 5	176. 7

2.10 Reference extraction methods

The conventional extraction methods for AG are reflux extraction with water and soaking or stirring extraction with water, and for DHQ, reflux extraction with hot water or ethanol solutions, and soaking or stirring extraction with ethanol. The extraction yields of AG and DHQ from the various extraction methods were compared with the optimal conditions of UAE-pretreated predicted by RSM: 40% ethanol solution, 50 min ultrasound extraction at 200 W and 1 : 18 solid-liquid ratio.

2.11 Pulping method and the physical properties of paper

To investigate the physical properties of wood pulp and paper, two kinds of pulping process were studied. One was the traditional Kraft pulping process, and the other, currently of high interest to researchers, was the high boiling solvent pulping process.

(1) Kraft pulping process (KP)

UAE-pretreated wood chips were mixed using 20% causticity, 25% sulfidity (sodium sulfide concentration), at a solid-liquid ratio of 1 : 3.5. The mixture was put into a ZQS1 electric cooking pot (Machinery Factory of Shanxi University of Science and Technology, Shanxi, China) then heated up to 140 ℃ at a pressure of 0.4 KPa, with a small release steam for 3 min to eliminate any false pressure. The mixture was then heated up to a maximum temperature of 170 ℃, at a pressure of 0.8 KPa with thermal retardation for 60 min, then a large release of steam was made until the pressure changed to zero. After the reaction, solids and liquids were separated and the mixture was then washed. After washing, the pulp was suitable for further processing into paper using a QJ1-B-Ⅱ sheet molding machine (Machinery Factory of Shanxi University of Science and Technology, Shanxi, China).

(2) High boiling solvent pulping process (HBSP)

UAE-pretreated (the optimal conditions predicted by RSM: 40% ethanol solution, 50 min ultrasound extraction at 200 W and 1 : 18 solid-liquid ratio) wood chips were mixed with a volume fraction of 80% ethanol with 10% acetic acid as the catalyst. The reaction temperature and pressure were 200 ℃ and 1.3 MPa, respectively. The reaction was performed using a T25FYX autoclave and FDK autoclave controller (Dalian Industry Autoclave Vessel Manufacturing Co. Ltd., Liaoning, China). The reaction continued for 6 h followed by solid-liquid separation. After washing, the pulp was suitable for further processing into paper using a QJ1-B-Ⅱ sheet molding machine (Machinery Factory of Shanxi University of Science and Technology, Shanxi, China).

(3) Determination of the physical properties of paper

To compare the physical properties of untreated pulp and its resulting paper and UAE-pretreated (the optimal conditions predicted by RSM: 40% ethanol solution, 50 min ultrasound extraction at 200 W and 1 : 18 solid-liquid ratio) pulp and its resulting paper, the parameters measured were the kappa number (according to GB/T 1546—1989), basis weight, apparent density, tensile strength (according to GB/T 453—1989), tearing strength (according to GB/T 455.1—1989), bursting strength (according to GB/T 454—1989) and folding strength (according to GB/T 454—1989). The instruments used in the tests were produced by the Changchun Small Testing Machine Co. Ltd. (Jilin, China) and consisted of a ZUS-4 paper thickness tester. a ZDNP-1 electronic paper bursting tester, a ZLD-300 electronic paper tensile tester, a ZSE-1000 paper tear tester and a YQ-Z-31 MIT folding endurance tester.

2.12 Statistical analysis method

Results were expressed as mean values ± SD (n53). Statistical significance was determined using Microsoft Excel statistical fractions. The data of optimal experiment were analyzed using RSM software design 7.0. Differences at P, 0.05 were considered to be significant.

3 Results and Discussion

3.1 Factors affecting UAE

3.1.1 Extraction solvent

As is well known, AG dissolves in water and its solubility decreases as the volume fraction of ethanol solution increases. DHQ can dissolve in any volume fraction of ethanol solution. To extract AG and DHQ simultaneously, we investigated the volume fraction of ethanol used as an extraction solvent and also as a precipitation solvent. First, AG and DHQ were extracted using UAE with lower volume fraction of ethanol solution, then AG was precipitated by adding anhydrous ethanol, after being centrifuged at 10000 r · min^{-1}. The optimal volume fraction of ethanol solution for AG precipitation was selected through calculating the extraction yield of AG. From Table 1 (the original data, File S1), when the volume fraction of ethanol in the solution was less than 40% (40% ethanol as the precipitation solvent, the AG yield after being centrifuged was 25.37±0.75 mg · g^{-1}), there was no obvious precipitation of AG. With more than 50% ethanol in the solution, the yield of AG showed an upward trend with the AG solution showing a milky turbidity (the AG yield was 40.35±0.41 mg · g^{-1} when 50% ethanol as the precipitation solvent, and 58.48±0.79 mg · g^{-1} when 60% ethanol as the precipitation solvent). When the volume fraction of ethanol was more than 70%, a white flocculent precipitation occurred within a short time and the yield of AG also showed an upward trend (the AG yield was 141.59±1.47 mg · g^{-1} when 70% ethanol as the precipitation solvent). When the volume fraction of ethanol was more than 80% (the AG yield was 168.78±1.74 mg · g^{-1} when 80% ethanol as the precipitation solvent), the upward trend of AG yield was less pronounced (the AG yield was 172.87±1.76 mg · g^{-1} when 90% ethanol as the precipitation solvent), so an 80% ethanol solution was selected as the precipitation solvent (the precipitation mass by 80% ethanol reached 97.6% of the precipitation mass by 90% ethanol).

As can be seen in Fig. 2 (the original data, File S1), when the volume fraction of ethanol solution increased, the yield of DHQ increased, with a maximum value when the volume fraction of ethanol solution was 60%. However, when the volume fraction of ethanol was 40%, the precipitation of AG did not occur rapidly, and the yield of DHQ showed little change from the maximum value. Therefore, to extract AG and DHQ simultaneously, a 40% ethanol solution as the extraction solvent and an 80% ethanol solution as the precipitation solvent were selected.

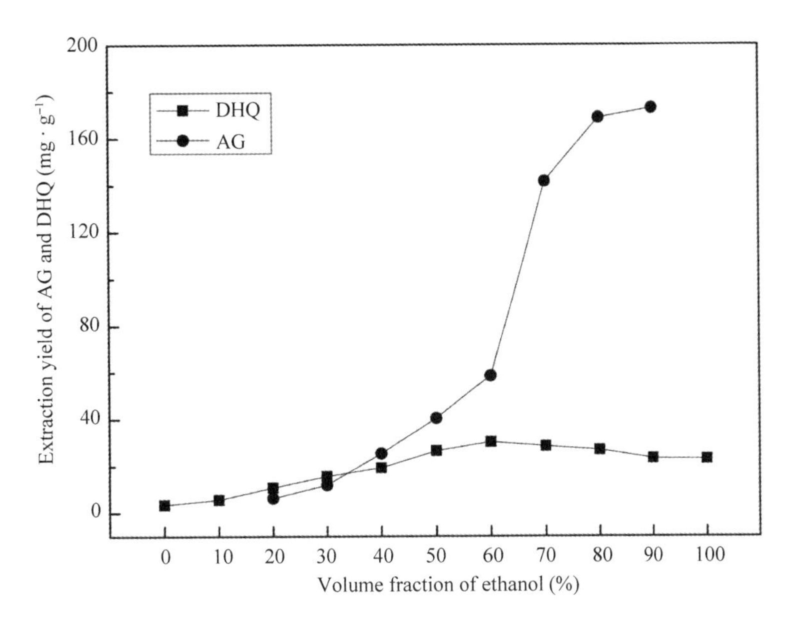

Fig. 2 Effect of volume fraction of ethanol on the extraction yields of AG and DHQ

3. 1. 2 Solid-liquid ratio

A high solvent content may cause complex procedures and unnecessary wastage and increase the energy consumption during recycling, while a low solvent content may lead to incomplete extraction. To evaluate the effect of the solid- liquid ratio, 30. 0 g of dried wood material was mixed with 40% ethanol solution as the extraction solvent, and then soaked for 8 h, the suspension was extracted 50 min by UAE at the power of 200 W. A series of extractions were carried out using different solid-liquid ratios (1 : 8-1 : 20 g · mL^{-1}). Fig. 3 (a) (the original data, File S1) indicates clearly that the extraction yield for AG increased with the increase in solvent volume, but at more than 1 : 18 (the yields of AG was 204. 66±11. 40 with 1 : 18 solid-liquid ratios), it was not significantly influenced by a further increase in the amount of solvent. The extraction yield of DHQ did not increase significantly when the solid-liquid ratios were more than 1 : 12 (the extraction yields of DHQ was 37. 71±0. 63 mg · g^{-1} with 1 : 12 solid-liquid ratios). Therefore, considering the yields of AG and DHQ, a range of solid-liquid ratios of 1 : 16-1 : 20 was selected and used in the further studies.

3. 1. 3 Soaking time

The role of infiltration in the extraction process is to make sure that the raw materials are fully soaked without consuming energy; this can help the dissolution of small molecules into the solution. The longer the time of infiltration, the better the soaking, but a long infiltration time may lead to moldy materials (if the solvent is water) and a long extraction cycle may lead to low extraction efficiency. In the present study, 30. 0 g dried wood chips were soaked with 40% ethanol for 0 h, 1 h, 2 h, 4 h, 8 h, 12 h or 24 h. The solid-liquid ratio was 1 : 18, the suspension was extracted 50 min by UAE at the power of 200 W, and the effect of soaking time was studied. Fig. 3 (b) (the original data, File S1) shows that the yields of AG and DHQ increased as the soaking time increased up to 8 h then remained almost unchanged after 8 h (the extraction yields of AG and DHQ were 198. 36±11. 13 and 46. 37±0. 33 mg · g^{-1}, respectively). To save time from a full infiltration, the soaking time used was 8 h for the further experiments.

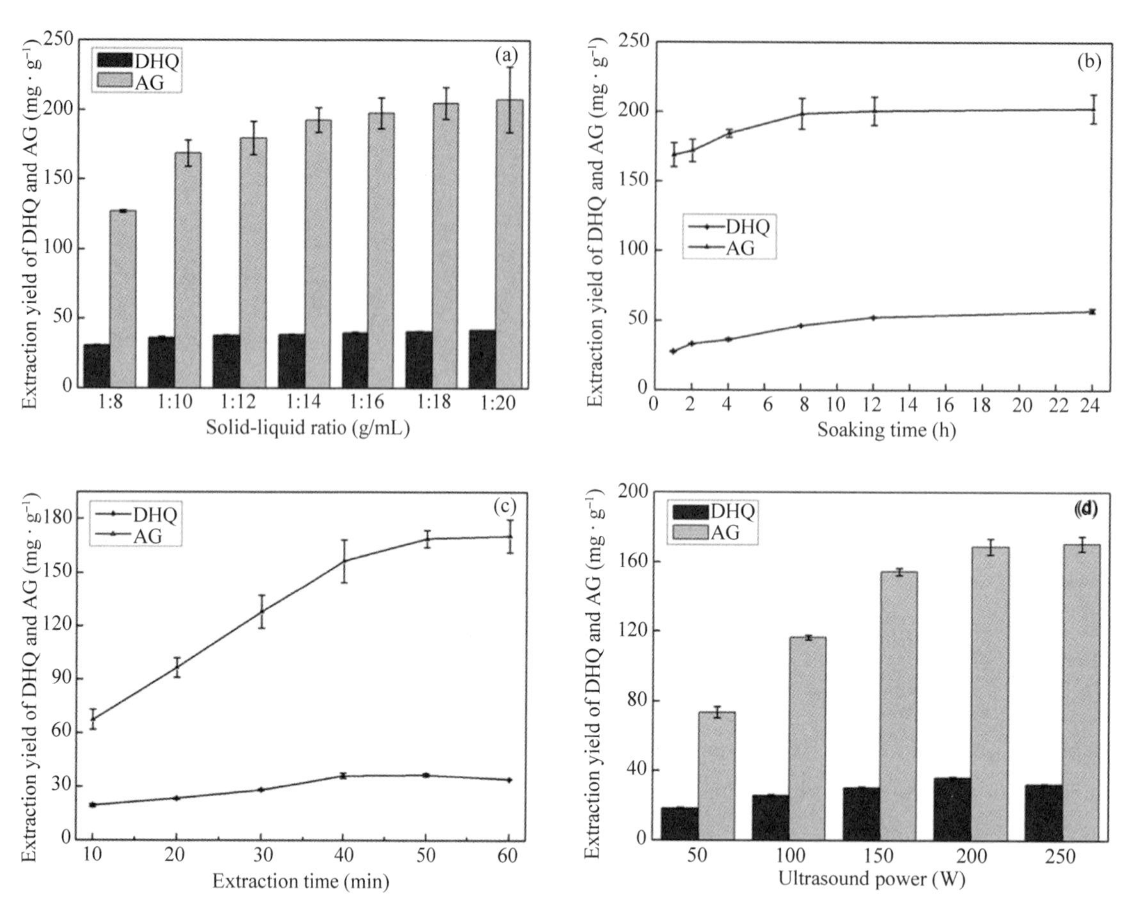

Fig. 3 Effect of single factors on the extraction yields of AG and DHQ. (a) Effect of solid-liquid ratio on the extraction yields of AG and DHQ. (b) Effect of soaking time on the extraction yields of AG and DHQ. (c) Effect of extraction time on the extraction yields of AG and DHQ. (d) Effect of ultrasound power on the extraction yields of AG and DHQ

3.2 Extraction dynamics of DHQ

To investigate the effect of extraction time on the yields of AG and DHQ using UAE, the process was performed in an ultrasound unit. 30.0 g of dried sample was mixed with 40% ethanol soaked for 8 h, and then extracted for 60 min. The power of UAE was 200 W with a solid-liquid ratio of 1 : 18, and the yields of AG and DHQ were tested every 10 min. Fig. 3(c) (the original data, File S1) shows that as the extraction time increased, the yields initially increased. Then after a UAE process for 40 min (the extraction yields of AG and DHQ were 156.32±11.94 and 36.00±1.53 mg · g^{-1}) and 50 min (the extraction yields of AG and DHQ were 168.79±4.81 and 36.52±0.96 mg · g^{-1}), respectively, the yields of DHQ and AG were almost unchanged. Therefore, a 40-60 min ultrasound treatment was selected as the optimal condition for extracting AG and DHQ.

3.3 Energy intensity of UAE

To examine the effect of ultrasound power on the extraction yields of AG and DHQ, 30.0 g of dried sample was mixed with 40% ethanol (the solid-liquid ratio was 1 : 18) soaked for 8 h, and then were carried out with a constant ultrasonic treatment time of 50 min at 50 W, 100 W, 150 W, 200 W and 250 W, respectively. Fig. 3(d) (the original data, File S1) indicates that the average extraction yields of AG and DHQ were significantly increased with the ultrasound power increasing. However, when the ultrasound power levels greater than 200 W (the extraction yields of AG and DHQ were 168.79±4.47 and 35.96±0.63 mg · g^{-1} at 200 W), the yield of DHQ was not increased (decreased slightly in the range of error). Therefore, as a high extraction yield and a low energy consumption were required, a range of 150–250 W ultrasound power was selected for further optimization experiments.

3.4 Extraction cycles

The extraction conditions of soaking time, solid-liquid ratio, ultrasonic time, and ultrasonic power are 8 h, 1 : 18 g · mL^{-1}, 50 min and 200 W, respectively. To optimize extraction cycles, extractions were carried out four times. From Fig. 4 (the original data, File S1), the total extraction yield of four cycles was defined as 100% (right-hand axis) and the extraction yields of AG and DHQ (left-hand axis) were compared with the total extraction yield. The extraction yield total for two cycles of AG reached 88.86% and the extraction yield total for two cycles of DHQ was 82.06% and the total for three cycles were more than 90% (97.71% for AG and 92.06% for DHQ). Considering these factors of time, solvent and energy consumption, an extraction process using three cycles seemed most reasonable.

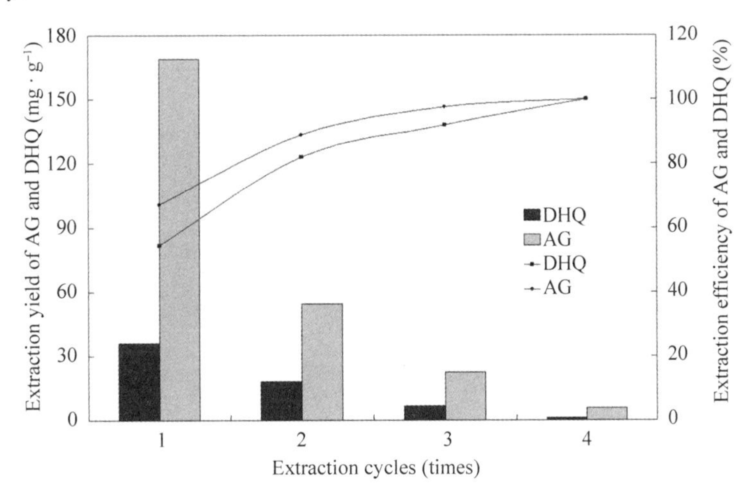

Fig. 4 Effect of extraction cycles on the extraction yields of AG and DHQ

3.5 Optimization of UAE conditions

To further study the interactions between factors, we optimized the extraction time, ultrasound power and solid-liquid ratio, using the yields of AG and DHQ as the response values.

In Table 3 (the original data, File S2), the Model F-values of 64.21 and 15.42 implied that the model was significant. There is only a 0.01% chance that a "Model F-Value" this large could occur by chance. If values of "Probability. *F*" are less than 0.0500, this indicates that the model terms are significant. The Probability. *F* value of "model" were , 0.0001 and 0.0008, respectively, It indicated that the experiment data were better fitting the model provided by RSM. In this case, For the yield of DHQ, *A* (the Probability. *F* value was , 0.0001), *B* (the Probability. *F* value was , 0.0001), *C* (the Probability. *F* value was,

0. 0001), *AC* (the Probability. *F* value was 0. 0094), *BC* (the Probability. *F* value was 0. 0094), *A2* (the Probability. *F* value was 0. 0038), and *B2* (the Probability. *F* value was , 0. 0001) were significant model terms. And for the yield of *AG*, *A* (the Probability. *F* value was 0. 0113), *B* (the Probability. *F* value was 0. 0003), *C* (the Probability. *F* value was 0. 0076), *A2* (the Probability. *F* value was 0. 0032), *B2* (the Probability. *F* value was 0. 0010), and *C2* (the Probability. *F* value was 0. 0286). If values were greater than 0. 1000 indicate that the model terms are not significant. If there are many insignificant model terms (not counting those required to support hierarchy), a reduction in the number of terms may improve the model. Table 4 (the original data, File S2) shows that the standard deviations of the model were 0. 14 and 3. 86. The lower the CV value, the better the stability. In the present study, the CV values of Y_1 and Y_2 were low at 0. 39% and 2. 24%, respectively, indicating a high level of reliability for the experimental operation. The R^2 values of 0. 9880 and 0. 9520 are in reasonable agreement with the adjusted R^2 values of 0. 9726 and 0. 8902, indicating a high level of reliability for 90% of the experimental data. "Adequacy Precision" measures the signal-to-noise ratio where a ratio greater than 4 is desirable; the ratios of 26. 785 and 9. 795 indicated an adequate signal. This model can therefore be used to explore the design space to find the optimal values for extraction.

Table 3 Test of significance for regression coefficient[a]

Source	*df*	Y_1 Sum of squares	Y_2 Sum of squares	Y_1 Mean square	Y_2 Mean square	Y_1 F-value	Y_2 F-value	Y_1 Pr. F	Y_2 Pr. F
A	1	0. 36	172. 98	0. 36	172. 98	18. 13	11. 63	0. 0038	0. 0113
B	1	5. 61	673. 45	5. 61	673. 45	281. 57	45. 27	,0. 0001	0. 0003
C	1	1. 12	204. 02	1. 12	204. 02	56. 45	13. 71	0. 0001	0. 0076
AB	1	2. 500610[23]	41. 60	2. 500610[23]	41. 60	0. 13	2. 80	0. 7336	0. 1384
AC	1	0. 25	51. 12	0. 25	51. 12	12. 54	3. 44	0. 0094	0. 1062
BC	1	0. 25	1. 56	0. 25	1. 56	12. 54	0. 11	0. 0094	0. 7553
A^2	1	0. 36	287. 45	0. 36	287. 45	18. 08	19. 32	0. 0038	0. 0032
B^2	1	3. 35	430. 58	3. 35	430. 58	168. 30	28. 95	,0. 0001	0. 0010
C^2	1	0. 019	112. 22	0. 019	112. 22	0. 96	7. 54	0. 3592	0. 0286
Model [b]	9	11. 52	2063. 82	1. 28	229. 31	64. 21	15. 42	,0. 0001	0. 0008
Residual	7	0. 14	104. 13	0. 020	14. 88	—	—	—	—
Lack of fit	3	0. 11	103. 83	0. 036	34. 61	4. 48	461. 47	0. 0908	,0. 0001
Pure Error	4	0. 032	0. 30	8. 000610[23]	0. 075	—	—	—	—
Corr Total	16	11. 66	2167. 95	—	—	—	—	—	—
Linear	3	7. 10	1050. 44	2. 37	350. 15	6. 75	4. 07	0. 0055	0. 0304
Quadratic	3	3. 92	919. 09	1. 31	306. 36	65. 5	20. 59	,0. 0001	0. 0008
Cubic	3	0. 11	103. 83	0. 036	34. 61	4. 48	461. 47	0. 0908	,0. 0001

[a]The results were obtained with the Design Expert 7. 0 software.

[b]*A* is extraction time (min), *B* is ultrasound power (W), *C* is solid-liquid ratio (*w/v*), and Y_1 is extraction yield of DHQ (mg · g^{-1}), Y_2 is extraction yield of AG (mg · g^{-1}).

The final extraction yields of DHQ (Y_1) and AG (Y_2) were found using RSM to be:

$Y1536.76+0.36A+0.78B+0.54C-0.025AB+0.050AC-0.12BC-0.38A2-0.81B2-0.28C2$

$Y25183.40+5.71A+9.55B+4.24C+3.22AB+4.45AC-0.13BC-9.70A2-8.68B2-5.85C2$

Where A is extraction time (min), B is ultrasound power (W), and C is solid- liquid ratio ($g \cdot mL^{-1}$), respectively.

Table 4 Credibility analysis of the regression equations

Index mark [a]	Y_1 is extraction yield of DHQ	Y_2 is extraction yield of AG
Std. Dev.	0. 14	3. 86
Mean	36. 17	172. 32
C. V. %	0. 39	2. 24
PRESS	1. 77	1661. 75
R-Squared	0. 9880	0. 9520
Adjust R-Squared	0. 9726	0. 8902
Predicted R-Squared	0. 8481	0. 2335
Adequacy Precision	26. 785	9. 795

[a] The results were obtained with the Design Expert 7. 0 software.

The response surfaces of the extraction yields of DHQ and AG were shown in Fig. 5 (the original data, File S2).

As indicated by the equation above, the conditions for the optimal point predicted by the software were: ultrasound extraction 52. 63 min at 200. 15 W and 1 : 18. 32 solid-liquid ratios with extraction yields for AG and DHQ of 183. 40 and 36. 76 $mg \cdot g^{-1}$, respectively.

3. 6 Verification test under optimum conditions

The verification tests (the original data, File S1) were performed three times under the optimal conditions predicted by RSM. 30. 0 g of dried sample was mixed with 40% ethanol (the solid-liquid ratio was 1 : 18) soaked for 8 h, and then were ultrasound extracted at 200 W for 50 min. The actual extraction yields of AG and DHQ were 179. 88 ± 10. 49 and 35. 96 ± 1. 66 $mg \cdot g^{-1}$, respectively, with correspond- ing errors of about 1. 92% and 2. 18%.

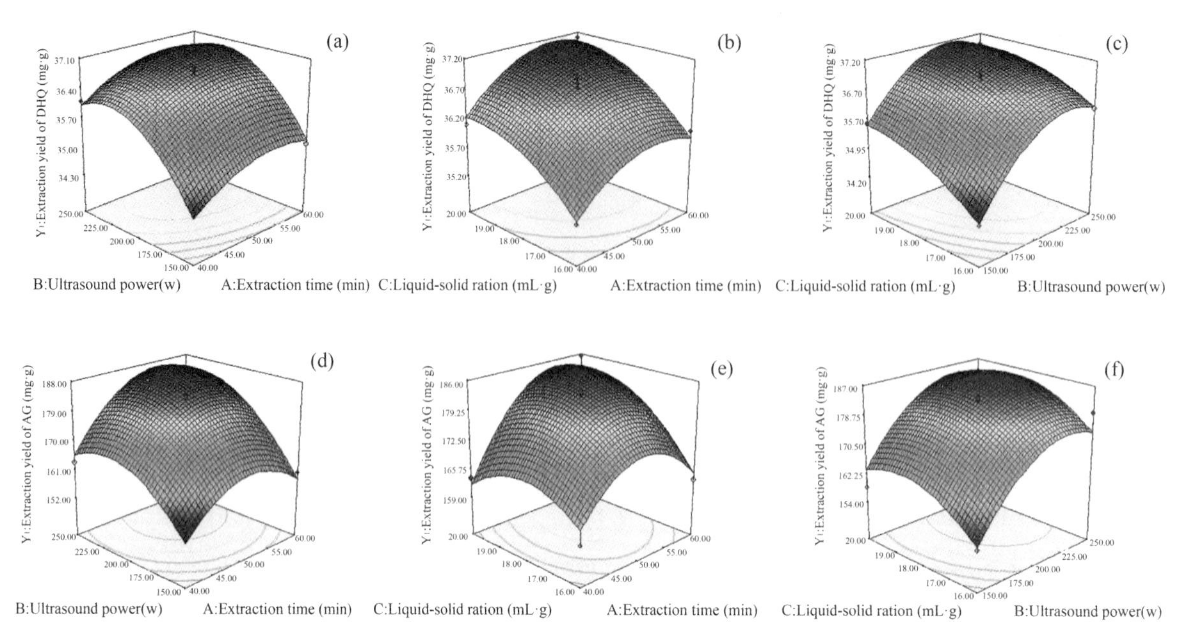

Fig. 5 Response surfaces. (a) Response surface of the extraction yields of DHQ, Y15f(A, B). (b) Response surface of the extraction yields of DHQ, Y15f(A, C). (c) Response surface of the extraction yields of DHQ, Y15f(B, C). (d) Response surface of the extraction yields of AG, Y25f(A, B). (e) Response surface of the extraction yields of AG, Y25f(A, C). (f) Response surface of the extraction yields of AG, Y25f(B, C)

3.7 Comparison of different extraction methods

Water is the most common and inexpensive solvent, therefore pure water is always selected as the co-solvent in various extraction processes, with AG being soluble in water and DHQ in hot water. Table 5 (the original data, File S1) shows the extraction yields of AG and DHQ for different reference extraction methods (n53). The extraction yields of AG and DHQ using soxhlet extraction for 12 h were the highest; 202.43±4.44 mg · g^{-1} for AG and 69.3±2.95 mg · g^{-1} for DHQ with 40% ethanol solution, and 223.43±4.88 mg · g^{-1} for AG and 63.4±2.29 mg · g^{-1} for DHQ with water. This extraction time was too long with higher energy consumption. The extraction yields of AG and DHQ using soaking at room temperature for 24 h were also high, reaching 164.22±4.28 mg · g^{-1} for AG and 56.6±1.94 mg · g^{-1} for DHQ, when the extraction solvent was 40% ethanol solution, and reaching 187.57±4.21 mg · g^{-1} for AG and 40.0±2.74 mg · g^{-1} for DHQ, when the extraction solvent was water. Although there was no energy consumption for this method, the extraction time was also too long and the extraction efficiency lower. Conversely, reflux extraction is a time-saving method, but consumes energy during the process of heating. The extraction yields of AG and DHQ with 40% ethanol solution reflux extraction for 2 h were 194.57±4.52 mg · g^{-1} for AG and 46.5±1.71 mg · g^{-1} for DHQ; the extraction yields of AG and DHQ with water reflux extraction for 2 h were 207.43±4.30 mg · g^{-1} for AG and 32.4±1.32 mg · g^{-1} for DHQ. The extraction yields of AG and DHQ using stirring extraction for 8 h with 40% ethanol solution at 50 ℃ were 200.32±4.27 mg · g^{-1} for AG and 60.7±2.15 mg · g^{-1} for DHQ, and using stirring extraction for 8 h with water at 50 ℃ were 218.42±5.31 mg · g^{-1} for AG and 58.5±1.13 mg · g^{-1} for DHQ, respectively. Compared with these extraction methods, the extraction yields of AG and DHQ using UAE were not the highest (183.40±4.26 mg · g^{-1} for AG and 36.8±1.21 mg · g^{-1} for DHQ with 40% ethanol, and

147. 48±3. 08 mg · g^{-1} for AG and 17. 6±0. 86 mg · g^{-1} for DHQ with), but it had the advantages of lower energy consumption and saving time. Moreover, when the solvent was 40% ethanol solution, it can obtain AG and DHQ simultaneously.

3. 8 Comparison of pulping processes and the physical properties of paper

Table 6 (the original data, File S1) shows that the pulp yield using UAE pretreatment was higher than that from untreated pulp and that the high-value active ingredients, AG and DHQ, could be obtained during this process. UAE not only increased the pulp yield but also created certain economic benefits. Moreover, the extraction of AG and DHQ using UAE pretreatment provides a convenient process for the subsequent pulp bleaching. The yield of HBSP was higher than that of KP, because the lignin removal during HBSP process is not as high compared with that during KP process. This can be explained by the kappa number of HBSP (77. 91±0. 06 for untreated and 77. 01±0. 06 for UAE- pretreated) being nearly three times higher than that for KP (27. 30±0. 13 for untreated and 26. 83±0. 08 for UAE-pretreated). However, compared with KP (the residual alkali value of untreated KP was 16. 30±0. 04, and that of UAE-pretreated KP was 16. 55±0. 06. It indicated that the alkali consumption was decreased through UAE pretreated), the natural advantage of the HBS pulping process is the recyclable solvent, which solves the problem of the high energy consumption for black liquor recovery and pollution of the environment during the process of alkaline pulping.

Table 5 Comparison of different extraction methods

Item	Raw material	Extraction method	Extraction yield of AG (mg · g^{-1}) (n53)	Extraction yield of DHQ (mg · g^{-1}) (n53)
A	Wood flour	Soxhlet extraction 12 h with 40% ethanol solution	202. 43±4. 44	69. 3±2. 95
B	Wood flour	Soxhlet extraction 12 h with water	223. 43±4. 88	63. 4±2. 29
C	Wood chips	Soak 24 h with 40% ethanol solution at room temperature	164. 22±4. 28	56. 6±1. 94
D	Wood chips	Soak 24 h with water at room temperature	187. 57±4. 21	40. 0±2. 74
E	Wood chips	Reflux extraction 2 h with 40% ethanol solution	194. 57±4. 52	46. 5±1. 71
F	Wood chips	Reflux extraction 2 h with water	207. 43±4. 30	32. 4±1. 32
G	Wood chips	40% ethanol solution stirring extraction 8 h at 50 ℃	200. 32±4. 27	60. 7±2. 15
H	Wood chips	Water stirring extraction 8 h at 50 ℃	218. 42±5. 31	58. 5±1. 13
I	Wood chips	UAE 50 min with 40% ethanol solution	183. 40±4. 26	36. 8±1. 21
J	Wood chips	UAE 50 min with water	147. 48±3. 08	17. 6±0. 86

Table 7 (the original data, File S1) shows that many physical properties paper were improved by UAE-pretreatment. The basis weight of paper was lower (85. 67±0. 64 and 82. 48±0. 55 g · cm^{-2} for paper from untreated KP and HBSP; 79. 94±0. 32 and 80. 25±0. 55 g · cm^{-2} for paper from UAE-pretreated KP and HBSP, respectively). The apparent densities of paper were nearly the same (0. 3273±0. 0016 and 0. 3085± 0. 0005 g · cm^{-3} for paper from untreated KP and HBSP; 0. 3272±0. 0034 and 0. 3054±0. 0014 g · cm^{-3} for paper from UAE-pretreated KP and HBSP, respectively). The tensile strengths were slightly higher (38. 69± 0. 07 and 37. 25±0. 07 N · m · g^{21} for paper from untreated KP and HBSP; 38. 75±0. 11 and 37. 90±0. 13 N · m · g^{-1} for paper from UAE-pretreated KP and HBSP, respectively). The tearing strengths were higher (14. 25±0. 08 and 14. 12±0. 06 mN · m^2 · g^{-1} for paper of untreated KP and HBSP; 16. 88±0. 08 and 16. 22

±0. 06 mN · m^2 · g^{-1} for paper of pretreated KP and HBSP, respectively). The bursting strengths were slightly higher (2. 17±0. 04 and 1. 87±0. 06 Kpa · m^2 · g^{-1} for paper from untreated KP and HBSP; 2. 25±0. 04 and 1. 93±0. 05 Kpa · m^2 · g^{-1} for paper from UAE-pretreated KP and HBSP, respectively). The folding strengths (8±1 and 6±1 times for paper from untreated KP and HBSP; 11±1 and 6±0 times for paper from UAE-pretreated KP and HBSP, respectively) of paper were higher. The reason for these changes is that during the process of UAE- pretreatment, compounds such as DHQ, the water-soluble AG, tannin and polyphenols, which are harmful to the pulping and papermaking processes, are removed. This improves the reaction efficiency of pulping and also the quality of the resulting paper.

Table 6 Technical analysis of pulp

Characters of pulping	Untreated KP	UAE-pretreated KP	Untreated HBSP	UAE-pretreated HBSP
Yield of pulp (%)	39. 60	41. 50	42. 37	44. 23
Kappa number (n53)	27. 30±0. 13	26. 83±0. 08	77. 91±0. 06	77. 01±0. 06
Residual alkali value (n53)(g · L^{-1})	16. 30±0. 04	16. 55±0. 06	—	—

Table 7 Physical properties of paper

Physical properties	Paper of untreated KP	Paper of UAE-pretreated KP	Paper of untreated HBSP	Paper of UAE-pretreated HBSP
Weight (g) (n53)	2. 69±0. 02	2. 51±0. 01	2. 59±0. 02	2. 52±0. 02
Basis weight (g · cm^{-2}) (n53)	85. 67±0. 64	79. 94±0. 32	82. 48±0. 55	80. 25±0. 55
Apparent density (g · cm^{-3}) (n53)	0. 3273±0. 0016	0. 3272±0. 0034	0. 3085±0. 0005	0. 3054±0. 0014
Tensile strength (N · m · g^{-1}) (n56)	38. 69±0. 07	38. 75±0. 11	37. 25±0. 07	37. 90±0. 13
Tearing strength (mN · m^2 · g^{-1}) (n56)	14. 25±0. 08	16. 88±0. 08	14. 12±0. 06	16. 22±0. 06
Bursting strength (Kpa · m^2 · g^{-1}) (n56)	2. 17±0. 04	2. 25±0. 04	1. 87±0. 06	1. 93±0. 05
Folding strength (times) (n56)	8±1	11±1	6±1	6±0

4 Conclusions

In summary, ultrasound-assisted extraction (UAE) was used for extracting arabinogalactan (AG) and dihydroquercetin (DHQ) simultaneously from Larix gmelinii wood, as a pretreatment for pulping and papermaking. Compared with untreated wood chips, the pulping characteristics and physical properties of the resulting paper were improved. The extraction parameters were optimized by Response Surface Methodology. The optimum conditions were three extraction cycles using 40% ethanol with 50 min for each cycle, ultrasound power at 200 W and a solid-liquid ratio of 1 : 18, leading to extraction yields for AG and DHQ of 179. 88±10. 49 and 35. 96±1. 66 mg · g^{-1}, respectively. The UAE pretreatment of wood chips not only improved the pulping characteristics and physical properties of paper, but also obtained substantial yields of the active compounds, AG and DHQ. The data from the present study will contribute to promoting the extraction of these valuable active compounds from larch wood.

Supporting Information

File S1. Original data of Table 1, Table 2, Table 5, Table 6, Table 7, and Fig. 2, Fig. 3, Fig. 4, and verification test. doi: 10. 1371/journal. pone. 0114105. s001 (XLS)

File S2. Original data of Table 3, Table 4, and Fig. 5.

doi: 10. 1371/journal. pone. 0114105. s002 (DX7)

Acknowledgments

We gratefully acknowledge the technical support from the College of Material Science and Engineering, Northeast Forestry University, and Key Laboratory of Forest Plant Ecology, Ministry of Education, Northeast Forestry University. We are thankful to two anonymous reviewers and editor for helpful comments.

Author Contributions

Conceived and designed the experiments: LY JY. Performed the experiments: CM WL. Analyzed the data: CM WL. Contributed reagents/materials/analysis tools: YZ JL. Wrote the paper: CM.

References

[1] Gernandt, David S, Aaron L. Internal Transcribed Spacer Region Evolution in Larix and Pseudotsuga (Pinaceae)[J]. Am. J. Bot., 1999, 86: 711-723.

[2] Gros-Louis M C, Bousquet J, Paques L E, et al. Species-diagnostic markers in Larix spp. based on RAPDs and nuclear, cpDNA, and mtDNA gene sequences, and their phylogenetic implications[J]. Tree Genetics & Genomes, 2005, 1: 50-63.

[3] Jie Z. Composition Study of Phenol Compounds of Chinese Larch Tree[J]. J. Cellulose Sci. and Technol, 2001, 9(4): 16-20.

[4] Chen S. Mechanism of Larch Kraft Pulping-Delignification Topochemistry and Dissolution of Polysaccharidcs[J]. Transactions of China. Pulp and Paper, 1989, 4: 61-69.

[5] Zhi L I. Research on Larch Kraft Pulp ECF Bleaching[J]. China Pulp. and Paper, 2012, 31(9): 6-9.

[6] Svetlana A M, Galina P A, Valentina I D, et al. Larch arabinogalactan as a perspective polymeric matrix for biogenic metals[J]. Chem Nat. Compd+, 2002, 2(7): 47-50.

[7] Medvedeva E N, Babkin V A, Ostroukhova L A. Arabinogalactan from larch-properties and usage perspectives (review)[J]. Chemical Plant Stock, 2003, 1(1): 27-37.

[8] Ernest, V, Groman, et al. Development of an immunoassay for larch arabinogalactan and its use in the detection of larch arabinogalactan in rat blood[J]. Carbohyd Res., 1997, 301(1-2): 69-76.

[9] Goellner E M, Utermoehlen J, Kramer R, et al. Structure of arabinogalactan from Larix laricina and its reactivity with antibodies directed against type-Ⅱ-arabinogalactans[J]. Carbohyd Polym, 2011, 86(4): 1739-1744.

[10] Marett R, Slavin J L. No long-term benefits of supplementation with arabinogalactan on serum lipids and glucose[J]. J. Am. Diet. Assoc., 2004, 104(4): 636-639.

[11] Josephson L, Groman E V, Chu J, et al. Targeting of therapeutic agents using polysaccharides. US Patent 5. 336. 506.

[12] Currier N L, Lejtenyi D, Miller S C. Effect over time of in-vivo administration of the polysaccharide arabinogalactan on immune and hemopoietic cell lineages in murine spleen and bone marrow[J]. Phytomedicine, 2003, 10(2-3): 145-153.

[13] Chen W, Wang W, Zhang H, Huang Q. Optimization of ultrasonic-assisted extraction of water-soluble polysaccharides from Boletus edulis mycelia using response surface methodology[J]. Carbohyd Polym, 2012, 87: 614-619.

[14] Sang, Mi, An, et al. Flavonoids, taxifolin and luteolin attenuate cellular melanogenesis despite increasing tyrosinase protein levels[J]. Phytother Res., 2008, 22(9): 1200-1207.

[15] Marozien A, Kliukien R, Arlauskas J, et al. Inhibition of phthalocyanine-sensitized photohemolysis of human erythrocytes by polyphenolic antioxidants: description of quantitative structure-activity relationships[J]. Cancer Lett., 2000, 157(1): 39-44.

[16] Sugihara N, Arakawa T, Ohnishi M, et al. Anti- and pro-oxidative effects of flavonoids on metal-induced lipid hydroperoxide-dependent lipid peroxidation in cultured hepatocytes loaded with α-linolenic acid[J]. Free Radic Biol. Med., 1999, 27(11-12): 1313.

[17] Chu S C, Hsieh Y S, Lin J Y. Inhibitory effects of flavonoids on Moloney murine leukemia virus reverse transcriptase activity[J]. J. Nat. Prod., 1992, 55(2): 179.

[18] Kawaii S, Tomono Y, Katase E, et al. Effect of citrus flavonoids on HL-60 cell differentiation[J]. Anticancer Res, 1999, 19(2A): 1261.

[19] Trouillas P, Fagnère C, Lazzaroni R, et al. A theoretical study of the conformational behavior and electronic structure of taxifolin correlated with the free radical-scavenging activity[J]. Food Chem., 2004, 88(4): 571-582.

[20] Pietarinen S P, Willför S M, Vikström F A, et al. Aspen Knots, a Rich Source of Flavonoids[J]. J. wood chem. technol, 2007, 26(3): 245-258.

[21] Peng J, Fan G, Chai Y, et al. Efficient new method for extraction and isolation of three flavonoids from Patrinia villosa Juss. By supercritical fluid extraction and high-speed counter-current chromatography[J]. J. Chromatogr A., 2006, 1102(1-2): 44-50.

[22] Ying W, Zu Y, Long J, et al. Enzymatic water extraction of taxifolin from wood sawdust of Larix gmelini (Rupr.) Rupr. and evaluation of its antioxidant activity[J]. Food Chem., 2011, 126(3): 1178-1185.

[23] Domini C, Vidal L, Cravotto G, et al. A simultaneous, direct microwave/ultrasound-assisted digestion procedure for the determination of total Kjeldahl nitrogen[J]. Ultrason Sonochem, 2009, 16(4): 564-569.

[24] Hu Y, Wang T, Wang M X, et al. Extraction of isoflavonoids from Pueraria by combining ultrasound with microwave vacuum[J]. Chem. Eng. Process, 2008, 47: 2256-2261.

[25] Iida Y, Tuziuti T, Yasui K, et al. Control of viscosity in starch and polysaccharide solutions with ultrasound after gelatinization[J]. Innov. Food Sci. Emerg Technol, 2008, 9(2): 140-146.

[26] Wang C H, Yang S Y, Zhao F J, et al. Ionic liquid-aqueous solution ultrasonic-assisted extraction of camptothecin and 10-hydroxycamptothecin from Camptotheca acuminata samara[J]. Chem. Eng. Process, 2012: 57-58, 59-64.

[27] Ma C H, Liu T T, Lei Y, et al. Study on ionic liquid-based ultrasonic-assisted extraction of biphenyl cyclooctene lignans from the fruit of Schisandra chinensis Baill. [J]. Analytica Chimica Acta, 2011, 689(1): 110-116.

[28] Liu T T, Ma C H, Sui X Y, et al. Preparation of shikonin by hydrolyzing ester derivatives using basic anion ion exchange resin as solid catalyst[J]. Ind. Crop Prod., 2012, 36(1): 47-53.

[29] Ultrasound-Assisted Extraction of Carnosic Acid and Rosmarinic Acid Using Ionic Liquid Solution from Rosmarinus officinalis[J]. Int. . J Mol. Sci., 2012, 13(12): 11027-11043.

[30] Schinor E C, Salvador M J, Turatti L C C, et al. Comparison of classical and ultrasound-assisted extractions of steroids and triterpenoids from three Chresta spp[J]. Ultrason Sonochem, 2004, 11(6): 415-421.

中文题目：作为制浆造纸预处理的超声辅助萃取落叶松阿拉伯半乳聚糖和二氢槲皮素

作者：马春慧，杨磊，李伟，岳金权，李坚，祖元刚

摘要：采用乙醇超声辅助提取(UAE)方法从落叶松中同时提取阿拉伯半乳聚糖(AG)和二氢槲皮素(DHQ)，作为制浆造纸的预处理。采用Box-Behnken实验设计，对提取过程各参数进行了优化以AG和DHQ的得率作为响应值。在最佳条件下(40%乙醇提取三次，每次超声提取时间50 min，超声

功率 200 W，料液比 1∶18，AG 和 DHQ 的产率分别为 183.4 和 36.76 mg·g^{-1}。经 UAE 预处理后的木片用于硫酸盐制浆(*KP*)和高沸醇溶剂制浆(*HBSP*)。预处理后的制浆得率(*HBSP*=44.23%，*KP*=41.50%)均高于未处理的制浆收率(*HBSP*=42.37%，*KP*=39.60%)，未经预处理的 kappa 值(*HBSP*=77.01，*KP*=26.83)低于预处理后的 kappa 值(*HBSP*=77.91，*KP*=27.30)。此外，还介绍了经预处理后的纸张特性优于未经处理的纸张特性：预处理后纸张定量(*HBSP*=85.67 g·cm^{-2}，*KP*=82.48 g·cm^{-2})比未处理的纸张定量高(*HBSP*=79.94 g·cm^{-2}，*KP*=80.25 g·cm^{-2})；并且纸张的抗张强度、撕裂度和耐折度均优于未经预处理的纸张。

关键词：超声辅助提取；阿拉伯半乳聚糖；二氢槲皮素；制浆造纸

影响未来的颠覆性技术——多元材料混合智造的3D打印*

李坚，许民，包文慧

摘要：阐述了3D打印基本概念、实体制造方法、国内外发展研究近况，对3D打印技术的优势与问题、选用的原材料与特点、3D打印技术与传统制造业关系进行讨论，简介3D打印技术市场占有率及认知度。着重分析3D打印技术与第三次工业革命的关系，并展望其发展趋势。预测：以木材加工剩余物、农作物秸秆等为原料的3D打印技术，将成为农林生物质及其废弃物高值利用的有效途径之一。

关键词：3D打印；制造流程；优势特点；第三次工业革命

美国麦肯锡全球研究院预测：3D打印是“影响未来的颠覆性技术”之一。

3D打印制造技术，是融合了计算机软件、材料、机械、控制、网络信息等多学科知识的系统性、综合性技术，是制造业领域的一项新兴技术；其依托信息技术、精密机械、材料科学等多学科尖端技术。其采用特殊材料，通过选择性黏结逐层叠加形成实体。

近年来，3D打印技术发展迅速，在各领域都取得长足进步，在航空航天、汽车制造、生物医疗、教学科研等领域发挥重要作用。当前，在发达国家大力倡导再工业化、再制造化的背景下，以3D打印技术为代表的数字化制造技术将引发第三次工业革命，对产品设计、制造工艺、制造设备及生产线、材料制备、相关工业标准、制造企业形态乃至整体制造体系，将产生全面、深刻的变革。当前，我国制造业大而不强，在信息化与工业化融合推进、走新型工业化道路进程中，应大力发展3D打印制造技术，对增强我国制造业创新能力、提升工艺制造能力、破解制造业发展与资源环境困局、培育新兴产业及调整产业结构等具有极其重要的战略意义。

林产工业作为一种传统产业，其与森林资源息息相关。中华人民共和国成立后的60余年，历经辉煌、平稳、低谷和复苏4个时期，面临新世纪产业革命和环境保护的双重压力，对于从事林业行业的生产者，如何提高生产效率获取最大利益、如何实现产品快速更新换代、如何能在多方压力下保证林业行业更加节能环保，这些都需要生产者从设计、生产、销售等其他各个环节去综合考虑的。3D打印技术的出现，在一定程度上这些问题也将会促使林业产业产生巨大变革。如果能使这项技术为林业生产企业服务，可为企业节约更多的时间成本；3D打印技术的增材加工方式，若能在木制品生产中大规模使用，也将比现有的减材加工方式节约大量原材料，减少制作过程中能耗。通过3D智能数字化技术的发展，将“中国制造”转变为“中国智造”，非常适合现在我国可持续发展环境友好的经济发展方针，具有跨时代的重要意义[1]。

* 本文摘自东北林业大学学报，2015，43(6)：1-9.

1 3D打印技术

1.1 3D技术概念

3D打印技术，最早于19世纪末起源于美国，20世纪80年代得到实现与发展。3D打印技术作为快速成型领域的一种新兴技术，是目前一种迅猛发展的潮流。近一段时间，3D打印技术吸引了国内外新闻媒体和社会公众的热切关注。英国《经济学人》杂志2011年2月刊载封面文章，对3D打印技术的发展做了介绍和展望，文章认为：3D打印技术未来的发展将使大规模的个性化生产成为可能，这将会带来全球制造业经济的重大变革。很多新闻媒体乐观地认为：3D打印产业将成为下一个具有广阔前景的朝阳产业[2]。

3D打印技术，是指通过连续的物理层叠加，逐层增加材料生成三维实体的技术，与传统的去除材料加工技术不同，因此又称添加制造。作为一种综合性应用技术，3D打印综合了数字建模技术、机电控制技术、信息技术、材料科学与化学等诸多方面的前沿技术知识，具有很高的科技含量[3]。3D打印又称三维打印，工业上称快速成型；最近，行业内部又统一更规范的专业术语——增材制造。3D打印，可以说是这项技术的“乳名”，“快速成型”可以说是它的“学名”，“增材制造”可以理解为它的专业术语[4]。

1.2 3D打印原理

3D打印基本原理，是断层扫描的逆过程。断层扫描，是把某个物品“切”成无数叠加的片；3D打印，则是逐片打印，然后叠加到一起，成为一个立体物体。3D打印机，是可以“打印”出真实3D物体的一种设备；功能上与激光成型技术一样，采用分层加工、叠加成型，即通过逐层增加材料生成3D实体，与传统的去除材料加工技术完全不同[5]。3D打印机是3D打印的核心装备，是集机械、控制及计算机技术等为一体的复杂机电一体化系统，主要由高精度机械系统、数控系统、喷射系统和成型环境等子系统组成。此外，新型打印材料、打印工艺、设计与控制软件等，也是3D打印技术体系的重要组成部分。称为“打印机”，是参照其技术原理，因为分层加工的过程与喷墨打印十分相似。

3D打印机，又称快速成型机；3D打印机可支持多种材料，较为普遍的有树脂、尼龙、石膏、塑料等可塑性较强的材料。3D打印机的精确度极高，即便是低档廉价的型号，也可以打印出模型中的大量细节；而且与铸造、冲压、蚀刻等传统方法相比，更能快速创建原型，特别是传统方法难以制作的特殊结构模型。通过3D打印获得一件物品，需要经历建模、分层、打印和后期处理4个主要阶段，具体流程如图1所示[6]。

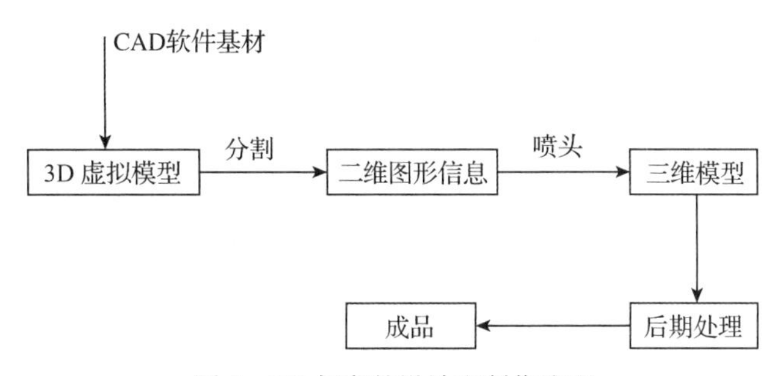

图1 3D打印的设计和制作流程

1.3 3D 打印实体制造方法

3D 打印技术，按照技术路径分为分层、叠加 2 个步骤。分层，是将设计者需要的实体，通过设计软件扫描并快速计算转换成表面网格结构；叠加，是采用各种不同类型材料，逐层堆垛成型的过程。2 个步骤所涉及的关键技术，即设计软件和成型技术。

在设计软件领域中离不开 STL 文件。STL 文件格式，是由 3D SYSTEMS 公司于 1988 年制定的一个接口协议，是一种为快速原型制造技术服务的三维图形文件格式。STL 文件在计算机图形应用系统中，主要通过三角形网格表述。其文件格式简单，只能描述三维物体的几何信息，不支持颜色材质等信息。STL 是最多快速原型系统所应用的标准文件类型。STL 文件一旦创建，3D 打印机将模型切“片”，存为一系列横截面的文件。STL 文件需要水密后，才可以进行三维打印。水密最好的解释，是无孔的有体积固体。虽然设计者设计的固体已经创建完成，但可能在模型中仍存在没有被留意的小孔。STL 文件也存在一定弊端，其在数字切片处理时，会删除一些设计细节，影响打印实体精度。为了更好处理复杂网格的设计文件，人们将 STL 文件升级，采用新标准 AMF(增材制造格式)。AMF 保留了 STL 格式的曲面网络结构，新增了可以处理不同颜色、不同类型材料、创建格子结构以及处理详细内部结构的功能。与之前的标准相比，更准确描述实体曲面。

3D 打印主流技术包括 FDM、SLA、SLS、LOM、DLP、3DP 等。

FDM(Fused Deposition Modeling)，熔融沉积成型。该工艺属于丝材挤出热熔成型，也称熔丝制造。技术原理：将直径约为 2mm 丝状热塑性材料通过喷头加热熔化；喷头底部带有微细喷嘴(直径为 0.2~0.6mm)，材料以一定压力挤喷出来，同时喷头沿水平方向移动，挤出的材料与前一个层面熔结在一起。一个层面沉积完成后，工作台垂直下降一个层厚度继续熔融沉积，直至完成完整实体造型。FDM 工艺使用两种材料：一种是成型材料；另一种是支撑材料，以防空腔或悬臂部分坍塌。由于 FDM 操作环境洁净、安全，没有产生毒气和化学污染，无须激光器等贵重元器件；原材料以卷轴丝形式提供，方便搬运更换。目前，市场上的桌面级 3D 打印机多使用这种工艺。但该技术产品成型后，表面粗糙、精度较低，还需后续处理。

SLA(Stereo Lithography Appearance)，光固化立体成型。即在树脂液槽中盛满透明、有黏性的液态光敏树脂，在紫外激光束的照射下会快速固化。成型开始时，光束按照预定截面轮廓要求进行扫描。扫描区域树脂固化，工作台下降让液面覆盖固化层，开始新的扫描。由于耗材为液态，SLA 制作的试件尺寸精度较高，适合做较小精细零件。但树脂类耗材价格相对偏高，力学性能、使用性能有局限性，SLA 自身设备价格、维护成本较高。工作环境要求苛刻。

SLS(Selective Laser Sintering)，选择性激光烧结。其利用粉末在激光照射下选择性烧结。首先铺一层粉末材料，将材料预热到接近熔化点，再使用高强度的 CO_2 激光器有选择地在该层截面上扫描，使粉末温度升至熔化点；其次烧结形成黏结；最后铺粉、烧结，直至完成整体模型。SLS 使用耗材相对广泛，成件效率高、时间短，无须支撑材料。但表面粗糙，需后处理工艺。

LOM(Laminated Object Manufacturing)，分层实体制造。利用激光或刀具切割薄层纸，然后通过热压或其他形式层层黏结，叠加获得三维实体零件。

DLP(Digital Light Processing)，数字光处理技术。采用高分辨率数字光处理器投影仪，固化液态光聚合物。

3DP(Three Dimensional Printing and Gluing)，三维打印黏结成型。该技术利用喷头喷黏结剂，选择性黏结粉末。层层叠加得到三维实物。

2 国内外研究发展现状

3D打印技术的发展已近30年，本文从专利申请、检测标准、企业概况、应用领域、科研发展等方面分析国内外发展现状。

专利检索：国外3D打印技术自诞生到1993年有80件专利，申请数量较低、分布较为分散。在1995年之前，该技术还处于萌芽期，专利申请量增长缓慢，主要申请人有麻省理工学院、3D公司等；在1996—1998年，专利申请处于平稳增长期[7]；1999—2003年，专利申请量呈快速增长趋势，2003年达到高峰，进入快速增长期。从专利占有比例看，美国、欧洲、日本，专利申请比例较大。随后，专利申请总量出现下滑趋势，技术发展进入相对成熟时期。国内，自20世纪90年代才开始涉足3D打印技术领域；1999年之前，专利申请量较少；2000—2007年，专利申请量稳步增长；2007年之后，专利申请量快速增长，直到2011年达到133件；说明我国还处于研发的快速成长阶段。国外，专利权人以企业为主，3D公司是该领域技术研发最活跃、技术水平领先者，也是进行产业化运作的主要企业。国外，重视企业合作、区域合作、协同创新的研发模式。通过企业之间的密切合作，不仅能使专利申请量提高，也使得企业互相取长补短，增强企业的技术创新能力，扩大在该技术领域的影响力和竞争力。国内，专利权人主要以高校、科研机构为主，华中科技大学、西安交通大学是国内最早相继进行研发的单位。我国缺少大型独立研发企业，且合作意识不强，企业之间缺少交流。相关企业依托高校产学研相结合方式获得研发动力。从专利技术研究内容看，国内研究，主要是基于分层实体制造、熔融沉积、光固化成型等技术研发及成型系统的研制。除此之外，还研究通过金属粉末、陶瓷、热塑性塑料等类型材料生成3D打印制品过程中涉及的材料配方优化及成型工艺、模型制作、成型设备等。国外公司主要从事的研究内容，有微滴喷射技术、全彩3D打印技术、熔融材料高分辨率选择性逐层喷射技术、光敏固化技术等方面。

标准制定：2002年，美国汽车工程师协会发布第一个增材制造标准，即宇航材料规范——AMS4999《退火Ti-6AL-4V钛合金激光沉积产品》，该标准的颁布是该技术在美国航空航天领域走向实际应用的重要标志[8]。2009年，美国材料与试验协会(ASTM)成立专门的增材制造技术委员会ASTM F-42；ASTM F- 42已经颁布4项标准，包括术语、文件格式等基础标准和产品标准等。2011年，国际标准化组织(ISO)也成立增材制造技术委员会ISO TC 261；目前，ISO TC 261进行的一项工作是制定ISO 17296《增材制造——快速技术(快速原型制造)》标准。此外，还有很多其他标准在制订计划中，例如：增材制造坐标系与命名标准、增材制造试验结果报告的惯例、网络结构术语、设计指南、文件格式标准规范等。我国增材制造技术标准的发展落后于国外，没能充分反映国内技术发展的水平；已经应用到产品领域的增材制造技术产品，均采用各企业的技术条件和规范；没有形成统一指标。由于缺少对沉积工艺过程的表征、控制和认证的规定，技术的大范围推广受到不同程度的制约，使已有的技术优势并没有迅速转化为产品优势和市场优势。因此，迫切需要开展增材制造技术的标准化工作。

企业发展：国外企业，在工艺技术、研发投入、人才基础、产业形态等领域都强于我国；企业规模比国内大，企业销售收入都在10亿人民币左右。美国3D Systems和Stratasys两家公司，占据全球3D行业绝大多数份额[9]。国内还没有一家企业收入过1亿人民币，超过5000万人民币的没有几家，大多数保持在两三千万人民币的水平。销售收入直接影响企业对研发经费的投入。目前，国内很多企业以直接或间接的方式与高校合作，利用其优质研究资源。而像北京殷华、湖南华曙高科，都是由海外归国团队建立，规模较小，产品技术处于低端。还有一些企业，采取与国外公司合作发展，如南京紫金立德电子有限公司、无锡飞而康快速制造有限公司。从企业运营模式看，国外企业采取设备销售与加工服务相结合的商业运作模式，有利于开拓市场，也有利于技术进一步完善。国内企业，基本上还采取传统制造模式，生产设备卖设备；只有北京隆源自动成型系统公司、武汉滨湖机电科技公司等少数几家企业，推行与加工服务相结合的运营模式。无论国外还是国内企业，3D 打印行业基本上还是各自为政，小而散的局面不利于形成产业链，也不利于技术发展推广。2012 年，我国率先成立中国 3D 打印技术产业联盟，而产业基地、基础配套设施也都随着国家政策扶持相继建立起来。

应用领域：目前，3D 打印技术的应用可分为 3 个领域：工业应用、家庭应用、生物医疗。其在新产品开发、个性化复杂产品定制、文化创意、教育培训等多方面，均有较好发展潜力[10]。在国家和地方的支持下，在全国建立 20 多个服务中心，设备用户遍布医疗、航空航天、汽车、军工、模具、电子电器、造船等行业，推动我国制造技术的发展。近 5 年，国内增材制造市场发展不大，主要还在工业领域应用，没有在消费品领域形成快速发展的市场。我国增材制造技术主要应用于模型制作，工业领域的金属零件直接制造技术也达到国际领先水平的研究与应用，例如，北京航空航天大学、西北工业大学、北京航空制造技术研究所制造出大尺寸金属零件，并应用在新型飞机研制过程中，显著提高了飞机研制速度[11]。生物 3D 打印，以活细胞、生物活性因子为基本成形单元，用于设计制造人体器官、植入细胞三维结构、药物开发、生物制药等领域。2013 年 2 月 22 日，美国康奈尔大学通过 3D 打印技术，帮助治疗先天性耳朵畸形或耳朵缺失人群直接打印耳朵。此外，他们还可以修复人体退化的脊椎间盘。

科学研究：我国自 20 世纪 90 年代初，在国家科技部等多部门持续支持下，西安交通大学、华中科技大学、清华大学、北京隆源公司等，在典型的成形设备、软件、材料等方面研究和产业化方面获得重大进展。随后，国内许多高校和研究机构也开展相关研究，如西北工业大学、北京航空航天大学、华南理工大学、南京航空航天大学、上海交通大学、大连理工大学、中北大学、中国工程物理研究院等单位，都在做探索性的研究和应用工作。我国研发出一批增材制造装备，在典型成型设备、软件、材料等方面的研究和产业化获得重大进展，到 2000 年，初步实现设备产业化接近国外产品水平，改变该类设备早期仰赖进口的局面。我国增材制造装备的部分技术水平，与国外先进水平相当；但在关键器件、成型材料、智能化控制和应用范围等方面，较国外先进水平落后。在增材的基础理论与成型微观机理研究方面，我国在一些局部点上开展了相关研究；但国外的研究，更基础、系统和深入。在工艺技术研究方面，国外是基于理论基础的工艺控制；而我国则更多依赖经验和反复的试验验证，导致我国增材制造工艺关键技术整体上落后于国外先进水平。

3 3D 打印的材料和特点

3D 打印材料是 3D 打印技术发展的重要物质基础；在某种程度上，材料的发展决定着 3D 打印能否有更广泛的应用。目前，3D 打印材料，主要包括工程塑料、光敏树脂、橡胶类材料、金属材料和陶瓷材料等；除此之外，彩色石膏材料、人造骨粉、细胞生物原料以及砂糖等食品材料，也在 3D 打印领域得到了应用[12]。3D 打印所用的这些原材料都是专门针对 3D 打印设备和工艺研发的。通常根据打印设备的类型及操作条件的不同以选取不同类型的材料，材料形态一般有粉末状、丝状、层片状、液体状等。以粉末状 3D 打印材料为例，其粒径大小由 1～100 μm 不等；为使粉末保持良好的流动性，一般要求粉末具有高球形度[13]。

(1)工程塑料。指被用作工业零件或外壳材料的工业用塑料，是强度、耐冲击性、耐热性、硬度及抗老化性均优的塑料。工程塑料是当前应用最广泛的一类 3D 打印材料，常见的有 Acrylonitrile

Butadiene Styrene(ABS)类材料、Polycarbonate(PC)类材料、尼龙类材料等[4]。

ABS材料是Fused Deposition Modeling(FDM，熔融沉积造型)快速成型工艺常用的热塑性工程塑料。具有强度高、韧性好、耐冲击等优点，正常变形温度超过90 ℃，可进行机械加工(钻孔、攻螺纹)、喷漆及电镀[14]。

PC材料是真正的热塑性材料。具备工程塑料的所有特性：高强度、耐高温、抗冲击、抗弯曲，可以作为最终零部件使用。使用PC材料制作的样件，可以直接装配使用，应用于交通工具及家电行业。PC材料的强度比ABS材料高出60%左右，具备超强的工程材料属性，广泛应用于电子消费品、家电、汽车制造、航空航天、医疗器械等领域[15]。

尼龙玻纤是一种白色粉末。与普通塑料相比，其拉伸强度、弯曲强度有所增强，热变形温度以及材料的模量有所提高，材料的收缩率减小；但表面变粗糙，冲击强度降低。材料热变形温度为110 ℃。主要应用于汽车、家电、电子消费品领域。

PC-ABS材料是一种应用最广泛的热塑性工程塑料。PC-ABS具备ABS的韧性和PC材料的高强度及耐热性，大多应用于汽车、家电及通信行业[16]。使用该材料配合Fortus设备，制作的样件强度比传统的FDM系统制作的部件强度高出60%左右；所以，使用PC-ABS能打印出包括概念模型、功能原型、制造工具及最终零部件等热塑性部件。

Polycarbonate-ISO(PC-ISO)材料是一种通过医学卫生认证的白色热塑性材料。具有很高的强度，广泛应用于药品及医疗器械行业，用于手术模拟、颅骨修复、牙科等专业领域。同时，因为具备PC的所有性能，也可用于食品及药品包装行业，做出的样品也可以作为概念模型、功能原型、制造工具及最终零部件使用。

Polysulfone(PSU)类材料是一种琥珀色的材料。热变形温度为189 ℃，是所有热塑性材料里面强度最高、耐热性最好、抗腐蚀性最优的材料，通常作为最终零部件使用，广泛用于航空航天、交通工具及医疗行业。PSU类材料能带来直接数字化制造体验，性能非常稳定，通过与Fortus设备的配合使用，可以达到令人惊叹的效果。

(2)光敏树脂。即Ultraviolet Rays(UV)树脂，由聚合物单体与预聚体组成，其中加有光(紫外光)引发剂(或称为光敏剂)。在一定波长的紫外光(250~300 nm)照射下，能立刻引起聚合反应，完成固化。光敏树脂一般为液态，可用于制作高强度、耐高温、防水材料。目前，研究光敏材料3D打印技术的主要有美国3D System公司和以色列Object公司[17]。

(3)橡胶类材料。具备多种级别弹性材料的特征，这些材料所具备的硬度、断裂伸长率、抗撕裂强度和拉伸强度非常适合于要求防滑或柔软表面的应用领域[18]。3D打印的橡胶类产品，主要有消费类电子产品、医疗设备以及汽车内饰、轮胎、垫片等。

(4)金属材料。近年来，3D打印技术逐渐应用于实际产品的制造，其中，金属材料的3D打印技术发展尤其迅速。在国防领域，欧美发达国家非常重视3D打印技术的发展，不惜投入巨资加以研究，3D打印金属零部件一直是研究和应用的重点。3D打印使用的金属粉末，一般要求纯净度高、球形度好、粒径分布窄、氧含量低。目前，应用于3D打印的金属粉末材料，主要有钛合金、钴铬合金、不锈钢、铝合金材料等，此外还有用于打印首饰用的金、银等贵金属粉末材料。

(5)陶瓷材料。具有高强度、高硬度、耐高温、低密度、化学稳定性好、耐腐蚀等优异特性，在航空航天、汽车、生物等行业有着广泛的应用。但由于陶瓷材料硬而脆的特点，使其加工成形尤其困难，特别是复杂陶瓷件需通过模具来成型，模具加工成本高、开发周期长，难以满足产品不断更新的需求[19]。3D打印用的陶瓷粉末，是陶瓷粉末和某一种黏结剂粉末所组成的混合物；由于黏结剂粉末的熔点较低，激光烧结时只是将黏结剂粉末熔化而使陶瓷粉末黏结在一起，在激光烧结之后，需将陶瓷制品放入温控炉中，在较高的温度下进行后处理；陶瓷粉末和黏结剂粉末的配比会影响到陶瓷零部件的性能。

(6)生物质材料。以木粉、农作物秸秆等生物质粉末为原料，与相适应的高分子聚合物粉末混合，制成木塑复合材料粉末。如将木粉与 PES(聚醚砜)粉末混合，通过选择性激光烧结获得成型件[20]。为了提高制件硬度，尚需对木塑复合材料进行增强处理，其中渗腊处理即可增加力学强度，又可改善制件外观。江苏锦禾高科技股份有限公司，采用秸秆、稻壳、淀粉等天然可再生植物纤维与高分子树脂经特殊工艺复合，形成新型绿色环保复合材料。广州优塑塑料有限公司，采用可降解壳聚糖改性的羟基丁酸脂作为 3D 打印耗材。

*7)其他 3D 打印材料。除了上面介绍的 3D 打印材料外，目前用到的还有彩色石膏材料、人造骨粉、细胞生物原料以及砂糖等材料。

4 3D 打印应用领域

3D 打印机的应用对象可以是任何行业，只要这些行业需要模型和原型[21]。目前，3D 打印技术已在工业设计、文化艺术、机械制造(汽车、摩托车)、航空航天、军事、建筑、影视、家电、轻工、医学、考古、雕刻、首饰等领域，都得到了应用。随着技术自身的发展，其应用领域将不断拓展。这些应用主要体现在以下 10 个方面[22]：

(1)设计方案评审。借助 3D 打印的实体模型，不同专业领域(设计、制造、市场、客户)的人员，可以对产品实现方案、外观、人机功效等进行实物评价。

(2)制造工艺与装配检验。3D 打印，可以较精确地制造出产品零件中的任意结构细节；借助 3D 打印的实体模型，结合设计文件，可有效指导零件和模具的工艺设计，或进行产品装配检验，避免结构和工艺设计错误。

(3)功能样件制造与性能测试。3D 打印的实体原型，本身具有一定的结构性能，同时利用 3D 打印技术可直接制造金属零件，或制造出熔(蜡)模；再通过熔模铸造金属零件，甚至可以打印制造出特殊要求的功能零件和样件等。

(4)快速模具小批量制造。以 3D 打印制造的原型作为模板，制作硅胶、树脂、低熔点合金等快速模具，可便捷地实现几十件到数百件零件的小批量制造。

(5)建筑总体与装修展示评价。利用 3D 打印技术可实现模型真彩及纹理特点的打印，可快速制造出建筑的设计模型，进行建筑总体布局、结构方案的展示和评价。

(6)科学计算数据实体可视化。计算机辅助工程、地理地形信息等科学计算数据，可通过 3D 彩色打印，实现几何结构与分析数据的实体可视化。

(7)医学与医疗工程。通过医学 CT 数据的三维重建技术，利用 3D 打印技术，制造器官、骨骼等实体模型，可指导手术方案设计，也可打印制作组织工程和定向药物输送骨架等。

(8)首饰及日用品快速开发与个性化定制。利用 3D 打印制作蜡模，通过精密铸造实现首饰和工艺品的快速开发和个性化定制。

(9)动漫造型评价。借助于动漫造型评价，可实现动漫等模型的快速制造，指导和评价动漫造型设计。

(10)电子器件的设计与制作。利用 3D 打印，可在玻璃、柔性透明树脂等基板上设计制作电子器件和光学器件，如 RFID、太阳能光伏器件、OLED 等。

5 3D 打印的优势与问题

5.1 优势

3D 打印核心优势，在于可以实现传统制造业难以解决的个性化、复杂化、高难度的制造难题；对于传统制造业，加工件形状复杂制造成本高；3D 打印的出现，可以解决部分高精尖、复杂零件制造难题。3D 打印可提供一些个性化定制服务，例如在模具模型、教育培训、文化创意、生物医疗等多领域，均可提供个性化定制服务[23]，并且随着时间的推移、技术的进步，可更广泛的应用到人们生产生活中。可以利用该技术解决航空发动机飞行叶片的复杂制造、在生物医疗领域进行人造器官的修复、在材料工程领域实现多组分材料的混合搭配以制造新材料。除此之外，该技术可提升设计创造空间。如果 3D 打印技术进入校园，有助于激发学生的创新性，提高学生的动脑动手能力。对产品开发，可快速实现设计制造，缩短研发周期，使新的产品更早上市。在能源利用方面，其可降低贵重资源的消耗，实现稀缺材料和其他资源的高效利用。超过 90%的原材料可以回收再利用，此技术具有节料、节能、环保的特点。这对节约一些大量应用于国民经济和国防工业的稀缺资源具有重要的战略意义。

5.2 目前存在的主要问题

(1)3D 打印的耗材。耗材，是目前制约 3D 打印技术广泛应用的关键因素。目前已研发的材料，主要有塑料、树脂和金属等；然而，3D 打印技术要实现更多领域的应用，需要开发更多的可打印材料。根据材料特点，深入研究加工、结构与材料之间的关系，开发质量测试程序和方法，建立材料性能数据的规范标准等。此外，在一些关键产业领域，寻找合适的材料也是一大挑战，例如：空客概念飞机的仿真结构，要求机身必须透明且有很高的硬度；为符合这些要求；需要研发新型复合材料。此外，目前对金属材料进行 3D 打印的需求尤为迫切，如工具钢、不锈钢、钛合金、镍基合金、银和金等，但目前这些打印技术尚未完全突破[24]。

(2)3D 打印机本身。据报道，世界上目前只有一种 3D 打印机能够同时打印出多种材料的产品。由于 3D 打印工艺发展还不完善，快速成型零件的精度和表面质量大多不能满足工程直接使用要求，只能作原型使用。3D 打印产品，由于采用叠加制造工艺，层与层之间连接得再紧密，目前也很难与传统锻件相媲美。

(3)3D 打印的价格。目前，3D 打印不具备规模经济的优势，价格方面的优势尚不明显。目前，1 kg 打印材料少则几百人民币，多则 4 万人民币左右；因此，3D 打印技术，在一段时间内还无法全面取代传统制造技术。但是，在单件小批量、个性化定制和网络社区化生产方面，对于大多数产品，不管打印 1 件还是 100 件，价格相差无几，因而 3D 打印具有无可比拟的优势。

(4)知识产权的保护。3D 打印技术的意义，不仅在于改变资本和工作的分配模式，而且也在于它能改变知识产权的规则。该技术的出现，使制造业的成功不再取决于生产规模，而取决于创意。然而，单靠创意也是很危险的，模仿者和创新者都能轻而易举地在市场上快速推出新产品，极有可能像当初的音乐领域一样面临盗版的威胁。

(5)3D 打印机的操作技能。3D 打印技术需要依靠数字模型进行生产，但是，普通用户学会使用计算机辅助设计工具(CAD)还有一定难度。随着社会发展，未来会有越来越多的人们学习并掌握这方面的技能，而且企业也会提供一些简单的产品数据库，用户不必学会 3D 设计技能就能制作模型，如同傻瓜相机的发展一样。

(6)政策方面。3D 打印技术的研发，需要大量的政府投入或产业界的资金支撑，如在医疗领域，可能会因缺少药品监管部门的许可，从而造成许多临床医疗产品应用的滞缓。

6 3D 打印与传统制造技术的关系

3D 打印技术与传统制造可以形成优势互补，两者相互依存，缺一不可。

所谓传统制造业，即指以劳动密集型与资金密集型为主导，以生产广泛面向市场的、不具备过高科学技术含量、附加值不高、满足基础设施和基础生活需要的产品的一类制造业的统称。传统制造技术经过数千年的积累与发展，已经在生产工艺、生产技术、材料应用等方面非常成熟，并形成配套完善、功能齐全、社会各界广泛认可的产业基础。但近些年来，随着全球资源的枯竭、劳动成本的攀升、经济萎靡，传统制造业所引发的问题及其自身的缺点越发明显。首先，利润较低——传统制造业制造成本逐年提高，而产品自身附加值并没有明显改变，致使利润降低，企业自身生存受到严重威胁。其次，创新性缺失——传统制造业对创新的重视与投入严重不足，缺乏核心技术，产业竞争力较差。最后，效率问题——传统制造业以高能耗、高污染、粗犷式、单一固定式的生产模式换取经济利益。在当前社会大背景下，这种生产模式显然与未来发展趋势背道而驰。

为改变传统制造业生产格局，使经济结构发生根本性转变；传统制造业产业升级转型，变得尤为重要。产业转型升级的关键，在于科技创新；以数字化、智能化、节能化，高科技含量为特征的 3D 打印技术，被认定为传统制造业创新的重要引擎。3D 打印技术相对传统制造技术，是一次重要的技术革命。它可以设计制造任何复杂、高难度的传统制造方式不能生产或生产加工周期很长的产品。除此之外，它节能环保，可以实现快捷、方便、低成本创新设计，解决传统制造所不能解决的技术难题。由于 3D 打印技术，目前成本较高、打印耗材种类较少、实际应用性不强、工艺水平还不成熟、产业规模较小，它并不具备传统制造业所擅长的大批量化、规模化、精细化生产；所以，它并不能取代传统制造技术。但 3D 打印技术，可与传统制造技术相结合，带动引领传统制造业的发展，相互补充，发挥各自优势。

7 3D 打印技术市场占有率与认知度

所谓市场占有率，即在目标区域内某种技术产品或其他，在交易总额中所占比例大小。3D 技术市场占有率，主要取决 3 方面因素：规模经济、竞争能力、产业化。本文仅从这 3 个方面分析 3D 打印技术市场占有率。

《国际增材制造行业发展报告》显示，2011 年，3D 打印技术全球直接产值 17. 14 亿美元；2012 年，全球市场规模 20 亿美元。3D 打印机销量同比上升 25%，其中：38%产自美国，10%产自中国。2013 年，3D 技术实现产值不足 40 亿美元。麦肯锡公司预测，到 2025 年，3D 打印对全球经济贡献值将为 2000 亿~6000 亿美元。目前，全球从事 3D 打印机研发的单位和生产商、材料商、配套服务商不足 100 家，国内大小企业仅三四十家。2012 年，国内 3D 打印设备销售和服务，约在 10 亿人民币的水平；2013 年，已突破 20 亿人民币；2014 年，有望达到 40 亿~50 亿人民币规模。由此可见，目前 3D 打印技术规模占制造业领域还是非常小的；但其具有广泛的发展空间。市场占有率小的一方面原因是 3D 打印技术与传统制造技术相比，不具备强劲竞争力；其自身还处于初级发展阶段，并不是替代性很强的技术，不能取代传统制造技术，更不是无所不能的技术。虽在一些领域得到很好的应用，但无法改变传统制造业格局。另一方面原因是 3D 打印技术难以找准市场定位，市场导向与用户需求并不明确，也造成 3D 打印技术有势无场的尴尬局面。

所谓认知度，是指一个社会组织被公众所认识、知晓的程度。3D打印技术，还没有像互联网那样渗透到人们生产生活的各个角落，也没有带来非常直接的变化和直观感受，所以一直得不到广泛关注和重视。社会各界，对3D打印技术的认知、市场的推广应用、材料、人才等方面，还存在明显不足。随着近些年的发展，用于设计和创意领域的桌面级3D打印机，已陆续走进校园和家庭。越来越多的年轻人对其产生浓厚兴趣，使更多的学生、设计人员进一步了解3D打印技术。为了让大家认知这项技术，国家应积极创造条件，让广大传统制造业企业和用户更直观接触到3D打印技术；而不仅仅是用户观望，商家仅停留在概念阶段炒作。3D打印技术产业创新中心的出现，可以向大家展示产品技术。而这个开放式的服务平台，也可以促使行业企业之间沟通交流、增进相互了解。2012年，美国总统奥巴马更是将3D打印作为重塑美国制造业的关键技术，由此，3D打印技术在社会公众中更引起较大兴趣和强烈反响。而公众，更应该通过学习、交流等媒介对3D打印技术做出科学客观的评价，增强对其先进性、重要性的认识。只有这样，才有利于技术的应用与发展。

8 3D打印与第三次工业革命的关系

第一次工业革命，始于18世纪的英国，它开创以机器代替手工劳动的时代。第二次工业革命，是指19世纪中期，电器开始代替机器，成为补充和取代以蒸汽机为动力的新能源。第三次工业革命，早在20世纪70年代末由西方提出；现今对第三次工业革命的认识与核心内容，还存在很多争议；但其特征，基本包括互联网、可再生能源、新材料、数字化、智能化、3D打印技术等。目前，3D打印技术与第三次工业革命的关系，主要有以下两种观点：第一种观点，即3D打印技术将引领第三次工业革命；第二种观点，是3D打印技术不会引发第三次工业革命。

《经济学人》《福布斯》《纽约时报》等杂志，宣称3D打印技术将推动“第三次工业革命”。美国经济学家杰里米·里夫金，在他的《第三次工业革命》书中预言，以互联网和新能源相结合为基础的新经济即将到来，成为世界的第三次工业革命。美国总统奥巴马，将3D打印技术作为引发制造业革命的一项战略举措。西方媒体，将3D技术誉为第三次工业革命的新技术；将3D打印技术认为是先进制造技术与生产方式变革的产物，对传统制造业是一种颠覆性变革，必将推动第三次工业革命的发展。

富士康总裁郭台铭公开表示：3D打印绝不等于第三次工业革命，因为它无法用于大量生产，不具有商业价值。世界3D打印技术产业联盟创始人罗军，认为3D技术能否推动引领第三次工业革命，还缺乏实例支撑。柏林工业大学3D实验室主任哈特穆特·施万特教授说：尽管3D打印技术对于科学和经济已有一个重要作用，并且赢得一个非常有活力的发展，设备和软件被不断研发更新，有越来越多的应用领域。但笔者认为：现阶段，说3D打印将带来第三次工业革命是夸张的，目前还没有人可以做这样的断言。《麻省理工技术评论》编辑大卫·罗特曼指出，关于通过3D打印技术彻底改变工业生产方式的结论，往往是由于对目前工业现实缺乏认识所造成的。哥本哈根未来研究学院（CIFS）的名誉主任约翰·彼得·帕鲁坦的一句话值得深思：我们的社会通常会高估新技术的可能性，同时却又低估它们的长期发展潜力。其实，该技术自诞生之时就饱受争议，技术的可实现性、生产工艺能否达到产品质量标准、技术昂贵、无法广泛应用等多项缺点，它能否引领第三次工业革命还有待观察[25]。

所谓革命，指推动事物发生根本变革，引起事物从旧质到新质的飞跃。从工业革命角度看，它的变革是从生产组织形式、劳动效率、产品价值、制造形式等多方面体现。以3D打印技术为代表的新兴制造技术，会对传统工业有着深远意义；对新型工业化的建设和促进传统产业的升级起到十分重要的引领作用；但由于3D打印技术还仅仅是一项非常基础前沿制造技术，还处于刚刚起步阶段，虽然有很大发展前景，但还不能对传统工业起到颠覆性作用[26]。中国3D打印技术产业联盟联席理事长、华中科技大学材料学院副院长史玉升教授指出：这种技术不是取代传统技术，而是跟传统技术相结合，形成良好的优势互补。中国工程院院士、中国工程物理研究院徐志磊研究员认为：3D打印是在第三次工业革命之内重要的一个环节，也是重要的领域[27]。中国工程院院士李培根认为：增材制造技术，将引发新一轮的科技革命[28]。显然短时间内，3D打印技术与第三次工业革命的关系还不能确定；而其做为一种新技术，对技术革命有重大影响的看法已经成为广泛共识。

9 3D打印技术在林业行业的应用预测

3D打印技术的魅力，在于它不需要在工厂操作，汽车小零件、灯罩、小提琴等小件物品，只需要一台类似台式计算机的小打印机即可，放在办公室或者房间的角落中[29]，人们只需要在打印过程中，控制特定的材料以及精密度[12]。因此，可以将这一技术引入林业行业。

(1)森林资源普查。国家每5年对森林资源做一次大的普查工作，以计算5年森林蓄积量的变化。可以利用扫描仪对森林进行现场扫描、收集数据信息；然后，利用3D技术复制打印出森林的详细信息资料，能够准确、快捷地获得森林地理地貌的详细信息。

(2)活立木勘探复原。针对原始森林中的珍贵树种进行扫描探测，及时了解活立木的生长情况，监控其材质变化。利用扫描仪进行内部精度扫描，准确定位掌握每颗树木的健康状况，发现问题及时补救处理。

(3)改变传统产业的生产模式。传统的木制品制作模式，是设计、画图、选料、锯削、刨削、胶结等生产方式，既浪费大量的优质资源，又导致加工工程高能耗、高污染、高噪声，严重影响工作人员的身心健康。如果将3D技术应用于木质家具和装饰领域，针对用户要求绘出效果图，按照一定比例打印出家具样品，并进行房间的布置和装饰，使用户置身其中体验感受未来家具和房间布局，依据自己的喜好改变和修改，能够减少很多纠纷和矛盾。

(4)异质复合新材料的开发研究。现有的熔融沉积打印(FDM)主要使用PLA和ABS。以此为聚合物，加入生物质粉料，先制作生物质复合丝状材料，通过界面改性和工艺技术创新，制备以生物质聚合物复合材料为原料的3D产品，并将聚合物种类扩展到PP、PE等通用塑料。改变现有木塑制品使用挤出成型单一的生产模式，将中低档木塑复合制品扩大到中高端市场，增加其使用范围和应用领域。另外，尝试将木质橡胶类复合材料应用到3D打印，真正实现集材料、功能、技术、创新多位一体的多领域共同发展、协同创新的生产模式。

总之，3D打印技术逐渐从打印物体外包造型过渡到打印物体的内部构成，最终发展到打印物体的高级功能和行为阶段。3D打印技术在世界上的发展速度已经开始加速，3D打印技术已初步形成一套体系，同时该技术可应用的领域也逐渐扩大，已涵盖产品设计、模具设计与制造、材料工程、医学研究、文化艺术、建筑工程等各个领域，前景远大。期望在不远的将来，3D技术能够应用在林业产业，并且不断提高打印技术和应用水平，持续推动林业产业快速、健康的发展。

10 发展及展望

3D打印技术诞生于20世纪80年代。在30多年的发展历史中，3D打印已应用到多个领域。近些年，全球资源日益匮乏、经济衰退，以美国为首的西方发达国家大力倡导“再工业化，再制造化”战略，提出数字化、智能化、新能源、新材料等关键制造技术为突破，以此来巩固提升制造业主导权。在这样背景下，具备以上特点的3D打印，即增材制造技术被广泛热议，被认定为第三次工业革命的重要因素、关键标志之一；而互联网、可再生能源、数字化、智能化，被认定为第三次工业革命的核心及主要特征。而以这些特征为主的工业革命，势必推动一些新型产业的诞生及发展，从而导致社会生产方式、制造模式、生产组织发生重大变革[30]。例如，大规模生产转向大规模个性化定制、工业化生产转向社会化生产，以3D打印为例的新型制造系统改变传统刚性生产系统，面向"柔性制造""批量定制"，赋予制造业更具包容性、灵活性、创新性的生产特点[31]。在此背景下，3D打印面临如下机遇与挑战。

（1）在大批量制造等方面，增材制造尚不具备取代传统工业制造的条件，不可能对现代制造产业产生颠覆性作用。增材制造也不是无所不能，但它势必会引发一场科技革命，它的主要作用是形成与传统批量制造互补的个性化制造模式。

（2）增材制造技术的发展，将削弱我国劳动力成本的优势，弱化中国制造[32]。我国企业应有强烈的危机意识，针对我国产业大而不强、设计创新能力薄弱、以量取胜的特点，应以此为机提升研发能力，抓住第三次工业革命的历史机遇，促进产业由大变强，推动我国由工业化大国向工业化强国的转变。

（3）政府制定相关政策法规，给予相关产业积极扶植。普及增材制造相关知识，加强教育培训、基础理论的研究，鼓励大学、实验室、企业加强研究和人才的培养，激发青少年对3D打印的兴趣爱好[3]。

（4）推动3D打印产业化，并组织制定发展路线图和中长期发展战略，开展基础研究，完善3D打印技术规范与标准制定。加大财税政策引导力度，加大对增材制造研发、产业化支持力度，适时筹建相关行业组织。

参考文献

[1]黄卫东. 如何理性看待增材制造(3D打印)技术[J]. 新材料产业，2013(8)：9-12.

[2][美]杰里米·里夫金. 第三次产业革命[J]. 张一萌，译. 国际参考研究，2013(6)：31-34.

[3]王雪莹. 3D打印技术与产业的发展及前景分析[J]. 中国高新技术企业，2013(26)：3-5.

[4]佚名. 解密3D打印[J]. 中国设计，2013，4(2)：15-19.

[5]陈新伟，三维喷绘机器人喷墨机理研究[D]. 天津：南开大学，2009.

[6]李青，王青. 3D打印：一种新兴的学习技术[J]. 远程教育杂志，2013(4)：29-35.

[7]刘红光，杨倩，刘桂锋，等. 国内外3D打印快速成型技术的专利情报分析[J]. 情报杂志，2013，32(6)：40-46.

[8]景绿路. 国外增材制造技术标准分析[J]. 航空标准化与质量，2013(4)：44-48，56.

[9]罗军. 中国3D打印的未来[M]. 北京：东方出版社，2014.

[10]Hod Lipson，Melba Kurman，赛迪研究院专家组. 3D打印从想象到现实[M]. 北京：中信出版社，2013.

[11]卢秉恒，李涤尘. 增材制造(3D打印)技术发展[J]. 机械制造与自动化，2013，42(4)：1-4.

[12]李涤尘，田小永，王永信，等. 增材制造技术的发展[J]. 电加工与模具，2012(增刊1)：20-22.

[13]Shen J. Material system for use in three-dimensional Printing：US Patent，7049363[P]. 2006-3-18.

[14]黄立本，张立基，赵旭涛. ABS树脂及其应用[M]. 北京：化学工业出版社，2001：312-313.

[15]贾玉珍，周春艳. 聚碳酸酯技术现状及发展趋势[J]. 化工文摘，2007(2)：46-49.

[16]李丽，王成国，李同生，等. 聚碳酸酯及聚碳酸酯合金导热绝缘高分子材料的研究[J]. 材料热处理学报，2007，28(2)：51-54.

[17]Avraham L, Petach T. Reverse thermal gels and the use thereof for rapid prototyping: US Patent, 6863859[P]. 2005-6-12.

[18]Objet Tango. 系列类橡胶材料[EB/OL]. [2013-11-09]. http://www. vx. com/news/dyc12013_ 6. html.

[19]张剑光，韩杰才，赫晓东，等. 制备陶瓷件的快速成型技术[J]. 材料工程，2001(6)：37-40.

[20]郭艳玲，姜凯译，辛宗生，等. 木粉/PES复合粉末选择性激光烧结成形及后处理技术研究[J]. 电加工与模具，2011(6)：29-32.

[21]Yang S F, Evans J R G. A dry powder jet printer for dispensing and combinatorial research[J]. Powder Technology, 2004, 142(2): 219-222.

[22]李小丽，马剑雄，李萍，等. 3D打印技术及应用趋势[J]. 自动化仪表，2014，35(1)：1-5.

[23]孙柏林. 试析"3D打印技术"的优点与局限[J]. 自动化技术与应用，2013，32(6)：1-6.

[24]杜宇雷，孙菲菲，原光，等. 3D打印材料的发展现状[J]. 徐州工程学院学报：自然科学版，2014，29(1)：20-24.

[25]王忠宏，李扬帆，张曼茵. 中国3D打印产业的现状及发展思路[J]. 经济纵横，2013(1)：90-93.

[26]吴怀宇. 3D打印三维智能数字化创造[M]. 北京：电子工业出版社，2014.

[27]徐志磊. 3D打印是第三次工业革命的重要环节[N]. 中国信息化周报，2013-09-30(5).

[28]李培根. 增材制造技术发展前景广阔[N]. 中国信息化周报，2013-07-22(5).

[29]王忠宏，李扬帆. 3D打印产业的实际态势、困境摆脱与可能走向[J]. 产业经济，2013(8)：29-36.

[30]左世全. 我国3D打印发展战略与对策研究[J]. 世界制造技术与装备市场，2014(5)：44-50.

[31]王忠宏. 我国3D打印产业现状及发展建议[N]. 中国经济时报，2012-10-19(5).

[32]王文涛，刘燕华. 3D打印制造技术发展趋势及对我国结构转型的影响[J]. 科技管理研究，2014(6)：22-25, 30.

英文题目：Impact of Future Disruptive Technologies—Multi-material 3D printing technology produced mixed

作者：Li Jian, Xu Min, Bao Wenhui

摘要：Describes the basic concepts of 3D printing, solid manufacturing methods, the development of research status on the advantages and problems of 3D printing technology, the choice of materials and features, 3D printing technology and traditional manufacturing industries to discuss relations Introduction of 3D printing technology market share rate and awareness. Analyzes the relationship between technology and 3D printing third industrial revolution, and the prospect of its development trend. Forecast: wood processing residues, straw and other raw materials for 3D printing technology, will become one of the effective ways to agriculture and forestry biomass and waste utilization of high value.

关键词：3D printing; Manufacturing processes; Benefits features; Industrial revolution

Two Solid-phase Recycling Method for Basic Ionic Liquid [C4mim] Ac by Microporous Resin and Ion Exchange Resin from *Schisandra chinensis* Fruits Extract*

Chunhui Ma, Yuangang Zu, Lei Yang, Jian Li

Abstract: In this study, two solid-phase recycling method for basic ionic liquid (IL) 1-butyl-3-methylimidazolium acetate ([C4mim] Ac) were studied through a digestion extraction system of extracting biphenyl cyclooctene lignans from Schisandra chinensis. The RP-HPLC detection method for [C4mim] Ac was established in order to investigate the recovery efficiency of IL. The recycling method of [C4mim] Ac is divided into two steps, the first step was the separation of lignans from the IL solution containing HPD 5000 macroporous resin, the recovery efficiency and purity of [C4mim] Ac achieved were 97.8% and 67.7%, respectively. This method cannot only separate the lignans from [C4mim] Ac solution, also improve the purity of lignans, the absorption rate of lignans in [C4mim] Ac solution was found to be higher (69.2%) than that in ethanol solution (57.7%). The second step was the purification of [C4mim] Ac by the SK1B strong acid ion exchange resin, an [C4mim] Ac recovery efficiency of 55.9% and the purity higher than 90% were achieved. Additionally, [C4mim] Ac as solvent extraction of lignans from S. chinensis was optimized, the hydrolysis temperature was 90 ℃ and the hydrolysis time was 2 h.

Keywords: Basic ionic liquid; Recovery; *Schisandra chinensis*; Biphenyl cyclooctene lignans; Macroporous-resin; Ion exchange resin;

1 Introduction

Since the 1990s, ionic liquids (ILs) have been increasingly investigated for use in many fields because of their excellent properties such as negligible vapor pressure, good thermal stability, wide liquid range, tunable viscosity, and miscibility with water and organic solvents. Additionally, various organic compounds have good solubility in ILs and are extractable. They have thus been used in analytical applications, organic syntheses[1,2], catalysis[3], and in separation science[4]. Some progress has been made in the field of IL-aided sample treatment techniques such as solid-liquid extractions[5], solid-phase microextractions[6], liquid-liquid extractions[7], liquid-phase microextractions[8], and aqueous two-phase system extractions[9]. ILs are promising solvents for the preparation of various active ingredients from medicinal plants such as essential oils[10,11], alkaloids[12,13], terpene lactones[4], polyphenolic compounds[14], and quinones[15]. Moreover, much research has been done on various catalytic reactions with an IL as a catalyst such as polymerization[16],

---

* 本文摘自 Journal of Chromatography B, 2015, 1(5): 976-977.

preconcentration[17], aza - Michael[18], hydrogenation[19], and other syntheses[20]. These reactions are characterized by a fast rate of reaction, high conversion rate, and exact reaction electivity. However, there is a few research has been done on IL detection and recycling methods. ILs are usually recycled in two ways: liquid-liquid extraction[21], or solid-phase extraction[22].

We mainly investigated a solid-phase extraction recycling process using macroporous resins and ion exchange resins. Moreover, a HPLC detection method for basic IL ([C4mim]Ac) was established in order to evaluate the recovery efficiency of ([C4mim]Ac). This method cannot only recycle ([C4mim]Ac) IL, also improve the purity of *Schisandra chinensis* (Turz.) Baill. (*S. chinensis*) lignans, We thus established a two solid-phase recycling method for [C4mim]Ac IL from *S. chinensis* fruits extract.

S. chinensis fruit is a popular herbal medicine and food additive[23], and has been extensively used in many countries[24]. Extensive studies have indicated that the major bioactive components of *S. chinensis* are free biphenyl cyclooctene lignans (FBCL)[25], such as schisandrin [S], schisantherin A [SA], deoxyschisandrin [DS] and γ-schisandrin [GS], which have a dibenzo cyclooctadiene type skeleton[25,26]. They have various positive biological effects such as antihepatotoxic, antioxidant, antitumor[27], detoxifying, anticarcinogenic[28], central nervous system maintenance, fatigue counteraction, increasing endurance, and anti-inflammatory effects[29], and allowing an improvement in the physical performance of sportspeople[30]. It has many pharmacological effects and can be used in the treatment of chronic cough, asthma, heart palpitations, spontaneous sweating, insomnia, forgetfulness, spermatorrhea, diabetes[31], and cancer because of its antiproliferative effects on tumor cells[32]. The hydroxyl group in FBCL can easily form ester bonds with organic acids in plants, and change into ester-bond biphenyl cyclooctene lignans (EBCL). So, there are two types of biphenyl cyclooctene lignans (FBCL and EBCL) in *S. chinensis* fruit. Hydrolysis can increase the FBCL content by the transformation of EBCL. The results of a previous study showed that the catalytic effect of FBCL was much better under alkaline conditions[33]. Therefore, three kinds of basic IL were selected as catalysts for the hydrolysis reaction of EBCL.

In this research, a two solid-phase recycling method of [C4mim] Ac IL from *S. chinensis* lignans solution was investigated, and the parameters of lignans extraction process with [C4mim] Ac IL a solvent were also optimized. Hope our preliminary experimental results can play an invaluable role in future research.

2 Experimental

2.1 Materials and chemicals

The fruit of *S. chinensis* was purchased from Sankeshu Medicinal Materials Market, Harbin, Heilongjiang Province, China. The dried fruit was powdered until a homogeneous size was obtained and then sieved (20-40 mesh) before use.

The standards of schisandrin [S], schisantherin A [SA], deoxyschisandrin [DS] and γ-schisandrin [GS] (98% purity) were purchased from the National Institute for the Control of Pharmaceutical and Biological Products (Beijing, China). The ionic liquids 1-butyl-3-methylimidazolium hydroxide ([C4mim] OH, 1-butyl- 3-methylimidazolium bisulfate ([C4mim]HSO_4), and 1-butyl-3- methylimidazolium acetate ([C4mim] Ac) were purchased from Chengjie (Shanghai, China) and used without further purification. Acetonitrile and acetic acid of HPLC grade were purchased from J&K Chemical Ltd. (Beijing,

China). Deionized water was purified using a Milli-Q Water Purification System (Millipore, MA, USA). The other solvents and chemicals used in this study were of analytical grade and purchased from Beijing Chemical Reagents Co. (Beijing, China). All the solutions prepared for HPLC were filtered through 0.45 μm membranes (GuangFu Chemical Reagents Co., Tianjin, China) before use. Ion exchange resins and macroporous resins were purchased from Guangfu Fine Chemical Research Institute (Tianjin, China).

2.2 Methods

2.2.1 HPLC analysis

The HPLC method used for *S. chinensis* lignans was reported by Ma et al. [5].

A HiQ sil-C18 reversed-phase column (4.6 mm 250 mm, 5 μm, KYA TECH) was used for the HPLC analysis of [C4mim]Ac, and acetonitrile-water-acetic acid (20 : 80 : 1, *v/v/v*) was used as the mobile phase with a 1.0 mL · min^{-1} flow rate, a 10-μL injection volume, and a 25 ℃ column temperature; The absorbance was mea- sured at 210 nm for the detection of [C4mim]Ac. The retention time was 6.6 min and the corresponding calibration curve for [C4mim] Ac was $Y_{IL} = 3.0\ 10^6 X + 16105$ ($r = 0.9999$). Good linearity was found in the range of 0.0156-4.0 mg · mL^{-1}.

Chromatograms for the four kinds of lignans (a), [C4mim]Ac (b) and the extraction sample (c) are shown in Fig. 1.

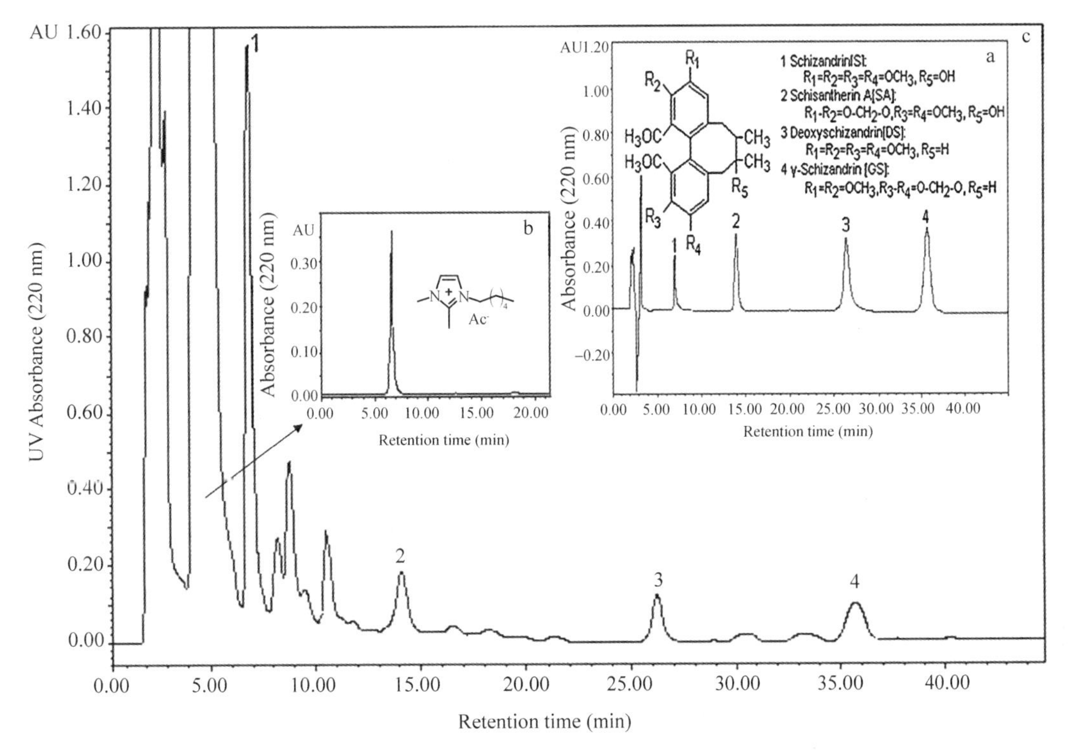

Fig. 1 RP-HPLC chromatogram of (a) the *S. chinensis* biphenyl cyclooctene lignans standards: 1, schisandrin [S]; 2, schisantherin A [SA]; 3, deoxyschisandrin [DS]; and 4, γ-schisandrin [GS]; (b) the basic ionic liquid-[C4mim] Ac; and (c) the heated hydrolysis sample

2.2.2 Basic ionic liquid heated digestion method

5.0 g of *S. chinensis* fruit were soaked in 50 mL IL aqueous solu tion for 4 h, and then the suspension was heated for extracting FBCL in a water bath. The concentration of the selected basic IL, the hydrolysis temperature and the hydrolysis time were systematically studied in this work. The solid-liquid ratio was optimized in our previous study. Each test was operated three times and the average value was recorded.

2.2.3 Reference methods of extracting FBCL

Pure water and 1.0 M sodium acetate[12,34,35] were selected as the reference solvent for extracting FBCL from *S. chinensis* fruit. The extraction experiments were operated under the optimized conditions except for solvent type. 5.0 g dried sample was mixed with 50 mL solvent, and the suspension was heated for 2 h The extract was filtrated through a 0.45 μm filter for HPLC analysis. Conventional reflux extraction (RE) was recognized as a more efficient extraction method for FBCL[36]. The conditions imposed on the conventional 80% ethanol RE were same as that shown above.

2.2.4 Separation of lignans and [C4mim]Ac by macroporous resin

Dynamic adsorption and desorption tests were run at 25 ℃ using the HPD 5000 (Supplementary Table S1) macroporous resin (dry weight) with the lignans solution obtained after hydrolysis. Adsorption step with macroporous resin is not only a purification step of lignans, is also a separation step of [C4mim] Ac from lignans solution. Most part of the lignans can be adsorbed on the macroporous resin, and the IL cannot be adsorbed washing with deionized water, while [C4mim] Ac was in the effluent. And then the dynamic desorption of lignans using an ethanol solution was performed[33]. Finally, the regeneration of the macroporous resin was evaluated as follows. The exhausted resin was washed with a 90% ethanol solution, a 1.0 M HCl solution, and deionized water until the pH was 7. And a 1.0 M NaOH solution, and deionized water until the pH was 7-8. Supplementary Table S1 related to this article can be found, in the online version, at http://dx.doi.org/10.1016/j.jchromb.2014.11.003.

2.2.5 Recycling of [C4mim] Ac by ion exchange resin

Dynamic ion exchange process: the cation of the IL (1-butyl-3-methylimidazolium) in the mixture was exchanged for other ions in the resin. The IL aqueous solution (the effluent in Section 2.2.4) was flowed through the fix-bed reactor and then strong cation exchanger cartridges were used for the ion exchange. Most of the FBCL in the effluent was allowed to flow out. And then the ion exchange resin was eluted with 0.1 M HCl, and [C4mim] Ac was flowed out. The system was eluted with 2.0 M HCl, finally 2.0 M NaOH and then with distilled water to convert the resin to its Na^{+} type. The selected resin was the strong acid gel-type ion exchange resin SK1B (Supplementary Table S1).

3 Results and discussion

3.1 Screening of the basic IL solution

Although several acid-base theories exist, most basic ILs have been used in a homogeneous system; the classification of alkaline ILs was according to the ionization theory of Arrhenius, Brönsted acid-base proton theory and the Lewis acid-base electron theory, respectively. Therefore, three kinds of ILs were selected containing the 1-alkyl-3-methylimidazolium cation. The selected ILs [C4mim] OH (Arrhenius base), [C4mim]HSO_4(Brönsted base) and [C4mim]Ac (Lewis base) were all water-soluble IL. And the neutral ionic liquid [C4mim] Br (water-soluble) and [C4mim] BF_4 (fat-soluble) are reference solvent, and the extraction yields of FBCL are 5.13 and 3.99, respectively. The water-soluble order of FBCL is S>SA>DS>GS, and the water-soluble of S is the strongest of all, moreover, the content of S is the highest in FBCL. So, the extraction yield of FBCL by water-soluble IL is higher. Results from the heated hydrolytic process using ILs were shown in Table 1.

From Table 1, the hydrolysis yield of FBCL with [C4mim] Ac (Lewis base) after the hydrolysis reaction was the highest (total FBCL was 7.51 mg · g^{-1}). The anion in IL-acetate, provided electron pair in the hydrolysis reaction, and made the bound state lignans into free state lignans. Therefore, [C4mim] Ac was selected for the hydrolytic process of *S. chinensis* lignans due to the significantly incremental change of FBCL. Catalyst amount is an important parameter in hydrolysis reaction. A higher [C4mim] Ac as catalyst loading allows the equilibrium to be reached sooner because of the increase in the total number of basic sites available for the reaction[37] (Table 2). Five copies of 50 mL IL solutions with different concentration were hydrolyzed at 90 ℃ in a water bath with stirring for 2.0 h. The concentration of [C4mim] Ac in the solutions were 0.2, 0.4, 0.6, 0.8, and 1.0 M, respectively. The optimum amount of [C4mim] Ac was found to be 1.0 M. At less than 1.0 M, the hydrolysis efficiency decreased significantly while at higher than 1.0 M, the viscosity of the IL solution increased resulting in a decrease in the reaction rate, made some difficulties in sampling and detection, and unnecessary waste.

Table 1 Heated extraction yield of free biphenyl cyclooctene lignans (FBCL) with basic ionic liquids

1.0 M IL	S (μg · g^{-1})	SA (μg · g^{-1})	DS (μg · g^{-1})	GS (μg · g^{-1})	Total (mg · g^{-1})
[C4mim] OH	7026.81	56.93	6.18	6.21	7.10
[C4mim] HSO_4	5429.85	104.06	4.43	6.78	5.55
[C4mim] Ac	7336.86	144.57	13.33	15.09	7.51
[C4mim] Br	4938.18	165.33	14.01	8.65	5.13
[C4mim] BF_4	3874.64	93.44	9.37	9.88	3.99

Table 2 Heated extraction yield of free biphenyl cyclooctene lignans (FBCL) with [C4mim] Ac of different concentration

[C4mim] Ac (M)	S (μg · g^{-1})	SA (μg · g^{-1})	DS (μg · g^{-1})	GS (μg · g^{-1})	Total (mg · g^{-1})
0.2	3964.11	23.37	1.95	4.79	3.99
0.4	4542.92	38.33	5.59	8.11	4.59
0.6	5118.24	56.23	8.47	10.71	5.19
0.8	6791.68	63.98	10.77	12.85	6.88
1.0	7336.86	144.57	13.33	15.09	7.51

3.2 Optimization of the hydrolytic process with [C4mim] Ac

For the ionic liquid aqueous solution, the highest temperature of heating under normal pressure is less than 100 ℃. Once reach the boiling point of water, the temperature of the ionic liquid aqueous solution no longer rises. This experiment adopts the water bath heating, and the highest temperature of heating under normal pressure is less than 100 ℃. So, in order to measure accurately, the effect of temperature on the

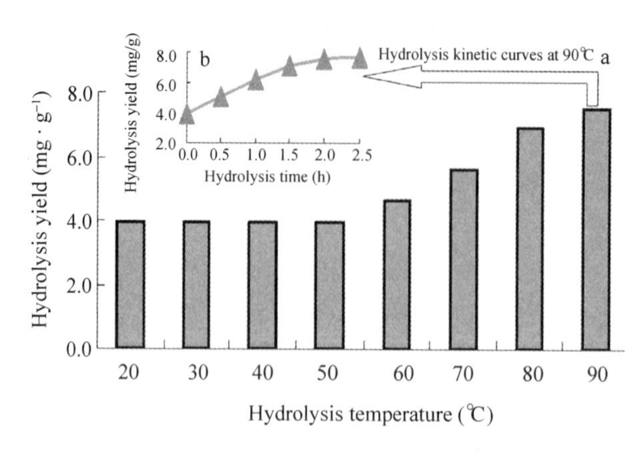

Fig. 2 Optimization of the hydrolysis conditions

hydrolysis of EBCL was studied from 20 to 90 ℃ under the normal pressure with [C4mim] Ac as the catalyst. If the temperature further increases within 100 ℃, the hydrolysis increment of FBCL is not obvious. The hydrolysis kinetic curves at 90 ℃ were also investigated. The experimental results are shown in Fig. 2. The hydrolysis efficiency improved as the hydrolysis temperature increased because the hydrolysis reaction is endothermic. From Fig. 2a, the hydrolysis reaction is endothermic reaction, and the hydrolysis reaction of EBCL was non spontaneous less than 50 ℃. However, above 90 ℃ under normal pressure, the hydrolysis efficiency of EBCL did not increase obviously, and this may be due to the high temperature caused by FBCL isomerization. Therefore, 90 ℃ was selected as the hydrolysis temperature for EBCL hydrolysis. Hydrolysis kinetic curves for EBCL in the presence of [C4mim] Ac were obtained at 90 ℃ (Fig. 2b). The hydrolysis efficiency of EBCL improved rapidly as the hydrolysis time increased before 1.5 h, after 1.5 h the efficiency improved slowly, and equilibrium was reached at about 2 h.

3.3 Comparison [C4mim] Ac heated digestion approach with reference methods

The reference methods include pure water extraction and 1.0 M sodium acetate solution extraction. Apparently, water is the most common and inexpensive solvent, therefore pure water is always selected as the co-solvent in various extraction processes. Aqueous solution of IL is used as an effective extraction solvent (in Table 3). [C4mim] Ac played a critical role on extraction yield of FBCL in IL- water solvent system. A comparison between IL and NaAc at an equimolar level indicated that the extraction yield of FBCL was mainly determined by the solvent effect of [C4mim] Ac and not determined by the salt effect or aqueous buffer effect. A comparison between the performance of NaAc solution and that of pure water showed that the extraction yield with NaAc (1.90 mg · g^{-1}) was lower than that with pure water (3.61 mg · g^{-1}), indicating that salt effect was not critical to extract FBCL. The most common method for FBCL extraction from *S. chinensis* is reflux extraction (RE) using 80% ethanol. The extraction yield by RE with 80% ethanol was 7.63 mg · g^{-1}.

3.4 Two solid-phase recycling method for [C4mim] Ac

3.4.1 Separation of lignans and [C4mim] Ac by HPD 5000 macroporous resin

Dynamic adsorption and desorption using a wet-packed (24.0 g of HPD5000 resin, BV 33 mL) fixed-bed column separator was done according to a previous report[33]. The results (Table 4) showed that 85.3% of [C4mim] Ac in the lignans solution was not adsorbed onto the macroporous resin surface, and most of the remaining 14.7% of [C4mim]Ac that was attached to the resin sur- face could be removed with water upon water washing. HPLC detection indicated that the cumulative outflow of [C4mim]Ac was more than 95% (97.8%). The purity of [C4mim]Ac after concentrating and drying under vacuum was 67.6%, and it was then dissolved in water and purified by an ion exchange resin. However, because of the IL presence in solution, the adsorption of lignans was much easier. The adsorption efficiency increased from 57.7% to 69.2%[33]. The fixed-bed column was then washed using an 80% ethanol solution, 63.6% of FBCL was eluted, and the purity of FBCL was 36.5%.

3.4.2 Recycling of [C4mim]Ac by SK1B ion exchange resin

A dynamic ion exchange process was carried out using a wet-packed (48.0 g of SK1B strong acid ion exchange resin, BV 80 mL) fixed-bed column reactor. The SK1B resin releases Na^+ and its exchange capacity is 2.0-2.2 meq/g; therefore, 48.0 g of ion exchange resin can exchange around 96.0 meq to 105.6 meq cations (viz. 96.0-105.6 mmol). The ion exchange capacity of 48.0 g SK1B resin can accommodate 100 mL of 1.0 M [C4mim]Ac solution. After [C4mim]Ac aqueous solution was added to the system, 70 mL 0.1 M HCl solution was added again to exchange with [C4mim]Ac. The exchanged solution was kept, and the system was washed until no white precipitates were present in the washed solution when one drop of $AgNO_3$ solution was added. The results (Table 4) showed that 65.2% of [C4mim] Ac was exchanged on SK1B resin, and upon 0.1 M HCl washing HPLC indicated that 55.9% of [C4mim] Ac was exchanged. The purity of [C4mim] Ac after concentrating and drying under vacuum was 91.2%.

Table 3 Reference extraction methods of free biphenyl cyclooctene lignans (FBCL)

Solvent	S (μg·g^{-1})	SA (μg·g^{-1})	DS (μg·g^{-1})	GS (μg·g^{-1})	Total (mg·g^{-1})
1.0 M [C4mim]Ac	7336.86	144.57	13.33	15.09	7.51
1.0 M NaAc	1873.86	29.19	-[a]	-	1.90
Pure water	3509.12	74.11	10.02	13.11	3.61
80%ethanol (RE)[b]	7268.64	278.24	55.45	31.21	7.63

a -: below the detection limit. b RE: reflux extracted for 2 h.

Table 4 Dynamic effect on separation of free biphenyl cyclooctene lignans (FBCL) from IL solution and [C4mim]Ac recycle

Dynamic process	[C4mim]Ac in effluent	Purity of [C4mim] Ac	FBCL in effluent	Purity of FBCL
Adsorption	85.3%	-a	30.8%	-
Water washing	12.5%	67.7%	-	-
Desorption	2.0%	-	63.6%	36.5%
Ion exchange	34.8%	-	-	-
Elution	55.9%	91.2%	-	-

a -: below thedetection limit.

4 Conclusions

In this research, two solid-phase recycling method for basic ionic liquid (IL) 1-butyl-3-methylimidazolium acetate ([C4mim]Ac) were studied through a digestion extraction system of extracting biphenyl cyclooctene lignans from *Schisandra chinensis* (Turz.) Baill. The RP-HPLC detection method for [C4mim] Ac was established in order to investigate the recovery efficiency of IL. The first step was the separation of IL and lignans solution with HPD 5000 macroporous resin. The recovery yields of FBCL and [C4mim] Ac were 63.6% and 97.8%. The purity of FBCL and [C4mim] Ac were found to be 36.5% and 67.7%, respectively. The second step was the purification of [C4mim] Ac by SK1B strong acid ion exchange resin, and [C4mim] Ac recovery yield was 55.9% with a purity of 91.2%.

Acknowledgements

The authors thank the Fundamental Research Funds for the Cen- tral Universities (2572014EY01) and Hei Long Jiang postdoctoral foundation for financial support.

References

[1] Cooper E R, Andrews C D, Wheatley P S, et al. Ionic liquids and eutectic mixtures as solvent and template in synthesis of zeolite analogues[J]. Nature, 2004, 430(7003): 1012-1016.

[2] Zhao D, Wu M, Kou Y, et al. Ionic liquids: Applications in catalysis[J]. Catal Today, 2002, 74(1-2): 157-189.

[3] Zhao H, Holladay J E, Brown H, et al. Metal chlorides in ionic liquid solvents convert sugars to 5-hydroxymethylfurfural[J]. Science, 2007, 316(5831): 1597-1600.

[4] Chi Y S, Zhang Z D, Liu Q S, et al. Microwave-assisted extraction of lactones from Ligusticum chuanxiong Hort. using protic ionic liquids[J]. Green Chem., 2011, 13(3): 666-670.

[5] Ma C H, Liu T T, Yang L, et al. Study on ionic liquid-based ultrasonic-assisted extraction of biphenyl cyclooctene lignans from the fruit of Schisandra chinensis Baill[J]. Anal. Chim. Acta., 2011, 689(1): 110-116.

[6] Meng Y, Pino V, Anderson J L. Role of counteranions in polymeric ionic liquid-based solid-phase microextraction coatings for the selective extraction of polar compounds[J]. Anal. Chim. Acta., 2011, 687(2): 141-149.

[7] Tzeng Y P, Shen C W, Yu T. Liquid-liquid extraction of lysozyme using a dye-modified ionic liquid[J]. J. Chromatogr A., 2008, 1193(1-2): 1-6.

[8] Zhang H, Chen X, Jiang X. Determination of phthalate esters in water samples by ionic liquid cold-induced aggregation dispersive liquid-liquid microextraction coupled with high-performance liquid chromatography[J]. Anal. Chim. Acta., 2011, 689(1): 137-142.

[9] Deive F J, Rodríguez A, Pereiro A B, et al. Ionic liquid-based aqueous biphasic system for lipase extraction[J]. Green Chem., 2011, 13(2): 390-396.

[10] Zhai Y, Sun S, Wang Z, et al. Microwave extraction of essential oils from dried fruits of Illicium verum Hook. f. and Cuminum cyminum L. using ionic liquid as the microwave absorption medium[J]. J. Sep. Sci., 2009, 32(20): 3544-3549.

[11] Ma C H, Liu T T, Yang L, et al. Ionic liquid-based microwave-assisted extraction of essential oil and biphenyl cyclooctene lignans from Schisandra chinensis Baill fruits[J]. J. Chromatogr A., 2011, 1218(48): 8573-8580.

[12] Cao X, Ye X, Lu Y, et al. Ionic liquid-based ultrasonic-assisted extraction of piperine from white pepper[J]. Anal Chim. Acta., 2009, 640(1-2): 47-51.

[13] Zhang L, Geng Y, Duan W, et al. Ionic liquid-based ultrasound-assisted extraction of fangchinoline and tetrandrine from Stephaniae tetrandrae[J]. J. Sep. Sci., 2009, 32(20): 3550-3554.

[14] Du F Y, Xiao X H, Luo X J, et al. Application of ionic liquids in the microwave-assisted extraction of polyphenolic compounds from medicinal plants[J]. Talanta, 2009, 78(3): 1177-1184.

[15] Wu K, Zhang Q, Liu Q, et al. Ionic liquid surfactant-mediated ultrasonic-assisted extraction coupled to HPLC: Application to analysis of tanshinones in Salvia miltiorrhiza bunge[J]. J. Sep. Sci., 2009, 32(23-24): 4220-4226.

[16] Sheldon R. Catalytic reactions in ionic liquids [J]. Chem Commun, 2001, 1(23): 2399-2407.

[17] M. German-Hernandeza, V. Pinoa, J. L. Andersonb, A. M. Afonsoa, J. Chromatogr. A 1227 (2012) 29-37.

[18] Qiao C Z, Zhang Y F, Zhang J C, et al. Activity and stability investigation of [BMIM][AlCl4] ionic liquid as catalyst for alkylation of benzene with 1-dodecene[J]. Appl. Catal A-Gen, 2004, 276(1-2): 61-66.

[19] Yang L, Xu L W, Zhou W, et al. Highly efficient aza-Michael reactions of aromatic amines and N-heterocycles catalyzed by a basic ionic liquid under solvent-free conditions[J]. Tetrahedron Lett., 2006, 47(44): 7723-7726.

[20] Gong K, Wang H L, Fang D, et al. Basic ionic liquid as catalyst for the rapid and green synthesis of substituted 2-amino-2-chromenes in aqueous media[J]. Catal. Commun., 2008, 9(5): 650-653.

[21] Turner M B, Spear S K, Huddleston J G, et al. Ionic liquid salt-induced inactivation and unfolding of cellulase from Trichoderma reesei[J]. Green Chem., 2003, 5(4): 443-447.

[22] Stepnowski P, Müller A, Behrend P, et al. Reversed-phase liquid chromatographic method for the determination of selected room-temperature ionic liquid cations[J]. J. Chromatogr A., 2003, 993(1-2): 173-178.

[23] Pharmacopoeia of the People's Republic of China, vol. I, Peoples Medicinal Publishing House, Beijng, 2010, pp. 61-62.

[24] Choi Y W, Takamatsu S, Khan S I, et al. Schisandrene, a dibenzocyclooctadiene lignan from Schisandra chinensis: Structure-antioxidant activity relationships of dibenzocyclooctadiene lignans[J]. J. Nat. Prod., 2006, 69(3): 356-359.

[25] Deng X, Chen X, Yin R, et al. Determination of deoxyschizandrin in rat plasma by LC-MS[J]. J. Pharmaceut Biomed, 2008, 46(1): 121-126.

[26] Kuo Y H, Kuo L M Y, Chen C F. Four new C19 homolignans, Schiarisanrins A, B, and D and cytotoxic schiarisanrin C, from Schizandra arisanensis[J]. J. Org. Chem., 1997, 62(10): 3242-3245.

[27] Park J Y, Shin H K, Lee Y J, et al. The mechanism of vasorelaxation induced by Schisandra chinensis extract in rat thoracic aorta[J]. J. Ethnopharmacol, 2009, 121(1): 69-73.

[28] Fu M, Sun Z H, Zong M, et al. Deoxyschisandrin modulates synchronized Ca^{2+} oscillations and spontaneous synaptic transmission of cultured hippocampal neurons[J]. Acta. Pharmacol Sin., 2008, 29(8): 891-898.

[29] Guo L Y, Hung T M, Bae K H, et al. Anti-inflammatory effects of schisandrin isolated from the fruit of Schisandra chinensis Baill[J]. Eur. J. Pharmacol, 2008, 591(1-3): 293-299.

[30] Peng J, Fan G, Qu L, et al. Application of preparative high-speed counter-current chromatography for isolation and separation of schizandrin and gomisin A from Schisandra chinensis[J]. J. Chromatogr A., 2005, 1082(2): 203-207.

[31] Wu H L, Wang H M. Sternocleidomastoid myocutaneous flap for reconstruction after resection of carcinoma in floor of mouth[J]. Zhejiang da xue xue bao Yi xue ban = Journal of Zhejiang University Medical sciences, 2005, 34(6): 574-577.

[32] Ma C H, Liu T T, Yang L, et al. Preparation of high purity biphenyl cyclooctene lignans from Schisandra extract by ion exchange resin catalytic transformation combined with macroporous resin separation[J]. J. Chromatogr B., 2011, 879(30): 3444-3451.

[33] Min H Y, Park E J, Hong J Y, et al. Antiproliferative effects of dibenzocyclooctadiene lignans isolated from Schisandra chinensis in human cancer cells[J]. Bioorg. Med. Chem. Lett., 2008, 18(2): 523-526.

[34] Du F Y, Xiao X H, Li G K. Application of ionic liquids in the microwave-assisted extraction of trans-resveratrol from Rhizma Polygoni Cuspidati[J]. J. Chromatogr A., 2007, 1140(1-2): 56-62.

[35] Huang T, Shen P, Shen Y. Preparative separation and purification of deoxyschisandrin and γ-schisandrin from Schisandra chinensis (Turcz.) Baill by high-speed counter-current chromatography[J]. J. Chromatogr A., 2005, 1066(1-2): 239-242.

[36] Lu Y, Ma W, Hu R, et al. Ionic liquid-based microwave-assisted extraction of phenolic alkaloids from the medicinal plant Nelumbo nucifera Gaertn[J]. J. Chromatogr A., 2008, 1208(1-2): 42-46.

[37] Ma C H, Yang L, Zu Y G, et al. A new approach to catalytic hydrolysis of ester-bound biphenyl cyclooctene lignans

from the fruit of Schisandra chinensis Baill by ion exchange resin[J]. Chem. Eng. Res. Des., 2012, 90(9): 1189-1196.

中文题目：从北五味子萃取物中分离［C4mim］Ac 离子液体的大孔树脂与离子交换树脂双固相回收方法

作者：马春慧，祖元刚，杨磊，李坚

摘要：本研究采用两种固相联用的方法从五味子萃取物中回收碱性离子液体(IL) 1-丁基-3-甲基咪唑([C4mim]Ac)。为了考察[C4mim]Ac 的回收效率，建立了[C4mim]Ac 的反相高效液相色谱(RP-HPLC)检测方法，第一步是利用大孔树脂 HPD 5000 从含 IL 的溶液中分离木脂素，[C4mim]Ac 的回收率和纯度分别为 97.8%和 67.7%。该方法不仅可以从[C4mim]Ac 溶液中分离出木脂素，而且可以提高木脂素的纯度，在[C4mim]Ac 溶液中木脂素的吸附率高达 69.2%。比乙醇溶液中多出 57.7%；第二步是利用强酸性离子交换树脂 SK1B 回收[C4mim]Ac，[C4mim]Ac 的回收效率为 55.9%，纯度高于 90%。优化了[C4mim]Ac 作为溶剂提取原位水解木脂素衍生物的工艺条件，水解温度 90 ℃，水解时间 2 h。

关键词：离子液体；双固相回收；大孔树脂；交换树脂

Poly (Vinyl Alcohol) Films Reinforced with Nanofibrillated Cellulose (NFC) Isolated from Corn Husk by High Intensity Ultrasonication*

Shaoliang Xiao, Runan Gao, LiKun Gao, Jian Li*

Abstract: This work was aimed at fabricating and characterizing poly(vinyl alcohol) films that was reinforced by nanofibrillated corn husk celluloses using a combination of chemical pretreatments and ultrasonication. The obtained nanofibrillated celluloses (NFC) possessed a narrow width ranging from 50-250 nm and a high aspect ratio (394). The crystalline type of NFC was cellulose Ⅰ type. Compared with the original corn husks, the NCF crystallinity and thermal stability increased due to the removal of the hemicelluloses and lignin. PVA films containing different NFC concentrations (0.5%, 1%, 3%, 5%, 7% and 9% *w/w*, dry basis) were examined. The 1% PVA/NFC reinforced films exhibited a highly visible light transmittance of 80%, and its tensile strength and the tensile strain at break were increased by 1.47 and 1.80 times compared to that of the pure PVA film, respectively. The NFC with high aspect ratio and high crystallinity is beneficial to the improvement of the mechanical strength and thermal stability.

Keywords: Nanofibriled cellulose (NFC); Corn husks; PVA; Light transmittance; Tensile strength

1 Introduction

Recently, many efforts have been made worldwide to develop environmentally friendly, low cost, renewable and biodegradable substitute materials from non-petroleum resources because the non-degradable plastics obtained from petroleum causes the environmental pollutions (Zhang et al., 2011). Nanofibrillated cellulose (NFC) is attractive reinforcing fillers in thermoplastic matrix materials due to their exceptional mechanical properties, such as low density, easy biodegradability, renewability, low cost, low thermal expansion and high strength (Dufresne, Cavaillé & Vignon, 1997; Fukuzumi et al., 2008; Habibi, Lucia & Rojas, 2010; Iwamoto et al., 2005; Jonoobi et al., 2010; Siqueira, Bras & Dufresne, 2010). These advantages make cellulose fibers to be a good reinforcing component for composite materials. The PVA/NFC reinforced films are transparent, light, flexible, biodegradable and easy to mold, which makes them excellent candidates for numerous applications including substrates for flexible displays, components for precision optical devices, packaging materials and automobile windows (Mihaela Jipa et al., 2012; Tang & Liu, 2008a; Yano et al., 2005).

---

* 本文摘自 Carbohydrate Polymers, 2016, 136: 1027.

As well-known agricultural and industrial wastes, corn husks offer an annual renewable, low cost, and copious source for natural cellulose fibers. With an annual yield of about 45 million tons worldwide. (Reddy & Yang, 2005), corn husks are large - volume solid wastes consisting of 38.2% cellulose, 44.5% hemicelluloses, 6.6% lignin, 1.9% protein, 2.8% ash and the rest being undetermined materials (Barl et al., 1991). Furthermore, the unique properties of corn husks cellulose fibers, such as, good pliability, outstanding strength, moderate durability, high elongation and ready biodegradability, would provide the corn husk products with a promising future (Reddy & Yang, 2005). Utilizing corn husks as a source for natural NFC will be significantly beneficial to reinforcement of polymer materials.

Ultrasonic technique has been recently used for the fabrication of NFC due to the simple, low cost, and environmentally friendly advantages. Individual NFC was prepared from varied cellulose resources, such as, prawn shell waste, BHKP, plant cellulose and bacterial cellulose (Cheng, Wang & Han, 2010; Li, Zhao, Huang & Liu, 2013; Lu et al., 2013; Tischer et al., 2010). NFC exhibits numerous advantages, including low density, exclusive renewability, high specific surface area, easy biodegradability, outstanding gas barrier property, and unlimited availability (Dufresne, Cavaillé & Vignon, 1997; Jonoobi et al., 2010; Siqueira, Bras & Dufresne, 2010). Moreover, NFC offers further means of modifying the cellulose biopolymer, including a highly reactive surface and the ability to make lighter and stronger materials with greater durability (Hubbe et al., 2008). NFC exhibits high strength and its hydroxyl groups offer reactive sites for chemical modification (Ebeling et al., 1999; Fakirov, Bhattacharyya & Shields, 2008; Fukuzumi et al., 2008; Nakagaito et al., 2009; Nakagaito & Yano, 2008). All of these outstanding properties reveal that NFC is beneficial to reinforced composites.

Poly (vinyl alcohol) (PVA) has been investigated as the potential matrix for fully biodegradable composites (Chakraborty, Sain & Kortschot, 2006; Cheng et al., 2007; Lu, Wang & Drzal, 2008). PVA is the most abundant synthetic water-soluble polymer with low cost, which has desirable properties such as water solubility, biocompatibility, and biodegradability, excellent chemical resistance, gas barrier properties, good thermostability, optical and physical properties (Guirguis & Moselhey, 2011; Mihaela Jipa et al., 2012; Ramaraj, 2007; Zhang et al., 2011). It has been studied as a polymeric matrix in the construction of nanocomposite films due to the improved versatility by incorporating nanostructures (Hubbe et al., 2008; Tang & Liu, 2008b).

In this paper, the NFC was separated from corn husks with the high intensity ultrasonication. The PVA/NFC films with different NFC contents were fabricated to examine the effect of NFC on the properties of the resulted materials. The NCF was characterized using the Scanning electron microscopy (SEM), X - ray diffraction (XRD), Fourier transform infrared (FTIR) and thermogravimetric analysis (TG). The mechanical properties, morphologies and optical properties of PVA/NFC composites were evaluated using the tensile test device, SEM and UV-Vis spectroscopy, respectively.

2 Materials and methods

2.1 Materials

Sixty-mesh corn husk powder after grinding and sieving was dried in a vacuum chamber at 60 ℃ for 24 h. The poly(vinyl alcohol) (PVA) ($DP = 1750 \pm 50$, 99 +% hydrolyzed), potassium hydroxide (KOH) hydrochloric acid (HCL), ethanol, benzene, acetic acid, sodium chlorite and other chemicals used were of analytical or reagent grades.

2.2 Preparation of the purification cellulose from corn husk

A brief information on the process for purification of cellulose from corn husk is presented in the supporting section.

2.3 Preparation of corn husk NFC and NFC foam

Based on the previous study (Jian et al., 2014), the purified corn husk cellulose was uniformly dispersed in distilled water at approximately 0.5% (*w/w*) solid content. About 450ml 0.5% aqueous dispersion was placed in an ultrasonic generator (JY99-IID, Scientz Technology, China) with a 1-cm^2-diameter titanium horn under a 50% duty cycle (i.e., a repeating cycle of 1-s ultrasonic treatment and 1-s shutdown). The ultrasonic process was performed at 60 kHz and conducted in an ice/water bath for 60 min with an output power of 1200 W, resulting in a suspension of NCF. In order to conduct further characterization tests, the NFC foam was fabricated successfully after the freeze-drying process.

2.4 Preparation of the PVA solution

The PVA solution [7% (*w/w*)] was prepared by dissolving PVA (7.0 g) in 100mL water with the magnetic stirring for 5 h at 90 ℃ until the PVA was completely dissolved. The PVA solution was sonicated in an ultrasonic bath for 10 min and then placed in a vacuum oven to remove bubbles. The prepared PVA solution appeared to be fully transparent.

2.5 Preparation of PVA/NFC reinforced films

Fig. 1 shows the preparation process of PVA/NFC reinforced films. Calculated volumes of the PVA solution were added to the NFC suspensions resulting in a final PVA concentration of 7% (*w/w*) and NFC concentrations (dry wt. basis) of 0.5% NCF/PVA, 1% NCF/PVA, 3% NCF/PVA, 5% NCF/PVA, 7% NCF/PVA, and 9% NCF/PVA. The PVA solution was cast in glass pane and dried at room temperature for 5 days. The average film thickness (accuracy of 0.1 mm ± 0.01) was obtained by measuring thicknesses at five random positions on each film. Films were then stored in a desiccator with a relative humidity of (65±5)% and a temperature of (25±2) ℃ for further tests.

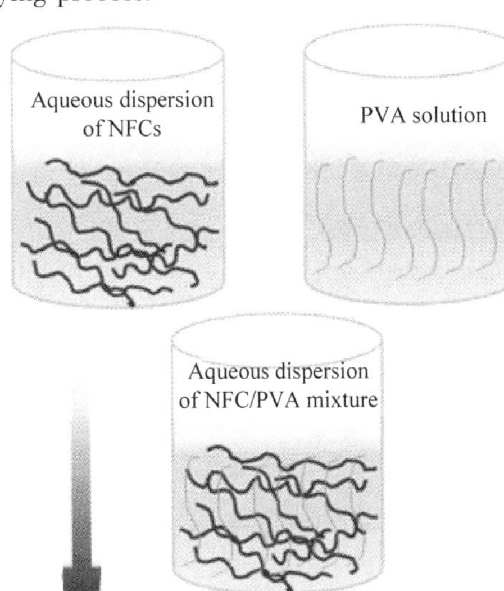

Fig. 1 Fabricated process for PVA/NFC films

2.6 Characterization

The morphology was characterized by SEM (FEI, Quanta 200) operating at 12.5kV. Crystalline structures were identified by the X-ray diffraction (XRD, Bruker D8 Advance, Germany) operating with Cu Kα radiation ($\lambda = 1.5418$ Å) at a scan rate (2θ) of 4° · min^{-1}, 40 kV, 40 mA ranging from 5° to 60°.

Chemical compositions were examined by the Fourier transform infrared spectroscopy (Magna-IR 560, Nicolet) in the range of 400-4000 cm^{-1} with a resolution of 4 cm^{-1}. The thermal stabilities were determined using the Instruments TGA Q500 (TA Instruments, USA) from 25 to 650 ℃ with a heating rate of 10 ℃ · min^{-1} in a N_2 atmosphere. The optical transmittance was measured from 400 to 700 nm using a UV-Vis spectrophotometer (UV-2600, Shimadzu, Kyoto, Japan). The mechanical properties of the PVA/NCF films were measured using a XLW (PC) auto tensile tester (Labthink Instruments Co., Ltd.).

3 Results and discussion

3.1 Morphology

Fig. 2a shows the microstructure of the surface of original materials. It can be clearly observed that unit cells were held together by lignin, hemicelluloses and extractives (waxes, oil, pectin, etc.), which were the cementing materials around the fiber-bundles. Moreover, it was also found the thick deposits on the surface (Reddy & Yang, 2005; Xiao et al., 2015). Fig. 2b shows the structure of the corn husk fibers after the chemical treatment. Different cell walls were well individualized but the microfibrils were still associated within the cell wall at this stage. The chemical-purified cellulose with clearer and smoother surface suggested that the hemicelluloses, lignin and extractives were mostly removed, which was favorable to the subsequent high intensity ultrasonic treatment because the ultrasonic power was concentrated for efficient defibrillation. Fig. 2c shows SEM micrographs of individualized cellulose microfibrils after the ultrasonic treatment. NCF was long filamentous and interconnected with each other. Fig. 2d shows the ultralight and high flexible foam (a bulk density of 4.56 mg · cm^{-3}). The obtained NFC possessed a high aspect ratio (394), which was helpful for the efficient reinforcement in composite materials (Liu et al., 2013; Siqueira et al., 2013).

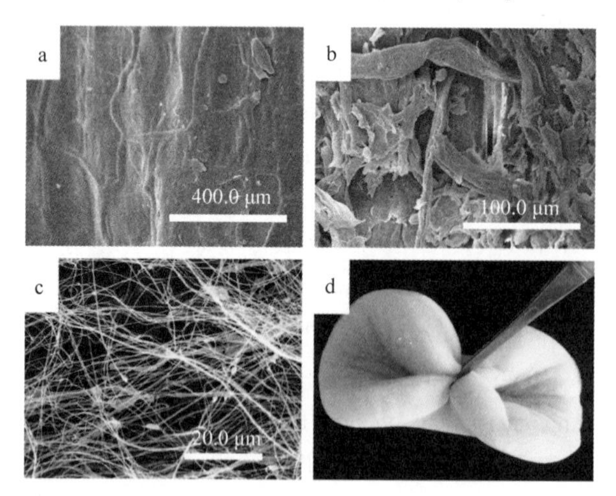

Fig. 2 SEM micrographs of (a) the raw corn husks, (b) the chemical-purified cellulose fibers and (c) the corn husks NFC; (d) the photograph of NCF foam

3.2 XRD analysis

The X-ray diffraction patterns of (a) the original corn husks, (b) corn husk NFC, (c) PVA films and

(d) 1% PVA/NFC films are shown separately in Fig. 3. It could be observed that (a) the original corn husks and (b) the corn husk NFC exhibited sharp diffraction peaks at around 16.3° (101) and 22.4° (002) (Cao et al., 2010), which were assigned to the typical crystalline lattice of cellulose Ⅰ (Isogai et al., 1989; Nishiyama et al., 2003). Cellulose crystallinity plays a key role in the determination of the mechanical and thermal properties. The crystallinity index (CrI) was calculated using Equation 1 (Segal et al., 1959):

$$CrI = 100 \times \frac{I_{002} - I_{am}}{I_{002}}$$

where I_{002} is the maximum intensity of the principal peak (002) lattice diffraction at 22.7° of 2θ for cellulose I, and I_{am} is the intensity of diffraction attributed to amorphous cellulose at 18° of 2θ for cellulose Ⅰ. The detailed data of I_{002} and I_{am} for the samples are listed in Table S1. The crystallinity of (b) the corn husk NFC reached 64.82%, which was increased by 28.85% as compared with (a) the original corn husks in insert stable. The increase in crystallinity was due to the removal of the most hemicelluloses which existed in the amorphous regions during the chemical pretreatment (Alemdar & Sain, 2008; Li et al., 2013). The pure PVA film (Fig. 3c) exhibited the diffraction peaks about 19.4° that assigned to the diffraction planes of 200, which was due to the strong intermolecular interaction between PVA chains through the inter- and intramolecular hydrogen bonding (Qian et al., 2001; Sun et al., 2014). The intensity of (d) 1% PVA/NFC films diffraction peaks has a small increase due to the addition of NFC. It was indicated that the diffractograms of the composite films displayed the superposition of the two components (Lu, Wang & Drzal, 2008; Qian et al., 2001).

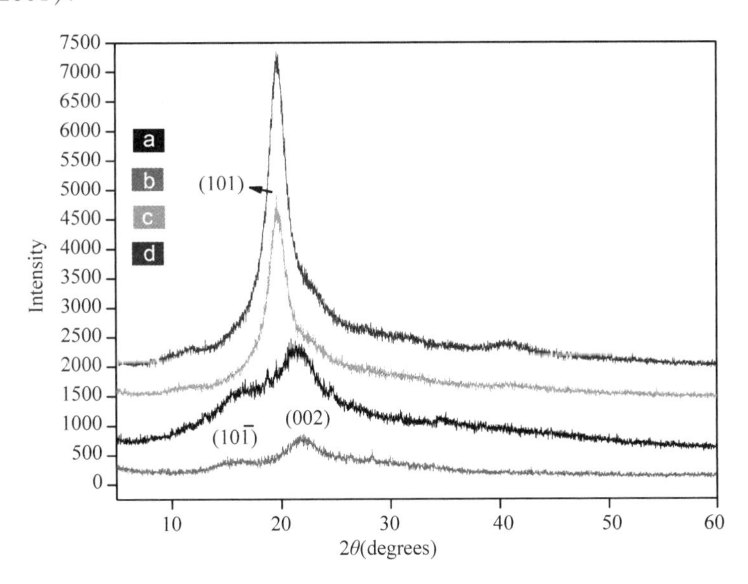

Fig. 3 X-ray diffraction patterns of (a) the original corn husks, (b) the corn husk NFC, (c) the pure PVA film and (d) the 1% PVA/NFC films

3.3 FTIR analysis

The FTIR spectroscopy was a nondestructive method for examining the physico-chemical properties of lignocellulosic materials and a highly effective means of investigating specific interactions between polymers. The FTIR spectra of (a) the original corn husks, (b) corn husk NFC, (c) pure PVA films and (d) 1% PVA/NFC films are shown in Fig. 4. Fig. 4 shows the FTIR spectra of (a) the original corn husks and (b) the corn husk NFC. The dominant absorptions at 3335 cm^{-1} and 2916 cm^{-1}, corresponding to stretching vibrations of OH and CH_2 groups, respectively, were the principal functional groups found in lignocellulosic materials. The sharp peak at 1734 cm^{-1} in the original corn husks was due to either theacetyl and uronic ester groups of the hemicelluloses or the esterlinkage of carboxylic group of the ferulic and p-coumeric acids of the lignin and/or hemicelluloses (Alemdar & Sain, 2008). The intensity of this peak was reduced significantly after the isolation process, which revealed a strong reduction of hemicelluloses and lignin in (b) the corn husk NFC. However, a small shoulder at 1734 cm^{-1} existed in the nanofiber sample, which indicated that there was a small amount of lignin in the cellulose. The major removal of lignin was confirmed by the reduced intensity of 1517 cm^{-1}, corresponding to aromatic C—O stretching mode for lignin and guaiacol ring of lignin in the corn husk NFC. This indicated that the lignin was well removed from the newly prepared corn husk cellulose nanofibers by the $NaClO_2$ treatment. The frequencies and assignments for (c) the pure PVA film were designated as follows: 3276 cm^{-1} for the stretching vibration of hydroxyl groups (O—H), 2915 cm^{-1} for the —CH_2 group stretching vibration, 1427 cm^{-1} for C—H bending, 1324 cm^{-1} for C—H wagging, 1145 cm^{-1} for the crystalline domains of PVA, 1085 cm^{-1} for C—O stretching, 827 cm^{-1} for rocking mode (Ahad, Saion & Gharibshahi, 2012; Sun et al., 2014). In contrast, the characteristic peaks of (d) 1% PVA/NFC films was nearly the same as that of (c) pure PVA film, which may be due to the small content of NFC.

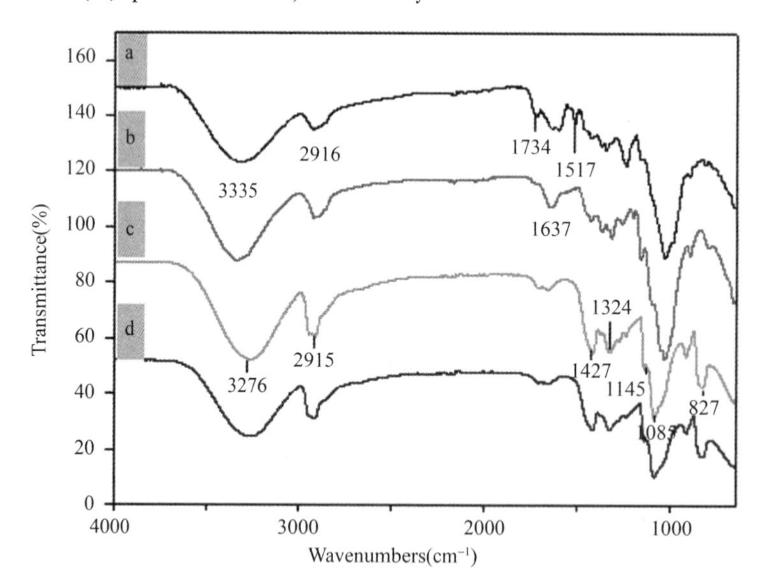

Fig. 4 FTIR spectra of (a) the original corn husks, (b) the corn husks NFC, (c) the pure PVA film and (d) the 1% PVA/NFC films

3.4 Thermal analysis

Fig. 5 shows TG and DTG curves of (a) the original corn husks, (b) the corn husk NFC, (c) the pure PVA film and (d) the 1% PVA/NFC films. Moreover, Table S2 shows the detailed information about the exact% weight loss at each temperature range of the sample. As shown in Fig. 5, two samples showed an initial peak below 200 ℃ due to the evaporation of the absorbed water or some low molecular materials. For the original corn husk, the DTG curves showed two main decomposition steps. The first weight loss stage at 200–320 ℃ was attributed to the hemicelluloses or pectin decomposition. The second weight loss stage at 320–400 ℃ was attributed to cellulose decomposition. Besides, lignin was the most difficult one to decompose among the three components. Its decomposition happened slowly under the whole temperature range from ambient to 650 ℃, and absence of its characteristic peak is due to the wide range of temperatures. For the corn husk NFC, there was only the severe decomposition step of cellulose at 320–400 ℃, which was attributed to the removal of hemecellulose and lignin during the chemical pretreatment process. The thermal degradation data (T_{on}), ($T_{50\%}$) and the peak degradation temperatures (T_{max}) are listed in Table S3. Compared with degradation temperatures (T_{on}, $T_{50\%}$ and T_{max}) of the original corn husk, the corn husk NFC presented a higher degradation temperature. It was revealed that the corn husk NFC had a better thermal stability, making it a promising candidate for the application in thermoplastic composites. For the composite films, three weight loss peaks were observed in Fig. 5c and d. The first region at a temperature range of 50–200 ℃ was due to the vaporization of H_2O molecules. In the second region, the major weight losses were observed in the range of 200–400 ℃, which was attributed to the degradation of side chain of PVA, dehydration accompanied by the formation of some volatile products and thermal degradation of the corn husk NFC (Liu et al., 2013; Lu, Wang & Drzal, 2008). The third region was observed at 400 – 500 ℃, which was attributed to the decomposition of main PVA chain (Liu et al., 2013). The decomposition of the side and main chain of PVA included polyene residues to yield a mixture of carbon and hydrocarbons, i. e., nalkanes, n-alkenes, and aromatic hydrocarbons (Mandal & Chakrabarty, 2014). Compared with the degradation temperatures (T_{on}, $T_{50\%}$ and T_{max}) of the pure PVA film, the thermal stability of 1% PVA/NFC films was slightly increased with the small addition of NFC due to the greater crystallinity and thermal stability of NFC.

Besides, as shown in Fig. S2 and Table S4, the detailed DSC analysis and discussion on the pure PVA film, PVA/NFC reinforced films and the corn husk NFC are provided in the supporting section.

3.5 Mechanical properties of PVA/NFC films

NFC was expected to improve mechanical properties of the PVA/NFC films. Fig. 6 shows the results of tensile strength and tensile strain at break of the pure PVA film and the PVA/NFC films with 0, 0.5, 1, 3, 5, 7 and 9% of NFC. The pure PVA film presented the tensile strength of 37.86 MPa and elongation of 186.6% at break. Compared with the pure PVA film, the tensile strength and tensile strain at break of the PVA/NFC films increased with the increase in NFC contents from 0.5% to 1%. Amongst all the films, 1% PVA/ NFC films showed the highest tensile strength of 55.56 MPa, and elongation of 335.8% at break, which were 1.47 and 1.80 times larger than that of the pure PVA film, respectively. This indicated that the good reinforcement and toughening effects were achieved via the addition of NFC to the PVA matrix. However, the tensile strength and tensile strain at break of the PVA/NFC films simultaneously decreased when the NFC contents were over 1%. It was probably that the dispersion of NFC decreased and the PVA/NFC films became fragile (Liu et al., 2013; Pereira et al., 2014). The aggregation of NCF in PVA caused local concentrations of stresses and tensile failure occurred easily during the testing stage.

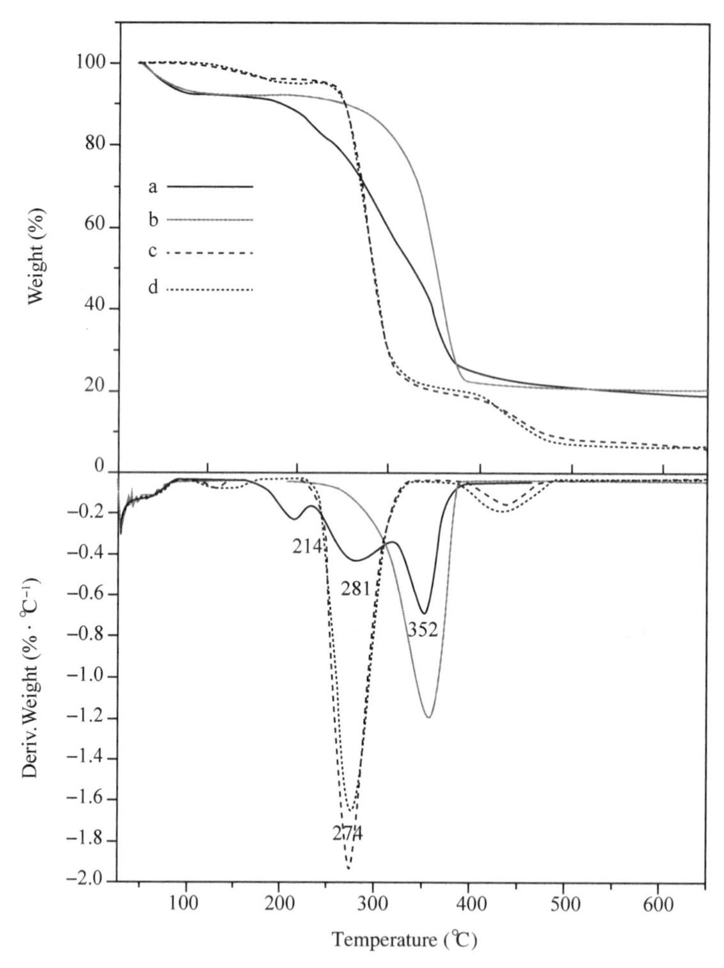

Fig. 5 TG and its derivative thermograms of (a) the original corn husks, (b) the corn husks NFC, (c) the pure PVA film and (d) the 1% PVA/NFC films

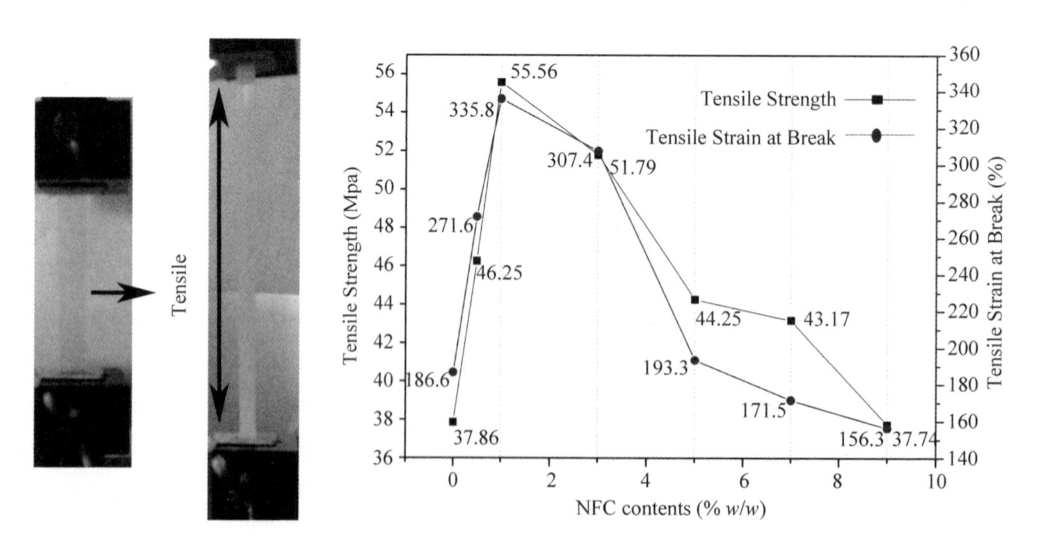

Fig. 6 Mechanical property curves for PVA/NFC films

3.6 Visual examination of PVA/NCF films

Fig. 7a shows digital images of the 1% PVA/NFC films. The background of the images can be clearly seen through the 1% PVA/NFC films, which demonstrated high transparency of the films. Fig. 7b shows the light

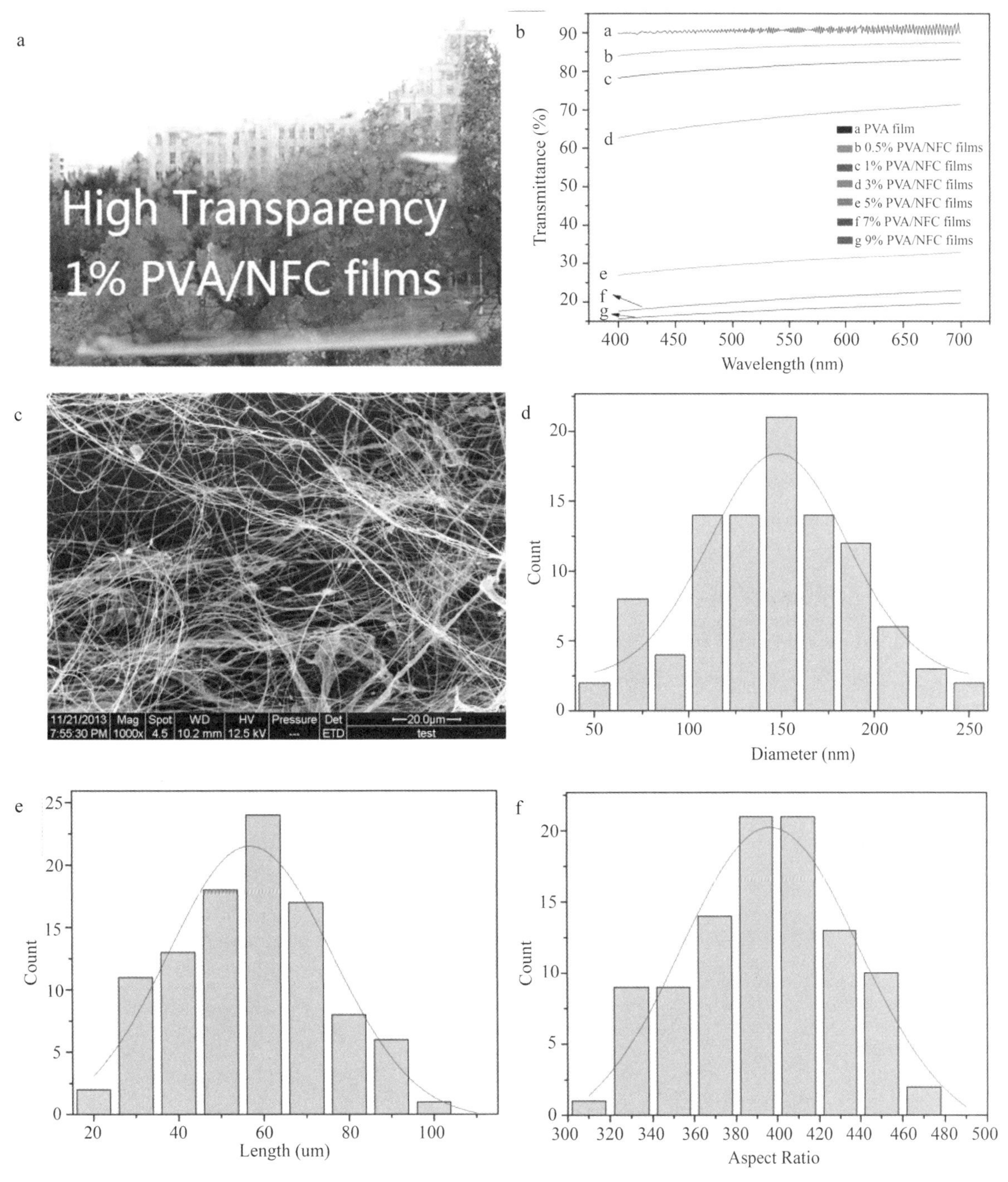

Fig. 7 (a) digital image of 1% PVA/NFC films, (b) transmittance property curves of the pure PVA film and its composite films with different contents of NFC, (c) SEM image of the corn husk NFC and (d) width, (e) Length and (f) aspect radio distributions of the corn husk NFC in (c), respectively

transmittance of the pure PVA film and PVA/NFC reinforced films with different NFC contents. The pure PVA film was transparent with nearly 90% light transmittance at a wavelength of 400 – 700 nm. The light transmittance of the PVA/NFC films decreased with NCF content increased. The light transmittance rates of the PVA/NFC films containing 0. 5%, 1%, 3%, 5%, 7% and 9% of NFC decreased by 85%, 80%, 65%, 30%, 20% and 15%, respectively. When the NCF content was 1%, the films were transparent and the transmittance was 80%. With the range of NFC content of 0. 5%–1%, less than 10% decreased in the transmittance was found for a wavelength ranging from 400–700 nm compared with the pure PVA film. However, a significant reduction in the transmittance was observed when the content of NFC was more than 1%. The size and distribution of nanofibers in matrix material were recognized as the key factors affecting the transparency and the transmittance of films. Figs. 7c and d show SEM image and the diameter distribution of the corn husk NFC. Moreover, Fig. 7 shows the length (L), width (D) and aspect ratio (L/D) distributions of the corn husk NFC measured by Nano measurer 1. 2 and OriginPro 8. 5 analysis. The detailed information was summarized in Table S5. The prepared NFC exhibited an excellent elongation due to the interconnected web with tiny fibrils and the high aspect ratio. Moreover, the prepared PVA/NFC reinforced films had superior toughness and strength due to the outstanding elongation (Sun et al., 2014; Yano et al., 2005). The light was actual a type of electromagnetic wave and may pass an object when its size was shorter than the light wavelength of 400–700nm. The SEM observation of NFC showed a uniform diameter. However, from the SEM images in Fig. S3, the dispersion of NFC became worse and worse with the increased content of NFC due to the stacking of NFC in certain locations in the matrix of PVA.

3.7 SEM micrographs of PVA/NFC films

Fig. 8a and c represent the surface of the pure PVA film and the 1% PVA/NFC films, respectively. The surface of the pure PVA film was smoother than that of the 1% PVA/NFC films. It was observed that the NFC

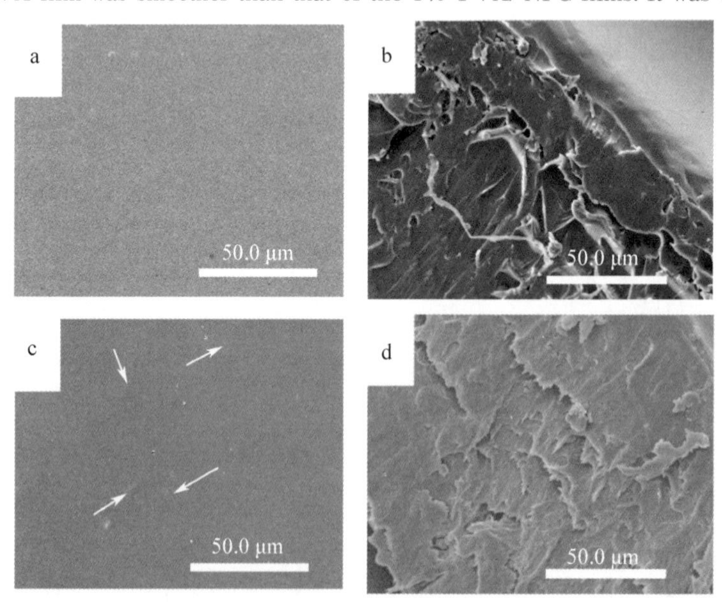

Fig. 8 SEM micrographs of (a) surface of the pure PVA film, (b) cross-section of the pure PVA film, (c) surface of 1% PVA/NFC film and (d) cross-section of 1% PVA/NFC film

was embedded in the PVA matrix on the surface of 1% PVA/NFC films. Fig. 8b and d show cross-section of the pure PVA film and the 1% PVA/NFC films by liquid nitrogen brittle fracture, respectively. In contrast, the cross-section of the 1% PVA/NFC films appeared to be more wavy respect to that of the pure PVA film. It was indicated that 1% PVA/NFC could be uniformly dispersed, well bonded and firmly embedded in the PVA matrix (Fortunati et al., 2013; Pereira et al., 2014). Moreover, refer to the results of the transmittance and mechanical property tests, it could be concluded that the NCF content of 1% was suitable for the preparation of PVA composites with excellent transparency and reinforcement.

4 Conclusions

In the present paper, NFC was successfully defibrillated from the corn husks by means of chemical pretreatments and ultrasonication. The obtained NFC with the mean diameter of 146.60 nm and length of tens microns possessed a relatively superior crystallinity and thermostability. Both NFC and PVA exhibited good compatibility due to the hydrogen bonding interaction. Simultaneously, the potential property improvement of PVA/NFC films depended on the degree of dispersion and the bonding ability between matrix and reinforcing phase. Interestingly, the tensile strength of PVA/NFC films increased from 37.86 MPa to 55.56 MPa, and the tensile strain at break increased from 186.6% to 335.8% at the NCF content of 1%. In addition, the PVA/NFC films presented high transparency. Moreover, the uniform dispersion and good compatibility of the PVA/NFC films play an important role in the improvement of the mechanical strength and light transmittance for the resulting products.

Acknowledgments

This project was supported financially by the National Natural Science Foundation of China (grant no. 31270590).

References

[1] Ahad N, Saion E, Gharibshahi E. Structural, thermal, and electrical properties of PVA-sodium salicylate solid composite polymer electrolyte[J]. J. Nanomater., 2012, 94.

[2] Alemdar A, Sain M. Isolation and characterization of nanofibers from agricultural residues-Wheat straw and soy hulls [J]. Bioresource Technol., 2008, 99(6), 1664-1671.

[3] Barl B, Biliaderis C G, Murray, et al. Combined chemical and enzymic treatments of corn husk lignocellulosics[J]. J. Sci. Food Agr., 1991, 56(2), 195-214.

[4] Cao Y, Li, H, et al. Structure and properties of novel regenerated cellulose films prepared from cornhusk cellulose in room temperature ionic liquids[J]. J. Appl. Polym. Sci., 2010, 116(1), 547-554.

[5] Chakraborty A, Sain M, Kortschot M. Reinforcing potential of wood pulp-derived microfibres in a PVA matrix. Holzforschung, 2006, 60(1), 53-58.

[6] Cheng Q, Wang S, Han Q. (2010). Novel process for isolating fibrils from cellulose fibers by high-intensityultrasonication. II. Fibril characterization[J]. J. Appl. Polym. Sci., 2010, 115(5), 2756-2762.

[7] Cheng Q, Wang S, Rials T G, et al. Physical and mechanical properties of polyvinyl alcohol and polypropylene composite materials reinforced with fibril aggregates isolated from regenerated cellulose fibers. Cellulose, 2007, 14(6), 593-602.

[8] Dufresne A, Cavaillé J Y, Vignon M R. Mechanical behavior of sheets prepared from sugar beet cellulose microfibrils

[J]. J. Appl. Polym. Sci., 1997, 64(6), 1185-1194.

[9] Virtanen S, Vuoti S, Heikkinen H, et al. High strength modified nanofibrillated cellulose-polyvinyl alcohol films [J]. Cellulose, 2014, 21(5): 3561-3571.

[10] Ebeling T, Paillet M, Borsali R, et al. Shear-induced orientation phenomena in suspensions of cellulose microcrystals, revealed by small angle X-ray scattering[J]. Langmuir, 2008, 15(19): 6123-6126.

[11] Fakirov S, Bhattacharyya D, Shields R. Nanofibril reinforced composites from polymer blends. Colloid. Surface. A., 2008, 313, 2-8.

[12] Fortunati E, Puglia D, Luzi F, et al. Binary PVA bio-nanocomposites containing cellulose nanocrystals extracted from different natural sources: Part I[J]. Carbohyd. Polym., 2013, 97(2), 825-836.

[13] Fukuzumi H, Saito T, Iwata T, et al. Transparent and high gas barrier films of cellulose nanofibers prepared by TEMPO-mediated oxidation. Biomacromolecules, 2008, 10(1), 162-165.

[14] Guirguis O W, Moselhey M T. Thermal and structural studies of poly (vinyl alcohol) and hydroxypropyl cellulose blends, Natural Science, 2012, 4(1): 57-67.

[15] Habibi Y, Lucia L A, Rojas O J. Cellulose nanocrystals: chemistry, self-assembly, and applications[J]. Chem. Rev., 2010, 110(6), 3479-3500.

[16] Hubbe M A, Rojas O J, Lucia L A, et al. Cellulosic nanocomposites: a review. BioResources, 2008, 3(3), 929-980.

[17] Isogai A, Usuda M, Kato T, et al. Solid-state CP/MAS carbon-13 NMR study of cellulose polymorphs. Macromolecules, 1989, 22(7), 3168-3172.

[18] Iwamoto S, Nakagaito A, Yano H, et al. Optically transparent composites reinforced with plant fiber-based nanofibers[J]. Appl. Phys. A, 2005, 81(6), 1109-1112.

[19] Jian L, Caichao W, Yun L, et al. Fabrication of cellulose aerogel from wheat straw with strong absorptive capacity. Front. Agric. Sci. Eng, 2014, 1(1), 46-52.

[20] Jonoobi M, Harun J, Mathew A P, et al. Mechanical properties of cellulose nanofiber (CNF) reinforced polylactic acid (PLA) prepared by twin screw extrusion[J]. Compos. Sci. Technol., 2010, 70(12), 1742-1747.

[21] Li W, Zhao X, Huang Z, et al. Nanocellulose fibrils isolated from BHKP using ultrasonication and their reinforcing properties in transparent poly (vinyl alcohol) films[J]. J. Polym. Res., 2013, 20(8), 1-7.

[22] Liu D, Sun X, Tian H, et al. Effects of cellulose nanofibrils on the structure and properties on PVA nanocomposites. Cellulose, 2013, 20(6), 2981-2989.

[23] Lu J, Wang T, Drzal L T. Preparation and properties of microfibrillated cellulose polyvinyl alcohol composite materials. Compos. Part A-Appl. S., 2008, 39(5), 738-746.

[24] Lu Y, Sun Q, She X, et al. Fabrication andcharacterisation of α-chitin nanofibers and highly transparent chitin films by pulsed ultrasonication[J]. Carbohyd. Polym., 2013, 98(2), 1497-1504.

[25] Mihaela Jipa I, Dobre L, Stroescu M, et al. Preparation and characterization of bacterial cellulose-poly (vinyl alcohol) films with antimicrobial properties. Mater. Lett., 2012, 66(1), 125-127.

[26] Nakagaito A N, Fujimura A, Sakai T, et al. Production of microfibrillated cellulose (MFC)-reinforced polylactic acid (PLA) nanocomposites from sheets obtained by a papermaking-like process[J]. Compos. Sci. Technol., 2009, 69(7), 1293-1297.

[27] Nakagaito A N, Yano H. The effect of fiber content on the mechanical and thermal expansion properties of biocomposites based on microfibrillated cellulose. Cellulose, 2008, 15(4), 555-559.

[28] Nishiyama Y, Sugiyama J, Chanzy H, et al. Crystal structure and hydrogen bonding system in cellulose Iα from synchrotron X-ray and neutron fiber diffraction[J]. J. Am. Chem. Soc., 2003, 125(47), 14300-14306.

[29] Pereira A L S, doNascimento D M, Morais J P S, et al. Improvement of polyvinyl alcohol properties by adding

nanocrystalline cellulose isolated from banana pseudostems[J]. Carbohyd. Polym., 2014, 112, 165-172.

[30] Pereira A, DMDNascimento, Filho M S, et al. Improvement of polyvinyl alcohol properties by adding nanocrystalline cellulose isolated from banana pseudostems[J]. Carbohyd. Polym., 2014, 112: 165-172.

[31] Qian X F, Yin J, Huang J C, et al. The preparation and characterization of PVA/Ag2S nanocomposite[J]. Mater. Chem. Phys., 2001, 68(1), 95-97.

[32] Ramaraj B. Crosslinked poly (vinyl alcohol) and starch composite films: Study of their physicomechanical, thermal, and swelling properties[J]. J. Appl. Polym. Sci., (2007), 103(2), 1127-1132.

[33] Reddy N, Yang Y. Properties and potential applications of natural cellulose fibers from cornhusks[J]. Green Chem., 2005, 7(4), 190-195.

[34] Siqueira G, Bras J, Dufresne A. Cellulosic bionanocomposites: a review of preparation, properties and applications [J]. Polymers, 2010, 2(4), 728-765.

[35] Siqueira G, Bras J, Follain N, et al. Thermal and mechanical properties of bio-nanocomposites reinforced by Luffa cylindrica cellulose nanocrystals[J]. Carbohyd. Polym., 2013, 91(2), 711-717.

[36] Sun X, Lu C, Liu Y, et al. Melt-processed poly (vinyl alcohol) composites filled with microcrystalline cellulose from waste cotton fabrics[J]. Carbohyd. Polym., 2014, 101, 642-649.

[37] Tang C, Liu H. Cellulosenanofiber reinforced poly (vinyl alcohol) composite film with high visible light transmittance. Compos. Part A-Appl. S., 2008, 39(10), 1638-1643.

[38] Tang C, Liu H. Cellulosenanofiber reinforced poly(vinyl alcohol) composite film with high visible light transmittance. Compos. Part A-Appl. S., 2008, 39(10), 1638-1643.

[39] Tischer P C F, Sierakowski M R, Westfahl Jr H, et al. Nanostructural reorganization of bacterial cellulose by ultrasonic treatment. Biomacromolecules, 2010, 11(5), 1217-1224.

[40] Xiao S, Gao R, Lu Y, et al. Fabrication and characterization of nanofibrillated cellulose and its aerogels from natural pine needles[J]. Carbohyd. Polym., 2015, 119, 202-209.

[41] Yano H, Sugiyama J, Nakagaito A N, et al. Optically transparent composites reinforced with networks of bacterial nanofibers[J]. Adv. Mater., 2005, 17(2), 153-155.

[42] Zhang W, Yang X, Li C, et al. Mechanochemical activation of cellulose and its thermoplastic polyvinyl alcohol ecocomposites with enhanced physicochemical properties[J]. Carbohyd. Polym., 2011, 83(1), 257-263.

[43] Liu D, Sun X, Tian H, et al. Effects of cellulose nanofibrils on the structure and properties on PVA nanocomposites. Cellulose, 2013, 20(6), 2981-2989.

[44] Lu J, Wang T, Drzal L T. Preparation and properties of microfibrillated cellulose polyvinyl alcohol composite materials. Compos. Part A-Appl. S., 2008, 39(5), 738 746.

[45] Mandal A, Chakrabarty D. Studies on the mechanical, thermal, morphological and barrier properties of nanocomposites based on poly (vinyl alcohol) and nanocellulose from sugarcane bagasse[J]. J. Ind. Eng. Chem., 2014, 20 (2), 462-473.

[46] Segal L, Creely J, Martin A, et al. An empirical method for estimating the degree of crystallinity of native cellulose using the X-ray diffractometer[J]. Text. Res. J., 1959, 29(10), 786-794.

[47] Sun X, Lu C, Liu Y, et al. Melt-processed poly (vinyl alcohol) composites filled with microcrystalline cellulose from waste cotton fabrics[J]. Carbohyd. Polym., 2014, 101, 642-649.

[48] Tang C, Liu H. Cellulosenanofiber reinforced poly (vinyl alcohol) composite film with high visible light transmittance. Compos. Part A-Appl. S., 2008, 39(10), 1638-1643.

[49] Yano H, Sugiyama J, Nakagaito A N, et al. Optically transparent composites reinforced with networks of bacterial nanofibers[J]. Adv. Mater., 2005, 17(2), 153-155.

中文题目：玉米苞皮纳米纤维增强可降解聚乙烯醇薄膜

作者：肖少良，高汝楠，高丽坤，李坚

摘要：本研究旨在利用化学预处理和超声波技术制备和表征玉米苞皮纳米纤丝化纤维素增强的聚乙烯醇(PVA)薄膜。所得的纳米纤丝化纤维素(NFCs)具有宽 50~250 nm 的窄宽度和高长宽比(394)。NFC 的结晶类型为纤维素 I 型。与原玉米苞皮相比，半纤维素和木质素的去除使 NFC 结晶度和热稳定性提高。考察了不同 NFC 浓度(0.5%、1%、3%、5%、7%和 9%，*w/w* 干基)的 PVA 薄膜。1% PVA/NFC 增强薄膜的可见光透过率为 80%，断裂时的拉伸强度和拉伸应变分别比纯 PVA 薄膜提高 1.47 倍和 1.80 倍。高长径比、高结晶度的 NFC 有利于提高机械强度和热稳定性。

关键词：纳米纤丝化纤维素(NFC)；玉米苞皮；聚乙烯醇；透光率；拉伸强度

Core-shell Composite of Wood-derived Biochar Supported MnO_2 Nanosheets for Supercapacitor Applications*

Caichao Wan, Yue Jiao, Jian Li

Abstract: Eco-friendly wood-derived biochar (WDB) was used as a substrate material to support sheet-like nano-MnO_2 via an easily-operated in-situ redox reaction between the biochar and $KMnO_4$. WDB was readily obtained by pyrolyzing wood waste of agriculture and industry. The MnO_2/WDB composite displays a core-shell structure and can be utilized as a free-standing and binder-free supercapacitor electrode. The MnO_2/WDB electrode has a moderate specific capacitance of 101 $F \cdot g^{-1}$, an excellent coulombic efficiency of 98%-100%, and a good cyclic stability with a capacitance retention of 85.0% after 10000 cycles, making it useful for supercapacitor applications. Moreover, it is expected that such porous inexpensive WDB can serve as a novel harmless substrate material to combine with other electrochemical active substances for the development of high-performance energy storage devices.

Keywords: Manganese oxide; Electrochemical properties; Carbon nanotubes; Graphene

1 Introduction

In response to the changing global landscape, energy has become a primary focus of the major world powers and scientific community. There is an increasing importance to develop and refine more efficient energy-conversion and storage devices. One class of such devices, called supercapacitors, have attracted increasing attention due to their ultrahigh power density, fast charge-discharge properties, long lifecycle, excellent reversibility (90%-95% or higher), widespread operation temperature, and their ability to bridge the power/energy gap between traditional dielectric capacitors and batteries/fuel cells.[1-2] The most extensively utilized electrode active materials for supercapacitors include carbon materials (typically like activated carbons, carbon nanofibers, carbon nanotubes, and graphene),[1,3,4] conducting polymers (typically like polyaniline, polypyrrole, and polythiophene),[5] and transition metal oxides/hydroxides (typically like RuO_2, CO_3O_4, NiO, MnO_2, $Ni(OH)_2$, and $CO(OH)_2$).[6] Among these materials, MnO_2 is particularly attractive because of its cost effectiveness, high specific capacitance (the theoretical value reaches up to 1370 $F \cdot g^{-1}$ based on a one-electron redox reaction per manganese atom),[7] environmental compatibility, and remarkable structural versatility.[8] Nevertheless, the poor electrical conductivity and redox kinetics of MnO_2 seriously hinder its electrochemical applications.[9,10] To address such intrinsic limitations and maximize utilization of MnO_2 pseudocapacitance, a common strategy is to integrate low-dimensional oxide materials with highly conductive substrates (like carbon materials).[11-13] This strategy creates a hybrid supercapacitor with a large

* 本文摘自 RSC Advances, 2016, 6: 64811-64817.

pseudocapacitance in addition to the capacitance provided by the electrical double-layer at the supporting electrode surface.[14] Also, this strategy contributes to acquiring both large active surface area and good electrical connection. Liu et al. coaxially coated ultrathin manganese oxide layers on a vertically aligned carbon nanofiber array via cathodic electrochemical deposition, and the core-shell nanostructure demonstrates high performance in maximum specific capacitance (365 $F \cdot g^{-1}$), specific energy (32.5 $W \cdot h \cdot kg^{-1}$) and specific power (6.216 $kW \cdot kg^{-1}$).[14] Recently, Wang et al. conformally coated α-MnO_2 nanowires onto the foamed metal foils supported 3D few-layer graphene/multi-walled carbon nanotube architecture by bath deposition, and the resultant hierarchical hybrid foam shows a high specific capacitance (1108.79 $F \cdot g^{-1}$) and power density (799.84 $kW \cdot kg^{-1}$) and a great capacitance retention (97.94%) after 13000 charge-discharge cycles.[15] Other typical carbon materials, such as activated carbon,[16] graphene oxide,[12] and carbon foam derived from melamine resin,[17] have also been used to combine with MnO_2 nanostructures with various morphologies for supercapacitor applications.

Wood, a naturally grown composite material of complex hierarchical cellular structure, has been always considered as an important natural renewable resource in the process of human being's development. Wood-derived biochar (WDB) is a carbonaceous solid residue produced from the thermal treatment of wood via oxygen-limited pyrolysis.[18] Compared with the biochar derived from other bioresources, WDB is dominant because of its large specific surface area[19] (as high as 683 $m^2 \cdot g^{-1}$) and abundant pore structure facilitating the fast penetration of electrolytes to allow rapid electron-transfer for charge storage and delivery.[20] Moreover, WDB is more readily available because of its fast simple preparation technique and cheap feedstock from wood waste of agriculture and industry, as compared to some aforementioned carbon materials. It is found that pure WDB supercapacitors (non-faradaic electric double-layer capacitance) display poor electrochemical activity, and the corresponding specific capacitances are only several tens of farad per gram in general.[19,21] When integrated with transition metal oxides[22] (like Cu_2O and CuO) or conducting polymers[23] (like polyaniline) with faradaic pseudo-capacitance, the hybrid supercapacitors show significantly improved electrochemical properties. So far, only several literatures studied WDB-based hybrid supercapacitors;[19,21-24] to the best of our-knowledge, there is still no report on the hybrid WDB/MnO_2 electrode. We now wish to report this electrode and its electrochemical properties.

MnO_2 can be synthesized using various techniques, such as thermal decomposition,[25] electrochemical deposition,[26] hydrothermal method,[27] and sol-gel method.[28] In the present work, we adopted a simple cost-effective chemical reduction method to convert potassium permanganate ($KMnO_4$) into MnO_2 by directly using the WDB as reducing agent, as illustrated in Fig. 1. This method helps to deposit MnO_2 nanosheets onto the surface of WDB and generate a core-shell structure. This MnO_2/WDB composite is expected to be used as a free-standing and binder-free electrode, which is beneficial to enhance rate capacitance performance and stability and reduce interface resistance.[29] The MnO_2/WDB composite was characterized by scanning electron microscopy (SEM), transmission electron microscopy (TEM), high-resolution TEM (HRTEM), selected area electron diffraction (SAED), energy dispersive X-ray spectroscopy (EDX), X-ray photoelectron spectroscopy (XPS), and X-ray diffraction (XRD). The electrochemical properties were studied through cyclic voltammograms (CV), galvanostatic charge-discharge (GCD), and electrochemical impedance spectroscopy (EIS) tests in a three-electrode configuration in 1 M Na_2SO_4 aqueous electrolyte.

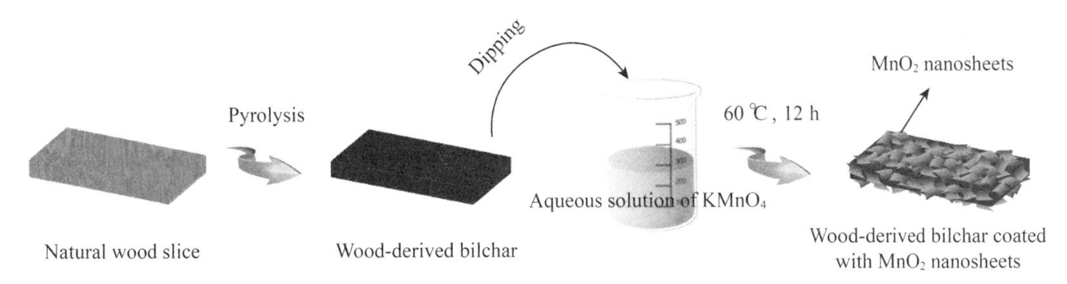

Fig. 1 Schematic illustration of the synthetic procedures for MnO_2/WDB

2 Experimental

2.1 Materials

Potassium permanganate ($KMnO_4$) was supplied by Kemiou Chemical Reagent Co. , Ltd. (Tianjin, China) and used as received. Wood waste of agriculture and industry was collected and cut into slices with a thickness of 1 mm. These wood slices were ultrasonically rinsed with distilled water for 30 min and dried at 60 ℃ for 24 h in a vacuum.

2.2 Synthesis of wood-derived biochar (WDB)

The biochar was synthesized by transferring the wood slice into a tubular furnace for pyrolysis under the protection of nitrogen. In a typical process, the wood slice was heated to 500 ℃ at a heating rate of 5 ℃ · min^{-1}, and this temperature was maintained for 1 h; then the sample was heated to 1000 ℃ at 5 ℃ · min^{-1} and held at this temperature for 2 h to allow for complete pyrolysis. Thereafter, the temperature decreased to 500 ℃ at 5 ℃ · min^{-1} and finally decreased naturally to the room temperature.

2.3 Synthesis of MnO_2/wood-derived biochar (MnO_2/WDB) composite

The as-prepared WDB (ca. 0. 07 g) was firstly immersed in the aqueous solution of $KMnO_4$(50 mL). The weight ratio of biochar to $KMnO_4$ was set as 4 : 1, which has been confirmed elsewhere to contribute to acquiring favorable electrochemical property for the composites consisting of MnO_2 and carbon. [30] The beaker containing the above mixture was then transferred into an oven and covered with a glass culture dish. The mixture was heated at 60 ℃ for 12 h. After the heating, the sample was rinsed with a great deal of distilled water and finally dried at 60 ℃ for 24 h in a vacuum.

2.4 Characterization

SEM observations were performed with a Hitachi S4800 SEM equipped with an EDX detector for element analysis. TEM and HRTEM observations and SAED were performed with a FEI, Tecnai G2 F20 TEM with a field-emission gun operating at 200 kV. XPS was carried out using a Thermo Escalab 250Xi XPS spectrometer-equipped with a dual X-ray source using Al-Kα. Deconvolution of the overlapping peaks was performed using a mixed Gaussian-Lorentzian fitting program (Origin 8. 5, Originlab Corporation). XRD spectroscopy was implemented on a Bruker D8 Advance TXS XRD instrument with Cu Kα (target) radiation (λ = 1. 5418 Å) at a scan rate (2θ) of 4° · min^{-1} and a scan range from 5° to 80°.

2.5 Electrochemical measurements

The electrochemical measurements were carried out on a CS350 electrochemical workstation (Wuhan CorrTest Instruments Co., Ltd., China) at room temperature in a three-electrode setup. The MnO_2/WDB composite (or WDB) directly served as the working electrode, and the area exposed to the electrolyte was about 0.5 cm^2. The mass of the exposed electrodes was around 2.48 mg for MnO_2/WDB and 1.88 mg for WDB, respectively. An Ag/AgCl electrode and a Pt wire electrode served as reference and counter electrodes, respectively. The electrolyte was 1 M Na_2SO_4 solution. CV curves were measured over the potential window from 0 to 0.8 V at different scan rates of 5, 10, 20 and 50 $mV \cdot s^{-1}$. GCD curves were measured in the potential range of 0-0.8 V at different current densities of 0.05, 0.1, 0.2, 0.5, 1, 2, 5 and 10 $A \cdot g^{-1}$. EIS measurements were carried out in the frequency range from 10^5 to 0.01 Hz with alternate current amplitude of 5 mV.

3 Results and discussion

3.1 Morphology observations and elemental analysis

Morphological characteristics of WDB before and after the deposition of MnO_2 were studied by SEM observations. As shown in Fig. 2a, WDB still maintained the feature structures of pristine wood after the pyrolysis treatment, and a porous structure with several pits whose diameters range from 2 to 4 μm can be observed. After WDB was dipped in the aqueous solution of $KMnO_4$, a redox reaction took place between $KMnO_4$ and WDB, which can be expressed as follows: [31]

$$4MnO_4^- + 3C + H_2O \rightarrow 4MnO_2 + CO_3^{2-} + 2HCO_3^- \quad (1)$$

This reaction results in the formation of MnO_2 on WDB, as indicated in Fig. 2b. It is clear that the initial smooth surface of WDB substrate was covered with an ocean of nano-MnO_2. According to a higher-magnification SEM image (160,000×) in Fig. 2c, the porous and cross-linked MnO_2 nanosheets are only a few nanometers in thickness and have numerous wrinkles and ripples. Such porous and ultrathin nanosheets can effectively improve the surface/interface area of MnO_2 nanocrystals and are expected to facilitate electrolyte diffusion among interspaces of MnO_2 nanosheets. [17] Fig. 2d shows the cross-section SEM image of MnO_2/WDB. The 3D honeycomb porous structure of the tracheid cell wall, which belongs to the structural characteristics of wood, can be clearly identified. The details on the surface features of MnO_2/WDB (the green-squared marker region in Fig. 2d) are presented in Fig. 2e. Similar to the results of Fig. 2b, the surface of WDB was encapsulated with dense MnO_2 layers (inset in Fig. 2e), which suggests that the material belongs to core-shell composite, i.e., the biochar serves as the core part, and the MnO_2 layers act as the shell part.

EDX analysis was used to obtain additional information about the elemental compositions. For WDB, only C, O and Au elements were detected (Fig. 2f). The Au element originates from the coating layer used for electric conduction during SEM observation. The low oxygen content of 5.3 wt% is attributed to the pyrolysis treatment that damaged the oxygen-containing groups. In contrast to WDB, apart from the common C, O and Au elements, the Mn and K elements were also detected for MnO_2/WDB. The strong Mn signals are assigned to the sheet-like nano-MnO_2 and indicate a high content (82.4 wt%) on the surface of MnO_2/WDB. The K signal might be derived from the $KMnO_4$ since there is always a possibility of K^+ co-existing in the MnO_2. [30,32]

In addition, the increased mass ratio of oxygen to carbon from 0.06 (WDB) to 0.76 (MnO_2/WDB) is associated with the generated core-shell structure, i.e., WDB coated with abundant MnO_2 nanosheets. The TEM image of MnO_2/WDB is presented in Fig. 2g, in which the wrinkled MnO_2 nanosheets are supported on WDB and no conspicuous agglomeration appeared. The HRTEM image of MnO_2/WDB reveals the presence of birnessite-type MnO_2 as shown in Fig. 2h; i.e., lattice fringes with spacings of around 0.257 and 0.213 nm agree well with the ($20\bar{1}$) and ($11\bar{2}$) lattice spacings. The corresponding SAED pattern exhibits a polycrystalline feature (inset in Fig. 2h) owing to the random orientation of different MnO_2 nanocrystals. Furthermore, the diffraction rings can be indexed to the (201), ($20\bar{1}$) and (203) planes of birnessite-type MnO_2.

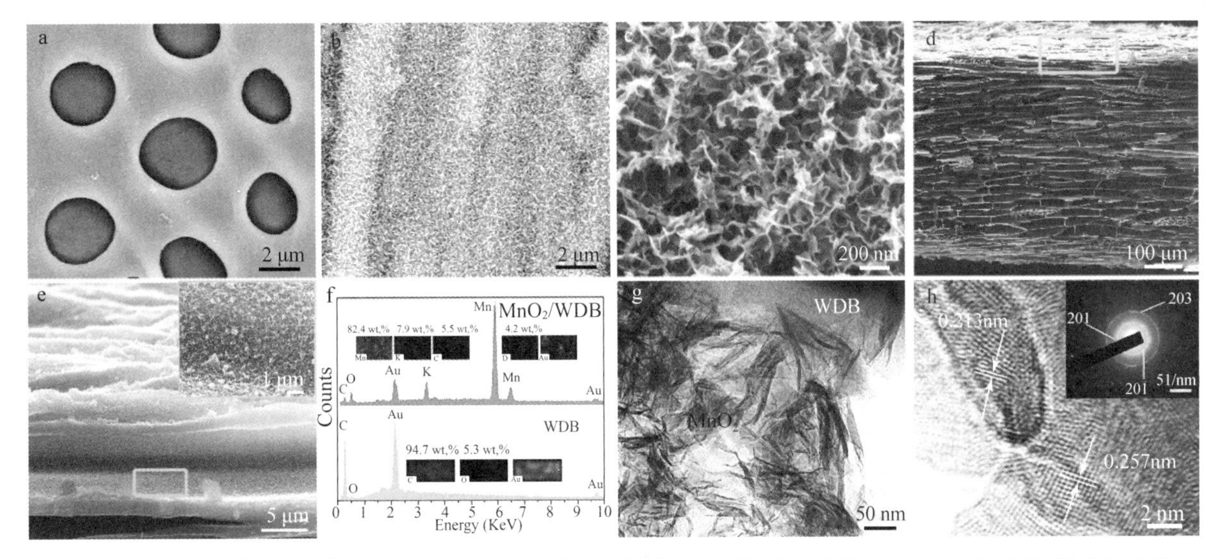

Fig. 2 (a) SEM image of WDB. (b) Low- and (c) high-magnification SEM images of MnO_2/WDB. (d) Cross-section SEM image of MnO_2/WDB. (e) Magnified image of the green-squared marker region in (d), and the inset is detail with enlarged scale of green marker in (e). (f) EDX patterns of WDB and MnO_2/WDB, and the insets show the corresponding elemental maps. (g) TEM and (h) HRTEM images of MnO_2/WDB, and the inset in (h) presents the corresponding SAED pattern

3.2 Crystal structure and chemical compositions

The crystal structures of WDB and MnO_2/WDB were characterized by XRD analysis. As shown in Fig. 3, WDB shows two broad diffraction peaks centered at 22.0° and 43.9°, which are ascribed to the (002) and (100) planes of amorphous carbon originated from the pyrolysis of wood.[33] For MnO_2/WDB, in addition to the peaks from WDB substrate, several new peaks located at around 12.4°, 36.4° and 65.6° can be observed, which correspond to the (001), (110) and ($31\bar{2}$) planes of birnessite-type MnO_2 crystalline phase (JCPDS no. 42-1317). The results are in accordance with the HRTEM and SAED analysis.

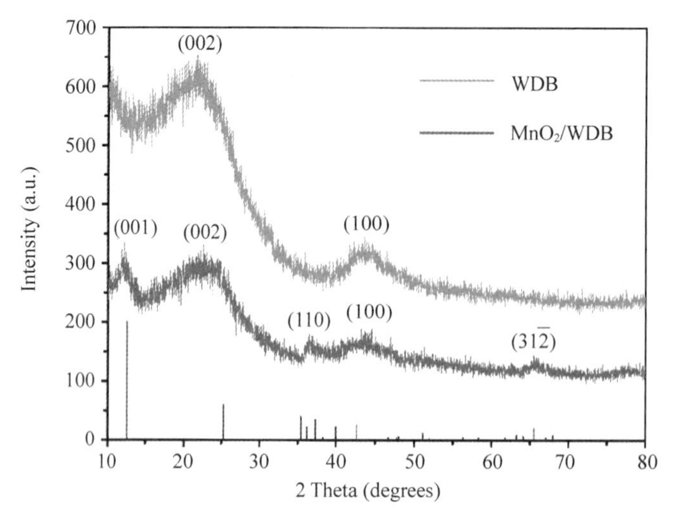

Fig. 3 XRD patterns of WDB and MnO_2/WDB, respectively. The bottom line is the standard JCPDS card no. 42-1317 for birnessite-type MnO_2

XPS analysis was carried out to investigate the oxidation state of manganese and the chemical bonding states of MnO_2/WDB. Fig. 4a shows the XPS survey spectrum of MnO_2/WDB. The XPS signals from elements C, O and Mn can be seen, agreeing well with the results of EDX analysis. The high-resolution Mn 2p spectrum is presented in Fig. 4b, where two strong peaks at 642. 7 and 654. 4 eV can be clearly identified, corresponding to the binding energy of Mn $2p_{3/2}$ and Mn $2p_{1/2}$, [34] respectively. Moreover, the spinning energy separation of 11. 7 eV between the Mn $2p_{3/2}$ and Mn $2p_{1/2}$ peaks is consistent with previously reported data in the spectrum of MnO_2, [30,35] suggesting that the predominant oxidation state is + 4. The high-resolution O 1s spectrum displays two peaks at 530. 2 and 532. 1 eV (Fig. 4c), which are assigned to Mn-O-Mn and Mn-O-H, [36] respectively.

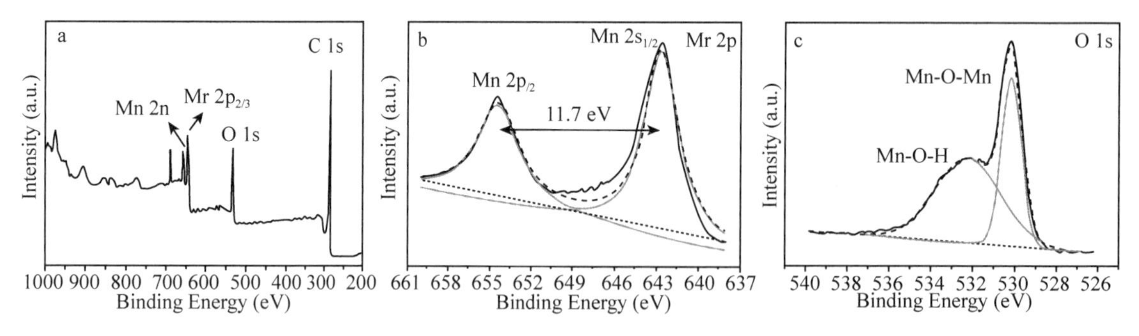

Fig. 4 XPS spectra of MnO_2/WDB: (a) wide scan survey spectrum; the core-level XPS signals of (b) Mn 2p and (c) O 1s

3. 3 Electrochemical properties

The electrochemical behaviors of WDB and MnO_2/WDB electrodes were investigated by CV, GCD, and EIS in a three-electrode system with 1 M Na_2SO_4 as the aqueous electrolyte. CV response of both WDB and MnO_2/WDB electrodes carried out at a scan rate of 10 mV · s^{-1} in the potential range of 0−0. 8 V is shown in

Fig. 5a. Obviously, the current density and the integration area of CV curve obtained from MnO_2/WDB electrode are much larger than those of WDB, indicating that the capacitance of the hybrid electrode is primarily from the pseudocapacitance of MnO_2 nanosheets. It is observed in Fig. 5b that the CV curves of WDB exhibit nearly symmetrical rectangular shape at scan rates of 5, 10, 20 and 50 mV · s^{-1}, indicating the low equivalent series resistance and the fast diffusion of electrolyte ions into the electrode.[37] WDB electrode (carbon as its main component) stores charges through electric double-layer capacitance, and thus expresses in the form of a rectangular shape of the voltammetry characteristics.[38] For MnO_2/WDB electrode (see Fig. 5c), a pair of weak redox peaks can be observed in the CV curves due to the reaction ($MnO_2+Na^++e^-=MnOONa$) of MnO_2 in the hybrid electrode. These CV curves are nearly rectangular, reflecting the good capacitive performance and excellent reversibility of MnO_2/WDB electrode. However, at the high scan rates (e. g., 50 mV · s^{-1}), the CV profile deviates from the rectangular shape, which might be attributed to the polarization of the desolvation process of the hydrated sodium ions and relatively low conductivity of MnO_2 in the hybrid electrode.[39]

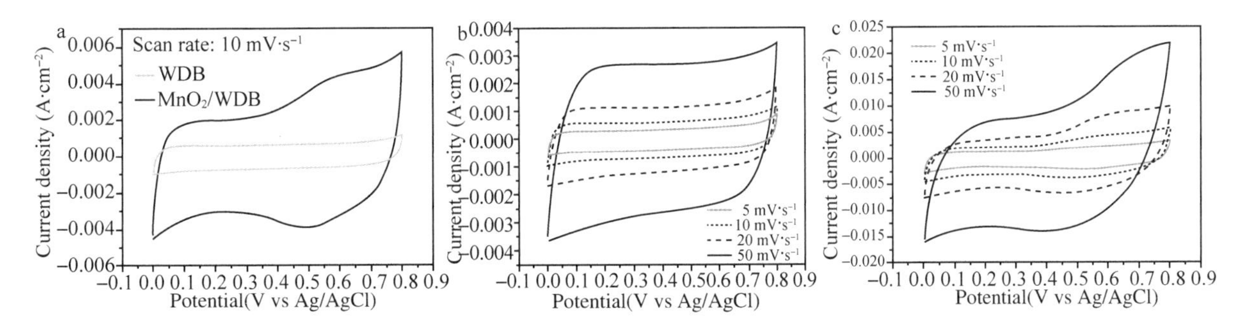

Fig. 5 (a) CV curves of WDB and MnO_2/WDB electrodes at a scan rate of 10 mV · s^{-1}. CV curves of (b) WDB and (c) MnO_2/WDB electrodes at various scan rates

Fig. 6a and b show the charge-discharge characteristics of WDB and MnO_2/WDB electrodes between 0 and 0. 8 V at various current densities of 0. 05–10 A · g^{-1}. During the charging and discharging steps, the charge curves of MnO_2/WDB electrode are almost symmetric to their corresponding discharge counterparts, revealing a good capacitive behavior and a highly reversible Faradic reaction between Na^+ and MnO_2, i. e., $MnO_2+Na^++e^-\leftrightarrow MnOONa$.[7,40] In addition, comparing the discharge curves of WDB and MnO_2/WDB electrodes, MnO_2/WDB electrode shows much longer discharge time, which is consistent with specific capacitance behavior since discharge time is directly proportional to the specific capacitance of electrode. The specific capacitance (C_m) was calculated from the GCD curves according to the following equation:[41]

$$C_m=\frac{I\times\Delta t}{\Delta V\times m} \tag{2}$$

where C_m is the specific capacitance (F · g^{-1}), I is the charge-discharge current (A), Δt is the discharge time (s), ΔV is the potential window (V), and m is the electrode mass (g). The specific capacitance was plotted versus discharge current in Fig. 6c. Because of the deposition of MnO_2, MnO_2/WDB electrode has a highest specific capacitance of 101 F · g^{-1} at a current density of 0. 05 A · g^{-1}, which is about five times of that of WDB (19 F · g^{-1}). The enhancement of capacitance is ascribed to the pseudocapacitance redox reactions of MnO_2. Nevertheless, the specific capacitance of 101 F · g^{-1} in this work is relatively lower, as compared with that of some recently reported composites consisting of MnO_2 and various carbon-based materials, such as carbon fiber cloth (684 F · g^{-1}),[42] graphene (328 F · g^{-1}),[43] and carbon papers (307 F · g^{-1}).[30] The specific capacitance of MnO_2/WDB electrode is expected to be further improved by the following modifications:

(1) A special oxidation of carbon for increasing the surface functionality (through chemical treatment,[44] electrochemical polarization,[45] plasma treatment[46]).

(2) Integrating the MnO_2/WDB electrode with some other types of electrochemical active substances (e. g., conducting polymers).[5,47,48]

(3) Further inserting electroactive particles of transition metals oxides (e. g., RuO_2, TiO_2, Cr_2O_3, and Co_3O_4) into the carbon material (namely WDB).[6]

(4) Minimizing the thickness of WDB substrate for reducing the weight of hybrid electrode.

These above modifications will be deeply studied in our future researches. Both electrodes show gradually decreased capacitance with the increase of current density. In contrast to WDB electrode (storing charges through electric double-layer capacitance), MnO_2/WDB electrode shows a more remarkable decrease of specific capacitance as the current density increased. The reduction is believed to be ascribed to the discrepant-insertion-deinsertion behaviors of Na^+ from the electrolyte to MnO_2.[12,27] At a low current density, the diffusion of ions from the electrolyte can gain access to almost all available pores of the electrode; nevertheless, with the increment of current density, the effective interaction between the ions and the electrode was significantly reduced, and thus a remarkable reduction in capacitance appeared. However, MnO_2/WDB electrode still retained 43% (a decrease from 101 F · g^{-1} to 43 F · g^{-1}) of its capacitance as the current density increased from 0.05 to 10 A · g^{-1}. In addition, the capacitance value is about three times higher than 10 F · g^{-1} of WDB electrode at 10 A · g^{-1}. These results confirm the superior capacitive characteristics of MnO_2/WDB electrode.

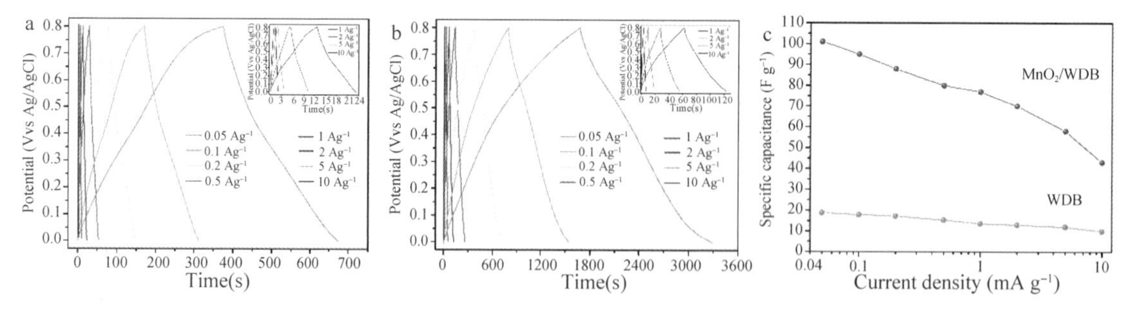

Fig. 6 GCD curves of (a) WDB and (b) MnO_2/WDB electrodes at various current densities. The insets show the enlarged images of the GCD curves at the high current densities. (c) Specific capacitance of WDB and MnO_2/WDB electrodes as a function of discharge current

EIS tests were carried out in a frequency range from 10^5 to 0.01 Hz to further evaluate the electrochemical behaviors of WDB and MnO_2/WDB electrodes. The EIS data were analyzed using Nyquist plots (see Fig. 7). The Nyquist plots can be divided into three parts: (1) a semicircle in the high-to-medium frequencies region indicating the intersection point at the real impedance axis (Z') representing the bulk solution resistance

(R_s) and the diameter of which represents the charge transport resistance (R_{ct}); (2) a straight line with a slope of 45° in the low-frequency range, corresponding to semi-infinite Warburg impedance due to the frequency dependence of ion diffusion/transport in the electrolyte; and (3) a vertical line at the very low frequencies, indicating the pure capacitive behavior.[35,49] In the high-frequency region (inset in Fig. 7), the both electrodes have almost identical R_s of ~2.0 Ω. However, the R_{ct} value of MnO_2/WDB electrode is obviously higher than that of WDB, due to the poor electrical conductivity of MnO_2 in the hybrid electrode. Such difference in R_{ct} is consistent with the prediction results of CV analysis. In the middle-frequency region, the projected length of Warburg curve on the Z' axis characterizes the ion penetration process.[50] The Warburg length of MnO_2/WDB electrode is shorter than that of WDB, which indicates that the hybrid electrode has a shorter ion-diffusion path. This is possibly because the compact covering layers of MnO_2 nanosheets increases the electrochemically active surface area accessed by electrolyte, thus promoting the electrolyte ions adsorption and shortening the migration pathways of ions during the charge-discharge processes.

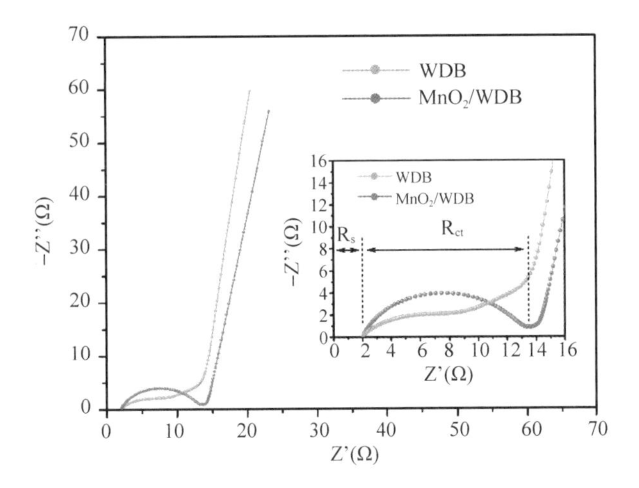

Fig. 7 Nyquist plots of WDB and MnO_2/WDB electrodes

The long-term cycle performances of WDB and MnO_2/WDB electrodes in a potential window of 0–0.8 V were measured by the consecutive GCD tests at a current density of 5 A · g^{-1}. In Fig. 8a, WDB electrode demonstrates good long-term cycle stability, retaining about 95.4% of its initial capacitance after 10000 cycles. Also, the Coulombic efficiency calculated from the ratio between charge capacitance and discharge capacitance remains a quite high value of 98% – 100% during the whole cycle. In contrast, MnO_2/WDB electrode shows slightly decreased capacitance retention of 85.0% after 10000 cycles, but still maintains a high Coulombic efficiency of more than 98% (Fig. 8b). The differences in electrochemical stability between WDB and MnO_2/WDB electrodes are possibly attributed to the different double-layer and pseudocapacitive contributions.[10] It is well-known that the double-layer capacitance only involves a charge rearrangement, while the pseudocapacitance is related to a chemical reaction. The double-layer capacitors generally have better electrochemical stability but lower specific capacitance as compared with those of pseudocapacitators.[51] In this work, WDB electrode (carbon as its main component), making more double-layer contribution compared to that of the hybrid electrode, has a significantly lower capacitance than the latter; however, its cycle stability was enhanced accordingly. For MnO_2/WDB electrode, the faster decrement in capacitance retention is attributed to the dissolution of active materials and mechanical faults caused by the ion insertion/extraction to the MnO_2 nanosheets. [10,52]

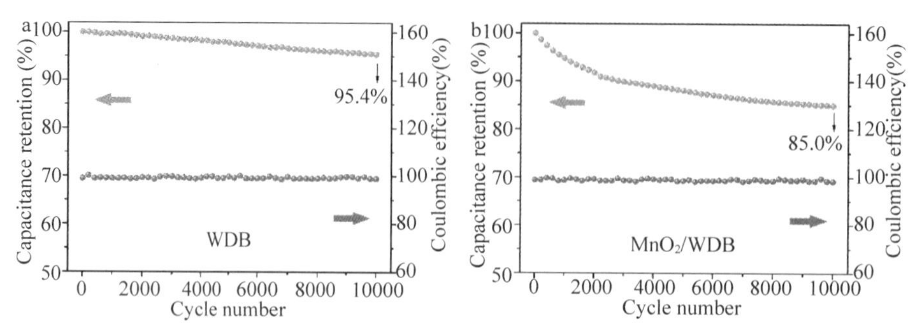

Fig. 8 Cycle performance and Coulombic efficiency of (a) WDB and (b) MnO_2/WDB electrodes measured at a current density of 5 A · g^{-1}

4 Conclusions

Wood-derived biochar was employed as a novel eco-friendly substrate to support MnO_2 nanosheets via a simple cheap in-situ redox reaction between the biochar and $KMnO_4$. This composite with a core-shell structure can serve as a free-standing and binder-free supercapacitor electrode, which shows a moderate specific capacitance of 101 F · g^{-1} at a current density of 0. 05 A · g^{-1}, an excellent coulombic efficiency of 98%-100%, and a favorable cyclic stability with a capacitance retention of 85. 0% after 10000 cycles. The specific capacitance is expected to be improved by integrating the electrode with some new electrochemical active substances (e. g. , conducting polymers), oxidation of carbon for increasing the surface functionality, or minimizing the thickness of WDB substrate, which will be further studied in our future researches. Perhaps wood-derived biochar is not the most suitable substrate material, but we try to give a new potential candidate for the development of high-performance energy storage devices.

Acknowledgements

This study was supported by the National Natural Science Foundation of China (grant nos. 31270590 and 31470584).

References

[1]Zhang LL, Zhao X S. Carbon-based materials as supercapacitor electrodes[J]. Chemical Society Reviews, 2009, 38 (9): 2520-2531.

[2]Wang G, Zhang L, Zhang J. A review of electrode materials for electrochemicalsupercapacitors[J]. Chemical Society Reviews, 2012, 41(2): 797-828.

[3]He S, Chen W. 3Dgraphene nanomaterials for binder-free supercapacitors: scientific design for enhanced performance [J]. Nanoscale, 2015, 7(16): 6957-6990.

[4]He S, Hou H, Chen W. 3D porous and ultralight carbon hybrid nanostructure fabricated from carbon foam covered by monolayer of nitrogen-doped carbon nanotubes for high performance supercapacitors[J]. Journal of Power Sources, 2015, 280: 678-686.

[5]Snook G A, Kao P, Best A S. Conducting-polymer-basedsupercapacitor devices and electrodes[J]. Journal of Power Sources, 2011, 196(1): 1-12.

[6]Zhi M, Xiang C, Li J, et al. Nanostructured carbon-metal oxide composite electrodes for supercapacitors: a review[J]. Nanoscale, 2013, 5(1): 72-88.

[7]Toupin M, Brousse T, Bélanger D. Charge storage mechanism of MnO_2 electrode used in aqueous electrochemical capacitor[J]. Chemistry of Materials, 2004, 16(16): 3184-3190.

[8]Thackeray MM, Rossouw M H, De Kock A, et al. The versatility of MnO_2 for lithium battery applications[J]. Journal of Power Sources, 1993, 43(1-3): 289-300.

[9]Peng Y, Chen Z, Wen J, et al. Hierarchical manganese oxide/carbon nanocomposites for supercapacitor electrodes [J]. Nano Lesearch, 2011, 4(2): 216-225.

[10]Hou Y, Cheng Y, Hobson T, et al. Design and synthesis of hierarchical MnO_2 nanospheres/carbon nanotubes/ conducting polymer ternary composite for high performance electrochemical electrodes [J]. Nano letters, 2010, 10(7): 2727-2733.

[11] Lee S W, Kim J, Chen S, et al. Carbon nanotube/manganese oxide ultrathin film electrodes for electrochemicalcapacitors[J]. ACS Nano, 2010, 4(7): 3889-3896.

[12]Chen S, Zhu J, Wu X, et al. Graphene oxide-MnO_2 nanocomposites for supercapacitors[J]. ACS nano, 2010, 4 (5): 2822-2830.

[13]He S, Zhang R, Zhang C, et al. Al/C/MnO_2 sandwich nanowalls with highly porous surface for electrochemical energy storage[J]. Journal of Power Sources, 2015, 299: 408-416.

[14]Liu J, Essner J, Li J. Hybrid supercapacitor based on coaxially coated manganese oxide on vertically aligned carbon nanofiber arrays[J]. Chemistry of Materials, 2010, 22(17): 5022-5030.

[15]Wang W, Guo S, Bozhilov K N, et al. Intertwined Nanocarbon and Manganese Oxide Hybrid Foam for High-Energy Supercapacitors[J]. Small, 2013, 9(21): 3714-3721.

[16]Khomenko V, Raymundo-Pinero E, Béguin F. Optimisation of an asymmetric manganese oxide/activated carbon capacitor working at 2V in aqueous medium[J]. Journal of Power Sources, 2006, 153(1): 183-190.

[17]He S, Chen W. High performancesupercapacitors based on three-dimensional ultralight flexible manganese oxide nanosheets/carbon foam composites[J]. Journal of Power Sources, 2014, 262: 391-400.

[18]Fang Q, Chen B, Lin Y, et al. Aromatic and hydrophobic surfaces of wood-derivedbiochar enhance perchlorate adsorption via hydrogen bonding to oxygen-containing organic groups[J]. Environmental Science & Technology, 2014, 48(1): 279-288.

[19]Taer E, Deraman M, Talib I A, et al. Physical, electrochemical and supercapacitive properties of activated carbon pellets from pre-carbonized rubber wood sawdust by CO_2 activation[J]. Current Applied Physics, 2010, 10(4): 1071-1075.

[20]Horng Y Y, Lu Y C, Hsu Y K, et al. Flexible supercapacitor based on polyaniline nanowires/carbon cloth with both high gravimetric and area-normalized capacitance[J]. Journal of Power Sources, 2010, 195(13): 4418-4422.

[21]Jiang J, Zhang L, Wang X, et al. Highly orderedmacroporous woody biochar with ultra-high carbon content as supercapacitor electrodes[J]. Electrochimica Acta, 2013, 113: 481-489.

[22]Teng S, Siegel G, Prestgard M C, et al. Synthesis and characterization of copper-infiltrated carbonized wood monoliths for supercapacitor electrodes[J]. Electrochimica Acta, 2015, 161: 343-350.

[23]Yu S, Liu D, Zhao S, et al. Synthesis of wood derived nitrogen-doped porous carbon-polyaniline composites for supercapacitor electrode materials[J]. RSC Advances, 2015, 5(39): 30943-30949.

[24]Taer E, Deraman M, Talib I A. Awitdrus, SA Hashmi and AA Umar[J]. Int. J. Electrochem. Sci., 2011, 6: 3301-3315.

[25]Kim S H, Kim S J, Oh S M. Preparation of layered MnO_2 via thermal decomposition of $KMnO_4$ and its electrochemical

characterizations[J]. Chemistry of Materials, 1999, 11(3): 557-563.

[26]Chou S, Cheng F, Chen J. Electrodeposition synthesis and electrochemical properties of nanostructured γ-MnO_2 films [J]. Journal of Power Sources, 2006, 162(1): 727-734.

[27] Subramanian V, Zhu H, Vajtai R, et al. Hydrothermal synthesis and pseudocapacitance properties of MnO_2 nanostructures[J]. The Journal of Physical Chemistry B., 2005, 109(43): 20207-20214.

[28]Wang X, Wang X, Huang W, et al. Sol-gel template synthesis of highly ordered MnO_2 nanowire arrays[J]. Journal of Power Sources, 2005, 140(1): 211-215.

[29]Li W, Zeng L, Yang Z, et al. Free-standing and binder-free sodium-ion electrodes with ultralong cycle life and high rate performance based on porous carbon nanofibers[J]. Nanoscale, 2014, 6(2): 693-698.

[30]He S, Hu C, Hou H, et al. Ultrathin MnO_2 nanosheets supported on cellulose based carbon papers for high-power supercapacitors[J]. Journal of Power Sources, 2014, 246: 754-761.

[31]Zhao X, Zhang L, Murali S, et al. Incorporation of manganese dioxide within ultraporous activated graphene for high-performance electrochemical capacitors[J]. ACS Nano, 2012, 6(6): 5404-5412.

[32]Ragupathy P, Park D H, Campet G, et al. Remarkable capacity retention of nanostructured manganese oxide upon cycling as an electrode material for supercapacitor[J]. The Journal of Physical Chemistry C, 2009, 113(15): 6303-6309.

[33]Sajitha E P, Prasad V, Subramanyam S V, et al. Synthesis and characteristics of iron nanoparticles in a carbon matrix along with the catalytic graphitization of amorphous carbon[J]. Carbon, 2004, 42(14): 2815-2820.

[34]Zhao X, Du Y, Li Y, et al. Encapsulation of manganese oxidesnanocrystals in electrospun carbon nanofibers as free-standing electrode for supercapacitors[J]. Ceramics International, 2015, 41(6): 7402-7410.

[35] Kim M, Hwang Y, Kim J. Graphene/MnO_2 - based composites reduced via different chemical agents for supercapacitors[J]. Journal of Power Sources, 2013, 239: 225-233.

[36]Lei Z, Shi F, Lu L. Incorporation of MnO_2-coated carbon nanotubes between graphene sheets as supercapacitor electrode[J]. ACS applied materials & interfaces, 2012, 4(2): 1058-1064.

[37]Yan J, Wei T, Shao B, et al. Electrochemical properties ofgraphene nanosheet/carbon black composites as electrodes for supercapacitors[J]. Carbon, 2010, 48(6): 1731-1737.

[38]Frackowiak E, Beguin F. Carbon materials for the electrochemical storage of energy in capacitors[J]. Carbon, 2001, 39(6): 937-950.

[39]LiH, He Y, Pavlinek V, et al. MnO_2 nanoflake/polyaniline nanorod hybrid nanostructures on graphene paper for high-performance flexible supercapacitor electrodes[J]. Journal of Materials Chemistry A, 2015, 3(33): 17165-17171.

[40]Bélanger D, Brousse T, Long J. Manganese oxides: battery materials make the leap to electrochemical capacitors[J]. The Electrochemical Society Interface, 2008, 17(1): 49.

[41]Lei Z, Zhang J, Zhao X S. Ultrathin MnO_2 nanofibers grown on graphitic carbon spheres as high-performance asymmetric supercapacitor electrodes[J]. Journal of Materials Chemistry, 2012, 22(1): 153-160.

[42]He S, Chen W. Application of biomass-derived flexible carbon cloth coated with MnO_2 nanosheets in supercapacitors [J]. Journal of Power Sources, 2015, 294: 150-158.

[43] Kim M, Hwang Y, Kim J. Graphene/MnO_2 - based composites reduced via different chemical agents for supercapacitors[J]. Journal of Power Sources, 2013, 239: 225-233.

[44]Jurewicz K, Frackowiak E. Modified carbon materials for electrochemical capacitors[J]. Molecular Physics Reports, 2000, 27: 38-45.

[45]Momma T, Liu X, Osaka T, et al. Electrochemical modification of active carbon fiber electrode and its application to double-layer capacitor[J]. Journal of Power Sources, 1996, 60(2): 249-253.

[46]Okajima K, Ohta K, Sudoh M. Capacitance behavior of activated carbon fibers with oxygen-plasma treatment[J]. Electrochimica Acta, 2005, 50(11): 2227-2231.

[47]Spitalsky Z, Tasis D, Papagelis K, et al. Carbon nanotube-polymer composites: chemistry, processing, mechanical and electrical properties[J]. Progress in Polymer Science, 2010, 35(3): 357-401.

[48]Peng C, Jin J, Chen G Z. A comparative study on electrochemical co-deposition and capacitance of composite films of conducting polymers and carbon nanotubes[J]. Electrochimica Acta, 2007, 53(2): 525-537.

[49]Sun L, Tian C, Li M, et al. From coconut shell to porous graphene-like nanosheets for high-power supercapacitors [J]. Journal of Materials Chemistry A., 2013, 1(21): 6462-6470.

[50]Weng Z, Su Y, Wang D W, et al. Graphene-cellulose paper flexible supercapacitors[J]. Advanced Energy Materials, 2011, 1(5): 917-922.

[51]Winter M, Brodd R J. What are batteries, fuel cells, and supercapacitors? [J]. Chemical Reviews, 2004, 104 (10): 4245-4270.

[52]Wu H B, Chen J S, Hng H H, et al. Nanostructured metal oxide-based materials as advanced anodes for lithium-ion batteries[J]. Nanoscale, 2012, 4(8): 2526-2542.

中文题目： 负载 MnO_2 纳米片的木质生物炭核壳复合材料基超级电容器电极

作者： 万才超，焦月，李坚

摘要： 环境友好型木质生物炭(WDB)作为基质材料，通过生物炭与高锰酸钾之间易于操作的原位氧化还原反应来负载片状纳米二氧化锰。WDB 易通过农业和工业废弃物热解制备获得。二氧化锰/WDB 复合材料具有核壳结构，可用作独立的无粘结剂超级电容器电极。二氧化锰/WDB 电极的比电容为 101 $F \cdot g^{-1}$，库仑效率为 98%~100%，循环稳定性好，循环 10000 次后的电容保留率为 85.0%，是超级电容器的有效材料。此外，这种多孔的廉价 WDB 有望作为一种新型无害的基底材料，与其他电化学活性物质结合，用于开发高性能的储能装置。

关键词： 锰氧化物；电化学性能；碳纳米管；石墨烯

Wood-derived Biochar Supported Polypyrrole Nanoparticles As A Free-standing Supercapacitor Electrode*

Caichao Wan, Jian Li

Abstract: Paulownia wood processing residues were adopted as both template and precursor for the synthesis of wood-derived biochar/polypyrrole (coded as WDB/PPy) composite via a simple cost-effective in-situ chemical oxidative polymerization method. WDB substrate was encapsulated with a plenty of PPy nanoparticles with diameters of a few hundred nanometers, and the resultant WDB/PPy composite can be utilized as a free-standing and binder-free supercapacitor electrode. This hybrid electrode has a high specific capacitance of 216 $F \cdot g^{-1}$ at 0.05 $A \cdot g^{-1}$, an excellent coulombic efficiency of more than 98%, and a favorable cyclic stability with 96.9% capacitance retention after 3000 cycles. Those results offer a low-cost eco-friendly design of electrode materials for supercapacitors. More importantly, this cheap and environmentally benign WDB substrate is expected to integrate with more types of electrochemical active substances to develop novel energy storage devices.

Keywords: High-performance supercapacitor; Activated carbon; Conducting-polymer; Energy storage; Composites

1 Introduction

Wood is the oldest material utilized by humans for construction after stone. Wood is known to have numerous favorable properties, such as renewability, low density, low thermal expansion, easy machining, desirable mechanical strength, and esthetically pleasing.[1,2] As a result, in modern times, wood is still widely applied in various fields like building industry, wooden furniture making, interior decoration, papermaking, bio-refinery, and industrial or household fuel. Wood can be easily transformed into carbon product (called as wood-derived biochar, WDB) via the pyrolysis of wood in oxygen-limited condition.[3] Compared with the biochar derived from other bioresources, WDB has more advantages such as large specific surface area[4] (as high as 683 $m^2 \cdot g^{-1}$) and abundant pore structure, which provide convenience for the combination with various guest substances. For instance, Karakoyun et al. synthesized environmentally benign hydrogel-WDB composites, which were used as adsorbents to remove phenol from aqueous environments.[5] Recently, Zhang et al. fabricated a magnetic WDB-based adsorbent with colloidal or nanosized γ-Fe_2O_3 embedded in porous biochar matrix.[6] This composite exhibited excellent ferromagnetic property and strong

* 本文摘自 RSC Advances, 2016, 6: 86006-86011.

sorption ability to aqueous arsenic. More recently, an engineered biochar, which was synthesized through direct pyrolysis of hickory wood pretreated with $KMnO_4$, displayed strong sorption ability to heavy metals including Pb(Ⅱ), Cu(Ⅱ), and Cd(Ⅱ) with maximum sorption capacities of 153.1, 34.2, and 28.1 $mg \cdot g^{-1}$, respectively.[7] Apart from these applications in environmental remediation, WDB-based composites are also useful in many fields such as biochar fertilizers,[8] catalyst,[9] direct carbon fuel cell,[10] microwave absorption,[11] and supercapacitors.[12]

Supercapacitors have been considered as one of the quite promising energy storage systems due to their ultrahigh power density, fast charge-discharge properties, long lifecycle, excellent reversibility (90%-95% or higher), and their ability to bridge the power/energy gap between traditional dielectric capacitors and batteries/fuel cells.[13-17] Researchers report that pure WDB supercapacitors (non-faradaic electric double-layer capacitance) display poor electrochemical activity, and the specific capacitances are only several tens of farad per gram in general.[4,12] It is expected to effectively improve the electrochemical properties of WDB-based supercapacitors by some special treatments, e.g., oxidation of carbon for increasing surface functionality through plasma treatment,[18] chemical treatment,[19-21] and electrochemical polarization.[22] Also, it is useful to enhance the electrochemical characteristics via the combination of WDB with some electroactive substances, such as transition metals oxides[15,23] (e.g., RuO_2 and MnO_2) and conducting polymers[24-26] [e.g., polyaniline (PANI) and polypyrrole (PPy)]. Especially, conducting polymers have been regarded as a class of momentous materials for the realization of high-performance supercapacitors due to their significantly higher pseudocapacitance than carbon materials, fast redox switching, high conductivity, and lightweight feature.[27] Conducting polymers have been widely integrated with various carbon materials (e.g., graphene,[28] carbon nanotube,[24,29] and carbon fibers[30]) to synthesize supercapacitors. It is worth to mention that the presence of carbon materials can enhance the mechanical stability of conducting polymers by restraining their considerable volume change during the repeated doping/dedoping process, which contributes to improving the cycling stability of supercapacitors.[31,32] PPy is one of typical conducting polymers, and applicable to develop a variety of electrochemical devices.[33] To the best of our knowledge, there is no report on the PPy/WDB hybrid electrode for supercapacitors. We now wish to study this electrode material and its electrochemical properties.

In the present work, we adopted a simple and cost-effective in-situ chemical oxidative polymerization method to deposit PPy nanoparticles onto WDB, as illustrated in Fig. 1. WDB was easily obtained from the pyrolysis of Paulownia wood processing residues collected from wood-working factory. Prior to the deposition of PPy, WDB was subjected to a concentrated nitric acid treatment to introduce various oxygen-containing groups onto WDB surface for the interaction with pyrrole monomers. The morphology and chemical compositions of the as-prepared PPy/WDB was characterized by scanning electron microscopy (SEM), energy dispersive X-ray spectroscopy (EDX), Fourier transform infrared spectroscopy (FTIR), and X-ray photoelectron spectroscopy (XPS). Moreover, this PPy/WDB composite is expected to serve as a free-standing and binder-free supercapacitor electrode, which is beneficial to enhance rate performance and reduce interface resistance. The electrochemical properties were studied through cyclic voltammograms (CV), galvanostatic charge-discharge (GCD), and electrochemical impedance spectroscopy (EIS) tests in a three-electrode configuration in 1 M Na_2SO_4 aqueous electrolyte.

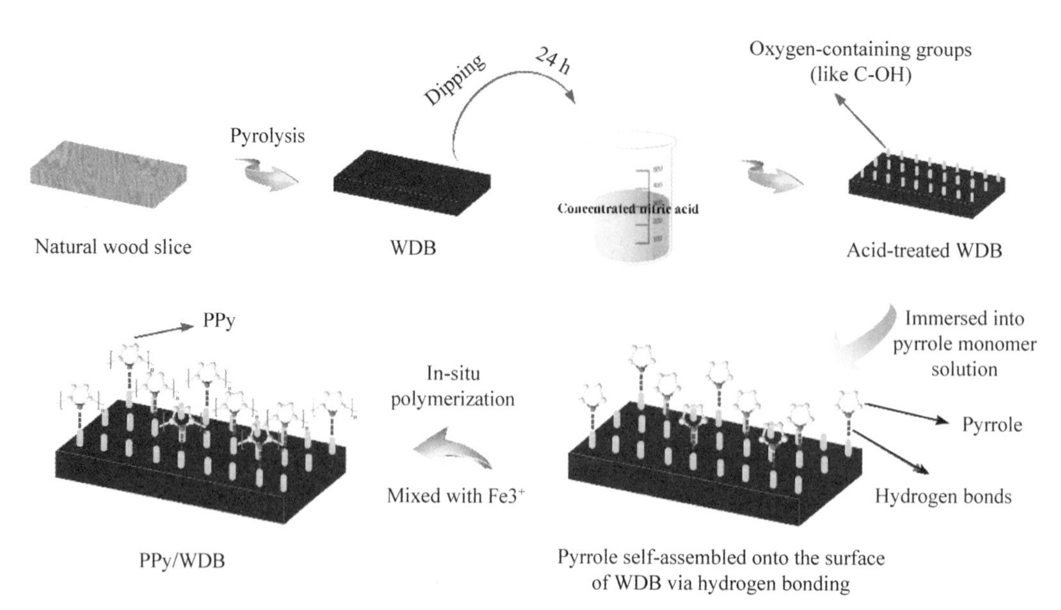

Fig. 1 Schematic illustration of the synthetic procedures for PPy/WDB

2 Experimental

2.1 Materials

Pyrrole, iron (Ⅲ) chloride hexahydrate ($FeCl_3 \cdot 6H_2O$), hydrochloric acid (HCl, 37%), and nitric acid (HNO_3, 65%) were supplied by Kemiou Chemical Reagent Co., Ltd. (Tianjin, China). Paulownia wood processing residues were collected from wood-working factory in China and cut into slices with a thickness of 1 mm. The wood slices were rinsed ultrasonically with distilled water for 30 min and dried at 60 ℃ for 24 h.

2.2 Synthesis of wood-derived biochar (WDB)

The biochar was synthesized by transferring the wood slice into a tubular furnace for pyrolysis under the protection of nitrogen. In a typical process, the wood slice was heated to 500 ℃ at a heating rate of 5 $℃ \cdot min^{-1}$, and this temperature was maintained for 1 h; then the sample was heated to 1000 ℃ at 5 $℃ \cdot min^{-1}$ and held at this temperature for 2 h to allow for complete pyrolysis. Thereafter, the temperature decreased to 500 ℃ at 5 $℃ \cdot min^{-1}$ and finally decreased naturally to the room temperature.

2.3 Synthesis of PPy/wood-derived biochar (PPy/WDB) composite

Prior to the deposition of PPy, WDB was treated with HNO_3(65%) for 24 h at room temperature to introduce various oxygen-containing groups onto WDB surface for the interaction with pyrrole monomers. The PPy was synthesized by the in-situ chemical oxidative polymerization of pyrrole using ferric ions as oxidizing agent according to the literature method with little modification.[34] The literature suggested some optimized synthetic protocols, which were also adopted in this work. In a typical process, WDB (ca. 0.07 g) was firstly immersed into pyrrole monomer solution (1 M, 50 mL) for 12 h with magnetic stirring at room temperature, enabling the pyrrole monomers to fully self-assemble onto the surface of WDB via hydrogen bonding. Thereafter, the in-situ polymerization of pyrrole was conducted for 3 h by slowly adding an aqueous solution of $FeCl_3 \cdot 6H_2O$. The molar ratio of ferric iron/pyrrole was 1 : 1. After the polymerization, the resultant PPy/WDB composite was dipped in 1 M HCl for 30 min, and then rinsed with a large amount of distilled water and dried in a vacuum oven at 50 ℃ for 24 h.

2.4 Characterizations

SEM observations were performed with a Hitachi S4800 SEM equipped with an EDX detector for element analysis. XPS was carried out using a Thermo Escalab 250Xi XPS spectrometer equipped with a dual X-ray source using Al-Kα. Deconvolution of the overlapping peaks was performed using a mixed Gaussian-Lorentzian fitting program (Origin 8.5, Originlab Corporation). FTIR spectrum was recorded by a Nicolet Nexus 670 FTIR instrument in the range of 500–4000 cm^{-1} with a resolution of 4 cm^{-1}.

2.5 Electrochemical measurements

The electrochemical measurements were carried out on a CS350 electrochemical workstation (Wuhan CorrTest Instruments Co., Ltd., China) at room temperature in a three-electrode setup. Because of the thin block shape, PPy/WDB (or WDB) can directly serve as working electrode without any binders. The mass of the electrodes is around 3.67 mg for PPy/WDB and 2.21 mg for WDB, respectively. An Ag/AgCl electrode and a Pt wire electrode served as reference and counter electrodes, respectively. The electrolyte was 1 M Na_2SO_4 solution. CV curves were measured over the potential window from −0.4 to 0.6 V at different scan rates of 5, 10, 20, 50 and 100 $mV \cdot s^{-1}$. GCD curves were measured in the potential range of −0.4~0.6 V at different current densities of 0.05, 0.1, 0.2, 0.5, 1.0 and 2.0 $A \cdot g^{-1}$. EIS measurements were carried out in the frequency range from 10^5 to 0.01 Hz with alternate current amplitude of 5 mV.

3 Results and discussion

The morphology characteristics of WDB before and after the deposition of PPy were studied by SEM observations. As shown in Fig. 2a, WDB remains the structure characteristics of wood (like vessels, rays and fibers). The rough surface is possibly due to the mechanical cutting or HNO_3 treatment. After the deposition of PPy, it is seen that WDB substrate was covered with dense nanoparticles (Fig. 2b), leading to the formation of a core-shell structure, i. e., WDB acts as the core part, and PPy nanoparticles serve as the shell part. The higher-magnification SEM images (Fig. 2c and d) show that these nanoparticles have diameters of a few hundred nanometers and were accumulated into micron-size aggregations. Fig. 2e displays the cross-section SEM image of PPy/WDB. The 3D honeycomb porous structure of the tracheid cell wall, which belongs to the structural characteristics of wood, can be clearly identified. Similar to the results of Fig. 2b-d, the surface of WDB was encapsulated with dense PPy layers, further suggesting that the material belongs to core-shell composite. In addition, it can be seen that the PPy nanoparticles were only in-situ polymerized on the surface of WDB and did not permeate into the inner of WDB (Fig. 2f). Fig. 2g shows the EDX spectrum of WDB, where several elements including carbon, oxygen, potassium, and nitrogen were detected. The C, O and K elements belong to the wood composition. The N element might be derived from the HNO_3 treatment. For PPy/WDB, apart from the above elements, the EDX spectrum shows the presence of Cl signals (Fig. 2h), which is originated from the dopant HCl. Moreover, the high nitrogen content (15.77 wt%) is attributed to the incorporated PPy, and the remarkably increased mass ratio of C/O from 6.4 (WDB) to 9.2 (PPy/WDB) is also related to the presence of PPy.

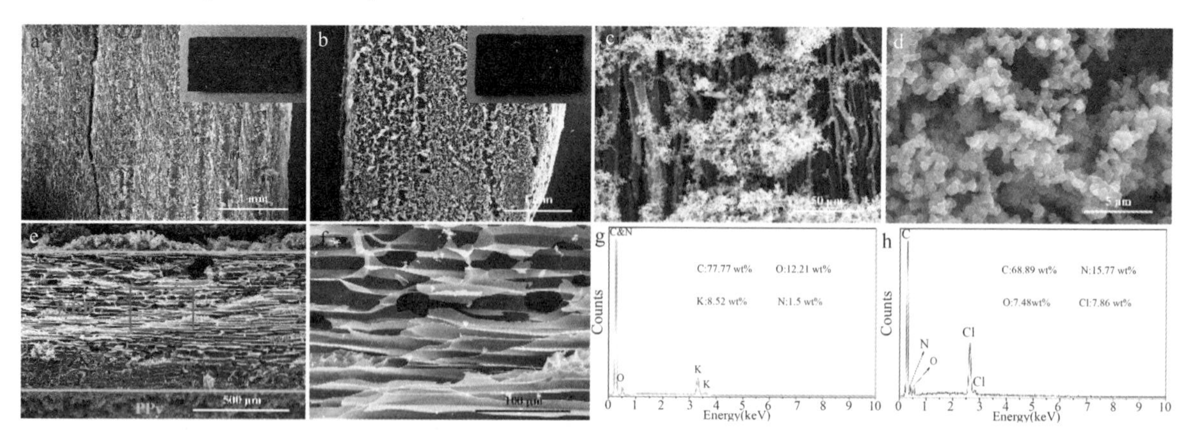

Fig. 2 (a) SEM image of WDB. (b-d) SEM images of PPy/WDB under different magnifications. The insets in (a) and (b) show the optical images of WDB and PPy/WDB, respectively. (e) Cross-section SEM image of PPy/WDB. (f) Magnified image of the red-squared marker region in (e). EDX spectra of (g) WDB and (h) PPy/WDB, respectively

The chemical compositions of PPy/WDB were investigated by FTIR analysis. As shown in Fig. 3a, the broad band at 3423 cm^{-1} is attributed to the O—H and N—H stretching vibrations,[35] and the bands at 1716 and 1151 cm^{-1} are assigned to the C═O stretching of carbonyl and the C—O antisymmetric bridge stretching,[34] indicating that the HNO_3 treatment successfully introduced oxygen-containing functional groups onto the surface of WDB. The bands at 1639 and 1618 cm^{-1} are due to the C═C stretching vibrations, in addition, the band at 1385 cm^{-1} is originated from the C—H deformation vibration.[36] The bands at 1495, 1190, and 937 cm^{-1} belong to the characteristic signals of PPy,[37] corresponding to the C—N stretching vibration in the pyrrole ring, the breathing vibration of the pyrrole ring, and the C—H out-of-plane ring deformation vibration, respectively. These characteristic signals suggest that PPy was successfully polymerized on the surface of WDB.

XPS spectra revealing the chemical bonding in PPy/WDB are shown in Fig. 3b-d. XPS survey spectrum displays four main peaks at binding energy (BE) of 532.4, 400.4, 285.4, and 198.5 eV (Fig. 3b) corresponding to O 1s, N 1s, C 1s, and Cl 2p, respectively. In detail, the N 1s spectrum can be deconvoluted into three main components at around 401.9, 400.3, and 398.6 eV (Fig. 3c), attributable to the positively charged nitrogen ($—N^+—$), amine (—NH—) group, and imine group (—N═),[38,39] respectively, which demonstrate the existence of PPy. The C 1s spectrum can be deconvoluted into four main components. As shown in Fig. 3d, the peaks at 288.8 eV (O—C═O), 286.9 eV (C—OH), and 284.7 eV (C—C or C—H) are derived from WDB,[40] consistent with the results of FTIR analysis. The peak at 286.1 eV (C—N) is another demonstration of the presence of PPy.

The electrochemical properties of WDB and PPy/WDB electrodes were studied by CV, GCD and EIS in a three-electrode system with 1 M Na_2SO_4 as the aqueous electrolyte. Fig. 4a displays the CV curves of supercapacitors based on WDB and PPy/WDB under the potential from −0.4 to 0.6 V at the scan rate of 5 $mV \cdot s^{-1}$. The CV curve of PPy/WDB is close to rectangular shape and displays symmetric current-potential characteristics, indicative of an ideal capacitive nature with ion response.[41] Different from common

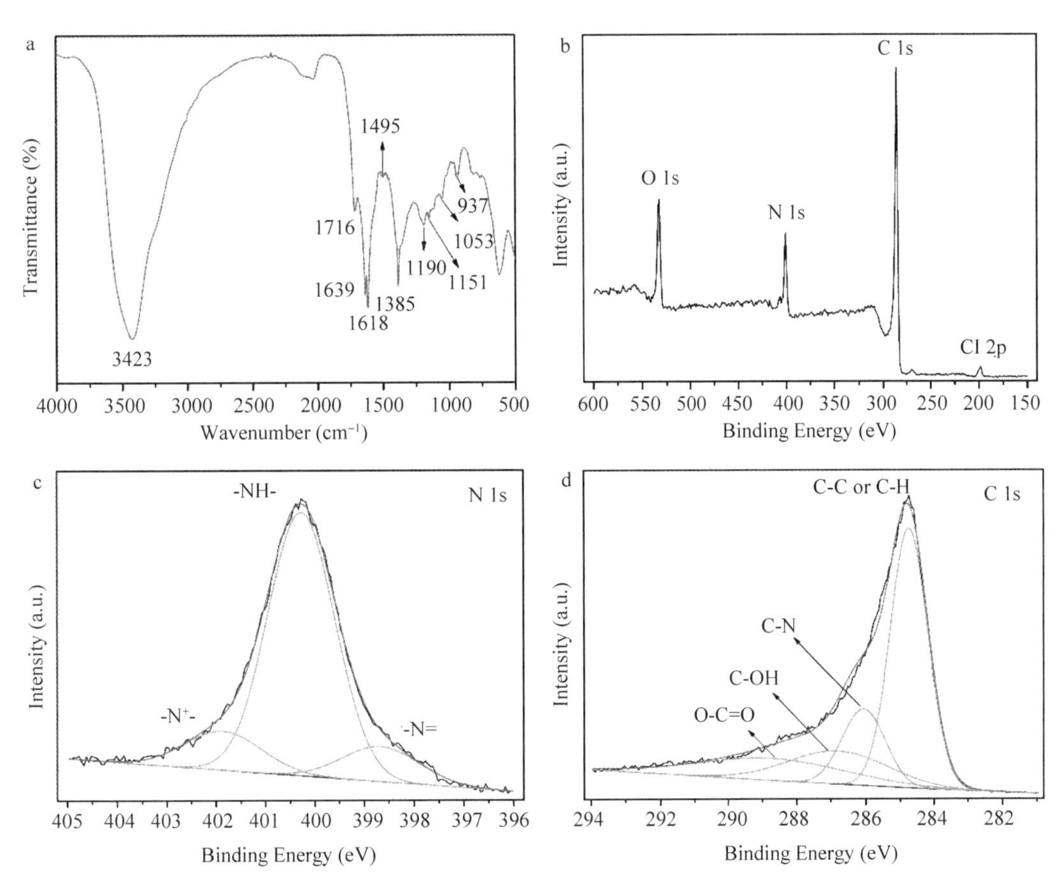

Fig. 3 (a) FTIR spectrum of PPy/WDB. (b) XPS survey spectrum and high resolution scan of (c) N 1s and (d) C 1s of PPy/WDB, respectively

rectangular CV curves for unmodified carbon, the CV profile for WDB after being treated with HNO_3 is heavily distorted from the rectangular shape, which is attributable to faradaic reactions that occur on the surface of WDB electrode, such as reaction between the carbonyl (C=O) and hydroxyl (C—OH) groups, i. e. , $C=O+H^++e^-\leftrightarrow C—OH$. [42,43] In addition, the current density and the integration area of CV curve obtained from PPy/WDB are much larger than those of WDB, indicating that the capacitance of the hybrid electrode is primarily from the pseudocapacitance of PPy. With the increase of scan rate from 5 to 100 $mV \cdot s^{-1}$, WDB maintains subtriangular CV curves, and the current response shows corresponding increases (Fig. 4b). In contrast, PPy/WDB remains a rectangular and symmetric CV shape at scan rate as high as 20 $mV \cdot s^{-1}$ (Fig. 4c), suggesting its favorable rate performance. However, at higher sweep rates (> 20 $mV \cdot s^{-1}$), the CV curves significantly deviate from the rectangular shape, which can be explained by entering into/ejecting and diffusion of counterions being too slow compared to the transfer of electrons in the electrode at high scan rates. [44]

GCD curves of WDB and PPy/WDB electrodes at various current densities (including 0.05, 0.1, 0.2, 0.5, 1.0 and 2.0 $A \cdot g^{-1}$) are presented in Fig. 5a and b, respectively. The charge curves of WDB and PPy/WDB electrodes are almost symmetric to their discharge counterparts, with slight curvature, indicating both double layer and pseudocapacitive contributions. [45] It is clear that the discharge time of PPy/WDB is

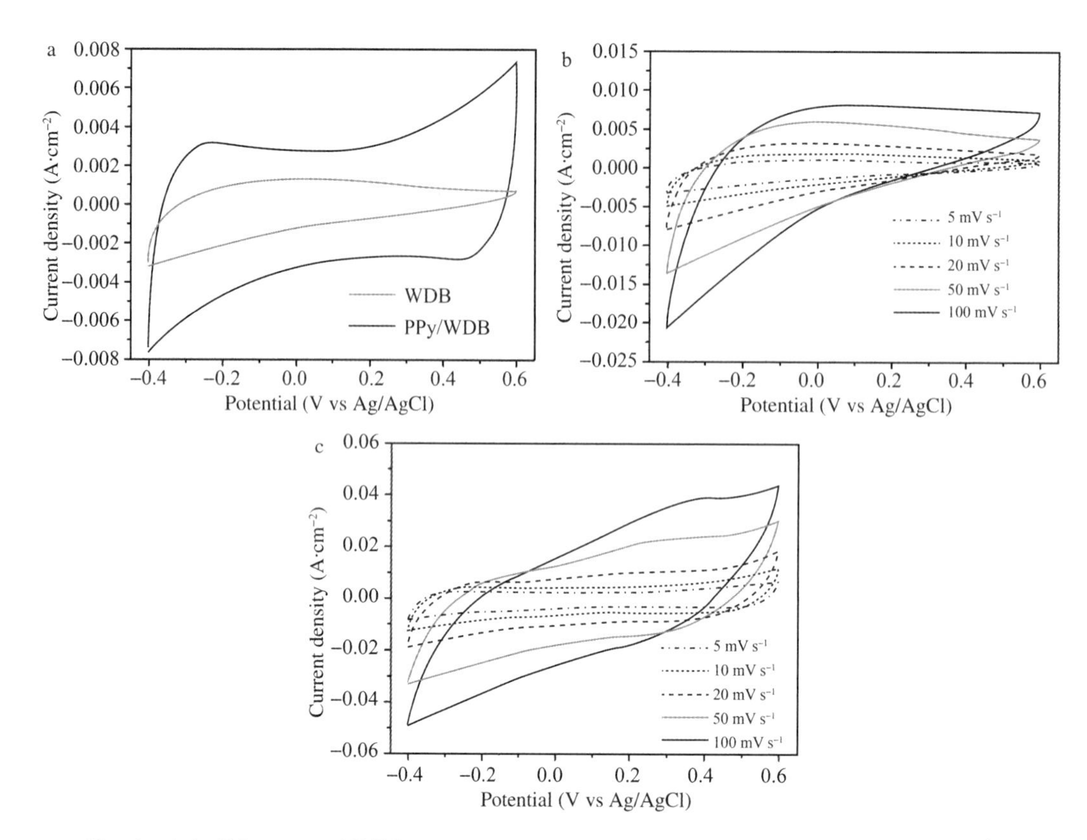

Fig. 4 (a) CV curves of WDB and PPy/WDB electrodes at the scan rate of 5 mV · s^{-1}. CV curves of (b) WDB and (c) PPy/WDB electrodes at various scan rates

much longer than that of WDB, suggesting the higher specific capacitance for PPy/WDB since discharge time is proportional to specific capacitance.

The specific capacitance (C_m) was calculated from the GCD curves according to the following equation:[46]

$$C_m = \frac{I \times \Delta t}{\Delta V \times m} \tag{1}$$

where C_m is the specific capacitance (F · g^{-1}), I is the charge-discharge current (A), Δt is the discharge time (s), ΔV is the potential window (V), and m is the electrode mass (g). The specific capacitance against the current density is plotted in Fig. 5c. PPy/WDB has a maximum specific capacitance of 216 F · g^{-1} at 0.05 A · g^{-1}, which is approximately three times of that of WDB (75 F · g^{-1}, 0.05 A · g^{-1}). This result confirms that the capacitance of the hybrid electrode is mainly derived from the pseudocapacitance of PPy, consistent with the CV analysis. In some previous literatures, a variety of carbon materials have been adopted to integrate with PPy for supercapacitor applications. In comparison, the capacitance value of PPy/WDB is significantly lower than that of previously reported PPY/multiwalled carbon nanotube composites (390 F · g^{-1}),[47] graphene oxide/PPy composites (424 F · g^{-1}),[48] and polypyrrole wrapped graphene hydrogels composites (375 F · g^{-1}),[49] but comparable to the value for multilayered nanoarchitecture of graphene nanosheets and PPy nanowires (165 F · g^{-1}),[50] graphene/PPy nanotube hybrid aerogel (253 F · g^{-1}),[51]

PPy-bonded air-plasma activated carbon nanotube ($264\ F \cdot g^{-1}$), [52] and PPy/sulfonated graphene composite films ($285\ F \cdot g^{-1}$). [53] Furthermore, both electrodes present gradually reduced capacitance with the increase of current density from 0.05 to $2\ A \cdot g^{-1}$, reflecting a much reduced electrochemically active surface area accessed by electrolyte. [41] Nevertheless, PPy/WDB electrode still remains a specific capacitance of $91\ F \cdot g^{-1}$ at the current density of $2.0\ A \cdot g^{-1}$ (42% of its maximum specific capacitance), much higher than $11\ F \cdot g^{-1}$ of WDB (15% of its maximum specific capacitance). The higher capacitance retention for PPy/WDB indicates more superior rate performance.

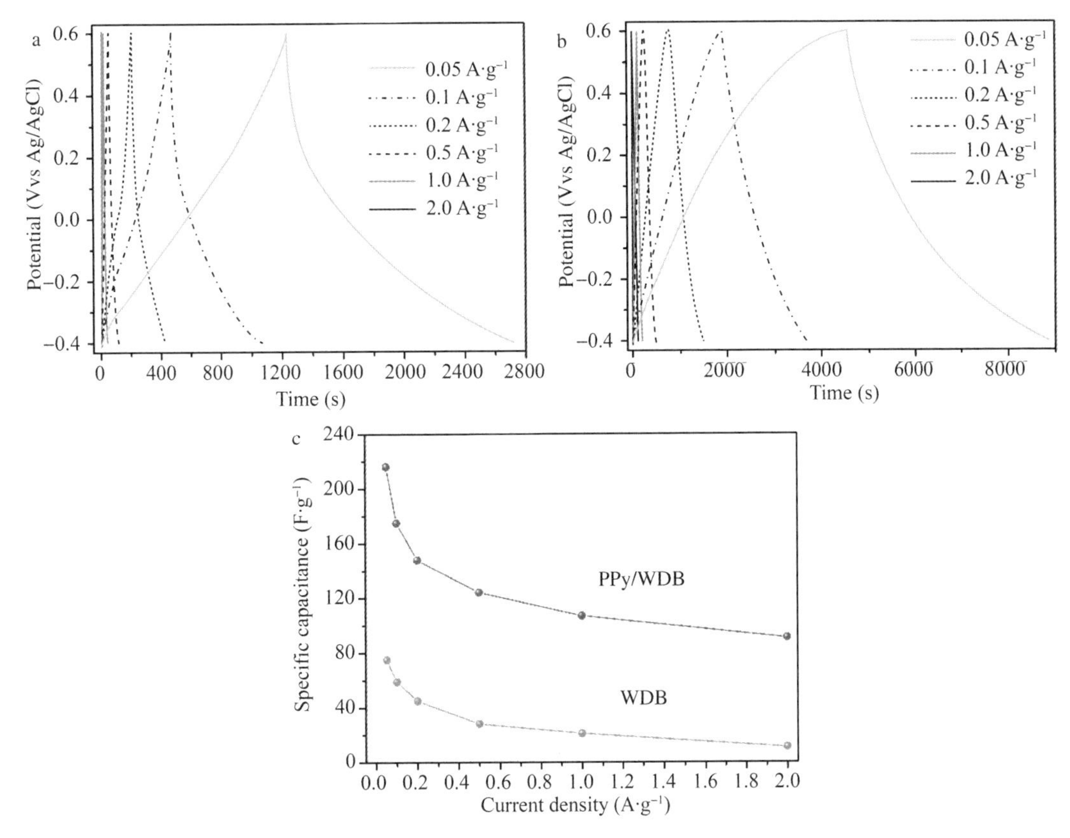

Fig. 5 GCD curves of (a) WDB and (b) PPy/WDB electrodes at various current densities. (c) Specific capacitance of WDB and PPy/WDB electrodes as a function of discharge current

EIS tests were carried out in a frequency range from 10^5 to 0.01 Hz to further evaluate the electrochemical behaviors of WDB and PPy/WDB electrodes. The EIS data were analyzed using Nyquist plots (see Fig. 6). The Nyquist plots can be divided into three parts: (1) a semicircle in the high-to-medium frequencies region indicating the intersection point at the real impedance axis (Z') representing the bulk solution resistance (R_s) and the diameter of which represents the charge transport resistance (R_{ct}); (2) a straight line with a slope of 45° in the low-frequency range, corresponding to semi-infinite Warburg impedance due to the frequency dependence of ion diffusion/transport in the electrolyte; and (3) a vertical line at the very low frequencies, indicating the pure capacitive behavior. [54-57] In the high-frequency region, both electrodes have almost identical R_s of ~1.5 Ω (inset in Fig. 6). Nevertheless, the R_{ct} value of PPy/WDB is ~1.2 Ω, much larger than that of WDB (ca. 0.25 Ω), which reveals that the hybrid electrode has higher charge transfer resistance since the incorporated PPy might have lower electrical conductivity as compared to that of WDB. Moreover, double-layer charging-discharging is not a charge-transfer reaction, and therefore double-layer formation is generally faster than pseudo-capacitive faradaic reaction, which is a diffusion-limited slower reaction.[58] The slower Faradaic reactions in PPy may be another reason for the higher R_{ct} value of PPy/WDB. The longer Warburg length of PPy/WDB suggests its larger Warburg resistance, which indicates greater variations in ion diffusion path lengths and increases obstruction of ion movement.[59] In the low-frequency region, PPy/WDB electrode exhibits a more vertical line than WDB electrode, demonstrating better capacitive behavior.[60]

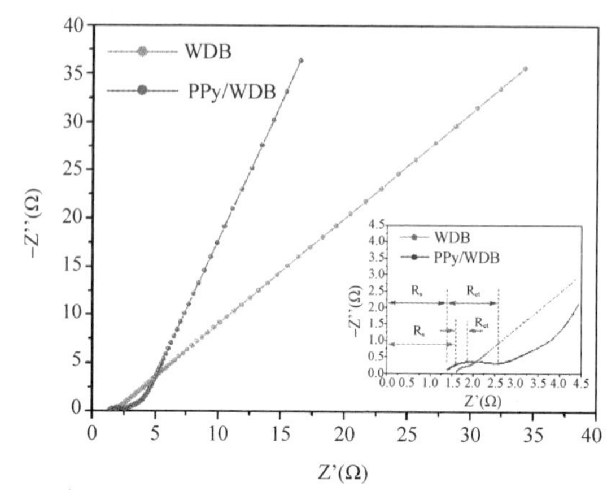

Fig. 6 Nyquist plots of WDB and PPy/WDB electrodes

Long cycle life is an important requirement for supercapacitors.[61] The cycle life test over 3000 cycles for PPy/WDB electrode was conducted by repeating the GCD test at the current density of 1.0 A · g^{-1}. The capacitance retention of PPy/WDB electrode as a function of cycle number is shown in Fig. 7. Obviously, PPy/WDB electrode has an excellent long-term electrochemical stability with 96.9% capacitance retention after 3000 cycles, which is possibly attributed to the presence of WDB preventing PPy from severe swelling and shrinkage during the charge-discharge process.[62] Furthermore, the Coulombic efficiency calculated from the ratio between discharge time and charged time remains a quite high value of more than 98% during the whole cycling.

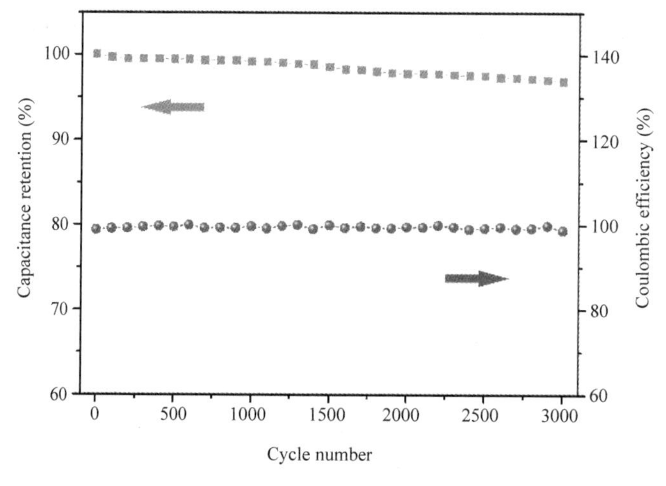

Fig. 7 Cycle performance and Coulombic efficiency of PPy/WDB electrode measured at the current density of 1 A · g^{-1}

4 Conclusions

The biochar derived from Paulownia wood processing residues was adopted as a substrate to support PPy nanoparticles via an easily-operated in-situ chemical oxidative polymerization method. The PPy nanoparticles with diameters of a few hundred nanometers were tightly adhered to the surface of WDB by hydrogen bonding. When used as a free-standing and binder-free supercapacitor electrode, PPy/WDB has a high specific capacitance of 216 $F \cdot g^{-1}$ at 0.05 $A \cdot g^{-1}$, which is around three times of that of WDB (75 $F \cdot g^{-1}$). PPy/WDB electrode shows favorable long-term electrochemical stability, which can keep 96.9% of the initial available capacitance after 3000 cycles. Also, the Coulombic efficiency remains a quite high value of more than 98% during the whole cycling. Thus this composite might be useful for the development of high-performance supercapacitors. Moreover, it is expected that this cheap eco-friendly WDB substrate can be integrated with other electrochemical active substances to develop more types of energy storage devices.

Acknowledgements

This study was supported by the National Natural Science Foundation of China (grant nos. 31270590 and 31470584) and the Fundamental Research Funds for the Central Universities (grant nos. 2572016AB22).

References

[1] Rowell R M. - WoodComposites [M]//Handbook of Wood Chemistry and Wood Composites. CRC Press, 2012: 338-429.

[2] Shmulsky R, Jones P D. Forest Products and Wood Science An introduction. Publish by A John Wiley & Sons [J]. 2011.

[3] Fang Q, Chen B, Lin Y, et al. Aromatic and hydrophobic surfaces of wood-derivedbiochar enhance perchlorate adsorption via hydrogen bonding to oxygen-containing organic groups[J]. Environmental science & technology, 2014, 48(1): 279-288.

[4] Taer E, Deraman M, Talib I A, et al. Physical, electrochemical and supercapacitive properties of activated carbon pellets from pre-carbonized rubber wood sawdust by CO_2 activation[J]. Current Applied Physics, 2010, 10(4): 1071-1075.

[5] Karakoyun N, Kubilay S, Aktas N, et al. Hydrogel-Biochar composites for effective organic contaminant removal from aqueous media[J]. Desalination, 2011, 280(1-3): 319-325.

[6] Zhang M, Gao B, Varnoosfaderani S, et al. Preparation and characterization of a novel magnetic biochar for arsenic removal[J]. Bioresource Technology, 2013, 130: 457-462.

[7] Wang H, Gao B, Wang S, et al. Removal of Pb (II), Cu (II), and Cd (II) from aqueous solutions by biochar derived from KMn04 treated hickory wood[J]. Bioresource Technology, 2015, 197: 356-362.

[8] Joseph S, Graber E R, Chia C, et al. Shifting paradigms: development of high-efficiencybiochar fertilizers based on nano-structures and soluble components[J]. Carbon Management, 2013, 4(3): 323-343.

[9] Kastner J R, Miller J, Geller D P, et al. Catalytic esterification of fatty acids using solid acid catalysts generated from biochar and activated carbon[J]. Catalysis Today, 2012, 190(1): 122-132.

[10] Elleuch A, Halouani K, Li Y. Investigation of chemical and electrochemical reactions mechanisms in a direct carbon fuel cell using olive wood charcoal as sustainable fuel[J]. Journal of Power Sources, 2015, 281: 350-361.

[11] Shaaban A, Se S M, Ibrahim I M, et al. Preparation of rubber wood sawdust-based activated carbon and its use as a filler of polyurethane matrix composites for microwave absorption[J]. New Carbon Materials, 2015, 30(2): 167-175.

[12] Jiang J, Zhang L, Wang X, et al. Highly orderedmacroporous woody biochar with ultra-high carbon content as

supercapacitor electrodes[J]. Electrochimica Acta, 2013, 113: 481-489.

[13]Vangari M, Pryor T, Jiang L. Supercapacitors: review of materials and fabrication methods[J]. Journal of Energy Engineering, 2013, 139(2): 72-79.

[14]Wang G, Zhang L, Zhang J. A review of electrode materials for electrochemicalsupercapacitors[J]. Chemical Society Reviews, 2012, 41(2): 797-828.

[15]Zhi M, Xiang C, Li J, et al. Nanostructured carbon-metal oxide composite electrodes for supercapacitors: a review [J]. Nanoscale, 2013, 5(1): 72-88.

[16]Yan J, Wang Q, Wei T, et al. Supercapacitors: Recent Advances in Design and Fabrication of Electrochemical Supercapacitors with High Energy Densities[J]. Advanced Energy Materials, 2014, 4(4): 1-43.

[17] Wang Q, Yan J, Fan Z. Carbon materials for high volumetric performancesupercapacitors: design, progress, challenges and opportunities[J]. Energy & Environmental Science, 2016, 9(3): 729-762.

[18] Lota G, Tyczkowski J, Kapica R, et al. Carbon materials modified by plasma treatment as electrodes for supercapacitors[J]. Journal of Power Sources, 2010, 195(22): 7535-7539.

[19]Jurewicz K, Frackowiak E. Modified carbon materials for electrochemical capacitors[J]. Molecular Physics Reports, 2000, 27: 38-45.

[20]Taer E , Deraman M , Talib I A , et al. Preparation of a Highly Porous Binderless Activated Carbon Monolith from Rubber Wood Sawdust by a Multi-Step Activation Process for Application in Supercapacitors[J]. International Journal of Electrochemical Science, 2011, 6(8): 3301-3315.

[21]Dobele G, Dizhbite T, Gil M V, et al. Production of nanoporous carbons from wood processing wastes and their use in supercapacitors and CO_2 capture[J]. Biomass and Bioenergy, 2012, 46: 145-154.

[22]Wang C C, Hu C C. Electrochemical catalytic modification of activated carbon fabrics by ruthenium chloride for supercapacitors[J]. Carbon, 2005, 43(9): 1926-1935.

[23]JiangH , Ma J , Li C . Mesoporous Carbon Incorporated Metal Oxide Nanomaterials as Supercapacitor Electrodes [J]. Advanced Materials, 2012, 24(30): 4197-4202.

[24]Peng C, Zhang S, Jewell D, et al. Carbon nanotube and conducting polymer composites for supercapacitors[J]. Progress in Natural science, 2008, 18(7): 777-788.

[25]Wu Q, Xu Y, Yao Z, et al. Supercapacitors based on flexible graphene/polyaniline nanofiber composite films[J]. ACS Nano, 2010, 4(4): 1963-1970.

[26]Yan J, Wang Q, Lin C, et al. Interconnected frameworks with a sandwiched porous carbon layer/graphene hybrids for supercapacitors with high gravimetric and volumetric performances[J]. Advanced Energy Materials, 2014, 4(13).

[27]Snook G A, Kao P, Best A S. Conducting-polymer-basedsupercapacitor devices and electrodes[J]. Journal of Power Sources, 2011, 196(1): 1-12.

[28] Gómez H, Ram M K, Alvi F, et al. Graphene - conducting polymer nanocomposite as novel electrode for supercapacitors[J]. Journal of Power Sources, 2011, 196(8): 4102-4108.

[29]Niu Z, Luan P, Shao Q, et al. A "skeleton/skin" strategy for preparing ultrathin free-standing single-walled carbon nanotube/polyaniline films for high performance supercapacitor electrodes[J]. Energy & Environmental Science, 2012, 5(9): 8726-8733.

[30]Cheng Q, Tang J, Ma J, et al. Polyaniline-coated electro-etched carbon fiber cloth electrodes for supercapacitors [J]. The Journal of Physical Chemistry C. , 2011, 115(47): 23584-23590.

[31]Mastragostino M, Arbizzani C, Meneghello L, et al. Electronically conducting polymers and activated carbon: Electrode materials in supercapacitor technology[J]. Advanced Materials, 1996, 8(4): 331-334.

[32] Salunkhe R R, Tang J, Kobayashi N, et al. Ultrahigh performance supercapacitors utilizing core-shell nanoarchitectures from a metal-organic framework-derived nanoporous carbon and a conducting polymer[J]. Chemical Science,

2016, 7(9): 5704-5713.

[33]Abdelhamid M E, O'Mullane A P, Snook G A. Storing energy in plastics: a review on conducting polymers & their role in electrochemical energy storage[J]. Rsc Advances, 2015, 5(15): 11611-11626.

[34] Xu J, Zhu L, Bai Z, et al. Conductive polypyrrole-bacterial cellulose nanocomposite membranes as flexible supercapacitor electrode[J]. Organic Electronics, 2013, 14(12): 3331-3338.

[35]Cho G, Fung B M, Glatzhofer D T, et al. Preparation and characterization of polypyrrole-coated nanosized novel ceramics[J]. Langmuir, 2001, 17(2): 456-461.

[36]Liu Y, Wang H, Zhou J, et al. Graphene/polypyrrole intercalating nanocomposites as supercapacitors electrode[J]. Electrochimica Acta., 2013, 112: 44-52.

[37]Yuan L, Yao B, Hu B, et al. Polypyrrole-coated paper for flexible solid-state energy storage[J]. Energy & Environmental Science, 2013, 6(2): 470-476.

[38]Tabačiarová J, Mičušík M, Fedorko P, et al. Study of polypyrrole aging by XPS, FTIR and conductivity measurements [J]. Polymer Degradation and Stability, 2015, 120: 392-401.

[39]Kasisomayajula S, Jadhav N, Gelling V J. In situ preparation and characterization of a conductive and magnetic nanocomposite of polypyrrole and copper hydroxychloride[J]. RSC Advances, 2016, 6(2): 967-977.

[40]Babu K F, Dhandapani P, Maruthamuthu S, et al. One pot synthesis of Polypyrrole silver nanocomposite on cotton fabrics for multifunctional property[J]. Carbohydrate Polymers, 2012, 90(4): 1557-1563.

[41]Lei Z, Shi F, Lu L. Incorporation of MnO_2-coated carbon nanotubes between graphene sheets as supercapacitor electrode[J]. ACS Applied Materials & Interfaces, 2012, 4(2): 1058-1064.

[42]M. Nagao, K. Kobayashi, Y. Yamamoto, and T. Hibino, J. Electrochem. Soc., 162, F410 (2015).

[43] Samuelsson R, Sharp M. The effect of electrode material on redox reactions ofquinones in acetonitrile [J]. Electrochimica Acta., 1978, 23(4): 315-317.

[44] Wang J P, Xu Y, Wang J, et al. High charge/discharge rate polypyrrole films prepared by pulse current polymerization[J]. Synthetic Metals, 2010, 160(17-18): 1826-1831.

[45]Li Z, Mi Y, Liu X, et al. Flexible graphene/MnO_2 composite papers for supercapacitor electrodes[J]. Journal of Materials Chemistry, 2011, 21(38): 14706-14711.

[46]He S, Chen W. 3Dgraphene nanomaterials for binder-free supercapacitors: scientific design for enhanced performance [J]. Nanoscale, 2015, 7(16): 6957-6990.

[47]Li X, Zhitomirsky I. Electrodeposition of polypyrrole-carbon nanotube composites for electrochemical supercapacitors [J]. Journal of Power Sources, 2013, 221: 49-56.

[48]Chang HH, Chang C K, Tsai Y C, et al. Electrochemically synthesized graphene/polypyrrole composites and their use in supercapacitor[J]. Carbon, 2012, 50(6): 2331-2336.

[49]Zhang F, Xiao F, Dong Z H, et al. Synthesis ofpolypyrrole wrapped graphene hydrogels composites as supercapacitor electrodes[J]. Electrochimica Acta, 2013, 114: 125-132.

[50] Biswas S, Drzal L T. Multilayered nanoarchitecture of graphene nanosheets and polypyrrole nanowires for high performance supercapacitor electrodes[J]. Chemistry of Materials, 2010, 22(20): 5667-5671.

[51]Ye S, Feng J. Self-assembled three-dimensional hierarchical graphene/polypyrrole nanotube hybrid aerogel and its application for supercapacitors[J]. ACS Applied Materials & Interfaces, 2014, 6(12): 9671-9679.

[52]Yang L, Shi Z, Yang W. Polypyrrole directly bonded to air-plasma activated carbon nanotube as electrode materials for high-performance supercapacitor[J]. Electrochimica Acta, 2015, 153: 76-82.

[53]Liu A, Li C, Bai H, et al. Electrochemical deposition of polypyrrole/sulfonated graphene composite films[J]. The Journal of Physical Chemistry C., 2010, 114(51): 22783-22789.

[54] Kim M, Hwang Y, Kim J. Graphene/MnO_2-based composites reduced via different chemical agents for

supercapacitors[J]. Journal of Power Sources, 2013, 239: 225-233.

[55]Weng Z, Su Y, Wang D W, et al. Graphene-cellulose paper flexible supercapacitors[J]. Advanced Energy Materials, 2011, 1(5): 917-922.

[56]He S, Hu C, Hou H, et al. Ultrathin MnO_2 nanosheets supported on cellulose based carbon papers for high-power supercapacitors[J]. Journal of Power Sources, 2014, 246: 754-761.

[57]He S, Hou H, Chen W. 3D porous and ultralight carbon hybrid nanostructure fabricated from carbon foam covered by monolayer of nitrogen-doped carbon nanotubes for high performance supercapacitors[J]. Journal of Power Sources, 2015, 280: 678-686.

[58]Sung J H, Kim S J, Lee K H. Fabrication of all-solid-state electrochemicalmicrocapacitors[J]. Journal of Power Sources, 2004, 133(2): 312-319.

[59]Zhang D, Zhang X, Chen Y, et al. Enhanced capacitance and rate capability ofgraphene/polypyrrole composite as electrode material for supercapacitors[J]. Journal of Power Sources, 2011, 196(14): 5990-5996.

[60]Ra E J, Raymundo-Piñero E, Lee Y H, et al. High power supercapacitors using polyacrylonitrile-based carbon nanofiber paper[J]. Carbon, 2009, 47(13): 2984-2992.

[61]Zhou J, Huang Y, Cao X, et al. Two-dimensional $NiCo_2O_4$ nanosheet-coated three-dimensional graphene networks for high-rate, long-cycle-life supercapacitors[J]. Nanoscale, 2015, 7(16): 7035-7039.

[62]Suematsu S, Oura Y, Tsujimoto H, et al. Conducting polymer films of cross-linked structure and their QCM analysis [J]. Electrochimica acta, 2000, 45(22-23): 3813-3821.

中文题目：木质生物炭负载聚吡咯纳米粒子作为自支撑的超级电容器电极

作者：万才超，李坚

摘要：采用泡桐木材加工残留物作为模板和前驱体，通过简单、经济的原位化学氧化聚合方法合成了木材衍生生物炭/聚吡咯(编码为 WDB/PPy)复合材料。WDB 基底被大量直径为几百纳米的 PPy 纳米粒子封装，并且所得的 WDB/PPy 复合材料可用作独立的、无黏结剂的超级电容器电极。该混合电极在 0.05 $A \cdot g^{-1}$ 时的比电容高达 216 $F \cdot g^{-1}$，库伦效率高达 98%以上，循环 3000 次后的循环稳定性良好，电容保持率为 96.9%。这些结果为超级电容器提供了一种低成本的环保电极材料设计。更重要的是，这种廉价、环保的 WDB 衬底有望与更多类型的电化学活性物质结合，开发出新型的能量存储器件。

关键词：高性能超级电容器；活性炭；导电聚合物；能量存储；复合材料

Durable, High Conductivity, Superhydrophobicity Bamboo Timber Surface for Nanoimprint Stamps*

Wenhui Bao, Daxin Liang, Ming Zhang, Yue Jiao,
Lijuan Wang, Liping Cai, Jian Li

Abstract: Superhydrophobic bamboo timber was fabricated by magnetron sputtering and nanoimprint stamps. Conductive copper film was deposited on the surface of bamboo timber, followed by transferring lotus leaf structure pattern to the Cu-bamboo timber surface. Modified by stearic acid, the as-prepared surface with lotus leaf structure showed superhydrophobic property with the water contact angle of 152°, and the sliding angle was only 5°. The coating showed excellent mechanical resistance, environmental stability and high conductivity. It was expected that this work could promote the applications of superhydrophobic and conductive bamboo products.

Keywords: Superhydrophobic; Magnetron sputtering; Bamboo timber; Conductive; Microcontact printing

1 Introduction

Bamboo is widely used as engineering and structural materials for various applications due to its numerous attractive properties including fast growth, good mechanical strength, and short rotation period[1]. However, bamboo materials also suffer some limitations due to its inherent hygroscopicity. To improve the water repellency of bamboo, surface coating and bulk treatment are two effective measures to impede hydroxyl groups of bamboo to absorb water. Water contact angle (CA) larger than 150° and sliding angle (SA) less than 5° are well-known as superhydrophobic surfaces, which is an effective measure for protecting exterior bamboo from deterioration[2,3].

In previous studies, the sol-gel process, chemical vapor deposition, phase separation, electrospinning, and layer-by-layer assembly were used to synthesize artificial superhydrophobic surfaces[4,5]. As a Physical Vapor Deposition (PVD) method, magnetron sputtering has become the process alternative for the deposition of a wide range of industrially important coatings. Compared with the traditional method, this method is simpler and easier to control, which works for substrates in any shape and does not need rigorous conditions[6].

In this paper, a new concept of the Cu-treated bamboo timber with lotus leaf microstructures composites was proposed. The first step is to deposit Cu films on the bamboo timber by radio frequency magnetron sputtering. As a widely used substance[7], copper (Cu) is a kind of inexpensive material[8], compared with the gold and silver. Due to its low electrical resistivity and good electro migration resistance, Cu is also recognized as an interconnect metal with multiple functions[9,10].

* 本文摘自 Progress in Natural Science: Materials International, 2017, 27(6): 669-673.

The second step is to fabricate a lotus-leaf-like morphology on Cu-bamboo timber by self-assembly method with the assistance of porous PDMS template[11,12]. Microcontact printing is a simple and inexpensive method of fabricating large quantities of microstructured materials[13]. Moreover, this stamping procedure can avoid the damage to the fragile cellulose-based materials, and then the surface modification of Cu films on bamboo timber surface was carried out by a self-assembly of stearic acid monolayer. At the same time, the morphology, chemical composition, crystallization structure, mechanical resistance and electric resistance properties of the Cu-bamboo timber hybrid were investigated in detail[14,15].

2 Experimental section

2.1 Materials

Purchased from the Shanghai Boyle Chemical Company Limited, the Stearic acid (97.0%), ethanol (95.0%), and Iron trichloride ($FeCl_3$) were of analytical grades. PDMS and curing agent were purchased from DOW Corning Corporation. Cu target was supplied by the Beijing Guanjinli New Material Company Limited. Bamboo was deforested in Zhejiang Province, China, and cut into specimens measuring $20\times20\times5$ mm^3 (length×width×height). After ultrasonically rinsing in ethanol and deionized water for 10 min, the bamboo timber specimens were oven-dried (24 h, 103 ℃±2 ℃) to a constant weight, and their weights were determined.

2.2 Preparation of Cu coating on the bamboo timber substrate

The target used was a copper (99.999% purity) plate with a diameter of 5 cm. The sputtering process was carried out in argon (Ar-99.995% purity) and the sputtering system was equipped with diffusion pump backed by a rotary pump to drive down the sputtering system to a base pressure of 2.1×10^{-2} Pa[16,17]. Cu films were deposited with a radio frequency magnetron sputtering system to improve the Cu thin films with high growth rate and low temperature deposition properties. The thicknesses of the Cu films were checked in situ with a quartz crystal monitor located near the bamboo timber substrate during the sputtering process[18]. The bamboo timber substrate was at a room temperature and the sputtering power was 200 W with a fixed target-substrate distance of 6 cm.

2.3 Preparation of PDMS template and fabrication of superhydrophobic Cu-bamboo

Firstly, PDMS and its catalyzer (mass ratio10 : 1) were poured into the surface of lotus leaf. The liquid PDMS was solidified after 24 h, and, a porous PDMS template was obtained by peeling off the lotus leaf[19]. 5 g $FeCl_3$ was added into 12 g ethanol and stirred for 15 min. Then, the glass slide was immersed in the $FeCl_3$ solution for 30 min and removed. Subsequently, the porous PDMS template was pressed on the wetted glass slide for 1 min. Secondly, the wetted porous PDMS template was covered on the Cu-bamboo timber under a certain pressure for 24 h. After the stress was released, a surface with lotus-leaf-like structure was obtained on the Cu-bamboo timber. Fig. 1 shows the procedure scheme of the lotus-leaf-like surface structures on the Cu-bamboo[20]. Finally, the rough Cu-bamboo timber with the lotus-leaf-like surface structure was immersed into a mixture of ethanol solution of 0.006 M stearic acid at 80 ℃ for 1 h. Afterward, the sample was rinsed with ethanol for ten minutes, and dried naturally in air while waiting for the further examination of its properties.

2.4 Characterization

The morphology structure of the as-prepared wood surface was observed using the scanning electron microscopy (SEM, FEIQUANTA200). The chemical composition of the treated or untreated bamboo was analyzed by the Fourier transform infrared spectroscopy (FTIR, Magna-IR 560, Nicolet). The water contact angles and sliding angles were measured with 5 μL deionized water droplet at room temperature using an optical contact angle meter (Hitachi, CA-A). The surface chemical state and structure of the treated samples were determined using the X-ray powder diffraction (XRD, Philiphs, PW 1840 diffractometer) operating with Cu-K radiation at a scan rate of 5°/min and accelerating voltage of 40 kV and applied current of 30 mA ranging from 5° to 80°.

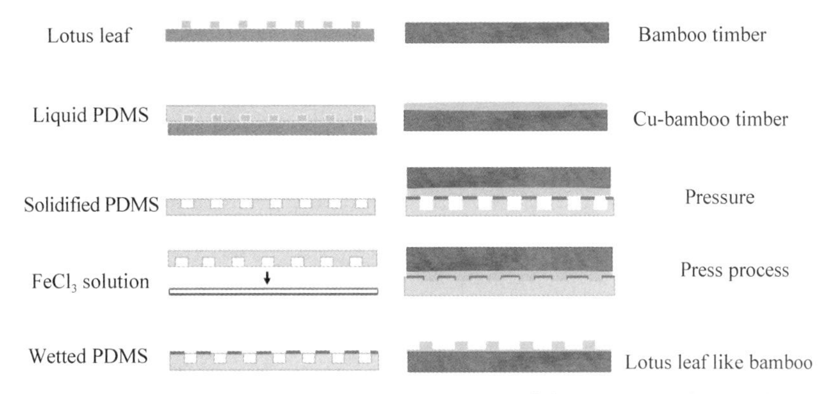

Fig. 1 The flow chart of the biomimetic preparation of lotus-leaf-like structure Cu-bamboo timber surfaces

3 Results and discussion

3.1 Surface morphology of the as-prepared bamboo timber

Fig. 2a shows the original bamboo timber with a clean and smooth surface. As shown in Fig. 2b, the surface morphology of Cu films deposited on a 30 μm thick layer appear to be smooth with little cracks, and Cu nanopaticles uniformly were deposited on the bamboo timber and filled into the pits of the cell wall structure. Fig. 2c shows that lotus leaves were mainly decided by the randomly distributed micrometer-scale papilla, which is essential for superhydrophobicity. The SEM images of PDMS template are shown in Fig. 2d. The morphology showed that the PDMS template was comprised of many pores that the structure of the lotus leaf was a negative surface. As exhibited in Fig. 2e, the morphology of the replicated lotus leaf Cu-bamboo surface was obtained by the combination of PDMS chemical etching. Comparing to Fig. 2c, the fabricated micrometer-scale structure of the Cu-bamboo surface was similar to the shapes of the papilla with the same lotus leaf. As observed[21], papillae had a diameter ranging from 3 μm to 5 μm and the distances among the adjacent papillae were between 10 μm to 15 μm. Fig. 2f shows the elements of carbon, oxygen, and copper from spectra of the Cu-bamboo timber. Carbon and oxygen stemmed from the bamboo substrate. A signal from Cu presented in the spectrum of the Cu-bamboo timber substrate, in which, the iron and chlorine almost undetectable, implying that the nanostructures were primarily caused by Cu.

3.2 Crystal morphology and chemical composition of the as-prepared surface

Fig. 3 shows the XRD patterns of the pristine bamboo, the Cu-bamboo timber and the lotus leaf Cu-

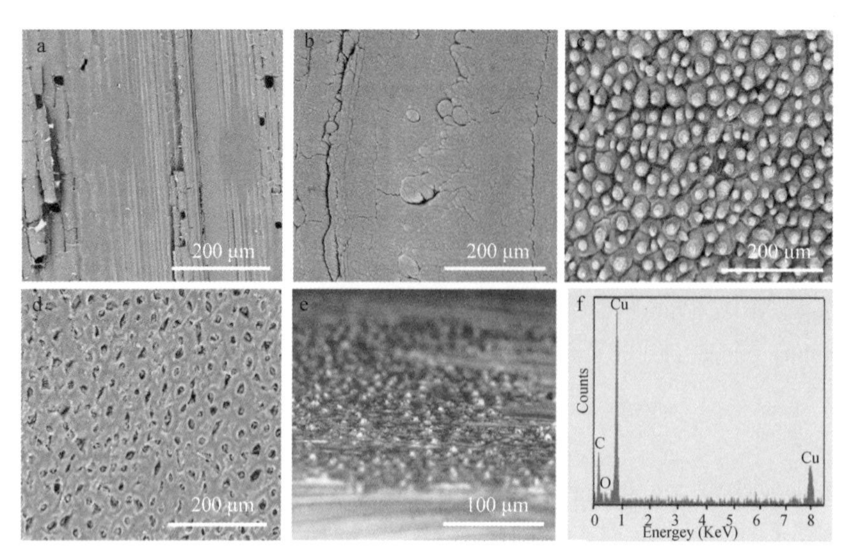

Fig. 2 SEM images of samples of (a) bamboo timber surface, (b) Cu-treated bamboo timber, (c) lotus leaf surface, (d) prepared PDMS template, (e) replicated lotus-leaf-like Cu-bamboo timber surface, and (f) the EDS spectrum of lotus-leaf-like Cu-bamboo timber surface

bamboo timber. As shown in Fig. 3, there was no peak expect for the characteristic peaks at 16° and 22°, belonging to the (101) and (002) crystal planes of the cellulose in the bamboo timber[22,23]. The XRD spectrum of the Cu-bamboo timber shows a strong diffraction peak attributed to the Cu (111) phase as observed at 43°, and the small diffraction peaks of 50. 43° and 74. 13° corresponded to Cu (200) and Cu (220), respectively, suggesting a highly oriented polycrystalline film in the Cu(111) direction. No other significant change peak was observed on the prepared Cu-bamboo timber with lotus leaf structure when the Cu-bamboo timber was covered by $FeCl_3$ of PDMS template. It can be deduced that the lotus leaf structure was formed on the Cu-treated bamboo timber surface according to reaction equation (1). The flux of cupric and ferrous ions went away from the Cu-bamboo timber surface.

$$Cu+2Fe^{3+} \rightarrow Cu^{2+}+2Fe^{2+} \tag{1}$$

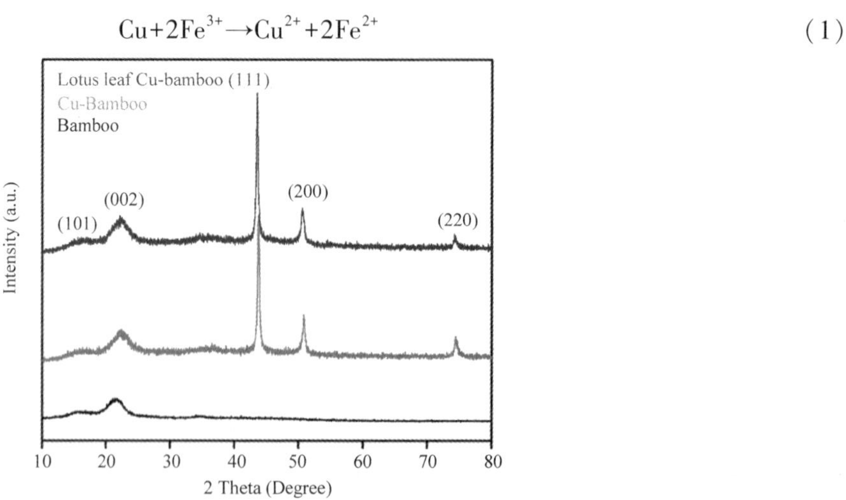

Fig. 3 XRD spectra of curve black line, red line and blue line correspond to the pristine bamboo timber, Cu-bamboo timber and Cu-bamboo timber with lotus leaf structure, respectively

To obtain a superhydrophobic surface, the stearic acid was used as a low surface energy material to modify the rough Cu-bamboo timber. In Fig. 4, the curves a and b show the FTIR spectra of the stearic acid and the superhydrophobic surface[24], respectively. It is well known that the spectrum of stearic acid exhibits the adsorption peaks at about 2913 cm^{-1} and 2847 cm^{-1} in the high frequency region, indicating the existence of a long-chain alkyl group on the surface. In the low frequency region, the carboxyl group (—COO) adsorption peak from stearic acid appears was at 1698 cm^{-1}, which was resulted from the free COO— band. Compared with the stearic acid spectrum, the modified Cu-bamboo timber surface exhibited the adsorption peaks at 1584 cm^{-1} and 1466 cm^{-1}. The two adsorption peaks may stem from asymmetric and symmetric stretches of COO group[25]. Copper nanoplates can be oxidized by dissolved stearic acid in ethanol provided copper oxidization[26] with an acidic environment. Cu^{2+} ions were released continuously from the copper nanoplates into the stearic acid solution according to reaction equation (2), while copper ions can be captured by coordination with stearic acid molecules, forming the copper carboxylate according to reaction equation (3). It can be deduced that hydrophobic groups were formed on the Cu-bamboo timber surface after the modification with stearic acid[27].

$$2Cu+O_2+4H^+ \rightarrow 2Cu^{2+}+2H_2O \tag{2}$$

$$Cu^{2+}+2CH_3(CH_2)_{16}COOH \rightarrow Cu[CH_3(CH_2)_{16}COO]_2+2H^+ \tag{3}$$

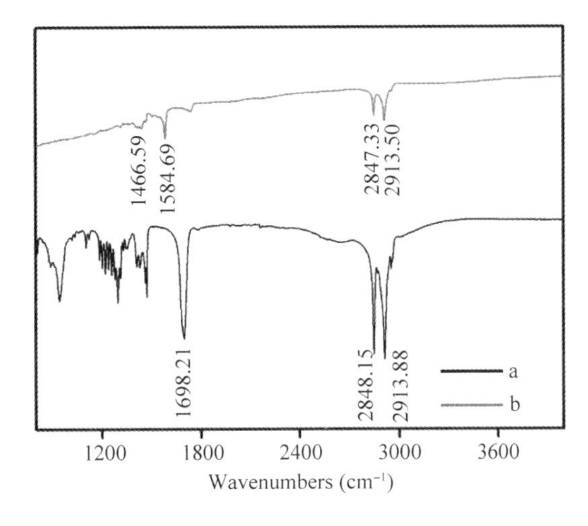

Fig. 4 FTIR spectra of (curve a) pure stearic acid powder, (curve b) modified Cu-bamboo timber

3.3 Wettability and electrical conductivity tests

To observe the surface wettability, the as-prepared samples were examined by CA measurements as shown in Fig. 5. The CA and SA were used to determine the liquid-repellent properties of the surfaces[28]. The pristine bamboo timber presented the CA of 44° (Fig. 5a), demonstrating the bamboo timber surface was hydrophilic. CA of the Cu-treated bamboo timber was 69°. Meanwhile, the Cu-bamboo timber surface with lotus leaf structure presents hydrophobicity had CA of 130° and SA of 15° (Fig. 5c). In Fig. 5d, after the Cu-bamboo timber surface with lotus leaf structure followed by the self-assembly of the stearic acid monolayer, the surface exhibited the excellent superhydrophobic with the CA of 152° and SA of 5°. The results indicated that the superhydrophobic properties of bamboo timber surfaces were significantly increased. The electrical conductivities of the as-prepared samples were examined by the multi-meter. The bamboo timber became a good electrical insulating material due to its inherent insulation. The conductivity of the superhydrophobic bamboo timber showed the excellent electrical conductivity with an electric resistance of (1.5±0.1)Ω[29]. Metal-based films exhibited conductivity and superhydrophobic properties, performing multiple functions simultaneously to improve the performance of products.

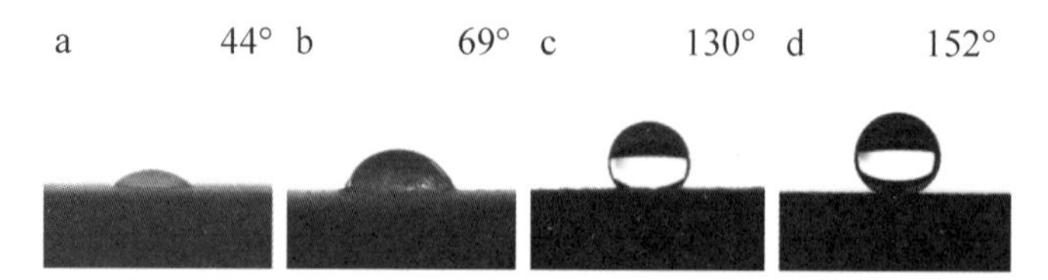

Fig. 5 Contact angle profiles of water droplets on different surfaces: (a) pristine bamboo timber; (b) Cu-bamboo timber surface with 30μm thick Cu layer; (c) Cu-bamboo timber surface with lotus leaf structure; (d) Cu-bamboo timber surface with lotus leaf structure treated by stearic acid

3.4 Mechanical, environmental durability of the superhydrophobic coatings

The mechanical stability of the as-prepared superhydrophobic bamboo timber was evaluated qualitatively by the sand abrasion experiments[30,31]. As shown in Fig. 6a, 20 g sand grains with a diameter ranging from 100 to 300 μm impinged on the surface from a height of 30 cm. After sand impact tests, CA of the as-prepared superhydrophobic bamboo timber surfaces were measured to assess the physical stability of the coatings as shown in Fig. 6c. This result indicated that the surface remained its superhydrophobic after the sand impact tests for 100 times. In order to investigate the thermal stability of the superhydrophobic bamboo timber[32], the as-prepared samples were immersed in a beaker at different temperatures and intensely magnetic stirred for 1 h (Fig. 6b). The results showed that the as-prepared samples without changing their superhydrophobicity, demonstrating that the samples could withstand hot washing.

Meanwhile, the electric resistance of the superhydrophobic bamboo timber did not show significant change, remained to be 1.5 Ω[33]. In Fig. 6d, the superhydrophobic bamboo timber surface resulted in a slight reduction in CA after exposing to the ambient condition for six months, which was reduced from 152° to 150.2°. This result proved that the high conductivity bamboo timber coated with the film showed good mechanical resistance and withstood thermal washing superhydrophobic property[34]. All results indicated that the superhydrophobic bamboo timber possessed excellent durable superhydrophobicity. Bamboo timber products with durable superhydrophobic properties could be attractive high-value-added products with great application future.

4 Conclusions

A simple technique for fabricating high electrical conductivity, superhydrophobic bamboo timber surface was developed. By combining the magnetron sputtering and nanoimprint, the CA and SA of the prepared superhydrophobic bamboo timber surfaces were 152° and 5°, respectively, with the electric resistance of 1.5 Ω ± 0.1 Ω. Meanwhile, the lotus leaf morphology on Cu-bamboo timber was fabricated, which showed a good mechanical stability. This method could also be applied for any shape of products. Moreover, this procedure can avoid the damage of the fragile cellulose-based materials. Based on the results, bamboo

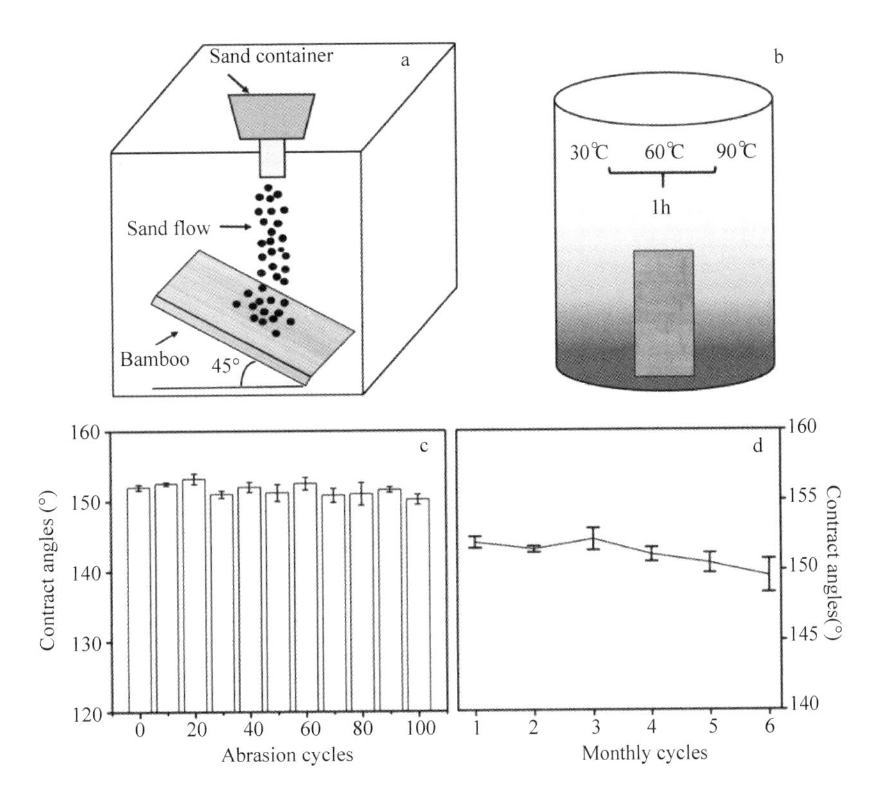

Fig. 6 (a) Mechanical resistance quantified by sand abrasion experiment. (b) Intense water stirring with different temperature for 1 h. (c) CA changes depending on the abrasion cycles. (d) CA changes depending on the monthly cycles

resources could be potentially used in a wider range of application.

Acknowledgements

This research was supported by the Fundamental Research Funds for the Central Universities (2572016AB64), Natural Science Foundation of Heilongjiang Province of China (LC201406) and the National Natural Science Foundation of China (31400497).

Reference

[1]Zhu H, Luo W, Ciesielski P N, et al. Wood-derived materials for green electronics, biological devices, and energy applications[J]. Chem Rev., 2016, 116: 9305-9374.

[2] Hill C A S. Wood modification: chemical, thermal and other processes, John Wiley & Sons, Inc., 2006.

[3]Wang S, Feng L, Jiang L. One-step solution-immersion process for the fabrication of stable bionic superhydrophobic surfaces[J]. Adv Mater., 2006, 18: 767-770.

[4]Xie Q D, Xu J, Feng L, et al. Facile creation of a super-amphiphobic coating surface with bionic microstructure[J]. Adv Mater., 2004, 16: 302-304.

[5]Dong H, Cheng M, Zhang Y, et al. Extraordinary drag-reducing effect of a superhydrophobic coating on a macroscopic model ship at high speed[J]. J Mater Chem A., 2013, 1: 5886.

[6] Kelly P J, Arnell R D. Control of the structure and properties of aluminum oxide coatings deposited by pulsed magnetron sputtering[J]. Vacuum., 1999, 17: 945.

[7]Yu X, Shen Z, Xu Z, Wang S. Fabrication and structural characterization of metal films coated on cenosphere particles by magnetron sputtering deposition[J]. Appl Surf Sci., 2007, 253: 7082-7088.

[8]Liu K, Jiang L. Metallic surfaces with special wettability[J]. Nanoscale., 2011, 3: 825-838.

[9]Yang F C, Guo J, Liu M M, et al. Design and understanding of a high-performance gas sensing material based on copper oxide nanowires exfoliated from a copper mesh substrate[J]. J Mater Chem A., 2015, 3: 20477-20481.

[10]Wang S, Feng L, Liu H, et al. Inorganic-organic hybrid photochromic materials[J]. Chemphyschem., 2005, 6: 1475-1478.

[11]Dai S, Ding W, Wang Y, et al. Fabrication of hydrophobic inorganic coatings on natural lotus leaves for nanoimprint stamps[J]. Thin Solid Films., 2011, 519: 5523-5527.

[12]Guan H, Han Z, Cao H, et al. Characterization of Multi-scale Morphology and Superhydrophobicity of Water Bamboo Leaves and Biomimetic Polydimethylsiloxane (PDMS) Replicas[J]. J Bionic eng., 2015, 12: 624-633.

[13]Gao J, Liu Y, Xu H, et al. Biostructure-like surfaces with thermally responsive wettability prepared by temperature-induced phase separation micromolding[J]. Langmuir., 2010, 26: 9673-9676.

[14]Nishimoto S, Bhushan B. Bioinspired self-cleaning surfaces with superhydrophobicity, superoleophobicity, and superhydrophilicity[J]. RSC Adv., 2013, 3: 671-690.

[15]Yang H, Deng Y. Preparation and physical properties of superhydrophobic papers[J]. J Colloid Interf Sci., 2008, 325: 588-593.

[16]Alkoy E M, Kelly P J. The structure and properties of copper oxide and copper aluminium oxide coatings prepared by pulsed magnetron sputtering of powder targets[J]. Vacuum., 2005, 79: 221-230.

[17]B. Purusottam Reddy, K. Sivajee Ganesh, O. M. Hussain et al. Growth, microstructure and supercapacitive performance of copper oxide thin films prepared by RF magnetron sputtering[J]. Appl Phys A., 2016, 122: 128-129.

[18]Mukherjee S K, Joshi L, Barhai P K. A comparative study of nanocrystalline Cu film deposited using anodic vacuum arc and dc magnetron sputtering[J]. Surf Coat Tech., 2011, 205: 4582-4595.

[19]Wang F, Li S, Wang L. Synergistic effect of mixed cationic/anionic collectors on flotation and adsorption of muscovite [J]. Colloid Surface A., 2016, 492: 181-189.

[20]Yuan Z, Wang X, Bin J, et al. A novel fabrication of a superhydrophobic surface with highly similar hierarchical structure of the lotus leaf on a copper sheet[J]. Appl Surf Sci., 2013, 285: 205-210.

[21]Yuan Z, Wang X, Bin J, et al. Controllable fabrication of lotus-leaf-like superhydrophobic surface on copper foil by self-assembly[J]. Appl Phys A., 2014, 116: 1613-1620.

[22]Yao Q, Wang C, Fan B, et al. One-step solvothermal deposition of ZnO nanorod arrays on a wood surface for robust superamphiphobic performance and superior ultraviolet resistance[J]. Sci Rep., 2016, 6: 35505.

[23]Gao L, Qiu Z, Gan W, et al. Negative oxygen ions production by superamphiphobic and antibacterial TiO_2/Cu_2O composite film anchored on wooden substrates[J]. Sci Rep., 2016, 6: 26055.

[24]Wang S, Shi J, Liu C, et al. Fabrication of a superhydrophobic surface on a wood substrate[J]. Appl Surf Sci., 2011, 257: 9362-9365.

[25]Li J, Wan H, Ye Y, et al. Chen. One-step process for the fabrication of superhydrophobic surfaces with easy repairability[J]. Appl Surf Sci., 2012, 258: 3115-3118.

[26]Meng H F, Wang S T, Xi J M, et al. Facile Means of Preparing Superamphiphobic Surfaces on Common Engineering Metals[J]. J Phys Chem C., 2008, 112: 11454-11458.

[27]Wang Q, Zhang B, Qu M, et al. Fabrication of superhydrophobic surfaces on engineering materialsurfaces with stearic acid[J]. Appl Surf Sci., 2008, 254: 2009-2012.

[28]Cassie A B D. Simulation of drag reduction in superhydrophobic microchannels based on parabolic gas-liquid interfaces[J]. Trans Faraday Soc., 1944, 40: 546-551.

[29]Jin C, Li J, Han S, et al. Sun. Silver mirror reaction as an approach to construct a durable, robust superhydrophobic surface of bamboo timber with high conductivity[J]. J Alloy Compd., 2015, 635: 300-306.

[30]Deng X, Mammen L, Butt H J, et al. Candle Soot as a Template for a Transparent Robust Superamphiphobic Coating [J]. Science., 2012, 335: 67-70.

[31]Gao L, Xiao S, Gan W, et al. Durable superamphiphobic wood surfaces from Cu_2O film modified with fluorinated alkyl silane[J]. RSC Adv., 2015, 5: 98203-98208.

[32]Jin C, Li J, Han S, et al. A durable, superhydrophobic, superoleophobic and corrosion-resistant coating with rose-like ZnO nanoflowers on a bamboo surface[J]. Appl Surf Sci., 2014, 320: 322-327.

[33]Li J, Sun Q, Yao Q, et al. Preparation of carbon-coated $MnFe_2O_4$ nanospheres as high-performance anode materials for lithium-ion batteries[J]. J Nanomater., 2015, 17: 173.

[34]Jin C, Li J, Han S, et al. A durable, superhydrophobic, superoleophobic and corrosion-resistant coating with rose-like ZnO nanoflowers on a bamboo surface[J]. Appl Surf Sci., 2014, 320: 322-327.

中文题目：采用纳米压印方法制备表面具有耐用、高导电率、超疏水性质的竹材

作者：包文慧，梁大鑫，张明，焦月，王立娟，蔡力平，李坚

摘要：通过磁控溅射和纳米压印方法制备超疏水竹材。在竹材表面磁控溅射沉积导电铜膜，然后将荷叶结构图案转印到竹木表面的 Cu 薄膜上。经硬脂酸改性修饰后，其表面呈现出超疏水性，接触角为 152°，滑动角为 5°。该涂层表现出优异的机械强度，环境稳定性和高导电性。预计这项工作可以促进超疏水和导电竹产品的应用。

关键词：超疏水；磁控溅射；竹材；导电；纳米压印

TiO_2 Microspheres Grown on Cellulose-based Carbon Fibers: Preparation, Characterizations and Photocatalytic Activity for Degradation of Indigo Carmine Dye*

Caichao Wan, Yue Jiao, Jian Li

Abstract: We describe a facile pathway to grow TiO_2 microspheres onto the surface of cellulose-based carbon fibers. These microspheres with diameters ranging from 0.65 to 1.45 μm show pure anatase phase. As an example of potential applications, the cellulose-based carbon fibers supported TiO_2 microspheres (coded as CCF/TMS) were employed as a photocatalyst to degrade indigo carmine dye in aqueous solution under UV radiation. The results show that the dark blue dye solution rapidly turned colorless within 40 min. Moreover, a faster decomposition rate of indigo carmine dye was acquired when using CCF/TMS, as compared to commercially available TiO_2 P25 and anatase TiO_2. In addition, CCF/TMS shows good shape stability during the whole photocatalycal reaction, which provides convenience for recovery processing. This work provides an eco-friendly, low-cost and high-efficiency photocatalyst for environmental remediation.

Keywords: TiO_2; Microspheres; Cellulose; Carbon fibers; Photocatalytic Activity.

1 Introduction

With the increasing emphasis on green chemistry, natural renewable biomaterials have received more and more attention from both scientific and industrial communities.[1,2] It is becoming more important to exploit new utilization of green bioresources in a great variety of frontier fields. Cellulose, the most abundant natural polymer on the earth, is a kind of typical carbohydrate polymer generated from repeating β-D-glucopyranose molecules that are covalently linked through acetal functions between the equatorial OH group of C4 and the C1 carbon atom (β-1, 4-glucan).[3] Cellulose is an important structural component of the primary cell wall of green plants, and the cellulose content varies in different bioresources. Natural bamboo fiber, as one of the typical cellulose fibers, comes from an abundant and renewable resource (namely bamboo) at low cost, which ensures a continuous fiber supply and a significant material cost saving. For natural bamboo fiber, cellulose constitutes the major portion (~73.83%), and other components include hemicellulose (~12.49%), lignin (~10.15%), pigment, tannin, pectin, etc.[4] The traditional applications of bamboo fiber involve multiple areas such as textile industry and papermaking. Owing to plentiful oxygen-containing groups on the surface, bamboo fiber is prone to interact with various nanomaterials by hydrogen bonding or electrostatic interaction,[5] contributing to acquiring high loading content and good dispersion of

* 本文摘自 Journal of Nanoscience and Nanotechnology, 2017, 17: 5525-5529.

guest substances. The major functions of the bamboo fiber-based composites are generally dependent on the incorporated guest materials, and the bamboo fiber serving as substrate can provide low density and good flexibility. For instance, recently, ZnO,[6] polylactic acid,[7] and TaC nanowires[8] have been carried out to integrate with bamboo fiber, and the resultant hybrids find potential applications in some advanced applications like antibacterial agents, thermal insulating materials, and supercapacitors. Motivated by these studies, it is meaning to further develop novel green bamboo fiber-based functional materials.

In this work, we adopted a mild easily-operated immersion-drying-pyrolysis method to grow TiO_2 microspheres (TMS) onto the surface of cellulose-based carbon fibers (CCF). The morphology, crystalline structure, and chemical compositions of the as-prepared CCF supported TMS (coded as CCF/TMS) were investigated. As an example of potential applications, CCF/TMS was used as a photocatalyst to degrade indigo carmine dye (typical organic dye pollutant). Moreover, its photocatalytic activity was compared to that of some typical TiO_2 photocatalysts, i. e., Degussa P25 and anatase TiO_2.

2 Experimental

2.1 Preparation of CCF/TMS

A TiO_2 sol was prepared referring to the literature method[9] with little modification. Briefly, 5 mL of titanium isopropoxide was added to ethanol (25 mL) with stirring in an ice bath. Another 25 mL of ethanol were mixed with water (0.5 mL) and 0.1 M HCl (0.5 mL), and the mixed solution was slowly poured into the titanium isopropoxide solution at 15 ℃. The dried bamboo fiber was dipped into this solution for 5 min. After dried for 1 h at 60 ℃, the sample was placed into a tube furnace to undergo a pyrolysis treatment under the protection of nitrogen. For the pyrolysis process, the temperature was firstly raised (50 ℃ $\cdot$ h^{-1}) to 400 ℃. Then the temperature was maintained at 400 ℃ for 6 h and finally cooled down (30 ℃ $\cdot$ h^{-1}) to room temperature.

2.2 Characterizations

Scanning electron microscope (SEM) observations were performed with a Hitachi S4800 SEM equipped with an Energy dispersive X-ray (EDX) detector for element analysis. X-ray photoelectron spectroscopy (XPS) was carried out using a Thermo Escalab 250Xi XPS spectrometer. X-Ray diffraction (XRD) spectroscopy was performed on a Bruker D8 Advance TXS XRD instrument. Thermal stability was determined using a thermogravimetry (TG) analyzer (TA, Q600) under a nitrogen atmosphere. Nitrogen adsorption-desorption tests were implemented using an accelerated surface area and porosimetry system (3H-2000PS2 unit, Beishide Instrument S&T Co., Ltd.).

2.3 Photocatalytic activities measurements

Usually, photocatalyst decontamination performance is evaluated by quantifying the photochemical degradation rate for either an actual pollutant or a model pollutant such as a dye with a specific absorption peak in the visible. In this work, the photocatalytic activity of CCF/TMS was evaluated by measuring the degradation rate of indigo carmine dye under UV irradiation, and the commercial Degussa P25 and the commercial powder-like anatase TiO_2 (coded as CPAT) were used for comparison. The amounts of P25 and CPAT were equal to the TiO_2 loading content of CCF/TMS determined by TG tests. In a typical photocatalytic experiment, the indigo carmine dye (50 mL, 0.05 g $\cdot$ L^{-1}) and CCF/TMS (ca. 0.4 g) were put into a glass dish and then stirred in the dark for 30 min to achieve adsorption equilibrium before irradiation. Thereafter, the dish was placed in a closed chamber with a UV source (mercury lamp, 120 W, and 365 nm). The distance between the dish and the UV light was around 5 cm. The dish was exposed to the mercury lamp for 40 min. The solution was collected at every time interval (10 min), and its absorbance was measured by a TU-1901 UV-vis spectrophotometer (Beijing Purkinje, China) at 610 nm. In addition, the photocatalytic experiments of P25 and CPAT followed the above procedures, and a centrifugation method was used to separate the powder-like photocatalysts from the solution.

3 Results and discussion

Fig. 1a presents the SEM image of CCF/TMS, where the surfaces of the fibers were coated with abundant particles (the white points). According to the enlarged SEM image shown in Fig. 1b, it is clear that these micro- and nano-scale particles exhibit spherical shape and are closely packed with each other. We drew the frequency distribution histogram of these microspheres (Fig. 1c). The average size (d) and standard deviation (σ) were calculated to be 1.11 μm and 0.22 μm, respectively. In addition, it can be seen that the experimental process also resulted in the generation of a small number of irregularly shaped particles with smaller sizes, indicating that these micron and submicron-scale spherical or nonspherical particles are formed by large quantities of small nanoparticles. Fig. 1d-h show the SEM image of CCF/TMS and the corresponding elemental mapping images, in which the carbon, oxygen, titanium, and gold elements were detected. Moreover, it is found that the microspheres consist of oxygen and titanium elements.

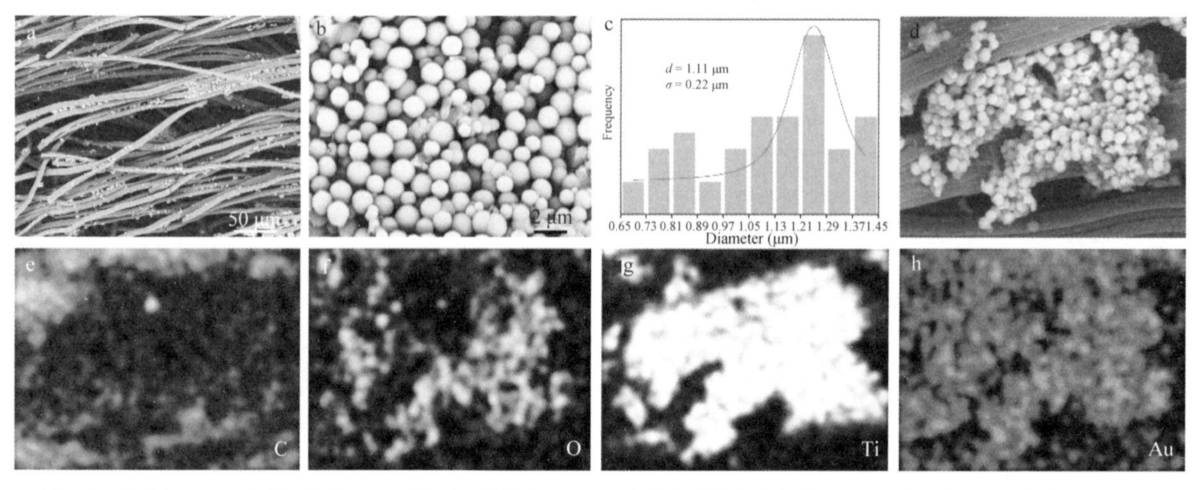

Fig. 1 (a) Low- and (b) high-magnification SEM images of CCF/TMS. (c) Diameter distribution of TMS. (d) SEM image of CCF/TMS with corresponding elemental mapping images of (e) C, (f) O, (g) Ti, and (h) Au

The crystalline structure was investigated by XRD analysis. As shown in Fig. 2a, the natural bamboo fiber shows typical cellulose I crystalline structure.[10] The diffraction peak at $2\theta = 15.8°$ is originated from the superposition of ($\bar{1}10$) and (110) planes, and the peak at $2\theta = 22.2°$ is corresponding to (200) plane, respectively. For CCF/TMS, the characteristic peaks of cellulose disappear due to the pyrolysis damaging the cellulose crystalline structure. Instead, some new strong peaks can be found and well indexed to the standard data of anatase TiO_2 (JCPDS card no. 21-1272). In addition, there is no indication of any other titanium dioxide crystalline phase (e.g., rutile or brookite TiO_2), suggesting a high phase purity of TiO_2 in the composite. The average crystallite size of the TiO_2 was calculated from the full width at half maximum of the strongest diffraction peak (101) using Scherrer equation:[11]

$$d = K\lambda/(\beta\cos\theta) \tag{1}$$

where λ is the X-ray wavelength, K is a constant taken as 0.89, θ is the angle of Bragg diffraction and β is the full width at half maximum. The average crystallite size of the TiO_2 is approximatively 9.2 nm.

XPS analysis was used to investigate the valence states and chemical bonding states of CCF/TMS. The bands at binding energies of 464.4 and 458.7 eV are attributed to the Ti $2p_{1/2}$ and Ti $2p_{3/2}$ spin-orbital splitting photoelectrons in Ti^{4+} (Fig. 2b),[12] respectively. In addition, the separation between these two bands was found 5.7 eV, consistent with presence of the normal state of Ti^{4+}. We also analyzed the O 1s core-level spectra of CCF/TMS. As shown in Fig. 2c, there are three peaks after fitting of the curve. The peak at 530.2 eV is attributed to the Ti-O linkages of TiO_2.[13] The peak at 531.5 eV may be caused by the O—H from the residual oxygen-containing groups of CCF or adsorbed water. Moreover, the peak at 533.4 eV is assigned to the C═O from CCF. The C 1s spectra (Fig. 2d) show the contributions of three components: the major peak at 284.7 eV is related to carbon atoms in C—C, C═C, and C—H bonds; the two weak peaks at 285.9 and 288.7 eV may be caused by the C—O and C═O,[14] respectively. The presence of these oxygen-containing functional groups provided anchors for the immobilization of the TiO_2.

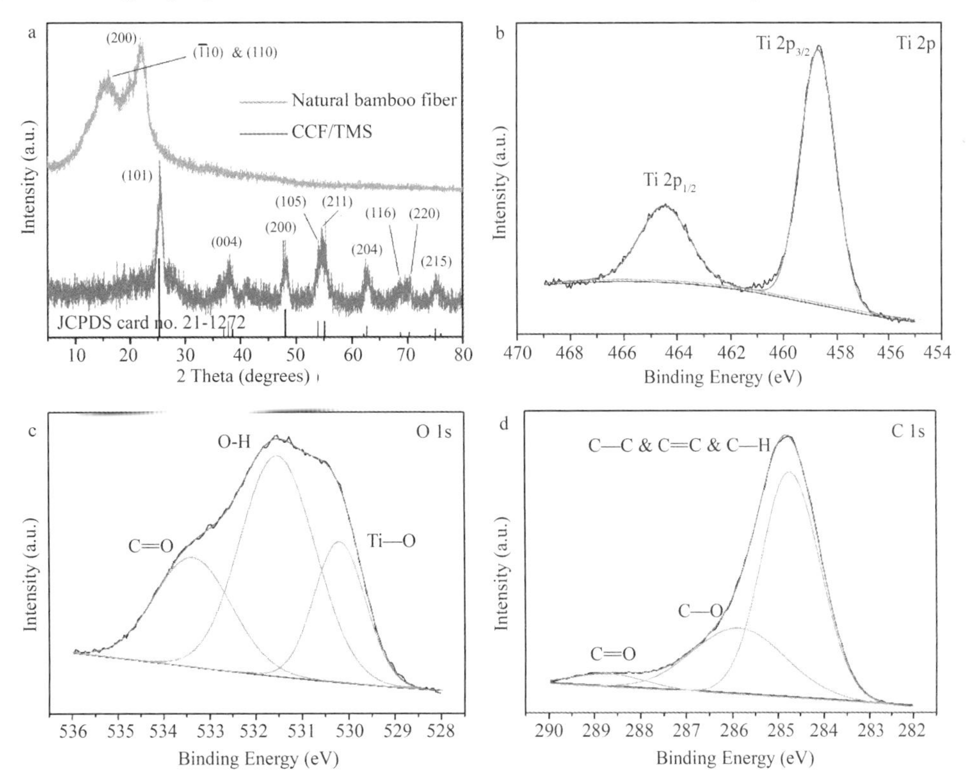

Fig. 2 (a) XRD patterns of the natural bamboo fiber and CCF/TMS, respectively. The bottom line is the standard JCPDS card no. 21-1272 for anatase TiO_2. High-resolution XPS spectra of CCF/TMS in (b) Ti 2p, (c) O 1s, and (d) C 1s regions

The loading content of TMS on the fibers was roughly calculated by the TG tests. As a contrast, the natural bamboo fiber without dipping the TiO_2 sol was subjected to the same pyrolysis treatment as described in the Experimental section, and the resultant CCF also underwent the TG test. As shown in Fig. 3a, CCF/TMS remained a higher weight of 89.6% at 700 ℃, 8.5% higher than that of CCF (81.1%). According to the differential thermogravimetry (DTG) curves shown in Fig. 3b, both the samples only exhibit a strong peak concerted at 525 ℃, possibly corresponding to the pyrolysis of residual thermally stable lignin.[15] Moreover, it is worth to mention that the carbothermal reduction between TiO_2 and carbon generally happens at a temperature higher than 1000 ℃.[16] Thus no weight loss associated with TiO_2 occurred at the temperature of 25–700 ℃. Therefore, the content of TMS is around 8.5% in the composite.

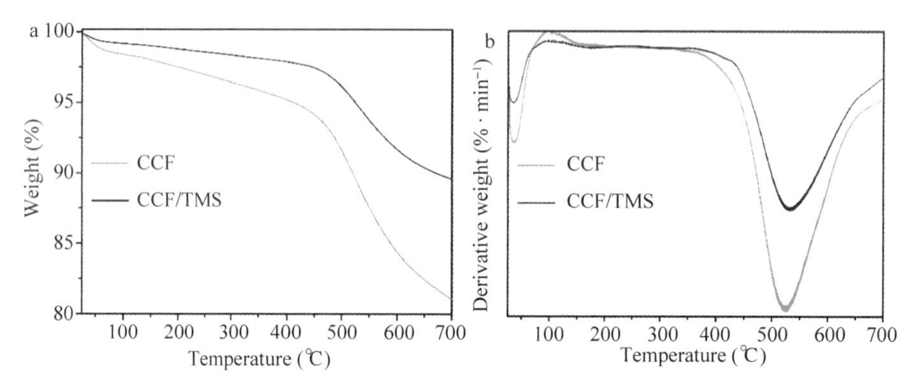

Fig. 3 (a) TG and (b) DTG curves of CCF and CCF/TMS, respectively

The nitrogen adsorption-desorption tests were carried out to determine the specific surface area of TMS (without combining with CCF), P25, and CPAT. As shown in Fig. 4, all the isotherms are identified as type Ⅳ, which is characteristic of mesoporous materials. The specific surface area was calculated over a relative pressure range of 0.05–0.30 from the multipoint Brunauer-Emmett-Teller (BET) method. TMS has a specific surface area of 103 $m^2 \cdot g^{-1}$, much higher than that of P25 (55 $m^2 \cdot g^{-1}$) and CPAT (11 $m^2 \cdot g^{-1}$), as shown in Table 1. The high specific surface area plays an important role in the photocatalytic activity.

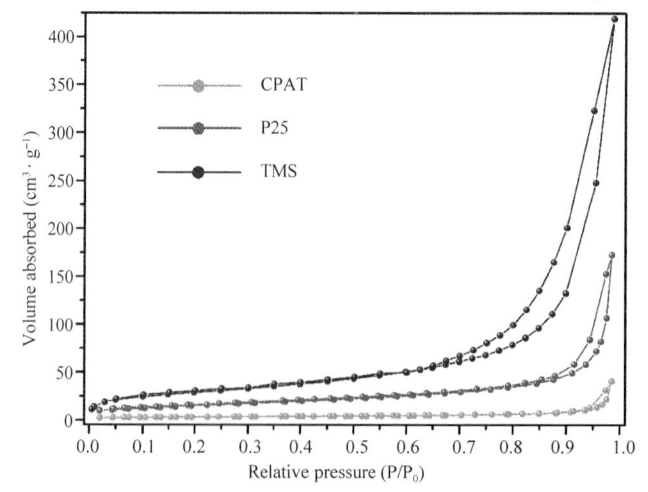

Fig. 4 Nitrogen adsorption-desorption isotherms of TMS, P25, and CPAT, respectively

Table 1 Characteristics of TMS, P25, and CPAT

	TMS	P25	CPAT
Crystallite size	9.2 nm	21 nm[a]	25 nm[a]
Surface area	103 $m^2 \cdot g^{-1}$	55 $m^2 \cdot g^{-1}$	11 $m^2 \cdot g^{-1}$
Crystal phase	Anatase (100%)	Anatase (80%) Rutile (20%)	Anatase (100%)

[a]The data were provided by the supplier

The photocatalytic activity of CCF/TMS was evaluated in the degradation reaction of indigo carmine dye in aqueous solution under UV radiation. Commercially available TiO_2 P25 and CPAT were used as photocatalytic references. The absorbance change (△ABS) represents the decomposition degree of indigo carmine dye. Large absolute value of △ABS means high decomposition. As shown in Fig. 5a, the negligible noncatalytic degradation of indigo carmine dye was observed by using CCF, indicating that CCF cannot decompose the dye molecules under UV irradiation. In contrast, the other three materials (including P25, CPAT and CCF/TMS) show obvious photocatalytic activity for the degradation of indigo carmine dye, with the reaction time prolonging from 0 to 40 min. Among these materials, CCF/TMS clearly shows the fastest increase in the absolute value of △ABS, suggesting that the decomposition rate of indigo carmine dye is fastest when using CCF/TMS. This speculation is consistent with the results of macroscopic observation shown in Fig. 5b. Generally, higher adsorption results in higher photocatalytic activity. Also, the larger specific surface area of TMS in CCF/TMS facilitates the absorption and utilization of UV light, which is essential for the photocatalytic degradation.[17] In addition, the crystalline size of the TiO_2 in CCF/TMS (9.2 nm) is much smaller than that of the P25 (ca. 21 nm) and CPAT (ca. 25 nm). It is commonly accepted that smaller crystalline size means more powerful redox ability due to the quantum-size effect.[18] Moreover, the smaller crystalline sizes are also beneficial for the separation of the photogenerated hole and electron pairs, which can slow the rate of e^--h^+ recombination and improve the photocatalytic activity.[19,20] Besides, the powder-like P25 and CPAT nanoparticles are easy to aggregate with each other due to their high surface energy, leading to the deterioration of photocatalytic activity. In contrast, the good dispersion of the TMS resulted from the presence of CCF contributes to obtaining the higher photocatalytic activity. Thus it is reasonable to explain a higher photocatalytic activity of CCF/TMS.

Moreover, according to the observations by naked eye, the cylindrical CCF/TMS sample maintained its well-defined shape without obvious shrinkage throughout the entire reaction process. The favorable shape stability provides convenience for recovery processing of the photocatalyst. In addition, this class of composite constructed by the fibers supported TMS can be weaved into large-sized cloth, which is possibly helpful for large-scale wastewater treatment. Also, this flexible and lightweight cloth is easily recovered. These projects will be deeply investigated in our future researches.

4 Conclusions

A mild easily-operated immersion-drying-pyrolysis approach was employed to grow anatase TMS onto the surface of CCF. The microspheres have an average size of 1.11 μm, and its content is approximately 8.5% in the composite measured by the TG tests. When evaluated as a photocatalyst, the composite can rapidly

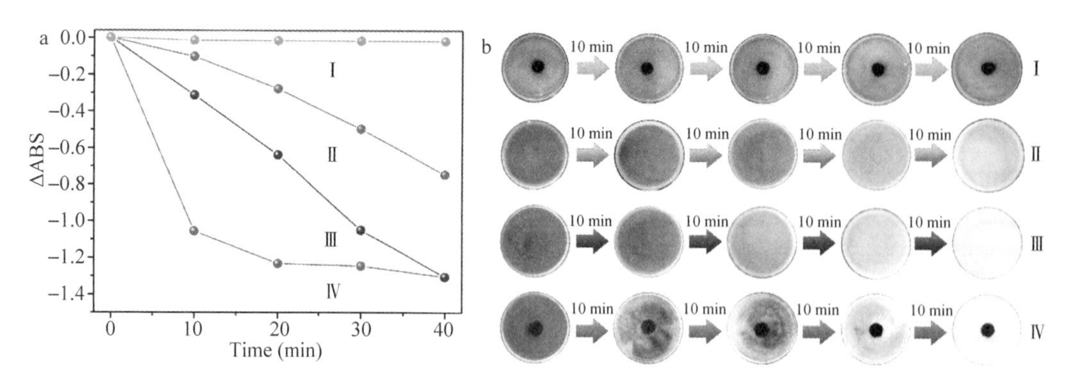

Fig. 5 (a) △ABS values and (b) color of indigo carmine dye for CCF (Ⅰ), CPAT (Ⅱ), P25 (Ⅲ) and CCF/TMS (Ⅳ) at various reaction times

decompose indigo carmine dye in aqueous solution under UV radiation within 40 min. Compared to the commercially available TiO_2 P25 and anatase TiO_2, a faster decomposition rate of indigo carmine dye was acquired when using CCF/TMS. In addition, CCF/TMS has good shape stability beneficial for recovery processing. To sum up, this composite with environmental friendliness, good economic effectiveness, and high photocatalytic activity might serve as a potential alternative for developing novel green photocatalyst.

Acknowledgements

This study was supported by the National Natural Science Foundation of China (grant nos. 31270590 and 31470584) and the Fundamental Research Funds for the Central Universities (grant nos. 2572016AB22).

References and Notes

[1] Mohanty A K. M. a. Misra and G. Hinrichsen, Biofibres, biodegradable polymers and biocomposites: an overview [J]. Macromol. Mater. Eng., 2000, 276(1): 1-24.

[2] Suh J K F, Matthew H W T. Application of chitosan-based polysaccharide biomaterials in cartilage tissue engineering: a review[J]. Biomaterials, 2000, 21(24): 2589-2598.

[3] Moon R J, Martini A, Nairn J, et al. Cellulose nanomaterials review: structure, properties and nanocomposites[J]. Chemical Society Reviews, 2011, 40(7): 3941-3994.

[4] Li L J, Wang Y P, Wang G, et al. Evaluation of properties of natural bamboo fiber for application in summer textiles [J]. Journal of Fiber Bioengineering and Informatics, 2010, 3(2): 94-99.

[5] He J, Kunitake T, Nakao A. Facile in situ synthesis of noble metal nanoparticles in porous cellulose fibers[J]. Chemistry of Materials, 2003, 15(23): 4401-4406.

[6] Zhang G, Morikawa H, Chen Y, et al. In-situ synthesis of ZnO nanoparticles on bamboo pulp fabric[J]. Materials Letters, 2013, 97: 184-186.

[7] Takagi H, Kako S, Kusano K, et al. Thermal conductivity of PLA-bamboo fiber composites[J]. Advanced Composite Materials, 2007, 16(4): 377-384.

[8] Tao X, Du J, Li Y, et al. TaC nanowire/activated carbon microfiber hybrid structures from bamboo fibers[J]. Advanced Energy Materials, 2011, 1(4): 534-539.

[9] Lakshmi BB, Dorhout P K, Martin C R. Sol-gel template synthesis of semiconductor nanostructures[J]. Chemistry of Materials, 1997, 9(3): 857-862.

[10] Han J, Zhou C, Wu Y, et al. Self-assembling behavior of cellulose nanoparticles during freeze-drying: effect of

suspension concentration, particle size, crystal structure, and surfacecharge [J]. Biomacromolecules, 2013, 14(5): 1529-1540.

[11]P. Scherrer and N. G. W. Gottingen, Math. -Pys. Kl, 1918, 2: 9.

[12]An G, Ma W, Sun Z, et al. Preparation oftitania/carbon nanotube composites using supercritical ethanol and their photocatalytic activity for phenol degradation under visible light irradiation[J]. Carbon, 2007, 45(9): 1795-1801.

[13]Hu C, Duo S, Liu T, et al. Low temperature facile synthesis ofanatase TiO_2 coated multiwalled carbon nanotube nanocomposites[J]. Materials Letters, 2010, 64(22): 2472-2474.

[14]Sevilla M, Fuertes A B. The production of carbon materials by hydrothermal carbonization of cellulose[J]. Carbon, 2009, 47(9): 2281-2289.

[15]Li M F, Sun S N, Xu F, et al. Ultrasound-enhanced extraction of lignin from bamboo (Neosinocalamus affinis): Characterization of the ethanol-soluble fractions[J]. Ultrasonics Sonochemistry, 2012, 19(2): 243-249.

[16]Berger L M, Langholf E, Jaenicke-Rößler K, et al. Mass spectrometric investigations on the carbothermal reduction of titanium dioxide[J]. Journal of Materials Science Letters, 1999, 18(17): 1409-1412.

[17]Herrmann J M. Heterogeneousphotocatalysis: fundamentals and applications to the removal of various types of aqueous pollutants[J]. Catalysis Today, 1999, 53(1): 115-129.

[18]Yu J C, Zhang L, Yu J. Directsonochemical preparation and characterization of highly active mesoporous TiO_2 with a bicrystalline framework[J]. Chemistry of Materials, 2002, 14(11): 4647-4653.

[19]Peng T, Zhao D, Dai K, et al. Synthesis of titanium dioxide nanoparticles with mesoporous anatase wall and high photocatalytic activity[J]. The Journal of Physical Chemistry B., 2005, 109(11): 4947-4952.

[20]Guo C, Ge M, Liu L, et al. Directed synthesis of mesoporous TiO_2 microspheres: catalysts and their photocatalysis for bisphenol A degradation[J]. Environmental Science & Technology, 2010, 44(1): 419-425.

中文题目：纤维素基碳纤维上生长的 TiO_2 微球：制备、表征及降解靛蓝胭脂红染料的光催化活性

作者：万才超，焦月，李坚

摘要：我们描述了一种在纤维素基碳纤维表面生长二氧化钛微球的简便方法。这些直径为 0.65-1.45 米的微球显示出纯锐钛矿相。举一个潜在应用的例子，以纤维素基碳纤维负载的二氧化钛微球（编码为 CCF/TMS）作为光催化剂，在紫外线辐射下降解水溶液中的靛蓝胭脂染料。结果表明，在 40 min 内，深蓝色染料溶液迅速变成无色，CCF/TMS 对靛蓝胭脂红染料的分解速度比市场上销售的二氧化钛 P25 和锐钛矿型二氧化钛更快。此外，CCF/TMS 在整个光催化反应过程中表现出良好的形状稳定性，为回收处理提供了方便。本研究为环境修复提供了一种环保、低成本、高效率的光催化剂。

关键词：二氧化钛；微球；纤维素；碳纤维；光催化活性

Flexible, Highly Conductive, and Free-standing Reduced Graphene oxide/Polypyrrole/Cellulose Hybrid Papers for Supercapacitor Electrodes*

Caichao Wan, Yue Jiao, Jian Li

Abstract: We report a facile scale-up process to fabricate a novel type of hybrid paper electrodes composed of reduced graphene oxide (RGO), polypyrrole (PPy) and cellulose. Through the optimization of preparation parameters (including the processes of in-situ polymerization of pyrrole and chemical reduction of graphene oxide by using $NaBH_4$), the fabricated hybrid paper acquired an extremely low sheet resistance of 1.7 Ω · sq^{-1}. Also, the hybrid paper possessed favorable mechanical flexibility and outstanding conductance stability. When the paper was evaluated as a free-standing and binder-free supercapacitor electrode, the unique construction (i. e., the cellulose fibers frame supporting interpenetrated RGO-PPy nanoarchitecture) endowed itself with high electrochemical activity, such as high areal capacitance of 1.20 F · cm^{-2} at 2 mA · cm^{-2} and good cyclic stability with a capacitance retention of 89.5% after cycling for 5000 times, which were tested in a three-electrode configuration. In addition, an all-solid-state laminated symmetric supercapacitor was prepared by assembling two pieces of hybrid paper electrodes and using a H_3PO_4/PVA gel as electrolyte. The solid-state supercapacitior has a high areal capacitance of 0.51 F · cm^{-2} at 0.1 mA · cm^{-2} and a high energy density of 1.18 mW · h · cm^{-3}. These results suggest that the hybrid paper is a promising electrode material and may be useful for the development of flexible high-performance and hand-held energy storage devices. More importantly, this work provides a good reference for the fabrication of other types of hybrid paper electrodes.
Keywords: Nanotube/cellulose composite paper; High power; Supercapacitors; Energy storage devices

1 Introduction

With the rapid consumption of non-renewable resources (such as fossil fuels) and the increasing seriousness of environmental pollution and greenhouse effect caused by exhaust emission of traditional internal combustion engines, it is extremely urgent to find eco-friendly high-performance energy-conversion and storage devices. Supercapacitors have been extensively considered as one of the quite promising energy storage systems due to the excellent reversibility (90%–95% or higher), [1,2] high power capability, [3,4] long cycle life ($>10^5$), [5-7] and safety of operations. Other advantages include high power density, [8-10] no maintenance, and widespread operation temperature. [11,12] Nowadays, the growing needs for stretchable, wearable and portable electronic gadgets like roll-up displays have motivated significant efforts to develop new forms of

* 本文摘自 Journal of Materials Chemistry A, 2017, 5(8): 3819–3831.

flexible ultrathin supercapacitors.[7,13-16] To meet these requirements, it firstly needs to make sure of the physical flexibility of various components of supercapacitors, especially for electrodes.

Papermaking technology is one of the four great inventions of ancient China, whose history dates back more than 2,000 years to the Han dynasty of China (206 BC-220 AD). In modern times, it is well-known that paper products are of special importance both in industries and in daily lives. Apart from widespread applications like information record, packaging, napkins, and decorative materials, paper recently has received great interests from some advanced fields. Thanks to the three-dimensional (3D) hierarchical porous fiber structures, abundant functional groups (e.g., hydroxyl), and hydrophilic property attributed to cellulose (major component), this kind of inexpensive, widely available and environmentally benign cellulose product provides an ideal platform for self-assembly of functional guest substances. This technique effectively integrates the given excellent properties of guest species and the unique physicochemical features of paper (e.g., flexibility), contributing to the exploitation of a multitude of novel functional materials. The guest species reported in recent literatures involve carbon nanotubes (CNTs),[17,18] nano-silver,[19] semiconductor oxides (Ga_2O_3-In_2O_3-ZnO),[20] Ti-O nanobelts ($Na_2Ti_3O_7$, $H_2Ti_3O_7$, TiO_2),[21] zinc oxide,[22] etc. The potential applications of these hybrids are eye-catching, such as electromagnetic interference shielding, antibacterial agents, field-effect transistors, photocatalysts, and strain and humidity sensors. Also, paper is an ideal candidate substrate for flexible supercapacitor applications since it can be easily incorporated with a large variety of electrochemically active substances like electronically conducting polymers (ECPs).[7,15,16,23,24] ECPs, such as polypyrrole (PPy), polyaniline (PANI), and poly-(3,4-ethylenedioxythiophene) (PEDOT), have been regarded as a class of momentous materials for the realization of high-performance supercapacitors on account of their significantly higher pseudocapacitance than carbon materials, fast redox switching, high conductivity, low cost, and lightweight feature.[25,26] Therefore, many investigations have been undertaken for supercapacitors fabricated from different ECPs or their derivatives. A simple low-cost "soak-polymerization" method can be carried out to effectively deposit PPy on common printing paper,[15] leading to the formation of a flexible and highly conductive paper possessing a low sheet resistance of $4.5\Omega \cdot sq^{-1}$. The solid-state supercapacitors fabricated by sandwiching a H_3PO_4/polyvinyl alcohol (PVA) membrane as the separator between the two PPy/paper electrodes achieved a high energy density of $1.0\ mW \cdot h \cdot cm^{-3}$ at a power density of $0.27\ W \cdot cm^{-3}$. Moreover, a core-sheath structured PPy/bacterial cellulose conductive nanocomposites[27] prepared by in-situ polymerization of self-assembled pyrrole have been reported to have an outstanding electrical conductivity as high as $77\ S \cdot cm^{-1}$, and a high mass specific capacitance hitting $316\ F \cdot g^{-1}$ at a current density of $0.2\ A \cdot g^{-1}$. Notwithstanding these fascinating merits of PPy/cellulose composites, the common poor mechanical stability of ECPs, i.e., the considerable volume change during the repeated doping/dedoping process (insertion/deinsertion of counter ions), gradually aggravates their conducting properties.[16,28,29] A swelling-shrinkage-cracking or breaking process always takes place. It is well-known that carbon-based materials (typically like activated carbons, CNTs, and graphene) have been widely employed to integrate with ECPs to compensate for the limitation of individual ECPs in electrochemical capacitors.[25,26,29,30] Although activated carbons are cheap electrode materials with high specific surface area giving a moderate capacity in supercapacitors, powder-like activated carbons are difficult to handle as electrode materials. Thus binder (e.g., poly(vinylidene fluoride) (PVDF) and poly (tetrafluoroethylene) (PTFE)) are always needed. It has been confirmed that the rate and capacitance performances are largely restricted by the addition of binder in the preparation of electrodes which can raise electrode resistance.[31] Also, the activated carbon electrodes usually cannot provide physical flexibility. Purified CNTs (i. e., without residual catalyst or amorphous carbon) generally have relatively low specific capacitance[32] typically from 15 to 80 $F \cdot g^{-1}$ with surface areas ranging from 120 to 400 $m^2 \cdot g^{-1}$. The observed contact resistance between CNT-based electrodes and current collector is another unfavourable factor.[33] In contrast, graphene-based supercapacitors have displayed superior specific capacitance values of more than 200 $F \cdot g^{-1}$ in aqueous electrolytes,[34,35] 120 $F \cdot g^{-1}$ in organic electrolytes,[36] and 75 $F \cdot g^{-1}$ in an ionic liquid.[37] Graphene, a flat monolayer of carbon atoms tightly packed into a two-dimensional (2D) honeycomb lattice, has become the most appealing carbon-based material for supercapacitor applications due to high chemical stability, high electrical and thermal conductivity, and high specific surface area (2630 $m^2 \cdot g^{-1}$).[38] Supercapacitors based on graphene and ECPs have already displayed a broad spectrum of brilliant properties, e. g., ultrahigh specific capacitance of 1046 $F \cdot g^{-1}$ (graphene/PANI),[39] high rate performance,[30,40] and prominent charge-discharge stability.[40,41] Consequently, self-assembly of ECPs and graphene sheets on the surface of paper substrate probably leads to an ultrathin, flexible, highly conductive, and free-standing supercapacitor electrodes with ideal electrochemical activity, which is the subject of this study.

In the present work, we have developed an easily-operated and scalable method to fabricate a type of reduced graphene oxide (RGO)/PPy/cellulose (RPC) hybrid papers, serving as flexible, ultrathin, highly conductive, and free-standing supercapacitor electrodes. The cellulose fibers in the RPC paper electrodes can effectively absorb electrolyte and act as electrolyte reservoirs to facilitate ion transport.[7] The free-standing and binder-free electrodes can serve as both electrodes and current collectors due to the excellent electrical conductivity, thereby simplifying the structure and reducing the cost of supercapacitors. RPC papers were prepared by the procedures in three steps: (1) in-situ polymerization of PPy on the surface of cellulose papers using ferric iron as oxidant; (2) immersion of the PPy/cellulose papers in the aqueous dispersion of GO; (3) reduction of GO by a nontoxic reductant namely sodium borohydride ($NaBH_4$). For acquiring high electrical conductivity of RPC papers, it is worth to decrease the sheet resistance of the precursor PPy/cellulose papers as far as possible. Therefore, the influences of two critical preparation parameters (i. e., the initial concentration of pyrrole solution and the molar ratio of ferric iron/pyrrole) were studied. In addition, different concentrations of GO dispersions were adopted to soak the PPy/cellulose papers to clarify the effects of RGO loading content on the electrochemical performance of synthesized RPC papers. As an example of potential applications in flexible solid-state supercapacitors, the RPC papers and H_3PO_4/PVA gel electrolyte were assembled into an all-solid-state symmetric supercapacitor (SSC), and its electrochemical property was studied.

2 Results and discussion

2.1 Effects of preparation parameters on sheet resistance of PPy/cellulose papers

Prior to the deposition of RGO, PPy was firstly self-assembled on the surface of cellulose paper via a facile soak-polymerization method. As shown in Fig. 1a, the cellulose paper was firstly soaked in the aqueous solution of pyrrole. Owing to the strong hydrogen-bond interaction between the hydroxyl groups of cellulose and the imine groups of pyrrole, plentiful pyrrole molecules were tightly immobilized on the paper surface.[42] Through the succeeding in-situ polymerization using Fe^{3+} as oxidant, the paper was gradually transformed from white to gray and eventually to black, suggesting the formation of PPy. For the sake of fabricating highly conductive RPC papers, the sheet resistance of precursor PPy/cellulose papers should be decreased as far as possible. Therefore, it is worth to optimize the crucial preparation parameters of PPy/cellulose papers, i. e., the initial concentration of pyrrole solution and the molar ratio of ferric iron/pyrrole. Fig. 1b-e illustrate the effects of the two aforementioned factors on the sheet resistance. It can be seen that the lowest sheet resistance (5. 1 $\Omega \cdot sq^{1}$) was achieved from the initial pyrrole concentration of 1. 35 M and the molar ratio of ferric iron/pyrrole of 1. The value is lower than that of previously reported Au/polyvinyl alcohol/polyaniline-coated paper (7 $\Omega \cdot sq^{1}$),[43] tin-doped indium oxide-coated paper (12 $\Omega \cdot sq^{1}$),[44] and carbon nanotube/cellulose composite paper (40 $\Omega \cdot sq^{1}$),[45] and comparable to the value for single-walled carbon nanotube/silver nanowire-coated paper (1 $\Omega \cdot sq^{1}$).[46] Furthermore, the corresponding electrical conductivity was calculated to be as high as 980 $S \cdot m^{-1}$, which is much higher than that of numerous conductive hybrid papers.[7,27,45]

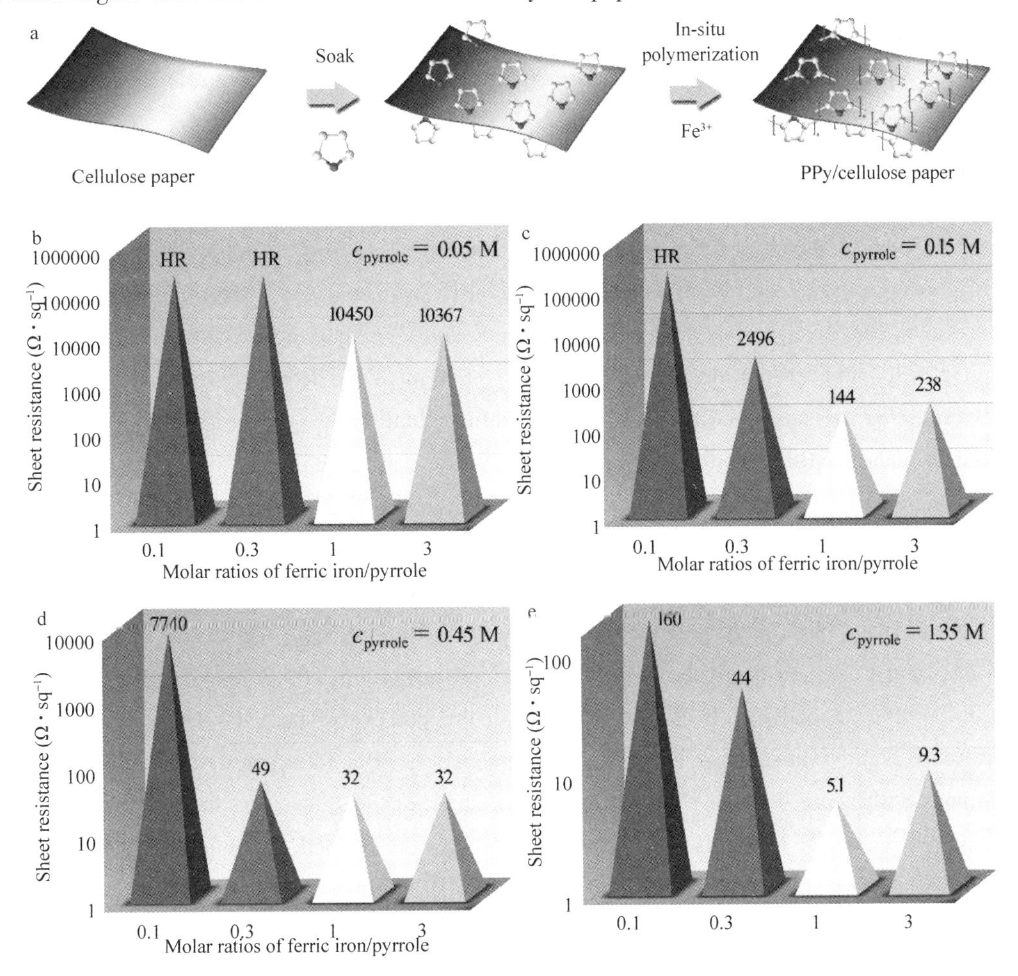

Fig. 1 (a) Schematic diagram of the fabrication of PPy/cellulose papers. Sheet resistance of PPy/cellulose papers fabricated from different initial pyrrole concentrations of 0. 05 M (b), 0. 15 M (c), 0. 45 M (d), and 1. 35 M (e), respectively. *HR* means the sheet resistance higher than the measuring range of the resistance tester (1.0×10^{-3}–1.9×10^{5} $\Omega \cdot sq^{1}$)

2.2 Characterizations of PPy/cellulose papers by SEM, EDX, FTIR and XPS

The PPy/cellulose paper displaying the lowest sheet resistance was characterized with scanning electron microscope (SEM), energy dispersive X-ray (EDX) spectroscopy, Fourier transform infrared (FTIR) spectroscopy, and X-ray photoelectron spectroscopy (XPS). For comparison, the untreated cellulose paper was also tested. It can be seen in Fig. 2b that the cellulose paper consists of typical interlaced fiber structures. After the in-site polymerization of pyrrole, the surface of fibers was tightly covered with a great number of nanoparticles (Fig. 2c), leading to the transformation of the cellulose paper surface color from white to black (Fig. 2a). Nevertheless, the porous fiber skeleton structure of the paper maintained well even after the coating. The higher-magnification SEM image in Fig. 2d shows that these well-distributed nanoparticles with the sizes of 150–450 nm were densely connected with each other, indicative of strong adhesion. The results of qualitative and quantitative elemental analysis on the control cellulose paper and the PPy/cellulose paper are presented in Fig. 2e and f, where the common elements of C and O mostly stemmed from the carbohydrate polymers. The Au element originated from the coating layer used for electric conduction during the SEM observation. Besides, the N element was detected and accounted for 12.6 wt.%, indicating the probable existence of PPy. Moreover, the notable increase in the mass ratio of C/O from 1.35 to 1.76 is another evidence for the presence of PPy.

The chemical compositions of the paper before and after the coating of PPy were characterized by FTIR (Fig. 2g). The cellulose paper exhibits the typical cellulose characteristic bands (for details, see Table S1 in the Supplementary Information).[47] However, these cellulose bands were disappeared in the FTIR spectrum of the hybrid paper. Instead, the characteristic signals of PPy can be clearly identified. The bands at around 1543 and 1454 cm^{-1} may be severally assigned to the C—C and C—N stretching vibrations in the pyrrole ring, and the band at 1303 cm^{-1} represents the C—H and C—N in-plane deformation modes.[15] The bands at 1172, 1038, and 782 are attributed to the N—C stretching vibration, C—H in-plane vibration of PPy ring, and C—H out-of-plane ring deformation, respectively.[27] The band at 902 cm^{-1} is associated with the =C—H out of plane vibration, indicating the polymerization of pyrrole.[48] These results confirm the fact that the paper substrate was almost covered by PPy.

XPS spectra revealing the chemical bonding in the PPy/cellulose paper are shown in Fig. 2h and i. XPS survey spectrum displays three main peaks at binding energy (BE) of 532, 400, and 285 eV corresponding to O 1s, N 1s, and C 1s, respectively. Regarding the high-resolution XPS N 1s spectrum, the three fitting peaks at 403.2, 400.0, and 397.8 eV are ascribed to the positively charged nitrogen (—N^+—), amine (—NH—) groups, and imine group (—N=), respectively, which further demonstrate the existence of PPy.[49] Moreover, the core levels of N 1s shift to higher binding energies, as compared with those of some previously reported pure PPy materials,[49,50] which is possibly due to both the shielding effect of PPy layers and the hydrogen bonding between the nitrogen lone pairs of PPy and the —OH groups of cellulose.

2.3 Morphologies and electrochemical properties of RPC papers

The preparation of RPC papers was carried out by integrating GO sheets with the PPy/cellulose paper with the lowest sheet resistance of 5.1 $\Omega \cdot sq^{1}$, and then reducing the GO component by using $NaBH_4$ which has already been confirmed elsewhere[51] as an effective reducing agent to acquire low-resistance RGO films. RPC

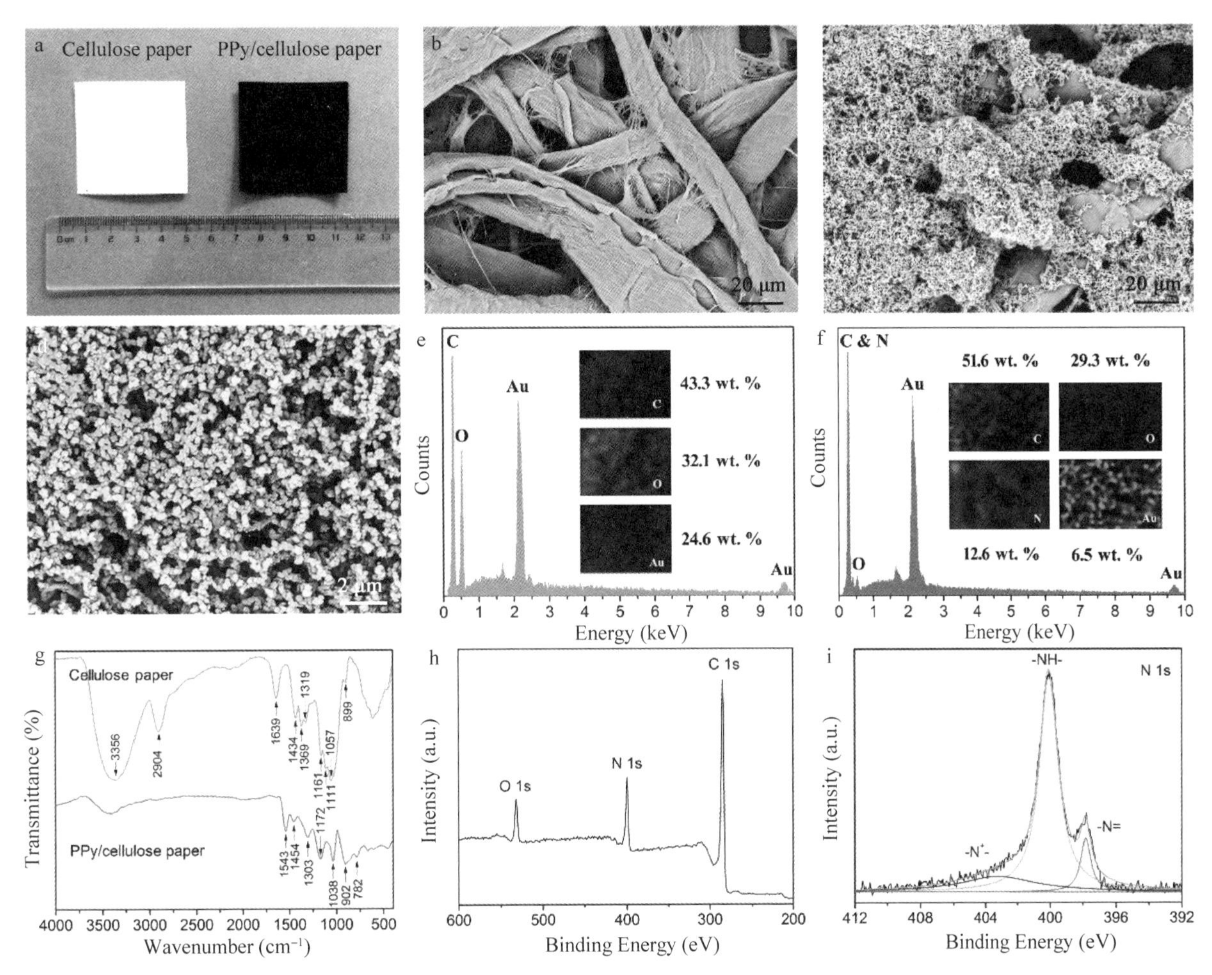

Fig. 2 (a) Optical images of the cellulose paper and the PPy/cellulose paper. (b) SEM image of the cellulose paper. (c) Low- and (d) high-magnification SEM images of the PPy/cellulose paper. EDX patterns of the cellulose paper (e) and the PPy/cellulose paper (f), and the insets provide the corresponding elemental maps and contents. (g) FTIR spectra of the cellulose paper and the PPy/cellulose paper. (h) XPS survey and (i) high-resolution XPS N 1s spectra of the PPy/cellulose paper, respectively

papers fabricated from the GO dispersions with different concentrations of 0.1, 0.5, and 2.5 $g \cdot L^{-1}$ were labeled as RPC-0.1, RPC-0.5, and RPC-2.5, respectively.

The morphologies of the RPC papers obtained by the SEM observations are shown in Fig. 3, where it can be clearly seen that the papers had been decorated with numerous thin sheet-like RGO with the size of several microns (the white circles signify the main distribution positions of RGO sheets). As shown in Fig. 3a-c, it is found that the proportion of RGO in the RPC papers obviously increases with the increasing initial concentration of GO dispersions. According to the results of thermogravimetric (TG) measurements (see Fig. S1 in the Supplementary Information), the RGO contents can be roughly calculated to be 0.94 wt.% for RPC-0.1, 3.41 wt.% for RPC-0.5, and 8.06 wt.% for RPC-2.5, respectively. In addition, unlike the simple deposition of RGO on the surface of PPy layers in the condition of the low-content RGO (see Fig. 3a), RGO sheets gradually penetrated into the porous PPy layers with the increasing RGO content (Fig. 3b and c), which contributes to acquiring stronger interactions and more superior interfacial bonding. The interactions are primarily derived from strong van der Waal's force and $\pi-\pi$ stacking interaction between conjugated backbones of PPy and RGO sheet. The high-magnification SEM images shown in Fig. 3d-f more clearly demonstrate the permeation trend of RGO sheets. Additionally, the incorporation of RGO gives rise to improvements in the conductive ability of the RPC papers. The sheet resistance of RPC-2.5 with 8.06 wt.% RGO content is 1.7 $\Omega \cdot sq^{-1}$, 0.51 and 0.64 times lower than that of RPC-0.5 (3.5 $\Omega \cdot sq^{-1}$) and RPC-0.1 (4.7 $\Omega \cdot sq^{-1}$) with the lower RGO contents of 3.41 wt.% and 0.94 wt.%. The results indicate remarkable negative correlation between the sheet resistance and the GO proportion, which might be explained as that the highly conductive and large-area RGO sheets participated in the construction of conductive network due to the good interpermeation between RGO and PPy layers in the condition of the high-content RGO. Also, the increase of RGO content leads to the increase of connections between RGO sheets and availability of a large number of charge carriers traveling through the entire network.

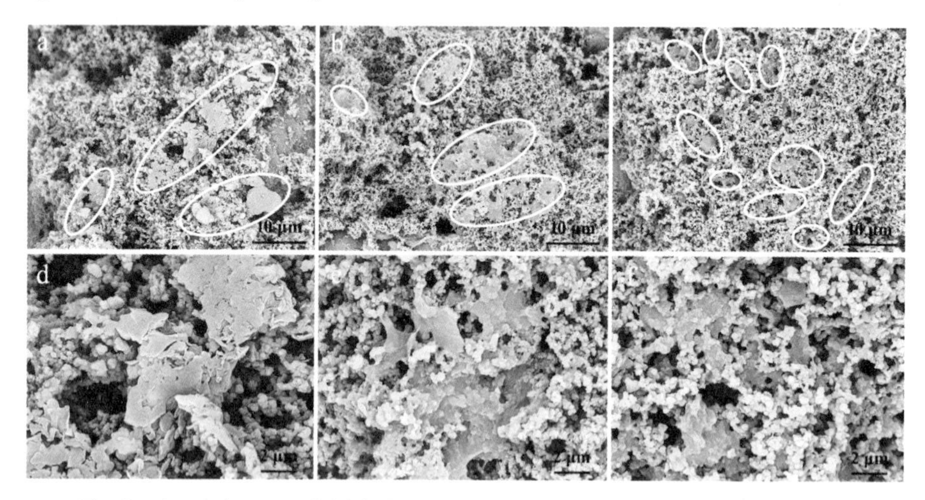

Fig. 3 (a-c) Low- and (d-f) high-magnification SEM images of RPC-0.1, RPC-0.5, and RPC-2.5, respectively. The white circles signify the main distribution positions of RGO sheets

Atomic force microscopy (AFM) characterization is one of the most direct methods of quantifying the degree of exfoliation to graphene level after the dispersion of the powder in a solvent. Fig. 4a depicts the typical AFM images of the as-prepared GO sheets. The thickness, measured from the height profile of the AFM image, is about 1.2 nm, which is consistent with the data reported in the literature,[52] suggesting that the formation of the single-layered GO. Thus the ultra-thin GO sheets were easy to penetrate into the porous PPy layer in the process of immersion accompanied with stirring and shaking. The transmission electron microscope (TEM) observations were executed to further study the structure of RPC papers. As shown in Fig. 4b, the TEM image presents that the RGO sheets were covered with abundant well-dispersed PPy nanoparticles, indicating good interface combination between them. Moreover, the higher-magnification TEM image (see Fig. 4c) further verifies that the RGO sheets have penetrated into the porous PPy layers, consistent with the results of SEM observations. Fig. 4d and e present the high-resolution XPS C 1s spectra of the GO/PPy/cellulose paper (obtained by dipping the PPy/cellulose paper in the 2.5 $g \cdot L^{-1}$ GO dispersion) before and after the reduction of GO. The deconvoluted C1s spectrum of the GO/PPy/cellulose paper (Fig. 4d) exhibits five Gaussian peaks with different binding energies, i. e. , the C—C at 284. 6 eV, C—N 285. 3 eV, C—OH 286. 8 eV, C =O 288. 0 eV, and O—C =O at 290. 0 eV.[53] Nevertheless, after the reduction of GO by $NaBH_4$, the C1s spectrum of the resultant (i. e. , RPC-2. 5) displays substantial decrease in peak intensity of the oxygen functionalities (Fig. 4e), corroborating that lessening of the oxygen functionality upon reduction. Moreover, the atomic ratio of C/O increases from 2. 7 to 6. 8 after the reduction, further demonstrative of the elimination of a large fraction of oxygenated groups on GO sheets by chemical reduction. The Raman spectra of the GO/PPy/cellulose paper before and after the reduction of GO was also studied (see Fig. 4f). In the case of the GO/PPy/cellulose paper, the characteristic D band at 1339 cm^{-1} represents the breathing vibrations of carbon atoms of dangling bonds in plane terminations of disordered and defected graphite, while the G band at 1567 cm^{-1} is assigned to the in-plane bond-stretching vibration of sp^2-bonded carbon atoms. The peaks at 952 and 1059 cm^{-1} belong to the characteristics peaks of PPy,[54] indicating the wrapping of PPy onto GO sheets and cellulose fibers. After the reduction of GO, the intensity ratio (I_D: I_G) increases notably, which may be attributed to the following reasons:[53]

(1) Increased number of sp^2 domains formed during the reduction process.

(2) Presence of unrepaired defects that remained after the removal of large amounts of oxygen-containing functional groups.

(3) Partially disordered crystal structure of in-situ formed RGO.

All these above results reveal the effective reduction of GO in the GO/PPy/cellulose paper by $NaBH_4$.

Fig. 4g displays the N_2 adsorption-desorption isotherms of the cellulose paper, PPy/cellulose paper, and RPC-2. 5. These isotherms can all be classified to type Ⅳ according to IUPAC classification, and the adsorption-desorption loops belong to the type H3, revealing the existence of slit-shaped pores. The Brunaure-Emmett-Teller (BET) specific surface area of RPC-2. 5 (9. 3 $m^2 \cdot g^{-1}$) is bigger than that of the cellulose paper (3. 4 $m^2 \cdot g^{-1}$) and the PPy/cellulose paper (6. 9 $m^2 \cdot g^{-1}$). Therefore, more active materials on RPC-2. 5 can contact with the electrolyte, which should enhance the specific capacitance of RPC-2. 5.

The electrochemical properties of the RPC papers were studied through cyclic voltammograms (CV), galvanostatic charge/discharge (GCD) and electrochemical impedance spectroscopy (EIS) measurements in a three-electrode configuration in 1 M NaCl aqueous electrolyte. Fig. 5 presents a comparative study of the CV curves for the control PPy/cellulose paper, RPC-0. 1, RPC-0. 5, and RPC-2. 5 at varied scan rates. Apparently, all the CV scans display a roughly rectangular mirror image with regard to the zero-current line at different scan rates in the potential window of -0. 2–0. 8 V. In addition, the CV curves became shuttle-like in shape when the scan rate reached 50 $mV \cdot s^{-1}$ and above, which can be interpreted as that the entering/ejecting and diffusion rates of counterions were slower than the transfer rate of electrons in the hybrid papers at the high scan rates[27]. The area of the closed CV loop of the PPy/cellulose paper is relatively smaller, implying a relatively low capacitance. In contrast, the areas of the CV curves of RPC papers are significantly larger due to the contribution of both double-layer and pseudo-capacitance of RGO and PPy. Also, it is obvious that the higher RGO proportion is beneficial to improve capacitance. With the increase of scan rate, RPC-2. 5 exhibits remarkably enlarged capacitive areas, as compared to the PPy/cellulose paper, implying that the RGO helped a lot in speeding up carriers' transportation along the PPy networking chains.[55]

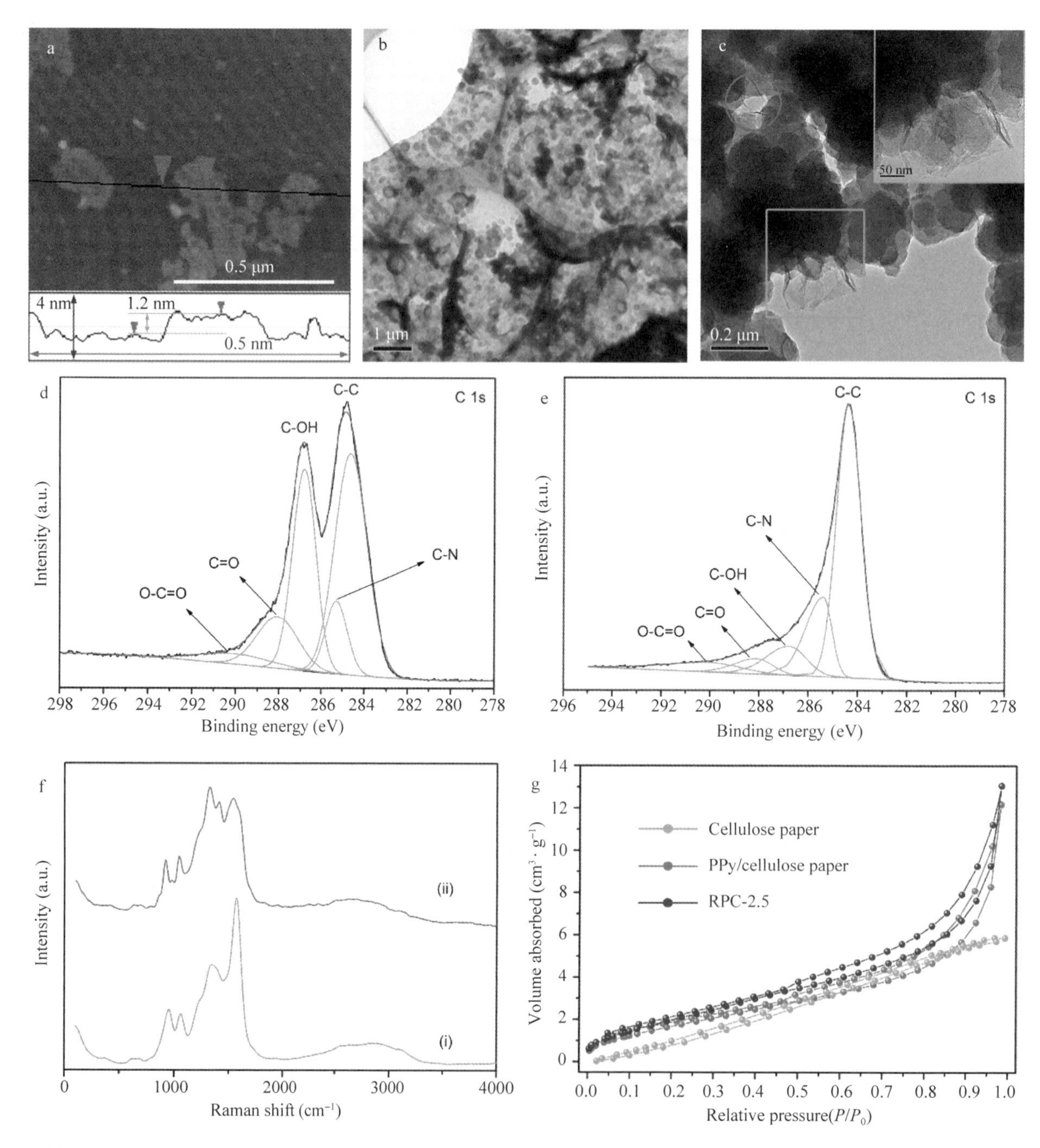

Fig. 4 (a) AFM analysis of GO nanosheets on mica surface with height profile. (b) Low- and (c) high-magnification TEM images of RPC-2. 5. The red and green circles in (c) signify the positions of RGO sheets, and the inset of (c) shows the enlarged image of the green-squared marker region. High-resolution XPS C 1s spectra of the GO/PPy/cellulose paper (d) and RPC-2. 5 (e). (f) Raman spectra of the GO/PPy/cellulose paper (i) and RPC-2. 5 (ii). (g) Nitrogen adsorption-desorption isotherms of the cellulose paper, PPy/cellulose paper, and RPC-2. 5

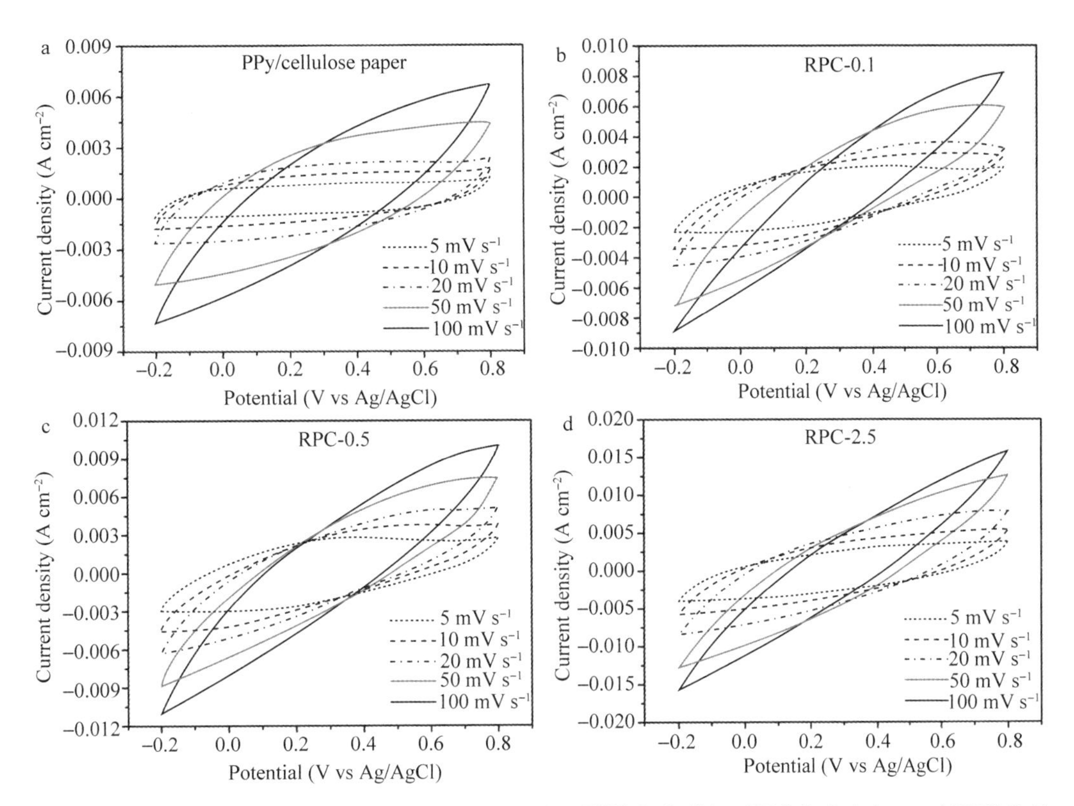

Fig. 5 CV curves of the PPy/cellulose paper (a), RPC-0. 1 (b), RPC-0. 5 (c), and RPC-2. 5 (d) at different scan rates of 5, 10, 20, 50, and 100 mV · s^{-1}, respectively

The electrochemical kinetics of electrode materials towards Na^+ can be evaluated by CV analysis. The total stored charge can be generally divided into three components, i. e. , the faradaic contribution from the Na^+ insertion process, the faradaic contribution from the charge transfer process with surface atoms (referred to as pseudocapacitance), and the nonfaradaic contribution from the double-layer effect. The latter two capacitive components cannot be separated. The currents (i) obey the power law:[56]

$$i = av^b \tag{1}$$

where v is the scan rate, and both a and b are adjustable parameters. The b-value can be determined by the slope of the log(v) − log(i) plots. In particular, the b-value of 0. 5 represents a total diffusion-controlled behavior, whereas 1. 0 indicates a capacitive process. For $b = 0.5$, the current is proportional to the square root of the sweep rate:[57]

$$i = nFAC \cdot D^{1/2} v^{1/2} (\alpha nF/RT) 1/2\pi^{1/2} \chi(bt) \tag{2}$$

where F is the Faraday constant, A is the surface area of the electrode materials, C^* is the surface concentration of the electrode material, D is the chemical diffusion coefficient, α is the transfer coefficient, n is the number of electrons involved in the electrode reaction, R is the molar gas constant, T is the temperature, and the function $\chi(bt)$ represents the normalized current for a totally irreversible system as indicated by the CV response. The current response in eq 2 is assumed to be diffusion controlled, indicating a faradaic insertion process.

For $b = 1.0$, the current is proportional to the sweep rate:[57]

$$i=vC_dA \quad (3)$$

where C_d is the capacitance.

Fig. 6a shows the log(v)-log(i) plots for RPC-2.5 electrode. The slope (b) takes a value between 0.5 (diffusion limited) and 1 (capacitive) at scan rates from 5 to 50 mV · s^{-1} for maximum current of positive-going potential sweep and from 5 to 20 mV · s^{-1} for maximum current of negative-going potential sweep, illustrating a mixed charge storage process involving both the intercalation of Na^+ and the pseudocapacitive storage. A remarkable decrease of slope can be found at scan rates above 50 mV · s^{-1} for positive-going potential sweep and above 20 mV · s^{-1} for negative-going potential sweep. The limitation to the rate capability should be ascribed to an increase of the ohmic contribution and/or diffusion constrains upon an ultra-fast scan rate.

The total capacitive contribution at a certain scan rate could be quantified on the base of separating the specific contribution from the capacitive and diffusion-controlled charge at a fixed voltage according to the following expression:[58]

$$i=k_1v+k_2v^{1/2} \quad (4)$$

$$\frac{i}{v^{1/2}}=k_1v^{1/2}+k_2 \quad (5)$$

where k_1v and $k_2v^{1/2}$ correspond to the current contributions from surface capacitive effects and the diffusion controlled insertion processes, respectively. k_1 and k_2 values can be obtained from the slope and the y-axis intercept point of a straight line (plotting the sweep rate dependence of the current at each fixed potential using eq 5), respectively. As shown in Fig. 6b, it is calculated that the slopes (k_1) range from -2.3×10^{-5} to -5.0×10^{-5} at the potentials of 0−0.6 V, indicating that the majority of diffusion contribution occurred in this potential range.

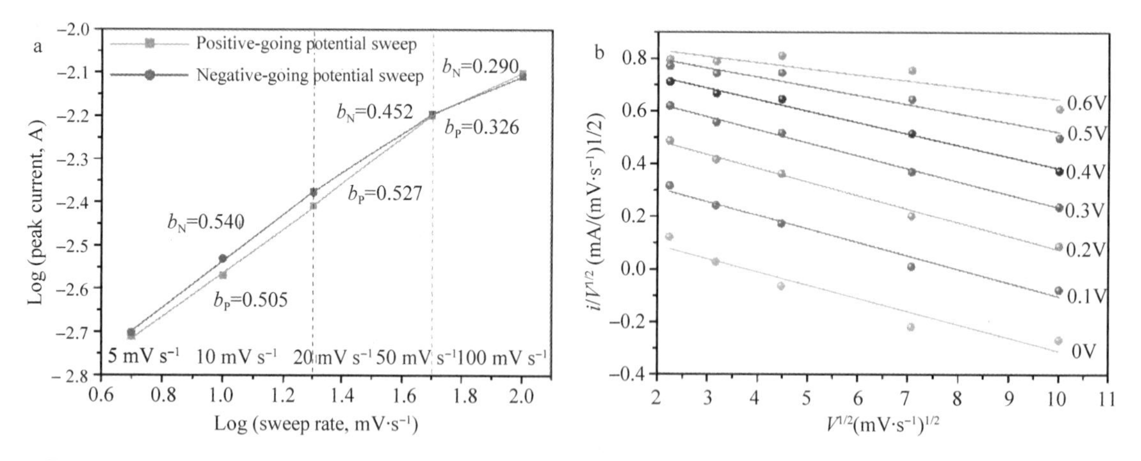

Fig. 6 (a) Determination of the b-value of RPC-2.5 using the relationship between maximum current and scan rate. (b) Use of eq 5 to analyze the voltammetric sweep data for RPC-2.5 at the different potentials

Fig. 7a-d illustrate the GCD traces of the control PPy/cellulose paper, RPC-0.1, RPC-0.5, and RPC-2.5, respectively. All the traces exhibit nearly linear and symmetric triangle shape with a slight curvature, indicating good electrochemical reversibility and charge-discharge properties and rapid I-V response. The internal resistance (IR) drop at the initial discharge can roughly reflect the internal resistance of electrode. As shown in Fig. 7a-d, RPC-2.5 has the minimal IR drop, suggesting a low internal resistance contributing to the reduction of amount of energy lost due to unwanted heat during charging/discharging processes. The areal capacitance at the different current densities was calculated by the equation: $C_A = (I \times \Delta t)/(\Delta V \times S)$, where C_A is the areal capacitance ($F \cdot cm^{-2}$), I is the charge-discharge current (A), Δt is the discharge time (s), ΔV is the potential window (V), and S is the superficial area of the working electrode (cm^{-2}). The areal capacitance of the PPy/cellulose paper, RPC-0.1, RPC-0.5, and RPC-2.5 was plotted *versus* discharge current in Fig. 7e. The areal capacitance increases with the RGO loading content, and the RPC-2.5 electrode exhibits the highest areal capacitance of about 1.20 $F \cdot cm^{-2}$ at a discharge current of 2 $mA \cdot cm^{-2}$, which is approximately 2.7, 2.3, and 1.6 times that of the control PPy/cellulose paper (0.44 $F \cdot cm^{-2}$), RPC-0.1 (0.52 $F \cdot cm^{-2}$), and RPC-0.5 (0.75 $F \cdot cm^{-2}$), respectively. Also, as a free-standing and binder-free electrode, the value is nearly one magnitude higher than most areal capacitances for graphene/PPy composites reported so far in literatures (see Table S2).[59-63] The significant enhancement in electrochemical performance of RPC-2.5, as compared to the others, is probably attributed to the following reasons: (1) RGO can offer high coverage available for the double-layer formation and charge-transfer reaction, and then improves the electrochemical performance; (2) RGO can provide enhanced electrode/electrolyte interface areas, which enable the electrochemical accessibility of electrolyte through the loosely packed PPy structure, facilitating rapid transport of the electrolyte ions in the electrode during charge/discharge processes and greatly reducing the diffusion length; (3) the incorporation of RGO can facilitate the oxidation or deoxidization of α-C or β-C atoms of PPy rings;[64] (4) the favorable interpermeation and intimate interfacial contact between RGO and PPy layers can reduce the diffusion and migration length, and thus improve the electrochemical utilization of PPy;[65] (5) the lower resistance results in the faster electron transport in the electrode; (6) the good synergistic effect between PPy and RGO. In addition, we have increased the concentration of GO to 5 $g \cdot L^{-1}$ to fabricate RPC paper electrode (coded as RPC-5). According to the GCD curves (see Fig. S2), the areal capacitance of RPC-5 was calculated to be 0.88 $F \cdot cm^{-2}$, 26.7% less than that of RPC-2.5 (1.20 $F \cdot cm^{-2}$). Moreover, it can be seen that the IR drop of RPC-5 is larger than that of RPC-2.5, indicating that RPC-5 has larger internal resistance of electrode. The lager internal resistance might be ascribed to the aggregation of RGO due to the high GO concentrations (see Fig. S3)

In addition, the capacitance decreases with the increasing current density. This is because of the low utilization of electroactive materials at high discharge current densities since the electrolyte ions were unable to enter into the inner structure of the active material, and only the outer active surface was utilized for charge storage. But RPC-2.5 still retains a high capacitance of 0.24 $F \cdot cm^{-2}$ (around 6 times that of the control PPy/cellulose paper) at the high current density of 10 $mA \cdot cm^{-2}$. The superior rate performance of RPC-2.5 is primarily ascribed to the short diffusion path of ions, enhanced active surface, increased electrical conductivity, and intimate interfacial contact between PPy and RGO, which can make higher materials utilization. Also, the interpenetrated RGO/PPy construction can provide fast charge carriers paths, allowing the trapped charges to easily shuttle in the electrode.[55] Furthermore, RPC-2.5 displays more outstanding charge-discharge stability. As shown in Fig. 7f, the capacitance retention of the PPy/cellulose paper is only about 48.8% after cycling for 5000 times. After the combination with RGO, the capacitance retention increases to 76.8%, 87.0%, and 89.5% for RPC-0.1, RPC-0.5, and RPC-2.5, respectively. Apparently, RPC-2.5 exhibits the most favorable capacitance retention ability, which might be explained as that the penetrated RGO sheets can act as the frameworks for sustaining PPy, thus preventing PPy chains from severely swelling and shrinking during the charge/discharge process. Compared with the others, RPC-2.5 is consequently more suitable for the pursuit of high-performance supercapacitors.

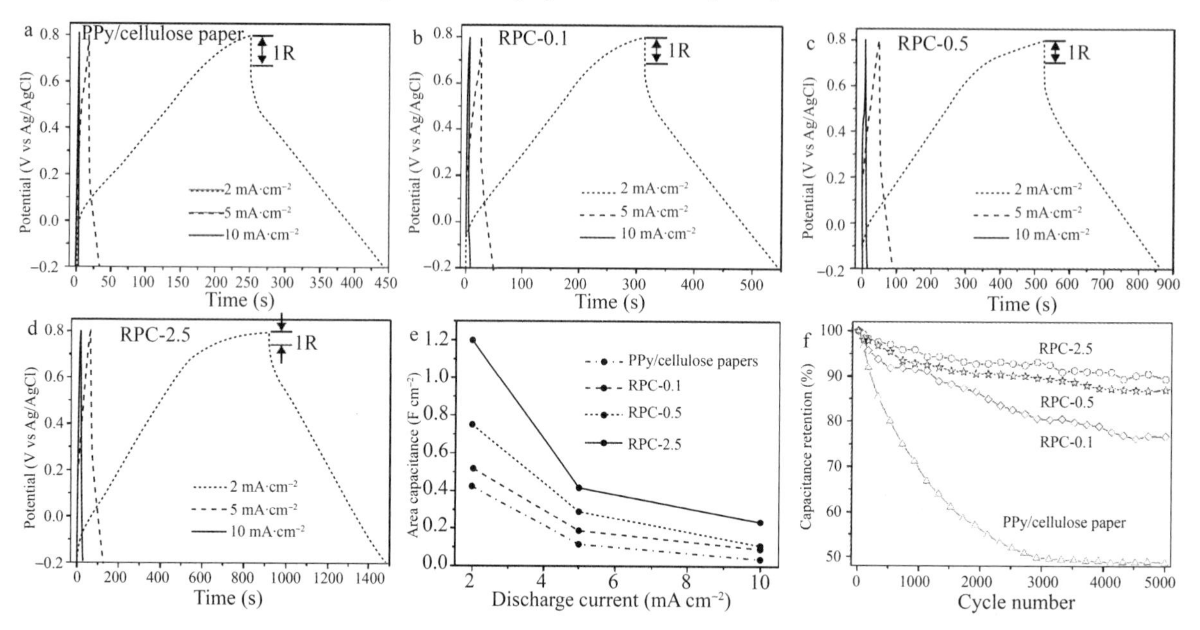

Fig. 7 GCD curves of the PPy/cellulose paper (a), RPC-0.1 (b), RPC-0.5 (c), and RPC-2.5 (d), respectively. (e) Areal capacitance of the PPy/cellulose paper, RPC-0.1, RPC-0.5, and RPC-2.5 at different current densities of 2, 5 and 10 mA · cm^{-2}. (f) Capacitance retention of the PPy/cellulose paper, RPC-0.1, RPC-0.5, and RPC-2.5 as a function of cycle number at the current density of 10 mA · cm^{-2}

The charge transfer and electrolyte diffusion in the electrode/electrolyte interface were evaluated by EIS. Fig. 8a presents a comparative study of the Nyquist plots for the PPy/cellulose paper and RPC-2.5. An equivalent circuit (inset of Fig. 8a) was used to fit the impedance curves, which is comprised of a solution resistance (R_s) arising from electrolyte, a contact interface resistance (R_c) from electrode/electrolyte interface, a constant phase element (CPE), and a Warburg diffusion resistance (W). [41] It can be seen in Fig. 8a that the plots are characterized by a semicircle over the high frequency range, a 45° Warburg region, and a linear pure capacitor part in the low frequency region. In the low-frequency region, RPC-2.5 exhibits a more vertical line than the PPy/cellulose paper, illustrating better capacitive behavior. [2] The knee frequencies of the PPy/cellulose paper and RPC-2.5 are 7.03 and 28.27 Hz. A higher knee frequency generally suggests a better rate performance, consistent with the GCD analysis. In the medium-frequency region, the projected length of the 45° Warburg region on the real impedance axis (Z') characterizes the ion penetration process. [7,35] A shorter Warburg length of RPC-2.5 (from 28.27 to 1.07 Hz) indicates a shorter ion-diffusion path. In the high-frequency region (Fig. 8b and c), the intersection of the semicircle with the real axis represents R_s value. [41] The intercept of RPC-2.5 (3.0 Ω) is slightly smaller than that of the PPy/cellulose paper (5.3 Ω), which reveals that the RPC-2.5 favors access of ions. Moreover, the Rc of RPC-2.5 (2.0 Ω) is far smaller than that of the PPy/cellulose paper (15.1 Ω), indicating that counterions pass the electrode/electrolyte interface into RPC matrix at a higher rate. Therefore, the EIS results reveal that the RPC-2.5 electrode has better ionic and electronic transport ability.

Fig. 8d presents the measurement of highly conductive RPC-2.5 serving as a connect component in circuit. As shown, a blue light-emitting diode (LED) can be lighted by a 4.5 V battery by using RPC-2.5 as a connecting lead. In addition, the influences of curvature on the electrical conductivity of flexible RPC-2.5 were investigated by monitoring the current change at a constant voltage. As shown in Fig. 8e, there is no significant difference in the electric current at the different bending states (state Ⅰ to Ⅴ, insets of Fig. 8e), which suggests that the conductance of flexible RPC-2.5 is hardly influenced by bending stress. Especially, the excellent conductance stability and flexibility play important roles in the production of stretchable or rollable supercapacitors. The tensile mechanical tests were performed to determine the mechanical property of RPC-2.5 and the cellulose paper. Fig. 8f shows their stress-strain curves. The co-presence of PPy and RGO effectively increases the tensile stress and elastic modulus by approximately 60% and 29% (from 3.0 MPa to 4.8 MPa, and from 126.3 MPa to 162.7 MPa). Although the presence of PPy possibly weakened the inter- and intramolecular hydrogen bonding of cellulose, the PPy filling in the nanopores of cellulose paper and the physical interlocking between them may strengthen the tensile property.[66] Also, the existence of RGO with excellent mechanical property (1 TPa Young's modulus and 130 GPa ultimate strength for monolayer graphene membranes)[67] is beneficial to improve the elastic modulus and tensile strength of the hybrid paper. Furthermore, it is of great significance to quantitatively characterize the mechanical flexibility of the papers. Stiffness (bending resistance) is an ideal parameter to reflect the flexibility of materials, and the higher stiffness indicates the worse flexibility. We used a Taber stiffness tester to measure the stiffness of the hybrid paper based on the standard Tappi T 489, i.e., deflecting the free end of vertically clamped specimen 15° from its center line. The cellulose paper has a stiffness of about 124 mN. The addition of PPy and RGO induced the improvement of stiffness. The stiffness of RPC-2.5 increased to 154 mN. It is well-known that paper stiffness is a function of both thickness and internal bonding. The deposition of PPy and RGO leads to the increase of paper thickness, and the strong bond strength between the fillers and the cellulose fibers is also responsible for the higher stiffness for RPC-2.5. Although the presence of fillers resulted in slight deterioration in the flexibility, RPC-2.5 still maintained favorable flexibility. As shown in Fig. 8g, RPC-2.5 can be easily bended and twisted.

2.4 All-solid-state flexible laminated SSC based on RPC-2.5

Flexible solid-state energy storage devices play important roles in realizing stretchable, wearable and portable electronic gadgets. Considering the security, a liquid electrolyte is inferior to its solid-state counterpart since robust encapsulation is needed to prevent liquid electrolyte leakage. In the present work, an all-solid-state flexible laminated SSC based on RPC-2.5 (coded as RPC-SSC) was fabricated by assembling two pieces of RPC-2.5 electrodes and using a H_3PO_4/PVA gel as electrolyte. This RPC-SSC device exhibits excellent flexibility, and the CV tests confirm that the bending stress has little influences on the capacitive behavior (Fig. 9a). Moreover, the CV curves for RPC-SSC maintain the quasi-rectangular shape at various scan rates from 5 to 100 $mV \cdot s^{-1}$ (see Fig. S4). The GCD curves of RPC-SSC shown in Fig. 9b are symmetric, with an approximatively linear relation of discharge/charge voltage versus time. These characteristics suggest an ideal

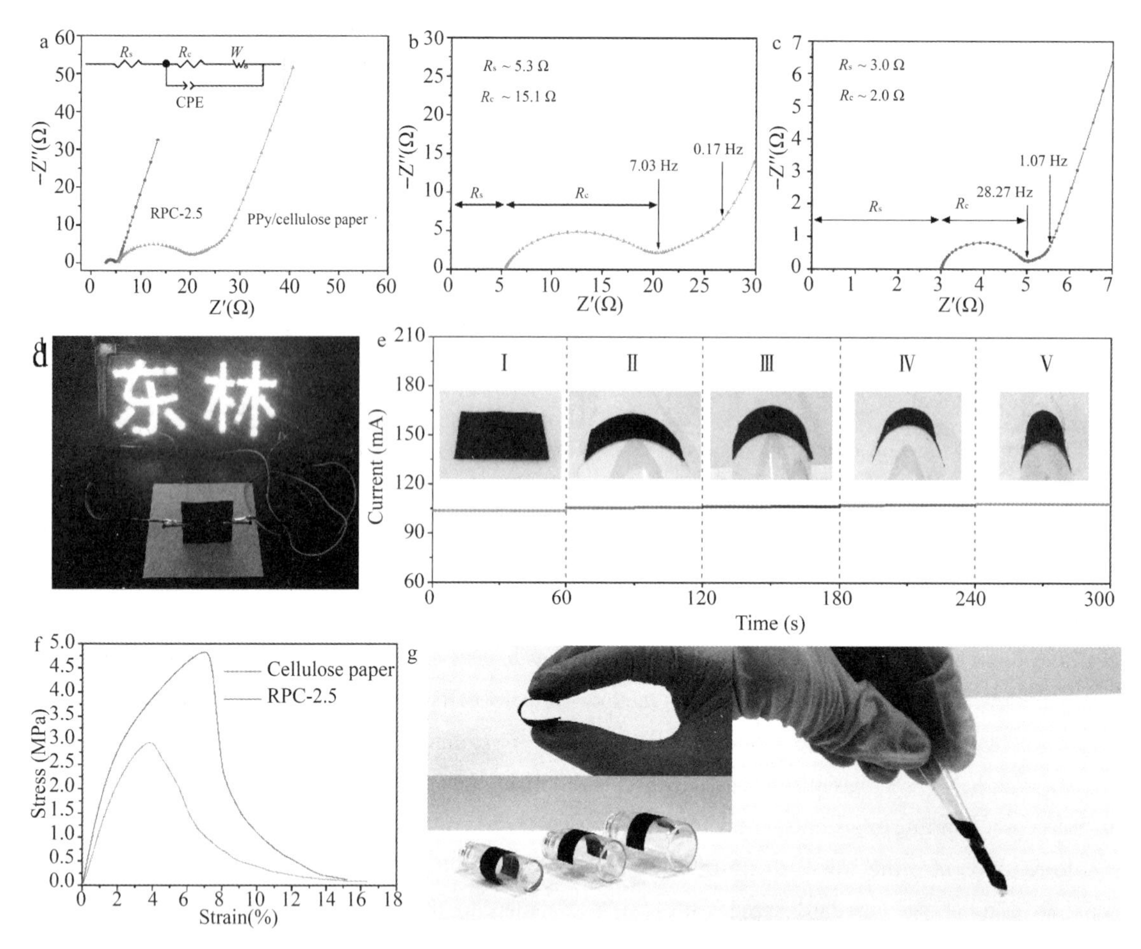

Fig. 8 (a) Nyquist plots of the PPy/cellulose paper and RPC-2. 5, and the inset shows the equivalent circuit used for EIS data fitting. (b-c) show their plots at high-frequency region with measurement of *R*s and *R*c. (d) RPC-2. 5 as an electrical connection to light a LED by a 4. 5 V battery. (e) Current-time curves of RPC-2. 5 bent with different curvatures under a constant voltage, and the upper insets labeled as Ⅰ, Ⅱ, Ⅲ, Ⅳ and Ⅴ represent the five different bending states. (f) Stress-strain curves of the cellulose paper and RPC-2. 5. (g) Photographs of RPC-2. 5 to reveal its flexibility

capacitance of the device. In addition, RPC-SSC exhibits the highest areal capacitance of 0. 51 F · cm^{-2} at a discharge current of 0. 1 mA · cm^{-2} (inset of Fig. 9b), and the corresponding volumetric capacitance was calculated to be about 8. 5 F · cm^{-3} (the thickness of the device is as low as 600 μm), which are larger than the values obtained from recent reports for other solid-state supercapacitors (see Table S3). [7,15,62,68-70]

The volumetric energy density (E) of RPC-SSC was calculated using the following equation:

$$E=\frac{1}{2}C(\Delta V)^2 \tag{6}$$

where E, C, and ΔV are the volumetric energy density, volumetric capacitance, and potential window, respectively.

The power density (P) of RPC-SSC was calculated from the ratio of energy density to discharge time as:

$$P = \frac{E}{t} \tag{7}$$

where E, t, and P are the volumetric energy density, discharge time, and volumetric power density, respectively. We compare our data with the data from other recently reported quasi/all-solid-state supercapacitors, as shown in Fig. 9c. The maximum volumetric energy density of RPC-SSC was calculated to be 1.18 mW · h · cm^{-3}, higher than values reported for other quasi/all-solid-state supercapacitors, such as CNTs-based SSC (0.008 mW · h · cm^{-3}, H_2SO_4/PVA),[71] single-walled CNT-based SSC (0.01 mW · h · cm^{-3}, H_3PO_4/PVA),[72] TiN-based SSC (0.05 mW · h · cm^{-3}, KOH/PVA),[70] carbon/MnO_2 fiber-based supercapacitors (0.22 mW · h · cm^{-3}),[73] TiO_2@MnO_2//TiO_2@C-based asymmetric supercapacitor (ASC, 0.30 mW · h · cm^{-3}, LiCl/PVA),[69] in-plane RGO-GO-RGO supercapacitor (0.43 mW · h · cm^{-3}),[74] VO_x//VN-based ASC (0.61 mW · h · cm^{-3}, LiCl/PVA),[68] and PPy/paper-based SSC (1.0 mW · h · cm^{-3}, H_3PO_4/PVA).[15] The superior energy density reveals that RPC-2.5 probably serves as a promising candidate electrode material for fabricating high-performance flexible supercapacitor devices.

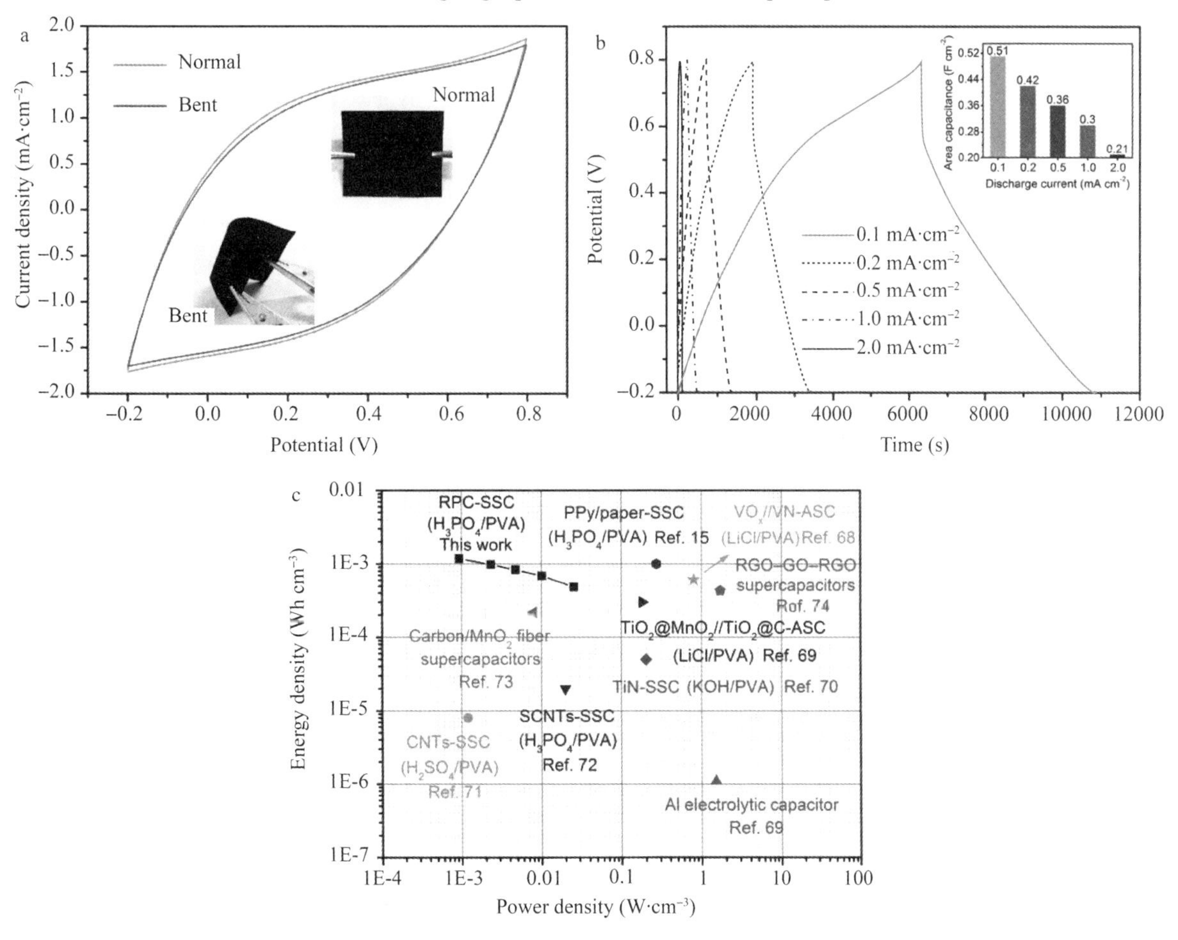

Fig. 9 (a) Comparison of CV curves at the scan rate of 5 mV · s^{-1} for flexible laminated RPC-SSC tested as normal and bent. (b) GCD curves of RPC-SSC at different current densities of 0.1, 0.2, 0.5, 1.0, and 2.0 mA · cm^{-2}, and the inset shows the corresponding areal capacitances. (c) Ragone plot of RPC-SSC, and the values reported for other supercapacitors were added for comparison

3 Conclusions

In summary, we have demonstrated an easily-operated scalable method to fabricate a type of hybrid RPC paper electrode. Through the optimization of preparation parameters, the as-prepared RPC-2.5 which was composed of cellulose fibers frame supporting interpenetrated RGO/PPy construction acquired an extremely low sheet resistance of 1.7 $\Omega \cdot sq^{-1}$. Also, this structure endowed RPC-2.5 with high areal capacitance of 1.20 $F \cdot cm^{-2}$ at a discharge current of 2 $mA \cdot cm^{-2}$, good cyclic stability with a capacitance retention of 89.5% after cycling for 5000 times, outstanding conductance stability, and favorable mechanical flexibility. Additionally, an all-solid-state flexible laminated SSC device fabricated with RPC-2.5 and H_3PO_4/PVA gel electrolyte achieved a high areal capacitance of 0.51 $F \cdot cm^{-2}$ at 0.1 $mA \cdot cm^{-2}$ and a superior energy density of 1.18 $mW \cdot h \cdot cm^{-3}$. These results indicate that the flexible, highly conductive, and free-standing RPC-2.5 holds great potential as a candidate electrode material for fabricating multifarious high-performance flexible and hand-held energy storage devices.

4 Experimental section

Fabrication of PPy/cellulose papers. All chemicals were of analytical grade and supplied by Tianjin Kemiou Chemical Reagent Co., Ltd. Cellulose papers (Fushun Minzheng Paper Co., Ltd., China) with the size of around 40 mm × 40 mm were soaked in four different concentrations of pyrrole monomer (0.05, 0.15, 0.45, and 1.35 M) for 2 h, respectively. Thereafter, the in-situ polymerization was carried out for 3 h by slowly adding an aqueous solution of iron (Ⅲ) chloride hexahydrate ($FeCl_3 \cdot 6H_2O$) as oxidant agent according to four different molar ratios of ferric iron/pyrrole (0.1, 0.3, 1, and 3). After the polymerization, the resulting sixteen groups of PPy/cellulose papers were dipped in 1 M HCl for 30 min, and then rinsed with distilled water. The clean hybrid papers were dried in a vacuum oven at 50 ℃ for 24 h.

Fabrication of GO. GO was fabricated by the modified Hummers' method.[75] Briefly, graphite powder (2 g) was firstly mixed with concentrated H_2SO_4 (80 mL) and sodium nitrate (4 g) with magnetic stirring in an ice bath. Secondly, potassium permanganate (8 g) was gently added to the suspension under vigorous agitation. The mixture was kept at 0 ℃ for 3 h, and then stirred for 2 h at 40 ℃. Thirdly, deionized water (200 mL) was poured into the aforementioned mixture, and the resulting solution was heated at 98 ℃ for 30 min. Finally, 30 ml H_2O_2 (30%) was added into the solution. After being rinsed with 5 wt.% HCl (200 mL) and a large amount of deionized water, the resultant graphite oxide was collected and immediately subjected to a facile ultrasonication treatment that exfoliates the graphite oxide into GO.

Fabrication of reduced graphene oxide/PPy/cellulose (RPC) papers. In order to synthesize RPC papers, the PPy/cellulose papers with the lowest sheet resistance were separately soaked for 2 h in different concentrations of GO dispersions (0.1, 0.5, and 2.5 $g \cdot L^{-1}$) which were prepared by one-hour ultrasonication (JY99-IID, Ningbo Scientz Biotechnology Co. Ltd., China) with an output power of 500 W in an ice/water bath. Subsequently, the GO component in these three kinds of GO/PPy/cellulose papers was reduced by adding an aqueous $NaBH_4$ solution (0.15 M), and the mixed solutions were stirred for 12 h. After repeated washing with deionized water for the removal of impurities, the RPC papers were dried at 50 ℃ for 24 h.

Fabrication of all-solid-state flexible laminated symmetric supercapacitor (SSC). To fabricate the flexible laminated SSC, the H_3PO_4/PVA gel electrolyte was firstly prepared by adding concentrated H_3PO_4(6 g) and PVA (6 g) in 60 mL deionized water with magnetic stirring at 85 ℃ until the solution became transparent. Thereafter, two pieces of RPC papers were dipped into the aforementioned gel electrolyte for 10 min and then pressed together, allowing the gel electrolyte layer on each electrode to combine into one thin separator. The as-prepared supercapacitor was finally transferred to a fume hood at room temperature for the evaporation of excess water.

Characterizations. The morphology was characterized by using a TEM (FEI, Tecnai G2 F20) and a SEM (Hitachi, S4800) equipped with an EDX detector for element analysis. AFM images were recorded using a Vecco Digital Instrument Nanoscope Ⅲ a Multimode AFM, with samples prepared by spin-coating GO suspension onto freshly exfoliated mica substrates. XPS analysis was carried out on a Thermo Escalab 250Xi system using a spectrometer with a dual Al Kα X-ray source. Deconvolution of the overlapping peaks was performed using a mixed Gaussian-Lorentzian fitting program (Origin 9.0, Originlab Corporation). FTIR spectra were recorded by a Nicolet Nexus 670 FTIR instrument in the range of 400–4000 cm^{-1} with a resolution of 4 cm^{-1}. Raman analysis was performed using a Raman spectrometer (Renishaw inVia, Germany) employing a helium/neon laser (633 nm) as the excitation source. Thermal stability was determined using a TG analyzer (TA, Q600) with a heating rate of 10 ℃ · min^{-1} from room temperature to 800 ℃ in a N_2 environment. Nitrogen adsorption-desorption tests were carried out at −196 ℃ by an accelerated surface area and porosimetry system (3H-2000PS2 unit, Beishide Instrument S&T Co. Ltd.). The stress-strain tests were performed on a tensile testing machine (EZ-LX, Shimadzu Corp., Kyoto, Japan) with a chuck distance of 15 mm and a tensile rate of 0.1 mm min^{-1} to obtain stress-strain curves. Samples were prepared by cutting strips from the papers of 4 cm × 3 cm (length × width). The stiffness was measured by a Taber stiffness tester (150-E, Taber Industries) according to the standard Tappi T 489, and the detailed test procedures were provided in the Supplementary Information. The sheet resistance (R) and electrical conductivity (σ) were measured at room temperature by a four-point probe resistivity/square resistance tester (KDB-1, Kunde Technology, China) using the following equation:

$$(\sigma,\ S \cdot m^{-1}) = \left(\frac{1}{\rho}\right) = \frac{1}{dR} \qquad (8)$$

where ρ is the resistivity, and d is the thickness of the sample.

Electrochemical measurements. The electrochemical measurements were conducted at room temperature in a three-electrode setup: RPC papers (or the PPy/cellulose papers) served as the working electrode, and an Ag/AgCl electrode and a Pt wire electrode served as reference and counter electrodes, respectively. The measurements were implemented in a 1 M NaCl aqueous electrolyte at room temperature. CV, GCD and EIS were measured using a CS350 electrochemical workstation (Wuhan CorrTest Instruments Co., Ltd.). CV tests were conducted over the potential window from −0.2 V to 0.8 V at different scan rates of 5, 10, 20, 50 and 100 mV · s^{-1}. GCD curves were tested in the potential range of −0.2~0.8 V at different current densities of 2, 5 and 10 mA · cm^{-2}. EIS measurements were carried out in the frequency range from 10^5 to 0.01 Hz with alternate current amplitude of 5 mV.

Acknowledgements

This study was supported by the National Natural Science Foundation of China (grant nos. 31270590 and

31470584) and the Fundamental Research Funds for the Central Universities (grant nos. 2572016AB22).

References

[1]Wang G, Zhang L, Zhang J. A review of electrode materials for electrochemicalsupercapacitors[J]. Chemical Society Reviews, 2012, 41(2): 797-828.

[2]Yan J, Fan Z, Wei T, et al. Fast and reversible surface redox reaction ofgraphene-MnO_2 composites as supercapacitor electrodes[J]. Carbon, 2010, 48(13): 3825-3833.

[3]Zhang LL, Zhao X S. Carbon-based materials as supercapacitor electrodes[J]. Chemical Society Reviews, 2009, 38(9): 2520-2531.

[4]Choi B G, Yang M H, Hong W H, et al. 3D macroporous graphene frameworks for supercapacitors with high energy and power densities[J]. ACS Nano, 2012, 6(5): 4020-4028.

[5]Lee J W, HallA S, Kim J D, et al. A facile and template-free hydrothermal synthesis of Mn_3O_4 nanorods on graphene sheets for supercapacitor electrodes with long cycle stability[J]. Chemistry of Materials, 2012, 24(6): 1158-1164.

[6]Wang J, Xu Y, Yan F, et al. Template-free prepared micro/nanostructured polypyrrole with ultrafast charging/discharging rate and long cycle life[J]. Journal of Power Sources, 2011, 196(4): 2373-2379.

[7]Weng Z, Su Y, Wang D W, et al. Graphene-cellulose paper flexible supercapacitors[J]. Advanced Energy Materials, 2011, 1(5): 917-922.

[8]Pech D, Brunet M, Durou H, et al. Ultrahigh-power micrometre-sized supercapacitors based on onion-like carbon[J]. Nature Nanotechnology, 2010, 5(9): 651-654.

[9]Liu C, Yu Z, Neff D, et al. Graphene-based supercapacitor with an ultrahigh energy density[J]. Nano Letters, 2010, 10(12): 4863-4868.

[10]Wu Z S, Parvez K, Feng X, et al. Graphene-based in-plane micro-supercapacitors with high power and energy densities[J]. Nature Communications, 2013, 4(1): 1-8.

[11]Masarapu C, Zeng H F, Hung K H, et al. Effect of temperature on the capacitance of carbon nanotube supercapacitors[J]. ACS Nano, 2009, 3(8): 2199-2206.

[12]Hung K, Masarapu C, Ko T, et al. Wide-temperature range operation supercapacitors from nanostructured activated carbon fabric[J]. Journal of Power Sources, 2009, 193(2): 944-949.

[13]He Y, Chen W, Li X, et al. Freestanding three-dimensionalgraphene/MnO_2 composite networks as ultralight and flexible supercapacitor electrodes[J]. ACS Nano, 2013, 7(1): 174-182.

[14]Chen L F, Huang Z H, Liang H W, et al. Flexible all-solid-state high-powersupercapacitor fabricated with nitrogen-doped carbon nanofiber electrode material derived from bacterial cellulose[J]. Energy & Environmental Science, 2013, 6(11): 3331-3338.

[15]Yuan L, Yao B, Hu B, et al. Polypyrrole-coated paper for flexible solid-state energy storage[J]. Energy & Environmental Science, 2013, 6(2): 470-476.

[16]Pushparaj V L, Shaijumon M M, Kumar A, et al. Flexible energy storage devices based on nanocomposite paper[J]. Proceedings of the National Academy of Sciences, 2007, 104(34): 13574-13577.

[17]Han J W, Kim B, Li J, et al. Carbon nanotube based humidity sensor on cellulosepaper[J]. The Journal of Physical Chemistry C, 2012, 116(41): 22094-22097.

[18]Fugetsu B, Sano E, Sunada M, et al. Electrical conductivity and electromagnetic interference shielding efficiency of carbon nanotube/cellulose composite paper[J]. Carbon, 2008, 46(9): 1256-1258.

[19]Tankhiwale R, Bajpai S K. Graft copolymerization onto cellulose-based filter paper and its further development as silver nanoparticles loaded antibacterial food-packaging material[J]. Colloids and Surfaces B: Biointerfaces, 2009, 69(2):

164-168.

[20]Fortunato E, Correia N, Barquinha P, et al. High-performance flexible hybrid field-effect transistors based on cellulose fiber paper[J]. IEEE Electron Device Letters, 2008, 29(9): 988-990.

[21]Wang Y, Du G, Liu H, et al. Nanostructured Sheets ofTiO Nanobelts for Gas Sensing and Antibacterial Applications [J]. Advanced Functional Materials, 2008, 18(7): 1131-1137.

[22]Gullapalli H, Vemuru V S M, Kumar A, et al. Flexible piezoelectric ZnO-paper nanocomposite strain sensor[J]. Small, 2010, 6(15): 1641-1646.

[23] Liu L, Niu Z, Zhang L, et al. Nanostructured graphene composite papers for highly flexible and foldable supercapacitors[J]. Advanced Materials, 2014, 26(28): 4855-4862.

[24]Perera S D, Patel B, Nijem N, et al. Vanadium oxide nanowire-carbon nanotube binder-free flexible electrodes for supercapacitors[J]. Advanced Energy Materials, 2011, 1(5): 936-945.

[25]Mastragostino M, Arbizzani C, Meneghello L, et al. Electronically conducting polymers and activated carbon: Electrode materials in supercapacitor technology[J]. Advanced Materials, 1996, 8(4): 331-334.

[26]Snook G A, Kao P, Best A S. Conducting-polymer-basedsupercapacitor devices and electrodes[J]. Journal of Power Sources, 2011, 196(1): 1-12.

[27] Xu J, Zhu L, Bai Z, et al. Conductive polypyrrole-bacterial cellulose nanocomposite membranes as flexible supercapacitor electrode[J]. Organic Electronics, 2013, 14(12): 3331-3338.

[28]Patrice S, Yury G. Materials for electrochemical capacitors[J]. Nature materials, 2008, 7(11): 845-54.

[29]Frackowiak E, Khomenko V, Jurewicz K, et al. Supercapacitors based on conducting polymers/nanotubes composites [J]. Journal of Power Sources, 2006, 153(2): 413-418.

[30] Gómez H, Ram M K, Alvi F, et al. Graphene-conducting polymer nanocomposite as novel electrode for supercapacitors[J]. Journal of Power Sources, 2011, 196(8): 4102-4108.

[31]He S, Hu C, Hou H, et al. Ultrathin MnO_2 nanosheets supported on cellulose based carbon papers for high-power supercapacitors[J]. Journal of Power Sources, 2014, 246: 754-761.

[32]Pandolfo A G, Hollenkamp A F. Carbon properties and their role in supercapacitors[J]. Journal of Power Sources, 2006, 157(1): 11-27.

[33]An K H, Kim W S, Park Y S, et al. Electrochemical properties of high-powersupercapacitors using single-walled carbon nanotube electrodes[J]. Advanced Functional Materials, 2001, 11(5): 387-392.

[34]Zhu Y, Murali S, Stoller M D, et al. Carbon-based supercapacitors produced by activation of graphene[J]. Science, 2011, 332(6037): 1537-1541.

[35]Wang Y, Shi Z, Huang Y, et al. Supercapacitor devices based on graphene materials[J]. The Journal of Physical Chemistry C., 2009, 113(30): 13103-13107.

[36]Lv W , Tang D M , He Y B , et al. Low-Temperature Exfoliated Graphenes: Vacuum-Promoted Exfoliation and Electrochemical Energy Storage[J]. Acs Nano, 2009, 3(11): 3730-3736.

[37] Vivekchand S R C, Rout C S, Subrahmanyam K S, et al. Graphene-based electrochemical supercapacitors[J]. Journal of Chemical Sciences, 2008, 120(1): 9-13.

[38]Allen M J, Tung V C, Kaner R B. Honeycomb carbon: a review of graphene[J]. Chemical Reviews, 2010, 110 (1): 132-145.

[39]Yan J, Wei T, Shao B, et al. Preparation of agraphene nanosheet/polyaniline composite with high specific capacitance [J]. Carbon, 2010, 48(2): 487-493.

[40]Zhang D, Zhang X, Chen Y, et al. Enhanced capacitance and rate capability ofgraphene/polypyrrole composite as electrode material for supercapacitors[J]. Journal of Power Sources, 2011, 196(14): 5990-5996.

[41] Biswas S, Drzal L T. Multilayered nanoarchitecture of graphene nanosheets and polypyrrole nanowires for high

performance supercapacitor electrodes[J]. Chemistry of Materials, 2010, 22(20): 5667-5671.

[42]Beneventi D, Alila S, Boufi S, et al. Polymerization of pyrrole on cellulose fibres using a $FeCl_3$ impregnation-pyrrole polymerization sequence[J]. Cellulose, 2006, 13(6): 725-734.

[43]Yuan L, Xiao X, Ding T, et al. Supporting Information (a) (b). 2012.

[44]Hu L, Zheng G, Yao J, et al. Transparent and conductive paper from nanocellulose fibers[J]. Energy & Environmental Science, 2013, 6(2): 513-518.

[45]Imai M, Akiyama K, Tanaka T, et al. Highly strong and conductive carbon nanotube/cellulose composite paper[J]. Composites Science and Technology, 2010, 70(10): 1564-1570.

[46]Hu L, Choi J W, Yang Y, et al. Highly conductive paper for energy-storage devices[J]. Proceedings of the National Academy of Sciences, 2009, 106(51): 21490-21494.

[47]Oh S Y, Yoo D I, Shin Y, et al. Crystalline structure analysis of cellulose treated with sodium hydroxide and carbon dioxide by means of X-ray diffraction and FTIR spectroscopy[J]. Carbohydrate Research, 2005, 340(15): 2376-2391.

[48]Omastova M, Trchová M, Kovářová J, et al. Synthesis and structural study of polypyrroles prepared in the presence of surfactants[J]. Synthetic metals, 2003, 138(3): 447-455.

[49]Wang H, Bian L, Zhou P, et al. Core-sheath structured bacterial cellulose/polypyrrole nanocomposites with excellent conductivity as supercapacitors[J]. Journal of Materials Chemistry A., 2013, 1(3): 578-584.

[50]Liu Y, Zhou J, Tang J, et al. Three-dimensional, chemically bondedpolypyrrole/bacterial cellulose/graphene composites for high-performance supercapacitors[J]. Chemistry of Materials, 2015, 27(20): 7034-7041.

[51]Shin H J, Kim KK, Benayad A, et al. Efficient reduction of graphite oxide by sodium borohydride and its effect on electrical conductance[J]. Advanced Functional Materials, 2009, 19(12): 1987-1992.

[52]Shen J, Hu Y, Shi M, et al. Fast and facile preparation of graphene oxide and reduced graphene oxide nanoplatelets [J]. Chemistry of Materials, 2009, 21(15): 3514-3520.

[53]Bose S, Kuila T, Uddin M E, et al. In-situ synthesis and characterization of electrically conductive polypyrrole/graphene nanocomposites[J]. Polymer, 2010, 51(25): 5921-5928.

[54]Liu Y C, Hwang B J. Identification of oxidizedpolypyrrole on Raman spectrum[J]. Synthetic Metals, 2000, 113(1-2): 203-207.

[55]Liu Y, Zhang Y, Ma G, et al. Ethylene glycol reducedgraphene oxide/polypyrrole composite for supercapacitor[J]. Electrochimica Acta, 2013, 88: 519-525.

[56]Chen C, Wen Y, Hu X, et al. Na^+ intercalationpseudocapacitance in graphene-coupled titanium oxide enabling ultra-fast sodium storage and long-term cycling[J]. Nature Communications, 2015, 6(1): 1-8.

[57]Wang J, Polleux J, Lim J, et al. Pseudocapacitive contributions to electrochemical energy storage in TiO_2(anatase) nanoparticles[J]. The Journal of Physical Chemistry C., 2007, 111(40): 14925-14931.

[58]Laskova B, Zukalova M, Zukal A, et al. Capacitive contribution to Li-storage in TiO_2(B) and TiO_2(anatase)[J]. Journal of Power Sources, 2014, 246: 103-109.

[59]Mini P A, Balakrishnan A, Nair S V, et al. Highly super capacitive electrodes made of graphene/poly (pyrrole)[J]. Chemical Communications, 2011, 47(20): 5753-5755.

[60]Zhou H, Han G, Xiao Y, et al. Facile preparation ofpolypyrrole/graphene oxide nanocomposites with large areal capacitance using electrochemical codeposition for supercapacitors[J]. Journal of Power Sources, 2014, 263: 259-267.

[61]Zhang J, Chen P, Oh B H L, et al. High capacitive performance of flexible and binder-freegraphene-polypyrrole composite membrane based on in situ reduction of graphene oxide and self-assembly[J]. Nanoscale, 2013, 5(20): 9860-9866.

[62]Ding X, Zhao Y, Hu C, et al. Spinning fabrication ofgraphene/polypyrrole composite fibers for all-solid-state, flexible fibriform supercapacitors[J]. Journal of Materials Chemistry A., 2014, 2(31): 12355-12360.

[63]Shi K, Zhitomirsky I. Electrophoretic nanotechnology of graphene-carbon nanotube and graphene-polypyrrole nanofiber composites for electrochemical supercapacitors[J]. Journal of colloid and interface science, 2013, 407: 474-481.

[64]Bose S, Kim N H, Kuila T, et al. Electrochemical performance of a graphene-polypyrrole nanocomposite as a supercapacitor electrode[J]. Nanotechnology, 2011, 22(29): 295202.

[65]Zhang H, Cao G, Wang W, et al. Influence of microstructure on the capacitive performance ofpolyaniline/carbon nanotube array composite electrodes[J]. Electrochimica Acta, 2009, 54(4): 1153-1159.

[66]Sasso C, Zeno E, Petit-Conil M, et al. Highly conducting polypyrrole/cellulose nanocomposite films with enhanced mechanical properties[J]. Macromolecular Materials and Engineering, 2010, 295(10): 934-941.

[67]Lee C, Wei X, Kysar J W, et al. Measurement of the elastic properties and intrinsic strength of monolayer graphene[J]. Science, 2008, 321(5887): 385-388.

[68]Lu X, Yu M, Zhai T, et al. High energy density asymmetric quasi-solid-state supercapacitor based on porous vanadium nitride nanowire anode[J]. Nano Letters, 2013, 13(6): 2628-2633.

[69]Lu X, Yu M, Wang G, et al. H-TiO_2@ MnO_2//H-TiO_2@ C core-shell nanowires for high performance and flexible asymmetric supercapacitors[J]. Advanced Materials, 2013, 25(2): 267-272.

[70]Lu X, Wang G, Zhai T, et al. Stabilized TiN nanowire arrays for high-performance and flexible supercapacitors[J]. Nano Letters, 2012, 12(10): 5376-5381.

[71]Kang Y J, Chung H, Han C H, et al. All-solid-state flexiblesupercapacitors based on papers coated with carbon nanotubes and ionic-liquid-based gel electrolytes[J]. Nanotechnology, 2012, 23(6): 065401.

[72]Kaempgen M, Chan C K, Ma J, et al. Printable thin film supercapacitors using single-walled carbon nanotubes[J]. Nano Letters, 2009, 9(5): 1872-1876.

[73]Xiao X, Li T, Yang P, et al. Fiber-based all-solid-state flexiblesupercapacitors for self-powered systems[J]. Acs Nano, 2012, 6(10): 9200-9206.

[74]Gao W, Singh N, Song L, et al. Direct laser writing of micro-supercapacitors on hydrated graphite oxide films[J]. Nature Nanotechnology, 2011, 6(8): 496-500.

[75]HummersJr W S, Offeman R E. Preparation of graphitic oxide[J]. Journal of the American Chemical Society, 1958, 80(6): 1339-1339.

中文题目：用于超级电容器电极的柔性、高导电性和独立的还原氧化石墨烯/聚吡咯/纤维素复合纸

作者：万才超，焦月，李坚

摘要：我们报道了一种由还原氧化石墨烯(RGO)、聚吡咯(PPy)和纤维素组成的新型复合纸电极的简易放大方法。通过优化制备参数(包括吡咯原位聚合和 $NaBH_4$ 化学还原氧化石墨烯的工艺)，制备的杂化纸具有极低的 1.7 $\Omega \cdot sq^{-1}$ 的薄层电阻。此外，杂化纸具有良好的机械柔韧性和优异的电导稳定性。当纸被评估为独立和无粘合剂的超级电容器电极时，独特的结构(即支持互穿 RGO-PPy 纳米结构的纤维素纤维框架)赋予其高电化学活性，例如在 2 $mA \cdot cm^{-2}$ 下 1.20 $F \cdot cm^{-2}$ 的高面积电容和良好的循环稳定性，在三电极配置中循环测试 5000 次后电容保持率为 89.5%。此外，通过组装两片复合纸电极并使用 H_3PO_4/PVA 凝胶作为电解质，制备了全固态叠层对称超级电容器。固态超级电容器在 0.1 $mA \cdot cm^{-2}$ 时具有 0.51 $F \cdot cm^{-2}$ 的高面积电容和 1.18 $mW \cdot h \cdot cm^{-3}$ 的高能量密度这些结果表明，复合纸是一种很有前途的电极材料，可用于开发其他高性能和手持式储能装置。更重要的是，这项工作为其他类型的复合型纸电极的制备提供了很好的参考。

关键词：纳米管/纤维素复合纸；高功率；超级电容器；储能装置

Multilayer Core-shell Structured Composite Paper Electrode Consisting of Copper, Cuprous Oxide and Graphite Assembled on Cellulose Fibers for Asymmetric Supercapacitors*

Caichao Wan, Yue Jiao, Jian Li

Abstract: An easily-operated and inexpensive strategy (pencil-drawing-electrodeposition-electro-oxidation) is proposed to synthesize a novel class of multilayer core-shell structured composite paper electrode, which consists of copper, cuprous oxide and graphite assembled on cellulose fibers. This interesting electrode structure plays a pivotal role in providing more active sites for electrochemical reactions, facilitating ion and electron transport and shorting their diffusion pathways. This electrode demonstrates excellent electrochemical properties with a high specific capacitance of 601 $F \cdot g^{-1}$ at 2 $A \cdot g^{-1}$ and retains 83% of this capacitance when operated at an ultrahigh current density of 100 $A \cdot g^{-1}$. In addition, a high energy density of 13.4 $W \cdot h \cdot kg^{-1}$ at the power density of 0.40 $kW \cdot kg^{-1}$ and a favorable cycling stability (95.3%, 8000 cycles) were achieved for this electrode. When this electrode was assembled into an asymmetric supercapacitor with carbon paper as negative electrode, the device displays remarkable electrochemical performances with a large areal capacitances (122 $mF \cdot cm^{-2}$ at 1 $mA \cdot cm^{-2}$), high areal energy density (10.8 $\mu W \cdot h \cdot cm^{-2}$ at 402.5 $\mu W \cdot cm^{-2}$) and outstanding cycling stability (91.5%, 5000 cycles). These results unveil the potential of this composite electrode as a high-performance electrode material for supercapacitors.

Keywords: Cuprous oxide; Cellulose; Graphite; Multilayer core-shell composites; Asymmetric supercapacitors

1 Introduction

Supercapacitors, one class of fascinating power sources, are designed to bridge the power/energy gap between traditional dielectric capacitors and batteries/fuel cells to generate fast-charging energy storage devices of intermediate specific energy[1-3]. Supercapacitors are known to have ultrahigh power density[4-6], long lifecycle[7,8], excellent reversibility[9] and widespread operation temperature[10]. Recently, the applications of supercapacitors in Airbus A380 planes and trolleybus (China Southern Locomotive and Rolling Stock Industry) have verified their safety and reliability. Supercapacitors can be classified into two categories on the basis of their energy storage mechanisms, i.e., electrochemical double-layer capacitors (EDLCs) and pseudocapacitors[11,12]. Pseudocapacitors, which store energy using fast and reversible Faradic reactions, theoretically have higher capacitance as compared to that of EDLCs, whose electron storage mechanism

* 本文摘自 Journal of Power Sources, 2017, 361: 122-132.

involves electrostatic adsorption and separation at the interface. Metal oxides/hydroxides, typically like RuO_2, MnO_2, NiO, $Ni(OH)_2$, Co_3O_4, $Co(OH)_2$, V_2O_5, SnO_2 and Fe_3O_4, are considered to be one of the best candidate materials for supercapacitors due to their high specific capacitance coupled with low resistance leading to a high specific power, which makes them very appealing in commercial applications[13-16]. In general, they provide higher energy density than conventional carbon materials and better electrochemical stability than conductive polymers. Among metal oxides/hydroxides, in contrast to well-known RuO_2 and MnO_2, cuprous (Ⅰ) oxide (Cu_2O) has not been paid extensive attention in the field of supercapacitors. However, Cu_2O holds unique charm due to its remarkable structural versatility[17,18], diversified synthetic strategy[19-21] and multitudinous applications (e. g., electro/photo-reduction of carbon dioxide[22] and solar-driven water splitting[23]). Most importantly, the theoretical capacitance of Cu_2O reaches up to 2247.6 $F \cdot g^{-1}$[24], predicting its promising applications in supercapacitors.

Papermaking technology is one of the four great inventions of ancient China, whose history dates back more than 2000 years to the Han dynasty of China (206 BC–220 AD). In recent years, cellulose paper-based electrodes have attracted much interest for the development of ultra-thin and ultra-lightweight supercapacitors[25,26]. Cellulose paper can act as a green, cost-effective, lightweight and flexible platform for supercapacitors by integration with various types of electrochemical active substances. Other critical characteristics that enable cellulose paper to serve as an ideal substrate for supercapacitors are the easily-functionalized compositions (e. g., cellulose with abundant hydroxyl groups) and the 3D porous fiber structures which can absorb electrolyte and act as electrolyte reservoirs to facilitate ion transport[27]. Nevertheless, cellulose paper is an insulator. To enhance the conductivity, cellulose paper is generally integrated with a variety of electrically conductive materials, e. g., carbon materials (typical like carbon nanotube[28] and graphene[27,29]), conductive polymers (typical like polypyrrole[30,31], polyaniline[32] and poly(3, 4-ethylenedioxythiophene)[33]), semiconductors[34] and metallic materials[35]. To date, numerous physicochemical techniques have been considered to deposit active substances onto multifarious substrates for different functions like powder metallurgy[36], plasma spraying[37], magnetron sputtering[38], electrodeposition[39] and vapor deposition[40]. Especially, electrodeposition is a traditional electrochemical method in preparing coatings with different functions. The first electrodeposition experiment dates back over 200 years when Luigi Valentino Brugnatelli electrodeposited gold upon silver[41]. Two typical approaches are involved in electrodeposition: the electrophoretic process which is based on the use of suspensions of particles, and the electrolytic process (also known as electroplating) which starts from solutions of metal salts[42]. Compared to other techniques, electrodeposition offers several advantages including (1) rapidity, (2) low cost, (3) good reproducibility, (4) high purity, (5) industrial applicability, (6) high deposition rates, (7) ability to produce coatings on widely differing substrates, (8) ability to form simple low-cost multilayers with sizes ranging from nanometer to micrometer, and (9) potential to overcome shape limitations or allows the production of free-standing parts with complex shapes[43]. Thus, electrodeposition provides a versatile convenient route to the realization of controlled coatings of composites.

Considering these above issues, in the present study, we designed a facile and cost-effective approach (i. e., pencil-drawing-electrodeposition-electro-oxidation) to prepare a free-standing and binder-free composite paper electrode with multilayer core-shell architecture, which was constructed with cellulose paper-supported graphite, Cu and Cu_2O nanostructure (coded as $Cu_2O/Cu/GCP$). The pencil-drawing was used to deposit graphite onto cellulose paper, and this easy method can endow cellulose paper with electrical conductivity for the subsequent electrodeposition of metallic Cu. The electrodeposited Cu was superficially electro-oxidized to highly electrochemically active Cu_2O by a constant-voltage polarization method, which has already been verified to effectively transform metals to their oxides or hydroxides[44,45]. This unique multilayer core-shell structure endows the electrode with excellent electrochemical properties. Additionally, an asymmetric supercapacitor device was designed and assembled by utilizing the Cu_2O/Cu/graphite complex as a positive electrode and carbon paper as a negative electrode, and the subjacent cellulose paper substrate was directly used as a separator. The electrochemical properties of this asymmetric supercapacitor device (labeled as $Cu_2O/Cu/GCP//CP$) were studied.

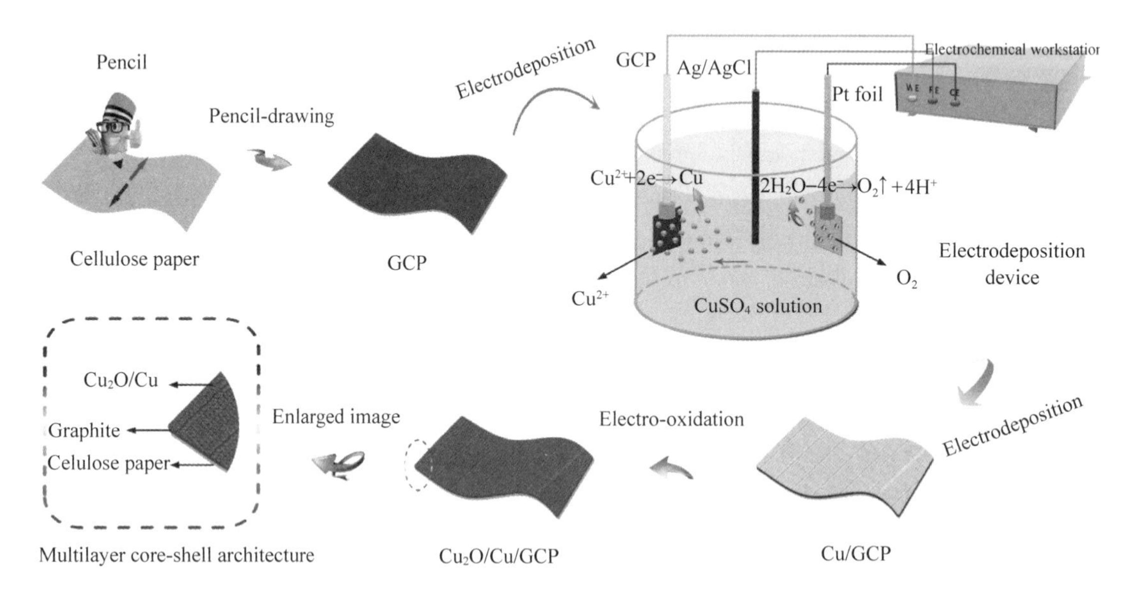

Fig. 1 Schematic illustration of the synthetic procedures for $Cu_2O/Cu/GCP$

2 Experimental section

2.1 Chemicals

High-purity chemicals including copper sulfate pentahydrate ($CuSO_4 \cdot 5H_2O$), sulfuric acid (H_2SO_4, 70%), hydrochloric acid (HCl, 37%) and potassium hydroxide (KOH) were purchased from Shanghai Aladdin Industrial Inc. (China) and used as received. Cellulose paper was supplied by Fuyang Special Paper Industry Co., Ltd. (Hangzhou, China). Polyethylene terephthalate (PET) film was provided by Dahua Plastic Industry Co., Ltd. (Hangzhou, China). Nickel foam was obtained from SXLZY Battery Material Co., Ltd. (Taiyuan, China). Carbon paper (TGP-H-060) was purchased from Toray International Inc. (Tokyo, Japan).

2.2 Preparation of GCP, Cu/GCP and $Cu_2O/Cu/GCP$

A schematic illustration for the synthesis strategy of the multilayered $Cu_2O/Cu/GCP$ electrode is shown in Fig. 1. A thin layer of graphite was firstly coated on the surface of cellulose paper through a fast, easy and eco-friendly pencil-drawing method, followed by the deposition of Cu layer on the graphite/cellulose paper (GCP) via cathodic electrodeposition, i. e., $Cu^{2+}+2e^{-}\rightarrow Cu$. Finally, the Cu_2O was in-situ grown on the surface of Cu layer though an electro-oxidation (constant-voltage polarization) process in a KOH solution. These procedures resulted in the formation of a quaternary composite with hierarchical multilayer core-shell architecture.

In detail, for the preparation of GCP, cellulose paper was firstly encapsulated with graphite by a simple fast pencil-drawing method, i. e., a common 8B pencil was used to draw on a piece of cellulose paper for five times, generating a uniform thin graphite coating. For the preparation of Cu/GCP, the electrodeposition experiment of Cu was carried out by an electrochemical workstation (CS350, Wuhan CorrTest Instruments Corp., Ltd.) with a three electrode configuration using a constant-current polarization method at −50 mA for 24 h. 0.6 M $CuSO_4$ aqueous solution was adjusted to pH = 2 with high-grade Cu sulfate and sulfuric acid. The GCP acted as the working electrode, a Pt foil (2 cm × 2 cm) was used as the counter electrode, and an Ag/AgCl electrode served as the reference electrode. After the electrodeposition, the Cu/GCP composite was washed with water and dried at 25 ℃. The synthesis of Cu_2O/Cu/GCP was also executed by an electrochemical workstation (CS350) with a three-electrode configuration, in which the as-prepared Cu/GCP acted as the working electrode, a Pt foil was used as the counter electrode, and an Ag/AgCl electrode served as the reference electrode. A constant-voltage polarization at 0.3 V was employed in a 6 M KOH solution to induce the in-situ growth of pyramid-shaped Cu_2O on the surface of Cu layer. The synthetic Cu_2O/Cu/GCP composite was rinsed with water and dried at 25 ℃.

2.3 Characterizations

The morphology and elemental compositions were analyzed by a scanning electron microscope (SEM, Hitachi S4800) equipped with an energy dispersive X-ray (EDX) detector, operating at an accelerating voltage of 20 kV. The samples were sputtered with gold before the observation. The thickness was measured from cross-section SEM observation.

X-ray photoelectron spectroscopy (XPS) analysis was carried out on a Thermo Escalab 250Xi system using a spectrometer with a dual Al Kα X-ray source. The base pressure of the instrument is about 1×10^{-9} Torr. The background contribution $B(E)$ (obtained by the Shirley method) caused by inelastic process was subtracted, and the curve-fitting was performed with a Gaussian-Lorentzian profile by a software (Origin 9.0, Originlab Corporation). The binding energies over the supported catalysts were calibrated using C 1s peak at 284.8 eV as reference. The instrument was calibrated using Au wire (Au $4f_{7/2}$ at 84.0 eV). XPS spectra were recorded at $\theta = 90°$ of X-ray sources.

X-ray diffraction (XRD) patterns were recorded on a Bruker D8 Advance TXS XRD instrument with Cu Kα (target) radiation ($\lambda = 1.5418$ Å) with the aid of Rietveld refinement using Code TOPAS3 at a scan rate (2θ) of $4°\ min^{-1}$ and a scan range from 10° to 80°. The accelerating voltage and the applied current were 40 kV and 30 mA, respectively.

Raman analysis was performed using a 50× objective on a Renishaw InVia Raman microscope equipped with an ultra-low noise charge-coupled device detector. The 633 nm line of a Helium-Neon laser was used for excitation, while a bandpass filter was used to reject the unwanted stray light. The maximum laser power available is 50 mW and the laser power at the sample can be adjusted by opening or closing an iris shutter. The spectrograph was equipped with an 1800-groove/mm grating, which offers a wavelength accuracy of 0.2 cm^{-1} with 633 nm excitation. All data acquisition and processing were executed using the WiRE 2.0 software package.

2.4 Electrochemical measurements

The electrochemical properties of GCP and Cu_2O/Cu/GCP were measured at room temperature using a CS350 electrochemical workstation in a three-electrode setup: the composite papers served as the working electrode, and an Ag/AgCl electrode and a Pt foil electrode served as the reference and counter electrodes, respectively. The electrolyte was 6 M KOH solution. Cyclic voltammetry (CV) and galvanostatic charge-discharge (GCD) curves were measured over the potential window from 0 to 0.4 V. Electrochemical impedance spectroscopy (EIS) measurements were carried out in the frequency range from 10^5 to 0.01 Hz with alternate current amplitude of 5 mV.

2.5 Calculations

The specific capacitance (C_m, $F \cdot g^{-1}$) was calculated from GCD curves by the following formula:

$$C_m = \frac{I \times \Delta t}{\Delta V \times m} \tag{1}$$

where I is the charge-discharge current (A), Δt is the discharge time (s), ΔV is the potential window (V) and m is the mass (g) of active material (Cu_2O). The energy density (E) was calculated from the equation:

$$E = \frac{1}{2} C_m (\Delta V)^2 \tag{2}$$

The power density (P) was calculated from the ratio of energy density to discharge time as:

$$P = \frac{E}{\Delta t} \tag{3}$$

The mass of Cu_2O was calculated according to the following steps. Firstly, the mass of the Cu_2O/Cu/GCP electrode was weighed (coded as m_1) by electronic balance. Secondly, this electrode was dipped into a 0.1 M HCl solution for about 10 min until no obvious color change occurred. During this step, the Cu_2O layer was removed by the HCl. The resultant was washed with a large amount of distilled water and completely dried at room temperature, and then weighed again (coded as m_2). Finally, the mass of Cu_2O attached on the electrode was calculated to be (m_1-m_2).

2.6 Fabrication of asymmetric supercapacitor based on Cu_2O/Cu/GCP electrode

For the assemble process of the asymmetric supercapacitor, the Cu_2O/Cu/graphite complex on the surface of Cu_2O/Cu/GCP electrode was used as the positive electrode and the carbon paper acted as the negative electrode, and the subjacent cellulose paper substrate was directly used as the separator. The two pieces of thin nickel foams were used as the current collectors. The electrodes, current collector and separator were sandwiched between two PET films. The assembled asymmetric supercapacitor device was immersed in a 6 MKOH aqueous solution for all of the electrochemical tests in a two-electrode system.

3 Results and Discussion

3.1 Analysis on morphology, crystal structure and chemical compositions

Fig. 2 compares the optic images, surface and cross-section morphologies, and elemental compositions of the cellulose paper, GCP, Cu/GCP and Cu_2O/Cu/GCP. The common cellulose paper has porous structure which consists of cross-linked 3D fibers network (Fig. 2a, e and i). In virtue of the pencil-drawing method, the paper turned from white to black (Fig. 2b) since a thin layer of graphitic sheets were uniformly covered on the paper surface (Fig. 2f and j) according to the SEM observation. Compared to the cellulose paper (Fig. 2m), the EDX spectrum of GCP shows that the presence of graphite coating leads to the significant enhancement of C/O ratio, and some other new elements (e. g., Mg, Al, Si, Ca and Fe) are derived from the clay component in the pencil (Fig. 2n).

After the electrodeposition of Cu on the GCP, the color of the hybrid paper turned purple red (Fig. 2c). The SEM images of Cu/GCP present that the Cu layer has a thickness of about 70 μm (Fig. 2g and k). In addition, its EDX spectrum confirms the existence of plentiful Cu composition (ca. 93.62 wt. %) on the surface of Cu/GCP (Fig. 2o). Regarding the electrodeposition process, this may proceed via two steps. Firstly, Cu^{2+} ions were weakly adsorbed on the cathode (i. e., GCP) surface with high surface coverage degree[46]. Secondly, Cu^{2+} ions were strongly adsorbed on the surface by Coulomb forces under the effect of applied electric field and fast got electron and then became pure Cu nucleus on the cathodic surface. The electropositive Cu^{2+} ions in the electrolyte continued migrating to the surface of cathode and were reduced to metallic Cu. According to the research by Grujicic and Pesic[47], the relatively high pH and Cu^{2+} concentration used in this paper tended to produce nuclei with a coarser texture and lower population density, which are consistent with the SEM observation (Fig. 2g).

The electro-oxidation treatment of Cu layer in the alkaline solution resulted in the formation of Cu_2O, and accordingly the color of the paper turned black (Fig. 2d). The operative chemical reactions are expected to be $Cu \rightarrow Cu^{+} + e^{-}$ and $2Cu^{+} + 2OH^{-} \rightarrow Cu_2O + H_2O$[48]. The SEM observation exhibits that the surface of the resultant (Cu_2O/Cu/GCP) became more homogeneous after the electro-oxidation (Fig. 2h and l). The higher-magnification SEM image reveals that the surface was constructed with abundant micro/nano-scale pyramid-shaped structure. As cuprous oxide (111) crystal plane has the highest density of oxygen atoms and the growth rate is small at low potential, morphology of Cu_2O depends on (111) crystal plane, leading crystal surface morphology to pyramid with four facets[49]. Such micro/nano-structures may effectively improve the surface/interface area of the composite and are expected to facilitate electrolyte diffusion among interspaces of the Cu/Cu_2O multiscale architecture. The EDX spectrum of Cu_2O/Cu/GCP presents the significant increase of oxygen content from 3.89 wt% to 7.09 wt% (Fig. 2p), verifying the oxidation of surficial Cu layer. In conclusion, these above morphological and elemental analyses indicate the generation of multilayer core-shell architecture for Cu_2O/Cu/GCP. The objectives of designing such multilayered paper electrode are listed as follows: (1) the cellulose paper serving as substrate provides lightweight features, and the fibrillar meshworks can absorb electrolyte and act as electrolyte reservoirs to facilitate ion transport; (2) the pencil-drawing method can easily deposit the thin layer of graphite on the cellulose paper and the conductivity of the paper was improved; (3) the Cu layer can offer electron "superhighways" for charge storage and delivery due to its outstanding electrical conductivity[50], and can serve as the starting material for the synthesis of Cu_2O; (4) the Cu_2O has an extremely high theoretical capacitance, and its pyramid-shaped micro/nano-scale morphology is beneficial to greatly enhance the electroactive surface area[51].

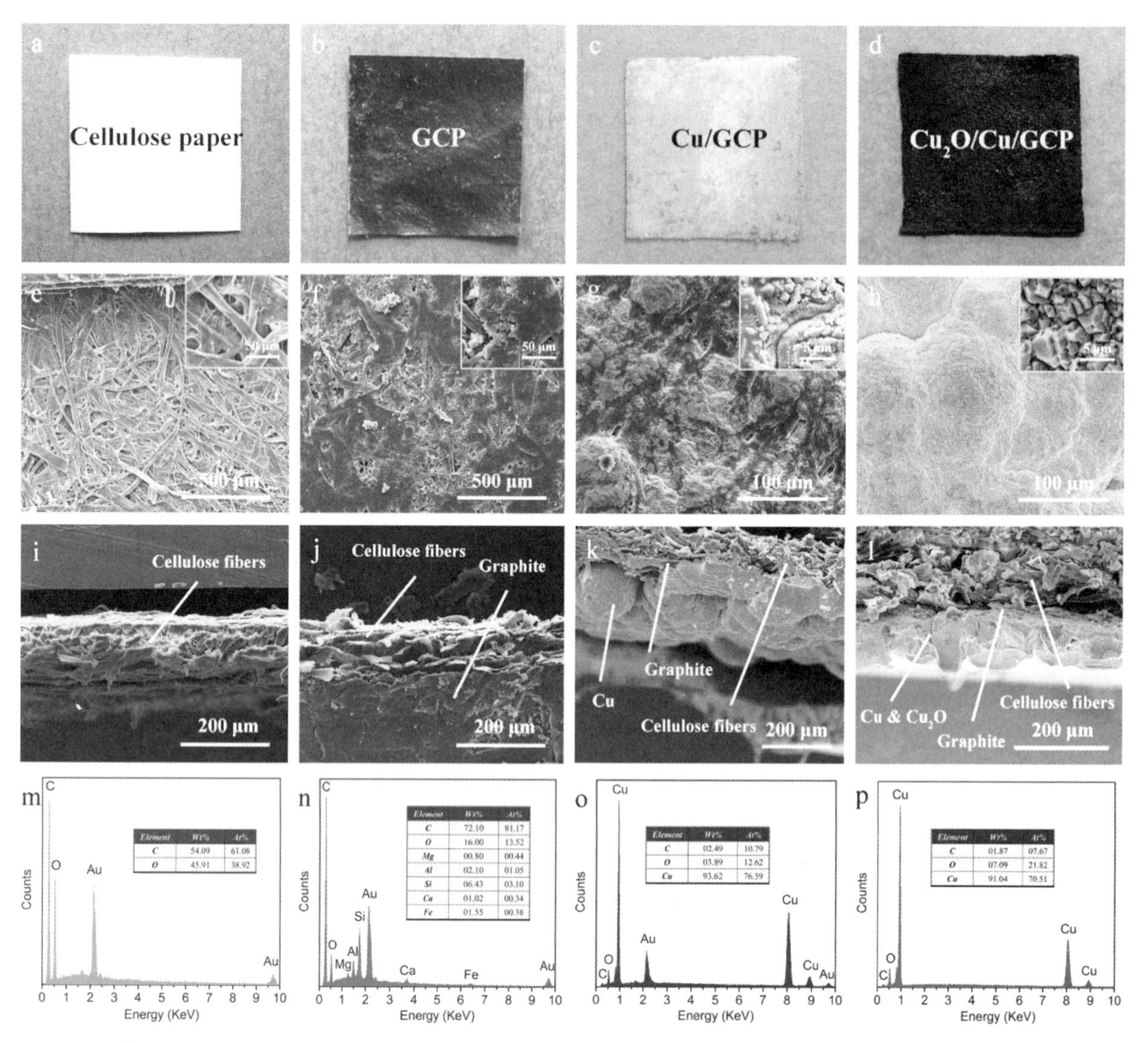

Fig. 2 Optic images of (a) cellulose paper, (b) GCP, (c) Cu/GCP and (d) Cu_2O/Cu/GCP. SEM images of (e) cellulose paper, (f) GCP, (g) Cu/GCP and (h) Cu_2O/Cu/GCP, and the insets in (e-h) show the corresponding enlarged images. Cross-section SEM images of (i) cellulose paper, (j) GCP, (k) Cu/GCP and (l) Cu_2O/Cu/GCP. EDX spectra of (m) cellulose paper, (n) GCP, (o) Cu/GCP and (p) Cu_2O/Cu/GCP

Raman spectroscopy is non-intrusive, non-destructive and particularly sensitive to the distinctive carbon signal of carbonaceous matters, and it is an ideal technique for the characterization of crystalline, nanocrystalline and amorphous carbons[52]. The Raman spectrum of GCP is displayed in Fig. 3a. The peak at 1345^{-1} is due to edge plane vibrations regarded as a disorder in graphite structure, which is referred as the

"D" peak[53]. The wavelength-dependent D peak is a characteristic of polycrystalline graphitic materials, which cannot be seen in single crystal graphite[54]. Another prominent Raman feature of graphite is a wavelength-independent "G" peak at 1580 cm^{-1}. Because visible excitation always resonates with the carbon π states, which makes sp^2 signals 50 – 250 times stronger than sp^3 vibrations, vibrations of sp^2 sites are dominated in the Raman spectra of carbonaceous materials[55]. Hence, Raman spectroscopy gives information about graphite component of pencil lead.

The crystal structure was studied by XRD analysis. As shown in Fig. 3b, the cellulose paper exhibits three peaks at 14.9°, 16.6 and 22.6°, which correspond to the ($\bar{1}10$), (110) and (200) planes of cellulose Ⅰ crystal structure[56], respectively. For GCP, the cellulose characteristic peaks still maintained, and two new peaks appeared at 26.6° and 54.7°, which can be indexed to (002) and (004) plane reflections of the graphite (JCPDS card no. 41-1487) contained in the pencil lead. For Cu/GCP, there are several new peaks appearing at 43.4°, 50.5° and 74.2°. All of these peaks can be successfully indexed to the (111), (200) and (220) plane reflections of metallic Cu (JCPDS card no. 04-0836). Moreover, the XRD signals of cellulose were completely shielded because of the existence of the superficial Cu cladding layer. After the electro-oxidation, some peaks related to Cu_2O (JCPDS card no. 05-0667) were generated at 36.4°, 42.3° and 61.4° for Cu_2O/Cu/GCP, indicative of the partial transformation from Cu to Cu_2O. To further investigate the chemical valences of Cu, XPS analysis has been executed for Cu_2O/Cu/GCP. The survey spectrum (Fig. 3c) suggests the presence of C, O and Cu and no obvious impurities were detected on the surface of Cu_2O/Cu/GCP, consistent with the EDX analysis. Fig. 3d displays the Cu 2p core-level spectrum, where the two strong peaks located at around 932.3 and 952.4 eV are related to the binding energy of Cu $2p_{3/2}$ and Cu $2p_{1/2}$, respectively, suggesting the existence of Cu^+ and Cu^0 species[57]. It is worth to mention that the Cu $2p_{3/2}$ binding energies of Cu_2O and Cu^0 are close to each other; the reported values are 932.5 and 932.6 eV[58], respectively. Therefore, Cu $2p_{3/2}$ XPS cannot differentiate between Cu^+ and Cu^0. In addition, two other small peaks with binding energy of 935.1 and 955.1 eV are corresponding to Cu $2p_{3/2}$ and Cu $2p_{1/2}$ in CuO or $Cu(OH)_2$[59], respectively, since a small amount of Cu^+ in Cu_2O was oxidized to Cu^{2+} in the air. The O 1s core-level spectrum can be fitted into three peaks (Fig. 3e). The binding energy at 530.4 eV represents the oxygen in the Cu_2O lattice further verifying the presence of Cu^+[58], and the peaks at 531.6 and 532.6 eV stand for the adsorbed OH groups [or OH groups in $Cu(OH)_2$] and molecular water on the Cu_2O/Cu/GCP surface[60], respectively.

3.2 Evaluation on electrochemical properties

Electrochemical properties of the Cu_2O/Cu/GCP electrode were investigated in a three-electrode system with 6 M KOH as the electrolyte. The GCD curves of Cu_2O/Cu/GCP electrodes, which were obtained from different time of electro-oxidation treatment of Cu/GCP in the KOH solution, are displayed in Fig. 4a. As shown, the discharge time significant increases as the electro-oxidation time prolonging, revealing the improvement in the specific capacitances of Cu_2O/Cu/GCP because the discharge time is proportional to the specific capacitance. According to Fig. 4b, with the increase of electro-oxidation time from 100 to 25600 s, the specific capacitance increases by 2.6 times from 160 to 583 $F \cdot g^{-1}$ due to the more transformation of Cu to Cu_2O. As the polarization time extended from 25600 to 51200 and eventually to 102400 s, the capacitance

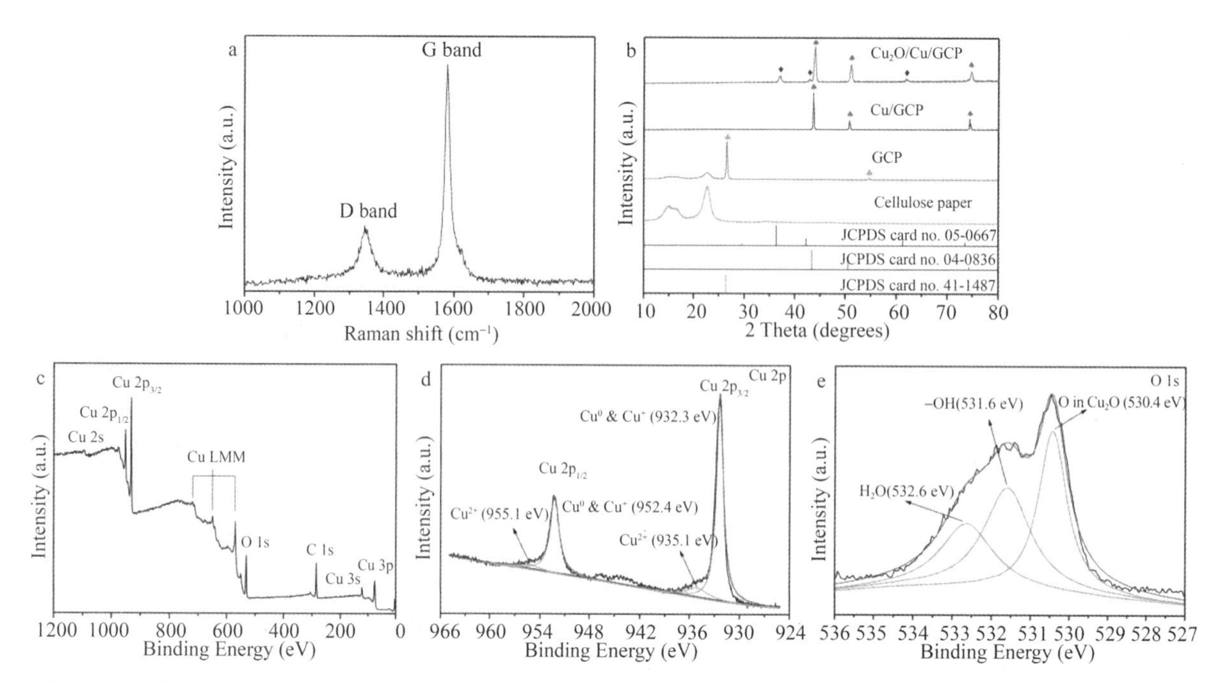

Fig. 3 (a) Raman spectrum of GCP. (b) XRD patterns of cellulose paper, GCP, Cu/GCP and Cu_2O/Cu/GCP. The bottom lines are the standard JCPDS cards no. 41-1487 for graphite, no. 04-0836 for Cu and no. 05-0667 for Cu_2O. (c) XPS survey spectrum and high resolution scan of (d) Cu 2p and (e) O 1s of Cu_2O/Cu/GCP, respectively

increments are not remarkable (approximatively 1.7% from 25600 to 51200s and 2.1% from 25600 to 102400 s), which may be ascribed to the almost complete conversion of the superficial layer of Cu to Cu_2O. The XRD analysis of Cu_2O/Cu/GCP verifies this speculation. As shown in Fig. S1, as the polarization time prolonging, the signals of Cu_2O gradually heighten. For Cu_2O/Cu/GCP-51200 and Cu_2O/Cu/GCP-102400 (Cu_2O/Cu/GCP-*x* represents the Cu_2O/Cu/GCP obtained from *x*-seconds electro-oxidation treatment of Cu/GCP), their intensities of Cu_2O signals are almost identical.

Fig. 4c shows the GCD curves of Cu_2O/Cu/GCP-102400 at various current densities from 2 to 100 A · g^{-1} with a voltage window of 0–0.4 V. As shown, both charge and discharge curves maintain good symmetry, indicative of the outstanding capacitance characteristics and electrochemical reversibility. The nonlinear feature of the charge-discharge behavior reflects the pseudocapacitive nature of the electrode[61]. Fig. 4d presents the values of the mass specific capacitances at various current densities based on the above discharge curves. A maximum capacitance value of 601 F · g^{-1} at 2 A · g^{-1} was achieved, which is higher than that of many Cu-based electrodes previously reported (see Table S1) such as 3D highly ordered nanoporous CuO (431 F · g^{-1})[62], Cu_2O-Cu$(OH)_2$-graphene (425 F · g^{-1})[63], rose rock-shaped Cu_2O-graphene (416 F · g^{-1})[64], CuO nanobelts grown on Cu substrate (392 F · g^{-1})[65], TiO_2/Cu_2O-Mn_3O_4(312 F · g^{-1})[66], and Cu_2O/CuO/reduced graphene oxide (173 F · g^{-1})[67]. The gradual decrease of capacitance with an increase of current density may be attributed to the incremental IR drop and insufficient active material involved in the redox reactions at higher current densities[68]. The specific capacitance of Cu_2O/Cu/GCP-102400 at 50 A · g^{-1} and 100 A · g^{-1} is as much as 88% and 83% of that at 2 A · g^{-1}. The high capacitance retention rates clearly demonstrate the excellent rate capability of the electrode under high current densities, which is important for supercapacitors in practical applications. In addition, we measured the electrochemical property of the GCP electrode (see Fig. S2). The GCP has a capacitance of only 1.5 F · g^{-1} at 2 A · g^{-1} (calculated based on its GCD curve), indicating that the capacitance of Cu_2O/Cu/GCP primarily originates from the pseudocapacitance of Cu_2O.

The energy density and power density, which were calculated from the GCD curves, are key factors for practical supercapacitor applications. As shown in Fig. 4e, the Cu_2O/Cu/GCP-102400 electrode exhibits a high energy density, from 13.4 W · h · kg^{-1} at 0.40 kW · kg^{-1} to 11.1 W · h · kg^{-1} at 20 kW · kg^{-1}, better than that of some reported Cu-based electrodes, e. g., carbon nanofibers/CuO nanorod arrays (8.8 W · h · kg^{-1}, 0.20 kW · kg^{-1})[69], CuO nanoflowers/carbon fiber fabric (10.1 W · h · kg^{-1} at 0.31 kW · kg^{-1})[70], $NiCo_2O_4$/Cu-based nanowires (12.6 W · h · kg^{-1} at 0.34 kW · kg^{-1})[71]. Fig. 4f displays the CV curves of Cu_2O/Cu/GCP-102400 at various scan rates from 5 to 100 mV · s^{-1}. The shape of the CV curves is not obviously affected by the increasing scan rate, further suggesting favorable rate capability and reversibility of the electrode. In addition, none of them show a quasi-rectangular shape, revealing that the charge storage capacity of the electrodes originates from the Faradic redox reactions. In an alkaline electrolyte, the redox behaviors of the Cu_2O/Cu/GCP electrode are mainly attributed to the oxidation of Cu^+ to Cu^{2+} and the reduction of Cu^{2+} to Cu^+. The possible reactions of Cu_2O in alkaline electrolytes are presented as follows[72,73]:

$$\frac{1}{2}Cu_2O+OH^- \Leftrightarrow CuO+\frac{1}{2}H_2O+e^- \quad (4)$$

$$\frac{1}{2}Cu_2O+\frac{1}{2}H_2O+OH^- \Leftrightarrow Cu(OH)_2+e^- \quad (5)$$

$$CuOH+OH^- \Leftrightarrow CuO+H_2O+e^- \quad (6)$$

$$CuOH+OH^- \Leftrightarrow Cu(OH)_2+e^- \quad (7)$$

The superior electrochemical behaviors of the Cu_2O/Cu/GCP electrode may be due to the following reasons: (1) the cellulose paper with high porous structure can absorb electrolyte and act as a electrolyte reservoir to facilitate ion transport; (2) the highly conductive Cu layer can provide electron "superhighways" for charge storage and delivery; (3) the Cu_2O nanostructure directly growing on the Cu layer increases the electrochemical active sites for the reversible redox reactions; (4) the Cu_2O has an extremely high theoretical capacitance.

EIS spectrum of Cu_2O/Cu/GCP was tested in the range of 0.01 – 10^5 Hz to further understand its capacitance behaviors. The Nyquist plot of the electrode is shown in Fig. 4g. The solution resistance (the real-axis intercept) is as low as 0.63 Ω. Moreover, no obvious semicircle, which reflects the charge-transfer resistance in the electrode, was observed at the high frequency region, revealing that the charge transfer rate is very fast and the electrode resistance is small[74]. Long cycle life is also an important requirement for supercapacitors. The cycle life test over 8000 cycles for Cu_2O/Cu/GCP-102400 was conducted by repeating the GCD tests at the current density of 50 A · g^{-1}. As shown in Fig. 4h, the electrode has an excellent long-term electrochemical stability with 95.3% capacitance retention after 8000 cycles. Moreover, the coulombic efficiency, which was calculated from the ratio between discharge time and charged time, remains a quite high value of 97.3%–99.6% during the whole cycling.

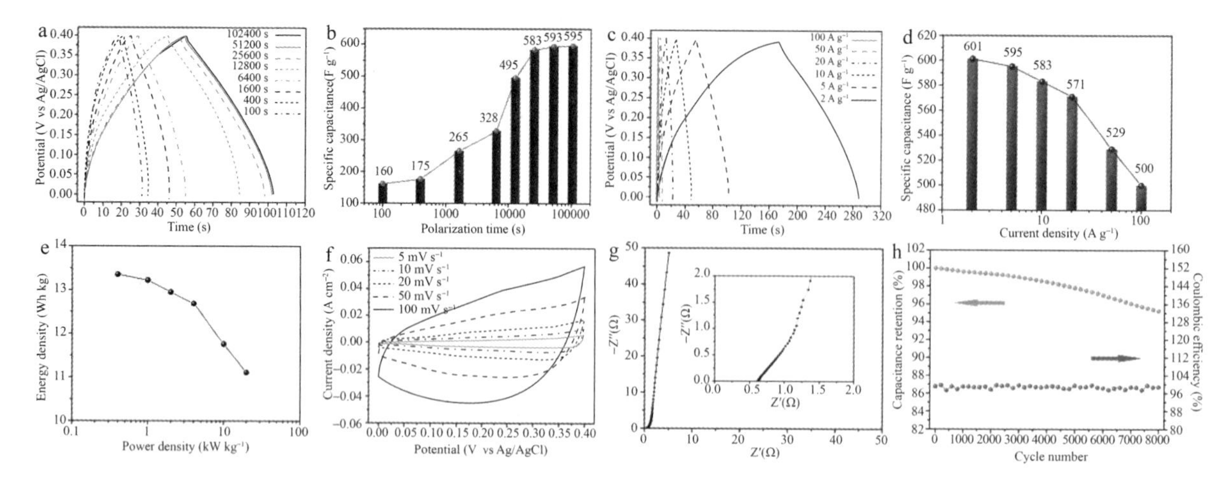

Fig. 4 (a) GCD curves and (b) specific capacitances of Cu_2O/Cu/GCP electrodes at the current density of 5 A · g^{-1} obtained from different time of electro-oxidation treatment of Cu/GCP. (c) GCD curves and (d) specific capacitances of Cu_2O/Cu/GCP-102400 at various current densities. (e) Energy density with respect to power density of Cu_2O/Cu/GCP-102400. (f) CV curves of Cu_2O/Cu/GCP-102400 at different scan rates. (g) Nyquist diagram of Cu_2O/Cu/GCP-102400 and the inset shows the enlarged image at the high frequency region. (h) Cycling performance and coulombic efficiency of Cu_2O/Cu/GCP-102400 measured at the current density of 50 A · g^{-1}

3. 3 Assembling asymmetric supercapacitor

For the sake of further studying the advantages of this interesting electrode design for practical applications, we assembled an asymmetric supercapacitor device with a "sandwiched" structure, which consists of two electrodes (i. e., the Cu_2O/Cu/graphite complex on the Cu_2O/Cu/GCP-102400 surface as the positive electrode and carbon paper as the negative electrode), a separator (the cellulose paper substrate), two current collectors (nickel foam) and two PET films, as shown in Fig. 5a. Due to the presence of carbon paper, the total potential window can be expressed as the sum of the potential range for the carbon paper and the Cu_2O/Cu/GCP electrode. Therefore, the potential window can be expanded to 0. 8 V in 6 M KOH aqueous electrolyte (see Fig. S3). The CV profiles (Fig. 5b) were very similar and quite different from the ideal rectangular shape for double-layer capacitance, revealing that the capacitance of the Cu_2O/Cu/GCP//CP asymmetric supercapacitor mainly originates from Faradic redox reactions[75]. The anodic peaks are primarily attributed to the oxidation of Cu_2O and/or CuOH to both CuO and/or $Cu(OH)_2$, while the broad cathodic peaks can be mainly due to the reduction of CuO and/or $Cu(OH)_2$ to Cu_2O and/or CuOH[76]. In addition, the broad peak feature may be because of the overlap of the redox signals.

The GCD curves of the Cu_2O/Cu/GCP//CP asymmetric supercapacitor at different current densities in the potential range of 0 – 0. 8 V are displayed in Fig. 5c. Their nonlinear shape further indicates that the effective storage of electrical energy is derived from the Faradic reactions[77], which is in agreement with the result of the CV curves. The Cu_2O/Cu/GCP//CP supercapacitor gives the highest areal capacitance value of 122 mF · cm^{-2} at the current density of 1 mA · cm^{-2}, and still kept 36 mF · cm^{-2} at a high current density of 16 mA · cm^{-2} (Fig. 5d). And the supercapacitor has a maximum areal energy density of 10. 8 μW · h · cm^{-2}

at the power density of 402.5 μW · cm^{-2} and still remained 3.2 μW · h · cm^{-2} at 6400 μW · cm^{-2} (Fig. 5e). The values are comparable to or higher than values recently reported for symmetric/asymmetric supercapacitors (coded as SSC and ASC, see Table S2), such as graphene/cellulose paper-based SSC (5.8 μW · h · cm^{-2} at 20 μW · cm^{-2})[27], RGO/MnO_2//RGO-based ASC (35.1 μW · h · cm^{-2} at 37.5 μW · cm^{-2})[78], graphene gel-based SSC (2.7 μW · h · cm^{-2} at 369.8 mW · cm^{-2})[79], RGO + CNT @ carboxymethyl cellulose-based SSC (3.8 μW · h · cm^{-2} at 20 μW · cm^{-2})[80], RGO/PPy-based SSC (61.4 μW · h · cm^{-2} at 10 mW · cm^{-2})[81], gallium nitride/graphite-based SSC (7.4 μW · h · cm^{-2} at 250 μW · cm^{-2})[82] and TiN@C nanotubes-based SSC (2.3 μW · h · cm^{-2} at 809 μW · cm^{-2})[83]. These favorable electrochemical properties are capable to meet requirement of certain applications, e.g., storing electrical power generated from thermoelectric and ambient airflow[84]. The Nyquist diagram of the Cu_2O/Cu/GCP//CP supercapacitor is presented in Fig. 5f, where the low intercept at Z real axis (ca. 1.1 Ω) at the high frequency region is related to the low solution resistance. Similarly, the assembled device displays no obvious semicircle at the high frequency region, which reflects the low electrode resistance. At the low frequency region, the experimental slope is close to 90° which suggests a predominantly capacitive behavior of the device[85]. Electrochemical stability of the Cu_2O/Cu/GCP//CP supercapacitor was investigated under continuous charge-discharge tests at a high current density of 8 mA · cm^{-2} for 5000 cycles. The supercapacitor is able to retain 91.5% of its original capacitance at the end of this cycling process (Fig. 5g), indicative of good cycling stability. Moreover, the stability property of our device is better than that of some recently reported Cu-based supercapacitors, such as Cu_2O-$Cu(OH)_2$-graphene/stainless steel-based SSC[63] (87%, 2000 cycles) and Cu_2O/$CuMoO_4$//activated carbon-based ASC[86] (86.6%, 3000 cycles). Additionally, in the whole cycling process, a high coulombic efficiency of 95.3%-99.8% was acquired, reflecting the high utilization rate of electric energy.

4 Conclusions

In summary, several unique merits of the Cu_2O/Cu/GCP electrode designed in this work make it a promising candidate as a free-standing and binder-free electrode material for energy storage devices. These characteristics include: (1) the easy and cost-effective synthesis approach (including pencil-drawing, electrodeposition and electro-oxidation) helps to improve the application value of the electrode; (2) the unique multilayer core-shell structure is able to obtain ideal electrochemical activity, e.g., the cellulose paper acting as electrolyte reservoir to facilitate ion transport, the conductive Cu layer providing electron superhighways for charge storage and delivery, and the Cu_2O with high electrochemical activity offering high pseudocapacitance; (3) the electrode displays a high capacitance value of 601 F · g^{-1} at 2 A · g^{-1}, a large energy density of 13.4 Wh · kg^{-1} at 0.40 kW · kg^{-1}, a high capacitance retention of 95.3% after 8000 cycles and ideal rate capability; (4) the assembled asymmetric supercapacitor device also possesses excellent electrochemical performances, such as large areal capacitances of 122 mF · cm^{-2} at 1 mA · cm^{-2}, high areal energy density of 10.8 μW · h · cm^{-2} at 402.5 μW · cm^{-2} and outstanding cycling stability (91.5%, 5000 cycles). Consequently, this work makes some contributions to the advancement in the energy storage devices and provides a good reference for the fabrication of other types of hybrid paper electrodes.

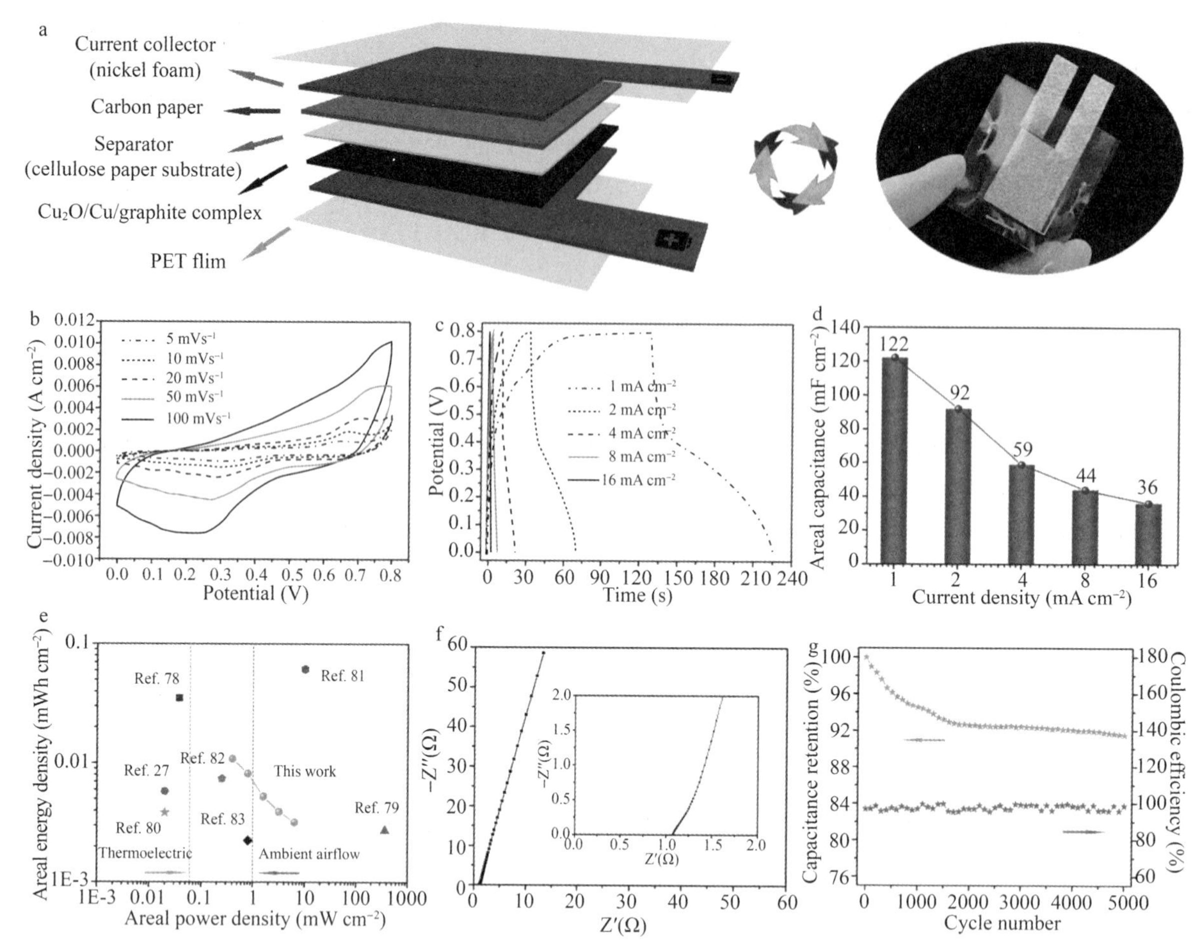

Fig. 5 (a) Schematic diagram (left) and digital photograph (right) of the assembled Cu_2O/Cu/GCP//CP asymmetric supercapacitor. (b) CV curves of the Cu_2O/Cu/GCP//CP supercapacitor at various scan rates. (c) GCD curves and (d) areal capacitances of the Cu_2O/Cu/GCP//CP supercapacitor at various current densities. (e) Areal energy density with respect to the areal power density of the Cu_2O/Cu/GCP//CP supercapacitor. (f) Nyquist diagram of the Cu_2O/Cu/GCP//CP supercapacitor and the inset shows the enlarged image at the high frequency region. (g) Cycling performance and coulombic efficiency of the Cu_2O/Cu/GCP//CP supercapacitor measured at the current density of 8 mA · cm^{-2}

Acknowledgements

This study was supported by the National Natural Science Foundation of China (grant no. 31270590 and 31470584) and the Fundamental Research Funds for the Central Universities (grant no. 2572016AB22).

References

[1] Wang G, Zhang L, Zhang J. A review of electrode materials for electrochemicalsupercapacitors[J]. Chemical Society Reviews, 2012, 41(2): 797-828.

[2] Chen H, Cong T N, Yang W, et al. Progress in electrical energy storage system: A criticalreview[J]. Progress in Natural Science, 2009, 19(3): 291-312.

[3]Winter M, Brodd R J. What are batteries, fuel cells, and supercapacitors? [J]. Chemical Reviews, 2004, 104(10): 4245-4270.

[4]Pech D, Brunet M, Durou H, et al. Ultrahigh-power micrometre-sized supercapacitors based on onion-like carbon [J]. Nature Nanotechnology, 2010, 5(9): 651-654.

[5]Chen L F, Huang Z H, Liang H W, et al. Bacterial-cellulose-derived carbonnanofiber@ MnO_2 and nitrogen-doped carbon nanofiber electrode materials: an asymmetric supercapacitor with high energy and power density[J]. Advanced Materials, 2013, 25(34): 4746-4752.

[6]Fan Z, Yan J, Wei T, et al. Asymmetricsupercapacitors based on graphene/MnO_2 and activated carbon nanofiber electrodes with high power and energy density[J]. Advanced Functional Materials, 2011, 21(12): 2366-2375.

[7]Li G, Li W, Xu K, et al. Sponge-like $NiCo_2O_4$/MnO_2 ultrathin nanoflakes for supercapacitor with high-rate performance and ultra-long cycle life[J]. Journal of Materials Chemistry A., 2014, 2(21): 7738-7741.

[8]Shen B, Zhang X, Guo R, et al. Carbon encapsulated RuO_2 nano-dots anchoring on graphene as an electrode for asymmetric supercapacitors with ultralong cycle life in an ionic liquid electrolyte[J]. Journal of Materials Chemistry A., 2016, 4 (21): 8180-8189.

[9]Jagadale A D, Guan G, Li X, et al. Ultrathin nanoflakes of cobalt-manganese layered double hydroxide with high reversibility for asymmetric supercapacitor[J]. Journal of Power Sources, 2016, 306: 526-534.

[10]Hung K, Masarapu C, Ko T, et al. Wide-temperature range operation supercapacitors from nanostructured activated carbon fabric[J]. Journal of Power Sources, 2009, 193(2): 944-949.

[11]Simon P, Gogotsi Y, Dunn B. Where do batteries end and supercapacitors begin? [J]. Science, 2014, 343(6176): 1210-1211.

[12]Patrice S, Yury G. Materials for electrochemical capacitors[J]. Nature Materials, 2008, 7(11): 845-854.

[13]Faraji S, Ani F N. Microwave-assisted synthesis of metal oxide/hydroxide composite electrodes for high power supercapacitors-a review[J]. Journal of Power Sources, 2014, 263: 338-360.

[14]Zhang Y, Feng H, Wu X, et al. Progress of electrochemical capacitor electrode materials: A review [J]. International Journal of Hydrogen Energy, 2009, 34(11): 4889-4899.

[15]Jiang J, Li Y, Liu J, et al. Recent advances in metal oxide-based electrode architecture design for electrochemical energystorage[J]. Advanced Materials, 2012, 24(38): 5166-5180.

[16]Wu Z S, Zhou G, Yin L C, et al. Graphene/metal oxide composite electrode materials for energy storage[J]. Nano Energy, 2012, 1(1): 107-131.

[17]Kuo C H, Huang M H. Morphologically controlled synthesis of Cu_2O nanocrystals and their properties[J]. Nano Today, 2010, 5(2): 106-116.

[18]Huang W C, Lyu L M, Yang Y C, et al. Synthesis of Cu_2O nanocrystals from cubic to rhombic dodecahedral structures and their comparative photocatalytic activity [J]. Journal of the American Chemical Society, 2012, 134 (2): 1261-1267.

[19]Yu H, Yu J, Liu S, et al. Template-free hydrothermal synthesis of CuO/Cu_2O composite hollow microspheres[J]. Chemistry of Materials, 2007, 19(17): 4327-4334.

[20]Zhang L, Jing D, Guo L, et al. In situ photochemical synthesis of Zn-doped Cu_2O hollow microcubes for high efficient photocatalytic H2 production[J]. ACS Sustainable Chemistry & Engineering, 2014, 2(6): 1446-1452.

[21]Wang Q, Jia Y, Wang M, et al. Synthesis of Cu_2O nanotubes with efficient photocatalytic activity by electrochemical corrosion method[J]. The Journal of Physical Chemistry C., 2015, 119(38): 22066-22071.

[22]Janáky C, Hursán D, Endrodi B, et al. Electro-and photoreduction of carbon dioxide: The twain shall meet at copper oxide/copper interfaces[J]. ACS Energy Letters, 2016, 1(2): 332-338.

[23]Luo J, Steier L, Son M K, et al. Cu_2O nanowire photocathodes for efficient and durable solar water splitting[J]. Nano

Letters, 2016, 16(3): 1848-1857.

[24]Xu P, Liu J, Liu T, et al. Preparation of binder-free CuO/Cu_2O/Cu composites: A novel electrode material for supercapacitor applications[J]. RSC Advances, 2016, 6(34): 28270-28278.

[25]Dong L, Xu C, Li Y, et al. Flexible electrodes and supercapacitors for wearable energy storage: a review by category [J]. Journal of Materials Chemistry A. 2016, 4(13): 4659-4685.

[26]Zhang Y Z, Wang Y, Cheng T, et al. Flexiblesupercapacitors based on paper substrates: a new paradigm for low-cost energy storage[J]. Chemical Society Reviews, 2015, 44(15): 5181-5199.

[27]Weng Z, Su Y, Wang D W, et al. Graphene-cellulose paper flexible supercapacitors[J]. Advanced Energy Materials, 2011, 1(5): 917-922.

[28]Kang Y J, Chun S J, Lee SS, et al. All-solid-state flexible supercapacitors fabricated with bacterial nanocellulose papers, carbon nanotubes, and triblock-copolymer ion gels[J]. ACS Nano, 2012, 6(7): 6400-6406.

[29]Ma L, Liu R, Liu L, et al. Facile synthesis of $Ni(OH)_2$/graphene/bacterial cellulose paper for large areal mass, mechanically tough and flexible supercapacitor electrodes[J]. Journal of Power Sources, 2016, 335(Dec. 15): 76-83.

[30]Yuan L, Yao B, Hu B, et al. Polypyrrole-coated paper for flexible solid-state energy storage[J]. Energy & Environmental Science, 2013, 6(2): 470-476.

[31]Wan C, Jiao Y, Li J. Flexible, highly conductive, and free-standing reducedgraphene oxide/polypyrrole/cellulose hybrid papers for supercapacitor electrodes[J]. Journal of Materials Chemistry A., 2017, 5(8): 3819-3831.

[32]Wang X, Gao K, Shao Z, et al. Layer-by-Layer assembled hybrid multilayer thin film electrodes based on transparent cellulose nanofibers paper for flexible supercapacitors applications[J]. Journal of Power Sources, 2014, 249: 148-155.

[33]Wang Z, Tammela P, Huo J, et al. Solution-processed poly (3, 4-ethylenedioxythiophene) nanocomposite paper electrodes for high-capacitance flexible supercapacitors[J]. Journal of Materials Chemistry A., 2016, 4(5): 1714-1722.

[34]Manekkathodi A, Lu M Y, Wang C W, et al. Direct growth of aligned zinc oxide nanorods on paper substrates for low-cost flexible electronics[J]. Advanced Materials, 2010, 22(36): 4059-4063.

[35]Koga H, Nogi M, Komoda N, et al. Uniformly connected conductive networks on cellulose nanofiber paper for transparent paper electronics[J]. NPG Asia Materials, 2014, 6(3): e93-e93.

[36]Schubert T. Ciupiński Ł, Zieliński W, Michalski A, Weißgärber T, Kieback B. Interfacial characterization of Cu/diamond composites prepared by powder metallurgy for heat sink applications[J]. Scripta Materialia, 2008, 58(4): 263-266.

[37]Hui R, Wang Z, Kesler O, et al. Thermal plasma spraying for SOFCs: Applications, potential advantages, and challenges[J]. Journal of Power Sources, 2007, 170(2): 308-323.

[38]Legnani C, Vilani C, Calil V L, et al. Bacterial cellulose membrane as flexible substrate for organic light emitting devices[J]. Thin Solid Films, 2008, 517(3): 1016-1020.

[39]Lu L, Shen Y, Chen X, et al. Ultrahigh strength and high electrical conductivity in copper[J]. Science, 2004, 304 (5669): 422-426.

[40]Reina A, Jia X, Ho J, et al. Large area, few-layer graphene films on arbitrary substrates by chemical vapor deposition[J]. Nano Letters, 2009, 9(1): 30-35.

[41]Bozzini B, Kourousias G, Gianoncelli A. In situ observation of dynamic electrodeposition processes by soft x-ray fluorescence microspectroscopy and keyhole coherent diffractive imaging[J]. Journal of Physics D: Applied Physics, 2017, 50 (12): 124001.

[42]Besra L, Liu M. A review on fundamentals and applications of electrophoretic deposition (EPD)[J]. Progress in Materials Science, 2007, 52(1): 1-61.

[43]Gurrappa I, Binder L. Electrodeposition of nanostructured coatings and their characterization-a review[J]. Science & Technology of Advanced Materials, 2008, 9(4): 043001.

[44]Yi Q, Zhang J, Huang W, et al. Electrocatalytic oxidation of cyclohexanol on a nickel oxyhydroxide modified nickel electrode in alkaline solutions[J]. Catalysis Communications, 2007, 8(7): 1017-1022.

[45] Dong C, Hua Z, Kou T, et al. Three-Dimensional Cu Foam-Supported Single Crystalline Mesoporous Cu_2O Nanothorn Arrays for Ultra-Highly Sensitive and Efficient Nonenzymatic Detection of Glucose[J]. ACS Applied Materials & Interfaces, 2015, 7(36): 20215-20223.

[46]Mohamed A M A, Ahmad Y H. Electrodeposition of nanostructured Nickel-Ceramic composite coatings: A review [J]. International Journal of Electrochemical Science, 2014, 9(4): 1942-1963.

[47]Grujicic D, Pesic B. Electrodeposition of copper: The nucleation mechanisms[J]. Electrochimica Acta, 2002, 47 (18): 2901-2912.

[48] Li Y, Chang S, Liu X, et al. NanostructuredCuO directly grown on copper foam and their supercapacitance performance[J]. Electrochimica Acta, 2012, 85: 393-398.

[49]Jiang X, Zhang M , Shi S , et al. Microstructure and optical properties of nanocrystalline Cu_2O thin films prepared by electrodeposition[J]. Nanoscale Research Letters, 2014, 9(1): 219.

[50]Feng J X , Li Q , Lu X F , et al. Flexible symmetrical planar supercapacitors based on multi-layered MnO_2/Ni/graphite/paper electrodes with high-efficient electrochemical energy storage[J]. Journal of Materials Chemistry A. , 2014, 2 (9): 2985-2992.

[51]Dong C, Wang Y, Xu J, et al. 3D binder-free Cu_2O@ Cu nanoneedle arrays for high-performance asymmetric supercapacitors[J]. Journal of Materials Chemistry A. , 2014, 2(43): 18229-18235.

[52]Ferrari A C, Robertson J. Interpretation of Raman spectra of disordered and amorphouscarbon[J]. Physical Review B, 2000, 61(20): 14095-14107.

[53]Yao B, Yuan L, Xiao X, et al. based solid-statesupercapacitors with pencil-drawing graphite/polyaniline networks hybrid electrodes[J]. Nano Energy, 2013, 2(6): 1071-1078.

[54]Navratil R , Kotzianova A , Halouzka V , et al. Polymer lead pencil graphite as electrode material: Voltammetric, XPS and Raman study[J]. Journal of Electroanalytical Chemistry, 2016: 152-160.

[55] Ferrari A C. Raman spectroscopy ofgraphene and graphite: Disorder, electron-phonon coupling, doping and nonadiabatic effects[J]. Solid State Communications, 2007, 143(1-2): 47-57.

[56]Wan C, Li J. Facile synthesis of well-dispersedsuperparamagnetic γ-Fe_2O_3 nanoparticles encapsulated in three-dimensional architectures of cellulose aerogels and their applications for Cr (VI) removal from contaminated water[J]. ACS Sustainable Chemistry & Engineering, 2015, 3(9): 2142-2152.

[57] Peng L, Hensen E. Highly efficient and robust Au/$MgCuCr_2O_4$ catalyst for gas-phase oxidation of ethanol to acetaldehyde. [J]. Journal of the American Chemical Society, 2013, 135(38): 14032 14035.

[58]Teo J J, Chang Y, Zeng H C. Fabrications of hollow nanocubes of Cu_2O and Cu via reductive self-assembly of CuO nanocrystals[J]. Langmuir, 2006, 22(17): 7369-7377.

[59]Jiang P, Prendergast D, Borondics F, et al. Experimental and theoretical investigation of the electronic structure of Cu_2O and CuO thin films on Cu (110) using x-ray photoelectron and absorption spectroscopy[J]. The Journal of Chemical Physics, 2013, 138(2): 024704.

[60][1]Rong X, Hua C Z . Self-generation of tiered surfactant superstructures for one-pot synthesis of Co_3O_4 nanocubes and their close- and non-close-packed organizations. [J]. Langmuir the Acs Journal of Surfaces & Colloids, 2004, 20(22): 9780-9790.

[61]Deng M J, Song C Z, Ho P J, et al. Three-dimensionally ordered macroporous Cu_2O/Ni inverse opal electrodes for electrochemical supercapacitors[J]. Physical Chemistry Chemical Physics, 2013, 15(20): 7479-7483.

[62]Moosavifard S E, El-Kady M F, Rahmanifar M S, et al. Designing 3D highly ordered nanoporous CuO electrodes for high-performance asymmetric supercapacitors[J]. ACS Applied Materials & Interfaces, 2015, 7(8): 4851-4860.

[63] Ghasemi S, Jafari M, Ahmadi F. $Cu_2O-Cu(OH)_2$ -graphene nanohybrid as new capacitive material for high performance supercapacitor[J]. Electrochimica Acta, 2016: 225-235.

[64] Zhang W, Yin Z, Chun A, et al. Rose rock - shapednano Cu_2O anchored graphene for high - performance supercapacitors via solvothermal route[J]. Journal of Power Sources, 2016, 318: 66-75.

[65]Zhang X, Yu L, Wang L, et al. High electrochemical performance based on ultrathin porousCuO nanobelts grown on Cu substrate as integrated electrode[J]. Physical Chemistry Chemical Physics, 2013, 15(2): 521-525.

[66]Sun D, Aref A A, Wang B, et al. Design and synthesis of hierarchical TiO_2 micro-nano spheres/$Cu_2O-Mn_3O_4$ nanoflakes composite for high performance electrochemical electrodes[J]. Journal of Alloys and Compounds, 2016, 688: 561-570.

[67]Wang K, Dong X, Zhao C, et al. Facile synthesis of Cu_2O/CuO/RGO nanocomposite and its superior cyclability in supercapacitor[J]. Electrochimica Acta, 2015, 152: 433-442.

[68]Yu X, Lu B, Xu Z. Super long-life supercapacitors based on the construction of nanohoneycomb-like strongly coupled $CoMoO_4$-3D graphene hybrid electrodes[J]. Advanced Materials, 2014, 26(7): 1044-1051.

[69]Moosavifard S E, Shamsi J, Fani S, et al. Facile synthesis of hierarchical CuO nanorod arrays on carbon nanofibers for high-performance supercapacitors[J]. Ceramics International, 2014, 40(10): 15973-15979.

[70]Xu W , Dai S , Liu G , et al. CuO Nanoflowers growing on Carbon Fiber Fabric for Flexible High-Performance Supercapacitors[J]. Electrochimica Acta, 2016, 203: 1-8.

[71]Kuang M, Zhang Y X, Li T T, et al. Tunable synthesis of hierarchical $NiCo_2O_4$ nanosheets-decorated Cu/CuO_x nanowires architectures for asymmetric electrochemical capacitors[J]. Journal of Power Sources, 2015, 283: 270-278.

[72] Li Y, Chang S, Liu X, et al. NanostructuredCuO directly grown on copper foam and their supercapacitance performance[J]. Electrochimica Acta, 2012, 85: 393-398.

[73]Zhao B , Liu P , Zhuang H , et al. Hierarchical self-assembly of microscale leaf-like CuO on graphene sheets for high-performance electrochemical capacitors[J]. Journal of Materials Chemistry A. , 2012, 1(2): 367-373.

[74]Iqbal N, Wang X, Babar A A, et al. Highly flexible $NiCo_2O_4$/CNTs doped carbon nanofibers for CO_2 adsorption and supercapacitor electrodes[J]. Journal of Colloid and Interface Science, 2016, 476: 87-93.

[75]Rath P C, Patra J, Saikia D, et al. Highly enhanced electrochemical performance of ultrafine CuO nanoparticles confined in ordered mesoporous carbons as anode materials for sodium-ion batteries[J]. Journal of Materials Chemistry A, 2016, 4(37): 14222-14233.

[76]Lu Y, Qiu K, Zhang D, et al. Cost-effective CuO nanotube electrodes for energy storage and non-enzymatic glucose detection[J]. RSC Advances, 2014, 4(87): 46814-46822.

[77] Qian T, Zhou J, Xu N, et al. On - chip supercapacitors with ultrahigh volumetric performance based on electrochemically co-deposited CuO/polypyrrole nanosheet arrays[J]. Nanotechnology, 2015, 26(42): 425402.

[78]Sumboja A, Foo C Y, Xu W, et al. Large Areal Mass, Flexible and Free-Standing Reduced Graphene Oxide/ Manganese Dioxide Paper for Asymmetric Supercapacitor Device[J]. Advanced Materials, 2013, 25(20): 2809-2815.

[79]Maiti U N, Lim J, Lee K E, et al. Three-dimensional shape engineered, interfacial gelation of reduced graphene oxide for high rate, large capacity supercapacitors[J]. Advanced Materials, 2014, 26(4): 615-619.

[80]LiangK , Huang T , Zheng B , et al. Coaxial wet-spun yarn supercapacitors for high-energy density and safe wearable electronics[J]. Nature Communications, 2014, 5(1): 3754.

[81]Yang C, Zhang L, Hu N, et al. Reducedgraphene oxide/polypyrrole nanotube papers for flexible all-solid-state supercapacitors with excellent rate capability and high energy density[J]. Journal of Power Sources, 2016, 302: 39-45.

[82]Wang S, Zhang L, Sun C, et al. Gallium nitride crystals: novelsupercapacitor electrode materials[J]. Advanced Materials, 2016, 28(19): 3768-3776.

[83]Sun P, Lin R, Wang Z, et al. Rational design of carbon shell endows nanotube based fiber

supercapacitors with significantly enhanced mechanical stability and electrochemical performance[J]. Nano Energy, 2016: 432-440.

[84]Paradiso J A, Starner T. Energy scavenging for mobile and wireless electronics[J]. IEEE Pervasive Computing, 2005, 4(1): 18-27.

[85]He S, Hu C, Hou H, et al. Ultrathin MnO_2 nanosheets supported on cellulose based carbon papers for high-power supercapacitors[J]. Journal of Power Sources, 2014, 246: 754-761.

[86]Du D, Lan R, Xu W, et al. Preparation of a hybrid $Cu_2O/CuMoO_4$ nanosheet electrode for high-performance asymmetric supercapacitors[J]. Journal of Materials Chemistry A., 2016, 4(45): 17749-17756.

中文题目：铜、氧化亚铜和石墨在纤维素纤维上组装多层核壳结构复合纸电极用于非对称超级电容器

作者：万才超，焦月，李坚

摘要：本文提出了一种操作简单、成本低廉的方法(铅笔-绘制-电沉积-电氧化)合成了一种由铜、氧化亚铜和石墨组装在纤维素纤维上的新型多层核壳结构复合纸电极。这种有趣的电极结构在为电化学反应提供更多的活性位点、促进离子和电子的传输以及缩短它们的扩散途径方面起着关键作用。该电极表现出优异的电化学性能，在 2 $A \cdot g^{-1}$ 下的比电容为 601 $F \cdot g^{-1}$，在 100 $A \cdot g^{-1}$ 的超高电流密度下仍保持 83%的比电容。此外，当功率密度为 0.40 $kW \cdot kg^{-1}$ 时，该电极的能量密度为 13.4 $W \cdot h \cdot kg^{-1}$，循环稳定性为 95.3%，循环次数为 8000 次。当该电极被组装成用碳纸作为负电极的非对称超级电容器时，该器件显示出非凡的电化学性能，拥有较大面积电容(1 $mA \cdot cm^{-2}$ 时为 122 $mF \cdot cm^{-2}$)、较高面能量密度(402.5 $\mu W \cdot cm^{-2}$ 时为 10.8 $\mu W \cdot h \cdot cm^{-2}$)和优异的循环稳定性(91.5%，5000 次循环)。这些结果都揭示了这种复合电极材料作为高性能超级电容器电极材料的潜力。

关键词：氧化亚铜；纤维素；石墨；多层核壳复合材料；非对称超级电容器

A Cellulose Fibers-supported Hierarchical Forest-like Cuprous Oxide/Copper Array Architecture As Flexible and Free-standing Electrodes for Symmetric Supercapacitors*

Caichao Wan, Yue Jiao, Jian Li

Abstract: Herein, we for the first time develop two facile and fast steps (including magnetron sputtering and electro-oxidation) to grow hierarchical forest-like Cu_2O/Cu array architecture onto the three-dimensional fibers framework of cellulose paper. The Cu rods serve as trunk and the oxidation product (Cu_2O) acts as branches. When utilized as a flexible and free-standing electrode, the unique architecture made full use of large interfacial area from the hierarchical multi-scale structure of the forest-like array, numerous channels for rapid diffusion of electrolyte ions from the porous fibers skeleton of hydrophilic cellulose paper, and fast electron transport and high electrochemical activity from the Cu_2O/Cu complex. These merits endowed the electrode with high specific capacitance of 915 $F \cdot g^{-1}$ (238 $mF \cdot cm^{-2}$) at 3.8 $A \cdot g^{-1}$, large specific energy of 53.7 $W \cdot h \cdot kg^{-1}$ at 1.25 $kW \cdot kg^{-1}$, superior rate capability, and excellent cycling stability with a capacitance retention of 91.7% after cycling 10000 times. More importantly, an easy interesting strategy was proposed to assemble a symmetric supercapacitor based on the Cu_2O/Cu/cellulose hybrid paper, that is, growing the forest-like Cu_2O/Cu array onto the two surfaces of cellulose paper. The device delivered a high specific capacitance of 409 $F \cdot g^{-1}$ (213 $mF \cdot cm^{-2}$) at 1.9 $A \cdot g^{-1}$, a superior specific energy of 24.0 $W \cdot h \cdot kg^{-1}$ at 0.625 $kW \cdot kg^{-1}$ and good cycling stability (90.2% capacitance retention after 10000 cycles). These fascinating results unveil the potential of the hybrid paper as high-performance electrode materials for flexible energy storage devices and portable electronics.

Keywords: High-performance supercapacitors; Energy storage

1 Introduction

To meet the rapidly increasing requirements for portable and wearable electronics in recent years, development of flexible energy storage devices becomes an emerging field.[1-4] Supercapacitors, a significant class of energy storage equipment, have higher power densities and higher energy densities compared to batteries and conventional capacitors, respectively, which are regarded as promising candidates for power devices in future generation.[5-7] Recent researches on flexible supercapacitors can be divided into three classes: fiber-like,

* 本文摘自 Journal of Materials Chemistry A, 2017, 5: 17267-17278.

paper-like and three-dimensional (3D) porous electrodes (or assembled supercapacitor devices).[8,9] Especially, paper-like flexible supercapacitors, which are also called planar supercapacitors, are expected to be utilized on electronic screens of foldable mobile phones, digital cameras, laptops and televisions. Electrochemically active materials in the form of powders can be mixed with binders to form a homogeneous paste and are then coated on various electrically conductive metal foils (such as flexible Al, Cu and Ni foils).[10] Also, these electrochemically active substances can be directly adsorbed, deposited or grown on the substrates via a variety of physicochemical techniques.[11-14] Apart from these metal substrates, recently, cellulose paper-based supercapacitors have been paid increasing attention.[15,16] Cellulose paper, serving as the substrate of supercapacitors, not only possesses bright economic and environmental advantages, but also can act as an electrolyte reservoir to facilitate ion transport due to its cross-linked porous fibers network structure and strong hydrophilic property.[17] Nevertheless, cellulose paper is an insulator. To enhance conductivity, cellulose paper is generally integrated with multifarious conductive materials (e. g., carbon materials, conductive polymers and metals) by virtue of some surface modification methods, e. g., magnetron sputtering,[18] electrodeposition,[19] atomic layer deposition,[20] and electroless plating.[21] Among these techniques, magnetron sputtering is a promising sputtering-based ionized physical vapor deposition technique and is already making its way to industrial applications. In the process of magnetron sputtering, inert gas atoms (commonly Ar) are ionized and accelerated due to the potential difference between the negatively biased target (cathode) and anode, and the interactions of the ions with the target surface result in ejection (sputtering) of atoms which condensate on a substrate and eventually generate a membrane.[22,23] The primary merits of magnetron sputtering can be summarized as follows: (1) high deposition rates, (2) ease of sputtering any metals, alloys or compounds, (3) high-purity films, (4) high adhesion of films, (5) outstanding coverage of steps and small features, (6) ability to coat heat-sensitive substrates, (7) ease of automation, and (8) excellent uniformity on large-area substrates.[24] Recent literatures have employed magnetron sputtering technique to encapsulate planar cellulose products with miscellaneous active materials for achieving different functions. For instance, Legnani et al. employed radio frequency magnetron sputtering to deposit indium tin oxide thin films onto bacterial cellulose membrane for the fabrication of organic light emitting diodes.[25] Fortunato et al. produced hybrid flexible field-effect transistors using cellulose fiber-based paper as the dielectric layer and a semiconductor oxide, deposited by radio frequency magnetron sputtering, as the channel layer.[26] Moreover, Lv et al. functionalized bacterial cellulose with Cu and Al_2O_3 by magnetron sputtering for endowing it with electromagnetic shielding property and improving its hydrophobic, mechanical and thermal properties.[27]

Motivated by these impressed researches, it is interesting to use magnetron sputtering to integrate cellulose paper with electrochemically active substances for exploitation of novel flexible and self-supported supercapacitor electrodes. In this paper, we for the first time synthesized hierarchical forest-like Cu_2O/Cu array architecture, which was grown on the 3D fibers framework of cellulose paper, via two fast and facile steps (magnetron sputtering and electro-oxidation). This ternary composite paper (labeled as Cu_2O/Cu/cellulose hybrid paper) acted as a flexible free-standing supercapacitor electrode and delivered a high specific capacitance of 915 $F \cdot g^{-1}$ (238 $mF \cdot cm^{-2}$), a high specific energy of 53.7 $W \cdot h \cdot kg^{-1}$ at a power density of 1.25 $kW \cdot kg^{-1}$ and 28.9 $W \cdot h \cdot kg^{-1}$ at a high power density of 40 $kW \cdot kg^{-1}$, and excellent cycling performance. More importantly, we assembled a symmetric supercapacitor device by growing the forest-like Cu_2O/Cu array onto the two surfaces of cellulose paper. The cellulose paper was directly used as a separator, and the two-sided Cu_2O/Cu complex served as the positive and negative electrodes, respectively. The device also features some favorable electrochemical properties like a high specific capacitance of 409 $F \cdot g^{-1}$ (213 $mF \cdot cm^{-2}$) and a superior specific energy of 24.0 $W \cdot h \cdot kg^{-1}$ at 0.625 $kW \cdot kg^{-1}$. In addition, the device displayed high capacitance retention of 90.2% after 10000 cycles. These outstanding electrochemical activities indicate its promising future in applications for energy storage.

2 Results and discussion

A schematic illustration for the synthesis strategy of the cellulose fibers-supported hierarchical forest-like Cu_2O/Cu array architecture is illustrated in Fig. 1a. A thin layer of metallic Cu array was firstly homogeneously deposited onto the 3D fibers framework of cellulose paper by magnetron sputtering. It is well-known that electro-oxidation method is a simple, fast and effective method to transform metals to their oxides or hydroxides.[28,29] Several literatures have successfully adopted cyclic voltammetry (CV) electro-oxidation method to transform Cu to Cu_2O or CuO.[30-32] In the present paper, the Cu/cellulose paper was also electro-oxidized by CV method to transform the superficial metallic Cu into the highly electrochemically active Cu_2O whose theoretical capacitance reaches up to 2075 $F \cdot g^{-1}$ (the calculation process is available in ESI). The electro-oxidation induced the formation of interlaced spinous Cu_2O on the surface of Cu rod array, that is, the rod-like metallic Cu serves as trunk and the oxidation product (namely Cu_2O) acts as branches, as schematically shown in Fig. 1b. This unique architecture takes advantage of a large interfacial area from the hierarchical multi-scale structure of the forest-like array, numerous channels for rapid diffusion of electrolyte ions from the porous 3D fibers skeleton of hydrophilic cellulose paper, and fast electron transport and high electrochemical activity from the Cu_2O/Cu complex.[33,34]

As shown in Fig. S1, distinct differences in surface color exist among the cellulose paper, Cu/cellulose paper and Cu_2O/Cu/cellulose hybrid paper, which suggest remarkable differences in micromorphology and chemical compositions. It is obvious that by comparing scanning electron microscope (SEM) images in Fig. 2a-b and Fig. 2d-e, the surface of 3D fibers framework of the cellulose paper (see also Fig. S2) was encapsulated with dense rod-like metallic Cu array with a height of approximately 15 μm after the magnetron sputtering. In addition, according to the higher-magnification SEM image aiming at the surface structure of the Cu/cellulose paper (see the inset in Fig. 2d), it is clear that these Cu rods have the diameters of 90–220 nm. Through the electro-oxidation, plentiful cross-linked Cu_2O branches were directly grown on the surface of Cu rods (see Fig. 2g and h), leading to the formation of the forest-like Cu_2O/Cu array architecture, i.e., the Cu and Cu_2O serve as trunk and branches, respectively. On the basis of the inset in Fig. 2g, the higher-magnification SEM image shows that the numerous cross-linked branch-like architectures have the lengths of 1–4 μm, and their diameters are in the range of 0.2–1 μm. Energy dispersive X-ray (EDX) was executed to analyze the elemental compositions. Compared with the cellulose paper (Fig. 2c), both the EDX spectra of the Cu/cellulose paper and Cu_2O/Cu/cellulose hybrid paper display the plentiful presence of Cu element (Fig. 2f and i). Furthermore, as compared to the Cu/cellulose paper, the intensity of oxygen signal improved significantly for the Cu_2O/Cu/cellulose hybrid paper because of the oxidization of Cu into Cu_2O. In addition, the self-supported Cu_2O/Cu/cellulose hybrid paper possesses good flexibility. As shown in Fig. 2j, the hybrid paper can be easily bended, folded and twisted. Its flexibility provides convenience for the development of flexible energy storage devices. To determine the conductivity of the Cu_2O/Cu/cellulose hybrid paper, we tested the electrical conductivity of the hybrid paper through the four-electrode method. The hybrid paper displays a high electrical conductivity of 3.6×10^4 S · m^{-1}, which is lower than that of metallic Cu (ca. 5.7×10^7 S · m^{-1} at 20 ℃), because of the presence of weakly conductive Cu_2O. However, the value is still higher than that of many electrical conductors, such as pristine graphite[35] (2500 S · m^{-1}) and conductive polymers[36] (10–40000 S · m^{-1}).

For the investigation of the fine structure of Cu_2O branches, the high-magnification SEM images are shown in Fig. 3a and b, where the branches were composed of plentiful nanowire-like structure with the diameters of approximately 25 nm. These nanowires aggregated together in parallel, and this self-assembled process led to the generation of the multiscale two-dimensional (2D) sheet-like and 3D branch-like structures. The transmission electron microscopy (TEM) images in Fig. 3c and d further verify that the Cu_2O branches were constructed with smaller structure and possessed rough marginal structure. In addition, the high-resolution TEM (HRTEM) image of the blue-circled area at the edge of branches is shown in Fig. 3e, in which the measured interplanar spacing of 2.49 Å for the well-defined lattice fringes is consistent well with the (111) plane of Cu_2O. Further analysis of the selected area electron diffraction (SAED) shows that the as-obtained branches are polycrystalline, and the diffraction rings corresponds to the lattice spacings of Cu_2O phase.

Regarding the possible formation mechanism of Cu_2O branches, according to the previous researches, the operative chemical reactions are expected to be $Cu \rightarrow Cu^+ + e^-$ and $2Cu^+ + 2OH^- \rightarrow Cu_2O + H_2O$.[37,38] The specific sites where oxidation took place located near defects (e. g., surface deformation and pits) on the surface of Cu rods. The formation of Cu_2O nanostructure is dependent dominantly on the applied potential.[29] At the mild applied potential (i. e., −0.4 to 0.4 V), the migration of fresh cuprous ions from the interior is expected to be slow and the reaction zone was confined to the surface and near-surface regions. Owing to the existence of a large number of hydroxyl ions in the electrolyte, these fresh cuprous ions reacted quickly and continuously with hydroxyl ions leading to the formation of well-developed Cu_2O nanowire configurations protruding nearly perpendicular to the surface of Cu. These nanowires then coalesced together to form 2D sheets and 3D branches.[39]

The crystal structure was investigated by X-ray diffraction (XRD). As can be seen by comparing Fig. 4a and b, the coverage of Cu array led to the disappearance of the characteristic peaks of cellulose Ⅰ crystal structure, and the three peaks at 2θ of 43.8°, 50.9° and 74.6° corresponding to the (111), (200) and (220) planes of Cu (JCPDS no. 04-0836) were detected. Furthermore, the weak (111) peak of Cu_2O was highlighted in Fig. 4c, verifying the partial conversion of Cu to Cu_2O due to the electro-oxidation. X-ray photoelectron spectroscopy (XPS) was used to analyze the chemical valences of Cu element in the Cu_2O/Cu/cellulose hybrid paper. Fig. 4d displays the Cu 2p XPS spectrum, where the two peaks at 932.5 and 952.3 eV correspond to the Cu $2p_{3/2}$ and Cu $2p_{1/2}$ peaks of Cu^+ and Cu^0 species, respectively.[40] It is worth to mention that the Cu $2p_{3/2}$ binding energies of Cu_2O and Cu^0 are close to each other; the reported values are 932.5 and 932.6 eV, respectively.[41] Therefore, Cu $2p_{3/2}$ XPS cannot differentiate between Cu^+ and Cu^0. On the other hand, two small peaks with binding energy of 933.5 and 953.2 eV and their shake-up peaks at 942.7 and 962.2 eV are assigned to the characteristic peaks of Cu^{2+},[42] revealing that a small amount of Cu^+ was oxidized to Cu^{2+} in the air. The O 1s XPS spectrum can be deconvoluted into two peaks (Fig. 4e). The peak at 530.6 eV represents the oxygen in Cu_2O lattice,[43] further confirming the presence of Cu^+. The peak at 531.6 stands for the adsorbed OH groups.[44] Hence, XPS and XRD results indicate the primary Cu_2O, Cu and cellulose components in the hybrid paper. The Brunauer-Emmett-Teller (BET) surface area was calculated based on the N_2 adsorption-desorption isotherms,[45] as shown in Fig. 4f. The co-existence of multiscale Cu rods and Cu_2O branches endows the Cu_2O/Cu/cellulose hybrid paper with higher BET surface area of 16.6 $m^2 \cdot g^{-1}$, when compared to that of the cellulose paper (3.4 $m^2 \cdot g^{-1}$) and Cu/cellulose paper (13.2 $m^2 \cdot g^{-1}$). According to these above data and the loading mass of Cu_2O in the hybrid paper, the surface area of Cu_2O component can be roughly calculated to be as high as 261.5 $m^2 \cdot g^{-1}$ (the calculation process is given in ESI). The large surface area helps to provide more electrochemically active sites accessed by electrolyte ions, promoting the ions adsorption and shortening the ions migration pathway during the reaction process.[46] The pore size distribution of the Cu_2O/Cu/cellulose hybrid paper was calculated from the N_2 adsorption-desorption isotherm using the Barrett-Joyner-Halenda (BJH) method (Fig. S3).[47] The pore sizes range from 3.1 to 193.8 nm, indicating that the hybrid paper consists of mesopores (2-50 nm) and macropores (> 50 nm). Such a pore size distribution is favorable for electrolyte penetration, allowing reaction species to quickly access the electrode.[48]

Table 1 Comparison of the specific capacitance, specific energy, cycling stability, BET surface area and electrical conductivity of some Cu_2O-based supercapacitior electrodes prepared by various strategies

Electrode materials	Synthesis method	Maximum specific capacitance ($F \cdot g^{-1}$)	Maximum specific energy ($W \cdot h \cdot kg^{-1}$)	Cycling stability	BET surface area ($m^2 \cdot g^{-1}$)	Electrical conductivity ($S \cdot m^{-1}$)	Refs
Cu_2O microspheres	Polyol reduction method	144 (0.1 $A \cdot g^{-1}$)	-	99.3% (100 cycles, 0.1 $A \cdot g^{-1}$)	4.3	-	[51]
Cu_2O microcubes	Hydrothermal methods	660 (1 $A \cdot g^{-1}$)	14.6 (0.2 $kW \cdot kg^{-1}$)	80% (1000 cycles, 5 $A \cdot g^{-1}$)	7.39	0.273×10^{-3}	[50]
Rose rock-shaped nano-Cu_2O anchored graphene	Solvothermal route	416 (1 $A \cdot g^{-1}$)	42 (1.5 $kW \cdot kg^{-1}$)	86% (1000 cycles, 1 $A \cdot g^{-1}$)	-	-	[52]
3D ordered macroporous Cu_2O/Ni inverse opals	Electrodeposition	502 (10 $mV \cdot s^{-1}$)	26.5 (1.2 $kW \cdot kg^{-1}$)	92% (500 cycles, 10 $mV \cdot s^{-1}$)	-	-	[53]
Cu_2O@ Cu nanoneedle arrays	Anodization electro-oxidation	510.2 (2.9 $A \cdot g^{-1}$)	-	89% (10000 cycles, 14.5 $A \cdot g^{-1}$)	5.36	1.770×10^6	[54]
Nanoplatelet-structured Ni/Cu/Cu_2O/CuO/ZnO	Chemical precipitation & annealing	241 (50 $mV \cdot s^{-1}$)	32 (0.55 $kW \cdot kg^{-1}$)	-	-	-	[55]
Cu_2O/CuO/Co_3O_4 core-shell nanowires	Chemical deposition & calcination	318 (0.5 $A \cdot g^{-1}$)	-	80% (3000 cycles, 5 $A \cdot g^{-1}$)	-	-	[56]
Cu_2O-Cu$(OH)_2$ nanoflakes/graphene/stainless steel	Electrodeposition	425 (5 $A \cdot g^{-1}$)	-	-	-	-	[57]

(continued)

Electrode materials	Synthesis method	Maximum specific capacitance ($F \cdot g^{-1}$)	Maximum specific energy ($W \cdot h \cdot kg^{-1}$)	Cycling stability	BET surface area ($m^2 \cdot g^{-1}$)	Electrical conductivity ($S \cdot m^{-1}$)	Refs
Graphene/polypyrrole/Cu_2O – $Cu(OH)_2$/Ni foam	Electrodeposition & electrochemical reduction	997 (10 $A \cdot g^{-1}$)	–	90% (2000 cycles, 10 $A \cdot g^{-1}$)	–	–	[58]
Cu_2O/Cu/cellulose hybrid paper	Magnetron sputtering & electro-oxidation	915 (3.8 $A \cdot g^{-1}$)	53.7 (1.25 $kW \cdot kg^{-1}$)	91.7% (10000 cycles, 61.5 $A \cdot g^{-1}$)	16.6	3.6×10^4	This work

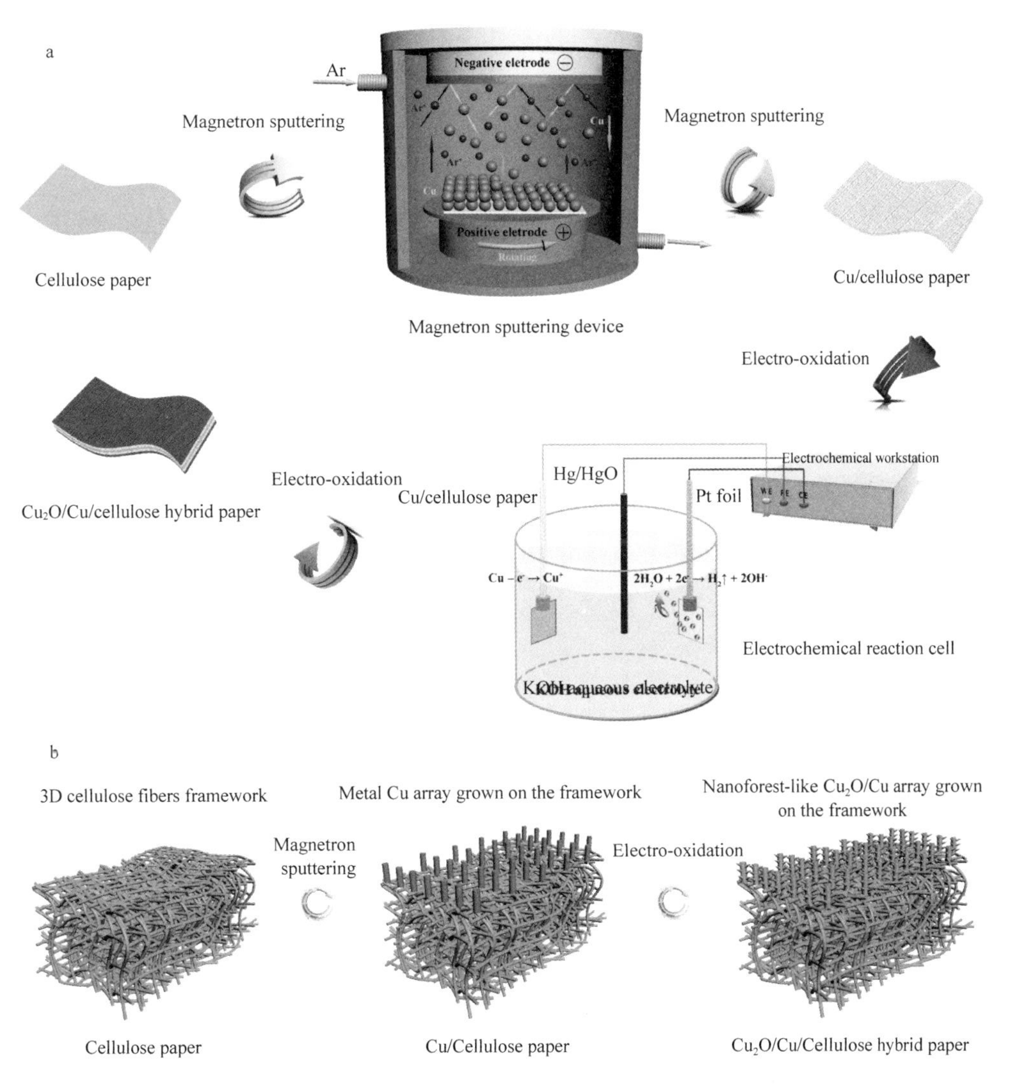

Fig. 1 Schematic illustration for (a) the synthesis of the Cu_2O/Cu/cellulose hybrid paper and (b) the formation of the cellulose fibers-supported hierarchical forest-like Cu_2O/Cu array architecture

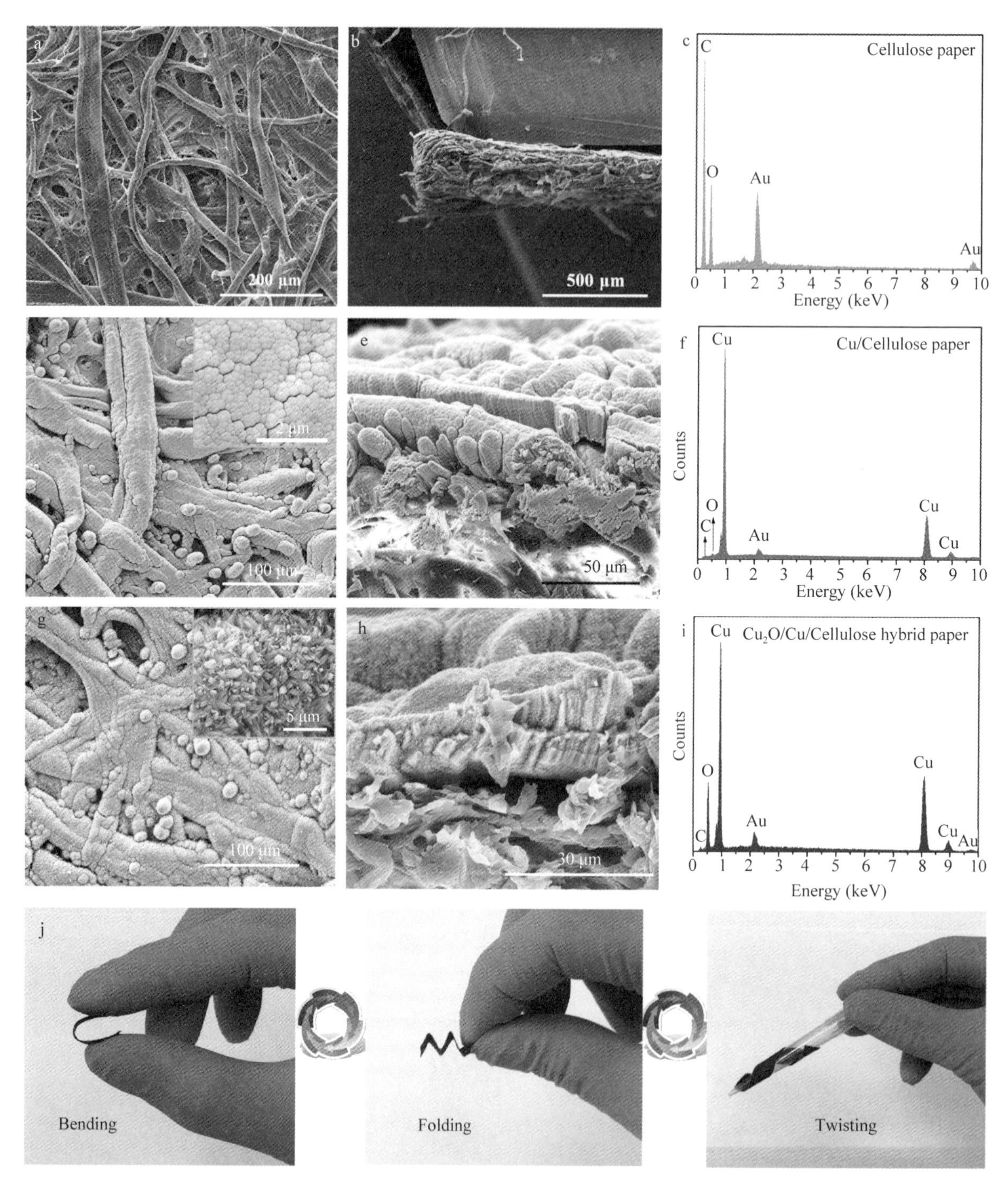

Fig. 2 (a) SEM image, (b) cross-section SEM image and (c) EDX spectrum of the cellulose paper. (d) SEM image, (e) cross-section SEM image and (f) EDX spectrum of the Cu/cellulose paper, and the inset in (d) shows the corresponding higher-magnification SEM image. (g) SEM image, (h) cross-section SEM image and (i) EDX spectrum of the Cu_2O/Cu/cellulose hybrid paper, and the inset in (g) shows the corresponding higher-magnification SEM image. (j) Photographs of the Cu_2O/Cu/cellulose hybrid paper to reveal its flexibility

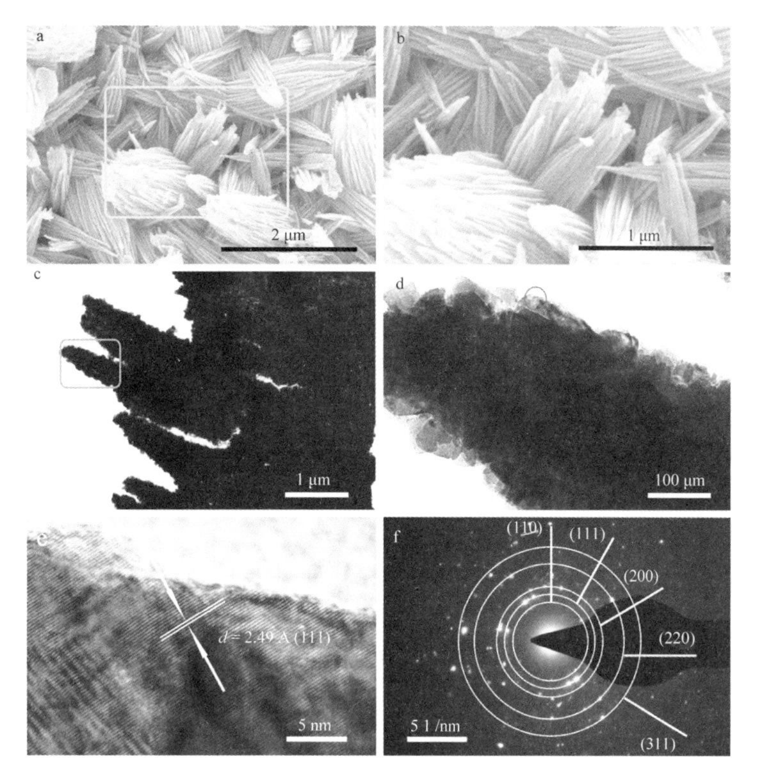

Fig. 3 (a) High-magnification SEM images (80000 x) of the Cu_2O branches. (b) Magnified image (160 000 x) of the green-squared marker region in (a). (c) TEM images of the Cu_2O branches, and (d) shows the magnified image of the red-squared marker region in (c). (e) HRTEM image and (f) SAED pattern of the blue-circled area in (d)

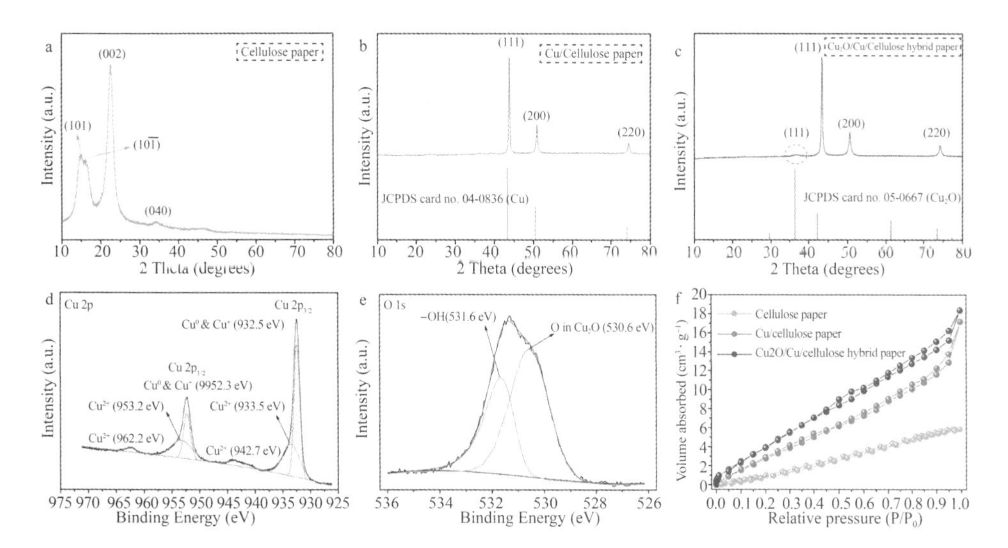

Fig. 4 XRD patterns of the cellulose paper (a), Cu/cellulose paper (b), and Cu_2O/Cu/cellulose hybrid paper (c). The bottom lines in (b) and (c) are the standard JCPDS cards no. 04-0836 for Cu and no. 05-0667 for Cu_2O, respectively. High-resolution XPS spectra of (d) Cu 2p and (e) O 1s core levels of the Cu_2O/Cu/cellulose hybrid paper. (f) N_2 adsorption-desorption isotherms of the cellulose paper, Cu/cellulose paper and Cu_2O/Cu/cellulose hybrid paper

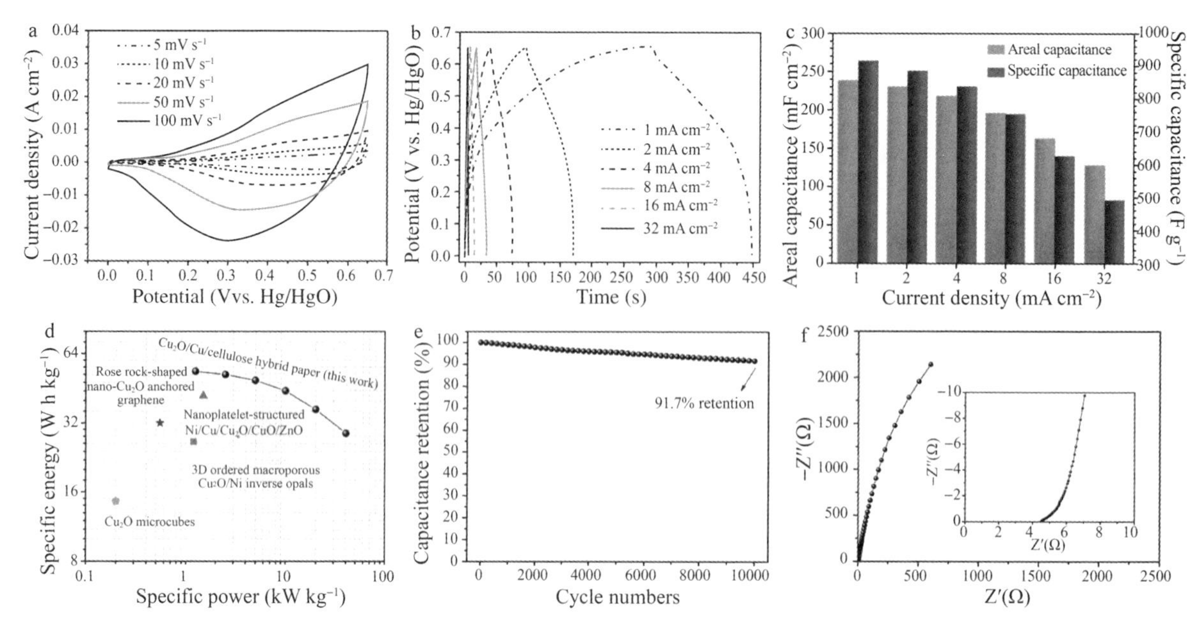

Fig. 5 (a) CV curves of the Cu_2O/Cu/cellulose hybrid paper electrode at various scan rates. (b) GCD curves and (c) areal and specific capacitances of the Cu_2O/Cu/cellulose hybrid paper electrode at various current densities. (d) Specific energy with respect to the specific power (Ragone plot) of the Cu_2O/Cu/cellulose hybrid paper electrode compared with data of other Cu_2O-based electrodes in the literatures. (e) Cycling performance of the Cu_2O/Cu/cellulose hybrid paper electrode measured at a current density of 16 mA · cm^{-2}. (f) Nyquist plot of the Cu_2O/Cu/cellulose hybrid paper electrode

To evaluate the application potentials of the cellulose fibers-supported hierarchical forest-like Cu_2O/Cu array architecture as electrochemical capacitors, a series of electrochemical tests were conducted in a three-electrode configuration. The CV curves of the Cu_2O/Cu/cellulose hybrid paper electrode at various scan rates are presented in Fig. 5a. These CV profiles are quite different from ideal rectangular shape for double-layer capacitance. This is primarily attributable to the Faradaic redox reactions of Cu(Ⅰ)/Cu(Ⅱ) transition, which are expressed as follows:[49]

$$\frac{1}{2}Cu_2O+OH^- \Leftrightarrow CuO+\frac{1}{2}H_2O+e^- \tag{1}$$

$$\frac{1}{2}Cu_2O+\frac{1}{2}H_2O+OH^- \Leftrightarrow Cu(OH)_2+e^- \tag{2}$$

$$CuOH+OH^- \Leftrightarrow CuO+H_2O+e^- \tag{3}$$

$$CuOH+OH^- \Leftrightarrow Cu(OH)_2+e^- \tag{4}$$

The XPS analysis of the Cu_2O/Cu/cellulose hybrid paper after the electrochemical tests confirms the plentiful formation of Cu^{2+} (see Fig. S4). In addition, the potentials of the cathodic peaks shifted towards negative direction with the increasing scan rate, which is associated with the internal resistance of the electrode and further demonstrates the pseudocapacitive characteristics.[50]

The galvanostatic charge-discharge (GCD) curves of the Cu_2O/Cu/cellulose hybrid paper electrode at various current densities are exhibited in Fig. 5b. The nonlinear feature also supports the pseudocapacitive nature of the hybrid electrode. Moreover, the charging curves are almost symmetrical with their corresponding discharge counterparts, which is indicative of the good reversibility. The areal and specific capacitances as a function of the discharge current density were plotted in Fig. 5c. A maximum capacitance value of 238 $mF \cdot cm^{-2}$ (915 $F \cdot g^{-1}$) was achieved at a current density of 1 $mA \cdot cm^{-2}$ (3.8 $A \cdot g^{-1}$). When the current density increased to 32 $mA \cdot cm^{-2}$ (123.1 $A \cdot g^{-1}$), the electrode still maintained the capacitance of 128 $mF \cdot cm^{-2}$ (492 $F \cdot g^{-1}$), as much as 54% of that at 1 $mA \cdot cm^{-2}$, which reveals its excellent rate capability. For the sake of comparison, we summarized recently reported Cu_2O-based supercapacitor electrodes, which were prepared by various approaches such as polyol reduction method, hydrothermal methods, chemical deposition-calcination and electrodeposition, as listed in Table 1. It is observed that the maximum specific capacitance of the Cu_2O/Cu/cellulose hybrid paper electrode is remarkably higher than or comparable with that of the unary Cu_2O electrodes like Cu_2O microspheres[51] (144 $F \cdot g^{-1}$) and microcubes[50] (660 $F \cdot g^{-1}$), the binary Cu_2O-based electrodes like rose rock-shaped nano-Cu_2O anchored graphene[52] (416 $F \cdot g^{-1}$), 3D ordered macroporous Cu_2O/Ni inverse opals[53] (502 $F \cdot g^{-1}$) and Cu_2O@Cu nanoneedle arrays[54] (510.2 $F \cdot g^{-1}$), and the polynary Cu_2O-based electrodes like nanoplatelet-structured Ni/Cu/Cu_2O/CuO/ZnO[55] (241 $F \cdot g^{-1}$), Cu_2O/CuO/Co_3O_4 core-shell nanowires[56] (318 $F \cdot g^{-1}$), Cu_2O-$Cu(OH)_2$ nanoflakes/graphene/stainless steel[57] (425 $F \cdot g^{-1}$) and graphene/polypyrrole/Cu_2O-$Cu(OH)_2$/Ni foam[58] (997 $F \cdot g^{-1}$). Besides the capacitances, energy density and power density are also crucial factors for supercapacitor applications. Although three-electrode configuration is not representative for a real device, these energy and power data still have certain guiding significance for the development of practical devices. The Cu_2O/Cu/cellulose hybrid paper electrode displays the high specific energies, from 53.7 $W \cdot h \cdot kg^{-1}$ at 1.25 $kW \cdot kg^{-1}$ to 28.9 $W \cdot h \cdot kg^{-1}$ at 40 $kW \cdot kg^{-1}$, better than those of the above Cu_2O-based electrodes (see Table 1 and Fig. 5d), i.e., Cu_2O microcubes[50] (14.6 $W \cdot h \cdot kg^{-1}$ at 0.2 $kW \cdot kg^{-1}$), 3D ordered macroporous Cu_2O/Ni inverse opals[53] (26.5 $W \cdot h \cdot kg^{-1}$ at 1.2 $kW \cdot kg^{-1}$), nanoplatelet-structured Ni/Cu/Cu_2O/CuO/ZnO[55] (32 $W \cdot h \cdot kg^{-1}$ at 0.55 $kW \cdot kg^{-1}$) and rose rock-shaped nano-Cu_2O anchored graphene[52] (42 $W \cdot h \cdot kg^{-1}$ at 1.5 $kW \cdot kg^{-1}$). Additionally, the Cu_2O/Cu/cellulose hybrid paper electrode possesses superior cycling stability, and only an 8.3% fall in the initial capacitance was found after 10000 cycles (Fig. 5e), which is also better than that of the most Cu_2O-based electrodes as listed in Table 1. In consequence, these comparisons not only display the more superior capacitance, specific energy and power and cycling performances for the Cu_2O/Cu/cellulose hybrid paper electrode, but also demonstrate that the synthesis strategy (i.e., magnetron sputtering and electro-oxidation) in this paper has more advantages.

The reasons for these favorable electrochemical behaviors of the Cu_2O/Cu/cellulose hybrid paper can be summed up as follows: (1) the cellulose paper with highly porous structure and good wettability can absorb electrolyte and serve as an electrolyte storage unit to facilitate ion transport;[17] (2) the hierarchical multi-scale structure of the forest-like array significantly increases the electrochemical active sites for the reversible redox reactions; (3) the highly conductive Cu rods array provides electron "superhighways" for charge storage and delivery;[59] (4) the generated Cu_2O branches have extremely high theoretical capacitance of 2075 $F \cdot g^{-1}$.

The Nyquist plot of the Cu_2O/Cu/cellulose hybrid paper electrode is provided in Fig. 5f. The equivalent series resistance (the real-axis intercept) is as low as approximately 4.5 Ω. Furthermore, the width of semicircle in the high-frequency region is indicative of the charge-transfer resistance. The Faradaic interfacial charge-transfer resistance is a well-known limiting factor of power density of pseudocapacitors.[60] The electrode has a remarkably low charge-transfer resistance, which indicates that the electrode favors access of ions due to the good wettability of cellulose fibers and helps to improve the power density of the pseudocapacitance.

Interestingly, the cellulose paper not only can serve as a 3D porous framework to support the forest-like Cu_2O/Cu array architecture or as an electrolyte storage unit to facilitate ion transport, but also can be directly used an electrical insulating separator to assemble supercapacitor device. For instance, in this paper, a symmetric supercapacitor was easily obtained by growing the forest-like Cu_2O/Cu array structure onto the two surfaces of cellulose paper. The Cu_2O/Cu complex on the both surfaces served as the positive and negative electrodes, respectively. The schematic diagram of this Cu_2O/Cu/cellulose hybrid paper-based symmetric supercapacitor device is shown in Fig. 6a. The device was soaked in a 1 M KOH solution for all of the electrochemical measurements in a two-electrode system.

Fig. 6b shows the CV curves of the device at different scan rates. For each curve, a typical pair of anodic and cathodic signals is clearly visible, indicating that the reversible and continuous faradaic reactions of Cu oxide were involved during the charging and discharging process. For investigating the contribution of the current collectors (namely platinum sheet) to the capacitance of the device, we tested the CV curves of the platinum sheet (see Fig. S5). The area enclosed within the CV curves, which corresponds to the energy storage capability, of the device is notably much larger than that of the platinum sheet. This result suggests that the capacitance contribution of the platinum sheet is negligible. The GCD curves of the device at different current densities are shown in Fig. 6c. During charging of the device, the positively charged electrode got oxidized to form CuO or $Cu(OH)_2$ at the outer layer of positive electrode. While at the negative electrode, K^+ ions were physically adsorbed on the surface of the electrode without Faradaic reactions. The discharge curve contains two regions, i. e., a fast potential IR drop followed by slow potential decay. The IR drop originated from the internal resistance of the device (including electrical resistance and solution resistance).[61] Fig. S6 displays the variation of areal and specific capacitances of the device with discharge current. A maximum specific capacitance of 409 $F \cdot g^{-1}$ (213 $mF \cdot cm^{-2}$) was achieved at a current density of 1.9 $A \cdot g^{-1}$, which are higher than some Cu_2O-based symmetric/asymmetric supercapacitors reported previously (see Table S1), for instance, Cu_2O@Cu nanoneedle arrays//active carbon asymmetric supercapacitor[54] (77 $F \cdot g^{-1}$), rose rock-shaped nano-Cu_2O anchored graphene symmetric supercapacitor[52] (92 $F \cdot g^{-1}$), Cu_2O-$Cu(OH)_2$ nanoflakes/graphene/stainless steel symmetric supercapacitor [57] (104 $F \cdot g^{-1}$), and graphene/polypyrrole/Cu_2O–$Cu(OH)_2$/Ni foam symmetric supercapacitor[58] (225 $F \cdot g^{-1}$). Ragone plot was plotted based on the GCD data, as shown in Fig. 6d. The device has a high specific energy of 24.0 $W \cdot h \cdot kg^{-1}$ at a specific power of 0.625 $kW \cdot kg^{-1}$, while maintains a specific energy of 2.5 $W \cdot h \cdot kg^{-1}$ at a high specific power of 10 $kW \cdot kg^{-1}$. These energy and power results are also comparable to or higher than those of the above Cu_2O-based symmetric/asymmetric supercapacitors[52,54,56–58] (12–35.6 $W \cdot h \cdot kg^{-1}$ and 0.162–8 $kW \cdot kg^{-1}$). Fig. 6e presents the excellent cycling performance of the device tested by continuous charge-discharge at 16 $mA \cdot cm^{-2}$ (30.8 $A \cdot g^{-1}$). The capacitance retained 90.2% of its initial value even after 10000 cycles.

It is noteworthy that, besides the Cu_2O/Cu/cellulose hybrid paper-based symmetric supercapacitor reported in this work, other types of symmetric/asymmetric supercapacitor devices with wider potential range may also be easily developed by depositing multifarious electrochemically active materials (e. g., metal oxides, conductive polymers and carbon materials) onto the two surface of cellulose paper via various physicochemical techniques like vapor deposition[62,63] and vacuum filtration[64,65]. This strategy is expected to be further investigated in our future study.

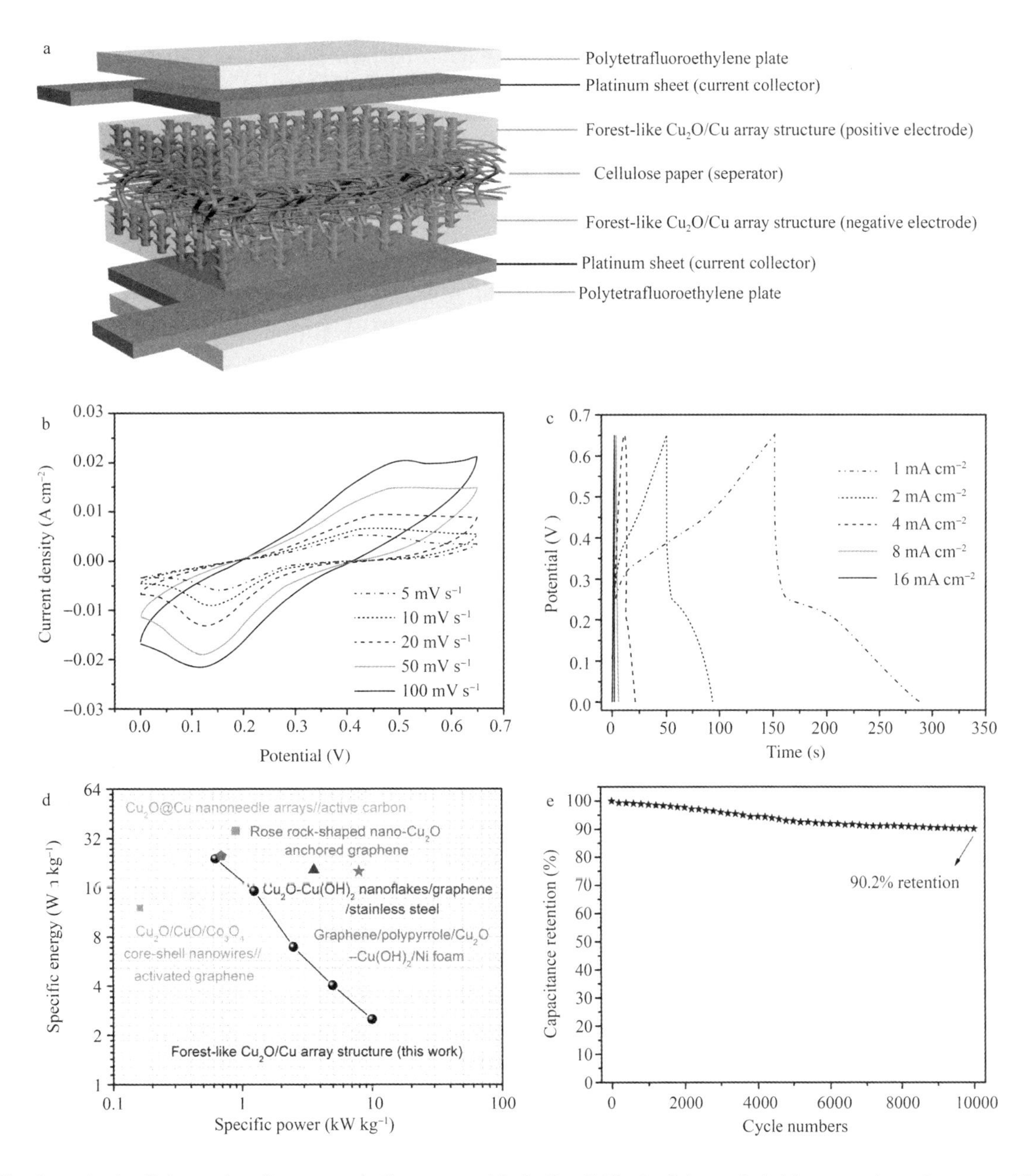

Fig. 6 (a) Schematic diagram of the assembled Cu_2O/Cu/cellulose hybrid paper-based symmetric supercapacitor device. (b) CV curves of the device at various scan rates. (c) GCD curves of the device at various current densities. (e) Ragone plot of the device compared with data of some Cu_2O-based symmetric/asymmetric supercapacitors in the literatures. (e) Cycling performance of the device measured at a current density of 16 mA · cm^{-2}

3 Conclusions

In summary, the hierarchical forest-like Cu_2O/Cu array architecture was grown onto the 3D fibers framework of flexible cellulose paper via two rapid simple steps including magnetron sputtering and electro-oxidation. This free-standing and flexible hybrid paper electrode achieved the high specific capacitance of 915 $F \cdot g^{-1}$ (238 $mF \cdot cm^{-2}$), high specific energy of 53.7 $W \cdot h \cdot kg^{-1}$ at 1.25 $kW \cdot kg^{-1}$, and outstanding cycling stability and rate capability, which are attributable to the synergistic contributions from the three components (namely strongly hydrophilic cellulose fibers, highly conductive Cu and highly electrochemically active Cu_2O), together with the merits of the multi-scale forest-like array structure providing large reaction area. More significantly, a symmetric supercapacitor was easily constructed by growing the forest-like Cu_2O/Cu array onto the two surfaces of cellulose paper. The high specific capacitance of 409 $F \cdot g^{-1}$ (213 $mF \cdot cm^{-2}$), superior specific energy of 24.0 $W \cdot h \cdot kg^{-1}$ at 0.625 $kW \cdot kg^{-1}$, and excellent cycling stability (90.2% capacitance retention after 10000 cycles) were achieved for the device. We believe that these encouraging results not only suggest the possibility to engineer the flexible hybrid paper with forest-like array structure into promising electrodes, but demonstrate a novel and easily-operated approach to develop hybrid electrode architectures for flexible portable electronics and energy storage devices.

4 Experimental section

4.1 Magnetron sputtering process

A thin layer of metallic Cu array was sputtered on the surface of cellulose paper by using a magnetron sputter (VTC-600-2HD, Shenyang Kejing Auto-instrument Co., Ltd, China). A Cu target (purity: 99.99%; diameter: 50 mm) was mounted on the cathode, while the cellulose paper was placed on the anode facing the target. The distance between the target and the substrate was 60 mm. Initially, the chamber was pumped to a pressure of 3×10^{-3} Pa before introduction of sputtering Ar gas (purity: 99.99%). The flow of Ar was set to 11 sccm. The Cu particles were sputtered on the cellulose paper by direct current magnetron sputtering (power: 100 W) with rotating speed of 90 rpm to achieve uniform deposition. To avoid substrate deformation and diffusion movement of the sputtered Cu particles at high temperature, water-cooling was applied in the whole sputtering process to control the temperature of substrate.

4.2 Electro-oxidation process

The electro-oxidation was executed by an electrochemical workstation (CS350, Wuhan CorrTest Instruments Corp., Ltd.) with a three-electrode configuration, in which the as-prepared Cu/cellulose paper acted as the working electrode, a Pt foil (2 cm × 2 cm) was used as the counter electrode, and an Hg/HgO electrode served as the reference electrode. The electro-oxidation process was conducted according to the literature method.[66] Briefly, the Cu_2O was formed by employing CV in the potential range from −0.4 to 0.4 V vs. Hg/HgO at the scan rate of 5 $mV \cdot s^{-1}$ for 1 segment in a 1 M KOH solution.

4.3 Characterizations

The morphology was characterized by using a SEM (Hitachi, S4800) equipped with an EDX detector for elemental analysis. TEM and HRTEM observations and SAED were performed with a FEI, Tecnai G2 F20

TEM with a field-emission gun operating at 200 kV. N_2 adsorption-desorption measurements were implemented at −196 ℃ using an accelerated surface area and porosimetry system (3H-2000PS2 unit, Beishide Instrument S&T Co., Ltd). XRD analysis was implemented on a Bruker D8 Advance TXS XRD instrument with Cu Kα (target) radiation (λ = 1.5418 Å) at a scan rate (2θ) of 4° min^{-1} and a scan range from 10 to 80°. XPS analysis was carried out on a Thermo Escalab 250Xi system using a spectrometer with a dual Al Ka X-ray source. Deconvolution of the overlapping peaks was performed using a mixed Gaussian-Lorentzian fitting program (Origin 9.0, Originlab Corporation). The sheet resistance (R) and electrical conductivity (σ) were measured by a H7756 four-point probe resistivity meter (Beijing Heng Odd Instrument Co., Ltd) at room temperature using the following equation:

$$(\sigma,\ S\cdot m^{-1}) = (\frac{1}{\rho}) = \frac{1}{dR} \tag{5}$$

where ρ is the electrical resistivity, and d is the thickness of the sample. The electrochemical properties were investigated at room temperature using a CS350 electrochemical workstation in a three-electrode setup: the composite papers served as the working electrode, and an Hg/HgO electrode and a Pt foil electrode served as reference and counter electrodes, respectively. The exposed geometric area of the working electrode is equal to 1 cm^2, and the mass loading of Cu_2O per area was calculated to be 0.26 mg · cm^{-2} (the calculation process is available in ESI). The electrolyte was 1 M KOH solution. CV and GCD curves were measured over the potential window from 0 to 0.65 V. Electrochemical impedance spectroscopy (EIS) measurements were carried out in the frequency range from 10^5 to 0.01 Hz with alternate current amplitude of 5 mV.

5 Calculations

The areal capacitance (C_S) was calculated from GCD curves by the following formula: [67]

$$C_S = \frac{I\times\Delta t}{\Delta V\times S} \tag{6}$$

where I is the charge-discharge current, Δt is the discharge time, ΔV is the potential window, and S is the electrode area. Similarly, the specific capacitance (C_m) can be calculated by the following formula:

$$C_m = \frac{I\times\Delta t}{\Delta V\times m} \tag{7}$$

where m is the mass of Cu_2O. The specific energy (E) and specific power (P) were calculated according to the following equations: [57]

$$E = \frac{1}{2}C_m(\Delta V)^2 \tag{8}$$

$$P = \frac{E}{\Delta t} \tag{9}$$

5.1 Fabrication of symmetric supercapacitor device

A symmetric supercapacitor was obtained by growing the forest-like Cu_2O/Cu array structure onto the two surfaces of cellulose paper. The Cu_2O/Cu complex on the both surfaces of cellulose paper served as the positive and negative electrodes, respectively, and the cellulose paper was used as a separator. Two pieces of thin platinum sheet were utilized as current collectors. These electrodes, separator and current collectors were sandwiched between two polytetrafluoroethylene plates. The whole device was clamped by a high-strength plastic clamp. The device was soaked in a 1 M KOH solution for all of the electrochemical measurements in a two-electrode system. The electrochemical properties of the device were also calculated according to the equations 6–8. Especially, the *m* is the total mass of Cu_2O in the positive and negative electrodes.

Acknowledgements

This study was supported by the National Natural Science Foundation of China (grant no. 31270590 and 31470584) and the Fundamental Research Funds for the Central Universities (grant no. 2572016AB22).

References

[1] Lee S Y, Choi K H, Choi W S, et al. Progress in flexible energy storage and conversion systems, with a focus on cable-type lithium-ionbatteries[J]. Energy & Environmental Science, 2013, 6(8): 2414–2423.

[2] Nyholm L, Nystrm G, Mihranyan A, et al. Toward Flexible Polymer and Paper-Based Energy Storage Devices[J]. Advanced Materials, 2011, 23(33): 3751–3769.

[3] Wang X, Lu X, Liu B, et al. Flexible energy-storage devices: design consideration and recentprogress[J]. Advanced Materials, 2014, 26(28): 4763–4782.

[4] Li L, Wu Z, Yuan S, et al. Advances and challenges for flexible energy storage and conversion devices and systems [J]. Energy & Environmental Science, 2014, 7(7): 2101–2122.

[5] Simon P, Gogotsi Y, Dunn B. Where do batteries end and supercapacitors begin? [J]. Science, 2014, 343(6176): 1210–1211.

[6] Winter M, Brodd R J. What Are Batteries, Fuel Cells, and Supercapacitors? [J]. Chemical Reviews, 2004, 104 (10): 4245–4270.

[7] Wang G, Zhang L, Zhang J. A review of electrode materials for electrochemicalsupercapacitors[J]. Chemical Society Reviews, 2012, 41(2): 797–828.

[8] Dong L, Xu C, Li Y, et al. Flexible electrodes and supercapacitors for wearable energy storage: a review by category [J]. Journal of Materials Chemistry A, 2016, 4(13): 4659–4685.

[9] Pérez–Madrigal MM, Edo M G, Alemán C. Powering the future: application of cellulose–based materials for supercapacitors[J]. Green Chemistry, 2016, 18(22): 5930–5956.

[10] Dubal D P, Kim J G, Kim Y, et al. Supercapacitors based on flexible substrates: an overview[J]. Energy Technology, 2014, 2(4): 325–341.

[11] Beidaghi M, Gogotsi Y. Capacitive energy storage in micro-scale devices: recent advances in design and fabrication of micro-supercapacitors[J]. Energy & Environmental Science, 2014, 7(3): 867–884.

[12] Xia X, Tu J, Zhang Y, et al. High–quality metal oxide core/shell nanowire arrays on conductive substrates for electrochemical energy storage. ACS Nano 6, 5531-5538 (2012)[J]. Energy Environ. Sci., 2012, 11: 2124–2133.

[13] Zhang G Q. HBwu, HE Hoster, MB Chan–Park, XW Lou[J]. Energy Environ. Sci, 2012, 5: 9453–9456.

[14] Peng S, Li L, Hao B W, et al. Controlled Growth of $NiMoO_4$ Nanosheet and Nanorod Arrays on Various Conductive Substrates as Advanced Electrodes for Asymmetric Supercapacitors[J]. Advanced Energy Materials, 2015, 5(2): 1–7.

[15] Zhang Y Z, Wang Y, Cheng T, et al. Flexiblesupercapacitors based on paper substrates: a new paradigm for low-cost energy storage[J]. Chemical Society Reviews, 2015, 44(15): 5181–5199.

[16] Hu L, Cui Y. Energy and environmental nanotechnology in conductive paper andtextiles[J]. Energy & Environmental Science, 2012, 5(4): 6423–6435.

[17] Weng Z, Su Y, Wang D W, et al. Graphene–cellulose paper flexible supercapacitors[J]. Advanced Energy

Materials, 2011, 1(5): 917-922.

[18] Hu L, Zheng G, Yao J, et al. Transparent and conductive paper from nanocellulose fibers[J]. Energy & Environmental Science, 2013, 6(2): 513-518.

[19] Yao B, Yuan L, Xiao X, et al. based solid-statesupercapacitors with pencil-drawing graphite/polyaniline networks hybrid electrodes[J]. Nano Energy, 2013, 2(6): 1071-1078.

[20] Jur J S, Sweet III W J, Oldham C J, et al. Atomic layer deposition of conductive coatings on cotton, paper, and synthetic fibers: conductivity analysis and functional chemical sensing using "all-fiber" capacitors[J]. Advanced functional materials, 2011, 21(11): 1993-2002.

[21] Dinderman M A, Dressick W J, Kostelansky C N, et al. Electroless Plating of Iron onto Cellulose Fibers[J]. Chemistry of Materials, 2006, 18(18): 4361-4368.

[22] Kelly P J, Arnell R D. Magnetron sputtering: a review of recent developments and applications[J]. Vacuum, 2000, 56(3): 159-172.

[23] Musil J, Baroch P, Vlcek J, et al. Reactive magnetron sputtering of thin films: present status and trends[J]. Thin Solid Films, 2005, 475(1-2): 208-218.

[24] Sarakinos K, Alami J, Konstantinidis S. High power pulsed magnetron sputtering: A review on scientific and engineering state of the art[J]. Surface & Coatings Technology, 2010, 204(11): 1661-1684.

[25] Legnani C, Vilani C, Calil V L, et al. Bacterial cellulose membrane as flexible substrate for organic light emitting devices[J]. Thin Solid Films, 2008, 517(3): 1016-1020.

[26] Fortunato E, Correia N, Barquinha P, et al. High-Performance Flexible Hybrid Field-Effect Transistors Based on Cellulose Fiber Paper[J]. IEEE Electron Device Letters, 2008, 29(9): 988-990.

[27] Lv P, Xu W, Li D, et al. Metal-based bacterial cellulose of sandwich nanomaterials for anti-oxidation electromagnetic interference shielding[J]. Materials & Design, 2016, 112: 374-382.

[28] Yi Q, Zhang J, Wu H, et al. Electrocatalytic oxidation of cyclohexanol on a nickel oxyhydroxide modified nickel electrode in alkaline solutions[J]. Catalysis Communications, 2007, 8(7): 1017-1022.

[29] Singh D P, Neti N R, Sinha A S K, et al. Growth of different nanostructures of Cu_2O (nanothreads, nanowires, and nanocubes) by simple electrolysis based oxidation of copper[J]. The Journal of Physical Chemistry C, 2007, 111(4): 1638-1645.

[30] Xiao X, Wang M, Li H, et al. Non-enzymatic glucose sensors based on controllable nanoporous gold/copper oxide nanohybrids[J]. Talanta, 2014, 125: 366-371.

[31] Dong C, Zhong H, Kou T, et al. Three-dimensional Cu foam-supported single crystalline mesoporous Cu_2O nanothorn arrays for ultra-highly sensitive and efficient nonenzymatic detection of glucose[J]. ACS applied materials & interfaces, 2015, 7(36): 20215-20223.

[32][1] Babu T G S, Ramachandran T, Nair B. Single step modification of copper electrode for the highly sensitive and selective non-enzymatic determination of glucose[J]. Microchimica Acta, 2010, 169(1-2): 49-55.

[33] Wang J, Zhang X, Wei Q, et al. 3D self-supportednanopine forest-like Co_3O_4@ $CoMoO_4$ core-shell architectures for high-energy solid state supercapacitors[J]. Nano Energy, 2016, 19: 222-233.

[34] Wan C, Jiao Y, Li J. Flexible, highly conductive, and free-standing reducedgraphene oxide/polypyrrole/cellulose hybrid papers for supercapacitor electrodes[J]. Journal of Materials Chemistry A, 2017, 5(8): 3819-3831.

[35] Stankovich S, Dikin D A, Piner R D, et al. Synthesis of graphene-based nanosheets via chemical reduction of exfoliated graphite oxide[J]. Carbon, 2007, 45(7): 1558-1565.

[36] Snook G A, Kao P, Best A S. Conducting-polymer-basedsupercapacitor devices and electrodes[J]. Journal of Power Sources, 2011, 196(1): 1-12.

[37] Patake V D, Joshi S S, Lokhande C D, et al. Electrodeposited porous and amorphous copper oxide film for application

in supercapacitor[J]. Materials Chemistry & Physics, 2009, 114(1): 6-9.

[38] Li Y, Chang S, Liu X, et al. NanostructuredCuO directly grown on copper foam and their supercapacitance performance[J]. Electrochimica Acta, 2012, 85: 393-398.

[39] Wu X, Bai H, Zhang J, et al. Copper hydroxide nanoneedle and nanotube arrays fabricated by anodization of copper [J]. The Journal of Physical Chemistry B, 2005, 109(48): 22836-22842.

[40] Zhang Z, Wang P. Highly stable copper oxide composite as an effective photocathode for water splitting via a facile electrochemical synthesisstrategy[J]. Journal of Materials Chemistry, 2012, 22(6): 2456-2464.

[41] Teo J J, Chang Y, Zeng H C. Fabrications of hollow nanocubes of Cu_2O and Cu via reductive self-assembly of CuO nanocrystals[J]. Langmuir, 2006, 22(17): 7369-7377.

[42] Wang K, Dong X, Zhao C, et al. Facile synthesis of Cu_2O/CuO/RGO nanocomposite and its superior cyclability in supercapacitor-ScienceDirect[J]. Electrochimica Acta, 2015, 152: 433-442.

[43] Jiang P, Prendergast D, Borondics F, et al. Experimental and theoretical investigation of the electronic structure of Cu_2O and CuO thin films on Cu (110) using x-ray photoelectron and absorption spectroscopy[J]. The Journal of Chemical Physics, 2013, 138(2): 024704.

[44] Zhong J H, Li G R, Wang Z L, et al. Facile electrochemical synthesis of hexagonal Cu_2O nanotube arrays and their application[J]. Inorganic Chemistry, 2011, 50(3): 757-763.

[45] Gelb L D, Gubbins K E. Characterization of Porous Glasses: Simulation Models, Adsorption Isotherms, and the BrunauerEmmettTeller Analysis Method[J]. Langmuir, 1998, 14(8): 2097-2111.

[46] Chmiola J, Yushin G, Dash R, et al. Effect of pore size and surface area of carbide derived carbons on specific capacitance[J]. Journal of Power Sources, 2006, 158(1): 765-772.

[47] Villarroel-Rocha J, Barrera D, Sapag K. Introducing a self-consistent test and the corresponding modification in the Barrett, Joyner and Halenda method for pore-size determination[J]. Microporous and Mesoporous Materials, 2014, 200: 68-78.

[48] Saliger R, Fischer U, Herta C, et al. High surface area carbon aerogels for supercapacitors[J]. 1998, 225(none): 0-85.

[49] Zhao B, Liu P, Zhuang H, et al. Hierarchical self-assembly of microscale leaf-like CuO on graphene sheets for high-performance electrochemical capacitors[J]. Journal of Materials Chemistry A., 2013, 1(2): 367-373.

[50] Kumar R, Rai P, Sharma A. Facile synthesis of Cu_2O microstructures and their morphology dependent electrochemical supercapacitor properties[J]. Rsc. Advances, 2016, 6(5): 3815-3822.

[51] Chen L, Zhang Y, Zhu P, et al. Copper salts mediated morphological transformation of Cu_2O from cubes to hierarchical flower-like or microspheres and their supercapacitors performances[J]. Scientific reports, 2015, 5(1): 1-7.

[52] Zhang W, Yin Z, Chun A, et al. Rose rock-shapednano Cu_2O anchored graphene for high-performance supercapacitors via solvothermal route[J]. Journal of Power Sources, 2016, 318: 66-75.

[53] Deng M J, Song C Z, Ho P J, et al. Three-dimensionally ordered macroporous Cu_2O/Ni inverse opal electrodes for electrochemical supercapacitors[J]. Physical Chemistry Chemical Physics, 2013, 15(20): 7479-7483.

[54] Dong C, Wang Y, Xu J, et al. 3D binder-free Cu_2O@ Cu nanoneedle arrays for high-performance asymmetric supercapacitors[J]. Journal of Materials Chemistry A., 2014, 2(43): 18229-18235.

[55] Fuku X, Kaviyarasu K, Matinise N, et al. Punicalagin Green Functionalized Cu/Cu_2O/ZnO/CuO Nanocomposite for Potential Electrochemical Transducer and Catalyst[J]. Nanoscale Research Letters, 2016, 11(1): 386.

[56] Kuang M, Li T T, Chen H, et al. Hierarchical Cu_2O/CuO/Co_3O_4 core-shell nanowires: synthesis and electrochemical properties[J]. Nanotechnology, 2015, 26(30): 304002.

[57] Ghasemi S, Jafari M, Ahmadi F. Cu_2O-$Cu(OH)_2$-graphene nanohybrid as new capacitive material for high performance supercapacitor[J]. Electrochimica Acta, 2016, 210: 225-235.

[58] Asen P, Shahrokhian S. A high performance supercapacitor based on graphene/polypyrrole/Cu_2O—Cu (OH)$_2$ ternary nanocomposite coated on nickel foam[J]. The Journal of Physical Chemistry C., 2017, 121(12): 6508-6519.

[59] Feng J X, Li Q, Lu X F, et al. Flexible symmetrical planar supercapacitors based on multi-layered MnO_2/Ni/graphite/paper electrodes with high-efficient electrochemical energy storage[J]. Journal of Materials Chemistry A., 2014, 2(9): 2985-2992.

[60] Liang K, Tang X, Hu W. High-performance three-dimensional nanoporous NiO film as a supercapacitor electrode [J]. Journal of Materials Chemistry, 2012, 22(22): 11062-11067.

[61] Guo Q, Zhou X, Li X, et al. Supercapacitors based on hybrid carbon nanofibers containing multiwalled carbon nanotubes[J]. Journal of Materials Chemistry, 2009, 19(18): 2810-2816.

[62] Qaiser A A, Hyland M M, Patterson D A. Control of polyaniline deposition on microporous cellulose ester membranes by in situ chemical polymerization. [J]. Journal of Physical Chemistry B., 2009, 113(45): 14986-14993.

[63] Ummartyotin S, Manuspiya H. An overview of feasibilities and challenge of conductive cellulose for rechargeable lithium based battery[J]. Renewable and Sustainable Energy Reviews, 2015, 50: 204-213.

[64] Kang Y J, Chun S J, Lee SS, et al. All-solid-state flexible supercapacitors fabricated with bacterial nanocellulose papers, carbon nanotubes, and triblock-copolymer ion gels. [J]. Acs Nano, 2012, 6(7): 6400-6406.

[65] Li S, D Huang, Zhang B, et al. Flexible Supercapacitors Based on Bacterial Cellulose Paper Electrodes[J]. Advanced Energy Materials, 2014.

[66] Dong C, Bai Q, Cheng G, et al. Flexible and ultralong-life cuprous oxide microsphere-nanosheets with superior pseudocapacitive properties[J]. RSC Advances, 2015, 5(8): 6207-6214.

[67] Yang P, Yong D, Lin Z, et al. Low-Cost High-Performance Solid-State Asymmetric Supercapacitors Based on MnO_2 Nanowires and Fe_2O_3 Nanotubes[J]. Nano Letters, 2014, 14(2): 731-736.

中文题目：一种纤维素纤维负载分层森林状氧化亚铜/铜阵列结构，作为对称超级电容器的柔性独立电极。

作者：万才超，焦月，李坚

摘要：在此，我们首次开发出两种简便快速的步骤(磁控溅射和电氧化)用在纤维素纸的三维纤维框架上生长层次化森林状 Cu_2O/Cu 阵列结构。在此结构中，铜棒作为主干，氧化产物(Cu_2O)作为分支。当用作柔性独立式电极时，这种独特的结构充分利用森林状阵列的分层多尺度结构的大界面面积、用于电解质离子从亲水性纤维素纸的多孔纤维骨架快速扩散的大量通道，以及 Cu_2O/铜复合物的快速电子传输和高电化学活性等优势使电极在 3.8 A·g^{-1} 时具有 915 F·g^{-1}(238 mF·cm^{-2})的高比电容，在 1.25 kW·kg^{-1} 时具有 53.7 W·h·kg^{-1} 的大比能量，优异的倍率性能和在循环 10000 次后的电容保持率仍为 91.7%的良好的循环稳定性。更重要的是，在 Cu_2O/Cu/纤维素复合纸的基础上，提出了一种简单有趣的策略来组装对称超级电容器，即在纤维素纸的两个表面上生长森林状的 Cu_2O/Cu 阵列，该结构在 1.9 A·g^{-1} 下的比电容为 409 F·g^{-1}(213 mF·cm^{-2})，在 0.625 kW·kg^{-1} 下的比能量为 24.0 W·h·kg^{-1}，且循环稳定性好(10000 次循环后的电容保持率为 90.2%)。这些结果揭示了复合纸作为高性能电极材料用于柔性储能装置和便携式电子设备的潜力。

关键词：高性能的超级电容器；能量储存

Scalable Synthesis and Characterization of Free-standing Supercapacitor Electrode Using Natural Wood As A Green Substrate to Support Rod-shaped Polyaniline*

Yue Jiao, Caichao Wan, Jian Li

Abstract: Natural wood slice was used as a green substrate to support rod-shaped polyaniline via a scalable easily-operated immersion-oxidative polymerization-freeze drying pathway. The scanning electron microscopy observations show that the wood surface was densely covered with plentiful polyaniline nanorods with diameters of 31-72 nm and lengths of 240-450 nm. The analysis of Fourier transform infrared spectroscopy provides further evidence of polyaniline coating onto the wood substrate. Moreover, the analysis of X-ray photoelectron spectroscopy indicates a strong hydrogen bonding between the nitrogen lone pairs (N) of polyaniline and the—OH groups of wood, which plays an important role in the interface bonding. This core-shell composite can serve as a free-standing supercapacitor electrode, which shows a high specific capacitance of 304 $F \cdot g^{-1}$ at 0.1 $A \cdot g^{-1}$, a high coulombic efficiency of 93%-100%, and a moderate cyclic stability with a capacitance retention of 72.3% after 5000 cycles. These make the nature wood a good alternative green substrate to develop novel eco-friendly energy storage devices.

Keywords: Wood; Polyaniline; Supercapacitors; Oxidative polymerization; Core-shell structure; Composites

1 Introduction

Electronically conducting polymers (ECPs) have been considered as promising materials for the realization of high-performance supercapacitors[1-3]. Compared with an electrochemical double-layer capacitor (EDLC), a pseudocapacitor containing ECPs generally stores a greater amount of capacitance per gram, because its charge process happens to the whole polymer mass and not only to the surface, as in the case of EDLC. Moreover, ECPs typically have high conductivities in the charged states and fast charge-discharge processes[4], which are conducive to developing electrochemical devices with low equivalent series resistance and high specific energy and power. As one of the main ECPs used for supercapacitors, polyaniline (PANI) has a history of more than 150 years. But until the discovery of switching characteristics between a conductor and an insulator under certain experimental conditions (in the early 1980s)[5,6], PANI has not captured the intense attention of the scientific community. PANI exists in a variety of forms that differ in physicochemical properties. The most common green protonated emeraldine has conductivity on a semiconductor level of the order of 10^0 $S \cdot cm^{-1}$, many orders of magnitude higher than that of common polymers ($<10^{-9}$ $S \cdot cm^{-1}$) but

* 本文摘自 Journal of Materials Science: Materials in Electronics, 2017, 28(3): 2634-2641.

lower than that of typical metals ($>10^4$ S · cm^{-1})[7]. Other main oxidation states include leucoemeraldine, blue protonated pernigraniline, and blue emeraldine base. The synthesis of PANI is usually implemented by oxidizing aniline monomer either chemically or electrochemically to form a cation radical followed by coupling to form dications, and the repetition leads to the polymer[8,9]. Besides good electrical conductivity and fast charge-discharge ability, PANI has reversible insulator-to-metal transitions[7] and electrochromic behavior (dependent on its oxidation state and pH)[10] as well as good stability in presence of air and humidity. Therefore, PANI has a good development prospect in numerous advanced fields like antistatic coating[11], chemical sensors[12], gas separation membranes[13], rechargeable batteries[14], and electrochromic devices[10]. In supercapacitors, PANI has been widely employed to integrate with various substrates to fabricate high-performance electrode materials. These substrates include bacterial cellulose[15], carbon cloth[16], carbon nanofiber[17], fluorine doped tin oxide[18], graphene hydrogel films[19], nickel[20], and polyvinyl alcohol[21].

Wood is the oldest material utilized by humans for construction after stone. In modern times, wood has been widely applied in various fields like building industry, wooden furniture making, interior decoration, papermaking, bio-refinery, and industrial or household fuel. Apart from these common applications, wood has also been employed as a substrate material to combine with a variety of nanomaterials for some advanced applications[22-25], because of its renewability, porous structure, low density, low thermal expansion, good workability, and desirable mechanical strength. Our previous researches also confirm that wood can serve as a good substrate to support Nano-ZrO_2[26] and graphene sheets[27]. Compared with the aforementioned substrates for supercapacitor applications, wood is undoubtedly cheaper, eco-friendly and widely available. Also, wood can be directly used without any complex costly purification and preparation technologies. Wood is known to have hierarchically porous structure, which facilitates the fast penetration of electrolytes to allow rapid electron-transfer for charge storage and delivery[28]. Also, its fibrous structures can substantially absorb electrolyte and act as electrolyte reservoirs to facilitate ion transport[29]. Wood has abundant oxygen-containing groups on its surface due to its main compositions including cellulose, hemicellulose and lignin. Thus hydrogen bonding is expected to occur between the nitrogen lone pairs (N) of aniline (or PANI) and the —OH groups of wood, which is believed to be beneficial to acquire good interface bonding. Considering that recent literatures[28,30] pay little attention to combining natural wood with electrochemical active materials for supercapacitor applications, it is meaningful to synthesize PANI/wood electrode as a typical example of wood-based electrodes and investigate its electrochemical properties.

In the present work, we adopted natural wood as a green substrate material to support rod-shaped polyaniline. The PANI/wood composite was synthesized by a scalable, but simple and efficient, approach that consists of immersion in the solution of aniline monomers, in-situ oxidative polymerization, and freeze drying. The as-prepared composite was characterized by scanning electron microscopy (SEM), energy dispersive X-ray spectroscopy (EDX), Fourier transform infrared spectroscopy (FTIR), and X-ray photoelectron spectroscopy (XPS). The electrochemical properties were studied through cyclic voltammograms (CV) and galvanostatic charge/discharge (GCD) tests in a three-electrode configuration in 1 M H_2SO_4 aqueous electrolyte. Moreover, a possible growth mechanism of PANI on the wood was proposed.

2 Materials and methods

2.1 Materials

All chemicals of analytical grade were supplied by Kemiou Chemical Reagent Co., Ltd. (Tianjin, China) and used as received. Natural wood slices without any chemical treatment were cut to 55 mm^3 × 55 mm^3 × 1 mm^3. The slices were ultrasonically rinsed with distilled water for 30 min and dried at 60 ℃ for 24 h in a vacuum.

2.2 Synthesis of the PANI/wood composite

The PANI/wood composite was prepared referring to the literature method[15] with little modification. The literature suggested some optimized synthetic protocols, which were also adopted in our work. In a typical process, the dried wood slice (ca. 0.75g) was immersed into aniline monomer solution (100 mL) for 24 h with magnetic stirring at room temperature, enabling the aniline monomers to fully self-assemble onto the surface of wood via hydrogen bonding. The mass ratio of wood/aniline was 1∶10. After the immersion, the above system was cooled to 0 in an ice-water bath. Then the mixture of oxidant (ammonium peroxide sulfate, 0.08 mol) and dopant (hydrochloric acid, 1 M, 96 mL) was dropwise added into the system and kept at 0–10 ℃ for 4 h. Finally, the wood slice was successively rinsed with distilled water and tert-butyl alcohol, and freeze-dried for 24 h at 25 Pa.

2.3 Characterization

SEM observation was performed with a Hitachi S4800 SEM equipped with an EDX detector for element analysis. XPS was carried out using a Thermo Escalab 250Xi XPS spectrometer equipped with a dual X-ray source using Al-Kα. Deconvolution of the overlapping peaks was performed using a mixed Gaussian-Lorentzian fitting program (Origin 9.0, Originlab Corporation). FTIR were recorded by a Nicolet Nexus 670 FTIR instrument in the range of 400–4000 cm^{-1} with a resolution of 4 cm^{-1}.

2.4 Electrochemical measurements

The electrochemical measurements were carried out on a CS350 electrochemical workstation (Wuhan CorrTest Instruments Co., Ltd., China) at room temperature in a three-electrode setup. The PANI/wood composite served as the working electrode. An Ag/AgCl electrode and a Pt wire electrode served as reference and counter electrodes, respectively. The electrolyte was 1 M H_2SO_4 solution. CV curves were measured over the potential window from −0.2 to 1.0 V at different scan rates of 5, 10, 20 and 50 mV · s^{-1}. GCD curves were measured in the potential range of 0–0.8 V at different current densities of 0.1, 0.2, 0.5 and 1.0 A · g^{-1}.

3 Results and discussion

3.1 Morphology observations and elemental analysis

Morphological characteristics before and after PANI was polymerized onto the wood were studied by SEM observations. Fig. 1a shows obviously that the native wood substrate is a porous material, and the diameters of its pits range from 3 to 7 μm. After PANI was polymerized on the wood, it can be seen that the wood surface was densely covered with a great number of nanomaterials (Fig. 1b), leading to the transformation of the wood surface color from light yellow to black (insets in Fig. 1a, b). According to a higher-magnification SEM image (160, 000×) shown in Fig. 1c, it is clear that these PANI nanorods have diameters of 31−72 nm and lengths of 240−450 nm, and are randomly intertwined with each other. EDX analysis was carried out to obtain additional information about the elemental compositions. As shown in Fig. 1d, apart from the common C, O and Au (originated from the coating layer used for electric conduction during the SEM observation), elements including N and S were detected from the EDX pattern of the PANI/wood. The N and S elements are derived from the PANI coating and residual chemicals from ammonium peroxide sulfate, respectively. Moreover, the increased mass ratio of carbon to oxygen from 2. 6 (the native wood) to 6. 2 (the PANI/wood) is associated with the formed core-shell structure, i. e. , the wood coated with plentiful rod-shaped PANI.

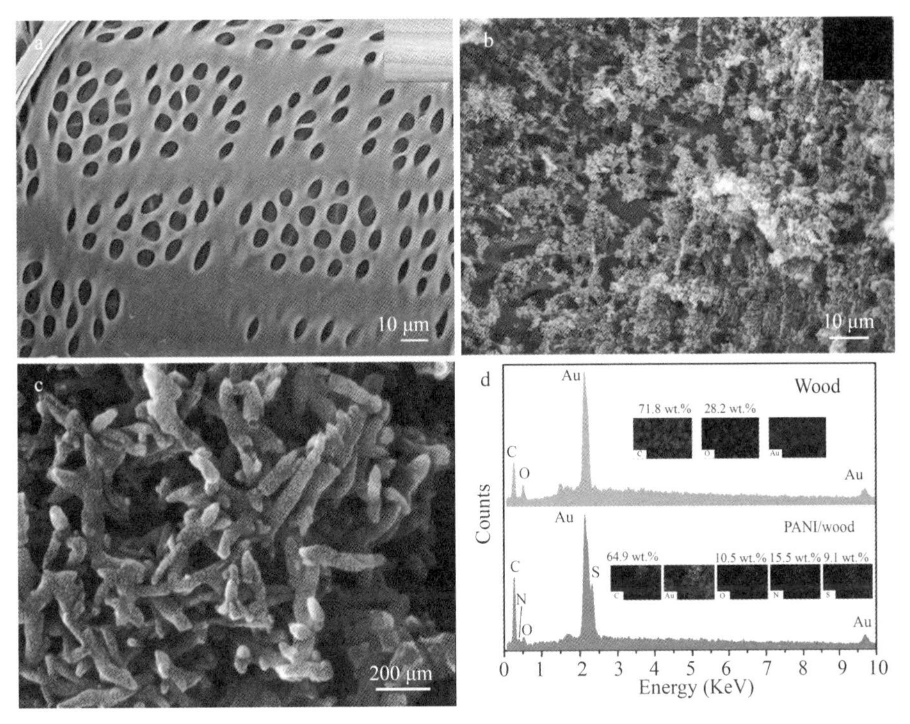

Fig. 1 (a) SEM image of the native wood. (b) Low- and (c) high-magnification SEM images of the PANI/wood composite. The insets in (a, b) present the surface color. (d) EDX patterns of the native wood and the PANI/wood composite, and the insets shows the elemental maps

3.2 FTIR and XPS analysis

The surface chemical compositions of the native wood and the PANI/wood composite were investigated by FTIR analysis. As shown in Fig. 2, in the case of the native wood, the broad adsorption band centered at 3341 cm^{-1} is attributed to the O—H stretching. The band at 2895 cm^{-1} is assigned to the C—H stretching. In addition, some characteristic bands related to the main compositions of wood (i. e. , cellulose, hemicellulose and lignin) can be clearly identified. The bands at 1423 cm^{-1}, 1032 cm^{-1} and 897 cm^{-1} are related to cellulose structure[31]; the band at 1506 cm^{-1} belongs to the aromatic skeletal vibrations of lignin[32]; the band at 1731 cm^{-1} is attributed to the C=O stretching in unconjugated ketones, carbonyls and in ester groups of hemicellulose[33]. The main signals of the wood and their corresponding assignments are provided in Table

1. In the case of the PANI/wood composite, almost no bands of wood are observed, suggesting that the deposited PANI nanorods on the wood surface cover the characteristic structures of wood. In addition, the new bands at 1564 and 1484 cm^{-1} correspond to the C =C stretching deformation of quinoid ring and benzenoid ring[34], respectively. The bands at 1294 and 793 cm^{-1} arise from the C—N in-plane ring bending modes and the aromatic C—H out of plane bending vibration[35], respectively. These characteristic bands further conform that the surface of wood was sufficiently coated with PANI.

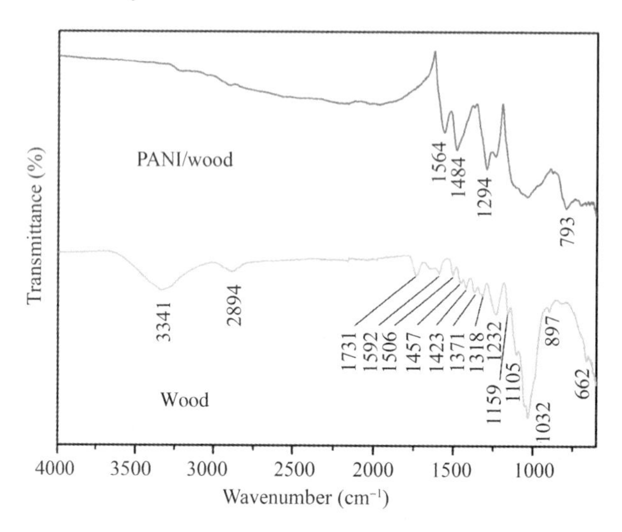

Fig. 2 FTIR spectra of the native wood and the PANI/wood composite, respectively

Table 1 Frequencies (cm^{-1}) of the main signals of the native wood and the PANI/wood composite

Absorption band (cm^{-1})	Assignment	Sample
3341	O—H stretching	Wood
2894	C—H stretching	Wood
1731	C=O stretching in unconjugated ketones, carbonyls and in ester groups	Wood
1592	Aromatic skeletal vibrations plus C=O stretching	Wood
1506	Aromatic skeletal vibrations	Wood
1457	CH_2 of pyran ring symmetric scissoring; OH plane deformation vibration	Wood
1423	Aromatic skeletal vibrations combined with C—H in plane deformation; CH_2 symmetric bending	Wood
1371	CH deformation vibration	Wood
1318	Condensation of guaiacyl unit and syringyl unit, syringyl unit and CH_2 bending stretching	Wood
1232	OH plane deformation, also COOH	Wood
1159	C—O—C stretching in pyranose rings, C=O stretching in aliphatic groups	Wood
1105	Ring asymmetric valence vibration	Wood
1032	Aromatic C—H in plane deformation; plus C—O deformation in primary alcohols; plus C=O stretch (unconjugated)	Wood
897	Anomere C-groups, C_1-H deformation, ring valence vibration	Wood

(continued)

Absorption band (cm^{-1})	Assignment	Sample
662	C—OH out-of-plane bending mode	Wood
1564	C=C stretching deformation of quinoid ring	PANI/wood
1484	C=C stretching deformation of benzenoid ring	PANI/wood
1294	C—N in-plane ring bending modes	PANI/wood
793	aromatic C—H out of plane bending vibration	PANI/wood

The presence of various species on the surface of the PANI/wood composite was studied by deconvoluting the O 1s, C 1s and N 1s core-level XPS spectra. Fig. 3b shows the oxygen XPS spectrum of the composite, in which there are distinct changes when compared to the same spectra for the native wood (Fig. 3a). A shoulder peak at 531.3 eV is observed, associated directly with C—OH bonding to N[36]. Moreover, the shift in the O 1s peak relating to C—OH bonds, from 532.8 to 531.3 eV for the native wood and the PANI/wood, further confirms the presence of a chemical interaction between OH and N. The shift is because the hydrogen bonding between the nitrogen lone pairs of PANI and the hydroxyl groups of wood decreases the binding energy of the oxygen from which the attached hydrogen atoms are being delocalized[36,37]. The C 1s spectrum of the native wood can be deconvoluted into three subpeaks at binding energies of 284.6, 286.2 and 287.4 eV (Fig. 3c), corresponding to C—H/C—C, C—O and C═O [38], respectively. The PANI coating introduced an additional subpeak at 285.1 eV shown in Fig. 3d, which corresponds to C—C/C—N/C═N. The N 1s spectrum of the PANI/wood was deconvoluted into four component peaks at 398.5 eV (—N═), 399.6 eV (—NH—), 400.8 eV ($—N^+—$), and 403.2 eV ($—N^+—$)[39,40] (Fig. 3e). The protonated nitrogen ($—N^+—$) preferentially derives from the imine nitrogen of PANI due to its higher dissociation constant as compared to that of amine ($pK_a=5.5$ for $-NH^+=$ versus 2.5 for $-NH_2$)[41,42]. In addition, the appearance of multiple peaks for positively charged nitrogen is related to localization/delocalization of the positive charge on nitrogen due to the variable association of doping Cl^- ions[38,43,44]. The delocalized nitrogen in protonated imine corresponds to the lower-BE peak and the more localized positively charged nitrogen results in the higher-BE peak.

3.3 Schematic diagram for the preparation of the PANI/wood composite

As shown in Fig. 4a, the PANI/wood composite was prepared in three steps: (1) dipping the native wood slice into the aqueous solution of aniline monomers; (2) adding oxidant [$(NH_4)_2S_2O_8$)] and dopant (HCl); (3) washing and freeze-drying the wood slice. During the dipping process, hydrogen bonding is expected to occur between the nitrogen lone pairs (N) of aniline and the —OH groups of wood substrate, allowing the aniline monomers to fully self-assemble onto the surface of wood. With the addition of the oxidant, the in-situ oxidative polymerization of aniline took place along with doping by HCl. The detailed polymerization mechanism of aniline can consult some previous literatures[45,46]. The final freeze-drying treatment worked by freezing the material and then reducing the surrounding pressure to allow the frozen liquids in the material to sublimate directly from the solid phase to the gas phase. The freeze-drying method was employed to maintain the morphology of PANI before and after the drying process.

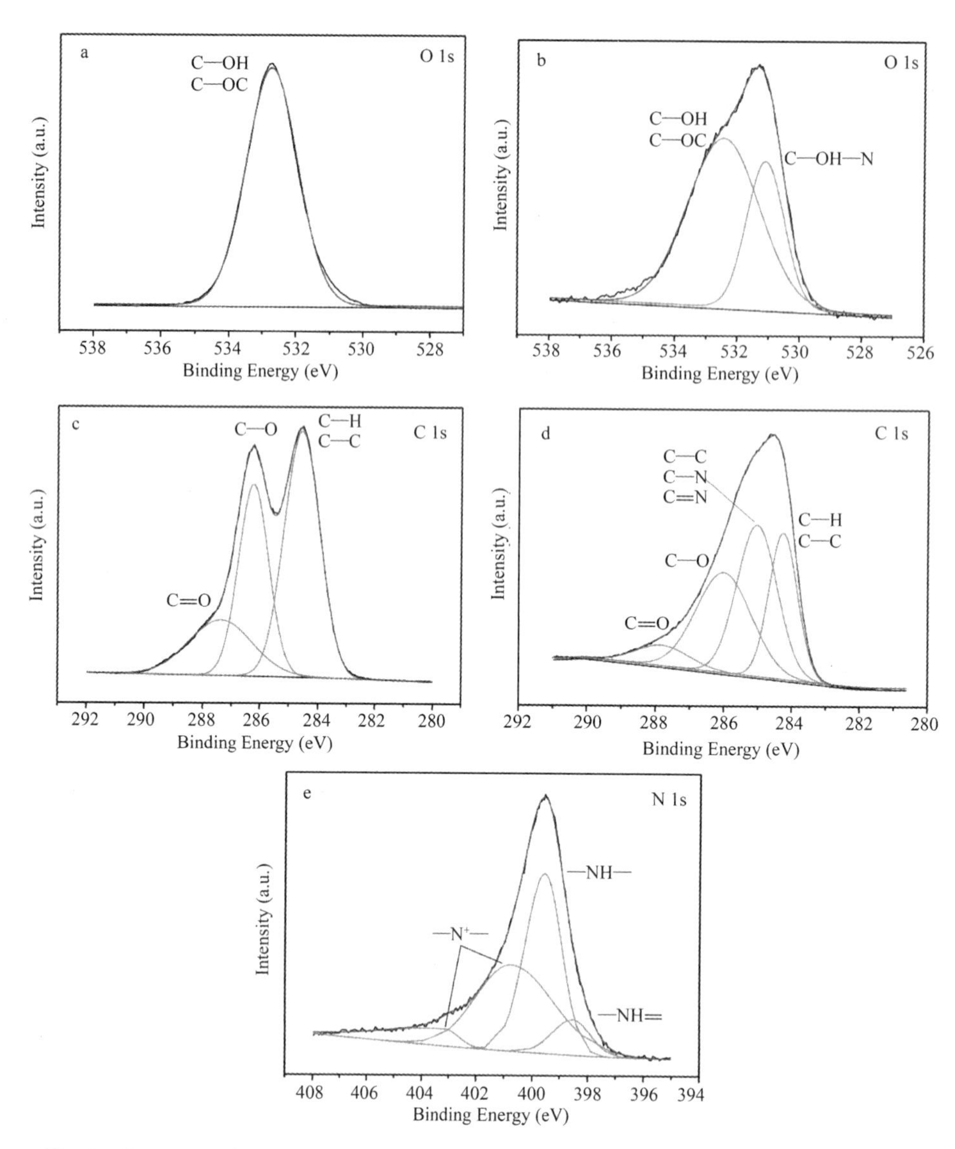

Fig. 3 O 1s and C 1s core-level XPS spectra of (a, c) the native wood and (b, d) the PANI/wood composite, respectively. (e) N 1s core-level XPS spectrum of the PANI/wood composite

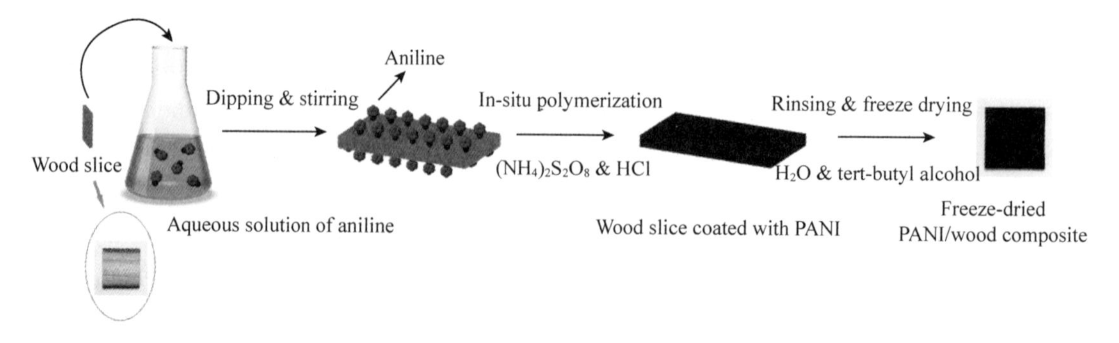

Fig. 4 Schematic diagram for the preparation of the PANI/wood composite

3.4 Electrochemical properties of the PANI/wood composite

The electrochemical properties of the PANI/wood composite were studied through CV and GCD measurements in a three-electrode configuration in 1 M H_2SO_4 aqueous electrolyte. Fig. 5a shows the CV curves in the potential of -0.2–1.0 V at scan rates of 5, 10, 20 and 50 mV · s^{-1}. It is seen that all the curves have two pairs of oxidation and reduction peaks. The peaks C_1/A_1 are ascribed to the redox transition between a semiconducting state (leucoemeraldine form) and a conducting state (polaronic emeraldine form), and the peaks C_2/A_2 are related to the Faradaic transformation of emeraldine-pernigraniline[47]. In addition, we can observe that the cathodic peaks (C_1/C_2) shifted positively and the anodic peaks (A_1/A_2) shifted negatively with the increasing scan rate from 5 to 50 mV · s^{-1}. The internal resistance of the PANI/wood electrode might be responsible for these shifts. Fig. 5b presents the anodic and cathodic peak currents as a function of scan rate. Apparently, there is an approximately linear relationship between anodic (or cathodic) peak current and scan rate, which indicates fast redox process of PANI[35].

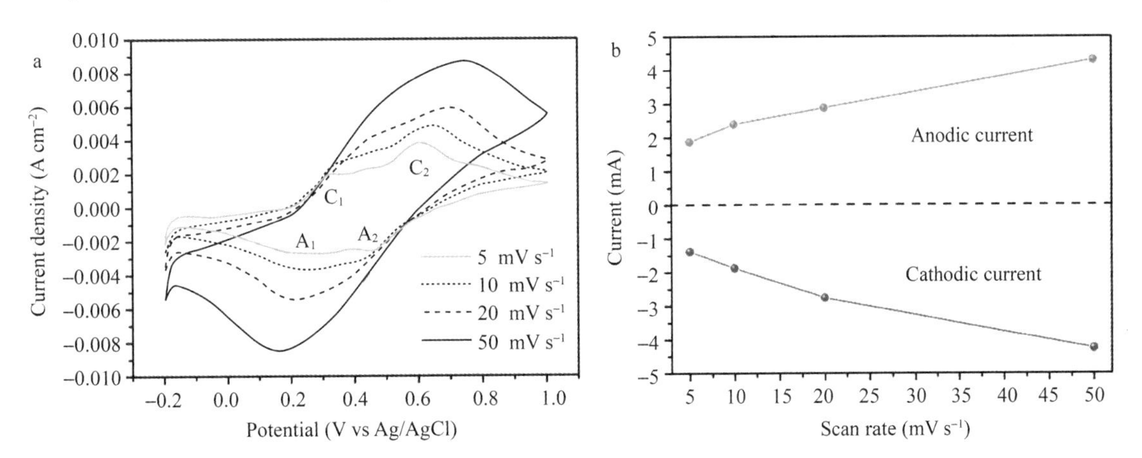

Fig. 5 (a) CV curves of the PANI/wood electrode at scan rates of 5, 10, 20 and 50 mV · s^{-1}. (b) Anodic and cathodic peak currents of the PANI/wood electrode against the CV scan rate

The GCD curves of the PANI/wood electrode at current densities of 0.1, 0.2, 0.5 and 1.0 A · g^{-1} are presented in Fig. 6a. The drop at the initial discharge originates from the internal resistance of the electrode. All the curves aren't ideal straight line, suggesting the process of a faradic reaction. The specific capacitance at the different current densities was calculated as follows:

$$C_m = \frac{I \times \Delta t}{\Delta V \times m} \quad (1)$$

where C_m is the specific capacitance (F · g^{-1}), I is the charge-discharge current (A), Δt is the discharge time (s), ΔV is the potential window (V), and m is the mass of active material (g). Besides, high-rate dischargeability (HRD) was used to study the power property of the electrode. The HRD was calculated as follows:

$$HRD\ (\%) = \frac{C_d}{C_1} \times 100 \quad (2)$$

where C_d and C_1 are the discharge capacity of electrode at a certain current density and 0.1 A · g^{-1}, respectively. The specific capacitance and HRD of the PANI/wood electrode were plotted versus discharge current in Fig. 6b. As shown, the electrode has a high specific capacitance of 304 $F\cdot g^{-1}$ at a current density of 0.1 $A\cdot g^{-1}$. In addition, the capacitance significantly decreased with the increasing current density. The HRD decreased to 92.2%, 37.3% and 12.9% as the current density increased to 0.2, 0.5 and 1.0 $A\cdot g^{-1}$, respectively. Thus the PANI/wood electrode is more suitable to be utilized in lower current density than 0.2 $A\cdot g^{-1}$.

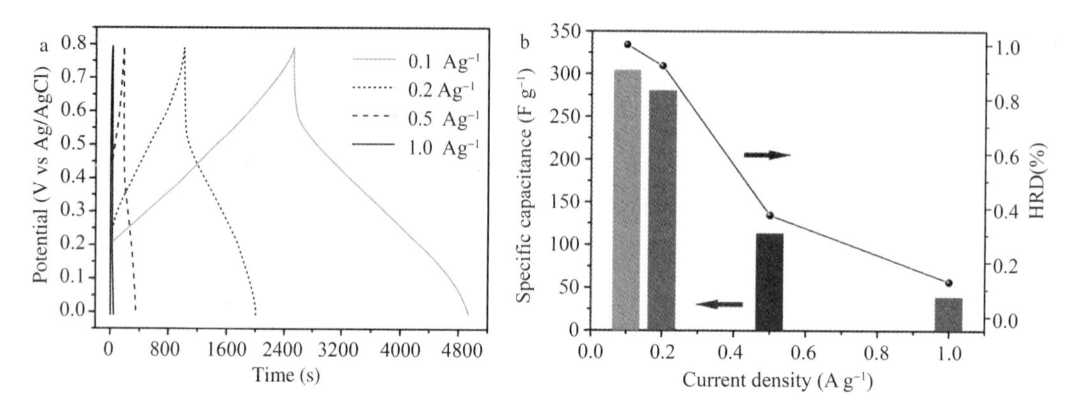

Fig. 6 (a) GCD curves and (b) specific capacitance and HRD of the PANI/wood electrode at different current densities of 0.1, 0.2, 0.5 and 1.0 $A\cdot g^{-1}$

The cycling performance of the PANI/wood electrode was studied at a current density of 0.2 $A\cdot g^{-1}$ in 1 M H_2SO_4 aqueous electrolyte. It is found that the electrode retained 72.3% of initial capacitance after 5000 cycles (Fig. 7). The capacitance retention can be improved by integrating the PANI/wood electrode with carbon materials (typically like carbon nanofibers, carbon nanotubes and graphene)[48,49] or transition metal oxides (typically like RuO_2, NiO and MnO_2)[50,51], which will be studied in our future researches. Moreover, the coulombic efficiency which is the ratio of charge capacitance and discharge capacitance, exhibits retention of as high as 93%-100%, indicating a good electrochemical reversibility.

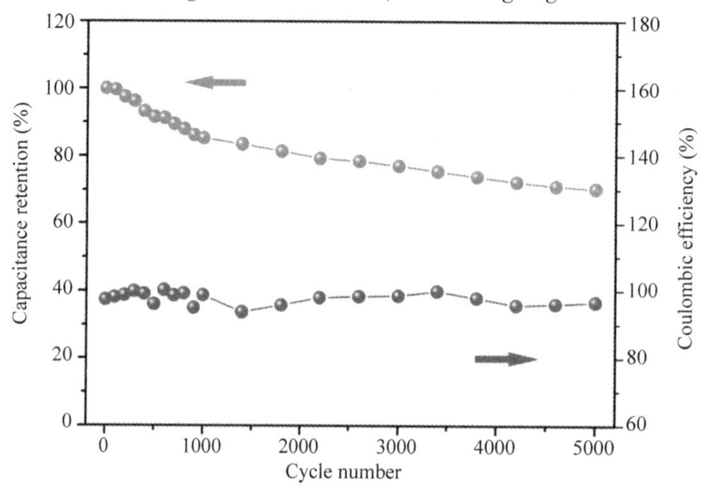

Fig. 7 Charge/discharge cycle of the PANI/wood electrode at a current density of 0.2 $A\cdot g^{-1}$ (left axis: capacitance retention, right axis: coulombic efficiency)

4 Conclusions

A scalable and facile immersion-oxidative polymerization-freeze drying approach was implemented to grow rod-shaped PANI on the natural wood slice. A strong hydrogen bonding occurred between the nitrogen lone pairs (N) of polyaniline and the —OH groups of wood. This core-shell composite can be utilized as a free-standing supercapacitor electrode. The electrode has a high specific capacitance of 304 $F \cdot g^{-1}$ at a current density of 0. 1 $A \cdot g^{-1}$, a high coulombic efficiency of 93% - 100% during the charge-discharge processes, and a moderate cyclic stability with a capacitance retention of 72. 3% after 5000 cycles. The cyclic stability is expected to be improved by integrating the electrode with carbon materials or metal oxides, which will be further studied in our future researches. In summary, this work suggests that the nature wood can act as a potential green substrate to develop novel and eco-friendly energy storage devices.

Acknowledgments

This study was supported by the National Natural Science Foundation of China (grant nos. 31270590 and 31470584) and the Fundamental Research Funds for the Central Universities (grant nos. 2572016AB22).

References

[1] Rudge A, Davey J, Raistrick I, et al. Conducting polymers as active materials in electrochemical capacitors[J]. Journal of Power Sources, 1994, 47(1-2): 89-107.

[2] Snook G A, Kao P, Best A S. Conducting-polymer-basedsupercapacitor devices and electrodes[J]. Journal of Power Sources, 2011, 196(1): 1-12.

[3] Mastragostino M, Arbizzani C, Soavi F. Conducting polymers as electrode materials in supercapacitors[J]. Solid State Ionics, 2002, 148(3-4): 493-498.

[4] Handbook of conductingpolymers[M]. CRC Press, 1997.

[5] Negi Y S, Adhyapak P V. Development in polyaniline conducting polymers[J]. Journal of Macromolecular Science, Part C: Polymer Reviews, 2002, 42(1): 35-53.

[6] Molapo K M, Ndangili P M, Ajayi R F, et al. Electronics of conjugated polymers (Ⅰ): polyaniline[J]. 2012.

[7] Stejskal J, Gilbert R G. Polyaniline. Preparation of a conducting polymer (IUPAC technical report)[J]. Pure and Applied Chemistry, 2002, 74(5): 857-867.

[8] Bhadra S, Khastgir D, Singha N K, et al. Progress in preparation, processing and applications of polyaniline[J]. Progress in Polymer Science, 2009, 34(8): 783-810.

[9] Blinova N V, Stejskal J, Trchová M, et al. Polyaniline and polypyrrole: A comparative study of the preparation[J]. European Polymer Journal, 2007, 43(6): 2331-2341.

[10] Lacroix J C, Kanazawa K K, Diaz A. Polyaniline: A very fast electrochromic material[J]. Journal of the Electrochemical Society, 1989, 136(5): 1308.

[11] Ohtani A, Abe M, Ezoe M, et al. Synthesis and properties of high-molecular-weight soluble polyaniline and its application to the 4MB-capacity barium ferrite floppy disk's antistatic coating[J]. Synthetic Metals, 1993, 57(1): 3696-3701.

[12] Huang J, Virji S, Weiller B H, et al. Polyaniline nanofibers: facile synthesis and chemical sensors[J]. Journal of the American Chemical Society, 2003, 125(2): 314-315.

[13] Illing G, Hellgardt K, Wakeman R J, et al. Preparation and characterisation of polyaniline based membranes for gas separation[J]. Journal of Membrane Science, 2001, 184(1): 69-78.

[14] MacDiarmid A G, Yang L S, Huang W S, et al. Polyaniline: Electrochemistry and application to rechargeable batteries[J]. Synthetic Metals, 1987, 18(1-3): 393-398.

[15]Wang H, Zhu E, Yang J, et al. Bacterial cellulosenanofiber-supported polyaniline nanocomposites with flake-shaped morphology as supercapacitor electrodes[J]. The Journal of Physical Chemistry C., 2012, 116(24): 13013-13019.

[16]Horng Y Y, Lu Y C, Hsu Y K, et al. Flexible supercapacitor based on polyaniline nanowires/carbon cloth with both high gravimetric and area-normalized capacitance[J]. Journal of Power Sources, 2010, 195(13): 4418-4422.

[17]Jang J, Bae J, Choi M, et al. Fabrication and characterization of polyaniline coated carbon nanofiber for supercapacitor[J]. Carbon, 2005, 43(13): 2730-2736.

[18]Deshmukh P R, Shinde N M, Patil S V, et al. Supercapacitive behavior of polyaniline thin films deposited on fluorine doped tin oxide (FTO) substrates by microwave-assisted chemical route[J]. Chemical Engineering Journal, 2013, 223 (Complete): 572-577.

[19]Wang Y, Yang X, Qiu L, et al. Revisiting the capacitance of polyaniline by using graphene hydrogel films as a substrate: the importance of nano-architecturing[J]. Energy & Environmental Science, 2013, 6(2): 477-481.

[20]Girija T C, Sangaranarayanan M V. Analysis of polyaniline-based nickel electrodes for electrochemical supercapacitors [J]. Journal of Power Sources, 2006, 156(2): 705-711.

[21]Patil D S, Shaikh J S, Dalavi D S, et al. Chemical synthesis of highly stable PVA/PANI films for supercapacitor application[J]. Materials Chemistry and Physics, 2011, 128(3): 449-455.

[22]Okahisa, Yoshi da, Miyaguchi, et al. Optically transparent wood-cellulose nanocomposite as a base substrate for flexible organic light-emitting diode displays[J]. Composites Sci. Technol, 2009.

[23]Franziska, Weichelt, Rico, et al. ZnO-Based UV Nanocomposites for Wood Coatings in Outdoor Applications[J]. Macromolecular Materials & Engineering, 2010.

[24]Teng S, Siegel G, Prestgard M C, et al. Synthesis and characterization of copper-infiltrated carbonized wood monoliths for supercapacitor electrodes[J]. Electrochimica Acta., 2015, 161: 343-350.

[25]Nagasawa C, Kumagai Y, Urabe K, et al. Electromagnetic shielding particleboard with nickel-plated wood particles [J]. Journal of Porous Materials, 1999, 6(3): 247-254.

[26]Wan C, Lu Y, Sun Q, et al. Hydrothermal synthesis of zirconium dioxide coating on the surface of wood with improved UVresistance[J]. Applied Surface Science, 2014, 321: 38-42.

[27]Wan C, Jiao Y, Li J. In situ deposition ofgraphene nanosheets on wood surface by one-pot hydrothermal method for enhanced UV-resistant ability[J]. Applied Surface Science, 2015, 347: 891-897.

[28]Lv S, Fu F, Wang S, et al. Novel wood-based all-solid-state flexible supercapacitors fabricated with a natural porous wood slice and polypyrrole[J]. RSC Advances, 2015, 5(4): 2813-2818.

[29]Weng Z, Su Y, Wang D W, et al. Graphene-cellulose paper flexible supercapacitors[J]. Advanced Energy Materials, 2011, 1(5): 917-922.

[30]S. Lv, F. Fu, S. Wang, J. Huang, L. Hu, Electron. Mater, Let, 11: 633.

[31]Lan W, Liu C F, Yue F X, et al. Ultrasound-assisted dissolution of cellulose in ionic liquid[J]. Carbohydrate Polymers, 2011, 86(2): 672-677.

[32]Schwanninger M, Rodrigues J C, Pereira H, et al. Effects of short-time vibratory ball milling on the shape of FT-IR Spectra of wood and cellulose[J]. Vibrational Spectroscopy., 2004, 36(1): 23-40.

[33]Chen H, Ferrari C, Angiuli M, et al. Qualitative and quantitative analysis of wood samples by Fourier transform infrared spectroscopy and multivariate analysis[J]. Carbohydrate Polymers, 2010, 82(3): 772-778.

[34]Yelil Arasi A, Juliet L J J, Sundaresan B, et al. The structural properties of Poly(aniline)-Analysis via FTIR spectroscopy[J]. Spectrochimica Acta Part A., 2009.

[35]Mi H, Zhang X, Yang S, et al. Polyaniline nanofibers as the electrode material for supercapacitors[J]. Materials Chemistry and Physics, 2008, 112(1): 127-131.

[36]Kelly F M, Johnston J H, Borrmann T, et al. Functionalised hybrid materials of conducting polymers with individual fibres of cellulose[J]. 2007: 5571-5577.

[37]Wagner C D, Zatko D A, Raymond R H. Use of the oxygen KLL Auger lines in identification of surface chemical

states by electron spectroscopy for chemical analysis[J]. Analytical Chemistry, 1980, 52(9): 1445-1451.

[38]Qaiser A A, Hyland M M, Patterson D A. Surface and charge transport characterization of polyaniline-cellulose acetate composite membranes[J]. The Journal of Physical Chemistry B., 2011, 115(7): 1652-1661.

[39]Han M G, Cho S K, Oh S G, et al. Preparation and characterization ofpolyaniline nanoparticles synthesized from DBSA micellar solution[J]. Synthetic Metals, 2002, 126(1): 53-60.

[40]Neoh K G, Pun M Y, Kang E T, et al. Polyaniline treated with organic acids: doping characteristics and stability[J]. Synthetic Metals, 1995, 73(3): 209-215.

[41]Tan S, Belanger D. Characterization and transport properties ofNafion/polyaniline composite membranes[J]. The Journal of Physical Chemistry B, 2005, 109(49): 23480-23490.

[42]Nagarale R K, Gohil G S, Shahi V K, et al. Preparation and electrochemical characterization of cation-and anion-exchange/polyaniline composite membranes[J]. Journal of colloid and interface science, 2004, 277(1): 162-171.

[43]Wei X L, Fahlman M, Epstein A J. XPS study of highly sulfonated polyaniline[J]. Macromolecules, 1999, 32(9): 3114-3117.

[44]Yue J, Epstein A J. XPS study of self-doped conducting polyaniline and parent systems[J]. Macromolecules, 1991, 24(15): 4441-4445.

[45]Tang S J, Wang A T, Lin S Y, et al. Polymerization of aniline under various concentrations of APS andHCl[J]. Polymer Journal, 2011, 43(8): 667-675.

[46]Gospodinova N, Terlemezyan L. Conducting polymers prepared by oxidative polymerization: polyaniline[J]. Progress in Polymer Science, 1998, 23(8): 1443-1484.

[47] Wang D W, Li F, Zhao J, et al. Fabrication ofgraphene/polyaniline composite paper via in situ anodic electropolymerization for high-performance flexible electrode[J]. ACS Nano, 2009, 3(7): 1745-1752.

[48]Zhang J, Zhao X S. Conducting polymers directly coated on reducedgraphene oxide sheets as high-performance supercapacitor electrodes[J]. The Journal of Physical Chemistry C., 2012, 116(9): 5420-5426.

[49]Al-Saleh M H, Sundararaj U. A review of vapor grown carbon nanofiber/polymer conductive composites[J]. Carbon, 2009, 47(1): 2-22.

[50]Gangopadhyay R, De A. Conducting polymer nanocomposites: a brief overview[J]. Chemistry of Materials, 2000, 12(3): 608-622.

[51]Hou Y, Cheng Y, Hobson T, et al. Design and synthesis of hierarchical MnO_2 nanospheres/carbon nanotubes/conducting polymer ternary composite for high performance electrochemical electrodes. [J]. Nano Letters, 2010, 10(7): 2727.

中文题目：以天然木材为绿色基材支撑棒状聚苯胺的自立式超级电容器电极的可伸缩合成与表征

作者：焦月，万才超，李坚

摘要：以天然木片为绿色基材，通过可扩展、易操作的浸泡-氧化聚合-冷冻干燥过程支撑棒状聚苯胺。扫描电镜观察表明，木材表面密布着大量的聚苯胺纳米棒，直径为31~72 nm，长度为240~450 nm。傅立叶变换红外光谱分析进一步证明了聚苯胺涂层在木材表面的存在。此外，X射线光电子能谱分析表明，聚苯胺的氮孤对与木材的-OH基团之间存在较强的氢键，这在界面键合中起着重要的作用。这种核壳复合材料可以作为独立的超级电容器电极，在0.1 $A \cdot g^{-1}$下表现出304 $F \cdot g^{-1}$的高比电容，93%~100%的高库仑效率，以及中等的循环稳定性，5000次循环后的容量保持率为72.3%。这些都使得天然木材成为开发新型环保储能装置的良好替代绿色基材。

关键词：木材；聚苯胺；超级电容器；氧化聚合；核壳结构；复合材料

Anatase TiO_2/Cellulose Hybrid Paper: Synthesis, Characterizations, and Photocatalytic Activity for Degradation of Indigo Carmine Dye*

Yue Jiao, Caichao Wan, Jian Li

Abstract: We report a facile easy method to deposit anatase titania (TiO_2) on cellulose paper. The anatase TiO_2/cellulose paper (ATCP) was characterized by scanning electron microscopy, transmission electron microscope, energy dispersive X-ray spectrometer, X-ray diffraction, Fourier transform infrared spectroscopy, X-ray photoelectron spectroscopy, and thermogravimetric analysis. This hybrid paper with the anatase TiO_2 content of around 13.86 wt.% can serve as an eco-friendly flexible photocatalyst, which can rapidly degrade blue indigo carmine dye into colorless within 30 min under UV radiation. Moreover, compared to commercially available TiO_2 P25 and anatase TiO_2 powder, a faster decomposition rate of indigo carmine dye was acquired when using ATCP. These results suggest that this hybrid paper might be useful in the treatment of organic dye wastewater.

Keywords: Cellulose; Anatase TiO_2; Photocatalysts; Composites

1 Introduction

Organic dyes are widely used in textile, paper making, printing, coating and many other fields. In the process of industrial production, some organic dye pollutants are often generated and directly released into rivers and lakes. These organic dye pollutants are constantly accumulated in water, soil and atmosphere, and their oxidation, hydrolysis or other chemical reactions can produce toxic metabolites[1]. Therefore, prior to industrial wastewater discharge, it is necessary to achieve effective removal of organic dye pollutants from the wastewater. Many methods have been applied to treat with organic dyes in wastewater, such as physical absorption (like activated carbon)[2,3], biodegradation technology[4,5], and photocatalysis technology[6]. Especially, photocatalysis is an effective, inexpensive and eco-friendly method to decompose organic dyes. Titania (TiO_2), a typical cheap non-toxic photocatalyst, mainly exists in four distinct crystallographic phases, i.e., anatase (tetragonal, space group $I4_1/amd$), rutile (tetragonal, space group $P4_2/mnm$), brookite (orthorhombic, space group $Pbca$), and TiO_2(B) (monoclinic, space group $C2/m$)[7]. It is believed that anatase phase has more superior photocatalytic activity within these four kinds of polymorphs. Because of some intractable problems like aggregation effects due to high surface energy and difficult recovery processing, nano-TiO_2 is generally combined with various substrate materials (like

* 本文摘自 Functional Materials Letters, 2017, 10(03): 1750018.

aerogels[8,9], fibers[10] and membranes[11]) prior to utilization. Motivated by these researches, we would like to develop a new flexible hybrid paper photocatalyst by depositing anatase TiO_2 on cellulose paper. Especially, the abundant hydroxyl groups of cellulose are beneficial to interact with and immobilize TiO_2 nanoparticles.

Herein, we employed a mild simple pathway to synthesize anatase TiO_2/cellulose paper (coded as ATCP) composite, and its morphology, crystalline structure, and chemical compositions were investigated. As an example of potential applications, ATCP was used as a photocatalyst to degrade indigo carmine dye (typical organic dye pollutant). Also, its photocatalytic activity was compared to that of Degussa P25 and anatase TiO_2 powder.

2 Materials and methods

2.1 Preparation of ATCP

All reagents were commercially available and reagent grade. Anatase TiO_2 nanosol was firstly prepared according to the method reported by Wu et al.[12]. Briefly, absolute ethyl alcohol (20 mL) and tetrabutyl titanate (5 mL) were mixed homogeneously with magnetic stirring for 30 min at the room temperature; then the mixture was added dropwise into 0.04 M HNO_3 solution (200 mL) with continuous magnetic stirring; finally, the solution was kept stirring for 96 h. The cellulose paper was dipped into the freshly prepared anatase TiO_2 nanosol for 3 h. After the immersion treatment, the nanosol with the paper was heated at 60 ℃ for 5 min in a thermostatic water bath and subsequently cured at 100 ℃ for 5 min to complete the formation of nano-TiO_2. Finally, the paper was washed several times with distilled water and then dried at 50 ℃ for 5 h. In addition, the anatase TiO_2 powder was prepared by the slow solvent evaporation of the anatase TiO_2 nanosol at 60 ℃ for 96 h.

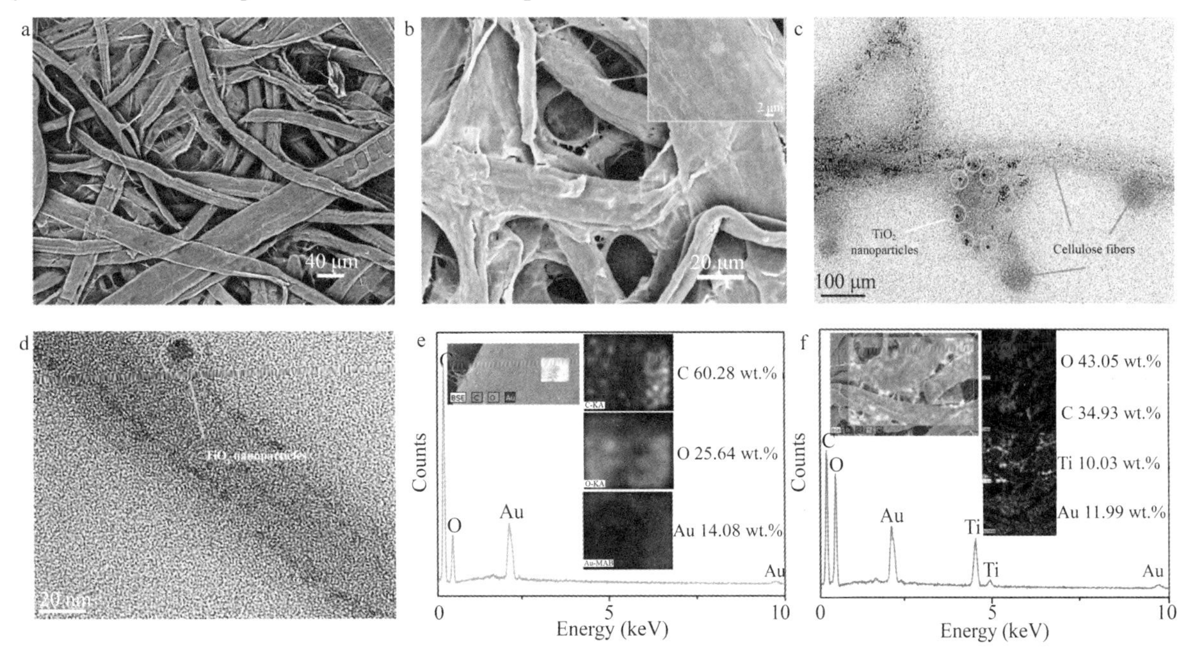

Fig. 1 SEM images of the cellulose paper (a) and ATCP (b), respectively, and the inset in (b) shows the corresponding enlarged image. (c and d) TEM images of ATCP under different magnification. EDX spectra of the cellulose paper (e) and ATCP (f), respectively. The insets in (e) and (f) show the corresponding elemental mapping images

2.2 Characterizations

Microstructure was observed by transmission electron microscope (TEM, FEI, and Tecnai G2 F20) and scanning electron microscopy (SEM, Quanta 200) equipped with an energy-dispersive X-ray (EDX) spectrometer for elemental analysis. Crystalline structure was investigated by X-ray diffraction (XRD, D/MAX 2200). Fourier transform infrared spectra (FTIR) were recorded on a FTIR instrument (Nicolet 6700, USA). X-ray photoelectron spectra (XPS) were recorded using a Thermo ESCALAB 250Xi XPS spectrometer. Thermogravimetric (TG) analysis was performed with a synchronous thermal analyzer (SDT-Q600, USA) under a nitrogen atmosphere.

2.3 Photocatalytic activities measurements

The photocatalytic activity of ATCP was evaluated by measuring the degradation rate of indigo carmine dye. The commercially available Degussa P25, anatase TiO_2 powder, and cellulose paper were used for comparison. The amount of anatase TiO_2 and P25 was equal to the TiO_2 loading of ATCP determined by the TG tests. The indigo carmine dye (100 mL, 6.43×10^{-5} mol · L^{-1}) and ATCP with a diameter of about 70 mm were put into a glass dish and then stirred in the dark for 30 min to achieve adsorption equilibrium. Thereafter, the dish was placed in a closed chamber with a UV source (mercury lamp, 120 W, and 365 nm). The dish was exposed to the mercury lamp. The solution was collected at every time interval (10 min), and its absorbance was measured by a TU-1901 UV-vis spectrophotometer (Beijing Purkinje, China) at 610 nm. The photocatalytic experiments of anatase TiO_2 powder and P25 followed the above processes, and an additional centrifugation method was used to separate the anatase TiO_2 and P25 powder from the solution. The degradation rate (η) of indigo carmine dye was calculated according to the following equation:

$$\eta = \frac{A_0 - A_t}{A_0} \times 100\% \qquad (1)$$

where A_0 and A_t are the initial absorbance and the absorbance at time t, respectively.

3 Results and discussion

It is seen in Fig. 1a that the interlaced fiber structures are obviously visible for the cellulose paper. For ATCP, it is hard to distinguish the synthetic TiO_2 from the surface of fibers (Fig. 1b). Even higher-magnification SEM image still could not identify the TiO_2 particles (the inset in Fig. 1b), possibly due to their extremely small size. Therefore, TEM observations were carried out. As shown in Fig. 1c and d, it is clear that plentiful nanoparticles were tightly anchored to the surface of cellulose fibers, and display good dispersion, indicating that the presence of cellulose fibers contributes to reducing particle aggregation. Moreover, it can be seen that the size of TiO_2 particles ranges from 2.3 to 9.2 nm. The main elements of the cellulose paper are shown in Fig. 1e, where C, O and Au elements were detected by EDX analysis. The Au peaks are originated from the coating layer used for electric conduction during the SEM observation. Apart from these elements, Ti element was detected in the EDX spectrum of ATCP and accounts for 10.03 wt. % (Fig. 1f). Moreover, the content of O element remarkably increases from 25.64 wt. % (cellulose paper) to 43.05 wt. % (ATCP), which is related to the presence of TiO_2.

As shown in Fig. 2a, the cellulose paper shows a typical cellulose I crystalline structure[13]. The diffraction peaks at around 15. 8° is originated from the superposition of (101) and (10$\bar{1}$) planes, and the peak at 22. 1° is derived from the (002) plane. For ATCP, the characteristic peaks of cellulose Ⅰ crystal structure can be clearly identified, which suggests that the dipping treatment cannot change the crystal form of cellulose. In addition, ATCP shows some new peaks (2θ = 25. 28°, 37. 80°, 48. 04°, 53. 89°, 55. 06°, 62. 68°, and 68. 76°), which are well indexed to the standard data of anatase TiO_2 (JCPDS card no. 21-1272). These results reveal that the anatase TiO_2 was successfully synthesized and deposited on the cellulose paper. Fig. 2b presents the FTIR spectra of the cellulose paper and ATCP. The band at 3309 cm^{-1} attributed to the stretching vibration of hydroxyl groups was obviously weakened in the spectrum of ATCP and shifted to a lower wavenumber of 3280 cm^{-1}, which is due to the interaction between the hydroxyl groups of cellulose and the TiO_2 [11]. This interaction led to the immobilization of TiO_2 on the surface of cellulose.

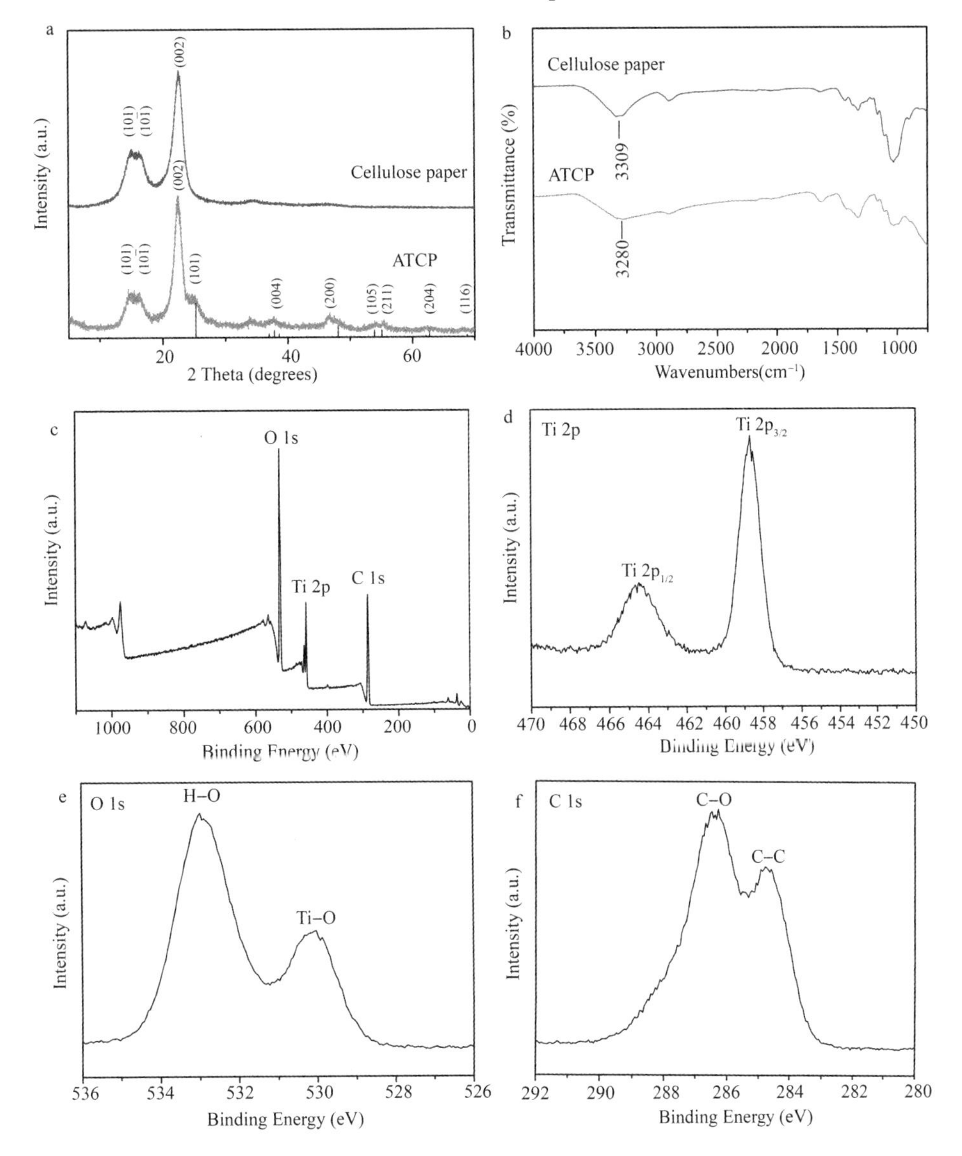

Fig. 2 (a) XRD patterns and (b) FTIR spectra of the cellulose paper and ATCP, respectively. XPS spectra of ATCP: (c) survey scan, (d) Ti 2p, (e) O 1s, and (f) C 1s

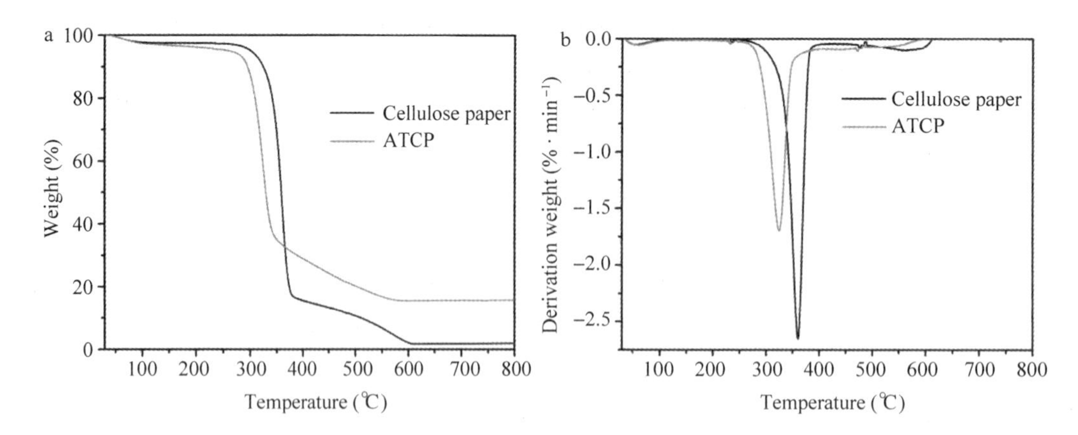

Fig. 3 (a) TG and (b) DTG curves of the cellulose paper and ATCP, respectively

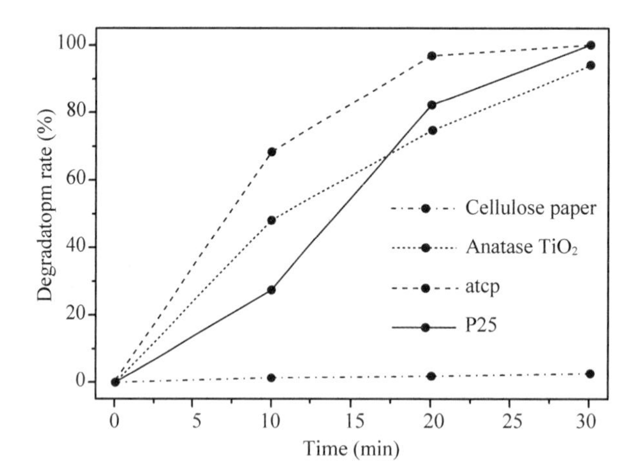

Fig. 4 Degradation rate of indigo carmine dye against irradiation time

The XPS survey spectrum of ATCP is shown in Fig. 2c. The Ti, O, and C elements were detected, consistent with the results of EDX analysis. The spectrum of Ti 2p is shown in Fig. 2d, where the binding energies of Ti $2p_{3/2}$ and $2p_{1/2}$ are centered at 458. 7 and 464. 6 eV, respectively. This clearly demonstrates that the Ti element is in oxidation state Ⅳ, corresponding to Ti^{4+}(TiO_2)[14]. In addition, the separation between these two bands is 5. 9 eV, in accordance with the presence of normal state of Ti^{4+}. In Fig. 2e, the O 1s spectrum can be fitted to two peaks at binding energies of 532. 9 and 530. 0 eV, which are attributed to the H—O and Ti—O, respectively. The C 1s spectrum can also be fitted to two peaks located at 284. 8 and 286. 0 eV, corresponding to C—C and C—O, respectively.

The TG and DTG curves of the cellulose paper and ATCP are presented in Fig. 3a and b. Both the samples exhibit only a strong peak at around 300-400 ℃, corresponding to the pyrolysis of cellulose[15]. In addition, the peak of ATCP is centered at 325 ℃, 24 ℃ lower than that of the cellulose paper. The relatively inferior thermal stability of ATCP might be due to catalytic character of TiO_2, the loosening of molecular chains in crystalline regions of cellulose resulted from infusion of TiO_2 during the impregnation process, and the damage of cellulose chains caused by HNO_3 in the sol[16,17]. In addition, ATCP remained a higher weight of 15. 82% at 800 ℃, and the cellulose paper remained a weight of only 1. 96% at 800 ℃, indicating that the content of

TiO_2 is approximately 13. 86 wt. % in ATCP.

The photocatalytic activity of ATCP was evaluated in the degradation reaction of indigo carmine dye in aqueous solution under UV radiation. Commercially available TiO_2 P25 was used as a photocatalytic reference. For comparison, the photocatalytic activities of cellulose paper and anatase TiO_2 powder were also measured. Before the UV radiation, the adsorption of dye in the dark by ATCP and P25 was measured. The results show that the absorbance of dye decreased from 0. 633 to 0. 583 for ATCP, while for P25, the absorbance of dye only decreased to 0. 609. The lower absorbance indicates that ATCP has stronger adsorption capacity for the dye than P25. The stronger adsorption capacity might be primarily attributed to the presence of cellulose paper. In addition, it is worth to mention that P25 is a porous material and has a specific surface area of about 55 $m^2 \cdot g^{-1}$ (the data is provided by the supplier), responsible for the decrease of absorbance. For the photocatalytic degradation tests, as shown in Fig. 4, the negligible noncatalytic degradation of indigo carmine dye was observed by using cellulose paper, indicating that cellulose cannot decompose the dye molecules under UV irradiation. In addition, both ATCP and P25 display 100% of degradation rate of indigo carmine dye within 30 min, indicating good photocatalytic activity. As compared to P25, the indigo carmine dye obviously decomposed more rapid when using ATCP, suggesting that ATCP has more superior photocatalytic property. For the anatase TiO_2 powder, a lower degradation rate of 93. 8% was obtained after the irradiation, indicative of worse photocatalytic activity.

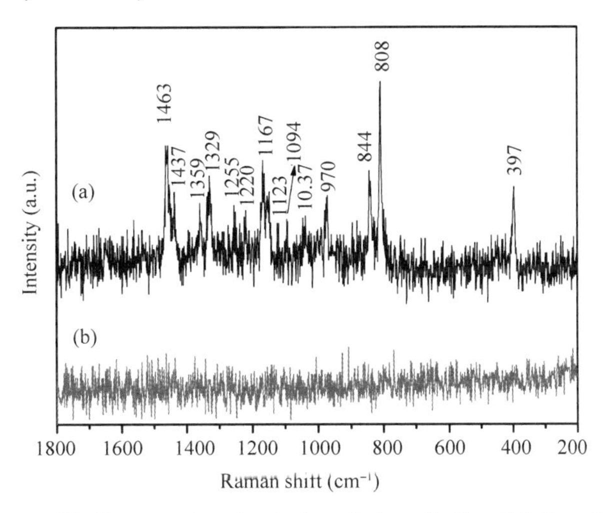

Fig. 5 Raman spectra of indigo carmine dye before (a) and after (b) the photocatalytic reaction

It is commonly accepted that smaller size represents more powerful redox ability due to the quantum-size effect[18]. Moreover, the smaller sizes are beneficial for the separation of the photogenerated hole and electron pairs, which can slow the rate of $e^- - h^+$ recombination and improve the photocatalytic activity[19,20]. The size of TiO_2 in ATCP is in the range of 2. 3–9. 2 nm, much smaller than that of the P25 (ca. 21 nm, provided by the supplier). For anatase TiO_2 powder obtained by the slow solvent evaporation of anatase TiO_2 sol at 60 ℃, the long drying process (ca. 96 h) and high temperature contribute to obtaining larger-size TiO_2 (Ostwald ripening)[21], as compared to that of the TiO_2 in ATCP. Therefore, the smaller size of TiO_2 in ATCP is responsible for the stronger photocatalytic activity. In addition, the powder-like P25 and anatase TiO_2 are easy to aggregate with each other in the aqueous solution due to their high surface energy, leading to the deterioration of photocatalytic activity. In contrast, the good dispersion of TiO_2 on the surface of cellulose contributes to obtaining the higher photocatalytic activity. Therefore, it is reasonable to explain the higher photocatalytic activity of ATCP.

For further demonstrating the mineralization effect, the Raman spectra of the indigo carmine dye were studied before and after the photocatalytic reaction. As shown in Fig. 5, before the photocatalytic tests, the Raman spectrum of the dye displays many peaks, which are related to the feature structure of indigo carmine. For instance, the band at 1463 cm^{-1} is attributed to the N—C stretching and to the in-plane C—H bending modes[22]. The band at 1437 cm^{-1} is attributed to the C—C and N—C ═C—N stretching contributions, the in-plane N—H bending mode, and the in-plane C—H bending mode of the phenyl rings[23]. The band at 1329 cm^{-1} is due to the in-plane C—H and N—H bending modes[24]. The band at 1255 cm^{-1} corresponds to the in-plane C—H and C ═O bending modes, while the band at 1220 cm^{-1} is related to the breathing modes of the phenyl rings[25]. The band at 1167 cm^{-1} corresponds to the asymmetrical O12-S10-O13/O12'-S10'-O13' stretching modes and to the in-plane C—C bending modes of the phenyl rings. The symmetrical O12-S10-O13/O12'-S10'-O13' stretching contributions are observed at 1123 cm^{-1}. The band at 1094 cm^{-1} corresponds to the out-of-plane bending modes of both O14-Na11/14'-Na11' bonds[23]. The out-of-plane C—H deformation mode of the phenyl rings is observed at 1037 cm^{-1}. The band at 970 cm^{-1} is assigned to the out-of-plane C—H bending mode and to the O12-S10-O13/O12'-S10'-O13' deformation modes. The peak at 844 cm^{-1} is due to the C5-S10/C5'-O10' stretching modes and to the C—C out-of-plane bending mode of the five-membered rings[26]. However, after the photocatalytic reaction, all of these bands disappeared and cannot be detected by the Raman spectroscopy, suggesting that the dye molecules had been almost completely decomposed. Also, this result suggests the good photocatalytic activity of ATCP.

4 Conclusions

We adopted a facile method to deposit anatase TiO_2 on the cellulose paper. This cheap eco-friendly and flexible hybrid paper can serve as a high-performance photocatalyst. This hybrid paper with the anatase TiO_2 content of around 13.86 wt.% can degrade blue indigo carmine dye into colorless within 30 min under UV radiation. In addition, the hybrid paper shows a more superior photocatalytic activity for the indigo carmine dye degradation than that of commercially available TiO_2 P25 and anatase TiO_2 powder. Thus this green high-efficiency hybrid paper may be useful in organic dye wastewater treatment.

Acknowledgement

This study was supported by the National Natural Science Foundation of China (grant nos. 31270590 and 31470584).

References

[1] Robinson T, McMullan G, Marchant R, et al. Remediation of dyes in textile effluent: a critical review on current treatment technologies with a proposed alternative[J]. Bioresource Technology, 2001, 77(3): 247-255.

[2] Namasivayam C, Kavitha D. Removal of Congo Red from water by adsorption onto activated carbon prepared from coir pith, an agricultural solid waste[J]. Dyes and Pigments, 2002, 54(1): 47-58.

[3] Luo X, Zhang L. High effective adsorption of organic dyes on magnetic cellulose beads entrapping activated carbon.

[J]. Journal of Hazardous Materials, 2009, 171(1-3): 340-347.

[4]Aksu Z. Application of biosorption for the removal of organic pollutants: a review[J]. Process Biochemistry, 2005, 40(3-4): 997-1026.

[5]Banat I M, Nigam P, Singh D, et al. Microbialdecolorization of textile-dyecontaining effluents: a review[J]. Bioresource Technology, 1996, 58(3): 217-227.

[6]Han F, Kambala V, Srinivasan M, et al. Tailored titanium dioxide photocatalysts for the degradation of organic dyes in wastewater treatment: A review[J]. Applied Catalysis A General, 2009, 359(1-2): 25-40.

[7]Lusvardi V S, Barteau M A, Farneth W E. The effects of bulk titania crystal structure on the adsorption and reaction of aliphatic alcohols[J]. Journal of Catalysis, 1995, 153(1): 41-53.

[8]Zeng J, Liu S, Cai J, et al. TiO_2 immobilized in cellulose matrix for photocatalytic degradation of phenol under weak UV light irradiation[J]. The Journal of Physical Chemistry C., 2010, 114(17): 7806-7811.

[9]Jiao Y, Wan C, Li J. Room-temperature embedment ofanatase titania nanoparticles into porous cellulose aerogels[J]. Applied Physics A, 2015, 120(1): 341-347.

[10]Chauhan I, Mohanty P. In situ decoration of TiO_2 nanoparticles on the surface of cellulose fibers and study of their photocatalytic and antibacterial activities[J]. Cellulose, 2015, 22(1): 507-519.

[11]Lin Y, Wu G S, Yuan X Y, et al. Fabrication and optical properties of TiO_2 nanowire arrays made by sol-gel electrophoresis deposition into anodic alumina membranes[J]. Journal of Physics: Condensed Matter, 2003, 15(17): 2917.

[12]Wu D, Long M, Zhou J, et al. Synthesis and characterization of self-cleaning cotton fabrics modified by TiO_2 through a facile approach[J]. Surface and Coatings Technology, 2009, 203(24): 3728-3733.

[13]Tsuboi M. Infrared spectrum and crystal structure of cellulose[J]. Journal of Polymer Science, 1957, 25(109): 159-171.

[14]Diebold U. The surface science of titaniumdioxide[J]. Surface science reports, 2003, 48(5-8): 53-229.

[15]Shafizadeh F, Fu Y L. Pyrolysis of cellulose[J]. Carbohydrate Research, 1973, 29(1): 113-122.

[16]H Wang, Zhong W, Xu P, et al. Polyimide/silica/titania nanohybrids via a novel non-hydrolytic sol-gel route Science Direct[J]. Composites Part A: Applied Science and Manufacturing, 2005, 36(7): 909-914.

[17]Yu Q, Wu P, Xu P, et al. Synthesis of cellulose/titanium dioxide hybrids in supercritical carbon dioxide[J]. Green Chemistry, 2008, 10(10): 1061-1067.

[18]Yu J C, Zhang L, Yu J. Directsonochemical preparation and characterization of highly active mesoporous TiO_2 with a bicrystalline framework[J]. Chemistry of Materials, 2002, 14(11): 4647-4653.

[19]Peng T, Zhao D, Dai K, et al. Synthesis of titanium dioxide nanoparticles with mesoporous anatase wall and high photocatalytic activity[J]. The journal of physical chemistry B., 2005, 109(11): 4947-4952.

[20]Guo C, Ge M, Liu L, et al. Directed synthesis of mesoporous TiO_2 microspheres: catalysts and their photocatalysis for bisphenol A degradation[J]. Environmental Science & Technology, 2010, 44(1): 419-425.

[21]Voorhees P W. The theory of Ostwald ripening[J]. Journal of Statistical Physics, 1985, 38(1): 231-252.

[22]Rode J E, Raczyńska E D, Górnicka E, et al. Low inversion energy barrier of cytisine NH group-an explanation for the FT-IR bands splitting[J]. Journal of Molecular Structure, 2005, 749(1-3): 51-59.

[23] Peica N, Kiefer W. Characterization of indigo carmine with surface-enhanced resonance Raman spectroscopy (SERRS) using silver colloids and island films, and theoretical calculations[J]. Journal of Raman Spectroscopy: An International Journal for Original Work in all Aspects of Raman Spectroscopy, Including Higher Order Processes, and also Brillouin and Rayleigh Scattering, 2008, 39(1): 47-60.

[24]Vandenabeele P, Bodé S, Alonso A, et al. Raman spectroscopic analysis of the Maya wall paintings in Ek'Balam, Mexico[J]. Spectrochimica Acta Part A: Molecular and Biomolecular Spectroscopy, 2005, 61(10): 2349-2356.

[25]AnnaBaran, AndreaFiedler, HartwigSchulz, et al. In situ Raman and IR spectroscopic analysis of indigo dye[J].

Analytical Methods, 2010, 2(9): 1372-1376.

[26] Shadi I T, Chowdhry B Z, Snowden M J, et al. Semi-quantitative analysis of indigo carmine, using silver colloids, by surface enhanced resonance Raman spectroscopy (SERRS) [J]. Spectrochimica Acta Part A: Molecular and Biomolecular Spectroscopy, 2003, 59(10): 2201-2206.

中文题目：锐钛矿型 TiO_2/纤维素复合纸的合成、表征及其光催化降解靛红染料的研究

作者：焦月，万才超，李坚

摘要：报道了一种在纤维素纸上沉积锐钛矿型二氧化钛(TiO_2)的简便方法。采用扫描电子显微镜、透射电子显微镜、X射线能谱仪、X射线衍射、傅立叶变换红外光谱、X射线光电子能谱和热重分析等手段对锐钛矿型 TiO_2/纤维素纸(ATCP)进行了表征。该杂化纸的锐钛矿型 TiO_2 含量在13.86 wt.%左右，可作为一种环保型柔性光催化剂，在紫外光照射下，可在30 min内将蓝色靛蓝胭脂红快速降解成无色溶液。此外，与市售的 TiO_2 P25和锐钛矿型 TiO_2 粉末相比，ATCP对靛红染料的分解速度更快。这些结果表明，该杂化纸可用于有机染料废水的处理。

关键词：纤维素纸；锐钛矿型二氧化钛；光催化剂；复合材料

High Performance, Flexible, Solid-State Supercapacitors Based on A Renewable and Biodegradable Mesoporous Cellulose Membrane*

Dawei Zhao, Chaoji Chen, Qi Zhang, Wenshuai Chen, Shouxin Liu,
Qingwen Wang, Yixing Liu, Jian Li, Haipeng Yu

Abstract: A flexible and transparent renewable mesoporous cellulose membrane (mCel-membrane) featuring uniform mesopores of ~24.7 nm and high porosity of 71.78% was prepared via a simple solution phase inversion process. KOH-saturated mCel-membrane as a polymer electrolyte demonstrated a high electrolyte retention of 451.2%, a high ionic conductivity of 0.325 $S \cdot cm^{-1}$, and excellent mechanical flexibility and robustness. A quasi-solid-state electric double layer capacitor (EDLC) using activated carbon as electrodes, KOH-saturated mCel-membrane as polymer electrolyte exhibited a high capacitance of 120.6 $F \cdot g^{-1}$ at 0.5 $A \cdot g^{-1}$, and long cycling life of 10000 cycles with 84.7% capacitance retention. Moreover, a highly integrated planar-type micro-supercapacitor (MSC) can be facilely fabricated by directly depositing the electrode materials on the mCel-membrane-based polymer electrolyte without using complicated devices. The resulting MSC exhibited a high areal capacitance of 153.34 $mF \cdot cm^{-2}$ and volumetric capacitance of 191.66 $F \cdot cm^{-3}$ at 10 $mV \cdot s^{-1}$, representing one of the highest values among all carbon-based MSC devices. Another prominent advantage of this kind of solid-state EDLC and MSC lies in the high biodegradability of both the mCel-membrane-based electrolyte and activated carbon electrode that would not tax our environment. These findings suggest that the developed renewable flexible mesoporous cellulose membrane holds great promise in the practical applications of flexible solid-state portable energy storage devices that are not limited to supercapacitors.

Keyword: Cellulose; Flexible; Membranes; Polymer electrolytes; Solid-state supercapacitors

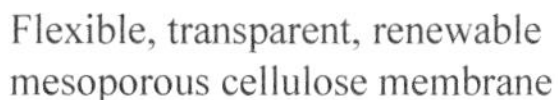
Flexible, transparent, renewable mesoporous cellulose membrane

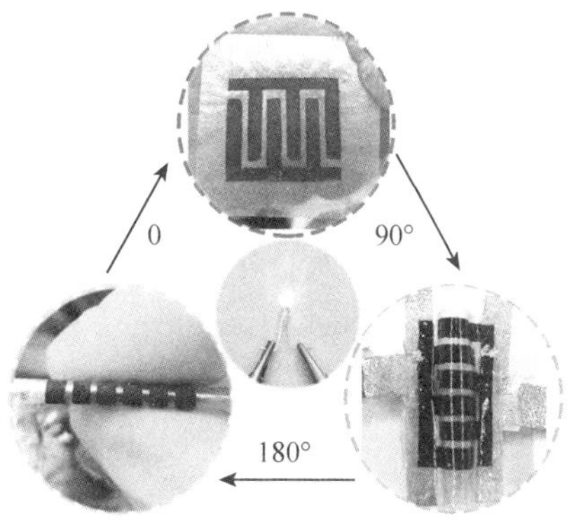

* 本文摘自 Advanced Energy Materials, 2017, 7(18): 1700739.

1 Introduction

Supercapacitors(SCs) and micro-supercapacitors(MSCs) have received increasing attention recently in the development of energy storage systems, particularly in systems that are low cost, miniaturized, light-weight and flexible. Such energy storage systems are in urgent need to meet the growing demand for portable consumer electronic devices.[1] The performance of SCs and MSCs depends on several components, including the electrode, electrolyte and separator. Much attention has been given to electrode materials and structures, in attempt to improve electrochemical performance.[2-6] However, much less attention has been given to the electrolyte and separator.

The electrolyte reportedly plays an important role in determining the maximum operating voltage, lifetime and safety of SCs.[7] Conventional liquid electrolytes include organic and aqueous electrolytes. The former is easy to catch fire thus brings safety issues, while the latter has narrow potential window that may lower the energy density of the SCs. Recently, polymer electrolytes have attracted increasing attention due to their ability to form thin films, their high ionic conductivity, wide electrochemical window, and service safety.[8-11] Incorporating the electrolyte and separator into a single polymer membrane has promoted the development of energy storage systems for portable and wearable electronics.[12-14] Much effort has been given to preparing high-performance gel polymer electrolytes[15] and porous polymer electrolytes.[16-19] For example, a zwitterionic gel electrolyte of propylsulfonate dimethylammonium propylmethacrylamide exhibited good water retention capability, which led to a high capacitance and cycling stability.[15] A microporous polymer electrolyte from a poly(ethylene glycol)-grafted poly(arylene ether ketone) composite membrane matrix incorporating a chitosan-based $LiClO_4$ gel was prepared.[18] The polymer electrolyte exhibited good mechanical properties and low leakage, and high cycling stability when incorporated into a solid-state electric double layer capacitor(EDLC). A graphene oxide-doped ion gel electrolyte exhibited a high ionic conductivity.[19] This gel electrolyte was then incorporated into an all-solid-state SC, which exhibited high capacitance performance and good cycling stability. All these works have demonstrated the effectiveness of the polymer electrolytes in EDLCs and all-solid-state SCs. However, the conventional polymer electrolytes are generally neither renewable nor biodegradable, or involve complicated preparation process, which have restricted their large-scale application. Achieving mechanical flexibility and environmental benignity of electrolytes, while retaining their good electrochemical performance, remains a challenge. A simple method for preparing versatile polymer electrolytes with high ionic conductivity, renewability, biodegradability and good mechanical properties is highly desirable but also challenging.

Renewable polymer solutes and appropriate solvents are essential for fabricating porous polymer electrolyte membrane. Among the numerous optional polymer solutes, cellulose is the most abundant natural biopolymer on earth. It is environmentally friendly and biodegradable, with high chemical and thermal stability, high mechanical performance, and flexibility. These properties have resulted in cellulose (especially cellulose acetate, methyl cellulose, cellulose nanofibers and cellulose-based composites) being widely used as separation membranes in energy storage devices.[20-23] However, cellulose does not readily dissolve in common organic solvents, so can be difficult to directly process into nanoporous membranes.[24] The phase-inversion method is typically the most efficient and techno-economic method for preparing cellulosic nanoporous membranes.[25-27] Nevertheless, the facile fabrication of the nanoporous cellulose electrolyte membrane with high ionic conductivity is not straight-forward. Detailed studies to in-depth reveal its performance in solid-state SCs and flexible MSCs are also currently necessary.

Here, a flexible and transparent renewable mesoporous cellulose membrane (hereafter referred to as mCel-membrane) was fabricated via a phase-inversion method with an IL as the solvent. Cellulose can be easily dissolved in the IL of 1-butyl-3-methylimidazolium chloride ([Bmim]Cl), and then regenerated into a film with pure water as a non-solvent. Two polytetrafluoroethylene (PTFE) millipore membranes were used as templates, to form mesopores during the drying process. The resulting mCel-membrane exhibited good flexibility, high light transmittance, good mechanical properties, and high thermal stability. A quasi-solid-state EDLC using potassium hydroxide (KOH)-saturated mCel-membrane as the polymer electrolyte and activated carbon as the electrodes demonstrated excellent electrochemical performance with high capacitance and long cycling life. Moreover, flexible MSCs could be facilely fabricated by directly depositing the activated carbon electrodes on the mCel-membrane in different configurations, demonstrating the great potential of our developed cellulose-based electrolyte membrane in flexible solid-state energy storage devices that are not limited to SCs.

2 Results and Discussion

2.1 Preparation and properties of the mCel-membrane

Cellulose was dissolved in [Bmim]Cl IL, which yielded a homogenous solution (denoted as Cel/IL) after stirring for 18 min (Fig. S1, Supporting Information). The good solubility of cellulose in [Bmim]Cl was attributed to the formation of hydrogen bonding interactions between the hydroxyl protons of cellulose and chloride anions/imidazolium cations in [Bmim]Cl.[28] Therefore, the dissolution of cellulose in IL was considered to be a physical process, and IL was used as a non-derivatizing solvent for regenerating the cellulose film.[29] The Cel/IL was viscous, and its viscosity decreased with increasing temperature up to 90 ℃, at which point the viscosity reached a plateau of approximately 1500 mPa s. From this result, 85 ℃ was selected as the optimum temperature for Cel/IL treatment. When the Cel/IL gel was immersed in distilled water, [Bmim]Cl was substituted by water which gradually diffused from the surface into the bulk of the gel. This spontaneous molecular self-assembly process resulted in cellulose chains rearranging, which formed an expanded white regenerated cellulose hydrogel. After solidifying and drying, a freestanding mCel-membrane was obtained.

The mCel-membrane was highly transparent with a high light transmittance of 91.1% at 600 nm (Fig. 1a and 1d). The mCel-membrane exhibited a super apparent flexibility, which could be processed into different sizes and shapes, even folded into various forms. High-resolution scanning electron microscopy (SEM) images indicated that the mCel-membrane had a macrovoid-free and three-dimensionally interconnected porous structure (Fig. 1b and 1c). The morphology of the mCel-membrane surface was very smooth. Abundant uniformly-distributed mesopores with diameters of 5-30 nm (most probable aperture of 24.7 nm) were observed (Fig. S2 in Supporting Information). The porosity of the mCel-membrane was 71.78%, which was higher than those of nonwoven polypropylene membrane (NKK-MPF30AC) (51.08%) and cellulosic separator paper (NKK-TF4030) (29.68%) (Fig. 1g). Although the crystal form of the mCel-membrane was converted to cellulose II, the ultimate tensile stress and Young's modulus of the mCel-membrane remained at 171.5 MPa and 8.93 GPa, respectively (Fig. S1d, Supporting Information). The initial thermal degradation temperature of the mCel-membrane was 275 ℃ (Fig. 1e), which was higher than those of most polymer membrane separators, such as the poly(ethylene glycol)-grafted poly(arylene ether ketone) (PAEK-g-PEG) membrane (160 ℃) and poly(aryl ether sulfone) membrane (200 ℃)[18,30].

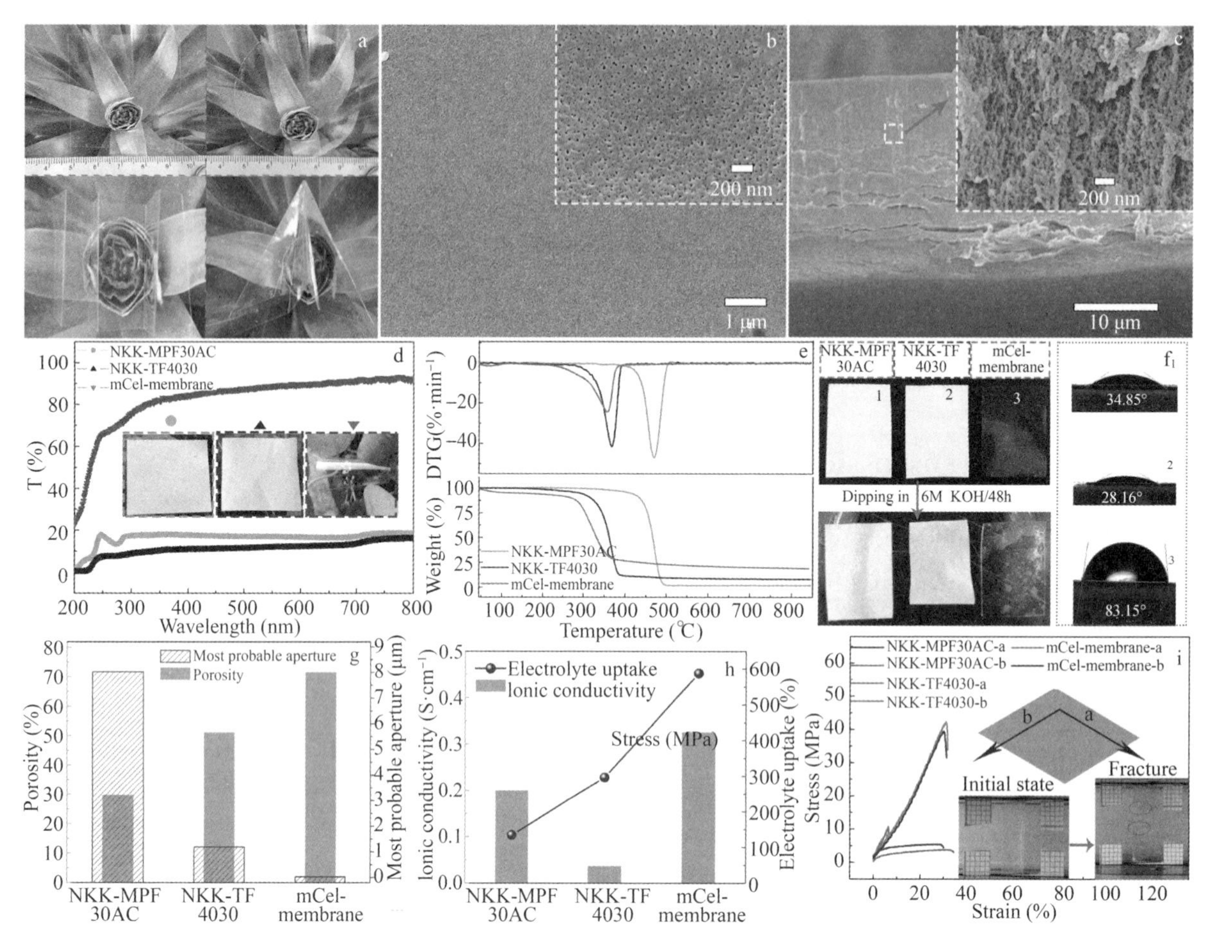

Fig. 1 Morphology and properties of the mCel-membrane. NKK-MPF30AC and NKK-TF4030 were also investigated for comparison. (a) Optical photographs show the transparency, flexibility and tailorability of the mCel-membrane. (b) Surface and. (c) cross-sectional SEM images. (d) Light transmittance (T%). Insets show optical images of the NKK-MPF30AC, NKK-TF4030, and mCel-membrane. (e) TG and its derivative TG (DTG) curves. (f) Static contact angles for a droplet of 6 M KOH solution, and absorption-swelling characteristics. (g) Porosity and most probable aperture size. (h) Ionic conductivity and electrolyte uptake. (i) Tensile stress-strain curves

The static contact angle of the mCel-membrane was 83.15° for a droplet 6 M KOH solution. This contact angle value resulted from the strong capillary tension of the smooth surface (Fig. 1f). After soaking in 6 M KOH solution for 48 h, the electrolyte uptake and ionic conductivity of the mCel-membrane were 587.5 wt.% and 0.325 $S \cdot cm^{-1}$, respectively (Fig. 1h). After 48 h of desorption at room temperature, the electrolyte retention of the mCel-membrane was 451.2 wt.% (Fig. S3, Supporting Information). The tensile strengths of the soppy mCel-membrane in the A and B directions were 42.98 MPa and 40.22 MPa, respectively. These were far higher than those of NKK-MPF30AC and NKK-TF4030 (Fig. 1i). In summary, these results suggested that the mCel-membrane with abundant mesopores, high ionic conductivity, excellent electrolyte retention, and robust mechanical properties had potential as an electrolyte/separator in energy storage devices.

2. 2 KOH saturated mCel-membrane as the polymer electrolyte in a quasi-solid-state EDLC

A quasi-solid-state EDLC was fabricated using the mCel-membrane/KOH as a polymer electrolyte, to achieve a portable energy storage system. The quasi-solid-state EDLC was fabricated by encapsulating the EDLC with EVA hot-melted glue. EDLCs containing NKK-MPF30AC, NKK-TF4030 and KOH/polyvinyl alcohol(PVA) as polymer electrolytes were prepared similarly for comparison. Their electrochemical properties were measured under typical atmospheric conditions. Cyclic voltammetry(CV) curves of these EDLCs at scan rates of 10, 50 and 300 $mV \cdot s^{-1}$ are shown in Fig. 2a-c, respectively. The CV curves had similar profiles in wide potential window of 0-1. 0 V. The integrated area of the CV curve of mCel-membrane-based EDLC was larger than those of the NKK-MPF30AC, NKK-TF4030 and PVA/KOH-based EDLCs (Fig. S4-S6, Supporting Information). This indicated better equivalent EDLC behavior and a higher specific capacitance of the quasi-solid-state EDLC with a mCel-membrane/KOH electrolyte. Galvanostatic charge/discharge (GCD) profiles measured at current densities of 1, 5 and 10 $A \cdot g^{-1}$ are shown in Fig. 2d-f, respectively. The mCel-membrane-based EDLC exhibited a longer discharge time and a lower voltage drop than the other three EDLCs. For example, the discharge time of the mCel-membrane-based EDLC was 2. 3 s at a current density of 10 $A \cdot g^{-1}$. This was 1. 9 times higher than that of the PVA/KOH-based EDLC(1. 2 s), 2. 3 times higher than that of the NKK-MPF30AC-based EDLC(1. 0 s), and 3. 2 times higher than that of the NKK-TF4030-based EDLC(0. 7 s). Nyquist plots were obtained by the alternating current impedance method from 0. 01 to 105 Hz. The intercept with the real axis at high frequencies indicated the series resistance of the EDLC(Fig. 2g). The equivalent series resistance of the mCel-membrane-based EDLC was 0. 41 Ω, which was smaller than those of the PVA/KOH-based EDLC (0. 691 Ω), NKK-MPF30AC-based EDLC (0. 892 Ω), and NKK-TF4030-based EDLC(4. 949 Ω).

All three EDLCs exhibited good specific capacitances and rate capabilities at current densities of 0. 5−10 $A \cdot g^{-1}$(Fig. 2h). The specific capacitance values for the mCel-membrane-based EDLC were 1. 4-1. 7 times higher than those of the PVA/KOH, NKK-MPF30AC, and NKK-TF4030-based EDLCs. The value of 120. 6 $F \cdot g^{-1}$ at 0. 5 $A \cdot g^{-1}$ for the mCel-membrane-based EDLC was also superior to reported values of 92. 79 $F \cdot g^{-1}$ for poly(aryl ether sulfone) in 6 M KOH,[30] 93 $F \cdot g^{-1}$ for polyethylene oxide in 6 M KOH,[31] and 118. 63 $F \cdot g^{-1}$ for the PAEK/PAEK-g-PEG/LiClO4 gel electrolyte.[18] The capacitance of the mCel-membrane-based EDLC was 81. 4 $F \cdot g^{-1}$ at 10 $A \cdot g^{-1}$, which was equivalent to 67. 5% of the capacitance of 120. 6 $F \cdot g^{-1}$ at 0. 5 $A \cdot g^{-1}$, indicating good rate capability. 84. 7% of the initial capacitance of the mCel-membrane-based EDLC was retained after 10, 000 cycles(Fig. 2i). This retention was considerably higher than those of the EDLCs containing PVA/KOH(65. 18%), NKK-MPF30AC(34. 73%), and NKK-TF4030 (4. 56%) electrolytes. Visual inspection of the mCel-membrane after 10, 000 cycles showed that the mCel-membrane remained intact and usable. When the same mCel-membrane was reused in a newly assembled EDLC, the CV, GCD and electrochemical impedance spectroscopy(EIS) curves exhibited similar profiles to those of the initial EDLC. The voltage drop of the newly assembled EDLC was 0. 037 V(Fig. S7, Supporting

Information). In summary, these results demonstrated that the quasi-solid-state mCel-membrane-based EDLC exhibited excellent electrochemical performance, which was superior to that of the PVA/KOH-based EDLC. This was attributed to: (1) the homogeneous structure and abundant active hydroxyl groups imparted high electrolyte retention and excellent mechanical properties on the mCel-membrane, which improved the ionic conductivity and capacitance cycling stability; (2) the high porosity and nanostructure provided abundant ion transport pathways, and abundant electrolyte storage space. This improved the rate capability and reduced the voltage drop.

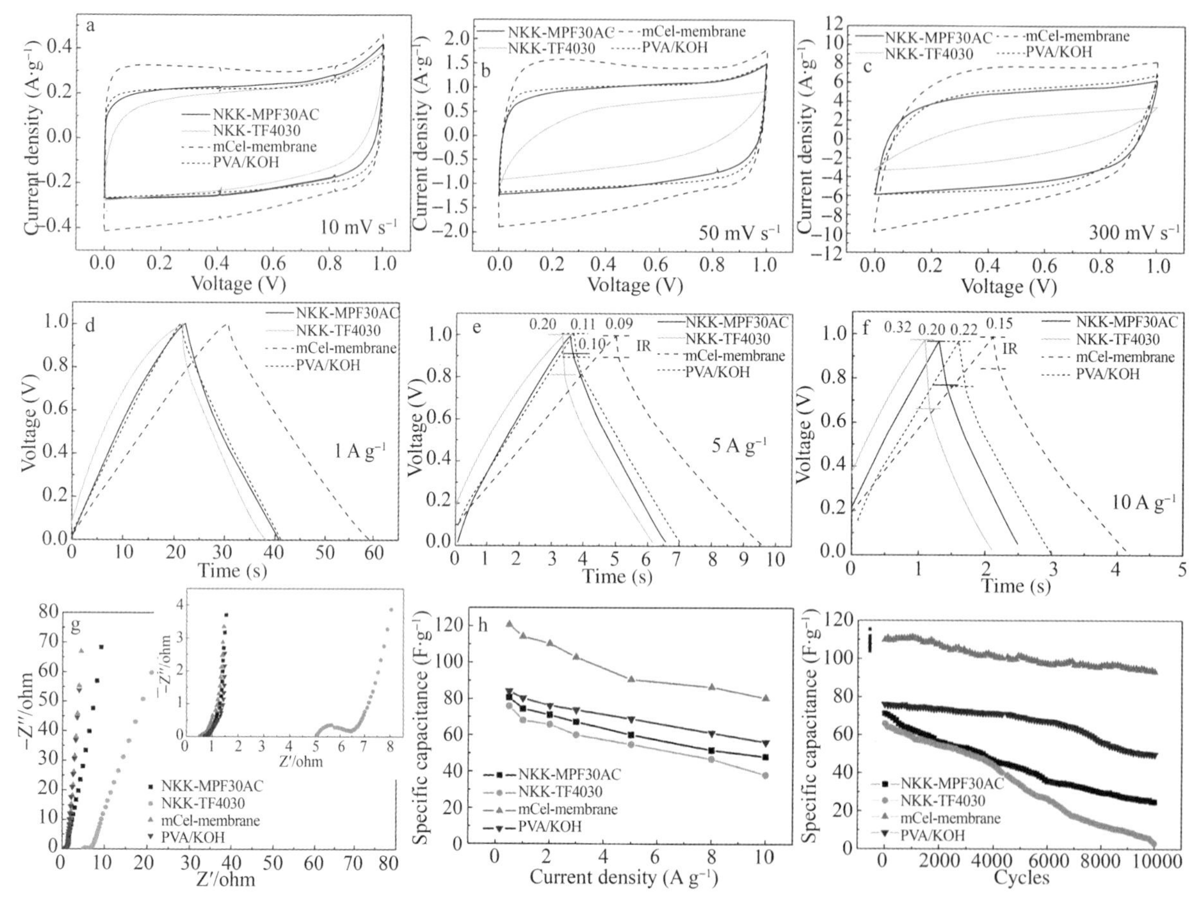

Fig. 2 Electrochemical properties of the quasi-solid-state mCel-membrane-based EDLC. Similar quasi-solid-state EDLCs containing NKK-MPF30AC, NKK-TF4030 and PVA/KOH as electrolyte were also tested for comparison. (a)-(c) CV curves measured at scan rates of 10, 50 and 300 $mV \cdot s^{-1}$, respectively. (d)-(f) GCD profiles measured at current densities of 1, 5 and 10 $A \cdot g^{-1}$, respectively. (g) Nyquist plots. (h) Specific capacitances measured at different current densities. (i) Cycling performance measured at a current density of 2 $A \cdot g^{-1}$

The lifetime stability of the quasi-solid-state mCel-membrane-based EDLC was evaluated by measuring its electrochemical performance after different time intervals. The CV and GCD curves in the scan potential windows of 0-0.5 V, 0-0.8 V and 0-1.0 V maintained their quasi-rectangular and symmetrical triangular profiles, respectively, 24 h after EDLC preparation (Fig. 3a, b). This suggested that the EDLC was adaptable to the wide working potential window. The profiles of the CV curves at a scan rate of 50 $mV \cdot s^{-1}$

were also well retained after different time intervals (Fig. 3c). Specific capacitances were calculated from the GCD profiles at a current density of 1 A · g^{-1}. The specific capacitance decreased by 1.3% after 24 h, 6.4% after 72 h, and 11.7% after 96 h (Fig. 3d). These results demonstrated the long lifetime stability of the mCel-membrane-based EDLC.

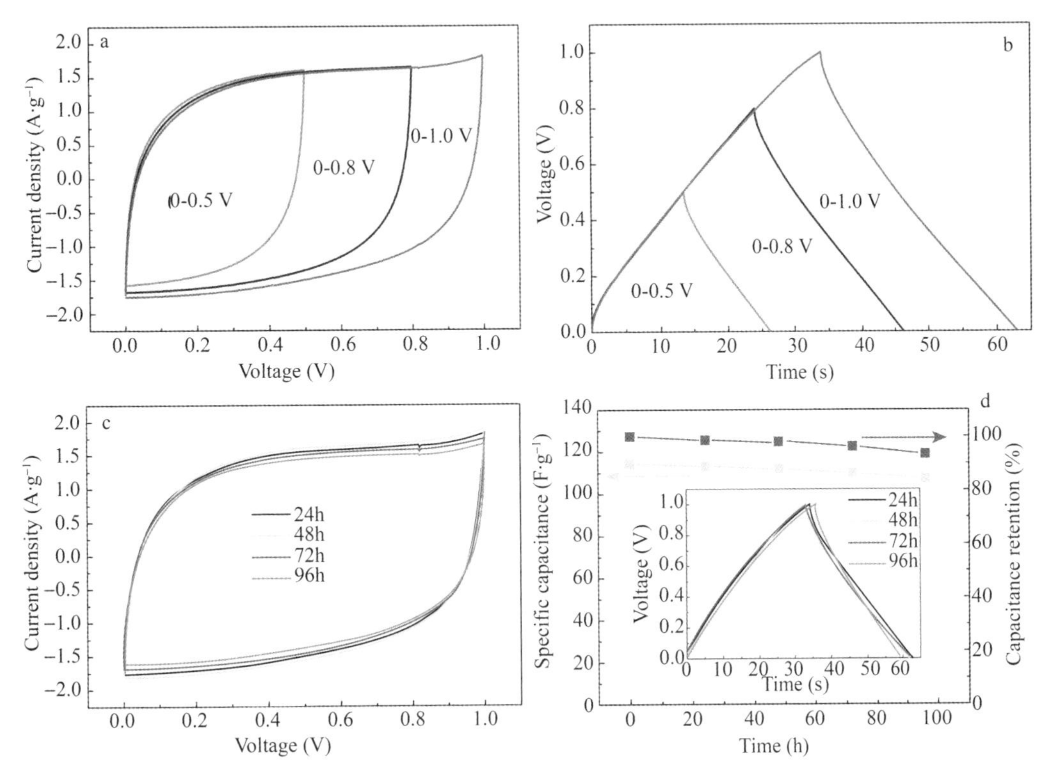

Fig. 3 Lifetime stability of the quasi-solid-state mCel-membrane-based EDLC. (a) CV curves measured in different scan potential windows, and at a scan rate of 50 mV · s^{-1}. (b) GCD profiles measured in different scan potential windows, and at a current density of 1 A · g^{-1}. (c) CV curves at 50 mV · s^{-1}, measured at different time intervals. (d) Specific capacitances and GCD profiles (at a current density of 1A · g^{-1}) calculated at different time intervals

The energy density and power density are important parameters for comparing the performance of EDLCs. The maximum energy density of the mCel-membrane-based EDLC was 4.37 Wh · kg^{-1} and power density being 248.79 W · kg^{-1} at a current density of 0.5A · g^{-1} (Fig. 4a). This value was 1.56 times higher than that of the NKK-MPF30AC-based EDLC, 1.66 times higher than that of the NKK TF4030-based EDLC, and 1.50 times higher than that of the PVA/KOH-based EDLC. At a current density of 10A · g^{-1}, the energy density of the mCel-membrane-based EDLC was 2.94 Wh · kg^{-1} at a power density of 5030.8 W · kg^{-1}. Two mCel-membrane-based EDLCs could be connected in series or in parallel, to obtain a device that outputted approximately twice the potential, or twice the current density and charge-discharge time of those of the single EDLC (Fig. 4b, c). The connected mCel-membrane-based EDLCs were stable in different voltage windows, which indicated their good practical prospects (Fig. S8, Supporting Information). As shown in Fig. 4d and Movie S1 in Supporting Information, the 2.0 V light-emitting diodes (LEDs) were powered by two mCel-membrane-based EDLCs in series connection for a few minutes.

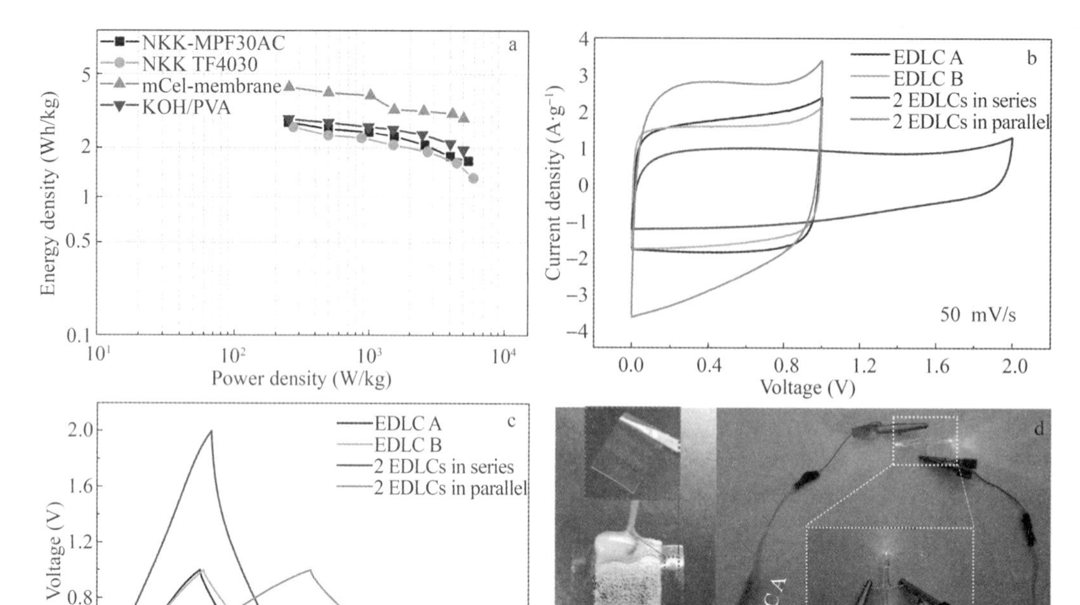

Fig. 4 Application performance of the quasi-solid-state mCel-membrane-based EDLC. (a) Ragone plots. (b) CV curves and (c) GCD profiles of two mCel-membrane-based EDLCs connected in series and in parallel, at a scan rate of 50 mV · s^{-1} and current density of 1A · g^{-1}. (d) Photographs of the LEDs powered by two mCel-membrane-based EDLCs connected in series

2.3 Flexible planar-type mCel-membrane-based MSC

The straightforward preparation and favorable properties of the mCel-membrane suggested its potential in flexible planar-type MSCs. During the preparation of the mCel-membrane, the adhesive properties of Cel/IL because of intermolecular forces and electrostatic interactions were sufficient to adhere to and retain electrodes for preparing a planar-type MSC. Fig. 5a shows that with the aid of a PTFE mask, activated carbon powder was directly deposited on the blank surface of the Cel/IL. Subsequent immersion in water and drying in an oven yielded a free-standing mCel-membrane-based MSC. This patterning technique here is free oF · glue or other additives, making it lower cost, simpler, and more versatile than many other patterning techniques such as photolithography,[32] ink-printing,[33] laser scribing or writing,[34] and plasma-scanning.[35] Different configurations of interdigitated electrodes could be fabricated on the mCel-membrane-based MSC, which demonstrated the good scalability of the procedure (Fig. 5b). The MSC exhibited excellent flexibility, and could be bent to 180° without compromising its structural integrity (Fig. S9, Supporting Information). SEM images showed that the activated carbon finger electrode was highly integrated with the mCel-membrane (Fig. S10, Supporting Information). No obvious potholes or interfacial defects were observed (Fig. S11, Supporting Information).

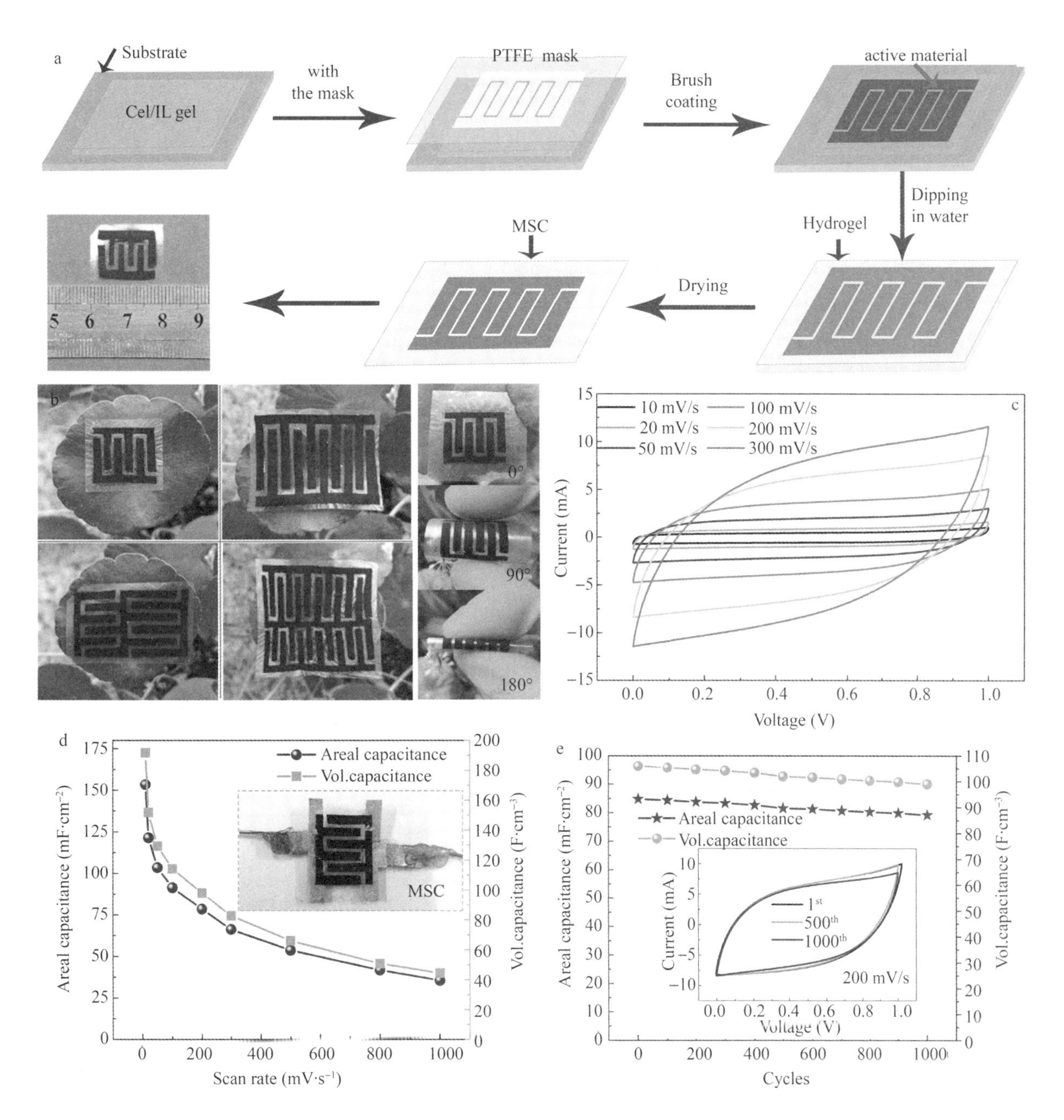

Fig. 5 Electrochemical properties of the mCel-membrane-based MSC. (a) Schematic diagram of the preparation. (b) Optical photographs of MSCs with different configuration patterns, and in bent states. (c) CV curves measured at scan rates of 10–300 mV · s^{-1}. (d) Areal and volumetric capacitances measured at different current densities. (e) Cycling performance measured at a scan rate of 200 mV · s^{-1}

The electrochemical properties of the mCel-membrane-based MSC were investigated under typical atmospheric conditions. CV curves were measured at scan rates of 10–300 mV · s^{-1} (Fig. 5c and Fig. S12 in Supporting Information). These exhibited symmetrical quasi-rectangular profiles at high scan rates. GCD profiles exhibited triangular profiles without a voltage drop, at a high current density of 1.0 A · cm^{-3} (Fig. S13, Supporting Information). In different potential windows, the CV and GCD curves also exhibited rectangular and symmetrical triangular profiles, respectively. The mCel-membrane-based MSC exhibited an areal capacitance of 153.34 mF · cm^{-2}, and a volumetric capacitance of 191.66 F · cm^{-3}, at 10 mV · s^{-1} (Fig. 5d). These values were far higher than those of some carbon electrode-based MSCs (see Table S1, Supporting Information). After 1000 cycles, the areal capacitance and volumetric capacitance slightly decreased from 84.8 to 79.2 mF · cm^{-2} and from 106 to 99 F · cm^{-3} (93.4% retention), respectively (Fig. 5e). The profile of the 1000th CV curve largely overlapped with those of the 1st and 500th CV curves (Fig. 5e inset), demonstrating the excellent cycling stability of the MSC. When bent to angles of 0°, 90° or 180°, the MSC did not show any obvious distortion in the CV curves or degradation in its capacitance (Fig. 6a). This suggested the potential of the MSC in flexible wearable devices. Ragone plots in Fig. 6b showed that the MSC delivered a volumetric energy density of 6.655 mWh · cm^{-3} at a power density of 0.2395 W · cm^{-3}, and 1.545 mWh · cm^{-3} at a power density of 5.561 W · cm^{-3}. These values were superior to those of some carbon-based MSCs.[36-39] Two MSCs connected in series exhibited twice the operating voltage of the single MSC. Two connected in parallel exhibited approximately twice the current and discharge time of the single MSC (Fig. 6c). Either in a plane or in a curved state, the MSC could power a digital watch for an extended period (Fig. 6d and Movie S2 in Supporting Information). The MSC retained its flexibility and integrity after electrochemical testing, demonstrating its robust application characteristics (Fig. S14 in Supporting Information).

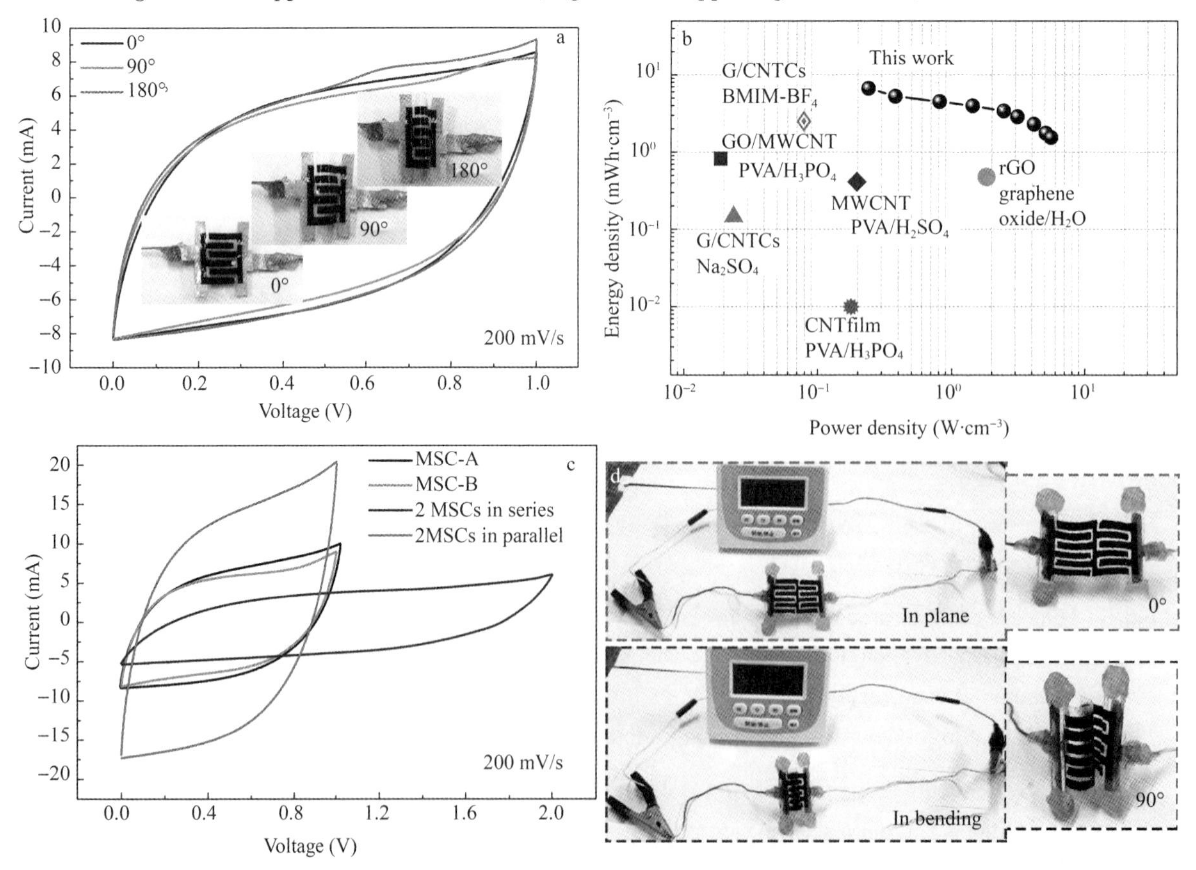

Fig. 6 (a) CV curves of the mCel-membrane-based MSC in bent states. (b) Ragone plots of the mCel-membrane-based MSC. (c) CV curves of two MSCs in series connection and in parallel connection. (d) Optical photographs of a digital watch powered by the mCel-membrane-based MSC

3 Conclusions

In summary, a transparent, flexible and robust mesoporous cellulose membrane was prepared by a simple dissolution-regeneration process. The mCel-membrane/KOH polymer electrolyte exhibited a high electrolyte retention of 451.2 wt. %, and a high ionic conductivity of 0.325 S · cm^{-1}. Symmetric quasi-solid-state EDLC comprising of mCel-membrane/KOH polymer electrolyte and active carbon electrode exhibited excellent electrochemical performance, demonstrated by a high capacitance of 120.6 F · g^{-1}, a high rate capability and excellent cyclability with 84.7% retention after 10, 000 cycles. The maximum energy density of the quasi-solid-state mCel-membrane-based EDLC was 1.5 – 1.66 times higher than those of the NKK-MPF30AC, NKK TF4030 and PVA/KOH-based EDLCs. Moreover, a solid-state MSC can be facilely prepared by directly depositing carbon electrode on the mCel-membrane. The highly integrated mCel-membrane-based MSC exhibited a high areal capacitance of 153.34 mF · cm^{-2} and volumetric capacitance of 191.66 F · cm^{-3} at 10 mV · s^{-1}, representing one of the highest values among all carbon-based MSCs. It also exhibited a maximum volumetric energy density of 6.655 mWh · cm^{-3} at the power density of 0.2395 W · cm^{-3}, and a maximum power density of 5.561 W · cm^{-3} at the energy density of 1.545 mWh · cm^{-3}. These results offer a renewable, transparent, flexible and biodegradable cellulose-based mesoporous quasi-solid-state electrolyte with versatile electrochemical performance in EDLCs and MSCs, holding great promise for solid-state biodegradable energy storage applications.

4 Experimental Section

Materials: Cellulose with an α-cellulose content of approximately 90% was purchased from Haojia Cellulose Co., Ltd. (Tianjin, China). The polymerization degree of the cellulose was 1484, with an average molecular weight of 240 kDa. NKK-MPF30AC and NKK-TF4030 were purchased from Nippon Kodoshi Corporation (Kochi, Japan). Activated carbon (porous volume: 1.0-1.2 cm^3 · g^{-1}, specific surface area: 2000 ± 100 m^2 · g^{-1}) and ethylene vinyl acetate (EVA) were purchased from Confluence Investment Management Co., Ltd. (Shanghai, China). [Bmim]Cl, KOH and PVA (molecular weight: 88 kDa) were purchased from Aladdin Chemistry Co., Ltd. (Shanghai, China). All chemicals were of analytical grade and used as received.

Preparation of mCel-membrane: [Bmim]Cl (50 g, 0.286 M) was added to a three-necked flask (250 mL) with a condenser, which was heated in an oil-bath at 85 ℃ for 10 min. Cellulose (2.08 g, 4 wt. %) was added to the flask, which was stirred until a homogeneous transparent solution was obtained. The Cel/IL was placed in a vacuum oven (0.01 MPa) at 85 ℃ for 6 h, to remove bubbles and decrease the viscosity. The obtained Cel/IL gel was cast on a silicon wafer (diameter > 40 mm), using a VTC-100 vacuum spin coater (Kejing Instrument Co. Ltd., Shenyang, China). After soaking in distilled water for 1 h, a hydrogel of the mCel-membrane was obtained. The hydrogel was sandwiched between two PTFE 0.1-μm millipore membranes (Hongqi Filter Equipment Co., Ltd., Haining, China), and the resulting structure was then dried at 60 ℃ for 12 h to generate the mCel-membrane.

Fabrication of quasi-solid-state mCel-membrane-based EDLC: The working electrode was prepared with 80 wt. % activated carbon, 10 wt. % Ketjen black and 10 wt. % PTFE, and was dried in an oven at 60 ℃ for

8 h. The electrodes and mCel-membrane were immersed in 6 M KOH aqueous solution for 48 h, and excess electrolyte was removed from the surface using absorbent tissue paper. The mCel-membrane-based EDLC was assembled in a sandwich-like configuration of electrode//mCel-membrane/KOH//electrode. The quasi-solid-state mCel-membrane-based EDLC was encapsulated using EVA hot-melted glue under an ambient atmosphere. Similar EDLCs containing activated carbon electrodes and NKK-MPF30AC, NKK-TF4030, or PVA/KOH gel as electrolytes were also prepared for comparison. The PVA/KOH polymer electrolyte was prepared by dissolving PVA powder(3 g) and KOH(3 g) in distilled water(30 mL). The mixture was heated at 80 ℃ under vigorous stirring, until it became homogeneously clear.

Fabrication of solid-state mCel-membrane-based MSC: To prepare the planar-type MSC, custom-made PTFE masks were used to define the required MSC patterns. The Cel/IL gel was cast on a silicon wafer, which was then placed in an oven at 90 ℃ for 45 min. With the aid of the PTFE masks, the active carbon electrode material was directly deposited on the blank surfaces of the Cel/IL. It was then immersed in distilled water for approximately 6 h, which formed a hydrogel state MSC. After drying at 60 ℃ for 8 h, a highly integrated planar-type MSC (hereafter referred to as mCel-membrane-based MSC) was obtained. Prior to electrochemical testing, the mCel-membrane-based MSC was impregnated in 6 M KOH solution for 48 h. Excess electrolyte on the MSC surface was then removed using absorbent tissue paper.

Characterizations: The light transmittance of the sample was measured using a TU-1901 spectrophotometer (Purkinje General Instrument Co., Beijing, China), with a wavelength range of 400-800 nm. The SEM micro-morphology of the sample was characterized using a JSM-7500F microscope (Hitachi, Tokyo, Japan), at an operating voltage of 10 kV. Thermogravimetric (TG) analysis was performed using a STA 6000 analyzer (Perkin Elmer Inc., Waltham, MA, USA) in the temperature range of 45-850 ℃, at a heating rate of 20 ℃ · min^{-1}, under a nitrogen atmosphere (40 mL · min^{-1}). The aperture of the film was measured by a 3H-2000PB type bubble pressure filter pore size analyzer (Beishide Instrument-ST Co., Beijing, China). The wetting liquid was anhydrous ethanol. Tensile tests were performed at room temperature using an Instron 5569 universal testing machine (Instron Corp., Canton, MA, USA), at a cross-head speed of 1 mm · min^{-1}. The sample was cut into 15 mm ×5 mm sized pieces before tensile testing.

Electrochemical testing of the EDLCs and MSCs: The electrochemical properties of the EDLCs and MSCs were investigated using a CHI 660D electrochemical workstation (Chenhua Instrument Co., Shanghai, China). CV, GCD and EIS measurements were performed. The bulk resistance was investigated from EIS measurements, with alternating current potential amplitude of 10 mV, and frequency range from 1 Hz to 100 kHz, at room temperature. The specific capacitance and energy density of the MSCs were estimated from CV measurements using equations in reference[40].

Acknowledgements

The authors kindly acknowledge the joint support by the National Natural Science Foundation of China (No. 31622016; 31670583), and the Natural Science Foundation of Heilongjiang Province of China (No. JC2016002).

References

[1] Yan J, Wang Q, Wei T, Fan, Z J. Recent advances in design and fabrication of electrochemical supercapacitors with

high energy densities[J]. Adv. Energy Mater., 2014, 4(4): 1300816.

[2]Fu K K, Cheng J, Li T, Hu L B. Flexible batteries: from mechanics to devices[J]. ACS Energy Lett., 2016, 1(5): 1065-1079.

[3]Chen C J, Zhang Y, Li Y J, Dai J Q, Song J W, Yao Y G, Gong Y H, Kierzewski I, Xie J, Hu L B. All-wood, low tortuosity, aqueous, biodegradable supercapacitors with ultra-high capacitance[J]. Energy Environ. Sci., 2017, 10(2): 538-545.

[4]Yan J, Wang Q, Lin C P, Wei T, Fan, Z J. Interconnected frameworks with a sandwiched porous carbon layer/graphene hybrids for supercapacitors with high gravimetric and volumetric performances[J]. Adv. Energy Mater., 2014, 4(13): 1400500.

[5]Yu G H, Hu L B, Liu N A, Wang H L, Vosgueritchian M, Yang Y, Cui Y, Bao Z A. Enhancing the supercapacitor performance of graphene/MnO_2 nanostructured electrodes by conductive wrapping[J]. Nano Lett., 2011, 11(10): 4438-4442.

[6]Zhu J, Tang S C, Wu J, Shi X L, Zhu B G, Meng X K. Wearable high-performance supercapacitors based on silver-sputtered textiles with FeCo2S4-NiCo2S4 composite nanotube-built multitripod architectures as advanced flexible electrodes[J]. Adv. Energy Mater., 2017, 7(2): 1601234.

[7]Yu ZN, Tetard L, Zhai L, Thomas J. Supercapacitor electrode materials: nanostructures from 0 to 3 dimensions[J]. Energy Environ. Sci., 2015, 8(3): 702-730.

[8]Zhong C, Deng Y D, Hu W B, Qiao J L, Zhang L, Zhang J J. A review of electrolyte materials and compositions for electrochemical supercapacitors[J]. Chem. Soc. Rev., 2015, 44(21): 7484-7539.

[9]WangY, Zhong W H, Tyler S, Allen E, Li B. A particle-controlled, high-performance, gum-like electrolyte for safe andfFlexible energy storage devices[J]. Adv. Energy Mater., 2015, 5(2): 1400463.

[10]Moon W G, Kim G P, Lee M, Song H D, Yi J. A biodegradable gel electrolyte for use in high-performance flexible supercapacitors[J]. ACS Appl. Mater. Interfaces, 2015, 7(6): 3503-3511.

[11]Lu X H, Yu M H, Wang G M, Tong Y X, Li Y. Flexible solid-state supercapacitors: design, fabrication and applications[J]. Energy Environ. Sci., 2014, 7(7): 2160-2181.

[12]Feng J, Sun X, Wu C Z, Peng L L, Lin C W, Hu S L, Yang J L, Xie Y. Metallic few-layered VS2 ultrathin nanosheets: high two-dimensional conductivity for in-plane supercapacitors[J]. J. Am. Chem. Soc., 2011, 133(44): 17832-17838.

[13]Tarascon J M, Armand M. Issues and challenges facing rechargeable lithium batteries[J]. Nature 2001, 414(6861): 359-367.

[14]Meng C Z, Liu C H, Chen L Z, Hu C H, Fan S S. HighlyfFlexible and all-solid-state paper like polymer supercapacitors[J]. Nano Lett., 2010, 10(10): 4025-4031.

[15]Peng X, Liu H L, Yin Q, Wu J C, Chen P Z, Zhang G Z, Liu G M, Wu C Z, Xie Y. A zwitterionic gel electrolyte for efficient solid-state supercapacitors[J]. Nat. Commun., 2016, 7: 11782.

[16]Liao H Y, Hong H Q, Zhang H Y, Li Z H. Preparation of hydrophilic polyethylene/methylcellulose blend microporous membranes for separator of lithium-ion batteries[J]. J. Membrane Sci., 2016, 498: 147-157.

[17]Lu W J, Yuan Z Z, Zhao Y Y, Zhang H Z, Zhang H M, Li X F. Porous membranes in secondary battery technologies[J]. Chem. Soc. Rev., 2017, 46(8): 2199-2236.

[18]Li S H, Huan, D K, Zhang B Y, Xu X B, Wang M K, Yang G, Shen Y. Flexible supercapacitors based on bacterial cellulose paper electrodes[J]. Adv. Energy Mater. 2014, 4(10): 1301655.

[19]Yang X, Zhang F, Zhang L, Zhang T F, Huang Y, Chen Y S. A high-performance graphene oxide-doped ion gel as gel polymer electrolyte for all-solid-state supercapacitor applications[J]. Adv. Funct. Mater., 2013, 23(26): 3353-3360.

[20]Zhu H L, Luo W, Ciesielski P N, Fang Z Q, Zhu J Y, Henriksson G, Himmel M E, Hu L B. Wood-derived materials for green electronics, biological devices, and energy applications[J]. Chem. Rev., 2016, 116(16): 9305-9374.

[21] Xiao S Y, Yang Y Q, Li M X, Wang F X, Chang Z, Wu Y P, Liu X. A composite membrane based on a biocompatible cellulose as a host of gel polymer electrolyte for lithium-ion batteries[J]. J. Power Sour., 2014, 270: 53-58.

[22] Nyholm L, Nystrom G, Mihranyan A, Stromme M. Toward flexible polymer and paper-based energy Storage devices [J]. Adv. Mater., 2011, 23(33): 3751-3769.

[23] Chun S J, Choi E S, Lee E H, Kim, J H, Lee S Y, Lee S Y. Eco-friendly cellulose nanofiber paper-derived separator membranes featuring tunable nanoporous network channels for lithium-ion batteries[J]. J. Mater. Chem., 2012, 22 (32): 16618-16626.

[24] Li Y N, Liu Y Z, Chen W S, Wang Q W, Liu Y X, Li J, Yu H P. Facile extraction of cellulose nanocrystals from wood using ethanol and peroxide solvothermal pretreatment followed by ultrasonic nanofibrillation[J]. Green Chem., 2016, 18 (4): 1010-1018

[25] Zhou J P, Zhang L, Shu H. Cellulose microporous membranes prepared from NaOH/urea aqueous solution[J]. J. Membrane Sci., 2002, 210(1): 77-90.

[26] Zhu S D, Wu Y X, Chen Q M, Yu Z N, Wang C W, Jin S W, Ding Y G, Wu G. Dissolution of cellulose with ionic liquids and its application: a mini-review[J]. Green Chem., 2006, 8(4): 325-327.

[27] Liu Y Z, Chen W S, Xia Q Q, Guo B T, Wang Q W, Liu S X, Liu Y X, Li J, Yu H P. Efficient cleavage of lignin-carbohydrate complexes and ultrafast extraction of lignin oligomers from wood biomass by microwave-assisted treatment with deep eutectic solvent[J]. ChemSusChem, 2016, 10(8): 1692-1700.

[28] Lu B L, Xu A R, Wang J J. Cation does matter: how cationic structure affects the dissolution of cellulose in ionic liquids[J]. Green Chem., 2014, 16(3): 1326-1335.

[29] Rabideau BD, Ismail AE. The effects of chloride binding on the behavior of cellulose-derived solutes in the ionic liquid 1-butyl-3-methylimidazolium Chloride[J]. J. Phys. Chem. B, 2012, 116(32): 9732-9743.

[30] Huo P F, Zhang S L, Zhang X R, Geng Z, Luan J S, Wang G B. Quaternary ammonium functionalized poly(aryl ether sulfone)s as separators for supercapacitors based on activated carbon electrodes[J]. J. Membrane Sci., 2015, 47: 562-570.

[31] Lewandowski A, Zajder M, Beguin F. Supercapacitor based on activated carbon and polyethylene oxide-KOH-H2O polymer electrolyte[J]. Electrochim. Acta, 2001, 46(18): 2777-2780.

[32] Qi D P, Liu Z Y, Liu Y, Leow W R, Zhu B W, Yang H, Yu J C, Wang W, Wang H, Yin S Y, Chen X D. Suspended wavy graphene microribbons for highly stretchable microsupercapacitors [J]. Adv. Mater., 2015, 27 (37): 5559-5566.

[33] Zhang Q, Huang L, Chang Q H, Shi W Z, Shen L, Chen Q. Gravure-printed interdigital microsupercapacitors on a flexible polyimide substrate using crumpled graphene ink[J]. Nanotechnology, 2016, 27(10): 105401.

[34] El-Kady M F, Kaner R B. Scalable fabrication of high-power graphene micro-supercapacitors for flexible and on-chip energy storage[J]. Nat. Commun., 2013, 4: 1475.

[35] Liu L, Ye D, Yu Y, Liu L, Wu Y. Carbon-based flexible micro-supercapacitor fabrication via mask-free ambient micro-plasma-jet etching[J]. Carbon, 2017, 111: 121-127.

[36] Kim S K, Koo H J, Lee A, Braun P V. Selective wetting-induced micro-electrode patterning for flexible micro-supercapacitors[J]. Adv. Mater., 2014, 26(30): 5108-5112.

[37] Wen F S, Hao C X, Xiang J Y, Wang L M, Hou H, Su Z B, Hu W T, Liu Z Y. Enhanced laser scribed flexible graphene-based micro-supercapacitor performance with reduction of carbon nanotubes diameter[J]. Carbon, 2014, 75: 236-243.

[38] Gao W, Singh N, Song L, Liu Z, Reddy, A L M, Ci L J, Vajtai R, Zhang Q, Wei B Q, Ajayan P M. Direct laser writing of micro-supercapacitors on hydrated graphite oxide films[J]. Nat. Nanotechnol., 2011, 6(8): 496-500.

[39] Lin J, Zhang C G, Yan Z, Zhu Y, Peng Z W, Hauge R H, Natelson D, Tour J M. 3-Dimensional graphene carbon

nanotube carpet-based microsupercapacitors with high electrochemical performance[J]. Nano Lett., 2013, 13(1): 72-78.

[40] Wu Z S, Zheng Y J, Zheng S H, Wang S, Sun C L, Parvez K, Ikeda T, Bao X H, Mullen K, Feng X L. Stacked-layer heterostructure films of 2D thiophene nanosheets and graphene for high-rate all-solid-state pseudocapacitors with enhanced volumetric capacitance[J]. Adv. Mater., 2017, 29(3): 1602960.

中文题目： 基于可再生、可生物降解介孔纤维素膜的高性能柔性固体超级电容器

作者： 赵大伟，陈朝吉，张奇，陈文帅，刘守新，王清文，刘一星，李坚，于海鹏

摘要： 通过一种简便、可扩展的溶液-相转化工艺制备了一种柔性、透明、可再生的介孔纤维素膜(mCel 膜)，其介孔均匀度为 24.7 nm，孔隙率为 71.78%。KOH 饱和的 mCel 膜作为聚合物电解质具有高的电解质保留率(451.2%)、高的离子电导率(0.325 $S \cdot cm^{-1}$)以及优异的机械柔性和鲁棒性。以活性炭为电极，以 KOH 饱和的 mCel 膜为聚合物电解质的固态双电层电容器(EDLC)在 1.0 $A \cdot g^{-1}$ 时具有 110 $F \cdot g^{-1}$ 的高电容，并具有较长的循环寿命，可循环 10 000 次，电容保持率为 84.7%。此外，不需要使用复杂的器件，直接将电极材料沉积在 mCel 膜基聚合物电解质上，可以方便地制备高度集成化的平面型微超级电容器。在 10mV $\cdot s^{-1}$ 下，得到的 MSC 的面积电容为 153.34 $mF \cdot cm^{-2}$，体积电容为 191.66 $F \cdot cm^{-3}$，是所有碳基 MSC 器件中最高的电容之一。这些发现表明，开发的可再生、柔性、介孔纤维素膜在柔性、固态、便携式储能设备的实际应用中具有很大的前景，而不仅仅局限于超级电容器。

关键词： 纤维素；柔性电子元件；膜；聚合物电解质；固态超级电容器

A High-performance, All-textile and Spirally Wound Asymmetric Supercapacitors Based on Core-sheath Structured MnO_2 Nanoribbons and Cotton-derived Carbon Cloth*

Caichao Wan, Yue Jiao, Daxin Liang, Yiqiang Wu, Jian Li

Abstract: An increasing emphasis on green chemistry and high-efficient utilization of natural resourcesraises more demands for facile, rapidand cost-effective approaches for preparation of energy-storage equipment. Herein, we demonstrate a simple, fast and cheap approach to create a core-sheath structured textile electrode based on cotton-derived carbon cloth (core) and MnO_2 nanoribbons (sheath). A good interface bonding between the in-situ grown MnO_2 and the carbon fibers aids electron transfer. The abundant MnO_2nanoribbons increase the electrochemically active areas accessed by electrolyte ions and the porous carbon cloth serves as an electrolytereservoir to shorten ion-diffusion path and facilitate efficient infiltrationof electrolyte ions. Because of these advantages, this textile electrode exhibits a high areal capacitance of 202 $mF \cdot cm^{-2}$. In addition, the flexible electrode is assembled into an all-textile and spirally woundasymmetric supercapacitor with an excellent electrochemical activity, like a high areal energy density of 30.1 $\mu W \cdot h \cdot cm^{-2}$ at 0.15 $mW \cdot cm^{-2}$ and an excellent capacitance retention of 87.7% after 5000 cycles. Another meritorious contribution is the synthetic strategy realizing a more direct, eco-friendly and efficientutilization method of cellulose resource, which avoids pollutionsfrom cellulose purification or pretreatment.

Keywords: Cotton; Carbon cloth; MnO_2 nanoribbons; Core-sheath structure; Spirally wound supercapacitors; In-situ redox.

1 Introduction

Promptly growing energy consumptioncoupled with critical issues of climate changehas become an unquestionable threat for the planet and the quality of human life[1]. Also, these issues have motivated significant efforts to develop multitudinous novel forms of sustainable and renewable energy-conversion and storage devices[2,3]. Among various novelenergy storage systems, supercapacitors are of special importance due to theirfascinating capability ofbridging the power/energy gapbetween traditional dielectric capacitors and batteries/fuelcells[4]. In addition, their highpower density, fast rates of charge-discharge, reliable cycling feature and safe operation are also highly regarded[5].

Textile supercapacitors are an important category of flexible supercapacitors and have attracted widespread

---

* 本文摘自 Electrochimica Acta, 2018, 285: 262-271.

attention because of their potential to develop green flexibleand wearableelectronicsfor some advanced future applications like high-tech sportswear, health monitoring systems and military camouflages[6-8]. For instance, Yuksel and Unalan[9] reportedflexible solid-state textile-based supercapacitors with ternary nanocomposite electrodes containing manganese oxide(MnO_2), single-walled carbon nanotubes and polyaniline or poly(3, 4-ethylenedioxythiophene) – poly (styrenesulfonate), which were layer-by-layer deposited onto cotton substrates. The measured maximum power and energy density and maximum capacitance were 746.5 W · kg^{-1}, 66.4 Wh · kg^{-1} and 294 F · g^{-1}, respectively. Laforgue[10] synthesized poly (3, 4-ethylenedioxythiophene) nanofibermats by the combination of electrospinning and vapor-phase polymerization, and the mats were then incorporated into all-textile flexiblesymmetricsupercapacitors (SSCs) whose maximum capacitance reached 20 F · g^{-1} at the current density of 2 mA · cm^{-2}. Meng et al. [11] preparedan all-graphene core-sheath fiber composed of graphene fiber core with a sheath of three-dimensional(3D) graphene networks via the combined method of hydrothermal and electrolytic routes. This as-prepared fiber supercapacitor was woven into a textile for wearable electronics. Besides aforesaid materials, some othertypes of carbon materials[12,13], conductive polymers[14] or metallic oxides/hydroxides[15,16] were also utilized to develop textile supercapacitors. However, it needs to be pointed out that the cost of some electrochemical active substancesused for the preparation of textile supercapacitors is prohibitive. More seriously, a number ofsynthetic pathwaysrequire eco-unfriendly chemicals, specific equipment or elaborate procedures, which create a barrier to their large-scale industrial promotion. As a result, it is significative to explore faster, cheaper and simpler approaches todevelop high-performance textile supercapacitors.

Cellulose is the most abundant nature polymer on Earth and consists of a linear chain of several hundred to many thousands of $\beta(1\rightarrow4)$ linked D-glucose units(see Fig. S1 in SI). As a main structural component of the primary cell wall of green plants and algae, cellulose has widely served as the feedstock of papermaking, skincare products, biofuel, textile industry and building products. In addition, owing to good flexibility, abundant pores and effortlessconversion to conductive carbon counterparts, cellulose products (e.g., 1D fibers, 2D films and 3D hydrogels or aerogels) have been verified experimentally as promising candidates for development of environmentally benign energy storage devices (like supercapacitors[17-19], lithium-ion battery[20,21], zinc-air battery[22,23] and solar cells[24,25]). Actually, plant-derived cellulose is usually found in a mixture with lignin, hemicellulose, pectin and other substances. Therefore, some purification methods of cellulose are generally involved. However, common chemical purificationconsumes a good deal of chemicals (like benzene)[26], and greener physical purification (like steam explosion) usually needs energy-extensive consumption[27]. Recently, bacterial cellulose serving asa substrate for energy storage has also been highlighted[17,28], but its production is still complex and lengthy. Undoubtedly, more direct, facile and green utilization ways of cellulose resource are high-profile.

Considering these issues, herein, we report aneasily-operated and rapid approachto prepare a flexible, high-performance and self-supported textile electrode. For achieving high-efficientutilization of cellulose resource, common jean cloth(100% cotton) was directly used as the feedstock to undergo a pyrolysis treatment for the conversion to its carbon counterpart(namely cotton-derived carbon cloth, coded as CDCC). CDCC was subsequently used as a flexible and free-standing substrate for the in-situ growth of MnO_2nanoribbons via a simple immersion treatment in the solution of potassium permanganate ($KMnO_4$) with strong oxidizing property. The as-prepared MnO_2/CDCC with a core-sheath structure served as the positive electrode and was assembled into an all-textile and spirally wound asymmetric supercapacitor(ASC) using CDCC as the negative electrode and thin cotton woven as the separator. Results of the electrochemical evaluation demonstrate that both the electrode and the ASC possess favorable electrochemical properties.

2 Experimental Section

2.1 Materials

Common old jean (100% cotton) served as raw material of CDCC after being washed repeatedly withdistilled water and ethyl alcohol and then dried at room temperature. Chemicals including NaOH, $KMnO_4$ and Na_2SO_4 were supplied by Shanghai Aladdin Industrial Inc. (China). Nickel foam was obtained from SXLZY Battery Material Co., Ltd. (Taiyuan, China).

2.2 Preparation of cotton-derived carbon cloth(CDCC)

The preparation of CDCC was carried out by transferring the clean and driedjeaninto a tubular furnace for pyrolysis under a flow ofnitrogen. The samples were heated to 1000 ℃ at a heating rate of 5 ℃ · min^{-1}, and this temperature was maintained for 1 h to allow for complete pyrolysis; subsequently, the furnace decreased naturally to the room temperature.

2.3 Preparation of MnO_2/cotton-derived carbon cloth(MnO_2/CDCC)

The as-prepared CDCC was immersed in the aqueous solution of $KMnO_4$. The weight ratio of CDCC to $KMnO_4$ was set as 4 : 1, which has been already confirmed elsewhere to contribute to acquiring good electrochemical property of the compositecontaining MnO_2 and carbon[29]. A beaker containing the mixture was transferred into an oven and covered with aglass culture dish. The mixture was heatedat 60 ℃ for 12 h. After the heating, the sample was rinsed with a great deal of distilled waterand finallydried at 60 ℃ for 24 h in a vacuum.

2.4 Fabrication of all-textile and spirallywoundasymmetric supercapacitor(ASC)

Anall-textile and spirallywound ASC was assembled by using MnO_2/CDCC as the positive electrode, CDCC as the negativeelectrode, thin cotton woven(i. e., the jean cloth) as the separator and nickel foam as the current collector. The MnO_2/CDCC and CDCC electrodes attached by the nickel foams were separated with the cotton woven, to form a "sandwich" structure. This sandwich-likeslice was rolled up tightly and dipped in a 1 M Na_2SO_4 aqueous electrolytefor all of the electrochemical tests.

2.5 Characterizations

The morphology was observedusing a scanning electron microscope(SEM, Hitachi S4800) equipped with an energy dispersive X-ray (EDX) detector for elemental analysis. Observation of transmission electron microscope(TEM) and high-resolution TEM(HRTEM) was performed with a FEI, Tecnai G2 F20 TEM with a field-emission gun operating at 200 kV. X-ray diffraction (XRD) analysis was implemented on a Bruker D8 Advance TXS XRD instrument with Cu Kα(target) radiation(λ = 1.5418 Å) at a scan rate(2θ) of 4° min^{-1} and a scan range from 5° to 90°. X-ray photoelectron spectroscopy(XPS) was carried out on a Thermo Escalab 250Xi system using a spectrometer with a dual Al Ka X-ray source. Deconvolution of the overlapping peaks was performed using a mixed Gaussian-Lorentzian fitting program (Origin 9.0, Originlab Corporation). Thermalstability was determined using a thermogravimetry (TG) analyzer (TA, Q600) under a nitrogen atmosphere.

2.6 Electrochemical characterizations

The electrochemical performancesof the self-supported MnO_2/CDCC electrode were studied at room temperature by using a CS350 electrochemical workstation (Wuhan CorrTest Instruments Co., Ltd., China) in a three-electrode setup: MnO_2/CDCCacted as the working electrode, and an Ag/AgCl electrode and a Pt foil electrode served as reference and counter electrodes, respectively. The exposed geometric area of the working electrode is equal to 2 cm^2. The electrolyte was 1 M Na_2SO_4 solution. Cyclic voltammetry (CV) and galvanostatic charge-discharge (GCD) curves were tested over the potential window from 0 to 0.8 V. Electrochemical impedance spectroscopy (EIS) measurements were carried out in the frequency range from 10^5 to 0.01 Hz with alternate current amplitude of 5 mV. For the assembled ASC, the device was soaked in 1 M Na_2SO_4 solution for all of the electrochemical tests in a two-electrode system.

2.7 Calculations

For the CDCC and MnO_2/CDCC electrodes and the assembled ASC, the gravimetric (C_m, $F \cdot g^{-1}$) and areal (C_s, $F \cdot cm^{-2}$) specific capacitances were calculated from their GCD curves at different current densities according to following equations:

$$C_m = I\Delta t/m\Delta V \ or \ C_s = I\Delta t/s\Delta V \tag{1}$$

where I(A) is the discharge current, Δt(s) is the discharge time, m(g) is the mass ofelectrodes, s (cm^2) is the specific area of the electrodes, and ΔV(V) is the operation discharge voltage window (excluding IR drop). The areal densities of MnO_2/CDCC and CDCC are 8.1 and 6.4 $mg \cdot cm^{-2}$, respectively. The areal energy density (E_s) and power density (P_s) were evaluated on the basis of capacitance values shown as follow:

$$E_s = \frac{1}{2} C_s (\Delta V)^2 / 3600 \tag{2}$$

$$P_s = \frac{E_s}{\Delta t} \times 3600 \tag{3}$$

where C_s($F \cdot cm^{-2}$) isthe areal specific capacitance, ΔV(V) is the discharging voltage and Δt(s) is the discharging time.

3 Results and discussion

3.1 Graphical illustration of the design concept of MnO_2/CDCC

For the purpose of realizing more high-efficient and green use of cellulose resource and seeking a more facile and economicalsynthesis process of textile supercapacitors, we design an easily-operated and cost-effective combined technique of pyrolysis and immersion to fabricate a high-performance and free-standing MnO_2/CDCC electrode. As illustrated in Fig. 1, a piece of flexible cotton woven was firstlypyrolyzed into the electrically conductive CDCC under the protection of nitrogen. After the pyrolysis, CDCC was dipped into the solution of $KMnO_4$ with strong oxidizing property, leading to the occurrence of a redox reaction between carbon and $KMnO_4$shown as follow[30]:

$$4MnO_4^- + 3C + H_2O \rightarrow 4MnO_2 + CO_3^{2-} + 2HCO_3^- \quad (4)$$

This reaction resulted in the in-situ information of plentiful nanoribbon-likeMnO_2 on the surface of carbon fibers of CDCC and hence generated a core-sheath structure. MnO_2 is a promising pseudocapacitive material with high theoretical capacitance(ca. 1400 F · g^{-1})[31]. Also, MnO_2 has abundant reserves in the nature, low price, non-toxicity and broad operating potential window in mild electrolyte. Nevertheless, MnO_2 suffers fromlow electrical conductivity (10^{-6} – 10^{-5} S · cm^{-1}), low ionic diffusion constant and structural susceptibility[32]. In consequence, in the present paper, the integration of MnO_2 with the electrically conductive CDCC may be helpful to overcome these drawbacks. Moreover, a delicatecore-sheath structure is beneficial to bring the electrochemical activity of the hierarchicalmulti-scale structureinto full play. In addition, the textile electrode of MnO_2/CDCC has excellentflexibility(see Fig. S2 in SI)and is readily assembled into anall-textile and spirally wound ASC device.

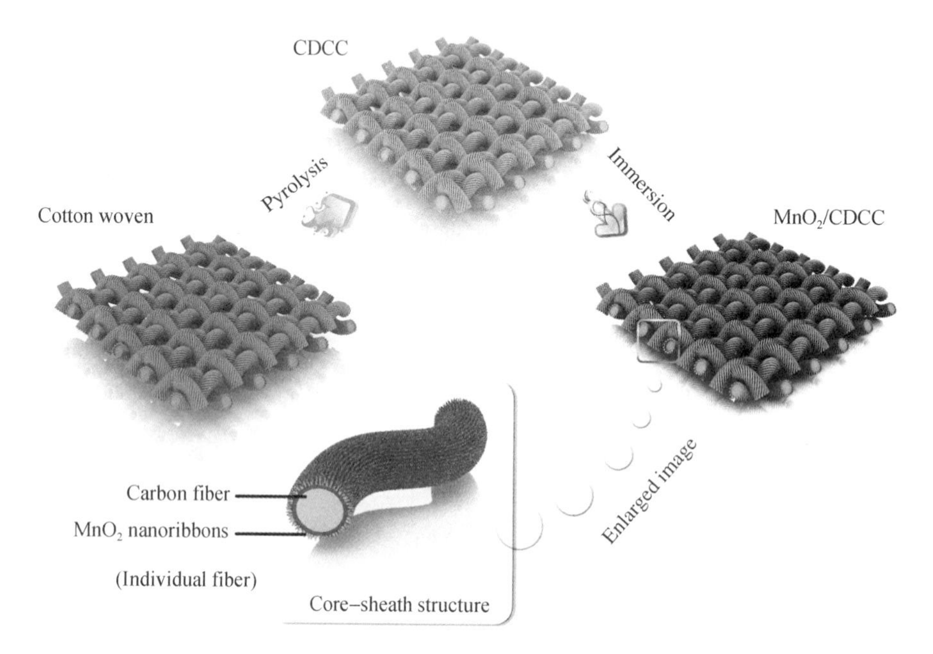

Fig. 1 Graphical illustration of the design concept of the core-sheath structured MnO_2/CDCC textile electrode

3.2 Microstructure, elemental components, crystal structure and chemical valence

By comparing Fig. 2a–c and d–f, we can see that the hierarchical braidingstructurecomposed of textile fibers was retained before and after the pyrolysis and immersion, while the diameter of fibers decreased from 12–17 μm to 3–8 μm. Meanwhile, the pyrolysisexpanded the distance between adjacent fibers and thus increased the sizes of the formed pores. In terms of supercapacitors, Weng et al.[33] demonstrated that the fibers can significantly absorb electrolyte and the porous architecture can serve as an electrolytereservoir to facilitate ion transport. By comparing Fig. 2g and h, we can find that the smooth surface of the cotton woven turned into a relatively rough surface of CDCC because of the exfoliation of cellulose compositions due to the pyrolysis. The higher surface roughness may increase the contact and reaction areas between CDCC and MnO_4^-, contributing to getting a high loading content and good interface bonding. In Fig. 2i, the fiber surface was encapsulated with a dense layer of MnO_2, whichsuggests that the material belongs to a core-sheath composite,

i. e. , the CDCC serves as the core part and the MnO_2 layer acts asthe sheath part. According to the higher-magnification SEM image in the *inset* of Fig. 2i, the formed nano-MnO_2 is curly and ribbon-like.

For EDX analysis in the *insets* in Fig. 2a−c, a sharp increase of C/O mass ratio from 1. 14 (cotton woven) to 8. 14 (CDCC) was found after pyrolysis. The introduction of MnO_2 induced a decrease to 3. 02 (MnO_2/CDCC). Besides, the strong characteristicsignals of Mn with a mass fraction of 24. 3% were detected, revealing the in-situ formation of MnO_2.

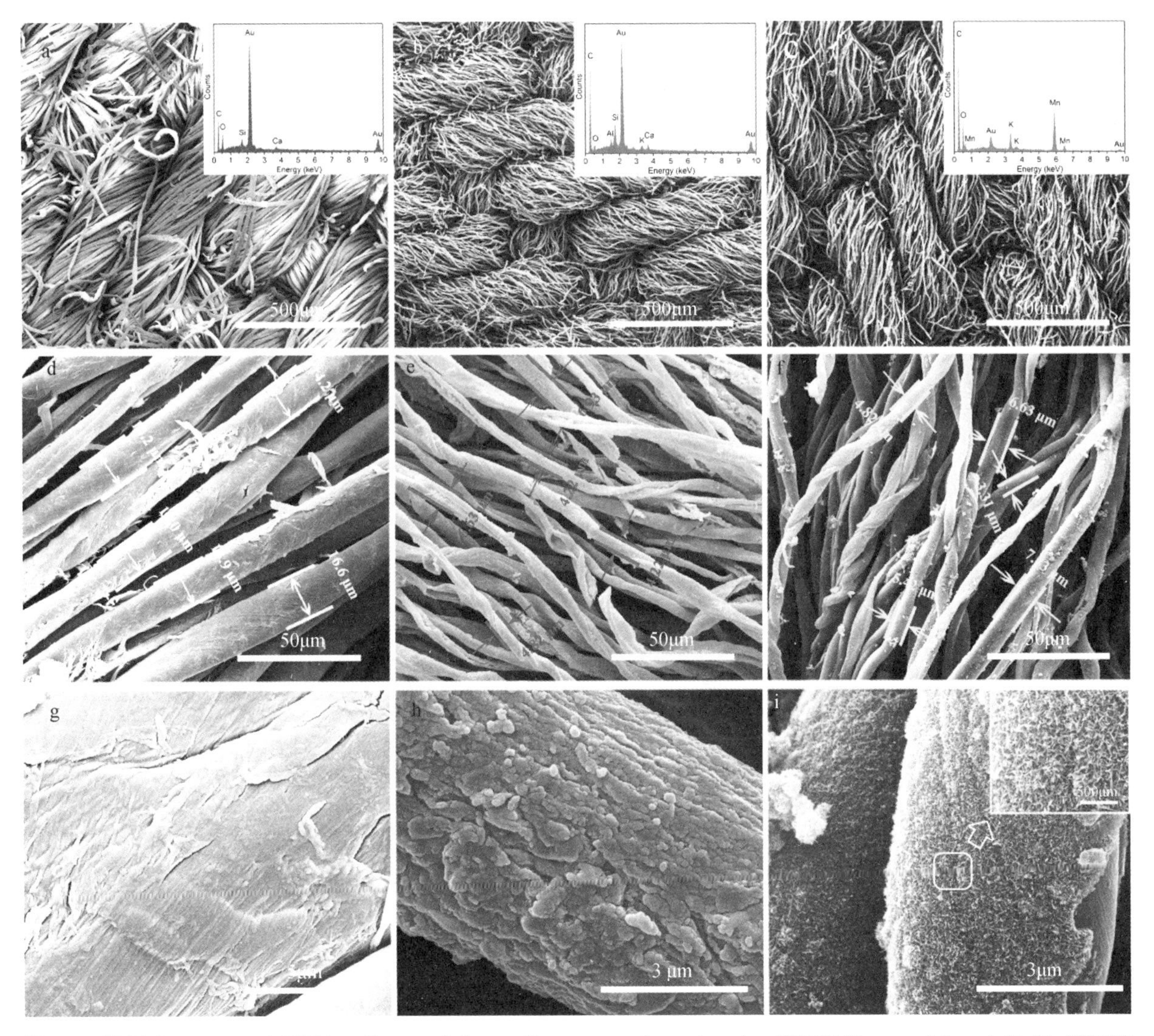

Fig. 2 SEM images and EDX patterns of the cotton woven (a, d, g), CDCC (b, e, h) and MnO_2/CDCC (c, f, i): (a−c) overall micromorphology, (d−f) measurement of fiber dimensions and (g−h) surface observation of individual fiber. *Insets* of a−c show the corresponding EDX patterns. *Inset* of i presents the enlarged image of the green range in i

For further investigating the micromorphology of MnO_2/CDCC, TEM observations were carried out. As illustrated in Fig. 3a, the underlying carbon substrate supports a large number of well-distributed MnO_2 nanoribbons. From the higher-magnification TEM image in Fig. 3b, we can find that the length and width of these nanoribbons on the carbon substrate are in the range of 1.9 – 2.8 nm and 11.3 – 21.5 nm, respectively. The HRTEM image of MnO_2/CDCC indicates the presenceof birnessite-type MnO_2 as shown in Fig. 3c; i. e., lattice fringes with spacings of approximately 0.347, 0.258 and 0.226 nm agree well with the(002), $(20\bar{1})$ and(201) lattice spacings, respectively. In addition, the analysis of XRD also confirms the existence of birnessite-type MnO_2. As seen in Fig. 3d, the XRD peaks at 2θ = 12.6°, 25.4°, 42.3° and 65.7° correspond to(001), (002), $(11\bar{2})$ and $(31\bar{2})$ planes of birnessite-type MnO_2($K_xMn_2O_4 \cdot 1.5H_2O$, JCPDS card no. 42-1317), respectively.

The possible formation mechanism of the MnO_2 nanoribbons might beattributed to the crimping of flakes by a "rolling mechanism"[34]. Graphite-type potassium manganese oxides have layered structures containing 2D sheets of edge-shared MnO_6 octahedra. These sheets are separated by potassium ions and/or water, which can be topologically extracted from or inserted between the sheets in either aqueous or organic phases[35]. As a result, they are ideal candidates for preparing many other manganese oxides. MnO_6 octahedral nuclei were initially combinedinto 1D matter with a thickness of few atomic layers, and hence the layer-structured 1D matter tended to curl and restack to form porous nanosheets[36,37]. Through rolling of the nanosheets, the ribbon-like MnO_2 was eventually generated. The graphical illustration of the possible mechanism was shown in Fig. 4.

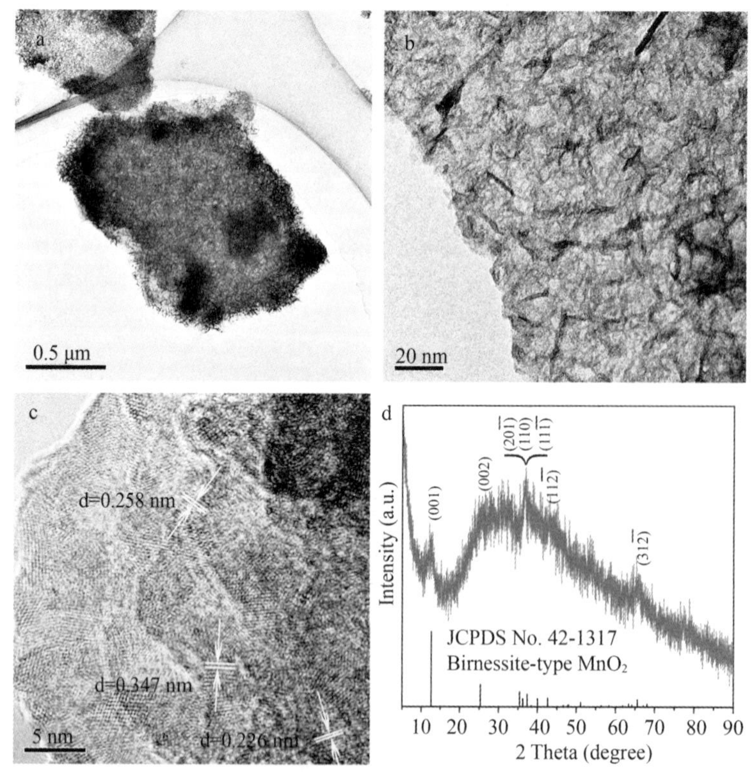

Fig. 3 (a, b) TEM images at different magnifications, (c) HRTEM image and (d) XRD pattern of MnO_2/CDCC, respectively. The bottom line is the standard JCPDS card no. 42-1317 for birnessite-type MnO_2

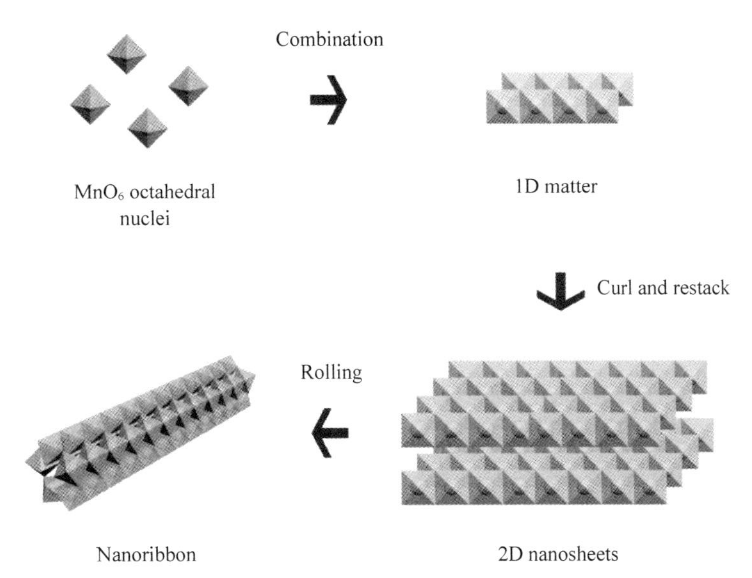

Fig. 4 Graphical illustration of thepossible mechanism of the ribbon-like MnO_2

XPS is a powerful tool which provides much useful information of the oxidation stateof manganese and the chemical bonding state of MnO_2/CDCC. The high-resolution Mn 2pspectrum shows two peaks at 653. 9 and 642. 3 eV (Fig. 5a), which are assigned tothe binding energy of Mn $2p_{1/2}$ and Mn $2p_{3/2}$, respectively. Besides, the spinenergy separation of 11. 6 eV betweenthe Mn $2p_{1/2}$ andMn $2p_{3/2}$ peaks is consistent with previouslyreported data for MnO_2[38], revealing that thepredominant oxidation state is+4. The high-resolution O1s spectrum is displayed in Fig. 5b, where two peaks at 531. 3 and 529. 7 eV are corresponding to Mn−O−H and Mn−O−Mn[39], respectively. This result further verifies the composition of this sheath is MnO_2. TG analysis was employed to roughly determine the loading content of MnO_2 in MnO_2/CDCC, i. e., subtracting the residual quantity of CDCC at 750 ℃ from that of MnO_2/CDCC. The slight weight loss, as shown in Fig. S3, might be ascribed to the evaporation or decomposition of residual oxygen-containing matter. After calculation, the content of the MnO_2sheath is about 17. 1%.

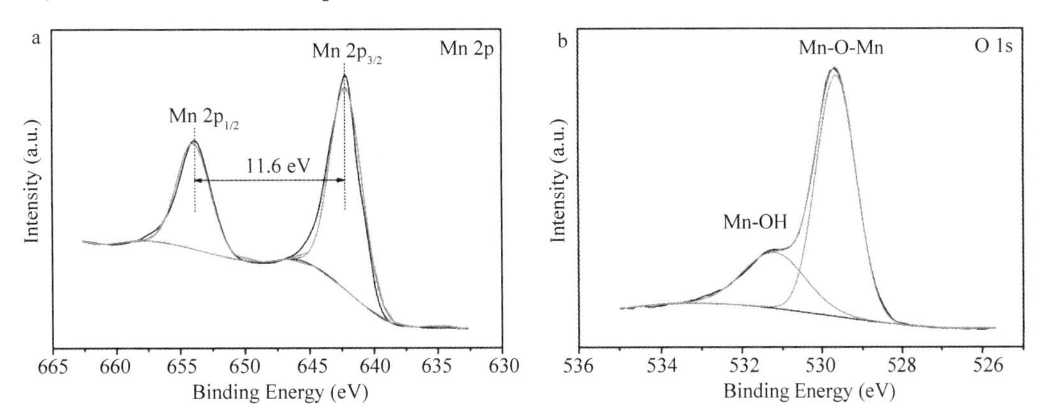

Fig. 5 XPS spectra of MnO_2/CDCC: thecore-level XPS signals of(a)Mn 2p and(b)O 1s

3.3 Electrochemical performances of MnO_2/CDCC in a three-electrode system

The electrochemical properties of the CDCC and MnO_2/CDCC electrodewere studied in a three-electrode configuration using 1M Na_2SO_4 solution as the electrolyte within the potentialrange of 0-0.8 V. As shown in Fig. 6a, the CV curves of MnO_2/CDCC at various scan rates of 5-100 mV · s^{-1}exhibited quasi-rectangular shapes with nearly mirror-image symmetry and displayed arapid current response to voltage reversal at each end potentialeven at the high scan rate of 100 mV · s^{-1}, which reflect an ideal electrochemicalcapacitive behavior[40]. In addition, the charge curves of MnO_2/CDCC are almost symmetric to the corresponding discharge counterparts(Fig. 6b), further revealing a good capacitive behavior and a highlyreversible faradic reaction between Na^+ and MnO_2, i.e., $MnO_2+Na^++e^-\leftrightarrow MnOONa$[38]. Fig. 6c presents the variation of areal and gravimetric specific capacitances of MnO_2/CDCC with the increase of current density. It is observed that the electrode showed a highest areal capacitance of 202 mF · cm^{-2} at the current density of 0.1 mA · cm^{-2}, which is higher than that of many MnO_2-based electrodes ever reported, such as the MnO_2 nanowire/ZnO nanorod array(41.5 mF · cm^{-2})[41], agarose gel-wrapped MnO_2(52.55 mF · cm^{-2})[42] and carbon-modified MnO_2 nanosheet array(143 mF · cm^{-2})[43]. In addition, MnO_2/CDCC exhibited gradually decreasedcapacitance with the increase of current density, which may be attributed to discrepant insertion-deinsertion behaviors of Na^+ from theelectrolyte to MnO_2[44,45]. The higher current densities significantly reduced effective interactions between the MnO_2 and ions, thus resulting in a remarkable decrease in the capacitance of MnO_2/CDCC.

Different from common rectangular CV curves of highly purifiedcarbon, CV profiles of CDCC is severely distorted from rectangular shape(Fig. 6d), which ispossibly due to faradaic reactions that occur on the surface of CDCC, like the reaction between carbonyl(C=O)andhydroxyl(C—OH)groups, i.e., $C=O+H^++e^-\leftrightarrow C—OH$[46]. Similar results can be found in other oxygen-containing carbon materials like HNO_3-activated biochar[47,48]. The GCD curves of CDCC have almost symmetric shape and slight curvature(Fig. 6e), which indicate synergistic effects of double-layer capacitance and pseudocapacitance. Fig. 6f presents the rate capability of CDCC. When the current density increased by 20 times(from 0.1 to 2 mA · cm^{-2}), the CDCC electroderetained only 2.5% of its originalcapacitance(a decrease from 119 mF · cm^{-2} to 3 mF · cm^{-2}). In comparison, the core-sheath structured MnO_2/CDCC electrode exhibited significantly highercapacitance retention of 37.3% (see Fig. 6c), suggesting its more superior rate capability. The better rate capability of MnO_2/CDCCmay be attributed to three main causes:

(1)The presence of abundant nano-scaled MnO_2 ribbons(serving as the sheath on the surface of carbon fibers) is beneficial to increase the electrochemically active surface areaaccessed by electrolyte ions, thus promoting the ionsadsorption and shortening the migration pathways of ionsduring the charge-discharge processes[49].

(2) The hierarchicallyporous CDCC substrate acted as an electrolytereservoir, which is beneficial to shortenthe ion-diffusion path and accelerate the efficient infiltrationof electrolyte ions. Besides, the conductive 3D carbon fiber networksensured the fast collect and transfer of electrons. The EIS result(see Fig. S4 in SI) demonstrates this conjecture. The diameter of the semicircle in the high frequency region(reflecting charge-transfer resistance)increased with the addition of MnO_2, which indicates that the CDCC with goodelectrical conductivity canreduce the charge-transfer resistance andthus improved the power density of the hybrid

electrode and even the assembled ASC;

(3) The two-step synthesizedcore-sheath nanostructuresaided the electron transferbetween the MnO_2 nanoribbons and carbon fibers of CDCC due to the good interface connection between them [50]. This self-supported structure also avoided the use of a polymer binder or conductive additive in the preparation of electrode, which adds extra contact resistance.

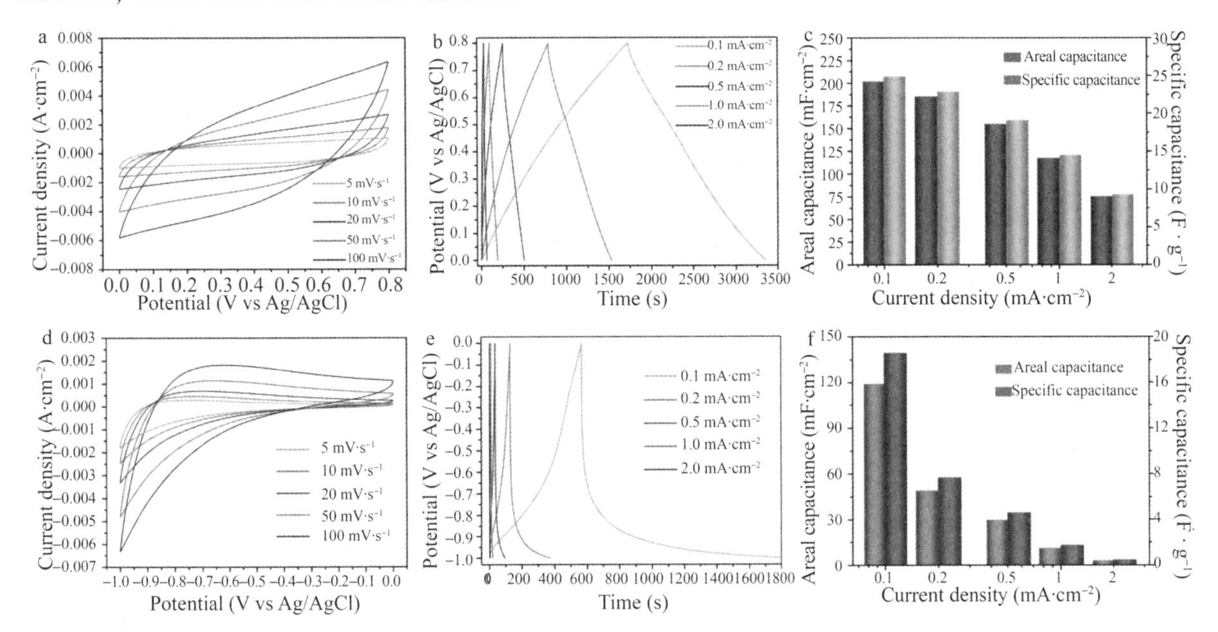

Fig. 6(a-c) Electrochemical properties of the MnO_2/CDCC electrode: (a) CV curves at various scan rates, (b) GCD profiles at different current densities and (c) rate capability; (d-f) Electrochemical properties of the CDCC electrode: (a) CV curves at various scan rates, (b) GCD profiles at different current densities and (c) rate capability

3.4 Electrochemical performances of CDCC//MnO_2/CDCC ASC in a two-electrode system

For further investigating the potential of the core-sheath structured MnO_2/CDCC in practical applications, anall-textile and spirally wound ASC was assembled as shown in Fig. 7a. The MnO_2/CDCC electrode and CDCC electrode acted as the positive electrode and negativeelectrode, respectively, and the two electrodes were separated with a piece of thin cotton woven to form a "sandwich" structure. Besides, two conductive nickel foamswas used as current collectors and severally attached tothe two electrodes. This five-layer architecture was finally rolled up and immersedin anaqueous electrolyte of 1 M Na_2SO_4 for electrochemical experiments. In addition, it is worth mentioning that, as compared with SSCs, ASCs generally feature both higher energy andpower densities because of theintegration of opposite operatingpotentialwindow of positiveand negative electrodes which extends the operating voltage range of ASCs. Therefore, ASCs have attracted a huge amount of interest already.

Fig. 7b presents the CV curves of the all-textile and spirally wound ASC at various voltage windows. It can be seen that the voltage window can be extended from 0.8 V to 1.5 V and no obvious distortion appeared for these CV curves as the current density increased. The CV profiles of the ASC at different scan rates (0-1.5 V) are shown in Fig. 7c, where these curves are obviously deviated from the ideal rectangular shape. This phenomenon reflects the pseudocapacitive feature and is primarily ascribed to the faradaic redoxreactions between Na^+ and MnO_2 and oxygen-containing groups of CDCC. In addition, for eliminating the interference of the current collector(namely nickel foam), we also measured the CV curves of the nickel foam in a three-electrode system and 1 M Na_2SO_4 solution. As shown in Fig. S5, the area enclosed within the CVcurve of the nickel foam, which is directly related to the energy storage capability, is remarkably smaller than that of the MnO_2/CDCC electrode. The result demonstrates the negligiblecontribution of the nickel foam to the capacitance of the ASC device.

The GCD curves of the ASC device at variouscurrent densities ranging from 0.2 to 10mA · $cm^{-2}$are shown in Fig. 7d. The nonlinear shape further reveals that the storage of electrical energy is mainly derived fromthe Faradic reactions, which is consistent with the result of CV analysis. This CDCC//MnO_2/CDCC ASC presented the highest areal capacitance of 96.3 mF · cm^{-2} at the current density of 0.2 mA · cm^{-2} and still kept 18.7 mF · cm^{-2} as the current density increased by 50 times to 10 mA · cm^{-2}(see Fig. 7e). The Ragone plotwas plotted based on the GCD data. As shown in Fig. 7f, the ASC device showed a high areal energy density of 30.1 μW · h · cm^{-2} at the areal power density of 0.15 mW · cm^{-2}, while maintaining a areal energy density of 5.8 μW · h · cm^{-2} at a higher areal power density of 7.5 mW · cm^{-2}. These energy and power densities are evidently more superior to those of some recently reported ASCs or SSCs, such as $Mn_3(PO_4)_2 \cdot 3H_2O$–graphene nanosheets SSC(0.17 μW · h · cm^{-2} at ~ 11 μW · cm^{-2})[51], gallium nitride/graphite paper SSC (7.4 μW · h · cm^{-2} at ~ 0.25 mW · cm^{-2})[52], polypyrrole/graphene oxide SSC(12.9 μW · h · cm^{-2} at 954.3 μW · cm^{-2})[53], carbon quantum dots reinforced polypyrrole nanowire SSC(26.5 μW · h · cm^{-2} at 0.2 mW · cm^{-2})[54], RuO_2/PEDOT: PSS//PEDOT: PSS ASC(0.053 μW · h · cm^{-2} at 147 μW · cm^{-2})[55] and Cu_2O/Cu/graphite/cellulose paper//carbon paper ASC(10.8 μW · h · cm^{-2} at 402.5 μW · cm^{-2})[56], as presented in Fig. 7f and Table 1. These comparisons indicate a favorable characteristic of high power and energy densities for our assembledall-textile and spirally wound ASC device.

The Nyquistdiagram of the ASC is shown in Fig. 7g, where the small intercept at Z real axis represents the low solution resistance(ca. 7.5 Ω) manifesting the favorable conductivity of the electrolyte. As mentioned above, the diameter of the semicircle in the high frequency region is concerned with the charge-transfer resistance. It is seen that this resistance is as low as 4.5 Ω, because of the good interface connection between the MnO_2 nanoribbons and carbon fibers. Besides, charge-transfer resistance, also called Faraday resistance, is a crucial restriction factor forthe power density of supercapacitors[57]. Therefore, the low charge-transfer resistance is one of the significant causes for the high power and energy densities of the CDCC//MnO_2/CDCC ASC. In addition, we can observe that the slope of the straight line in the lowfrequencyregion is larger than that of the 45° straight line, representing typical capacitor behavior. The cycling stability of the ASC was studied by repeating the GCD process for 5000 times at a current density of 5 mA · cm^{-2}. It is seen in Fig. 7h that the ASC could maintain 87.7% of its original capacitance after 5000 cycles, which is comparable to or even higher than the value of these SSCs or ASCs as listed in Table 1. The resultindicates an excellent cycling performance of the ASC device.

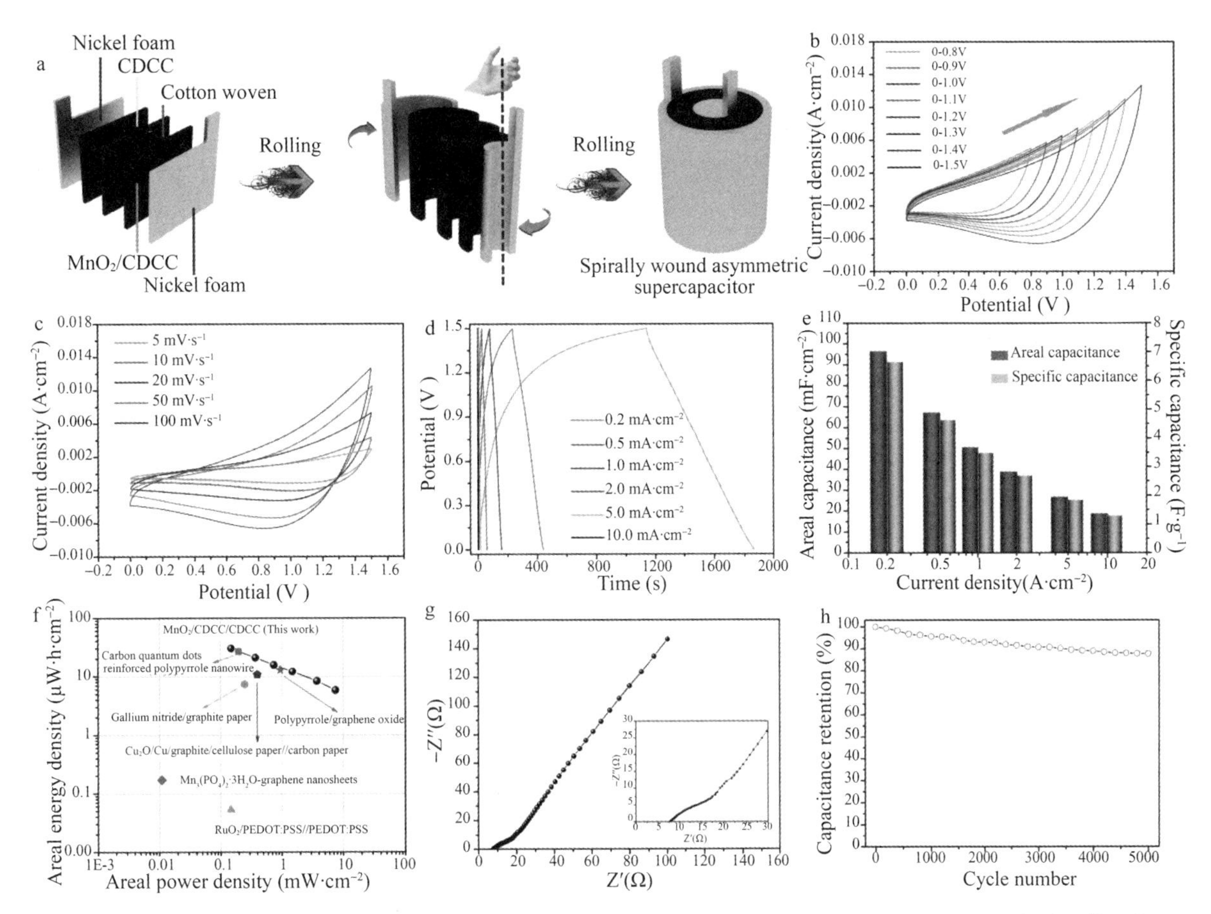

Fig. 7 (a) Primary components and assembly process of the all-textile and spirally wound CDCC//MnO_2/CDCC ASC. (b-h) Electrochemical performance of theCDCC//MnO_2/CDCC ASC: (b) CV curves within different potential ranges, (c) CV curves at various scan rates, (d) GCD profiles at different current densities, (e) rate capability, (f) Ragone plot of the device compared with data of otherSSCs or ASCs in the literature, (g) Nyquist plot and (h) cycling performance

Table 1 Comparison of areal energy density, areal capacitance and cycling stability of the CDCC//MnO2/CDCC ASC with other recently reported ASCs or SSCs

Electrodes	Type	Maximum areal energy density ($\mu W \cdot h \cdot cm^{-2}$)	Areal capacitance ($mF \cdot cm^{-2}$)	Cycling stability	Potential range(V)	Electrolyte	Refs
$Mn_3(PO_4)_2 \cdot 3H_2O$-graphene nanosheets	SSC	0.17 (~11 $\mu W \cdot cm^{-2}$)	40 (0.13 $mA \cdot cm^{-2}$)	~100% (2000 cycles)	0-0.35	PVA/KOH	51
Gallium nitride/graphite paper	SSC	7.4(~0.25 $mW \cdot cm^{-2}$)	53.2(0.5 $mA \cdot cm^{-2}$)	~100%(10000 cycles)	0-1	H_2SO_4/PVA	52
Polypyrrole/graphene oxide	SSC	12.9(954.3 $\mu W \cdot cm^{-2}$)	152(10 $mV \cdot s^{-1}$)	88.3%(10000 cycles)	-0.5-0.5	1 M KCl	53
Carbon quantum dots reinforced polypyrrole nanowire	SSC	26.5(0.2 $mW \cdot cm^{-2}$)	248.5(0.2 $mA \cdot cm^{-2}$)	85.2%(5000 cycles)	0-0.8	1 M KCl	54
RuO_2/PEDOT: PSS//PEDOT: PSS[a]	ASC	0.053(147 $\mu W \cdot cm^{-2}$)	1.06(50 $mV \cdot s^{-1}$)	93%(6000 cycles)	0-1.2	PVA/H_2SO_4	55
Cu_2O/Cu/graphite/cellulose paper//carbon paper	ASC	10.8(402.5 $\mu W \cdot cm^{-2}$)	122(1 $mA \cdot cm^{-2}$)	91.5%(5000 cycles)	0-0.8	6 M KOH	56
MnO_2/CDCC//CDCC	ASC	30.1(0.15 $mW \cdot cm^{-2}$)	96.3(0.2 $mA \cdot cm^{-2}$)	87.7%(5000 cycles)	0-1.5	1 M Na_2SO_4	This work

[a]PEDOT: poly(3, 4-ethylene-dioxythiophene), PSS: poly(styrene-4-sulfonate).

4 Conclusions

In summary, we have demonstrated a simple and rapid approach (i. e., pyrolysis and immersion) to prepare a core-sheath structured electrode based on MnO_2 nanoribbons and CDCC. The ribbon-like nano-MnO_2 was well dispersed on the surface of carbon fibers of CDCC by virtue ofan in-situredox reaction between $KMnO_4$ and CDCC during the immersion process. This synthetic strategyprimarilycreates three merits, namely (1) this good interface connection between MnO_2 nanoribbons (sheath part) and CDCC (core part) aided the electron transfer; (2) abundant nanostructured MnO_2 increased the electrochemically active surface areaaccessed by electrolyte ions; (3) the hierarchically porous CDCC substrate served as an electrolytereservoir to shortenthe ion-diffusion path and accelerate the efficient infiltrationof electrolyte ions. By taking these advantages, the electrode showed a high areal capacitance of 202 $mF \cdot cm^{-2}$ at 0.1 $mA \cdot cm^{-2}$. In addition, the CDCC and MnO_2/CDCC electrodes were assembled into an all-textile and spirally wound CDCC//MnO_2/CDCC ASC, which possessed a high areal energy density of 30.1 $\mu W \cdot h \cdot cm^{-2}$ at 0.15 $mW \cdot cm^{-2}$, an excellent cycling stability (87.7% after 5000 cycles) and a moderate areal capacitance of 96.3 $mF \cdot cm^{-2}$ at 0.2 $mA \cdot cm^{-2}$. More importantly, this synthetic strategy achieved a more direct and green utilization approach of cellulose resource and avoided pollutions derived from pretreatment or purification of cellulose. It is anticipated that such an energy storage device is cheap, easilyproduced and eco-friendly and thus may have potential values in numerous emerging fields like novelportable electronics.

Acknowledgements

This study was supported by the National Natural Science Foundation of China (grant no. 31470584) and Overseas Expertise Introduction Project for Discipline Innovation, 111Project (No. B08016).

Appendix A. Supplementary data

Please find the Supplementary data in the attachment.

References

[1] McMichael A J, Powles J W, Butler C D, et al. Food, livestock production, energy, climate change, and health[J]. The Lancet, 2007, 370(9594): 1253-1263.

[2] Chu S, Cui Y, Liu N. The path towards sustainable energy[J]. Nature Materials, 2017, 16(1): 16-22.

[3] Liu J, Kopold P, van Aken P A, et al. Energy storage materials from nature through nanotechnology: a sustainable route from reed plants to a silicon anode for lithium-ion batteries[J]. Angewandte Chemie, 2015, 127(33): 9768-9772.

[4] Yu Z, Tetard L, Zhai L, et al. Supercapacitor electrode materials: nanostructures from 0 to 3 dimensions[J]. Energy & Environmental Science, 2015, 8(3): 702-730.

[5] Wang G, Zhang L, Zhang J. A review of electrode materials for electrochemicalsupercapacitors[J]. Chemical Society Reviews, 2012, 41(2): 797-828.

[6] Liu J, Kopold P, van Aken P A, et al. Energy storage materials from nature through nanotechnology: a sustainable route from reed plants to a silicon anode for lithium-ion batteries[J]. Angewandte Chemie, 2015, 127(33): 9768-9772.

[7] Jost K, Dion G, Gogotsi Y. Textile energy storage in perspective[J]. Journal of Materials Chemistry A, 2014, 2(28): 10776-10787.

[8] Hu L, Pasta M, LaMantia F, et al. Stretchable, porous, and conductive energy textiles[J]. Nano letters, 2010, 10

(2): 708-714.

[9] Yuksel R, Unalan H E. Textile supercapacitors-based on MnO_2/SWNT/conducting polymer ternary composites[J]. International Journal of Energy Research, 2015, 39(15): 2042-2052.

[10] Laforgue A. All-textile flexible supercapacitors using electrospun poly (3, 4-ethylenedioxythiophene) nanofibers[J]. Journal of Power Sources, 2011, 196(1): 559-564.

[11] Meng Y, Zhao Y, Hu C, et al. All-graphene core-sheath microfibers for all-solid-state, stretchable fibriform supercapacitors and wearable electronic textiles[J]. Advanced Materials, 2013, 25(16): 2326-2331.

[12] Jost K, Stenger D, Perez C R, et al. Knitted and screen printed carbon-fiber supercapacitors for applications in wearable electronics[J]. Energy & Environmental Science, 2013, 6(9): 2698-2705.

[13] Zhang D, Miao M, Niu H, et al. Core-spun carbon nanotube yarn supercapacitors for wearable electronic textiles [J]. ACS Nano, 2014, 8(5): 4571-4579.

[14] Kong D, Ren W, Cheng C, et al. Three-dimensional $NiCo_2O_4$@ polypyrrole coaxial nanowire arrays on carbon textiles for high-performance flexible asymmetric solid-state supercapacitor[J]. ACS applied materials & interfaces, 2015, 7 (38): 21334-21346.

[15] Shakir I, Ali Z, Bae J, et al. Layer by layer assembly of ultrathin V_2O_5 anchored MWCNTs and graphene on textile fabrics for fabrication of high energy density flexible supercapacitor electrodes[J]. Nanoscale, 2014, 6(8): 4125-4130.

[16] Javed M S, Zhang C, Chen L, et al. Hierarchical mesoporous $NiFe_2O_4$ nanocone forest directly growing on carbon textile for high performance flexible supercapacitors[J]. Journal of Materials Chemistry A, 2016, 4(22): 8851-8859.

[17] Long C, Qi D, Wei T, et al. Nitrogen-doped carbon networks for high energy densitysupercapacitors derived from polyaniline coated bacterial cellulose[J]. Advanced Functional Materials, 2014, 24(25): 3953-3961.

[18] Chen W, Yu H, Lee S Y, et al. Nanocellulose: a promising nanomaterial for advanced electrochemical energy storage [J]. Chemical Society Reviews, 2018, 47(8): 2837-2872.

[19] Gui Z, Zhu H, Gillette E, et al. Natural Cellulose Fiber as Substrate for Supercapacitor[J]. ACS Nano, 2013, 7 (7): 6037-6046.

[20] Wang B, Li X, Luo B, et al. Pyrolyzed bacterial cellulose: a versatile support for lithium ion battery anode materials [J]. Small, 2013, 9(14): 2399-2404.

[21] Kim J M, Guccini V, Seong K, et al. Extensively interconnected silicon nanoparticles via carbon network derived from ultrathin cellulose nanofibers as high performance lithium ion battery anodes[J]. Carbon, 2017, 118: 8-17.

[22] Liang H W, Wu Z Y, Chen L F, et al. Bacterial cellulose derived nitrogen-doped carbonnanofiber aerogel: an efficient metal-free oxygen reduction electrocatalyst for zinc-air battery[J]. Nano Energy, 2015, 11: 366-376.

[23] Lu Y, Ye G, She X, et al. Sustainable route for molecularly thin cellulosenanoribbons and derived nitrogen-doped carbon electrocatalysts[J]. ACS Sustainable Chemistry & Engineering, 2017, 5(10): 8729-8737.

[24] Bella F, Galliano S, Falco M, et al. Approaching truly sustainable solar cells by the use of water and cellulose derivatives[J]. Green Chemistry, 2017, 19(4): 1043-1051.

[25] Neo C Y, Ouyang J. Ethyl cellulose and functionalized carbon nanotubes as a co-gelator for high-performance quasi-solid state dye-sensitized solar cells[J]. Journal of Materials Chemistry A, 2013, 1(45): 14392-14401.

[26] Johar N, Ahmad I, Dufresne A. Extraction, preparation and characterization of cellulose fibres and nanocrystals from rice husk[J]. Industrial Crops and Products, 2012, 37(1): 93-99.

[27] Kaushik A, Singh M, Verma G. Green nanocomposites based on thermoplastic starch and steam exploded cellulose nanofibrils from wheat straw[J]. Carbohydrate Polymers, 2010, 82(2): 337-345.

[28] Chen L F, Huang Z H, Liang H W, et al. Bacterial-cellulose-derived carbon nanofiber@ MnO_2 and nitrogen-doped carbon nanofiber electrode materials: an asymmetric supercapacitor with high energy and power density[J]. Advanced materials, 2013, 25(34): 4746-4752.

[29] Zhang Y X, Huang M, Li F, et al. One-pot synthesis of hierarchical MnO_2-modified diatomites for electrochemical capacitor electrodes[J]. Journal of Power Sources, 2014, 246: 449-456.

[30] Jin X, Zhou W, Zhang S, et al. Nanoscale microelectrochemical cells on carbon nanotubes[J]. Small, 2007, 3(9): 1513-1517.

[31] Lu X, Yu M, Wang G, et al. H-TiO_2@ MnO_2//H-TiO_2@ C core-shell nanowires for high performance and flexible asymmetric supercapacitors[J]. Advanced Materials, 2013, 25(2): 267-272.

[32] Zhang Y X, Huang M, Li F, et al. One-pot synthesis of hierarchical MnO_2-modified diatomites for electrochemical capacitor electrodes[J]. Journal of Power Sources, 2014, 246: 449-456.

[33] Weng Z, Su Y, Wang D W, et al. Graphene-cellulose paper flexible supercapacitors[J]. Advanced Energy Materials, 2011, 1(5): 917-922.

[34] Li Y D, Li X L, He R R, et al. Artificial lamellar mesostructures to WS_2 nanotubes[J]. Journal of the American Chemical Society, 2002, 124(7): 1411-1416.

[35] Moreo A, Yunoki S, Dagotto E. Phase separation scenario for manganese oxides and related materials[J]. Science, 1999, 283(5410): 2034-2040.

[36] Zhang H T, Chen X H, Zhang J H, et al. Synthesis and characterization of one-dimensional $K_{0.27}MnO_2 \cdot 0.5H_2O$ [J]. Journal of Crystal Growth, 2005, 280(1-2): 292-299.

[37] Li Y, Yang X Y, Feng Y, et al. One-dimensional metal oxide nanotubes, nanowires, nanoribbons, and nanorods: synthesis, characterizations, properties and applications[J]. Critical Reviews in Solid State and Materials Sciences, 2012, 37(1): 1-74.

[38] Yuan L, Lu X H, Xiao X, et al. Flexible solid-statesupercapacitors based on carbon nanoparticles/MnO_2 nanorods hybrid structure[J]. ACS Nano, 2012, 6(1): 656-661.

[39] Lei Z, Shi F, Lu L. Incorporation of MnO_2-coated carbon nanotubes between graphene sheets as supercapacitor electrode[J]. ACS applied materials & interfaces, 2012, 4(2): 1058-1064.

[40] Chen C, Zhang Y, Li Y, et al. All-wood, low tortuosity, aqueous, biodegradablesupercapacitors with ultra-high capacitance[J]. Energy & Environmental Science, 2017, 10(2): 538-545.

[41] Li S, Wen J, Mo X, et al. Three-dimensional MnO_2 nanowire/ZnO nanorod arrays hybrid nanostructure for high-performance and flexible supercapacitor electrode[J]. Journal of Power Sources, 2014, 256(15): 206-211.

[42] Park S, Nam I, Kim G P, et al. Hybrid MnO_2 film with agarose gel for enhancing the structural integrity of thin film supercapacitor electrodes[J]. ACS applied materials & interfaces, 2013, 5(20): 9908-9912.

[43] Huang Y, Li Y, Hu Z, et al. A carbon modified MnO_2 nanosheet array as a stable high-capacitance supercapacitor electrode[J]. Journal of Materials Chemistry A, 2013, 1(34): 9809-9813.

[44] Chen S, Zhu J, Wu X, et al. Graphene oxide-MnO_2 nanocomposites for supercapacitors[J]. ACS nano, 2010, 4(5): 2822-2830.

[45] Wan C, Jiao Y, Li J. Flexible, highly conductive, and free-standing reducedgraphene oxide/polypyrrole/cellulose hybrid papers for supercapacitor electrodes[J]. Journal of Materials Chemistry A, 2017, 5(8): 3819-3831.

[46] Kobayashi K, Nagao M, Yamamoto Y, et al. Rechargeable PEM fuel-cell batteries using porous carbon modifiedwith carbonyl groups as anode materials[J]. Journal of The Electrochemical Society, 2015, 162(8): F868-F877.

[47] Wan C, Li J. Wood-derivedbiochar supported polypyrrole nanoparticles as a free-standing supercapacitor electrode [J]. RSC Advances, 2016, 6(89): 86006-86011.

[48] Jiang J, Zhang L, Wang X, et al. Highly orderedmacroporous woody biochar with ultra-high carbon content as supercapacitor electrodes[J]. Electrochimica Acta, 2013, 113: 481-489.

[49] Peng L, Peng X, Liu B, et al. Ultrathin two-dimensional MnO_2/graphene hybrid nanostructures for high-performance, flexible planar supercapacitors[J]. Nano Letters, 2013, 13(5): 2151-2157.

[50] Zhang G Q, Wu H B, Hoster H E, et al. Single-crystalline $NiCo_2O_4$ nanoneedle arrays grown on conductive substrates as binder-free electrodes for high-performance supercapacitors[J]. Energy & Environmental Science, 2012, 5(11): 9453-9456.

[51] Yang C, Dong L, Chen Z, et al. High-performance all-solid-statesupercapacitor based on the assembly of graphene and manganese (II) phosphate nanosheets[J]. The Journal of Physical Chemistry C, 2014, 118(33): 18884-18891.

[52] Wang S, Sun C, Shao Y, et al. Self-SupportingGaN Nanowires/Graphite Paper: Novel High-Performance Flexible Supercapacitor Electrodes[J]. Small, 2017, 13(8): 1603330.

[53] Zhou H, Han G, Xiao Y, et al. Facile preparation ofpolypyrrole/graphene oxide nanocomposites with large areal capacitance using electrochemical codeposition for supercapacitors[J]. Journal of Power Sources, 2014, 263: 259-267.

[54] Jian X, Li J, Yang H, et al. Carbon quantum dots reinforced polypyrrole nanowire via electrostatic self-assembly strategy for high-performance supercapacitors[J]. Carbon, 2017, 114: 533-543.

[55] Zhang C J, Higgins T M, Park S H, et al. Highly flexible and transparent solid-statesupercapacitors based on RuO_2/PEDOT: PSS conductive ultrathin films[J]. Nano Energy, 2016, 28: 495-505.

[56] Wan C, Jiao Y, Li J. Multilayer core-shell structured composite paper electrode consisting of copper, cuprous oxide and graphite assembled on cellulose fibers for asymmetric supercapacitors[J]. Journal of Power Sources, 2017, 361: 122-132.

[57] Bao L, Zang J, Li X. Flexible Zn_2SnO_4/MnO_2 core/shell nanocable-carbon microfiber hybrid composites for high-performance supercapacitor electrodes[J]. Nano Letters, 2011, 11(3): 1215-1220.

中文题目：一种基于核鞘结构 MnO_2 纳米带和棉基碳布的高性能、全纺织和螺旋线性的非对称超级电容器

作者：万才超，焦月，梁大鑫，吴义强，李坚

摘要：随着对绿色化学和自然资源高效利用的日益重视，对制备更加简便、快速和成本效益更高的储能设备提出了更多要求。在这里，我们展示了一种简单、快速并廉价的方法来制备基于棉花基碳布(CDCC，核)和二氧化锰纳米带(鞘)的核鞘结构的纺织电极。而原位生长的二氧化锰与 CDCC 碳纤维之间良好的界面结合有助于电子转移。丰富的二氧化锰纳米结构增加了电解质离子进入的电化学活性区域，多孔 CDCC 充当电解质储库以缩短离子扩散路径并促进电解质离子的有效渗透。基于这些优点，二氧化锰/CDCC 电极显现出 202 $mF \cdot cm^{-2}$ 的高面积比电容。此外，柔性电极被组装成全纺织品和螺旋线性的非对称超级电容器，具有优异的电化学活性，在 0.15 $mW \cdot cm^{-2}$ 时具有 30.1 $\mu W \cdot h \cdot cm^{-2}$ 的高面能密度，循环 5000 次后电容保持率可达 87.7%。值得一提的是，该合成策略实现了更直接、更环保、更高效的纤维素资源利用途径，避免了纤维素净化或预处理带来的污染。

关键词：棉花；碳布；二氧化锰纳米带；核鞘结构；螺旋线性超级电容器；原位氧化还原

A Geologic Architecture System-inspired Micro-/Nano-heterostructure Design for High-performance Energy Storage*

Caichao Wan, Yue Jiao, Daxin Liang, Yiqiang Wu, Jian Li

Abstract: Nature-inspired strategies are extensively proposed as novel and effective routines to address challenges for eco-friendly and high-performance energy storage devices with high energy/power density and long cycling life. Inspired by synergistic functions and integrated form of a geologic architecture system (i. e., "ground-mountain-vegetation"), here we create a novel and hierarchical cellulose-supported Co@ $Co(OH)_2$ heterostructure based on a facile combined method of magnetron sputtering and electrooxidation. Thanks to the synergistic effects of this multiscale structure (i. e., the storage capacity of cellulose substrate ("ground") for electrolyte ions, electron superhighway supplied by interlayered metallic Co ("mountain"), and ultra-high electrochemical activity and mechanical stability of in-situ grown and quasi-honeycomb $Co(OH)_2$ ("vegetation") with large surface area, the composite displays a high specific capacitance ($642\ mF \cdot cm^{-2}$/ $958\ F \cdot g^{-1}$ at $2\ mA \cdot cm^{-2}$), excellent rate performance and outstanding cycling stability (only 2.1% loss after 10000 cycles), which are significantly superior to those of other microstructure designs of $Co(OH)_2$-based electrodes. Also, the assembled asymmetric supercapacitor exhibits highly competitive energy/power density ($166\ \mu W \cdot h \cdot cm^{-2}$ at $1.5\ mW \cdot cm^{-2}$) and excellent cycling stability. Combined with the outstanding electrochemical properties, facile synthesis technology, environmental friendliness and low cost, this ingenious nature-inspired composite holds great promise for green high-performance energy storage devices.
Keywords: Functional materials; Heterostructure; Nature-inspired design; Supercapacitors

1 Introduction

Since ancient times, the natural world always acts as the headspring of technological thoughts, engineering principles and significant inventions in the human's civilization.[1-3] This is primarily because human intelligence is not only restricted to observing and understanding nature, but also learning from nature by duplicating or mimicking the components, structures and systems existing in the nature world. As a result, a series of helpful tools or apparatuses (like hacksaw-mantis, radar-bat and gyrotron-fly) have been invented. Over the past several decades of development of nanoscience, the study on nature-inspired materials has expanded to micro-/nanoscale, which has facilitated new breakthroughs on the design of micro-/nano-structural materials with attractive properties.[2,4-9] For instance, directional water collection material,[10]

* 本文摘自 Advanced Energy Materials, 2018, 8(33): 1802388.

antireflection film,[11,12] and superhydrophobic surface[13] have been designed and prepared by mimicking the microstructures of spider silk, transparent wings of glasswing butterfly/moth eye, and lotus leaf, respectively.

In the field of electrochemical energy storage, natural structures (like bamboo,[14] flower,[15,16] forest[17,18] and honeycomb[19,20]) have been and continue to be intriguing as models for moulding similar micromorphologies of electrochemically active materials since these multiscale hierarchical structures exert enormous functions on electrochemical reactions (e. g., increasing reaction area[21], providing fast electron paths[22] and reducing diffusion path of electrolyte ions[23]). Especially, these active substances usually have a hybrid heterostructure, which is beneficial to obtain a high power/energy density and good cycling stability; taking an example of binary materials, the common integrated form is "metals or alloys/carbon materials/conducting polymers"-"carbon materials/metallic oxides (or hydroxides)".[24-28] In addition, besides high-performance active materials, a flexible, porous and self-supporting substrate is also crucial for developing a free-standing electrode without conductive or adhesive agents. Recently, an increasing emphasis on green chemistry has promoted the attention of people on the utilization of green resources for design and preparation of energy storage equipment.[29-31] Especially, cellulose, the most abundant natural polymer on the earth, has been placed great expectations on account that different dimensions of cellulose products (including fibers, films, hydrogels or aerogels) have displayed fascinating peculiarities like high mechanical strength, good flexibility, high porosity, high transparency and excellent chemical reactivity.[32,33] In consequence, these features make them suitable candidates as flexible and porous building blocks to support or encapsulate both of conductive and electrochemically active substances.

Mountain, a stable and mechanically strong structure with relatively homogeneous components, is a representative product of geological processes. More importantly, the mountain and its underlying ground and surface vegetation construct a stable and harmonious ecosystem. For withstanding the forces of nature, taking rainstorm as an example, the porous soil can act as a water reservoir to restrict the movement of water, and the solid and high-density mountain provides strength and stability, and the good interaction between mountain and its surface vegetation also contributes to reducing water erosion. Here, inspired by the synergistic functions and integrated form of this geologic architecture system (namely "ground-mountain-vegetation"), we create a novelhierarchical cellulose-supported Co@ $Co(OH)_2$ heterostructure for high-performance energy storage. It is worth mentioning that $Co(OH)_2$ is a promising electrode material for supercapacitors due to its high theoretical capacitance ($3460\ F \cdot g^{-1}$) and long cyclability, which are attributed to the intrinsic structural similarity between $Co(OH)_2$ and corresponding charged phase CoOOH sparingunnecessary and unfavourable massive atom rearrangements for the phase transformation and the battery-mimic mechanism (a key factor to achieve high energy densities).[34-36] The synthesis of this free-standing composite proceeded via a mild combined routine of magnetron sputtering and electrooxidation (as illustrated in Fig. 1). To our knowledge, the researches on biomass-based electrodes synthesized based on the magnetron sputtering-electrooxidation method are very scarce.

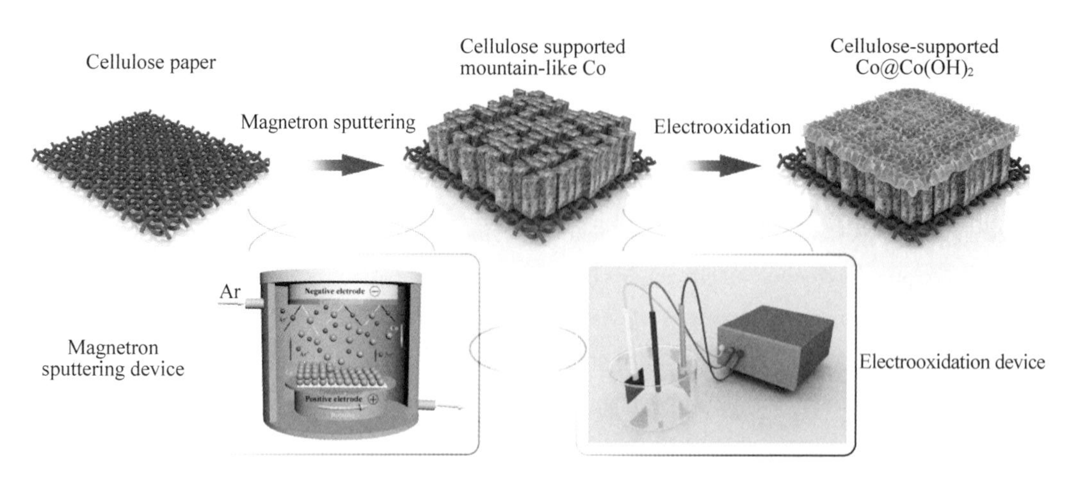

Fig. 1 Schematic illustration of the synthesis of the free-standing cellulose-supported Co@Co(OH)$_2$ heterostructure

Such a design has the following philosophies(as shown in Fig. 2a): (1)a piece of common cellulose paper serves as the cellulose substrate for displaying universality of the preparation method. Similar to the role of ground, the hydrophilic cellulose fibers (as illustrated in Fig. S1 in SI) are expected to significantly absorb electrolyte and their cross-linked pores can serve as electrolyte reservoirs to facilitate ion transport(Fig. 2b). [37] Also, the presence of cellulose donates the flexibility and lightweight characteristic to the composite(Fig. S2); (2) the sputtered mountain-like Co component with a height of ~ 40. 3 μm (Fig. 2c-e) brings rigidity and mechanical strength to the system. Besides, the presence of high-purity metallic Co endows the composite with a high electrical conductivity of ~ 5.1×10^2 S · cm^{-1} (the calculation and comparisonsare available in SI) and consequentlyplays a key role as electron "superhighway" for charge delivery; (3) the quasi-honeycomb nanoarchitecture of Co(OH)$_2$, which is constituted by the interconnected nanosheets with a thickness and width of around 12 - 20 nm and 300 - 500 nm (Fig. 2f and g), is in-situ grown on the surface of Co by the electrooxidation. Therefore, the Co(OH)$_2$ nanoarchitecture has a perfect interfacial bonding with the internal Co and possesses a large specific surface area of 318 m^2 · g^{-1}(see Fig. S3 and the calculation is available in SI), increasing the electrochemically active sites for redox reactions. Considering these strategies, we believe that this nature-inspired design is suitable for high-performance energy storage.

The morphology of the cellulose-supported Co @ Co (OH)$_2$ heterostructure was further studied by transmission electron microscope (TEM) observations. We can find that the mountain-like Co consists of numeroussubrotund nanoparticles with a size of ~ 5 nm(Fig. 2i) and the Co(OH)$_2$nanosheets directly attach to the Co surface (Fig. 2h), again reflecting their good interface bonding. In addition, the high-resolution TEM (HRTEM) images(as seen in Fig. 2j and k), corresponding to the marked areas in Fig. 2h and i, not only show the existence of Co, but also demonsrate the co-presence of α-and β-Co (OH)$_2$ crystal formsin the Co (OH)$_2$nanoarchitecture, consistent with the results of the X-ray diffraction(XRD) analysis(Fig. S4). As for the chemical valences of Co, the X-ray photoelectron spectroscopy(XPS) analysis demonstrates the main existence of Co^{2+} accompanied bya small quantity of Co^{3+}in the composite. Please find more detailed discussions on XRD and XPS in SI (Fig. S4 and S5). Based on the above-mentionedcharacterization results, the possible synthesis mechanisms of the "ground-mountain-vegetation" structure are proposedin SI(Fig. S6–S8).

Fig. 2 Nature-inspired design philosophies and microstructure of the cellulose-supported Co @ Co (OH)$_2$ composite: (a) schematic illustration of the nature-inspired nanostructure design; scanning electron microscope(SEM) images of(b) the cellulose paper and(c) the cellulose-supported Co; cross-section SEM images of(d and e) the cellulose-supported Co and (f) the cellulose-supported Co @ Co (OH)$_2$; (g) high-magnification SEM image of the quasi-honeycomb Co(OH)$_2$; (h, i) TEM and(j, k) HRTEM images of the cellulose-supported Co@Co(OH)$_2$. Insets of *c* and *g* show the corresponding enlarged images

The electrochemical performances of the free-standing cellulose-supported Co @ Co (OH)$_2$ composite electrode without any binder and conductive additives were studied using a conventional three-electrode system, as illustrated in Fig. 3a. Fig. 3b presents the cyclic voltammetry (CV) curves of the electrode at the various scan rates of 5-100 mV · s^{-1} over a potential window of 0 to 0. 5 V(vs Hg/HgO). The existence of apparent redox peaks is a characteristic of redox insertion reactions, which is derived from the 3D absorption of electroactive species into the bulk solid electrode material. [38] In addition, we observe that the potential difference between the anodic and cathodic peaks is far higher than the theoretical value (0. 058 V) for a reversible, single-electron-transfer redox process in solution.[39] Thus two plausible reactions may take place in the process of potential sweep, which are presented as follows:[15]

$$Co(OH)_2+OH^- \leftrightarrow CoOOH+H_2O+e^- \quad (1)$$

$$CoOOH+OH^- \leftrightarrow CoO_2+H_2O+e^- \quad (2)$$

Besides the famous reaction 1, the second redox reaction has also been discussed in numerous works and is associated with many factors (e. g., specific surface area).[39,40] Moreover, this redox reaction is also easily affected by the interlayer distance between adjacentCo ($OH)_2$ single sheets. The oxidation process toCoOOH is believed to dominate. The XPS spectrum of the composite after the electrochemical tests also demonstrates the noticeable increase of the proportion of Co^{3+} (Fig. S9). Moreover, apart from the slight shift of the redox peaks with the increasing scan rate, the shape of the CV curves remains unchanged. This result reveals that the unique "ground-mountain-vegetation" structure and quasi-honeycombCo($OH)_2$ are conducive to the rapid redox reactions and beneficial to acquire a superior rate performance.

To study the electrochemical capacitive property of the composite electrode, the galvanostatic charge-discharge (GCD) tests were implemented at the different current densities ranging from 2 to 50 mA · cm^{-2} (3.0–74.6 A · g^{-1}). As shown in Fig. 3c, the shape of the GCD curves reflects the feature of pseudocapacitance, agreeing well with the results of CV test. Also, no obvious IR drop appears, suggesting again that the "ground-mountain-vegetation" structure possesses rapid I-V response ability and low internal resistance. Fig. 3d displays the rate performance of the composite electrode (areal or gravimetric specific capacitances*versus* current densities). The maximum specific capacitance (642 mF · cm^{-2}/958 F · g^{-1}) of the electrode is achieved at 2 mA · cm^{-2} (3.0 A · g^{-1}), much higher than the values of other morphologies of Co($OH)_2$-based electrodes, e. g., nanowall (205 F · g^{-1}),[41] nanoflower (405 F · g^{-1}),[15] nanowire arrays (422.36 F · g^{-1}),[42] interlaced sheets (693.8 F · g^{-1})[43] and nanocones (729 F · g^{-1}),[44] as summarized in Table S1. In addition, when the current density increases from 2 to 50 mA · cm^{-2}, a capacitance retention of 57.6% (552 F · g^{-1} at 50 mA · cm^{-2}) is acquired and reveals the excellent rate capability of our composite electrode. Moreover, a negligible loss of capacitance after 10000 cycles (~2.1% loss) reflects its outstanding cycling stability (Fig. 3e). As compared to other Co($OH)_2$-based electrodes (Table S1), our cycling stability is dominated and possibly due to the porous and fluffy quasi-honeycomb structure mitigating the mechanical degradation and volume expansion (Fig. 3g).[19] Besides, the SEM image of the electrode after 10000 cycles verifies that the quasi-honeycomb structure was well maintained (see the inset of Fig. 3e), indicating its good shape stability.

Based on the characterization results, we consider that the excellent electrochemical behaviors of the cellulose-supported Co@ Co($OH)_2$ composite electrode are primarily attributable to the synergistic functions of the "ground–mountain–vegetation" components, i. e., (1) the strongly hydrophilic and porous cellulose paper ("ground") can adsorb and store electrolyte and thus facilitates the ion migration;[37] (2) the highly conductive mountain-like Co acts as the electron superhighway for charge delivery, and the perfect interfacial bonding between the Co and the in-situ grown Co($OH)_2$ ("vegetation") minimizes the interface resistance and also facilitates charge transfer reactions.[45] Especially, the inexistence of semicircle in the high-frequency region of the Nyquist plot of the cellulose-supported Co@ Co($OH)_2$ further demonstrates the ultralow charge-transfer resistance (as shown in Fig. 3f); (3) the Co($OH)_2$ displays a quasi-honeycomb and vertically oriented nanostructure with a large surface area of 318 $m^2 \cdot g^{-1}$ and good hydrophilicity (Fig. S1), which is believed to significantly increase the contact area between the electrolyte and the electrode, shorten the ion diffusion length and promote the ion diffusion rate that are closely related to rate performance (Fig. 3g). [46-48] Apart from these merits from the unqiue nature-inspired structure, according to the research by Deng et al., [35] both $Co(OH)_2$ and CoOOH have the same layered cobalt structureand large interlayer spacing that facilitates the intercalation of H^+ species. Furthermore, from the point of supercapacitors, the structural similarity sparesmassive structural changes for this phase transformation, which contributes tohigh pseudocapacitance and long cycling life. From the ponit of batteries, the H^+ intercalation/deintercalation for energy store/release is simialr to the Li^+ intercalation/de-intercalation of lithium battery. Therefore, this battery-mimic mechanism of $Co(OH)_2$ also plays an important role in achieving high electrochemical properties.

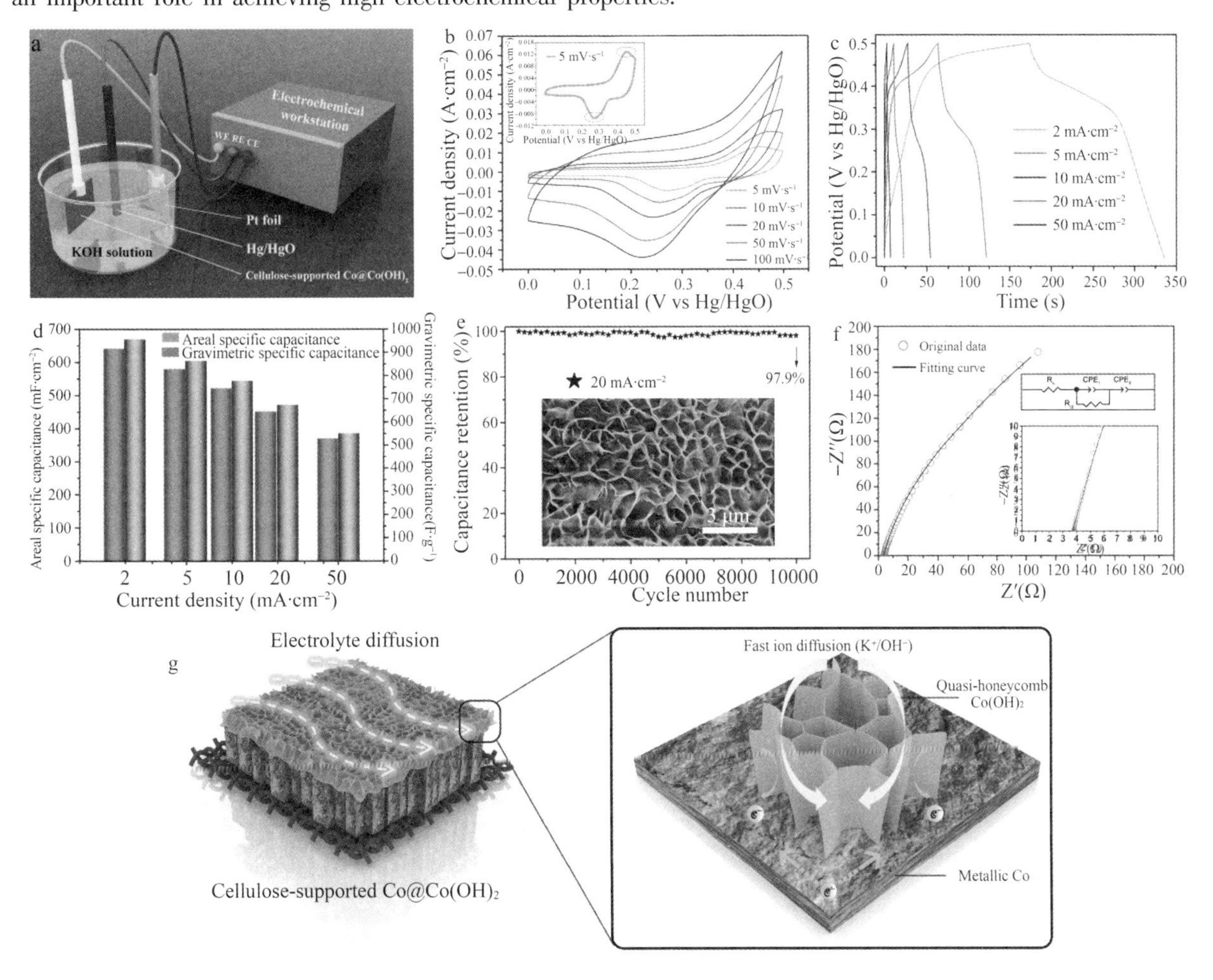

Fig. 3 Electrochemical properties of the free-standing cellulose-supported Co @ Co (OH)$_2$ compositeelectrode: (a) schematic diagram of the electrochemical testing apparatus using a conventional three-electrode system; (b) CV curves at the scan rates of 5–100 $mV \cdot s^{-1}$; (c) GCD curves at the current densities of 2–50 $mA \cdot cm^{-2}$; (d) rate performance; (e) cycling stability at the current density of 20 $mA \cdot cm^{-2}$, and the inset shows the micromorphology of $Co(OH)_2$ after the 10000 cycles; (f) Nyquist plot (the electrical equivalent circuitwith solution resistance R_s, charge transfer resistance R_{ct} and two constant phase elements (CPE) is compatible with the Nyquist diagram); (g) schematic diagram of the functions of the quasi-honeycomb structure on the electrochemical reactions

An asymmetric supercapacitor (ASC) device was assembled to evaluate the potential of the cellulose-supported Co@Co$(OH)_2$ composite for practical applications. The schematic diagram (Fig. 4a) depicts the fabrication of ASC with the Co@Co$(OH)_2$ heterostructure as a positive electrode and the underlying cellulose substrate as a separator. A piece of carbon cloth (CC), which was obtained by pyrolyzing a piece of common cotton cloth, acts as a negative electrode with a moderate capacitance of 219 mF · cm^{-2} (34 F · g^{-1}) at 2 mA · cm^{-2}. Its morphology and detailed electrochemical evaluation in a three-electrode cell are available in SI (Fig. S10 and S11). Based on the CV (Fig. 4b), the CC electrode was measured within a stable potential window of −1 ~ 0 V vs Hg/HgO while that of the cellulose-supported Co@Co$(OH)_2$ electrode was tested from 0 to 0.5 V, suggesting that such two electrodes combination as ASC may afford a device with 1.5 V operation voltages in 6M KOH as electrolyte.

Fig. 4c presents the CV curves of the Co@Co$(OH)_2$//CC ASC at the various scan rates. As expected, the ASC exhibits a favorable stability in the voltage range of 0−1.5 V (Fig. S12), reflecting the good maching effects between the positive and negative electrodes. Moreover, two pairs of redox peaks are obviously identified (see the inset in Fig. 4c) and are corresponding to the redox reactions of $Co^{2+} \leftrightarrow Co^{3+} \leftrightarrow Co^{4+}$. Also, these curves behave similarly in shape and the current density increases with the increase of scan rate from 5 to 100 mV · s^{-1}, revealing the fast charge/discharge and good reversibility. To further evaluate the rate capability of the ASC, we tested its GCD curves at various current densities. As shown in Fig. 4d, both charge and discharge curves remain a good symmetry at such a high operation voltage of 1.5 V, again demonstrating the outstanding reversibility and good coulombic efficiency (i.e., the ratio of discharging and charging capacitances). The non-linear shapes are because of the pseudocapacitance arising out from the redox reactions within this voltage range. In addition, the ASC displays a high areal specific capacitance of 530 mF · cm^{-2} (~75 F · g^{-1} based on the total mass of active materials in both positive and negative electrodes) at the initial discharge current density of 2 mA · cm^{-2} and still remains 333 mF · cm^{-2} (~47 F · g^{-1}) even at a high current density of 50 mA · cm^{-2} with an excellent rate capability of 62.8% (Fig. 4e). The maximum volume capacitance of the ASC with a thickness of ~460 μm can reach up to 11.5 F · cm^{-3} at 2 mA · cm^{-2} (43.5 mA · cm^{-3}).

The Ragone plot relative to the energy and power densities for the Co@Co$(OH)_2$//CC ASC is shown in Fig. 4f. The ASC stores the maximum areal energy density of 166 μW · h · cm^{-2} (~23.5 W · h · kg^{-1}) with an areal power density of 1.5 mW · cm^{-2} (~212 W · kg^{-1}) at the current density of 2 mA · cm^{-2} and still maintains 104 μW · h · cm^{-2} (~14.7 W · h · kg^{-1}) with a maximum areal power density of 37.5 mW · cm^{-2} (~5304 W · kg^{-1}) at 50 mA · cm^{-2}. The attained energy densities of our ASC are competitive or superior to those of previously reported metallic oxide or hydroxide-based ASCs (see Fig. 4f and Table S2 in SI).[49-53] The maximum energy and power density based on the volume of the ASC are 3.6 mW · h · cm^{-3} and 815 mW · cm^{-2}, respectively. Moreover, the Co@Co$(OH)_2$//CC ASC retains 91.6% of the original capacitance after recycling the ASC 10000 times at a large current density of 20 mA · cm^{-2} (Fig. 4g), revealing its remarkable cycling stability. The corresponding Nyquist plot is shown in the inset of Fig. 4g, where a low charge-transfer resistance (R_c) of ~0.5 Ω reflects the fast charge transfer and low internal resistance. This feature helps to enhance the power density of the ASC. Combined with the meritsof excellent electrochemical properties, facile synthesis technology, low cost and environmental friendliness, such an ingenious geologic architecture system-inspired composite may act as a promising candidate for energy storage applications.

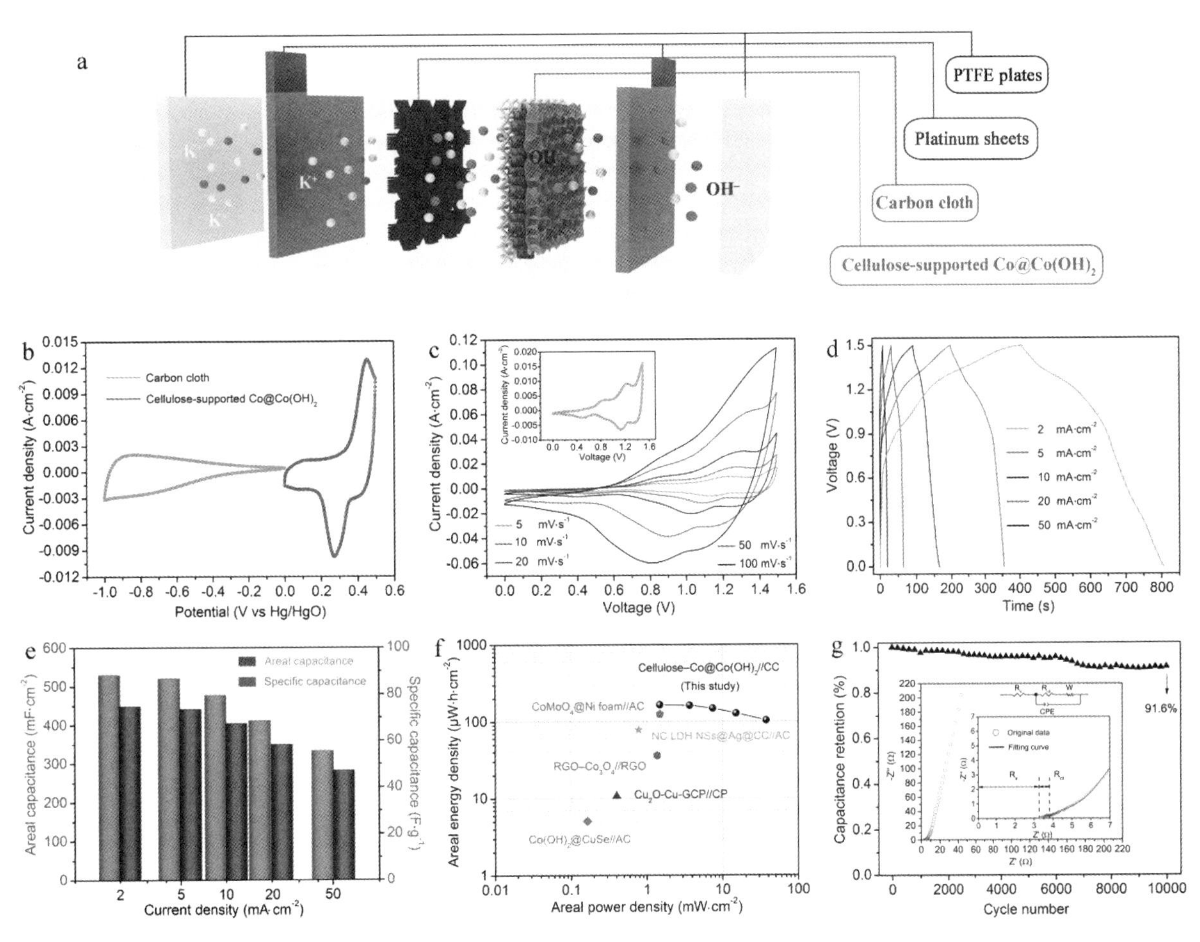

Fig. 4 Electrochemical properties of the Co@Co(OH)$_2$//CC ASC: (a) schematic diagram of the assembled ASC device; (b) CV curves of the Co@Co(OH)$_2$ electrode (positive) and CC electrode (negative) at 5 mV cm^{-2} in a three-electrode cell; (c) CV curves of the ASC at the scan rates of 5–100 mV · s^{-1}; (d) GCD curves of the ASC at the current densities of 2–50 mA · cm^{-2}; (e) rate performance of the ASC; (f) Ragone plot of the ASC compared with data of some other ASCs in the literature; (g) cycling stability of the ASC at the current density of 20 mA · cm^{-2}, and the inset shows the Nyquist plot (W: Warburg diffusion resistance)

In summary, a novel and nature-inspired design concept, i. e., mimicking and duplicating the "ground-mountain-vegetation" geologic architecture system, is purposed to construct a free-standing cellulose-supported Co@Co(OH)$_2$ composite based on a mild combined technology of magnetron sputtering and electrooxidation. By taking advantage of the unique structural merits of the three components (i. e., the highly hydrophilic and porous cellulose (ground), highly conductive Co (mountain), and quasi-honeycomb Co(OH)$_2$ (vegetation) with large surface area and high electrochemical activity as well as good mechanical stability), a high specific capacitance (642 mF · cm^{-2}/958 F · g^{-1} at 2 mA · cm^{-2}), excellent rate performance and outstanding cycling stability (only 2.1% loss after 10000 GCD tests) are achieved for the composite electrode. In addition, the assembled Co@Co(OH)$_2$//CC ASC presents a highly competitive energy/power density (166 μW · h · cm^{-2} at 1.5 mW · cm^{-2}) and excellent cycling stability among ever-reported metallic oxide or hydroxide-based ASCs. More importantly, our purposed synthesis strategy can also be applied to build other similar heterostructures of "$X-X_nO_m$ or $X(OH)_y$" (X= Ni, Cu, V…) on porous flexible substrates, expanding a promising pathway for developing high-performance energy storagedevices that are not restricted to supercapacitors.

Supporting Information

Supporting Information is available from the Wiley Online Library or from the author.

Acknowledgements

C. W. and Y. J. contributed equally to this work. This study was supported by the National Natural Science Foundation of China (grant no. 31530009 and 31470584) and Overseas Expertise Introduction Project for Discipline Innovation, 111 Project (No. B08016).

References

[1] Sun T, Qing G, Su B, et al. Functionalbiointerface materials inspired from nature[J]. Chemical Society Reviews, 2011, 40(5): 2909-2921.

[2] Zhang Y, Peng J, Li M, et al. Bioinspired supertough graphene fiber through sequential interfacial interactions[J]. ACS Nano, 2018, 12(9): 8901-8908.

[3] Liu K, Tian Y, Jiang L. Bio-inspired superoleophobic and smart materials: design, fabrication, and application[J]. Progress in Materials Science, 2013, 58(4): 503-564.

[4] Wegst U G K, Bai H, Saiz E, et al. Bioinspired structural materials[J]. Nature Materials, 2015, 14(1): 23-36.

[5] Yao H B, Fang H Y, Wang X H, et al. Hierarchical assembly of micro-/nano-building blocks: bio-inspired rigid structural functional materials[J]. Chemical Society Reviews, 2011, 40(7): 3764-3785.

[6] Yao H B, Ge J, Mao L B, et al. 25th anniversary article: artificial carbonate nanocrystals and layered structural nanocomposites inspired by nacre: synthesis, fabrication and applications[J]. Advanced Materials, 2014, 26(1): 163-188.

[7] Neagu D, Tsekouras G, Miller D N, et al. In situ growth of nanoparticles through control of non-stoichiometry[J]. Nature Chemistry, 2013, 5(11): 916-923.

[8] Zhang W, Chen M, Theil Kuhn L, et al. Electrochemistry unlocks wettability: epitaxial growth of oxide nanoparticles on rough metallic surfaces[J]. ChemElectroChem, 2014, 1(3): 520-523.

[9] Zhang W, Zheng W. Exsolution-Mimic Heterogeneous Surfaces: Towards Unlimited Catalyst Design [J]. ChemCatChem, 2015, 7(1): 48-50.

[10] Zheng Y, Bai H, Huang Z, et al. Directional water collection on wetted spider silk[J]. Nature, 2010, 463(7281): 640-643.

[11] Siddique R H, Gomard G, Hölscher H. The role of random nanostructures for the omnidirectional anti-reflection properties of the glasswing butterfly[J]. Nature Communications, 2015, 6(1): 1-8.

[12] Tan G, Lee J H, Lan Y H, et al. Broadband antireflection film with moth-eye-like structure for flexible display applications[J]. Optica, 2017, 4(7): 678-683.

[13] Liu K, Yao X, Jiang L. Recent developments in bio-inspired specialwettability[J]. Chemical Society Reviews, 2010, 39(8): 3240-3255.

[14] Sun Y, Sills R B, Hu X, et al. A bamboo-inspired nanostructure design for flexible, foldable, and twistable energy storage devices[J]. Nano letters, 2015, 15(6): 3899-3906.

[15] Wang R, Yan X, Lang J, et al. A hybrid supercapacitor based on flower-like $Co(OH)_2$ and urchin-like VN electrode materials[J]. Journal of Materials Chemistry A, 2014, 2(32): 12724-12732.

[16] Yu D, Pang Q, Gao Y, et al. Hierarchical flower-like VS_2 nanosheets-A high rate-capacity and stable anode material for sodium-ion battery[J]. Energy Storage Materials, 2018, 11: 1-7.

[17] Wang J, Zhang X, Wei Q, et al. 3D self-supportednanopine forest-like Co_3O_4@ $CoMoO_4$ core-shell architectures for high-energy solid state supercapacitors[J]. Nano Energy, 2016, 19: 222-233.

[18] Wan C, Jiao Y, Li J. A cellulose fibers-supported hierarchical forest-like cuprous oxide/copper array architecture as a flexible and free-standing electrode for symmetricsupercapacitors[J]. Journal of Materials Chemistry A, 2017, 5(33): 17267-17278.

[19] Xiao K, Xia L, Liu G, et al. Honeycomb-like $NiMoO_4$ ultrathin nanosheet arrays for high-performance electrochemical energy storage[J]. Journal of Materials Chemistry A, 2015, 3(11): 6128-6135.

[20] Puthusseri D, Aravindan V, Madhavi S, et al. 3D micro-porous conducting carbon beehive by single step polymer carbonization for high performance supercapacitors: the magic of in situ porogen formation[J]. Energy & Environmental Science, 2014, 7(2): 728-735.

[21] Zhu Y, Murali S, Stoller M D, et al. Carbon-based supercapacitors produced by activation of graphene[J]. Science, 2011, 332(6037): 1537-1541.

[22] Wang H, Wang D, Deng T, et al. Insight intographene/hydroxide compositing mechanism for remarkably enhanced capacity[J]. Journal of Power Sources, 2018, 399: 238-245.

[23] Shi X, Zhang W, Wang J, et al. $(EMIm)^+(PF6)^-$ Ionic Liquid Unlocks Optimum Energy/Power Density for Architecture of Nanocarbon-Based Dual-Ion Battery[J]. Advanced Energy Materials, 2016, 6(24): 1601378.

[24] Wang G, Zhang L, Zhang J. A review of electrode materials for electrochemicalsupercapacitors[J]. Chemical Society Reviews, 2012, 41(2): 797-828.

[25] Yu Z, Tetard L, Zhai L, et al. Supercapacitor electrode materials: nanostructures from 0 to 3 dimensions[J]. Energy & Environmental Science, 2015, 8(3): 702-730.

[26] Liu Y, He K, Chen G, et al. Nature-Inspired Structural Materials for Flexible Electronic Devices[J]. Chemical Reviews, 2017: 12893.

[27] Peng S, Li L, Wu H B, et al. Controlled growth of $NiMoO_4$ nanosheet and nanorod arrays on various conductive substrates as advanced electrodes for asymmetric supercapacitors[J]. Advanced Energy Materials, 2015, 5(2): 1401172.

[28] Wang H X, Zhang W, Drewett N E, et al. Unifying miscellaneous performance criteria for a prototype supercapacitor via Co (OH) 2 active material and current collector interactions[J]. Journal of Microscopy, 2017, 267(1): 34-48.

[29] Irimia-Vladu M. "Green" electronics: biodegradable and biocompatible materials and devices for sustainable future [J]. Chemical Society Reviews, 2014, 43(2): 588-610.

[30] Ying Y, Feng W. Naturally derived nanostructured materials from biomass for rechargeable lithium/sodium batteries [J]. Nano Energy, 2015, 17: 91-103.

[31] Zhang L, Liu Z, Cui G, et al. Biomass-derived materials for electrochemical energystorages[J]. Progress in Polymer Science, 2015, 43: 136-164.

[32] Chen W, Yu H, Lee S Y, et al. Nanocellulose: a promising nanomaterial for advanced electrochemical energy storage [J]. Chemical Society Reviews, 2018, 47(8): 2837-2872.

[33] Dutta S, Kim J, Ide Y, et al. 3D network of cellulose-based energy storage devices and related emerging applications [J]. Materials Horizons, 2017, 4(4): 522-545.

[34] Li H B, Yu M H, Wang F X, et al. Amorphous nickel hydroxidenanospheres with ultrahigh capacitance and energy density as electrochemical pseudocapacitor materials[J]. Nature Communications, 2013, 4(1): 1-7.

[35] Deng T, Zhang W, Arcelus O, et al. Atomic-level energy storage mechanism of cobalt hydroxide electrode for pseudocapacitors[J]. Nature Communications, 2017, 8(1): 1-9.

[36] Simon P, Gogotsi Y, Dunn B. Where do batteries end and supercapacitors begin? [J]. Science, 2014, 343(6176):

1210-1211.

[37]Weng Z, Su Y, Wang D W, et al. Graphene-cellulose paper flexible supercapacitors[J]. Advanced Energy Materials, 2011, 1(5): 917-922.

[38]Jiang J, Li Y, Liu J, et al. Recent advances in metal oxide-based electrode architecture design for electrochemical energystorage[J]. Advanced Materials, 2012, 24(38): 5166-5180.

[39]Cao L, Xu F, Liang Y Y, et al. Preparation of the novel nanocomposite $Co(OH)_2$/ultra-stable Y zeolite and its application as a supercapacitor with high energy density[J]. Advanced Materials, 2004, 16(20): 1853-1857.

[40]Chang J K, Wu C M, Sun I W. Nano-architectured $Co(OH)_2$ electrodes constructed using an easily-manipulated electrochemical protocol for high-performance energy storage applications[J]. Journal of Materials Chemistry, 2010, 20(18): 3729-3735.

[41]Ramadoss A, Kim S J. Enhanced supercapacitor performance using hierarchical TiO_2 nanorod/$Co(OH)_2$ nanowall array electrodes[J]. Electrochimica Acta, 2014, 136: 105-111.

[42]Jiang J, Liu J, Ding R, et al. Large-scale uniform α-$Co(OH)_2$ long nanowire arrays grown on graphite as pseudocapacitor electrodes[J]. ACS Applied Materials & Interfaces, 2011, 3(1): 99-103.

[43]Zhao C, Wang X, Wang S, et al. Synthesis of $Co(OH)_2$/graphene/Ni foam nano-electrodes with excellent pseudocapacitive behavior and high cycling stability for supercapacitors[J]. International journal of hydrogen energy, 2012, 37(16): 11846-11852.

[44]Lei W, Zhi H D, Zheng G W, et al. Layered α-Co(OH) Nanocones as Electrode Materials for Pseudocapacitors: Understanding the Effect of Interlayer Space on Electrochemical Activity[J]. Advanced Functional Materials, 2013, 23(21): 2758-2764.

[45]Sivanantham A, Ganesan P, Shanmugam S. Bifunctional Electrocatalysts: Hierarchical $NiCo_2S_4$ Nanowire Arrays Supported on Ni Foam: An Efficient and Durable Bifunctional Electrocatalyst for Oxygen and Hydrogen Evolution Reactions [J]. Advanced Functional Materials, 2016, 26(26): 4660-4660.

[46]Deng T, Lu Y, Zhang W, et al. Inverted Design for High-Performance Supercapacitor Via $Co(OH)_2$-Derived Highly Oriented MOF Electrodes[J]. Advanced Energy Materials, 2018, 8(7): 1702294.

[47]Yoon Y, Lee K, Kwon S, et al. Vertical alignments ofgraphene sheets spatially and densely piled for fast ion diffusion in compact supercapacitors[J]. Acs Nano, 2014, 8(5): 4580-4590.

[48]Xiong G, Hembram K, Reifenberger R G, et al. MnO_2-coated graphitic petals for supercapacitor electrodes[J]. Journal of Power Sources, 2013, 227(1): 254-259.

[49]Gong J, Tian Y, Yang Z, et al. High-performance flexible all-solid-state asymmetric supercapacitors based on vertically aligned CuSe@ $Co(OH)_2$ nanosheet arrays[J]. The Journal of Physical Chemistry C, 2018, 122(4): 2002-2011.

[50]Wan C, Jiao Y, Li J. Multilayer core-shell structured composite paper electrode consisting of copper, cuprous oxide and graphite assembled on cellulose fibers for asymmetric supercapacitors[J]. Journal of Power Sources, 2017, 361: 122-132.

[51]Ghosh D, Lim J, Narayan R, et al. High energy density all solid state asymmetric pseudocapacitors based on free standing reduced graphene oxide-Co_3O_4 composite aerogel electrodes[J]. ACS Applied Materials & Interfaces, 2016, 8(34): 22253-22260.

[52]Sekhar S C, Nagaraju G, Yu J S. Conductive silver nanowires-fenced carbon cloth fibers-supported layered double hydroxide nanosheets as a flexible and binder-free electrode for high-performance asymmetric supercapacitors[J]. Nano Energy, 2017, 36: 58-67.

[53]Veerasubramani G K, Krishnamoorthy K, Kim S J. Improved electrochemical performances of binder-free $CoMoO_4$ nanoplate arrays@ Ni foam electrode using redox additive electrolyte[J]. Journal of Power Sources, 2016, 306: 378-386.

中文题目：地质结构系统启发用于高性能能量存储的微/纳米异质结构设计

作者：万才超，焦月，梁大鑫，吴义强，李坚

摘要：受自然启发提出一种新颖而有效的方法，以应对具有高能量/功率密度和长循环寿命的环保型高性能储能设备的挑战。受地质建筑系统(即地面-山脉-植被)的协同作用和综合形式的启发，本文基于磁控溅射和电氧化的简易组合方法，创建了一种新颖的、分层的纤维素负载 Co@Co(OH)$_2$ 异质结构。由于这种多尺度结构的协同效应(即纤维素基质(地面)对电解质离子的储存能力、由层间金属钴(山脉)提供的电子高速公路以及原位生长和准蜂窝状大表面积 Co(OH)$_2$(植被)的超高电化学活性和机械稳定性)，该复合材料显示出高比电容(2 mA·cm^{-2} 时为 642 mF·cm^{-2}/958 F·g^{-1})、优异的倍率性能和良好的循环稳定性(10000 次循环后仅有 2.1%的损失)，明显优于其他 Co(OH)$_2$ 基电极的微结构设计。此外，组装的非对称超级电容器具有较强的能量/功率密度(1.5 mW·cm^{-2} 时为 166 μW·h·cm^{-2})和出色的循环稳定性。结合其突出的电化学性能、简便的合成技术、环保低成本的优势，这种受自然启发的巧妙的复合材料为绿色高性能储能设备带来了巨大的希望。

关键词：功能材料；异质结构；受自然启发的设计；超级电容器

Mussel Adhesive-inspired Design of Superhydrophobic Nanofibrillated Cellulose Aerogels for Oil/Water Separation*

Runan Gao, Shaoliang Xiao, Wentao Gan, Qi Liu, Hassan Amer, Thomas Rosenau, Jian Li, Yun Lu

Abstract: Inspiration from biomimetics-the chemistry of mussel adhesives-was taken for imparting superhydrophobic properties to nanofibrillated cellulose (NFC) matrices. Apolydopamine (PDA) surface coating was introduced, acting as an anchor between the NFCscaffold and octadecylamine (ODA): PDA is coated onto the NFC scaffolds by its adhesiveproperties and the ODA is successfully attached to the PDA by a Schiff base reaction. Theultralow density of 6. 04 mg · cm^{-3}combined with the high contact angle of 152. 5° endows thecomposite aerogel with superb buoyancy and excellent oil/water separation selectivity. Oil can be rapidly absorbed from a mixture of oil and water. In addition, the modified aerogel cantake up a wide range of organic solvents, the maximum absorption capacity reaching up to 176 g · g^{-1}, depending on the density of the liquids. The novel superhydrophobic aerogel showsgreat potential as adsorber for oil and solvent spills and as oil-water separator.

Keywords: Nanofibrillated cellulose; Polydopamine; Octadecylamine; Aerogel; Superhydrophobic; Oil absorption

1 Introduction

Oil spills have become a major concern with regard to both marine and terrestrial pollution, and have posed a huge burden to economy and ecology. [1] The catastrophic oil spills of the lastdecades have called forth various strategies for oil leakage problem, such as filtration, mechanical extraction, chemical degradation, bioremediation and sorbents. [2-3] Among thesemethods, the sorbent stands out due to their facile operability, significant capacity for oilcapture, and high cost-efficiency. [4-6] Such sorbents are not only important on a large scale, such as in supertanker accidents, but also in chemical labs, production sites and even households.

Nanocellulose-based aerogels have attracted much attention for use in oil/waterseparation. [7-10] Compared with synthetic polymer aerogels, the abundant and sustainable nanofibrillated cellulose (NFC) aerogels stick out because of their environmentally friendliness and biocompatibility. They have large specific area and high porosities, which renders them most promising for renewable absorbent materials. [11-12] However, the intrinsic

* 本文出自 ACS Sustainable Chem. Eng, 2018, 6(7): 9047-9055.

hydrophilicity of cellulose poses an obstacle for its application in oil/water separation.[13] Nowadays, various methods have been developed to modify the surface of cellulose nanofibers in order to alter and tune the wettability of cellulose-based aerogels without significant intake of chemicals and without impairing density and pore characteristics. Chemical vapor deposition(CVD) technique is widely applied in hydrophobic modification of cellulose-based aerogels. For example, with the CVD method, the methyltrimethoxysilane(MTMS)-coated cellulose aerogel has a contact angle of 145° and an oil absorption capacityof 24.4 g.[14] Atomic layer deposition(ALD),[15] sol-gel[16] and the cool plasma technique[17] were also employed to alter the wettability of NFC aerogels. These methods imply use of toxic modifiers or precursors, specific equipment, complicated processes and result in non-uniformstructures, making large-scale application of NFC aerogels difficult. A scalable strategy forfacile and green modification of cellulose nanofibers, which renders NFC aerogels morefavorable for widespread application, is still needed.

Bionic-also called biomimetic-approaches provide an alternative tactics for material design and construction, trying to learn from nature and to transfer natural design into technical principles. In nature, mussels can tightly attach to virtually all types of substrates, even on blank surfaces under wet conditions, on Teflon or glass or metals by secreting adhesive proteins.[18] This robust adhesion is effected by 3,4-dihydroxy-L-phenylalanine(DOPA) and lysine amino acids.[19-20] Further studies revealed that dopamine, a mimic of DOPA that contains both the catechol and the amine groups, can be oxidized and polymerized under very mild basic conditions to form to polydopamine(PDA) coatings onvarious solid surfaces.[21] The different functional groups incorporated in PDA, such asphenolic hydroxyl, ortho-quinone(from catechol oxidation), amine and imine, can serve aslinks to surfaces and as covalent anchor groups to desired molecules.[22-23] In addition, PDA exhibits excellent chemical reactivity toward various thiol-or amine-containing molecules.[24-25] The mussel adhesive-inspired surface chemistry has opened new routes to diverse hybrid materials with application potential in a wide range of technology fields, suchas chemical, biological, energy, sensing and environmental applications.

Recently, mussel adhesive chemistry has also inspired fabrication of functional nanocellulose materials. For instance, Ag-PDA functionalized cellulose nanocrystals (CNC) weredeveloped for antibacterial activity and reduction of 4-nitrophenol,[26-27] and PDA-modified CNC was used for Fe^{3+} detection.[28] In this work, we employ the PDA interlayer as a mediator to bridge the hydrophilic NFC and the desired hydrophobic octadecylamine(ODA) molecules, thus endowing the NFC surface with hydrophobicity. Simply by dipping NFC intoa dopamine/ODA emulsion and subsequent freeze-drying, composite aerogels with favorable characteristics of superhydrophobicity, ultralow density and high porosity are obtained(Fig. 1). They show great oil/water absorption selectivity and can absorb a wide range oforganic solvents. While according to the CVD method, the hydrophobic reagent generally hasan uneven distribution, the method presented here allow full contact between the NFC scaffolds and the modifiers, which leads to uniform structures of the composite aerogels andkeeps its original porous structure intact. In addition, tedious procedures and specific equipment, which are for instance necessary in ALD or cool plasma techniques, are avoidedin this work. This green and facile method may in fact provide an easy and scalable way for NFC-based oil absorbent preparations.

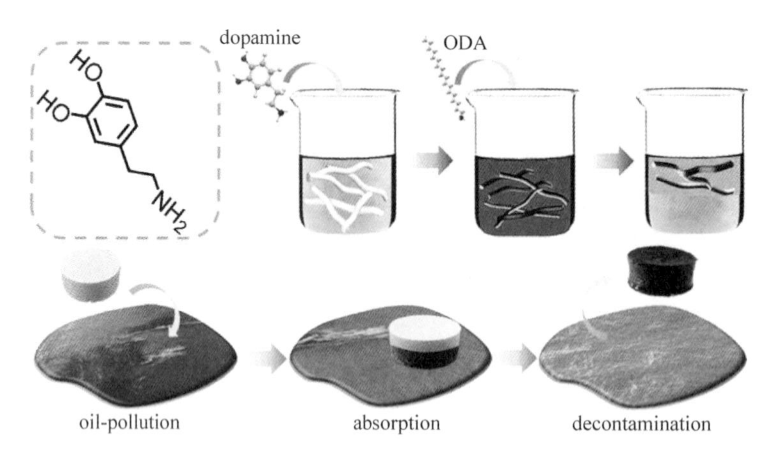

Fig. 1 Schematic illustration of the facile fabrication of mussel adhesive-inspired, superhydrophobic NFC-based aerogels for highly efficient oil/water separation

2 Experimental section

2.1 Materials

Larix gemlini NFC suspension was prepared according to our previously reported method. [29] Dopamine hydrochloride, octadecylamine, Tris (hydroxymethyl) aminomethane (Tris), oilred, and methylene blue were purchased from Sigma-Aldrich Co. , Ltd. Lubricating oil was purchased from the Si Fang special oil factory(Beijing China). Other chemicals were used asreceived.

2.2 Fabrication of NFC and PDA-anchored NFC aerogel

The NFC suspension(0. 1%) was poured into a dialysis bag and the solvent replaced with *tert*-butanol for 2 h. This replacement process was repeated twice, and the sample was placedinto molds. NFC aerogels were obtained after the freeze-drying(−20 ℃) for 12 h. PDA-anchored NFC aerogels(designated as PDA@ NFC): 0. 02 g dopamine was immersed in0. 1% NFC-Trissuspension(pH=8. 5, 100 mL) and stirred at room temperature for 24 h. The sample was subjected to solvent exchange with *tert*-butanol and freeze-dried(−20 ℃).

2.3 Preparation of the superhydrophobic aerogels

ODA(0. 5 g) was dispersed into Tris buffer(pH=8. 5, 100 mL) and ultrasonicated for 3 min to obtain an emulsion. Dopamine hydrochloride(0. 02 g) and the prepared ODA emulsion were added to the Larix gmelini NFC-Tris suspension (0. 1%, 100 mL, pH = 8. 5). The mixture was stirred at room temperature for 24 h. Formation of two phases showed completion of the process. The composite phase was rinsed with distilled water and then the water was replaced with tert-butanol, followed by freeze-drying (-20℃), the superhydrophobic aerogels(designated as ODA-PDA@ NFC) were obtained. ODA-admixed NFC aerogels for comparison were prepared by the above method just without adding dopamine hydrochloride.

2.4 Characterization

Morphologies of NFC and composite aerogels were observed by a Hitachi JSM-7500 F scanning electron microscope(SEM). Samples were sprayed with a layer of gold before hand to improve conductivity. The

Fourier transform infrared spectroscopy(FTIR) spectra of samples were examined on a Nicolet Nexus 670 FTIR instrument in the range of 4000 – 600 cm^{-1}. The chemical composition on surface of NFC scaffolds and composite fiber were determined by a Thermo Scientific K-Alpha X-ray Photoelectron Spectrometer(XPS). N_2 adsorption-desorption isotherms were recorded on a BK112 specific surface area and poresize analyzer. The specific surface area was determined according to theBrunauer-Emmett-Teller method. Pore size distribution was estimated according to the BJH approach. The bulk density of aerogels was calculated on the basis of the physical dimensions and weights of the samples. Porosity of aerogels was evaluated as

$$\rho_a(\%) = \left(1 - \frac{\rho_a}{\rho_s}\right) \times 100$$

where ρ_a is the bulk density of the NFC or the composite aerogel, and ρ_s is the bulk density of solid scaffold (NFC or composite NFC). Bulk densities of PDA@NFC and ODA-PDA@NFC were estimated as follow:

$$\rho_s = \frac{1}{\frac{W_{cellulwse}}{\rho_{cellulose}} + \frac{W_{PDA}}{\rho_{PDA}} + \frac{W_{ODA}}{\rho_{ODA}}}$$

Here $W_{cellulose}$ is the weight fraction of cellulose, W_{PDA} is the weight fraction of PDA, and w_{ODA} is the weight fraction of ODA. The bulk density of cellulose ($\rho_{cellulose}$) is taken as 1600 mg · cm^{-3}, while the bulk density of PDA(ρ_{PDA}) is assumed to be 1400 mg · cm^{-3}, and that of ODA(ρ_{ODA}) is taken as 862 mg · cm^{-3}.

Water contact angle(WCA) measurements were carried out with a Data Physics OCA 20 video contact angle measurement instrument (Data Physics). The stress-strain curves of aerogel materials were measured and recorded by a Shimadzu AG-A10T universal mechanical testing system. Cylinder aerogels with a size of 40 mm × 25 mm were compressed with a speed of 1 mm · min^{-1} to 50% of its original length.

Oil and organic solvents adsorption capacities of ODA-PDA@NFC. Modified aerogels were immersed in 40 mL of oil or organic solvents. Sorbents were taken out after completely wetting and redundant solvents were extracted with filter paper. The practical absorption capacity (C_p) (w/w) of ODA-PDA @ NFC was calculated by:

$$C_p(w/w) = \frac{m_1 - m_0}{m_0}$$

where m_0 and m_1 are the weights of aerogel before and after absorption, respectively. The theoretical mass-based absorption capacity of aerogels was calculated by:

$$C_m(w/w) = \text{porosity} \times \rho_{liquid} / \rho_{aeroge}$$

The theoretical volume-based absorption capacity of aerogels was calculated by:

$$C_v(v/w) = \frac{\rho_{orosity}}{\rho_{aerogel}}$$

3 Results and discussion

"Switching" the medium compatibility of the aerogel by the ODA-PDA modifier is demonstrated in Fig. 2 (a-c). When an NFC aerogel encountered a water or oil droplet, both liquids are promptly taken up, causing partial collapse in the water case. After the modification, ODA-PDA@NFC still absorbed oil droplets rapidly, but blocked water and aqueous solutions (milk and tea as examples in the picture), which were staying outside the aerogel and not penetrating into inner structures. Upon immersing NFC aerogel into water, it was deformed immediately. Once saturated with water, it sank to the bottom of the vessel. Conversely, the ODA-PDA@ NFC aerogel floated on water due to its lightweight and superhydrophobicity. If submerged into water by external force, then it was not wetted and immediately rose to the surface when the external force was withdrawn(Supporting Information, SI, Video S1). Its outstanding floatability renders ODA-PDA@ NFC a promising absorbent for oil spills in bulk water.

The morphology of the samples was illustrated by SEM images. Fig. 2(d) shows the microstructure of a NFC aerogel. Intertwining of the fibrils, forming the porous three-dimensional (3D) network structure, can nicely be seen. Coating with PDA broadened the fibrils and roughened the surface. The 3D network became denser and the pore structure shrunk[Fig. 2(e)]. It was obviously that ODA-PDA fully covered the NFC scaffolds, shaped the sheet-like fibrous structure and generated irregular roughness on NFC surfaces. When only ODA was mixed with NFC, it dispersed unregularly in the NFC network and jammed the pores. The characteristic pore structure of aerogel material was canceled out(see Fig. S1). The SEM result indicated that PDA can orient the ODA molecules on the NFC's surface in a way that the 3D network is kept, and porous structure as well as low density are maintained.

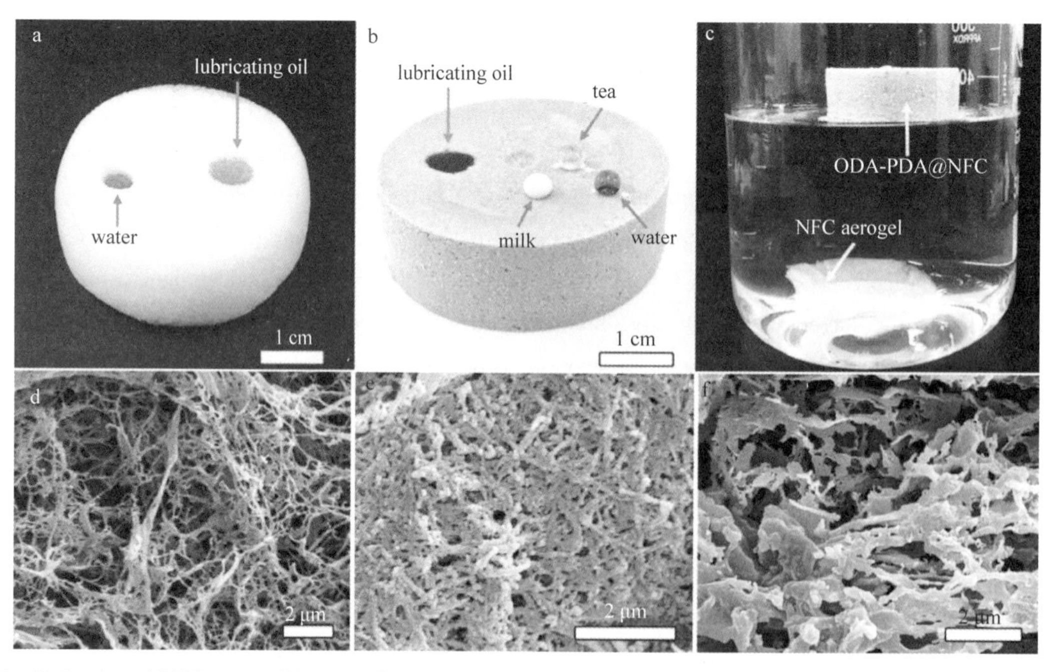

Fig. 2 Behavior of NFC aerogel(a) and ODA-PDA@ NFC aerogel(b) towards different liquids andin bulk water (c), SEM images of microstructure of NFC(d), PDA@NFC(e) and ODA-PDA@NFC(f)

Generally, the PDA coating can easily adhere on various substrates through the oxidative self-polymerization of dopamine molecules under basic conditions. Dopamine is first transformed into intermediate products, such as 5,6-dihydroxyindole(DHI), then further oxidatively polymerized into the final PDA.[15] The combination of multiple functional groups in PDA, such as hydroxyl groups, indole motifs, amino groups, catechol, or quinone moieties, causes the strong adhesion capability of PDA to various substrates.[30] When ODA molecules are involved in copolymerization(Fig. 3), it is suggested that the amine moieties of ODA react with the intermediates and products of oxidative dopamine polymerization by Schiff base formation or/and Michael addition.[31] In this work, the formation of C=N structures with simultaneous depletion of carbonyl (quinone) structures in the PDA-ODA product (Fig. 4), as seen by FTIR and XPS, confirms the formation of Schiff bases (C=N).

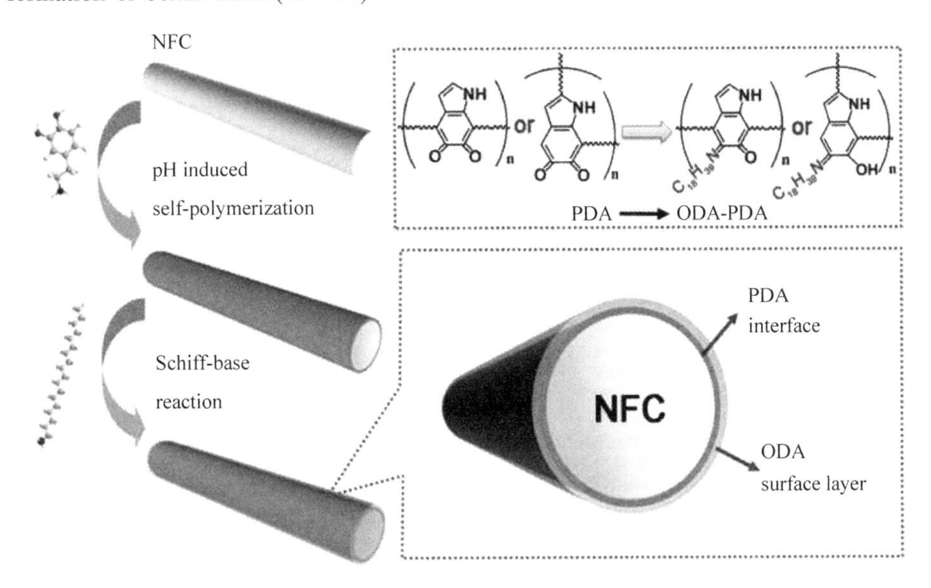

Fig. 3 Schematic illustration of the fabrication mechanism of the musseladhesive-inspired, superhydrophobic NFC-based aerogels

Fig. 4(a) shows the FTIR spectra of NFC and modified aerogels. The former showed the characteristic broad band in the region of 3600-3000 cm^{-1}, originating from the stretching vibration of H-bonded —OH groups in cellulose molecules, and other characteristic peaks of cellulose. The peak at 2900 cm^{-1} was attributed to C—H stretching vibration,[32] those at 1428 and 1370 cm^{-1} to asymmetric CH_2 wagging and bending,[33] and the one at 897 cm^{-1} to anomeric carbon (C1) deformation. The prominent peak at 1050 cm^{-1} is associated with the C—O—C pyranose ring (antisymmetric in-phase) stretching vibration.[34] After coating with PDA, the band around 3300 cm^{-1} was broadened, indicating functional groups of PDA forming new hydrogen bonds with —OH groups of NFC. Moreover, the characteristic peak at 1508 cm^{-1} was attributed to the N—H shearing vibration of PDA,[35] which was in agreement with the FTIR spectra of PDA and dopamine monomer (Fig. S2). For ODA-PDA@ NFC, the broad band assigned to —OH groups was weaker, and a peak at 3326 cm^{-1} attributed to the stretching vibration of N—H appeared. Two significant peaks at 2916 and 2849 cm^{-1} were designated to stretching vibrations of —CH_3 and —CH_2— of ODA.[36] The band at 1645 cm^{-1} was attributed to the C=N stretching vibrations of Schiff base reaction products between PDA and ODA; it cannot be found in the spectrum of pure ODA.

The surface compositions of samples were further studied by XPS. Surface composition changes before and after the modification are clearly presented in survey spectra (Fig. 4b). The characteristic peak of the N element appeared both after coating NFC with PDA and with the mixture of PDA and ODA. The N/C ratio of PDA@ NFC is 0.1 (see the atomic composition listed in Table S1), slightly lower than the theoretic value of 0.125 of dopamine,[37] indicating the NFC scaffolds to be largely covered with PDA. Moreover, after ODA was deposited on the surface of scaffolds, the intensities of the C 1s peak remarkably increased while the intensity of O 1s sharply decreased. The N/C ratio of ODA-PDA@ NFC is 0. 053, which is very close to the theoretical value of ODA (0. 055). [31] This indicates that most hydroxyl groups were hidden from the surface, which instead presents ODA molecules almost exclusively. To further prove this, high-resolution XPS C 1s and N 1s spectra were examined. Fig. 4c presents the C 1s core-level spectra of aerogel before and after the modification. For pure NFC aerogels, curve-fitted peaks at 284. 6, 286. 7, 288. 3, and 289. 2 eV were designated to C—C, C—O, C=O, and O—C=O moieties, respectively. [38] In the high-resolution C 1s spectra of PDA @ NFC, there were three characteristic contributions at 284. 5, 285. 8, and 287. 7 eV assigned to CHx/C—NH_2, C—O, and C=O species, respectively.

The N 1s spectra provided crucial evidence for the successful deposition of PDA on NFC. As shown in Fig. 4d, the N 1s region was composed of two peaks from primary(R—NH_2, 401. 6 eV) and secondary(R—NH—R, 399. 6 eV) amine functionalities. [40] The primary amine was associated with self-assembled $(dopamine)_2$/DHI trimer and the secondary amine was associated with both intermediate species(DHI) and polydopamine. In the case of ODA-PDA@ NFC, peaks from carbon-oxygen bonds almost disappeared and the

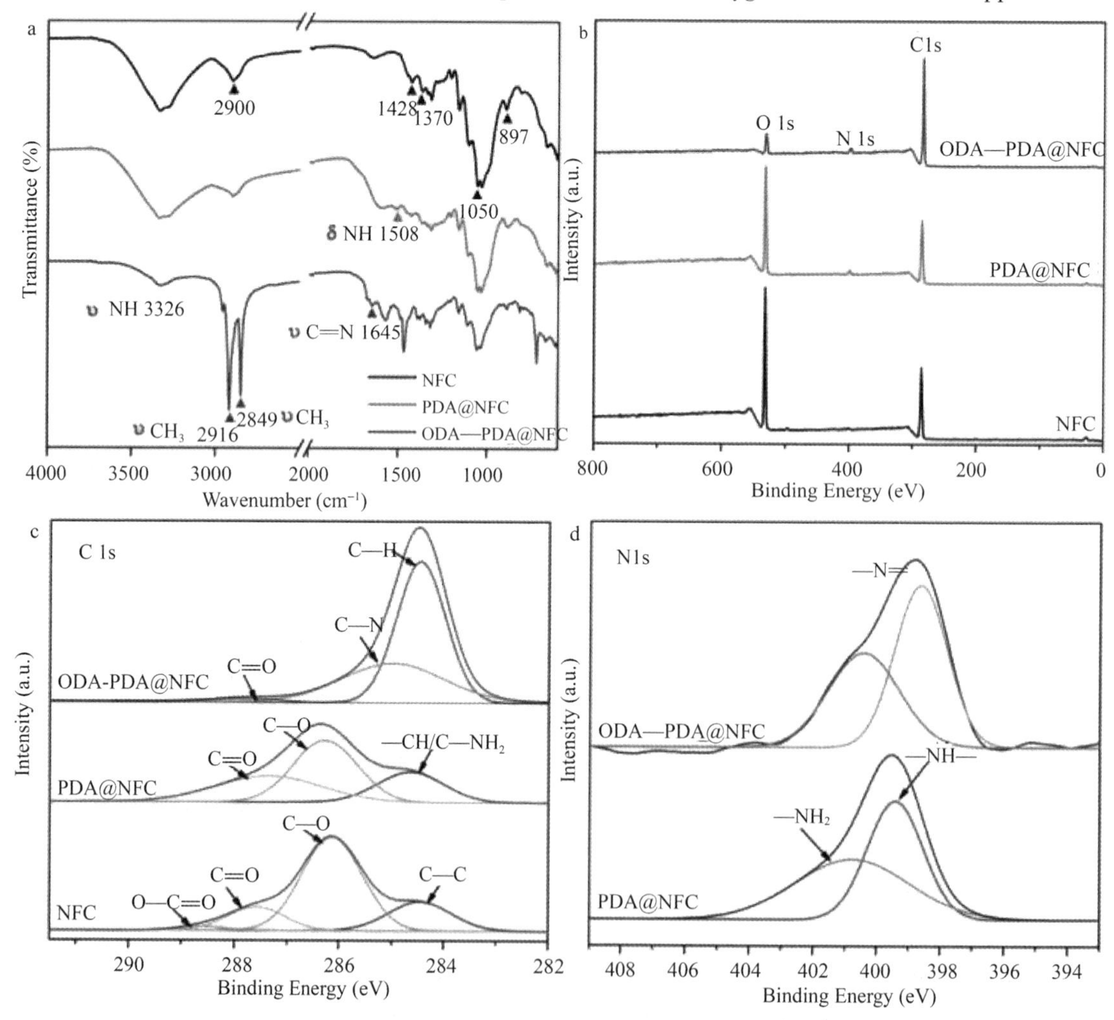

Fig. 4 (a)FTIR spectra of NFC and composite aerogels, (b)survey XPS spectra of NFCand composite aerogels, (c-d)high-resolution XPS spectra(C1s and N1s, respectively)ofNFC and composite aerogels

peak attributed to C—H and C—N at 284.6 and 285.5 eV, respectively, intensified. In the N 1s region, a new intense signal at 398.8 eV appeared, assigned to tertiary (—N═) imine moieties, implying the formation of Schiff bases (see above). This result was consistent with that from FTIR characterization and further supported the successful anchoring of ODA on the NFC-PDA scaffolds.

The specific surface area and pore structure of the different aerogel specimens were examined by N_2 adsorption-desorption (Fig. 5). The presence of H3 type hysteresis demonstrated their mesoporous structure.[41] The NFC aerogel had a large specific surface area of 186.5 $m^2 \cdot g^{-1}$ and a pore volume of 0.76 $cm^3 \cdot g^{-1}$, respectively. As the functional molecules deposited on the NFC, the density of aerogel material increased. Inversely, the specific surface area and pore volume of composite aerogels decreased. However, compared with other reported hydrophobic modified NFC aerogels, ODA-PDA@ NFC possessed relatively low density (6.0 $mg \cdot cm^{-3}$) and high specific surface area (93.1 $m^2 \cdot g^{-1}$).[13,16] Using tert-butanol as the solvent helped to maintain the high specific area. The fast crystallization of tert-butanol during freezing leads to small pores and fine interconnected structure.[42] When use water as the solvent, large ice crystals formed during freezing will press the network flat and generate large pores with honeycomb structures, leading to a low specific surface area and large pore structure of the composite aerogel. As for ODA mixed NFC aerogel, the specific surface area and the pore volume was only 22.07 $m^2 \cdot g^{-1}$ and 0.07 $cm^3 \cdot g^{-1}$, respectively. As revealed in Fig. S1, in the absence of the mediating PDA interlayer the ODA molecules were disordered and irregularly dispersed between NFC scaffolds, jamming its pore structure and in this way causing a sharp decline in both specific surface area and pore volume. Detailed information about density, specific surface area and pore volume is listed in Table 1.

Table 1 Bulk density, specific surface area and pore volume of the studied NFC and composite aerogels

Sample	$\rho_a$etogel($mg \cdot cm^{-3}$)	BET specific surface aren($m^2 \cdot g^{-1}$)	Pore diameter(nm)	Pore volume($cm^3 \cdot g^{-1}$)
NFC	2.57	186.50	10.92	0.76
PDA@ NFC	3.04	178.80	10.34	0.70
ODA-PDA@ NFC	6.04	93.08	8.04	0.37
ODA mixed NFC	7.37	22.07	5.73	0.07

Upon water contact angle (WCA) measurements, it was hard to place a drop of water on the surface of ODA-PDA@ NFC. When the microsyringe was withdrawn, the water droplet lifted with it (SI Video S2), a nice visualization of the water repellency. In Fig. 6b it is shown that water droplets rolled off the support very easily as soon as it was tilted to a certain angle (SI Video S3). The ceraceous (wax-like) ODA molecules anchored on the NFC surface evidently decreased the surface energy and sealed off the hydrophilic hydroxyl groups quite efficiently. Moreover, the porous and rough structure of aerogel traps air below the water drops to form air pockets, i. e., a Cassie-Baxter wetting regime. Since the drops "sit" partially on the air, hydrophobicity is strengthened.[43] The ODA-PDA@ NFC showed superhydrophobic characteristics and possessed a high contact angle of 152.5° (see SI Fig. S4). Such superhydrophobic aerogels exhibit neither water-wetting nor water-absorbing characteristics, however, they still absorb oil and some organic solvents.

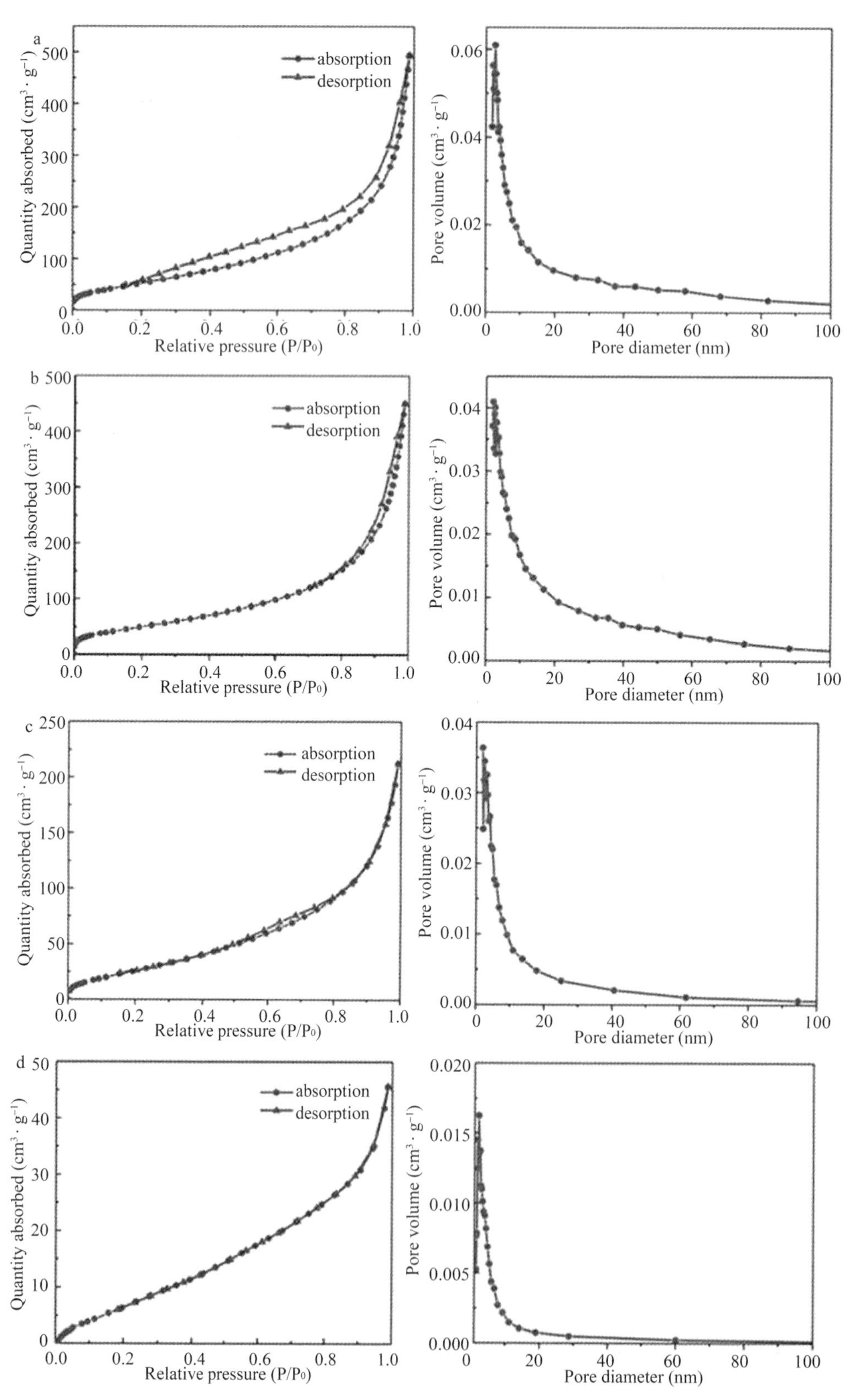

Fig. 5 N_2 adsorption-desorption isotherms of the aerogel samples：(a)NFC，(b)PDA@NFC，(c)ODA-PDA@NFC and(d)ODA mixed NFC

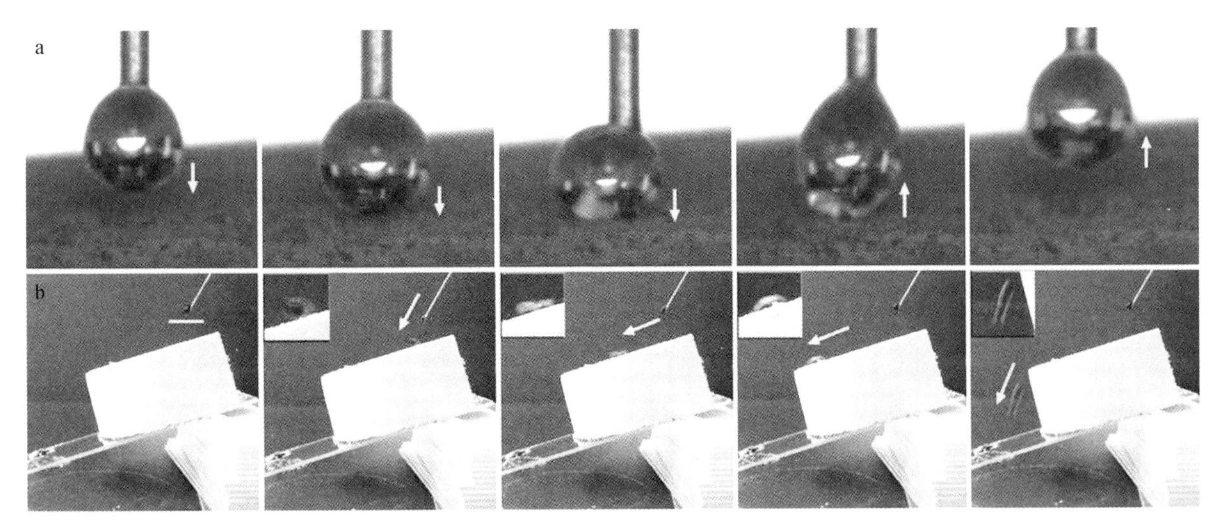

Fig. 6 Water repellency(a) and self-cleaning characteristic(b) of ODA-PDA@NFC

As shown in Fig. 7, the composite aerogel can rapidly collect oil from the surface of water. To separate organic solvents with higher density than the aqueous phase-taking chloroform as an example-a great amount of "air pockets" on the surface of the aerogel formed a water barrier surrounding the aerogel, blocking the water, but enabling the solvent to be absorbed. The mechanical integrity of the aerogel remained unchanged. The NFC aerogel, by contrast, was thoroughly wetted and saturated with water, which incapacitated it for further interaction with the chloroform. Compared with ODA-PDA@NFC, the NFC aerogel was more fragile and friable and very easy to be disintegrated after saturation with water (Fig. 7c). This can be verified by the mechanical testing. The mechanical performance of aerogels was proved after the modification (Fig. S3). No apparent linear elastic deformation was observed for both NFC and composite aerogels. The compressive stress of aerogels at 50% strain dramatically increased from 0.43 kPa (NFC) to 2.6 kPa (ODA-PDA@NFC). The shape recovery of ODA-PDA@NFC increased by 12% compared to that of NFC. ODA covered on the scaffolds weakened hydrogen bonds inside the interconnected network, which allows the composite aerogel to recover.[44] The improvement in mechanical performance rendered the aerogel more favorable as an absorbent.

Absorption performance of different composite samples for some organic solvents was investigated. The mass-based absorption capacity, to a large extent, depended on the solvent density [Fig. 8(a, c)]. The maximum absorption capacities of ODA-PDA@NFC for different oil and organic solvents were 83 to 176 $g \cdot g^{-1}$. Compared with cellulosic aerogels modified according to the CVD method, the value reported in our work is higher than that of sorbents with large densities (24−95 $g \cdot g^{-1}$), and lower than that with low densities (88−356 $g \cdot g^{-1}$) (Table S2). Furthermore, ODA-PDA@NFC has a significantly better absorption performance than synthetic polymer materials, such as nickel foam (3.5 g), polydimethysiloxane sponges (4.3 g), polyurethane sponges (25−87 $g \cdot g^{-1}$), or melamine sponges (60−150 $g \cdot g^{-1}$). The volume-based absorption capacities [Fig. 8(b)] reveal the absorption performance of the material itself. Approximately 72% of the pore volume was enabled to withhold the organic solvents. However, there is an exception when taking up the dichloromethane and only 64% of the pore volume was filled up. This may be due to high volatility of dichloromethane, which makes it difficult for ODA-PDA@NFC to hold the solvent. The average 28% of pores that remained inaccessible to the solvent are thought to be occupied with air bubble or failed to keep the solvent when the composite aerogel was removed from the liquid phase due to unfavorable geometries. The pore volume of aerogels plays a crucial role in absorbing process as revealed by Arie et al. In this work, pore size and density of materials were shown to have also an apparent effect on the practical absorption capacities. PDA@ NFC with much higher pore volume, which is supposed to have a double volume-based absorption capacity (328.3 $mL \cdot g^{-1}$) compared to ODA-PDA@ NFC(164.7 $mL \cdot g^{-1}$), is only 0.47 times higher in absorption performance. In fact, only 54% of the pores of PDA@ NFC were occupied by solvent. This may be caused by the lower density and larger pores of PDA@ NFC as depicted in Fig. 5 and Table 1. The result indicates that denser aerogels with smaller pores are more capable of holding the absorbed solvents, which is consistent with previous work[13,45] using the NFC as the matrix.

For ODA-ad mixed NFC foam aerogel, the mass-based absorption capacities (27 – 45.6 $g \cdot g^{-1}$) were

Fig. 7 Separation the lubricating oil(a) and chloroform(b) from the water phase withODA-PDA@NFC, (c) unsuccessful attempt to collect chloroform with neat NFC aerogel, forcomparison

significantly smaller. In addition, there was no predictable pattern in volume-basedabsorption performance (Fig. S5). This may be not only due to the low pore volume butalso the chaotic pore structure. Without adding PDA, random dispersion of ODA may seal offpores from both the outside and inside and cut off the chance for liquids to get through. Thecomparison of the different aerogels confirmed the orientation feature of PDA and provedthat PDA rendered the composite aerogels more favorable for absorbing purposes.

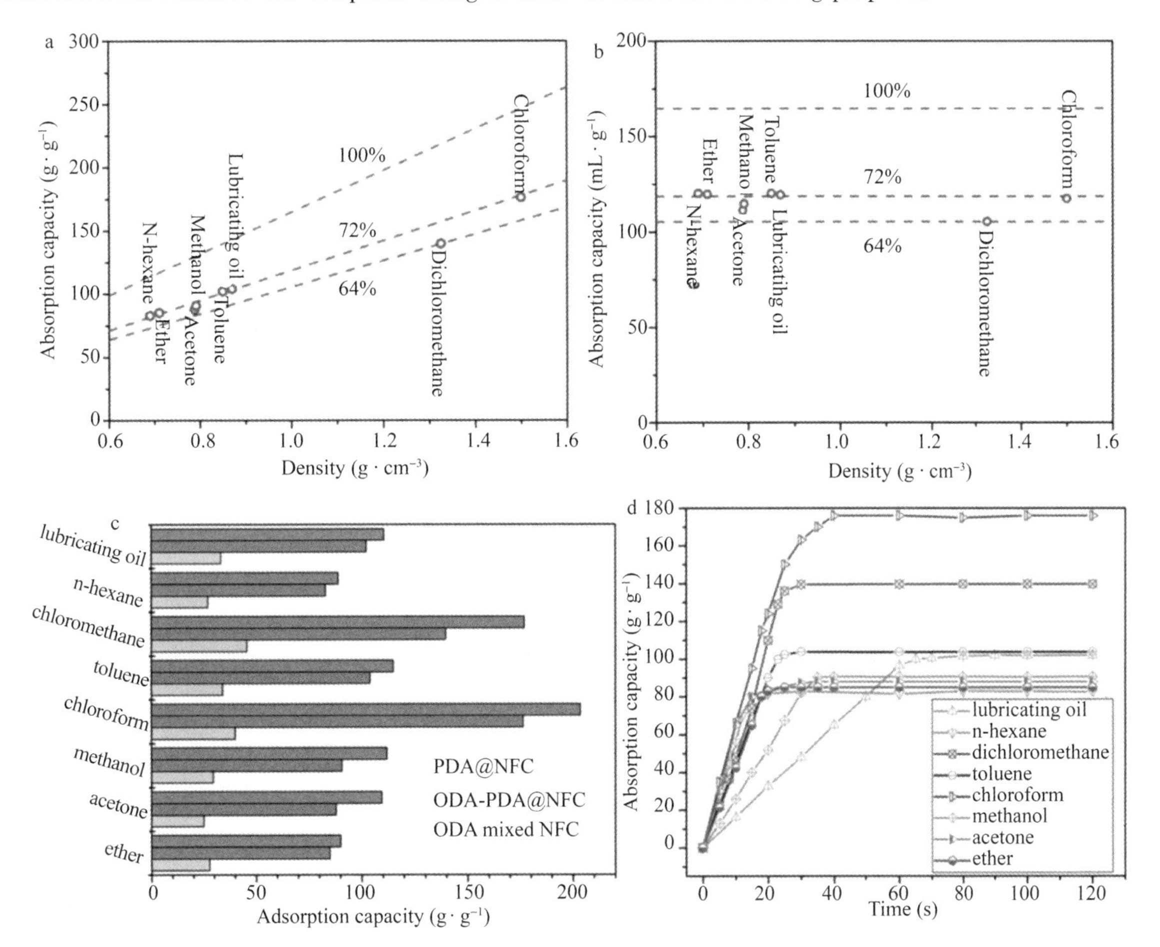

Fig. 8 Absorption capacities of ODA-PDA@NFC for various organic solvents: (a) mass-based, (b) volume-based, (c) absorption capacity contrast between various compositesamples, (d) kinetics of absorption of organic solvents to ODA-PDA@NFC

The curves of the absorption kinetics of different organic solvents to ODA-PDA@NFC are shown in Fig. 8d. In addition to above-mentioned factors, the absorption capacity was also related to the viscosity. For high-viscosity lubricating oil, it took 80 s to reach the absorption equilibrium, while for low-viscosity solvents, such as n-hexane, ether, and acetone, the absorption equilibrium was reached within 25 s. The open pore structure is more accessible to organic solvents with low viscosity, thereby faster establishing the absorption equilibrium. It was evident that ODA-PDA@NFC efficiently absorbed a variety of organic solvents.

An absorption-squeezing test was preformed to evaluate the reusability of the composite aerogels (Fig. S6). Taking the lubricating oil as example, the absorption capacity was reduced to half after the first test cycle and remained constant until the structure collapsed at the forth squeezing process. However, this performance of ODA-PDA@ NFC is less satisfactory compared with 30 usage cycles of microfibrillated cellulose aerogels reported by Wang et al.

4 Conclusions

We have developed a facile and reproducible method to fabricate novel superhydrophobic NFC aerogels. The mussel adhesive-inspired polydopamine(PDA) coating was evenly spun on the NFC scaffolds and worked as a mediator or compatibilizer between cellulose and the hydrophobizer, which in our case was octadecylamine(ODA) grafted on the scaffolds by Schiff base reaction. The obtained ODA-PDA@ NFC aerogel had a high contact angle of 152.5° and can collect oil and a wide range of organic solvents from aqueous phases. The maximum absorption capacity reached as much as 176 $g \cdot g^{-1}$, depending on density and viscosity of the target liquids. We currently work on the reusability of ODA-PDA@ NFC, but already at the present stage we can state that the ODA-PDA@ NFC aerogels are promising as absorbent materials for handling oil and solvent spill challenges.

References

[1] Cheng Y, Li X, Xu Q, et al. SAR observation and model tracking of an oil spill event in coastal waters[J]. Mar Pollut Bull, 2011, 62(2): 350-363.

[2] Sai H, Fu R, Xing L, et al. Surface modification of bacterial cellulose aerogels' web-like skeleton for oil/water separation[J]. ACS Appl. Mater. Inter., 2015, 7: 7373-7381.

[3] Almasian A, Jalali L, Fard C, et al. Surfactant grafted PDA-PAN nanofiber: Optimization of synthesis, characterization and oil absorption property[J]. Chem. Eng. J., 2017, 326: 1232-1241.

[4] Maleki H, Recent advances in aerogels for environmental remediation applications: A review[J]. Chem. Eng. J., 2016, 300: 98-118.

[5] Lu Y, Liu H, Gao R, et al. Coherent-Interface-Assembled Ag_2O-Anchored Nanofibrillated Cellulose Porous Aerogels for Radioactive Iodine Capture[J]. ACS Appl. Mater. Inter., 2016, 8: 29179-29185.

[6] Chen C, Song J, Zhu S, et al. Scalable and Sustainable Approach toward Highly Compressible, Anisotropic, Lamellar Carbon Sponge[J]. Chem, 2018, 4: 544-554.

[7] ZhouS, Liu P, Wang M, et al. Sustainable, Reusable, and Superhydrophobic Aerogels from Microfibrillated Cellulose for Highly Effective Oil/Water Separation[J]. ACS Sustain. Chem. Eng., 2016, 4: 6409-6416.

[8] Sun F, Liu W, Dong Z, et al. Underwater superoleophobicity cellulose nanofibril aerogel through regioselective sulfonation for oil/water separation[J]. Chem. Eng. J., 2017, 330: 774-782.

[9] Wang S, Peng X, Zhong L, et al. An ultralight, elastic, cost-effective, and highly recyclable superabsorbent from microfibrillated cellulose fibers for oil spillage cleanup[J]. J. Mater. Chem. A, 2015, 3: 8772-8781.

[10] Štefelová J, Slovák V, Siqueira G, et al. Drying and Pyrolysis of Cellulose Nanofibers from Wood, Bacteria, and Algae for Char Application in Oil Absorption and Dye Adsorption[J]. ACS Sustain. Chem. Eng, 2017, 5: 2679-2692.

[11] Song J, Chen C, Zhu S, et al. Processing bulk natural wood into a high-performance structural material[J]. Nature, 2018, 554: 224.

[12] Chen C, Zhang Y, Li Y, et al. All-wood, low tortuosity, aqueous, biodegradable supercapacitors with ultra-high capacitance[J]. Energ. Environ. Sci., 2017, 10: 538-545.

[13] MulyadiA, Zhang Z, Deng Y. Fluorine-Free Oil Absorbents Made from Cellulose Nanofibril Aerogels[J]. ACS Appl.

Mater. Inter. , 2016, 8: 2732-2740.

[14]Nguyen S, T Feng, J Le, et al. Cellulose aerogel from paper waste for crude oil spill cleaning[J]. Ind. Eng. Chem. Res. , 2013, 52: 18386-18391.

[15]Korhonen T, Kettunen M, Ras A, et al. Hydrophobic nanocellulose aerogels as floating, sustainable, reusable, and recyclable oil absorbents[J]. ACS Appl. Mater. Inter. , 2011, 3(6): 1813-1816.

[16]Zhang Z, Sebe G, Rentsch D, et al. Ultralightweight and Flexible Silylated Nanocellulose Sponges for the Selective Removal of Oil from Water[J]. Chem. Mater. , 2014, 26: 2659-2668.

[17]Lin R, Li A, Zheng T, et al. Hydrophobic and flexible cellulose aerogel as an efficient, green and reusable oil sorbent [J]. RSC Adv, 2015, 5: 82027-82033.

[18]Liu Y, Ai K, Lu L. Polydopamine and Its Derivative Materials: Synthesis and Promising Applications in Energy, Environmental, and Biomedical Fields[J]. Chem. Rev. , 2014, 114: 5057-5115.

[19] Waite H, Qin X. Polyphosphoprotein from the adhesive pads of Mytilus edulis [J]. Biochem, 2001, 40: 2887-2893.

[20]Ryu H, Messersmith B, Lee H. Polydopamine Surface Chemistry: A Decade of Discovery[J]. ACS Appl. Mater. Inter. , 2018, 10: 7523-754.

[21]Lee H, Dellatore M, Miller M, et al. Mussel-Inspired Surface Chemistry for Multifunctional Coatings[J]. Science, 2007, 318: 426-430.

[22]LiS, Chu L, Gong X, et al. Hydrogen Bonding Controls the Dynamics of Catechol Adsorbed on a TiO_2(110) Surface [J]. Science, 2010, 328: 882-884.

[23]WangJ, TahirN, Kappl M, et al. Influence of BindingSite Density in Wet Bioadhesion[J]. Adv. Mater. , 2008, 20: 3872-3876.

[24]ZhuQ, Pan Q. Mussel-Inspired Direct Immobilization of Nanoparticles and Application for Oil-Water Separation[J]. ACS Nano, 2014, 8: 1402-1409.

[25] Huang S. Mussel - Inspired One - Step Copolymerization to Engineer Hierarchically Structured Surface with Superhydrophobic Properties for Removing Oil from Water[J]. ACS Appl. Mater. Inter. , 2014, 6: 17144-17150.

[26]Shi Z, Tang J, ChenL, et al. Enhanced colloidal stability and antibacterial performance of silver nanoparticles/ cellulose nanocrystal hybrids[J]. J. Mater. Chem. B, 2015, 3: 603-611.

[27]TangJ, Shi Z, Berry R, et al. Mussel-Inspired Green Metallization of Silver Nanoparticles on Cellulose Nanocrystals and Their Enhanced Catalytic Reduction of 4-Nitrophenol in the Presence of β-Cyclodextrin[J]. Ind. Eng. Chem. Res. , 2015, 54: 3299-3308.

[28]HanY, Wu X, Zhang X, et al. Dual Functional Biocomposites Based on Polydopamine Modified Cellulose Nanocrystal for Fe^{3+}-Pollutant Detecting and Autoblocking[J]. ACS Sustain. Chem. Eng. , 2016, 4: 5667-5673.

[29]Xiao S, Gao R, Lu Y, et al. Fabrication and characterization of nanofibrillated cellulose and its aerogels from natural pine needles[J], Carbohyd. Polym. , 2015, 119: 202-209.

[30]Josep S, Javier S, Felix B, et al. Catechol-based biomimetic functional materials. Adv. Mater. , 2013, 25: 653-701.

[31]Chen S, Cao Y, FengJ. Polydopamine as an Efficient and Robust Platform to Functionalize Carbon Fiber for High-Performance Polymer Composites[J]. ACS Appl. Mater. Inter. 2014, 6: 349-356.

[32] Ram B, Chauhan G. New spherical nanocellulose and thiol-based adsorbent for rapid and selective removal of mercuric ions[J]. Chem. Eng. J. , 2018, 331: 587-596.

[33]Das K, Ray D, Bandyopadhyay N, et al. Study of the Properties of Microcrystalline Cellulose Particles from Different Renewable Resources by XRD, FTIR, Nanoindentation, TGA and SEM[J]. J Polym. Environ. , 2010, 18 (3) : 355-363.

[34]Maiti S, Jayaramudu J, Das K, et al. Preparation and characterization of nano-cellulose with new shape from different

precursor[J]. Carbohyd. Polym, 2013, 98: 562-567.

[35]Zhu L, Lu Y, Wang Y, et al. Preparation and characterization of dopamine-decorated hydrophilic carbon black[J]. Appl. Surf. Sci., 2012, 258: 5387-5393.

[36]Oribayo O, Feng X, Rempel G L, et al. Modification of formaldehyde-melamine-sodium bisulfite copolymer foam and its application as effective sorbents for clean up of oil spills[J]. Chem. Eng. Sci., 2017, 160: 384-395.

[37]Lee H, Rho J, Messersmith P. Facile Conjugation of Biomolecules onto Surfaces via Mussel Adhesive Protein Inspired Coatings[J]. Adv. Mater., 2009, 21: 431-434.

[38]Belgacem M, Czeremuszkin G, Sapieha S, et al. Surface characterization of cellulose fibres by XPS and inverse gas chromatography[J]. Cellulose, 1995, 2(3): 145-157.

[39]Zangmeister R, Morris T, Tarlov M. Characterization of polydopamine thin films deposited at short times by autoxidation of dopamine[J]. Langmuir, 2013, 29(27): 8619-8628.

[40]Clark B, Gardella A, Schultz T, et al. Solid-state analysis of eumelanin biopolymers by electron spectroscopy for chemical analysis[J]. Anal. Chem., 1990, 62: 949-956.

[41]Gao R, Lu Y, Xiao S, et al. Facile Fabrication of Nanofibrillated Chitin/Ag2O Heterostructured Aerogels with High Iodine Capture Efficiency[J]. Sci. Rep., 2017, 7: 4303.

[42] Petersson L, Kvien I, Oksman K. Structure and thermal properties of poly(lacticacid)/cellulose whiskers nanocomposite materials[J]. Compos. Sci. Technol., 2007, 67: 2535-2544.

[43]Bormashenko E, Bormashenko Y, Stein T, et al. Why do pigeon feathers repel water? Hydrophobicity of pennae, Cassie-Baxter wetting hypothesis and Cassie-Wenzel capillarity-induced wetting transition[J]. J. Colloid Interf Sci, 2007, 311: 212-216.

[44]Zhang W, Zhang Y, Lu C, et al. Aerogels from crosslinked cellulose nano/micro-fibrils and their fast shape recovery property in water[J]. J. Mater. Chem., 2012, 22: 11642-11650.

[45]Feng J, Hsieh Y. Super water absorbing and shape memory nanocellulose aerogels from TEMPO-oxidized cellulose nanofibrils via cyclic freezing-thawing[J]. J. Mater. Chem. A, 2013, 2: 350-359.

Synopsis: Bio-inspired superhydrophobic nanofibrillated nanocellulose aeroegls show highoil/water separation efficiency.

中文题目：贻贝粘附性启发的超疏水纳米纤丝化纤维素气凝胶用于油水分离

作者：高汝楠，肖少良，甘文涛，刘琦，Hassan Amer，Thomas Rosenau，李坚，卢芸

摘要：受到贻贝粘附性的启发，仿生设计了具有超疏水特性的纳米纤丝化纤维素(NFC)气凝胶材料。首先在NFC表面制备将聚多巴胺(PDA)涂层，并以该涂层为桥梁链接十八胺(ODA)。PDA通过自聚合以及粘附特性附着在NFC表面，然后通过席夫碱反应将ODA接枝到复合骨架上。超低密度和高疏水角152.5°赋予了复合气凝胶材料高效的选择性油水分离功能。此外，该材料可以吸附一些列有机溶剂，吸附量高达176 g·g^{-1}。该新型超疏水气凝胶材料在油污吸附领域展现出强劲的应用潜力。

关键词：纳米纤丝化纤维素；聚多巴胺；十八胺；气凝胶；超疏水；油水分离

Biomorphic Carbon-Doped TiO_2 for Photocatalytic Gas Sensing with Continuous Detection of Persistent Volatile Organic Compounds*

Likun Gao, Wentao Gan, Qiu Zhe, Guoliang Cao, Xianxu Zhan,
Tiangang Qiang, Jian Li

Abstract: How to protect the sensitivity of gas sensing system in the case of persistent leakage in a closed environment? The aim of this study is to combine the gas sensing property of semiconductor with its photocatalytic performance, which may be viable alternative to give recovery time to gas sensors in the closed environment. By using *Papilioparis* butterfly wings as biotemplates, we herein demonstrate a facile way to synthesize biomass carbon dopedTiO_2 with the replication of quasi-honeycomb scales structures, which is beneficial to highest specific area ($85.27 m^2 \cdot g^{-1}$) in comparison with pure TiO_2 and Ag-doped TiO_2. The biomorphicC/TiO_2 exhibit not only excellent responses to benzene and dimethylbenzene vapors at 300 ℃ operating temperature superior to that of Ag-doped TiO_2, but also have excellent sensitivity to visible light. Furthermore, the multifunctional biomorphic C/TiO_2, used as safe concentration detectors, could determine vapors concentrations by gas sensing response values. Adoptingilluminating the photocatalysts in the closed environment, the responses to dimethylbenzene and benzenehave almost no changes along with continuous vapors injection. This work shows a good example for exploring the integrated application of semiconductor materials and improving the gas sensors lifetime.

Keywords: Butterfly wings; C/TiO_2; Gas sensor; Photocatalytic; VOCs

1 Introduction

Semiconductor gas sensors have been widely used for environmental monitoring in industrial production. Most attention of its researches is paidto lower the operating temperatures for gas detection and sensing[1-2]. However, in the practical application, the sensitivity of gas sensing system would be destroyed in the case of persistent leakage in a closed environment, which seriously affects the safety of rescue. That is, in the absence of ventilation, the gas sensors without recovery time have the sensitivity of gradual reduction. It is well known that semiconductor materials could be used as photocatalysts under suitable light irradiation and gas sensors at optimal operating temperature. Herein, we design a multifunctional material possessing both detecting and eliminating properties for hazardous compounds. Therefore, in such a closed environment, after gas sensing detecting for a while, illuminating the photocatalysts with suitable light is viable alternative to give

* 本文摘自 ACS Applied Nano Materials, 2018, 1(4): 1766-1775.

recovery time to gas sensors, which would significantlyreduce pernicious hurts to the human and improve device lifetime.

Moreover, recently, environment pollutants have drawn great attention due to their adverse effects on human health, especially the indoor volatile organic compounds(VOCs)pollutions. After years of effort, VOCs photocatalytic degradation is recognized as an effective method because it could convert solar energy to any other energy we need, and it would not produce secondary pollution in degradation process[3,4]. Moreover, due to the low-cost and easy-produce advantages of photocatalysts, the degradation of pollutants using photocatalysts is considered as one of the ideal ways to solve the pollution problems[5-7].

TiO_2 with bandgap of 3.2 eV is an important n-type semiconducting metal oxide, which has attracted a great dealof attention, due to its photocatalytic applications[8-11]. It is wellacceptable that photocatalytic process is based on the mechanisminvolving electron-hole separation and electron-hole recombination with light irradiation on semiconducting oxide[12,13]. For semiconductor photocatalysts, two aspects including structures and compositions are carried on to enhance photocatalytic activity and energy efficiency of photocatalysts[14,15]. According to these, reducing grain size, building hierarchical structure and doping with metal or nonmetal atoms are used widely for developing high active photocatalysts capable of using solar light to eliminate hazardous compounds. The grain size directly affects the specific surface area, that is, the smaller the grain size is, the larger specific surface area will be[16]. Then, the migration distance of electrons and holes from the photocatalysts interior to the photocatalysts surfaces would decrease, while the electron-hole recombination would also reduce. It is also considered that in the photocatalytic performance of some photocatalysts, hierarchical structures can enhance capture efficiency of incident light and provide more active sites[17,18]. In view of the modification of TiO_2 sensitive to visible light, one approach is to substitute Cu, Ni, or Fe etc. for Ti[19-21], and another is to introduce oxygen vacancy in TiO_2 for forming Ti^{3+} site[22-24].

Biomass carbon originating from nature has been found that could provide a convenient and renewable source for fabricating novel functional material with special structures and enhanced performances. For example, Sond et al. have synthesized seaweed carbon for enhanced electrocatalytic activity in the oxygen reduction reaction in alkaline medium compared with the commercial Pt catalyst[25]. He et al. have prepared cyanobacteriatemplated titania photocatalyst with higher visible-light photocatalytic performance than commercial Degussa P25 for Rhodamine B degradation[15]. Moreover, butterfly wings with special architectures of scales present charming properties, such as structural color[26,27], superhydrophobicity[28,29] and high selectivity to vapors[30,31]. Many scientists have applied the wings as biotemplates to replicate the 3D porous structures, which is beneficial to both the chemical vapor sensing properties and the photocatalytic degradation process.

In this work, we utilized *Papilioparis* butterfly wings as biotemplates to prepare a biomorphic C/TiO_2 through a simple hydrothermal synthesis and a preferred calcination process. The biomorphic C/TiO_2 composites were used as gas sensors at 300 ℃ operating temperature and photocatalysts for the degradation of benzene and dimethylbenzene vapors under visible light irradiation. In the case of persistent vapors leakage, combining with photocatalytic degradation can keep the gas sensing system sensitive. Furthermore, the safe gas concentrations of benzene and dimethylbenzene were tested by gas sensing system after photodegradation. Additionally, the biomorphic C/TiO_2 exhibited enhanced gas sensing properties relative to the state of metal Ag doped TiO_2, which demonstrates that biomorphic C/TiO_2 is a promising alternative to costly Ag-modified TiO_2 for gas sensors. Our study successfully offers a promisingfabrication strategy for the rational design of comparativelycheap, high performance multifunctional materials and then facilitates their practical applications inenvironmental issues, especially indoor air pollutions.

2 Materials and methods

2.1 Synthesis

The *Papilioparis* butterfly wings were borrowed from Shanghai Natural WildInsect Kingdom Co., Ltd. *Papilioparis* butterfly wings were pretreatedand washed with anhydrous ethanol, and dried in air for 12 h. All chemicals supplied by Shanghai Boyle Chemical Company were of analytical reagent-grade quality and used without further purification. Deionized water was used throughout the study.

The TiO_2 precursor was synthesized according to a procedure described in our previous paper[32]. In typical process, the ammonium fluorotitanate(0.4 M) and boric acid(1.2 M) were dissolved in the distilled water in a 500 mL glass containerat a room temperature with vigorous magnetic stirring. Asolution of 0.3 M hydrochloric acid was added until the pHvalue reached approximately 3. The pretreated butterfly wings fixed in the glass slides were immersed into TiO_2 precursor and transferred into a 100 mL Teflon container. The system was placed in an autoclave andkept at 90 ℃ for 5 h. After that, the treated butterfly wings were taken out, washed with deionized water, and dried overnight at room temperature. Thus, the TiO_2-treated wings were obtained. The hierarchical structure replication was realized through calcineingthe TiO_2-treated wings in air at 550 ℃ for 3 h. It is worthwhile to note that the brilliant blue regions in the butterfly wings were removed before calcination. Finally, the *Papilioparis*-carbon-TiO_2 air (denoted as C-T) was produced. The weight changes after calcination between the TiO_2-treated wings and the pure TiO_2 was calculated. Thus the excess weight was ascribed to the biomass C, and the dopant content of biomass C is approximately 6%. For comparison, the pure TiO_2 powder without *Papilioparis* butterfly wings was prepared using the same hydrothermal synthesis and calcination in air at 550 ℃ for 3 h (denoted as pure T). For illustrating the effect of biomass carbon, the noble metal (Ag) with the content of 6% doped TiO_2 powder without *Papilioparis* butterfly wings was prepared through a hydrothermal synthesis used as above and a mirror reaction, as well as calcination in air at 550 ℃ for 3 h (denoted as Ag-T). The mirror reaction was carried out according to our previous paper[33].

2.2 Characterizations

The morphologies and microstructures were characterized by field-emission scanning electron microscopy (FE-SEM, JSM-7500F, JEOL, Tokyo, Japan) operating at 12.5 kVin combination with energy dispersive spectroscopy (X-Max, Oxford Instruments, Abingdon, Oxfordshire, UK). The crystal structure of the as-prepared product was investigated by X-ray diffraction (D8Advance, Bruker, Billerica, MA, USA) with Cu Kα radiation of wavelength $\lambda = 1.5418$ Å, usinga step scan mode with the step size of 0.02° and a scan rate of 4° · min^{-1}, at 40 kV and 40 mA ranging from 5° to 80°. Further evidence for the composition of the product was inferred from the X-ray photoelectronspectroscopy (XPS, Thermo ESCALAB 250XI, USA), using an ESCA Lab MKII X-ray photoelectron spectrometer with Mg-Kα X-rays as the excitationsource. Specific surface areas of the prepared products were measured by the Brunauer-Emmett-Teller(BET) method based on N_2 adsorption at the liquidnitrogen temperature using a 3H-2000PS2 unit (Beishide Instrument S&T Co., Ltd). Optical properties were characterized by the UV-vis diffuse reflectance spectroscopy (TU-190, Beijing Purkinje General Instrument Co., Ltd., Beijing, China) equipped with an integrating sphere attachment, which $BaSO_4$ was the reference. Photoluminescence (PL) emission spectra were used to investigate the fate of photogenerated electrons and holes in the sample, and were recorded on a FluoroMax 4 fluorescence spectrometer (HORIBA Jobin Yvon Company, France). The excitation wavelength was 350 nm with the scanning speed of 600 nm · min^{-1}. The widths of both excitation slit and emission slit were 10 nm.

2.3 Gas sensor fabrication and measurements

Gas sensing of the samples was performed in a WS-30Astatic gas-sensing system (Zhengzhou Wei-Sheng Electronics Technology Co., Ltd., Henan, P. R. China) with the export voltage of 5 V. The sensors were fabricated as follows. Firstly, the samples were grinded with ethanol in an agate mortar to form the paste. A Ni-Cr heating wire was placed inside the component as a resistor to adjust the whole operating temperature of the gas sensors. Subsequently, the ceramic chip was pasted to the conductive pedestal and welded together with wire. All the sensors were aged at 300 ℃ for 3 days toimprove their stability. The measurement was processed by a static process in a test chamber (320 mm × 320 mm × 250 mm). The desired gases concentrationwas obtained byevaporating the certain volume of liquidsthrough a heater in the testing chamber. Then, the gas sensing measurement was set out. The response sensitivity, S, is determined as the ratio, $R_a/R_g \times 100\%$, where R_a is the resistance in airand R_g is the resistance in the tested gas atmosphere.

2.4 Tests of continuous detection of persistent VOCs vapors leakage using photocatalytic gas sensing system

The experiments were performed inthe obturator of the gas sensing system. Firstly, 0.5 g catalyst powders were uniformly spun on five glass slides and used in the photocatalytic degradation process under visible light irradiation. A 500 W Xe lamp was applied as an artificial solar lightsource, which was positioned on the top of the samples and outside the obturator. In the case of visible light (>420 nm) irradiation, a UV cutoff filter wasused.

For measuring gases concentrations, the 500 ppm dimethylbenzeneor benzene was injected into the obturator and thegas sensing system was set out for 60 seconds. Then, the photocatalytic degradation was performed by transporting the gas acrossthe samples continuously when Xe lamp was turned on, the process lasted for 30 minutes, and after each photodegradation process, the gas sensing measurement was carried for 60 seconds. For exploring the light effect on sensors, a baffle was settled in front of the sensors to avoid light irradiation, and the voltage-time curves from gas-sensing system were collected without photocatalysts after irradiation for 30 minutes.

For the continuous detection of persistent VOCs vapors leakage in the closed environment, 200 ppm dimethylbenzene or benzene was continuously injected after 60 secondsdetection by gas sensing system. To prove the effect of photocatalytic action, the photocatalysts are illuminated for 60 minutes under visible light during the interval time before new vapor is injected. For comparison, the same process is made without irradiation.

3 Results and discussion

As shown in Fig. 1a, the *Papilioparis* butterfly wings appear mostly black (the drawn blue region) and some brilliant blue (the drawn yellow region). There are abundant metallic green scales on their surfaces. Through the SEM observations, it could be seen that the wings have two kinds of scales structures, the surface scales and the basal scales. The metallic green scales are caused by the surface scales structure (Fig. 1b). The high magnification image in the inset of Fig. 1b shows that the surface scales consist of the parallel ridges andthe periodic latticework ofquasi-honeycomblike pores. The wide between the two adjacent ridges is about 3.66 μm and the average diameter of the pores is about 1.27 μm. Fig. 1c reveals that the different colors between the black region and the brilliant blue region are caused by the different arrangement of surface scales and basal scales. In the drawn yellow region of the wings, the basal scales covered onto the surface scales. After hydrothermal treating with TiO_2 (Fig. 1d), the uniform nanoparticlessettled onto the ridges and covered quasi-honeycomb like structures. From the high magnification image in the inset of Fig. 1d, it could be calculated that the average diameter of the nanoparticles is about 32.5 nm. An extremely good replication of the fine detail of the original scale structures is clearly seen in the*Papilioparis*-carbon-TiO_2 air(C-T) (Fig. 1e). The nanoscale rib ridges of the wings were both maintained in the sample after calcination in air at elevated temperature. The longitudinal ridges supported by cross-ribs are approximately 2.59 μm apart as shown in Fig. 1f. After calcination, the average diameter of the pores is approximately 85.1 nm as shown in the inset of Fig. 1f. This represents a slight shrinkage of scales structure during the calcinations. The particles became smaller and more homogeneous after calcination, and the average diameter of the TiO_2 particles is 29.87 nm. The element mapping results in Fig. 1g-i show that C, O and Ti homogenously distribute in the sample. The C distribution in element mapping is ascribed to the biomass carbon originating from the wings. For better illustrating the effect of biomass C and scales structure, the comparison was carried by preparing metal Ag doped TiO_2 (Ag-T). As shown in Fig. 1j, the particles of the Ag-T agglomerate as stones shapes of different sizes. Therefore, biomass carbon may act as a preferred support instead of metal Ag, to prevent TiO_2 particles from agglomeration. Additionally, EDS spectra of the samples at different stages (original butterfly wings, TiO_2-treated wings, C-T and Ag-T) are presented in Fig 1k-n. We can observe that the carbon content in the samples decreased in the process of preparing C-T. However, after calcination in air at 550 ℃ for 3 h (Fig 1m), there is still remaining carbon element.

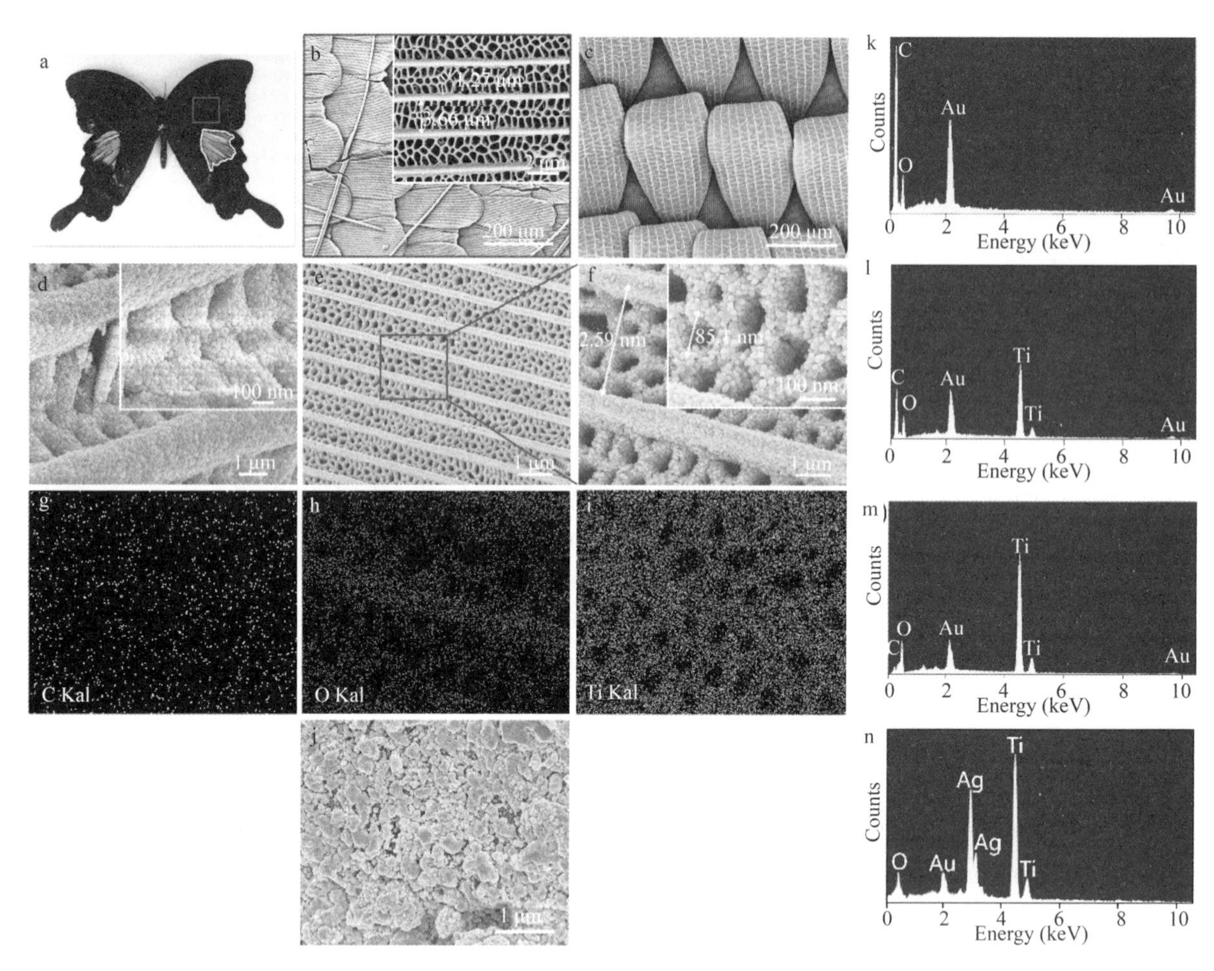

Fig. 1 (a) An overall view of the characterization of a *Papilioparis* butterfly. SEM images of (b) the drawn blue regions and (c) the drawn yellow regions in original butterfly wings, and (d) TiO_2-treated wings. (e) SEM image and (f) the high magnification images of C-T. (g-i) The elementmappings of C-T. (j) SEM image of Ag-T. (k-n) EDS spectra of original butterfly wings, TiO_2-treated wings, C-T and Ag-T

XRD patterns of the obtained samples are compared in Fig. 2. There appear three main peaks at 15.9°, 19.2° and 23.0° on the XRD pattern of the original wings in Fig. 2a. The 19.2° peak also appears in the TiO_2-treated wings (Fig. 2b), thusit can be considered that the chemicalcomponents of the original wings remain intact in the hydrothermal process. As shown in Fig. 2b-e, all the samples present the peaks of the pure anatase TiO_2 (JCPDS card no. 21-1272)[34]. This demonstrates the formation of TiO_2 crystallites in these samples. Besides, the additional peaks in Fig. 2e oftheAg-T are ascribed to the planes of Ag crystal (JCPDS card no. 04-0783)[35]. The average grain sizes of the TiO_2-treated wings, the C-T, the pure T and the Ag-T calculated via the Scherrer formula are 14.8 nm, 10.7 nm, 16.6 nm and 23.1 nm, respectively. Thisindicates that the existence of biomass carbonfavors the formation of TiO_2 with a smallest crystalline size, and the biomass carbon acted as a support assists the replication of butterfly wings, which suppresses thegrain growth. This could be ascribed to the effect of CO that was produced in the calcination process. CO could be easily obtained by oxidizing the carbon during calcination in air. And in the report of

Colón *et al.*, the existence of carbon in the samples is advantageous to decrease the crystallite sizes and increase the surface area of the samples, which is benefited to photocatalysts and gas sensors. Moreover, we have study the TiO_2 and WO_3 fabricated by wood fibers as templates, the results indicated that the carbon existed in the samples could decrease the crystallite sizes in the calcination process. It is worth noting that Fig. 2 fshows a magnified view of (101) peaks of C-T and pure T, which the curve of C-T exhibits a slightshiftof 0.14° to lower angles compared with that of pure T. This may be attributed to the lattice expansion dueto the incorporation of carbon atoms in the C-T, and alsoproves that biomass carbon may have a certain influence on the $d(101)$ spacing of the TiO_2.

The nitrogen adsorption-desorption performances are measured to illustrate the porous structures of C-T, pure T and Ag-T. As shown in Fig. 2g, all the samples have isotherms of type Ⅳ and H_3 type hysteresis loops in the relative pressure P/P_0 ranging from 0.18 to 1.0, indicating the presences of mesopores (2−50 nm) and slit-like pores. The pore size distribution of the C-T exhibits a broadened pore size range (Fig. 2h). The S_{BET} and average pore width of all samples are given in Table 1. The TiO_2 prepared by doping with Ag has alow surface area of 43.6 $m^2 \cdot g^{-1}$. When the butterfly wings are used as biotemplates, the surface area of the resulting C-T is increased to 85.3 $m^2 \cdot g^{-1}$. Both C-T and Ag-T have larger surface areas thanpure T (S_{BET} = 30.9 $m^2 \cdot g^{-1}$) due to the modification by doping C or Ag. In particular, the C-T have the largest pore volume of 0.14 $cm^3 \cdot g^{-1}$, which is ascribed to the replication of the quasi-honeycomb like and porous scales structure. Thus, the presence of the biomass C and the replicated scales structure in the C-T greatlyincrease the specific surface areas and broaden therange of mesopores sizes, which would provide more surface active sites for photocatalysis. In addition, the hollow pores structure provides channels for gas diffusion, and the mesoporous walls between TiO_2 nanoparticles promote the gas to better react with the inner particles. Hence, the replicated scales structure also play a crucial role on the improvement of gas sensitivity.

Table 1 The structure parameters of C-T, T and Ag-T

Sample	BET surface area ($m^2 \cdot g^{-1}$)	Pore size (nm)	Pores volume ($cm^3 \cdot g^{-1}$)
C-T	85.3	12.7	0.14
Pure T	30.9	17.3	0.06
Ag-T	43.6	10.6	0.05

The typical procedure for the fabrication of biomorphic C/TiO_2 with replicatedhierarchical structure of butterfly wings is depicted in Fig. 3a. Combining the necessary calcination process of preparing a gas sensor, the C-T was designed to achieve by calcineingthe TiO_2-treated wings in air at 550 ℃ for 3 h. Additionally, according to the above results, it can be concluded that the existence of biomass carbon in the calcination process is advantageous to decrease the grain sizes and increase the surface area of the samples. TheC-T with scales structure was mixedwith ethanol until a viscous state was formed, and then uniformly put on the surface of ceramic tube. Finally, the C-T sensors were fabricated with a Ni-Cr heating wire placed inside the component as a resistor to adjust the operating temperature (Fig. 3b).

In order to estimate the gas sensing capability to the two kinds of common indoor volatile organic compounds (VOCs), that is, benzene and dimethylbenzene, the gas sensing tests of C-T and Ag-T were carried out. Fig. 3c shows the operating temperaturedependences of benzene and dimethylbenzene (100 ppm)

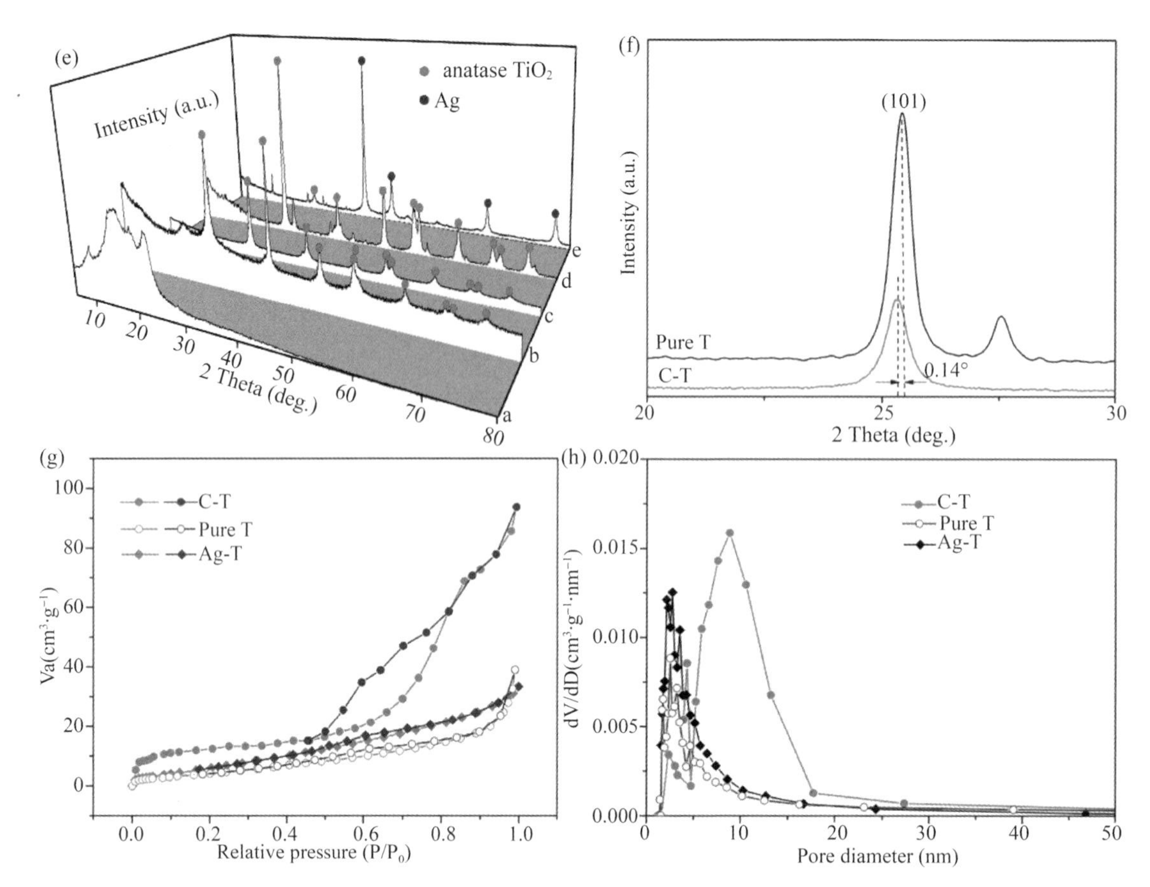

Fig. 2 XRD patterns of(a)original butterfly wings, (b)TiO_2-treated wings, (c)C-T, (d)pure T, and(e)Ag-T. (f)Magnifiedpeaks of(101)planes of C-T and pure T in the range from 20° to 30°. (g) Nitrogen adsorption-desorption isotherms and(h)the corresponding pore size distribution curves of C-T, pure T and Ag-T

response of the sensors based on C-T and Ag-T. The C-T sensors and the Ag-T sensors show dimethylbenzene responses as high as 10.4 and 4.1, and benzene responses as high as 5.0 and 4.2at the corresponding optimaloperating temperature(300 ℃), respectively. Thus, for excluding the effect of operating temperature on gas sensing properties, the tests should be operated at optimal operating temperature(300 ℃). It is also obvious that benzene and dimethylbenzene responsesof C-T sensors significantly exceed that of Ag-T sensors. This illustrates that in the case of same operating temperature and the same dopant content, the sensors fabricated by biomass C doped TiO_2 possess better gas sensitivities to benzene and dimethylbenzene. That is, for improving the gas sensing properties of semiconductor sensors, it is in favor of doping with biomass C rather than metal Ag.

Fig. 3d displays the real-time sensing response to benzene and dimethylbenzene of different concentrations (at 300 ℃), respectively. All the sensors show reversible responses to benzene and dimethylbenzene with rapid response and recovery capabilities, while the responses of C-T sensors to benzene and dimethylbenzene are higher. Fig. 3e shows the response variation of the sensors exposed to benzene and dimethylbenzene of different concentrations (at 300 ℃), respectively. Compared with the responses of metal Ag doped TiO_2

sensors to benzene and dimethylbenzene, the corresponding values of biomass C doped TiO_2 with biomorphic structure is superior. In 1991, Yamazoe et al. has proposed that the sensing characteristic of a semiconductor gas sensor is deeply related with the grain size of the semiconductor(D) and the thickness of the surface space-charge layer(L)[36]. And the thickness of the surface space-charge layer is determined by the surface charge and the Debye length(L_d). For n-type semiconductors, the electrons concentration is more dominant than the holes concentration. As the smallest grain size of 10.7 nm for TiO_2 in C-T is comparable to $2L_d$(as suggested from the reported L_d value, 3.5 nm[37]), the results that C-T sensor shows highest sensitivity to the gases is in line with the above reports. Meanwhile, in the case of replicating with scales structure of the wings, the quasi-honeycomb like pores between the parallel ridges provide more chances for the gas to diffuse into the sensors interior. Based on the above results, it could be concluded that the high response sensitivities are attributed to the small grain size of TiO_2 nanoparticles and the replicated porous scales structure.

It can also be observed that the high responses of C-T sensorsincrease nearly linearly with the concentrations increases in both benzene and dimethylbenzene in the case of the vapors concentrations below 100 ppm. For C-T sensors(operating temperature was 300 ℃), the relations equations between the responses (S) and the gas concentrations(C) are depicted as follows:

$$S = 0.053C + 0.954 \text{(for dimethylbenzene gas)}$$

$$S = 0.043C + 1.420 \text{(for benzene gas)}$$

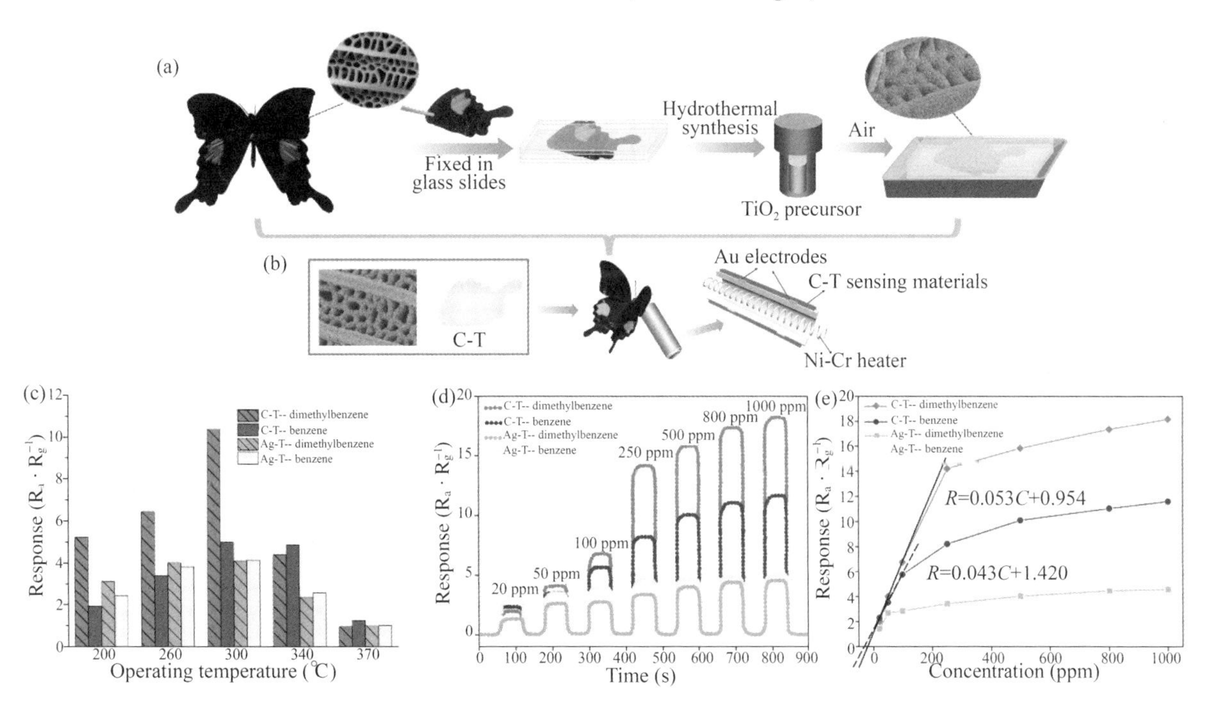

Fig. 3 Illustration depicting the synthesis process of (a) biomorphic C/TiO_2 replica with hierarchical structure of butterfly wings and (b) C-T sensors. (c) Temperature dependences of gas responses of C-T sensors airand Ag-T sensors, respectively (vapor concentration was 100 ppm, relative humidity (RH), 10%). (d) Real-time sensing responses and (e) response variations of C-T sensors and Ag-T sensors to benzene and dimethylbenzene of different concentrations, respectively (operating temperature was 300 ℃, relative humidity (RH), 10%)

To confirm the existence form of C and its effect to support the mechanisms of photocatalytic property and gas sensing performance, XPS is employed to get the elements information in C-T compared with that in pure T, and the changes of elements after irradiation for 60s. As shown in Fig. 4a, the XPS spectrum of Ti $2p_{3/2}$ in pure T is located at 458. 5 eV, while there is a red-shift of 0. 4 eV in C-T. With biomass C doped, the full width at half-maximum(FWHM) of the Ti $2p_{3/2}$ peak centered at 458. 9 eV increases, indicating the creation of Ti^{3+}[38]. In general, tosatisfy the requirement of charge equilibrium, the oxygen vacancies must exist around Ti^{3+} in C-T, and thereby the surface oxygen vacancies are formedon thetitania surface[39-41].

The peaks at 458. 6 eV and 464. 1 eV clearly indicate the presence of Ti^{3+}, and the peaks at 459. 0 eV and 464. 9 eV are assigned to Ti^{4+}, which the Ti^{3+}/Ti^{4+} ratio is about 58% calculated from the peak areas. Both the sensing mechanism and the photocatalytic mechanism of the present C-T can be attributed to a partial reduction of Ti^{4+} in the titania film to Ti^{3+}. Moreover, the Ti 2p spectrum of C-T after irradiation is simulated with Gaussian simulation as shown in Fig. 4b. The peakat 457. 3 eV with FEHM of 3. 3 eV is attributed to Ti^{3+}, and the peak at 459. 3 eV with FEHM of 1. 24 eV is attributed to Ti^{4+}, respectively, which the peak ofTi^{3+} became much broader than Ti^{3+} peaks in pure T and C-T without irradiation. TheTi^{3+}/Ti^{4+} ratio is increased to 75% after visible light irradiation. This also verifies that Ti^{3+} does exist in C-T, and the increase of $Ti^{3+}$when it is irradiated under visible light.

From the C1s XPS spectra in Fig. 4c, for C-T, peaksat 284. 2, 286. 3, and 288. 8 eVare found. The peak at 284. 2 eV that is also observed in C-T after irradiation should be ascribed to the ambient organic impurities, while the peaks at 286. 3 eV and 288. 8 eVsuggest the presence of carbonate species. A XPS peak assignedto the carbonate species was also observed after visible light irradiation. C 1s peak ascribed to Ti-C bond with a low binding energy of 281. 8 eV cannot be found in the sample[24]. Therefore, the biomass carbon exists in form of carbonate as interstitial dopant. The O1s spectra for the C-T can befitted into twopeakswith binding energies of about 529. 6 eV and 530. 1 eV(Fig. 4d). The first peak arises from lattice oxygen(Ti-O), and the latter peakis attributed to the oxygen in the carbonate[C-OOR(H), C-OR(H), C=O]. After irradiation, the binding energy of the peak drifts by 0. 9 eV, which is related with the creation of oxygen vacancies as a result of the carriers capture excited by visible light.

For observing light sensitivities of each sample, the UV-vis absorption spectra and the corresponding band gapscalculated by Tauc plot $[(\alpha h\nu)^{1/2} = A(h\nu - E_g)]$ are presented in Fig. 4e and f. As shown in Fig. 4e, compared with pure T, the main absorption edge of C-T shifts to higher wavenumber. This suggests that doping with biomass Ccan indeed extend the absorbance into the visible light range. It can also be seen from the band gap energy in Fig. 4f. The band gapenergy of pure T is measured to be about 3. 15 eV, and the band gapenergy of C-T is about 2. 25 eV. The maximum band-gap narrowing of 0. 9 eV iscomparable with the value of 0. 14eV observed for carbon-doped TiO_2 reported by Sakthivel et al.[42] Therefore, the C-T powders are subsequently used as photocatalysts in the photodegradation process.

From the PL spectra of pure Tand C-T in Fig. 4g, The PLemission intensity of C-T at 390 nm is lower than that of pure T, which indicated that therecombination rate of photogenerated electrons and holes is inhibited considerably in C-T. This is due to theformation of oxygen vacancies, which actually served as electron capture traps, andhence separated the charge carriers and reduced therecombination significantly.

Based on the results presented above, a possible mechanism for photodegradation under visible light

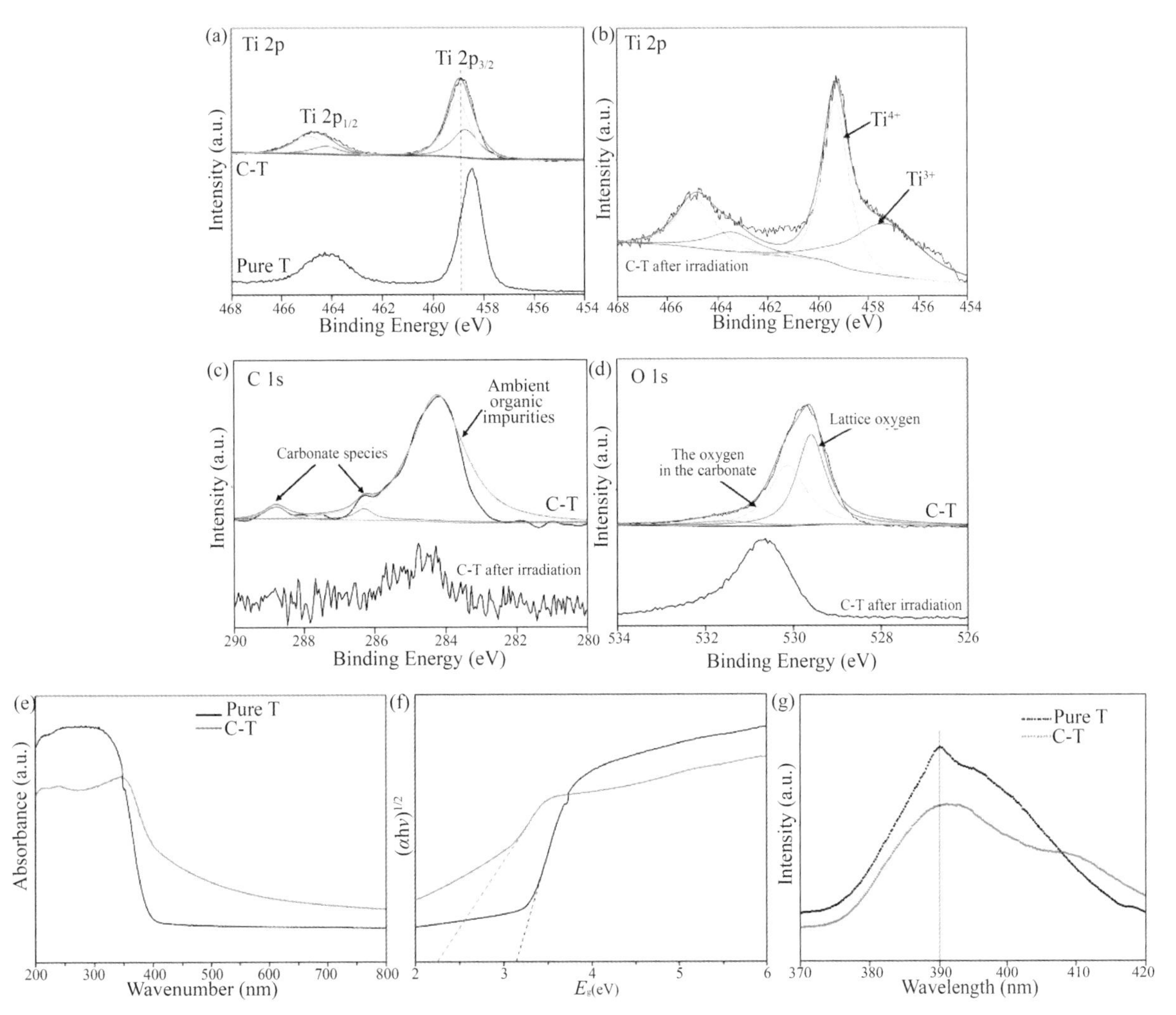

Fig. 4 (a) Ti 2p XPS spectra of pure T and C-T, (b) the peaking-fitting result of Ti 2p XPS spectrum of C-Tafter visible light irradiation for 60s, (c) C 1s and (d) O1s XPS spectra of C-T and C-T after irradiation, respectively. (e) UV-vis absorption spectra of pure T and C-T, and (f) the corresponding evaluation of the optical band gap using the Tauc plot. (g) PL spectra of pure T and C T

irradiation and gas sensing performances of VOCs (benzene and dimethylbenzene) is illustrated in Scheme 1. The enhanced photocatalytic performance of C-T under visible light is attributed to the formation of oxygen vacancy state (V_O) in the biomass C doped TiO_2 between the valenceand the conduction bands in the TiO_2 band structure. In the degradation of VOCs, the photoexcited carriers are produced under visible light irradiation. The superoxide radical ($\cdot O_2^-$) is produced by the oxidation reaction between the electrons and O_2. The generated $\cdot O_2^-$ after protonation would further react with H_2O_2 and form thehydroxyl radical ($\cdot OH$). Meanwhile, the holes react with hydroxyl groups derived from the adsorbed H_2O and produce more $\cdot OH$ radicals. The $\cdot OH$ radicals play the key role in degradation of benzene and dimethylbenzene, which would react with pollutants and transform them into CO_2 and H_2O (eqn. 2 and 3).

$$\text{C-T} + h\nu \rightarrow e^- + h^+ \tag{1}$$

$$30 \cdot OH + C_6H_6 \rightarrow 6CO_2 + 18H_2O \quad (2)$$

$$42 \cdot OH + C_8H_{10} \rightarrow 8CO_2 + 26H_2O \quad (3)$$

For pure stoichiometric TiO_2, there are no electrons on its conduction band, lowest unoccupied molecular orbital (the bottom of conduction band) is ascribed to Ti^{4+} 3d orbital electrons, while highest occupied molecular orbital (the top of valence band) is ascribed to O^{2-} 2p orbital electrons. The energy level distance (i. e. band gap) between the conduction band and the valence band full of electrons is 3.2 eV. Therefore, pure stoichiometric TiO_2 exhibits high resistance at room temperature[43,44]. However, with the raise of temperature, to satisfy the requirement of thermal equilibrium condition, the oxygen vacancies defects (V_O) and the electrons would be produced in TiO_2 according to eqn. 4. The V_O represents an empty position originating from the removals of O^{2-} in the lattice.

$$2Ti^{4+} + O^{2-} \rightarrow 4Ti^{4+} + 2e^-/V_O + 0.5O_2 \quad (4)$$

Under operating temperature conditions, the electrons from sensors are trapped by the adsorbed oxygen to form the chemisorbed oxygen ions (O_2^-, O^- and O^{2-}) on the surfaces, while the oxygen is mainly in the form of O^- when the operating temperature is 300 ℃.

As shown in Scheme. 1, when C-T sensor is exposed to the reducing gases (benzene or dimethylbenzene gases), the adsorbed O^- species would react with the gases molecules, and the reduction removes charged oxygen ion species from the oxide surface and that decreases the depleting layer width, which, in turn, leads to a wider conductive channel and higher conductivity. As a result, C-T sensors exhibit excellent gas sensitivity.

Moreover, both the porous architecture and the grain size are responsible for the sensing capability. While C-T replicated the well-organized porous frameworks of the butterfly wings, the connective hollow pores provide abundant channels for gas diffusion, and then the gas could contact with the inner grains swimmingly through molecular diffusion. Thus, more reactions between the gases and the adsorbed O^- species on the grains surfaces would give rise to the increase of response sensitivities.

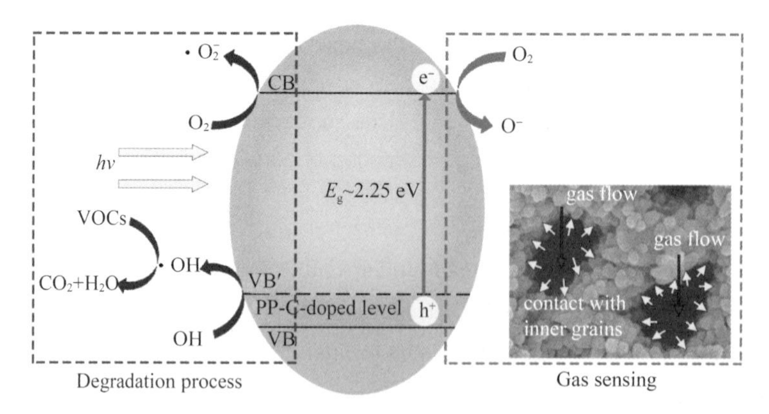

Fig. 5 The mechanism proposed for the photocatalytic degradation and gas sensing performances of C-T

The work is not to study the light irradiation effect on the response behavior of C-TiO_2 sensor to VOCs, but is to study the light irradiation effect on VOCs concentration. The VOCs concentration without continuous increase can make the C-TiO_2 sensor keep sensitive. That is, the photocatalysis can indirectly protect the sensitivity of gas sensors in the case of persistent leakage in a closed environment. Therefore, for excluding the light effect on semiconductor sensors, a baffle was settled in front of the sensors to avoid light irradiation, and the test was carried out without photocatalysts. In Fig. 5a, the voltage-time curves of PP-C-TiO_2 air sensor to 500 ppm benzene and dimethylbenzene(the blue region and the pink region)was presented. In the test without light irradiation(black lines), it could be seen that voltage-time curves had no changes. And during the test with light irradiation, the light was set up after 120s and meanwhile the gas sensing system was paused, the irradiation lasted for 30 minutes, finally after 30 minutes irradiation, the gas sensing tests were continuous. In the test with irradiation (red lines), we could observe that for the vapors, the voltage response values decrease to initial values after 30 minutes irradiation. Therefore, the results indicated that the light make no difference on gas responses due to the no changes of voltage-time curves without light irradiation(black lines). Moreover, Fig. 5b presents the voltage-time curves of PP-C-TiO_2 air sensor to residual gases after irradiation with photocatalysts for 30 minutes, and the corresponding responses. Based on the results of Fig. 3 that C-T sensor shows highest sensitivity to gases, we focus on the calculation of residual gas concentration after photodegradation using C-T sensor. We can get the residual gas concentration based on the corresponding responsesaccording to the relations between the responses and the gas concentrations given in Fig. 3e. In Fig. 5b, the responses of the sensor to the residual dimethylbenzene and benzene are 2.6 and 2.5, respectively. It could be calculated that the residual dimethylbenzene and benzene are 31.1 ppm and 25.1 ppm, respectively.

The continuous detection of persistent VOCs vapors leakage was carried on through injecting 200 ppm dimethylbenzene or benzene after 60 minutes irradiation(Fig. 5d), which was performed in the same method without irradiation(Fig. 5c). It is worth noting that the irradiation lasted for 60 minutes is to ensure complete degradation of vapors. Obviously, as shown in Fig. 5c, in the case of without irradiation, the responses to dimethylbenzene and benzenedecrease from 2.84 to 1.02 and from 1.57 to 1.08, respectively, along with the continuous vapors injection. This better verifies that the sensitivity of gas sensing system would be destroyed when there is persistent vapors leakage in practical application. To address the problem, we try to adopt the photocatalytic technology in gas sensing detection. In the case of with irradiation(Fig. 5d), the responses to dimethylbenzene and benzenehave almost no changes along with the continuous vapors injection.

4 Conclusions

In summary, biomass C doped TiO_2 with hierarchical structures was fabricated by a facile hydrothermal synthesis and a calcination process, which enables the control of crystal phase and morphology, and the dopant of carbon originating from *Papilioparis* butterfly wings. The perfect replication of quasi-honeycomb scales structures in C-Tis responsible for the high special surface area and the small grain size. The biomass carbon in form of carbonate has been well introduced into the products as interstitial dopant, and the dopant content of biomass C was approximately 6%. The effects of the biotemplates and the biomass C dopant on structures and properties were investigated in comparison with the state of the Ag nanoparticles doped TiO_2. From the SEM, XRD and BET results, it could be concluded that the existence of biomass C can suppress the grain growth, prevent TiO_2 particles from agglomeration, and enhance the special surface area up to 85.27 $m^2 \cdot g^{-1}$. These

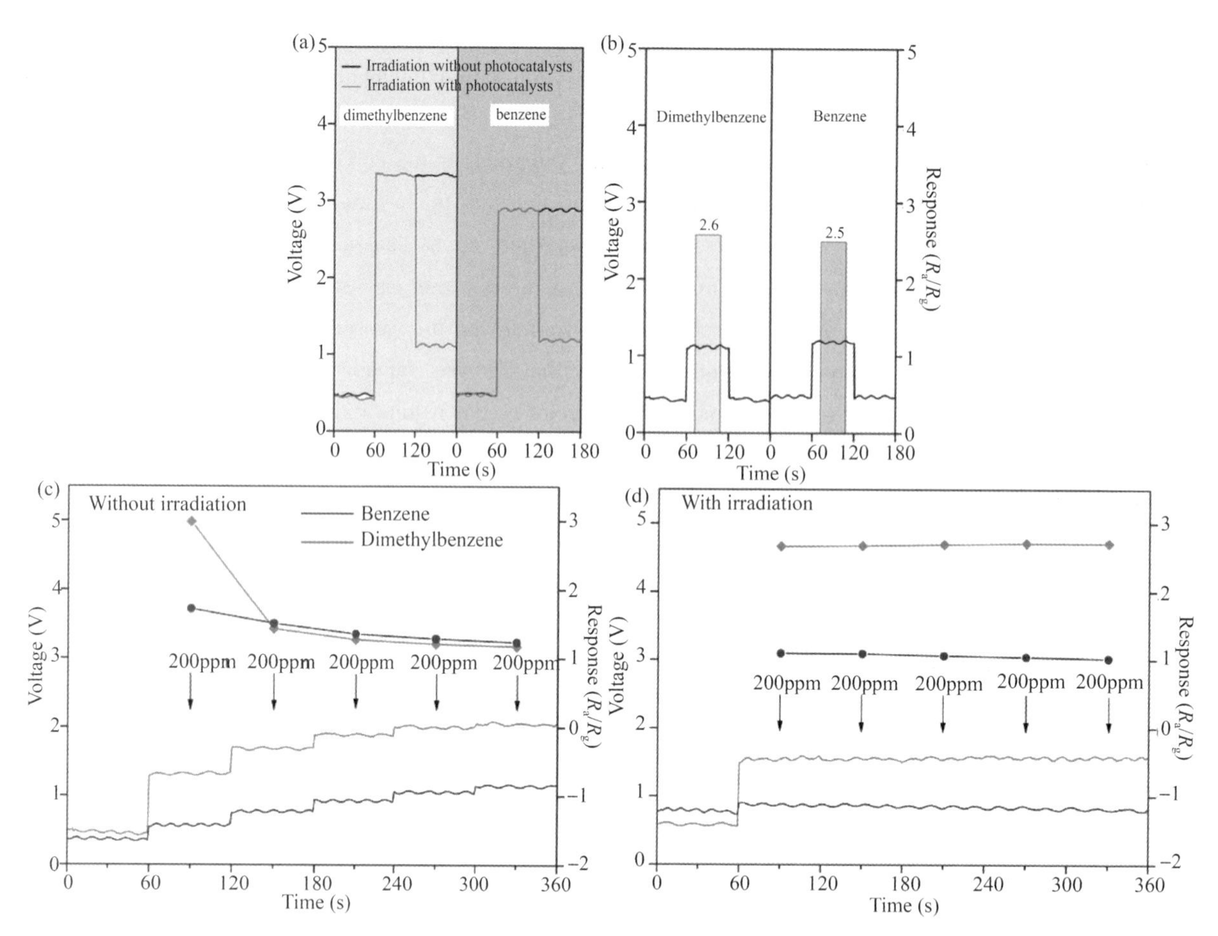

Fig. 6 (a) Voltage-time curvesof C-T sensor to 500 ppm benzene and dimethylbenzene before irradiation, after irradiation without photocatalysts, and after irradiation with photocatalysts for 30 minutes, respectively. (b) Voltage-time curvesof C-T sensor toresidual gases after irradiation with photocatalysts for 30 minutes, and the corresponding responses. (c, d) Voltage-time curves and response variations of C-T sensor during continuous detection of persistent injection of 200 ppm benzene and dimethylbenzene without irradiation and with irradiation, respectively(Operating temperature was 300 ℃, relative humidity(RH) was 10%)

illustrate that the doping of biomass carbon is a promising alternative to the costly doping ofmetal Ag. Moreover, the XPS results proved the creation of Ti^{3+} ions and the formation of surface oxygen vacancies on the titania surfacewhen doped with the biomass C, as well as the increase of Ti^{3+} in C-Twhen it is irradiated under visible light. Oxygen vacancydefectsplay the key role in both sensing and photocatalytic properties. Oxygen vacancy defects would act as electron scavengers in the photocatalysis, which separate the charge carriers and reduce the electron-hole recombination. For photocatalytic process, the longer separation time is favorable for producing more superoxide radical($\cdot O_2^-$) and hydroxyl radical($\cdot OH$), which would react with pollutants and transform them into CO_2 and H_2O. For gas sensing properties, the operating temperature provides condition for TiO_2 to produce oxygen vacancydefects and electrons. Chemisorbed oxygen ions(O^-) are formed by reaction between adsorbed oxygen molecules and electrons. When the sensors were exposed to reductive gas vapors (benzene and dimethylbenzene), the vapors could react with O^- ions, which is accompanied by the increasing electron concentration. As a result, the C-T sensors present high gas responses due to the decrease of resistance.

Furthermore, we imitated a safe vapor concentration detecting, which the as-prepared biomorphic C/TiO_2 was used as both gas sensors and photocatalysts. The vapors concentrations can be estimated by the gas sensing response values according to the derived relations equations. In the process of gas sensing detection, we found that the sensitivity of gas sensing system would be destroyed when there is persistent vapors leakage in the closed environment. Thus, the photocatalytic technology is adopted in gas sensing detection. The results illustrated that in the case of with irradiation the responses to dimethylbenzene and benzene have almost no changes along with the continuous vapors injection. That is, the photocatalytic action could play a protective role in the gas sensing system.

The aim of the study is to present the effect of biomass carbon from butterfly wings. We tried to compare the effect of biomass C dopant and metal dopant. Because biomass C is more environmental-friendly and more renewable than metals. And we tried to explain that the materials with biomass C doped could have higher performances(such as photocatalysis and gas sensors) than that with metal dopant. Among the metals that were widely used in the modification of TiO_2, Ag was more easy and cheaper to synthetize. However, it would make much more sense to compare two carbon-doped TiO_2 samples. Therefore, further study will be taken on to explore the effects of different carbon resource and different templates.

Acknowledgments

This work was financially supported by the National Natural Science Foundation of China (grant no. 31470584) and the Fundamental Research Funds for the Central Universities(grant no. 2572017AB08).

References

[1]Prades J D, Jiménez-Díaz, R, Hernandez-Ramirez, F, et al. Equivalence between Thermal and Room Temperature UV Light-Modulated Responses of Gas Sensors Based on Individual SnO_2 Nanowires[J]. Sensor. Actuat. B-Chem., 2009, 140: 337-341.

[2]Wang X F, Ma W, Jiang F, et al. Prussian Blue Analogue Derived Porous $NiFe_2O_4$ Nanocubes for Low-Concentration Acetone Sensing at Low Working Temperature[J]. Chem. Eng. J., 2018, 338: 504-512.

[3]Chaturvedi S, Dave P N. Environmental Application of Photocatalysis[J]//Materials Science Forum. Trans Tech Publicationsv, 2013, 734: 273-294.

[4]Ma S, Reish M E, Zhang Z, et al. Anatase-Selective Photoluminescence Spectroscopy of P25 TiO_2 Nanoparticles: Different Effects of Oxygen Adsorption on the Band Bending of Anatase[J]. J. Phys. Chem. C., 2017, 121: 1263-1271.

[5]Lyulyukin M N, Kolinko P A, Selishchev D S, et al. Hygienic Aspects of TiO_2-Mediated Photocatalytic Oxidation of Volatile Organic Compounds: Air purification analysis using a total hazard index[J]. Appl. Catal. B-Environ., 2018, 220: 386-396.

[6] Selishchev D S, Kolobov N S, Pershin A A, et al. TiO_2 Mediated Photocatalytic Oxidation of Volatile Organic Compounds: Formation of CO as a Harmful By-Product[J]. Appl. Catal. B-Environ., 2017, 200: 503-513.

[7]Maira A J, Yeung K L, Soria J, et al. Gas-Phase Photo-Oxidation of Toluene Using Nanometer-Size TiO_2 Catalysts [J]. Appl. Catal. B-Environ., 2001, 29: 327-336.

[8] Wang Z, Peng X, Huang C, et al. CO Gas Sensitivity and Its Oxidation over TiO_2 Modified by PANI under UV Irradiation at Room Temperature[J]. Appl. Catal. B-Environ., 2017, 219: 379-390.

[9]Peng X, He Z, Yang K, et al. Correlation between Donating or Accepting Electron Behavior of the Adsorbed CO or H_2 and Its Oxidation over TiO_2 under Ultraviolet Light Irradiation[J]. Appl. Surf. Sci., 2016, 360: 698-706.

[10]Lui G, Liao J Y, Duan A, et al. Graphene-Wrapped Hierarchical TiO_2 Nanoflower Composites with Enhanced Photocatalytic Performance[J]. J. Mater. Chem. A, 2013, 1: 12255-12262.

[11]Yang H G, Sun C H, Qiao S Z, et al. Anatase TiO_2 Single Crystals with A Large Percentage of Reactive Facets[J]. Nature, 2006, 453: 638.

[12]Carp O, Huisman C L, Reller A. Photoinduced Reactivity of Titanium Dioxide[J]. Prog. Solid State Ch., 2004, 32: 33-177.

[13]Hoffmann M R, Martin S T, Choi W, et al. Environmental Applications of Semiconductor Photocatalysis[J]. Chem. Rev., 1995, 95: 69-96.

[14]Shi M, Wei W, Jiang Z, et al. Biomass-Derived Multifunctional TiO_2/Carbonaceous Aerogel Composite as A Highly Efficient Photocatalyst[J]. Rsc. Adv., 2016, 6: 25255-25266.

[15]He J, Zi G, Yan Z, et al. Biogenic C-Doped Titania Templated by Cyanobacteria for Visible-Light Photocatalytic Degradation of Rhodamine B[J]. J. Environ. Sci-China, 2014, 26: 1195-1202.

[16]Yin C, Zhu S, Chen Z, et al. One Step Fabrication of C-Doped $BiVO_4$ with Hierarchical Structures for A High-Performance Photocatalyst under Visible Light Irradiation[J]. J. Mater. Chem. A, 2013, 1: 8367-8378.

[17]Yao F, Yang Q, Yin C, et al. Biomimetic Bi_2WO_6 with Hierarchical Structures from Butterfly Wings for Visible Light Absorption[J]. Mater. Lett., 2012, 77: 21-24.

[18]Zhang W, Zhang D, Fan T, et al. Novel Photoanode Structure Templated from Butterfly Wing Scales[J]. Chem. Mater., 2009, 21: 33-40.

[19]Pham T D, Lee B K. Advanced Removal of C. Famata in Bioaerosols by Simultaneous Adsorption and Photocatalytic Oxidation of Cu-Doped TiO_2/PU under Visible Irradiation[J]. Chem. Eng. J., 2016, 286: 377-386.

[20]Khan R, Kim T. J. Preparation and Application of Visible-LighResponsive Ni-Doped and SnO_2-Coupled TiO_2 Nanocomposite Photocatalysts[J]. J. Hazard. Mater., 2009, 163: 1179-1184.

[21]Zhu J, Ren J, Huo Y, et al. Nanocrystalline Fe/TiO_2 Visible Photocatalyst with A Mesoporous Structure Prepared via A Nonhydrolytic Sol-Gel Route[J]. J. Phys. Chem. C, 2007, 111: 18965-18969.

[22]Hu X, Zhang T, Jin Z, et al. Fabrication of Carbon-Modified TiO_2 Nanotube Arrays and Their Photocatalytic Activity [J]. Mater. Lett., 2008, 62: 4579-4581.

[23]Valentin C D, Pacchioni G, Selloni A. Theory of Carbon Doping of Titanium Dioxide[J]. Chem. Mater., 2005, 17: 6656-6665.

[24]Irie H, Watanabe Y, Hashimoto K. Carbon-Doped Anatase TiO_2 Powders as A Visible-Light Sensitive Photocatalyst [J]. Chem. Lett., 2003, 32: 772-773.

[25]Song M Y, Park H Y, Yang D S, et al. Seaweed-Derived Heteroatom-Doped Highly Porous Carbon as An Electrocatalyst for The Oxygen Reduction Reaction[J]. Chemsuschem, 2014, 7: 1755.

[26]Kolle M, Salgard-Cunha P M, Scherer M R, et al. Mimicking The Colourful Wing Scale Structure of The Papilio Blumei Butterfly[J]. Nat. Nanotechnol., 2010, 5: 511-515.

[27]Zhang D, Zhang W, Gu J, et al. Inspiration from Butterfly and Moth Wing Scales: Characterization, Modeling, and Fabrication[J]. Prog. Mater. Sci., 2015, 68: 67-96.

[28]Fang Y, Sun G, Bi Y, et al. Multiple-Dimensional Micro/Nano Structural Models for Hydrophobicity of Butterfly Wing Surfaces and Coupling Mechanism[J]. Sci. Bull., 2015, 60: 256-263.

[29]Liu F, Liu Y, Huang L, et al. Replication of Homologous Optical and Hydrophobic Features by Templating Wings of Butterflies Morpho Menelaus[J]. Opt. Commun., 2011, 284: 2376-2381.

[30]Wu W, Liao G, Shi T, et al. The Relationship of Selective Surrounding Response and The Nanophotonic Structures of

Morpho Butterfly Scales[J]. Microelectron. Eng., 2012, 95: 42-48.

[31]Zhang J, Liang Y, Mao J, et al. 3D Microporous Co_3O_4-Carbon Hybrids Biotemplated from Butterfly Wings as High Performance VOCs Gas Sensor[J]. Sensor. Actuat. B-Chem., 2016, 235: 420-431.

[32]Gao L, Zhan X, Lu Y, et al. pH-Dependent Structure and Wettability of TiO_2-Based Wood Surface[J]. Mater. Lett., 2015, 142: 217-220.

[33]Gao L, Lu Y, et al. Superhydrophobic Conductive Wood with Oil Repellency Obtained by Coating with Silver Nanoparticles Modified by Fluoroalkyl Silane[J]. Holzforschung, 2015, 70: 63-68.

[34]Gao L, Gan W, Zhe Q, et al. Preparation of Heterostructured WO_3/TiO_2 Catalysts from Wood Fbers and Its Versatile Photodegradation Abilities[J]. Sci. Rep., 2017, 7: 1102.

[35]Gao L, Gan W, Xiao S, et al. Enhancement of Photo-catalytic Degradation of Formaldehyde through Loading Anatase TiO_2 and Silver Nanoparticle Films on Wood Substrates[J]. Rsc. Adv., 2015, 5: 52985-52992.

[36]Yamazoe N. New Approaches for Improving Semiconductor Gas Sensors[J]. Sensor. Actuat. B-Chem., 1991, 5: 7-19.

[37]Wilson J. N, Idriss H. Effect of Surface Reconstruction of TiO_2(001) Single Crystal on The Photoreaction of Acetic Acid[J]. J. Catal., 2003, 214: 46-52.

[38]Du X, Wang Y, Mu Y, et al. A New Highly Selective H_2 Sensor Based on TiO_2/PtO-Pt Dual-Layer Films[J]. Chem. Mater., 2002, 33: 3953-3957.

[39]Jing L, Xin B, Yuan F, et al. Effects of Surface Oxygen Vacancies on Photophysical and Photochemical Processes of Zn-Doped TiO_2 Nanoparticles and Their Relationships[J]. J. Phys. Chem. B, 2006, 110: 17860-17865.

[40]Jiang X, Zhang Y, Jiang J, et al. Characterization of Oxygen Vacancy Associates within Hydrogenated TiO_2: A Positron Annihilation Study[J]. J. Phys. Chem. C, 2012, 116: 22619-22624.

[41]Liu C, Xian H, Jiang Z, et al. Insight into The Improvement Effect of The Ce Doping into The SnO_2 Catalyst for The Catalytic Combustion of Methane[J]. Appl. Catal. B-Environ., 2015, s 176-177: 542-552.

[42]Sakthivel S, Kisch H. Daylight Photocatalysis by Carbon-Modified Titanium Dioxide[J]. Angew. Chem. Int. Edit., 2003, 42: 4908-4911.

[43]Blumenthal R N, Baukus J, Hirthe W M. Studies of The Defect Structure of Nonstoichiometric Rutile, TiO_{2-x}[J]. J. Electrochem. Soc., 1967, 114: 172-176.

[44]Kofstad P. Nonstoichiometry, Diffusion, and Electrical Conductivity in Binary Metal Oxides[J]. New York: Wiley-Interscience., 1972, 155.

中文题目：仿生蝶翅结构制备生物质衍生碳掺杂 TiO_2 及光催化气敏系统的研究

作者：高丽坤，甘文涛，邱哲，曹国良，詹先旭，强添刚，李坚

摘要：当封闭环境中气体持续泄露时，如何保护气敏传感系统的灵敏性？本研究将半导体材料的气敏特性与光催化性能相结合，使之在封闭环境中为气敏元件的恢复时间提供可行性选择。以巴黎凤尾蝶蝶翅为生物质模板，成功仿生复制了蝶翅鳞片的蜂窝状结构，制备了生物质衍生碳掺杂的二氧化钛，相比于单纯的二氧化钛和银掺杂的二氧化钛，该样品具有较高的比表面积(85.27 $m^2 \cdot g^{-1}$)。相比于银掺杂的二氧化钛，仿生的碳/二氧化钛不仅在300 ℃的操作温度下对苯和二甲苯气体具有较高的灵敏度，而且对可见光具有优异的响应性。此外，该多功能性的仿生碳/二氧化钛作为安全浓度探测器，可通过气体响应值来确定气体的浓度。当在封闭环境中光照射光催化剂时，气敏元件对苯和二甲苯的灵敏度随着气体的不断注入几乎无变化。此研究为探索半导体材料的集成应用和提高气敏元件的使用寿命提供了新的思路。

关键词：蝶翅；碳/二氧化钛；气敏元件；光催化；挥发性有机化合物

Fabrication of A Superamphiphobic Surface on the Bamboo Substrate*

Wenhui Bao, Zhen Jia, Liping Cai, Daxin Liang, Jian Li

Abstract: The superamphiphobic-functionalized CuO microflowers/$Cu(OH)_2$ nanorod arrays hierarchicalstructure was prepared on the bamboo surface as a rough coating via a facile alkali assistant surface oxidation technique. Thereafter, the long-hydrophobic groups were grafted onto the bamboo to obtain superamphiphobic surfaces, which were super-repellent for water and hexadecane. The morphologies, microstructures, crystal structure, chemical compositions and states, as well as the hydrophobicity of the films on the bamboo substrates were analyzed by means of the scanning electron microscopy (SEM), X-ray diffraction (XRD), X-ray photoelectron spectroscopy (XPS) and water contact angle measurement. Additionally, the coating showed excellent environmentalstability and high conductivity. It is expected that the work mightexpand the application of the superamphiphobic bamboo products.

Keywords: Room-temperature; Wood surface; Energy; Superhydrophilicity

1 Introduction

Wettability and repellency are important properties of a solid surface from both fundamental research and practical application aspects. Surface energyand rough structure are the two key ingredients inoptimizing thewettability of a solid surface (Nishimoto and Bhushan, 2013). Inspired by the lotus leaf, superhydrophobic surfaces exhibit great water-repellent properties with watercontact angles of greater than 150°. The water-repellent surface on the biologic material is an effective measure to protect exteriordeterioration. However, the superhydrophobic surface does not have hydrophobic property when the surface tension of liquid is low, such ashexadecane ($27.5\ mN \cdot m^{-1}$). Based on this, superoleophobic surfacesare defined as that the resisting wetting of liquidshave the surface tension lower than $35\ mN \cdot m^{-1}$. Surfaces that exhibit both water-repellent and oil-repellent properties on the same substrate have attracted muchinterest because of potential practical applications, including the crude oil transfer, fluid power system, anti-fouling, anti-icing, anti-crawling, anti-corrosion materials (Xi et al., 2008), and the so-called superamphiphobicsurfaces.

Bambooproducts with superamphiphobic properties would begreatly appreciated by various high value-added products (Zhu et al., 2016). In a previous study, Gao et al. (2015) reported that a simple hydrothermal process with further hydrophobization was developed for fabricating durablesuperamphiphobic films of cuprous oxide (Cu_2O) microspheres on a wood substrate. Yao et al. (2016) demonstrated a solvothermal

---

* 本文摘自 European Journal of Wood and Wood Products, 2018, 76(6): 1595-1603.

depositionof ZnO nanorod arrays ona wood surface for robustsuperamphiphobic performanceand superior ultraviolet resistance. Tuominen et al. (2016) developed an overhang structure of the TiO_2 nanoparticles on the wood surface using the liquid flame spray(LFS) method. The so-called LFS is a thermal aerosol-based process utilized to deposit nanometre-sized metal and metal oxide particles. Teisala et al. (2018) fabricated a superamphiphobic and opticallytransparentnanoparticle coating on glass by LFS, and examined thesurface morphologyby varying the ratio of SiO_2 and TiO_2 inthe coating. Jin et al. (2014) developed a method to render bamboo surfaces superamphiphobic by combining control of ZnO nanostructures and fluoropolymer deposition while maintaining their corrosion resistance. However, the poor reproducibility, complicated procedure, high manufacturing costs of these methods have limited their practical uses.

Hence, this study aims at developing a simple and inexpensive method to fabricatesuperamphiphobicand conductive coatings on bamboo surfaces. As illustrated in Fig. 1, in the first step, Cu films were deposited on bamboo substrate via a radio frequency magnetron sputtering method. As a Physical Vapor Deposition (PVD) method, the magnetron sputtering has become the process alternativefor the deposition of a wide range of industrially important coatings (Kelly and Arnell, 1999). Examples include wear-resistant coatings (Burnett and Rickerby, 1987), low friction coatings (Cumings and Zettl, 2000), corrosion resistant coatings (Boden, 1972), decorative coatings and coatings with specific optical or electrical properties. Compared with the traditional approach, this method is simpler and easier to be controlled, which works for substrates in any shape and does not need rigorous conditions. Copper (Cu) is widely used in industrial applications (Zalba et al., 2003), and is a kind of inexpensive conductive material.

In the second step, the as-prepared sample were immersedinto an aqueous solution of sodium hydroxide and ammonium persulfate, the CuO microflowers/$Cu(OH)_2$ nanorod arrays hierarchicalstructure was prepared on the Cu-bamboo as a rough coating by an alkali assistant surface oxidation. $Cu(OH)_2$ is a well-known layered material, its magneticproperties are remarkably sensitive to the intercalation ofmolecular anions. CuO is a p-type semiconductor (Yang et al., 2015) with low resistance value, and is also applied to chemical sensors because of its efficientcatalytic activity, good electrochemical performance (Liu et al., 2016), andexcellent surface adsorption property (Tanvir et al., 2016). Followed by the immersion into perfluorocarboxylic acid [$CF_3(CF_2)_8COOH$], ethanol solution grafts long-hydrophobic group onto the surface to obtain a superamphiphobic surface (Liu and Jiang, 2011). In this study, the sand abrasion, sandpaper abrasion and compression testswere adoptedto explorethe mechanical resistance properties of the products.

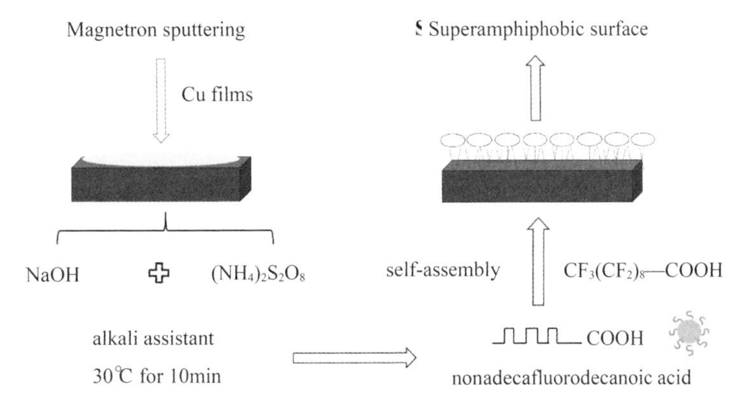

Fig. 1 Experimental strategy for the fabrication of superamphiphobic coatings on bamboo surface

2 Materials and methods

2.1 Materials

The Cu target was supplied by Beijing Guanjinli New Matrial Co., Ltd, China. The reagents used for CuO microflowers/$Cu(OH)_2$ nanorod arrays synthesis including NaOH(96.0%) and $(NH_4)_2S_2O_8$(98.0%), were purchased fromTianjin Kermel Chemical Reagent Manufacturing Co., Ltd, China. Deforested in Zhejiang Province, China, the bamboo was cut into specimens measuring 20mm×20mm×5 mm (length × width × height). After being ultrasonically rinsed in ethanol and deionized water for 10 min, the bamboo specimenswere oven-dried(24 h, 103 ± 2 ℃) to a constant weight, and their weights were determined.

2.2 Deposition of Cu coating on the bamboo substrate

The Cu films were depo sited with a radiofrequency magnetronsputtering system(Ogwu et al., 2005). The target used was a copper(99.999% purity) plate of 5cmdiameter. The sputtering was carried out in argon (Ar-99.995% purity), the sputtering powerwas 100W, and the sputtering system used a base pressureof 3×10^{-3}mba. The filmswere deposited on the 15μm thick Cu layer.

2.3 Fabrication of re-entrant structures and surface modification

Cu-treatedbamboo sampleswere immersedinto an aqueous solution of 2.5M NaOH and 0.12M $(NH_4)_2S_2O_8$ at room temperaturefor 15 min. After taking them out of the solution, the prepared samples were ultrasonically washed with distilledwater and ethanol for 5 min, followed by drying undernitrogen. Then they were immersed into a 0.01M perfluorocarboxylic acid[$CF_3(CF_2)_8COOH$] ethanol solution for 30 min and dried in air at room temperature.

2.4 Characterization

The morphology structure of the as-prepared bamboo samples was observed using the scanning electron microscopy(FEI Quanta 200, Eindhoven, Holland, 12.5 kV). The chemical composition of the treated and untreated bamboo was analyzed by the Fourier transform infrared spectroscopy (FTIR, Magna-IR 560, Nicolet) and X-ray photoelectron spectroscopy (XPS, ThermoFisher Scientific Company, K-Alpha). The surface chemical state and structure of thetreated samples were determined using the X-ray powder diffraction (XRD, Philiphs, PW 1840 diffractometer) operating with Cu-Kradiation at a scan rate of $4°\cdot min^{-1}$ and accelerating voltageof 40 kV and applied current of 30 mA ranging from 5° to 80°. The non-contact surface profilometry was used to calculate the surface morphology. The sample color was measured with an optical spectrophotometer(CM-2300d, Konica Minolta Sensing, Inc.), and the color change was evaluated using L, a, b values. Thetotal color changes(ΔE) of the sample coatings were calculatedusing the following equation:

$$\Delta E=\Delta a^2+\Delta b^2+\Delta L^2 \quad (1)$$

The contact angles and sliding angles were measured with 5μL deionized water and oil (hexadecane) droplet at room temperature using an optical contact angle meter(Hitachi, CA-A).

3 Results and discussion

3.1 Surface morphology and 3D scan image of the as-preparedbamboo sample

The surface morphologies of the pristinebamboo, the Cu-treated bamboo and the as-prepared samples that were immersed in the alkali assistant oxidation for 5 to 10 min are presented in Fig. 2. In Fig. 2a, the original bamboo surface displays a clean and smooth surface(Li et al. , 2015). After Cu films were deposited on the bamboo substrate by magnetron sputtering, Cu nanopaticles uniformly filled into the pits of the cell wall structure, as shown in Fig. 2b. The Cu-treated bamboo was immersed in the alkaline aqueous solution with $(NH_4)_2S_2O_8$ at room temperature for 5 min. Fig. 2 cshows that the $Cu(OH)_2$ was formed in the nanorod arrays standing upright, the wire-like morphology covered the Cu-treated bamboo substrate uniformly, and the metallic luster of the Cu-treated bamboo surface turnedgradually tofaint blue(Chen et al. , 2009b). With the increase in reaction time from 5 to 6 min, sporadic agglomeration structures appeared in the nanorod arrays, while the size of the nanorods remained unchanged (Fig. 2d). The agglomeration structure became larger and showed a microflower shapewhen immersed for 8 minutes(Fig. 2e). Finally, the microflowers densely stood with the nanorod arrays and formed re-entrant structures, and a sample ofblack film was obtained owing to the transformation of $Cu(OH)_2$ to CuO, as shown in Fig. 2f.

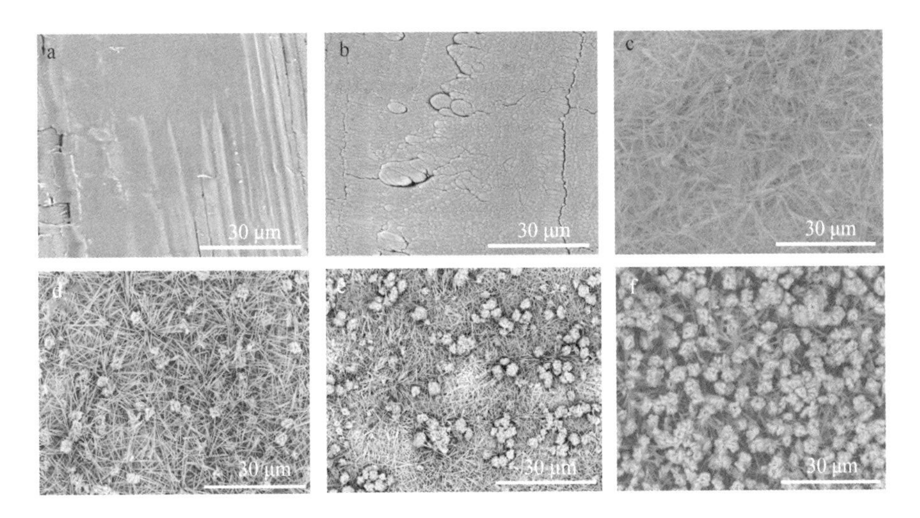

Fig. 2 (a and b) SEM images of the surface bamboo and Cu-treated bamboo. Surface morphologies after immersion in alkaline aqueous solution with(NH4)2S2O8 for 5min(c), 6min(d), 8min(e), 10min(f)

To prove that the hierarchical structure had three-dimension, the cross-section SEM image and 3D scan image were utilized. Fig. 3a displays the cross-section SEM image of the Cu-bambooafter the immersion in alkaline aqueous solution for 8min. The divergentnanorod arrays were distributed on the surface likebunches of straws. The nanorod arrays were mixed with microflower shape on the top. Fig. 3b-epresent 3D images of the as prepared samples. As shown in Fig. 3b, the Cu-bamboo surface was covered mostly by flat areas, and the defects existed on the surface due to the anisotropy of bamboo. Fig. 3c shows the small raised structure, representing the generation of the split in the vertical direction, which verified that the nanorod structures had a certain height relative to the horizontal plane. Fig. 3d shows a similar structure, and the difference is the more clear vertical gradient due to the more obvious raised structure. Likewise, Fig. 3e shows a more obvious staggered convex structure and the uniform distribution on the surface, indicating the formation of space structures.

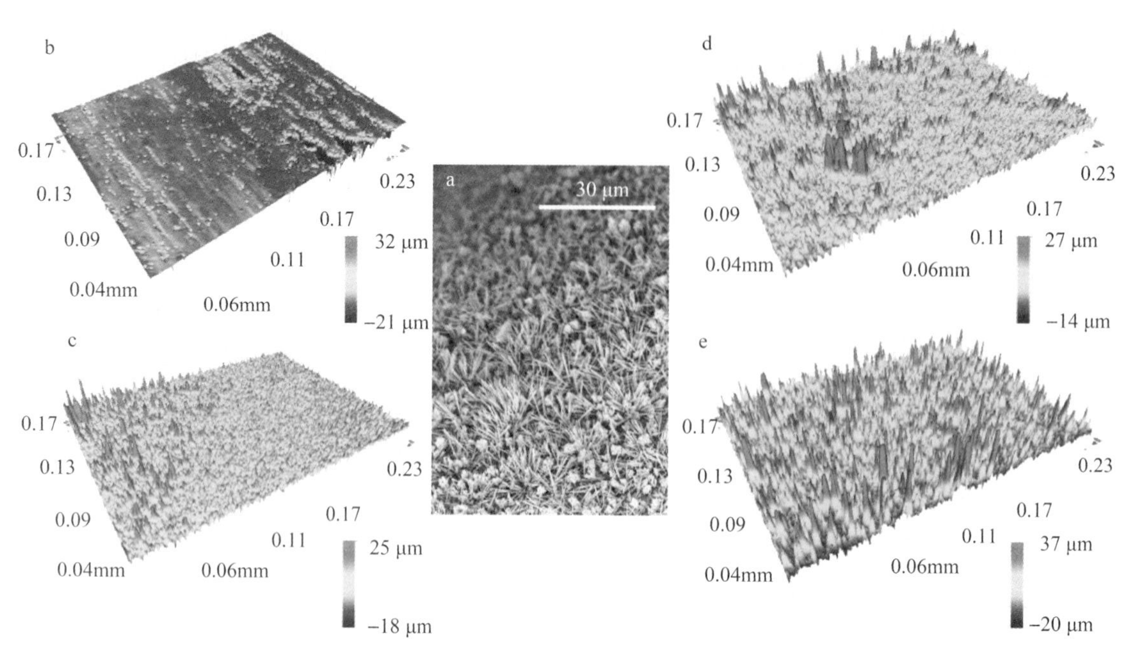

Fig. 3 (a) Cross-section SEM images of the sample. 3D scan image of the Cu-bamboo sample (b), and samples after immersion in alkaline aqueous solution with $(NH_4)_2S_2O_8$ for 5min (c), 8min (d), 10min (e)

3.2 Crystal morphology of the as-prepared bamboo sample and theoretical explanation

As shown in Fig. 4, no characteristic peaks were observed at 16° and 22° belonging to the (101) and (002) crystal planes of the cellulose in the bamboo. The XRD spectrum of the Cu-treated bamboo showed three main peaks at 43.30°, 50.43° and 74.13°, which could be attributed to the cubic copper phases with orientation planes (111), (200) and (220), respectively, and no other phases were detected (Mukherjee et al., 2011). The peaksmarked with rhombus can be readily indexed to orthorhombic-phase $Cu(OH)_2$ (JCPDS card No. 72-0140) and the diffractionpeaks marked with triangles can be indexed to monoclinic-phase CuO (JCPDS card No. 80-0076) (Chen et al., 2009a). The Cu-bamboo after alkali oxidation is shown in Fig. 3, illustrating the CuO microflowers / $Cu(OH)_2$ nanorodarrays hierarchical structure on Cu-treated bamboo (Zhu et al., 2012).

The formation ofhierarchical structures on Cu-treated bamboo involved inorganic polymerization reactions. $Cu(OH)_2$ proceeded through the condensation of Cu^{2+} and OH- (olationreaction), forming μ4-OH bridges (Zhang et al., 2003a). The OH-ligand acted as anucleophile and underwent a change in coordination onlywhen it switched from a terminal ligand in a monomer to a bridging ligand in a condensed species (Zhang et al., 2003b). Infact, the structure of $Cu(OH)_2$ was unstable against oxolation because oxygen atoms were either pentacoordinated or tricoordinated (Wang et al., 2003). Consequently, it was transformed to CuO by breaking the interplanarhydrogen bonds under certain pH conditions or long reaction time. The reaction processes were summarizedand are presented as follows (Wang et al., 2003).

$$Cu + 4OH^- + (NH_4)_2S_2O_8 \rightarrow Cu(OH)_2 + 2SO_{42-} + 2NH_3 + 2H_2O \quad (2)$$

$$Cu(OH)_2 \rightarrow CuO + H_2O \quad (3)$$

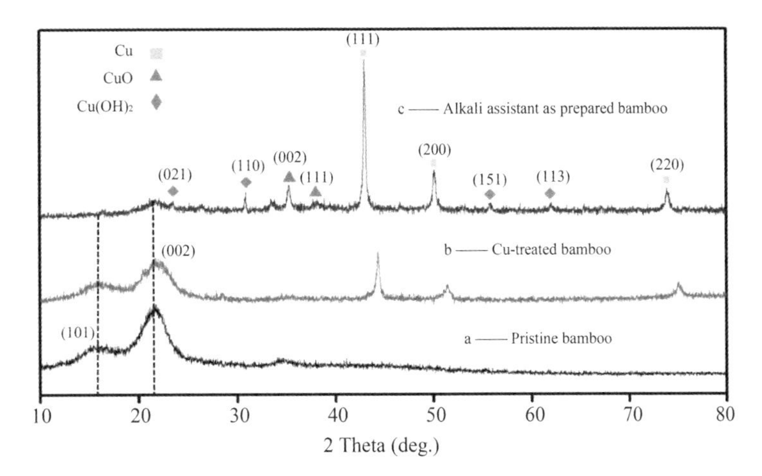

Fig. 4 XRD patterns of(a)the pristine bamboo, (b)the Cu-treated bamboo, and(c)the Cu-treated bamboo after immersion in an alkaline aqueous solution with$(NH_4)_2S_2O_8$ for 10 min

3.3 FTIR and XPS measurements

The chemical composition of as-prepared bamboo after the fluorinationwas investigated by FTIR and XPS measurements as shown in Fig. 5. As shown in Fig. 5a, the FTIR spectra indicated that the stretching vibration of free COOH band from perfluorocarboxylic acid at 1688.47 $cm^{-1}$was no longer present, and turned to the coordinated COO band at1681.14 cm^{-1}, and Cu^{2+} ions immediately coordinated with $CF_3(CF_2)_8COO$-ions toform $Cu_2[CF_3(CF_2)_8COO]_4$(Meng et al., 2008). In Fig. 5b, the more detailed information concerning the elemental and chemical state of the as-prepared bamboo is examined using XPS. The elements of the superamphiphobic bamboo surfacesafter the chemical modificationwere C, O, F, Cu andthe peak for F1s was detected at 689.08 eV(Fig. 5b). The XPS high-resolution C 1sspectrum showed that there were seven types of carbon in the samplebesides the carbon source from the detection. The analysis of the spectrum found that there were—CF_3, —CF_2, —COO and C1-C4 groups. The surface free energydecreased in the order of —CH_2>—CH_3>—CF_2>—CF_2H>—CF_3(Nishino et al., 1999). The —CF_3, —CF_2groups resulted in a surface with low surface energy, which was necessary forachieving the superoleophobicity (Xie et al., 2004). It was revealed that thefluorine polymers successfully modified the bamboo. C1-C4 components were the major lignocellulosic constituents in the bamboo(Xu et al., 2013). Bamboo is composed of cellulose, hemicellulose, lignin, and extracts(Liu et al., 2010). The C1(—O—C=O, 289.14eV)and C2(—O—C—O—, —C=O, 288.44eV) components were mainly from hemicellulose and the extracts. C3 (—O—C—, 286.72eV)was assigned to cellulose, and the C4(—C—H, —C—C, 285.02 eV) components were constituents of lignin.

3.4 Wettability and electrical conductivity tests

Fig. 6 demonstrates the contact angles for water and hexadecane of the original bamboo, Cu-treated bamboo, and the superamphiphobic bamboo(from left to right). The original bamboo presented the water contact angle (WCA) of 42° and the oil contact angle (OCA) of 0°, which demonstratedthat the pristine bamboo surface was hydrophilic and oleophilic (Wang et al., 2014). The Cu-treated bamboo presented hydrophobicity with WCA of 44° and OCA of 0°. The samemethod was used to evaluate samples after

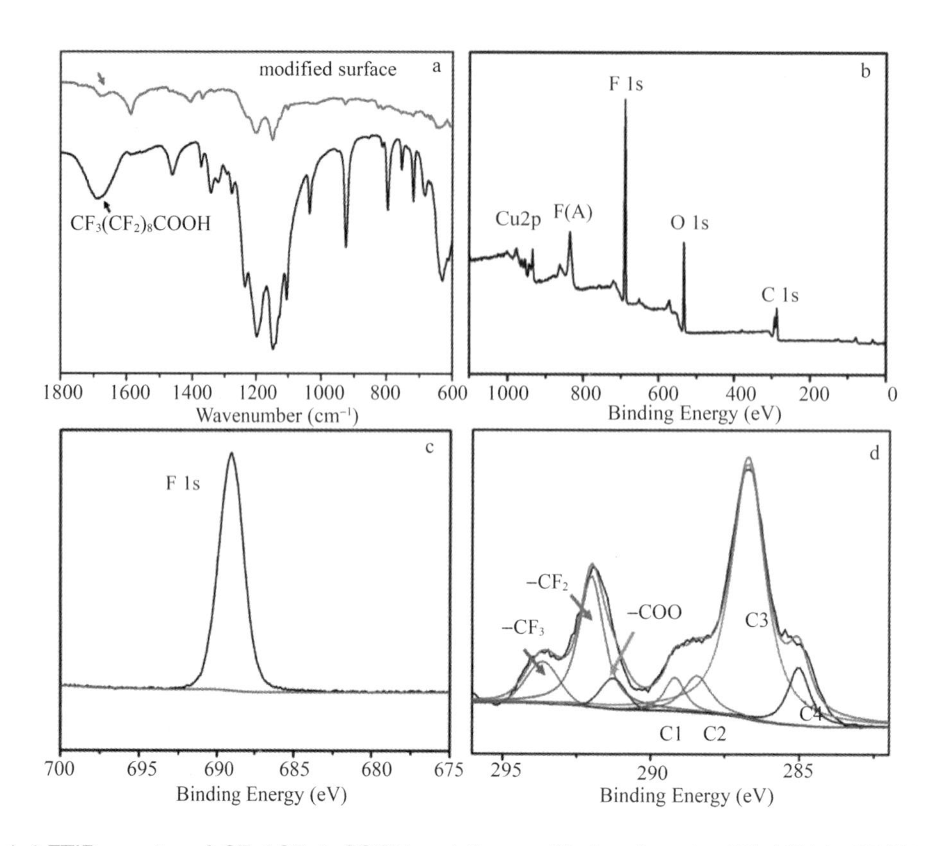

Fig. 5 (a) FTIR spectra of $CF_3(CF_2)_8COOH$ and the modified surface by $CF_3(CF_2)_8COOH$. (b) XPS survey of the surfaces after $CF_3(CF_2)_8COOH$ modification. (c) and (d) are the F1s survey spectrum and C1s peaks of XPS measurement

immersion in alkaline aqueous solution with $(NH_4)_2S_2O_8$ for 5min (c), 8min (d), 10min (e), and the results are listed in Table 1. Results showed that the hydrophobicity properties of different surface topography was similar and theoleophobicity propertiesincreased with the reaction time. The hydrophobicity of the sample after immersion in alkaline aqueous solution with $(NH_4)_2S_2O_8$ for 10min was raised to superamphiphobicity, while WCA and OCA reached 154° and 152°, respectively, and the roll-off angle (α) was less than 6° (Fig. S1, Supporting Information). The resultindicatedthat the microflower geometrical shape was also critical in establishing superoleophobicity (Fig. S2, Supporting Information). The hierarchical structure can retain more air to prevent oil droplets from penetrating.

Table 1 After hydrophobization, contact angles of water and oil on the Cu-bamboo surfacesafter immersion in the solutions of NaOH and $(NH_4)_2S_2O_8$ for 0~10min

Immersion time(min)	5	6	8	10
WCA	148°	154°	151°	154°
OCA	122°	135°	141°	152°

A simple method was developed for the measurement of the electrical resistance for the as-prepared samples by voltammetry (Jin et al., 2015). Dry bamboo is a good electrical insulation material due to its inherent insulation property. The dazzling lightin the picture implied that the Cu-treated bamboo had excellent electrical conductivity with an electricresistance of 0.5 ± 0.1Ω measured using a multimeter. After the hydrophobization, the conductivity of the superamphiphobic bamboo did not show any visible change, continuing tolight a bulb with an electricresistance of 2.2± 3.1Ω.

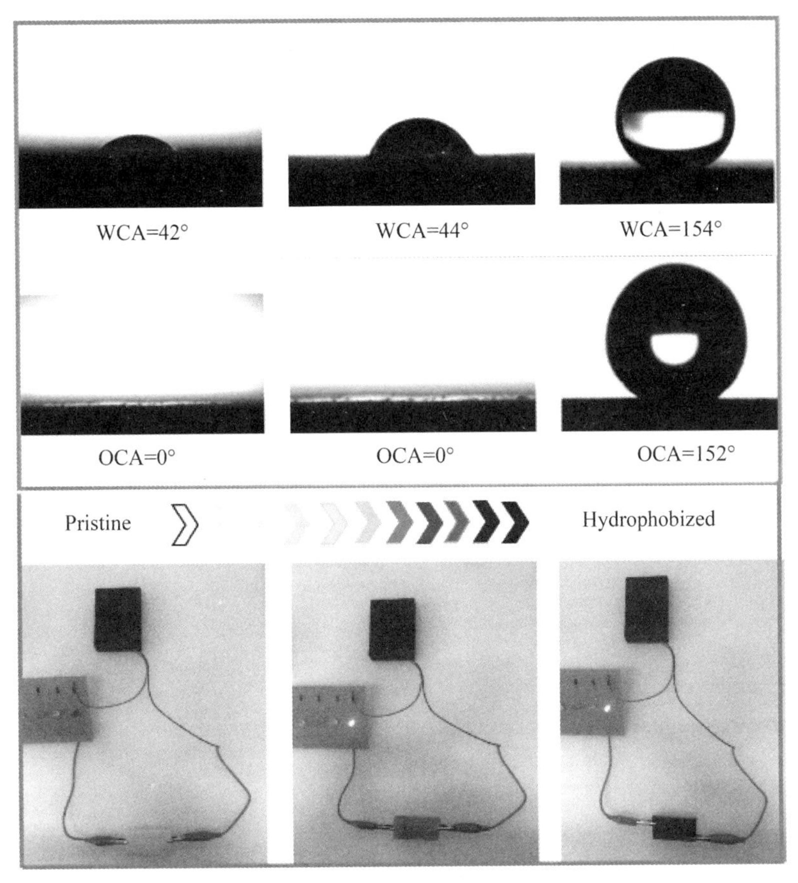

Fig. 6 Contact angle profiles and electrical conductivity of the original bamboo, the Cu-treated bamboo, andthe superamphiphobic bamboo

3.5 Color measurements and mechanical abrasion testsof the superamphiphobic bamboo surface

In general, surface coatinghas been recognized as an efficientstrategy in improving the properties of bamboo. However, the natural color texture of the bamboo surface may be affected by the coating. Therefore, the effect of the coating on the color of the bamboo was explored. The total color change of samples isshown in Table 2. The lightnessfactor L of the samplesshowed a decreasingtrend from 80.25 to 31.321, indicating that the surface color turned darker. A significant decline in b value from 24.373 to 3.116 was observed, implyingthe surface color changed from yellow to blue. The a values changed slightly. The total change of ΔE value varied from 0 to 53.459, corresponding to a greater color difference.

Table 2 Chromaticity indexes change of samples

Sample *Lab*	Color change						
	△*L*	△*a*	△*b*	△*E*			
Bamboo	80.25	5.128	24.373	0	0	0	0
Cu-treated bamboo	59.219	10.987	12.479	-21.01	5.848	-11.923	24.856
Superamphiphobic bamboo	31.321	1.568	3.116	-48.908	-3.57	-21.286	53.459

As well-known, the commercial bamboo-based products are required to be last longunder harsh weather conditions. To assess the mechanical resistance of the superamphiphobic bamboo, the sand abrasion(Deng et al. 2012)(a), sandpaper abrasion(Liu et al. 2013)(b), compression tests(Yao et al. 2016)(c)were used, as shown in Fig. 7. In Fig. 7a, 20 g sand grains with a diameterranging from 100 to 300 μm were impinged to the as-prepared surface from a height of 30 cm. Fig. 7b shows that the as-prepared sample with 100 g weight on the top was placed face-down on the sandpaper and moved 12 cm. In Fig. 7c, the dropletwas pressed by finger on the top with a certainpressure. After mechanical abrasiontests, the contact angles of the superamphiphobic bamboo surfaces were measured to estimate the mechanical stability of the coatings.

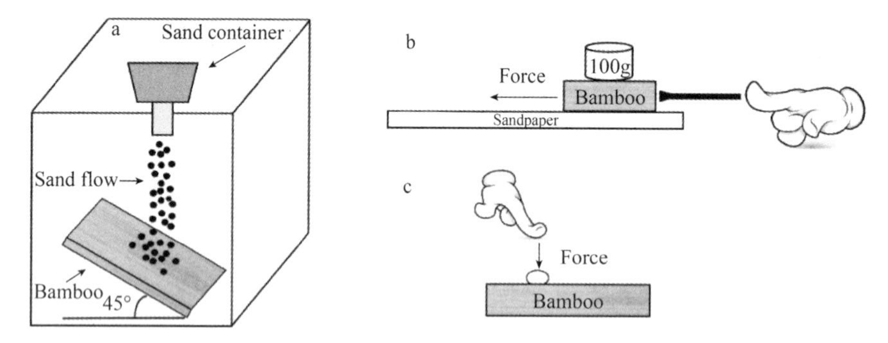

Fig. 7 (a)Schematic drawing of(a)sand abrasion test, (b)sandpaper abrasion test and(c)compression test

Fig. 8a demonstrates thatafter 100 cycles of sand abrasion tests, WCA and OCA on the as-prepared bamboo surface remained superamphiphobic with a WCA of 151.2°, and OCA of 150.4°, and α was less than 10°(Fig. 8c). Fig. 8b shows that, after being scratched by moving the sandpaper for 9 cm, the surface lost its super-repellency, with a WCA of 45° and OCA of 0°, and the α wasmore than 30°(Fig. 8d). Obviously, the re-entrant structures were particularly weak undervertical destroyable forces by the sandpaper abrasion. However, most of the removed surface structures can be recast. The good reparability isgreatly appreciatedas high-value-added products. In the compression experiment, the finger contact area remained superamphiphobic after being pressed with a finger force, demonstrating that the water and hexadecane did not penetratethe re-entrant structure composite interface under the compression. In addition, the superamphiphobic durability of as-prepared sampleswas evaluated after exposure to the ambient condition for six months, and the WCA and OCA were maintained larger than 150°. These results indicated that the superamphiphobic bamboo would not be ideal for outdoor applications because the micro-nano structure was easily damagedwhen the external force was greater than its own load.

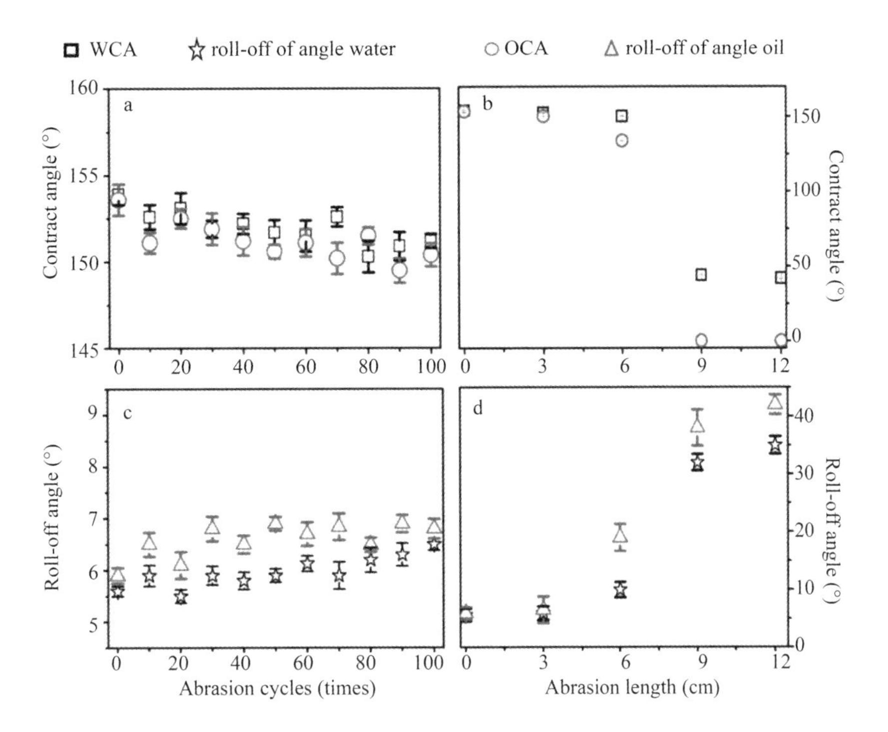

Fig. 8 (a and c) WCAs and roll-off angles of water changes depending on the sand abrasioncycles. (b and d) Effect of sandpaper abrasion on the OCAs and roll-off angles of oil at longer abrasion length

4 Conclusions

In summary, superamphiphobic and conductive films were deposited onto the surface of bamboo through the magnetron sputtering at alkaline and oxidative conditions. Thecoating of Cu films not only provided the construction with re-entrant structures but also imparted the metallic featureto the surface rendering the bamboo conductive property. The sand abrasion and compression tests showed that the film had a high mechanical resistance property. However, the sandpaper abrasion test showed that there-entrant structures were particularly weak under vertical destroyable forces. Consequently, the superamphiphobic bamboo surface is not suitable for outdoor environments. This study provides aneffective and convenient path to achieve novel bamboo products with micro-nano structure surfaces and large-scale engineering capacity.

Acknowledgements

This research was supported by The National Natural Science Foundation of China(31470584), Overseas Expertise Introduction Project for Discipline Innovation, 111 Project(No. B08016).

Reference

[1] Boden PJ. The mechanical properties of wear-resistant coatings: I: Modelling of hardness behaviour [J]. Thin Solid Films., 1987, 148: 41-50.

[2] Chen X, Kong L, Dong D, et al. Synthesis and characterization of superhydrophobic functionalized $Cu(OH)_2$ nanotube arrays on copper foil[J]. Appl Surf Sci., 2009, 255: 4015-4019.

[3] Chen XH, Kong LH, Dong D, et al. Fabrication of Functionalized Copper Compound Hierarchical Structure with Bionic Superhydrophobic Properties [J]. J. Phys Chem C., 2009, 113: 5396-5401.

[4] Cumings J, Zettl A. Low-Friction Nanoscale Linear Bearing Realized from Multiwall [J]. Carbon Nanotubes Science., 2000, 289: 602-604.

[5] Deng X, Mammen L, Butt HJ, et al. Candle Soot as a Template for a Transparent Robust Superamphiphobic Coating [J]. Science., 2012, 335: 67-70.

[6] Gao L, Xiao S, Gan W, et al. Durable superamphiphobic wood surfaces from Cu_2O film modified with fluorinated alkyl silane[J]. RSC Adv., 2015, 5: 98203-98208.

[7] Jin C, Li J, Han S, et al. A durable, superhydrophobic, superoleophobic and corrosion-resistant coating with rose-like ZnO nanoflowers on a bamboo surface[J]. Appl Surf Sci., 2014, 320: 322-327.

[8] Jin C, Li J, Han S, et al. Silver mirror reaction as an approach to construct a durable, robust superhydrophobic surface of bamboo timber with high conductivity[J]. J. Alloys Compd., 2015, 635: 300-306.

[9] Kelly PJ, Arnell RD. Magnetron sputtering: a review of recent developmentsand applications[J]. Vacuum., 1999, 56: 159-172.

[10] Li J, Sun Q, Han S, et al. Reversibly light-switchable wettability between superhydrophobicity and superhydrophilicity of hybrid ZnO/bamboo surfaces via alternation of UV irradiation and dark storage[J]. Prog Org Coat., 2015, 87: 155-160.

[11] Liu F, Wang S, Zhang M, et al. Improvement of mechanical robustness of the superhydrophobic wood surface by coating PVA/SiO2 composite polymer[J]. Appl Surf Sci., 2013, 280: 686-692.

[12] Liu K, Jiang L. Metallic surfaces with special wettability[J]. Nanoscale., 2011, 3: 825-838.

[13] Liu QS, Tong Z, Peng W, et al. Preparation and characterization of activated carbon from bamboo by microwave-induced phosphoric acid activation[J]. Industrial Crops & Products., 2011, 31: 233-238.

[14] Liu X, et al. Sensitive Room Temperature Photoluminescence-Based Sensing of H2S with Novel CuO-ZnO Nanorods [J]. ACS Appl Mater Interfaces., 2016, 8: 16379-16385.

[15] Meng HF, Wang ST, Xi JM, et al. Facile means of preparing superamphiphobic surfaces on common engineering metals[J]. J. Phys Chem C., 2018, 112: 11454-11458.

[16] Mukherjee SK, Joshi L, Barhai PK. A comparative study of nanocrystalline Cu film deposited using anodic vacuum arc and dc magnetron sputtering[J]. Surf Coat Technol., 2011, 205: 4582-4595.

[17] Nishimoto S, Bhushan B. Bioinspired self-cleaning surfaces with superhydrophobicity, superoleophobicity, and superhydrophilicity[J]. RSC Adv., 2013, 3: 671-690.

[18] Nishino T, Meguro M, Nakamae K, et al. The Lowest Surface Free Energy Based on $-CF_3$ Alignment [J]. Langmuir: the ACS journal of surfaces and colloids., 1999, 15: 4321-4323.

[19] Ogwu AA, Bouquerel E, Ademosu O, et al. An investigation of the surface energy and optical transmittance of copper oxide thin films prepared by reactive magnetron sputtering[J]. Acta Mater., 2005, 53: 5151-5159.

[20] Tanvir NB, Yurchenko O, Wilbertz C, et al. Investigation of CO_2 reaction with copper oxide nanoparticles for room temperature gas sensing[J]. J. Mater Chem A., 2016, 4: 5294-5302.

[21] Teisala H, Geyer F, Haapanen J, et al. Ultrafast Processing of Hierarchical Nanotexture for a Transparent Superamphiphobic Coating with Extremely Low Roll-Off Angle and High Impalement Pressure [J]. Adv Mater., 2018, 30: e1706529.

[22] Tuominen M et al. Superamphiphobic overhang structured coating on a biobased material[J]. Appl Surf Sci., 2016, 389: 135-143.

[23] Wang C et al. One-step synthesis of unique silica particles for the fabrication of bionic and stably superhydrophobic coatings on wood surface[J]. Adv Powder Technol., 2014, 25: 530-535.

[24] Wang WZ, Varghese OK, Ruan CM, et al. Synthesis of CuO and Cu_2O crystalline nanowires using $Cu(OH)_2$ nanowire templates[J]. J. Mater Res., 2013, 18: 2756-2759.

[25] Xi J, Feng L, Jiang L. A general approach for fabrication of superhydrophobic and superamphiphobic surfaces[J]. Appl Phys Lett., 2008, 92: 053102.

[26] Xie QD, Xu J, Feng L, et al. Facile creation of a super-amphiphobic coating surface with bionic microstructure[J]. Adv Mater., 2004, 16: 302-+.

[27] Xu G, Wang L, Liu J, et al. FTIR and XPS analysis of the changes in bamboo chemical structure decayed by white-rot and brown-rot fungi[J]. Appl Surf Sci., 2013, 280: 799-805.

[28] Yang FC, Guo J, Liu MM, et al. Design and understanding of a high-performance gas sensing material based on copper oxide nanowires exfoliated from a copper mesh substrate[J]. J Mater Chem A., 2015, 3: 20477-20481.

[29] Yao Q, Wang C, Fan B, et al. One-step solvothermal deposition of ZnO nanorod arrays on a wood surface for robust superamphiphobic performance and superior ultraviolet resistance[J]. Sci Rep., 2016, 6: 35505.

[30] Zalba B, Marín JMa, Cabeza LF, et al. Review on thermal energy storage with phase change: materials, heat transfer analysis and applications[J]. Appl Therm Eng., 2003, 23: 251-283.

[31] Zhang W, Wen X, Yang S. Controlled Reactions on a Copper Surface: Synthesis andCharacterization of Nanostructured Copper Compound Films [J]. Inorg Chem., 2003, 2: 5005-5014.

[32] Zhang WX, Wen XG, Yang SH, et al. Single-crystalline scroll-type nanotube arrays of copper hydroxide synthesized at room temperature[J]. Adv Mater., 2003, 15: 822.

[33] Zhu H et al. Wood-Derived Materials for Green Electronics, Biological Devices, and Energy Applications [J]. Chem Rev., 2016, 1160: 9305-9374.

[34] Zhu X, Zhang Z, Xu X, et al. Facile fabrication of a superamphiphobic surface on the copper substrate[J]. J. Colloid Interface Sci., 2012, 367: 443-449.

中文题目：超双疏竹材的制备

作者：包文慧，贾贞，蔡力平，梁大鑫，李坚

摘要：通过方便的碱辅助表面氧化技术，在竹表面上制备超双疏功能的CuO微花/ $Cu(OH)_2$ 纳米棒阵列分层结构。将长链疏水基团接枝到竹材表面以获得超双疏表面，该表面对水和十六烷具有超强排斥性。通过扫描电子显微镜(SEM)、X射线衍射(XRD)、X射线光电子能谱(XPS)及水接触角分析竹基材薄膜上的形貌、微观结构、晶体结构、化学组成和状态以及疏水性。此外，该涂层显示出优异的环境稳定性和高导电性。预计这项工作可能会扩大超双疏竹制品的应用。

关键词：超双疏；磁控溅射；导电；竹材

Nanocellulose: A Promising Nanomaterial for Advanced Electrochemical Energy Storage*

Wenshuai Chen, Haipeng Yu, Sangyoung Lee, Tong Wei, Jian Li, Zhuangjun Fan

Abstract: Nanocellulose has emerged as a sustainable and promising nanomaterial owing to its unique structures, superb properties, and natural abundance. Here, we present a comprehensive review of the current research activities that center on the development of nanocellulose for advanced electrochemical energy storage. We begin with a brief introduction of the structural features of cellulose nanofibers within the cell walls of cellulose resources. We then focus on a variety of processes that have been explored to fabricate nanocellulose with various structures and surface chemical properties. Next, we highlight a number of energy storage systems that utilize nanocellulose-derived materials, including supercapacitors, lithium-ion batteries, lithium-sulfur batteries, and sodium-ion batteries. In this section, the main focus is on the integration of nanocellulose with other active materials, developing films/aerogel as flexible substrates, and the pyrolyzation of nanocellulose to carbon materials and their functionalization by activation, heteroatom-doping, and hybridization with other active materials. Finally, we present our perspectives on several issues that need further exploration in this active research field in the future.
Keywords: Nanocellulose; Supercapacitors; Lithium – ion batteries; Lithium – sulfur batteries; Sodium – ion batteries

1 Introduction

Because of the rapid development of modern society, the fast-growing demand for energy because of worldwide power consumption has become a serious issue and will continue to be into the future. Interest in the sustainable development of advanced, low-cost, and environmentally friendly energy storage devices has been growing steadily.[1] Electrochemical energy storage systems in terms of supercapacitors and batteries (such as lithium (sodium)-ion batteries and lithium-sulfur batteries) have demonstrated great potential in powering portable electronics, electric vehicles, hybrid electric vehicles, and even huge energy-storage systems.[2-10] Except for performance and safety enhancements, the major remaining challenges for the future development of energy storage devices are the reduction of both production and overall device costs, the realization of flexible devices, the utilization of green and abundant raw materials, the realization of environmentally friendly processes, and the development of easily recyclable and up-scalable systems.[11-15] Among numerous candidates, cellulose-derived materials have obtained increasing attention as attractive components of various electrochemical energy storage devices.[16-28]

---

* 本文摘自 Chemical Society Reviews, 2018, 47(8): 2837-2872.

Cellulose is the most abundant renewable organic polymeron earth. The world production of biomass is estimated at approximately 150-200 billion metric tons a year, most of which is cellulose.[29,30] It is used in the form of wood, cotton, or other plant fibers as building materials, furniture, paper, clothing, and many otherapplications. Regardless of its source, cellulose is a high-molecular-weight homopolymer of $\beta-1$, 4-linked anhydro-D-glucose units in which every unit is corkscrewed 180° with respect to its neighbors, and the repeat segment is frequently taken to be a dimer of glucose, known as cellobiose. During biosynthesis, van der Waals forces and intermolecular hydrogen bonding promote the parallel stacking of cellulose chains to form nanosized elementary fibrils that further organize into larger fibrils.[31] Within these nanosized cellulose fibers, there are regions in which the cellulose chains are arranged in highly crystalline structures, as well as regions containing amorphous structures. The unique structure of these nanofibers, with advantageous mechanical properties and a low coefficient of thermal expansion, makes them ideal building blocks for advanced functional products.[32-36] Enormous efforts have been made to extract nanofibers from the cell walls of cellulose resources such as from higher plants, or the synthesis of nanofibers by bacteria, which gave birth to nanocellulose.[26,31,37-52] Interest in nanocellulose has been growing steadily and nanocellulose has been widely applied in optically transparent materials,[53-55] reinforced polymer nanocomposites,[32,33] biomimetic materials,[56,57] templates,[35,58] sensors,[59,60] and energy harvesters.[61,62] There are also much more potential application fields for nanocellulose such as in microfluidics channels, as cell cultivation substrate, or substrate for printed electronics. Among the multiple applications, the development of nanocellulose for energy storage (Fig. 1) has received increasing attention because nanocellulose displays advantages based on its intrinsic structures and properties, which can be broadly categorized into the following six groups (Fig. 2):

(1) Because nanocellulose has prospective mechanical properties including a high Young's modulus of 138 GPa[63] and an estimated strength of 2-3 GPa, it can be utilized to develop self-standing and high strength materials for electrodes and separators.

(2) Because it has reactive surfaces containing hydroxyl groups, it is possible and convenient for chemical modification and integration of nanocellulose with active materials. By changing the composite strategies and turning the ratios of active materials in the composites, the properties of the nanocellulose-based composites can be tailored to enhance their electrochemical performance for specific applications.

(3) Nanocellulose fibers with a high aspect ratio, such as those extracted from higher plant or secreted by bacteria form entangled web-like structures, can be utilized to fabricate strong film/aerogel substrates for furtherdevelopment of flexible energy storage devices.

(4) Because of the nanometer scale, high specific surface area, thermal stability, and easy processing, it is possible to fabricate thermally stable nanoporous separators with controlled pore structures, which can simultaneously electrically separate the electrodes and facilitate ionic transportation.

(5) Nanocellulose has desired constituents with relatively high carbon content. Therefore, it is an ideal precursor for fabricating carbon-based porous materials or carbon hybrid materials that can be further functionalized as high-performance carbon electrodes for energy storage devices.

(6) Last but not the least, nanocellulose with various structures and surface chemistry properties can be developed from different cellulose sources through various fabrication strategies. Thus, it is possible to develop multiple kinds of materials or devices motioned in the above 5 points by using various types of nanocellulose

and their hybrids as building blocks.

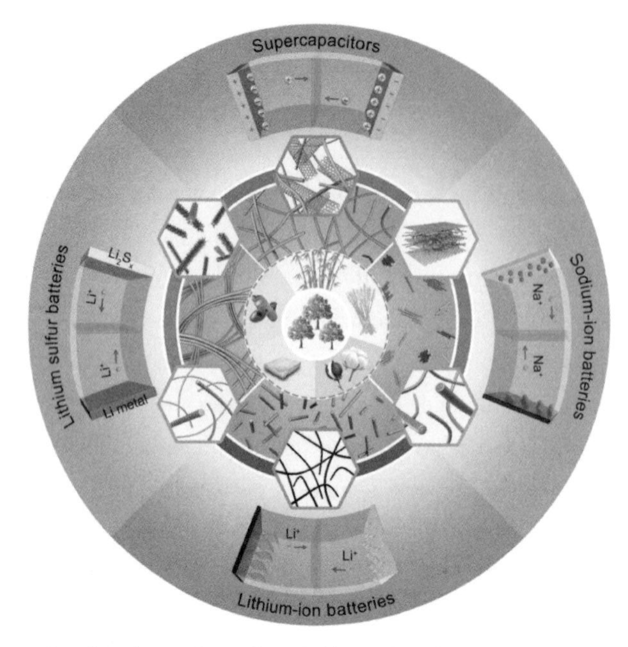

Fig. 1 Schematic diagram showing the main topics of this review from the cellulose resources, nanocellulose fabrication, structural and functional design to the energy storage applications

Thanks to the efforts from many research groups, a series of nanocellulose-derived materials have recently been developed for energy storage applications. Cellulose derivatives have also been confirmed as good candidates for the same purpose.[64-66] After processing the cellulose derivatives into nanoparticles, these novel materials that having advantages both of nano-sized cellulose and new chemical functions from the derivatives, will build up new opportunity for developing of new type of cellulose nanomaterial-based energy storage systems. Over the past few years, extensive reviews that have focused on different aspects in the field of lignocellulosic materials for energy storage have been published and provide great insights to researchers.[16,18-21,67-72] As a very promising nanomaterial, nanocellulose has been included but usually just summarized as a part of some reviews. Recently, several reviews summarized the utilization of nanocellulose and its derived materials for supercapacitors, LIBs and several energy conversion devices.[73-76] However, till now, no reviews have specifically focused on, and comprehensive summarized and comparison, the development of different types of nanocellulose for various energy storage systems. It is timely to have a review on this topic to summarize the recent advances and, more importantly, to provide a systematic understanding of the merits ofthe use of nanocellulose and its derived materialsfor electrodes, separators, and energy storage devices. This review focuses on the recent advances on putting nanocellulose to work for electrochemical energy storage research(Fig. 1). We begin with a brief introduction of the structural features of cellulose nanofibers within the cell walls of various kinds of cellulose resources. We then focus on a variety of methods that have been explored to fabricate various types of nanocellulose with different structures and surface chemistry properties from a variety of sources. Finally, we review the recent progress in the development of nanocellulose-derived materials for energy storage applications including supercapacitors, lithium-ion batteries

(LIBs), lithium-sulfur(Li-S) batteries, and sodium-ion batteries(NIBs).

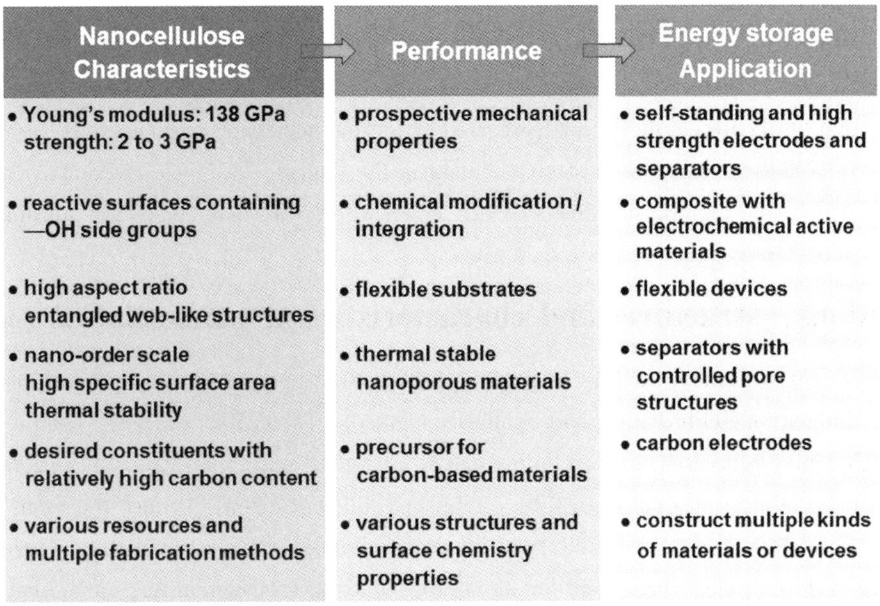

Nanocellulose Characteristics	Performance	Energy storage Application
• Young's modulus: 138 GPa strength: 2 to 3 GPa	• prospective mechanical properties	• self-standing and high strength electrodes and separators
• reactive surfaces containing —OH side groups	• chemical modification / integration	• composite with electrochemical active materials
• high aspect ratio entangled web-like structures	• flexible substrates	• flexible devices
• nano-order scale high specific surface area thermal stability	• thermal stable nanoporous materials	• separators with controlled pore structures
• desired constituents with relatively high carbon content	• precursor for carbon-based materials	• carbon electrodes
• various resources and multiple fabrication methods	• various structures and surface chemistry properties	• construct multiple kinds of materials or devices

Fig. 2 Summary of the structural and property advantages of nanocellulose for energy storage applications

2 Cellulose nanofibers in the cell walls

Cellulose nanofibers are mainly synthesized by higher plants, but also by algae, fungi, bacteria, and an animal, tunicate. During the biosynthesis, van der Waals and intermolecular hydrogen bonds between hydroxyl groups and oxygens of adjacent molecules promote parallel stacking of fully extended cellulose chains, forming elemental fibrils with cross sections ranging from 2 to 50 nm. Owing to differences in biosynthetic mechanisms, the structure of elementary fibrils differs depending on the source organism. The elementary fibrils are further assembled into larger fibrils within the cell walls. The intraand interchain hydrogen bonding network makes cellulose a relatively stable polymer, and gives the cellulose fibrils high axial stiffness.[31] As the skeletal component in cellulose sources, cellulose fibrils are organized in a cellular hierarchical structure. In combination with the accompanying substances such as hemicellulose, lignin, and pectin, this structure leads to extraordinary properties of native cellulose materials.

Fig. 3 illustrates the cellulose nanofibers in the cell walls of wood, bamboo, and cotton. The cellulose molecules organized in the cell walls in the form of nanofibers have characteristic orientations(helix angles), which vary depending on the type of plant sources and the cell wall layers(Fig. 3b, g, m). Because the cellulose nanofibers are embedded in matrix such as hemicellulose and lignin of wood, the high-magnification SEM observations of the samples are usually not clear and cannotreveal the nanostructure of cellulose fibers. When chemical purification methods are used to remove the matrix from wood, such as the removal of lignin using acidified sodium chlorite and the removal of a large amount of hemicellulose using potassium hydroxide(Fig. 3d), the nanofiber characteristics of the cellulose within the cell walls can be clearly revealed (Fig. 3e).[82] Similar results have also been obtained for chemically purified cellulose fibers of bamboo

(Fig. 3i, j, k),[80] cotton,[81] rice straw,[83] and potato tuber,[83] confirming that the higher plant cell walls are supported by high aspect ratio cellulose nanofibers. After nanofibrillation through various mechanical or chemical methods, these cellulose nanofibers are released from the cell walls, resulting in nanocellulose. It should be noted that the uniform nanofibers observed from SEM images in Fig. 3 is elemental fibril aggregates and single elemental fibrils are parallel align aggregated within the aggregates and cannot be clearly seen by the SEM observation.[84] For example, in most higher plants, the diameter of single elemental fibrils is reported to be ~3 nm.[85] Thus, the uniform nanofibers of 12-20 nm within the wood chemically purified cellulose can be recognized as the ~3 nm elemental fibril aggregates.[82,83]

3 Fabrications, structures and characteristics of nanocellulose

Nanocellulose with various structures and surface chemistry properties can be produced from various sources through different nanofibrillation and synthesis methods (Table 1).[77,82,86-91] On the basis of their structures and fabrication processes, nanocellulose can be classified into three main subcategories. So far, the nomenclature of nanocellulose has not been fully established. In this review, we use the terms nanofibrillated cellulose (NFC), cellulose nanocrystal (CNC) and bacterial cellulose (BC), because they are widely used in the literature. It should be noted that except for the above mentioned nanocellulose, other shapes of cellulose nanoparticles can also be developed from different precursors through various fabrication strategies. For example, spherical shaped cellulose nanoparticles were developed from a series of cellulose derivatives,[92-94] while ribbon shaped cellulose nanofibers were produced from cellulose-based precursors through electrospinning technique.[95,96] These cellulose nanoparticles also have potential to be developed for energy storage applications.

3.1 Nanofibrillated cellulose (NFC)

The fabrication of NFC was developed by Turbak and his coworkers at ITT Rayonier in the 1980s.[97,98] NFC was prepared by nanofibrillation of wood cellulose pulps in water using a high-pressure homogenizer. During the nanofibrillation process, cellulose slurry is pumped at high pressure through a spring-loaded valve assembly. The valve is opened and closed in a reciprocating motion, subjecting the cellulose to a large pressure drop of shearing and impact forces.[99] This combination of repeated mechanical forces promots the disintegration of cellulose fibers into nanosized fibers, which generates NFC. The NFC is long with a length greater than 2 μm and displays web-like entangled structures.

Besides the high-pressure homogenizer, other apparatus such as grinder[100,101] and high-intensity ultrasonicator[102,103] can also be directly utilized for nanofibrillation of the cellulose fibers for producing NFC, although the energy required is still large. To reduce the energy requirement and improve the nanofibrillation degree, removing the matrix from the cellulose sources is critical before nanofibrillation. The as-prepared cellulose pulps also need to be kept in the water-swollen state during the whole chemical purification process. Abe et al. reported the fabrication of NFC with a uniform width of 15 nm by using chemical treatment to remove lignin and a large amount of hemicellulose from wood, followed by nanofibrillation of the cellulose pulps using a grinder.[82] Using a similar process, NFC with a uniform width has been successfully produced from rice straw,[83] potato tuber,[83] and bamboo.[80] Enzymatic hydrolysis combined with mechanical nanofibrillation is another route to produce NFC. Pääkkö et al. prepared NFC using enzymatic hydrolysis

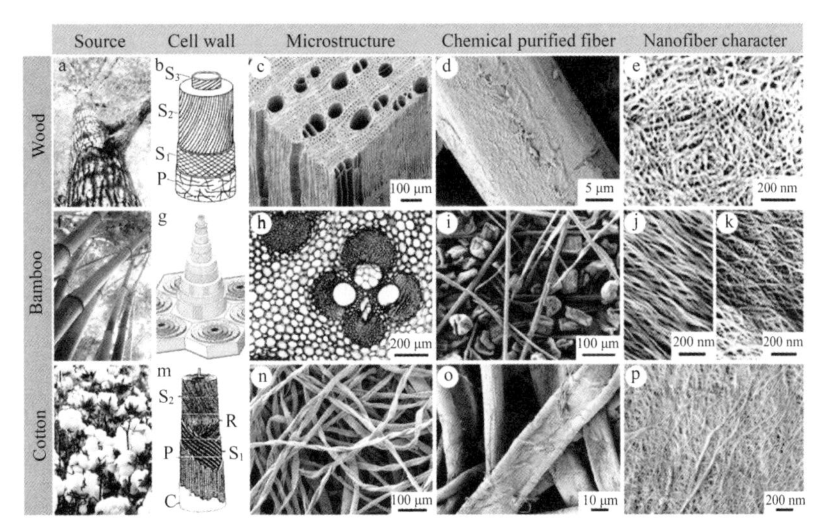

Fig. 3 The cellulose nanofibers in the cell walls of wood, bamboo, and cotton. From left to right are the(a, f and l)digital photos of cellulose sources, (b, g and m)cell wall models, 29, 77, 78(b)is reproduced with permission from ref. 77. Copyright 2010 American Chemical Society. (g)is reproduced with permission from ref. 78. Copyright 2015 Springer Nature. (m)is reproduced with permission from ref. 29. Copyright 2005 Wiley-VCH. SEM images of(c, h and n)raw materials, 79–81(c)is reproduced with permission from ref. 79. Copyright 2012 Scientia Silvae Sinicae. (h)is reproduced with permission from ref. 80. Copyright 2009 Springer. (n)is reproduced with permission from ref. 81. Copyright 2014 Springer. (d, i and o) chemical purified cellulose fibers after removal of the matrix, 80 – 82 (d) is reproduced with permission from ref. 82. Copyright 2007 American Chemical Society. (i)is reproduced with permission from ref. 80. Copyright 2009 Springer. (o) is reproduced with permission from ref. 81. Copyright 2014 Springer. (e, j, k and p) cellulose nanofibers within the chemically purified cellulose fibers. 80 – 82 (e) is reproduced with permission from ref. 82. Copyright 2007 American Chemical Society. (j and k)is reproduced with permission from ref. 80. Copyright 2009 Springer. (p)is reproduced with permission from ref. 81. Copyright 2014 Springer. The image(j) and(k) are the high-magnification FE-SEM images of fiber and parenchyma cells, respectively, as shown in image(i)

combined with mechanical shearing and high-pressure homogenization.[87] The resulting NFC mainly consisted of nanofibers with a width of 5–6 nm and nanofiber aggregates with a width of approximately 10–20 nm. Although high-quality NFC can be produced from most cellulose sources, the uniform nanofibrillation of cellulose fibers from raw materials with a high cellulose content such as cotton and flax[104] is difficult because a small amount of matrix can be removed during the chemical purification process, and there are strong interactions between adjacent cellulose nanofibers that have a small amount or no hemicellulose on the surfaces within the cell walls. Recently, Chen et al. reported a method to individualize NFC from raw cotton by chemical purification (removal of noncellulosic components) and pretreatment with a high-speed blender (breaking down the fiber structures) combined with high-pressure homogenization (nanofibrillation).[81] The resultant cotton NFC was found to have a uniform width of approximately 10–30 nm and high aspect ratios. Therefore, it can be confirmed that NFC with high aspect ratio and web-like entangled structures can be produced from any kind of higher plant sources.

Although the above-mentioned mechanical nanofibrillation methods can be used to fabricate high aspect ratio NFC, a large amount of nanofiber bundles are still existed.[105] To further improve the nanofibrillation degree and decrease the process energy consumption, chemical modification of the surface of cellulose pulps before nanofibrillation is effective. Isogai et al. pioneered the development of the TEMPO (2,2,6,6-tetramethylpiperidine-1-oxyl radical)-mediated oxidation method to produce uniform dispersion of individualized NFC.[51,88,106] Significant amounts of C_6 carboxylate groups are formed on each cellulose nanofiber surface. Electrostatic repulsion and/or osmotic effects working between the anionicallycharged cellulose nanofibers cause the formation of completely individualized NFC dispersed in water by a gentle mechanical disintegration treatment of TEMPO-oxidized cellulose fibers. The wood TEMPO-oxidized NFC is 3-4 nm in width and a few micrometers in length. However, the formation of sodium carboxylate groups from the C_6 primary hydroxyls of NFC surfaces leads to a significant decrease of the thermal stability of the NFC.[107,108] Another route toward the preparation of carboxylated NFC suspensions with individualized nanofibers was developed by Wågberg et al. by using high-pressure homogenization of carboxymethylated cellulose fibers followed by ultrasonication and centrifugation.[89]

The high aspect ratio and web-like entangled structures of NFC are beneficialfordeveloping flexible substrates such as films and aerogels, for integrating with active materials, or converting NFC to carbon materials to develop electrodes and separators for flexible and high strength energy storage devices.

3.2 Cellulose nanocrystal (CNC)

CNC has been produced using an acid hydrolysis approach from a range of cellulose sources. Disordered or paracrystalline regions of cellulose are preferentially hydrolyzed, whereas crystalline regions that have a higher resistance to acid attack remain intact. Thus, the CNC is short and displays lower aspect ratios and a larger relative crystallinity compared with that of NFC. Sulfuric acids have been extensively used for CNC preparation. Rånby first reported that colloidal suspensions of cellulose can be obtained by controlled sulfuric acid-catalyzed degradation of cellulose fibers.[109] Marchessault et al. demonstrated that colloidal suspension of sulfuric acid hydrolyzed CNC exhibited nematic liquid crystalline alignment.[110] The sulfuric acid reacts with the surface hydroxyl groups of cellulose to yield charged surface sulfate esters that promote dispersion of the individual CNC in water. However, the introduction of charged sulfate groups decreases the thermal stability of the CNC. The concentration of sulfuric acid in the hydrolysis reactions is usually controlled at a typical value of 65 wt%; this level is determined, on the one hand, by the requirement for efficient hydrolysis and sulfate ester formation, and, on the other hand, by the requirement to prevent the carbonization of cellulose by dehydration. The structures of CNC obtained by using similar sulfuric acid hydrolysis methods vary depending on the cellulose sources. For example, CNCs from cotton and Avicel have a length between 100 and 300 nm, whereas that from tunicin is several micrometers long and has a whisker-like morphology.[90] Hydrochloric acid was also used as a hydrolyzing agent for producing CNC. It can remove the amorphous area of cellulose but does not react with the surface hydroxyl groups of cellulose, resulting in CNC with large amount of CNC bundles. For the hydrochloric acid hydrolyzed cotton CNC,[86] the single CNC in the bundles were 100-400 nm long and 4-9 nm wide, with aspect ratios ranging from 11 to 100. The CNC displays high crystallinity and thermal stability with thermal degradation temperatures of 341.7 ℃. Thermally stable CNC can also be produced by other strong acid hydrolysis (such as phosphoric acid).

The highly crystalline CNC with high specific surface area are conducive for integrating with active materials or converting to carbon materials to develop electrodes with high specific surface areas. The chiral nematic structures derived from CNC are expected to be utilizedfor the development ofelectrodes with novel structures and properties for high-performance energy storage.

Table 1 Categories, fabrications, structures and characteristics of nanocellulose

Type	Sub-clarification	Schematic illustration	Source	Representative SEM/TEM/AFM image	Fabrication method	Surface chemical group	Ref.
NFC	Containing nanofiber bundles	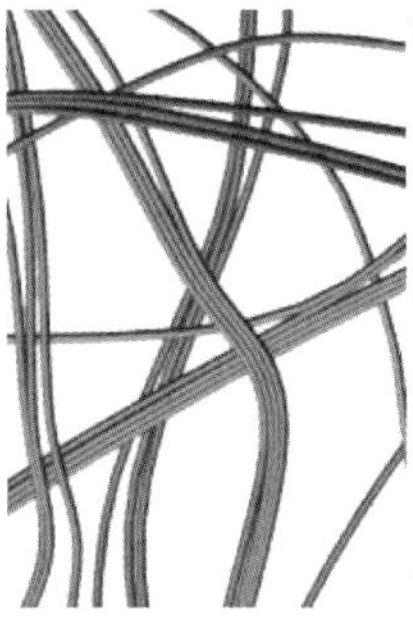	Wood	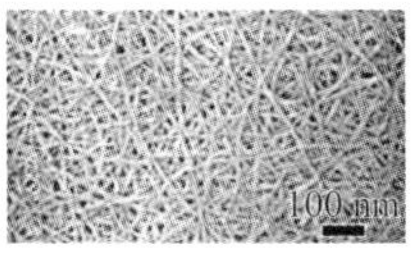 100 nm	Chemical purification +grinder	Hydroxyl groups	[82]
			Wood	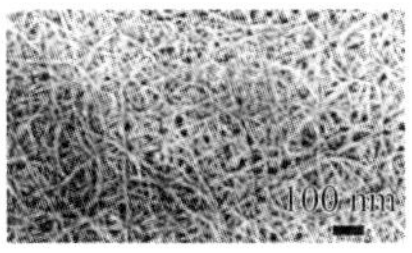 100 nm	Chemical purification +blender	Hydroxyl groups	[77]
			Wood	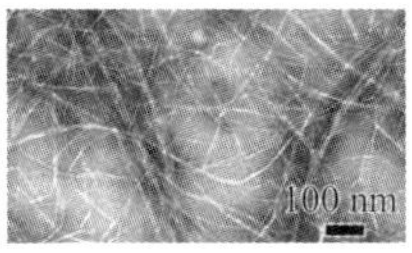 100 nm	Chemical purification +high intensity ultrasonicator	Hydroxyl groups	[86]
			Wood	 100 nm	Enzymatic hydrolysis+ refiner+ high-pressure homogenizer	Hydroxyl groups	[87]
	Individual nanofibers	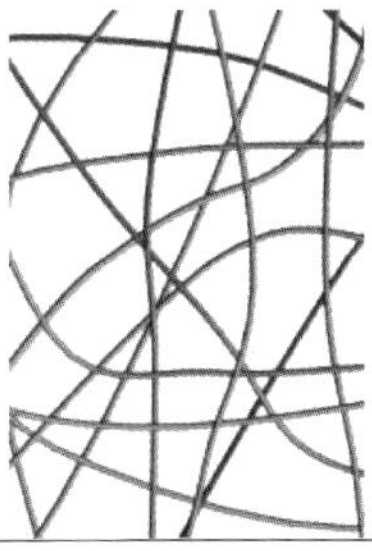	Wood	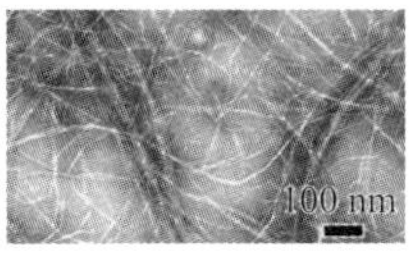 100 nm	TEMPO-mediated oxidation+magnetic stirring	Hydroxyl, carboxylate, aldehyde groups	[88]
			Wood	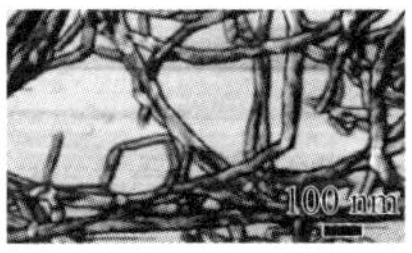 100 nm	Carboxymethylated+ high-pressure homogenizer	Hydroxyl, carboxyl groups	[89]
CNC	Containing CNC bundles		Cotton	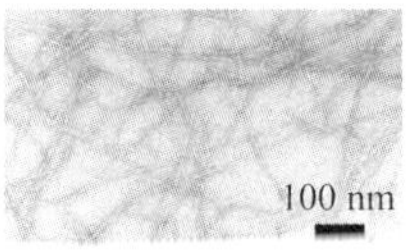 100 nm	Hydrochloric acid hydrolysis	Hydroxyl groups	[86]
	Individual CNC		Cotton	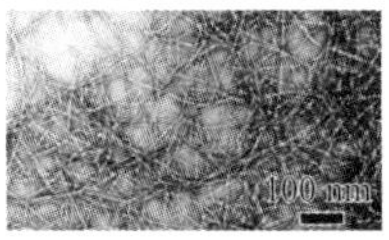 100 nm	Sulfuric acid hydrolysis	Hydroxyl, sulfate groups	[86]
			Tunicate	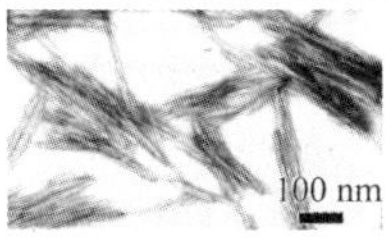 100 nm	Sulfuric acid hydrolysis	Hydroxyl, sulfate groups	[90]
BC	—	—	Coconut milk, sucrose	 100 nm	Secreted by *Acetobacter xylinum* FF-88	Hydroxyl groups	[91]

3.3 Bacterial cellulose (BC)

BC is another kind of high aspect ratio nanocellulose that is produced through a microbial fermentation process. BC has long been utilized as the raw material for *nata-de-coco*, an indigenous dessert food, for which one centimeter thick gel sheets fermented with coconut water are cut into cubes and immersed in sugar sirup.[111] BC is secreted extracellularly by certain bacteria belonging to the genera *Acetobacter*, *Agrobacterium*, *Alcaligenes*, *Pseudomonas*, *Rhizobium*, or *Sarcina*.[43,112] In contrast to NFC and CNC, which are mainly extracted from higher plant cellulose sources, BC is fabricated by a biotechnological assembly processes from low-molecular weight carbon sources, such as D-glucose.[44] The bacteria are cultivated in common aqueous nutrient media, and the BC is excreted as exopolysaccharide at the interface to the air, producing a thick gel composed of an interconnected 3D porous BC nanofiber network structure and ~99% water, called pellicle. The structure of the nanofiber network of BC can be controlled by changing the type of bacterial strain, the additives in the culture medium, the type and conditions of cultivation, and the drying processes in the post-processing stage.[113] There are apparent structural differences between BC and higher plant derived NFC such as wood NFC. BC is secreted as a ribbon-shaped fibril of around 100 nm in width and around 100 μm in length,[114] which is composed of 2–4 nm width nanofibrils. BC is free of functional groups (carbonyl, carboxyl) and is almost pure cellulose containing no lignin and other foreign substances.[111] Moreover, BC is characterized by very long polymer chain with polymerization up to 8000 and displays a distinct crystallinity of up to 90%. It should be noted that BC can also be utilized as a raw material to prepare NFC[106] and CNC.[115]

BC is high-purity cellulose with a web-like entangled structure and can be utilized to integrate with active materials or converted to carbon materials to develop flexible energy storage materials/devices. However, compared with NFC/CNC, which can be produced from abundant lignocellulosic resources, the cost of BC is relative high, although the industrial-scale production of BC has been achieved.

4 Supercapacitors

Supercapacitors, also known as ultracapacitors or electrochemical capacitors, offer a promising approach to meet the ever-increasing power demand owing to their high power density, superior rate capability, rapid charging/discharging rate, long cycle life, simple principles, fast dynamics of charge propagation and low maintenance cost.[116-118] Supercapacitors can be typically classified into two types: electrical double layer capacitors, which store energy through electrostatically accumulated charges at the electrode/electrolyte interface, and pseudocapacitors, which are based on rapid redox reactions at the electrodes for huge pseudocapacitance.

Nanocellulose can be integrated with active materials or converted to carbon materials as electrode materials for supercapacitors. It can also be processed as separator materials for supercapacitors.[119,120] The electrochemical performance of nanocellulose-derived materials, thatwere used as electrodes and tested under different conditionsfor supercapacitors, has been compared and summarized in Table 2.[121-139] The Ragone plot of supercapacitors fabricated from several typical nanocellulose-derived hybrid materials and carbon materials (including porous carbon materials and carbon hybrid materials) is shown in Fig. 4.[123,125,130-133,135-141] The electrochemical properties (such as energy density and power density) of those supercapacitors are effectively improved by integrating nanocellulose and nanocellulose-derived carbon materials with other active materials. Although several supercapacitors exhibit high energy density and power density, large space for spontaneously improving the energy density and power density is

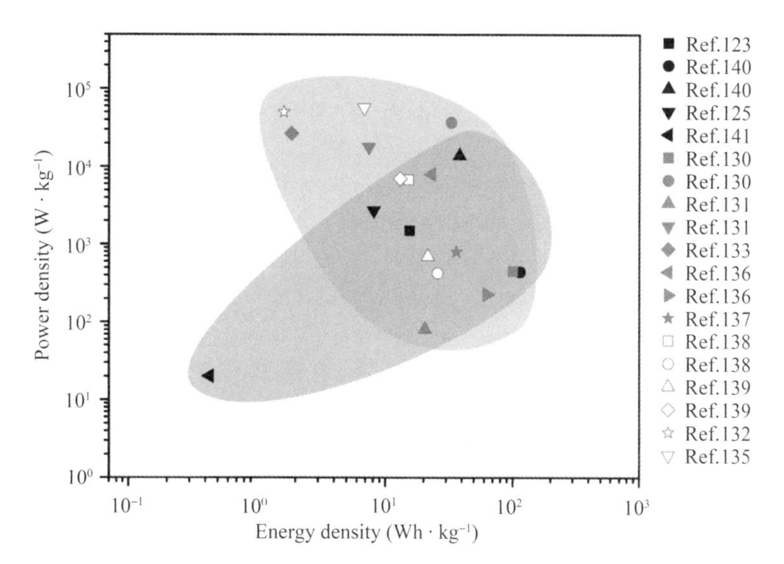

Fig. 4 Ragone plot of supercapacitors fabricated from several typical nanocellulose-derived hybrid materials and carbon materials. The black symbols within the light blue background region represent the supercapacitors fabricated from nanocellulose-derived hybrid materials. The red symbols within the light green background region represent the supercapacitors fabricated from nanocellulose-derived porous carbon materials and carbon hybrid materials. Data obtained from Ref. 123, 125, 130-133, and 135-141

still exist. In the ensuing sections, we discuss recent progress reported on the development of nanocellulose and its derived materials for supercapacitor applications.

4.1 Flexible supercapacitors based on NFC/BC-derived hybrid materials

Because of the high aspect ratio and web-like entangled structures, NFC/BC can form hybrids with active materials such as nanocarbon or conductive polymers to fabricate electrodes for flexible supercapacitors. The active materials can be integrated with NFC/BC usingdifferent composite approaches such as extrusion,[121] vacuum filtration[142] and freeze-drying,[125] resulting in the formation of various structures such as 1D fibers, 2D films/papers, and 3D aerogels/foams. Owing to the high aspect ratio and intrinsic mechanical strength, the introduction of NFC/BC into the hybrid materialsmakes the electrodes/supercapacitors flexible, which can store energy upon repeated bending, folding and/or compressing without a dramaticdecrease in the electrochemical performance.

4.1.1 Hybrids with nanocarbon

NFC/BC can be mixed with nanocarbonmaterials (Fig. 5) such as carbon nanotubes (CNTs) and graphene to form uniform solutions. 1D fibers can be produced from the composite solutions by wet spinning. Macrofiber with a uniform diameter of ~ 50 μm was fabricated by extrusion of NFC/single-walled CNT (SWCNT) suspensions in an ethanol coagulation bath (Fig. 5a) and drying in air under restricted conditions.[121] The SWCNTs were preferentially oriented along the axial direction of the macrofiber, whereas the NFC effectively prevented the aggregation of SWCNTs. The as-prepared nonwoven macrofiber mat wearable supercapacitors exhibited good electrochemical properties (Fig. 5b), outstanding tailorability, and damage reliability. The supercapacitors maintained 96.0% of their initial capacitance after 1500 bending cycles (Fig. 5c). After a fifth extreme deformation, the capacitance retention of the supercapacitors was approximately 93%.

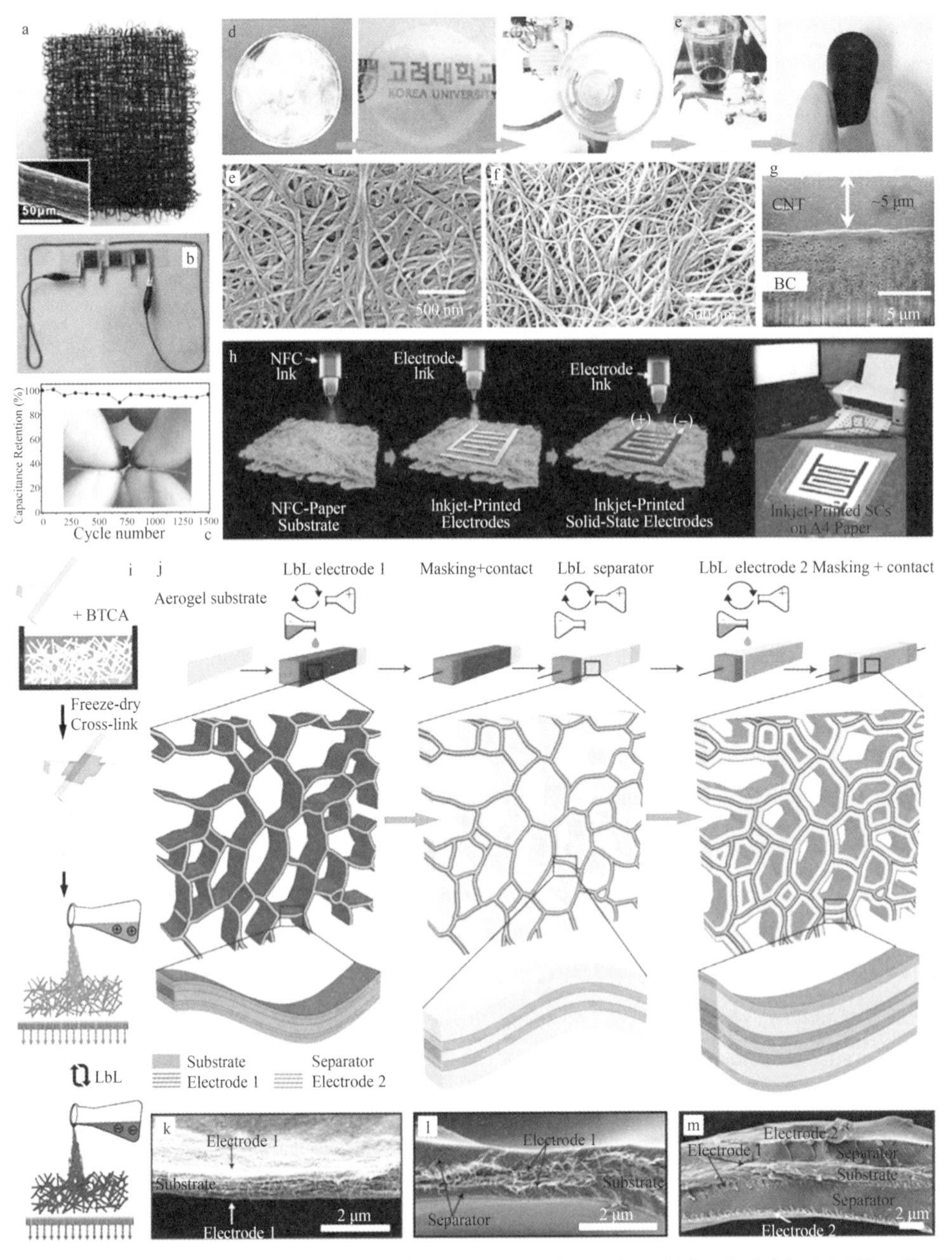

Fig. 5 The hybrid of NFC/BC with nanocarbon for supercapacitors. (a-c) The hybrid of NFC with SWCNT to fabricate 1D fibers:[121] (a) photograph of the wet NFC/SWCNT hybrid nonwoven macrofiber mat; (b) three nonwoven macrofiber mat wearable supercapacitors connected in series can illuminate a LED light; (c) the durability test of the nonwoven macrofiber mat wearable supercapacitor undergoing 1500 bending cycles. (a-c) are reproduced with permission from ref. 121. Copyright 2014 Royal Society of Chemistry. (d-h) The hybrid of NFC/BC with nanocarbon to fabricate 2D films/papers:[123,143] (d) fabrication process of BC papers coated with CNTs; SEM images of (e) a BC paper and (f) CNTs coated on a BC paper; (g) cross-sectional view of a BC/CNTs paper; (d-g) are reproduced with permission from ref. 123. Copyright 2012 American Chemical Society. (h) Schematic illustration of the stepwise fabrication procedure of the inkjet-printed supercapacitors using NFC mediated nanoporous mats as a primer layer. Reproduced from ref. 143. Copyright 2016 Royal Society of Chemistry. (i-m) The hybrid of NFC with nanocarbon to fabricate 3D aerogels:[141,144] (i) representations of the crosslinking in NFC, and aerogel construction and the LBL assembly on aerogels; Reproduced from ref. 144. Copyright 2013 Wiley-VCH. (j) schematics of the LBL process used to assemble 3D devices in an NFC aerogel and cross-section SEM images of (k) the first PEI/CNT electrode, (l) the PEI/CNT electrode with separator, and (m) the full device. Scale bars in (k-m) are 2 μm. (j-m) are Reproduced from ref. 141. Copyright 2015 Springer Nature

2D films/papers can be fabricated by mixing NFC/BC and nanocarbon solutions followed by filtration and drying. Films containing the covalent intercalation of graphene oxide (GO) and BC were fabricated by filtration of the reaction mixture followed by drying. [142] The composite films exhibited a tensile strength of 18.5 MPa and an elongation at break of 24%. With an electrical conductivity of 171 $S \cdot m^{-1}$, the composite films demonstrated specific capacitance of 160 $F \cdot g^{-1}$ at 0.4 $A \cdot g^{-1}$. 2D films/papers can also be fabricated by the deposition of nanocarbons on the NFC/BC films/papers. Kang et al. reported the preparation of 2D papers by the deposition of a CNT layer on the BC paper through a vacuum filtering process (Fig. 5d-g). [123] The CNTs were coated on and integrated with the BC paper through hydrogen bonding and van der Waals' interactions. Flexible supercapacitors fabricated by incorporating ionic liquid based polymer gels between two BC papers coated with CNTs showed a specific capacitance of 50.5 Fg^{-1}, aspecific energy density of 15.5 $mWhg^{-1}$, and a specific power density of 1.5 Wg^{-1} (measured at 1 Ag^{-1}). Owing to the good interfacial quality between the three different layers, the supercapacitor performance was generally maintained over 200 bending cycles to a bending radius of 3 mm. Recently, Lee et al. reported the fabrication of all-inkjet-printed, solid-state flexible supercapacitors on paper (Fig. 5h). [143] A NFC-mediated nanomat was inkjet-printed as a primer layer to enable high-resolution images. CNT-assisted photonic interwelding of Ag nanowires was introduced onto the CNT/activated carbonelectrodes to improve the electrical conductivity. The ([BMIM][BF_4]/ETPTA) solid-state electrolyte was applied to the inkjet-printed electrodes through inkjet printing, after which the UV-crosslinking process was performed. The inkjet-printed supercapacitors exhibited reliable electrochemical performance and good mechanical flexibility. Even after 1000 bending cycles with a bending radius of 2.5 mm, the supercapacitors still retained their structural integrity without impairing the cell capacitance.

3D aerogels/foams for supercapacitor can be fabricated through freeze-drying/critical point drying/supercritical drying of mixtures of NFC/BC and nanocarbons. The as-developed aerogels/foams exhibited web-like entangled structures, high porosity, low density, high conductivity, as well as the synergy effect of NFC/BC and nanocarbons, which makes them good electrodes for flexible supercapacitors. NFC/reduced GO (RGO) hybrid aerogels were prepared by supercritical CO_2 drying of the corresponding hydrogels. [124] Flexible supercapacitors were developed using the hybrid aerogel film as an electrode, and exhibited areal capacitance of 158 $mF \cdot cm^{-2}$, and a maximum areal energy density of 20 $\mu \cdot Wh \cdot cm^{-2}$ at an areal power density of 15.5 $mW \cdot cm^{-2}$. Aerogels derived from NFC/RGO/CNT were developed by freeze-drying of aqueous dispersion of NFC/GO nanosheets (GONSs)/CNT, followed by thermal reduction of the GONSs. [125] The supercapacitors assembled by the compressed NFC/RGO/CNT aerogel exhibited a specific capacitance of 252 $F \cdot g^{-1}$, an energy density of 8.1 mWh g^{-1} at a power density of 2.7 $W \cdot g^{-1}$, as well as good cyclic stability with more than 99.5% capacitance retention after 1000 charge/discharge cycles. Recently, Hamedi et al. pioneeredthe development of a layer-by-layer (LBL) assembly technology for coating functional materials onto NFC aerogel surfaces (Fig. 5i). [144] Prior to LBL assembly, covalent crosslinking of the NFC aerogels was conducted to avoid the disintegration of the NFC aerogels during the LBL assembly process. Active materials such as CNTs were assembled onto NFC aerogels by LBL, and enhanced compressive strength, super elasticity in the wet state, and elastic mechanoresponsive resistance were achieved. A SWCNT-functionalized NFC aerogel electrode showed a specific capacitance of 419 $F \cdot g^{-1}$ by accounting for the measured weight of the active LBL layer. Using similar covalently crosslinked NFC aerogels as scaffolds, CNT electrodes and separator were LBL deposited on the NFC aerogel surface, resulting in the assembly of 3D supercapacitors (Fig. 5j).[141] The pore walls of the NFC aerogel weregradually thickenedwhen the first electrode, the separator, and the second electrode were added(Fig. 5k-m). The 3D supercapacitor showed stable operation over 400 cycles with a capacitance of 25 $F \cdot g^{-1}$, and was fully functional even at compressions up to 75%.

Table 2 Examples of different nanocellulose-derived materials for supercapacitor electrodes reported in the literature

Classification	Sub-classification	Nanocellulose type	Nanocellulsoe process	Electrode	Electrolyte	Performance: Capacity/Energy density/Power density	Cycles	CR[a) [%]	Ref.
Flexible supercapacitors based on nanocellulose-derived hybrid materials	With nanocarbon	TEMPO-oxidized NFC	Hybrid with SWCNTs	NFC/SWCNT hybrid nonwoven macrofiber mat	H_3PO_4 - Poly-(vinyl alcohol) gel	Area capacitance is about 6.0 $mF \cdot cm^{-2}$ at 0.02 $mA \cdot cm^{-2}$; the energy density is 0.7 $\mu Wh \cdot cm^{-2}$ at a power density of 2.4 $mW \cdot cm^{-2}$	5000	97.0	[121]
		Wood TEMPO-oxidized NFC	Hybrid with MWCNTs	NFC/MWCNT hybrid aerogel film	H_2SO_4-Poly-(vinyl alcohol) gel	Specific capacitance is up to 178.0 $F \cdot g^{-1}$ at 5 $mV \cdot s^{-1}$; areal energy density: 20.0 $mWh \cdot cm^{-2}$; maximum power density: about 13.6 $mW \cdot cm^{-2}$	1000	99.9	[122]
		BC	Hybrid with CNTs	CNT-coated BC paper	Ionic liquid based polymer gel	Specific capacitance is 50.5 $F \cdot g^{-1}$ at 1 $A \cdot g^{-1}$; the energy density is 15.5 mWh g^{-1} at a power density of 1.5 W g^{-1}	5000	~99.5	[123]
		TEMPO-oxidized NFC	Hybrid with RGO	NFC/RGO hybrid aerogel film	H_2SO_4-Poly-(vinyl alcohol) gel	Specific capacitance is 203.0 $F \cdot g^{-1}$ at 0.7 $mA \cdot cm^{-2}$; the maximum areal energy density is 20.0 $\mu Wh \cdot cm^{-2}$ at an areal power density of 15.5 $mW \cdot cm^{-2}$	5000	99.1	[124]
		Wood NFC	Hybrid with RGO and CNTs	NFC/RGO/CNT hybrid aerogel film	H_2SO_4-Poly-(vinyl alcohol) gel	Areal capacitance can be 216.0 $mF \cdot cm^{-2}$; the areal energy density is 28.4 $\mu Wh \cdot cm^{-2}$ at an areal power density of 9.5 $mW \cdot cm^{-2}$	1000	>99.5	[125]

Table 2 continued

Classification	Subclassification	Nanocellulose type	Nanocellulsoe process	Electrode	Electrolyte	Performance: Capacity/Energy density/Power density	Cycles	CR[a) [%]	Ref.
Flexible supercapacitors based on nanocellulose-derived hybrid materials	With conductive polymer	Cladophora NFC with quaternary amine groups on the surface	Hybrid with PPy	NFC/PPy composite	2M NaCl	The volumetric capacitance is 122.0 F · cm^{-3} at 300 mA · cm^{-2}; the volumetric energy and power densities are 3.1 mWh · cm^{-3} and 3.0 W · cm^{-3}, respectively	5000	93.0	[126]
		Cladophora NFC	Hybrid with PEDOT	NFC/PEDOT composite paper	1M H_2SO_4	The volumetric energy and power densities are 1.1 mWh · cm^{-3} and 900.0 mW · cm^{-3}, respectively	15000	93.0	[127]
	With nanocarbon and conductive ymer	BC	Hybrid with PPy and MWCNTs	BC/PPy/MWCNTs hybrid membrane	2M LiCl	Areal capacitance can be 590.0 mF · cm^{-2} at 1 mA · cm^{-2}	5000	94.5	[128]
		BC	Hybrid with PANI and graphene	BC/PANI/graphene composite film	1M H_2SO_4	The areal capacitance is 1.9 F · cm^{-2} at 0.25 mA · cm^{-2}; the maximum energy density is about 0.2 mWh · cm^{-2} at a power density of about 0.1 mW · cm^{-2}	5000	~53.6	[129]
Supercapacitors based on nanocellulose-derived carbon materials	Porous carbon materials	BC	Pyrolyzation and activated by KOH	NiCoAl-layered double hydroxide as the positive electrode; activated BC-derived carbon as the negative electrode	6M KOH	The maximum energy density is 100.0 Wh · kg^{-1} at a power density of 451.0 W · kg^{-1}	10000	113.0	[130]
		BC	Hybrid with potassium citrate followed by pyrolyzation and acid washing	BC-derived carbon nanofiber-bridged porous carbon nanosheet	1M Na_2SO_4	The energy density is 20.4 Wh · kg^{-1} at a power density of 81.8 W · kg^{-1}	10000	94.8	[131]

Table 2 continued

Classifi-cation	Sub-classification	Nanocellulose type	Nanocellulsoe process	Electrode	Electrolyte	Performance: Capacity/Energy density/Power density	Cycles	CR[a)] [%]	Ref.
Supercapacitors based on nanocellulose-derived carbon materials	Porous carbon materials	BC	Pyrolyzation and activated by mixed gas of CO_2/argon; N-doped via a hydrothermal synthesis	Activated N-doped BC-derived carbon nanofiber	H_2SO_4-Poly-(vinyl alcohol) gel	Maximum energy density: 6.1 Wh · kg^{-1}; maximum power density: 390.5 kW · kg^{-1}	5000	~95.9	[132]
		BC	Hybrid with $NH_4H_2PO_4$ followed by pyrolyzation	N, P codoped BC-derived carbon nanofibers	2M H_2SO_4	The maximum power density is 186.0 kW · kg^{-1}	4000	-	[133]
	Carbon hybrid materials	BC	Hybrid with PPy and GO followed by pyrolyzation	N-doped BC-derived carbon nanofiber/RGO/BC	6M KOH/1M H_2SO_4	Areal maximum energy density: 0.1 mWh · cm^{-2} in KOH and 0.3 mWh · cm^{-2} in H_2SO_4; areal maximum power density: 27.0 mW · cm^{-2} in KOH and 37.5 mW · cm^{-2} in H_2SO_4	10000	~99.6	[134]
		BC	Pyrolyzation; coated with MnO_2; N-doped via a hydrothermal method	BC-derived carbon nano-fibers coated with MnO_2 as the positive electrode; N-doped BC-derived carbon nanofibers as the negative electrode	1M Na_2SO_4	The maximum energy density is 32.9 Wh · kg^{-1}; The maximum power density is 284.6 kW · kg^{-1}	2000	~95.4	[135]
		BC	Hybrid with PANI followed by pyrolyzation; coated with MnO_2; activated with KOH	N-doped BC-derived carbon coated with MnO_2 as the positive electrode; KOH activated N-doped BC-derived carbon as the negative electrode	1M Na_2SO_4	Specific capacitance is 113.0 F · g^{-1} at 2 mV · s^{-1}; the maximum energy density is 63.0 Wh · kg^{-1} at a power density of 227.0 W · kg^{-1}	5000	92.0	[136]

Table 2 continued

Classifi-cation	Sub-classification	Nanocellulose type	Nanocellulsoe process	Electrode	Electrolyte	Performance: Capacity/Energy density/Power density	Cycles	CR[a) [%]	Ref.
Supercapacitors based on nanocellulose-derived carbon materials	Carbon hybrid materials	BC	Hybrid with PANI followed by pyrolyzation; deposition of nickel-cobalt layered double hydroxide (Ni-Co LDH) nanosheets	N-doped BC-derived carbon nano-fiber@ LDH composites as the positive electrode; N-doped BC-derived carbon nanofibers as the neg-ative electrode	6M KOH	The maximum energy density is 36.3 Wh · kg^{-1} at a power density of 800.2 W · kg^{-1}	2500	89.3	[137]
		BC	Pyrolyzation; growing Ni_3S_2 nanopar-ticles via a hydrothermal method	BC-derived carbon nano-fiber coated Ni_3S_2 nano-particles as positive elec-trode; BC-derived carbon nanofiber as negative electrode	2M KOH	Specific capacitance is 56.6 F · g^{-1} at 1 A · g^{-1}; the energy density is 25.8 Wh · kg^{-1} at a power density of 425.0 W · kg^{-1}	2500	97.0	[138]
		BC	Dissolution/gelation/car-bonization; in-situ growth of NiS particles via a hydroth-ermal mathod	NiS/BC-derived carbon aerogel composite as positive elec-trode; BC-derived carbon aerogel as negative electrode	2M KOH	The energy density is ~21.5 Wh · kg^{-1} at a power density of 700.0 W · kg^{-1}	10000	~87.1	[139]

a) CR = capacitance retention.

4.1.2 Hybrids with conductive polymers

Conductive polymers such as polypyrrole (PPy),[145-150] poly aniline (PANI),[151] and poly (3, 4-ethylenedioxythiophene) (PEDOT)[127] are usually polymerized *in situ* and form a homogenous layer around NFC/BC, resulting in composite films/papers with high electrical conductivity. To improve the electrochemical performance, the composites are usually thick with large active mass loadings, which often sacrifice the electrode flexibility. Thus, improving the interfacial interaction between nanocellulose and the conductive polymers, wrapping the nanocellulose with a uniform and sufficiently thin conductive polymer layer, and decreasing the electrode thickness are critical for fabricating flexible electrodes. Wang et al. reported the introduction of quaternary amine groups on the surface of NFC to prepare NFC with cationic (c-NFC) surface charges(Fig. 6a).[126] After the polymerization of pyrrole on the c-NFC(Fig. 6b), the composites delivered high normalized gravimetric (127 F · g^{-1}) and volumetric (122 F · cm^{-3}) capacitances at 300 mA · cm^{-2}. Symmetrical supercapacitor using these composites as electrodes exhibited a volumetric energy and power densities of 3. 1 mWh · cm^{-3} and 3 W · cm^{-3} in aqueous electrolyte, respectively. The functionality of the devices was maintained even in different mechanically challenging states(Fig. 6c-e). Wang et al. also reported the fabrication of flexible NFC/PEDOT composite papers by chemical polymerization of 3,4-ethylenedioxythiophene at room temperature in the presence of NFC followed by vacuum filtration and drying (Fig. 6f).[127] A uniform PEDOT coating layer was formed on NFC (Fig. 6g) and the composite papers exhibited a surface area of 137 m^2 · g^{-1}, a sheet resistance of 1. 4 Ω^{-1}, and an active mass loading of 7. 3 mg · cm^{-2}. Symmetric NFC/PEDOT paper-based supercapacitors demonstrated high specific electrode capacitances(920 mF · cm^{-2}) and good cycling stability (93% capacity retention after 15000 cycles at 30 mA · cm^{-2}) in 1M H_2SO_4. Under harsh mechanical conditions involving the twisted and bent states, the device exhibited good stability (Fig. 6h-j) and retained 96. 5% of its initial specific capacitance after 1200 cycles. Moreover, PPy/BC composite membranes,[152,153] PPy/cobalt sulfide/BC compositemembranes,[154] PPy/nickel sulfide/BC composite membranes,[155] PPy/copper sulfide/BC composite membranes,[156] and PANI/silver NFC aerogels,[157] have all been fabricated as electrodes for developing flexible supercapacitors.

Moreover, PPy/BC composite membranes,[152,153] PPy/cobalt sulfide/BC composite membranes,[154] PPy/nickel sulfide/BC composite membranes,[155] PPy/copper sulfide/BC composite membranes,[156] and PANI/silver NFC aerogels,[157] have all been fabricated as electrodes for developing flexible supercapacitors.

4. 1. 3 Hybrids with nanocarbon and conductive polymers

These combinationsgive electrodes enhanced flexibility and mechanical robust ness, and maintained their capacitances.[129,158,159] A flexible BC/RGO/PPy composite film was prepared through an *in situ* polymerization and filtering method. The assembled symmetric supercapacitor delivered an areal capacitance of 1. 67 F · cm^{-2}, an areal energy density of 0. 23 mW h cm^{-2}, and a maximum power density of 23. 5 mW · cm^{-2}. 160 An NFC/GO/PPy composite paper was produced by straightforward polymerization of a PPy layer both on the NFC attached to the GO sheets and the GO itself.[161] The NFC reinforced the entire electrode yielding a robust and compact electrode that still contained a sufficient number of mesopores to allow the material to undergo fast PPy charge and discharge reactions. Symmetrical supercapacitors containing NFC/GO/PPy composite paper were flexible, with 485% capacitance retention over 16 000 cycles at 5 A · g^{-1} and high volumetric specific capacitance (301 F · cm^{-3} based on PPy).

BC/multiwalled CNTs (MWCNTs)/PANI compositefilms were produced by vacuum filtration of a MWCNTs layer onto BC paper followed by electrodeposition of PANI (Fig. 7a–h).[162] Owing to the porous structure and electrolyte absorption properties of the BC paper, the flexible BC/MWCNTs/PANI hybrid electrode exhibited high specific capacitance (656 F · g^{-1} at 1 A · g^{-1}) and good cycling stability with capacitance loss of less than 0. 5% after 1000 cycles at 10 A · g^{-1}. Moreover, the electrodes with high mass loading in the range of 7 – 12 mg · cm^{-2} were prepared by using BC/PPy nanofibers in combination with MWCNTs through a vacuum-filtering method (Fig. 7i–o).[128] The as-prepared symmetric supercapacitors exhibited a capacitance of 590 mF · cm^{-2} and good cycling stability with 94. 5% of capacitance retention after 5000 cycles.

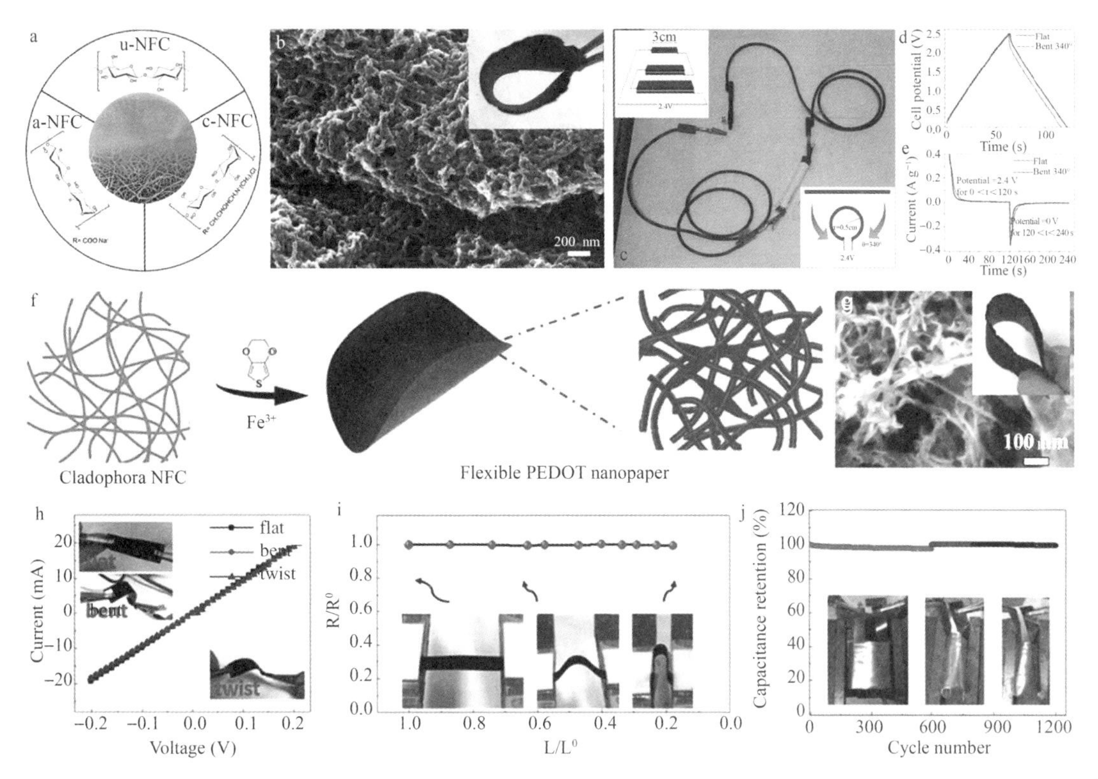

Fig. 6 The hybrid of NFC with conductive polymers for supercapacitors. (a-e) The hybrid of NFC with PPy:[126] (a) photograph of NFC paper and the molecular structure of unmodified and modified NFC; (b) SEM image of PPy/c-NFC composite; (c) photo of a red LED powered by three flexible PPy/c-NFC-based supercapacitors coupled in series for a bending angle of ~340°; (d) charge/discharge profiles for the in-series supercapacitors under flat and bending conditions for an applied current of ±20 mA as well as (e) the corresponding current responses owing to potential steps to 2.4 and 0 V, respectively. (a-e) are reproduced with permission from ref. 126. Copyright 2015 American Chemical Society. (f-j) The hybrid of NFC with PEDOT:[127] (f) schematic illustration of the manufacturing of PEDOT nanocomposite paper electrodes; (g) SEM image of PEDOT nanopaper; (h) conductivity of PEDOT nanopaper in various geometrical states; (i) resistance (R) of PEDOT nanopaper in different bending states normalized with respect to the resistance (R_0) of the flat PEDOT nanopaper, in which L_0 is the length of the flat paper and L is the distance between the two paper ends in the different bending states; (j) the cycling stability during 1200 cycles for the flexible device under different bending conditions, in which the insets show photographs of the device for the different test conditions. (f-j) are reproduced with permission from ref. 127. Copyright 2016 Royal Society of Chemistry

Overall, NFC/BC-derived hybrid electrodes hold great promise for flexible supercapacitors. By introducing active materials, the electrochemical performance of the hybrid electrodes can be increased, whereas the addition of NFC/BC can improve the mechanical and surface properties of the hybrid electrodes in several ways. The high aspect ratio NFC/BC formed strong entangled networks with the active materials, which made the hybrid electrodes flexible, bendable, foldable, and compressible without deterioration of the electrochemical performance. The ratio of active materials plays a critical role on the electrode performance. The interface interactions between the active materials and the NFC/BC should be optimized to maximize the synergistic effects and enhance the activity of supercapacitor and spontaneously maintain its mechanical

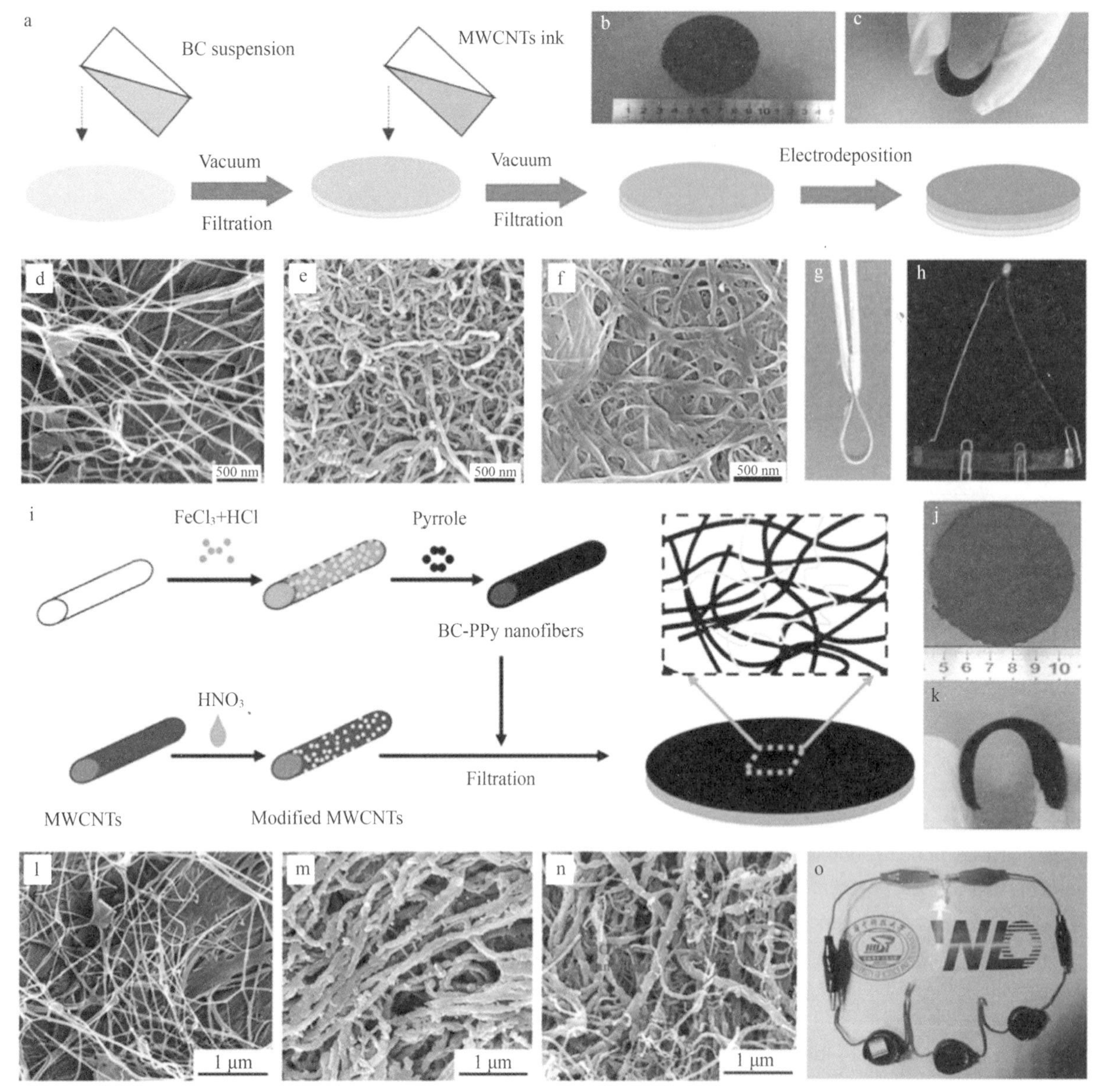

Fig. 7 The hybrid of BC with nanocarbon and conductive polymers for supercapacitors. (a–h) The hybrid of NFC with MWCNTs and PANI: 162 (a) schematic of the fabrication process of BC/MWCNTs/PANIx freestanding paper electrodes; (b, c) digital image of BC/MWCNTs paper and BC/MWCNTs/PANI paper electrode; SEM images of (d) BC paper, (e) BC/MWCNTs paper and (f) BC/MWCNTs/PANI paper electrode; the optical images of (g) flexible supercapacitor device based on BC/MWCNTs/PANI and (h) a red LED lighted by such devices. (a–h) are reproduced with permission from ref. 162. Copyright 2014 Wiley-VCH. (i–o) The hybrid of NFC with MWCNTs and PPy: 128 (i) schematic of the synthesis procedure of BC/MWCNTs/PPy hybrid membrane; optical images of (j) BC/MWCNTs/PPy membrane and (k) the bending state; SEM images of (l) BC membrane, (m) BC/PPy membrane, and (n) BC/MWCNTs/PPy hybrid membrane; (o) optical image of a red LED lit by three charged supercapacitors connected in series. (i–o) are reproduced with permission from ref. 128. Copyright 2014 Springer

strength and high flexibility.

4.2 CNC-derived hybrid materials

CNC, which has a higher specific surface area compared with that of NFC/BC, can be integrated with active materials in several ways to fabricate electrodes for supercapacitors. CNC can be integrated with conductive polymers such as PPy through electrodeposition methods.[163] A cotton CNC was subjected to TEMPO-mediated oxidation to convert the surface primary hydroxyls to carboxylate functionalities. Then, the TEMPO-oxidized CNC was electrodeposited with PPy to fabricate CNC/PPy nanocomposites. A symmetrical supercapacitor using CNC/PPy nanocomposites as the electrodes was constructed and tested at an operating voltage of 1 V. The specific energy density of the supercapacitor was 8.34 $W \cdot h \cdot kg^{-1}$ at 1 $mA \cdot cm^{-2}$. The supercapacitor retained 70% of its initial capacitance after cycling 10 000 times and 47% of its capacitance after an excessive 50000 cycles. The good cycling stability was mainly attributed to the stability of the individual electrodes owing to the highly porous structure of the CNC/PPy nanocomposites and the thin coating of PPy onto individual CNC, which facilitates the ion and solvent movements during the charge/discharge cycles and thus makes the composites with stable volumetrically. CNC can also be integrated with conductive polymers through *in situ* polymerization methods.[164] PPy was readily coated on the surface of TEMPO-oxidized CNC and formed a conductive shell on individual CNC through the *in situ* chemical polymerization of pyrrole. The 1D structure of the CNC not only facilitated the efficient transport of electrons along one controllable direction, but also provided a very low percolation concentration of CNC/PPy to achieve a conductive network. The electrical conductivity of the CNC/PPy composites approached 4 $S \cdot cm^{-1}$ with a capacitance of 248 $F \cdot g^{-1}$ at 0.01 $V \cdot s^{-1}$, retaining 90% capacitance at up to 0.1 $V \cdot s^{-1}$. To ensure a uniform PPy coating, PVP was used as a surface modifier through physical adsorption onto the CNC surface prior to the introduction of pyrrole.[165] PVP introduced hydrophobicity on the CNC surface, making it more favorable for the growth of the hydrophobic PPy shell, which acted as a steric stabilizer preventing further agglomeration of the nanoparticles. The CNC/PVP/PPy composites exhibited electronic conductivity of 36.9 $S \cdot cm^{-1}$, a specific capacitance of 322.6 $F \cdot g^{-1}$, and cycling stability with less than 9 and 13% capacitance loss after 1000 and 2000 cycles, respectively.

Because CNC is short with a low aspect ratio, CNC-derived bulk materials such as aerogels are usually mechanically weak[86] and easily broken when bent or compressed. Cranston et al. developed a chemical crosslinking method to produce CNC aerogels with tailorable mechanical performance and shape recovery abilities (Fig. 8a–e).[166] The CNC aerogels were fabricated based on the chemical crosslinking of aldehyde-modified CNCs and hydrazide-modified CNCs, which formed covalent hydrazone crosslinks immediately upon contact. These CNC aerogels acted as universal substrates for a variety of active nanoparticles, such as PPy nanofibers, PPy-coated CNTs, and spherical manganese dioxide nanoparticles during the aerogel assembly, offering enough accessible surface area of active material, which promoted charge storage. The CNC-active nanoparticle hybrid aerogels were strong and remained intact when compressed in both air and aqueous electrolytes. Symmetric coin-cell supercapacitors were assembled containing two hybrid aerogels separated by a porous polyethylene membrane immersed in saturated Na_2SO_4 aqueous electrolyte. The areanormalized specific capacitances at 2 $mV \cdot s^{-1}$ for PPy nanofibers, PPy-coated CNTs, and spherical manganese dioxide nanoparticle devices were 3.32, 2.42, and 2.14 $mF \cdot cm^{-2}$, respectively.

Using similar CNCs as building blocks, anaerogel-based current collector composed of crosslinked CNCs and MWCNTs was produced using bile acid as a dispersant (Fig. 8f–k).[167] PPybased electrodes with a mass loading up to 17.8 mg cm^{-2} were fabricated based on *in situ* polymerization of pyrrole inside the CNC/MWCNT aerogel. The CNC/MWCNT/PPy electrodes were flexible and had good compression stability at 80% compression, exhibited a capacitance of 2.1 F · cm^{-2} with a mass ratio of active material/current collector of 0.57. The electrodes and devices showed good capacitance retention at high charge/discharge rates and good cyclic stability and capacitance stability during compression and bending cycles.

Overall, compared with NFC/BC, highly crystalline CNC has a higher specific surface area to integrate with active materials. A uniformly, full, and thin coating of an active materials on individual CNC fibers is effective for facilitating the ion movements during the charge/discharge cycles. The CNC-derived bulk materials such as films and aerogels are fragile and easily broken when bent or compressed, mainly because of the low aspect ratio of CNCs, which limits their use in several practical applications such as for the development of flexible supercapacitors. Although several methods such as chemical crosslinking can be utilized to fabricate flexible CNC electrodes, the complicated synthesis routes and sacrificing of the specific surface area of the bulk materials is still a problem that should not be ignored. The low thermal degradation temperature of H_2SO_4-hydrolyzed CNCs hampers the thermal stability of the supercapacitors. Thus, the surface/interface properties and interactions of the CNCs and the active materials should be rationally designed and controlled to achieve uniform integration of active materials and CNCs with a high active material loading, and to make composites with high mechanical strength, flexibility, and thermal stability without decreasing the specific surface area of the bulk materials.

4.3 Nanocellulose-derived carbon materials

High temperature pyrolysis under an inert atmosphere can convert nanocellulose bulk materials into conductive carbon materials.

The carbon materials possessed an ultralow apparent density, high specific surface area, and high electric conductivity. Nanocellulose-derived porous carbon materials can be directly utilized as electrodes for supercapacitors. The electrochemical properties of the carbon materials can be further improved through activation treatment, doping with heteroatoms, and the formation of hybrids with active materials. Recently, nanocellulose-derived porous carbon materials and carbon hybrid materials have been frequently developed as the electrode materials for supercapacitors.

4.3.1 Porous carbon materials

NFC and BC were utilized as building blocks to prepare films/aerogels for pyrolysisand conversion to porous carbon aerogels (Fig. 9a and b).[168-170]

The as-fabricated carbon aerogels shrank in three dimen sions after the pyrolysis process and the 3D structure of NFC/BC aerogels was inherited. Although the carbon aerogels possessed ultra-low density, extraordinary compressibility, and high electric conductivity, they exhibited low electrochemical properties for supercapacitors. For example, the specific capacitance of the BC-derived carbon aerogelswas only 77.7 F · g^{-1} at 1 A · g^{-1},[135] mainly because of its relatively low surface area and wettability. Activation treatment can apparently increase the specific surface areas. The activation of BC-derived carbon aerogels with a mixed gas of CO_2/argon exhibited a high BET surface area of 699.7 m^2 · g^{-1} compared to that of 286.4 m^2 · g^{-1} for the

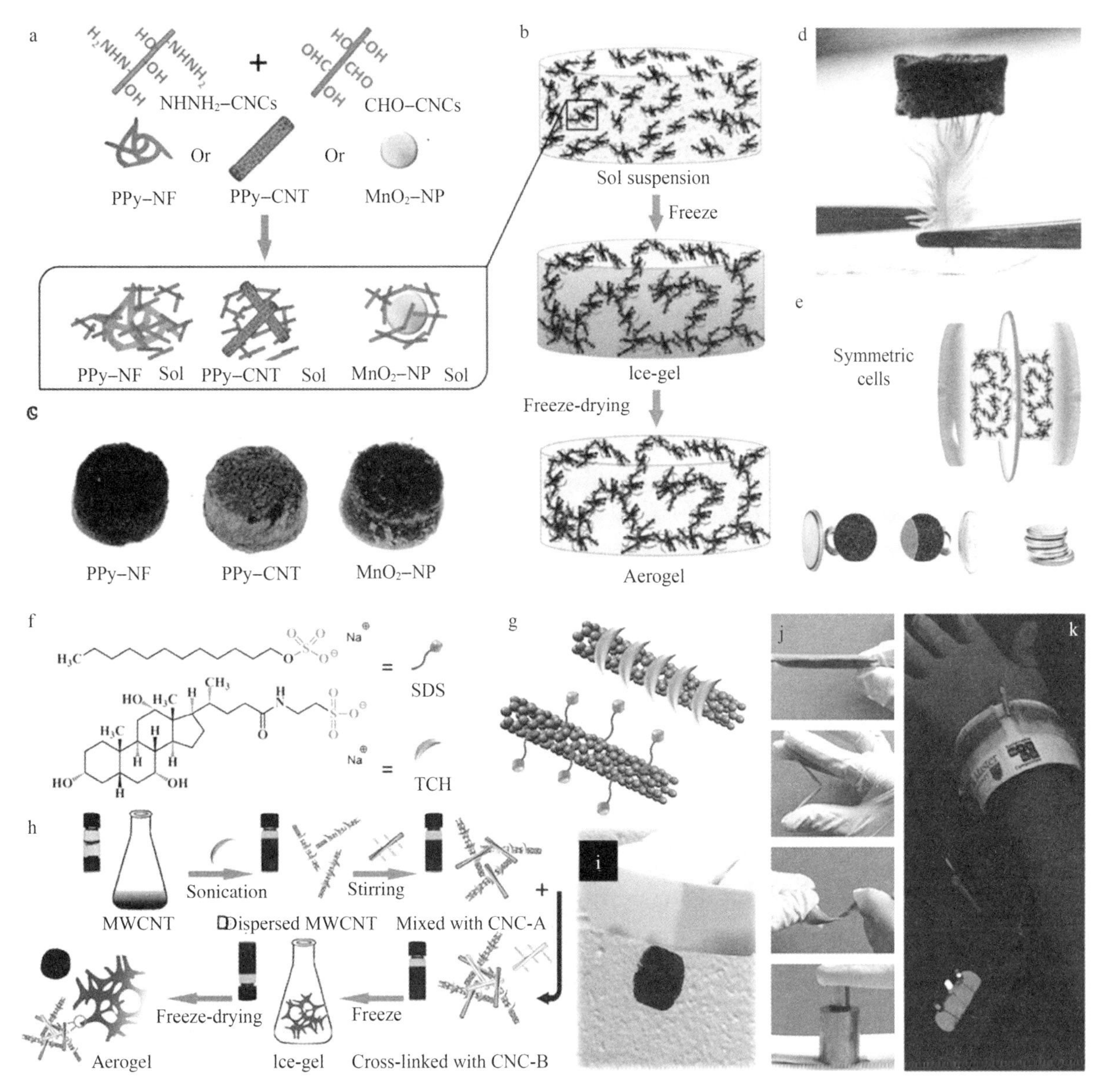

Fig. 8 CNC-derived hybrid materials for supercapacitors.[166,167] (a) Schematic representation of the aerogel components, including functionalized CNC and capacitive nanoparticles that form the initial suspension of crosslinked clusters. (b) The sol-gel process used to prepare aerogels. (c) Photograph of the final hybrid aerogels. (d) Photograph emphasizing the lightweight nature of a hybrid aerogel resting on top of a feather. (e) Schematic showing the fabrication of symmetric supercapacitor cells. (a-e) are reproduced with permission from ref. 166. Copyright 2015 Wiley-VCH. (f) Chemical structures and (g) MWCNT dispersion mechanisms of SDS and TCH. (h) Scheme of CNC/MWCNT aerogel fabrication method. (i) Aerogel with a density of 21 ± 3 mg · cm^{-3} attached to a weighing dish through static electricity. (j) Electrochemical supercapacitor cell: as fabricated and under bending, twisting, and partially compressing stress; (k) Wearable cells that are connected in series are able to power LED bulbs. (f-k) are reproduced with permission from ref. 167. Copyright 2016 Wiley-VCH

carbon aerogels without activation.[132] The CV curves of the activated carbon aerogel had roughly rectangular shapes and displayed almost mirror images, which demonstrated good capacitive behavior and reversibility. Activation of the carbon aerogels with potassium hydroxide[130,136,171] or potassium citrate[131] is another strategy to improve the specific surface areas, which ensures more effective electroactive sites for charge accommodation and thus improvement of the specific capacitance of the carbon aerogels. Potassium hydroxide was also found to play a critical role for facilitating the crosslinking of BC and acting as a hard template to generate macropores during the carbonization process. The obtained carbon aerogels had a 3D interconnected honeycomb-like hierarchical network structure (Fig. 9c),[130] exhibited a high specific surface area of 1533 $cm^2 \cdot g^{-1}$, and a high specific capacitance of 422 $F \cdot g^{-1}$ in 6 M KOH, as well as good rate capability with 73.7% of capacitance retention at 500 $mV \cdot s^{-1}$.

The electrochemical properties of porous carbon materials can be improved by doping with heteroatoms, becausethe heteroatom dopants can modulate the physical and chemical properties of the carbon material to generate more reactive sites. The doping with heteroatoms has been achieved mainly by three routes:

(1) Annealing of NFC/BC aerogels/foams with heteroatom-rich compounds. Because NFC/BC aerogels/foams are hydrophilic and exhibit high porosity, the heteroatom-rich compounds can easily fill with the pores of the aerogels/foams. After high temperature pyrolysis under an inert atmosphere, heteroatom doping was achieved. Various doped carbon aerogels were produced by Wu *et al.* through the pyrolysis of BC pellicles that had adsorbed or were dyed with different toxic organic dyes (Fig. 9d).[172] From the sample obtained from methylene blue ($C_{16}H_{18}C_1N_3S$), N and S atoms were codoped and homogeneously distributed within the carbon networks. When used as supercapacitor electrodes, the N, S dual-doped carbon aerogels exhibited better electrochemical properties compared with the raw carbon aerogels. In addition, P-doped, N, P-codoped, and B, P-codoped carbon aerogels were obtained by the pyrolysis of BC immersed in H_3PO_4, $NH_4H_2PO_4$, and H_3BO_3/H_3PO_4 aqueous solution, respectively.[133] The N, P-codoped carbon aerogels as electrodes for supercapacitors showed a maximum power density of 186.03 $kW \cdot kg^{-1}$. The high power density was mainly attributed to the codoping of N and P and the 3D porous nanofiber network of the carbon aerogels, which ensured good electrical conductivity while avoiding the use of conducting additives and polymer binders, which blocks the pores and increases contact resistance.

(2) Annealing of heteroatom – containing polymers and NFC/BC composite aerogels/foams. Heteroatomcontaining polymers such as PANI and PPy can form composites with NFC/BC aerogels/foams through *in situ* polymerization or electrodeposition methods. After carbonization, the heteroatomdoped carbon aerogels/foams were fabricated. BC was used as both a template and precursor for the synthesis of N-doped carbon networks through the carbonization of PANI-coated BC (Fig. 9e).[136] KOH activation was conducted to further optimize the pore structure and improve the specific surface area. N doping enhances the surface polarity, electric conductivity, and electrondonor tendency of carbon materials. The as-prepared N-doped porous carbon displayed a high specific capacitance of 296 $F \cdot g^{-1}$ at 2 $mV \cdot s^{-1}$, higher than that of raw BC-derived carbon (161 $F \cdot g^{-1}$). The N-doped carbon also exhibited excellent rate performance and cycling stability, which was ascribed to its unique con ductive networks and energy storage units. An interconnected, well-organized 3D conductive network is beneficial for electron transport, and a N-doped porous carbon network with high surface area can provide more active sites for energy storage and short ion diffusion length during the charge/discharge process.

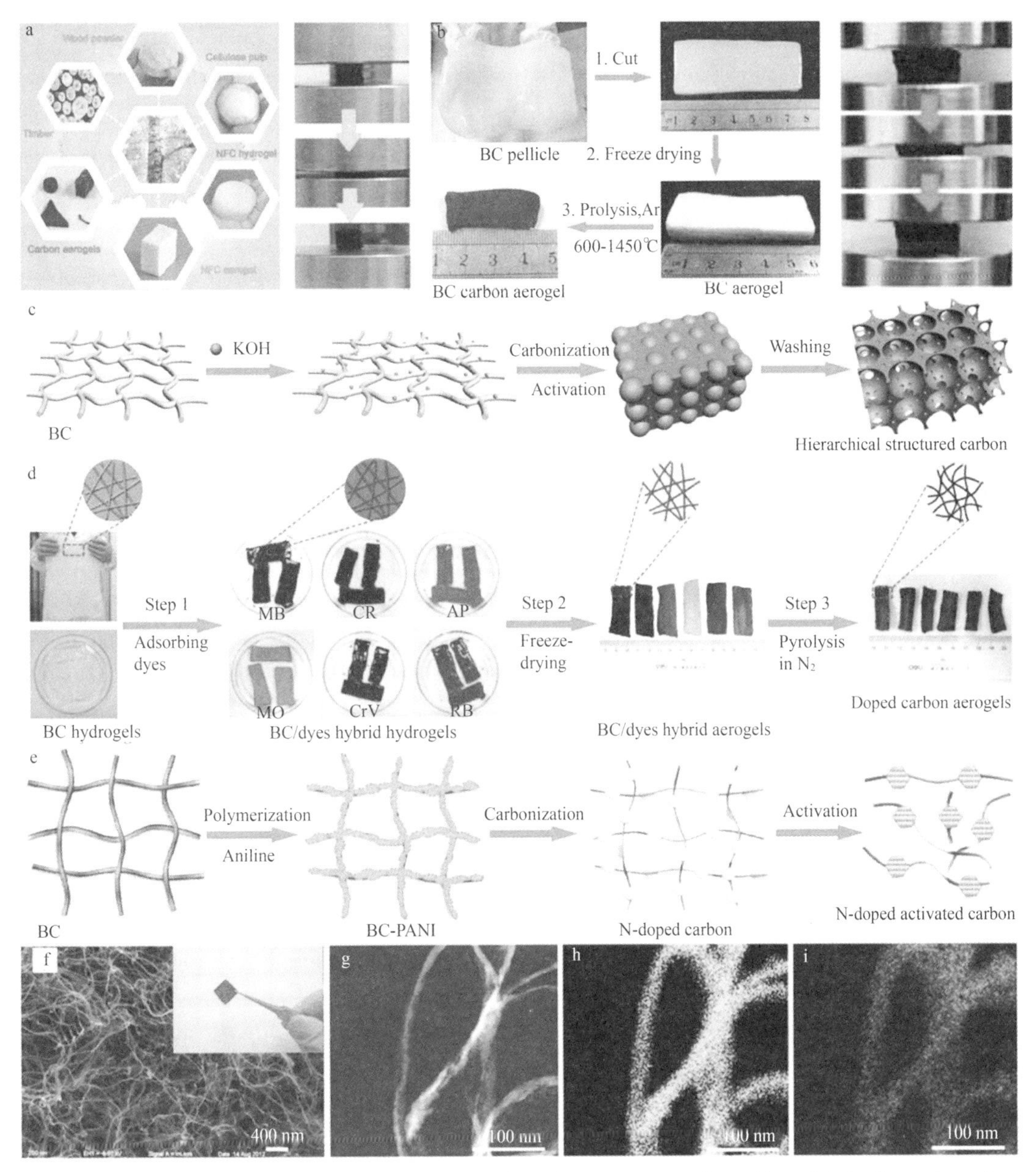

Fig. 9 NFC/BC-derived porous carbon materials for supercapacitors. The fabrication of carbon aerogels from (a) wood NFC[169] and (b) BC.[168] (a) is reproduced with permission from ref.[169]. Copyright 2016 Wiley-VCH. (a) is reproduced with permission from ref. 168. Copyright 2012 Springer Nature. The carbon aerogels show high flexibility. (c) Schematic illustration of the synthesis of hierarchically structured carbon from BC by a one-step carbonization/activation.[130] Reproduced with permission from ref. 130. Copyright 2016 Royal Society of Chemistry. (d) The fabrication process of doped carbon aerogels by annealing BC aerogels with different toxic organic dyes.[172] Reproduced with permission from ref. 172. Copyright 2014 Springer. (e) Schematic illustration of the synthesis of N-doped carbon by annealing BC/PANI composite aerogels followed by activation.[136] Reproduced with permission from ref. 136. Copyright 2014 Wiley-VCH. (f-i) The fabrication of N-doped carbon aerogels by hydrothermal treatment of BC-derived carbon aerogels and urea:[135] (f) SEM image and photograph (inset) of N-doped carbon aerogels; (g) typical EFTEM image of N-doped carbon aerogels; (h, i) corresponding elemental mapping images of (h) C and (i) N. (f-i) are reproduced with permission from ref. 135. Copyright 2013 Wiley-VCH

(3) Hydrothermal treatment of carbon aerogelswith heteroatom-containing molecules in solution. Because the NFC/BC-derived carbon aerogels have high porosity, an aqueous solution containing heteroatom molecules such as urea and ammonia can easily permeate through the carbon aerogels. N-Doped carbon aerogels were fabricated by Chen *et al.* using BC-derived carbon aerogels and urea through a hydrothermal reaction under mild conditions. [135] The as-prepared N-doped carbon aerogels still maintained a free-standing 3D pellicle with interconnected nanofibers and crosslinking pores (Fig. 9f−i). Owing to the high content of N doping and proper N species, the interconnected pore structure, and the high specific surface area of the carbon aerogels, the N-doped carbon aerogels exhibited a high specific capacitance of 173. 3 F · g^{-1}, which is apparently higher than that of BC-derived carbon aerogels (77. 7 F · g^{-1}). N-Doped carbon aerogels can also be produced using a hydro-thermal method with BC-derived carbon aerogels (after activation with a mixed gas of CO_2/argon) and aqueous ammonia. [132] An all-solid-state supercapacitor was fabricated by impregnating poly(vinyl alcohol)—H_2SO_4 gel electrolyte into the framework of N-doped carbon aerogels. The assembled supercapacitor exhibited good flexibility with a maximum power density of 390. 5 kW · kg^{-1} and stable cycling stability over 5000 cycles.

CNC can also be converted into porous carbon materials that can be further functionalized and utilized for supercapacitors. CNC has been used as both a carbon source and a template for the controlledgrowth of the N precursor to form melamine-formaldehyde (MF) coated CNCs (Fig. 10a). [173] The hybrid material was subjected to pyrolysis to fabricate N-doped carbon materials with micro-, meso-, and macropores (Fig. 10b). Optimal capacitances of 328. 5 F · g^{-1} at 0. 01 V s^{-1} and 352 F · g^{-1} at 5 A · g^{-1} were achieved for the carbon materials in 1 M H_2SO_4. The carbon materials also exhibited good cycling stability (less than 4. 6% loss after 2000 cycles) at 20 A · g^{-1}.

Another strategy for using CNC-derived porouscarbon for supercapacitors is the development of chiral nematic mesoporous carbon. [174,175] In water, suspensions of CNC organize into a chiral nematic phase that can be preserved upon slow evaporation, thereby resulting in chiral nematic films (Fig. 10c). [34] MacLachlan *et al.* reported that the evaporationinduced self-assembly of CNC with silica precursors can result in composite films with chiral nematic structures. Upon pyrolysis and etching of the silica, freestanding films of chiral nematic mesoporous carbon are obtained (Fig. 10d−g). [176] In a symmetrical capacitor with H_2SO_4 as the electrolyte, the mesoporous carbon films display near-ideal capacitor behavior with specific capacitance of 170 F · g^{-1} at 230 mA · g^{-1}.

CNC and NFC can also be integrated together to form porous carbon materials for supercapacitors (Fig. 10h). [177] For the hybrid films, the NFC formed a macroporous framework and the CNC was assembled around the NFC framework. By depositing a thin layer of Al_2O_3 as a protective layer by atomic layer deposition followed by etching away Al_2O_3 using HCl after carbonization, a freestanding carbon film with a hierarchical porous structure and a specific surface area of over 1200 m^2 · g^{-1} was produced. The two-level hierarchical porous structure derived from NFC and CNC facilitated fast ion transport in the film. When tested as an electrode material with a mass loading of 4 mg · cm^{-2} for supercapacitors, the carbon film exhibited a specific capacitance of 170 F · g^{-1}. Even at 50 A · g^{-1}, it still retained 65% of its original specific capacitance.

4. 3. 2 Carbon hybrid materials

To further improvethe specific capacitance of nano cellulose-derived carbon, carbon hybrid materials were

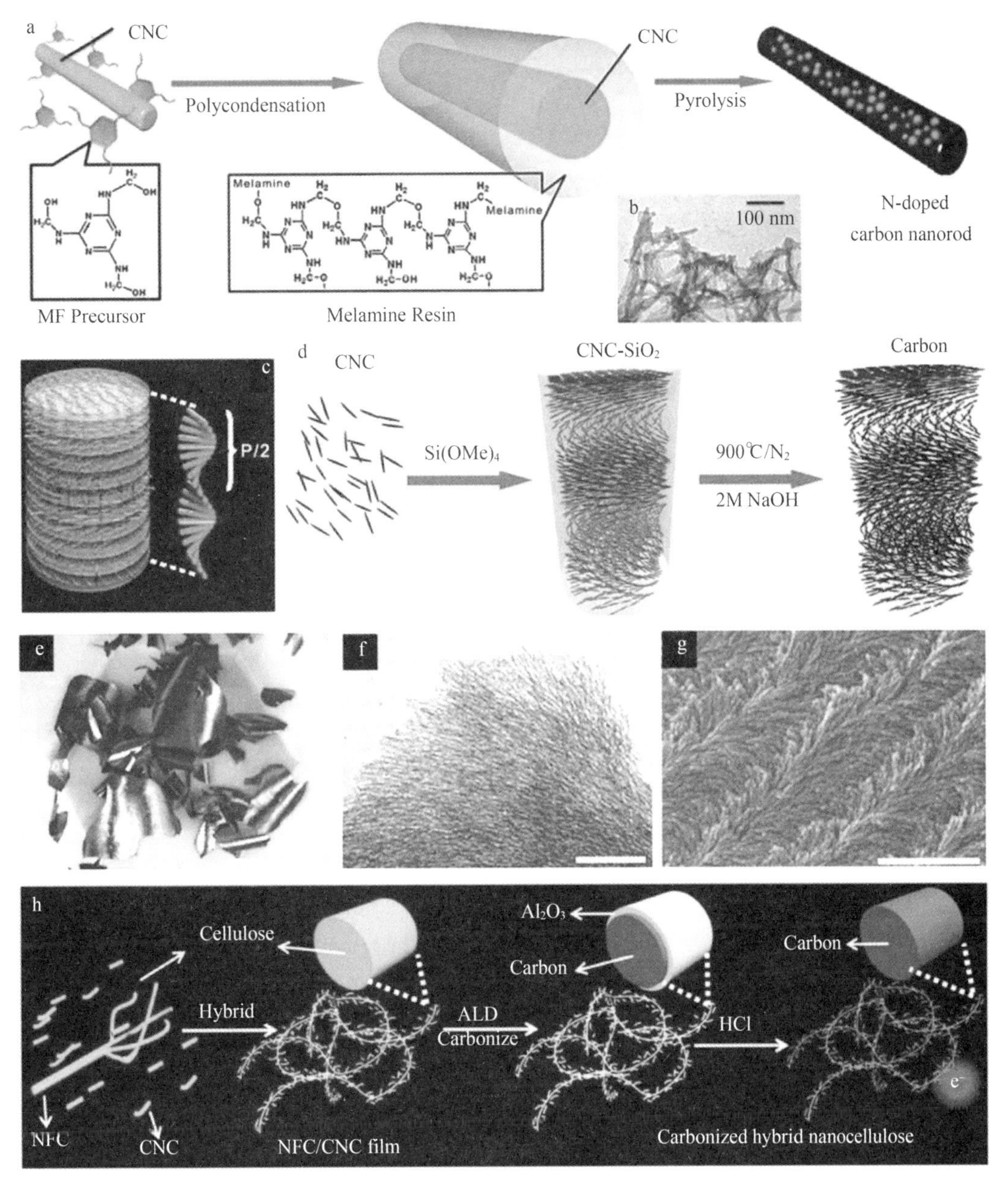

Fig. 10 CNC-derived porous carbon materials for supercapacitors. (a) Schematic of the synthesis of CNC-derived N-doped carbon nanorods.[173] (b) TEM image of the CNC-derived N-doped carbon nanorods.[173] (a, b) are reproduced with permission from ref. 173. Copyright 2015 Royal Society of Chemistry. (c) Schematic of the chiral nematic ordering present in CNC, along with an illustration of the half-helical pitch P/2 (~150–650 nm).[34] Reproduced with permission from ref. 34. Copyright 2010 Springer Nature. (d) Schematic illustration of the synthesis of chiral nematic mesoporous carbon by the evaporation induced self-assembly of CNC with silica precursors, followed by pyrolysis and etching of the silica.[176] (e) A photograph of a sample of the chiral nematic mesoporous carbon (scale bar: 2 cm).[176] (f) TEM image of chiral nematic mesoporous carbon (scale bar: 200 nm).[176] (g) SEM image of chiral nematic mesoporous carbon (scale bar: 500 nm).[176] (d-g) are reproduced with permission from ref. 176. Copyright 2011 Wiley-VCH. (h) Schematic illustration of the fabrication process for carbonized hybrid CNC and NFC.[177] Reproduced with permission from ref. 177. Copyright 2017 Springer

developed by integration with active materials such as activated carbon, nanocarbon, metal oxides, hydroxides, and sulfides, *etc*. The carbon hybrid materials can be prepared by direct pyrolyzation of the composites of nanocellulose and carbon materials. Ma et al. reported the preparation of N-doped carbon materials/RGO/BC composite paper.[134] The BC was exploited as both a substrate and a biomass precursor for N-doped carbon by pyrolysis. The one-step carbonization treatment not only fabricated the N-doped 3D nanostructured carbon composite materials but also forms the reduction of the GO sheets at the same time. The symmetric supercapacitor obtained using the composite paper as an electrode showed an areal capacitance of 810 mF · cm^{-2} in KOH and 920 mF · cm^{-2} in H_2SO_4, delivered an energy density of 0. 11 mW · h · cm^{-2} in KOH and 0. 29 mW h cm^{-2} in H_2SO_4, and a maximum power density of 27 mW · cm^{-2} in KOH and 37. 5 mW · cm^{-2} in H_2SO_4. Free-standing carbon nanofibers/activated carbon composite films were prepared by carbonization of the NFC/activated carbon composite films.[178] The carbon nanofibers are 'welded' on activated carbon particles and integrated into one piece of carbon. The network of carbonized NFC possessed better electron transport efficiency than the activated carbon particles. When tested as supercapacitor electrodes at commercial level mass loading, the composite films exhibited 2-times-slower capacitance fading at high current and 3-times-higher maximum power density than the bare activated carbon. The carbon hybrid materials can also be prepared by pyrolyzation of the composites of nanocellulose and the carbon precursors. Highly graphitized carbon aerogels were synthesized based on BC toughened lignin-resorcinol- formaldehyde aerogels.[179] The as-prepared carbon aerogels, consisting of a blackberry-like, core-shell structured, and highly graphitized carbon nanofibers, were able to undergo at least 20% reversible compressive deformation. Owing to the unique nanostructure and large mesopore population, the carbon materials exhibited high areal capacitances calculated using the total surface area and the mesopore surface area of 62. 2 and 78. 7 mF · cm^{-2}, respectively.

Carbon hybrid materials can be achieved by direct reaction between carbon framework and metal precursor under certain conditions. MnO_2-decorated carbon aerogels were fabricated by employing BC-derived carbon aerogels as a precursor. Based on the redox reaction $4MnO_4^- + 3C + H_2O$ $4MnO_2 + CO_3^{2-} + 2HCO_3^-$, MnO_4^- ions were reduced spontaneously to MnO_2 on the surface of BC-derived carbon aerogels by oxidizing the exterior carbon (Fig. 11a–d).[135] The carbon aerogel with high graphitization serves as a highly conductive matrix for fast ion and electron transport, whereas the ultrathin MnO_2 film that is strongly anchored on the surface of carbon aerogels can effectively shorten the diffusion length of electrolyte ions during the charge/discharge process and provide a large electrochemically active surface area for the fast reversible Faradic redox reactions. Asymmetric supercapacitor (ASC) fabricated by using the carbon aerogels/MnO_2 with a specific capacitance of 254. 6 F · g^{-1} as positive electrodes, the N-doped carbon aerogels with a specific capacitance of 173. 3 F · g^{-1} as negative electrodes, and Na_2SO_4 as aqueous electrolyte, worked very well at different voltages, even up to 2. 0 V. Such an ASC displayed a maximum energy and power density of 32. 91 W · h · kg^{-1} and 284. 63 kW · kg^{-1}, respectively. Fan et al. decorated N-doped BC-derived carbon with flower-like MnO_2(Fig. 11e–k).[136] The BC-derived carbon/MnO_2 exhibited a high specific capacitance of 273 F · g^{-1} at 2 mV s^{-1}. The as-assembled carbon/MnO_2//KOH activated N-doped carbon ASC exhibited a considerably high energy density of 63 W · h · kg^{-1} in 1 M Na_2SO_4. The ASC also exhibited good cycling performance with 92% specific capacitance retention after 5000 cycles. To further improve the specific capacitance, the nanocellulosederived carbon materials were used as a scaffold for deposition of metal hydroxyl and sulfide. Lai et al. reported the further deposition of ultrathin nickel-cobalt layered double hydroxide (Ni—Co LDH) nanosheets on N-doped BC-derived carbon nanofibers (Fig. 11 a and m).[137] The composite electrodes exhibited significantly enhanced specific capacitance (1949.5 F · g^{-1} at 1 A · g^{-1}), high capacitance retention of 54.7% even at 10 A · g^{-1}, and good cycling stability with 74.4% retention after 5000 cycles. ASCs assembled by using the N-doped carbonized BC@ Ni—Co LDH composites as positive electrode and N-doped carbonized BC as negative electrode exhibited an energy density of 36.3 W · h · kg^{-1} at a power density of 800.2 W · kg^{-1}. Zuo et al. reported the uniform decoration of the pore walls of BC-derived carbon aerogels with NiS particles through a hydrothermal process.[139] The ASCs were constructed utilizing NiS composite carbon aerogels and carbon aerogels as the positive and negative electrodes, respectively. Through the synergistic effect of the 3D porous structures and conductive networks derived from the carbon aerogel and the high capacitive performance of NiS, the ASC exhibited an energy density of B21.5 W · h · kg^{-1} at a power density of 700 W · kg^{-1}, as well as good cycle stability with B87.1% specific capacitance retention after 10000 cycles.

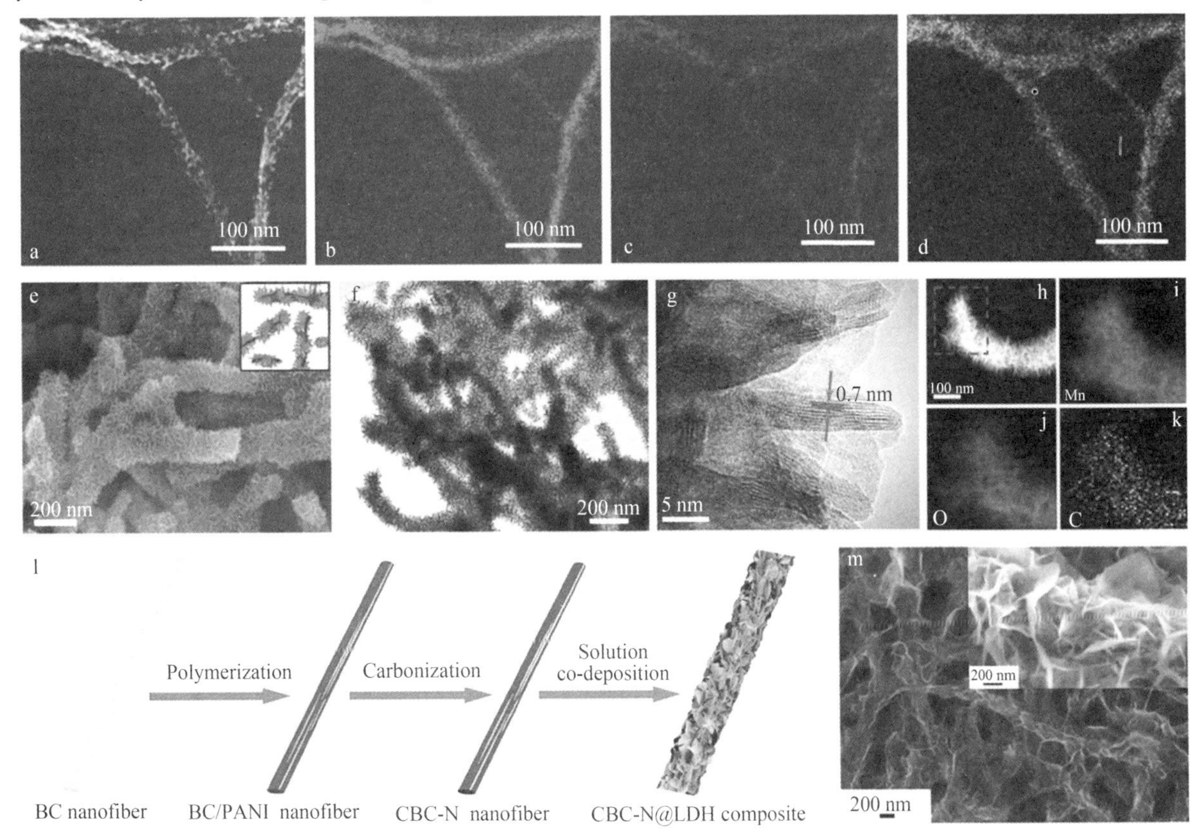

Fig. 11 Carbon hybrid materials for supercapacitors. (a-d) Hybrid BC-derived carbon aerogels with MnO_2:[135] STEM images of the BC-derived carbon aerogels/MnO_2 with corresponding elemental mapping images of (b) C, (c) Mn, and (d) O. (a-d) are reproduced with permission from ref. 135. Copyright 2013 Wiley-VCH. (e-k) Hybrid N-doped BC-derived carbon with MnO_2:[136] (e, f) SEM and TEM images of the N-doped BC-derived carbon/MnO_2 composite; (g) high resolution TEM image of the N-doped BC-derived carbon/MnO_2 composite; (h-k) STEM images of the N-doped BC-derived carbon/MnO_2 composite with corresponding elemental mapping images of (i) Mn, (j) O, and (k) C in the dashed square region of (h). (e-k) are reproduced with permission from ref. 136. Copyright 2014 Wiley-VCH. (1, m) Hybrid BC-derived carbon aerogels with CBC-N @ LDH:[137] (1) schematic for the preparation of CBC-N@LDH composites; (m) FESEM image of a CBC-N@LDH composite. (1, m) are reproduced with permission from ref. 137. Copyright 2016 Wiley-VCH

Overall, the pyrolyzation of nanocellulose-based bulk materials to carbon materials can obviously increase their specific surface area and electrical conductivity. After further functionalization by activation, heteroatom-doping, or the formation of hybrids with other active materials, nanocellulose-derived carbon electrodes showed good electrochemical performance. Controlling the morphologies, structures, and ratios of the doping heteroatoms/active materials of the nanocellulose-derived carbon materials would be the main challenge in the future. Surface/interface design on nanocellulose/nanocellulose-derived carbon and active materials/precursors is crucial for controlling the interface interactions during the preparation, pyrolyzation, activation, and etching process. The ultimate goal is to achieve well-defined bulk-, micro-, nano-, and even atom-scale structures of the carbon materials. The structure-property relationship of the novel developed nanocellulose-derived carbon materials should be better understand for maximum utilizing the advantageous of the carbon materials for high performance supercapacitors.

Table 3 Examples of different nanocellulose-derived materials for LIB electrodes reported in the literature

Nanocellulose type	Nanocellulsoe process	Electrode	Electrochemical performance: Capacity/Rate performance/Energy density/Columbic efficiency	Cyclic stability	Ref.
TEMPO-oxidized NFC	Hybrid with $LiFePO_4$ and Super-P carbon particles	NFC/$LiFePO_4$/Super-P carbon particle composite paper as cathode	The Coulombic efficiency for the samples dried at 110 ℃ is around 97% during the first ten cycles, whereas the samples stored at 170 ℃ produced efficiencies of around 99%	The cycling performance was 151 $m \cdot Ah \cdot g^{-1}$ at C/10 and 132 $m \cdot Ah \cdot g^{-1}$ at 1C for samples dried at 170 ℃	[184]
Cladophora NFC	Hybrid with silicon nanoparticles and CNTs	NFC/silicon nanoparticle/CNT composite paper as anode	The areal capacities is up to 2.5 mAh cm^{-2} and the specific capacities is 3200 $m \cdot Ah \cdot g^{-1}$ (based on the weight of the Si) corresponding to electrode specific capacities of up to 800 $m \cdot Ah \cdot g^{-1}$ at 200 $mA \cdot g^{-1}$; Coulombic efficiency: 49%	The capacity is 1840 $m \cdot Ah \cdot g^{-1}$ (based on the weight of the Si) after 100 cycles at 1 $A \cdot g^{-1}$	[185]
Wood carboxymethylated NFC	Hybrid with SWCNTs and silicon	NFC/SWCNT nanopaper aerogel deposited with a thin layer of silicon as anode	The first charge has a specific capacity of ~5000 $m \cdot Ah \cdot g^{-1}$ Si; the first discharge reveals the reversible capacity to be ~2100 $m \cdot Ah \cdot g^{-1}$; Coulombic efficiency: 43%	The capacity is 1200 $m \cdot Ah \cdot g^{-1}$ for 100 cycles in half-cells	[186]
TEMPO-oxidized NFC	Hybird with graphite, Super-P carbon, $LiFePO_4$, SiO_2	NFC/graphite/Super-P carbon composite layer as anode; NFC/$LiFePO_4$/Super-P carbon composite layer as cathode	An energy density of full paper battery is 188 $mW \cdot h \cdot g^{-1}$ at C/10; the Coulombic efficiency readings were consistent in the range 99.5 ~ 100%, with the exception of the first cycles for each new C rate	The reversible capacities is 146 $mA \cdot h \cdot g^{-1}$ $LiFePO_4$ at C/10 and 101 $mA \cdot h \cdot g^{-1}$ $LiFePO_4$ at 1 C	[187]
BC	Pyrolyzation	BC-derived carbon aerogel as anode	The initial discharge and charge capacities at 75 $mA \cdot g^{-1}$ were 797 and 386 $m \cdot Ah \cdot g^{-1}$, respectively	After 100 cycles, the charge capacity showed a slight decrease to 359 $m \cdot Ah \cdot g^{-1}$	[188]

Table 3 continued

Nanocellulose type	Nanocellulsoe process	Electrode	Electrochemical performance: Capacity/Rate performance/Energy density/Columbic efficiency	Cyclic stability	Ref.
BC	Pyrolyzation and activated by KOH	KOH activated BC-derived carbon nanofiber as anode	The initial discharge/charge delivers specific capacities of 1948.5 and 1091.4 m · Ah · g^{-1}, respectively, based on the total mass of the electrode	The electrode delivers a specific capacity of over 857.6 m · Ah · g^{-1} after 100 cycles at 100 mA · g^{-1} and retain a capacity of 325.4 m · Ah · g^{-1} when cycled at 4000 mA · g^{-1}	[189]
Wood TEMPO-oxidized NFC	Hybrid with GO followed by pyrolyzation	NFC/GO composite derived carbon microfiber as anode	The second cycle has an initial discharge capacity at 317.3 m · Ah · g^{-1}, with a Columbic efficiency of 62.2%	The discharge capacity remains 312 m · Ah · g^{-1} at the 63th cycle	[190]
BC	Pyrolyzation; in situ assembly of SnO_2 nanoparticles	BC-derived carbon loaded with SnO_2 as anode	The specific capacity is ca. 380 m · Ah · g^{-1} under 1A · g^{-1}, based on the total mass of the electrode; average Coulombic efficiency: over 98.5%	The reversible capacity is ca. 600 m · Ah · g^{-1} after 100 cycles under 100 mA · g^{-1}	[191]
BC	Hybrid with Fe_2O_3 followed by pyrolyzation	BC-derived carbon aerogel decorated with nano-Fe_3O_4 as anode	The discharge capacity is 410 m · Ah · g^{-1} at 1 A · g^{-1}	The electrode delivers a reversible capacity of 754 m · Ah · g^{-1} after 100 cycles at 100 mA · g^{-1}	[192]
BC	hybrid with MoS_2 followed by pyrolyzation	BC-derived carbon nanofiber loaded with MoS_2 nanoleaves	A reversible capacity is 935 m · Ah · g^{-1} at 0.1A · g^{-1}(267 m · Ah · g^{-1} at 4A · g^{-1}), with a Coulombic efficiency of 66%	The capacity is 581 m · Ah · g^{-1} after 1000 cycles at 1 A · g^{-1}	[193]

5 Lithium-ion batteries

Since the introduction by Sony Inc. in 1991, LIBs have emerged as the main power source for portable electronic devices because of their high energy density, large output voltage, appreciable lifespan, and environmentally benign operation.[180-183] A LIB mainly consists of four functional components: anode, cathode, electrolyte, and separator. In principle, the charging and discharging process of a LIB is realized through the insertion/deintercalation of Li ions transporting between the anode and the cathode material. During charging, an external electrical power source forces the current to pass in the reverse direction and enables the migration (diffusion) of Li ions from the cathode to the anode through the electrolyte. During the discharging process, Li ions move back to the anode through the nonaqueous electrolyte, carrying the current. Nanocellulose can be integrated with other active materials or converted to carbon materials to be further developed as the electrodes for LIBs. The electrochemical performance of nanocellulose-derived materials used as electrodes for LIBs has been compared and summarized in Table 3.[184-193] Nanocellulose-derived films/

papers can also be designed as separators for LIBs after carefully controlling the micro/nanostructures. In the following sections, we discuss recent advances related to the development of nanocellulose and its derived materials for LIBs.

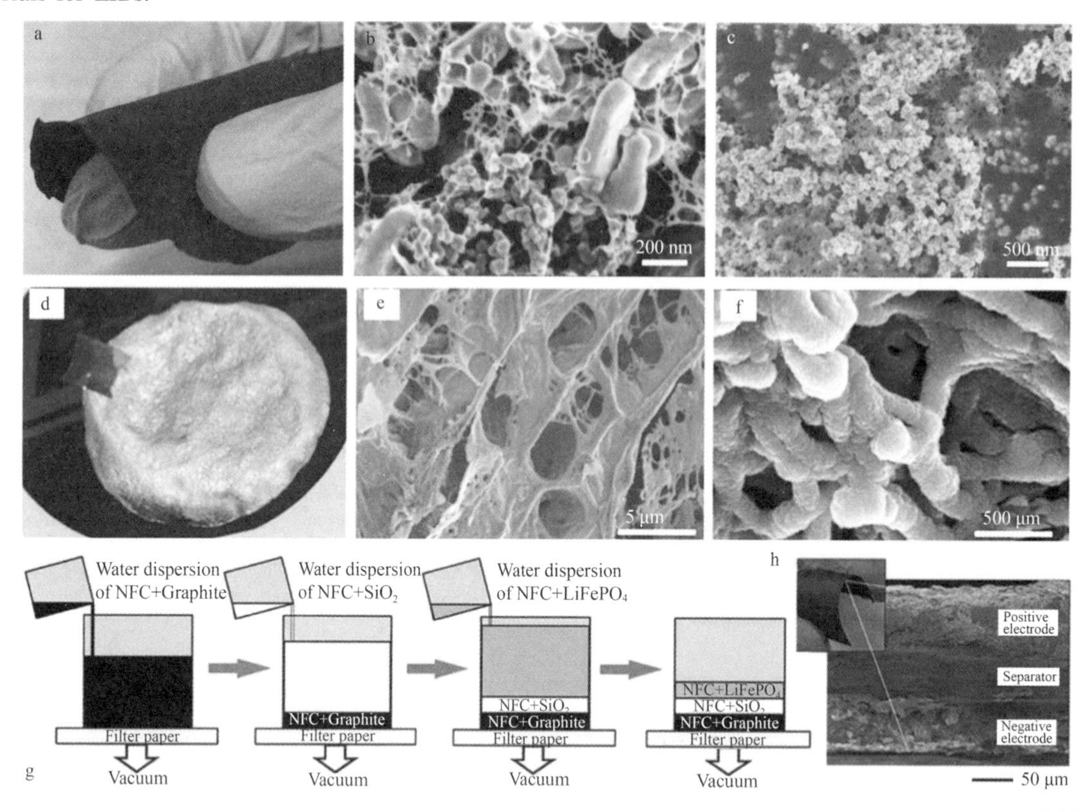

Fig. 12 Flexible LIBs based on NFC hybrid materials. (a-c) Fabrication of flexible positive electrodes:[184] (a) digital image of a flexible single-layered $LiFePO_4$ paper electrode; (b) SEM image showing the top-side of a $LiFePO_4$ paper electrode; (c) SEM image showing the carbon(bright particles)/NFC(dark surfaces) side of a bi-layered electrode paper. (a-c) are reproduced with permission from ref. 184. Copyright 2013 Springer. (d-f) Fabrication of Si-conductive nanopaper anode:[186] (d) digital image of conductive nanopaper coated with Si sitting on a silicon wafer; (e, f) SEM images of the surface morphology of the Si-conductive nanopaper. (d-f) are reproduced with permission from ref. 186. Copyright 2013 Springer. (g, h) Fabrication of single-paper LIB cells:[187] (g) an illustration of the sequential filtration steps during the production of a paper battery cell; (h) digital image of a paper battery and SEM image of a paper battery cross-section. (g, h) are reproduced with permission from ref. 187. Copyright 2013 Royal Society of Chemistry

5.1 Flexible LIBs based on NFC/BC hybrid materials

The high aspect ratio NFC/BC can be used as a flexible substrate and binder material to integrate with active materials to prepare paper/film electrodes for flexible LIBs. Flexible positive electrodes were developed by filtration of a water dispersion of NFC, $LiFePO_4$, and Super-P carbon particles (Fig. 12a-c).[184] The electrodes showed good mechanical properties both dry and when soaked with battery electrolyte. The capacities were 151 mA · h · g^{-1} at C/10 and 132 mA · h · g^{-1} at 1C for sample dried at 170 1C. Flexible negative electrodes have been obtained by integrating nanocellulose with active materials such as CNTs and silicon. Flexible paper anodes were prepared through a paper-making process using NFC, silicon nanoparticles and CNTs as the building blocks.[185] Owing to the uniform distribution and strong adhesion of the silicon nanoparticles to the 3D conductive CNT/NFC network, the electrodes exhibited good energy storage performances with specific capacities of up to 800 mA · h · g^{-1}(based on the weight of the whole electrode) and good cycling performances. Silicon can also be integrated with CNT/NFC network by deposition of a thin layer of silicon on the NFC/CNT nanopaper aerogels through a plasma-enhanced CVD method (Fig. 12d–f).[186] The open channels of the nanopaper aerogels allow for electrolyte accessibility to the surface of silicon. The aerogels performed well as anodes with a stable capacity of 1200 mA · h · g^{-1} for 100 cycles tested in half-cells.

Other active material such as $Li_4Ti_5O_{12}$ has also been integrated with NFC and CNT to prepared flexible paper electrodes. The hybrid films had a dual-layer structure consisting of $Li_4Ti_5O_{12}$/CNT/NFC and CNT/NFC.[194] The CNT/NFC layer with a thickness of approximately 10 mm was utilized as a current collector foil, which reduced the total mass of the electrode while keeping the same areal loading of the active materials. The flexible paper-electrodes showed high rate cycling performance, and the specific charge/discharge capacities were up to 142 mA · h · g^{-1} at 10C. TEMPO-oxidized NFC was used as a binder material to integrate with $Li_4Ti_5O_{12}$ and carbon fibers to fabricate flexible electrodes for LIBs.[195] Even a small amount of NFC (2 wt%) can be used to prepare the $Li_4Ti_5O_{12}$ electrode, which could meet the demands for high gravimetric energy density applications when a low C rate is acceptable. A full cell assembled using the $Li_4Ti_5O_{12}$/carbon fiber/NFC as a negative electrode and $LiFePO_4$/carbon fiber/NFC as a positive electrode exhibited a stable cycling performance with the specific capacity of approximately 125 mA · h · g^{-1} after 20 cycles at 0. 1C, and a high energy density of approximately 90 W · h · kg^{-1}(for $Li_4Ti_5O_{12}$-based batteries). Flexible single-paper LIBs were developed by Leijonmarck et al. through sequential filtration of water dispersions containing the negative electrode (consisting of graphite, Super-P carbon, and NFC), the separator (consisting of NFC and SiO_2), and the positive electrode (contained $LiFePO_4$, Super-P carbon, and NFC) (Fig. 12g and h).[187] The resulting paper LIB was thin and strong and had good cycling performances. The energy density of the full paper battery at C/10 was 188 mW · h · g^{-1}.

Overall, NFC/BC display great potential for the fabrication of flexible LIBs because of their high aspect ratio, web-like entangled structures, and high mechanical strength. However, most electrode materials reported so far have been synthesized simply by mixing the active materials with nanocellulose, which leads to relatively low interfacial interactions. Optimizing the interfaces between the NFC/BC and the active materials using covalent or noncovalent techniques should be considered for improving the interface interactions, to achieve electrodes/LIBs with appropriate and effective mass loading, high flexibility, mechanical strength, and good electrochemical performance. The integration of active materials into NFC/BC networks using other novel techniques such as ALD or CVD is also suggested.

5.2 Nanocellulose-derived carbon materials

Nanocellulose-derived carbon materials can be directly usedas electrodes for LIBs. The utilization of KOH activated BC-derived carbon aerogels as anodes for LIBs can deliver a specific capacity of over 857. 6 mA · h · g^{-1} after 100 cycles at 100 mA · g^{-1} and retain a capacity of 325. 4 mA · h · g^{-1} even cycled at 4 A · g^{-1}.[189]

The good lithium storage performance of the carbon aerogels can be attributed to its hierarchical micropore-mesopore structure and high surface area, which can enhance the contact area of the electrode-electrolyte, decrease the diffusion resistance of the lithium ions, shorten the diffusion length of the lithium ions, and provide a solid and continuous pathway for electron transport. To further improve the electrochemical

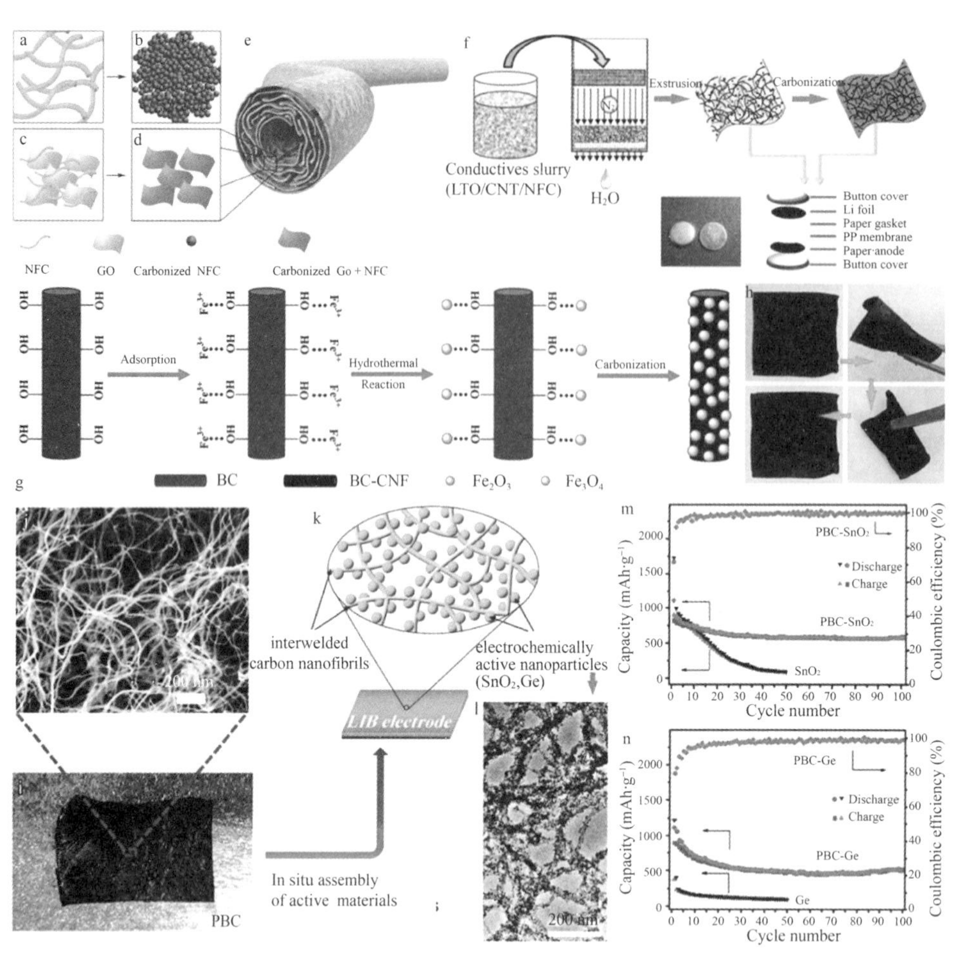

Fig. 13 Nanocellulose-derived carbon materials for LIBs. (a-e) The fabrication of NFC/GO-derived hybrid carbon fibers:[190] (a, b) schematic to show the morphology of NFC changed from fibers (before carbonization) to sphere particles after carbonization. In the carbonized (GO+NFC) microfiber, only sheets formed after carbonization as shown in (c, d); (e) is the structure of a carbonized (GO+NFC) microfiber. (a-e) are reproduced with permission from ref. 190. Copyright 2014 Wiley-VCH. (f) Schematic illustration of preparation the flexible single-layer paper electrode by using a papermaking method.[196] Reproduced with permission from ref. 196. Copyright 2016 American Chemical Society. (g, h) Fabrication of BC-derived carbon nanofiber/Fe_3O_4 composites:[192] (g) schematic illustration of the fabrication of BC-derived carbon nanofiber/Fe_3O_4 composites; (h) sequential photographs of the BC-derived carbon nanofiber/Fe_3O_4 composite during repeated bending and compressing demonstrating the excellent flexibility. (g, h) are reproduced with permission from ref. 192. Copyright 2015 Royal Society of Chemistry. (i-n) Fabrication of BC-derived carbon aerogel/active nanoparticle composites:[191] (i) the digital image of the carbon aerogel; (j) SEM image of the carbon aerogel; (k) *in situ* assembly of active nanoparticles into interwelded carbon nanofibers as LIB electrode; (l) TEM image of the carbon aerogel/SnO_2 composites; (m) comparison of cycling performances of carbon aerogel/SnO_2 composites and aggregated SnO_2 nanoparticles over 100 cycles at 100 mA · g^{-1}; (n) comparison of cycling performances of carbon aerogel/Gecomposites and aggregated Ge nanoparticles over 100 cycles at 100 mA · g^{-1}. (i-n) are reproduced with permission from ref. 191. Copyright 2013 Wiley-VCH

properties, active materials have been integrated with nanocellulose to develop composites for carbonization. Li et al. reported the fabrication of microfibers with a conductivity of 649 T 60 S · cm^{-1} through a wet-spinning method to prepare well-aligned NFC/GO hybrid fibers followed by carbonization (Fig. 13a-e).[190] GO acted as a template for NFC carbonization, which changed the morphology of the carbonized NFC from microspheres to sheets while improving the carbonization of NFC. The as-prepared fibers were demonstrated as lithium ion battery anodes, with a stable discharge capacity of 312 mA · h · g^{-1}. A carbon hybrid film was fabricated by integrating

NFC with $Li_4Ti_5O_{12}$ and CNTs through a papermaking method followed by carbonization (Fig. 13f).[196] Owing to the uniform distribution and immobilization of $Li_4Ti_5O_{12}$ into the interpenetrating carbonized NFC/CNT conductive network, the carbonized paper anode displayed favorable transport behavior of the lithium ion, and delivered a reversible discharge capacity of 160 $mA \cdot h \cdot g^{-1}$ and remained at 140 $mA \cdot h \cdot g^{-1}$ after 10C with a Coulombic efficiency of nearly 100%. Nanocellulose-derived carbon materials can be employed as versatile conductive porous scaffolds to support active anode materials. SnO_2 nanoparticles,[191] Ge nanoparticles,[191] amorphous Fe_2O_3,[197] Fe_3O_4 nanoparticles[192] (Fig. 13g and h) and MoS_2 nanoleaves,[193] have been successfully synthesized on nanocellulose-derived carbon scaffolds through an *in situ* growth method. Wang et al. demonstrated the utilization of BC-derived carbon aerogels as scaffold to support the *in situ* assembly of active SnO_2 and Ge nanoparticles (Fig. 13i–n).[191] The SnO_2 and Ge nanoparticles were uniformly immobilized onto the BC-derived carbon nanofiber surface. The carbon hybrid aerogels showed high specific capacity and good cycling stability. The BC-derived carbon-SnO_2 hybrid aerogels retained a reversible capacity of approximately 600 $mA \cdot h \cdot g^{-1}$ even after 100 cycles under 100 $mA \cdot g^{-1}$. In addition, the hybrid aerogel electrodes exhibited good rate capability. Even at 1 $A \cdot g^{-1}$, a specific capacity of approximately 380 $mA \cdot h \cdot g^{-1}$ was obtained on the basis of the total electrode weight, which is higher than the theoretical capacity of graphite. The high specific capacity and good cycling stability of the carbon hybrid aerogels were mainly attributed to the well-dispersed active nanoparticles, interconnected conductive carbon nanofibers creating efficient electron conduction pathways at the whole electrode scale, and numerous interconnected voids facilitating the diffusion of lithium ions.

In summary, the high specific surface area, high electrical conductivity and porous structure of the nanocellulose-derived carbon materials lead to high performances of the carbon electrodes for LIBs. Integration with active materials can further improve the electrochemical properties. More attention should be paid to the integration of nanocellulose-derived carbon and active materials with high theoretical capacity, such as SnO_2 and Ge, in a rational manner. The morphology and crystallization of active materials anchored on carbon materials should be carefully controlled during the fabrication process, and the interface interactions between these active components and electrolytes should also be considered. In addition, the construction and optimization of hierarchical micro-/mesoporous structures of nanocellulose-derived carbon materials through pore engineering techniques is expected to introduce additional and suitable positions for lithium storage and shorten the diffusion length for ion and charge transport.

5.3 Nanocellulose-derived separators

The separator in a LIB prevents the direct contact betweenthe positive and negative electrodes while serving as the electrolyte reservoir to enable the transportation of lithium ions between the two electrodes.[198] The separator is not directly involved in any cell reactions, but its structure and properties play a critical role in determining the LIB performance, including cycle life, safety, energy density, and power density by influencing the cell kinetics. Because nanocellulose-based paper/film is hydrophilic and exhibits excellent mechanical/thermal properties, utilizing the nanocellulose-derived paper/film as a separator for LIBs has gained more interest.

An NFC film was obtained from densely packed NFC that exhibited high strength and thermal stability, and was electrolytephilic. Therefore, the NFC separator imparted substantial improvement in the ionic conductivity,

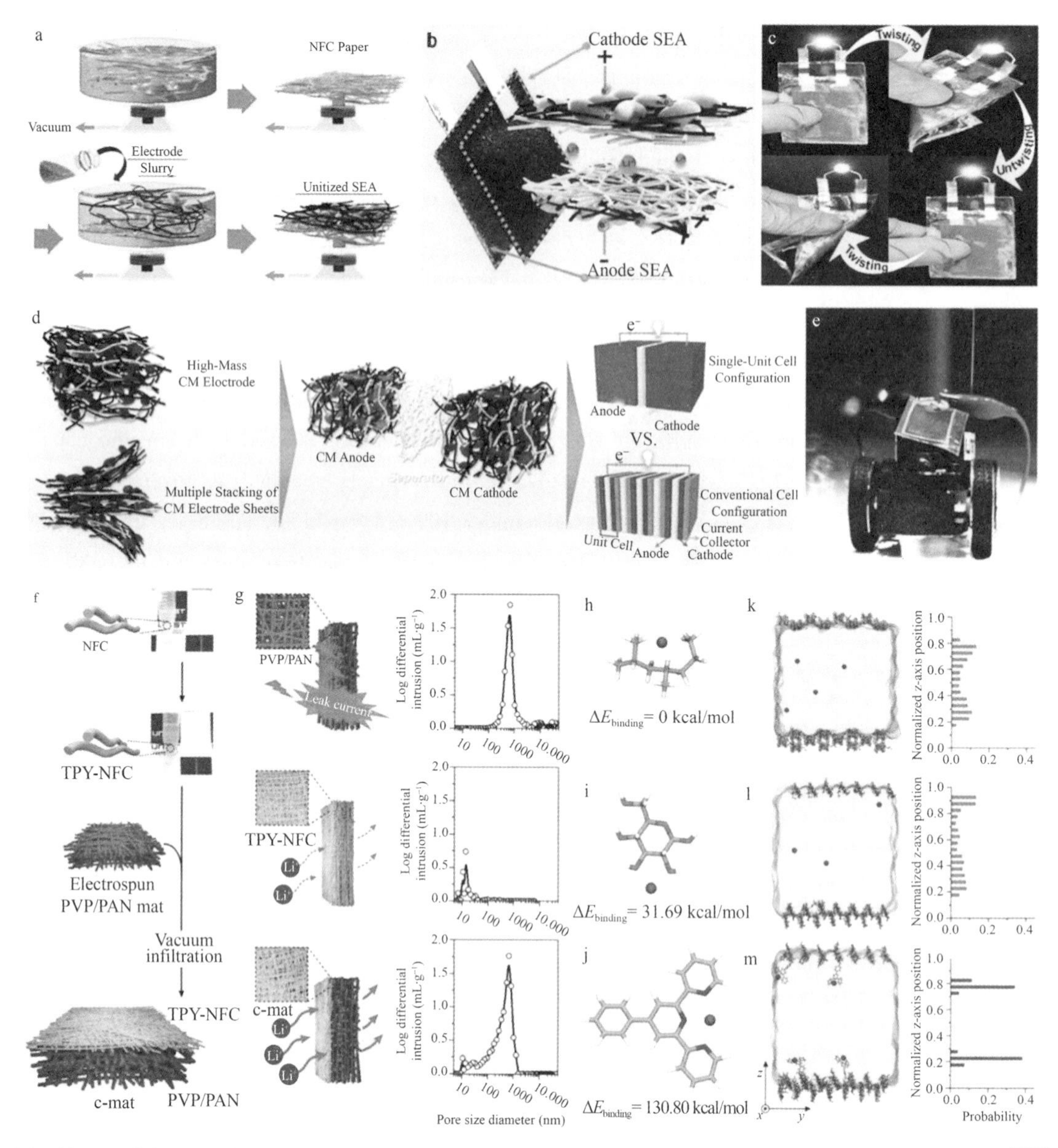

Fig. 14 Nanocellulose-derived separators for LIBs. (a-c) The fabrication of NFC/SWCNT-based SEAs:[203] (a) schematic representation of the overall fabrication procedure for unitized SEA; (b) a photograph of a *h*-nanomat full cell and a schematic illustration describing the electron transport pathways and electrolyte accessibility; (c) A photograph showing the electrochemical activity of a *h*-nanomat cell after the repeated twisting/untwisting deformation. (a-c) are reproduced with permission from ref. 203. Copyright 2014 American Chemical Society. (d, e) Development of hetero-nanonet rechargeable paper batteries using NFC separator:[204] (d) a conceptual representation of a single-unit cell and its contribution to the energy density of cells; (e) a photograph showing the operation of a mini toy car installed with the single-unit hetero-nanonet paper cell. (d, e) are reproduced with permission from ref. 204. Copyright 2015 Wiley-VCH. (f-m) Development of hierarchical/asymmetric porous membrane for smart battery separator:[205] (f) schematic depiction of the overall fabrication procedure of the separator; (g) schematic and pore size distribution of the PVP/PAN electrospun mat, TPY/NFC, and TPY/NFC on PVP/PAN mat; (h-m) Mn^{2+} ion binding capability of TPY-NFC. B3LYP/6-31G(d, p)-optimized geometries of (h) $Mn^{2+}(PP_3)$, (i) Mn^{2+}(Glc), and (j) Mn^{2+}(TPY) complexes with the relative binding energies from DFT calculations at B3LYP/6-311+G(2d, 2p) level, including BSSE corrections. Equilibrium morphologies after a 400 ns MD simulation at 333 K and normalized *z*-axis position probability of Mn^{2+} ions for the last 100 ns on the surface of (k) PP, (l) cellulose, and (m) TPY-cellulose. Violet beads represent the Mn^{2+}, and the ethylene carbonate solvent is shown as a transparent surface. (f-m) are reproduced with permission from ref. 205. Copyright 2016 American Chemical Society

electrolyte wettability, and thermal shrinkage, as compared with that of the commercialized PP/PE/PP separators.[199] The cladophora NFC separators have a typical thickness of approximately 35 mm, an average pore size of approximately 20 nm, a Young's modulus of 5.9 GPa, and an ionic conductivity of 0.4 mS · cm^{-1} after soaking into 1 M $LiPF_6$ ethylene carbonate: diethyl carbonate electrolyte.[200] The NFC separators were thermally stable at 150 1C and were electrochemically inert in the potential range between 0 and 5V *vs.* Li^+/Li. A $LiFePO_4/Li$ cell containing a NFC separator showed good cycling stability with 99.5% of capacity retention after 50 cycles at 0.2C. Because NFC separators were usually hard in providing highly porous structures owing to the densely packing of NFC, SiO_2 nanoparticles have been introduced as a NFC "disassembling" agent to facilitate the evolution of a more porous structure.[201] The porous structure can be tuned by varying the SiO_2 content in the NFC suspension. The as-prepared separator exhibited high ionic conduction, thus contributing to good cell performance. Wang et al. developed a bilayered nanocellulose-based separator, which contains a mesoporous insulating NFC layer and a redox-active PPy-containing support layer. The as-fabricated redox-active separator can be used to enhance the capacity of LIBs.[202]

Recently, Lee et al. pioneered the development of new kinds of NFC-based separators. A heterolayered mat battery was fabricated based on a unitized separator/electrode assembly (SEA) architecture (Fig. 14a–c).[203] The unitized SEAs consisted of NFC separators and solely SWCNT-netted electrode active materials ($LiFePO_4$ (cathode) and $Li_4Ti_5O_{12}$ (anode) powders were chosen as model systems). The NFC separator played a critical role in securing the tightly interlocked electrode-separator interface, whereas the SWCNTs exhibited multifunctional roles as electron conductive additives, binders, current collectors and also non-Faradaic active materials. The SEA allowed significant improvements in the mass loading of active materials, electron transport pathways, electrolyte accessibility, and misalignment-proof of separator/electrode interface. The batteries fabricated by stacking anode SEA and cathode SEA exhibit advances in the electrochemical performances, shape flexibility, and safety tolerance.

Using a similar NFC separator in batteries, the electrode was further modified by using NFC/MWCNT-intermingled heteronets embracing electrode active powders (NM electrodes). The multi-stackable NM electrodes were assembled with a NFC separator, leading to a single-unit hetero-nanonet paper cell (composed of one NM cathode/one NFC separator/one NM anode) (Fig. 14d and e),[204] which provided an increase in the energy density (226 W · h · kg^{-1} at 400 W · kg^{-1} for $L_iN_{i0.5}Mn_{1.5}O_4$/graphite system). The batteries were flexible and would be folded to fabricate paper crane batteries through an origami folding technique. By considering the unwanted byproducts in electrolytes, such as manganese ions (Mn^{2+}) dissolved from lithium manganese oxide electrode materials and hydrofluoric acid (HF) generated by side reactions between residual water and lithium salts, a heterolayered nanomat-based hierarchical/asymmetric porous separator was developed.[205] The separator was composed of a terpyridine (TPY)-functionalized NFC nanoporous thin mat as the top layer and an electrospun polyvinylpyrrolidone (PVP)/polyacrylonitrile (PAN) macroporous thick mat as the support layer (Fig. 14f–m). The hierarchical/asymmetric porous structure of the heterolayered nanomat was rationally designed by considering the trade-off between the leakage current and the ion transport rate. The TPY (to chelate Mn^{2+} ions) and PVP (to capture hydrofluoric acid)-mediated chemical functionalities bring a synergistic coupling in suppressing Mn^{2+}-induced adverse effects, thus enabling a substantial improvement in the high-temperature (60 1C) cycling performance (capacity retention B80% after

100 cycles) far beyond that attainable with conventional membrane technologies (5% for a commercial PP/PE/PP separator).

Overall, separators constructed with nanocellulose as building blocks display high wettability for the electrolyte, are extremely thin with strong mechanical strength, are electrochemically, structurally, and thermally stable, and have a porous structure, which makes them promising candidates for LIBs. Precisely controlling the pore structures and sizes is still challenging because the nanocellulose is usually densely packed within the thin separators. A porous structure with high tortuosity is expected to prevent the growth of dendritic lithium. Therefore, efficient pore engineering techniques are needed to adjust the pore structures and sizes as well as the porosity of the nanocellulose separators. The integration of nanocellulose membranes and other type of membranes such aselectrospun nonwoven mats should be considered to exploit the synergy effect of the beneficial influences of two or multiple layers.

6 Other batteries

6.1 Lithium-sulfur batteries

Li-S batteries have received increasing attention recently because they can offer high theoretical gravimetric (2500 $W \cdot h \cdot kg^{-1}$) and volumetric (2800 $W \cdot h \cdot L^{-1}$) energy density and 1 order of magnitude higher capacity (1675 $mA \cdot h \cdot g^{-1}$) than that of conventional LIBs.[9,10,206-210] In addition, elemental sulfur is environmentally friendly, cheap, and abundant, which makes Li-S batteries a particularly attractive and low-cost energy storage device. However, the practical application of Li-S batteries still faces several challenges. First, the volume changes of sulfur particles during electrochemical cycles lead to structure changes of the active materials and cause low capacity. Second, sulfur and Li_2S are both insulating, resulting in sluggish electrochemical kinetics. Third, polysulfides are easy to dissolve in the electrolyte, which is the main cause of the "shuttle phenomenon".[206,208] To solve the problems mentioned above, various efforts have been made to design and develop novel electrodes, separators, and electrolytes. Recently, several types of nanocellulose materials have been developed to construct high performance Li-S batteries, mainly based on their intrinsic structures and properties.

Nanocellulose can be integrated with active materials to fabricate hybrid electrodes for Li-S batteries. Owing to its high aspect ratio, entangled networks, and high abundance hydroxyl groups on the surfaces, NFC was developed as a flexible building block to fabricate freestanding and sandwich-structured cathode materials with high areal mass loading for long-life Li-S batteries (Fig. 15a-e).[211]

Sulfurwas impregnated in N-doped graphene and constructed as a primary active material, which was further welded in the NFC/CNT framework. Interconnected NFC/CNT layers on both sides of the active layer were uniquely synthesized to entrap polysulfide species and supply efficient electron transport. The synergistic effects of the physical encapsulation by carbonaceous materials (graphene and CNT/NFC fibers) and chemisorption for lithium polysulfides by chemical functionalization (hetero N-doping and hydroxyl groups) resulted in good electronic conductivity and suppression of polysulfide dissolution and migration. The rationally designed structure endowed the electrode with high specific capacity and good rate performance. The electrode with an areal sulfur loading of 8.1 $mg \cdot cm^{-2}$ exhibited an areal capacity of B8 $mA \cdot h \cdot cm^{-2}$ and an ultralow capacity fading of 0.067% per cycle over 1000 charge/discharge cycles at C/2 rate, whereas the average

Coulombic efficiency was approximately 97.3%, indicating good electrochemical reversibility.

Nanocellulose-derived carbon materials exhibit good performance as electrodes for Li-S batteries. Huang et al. reported the utilization of BC-derived carbon aerogels as a 3D framework for sulfur to fabricate a flexible sulfur cathode for Li-S batteries.[212] The as-prepared 3D carbon nanofiber networks exhibited high electrical conductivity and good mechanical stability. The macroporous structure of the carbon aerogels contributed to a high sulfur loading of 81 wt% through sulfur precipitation from the CS_2 solution. The sulfur species wrapped around the carbon nanofibers were well dispersed. Even at such a high loading, the S/carbon aerogel composite still contained sufficient free space to accommodate the volume expansion of sulfur during lithiation. An ultralight and thin BC-derived carbon aerogel interlayer was inserted between the sulfur cathode and separator to provide an extra conductive framework and somewhat adsorb the migrating polysulfides. The carbon aerogel interlayer can also act as an additional collector for sulfur and thus could prevent the over-aggregation of insulated sulfur on the cathode surface. With a thin carbon aerogel interlayer, the flexible S/carbon aerogel cathode exhibited a high discharge capacity (1134 mA h per g_{Sul} at 200 mA · g^{-1}), long-term cycle stability (700 mA h per g_{Sul} at 400 mA · g^{-1} over 400 cycles), and good rate capability. To further improve the electrochemical capability, heteroatom doping has attracted attention because it is an effective way to tailor the electronic and chemical properties of the carbon host. N is the most extensively studied heteroatom, which could improve the electron conductivity and strongly adsorb polysulfides, endowing a good rate capability and remarkable cycling stability. N-Doped carbon materials have been prepared from carbonization and activation (with KOH) the aerogels derived from BC hydrogels and its residual medium with enriched carbon and nitrogen. The S/carbon (with N-doped) composites were synthesized through a melt-diffusion method and used as the cathode material, which showed an initial discharge capacity of 1267 mA · h · g^{-1} and capacity retention of 995 mA · h · g^{-1} after 500 charge/discharge cycles at a rate of 0.1C with a Coulombic efficiency of 99.0%.[213] Recently, Li et al. developed a N, O-codoped carbon nanofiber aerogel by pyrolysis a BC aerogel in an ammonia atmosphere.[214] O was doped owing to the oxygenated species in the cellulose that remained in the aerogel after pyrolysis, whereas N was originated from the N-containing compounds left by culture media and secretions and the ammonia gas used during pyrolysis. The oxygen in the carbon nanofiber aerogel could further enhance the polysulfide retention capacity owing to improved hydrophilicity of the surface. The N, O-codoped carbon nanofiber aerogel was employed to a gel-based sulfur cathode, which simultaneously achieved both high sulfur content and high sulfur loading. With a sulfur loading of 6.4 mg cm^{-2} and a sulfur content of 90% in the whole electrode (including current collector) level, a capacity of 943 mA · h · g^{-1} was achieved, which corresponded to an areal capacity of 5.9 mA h cm^{-2}. CNC was also utilized as a carbon source for dual-atom doping to prepare carbon materials with hierarchical porous structures. N, S-Codoped carbons with a high pore volume and surface area were fabricated from the liquidcrystal-driven self-assembly of CNC coated with polyrhodanine followed by pyrolysis and etching of silica (silica precursors were added into the dispersion to provide a spacer between particles and avoid structural collapse upon carbonization) (Fig. 15f).[215] A synergistic effect from N and S heteroatoms modified the electron density distribution of the host substrate, leading to stronger polysulfide binding than that for non-doped or N single-doped carbons. *Ab initio* calculations based on density functional theory revealed the strong interactions between Li^+ and doped N atoms (Li-N) as well as the polysulfide anions and doped S atoms (S-

S) (Fig. 15g–j). The electrical conductivity of N, S-codoped carbon was also greatly improved compared with nondoped carbon, favoring high-rate kinetics. Using this dualdoped carbon as sulfur host, the sulfur electrode was able to deliver a high capacity of 1370 mA · h · g^{-1} at C/20 and charge/discharge for 1100 cycles at 2C with a very low capacity fading of 0.052% per cycle. This was attributed to the high pore volume and surface area of the host that can provide sufficient active sites for chemically adsorbed lithium polysulfides.

Nanocellulose-derived hybrid materials and carbonhybrid materials with a rational structure design can be used as separators or electrolytes for Li–S batteries. BC-derived carbon materials were mixed with TiO_2 to form a composite slurry and coated on one side of a separator.[216] The carbon materials served as an upper collector and supplied strong physical adsorption of polysulfides, whereas the TiO_2 can chemically adsorbed polysulfides. Thus, the carbon materials/TiO_2 modified separator was efficient restraining the shuttle effect of Li–S batteries. Cells with carbon materials/TiO_2 modified separator showed an initial discharge capacity of 1314 mA · h · g^{-1} at 0.2C, and the capacity retention was 1048.5 mA · h · g^{-1} after 50 cycles. During the rate test, Li–S cells delivered a discharge capacity of 537.1 mA · h · g^{-1} at 2C. Nair et al. prepared NFC-laden composite polymer electrolytes using a thermally induced polymerization process.[217] The NFC mainly acted as a reinforcing agent and provided an interpenetrated network for methacrylate-based polymer matrix to fabricate opaque and self-standing polymer electrolyte membranes. The activated membrane in the liquid electrolyte showed a high conductivity (41.2 mS · cm^{-1}) and a stable interface towards lithium metal. The final cell with a Li/activated polymer electrolyte separator/sulphur-activated carbon cell showed good cycling stability at above 700 mA · h · g^{-1} even at 1C rate and a Columbic efficiency of 499% at ambient temperature. The stable cycling profile was attributed to the reduction of the migration of polysulphide towards the anode by the entrapment of NFC in the polymer matrix.

Overall, nanocellulose-derived hybrid materials and carbon materials have the potential to be used as electrodes, separators and electrolytes for high performance Li–S batteries. However, challenges still exist such as how to achieve both high content and high loading of active materials in the same cathode while still preserving the flexible, high mechanical strength, and excellent electronic conductivity of the nanocellulose and nanocellulose-derived materials. To alleviate the polysulfide shuttle with a stable and persistent effect is still challenge for the nanocellulose-derived electrodes and separators. Thus, the rational designing and optimization of hierarchical porous structures and surface/interface properties of nanocellulose-derived hybrid materials and carbon materials is critical for constructing high-performance Li–S batteries. Progress in this direction is expected to be accomplished through a synergistic combination of theoretical calculations and experimental approaches.

6.2 Sodium-ion batteries

Over the last few years, NIBs have attracted increasing attention as an important source of energy storage devices largely owing to the fact that Na is the sixth most abundant element in the earth's crust and it is inexpensive and environmentally friendly for large-scale energy storage.[8,218–223] The working principle of NIBs is similar to that of LIBs. However, several good anode materials used in LIBs such as graphitic carbon materials exhibited poor storage properties for NIBs, mainly because the radius of Na^+ is B55% larger than that of Li^+; most materials do not possess host frameworks with larger interstitial spaces for Na ion insertion/extraction. Besides the kinetic issue, the larger Na^+ radius is also relevant for possible structural change during insertion or

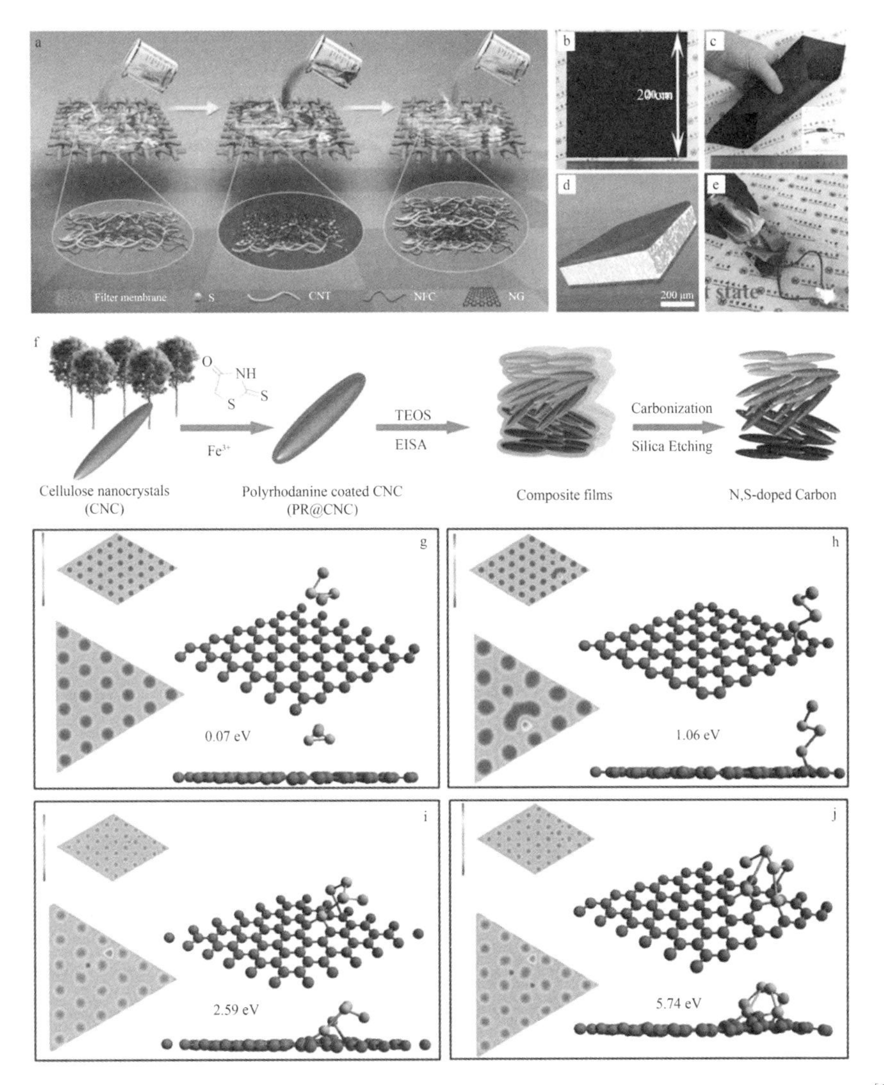

Fig. 15 Nanocellulose-derived materials for Li-S batteries. (a-e) NFC-derived materials:[211] (a) Schematic illustration of the freestanding electrode preparation using NFC as one of the building blocks; digital photographs of (b) the electrode and (c) its flexibility; (d) 3D XRM image of the electrode with 8. 1 mg of sulfur per cm^2. To get a stark contrast, sulfur in yellow, carbon in gray, and white in the outmost two layers and middle layer, respectively; (e) experimental demonstration showing that the prototype soft pack battery can light up three blue LEDs under bent states. (a-e) are reproduced with permission from ref. 211. Copyright 2017 Wiley-VCH. (f-j) CNC-derived materials:[215] (f) schematic illustration of the synthesis of CNC-derived N/S-doped carbon; Ab initio calculations illustrating the most stable Li_2S_2 binding configurations after full relaxation for carbon substrates (g) with no doping and doped with (h) 1N doping, (i) 1N+ 1S, and (j) 1N+ 2S per supercell (bottom right for side view). Insets on the top left are the 2D deformation charge distributions of the corresponding substrates only without Li_2S_2 (red for accepting electrons, blue for donating electrons), with the magnified versions (around the heteroatoms) shown on the bottom left. Gray, blue, purple, and yellow balls represent C, N, Li, and S atoms, respectively. (f-j) are reproduced with permission from ref. 215. Copyright 2015 Wiley-VCH

extraction. Thus, the development of appropriate electrode materials is critical for improving the performance of NIBs. Recently, nanocellulose-derived materials have gained interest for use as electrodes in NIBs.

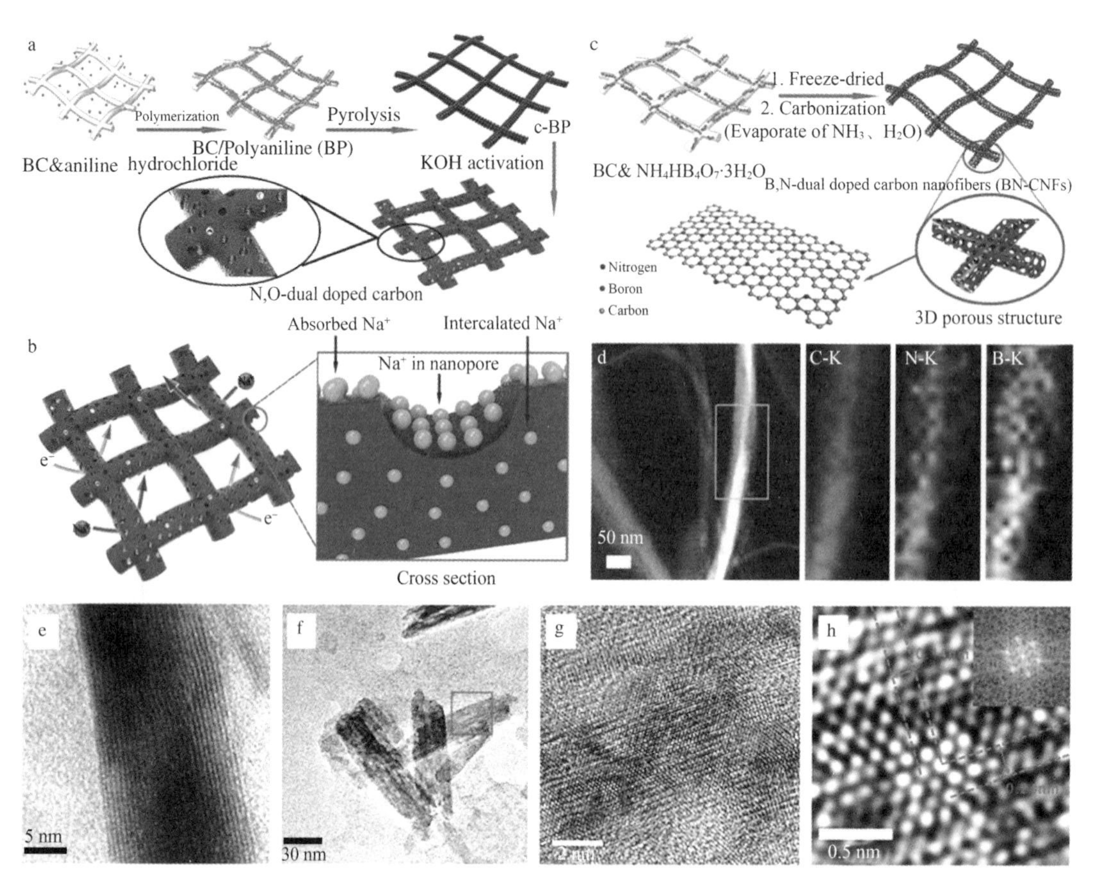

Fig. 16 Nanocellulose-derived materials for NIBs. (a, b) Fabrication of BC-derived N, O-dual doped carbon:[227] (a) schematic illustration of the fabrication process of the N, O-dual doped carbon network composite; (b) schematic illustration of the transportation process of Na-ion and electron in the N, O-dual doped carbon network composite. (a, b) are reproduced with permission from ref. 227. Copyright 2016 Wiley-VCH. (c, d) Fabrication of BC-derived B, N-dual doped carbon:[228] (c) schematic illustration of the fabrication process of B, N-dual doped carbon; (d) elemental mapping images of B, N-dual doped carbon electrode. (c, d) are reproduced with permission from ref. 228. Copyright 2016 Wiley-VCH. (e-h) Fabrication of CNC-derived carbon:[229] (e) HRTEM images of CNC; (f) TEM image of the CNC derived carbon (front-on); (g) higher magnification HRTEM image of the area contained in the red squared in (f); (h) higher magnification HRTEM image of the area contained in the red squared in (g) with the Fourier transform shown in the inset. (e-h) are reproduced with permission from ref. 229. Copyright 2017 Springer

Nanocellulose can be integrated with active materials to fabricate electrodes for NIBs. Li et al. reported the use of TEMPO-oxidized NFC as a dispersant to efficiently disperse and form a hybrid with 2D materials such as molybdenum disulfide (MoS_2) in aqueous solution.[224] The as-prepared NFC/MoS_2/CNTs composite films with a weight ratio of 16.5 : 67 : 16.5 can act as flexible electrodes for NIB anodes. The first cycle discharge capacity was 147 mA · h · g^{-1}, which was higher than that of the MoS_2/CNTs composite anode, even though it had a lower MoS_2 content. The Coulombic efficiency of the first cycle was 43.8% and increased to 89.7% by the third cycle.

Nanocellulose-derived carbon materials display advantages such as chemical and thermal stability, high electrical conductivity, as well as high Na-storage capacity for the development of anode materials for NIBs. Luo et al. reported the synthesis of carbon nanofibers directly from NFC through a pyrolysis process.[225] The unique morphology of the carbon nanofibers facilitated highly reversible ion migration and provided plenty of contact area between the electrode and the electrolyte. Thus, when utilized as an anode material for NIBs, the carbon nanofibers delivered not only a high reversible capacity of 255 $mA \cdot h \cdot g^{-1}$ at 40 $mA \cdot g^{-1}$ and a rate capability of 85 $mA \cdot h \cdot g^{-1}$ at 2000 $mA \cdot g^{-1}$, but also a good cycling stability of approximately 180 $mA \cdot h \cdot g^{-1}$ at 200 $mA \cdot g^{-1}$ over 600 cycles. To further improve the electrochemical performance, engineering the 3D porous carbon materials by heteroatom doping was conducted. Zhang et al. synthesized core-sheath structured carbon nanofiber@ N-doped porous carbon (CNF@ NPC) composite through carbonization of PPy coated BC.[226] The inner carbon nanofiber cores pyrolyzed from BC ensured the high electronic conductivity and the supply of a 3D matrix, whereas the outer NPC layers derived from PPy with a high surface area and an appropriate N doping improved the reversible capacity. When utilized as an anode material, the CNF@ NPC exhibited a high reversible specific capacity (240 $mA \cdot h \cdot g^{-1}$ at 100 $mA \cdot g^{-1}$ over 100 cycles), high rate performance (146.5 $mA \cdot h \cdot g^{-1}$ at 1 $A \cdot g^{-1}$), and good cycling stability (148.8 $mA \cdot h \cdot g^{-1}$ at 500 $mA \cdot g^{-1}$ over 400 cycles).

Comparedto single-atom doping, dual-atom doping in carbon could exploit the synergistic effects of the advanced features of the two heteroatoms, which takes advantage of the additionally created defect sites and electronic conductivity for sodium storage. Recently, Yu et al. synthesized N, O-dual doped carbon networks by using the pyrolysis of BC/PANI and subsequent activation with KOH (Fig. 16a and b).[227] BC acted as a 3D scaffold to deposit PANI but also formed a conductive carbon network during the carbonization, whereas the PANI formed large amounts of small N, O-dual doped carbon particles along the interconnected nanofibers *in situ*.

Theobtained carbon networks exhibited a high specific capacity (798 $mA \cdot h \cdot g^{-1}$), long cycle life (545 $mA \cdot h \cdot g^{-1}$ at 100 $mA \cdot g^{-1}$ after 100 cycles), and high rate capability (240 $mA \cdot h \cdot g^{-1}$ at 2 $A \cdot g^{-1}$ after 2000 cycles). The carbon materials with 3D well-defined porosity and N, O-dual doped induced active sites, contributing to the enhanced sodium storage. B, N-Dual doped carbon materials were developed by fully infiltrating $NH_4HB_4O_7$ H_2O into the BC pellicle followed by carbonization (Fig. 16c and d).[228] The B, N-dual doping created synergistic effects to enlarge the carbon layer spacing for Na^+ insertion, and improved the electrochemical activity as well as electronic conductivity. The obtained carbon materials delivered a specific charge capacity of 581 $mA \cdot h \cdot g^{-1}$ at 100 $mA \cdot g^{-1}$after 120 cycles and good cycling stability (277 $mA \cdot h \cdot g^{-1}$ at 10 $A \cdot g^{-1}$ after 1000 cycles). In addition to NFC/BC with a high aspect ratio, Zhu et al. reported the carbonization of short and high specific surface area CNCs at 1000 1C for use in NIBs.[229] During the carbonization, the CNC was transformed into porous carbon with an increased short-range ordered lattice and percolated carbon nanofiber (Fig. 16e–h), which exhibited a good structure and conductivity for NIB anodes. A starting capacity of 340 $mA \cdot h \cdot g^{-1}$ at 100 $mA \cdot g^{-1}$ was achieved, which maintained 88.2% capacity over 400 cycles.

Overall, nanocellulose has potential to be developed as an electrode material for NIBs. Nanocellulose can form a hybrid with active materials or be converted to 3D interconnected carbon materials and be further functionalized by doping with heteroatoms or integrated with active materials. Compared with LIBs, the larger size of sodium ions hampers the kinetics of the electrochemical reactions. Therefore, reversible electrodes derived from nanocellulose-derived materials need to possess large enough channels and/or interstitial sites. The challenge in the future is to design and construct novel hierarchical 3D fibrous porous structures with appropriate space for Na storage, short diffusion distance for ions, high Na^+ diffusion coefficients, and fast electron transport from various types of nanocellulose through multiple synthesis strategies.

7 Summary and outlook

The past few years have witnessed significant progress in the development of nanocellulose forelectrochemical energy storage owing to its exceptional structures and properties as well as its sustainability. In this review, we have summarized the recent advances in this area, with a particular focus on the structures and characteristics of nanocellulose within the cell walls of cellulose sources, the fabrication, structural and functional design of nanocellulose, and the application of the nanocellulose-derived materials for advanced energy storage systems. Thanks to the efforts of many research groups, a rich variety of nanocellulose materials with various structures and surface chemistry performance have been developed from various sources using multiple fabrication strategies (Table 1),[77, 82, 86-91] which provides plenty of building blocks for further development of high-performance Energy Storage Mater. and devices. Owing to the nanoscale dimension, high specific surface area, high relative crystallinity, reactive surfaces containing hydroxyl groups, entangled web-like structures, and advantageous mechanical and thermal properties, nanocellulose can be utilized for ① constructing self-standing and high strength electrodes or separators, ② integration with other electrochemical active materials, ③ developing flexible substrates/scaffolds for flexible devices, ④ designing multifunctional separators, ⑤ developing carbon-based porous materials and carbon hybrid materials, ⑥ constructing multiple kinds of materials or devices using various kinds of nanocellulose and their hybrids as building blocks. The nanocellulose-derived materials exhibit advantages for various electrochemical energy storage applications, including supercapacitors, LIBs, Li-S batteries, and NIBs.

Although nanocellulose has been successfully developed to construct a series ofAdv. Mater./devices for high performance energy storage, there are some issues that need to be addressed in the future.

(1) Large-scale fabrication. Although the fabrication of nanocellulose has become mature during the past 10 years and mass production of nanocellulose has been achieved by several research institutions and companies, the quality such as the structures, nanofibrillation degree, and surface chemical properties of nanocellulose is still quite different compared with those fabricated at a laboratory scale. Thus, it is still difficult to perform large-scale production of nanocellulose with high quality and uniformity, especially for the large-scale production of high aspect ratio and individualized NFC. We should pay more attention on developing and optimizing mechanical processing equipment such as large capacity high pressure homogenizer, beads mill, twin screw extruder, and high-intensity ultrasonicator to achieve the mass production of high quality nanocellulose for sustainable supply for energy storage development and other applications. Furthermore, limited by the preparation technology and apparatus such as vacuum filter, supercritical dryer, and tube furnace, it is still a big challenge to perform continuous and large-scale fabrication of nanocellulose-derived electrodes/separators with a large size for energy storage devices.

(2) Surface/Interface Engineering. Surfaces and interfaces play crucial roles in the fabrication and

performance of nanocellulose-derived Energy Storage Mater. /devices. Surface/interface engineering can induce novel physicochemical properties and strong synergistic effects for nanocellulose and nanocellulose-derived materials, providing new and efficient strategies to enhance the energy storage activities. Thus, we should exploit the surface atomic or molecular engineering (e. g. , heteroatom-doping, chemical modification) to directly utilization or chemical modification of the surface structures and chemistry of nanocellulose and nanocellulose-derived carbon materials, and further integrate the materials with other active materials through various interface engineering routes such as physical and chemical treatment related to weak (e. g. , hydrogen bonding, van der Waals interaction) or strong (e. g. , covalent bonding) effects. More alternative methodologies can be inspired from the success work of other similar materials such as polymeric materials.[230] We should develop surface/interface engineering technique to provide nanocellulose-derived Energy Storage Mater. /devices with tunable micro-, nanoand even atom-scale structures, tailorable physicochemical properties, excellent electronic conductivity, flexibility, advantageous mechanical properties, physical and chemical stability as well as high energy storage performance.

(3) Pore Engineering. The sizes and structures of the pores and the porosity of the nanocellulose-derived materials critically affect their performance. The pore size of materials can range from the nanometer to the micrometer scale, depending on the structure of nanocellulose and the active materials as well as the processing methods. For the electrodes, a hierarchical micropore-mesopore structure should be developed to improve the contact area of the electrode-electrolyte, decrease the diffusion resistance of electrolyte ions, and shorten the diffusion length of the electrolyte ions. For the separator, the porosity and pore size of the nanocellulose film should be designed to achieve the electrically isolate the electrodes but facilitate the ionic transport. However, it is still difficult to precisely control the pore structures and sizes of the nanocellulose-derived materials. Therefore, pore engineering that can tune the pore size, especially at the nanometerand even the atom-scale level is highly desirable. It is also suggested to get more inspiration from the research experience of other similar polymeric materials,[231,232] to optimize the pore engineering technique of nanocellulose-derived materials. In the future, we should design all possible pathways to maximize generate and control the porous structures and avoid the collapse of web-like structures of the nanocellulose-derived materials throughout the whole life cycles.

(4) Flexibility. Forthcoming smart electronics era necessitates the development of advanced flexible energy storage devices with environmentally friendly characteristics, lightweight, low cost, and advantageous energy storage performance. Nanocellulose meets all the above-mentioned critical requirements and can be utilized as a building block to construct flexible substrates/scaffolds or as carbon precursors to fabricate light and flexible energy storage devices with shape-conformability, aesthetic diversity, and advantageous mechanical properties. For the separators and electrodes derived from nanocellulose or nanocellulose composites, the mechanically robust and entangled networks of nanocellulose should be preserved during the whole fabrication and utilization process. For the nanocellulose-derived carbon electrodes, the hydrogen bond interaction between the nanocellulose should be removed and the interconnected network structures of the nanocellulose should be carefully preserved during carbonization, which would ensure robust mechanical properties with high flexibility of the carbon electrodes. Furthermore, excellent conductivity and a high capacity should be achieved for nanocellulose-derived electrodes by integration with active materials without

sacrificing the structural integrity and flexibility. It is also necessary to integrate the electrolytes/separators, electrode materials, current collectors, and packaging to maximize the synergistic effect of the composite structures and utilization of the flexibility derived from the nanocellulose entangled networks, which make the devices simultaneously exhibit high flexibility and high energy storage performance.

(5) New materials and new applications. As nanocellulose can be fabricated from various sources using different preparation methods, new materials/devices for energy storage should be explored using different types of nanocellulose and their combinations as building blocks, based on the sizes, structures, and surface chemical performance of nanocellulose. For example, optically transparent and flexible electrodes can be developed by the integration of TEMPO-oxidized NFC with transparent conductive materials. Chiral nematic mesoporous carbon electrodes can be developed by using sulfuric-acid-hydrolyzed CNC as a template. Investigation of new materials as well as the fabrication strategies could open up new opportunities for a wide range of applications. To date, nanocellulose and its derived materials have been widely used for supercapacitors and LIBs, however, a few works were reported for their application in Li-S batteries and NIBs. Moreover, some new types of energy storage systems such as Mg (Al, Mn)-ion batteries have been rarely explored. Considering that nanocellulose and its derived materials are mechanically robust and with structural, surface/interface chemistry adjustable, they can be readily used as a green material solution for newly emerging energy storage systems through customized control of their structure and characteristics.

Acknowledgments

This work was supported in part by the National Natural Science Foundation of China (No. 31400495; 51672055; 31770594), Young Elite Scientists Sponsorship Program by CAST (No. 2017QNRC001), Ten Thousand Talent Program (No. W02020249), Funds supported by the Fok Ying-Tong Education Foundation, China (No. 161025), and Fundamental Research Funds for the Central Universities. We would like to acknowledge Prof. Shigenori Kuga, Graduate School of Agricultural and Life Sciences, The University of Tokyo, Japan, for valuable suggestions and revision of the manuscript.

References

[1] Chu S, Majumdar A. Opportunities and challenges for a sustainable energy future[J]. Nature, 2012, 488(7411): 294-303.

[2] Dunn B, Kamath H, Tarascon J M. Electrical Energy Storage for the Grid: A Battery of Choices[J]. Science, 2011, 334(6058): 928-935.

[3] Ji L W, Lin Z, Alcoutlabi M, Zhang X W. Recent developments in nanostructured anode materials for rechargeable lithium-ion batteries[J]. Energy Environ. Sci., 2011, 4(8): 2682-2699.

[4] Choi N S, Chen Z H, Freunberger S A, et al., Challenges Facing Lithium Batteries and Electrical Double-Layer Capacitors[J]. Angew. Chem. -Int. Edit., 2012, 51(40): 9994-10024.

[5] Xu K. Nonaqueous liquid electrolytes for lithium-based rechargeable batteries[J]. Chem. Rev., 2004, 104(10): 4303-4417.

[6] Gogotsi Y. What Nano Can Do for Energy Storage[J]. ACS Nano, 2014, 8(6): 5369-5371.

[7] Palomares V, Serras P, Vıllaluenga I, Hueso K B, Carretero-Gonzalez J, Rojo T. Na-ion batteries, recent advances and present challenges to become low cost energy storage systems[J]. Energy Environ. Sci., 2012, 5(3): 5884-5901.

[8] Palomares V, Casas-Cabanas M, Castillo-Martinez E, Han M H, Rojo T. Update on Na-based battery materials. A

growing research path[J]. Energy Environ. Sci., 2013, 6(8): 2312-2337.

[9] Peng H J, Huang J Q, Cheng X B, Zhang Q. Review on high-loading and high-energy lithium-sulfur batteries[J]. Adv. Energy Mater., 2017, 7(24): 1700260.

[10] Fang R, Zhao S, Sun Z, Wang D W, Cheng H M, Li F. More reliable lithium-sulfur batteries: status, solutions and prospects[J]. Adv. Mater., 2017, 29(48): 1606823.

[11] Larcher D, Tarascon J M. Towards greener and more sustainable batteries for electrical energy storage[J]. Nat. Chem., 2015, 7(1): 19-29.

[12] Yang Z, Zhang J, Kintner-Meyer M C W, Lu XC, Choi DW, Lemmon J P, Liu J. Electrochemical energy storage for green grid[J]. Chem. Rev., 2011, 111(5): 3577-3613.

[13] Armstrong R C, Wolfram C, De Jong K P, Gross P, Lewis N S, Boardman B, Ragauskas A J, Martinez K E, Crabtree G, Ramana M V. The frontiers of energy[J]. Nat. Energy, 2016, 1(1): 1-8.

[14] Arico A S, Bruce P, Scrosati B, Tarascon J M, Schalkwijk W V. Nanostructured materials for advanced energy conversion and storage devices[J]. Nat. Mater., 2005, 4(5): 366-377.

[15] Simon P, Gogotsi Y. Materials for electrochemical capacitors[J]. Nat. Mater., 2008, 7(11): 845-854.

[16] Dutta S, Kim J, Ide Y, Kim J H, Hossain M S A, Bando Y, Yamauchi Y, Wu K C-W. 3D network of cellulose-based energy storage devices and related emerging applications[J]. Mater. Horizons, 2017, 4(4): 522-545.

[17] Ummartyotin S, Manuspiya H. An overview of feasibilities and challenge of conductive cellulose for rechargeable lithium-based battery[J]. Renew. Sust. Energ. Rev., 2015, 50: 204-213.

[18] Jabbour L, Bongiovanni R, Chaussy D, Gerbaldi C, Beneventi D. Cellulose-based Li-ion batteries: a review[J]. Cellulose, 2013, 20(4): 1523-1545.

[19] Wang Z, Tammela P, Strømme M, Nyholm L. Cellulose-based supercapacitors: material and performance considerations[J]. Adv. Energy Mater., 2017, 7(18): 1700130.

[20] Pérez-Madrigal M M, Edo M G, Alemán C. Powering the future: application of cellulose-based materials for supercapacitors[J]. Green Chem., 2016, 18(22): 5930-5956.

[21] Zheng G, Cui Y, Karabulut E, Wågberg L, Zhu H, Hu L. Nanostructured paper for flexible energy and electronic devices[J]. MRS Bull., 2013, 38(4): 320-325.

[22] Hu L, Cui Y. Energy and environmental nanotechnology in conductive paper and textiles[J]. Energy Environ. Sci., 2012, 5(4): 6423-6435.

[23] Nyholm L, Nyström G, Mihranyan A, Strømme M. Toward flexible polymer and paper-based energy storage devices [J]. Adv. Mater., 2011, 23(33): 3751-3769.

[24] Nguyen T H, Fraiwan A, Choi S. Based batteries: A review[J]. Biosens. Bioelectron., 2014, 54: 640-649.

[25] Sharifi F, Ghobadian S, Cavalcanti F R, Hashemia N. Based devices for energy applications[J]. Renew. Sust. Energ. Rev., 2015, 52: 1453-1472.

[26] Kim J H, Shim B S, Kim H S, Lee Y J, Jang D, Abas Z, Kim J. Review of nanocellulose for sustainable future materials[J]. Int. J. Precis Eng Manuf-Green Technol., 2015, 2(2): 197-213.

[27] Zhang Y Z, Wang Y, Cheng T, Lai W Y, Pang H, Huang W. Flexible supercapacitors based on paper substrates: a new paradigm for low-cost energy storage[J]. Chem. Soc. Rev., 2015, 44(15): 5181-5199.

[28] Yao B, Zhang J, Kou T, Song, Y, Liu T Y, Li Y. Paper-based electrodes for flexible energy storage devices[J]. Adv. Sci., 2017, 4(7): 1700107.

[29] Klemm D, Heublein B, Fink H P, Bohn A. Cellulose: fascinating biopolymer and sustainable raw material[J]. Angew. Chem. -Int. Edit., 2005, 44(22): 3358-3393.

[30] Balat M, Ayar G. Biomass energy in the world, use of biomass and potential trends[J]. Energy Sources, 2005, 27 (10): 931-940.

[31] Moon R J, Martini A, Nairn J, Simonsen J, Youngblood J. Cellulose nanomaterials review: structure, properties and nanocomposites[J]. Chem. Soc. Rev., 2011, 40(7): 3941-3994.

[32] Capadona J R, Shanmuganathan K, Tyler D J, Rowan S J, Weder C. Stimuli-responsive polymer nanocomposites inspired by the sea cucumber dermis[J]. Science, 2008, 319(5868): 1370-1374.

[33] Capadona J R, Van Den Berg O, Capadona L A, Schroeter M, Rowan S J, Tyler D J, Weder C. A versatile approach for the processing of polymer nanocomposites with self-assembled nanofibre templates[J]. Nat. Nanotechnol., 2007, 2(12): 765-769.

[34] Shopsowitz K E, Qi H, Hamad W Y, MacLachlan M J. Free-standing mesoporous silica films with tunable chiral nematic structures[J]. Nature, 2010, 468(7322): 422-425.

[35] Olsson R T, Samir M A S A, Salazar-Alvarez G, Belova L, Strom V, Berglund L A, Ikkala O, Nogues J, Gedde U W. Making flexible magnetic aerogels and stiff magnetic nanopaper using cellulose nanofibrils as templates[J]. Nat. Nanotechnol., 2010, 5(8): 584-588.

[36] Wicklein B, Kocjan A, Salazar-Alvarez G, Carosio F, Camino G, Antonietti M, Bergstrom L. Thermally insulating and fire-retardant lightweight anisotropic foams based on nanocellulose and graphene oxide[J]. Nat. Nanotechnol., 2015, 10(3): 277-283.

[37] Eichhorn S J. Cellulose nanowhiskers: promising materials for advanced applications[J]. Soft Matter, 2011, 7(2): 303-315.

[38] Lagerwall J P F, Schütz C, Salajkova M, Noh J, Park J H, Scalia G, Bergstrom L. Cellulose nanocrystal-based materials: from liquid crystal self-assembly and glass formation to multifunctional thin films[J]. NPG Asia Mater., 2014, 6(1): e80-e80.

[39] Habibi Y, Lucia L A, Rojas O J. Cellulose nanocrystals: chemistry, self-assembly, and applications[J]. Chem. Rev., 2010, 110(6): 3479-3500.

[40] Habibi Y. Key advances in the chemical modification of nanocelluloses[J]. Chem. Soc. Rev., 2014, 43(5): 1519-1542.

[41] Peng B L, Dhar N, Liu H L, Tam KC. Chemistry and applications of nanocrystalline cellulose and its derivatives: a nanotechnology perspective[J]. Can. J. Chem. Eng., 2011, 89(5): 1191-1206.

[42] Lavoine N, Desloges I, Dufresne A, Bras J. Microfibrillated cellulose-Its barrier properties and applications in cellulosic materials: A review[J]. Carbohydr. Polym., 2012, 90(2): 735-764.

[43] Siró I, Plackett D. Microfibrillated cellulose and new nanocomposite materials: a review[J]. Cellulose, 2010, 17(3): 459-494.

[44] Klemm D, Kramer F, Moritz S, Lindstrom T, Ankerfors M, Gray D, Dorris A. Nanocelluloses: a new family of nature-based materials[J]. Angew. Chem.-Int. Edit., 2011, 50(24): 5438-5466.

[45] Kalia S, Boufi S, Celli A, Kango S. Nanofibrillated cellulose: surface modification and potential applications[J]. Colloid Polym. Sci., 2014, 292(1): 5-31.

[46] Hamad W. On the development and applications of cellulosic nanofibrillar and nanocrystalline materials[J]. Can. J. Chem. Eng., 2006, 84(5): 513-519.

[47] Eichhorn S J, Dufresne A, Aranguren M, Marcovich N E, Capadona J R, Rowan S J, Weder C, Thielemans W, Roman M, Renneckar S, Gindl W, Veigel S, Keckes J, Yano H, Abe K, Nogi M, Nakagaito A N, Mangalam A, Simonsen J, Benight A S, Bismarck, A, Berglund L A, Peijs T. Current international research into cellulose nanofibres and nanocomposites[J]. J. Mater. Sci., 2010, 45(1): 1-33.

[48] Azizi Samir M A S, Alloin F, Dufresne A. Review of recent research into cellulosic whiskers, their properties and their application in nanocomposite field[J]. Biomacromolecules, 2005, 6(2): 612-626.

[49] Eyley S, Thielemans W. Surface modification of cellulose nanocrystals[J]. Nanoscale, 2014, 6(14): 7764-7779.

[50] Isogai A. Wood nanocelluloses: fundamentals and applications as new bio-based nanomaterials[J]. J. Wood Sci., 2013, 59(6): 449-459.

[51] Isogai A, Saito T, Fukuzumi H. TEMPO-oxidized cellulose nanofibers[J]. Nanoscale, 2011, 3(1): 71-85.

[52] Dufresne A. Nanocellulose: a new ageless bionanomaterial[J]. Mater. Today, 2013, 16(6): 220-227.

[53] Yano H, Sugiyama J, Nakagaito A N, Nogi M, Matsuura T, Hikita M, Handa K. Optically transparent composites reinforced with networks of bacterial nanofibers[J]. Adv. Mater., 2005, 17(2): 153-155.

[54] Nogi M, Iwamoto S, Nakagaito A N, Yano H. Optically transparent nanofiber paper[J]. Adv. Mater., 2009, 21(16): 1595-1598.

[55] Nogi M, Yano H. Transparent nanocomposites based on cellulose produced by bacteria offer potential innovation in the electronics device industry[J]. Adv. Mater., 2008, 20(10): 1849-1852.

[56] Svagan A J, Samir M A S A, Berglund L A. Biomimetic foams of high mechanical performance based on nanostructured cell walls reinforced by native cellulose nanofibrils[J]. Adv. Mater., 2008, 20(7): 1263-1269.

[57] Laaksonen P, Walther A, Malho J M, Kainlauri M, Ikkala O, Linder M B. Genetic engineering of biomimetic nanocomposites: diblock proteins, graphene, and nanofibrillated cellulose[J]. Angew. Chem.-Int. Edit., 2011, 50(37): 8688-8691.

[58] Korhonen J T, Hiekkataipale P, Malm J, Karppinen M, Ikkala O, Ras R H A. Inorganic hollow nanotube aerogels by atomic layer deposition onto native nanocellulose templates[J]. ACS nano, 2011, 5(3): 1967-1974.

[59] Rajala S, Siponkoski T, Sarlin E, Mettanen M, Pammo A, Juuti J, Rojas O J, Franssila S, Tuukkanen S. Cellulose nanofibril film as a piezoelectric sensor material[J]. ACS Appl. Mater. Interfaces, 2016, 8(24): 15607-15614.

[60] Mangayil R, Rajala S, Pammo A, Sarlin, E, Luo J, Santala V, Karp M, Tuukkanen, S. Engineering and characterization of bacterial nanocellulose films as low cost and flexible sensor material[J]. ACS Appl. Mater. Interfaces, 2017, 9(22): 19048-19056.

[61] Yao C, Hernandez A, Yu Y, Cai Z Y, Wang X D. Triboelectric nanogenerators and power-boards from cellulose nanofibrils and recycled materials[J]. Nano Energy, 2016, 30: 103-108.

[62] Chen B, Yang N, Jiang Q, Chen W S, Yang Y. Transparent triboelectric nanogenerator-induced high voltage pulsed electric field for a self-powered handheld printer[J]. Nano Energy, 2018, 44: 468-475.

[63] Sakurada I, Nukushina Y, Ito T. Experimental determination of the elastic modulus of crystalline regions in oriented polymers[J]. J. Polym. Sci., 1962, 57(165): 651-660.

[64] Nirmale T C, Karbhal I, Kalubarme R S, Shelke M V, Varma A J, Kale B B. Facile synthesis of unique cellulose triacetate based flexible and high-performance gel polymer electrolyte for lithium-ion batteries[J]. ACS Appl. Mater. Interfaces, 2017, 9(40): 34773-34782.

[65] Wan J, Zhang J, Yu J, Zhang J. Cellulose aerogel membranes with a tunable nanoporous network as a matrix of gel polymer electrolytes for safer lithium-ion batteries[J]. ACS Appl. Mater. Interfaces, 2017, 9(29): 24591-24599.

[66] Deng L, Young R J, Kinloch I A, Abdelkader A M, Holmes S M, De Haro-Del Rio D A, Eichhorn S J. Supercapacitance from cellulose and carbon nanotube nanocomposite fibers[J]. ACS Appl. Mater. Interfaces, 2013, 5(20): 9983-9990.

[67] White R J, Brun N, Budarin V L, Clark J H, Titirici M M. Always look on the "light" side of life: sustainable carbon aerogels[J]. ChemSusChem, 2014, 7(3): 670-689.

[68] Zhang L, Liu Z, Cui G, Chen L Q. Biomass-derived materials for electrochemical energy storages[J]. Prog. Polym. Sci., 2015, 43: 136-164.

[69] Titirici M M, White R J, Brun N, Budarin V L, Su D S, Monte F, Clark J H, MacLachlan M J. Sustainable carbon materials[J]. Chem. Soc. Rev., 2015, 44(1): 250-290.

[70] Zhu H, Luo W, Ciesielski P N, Fang Z Q, Zhu J Y, Henriksson G, Himmel M E, Hu L B. Wood-derived materials

for green electronics, biological devices, and energy applications[J]. Chem. Rev., 2016, 116(16): 9305-9374.

[71] Nirmale T C, Kale B B, Varma A J. A review on cellulose and lignin-based binders and electrodes: Small steps towards a sustainable lithium ion battery[J]. Int. J. Biol. Macromol., 2017, 103: 1032-1043.

[72] Chen L F, Feng Y, Liang H W, Wu Z Y, Yu S H. Macroscopic-scale three-dimensional carbon nanofiber architectures for electrochemical energy storage devices[J]. Adv. Energy Mater., 2017, 7(23): 1700826.

[73] Wu Z Y, Liang H W, Chen L F, Hu B C, Yu S H. Bacterial cellulose: a robust platform for design of three dimensional carbon-based functional nanomaterials[J]. Accounts Chem. Res., 2016, 49(1): 96-105.

[74] Hoeng F, Denneulin A, Bras J. Use of nanocellulose in printed electronics: a review[J]. Nanoscale, 2016, 8(27): 13131-13154.

[75] Du X, Zhang Z, Liu W, Deng Y L. Nanocellulose-based conductive materials and their emerging applications in energy devices-A review[J]. Nano Energy, 2017, 35: 299-320.

[76] Wang X, Yao C, Wang F, Li Z D. Cellulose-based nanomaterials for energy applications[J]. Small, 2017, 13(42): 1702240.

[77] Uetani K, Yano H. Nanofibrillation of wood pulp using a high-speed blender[J]. Biomacromolecules, 2011, 12(2): 348-353.

[78] Wegst U G K, Bai H, Saiz E, Tomsia A P, Ritchie R O. Bioinspired structural materials[J]. Nat. Mater., 2015, 14(1): 23-36.

[79] Sun Y X, Qi J Z, Yang G, Pang J Y, Du F G. Structural features and properties of Betula schmidtii entitled as rigidy and heavy wood[J]. Scientia Silvae Sinicae, 2012, 48(2): 180-186.

[80] Abe K, Yano H. Comparison of the characteristics of cellulose microfibril aggregates isolated from fiber and parenchyma cells of Moso bamboo (Phyllostachys pubescens)[J]. Cellulose, 2010, 17(2): 271-277.

[81] Chen W, Abe K, Uetani K, Yu H P, Yano H. Individual cotton cellulose nanofibers: pretreatment and fibrillation technique[J]. Cellulose, 2014, 21(3): 1517-1528.

[82] Abe K, Iwamoto S, Yano H. Obtaining cellulose nanofibers with a uniform width of 15 nm from wood[J]. Biomacromolecules, 2007, 8(10): 3276-3278.

[83] Abe K, Yano H. Comparison of the characteristics of cellulose microfibril aggregates of wood, rice straw and potato tuber[J]. Cellulose, 2009, 16(6): 1017-1023.

[84] Terashima N, Kitano K, Kojima M, Yoshida M, Yamamoto H, Westermark U. Nanostructural assembly of cellulose, hemicellulose, and lignin in the middle layer of secondary wall of ginkgo tracheid[J]. J. Wood Sci., 2009, 55(6): 409-416.

[85] Somerville C, Bauer S, Brininstool G, Facette M, Hamann T, Milne J, Osborne E, Paredez A, Persson S, Raab T, Vorwerk S, Youngs H. Toward a systems approach to understanding plant cell walls[J]. Science, 2004, 306(5705): 2206-2211.

[86] Chen W, Li Q, Wang Y, Yi X, Zeng J, Yu H P, Liu Y X, Li J. Comparative study of aerogels obtained from differently prepared nanocellulose fibers[J]. ChemSusChem, 2014, 7(1): 154-161.

[87] Pääkkö M, Ankerfors M, Kosonen H, Nykanen A, Ahola S, Osterberg M, Ruokolainen J, Laine J, Larsson P T, Ikkala O, Lindstrom T. Enzymatic hydrolysis combined with mechanical shearing and high-pressure homogenization for nanoscale cellulose fibrils and strong gels[J]. Biomacromolecules, 2007, 8(6): 1934-1941.

[88] Saito T, Kimura S, Nishiyama Y, Isogai A. Cellulose nanofibers prepared by TEMPO-mediated oxidation of native cellulose[J]. Biomacromolecules, 2007, 8(8): 2485-2491.

[89] Wagberg L, Decher G, Norgren M, Lindstrom T, Ankerfors M, Axnas K. The build-up of polyelectrolyte multilayers of microfibrillated cellulose and cationic polyelectrolytes[J]. Langmuir, 2008, 24(3): 784-795.

[90] Elazzouzi-Hafraoui S, Nishiyama Y, Putaux J L, Heux L, Dubreuil F, Rochas C. The shape and size distribution of crystalline nanoparticles prepared by acid hydrolysis of native cellulose[J]. Biomacromolecules, 2008, 9(1): 57-65.

[91] Ifuku S, Nogi M, Abe K, Handa K, Nakatsubo F, Yano, H. Surface modification of bacterial cellulose nanofibers for property enhancement of optically transparent composites: dependence on acetyl-group DS[J]. Biomacromolecules, 2007, 8(6): 1973-1978.

[92] Sharma P R, Varma A J. Functional nanoparticles obtained from cellulose: Engineering the shape and size of 6-carboxycellulose[J]. Chem. Commun., 2013, 49(78): 8818-8820.

[93] Sharma P R, Varma A J. Thermal stability of cellulose and their nanoparticles: Effect of incremental increases in carboxyl and aldehyde groups[J]. Carbohydr. Polym., 2014, 114: 339-343.

[94] Hornig S, Heinze T. Efficient approach to design stable water-dispersible nanoparticles of hydrophobic cellulose esters [J]. Biomacromolecules, 2008, 9(5): 1487-1492.

[95] Nosar M N, Salehi M, Ghorbani S, Beiranvand, S P, Goodarzi A, Azami, M. Characterization of wet-electrospun cellulose acetate based 3-dimensional scaffolds for skin tissue engineering applications: influence of cellulose acetate concentration [J]. Cellulose, 2016, 23(5): 3239-3248.

[96] Rojas O J, Montero G A, Habibi Y. Electrospun nanocomposites from polystyrene loaded with cellulose nanowhiskers [J]. J. Appl. Polym. Sci., 2009, 113(2): 927-935.

[97] Turbak A F, Snyder F W, Sandberg K R. Microfibrillated cellulose, a new cellulose product: properties, uses, and commercial potential[J]. J Appl Polym Sci Appl Polym Symp. 1983, 37(9): 815-827.

[98] Herrick F W, Casebier R L, Hamilton J K, Sandberg K R. Microfibrillated cellulose: morphology and accessibility [J]. Appl. Polym. Sci.: Appl. Polym. Symp., 1983, 37(CONF-8205234-Vol. 2).

[99] Nakagaito A N, Yano H. Novel high-strength biocomposites based on microfibrillated cellulose having nano-order-unit web-like network structure[J]. Appl. Phys. A, 2005, 80(1): 155-159.

[100] Iwamoto S, Nakagaito A N, Yano H. Nano-fibrillation of pulp fibers for the processing of transparent nanocomposites [J]. Appl. Phys. A, 2007, 89(2): 461-466.

[101] Iwamoto S, Nakagaito A N, Yano H, Nogi M. Optically transparent composites reinforced with plant fiber-based nanofibers[J]. Appl. Phys. A, 2005, 81(6): 1109-1112.

[102] Zhao H P, Feng X Q, Gao H. Ultrasonic technique for extracting nanofibers from Nat. Mater. [J]. Appl. Phys. Lett., 2007, 90(7): 073112.

[103] Cheng Q, Wang S, Rials T G. Poly (vinyl alcohol) nanocomposites reinforced with cellulose fibrils isolated by high intensity ultrasonication[J]. Compos. Pt. A-Appl. Sci. Manuf., 2009, 40(2): 218-224.

[104] Chen W, Yu H, Liu Y, Hai Y F, Zhang M X, Chen, P. Isolation and characterization of cellulose nanofibers from four plant cellulose fibers using a chemical-ultrasonic process[J]. Cellulose, 2011, 18(2): 433-442.

[105] Chen W, Li Q, Cao J, Liu Y X, Li J, Zhang J S, Luo S Y, Yu H P. Revealing the structures of cellulose nanofiber bundles obtained by mechanical nanofibrillation via TEM observation[J]. Carbohydr. Polym., 2015, 117: 950-956.

[106] Saito T, Nishiyama Y, Putaux J L, Vignon M, Isogai A. Homogeneous suspensions of individualized microfibrils from TEMPO-catalyzed oxidation of native cellulose[J]. Biomacromolecules, 2006, 7(6): 1687-1691.

[107] Fukuzumi H, Saito T, Okita Y, Isogai, A. Thermal stabilization of TEMPO-oxidized cellulose[J]. Polym. Degrad. Stabil., 2010, 95(9): 1502-1508.

[108] Fukuzumi H, Saito T, Iwata T, et al. Kumamoto, Y, Isogai A. Transparent and high gas barrier films of cellulose nanofibers prepared by TEMPO-mediated oxidation[J]. Biomacromolecules, 2009, 10(1): 162-165.

[109] Ranby B G. Fibrous macromolecular systems. Cellulose and muscle. The colloidal properties of cellulose micelles [J]. Discussions of the Faraday Society, 1951, 11: 158-164.

[110] Marchessault R H, Morehead F F, Walter N M. Liquid crystal systems from fibrillar polysaccharides[J]. Nature, 1959, 184(4686): 632-633.

[111] Iguchi M, Yamanaka S, Budhiono A. Bacterial cellulose-a masterpiece of nature's arts[J]. J. Mater. Sci., 2000,

35(2): 261-270.

[112]El-Saied H, Basta A H, Gobran R H. Research progress in friendly environmental technology for the production of cellulose products (bacterial cellulose and its application)[J]. Polym. -Plast. Technol. Eng., 2004, 43(3): 797-820.

[113]Gatenholm P, Klemm D. Bacterial nanocellulose as a renewable material for biomedical applications[J]. MRS Bull., 2010, 35(3): 208-213.

[114]Torres F G, Commeaux S, Troncoso O P. Biocompatibility of bacterial cellulose based biomaterials[J]. Journal of Functional Biomaterials, 2012, 3(4): 864-878.

[115]Roman M, Winter W T. Effect of sulfate groups from sulfuric acid hydrolysis on the thermal degradation behavior of bacterial cellulose[J]. Biomacromolecules, 2004, 5(5): 1671-1677.

[116] Wang Y, Song Y, Xia Y. Electrochemical capacitors: mechanism, materials, systems, characterization and applications[J]. Chem. Soc. Rev., 2016, 45(21): 5925-5950.

[117]Yan J, Wang Q, Wei T, Zhuang J F. Recent advances in design and fabrication of electrochemical supercapacitors with high energy densities[J]. Adv. Energy Mater., 2014, 4(4): 1300816.

[118]Wang F, Wu X, Yuan X, Liu Z C, Zhang Y, Fu L J, Zhu Y S, Zhou Q M, Wu Y P, Huang W. Latest advances in supercapacitors: from new electrode materials to novel device designs[J]. Chem. Soc. Rev., 2017, 46(22): 6816-6854.

[119]Torvinen K, Lehtimäki S, Keränen J T, Sievänen J, Vartiainen J, Hellén E, Lupo D, Tuukkanen S. Pigment-cellulose nanofibril composite and its application as a separator-substrate in printed supercapacitors[J]. Electron. Mater. Lett., 2015, 11(6): 1040-1047.

[120] Tuukkanen S, Lehtimäki S, Jahangir F, Eskelinen A P, Lupo D, Franssila S. Printable and disposable supercapacitor from nanocellulose and carbon nanotubes[C]. Proceedings of the 5th Electronics System-integration Technology Conference (ESTC). IEEE, 2014: 1-6.

[121]Niu Q, Gao K, Shao Z. Cellulose nanofiber/single-walled carbon nanotube hybrid non-woven macrofiber mats as novel wearable supercapacitors with excellent stability, tailorability and reliability[J]. Nanoscale, 2014, 6(8): 4083-4088.

[122]Gao K, Shao Z, Wang X, Zhang, Y H, Wang W J, Wang F J. Cellulose nanofibers/multi-walled carbon nanotube nanohybrid aerogel for all-solid-state flexible supercapacitors[J]. RSC Adv., 2013, 3(35): 15058-15064.

[123] Kang Y J, Chun S J, Lee S S, Kim B Y, Kim J H, Chung H, Lee, SY, Kim, W. All-solid-state flexible supercapacitors fabricated with bacterial nanocellulose papers, carbon nanotubes, and triblock-copolymer ion gels[J]. ACS nano, 2012, 6(7): 6400-6406.

[124]Gao K, Shao Z, Li J, Wang X, Peng X Q, Wang W J, Wang F J. Cellulose nanofiber-graphene all solid-state flexible supercapacitors[J]. J. Mater. Chem. A, 2013, 1(1): 63-67.

[125]Zheng Q, Cai Z, Ma Z, Gong S Q. Cellulose nanofibril/reduced graphene oxide/carbon nanotube hybrid aerogels for highly flexible and all-solid-state supercapacitors[J]. ACS Appl. Mater. Interfaces, 2015, 7(5): 3263-3271.

[126]Wang Z, Carlsson D O, Tammela P, Hua K, Zhang P, Nyholm L, Stromme M. Surface modified nanocellulose fibers yield conducting polymer-based flexible supercapacitors with enhanced capacitances[J]. ACS Nano, 2015, 9(7): 7563-7571.

[127] Wang Z, Tammela P, Huo J, Zhang P, Stromme M, Nyholm L. Solution-processed poly (3,4-ethylenedioxythiophene) nanocomposite paper electrodes for high-capacitance flexible supercapacitors[J]. J. Mater. Chem. A, 2016, 4(5): 1714-1722.

[128]Li S, Huang D, Yang J, Zhang B Y, Zhang X F, Yang G, Wang M K, Shen Y. Freestanding bacterial cellulose-polypyrrole nanofibres paper electrodes for advanced energy storage devices[J]. Nano Energy, 2014, 9: 309-317.

[129]Liu R, Ma L, Huang S, Mei J, Xu J, Yuan G H. Large areal mass, flexible and freestanding polyaniline/bacterial cellulose/graphene film for high-performance supercapacitors[J]. RSC Adv., 2016, 6(109): 107426-107432.

[130]Shan D, Yang J, Liu W, Yan J, Fan Z J. Biomass-derived three-dimensional honeycomb-like hierarchical structured

carbon for ultrahigh energy density asymmetric supercapacitors[J]. J. Mater. Chem. A, 2016, 4(35): 13589-13602.

[131] Jiang Y, Yan J, Wu X, Shan D D, Zhou Q H, Jiang L L, Yang D R, Fan Z J. Facile synthesis of carbon nanofibers-bridged porous carbon nanosheets for high-performance supercapacitors [J]. J. Power Sources, 2016, 307: 190-198.

[132] Chen L F, Huang Z H, Liang H W, Yao WT , Yu ZY, Yu S H. Flexible all-solid-state high-power supercapacitor fabricated with nitrogen-doped carbon nanofiber electrode material derived from bacterial cellulose[J]. Energy Environ. Sci., 2013, 6(11): 3331-3338.

[133] Chen L F, Huang Z H, Liang H W, Gao H L, Yu S H. Three-Dimensional Heteroatom-Doped Carbon Nanofiber Networks Derived from Bacterial Cellulose for Supercapacitors[J]. Adv. Funct. Mater., 2014, 24(32): 5104-5111.

[134] Ma L N, Liu R, Niu H J, Xing L X, Liu L, Huang Y D. Flexible and freestanding supercapacitor electrodes based on nitrogen-doped carbon networks/graphene/bacterial cellulose with ultrahigh areal capacitance [J]. ACS Appl. Mater. Interfaces, 2016, 8(49): 33608-33618.

[135] Chen L F, Huang Z H, Liang H W, Guan Q F, Yu S H. Bacterial-cellulose-derived carbon nanofiber@ MnO2 and nitrogen-doped carbon nanofiber electrode materials: an asymmetric supercapacitor with high energy and power density[J]. Adv. Mater., 2013, 25(34): 4746-4752.

[136] LongC l , Qi D , Wei T , Yan J , Jiang L L , Fan Z J. Nitrogen-doped carbon networks for high energy density supercapacitors derived from polyaniline coated bacterial cellulose[J]. Adv. Funct. Mater., 2014, 24(25): 3953-3961.

[137] Lai F L, Miao Y E, Zuo L Z, Lu H Y, Huang Y P, Liu T X. Biomass-Derived Nitrogen-Doped Carbon Nanofiber Network: A Facile Template for Decoration of Ultrathin Nickel-Cobalt Layered Double Hydroxide Nanosheets as High-Performance Asymmetric Supercapacitor Electrode[J]. Small, 2016, 12(24): 3235-3244.

[138] Yu W D, Lin W R, Shao X F, Hu Z X, Li R C, Yuan D S. High performance supercapacitor based on Ni3S2/carbon nanofibers and carbon nanofibers electrodes derived from bacterial cellulose [J]. J. Power Sources, 2014, 272: 137-143.

[139] Zuo L Z, Fan W, Zhang Y F, Huang Y P, Gao W, Liu T X. Bacterial cellulose-based sheet-like carbon aerogels for the in-situ growth of nickel sulfide as high-performance electrode materials for asymmetric supercapacitors[J]. Nanoscale, 2017, 9(13): 4445-4455.

[140] Zheng Q F, Kvit A, Cai Z Y, Ma Z Q, Gong S Q. A freestanding cellulose nanofibril-reduced graphene oxide-molybdenum oxynitride aerogel film electrode for all-solid-state supercapacitors with ultrahigh energy density [J]. J. Mater. Chem. A, 2017, 5(24): 12528-12541.

[141] Nystrom G, Marais A, Karabulut E, Wagberg L, Cui Y, Hamedi M M. Self-assembled three-dimensional and compressible interdigitated thin-film supercapacitors and batteries[J]. Nat. Commun., 2015, 6: 7259.

[142] Liu Y, Zhou J, Zhu EW, Tang J, Liu XH, Tang WH. Facile synthesis of bacterial cellulosefibres covalently intercalated with graphene oxide by one-step cross-linking for robust supercapacitors[J]. J. Mater. Chem. C, 2015, 3(5): 1011-1017.

[143] Choi K H, Yoo J, Lee C K, Lee S Y. All-inkjet-printed, solid-state flexible supercapacitors on paper[J]. Energy Environ. Sci., 2016, 9(9): 2812-2821.

[144] Hamedi M, Karabulut E, Marais A, Herland A, Nystrom G, Wagberg L. Nanocellulose Aerogels Functionalized by Rapid Layer-by-Layer Assembly for High Charge Storage and Beyond [J]. Angew. Chem.-Int. Edit., 2013, 52 (46): 12038-12042.

[145] Wang HH, Bian L Y, Zhou P P, Tang J, Tang W H. Core-sheath structured bacterial cellulose/polypyrrole nanocomposites with excellent conductivity as supercapacitors[J]. J. Mater. Chem. A, 2013, 1(3): 578-584.

[146] Wang Z H, Tammela P, Zhang P, Huo J X, Ericson F, Stromme M, Nyholm L. Freestanding nanocellulose-composite fibre reinforced 3D polypyrrole electrodes for energy storage applications [J]. Nanoscale, 2014, 6 (21):

13068-13075.

[147] Wang Z H, Tammela P, Zhang P, Stromme M, Nyholm L. Efficient high active mass paper-based energy-storage devices containing free-standing additive-less polypyrrole-nanocellulose electrodes[J]. J. Mater. Chem. A, 2014, 2(21): 7711-7716.

[148] Wang Z H, Tammela P, Zhang P, Stromme M, Nyholm L. High areal and volumetric capacity sustainable all-polymer paper-based supercapacitors[J]. J. Mater. Chem. A, 2014, 2(39): 16761-16769.

[149] Carlsson D O, Nystrom G, Zhou Q, Berglund L A, Nyholm L, Stromme M. Electroactive nanofibrillated cellulose aerogel composites with tunable structural and electrochemical properties[J]. J. Mater. Chem., 2012, 22(36): 19014-19024.

[150] Razaq A, Nyholm L, Sjodin M, Stromme M, Mihranyan A. Paper-based energy-storage devices comprising carbon fiber-reinforced polypyrrole-cladophora nanocellulose composite electrodes[J]. Adv. Energy Mater., 2012, 2(4): 445-454.

[151] Wang H, Zhu E, Yang J, et al. Zhou P P, Sun D P, Tang W H. Bacterial cellulose nanofiber-supported polyaniline nanocomposites with flake-shaped morphology as supercapacitor electrodes[J]. J. Phys. Chem. C, 2012, 116(24): 13013-13019.

[152] Wang F, Kim H J, Park S, et al. Kee, C D, Kim S J, Oh I K. Bendable and flexible supercapacitor based on polypyrrole-coated bacterial cellulose core-shell composite network[J]. Compos. Sci. Technol., 2016, 128: 33-40.

[153] Xu J, Zhu L G, Bai Z K, Liang G J, Liu L, Fang D, Xu W L. Conductivepolypyrrole-bacterial cellulose nanocomposite membranes as flexible supercapacitor electrode[J]. Org. Electron., 2013, 14(12): 3331-3338.

[154] Peng S, Xu Q, Fan LL, Wei C Z, Bao H F, Xu W L, Xu J. Flexible polypyrrole/cobalt sulfide/bacterial cellulose composite membranes for supercapacitor application[J]. Synth. Met., 2016, 222: 285-292.

[155] Peng S, Fan LL, Wei C Z, Bao H F, Zhang H W, Xu W L, Xu J. Polypyrrole/nickel sulfide/bacterial cellulose nanofibrous composite membranes for flexible supercapacitor electrodes[J]. Cellulose, 2016, 23(4): 2639-2651.

[156] Peng S, Fan LL, Wei C Z, Liu X H, Zhang H W, Xu W L, Xu J. Flexible polypyrrole/copper sulfide/bacterial cellulose nanofibrous composite membranes as supercapacitor electrodes[J]. Carbohydr. Polym., 2017, 157: 344-352.

[157] Zhang X d, Lin Z Y, Chen B, Zhang W, Sudhir S, Deng Y L. Solid-state flexible polyaniline/silver cellulose nanofibrils aerogel supercapacitors[J]. J. Power Sources, 2014, 246: 283-289.

[158] Yang C, Chen CC, Pan Y Y, Li S Y, Wang F, Li J Y, Li N N, Li X Y, Zhang Y Y, Li D G. Flexible highly specific capacitance aerogel electrodes based on cellulose nanofibers, carbon nanotubes and polyaniline[J]. Electrochim. Acta, 2015, 182: 264-271.

[159] Li NN, Li X Y, Yang C, Wang F, Li J Y, Wang H Y, Chen C C, Liu S N, Pan Y Y, Li D G. Fabrication of a flexible free-standing film electrode composed of polypyrrole coated cellulose nanofibers/multi-walled carbon nanotubes composite for supercapacitors[J]. RSC Adv., 2016, 6(89): 86744-86751.

[160] Ma L N, Liu R, Niu H J, Wang F, Liu L, Huang Y D. Freestanding conductive film based on polypyrrole/bacterial cellulose/graphene paper for flexible supercapacitor: large areal mass exhibits excellent areal capacitance[J]. Electrochim. Acta, 2016, 222: 429-437.

[161] Wang Z H, Tammela P, Stromme M, Nyholm L. Nanocellulose coupled flexible polypyrrole@graphene oxide composite paper electrodes with high volumetric capacitance[J]. Nanoscale, 2015, 7(8): 3418-3423.

[162] Li S H, Huang D K, Zhang B Y, Xu X B, Wang M K, Yang G, Shen Y. Flexible supercapacitors based on bacterial cellulose paper electrodes[J]. Adv. Energy Mater., 2014, 4(10): 1301655.

[163] Liew S Y, Walsh D A, Thielemans W. High total-electrode and mass-specific capacitance cellulose nanocrystal-polypyrrole nanocomposites for supercapacitors[J]. RSC Adv., 2013, 3(24): 9158-9162.

[164] Wu X Y, Chabot V L, Kim B K, Yu A P, Berry R M, Tam K C. Cost-effective and scalable chemical synthesis of conductive cellulose nanocrystals for high-performance supercapacitors[J]. Electrochim. Acta, 2014, 138: 139-147.

[165] Wu X Y, Tang J T, Duan Y C, Yu A P, Berry R M, Tam K C. Conductive cellulose nanocrystals with high cycling

stability for supercapacitor applications[J]. J. Mater. Chem. A, 2014, 2(45): 19268-19274.

[166] Yang X, Shi K Y, Zhitomirsky I, Cranston E D. Cellulose nanocrystal aerogels as universal 3D lightweight substrates for supercapacitor materials[J]. Adv. Mater., 2015, 27(40): 6104-6109.

[167] Shi K Y, Yang X, Cranston E D, Zhitomirsky I. Efficient lightweight supercapacitor with compression stability[J]. Adv. Funct. Mater., 2016, 26(35): 6437-6445.

[168] Liang H W, Guan Q F, Zhu Z, Song L T, Yao H B, Lei X, Yu S H. Highly conductive and stretchable conductors fabricated from bacterial cellulose[J]. NPG Asia Mater., 2012, 4(6): e19-e19.

[169] Chen W S, Zhang Q, Uetani K, Li Q, Lu P, Cao J, Wang Q W, Liu Y X, Li J, Quan Z C, Zhang Y S, Wang S F, Meng Z Y, Yu H P. Absorption Materials: Sustainable Carbon Aerogels Derived from Nanofibrillated Cellulose as High-Performance Absorption Materials (Adv. Mater. Interfaces 10/2016)[J]. Adv. Mater. Interfaces, 2016, 3(10): 1600004.

[170] Virtanen J, Pammo A, Keskinen J, Sarlin E, Tuukkanen S. Pyrolysed cellulose nanofibrils and dandelion pappus in supercapacitor application[J]. Cellulose, 2017, 24(8): 3387-3397.

[171] Wang X J, Kong D B, Wang B, Song Y, Zhi L J. Activated pyrolysed bacterial cellulose as electrodes for supercapacitors[J]. Sci. China-Chem., 2016, 59(6): 713-718.

[172] Wu Z Y, Liang H W, Li C, Hu B C, Xu XX, Wang Q, Chen J F, Yu S H. Dyeing bacterial cellulose pellicles for energetic heteroatom doped carbon nanofiber aerogels[J]. Nano Res., 2014, 7(12): 1861-1872.

[173] Wu X Y, Shi Z Q, Tjandra R, Cousins A J, Sy S, Yu A P, Berry R M, Tam K C. Nitrogen-enriched porous carbon nanorods templated by cellulose nanocrystals as high-performance supercapacitor electrodes[J]. J. Mater. Chem. A, 2015, 3(47): 23768-23777.

[174] Giese M, Blusch L K, Khan M K, MacLachlan M J. Functional materials from cellulose-derived liquid-crystal templates[J]. Angew. Chem. -Int. Edit., 2015, 54(10): 2888-2910.

[175] Kelly J A, Giese M, Shopsowitz K E, Hamad W Y, MacLachlan M J. The development of chiral nematic mesoporous materials[J]. Accounts Chem. Res., 2014, 47(4): 1088-1096.

[176] Shopsowitz K E, Hamad W Y, MacLachlan M J. Chiral nematic mesoporous carbon derived from nanocrystalline cellulose[J]. Angew. Chem. -Int. Edit., 2011, 50(46): 10991-10995.

[177] Li Z, Ahadi K, Jiang K, Ahvazi B, Li P, Anyia A O, Cadien K, Thundat T. Freestanding hierarchical porous carbon film derived from hybrid nanocellulose for high-power supercapacitors[J]. Nano Res., 2017, 10(5): 1847-1860.

[178] Li Z, Liu J, Jiang K, Thundat T. Carbonized nanocellulose sustainably boosts the performance of activated carbon in ionic liquid supercapacitors[J]. Nano Energy, 2016, 25: 161-169.

[179] Xu X, Zhou J, Nagaraju D H, Jiang L, Marinov V R, Lubineau G, Alshareef H N, Oh M. Flexible, highly graphitized carbon aerogels based on bacterial cellulose/lignin: Catalyst-free synthesis and its application in energy storage devices[J]. Adv. Funct. Mater., 2015, 25(21): 3193-3202.

[180] Li H, Wang Z, Chen L, Huang X J. Research on Adv. Mater. for Li-ion batteries[J]. Adv. Mater., 2009, 21(45): 4593-4607.

[181] Etacheri V, Marom R, Elazari R, Salitra, G, Aurbach D. Challenges in the development of advanced Li-ion batteries: a review[J]. Energy Environ. Sci., 2011, 4(9): 3243-3262.

[182] Li W, Song B, Manthiram A. High-voltage positive electrode materials for lithium-ion batteries[J]. Chem. Soc. Rev., 2017, 46(10): 3006-3059.

[183] Tang Y, Zhang Y, Li W, Ma B, Chen, XD. Rational material design for ultrafast rechargeable lithium-ion batteries [J]. Chem. Soc. Rev., 2015, 44(17): 5926-5940.

[184] Leijonmarck S, Cornell A, Lindbergh G, Wagberg, L. Flexible nano-paper-based positive electrodes for Li-ion batteries-Preparation process and properties[J]. Nano Energy, 2013, 2(5): 794-800.

[185] Wang Z, Xu C, Tammela P, Huo J X, Stromme M, Edstrom K, Gustafsson T, Nyholm L. Flexible freestanding

Cladophora nanocellulose paper based Si anodes for lithium-ion batteries[J]. J. Mater. Chem. A, 2015, 3(27): 14109-14115.

[186]Hu L, Liu N, Eskilsson M, Zheng G Y, McDonough J, Wagberg L, Cui Y. Silicon-conductive nanopaper for Li-ion batteries[J]. Nano Energy, 2013, 2(1): 138-145.

[187]Leijonmarck S, Cornell A, Lindbergh G, Wagberg L, Single-paper flexible Li-ion battery cells through a paper-making process based on nano-fibrillated cellulose[J]. J. Mater. Chem. A, 2013, 1(15): 4671-4677.

[188]Wang L, Schütz C, Salazar-Alvarez G, Titirici M M. Carbon aerogels from bacterial nanocellulose as anodes for lithium ion batteries[J]. RSC Adv., 2014, 4(34): 17549-17554.

[189]Wang W, Sun Y, Liu B, Wang S G, Cao M H. Porous carbon nanofiber webs derived from bacterial cellulose as an anode for high performance lithium ion batteries[J]. Carbon, 2015, 91: 56-65.

[190]Li Y, Zhu H, Shen F, Wan J Y, Han X G, Dai J Q, Dai H Q, Hu LB. Highly conductive microfiber of graphene oxide templated carbonization of nanofibrillated cellulose[J]. Adv. Funct. Mater., 2014, 24(46): 7366-7372.

[191]Wang B, Li X, Luo B, Yang J X, Wang X J, Song Q, Chen S Y, Zhi L J. Pyrolyzed bacterial cellulose: a versatile support for lithium ion battery anode materials[J]. Small, 2013, 9(14): 2399-2404.

[192]Wan Y, Yang Z, Xiong G, Luo H. A general strategy of decorating 3D carbon nanofiber aerogels derived from bacterial cellulose with nano-Fe_3O_4 for high-performance flexible and binder-free lithium-ion battery anodes[J]. J. Mater. Chem. A, 2015, 3(30): 15386-15393.

[193]Zhang F, Tang Y, Yang Y, Zhang X, Lee C S. In-situ assembly of three-dimensional MoS_2 nanoleaves/carbon nanofiber composites derived from bacterial cellulose as flexible and binder-free anodes for enhanced lithium-ion batteries[J]. Electrochim. Acta, 2016, 211: 404-410.

[194]Cao S, Feng X, Song Y, Xue X, Liu H J, Miao M, Fang J H, Shi, LY. Integrated fast assembly of free-standing lithium titanate/carbon nanotube/cellulose nanofiber hybrid network film as flexible paper-electrode for lithium-ion batteries[J]. ACS Appl. Mater. Interfaces, 2015, 7(20): 10695-10701.

[195]Lu H, Hagberg J, Lindbergh G, Cornell A. $Li_4Ti_5O_{12}$ flexible, lightweight electrodes based on cellulose nanofibrils as binder and carbon fibers as current collectors for Li-ion batteries[J]. Nano Energy, 2017, 39: 140-150.

[196]Cao S, Feng X, Song Y, Liu H J, Miao M, Fang J H, Shi L Y. In situ carbonized cellulose-based hybrid film as flexible paper anode for lithium-ion batteries[J]. ACS Appl. Mater. Interfaces, 2016, 8(2): 1073-1079.

[197] Huang Y, Lin Z, Zheng M, Wang T H, Yang J Z, Yuan F S, Lu X Y, Liu L, Sun D P. Amorphous Fe_2O_3 nanoshells coated on carbonized bacterial cellulose nanofibers as a flexible anode for high-performance lithium ion batteries[J]. J. Power Sources, 2016, 307: 649-656.

[198]Lee H, Yanilmaz M, Toprakci O, Fu K, Zhang X W. A review of recent developments in membrane separators for rechargeable lithium-ion batteries[J]. Energy Environ. Sci., 2014, 7(12): 3857-3886.

[199]Chun S J, Choi E S, Lee E H, Kim J H, Lee S Y, Lee S Y. Eco-friendly cellulose nanofiber paper-derived separator membranes featuring tunable nanoporous network channels for lithium-ion batteries[J]. J. Mater. Chem., 2012, 22(32): 16618-16626.

[200]Pan R, Cheung O, Wang Z, Tammela P, Huo J X, Lindh J, Edstrom K, Stromme M, Nyholm L. Mesoporous Cladophora cellulose separators for lithium-ion batteries[J]. J. Power Sources, 2016, 321: 185-192.

[201] Kim J H, Kim J H, Choi E S, Yu H K, Kim J H, Wu Q L, Chun S J, Lee S Y, Lee S Y. Colloidal silica nanoparticle-assisted structural control of cellulose nanofiber paper separators for lithium-ion batteries[J]. J. Power Sources, 2013, 242: 533-540.

[202] Wang Z H, Pan R J, Ruan C Q, Edstrom K , Stromme M, Nyholm L . Redox-Active Separators for Lithium-Ion Batteries[J]. Adv. Sci., 2018, 5(3): 1700663.

[203] Cho S J, Choi K H, Yoo J T, Kim J H, Lee Y H, Chun S J, Park S B, Choi D H, Wu Q L, Lee S Y, Lee S Y.

Heterolayered, one-dimensional nanobuilding block mat batteries[J]. Nano Lett. , 2014, 14(10): 5677-5686.

[204] Cho S J, Choi K H, Yoo J T, Kim J H, Lee Y H, Chun S J, Park S B, Choi D H, Wu Q L, Lee S Y, Lee S Y. Hetero-nanonet rechargeable paper batteries: toward ultrahigh energy density and origami foldability[J]. Adv. Funct. Mater. , 2015, 25(38): 6029-6040.

[205] Kim J H, Gu M, Lee D H, Kim J H, Oh Y S, Min S H, Kim B S, Lee S Y. Functionalized nanocellulose-integratedheterolayered nanomats toward smart battery separators[J]. Nano Lett. , 2016, 16(9): 5533-5541.

[206] Mai L Q, Wei Q L, Tian X C, Zhao Y L, AnQ Y. Nanowire electrodes for electrochemical energy storage devices [J]. Chem. Rev. , 2014, 114(23): 11828-11862.

[207] Peng H J, Huang J Q, Zhang Q. A review of flexible lithium - sulfur and analogous alkali metal - chalcogen rechargeable batteries[J]. Chem. Soc. Rev. , 2017, 46(17): 5237-5288.

[208]Seh Z W, Sun Y, Zhang Q, Cui Y. Designing high-energy lithium-sulfur batteries[J]. Chem. Soc. Rev. , 2016, 45 (20): 5605-5634.

[209] Yin Y X, Xin S, Guo Y G, Wan L J. Lithium - sulfur batteries: electrochemistry, materials, and prospects [J]. Angew. Chem. -Int. Edit. , 2013, 52(50): 13186-13200.

[210] Manthiram A, Fu Y, Chung S H, Zu C, Su Y S. Rechargeable lithium-sulfur batteries[J]. Chem. Rev. , 2014, 114(23): 11751-11787.

[211] Yu M P, Ma J S, Xie M, Song H Q, Tian F Y, Xu S S, Zhou Y, Li B, Wu D, Qiu H, Wang R M. Freestanding and Sandwich-Structured Electrode Material with High Areal Mass Loading for Long-Life Lithium-Sulfur Batteries[J]. Adv. Energy Mater. , 2017, 7(11): 1602347.

[212] Huang Y, Zheng M B, Lin Z X, Zhao B, Zhang S T, Yang J Z, Zhu C L, Zhang H, Sun D P, Shi Y. Flexible cathodes and multifunctional interlayers based on carbonized bacterial cellulose for high-performance lithium-sulfur batteries[J]. J. Mater. Chem. A, 2015, 3(20): 10910-10918.

[213] Quan Y T, Han D M, Feng Y H, Wang S J, Xiao M, Meng Y Z. Microporous carbon materials from bacterial cellulose for lithium-sulfur battery applications[J]. Int JElectrochem Sci, 2017, 12: 5984-5997.

[214] Li S Q, Mou T, Ren G F, Warzywoda J, Wei Z D, Wang B, Fan Z Y. Gel based sulfur cathodes with a high sulfur content and large mass loading for high-performance lithium-sulfur batteries[J]. J. Mater. Chem. A, 2017, 5(4): 1650-1657.

[215] Pang Q, Tang J T, Huang H, Liang X, Hart C, Tam K C, Nazar L F. A nitrogen and sulfur dual-doped carbon derived from polyrhodanine@ cellulose for advanced lithium-sulfur batteries[J]. Adv. Mater. , 2015, 27(39): 6021-6028.

[216] Li F, Wang G, Wang P, Yang J, Zhang K, Liu Y X, Lai Y Q. High-performance lithium-sulfur batteries with a carbonized bacterial cellulose/TiO_2 modified separator[J]. Journal of Electroanal. Chem. , 2017, 788: 150-155

[217] Nair J R, Bella F, Gerbaldi C. Nanocellulose-laden composite polymer electrolytes for high performing lithium-sulphur batteries[J]. Energy Storage Mater. , 2016, 3: 69-76.

[218] Kim SW, Seo D H , Ma X H, Ceder G, Kang K. Electrode materials for rechargeable sodium-ion batteries: potential alternatives to current lithium-ion batteries[J]. Adv. Energy Mater. , 2012, 2(7): 710-721.

[219] Che H Y, Chen S L, Xie Y Y, Wang H, Amine K, Liao X Z, Ma Z F. Electrolyte design strategies and research progress for room-temperature sodium-ion batteries[J]. Energy Environ. Sci. , 2017, 10(5): 1075-1101.

[220] Luo W, Shen F, Bommier C, Zhu H L, Ji X L, Hu, L B. Na-ion battery anodes: materials and electrochemistry [J]. Accounts Chem. Res. , 2016, 49(2): 231-240.

[221] Kim H, Kim H, Zhang D, Lee M H, Lim K , Yoon G, Kang K. Recent progress in electrode materials for sodium-ion batteries[J]. Adv. Energy Mater. , 2016, 6(19): 1600943.

[222] Pan H L, Hu Y S, Chen L Q. Room-temperature stationary sodium-ion batteries for large-scale electric energy storage[J]. Energy Environ. Sci. , 2013, 6(8): 2338-2360.

[223] Hwang J Y, Myung S T, Sun Y K. Sodium-ion batteries: present and future[J]. Chem. Soc. Rev., 2017, 46(12): 3529-3614.

[224] Li YY, Zhu H L, Shen F, Wan J Y, Lacey S, Fang Z Q, Dai H Q, Hu L B. Nanocellulose as green dispersant for two-dimensional energy materials[J]. Nano Energy, 2015, 13: 346-354.

[225] Luo W, Schardt J, Bommier C, Wang B, Razink J, Simonsen J, Ji X L. Carbon nanofibers derived from cellulose nanofibers as a long-life anode material for rechargeable sodium-ion batteries[J]. J. Mater. Chem. A, 2013, 1(36): 10662-10666.

[226] Zhang Z A, Zhang J, Zhao XX, Yang F H. Core-sheath structured porous carbon nanofiber composite anode material derived from bacterial cellulose/polypyrrole as an anode for sodium-ion batteries[J]. Carbon, 2015, 95: 552-559.

[227] Wang M, Yang ZZ, Li W H, Gu L, Yu Y. Superior sodium storage in 3D interconnected nitrogen and oxygen dual-doped carbon network[J]. Small, 2016, 12(19): 2559-2566.

[228] Wang M, Yang Y, Yang ZZ, Gu L, Chen Q W, Yu Y. Sodium-ion batteries: Improving the rate capability of 3D Interconnected carbon nanofibers thin film by boron, nitrogen dual-doping[J]. Adv. Sci., 2017, 4(4): 1600468.

[229] Zhu H L, Shen F, Luo W, Zhu S Z, Zhao M H, Natarajan B, Dai J Q, Zhou L H, Ji X L, Yassar R S, Li T, Hu L B.

Low temperature carbonization of cellulose nanocrystals for high performance carbon anode of sodium-ion batteries[J]. Nano Energy, 2017, 33: 37-44.

[230] Zoppe J O, Ataman N C, Mocny P, Wang J, Moraes J, Klok H A. Surface-initiated controlled radical polymerization: state-of-the-art, opportunities, and challenges in surface and interface engineering with polymer brushes[J]. Chem. Rev., 2017, 117(3): 1105-1318.

[231] Hsueh H Y, Yao C T, Ho R M. Well-ordered nanohybrids andnanoporous materials from gyroid block copolymer templates[J]. Chem. Soc. Rev., 2015, 44(7): 1974-2018.

[232] Zhang A, Zhang Q, Bai H, Li L, Li, J. Polymeric nanoporous materials fabricated with supercritical CO_2 and CO_2-expanded liquids[J]. Chem. Soc. Rev., 2014, 43(20): 6938-6953.

中文题目: 纳米纤维素:极具潜力的先进电化学能量储存材料

作者: 陈文帅,于海鹏,Sang-Young Lee,魏彤,李坚,范壮军

摘要: 因具有独特结构、优越性能以及天然丰富性,纳米纤维素已成为一种新兴的、极具潜力的可持续纳米材料。本文中,我们对以利用纳米纤维素为基本构筑单元开发电化学储能系统的前沿研究进展进行了总结和论述。全文首先简要介绍了细胞壁内纤维素纳米纤维的结构特征,并详细论述了具有不同结构和表面化学性质的纳米纤维素的主要制备策略。在此基础上,我们重点介绍了利用纳米纤维素及其衍生材料开发的超级电容器、锂离子电池、锂硫电池和钠离子电池等能量储存系统。本部分主要讨论纳米纤维素与活性材料的集成整合、开发膜/气凝胶用于柔性基底、热解成碳,以及通过活化、杂原子掺杂和与活性材料复合等策略对纳米纤维素衍生碳材料的功能化开发。最后,我们提出了这一新兴研究领域在今后需要重点关注的研究工作。

关键词: 纳米纤维素;超级电容器;锂离子电池;锂-硫电池;钠离子电池

Carbon Fibers Encapsulated with Nano-copper: A Core-shell Structured Composite for Antibacterial and Electromagnetic Interference Shielding Applications*

Yue Jiao, Caichao Wan, Wenbo Zhang, Wenhui Bao, Jian Li

Abstract: A facile and scalable two-step method (including pyrolysis and magnetron sputtering) is created to prepare a core – shell structured composite consisting of cotton-derived carbon fibers (CDCFs) and nano-copper. Excellent hydrophobicity (water contact angle = 144°) and outstanding antibacterial activity against *Escherichia coli* and *Staphylococcus aureus* (antibacterial ratios of >92%) are achieved for the composite owing to the composition transformation from cellulose to carbon and nano-size effects as well as strong oxidizing ability of oxygen reactive radicals from interactions of nano-Cu with sulfhydryl groups of enzymes. Moreover, the core-shell material with high electrical conductivity induces the interfacial polarization loss and conduction loss, contributing to a high electromagnetic interference (EMI) shielding effectiveness of 29.3 dB. Consequently, this flexible and multi-purpose hybrid of nano-copper/CDCFs may be useful for numerous applications like self-cleaning wall cladding, EMI shielding layer and antibacterial products.

Keywords: Carbon fibers; Core-shell structure; Magnetron sputtering; Antibacterial materials; Electromagnetic interference shielding

1 Introduction

Recently, rapid consumption of non-renewable resources (like petrochemical resources) and increasing seriousness of environmental pollutions have prompted researchers to pay more attention to the utilization of green and renewable biomass resources. Thus, biomass-based functional materials are attracting increasing interest from research and industrial circles. For instance, some biomass materials (such as wheat straw, wood and bamboo fibers) are directly combined with nanomaterials for specific applications like water purification and energy storage[1-5]. Biomass materials can also be transformed into their corresponding carbon counterparts for the development of various electroresponse products[6,7]. Moreover, some nano-components or novel reconstituted materials from biomass resource (like 1D cellulose nanofibrils[8], 2D cellulose films[9,10] and 3D cellulose hydrogel or aerogels[11-14]) are used as templates to support multifarious guest substances for the creation of novel and eco-friendly functional composites[15]. In the process of preparation, numerous physicochemical methods (e.g., hydrothermal method[16], vapor phase polymerization[17], electrodeposition[18,19] and atomic layer deposition[20,21]) are involved. However, some studies generally involve

---

* 本文摘自 Nanomaterials, 2019, 9(3): 460.

high consumption of energy and chemicals during the separation/disassembly of raw materials and complicated or low-precision synthetic methods, which seriously restrict the practical applications and bulk production of these biomass-based functional materials. Therefore, it is of significance to screen easily available and cheap raw materials as well as mild and scalable synthetic methods.

Amongst a variety of physicochemical techniques, magnetron sputteringis a powerful sputtering-based ionized physical vapor deposition technique and is already making its way to industrial applications[22]. Its deposition is achieved by rapidly colliding ionized inert gas atoms(commonly Ar) with the surface of negatively biased target under high electric field and thus inducing the ejection(sputtering) of atoms which condenses on a substrate and eventually generates a membrane[23]. Magnetron sputtering has numerous advantages, like high purity, high adhesion, high deposition rate, excellent uniformity on large-area substrates, ease of automation, ease of sputtering any metals, alloys or compounds and extensive applicability(like heat-sensitive substrates)[24]. The deposition of metallic materials by magnetron sputtering may improve the electrical conductivity, hydrophobicity and electroresponse of biomass materials, thus expanding their potential application areas[25,26]. Besides, compared to other methods that deposit metals onto biomass materials (like electrodeposition, electroless plating and chemical reduction method), magnetron sputtering method has stronger capability to accurately control the thickness, homogeneousness and purity of deposited layer[27]. Therefore, magnetron sputtering is an ideal technique for the surface modification of natural biomass.

In this work, we adopted a low-cost and widely available biomass material (i. e., cotton cloth) as raw materials. For the purpose of extending the application scopes, the cotton fibers were firstly pyrolyzed into a conductive and hydrophobic cotton-derived carbon fibers(coded as CDCFs). A thin layer of nano-copper was then deposited on the surface of CDCFs via the facile magnetron sputtering, resulting in the generation of core–shell structured composite. As a typical application example, the nano-Cu/CDCFs composite serves as a multi-purpose electromagnetic interference(EMI) shielding material with favorable flexibility, antibacterial activity and hydrophobic nature. Also, the synergistic effects of nano-Cu and CDCFs compositions on the above properties of the composite were analyzed.

2 Experimental section

2.1 Materials

Common old jean(100% cotton) was employed as the feedstock of CDCFs after being washed repeatedly with distilled water and ethyl alcohol and then dried at room temperature. A Cu target(purity: 99.99%) with a diameter of 50 mm was purchased from Shenyang Kejing Auto-instrument Co., Ltd, China. *Escherichia coli* (*E. coil*, ATCC 25922) and *Staphylococcus aureus* (*S. aureus*, ATCC 6538) were supplied by Guangdong Detection Center of Microbiology, China. Other chemicals were provided by Kemiou Chemical Reagent Co. Ltd. (Tianjin, China) and used as received.

2.2 Preparation of nano-Cu/CDCFs composite

The preparation of nano-Cu/CDCFs composite is primarily based on two processes, i. e., pyrolysis and magnetron sputtering. First, the clean and dried jean was transferred into a tubular furnace for pyrolysis under the protection of nitrogen. The sample was heated to 1000 ℃ at a heating rate of 5 ℃ · min^{-1}, and this temperature was maintained for 1 h to allow for complete pyrolysis; subsequently, the furnace decreased naturally to the room temperature and the following CDCFs were obtained. Second, the nano-Cu shell was deposited on the surface of CDCFs using magnetron sputter deposition technology with a DC sputter source (VTC-600-2HD, Shenyang Kejing Auto-instrument Co., Ltd, China), where the CDCFs and Cu target were placed on the anode and cathode with a distance of 60 mm between them, respectively. Regarding the sputtering process, we firstly pumped the chamber to a pressure of 3×10^{-3} Pa and then Argon was used as a sputtering gas and slowly introduced to the chamber with a flow rate of 11 sccm. With a target power of 100 W and rotating speed of 20 rpm, a homogeneous sputtering of Cu was achieved on the surface of CDCFs. With 40 ℃ maximum due to the water-cooling, the thermal stress of the substrate is on a very low level contributing to preventing the deformation and diffusion movement of the deposited Cu. The deposition thickness of Cu was set as 50 nm and the deposition was performed twice on the two sides of carbon cloth.

2.3 Characterizations

Morphology observations were performed on a scanning electron microscope (SEM, Hitachi S4800) equipped with an energy dispersive X-ray (EDX) detector. Crystal structure was analyzed by X-ray diffraction (XRD, Bruker D8 Advance TXS) with Cu Kα (target) radiation ($\lambda = 1.5418$ Å). The scan rate and scan range were set as 4° min^{-1} and 10°–80°, respectively. Water contact angle (WCA) tests were performed on a contact angle analyzer (JC2000C). Electrical conductivity was tested using a four-point probe resistivity/square resistance tester (KDB-1, Kunde Technology Company Ltd., China).

2.4 Antibacterial activity studies

Antibacterial activity studies were conducted using a shake flask method[28]. *E. coli* and *S. aureus* were used as the models of Gram-negative and Gram-positive bacteria for the tests. For preparing bacteria suspensions, the bacteria were grown in Luria Broth (LB) growth solutions for 18 h at 37 ℃. A colony was lifted off with a platinum loop, placed in 30 mL of nutrient broth, and incubated with shaking for 18 h at 37 ℃. After washed twice with phosphate buffer saline (PBS, pH = 7.4), they were resuspended in PBS to yield $1.0\times10^5-1.5\times10^5$ colony forming unit CFU · mL^{-1}. By measuring the absorbance of cell suspension, the bacterial cell concentration can be estimated[29]. To evaluate the antimicrobial properties of the cotton fibers, CDCFs and nano-Cu/CDCFs, 1 cm× 1 cm of the sample was immersed into a falcon tube containing 5.0 mL of 1.0×10^{-3} M PBS culture solution with a cell concentration of $1.0\times10^5-1.5\times10^5$ CFU · mL^{-1}. The falcon tube was then shaken at 200 rpm on a shaking incubator at 25 ℃ for 24 h. After shaking vigorously to detach adhered cells from the sample surfaces, the solution was serially diluted, and then 0.1 mL of each diluent was spread onto the agar plates. Viable microbial colonies were counted after incubating the plates for 18 h at 37 ℃. In addition, a blank control experiment was also conducted following the same method while any tested materials were not added into the falcon tube containing PBS culture solution and bacterial cell.

2.5 EMI shielding effectiveness studies

EMI shielding effectiveness was measured with the samples of dimension of 22.9 mm × 10.2 mm × 2 mm to fit waveguide sample holder using a PNA-X net-work analyzer (N5244a) at the frequency range of 8.2–12.4 GHz (X-band). S-parameters connect the input and output circuit quantities using the reflection and transmission parameters normally adopted in microwave analysis. By means of such parameters it is possible to determine the EMI shielding effectiveness due to reflection or absorption.

3 Results and discussion

3.1 Schematic diagram for preparation of nano-Cu/CDCFs composite

For the sake of seeking easily available and cheap biomass feedstock and developing mild and scalable methods for the synthesis of novel and high-performance functional products, as illustrated in Fig. 1, we chose disused jean cloth (100% cotton) as raw material and its main composition (namely cellulose) is easily transformed into the corresponding carbon material possessing new functions (like electrical conduction and hydrophobicity). The following magnetron sputtering process was conducted through the atom ejection of target materials due to the collision of high-speed Ar^+, resulting in the deposition of Cu on the surface of fibers. Therefore, this simple and easily scalable two-step method could generate the core-shell structured nano-Cu/CDCFs composite.

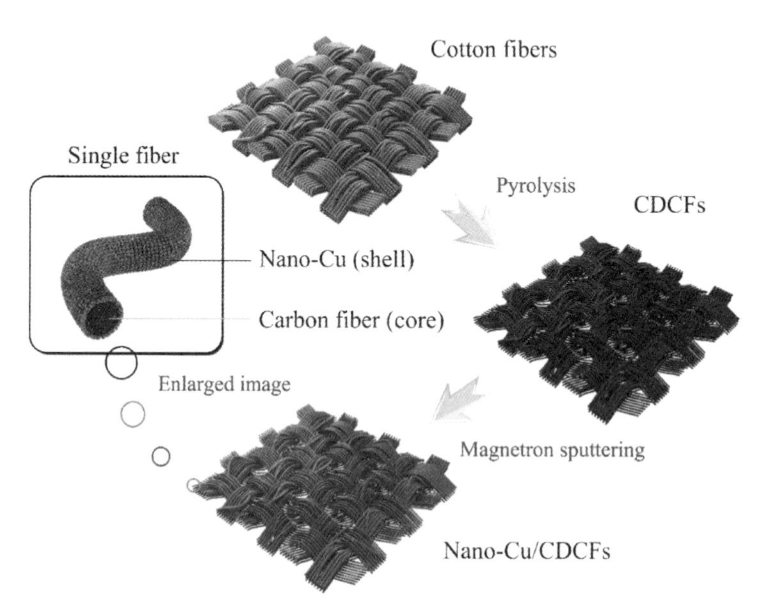

Fig. 1 Graphical illustration of the design concept of the core-shell structured nano-Cu/CDCFs composite

3.2 Morphology observations, elemental analysis and crystal structure

The changes of morphology and elemental compositions derived from the treatments of pyrolysis and magnetron sputtering were analyzed by SEM and EDX. By comparing Fig. 2a and d, we can find that the pyrolysis causes a reduction in the average size of fibers from 12.96 μm (cotton fibers) to 5.66 μm (CDCFs), while the fibers still tightly intertwine with each other after the pyrolysis ensuring the structural integrity of carbon cloth. Moreover, the fiber surface becomes rougher (Fig. 2b and e) after the pyrolysis owing to the thermal decomposition of most oxygen-containing compositions. Also, the significantly increased C/O ratio from 1.3 to 8.1 based on the EDX patterns (Fig. 2b and e) indicates the transformation from cellulose to carbon. Through the subsequent magnetron sputtering, the diameters and entangled state of fibers are almost unchanged (Fig. 2g), while the surface of CDCFs was coated with many particles with a size of dozens of nanometers(Fig. 2h), resulting in the formation of a core-shell structured composite, i. e., the CDCF serves as the core part and the superficial nano-layer acts as the sheath part. The thickness of nano-Cu/CDCFs is about 0.2 mm and its density is around 0.317 g · cm^{-3}, similar to that of pure CDCFs(ca. 0.315 g · cm^{-3}). Moreover, Cu signals were detected in the EDX pattern of the composite(Fig. 2i), revealing that Cu element is one of the main elemental compositions of nano-layer. These results demonstrate that the two-step method successfully generated the core-shell structured composite. Besides, Au peaks in these EDX patterns were originated from the coating layer used for electrical conduction during the SEM observation.

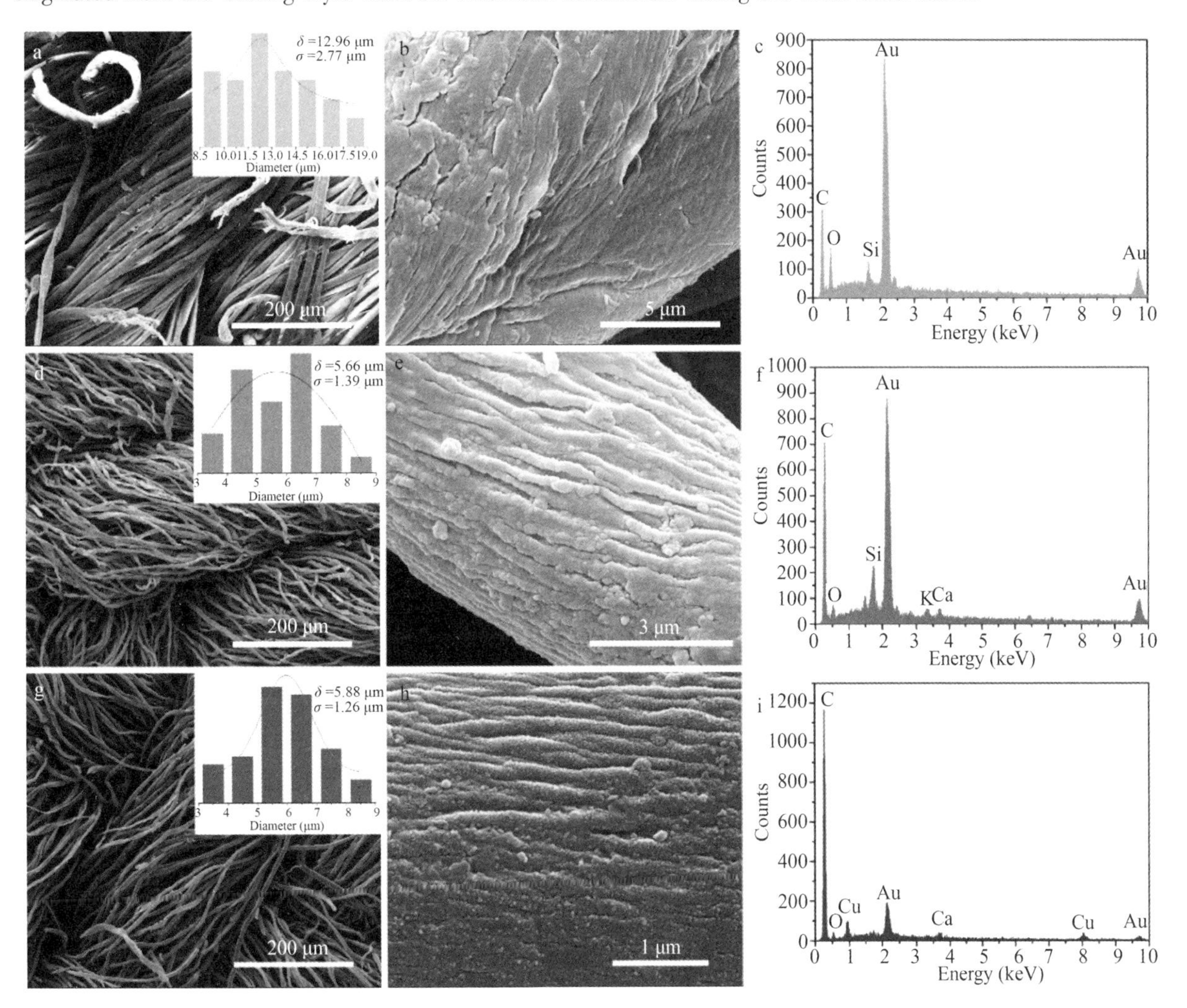

Fig. 2 SEM images and EDX patterns ofthe cotton fibers (a-c), CDCFs (d-f) and nano-Cu/CDCFs (g-i): (a, d, g) distribution state of fibers and insets show the corresponding diameter distribution of fibers; (b, e, g) surface observation of single fiber; (c, f, i) EDX patterns

The cross-section SEM image of nano-Cu/CDCFs is exhibited in Fig. 3a, where a core-shell structure can be clearly identified. XRD analysis was carried out to study the changes of crystal structure before and after the pyrolysis and magnetron sputtering. From Fig. 3b, cotton fiber(mainly consisting of cellulose) displays a typical cellulose Ⅰ crystal structure with peaks at 15.3°, 17.0°, 23.2° 20.9° and 34.6°, corresponding to the planes of (101), (10$\bar{1}$), (021), (002) and (040), respectively[30]. These characteristic peaks disappear after the pyrolysis while two broad peaks centered at around 24.1° and 43.3° appear in the XRD pattern of CDCFs, which are related to the(002)and(100)planes of graphite and suggest the generation of amorphous carbon[31]. The result agrees well with that of EDX analysis. After the deposition of Cu nanoparticles by magnetron sputtering, a new peak at 43.3° is detected and related to the(111)plane of Cu(JCPDS No. 04-0836). In addition, the (002) diffraction shifts towards a lower angle after the magnetron sputtering, indicative of a decrease in theorder of crystallinity in carbon materials[32].

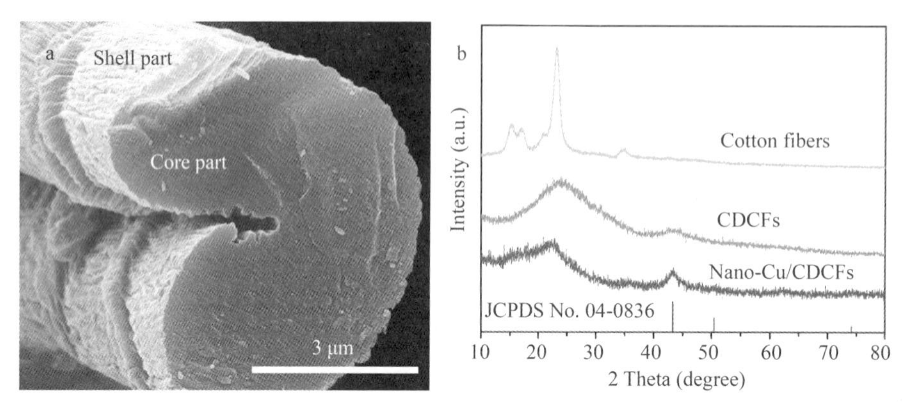

Fig. 3 (a) Cross-section SEM image of nano-Cu/CDCFs. (b) XRD patterns of the cotton fibers, CDCFs and nano-Cu/CDCFs, and the bottom line is the standard JCPDS card no. 04-0836 for Cu

3.3 Hydrophobic property and antibacterial activity

Hydrophobic property plays a crucial role in the self-cleaning ability of materials[33]. The influences of the treatments of pyrolysis and magnetron sputtering on the hydrophobic property were studied by WCA tests. As shown in Fig. 4a and d, the cotton fibers can readily absorb water drop once the drop contacts its surface, indicative of favorable hydrophilicity of the cotton. By contrast, the pyrolyzed carbon fibers can stably support a water drop on its surface with a WCA value of 118° (Fig. 4b and e), suggesting the formation of hydrophobicity owing to the pyrolysis. After the coating of nano-Cu, the core–shell material shows an improved hydrophobic property with a higher WCA value of 144°(Fig. 4c and f). Furthermore, comparing Fig. 4g and h, it is clear that the water drop quickly tumbles from the surface of nano-Cu/CDCFs as soon as the drop touches the surface, further demonstrating its excellent water repellency. This dropping process of water mixed with various smudges (like dust and mucus) rises an important self-cleaning function. In addition, we also calculated some characteristic parameters including the work of adhesion (W_a), the coefficient of spreading (S) and the work of wetting (W_w) via the equations, i.e., $W_a = \gamma_{LG}(1+\cos\theta)$, $S = \gamma_{LG}(\cos\theta - 1)$ and $W_w = \gamma_{LG}\cos\theta$ (γ_{LG} is the interfacial tension between liquid and gas, $\theta$is the water contact angle)[34], respectively. In this paper, θ is 144° and γ_{LG} is 72.75 mN · m^{-1}; thus, W_a, S and W_w are calculated as 13.9 mN · m^{-1}, −131.6 mN · m^{-1} and −58.9 mN · m^{-1}, respectively. The negative values of S and W_w reveal that the water cannot be spread on the surface of nano-Cu/CDCFs, i.e., excellent hydrophobicity.

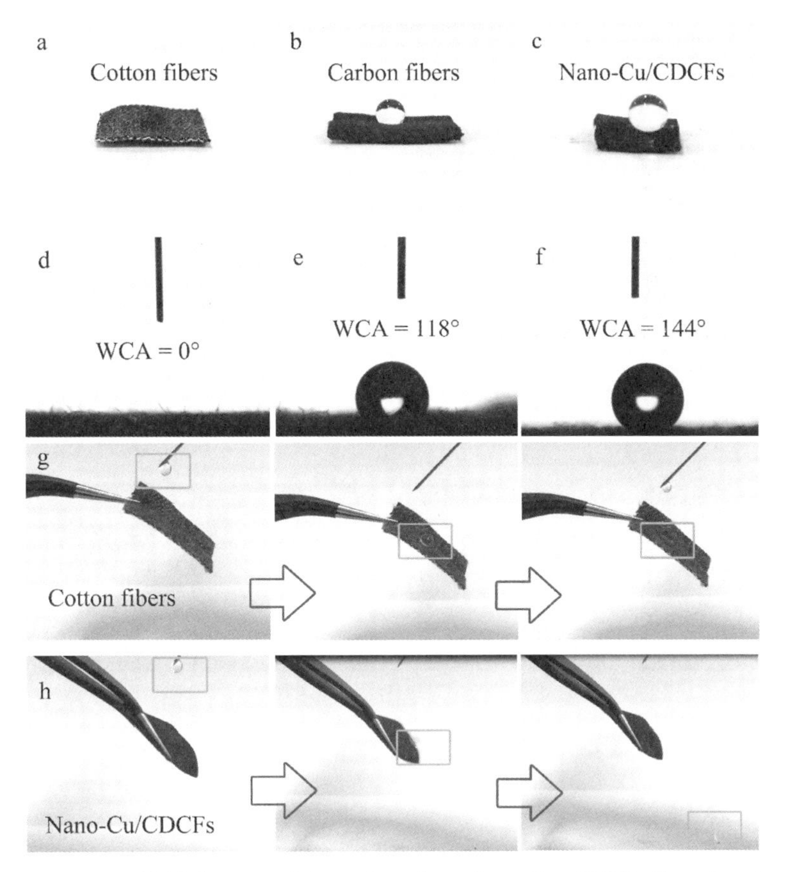

Fig. 4 Hydrophobic property and WCA tests of the cotton fibers(a, d), CDCFs(b, e)and nano-Cu/CDCFs(c, f), respectively. Photographs of water drop contacting the surfaces of the cotton fibers(g)and nano-Cu/CDCFs(h)

The discovery of antibacterial effects of Cu can be traced back to ancient civilizations that used to employ different forms of Cu compounds to treat several afflictions and to maintain hygiene. However, the reports on the antibacterial properties of Cu-containing nanomaterials are not abundant. Herein, theantibacterial activityof the cotton fibers, CDCFs and nano-Cu/CDCFs was tested via a shake flask method. *E. coli* and *S. aureus*were chosen as the models of Gram-negative and Gram-positive bacteria for the tests. As shown in Fig. 5, the surface concentrations of live *E. coli and S. aureus*on the cotton fibers(ca. 6.7×10^{10} and 5.5×10^{10}CFU · cm^{-2})are close to those of blank control(ca. 6.8×10^{10} and 5.6×10^{10} CFU · cm^{-2}), indicative of a negligible antibacterial activity of the cellulose material(i. e., cotton fibers). The phenomenon is consistent with that of our previous study[35]. Through introducing CDCFs, the surface concentrations of live *E. coli* and *S. aureus* decline to 2.5×10^{10} and 1.2×10^{10} CFU · cm^{-2}, respectively. The introduction of nano-Cu/CDCFs further reduces the concentrations to 5.2×10^{9} and 0 CFU · cm^{-2}, respectively, revealing an excellent antibacterial activity of nano-Cu/CDCFs with 92.35% and 100% of antibacterial ratios for *E. coli* and *S. aureus*. The origins of the antibacterial activity of nano-Cu/CDCFs may mainly be concluded to the two points: (1)the interaction of nano-Cu with sulfhydryl groups of enzymes is one of the possible protein oxidation routes resulting in the generation of oxygen reactive radicals that eventually causes irreparable damage like oxidation of proteins, cleavage of DNA and RNA molecules, and membrane damage due to lipid peroxidation[36,37]; (2)the direct cell contact with CDCFs may impact cellular membrane integrity, metabolic activity and morphology of bacteria[38].

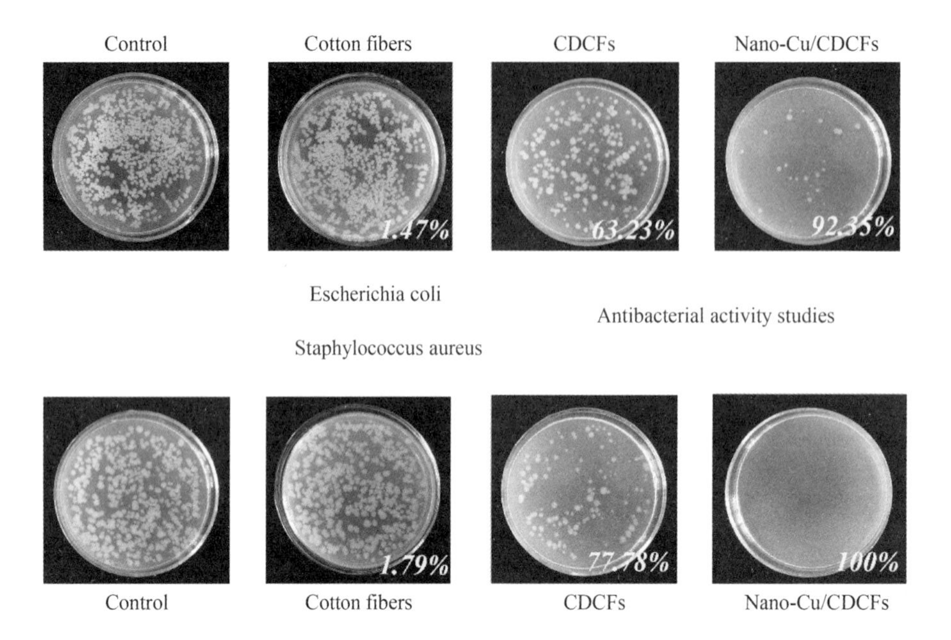

Fig. 5 Antibacterial activity studies of the cotton fibers, CDCFs and nano-Cu/CDCFs using a shake flask method

3.4 EMI shielding properties

Nowadays, the widespread use of high-power machines(such as base station, radar and some household appliances like induction cooker) has caused serious electromagnetic pollution, which has a strong adverse impact on the human health and normal operation of equipment. Especially, electromagnetic pollution has been listed as the fifth major pollution sources on the earth, following atmospheric pollution, water pollution, solid waste pollution and noise pollution. Therefore, it is urgent to develop high-performance, cheap and easily produced EMI shielding products.

EMI shielding property is evaluated by shielding effectiveness expressed in decibels (dB) over the frequency range of 8.2–12.4 GHz(X-band). A higher decibel level reveals less energy transmitted through shielding materials. The total shielding effectiveness(SE_{total}) can be expressed as[39]:

$$SE_{total}(dB) = 10\log\frac{P_i}{P_t} = SE_A + SE_R + SE_M \tag{1}$$

where P_i and P_t are the incident and transmitted electromagnetic power, respectively. SE_R and SE_A are the shielding effectiveness from reflection and absorption, respectively. SE_M is multiple reflection effectiveness inside the material, which can be negligible when $SE_{total} > 10$ dB. Besides, SE_R and SE_A can be described as[40]:

$$SE_R = -10\log(1-R) \tag{2}$$

$$SE_A = -10\log[T/(1-R)] \tag{3}$$

where R and T are reflected power and transmitted power, respectively.

The shielding effectiveness of the cotton fibers, CDCFs and nano-Cu/CDCFs within the frequency scope of 8. 2–12. 4 GHz is presented in Fig. 6. As seen in Fig. 6a, the shielding effectiveness of the cotton fibers is negligible. The low SE_{total} value of 0. 7 dB is ascribed to its ignorable magnetic permeability and electrical conductivity which are both decisive for EMI shielding effectiveness. In addition, CDCFs exhibit an obviously higher electrical conductivity of 4. 57 S · cm^{-1} than that of the cotton fibers(< 5×10^{-6} S · cm^{-1}). As a result, the maximum SE_{total} value of CDCFs can reach 18. 9 dB(Fig. 6b), close to the requirement for the commercial EMI shieldingapplications(>20 dB). Moreover, the contributions from SE_R and SE_A account for 50. 3% and 49. 7% at 8. 2 GHz, respectively, indicating that the EMI shielding action due to absorption is close to that due to reflection. The introduction of nano-Cu leads to the increase of electrical conductivity of nano-Cu/CDCFs to 20. 3 S · cm^{-1}. Moreover, the surface resistivity of CDCFs andnano-Cu/CDCFs is calculated as ~ 10. 9 Ω · sq^{-1} and ~ 2. 5 Ω · sq^{-1}, respectively, according to the equation, i. e., $\sigma = 1/\rho = 1/(dR)$ (σ is the electrical conductivity, d is the thickness of the sample, R is the sheet resistance and ρ is the electrical resistivity). Besides, a positive correlation between electrical conductivity and EMI shielding effectiveness has been already verified[41]. The maximum SE_{total} value reaches up to 29. 3 dB(Fig. 6b), comparable to or even higher than that of many EMI shielding materials like scoured canvas fabric/polyaniline(13 dB)[42], nickel-plated multiwalled carbon nanotubes/high-density polyethylene composites (12 – 16 dB)[43], carbonyliron powder-carbon fiber cloth/epoxy resin(12–47 dB)[44], neat carbon nanofiber networks(17–18 dB)[45], d-$Ti_3C_2T_x$/cellulose nanofiber composite paper(21 – 26 dB)[46] andcarbon nanofiber – graphene nanosheet networks(25–28 dB)[45]. In addition, specific shielding effectiveness(*SSE*) is derived to compare the effectiveness of shieldingmaterials taking into account the density. Mathematically, *SSE* can be obtained by dividing the SE_{total} by density of material. Also, to account for thethickness contribution, absolute effectiveness (SSE_t) is introduced and calculated by dividing the *SSE* by thickness of material. As show in Table 1, the *SSE* and SSE_t values of nano-Cu/CDCFsare about 92 dB cm^3 · g^{-1} and 4621 dB cm^2 · g^{-1}, respectively, comparable with those of these above composites. Furthermore, our synthetic method is relatively simpler and more easily scalable and the composite also has good cost effectiveness and environmental friendliness(free of harmful substances). Inspired by the works reporting Cu-clad carbon fiber nonwoven fabrics[47] and MXene-Graphene-PVDF composite[48] with higher EMI shielding efficiencies, the aim of our future research is to seek better parameters to increase the thickness of Cu composition and also reduce the sputtering time as much as possible, for the sake of preparing more superior EMI shielding property of nano-Cu/CDCFs.

The synergistic effects of this core – shell structured nano-Cu/CDCFs are responsible for its good EMI shielding property, as illustrated in Fig. 7. For CDCFs (core), in the process of electromagnetic wave propagation, the multi-scaled reticulated conductive structure contributes to the occurrence of time-varying electromagnetic-field-induced currents and long-range induced currents. The presence of these currents caused an electric-thermal conversion and rapid decay of massive incident wave[49]. Other effects like dielectric relaxation also caused the attenuation of electromagnetic wave[50]. For nano-Cu(shell), it is well-known that copper is one of the most reliable materials in EMI shielding because it is highly effective in attenuating magnetic and electrical waves. In this core–shell material, the existence of interface between the nano-Cu and the CDCFs would cause interfacial polarization loss under an electromagnetic field and the formed conductive Cu layer could induce conduction loss[51,52].

The contribution from SE_A accounts for 61. 1% at 8. 2 GHz, much higher than that from SE_R(38. 9%).

The higher contribution from SE_A can be explained by the equations of(4) and(5)[53]:

$$SE_A = 20d\sqrt{\frac{\mu_r \omega \sigma_{AC}}{2}} \cdot \log e \tag{4}$$

$$SE_R = 10\log\left(\frac{\sigma_{AC}}{16\omega\mu_r\varepsilon_0}\right) \tag{5}$$

where d is the thickness of the shield, μ_r is the magnetic permeability, ω is the angular frequency, σ_{AC} is the frequency dependent conductivity and ε_0 is the permittivity of the free space. Obviously, dependence of SE_A and SE_R on conductivity and permeability indicates that the material having higher conductivity and magnetic permeability can achieve better absorption properties. This absorption-dominant EMI shielding mechanism ofnano-Cu/CDCFs is beneficial to alleviate secondary radiation and considered as a more attractive alternative for the fabrication of electromagnetic radiation protection products.

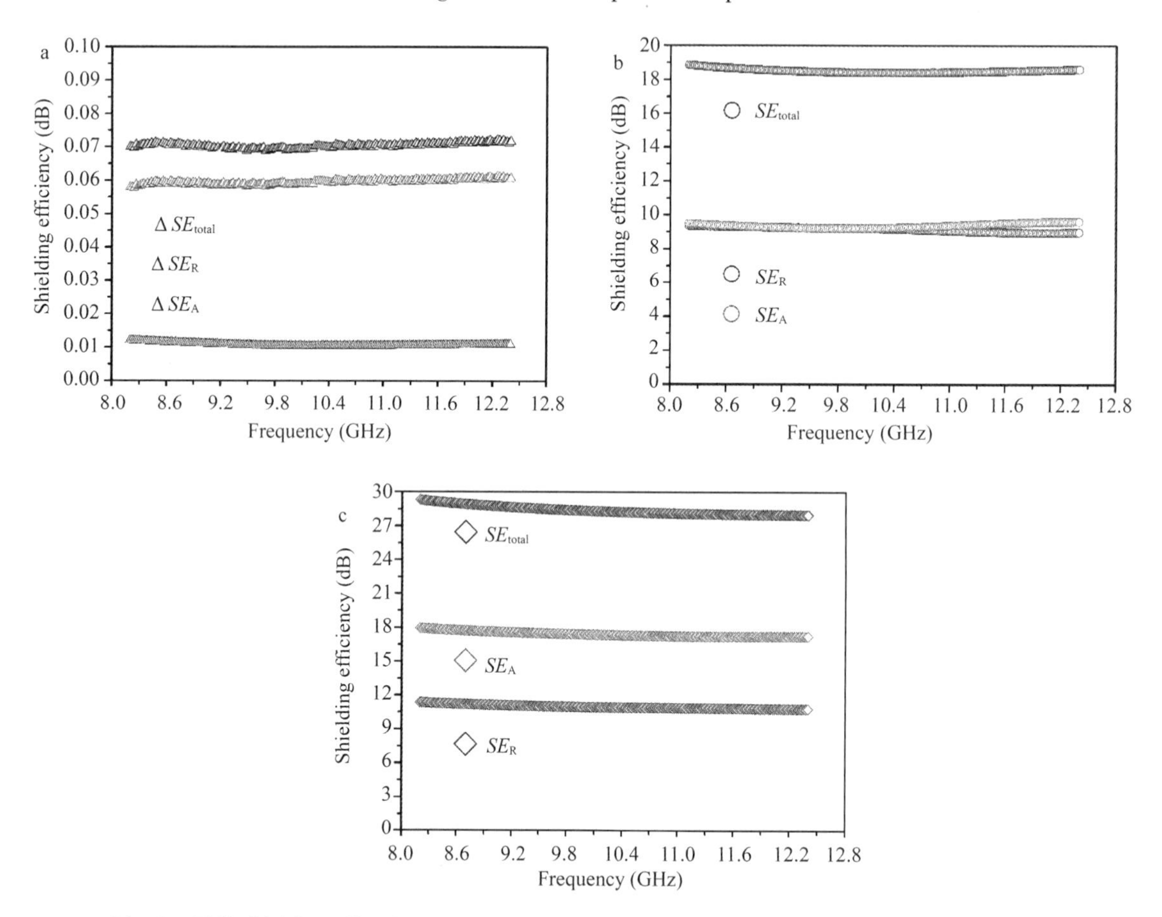

Fig. 6 EMI shielding effectiveness(including SE_{total}, SE_R and SE_A) of(a) the cotton fibers, (b) CDCFs and(c) nano-Cu/CDCFs, respectively

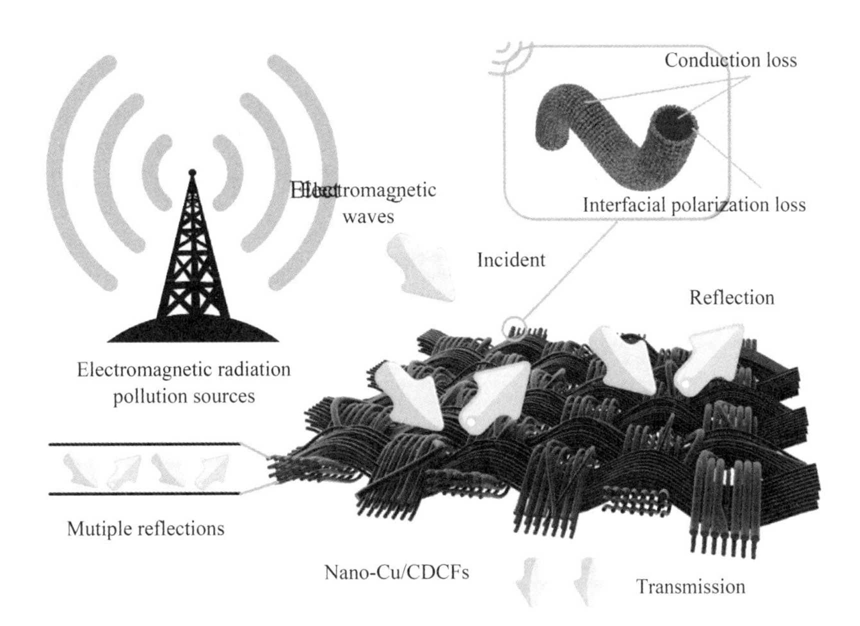

Fig. 7 Graphical illustration of the EMI shielding mechanism of the core-shell structured nano-Cu/CDCFs composite

Table 1 Comparison of EMI shielding properties of composites

Composites	MaximumSE_{total}(dB)	SSE(dB cm^3 · g^{-1})	SSE_t(dB cm^2 · g^{-1})	Ref.
Scoured canvas fabric/polyaniline	13	-	-	[42]
Nickel-plated multiwalled carbon nanotubes/ high-density polyethylene composites	16	-	-	[43]
Carbonyliron powder-carbon fiber cloth/ epoxy resin	47	-	-	[44]
Neat carbon nanofiber networks	18	180	6667	[45]
d-$Ti_3C_2T_x$/cellulose nanofiber composite paper	25.8	12.4	2647	[46]
Carbon nanofiber-graphene nanosheet networks	28	280	10370	[45]
Nano-Cu/CDCFs	29.3	92	4621	This work

4 Conclusions

An easily-operated and scalable two-step method (pyrolysis and magnetron sputtering) is developed to create a green and core-shell structured composite of nano-Cu/CDCFs. The flexible composite shows numerous alluring properties like excellent hydrophobic property (WCA = 144°), outstanding antibacterial activity against *E. coli* and *S. aureus* (antibacterial ratios of>92%), and good EMI shielding ability with a high SE_{total} value of 29.3 dB and absorption-dominant shielding mechanism, due to the physicochemical properties, nano-size effect and synergistic effects of the two components. In conclusion, this multi-purpose eco-friendly biomass-based product is expected to find applications in many fields, e.g., self-cleaning wall cladding, waterproof layer, antibacterial agents and EMI shielding case.

Acknowledgements

This study was supported by the National Natural Science Foundation of China(grant no. 31470584), the Fundamental Research Funds for the Central Universities(grant no. 2572018AB09), and the Youth Scientific Research Foundation, Central South University of Forestry and Technology(grant no. QJ2018002A).

Data Availability

The raw/processed data required to reproduce these findings cannot be shared at this time as the data also forms part of an ongoing study.

References

[1]Han R, Zhang L, Song C, et al. Characterization of modified wheat straw, kinetic and equilibrium study about copper ionand methylene blue adsorption in batch mode[J]. Carbohydrate Polymers, 2010, 79(4): 1140-1149.

[2]Li D, Zhu F Z, Li J Y, et al. Preparation and characterization of cellulose fibers from corn straw as natural oilsorbents [J]. Industrial & Engineering Chemistry Research, 2013, 52(1): 516-524.

[3]Lv S, Fu F, Wang S, et al. Novel wood-based all-solid-state flexible supercapacitors fabricated with a natural porous wood slice and polypyrrole[J]. RSC advances, 2015, 5(4): 2813-2818.

[4]Jiao Y, Wan C, Li J. Scalable synthesis and characterization of free-standingsupercapacitor electrode using natural wood as a green substrate to support rod-shaped polyaniline[J]. Journal of Materials Science: Materials in Electronics, 2017, 28(3): 2634-2641.

[5]Zhou Q, Gong W, Xie C, et al. Removal of Neutral Red from aqueous solution by adsorption on spent cottonseed hull substrate[J]. Journal of Hazardous Materials, 2011, 185(1): 502-506.

[6]Chen C, Zhang Y, Li Y, et al. All-wood, low tortuosity, aqueous, biodegradable supercapacitors with ultra-high capacitance[J]. Energy Environ Sci, 2017, 10(2): 538-545.

[7]Wang L, Gao B, Peng C, et al. Bamboo leaf derived ultrafine Si nanoparticles and Si/C nanocomposites for high-performance Li-ion battery anodes[J]. Nanoscale, 2015, 7(33): 13840-13847.

[8]Khalil H P S A, Bhat A H, Yusra A F I. Green composites from sustainable cellulose nanofibrils: A review[J]. Carbohydrate Polymers, 2012, 87(2): 963-979.

[9]Yang Q, Fukuzumi H, Saito T, et al. Transparent cellulose films with high gas barrier properties fabricated from aqueous alkali/urea solutions[J]. Biomacromolecules, 2011, 12(7): 2766-2771.

[10]Kim J H, Kim J H, Choi E S, et al. Colloidal silica nanoparticle-assisted structural control of cellulosenanofiber paper separators for lithium-ion batteries[J]. Journal of Power Sources, 2013, 242: 533-540.

[11]Wang Q, Cai J, Zhang L, et al. A bioplastic with high strength constructed from a cellulose hydrogel by changing the aggregated structure[J]. Journal of Materials Chemistry A, 2013, 1(22): 6678-6686.

[12]Wan C, Li J. Facile synthesis of well-dispersedsuperparamagnetic γ-Fe_2O_3 nanoparticles encapsulated in three-dimensional architectures of cellulose aerogels and their applications for Cr (VI) removal from contaminated water[J]. ACS Sustainable Chemistry & Engineering, 2015, 3(9): 2142-2152.

[13] Wan C, Li J. Cellulose aerogels functionalized withpolypyrrole and silver nanoparticles: In-situ synthesis, characterization and antibacterial activity[J]. Carbohydrate Polymers, 2016, 146: 362-367.

[14]Lu Y, Liu H, Gao R, et al. Coherent-interface-assembled Ag_2O-anchored nanofibrillated cellulose porous aerogels for radioactive iodine capture[J]. ACS applied materials & interfaces, 2016, 8(42): 29179-29185.

[15]Moon R J, Martini A, Nairn J, et al. Cellulose nanomaterials review: structure, properties and nanocomposites[J]. Chemical Society Reviews, 2011, 40(7): 3941-3994.

[16] Wan C, Li J. EmbeddingZnO nanorods into porous cellulose aerogels via a facile one-step low-temperature hydrothermal method[J]. Materials & Design, 2015, 83: 620-625.

[17] Shi Z, Gao H, Feng J, et al. In situ synthesis of robust conductive cellulose/polypyrrole composite aerogels and their potential application in nerve regeneration[J]. Angewandte Chemie International Edition, 2014, 53(21): 5380-5384.

[18] Yin Y, Huang R, Zhang W, et al. Superhydrophobic-superhydrophilic switchable wettability via TiO_2 photoinduction electrochemical deposition on cellulose substrate[J]. Chemical Engineering Journal, 2016, 289: 99-105.

[19] Zhang X, Lin Z, Chen B, et al. Solid-state flexiblepolyaniline/silver cellulose nanofibrils aerogel supercapacitors[J]. Journal of Power Sources, 2014, 246: 283-289.

[20] Kemell M, Pore V, Ritala M, et al. Atomic layer deposition in nanometer-level replication of cellulosic substances and preparation of photocatalytic TiO_2/cellulose composites[J]. Journal of the American Chemical Society, 2005, 127(41): 14178-14179.

[21] Korhonen J T, Hiekkataipale P, Malm J, et al. Inorganic hollow nanotube aerogels by atomic layer deposition onto native nanocellulose templates[J]. ACS nano, 2011, 5(3): 1967-1974.

[22] Kelly P J, Arnell R D. Magnetron sputtering: a review of recent developments and applications[J]. Vacuum, 2000, 56(3): 159-172.

[23] Wan C, Jiao Y, Liang D, et al. A Geologic Architecture System-Inspired Micro-/Nano-Heterostructure Design for High-Performance Energy Storage[J]. Advanced Energy Materials, 2018, 8(33): 1802388.

[24] Sarakinos K, Alami J, Konstantinidis S. High power pulsed magnetron sputtering: A review on scientific and engineering state of the art[J]. Surface and Coatings Technology, 2010, 204(11): 1661-1684.

[25] Alexeeva O K, Fateev V N. Application of the magnetron sputtering for nanostructured electrocatalysts synthesis[J]. International Journal of Hydrogen Energy, 2016, 41(5): 3373-3386.

[26] Wang Q, Xiao S, Shi S Q, et al. Self-bonded natural fiber product with high hydrophobic and EMI shielding performance via magnetron sputtering Cu film[J]. Applied Surface Science, 2019, 475(1): 947-952.

[27] Wan C, Jiao Y, Li J. A cellulose fibers-supported hierarchical forest-like cuprous oxide/copper array architecture as a flexible and free-standing electrode for symmetricsupercapacitors[J]. Journal of Materials Chemistry A, 2017, 5(33): 17267-17278.

[28] Daoud W A, Xin J H, Zhang Y H. Surface functionalization of cellulose fibers with titanium dioxide nanoparticles and their combined bactericidal activities[J]. Surface Science, 2005, 599(1-3): 69-75.

[29] Mi L, Licina G A, Jiang S. Nonantibiotic-Based Pseudomonas aeruginosa Biofilm Inhibition with Osmoprotectant Analogues[J]. Acs Sustainable Chemistry & Engineering, 2014, 2(10): 2448-2453.

[30] Nishiyama Y, Langan P, Chanzy H. Crystal structure and hydrogen-bonding system in cellulose Iβ from synchrotron X-ray and neutron fiber diffraction[J]. Journal of the American Chemical Society, 2002, 124(31): 9074-9082.

[31] Popov V, Orlova T, Magarino E, et al. Specific features of electrical properties of porous biocarbons prepared from beech wood and wood artificial fiberboards[J]. Physics of the Solid State, 2011, 53(2): 276-283.

[32] Dandekar A, Baker R T K, Vannice M A. Characterization of activated carbon, graphitized carbon fibers and synthetic diamond powder using TPD and DRIFTS[J]. Carbon, 1998, 36(12): 1821-1831.

[33] Bhushan B, Jung Y C. Natural and biomimetic artificial surfaces for superhydrophobicity, self-cleaning, low adhesion, and drag reduction[J]. Progress in Materials Science, 2011, 56(1): 1-108.

[34] Kim M J, Kim Y K, Kim K H, et al. Shear bond strengths of various luting cements to zirconia ceramic: Surface chemical aspects[J]. Journal of Dentistry, 2011, 39(11): 795-803.

[35] Wan C, Jiao Y, Sun Q, et al. Preparation, characterization, and antibacterial properties of silver nanoparticles embedded into celluloseaerogels[J]. Polymer Composites, 2016, 37(4): 1137-1142.

[36] Longano D, Ditaranto N, Cioffi N, et al. Analytical characterization of laser-generated copper nanoparticles for

antibacterial composite food packaging[J]. Analytical and Bioanalytical Chemistry, 2012, 403(4): 1179-1186.

[37]Peña MM O, Koch K A, Thiele D J. Dynamic regulation of copper uptake and detoxification genes in Saccharomyces cerevisiae[J]. Molecular and Cellular Biology, 1998, 18(5): 2514-2523.

[38]Kang S, Herzberg M, Rodrigues D F, et al. Antibacterial effects of carbon nanotubes: size does matter! [J]. Langmuir, 2008, 24(13): 6409-6413.

[39]Wan C, Jiao Y, Qiang T, et al. Cellulose-derived carbon aerogels supported goethite (α-FeOOH) nanoneedles and nanoflowers for electromagnetic interference shielding[J]. Carbohydrate polymers, 2017, 156: 427-434.

[40]Wan C, Li J. Synthesis and electromagnetic interference shielding of cellulose-derived carbon aerogels functionalized with α-Fe2O3 and polypyrrole[J]. Carbohydrate Polymers, 2017, 161: 158-165.

[41]Li N, Huang Y, Du F, et al. Electromagnetic interference (EMI) shielding of single-walled carbon nanotube epoxycomposites[J]. Nano letters, 2006, 6(6): 1141-1145.

[42]Akşit A C, Onar N, Ebeoglugil M F, et al. Electromagnetic and electrical properties of coated cotton fabric with barium ferrite doped polyaniline film[J]. Journal of Applied Polymer Science, 2009, 113(1): 358-366.

[43]Yim Y J, Rhee K Y, Park S J. Electromagnetic interference shielding effectiveness of nickel-plated MWCNTs/high-density polyethylene composites[J]. Composites Part B: Engineering, 2016, 98: 120-125.

[44]HuT , Wang J , Wang J . Electromagnetic interference shielding properties of carbon fiber cloth based composites with different layer orientation[J]. Materials Letters, 2015, 158(1): 163-166.

[45] Song W L, Wang J, Fan L Z, et al. Interfacialengineering of carbon nanofiber-graphene-carbon nanofiber heterojunctions in flexible lightweight electromagnetic shielding networks[J]. ACS Applied Materials & Interfaces, 2014, 6 (13): 10516-10523.

[46]Cao W T, Chen FF, Zhu Y J, et al. Binary strengthening and toughening of MXene/cellulose nanofiber composite paper with nacre-inspired structure and superior electromagnetic interference shielding properties[J]. ACS Nano, 2018, 12(5): 4583-4593.

[47]Lee J, Liu Y, Liu Y, et al. Ultrahigh electromagnetic interference shielding performance of lightweight, flexible, and highly conductive copper-clad carbon fiber nonwoven fabrics[J]. Journal of Materials Chemistry C, 2017, 5(31): 7853-7861.

[48]Raagulan K, Braveenth R, Jang H J, et al. Electromagnetic shielding by MXene-graphene-PVDF composite with hydrophobic, lightweight and flexible graphene coated fabric[J]. Materials, 2018, 11(10): 1803.

[49]Song Q, Ye F, Yin X, et al. Carbon nanotube-multilayered graphene edge plane core-shell hybrid foams for ultrahigh-performance electromagnetic-interference shielding[J]. Advanced Materials, 2017, 29(31): 1701583.

[50]Micheli D, Vricella A, Pastore R, et al. Synthesis and electromagnetic characterization of frequency selective radar absorbing materials using carbon nanopowders[J]. Carbon, 2014, 77: 756-774.

[51]Zhao Z, Zheng W, Yu W, et al. Electrical conductivity of poly (vinylidene fluoride)/carbon nanotube composites with a spherical substructure[J]. Carbon, 2009, 47(8): 2118-2120.

[52]Zhao B, Park C B. Tunable electromagnetic shielding properties of conductive poly (vinylidene fluoride)/Ni chain composite films with negative permittivity[J]. Journal of Materials Chemistry C, 2017, 5(28): 6954-6961.

[53]Ohlan A, Singh K, Chandra A, et al. Microwave absorption behavior of core-shell structured poly (3, 4-ethylenedioxy thiophene)-barium ferrite nanocomposites[J]. ACS Applied Materials & Interfaces, 2010, 2(3): 927-933.

中文题目：纳米铜包裹碳纤维：抗菌和电磁屏蔽用核壳结构复合材料

作者：焦月，万才超，张文博，包文慧，李坚

摘要：提出了一种简便、可扩展的两步法(包括热解和磁控溅射)来制备由棉基碳纤维(CDCF)和纳米铜组成的核壳结构复合材料。该复合材料具有优异的疏水性(水接触角=144°)和出色的针对大肠杆菌和金黄色葡萄球菌的抗菌活性(抑菌率>92%)，这是由于其成分从纤维素转变为碳和纳米级效

果，以及纳米铜与酶的巯基相互作用，使氧反应性自由基具有很强的氧化能力。此外，具有高电导率的核-壳材料会引起界面极化损耗和传导损耗，从而导致 29.3 dB 的高电磁干扰(EMI)屏蔽效果。因此，这种柔性多用途纳米铜/CDCFs 的杂化物可用于许多应用，例如自清洁壁板，EMI 屏蔽层和抗菌产品。

关键词：碳纤维；核壳结构；磁控溅射；抗菌材料；电磁屏蔽

Polyaniline-polypyrrole Nanocomposites Using A Green and Porous Wood As Support for Supercapacitors*

Jian Li, Yue Jiao

Abstract: Wood is an ideal type of support material whose porous structure and surface functional groups are benefificial for deposition of various guest substances for different applications. In this paper, wood is employed as a porous support, combined with two kinds of conductive polymers [i. e., polyaniline (PANI) and polypyrrole (PPy)] using an easy and fast liquid polymerization method. Scanning electron microscope observations indicate that the PANI-PPy complex consists of nanoparticles with a size of ~ 20 nm. The interactions between oxygencontaining groups of the wood and the nitrogen composition of PANI-PPy were verified by Fourier transform infrared spectroscopy. The self-supported PANI-PPy/ wood composite is capable of acting as a free-standing supercapacitor electrode, which delivers a high gravimetric specifific capacitance of 360 $F \cdot g^{-1}$ at 0.2 $A \cdot g^{-1}$.

Keywords: Wood; Polypyrrole; Polyaniline; Supercapacitors; Nanocomposites

1 Introduction

Supercapacitors are a novel class of electrochemical energy storage systems, which have the ability to bridge the power/energy gap between traditional dielectric capacitors and batteries/fuel cells[1]. It is known that supercapacitors universally possess ultrahigh power densities[2], fast charge-discharge properties[3], excellent reversibility[4], a long lifecycle[5] and wide operation temperature[6]. According to the difference in energy storage mechanisms, supercapacitors can be classified into two categories: electrical double-layer capacitor (EDLC) materials (such as carbon materials) and pseudocapacitive materials (e. g., transition-metal oxides or hydroxides and conductive polymers)[7]. For EDLCs, the capacitance is due to electrostatic charge separation at the interface between the electrode and the electrolyte; while for pseudocapacitors, the capacitance relies on fast and reversible faradaic redox reactions to store charges[8]. In general, pseudocapacitive materials deliver higher capacitance and energy density than those of EDLC materials. Conductive polymers [for instance polyaniline (PANI), polypyrrole (PPy) and polythiophene (PTh)] are a class of important pseudocapacitive materials[9]. Supercapacitor electrodes prepared with these conductive polymers have already exhibited numerous merits, such as high electrochemical activity and conductivity, good cost-effectiveness and easy of synthesis[10-13]. Nevertheless, because of their poor mechanical strength, difficulty in processing and handling, ease of agglomeration and low porosity, conductive polymers are widely combined with various host materials, e. g., carbon aerogels[14], cellulose

* 本文摘自 Front. Agr. Sci. Eng., 2019, 6(2): 137-143.

nanofibrils[15] and cellulose paper[16].

Wood is a green and ideal matrix with many fascinating properties, for instance, renewability, low density and thermal expansion, ease of machining, and good mechanical strength. Wood has been combined with various organic or inorganic substances (such as ZrO_2[17], graphene[18] and polyvinyl alcohol[19]) for multifarious applications (such as ultraviolet resistant catalysts and reinforcing materials). In the field of supercapacitors, wood has abundant direct channels with low tortuosity along the tree-growing direction, which have been demonstrated to allow fast transport of electrolyte ions[20]. Moreover, in our previous research, we have integrated wood with PANI via a method of oxidative polymerization and this binary composite shows a moderate specific capacitance of 304 $F \cdot g^{-1}$ at 0.1 $A \cdot g^{-1}$[21]. However, the value can be further improved by introducing new electrochemically active components.

Herein, natural wood was used as a green and nanoporous support to combine with PANI-PPy nanocomposites by virtue of a liquid-phase synthesis method. The micromorphology and chemical bonds of the asprepared wood-supported PANI-PPy hybrid (coded as PANI-PPy/wood) was studied by scanning electronmicroscopy (SEM) and Fourier transform infrared spectrum (FTIR) analysis, respectively. In addition, theelectrochemical performances of PANI-PPy/wood werestudied in a three-electrode configuration via cyclicvoltammetry (CV), galvanostatic charge-discharge (GCD) and electrochemical impedance spectroscopy (EIS) methods.

2 Materials andmethods

2.1 Materials

Pyrrole, iron(Ⅲ) chloride hexahydrate ($FeCl_3 \cdot 6H_2O$), hydrochloric acid (HCl, 37%), aniline and ammonium peroxide sulfate, tert-butyl alcohol and absolute ethyl alcohol were supplied by Kemiou Chemical Reagent Co., Ltd. (China). Paulownia wood processing residues were collected from a wood-working factory in China (Linwei Wood Industry, Caoxian County in Shandong province) and cut into slices with a thickness of ~ 1 mm. These wood slices were rinsed ultrasonically with absolute ethyl alcohol and distilled water for 30 min and dried at 60 ℃ for 24 h.

2.2 Deposition of PANI on wood support

The deposition of PANI on the wood support was performed by the method described in[21]. Briefly, the wood slices were immersed in a 100 mL aqueous solution of aniline monomer for 24 h at room temperature. The mass ratio of wood to aniline was set as 1 : 10. Subsequently, a mixture of oxidant (ammonium peroxide sulfate, 0.08 mol) and dopant (hydrochloric acid, 1 $mol \cdot L^{-1}$, 96 mL) was dropwise added into the above solution precooled to 0 ℃. The reaction was allowed to proceed for 4 h at 0-10 ℃. Finally, the resultant PANI/ wood slice was rinsed with a large amount of distilled water.

2.3 Deposition of PPy on PANI/wood

The deposition of PPy on PANI/wood was conducted according to the method in our previous report[11]. Typically, PANI/wood was soaked in 100 mL water solution of pyrrole monomer (1.35 $mol \cdot L^{-1}$) for 2 h andthen slowly mixed with an aqueous solution of $FeCl_3 \cdot 6H_2O$ (oxidant). The molar ratio of ferric iron/ pyrrole was set as 1 : 1. After the polymerization, the asprepared PANI-PPy/wood was washed with distilled water and tert-butyl alcohol in sequence and then underwent a freeze-drying process for 24 h at 25 Pa.

2.4 Characterization

The morphology was characterized by an SEM(Hitachi, S4800) equipped with an energy dispersive X-ray (EDX) detector for elemental analysis. FTIR spectra were recorded using a Nicolet Nexus 670 FTIR instrument in the range of 500-4000 cm^{-1} with a resolution of 4 cm^{-1}. Electrochemical tests were conducted in a three-electrode setup with PANI-PPy/wood as the working electrode and an Ag/AgCl electrode and a Pt wire electrode as reference and counter electrodes, respectively and 1 mol · L^{-1} HCl aqueous solution electrolyte. CV, GCD and EIS tests were implemented using a CS350 electrochemical workstation (Wuhan CorrTest Instruments Co., Ltd). CV tests were conducted over the potential window from −0.2 V to 1.0 V at different scan rates of 5 mV · s^{-1}, 10 mV · s^{-1}, 20 mV · s^{-1} and 50 mV · s^{-1}. GCD curves were tested in the potential range of 0−0.8 V at different current densities of 0.2 A · g^{-1}, 0.5 A · g^{-1}, 1 A · g^{-1} and 2 A · g^{-1}. EIS measurements were carried out in the frequency range of 105 Hz to 0.01 Hz with an alternate current amplitude of 5 mV.

3 Results and discussion

3.1 Schematic diagram for the synthesis of PANI-PPy/wood

Wood is a good porous support material for the growth of various guest substances. The abundant oxygen-containing functional groups(such as hydroxyl groups and carboxyl groups) on its surface have the ability to fix ions and small molecules and guide theirin situ growth or polymerization. In this paper, based on this strategy, we used wood as host material; through a facile liquid polymerization approach, the PANI and PPy nanoparticles were successively deposited onto the surface of the wood matrix, resulting in the generation of a self-supported ternary composite, i. e., PANI-PPy/wood(Fig. 1).

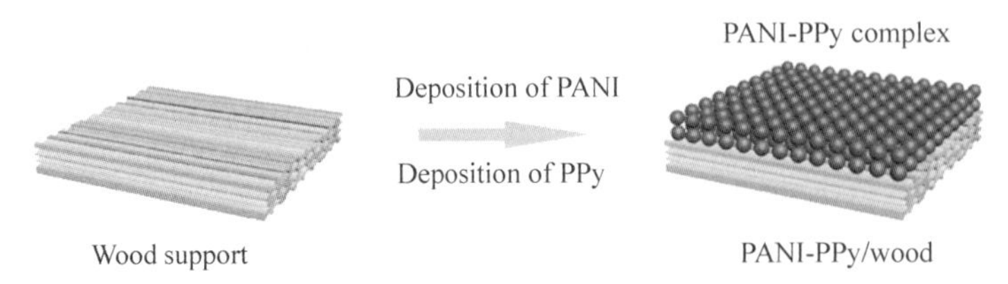

Fig. 1 Schematic diagram for the synthesis of PANI-PPy/wood

3.2 Micromorphology and elemental analysis

The micromorphology changes in the wood support before and after the deposition of PANI and PPy were studied by SEM observations. The structure of wood is presented in Fig. 2(a), where wood vessels can be clearly identified. After the deposition of PANI and PPy, the characteristic structure of wood disappeared; instead, a large number of nano-scale particles appeared on the surface of the wood support[Fig. 2(b)]. Based on the higher-magnification SEM image in Fig. 2(c), we found that these nanoparticles have a size of~20 nm. The tight connection of these particles is expected to improve the structural stability of the composite during the charge-discharge process. Also, this connection may be beneficial for the transport of electrons. The elemental compositions of the wood and PANI-PPy/wood were investigated by EDX analysis. As seen in

Fig. 2(d), the elements including C, O and Au were detected in the wood. The Au element in particular came from the coating layer used for electric conduction during the SEM observation. For the PANI-PPy/wood, a new N element signal was detected derived from the PANI and PPy and a Cl signal derived from the reactants ($FeCl_3$ or and HCl).

Fig. 2 SEM images of the wood support(a) and PANI-PPy/wood(b, c); (d) EDX patterns of the wood support and PANI-PPy/wood

3.3 FTIR analysis

The chemical composition of the wood support and PANI-PPy/wood were studied by FTIR analysis. As seen in Fig. 3, for the wood support, the wide band at 2990-3720 cm^{-1} is attributed to the O—H stretching of polymeric compounds, while the bands at 2920 cm^{-1} and 2857 cm^{-1} are derived from the alkane C—H asymmetric and symmetric stretching vibrations[22], respectively. The band at 1734 cm^{-1} belongs to the C=O stretching in unconjugated ketone, carbonyl and aliphatic groups (xylan) and the strong band at 1641 cm^{-1} is due to the bending mode of the absorbed water[23]. The bands at 1420 cm^{-1}, 1163 cm^{-1} and 1066 cm^{-1} are attributed to the cellulose structure, correspond to the CH_2 symmetric bending, C—O antisymmetric stretching and C—O—C pyranose ring skeletal vibration[24], respectively. The band at 1261 cm^{-1} originates from the C—O symmetric stretching.

For PANI-PPy/wood, the band position of the O—H stretching moves to the lower wavenumber from 3441 cm^{-1} compared to 3423 cm^{-1} in wood, possibly due to the interaction between the oxygen-containing groups of wood and the nitrogen composition of PANI-PPy. The strong interaction between them helps to enhance the adhesion effect of active materials on the wood, possibly contributing to the strengthening of electrochemical stability. The bands at 1541 cm^{-1} and 1458 cm^{-1} may be assigned to the C—C and C—N stretching vibrations in the pyrrole ring, respectively. In addition, the bands at 1031 cm^{-1} and 774 cm^{-1} are due to the N—H in-plane vibration of the PPy ring and C—H out-of-plane ring deformation[25], respectively. Additionally, the C—H stretching vibration of aromatic conjugation of PANIdisplays adsorption bands at 1375 cm^{-1} and 1163 cm^{-1}[26]. Moreover, the band at 1298 cm^{-1} arises from the C—N in-plane ring bending mode of PANI. The results of FTIR analysis thus suggest the successful synthesis of PANI and PPy on the surface of the wood host.

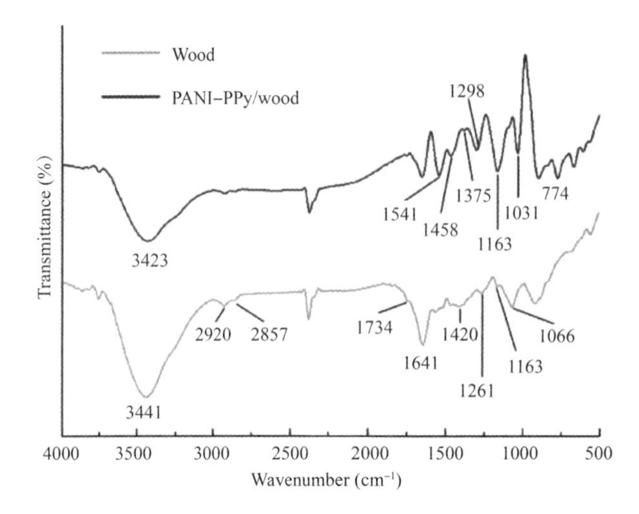

Fig. 3 FTIR spectra of the wood support and PANI-PPy/wood

3.4 Electrochemical analysis

The PANI-PPy/wood can serve as a free-standing electrode and its electrochemical properties were studied by CV, GCD and EIS in a three-electrode system, using a 1 mol · L^{-1} HCl aqueous solution. The CV curves of the PANI-PPy/wood electrode at different scan rates are displayed in Fig. 4. At scan rates of 5 and 10 mV · s^{-1}, the CV plots display quasi-rectangular shapes, whereas when the scan rate reached 20 mV · s^{-1} and above, the CV curves became shuttle-like in shape, which can possibly be attributed to the slower entering/ejecting and diffusion rates of counterions compared to the transfer rate of electrons in the electrode at the high scan rates[11].

According to the GCD traces (Fig. 5) at different current densities, the obvious curvature reflects the process of a faradic reaction. In addition, during the charging and discharging step, the charge curves of the PANI-PPy/wood electrode are almost symmetric to their corresponding discharge counterparts, suggesting good capacitive behavior and highly reversible electrochemical reactions. Moreover, GCD is a conventional method to calculate the capacitance value of supercapacitor electrodes. The gravimetric (C_m, F · g^{-1}) specific capacitances were calculated from these GCD curves at different current densities based on the following equations[27]:

$$C_m = I\Delta t \ / \ m\Delta V \tag{1}$$

where I(A) is the discharge current, Dt(s) is the discharge time, m(g) is the mass of active materials (the areal density of PANI-PPy on the electrode is ca. 10 mg · cm^{-2}) and DV(V) is the operation discharge voltage window. The result displays that the PANI-PPy/wood electrode achieves the highest value of 360 F · g^{-1} at 0.2 A · g^{-1}, which is higher than that of some congeneric products in the literature, such as core-sheath structured bacterial cellulose/polypyrrole nanocomposites (316 F · g^{-1} at 0.2 A · g^{-1})[28], bacterial cellulose nanofiber-supported polyaniline nanocomposites (273 F · g^{-1} at 0.2 A · g^{-1})[29] and PANI/wood (280 F · g^{-1}

at 0.2 $A \cdot g^{-1}$)[21]. With the increase of current densities from 0.5 $A \cdot g^{-1}$ to 2.0 $A \cdot g^{-1}$, the specific capacitance value decreased to 259 $F \cdot g^{-1}$, 180 $F \cdot g^{-1}$ and 77 $F \cdot g^{-1}$, as shown in Fig. 6. The decline is associated with the low utilization of electroactive materials at high discharge current densities since the electrolyte ions cannot enter into the inner structure of the active material, and only the outer active surface was utilized for charge storage[30].

EIS was conducted to evaluate the charge transfer and electrolyte diffusion in the electrode/electrolyte interface. Fig. 7 shows the Nyquist plot for the PANI-PPy/wood electrode. As illustrated, the curve has a characteristic semicircle over the high frequency range, a 45° Warburg region, and a linear pure capacitor part in the low frequency region. In the high-frequency region, the intersection of the semicircle with the real axis reveals the solution resistance value (i. e., 1.1Ω), indicative of a high electrical conductivity of the electrolyte. The diameter of the semicircle represents the charge transport resistance (ca. 6.6Ω). In the medium-frequency region, the projected length of the 45° Warburg region on the real impedance axis characterizes the ion penetration process. This short Warburg length indicates a short ion-diffusion path. In the low frequency region, the slope of the linear part is low, suggesting a diffusion-controlled behavior[31].

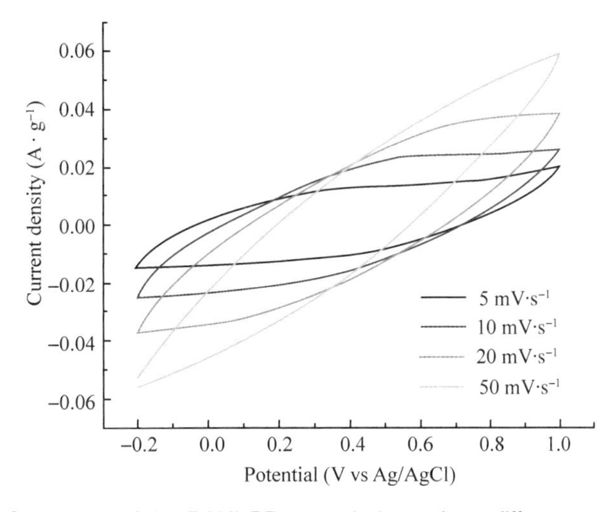

Fig. 4 CV curves of the PANI-PPy/wood electrode at different scan rates

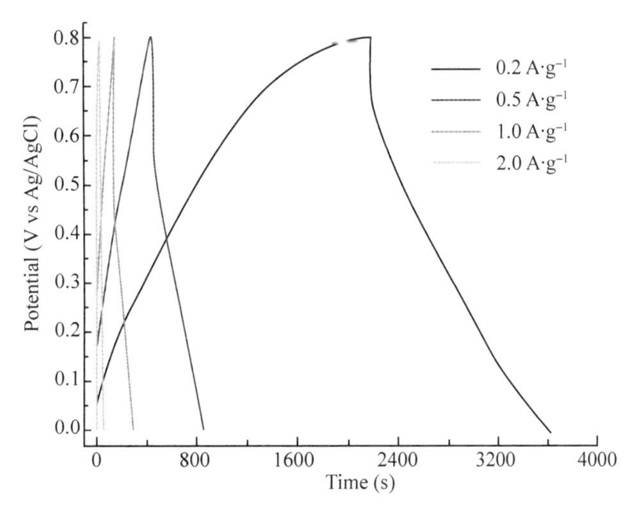

Fig. 5 GCD curves of the PANI−PPy/wood electrode at different current densities

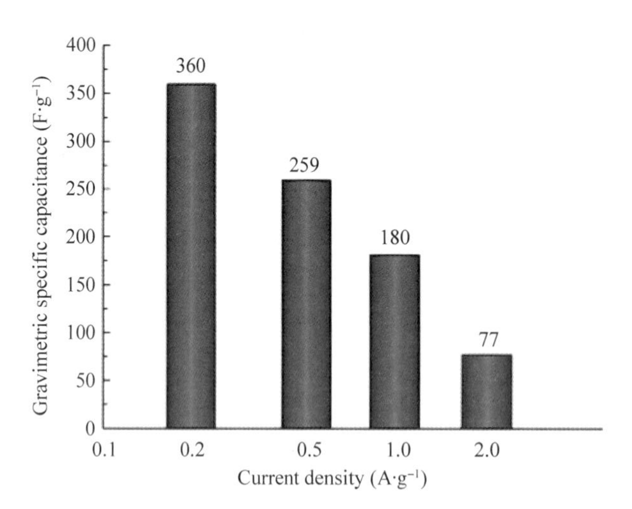

Fig. 6 Gravimetric specifific capacitances of the PANI-PPy/wood electrode at different current densities

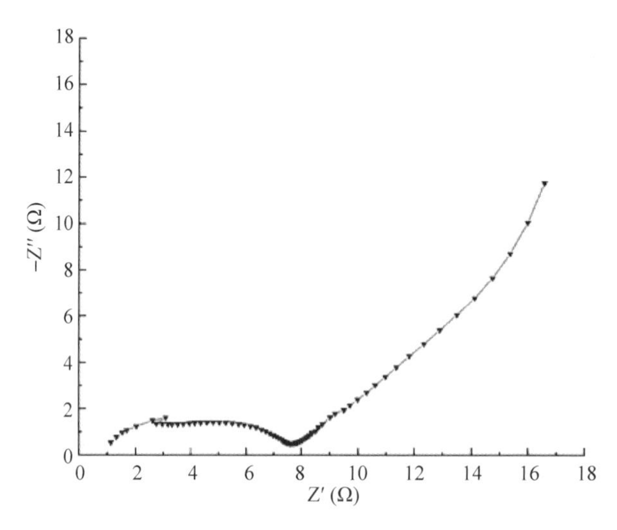

Fig. 7 Nyquist plots of the PANI-PPy/wood electrode

4 Conclusions

The electrochemically active substances, PANI and PPy, were deposited in sequence onto the surface of a wood support using a liquid-phase polymerization method. The PANI-PPy mixed particles have a diameter of around 20 nm according to SEM observations. EDX and FTIR analysis also reflects the characteristics of both PANI and PPy compositions and demonstrates their successful synthesis on the wood surface. Furthermore, this self-supported PANI-PPy/wood composite has the ability to serve as a free-standing electrode, which shows a high gravimetric specific capacitance of 360 $F \cdot g^{-1}$ at 0.2 $A \cdot g^{-1}$ superior to some congeneric products.

Acknowledgments

This work was financially supported by The NatiThis study was supported by the Fundamental Research Funds for the Central Universities (2572018AB09) and the National Natural Science Foundation of China (31470584).

Compliance with ethics guidelines

Jian Li and Yue Jiao declare that they have no conflicts of interest or financial conflicts to disclose.

This article does not contain any studies with human or animal subjects performed by the any of the authors.

References

[1]Wang G P, Zhang L, Zhang J J. A review of electrode materials for electrochemical supercapacitors[J]. Chem Soc Rev, 2012, 41(2): 797-828.

[2]Du C S, Pan N. High power density supercapacitor electrodes of carbon nanotube films by electrophoretic deposition [J]. Nanotechnology, 2006, 17(21): 5314-5318.

[3]Yan J, Fan Z J, Wei T, Qian W Z, Zhang M L, Wei F. Fast and reversible surface redox reaction of graphene-MnO_2 composites as supercapacitor electrodes[J]. Carbon, 2010, 48(13): 3825-3833.

[4]Hu C C, Chen J C, Chang K H. Cathodic deposition of $Ni(OH)_2$ and $Co(OH)_2$ for asymmetric supercapacitors: importance of the electrochemical reversibility of redox couples[J]. J Power Sources, 2013, 221: 128-133.

[5]Wan C C, Jiao Y, Li J. A cellulose fibers-supported hierarchical forest like cuprous oxide/copper array architecture as a flexible and free standing electrode for symmetric supercapacitors[J]. J Mate Chem A: Materials for Energy and Sustainability, 2017, 5(33): 17267-17278.

[6]Hung K S, Masarapu C, Ko T, et al. Wide-temperature range operation supercapacitors from nanostructured activated carbon fabric[J]. J Power Sources, 2009, 193(2): 944-949.

[7]Zhi M J, Xiang C C, Li J T, et al. Nanostructured carbon-metal oxide composite electrodes for supercapacitors: a review[J]. Nanoscale, 2013, 5(1): 72-88.

[8] Zhong C, Deng Y, Hu W B, et al. A review of electrolyte materials and compositions for electrochemical supercapacitors[J]. Chem Soc Rev, 2015, 44(21): 7484-7539.

[9]Snook G A, Kao P, Best A S. Conducting-polymer-based supercapacitor devices and electrodes[J]. J Power Sources, 2011, 196(1): 1-12.

[10]Zhang J T, Zhao X S. Conducting polymers directly coated on reduced graphene oxide sheets as high-performance supercapacitor electrodes[J]. J Phys Chem C, 2012, 116(9): 5420-5426.

[11]Wan C C, Jiao Y, Li J. Flexible, highly conductive, and free-standing reduced graphene oxide/polypyrrole/cellulose hybrid papers for supercapacitor electrodes[J]. J Mater Chem A: Materials for Energy and Sustainability, 2017, 5(8): 3819-3831.

[12]Wan C C, Li J. Wood-derived biochar supported polypyrrole nanoparticles as a free-standing supercapacitor electrode [J]. RSC Advances, 2016, 6(89): 86006-86011.

[13]Li J S, Lu W B, Yan Y S, Chou T W. High performance solid-state flexible supercapacitor based on Fe_3O_4/carbon nanotube/polyaniline ternary films[J]. J Mater Chem A: Materials for Energy and Sustainability, 2017, 5(22): 11271-11277.

[14] Wan C C, Li J. Synthesis and electromagnetic interference shielding of cellulose - derived carbon aerogels functionalized with α-Fe_2O_3 and polypyrrole[J]. Carbohyd Polym, 2017, 161: 158-165.

[15]Tian J, Peng D F, Wu X, Li W, Deng H B, Liu S. Electrodeposition of Ag nanoparticles on conductive polyaniline/

cellulose aerogels with increased synergistic effect for energy storage[J]. Carbohyd Polym, 2017, 156: 19-25.

[16] Zhang Y Z, Wang Y, Cheng T, Lai W Y, Pang H, Huang W. Flexible supercapacitors based on paper substrates: a new paradigm for low-cost energy storage[J]. Chem. Soc. Rev., 2015, 44 (15): 5181-5199.

[17] Wan C C, Lu Y, Sun Q F, Li J. Hydrothermal synthesis of zirconium dioxide coating on the surface of wood with improved UV resistance[J]. Appl. Surf. Sci., 2014, 321: 38-42.

[18] Wan C C, Jiao Y, Li J. In situ deposition of graphene nanosheets on wood surface by one-pot hydrothermal method for enhanced UV resistant ability[J]. Appl. Surf. Sci., 2015, 347: 891-897

[19] Bana R, Banthia A K. Green composites: development of poly(vinyl alcohol)-wood dust composites[J]. Polym-plast Technol, 2007, 46(9): 821-829.

[20] Chen C J, Zhang Y, Li Y J, Dai J Q, Song J W, Yao Y G, Gong Y H, Kierzewski I, Xie J, Hu L B. All-wood, low tortuosity, aqueous, biodegradable supercapacitors with ultra-high capacitance[J]. Energ Environ Sci, 2017, 10(2): 538-545.

[21] Jiao Y, Wan C C, Li J. Scalable synthesis and characterization of freestanding supercapacitor electrode using natural wood as a green substrate to support rod-shaped polyaniline[J]. J Mater Sci-Mater El, 2017, 28(3): 2634-2641.

[22] Wan C C, Lu Y, Jiao Y, Jin C D, Sun Q F, Li J. Ultralight and hydrophobic nanofibrillated cellulose aerogels from coconut shell with ultra-strong adsorption properties[J]. J. Appl. Polym. Sci., 2015, 132(24): 42037.

[23] Tjeerdsma B F, Militz H. Chemical changes in hydrothermal treated wood: FTIR analysis of combined hydrothermal and dry heat-treated wood[J]. Holz als Roh-und Werkstoff, 2005, 63(2): 102-111.

[24] Oh S Y, Yoo D I, Shin Y S, Seo G. FTIR analysis of cellulose treated with sodium hydroxide and carbon dioxide[J]. Carbohyd Res, 2005, 340(3): 417-428.

[25] Xu J, Zhu L G, Bai Z K, Liang G J, Liu L, Fang D, Xu W L. Conductive polypyrrole-bacterial cellulose nanocomposite membranes as flexible supercapacitor electrode[J]. Org. Electron, 2013, 14(12): 3331-3338.

[26] Xu J Q, Zhang Y Q, Zhang D Q, Tang Y M, Cang H. Electrosynthesis of PANI/PPy coatings doped by phosphotungstate on mild steel and their corrosion resistances[J]. Prog Org Coat, 2015, 88: 84-91.

[27] Wan C C, Yue J, Jian L. Core-shell composite of wood-derived biochar supported MnO2 nanosheets for supercapacitor applications[J]. RSC Advances, 2016, 6(69): 64811-64817.

[28] Wang H H, Bian L Y, Zhou P P, Tang J, Tang W H. Core-sheath structured bacterial cellulose/polypyrrole nanocomposites with excellent conductivity as supercapacitors[J]. J Mater Chem A: Materials for Energy and Sustainability, 2013, 1(3): 578-584.

[29] Wang H H, Zhu E W, Yang J Z, Zhou P P, Sun D P, Tang W H. Bacterial cellulose nanofiber-supported polyaniline nanocomposites with flake-shaped morphology as supercapacitor electrodes[J]. J Phys Chem C, 2012, 116(24): 13013-13019.

[30] Chen H C, Jiang J J, Zhang L, Wan H Z, Qi T, Xia D D. Highly conductive NiCo2S4 urchin-like nanostructures for high-rate pseudocapacitors[J]. Nanoscale, 2013, 5(19): 8879-8883.

[31] Zhao T, Jiang H, Ma J. Surfactant-assisted electrochemical deposition of α-cobalt hydroxide for supercapacitors[J]. J Power Sources, 2011, 196(2): 860-864.

中文题目：绿色多孔木材基聚苯胺-聚吡咯纳米复合材料超级电容器的制备

作者：李坚，焦月

摘要：木材是一种理想的载体材料，其多孔结构和表面官能团有利于各种客体物质在不同应用中的沉积。本文以木材为多孔载体，与两种导电聚合物[聚苯胺(PANI)和聚吡咯(PPy)]进行简单、快速的液相聚合。扫描电镜观察表明，PANI-PPy复合物的粒径约为20 nm。用傅里叶变换红外光谱证实了木材含氧基团与 PANI-PPy 的氮组分之间的相互作用。自支撑的 PANI-PPy/木材复合材料可以作为独立的超级电容器电极，在 0.2 $A \cdot g^{-1}$ 下具有高达 360 $F \cdot g^{-1}$ 的高质量比电容。

关键词：木材；聚吡咯；聚苯胺；超级电容器；纳米复合材料

木质基新型能量存储与转换材料研究进展*

李坚，焦月

摘要：木质资源是自然界中取之不尽、用之不竭的绿色资源，具有储量丰富、可再生、可生物降解和碳中性等特点。以木材为代表的各向异性木质资源可以直接作为基质材料来诱导微纳米功能单元的沉积和生长，进而发展各向异性的能量存储和转换材料。此外，以木质资源为原料，通过拆解和重组可以得到具有大比表面积和高反应活性的一维纳米纤丝、二维纳米薄膜和三维凝胶等材料。这些重组材料也可以作为基质与微纳米功能单元复合，从而发展绿色新型的木质基能量存储和转换材料。分析近年来木质资源在超级电容器、锂离子电池、催化析氢和太阳能电池等能量存储与转换材料领域中的国内外研究进展，并展望其未来发展方向，旨在为木质资源的功能化和高值化利用提供新的思路。

关键词：木质资源，可再生能源，能量存储与转换，超级电容器，锂离子电池，催化析氢

随着世界工业的快速发展，化石资源的过度使用引发了环境污染和能源危机等诸多问题，目前研发绿色新型能量存储与转换器件是解决以上难题最有效的方法之一。新型的能量存储与转换器件（如超级电容器、锂电池和太阳能电池等），由于具有超高性能、长使用寿命及良好的安全性等优势得到了研究人员的广泛关注[1-3]，材料的选用则是超级电容器、锂电池、太阳能电池和催化析氢等领域的研究重点，开发新型的复合组分和合理设计材料的微观结构是实现器件性能提高、制备成本降低、安全性增强等目标的关键途径。

木质资源是一种取之不尽、用之不竭的天然资源，它含有多种组织结构、细胞形态、孔隙结构和化学组分，是结构层次分明、构造高度有序的聚合物基天然复合材料[3-5]。木质资源具有天然的分级多孔结构，孔径尺寸从毫米到纳米，可为离子的快速传输提供通道；此外，木质资源具有可再生、可生物降解及碳中性等特点，是制备绿色新型能量存储与转换器件中电极、电解质、催化剂和隔膜等的理想材料[6-8]。近年来，许多研究利用木质资源天然的结构与理化属性，结合纳米复合技术、界面修饰技术等处理方法，基于增材制造和减材制造两大方向实现了木质资源在能量存储和转换领域的高值化应用。综述近年来木质资源在新型绿色能量存储与转化领域的研究进展，重点介绍木质资源在超级电容器、锂电池、催化析氢、太阳能电池等方向的应用，并展望其未来发展方向，旨在为木质资源的功能化和高值化利用提供新的思路。

1 木质资源在超级电容器方向的典型应用

超级电容器是介于传统电容器和充电电池之间的一种新型储能装置，它既具有电容器快速充放电的特性，同时又具有电池的储能特性。根据电荷存储机制不同，可将电容器分为两类：双电层超级电容器和赝电容超级电容器。双电层超级电容器的电荷存储依赖于活性电极材料表面电解液离子的静电吸附，赝电容超级电容器依赖于可逆的氧化还原反应实现电荷存储。

* 本文摘自森林与环境学报，2020，40(4)：337-346.

木质资源在超级电容器领域的应用可以分为两个方向，即减材制造和增材制造。减材制造是指直接以木质原料为基质，通过生物、物理、化学等处理对木质原料的成分进行脱除或转化以及对结构进行定向调控，随后与微纳米活性单元复合，制备木质基复合材料；增材制造是将已经纳米化的木质原料通过重组得到大比表面积和高反应活性的一维纳米纤丝、二维纳米薄膜和三维凝胶等，在重组的过程中或者形成重组材料后与微纳米活性单元复合，以制备电极、隔膜等超级电容器的组件。

木材是减材制造法制备超级电容器电极材料的典型木质材料，木材具有分层多孔的框架结构，是支撑电极活性材料(如：金属氧化物/氢氧化物、导电聚合物等)的理想基底/支架[9-11]；此外，木材分层次的多孔结构有利于电极与电解液的接触，其连续的低弯曲度孔道结构可促进离子的迁移并提供大量的表面位点来负载大量的活性物质。CHEN et al.[12]将木材热解制备的三维多孔导电木炭作为支架，在其表面生长 MnO_2 纳米片，利用木炭结构的大比表面积实现 MnO_2 的高密度负载(>75 mg · cm^{-2})；全木质非对称超级电容器组装图解及性能分析如图 1 所示。CHEN et al.[12]设计的全木质非对称超级电容器采用木炭作为阳极、木质薄膜作为隔膜、MnO_2/木炭作为阴极组装全木质的非对称超级电容器；该超级电容器厚度仅 1 mm；由于其独特的结构、低弯曲度和高活性物质负载量等特点，该器件的面电容高达 3.6F · cm^{-2}，在功率密度为 1044 mW · cm^{-2} 时，能量密度可达 1.6 mW · h · cm^{-2}，且具有较长的循环寿命。

YANG et al.[13]利用聚合物辅助金属沉积法在拟提供柔性和存储离子的棉织物表面负载金属镍，随后通过真空抽滤法负载多壁碳纳米管和还原氧化石墨烯，制备了三维多孔的多壁碳纳米管/还原氧化石墨烯/金属镍复合织物电极；该电极可提供大量的连接通道和表面积用于离子快速扩散和吸收；组装的全固态棉织物基超级电容器可以便捷地嵌入到衣物中，可发展为镶嵌便携式健康监控设备、电子娱乐设备、运动记录设备等电子器件的新型服装；YANG et al.[13]组装了一个 10 cm^2 的全固态棉织物基超级电容器。该超级电容器器件在扫描速率 10~100 mV · s^{-1} 范围内呈现出近乎矩形的循环伏安曲线；不同电流密度下的恒电流充放电曲线具有标准的三角形形状表明其良好的库仑效率(在 20 mA · cm^{-2} 可达到 99%)；当使用十层多壁碳纳米管/还原氧化石墨烯/金属镍复合织物做电极时，器件的电容在 20 mA · cm^{-2} 时可达到 2.7 F · cm^{-2}；由该电极组装的全固态超级电容器器件在超过 10000 次循环测试后没有发生明显电容衰减。

随着纳米技术的快速发展，将木质资源纳米化拆解得到的纳米材料在重组的过程中或者形成重组材料后与微纳米活性单元复合来制备超级电容器电极材料，符合增材制造的理念。木质材料经化学预处理脱除木质素、半纤维素等物质后可得到亲水的纤维素；纤维素已被广泛应用于制备超级电容器电极材料。ZHANG et al.[14]通过 2, 2, 6, 6-四甲基哌啶氧化物(2, 2, 6, 6-tetramethylpiperidine-1-oxyl, TEMPO)法从硬木纸浆中提取纤维素纤维，以柠檬酸-Fe^{3+}复合物为氧化剂控制聚吡咯沉积量，制备了具有三维多孔结构的柔性纤维素纤维/还原氧化石墨烯/聚吡咯气凝胶电极；以纤维素纤维/还原氧化石墨烯/聚吡咯气凝胶膜为电极、聚乙烯醇/H_2SO_4 凝胶为电解质和隔膜制备柔性全固态超级电容器；由于该电极具有多孔结构、高导电性和良好的润湿性，组装的超级电容器展示了优异的电化学性能。该电极在电流密度为 0.25 mA · cm^{-2} 时其面积比电容最高可达 720 mF · cm^{-2}，且经 2000 次循环后电容保持率为 95%；该器件最大能量密度为 60.4 μW · h · cm^{-2}，并在不同弯曲条件下电容仍能保持稳定，表明其在柔性电子元件领域具有较大的应用前景。

WAN et al.[15]采用直流磁控溅射、循环伏安电氧化等现代微纳米制造技术，仿生自然界中的“土壤-山体-植被”地质生态系统，构筑超强储能特性的“功能集聚体”，其设计原理和电化学性能分析如图 2 所示。其中，仿生土壤即纤维素网络，兼具保水、亲水、柔性等功能(促进离子迁移)；仿生山体即金属钴兼具刚性、电子高速传输、界面黏附等功能(促进电子传输);仿生植被即垂直取向的蜂巢状氢氧化钴,兼具电化学高分辨响应、赝电容存储、离子协助扩散、缓和脱嵌应力等功能(促进电化学反应)。该仿生超级电容器的能量密度高达 166 $\mu W \cdot h \cdot cm^{-2}$,在高电流密度下经过 10000 次连续充放电测试,其比电容值仅损失 8.4%,性能优于诸多同类产品。将仿生天然地质结构系统所构筑的纤维素-Co@ $Co(OH)_2$ 复合材料与碳布进行组装,构建非对称的超级电容器元器件,电压窗口可扩展至 0~1.5 V。在两电极体系下,该器件展现出诸多优于同类产品的电化学特性,能量密度可达 166 $\mu W \cdot h \cdot cm^{-2}$,功率密度可达 37.5 $mW \cdot cm^{-2}$,具有优异的循环稳定性(电容保持率 91.6%,10000 次循环)以及良好的倍率特性。

除了制备电极材料,木质资源还可以用于制备超级电容器隔膜。超级电容器隔膜一般要求为电子的绝缘体和离子的良导体,且具有化学稳定性好、吸液和保液性强、隔离性能好、机械强度高和柔韧性好等特点。纤维素大分子的自组装体具有丰富的孔结构、优异的绝缘性、强亲水性、良好的柔性和力学强度,是制备超级电容器隔膜的理想选择。制备纤维素膜的常用方法主要包括 3 种[16-17]:①通过

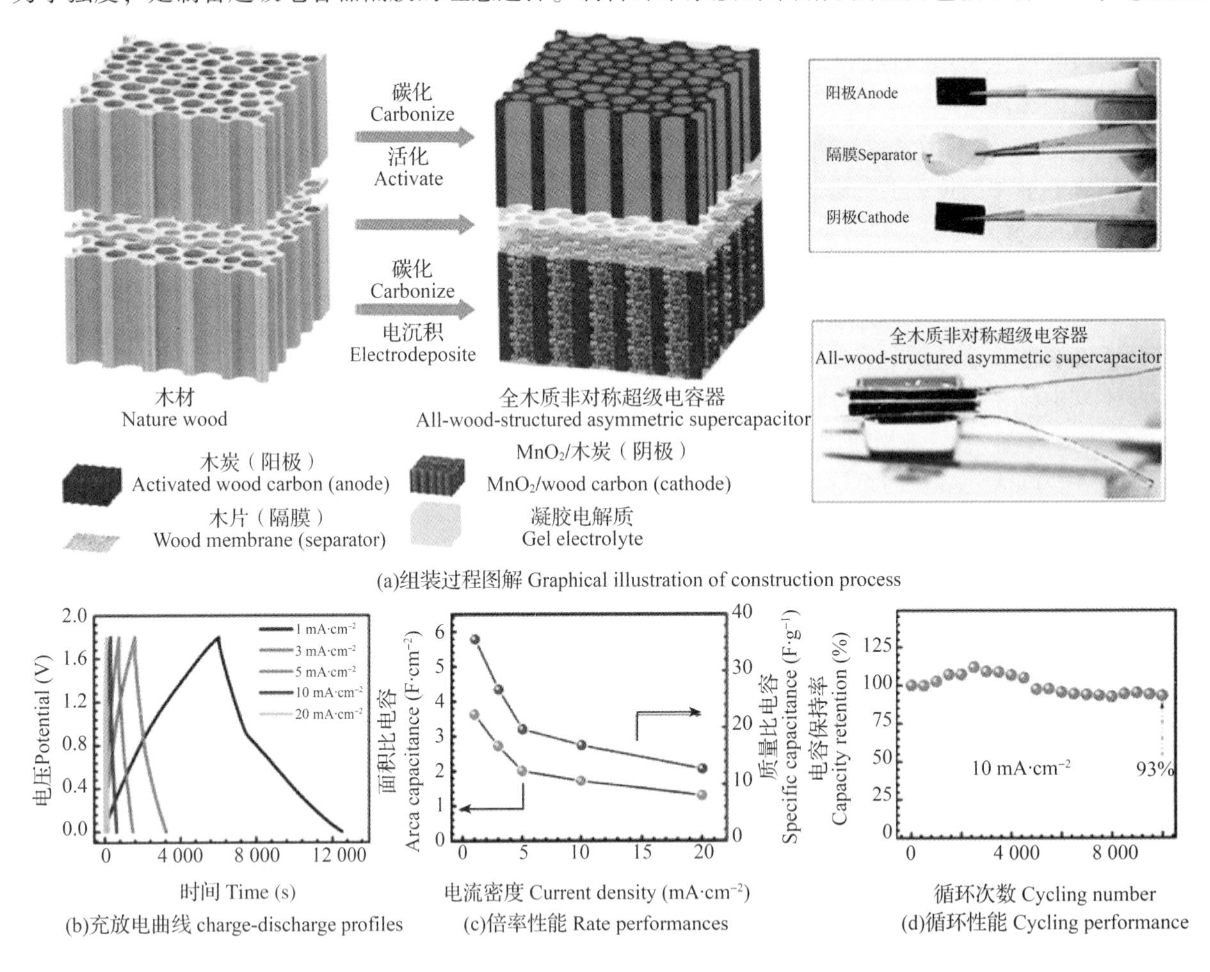

图 1 全木质非对称超级电容器的组装和电化学性能

Fig. 1 Schematic diagram of the preparation of a wooden asymmetric supercapacitor, and its electrochemical properties

注:图片来源于参考文献[12]。Note: the pictures are from references[12].

高压均质法、高频脉冲超声法等物理处理结合化学预处理或酶预处理等前期处理，实现木质资源的纳米化，随后通过超分子作用实现重组成膜；②通过溶解实现木质资源的纳米化，随后通过加入抗溶剂干扰溶解平衡，促使纤维素分子重组成膜；③采用酯化、醚化等修饰处理或氧化开环处理，促使纤维

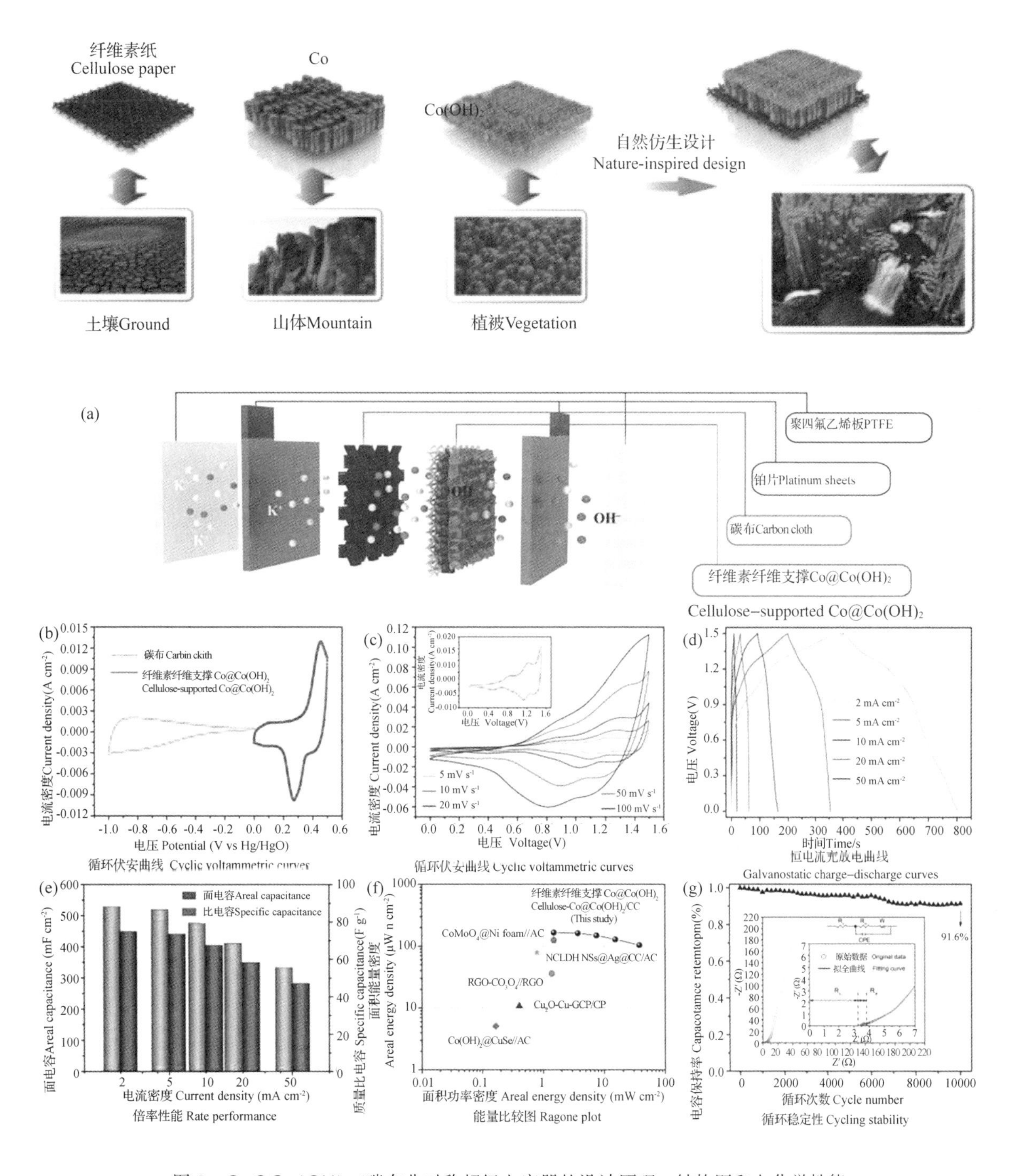

图 2　Co@Co(OH)$_2$/碳布非对称超级电容器的设计原理、结构图和电化学性能

Fig. 2　Schematic diagrams of the nature-inspired design philosophy, structure, and the electrochemical properties of the Co@ Co(OH)$_2$/cotton-derived carbon cloth asymmetricsupercapacitor

注：图片来源于参考文献[15]。Note: the pictures are from references[15].

素衍生物直接溶解在水中，再经由溶剂蒸发法制得透明膜材料。将纤维素膜直接浸渍在电解质中或与电解质混合后重组成膜即可得到纤维素基超级电容器隔膜。ZHAO et al.[18]通过简便、可控的固相转化法制备了具良好柔韧性、透明性和可再生的介孔纤维素膜，并利用饱和 KOH 电解质、介孔纤维素隔膜与碳材料电极组成固态双电层超级电容器；在电流密度为 1.0 A・g^{-1} 时，该器件的电容为 110 F・g^{-1}，且循环 10000 次后电容保留率可达 84.7%。另外，在不使用复杂设备的情况下，将电极材料直接沉积在介孔纤维素基聚合物电解质上，可方便地制备出高度集成的平面微型超级电容器，其面电容可达 53.34 mF・cm^{-2}。

2 木质资源在锂电池方向的典型应用

锂电池可分为锂金属电池和锂离子电池(lithium ion battery，LIBs)，其中 LIBs 可根据采用的正极材料类型进行细分，可分为锰酸锂电池、钴酸锂电池、磷酸锂电池等。典型的完整 LIBs 包括正极、负极、隔膜和 Li^+导电电解质。LIBs 电极制备过程通常是将活性电极材料与黏结剂和导电剂混合在有机溶剂中形成浆料，然后将浆料浇注在金属集流器(如铝或铜箔)上形成电极。然而，这种传统方式仍存在一些缺点，如：添加剂和集流器对电容没有贡献，从而降低了能量密度；机械稳定性差导致填料密度低，易于从集流器上脱落。另外，从环境可持续和生态友好性出发，应研发更加绿色的锂电池设备。木质资源及其衍生物由于独特的结构和高环保性，是制备 LIBs 电极的理想材料[19-20]。

为解决 LIBs 在充放电过程中电容快速衰减的问题，BAO et al.[21]利用静电纺丝工艺制备了具有双功能核-壳结构的新型正极材料，纤维素-炭黑复合材料的合成示意图和结构图如图 3(a)和图 3(b)所示；硫掺杂的介孔碳为材料的核部分，纤维素纤维为材料的壳部分。纤维素作为外壳材料其具有较高的电导率，可促进离子迁移；纤维素外壳的高韧性能够承受循环过程中正极体积的变化，抑制多硫

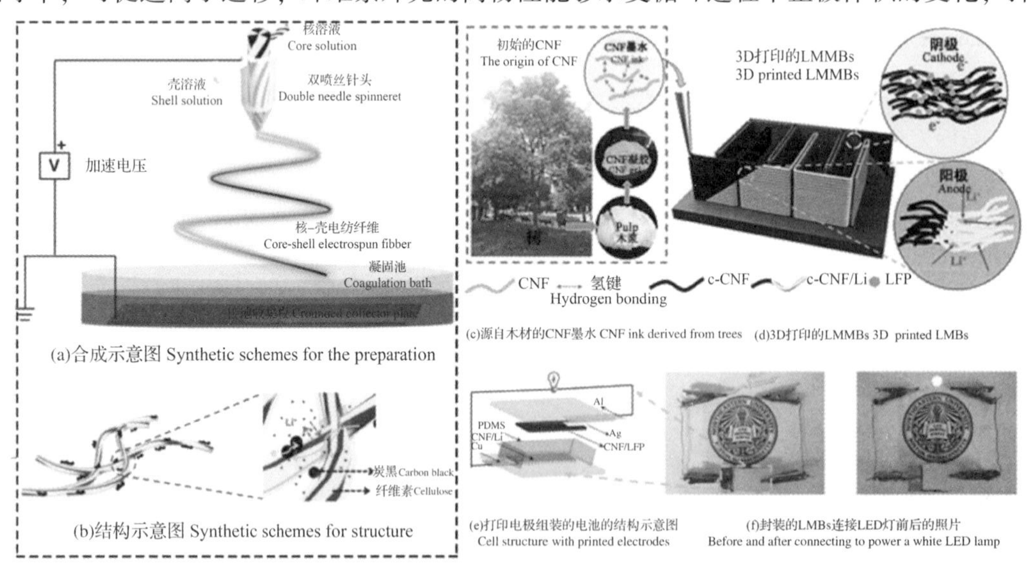

图 3 纤维素/炭黑复合材料的合成图及结构图和 CNF 墨水 3D 打印的 LMBs

Fig. 3 Schematic diagrams of the preparation and the structure of the cellulose /carbon black composite, and CNF ink 3D printed LMBs

注：图片(a)和(b)来源于参考文献[21]，图片(c)-(f)来源于参考文献[22]。

Note: the pictures (a) and (b) are from references[21], the pictures(c)-(f) are from references[22].

的穿梭效应；将该正极材料组装成电池后进行测试，其首次放电比容量高达 1200 mA·h·g^{-1}；经过 300 次循环后，容量仍保持在 660 mA·h·g^{-1}，且库伦效率高达 99%；通过计算纤维素中空纤维中的冯米斯应力分布，验证了柔软的纤维素材料比刚性壳体能更好地限制和捕捉硫原子也减少了应力集中和结构破坏。CAO et al.[22]首次通过 3D 打印技术，使用纤维素纳米纤维(cellulose nanofiber，CNF)制备高性能的锂金属电池。CNF 水分散液可以作为 3D 打印“墨水”的理想选择[图 3(c)]；采用 3D 打印和热解技术制备的纤维素碳纳米纤维/磷酸铁锂(LFP)正极材料[图 3(c)-(f)]；打印出的电极内部的多孔结构具有很高的离子可及性，可以有效抑制不均匀锂沉积/剥离形成的枝晶；采用 3D 打印的锂正极和 $LiFePO_4$ 负极组装的电池，在充放电速率为 10 C 时，其容量可达 80 mA·h·g^{-1}，循环 3000 次后容量保持率为 85%。

木材具有层次结构分明、机械强度大等特点，在惰性气体下经高温碳化后具有较高的导电性；木材的这些特性可以显著地促进离子和电子的快速传输，使得木材衍生的碳材料成为 LIBs 理想的负极材料。ZHANG et al.[23]使用 3D 高孔隙率的碳化木(孔隙率为 73%)导电骨架作为负载锂的基质，利用木材天然的低弯曲率和高孔隙率的特点，将熔融的金属 Li 注入到木材的直管中，为了增加碳化木与锂之间的相容性，在注入熔融的金属锂之前，在碳化木表面涂一层薄薄的 ZnO；将熔融的锂金属快速地注入到木材的管道中，制备得到 Li /碳化木电极。相比由单一锂金属电极组装的对称电池，Li /碳化木电极展现出更加稳定的剥离/电镀特性、更好的循环稳定性(在 3 mA·cm^{-2} 时为 150 h)和更低的过电位(在 3 mA·cm^{-2} 时为 90 mV)。YANG et al.[24]以木材为原料制备了 MnO/木炭复合材料，CHEN et al.[25]用 MnO/木炭复合材料作锂电池的负极[图 4(a)]；MnO 纳米材料在木炭纤维表面均匀分布，MnO 和木炭之间的协同效应促进了复合材料的电荷存储[图 4(b)~(e)]；MnO/木炭复合材料电极在扫描速率为 0.1 mV·s^{-1} 和电压范围为 0.01~3.00 V 时循环伏安曲线的前 4 圈如图 4(f)所示；在 3.00~0.01 V 范围的阴极扫描下，在 1.40 V 处出现的峰归因于在煅烧过程中 MnO_2 的不完全氧化导致 Mn^{3+} 或 Mn^{4+} 还原为 Mn^{2+}；在 0.01~3.00 V 的阳极扫描下，在 1.0~1.5 V 的电压范围内电极呈现出 MnO 到 Mn^{2+} 氧化的相关信号；图 4(g)显示在电流密度为 0.1 A·g^{-1} 时电极获得 952 mA·h·g^{-1} 的放电容量，且经 100 次充放电后仍具有稳定的循环能力，库仑效率稳定在 99%左右；图 4(h)进一步研究了 MnO/木炭纳米复合材料的倍率性能和不同电流密度下的循环稳定性，首先以 0.1 A·g^{-1} 的电流密度对电池进行测试，然后将充放电的电流密度依次增大至 0.2、0.5、1.0 和 3.0 A·g^{-1}；经过 10 次循环后对应的平均充电容量分别为 780、674、571、386 和 275 mA·h·g^{-1}；当电流密度重新降至 0.1 A·g^{-1} 时，平均容量恢复至 604.0 mA·h·g^{-1}，表明其具有良好的倍率性能和循环稳定性。

与超级电容器相似，常规的锂电池的隔膜主要起隔离正负极、防止短路和提供锂离子通道的作用，本身不参加任何化学或电化学反应。普通的隔膜对电池容量没有任何贡献，然而它却占据了锂离子电池 20%左右的体积。WANG et al.[26]制备了包含多孔氧化还原活性层的木质基双层纤维素隔膜，显著增强了 LIBs 的电化学性能。该柔性氧化还原活化隔膜是以绝缘的介孔纤维素纤维为基质，在其表面负载导电且具有高电化学活性的聚吡咯层；前者为电极和聚吡咯-纳米纤维素提供必要的绝缘性，后者增强了纳米纤维素层机械强度，同时高电化学活性的聚吡咯通过离子脱嵌机理为电池提供额外容量；利用合成的隔膜代替传统的聚乙烯隔膜，并以 $LiFePO_4$ 阴极和锂金属阳极构成的 LIBs 的容量从 0.16 mA·h 增加到 0.276 mA·h。

为了提高锂电池的能量密度，降低非电化学活性成分的比例并制备厚电极是一种有效的方法。然而，普通的厚电极离子扩散能力较差，且易变形导致电极开裂并脱离集流器。CHEN et al.[25]受木材独特的平行通道结构的启发，将木材直接碳化，使其成为具有高导电性、多孔性、轻质和低弯曲率的碳骨架，用作多通道木炭集流器，得益于碳骨架独特的多通道结构，厚度 800 μm 的碳骨架可负载 60 mg · cm^{-2} 的 $LiFePO_4$(LFP)，进而制备成 LFP-碳骨架三维电极。该 LFP-碳骨架三维电极的容量可达 7.6 mA · h · cm^{-2}，且机械性能显著增强；与此同时，LFP-碳骨架三维电极具有良好的循环稳定性和容量保持率。

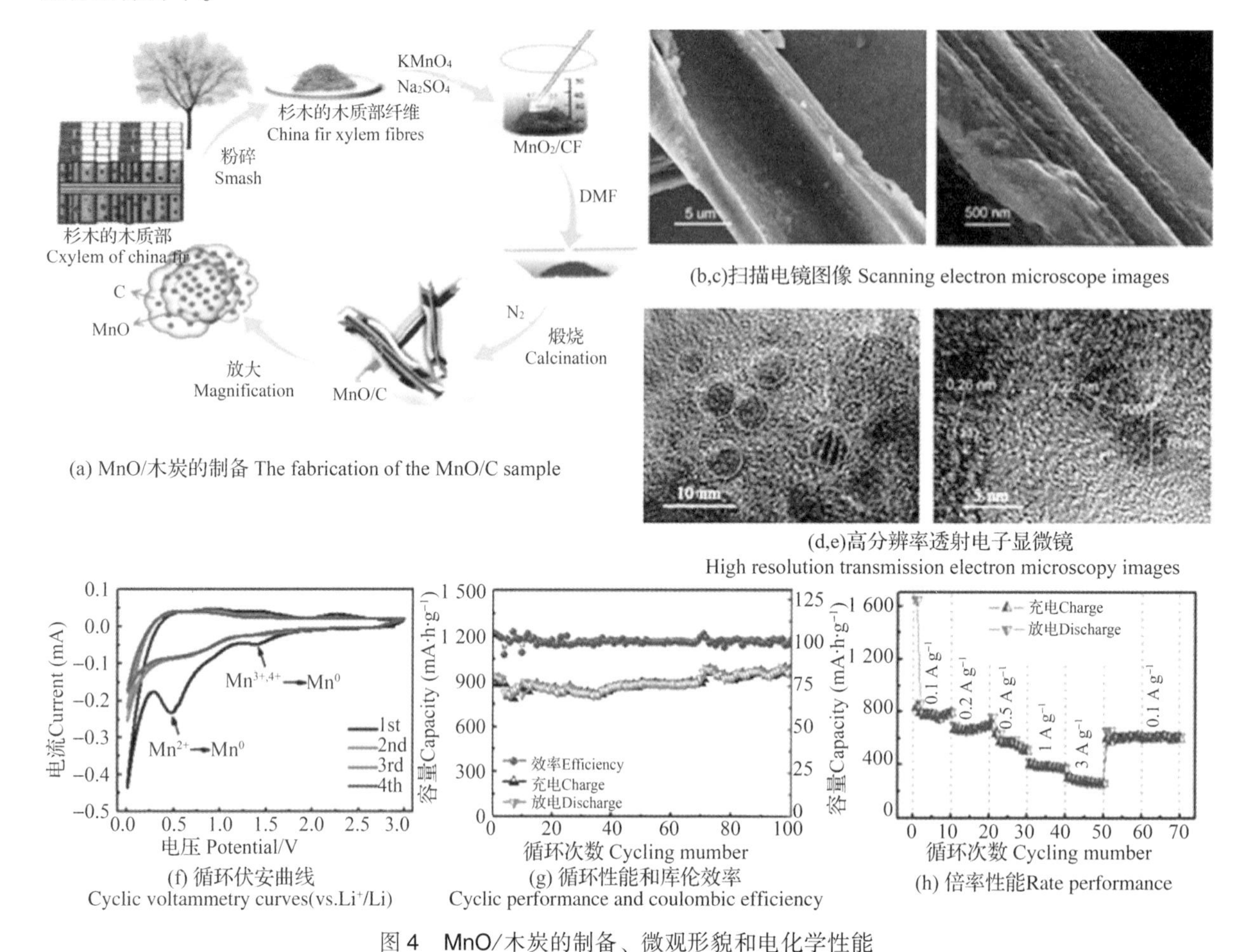

图 4 MnO/木炭的制备、微观形貌和电化学性能

Fig. 4 Schematic diagrams of the fabrication, micromorphology and electrochemical properties of the MnO/C sample

注：图片来源于参考文献[25]。Note：the pictures are from references[25].

3 木质资源在催化析氢方向的典型应用

氢能是一种高效的可再生清洁能源，被认为是未来最有可能代替化石燃料的可再生化学燃料之一。电催化析氢是水裂解过程的阴极反应，也是获得高纯度氢气并实现可持续分布式存储的重要途径。催化剂性能是催化析氢过程中最关键的一环，催化剂性能的提高主要包括 3 个途径[27-28]：①构建层次多孔结构；②提高比表面积；③增加活性位点浓度和杂原子掺杂。木质资源及其衍生物由于其独特的结构、环保性和可持续性是制备催化剂和催化剂载体的理想材料。下文分别介绍了木质材料及其衍生物在催化析氢中作为催化剂模板、载体或原料的应用。

$MoSe_2$ 自身活性位点有限且导电性差，因此，其应用受到限制。LAI et al.[29] 利用细菌纤维素衍生的纤维素碳纳米纤维(carbonized bacterial cellulose，CBC)为基底诱导 $MoSe_2$ 纳米片沿纤维径向定向生长，$MoSe_2$ 聚集体的合成示意图如图 5(a)和图 5(b)所示。CBC 的参与不仅抑制了 $MoSe_2$ 纳米片的团聚，同时提升其导电性和活性位点数量；由一维纳米纤维诱导生成的 $MoSe_2$ 纳米片呈现薄层特性(5~8 层，层间距 0.62 nm)，使得纳米片能充分与电解液离子接触，显著提升催化产氢的活性位点数目；$MoSe_2$ 和 CBC 之间的协同效应有利于复合材料表现出优异的电催化产氢性能；在所测的样品中，CBC/MoSe2-2复合材料的阴极电流密度最大，在 300 mV 的过电位下其电流密度为 87 $mA \cdot cm^{-2}$，分别为 $MoSe_2$ 聚集体和 CBC 的 15 和 174 倍[图 5(c)]，说明其具有更强的析氢能力；对不同样品的塔菲尔图[图 5(d)]进行分析，研究表明与单组分的 CBC(112 $mV \cdot dec^{-1}$)和 $MoSe_2$ 聚集体(105 $mV \cdot dec^{-1}$)相比，CBC / $MoSe_2$ 复合材料的塔菲尔斜率更小(55 $mV \cdot dec^{-1}$)，表明其更快的析氢速率；CBC / $MoSe_2$ 复合材料在析氢过程中的电化学阻抗谱，如图 5e 所示；CBC / $MoSe_2$ 复合材料的电荷转移电阻明显低于 $MoSe_2$ 聚集体，这是由于其具有三维网络结构和较大的比表面积，可实现更快的电子传输。为了研究酸性条件下 CBC / $MoSe_2$ 复合材料的耐久性，在扫描速率为 100 $mV \cdot s^{-1}$ 的条件下，进行了 1000 次循环伏安测试，电势范围为-0.4~0.2 V。1000 次循环前后的线性扫描伏曲线之间的差异可以忽略不计，证明复合材料是一种能够承受酸腐蚀的高稳定催化剂[图 5(f)]。

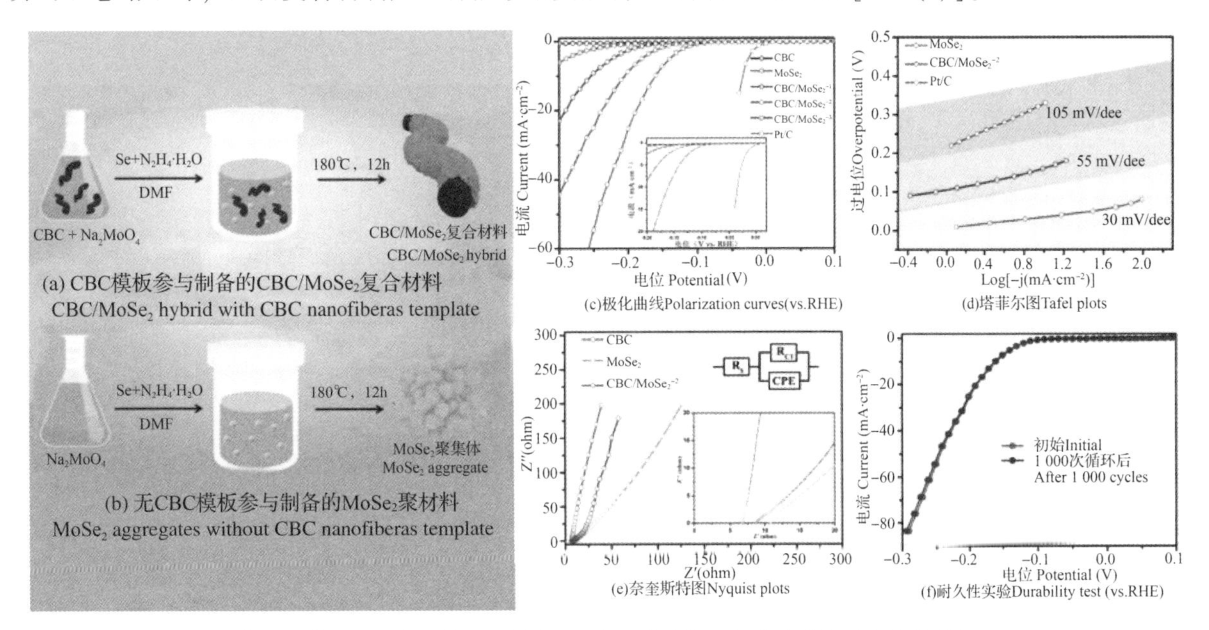

图 5　CBC/$MoSe_2$ 复合材料和 $MoSe_2$ 聚集体的合成示意图和电化学性能

Fig. 5　Schematic diagrams of the preparation and electrochemical properties of the CBC/$MoSe_2$hybrid and $MoSe_2$aggregates

注：图片来源于参考文献[29]。Note：the pictures are from references[29].

MULYADI et al.[30] 采用溶剂热碳化和热解工艺，将纤维素纳米纤丝转变为具有三维结构的氮磷硫掺杂的碳网络，用作无金属析氢催化剂，其制备原理如图 6(a)所示。该催化剂避免了贵金属元素的使用，降低了成本。大多数的石墨烯基催化剂仅对氧还原反应(oxygen reduction reaction，ORR)具有良好的促进作用，而氮磷硫掺杂的碳基无金属催化剂对析氢反应(hydrogen evolution reaction, HER)和 ORR 均表现出良好的催化效果。掺杂的碳基材料在初始电位 233 mV 具有 HER 性能(与可逆氢电极的电极电势相比), 331 mV(与可逆氢电极的电极电势相比)的电流密度为 10 mA · cm^{-2}, 塔菲尔斜率为 99 mV · dec^{-1}[图 6(b)]; 将同样的材料用于 ORR, 它的初始电位相比于商用的 Pt /C 低 10 mV, 阴极峰值为 0. 84 V(与可逆氢电极的电极电势相比); 氮磷硫掺杂提高了纤维素碳纤维网络的表面粗糙度和电子传导速度, 增加了反应位点数目, 提高了该催化剂的催化活性。

木质资源中含有的碳水化合物是制备清洁燃料的良好原料。它不仅能转化为氢气, 还能生产出乙醇。在室温下, 太阳能驱动的光催化木质纤维素制氢是实现这一目标的绿色途径。WAKERLEY et al.[31]模拟光合作用, 在光照条件下, 以半导体硫化镉基量子点为催化剂, 催化浸泡于碱性水溶液中的纤维素、半纤维素和木质素制氢; 其中, CdS /CdO_x 催化剂充当"光合作用线粒体"的功能; 在该体系下, 除木材外, 未经处理的打印纸、甘蔗渣、锯屑等木质原料或制品也呈现出一定的析氢现象。

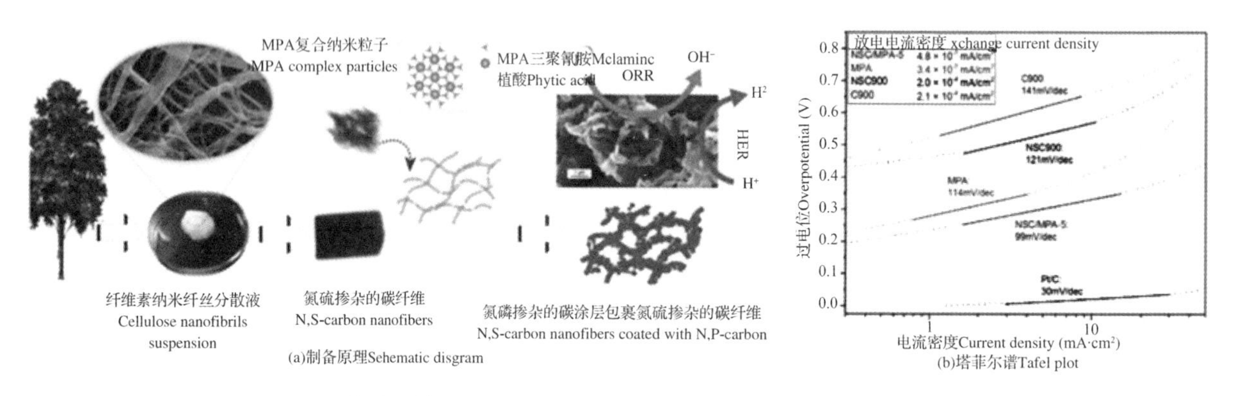

图 6　氮磷硫掺杂的碳基无金属碳催化剂的制备原理和塔菲尔谱图

Fig. 6　Schematic diagram of the preparation of N-, P-, and S-doped carbon-based metal-free catalysts, and the corresponding Tafel plots

注: 图片来源于参考文献[30]。Note: the pictures are from references[30].

4　木质资源在太阳能电池方向的典型应用

太阳能电池是通过光电效应或者光化学效应直接或间接把太阳辐射能转化成电能的装置。太阳能资源取之不尽, 用之不竭。太阳能电池具有诸多优点[32-33], 如绿色环保, 不受制于地域、海拔等因素应用范围广, 无机械转动部件, 操作、维护简单, 运行稳定可靠, 体积小且轻, 便于运输和安装, 建设周期短。现代太阳能电池可分为 3 种类型[34]: 染料敏化太阳能电池是以低成本的纳米 TiO_2 和光敏染料为主要原料, 模拟自然界中植物利用太阳能进行光合作用, 将太阳能转化为电能的一种装置; 量子点敏化太阳能电池, 是以量子点为敏化剂的太阳能电池, 与染料相比, 量子点光谱吸收范围更广, 具有更大的消光系数和光化学稳定性; 钙钛矿太阳能电池, 是利用钙钛矿型的有机金属卤化物半导体作为吸光材料的太阳能电池。木质资源及其衍生物由于其独特的多孔结构、亲水性、可再生和可降解性等, 可用于制备太阳能电池的多个组成部分。BRISCOE et al.[35]将几丁质、壳聚糖和葡萄糖等经水热碳化所制备的碳量子点与半导体 ZnO 纳米棒相结合, 制备了碳量子点敏化太阳能电池(图 7),

器件的性能与碳量子点上的官能团紧密相关；其中，壳聚糖和几丁质衍生物结合制备的碳量子点敏化太阳能电池器件光伏电池效率最高可达 0.077%；相对于一些染料敏化太阳能电池和半导体敏化太阳能电池，碳量子点敏化太阳能电池活性更高且更绿色环保。

太阳能电池通常在玻璃或塑料上进行组装，然而这两种方法制得的产品不易回收。纤维素纳米材料是一种新兴的高附加值纳米材料，可从植物中提取，具有储量丰富、可再生和可持续等特点。从木质资源中分离而来的环境友好的纤维素纳米纤维，可以重组制备高透明性的膜材料。ZHOU et al.[36] 以纳米纤维素(cellulose nanocrystal，CNC)为基质，将太阳能电池的组分组装在高透明的 CNC 基质上，基于 CNC 基板组装的太阳能电池器件结构示意图和宏观照片如图 8(a)和图 8(b)所示。这种太阳能电池器件可以很容易在使用寿命结束后进行回收处理，具有良好的整流性，且光伏电池效率为 2.7%；在 CNC 基板上组装的高效且易于回收的太阳能电池有望成为绿色、可持续的光电转换器件。利用咪唑啉类离子液体制成的凝胶聚合物电解质在染料敏化太阳能电池中的应用受到广泛关注。KHANMIRZAEI et al.[37] 将羟丙基纤维素添加入碘化钠(NaI)、离子液体 1-甲基-3-丙基碘化咪唑鎓(MPII)、碳酸乙烯酯和碳酸丙烯酯混合物中，可制备不具挥发性的凝胶聚合物电解质(HNaP)，用于制备染料敏化太阳能电池；其离子电导率最高可达 7.37×10^{-3} S·cm^{-1}；对制备的染料敏化太阳能电池的光电流密度-电池电位(J-V)特性进行分析，含有质量分数为 100% 的 MPII 离子液体的凝胶聚合物电解质具有最佳性能，其能量转换效率可达 5.79%，且短路电流密度、开路电压和填充率分别为 13.73 mA·cm^{-2}、610 mV 和 69.1%。

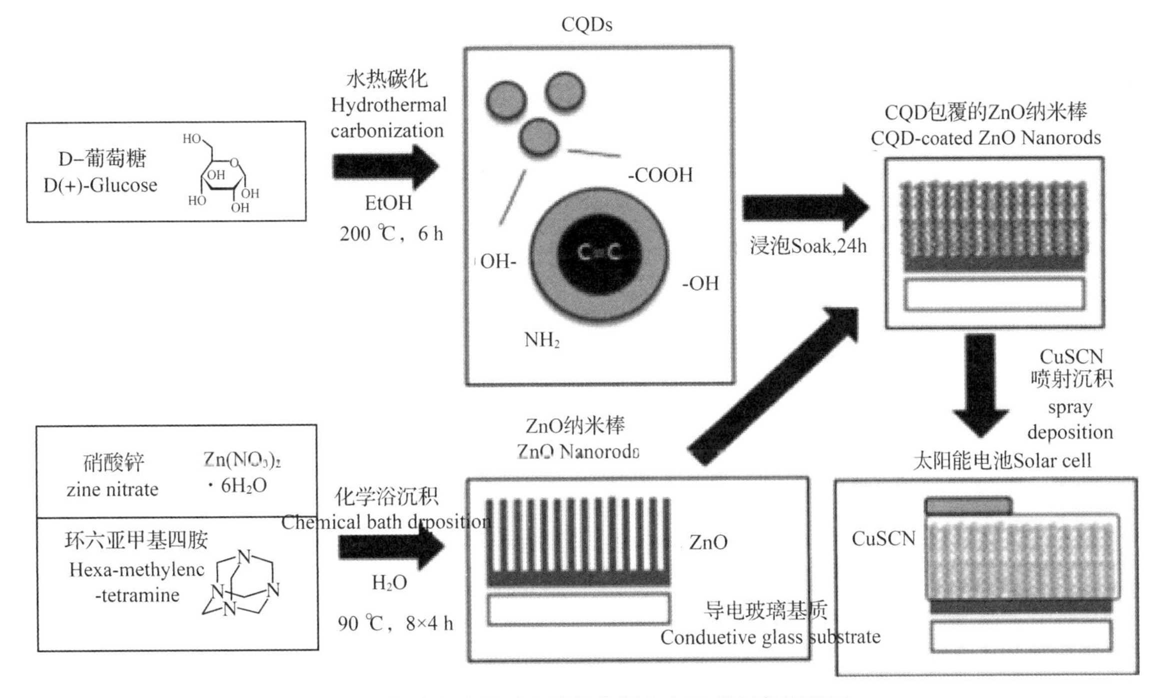

图 7 木质基碳量子点敏化太阳能电池的制备流程图

Fig. 7 Fabrication flow chart of wood-based carbon quantum dot sensitized solar cells

注：图片来源于参考文献[35]。Note：the pictures are from references[35].

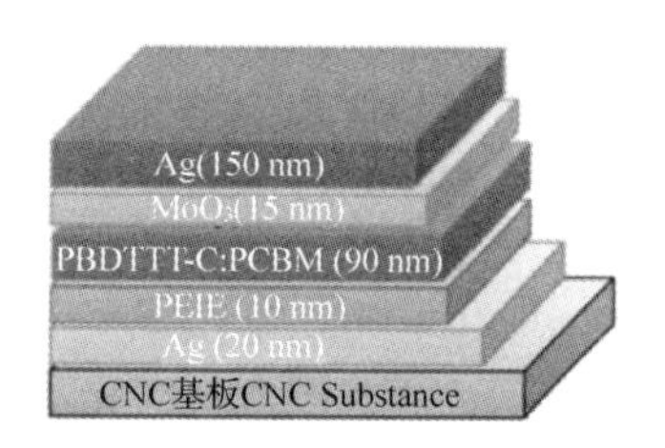

(a)结构示意图 Schematic diagram

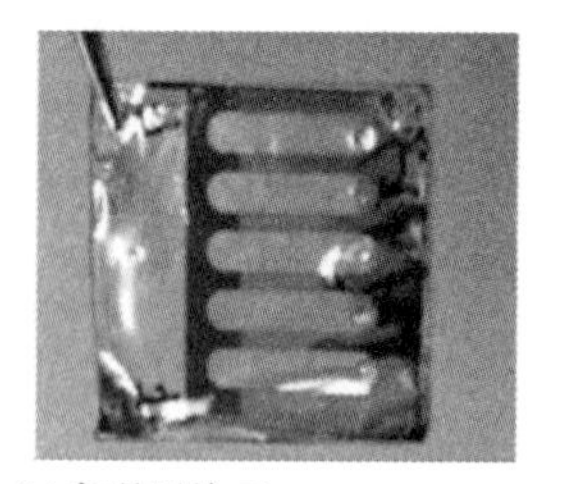
(b)宏观照片 Photomacrograph

图8　基于纳米纤维素基板组装的太阳能电池器件

Fig. 8　Solar cell device based on CNC substrate assembly

注：图片来源于参考文献[37]。Note：the pictures are from references[37].

5　发展趋势

随着自然资源的耗竭和环境的恶化，人们越来越注重可持续发展，木质材料及其衍生物由于具有环境友好性、独特的结构特性和较高的反应活性，基于增材制造和减材制造两大方向在超级电容器、锂离子电池、催化析氢和太阳能电池等能量存储与转换材料领域推广应用前景。木质材料及其衍生物的天然分层多孔结构的充分利用、表界面化学性质的深入分析以及微观结构与性能之间的构效关系分析等均为当前研究的重点。在未来的研究中，应从降低木质材料成本、提高性能、扩大应用和有效利用廉价木材产品等方面继续深入研究。努力寻找有效的提纯方法降低生产成本，开发合适的预处理溶剂体系、可回收利用的催化剂、减少化学药品及水的消耗；对木质材料及其衍生物的表面进行功能化处理，在实现高值化利用的同时更要注重遵循绿色化学的原则，避免二次污染；系统地探明木质基材料分级多孔结构与性能的关系，实现在一维纳米纤丝、二维纳米薄膜和三维凝胶等不同维度的木质基质合理设计和表面微纳米结构的精确控制；广泛挖掘材料的“一剂多效”特性，发展新型木质基能量存储与转化材料。

参考文献

[1]商洪涛，岳立平，杨献奎，等. 超级电容器结构及应用发展概述[J]. 化学工程与装备，2014(9)：177-179.

[2]邹群，楼台芳，王银平，等. 析氢催化电极的研究现状[J]. 材料保护，2002，35(3)：11-14.

[3]李坚，孙庆丰. 大自然给予的启发：木材仿生科学刍议[J]. 中国工程科学，2014，16(4)：4-12.

[4]李坚. 木材的生态学属性：木材是绿色环境人体健康的贡献者[J]. 东北林业大学学报，2010，38(5)：1-8.

[5]李坚. 拓展生物质利用空间助推增材制造绿色发展[J]. 科技导报，2016，34(19)：13.

[6]Wang G P，Zha L，Zhang J J. A review of electrode materials for electrochemical supercapacitors[J]. Chem. Soc. Rev.，2012，41(2)：797-828.

[7]Sun Y M，SILLS R B，Hu X L，et al. A bamboo-inspired nanostructure design for flexible，foldable，and twistable energy storage devices[J]. Nano Letters，2015，15(6)：3899-3906.

[8]李坚，焦月，万才超. 负载 $Ni(OH)_2$/NiOH 微球碳气凝胶的水热合成与储能应用[J]. 森林与环境学报，2018，38(3)：257-264.

[9]Wang R T，Yan X B，Lang J W，et al. A hybrid supercapacitor based on flower-like $Co(OH)_2$ and urchin-like VNelectrode materials[J]. J. Mater Chem. A，2014，2(32)：12724-12732.

[10]Wan C C，Jiao Y，Li J. A cellulose fibers-supported hierarchical forest-like cuprous oxide /copper array architecture

as a flexible and free-standing electrode for symmetric supercapacitors[J]. J. Mater Chem. A, 2017, 5(33): 17267-17278.

[11]刘铎. 木竹材衍生碳基超级电容器电极材料的制备及其性能研究[D]. 杭州: 浙江农林大学, 2016.

[12]Chen C J, Zhang Y, Li Y J, et al. All-wood, low tortuosity, aqueous, biodegradable supercapacitors with ultra-high capacitance[J]. Energ. Environ. Sci., 2017, 10(2): 538-545.

[13]Yang Y, Huang Q Y, Niu L Y, et al. Waterproof, ultrahigh areal-capacitance, wearable supercapacitor fabrics[J]. Adv. Mater, 2017, 29(19): 1606679.

[14]Zhang Y H, Shang Z, Shen M X, et al. Cellulose nanofibers/reduced graphene oxide /polypyrrole aerogel electrodes for high-capacitance flexible all-solid-state supercapacitors[J]. ACS Sustain Chem. Eng., 2019, 7(13): 11175-11185.

[15]Wan C C, Jiao Y, Liang D X, et al. A geologic architecture system-inspired micro-/nano- heterostructure design for high-performance energy storage[J]. Adv. Energy Mater, 2018, 8(33): 1802388.

[16]Xu Q, Chen L F. Characterization of cellulose film prepared from zinc-cellulose complexes[J]. Biomass and Bioenergy, 1994, 6(5): 415-417.

[17]OH Y S, KIM W C, JUNG J C. Method for preparing cellulose-based film and cellulose-based film: 8883056[P]. 2014-11-11.

[18]Zhao D W, Chen C J, Zhang Q, et al. High performance, flexible, solid-state supercapacitors based on a renewable and biodegradable mesoporous cellulose membrane[J]. Adv. Energy Mater, 2017, 7(18): 1700739.

[19]吴宇平. 锂离子电池: 应用与实践[M]. 北京: 化学工业出版社, 2004.

[20]陈立泉. 锂离子电池正极材料的研究进展[J]. 电池, 2002, 32(S1): 6-8.

[21]Bao L H, LI X D. Towards textile energy storage from cotton T-shirts[J]. Adv. Mater, 2012, 24(24): 3246-3252.

[22]Cao D X, Xing Y J, TANT R ATIAN K, et al. 3D printed high-performance lithium metal microbatteries enabled by nanocellulose[J]. Adv. Mater, 2019, 31(14): 1807313.

[23]Zhang Y, Luo W, Wang C W, et al. High-capacity, low-tortuosity, and channel-guided lithium metal anode[J]. P. Natl. Acad. Sci. USA, 2017, 114(14): 3584-3589.

[24]Yang C X, Gao Q M, Tian W Q, et al. Superlow load of nanosized MnO on a porous carbon matrix from wood fibre with superior lithium ion storage performance[J]. J. Mater Chem. A, 2014, 2(47): 19975-19982.

[25]Chen C J, Zhang Y, Li Y J, et al. Highly conductive, lightweight, low-tortuosity carbon frameworks as ultrathick 3D current collectors[J]. Adv. Energy Mater, 2017, 7(17): 1 700595.

[26]Wang Z H, Pang J, Ruan C Q, et al. Redox-active separators for lithium-ion batteries[J]. Adv. Sci., 2018, 5(3): 1700663.

[27]吴玉琪, 吕功煊, 李树本. CoOx 改性 TiO_2 光催化剂的制备、优化及其光催化分解水析氢性能研究[J]. 无机化学学报, 2005, 21(3): 309-314.

[28]乔乔. 碳基非铂复合催化剂的电催化析氢性能的研究[D]. 秦皇岛: 燕山大学, 2015.

[29]Lai F L, Yong D Y, Ning X L, et al. Bionanofiber assisted decoration of few-layered MoSe2 nanosheets on 3D conductive networks for efficient hydrogen evolution[J]. Small, 2017, 13(7): 1602866.

[30]Mulyadi A, Zhang Z, Dutze R M, et al. Facile approach for synthesis of doped carbon electrocatalyst from cellulose nanofibrils toward high-p. erformance metal-free oxygen reduction and hydrogen evolution[J]. Nano Energy, 2017, 32: 336-346.

[31]Wakerley D W, Kuehnel M F, OrchardD K L, et al. Solar-driven reforming of lignocellulose to H2 with a CdS / CdO_x photocatalyst[J]. Nature Energy, 2017, 2(4): 17021.

[32]刘佳琦. 钙钛矿薄膜的制备及性能表征[D]. 北京: 北京交通大学, 2015.

[33]GONZLEZ-PEDRO V, Xu X Q, MORA-SER I, et al. Modeling high-efficiency quantum dot sensitized solar cells [J]. ACS Nano, 2010, 4(10): 5783-5790.

[34]杨林, 左智翔, 于凤琴, 等. 钙钛矿太阳能电池的研究进展[J]. 化工技术与开发, 2015(9): 40-45.

[35] Briscoe J, Marinovic A, Sevilla M, et al. Biomass-derived carbon quantum dot sensitizers for solid-state nanostructured solar cells[J]. Angew. Chem. Int. Ed., 2015, 54(15): 4463-4468.

[36]Zhou Y H, FUENTES-HERNANDEZ C, KHAN T M, et al. Recyclable organic solar cells on cellulose nanocrystal substrates[J]. Sci Rep-uk, 2013, 3: 1536.

[37]Khanmirzaei M H, Ramesh S, Ramesh K. Hydroxypropyl cellulose based non-volatile gel polymer electrolytes for dye-sensitized solar cell applications using 1-methyl-3-propylimidazolium iodide ionic liquid[J]. Sci. Rep-uk., 2015, 5: 18056.

英文题目: Novel Wood-based Energy Storage and Conversion Materials

作者: Li Jian, Jiao Yue

摘要: Wood is an inexhaustible resource with many advantages, such as high abundance, renewability, biodegradability, and carbon balance. Anisotropic wood can serve as a substrate to facilitate the deposition and growth of micro /nano functional units for the development of anisotropic energy storage and conversion materials. Moreover, wood can serve as a raw material for preparing novel products with large surface areas and high reaction activities, such as 1D nanofibrils, 2D nanofilms, and 3D gels, through disassembly and reassembly. These reassembled materials can also act as substrates that can be combined with micro /nano functional units for the development of green novel energy storage and conversion materials. In this paper, the recent research on wood-based energy storage and conversion materials, including supercapacitors, lithium-ion batteries, water splitting, and solar cells, is summarized, and the direction of future development is proposed by providing new ideas for the applications of this high-value resource.

关 键 词: Wood resources; Renewable energy; Energy storage and conversion; Supercapacitors; Lithium-ion batteries; Water splitting

A Holocellulose Framework with Anisotropic Microchannels for Directional Assembly of Copper Sulphide Nanoparticles for Multifunctional Applications*

Caichao Wan, Yue Jiao, Wenyan Tian, Luyu Zhang, Yiqiang Wu, Jian Li, Xianjun Li

Abstract: A holocellulose framework (HCFW) with anisotropic microchannels, which is synthesized via a facile top-down strategy, is demonstrated to be an ideal anisotropic substrate to guide the directional assembly of CuS nanoparticles (CuS NPs) along the walls of microchannels. The oxygen-containing groups on the surface of HCFW and its abundant low tortuosity of channels contribute to the high loading (ca. 22.7 wt. %) and homogeneous dispersion of CuS NPs as well as the fast outdiffusion of cytotoxic substances [such as Cu(Ⅰ)/Cu(Ⅱ)], hence endowing the composite with an outstanding antibacterial activity which achieves approximatively 100% of growth inhibition ratios for *Escherichia coli* and *Staphylococcus aureus* and extremely low minimal inhibitory concentrations of 8 μg · mL^{-1} for these both bacteria. For drug delivery application, the release of doxorubicinhydrochloride (DOX) from CuS NPs@HCFW is well regulated by solution pH. A maximum cumulative release capacity of DOX (78.3%) is achieved at pH of 2.2. The cancer therapy capability of DOX-loaded CuS NPs@HCFW is also validated *in vivo*. Moreover, CuS NPs@HCFW has a strong antibiotic removal ability (~100% complete removal of ciprofloxacin within 120 min) by virtue of adsorption and catalytic oxidation in the presence of potassium peroxymonosulfate. These inspiring results not only indicate the high potential of CuS NPs@HCFW for antibacterial, pH-responsive drug delivery and antibiotic removal applications, but also provide a new sight for the development of novel anisotropic composites by using anisotropic HCFW for more extensive uses.

Keywords: Anisotropic structure; Directional assembly; Copper sulphide; Antibacterial materials; Drug delivery; Antibiotics.

1 Introduction

Nowadays, the common pursuit for eco-friendly functional composites has stimulated a variety of studies developing green, cost-effective and easily available substrates for the integration with functional components. Some natural polymers (like cellulose and chitosan), which extensively exist in plants and/or animals, have attracted considerable attention of researchers because of their natural abundance, excellent biodegradability, high modifiability and low cost[1-3]. As a result, these biomacromolecules are usually transformed into different dimensions of supramolecular assemblies (such as fibers[4,5], membranes[6,7],

---

* 本文摘自 Chemical Engineering Journal, 2020, 393: 124637.

paper[8-10], hydrogels[11,12] and aerogels[13,14]) by virtue of a combined top-down and down-top pathway, namely firstly disassembling nano-scaled biomacromolecules from biomass feedstocks and then reassembling them into macroscopical materials. Taking 3D cellulose gels as examples, traditionally, cellulose gels are prepared by two steps including(1)mechanical or chemical disintegration of cellulose nanofibers from biomass resources and(2)assembly of 3D cross-linked nano-network using nanofibers as building units[3]. However, it is noted that the mechanical or chemical disintegration treatments generally require a large quantity of energy and/or chemicals and also cause pollution. More seriously, the succedent self-assembly is slow and random, eventually causing an isotropic structure. Also, it still remains challenging for the construction of an anisotropic structure dependent on the down-top approach.

Compared to the down-top process, the top-down strategy(i. e., directly transforming existing natural materials into anisotropic substrates by physicochemical modifications) is obviously faster and greener. Wood, naturally grown anisotropic material of complex hierarchical cellular structure, has always been considered as an important renewable substrate in the process of preparing nanocomposites[15]. In recent years, wood-based nanotechnologies have achieved notable progress[16]. The abundant vertically aligned channels have the ability to serve as microreactors to guide the directional growth of nanomaterials. Taking advantage of the unique structure that naturallyoccurred in wood, wood-based advanced materials have been widely applied in many fields. For instance, in the field of transparent materials, Zhu et al. [17] report anisotropic transparent wood by removing thelight-absorbing components in basswood, followed by infiltration with refractive index matchingepoxy resin. In the field of supercapacitors, Chen et al. [18] demonstrate that electrodepositing MnO_2 onto the open channel walls of wood carbon along the growth direction can acquire ultra-high electrochemical activity due to the high loading, high electrical conductivity and low tortuosity. Similarly, high porosity, low tortuosity, and disordered amorphous feature of carbonized wood has been verified as lithium metal host in lithium-ion battery[19,20]. In the area of water treatment, through decorating 3D wood membrane with high-density palladium (Pd) particles, the long Pd-decorated channels facilitate bulk treatment as water flows through the entire mesoporous wood membrane[21]. Moreover, wood-based nanotechnologies have also been applied in the other important areas such as solar cells[22], super-strongmaterials[23] and high-efficiency solar steam toward water desalination[24]. In our previous works, we also combine nano-scaled conductive polymers and MnO_2 with wood for electrochemical application[25,26]. It is known that the main components of wood cell wall are cellulose, hemicellulose and lignin as well as a quite small quantity of extractives. Previous studies indicate that the delignification can significantly enhance the porosity of wood and promote the permeation of modifiers[27,28]. Consequently, a radially cutting delignified wood(also called as "holocellulose") framework with abundant microchannels has the ability to serve as a natural, anisotropic and easily available substrate to create novel anisotropic functional composites. Also, reports on the holocellulose framework (HCFW)-based composites are quite rare.

In this study, we demonstrate that HCFW can be used as a novel and ideal anisotropic substrate to guide the assembly of copper sulphide nanoparticles(labeled as CuS NPs) along its abundant microchannels. The numerous oxygen-containing functional groups on the surface of HCFW can effectively immobilize CuS NPs, which is beneficial to acquire high loading(~22.7 wt.%) and good distribution of CuS NPs as well as strong interfacial adhesion(crucial for the nano-size effects). More importantly, with the help of the hierarchical anisotropic channel structure with low tortuosity (guiding the rapid outdiffusion of Cu(Ⅰ)/Cu(Ⅱ) ions and enrichment effects) and the nano-size effects of CuS NPs, an outstanding antibacterial property of CuS NPs@ HCFWagainst the gram-negative *Escherichia coli* (*E. coli*) and gram-positive *Staphylococcus aureus* (*S. aureus*) are demonstrated qualitatively and quantitatively. The cytotoxicity in the keratinocytes (HaCaTcells) is studied by the 3-(4, 5-Dimethylthiazol-2-yl)-2, 5-diphenyltetrazolium bromide (MTT) assay and the inverted fluorescence microscope. For the application of controlled drug delivery, the effect of pH on the release of doxorubicinhydrochloride (DOX) from CuS NPs@ HCFW is analyzed. The cancer therapy capability of DOX-loaded CuS NPs@ HCFW is also examined *in vivo*. In addition, for another application (i. e., removal of antibiotics from aquatic environment), ciprofloxacin is selected as the model of antibiotics, and the influences of adsorption and catalytic oxidation functions of CuS NPs@ HCFW on the ciprofloxacin removal are studied in the presence of potassium peroxymonosulfate (PMS).

2 Materials and methods

2.1 Materials

Basswood was purchased from Linwei Wood Industry (China). Contents of components (namely cellulose, hemicellulose and lignin) in the basswood sample were determined to be 53.8%, 24.3% and 18.6%, respectively, by virtue of a classical chemical titration method[29]. The averagedeviation of the titration is less than±0.5 wt.%. The detailed measured processes are available in SI. Before use, the wood was successively rinsed ultrasonically with ethyl alcohol and distilled water and then dried at 60 ℃ for 24 h. Chemicals including sodium hydroxide (NaOH), sodium sulfite (Na_2SO_3), hydrogen peroxide (H_2O_2), copper sulfate pentahydrate ($CuSO_4 \cdot 5H_2O$), sodium sulphide (Na_2S), doxorubicinhydrochloride (DOX), sodium phosphate dibasic anhydrous (Na_2HPO_4), sodium dihydrogen phosphate anhydrous (NaH_2PO_4), dimethyl sulfoxide (DMSO), MTT, calcein acetoxymethyl ester (AM) and propidium iodide (PI), PMS, ciprofloxacin, tert-butyl alcohol and absolute ethyl alcoholwere purchased from Shanghai Aladdin Industrial Corporation and used without further purification.

2.2 Synthesis of holocellulose framework (HCFW)

The synthesis of HCFW was conducted according to the following steps: (1) the clean and dried basswood was dipped in a mixed solution of NaOH ($2.5\ mol \cdot L^{-1}$) and Na_2SO_3 ($0.4\ mol \cdot L^{-1}$) at 100 ℃ for 5 h; (2) the resultant was washed with plenty of distilled water to remove residual chemicals and then added into a solution of H_2O_2 ($2.5\ mol \cdot L^{-1}$) at 100 ℃ for 9 h; (3) the resultant HCFW was repeatedly rinsed by distilled water for the removal of impurities.

2.3 Synthesis of CuS NPs@HCFW composite

The HCFW sample was dipped into a certain concentration of $CuSO_4$ solution (50 mL), and then the mixed system was gently shakenfor3h at room temperature. Subsequently, the system was subjected to an ultrasonic treatment for 60 min with an output power of 80 W. After that, the system was shaken again for 1 h to ensure that Cu^{2+} ions adequately accessed into the channels of HCFW. Thereafter, the Cu(Ⅱ)@ HCFW wastransferred to 50 mL solution of Na_2S ($0.1\ mol \cdot L^{-1}$). Monitoring the color change of HCFW until the color didn't change (from white to black). The as-prepared CuS NPs@ HCFW composite was repeatedly washed with distilled water to remove impurities and finally underwent a tert-butyl alcohol freeze-drying treatment to remove most liquids.

2.4 Characterizations

Scanning electron microscope(SEM) observation was performed with a Hitachi S4800 SEM equipped with an energy dispersive X-ray (EDX) detector for element analysis. Transmission electron microscope (TEM) and high-resolution TEM (HRTEM) observations and selected area electron diffraction (SAED) were performed with a FEI, Tecnai G2 F20 TEM. Fourier transform infrared spectra (FTIR) were recorded on a Nicolet Nexus 670 FTIR instrument in the range of 520-4000 cm^{-1} with a resolution of 4 cm^{-1}. X-ray photoelectron spectra (XPS) was recorded using a ThermoEscalab 250Xi XPS spectrometer. Deconvolution of the overlapping peaks was performed via a mixed Gaussian-Lorentzian fitting program (Origin 9.0, Originlab Corporation). Thermal stability was analyzed by a thermogravimetric (TG) analyzer (TA, Q600) with a heating rate of 10 ℃ · min^{-1} in a N_2 environment from room temperature to 700 ℃. It is noted that these analyses of SEM, TEM, XPS, FTIR and TG all aim at the CuS NPs @ HCFW prepared using 0.1 M $CuSO_4$ unless specifically indicated otherwise. UV-vis absorption spectra were recorded by aUV-vis spectrophotometer (UV-6000PC, Shanghai Metash Instruments Co., Ltd.).

2.5 Antibacterial activity studies

Antibacterial activity was studied against *E. coli* as the model Gram-negative bacteria and *S. aureus*as the model Gram-positive bacteria. Qualitative and quantitative tests on CuS NPs@ HCFW were performed by three methods (i.e., shake flask method, sticking membrane method, and determination of minimal inhibitory concentration (MIC)), respectively. Prior to these tests, all the samples were sterilized in an autoclave at 121 ℃ for 40 min and subsequently rinsed with sterile distilled water. Each test was performed in triplicate.

2.5.1 Shake flask method

For preparing bacteria suspensions, the bacteria were grown in Luria Broth (LB) growth solutions for 18 h at 37 ℃. A colony was lifted off with a platinum loop, placed in 30 mL of nutrient broth, and incubated with shaking for 18 h at 37 ℃. After washed twice with phosphate buffer saline (PBS, pH = 7.4), they were resuspended in PBS to yield 1.0×10^5-1.5×10^5 colony forming unit (CFU) mL^{-1}. By measuring the absorbance of cell suspensions, the concentration of bacterial cell can be estimated[30,31]. To evaluate the antibacterial property, around 10 mm × 10 mm × 2 mm of the sterilized sample was dipped into a falcon tube containing 5.0 mL of 1.0×10^{-3} M PBS culture solution with a cell concentration of 1.0×10^5-1.5×10^5 CFU mL^{-1}. The falcon tube was then shaken at 200 rpm on a shaking incubator at 25 ℃ for 24 h. After shaking vigorously to detach adhered cells from the sample surface, the suspension was diluted 10^5 times, and then 0.1 mL of the diluent was spread onto an agar plate. Viable microbial colonies were counted after incubating the plate for 18 h at 37 ℃.

2.5.2 Sticking membrane method

Bacteria suspension with a concentration of 1.0×10^5-1.5×10^5 CFU mL^{-1} was prepared according to the method in 2.5.1. Then, around 20 mm × 20 mm × 2 mm of the sterilized sample was placed in a sterilized plate, and 0.2–0.5 mL of bacteria suspension was subsequently dropped onto the sample surface. After that, the sample was uniformly covered with a thin and transparent sterilized membrane, completely removing bubbles for the sufficient contact between the sample and the bacteria suspension. Then the plate was incubated at 37 ℃ for 24 h at 90% relative humidity. After the incubation, the sample and the covering filmwere washed thrice by using 20 mL eluent per time; later, the eluent was diluted in a conical flask. The diluted eluent was vaccinated onto a culture plate, and the nutrient agar medium at 45 - 55 ℃ was then poured into the plate. After the solidification, they were put in an incubator at 37 ℃ for 24 h. The resulting product was used for colony count.

2.5.3 Determination of MIC

The MIC value represents the lowest concentration ($\mu g \cdot mL^{-1}$) of the antibacterial agent that prevents visible growth of microorganisms compared with the growth in the control plate. The determination of MIC followed the agar dilution way. Firstly, the sterilized sample was ground into powder and dispersed in sterile PBS. Different concentrations of suspensions were subsequently added at twofold concentrations (0.125 - 64$\mu g \cdot mL^{-1}$) to sterilized Mueller-Hinton agar (tempered to 50 ℃) prior to pouring plates. After the solidification, 10 μL (ca. 10^4 CFU) of each prepared culture was spotted (five spots per plate) on the agar surfaces. All test plates were incubated at 37 ℃ for 24 h. Inoculated agar plates without adding any samples were used as positive controls.

2.6 Cytotoxicity studies

Determination of cytotoxicity (represented in term of cell viability) was conducted by using the MTT assay in HaCaT cells and inverted fluorescence microscope. At first, HaCaT cells were seeded into a 96-well plate (5×10^3 cells per well) andincubated for 24 h at 37 ℃ under 5% CO_2. After that, Ha CaT cells were incubated with HCFW or CuS NPs@ HCFW for 24 h. On the other hand, as untreated controlsamples, some wells that only contained the cell medium were also prepared. After washing with PBS and treating with MTT (200 μL, 0.5 $mg \cdot mL^{-1}$) for 4 h, 150 μL DMSO were added into the cells and theabsorbance of the mixture in each well was recorded by using a multimodemicroplate reader. Three replicates were prepared for eachtreatment group. In addition, calcein AM and PI were used to stainlive and dead cells, then the luminescence images were acquired.

2.7 DOX loading and release studies

2.7.1 DOX loading

DOX loading was carried out by immersing a piece of CuS NPs@ HCFW (30 mm × 30 mm × 2 mm, ~310 mg) in 20 mL of PBS (10 mM, pH = 7.4) containing 20 mg of DOX. The mixture was gently stirred at room temperature for36 h in the dark to reach the equilibrium state. Theamount of loaded DOX forCuS NPs@ HCFW was determined by a UV-Visspectrophotometer at 490 nm. The DOX loading capacity was determined by the following equation:

$$DOX \text{ loading capacity } (mg \cdot mg^{-1}) = \frac{W_{\text{Initial DOX}} - W_{\text{DOX after loading}}}{W_{\text{CuS NPs@ HCFW}}} \quad (1)$$

where $W_{\text{Initial DOX}}$ and $W_{\text{DOX after loading}}$ are the amount of DOX in the initial solution andthe solution after loading, respectively. $W_{\text{CuS NPs@ HCFW}}$ is the weight of CuS NPs@ HCFW.

2.7.2 DOX release

DOX release studies of DOX-loaded CuS NPs@ HCFWwere conducted in different pH values (7.4, 4.5, and 2. 2) of PBS. The samples were resuspended in these PBS and then stirred at 37 ℃ for DOX release. At the selected time interval (1 hour), the supernatant was collected and determined by a UV-Vis spectrophotometer at 490 nm. After that, the collected supernatant was poured back to the release system. The results were presented in termof cumulative release capacity (%) as a function of time by using thefollowing equation:

$$\text{Cumulative release capacity (\%)} = \frac{W_t}{W_I} \tag{2}$$

where W_I is the amount of DOX loaded onto the CuS NPs@ HCFW composite and W_t is the amount of DOX released from the composite at time t.

2.8 In vivo therapy studies

For the establishment of tumor-bearing mouse models, the axillary fossa of mice was subcutaneously injected with H22 tumor cells (5×10^6 in 50 μL PBS). When the tumor volume grew to approximately 30-50 mm^3, the tumor-bearing mice were randomly divided into three groups and were injected with PBS, DOX ($1.0\ mg \cdot kg^{-1}$) or DOX-loaded CuS NPs@ HCFW ($2.0\ mg \cdot kg^{-1}$) through an intratumor injection individually. Besides, for the mice injected with DOX-loaded CuS NPs@ HCFW, an 808 nm laserwith a power density of $0.76\ W \cdot cm^{-2}$ was subsequentlyused to perpendicularly irradiate theinjected area for 5 min. For eachmouse, the intratumor injection was conducted once everyother day, while their body weight was monitoredevery day. All tests obey the Animal Management Rules of the Ministry of Healthof the People's Republic of China (Document no. 55, 2001).

2.9 Adsorption and catalytic oxidation of ciprofloxacin

For the adsorption and catalytic oxidation experiment of ciprofloxacin, a piece of CuS NPs@ HCFW or HCFW (10 mm × 10 mm × 2 mm) and 0.5 g of PMS were added into 50 mL ciprofloxacin solution with a concentration of 20 ppm under amagnetic stirring. After aspecified time, a small quantity of ciprofloxacinsupernatant was taken out and its concentration was determined by a UV-Vis spectrophotometer at 275 nm. A control group without adding CuS NPs@ HCFW or HCFW was set. In addition, for comparison, an individual adsorption test without adding PMS was also conducted, and the change of ciprofloxacin concentration was monitored.

3 Results and discussion

3.1 Schematic diagram for the preparation of CuS NPs@HCFWcomposite

Schematic diagram for the preparation of CuS NPs@ HCFW is illustrated in Fig. 1, where a facile rapid top-down strategy is conducted to fabricate an anisotropic green substrate (i. e., HCFW) with the help of delignification of natural basswood. Compared to traditional down-top method, this top-down fabrication eliminates the complicated and time-consuming steps of nanofiber isolation and reassembly. Moreover, previous researches have demonstrated that the lignin removal effectively increases the porosity and chemical accessibility of wood[27,28], which is beneficial for the access, fixation and reaction of guest substances. Because of the presence of abundant surface oxygen-containing functional groups on HCFW, the strong electro-static interactions (like ion-dipole interaction) occur between the electron-rich oxygen atoms of polar hydroxyl and carboxyl groups of HCFW and the electropositive copper cations, resulting in the immobilization of copper cations on HCFW. After introducing sulfur anions, plentiful CuS nanonuclei are quickly generated and grown at the original ionized selective sites of HCFW, resulting in a high-density loading and high dispersion of CuS NPs on HCFW. Photographs of the basswood and CuS NPs@HCFW samples are shown in Fig. S1, where the color of the basswood turned from brown to black after depositing CuS NPs.

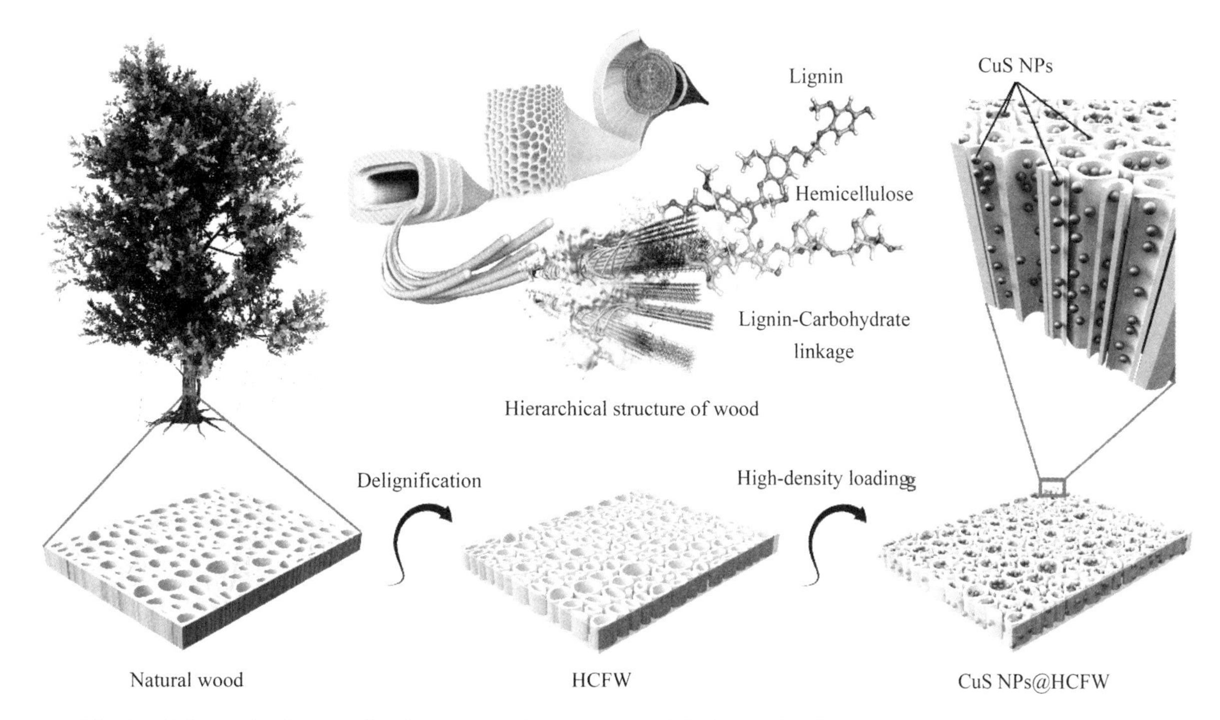

Fig. 1 Schematic diagram for the preparation strategy of CuS NPs@HCFW (the hierarchical structure of wood is reprinted with permission from Ref. [32])

3.2 Chemical compositions and thermal property

The FTIR spectra of the natural wood, HCFW and CuS NPs@HCFW are displayed in Fig. 2a. For the natural wood, the bands at 3338, 2925 and 2853 cm^{-1} are assigned to the stretching vibration of O—H, asymmetric stretching vibration of C—H and symmetric stretching vibration of C—H[33], respectively. The strong bands at 1737, 1593 (1505) and 1457 cm^{-1} are due to the stretching vibration of C=O, aromatic skeletal vibrations and bending vibration of C—H & aromatic skeletal vibrations[34], respectively. Other strong signals at 1421, 1369, 1236 and 1158 cm^{-1} originate from the scissoring/bending vibration of CH_2, bending vibration of C—H, stretching vibration of Ar—O and stretching vibration of C—O—C[35], respectively. The band at 897 cm^{-1} is assigned to the anomer C-groups, C_1—H deformation and ring valence vibration[36]. The attributions of these FTIR characteristic bands are summarized in Table 1. After the chemical treatment (i. e., the delignification), a significant decline occurs in the intensity of lignin characteristic bands, such as 1737, 1593 and 1236 cm^{-1} corresponding to the C=O stretching vibration, aromatic skeletal vibrations and stretching vibration of Ar—O, respectively. This result indicates the removal of a large portion of lignin and the formation of holocellulose after the delignification. Moreover, after the assembly of CuS NPs, a shoulder peak at 599 cm^{-1} appears in the FTIR spectrum of CuS NPs@ HCFW, corresponding to the Cu-S stretching modes[37].

Table 1 FTIR characteristic bands and their assignments of the natural wood, HCFW and CuS NPs@HCFW

Absorption band(cm^{-1})	Assignments
3338	Stretching vibration of O-H
2925	Asymmetric stretching vibration of C-H
2853	Symmetric stretching vibration of C-H
1737	Stretching vibration of C=O(hemicellulose, lignin)
1593	Aromatic skeletal vibrations(lignin)
1505	Aromatic skeletal vibrations(lignin)
1457	Bending vibration of C-H(lignin), aromatic skeletal vibrations(lignin)
1421	CH_2 scissoring(cellulose), bending vibration of CH_2(lignin)
1369	Bending vibration of C-H(cellulose, hemicellulose)
1236	Stretching vibration of Ar-O(lignin)
1158	Stretching vibration of C-O-C(cellulose, hemicellulose)
897	Anomer C-groups, C_1-H deformation, ring valence vibration
599	Cu-S stretching modes(CuS NPs)

The surface chemical compositions of CuS NPs@ HCFW are further confirmed by XPS analysis, and the results are given in Fig. 2b and c. The high-resolution XPS spectrum of Cu 2p in Fig. 2b shows two peaks at 932. 0 and 952. 1 eV which belong to Cu $2p_{3/2}$ and Cu $2p_{1/2}$, respectively. The two signals separated by 20. 1 eV are essentially identical binding energies for the Cu 2p orbital in accord with Cu(Ⅱ)in CuS[38]. Moreover, for the core-level XPS signals of S 2p(Fig. 2c), it is obvious that the presence of three fitting peaks at 161. 8, 163. 1 and 168. 8 eV are attributed to the S^{2-} state in CuS[39]. In addition, the fitting peaks at ~284. 5, 286. 2 and 288. 3 eV in the high-resolution XPS spectrum of C 1s(see Fig. S2 in SI) are related to the C—C/C—H, C—OH and O—C—O[40], respectively; besides, the peaks at ~531. 2 and 532. 6 eV in the high-resolution XPS spectrum of O 1s(Fig. S3) are assigned to the C—OH and C—OC, respectively. These C and O signals are primarily derived from the HCFW component. All of the aforesaid analyses demonstrate the successful growth of CuS NPs onto HCFW.

The thermal property of HCFW and CuS NPs @ HCFW is investigated by TG and derivative thermogravimetric(DTG) analyses. As seen in Fig. 2d, the residual weight at 700 ℃ is 3. 84 wt. % for HCFW and 17. 45 wt. % for CuS NPs@ HCFW, respectively. From the DTG curve of CuS NPs@ HCFW(see the inset of Fig. 2d), the small initial drops occurring before 150 ℃ is due to the evaporation of retained moisture; the peak centered at 264. 1 ℃ is ascribed to the decomposition reaction of CuS to Cu_2S(namely $2CuS = Cu_2S + S$); the peak centered at 321. 0 ℃ corresponds to the thermal degradation of holocellulose[3]; the shoulder peak at 355. 8 ℃ might be attributed to the reaction between the pyrolytic carbon and the generated elemental sulfur. Moreover, it is well-known that Cu_2S possesses high thermal stability with a high melting point of 1130 ℃. Therefore, the main components in the residual of CuS NPs@ HCFW at 700 ℃ may be carbon and Cu_2S. As a result, the content of CuS in CuS NPs@ HCFW can be roughly calculated as(17. 45%−3. 84%)/[(32. 065 + 63. 546)/(32. 065+ 63. 546×2)] = 22. 7 wt. %. Such a high value suggests the high-density loading of CuS NPs on the HCFW substrate.

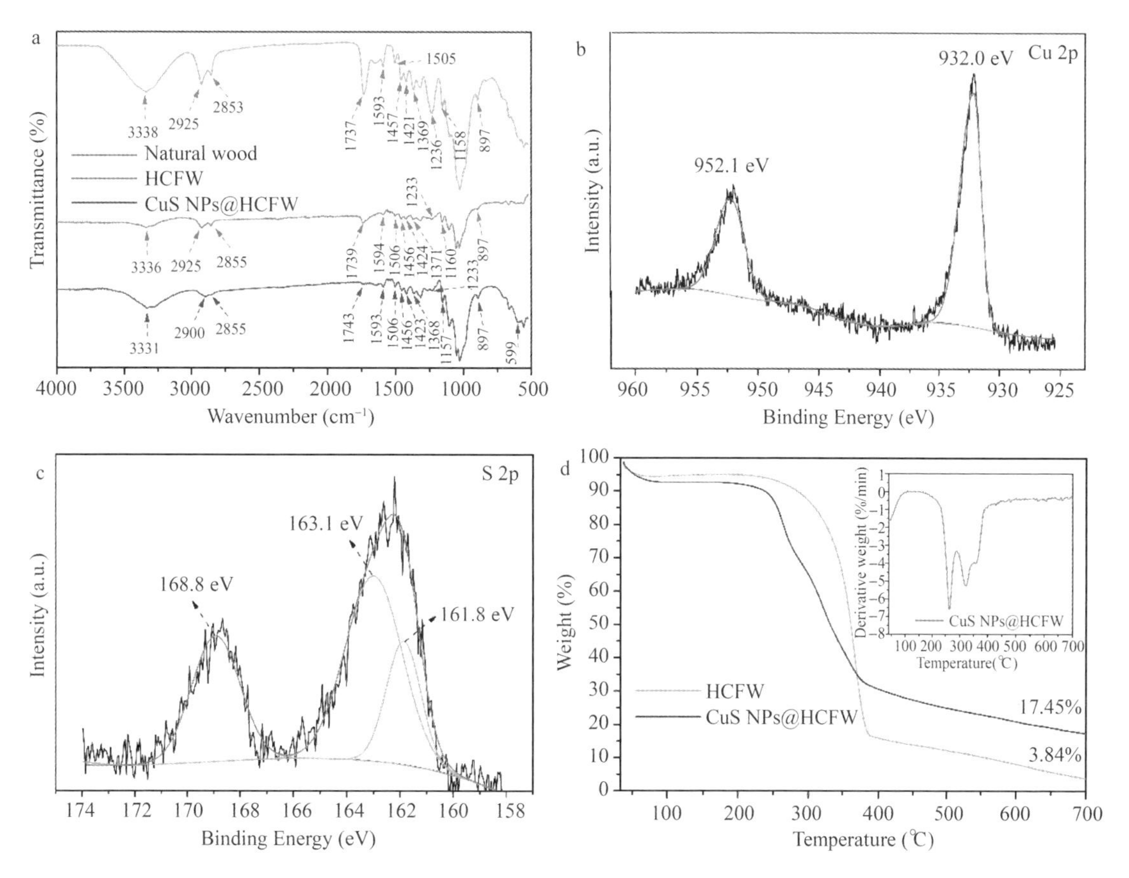

Fig. 2 FTIR, XPS and TG analyses. (a) FTIR spectra of the natural wood, HCFW and CuS NPs@HCFW; XPS spectra of CuS NPs@HCFW: the core-level XPS signals of(b) Cu 2p and(c) S 2p; (d) TG curves of HCFW and CuS NPs@HCFWand the inset shows the DTG curve of CuS NPs@HCFW

3. 3 Morphology observations and elemental analysis

Fig. 3a and c presents the macrophotographs of the basswood before and after the delignification treatment. For studying the effect of the treatment on the dimension, we selected large thickness of basswood samples[around 20 mm(length) × 20 mm(width) × 10 mm(thickness)]. As shown, the brown natural wood becomes totally white after the delignification, and the bulk density deceases from 0. 40 to 0. 26 g · cm^{-3} while the volume is almost unchanged, which indicates the high structural rigidity of basswood. For the micromorphology, the natural wood displays a three-dimensional(3D) porous structure with lumina of dozens of micrometers in diameter(Fig. 3b). The removal of lignin composition results in an evolution from the irregular hexagonal lumina to the sheet-like structure (as shown in Fig. 3d), while the anisotropic structure is retained. Besides, from the cross-sectional and longitudinal SEM images of CuS NPs@ HCFW(Fig. 3e-g), it is clear that the walls of HCFW are loaded with plentiful nanoparticles with sizes of 100 − 170 nm. The corresponding TEM image in Fig. 3i suggests that these secondary particles are composed of smaller primary particles(about 8-11 nm in diameter). Moreover, the element mapping images(Fig. 3h) further demonstrate that CuS NPs are homogeneously and densely loaded onto the walls of HCFW by the simple fast chemical precipitation method. The HRTEM image of CuS NPs@ HCFW suggests the existence of CuS (as shown in Fig. 3j), i. e., lattice fringes with spacing of ~ 0. 325 nm agreeing well with the (101) lattice spacing of CuS. Moreover, the corresponding SAED pattern shows uniform ring patterns typical of nanocrystalline materials, and is exactly consistent with the lattice spacings of CuS(see the inset in Fig. 3j).

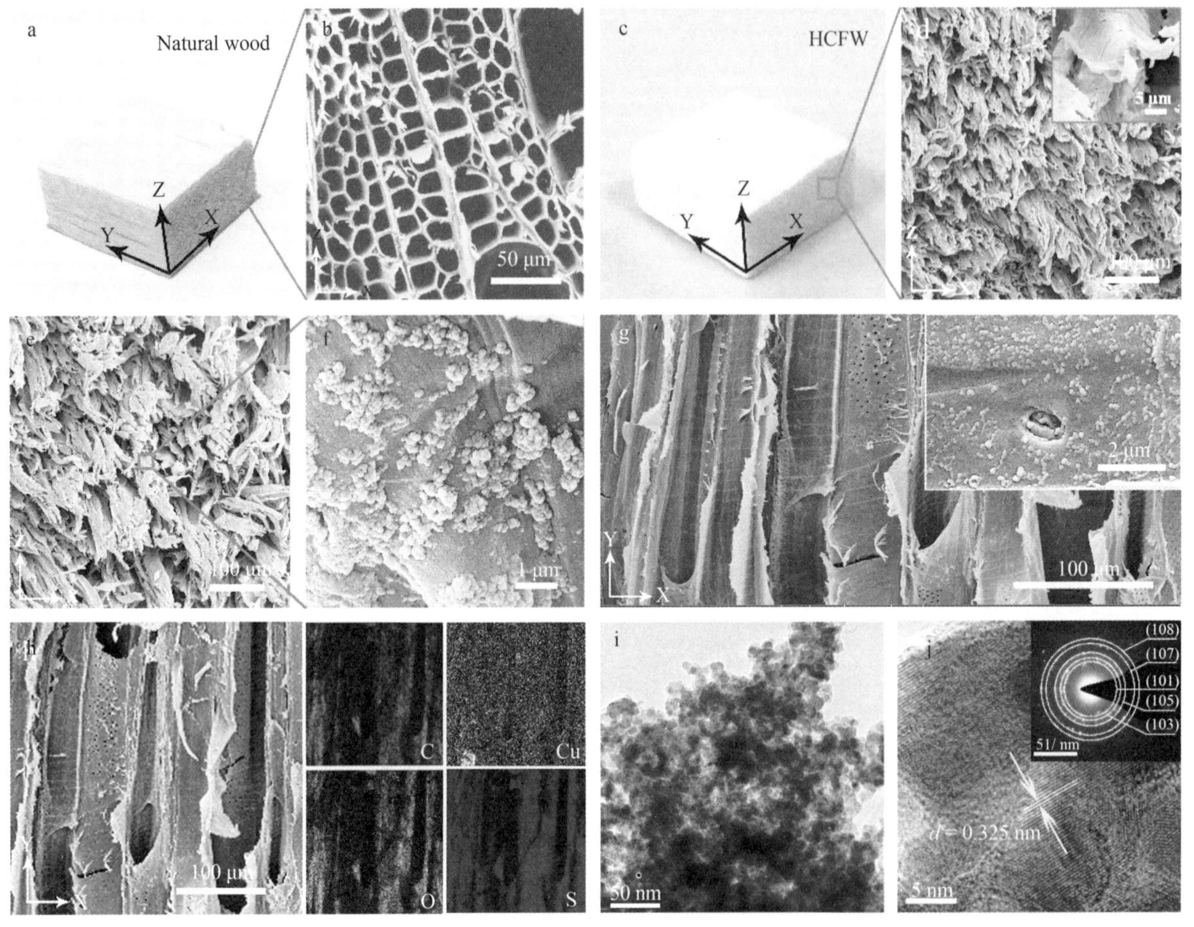

Fig. 3 Morphology observations and elemental analysis. (a) Digital and (b) cross-sectional SEM images of the natural wood(in the XZ plane); (c) digital and(d) cross-sectional SEM images of HCFW(in the XZ plane), and the inset in *d* is the enlarged image; (e) cross-sectional SEM image of CuS NPs@HCFW(in the XZ plane) and(f) is the enlarged image of the red area in *e*; (g) longitudinal SEM image of CuS NPs@HCFW(in the XY plane) and the inset is the enlarged image; (h) Element mapping images of C, O, Cu, S and their integration; (i) TEM and (j) HRTEM images of CuS NPs@HCFW, and the inset in *j* is the SAED pattern

3.4 Qualitative and quantitative analyses of antibacterial activity

Discovery of antibacterial activity of Cu can be traced back to ancient civilizations that used to employ different forms of Cu compounds to treat infections and to maintain hygiene. However, reports on the antibacterial property of Cu-containing nanomaterials are not abundant. In the present study, theantibacterial activityof CuS NPs@ HCFW is evaluated qualitatively and quantitatively via three pathways(i. e., shake flask method, sticking membrane methodand determination of MIC). *E. coli* and *S. aureus*are chosen as the models of Gram-negative and Gram-positive bacteria for these experiments. Our previous studies and some literatures had verified the negligible antibacterial activity of cellulose and hemicellulose[41-43]; hence, we do not perform the antibacterial tests of HCFW in the current study. For the shake flask method, bacterial inhibition ratio (%), which is calculated by dividing the surface concentrations of live *E. coli and S. aureus* after mixing with samples by those of blank control, is used to evaluate the antibacterial activity. Influences of the concentration of initial reactant $CuSO_4$ solution(including 0. 02, 0. 1 and 0. 5 M)on the antibacterial activity of CuS NPs@ HCFW are studied(the composite is labeled as CuS NPs@ HCFW-*X*, *X* means the concentration). As shown in Fig. 4a and b, thesurface concentrations of live *E. coli and S. aureus*after mixing with CuS NPs@ HCFW-0. 02(namely 2.3×10^{10} and 1.6×10^{10} CFU mL^{-1}) are close to those of blank control(ca. 2.6×10^{10} and 1.8×10^{10} CFU mL^{-1}). When the concentration of $CuSO_4$ solution increases to 0. 1 M, the acquired surface concentrations of live *E. coli* and *S. aureus*sharply decline to 2.1×10^{7} and 1.8×10^{8} CFU mL^{-1}, respectively, indicative of the more outstanding antibacterial property of CuS NPs @ HCFW-0. 1 with approximatively 100% (namely 99. 9% and 99. 0%) of inhibition ratios for *E. coli* and *S. aureus* (see Fig. 4c and d). However, when the concentration further increases to 0. 5 M, the inhibition ratios of *E. coli* and *S. aureus* significantly decrease to 84. 2% and 75. 6% due to the increased surface concentrations of live *E. coli* and *S. aureus* (ca. 4.1×10^{9} and 4.4×10^{9} CFU mL^{-1}), respectively, revealing the deterioration of antibacterial performance possibly owing to the drastic agglomeration of CuS NPs in the composite(Fig. S4). As a result, the comparisons indicate that 0. 1 M of $CuSO_4$ is the optimum concentration for the preparation of CuS NPs@ HCFW with high antibacterial activity. In addition, for the individual CuS NPs whose synthesis process excludes the HCFW substrate, the surface concentrations of live *E. coli and S. aureus* reach 5.4×10^{9} and 7.6×10^{9} CFU mL^{-1}, respectively. The corresponding inhibition ratios for *E. coli* and *S. aureus* are 79. 2% and 57. 8%(Fig. 4c and d), respectively, 20. 7% and 41. 6% lower than those of CuS NPs@ HCFW-0. 1. The lower inhibition ratios may also be related to the agglomeration of CuS NPs. Moreover, the result also demonstrates the significance of the HCFW substrate in the high antibacterial activity of CuS NPs @ HCFW-0. 1.

For potential application examples, CuS NPs@ HCFW is expected to be an eco-friendly and sticky antibacterial paste, which is easily used by sticking the antibacterial paste on the surface of protected products (as indicated by Fig. 5a). Based on this, we employ a sticking membrane method(a suitable method for the determination of antibacterial property of sheet-like materials) to further evaluate the antibacterial property of CuS NPs@ HCFW-0. 1. As shown in Fig. 5b-d, the results verify the excellent antibacterial property of CuS NPs@ HCFW-0. 1 with high inhibition ratios of approximatively 100%(ca. 99. 8% and 98. 5%)for both *E. coli* and *S. aureus*. Therefore, the tests suggest that CuS NPs@ HCFW composite holds good potential as convenient antibacterial pastes.

Apart from the application as the membrane-like form, CuS NPs@ HCFW can also be used as a powder-like form, and the corresponding antibacterial property is suitable to be evaluated through the method of determination of MIC. MIC is defined as the lowest concentration of the antibacterial agent preventing visible growth of microorganisms under defined conditions. As seen in Table 2, the MIC values of CuS NPs@ HCFW-0. 1 is 8 $\mu g\cdot mL^{-1}$ for *E. coli* and 8 $\mu g\cdot mL^{-1}$ for *S. aureus*, respectively. Visible growth would be observed

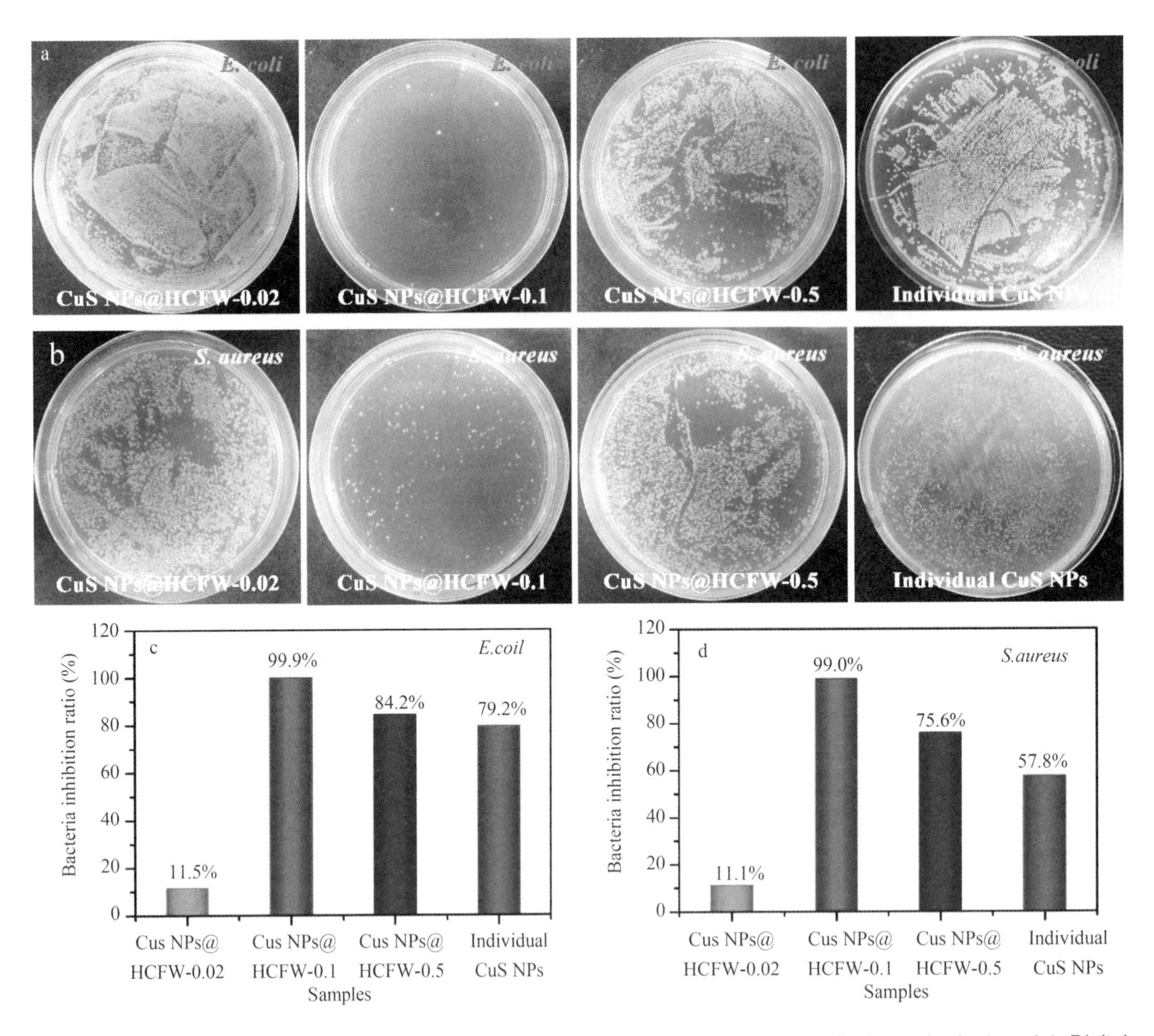

Fig. 4 Antibacterial activity towards *E. coli* and *S. aureus*determined byshake flask method. (a, b) Digital photographs presenting growth inhibition effects of individual CuS NPs and CuS NPs @ HCFW prepared using different concentrations of $CuSO_4$ solution(0.02, 0.1 and 0.5 M); (c, d)corresponding bacteria inhibition ratios

when the lower concentrations are used, suggesting no inhibitory effects on the growth of microorganisms under these circumstances. Such values are obviously lower than those of many recently reported cellulose/wood-based antibacterial agents, for instance, hydroxyethyl cellulose-grafted poly(ionic liquid)(10 and 19 μg · mL^{-1})[44], ZnO/bacterial cellulose(35 μg · mL^{-1})[45], cellulose nanofibers-titania-phosphomycin(>100 and 10 μg · mL^{-1})[46], polyrhodanine-coated cellulose nanocrystals(500-1000 μg · mL^{-1})[47], cellulose/γ-Fe_2O_3/Ag(1024 and 512 μg · mL^{-1})[48], lysozyme-conjugated nanocellulose (1000 and 1000 μg · mL^{-1})[49], allicin-conjugated nanocellulose(1000 and 1000 μg · mL^{-1})[49], *Azadirachta indica*(2000 and 1000 μg · mL^{-1})[50] and *Cinnamomum cassia*(2640 and 1320 μg · mL^{-1})[51], as listed in Table 3. The comparisons further verify the strong antibacterial activity of CuS NPs@ HCFW.

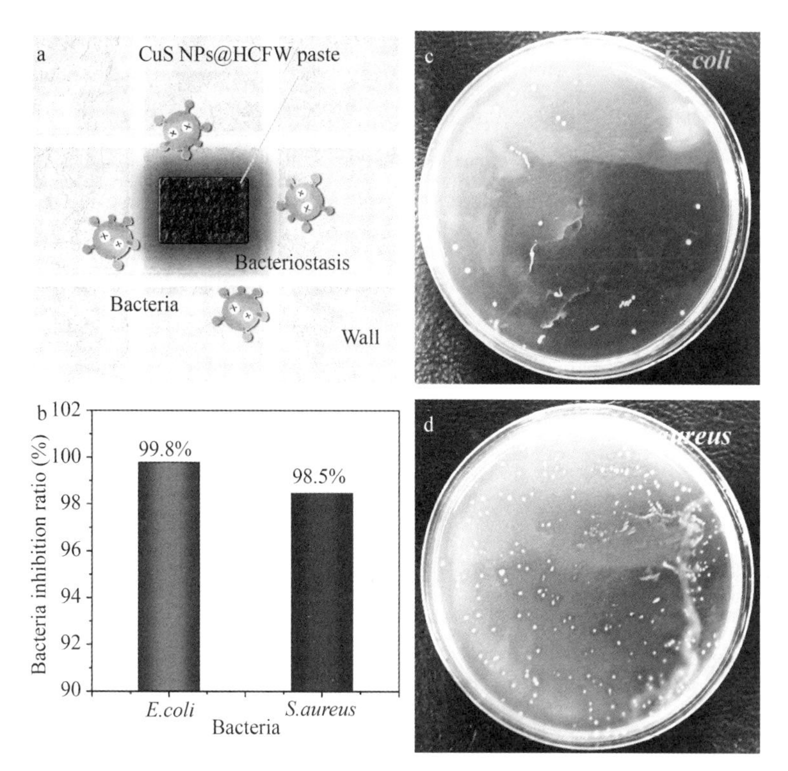

Fig. 5 Antibacterial property of CuS NPs @ HCFW-0. 1 towards *E. coli* and *S. aureus* determined by sticking membrane method. (a) Schematic diagram for the application of antibacterial paste; (b) inhibition ratios; (c, d) digital photographs presenting growth inhibition effects

Table 2 Determination of MIC of CuS NPs@HCFW-0. 1 for the studied strains

Concentration(μg · mL^{-1})	0. 125	0. 25	0. 5	1	2	4	8	16	32	64
E. coli	+	+	+	+	+	+	-	-	-	-
S. aureus	+	+	+	+	+	+	-	-	-	-
+ visible growth										
- invisible growth										

Table 3 Comparison of MIC(μg · mL^{-1}) of CuS NPs@HCFW-0. 1 with other recently reported cellulose/wood-based antibacterial agents

Antibacterial agents	*E. coli*	*S. aureus*	Ref.
CuS NPs@ HCFW-0. 1	8	8	This work
Hydroxyethyl cellulose-grafted poly(ionic liquid)	10	19	[44]
ZnO/bacterial cellulose	35	-	[45]
Cellulose nanofibers-titania-phosphomycin	>100	10	[46]
Polyrhodanine-coated cellulose nanocrystals	500-1000	-	[47]
Cellulose/γ-Fe_2O_3/Ag	1024	512	[48]
Lysozyme-conjugated nanocellulose	1000	1000	[49]
Allicin-conjugated nanocellulose	1000	1000	[49]
Azadirachta indica	2000	1000	[50]
Cinnamomum cassia	2640	1320	[51]

Regarding the possible antibacterial mechanism of CuS NPs@ HCFW, copper, a redox active transition metal, can exist in two redox states(i. e., oxidized cupric and reduced cuprous species). A long-standing hypothesis, that copper reacts with endogenous H_2O_2 of bacteria to generate hydroxyl radicals ina process analogous to the Fenton reactions, has been verified experimentally[52,53]. As a result, the bacteriostatic action of CuS NPs may primarily originate from the direct interaction of Cu(Ⅱ)species with the cell components, leading to the reduction of Cu(Ⅱ)in CuS NPs to Cu(Ⅰ)according to the Fenton reactions(the latter is much more toxic to *E. coli* and *S. aureus* than the previous one[54]). Ascorbic acid, glutathione and other amino acids from intracellular proteins of bacteria are ready to chelate Cu(Ⅰ) ions so that the proteins fail to associate with DNA, which is primarily responsible for the inactivation of *E. coli* and *S. aureus*. In addition, the significantquantity of reactive oxygen species(like OH ·)must be rendereddirectly from the surface defect sites in nanocrystalline CuS[55]. However, the agglomeration of CuS NPs(like in the case of CuS NPs@ HCFW-0. 5, see Fig. S4)would reduce the sites and thus declines the amount of reactive oxygen species and Cu(Ⅰ), leading to the deterioration of antibacterial activity.

The expected functions of HCFW in the process of bacterial growth inhibition can be primarily summarized into three points(as illustrated in Fig. 6):

(1)the abundant oxygen-containing functional groups on the surface of the anisotropic HCFW ensure the high loading(ca. 22. 7 wt. % as indicated by the TG analysis in Fig. 2d)and good distribution of CuS NPs within a certain concentration range(as indicated by SEM observations in Fig. 3f-i). These roles are helpful to acquire high nano-size effects and plentiful reactive oxygen species and Cu(Ⅰ)(because of the Fenton reactions between Cu(Ⅱ)species and cell components), contributing to the superior antibacterial activity;

(2)HCFW is originated from natural wood and inherits its unique anisotropic feature with many open channels. More importantly, these microchannels with low tortuosity would accelerate the ion transfer process, which has been demonstrated as a positive contribution for performance improvement[17,56,57]. The cytotoxic Cu(Ⅰ)and O_2^- · or hydroxyl radicals(OH ·), which are generated by the Fenton reactions between Cu(Ⅱ) species and the cell components, are outdiffused along the abundant channels of HCFW to the surrounding area of CuS NPs@ HCFW for wider areas of bacteriostasis. Hence, the low tortuosity of channel structure of HCFW is expected to contribute to the faster and more efficient diffusion of cytotoxic substances, helpful to enhance the antibacterial activity of CuS NPs@ HCFW;

(3)the abundant micrometer-scale channels from the anisotropic feature of HCFW have enrichment effects because of the capillary action. Therefore, the increase of bacterial concentration in the vicinity of CuS NPs@ HCFW due to the enrichment effects would be useful for the more efficient bacteriostasis.

3.5 Cytotoxicity, pH-responsive drug release and in vivo anti-tumor effects

When CuS NPs@ HCFW is used as antibacterial paste or other product for external use, it is easy for human skin to touch it. Thus, it is meaningful to study the cytotoxicity of CuS NPs@ HCFW. Human skin consists of epidermis anddermis, with the epidermis constituting the exterior layer andprimarily comprised of keratinocytes. As a result, monolayer cultures of HaCaTcells(a kind of typical keratinocytes)are chosen as the reconstructed epidermis model, for evaluatingcytotoxicity. HaCaTcells stained with calcein-AM and PI were incubated with CuS NPs@ HCFWor HCFW. This staining method is fluorescence based for live(green)and/or dead cells(red)detection with two probes, indicating cellular activities andplasma membrane integrity. As seen

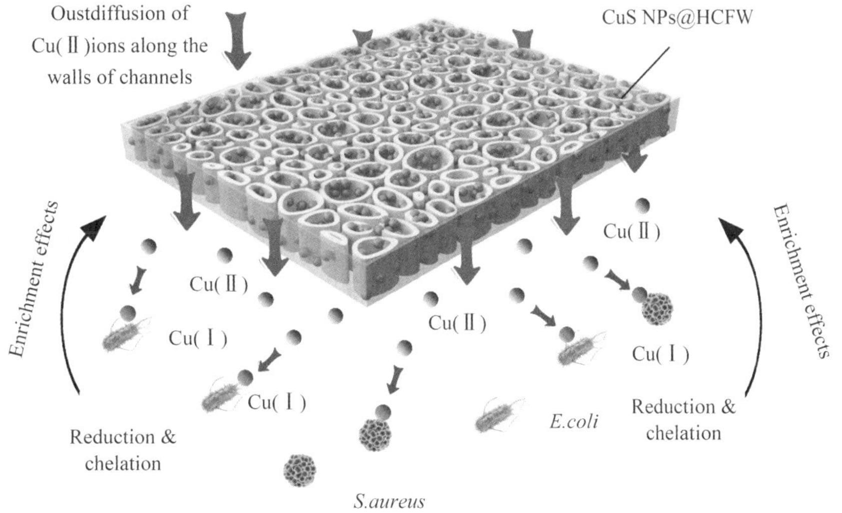

Fig. 6 Schematic diagram for the possible antibacterial mechanism of CuS NPs@HCFW

in Fig. 7a-c, CuS NPs@ HCFW displays a distinctly stronger cytotoxicity than that of HCFW since the proportion of red dead cells in the case of CuS NPs@ HCFW is far higher than that in the case of HCFW. The result is well consistent with the quantitative MTTassay result(Fig. 7d), reflecting the smaller cell viability of 40.9% for CuS NPs@ HCFW than that of HCFW(76.6%). Therefore, it should try to avoid direct skin-contact in the course of using CuS NPs@ HCFW.

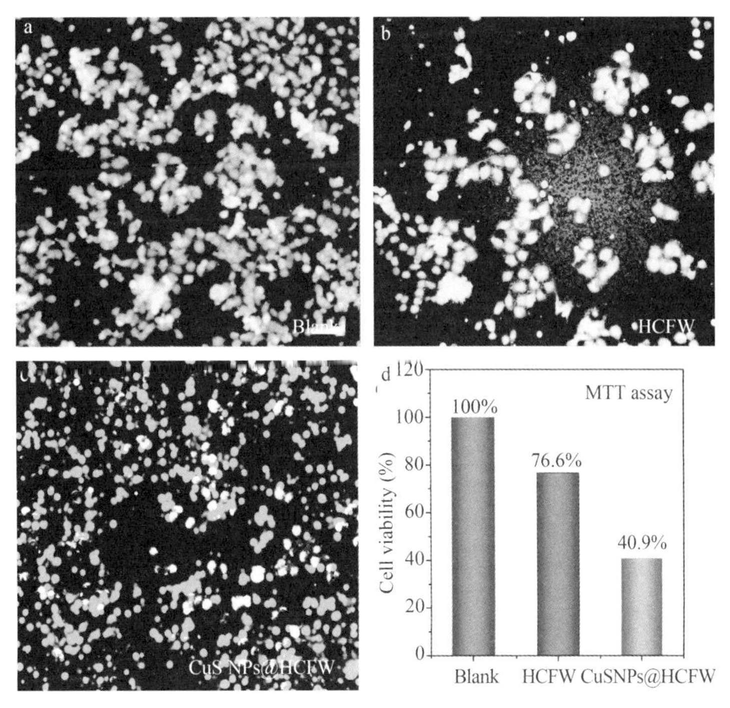

Fig. 7 Cytotoxicity of CuS NPs@HCFW. (a-c) Calcein-AM and PI double staining assays ofHaCaT cells without any treatment(a), incubated with HCFW(b), and incubated with CuS NPs@HCFW(c); (d) viability of HaCaT cells incubated with HCFW or CuS NPs@HCFW

For the pH-responsive drug release application of CuS NPs@ HCFW, DOX, a typical and common chemotherapeutic drug for cancer treatment, is selected as a model drug. UV-vis absorption spectrum of DOX presents the characteristic peaks of DOX at 232, 254, 292, and 490 nm (Fig. S5)[58]. The DOX release behavior of CuS NPs@ HCFWat different pH(2.2, 4.5 and 7.4) is investigated, and the curve of cumulative release capacity verses time for DOX-loaded CuS NPs @ HCFW is presented in Fig. 8. As shown, the cumulative release capacity of DOX at pH of 7.4 is very small (only 2.3%) after 16 h since theinteraction between positively charged DOX and negatively charged CuS is quite strong at this pH value according to the research by Wang et al.[59]. The electrostatic interaction is a primary factor forthe integration between DOX and CuS. As the pH value decreases from 7.4 to 4.5 and 2.2, the CuS NPs on the surface of CuS NPs@ HCFWare protonated, causing weakening of the electrostatic interaction. As a result, the cumulative release capacity of DOX increases to 31.0% at pH of 4.5 and further increases to 78.3% at pH of 2.2. As a result, the release of DOX from CuS NPs@ HCFW can be well regulated by solution pH. Most importantly, the pH-dependent stability not only avoids undesired release during the transportation of DOX inblood circulation, but can increases the effective release of DOX within tumor cells since most tumor tissues have lower extracellularpH values than that of normaltissues and bloodstream.

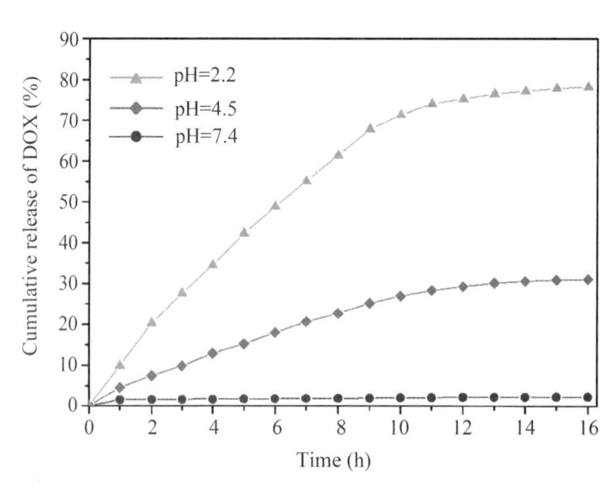

Fig. 8 In vitro release profiles of DOX from CuS NPs@HCFW underdifferent pH conditions

To examine the ability of CuS NPs@ HCFW for practical biomedical application, an *in vivo* experiment on the therapeutic efficacy of DOX-loaded CuS NPs@ HCFW for H22 tumor-bearing mice was conducted. The H22 tumor-bearing mice(n = 18) were injected with PBS (control group), DOX or DOX-loaded CuS NPs@ HCFW via an intratumor injection individually once every other day. Moreover, the tumors of DOX-loaded CuS NPs@ HCFW group were also irradiated by near-infrared laser for 5 min (808 nm, 0.76 W · cm^{-2}), as illustrated in Fig. 9a. The body weight change of mice is an important indicator to evaluate the toxicity of injectants since the high toxicity always causes a remarkable weight loss. Fig. 9b presents weight changes of mice given PBS, DOX or DOX-loaded CuS NPs @ HCFW, respectively. In allgroups, the insignificant changes in body weightduring the 9-day observation period indicate the lowsystemic toxicity. Apart from the body weight of mice, the tumor sizes grew indifferent groups also reflect the health condition ofmice. The tumor inhibitionratio was calculated by the equation:

$$\text{Inhibition ratio}(\%) = \frac{V_t - V_c}{V_c} \tag{3}$$

where V_c and V_t represent the average tumor volumefor the control group and treatment groups, respectively. For the mice treated by DOX or DOX-loaded CuS NPs@ HCFW, the corresponding tumor inhibition rates are 48.3% and 64.6% (Fig. 9c), respectively. The higher anti-tumoractivity of DOX-loaded CuS NPs@ HCFW can be ascribed to the synergistic effects between chemotherapy and photothermaltherapy (CuS) with the near-infrared laser irradiation. The results not only demonstrate that DOX-loaded CuS NPs@ HCFW is a powerful tool for the integrated chemotherapy and photothermal therapyof cancer *in vivo*, but also indicate that the use of DOX-loaded CuS NPs@ HCFW can reduce the drug dose (like DOX) with the help of near-infrared laser irradiation, which is beneficial to reduce the toxic sideeffect of drugs.

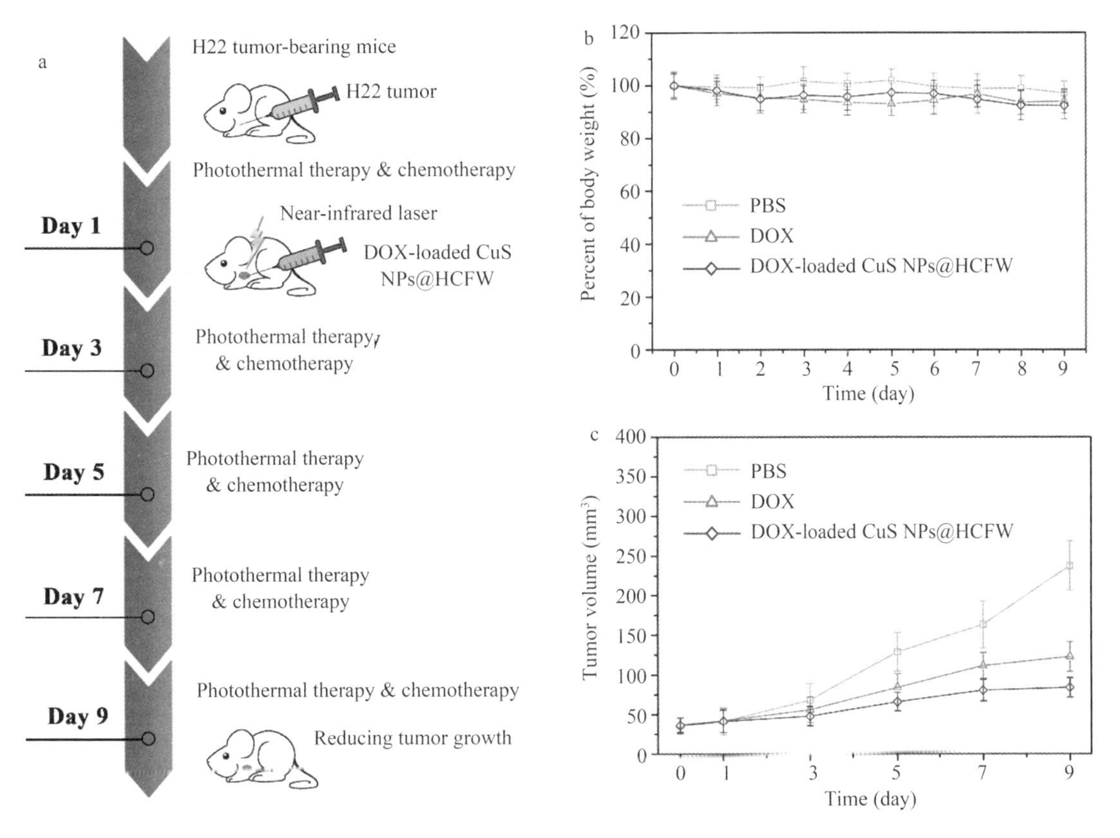

Fig. 9 *In vivo* anti-tumor effects of various formulations in the H22 tumor-bearing mouse models. (a) Experimental scheme to evaluate the anti-tumor effect of DOX-loaded CuS NPs@HCFW; (b) body weight changes and (c) tumorsize changes of tumor-bearing mice during 9 daystreatment of PBS, DOX or DOX-loaded CuS NPs@HCFW & near-infrared laser

3.6 Removal of ciprofloxacinby adsorption and catalytic oxidation

Another interesting application of CuS NPs @ HCFW is to remove ciprofloxacin (a kind of typical antibiotics) by adsorption and catalytic oxidation in the presence of PMS. In recent years, antibiotics have become one of the most important water contaminants. Antibiotics easily enterthe environment through urine and feces and are difficult to biodegrade, causing resistance in bacterialpopulations and deactivating antibiotics in the treatment of diseases[60]. Fig. 10 shows the removal ability of CuS NPs@ HCFW towards ciprofloxacin. Only by virtue of adsorption, HCFW delivers a ciprofloxacin removal proportion of 32% after 120 min(Fig. 10a). By contrast, CuS NPs@ HCFW achieves a remarkable enhancement to 54% of ciprofloxacin removal, which is mainly ascribed to the coordination interaction occurring between the donor N atoms of ciprofloxacin and Cu atoms of CuS NPs[61].

After introducing catalytic oxidation effect in the presence of PMS, the removal kinetics of ciprofloxacin is shown in Fig. 10b. An improved ciprofloxacin removal proportion of 73% is acquired for individual HCFW after 120 min. The production of ·OH withstrong oxidizing property, due to the primary alcohol oxidation reaction of cellulose by PMS[62], is possibly responsible for the improvement. In the case of CuS NPs @ HCFW, the ciprofloxacin in the aqueous solution is almost completely removed (~99.6%). Theabsorption kinetics data are fitted by the pseudo-second-order kinetic model, which is expressed as:

$$\frac{t}{Q_t}=\frac{t}{Q_e}+\frac{1}{K_2Q_e^2} \tag{4}$$

where Q_e and Q_t($mg \cdot g^{-1}$) are the removal amountat equilibrium and at time t, respectively, and K_2 is the absorption rateconstant($mg \cdot g^{-1} \cdot min^{-1}$). In Fig. 10c, it is clear that the experimental data fit the model well, with high correlation coefficient(R^2) values of 0.995–0.998. Furthermore, on the basis of the eq 3, the K_2 value of CuS NPs@ HCFW is 4.59 $mg \cdot g^{-1} \cdot min^{-1}$, which is 1.33 times higher than that of HCFW (1.97 $mg \cdot g^{-1} \cdot min^{-1}$). The result demonstrates the faster ciprofloxacin removal ability of CuS NPs@ HCFW. The possible catalytic oxidation mechanism is summarized as follows[63]:

$$\equiv Cu(II) + HSO_5^- \rightarrow \equiv Cu(III) + SO_4^{2-} + \cdot OH \tag{5}$$

$$\equiv Cu(II)-OH^- + HSO_5^- \rightarrow \equiv Cu(II)-(OH)OSO_3^- + OH^- \tag{6}$$

$$\equiv Cu(II)-(OH)OSO_3^- \rightarrow \equiv Cu(III)-OH^- + SO_4^- \cdot \tag{7}$$

$$\cdot OH + SO_4^- \cdot + ciprofloxacin \rightarrow Multiple\ steps \rightarrow CO_2 + H_2O + SO_4^{2-} \tag{8}$$

This catalytic oxidation mechanism of ciprofloxacin by using CuS NPs@ HCFW in thepresence of PMS is illustrated in Fig. 10d. The Cu(II) in CuS NPs@ HCFW reacts with HSO_5^- in PMS to form hydroxyl radicals (·OH) with a strong oxidizing property(eq 4). Moreover, the $\equiv Cu(II)-OH^-$, which is generated by the H_2O adsorption on CuS NPs, also reacts with HSO_5^-, leading to the generation of $\equiv Cu(II)-(OH)OSO_3^-$ and subsequent decomposition into sulfate radicals $SO_4^- \cdot$ (namely eq 5 and 6). The strong oxidizing property of ·OH and $SO_4^- \cdot$ play an important role in the degradation of ciprofloxacin(eq 7).

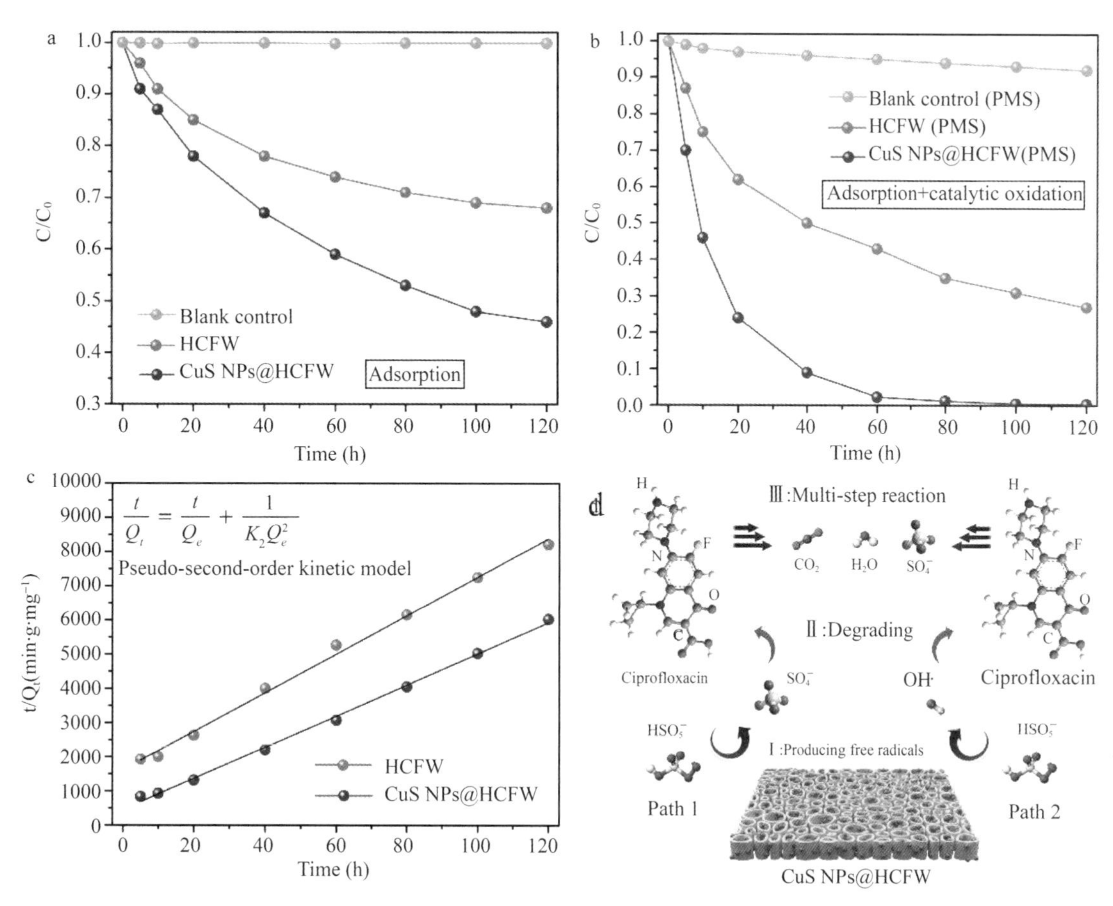

Fig. 10 Removal ability of CuS NPs@HCFW towards ciprofloxacin. (a, b) Removal kinetics of ciprofloxacin by adsorption(a) and adsorption & catalytic oxidation(b); (c) fitting theexperimental data in *b* to the pseudo-second-order kinetic model; (d) schematic diagram for the possible catalytic oxidation mechanism of ciprofloxacin by CuS NPs@HCFW in thepresence of PMS

4 Conclusions

In summary, we demonstrate that anisotropic HCFW with abundant microchannels can serve as an ideal substrate to guide the directional assembly of CuS NPs along the walls of microchannels by virtue of a facile and fast chemical precipitation method. The high loading(~22. 7 wt. %) and homogeneous dispersion of CuS NPs ensure the high nano-size effects. The low tortuosity of channel structure is beneficial for the outdiffusion of cytotoxic substances. As a result, CuS NPs@ HCFW possesses a superior antibacterial activity achieving approximatively 100% of inhibition ratios for *E. coli* and *S. aureus* and extremely low MIC values of 8 μg · mL^{-1} for both *E. coli* and *S. aureus*. The direct interaction of Cu(Ⅱ) species in CuS with the cell components of bacterial is possibly responsible for the strong bacterial inactivation. The potential enrichment function of anisotropic microchannelsis also expected to reinforce bacteriostasis. In addition, CuS NPs @ HCFW has cytotoxicity in the HaCaTkeratinocytes with a cell viability of 40. 9%. For the drug delivery application, the release of DOX from CuS NPs@ HCFW is well regulated by solution pH, and a high cumulative release capacity of DOX(78.3%) is achieved at pH of 2.2. We validate the cancer therapy capability of DOX-loaded CuS NPs@ HCFW *in vivo* with a high tumor inhibition rate of 64.6%. In addition, for the ciprofloxacin removal application, a strong and fast removal ability(~100% complete removal within 120 min) is achieved for CuS NPs@ HCFW, with the help of adsorption and catalytic oxidation in the presence of PMS. The results verify its high potential for antibacterial, pH-responsive drug delivery and antibiotic removal applications.

Acknowledgements

This study was supported by the National Natural Science Foundation of China(grant nos. 31901249, 31890771 and 31530009), the Scientific Research Foundation of Hunan Provincial Education Department (grant no. 18B180) and the Youth Scientific Research Foundation of Central South University of Forestry and Technology(grant no. QJ2018002A).

References

[1]Kim J H, Lee D, Lee Y H, et al. Nanocellulose for energy storage systems: beyond the limits of synthetic materials [J]. Advanced Materials, 2019, 31(20): 1804826.

[2] Ling S, Chen W, Fan Y, et al. Biopolymernanofibrils: Structure, modeling, preparation, and applications[J]. Progress in Polymer Science, 2018, 85: 1-56.

[3]Wan C, Jiao Y, Wei S, et al. Functionalnanocomposites from sustainable regenerated cellulose aerogels: A review[J]. Chemical Engineering Journal, 2019, 359: 459-475.

[4]Koga H, Nogi M, Isogai A. Ionic liquid mediated dispersion and support of functional molecules on cellulose fibers for stimuli-responsive chromic paper devices[J]. ACS Applied Materials & Interfaces, 2017, 9(46): 40914-40920.

[5]Wan C, Jiao Y, Li J. A cellulose fibers-supported hierarchical forest-like cuprous oxide/copper array architecture as a flexible and free-standing electrode for symmetricsupercapacitors[J]. Journal of Materials Chemistry A, 2017, 5(33): 17267-17278.

[6]Zheng H, Li W, Li W, et al. Uncovering the circular polarization potential of chiral photonic cellulose films for photonic applications[J]. Advanced Materials, 2018, 30(13): 1705948.

[7]Li M C, Wu Q, Song K, et al. Chitinnanofibers as reinforcing and antimicrobial agents in carboxymethyl cellulose films: Influence of partial deacetylation[J]. ACS Sustainable Chemistry & Engineering, 2016, 4(8): 4385-4395.

[8]Zhang Y, Li L, Zhang L, et al. In-situ synthesizedpolypyrrole-cellulose conductive networks for potential-tunable foldable power paper[J]. Nano Energy, 2017, 31: 174-182.

[9]Booshehri A Y, Wang R, Xu R. Simple method of deposition of CuO nanoparticles on a cellulose paper and its antibacterial activity[J]. Chemical Engineering Journal, 2015, 262: 999-1008.

[10]Wan C, Jiao Y, Li J. Multilayer core-shell structured composite paper electrode consisting of copper, cuprous oxide and graphite assembled on cellulose fibers for asymmetricsupercapacitors[J]. Journal of Power Sources, 2017, 361: 122-132.

[11]Xu M, Huang Q, Wang X, et al. Highly tough cellulose/graphene composite hydrogels prepared from ionic liquids [J]. Industrial Crops and Products, 2015, 70: 56-63.

[12]Wang Q, Guo J, Wang Y, et al. Creation of the tunable color light emission of cellulose hydrogels consisting of primary rare-earth compounds[J]. Carbohydrate Polymers, 2017, 161: 235-243.

[13]Lu Y, Liu H, Gao R, et al. Coherent-interface-assembled Ag_2O-anchored nanofibrillated cellulose porous aerogels for radioactive iodine capture[J]. ACS Applied Materials & Interfaces, 2016, 8(42): 29179-29185.

[14]Ren W, Gao J, Lei C, et al. Recyclable metal-organic framework/cellulose aerogels for activating peroxymonosulfate to degrade organic pollutants[J]. Chemical Engineering Journal, 2018, 349: 766-774.

[15]Lu LL, Lu Y Y, Xiao Z J, et al. Wood-inspired high-performance ultrathick bulk battery electrodes[J]. Advanced Materials, 2018, 30(20): 1706745.

[16]Jiang F, Li T, Li Y, et al. Wood-based nanotechnologies towardsustainability[J]. Advanced Materials, 2018, 30(1): 1703453.

[17]Zhu M, Song J, Li T, et al. Highly anisotropic, highly transparent woodcomposites[J]. Advanced Materials, 2016, 28(26): 5181-5187.

[18]Chen C, Zhang Y, Li Y, et al. All-wood, low tortuosity, aqueous, biodegradablesupercapacitors with ultra-high capacitance[J]. Energy & Environmental Science, 2017, 10(2): 538-545.

[19]Zhang Y, Luo W, Wang C, et al. High-capacity, low-tortuosity, and channel-guided lithium metal anode[J]. Proceedings of the National Academy of Sciences, 2017, 114(14): 3584-3589.

[20]Jin C, Sheng O, Lu Y, et al. Metal oxide nanoparticles induced step-edge nucleation of stable Li metal anode working under an ultrahigh current density of 15 mA cm^{-2}[J]. Nano Energy, 2018, 45: 203-209.

[21]Chen F, Gong A S, Zhu M, et al. Mesoporous, three-dimensional wood membrane decorated with nanoparticles for highly efficient water treatment[J]. Acs Nano, 2017, 11(4): 4275-4282.

[22]Jia C, Li T, Chen C, et al. Scalable, anisotropic transparent paper directly from wood for light management in solar cells[J]. Nano Energy, 2017, 36: 366-373.

[23]Song J, Chen C, Zhu S, et al. Processing bulk natural wood into a high-performance structural material[J]. Nature, 2018, 554(7691): 224-228.

[24]Zhu M, Li Y, Chen G, et al. Tree-inspired design for high-efficiency waterextraction[J]. Advanced Materials, 2017, 29(44): 1704107.

[25]Wan C, Jiao Y, Li J. Core-shell composite of wood-derived biochar supported MnO_2 nanosheets for supercapacitor applications[J]. RSC Advances, 2016, 6(69): 64811-64817.

[26]Jiao Y, Wan C, Li J. Scalable synthesis and characterization of free-standingsupercapacitor electrode using natural wood as a green substrate to support rod-shaped polyaniline[J]. Journal of Materials Science: Materials in Electronics, 2017, 28(3): 2634-2641.

[27]Stone J E, Scallan A M. Effect of component removal upon the porous structure of the cell wall of wood[C] Journal of Polymer Science Part C: Polymer Symposia. New York: Wiley Subscription Services, Inc., A Wiley Company, 1965, 11(1): 13-25.

[28]Junior C S, Milagres A M F, Ferraz A, et al. The effects of lignin removal and drying on the porosity and enzymatic hydrolysis of sugarcane bagasse[J]. Cellulose, 2013, 20(6): 3165-3177.

[29]Qi W Y, Hu C W, Li G Y, et al. Catalytic pyrolysis of several kinds of bamboos over zeoliteNaY[J]. Green Chemistry, 2006, 8(2): 183-190.

[30]Wei D, Li Z, Wang H, et al. Antimicrobial paper obtained by dip-coating with modified guanidine-based particle aqueousdispersion[J]. Cellulose, 2017, 24(9): 3901-3910.

[31]Mi L, Licina G A, Jiang S. Nonantibiotic-based Pseudomonas aeruginosa biofilm inhibition with osmoprotectant analogues[J]. ACS Sustainable Chemistry & Engineering, 2014, 2(10): 2448-2453.

[32]Nishimura H, Kamiya A, Nagata T, et al. Direct evidence for α ether linkage between lignin and carbohydrates in wood cell walls[J]. Scientific Reports, 2018, 8(1): 6538.

[33]Abidi N, Cabrales L, Haigler C H. Changes in the cell wall and cellulose content of developing cotton fibers investigated by FTIR spectroscopy[J]. Carbohydrate Polymers, 2014, 100(2): 9-16.

[34]Ferraz A, Baeza J, Rodriguez J, et al. Estimating the chemical composition of biodegraded pine and eucalyptus wood by DRIFT spectroscopy and multivariate analysis[J]. Bioresource Technology, 2000, 74(3): 201-212.

[35]Shen D K, Gu S, Luo K H, et al. The pyrolytic degradation of wood-derived lignin from pulping process[J].

Bioresource technology, 2010, 101(15): 6136-6146.

[36]Cao X, Sun S, Peng X, et al. Rapid synthesis of cellulose esters by transesterification of cellulose with vinyl esters under the catalysis of NaOH or KOH in DMSO[J]. Journal of agricultural and food chemistry, 2013, 61(10): 2489-2495.

[37]Mathew, X, Sanchez-Mora, et al. Synthesis of CuS nanoparticles by a wet chemical route and their photocatalytic activity[J]. Journal of nanoparticle research: An interdisciplinary forum for nanoscale science and technology, 2015, 17(7).

[38]Nekouei F, Nekouei S, Kargarzadeh H. Enhanced adsorption and catalytic oxidation of ciprofloxacin on hierarchical CuS hollow nanospheres@ N-doped cellulose nanocrystals hybrid composites: kinetic and radical generation mechanism studies [J]. Chemical Engineering Journal, 2018, 335: 567-578.

[39]Saeed R M Y, Bano Z, Sun J, et al. CuS-functionalized cellulose based aerogel as biocatalyst for removal of organic dye[J]. Journal of Applied Polymer Science, 2019, 136(15): 47404.

[40]Liu P, Huang Y, Yan J, et al. Construction of CuS Nanoflakes Vertically Aligned on Magnetically Decorated Graphene and Their Enhanced Microwave Absorption Properties[J]. Acs Appl Mater Interfaces, 2016, 8(8): 5536-5546.

[41]Jiao Y, Wan C, Zhang W, et al. Carbon fibers encapsulated withnano-copper: a core-shell structured composite for antibacterial and electromagnetic interference shielding applications[J]. Nanomaterials, 2019, 9(3): 460.

[42]Wan C, Jiao Y, Sun Q, et al. Preparation, characterization, and antibacterial properties of silver nanoparticles embedded into celluloseaerogels[J]. Polymer Composites, 2016, 37(4): 1137-1142.

[43]Khan B A, Chevali V S, Na H, et al. Processing and properties of antibacterial silver nanoparticle-loaded hemp hurd/poly (lactic acid) biocomposites[J]. Composites Part B: Engineering, 2016, 100: 10-18.

[44]Joubert F, Yeo R P, Sharples G J, et al. Preparation of an antibacterial poly (ionic liquid) graft copolymer of hydroxyethyl cellulose[J]. Biomacromolecules, 2015, 16(12): 3970-3979.

[45]Wang P, Zhao J, Xuan R, et al. Flexible and monolithic zinc oxide bionanocomposite foams by a bacterial cellulose mediated approach for antibacterial applications[J]. Dalton Transactions, 2014, 43(18): 6762-6768.

[46]Galkina O L, Önneby K, Huang P, et al. Antibacterial and photochemical properties of cellulose nanofiber-titania nanocomposites loaded with two different types of antibiotic medicines[J]. Journal of Materials Chemistry B, 2015, 3(35): 7125-7134.

[47]Tang J, Song Y, Tanvir S, et al. Polyrhodanine coated cellulose nanocrystals: a sustainable antimicrobial agent[J]. ACS Sustainable Chemistry & Engineering, 2015, 3(8): 1801-1809.

[48]Maleki A, Movahed H, Paydar R. Design and development of a novel cellulose/γ-Fe_2O_3/Ag nanocomposite: a potential green catalyst and antibacterial agent[J]. RSC advances, 2016, 6(17): 13657-13665.

[49]Jebali A, Hekmatimoghaddam S, Behzadi A, et al. Antimicrobial activity of nanocellulose conjugated with allicin and lysozyme[J]. Cellulose, 2013, 20(6): 2897-2907.

[50]Fabry W, Okemo P O, Ansorg R. Antibacterial activity of East African medicinal plants[J]. Journal of ethnopharmacology, 1998, 60(1): 79-84.

[51]Alzoreky N S, Nakahara K. Antibacterial activity of extracts from some edible plants commonly consumed in Asia [J]. International Journal of Food Microbiology, 2003, 80(3): 223-230.

[52]Lloyd D R, Phillips D H. Oxidative DNA damage mediated by copper (II), iron (II) and nickel (II) Fenton reactions: evidence for site-specific mechanisms in the formation of double-strand breaks, 8-hydroxydeoxyguanosine and putativeintrastrand cross-links[J]. Mutation Research/Fundamental and Molecular Mechanisms of Mutagenesis, 1999, 424(1-2): 23-36.

[53]Passos W, Scarpellini M, Drago V, et al. Phosphate diester hydrolysis and DNA damage promoted by new cis-aqua/hydroxy copper(II) complexes containing tridentate imidazole-rich ligands. [J]. Inorganic Chemistry, 2003, 42(25): 8353-8365.

[54]Park H J, Nguyen TT M, Yoon J, et al. Role of reactive oxygen species in Escherichia coli inactivation by cupric ion

[J]. Environmental science & technology, 2012, 46(20): 11299-11304.

[55] Christopher, Rensing, Gregor, et al. Escherichia colimechanisms of copper homeostasis in a changing environment [J]. FEMS Microbiology Reviews, 2003, 27(2-3): 197-213.

[56] Billaud J, Bouville F, Magrini T, et al. Magnetically aligned graphite electrodes for high-rate performance Li-ion batteries[J]. Nature Energy, 2016, 1(8): 1-6.

[57] Sander J S, Erb R M, Li L, et al. High-performance battery electrodes via magnetic templating[J]. Nature Energy, 2016, 1(8): 1-7.

[58] Liao H, Liu H, Li Y, et al. Antitumor efficacy of doxorubicin encapsulated withinPEGylated poly (amidoamine) dendrimers[J]. Journal of applied polymer science, 2014, 131(11).

[59] Wang Y, Xiao Y, Zhou H, et al. Ultra-high payload of doxorubicin and pH-responsive drug release inCuS nanocages for a combination of chemotherapy and photothermal therapy[J]. RSC advances, 2013, 3(45): 23133-23138.

[60] Carvalho I T, Santos L. Antibiotics in the aquatic environments: a review of the European scenario[J]. Environment international, 2016, 94: 736-757.

[61] Nekouei F, Nekouei S, Kargarzadeh H. Enhanced adsorption and catalytic oxidation of ciprofloxacin on hierarchical CuS hollow nanospheres@ N-doped cellulose nanocrystals hybrid composites: kinetic and radical generation mechanism studies [J]. Chemical Engineering Journal, 2018, 335: 567-578.

[62] Ruan C Q, Stromme M, Mihranyan A, et al. Favored surface-limited oxidation of cellulose with Oxone in water[J]. RSC advances, 2017, 7(64): 40600-40607.

[63] Zhang T, Zhu H, Croue J P. Production of sulfate radical from peroxymonosulfate induced by a magnetically separable $CuFe_2O_4$ spinel in water: efficiency, stability, and mechanism[J]. Environmental science & technology, 2013, 47(6): 2784-2791.

中文题目：一种用于定向组装多功能硫化铜纳米粒子的具有各向异性微通道的全纤维素框架

作者：万才超，焦月，田文燕，张陆雨，吴义强，李坚，李贤军

摘要：通过简单的自上而下的方法合成的具有各向异性微通道的全纤维素框架(HCFW)是一种理想的各向异性基底，可以引导 CuS 纳米粒子(CuS NPs)沿微通道壁定向组装。HCFW 表面的含氧基团及其丰富的低弯度曲度通道有助于提高 CuS NPs 的负载量(约 22.7 wt.%)和粒子的均匀分散以及细胞毒性物质(如 Cu(Ⅰ)/Cu(Ⅱ))的快速扩散。因此，该复合材料具有出色的抗菌活性，对大肠杆菌和金黄色葡萄球菌的生长抑制率接近 100%，对这两种细菌的最低抑制浓度为 8 $\mu g \cdot mL^{-1}$。在药物输送方面，盐酸阿霉素(DOX)从 CuS NPs@ HCFW 的释放受溶液 pH 的调控。pH 为 2.2 时，DOX 的最大累积释放量为 78.3%。在体内也验证了负载 DOX 的 CuS NPs@ HCFW 的癌症治疗能力。此外，CuS NPs@ HCFW 在过硫酸钾存在的情况下通过吸附和催化氧化，具有较强的抗生素去除能力(在 120 min 内可 100%完全去除环丙沙星)。这些令人振奋的结果不仅表明了 CuS NPs@ HCFW 在抗菌、pH 响应药物输送和抗生素去除方面的巨大潜力，而且为利用各向异性 HCFW 开发新型各向异性复合材料提供了新视角。

关键词：各向异性结构；定向组装；硫化铜；抗菌材料；药物输送；抗生素

Ultra-high Rate Capability of Nanoporous Carbon Network@ V_2O_5 Sub-micron Brick Composite As A Novel Cathode Material for Asymmetric Supercapacitors*

Yue Jiao, Caichao Wan, Yiqiang Wu, Jingquan Han, Wenhui Bao,
He Gao, Yaoxing Wang, Chengyu Wang, Jian Li

Abstract: A green biomass-derived nanoporous carbon network (NCN) has been prepared and integrated with V_2O_5 submicron bricks (SMBs). The large surface area and high pore volume of the NCN can not only provide abundant sites for electrochemical reactions but also stabilize the structure of the V_2O_5 SMBs. The NCN@ V_2O_5 SMB composite, acting as a novel cathode material, delivers a high areal capacitance of 786 $mF \cdot cm^{-2}$ at 0. 2 $mA \cdot cm^{-2}$ and superior cycling stability with 89. 5% capacitance retention after 5000 cycles. Besides, the electrode achieves an ultra-high rate capability (82% capacitance retention as the current density increases from 0. 2 to 5 $mA \cdot cm^{-2}$) since the contribution from the non-diffusion-controlled process is estimated to be as high as 95. 5%-98. 5% according to the kinetic analysis. Furthermore, the micropores are more favorable than the mesopores at lower current densities (0. 2-2 $mA \cdot cm^{-2}$), while the contribution of the external surface area becomes more significant for current densities higher than 2 $mA \cdot cm^{-2}$. Moreover, an asymmetric supercapacitor assembled using this cathode and the NCN anode shows superior electrochemical properties, such as wide operating voltage, long cycle life and large energy density (72. 2 $\mu W \cdot h \cdot cm^{-2}$). Their excellent electrochemical features and good eco-friendliness confirm the potential of the NCN@ V_2O_5.
Keywords: Cathode material; Nanoporous carbon network; Ultra-high rate capability

1 Introduction

The rapid consumption of petrochemical resources accompanied by serious environmental concerns has promoted the development of novel green and regenerative energy storage and conversion systems[1-3]. Particularly, supercapacitors have shown great promise as a new type of energy storage device because of their unique merits such as high power density, long cycle stability, wide operation temperature and low cost of maintenance[4-6]. Electrode materials are the most crucial components of supercapacitors because they significantly determine the performance of supercapacitors. The state-of-the-art theory and technology primarily focus on three classes of electrode materials, namely carbon materials (such as porous carbon[7,8] and graphene[9]), metallic compounds (such as $Ni(OH)_2$[10] and MnO_2[11]) and conductive polymers (such as polypyrrole[9,12] and polyaniline[13]). So far, metallic compounds have generally delivered the highest

* 本文摘自 Nanoscale, 2020, 12(45): 23213-23224.

capacitance and energy density due to their faradaic mechanism that is dependent on redox reactions. However, this class of metalbased electrochemically active materials commonly suffer from poor cycling stability and low rate capability because of their large volume changes in the charging-discharging process and diffusion limitation in relation to inner-pore ion transport[14,15]. A widely accepted solution is to integrate metallic compounds into carbon materials by an electronic double-layer capacitive (EDLC) mechanism[7,14,16]. The porous structure and high porosity of carbon materials play a crucial role in suppressing the volume change and shortening of the diffusion distance of the electrolyte ions. A bimodal porosity containing micropores and either meso-or macro-pores is currently believed to be ideal as the actual energy storage occurs predominantly in the smaller micropores where the bulk of the surface area lieswhile the larger pores provide the fast mass-transport of electrolytes to and from the micropores[17].

Among the various metallic compounds, vanadium oxides are of particular interest due to the high theoretical capacitance (1060 $F \cdot g^{-1}$ for pseudocapacitive nano-size V_2O_5 (extrinsic pseudocapacitive materials), the calculation is available in the ESI), high energy density, high abundance, easy synthesis and various morphologies(such as nanonetworks and nanoribbons)[18-20]. Among vanadium oxides, vanadium pentoxide(V_2O_5) is the most stable oxide in the V-O system; besides, its high oxidation state promotes the feasibility of storing multiple electrons per formula unit (+5, +4, +3 and +2 are electrochemically accessible). In addition, V_2O_5 has the ability to generate a layered structure with strong chemical bonds in-plane but weak out-of-plane bonding, allowing the intercalations of electrolyte ions accompanied by large free energy contributing to high electrochemical activity[21]. Magnetron sputtering is an effective way to create high-purity metal-and metal oxide-based nanomaterials for use in supercapacitors. Magnetron sputtering is a kind of physical vapor deposition method which depends on the collision process between the injected gas and targets. Because of this collision, the target atoms near the surface gain sufficient momentum to move outward, and hence they are sputtered out. The merits of magnetron sputtering include fast deposition rate, lowdamage to the film, high interface bonding between the sputtering films and substrates, high purity and controllable thickness[22]. Kumar et al. deposited α-MnO_2 nanorods on a current collector(Ni coated aluminum oxide) by the reactive directcurrent sputtering technique. The assembled binder-free symmetric supercapacitor device delivered a high areal specific capacitance(112.6 $mF \cdot cm^{-2}$) and energy density(4.2 $W \cdot h \cdot kg^{-1}$)[23]. Wei et al. prepared CrN thin films by reactive directcurrent magnetron sputtering from a metallic Cr target using Ar and N_2 gases as the sputtering and reactive gases. The symmetric devices, assembled with a pair of CrN thin film electrodes, deliver a high energy density of 8.2 $mW \cdot h \cdot cm^{-3}$ at the power density of 0.7 $W \cdot cm^{-3}$ along with outstanding cycling stability[24]. However, the low electrical conductivity(10^{-2}–10^{-3} $S \cdot cm^{-1}$)[25] and unsatisfactory electrochemical stability are the main challenges of V_2O_5 in supercapacitor applications. Carbon materials, including graphene and carbon nanotubes (CNTs), have been utilized to improve the ionic and electronic transfer ability and cycling stability[26,27]. However, considering the requirement for cost reduction and green chemistry, it is still important to develop new, cheap and eco-friendly porous carbon materials. Apart from integration with EDLC materials, reducing the size of V_2O_5 materials to the nanoscale is another tactic to improve the electrochemical activity of V_2O_5 because nanomaterials can supply more electrochemically active sites. Also, some faradaic electrode materials demonstrate quite rapid charging rates and much reduced voltage polarizations when their size is reduced to the nanoscale, which is helpful for the improvement of rate capability[28]. Therefore, it is of significance to develop a fast and precise pathway to prepare nanostructured V_2O_5.

To make better use of V_2O_5 in the field of supercapacitors, in this paper, we demonstrate a fast and precise combined strategy of magnetron sputtering and calcination to prepare a layered structure of V_2O_5 sub-micron bricks(SMBs). For the sake of stabilizing the nanostructure and improving the electroconductibility of V_2O_5 SMBs, an eco-friendly cellulosederived nanoporous carbon network(NCN) was developed and integrated with V_2O_5 by virtue of a fast, easily operated ballmilling process. The as-prepared NCN@ V_2O_5 sub-micron brick(coded as NCN@ V_2O_5 SMBs) composite, which serves as a novel cathode material, delivers a high areal capacitance of 786 mF · cm^{-2} at the current density of 0.2 mA · cm^{-2} and excellent cycling stability with an 89.5% capacitance retention after 5000 cycles. An ultra-high rate capability of 82% was achieved as the current density increases 25 times from 0.2 to 5 mA · cm^{-2}. We were encouraged as the asymmetric supercapacitor(ASC) device, which is assembled by using this cathode and the NCN anode, shows superior electrochemical properties, e.g., wide operating voltage(0-1.6 V), large energy density(72.2 μW · h · cm^{-2} at the power density of 0.16 W · cm^{-2}) and long cycle life.

2 Materials andmethods

2.1 Materials

The chemicals tert-butyl alcohol, hydrochloric acid (HCl), potassium sulfate (K_2SO_4), sodium hydroxide(NaOH) and polyethylene glycol-4000(PEG-4000) were purchased from Shanghai Aladdin Industrial Inc. (China). Bamboo fiber was supplied by Zhejiang Mingtong Textile Technology Co. Ltd and further washed several times with distilled water and dried at 60 ℃ for 24 h. The dried bamboo fiber was directly used as the cellulose feedstock without further purification. Superfine acetylene black, nickel (Ni) foam and polytetrafluoroethylene (PTFE) binder were obtained from China New Metal Materials Technology Co. Ltd. PTFE plates were supplied by Dahua Plastic Industry Co., Ltd(Hangzhou, China). A vanadium (V) sputtering target(purity: 99.95%) was bought from China New Metal Materials Technology Co. Ltd.

2.2 Preparation of V_2O_5 SMBs

V_2O_5 SMBs were prepared through a combined method of magnetron sputtering and calcination. For the magnetron sputtering process, a thin layer of metallic V(the thickness was set as 500 nm) was sputtered on the surface of common filter paper by using a magnetron sputter (VTC-600-2HD, Shenyang Kejing Auto-instrument Co. Ltd, China). A V target(99.95% purity) was mounted on the cathode, and the filter paper was placed on the anode facing the target. The distance between the target and substrate was 60 mm. Initially, the chamber was pumped to a pressure of 1×10^{-3} Pa prior to the introduction of the sputtering gas, Ar (99.99% purity). The flow of Ar was 11 sccm. The V particles were sputtered on the filter paper through direct current magnetron sputtering(power: 135 W) with a rotation speed of 20 rpm to achieve homogeneous deposition. To avoid substrate deformation and diffusion movement of the sputtered V particles at high temperature, water-cooling was applied in the whole sputtering process to control the temperature of the substrate. For the calcination process, the filter paper/V composite was first transferred into a muffle furnace. Then the composite was heated to 600 ℃ in air at a heating rate of 5 ℃ · min^{-1} and maintained at this temperature for 30 min. After that, the composite was naturally cooled to room temperature.

2.3 Preparation of the NCN

The NCN was prepared by pyrolyzing its precursor (namely a porous cellulose network) in a nitrogen atmosphere. The synthesis of the porous cellulose network primarily involves three procedures: (1) dissolving the bamboo fiber in a solution of NaOH and PEG-4000; (2) regenerating the cellulose solution in a 1 v% HCl solution; (3) freeze-drying the resultant cellulose hydrogel after solution replacement with tertiary butanol. Details are available in our recent reports[29,30]. The following pyrolysis was performed by first heating the porous cellulose network to 500 ℃ at a heating rate of 5 ℃ · min^{-1} and retaining the material at 500 ℃ for 1 h. Then, the material was further heated to 1000 ℃ at 5 ℃ · min^{-1} and kept at this temperature for 2 h. Finally, the material was cooled to 500 ℃ at a cooling rate of 5 ℃ · min^{-1}. After these procedures, the material was naturally cooled to room temperature.

2.4 Preparation of the NCN@ V_2O_5 SMBs

The NCN and V_2O_5 SMBs were homogeneously blended by ball-milling with a mass ratio of 1 : 4. The ball-milling was carried out by using a planetary ball-mill (Changsha Deco Equipment Co. Ltd, China) with a rotation speed of 250 r min^{-1} using a 50 mL zirconia vial with zirconia balls (10 mm in diameter) under an argon atmosphere. The ball-to-powder ratio was 40 : 1, and the milling time was 1 h.

2.5 Fabrication of the cathode, anode and ASC device

The NCN@ V_2O_5 SMBs, acetylene black and PTFE were mixed in a mass ratio of 85 : 10 : 5 and then ground in alcohol. The plain NCN powder possesses an electrical conductivity of 0.26 S · m^{-1} and the NCN@ V_2O_5 SMB composite displays an electrical conductivity of 0.012 S · m^{-1}. It is noted that the addition of superfine acetylene black with a high electrical conductivity of ~400 S · m^{-1} is helpful in ensuring high inter-particle electric conductivity (like tunnel current), notwithstanding that NCN is also conductive. After being mixed with superfine acetylene black, the NCN@ V_2O_5 SMB/superfine acetylene black mixture has an electrical conductivity of 6.85 S · m^{-1}, approximately 570 times higher than that of the NCN@ V_2O_5 SMBs. The resultant slurry was pasted and pressed onto a Ni foam substrate at 10 MPa and then dried at 80 ℃ overnight, which would be the cathode. Plain NCN was also mixed with acetylene black and PTFE, and the mixture was pressed onto a Ni foam following the above method. The NCN electrode served as the anode. The exposed geometric area of these electrodes equals to 1 × 1 cm^2. For the assembly of an ASC device, a piece of common and thin cellulose paper acts as a separator. These electrodes and separator were sandwiched between two PTFE plates and the whole ASC device was clamped with a high-strength plastic clamp.

2.6 Characterization

The micromorphology was observed with a scanning electron microscope (SEM, Hitachi S4800) equipped with an energy-dispersive X-ray (EDX) detector for elemental analysis. Transmission electron microscopy (TEM) was performed with a FEI, Tecnai G2 F20 TEM with a field-emission gun operating at 200 kV. X-ray diffraction (XRD) analysis was implemented on a Bruker D8 Advance TXS XRD instrument with Cu Kα (target) radiation (λ = 1.5418 Å) at a scan rate (2θ) of 4° min^{-1} and a scan range from 5° to 90°. X-ray photoelectron spectroscopy (XPS) was carried out using a Thermo Escalab 250Xi system using a spectrometer with a dual Al Ka X-ray source. Deconvolution of the overlapped peaks was achieved by using a mixed Gaussian-Lorentzian fitting program(Origin 9.0, Originlab Corporation). N_2 adsorption-desorption tests were performed at −196 ℃ on an accelerated surface area and porosimeter system(3H-2000PS2 unit, Beishide Instrument S&T Co. Ltd). Electrical conductivity was measured by using a H7756 four-point probe resistivity meter with a testing range of $10^{-5}-10^{5}$ S · cm^{-1}(Heng Odd Instrument Co. Ltd, China).

2.7 Electrochemical characterization

The electrochemical performance of the NCN@ V_2O_5 SMB electrode was studied by using a CS350 electrochemical workstation(Wuhan CorrTest Instruments Corp. Ltd, China)in a three-electrode system: NCN @ V_2O_5 SMBs acted as the working electrode, and an Ag/AgCl electrode and a Pt foil served as the reference and counter electrodes, respectively. The electrolyte was 0.5 M K_2SO_4 solution. Cyclic voltammetry(CV) and galvanostatic charge-discharge (GCD) plots were measured over the potential window of 0 to 1 V. Electrochemical impedance spectroscopy(EIS) tests were carried out in the frequency range from 105 to 0.01 Hz with the alternate current amplitude of 5 mV. The exposed geometric area of the NCN@ V_2O_5 SMB working electrode was equal to 1 cm× 1 cm, and the mass loading of the electrochemically active substances (namely NCN and V_2O_5 SMBs) per area was ~2.5 mg · cm^{-2}. For comparison, the individual NCN electrode and individual V_2O_5 SMB electrode were also prepared. These materials also were subjected to the ball-milling treatment before mixing with PTFE and acetylene black. Besides, their mass loading of active substance per area was also controlled to be ~2.5 mg · cm^{-2}. In addition, the NCN was also used to prepare the anode by mixing ball-milled NCN with PTFE and acetylene black, and the working potential window was−0.8−0 V. Its mass loading of the electrochemically active substance(namely NCN) per area was ~1.8 mg · cm^{-2}. For the NCN@ V_2O_5 SMB//NCN ASC device, it was immersed in an aqueous electrolyte of 0.5 M K_2SO_4 for all the electrochemical experiments in a two-electrode system. The geometric area of the ASC was 1 × 1 cm^2, and the thickness was around 280 μm.

2.8 Calculations

For the NCN, V_2O_5 SMB and NCN@ V_2O_5 SMB electrodes and the assembled ASC device, the gravimetric(C_m, F · g^{-1}) and areal(C_s, F · cm^{-2}) specific capacitances were calculated based on the GCD curves at different current densities according to the following equations:[9]

$$C_m = I\Delta t/m\Delta V \text{ or } C_s = I\Delta t/s\Delta V \tag{1}$$

where I(A) is the discharge current, Δt(s) is the discharge time, m(g) is the mass of active materials, s is the specific area(cm^2), and ΔV(V) is the operation discharge potential/voltage window. For the ASC, m (g) is the total mass of active materials for both the anode and cathode electrodes. The areal(E_s) or gravimetric (E_g) energy density and areal(P_s) or gravimetric(P_g) power density were evaluated based on the capacitance values shown below:[7]

$$E_s = C_s(\Delta V)^2/3600 \text{ or } E_m = C_m(\Delta V)^2/3600 \tag{2}$$

$$P_s = \times 3600 \text{ or } P_m = \times 3600 \tag{3}$$

3 Results and discussion

The schematic diagram for the synthesis of NCN@ V_2O_5 SMBs is illustrated in Fig. 1a. There are three major steps involved in the synthesis, i. e., (1) magnetron sputtering of high-purity nano-V, (2) the oxidization of nano-V to V_2O_5 SMBs at hightemperatures in air; and (3) integrating the V_2O_5 SMBs into the NCN with the help of ball-milling treatment. It is worth mentioning that ball-milling is a cost-efficient and environmentally friendly physical processing method that has been proved to be capable of achieving milling accuracy as small as 100 nm. The micromorphology of NCN, V_2O_5 SMBs and NCN @ V_2O_5 SMBs was observed by SEM. As seen in Fig. 1b, NCN possesses a cross-linked and 3D porous network structure, and the enlarged image (the inset of Fig. 1b) clearly confirms the existence of mesopores. From Fig. 1c, it is clear that the brick-like nano-V_2O_5 with a thickness of 210−490 nm and a width of 1. 1−2. 7 μm is successfully formed through the magnetron sputtering and calcination. Also, the lamellar structure of V_2O_5 can be identified (blue circles) and is expected to supply more space for the electrochemical reactions of the ions. After the ball-milling treatment, the porous NCN is dispersed on the surface of V_2O_5 SMBs while the shape and size of V_2O_5 are well kept, as shown in Fig. 1d. The coating of NCN is believed to stabilize the structure of V_2O_5 SMBs and facilitate the electron transport[14]. In addition, the EDX analysis of the NCN shows a strong C signal and an extremely weak O signal (see the inset of Fig. 1d), which are ascribed to the high-temperature pyrolysis treatment. Compared with that of the NCN, the EDX pattern of NCN@ V_2O_5SMBs clearly demonstrates the existence of V element. TEM was performed to further observe the combination between V_2O_5 SMBs and NCN. As shown in Fig. 1e and f, V_2O_5 SMBs are densely covered with the network-like NCN; besides, from the enlarged image of the V_2O_5 SMB surface, thin sheet-like NCN can be identified and it is homogeneously anchored on V_2O_5. In addition, the high-resolution TEM (HRTEM) image of the yellow-circled area at the edge of the V_2O_5 SMBs is presented in Fig. 1g, in which the measured interplanar spacing of 0. 199 nm for the well-defined lattice fringes is well consistent with the (411) plane of V_2O_5. For these thin NCN sheets (red-circled area), the HRTEM image shows no obvious long range-ordered structure, which suggests an amorphous nature.

The crystal structure of the NCN@ V_2O_5 SMBs was studied by XRD analysis. All the diffraction peaks can be indexed to the orthorhombic V_2O_5 phase with the lattice parameters of a = 11. 516 Å, b = 3. 5656 Å and c = 4. 3727 Å (JCPDS no. 41-1426) (Fig. 2a). The elemental composition and chemical valence of the NCN @ V_2O_5 SMBs were determined by XPS analysis. The XPS survey spectrum reveals that the composite completely consists of C, O and V elements (Fig. 2b), which is in agreement with the results of the EDX analysis. In the V 2p corelevel XPS spectrum (Fig. 2c), there are two main peaks with binding energies of 517. 01 and 524. 55 eV that correspond to the V $2p_{3/2}$ and V $2p_{1/2}$ of V^{5+} orbitals. 31 The results indicate the formation of a V_2O_5 phase. Moreover, there are also two weak peaks at 515. 67 and 522. 88 eV, which are associated with the V $2p_{3/2}$ and V $2p_{1/2}$ of V^{4+} orbitals, 32 indicating the presence of a small quantity of VO_2 in the vanadium oxides. The O 1s spectrum (Fig. 2d) with fitting peak positions at the binding energies of 529. 87, 531. 52, and 532. 82 eV is attributed to O (1s) of V_2O_5, −OH, and H_2O molecules, 33 respectively. The pore structure was studied by N_2 adsorption-desorption experiments, and some characteristic parameters are calculated according to the Brunauer-Emmett-Teller (BET) and Barrett- Joyner-Halenda (BJH) methods. As seen in Fig. 2e, the NCNhas a type-IV adsorption isotherm based on the IUPAC classification, and the hysteresis loop between adsorption and desorption isotherms is due to the capillary condensation occurring in the mesopores. The NCN achieves a high BET surface area of 459 $m^2 \cdot g^{-1}$ and pore volume of 1. 0 $cm^3 \cdot g^{-1}$. The

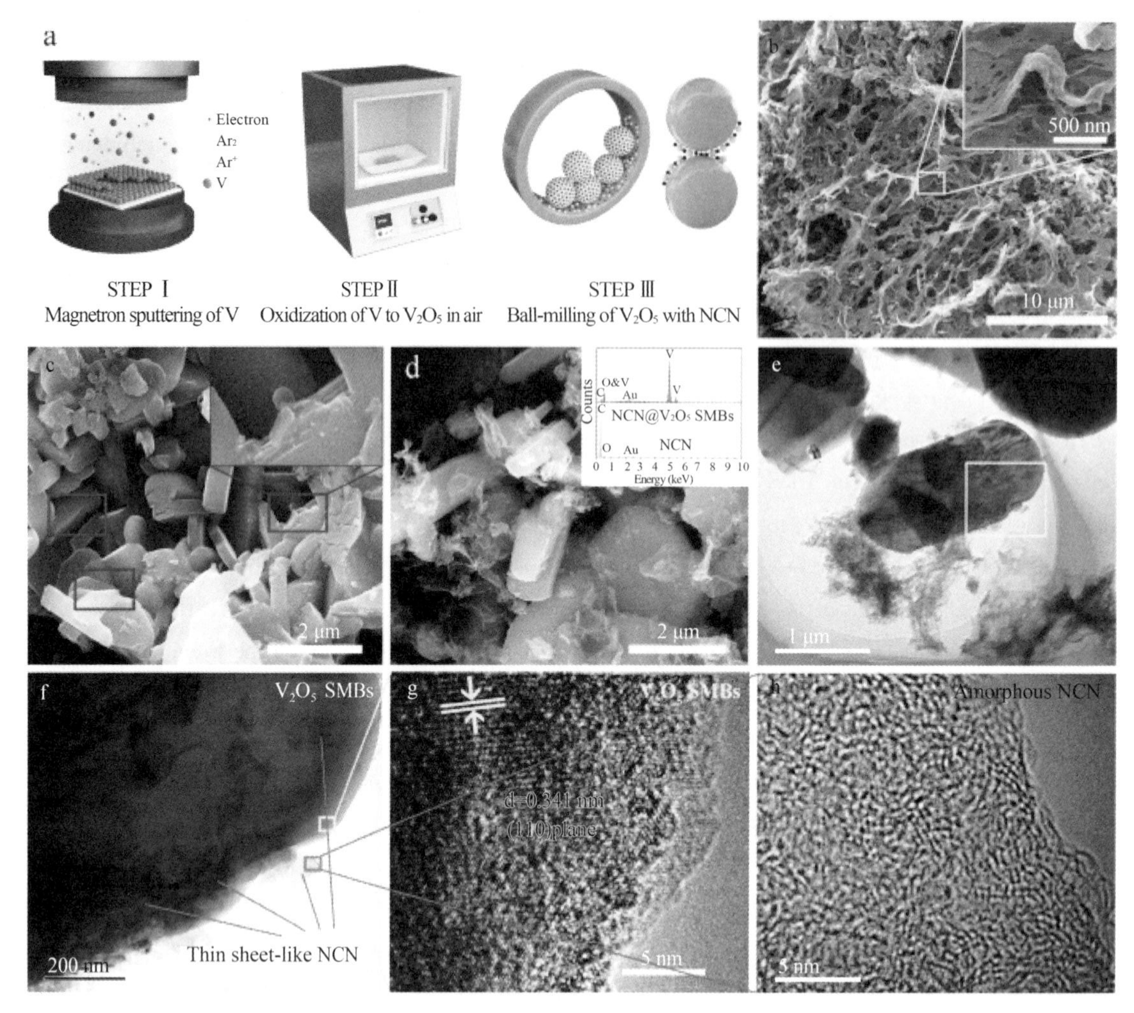

Fig. 1 Synthesis strategy, microstructure and elemental analysis of NCN@V_2O_5 SMBs. (a) Schematic diagram for the synthesis of NCN@V_2O_5 SMBs; SEM images of (b) NCN, (c) V_2O_5 SMBs and (d) NCN@V_2O_5 SMBs (the inset shows the EDX patterns); (e and f) TEM images of NCN@V_2O_5, and f is the enlarged image of the area within the yellow frame in e; (g and h) HRTEM images of the yellow and red frames in f

introduction of the NCN significantly improves the BET surface area of the NCN@ V_2O_5 SMBs from 1.58 $m^2 \cdot g^{-1}$ (individual V_2O_5 SMBs) to 19.5 $m^2 \cdot g^{-1}$. Furthermore, the pore volume of the NCN@ V_2O_5SMBs also increases from 0.013 $cm^3 \cdot g^{-1}$ (individual V_2O_5 SMBs) to 0.089 $cm^3 \cdot g^{-1}$, as summarized in Table S1. The larger surface area and higher pore volume can provide more available active sites for electrochemical reactions. Moreover, for all samples, most peaks are located within the scope of 1-10 nm (Fig. 2f), suggesting that these materials are primarily made up of mesopores and micropores.

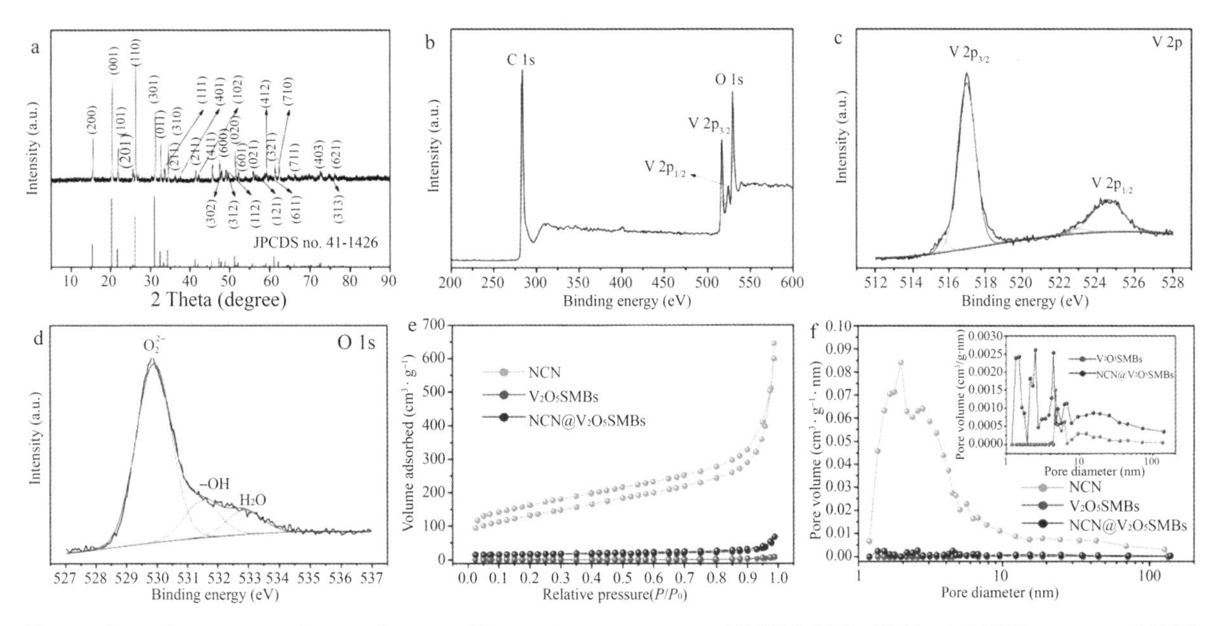

Fig. 2 Crystal structure, elemental composition and pore structure of NCN@ V_2O_5 SMBs. (a) XRD patterns of NCN @ V_2O_5 SMBs; (b) XPS survey spectrum and (c) V 2p and (d) O 1s core-level XPS spectra of NCN@V_2O_5 SMBs; (e) N_2 adsorption-desorption isotherms and (f) pore size distributions of NCN, V_2O_5 SMBs and NCN@V_2O_5 SMBs

The electrochemical properties of the NCN@ V_2O_5 SMB electrode were investigated by CV, GCD and EIS in a three-electrode system. The CV curves of the NCN@ V_2O_5 SMBs, individual V_2O_5 SMBs and individual NCN electrodes were tested and compared at a scan rate of 100 mV · s^{-1} within the positive potential window of 0-1V (vs. Ag/AgCl). As shown in Fig. 3a, the area enclosed by the CV curve of the NCN@ V_2O_5 SMBs is much higher than that of individual V_2O_5 SMB or NCN electrodes, revealing the higher charge storage ability of the NCN@ V_2O_5 SMBs because of the synergistic effect of both components. Moreover, the CV curves of the NCN @ V_2O_5 SMB electrode at different scan rates in the range of 5-100 mV · s^{-1} are shown in Fig. 3b, in which the well-preserved CV curve even at a high rate of 100 mV · s^{-1} reveals its potential good rate capability. A pair of broad redox peaks is clearly identified (the inset of Fig. 3b) and mainly attributed to the faradaic charge transfer reaction: [34]

$$V_2O_5+xK^++xe^+\leftrightarrow K_xV_2O_5 \quad (4)$$

where x is the mole fraction of inserted K^+ ions. In addition, based on the fitting result of XPS (Fig. 2c), there is still a small fraction of VO_2 (ca. 5.2%). For VO_2, the charge storage mechanism in the neutral K_2SO_4 electrolyte involves incorporation of the electrolyte cation K^+:

$$VO_2+xK^++xe^+\leftrightarrow K_xVO_2 \quad (5)$$

The redox reaction of VO_2 also contributes to the electrochemical energy storage of the composite NCN@ V_2O_5 SMB electrode. Charge-storage kinetics of electrode materials towards K^+ was studied by CV analysis. The analysis was performed regarding the behavior of anodic peak currents byassuming that the current (i) obeys a power-law relationship with the scan rate (ν): [35]

$$i=a\nu^b \quad (6)$$

It was proposed that by calculating the value of b, the diffusion-controlled contribution and capacitive contribution could be quantitatively determined. The b-value was determined by the slope of log(v)-log(i) plots, as illustrated in Fig. 3c. In particular, the b-value of 0.5 represents a total diffusion-controlled behavior, whereas 1.0 reveals a capacitive process. The b value of the NCN@ V_2O_5 SMB electrode was calculated to be as high as 0.88, indicating that the majority of currents at the peak potential were capacitive. The result confirms that the capacitive behavior dominates the reaction of the NCN@ V_2O_5 SMBs with fast kinetics. To further study the capacitance contribution of the NCN@ V_2O_5 SMBs, the contributions from the surface faradaic reaction and the insertion reaction were calculated at different scan rates based on eqn(6):[35]

$$I(V) = k_1 v + k_2 v^{1/2} \tag{6}$$

where $k_1 v$ and $k_2 v^{1/2}$ denote the contributions from the surface reaction (pseudocapacitance and double-layer capacitance) and the insertion reaction (diffusion-controlled process), respectively. By determining k_1 and k_2, we can distinguish the fractions arising from the surface and insertion processes at a specific potential. At scan rates in the range of 5–50 $mV \cdot s^{-1}$, the contribution resulting from non-diffusion-controlled process (i.e., surface reaction) is estimated to be as high as 95.5%–98.5% (see Fig. 3d). The result is consistent with that of b-value analysis.

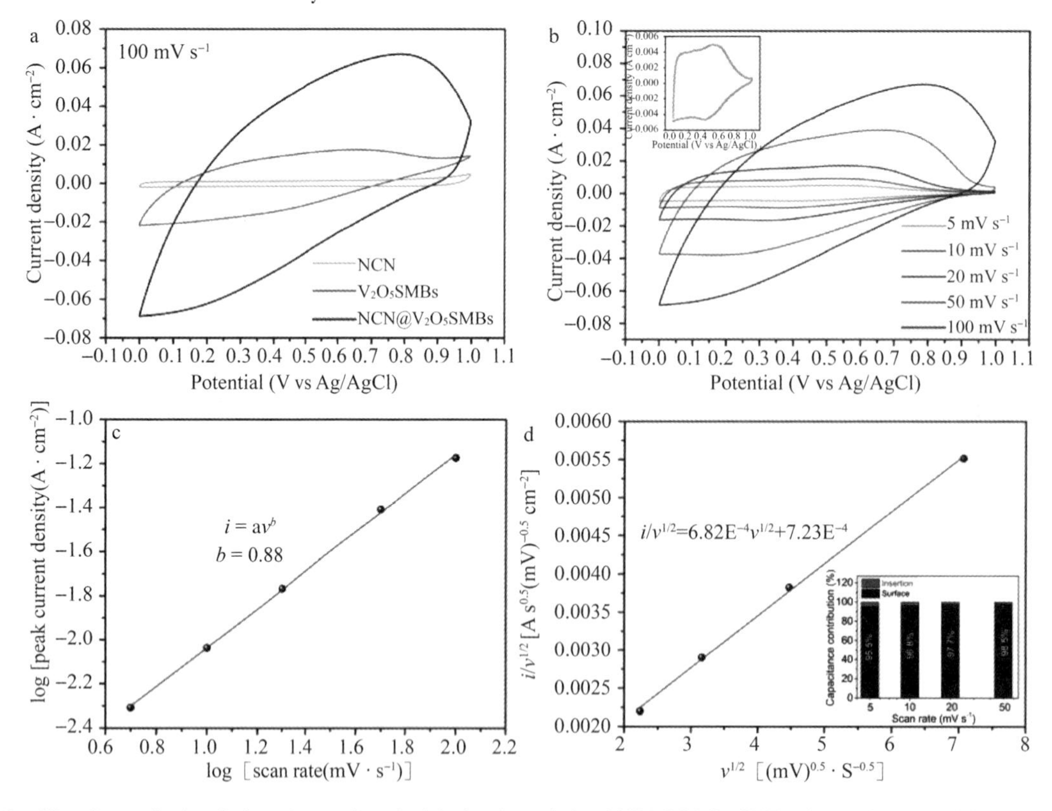

Fig. 3 Kinetic analysis of the electrochemical behavior of the NCN@V_2O_5SMB electrode tested in a positive potential range in a three-electrode system. (a) CV curves of NCN, V_2O_5 SMBs and NCN@V_2O_5 SMBs at 100 $mV \cdot s^{-1}$; (b) CV curves of NCN@V_2O_5 SMBs at scan rates of 5-100 $mV \cdot s^{-1}$, and the inset shows the enlarged image of CV curve at 5 $mV \cdot s^{-1}$; (c) determination of b value using the relationship between the peak current and the scan rate according to the voltammograms in b; (d) $i/v^{1/2}$ vs. $v^{1/2}$ plot using the anodic peak current, and the inset shows contributions from surface and insertion reactions at different scan rates

To distinguish the functions between micropores and external surfaces on rate capability, contributions of micropore and external surface area to the capacitance were determined using different mass ratios of the NCN @ V_2O_5 SMBs(m_{MCN} : m_{V2O5} = 1 : 4, 1 : 6 and 1 : 8). Their specific surface area and specific capacitance at different current densities were tested. According to the literature, 36, 37 the contribution of the external surface area(S_{ext}) to the total capacitance(C_{total}) can be estimated based on eqn(7):

$$C_{total} = C_{micro}S_{micro} + C_{ext}S_{ext} \quad (7)$$

where C_{micro} and C_{ext} are the capacitances per unit micropore and external surface area, respectively. Moreover, eqn(7) can be reorganized into eqn(8):

$$\frac{C_{total}}{S_{ext}} = C_{micro}\frac{S_{micro}}{S_{ext}} + C_{ext} \quad (8)$$

If the plot of C_{total}/S_{ext} versus S_{micro}/S_{ext} is linear, the C_{micro} and C_{ext} can be determined by the slope and intercept, respectively. As seen in Fig. 4a, there is a linear fit between C_{total}/S_{ext} and S_{micro}/S_{ext} by using the least-squares method at different current densities, and the slope decreases as the current density increases. By virtue of the slope and intercept, the surface capacitances from the micropores and external surfaces are calculated and plotted in Fig. 4b. As the current density increases from 0. 2 to 5 mA · cm^{-2}, the surface capacitances from the external surfaces gradually increase from 21. 2% to 73. 8% while the surface capacitances from the micropores decrease from 78. 8% to 26. 2% . Moreover, it is noted that the capacitances from the external surfaces and the micropores are almost identical at the current density of 2 mA · cm^{-2}. These results indicate that micropores are more favorable than mesopores at lower current densities(0. 2-2 mA · cm^{-2}), while the contribution of the external surface area becomes more significant than that of micropore surface area for current densities higher than 2 mA · cm^{-2}. As a result, it is deduced that if the diameter of mesopores increases and/or the smaller ion size of electrolytes are used, the contribution of the mesopore surface area to capacitance may increase at high current densities but decrease at low current densities.

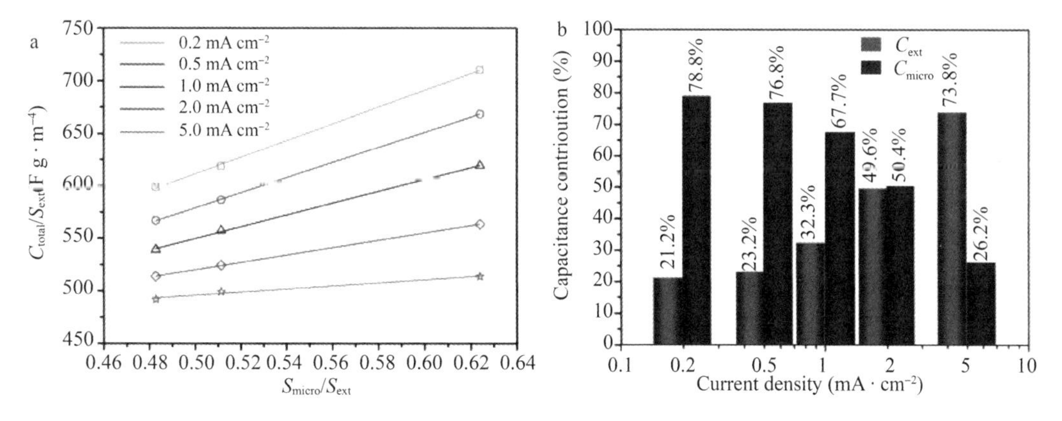

Fig. 4 Determination of capacitance contributions from micropores and external surfaces of NCN @ V_2O_5 SMBs. (a) Ratio of specific capacitance to the external surface area against the ratio of the micropore surface area to external surface area; (b) capacitance contribution ratios of the micropore and external surface area

Fig. 5a shows the GCD curves of the NCN@ V_2O_5 SMB electrode at different current densities over the range of 0. 2–5 mA · cm^{-2}. The almost triangular and symmetric shape of the GCD curves indicates an ideal capacitance feature of the composite; also, the slight distortion from linearity reflects the pseudocapacitive characteristic. The maximum areal capacitance of 786. 4 mF · cm^{-2}(314. 6 F · g^{-1}) is achieved for the NCN@ V_2O_5 SMBs at the current density of 0. 2 mA · cm^{-2}(Fig. 5b), higher than that of some recently reported V_2O_5-based electrodes, such as Mo-doped V_2O_5 thin film(175 mF · cm^{-2}), 38 V_2O_5 nanorods/stainless steel (337. 6 mF · cm^{-2})[39], V_2O_5-reduced graphene oxide(382. 0 mF · cm^{-2})[40], carbon-coated flowery V_2O_5 (417. 0 mF · cm^{-2})[41], 3D N-doped carbon nanofibers/ V_2O_5 aerogels(476. 1 mF · cm^{-2})[42] and V_2O_5-polyaniline(664. 5 mF · cm^{-2})[43], as listed in Table S2. Rate capability is also a crucial parameter for supercapacitors. High rate capability means ultrafast charge-discharge characteristics along with excellent capacitance and power density. As summarized in Table S2, these V_2O_5-based electrodes only deliver 16. 2%–63. 8% of the capacitance retention ratio when the current density increases 2–20 times. In contrast, in this study, the as-prepared NCN@ V_2O_5 SMB electrode distinctly has a more superior rate capability, which remains at 85. 8% and 82. 1% capacitance retention ratios, respectively, as the current density increases 10 times and 25 times. Surprisingly, when the current density increases 250 times to a higher value of 50 mA · cm^{-2}, a high retention ratio of 61. 7% is still achieved(485. 0 mF · cm^{-2}) (Fig. S1), indicating that the electrode can be applied in the case of large current density also. In addition, the commercially available V_2O_5 (coded as CA V_2O_5) particles, which were purchased from Aladdin Industrial Corporation (Shanghai, China), were used to prepare electrodes following the above method. As shown in Fig. S2a and b, the CA V_2O_5 electrode shows a maximum areal specific capacitance of 90. 6 mF · cm^{-2} at 0. 2 mA · cm^{-2}, only 11. 5% of the value for the NCN@ V_2O_5 SMBs. When the current density increases from 0. 2 to 5 mA · cm^{-2}, the CA V_2O_5 electrode only achieves 60. 1% of capacitance retention, distinctly lower than that of the NCN@ V_2O_5 SMBs(82. 1%). Moreover, the CA V_2O_5 particles were also mixed with NCN by ball-milling, and the NCN@ CA V_2O_5 electrode has a maximum areal specific capacitance of 177. 9 mF · cm^{-2} and a capacitance retention ratio of 72. 0% as the current density increases 25 times(Fig. S2a and b). Clearly, these values are lower than those of the NCN@ V_2O_5 SMBs. As a result, these comparisons indicate the more superior electrochemical properties of the NCN@ V_2O_5 SMBs.

The evaluation of the cycling stability of the NCN@ V_2O_5SMB electrode shows that it delivers an excellent cycling stability with an 89. 5% capacitance retention ratio after 5000 cycles(Fig. 5c). Also, nearly 100% of coulombic efficiency(97. 8% –99. 9%) is obtained in the whole cycling process, revealing the high energy utilization efficiency. The Nyquist plots of the NCN, V_2O_5 SMBs and NCN@ V_2O_5 SMBs are shown in Fig. 5d, where the semicircle width of the NCN@ V_2O_5 SMBs(0. 9 Ω) is close to that of the NCN(0. 5 Ω) but lower than that of the V_2O_5 SMBs(2. 0 Ω). The result indicates the low internal resistance of the NCN@ V_2O_5 SMBs. In addition, in the low-frequency region, the NCN@ V_2O_5 SMB electrode displays a straight line whose slope is larger than that of the others, which indicates the lower ion diffusion resistance and better capacitive behavior[21,44].

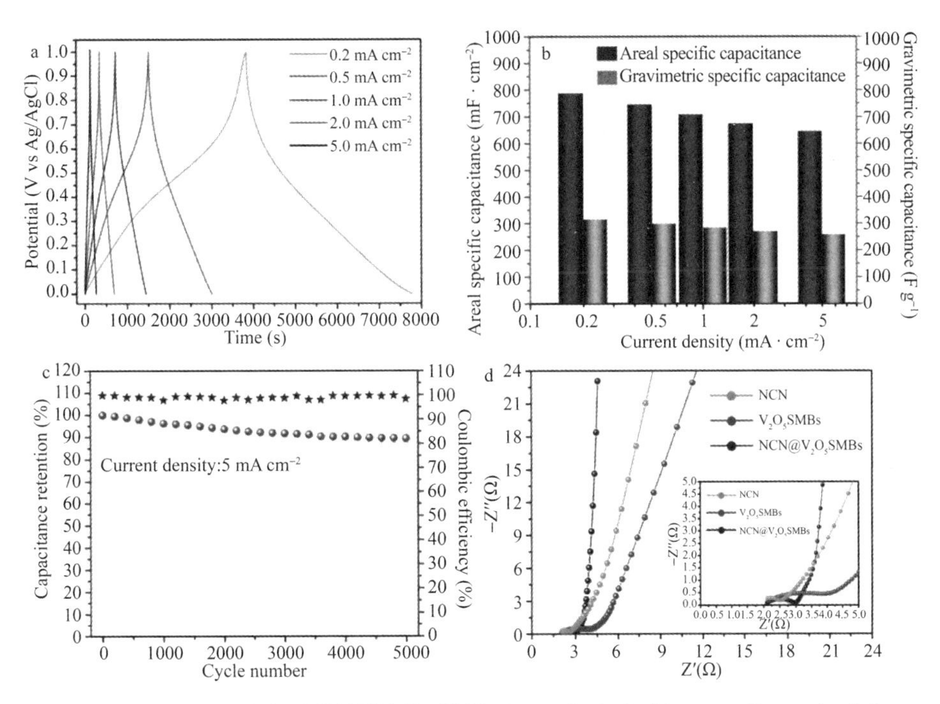

Fig. 5 Electrochemical properties of the NCN@V_2O_5 SMB electrode tested in a positive potential range in a three-electrode system. (a) GCD curves at current densities of 0.2–5 mA · cm^{-2} (b) rate performance; (c) cycling stability and coulombic efficiency at 5 mA · cm^{-2}; (d) Nyquist plots and the inset shows the enlarged images in the high-frequency region

The possible reasons for the outstanding electrochemical properties are concluded into three points: (1) the high theoretical capacitance of the V_2O_5 SMBs is due to their lamellar structure that can provide more space for the intercalation and electrochemical reactions of the ions, which is beneficial for achieving high electrochemical activity[21,44,45]; (2) large surface area and high pore volume of the NCN not only provides abundant and available sites for electrochemical reactions and serves as a reservoir to shorten the diffusion path of the ions[17] but also stabilizes the structure of V_2O_5SMBs and promotes electron transport (as illustrated in Fig. 6); (3) the synergistic effects of the EDLC and pseudocapacitance mechanisms significantly improve the electrochemical performance.

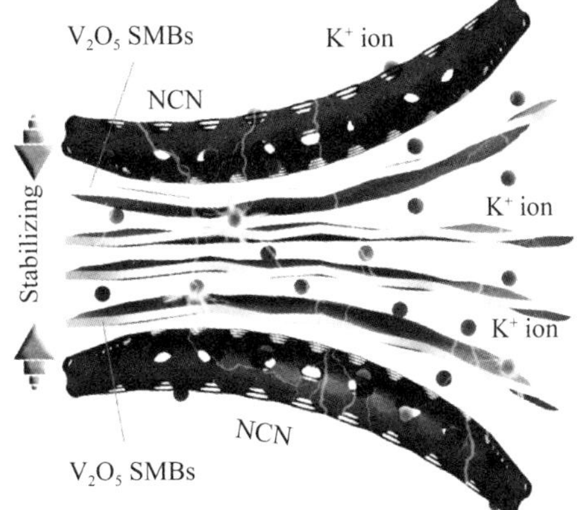

Fig. 6 Schematic diagrams of functions of NCN and V_2O_5 SMB components on the electrochemical reactions

So far, carbon materials are the most common anode materials for electrochemical energy storage. Therefore, in this work, the electrochemical property of the NCN acting as the anode is also evaluated in a negative potential range for the subsequent assembly of an asymmetrical supercapacitor(ASC) device. The CV curves of the NCN at different scan rates over 5-100 mV · s^{-1} are given in Fig. 7a. The heavy distortion from the rectangular shape at a high scan rate is attributable to the faradaic reactions occurring on the surface of the NCN electrode, such as the reactions between the carbonyl (C=O) and hydroxyl (C—OH) groups, i. e., $C{=}O + H^+ + e^- \rightarrow C{-}OH$. 46 Based on the GCD curves(see Fig. 7b), the NCN electrode delivers the maximum areal specific capacitance of 144. 8 mF · cm^{-2}(80. 5 F · g^{-1}) at 0. 2 mA · cm^{-2}. An inferior rate capability with 23. 3% capacitance retention is obtained when the current density increases from 0. 2 mA · cm^{-2} to 5 mA · cm^{-2}(Fig. 7c). However, the NCN possesses a superior cycling stability that retains 91. 5% of the initial capacitance after 5000 cycles(Fig. 7d), which is the typical characteristic of EDLC materials.

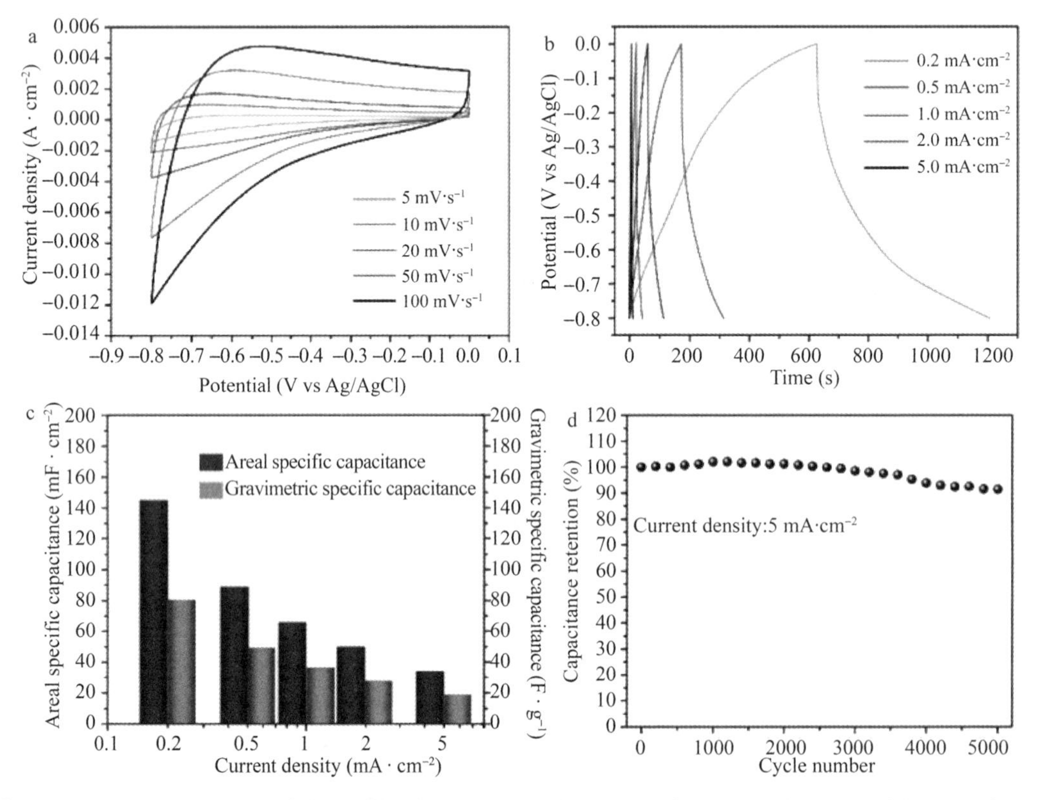

Fig. 7 Electrochemical properties of the NCN electrode (anode) tested in a negative potential range in a three-electrode system. (a) CV curves at scan rates of 5-100 mV · s^{-1}; (b) GCD curves at current densities of 0. 2-5 mA · cm^{-2}; (c) rate performance; (d) cycling stability and coulombic efficiency at 5 mA · cm^{-2}

An ASC device is assembled to study the potential of the NCN@ V_2O_5 SMBs for practical applications. The schematic diagram(Fig. 8a) depicts the fabrication of the ASC with the NCN@ V_2O_5 SMBs as the cathode and the individual NCN as the anode. The CV curves of the ASC device in different voltage windows(from 0-0. 8 V to 0-1. 6 V) are shown in Fig. 8b. It is seen that the stable voltage window of the ASC can be extendedto 1. 6 V. Moreover, the quasi-rectangular CV curves of the ASC with a pair of redox peaks at various scan rates indicate that the capacitance comes from both the EDLC and faradaic pseudocapacitance (Fig. 8c). Fig. 8d shows the GCD curves of the ASC at difffferent current densities. Both charge and discharge curves remain at a good symmetry at an operating voltage as high as 1.6 V, verifying its outstanding reversibility and coulombic effiffifficiency. The ASC device has an areal specific capacitance of 203.0 mF · cm^{-2} at 0.2 mA · cm^{-2} and remains at 108.4 mF · cm^{-2} at 5 mA · cm^{-2}. From the Ragone plot (Fig. 8e), the ASC exhibits a maximum areal energy density of 72.2 μW · h · cm^{-2} at a power density of 0.16 mW · cm^{-2} and an areal energy density of 38.5 μW · h · cm^{-2} at a maximum power density of 4 mW · cm^{-2}. The attained energy densities are superior to those of numerous recently reported V_2O_5-based ASCs or symmetrical supercapacitors (SSCs), such as V_2O_5 thin film SSC (0.68 μW · h · cm^{-2})[47], reduced graphene oxide (rGO)/ V_2O_5 SSC (3.30 μW · h · cm^{-2}),[48] V_2O_5/CNTs SSC (5.15 μW · h · cm^{-2})[49], V_2O_5/ PEDOT SSC (11.0 μW · h · cm^{-2})[50], V_2O_5/rGO//activated carbon ASC (14.8 μW · h · cm^{-2})[51], and $TiO_2-V_2O_5$//activated carbon ASC

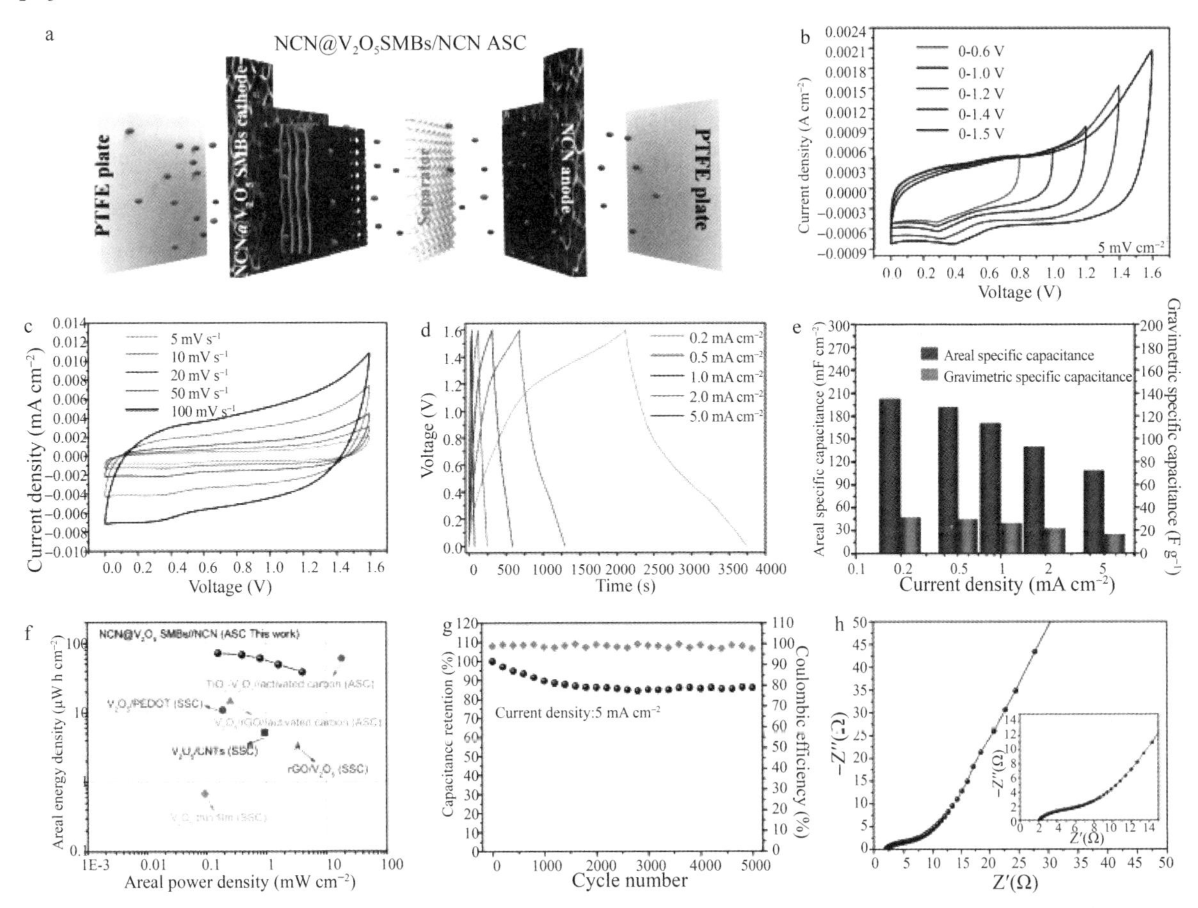

Fig. 8 Electrochemical properties of NCN@V_2O_5 SMBs//NCN ASC in a two-electrode system. (a) Schematic diagram of assembled ASC device; (b) CV curves in different voltage windows at a scan rate of 5 mV · s^{-1}; (c) CV curves at scan rates in the range of 5-100 mV · s^{-1}; (d) GCD curves at current densities over 0.2–5 mA · cm^{-2}; (e) rate performance; (f) Ragone plot of the ASC compared with the data of some other SSCs or ASCs in the literature; (g) cycling stability and coulombic efficiency at 50 mA · cm^{-2}; (h) Nyquist plots (the inset shows the enlarged image in the high-frequency region)

(~60. 54 μW · h · cm^{-2}),[52] as shown in Fig. 8f and Table 1. In addition, the thickness of the ASC device is ~280 μm. Thus, the volumetric specific capacitances of the device are in the range of 3. 87–7. 25 F · cm^{-3} at current densities over 7. 14–178. 57 mA · cm^{-3}. Moreover, the device has the maximum volumetric energy density of 2. 58 mW · h · cm^{-3} at the volumetric power density of 5. 7 mW · cm^{-3} and still maintains 1. 38 mW · h · cm^{-3} at the maximum volumetric power density of 142. 9 mW · cm^{-3}. The mass loading of active substance per area is 4. 3 mg · cm^{-2}. As a result, the ASC device achieves a maximum gravimetric specific capacitance of 47. 2 F · g^{-1} at 0. 047 A · g^{-1} and a maximum gravimetric energy density of 16. 8 W · h · kg^{-1} at the power density of 37. 2 W · kg^{-1}. Moreover, the gravimetric energy density of the NCN@ V_2O_5 SMBs// NCN ASC is higher than that of some recent V_2O_5-based ASCs, for instance, V_2O_5/rGO//activated carbon (7. 4 W · h · kg^{-1})[51], C@ V_2O_5nanorods//activatedcarbon (9. 4 W · h · kg^{-1})[53] and V_2O_5-rGO//rGO (13. 3 W · h · kg^{-1}). The comparison demonstrates the performance superiority of the NCN@ V_2O_5 SMBs// NCN ASC.

With respect to cycling stability, the ASC device retains 86. 4% of the initial capacitance after the first 2500 cycles, and then the capacitance retention ratio tends to be stable for the next 2500 cycles (see Fig. 8g). Eventually, the ASC device delivers good cycling stability with an 86. 2% capacitance retention ratio after 5000 cycles. Also, nearly 100% of coulombic efficiency is achieved in the whole cycling process. The Nyquist plot of the ASC is given in Fig. 8h, where the corresponding solution resistance and charge transfer resistance are as low as 2. 0 Ω and 3. 0 Ω, which are expected to enhance electron and ion transfer ability and contribute to achieving high power density.

Table 1 Comparison of areal energy density and other parameters of the NCN@V_2O_5 SMBs//NCN ASC with other recently reported V_2O_5-based ASCs or SSCs

Devices	Type	Maximum areal energy density (μW · h · cm^{-2})	Areal capacitance (mF · cm^{-2})	Voltage range (V)	Electrolyte	Ref.
V_2O_5 thin film	SSC	0. 68(95 μW · cm^{-2})	9. 7(10 mV · s^{-1})	0–1. 0 V	PVA/KOH	47
rGO/ V_2O_5	SSC	3. 30(3. 30 mW · cm^{-2})	24(1 mV · s^{-1})	0–1. 0 V	PVA/H_3PO_4	48
V_2O_5/CNTs	SSC	5. 15(0. 95 mW · cm^{-2})	34. 8(0. 1A · g^{-1})	0–0. 8 V	PVA/LiCl	49
V_2O_5/PEDOT	SSC	11. 0(0. 19 mW · cm^{-2})	240(0. 5 mA · cm^{-2})	0–0. 8 V	PVA/LiCl	50
V_2O_5/rGO//activated carbon	ASC	14. 8(0. 254 mW · cm^{-2})	106(1. 0 mA · cm^{-})	0–1. 0 V	0. 5 M K_2SO_4	51
TiO_2–V_2O_5//activated carbon	ASC	~60. 54(~17. 82 mW · cm^{-2})	181. 5(2 mV · s^{-1})	0–1. 3 V	1. 0 M Na_2SO_4	52
NCN@ V_2O_5 SMBs//NCN	ASC	72. 2(0. 16 mW · cm^{-2})	203. 0(0. 2 mA · cm^{-2})	0–1. 6 V	0. 5 M K_2SO_4	This work

4 Conclusions

In summary, we have demonstrated that the NCN@ V_2O_5 SMB composite can act as a novel and high-performance cathode material, which achieves a high areal capacitance of 786 mF · cm^{-2} at 0. 2 mA · cm^{-2} and excellent cycling stability with 89. 5% capacitance retention after 5000 cycles. Moreover, an ultrahigh rate capability (82% capacitance retention) is achieved as the current density increases 25 times from 0. 2 to 5 mA · cm^{-2}, verifying that the NCN @ V_2O_5 SMB electrode can be used in the case of large current density. The kinetic analysis reveals that the contribution from the non-diffusion-controlled process is estimated to be as high as 95. 5%-98. 5%, which plays a key role in the high rate capability. At lower current densities (0. 2-2 mA · cm^{-2}), the micropores are more favorable than the mesopores, while the contribution of the external surface area becomes more significant for current densities higher than 2 mA · cm^{-2}. In addition, the assembled NCN@ V_2O_5 SMBs//NCN ASC device shows a high energy density of 72. 2 μW · h · cm^{-2} at a power density of 0. 16 mW · cm^{-2}, superior to that of many V_2O_5-based ASCs or SSCs. The ASC also has a moderate areal specific capacitance and favorable cycling stability. By virtue of these electrochemical features and favorable environmental compatibility, the novel NCN@ V_2O_5 SMB composite demonstrates its potential for applications in supercapacitors.

Acknowledgments

This study was supported by the National Natural Science Foundation of China(grant no. 31901249), the Young Elite Scientists Sponsorship Program by CAST(grant no. 2019QNRC001) and the Scientific Research Foundation of Hunan Provincial Education Department(grant no. 18B180).

Compliance with ethics guidelines

There are no conflicts to declare.

References

[1]Yang Z H, Zhang J L, Kintner-Meyer M C W, et al. Electrochemical Energy Storage for Green Grid[J]. Chem. Rev., 2011, 111(5): 3577-3613.

[2]Zhang L X, Liu Z H, Cui G L, et al. Biomass-derived materials for electrochemical energy storages[J]. Prog. Polym. Sci., 2015, 43: 136-164.

[3]Wang H, Yang Y, Guo L. Nature-Inspired Electrochemical Energy-Storage Materials and Devices[J]. Adv. Energy Mater, 2017, 7(5): 1601709.

[4] Zhong C, Deng Y, Hu W B, et al. A review of electrolyte materials and compositions for electrochemical supercapacitors[J]. Chem. Soc. Rev., 2015, 44(21): 7484-7539.

[5]Liang J, Jiang C Z, Wu W. Towards fiber-, paper- and foam-like flexible solid-state supercapacitors: Electrode materials and device design[J]. Nanoscale, 2019, 11(15): 7041-7061.

[6]Wan C C, Jiao Y, Liang D X, et al. A Geologic Architecture System-Inspired Micro-/Nano-Heterostructure Design for High-Performance Energy Storage[J]. Adv. Energy Mater, 2018, 8(33): 1802388.

[7]Wan C C, Jiao Y, Bao W H, et al. Self-stacked multilayer FeOCl supported on a cellulose-derived carbon aerogel: A new and high-performance anode material for supercapacitors[J]. J. Mater Chem. A, 2019, 7(16): 9556-9564.

[8]Zhang S, Li D H, Chen S, et al. Highly Stable Supercapacitors with MOF-derived Co_9S_8/Carbon Electrodes for High Rate Electrochemical Energy Storage[J]. J. Mater Chem. A, 2017, 5(24): 12453-12461.

[9]Wan C C, Jiao Y, Li J. Flexible, highly conductive, and free-standing reduced graphene oxide/polypyrrole/cellulose hybrid papers for supercapacitor electrodes[J]. J. Mater Chem. A, 2017, 5(8): 3819-3831.

[10]YanJ, Fan Z J, Sun W, et al. Advanced Asymmetric Supercapacitors Based on $Ni(OH)_2$/Graphene and Porous Graphene Electrodes with High Energy Density[J]. Adv. Funct. Mater, 2012, 22(12)2632-2641.

[11]Wan C C, Jiao Y, Liang D X, et al. A high-performance, all-textile and spirally wound asymmetric supercapacitors based on core-sheath structured MnO_2 nanoribbons and cotton-derived carbon cloth[J]. Electrochim Acta., 2018, 285: 262-271.

[12]Jurewicz K, Delpeux S, Bertagna V, et al. Determination of vanadium valence in hydrated compounds[J]. Chem.

Phys. Lett. , 2001, 347: 36-40.

[13]Oh J K, Kim Y K, Lee J S, et al. Highly Porous Structured Polyaniline Nanocomposite for Free-sized and Flexible High-Performance Supercapacitor[J]. Nanoscale, 2019, 11(13): 6462-6470.

[14]Jiang H, Ma J, Li C Z. Mesoporous Carbon Incorporated Metal Oxide Nanomaterials as Supercapacitor Electrodes[J]. Adv. Mater, 2012, 24(30): 4197-4202.

[15]He Y-B, Li G-R, Tong Y X, et al. Single-crystal ZnO nanorod/amorphous and nanoporous metal oxide shell composites: Controllable electrochemical synthesis and enhanced supercapacitor performances[J]. Energy Environ. Sci. , 2011, 4(4): 1288-1292.

[16]Wang H Y, Deng J, Chen Y Q, et al. Hydrothermal synthesis of manganese oxide encapsulated multiporous carbon nanofibers for supercapacitors[J]. Nano Res. , 2016, 9(9): 2672-2680.

[17]Zhai Y P, Dou Y Q, Zhao D Y, et al. ChemInform Abstract: Carbon Materials for Chemical Capacitive Energy Storage [J]. Adv. Mater, 2011, 23(2): 4828-4850.

[18]Ghosh M, Vijayakumar V, Soni R, et al. Rationally designed self-standing V_2O_5 electrode for high voltage non-aqueous all-solid-state symmetric (2. 0 V) and asymmetric (2. 8 V) supercapacitors[J]. Nanoscale, 2018, 10(18): 8741-8751.

[19]Deng M-J, Yeh L-H, Lin Y-H, et al. 3D Network V_2O_5 Electrodes in a Gel Electrolyte for High-Voltage Wearable Symmetric Pseudocapacitors[J]. ACS Appl. Mater Interfaces, 2019, 11(33): 29838-29848.

[20]Qu Q T, Zhu Y S, Gao X W, et al. Core-Shell Structure of Polypyrrole Grown on V_2O_5 Nanoribbon as High Performance Anode Material for Supercapacitors[J]. Adv. Energy Mater, 2012, 2(8): 950-955.

[21] Guo C X, Yilmaz G, Chen S S, et al. Hierarchical nanocomposite composed of layered V_2O_5/PEDOT/MnO_2 nanosheets for high-performance asymmetric supercapacitors[J]. Nano Energy, 2015, 12: 76-87.

[22] Sarakinos K, Alami J, Konstantinidis S. High power pulsed magnetron sputtering: A review on scientific and engineering state of the art[J]. Surf Coat Technol. , 2010, 204(11): 1661-1684.

[23]Kumar A, Sanger A, Kumar A, et al. Sputtered Synthesis of MnO_2 Nanorods as Binder Free Electrode for High Performance Symmetric Supercapacitors[J]. Electrochim Acta. , 2016, 222: 1761-1769.

[24]WeiB B, Liang H F, Zhang D F, et al. CrN Thin Film Prepared by Reactive Magnetron Sputtering for Symmetric Supercapacitors[J]. J. Mater Chem. A, 2017, 5(6): 2844-2851.

[25]Coustier F, Hill J, Owens B B, et al. Silver-Doped Vanadium Oxides as Host Materials for Lithium Intercalation[J]. J. Electrochem. Soc. , 1999, 146: 1355-1360.

[26]Zhang J T, Jiang J W, Li H L, et al. A high-performance asymmetric supercapacitor fabricated with graphene-based electrodes[J]. Energy Environ. Sci. , 2011, 4(10): 4009-4015.

[27]Afzal A, Abuilaiwi F A, Habib A, et al. Polypyrrole/carbon nanotube supercapacitors: Technological advances and challenges[J]. J. Power Sources, 2017, 352: 174-186.

[28]AugustynV, Simon P, Dunn B. Pseudocapacitive oxide materials for high-rate electrochemical energy storage[J]. Energy Environ. Sci. , 2014, 7(5): 1597-1614.

[29]Wan C C, Jiao Y, Wei S, et al. Functional nanocomposites from sustainable regenerated cellulose aerogels: A review [J]. Chem. Eng. J. , 2019, 359: 459-475.

[30]Wan C C, Li J. Facile Synthesis of Well-Dispersed Superparamagnetic γ-Fe_2O_3 Nanoparticles Encapsulated in Three-Dimensional Architectures of Cellulose Aerogels and Their Applications for Cr(VI) Removal from Contaminated Water[J]. ACS Sustainable Chem. Eng. , 2015, 3(9): 2142-2152.

[31]Zhang H, Xie A J, Wang C P, et al. Bifunctional Reduced Graphene Oxide/V_2O_5 Composite Hydrogel: Fabrication, High Performance as Electromagnetic Wave Absorbent and Supercapacitor[J]. Chem. Phys. Chem. , 2014, 15(2): 366-373.

[32]Bondarenka V, Grebinskij S, Mickevičius S, et al. Determination of vanadium valence in hydrated compounds[J]. J.

Alloys. Compd. , 2004, 382: 239-243.

[33] WangJ-G, Liu H Y, Liu H Z, et al. Interfacial Constructing Flexible V_2O_5@ Polypyrrole Core-Shell Nanowire Membrane with Superior Supercapacitive Performance[J]. ACS Appl. Mater Interfaces, 2018, 10(22): 18816-18823.

[34] Saravanakumar B, Purushothaman K K, Muralidharan G. Interconnected V_2O_5 Nanoporous Network for High-Performance Supercapacitors[J]. ACS Appl. Mater Interfaces, 2012, 4(9): 4484-4490.

[35] Wang J, Polleux J, Lim J, et al. Pseudocapacitive Contributions to Electrochemical Energy Storage in TiO_2(Anatase) Nanoparticles[J]. J. Phys. Chem. C, 2007, 111(40): DA14925-14931.

[36] Shi H. Activated Carbons and Double Layer Capacitance[J]. Electrochim Acta, 1996, 41(10): 1633-1639.

[37] Gryglewicz G, Machnikowski J, Lorenc-Grabowska E, et al. Effect of Pore Size Distribution of Coal-Based Activated Carbons on Double Layer Capacitance[J]. Electrochim Acta, 2005, 50(5): 1197-1206.

[38] Prakash N G, Dhananjaya M, Reddy B P, et al. Molybdenum doped V_2O_5 Thin Films electrodes for Supercapacitors [J]. Mater. Today: Proc. , 2016, 3(10): 4076-4081.

[39] Balamuralitharan B, Cho I. -H, Bak J-S, et al. V_2O_5 nanorod electrode material for enhanced electrochemical properties by facile hydrothermal method for supercapacitor applications[J]. New J. Chem. , 2018, 42(14): 11862-11868.

[40] Foo C Y, Sumboja A, Tan D J H, et al. Flexible and Highly Scalable V2O5-rGO Electrodes in an Organic Electrolyte for Supercapacitor Devices[J]. Adv. Energy Mater, 2014, 4(12): 1400236.

[41] Balasubramanian S, Purushothaman K K. Carbon Coated Flowery V_2O_5 Nanostructure as Novel Electrode Material for High Performance Supercapacitors[J]. Electrochim. Acta, 2015, 186: 285-291.

[42] Sun W, Gao G H, Zhang K, et al. Self-assembled 3D N-CNFs/V_2O_5 aerogels with core/shell nanostructures through vacancies control and seeds growth as an outstanding supercapacitor electrode material[J]. Carbon, 2018, 132: 667-677.

[43] Bai M-H, Liu T-Y, Luan F, et al. Electrodeposition of vanadium oxide-polyaniline composite nanowire electrodes for high energy density supercapacitors[J]. J. Mater Chem. A, 2014, 2(28): 10882-10888.

[44] Ktz R, Carlen M. Principles and applications of electrochemical capacitors[J]. Electrochim Acta, 2000, 45(15-16): 2483-2498.

[45] Sathiya M, Prakash A S, Ramesha K, et al. V_2O_5-Anchored Carbon Nanotubes for Enhanced Electrochemical Energy Storage[J]. J. Am. Chem. Soc. , 2011, 133: 16291-16299.

[46] KobayashiK, Nagao M, Yamamoto Y, et al. Rechargeable PEM Fuel-Cell Batteries Using Porous Carbon Modified with Carbonyl Groups as Anode Materials[J]. J. Electrochem. Soc. , 2015, 162(8): F868-F877.

[47] Ramasamy V, Premkumar J, Pitchai R, et al. Robust, Flexible and Binder Free Highly Crystalline V_2O_5 Thin Film Electrodes and Their Superior Supercapacitor Performances[J]. ACS Sustainable Chem. Eng. , 2019, 7(15): 13115-13126.

[48] Boruah B D, Nandi S, Misra A. Layered Assembly of Reduced Graphene Oxide and Vanadium Oxide Heterostructure Supercapacitor Electrodes with Larger Surface Area for Efficient Energy Storage Performance[J]. ACS Appl. Energy Mater, 2018, 1(4): 1567-1574.

[49] YilmazG, Guo C X, Lu X M. High Performance Solid-State Supercapacitors Based on V_2O_5/Carbon Nanotube Composites[J]. Chem. Electro. Chem. , 2016, 3(1): 158-164.

[50] Qi R J, Nie J H, Liu M Y, et al. Stretchable V_2O_5/PEDOT Supercapacitors: A Modular Fabrication Process and Charging with Triboelectric Nanogenerators[J]. Nanoscale, 2018, 10(16): 7719-7725.

[51] Saravanakumar B, Purushothaman K K, Muralidharan G. Fabrication of two-dimensional reduced graphene oxide supported V_2O_5 networks and their application in supercapacitors[J]. Mater Chem. Phys. , 2016, 170: 266-275.

[52] Ray A, Roy A, Sadhukhan P, et al. Electrochemical properties of TiO_2-V_2O_5 nanocomposites as a high performance Supercapacitors electrode material[J]. Appl. Surf. Sci. , 2018, 443: 581-591.

[53] Saravanakumar B, Purushothaman K K, Muralidharan G. High performance supercapacitor based on carbon coated V_2O_5 nanorods[J]. J. Electroanal. Chem. , 2015, 758: 111-116.

中文题目：纳米多孔碳网络@ V_2O_5 亚微米砖复合材料的制备及其高倍率性能研究

作者：焦月，万才超，吴义强，韩景泉，包文慧，高鹤，王耀星，王成毓，李坚

摘要：制备了一种绿色生物质基纳米多孔碳网络(NCN)，并与 V_2O_5 亚微米砖(SMBs)集成。NCN的大比表面积和高孔容不仅为电化学反应提供了丰富的位点，而且稳定了 V_2O_5 SMBs 的结构。NCN@ V2O5 SMB 复合材料作为一种新型的阴极材料，在 0.2 mA · cm^{-2} 时具有较高的面积电容，达到 786 mF · cm^{-2}，循环 5000 次后具有良好的循环稳定性，电容保持率为 89.5%。此外，根据动力学分析，非扩散控制过程的贡献估计高达 95.5%-98.5%，使得电极具有超高的倍率性能(当电流密度从 0.2 增加到 5 mA · cm^{-2} 时，电容保留率为 82%)。在较低的电流密度(0.2-2 mA · cm^{-2})下，微孔比介孔更有利，而当电流密度大于 2 mA · cm^{-2} 时，外表面积的贡献更为显著。此外，用该阴极和 NCN 阳极组装的非对称超级电容器具有工作电压宽、循环寿命长、能量密度大(72.2 μW · h · cm^{-2})等优异的电化学性能。其优异的电化学性能和良好的生态友好性证实了 NCN@ V_2O_5 SMBs 作为超级电容器的潜力。

关键词：阴极材料；纳米多孔碳网络；超高速率容量

Wood-Derived Carbon Materials and Light-Emitting Materials*

Wei Li, Zhijun Chen, Haipeng Yu, Jian Li, Shouxin Liu

Abstract: Wood is a sustainable and renewable material that naturally has a hierarchical structure. Cellulose, hemicellulose, and lignin are the three main components of wood. The unique physical and chemical properties of wood and its derivatives endow them with great potential as resources to fabricate advanced materials for use in bioengineering, flexible electronics, and clean energy. Nevertheless, comprehensive information on wood-derived carbon andlight-emitting materials is scarce, although much excellent progress has been made in this area. Here, the unique characteristics of wood-derived carbon and light-emitting materials are summarized, with regard to the fabrication principles, properties, applications, challenges, and future prospects of wood-derived carbon and light-emitting materials, with the aim of deepening the understanding and inspiring new ideas in the area of advanced wood-based materials.

Keywords: Carbon materials; Light-emitting materials; Wood; Wood derivatives

1 Introduction

As a traditional material, wood has contributed hugely to the development of human society.[1,2] Wood is widely used in homes, heating, furniture, and paper.[3] Resource shortages have led to wood drawing increased attention because it is an environ-mentally friendly material. From a chemical viewpoint, wood is a natural composite material that consists of cellulose (40%-45% by weight), hemicellulose (20%-35% by weight), and lignin (10%-30% by weight).[4] Cellulose and hemicellulose are types of polysaccharides, which are easily carbonized to a controlled degree.[5-7] Considering its natural hierarchical structure, wood with a high polysaccharide content is advantageous for developing carbon materials with regular morphology.[8-10] Lignin is a heterogeneous amorphous polymer that constitutes a large proportion of the cell wall of wood, making it the second most abundant biomass on Earth after cellulose.[11,12] Lignin has interesting self-association and fluorescence emission properties, which have facilitated its use as self-assembled light-emitting nanomaterials.[13] Additionally, both polysaccharides and lignin possess abundant hydroxyl moieties, which allow these wood-derived components to readily undergo chemical modification. These inherent advantages of wood make it attractive as a raw material to fabricate advanced carbon and light-emit-ting materials.

Although substantial progress has been achieved in the development of wood-derived advanced carbon and

* 本文摘自 Applied Surface Science, 2015, 332: 565-572.

light-emitting materials, there have been few systematic reviews of this research area. Our goal here is to provide a critically selected overview of recent progress in this field. Typical preparation strategies, properties, and applications of recent wood-derived carbon materials and light-emitting mate-rials are introduced (Fig. 1). Specifically, wood-derived carbon spheres, carbon sponges, and carbon fibrils derived from wood are described. Notably, traditional wood-derived activated carbon will not be covered in this review as there are other systematic reviews in this area.[14,15] We would focus on discussion of preparation strategies, formation mechanism and morphology tuning of these carbon spheres, carbon sponges, and carbon fibrils. The preparation strategies and structure-property relationships of these materials are discussed in detail. Wood-derived light-emitting materials including fluorescent carbon dots (CDs), materials that resist aggregation-caused quenching (ACQ) (denoted as anti-ACQ materials), and circularly polarized luminescence (CPL) materials are then introduced, including their fabrication principles, structures, and optical proper-ties. Particularly, we would give a detailed discussion about the inherent priorities of wood-derived light-emitting materials. The applications of these wood-derived carbon and light-emitting materials including adsorption, bioimaging, chemical sensing, solar steam generation, and electronic devices are then summarized. Finally, the challenges and future perspectives of wood-based advanced materials are discussed.

2 Wood-Derived Carbon Materials

Wood-derived components could be converted to various carbon materials with a certain morphology, including sphere, sponges and fibrils.[16] Interestingly, the morphology of these carbon materials could be well tuned by controlling the car-bonization temperature, heating speed, concentration of precursors and additives.[17,18] While, morphology of carbon materials directly derived from natural wood was not so easy to be tuned, attributed to their inherent difficult microcrystalline preferred orientation. In this part, we would focus on discus-sion of the preparation methods, formation mechanism and morphology tuning on wood-derived carbon materials with various morphology.

2.1 Carbon Spheres

Wood-derived polysaccharides can be converted to carbon spheres (CSs) via hypothermal carbonization (HTC).[19] During the HTC, polysaccharide chains first hydrolyzed into oligomers and glucose, followed by dehydration and fragmentation into soluble products like hydroxymethylfurfural (HMF).[20,21] Subsequent condensation or polymerization reactions led to the formation of soluble polymers. The final carbon structures were obtained through a switch from a carbonaceous polyfuran to a carbon network of increasingly enlarging aromatic domains (Fig. 2a).[22] The reason that HTC of carbohydrates leads to spherical morphology is probably because the small nuclei formed by the polymerization/polycondensation of HMF decomposed from carbohydrates and these nuclei grow into spheres with a certain size according to the LaMer model.[23] In 2009, Sevilla and Fuertes[24] prepared CSs with particle sizes of 0.2–1.5 μm by the HTC of cellulose in the temperature range of 200–250 ℃. Titirici's group also investigated the transformation mechanism of cellulose into spherical hydrochars. Our group reported the preparation of CSs using carboxymethyl cellulose (CMC) as a carbon precursor. As shown in Fig. 2b, the CSs prepared via HTC of CMC at 210 ℃ had a smooth surface and dispersed uniformly with a narrow diameter distribution of 2–3 μm.[25] Our group also reported a cage-like ordered mesoporous CSs using CMC as a carbon pre-cursor and F127 as a soft template (Fig. 2c). By varying the HTC reaction time at 210 ℃, the mesoporous structure of CS changed from stripe-like hexagonal to mixed hexagonal and cubic to cubic.[26] Additionally, the glucose which derived from wood is a simple monosaccharide that has been used extensively as a starting material to prepare CSs through HTC.[27] For example, Sun et al.[28] prepared CSs from glucose by HTC at 160–180 ℃. The diameters of the CSs were tuned from 200 to 1500 nm by changing the HTC duration, temperature, and glucose concentration. Besides these CSs micellar sphere, hollow carbon sphere could also be prepared from wood-derived polysaccharide. Ikeda and co-workers used silicon dioxide as a hard template to prepare hollow CSs with porous walls from hydrothermally treated glucose.[29] The hollow CSs inversely replicated the hard templates and exhibited a cavity with a diameter of 170 nm and a carbon shell with a thickness that could be controlled by varying the glucose concentration. Wood-derived polysaccharide could also be converted to functional carbon composites together with metal ions or other organic polymers. Carbon-coated noble metal core-shell colloids were prepared by introducing noble metal ions [i. e., Au(Ⅰ) and Ag(Ⅰ)] during HTC (Fig. 2c).[28] Zhang's group recently proposed a method to fabricate of N/S co-doped carbon microspheres with an expanded interlayer by pyrolyzing cellulose/polyaniline composite microspheres containing dodecylbenzene sulfonic acid. The microspheres before carbonization were prepared from cellulose/polyaniline solution via an emulsion method, as depicted in Fig. 2e. The fabricated carbon microspheres possessed a rough surface with interconnected porous nano-walls and showed high performance as anode materials for sodium-ion batteries.[30]

2. 2 Carbon Sponges

Wood-derived carbon sponges possess low density, desirable conductivity, and interconnected transport pathways, attributed to natural anisotropic channels of wood.[31] The 3D hierarchical porous structure of wood with open and elongated microchannels has excellent advantages for fast charge storage and transport, which can realize low tortuosity and high ion/ electron diffusivity. For example, Lu and co-workers duplicated the vertical microchannel structures of wood into a free-standing bulk $LiCoO_2$ cathode with a thickness of up to 1 mm through sol-gel infiltration and calcinations process, as graphically illustrated in Fig. 3a.[8] As shown in Fig. 3b, in a traditional $LiCoO_2$ electrode, the active $LiCoO_2$ particles are connected by conductive carbon black and polymer binders to form an architecture which contain poor electrolyte diffusion and long lithium-ion transport pathways. On the contrary, the microchannel structures in the wood inspired $LiCoO_2$ cathode lowered the tortuosity, which facilitated electrolyte diffusion and shortened the lithium-ion transport distance. As a result, the ultrathick $LiCoO_2$ cathode with continuous microchannels delivered a high areal capacity of 22.7 $mAh \cdot cm^{-2}$ in a dynamic discharge test, which far exceeded the areal loading capacity of a commercial $LiCoO_2$ cathode.

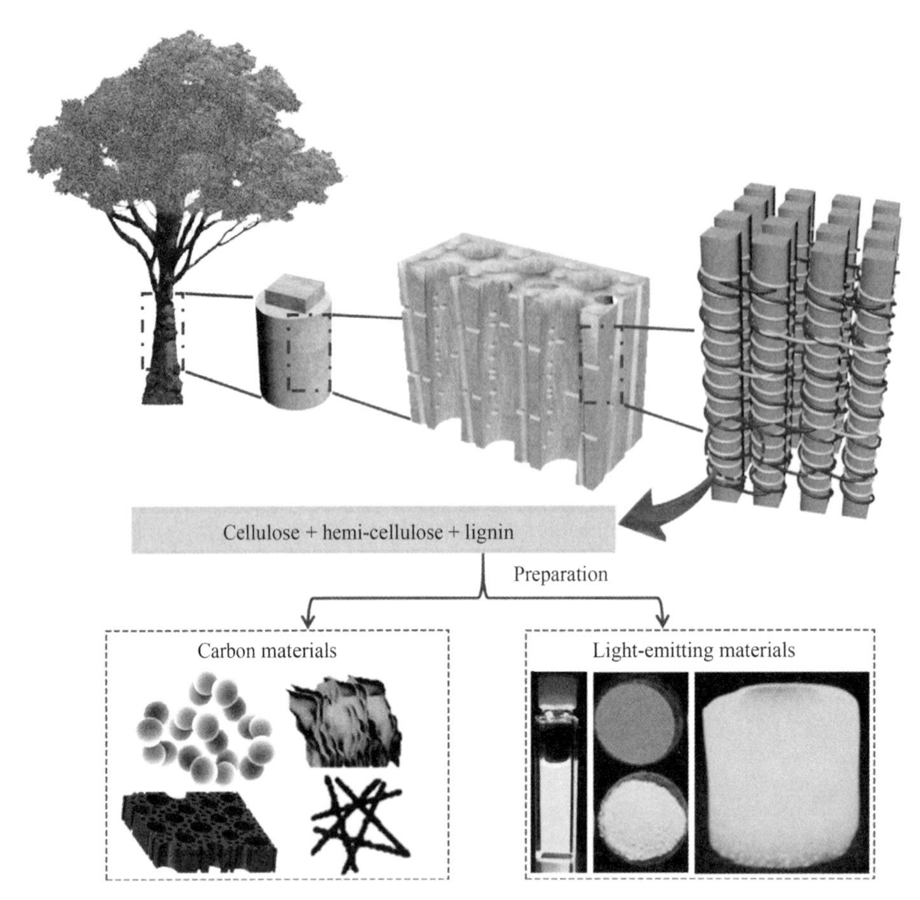

Fig. 1 Schematic illustration of the preparation of carbon and light-emitting materials from wood-derived components

Carbonized and activated wood were used as a substrate by loading ruthenium nanoparticles in its porous microchannels to construct a cathode material for lithium-oxygen batteries.[32] As illustrated in Fig. 3c, the aligned microchannels accelerated oxygen gas diffusion through the whole ultrathick electrode. The electrolyte could wet the plentiful hierarchical pores of the microchannel walls. Along the microchannel direction, the thin electrolyte layers formed on the walls ensured continuous pathways for fast lithium-ion transport. Wood-derived carbon sponges are also desirable to construct compressible materials. Chen et al.[9] directlyconverted wood into a highly compressible wood carbon sponge(WCS). As shown in Fig. 3d, nucleophilic sodium sulfite in alkaline solution promoted the sulfonation of lignin, hemicellulose, and part of the cellulose in the wood sample, leading to their removal. The thin cell walls became porous and even broke, whereas the thicker rays survived(step 1). Lignin and hemicelluloses were removed through hydrogen peroxide treatment which leads to the breakage of the thin cell walls(step 2). The cell walls can form in a unique lamellar structure with multiple stacked and connecting arched layers during freeze drying. This lamellar structure exhibited remarkable mechanical compressibility,

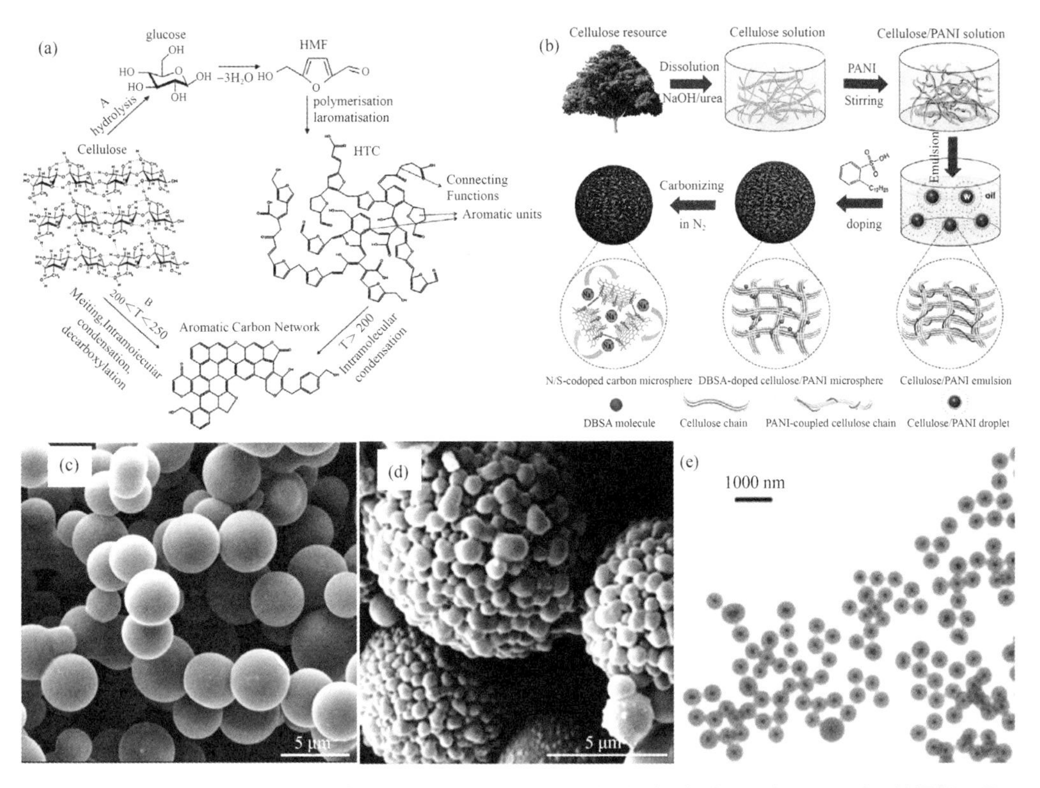

Fig. 2 a) Conversion of cellulose into a furan-rich aromatic network via hydrothermal conversion (HTC). Reproduced with permission.[22] Copyright 2012, Royal Society of Chemistry. b) Illustration of the formation process of a cellulose-derived carbon microsphere composite material. Reproduced with permission.[26] Copyright 2015, Elsevier. c) Smooth carbon spheres (CSs) prepared from carboxymethylcellulose (CMC) via HTC. Reproduced with permission.[25] Copyright 2004, Wiley-VCH. d) Cage-like CSs prepared from CMC via the combination of HTC and soft templating. Reproduced with permission.[26] Copyright 2015, Elsevier. e) Carbon spheres with an Au core prepared from glucose. Reproduced with permission.[28] Copyright 2016, Wiley-VCH

fatigue resistance, and pressure sensitivity after carbonization. Based on a WCS, Huang and co-workers established a solution treatment and carbonization method to convert rigid nonconductive bulk wood into flexible and conductive composites.[33] As shown in Fig. 3e, the bleached trunk was carbonized in an inert atmosphere to achieve high conductivity. Polydimethylsiloxane was infiltrated into the channels of the carbonized wood to improve the stability, elasticity, and flexibility of the resulting WCS. In this manner, WCS showed that high sensitivity (10.74 kPa^{-1}) was obtained over a wide linear region (100 kPa, $R^2 = 99\%$). The sensors achieved accurate human physiological signal monitoring, such as real-time respiration detection and epidermal pulse measurement.

2.3 Carbon Fibrils

The first carbon fibrils obtained from sustainable biomass fibers, such as cotton, linen threads, wood splints and paper.[34] Considerable research on sustainable precursors was conducted in the 1950s-70s, but difficulties such as low yields and high production costs limited their utilization at that time. Interest in producing carbon fibers from sustainable resources such as cellulose, lignin, and liquefied wood is again increasing because of the oil and energy shortage.[35,36] To produce carbon fibers from cellulose fibers with high carbon yield, three general options have been developed for this purpose include using flame retardants, slow heating rates and active

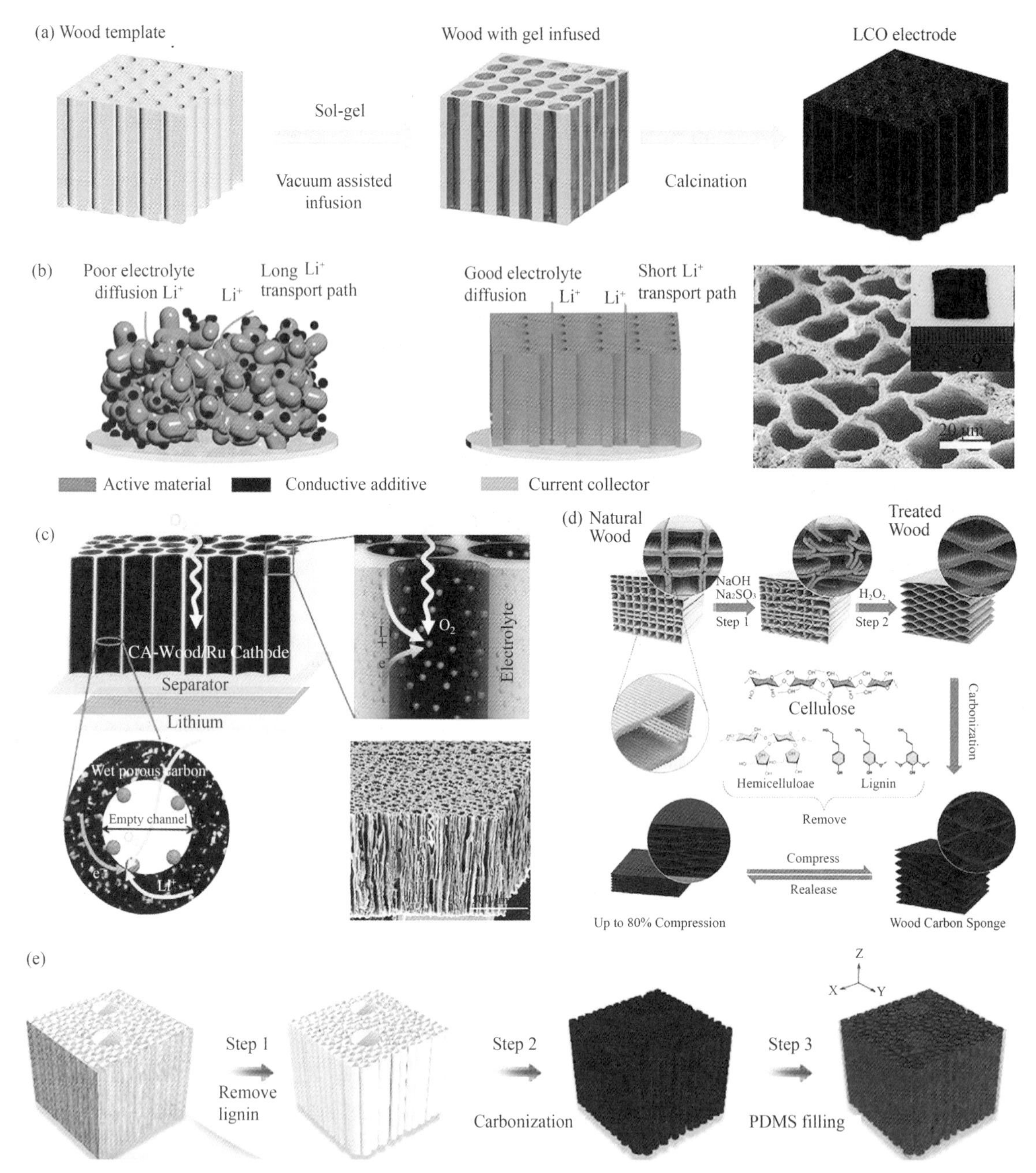

Fig. 3 a) Illustration of the fabrication procedure of an ultrathick $LiCoO_2$ cathode by wood templating. b) Illustrations of a traditional electrode with poor electrolyte diffusion and long lithium-ion transport path and wood-inspired electrode with vertical channels to shorten the lithium-ion transport path and cross-sectional scanning electron microscopy image of an ultrathick $LiCoO_2$ cathode. a, b) Reproduced with permission.[8] Copyright 2017, Wiley-VCH. c) Schematic diagram of a lithium-oxygen battery with a CA-wood/Ru cathode. Reproduced with permission.[32] Copyright 2017, Wiley-VCH. d) Graphical illustration of the design and fabrication process of a highly compressible WCS. Reproduced with permission.[9] Copyright 2018, Cell Press. e) Synthetic route to flexible and conductive carbon/silicone composites. Reproduced with permission.[33] Copyright 2018, Wiley-VCH

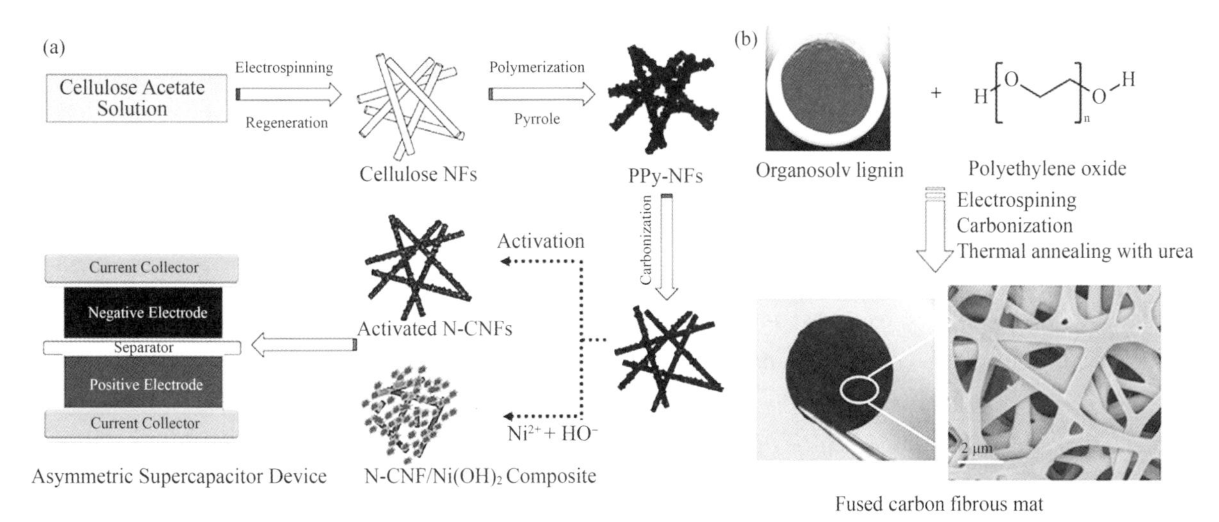

Fig. 4 a) Schematic of the preparation of carbonized fibrils from cellulose. Reproduced with permission.[38] Copyright 2015, American Chemical Society. b) Schematic of the preparation of carbonized fibrils from lignin. Reproduced with permission.[43] Copyright 2013, American Chemical Society

pyrolysis atmosphere.[37] Nitrogen-functionalized carbon nanofibers (N-CNFs) were prepared as a high-performance supercapacitor electrode material using a multistep method (Fig. 4a).[38] They first electrospun cellulose acetate (CA) into uniform nanofibers. The as-prepared nanofibers were immersed in an alkaline solution to convert CA into cellulose. Subsequently, polypyrrole was coated onto the cellulose nanofibers and then direct carbonization of the polypyrrole-coatednanofibers to form carbon nanofibers was conducted. Electrodes prepared using the N-CNFs showed high specific capacitance and excellent cycling stability. Besides cellulose, lignin has also been used as a raw material to fabricate carbonized fibers.[39-41] Otani et al.[42] described several methods to form carbonized fibers from lignin. A nitrogen-doped freestanding fused carbon fibrous mat was fabricated from a lignin/polyethylene oxide blend via electrospinning method followed by carbonization and thermal annealing in the presence of urea (Fig. 4b). The electrochemical properties of the carbon fibers were investigated as an anode in a lithium-ion battery.[43] Although lignin is sustain-able and cheap as a raw material to prepare carbon fibrils, further effort still need to be expended to improve its purity and narrow its molecular weight distribution to produce high-quality carbonized fibrils.[42] Carbonized fibrils can also be prepared from liquefied wood.[44-46] Ma and co-workers prepared carbon fibers from liquefied wood.[47] Characterization of the fibers by Fourier transform infrared spectroscopy and X-ray diffraction indicated that the apparent crystallite size and layer-plane length parallel to the fiber axis gradually increased during carbonization. Theas-obtained carbon fibers showed a maximum tensile strength of 1.7 GPa. Preparation of carbonized fibrils from liquefied wood did not require separation of cellulose and lignin from wood cell walls, which simplified their preparation compared to thatof carbonized fibrils made from cellulose and lignin. However, the as-obtained carbonized fibrils contained numerous structural defects and relatively compromised mechanical performance because of their multiple and unpurified components.

2.4 Carbon Aerogels

Carbon aerogels have unique 3D networks and thus exhibit many fascinating physical properties.[48,49] As a result, lots of technical methods, such as in situ self-assembly, freeze casting, the sol-gel method and 3D printing, were developed for fabricating carbon aerogels.[50] Wood-derived carbon aerogels show advantages over traditional carbon aerogels in terms of compressibility and eco-friendliness. Recently, Freeze casting has been well developed for preparing carbon aerogels with anisotropic structures. Zhuo et al.[51] prepared a lamellar carbon aerogel via mixing, directional freezing, freeze drying, and carbonization using cellulose nanocrystals (CNCs), glucose, and graphene oxide as precursors (Fig. 5a). The arch-shaped lamellae (Fig. 5b) sustained large out-of-plane deformation without leading to plastic deformation due to their small material strain. And then it recovered to their original shape immediately after the compression force was removed. Additionally, Peng and co-workers fabricated a highly compressible wave-shaped carbon nanotube (CNT)/reduced graphene oxide (rGO)-carbon nanofiber (CNF) carbon aerogel by directional freeze casting (Fig. 5c).[52] The CNFs can enhance the interaction of rGO with CNTs and the CNTs produced strong yet flexible wave-shaped layers. As shown in Fig. 5d, the CNT/rGO-CNF aerogel could be highly compressed and then recovered its original height at high strain. Based on the continuous, oriented, and CNF-strengthened wave-shaped carbon layers, the aerogel demonstrated a stable structure with excel-lent mechanical compressibility and elasticity. In addition to freeze casting, the sol-gel method is also a common approach to prepare carbon aerogels. For example, Alatalo et al.[53] reported the hydrothermal synthesis of nitrogen-rich carbon materials from cellulose. These materials were designed for the removal of metal contaminants from water. Sulfuric acid was added to the reaction mixture because dehydration of cellulose can be catalyzed at low pH. In the reaction, ovalbumin played two roles, that is, it acted both as structure directing agentand natural nitrogen source. Thermally induced aqueous gels were formed above the denaturation temperature, while its can achieve co-condensation with HMF during the initial formation of the carbonaceous network via Maillard reactions. The same group also used a similar method to prepare a cellulose carbon aerogel for electrocatalytic oxygen reduction from soy protein.[54] A limitation of both freeze casting and the sol-gel method is that they require a time-consuming freeze-drying process. To overcome this limitation, Li and co-workers developed the homogenization-filtration method, which involves drying at ambient temperature.[55] Using their method, A novel pressure-sensitive and conductive carbon aerogels were fabricated by pyrolysis of cellulose aerogels without any deformation through directly dried in an oven. The resulting aerogels possessed an ultralow density of 4.3 $mg \cdot cm^{-3}$, high compressibility of 80%, high electrical conductivity of 0.47 $S \cdot cm^{-1}$, and high absorbency of 80-161 $g \cdot g^{-1}$ for oils and organic liquids.

3 Wood-Derived Light-Emitting Materials

Light-emitting materials display light emission after being excited by photons, electrons, or chemicals.[56-62] Most wood-derived light-emitting materials are photoluminescent, suggesting that they will emit fluorescence or phosphorescence upon light irra-diation.[63-65] The main focus of this section is fluorescent and phosphorescent wood-derived materials including fluorescent CDs, anti-ACQ materials, and CPL materials.

3.1 Fluorescent Carbon Dots Prepared fromWood-Derived Components

CDs are novel luminescent nano-size carbon materials that are less than 10 nm in size.[66,67] CDs were first reported in 2004 as a byproduct in the synthesis of carbon nanotubes.[68] In 2006, Sun et al.[69]

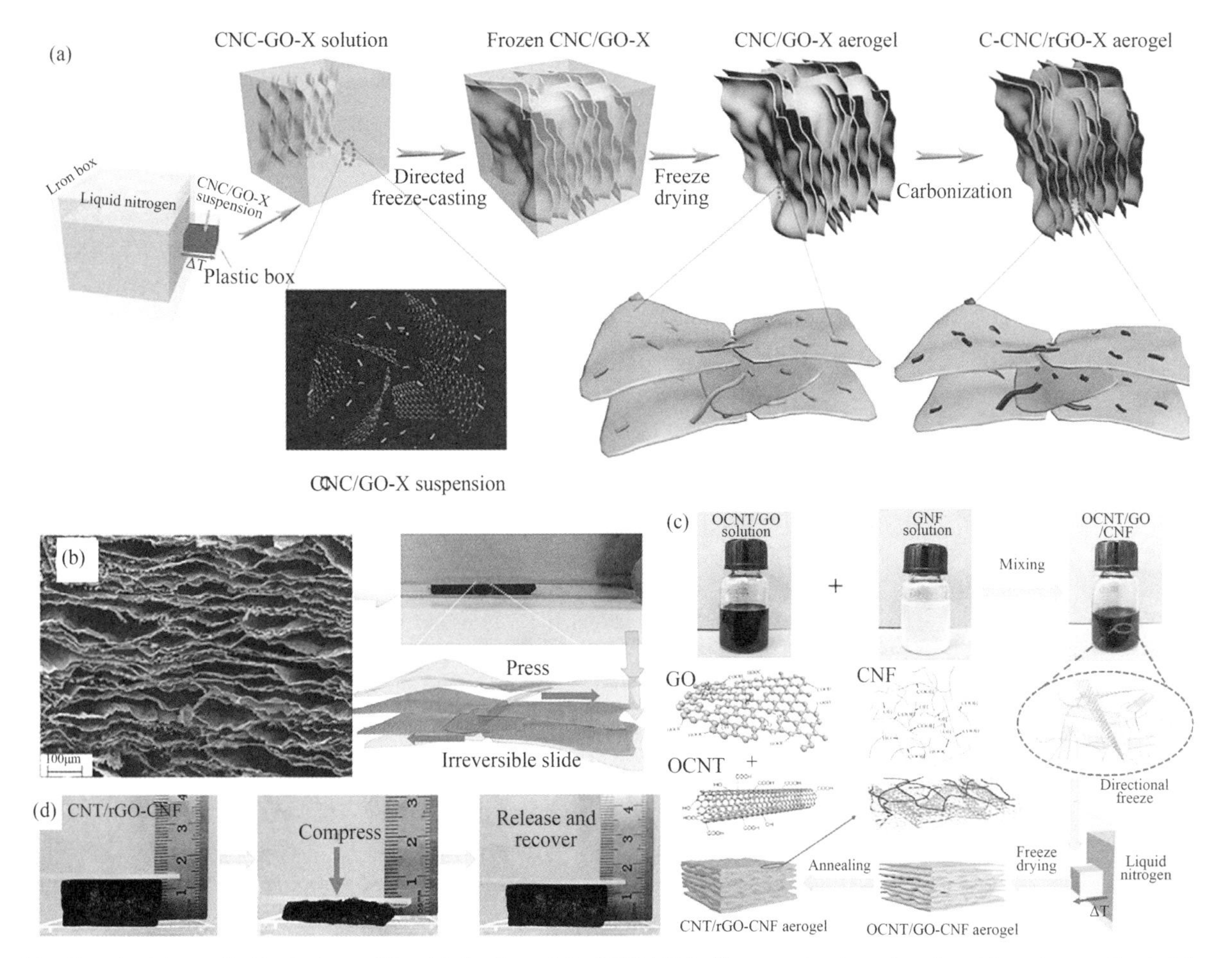

Fig. 5 a) Schematic illustration of the fabrication of C-CNC/rGO-X aerogels via a directional freezing method. b) Digital photographs and sche- matic illustrations of the supercompression and elasticity mechanisms of the C-CNC/rGO-glu2 carbon aerogel. a, b) Reproduced with permission.[51] Copyright 2018, Wiley-VCH. c) Schematic illustration of the fabrication of the CNT/rGO-CNF aerogel. d) Digital photographs of the compression and recovery of CNT/rGO-CNF. c, d) Reproduced with permission.[52] Copyright 2018, Royal Society of Chemistry

reported to prepare luminescent CDs using laser ablation and surface passivation. Since then, a number of studies investigating the fascinating properties of CDs have been carried out.[70-72] Compared with semiconductor quantum dots,[73,74] upconversion nanoparticles,[75,76] and organic dyes,[77,78] photoluminescent CDs prepared from wood-derived biomass show many advantages, including sustainability, easy preparation, and favorable optical properties.[79] Additionally, compared to other biomass materials, wood-derived polysaccharide was easy carbonized and fluorescent carbon dots can be easily prepared via low temperature hydrothermal carbonization method. Another wood-derived component, lignin, had inherent fluorescence. Thus, using lignin to prepare carbon dots can be simply via molecular aggregation. Attributed to these reasons, wood-derived carbon dots were a hot topic in past years. Ding and co-workers produced single-crystalline graphene quantum dots (GQDs) from lignin biomass via a two-step method including oxidative cleavage and aromatic fusion (Fig. 6a).[80] The as-prepared GQDs contained a honeycomb graphene network and were one to three atomic layers thick. Interestingly, the GQDs showed not only bright fluorescence, but also upconversion properties. Additionally, the GQD sexhibited long-term photostability, water solubility, and biocompatibility. Niuet al.[81] prepared fluorescent CDs using a novel molecular aggregation method(Fig. 6b). The self-assembly of lignin molecules in ethanol solution can by tuned by regulating their concentration. Accordingly, they finely tuned the lignin concentration to produce fluorescent CDs with a size of 5 - 10 nm. The as-prepared CDs demonstrated interesting excitation-dependent downconversion emission and near-infrared-excited upconversion emission.

Cellulose and their derivatives have also been converted to fluorescent CDs via an HTC method. Our group prepared fluorescent nitrogen-containing carbon nanodots(CNDs) via the one-pot HTC of CMC using and urea (Fig. 6c).[24] The CNDs possessed diameters of 2-8 nm and demonstrated pH-sensitive fluorescence. Then CNDs were used as a photosensitizer to assist photocatalytic degradation of methylene blue(MB) and phenol by titanium dioxide(TiO_2) under visible-light irradiation. We also converted CNCs into a CD-doped fluorescent hydrogel in situ using a hydrothermal method (Fig. 6d).[82] Hydrothermal treatment triggered molecular crosslinking and carbonization of CNCs, leading to simultaneous formation of the hydrogel andfluorescent CDs. The CDs generated in situ had diameters of 2-6 nm. The CD-decorated hydrogel exhibited a broad spectral response and high fluorescence stability at various pH values. Although numerous achievements have been obtained in the preparation of fluorescent CDs from wood-derived biomass, their emission mechanism needs to be clarified and their excita-tion/emission wavelength should be lengthened in the future.

3.2 Cellulose/Lignin-Based Anti-ACQMaterials

Organic fluorescent chromophores show great potential for use in biomedical applications and organic light-emitting devices.[83-85] However, most of these chromophores lose their fluorescent emission when they aggregate via π-π stacking, which compromises their performance in these applications.[86,87] Several methods have been proposed to conquer this challenge, such as designing molecules with aggregation-induced emission (AIE),[88,89] introducing moieties with high steric bulk,[90,91] and dispersing the chromophores in a matrix.[92,93] Wood-derived components such as cellulose and lignin have shown great promise for the preparation of chromophores without ACQ. Integrating chromophores with cellulose and using the inherent aggregation-induced emission of cellulose/lignin are currently the two main strategies employed to prepare anti-ACQ materials.

Tian et al.[94] first demonstrated the covalent grafting of an ACQ chromophore to cellulose to prepare an anti-ACQ solid emission material(Fig. 7a). The synergy between the luminogen anchoring and diluting effects of the cellulose skeleton efficiently inhibited their aggregation and self-quenching. After that, the same group designed a cellulose-based anti-ACQ ratiometric fluorescent material using cellulose as a backbone (Fig. 7b).[95] Fluorescein isothiocyanate (FITC), which is a biogenic-amine indicator that exhibits green emission, and protoporphyrin IX(PpIX) as a red-emitting internal reference were covalently bonded onto CA chains to form CA-FITC and CA-PpIX, respectively, and remove their ACQ properties. Ratio-metric fluorescent materials were obtained by mixing CA-FITC, the cellulose derivatives and CA-PpIX. The fluorescence color was finely tuned by the mixing ratio. As-prepared ratiometric fluorescent materials exhibited fast and reversible responses to biogenic amines and also possessed high processability.

Li et al.[96] also combined an ACQ chromophore and cellulose via non-covalent hydrogen bonding to prepare an anti-ACQ solid fluorescent material(Fig. 7c). The hydroxyl moieties of cellulose formed hydrogen

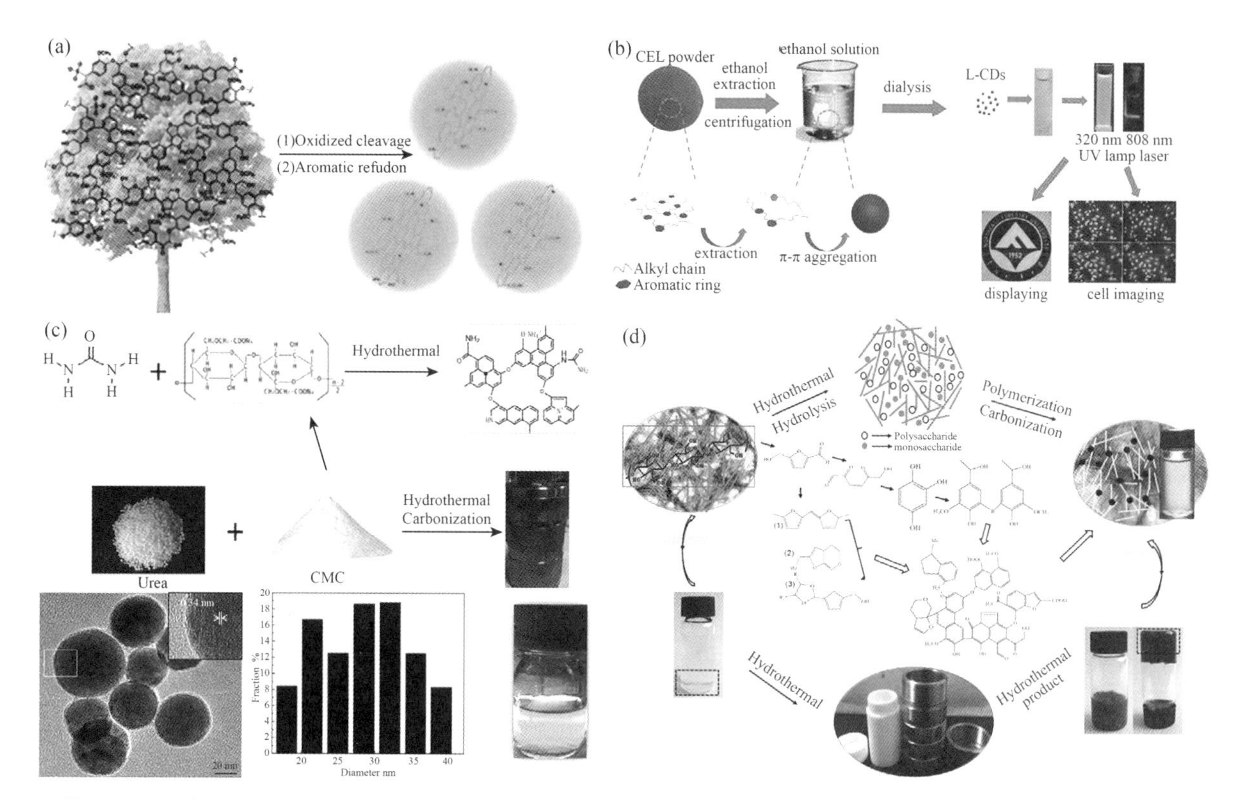

Fig. 6 a) Schematic illustration of the preparation of lignin-based carbon dots (CDs) via hydrothermal carbonization. Reproduced with permission.[80] Copyright 2018, Royal Society of Chemistry. b) Preparation of lignin-based CDs via molecular aggregation and a fluorescence image of their emission. Reproduced with permission.[81] Copyright 2017, American Chemical Society. c) Preparation of cellulose derivatives-based CDs and a fluorescence image of their emission. Reproduced with permission.[25] Copyright 2015, Elsevier. d) Preparation of a cellulose derivative-based CD/hydrogel hybrid and a fluorescence image showing its emission. Reproduced with permission.[82] Copyright 2017, Elsevier

bonds with guest ACQ coumarin chromophores. As a result, π-π stacking of ACQ chromophore molecules was strongly suppressed, which was attributed to the steric effect of cellulose. In such a manner, cellulose successfully converted an ACQ fluorescent compound to a material exhibiting stable emission behavior. These reportsdemonstrated that grafting ACQ chromophores to cellulose via covalent or noncovalent bonding is an efficient strategy to pre-pare anti-ACQ materials with attractive emission properties.

In fact, using cellulose for preparation of solid-state fluorescent materials was general for all aggregation-caused quenching fluorescent dyes with hydroxyl moieties. Most of the fluorescence quenching for the solid state of these dyes was attributed to the π−π stacking. However, all these dyes with hydroxyl moieties can form hydrogen bond with cellulose. As a result, cellulose served as a space to separate dyes for preventing their π−π stacking and led to a solid fluorescence emission, which was promising for fabricating cheap and sustainable solid fluorescent materials in wide range.

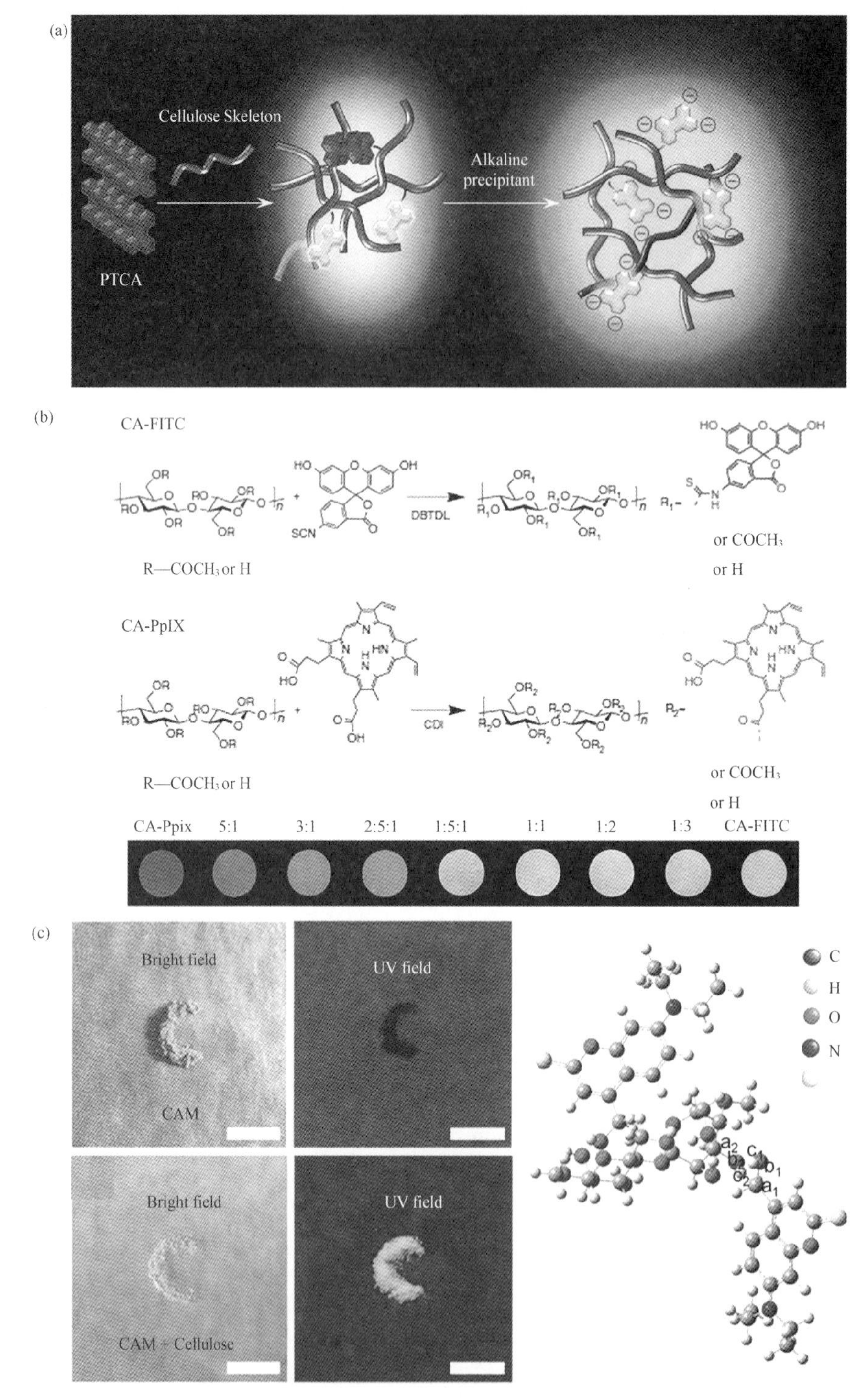

Fig. 7 a–c) Schematic illustrations of cellulose-based anti-ACQ perylene tetracarboxylic acid materials (a), CA-FITC and CA-PpIX materials (b), and coumarin materials (c). a) Reproduced with permission.[94] Copyright 2016, Wiley-VCH. b) Reproduced under the terms of the CC-BY Creative Commons Attribution 4.0 International License (http://creativecommons.org/licenses/by/4.0/).[95] Copyright 2019, The Authors, published by Springer Nature. c) Reproduced with permission.[96] Copyright 2019, American Chemical Society

Designing AIE molecules is another strategy to develop anti-ACQ materials. Generally, AIE molecules are prepared via com-plicated and toxic organic syntheses, which might hinder their practical applications. Xue et al.[97] reported that sulfonated alkali lignin displayed AIE active fluorescence. Clustering of carbonyl groups and restriction of intramolecular rotation effects together contributed to the AIE activity of sulfonated alkali lignin. Our group reported the AIE fluorescence of cellulolytic enzyme lignin(Fig. 8a).[98] Addition of ethanol to lignin basic aqueous solution(the fraction of ethanol was increased from 0% to 90%), which promoted the aggregationof lignin molecules, triggered intensified fluorescence. The proposed mechanism of this AIE emission involved the formation of J-aggregates during lignin aggregation. Interestingly, we found that carboxymethylated nanocellulose(C—CNC) without aromatic structure also showed fluorescence(Fig. 8b).[99] C—CNC was nonluminescent in diluted solution, but has strong emission when aggregated in nanosuspensions. A den-sity functional theory calculation confirmed that the luminescence of C—CNC originated from the through-space conjugation of O and C =O of C—CNC. Yuan's group investigated the photo-luminescence of microcrystalline cellulose, 2-hydroxyethyl cellulose, hydroxypropyl cellulose, and CA (Fig. 8c).[100] While the first three materials showed bright emission and distinct room-temperature phosphorescence, CA displayed relatively low intensity emission without noticeable room-temperature

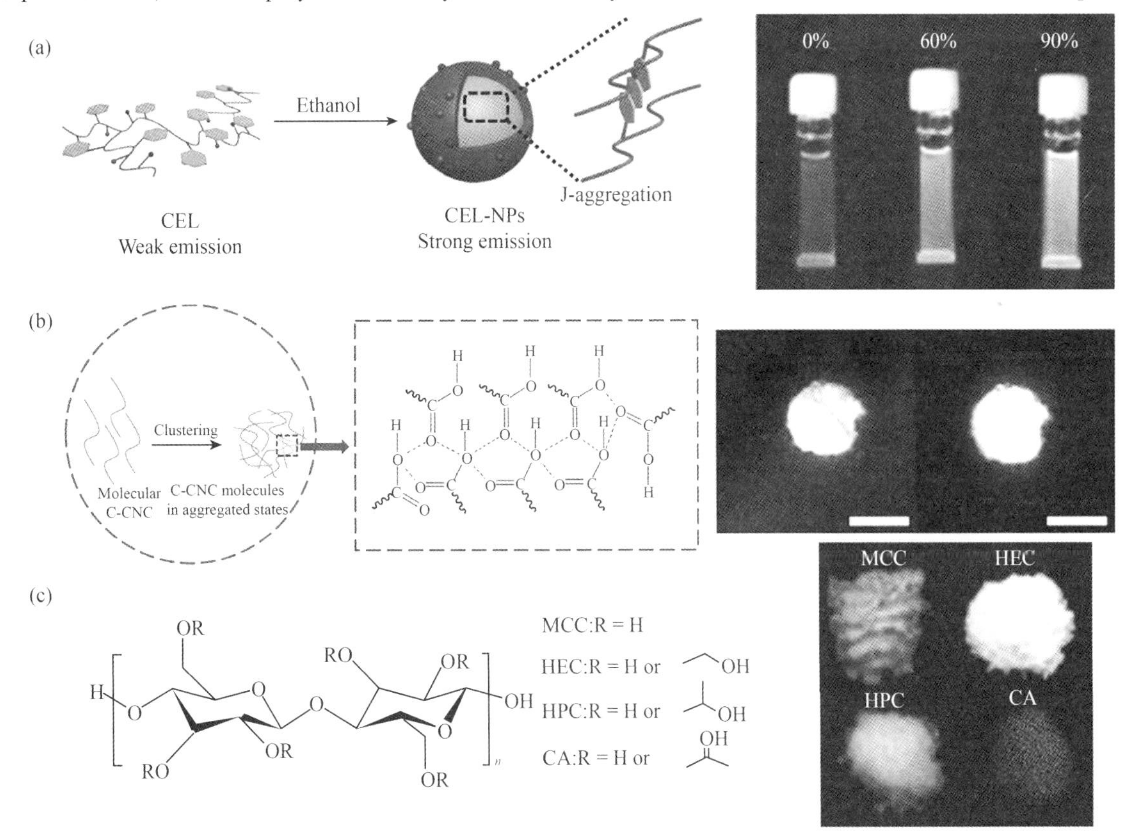

Fig. 8 a) Schematic illustration of AIE fluorescence of lignin; b) AIE fluorescence of C-CNC under UV light; c) clustering-induced emission of cel- lulose derivatives under UV light. a) Reproduced with permission.[98] Copyright 2018, American Chemical Society. b) Reproduced under the terms of the CC-BY Creative Commons Attribution 4. 0 International License(http: //creativecommons. org/licenses/by/4. 0/).[99] Copyright 2019, The Authors, published by Frontiers. c) Reproduced with permission.[100] Copyright 2019, Springer Nature phosphorescence. The emission behavior of the materials was explained in terms of the clustering-triggered emission mechanism and conformation rigidification.

3.3 Cellulose-Based CPLMaterials

CPL has drawn a great deal of interest because of its wide application prospects in 3D optical displays, enantioselective synthesis, chiral recognition, and optical information storage devices.[101-107] Inspired by

crustaceans, solid chiral photonic films were pre-pared via the self-assembly of CNCs.[108] The films could selectively transmit right-handed circularly polarized light and reflect left-handed circularly polarized light because of the left-handed photonic crystal structure. As a result, integrating normal fluorescent chromophores such as organic chromophores, CDs, and inorganic phosphors into chiral photonic cellulose films is a convenient and efficient way to prepare CPL materials.

Ikai and co-workers reported cellulose-based materials that exhibited CPL in solution (Fig. 9b).[109] A series of fluorescent cellulose derivatives containing fluorescent pyrene-basedp-conjugated groups were obtained from microcrystalline cellulose through carbamoylation and subsequent cross-coupling reactions. These chiroptical cellulose derivatives showed greenish CPL with dissymmetry factors (glum) larger than 3.0×10^{-3}. The efficient CPL of the cellulose derivatives wasattributed to the helical structure of the cellulose

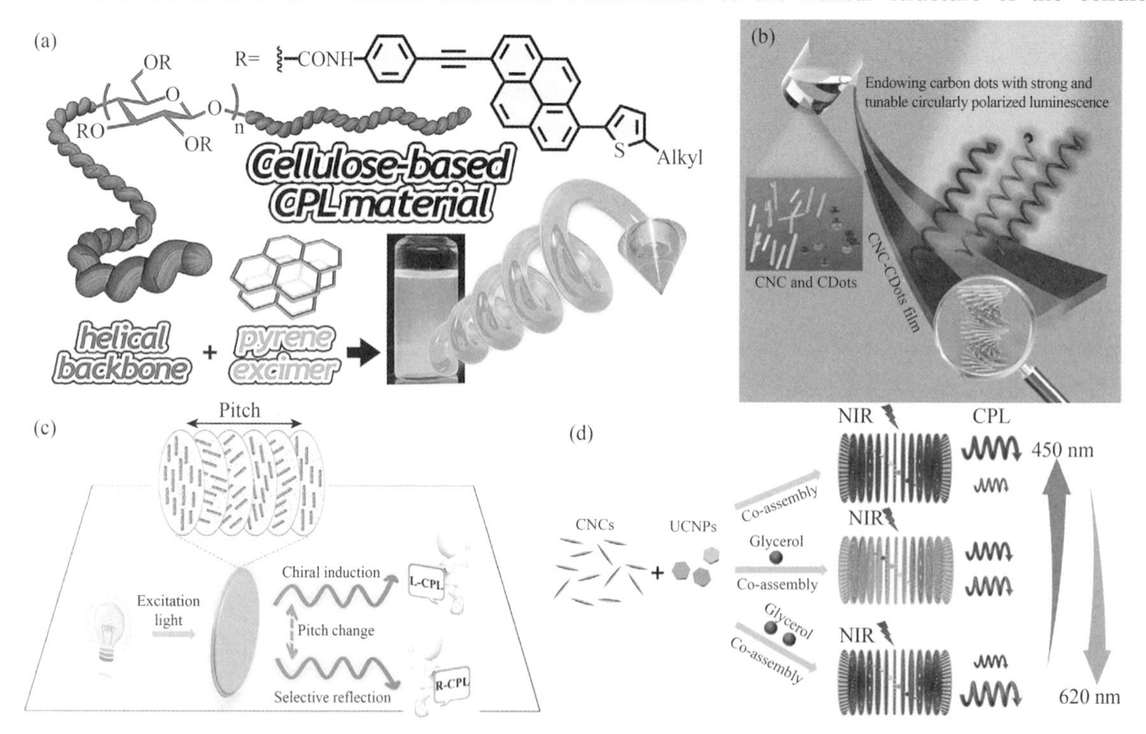

Fig. 9 a) Cellulose-based CPL materials with pyrene chromophores. Reproduced with permission.[109] Copyright 2017, Elsevier. b) Schematic illus-tration of the evaporation-induced cooperative assembly of CNC-derived CDs exhibiting PBG-based right-handed CPL with precise handedness and tunable chiroptical activity. Reproduced with permission.[111] Copyright 2018, Wiley-VCH. c) Illustration of the generation of full-color and switchable CPL from achiral dyes assembled in cholesteric cellulose films. Reproduced with permission.[112] Copyright 2019, Royal Society of Chemistry. d) Tunable UC-CPL emission (450 and 620 nm) with a tailored glum that was obtained from chiral photonic films with a tunable PBG. Reproduced with permis-sion.[114] Copyright 2019, American Chemical Society backbone and intramolecular excimer formation of the pyrene units. Xu's group incorporated several luminophores into photonic cellulose films that displayed CPL that could be tuned by simply changing the photonic bandgap(PBG) and selecting different luminophores ranging from molecular to nanometer scale. [110] After that, the same group developed tunable CPL materialsusing CDs and chiral photonic cellulose films. [111] CPL strength and | g | values could be tuned by changing helical super-structures, CD loading amount, and irradiation wavelength(Fig. 9b).

He et al. [112] reported full-color and mirror-image CPL from blending achiral fluorescent and cholesteric cellulose films(Fig. 9c). The CPL direction was easily switched by changing the helical pitch of photonic films. Left-handed CPL with glum of 10^{-2} and right-handed CPL with glum of 10^{-1} were ascribed to the chiral induction effect and selective reflection effect, respectively. Our group reported a CPL-active nanocomposite with tun-able right-handed CPL and tailored glum by assembling quantum dots in photonic CNC films. [113] Because of their passive CPL emission, the photonic films were developed as optical labels to encode and decode the hidden information in a QR code. We also achieved tunable upconverted circularly polarized luminescence(UC-CPL) by incorporating multiple emissive upconversion nanoparticles into CNC-based chiral photonic films with tun-able PBGs(Fig. 9d). [114] A tailored glum with tunable UC-CPL emission at 450 and 620 nm was realized by the targeted control of the PBGs within the photonic films. Interestingly, humidity-responsive UC-CPL with glum that ranged from -0. 156 to -0. 033 were attained in glycerol composite photonic films because the PBG was sensitive to the relative humidity of the environment.

4 Applications of Wood-DerivedAdvanced Materials

4.1 Hosting Matrix for GuestMaterials

Inheriting from hierarchical porous structure of natural wood, wood-derived carbon showed great potential in hosting guest materials for oil, metal, gas, dyes and cations, which contributed its applications in adsorption and pollution control. Hu ' s group devised a top-down strategy to produce a wood-derived solar-heated carbon adsorbent with aligned channels for fast crude oil adsorption. The wood-derived adsorbent was prepared by carbonization and then coated with a layer of hexamethyldisilazane-treated silicon dioxide to enhance its hydrophobic and oleophilic properties. The resulting material exhibited fast capillary-driven oil adsorption behavior with lowflow resistance, which was attributed to the inherited wood porous structure with low tortuosity. As shown in Fig. 10a, the unique porous structure of the wood carbon adsorbent also enabled efficient solar absorption, effectively trapping over 99% of the incident irradiation. The low surface tension and viscosity of the oil heated by the absorbed light and decreased dynamic contact angle of the heated oil at the inner surface of the wood carbon channels promoted fast oil adsorption. More-over, efficient solar-thermal conversion and excellent thermal management of the adsorbent also contributed to the rapid oil adsorption, reaching a fast adsorption rate of 1550 mL m^{-2} in 30 s under one-sun irradiation. [115] Carbonized wood can also be used as a 3D substrate to support metals. Xia et al. [116] fabricated constructed a novel 3D wood-derived membrane with iron manganese oxide nanosheets (Fe-Mn-O NSs) on the inherited open microchannels of carbonized wood for watertreatment(Fig. 10b). The uniformly dispersed Fe-Mn-O NSs throughout the aligned microchannels exposed numerous active sites that contacted with contaminants in wastewater. Moreover, the diffusion of pollutants was facilitated by the open microchannels, numerous hierarchical pores, and

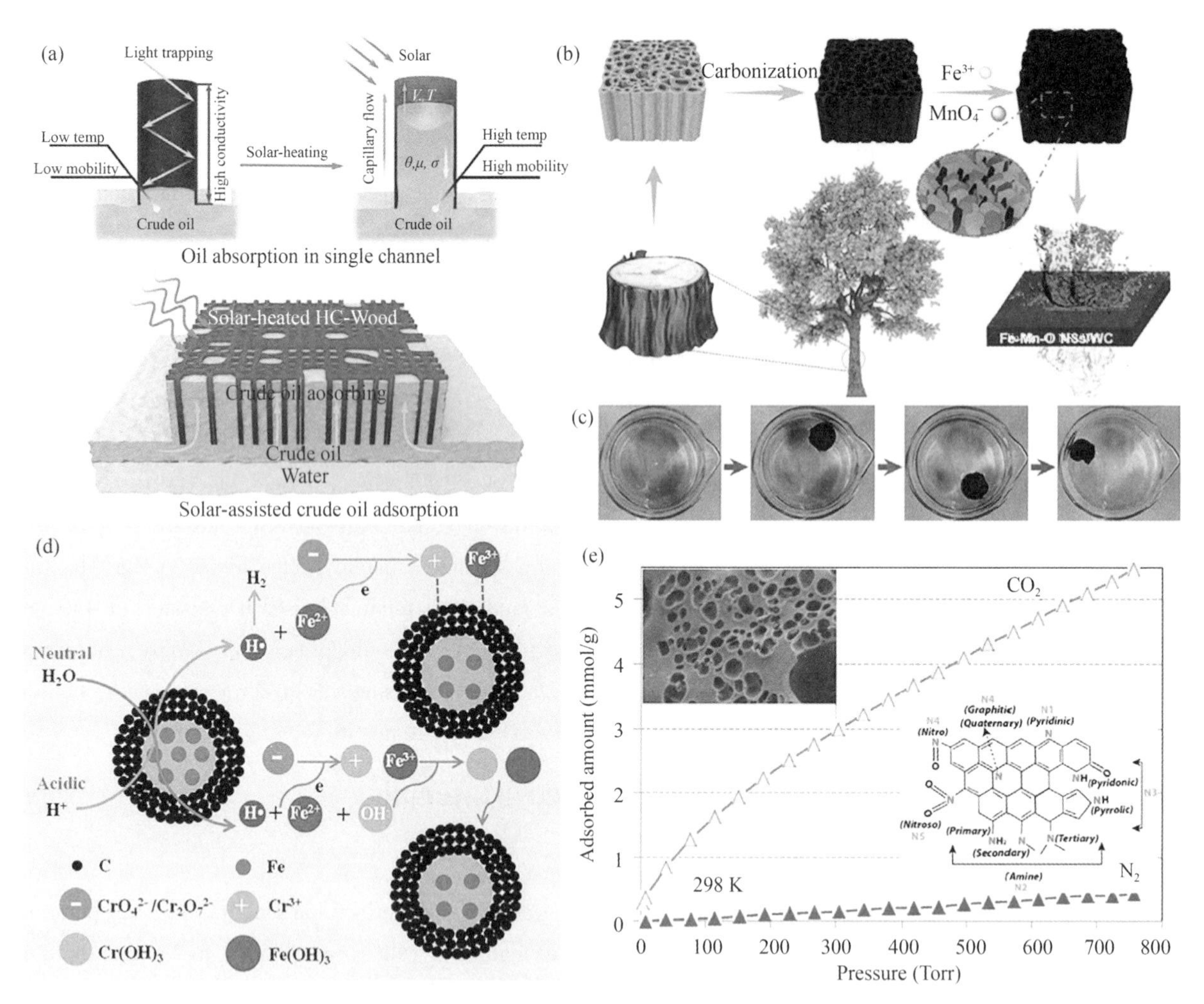

Fig. 10 a) Graphical illustration of a solar-assisted carbon adsorbent showing the process of oil adsorption. Reproduced with permission.[115] Copyright 2019, Wiley-VCH. b) Schematic illustration of the synthesis of an Fe-Mn-O NS/wood carbon membrane for water treatment. Reproduced with permission.[116] Copyright 2019, Elsevier. c) Photographs showing the absorption of heptane (stained with Sudan Ⅲ) by a piece of carbon aerogel prepared from nanofibrillated cellulose. Reproduced with permission.[118] Copyright 2016, Wiley-VCH. d) Mechanism of Cr(Ⅵ) removal by a cellulose-derived magnetic carbon adsorbent. Reproduced with permission.[119] Copyright 2014, Royal Society of Chemistry. e) SEM image of a lignin-derived carbon adsorbent (scale bar: 200 nm), schematic of the nitrogen functionality on carbon surface, and the CO_2 and N2 adsorption isotherms of the adsorbent at 298 K. Reproduced with permission.[122] Copyright 2017, Elsevier

perforation plates inside the wood carbon structure. The unique mesoporous wood carbon structure enabled favorable fluid dynamics as wastewater passed through the 3D membrane. As a result, a high MB removal efficiency of ca. 99.8% was achieved by the Fe-Mn-O NS/wood carbon membrane. Sun and co-workers reported a cellulose-derived carbon adsorbent with honeycomb-like microstructure and hierarchical pores for the removal of methyl orange.[117] Different temperatures and amounts of cellulose were used to tune the porous structure and surface properties of the adsorbent, which were two important parameters influencing its adsorption behavior in aqueous solution. The prepared carbon adsorbent that was thermally treated at 800 ℃

possessed a high specific surface area of around 1259. 4 $m^2 \cdot g^{-1}$, total pore volume of 2. 7 $cm^3 \cdot g^{-1}$, and high methyl orange adsorption capacity of 337. 8 $mg \cdot g^{-1}$. Chen et al. [118] used wood nanofibrillated cellulose to fabricate carbon aerogels for the adsorption of oils and organic solvents from water. The synthesized carbon aerogel with low density, high porosity, good flexibility, and excellent hydrophobic and oleophilic characteristics showed high absorption capacity even under harsh conditions. As shown in Fig. 10c, an organic pollutant was easily removed by a piece of the carbon aerogel. By combining the carbon aerogel with a composite sponge and vacuum pump, continuous and effective adsorption of contaminants from water was achieved. Qiuet al. [119] fabricated cellulose-derived magnetic mesoporous carbon-iron nanoadsorbents for the removal of Cr(Ⅵ). During the pyrolysis of cellulose, reductive intermediates like H_2, CO, and carbon were generated from its decomposition, which reduced Fe^{3-} to zero-valence iron. Fast Cr(Ⅵ) removal rates were achieved in acidic solutions because the magnetic carbon adsorbents tended to remove $HCrO^{4-}$ rather than CrO_4^{2-}. The mesopores in the adsorbents improved the diffusion of Cr(Ⅵ) inside the particles. As shown in Fig. 10d, under acidic conditions, zero-valence iron reacted with protons and produced reductive intermediates, which then reduced the Cr(Ⅵ) to Cr(Ⅲ). The positively charged Cr(Ⅲ) was adsorbed onto the surface of the magnetic carbon nanoadsorbent by electrostatic attraction. As a result, most of the Cr(Ⅵ) in a solution was removed by the magnetic carbon nanoadsorbent after 10 min at pH 1. 0. The used nanoadsorbent could be easily separated from the solution using a permanent magnet.

Wood-derived carbon materials are also promising adsorbents for the removal of gases because of their excellent textural properties, tunable porosity, and low cost. [120,121] Dipendu and co-workers produced nitrogen-doped hierarchical porous carbon adsorbents from lignin by KOH and NH_3 activation. The adsorbents possessed a high specific surface area of 1631-2922 $m^2\ g^{-1}$, which increased their ability to capture CO_2. [122] As shown in Fig. 10e, the fabricated carbon adsorbents contained patterned holes with diameters of 200-500 nm. The patterned holes combined with micropores and some mesopores endowed the carbon adsorbents with hierarchical porosity. Meanwhile, NH_3 activation introduced differentnitrogen species onto the carbon surface. Through the Lewis acid-base interactions between CO_2 and pyridinic nitrogen, along with hydrogen bonding between CO_2 and the pyridine group, an excellent CO_2 adsorption capacity of 5. 48 $mmol \cdot g^{-1}$ was achieved at 298 K and pressure of up to 760 torr(1 bar). Moreover, the carbon adsorbents showed good selectivity for CO_2 over N_2, making them suitable for highly efficient capture of CO_2 from N_2.

4. 2 Microwave Absorption

To remove and reduce electromagnetic(EM) pollution, porous carbon(PC) materials show great potential for EM wave absorption owing to their low density, high surface area, and excellent dielectric loss ability. [123] Biomass sources are a sustainable, low-cost material with hierarchical structure and unique nanostructure which make them become a potential EM wave absorption materials. Additionally, using biomass as raw materials to fabricate carbon-based EM wave absorption materials is an eco-friendly and promising route. Specifically, porous carbon can be prepared through a mild thermal treatment process which are beneficial for enhancing EM wave absorption due to the preserving the elaborate periodic porous microstructure and microtubular channels. [124,125] Much current research has been dedicated to the exploitation of biomass-derived PC in EM absorption, including optimizing the pore size, enlarging the surface area, and building multicomponent. [126-128] Fig. 11a shows a schematic representation of biomass-derived PC for electromagnetic

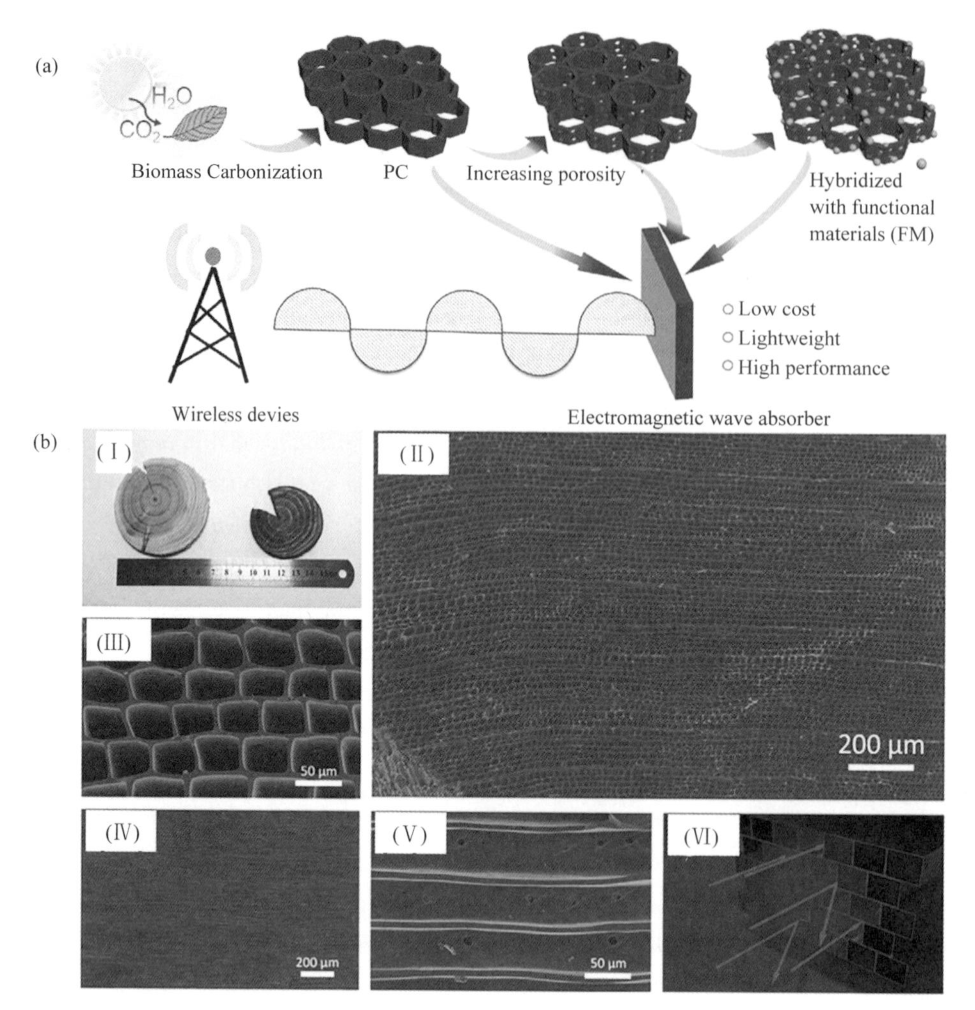

Fig. 11 a) Schematic representation of biomass-derived porous carbon for electromagnetic absorption application, showing the fabrication process of porous carbon. Reproduced under the terms of the CC-BY Creative Commons Attribution 4.0 International License (http://creativecommons.org/licenses/by/4.0/).[129] Copyright 2019, The Authors, published by Springer Nature. b) Images of the porous biomass-pyrolyzed carbon material (Ⅰ-Ⅵ); digital photograph of wood (left) and porous biomass-pyrolyzed carbon (right) (Ⅰ); SEM images of radial section of porous biomass-pyrolyzed carbon (Ⅱ, Ⅲ); SEM images of the axial section of porous biomass-pyrolyzed carbon (Ⅳ, Ⅴ); schematic representation of PBPC microwave absorption mechanism(Ⅵ). Reproduced with permission.[134] Copyright 2017, Elsevier

absorption applications, which demonstrates a basic synthetic route for designing novel wave-absorption materials from biomass resources.[129] The PC materials were prepared via a one-step pyrolysis method from a wide range of biomass resources such as walnut shells,[130] bamboo,[131] peanut shells,[132] apricot pits,[133] and wood.[134] As a natural material, wood is pervasively used as a structural material owing to its highly-ordered microstructure.[135] Xi et al. reported the production of a porous biomass-pyrolyzed carbon(PBPC) with orderly parallel channel structures(Fig. 11b, Ⅰ-Ⅴ) by thermal annealing of fir wood, and demonstrated its micro-wave absorption (MA) performance parallel to the direction of the channels in the range 2-18 GHz. Owing to its highly-oriented pore structure arrangement, only minimal reflection of an incident microwave occurred on the side walls of channel, while most microwaves entered the structure via the channels(Fig. 11b

Ⅵ). The PCBC samples exhibited different attenuation capacities as a function of EM energy under different carbonation temperatures. It was found that the RL values of the PCBC materials prepared at 680 ℃ (PCBC-680) are much higher than those of other samples. The maximum RL value reaches -68. 3 dB with a broad frequency band width of 6. 13 GHz at a thickness of 4. 28 mm. [134] Therefore, the thermaltreatment condition is a significant factor for enhancing the microwave adsorption property. The templating method and activation method are two other strategies for controlling the pore structure of PC to achieve remarkable microwave adsorption properties. Furthermore, the combination of bio-inspiration, chemical synthesis and nanotechnology is expected in the near future to generate an increase in nanostructuredmaterials from wood-derived sources for microwave absorption applications. [129]

4. 3 Energy Storage and Conversion

Fabrication of porous materials is important for development of energy devices including supercapacitors and cell cathodes. [136-138] The efficiency of these energy devices is deter-mined by the electrical double layers structure of the electrode electrolyte interface. [139,140] Wood and its derivatives had natural porous structures, which enabled them as excellent sources for energy storage and conversion materials. [141,142] Huang et al. [143] prepared porous carbon fibers for use in supercapacitors by KOH activation of liquefied wood sawdust (Fig. 12a). Increasing the activation time and the KOH/fiber ratio helped to ensure thorough activation with further developed porosity that consisted of large micropores to 2-5 nm small mesopores. The optimized sample presented a specific capacitance of 225 F · g^{-1}. Moreover, the supercapacitor showed good cyclingstability with

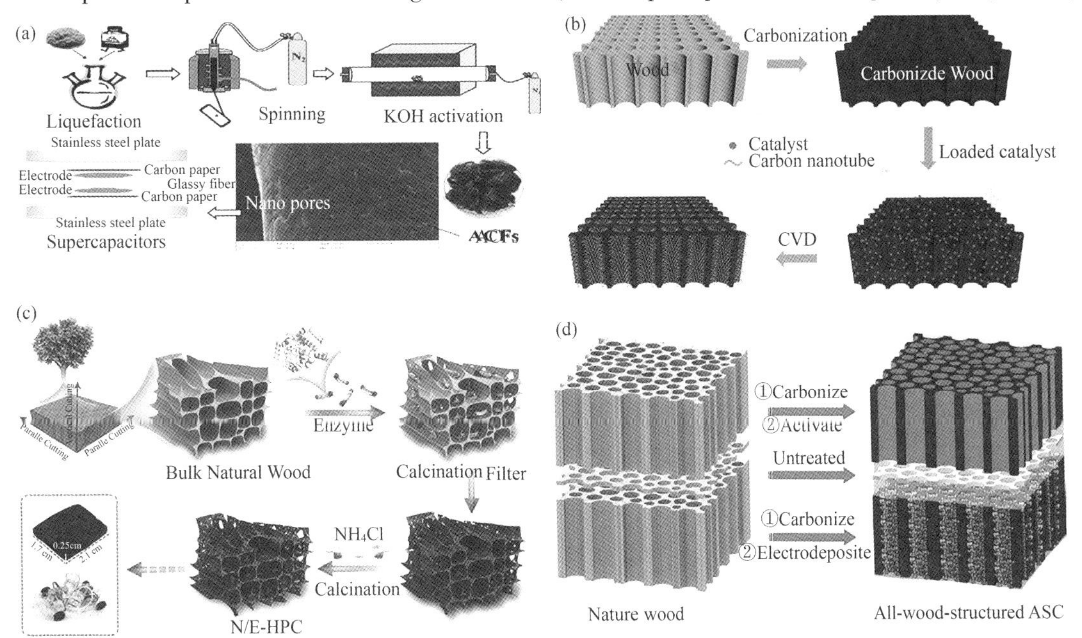

Fig. 12 a-d) Schematic illustrations of: a) an activated carbon fiber-based supercapacitor; b) a CNT-containing carbonized wood-based high-performance supercapacitor; c) hierarchically porous carbon plates derived from wood as bifunctional ORR/OER electrodes; and d) an all-wood, low-tortuosity, aqueous, biodegradable supercapacitor with ultrahigh capacitance. a) Reproduced with permission. [143] Copyright 2016, American Chemical Society. b) Reproduced with permission. [144] Copyright 2019, Elsevier. c) Reproduced with permission. [10] Copyright 2019, Wiley-VCH. d) Reproduced with permission. [146] Copyright 2017, Royal Society of Chemistry

94.2% capacitance retention after 10000 charge-discharge cycles. Wood-derived carbon have been used as cathodes. In order to increase the specific surface area of wood-derived carbon without affecting its conductivity, Wu and co-workers prepared CNTs on the inner wall of tracheids in wood carbon slices catalyzed by nickel nanoparticles (Fig. 12b).[144] The as-prepared carbonized carbon showed a specific surface area of 537.9 $m^2 \cdot g^{-1}$ and an enhanced volumetric capacitance of 76.5 $F \cdot cm^{-3}$. The energy density of the all-solid-state super-capacitor was 39.8 $Wh \cdot kg^{-1}$ and it showed 96.2% capacitance retention after 10000 charge-discharge cycles. Peng et al.[10] used enzymes to partially hydrolyze the cellulose in raw wood and prepare nanoporous wood. Hydrolysis helped to expose the inner parts of the wood to ensure sufficient nitrogen doping during the subsequent carbonation process (Fig. 12c). The obtained carbon exhibited excellent oxygen reduction reaction (ORR) and oxygen evolution reaction (OER) catalytic activity. When the as-fabricated carbonized wood was applied as a metal-free cathode in a zinc-air battery, a specific capacity of 801 mAh/g and energy density of 955 $Wh \cdot kg^{-1}$ were achievedalong with long-term stability of 110 h (Fig. 12c). Chen and co-workers directly carbonized natural wood to prepare a higher conductive, lightweight, and lower tortuosity carbon frame-work as an ultrathick 3D current collector.[145] Attributed to the unique multichanneled carbon framework, an ultrathick 3D lithium iron phosphate electrode with a thickness of 800 μm and higher active material mass loading of 60 $mg \cdot cm^{-2}$ delivered a rational capacity of 7.6 $mAh \cdot cm^{-2}$, long cycling life span, low deformability, and enhanced mechanical properties. Chen et al.[146] constructed an asymmetric supercapacitor (ASC) based on an activated carbonized wood carbon anode, wood membrane separator, and MnO_2/wood carbon (MnO_2@ WC) cathode (Fig. 12d). Because of its structural virtues, the all-wood-structured ASC realized higher areal mass loadings, higher energy density of 1.6 mWh/cm^2, and a maximum power density of 24 W/cm^2, representing the highest areal energy densities and mass loading of previous reported on MnO_2-based supercapacitors.

4.4 Solar Steam GenerationDevices

Solar steam generation techniques show promise in wastewater treatment and desalination[147-152] In particular, carbonized wood-based solar steam generation devices (CW-SSGDs) demonstrate great potential, attributed to sustainability and easily preparation.[153-158] Moreover, attributed to the inherent naturalstructural advantages, carbonized wood-based solar steam generation devices have nice anti-salty accumulation capacity, which enables their long term high performance during the evaporation. Additionally, carbonized wood-based solar steam generation devices can be prepared in a large scale at low costs. CW-SSGDs typically have two layers, a photothermal layer and a wood matrix.[159-163] Current CW-SSGDs can be divided into two types: those with surficial carbonization and those with complete carbonization. First, CW-SSGDs based on surficially carbonized wood are introduced. Hu's group first reported the use of surficially carbonized wood for solar steam generation (Fig. 13a).[164] This design offers several advantages, including efficient water transport and evaporation in the wood lumens, strong light absorption (≈99%) in the open wood channels and low thermal conductivity. The same group used surficially carbonized wood with a bimodal porous structure as a solar vapor generator, which efficiently solved the problem of salt blocking generally found in solar steam generation.[165] Taking advantage of the inherent and interconnected bimodal porous microstructure, efficient transport between the micro-and macrochannels led to quick replenishment of surface-vaporized brine, which efficiently prevented the salty accumulation. Zhou and co-workers developed CW-SSGDs with surficial carbonization that

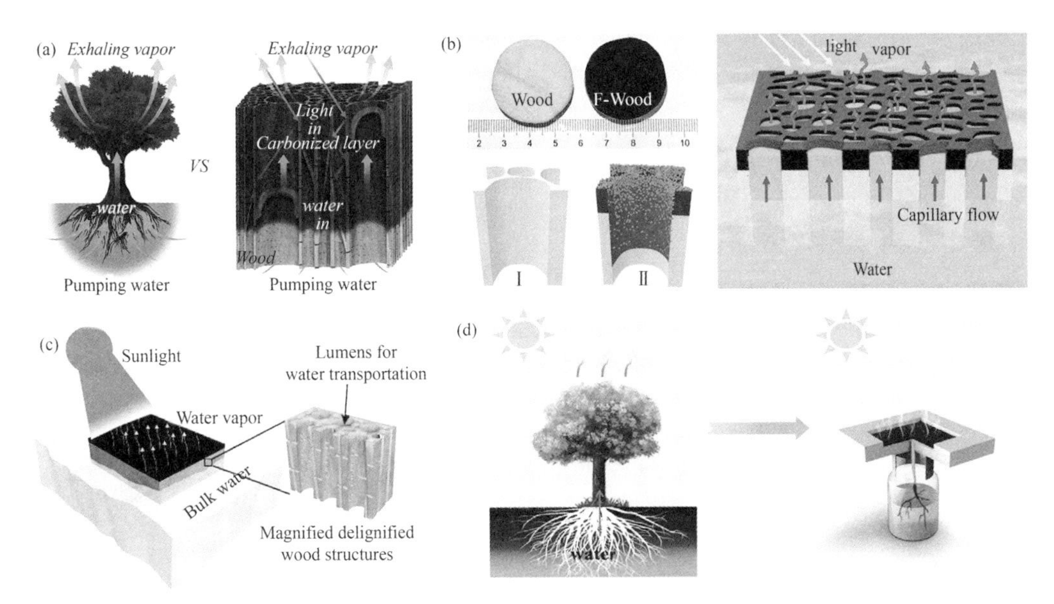

Fig. 13 a–d) Schematic illustrations of CW-SSGDs based on: a–c) surficially carbonized wood and d) completely carbonized wood. a) Repro-duced with permission.[164] Copyright 2017, Wiley -VCH. b) Reproduced with permission.[166] Copyright 2017, American Chemical Society. c) Repro duced with permission.[167] Copyright 2019, American Chemical Society. d) Reproduced with permission.[168] Copyright 2019, Royal Society of Chemistry

showed a solar-thermal efficiency of ≈72% at one standard solar irradiation (1 kW · m^{-2}) (Fig. 13b).[166] Our group used deep eutectic solvents to delignify surficial carbonized wood for preparing a high-performance CW-SSGD (Fig. 13c). Wooddelignification enhanced its water transportation capacity, whilst lowing thermal conductivity. The improved properties of the delignified wood showed an evaporation rate as high as 1.3 kg · m^{-2}/h and a steam generation efficiency of 89% under one-sun irradiation (100 mW · cm^{-2}).[167] Second, CW-SSGDs based on completely carbonized wood are described. Motivated by water transport and transpiration of trees, Zhao's group designed a transpiration system.[168] In this design, hydrophilic air laid paper acted as the root, polyethylene foam was used as the soil, and completely carbonized wood acted as the photo-thermal convertor (Fig. 13d). The as-prepared CW-SSGD achieved a conversion efficiency of 91.3% under one-sun irradiation (100 mW · cm^{-2}). Qiuet al.[169] reported novel porous 3D carbon foams (CFs) via alkali activation of naturally abundant wood carbon for solar steam generation. Their rough surface enabled the CFs to harvest more than 97% of the energy of the solar spectrum. The hydrophilic microchannels of the CFs facilitated their efficiency in heat localization and transportation. Benefiting from these advantages, the steam generation efficiency of the CFs reached as high as 80.1% under one-sun irradiation.

4.5 Catalysis

Generally, porous carbon was widely used as catalyst or catalyst support.[170] Our group synthesized conductive CSs and water soluble fluorescent CNDs by the one-pot HTC method using CMC as raw materials, and then used the CNDs as a photosensitizer in a CND/TiO_2 system for MB and phenoldegradation under visible-light irradiation.[24] As shown in Fig. 14a, CNDs with abundant oxygen-containing groups were sensitized through photon absorption upon irradiation by visible light, allowing them to act as a photosensitizer of TiO_2. That is, photons with adequate energy excited electrons in the highest occupied molecular orbital of the CNDs and induced their transfer to the lowest unoccupied molecular orbital of the CNDs. Free electrons as well energy were then transferred to the conduction band of TiO_2. These electrons then reacted with oxidants (O_2) adsorbed on TiO_2 to produce active oxygen radicals($O_2^{-\cdot}$), which degraded MB into small molecules. The spectral response range of the CND/TiO_2 composite extended to the visible region and 97. 1% MB degradation

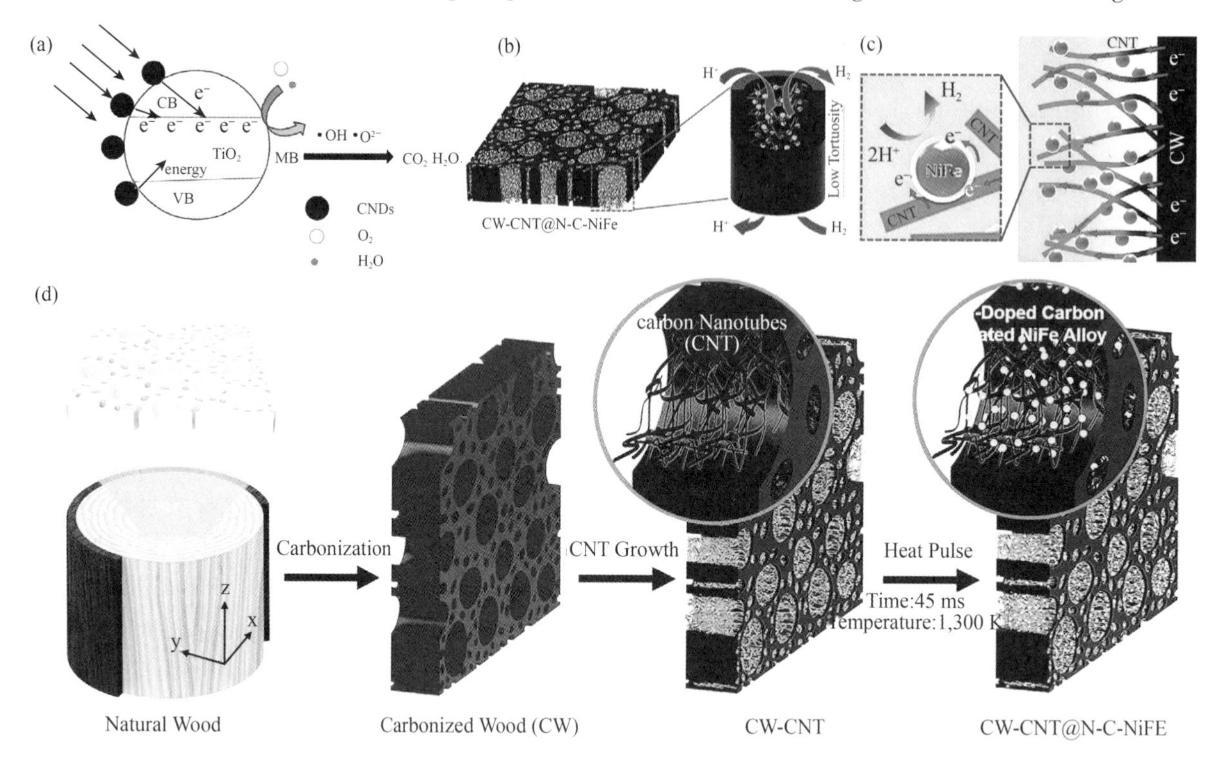

Fig. 14 a) The catalytic mechanism of CND/TiO_2 under visible-light irradiation. Reproduced with permission.[25] Copyright 2015, Elsevier. b) Schematic illustration showing the open and low-tortuosity structure of the CW-CNT@N-C-NiFe electrode. c) Schematic illustration of the hydrogen evolution reaction on the N-C-NiFe nanoparticles. Electrons travel along the CNT to the N-C-NiFe nanoparticles for the reduction of protons d) Schematic representation of the fabrication process of the CW-CNT@N-C-NiFe electrode. b–d) Reproduced with permission.[171] Copyright 2018, Wiley-VCH

was achieved under visible-light irradiation. The direct conversion the natural wood into nanostructured catalysts is advantageous because of the ease and cost-effectiveness of such a process as well as the ready exploitation of the natural hierarchical structure of wood. For example, Peng et al.[10] prepared metal-free mesoporous nitrogen-doped carbons containing abundant hierarchical pores through facile carbonation of wood via enzyme-catalyzed hydrolysis of cellulose microfibers. The resulting materials contained a large number of preformed micropores and were suitable to catalyze the ORR and OER. To fabricate this catalyst material, the cellulose in wood was selectively hydrolyzed, which lead to formation of nanopores. This can maximize the exposure of the catalytic active sites for accelerating the reactions. The abundant active sites and the hierarchical porous structure facilitated mass diffusion, leading to the catalyst exhibiting superb ORR and OER performance in alkaline electrolyte. A self-supported wood-based carbon framework was designed which combinedwith CNTs and nitrogen-doped few-graphene-layer-encapsulated NiFe alloy (N-C-NiFe) nanoparticles for hydrogen generation, as graphically illustrated in Fig. 14b–d. Because of the open and low-tortuosity wood structure, the electrolyte easily permeated into the porous framework of the CW-CNT@ N-C-NiFe catalyst during the hydrogen evolution reaction, and the generated hydrogen gas on the catalyst surface was readily released from the microchannels without blocking the mass transfer pathway (Fig. 14b). In Addition, the core-shell N-C-NiFe nanoparticles were anchored on the CNTs, which led to forming electron "highways" that enabled fast charge transfer on the surface of the N-C-NiFe nanoparticles, resulting in superior cycling durability and high electrocatalytic activity in hydrogen evolution (Fig. 14c). The CWCNT @ N-C-NiFe electrode exhibited impressive electrochemical performance toward hydrogen evolution with a small Tafel slope of 52.8 mV dec^{-1} and an overpotential of 179 mV at 10 mA · cm^{-2} with favorable long-term cycling stability.[171]

4.6 Bioimaging and Sensing

Because of their sustainability, biocompatibility, and safety,[172] wood-derived fluorescent materials have been widely applicable in bioimaging.[173-175] Zhang's group developed a rare-earth-doped phosphor-containing cellulose hydrogel for in vivo imaging(Fig. 15a).[176] The cellulose hydrogel emitted intense green fluorescence with a long-lasting afterglow under UV irradiation. Guo et al.[177] prepared biocompatible photoluminescent hybrid materials composed of CDs and CNCs(Fig. 15b). Amino-functionalized CDs were grafted to CNCs via carbodiimide-assisted coupling chemistry. The as-prepared hybrid fluorescent materials showed enhanced cytocompatibility, cellular association, and internalization compared with those of CNCs. The fluorescent hybrid materials were successfully used to image RAW 264.7 cell lines. Besides these cellulose-based fluorescent materials, lignin fluorescence has also showed potential in bioimaging. Our group used fluorescent lignin carbon dots (L-CDs) for cellular imaging via one-photon and two-photon excitation.[81] L-CDs were prepared from cellulolytic enzyme lignin using a green, simple, and easy-to-operate molecular aggregation method. The L-CDs showed high biocompatibility and emitted fluorescence in the cytoplasm of HeLa cells upon both one-and two-photon excitation. Wood-derived fluorescent materials also demonstrate great potential for chemical sensing of guest gas species. Wu et al.[178] covalently grafted fluorescent CDs on cellulose nanofibrils in an aerogel matrix to prepare a fluorescent aerogel(Fig. 15c). The as-prepared aerogel exhibited bright blue fluorescence with a high fluorescence quantum yield of 26.2% under UV radiation. The aerogel also displayed high sensitivity and selectivity in the recognition of NO_x and aldehyde species, and detected glutaraldehyde at ultralow concentrations (ppm) in water. Our group also demonstrated the use of the inherent fluorescence of lignin nanoparticles tosense formaldehyde vapor.[98] Lignin nanoparticles showed turn-on fluorescence when exposed to formaldehyde vapor. The detection limit(3 s/K, where s is the standard deviation of the blank signal and K is 17.8) was 8 μm. The following sensing mechanism for formaldehyde vapor was proposed. Formaldehyde vapor was first reacted with lignin nanoparticles. As a result, hydroxyl moieties were formed at the aromatic rings. The newly generated hydroxyl moieties promoted J-aggregation and decreased the motion of the aromatic rings, which enhanced the AIE of the lignin nanoparticles. Xiong and co-workers repotted to prepare fluorescent lignin nanoparticles using 1-pyrenebutyric acid as fluorescent reagent.[179] The fluorescence probe 1-pyrenebutyric acid (PBA) was grafted onto the lignin surface by amidation. Oxygen quenched the fluorescence of the lignin-PBA probe. Furthermore, with increasing PBA concentration, the sensitivity of the probe to oxygen increased.

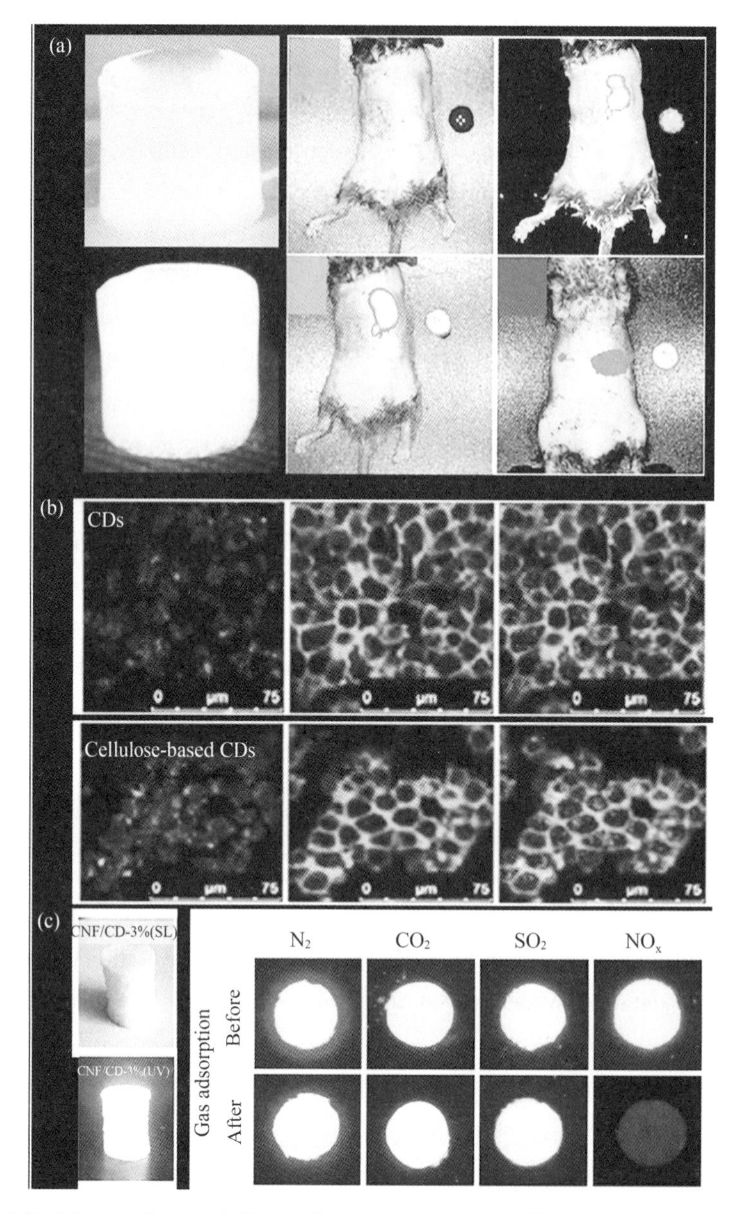

Fig. 15 a) Bright-field images (upper left) and fluorescence images (lower left) of the hydrogel. A mouse after injection of CPH5 under the skin and in the stomach. Reproduced with permission.[176] Copyright 2017, American Chemical Society. b) Confocal fluorescence microscopy images of RAW264.7 macrophage cells. Reproduced with permission.[177] Copyright 2014, Royal Society of Chemistry. c) Bright-field images (upper left) and fluorescence images (lower left) of a CNF/CD aerogel. Fluorescence behavior of the fabricated CNF/CD-3% aerogel under UV light before and after gas adsorption (pure N_2, CO_2, SO_2, and NOx) (right). Reproduced with permission.[178] Copyright 2019, American Chemical Society

5 Conclusions

Wood and wood-derived components including cellulose, hemicellulose, and lignin not only are biocompatible and earth abundant but also have nature-provided intrinsic advantages for potentially transformative device performance. Because of these inherent advantages, carbon materials and light-emittingmaterials derived from wood are emerging as attractive solutions to a range of technological challenges. Although considerable progress has been achieved in these areas, there are still several challenges that currently limit the practical use of advanced wood-derived carbon materials and light-emitting materials, especially in emerging applications in the fields of energy and biomedicine. These challenges facing wood-derived carbon materials include the limited fundamental under-standing of the carbonization process of wood-derived components and the need to develop a strategy for targeted conversion of wood-derived components to carbon materials with regular morphology. Moreover, it would be ideal that developing a facile method to dynamically tune the morphology of wood-derived carbon materials ranging from fibrils, sphere to foams in the future. The challenges facing wood-derived light-emitting mate-rials include the need to red-shift the emission wavelength of wood-derived photoluminescent materials to near-infrared optical window. The long optical window enabled its penetration in deep tissue, which facilitated their applications as bio-imaging and therapeutic reagents both in vitro and in vivo. It is also desirable to develop more sustainable anti-ACQ fluorescent materials and CPL materials with stimuli-responsive tunable properties. In the future, more natural dyes, as replacement of available petrol-based dyes, might be employed for preparing anti-ACQ materials together with cellulose. In this manner, pure sustainable solid emission materials might be obtained. For CPL materials, stimuli-sensitive components, such as thermo, light or redox-responsive molecules/polymers, might also be assembled in cellulose chiral films for producing a CPL materials with a dynamic responsive to external stimuli. To date, both of these types of wood-derived materials have mostly been produced from wood-derived components rather than raw wood. Thus, it is also necessary to design a low-cost and green method to separate cellulose, hemicellulose, and lignin fromwood cell walls. We believe that joint worldwide effort will conquer these challenges and promote the further development of wood-derived carbon and light-emitting materials, which will greatly improve our everyday life.

Acknowledgements

W. L. and Z. C. contributed equally to this work. This work was financially supported by the National Natural Science Foundation of China(31890773, 31971601).

Conflict of Interest

The authors declare no conflict of interest.

References

[1]Zhu M, Song J, Li T, et al. Highly anisotropic, highly transparent wood composites[J]. Adv. Mater., 2016, 28(26): 5181-5187.

[2]Berglund L A, Burgert I. Bioinspired wood nanotechnology for functional materials[J]. Adv. Mater., 2018, 30(19): 1704285.

[3]Zhu H, Luo W, Ciesielskl P N, et al. Wood-derived materials for green electronics, biological devices, and energy

applications[J]. Chem. Rev., 2016, 116(19): 12650.

[4]Li T, Zhu M, Yang Z, et al. Wood composite as an energy efficient building material: guided sunlight transmittance and effective thermal insulation[J]. Adv. Energy Mater., 2016, 6(22): 1601122.

[5]Vahayal L, Wilson P, Prabhakaran K. Waste to wealth: Lightweight, mechanically strong and conductive carbon aerogels from waste tissue paper for electromagnetic shielding and CO_2 adsorption[J]. Chem. Eng. J., 2020, 381: 122628.

[6]E L, Li W, Ma C, et al. An ultra-lightweight recyclable carbon aerogel from bleached softwood kraft pulp for efficient oil and organic absorption[J]. Mater. Chem. Phys., 2018, 214: 291-296.

[7]Vithanage M, Herath I, Joseph S, et al. Interaction of arsenic with biochar in soil and water: A critical review[J]. Carbon, 2017, 113: 219-230.

[8]Lu L L, Lu Y Y, Xiao Z J, et al. Wood-inspired high-performance ultrathick bulk battery electrodes[J]. Adv. Mater., 2018, 30(20): 1706745.

[9]Chen C, Song J, Zhu S, et al. Scalable and sustainable approach toward highly compressible, anisotropic, lamellar carbon sponge[J]. Chem, 2018, 4(3): 544-554.

[10]Peng X, Zhang L, Chen Z, et al. Hierarchically porous carbon plates derived from wood as bifunctional ORR/OER electrodes[J]. Adv. Mater., 2019, 31(16): 1900341.

[11]Jia H, Sun N, Dirican M, et al. Electrospun kraft lignin/cellulose acetate-derived nanocarbon network as an anode for high-performance sodium-ion batteries[J]. ACS Appl. Mater. Interfaces, 2018, 10(51): 44368-44375.

[12]Simon C, Lion C, Spriet C, et al. One, two, three: A bioorthogonal triple labelling strategy for studying the dynamics of plant cell wall formation in vivo[J]. Angew. Chem., Int. Ed., 2018, 57(51): 16665-16671.

[13]Zhang B, Liu Y, Ren M, et al. Sustainable synthesis of bright green fluorescent nitrogen-doped carbon quantum dots from alkali lignin[J]. ChemSusChem, 2019, 12(18): 4202-4210.

[14]Danish M, Ahmad T. A review on utilization of wood biomass as a sustainable precursor for activated carbon production and application[J]. Renewable Sustainable Energy Rev., 2018, 87: 1-21.

[15]Ao W, Fu J, Mao X, et al. Microwave assisted preparation of activated carbon from biomass: A review[J]. Renewable Sustainable Energy Rev., 2018, 92, 958-979.

[16]Zhao X, Chen H, Kong F, et al. Fabrication, characteristics and applications of carbon materials with different morphologies and porous structures produced from wood liquefaction: A review[J]. Chem. Eng. J., 2019, 364: 226-243.

[17]Zhang J, Terrones M, Park C R, et al. Carbon science in 2016: Status, challenges and perspectives[J]. Carbon, 2016, 98: 708-732.

[18]Bazaka K, Jacob M V, Ostrikov K. Sustainable life cycles of natural-precursor-derived nanocarbons[J]. Chem. Rev., 2016, 116(1): 163-214.

[19]Peterson A A, Vogel F, Lachance R P, et al. Thermochemical biofuel production in hydrothermal media: A review of sub- and supercritical water technologies[J]. Energy Environ. Sci., 2008, 1(1), 32-65.

[20]Deng J, Xiong T Y, Xu F, et al. Effects of Cellulose, Hemicellulose, and Lignin on the Structure and Morphology of Porous Carbons[J]. ACS Sustainable Chem. Eng., 2016, 4(7): 3750-3756.

[21]Chheda J N, Huber G W, Dumesic J A. Liquid-phase catalytic processing of biomass-derived oxygenated hydrocarbons to fuels and chemicals[J]. Angew. Chem., Int. Ed., 2007, 46(38): 7164-7183.

[22]Titirici M M, White R J, Falco C, et al. Black perspectives for a green future: hydrothermal carbons for environment protection and energy storage[J]. Energy Environ. Sci., 2012, 5(5): 6796-6822.

[23]Zhang P, Qiao Z A, Dai S. Recent advances in carbon nanospheres: synthetic routes and applications[J]. Chem. Commun., 2015, 51(45): 9246-9256.

[24]Sevilla M, Fuertes A B. The production of carbon materials by hydrothermal carbonization of cellulose[J]. Carbon, 2009, 47(9): 2281-2289.

[25]Wu Q, Li W, Tan J, et al. Hydrothermal carbonization of carboxymethylcellulose: One-pot preparation of conductive carbon microspheres and water-soluble fluorescent carbon nanodots[J]. Chem. Eng. J., 2015, 266: 112-120.

[26]Wu Q, Li W, Wu Y, et al. Effect of reaction time on structure of ordered mesoporous carbon microspheres prepared from carboxymethyl cellulose by soft-template method[J]. Ind. Crops Prod., 2015, 76: 866-872.

[27] Falco C, BAccile N, Titirici M M. Morphological and structural differences between glucose, cellulose and lignocellulosic biomass derived hydrothermal carbons[J]. Green Chem., 2011, 13(11): 3273-3281.

[28]Sun X M, Li Y D. Colloidal carbon spheres and their core/shell structures with noble-metal nanoparticles[J]. Angew. Chem., Int. Ed., 2004, 43(5): 597-601.

[29]Ikeda S, Tachi K, Ng Y H, et al. Selective adsorption of glucose - Derived carbon precursor on amino-functionalized porous silica for fabrication of hollow carbon spheres with porous walls[J]. Chem. Mater., 2007, 19(17): 4335-4340.

[30]Xu D, Chen C, Xie J, et al. A Hierarchical N/S-co doped carbon anode fabricated facilely from cellulose/polyaniline microspheres for high-performance sodium-ion batteries[J]. Adv. Energy Mater., 2016, 6(6): 1501929.

[31]Cheong J Y, Benker L, Zhu J, et al. Generalized and feasible strategy to prepare ultra-porous, low density, compressible carbon nanoparticle sponges[J]. Carbon, 2019, 154: 363-369.

[32]Song H, Xu S, Li Y, et al. Hierarchically porous, ultrathick, "breathable" wood-derived cathode for lithium-oxygen batteries[J]. Adv. Energy Mater., 2018, 8(4): 1701203.

[33]Huang Y, Chen Y, Fan X, et al. Wood derived composites for high sensitivity and wide linear-range pressure sensing [J]. Small, 2018, 14(31): 1801520.

[34]Edison T A. Electric -Lamp. US223898, 1880.

[35]Byrne N, Setty M, Blight S, et al. Cellulose-derived carbon fibers produced via a continuous carbonization process: Investigating precursor choice and carbonization conditions[J]. Macromol. Chem. Phys., 2016, 217(22): 2517-2524.

[36]Dumanli A G, Windle A H. Carbon fibres from cellulosic precursors: a review[J]. J. Mater. Sci., 2012, 47(10): 4236-4250.

[37]Frank E, Steudle L M, Ingildeev D, et al. Carbon fibers: precursor systems, processing, structure, and properties [J]. Angew. Chem., Int. Ed., 2014, 53(21): 5262-5298.

[38]Cai J, Niu H, Li Z, et al. High-performance supercapacitor electrode materials from cellulose-derived carbon nanofibers[J]. ACS Appl. Mater. Interfaces 2015, 7(27): 14946-14953.

[39]Mainka H, Tager O, Korner E, et al. Lignin - an alternative precursor for sustainable and cost-effective automotive carbon fiber[J]. J. Mater. Res. Technol., 2015, 4(3): 283-296.

[40]Xu X, Zhou J, Jiang L, et al. Lignin-based carbon fibers: Carbon nanotube decoration and superior thermal stability [J]. Carbon, 2014, 80: 91-102.

[41]Li Q, Xie S, Serem W K, et al. Quality carbon fibers from fractionated lignin[J]. Green Chem., 2017, 19(7): 1628-1634.

[42]Baker D A, Rials T G. Recent advances in low-cost carbon fiber manufacture from lignin[J]. J. Appl. Polym. Sci., 2013, 130(2): 713-728.

[43]Wang S X, Yang L, Stubbs L P, et al. Lignin-derived fused electrospun carbon fibrous mats as high-performance anode materials for lithium-ion batteries[J]. ACS Appl. Mater. Interfaces 2013, 5(23): 12275-12282.

[44]Liu Z, Huang Y, Zhao G. Preparation and characterization of activated carbon fibers from liquefied wood by $ZnCl_2$ activation[J]. BioResources, 2016, 11(2): 3178-3190.

[45]Huang Y, Ma E, Zhao G. Thermal and structure analysis on reaction mechanisms during the preparation of activated carbon fibers by KOH activation from liquefied wood-based fibers[J]. Ind. Crops Prod., 2015, 69: 447-455.

[46]Ma X, Zhang F, Zhu J, et al. Preparation of highly developed mesoporous activated carbon fiber from liquefied wood using wood charcoal as additive and its adsorption of methylene blue from solution[J]. Bioresour. Technol. 2014, 164: 1-6.

[47]Ma X, Zhao G. Preparation of carbon fibers from liquefied wood[J]. Wood Sci. Technol., 2010, 44(1): 3-11.

[48]Sun J, E L, Ma C, et al. Fabrication of three-dimensional microtubular kapok fiber carbon aerogel/RuO_2 composites for supercapacitors[J]. Electrochim. Acta, 2019, 300, 225-234.

[49]Sun J, Li W, E L, et al. Ultralight carbon aerogel with tubular structures and N-containing sandwich-like wall from kapok fibers for supercapacitor electrode materials[J]. J. Power Sources, 2019, 438: 227030.

[50]Jiang S, Agarwal S, Greiner A. Low-density open cellular sponges as functional materials[J]. Angew. Chem., Int. Ed., 2017, 56(49): 15520-15538.

[51]Zhuo H, Hu Y, Tong X, et al. A supercompressible, elastic, and bendable carbon aerogel with ultrasensitive detection limits for compression strain, pressure, and bending angle[J]. Adv. Mater., 2018, 30(18): 1706705.

[52]Peng X, Wu K, Hu Y, et al. A mechanically strong and sensitive CNT/rGO-CNF carbon aerogel for piezoresistive sensors[J]. J. Mater. Chem. A, 2018, 6(46): 23550-23559.

[53]Alatalo S, Pileidis F, Makila E, et al. Versatile cellulose-based carbon aerogel for the removal of both cationic and anionic metal contaminants from water[J]. ACS Appl. Mater. Interfaces, 2015, 7(46): 25875-25883.

[54]Alatalo S, Qiu K P, Preuss K, et al. Soy protein directed hydrothermal synthesis of porous carbon aerogels for electrocatalytic oxygen reduction[J]. Carbon, 2016, 96: 622-630.

[55]Li L, Hu T, Sun H, et al. Pressure-sensitive and conductive carbon aerogels from poplars catkins for selective oil absorption and oil/water separation[J]. ACS Appl. Mater. Interfaces, 2017, 9(21): 18001-18007.

[56]Jiang M, Kwok R T K, Li S, et al. A simple mitochondrial targeting AIEgen for image-guided two-photon excited photodynamic therapy[J]. J. Mater. Chem. B, 2018, 6(17): 2557-2565.

[57]Chen P, Wang J, Niu L, et al. Carbazole-containing difluoroboron beta-diketonate dyes: two-photon excited fluorescence in solution and grinding-induced blue-shifted emission in the solid state[J]. J. Mater. Chem. C, 2017, 5(47): 12538-12546.

[58]Zhang W, Peng L, Liu J, et al. Controlling the cavity structures of two-photon-pumped perovskite microlasers[J]. Adv. Mater., 2016, 28(21): 4040-4046.

[59]Chen Z, He S, Butt H, et al. Photon upconversion lithography: patterning of biomaterials using near-infrared light [J]. Adv. Mater., 2015, 27(13): 2203.

[60]Chen Z, Sun W, Butt H, et al. Upconverting-nanoparticle-assisted photochemistry induced by low-intensity near-infrared light: how low can we go? [J]. Chem. - Eur. J., 2015, 21(25): 9165-9170.

[61]Lederhose P, Chen Z, Müller R, et al. Near-infrared photoinduced coupling reactions assisted by upconversion nanoparticles[J]. Angew. Chem., Int. Ed., 2016, 55(40): 12195-12199.

[62]Chen Z, Xiong Y, Etchenique R, et al. Manipulating pH using near-infrared light assisted by upconverting nanoparticles[J]. Chem. Commun., 2016, 52(97): 13959-13962.

[63]Li M, Li X, Xiao H N, et al. Fluorescence sensing with cellulose-based materials[J]. ChemistryOpen, 2017, 6 (6): 685-696.

[64]Zhang Y, Halidan M, Zhang B. Preparation of cellulose-based fluorescent carbon nanoparticles and their application in trace detection of Pb(II)[J]. RSC Adv. 2017, 7(5): 2842-2850.

[65]Zeng R, Deng H, Xiao Y, et al. Cross-linked graphene/carbon nanotube networks with polydopamine "glue" for flexible supercapacitors[J]. Compos. Commun., 2018, 10: 73-80.

[66]Liu J, Wang N, Yu Y, et al. Carbon dots in zeolites: A new class of thermally activated delayed fluorescence materials with ultralong lifetimes[J]. Sci. Adv., 2017, 3(5): e1603171.

[67]Wang B, Yu Y, Zhang H, et al. Carbon dots in a matrix: energy-transfer-enhanced room-temperature red phosphorescence[J]. Angew. Chem. Int. Ed., 2019, 58(51): 18614.

[68]Xu X, Ray R, Gu Y, et al. Electrophoretic analysis and purification of fluorescent single-walled carbon nanotube

fragments[J]. J. Am. Chem. Soc., 2004, 126(40): 12736-12737.

[69]Sun Y, Zhou B, Lin Y, et al. Quantum-sized carbon dots for bright and colorful photoluminescence[J]. J. Am. Chem. Soc., 2006, 128(24): 7756-7757.

[70]Wang C, Xu Z, Cheng H, et al. A hydrothermal route to water-stable luminescent carbon dots as nanosensors for pH and temperature[J]. Carbon, 2015, 82: 87-95.

[71]Bao L, Zhang Z, Tian Z, et al. Electrochemical tuning of luminescent carbon nanodots: from preparation to luminescence mechanism[J]. Adv. Mater., 2011, 23(48): 5801-5806.

[72]Miao P, Han K, Tang Y, et al. Recent advances in carbon nanodots: synthesis, properties and biomedical applications[J]. Nanoscale, 2015, 7(5): 1586-1595.

[73]Cayuela A, Soriano M L, Carrillo-Carrión C, et al. Semiconductor and carbon-based fluorescent nanodots: the need for consistency[J]. Chem. Commun., 2016, 52(7): 1311-1326.

[74]Hepp S, Jetter M, Portalupi S L, et al. Semiconductor quantum dots for integrated quantum photonics[J]. Adv. Quantum Technol., 2019, 2(9): 1900020.

[75]Nsubuga A, Zarschler K, Sgarzi M, et al. Towards utilising photocrosslinking of polydiacetylenes for the preparation of "Stealth" upconverting nanoparticles[J]. Angew. Chem., Int. Ed., 2018, 57(49): 16036-16040.

[76]Pickel A, Teitelboim A, Chan E, et al. Apparent self-heating of individual upconverting nanoparticle thermometers [J]. Nat. Commun., 2018, 9: 4907.

[77]Jogela J, Uri A, Palsson L O, et al. Almost complete radiationless energy transfer from excited triplet state of a dim phosphor to a covalently linked adjacent fluorescent dye in purely organic tandem luminophores doped into PVA matrix[J]. J. Mater. Chem. C, 2019, 7(22): 6571-6577.

[78]Gartzia-Rivero L, Leiva C R, Sanchez-Camerero E M, et al. Chiral microneedles from an achiral bis(boron dipyrromethene): spontaneous mirror symmetry breaking leading to a promising photoluminescent organic material [J]. Langmuir, 2019, 35(14): 5021-5028.

[79]Sahu S, Behera B, Maiti T K, et al. Simple one-step synthesis of highly luminescent carbon dots from orange juice: application as excellent bio-imaging agents[J]. Chem. Commun., 2012, 48(70): 8835-8837.

[80]Ding Z, Li F, Wen J, et al. Gram-scale synthesis of single-crystalline graphene quantum dots derived from lignin biomass[J]. Green Chem., 2018, 20(6): 1383-1390.

[81]Niu N, Ma Z, He F, et al. Preparation of carbon dots for cellular imaging by the molecular aggregation of cellulolytic enzyme lignin[J]. Langmuir, 2017, 33(23): 5786-5795.

[82]Li W, Chun S, Li Y, et al. One-step hydrothermal synthesis of fluorescent nanocrystalline cellulose/carbon dot hydrogels[J]. Carbohydr. Polym., 2017, 175: 7-17.

[83]Munter R, Kristensen K, Pedersbaek D, et al. Dissociation of fluorescently labeled lipids from liposomes in biological environments challenges the interpretation of uptake studies[J]. Nanoscale, 2018, 10(48): 22720-22724.

[84]Tang L, Ji R, Cao X, et al. Deep ultraviolet photoluminescence of water-soluble self-passivated graphene quantum dots[J]. ACS Nano, 2012, 6(6): 5102-5110.

[85]Li W, Zhang Z, Kong B, et al. Simple and green synthesis of nitrogen-doped photoluminescent carbonaceous nanospheres for bioimaging[J]. Angew. Chem., Int. Ed., 2013, 52(31): 8151-8155.

[86]Shen C, Wang J, Cao Y, et al. Facile access to B-doped solid-state fluorescent carbon dots toward light emitting devices and cell imaging agents[J]. J. Mater. Chem. C, 2015, 3(26): 6668-6675.

[87]Chen P, Chen Y, Hsu P C, et al. Photoluminescent organosilane-functionalized carbon dots as temperature probes [J]. Chem. Commun., 2013, 49(16): 1639-1641.

[88]Kingshuk D, Jampani S, Beeraish B, et al. Aggregation-induced emission active donor-acceptor fluorophore as a dual sensor for volatile acids and aromatic amines[J]. ACS Appl. Mater. Interfaces, 2019, 11(51): 48249-48260.

[89]Zhang Y, Zhu W, Huang X, et al. Supramolecular aggregation-induced emission gels based on Pillar[5]arene for ultrasensitive detection and separation of multianalytes[J]. ACS Sustainable Chem. Eng., 2018, 6(12): 16597-16606.

[90]Ishi-i T, Tanaka H, Youfu R, et al. Mechanochromic fluorescence based on a combination of acceptor and bulky donor moieties: tuning emission color and regulating emission change direction[J]. New J. Chem., 2019, 43(13): 4998-5010.

[91]Xue P, Sun J, Chen P, et al. Strong solid emission and mechanofluorochromism of carbazole-based terephthalate derivatives adjusted by alkyl chains[J]. J. Mater. Chem. C, 2015, 3(16): 4086-4092.

[92]Liu K, Zhang J, Xu L, et al. Film-based fluorescence sensing: a "chemical nose" for nicotine[J]. Chem. Commun., 2019, 55(84): 12679-12682.

[93]Zhang Y, Li Q, Guo P, et al. Fluorescence-enhancing film sensor for highly effective detection of Bi^{3+} ions based on SiO_2 inverse opal photonic crystals[J]. J. Mater. Chem. C, 2018, 6(27): 7326-7332.

[94]Tian W, Zhang J, Yu J, et al. Cellulose-based solid fluorescent materials[J]. Adv. Opt. Mater., 2016, 4(12): 2044-2050.

[95]Jia R, Tian W, Bai H, et al. Amine-responsive cellulose-based ratiometric fluorescent materials for real-time and visual detection of shrimp and crab freshness[J]. Nat. Commun., 2019, 10: 795.

[96]Li M, An X, Jiang M, et al. "Cellulose Spacer" Strategy: anti-aggregation-caused quenching membrane for mercury ion detection and removal[J]. ACS Sustainable Chem. Eng., 2019, 7(18): 15182-15189.

[97]Xue Y, Qiu X, Wu Y, et al. Aggregation-induced emission: the origin of lignin fluorescence[J]. Polym. Chem., 2016, 7(21): 3502-3508.

[98]Ma Z, Liu C, Niu N, et al. Seeking brightness from nature: J-Aggregation-induced emission in cellulolytic enzyme lignin nanoparticles[J]. ACS Sustainable Chem. Eng., 2018, 6(3): 3169-3175.

[99]Li M, Li X, An X, et al. Clustering-triggered emission of carboxymethylated nanocellulose[J]. Front. Chem., 2019, 7: 447.

[100]Du L, Jiang B, Chen X, et al. Clustering-triggered emission of cellulose and its derivatives[J]. Chin. J. Polym. Sci., 2019, 37(4): 409-415.

[101]Lee D M, Song J W, Lee Y J, et al. Control of circularly polarized electroluminescence in induced twist structure of conjugate polymer[J]. Adv. Mater., 2017, 29(29): 1700907.

[102]Srivastava A K, Zhang W L, Schneider J, et al. Photoaligned nanorod enhancement films with polarized emission for liquid-crystal-display applications[J]. Adv. Mater., 2017, 29(33): 1701091.

[103]Kawasaki T, Sato M, Ishiguro S, et al. Enantioselective synthesis of near enantiopure compound by asymmetric autocatalysis triggered by asymmetric photolysis with circularly polarized light[J]. J. Am. Chem. Soc., 2005, 127(10): 3274-3275.

[104]Yang G, Zhu L, Hu J, et al. Near-infrared circularly polarized light triggered enantioselective photopolymerization by using upconversion nanophosphors[J]. Chem. - Eur. J., 2017, 23(33): 8032-8. 38.

[105]Carr R, Evans N H, Parker D. Lanthanide complexes as chiral probes exploiting circularly polarized luminescence [J]. Chem. Soc. Rev., 2012, 41(23): 7673-7686.

[106]Imai Y, Nakano Y, Kawai T, et al. A smart sensing method for object identification using circularly polarized luminescence from coordination-driven self-assembly[J]. Angew. Chem., Int. Ed., 2018, 57(29): 8973-8978.

[107]Wagenknecht C, Li C M, Reingruber A, et al. Experimental demonstration of a heralded entanglement source[J]. Nat. Photonics, 2010, 4(8): 549-552.

[108]Fernandes S N, Lopes L F, Godinho M H. Recent advances in the manipulation of circularly polarised light with cellulose nanocrystal films[J]. Curr. Opin. Solid State Mater. Sci., 2019, 23(2): 63-73.

[109]Ikai T, Kojima Y, Shinohara K I, et al. Cellulose derivatives bearing pyrene-based pi-conjugated pendants with

circularly polarized luminescence in molecularly dispersed state[J]. Polymer, 2017, 117: 220-224.

[110]Zheng H, Li W, Wang X, et al. Uncovering the circular polarization potential of chiral photonic cellulose films for photonic applications[J]. Adv. Mater., 2018, 30(13): 1705948.

[111]Zheng H, Ju B, Wang X, et al. Circularly polarized luminescent carbon dot nanomaterials of helical superstructures for circularly polarized light detection[J]. Adv. Opt. Mater., 2018, 6(23): 1801246.

[112]He J, Bian K, Li N, et al. Generation of full-color and switchable circularly polarized luminescence from nonchiral dyes assembled in cholesteric cellulose films[J]. J. Mater. Chem. C, 2019, 7(30): 9278-9283.

[113]Xu M, Ma C, Zhou J, et al. Assembling semiconductor quantum dots in hierarchical photonic cellulose nanocrystal films: circularly polarized luminescent nanomaterials as optical coding labels[J]. J. Mater. Chem. C, 2019, 7(44): 13794-13802.

[114]Li W, Xu M, Ma C, et al. Tunable upconverted circularly polarized luminescence in cellulose nanocrystal based chiral photonic films[J]. ACS Appl. Mater. Interfaces, 2019, 11(26): 23512-23519.

[115]Kuang Y, Chen C, Chen G, et al. Bioinspired solar-heated carbon absorbent for efficient cleanup of highly viscous crude oil[J]. Adv. Funct. Mater., 2019, 29(16): 1900162.

[116]Xia H, Zhang Z, Liu J, et al. Novel Fe-Mn-O nanosheets/wood carbon hybrid with tunable surface properties as a superior catalyst for Fenton-like oxidation[J]. Appl. Catal., B, 2019, 259: 118058.

[117]Sun B, Yuan Y, Li H, et al. Waste-cellulose-derived porous carbon adsorbents for methyl orange removal[J]. Chem. Eng. J., 2019, 371, 55-63.

[118]Chen W, Zhang Q, Uetani K, et al. Sustainable carbon aerogels derived from nanofibrillated cellulose as high-performance absorption materials[J]. Adv. Mater. Interfaces, 2016, 3(10): 1600004.

[119]Qiu B, Gu H, Yan X, et al. Cellulose derived magnetic mesoporous carbon nanocomposites with enhanced hexavalent chromium removal[J]. J. Mater. Chem. A, 2014, 2(41): 17454-17462.

[120]Singh G, Lakhi K S, Sil S, et al. Biomass derived porous carbon for CO_2 capture[J]. Carbon, 2019, 148, 164-186.

[121]Supanchaiyamat N, Jetsrisuparb K, Knijnenburg J T N, et al. Lignin materials for adsorption: Current trend, perspectives and opportunities[J]. Bioresour. Technol., 2019, 272, 570-581.

[122]Saha D, Bramer S E V, Orkoulas G, et al. CO_2 capture in lignin-derived and nitrogen-doped hierarchical porous carbons[J]. Carbon 2017, 121, 257-266.

[123]P. Chen, L. Wang, G. Wang, et al. Energy Environ. Sci. 2014, 7, 4095.

[124]H. Xu, X. Yin, M. Zhu, et al. Cheng, ACS Appl. Mater. Interfaces 2017, 9, 6332.

[125]Fang J, Liu T, Chen Z, et al. A wormhole-like porous carbon/magnetic particles composite as an efficient broadband electromagnetic wave absorber[J]. Nanoscale, 2016, 8(16): 8899-8909.

[126]Lu S, Jin M, Zhang Y, et al. Chemically exfoliating biomass into a graphene-like porous active carbon with rational pore structure, good conductivity, and large surface area for high-performance supercapacitors[J]. Adv. Energy Mater., 2018, 8(11): 1702545.

[127]Cao J, Zhu C, Aoki Y, et al. Starch-derived hierarchical porous carbon with controlled porosity for high performance supercapacitors[J]. ACS Sustainable Chem. Eng., 2018, 6(6): 7292-7303.

[128]Xie A, Dai J, Chen X, et al. Ultrahigh adsorption of typical antibiotics onto novel hierarchical porous carbons derived from renewable lignin via halloysite nanotubes-template and in-situ activation[J]. Chem. Eng. J., 2016, 304, 609-620.

[129]Zhao H, Cheng Y, Liu W, et al. Biomass-derived porous carbon-based nanostructures for microwave absorption [J]. Nano-Micro Lett., 2019, 11(1): 24.

[130] Gao S, An Q, Xiao Z, et al. Significant promotion of porous architecture and magnetic Fe3O4 NPs inside honeycomb-like carbonaceous composites for enhanced microwave absorption[J]. RSC Adv., 2018, 8(34): 19011-19023.

[131] Gong Y, Li D, Luo C, et al. Highly porous graphitic biomass carbon as advanced electrode materials for supercapacitors[J]. Green Chem., 2017, 19(17): 4132-4140.

[132]Lv W, Wen F, Xiang J, et al. Peanut shell derived hard carbon as ultralong cycling anodes for lithium and sodium batteries[J]. Electrochim. Acta, 2015, 176, 533-541.

[133]Zhu Y, Chen M, Li Q, et al. A porous biomass-derived anode for high-performance sodium-ion batteries[J]. Carbon, 2018, 129: 695-701.

[134]Xi J, Zhou E, Liu Y, et al. Wood-based straightway channel structure for high performance microwave absorption [J]. Carbon, 2017, 124: 492-498.

[135]Lv H, Ji G, Liang X, et al. A novel rod-like MnO_2@ Fe loading on graphene giving excellent electromagnetic absorption properties[J]. J. Mater. Chem. C, 2015, 3(19): 5056-5064.

[136]Xu P, Gao Q, Ma L, et al. A high surface area N-doped holey graphene aerogel with low charge transfer resistance as high-performance electrode of non-flammable thermostable supercapacitors[J]. Carbon, 2019, 149: 452-461.

[137]Ma L, Liu J, Lv S, et al. Scalable one-step synthesis of N, S co-doped graphene-enhanced hierarchical porous carbon foam for high-performance solid-state supercapacitors[J]. J. Mater. Chem. A, 2019, 7(13): 7591-7603.

[138]Wan C, Jiao Y, Bao W, et al. Self-stacked multilayer FeOCl supported on a cellulose-derived carbon aerogel: a new and high-performance anode material for supercapacitors[J]. J. Mater. Chem. A, 2019, 7(16): 9556-9564.

[139]Cetinkaya T, Dryfe R A W. Electrical double layer supercapacitors based on graphene nanoplatelets electrodes in organic and aqueous electrolytes: Effect of binders and scalable performance[J]. J. Power Sources, 2018, 408: 91-104.

[140]Li J, Wang N, Tian J, et al. Cross-coupled macro-mesoporous carbon network toward record high energy-power density supercapacitor at 4 V[J]. Adv. Funct. Mater. 2018, 28(51): 1806153.

[141] Lee Y, Lee J, Kim D, et al. Mussel-inspired surface functionalization of porous carbon nanosheets using polydopamine and Fe^{3+}/tannic acid layers for high-performance electrochemical capacitors[J]. J. Mater. Chem. A, 2017, 5 (48): 25368-25377.

[142]Shao R, Niu J, Liang J, et al. Mesopore- and macropore-dominant nitrogen-doped hierarchically porous carbons for high-energy and ultrafast supercapacitors in non-aqueous electrolytes[J]. ACS Appl. Mater. Interfaces, 2017, 9(49): 42797-42805.

[143]Huang Y, Peng L, Liu Y, et al. Highly sensitive protein detection based on smart hybrid nanocomposite-controlled switch of DNA polymerase activity[J]. ACS Appl. Mater. Interfaces, 2016, 8(41): 28202-28207.

[144]Wu C, Zhang S, Wu W, et al. Carbon nanotubes grown on the inner wall of carbonized wood tracheids for high-performance supercapacitors[J]. Carbon, 2019, 150: 311-318.

[145]Chen C, Zhang Y, Li Y, et al. Highly conductive, lightweight, low-tortuosity carbon frameworks as ultrathick 3D current collectors[J]. Adv. Energy Mater. 2017, 7(17): 1700595.

[146]Chen C, Zhang Y, Li Y, et al. All-wood, low tortuosity, aqueous, biodegradable supercapacitors with ultra-high capacitance[J]. Energy Environ. Sci., 2017, 10(2): 538-545.

[147]Ding H, Peng G, Mo S, et al. Ultra-fast vapor generation by a graphene nano-ratchet: a theoretical and simulation study[J]. Nanoscale, 2017, 9(48): 19066-19072.

[148]Cao S, Jiang Q, Wu X, et al. Advances in solar evaporator materials for freshwater generation[J]. J. Mater. Chem. A, 2019, 7(42): 24092-24123.

[149]Liu H, Huang Z, Liu K, et al. Interfacial solar-to-heat conversion for desalination[J]. Adv. Energy Mater., 2019, 9(21): 1900310.

[150]Yu Z, Cheng S A, Li C, et al. Enhancing efficiency of carbonized wood based solar steam generator for wastewater treatment by optimizing the thickness[J]. Sol. Energy, 2019, 19: 3434-441.

[151]Zha X, Zhao X, Pu J, et al. Flexible anti-biofouling MXene/cellulose fibrous membrane for sustainable solar-driven

water purification[J]. ACS Appl. Mater. Interfaces, 2019, 11(40): 36589-36597.

[152]Huang W, Hu G, Tian C, et al. Nature-inspired salt resistant polypyrrole-wood for highly efficient solar steam generation[J]. Sustain. Energ. Fuels, 2019, 3(11): 3000-3008.

[153]Wang Z, Yan Y, Shen X, et al. A wood-polypyrrole composite as a photothermal conversion device for solar evaporation enhancement[J]. J. Mater. Chem. A, 2019, 7(36): 20706-20712.

[154]Liu S, Huang C, Huang Q, et al. A new carbon-black/cellulose-sponge system with water supplied by injection for enhancing solar vapor generation[J]. J. Mater. Chem. A, 2019, 7(30): 17954-17965.

[155]Hou Q, Xue C, Li N, et al. Self-assembly carbon dots for powerful solar water evaporation[J]. Carbon, 2019, 149: 556-563.

[156]Kuang Y, Chen C, He S, et al. A high-performance self-regenerating solar evaporator for continuous water desalination[J]. Adv. Mater., 2019, 31(23): 1900498.

[157]He Y, Li H, Guo X, et al. Delignifiedwood-based highly efficient solar steam generation device via promoting both water transportation and evaporation[J]. BioResources, 2019, 14(2): 3758-3767.

[158]Guo D, Yang X. Highly efficient solar steam generation of low-cost TiN/bio-carbon foam[J]. Sci. China Mater., 2019, 62(5): 711-718.

[159]Liu S, Huang C L, Luo X, et al. Performance optimization of bi-layer solar steam generation system through tuning porosity of bottom layer[J]. Appl. Energy, 2019, 239: 504-513.

[160]Wang M, Wang P, Zhang J, et al. A ternary Pt/Au/TiO_2-decorated plasmonic wood carbon for high-efficiency interfacial solar steam generation and photodegradation of tetracycline[J]. ChemSusChem, 2019, 12(2): 467-472.

[161]Luo X, Huang C, Liu S, et al. High performance of carbon-particle/bulk-wood bi-layer system for solar steam generation[J]. Int. J. Energy Res., 2018, 42(15): 4830-4839.

[162]Liu F, Zhao B, Wu W, et al. Low cost, robust, environmentally friendly geopolymer-mesoporous carbon composites for efficient solar powered steam generation[J]. Adv. Funct. Mater. 2018, 28(47): 1803266.

[163]Li C, Yu J, S. Xue S, et al. Wood-inspired multi-channel tubular graphene network for high-performance lithium-sulfur batteries[J]. Carbon, 2018, 139: 522-530.

[164]Zhu M, Li Y, Chen G, et al. Tree-inspired design for high-efficiency water extraction[J]. Adv. Mater., 2017, 29(44): 1704107.

[165]He S, Chen C, Kuang Y D, et al. Nature-inspired salt resistant bimodal porous solar evaporator for efficient and stable water desalination[J]. Energy Environ. Sci., 2019, 12(5): 1558-1567.

[166]Xue G, Liu K, Chen Q, et al. Robust and low-cost flame-treated wood for high-performance solar steam generation[J]. ACS Appl. Mater. Interfaces, 2017, 9(17): 15052-15057.

[167]Chen Z, Dang B, Luo X, et al. Deep eutectic solvent-assisted in situ wood delignification: A promising strategy to enhance the efficiency of wood-based solar steam generation devices[J]. ACS Appl. Mater. Interfaces, 2019, 11(29): 26032-26037.

[168]Liu P, Miao L, Deng Z, et al. A mimetic transpiration system for record high conversion efficiency in solar steam generator under one-sun[J]. Mater. Today Energy, 2018, 8: 166-173.

[169]Qiu P, Liu F, Xu C, et al. Porous three-dimensional carbon foams with interconnected microchannels for high-efficiency solar-to-vapor conversion and desalination[J]. J. Mater. Chem. A, 2019, 7(42): 13036-13042.

[170]Zhang X, Jiang M, Niu N, et al. Natural-product-derived carbon dots: from natural products to functional mterials[J]. ChemSusChem, 2018, 11(1): 11-24.

[171]Li Y, Gao T, Yao Y, et al. In situ "chainmail catalyst" assembly in low-tortuosity, hierarchical carbon frameworks for efficient and stable hydrogen generation[J]. Adv. Energy Mater., 2018, 8(25): 1801289.

[172]Dong S, Roman M. Fluorescently labeled cellulose nanocrystals for bioimaging applications[J]. J. Am. Chem.

Soc., 2007, 129(45): 13810.

[173]Fan X, Yu H, Wang D, et al. Designinghighly luminescent cellulose nanocrystals with modulated morphology for multifunctional bioimaging materials[J]. ACS Appl. Mater. Interfaces, 2019, 11(15): 48192-48201.

[174]Zhang Y, X. Ma X, Gan L, et al. Fabrication of fluorescent cellulose nanocrystal via controllable chemical modification towards selective and quantitative detection ofCu(II) ion[J]. Cellulose, 2018, 25(10): 5831-5942.

[175]Li Y P, Zhang J M, Guo Y Z, et al. Cellulosic micelles as nanocapsules of liposoluble CdSe/ZnS quantum dots for bioimaging[J]. J. Mater. Chem. B, 2016, 4(39): 6454-6461.

[176]Wang Z, Fan X, He M, et al. Construction of cellulose-phosphor hybrid hydrogels and their application for bioimaging[J]. J. Mater. Chem. B, 2014, 2(43): 7559-7566.

[177]Guo J, Liu D, Filpponen I, et al. Photoluminescent hybrids of cellulose nanocrystals and carbon quantum dots as cytocompatible probes for in vitro bioimaging[J]. Biomacromolecules, 2017, 18(7): 2045-2055.

[178]Wu B, Zhu G, Dufresne A, et al. Fluorescent aerogels based on chemical crosslinking between nanocellulose and carbon dots for optical sensor[J]. ACS Appl. Mater. Interfaces, 2019, 11(17): 16048-16058.

[179]Xiong F, Han Y, Li G, et al. Synthesis and characterization of renewable woody nanoparticles fluorescently labeled by pyrene[J]. Ind. Crops Prod., 2016, 83: 663-669.

中文题目：木材衍生炭材料和发光材料

作者：李伟，陈志俊，于海鹏，李坚，刘守新

摘要：木材是一种天然的具有多级结构的、可持续和可再生的材料。纤维素、半纤维素和木质素是木材的三大组分。木材及其衍生物独特的物理化学性质，使其在生物工程、柔性电子和清洁能源等先进材料中具有巨大的应用潜力。然而，尽管在木材衍生碳材料和发光材料已经有了很多研究进展的报道，但关于其相关的系统的综述却鲜有报道。这里，我们对木材衍生炭材料和发光材料的特性进行了总结，并对其制备原理、性质、应用、挑战和未来前景进行了详细的总结和展望，为进一步深化了解和启发木质基新材料研究领域提供新思路。

关键词：炭材料；发光材料；木材；木材衍生物

Multifunctional Reversible Self-assembled Structures of Cellulose-derived Phase Change Nanocrystals*

Yonggui Wang, Zhe Qiu, Zhen Lang, Yanjun Xie, Zefang Xiao, Haigang Wang, Daxin Liang, Jian Li, Kai Zhang

Abstract: Owing to advantageous properties attributed to well-organized structures, multifunctional materials with reversible hierarchical and highly ordered arrangement in solid-state assembled structures have drawn tremendous interest. However, such materials rarely exist. Based on the reversible phase transition of phase change materials (PCMs), phase change nanocrystals (C18-UCNCs) were presented herein, which are capable of self-assembling into well-ordered hierarchical structures. C18-UCNCs have a core-shell structure consisting of a cellulose crystalline core that retains the basic structure and a soft shell containing octadecyl chains that allow phase transition. The distinct core-shell structure and phase transition of octadecyl chains allow C18-UCNCs to self-assemble into flaky nano/microstructures. These self-assembled C18-UCNCs exhibit efficient thermal transport and light-to-thermal energy conversion, and thus are promising for thermosensitive imaging. Specifically, flaky self-assembled nano/microstructures with manipulable surface morphology, surface wetting, and optical properties are thermoreversible and show thermally induced self-healing properties. By using phase change nanocrystals as a novel group of PCMs, reversible self-assembled multifunctional materials can be engineered. This study proposes apromising approach for constructing self-assembled hierarchical structures by using phase change nanocrystals andthereby significantly expands the application of PCMs.

Keywords: Cellulose; Phase change nanocrystal; Core-shell structure; Thermal imaging; Thermal reversible

Abbreviations:

AFM, atomic force microscopy;

C18-UCNCs, phase change nanocrystals by immobilizing 1-octadecanethiol onto UCNCs;

$C18_{0.5}$-UCNCs or $C18_{0.75}$-UCNCs, partially 1-octadecanethiolated UCNCs by immobilizing low ratio of 1-Octadecanethiol(0.5 or 0.75 mol per mol C=C);

CNCs, cellulose nanocrystals;

DS, the average degree of substitution;

DSC, differential scanning calorimetry;

G+C18-UCNCs, the glass slide coated withC18-UCNCs;

G+stearic acid, the glass slide coated with stearic acid;

---

* 本文出自 Advanced Materials, 2021, 33: 2005263.

MCC, microcrystalline cellulose;

PCMs, phase change materials;

POM, polarized optical microscopy;

RT, room temperature;

SAXS, small-angle X-ray scattering;

SEM, scanning electron microscopy;

TEM, transmission electron microscopy;

THF, tetrahydrofuran;

UCNCs, undecenoated cellulose nanocrystals;

Dry-state structures engineered using the precise positional arrangement of molecules or nanoparticles at different length scales have attracted great interest for functional materials.[1] In stimuli-responsive materials, ordered molecular arrangements can give rise to novel structures and functionalities in response to the environmental changes.[2] For example, phase change materials (PCMs) are temperature-responsive materials that can reversibly store and release abundant amounts of thermal energy, when they change their molecular arrangements from one physical state to another.[3] Organic PCMs, such as polyethylene glycol, fatty alcohols, paraffin, and fatty acids, generally exhibit large latent heat and solid-liquid phase transitions, rendering them highly suitable for the preparation of thermal energy storage materials.[4,5]. These phase changes are governed by diverse intermolecular interactions, such as van der Waals forces, dipole-dipole interactions, and hydrogen bonding.[6] Adjusting these interactions during phase transitions allows control not only over thermal energy but self-assembled structures as well via molecular arrangement. Some solid-liquid PCMs, including stearic acid and alkylated comb-like polymers, have been used to create distinct functional nanostructures.[7,8] However, the obtained structures are generally irreversible owing to the high melting temperature and ease of deformability.

Substantial efforts have been devoted to stabilizing solid-liquid PCMs to prevent leakage during melting. Conventional techniques include PCM encapsulation in a core material,[9] porous adsorption,[10] and physical/chemical bonding in micro-or macroscale molecular networks to form composites.[11] For these stabilized PCMs, the mobility and arrangement of their molecules are generally restricted to a small area, and only crystalline structures can be reversibly changed by varying the temperature.[12] However, the overall morphologies and shapes of PCMs mainly depend on stabilizing substrates (e. g., silica shells, metal-organic frameworks, porous graphene and coaxial fibers),[13,14] to obtain PCMs in micro/nanoparticles, porous tubes, foams, bulk sheets, and fibers.

By contrast, the construction of ordered hierarchical nano/microstructures by using nanoparticles still presents a challenge and rarely obtains reversible structures. In the current study, a strategy was developed to prepare phase change nanocrystals with self-stabilized shapesby immobilizing 1-octadecanethiol onto undecenoated cellulose nanocrystals (UCNCs) (Fig. 1a), resulting in core-shell C18-UCNCs. These phase change nanocrystals not only possess efficient thermal transport and conversionproperties, but can also self-assemble into solid-state thermoreversible hierarchical structures. Different than conventional PCM molecules requiring structuring techniques, immobilized octadecyl chains in the soft shell of C18-UCNCs can undergo a phase change because of the feasibility of crystallization, while their mobility is strongly confined owing to the presence of CNCs as inert cores. Molecular interactions between the C18-UCNCs shells ultimately lead to anisotropic flaky nanostructures to which complete reversibility and thermoreversible properties are attributed.

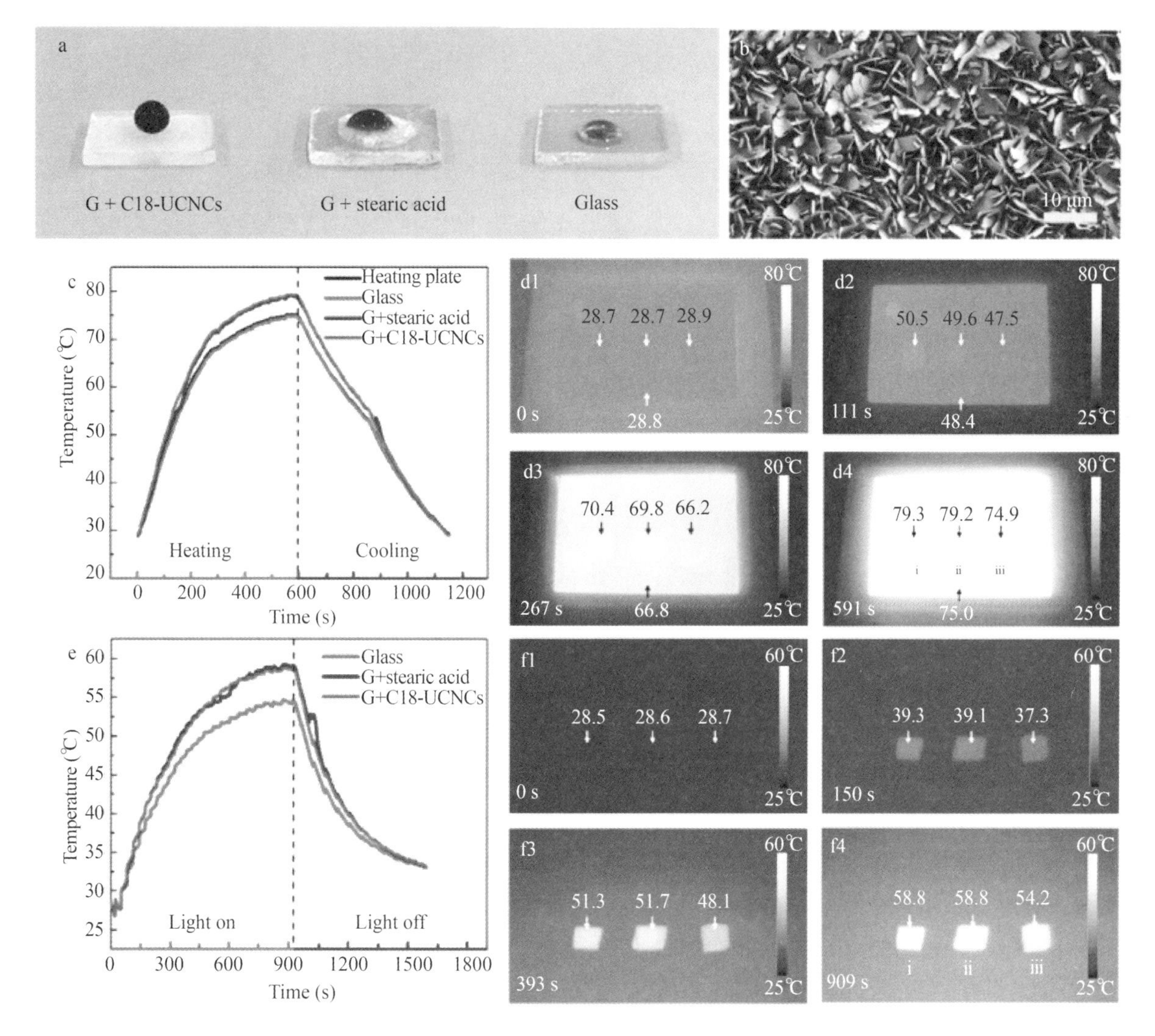

Fig. 1 Synthesis and characterization of C18-UCNCs as phase change nanocrystals. (a) Schematic of the synthesis of C18-UCNCs from UCNCs via the thiol-ene reaction with 1-octadecanethiol; (b1-d1) AFM images of (b1) CNCs, (c1) UCNCs, and (d1) C18-UCNCs; (b2-d2) TEM images of (b2) CNCs, (c2) UCNCs, and (d2) C18-UCNCs; (e) Solid-state ^{13}C NMR spectra of CNCs, UCNCs, and C18-UCNCs; (f) DSC curves of C18-UCNCs asthe 1st and 10th thermal cycle

CNCs wereextracted from microcrystalline cellulose (MCC) by hydrochloric acid hydrolysis.[15] Obtained CNCs formed stable suspensions in water (Fig. S1). The morphology of the CNCs was characterized by atomic force microscopy (AFM) and transmission electron microscopy (TEM) (Fig. 1b), which showed typically rigid rod-shaped CNCs. CNCs had an average length of 197±73 nm and an average width of 12±4 nm, as revealed by TEM. UCNCs were synthesized by modifying CNCs with 10-undecenoyl chloride in pyridine under heterogeneous reaction conditions (Fig. 1a).[16] After surface esterification, obtained UCNCs were well dispersed in tetrahydrofuran(THF)(Fig. S1), retaining their rod-shaped form, as revealed in the AFM and TEM images(Fig. 1c). The solid-state ^{13}C NMR spectrum of UCNCs showed typical signals at 114.9 ppm (C17) and 138.9 ppm (C16) ascribed to carbons in terminal olefin groups. The signals between 20 and 40 ppm and those near 173 ppm were attributed to alkane and carbonyl carbons in the group with 10-undecenoyl, respectively(Fig. 1e).[17] The average degree of substitution(DS) ascribed to 10-undecenoyl groups in UCNCs was approximately 0.9, as determined by elemental analysis. Thus, more than two-thirds of the hydroxyl group on cellulose chains remained unreacted. The heterogeneous esterification of CNCs generally begins from the particle surface, extending to the interior.[18] The intact hydroxyl groups should mainly be located in the core of the UCNCs. The presence of the unreacted cellulose core within the UCNCs was confirmed by solid-state ^{13}C NMR spectroscopy, which detected the signal at 84 ppm (C4) attributed to crystalline cellulose (Fig. S2).[19]

Chemicals containing straight long-chain alkanes are widely used as organic PCMs.[5] To prepare C18-UCNCs as phasechange nanocrystals with phasechange properties, 1-octadecanethiol was grafted in the shell layer ofUCNCs containing terminal vinyl groups via thiol-ene reaction(Fig. 1a).[20] Compared with those of UCNCs, the peaks related to vinyl groups at 113.9 and 138.2 ppm in the ^{13}C NMR spectrum of C18-UCNCs disappeared after the thiol-ene reaction, while a new signal attributed to the CH_3-group appeared at 15.0 ppm (Fig. 1e). These results confirmed the successful introduction of 1-octadecanethiol groups. For C18-UCNCs with long-chain alkanes, the nanocrystals tend to assemble into aggregates(Fig. 1d). Compared with UCNCs, C18-UCNCs showed a similar crystalline structure, as observed from the solid-state ^{13}C NMR spectrum, indicating that the cellulose crystalline core was retained(Fig. 1a and Fig. 2). Thus, the C18-UCNCs consist of a highly esterified phase change soft shell and an intactcrystalline cellulose core.

As a crucial factor, the phase change enthalpy of C18-UCNCs as phase change nanocrystalswas investigated using differential scanning calorimetry (DSC) (Fig. 1f). In comparison to pristine 1-octadecanethiol with a melting point(T_m) at approximately 31 ℃ and a crystallization temperature(T_c) at about 26 ℃.[21] C18-UCNCswithoctadecyl chainsin the shell had higher T_m and T_c at nearly 60 ℃ and 54 ℃, respectively(Fig. 1f). This change was related to the presence of the octadecyl chains with a high density in the shell, which promoted the formation of crystalline regionsby restricted chains.[22] C18-UCNCs showed excellent thermal properties with a latent heat ofmelting(H_m) and crystallization(H_c) of about 40.8 $J \cdot g^{-1}$ and 41.5 $J \cdot g^{-1}$, respectively (Fig. 1f). Notably, C18-UCNCs show excellent energy storage and release stability, and no obvious change of temperature and latent heat after 10 phase change cycles.

Thermal transport and conversion are considered as important properties for PCMs. Based on the enhanced crystallization of surface-attached octadecyl chains, the thermal transport under heating and the light-to-thermal energy conversion under irradiation of C18-UCNCs were examined. The glass slide coated withC18-UCNCs(referred to as G+C18-UCNCs)wascompared with the glass slide coated with stearic acid(referred to as G+stearic acid) and a pure glass slide. C18-UCNCs formed a layer of nanoflakeson the surface exhibiting superhydrophobicity(Fig. 2a-b) (further investigated in the subsequent section). By contrast, stearic acid formed no such nanoflakes on the glass surfaceand instead formed only rough aggregates as crystallized stearic acid(Fig. S3). Fig. 2c presents the temperature development curves of the three samples during heating and cooling on the plate(Fig. S4a). When the temperature was set to 80 ℃ and the heater was turned on, the

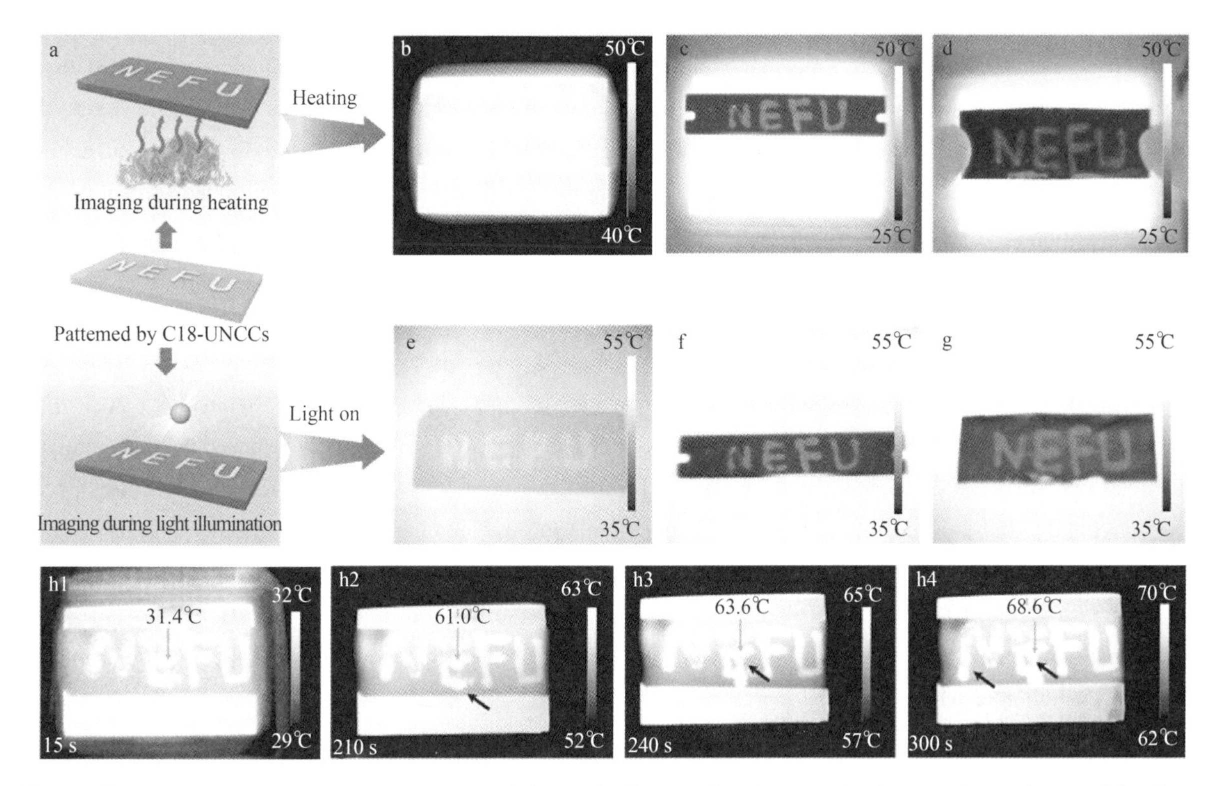

Fig. 2 The thermal transport and conversion of C18-UCNCs. (a) Dyed water droplets on the surfaces of G+C18-UCNCs, G+stearic acid, and glass; (b) SEM image of the surface morphology of G+C18-UCNCs; (c) Time-dependent temperature evolution curves of G+C18-UCNCs, G+stearic acid, and glass during heating and cooling; (d) Infrared images showing temperature variations during heating(c); (e) Light-to-thermal energy conversion curves of G+C18-UCNCs, G+stearic acid, and glass; (f) Infrared images showing the temperature variations of simulated sunlight illumination of(e). Samples i, ii, and iii in(d)and(f)represent the samples of G+C18-UCNCs, G+stearic acid, and glass, respectively. The numbers above or below the arrows in(b)and(d)denote the temperatures(℃)of the corresponding areas

temperatures of G+C18-UCNCs, G+stearic acid, and the glass increased from ~28 ℃ and remained constant at 79.3 ℃, 79.2 ℃, and 74.9 ℃, respectively(Fig. 2c and Movie S1) Stearic acid containing octadecyl chains is recommended as a promisingPCM because of the high latent heat storage density and energy preservation ability, which make it more suitable for thermal energy storageapplications.[23] The temperature of G+C18-UCNCs increased at almost the same rate as that of G+stearic acid, while faster than that of glass, despite the considerably higher thermal conductivity of the glass slide(Fig. 2c and Fig. S5). This finding may be related to the grafted octadecyl chains in the shell of C18-CNCs. Their presence with a high density formed a phase change shell that facilitated thermal transport and conversion performance. During heating or cooling, no thermal storage plateau was observed for G+C18-UCNCs, unlike stearic acid, which should be in turn attributed to the extremely low absolute amount of the octadecyl chains in the C18-UCNCs shell.

Tointuitively and visually observe the change in surface transient temperature on the plate during heating and cooling, the thermal infrared images of G+C18-UCNCs, G+stearic acid, and glass were recorded (Fig. 2d). It is notable that the temperature of G+C18-UCNCs was similar to that of G+stearic acid and consistently higher than those of glass and the plate (Fig. 2d and Movie S1). This difference should be attributed to the efficient thermal conversion and preservation of C18-UCNCs, which clearly delayed direct energy transport to the surrounding air and thus led to increased temperature in the area of the plate covered by C18-UCNCs. By contrast, the plate surface without the C18-UCNCs coverage orcoveredby glass had a lower temperature due to faster energy transport to the air. Moreover, the leakage test was conducted to evaluate the thermal stability of C18-UCNCs. Photographs and thermal infrared images of G+C18-UCNCs and G+stearic acid before and after heat treatment at 80 ℃ for 5 min are presented in Fig. S4, Fig. S6 and Movie S2. The stearic acid on the surface of G+stearic acid completely melted into liquid, acquiring the ability to flow, whereas C18-UCNCs on the G+C18-UCNCs surface showed no such leakage. Moreover, C18-UCNCs also demonstrated considerable light-to-thermal energy conversion under light irradiation by using a solar light simulator(Fig. S7). Samples of G+C18-UCNCs, G+stearic acid, and glass were exposed to the simulated solar light and then cooled to room temperature (RT) after the light was turned off. Time-temperature curves were generated, and thermal infrared images of the samples were obtained using an infrared thermal camera (Fig. 2e-f and Movie S3). Similar to that of G+stearic acid, the temperature of G+C18-UCNCs increased under light irradiation at a rate faster than that of glass during the illustrating process. G+C18-UCNCs reached the highest temperature(58. 8 ℃) at 909 s, which was 4. 6 ℃ higher than that of the glass. This result further indicated the efficient light-to-thermal conversion of C18-UCNCs.

The efficient thermal preservation and light-to-thermal conversion of the self-assembled C18-UCNCs suggest that C18-UCNCs show great potential for application in thermosensitive imaging. Imaging via heating or illumination is illustrated in Fig. 3a. During heating or under light irradiation, patterned areas formed by C18-UCNCs should significantly increase their temperatures, which can be captured with a thermal imaging camera. We fabricated patterned letters of "NEFU" with C18-UCNCs on three substrates as glass, steel, and foil paper(Fig. S8a-c). By heating the patterned glass slide on the plate at different temperatures, the efficient thermal transport and preservation of C18-UCNCs result in the higher temperatures of the patterned areas and consequently, brighter patterns in the infrared images(Fig. S8, Fig. S9 and Fig. 3b). In addition, the "NEFU" patterns with higher temperatures than those of the surrounding area on the steel blade and foil paper also became apparent during heating at 50 ℃ (Fig. 3c–d). Under simulated sunlight irradiation, all C18-UCNCs patterns on glass, steel, and foil paper could be captured with a thermal imaging camera, revealing high-quality imaging(Fig. 3e–g). The C18-UCNCs showed significantly higher thermal stability than that of molecular PCM stearic acid, as reflected in the patterned letters of "NE" formed by stearic acid. The letters "FU" were formed by C18-UCNCs(Fig. 3h). During the heat treatment at the set temperature (80 ℃), the letters "NE" were gradually damaged by the melting of stearic acid. By contrast, the letters "FU" remained unchanged throughout because of the presence of the cellulose crystalline core, indicating excellent thermal stability for C18-UCNCs (Fig. 3h). Therefore, C18-UCNCsrepresent promising candidates for thermosensitive imaging.

The unique flaky structures were observed on the surface of G+C18-UCNCs(Fig. 2b). To understand the mechanisms underlying the self-assembly of C18-UCNCs into flaky structures, self-standing films were formed by solvent casting from the THF suspensions of C18-UCNCs on Teflon substrates(Fig. 4a and Fig. 11a). The stress-strain curve fromthe tensile test in Fig. S11bshows a fracture strength of around 9. 71 MPa for C18-

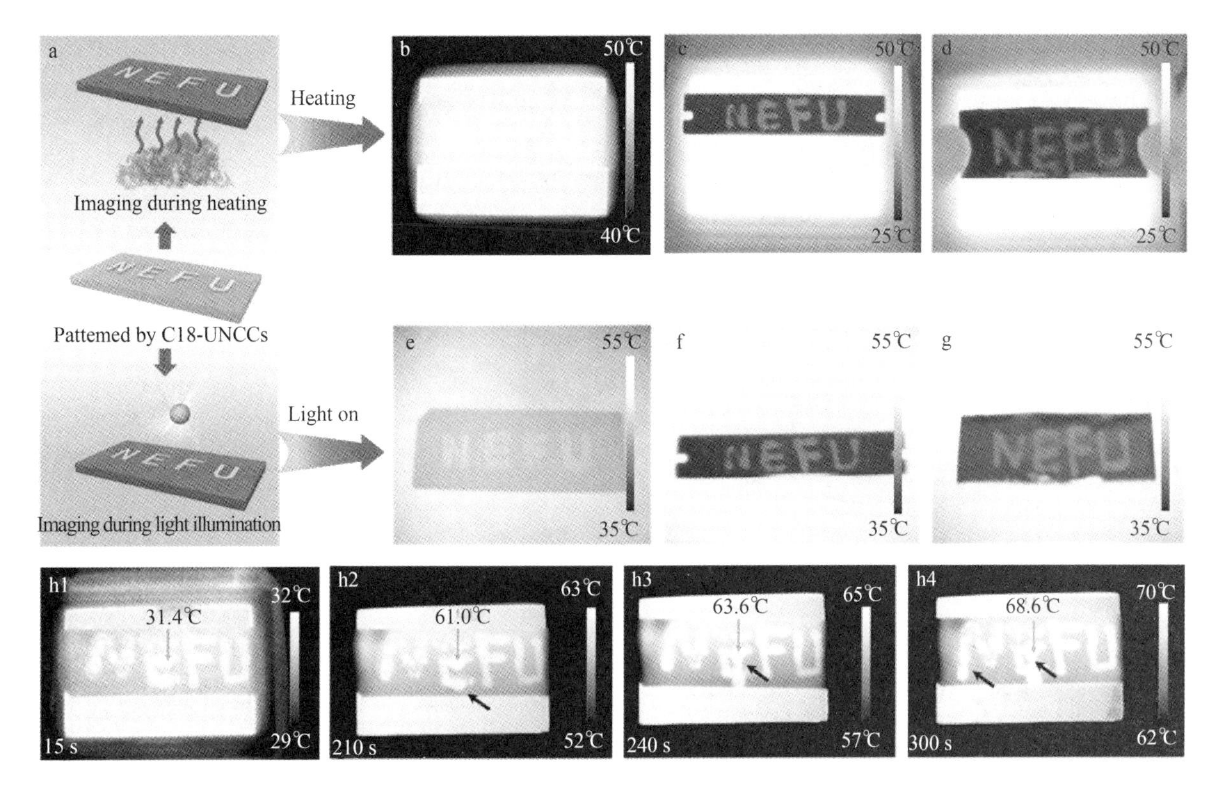

Fig. 3 Thermosensitive imaging with C18-UCNCs. (a) Schematic illustration for the thermosensitive imaging of surfaces containing C18-UCNC patterns of "NEFU" during heating or under light illumination; (b-d) Infrared images showing the temperature variations of (b) a patterned glass slide, (c) a patterned steel blade, and (d) a sheet of patterned foil paper on a heating plate at the set temperature of 50 ℃; (e-g) Infrared images showing the temperature variations of (e) a patterned glass slide, (f) a patterned steel blade, and (g) a sheet of patterned foil paper under simulated sunlight illumination; (h) Time-dependent infrared images showing temperature variations of the patterned glass slide on a heating plate with a tilt angle of 45° during heating from RT to the set temperature of 80 ℃. The patterned letters "NE" are formed by stearic acid, and the letters "FU" are formed using C18-UCNCs. The red arrows in (h) indicate the temperature of the corresponding area. Black arrows in (h1, h2, and h3) show the flow and leakage of stearic acid caused by melting

UCNCs films. After the evaporation of THF, peeled-off C18-UCNCs films became nontransparent and superhydrophobic, because flaky self-assembled structures were formed on both surfaces of the films (Fig. 4a-b). Irregular flakes with thicknesses of about 150 nm were organized into a flowerlike porous morphology on the film surface (Fig. 4a-b and Table S1). By contrast, the UCNCs suspensionsand pure 1-octadecanethiol solution in THFdid not form self-assembled flaky structures (Fig. S12a-b), but only smooth surface or random aggregates under similar conditions. In addition, polymeric cellulose derivative, which was obtained after the introduction of 1-octadecanethiol into backbone of cellulose 10-undecenoyl ester[17] with the complete derivatization of hydroxy groups (DS = 3) via thiol-ene reaction, also did not form similar self-assembled structure (Fig. S12c). Furthermore, the density of octadecyl chains in the shell of UCNCs was altered to examine the effect of theoctadecyl chains on theself-assembly process. The surfaces of $C18_{0.5}$-UCNCs and $C18_{0.75}$-UCNCs with low densities of octadecyl chain, which were synthesized by partially immobilizing 1-

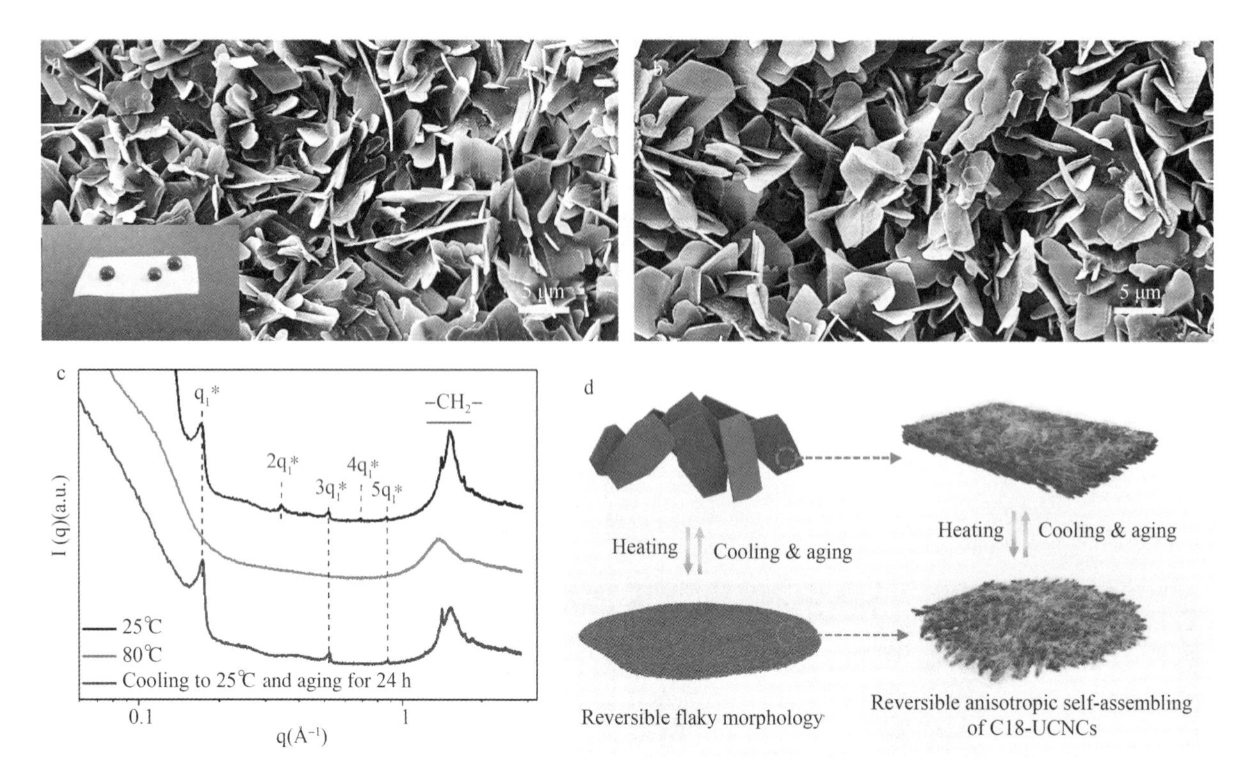

Fig. 4 Flaky self-assembled structures by C18-UCNCs. (a-b) SEM images of flaky self-assembled structures on (a) the top surface and (b) the bottom surface (the surface in contact with Teflon during film formation) of the as-prepared C18-UCNCs film. The inset in (a) shows the dyed water droplets on the surface of the C18-UCNCs film. (c) SAXS curves of a C18-UCNCs film measured under different conditions: as-prepared films at 25 ℃, heated at 80 ℃ and cooled from 80 ℃ to 25 ℃ and aging for 24 h. (d) Schematic of the formation of a thermoreversible flaky structure

octadecanethiol (0.5 or 0.75 mol per mol C=C) throughthe thiol-ene reaction, did not contain fully covered flaky structures, and were only covered by sparsely distributed micro-lamella clusters (Fig. S12d-e). Resulting surfaces showed a lower hydrophobicity in comparison to those of C18-UCNCs. Thus, the presence of phasechange octadecyl chains of a high density within the shell of C18-UCNCs that are at the same time immobilized in a comfined surrounding on crystalline cellulose coresshould be the determining factors for the formation of hierarchical self-assembled structures. Moreover, the THF suspension of stearoylated CNCs with DS of 1.2, which was obtained after surface esterification of CNCs using stearoyl chloride, only formedtransparent films and no self-assembled flaky structure was observed (Fig. S12f). Therefore, the undecenoyl chains are also critical for the formation of the thermoreversible flaky structures, which guarantee the efficient flexibility and mobility for the ordering assembly of the octadecyl groups in dry state.

Alkyl side chains with a critical length of longer than 12 carbon atoms can form crystalline structures.[24] Thus, the formation of the flaky structures should be attributed to the ordered organization of C18-UCNCs by molecular interactions between octadecylchains, which led to the crystalline structure within the flaky units as confirmed with small-angle X-ray scattering curves (SAXS) (Fig. 4c). Within the SAXS curve of dried C18-UCNCs at RT, the intermediate q range of 0.1 -1 $Å^{-1}$ was dominated by several diffraction peaks, representing the existence of a lamellar morphology with a periodicityof 3. 7 nm. Previous studies revealed that the length of stretched octadecyl chainswas around 2. 0 nm. [8] Therefore, this periodicity in dried C18-UCNCs films suggests that the crystalline domains consisted of head-to-head packedoctadecyl chains belonging to different backbones. Furthermore, the scattering peaks in the wide-angle region($1\ \text{Å}^{-1} < q < 2\ \text{Å}^{-1}$) corresponded to the stacking distance between the octadecylmoieties. The sharp peak at $1.42\ \text{Å}^{-1}$ ascribed to a d-spacing of 0. 44 nm was in accordance with the distance between highlyorganized octadecyl chains. After heating at 80 ℃, C18-UCNCs became isotropic due to the disassociation of the crystalline domains of packed octadecyl chains (Fig. 4d). SAXS analysis exhibited only a broad peak from random scattering of the dissociated octadecyl chains(Fig. 4c). By decreasing the temperature to 25 ℃ and aging for further 24 h, C18-UCNCs regenerated nanoflakes with thecrystalline texture. The SAXS curve exhibitedfeatured peaks reflecting a lamellar structure, revealing the reversibility of this self-assembled structure(Fig. 4c-d).

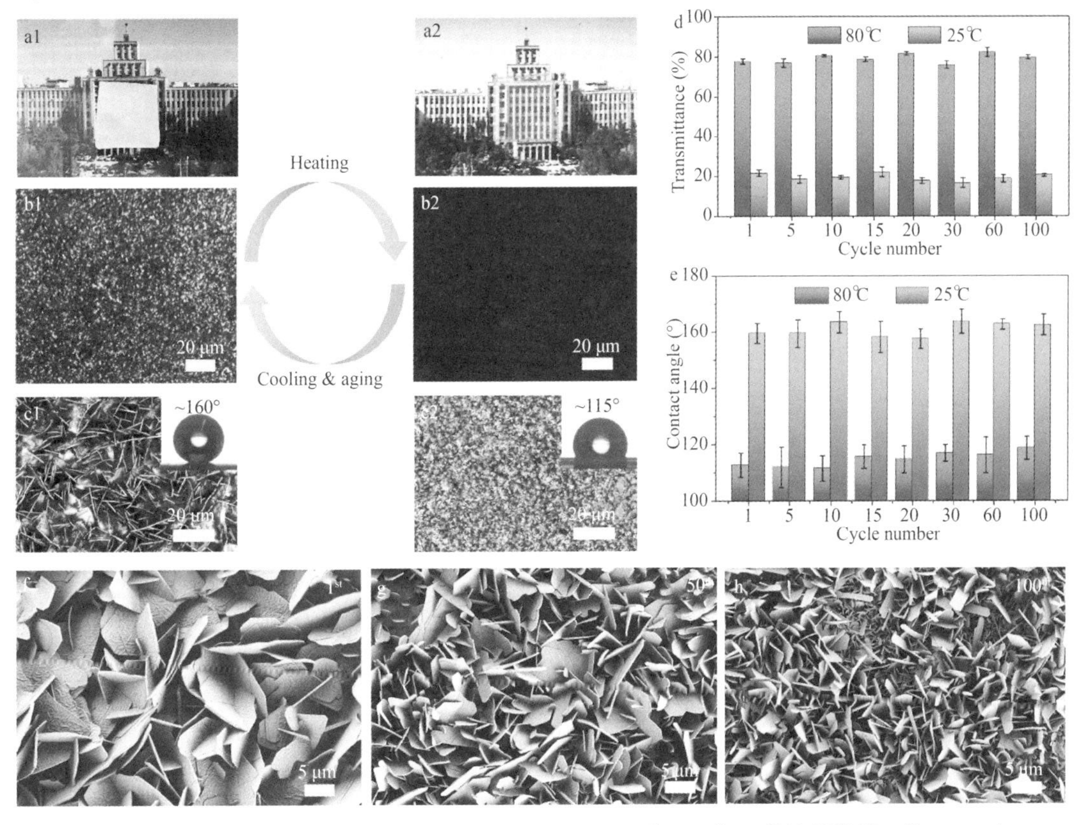

Fig. 5 Thermoreversible self-assembled flaky structures on self-standing C18-UCNCs films under repeated heating at 80 ℃ and cooling with subsequent aging at RT. (a) Photos of a self-standing film of C18-UCNCs at (a1) RT and (a2) 80 ℃; (b) POM images of a C18-UCNCs film at (b1) RT and (b2) 80 ℃; (c) Laser scanning microscopy (LSM) images of the surface morphology of the C18-UCNCs film at (c1) RT and (c2) 80 ℃. The insets in (c1) and (c2) show the static water contact angles on the corresponding surfaces; (d) Transparency of C18-UCNCs films under repeated heating at 80 ℃ and cooling with subsequent aging at 25 ℃ for 100 cycles; (e) Water contact angles on the surface of C18-UCNCs films under repeated heating at 80 ℃ and cooling with subsequent aging at 25 ℃ for 100 cycles; (f-h) SEM images of flaky structures on self-standing C18-UCNCs films after cooling to RT with subsequent aging for>24 h from (d) the first heat treatment at 80 ℃; (e) 50th heat treatment at 80 ℃; and (f) 100th heat treatment at 80 ℃

To elucidate the thermoreversibility of the self-assembled flaky structures, diverse features were investigated, which include optical properties, wetting ability, and morphology. Owing to the crystalline packing of octadecyl moieties between different C18-UCNCs, the opaque C18-UCNCs films at RT exhibited apparent birefringence, as determined from their polarized optical microscopy (POM) spectra (Fig. 5a-b). Different from homogenous molecular crystals with uniform crystalline patterns, the long alkane chains only crystallized in the soft shells of C18-UCNCs, undergoing microphase separation to form flaky structures. Owing to the thermoreversibilityof such crystalline packing by octadecyl moieties, these self-assembled flaky structures endowed obtained C18-UCNCs films with thermoreversible properties. The nontransparent C18-UCNCsfilms at RT became transparent after heating to 80 ℃ and *vice versa* after cooling with accompanied aging for several hours at RT(Fig. 5a and Fig. 5d). Similarly, the surfaces fully covered with ordered flaky structures turned considerably smoother with uniform morphologies after storage at 80 ℃ for several seconds (Fig. 5c and Fig. S13). This change was accompanied by an increase in visible transparency to ~ 80% and a decrease in the static water contact angle to ~ 115° (Fig. 5e). After cooling down by further aging at RT, flaky structures were regenerated, accompanied by the recovery of opacity, birefringence, and superhydrophobicity (Fig. 5a-e). Moreover, the self-assembled flakes exhibited excellent thermoreversibility and maintained similar morphologies with almost the same thickness (Table S1). No significant fatigue was observed in the transparency and wetting properties(Fig. 5d-e). On the contrary, the density of the porous flaky structures gradually increased after 50 and 100 heating-cooling cycles(Fig. 5f-h).

These macroscopic thermoreversible hierarchical structures have not been observed in commonly used PCM systems, which generally undergo melting and crystallization within trapped areas by using stable agents and do not affect changes in macroscopic morphologies.[13,25] Compared with conventional PCMs, the immobilization of the phasechangeoctadecylmoieties in the soft shell of C18-UCNCs induced the mobility of the whole shell during the phase change of the octadecyl chains. When the temperature exceeded the melting temperature of crystallized octadecyl chains (e. g. at 80 ℃) (Fig. 1f), the flakes disintegrated owing to the relaxation and movement of disentangled octadecyl chains in the shell. This process produced a homogeneous non-ordered structurein the films (Fig. 4d, Fig. 5b-c and Movie S4). Unlike the melting of commonly used solid–liquid PCMs, the intact crystalline cellulose core is considerably more inert and retained the film stability during the phase transitions. The absorbed latent heat during the melting of the octadecyl chains released while cooling down to RT induced the phase separation of the relaxed shell layer and the reorganization of the C18-UCNCs at RT. The whole transition process in dry state caused a slow assemblyof the octadecyl chains to form ordered structures. The temperature dropped to RT within 5 min, while the recovery of the flaky structures took much longer time (Fig. S14a). The formation of self-assembled flaky structures occurred immediately after cooling down and grew gradually, which was visualized by a digital microscope with a large depth of field(Fig. S14b and Movie S5). After 30 minutes(Fig. S14b4), the surface was already fully covered with the self-assembled structures with larger populations compared to the image after 1 min in Fig. S14b1. With the re-assembling of the flaky structures, the static water contact angle on the surface of these films already increased to higher than 150° within 1 hour (Fig. S14c). After 3 hours aging, the visible transparency of the films decreased to ~ 15%. With efficient aging time, the flaky structures gradually reassembled on and fully covered the surface of the C18-UCNCs films via the crystallization of octadecyl chains.

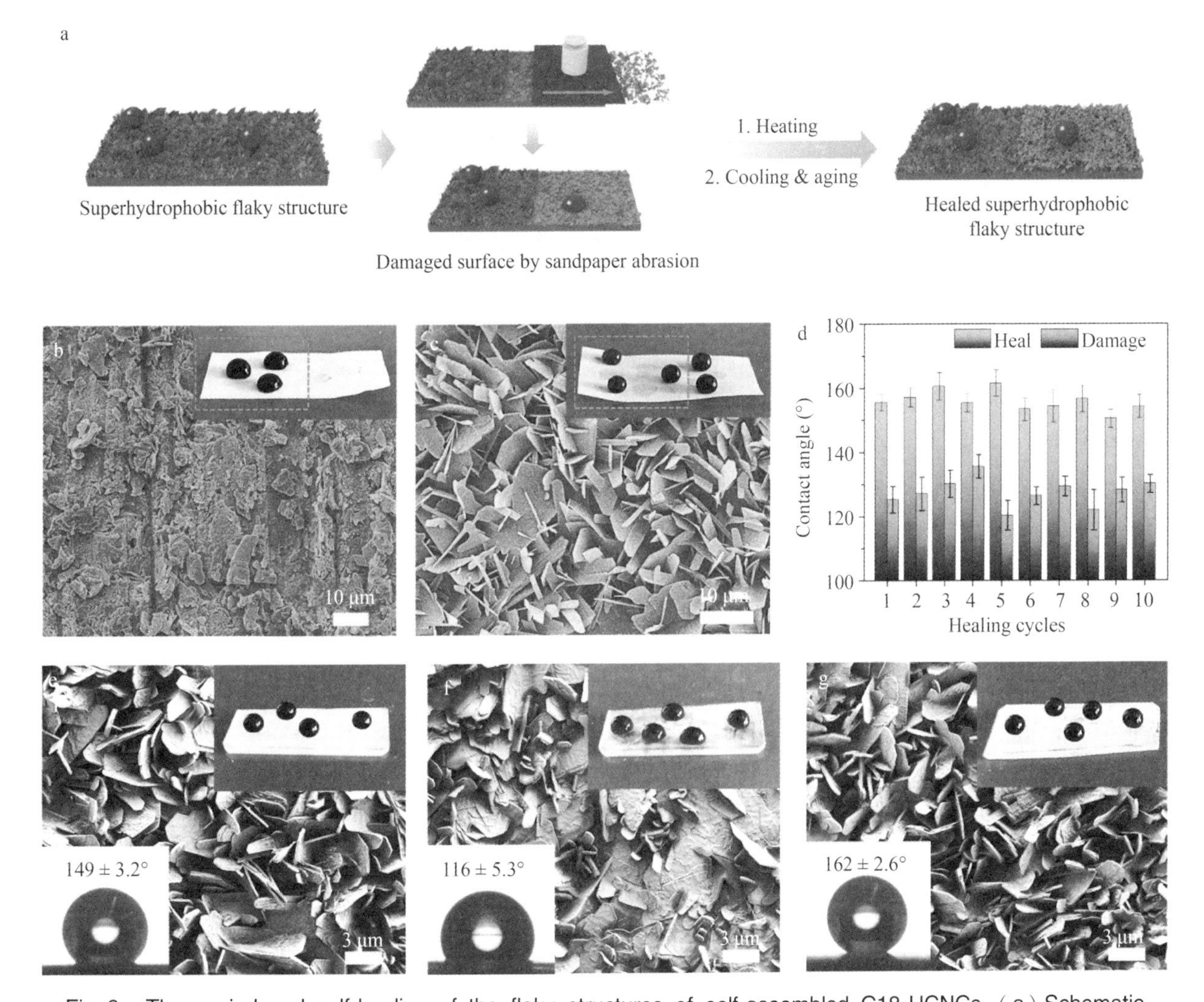

Fig. 6 Thermo-induced self-healing of the flaky structures of self-assembled C18-UCNCs. (a) Schematic illustration of the self-healing process; (b-c) SEM images of the flaky structures on self-standing C18-UCNCs films: (b) after the 1st damage by sandpaper abrasion, and (c) after the 1st healing by heating at 80 ℃ for 30 min, cooling down, and aging at RT for 24 h; (d) Changes in static water contact angles on the C18-UCNCs films as the function of the number of abrasion-healing cycles; (e-g) SEM images of the surface morphologies of the glass slides coated with C18-UCNCs under different conditions: (e) the as-prepared sample, (f) the sample damaged by scratching, (g) the sample healed by heating at 80 ℃ for 30 min, cooling down and aging at RT for 24 h. Insets in (b) and (c) show dyed water droplets on the C18-UCNCs film surface, and the red areas show the damaged area and corresponding healed area. The insets in (e-g) at top right show dyed water droplets on the corresponding surfaces, and those at bottom left show static water contact angles on the corresponding surfaces

To investigate the mechanical durability of the assembled structures, anabrasion test was conducted on the surface of C18-UCNCs films. As shown in Fig. 6a and Fig. S15, the C18-UCNCs films were abraded using 800 mesh sandpaper with a 50 gloading. After the abrasion, the obvious white powder on the sandpaper indicated significant damage of the flaky structures, revealing limited mechanical stability of the assembled flaky structures. The damage turned the superhydrophobic surface less hydrophobic with a static water contact angle of about 115°(Fig. 6b and Fig. 6d). Thanks tothethermoreversibility of these self-assembled flakes, the C18-UCNCs film surface was endowed with self-healing properties. After heating at 80 ℃ for 30 min, cooling down, and aging at RTfor 24 h, the original superhydrophobicity was restored, and distinct hierarchical flaky structures were regenerated on the damaged surfaces(Fig. 6c-d). As shown in Fig. 6d and Fig. S16, the C18-UCNCs films restored their original topographic features with maintained superhydrophobicity even after 10 abrasion-healing cycles. Thus, the self-assembled flaky structures on C18-UCNCs films showed excellent self-healing properties. The crystallization of the octadecyl chains after heating and cooling could induce the exposed C18-UCNCs to organize and assemble into flaky structures.

The formation of hierarchical flaky structures on flat substrates was further explored using C18-UCNCs. The THF suspension of C18-UCNCs was coated onto the glass slide and filter paper for surface functionalization. The treated glass slide and filter paper were transformed from being hydrophilic to almost superhydrophobic(Fig. 6e and Fig. S17a). The enhanced hydrophobicity was attributed to the formation of the flaky structure consisting of C18-UCNCs on the surface. This hierarchical structure could trap microscopic air pockets beneath water drops, producing superhydrophobic surfaces [26]. Specifically, the morphology and superhydrophobicity of the C18-UCNCs-coated glass slide and filter paper also exhibited thermo-induced self-healing properties. The physically damaged flaky structures on the surfaces were healed by heat treatment, together with the recovery of the surface morphology and superhydrophobicity (Fig. 6f-g and Fig. S17b-c). Thus, the formation of the C18-UCNCs flaky structures on substrates via coating represents a facile and effective method of superhydrophobizing materials with self-healing properties.

In conclusion, phase change nanocrystals C18-UCNCs with core-shell structures as a stabilized PCM was successfully synthesized by immobilizing 1-octadecanethiol in the shell layer of UCNCs. Duringthe evaporation from its THF suspension, C18-UCNCs could self-assemble into well-ordered flaky hierarchical structures on other substrates or on the surface of self-standing films formed by themselves. This ability is attributed to the synergistic phase segregation and crystallization of octadecyl chains in the shell of C18-UCNCs, leading to lamellar crystalline structures with 3. 7 nm periodicity. The phase transition of the C18-UCNCsshell conferred on the flaky structures superior thermo-induced self-healing and thermoreversible properties, including the morphology, birefringence, superhydrophobicity, and transparency of the material. Moreover, C18-UCNCs showed efficient thermal transport and light-to-thermal energy conversion, largely expanding their potential uses in various fields, such as smart surfaces, temperature-preserving coatings, and thermosensitive imaging.

Supporting Information

Supporting Information is available from the Wiley Online Library or from the author.

Acknowledgements

Y. W. thanks National Natural Science Foundation of China (31890774, 31890770) and China Postdoctoral Science Foundation (2018M640286) for the financial support. K. Z. thanks German Research Foundation(DFG) with the project number of ZH546/2-1 for the financial support.

References

[1] Cui Y, Wei Q, Park H, et al. Nanowirenanosensors for highly sensitive and selective detection of biological and chemical species[J]. Science, 2001, 293(5533): 1289-1292. b) Wang Z L, Song J H. Piezoelectric nanogenerators based on zinc oxide nanowire arrays[J]. Science, 2006, 312(5771): 242-246. c) Peer D, Karp J M, Hong S, et al. Nanocarriers as an emerging platform for cancer therapy[J]. Nat. Nanotechnol., 2007, 2(12): 751-760. d) Tian Q W, Tang M H, Sun Y G, et al. Hydrophilic flower-like CuS superstructures as an efficient 980 nm laser-driven photothermal agent for ablation of cancer cells[J]. Adv. Mater., 2011, 23(31): 3542-3547. e) Bao Z H, Weatherspoon M R, Shian S, et al. Chemical reduction of three-dimensional silica micro-assemblies into microporous silicon replicas[J]. Nature, 2007, 446(7132): 172-175. f) Lim B, Jiang M J, Camargo P H C, et al. Pd-Pt bimetallic nanodendrites with high activity for oxygen reduction[J]. Science, 2009, 324(5932): 1302-1305. g) King'ondu C K, Iyer A, Njagi E C, et al. Light-assisted synthesis of metal oxide heirarchical structures and their catalytic applications[J]. J. Am. Chem. Soc., 2011, 133(12): 4186-4189.

[2] Zha R H, Vantomme G, Berrocal J A, et al. Photoswitchable nanomaterials based on hierarchically organized siloxane oligomers[J]. Adv. Funct. Mater. 2018, 28(1): 1703952.

[3] Kenisarin M M. Thermophysical properties of some organic phase change materials for latent heat storage. A review[J]. Sol. Energy, 2014, 107: 553-575.

[4] a) Oró E, De Gracia A, Castell A, et al. Review on phase change materials (PCMs) for cold thermal energy storage applications[J]. Appl. Energy, 2012, 99: 513-533. b) Sharma A, Tyagi V V, Chen C R, et al. Review on thermal energy storage with phase change materials and applications[J]. Renew. Sust. Energ. Rev., 2009, 13(2): 318-345.

[5] Prajapati D G, Kandasubramanian B. Biodegradable polymeric solid framework-based organic phase-change materials for thermal energy storage[J]. Ind. Eng. Chem. Res., 2019, 58(25): 10652-10677.

[6] Han G G D, Li H S, Grossman J C. Optically-controlled long-term storage and release of thermal energy in phase-change materials[J]. Nat. Commun., 2017, 8(1): 1-10.

[7] Elschner T, Lüdecke C, Kalden D, et al. Zwitterionic cellulose carbamate with regioselective substitution pattern: a coating material possessing antimicrobial activity[J]. Macromol. Biosci., 2016, 16(4): 522-534.

[8] Zhang K, Geissler A, Chen X, et al. Polymeric flower-like microparticles from self-assembled cellulose stearoyl esters [J]. ACS Macro Lett., 2015, 4(2): 214-219.

[9] a) Sarı A, Alkan C, Bilgin C. Micro/nano encapsulation of some paraffin eutectic mixtures with poly (methyl methacrylate) shell: Preparation, characterization and latent heat thermal energy storage properties[J]. Appl. Energy, 2014, 136: 217-227. b) Li Y, Yu S, Chen P, et al. Cellulose nanofibers enable paraffin encapsulation and the formation of stable thermal regulation nanocomposites[J]. Nano Energy, 2017, 34: 541-548.

[10] a) Chen L J, Zou R Q, Xia W, et al. Electro-and photodriven phase change composites based on wax-infiltrated carbon nanotube sponges[J]. ACS Nano, 2012, 6(12): 10884-10892. b) Chen X, Gao H Y, Yang M, et al. Highly graphitized 3D network carbon for shape-stabilized composite PCMs with superior thermal energy harvesting[J]. Nano energy, 2018, 49: 86-94.

[11] a) Wang Y, Liu Z, Zhang T, et al. Preparation and characterization of graphene oxide-grafted hexadecanol composite phase-change material for thermal energy storage[J]. Energy Technol., 2017, 5(11): 2005-2014; b) Royon L, Guiffant G, Flaud P. Investigation of heat transfer in a polymeric phase change material for low level heat storage[J]. Energy Convers. Manag., 1997, 38(6): 517-524.

[12] a) Alva G, Lin Y, Fang G. Synthesis and characterization of chain-extended and branched polyurethane copolymers as form stable phase change materials for solar thermal conversion storage[J]. Sol. Energy Materials and Solar Cells, 2018, 186: 14-28; b) Lian Q S, Li K, Sayyed A A S, et al. Study on a reliable epoxy-based phase change material: facile preparation, tunable properties, and phase/microphase separation behavior [J]. J. Mater. Chem., 2017, 5(28): 14562-14574.

[13] Tang J, Chen X Y, Zhang L G, et al. Alkylated meso-macroporous metal-organic framework hollow tubes as nanocontainers of octadecane for energy storage and thermal regulation[J]. Small, 2018, 14(35): 1801970.

[14] a) Zhou M, Lin T Q, Huang F Q, et al. Highly conductive porous graphene/ceramic composites for heat transfer and thermal energy storage[J]. Adv. Funct. Mater., 2013, 23(18): 2263-2269; b) Li S Q, Wang H X, Mao H Q, et al. Light-to-thermal conversion and thermoregulated capability of coaxial fibers with a combined influence from comb-like polymeric phase change material and carbon nanotube[J]. ACS Appl. Mater. Interfaces, 2019, 11(15): 14150-14158.

[15] Yu H Y, Qin Z Y, Liang B L, et al. Facile extraction of thermally stable cellulose nanocrystals with a high yield of 93% through hydrochloric acid hydrolysis under hydrothermal conditions[J]. J. Mater. Chem. A, 2013, 1(12): 3938-3944.

[16] Wang Y G, Groszewicz P B, Rosenfeldt S, et al. Thermoreversible self-assembly of perfluorinated core-coronas cellulose-nanoparticles in dry state[J]. Adv. Mater., 2017, 29(43): 1702473.

[17] Wang Y G, Heinze T, Zhang K. Stimuli-responsive nanoparticles from ionic cellulose derivatives[J]. Nanoscale, 2016, 8(1): 648-657.

[18] Berlioz S, Molina-Boisseau S, Nishiyama Y, et al. Gas-phase surface esterification of cellulose microfibrils and whiskers[J]. Biomacromolecules, 2009, 10(8): 2144-2151.

[19] Fumagalli M, Ouhab D, Boisseau S M, et al. Versatile gas-phase reactions for surface to bulk esterification of cellulose microfibrils aerogels[J]. Biomacromolecules, 2013, 14(9): 3246-3255.

[20] Wang J X, Zhang K. Modular adjustment of swelling behaviors of surface-modified solvent-responsive polymeric nanoparticles based on cellulose 10-undecenoyl ester[J]. J. Phys. Chem. C, 2018, 122(13): 7474-7483.

[21] Lian Q S, Li K, Sayyed A A S, et al. Study on a reliable epoxy-based phase change material: facile preparation, tunable properties, and phase/microphase separation behavior[J]. J. Mater. Chem. A, 2017, 5(28): 14562-14574.

[22] a) Fu X W, Xiao Y, Hu K, et al. Thermosetting solid-solid phase change materials composed of poly (ethylene glycol)-based two components: flexible application for thermal energy storage[J]. Chem. Eng. J., 2016, 291: 138-148; b) Sari A, Alkan C, Biçer A, Karaipekli A. Synthesis and thermal energy storage characteristics of polystyrene-graft-palmitic acid copolymers as solid-solid phase change materials[J]. Sol. Energy Mater Sol. Cells, 2011, 95(12): 3195-3201.

[23] Venkataraman S, Lee A L Z, Tan J P K, et al. Functional cationic derivatives of starch as antimicrobial agents[J]. Polym. Chem., 2019, 10(3): 412-423.

[24] a) Chabinyc M L, Toney M F, Kline R J, et al. X-ray scattering study of thin films of poly(2, 5-bis(3-alkylthiophen-2-yl)thieno[3, 2-b]thiophene)[J]. J. Am. Chem. Soc., 2007, 129(11): 3226-3237. b) Sealey J E, Samaranayake G, Glasser W G. Novel cellulose derivatives. 4. Preparation and thermal analysis of waxy esters of cellulose[J]. J. Polym. Sci. Pt. B-Polym. Phys., 1996, 34(9): 1613-1620. c) Vaca-Garcia C, Gozzelino G, Glasser W G, et al. Dynamic mechanical thermal analysis transitions of partially and fully substituted cellulose fatty esters[J]. J. Polym. Sci. Pt. B-Polym. Phys., 2003, 41(3): 281-288. d) Chen X, Zheng N, Wang Q, et al. Side-chain crystallization in alkyl-substituted cellulose esters and hydroxypropyl cellulose esters[J]. Carbohydr. Polym., 2017, 162: 28-34.

[25] Liu C, Xu Z, Song Y, et al. A novel shape-stabilization strategy for phase change thermal energy storage[J]. J. Mater. Chem. A, 2019, 7(14): 8194-8203.

[26] a) Feng L, Li S H, Zhu D B. Super-hydrophobic surfaces: from natural to artificial[J]. Adv. Mater., 2002, 14(24): 1857-1860. b) Geissler A, Chen L, Zhang K, et al. Superhydrophobic surfaces fabricated from nano- and microstructured cellulose stearoyl esters[J]. Chem. Commun., 2013, 49(43): 4962-4964. c) Deng X, Mammen L, Zhao Y, et al. Transparent, thermally stable and mechanically robust superhydrophobic surfaces made from porous silica capsules[J]. Adv. Mater., 2011, 23(26): 2962-2965. d) Hozumi A, Cheng D F, Yagihashi M. Hydrophobic/superhydrophobic oxidized metal surfaces showing negligible contact angle hysteresis[J]. J. Colloid Interface Sci., 2011, 353(2): 582-587.

中文题目： 纤维素衍生物相变纳米晶体的多功能可逆自组装结构

作者： 王永贵，邱哲，郎真，谢延军，肖泽芳，王海刚，梁大鑫，李坚，张凯

摘要： 由于具有良好组织结构的优点，在固态组装结构中具有可逆分层和高度有序排列的多功能材料引起了人们的极大兴趣。然而，这样的材料很少存在。基于相变材料（PCMs）的可逆相变特性，提出了能够自组装成有序层次结构的相变纳米晶体（C18-UCNCs）。C18-UCNCs 具有核壳结构，由保持基本结构的纤维素晶体核和包含允许相变的十八烷基链的软壳组成。独特的核壳结构和十八烷基链的相变使得 C18-UCNCs 能够自组装成片状的纳米/微结构。这些自组装的 C18-UCNCs 具有高效的热传输和光到热能转换能力，因此有望用于热敏成像。具体来说，片状自组装纳米/微结构具有可操作的表面形貌、表面润湿和光学性能，是热可逆的，并表现出热诱导自愈合性能。利用相变纳米晶体作为一类新型的相变材料，可制备出可逆自组装多功能材料。本研究提出了一种利用相变纳米晶体构建自组装层次结构的方法，从而极大地扩展了相变材料的应用。

关键词： 纤维素；核-壳结构；相变纳米晶体；热成像；热可逆性

Magnetic-driven 3D Cellulose Film for Improved Energy Efficiency in Solar Evaporation*

Wentao Gan, Yaoxing Wang, Shaoliang Xiao, Runan Gao, Ying Shang,
Yanjun Xie, Jiuqing Liu, Jian Li

Abstract: The architecture of cellulose nanomaterials is definitizedby random depositionand cannotchangein response to shifting application requirements. Herein, we present a magnetic field-controlled cellulose filmderived from wood that exhibitsgreat magnetic properties and reliable tunability enabled by incorporated Fe_3O_4nanoparticles andcellulose nanofibers (CNF) with large length-diameter ratio. Fe_3O_4 nanoparticles are dispersed in suspensions of CNF so as to enhance the magnetic response. The plane magnetic CNFcan be processedto form athree-dimensional(3D)flower-like structurealong the magnetic induction line after applying an external magnet. Inspired by the fluidic transport innatural flower, a bilayer structurewascreated by using the 3D flower-like film as the solar energy receiver and the natural wood as the water pathway in solar-derived evaporation system. Comparing with planar cellulose filmdecorated withFe_3O_4, the 3D structure design can greatly improve the evaporation rate from 1.19 $kg \cdot m^{-2} \cdot h^{-1}$ to 1.39 $kg \cdot m^{-2} \cdot h^{-1}$and theefficiency from 76.9% to 90.6% under 1 sun. Finite element molding further reveals that the 3D-structural top layer is beneficial for the formation of gradient temperature profile and the improvement ofthe energy efficiencythrough the reduction of thermal radiation. The magneticallycontrolledfabrication represents a promising strategy for designing cellulosenanomaterials with complicated structure and controllable topography, whichhas a wide spectrum of applications in energy storage devices and water treatment.

Keywords: Bionics; Magnetic film; Cellulose nanofibers; Stimuli-response; Solar desalination

1 Introduction

Solar evaporation is regarded as a brillianttechnologyto mitigate the global water crisis because of the abundant and renewable solar energy resources.[1-3] With the development of novel photothermal materials and energy management system in recent years, the solar steam generation technologies are not limited to steam and clean water production from seawater and underground water. Many other applications based on the photo-to-thermal conversion have been exploited in fields of electricity generation, fuel production and steam sterilization.[3-5] It is imperative requirement to develop solar vapor generators with a high energy conversion efficiency in practical applications.

Usually, solar evaporation system is made up bythe absorber withlight-to-heat conversionefficiency and the substrate with thermal insulation as well as water transport function.[6-8] Photothermal materials range from

* 本文出自 ACS Applied Materials & Interfaces, 2021, 13: 7756-7765.

plasmonic metal nanoparticles, semiconductors, conductive polymer, to carbon-based materials with broad light absorption and well-designed architecturehave been investigated for efficient water evaporation.[9-12] Although theyexhibit excellent light-to-heat conversions, the utilization cost, complex fabrication, and fragile nanostructurestill need improvement. As for the supporting substrate, itcan pump water into the evaporation layer for continuous steam generation and reduce the heat loss from the top light absorber to the bulk water. Synthetic polymers such as polyurethane (PU) and polystyrene (PS) are widely used substrates due to their low cost and good thermal insulation.[13,14] Nevertheless, the synthetic polymers always suffer limitations due to the non-biodegradabilityand the concerns to the environment.[15,16] In terms ofsustainability and environmental friendliness, there is a compelling interest todevelop renewable biomass materials such as wood, cellulose aerogel, and bamboo for solar steam generation.[17-22]

As a kind of important renewable resource on earth, cellulose nanofibrils in wood possess the transportationof moisture. From the material point of view, cellulose nanomaterials are highly promising for solar evaporatorsvia various structural designs, such as bottom-up assembly of 2D cellulose paper, 3D porous celluloseaerogel and wood with intrinsic hierarchical structure.[23-26] Interconnected porous structure and low thermal conductivity of cellulose nanomaterials can improve the light absorption and enhance heat localization when the photothermal materials loaded onto thesesubstrates.[27,28] To improve the evaporation rate and light-to-vapor energy efficiency, current research attentions have shifted to improving configuration design through surface engineering of light absorber from a conventional planar architecture to a 3D structure.[29-32] By using 3D cellulose-based aerogels, the evaporation rate and energy conversion efficiency have been dramatically improveddue to the increased evaporationsurface areaand simultaneousenergy harvestingfrom the solar irradiation, bulk water and convective air flow from surroundings.[33-35] Although great advantages have been made in utilization of various3D cellulose nanomaterials in solar evaporation, the architecture of cellulose-based materials is definitizedby random deposition and their fabrication always requires high-energy consumption and complex polymerization processes. It is still a challenge to realize the structural control of cellulose nanomaterials through a convenient method.

In this work, we proposed a magnetic field assistantstrategy to construct the 3D flower-like cellulose nanomaterials that exhibit non-contacting control and reliable tunability enabled by Fe_3O_4 with great magnetic properties and CNF with large length-diameter ratio. The CNFwasfabricated by the chemical pretreatment combined with high frequency ultrasonic treatment from wood sawdust. To enhance the magnetic responses, Fe_3O_4 nanoparticles were added to the CNFsuspension, therefore cellulosenanomaterialscan be actuated by the external magnet and furtherprocessing into 3D architecture. After applying an external magnetic field of 200 mT, the magnetic suspension formed a 3D flower-like structurealong the magnetic induction line on the nonwovens. Benefitingfromthe large amount of hydroxyl groups and highsurface area of CNF, a stable interwoven structure of magnetic nanoparticles and CNF in the petal was created after the suspension wastotally dried (Fig. 1a). As a narrow bandgap semiconductor, magnetic Fe_3O_4 nanoparticles can generate electron-hole pairs and convert the extra energy of solar light into heat. Inspired by thefluidic transport innatural flower, we propose a unique solar energy receiver made of 3D flower-like cellulose film with consideration in the magnetic response, easy of fabrication, enlarged surface areaand high light absorption. The 3D flower-like cellulose film can efficiently absorb the solar illumination causing the increasement of temperatureon the top

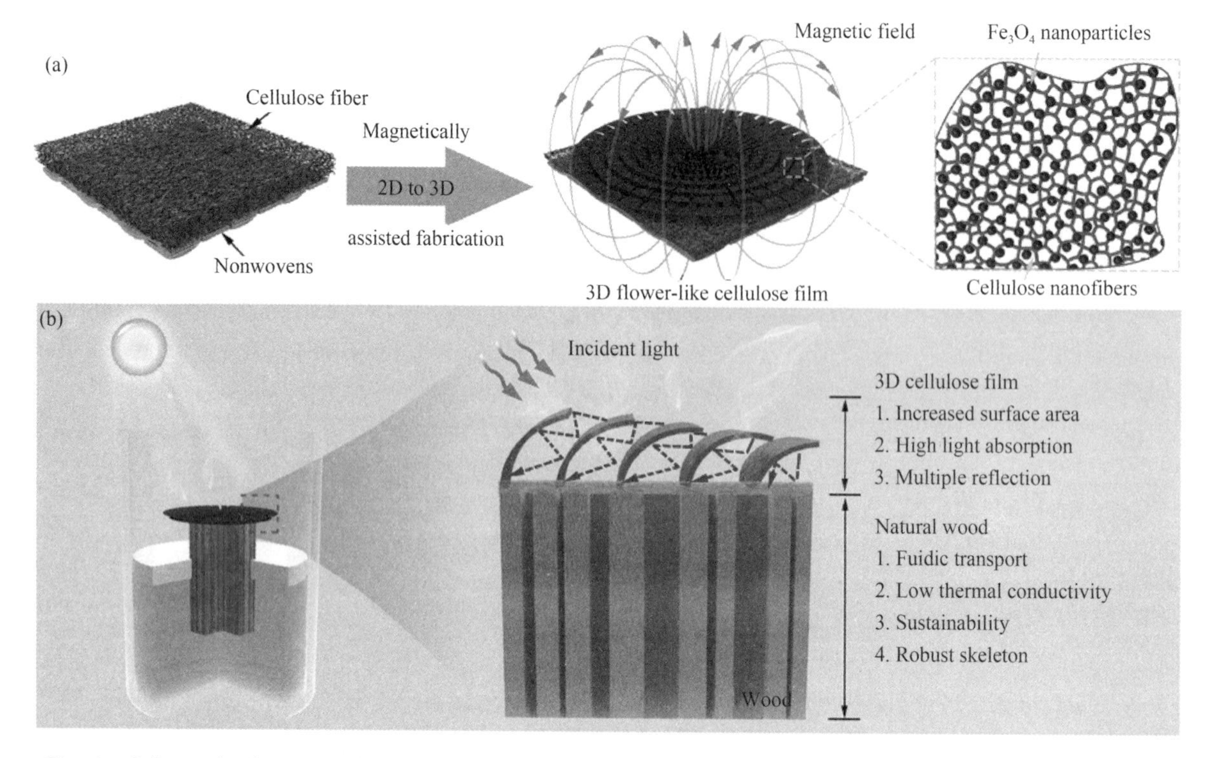

Fig. 1 Schematic demonstrating the working principle of 3D flower-like cellulose film as an efficient solar steam generation material. (a) The flower-like structure design of the magnetic cellulose gel actuated by external magnetic field. (b) Artificial solar steam generator with a bilayer structure. The top flower-like cellulose layer acts as a 3D light receiver and can effectively absorb sunlight. The bottom wood is hydrophilic, which promotes rapid water transport for continuous solar steam operation

surface. Meanwhiles, as the most abundant renewable material on the earth, thesustainable and low-cost wood is used as a robust substrate in our solar steam generation system, which can help pump water upwards during continuous water evaporation (Fig. 1b). Thus, the 3D structure design provides a high water evaporation rate of 1.39 $kg \cdot m^{-2} \cdot h^{-1}$ and an energy efficiency of 90.6% at 1 Sun, which outperform most reported photothermal materials (Fig. S1).[4,13,36-40] Compare to 2D structure, 3D cellulose film provides an ideal template for enhancing water evaporation rates and improving solar steam efficiency, owing to the enlarged water/air interfacefor steam escape, the efficient energyconversionand good capability of reabsorbing lost energy caused by multiple reflection. The demonstrated cellulose film with instantaneous response, controllabletopography, and efficient light absorption unfoldsoutstanding capacity of light-to-heat conversion in wide applications of energy storage devices and environmental protection.

2 Results and discussion

CNF with high aspect ratio and abundant hydroxy groups is a promising building-block to construct composite film materials. As shown in Fig. 2a, the translucent CNF film was obtained after air drying. For the control sample, the cracking and detachment phenomenonare observed in magnetic nanoparticles, demonstrating the fragile nanostructure and low interface bonding forces between the neighboring Fe_3O_4

nanoparticles(Fig. 2b). After the equal mixture of CNF suspensionand magnetic nanoparticles, the magnetic cellulose film with homogeneously distributed magnetic nanoparticles shows opaque characteristic, revealing the structure stability and light-absorbing performance (Fig. 2c). Transmission electron microscopy (TEM) images were taken to confirm the morphology and microstructures of as-prepared Fe_3O_4@ CNF composites. The TEM image show that the spherical Fe_3O_4 nanoparticles with average size of ~ 200 nmis well mixed with CNF (Fig. 2d). At highmagnification in Fig. 2e, the magnetic nanoparticles were anchoredon the surface CNF. The obvious peak attributed to O-H stretching vibrationswas observed in the FTIR spectrum of magnetic film, demonstrating the magnetic nanoparticles and cellulose nanofibers were bound together through hydrogen bonds(Fig. S2). Their anchoringon the cellulose nanofibersreduces the propensity of nanoparticles slipping from the substrate after stirring in waterfor 60 mins(Fig. S3), demonstrating the improved structure stability of the composite material. A high-resolution transmission electron microscopy (HRTEM) image of a Fe_3O_4 nanoparticle(Fig. 2f) shows well-resolved lattice fringes with interplanar distance of 0. 285 nm, demonstrating the(220) plane of Fe_3O_4. The selective area electron diffraction(SAED) pattern exhibits several bright discrete spots, which are well corresponded to the octahedral Fe_3O_4 crystals. [41,42] Fig. 2g depicts XRD patterns of the CNFfilm and magnetic film. The diffraction peaks appeared at 30. 6°, 35. 8°, 43. 5°, 53. 9°, 57. 4°, and 63. 1° attributed to the(220), (311), (400), (422), (511), and(440) lattice planes of spinel phase of Fe_3O_4(JCPDS 19-0629). The peaks centered at 14. 9° and 22. 4° in CNFfilm represent the typical crystalline region of the cellulose Ⅰ. [43,44]

The resulting microstructure of the Fe_3O_4@ CNF composite caused a dramatic improvement in magnetic properties. Fig. 2h displays the room-temperature magnetization curves of magnetic filmand cellulose film in the presence of the applied magnetic field ranging from −20000 to +20000 Oe. CNF film exhibits appearance of straight line close to zero, revealing the nonmagnetic properties as expected. Interestingly, the magnetic film shows ferromagnetic behaviors with the saturation magnetization of 78 emu g^{-1}as evident from its substantial hysteresis loop in magnetization curves. Photograph in Fig. 2h also reveals that the magnetic Fe_3O_4@ CNF film can be easily lifted and actuated by an external magnet. Such sensitive magnetic responseisimportant to realize the non-touching control of the surface topographyofcellulose nanomaterials. Additionally, the mechanical stability of cellulose nanomaterials also should be taken into consideration owning to the weak interface bonding force as shown in Fig. 2b. We conducted tensile tests to evaluate the material' s mechanical performance (Fig. 2h). The tensile strength of cellulose film is 238 MPa, indicating the good mechanical properties. After the addition of Fe_3O_4 nanoparticles, the tensile strength of magnetic film still reaches to 77. 6 MPa, which could be explained by the embedded Fe_3O_4 reduced the interfacial hydrogen bonding between the cellulose nanofibers(Fig. S4). Compare with the fragile magnetic nanoparticles, the cellulose nanofibers not only improve the film-making capacitybecause of the high aspect ratio, but also provide more anchoring sites for magnetic nanoparticles, which are critical to enhancing the structural stability and mechanical properties of magnetic film.

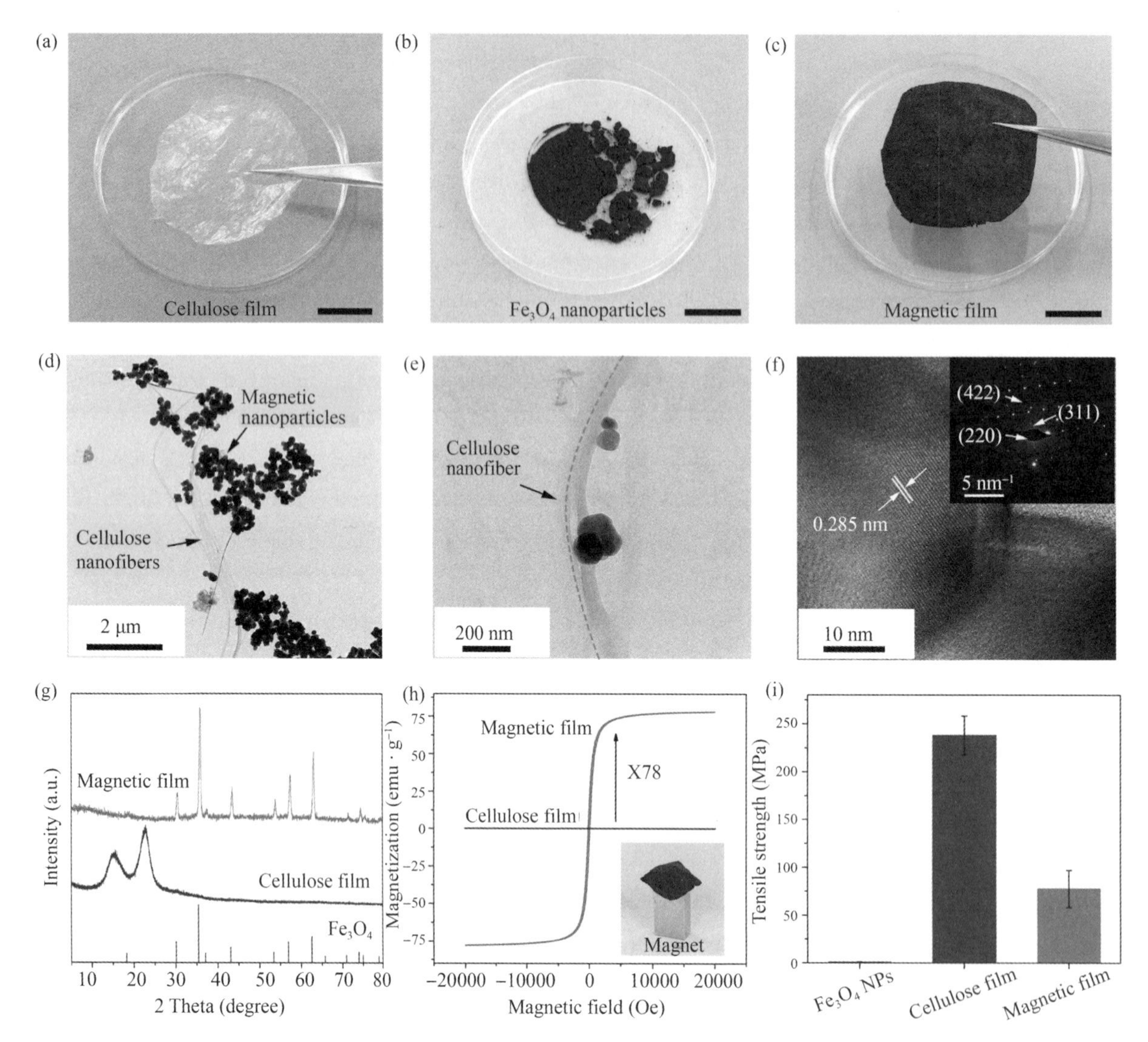

Fig. 2 Photographs of(a)CNFfilm, (b)magnetic nanoparticles, and(c)magnetic film. Scale bar: 2 cm. (d) and(e) TEM observation of the magnetic nanoparticles and CNFs. (f) High resolution TEM image of the magnetic nanoparticles. Inset: the corresponding SAED pattern. (g)XRD patterns of magnetic film, cellulose film and Fe_3O_4 nanoparticles. (h)Magnetization curve of cellulose film and magnetic film at room temperature, inset shows that the magnetic film can be easily actuated by a bar magnet. (i)Corresponding tensile strength of the Fe_3O_4 nanoparticles(NPs), cellulose film and magnetic film

To actuate the Fe_3O_4@ CNFnanomaterialsbased on the strong magnetic response tothe external magnetic field, a neodymium-iron-boron(NdFeB)permanent magnet with the diameter of 40 mm and the magnetic field intensity of 200 mT is placed below a nonwoven to assemble the cellulose nanomaterials into 3D flower-like structure. The magnetic field of permanent magnet wassimulated and the bending behavior of the cellulose nanomaterials wasanalyzed. As shown in Fig. 3a, the magnetic field intensity is increased from the center to the edge of magnet. Under the influence of magnetic field, Fe_3O_4@ cellulose nanomaterials gradually bended along the direction of magnetic induction line and reached to the maximum curving degree, therefore a flower-like structurewas fabricated after followed solidification for 6 h at 60 ℃. Using the simulated angles between magnetic induction line and magnet surface, we canpredict the bending degrees of cellulose nanomaterials and further control the surface topography(Fig. 3b). As shown in Fig. S5, the angle changes between magnetic induction line and magnet surface are normally distributed, the maximum angle reaches to 90°at the center of magnetic field, and the minimum angle is 11° at the edge of permanent magnet. The cross-sectional photographs of cellulose film after magnetic field actuation also shows the 3D flower-like structure(Fig. S6). Our experimental results exhibited the similar tendency of angle changes, demonstrating that we can control the distribution of the cellulosic petals through a tunable magnetic field, which provide an experimental backing to the simulated bending behaviors(Fig. S7).

In the absence of a magnetic field, the magnetic cellulose suspension behaves as a liquid wherein the suspended cellulose nanofibers and Fe_3O_4 nanoparticles are randomly distributed, and the suspension freely flows to form aplatform when it wasdeposited on the nonwoven(Fig. 3c). After a magnetic field was applied, the suspended cellulose nanomaterials aligned into microplatesalong the field lines and formed a 3D flower-like structure(Fig. 3d). Different from the self-arrangement of cellulose nanomaterials, the fluidity of cellulose suspensionwas decreased until reaching a fixed flower-like structuredue to the alignmentof Fe_3O_4 nanoparticles in cellulose suspensionalong magnetic induction line(Fig. 3e). The optical micrographs obviously exhibit that the magnetic Fe_3O_4@ cellulose composites are orientedand tightly stacked together along the magnetic field direction(Fig. 3f and Fig. S8). As we discussed above, the magnetic nanoparticles can be anchored on the surface of CNF due to their highaspect ratioand existedhydroxyl groups. After totally dried, the interconnected CNFnetwork and magnetic nanoparticles aggregation can be directly observed in SEM images, resulting in a flower petal with porous structure(Fig. 3g and Fig. S9). The 3D flower-like magnetic film caused by the assembly of Fe_3O_4 and CNF enables tunable topography, magnetic response, and increased surface area, which has tremendouspotentialin solar steam generation, photochemical catalysisand energy storages.

We made a solar steam generator constructed by 3D flower-like magnetic film as the energy receiver, natural wood as the water supporter and polystyrene (EPS) foam as the thermal insulator(Fig. 4a). The geometry ofour solar steam generator is optimized for minimizing the heat loss which including heat conduction and radiation. The natural wood avoids the direct contact between the cellulose nanomaterials and bulk water, resulting in the minimization of heat conduction loss. Based on numerouspetals of the 3D flower-like magnetic film, multiple reflections can be occurred between neighboring petals, which effectively reabsorb the lost energy fromheat radiation. Besides the enlarged surface areas under projecting light for water evaporation, 3D magnetic film also offers substantially enhanced absorption capacity compared to those of 2D absorbers. As shown in Fig. 4b, the nonwoven substratewithout magnetic cellulose coating hasonly 10% absorption of solar energy. After magnetic nanoparticles treatment, the absorptioncapacity dramatically increased to 93%, which is a result of the increased light absorption of Fe_3O_4 coating. After actuating the magnetic cellulosic suspension to a 3D structure, the optical absorption capacity further increases up to 97%. This enhancement could be attributed to the internal multi-reflection caused by the 3D petals as illustrated in Fig. 1b. There is only a single reflection occurring on the planar cellulose film, while the multiple reflections existed in the 3D cavities of the bended structure. An infrared camera is used to carefully investigate heat behaviorsofcellulose nanomaterials under 1 sun illumination. The typical Fe_3O_4 coated woodisused as the controlsample(Fig. S10), the average

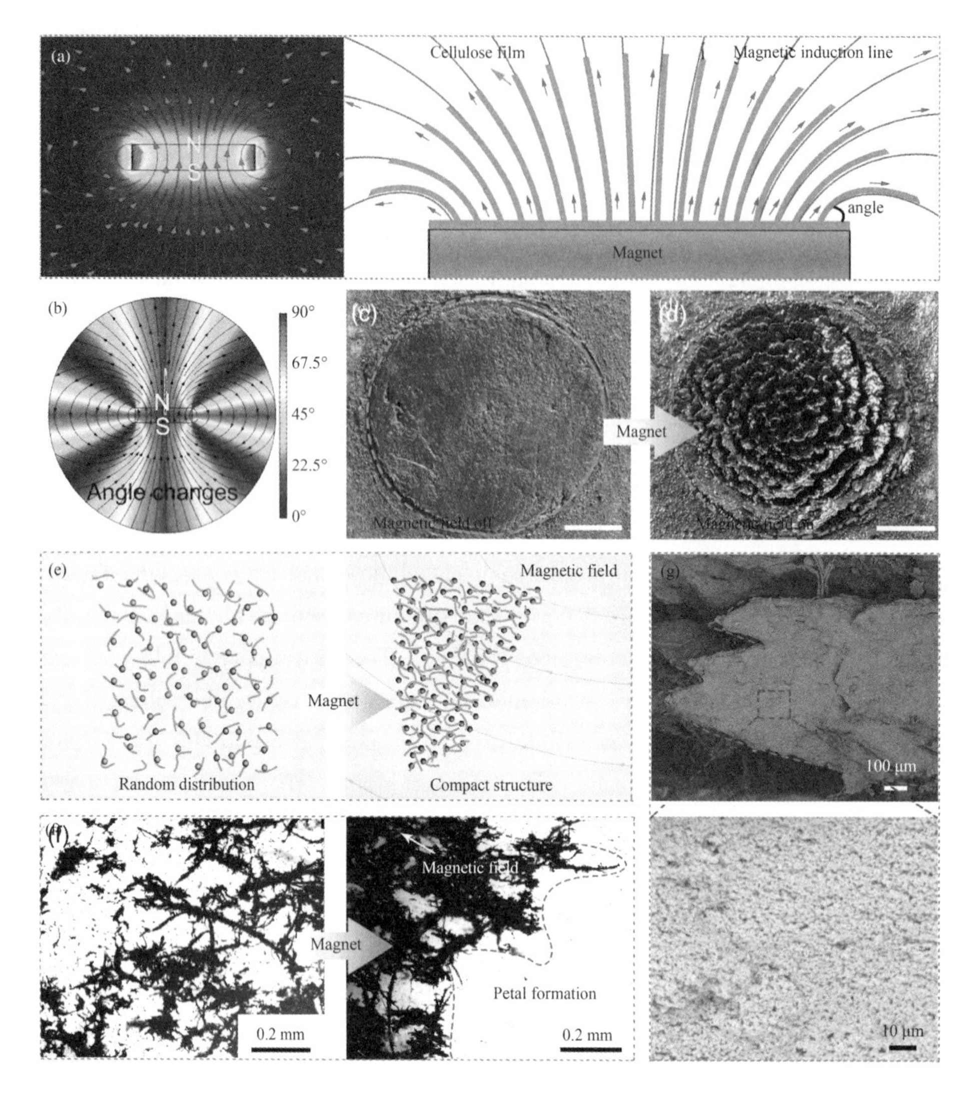

Fig. 3 The formation of flower-like magnetic film. (a) Simulation of magnetic field and cellulose film bends into alignment with the magnetic induction line. (b) Simulation of the angle between magnetic induction line and magnet surface. (c) Photograph of the magnetic cellulose suspension forming a liquid pool in the absence of a magnetic field. (d) Photograph of the magnetic cellulose suspension forming an orderedflower-like film in the presence of a magnetic field. Scale bar: 1 cm. (e) Schematic of the magnetic cellulose suspension with random distribution in the absence of a magnetic field, and the magnetic cellulose suspension with compact structure in the presence of a magnetic field. (f) Optical microscope photographs show the random distribution of cellulose fibers and magnetic nanoparticles in the absence of a magnetic field, and the petal with compact structure formation in the presence of a magnetic field. (g) SEM images of the petal in flower-like cellulose film, demonstrating the nanopores on the surface, which benefits to increase the surface area

temperaturesof the as-prepared samples are plotted in Fig. 4c as a function of reaction time. Typical infrared photographs of the surfaces in 2D and 3D flower-like magnetic film are also displayed in Fig. 4d and 4e. The temperatures of as-prepared samples are around 25 ℃ during the initial stage. Once the light is turned on, rapid increasement in average temperature are observed. According to the increasing trend in temperature, ≈ 360s is needed for all samples to obtain quasi-steady states. The average temperature of the surface in magnetic wood, 2D and 3D cellulose film reached ≈41, 43 and 45 ℃ after 1 h illumination, respectively. The increased absorption of the Fe_3O_4 coated wood and 2D cellulose filmresulted in a high surface temperature. As for the 3D structure, the petals on the surface can enhance the absorption and suppress the radiation losses by multi-reflection of solar light.

To systematically evaluate the performance of as-prepared solar steam system, the evaporation rates and the solar-to-vapor conversion efficiencies are accurately examined by recording the weight changes under 1 sun solar illumination (1 $kW \cdot m^{-2}$). The time-dependent mass changes curves under various conditions are plotted in Fig. 4f. The evaporation rates are calculated from the slope of the curves. As shown in Fig. S11, the utilization of 2D cellulose film promotes the maximum evaporation rate within 1 hour to 1. 19 $kg \cdot m^{-2} \cdot h^{-1}$ from 0. 49 $kg \cdot m^{-2} \cdot h^{-1}$of pure water and 1. 01 $kg \cdot m^{-2} \cdot h^{-1}$ of typical Fe_3O_4 coated wood. After magnetic field stimulation, the enhanced surface areas of flower-like structure under projecting light further improve maximum water evaporation rate to 1. 39 $kg \cdot m^{-2} \cdot h^{-1}$ within 1 hour. Additionally, we carried out six-hours illumination tests to investigate the stability of as-prepared solar steam systems. After continuous evaporation for 6 hours, the evaporation rates of 3D cellulose film were slightly decreased due to the accumulated salts (Fig. S12 and S13). These deposited salts on the surface of 3D cellulose film can affect the adsorption of incident solar energy and clog the water path.[45] However, the deposited salts on the surface of 3D cellulose film can be quickly removed in the distilled water (Fig. S14). Benefiting to the excellent magnetic performance, we can collect all of magnetic cellulose materials together and rearrange them into 3D structure conveniently. No salt was observed on the surface of the rearranged 3D cellulose film, resulting in a stable evaporation rate during the 10 cycles tests (Fig. S13 and S14). The above results demonstrate that the magnetic field assisted method is an alternative reusable strategy to improve the cycling performance of 3D cellulose-based evaporators.

We further calculated solar-thermal conversion efficiencies of Fe_3O_4 coated wood, 2D and 3D cellulo3cfilm(Fig. S15). Consistent with the trend of water evaporation rates, the 3D flower-like cellulose film exhibits the highest energy conversion efficiency of 90. 6%, followed by 2D film and Fe_3O_4 coated wood that have the efficiency of 76. 9% and 65. 6%, respectively. These results indicate that our 3D flower-like cellulose film can achieve fast water evaporation as well as high light-to-heat conversion efficiency, which is superior to the reported biomass-based solar steam generators (Fig. 4g and Fig. S16).[7,17,19-22,28,45-52] In addition, the influence of the heights between 3D cellulose film and bulk water on the energy conversion efficiencies was evaluated. As shown in Fig. S17, a 0. 97 cm and 2. 05 cm water transport distance in natural wood can be found in the first 1 minute and 5 minutes, respectively, demonstrating the natural wood can efficiently pump the water to the height of 2. 05 cm. Limited by the water transport of natural wood, the evaporation rate and efficiency of solar evaporators were greatly restricted when the heights between the 3D cellulose film and the water surface as high as 3 cm(Fig. S18). As demonstrated by Zhu and many other groups,[33-35,53] 3D structured evaporators with long and efficient water transport pathway can enhance the energy harvesting from the ambient environmentand improve the energy conversion efficiencies. Comparing with the cellulose-based generator close to the bulk water, the 3D cellulose solar generator with the heights of 1 cm and 2 cm can achieve fast evaporation rate of 1.39 and 1.41 kg · m^{-2} · h^{-1}, corresponding to the energy conversion efficiencies of 90.6 and 91.9%, respectively (Fig. S18). The desalination effects of the 3D cellulose film were also evaluated using artificial seawater containing NaCl, KCl, $CaCl_2$ and $MgCl_2$. The ion concentrations of Na^+, Mg^{2+}, Ca^{2+}, and K^+ in the desalinated water were significantly decreased from 26730, 5510, 1150 and 720 ppm to 1.299, 0.292, 0.485 and 0.236 ppm, respectively. These ion concentrations are much lower than the ion concentrations of World Health Organization (WHO) and U. S. Environmental Protection Agency (EPA) for drinkable water (Fig. S19).[33] Furthermore, the outstanding evaporation rates and energy conversion efficiency of 3D flower-like film also can be achieved in many other hydrophilic substrates such as cotton and sponge, indicatingthat the magnetic responsive cellulose filmis auniversal materialfor solar steam system to enhance the light-to-heat conversion efficiency (Fig. 4h).

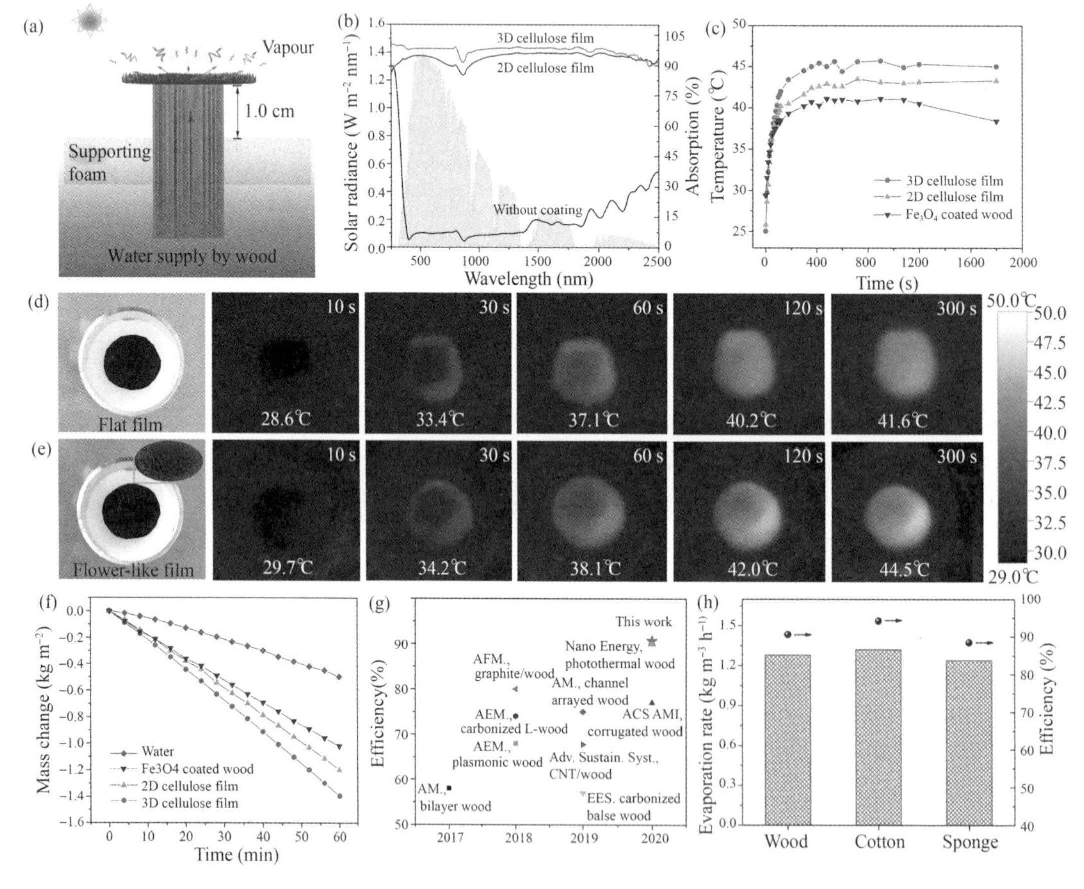

Fig. 4 Solar steam generation of the cellulose nanomaterials under 1 sun illumination. (a) Schematic of the 3D flower-like cellulose-based generator. (b) Solar spectral irradiance (AM 1.5 G) (gray, left side axis) and absorption (red and black, right side axis) of nonwoven, 2D and 3D cellulose films. (c) Average temperatures of surfaces for magnetic wood, 2D magnetic cellulose film and 3D flower-like cellulose film as a function of time. Infrared photos of (d) 2D cellulose film and (e) 3D flower-like film. The photographs, in order from left to right, correspond to t = 0, 30, 60, 120, and 300 s after illumination. (f) Mass changes of water over time under 1 sun illumination. (g) Comparisons of solar steam efficiency of 3D flower-like cellulose-based generator with other wood-based solar steam systems. (h) Evaporation rate and solar steam efficiency of 3D flower-like film using in wood, cotton and sponge substrate

To offer a thermodynamicunderstanding of the superb light-to-heat conversion efficiency and solar steam generating performance of 3D flower-like cellulose film, we carried out finite element modeling under two different conditions(with and without fluid transport) to reveal the heat transferring behavior of the as-prepared 2D and 3D cellulose film under 1 sun illumination (detailed in the Supporting Information). The 3D topography of cellulose film helps improve the efficiency of light-to-heat conversion through reducing the thermal radiation, which in turn plays a key role in the efficient solar steam generator. Fig. 5a compares the temperature distribution of the simulation models in 2D and 3D cellulose film after 1 sun illumination for 5 mins and clearly demonstrates the critical role of surface topography of 3D cellulose film on the heat transferring. In planar magnetic film, the temperaturedistributesuniformly on its surface and reaches to the saturation value of 72.18 ℃ after 5 mins calculation. Obviously, the 3D magnetic film with flower-like structure shows fast heat transferring behaviors in the central area, then reaches to the saturation value of 72.25 ℃ after 5 mins illumination, which is similar to our experimental results (Fig. S20 and Fig. S21). Meanwhile, a gradually changed temperature profile of 3D structural magnetic film can be observedin modeling(Fig. 5b). The curves Ⅰ and Ⅱ represent the top and bottom surface of the temperature changes in 3D cellulose film. Different from the average temperature distribution in 2D cellulose film, the temperatures ofsurface Ⅰ are higher than those of surface Ⅱ in 3D flower-like film. Although the bended magnetic cellulose petals increased the surface area exposed to incident light, it impeded the direct sun illumination to surface Ⅱ at the initial stage. In actual case of water evaporation, the temperatures of the bottom surface Ⅱ are always lower than those of top surface Ⅰ because of the cold water transport. The fast light absorption in surface Ⅰ and temperature gradient formation in 3D cellulose film can reduce the heat loss caused by heat conduction from the surface Ⅱ to the bulk waterthusimprove the energy efficiency. The above results offer a thermodynamic understanding of the 3D structure in heat transferring without fluidic transport, which is shown to be pivotal for the enhanced light-to-heat conversion of the 3D flower-like cellulosefilm.

To this end, we next modeled the heat transferring behavior of the solar steam generator with fluidic transport composed by 3D cellulose film and compare it with that of 2D cellulose film (Fig. S22). Fig. 5c depicts the finite element modeling of 3D cellulose film, natural wood, EPS foam and water. With continuous energy input (1 sun) for 5 mins, the surface temperatures of 3D cellulose film are higher than those of wood and water due to the excellent light absorption. Fig. S13 exhibits the calculated temperature changes as the function of reaction time of 2D and 3D cellulose film. The temperatures of 2D and 3D cellulose film increase dramatically and remain relatively stable to 42 ℃ and 45 ℃, respectively, which agree well with the experimental results. Meanwhile, the steady-state heat loss of heat convection, radiation and conduction are further calculated, as shown in Fig. 5c. For the 2D magnetic film, P_{conv}, P_{rad}, and P_{cond} are simulated to be 4.7, 118.6, and 12 $W \cdot m^{-2}$, respectively (Fig. 5d). Interestingly, the steady-state heat loss of P_{conv}, P_{rad}, and P_{cond} of 3D magnetic film are 5.8, 42.3, and 17.1 $W \cdot m^{-2}$, repsectively. As a result, ~13.6% of the incident energy was lost for 2D structure, whereas only ~ 6.5% heat loss was simulated by the 3D structure. This decrement can be attributed to the 3D flower-like structure with enhanced internal light reflection, which can effectively reabsorb the lost energy of heat radiation. Furthermore, the evaporation rates can be simulated from the concentration variation of water in the modeling. As the illumination time was prolonged, the evaporation rates of 3D magnetic film were obviously higher than that of 2D magnetic film

(Fig. 5e). That is to say, the solar steam generator constructed by 3D flower-like film can obviouslyimprove theenergy efficiency and fluidic evaporation rate.

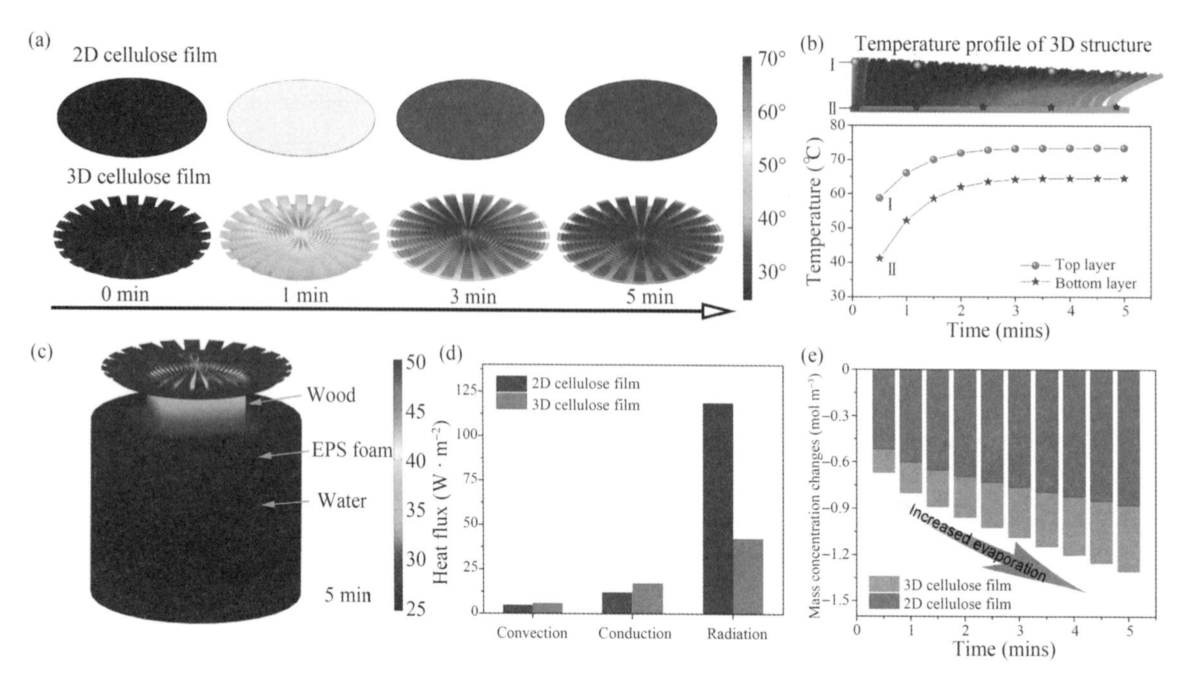

Fig. 5 Thermodynamicunderstanding from modeling. (a) Finite element modeling of the surface temperature distribution of 2D and 3D cellulose film under 1 sun illumination. (b) Temperature profiles of the surface Ⅰ and Ⅱ of 3D cellulose film under 1 sun illumination during 5 mins calculation. (c) The simulation model of solar steam generator made by 3D cellulose film, natural wood, EPS foam and water. (d) Finite element modeling of heat losses of 2D and 3D cellulose film. (e) The comparison of mass concentration changes of 2D and 3D cellulose film as a function of simulated time

3 Conclusions

In this work, we demonstrated a magnetically assistedstrategy towardscontrollablefabrication of the 3D flower-like cellulose material via a film-make process. The cellulose nanofibers derived from natural wood acts as thescaffold and reinforcement agent, which provides large number of anchoring sites for magnetic Fe_3O_4 nanoparticles. Different to the deposition for functional cellulose film, under the influence of an external magnetic field, the magnetic nanoparticleswere ordered along the magnetic induction line, resulting inthe petals with the compactly stacked cellulose nanofibers. Meanwhile, the bending degrees of magnetic cellulose petals are regulated in a remote fashion via an external magnetic field dynamically. In addition to enlarged surface areas of 3D structure, the flower-like cellulose film exhibits enhanced optical absorption of 97% on average from 250 to 2500 nm compared to conventional planar structures. As a result, the evaporation rate and efficiency of the solar steam generator constructed by 3D flower-like magnetic film and natural wood under a 1 sun irradiation are 1.39 $kg \cdot m^{-2} \cdot h^{-1}$ and 90.6%, respectively, which are surpassed than most reported 2D planar photothermal materials. The combined finite element modeling reveals that the 3D flower-like structure can decrease energy loss from heat radiation and enables the fast and effective solar steam evaporation. The magnetically actuated design offers the cellulose film many advances, including non-contacting control, reliable tunability, mechanical stability, instantaneous magnetic response, great light absorptionand outstanding light-to-heat conversion effectiveness. This concept to adjust the topography and distribution of cellulose film provides a general and powerful strategy for designing cellulose nanomaterials with complicated structure in response to shifting application requirements, and it would benefit a wide range of fields such as water treatment, energy storage, microrobot, and intelligent devices.

4 Experimental section

4.1 Materials and Chemicals

Poplar sawdust used for this experiment was derived from poplar wood. Sodium chlorite, acetic acid and Fe_3O_4 nanoparticles were purchased from Shanghai Aladdin Biochemical Technology Co., Ltd. The magnet with the diameter of 40 mm was purchased from Dongguan Jusheng Magnet Co. Ltd.

4.2 Preparation of cellulose nanofibers

Poplar sawdust(5g) was immersed in 0.55 M $NaClO_2$ solution, and the pH was adjusted to 5 by adding acetic acid. The mixture was then heated to 100 ℃ until the sawdust became totally white. After filtration, the white sawdust was dispersed in distilled water(500 mL), and the dispersion was performed at 25 kHz with a common ultrasonic generator(SB-5200DT, Ningbo Scientz Biotechnology Co., Ltd., China) with 1 kW to fabricatethe cellulose nanofibers. Finally, cellulose nanofibers were obtained by collecting the supernatant after centrifugation at 8000 rpm for 5 minutes.

4.3 Fabrication of 3D flower-like magnetic filmfor solar steam generation

To fabricate the Fe_3O_4/cellulose nanomaterials, the Fe_3O_4 nanoparticles was added to cellulosic suspension with a ratio of 1 : 1 (w/w) and thoroughly mixed. After external magnetic field operation, the flower-like magnetic film was dried for 6 h at 60 ℃ in a vacuum drying oven. A balsa wood was cut along the longitudinal direction to form wood blocks(2 cm×2 cm×3.5 cm). The 3D flower-like magnetic film and natural wood was stacked together and placed in a beaker(100 mL) with 50 mL of saline water(NaCl, 3.5 wt%) inside, the polystyrene(EPS) foam was used as the insulation layer to avoiddirectly contact between the magnetic film and bulk water. The height between the magnetic film and the bulk water is keep at 1.0 cm. Then the device was irradiated by a solar simulator(CEL-PE300L, Beijing zhongjiaojinyuan Technology Co. Ltd) at room temperature ~ 25 ℃ and an environmental humidity of ~ 60%. The mass changes were monitored by an electronic analytical scale(FA2004, Shanghai Sunny Hengping Scientific Instrument Co., Ltd.) and real time recorded by a video camera. The measured evaporation rates of as-prepared samples under the dark condition were subtracted. The solar energy conversion efficiency(η) of the evaporation system can be calculated by the following equation:

$$\eta = \frac{m(h_{LV} + C_m \Delta T)}{C_{opt} q_i} \tag{1}$$

where m is the water evaporation rate($kg \cdot m^{-2} \cdot h^{-1}$), h_{LV} is the liquid-vapor phase change enthalpy, normally taken to be 2256.0 $kJ \cdot kg^{-1}$, C_m is the specific heat capacity of water(4.2 $kJ \cdot kg^{-1}℃^{-1}$), ΔT is the temperature change from the initial temperature to the final temperature, q_i is the energy input of the incident light, and C_{opt} is the optical concentration. [45-50]

5 Characterization

The microscopic morphology of materials was characterized by optical microscope (Mingmei, MJ31), scanning electron microscope (SEM, HITACHI TM3030), and transmission electron microscope (TEM, FEI Talos F200X). X-ray diffraction analysis (XRD, SHIMADZU 6100) was conducted at scan rate of 4° · min^{-1} from 5° to 80°. The absorption spectrum was measured by a UV-Vis Spectrometer SARY 300 from 250-2500nm. Magnetic properties of materials with the mass of 40.7 mg were determined using a superconducting quantum interference device (MPMS XL-7) at room temperature. The thermal conductivity of magnetic film was measured at room temperature using the transient plane source method (Hot Disk TPS2500S, Sweden). The special heat capacity of magnetic film in the temperature ranges from 25 ℃ to 60 ℃ was collected by differential scanning calorimetry (DSC, TA Q20). The mechanical tensile tests of cellulose nanomaterials were performed on a universal testing machine (Suns, UTM2503) with loading speed of 2 mm min^{-1}. The change of temperature wasrecorded via an infrared thermal imager (Testo 869) in real time. The ion concentrations of the condensed water were measured by inductively coupled plasma-mass spectrometry (ICP-MS, ICS-5000).

References

[1] Ghasemi H, Ni G, Marconnet A M, et al. Solar steam generation by heat localization[J]. Nat. Commun., 2014, 5: 4449.

[2] Xu N, Hu X Z, Xu W C, et al. Mushrooms as efficient solar steam-generation devices[J]. Adv. Mater., 2017, 29 (28): 1606762.

[3] Chen C J, Kuang Y D, Hu L B. Challenges and opportunities for solar evaporation[J]. Joule, 2019, 3(3): 683-718.

[4] Yang P H, Liu K, Chen Q, et al. Solar-driven simultaneous steam production and electricity generation from salinity [J]. Energy Environ. Sci., 2017, 10(9): 1923-1927.

[5] Neumann O, Feronti C, Neumann A D, et al. Compact solar autoclave based on steam generation using broadband light-harvesting nanoparticles[J]. Proc. Natl. Acad. Sci. U. S. A., 2013, 110(29): 11677-11681.

[6] Yang Y, Zhao R Q, Zhang T F, et al. Graphene-based standalone solar energy converter for water desalination and purification[J]. ACS Nano, 2018, 12(1): 829-835.

[7] Zhu M W, Li Y J, Chen F J, et al. Plasmonic wood for high-efficiency solar steam generation[J]. Adv. Energy Mater., 2018, 8(4): 1701028.

[8] Hong S H, Shi Y, Li R Y, et al. Nature-inspired, 3D origami solar steam generator toward near full utilization of solar energy[J]. ACS Appl. Mater. Interfaces, 2018, 10(34): 28517-28524.

[9] Zhou L, Tan Y L, Ji D X, et al. Self-assembly of highly efficient, broadband plasmonic absorbers for solar steam generation[J]. Sci. Adv., 2016, 2(4): e1501227.

[10] Mu P, Bai W, Fan Y K, et al. Conductive hollow kapok fiber-PPy monolithic aerogels with excellent mechanical robustness for efficient solar steam generation[J]. J. Mater. Chem. A, 2019, 7(16): 9673-9679.

[11] Hu X Z, Xu W C, Zhou L, et al. Tailoring graphene oxide-based aerogels for efficient solar steam generation under one sun[J]. Adv. Mater., 2017, 29(5): 1604031.

[12] Ding D D, Huang W C, Song C Q, et al. Non-stoichiometric MoO_3-x quantum dots as a light-harvesting material for

interfacial water evaporation[J]. Chem. Commun., 2017, 53(50): 6744-6747.

[13] Wang G, Fu Y, Guo A K, et al. Reduced graphene oxide-polyurethane nanocomposite foam as a reusable photoreceiver for efficient solar steam generation[J]. Chem. Mater., 2017, 299(13): 5629-5635.

[14] Elsheikh A H, Sharshir S W, Ali M K A, et al. Thin Film technology for solar steam generation: A new dawn[J]. Sol. Energy, 2019, 177: 561-575.

[15] Haward M. Plastic Pollution of the world's seas and oceans as a contemporary challenge in ocean governance[J]. Nat. Commun., 2018, 9: 1-3.

[16] Singh N, Hui D, Singh R, et al. Recycling of plastic solid waste: A state of art review and future applications[J]. Composites, Part B, 2017, 115: 409-422.

[17] Zhang Q, Li L, Jiang B, et al. Flexible and mildew-resistant wood-derived aerogel for stable and efficient solar desalination[J]. ACS Appl. Mater. Interfaces, 2020, 12(25): 28179-28187.

[18] Zhang P P, Liao Q H, Yao H Z, et al. Direct solar steam generation system for clean water production[J]. Energy Storage Mater., 2019, 18: 429-446.

[19] He F, Han M C, Zhang J, et al. A Simple, Mild and Versatile Method for Preparation of photothermal woods toward highly efficient solar steam generation[J]. Nano Energy, 2020, 71: 104650.

[20] Jang H, Choi J, Lee H, et al. Corrugated wood fabricated using laser-induced graphitization for salt-resistant solar steam generation[J]. ACS Appl. Mater. Interfaces, 2020, 12(27): 30320-30327.

[21] Wang Y L, Liu H, Chen C J, et al. All natural, high efficient groundwater extraction via solar steam/vapor generation [J]. Adv. Sustainable Syst., 2019, 3(1): 1800055.

[22] Li Z T, Wang C B, Lei T, et al. Arched bamboo charcoal as interfacial solar steam generation integrative device with enhanced water purification capacity[J]. Adv. Sustainable Syst., 2019, 3(4): 1800144.

[23] Fu Q L, Chen Y, Sorieul M. Wood-based flexible electronics[J]. ACS Nano, 2020, 14(3): 3528-3538.

[24] Garemark J, Yang X, Sheng X, et al. Top-Down Approach making anisotropic cellulose aerogels as universal substrates for multi-functionalization[J]. ACS Nano, 2020, 14(6): 7111-7120

[25] Gan W T, Chen C J, Kim H T, et al. Single-digit-micrometer thickness wood speaker[J]. Nat. Commun., 2019, 10: 5084.

[26] Fu Q L, Ansari F, Zhou Q, et al. Wood nanotechnology for strong, mesoporous, and hydrophobic biocomposites for selective separation of oil/water mixtures[J]. ACS Nano, 2018, 12(3): 2222-2230.

[27] Cao S S, Rathi P, Wu X H, et al. Cellulose nanomaterials in interfacial evaporators for desalination: A "natural" choice[J]. Adv. Mater., 2020, 33(28): 2000922.

[28] Jiang F, Liu H, Li Y, et al. Lightweight, mesoporous, and highly absorptive all-nanofiber aerogel for efficient solar steam generation[J]. ACS Appl. Mater. Interfaces, 2018, 10(1): 1104-1112.

[29] Li W B, Li Z, Bertelsmann K, et al. Portable low-pressure solar steaming-collection unisystem with polypyrrole origamis[J]. Adv. Mater., 2019, 31(29): 1900720.

[30] Shi Y, Li R Y, Jin Y, et al. A 3D photothermal structure toward improved energy efficiency in solar steam generation [J]. Joule, 2018, 2(6): 1171-1186.

[31] Li X Q, Lin R X, Ni G, et al. Three-dimensional artificial transpiration for efficient solar waste-water treatment[J]. Natl. Sci. Rev., 2018, 5(1): 70-77.

[32] Sun P, Zhang W, Zada I, et al. 3D-structured carbonized sunflower heads for improved energy efficiency in solar steam generation[J]. ACS Appl. Mater. Interfaces, 2019, 12(2): 2171-2179.

[33] Storer D P, Phelps J L, Wu X, et al. Graphene and rice-straw-fiber-based 3d photothermal aerogels for highly efficient solar evaporation[J]. ACS Appl. Mater. Interfaces, 2020, 12(13): 15279-15287.

[34] Wang Y, Wu X, Yang X F, et al. Reversing heat conduction loss: Extracting energy from bulk water to enhance solar

steam generation[J]. Nano Energy, 2020, 78: 105269.

[35]Wang Y D, Wu X, Gao T, et al. Same materials, bigger output: a reversibly transformable 2D-3D photothermal evaporator for highly efficient solar steam generation[J]. Nano Energy, 2020, 79: 105477.

[36]Chen R, Zhu K H, Gan Q M, et al. Interfacial solar heating by self-assembled Fe_3O_4@C film for steam generation [J]. Mater. Chem. Front., 2017, 1(12): 2620-2626.

[37]Xu W C, Hu X Z, Zhuang S D, et al. Flexible and salt resistant janus absorbers by electrospinning for stable and efficient solar desalination[J]. Adv. Energy Mater., 2018, 8(14): 1702884.

[38]Li X Q, Xu W C, Tang M Y, et al. Graphene oxide-based efficient and scalable solar desalination under one sun with a confined 2D water path[J]. Proc. Natl. Acad. Sci. U. S. A., 2016, 113(49): 13953-13958.

[39]Li R Y, Zhang L B, Shi L, et al. MXene Ti_3C_2: an Effective 2D light-to-heat conversion material[J]. ACS Nano, 2017, 11(4): 3752-3759.

[40]Liu Z J, Song H M, Ji D X, et al. Extremely cost-effective and efficient solar vapor generation under nonconcentrated illumination using thermally isolated black paper[J]. Glob Chall, 2017, 1(2): 1600003.

[41]Gan W T, Gao L K, Xiao S L, et al. Magnetic wood as an effective induction heating material: Magnetocaloric effect and thermal insulation[J]. Adv. Mater. Interfaces, 2017, 4(22): 1700777.

[42]Wu N N, Liu C, Xu D M, et al. Enhanced electromagnetic wave absorption of three-dimensional porous Fe_3O_4/C composite flowers[J]. ACS Sustainable Chem. Eng., 2018, 6(9): 12471-12480.

[43]Gao R N, Xiao S L, Gan W T, et al. Mussel Adhesive-inspired design of superhydrophobic nanofibrillated cellulose aerogels for oil/water separation[J]. ACS Sustainable Chem. Eng., 2018, 6(7): 9047-9055.

[44]Guo R Q, Zhang L X, Lu Y, et al. Research progress of nanocellulose for electrochemical energy storage: A Review [J]. J. Energy Chem., 2020, 51: 342-361.

[45]He S M, Chen C J, Kuang Y D, et al. Nature-inspired salt resistant bimodal porous solar evaporator for efficient and stable water desalination[J]. Energy Environ. Sci., 2019, 12(5): 1558-1567.

[46]Zhu M W, Li Y J, Chen G, et al. Tree-inspired design for high-efficiency water extraction[J]. Adv. Mater., 2017, 29(44): 1704107.

[47]Liu H, Chen C J, Chen G, et al. High-performance solar steam device with layered channels: Artificial tree with a reversed design[J]. Adv. Energy Mater., 2018, 8(8): 1701616.

[48]L T, Liu H, Zhao X P, et al. Scalable and highly efficient mesoporous wood-based solar steam generation device: Localized heat, rapid water transport[J]. Adv. Funct. Mater., 2018, 28(16): 1707134.

[49]Kuang Y D, Chen C J, He S M, et al. A high-performance self-regenerating solar evaporator for continuous water desalination[J]. Adv. Mater., 2019, 31(23): 1900498.

[50]Chen S, Sun Z Y, Xiang W L, et al. Plasmonic wooden flower for highly efficient solar vapor generation[J]. Nano Energy, 2020, 76: 104998.

[51]Zhang H T, Li L, Jiang B, et al. Highly thermally insulated and superhydrophilic corn straw for efficient solar vapor generation[J]. ACS Appl. Mater. Interfaces, 2020, 12(14): 16503-16511.

[52]Guan Q F, Han Z M, Ling Z C, et al. Sustainable wood-based hierarchical solar steam generator: A biomimetic design with reduced vaporization enthalpy of water[J]. Nano Lett., 2020, 20(8): 5699-5704.

[53]Li J L, Wang X Y, Lin Z H, et al. Over 10 kg m^{-2} h^{-1} evaporation rate enabled by a 3D interconnected porous carbon foam[J]. Joule, 2020, 4(4): 928-937.

中文题目：磁驱动三维纤维素薄膜提高太阳能蒸发的能源效率

作者：甘文涛，王耀星，肖少良，高汝楠，尚莹，谢延军，刘九庆，李坚

摘要：纤维素纳米材料的结构是随机沉积的，不会因应用需求的变化而改变。在此，我们提出了

一种磁场控制的木材纤维素薄膜，该薄膜具有良好的磁性和可靠的可调性，通过加入 Fe_3O_4 纳米颗粒和长径比大的纤维素纳米纤维(CNF)。将 Fe_3O_4 纳米粒子分散在 CNF 的悬浮液中以增强磁性。平面磁性 CNF 在外加磁体后可沿磁感应线加工成三维花状结构。受天然花卉中的流体传输的启发，利用 3D 花卉状薄膜作为太阳能接收器，天然木材作为太阳能蒸发系统中的水通道，创造了双层结构。与 Fe_3O_4 修饰的平面纤维素膜相比，该三维结构设计可使蒸发率从 1.19 $kg \cdot m^{-2} \cdot h^{-1}$ 提高到 1.39 $kg \cdot m^{-2} \cdot h^{-1}$，蒸发率从 76.9%提高到 90.6%。有限元模型进一步揭示了三维结构顶层有利于形成梯度温度分布，并通过降低热辐射来提高能源效率。磁控制备技术是设计具有复杂结构和可控形貌的纤维素纳米材料的一种很有前途的方法，在储能器件和水处理方面有着广泛的应用。

关键词：仿生学；磁膜；纤维素纳米纤维；刺激效应；太阳能海水淡化

打造林业人才的四个注重*

习近平总书记在两院院士大会上指出，人才是创新的第一资源，一切创新成果都是人做出来的，我国要建设世界科技强国，关键是要建设一支规模宏大、结构合理、素质优良的创新人才队伍。

作为中华民族永续发展千年大计的生态文明建设的主力军，林业人才在建设世界科技强国的大潮中不可或缺，他们使命光荣、责任重大。我认为，培植好林业人才成长的沃土，打造一批有理想、有本领、有担当的林业人才，就要注重启发青年的好奇心和想象力、注重培养青年的行动力、注重培育青年的坚定信念、注重引导青年服务于经济社会。

一、注重启发青年的好奇心、想象力

爱因斯坦曾说过："想象力比知识更重要，因为知识是有限的，而想象力概括了世界上的一切，推动着社会进步，是知识进步的源泉。"

在科学研究中，提出问题，往往比解决问题更为重要。一切创新产品都是想象力的具化。国家与国家之间的差距，便是想象力之间的差距，这一现象也被经济学家称为"想象力差额"。所以，一个国家竞争力的培养，追根到底就是对"想象力"的培养。作为培育青年人才、开展科学研究的高等学校，不应该只成为向青年传播纯粹知识的场所，更应该注重激发青年的好奇心和想象力。

近年来，随着教育的功利化追求，教育成为升学、就业的标准路径。考试标准答案的唯一性，使学生的思维训练陷入了求同、求一的死胡同。个性张扬只能换来排名的落后、奇思妙想与考试的评价格格不入，这些都泯灭了本应属于青年人的激情和灵气，在一定程度上扼杀了青年的想象力和好奇心。虽然近几年推行的高校自主招生力求突破这样的限制，但由于体制机制尚不健全，自主招生难免受到权力、金钱、关系等各方面人为因素的干扰，成为滋生职务犯罪的温床，也在一定程度上弱化了青年的奋斗激情。

2009 年教育进展国际评估组织对全球 21 个国家进行的调查显示，中国孩子的计算能力排名世界第一，想象力却排名倒数第一。在中国学生中，认为自己有好奇心和想象力的，只有 4.7%，而希望培养想象力和创造力的只占 14.9%。

中国并不缺乏自主研发的实力，缺乏的是具有创新思维的人才。国家可以投入巨资、搭建平台，推动科学研究，但是人的创造力是无法通过激励机制来促进的，它与人的思维方式息息相关。回望我们的每一个教育环节，基本都是在储备知识和技术，而对想象力的储备却是最大的稀缺。

国家教育部门应该下大力气改变社会的价值导向、教育理念、教育内容、评价机制，恢复青少年已经丧失的想象的热情和天生的好奇心，用开放的思想，给孩子们留出想象的空间和时间，鼓励他们开拓思路，让他们在充满青春活力的时期，敢于冒险、自由思考。好奇心、想象力需要悠闲自在、无拘无束、无忧无虑的氛围，作为教育工作者，我们就要尽量营造这样的氛围，让众多青年可以在宽松

---

* 本文摘自《百名院士谈建设科技强国》，北京：人民出版社，第 623-628 页，2019.

的环境中碰撞思想、相互启发，全面提升想象力。

二、注重培养青年的行动力

如果说想象力是梦想的起点，那么将想象力具化成真正的创新成果，需要的则是行动力。

习近平在北京大学师生座谈会上指出，“学到的东西，不能停留在书本上，不能只装在脑袋里，而应该落实到行动上，做到知行合一、以知促行、以行求知”。

当代青年肩负着建设“两个一百年”的历史使命，在建设世界科技强国的进程中，广大青年或许在短时间内还不能承担起宏伟、重大的科研攻坚任务，但天下大事必做于细、天下难事必做于易，社会进步固然体现在那些宏大事件上，却更体现在具体而微的进步上。只有广大青年以更强的行动力去参与社会建设，才能够聚沙成塔，以个体的进步推动国家的进步。

青年有行动力，国家才有活力。现在90后已经登上历史舞台、成为干事创业的中坚力量，他们从小生活在条件相对优越的环境，他们没有太多的历史包袱，可以轻装上阵去筑梦、逐梦。虽然他们眼界宽阔、思维活跃，但在一些人中的确存在好高骛远、作风漂浮等情况。

作为教育工作者，培养青年的行动力，就要引导他们树立起强烈的责任意识和进取精神，克服夸夸其谈的毛病，鼓励他们从小事做起，从点滴做起，积小胜为大胜。同时我们也要为青年创新实践搭建更为广阔的舞台，为青年塑造人生提供更丰富的机会，为青年建功立业创造更有利的条件，建立激发青年创新创造的机制，让青年人展现出更多的新气象，成为建设科技强国的生力军。

三、注重培育青年的坚定信念

习近平在多个场合鼓励青年要励志，要立鸿鹄之志。林业作为艰苦行业，更需要众多不怕困难、信念执着、顽强拼搏、永不气馁的奋斗者。

林业是艰苦和冷门的代名词，很少受到高考学子的青睐。但林业却是国民经济、社会发展时刻不可或缺的产业。党的十八大把生态文明建设纳入“五位一体”总体布局，将生态文明建设放在突出位置。林业是建设生态文明的关键领域和主要阵地，历史和现实反复证明，森林兴则生态兴，生态兴则文明兴。建设生态文明，对林业人才的素质和能力提出了更高的要求，但是现在林业人才普遍存在储备严重不足、教育教学方法和手段严重落后、教育资源严重不均衡、偏远林区知识更新严重滞后等制约着林业人才供给和林业事业健康发展的严重问题。

现在高等教育正由“稀缺资源”向“选择资源”的竞争时代迈进，作为林业高等学校，由于办学资源有限，面临着来自综合性大学的众多挑战。由于林业基层岗位对毕业生吸引力不足，毕业生到林业基层岗位难，在基层岗位稳定发展更难。

功以才成，业由才广。为了让涉林毕业生可以在林区留得下、用得住，一方面国家要努力营造良好的林业基层环境，出台有利于林业人才成长的体制机制，另一方面还要对培养林业人才的高等学校在“双一流”建设等方面给予一定的政策倾斜，让林业高等学校可以强化办学特色，聚焦重点和优势。

此外，教育工作者还要注重培育青年树立愿意为祖国林业事业不断奋斗的坚定信念。习近平指出：“人世间没有一帆风顺的事业。综观世界历史，任何一个国家、一个民族的发展，都会跌宕起伏甚至充满曲折。”只有理想信念坚定的立志者，才会以不畏艰险、攻坚克难的勇气，以昂扬向上、奋发有为的锐气，不断把中华民族伟大复兴事业推向前进。

四、注重引导青年服务于经济社会

科学研究的价值，体现在对知识、真理的追求，也要靠服务经济社会发展、增进人民群众福祉的实效来检验。

党的十九大提出了新时代坚持和发展中国特色主义的战略任务，描绘了把我国建成社会主义现代化强国的宏伟蓝图。习近平总书记指出，实现中华民族伟大复兴的中国梦，必须具有强大的科技实力和创新能力。

虽然林业行业的科技成果不像信息、生命、制造、空间、海洋等领域的前沿技术可以直接拓展人类生存发展的新疆域，但是林业的科技成果却可以为国家节约森林资源，从高效利用木材的角度来说，提高林业科学技术，就等于不植树地造林。目前，我国低成本资源和要素投入形成的驱动力明显减弱，需要依靠更多更好的科技创新为经济发展注入新动力。为此，我们应注重引导林业青年人才，在进行科学研究时，要紧密联系社会需求，让科研成果可以服务于经济社会。

习近平曾经例举了清政府组织传教士们绘制《皇舆全览图》的例子，他指出，虽然这项科学水平空前的《皇舆全览图》走在了世界前列，但是因为它长期被作为密件收藏内府，所以并没有对经济社会发展起到作用。习近平指出，科学技术必须同社会发展相结合，要打通从科技强到产业强、经济强、国家强的通道，把创新驱动的新引擎全速发动起来。

科学技术是世界性的、时代性的，发展科学技术必须具有全球视野。当前我们既面临千载难逢的历史性机遇，又面临必须克服差距拉大风险的严峻挑战。习近平指出，形势逼人、挑战逼人、使命逼人，我们必须把握大势、抢占先机，直面问题、迎难而上，瞄准世界科技前沿，不断聚焦经济社会发展主战场，努力把论文写在祖国的大地上，将科技成果应用在实现现代化的伟大事业中。

关于深化对外合作交流的政策建议*

李坚　甘文涛

摘要：加强对外学术合作与交流，跟踪全球科技前沿，鼓励自主创新，实现我国科技从“跟跑者”到“领跑者”的角色转变是科技强国战略的重要组成部分。针对目前某些科技发达国家的限制，对华学术交流和对外合作交流中存在的合作对象较窄、合作项目较少的问题，课题组建议：打破时空局限，完善数字化教学机制，鼓励开展云端学术论坛；拓宽对外交流合作对象范围，让科技合作服务国家重大发展战略；立足自主创新，树立中国期刊品牌，提升中国学术圈影响力。

国家之间的学术交流是促进科学发展的重要推动力。通过学术交流与合作，高校可以培养具有国际化视野的学生，科研院所可以攻克国际难题，高新企业可以推动科技产品更新换代。然而，目前重大科学问题与工程技术难题往往涉及的问题复杂，单一学科难以解决，因而更加需要来自不同学科领域的专家组成科研团队，共同完成攻关任务。基于全球信息化高速发展的形势，加强国际合作，深化对外学术交流，是提高我国自主创新能力，实现建设世界科技强国奋斗目标的关键。

一、对外合作交流中存在的问题

为扩大对外学术合作，提升科技创新能力，国家制定了一系列发展政策，从学术交流、技术合作，到联合办学、共同建设实验室或者研究中心等，吸引了国外大量优秀的专家、教授来华工作或进行技术指导，取得了丰硕的成果。近年来，某些国家政府却以防止外部势力攫取科研成果为名，阻碍中外科学家之间的正常学术交流，缩减中国访问学者数量，希望从人才源头上遏制我国高新科技产业发展。为此，不少知名学者都遭遇了不公正的对待与审查，极大地增加了国外科技人才在工作和生活方面的压力，迫使他们不得不放弃与国内的合作关系。短期内这样的限制措施对我国人才强国战略有一定影响，但从长期发展来说，这必将刺激我国对外合作政策的改变，加强我国对全球顶尖人才的吸引力度和本土人才自主创新能力的培养力度。

反思这一现象的根源，除了中国崛起的速度令世界侧目之外，也体现了我国对外交流合作对象较窄和合作项目较少的问题。以国家公派留学生项目为例，美国、英国、澳大利亚都是主流的留学目的国，不可否认，它们拥有先进的学术思想和科学技术，然而，世界上以东欧、中欧为代表的其他国家尽管在整体科技水平上落后于美英等发达国家，但在某些领域同样具有知名的高校和卓越的工业技术，如波兰的弗罗茨瓦夫大学、捷克的机械工业，与这些国家合作不仅成本较低，还可以相对容易获得关键技术。另一方面，我国引进人才的政策和评价体系仍然不够灵活，合作办学、项目申请在一定

---

* 本文摘自教育部科学技术委员会《专家建议》第19期(总第323期)，2020年4月21日.

程度上仍然存在行政主导的因素，因而难吸引国际知名专家来国内工作和全球科技巨头与国内市场合作研发等问题。基于以上背景和对外合作中存在的若干问题，完善对外合作交流政策，提升高校和科研院所的自主创新能力，对实现科技强国目标具有重要意义。

二、深化对外合作交流的政策建议

深化对外合作交流关键在于实现“走出去”和“引进来”，既让更多科研工作者走出过国门，又要将国外优秀的人才和高新科技引进来，为此，我们提出如下建议。

第一，打破时空局限，完善数字化教学机制，鼓励开展云端学术论坛

1)完善数字化教学机制

高校要培育国际化人才，提高自主创新能力，最直接的培养方式就是课程设置和教学管理。近年来，在推进高校的国际化进程中，国内高校与国际知名高校强强联合，共建办学的模式取得了一定成效，但在联合办学模式下，培养的学生大部分都选择出国留学，少有毕业生在国内继续深造，反映出联合办学模式仅面向部分有出国留学意向的学生，高校国际化进程的深度和广度还有待提高。深化高校之间联合办学模式，需要解决两个核心问题：一、组建高质量的教师团队；二、推广多元化学习课程，让大部分学生参与其中。

2020 年，全国人民抗击新型冠状病毒的战役，再次展现了中华民族精诚团结、守望相助的伟大民族精神。疫情期间，众多高校都改变了传统授课方式，许多在线课程、学术讲座、科学研讨会如雨后春笋般涌现。通过这些专家、教授和学生们的共同努力，营造了良好的云端学术环境，充分体现了学术自由的思想。科学技术是第一生产力，21 世纪，信息化、数字化就是加快科学技术更新换代的引擎。充分利用好网络信息技术，完善外聘专家在线教学和网络授课方式，构建以学生为中心，以外籍专家为主导，以信息技术为支撑的在线学习系统，可以打破时空局限，缩减时间成本，加快知识传播速度，助力我国科技实现弯道超车。

因此，建议高校根据自身办学特色和专业需求，联合国际知名高校或引进国外专家，签署网上教学协议，开设专门的国际网络课程，确保学生能够听到专家讲解的专业知识，能够追踪本学科国际前沿发展动态，国际最新的研究成果。数字化教学模式不仅能让外籍专家们更加公开、透明地传播学术思想，减轻他们与国内合作的政治负担，同时也简化了联合教学流程，让专家们更加便捷地融入国内开展教学工作。

推广这一新的教学方法，国家政策上的支持和稳定的数字化传输技术十分重要。国家制定政策，增设联合国际知名高校和聘请外籍专家参与共建一流大学的项目、合作开发精品线上课程、探索全球中英文基础课等方式，支持远程互动教学新模式。通过不断改善数字化传输技术，在本科教育中融入专业化知识、国际化视角，健全教学评价机制，才能充分调动国外专家参与共建国际一流高校的积极性，使各大高校充满国际化的学术氛围。

2)鼓励开展云端学术论坛

学术论坛作为全世界科研工作者合作与交流的重要方式，其包含了前瞻性、针对性、互动性、启发性等特征。在学术论坛中，科研工作者本着自由探索、交叉通融、开拓创新的理念进行相互交流，既可以了解领域前沿，又可以分享研究成果，启发科研思路，推动技术革新。但举办一场学术报告，从准备到顺利闭幕，需要多部门协调联动，从资金预算到场地评估，都需要消耗大量时间、财力和物力。

随着网络信息技术对学术领域地全面渗透，学术期刊开放存储、平台化交流、新媒体等以互联网为媒介，学习和科研交流方式的出现，极大地提高了学者们信息交换的速度。顺应时代发展，利用互联网技术，鼓励开展云端学术论坛，跨越时空，邀请国外专家分享新近研究成果、点评热点，可以帮助科研工作者快速拓展学术空间，提升学术视野，有效进行全方位、系统性的科学思考和开放式创新研究。

组织云端学术会议需要建立在科学家已有学术网络基础上，因此，构建高校各学科之间、国内各学术协会与国外各学术机构之间的科研网络圈至关重要。此外，云端学术会议需要积极争取多方支持，精心设计国际热点议题或企业难题，依托各高校各科研单位丰富的专家资源，联合国际顶级期刊和科技企业，高效利用视频、音频、数据编码、新媒体等网络信息发布和传递机制开展学术交流活动，并结合高校、企业和学术期刊的三方优势，将开放式学术交流思维有效融入云端学术会议，才能启发众多科研工作者，切实提高学术交流效果。

总而言之，基于网络信息技术的高速发展，不管是在对外合作中完善在线教学机制，还是开展云端学术交流活动，其普适性、时效性、便捷性等优点是显而易见的，但以网络为载体进行学术交流的合法性，尤其是对知识产权的保护还需要健全的法律制度。只有形成法律制度规范，保证网络学习与学术交流制度在组织、政策、经费管理方面的合理性，才能树立网络学术交流制度的规范性和权威性。只有在尊重和保障国内外科研工作者知识产权的前提下，专家和学生才能更加高效地发挥出网络学术交流的优势。

第二，拓宽对外合作交流对象范围，让科技合作服务国家重大发展战略

目前，我国科技和经济高速发展，“一带一路”国家发展战略倡议与沿线国家开展更大范围、更高水平和更深层次的合作。以此为契机，践行人类命运共同体理念，深化与“一带一路”签约国家的学术交流和科技合作具有重要的战略意义。虽然沿线各国经济发展水平各有差异，但我们也绝不能轻视，这里包含了大量互利共赢的机会。在合作对象的选择上，可在对沿线国家的科技优势、企业文化、国家政策及资源分布进行系统分析的基础上，合理设计合作议题和合作项目，分批次派遣相关专业留学生访问交流，高薪聘请当地优秀的学者回国任教，从学术交流，到技术研发、共建实验室和合作创办高科技企业等方式逐步深化合作层次，推动科技创新。通过扩大对外交流的合作对象，既可以打破某些发达国家的科技出口限制，又可以充分吸收其他国家先进的科学知识和技术，彰显大国实力，提高中国高校和企业的国际知名度，拓宽国际学术合作交流的渠道。

第三，立足自主创新，树立中国期刊品牌，提升中国学术圈影响力

一流的学术刊物对众多学者的吸引力和在学术圈的号召力不言而喻。创办一批一流的国内学术期刊能够吸引大量的科技人才发表最优秀的研究成果。利用学术期刊的品牌效应，提升中国学术的国际地位，也应是我国立足于自主创新，引进顶尖人才和先进科技的关键步骤。虽然一流的学术刊物与高质量科技成果之间不能画等号，但至少具有一定程度的正相关。如今，在学术论文的创造上，我国已经成为了真正的论文发表大国，质量和数量都有了极大的提升，但缺乏一流的国内学术期刊，大量优秀的学术成果都会优先发表在国外期刊。其中除了需要完善国内人才晋升机制和学术成果评价体系外，政府也应该加大财政支持力度，加快在关键学科重点建设优秀国内学术期刊的步伐。通过组建优秀的编辑团队，严格控制好稿源质量，透明化审稿流程，创办一批能够吸引顶尖科学家投稿发表的世界顶尖学术期刊。同时，高校和科研院所应该加强宣传，为国内期刊树立品牌形象。这主要涉及期刊上发表的高水平文章和重要的研究成果应该借助国际学术会议、国外媒体、互联网在世界范围内大力推广等宣传措施。

需要指出的是，不管世界局势如何变化，深化对外合作交流的核心都应该以受教育者为本，以培养具有国际化思维的学术人才和世界顶尖的学术大师为目标，以合作交流带动创新驱动，以科技发展带动社会进步。高校和科研院所应坚定国家既定对外合作交流政策不动摇，辅以外界信息技术革新和自主创新能力的提高，国际化学生培养模式，高效化学术交流模式，让更多国外的科技人才和先进技术走向国内，共建国家的发展和繁荣之路。

大爱无疆，铿锵的中国力量*

李坚

抗击突如其来的新冠病毒流行，是一场没有硝烟的战争。病毒的传播，没有国界，不分种族，对人类健康和生命的危害巨大。世界卫生组织已先后宣告 Covid-19 是国际公共卫生突发事件，是一种世界大流行的疾病。

中国人民在中国共产党的领导下，万众一心，团结抗疫，仅仅用了两个多月的时间，就有效地抑制了疫情的蔓延，取得了阶段性的重大胜利。其根本原因在于，中国共产党与中国人民心连心，始终坚持以人民利益为宗旨，亲民爱民，将保护人民的生命安全放在第一位。全国人民风雨同舟，守望相助，众志成城，共渡难关。中国政府和中国人民用大爱点燃了中国速度、中国智慧和中国力量！

一、超强的凝聚力

在以习近平同志为核心的党中央英明领导下，战“疫”进程中，全国一盘棋，党政军民学、东西南北中一体行动，形成了强劲的凝聚共识、凝聚智慧、凝聚力量。14 亿中国人民团结一心的凝聚力，构筑了攻克难关的坚强有力的精神防线，这是中国力量的重要元素。

二、卓越的执行力

2019 年 10 月 28~31 日，在北京召开了中国共产党第十九届中央委员会第四次全体会议。会议指出，中国特色社会主义制度是党和人民在长期实践探索中形成的科学制度体系，我国国家治理一切工作和活动都依照中国特色社会主义制度展开，我国国家治理体系和治理能力是中国特色社会主义制度及其执行能力的集中体现。我国在短期内能取得抗击新冠肺炎斗争的阶段性胜利，已经充分彰显了新时代中国政治制度的优越性，并向全世界展现了中国共产党无以伦比的执行力！这也是西方某些国家的政客永远不能理解的，是他们永远抄不来的作业。

疫情防控斗争实践证明，中国共产党和我国社会主义制度、我国国家治理体系具有强大的生命力，其治理能力排山倒海，能够战胜任何艰难险阻！

在这次战“疫”中，充分发挥了党集中统一领导这一最大的政治优势，以及“集中力量办大事”这一显著的制度优越性，万众一心，共克时艰，实现了中国最早预警、最早防疫、最早取得抗击疫情阻击战和歼灭战的阶段性胜利。

中国共产党始终坚持以人民为中心，将广大人民群众的生命安全和身体健康放在第一位！亲民爱

---

* 本文摘自《担当》，北京：人民出版社，303-312，2021.

民的思想和精神，已牢牢扎根于广大共产党员心中。哪里有疫情、哪里疫情严重，只要党中央发出指令，全国各界马上就一呼百应，从速调集全国最优秀的医护人员、最急需的设备和资源，全力以赴投入疫情防控第一线。在中国共产党的领导下，集中多方面优势，构筑起了最严密的疫情防控体系。实践证明，在这场没有硝烟的战争中，14亿中国人民心心相印、同舟共济，就没有战胜不了的艰难险阻，就没有不可跨越的道道难关，这才是英雄中国！

三、理念的辐射力

习近平总书记在2018年5月召开的两院院士大会上郑重指出："深度参与全球科技治理，贡献中国智慧，着力推动构建人类命运共同体……共同应对未来发展、粮食安全、能源安全、人类健康、气候变化等人类共同挑战，在实现自身发展的同时惠及其他更多国家和人民，推动全球范围平衡发展。"在这场全人类面临的抗击疫情的共同挑战中，我国始终秉持人类命运共同体的理念，及时地向世界卫生组织通报疫情，积极地毫无保留地分享抗疫经验，并主动向塞尔维亚、柬埔寨、意大利等数十个国家派遣了专家医疗队，身临其境传授中国的抗疫经验。有的医护人员是刚从援鄂战"疫"中返回的，尚未休息就奔赴新的抗疫前线，得到了所在国家的热烈欢迎和由衷感谢。譬如，当中国医护专家到达塞尔维亚首都机场时，总统先生亲自在机场迎接，并亲吻了中国国旗，流出了热泪……场面令人难忘！另外，中国主动向世界200多个国家和组织提供力所能及的医护专用物资与设备援助，诸如口罩、检测试剂盒、呼吸机等。为了救助世界贫困地区开展抗疫斗争，中国向世界卫生组织分两批捐助资金5000万美元，彰显了负责任大国的担当。

四、榜样的影响力

在这场突如其来的疫情防控中，以钟南山院士为代表的中国工程院院士首当其冲，以榜样的力量，与时间赛跑，与疫情斗争，展现了为国担当的使命和情怀，其榜样的影响力无穷无尽！疫情是一场大考，更是对人类智慧和力量的考验。

钟南山院士首先提出"新冠病毒肺炎肯定人传人"，号召全国人民做到早发现、早隔离、早治疗，要戴口罩、勤洗手、多通风、少去人群密集的地方……他在国家卫健委和相关省(市)卫健委的工作会议上，都及时地、有针对性地提出了自己的见解、预测和主张，在每一节点上都发挥了重要作用。钟南山院士的发言是最有可信度的。老百姓说，有一种信任叫作"钟南山"，有一种戒律叫作"宅在家里"。2020年2月18日，在广东省政府召开的疫情防控新闻发布会上，钟南山院士终于露出了久违的笑容。他的笑容，也感染了无数网友，让大家看到了疫情防控胜利的曙光。相信在不久的将来，疫情彻底结束后，钟南山院士和所有人都能酣畅淋漓地笑起来。

钟南山院士已经84岁高龄了，却始终坚守在疫情防控的第一线。1月18日，他从广州乘高铁去武汉时，由于长期疲劳，在座位上坐着坐着就睡着了。谁若能亲眼看到这一幕，都将铭记心田、热泪盈眶……百姓对钟南山院士的评价是：民族脊梁，国士无双，中国的铮铮铁汉！他因敢医敢言、敢于讲实话、敢于肩挑大义，而受人尊重。他有院士的专业，有战士的勇猛，更有国士的担当，是最美的逆行者。

李兰娟院士首先提出"武汉封城刻不容缓"。1月20日，李克强总理主持召开国务院常务会议，李兰娟院士和钟南山院士都参加了会议。1月22日，李兰娟院士向国家建议，武汉必须严格封城；

次日，武汉这样一座1000万人口的城市被瞬间封城。

李兰娟院士敢为人先，在1966年创建了“李氏人工肝支持系统”，使急性、亚急性重型肝炎治愈好转率由11.9%上升至78.9%。

新冠肺炎疫情来临，她于2月1日(农历大年初八)再赴武汉。她和她的团队特别注重研究有关药物在治疗新冠肺炎上的疗效，特别注重拯救重症患者的生命。

李兰娟院士在一次接受记者采访时，说了这样一段话：“这次疫情结束后，希望国家逐步给年轻一代树立正确的人生导向和正确的人生观！把高薪、高福利、高地位留给德才兼备的科研、军事技术人员，让孩子们明白真正偶像的含义！”只有少年强，才国家强，为祖国未来发展培养国之栋梁！这些话发自肺腑，令人深省。

王辰院士首先提出“建立方舱医院”。2020年2月1日，王辰院士和中日友好医院医疗队赶赴武汉，经实地走访考察后指出，武汉最紧迫的任务，是如何解决新冠病毒的社会传播和扩散问题，首先要迅速地把确诊的轻症患者全部收治在医院里，以避免造成新的传染源。王辰院士建议应收尽收、应治尽治。上述建议得到了武汉相关部门的快速响应。48小时后，首批3座方舱医院开舱，共有4000多个床位，成为隔离在家、孤立无援患者保护生命的“绿色通道”，同时也阻绝了这些患者引发新的传染。

陈薇院士自主研发检测试剂和新冠病毒疫苗。从2020年1月30日开始，她和她的团队转入帐篷式移动检测实验室，应用自主研发的检测试剂盒，配合核酸自动提取技术，使核酸检测时间大大缩短。陈薇院士坚定地表示，力争在最短的时间内，将正在研制的重组新冠病毒疫苗推向临床、实际应用。这将向全世界彰显中国智慧，对拯救人类生命具有极其重大的意义。

张伯礼院士提出轻、重症患者分开治疗，征用学校、酒店等空间作为隔离场所。在中央召开的有关会议上，他提出，必须立即对新冠肺炎患者分类分层管理、集中隔离。对确诊的患者要按轻症和重症分开治疗，并且给患者普遍服用中药，用“大水漫灌”方式达到了早期干预的目的，一些轻症患者退烧了，取得了良好的效果。

此外，其他中国工程院院士针对新冠病毒的检测、防控进行了大量研究，研究成果对疫情的缓解和抑制起到了重要作用。

习近平总书记指出：“院士是我国科学技术界、工程技术界的杰出代表，是国家的财富、人民的骄傲、民族的光荣。”在这次疫情中，以钟南山、李兰娟、王辰、陈薇和张伯礼为代表的诸多院士日以继夜的辛勤劳动，换来了中国百姓的平平安安。他们的大爱点燃了中国智慧和中国力量。他们在疫情大考中的成绩，充分证明了习近平总书记对院士英明伟大的论断。

综上所述，凝聚力、执行力、辐射力、影响力和战斗力的汇聚融合，彰显了中国的震撼力。这一力量所向披靡，无坚不摧！在它面前，那些心怀敌意、掠夺成性、草菅人命、抹黑中国、善于巧舌鼓噪的国际政客永远苍白无力，自享由他们自己酿造的那“混乱的灾难”吧！那些政客已经把天灾变成了人祸，建议他们把用于讹诈、威胁、“甩锅”所花费的大量时间，好好用在挽救他们所在国家的宝贵生命上！

五、医护的战斗力

“病毒最怕的是团结一致的中国人！”武汉疫情发生后，中央一声令下，来自全国东南西北中的数万名医护人员日夜兼程来到武汉，一批批防护物资和专用设备源源不断地驰援湖北；火神山医院从除夕前夜开始准备，7000余名工人苦战在工地上，历时10天10夜，拔地而起。请问，世界上除了伟大的中国共产党，还有谁能做到？全国优秀的医护人员爱心汇聚，同心抗疫，昼夜奋战，与时间赛跑，同病毒抗争，全身心地投入到抗击疫情的总体战、阻击战中，涌现出大量的先进事迹和模范人物。下面举几个典型事例。

李文亮：疫情“吹哨人”。2月7日，李文亮医生去世。他因最早向外界发出预警，而被称为疫情“吹哨人”。由于在一线工作，李文亮不幸感染新冠病毒。治疗期间，他还想着尽快康复，因为“疫情还在扩散，不想当逃兵，恢复以后还是要上一线”。

彭银华：推迟婚礼抗疫情。农历正月初八，彭银华原定在这一天与心爱的人举行婚礼。他工作的医院考虑到这种情况，一开始并没有安排他在春节期间值班。但他得知疫情状况后，主动推迟婚礼，奔赴抗疫一线，后因感染新冠病毒不幸去世，年仅29岁。

黄文军：岂因祸福避趋之。疫情发生后，湖北省孝感市中心医院医生黄文军写下请战书：“苟利国家生死以，岂因祸福避趋之。我申请去隔离病房，共赴国难，听从组织安排！”之后，他一直奋战在抗击新冠肺炎一线，不幸感染，英勇殉职！

肖俊：坚守一线不后退。年已50岁的肖俊，大学一毕业，便在武汉市红十字会医院工作，已有29年。疫情发生后，肖俊一直坚守抗疫一线。白天工作了一天，晚上还值夜班，有时是24小时连班。他不幸感染新冠病毒，经抢救无效去世。

张军浩：连续奋战不松懈。张军浩，57岁，湖北省黄冈市疾病预防控制中心医生，疫情发生后，曾连续16天奋战在新冠肺炎疫情防控一线，因劳累过度，在2020年2月9日突发心梗去世。

朱峥嵘：藏起住院通知书，坚守岗位。年前，江苏省启东市南阳镇社区卫生服务中心医生朱峥嵘已经被查出患大动脉炎，需要住院治疗。新冠肺炎疫情发生后，朱峥嵘藏起了住院通知书，坚守一线，连续工作。从农历除夕开始，整整20多天，他始终奋战在疫情防控一线，直到最后倒下，年仅48岁。

在这场战“疫”中，从医生到医院院长，从护士到护士长，坚守岗位，日日夜夜，创造出许许多多保护人民生命的奇迹，将永远铭刻在全中国人民心中！

附录一

青年科技工作者“肩挑”科技强国的使命

——访中国工程院院士李坚

《科技导报》：科学家承载着国家和民族创新发展的历史、现在和未来。进入新时代，中国科学家精神赋予了哪些新的内涵和时代特征？

李坚：科学家承载着国家和民族创新发展的历史、现在和未来。马克思说：“科学绝不是一种自私自利的享乐，有幸能够致力于科学研究的人，首先应该拿自己的学识为人类服务”。

科学家们致力于探索未知、发现真理，承担着发展先进的科学技术、改造世界、造福人类的重任。“实事求是、追求真理”是科学家的精神基础，也是新时代科学家精神的本质特征。钱学森、邓稼先、袁隆平等老一辈科学家敢于担当，甘于奉献、不畏艰难、勇攀高峰，他们用实际行动诠释了科学家精神。

我国进入了中国特色社会主义新时代，科技事业蓬勃发展，实现了伟大的变革，与新时代的战略任务紧密相连。那么，新时代中国科学家精神在“求真、务实”的基础上，更赋予了“爱国自信、协同创新、精神传承”的新内涵和时代特征，也是社会主义核心价值观的辩证体现。“科学无国界，科学家是有祖国的”。钱学森、黄大年等科学家的爱国事迹家喻户晓，尽人皆知，用实际行动书写了“心有大我，至诚报国”的人生篇章。

科技是“国之利器”，科技创新能力是当今最值得重视的国家核心竞争力。习近平总书记指出：“创新是一个民族进步的灵魂，是一个国家兴旺发达的不竭动力，也是中华民族最深沉的民族禀赋。”科学家们的创新意识、创新能力关乎国家发展和未来命运。新时代的科学家要在创新驱动发展上有大作为，争做科技创新的领头雁。

我国实现科技强国的伟大梦想，归根结底要靠青年人。科学家要秉承甘为人梯、奖掖后学的育人精神，努力传承“爱国精神、求实精神、奉献精神、协同精神”，争做提携后学的引路人，引导青年一代创新发展，建功新时代，为中华民族的伟大复兴贡献青春和智慧。

《科技导报》：怎样引导广大青年科技工作者扣好科研生涯的第一粒扣子？一方面，青年科技工作者应该怎样处理、看待科学研究事业；另一方面，国家应采取怎样的措施或政策鼓励、支持、引导广大青年工作者全身心投身于科技创新事业？

李坚：青年是祖国的前途、民族的希望、创新的未来。青年一代有理想、有本领、有担当，科技就有前途，创新就有希望。尤其对青年科技工作者而言，他们是创新的主要源泉，是国家科技发展的希望，因此引领好他们扣好科研生涯的第一粒扣子尤为重要。

我认为，作为青年科技工作者首先要树立正确的社会主义核心价值观，习总书记说："青年的价值取向决定了未来整个社会的价值取向，而青年又处在价值观形成和确立的时期，抓好这一时期的价值观养成具有划时代意义。这就像穿衣服扣扣子一样，如果第一粒扣子扣错了，剩余的扣子都会扣错。人生的扣子从一开始就要扣好。"凿井者，起于三寸之坎，以就万仞之深。"青年要从现在做起、从自己做起，使社会主义核心价值观成为自己的基本遵循，并身体力行大力将其推广到全社会去。"

其次，要引领好青年科技工作者们有"沉下气、静下心，守得住寂寞、坐得住冷板凳"和"十年磨一剑"的决心和信心，在科研工作中要遵守实事求是、求真务实、创新发展等系列学术规范。

第三，更要"以识才的慧眼、爱才的诚意、用才的胆识、容才的雅量、聚才的良方，放手使用优秀青年人才，为青年人才成才铺路搭桥，让他们成为有思想、有情怀、有责任、有担当的社会主义建设者和接班人。"作为青年科技人才的培养，要注重启发青年的好奇心、想象力、注重培养青年的行动力、注重培养青年的坚定信念、注重引导青年服务于经济社会。我们这一辈科研工作者要崇德向善，甘当人梯，助推他们追求真理，锐意创新，追逐梦想，德技双馨，成为新时代砥砺前行的生力军。

附录二

论文、著作、专利名录（2012—2021）

一、论文

1. Nguyen Thi Thanh Hien, **Li Jian***, Li Shujun. Effects of water-borne rosin on the fixation and decay resistance of copper-based preservative treated wood. BioResources, 2012, 7(3): 3573-3584
2. Lili Sun, **Jian Li**, Lijuan Wang. Electromagnetic interference shielding material from electroless copper plating on birch veneer. Wood Science & Technology, 2012, 46: 1061-1071
3. Hao Feng, **Jian Li**, Lijuan Wang. The removal of reactive red 228 dye from aqueous solutions by chitosan-modified flax shive. BioResources, 2012, 7(1): 624-639
4. Shuliang Wang, Chengyu Wang, Changyu Liu, Ming Zhang, Hua Ma, **Jian Li**. Fabrication of superhydrophobic spherical-like α-FeOOH films on the wood surface by a hydrothermal method. Colloids and Surfaces A: Physicochemical and Engineering Aspects, 2012, 403: 29-34
5. Ming Zhang, Shuliang Wang, Chengyu Wang, **Jian Li**. A facile method to fabricate superhydrophobic cotton fabrics. Applied Surface Science, 2012, 261: 561-566
6. Ming Zhang, Chengyu Wang, Shuliang Wang, Yunling Shi, **Jian Li**. Fabrication of coral-like superhydrophobic coating on filter paper for water-oil separation. Applied Surface Science, 2012, 261: 764-769
7. 赵骏衡，金海兰，**李坚**. 改善造纸污泥脱水性条件对 CST 的影响. 中国造纸，2013，32(11)：42-45
8. Nguyen Thi Thanh Hien, Li Shujun, **Li Jian***. The combined effects of copper sulfate and rosin sizing agent treatment on some physical and mechanical properties of poplar wood. Construction and Building Material, 2013, 40: 33-39
9. Ming Zhang, Chengyu Wang, Shuliang Wang, **Jian Li**. Fabrication of superhydrophobic cotton textiles for water-oil separation based on drop-coating route. Carbohydrate Polymers, 2013, 97(1): 59-64
10. Feng Liu, Shuliang Wang, Ming Zhang, Miaolian Ma, Chengyu Wang, **Jian Li**. Improvement of mechanical robustness of the superhydrophobic wood surface by coating PVA/SiO_2 composite polymer. Applied Surface Science, 2013, 280: 686-692
11. Lijuan Wang, **Jian Li**. Electromagnetic shielding wood-based materials from a novel electroless copper plating process. BioResources, 2013, 8(3): 3414-3425
12. Bin Hui, **Jian Li**, Lijuan Wang. Preparation of EMI shielding and corrosion-resistant composite based on electroless Ni-Cu-P coated wood. BioResources, 2013, 8(4): 6097-6110
13. Lijuan Wang, **Jian Li**. Adsorption of C. I. Reactive Red 228 dye from aqueous solution by modified cellulose from flax shive: Kinetics, equilibrium, and thermodynamics. Industrial Crops and Products,

2013, 42: 153-158

14. Lijuan Wang, **Jian Li**. Removal of methylene blue from aqueous solution by adsorption onto crofton weed stalk. BioResources, 2013, 8(2): 2521-2536
15. Yue Qi, **Jian Li**, Lijuan Wang. Removal of Remazol Turquoise Blue G-133 from aqueous medium using functionalized cellulose from recycled newspaper fiber. Industrial Crops and Products, 2013, 50: 15-22
16. Nguyen Thi Thanh Hien, Li Shujun, **Li Jian**, Liang Tao. Micro-distribution and fixation of a rosin-based micronized-copper preservative in poplar wood. International Biodeterioration & Biodegradation, 2013, 83: 63-70
17. **李坚***, 孙庆丰. 大自然给予的启发——木材仿生科学刍议. 中国工程科学, 2014, 16(4): 4-12
18. 万才超, 卢芸, 孙庆丰, **李坚***. 新型木质纤维素气凝胶的制备、表征及疏水吸油性能. 科技导报, 2014, 32(4-5): 79-85
19. Wei Li, Qiong Wu, Xin Zhao, Zhanhua Huang, Jun Cao, **Jian Li***, Shouxin Liu. Enhanced thermal and mechanical properties of PVA composites formed with filamentous nanocellulose fibrils. Carbohydrate Polymers, 2014, 113: 403-410
20. Caichao Wan, Yun Lu, Qingfeng Sun, **Jian Li***. Hydrothermal synthesis of zirconium dioxide coating on the surface of wood with improved UV resistance. Applied Surface Science, 2014, 321: 38-42
21. Chunhui Ma, Wei Li, Yuangang Zu, Lei Yang, **Jian Li***. Antioxidant properties of pyroligneous acid obtained by thermochemical conversion of schisandra chinensis baill fruits. Molecules, 2014, 19: 20821-20838
22. Chunhui Ma, Lei Yang, Wei Li, Jinquan Yue, **Jian Li***, Yuangang Zu. Ultrasound-assisted extraction of arabinogalactan and dihydroquercetin simultaneously from Larix gmelinii as a pretreatment for pulping and papermaking. Plos One, 2014, 12(2): 1-21
23. Bin Hui, **Jian Li**, Qi Zhao, Tie qiang Liang, Lijuan Wang. Effect of $CuSO_4$ content in the plating bath on the properties of composites from electroless plating of Ni-Cu-P on birch veneer. BioResources, 2014, 9(2): 2949-2959
24. Bin Hui, **Jian Li**, Lijuan Wang. Electromagnetic shielding wood-based composite from electroless plating corrosion-resistant Ni-Cu-P coatings on *Fraxinus mandshurica* veneer. Wood Science and Technology, 2014, 48: 961-979
25. Dongying Hu, Peng Wang, **Jian Li**, and Lijuan Wang. Functionalization of microcrystalline cellulose with N, N-dimethyldodecylamine for the removal of Congo Red dye from an aqueous solution. BioResources, 2014, 9(4): 5951-5962
26. Wenshuai Chen, Qing Li, Youcheng Wang, Xin Yi, Jie Zeng, Haipeng Yu, Yixing Liu, **Jian Li**. Comparative study of aerogels obtained from differently prepared nanocellulose fibers. ChemSusChem, 2014, 7(1): 154-161
27. **李坚***, 高丽坤. 光控润湿性转换的抑菌性木材基银钛复合薄膜. 森林与环境学报, 2015, 35(3): 193-198
28. 李坚*, 许民, 包文慧. 影响未来的颠覆性技术: 多元材料混合智造的3D打印. 东北林业大学学报, 2015, 43(6): 1-9
29. 代林林, 李伟, 曹军, **李坚**, 刘守新. 纳米纤维素手性向列液晶相结构的形成、调控及应用. 化学进展, 2015, 27(7): 861-869
30. Likun Gao, Xianxu Zhan, Yun Lu, **Jian Li***, Qingfeng Sun. pH-dependent structure and wettability of

TiO_2-based wood surface. Materials Letters, 2015, 142: 217-220

31. Likun Gao, Wentao Gan, Shaoliang Xiao, Xianxu Zhan, **Jian Li***. Enhancement of photo-catalytic degradation of formaldehyde through loading anatase TiO_2 and silver nanoparticle films on wood substrates. RSC Advances, 2015, 5(65): 52985-52992
32. Likun Gao, Shaoliang Xiao, Wentao Gan, Xianxu Zhan, **Jian Li***. Durable superamphiphobic wood surfaces from Cu_2O film modified with fluorinated alkyl silane. RSC Advances, 2015, 5: 98203-98208
33. Likun Gao, Yun Lu, Xianxu Zhan, **Jian Li***, Qingfeng Sun. A robust, anti-acid, and high-temperature-humidity-resistant superhydrophobic surface of wood based on a modified TiO_2 film by fluoroalkyl silane. Surface & Coatings Technology, 2015, 262: 33-39
34. Wentao Gan, Likun Gao, Qingfeng Sun, Yun Lu, **Jian Li***. Multifunctional wood materials with magnetic, superhydrophobic and anti-ultraviolet properties. Applied Surface Science, 2015, 332: 565-572
35. Wentao Gan, Ying Liu, Likun Gao, Xianxu Zhan, **Jian Li***. Growth of $CoFe_2O_4$ particles on wood template using controlled hydrothermal method at low temperature. Ceramics International, 2015, 41(10): 4876-14885
36. Wentao Gan, Likun Gao, Xianxu Zhan, **Jian Li***. Hydrothermal synthesis of magnetic wood composites and improved wood properties by precipitation with $CoFe_2O_4$/hydroxyapatite. RSC Advances, 2015, 5(57): 45919-45927
37. Caichao Wan, Yun Lu, Yue Jiao, Jun Cao, Qingfeng Sun, **Jian Li***. Cellulose aerogels from cellulose-NaOH/PEG solution and comparison with different cellulose contents. Materials Science and Technology, 2015, 31(9): 1096-1102
38. Caichao Wan, Yun Lu, Yue Jiao, Jun Cao, Qingfeng Sun, **Jian Li***. Preparation, characterization and oil adsorption properties of cellulose aerogels from four kinds of plant materials via a NaOH/PEG aqueous solution. Fibers and Polymers, 2015, 16(2): 302-307
39. Caichao Wan, Yun Lu, Yue Jiao, Chunde Jin, Qingfeng Sun, **Jian Li***. Ultralight and hydrophobic nanofibrillated cellulose aerogels from coconut shell with ultrastrong adsorption properties. Journal of Applied Polymer Science, 2015, 132(24): 42037
40. Caichao Wan, Yun Lu, Yue Jiao, Chunde Jin, Qingfeng Sun, **Jian Li***. Fabrication of hydrophobic, electrically conductive and flame-resistant carbon aerogels by pyrolysis of regenerated cellulose aerogels. Carbohydrate Polymers, 2015, 118: 115-118
41. Caichao Wan, Yun Lu, Chunde Jin, Qingfeng Sun, **Jian Li***. Thermally induced gel from cellulose/NaOH/PEG solution: preparation, characterization and mechanical properties. Applied Physics A, 2015, 119(1): 45-48
42. Caichao Wan, Yun Lu, Chunde, Jin Qingfeng Sun, **Jian Li***. A facile low-temperature hydrothermal method to prepare anatase titania/cellulose aerogels with strong photocatalytic activities for rhodamine b and methyl orange degradations. Journal of Nanomaterials, 2015, 2015: 717016
43. Caichao Wan, Yun Lu, Yue Jiao, Jun Cao, Qingfeng Sun, **Jian Li***. Preparation of mechanically strong and lightweight cellulose aerogels from cellulose-NaOH/PEG solution. Journal of Sol-Gel Science and Technology, 2015, 74(1): 256-259
44. Caichao Wan, Yue Jiao, **Jian Li***. In situ deposition of graphene nanosheets on wood surface by one-

pot hydrothermal method for enhanced UV-resistant ability. Applied Surface Science, 2015, 347: 891-897
45. Caichao Wan, **Jian Li***. Embedding ZnO nanorods into porous cellulose aerogels via a facile one-step low-temperature hydrothermal method. Materials & Design, 2015, 83: 620-625
46. Caichao Wan, **Jian Li***. Synthesis of well-dispersed magnetic $CoFe_2O_4$ nanoparticles in cellulose aerogels via a facile oxidative co-precipitation method. Carbohydrate Polymers, 2015, 134: 144-150
47. Caichao Wan, **Jian Li***. Facile synthesis of well-dispersed superparamagnetic γ-Fe_2O_3 nanoparticles encapsulated in three-dimensional architectures of cellulose aerogels and their applications for Cr (VI) removal from contaminated water. ACS Sustainable Chemistry & Engineering, 2015, 3(9): 2142-2152
48. Yue Jiao, Caichao Wan, **Jian Li***. Room-temperature embedment of anatase titania nanoparticles into porous cellulose aerogels. Applied Physics A, 2015, 120(1): 341-347
49. Shaoliang Xiao, Runan, Gao, Yun Lu, **Jian Li***, Qingfeng Sun. Fabrication and characterization of nanofibrillated cellulose and its aerogels from natural pine needles. Carbohydrate Polymers, 2015, 119: 202-209
50. Chunhui Ma, Yuangang Zu, Lei Yang, **Jian Li***. Two solid-phase recycling method for basic ionic liquid [C4mim]Ac by macroporous resin and ion exchange resin from Schisandra chinensis fruits extract. Journal of Chromatography B, 2015, 976-977: 1-5
51. Caichao Wan, **Jian Li***. Cellulose aerogels based on a green NaOH/PEG solution: Preparation, characterization and influence of molecular weight of PEG. Fibers and Polymers, 2015, 16(6): 1230-1236
52. Likun Gao, Yun Lu, Jun Cao, **Jian Li**, Qingfeng Sun. Reversible photocontrol of wood-surface wettability between superhydrophilicity and superhydrophobicity based on a TiO_2 Film. Journal of Wood Chemistry and Technology, 2015, 35(5): 365-373
53. Bin Hui, Guoliang Li, Guanghui Han, Yingying Li, Lijuan Wang, **Jian Li**. Fabrication of magnetic response composite based on wood veneers by a simple in situ synthesis method. Wood Science and Technology, 2015, 49(4): 755-767
54. Feng Liu, Zhengxin Gao, Deli Zang, Chengyu Wang, **Jian Li**. Mechanical stability of superhydrophobic epoxy/silica coating for better water resistance of wood. Holzforschung, 2015, 69(3): 367-374
55. Wenshuai Chen, Qing Li, Jun Cao, Yixing Liu, **Jian Li**, Jiangshuai Zhang, Shuiyang Luo, Haipeng Yu. Revealing the structures of cellulose nanofiber bundles obtained by mechanical nanofibrillation methods via TEM observation. Carbohydrate Polymers, 2015, 117: 950-956
56. Wang Hui, Nguyen Thi Thanh Hien, Li, Shujun, Liang Tao, Zhang Yuanyuan, **Li Jian**. Quantitative structure-activity relationship of antifungal activity of rosin derivatives. Bioorganic & Medical Chemistry Letters, 2015, 25(2): 347-354
57. Hailan Jin, Junhyung Cho, Takayuki Okayam, Lihui Chen, **Jian Li**. Effects of carboxymethyl cellulose sodium addition on properties of beaten recycled reed pulps and the handsheets. SEN'I GAKKAISI, 2015, 71(9): 291-296
58. Hailan Jin, Kosei Watanabe, Hajime Ohtani, **Jian Li**, Takayuki Okayama. Effects of internal addition of bulking agents on properties of recycled handsheet. SEN'I GAKKAISI, 2015, 71(6): 201-206
59. **李坚**. 大自然的启发——木材仿生与智能响应. 科技导报, 2016, 34(19): 卷首语
60. 高丽坤, **李坚***. 木材基银钛复合薄膜的制备及其可见光降解甲醛. 科技导报, 2016, 34: 127-131

61. **李坚**[*], 张明, 强添刚. 特殊润湿性油水分离材料的研究进展. 森林与环境学报, 2016, 36(3): 257-265
62. 代林林, 李伟, 曹军, **李坚**, 刘守新. 湿敏手性纳米纤维素薄膜的自组装制备及表征. 高分子材料科学与工程, 2016, 32(8): 115-119
63. 高鹤, 梁大鑫, **李坚**. 纤维素气凝胶材料研究进展. 科技导报, 2016, 34(19): 138-142
64. 高鹤, 梁大鑫, **李坚**, 庞广生, 方振兴. 纳米 TiO_2-ZnO 二元负载木材的制备及性质. 高等学校化学学报, 2016, 37(06): 1075-1081
65. Shaoliang Xiao, Runan Gao, Likun Gao, **Jian Li**[*]. Poly (vinyl alcohol) films reinforced with nanofibrillated cellulose (NFC) isolated from corn husk by high intensity ultrasonication. Carbohydrate Polymers, 2016, 136: 1027-1034
66. Likun Gao, Yun Lu, **Jian Li**[*], Qingfeng Sun. Superhydrophobic conductive wood with oil repellency obtained by coating with silver nanoparticles modified by fluoroalkyl silane. Holzforschung, 2016, 70 (1): 63-68
67. Likun Gao, Wentao Gan, Shaoliang Xiao, Xianxu Zhan, **Jian Li**[*]. A robust superhydrophobic antibacterial Ag-TiO_2 composite film immobilized on wood substrate for photodegradation of phenol under visible-light illumination. Ceramics International, 2016, 42: 2170-2179
68. Likun Gao, Zhe Qiu, Wentao Gan, Xianxu Zhan, **Jian Li**[*], Tiangang Qiang. Negative oxygen ions production by superamphiphobic and antibacterial TiO_2/Cu_2O composite film anchored on wooden substrates. Scientific Reports, 2016, 6: 26055-26064
69. Caichao Wan, Yue Jiao, **Jian Li**[*]. Core-shell composite of wood-derived biochar supported MnO_2 nanosheets for supercapacitor applications. RSC Advances, 2016, 6(69): 64811-64817
70. Caichao Wan, **Jian Li**[*]. Wood-derived biochar supported polypyrrole nanoparticles as a free-standing supercapacitor electrode. RSC Advances, 2016, 6(89): 86006-86011
71. Wentao Gan, Likun Gao, Wenbo Zhang, **Jian Li**[*], Liping Cai. Removal of oils from water surface via useful recyclable $CoFe_2O_4$/sawdust composites under magnetic field. Materials & Design, 2016, 98: 194-200
72. Wentao Gan, Likun Gao, Wenbo Zhang, **Jian Li**[*], Xianxu Zhan. Fabrication of microwave absorbing $CoFe_2O_4$ coatings with robust superhydrophobicity on natural wood surfaces. Ceramics International, 2016, 42(11): 13199-13206
73. Wentao Gan, Likun Gao, Ying Liu, Xianxu Zhan, **Jian Li**[*]. The magnetic, mechanical, thermal properties and UV resistance of $CoFe_2O_4/SiO_2$-coated film on wood. Journal of Wood Chemistry and Technology, 2016, 36(2): 94-104
74. Wentao Gan, Likun Gao, Xianxu Zhan, **Jian Li**[*]. Preparation of thiol-functionalized magnetic sawdust composites as an adsorbent to remove heavy metal ions. RSC Advances, 2016, 6(44): 37600-37609
75. Caichao Wan, **Jian Li**[*]. Incorporation of graphene nanosheets into cellulose aerogels: enhanced mechanical, thermal, and oil adsorption properties. Applied Physics A, 2016, 122(2): 105
76. Caichao Wan, **Jian Li**[*]. Cellulose aerogels functionalized with polypyrrole and silver nanoparticles: In-situ synthesis, characterization and antibacterial activity. Carbohydrate Polymers, 2016, 146: 362-367
77. Caichao Wan, **Jian Li**[*]. Graphene oxide/cellulose aerogels nanocomposite: Preparation, pyrolysis, and application for electromagnetic interference shielding. Carbohydrate Polymers, 2016, 150: 172-179
78. Caichao Wan, Yue Jiao, Qingfeng Sun, **Jian Li**[*]. Preparation, characterization, and antibacterial

properties of silver nanoparticles embedded into cellulose aerogels. Polymer Composites, 2016, 37(4): 1137-1142

79. Caichao Wan, Yue Jiao, **Jian Li***. Influence of pre-gelation temperature on mechanical properties of cellulose aerogels based on a green NaOH/PEG solution—a comparative study. Colloid and Polymer Science, 2016, 294(8): 1281-1287
80. Yue Jiao, Caichao Wan, **Jian Li***. Synthesis of carbon fiber aerogel from natural bamboo fiber and its application as a green high-efficiency and recyclable adsorbent. Materials & Design, 2016, 107: 26-32
81. Yue Jiao, Caichao Wan, Tiangang Qiang, **Jian Li***. Synthesis of superhydrophobic ultralight aerogels from nanofibrillated cellulose isolated from natural reed for high-performance adsorbents. Applied Physics A, 2016, 122(7): 686
82. Miao Yu, **Jian Li**, Lijuan Wang. Preparation and characterization of magnetic carbon aerogel from pyrolysis of sodium carboxymethyl cellulose aerogel crosslinked by iron trichloride. Journal of Porous Materials, 2016, 23(4): 997-1003
83. Yanna Li, Yongzhuang Liu, Wenshuai Chen, Qingwen Wang, Yixing Liu, **Jian Li**, Haipeng Yu. Facile extraction of cellulose nanocrystals from wood using ethanol and peroxide solvothermal pretreatment followed by ultrasonic nanofibrillation. Green Chemistry, 2016, 18(4): 1010-1018
84. Wenshuai Chen, Qi Zhang, Kojiro Uetani, Qing Li, Ping Lu, Jun Cao, Qingwen Wang, Yixing Liu, **Jian Li**, Zhichao Quan, Yongshi Zhang, Sifan Wang, Zhenyu Meng, Haipeng Yu. Sustainable carbon aerogels derived from nanofibrillated cellulose as high-performance absorption materials. Advanced Materials Interfaces, 2016, 3(10): 1600004
85. Yao Tan, Yongzhuang Liu, Wenshuai Chen, Yixing Liu, Qingwen Wang, **Jian Li**, Haipeng Yu. Homogeneous dispersion of cellulose nanofibers in waterborne acrylic coatings with improved properties and unreduced transparency. ACS Sustainable Chemistry & Engineering, 2016, 4(7): 3766-3772
86. Hailan Jin, Junhyung Cho, Takayuki Okayam, Lihui Chen, **Jian Li**. Effects of ultrasonic cell crushing apparatus on the properties of recycling bamboo pulp handsheets. Journal of Fiber Science and Technology, 2016, 72(2): 44-48
87. 高丽坤，**李坚***. TiO_2/Cu_2O 膜覆木材的表面及负氧离子释放特性. 功能材料, 2017, 4: 04116-04121
88. **李坚***，甘文涛. 趋磁性木材的制备与多功能化修饰. 森林与环境学报, 2017, 37(3): 257-265
89. Likun Gao, Wentao Gan, Guoliang Cao, Xianxu Zhan, Tiangang Qiang, **Jian Li***. Visible-light activate Ag/WO_3 films based on wood with enhanced negative oxygen ions production properties. Applied Surface Science, 2017, 425: 889-895
90. Wenhui Bao, Daxin Liang, Ming Zhang, Yue Jiao, Lijuan Wang, Liping Cai, **Jian Li***. Durable, high conductivity, superhydrophobicity bamboo timber surface for nanoimprint stamps. Progress in Natural Science: Materials International, 2017, 27(6): 669-673
91. Wenhui Bao, Ming Zhang, Zhen Jia, Liping Cai, Daxin Liang, **Jian Li***. Cu thin films on wood surface for robust superhydrophobicity by magnetron sputtering treatment with perfluorocarboxylic acid. European Journal of Wood and Wood Products, 2017, 77(1): 115-123
92. Likun Gao, Wentao Gan, Guoliang Cao, Xianxu Zhan, Tiangang Qiang, **Jian Li***. Fabrication of biomass-derived C-doped Bi_2WO_6 templated from wood fibers and its excellent sensing of the gases containing carbonyl groups. Colloids and Surfaces A: Physicochemical and Engineering Aspects, 2017,

529: 487-494
93. **Jian Li***, Likun Gao, Wentao Gan. Bioinspired C/TiO_2 photocatalyst for rhodamine B degradation under visible light irradiation. Frontiers of Agricultural Science and Engineering, 2017, 4(4): 459-464
94. Yingying Li, Bin Hui, Guoliang Li, **Jian Li***. Fabrication of smart wood with reversible thermoresponsive performance. Journal of Materials Science, 2017, 52(13): 7688-7697
95. Wentao Gan, Shaoliang Xiao, Likun Gao, Runan Gao, **Jian Li***, Xianxu Zhan. Luminescent and transparent wood composites fabricated by poly (methyl methacrylate) and γ-Fe_2O_3@YVO_4: Eu^{3+} nanoparticle impregnation. ACS Sustainable Chemistry & Engineering, 2017, 5(5): 3855-3862
96. Wentao Gan, Likun Gao, Shaoliang Xiao, Runan Gao, Wenbo Zhang, **Jian Li***, Xianxu Zhan. Magnetic wood as an effective induction heating material: Magnetocaloric effect and thermal insulation. Advanced Materials Interfaces, 2017, 4(22): 1700777
97. Wentao Gan, Likun Gao, Shaoliang Xiao, Wenbo Zhang, Xianxu Zhan, **Jian Li***. Transparent magnetic wood composites based on immobilizing Fe_3O_4 nanoparticles into a delignified wood template. Journal of Materials Science, 2017, 52(6): 3321-3329
98. Wentao Gan, Likun Gao, Xianxu Zhan, **Jian Li***. Removal of Cu^{2+} ions from aqueous solution by amino-functionalized magnetic sawdust composites. Wood Science and Technology, 2017, 51(1): 207-225
99. Wentao Gan, Ying Liu, Xianxu Zhan, **Jian Li***. Magnetic property, thermal stability, UV-resistance, and moisture absorption behavior of magnetic wood composites. Polymer Composites, 2017, 38(8): 1646-1654
100. Caichao Wan, Yue Jiao, **Jian Li***. TiO_2 microspheres grown on cellulose-based carbon fibers: preparation, characterizations and photocatalytic activity for degradation of indigo carmine dye. Journal of Nanoscience and Nanotechnology, 2017, 17(8): 5525-5529
101. Caichao Wan, Yue Jiao, Tiangang Qiang, **Jian Li***. Cellulose-derived carbon aerogels supported goethite (α-FeOOH) nanoneedles and nanoflowers for electromagnetic interference shielding. Carbohydrate Polymers, 2017, 156: 427-434
102. Caichao Wan, Yue Jiao, Tiangang Qiang, **Jian Li***. Synthesis and electromagnetic interference shielding of cellulose-derived carbon aerogels functionalized with α-Fe_2O_3 and polypyrrole. Carbohydrate Polymers, 2017, 161: 158-165
103. Caichao Wan, Yue Jiao, **Jian Li***. Flexible, highly conductive, and free-standing reduced graphene oxide/polypyrrole/cellulose hybrid papers for supercapacitor electrodes. Journal of Materials Chemistry A, 2017, 5(8): 3819-3831
104. Caichao Wan, Yue Jiao, **Jian Li***. Multilayer core-shell structured composite paper electrode consisting of copper, cuprous oxide and graphite assembled on cellulose fibers for asymmetric supercapacitors. Journal of Power Sources, 2017, 361: 122-132
105. Caichao Wan, Yue Jiao, **Jian Li***. A cellulose fibers-supported hierarchical forest-like cuprous oxide/copper array architecture as a flexible and free-standing electrode for symmetric supercapacitors. Journal of Materials Chemistry A, 2017, 5(33): 17267-17278
106. Yue Jiao, Caichao Wan, **Jian Li***. Scalable synthesis and characterization of free-standing supercapacitor electrode using natural wood as a green substrate to support rod-shaped polyaniline. Journal of Materials Science: Materials in Electronics, 2017, 28(3): 2634-2641

107. Yue Jiao, Caichao Wan, **Jian Li***. Anatase TiO_2/cellulose hybrid paper: Synthesis, characterizations, and photocatalytic activity for degradation of indigo carmine dye. Functional Materials Letters, 2017, 10 (03): 1750018

108. **Jian Li***, Yue Jiao, Caichao Wan. Synthesis of $MnFe_2O_4$/cellulose aerogel nanocomposite with strong magnetic responsiveness. Frontiers of Agricultural Science and Engineering, 2017, 4(1): 116-120

109. Runan Gao, Yun Lu, Shaoliang Xiao, **Jian Li***. Facile Fabrication of nanofibrillated chitin/Ag_2O heterostructured aerogels with high iodine capture efficiency. Scientific Reports, 2017, 7: 4303

110. Miao Yu, Yingying Han, **Jian Li**, Lijuan Wang. One-step synthesis of sodium carboxymethyl cellulose-derived carbon aerogel/nickel oxide composites for energy storage. Chemical Engineering Journal, 2017, 324: 287-295

111. Miao Yu, **Jian Li**, Lijuan Wang. KOH-activated carbon aerogels derived from sodium carboxymethyl cellulose for high-performance supercapacitors and dye adsorption. Chemical Engineering Journal, 2017, 310: 300-306

112. Miao Yu, Yingying Han, **Jian Li**, Lijuan Wang. CO_2-activated porous carbon derived from cattail biomass for removal of malachite green dye and application as supercapacitors. Chemical Engineering Journal, 2017, 317: 493-502

113. Likun Gao, Wentao Gan, Zhe Qiu, Xianxu Zhan, Tiangang Qiang, **Jian Li**. Preparation of heterostructured WO_3/TiO_2 catalysts from wood fibers and its versatile photodegradation abilities. Scientific Reports, 2017, 7(1): 1-13

114. Niu Na, Ma Zhuoming, He Fei, Li Shujun, **Li Jian**, Liu Shouxin, Yang Piaoping. Preparation of carbon dots for cellular imaging by the molecular aggregation of cellulolytic enzyme lignin. Langmuir, 2017, 33(23): 5786-5795

115. Yongzhuang Liu, Wenshuai Chen, Qinqin Xia, Bingtuo Guo, Qingwen Wang, Shouxin Liu, Yixing Liu, **Jian Li**, Haipeng Yu. Efficient cleavage of lignin-carbohydrate complexes and ultrafast extraction of lignin oligomers from wood biomass by microwave-assisted treatment with deep eutectic solvent. ChemSusChem, 2017, 10(8): 1692-1700

116. Dawei Zhao, Qi Zhang, Wenshuai Chen, Xin Yi, Shouxin Liu, Qingwen Wang, Yixing Liu, **Jian Li**, Xianfeng Li, Haipeng Yu. Highly flexible and conductive cellulose-mediated PEDOT: PSS/MWCNT composite films for supercapacitor electrodes. ACS Applied Materials & Interfaces, 2017, 9: 13213-13222

117. Dawei Zhao, Chaoji Chen, Qi Zhang, Wenshuai Chen, Shouxin Liu, Qingwen Wang, Yixing Liu, **Jian Li**, Haipeng Yu. High performance flexible solid-state supercapacitors based on a renewable and biodegradable mesoporous cellulose membrane. Advanced Energy Materials, 2017, 7: 1700739

118. Yongzhuang Liu, Bingtuo Guo, Qinqin Xia, Juan Meng, Wenshuai Chen, Shouxin Liu, Qingwen Wang, Yixing Liu, **Jian Li**, Haipeng Yu. Efficient cleavage of strong hydrogen bonds in cotton by acidic deep eutectic solvents and fabrication of cellulose nanocrystals with high yields. ACS Sustainable Chemistry & Engineering, 2017, 5(9): 7623-7631

119. Bingtuo Guo, Yongzhuang Liu, Qi Zhang, Fengqiang Wang, Qingwen Wang, Yixing Liu, **Jian Li**, Haipeng Yu. Efficient flame-retardant and smoke-suppression properties of Mg-Al layered double hydroxide nanosheets on a wood substrate. ACS Applied Materials & Interfaces, 2017, 9(27): 23039-23047

120. Wei Li, Sichun Wang, Ying Li, Chunhui Ma, Zhanhua Huang, Chunsheng Wang, **Jian Li**, Zhijun Chen, Shouxin Liu. One-step hydrothermal synthesis of fluorescent nanocrystalline cellulose/carbon dot hydrogels. Carbohydrate Polymers, 2017, 175: 7-17
121. Chunhui Ma, Yuangang Zu, **Jian Li**, Wei Li, Shouxin Liu. Dynamic distribution and biological storage analysis of procyanidins, arabinogalactan and dihydroquercetin in different parts of Larix gmelinii. Industrial Crops and Products, 2017, 95: 324-331
122. **李坚***, 焦月, 万才超. 负载 $Ni(OH)_2/NiOOH$ 微球碳气凝胶的水热合成与储能应用. 森林与环境学报, 2018, 38(3): 257-264
123. 王引航, 李伟, 罗沙, 刘守新, 马春慧, **李坚**. 离子液体固载型功能材料的应用研究进展. 化学学报, 2018, 76: 85-94
124. Yingying Li, Bin Hui, Gaoling Li, **Jian Li***. Facile one-pot synthesis of wood based bismuth molybdate nano-eggshells with efficient visible-light photocatalytic activity. Colloids and Surfaces A: Physicochemical and Engineering Aspects, 2018, 556: 284-290
125. Wenhui Bao, Zhen Jia, Liping Cai, Daxin Liang, **Jian Li***. Fabrication of a superamphiphobic surface on the bamboo substrate. European Journal of Wood and Wood Products, 2018, 76(6): 1595-1603
126. Likun Cao, Wentao Gan, Zhe Qiu, Guoliang Cao, Xianxu Zhan, Tiangang Qiang, **Jian Li***. Biomorphic carbon-doped TiO_2 for photocatalytic gas sensing with continuous detection of persistent volatile organic compounds. ACS Applied Nano Materials, 2018, 1(4): 1766-1775
127. Yingying Li, **Jian Li***. Fabrication of reversible thermoresponsive thin films on wood surfaces with hydrophobic performance. Progress in Organic Coatings, 2018, 119: 15-22
128. Yingying Li, Bin Hui, Miao Lv, **Jian Li***. Inorganic-organic hybrid wood in response to visible light. Journal of Materials Science, 2018, 53(5): 3889-3898
129. Caichao Wan, Yue Jiao, Daxin Liang, Yiqiang Wu, **Jian Li***. A high-performance, all-textile and spirally wound asymmetric supercapacitors based on core-sheath structured MnO_2 nanoribbons and cotton-derived carbon cloth. Electrochimica Acta, 2018, 285: 262-271
130. Caichao Wan, Yue Jiao, Daxin Liang, Yiqiang Wu, **Jian Li***. A geologic architecture system-inspired micro-/nano-heterostructure design for high-performance energy storage. Advanced Energy Materials, 2018, 8(33): 1802388
131. Yue Jiao, Caichao Wan, Wenhui Bao, He Gao, Daxin Liang, **Jian Li***. Facile hydrothermal synthesis of Fe_3O_4@cellulose aerogel nanocomposite and its application in Fenton-like degradation of Rhodamine B. Carbohydrate Polymers, 2018, 189: 371-378
132. Yue Jiao, Caichao Wan, **Jian Li***. Hydrothermal synthesis of SnO_2-ZnO aggregates in cellulose aerogels for photocatalytic degradation of Rhodamine B. Cellulose Chemistry and Technology, 2018, 52(1-2): 141-145
133. Runan Gao, Shaoliang Xiao, Wentao Gan, Qi Liu, Hassan Amer, Thomas Rosenau, **Jian Li***, Yun Lu. Mussel adhesive-inspired design of superhydrophobic nanofibrillated cellulose aerogels for oil/water separation. ACS Sustainable Chemistry & Engineering, 2018, 6: 9047
134. Jie Shen, Chang Cui, **Jian Li** and Lijuan Wang. In situ synthesis of a silver-containing superabsorbent polymer via a greener method based on carboxymethyl celluloses. Molecules, 2018, 23(10): 2483
135. Lele Cao, Tieqiang Liang, Xipeng Zhang, Wenbo Liu, **Jian Li**, Lijuan Wang. In-situ pH-sensitive fibers via the anchoring of bromothymol blue on cellulose grafted with hydroxypropyltriethylamine groups

via adsorption. Polymers, 2018, 10(7): 709

136. He Ting, Wang Hui, Chen Zhijun, Liu Shouxin, **Li Jian**, Li Shujun. Natural Quercetin AIEgen composite film with antibacterial and antioxidant properties for in situ sensing of Al^{3+} residues in food, detecting food spoilage, and extending food storage times. ACS Applied Bio Materials, 2018, 1: 636-642
137. Niu Na, Zhang Zhe, Gao Xi, Chen Zhijun, Li Shujun, **Li Jian**. Photodynamic therapy in hypoxia: Near-infrared-sensitive, self-supported, oxygen generation nano-platform enabled by upconverting nanoparticles. Chemical Engineering Journal, 2018, 352: 818-827
138. Zhang Xinyue, Wang Hui, Ma Chunhui, Niu Na, Chen Zhijun, Liu Shouxin, **Li Jian**, Li Shujun. Seeking value from biomass materials: Preparation of coffee bean shells-derived fluorescent carbon dots via molecular aggregation for antioxidation and bioimaging applications. Materials Chemistry Frontiers, 2018, 2: 1269-1275
139. Ma Zhuoming, Liu Chen, Niu Na, Chen Zhijun, Li Shujun, Liu ShouXin, **Li Jian**. Seeking brightness from nature: J-aggregation-induced emission in cellulolytic enzyme lignin nanoparticles. ACS Sustainable Chemistry & Engineering, 2018, 6(5): 3169-3175
140. Luo Xiongfei, Chen Zhijun, Li Shujun, Liu Shouxin, **Li Jian**. Preparation of a smart and portable film for in situ sensing of iron microcorrosion. ACS Applied Materials & Interfaces, 2018, 10(5): 4981-4985
141. He Ting, Niu Na, Chen Zhijun, Li Shujun, Liu Shouxin, **Li Jian**. Novel quercetin aggregation-induced emission luminogen (AIEgen) with excited-state intramolecular proton transfer for in vivo bioimaging. Advanced Functional Materials, 2018, (11): 1706196
142. Zhang Xinyue, Jiang Mingyue, Niu Na, Chen Zhijun, LiShujun, Liu Shouxin, **Li Jian**. Natural-product-derived carbon dots: from natural products to functional materials. ChemSusChem, 2018, 11(1): 11-24
143. Haiyue Yang, Yazhou Wang, Qianqian Yu, Guoliang Cao, Xiaohan Sun, Rue Yang, Qiong Zhang, Feng Liu, Xin Di, **Jian Li**, Chengyu Wang, Guoliang Li. Low-cost, three-dimension, high thermal conductivity, carbonized wood-based composite phase change materials for thermal energy storage. Energy, 2018, 159: 929-936
144. Wenshuai Chen, Haipeng Yu, Sang-Young Lee, Tong Wei, **Jian Li**, Zhuangjun Fan. Nanocellulose: a promising nanomaterial for advanced electrochemical energy storage. Chemical Society Reviews, 2018, 47: 2837-2872
145. Qinqin Xia, Yongzhuang Liu, Juan Meng, Wanke Cheng, Wenshuai Chen, Shouxin Liu, Yixing Liu, **Jian Li**, Haipeng Yu. Multiple hydrogen bond coordination in three-constituent deep eutectic solvent enhances lignin fractionation from biomass. Green Chemistry, 2018, 20: 2711-2721
146. Xuehua Liu, Rue Yang, Mincong Xu, Chunhui Ma, Wei Li, Yu Yin, Qiongtao Huang, Yiqiang Wu, **Jian Li**, Shouxin Liu. Hydrothermal synthesis of cellulose nanocrystal-grafted-acrylic acid aerogels with superabsorbent properties. Polymers, 2018, 10: 1168
147. Mingcong Xu, Rue Yang, Qiongtao Huang, Xin Zhao, Chunhui Ma, Wei Li, **Jian Li**, Shouxin Liu. Preparation and characterization of acetylated nanocrystalline cellulose-reinforced polylactide highly regular porous films. BioResources, 2018, 13(4): 8432-8443
148. Mingcong Xu, Wei Li, Chunhui Ma, Haipeng Yu, Yiqiang Wu, Yonggui Wang, Zhijun Chen, **Jian Li**

and Shouxin Liu. Multifunctional chiral nematic cellulose nanocrystals/glycerol structural colored nanocomposites for intelligent responsive Films, photonic inks and iridescent coatings. Journal of Materials Chemistry C, 2018, 6: 5391-5400

149. Qianyun Ma, Lele Cao, Tieqiang Liang, **Jian Li**, Lucian Lucia, Lijuan Wang. Active tara gum/PVA blend films with curcumin - loaded CTAC brush - TEMPO - oxidized cellulose nanocrystals. ACS Sustainable Chemistry & Engineering, 2018, 6(7): 8926-8934

150. Tieqiang Liang, Guohou Sun, Lele Cao, **Jian Li**, Lijuan Wang. Rheological behavior of film-forming solutions and films properties from *Artemisia sphaerocephala Krasch*. gum and purple onion peel extract. Food hydrocolloids, 2018, 82: 124-134

151. Miao Yu, Yingying Han, **Jian Li**, Lijuan Wang. Magnetic carbon aerogel pyrolysis from sodium carboxymethyl cellulose/sodium montmorillonite composite aerogel for removal of organic contamination. Journal of Porous Materials, 2018, 25(3): 657-664

152. Miao Yu, Yingying Han, **Jian Li**, Lijuan Wang. Polypyrrole-anchored cattail biomass-derived carbon aerogels for high performance binder - free supercapacitors. Carbohydrate Polymers, 2018, 199: 555-562

153. Miao Yu, Yingying Han, **Jian Li**, Lijuan Wang. Magnetic N - doped carbon aerogel from sodium carboxymethyl cellulose/collagen composite aerogel for dye adsorption and electrochemical supercapacitor. International Journal of Biological Macromolecules, 2018, 115: 185-193

154. Miao Yu, Yingying Han, **Jian Li**, Lijuan Wang. Three - dimensional porous carbon aerogels from sodium carboxymethyl cellulose/poly(vinyl alcohol) composite for high-performance supercapacitors. Journal of Porous Materials, 2018, 25(6): 1679-1689

155. **李坚***, 李莹莹. 木质仿生智能响应材料的研究进展. 森林与环境学报, 2019, 39(4): 337-343

156. 石江涛, **李坚**. 基于代谢组学分析技术的木材形成机理研究. 西南林业大学学报(自然科学), 2019, 39(1): 1-8

157. **李坚**. 打造林业人才的四个注重.《百名院士谈建设科技强国》, 北京: 人民出版社, p623-628, 2019

158. YingyingLi, Gaoliang Li, **Jian Li***. Photoresponsive wood-based composite fabricated by a simple drop-coating procedure. Wood Science and Technology, 2019, 53(1): 211-226

159. YingyingLi, Runan Gao, **Jian Li***. Energy saving wood composite with temperature regulatory ability and thermoresponsive performance. European Polymer Journal, 2019, 118: 163-169

160. YingyingLi, Likun Gao, Yingtao Liu, **Jian Li***. Structurally colored wood composite with reflective heat insulation and hydrophobicity. Journal of Wood Chemistry and Technology, 2019, 39(6): 454-463

161. Caichao Wan, Yue Jiao, Wenhui Bao, He Gao, Yiqiang Wu, **Jian Li***. Self-stacked multilayer FeOCl supported on a cellulose-derived carbon aerogel: a new and high-performance anode material for supercapacitors. Journal of Materials Chemistry A, 2019, 7(16): 9556-9564

162. Yue Jiao, Caichao Wan, Wenbo Zhang, Wenhui Bao, **Jian Li***. Carbon fibers encapsulated with nano-copper: a core-shell structured composite for antibacterial and electromagnetic interference shielding applications. Nanomaterials, 2019, 9(3): 460

163. **Jian Li***, Yue Jiao. Polyaniline-polypyrrole nanocomposites using a green and porous wood as support for supercapacitors. Frontiers of Agricultural Science and Engineering, 2019, 6(2): 137-143

164. Caichao Wan, Yue Jiao, Song Wei, Xianjun Li, Wenyan Tian, Yiqiang Wu, **Jian Li***. Scalable top-to-bottom design on low tortuosity of anisotropic carbon aerogels for fast and reusable passive capillary absorption and separation of organic leakages. ACS Applied Materials & Interfaces, 2019, 11(51): 47846-47857

165. Chen Zhijun, Dang Ben, Luo Xiongfei, Li Wei, **Li Jian**, Yu Haipeng, LiShouxin, Li Shujun. Deep eutectic solvent-assisted in situ wood delignification: a promising strategy to enhance efficiency of wood-based solar steam generation devices. ACS Applied Materials & Interfaces, 2019, 11(29): 26032-26037

166. Wang Ping, Liu Chen, Tang Weiqiang, Ren Shixue, Chen Zhijun, Guo Yuanru, Rahele Rostamian, Zhao Shuangliang, **Li Jian**, Liu Shouxin, Li Shujun. Molecular glue strategy: Large-scale conversion of clustering-induced emission luminogen to carbon dots. ACS Applied Materials & Interfaces, 2019, 11: 19301-19307

167. Luo Xiongfei, Ma Chunhui, Chen Zhijun, Zhang Xinyue, Niu Na, **Li Jian**, Liu Shouxin, Li Shujun. Biomass-derived solar-to-thermal materials: promising energy absorbers to convert light to mechanical motion. Journal of Materials Chemistry A, 2019, 7: 4002

168. Haiyue Yang, Weixiang Chao, Xin Di, Zhaolin Yang, Tinghan Yang, Qianqian Yu, Feng Liu, **Jian Li**, Guoliang Li, Chengyu Wang. Multifunctional wood based composite phase change materials for magnetic-thermal and solar-thermal energy conversion and storage. Energy Conversion and Management, 2019, 200: 112029

169. Haiyue Yang, Weixiang Chao, Siyuan Wang, Qianqian Yu, Guoliang Cao, Tinghan Yang, Feng Liu, Xin Di, **Jian Li**, Chengyu Wang, Guoliang Li. Self-luminous wood composite for both thermal and light energy storage. Energy Storage Materials, 2019, 18: 15-22

170. Qi Zhang, Chaoji Chen, Wenshuai Chen, Xiaoyu Guo, Shouxin Liu, Qingwen Wang, Yixing Liu, **Jian Li**, Haipeng Yu. Nanocellulose enabled all-nanofiber, high performance supercapacitor. ACS Applied Materials & Interfaces, 2019, 11(6): 5919-5927

171. Yue Ma, Qinqin Xia, Yongzhuang Liu, Wenshuai Chen, Shouxin Liu, Qingwen Wang, **Jian Li**, Haipeng Yu. Production of nanocellulose using hydrated deep eutectic solvent combined with ultrasonic treatment. ACS Omega, 2019, 4(5): 8539-8547

172. Yushu Wang, Jialiang Wang, Shengjie Ling, Haiwei Liang, Ming Dai, Lulu Bai, Qing Li, Zhuangjun Fan, Sang-Young Lee, Haipeng Yu, Shouxin Liu, Qingwen Wang, Yixing Liu, **Jian Li**, Tianmeng Sun, Wenshuai Chen. Nanoparticle separation: wood-derived nanofibrillated cellulose hydrogel filters for fast and efficient separation of nanoparticles. Advanced Sustainable Systems, 2019, 3(9): 1970019

173. Mingcong Xu, Chunhui Ma, Jin Zhou, Yushan Liu, Xueyun Wu, Sha Luo, Wei Li, Haipeng Yu, Yonggui Wang, Zhijun Chen, **Jian Li**, Shouxin Liu. Assembling semiconductor quantum dots in hierarchical photonic cellulose nanocrystal films: circularly polarized luminescent nanomaterials as optical coding labels. Journal of Materials Chemistry C, 2019, 7: 13794-13802

174. Xuehua Liu, Mingcong Xu, Bang An, Zhenwei Wu, Rue Yang, Chunhui Ma, Qiongtao Huang, Wei Li, **Jian Li**, Shouxin Liu. A facile hydrothermal method fabricated robust and ultralight weight cellulose nanocrystal-based hydro/aerogels for metal ions removal. Environmental Science and Pollution Research, 2019, 26(25): 25583-25595

175. Wei Li, Mingcong Xu, Chunhui Ma, Yushan Liu, Jin Zhou, Zhijun Chen, Yonggui Wang, Haipeng

Yu, **Jian Li**, Shouxin Liu. Tunable upconverted circularly polarized luminescence in cellulose nanocrystal based chiral photonic films. ACS Applied Materials & Interfaces, 2019, 11: 23512-23519

176. Caichao Wan, Yue Jiao, Song Wei, Liyu Zhang, Yiqiang Wu, **Jian Li**. Functional nanocomposites from sustainable regenerated cellulose aerogels: a review. Chemical Engineering Journal, 2019, 359: 459-475
177. Zhe Qiu, Zefang Xiao, Likun Gao, **Jian Li**, Haigang Wang, Yonggui Wang, Yanjun Xie. Transparent wood bearing a shielding effect to infrared heat and ultraviolet via incorporation of modified antimony-doped tin oxide nanoparticles. Composites Science and Technology, 2019, 172: 43-48
178. Wenbo Che, Zefang Xiao, Zheng Wang, **Jian Li**, Haigang Wang, Yonggui Wang, Yanjun Xie. Wood-based mesoporous filter decorated with silver nanoparticles for water purification. ACS Sustainable Chemistry & Engineering, 2019, 7(5): 5134-5141
179. Lele Cao, Qianyun Ma, Tieqiang Liang, Guohou Sun, Wenrui Chi, Cijian Zhang, **Jian Li**, Lijuan Wang. A semen cassia gum-based film with visual-olfactory function for indicating the freshness change of animal protein-rich food. International Journal of Biological Macromolecules, 2019, 133: 243-252
180. Guohou Sun, Wenrui Chi, Cijian Zhang, Shiyu Xu, **Jian Li**, Lijuan Wang. Developing a green film with pH-sensitivity and antioxidant activity based on κ-carrageenan and hydroxypropyl methylcellulose incorporating Prunus maackii juice. Food hydrocolloids, 2019, 94: 345-353
181. Lele Cao, Guohou Sun, Cijian Zhang, Wenbo Liu, **Jian Li**, Lijuan Wang. An intelligent film based on cassia gum containing bromothymol blue-anchored cellulose fibers for real-time detection of meat freshness. Journal of Agricultural and Food Chemistry, 2019, 67(7): 2066-2074
182. Tieqiang Liang, Guohou Sun Lele Cao, **Jian Li**, Lijuan Wang. A pH and NH_3 sensing intelligent film based on Artemisia sphaerocephala Krasch. gum and red cabbage anthocyanins anchored by carboxymethyl cellulose sodium added as a host complex. Food hydrocolloids, 2019, 87: 858-868
183. Miao Yu, Yingying Han, Yao Li, **Jian Li**, Lijuan Wang. Improving electrochemical activity of activated carbon derived from popcorn by $NiCo_2S_4$ nanoparticle coating. Applied Surface Science, 2019, 463: 1001-1010
184. Houjuan Qi, Min Teng, Miao Liu, Shouxin Liu, **Jian Li**, Haipeng Yu, Chunbo Teng, Zhanhua Huang, Hu Liu, Qian Shao, Ahmad Umar, Tao Ding, Qiang Gao, Zhanhu Guo. Biomass-derived nitrogen-doped carbon quantum dots: highly selective fluorescent probe for detecting Fe^{3+} ions and tetracyclines. Journal of Colloid and Interface Science, 2019, 539: 332-341
185. Cai Shi, Houjuan Qi, Rongxiu Ma, Zhe Sun, Lidong Xiao, Guangbiao Wei, Zhanhua Huang, Shouxin Liu, **Jian Li**, Mengyao Dong, Jincheng Fan, Zhanhu Guo. N, S-self-doped carbon quantum dots from fungus fibers for sensing tetracyclines and for bioimaging cancer cells. Materials Science & Engineering C, 2019, 105: 110132
186. Houjuan Qi, Xiaodi Ji, Cai Shi, Rongxiu Ma, Zhanhua Huang, Minghui Guo, **Jian Li**, Zhanhu Guo. Bio-templated 3D porous graphitic carbon nitride hybrid aerogel with enhanced charge carrier separation for efficient removal of hazardous organic pollutants. Journal of Colloid and Interface Science, 2019, 556: 366-375
187. **李坚***, 焦月. 木质基新型能量存储与转换材料研究进展. 森林与环境学报, 2020, 40(4): 337-346
188. 黄睿, 韦双颖, 王成毓, 王砥, **李坚**. 硅烷改性铽配合物修饰木质基材料的耐光老化性能. 林业

工程学报, 2020, 5(5): 35-39

189. 陈志俊, 高鹤, 李伟, 李淑君, 刘守新, **李坚**. 生物质基光学材料研究进展. 林业工程学报, 2020, 5: 1-12

190. 王梦竹, 方颖, 李勍, 白璐璐, 于海鹏, 刘守新, **李坚**, 陈文帅. 植物细胞壁纳米结构与纳米纤维素的化学纯化处理结合机械解纤法制备. 高分子学报, 2020, 51(6): 586-597

191. 刘学, 李淑君, 刘守新, **李坚**, 陈志俊. 木质素纳米颗粒的制备及其功能化应用研究进展. 生物质化学工程, 2020, 54(5): 53-65

192. Wei Li, Zhijun Chen, Haipeng Yu, **Jian Li***, Shouxin Liu. Wood-derived carbon materials and light-emitting materials. Advanced Materials, 2020: 2000596

193. He Gao, Mingming Yang, Ben Dang, Xiongfei Luo, Shouxin Liu, Shujun Li, Zhijun Chen, **Jian Li***. Natural phenolic compound-iron complexes: sustainable solar absorbers for wood-based solar steam generation devices. RSC Advances, 2020, 10: 1152-1158

194. Jiao Yue, Wan Caichao, Wu Yiqiang, Han Jingquan, Bao Wenhui, Gao He, Wang Yaoxing, Wang Chengyu, **Li Jian***. Ultra-high rate capability of nanoporous carbon network@ V_2O_5 sub-micron brick composite as a novel cathode material for asymmetric supercapacitors. Nanoscale, 2020, 12: 23213-23224

195. Caichao Wan, Yue Jiao, Wenyan Tian, Luyu Zhang, Yiqiang Wu, **Jian Li***, Xianjun Li. A holocellulose framework with anisotropic microchannels for directional assembly of copper sulphide nanoparticles for multifunctional applications. Chemical Engineering Journal, 2020, 393: 124637

196. Mingcong Xu, Xueyun Wu, Yang Yang, Chunhui Ma, Wei Li, Haipeng Yu, Zhijun Chen, **Jian Li**, Kai Zhang, Shouxin Liu. Designing hybrid chiral photonic films with circularly polarized room-temperature phosphorescence. ACS Nano, 2020, 14: 11130-11139

197. Wenxin Zhu, Wei Huang, Wenhua Zhou, Zhe Qiu, Zheng Wang, Haojun Li, Yonggui Wang, **Jian Li**, Yanjun Xie. Sustainable and antibacterial sandwich-like Ag-Pulp/CNF composite paper for oil/water separation. Carbohydrate Polymers, 2020, 245: 116587

198. Li Xiaoning, Li Meng, Yang Mengqing, Xiao Huining, Wang Lidong, Chen Zhijun, Liu Shouxin, **Li Jian**, Li Shujun, Tony D. James. "Irregular" aggregation-induced emission luminogens. Coordination Chemistry Reviews, 2020, 418: 213358

199. Kütahya Ceren, Wang Ping, Li Shujun, Liu Shouxin, **Li Jian**, Chen Zhijun, Strehmel Bernd. Carbon dots as promising green photocatalyst for free radical and ATRP-Based radical photopolymerization with blue LEDs. Angewandte Chemie International Edition, 2020, 59: 3166-3171

200. Jiang Mingyue, Liu Xue, Chen Zhijun, **Li Jian**, Liu Shouxin, Li Shujun. Near-infrared-detached adhesion enabled by upconverting nanoparticles. iScience, 2020, 23: 100832

201. Zhe Qiu, Shuo Wang, Yonggui Wang, **Jian Li**, Zefang Xiao, Haigang Wang, Daxin Liang, Yanjun Xie. Transparent wood with thermo-reversible optical properties based on phase-change material. Composites Science and Technology, 2020, 200: 108407

202. Wei Huang, Xiangyu Tang, Zhe Qiu, Wenxin Zhu, Yonggui Wang, You-Liang Zhu, Zefang Xiao, Haigang Wang, Daxin Liang, **Jian Li**, and Yanjun Xie. Cellulose-based superhydrophobic surface decorated with functional groups showing distinct wetting abilities to manipulate water harvesting. ACS Applied Materials & Interfaces, 2020, 12: 40968-40978

203. Yonggui Wang, Zhe Qiu, Zhen Lang, Yanjun Xie, Zefang Xiao, Haigang Wang, Daxin Liang, **Jian**

Li, and Kai Zhang*. Multifunctional reversible self-assembled structures of cellulose-derived phase-change nanocrystals. Advanced Materials, 2020: 2005263

204. Lele Cao, Huiyan Feng, Fansong Meng, **Jian Li**, Lijuan Wang. Fabrication of a high tensile and antioxidative film via a green strategy of self-growing needle-like quercetin crystals in cassia gum for lipid preservation. Journal of Cleaner Production, 2020, 266: 121885
205. Wenrui Chi, Lele Cao, Guohou Sun, Fansong Meng, Cijian Zhang, **Jian Li**, Lijuan Wang. Developing a highly pH-sensitive κ-carrageenan-based intelligent film incorporating grape skin powder via a cleaner process. Journal of Cleaner Production, 2020, 244: 118862
206. Lele Cao, Tingting Ge, Fansong Meng, Shiyu Xu, **Jian Li**, Lijuan Wang. An edible oil packaging film with improved barrier properties and heat sealability from cassia gum incorporating carboxylated cellulose nano crystal whisker. Food hydrocolloids, 2020, 98: 105251
207. Danyang Wang, Ruoting Liu, Yanjun Xie, **Jian Li**, Lijuan Wang. Fabrication of a laminated felt-like electromagnetic shielding material based on nickel-coated cellulose fibers via self-foaming effect in electroless plating process. International Journal of Biological Macromolecules, 2020, 154: 954-961
208. Cai Shi, Houjuan Qi, Zhe Sun, Keqi Qu, Zhanhua Huang, **Jian Li**, Mengyao Dong, Zhanhu Guo. Carbon dot-sensitized urchin-like Ti^{3+} self-doped TiO_2 photocatalysts with enhanced photoredox ability for highly efficient removal of Cr^{6+} and RhB. Journal of Materials Chemistry C, 2020, 8: 2238
209. Haiyue Yang, Siyuan Wang, Xin Wang, Weixiang Chao, Nan Wang, Xiaolun Ding, Feng Liu, Qianqian Yu, Tinghan Yang, Zhaolin Yang, **Jian Li**, Chengyu Wang, Guoliang Li. Wood-based composite phase change materials with self-cleaning superhydrophobic surface for thermal energy storage. Applied Energy, 2020, 261: 114481
210. Dawei Zhao, Ying Zhu, Wanke Cheng, Guangwen Xu, Qingwen Wang, Shouxin Liu, **Jian Li**, Chaoji Chen, Haipeng Yu, Liangbing Hu. A dynamic gel with reversible and tunable topological networks and performances. Matter, 2020, 2(2): 390-403
211. Caichao Wan, Yue Jiao, Xianjun Li, Wenyan Tian, **Jian Li**, Yiqiang Wu. Multi-dimensional and level-by-level assembly strategy on flexible and sandwich-type nanoheterostructures for high-performance electromagnetic interference shielding. Nanoscale, 2020, 12(5): 3308-3316
212. Song Wei, Caichao Wan, Yue Jiao, Xianjun Li, **Jian Li**, Yiqiang Wu. 3D nanoflower-like $MoSe_2$ encapsulated with hierarchically anisotropic carbon architecture: a new and free-standing anode with ultra-high areal capacitance for asymmetric supercapacitors. Chemical Communications, 2020, 56: 340-343
213. Wei Yifan, Fu Zhengquan, Zhao Hao, Liang Ruiqi, Wang Chengyu, Wang Di, **Li Jian**. Preparation of PVA fluorescent gel and luminescence of europium sensitized by terbium (III). Polymers, 2020, 12 (4): 893
214. Yue You, Keqi Qu, Cai Shi, Zhe Sun, Zhanhua Huang, **Jian Li**, Mengyao Dong, Zhanhu Guo. Binder-free CuS/ZnS/sodium alginate/rGO nanocomposite hydrogel electrodes for enhanced performance supercapacitorsInternational Journal of Biological Macromolecules, 2020, 162: 310-319
215. **李坚**，甘文涛. 关于深化对外合作交流的政策建议. 专家建议，2020年第19期(总第323期)
216. **李坚**. 大爱无疆，铿锵的中国力量.《担当》，北京：人民出版社，p303-312，2021
217. **李坚**.《木材科学与技术》刊首寄语. 木材科学与技术. 2021，35(1)：刊首语
218. Wentao Gan, Yaoxing Wang, Shaoliang Xiao, Runan Gao, Ying Shang, Yanjun Xie*, Jiuqing Liu*,

and**Jian Li**[*]. Magnetically driven 3D cellulose film for improved energy efficiency in solar evaporation. ACS Applied Materials & Interfaces, 2021, 13: 7756-7765

219. Yang Haiyue, Liu Yushan, **Li Jian**, Wang Chengyu, Li Yudong. Full-wood photoluminescent and photothermic materials for thermal energy storage. Chemical Engineering Journal, 2021, 403: 126406
220. Cijian Zhang, Guohou Sun, **Jian Li**, Lijuan Wang. A green strategy for maintaining intelligent response and improving antioxidant properties of kappa-carrageenan-based film via cork bark extractive addition. Food Hydrocolloids, 2021, 113: 106470
221. Liu Ruoting, Li Tingting, Xu Jin, Zhang Tongcheng, Xie Yanjun, **Li Jian**, Wang Lijuan. Sandwich-structural Ni/Fe_3O_4/Ni/cellulose paper with a honeycomb surface for improved absorption performance of electromagnetic interference. Carbohydrate Polymers, 2021, 260(15): 117840
222. Tongcheng Zhang, Dong Lv, Ruoting Liu, Danyang Wang, Tingting Li, Jin Xu, Yanjun Xie, **Jian Li**, Lijuan Wang. Developing a superhydrophobic absorption-dominated electromagnetic shielding material by building clustered Fe_3O_4 nanoparticles on copper-coated cellulose paper. ACS Sustainable Chemistry & Engineering, 2021, 9(19): 6574-6585
223. Zhao Xinpeng, Huang Caoxing, Xiao Daming, Wang Ping, Luo Xiongfei, Liu Wenbo, Liu Shouxin, **Li Jian**, Li Shujun, Chen Zhijun. Melanin-inspired design: preparing sustainable photothermal materials from lignin for energy generation. ACS Applied Materials & Interfaces, 2021, 13: 7600-7607
224. Wang Ping, Zheng Dongxiao, Liu Shouxin, Luo Mengkai, **Li Jian**, Shen Shen, Li Shujun, Zhu Liangliang, Chen Zhijun. Producing long afterglow by cellulose confinement effect: A wood-inspired design for sustainable phosphorescent materials. Carbon, 2021, 171: 946-952
225. Kütahya Ceren, Zhai Yingxiang, Li Shujun, Liu Shouxin, **Li Jian**, Strehmel Veronika, Chen Zhijun, Strehmel Bernd. Distinct sustainable carbon nanodots enable free radical photopolymerization, photo-ATRP and photo-CuAAC chemistry. Angewandte Chemie International Edition, 2021, 133(19): 11078-11087
226. Ge Min, Huang Xun, Ni Jiaxin, Han Youqi, Zhang Chunlei, Li Shujun, Cao Jun, **Li Jian**, Chen Zhijun, Han Shiyan. One-step synthesis of self-quenching-resistant biomass-based solid-state fluorescent carbon dots with high yield for white lighting emitting diodes. Dyes and Pigments, 2021, 185: 108953
227. Min Ge, Youqi Han, Jiaxin Ni, Yudong Li, Shiyan Han, Shujun Li, Haipeng Yu, Chunlei Zhang, Shouxin Liu, **Jian Li**, Zhijun Chen. Seeking brightness from nature: Sustainable carbon dots-based AIEgens with tunable emission wavelength from natural rosin. Chemical Engineering Journal, 2021, 413: 127457
228. Meng Fansong, Zhang Cijian, **Li Jian**, Wang Lijuan. Self-assembling crystals of an extract of Flos Sophorae Immaturus for improving the antioxidant, mechanical and barrier properties of a cassia gum film. International Journal of Biological Macromolecules, 2021, 167: 1281-1289
229. Zhang Zhihui, Zhao Zhengdong, Lu Yujia, Wang Di, Wang Chengyu, **Li Jian**. One-step synthesis of Eu^{3+}-modified cellulose acetate film and light conversion mechanism. Polymers, 2021, 13(1): 113
230. Li Ming, Huang Rui, Fu Zhengquan, Wang Di, Wang Chengyu, **Li Jian**. Multi-functional luminescent coating for wood fabric based on silica sol-gel approach. Polymers, 2021, 13(1): 127
231. Li Ming, Wang Chengyu, Wang Di, **Li Jian**. Structure-dependent photoluminescence of europium (III) coordination oligomeric silsesquioxane: Synthesis and mechanism. ACS omega, 2021, 6(1):

227-238

232. Wang Mengzhu, Sun Tianmeng, Wan Dehui, Dai Ming, Ling Shengjie, Wang Jialiang, Liu Yuqiu, Fang Ying, Xu Shuhan, Yeo Jingjie, Yu Haipeng, Liu Shouxin, Wang Qingwen, **Li Jian**, Yang Ya, Fan Zhuangjun, Chen Wenshuai. Solar-powered nanostructured biopolymer hygroscopic aerogels for atmospheric water harvesting. Nano Energy, 2021, 80: 105569
233. Lulu Bai, Qing Li, Ya Yang, Shengjie Ling, Haipeng Yu, Shouxin Liu, **Jian Li**, Wenshuai Chen. Biopolymer nanofibers for nanogenerator development. Research, 2021: 1843061
234. Zhe Sun, Keqi Qu, Yue You, Zhanhua Huang, Shouxin Liu, **Jian Li**, Qian Hu, Zhanhu Guo. Overview of cellulose-based flexible materials for supercapacitors. Journal of Materials Chemistry A, 2021, 9(12): 7278-7300

二、著作

235. **李坚**(主编). 木材科学(第三版)/全国高等院校研究生规划教材. 北京: 科学出版社, 2014
236. 陈文帅, 于海鹏, 李勍, 刘一星, **李坚**. 纳米纤维素机械法制备和应用基础. 北京: 科学出版社, 2014
237. 卢芸, **李坚**, 等. 生物质纳米材料与气凝胶. 北京: 科学出版社, 2015
238. 马春慧, 祖元刚, **李坚**. 落叶松化学成分分析及高附加值产品加工利用. 北京: 科学出版社, 2016
239. 邢东, 王立娟, **李坚**. 木材热改性处理. 北京: 科学出版社, 2016
240. 李伟, 刘守新, **李坚**. 纳米纤维素的制备及功能化应用基础. 北京: 科学出版社, 2016
241. **李坚**(主编). 生物质复合材料学(第二版)/"十二五"普通高等教育本科国家级规划教材. 北京: 科学出版社, 2017(获首届黑龙江省教材建设奖特等奖, 黑教联〔2020〕47号)
242. **李坚**, 孙庆丰, 王成毓, 李伟, 高丽坤, 甘文涛. 木材仿生智能科学引论. 北京: 科学出版社, 2018
243. Yun Lu, Runan Gao, Shaoliang Xiao, Yafang Yin, Qi Liu, **Jian Li**, Biobased Biobased Aerogels: Polysaccharide and Protein-based Materials. Chapter 5: Cellulose Based Aerogels: Processing and Morphology, RSC Publishing, 2018
244. **李坚**, 万才超, 焦月. 纤维素气凝胶基多功能纳米复合材料 北京: 科学出版社, 2019
245. **李坚**(主编). 木材波谱学(第二版)/全国高等院校研究生规划教材. 北京: 科学出版社, 2020
246. **李坚**, 孙庆丰, 陈志俊, 等. 木竹材仿生与智能响应. 北京: 科学出版社, 2020

三、专利

247. **李坚**, 苏文强, 杨冬梅, 李淑君. 从怀槐中提取抑制木材腐朽菌提取物的方法. 专利号: ZL200910307616. 2, 2012
248. **李坚**, 王成毓. 仿生矿化法原位制备双疏性木材/碳酸钙复合材的方法. 专利号: ZL201010148727. 6, 2012
249. 李淑君, **李坚**, 梁涛, 金燕, 孙月正, 张明鑫, 王前. 松香-铜复合木材防腐剂的制备方法. 专利号: ZL201010561000. 0, 2013
250. 王立娟, 梁铁强, **李坚**. 一种环保型木材单板的染色方法. 专利号: ZL201310303118. 7, 2014

251. **李坚**，甘文涛，詹先旭，高丽坤，万才超，孙庆丰，卢芸．一种木材表面原位生长磁性纳米 Fe_3O_4 的方法，专利号：ZL 201310703641．9，2014
252. **李坚**，孙庆丰，高丽坤，卢芸，万才超，甘文涛．一种木材表面亲疏可逆开关的制备方法．专利号：ZL201310703642．3，2015
253. **李坚**，万才超，焦月，包文惠．一种基于碳材料、聚吡咯和 α-Fe_2O_3 三种成分的电磁屏蔽材料的制备方法．专利号：ZL201610990084．7，2016
254. **李坚**，万才超，焦月．一种提高纤维素气凝胶的力学强度的方法．专利号：ZL201610560600．2，2016
255. **李坚**，万才超，焦月．一种以柔性纤维为原料的碳纤维气凝胶循环吸附剂的制备方法．专利号：ZL201610560599．3，2016
256. **李坚**，甘文涛，高丽坤，詹先旭，万才超，肖少良，高汝楠．一种用于吸收电磁波的磁性木材的制备方法，专利号：ZL 201610251714．9，2016
257. **李坚**，高丽坤，詹先旭，甘文涛，肖少良．一种在紫外光照射下释放负氧离子的木材基材料．专利号：ZL201610153155．8，2017
258. **李坚**，李莹莹，李国梁，惠彬，程婉可．一种改性木基质可逆温致变色复合材料的制备方法．专利号：ZL201610528367．X，2017
259. **李坚**，甘文涛，高丽坤，詹先旭，肖少良，高汝楠．一种荧光透明磁性木材的制备方法．专利号：ZL 201610899860．2，2017
260. **李坚**，甘文涛，高丽坤，詹先旭，肖少良，高汝楠．一种用于磁感应加热磁性木材的制备方法．专利号：ZL 201710364026．8，2017
261. **李坚**，李莹莹，李国梁．一种快速响应的持久型无机-有机杂化光敏木材的制备方法．专利号：ZL201710520521．3，2018
262. **李坚**，李莹莹，李国梁，惠彬．一种有机-无机杂化光响应木材的制备方法．专利号：ZL201710520522．8，2018
263. 梁大鑫，高鹤，包文慧，焦月，王成毓，**李坚**．无污染提高木材耐侯性能和双疏性能的处理方法．专利号：ZL201610891199．0，2018
264. 王立娟，曹乐乐，冯昊，**李坚**．一种 pH 响应变色纤维素纤维的制备方法．专利号：ZL201711292045．0，2019
265. **李坚**，李莹莹，李国梁，惠彬．一种正向开关温致可逆变色复合薄膜的制备方法．专利号：ZL201610528334．5，2019
266. **李坚**，李莹莹，惠彬，高立坤，高汝楠．一种光温双重响应木质储能材料的制备方法．专利号：ZL201810920330．0，2019
267. **李坚**，李莹莹，高汝楠．一种生长在木材表面的环境友好型蛋壳状光催化剂的制备方法．专利号：ZL201711454976．6，2019
268. **李坚**，高汝楠，肖少良，甘文涛．一种超疏水吸油纳米纤维素气凝胶材料的制备方法．专利号：ZL201710436369．0，2019
269. 李伟，刘守新，徐明聪，马春慧，黄琼涛，吴义强，**李坚**．一种手性向列纤维素纳米晶体-丙三醇复合薄膜及其制备方法和应用．专利号：ZL201810119604．6，2019
270. 李伟，刘守新，刘学华，马春慧，罗沙，**李坚**．一种纳米纤维素凝胶及其制备方法和应用．专利号：ZL201810119073．0，2019